Mathematical Formulas*

Quadratic Formula

If $ax^2 + bx + c = 0$, then $x = \dfrac{-b \pm \sqrt{b^2 - 4ac}}{2a}$.

Properties of Simple Geometric Objects

Triangle of base b and height h	Area $= \frac{1}{2}bh$	
Circle of radius r	Circumference $= 2\pi r$	Area $= \pi r^2$
Sphere of radius r	Surface area $= 4\pi r^2$	Volume $= \frac{4}{3}\pi r^3$
Cylinder of radius r and height h	Area of curved surface $= 2\pi rh$	Volume $= \pi r^2 h$

Trigonometry

Law of sines: $\dfrac{a}{\sin\alpha} = \dfrac{b}{\sin\beta} = \dfrac{c}{\sin\gamma}$

Law of cosines: $c^2 = a^2 + b^2 - 2ab\cos\gamma$

$\sin\theta = \dfrac{1}{\csc\theta}$; $\cos\theta = \dfrac{1}{\sec\theta}$; $\tan\theta = \dfrac{1}{\cot\theta}$

$\sin(90° - \theta) = \cos\theta$; $\cos(90° - \theta) = \sin\theta$; $\tan(90° - \theta) = \cot\theta$

$\sin^2\theta + \cos^2\theta = 1$

$\tan\theta = \dfrac{\sin\theta}{\cos\theta}$

$\sin 2\theta = 2\sin\theta\cos\theta$

$\cos 2\theta = \cos^2\theta - \sin^2\theta = 2\cos^2\theta - 1 = 1 - 2\sin^2\theta$

$\sin(\alpha \pm \beta) = \sin\alpha\cos\beta \pm \cos\alpha\sin\beta$

$\cos(\alpha \pm \beta) = \cos\alpha\cos\beta \mp \sin\alpha\sin\beta$

Expansions

$$(1 \pm x)^n = 1 \pm \frac{nx}{1!} + \frac{n(n-1)x^2}{2!} + \cdots \quad (x^2 < 1)$$

$$(1 \pm x)^{-n} = 1 \mp \frac{nx}{1!} + \frac{n(n+1)x^2}{2!} + \cdots \quad (x^2 < 1)$$

$$\sin x = x - \frac{x^3}{3!} + \frac{x^5}{5!} - \cdots$$

$$\cos x = 1 - \frac{x^2}{2!} + \frac{x^4}{4!} - \cdots$$

$$e^x = 1 + x + \frac{x^2}{2!} + \cdots$$

*There is a more comprehensive list in App. D.

UNIVERSITY PHYSICS

Publisher's Note

University Physics is available as a standard length casebound text, as an extended casebound text, or as three paperbound separates.

Binding Options	Description	ISBN
University Physics, Standard Version	Full-length text	0-697-05884-0
University Physics, Extended Version	Full-length text with modern physics	0-697-27337-7
University Physics, Volume I Mechanics and Thermodynamics	Chapters 1–21	0-697-29949-X
University Physics, Volume II Electricity & Magnetism and Optics	Chapters 22–39	0-697-29950-3
University Physics, Volume III Modern Physics	Chapters 40–46	0-697-29951-1

UNIVERSITY PHYSICS

Jeff Sanny & William Moebs

Loyola Marymount University

Wm. C. Brown Publishers

Dubuque, IA Bogota Boston Buenos Aires Caracas Chicago
Guilford, CT London Madrid Mexico City Sydney Toronto

Book Team

Editor *Megan Johnson*
Developmental Editor *Tom Riley*
Publishing Services Coordinator *Julie Avery Kennedy*

Wm. C. Brown Publishers

A Times Mirror Company

Cover photo: Telegraph Colour Library/FPG International Corp.
Production by Lachina Publishing Services, Inc.
Interior design by Tara Bazata

The credits section for this book begins on page C-1 following the answer section and is considered an extension of the copyright page.

Library of Congress Catalog Card Number: 94-70033

ISBN 0-697-05884-0   Volume 1 0-697-29949-X
Volume 2 0-697-29950-3
Volume 3 0-697-29951-1

Printed in the United States of America by Times Mirror Higher Education Group, Inc., 2460 Kerper Boulevard, Dubuque, IA 52001

10 9 8 7 6 5 4 3 2 1

With love to our parents, wives,
and children

BRIEF CONTENTS

PART V

Electromagnetic Waves and Optics

PART VI

Modern Physics

CONTENTS

PART I

MECHANICS

PART II

MECHANICAL WAVES

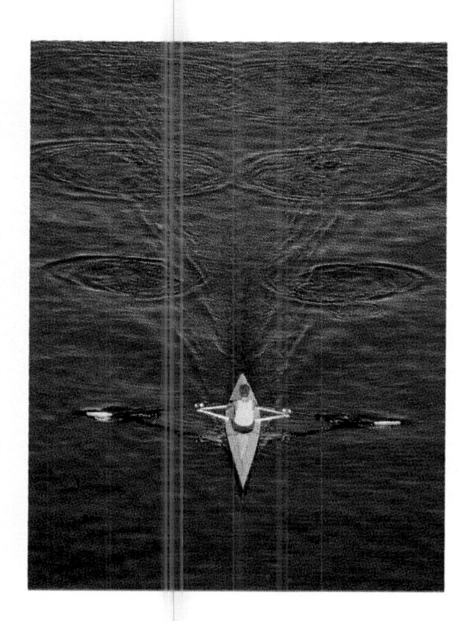

PART III

THERMODYNAMICS

PART IV

Electricity and Magnetism

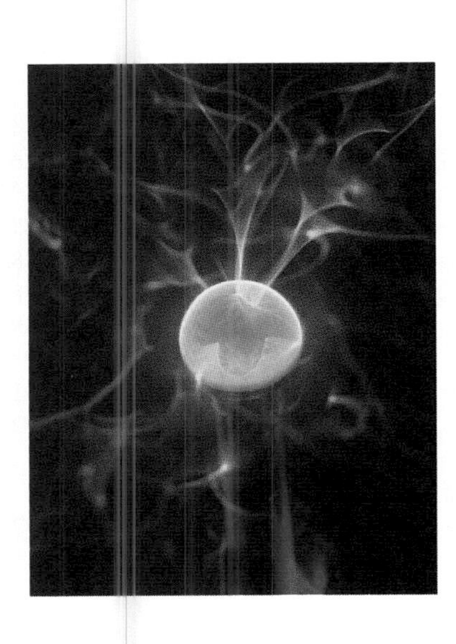

PART V

ELECTROMAGNETIC WAVES AND OPTICS

PART VI

MODERN PHYSICS

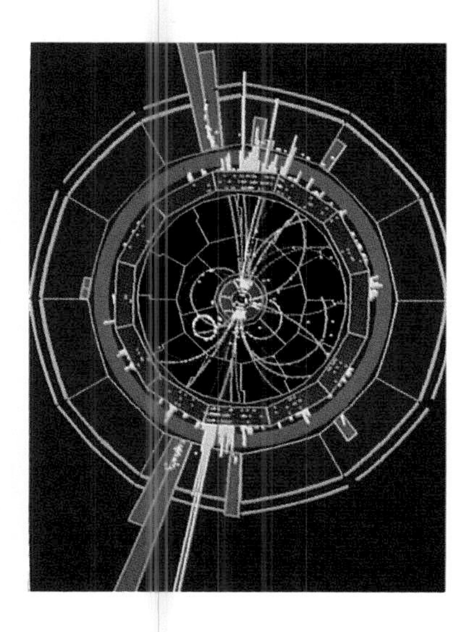

PREFACE

Altogether, we have taught physics at the university level for over forty years. Perhaps the most common problem we have encountered in calculus-based, general physics courses is that a large amount of challenging material has to be taught in a relatively short period of time to a student population that often has little preparation for what is to come. We have attempted to overcome this difficulty by writing a textbook in a style that is direct and easy to read while using mathematics that is not beyond what should be common background of the students at this level. However, we do not feel that we have compromised the depth of the physics, and all of the standard material is presented. Our goal is to provide a clear and concise presentation that will make it easier for students to (a) learn the basic principles of physics, (b) be able to apply these principles to realistic situations, and (c) eventually develop a sense of logic and intuition that will benefit them in whatever profession they choose.

The textbook covers seven main topics over thirty-nine chapters. In order of appearance, they are Newtonian mechanics (fourteen chapters), fluids (one chapter), wave motion (two chapters), heat and thermodynamics (four chapters), electricity and magnetism (thirteen chapters), light and optics (four chapters), and special relativity (one chapter). There is also an extended version of this textbook. It is identical to this one, except it includes several chapters on modern physics.

The book is intended for students who are taking calculus concurrently with their physics courses. We assume that the students start calculus at the same time that they take their first physics course, and that they continue with calculus through the normal three-semester sequence taken by almost all science and engineering students. To understand the essential material of the first eleven chapters, students have to know how to differentiate simple polynomials, but they do not have to be familiar with integration. These chapters include Sections, Examples, and Drill Problems that use integration, but they can all be skipped without any loss of continuity. So that they can be easily identified, their titles are shaded light blue, as are the numbers of problems at the ends of these chapters that require the use of integration. After Chapter 11, integration is gradually added to the text, until by the time electric fields are covered, it is assumed that the students can integrate most simple functions.

This textbook emphasizes the skill of problem solving. Like most physicists, we feel that students do not really understand a particular theory until they can solve problems related to it. To help students develop these problem-solving skills, we have included step-by-step techniques that can be used to solve certain classes of problems. For example, we have described techniques consisting of five steps for applying Newton's second law, six steps for using the work-energy theorem, and five steps for applications of Gauss' law for electric fields. Immediately after these techniques are described, they are carefully illustrated in Examples, which are followed by Drill Problems in which students can practice applying what they have just learned. Throughout the textbook these techniques are presented as highlighted lists of steps that the students may easily follow.

When introducing a new principle, we usually present a simple special case before proceeding to the general theoretical development of the principle. For example, we discuss pushing a toy chest across a rough floor before we investigate the work-energy theorem. We believe that this approach helps students develop the physical intuition that makes both the mathematics and the philosophical implications of a general theory much easier to comprehend. In our opinion, the reverse approach—that is, the presentation of a general theory before any of its special cases—often leads to confusion and frustration for most students. After all, the brilliant scientists who discovered the many beautiful theories of physics usually did so by proceeding from special cases to general theories.

Optional Sections

The material is organized so that instructors can, with considerable flexibility, choose the particular subjects they think should be covered in their courses. To help instructors make these choices, we have designated with asterisks (*) those sections that can be omitted with effectively no loss of continuity. Examples of topics presented in asterisked sections are applications of integration in the first eleven chapters, rolling motion, elastic collisions, angular impulse, various applications of the magnetic force law and of electromagnetic induction, mutual inductance, and multiple-slit interference.

Previews and Summaries

At the beginning and end of the body of each chapter are, respectively, a Preview and a Summary. These two sections are closely matched. The Preview is a listing and brief description of the topics in the chapter the students are about to study. The Summary presents a succinct review of those same topics along with the main equations that are used to describe the topics.

Supplements

Some chapters include Supplements. In contrast to the Appendices, which contain useful reference material at the end of the textbook, the Supplements are sections attached to certain chapters. A Supplement generally contains a derivation or a proof that is so involved mathematically that it detracts from the continuity of the discussion. Of course, we do not feel that an important proof/derivation should be ignored simply because it is too complex; hence we have compromised by using Supplements. Hopefully, a student may now follow the physical arguments used in developing a particular theory without getting too distracted by the complicated mathematics of a derivation. After gaining some insight into the situation, the student can then consult the Supplement for the mathematical details. Supplements are also occasionally used to provide a parallel treatment of a particular topic. For example, the potential energy functions associated with the gravitational force and the spring force are obtained by integration in the body of Chapter 7. But a student not yet familiar with the integral may follow, without any loss of continuity, an alternate derivation of the potential energy functions given in Supplement 7.1. After having acquired the necessary mathematical tools at a later time, the student may then return and study the more advanced treatment. Another instance where we use a Supplement to provide a parallel treatment of a topic occurs in Chapter 35 ("Electromagnetic Waves and the Nature of Light"), the first of the optics chapters. The opening section is a discussion of the propagation of electromagnetic waves and draws heavily from Maxwell's equations. At institutions where optics is covered before Maxwell's equations, this arrangement would not be appropriate. Hence, we provide an alternate discussion in a Supplement. There, the properties of electromagnetic waves are described assuming the student has not yet studied electricity and magnetism. The student reading this Supplement arrives at the same results as one reading the opening section of Chapter 35, using equations that are numbered identically to those in the section.

Examples and Drill Problems

Each chapter contains many Examples that illustrate how the physical principles can be used to analyze systems—or, putting it more succinctly, they illustrate how to solve problems. The Examples are often grouped in sets of three or more, each illustrating a different aspect of the physical principle under consideration. Examples are often immediately followed by Drill Problems. Most of these allow the students to determine whether or not they can do a calculation similar to that presented in the preceding examples. Doing these Drill Problems will give students immediate feedback to help them determine their progress and understanding of the material just covered. Another advantage of the Drill Problems is their placement. In their reading, the students arrive naturally at these problems and do not have to search for particular problems at the appropriate level at the end of the chapter. Finally, the Drill Problems are also suitable for homework assignment.

Questions and Problems

In addition to the Examples and Drill Problems, there are many Questions (~900) and Problems (~2200) at the end of the chapters for the student to work on. These also test the student's understanding of the material presented in the chapter. The problems are grouped according to topic, which, in most cases, is equivalent to grouping according to the chapter section. The problems are ordered in each group according to our best estimate of their degree of difficulty. Placed at the end of each chapter's problem set is a group of General Problems. These problems are usually more difficult than the earlier problems of the chapter and often require the application of more than one idea to solve.

Organization

Although this textbook covers the standard material on mechanics, waves, thermodynamics, electricity and magnetism, and optics, the organization of the material is, in many ways, different from that of any other physics textbook now available. These changes are made not only because we feel they give a more logical flow to the presentation of topics, but also because they address a tendency that we have observed in our students to treat the many facets of physics as separate entities rather than a related, glorious whole. We think this is detrimental to understanding. If students can get a better overall view of general physics with its numerous interrelated parts and draw analogies, then their understanding will be enhanced. Among these organizational differences are the following:

- The use of vectors is integrated into the text from the very beginning. The student is reminded of the vector nature of kinematic quantities such as displacement, velocity, etc., even in the discussion of one-dimensional motion.
- Newton's law of gravitation is included in the presentation of Newton's laws of motion in Chapters 5 and 6. The instructor is then able to use this important law when explaining weight and mass along with the constancy of g. The concept of a field is also introduced at this point in connection with the gravitational force.
- Rotational dynamics is presented before momentum and angular momentum are introduced. This topic is treated as an extension of linear dynamics, with a continued emphasis on analyzing forces and drawing free-body diagrams. In many other textbooks, rotational dynamics and angular momentum are mixed together in a way that tends to make both subjects confusing

for most students. Also, with the separation of rotational dynamics and angular momentum, we are able to develop momentum and angular momentum in successive chapters with very similar arguments. This should help students understand the many similarities between these two physical quantities. Our approach is also beneficial to instructors using the quarter system, where it is often necessary due to lack of time to cover linear momentum in a very hurried way in the last week of the first quarter, depriving this important topic of the attention it deserves. The organization of this textbook brings out rotational dynamics more naturally as an extension of linear motion and provides an 11-chapter package that can be covered in the first quarter.

- In the chapters on electricity and magnetism, the early emphasis is on the physics of the electric and magnetic fields, culminating in Maxwell's equations. These equations are developed before capacitance, inductance, and *RC* and *RL* circuits are introduced. With this approach, we are able to demonstrate much better the similarities and differences between the electric field and the magnetic field. Our comparison continues with a chapter on capacitance and inductance, followed by one on *RC* and *RL* circuits. A further advantage of this approach is that the treatment of capacitance occurs later than in other textbooks. As a result, dc circuits are covered sooner, allowing the laboratory part of the course to start using circuit theory earlier in the term. This should be helpful for scheduling experiments, since so many of them involve electric circuits.

Ancillaries

An extensive teaching package supports the use of *University Physics*. The authors and the publisher have worked closely with your colleagues to develop these materials.

For the Instructor

Instructor's Solutions Manual You can find thoroughly worked-out solutions to every problem in the textbook in this manual. This thorough supplement was prepared by Robert Cole, University of Southern California, Thurman Kremser, Albright College, and Bo Lou, Ferris State University. Professor Lou also typeset this manual. Each solution was checked at least six times for accuracy—including a painstaking check by the authors of the main text.

Answer Booklet A complete set of just answers to every end-of-chapter problem is available on disk or in a handy 6×9 booklet to slip into your briefcase.

Instructor's Resource Manual with Test Item File This resource manual package includes selected transparency masters, suggested syllabi, and hints for handling common student difficulties. The Test Item File is a set of over 4,000 classroom-tested exam questions. These quantitative questions are grouped around a common principle, often around a single diagram. Test items are also available on disk, either with the test-generating software MicroTest for Macintosh or Windows, or in Microsoft Word for Macintosh or Windows.

Wm. C. Brown Publishers' Physics and Astronomy Electronic Image Bank This collection has nearly 2,000 pieces of line-art, photos, and diagrams—500 from this text alone—on a single CD-ROM. This Wm. C. Brown exclusive CD-ROM employs an accessible program that enables you to move quickly among the images and create your own multimedia presentation. Contact your Wm. C. Brown representative for details.

Transparency Set A set of 120 full-color transparencies is available to adopters of this text.

Physics Videodisc Library Over 600 lab demonstrations are available from the Wm. C. Brown Publishers' videodisc library. These professionally produced experiments are considered the best available lab demonstrations on the market. Please call your Wm. C. Brown sales representative for more information.

For the Student

Electronic Physics Solutions Guide This exclusive Wm. C. Brown Publishers' CD-ROM tutorial walks the student through twenty-five typical problems in mechanics and E&M. Step-by-step instruction, extensive voice-over, 3-D animation, and full-motion video help students visualize and internalize the problem-solving strategies they need to succeed in your course. Contact your Wm. C. Brown sales representative for details.

Student Solutions Manual Solutions to approximately twenty percent of the problems in the textbook can be found in this helpful manual. Each solution is completely worked out for the student.

Student Study Guide An excellent resource for the student, this extensive guide reinforces the concepts presented in the text and provides the student with sample at-home quizzes and problem-solving techniques.

Acknowledgments

At our own institution, there are several people that we are particularly indebted to: Mary-Margret Grady, who transcribed the initial, ancient draft of our project (on a typewriter!); Erianne Aichner, a wonderful secretary and supportive friend; Fr. Hanford Weckbach, S.J., who was always able to provide us with answers on how anything worked when we had none; and Joshua Crayton, our laboratory technician and, in his words, "facsimile engineer."

Our greatest thanks go naturally to our families, who have been our strength during this long project. To our parents, who instilled an interest in science in us; to our wives, Dolores and Virginia, and to our children—thank you for everything.

Faculty members and graduate students at several institutions were involved in the extensive accuracy checking done at every stage of the production process, including physicists at Albright College, Austin Community College, Drexel University, Ferris State University, Kirkwood Community College, Loyola Marymount University, Oklahoma State University,

University of Iowa, University of Southern California, University of Texas, and University of Utah. Thanks to all those individuals who contributed to the accuracy of the text and supplementary materials.

Roman Basko
Ivar Christopher
Robert Cole
Jamie Cooney
Ronald Cosby
George Dixon
Jay Happel
Richard Haracz
Timothy Johnson
Adam Johnston
Thurman Kremser
Philip Moore
Michael Ottinger
Jennifer Siders
Christopher Stigers
Joel Walter

We would also like to thank all those at Wm. C. Brown Publishers who brought this project to fruition, especially Vice-President and Publisher Jeff Hahn, Acquisitions Editor Megan Johnson, Developmental Editor Tom Riley, and Publishing Services Coordinator Julie Kennedy. Special thanks as well go to Jeff Lachina and his staff at Lachina Publishing Services for producing a marvelously clean text.

Reviewers

Our appreciation goes out to all those who offered their suggestions for the manuscript in its several iterations.

B.N. Narahari Achar
University of Memphis

Arun Bansil
Northeastern University

Robert Cole
University of Southern California

George Dixon
Oklahoma State University

Fred Domann
University of Wisconsin-Platteville

Harry D. Downing
Stephen F. Austin State University

F. Paul Esposito
University of Cincinnati

Edward F. Gibson
California State University, Sacramento

Richard D. Haracz
Drexel University

Marcus Price
University of New Mexico

Charles W. Scherr
University of Texas at Austin

Erwin Selleck
SUNY College of Technology at Canton

Javier Torner
California State University

The Learning System

CHAPTER-OPENING PREVIEWS AND OUTLINES

Music is the superposition of many harmonic waves.

CHAPTER 17 Superposition of Waves

Preview

In this chapter we investigate what happens when there is more than one wave in a particular region of space. The main topics to be discussed here include the following:

1. **Principle of superposition.** This principle describes how waves are added.
2. **Superposition of two harmonic waves of the same frequency.** The superposition principle is used to find the sum of two harmonic waves of the same frequency. Constructive and destructive interference are defined.
3. **Beats.** The superposition principle is applied to two harmonic waves of different frequency. The concept of beats is introduced.
4. **Boundary conditions.** When a wave encounters the boundary between two media, it is partially reflected and partially transmitted. How the reflected and transmitted waves are related to the incident wave is determined.
5. **Standing waves.** We see how two waves of the same frequency that are moving in opposite directions combine to form a standing wave. Properties of the standing wave are considered.
6. **Standing waves on a string.** We determine the conditions necessary for the production of a standing wave on a string fixed at both ends.
7. **Standing waves in a tube.** We determine the conditions necessary for the production of standing waves in open and closed tubes.

*8. **Music and musical instruments.** We see how standing waves on strings and in tubes are related to the production of musical notes.

307

Each chapter begins with a preview and outline. These allow students to tell at a glance how a chapter is organized and what major topics have been included in the chapter.

The Learning System

IN-CHAPTER EXAMPLES

terminates at a charge; hence there are $2N$ lines emanating from the two charges. At very large distances, these lines appear to be directed radially from the charge configuration. In other words, it appears as though the $2N$ field lines were coming from a single charge $+2q$. In Fig. 23-8*c* $3N$ lines emerge from the $+3q$ charge, while N lines terminate at the $-q$ charge. The remaining number, $2N$, emanate outward. Once again, at large distances from the charges, the field lines resemble those from a single $+2q$ charge. Hence we see that far away from a group of charges, the field lines (and thus the electric field) appear as though they are due to a single charge that is the sum of all the charges in the group. Try applying this result to the two equal and opposite charges of Fig. 23-8*a*. Can you picture what happens to the field lines (and the electric field) at progressively greater distances away from the charges?

DRILL PROBLEM 23-2

Do field lines ever cross? (*Hint:* Consider the definition of a field line.)

23-4 Calculating the Electric Field

The following examples illustrate how the electric fields of various charge distributions are determined. In calculating **E**, you should remember that (1) the electric field of a positive charge is directed away from that charge, while the electric field of a negative charge is directed toward the charge, and (2) electric fields are vectors and must be added accordingly. For a group of discrete charges, the net field is simply the vector sum of the fields of the individual charges (Eq. 23-3). For a continuous charge distribution, you must perform an integration (Eqs. 23-4) to find the net field. The continuous charge distributions we consider generally have some spatial symmetry, and as a result, the integral is simplified. Such continuous distributions are analyzed in Examples 23-5 through 23-9.

EXAMPLE 23-3 The Electric Field of Two Identical Point Charges

Two positive point charges q are separated by a distance $2a$. (See Fig. 23-9.) Find the electric field due to the two charges at P and at P′.

SOLUTION (*a*) At P, the electric field of the left charge points in the positive x direction, the field of the right charge points in the negative x direction, and the magnitude of each field equals $q/4\pi\epsilon_0 a^2$. The resultant electric field at P is therefore zero.

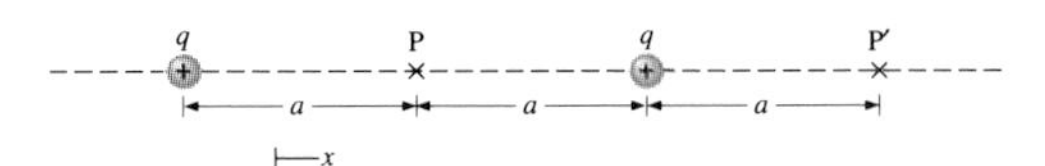

FIGURE 23-9 Two point charges.

(*b*) At P′, the electric fields due to the left and right charges are $[q/4\pi\epsilon_0(3a)^2]\mathbf{i}$ and $(q/4\pi\epsilon_0 a^2)\mathbf{i}$, respectively. The resultant electric field at P′ is then

$$\mathbf{E} = \frac{1}{4\pi\epsilon_0}\left[\frac{q}{(3a)^2}\mathbf{i} + \frac{q}{a^2}\mathbf{i}\right] = \frac{5}{18\pi\epsilon_0}\frac{q}{a^2}\mathbf{i}.$$

DRILL PROBLEM 23-3

Suppose the right particle of Fig. 23-9 is replaced by a particle of charge $-q$. Now what are the electric fields at P and at P′?
ANS. $(q/2\pi\epsilon_0 a^2)\mathbf{i}$; $(-2q/9\pi\epsilon_0 a^2)\mathbf{i}$.

EXAMPLE 23-4 The Electric Field of Two Point Charges

(*a*) Point charges $q_1 = 6.0 \times 10^{-9}$ C and $q_2 = -4.0 \times 10^{-9}$ C are placed 10.0 cm apart, as shown in Fig. 23-10*a*. Calculate the electric field due to these two charges at P and at P′. Point P is between the charges, and P′ and the two charges are at the corners of an equilateral triangle. (*b*) A third point charge $q_3 = 3.0 \times 10^{-9}$ C is placed on the positive y axis 4.0 cm above q_1. (See Fig. 23-10*b*.) Now what is the electric field at P′?

SOLUTION (*a*) At P, the field due to q_1 points in the positive x direction and has magnitude

$$E_1 = (9.0 \times 10^9\ \text{N}\cdot\text{m}^2/\text{C}^2)\frac{6.0 \times 10^{-9}\ \text{C}}{(0.040\ \text{m})^2} = 3.4 \times 10^4\ \text{N/C}.$$

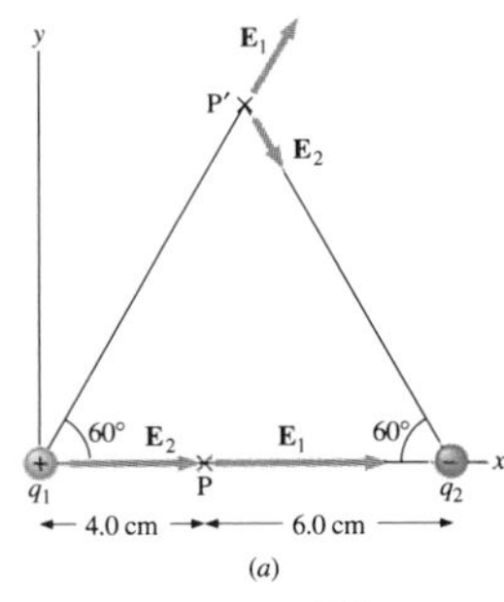

(*a*)

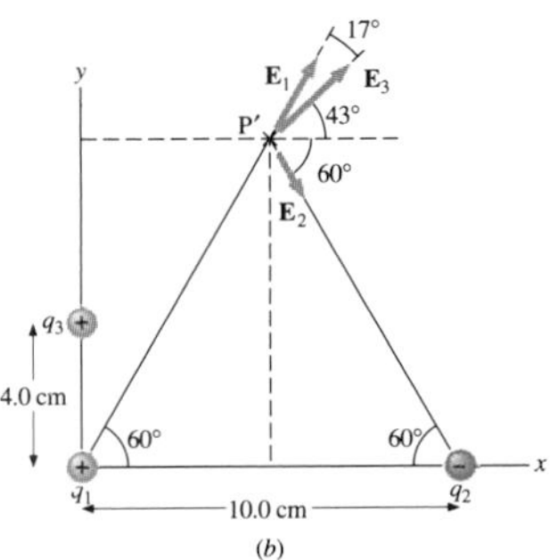

(*b*)

FIGURE 23-10 Simple charge distributions.

In-chapter examples help the student walk through typical physics problems, showing them in detail how a solution is derived.

THE LEARNING SYSTEM

DRILL PROBLEMS

EXAMPLE 18-3 **STANDARD TEMPERATURE AND PRESSURE**

(*a*) What volume does 1.00 mol of an ideal gas occupy at a pressure of 1 atm and a temperature of 0°C (called *standard temperature and pressure*, or *STP*)? (*b*) If this ideal gas is air, what is its density?

SOLUTION (*a*) With $T = 273$ K, we have from the ideal-gas law,

$$V = \frac{nRT}{p} = \frac{(1.00 \text{ mol})(8.31 \text{ J/mol}\cdot\text{K})(273 \text{ K})}{1.01 \times 10^5 \text{ N/m}^2}$$
$$= 22.4 \times 10^{-3} \text{ m}^3 = 22.4 \text{ L}.$$

This is the approximate volume occupied by 1 mol of *any dilute gas* at STP.

(*b*) The average molecular mass of air is 29.0 g/mol, so the mass of 1.00 mol of air is $M = 29.0$ g $= 0.0290$ kg. The density of the air at STP is then

$$\rho = \frac{m}{V} = \frac{0.0290 \text{ kg}}{22.4 \times 10^{-3} \text{ m}^3} = 1.29 \text{ kg/m}^3.$$

EXAMPLE 18-4 **OXYGEN AS AN IDEAL GAS**

A dilute O_2 gas is confined to a rigid container of volume 1.50 m^3 at a pressure of 0.400 atm and a temperature of 40°C. (*a*) If the temperature is raised to 100°C, what is the pressure of the gas? (*b*) What is the mass of the gas?

SOLUTION (*a*) Since both the number of moles and the volume of the gas are constant, we have from the ideal-gas law

$$\frac{p_1}{T_1} = \frac{p_2}{T_2} = \text{constant},$$

so

$$p_2 = p_1 \frac{T_2}{T_1}.$$

Here the initial pressure and temperature are $p_1 = 0.400$ atm and $T_1 = (40 + 273)$ K $= 313$ K, and the final temperature is $T_2 = (100 + 273)$ K $= 373$ K; thus the final pressure is

$$p_2 = p_1 \frac{T_2}{T_1} = (0.400 \text{ atm})\left(\frac{373 \text{ K}}{313 \text{ K}}\right) = 0.477 \text{ atm}.$$

(*b*) The number of moles of the gas can be calculated from the ideal-gas law. With $p_1 = 0.400$ atm $= 4.04 \times 10^4$ N/m^2, we have

$$n = \frac{p_1 V_1}{RT_1} = \frac{(4.04 \times 10^4 \text{ N/m}^2)(1.50 \text{ m}^3)}{(8.31 \text{ J/mol}\cdot\text{K})(313 \text{ K})} = 23.3 \text{ mol}.$$

The molecular mass of oxygen is 0.0320 kg/mol, so the mass of the gas in the container is

$$(0.0320 \text{ kg/mol})(23.3 \text{ mol}) = 0.746 \text{ kg}.$$

DRILL PROBLEM 18-2

If dilute nitrogen gas at 1.00 atm pressure is at the same temperature as boiling water at 1.00 atm pressure, what is the density of the gas?
ANS. 0.912 kg/m^3.

DRILL PROBLEM 18-3

An ideal gas is confined to a volume of 4.0 L with a pressure of 5.0 atm. If the gas expands to a volume of 10.0 L without any change in its temperature, what is its new pressure?
ANS. 2.0 atm.

DRILL PROBLEM 18-4

The stopcock on a container is opened and a portion of the ideal gas it contains is released. When the remaining gas returns to its original temperature, its pressure is found to have dropped by a factor of 3. What fraction of the gas is released?
ANS. 2/3.

18-4 A STATISTICAL INTERPRETATION OF AN IDEAL GAS

The properties of an ideal gas can be understood in terms of a simple statistical model. With this model, we can determine averages of certain mechanical properties of the molecules and then relate these averages to measurable macroscopic properties of the gas. The model is based on the following assumptions:

1. The gas is in thermal equilibrium with its environment.
2. The molecules of the gas move randomly and obey Newton's laws. The gas can be pictured as a very large number of point masses that collide elastically as they move randomly throughout the container. (See Fig. 18-4.)

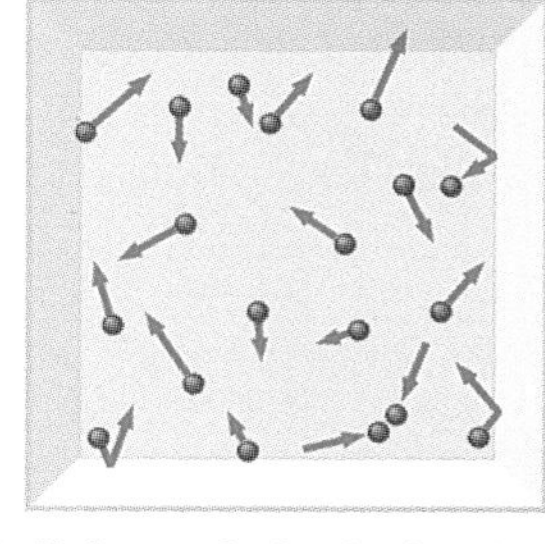

FIGURE 18-4 An ideal gas can be thought of as a large number of point masses moving randomly throughout the container and colliding elastically with each other and the walls.

3. The total number of molecules is large enough that averages of molecular dynamic variables such as speed are meaningful physically.
4. The total volume of the molecules is negligible compared with the volume that the gas occupies.
5. Intermolecular collisions and collisions between molecules and the walls of the container are elastic and occur over time

Drill problems follow examples to give the student *an immediate opportunity* to work a problem on their own. Brief answers are given to these problems.

The Learning System

OPTIONAL SECTIONS

*37-3 The Eye

A sketch of the eye is shown in Fig. 37-14. Light enters the eye through an opening called the pupil, whose size is controlled by the iris. Through refractions at the cornea (the front surface membrane) and at the crystalline lens, this light forms an image on the retina. The image is detected by optical receptors, known as rods and cones, at the retina, and information is transmitted to the brain by the optic nerve. The rods are more sensitive than the cones, and they are responsible for sight in very dim light. However, the rods do not distinguish color, and the images they send to the brain are not very sharp. The cones, discussed earlier in Sec. 35-2, are responsible for color vision, and they produce the sharp images seen in bright light. The region between the cornea and the lens is filled with a liquid called the aqueous humor, and a thin jelly called the vitreous humor fills the volume between the lens and the retina. The indices of refraction of the humors are very close to that of water, while the crystalline lens has a refractive index of about 1.44. Because the refractive indices of the humors and the lens are almost identical, most of the refraction of the light entering the eye occurs at the cornea, whose index of refraction is about 1.38. Fine adjustments in focusing are made by the lens, as its shape (and hence its focal length) is changed by ciliary muscles located around the edge of the lens. This process is known as *accommodation*.

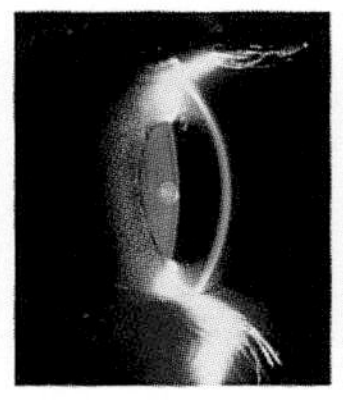

(*a*) A side view of the human eye. (*b*) Rods and cones magnified by a scanning electron microscope.

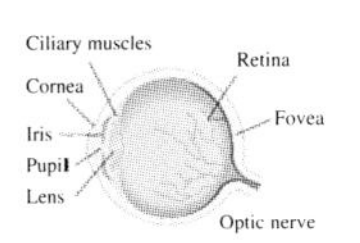

FIGURE 37-14 The human eye.

The extremes of the range of the object distance for which a clear image is formed on the retina are called the *far point* and the *near point*. The far point of the normal eye (corresponding to fully relaxed ciliary muscles) is at infinity. The near point depends on the eye's ability to accommodate. We generally use 25 cm, a normal reading distance, as the near point of the normal eye in calculations. Since the lens loses flexibility with age, our near points increase as we get older. A typical near point of the healthy eye of a 20-year-old is around 25 cm. This increases to about 50 cm when a person reaches 40 years old. The gradual recession of the near point is called *presbyopia*.

The most common defects in vision are caused by distortions in the shape of the eyeball. A normal eye focuses a distant object by relaxing the ciliary muscles. Now if the eyeball happens to be too long from front to back, the image is produced in front of the retina. This condition is known as *myopia* (or more colloquially, nearsightedness). The far point of a myopic eye is not at infinity, and in fact, is often quite close to the eye. If the eyeball is too short, a sharp image is formed behind the retina. This condition is called *hyperopia* (farsightedness). In this case, the image can still be focused on the retina through accommodation. However, the limited range of accommodation prevents the eye from focusing objects that are at the normal near point. The near point of a hyperopic eye is therefore more distant than that of a normal eye. Finally, when the cornea is not spherical, line objects in one direction are focused in a different plane than line objects in another direction. This condition is known as *astigmatism*.

Many vision defects are remedied with corrective lenses such as eyeglasses or contact lenses (Figs. 37-15 and 37-16). For myopia, a diverging lens is used to form the image of a distant object at the eye's far point. For example, if the far point of an eye is 300 cm, the corrective lens must focus objects at infinity

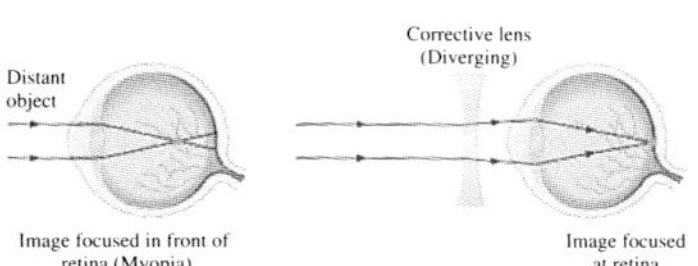

FIGURE 37-15 Myopia and its correction with a diverging lens.

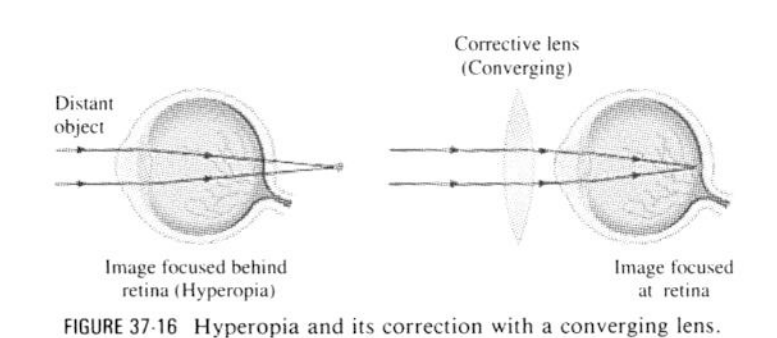

FIGURE 37-16 Hyperopia and its correction with a converging lens.

ither a presbyopic or a hy-
:han the accepted distance
/ith a converging lens that
nt from an object located
ding distance). To correct
directions must be focused
with corrective lenses that

G Vision Defects

)0 cm. What is the focal length
·ect this defect? (*b*) The near
Vhat is the focal length of the
this defect?

; must produce an image at
, so from the lens equation,

:ngth of

−60 cm of an object at 25 cm.

ing lens is

If a hyperopic eye has a near point of 100 cm, what is the focal length of the corrective lens needed so that objects at 25 cm can be viewed comfortably?
ANS. +33 cm.

The closer an object is to us, the larger it appears to be. Its apparent size is determined by the size of its image on the retina, or equivalently, by the number of rods and cones stimulated by the light. As shown in Fig. 37-17, the size of the image on the retina depends on the angle subtended at the eye by the object. In fact, since the image distance (the distance from the lens to the retina) remains constant, *the size of the image is directly proportional to the angle that the object subtends at the eye*. The largest focused image of a given object formed on the retina by the unaided eye is therefore produced when the subtended angle is largest, that is, when the object is at the near point. With 25 cm as the distance of the near point from the eye, the maximum angle an object of height y (in centimeters) can subtend at the eye is

$$\theta_N \approx \frac{y}{25\ \text{cm}}.$$

To increase the size of the image on the retina, we use an optical device such as a microscope or a telescope to view the object. The image produced by such a device must be far enough from the eye (>25 cm) in order to be seen clearly. If this image subtends an angle θ at the eye, then by definition, the **angular magnification** M of the optical device is

$$M = \frac{\theta}{\theta_N}. \tag{37-7}$$

The angular magnification of an optical device depends on its geometry and the properties of its components. In the following sections, we consider three such devices.

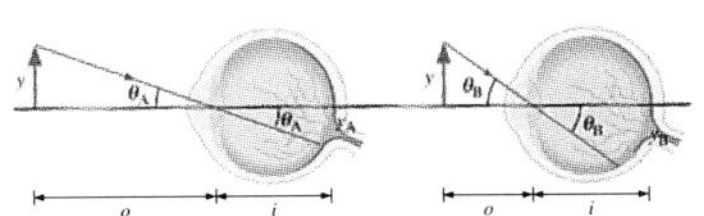

FIGURE 37-17 The apparent size of an object is determined by the size of its image on the retina. This is proportional to the angle that the object subtends at the eye.

*37-4 The Simple Magnifier

The simple magnifier, which is a single converging lens, is a basic optical device used to magnify an image on the retina. When an object is placed at the focal point of the lens of Fig. 37-18*a*, the eye perceives the erect, virtual image formed by the lens. Since this image is at infinity, it can be easily focused

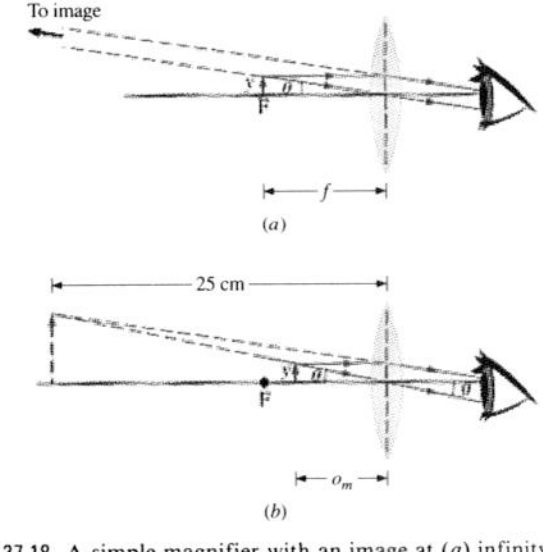

FIGURE 37-18 A simple magnifier with an image at (*a*) infinity and (*b*) the near point (see Example 37-5).

Optional sections in selected chapters elaborate on topics of special interest in that chapter and allow for abbreviated coverage where necessary.

The Learning System

END-OF-CHAPTER SUMMARIES

SUMMARY

1. **Types of waves**
   (*a*) Mechanical waves travel through a medium by disturbing the particles of the medium. An electromagnetic wave can travel through a vacuum. The waves associated with elementary particles are called matter waves.
   (*b*) A transverse wave displaces the medium perpendicular to its direction of propagation. A longitudinal wave displaces the medium parallel to its direction of propagation.
2. **Traveling waves**
   A one-dimensional traveling wave can be represented by
   $$y(x, t) = F(x \pm vt),$$
   where v is the speed of the wave.
3. **Wave speed**
   (*a*) The speed of a wave on a string is
   $$v = \sqrt{\frac{F}{\mu}},$$
   where F is the tension in the string and μ is the linear mass density of the string.
   (*b*) The speed of a sound wave in a bulk medium is
   $$v = \sqrt{\frac{B}{\rho}},$$
   where B is the bulk modulus of the medium through which the sound wave is moving and ρ is the density of the medium.
4. **Harmonic waves**
   (*a*) The harmonic wave can be represented by
   $$y(x, t) = A \sin(kx \pm \omega t + \phi),$$
   where A is the amplitude, k the wave number, ω the angular frequency, and ϕ the phase constant of the wave.
   (*b*) The wave number and angular frequency are related to the wavelength λ and frequency f by
   $$k = \frac{2\pi}{\lambda} \quad \text{and} \quad \omega = 2\pi f.$$
   (*c*) The speed of the harmonic wave is
   $$v = f\lambda = \frac{\omega}{k}.$$
5. **Energy transport in harmonic waves**
   The average power transmitted by a mechanical harmonic wave is proportional to the product of the square of the amplitude and the square of the frequency.
6. **Circular and spherical waves**
   (*a*) Two-dimensional waves emanating from point sources are represented by circular wavefronts spaced a distance λ apart. Three-dimensional waves from a point source are represented by spherical wavefronts spaced a distance λ apart.
   (*b*) The intensity of a spherical wave is the average power transmitted by the wave across a unit area normal to the direction of propagation of the wave. The intensity of a spherical wave decreases with the inverse square of the distance from the source.
7. **Plane-wave approximation**
   Far from a source of spherical waves, small sections on a spherical wavefront can be approximated by planes.

*8. **Sound intensity in decibels**
   If the intensity of sound in watts per square meter is I, then the intensity level β (in decibels) is given by
   $$\beta = 10 \log\left(\frac{I}{I_0}\right),$$
   where the base of the logarithm is 10 and $I_0 = 10^{-12}$ W/m^2.

*9. **Doppler effect**
   (*a*) When an observer (O) is moving toward or away from a stationary source (S) of waves of frequency f_S, the frequency detected by the observer is
   $$f_O = \left(1 \pm \frac{v_O}{v}\right) f_S,$$
   where the positive sign corresponds to the observer approaching and the negative sign to the observer receding from the source.
   (*b*) If the observer is stationary and the source is moving,
   $$f_O = \left(\frac{1}{1 \mp v_S/v}\right) f_S,$$
   where the positive sign corresponds to the source moving away from the observer and the negative sign to the source approaching the observer.

At the end of each chapter a summary is given to help underscore the key topics in each chapter and to help facilitate a student's review of that chapter.

The Learning System

CHAPTER SUPPLEMENTS

Supplement 16-1 Relationship Between the Harmonic Displacement Wave and the Harmonic Pressure Wave for Sound

In this supplement we will derive the relationship between the harmonic displacement wave and the harmonic pressure wave for sound. Figure 16-25 shows a harmonic displacement wave moving through air contained in a long tube of cross-sectional area $\mathcal{A}$. The mathematical representation of this wave is given by Eq. (16-13*a*). At a time t, the displacement of the air at $x + \Delta x$ is (assuming $\phi = 0$)

$$s(x + \Delta x, t) = s_0 \sin [k(x + \Delta x) - \omega t],$$

while its displacement at x is

$$s(x, t) = s_0 \sin (kx - \omega t).$$

Before the wave moved down the tube, the volume of the air between x and $x + \Delta x$ was $V = \mathcal{A}\,\Delta x$. At the time t, the displacement wave causes the volume of this same air to change by an amount

$$\Delta V = \mathcal{A}[s(x + \Delta x, t) - s(x, t)]. \qquad (i)$$

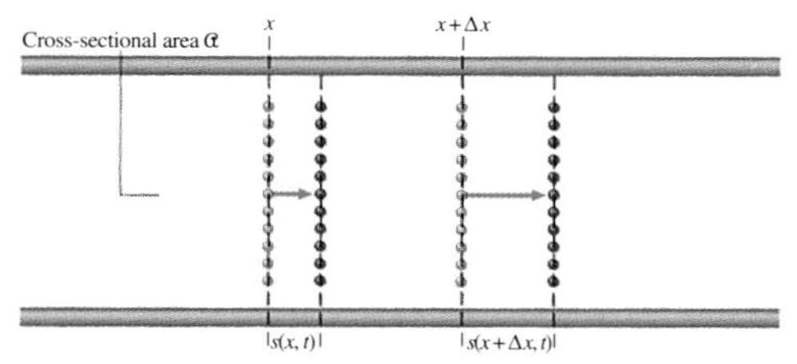

FIGURE 16-25 The change in volume of a given section of air as a harmonic displacement wave passes through the air. At a time t, the wave has displaced the molecules at x an amount $s(x, t)$ and the molecules at $x + \Delta x$ an amount $s(x + \Delta x, t)$.

Using the trigonometric identity $\sin (\alpha + \beta) = \sin \alpha \cos \beta + \cos \alpha \sin \beta$, we can write

$$s(x + \Delta x, t) = s_0 [\sin (kx - \omega t) \cos (k\,\Delta x) + \cos (kx - \omega t) \sin (k\,\Delta x)].$$

But Δx, and therefore $k\,\Delta x$, is small, so $\cos (k\,\Delta x) \simeq 1$ and $\sin (k\,\Delta x) \simeq k\,\Delta x$; thus

$$s(x + \Delta x, t) = s_0 [\sin (kx - \omega t) + (k\,\Delta x) \cos (kx - \omega t)].$$

Now from Eq. (*i*),

$$\Delta V = \mathcal{A} s_0 [\sin (kx - \omega t) + (k\,\Delta x) \cos (kx - \omega t) - \sin (kx - \omega t)],$$

which simplifies to

$$\Delta V = s_0 k \mathcal{A}\,\Delta x \cos (kx - \omega t). \qquad (ii)$$

From the definition of the bulk modulus (Eq. 11-9),

$$\Delta V = -\frac{\Delta p V}{B} = -\frac{\Delta p \mathcal{A}\,\Delta x}{B}.$$

Combining this with Eq. (*ii*) gives

$$s_0 k \mathcal{A}\,\Delta x \cos (kx - \omega t) = -\frac{\Delta p \mathcal{A}\,\Delta x}{B},$$

which when solved for Δp leaves

$$\Delta p = -s_0 k B \cos (kx - \omega t).$$

Finally, using $v = \sqrt{B/\rho}$, we have the expression for the pressure wave:

$$\Delta p = -\rho k v^2 s_0 \cos (kx - \omega t).$$

Notice that the amplitude of this wave is $\Delta p_0 = \rho k v^2 s_0$, which is in agreement with Eq. (16-14).

Questions

16-1. Discuss the differences and similarities between a wave on a string and a sound wave propagating in a fluid in a long tube.

16-2. If $F(x - vt)$ describes a pulse traveling in the positive x direction, what does $-F(x - vt)$ describe?

16-3. Does the speed of a pulse traveling along a wire depend on the wire's diameter?

16-4. What happens to the speed of a wave pulse on a string if we (*a*) increase the length of the string? (*b*) decrease the tension? (*c*) wrap tape around the string?

16-5. What happens to the speed of a wave pulse in a long tube of air if we (*a*) pump more air into the tube? (*b*) replace some of the air with helium? (*c*) increase the diameter of the tube while keeping the air density constant?

16-6. Would air in a narrow tube better approximate a one-dimensional medium for high- or low-frequency sound waves? Explain.

16-7. A popular lecture demonstration is performed by placing a ringing bell inside a jar attached to a vacuum pump. When the air is pumped from the jar, the bell can no longer be heard. Why? If the electrical power operating the bell is turned off, will the bell keep vibrating longer when there is air in the jar or when the air is evacuated?

16-8. The densities of most solids are more than 1000 times that of air; yet the speed of sound in a solid is usually greater than its speed in air. What does this tell you about the bulk moduli of solids as compared with that of air?

16-9. Is it possible to have a transverse or longitudinal wave if the vibrational motion of its source is not simple harmonic?

16-10. A string is attached to a vibrator of constant frequency. How would the wavelength of the harmonic wave produced be affected if the string were twice as thick?

16-11. Suppose that the vibrator of the previous question is set to a new frequency while oscillating at the same amplitude. Which of the following properties of the harmonic wave it produces would be affected? (*a*) frequency; (*b*) wavelength; (*c*) speed; (*d*) wave number; (*e*) amplitude; (*f*) period.

Chapter supplements offer a more rigorous discussion of additional topics that may not be part of the typical syllabus, allowing flexible and in-depth presentation of the course material.

The Learning System

END-OF-CHAPTER QUESTIONS

196 PART I Mechanics

11-8. What force **F** is needed to lift the 100-N weights upward at constant velocity for each of the pulley systems shown in the accompanying figure? Ignore the weights of the pulleys and ropes.

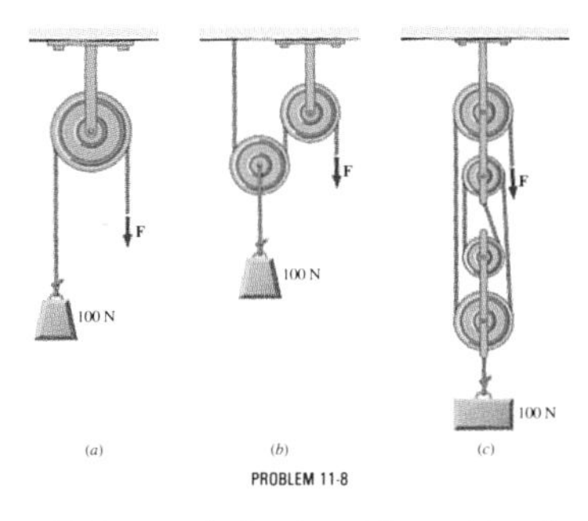

PROBLEM 11-8

11-9. The accompanying figure shows a pulley system being used to lift a 2000-kg safe up the incline. What minimum force must the men apply to move the safe up the incline? Assume that friction on the incline is negligible.

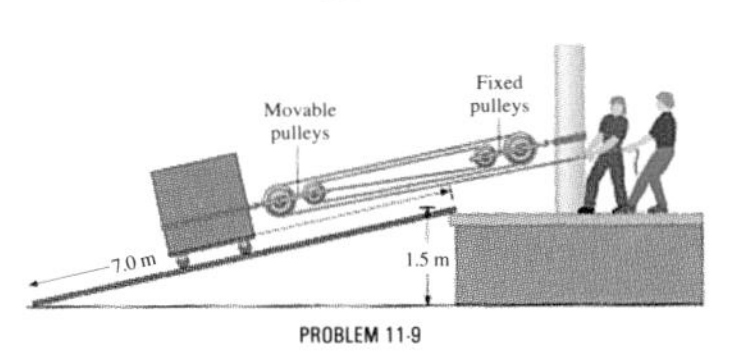

PROBLEM 11-9

11-10. A device used to lift large loads is shown in the accompanying figure. What force **F** is needed to lift the 600-kg load? Ignore the weights of the pulleys and rope.

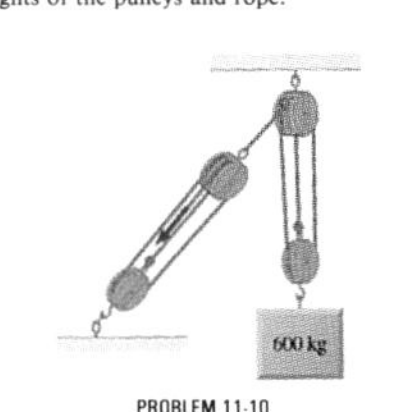

PROBLEM 11-10

11-11. What force **F** is needed to lift the 100-N weight shown in the accompanying figure at constant velocity? Ignore the weights of the pulleys.

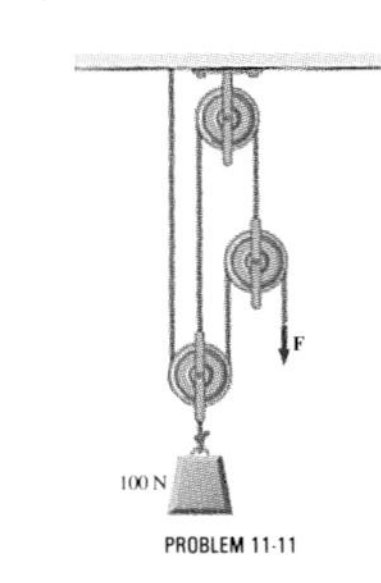

PROBLEM 11-11

11-12. What force must be applied at P to keep the structure shown in the accompanying figure in equilibrium? The weight of the structure is negligible.

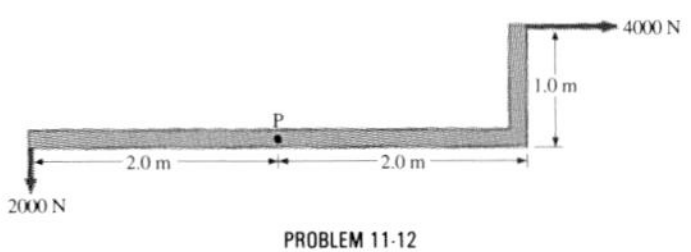

PROBLEM 11-12

11-13. Is it possible to apply a force at R to keep the structure shown in the accompanying figure in equilibrium? The weight of the structure is negligible.

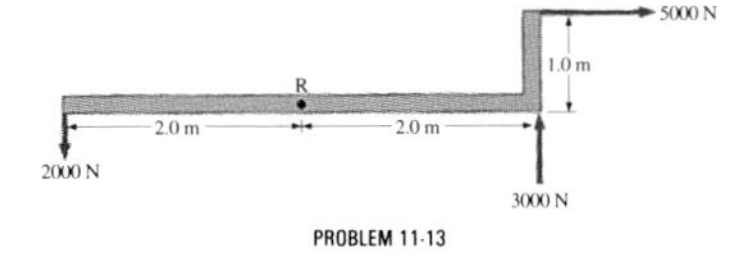

PROBLEM 11-13

11-14. The structure shown in the accompanying figure is supported at P. If it is in equilibrium, what is the magnitude of **F** and what

724 PART V Electromagnetic Waves and Optics

38-56. Find the radius of a star's image on the retina of an eye if its pupil is open to 0.65 cm and the distance from the pupil to the retina is 2.8 cm. Assume that $\lambda = 550$ nm.

38-57. *Radio telescopes* are telescopes used for the detection of radio emission from space. Because radio waves have much longer wavelengths than visible light, the diameter of a radio telescope must be very large to provide good resolution. For example, the radio telescope in Arecibo, Puerto Rico (see the accompanying figure), has a diameter of 300 m and can be tuned to wavelengths as low as 4.0 cm. At this wavelength, what is the minimum angular separation of two stars that can be resolved by the telescope?

PROBLEM 38-57

38-58. A spy satellite orbits the earth at a height of 180 km. What is the minimum diameter of the objective lens in a telescope that must be used to resolve columns of troops marching 2.0 m apart? Assume that $\lambda = 550$ nm.

Diffraction Grating

38-59. How many lines per centimeter must be ruled on a 5.0-cm-wide grating if it is just able to resolve the two wavelengths 415.724 and 415.744 nm in first order?

38-60. Yellow sodium light consists of two closely spaced wavelengths, $\lambda_1 = 589.00$ nm and $\lambda_2 = 589.59$ nm. If this light falls normally on a diffraction grating with 5000 rulings per centimeter, how wide must the grating be to just resolve the two fringes in second order?

38-61. A grating with 5000 rulings per centimeter is used to produce two sets of fringes with $\lambda_1 = 570.720$ nm and $\lambda_2 = 570.730$ nm. (*a*) What is the highest order in the diffraction pattern of these fringes? (*b*) For this order, how wide must the grating be so that the two fringes are just resolved?

38-62. A diffraction grating produces a second maximum that is 8.97 cm from the central maximum on a screen 2.0 m away. If the grating has 600 rulings per centimeter, what is the wavelength of the light that produces the diffraction pattern?

38-63. A grating with 4000 lines per centimeter is used to diffract light that contains all wavelengths between 400 and 650 nm. How wide is the first-order spectrum on a screen 3.0 m from the grating?

38-64. A diffraction grating with 2 to measure the wavelengths e tube. (*a*) At what angles wil first-order blue fringes of wa maxima of two other first-c found at $\theta_1 = 0.0972$ rad a wavelengths of these fringes

38-65. For white light (400 nm < λ diffraction grating, show tha tra overlap no matter what t

38-66. How many complete orders λ < 700 nm) can be produced tains 5000 rulings per centim

General Problems

38-67. In Young's double-slit experi fringe is 1.80 cm from the ce positions does the intensity c its maximum value?

38-68. White light falls on two nar The interference pattern is o (*a*) What is the separation bet ($\lambda = 700$ nm) and violet light nearest the central maximum ($\lambda = 600$ nm) coincide with a tify the order for each maxi

38-69. (*a*) Show that in the *N*-slit in maxima are located approxim $\pm 5\pi/2N, \ldots$ (*b*) Show that tensity of a first subsidiary m that of a principal maximum a second subsidiary maxim maximum.

38-70. Different filters are placed i thereby producing monochron The light falls normally on a with a thin film of water. T reach a minimum at $\lambda = 400$ at $\lambda = 700$ nm. What is the t

38-71. A thin layer of transparent plastic ($n = 1.60$) is deposited on a glass ($n = 1.50$) surface. When illuminated from above by light ($\lambda = 589$ nm) from a sodium lamp, the surface appears dark. What are the two smallest values possible for the thickness of the plastic layer?

38-72. The sun is 1.5×10^{11} m from the earth and has a mean radius of 7.0×10^8 m. Does the sun qualify as a good point source for a pair of slits separated by 0.10 mm?

38-73. Microwaves of wavelength 10 mm fall normally on a metal plate that contains a slit 25 mm wide. (*a*) Where are the first minima of the diffraction pattern? (*b*) Would there be minima if the wavelength were 30 mm?

38-74. (*a*) By differentiating Eq. (38-17), show that the higher-order maxima of the single-slit diffraction pattern occur at values of β that satisfy $\tan\beta = \beta$. (*b*) Plot $y = \tan\beta$ and $y = \beta$ versus β and find the intersections of these two curves. What information do they give you about the locations of the maxima? (*c*) Convince yourself that these points do not occur exactly at $\beta = (n + 1/2)\pi$, where $n = 0, 1, 2, \ldots$, but are quite close to these values.

A large number of problems are offered at the end of each chapter. Beginning with questions that are separated by section and graded in difficulty, the problems sets end with general questions that require the student to combine concepts and apply knowledge from different sections of the book to solve the problem.

COLOR KEY

Vectors

- General vector
- Resultant vector
- Components
- Position (**r**)
- Displacement(Δ**r**)
- Velocity (**v**)
- Acceleration (**a**)
- Force (**F**)
- Electric field (**E**)
- Magnetic field (**B**)
- Other vectors

Electricity and Magnetism

- Electric current
- Electric field lines
- Equipotential lines
- Magnetic field lines
- Electric charge (+)
- Electric charge (–)

Electric Circuit Symbols

- Wire
- Resistor
- Capacitor
- Inductor
- Battery
- Ground

Optics

- Light rays
- Object
- Real image
- Virtual image

Other

- Path of a moving object
- Direction of motion

UNIVERSITY PHYSICS

PART I

MECHANICS

Photo courtesy of Ringling Bros. and Barnum & Bailey Combined Shows, Inc.

Stonehenge, thought to be an ancient observatory, was constructed nearly 4,000 years ago on Salisbury Plain in southern England.

CHAPTER 1 INTRODUCTION

PREVIEW

In this chapter the basic units used in the measurement of physical properties are introduced. Nearly all the quantities you will encounter in your study of physics are expressed in these units.

1. **Standards for length, mass, and time**. The International System (SI) of units is presented. The standards of length (meter), mass (kilogram), and time (second) are discussed.
2. **Dimensional analysis and the conversion of units**. Dimensional analysis is a useful method for determining if an equation is incorrect. Conversions among the basic sets of units are illustrated.
3. **Significant figures**. The concept of significant figures, which represent how accurately a quantity is known, is discussed.

Physics is the study of the laws of nature. With an understanding of these laws, the physicist is able to investigate almost any phenomenon in the natural world. Many physicists engage in research on the properties of and interactions among elementary particles, atoms, and molecules in order to acquire a basic knowledge of the behavior of matter. Yet many others are active in such diverse areas as electronics, biology, acoustics, astrophysics, optics, and medical physics. The areas of research open to the physicist are almost endless.

Most engineering advances are based on applications of the laws of physics. Consequently, the engineer must also have a clear understanding of physics. In fact, the engineer's work frequently cannot be distinguished from that of the physicist. The two professions are so intertwined that physicists and engineers often work side by side, using the same experimental and theoretical methods, reading the same books and journals, in order to solve the same problems.

The work of physicists is generally categorized as either experimental or theoretical. The *experimental physicist* accumulates quantitative data by doing controlled experiments that usually focus on one well-defined aspect of nature—for example, how acceleration is related to force. The *theoretical physicist*, on the other hand, attempts to determine the behavior of physical systems by doing calculations whose end products are numerical predictions. Theory and experiment are closely intertwined. Theoretical physicists often use the information furnished by experiments to develop a theory that explains the behavior of all physical phenomena of a certain type. For example, Isaac Newton originally devised his theories of motion and gravitation to explain the data concerning the orbits of planets that had been accumulated by Tycho Brahe and analyzed by Johannes Kepler. As it turned out, this theory explains all motion in the macroscopic world. But theory can also precede experiment. A well-known example is Albert Einstein's prediction in 1915 that light should be deflected by a gravitational field. This prediction was confirmed by astronomers in 1919 when they observed the deflection of sunlight during a total solar eclipse.

The successful interplay of experiment and theory eventually leads to the development of a physical theory expressed in terms of mathematical laws. This theory can then be used to make predictions about an entire class of phenomena. For example, Newton's three laws of mechanics describe the motion of all bodies of macroscopic size moving at ordinary speeds (much less than the speed of light). These laws can be used to calculate something as simple as the path of a batted baseball or as complicated as the motion of a space probe through the solar system or the stresses in the beams of a large bridge.

However, all known theories do have boundaries outside of which they are not applicable. For example, Newton's laws cannot be used to describe the motion of atomic and subatomic particles; the behavior of such particles is determined by the laws of quantum mechanics. Someday there might be a "theory of theories" that describes all phenomena. But at this point in history, there are different (though not completely unrelated) theories, each of which describes a certain class of phenomena.

Many of these important basic theories are considered in this textbook. You will be studying the laws of motion, fluid mechanics, wave motion, thermodynamics, electricity and magnetism, optics, and Einstein's theory of special relativity. A short introduction to quantum mechanics is also presented. Just as important as an understanding of these theories is the ability to use or apply them. This skill is given special emphasis in this textbook. There are hundreds of examples and thousands of problems, most of which are a direct application of some basic law of physics.

1-1 Standards of Measurement

To be truly useful, the language of measurement must be universal. One second must represent exactly the same time interval in Japan as it does in Brazil. To ensure uniform interpretation of measurements, the application of standards is essential.

Since 1889 the General Conference on Weights and Measures, composed of representatives from all major countries, has defined the standards for the basic units. The metric system of units established by the conference is known as the *International System*, abbreviated *SI* because of its French translation, *Système International*. The units of length, time, and mass in this system are the *meter* (*m*), *second* (*s*), and *kilogram* (*kg*), respectively. Other defined SI units are those for temperature (the *kelvin*, *K*), electric current (the *ampere*, *A*), amount of a substance (the *mole*, *mol*), and luminous intensity (the *candela*, *cd*). The units of all other physical quantities are a combination of these basic seven and are known as *derived units*. For example, the mass density, or mass per unit volume, of a body is expressed in kilograms per meter cubed (kg/m^3), and velocity is given in meters per second (m/s).

Standard for Length

Prior to 1960, the length standard was the distance between two lines on a platinum-iridium bar stored at the International Bureau of Weights and Measures near Paris. This standard had serious inadequacies, for the accuracy with which the marks could be scribed on the bar was limited and the bar itself could easily be destroyed in a catastrophe. An improved standard for the meter was developed and accepted in 1961. It was based on optical interference techniques, which can be used to determine distances in terms of the wavelength of the radiation emitted in a specific atomic transition. The meter was defined as 1,650,763.73 times the wavelength of a particular orange radiation emitted by krypton atoms. (See Fig. 1-1.) This number was chosen by measuring the length of the platinum-iridium bar in terms of the wavelength of the krypton radiation, thereby making the two standards consistent. Since krypton is readily available, this wavelength standard was accessible and also reproducible to better accuracy than the bar standard.

Because the wavelength of the krypton spectral line does vary slightly, the precision of this standard was limited to approximately one part in 10^9. In 1983 a more accurate length standard was adopted. First, the speed of light was fixed at exactly

$$c = 2.99792458 \times 10^8 \text{ m/s};$$

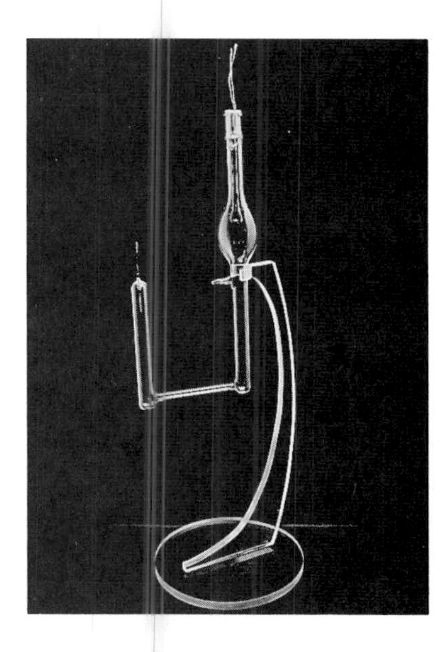

FIGURE 1-1 The wavelength of the radiation emitted from a krypton lamp was used as a length standard.

then the meter was defined as *the length that a light beam travels in a time interval of (1/2.99792458) s.*

Now this is simply a definition. In practice, length measurements are made with this standard using the stabilized laser, which produces light whose wavelength fluctuates less than two parts in 10^{12}. The time required for the laser to emit one complete wave of the radiation is measured and then multiplied by the speed of light to yield the wavelength. Since the time standard (to be discussed below) is accurate to one part in 10^{13}, the wavelength itself is precise to roughly one part in 10^{12}. This value can then be used to determine other lengths using interferometric techniques similar to those employed with the krypton standard. At present, distances of a few millimeters can be measured to within 10^{-12} m.

Standard for Mass

The standard for the SI mass unit, the kilogram, is a platinum-iridium cylinder that is stored at the International Bureau outside Paris. Copies of the standard are available in many countries. (See Fig. 1-2.) With very precise measurements, masses can now be compared to the standard with an accuracy of a few parts in 10^9.

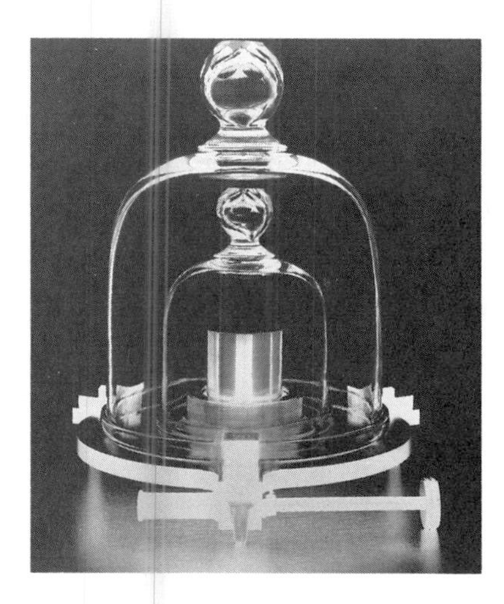

FIGURE 1-2 A duplicate of the platinum-iridium kilogram cylinder is stored in a double bell jar at the National Bureau of Standards in Washington, D.C.

At present there isn't an atomic standard for mass that is as accurate as the platinum-iridium cylinder. Although the masses of different atoms can be compared very precisely, the number of atoms in a material can be counted with an accuracy of roughly only one part in 10^6. Consequently, an atomic-mass determination of the mass of a sample is limited to this same accuracy.

Standard for Time

The SI time unit was originally based on the mean solar day, which is the time interval averaged over one year between successive passages of the sun across a particular meridian. The second was defined as (1/60)(1/60)(1/24) of the mean solar day. As the rotational rate of the earth does vary, this definition was not very precise.

The time standard was improved considerably in 1967 when it was redefined in terms of the transition of the cesium-133 atom between its two lowest energy states. When the atom is in the lower energy state, it can jump to the higher state by absorbing electromagnetic radiation whose frequency is measurable to an accuracy of one part in 10^{13}. The second is *the time required for 9,192,631,770 oscillations of this radiation.* It's interesting to note that an accuracy of one part in 10^{13} corresponds to a loss or gain of no more than a second in 300,000 years! The cesium "atomic clock" at the National Bureau of Standards in Washington, D.C., is shown in Fig. 1-3.

Precise time signals that are coordinated with the cesium clocks at the National Bureau of Standards are transmitted by radio station WWV, Fort Collins, Colorado. These time signals

FIGURE 1-3 The cesium atomic clock at the National Bureau of Standards.

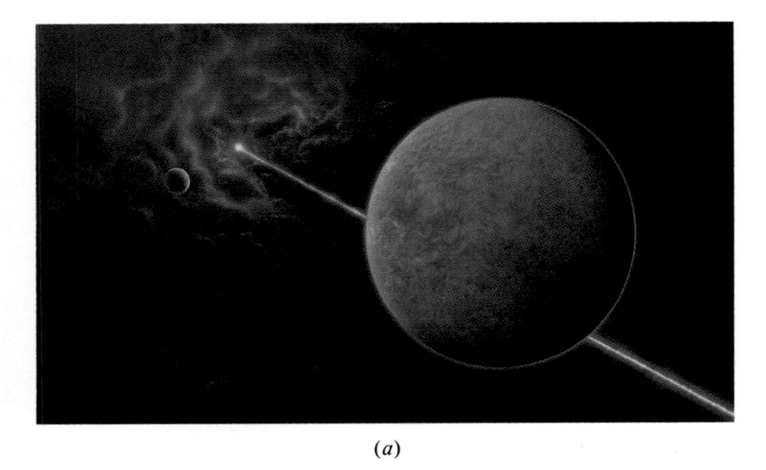

(*a*)

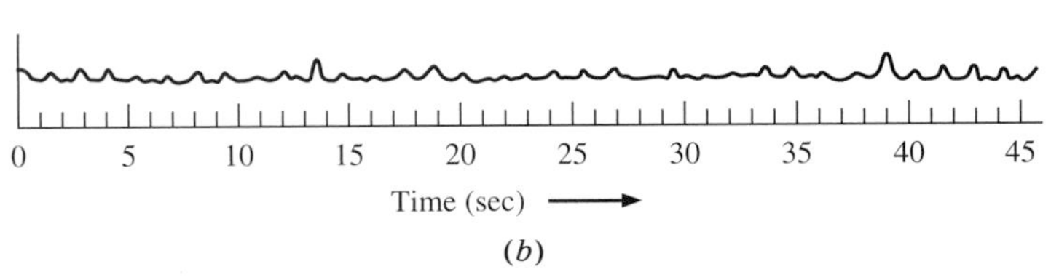

(*b*)

(*a*) An artist's rendition of a pulsar; (*b*) A sketch of the chart recording of the intensity of the radio emission from the first pulsar discovered (CP1919). The time intervals between the emissions are extremely regular (1.3373011 s). Pulsars are therefore naturally occurring clocks.

are broadcast at the shortwave frequencies of 2.5, 5, 10, 15, and 20 MHz (megahertz) and are received worldwide.

Most countries now employ the SI units as a basis for trade and commerce. These countries also use *cgs units*, which express length, mass, and time in *centimeters* (*cm*), *grams* (*g*), and *seconds* (*s*), respectively. The cgs units are defined in terms of SI units by the following *exact* relationships:

$$1 \text{ centimeter} = 10^{-2} \text{ meter}$$
$$1 \text{ gram} = 10^{-3} \text{ kilogram.}$$

A few countries (including the United States) still use a system of units known as the *British system*. Here, length is measured in *feet* (*ft*), mass in *slugs* (*slug*), and time in *seconds*. The British unit for length is defined in terms of the corresponding SI unit by the following *exact* relationship:

$$1 \text{ foot} = 0.3048 \text{ meter.}$$

However, the slug is not defined directly in terms of its corresponding SI unit, the kilogram. It turns out that the British force unit*, the pound, is defined exactly in terms of the SI force unit, the newton; then the slug is defined in terms of the pound by an exact relationship. Of course, the slug can be related to the kilogram. To roughly one part in a million (which is certainly good enough for us),

$$1 \text{ slug} = 14.59390 \text{ kilograms.}$$

Except for some applications in mechanics, we will be using SI units throughout this book.

1-2 Dimensional Analysis and the Conversion of Units

A measurement of length, mass, or time is said to have the *dimension* length L, mass M, or time T, respectively. For example, the distance d between two points has the dimension $[d] = L$; a time interval Δt has the dimension $[\Delta t] = T$; an area A, which is a length squared, has the dimensions $[A] = L^2$; and a speed v, which is a length per unit time, has the dimensions $[v] = L/T$. Associated with any dimension is a unit. The meter and the mile (mi) are length units, the second and the hour (h) are time units, and the meter per second and the mile per hour are speed units. There are also dimensionless quantities (often called *pure numbers*). Some examples of these are angles and their trigonometric functions such as the sine or cosine, exponentials, and logarithms.

Dimensions and units obey the simple rules of algebra. By multiplying a speed with dimensions L/T by a time with dimension T, we obtain a distance with dimension L:

$$\frac{L}{T}T = L.$$

In terms of SI units,

$$\frac{m}{s}s = m.$$

Every term in an equation must have the same dimensions. As an example, consider the equation (to be discussed in Chap. 3)

$$x = v_0 t + \tfrac{1}{2}at^2,$$

where x is a position with dimension $[x] = L$, and v_0, a, and t are, respectively, initial velocity with dimensions $[v_0] = L/T$, acceleration with dimensions $[a] = L/T^2$, and time with dimension $[t] = T$. Since

$$[v_0 t] = \frac{L}{T}T = L \quad \text{and} \quad \left[\frac{1}{2}at^2\right] = \frac{L}{T^2}T^2 = L,$$

the dimension of every term is the same.

Dimensional analysis can tell us when a relationship is incorrect. For example, the equation

$$x = v_0 t^2 + at$$

must be *incorrect* because all of its terms are *not the same dimensionally*. The dimensions of the three terms in this equation are L, $(L/T)(T^2) = LT$, and $(L/T^2)(T) = L/T$, respectively. You should be aware, however, that not all equations that satisfy a dimensional check are necessarily physically correct. The

*Force and its units are defined in Chap. 5. The exact relationship that defines the British force unit in terms of the SI force unit is 1 lb = 4.448221615260 N (newtons). The slug is defined exactly by 1 slug = (4.448221615260/0.3048) kg.

Conversion from feet to meters.

equations $x = v_0t + at^2/2$, $x = 2v_0t$, and $x = v_0t + at^2$ are all consistent dimensionally; yet only the first of these is physically correct for the case of constant acceleration.

Frequently a quantity must be converted from one set of units to another. This can be done by treating the units as algebraic symbols. As an example, suppose a distance measurement of 1 mi must be converted to inches (in.). Knowing that there are 5280 ft in a mile and 12 in. in a foot, we can make the conversion as follows:

$$1\text{ mi} = (1\ \cancel{\text{mi}})\left(5280\,\frac{\cancel{\text{ft}}}{\cancel{\text{mi}}}\right)\left(12\,\frac{\text{in.}}{\cancel{\text{ft}}}\right) = 63{,}360\text{ in.}$$

Notice how the various units cancel, leaving the final result in inches.

Some of the frequently used conversion factors are given in Table 1-1. There is a more extensive table in App. A at the end of the book. Although the conversion factors are written as equalities, they are not "equal" in the normal sense. The statement 2.54 cm = 1 in. does not mean that 2.54 = 1. Instead, it tells us that a distance measured to be 2.54 cm with a meterstick is found to be 1 in. with a yardstick. You can think of the conversion factors as ratios that are equal to 1; that is, 1 = 2.54 cm/1 in., 1 = 60 s/1 min, etc. A change of units is equivalent to the multiplication of a quantity by the appropriate unit ratios. For example, one day can be converted to seconds by multiplying by 1 three times:

$$\begin{aligned}1\text{ day} &= (1\text{ day})(1)(1)(1)\\ &= (1\ \cancel{\text{day}})\left(24\,\frac{\cancel{\text{h}}}{\cancel{\text{day}}}\right)\left(60\,\frac{\cancel{\text{min}}}{\cancel{\text{h}}}\right)\left(60\,\frac{\text{s}}{\cancel{\text{min}}}\right)\\ &= 86{,}400\text{ s.}\end{aligned}$$

EXAMPLE 1-1 Mass Density

If a homogeneous substance has a mass m and a volume V, its mass density ρ is by definition

$$\rho = \frac{m}{V}. \tag{1-1}$$

The mass density of mercury is 13.6 g/cm^3. What is this mass density in kilograms per cubic meter?

TABLE 1-1 Common Conversion Factors

Length	
1 m	= 100 cm = 1000 mm
1 km	= 1000 m
1 in.	= 2.54 cm
1 ft	= 0.3048 m
1 mi	= 1.609 km
1 mi	= 1760 yd = 5280 ft
1 ft	= 12 in.
1 yd	= 3 ft
Mass	
1 kg	= 1000 g
1 slug	= 14.59 kg
Time	
1 yr	= 365.24 days = 3.1557 × 10^7 s
1 min	= 60 s
1 h	= 60 min

SOLUTION From Table 1-1, 1 kg = 1000 g and 1 m = 100 cm, so

$$\rho = \left(13.6\,\frac{\text{g}}{\text{cm}^3}\right)\left(10^{-3}\,\frac{\text{kg}}{\text{g}}\right)\left(100\,\frac{\text{cm}}{\text{m}}\right)^3 = 1.36\times10^4\,\frac{\text{kg}}{\text{m}^3}.$$

Notice that we have to convert cubic centimeters to cubic meters, which makes us *cube* the term 100 cm/m. Ignoring powers of units is a common mistake that you should avoid.

EXAMPLE 1-2 The Radian

The angle θ of Fig. 1-4 is given in radians (rad) by the ratio of the circle's arc length s to the radius r: $\theta = s/r$. What are the dimensions of θ?

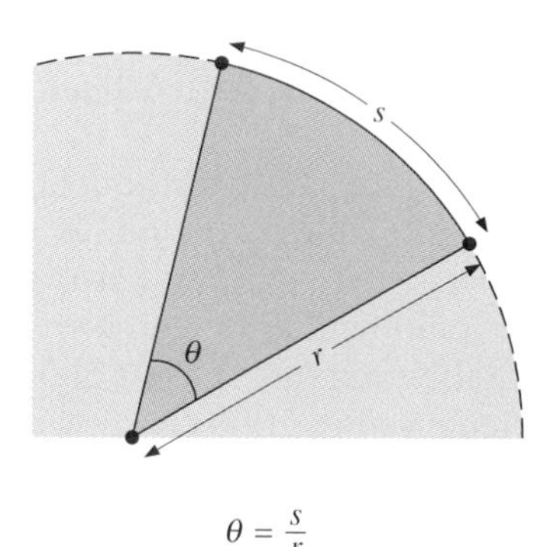

FIGURE 1-4 Angle θ expressed in radians is the ratio of the arc length s to the radius r.

SOLUTION Since both s and r have the dimension length, the dimensions of θ are

$$[\theta] = \frac{L}{L} = 1;$$

that is, the angle is dimensionless. Because the radian is dimensionless, it can be included or omitted in the dimensional analysis of an equation. We use this unit only as a reminder that an angular measurement is in radians rather than degrees (π rad = 180°).

DRILL PROBLEM 1-1

What is a speed of 30 mi/h in feet per second?
ANS. 44 ft/s.

DRILL PROBLEM 1-2

The density of water at 3.98°C is 1.000 g/cm^3. Express this density in kilograms per cubic meter and slugs per cubic foot.
ANS. 1000 kg/m^3; 1.941 slug/ft^3.

DRILL PROBLEM 1-3

Let x, v, a, m, and t have dimensions L, L/T, L/T^2, M, and T, respectively. Which of the following equations are dimensionally correct? (*a*) $v = at$; (*b*) $x = at$; (*c*) $v^2 = 2ax$; (*d*) $v = at^2$.
ANS. (*a*) and (*c*).

1-3 Significant Figures

All physical measurements are limited in accuracy by a variety of factors such as the quality of the equipment used, the number of measurements taken, and how well external factors such as temperature and pressure can be controlled. An important part of any experiment is an accurate estimate of how precise a particular measurement is. Furthermore, the accuracy to which a measured quantity is determined must somehow be contained in the way the value of that quantity is presented.

There are methods that can be used to estimate the accuracy of a particular experiment. These methods allow us to express a measurement as a number with some uncertainty. For example, we might say that we have measured the length and width of a rectangular floor to be 1120 ± 3 cm and 643 ± 3 cm, respectively. These two numbers indicate that our measurements allow us to state only that the length lies somewhere between 1117 and 1123 cm and the width somewhere between 640 and 646 cm.

When we present the value of a physical quantity without an error, we represent the accuracy of that value by the number of *significant figures* we use to express it. In stating that the width of the floor is 643 cm, we are claiming that we know this measurement to three significant figures and that the last digit (the 3) is uncertain. Similarly, the length of the floor is known to four significant figures with some uncertainty in the 0. Zeros in the middle or at the end of a number are just as important as any other digit and should therefore be counted as significant figures. For example, 50,300 has five significant figures. If the two zeros at the end were not meaningful, we would express this number as 5.03×10^4. Zeros at the beginning of a number are present only to locate the decimal point and are *not* significant figures. Therefore the number 0.0003405 has four significant figures. We can avoid these insignificant zeros by expressing numbers in scientific notation, such as $0.0003405 = 3.405 \times 10^{-4}$.

What is the area of our floor? With our measurement uncertainties in the length and width, we can only say that the area is approximately (643 cm)(1120 cm) = 720,160 cm^2 with a lower bound of (640 cm)(1117 cm) = 714,880 cm^2 and an upper bound of (646 cm)(1123 cm) = 725,458 cm^2. Notice that with these limits we cannot even estimate the third digit of the value of the area to any better than ± 5. If we must estimate this third digit, we surely have no idea what the fourth, fifth, and sixth digits are. Consequently, we should only express the area to three significant figures, or as 7.20×10^5 cm^2.

This example illustrates the rule for determining the number of significant figures in a product or a quotient: *When multiplying or dividing several numbers, you should state the answer using the same number of significant figures as found in the term with the least number of significant figures.* In our example, we multiplied a quantity with four significant figures by one with three significant figures; the result must then be expressed to three significant figures only.

A touch of common sense must be used when applying this rule. As an example, consider two floors, one whose length and width are 18.6 and 9.7 m, respectively, the other whose dimensions are 18.6 by 10.1 m. Strict adherence to the rule requires that we express the areas of the two floors as 1.8×10^2 m^2 and 188 m^2. However, contrary to the way they are written, these two areas are known to almost the same accuracy. The fact is that in this case we are justified in adding a third significant figure to the first area and writing it as 180 m^2.

There is also a general rule for determining significant figures when numbers are added or subtracted. This rule is illustrated by the following summation:

```
2347.56??
  53.9521
 300.2???
---------
2701.7???
```

Here the question marks in the numbers being added indicate that we do not know what digits belong in those locations. Obviously, then, we don't know what digits belong in the designated spots in the sum—which must therefore be represented as 2701.7. We can now state the general rule for addition and subtraction: *When adding or subtracting several numbers, you should state the answer using the same number of decimal places as that quantity with the least number of decimal places.*

In a calculation involving multiple arithmetic operations, one or two extra (or insignificant) digits are usually carried at each step. The final result is then reduced to the proper number of significant figures after the last calculation. This avoids an accumulation of rounding errors, which can lead to a rather inaccurate final result, even when that result is stated to the proper number of significant figures.

To avoid questions about significant figures and errors that may obscure explanations of the basic physics, we adopt a rather pragmatic approach to using the proper number of significant figures in this textbook: *The values of all quantities are assumed to be precise enough to give all calculated results to either two or three significant figures, where the "two or three" depends on the particular application.*

Now there are terms in equations that do not come from measurements with inherent inaccuracies. They represent exact quantities and therefore play no part in decisions about significant figures. Fractions, usually exponents, and often conversion factors are terms of this type. For example, in the equation

$$d = \tfrac{1}{2}(9.80\ \text{m/s}^2)(2.00\ \text{s})^2(1\ \text{ft}/0.3048\ \text{m}) = 64.3\ \text{ft},$$

the fraction 1/2, the exponent 2, and the conversion factor 1 ft/0.3048 m are exact.

EXAMPLE 1-3 VOLUME OF A RIGHT CIRCULAR CONE

The volume V of a right circular cone is given by $V = \pi r^2 h/3$, where r is the radius of the base and h is the altitude (the distance from the base to the apex). A student measures r and h and finds that $r = 5.53$ cm and $h = 17.56$ cm. Calculate the volume of the cone to the proper number of significant figures.

SOLUTION The value of π can be found to an arbitrarily large number of significant figures. Thus the volume is limited in accuracy by the measurements of the radius and altitude, in this case by the measurement to the three significant figures of r. The volume is then

$$V = \frac{\pi(5.53\ \text{cm})^2(17.56\ \text{cm})}{3} = 562\ \text{cm}^3.$$

DRILL PROBLEM 1-4

A metal piece of mass 457.9 g is machined, after which its mass is found to be 456.2 g. Calculate the mass lost in the machining to the proper number of significant figures.
ANS. 1.7 g.

DRILL PROBLEM 1-5

The lengths of the base and height of a thin rectangle are measured to be 20.005 and 0.032 cm, respectively. Calculate the area of the rectangle to the proper number of significant figures.
ANS. 0.64 cm^2.

SUMMARY

1. **Standards for length, mass, and time**
   The standards for the three basic units of the International System (SI) of units are given below:
   (*a*) The meter is the length that a light beam travels in a time interval of (1/2.99792458) s.
   (*b*) The kilogram is the mass of a platinum-iridium cylinder stored at the International Bureau of Weights and Measures.
   (*c*) The second is the time required for 9,192,631,770 oscillations of the radiation absorbed or emitted when a cesium-133 atom makes a transition between its two lowest energy states.
2. **Dimensional analysis and the conversion of units**
   All the terms in a correct equation must have the same dimensions. However, an equation that satisfies a dimensional check may not be a valid representation of a physical phenomenon. Conversions among the SI, cgs, and British units are given in Table 1-1.
3. **Significant figures**
   (*a*) When multiplying or dividing several numbers, you should state the answer using the same number of significant figures as that term with the least number of significant figures.
   (*b*) When adding or subtracting several numbers, you should state the answer using the same number of decimal places as that quantity with the least number of decimal places.

QUESTIONS

1-1. An inscription on the wall of an ancient temple reads, "Twenty paces to the south, thirty paces to the west, dig." What is the length standard in this message? Is this a good length standard?

1-2. Discuss the properties that a phenomenon should have in order that we may base a time standard on it.

1-3. Suggest some natural phenomena that would serve as reasonably accurate timekeepers.

1-4. Suppose you could communicate by radio with scientists from another planetary system. With the help of your explanations, could they reproduce our old standards for length and time? What about our modern standards? Could they reproduce the modern standard for mass?

1-5. Discuss the advantages that the SI system of units has over the British system.

1-6. Why do you think it was necessary to specify the temperature for the platinum-bar length standard? Do we have to specify the temperature for the contemporary length standard? for the mass standard?

1-7. Could we use length, time, and density in place of length, mass, and time as standards? What about length, mass, and density? or length, mass, and speed?

1-8. Would a pulse rate be a good mechanism to use for measuring time intervals? If, shortly after some strenuous exercise, you used your pulse rate to estimate a time interval, would your estimate be high or low?

1-9. Obstetricians generally describe a woman's term of pregnancy as 40 weeks rather than 9 months. What advantage(s) does the "week" unit have over the "month" unit?

1-10. Car speedometers are generally calibrated in both miles per hour and kilometers per hour. Verify that the conversions are accurate on the speedometer of a car to which you have access.

1-11. Consider the equation $x = y + z$. If y and z both have the dimensions of ML/T^2, what are the dimensions of x? What are the units of x in the SI system? cgs system? British system?

1-12. If the dimensions of y were different from the dimensions of z in the equation of Ques. 1-11, could we determine a dimension for x? Would the equation be physically meaningful?

1-13. In the equation $x = y/z$, is it necessary for y and z to have the same dimensions for the equation to be meaningful? If y and z do have the same dimensions, what would the dimensions of x be?

1-14. Does the value of the angle subtended by an arc of a circle depend on the unit of length used for the arc length and radius of the circle?

1-15. Can you suggest a way to estimate the diameter of an atom?

PROBLEMS

Measurement, Dimensional Analysis, and the Conversion of Units

1-1. Consider the equation $y = mt + b$, where y and t are variables with the dimensions of length and time, respectively, and m and b are constants. What are the dimensions of m and b? What are the units of these constants in the SI and in the cgs systems?

1-2. The formula for the kinetic energy of a particle is $mv^2/2$, where m and v are the mass and speed of the particle. What are the dimensions of kinetic energy? What are its SI units?

1-3. Show that the formula $v^2 = 2ax$ is dimensionally correct. The variables v and x represent the speed of a particle and the distance that it has traveled; a is its acceleration.

1-4. Suppose that the distance traveled is related to the time by $x = pt^3$. What are the dimensions of p?

1-5. The formula for the period of a simple pendulum is $T = 2\pi\sqrt{l/g}$, where l is a length and g is an acceleration. Show that this formula is dimensionally correct.

1-6. What is your height in meters? in centimeters?

1-7. Astronomers usually measure astronomical distances in light-years. By definition, *one light-year is the distance that light travels in one year*. Given that the speed of light is 3.00×10^8 m/s, what is one light-year in meters?

1-8. What is the height in light-years of a person 6.00 ft tall? (See Prob. 1-7.)

1-9. A pressure of 1 atmosphere (atm) is also expressed as 76.0 cm of mercury. What is 1 atm in inches of mercury?

1-10. Which is longer, the record time for running 100 yd or 100 m? for running 1.000 mi or 1500 m?

1-11. A speed-limit sign on a Canadian road reads "80 km/h." What is this limit in miles per hour?

1-12. Use the thickness of your book and the number of pages in it to estimate the thickness of one page.

1-13. In baseball, the distance between the bases is exactly 90 ft. What is the distance in meters between home plate and second base?

1-14. When a woman turns 100 years old, how many seconds has she lived?

1-15. (*a*) How many degrees are equivalent to 0.500 rad? (*b*) How many radians are there in an angle of 270°? Express your answer in terms of π.

1-16. Express an angle of 15.65° in (*a*) minutes and (*b*) seconds. (*Note:* 1 degree = 60 minutes and 1 minute = 60 seconds.)

1-17. Both the sun and the full moon subtend angles of about 0.50° to a viewer on the earth. Estimate the solar and lunar diameters using 1.5×10^{11} m as the sun-earth distance and 3.8×10^8 m as the moon-earth distance.

1-18. The mass density of mercury is 13.6×10^3 kg/m^3. Express this density in slugs per cubic feet.

1-19. Assuming that the human body is composed primarily of water of mass density 1.0 g/cm^3, estimate the volume in cubic meters of a 70-kg person.

1-20. An acre, which is 43,560 ft^2, is an area unit often used to describe the size of a farm. What is the size of a 600-acre farm in square miles?

1-21. A football field is 120 yd long and 53 yd wide. What is the size of a football field in acres?

1-22. Suppose you are paying $1.20 per gallon for gasoline when the United States makes the switch to the SI system of units for commerce. If you find the price of gasoline is now $0.35 per liter, are you paying more than before the switch was made?

1-23. The radius and mass of the sun are approximately 7.0×10^8 m and 2.0×10^{30} kg, respectively. What is the mass density of the sun in grams per cubic centimeter?

1-24. Assume that your body is all water. Estimate the number of molecules in it.

1-25. The mass density of air is approximately 1.2 kg/m^3. Estimate the mass of the air in your classroom.

1-26. The earth is approximately spherical, with a radius of 6.4×10^3 km and a mean mass density of 5.5×10^3 kg/m^3. (*a*) Use this information to calculate the mass of the earth. (*b*) Assuming that the average atomic mass of the earth is 30 g/mol, estimate the number of molecules in the earth.

1-27. The proton is sometimes represented as a sphere of radius 1.5×10^{-15} m with a mass of 1.67×10^{-27} kg. Estimate the mass density of a proton and compare it with the mass density of water, which is 1000 kg/m^3.

Significant Figures

1-28. How many significant figures are there in the following measurements? (*a*) 72 km; (*b*) 0.009345 m; (*c*) 9.3004500 cm; (*d*) 4.7450×10^3 s.

1-29. Express the following calculations to the proper number of significant figures: (*a*) (1.27 m)(2.5765 m); (*b*) 3.2576 ft + 1.2 ft; (*c*) $4\pi(2.72\text{ cm})^2$; (*d*) $3(1.2543 \times 10^{-4}\text{ kg})/4\pi(2.0\text{ m})^3$.

1-30. The length of a very thin rectangle is found with a meterstick to be 94.5 cm, and its width is measured with a vernier caliper to be 1.384 cm. Determine the perimeter and area of the rectangle.

1-31. What is the volume of a sphere of radius 0.0703 m?

1-32. The mass of a steel ball bearing is measured to be 43.82 g. If the mass density of the steel is 7.86 g/cm^3, what is the radius of the ball bearing?

General Problems

1-33. Given that the radius of the earth is 6.37×10^6 m, how far must a ship travel along the equator to cover 1.00 minute of arc? (See Prob. 1-16.) What is this distance in feet? in miles? (*Note:* By definition, this distance is 1 nautical mile.)

1-34. Prior to the development of modern electronic timers, the time required to complete a race in a track meet was determined by a person holding a stopwatch at the finish line. The timer started the watch when he saw the puff of smoke from the starter's pistol and stopped it when the winner crossed the finish line. Suppose that rather than using the sighting of the puff of smoke to start the watch, the timer used the sound from the firing of the gun. Assuming that the speed of sound is 340 m/s, by how much would the measurement of the winning time in a 100-m dash be in error? If the actual winning time was 10.2 s, what would be the percentage error in the measurement?

1-35. Suppose you are looking across the ocean at the horizon with your eyes a distance h above sea level, as represented in the figure. Your eyes are at B; the horizon is at C, which is a distance s away from A; and the radius of the earth is R. (*a*) Show that $\phi = \theta/2$. (*b*) Since ϕ is very small, h is approximately equal to $s\phi$. Use this to show that $\theta \approx \sqrt{2h/R}$. (*c*) Use this approximation to show that the distance that the horizon is away from you is given by $s \approx \sqrt{2hR}$. (*d*) If h is in feet, show that s in miles is given by $s \approx \sqrt{1.5h}$. (*e*) If you were standing on the deck of a ship with your eyes 50 ft above sea level, how far away would the horizon be?

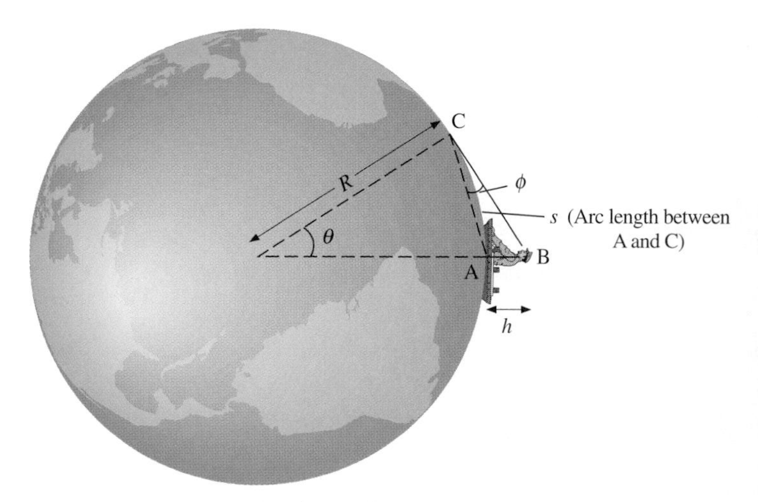

(Angles greatly exaggerated)

PROBLEM 1-35

The displacement vector involves both a distance and direction.

CHAPTER 2 VECTORS

PREVIEW

This chapter is an introduction to vectors and their properties. The main topics to be discussed here are the following:

1. **Displacement**. Displacement is used as a vector prototype for illustrating general vector properties.
2. **Vector arithmetic**. The equality of vectors, the addition and subtraction of vectors, the commutative and associative laws for vectors, and multiplication of a vector by a scalar are considered.
3. **Components of a vector**. Vectors are expressed in terms of their projections, or components, along defined coordinate axes. The process of determining these projections is known as vector resolution.
4. **Scalar and vector products**. The multiplication of vectors is discussed.

Many of the physical quantities we are familiar with can be completely specified by a single number and the appropriate unit. For example, our normal body temperature is 37°C, a class period lasts 50 min, and a pair of shoes might cost $45. A quantity that can be specified in this manner is called a **scalar**. Scalars obey standard addition and subtraction operations. A person whose body temperature is 2°C above normal has a temperature of 37°C + 2°C = 39°C. Similarly, a class ending 10 min early lasts 50 min − 10 min = 40 min.

There are other quantities, however, that require not just a number and a unit for specification, but also a direction. Such quantities are called **vectors**. In this chapter, we will consider how vectors are represented, added, subtracted, and multiplied. The role of vectors in physics is a significant one, for many important physical quantities are vectors. Some examples include displacement (which we will use as a vector prototype), velocity, acceleration, force, and electric and magnetic fields.

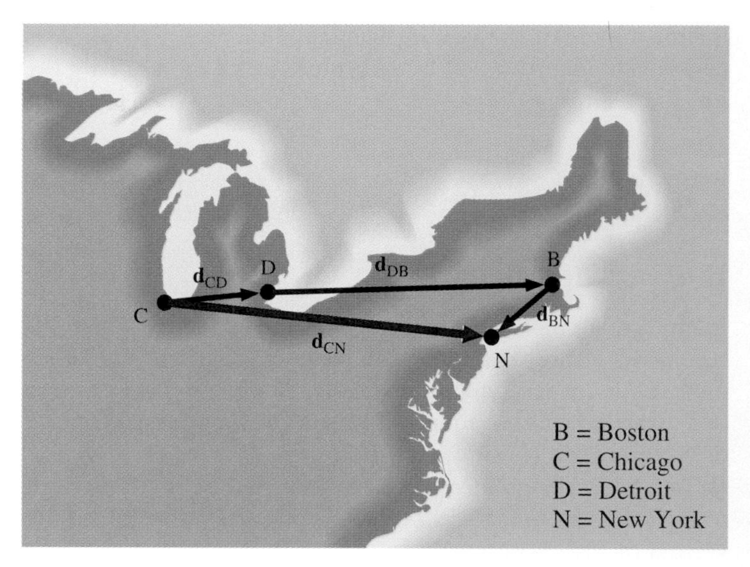

FIGURE 2-2 Displacements between Chicago and Detroit, Detroit and Boston, Boston and New York City, and Chicago and New York City are represented by $\mathbf{d}_{CD}$, $\mathbf{d}_{DB}$, $\mathbf{d}_{BN}$, and $\mathbf{d}_{CN}$, respectively.

2-1 Introduction to Vectors

Suppose you tell a friend with whom you are on a camping trip that you've discovered a terrific fishing hole 2 km from your tent. It's unlikely that he would be able to find the hole easily unless you also tell him the direction in which he has to walk to reach it. For example, you might say the fishing hole is 2 km northeast of your campsite. The key here is you have to provide *two* pieces of information: the distance, or *magnitude*, of the walk he has to make and the *direction* in which he has to travel.

The general term used to describe a *change in position* such as the trip from the tent to the fishing hole is **displacement**. This is one example of a vector. If you walk from the tent (point A) to the hole (point B), as shown in Fig. 2-1, the vector that represents your displacement is given by the straight line drawn between A and B. The fact that the displacement is directed from A to B (rather than from B to A) is designated by the arrowhead. The *actual* path taken in going from A to B does not have to be the straight line joining these points. You could have traveled along any of the curved paths shown—but your displacement vector is *defined* to be the straight line. Displacement is independent of the path traveled.

The addition of vectors is illustrated by the displacements made during the following trip. Imagine traveling from Chicago to Detroit, then leaving Detroit for Boston, and finally going from Boston to New York City. Your *net* displacement is the straight line from Chicago to New York City (see Fig. 2-2). This is equivalent to the displacement from Chicago to Detroit plus the displacement from Detroit to Boston plus the displacement from Boston to New York City.

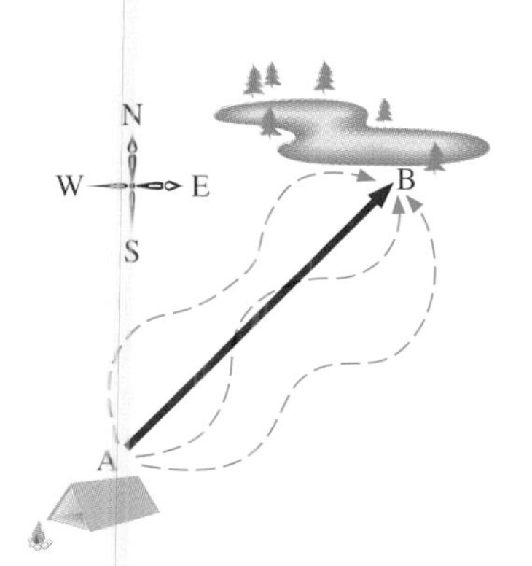

FIGURE 2-1 The displacement from A (the campsite) to B (the fishing hole) is the straight line drawn between A and B. This displacement is independent of the paths (the curved lines) that might actually be taken between the two locations.

Notice that the distances between the various cities on the trip do *not* add to give the distance from the starting point, Chicago, to the final destination, New York City. The distances between Chicago and Detroit, between Detroit and Boston, and between Boston and New York City are 270, 700, and 210 mi, respectively. These add to 1180 mi, while the length of the straight line from Chicago to New York City is 800 mi.

We will designate vectors by boldface type. For example, we might represent the individual displacements between the cities of Fig. 2-2 as $\mathbf{d}_{CD}$, $\mathbf{d}_{DB}$, and $\mathbf{d}_{BN}$. The net displacement $\mathbf{d}_{CN}$ is then given by

$$\mathbf{d}_{CN} = \mathbf{d}_{CD} + \mathbf{d}_{DB} + \mathbf{d}_{BN}.$$

Graphically, a vector is represented in a chosen coordinate system by a line drawn with an arrowhead denoting its direction. (See Fig. 2-3.) The length of that line must be proportional to

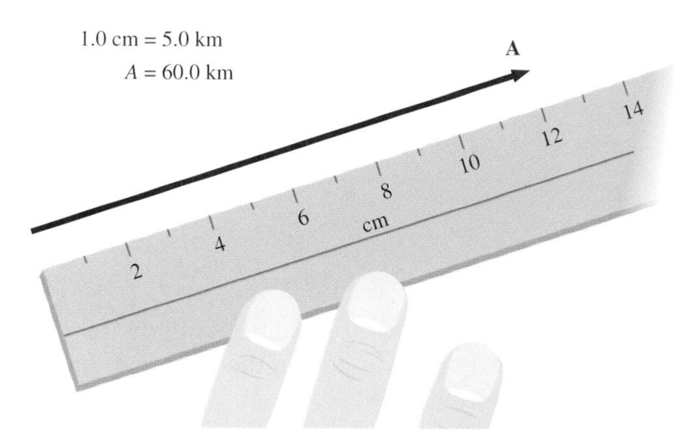

FIGURE 2-3 The vector **A** points in the direction shown. The length of the vector is scaled as indicated.

the magnitude of the vector. The magnitude of a vector **A** is designated either by |**A**| (the absolute value of **A**) or by the italicized letter A.

2-2 Vector Arithmetic

Equality of Vectors

Figure 2-4 shows two displacements of magnitude 300 km directed 45° north of east. One starts in Chicago and the other in St. Louis. Since displacement is a *change* in position, these two displacements are equal. This illustrates a general property of vectors: *On a graph with a given scale, all vectors that have the same length and direction are mathematically equal, even though they are at different locations.* For example, all of the vectors of Fig. 2-5 are equal.

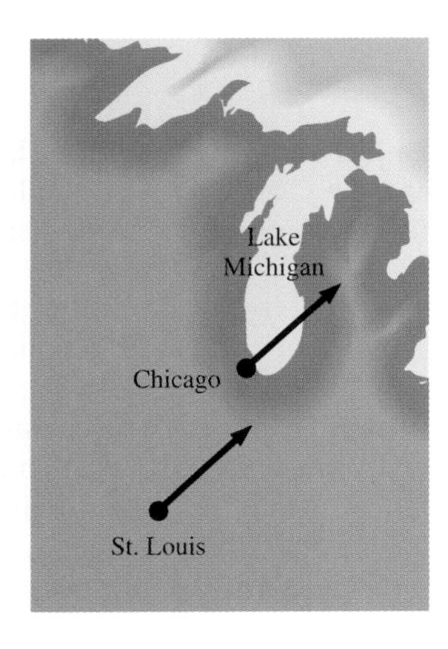

FIGURE 2-4 Displacements of the same magnitude and direction are equal even though their origins and destinations are different.

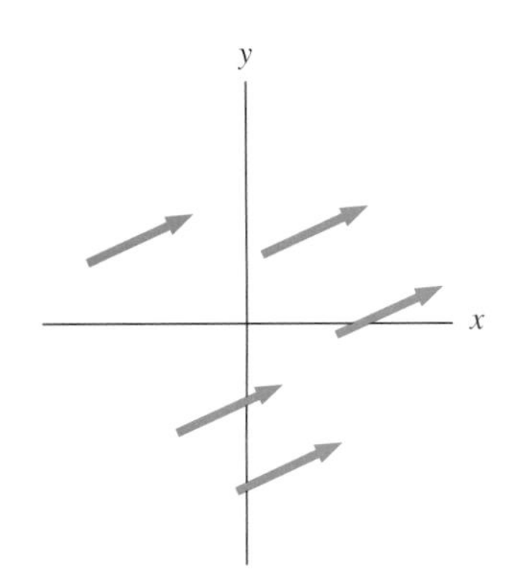

FIGURE 2-5 The vectors shown are all equal because they have the same direction and magnitude.

Addition of Two Vectors

We have already discussed vector addition with the help of displacement vectors. In general, to add the vectors **A** and **B**, we carry out the following steps:

1. Choose a scale for the magnitudes of the vectors. For example, 3.0 cm on a graph might represent 60 km.

Vectors are an important part of navigation.

2. Draw vector **A** to scale and in the appropriate direction.
3. Starting at the head of **A**, draw **B** to scale and in the appropriate direction.
4. The resultant vector **R** = **A** + **B** is then the vector drawn from the tail of **A** to the head of **B**. (See Fig. 2-6.)

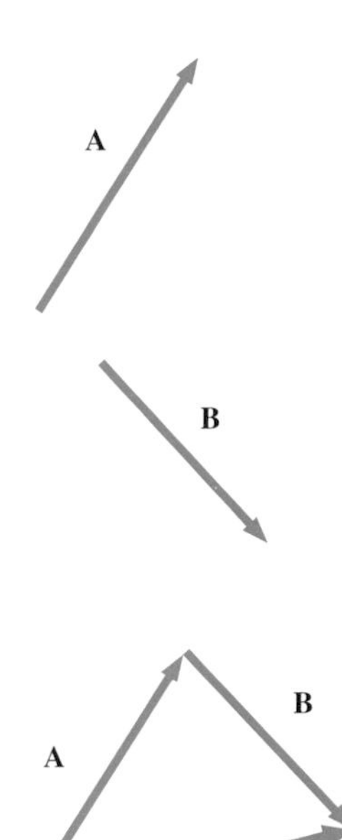

FIGURE 2-6 The sum of two vectors: **R** = **A** + **B**.

Addition of Three or More Vectors

By definition, **A** + **B** + **C** = (**A** + **B**) + **C**. The application of this rule is illustrated in Fig. 2-7*a*. Notice that the addition rule for two vectors is used twice—once on **A** and **B** to get **A** + **B** ($\mathbf{R}_1$ in the figure), and then on $\mathbf{R}_1$ and **C** to obtain **R** = **A** + **B** + **C**. Fortunately, you don't have to go through these steps

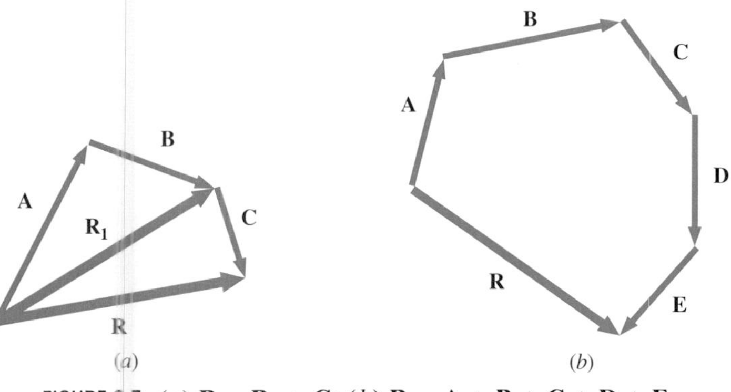

FIGURE 2-7 (*a*) $\mathbf{R} = \mathbf{R}_1 + \mathbf{C}$; (*b*) $\mathbf{R} = \mathbf{A} + \mathbf{B} + \mathbf{C} + \mathbf{D} + \mathbf{E}$.

each time you add three or more vectors. You just need to lay out tail-to-head the vectors that you are adding and then draw the resultant vector **R** from the tail of the first vector to the head of the last. An example of the addition of five vectors is shown in Fig. 2-7*b*.

Commutative and Associative Laws

The commutative law is expressed by

$$\mathbf{A} + \mathbf{B} = \mathbf{B} + \mathbf{A}, \tag{2-1}$$

and the associative law by

$$(\mathbf{A} + \mathbf{B}) + \mathbf{C} = \mathbf{A} + (\mathbf{B} + \mathbf{C}). \tag{2-2}$$

The validity of these laws is established in Fig. 2-8. The laws allow vectors to be added in any order; that is, $\mathbf{A} + \mathbf{B} + \mathbf{C} = \mathbf{A} + \mathbf{C} + \mathbf{B} = \mathbf{B} + \mathbf{C} + \mathbf{A}$, etc.

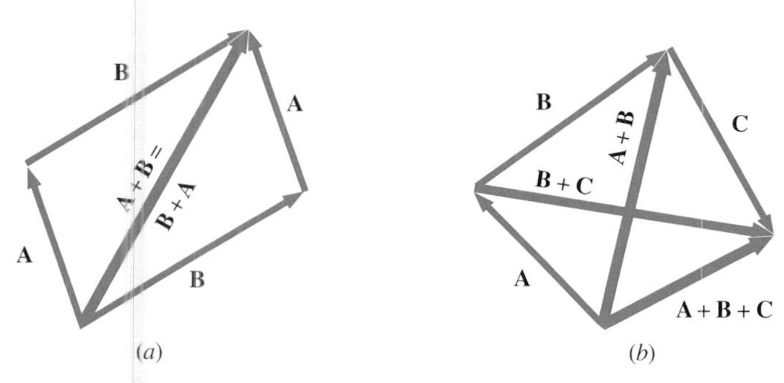

FIGURE 2-8 (*a*) The commutative law and (*b*) the associative law.

Multiplication of a Vector by a Scalar

The vector $k\mathbf{A}$ is defined to be a vector of magnitude $|k||\mathbf{A}|$ whose direction is along **A** if $k > 0$ and opposite to **A** if $k < 0$. This rule is illustrated in Fig. 2-9.

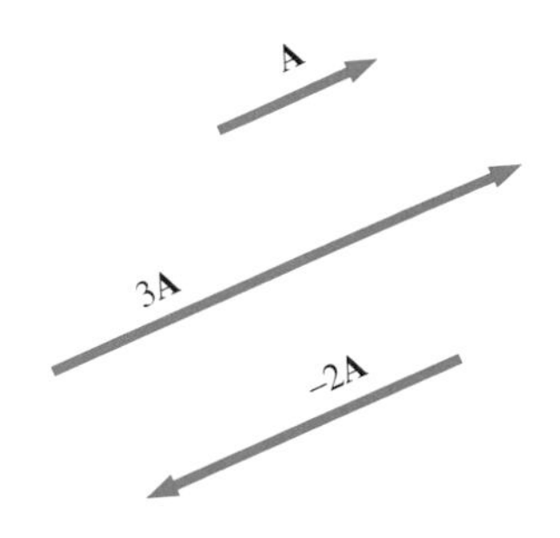

FIGURE 2-9 The multiplication of a vector by a scalar.

Subtraction of Vectors

The vector difference is defined in terms of the vector sum as follows:

$$\mathbf{A} - \mathbf{B} = \mathbf{A} + (-1)\mathbf{B}. \tag{2-3}$$

This operation is illustrated in Fig. 2-10.

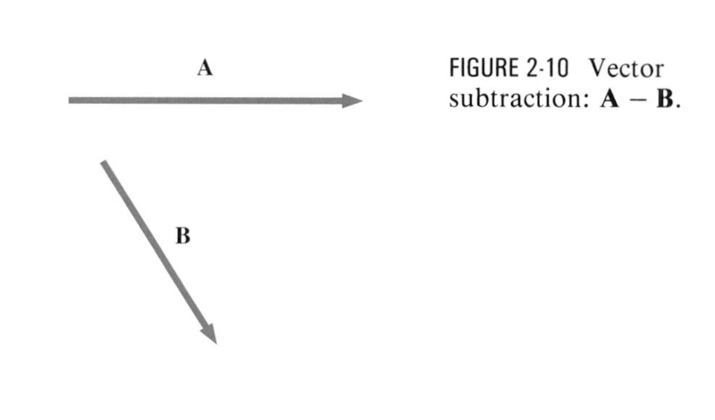

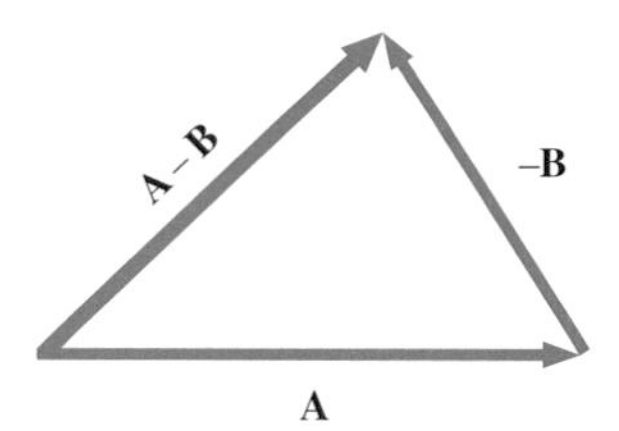

FIGURE 2-10 Vector subtraction: $\mathbf{A} - \mathbf{B}$.

EXAMPLE 2-1 **Geometrical Constructions of Vector Sums**

Choose a convenient scale; then with a protractor and a ruler construct $\mathbf{A} + \mathbf{B}$, $\mathbf{A} - \mathbf{B}$, and $\mathbf{A} - 3\mathbf{B} + \mathbf{C}$ for the vectors given in Fig. 2-11*a*.

SOLUTION Geometrical constructions of the vector sums are shown in Fig. 2-11*b*, *c*, and *d*.

DRILL PROBLEM 2-1

Choose a convenient scale; then with a protractor and a ruler construct $\mathbf{A} + \mathbf{B}$, $\mathbf{A} - \mathbf{B}$, and $\mathbf{A} + 2\mathbf{B} - \mathbf{C}$ for the vectors of Fig. 2-12.
ANS. 5.9 at −8.2°; 16.2 at 49°; 28.2 at 291°.

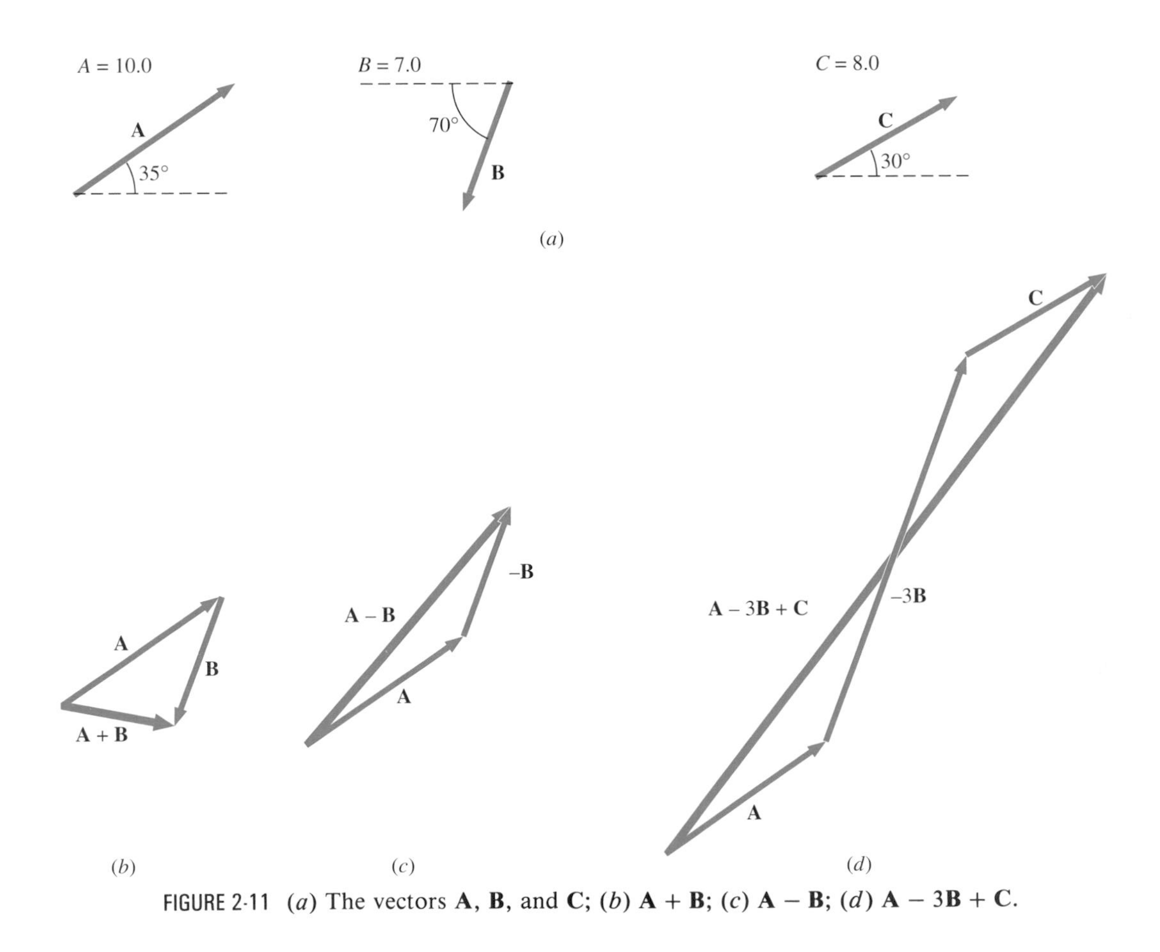

FIGURE 2-11 (*a*) The vectors **A**, **B**, and **C**; (*b*) **A** + **B**; (*c*) **A** − **B**; (*d*) **A** − 3**B** + **C**.

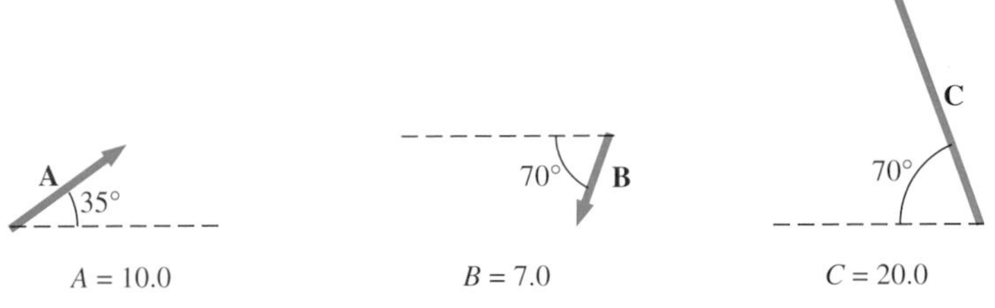

FIGURE 2-12 The vectors of Drill Prob. 2-1.

2-3 Components of a Vector

Vectors are usually described in terms of their *projections* onto the axes of a coordinate system. For example, if someone asks you for directions to a particular location, you might say that it is 40 km to the east and 30 km to the north rather than 50 km at 37° north of east. The coordinate system we most frequently use is the *rectangular coordinate system* with its three perpendicular axes denoted as the x, y, and z axes. Figure 2-13 shows the projections of the planar vector **A** onto the x and y axes. These projections are called the *rectangular components* of vector **A**, where A_x is the x component and A_y is the y component. In terms of the magnitude and direction of the vector itself, A_x and A_y are

$$A_x = A\cos\theta \tag{2-4a}$$

$$A_y = A\sin\theta, \tag{2-4b}$$

where θ is the angle between **A** and the positive x axis.*

The components can be either positive or negative, depending on the direction of the vector and the particular set of coordinate axes used. In Fig. 2-13, both A_x and A_y are positive; however, in the coordinate system of Fig. 2-14, A_x is negative while A_y is positive.

We can express any vector in terms of its components with the help of unit vectors. A **unit vector** is a vector whose magnitude is unity; its sole function is to specify direction. While there are an infinite number of possible unit vectors, we are interested at this point only in those along the x, y, and z axes; these unit vectors are mutually perpendicular and are denoted by **i**, **j**, and **k**, respectively. (See Fig. 2-15.) So in Fig. 2-13, $A_x\mathbf{i}$ is a vector of magnitude A_x, which is parallel to the x axis, and

*Be aware that Eqs. (2-4) are valid *only* if θ is the angle between **A** and the positive x axis! In some cases, θ may represent the angle between **A** and the positive y axis. The components would then be given by $A_x = A\sin\theta$ and $A_y = A\cos\theta$.

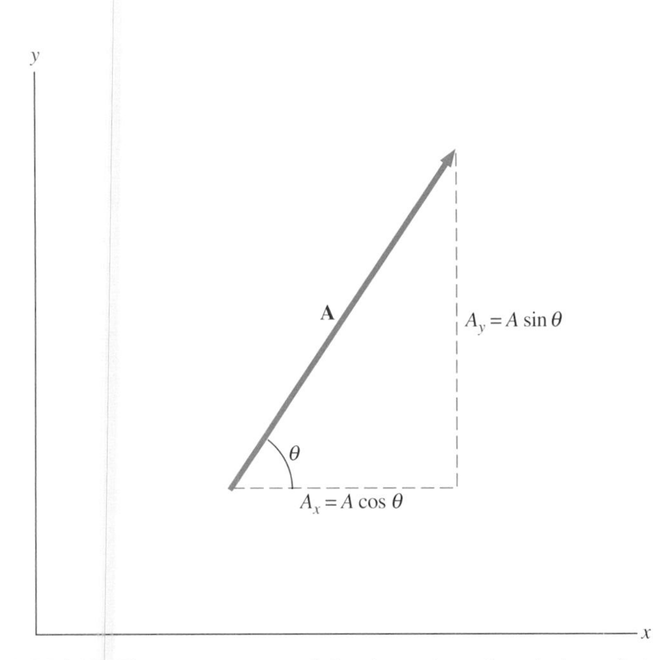

FIGURE 2-13 The components of **A**: $A_x = A\cos\theta$ and $A_y = A\sin\theta$.

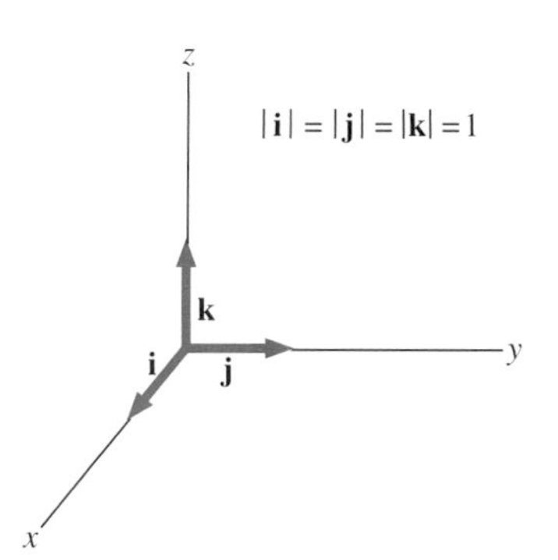

FIGURE 2-15 The unit vectors **i**, **j**, and **k**.

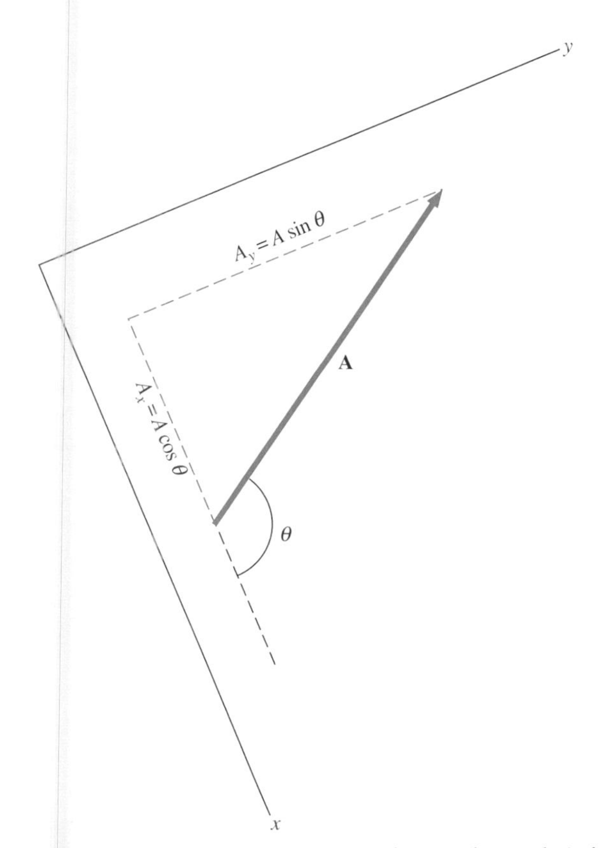

FIGURE 2-14 In this coordinate system, A_x is negative and A_y is positive.

$A_y\mathbf{j}$ is a vector of magnitude A_y, which is parallel to the y axis. The addition of these two vectors then gives us **A** (see Fig. 2-16):

$$\mathbf{A} = A_x\mathbf{i} + A_y\mathbf{j}. \tag{2-5}$$

Vector **A** has been *resolved* into its rectangular components.

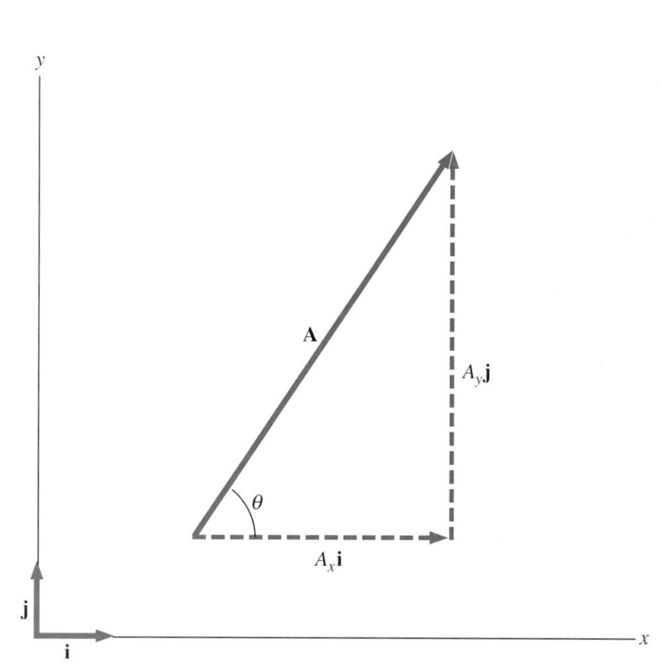

FIGURE 2-16 $\mathbf{A} = A_x\mathbf{i} + A_y\mathbf{j}$.

We have just calculated the components of a vector whose magnitude and direction are given. Now let's do the reverse. Suppose that A_x and A_y are known—how do we find A and θ? With the help of Fig. 2-16 and the Pythagorean theorem, we obtain

$$A = \sqrt{A_x^2 + A_y^2}, \tag{2-6}$$

and

$$\theta = \tan^{-1}\frac{A_y}{A_x}, \tag{2-7}$$

where θ is the angle between **A** and the positive x axis.

Since vector addition is associative, it is easy to add vectors when their components are known. If $\mathbf{A} = A_x\mathbf{i} + A_y\mathbf{j}$ and $\mathbf{B} = B_x\mathbf{i} + B_y\mathbf{j}$, then

$$\mathbf{R} = \mathbf{A} + \mathbf{B} = (A_x + B_x)\mathbf{i} + (A_y + B_y)\mathbf{j}.$$

Thus the rectangular components of **R** are

$$R_x = A_x + B_x \tag{2-8a}$$

and

$$R_y = A_y + B_y; \qquad (2\text{-}8b)$$

that is, the sum of the components in a particular direction is equal to the component of the resultant vector in that direction. This rule also holds in three dimensions, in which case we must include the equation

$$R_z = A_z + B_z. \qquad (2\text{-}8c)$$

Vector arithmetic is done much more efficiently by resolving vectors into their components. While the geometric approach provides insight into the properties of vectors, it is often cumbersome to apply in physical situations. We will generally analyze problems involving vector quantities in terms of their components.

EXAMPLE 2-2 VECTOR COMPONENTS

Express the displacement vectors **A**, **B**, and **C** of Fig. 2-17 in terms of their rectangular components.

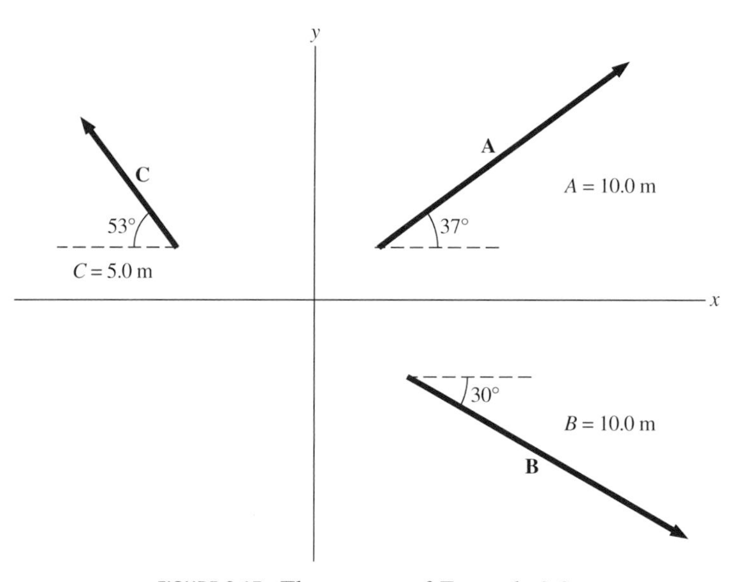

FIGURE 2-17 The vectors of Example 2-2.

SOLUTION Using trigonometry, we find

$$A_x = (10.0 \text{ m}) \cos 37° = 8.0 \text{ m},$$
$$A_y = (10.0 \text{ m}) \sin 37° = 6.0 \text{ m};$$
$$B_x = (10.0 \text{ m}) \cos 30° = 8.7 \text{ m},$$
$$B_y = -(10.0 \text{ m}) \sin 30° = -5.0 \text{ m};$$
$$C_x = -(5.0 \text{ m}) \cos 53° = -3.0 \text{ m},$$
$$C_y = (5.0 \text{ m}) \sin 53° = 4.0 \text{ m}.$$

The three vectors are therefore

$$\mathbf{A} = (8.0\mathbf{i} + 6.0\mathbf{j}) \text{ m};$$
$$\mathbf{B} = (8.7\mathbf{i} - 5.0\mathbf{j}) \text{ m};$$
$$\mathbf{C} = (-3.0\mathbf{i} + 4.0\mathbf{j}) \text{ m}.$$

EXAMPLE 2-3 A SKIER'S DISPLACEMENT

Starting at a ski lodge, a cross-country skier goes 5.0 km to the north, then 3.0 km to the west, and finally 4.0 km to the southwest. (*a*) Using a rectangular coordinate system with the origin at the lodge, the *x* axis pointing east, and the *y* axis pointing north, determine the displacement of the skier for the entire journey. (*b*) What fourth displacement would be required for the skier to return to the lodge?

SOLUTION (*a*) The three displacements that the skier makes are shown in Fig. 2-18. For the chosen coordinate system, these displacements are

$$\mathbf{r}_1 = 5.0\mathbf{j} \text{ km}, \qquad \mathbf{r}_2 = -3.0\mathbf{i} \text{ km},$$

and

$$\mathbf{r}_3 = (-4.0 \cos 45°\mathbf{i} - 4.0 \sin 45°\mathbf{j}) \text{ km} = (-2.8\mathbf{i} - 2.8\mathbf{j}) \text{ km}.$$

The net displacement **R** is

$$\mathbf{R} = \mathbf{r}_1 + \mathbf{r}_2 + \mathbf{r}_3 = (-5.8\mathbf{i} + 2.2\mathbf{j}) \text{ km}.$$

The magnitude and the direction of **R** are then

$$R = \sqrt{(5.8)^2 + (2.2)^2} \text{ km} = 6.2 \text{ km}$$

and

$$\theta = \tan^{-1} \frac{2.2 \text{ km}}{-5.8 \text{ km}} = 159°.$$

The net displacement of the skier is therefore 6.2 km in a direction 21° north of west.

(*b*) When the skier returns to the lodge, her overall displacement is zero. Hence the displacement $\mathbf{r}_4$ required for her return is

$$\mathbf{r}_4 = -\mathbf{R} = (5.8\mathbf{i} - 2.2\mathbf{j}) \text{ km}.$$

This displacement is 6.2 km in a direction 21° south of east.

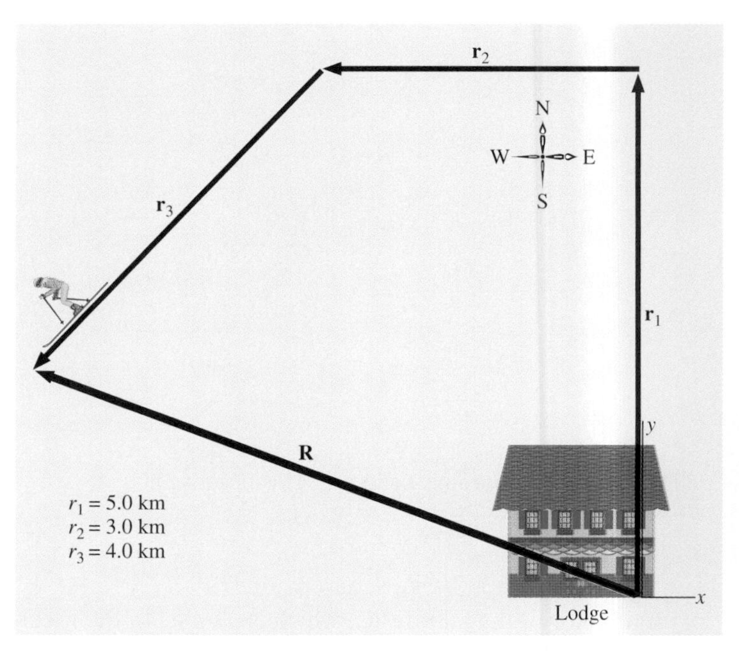

FIGURE 2-18 The displacements of the skier from the lodge are $\mathbf{r}_1$, $\mathbf{r}_2$, and $\mathbf{r}_3$.

EXAMPLE 2-4 Finding the Net Displacement

For the displacement vectors of Fig. 2-17, calculate $\mathbf{R} = \mathbf{A} - 3\mathbf{B} + 2\mathbf{C}$.

SOLUTION From Eqs. (2-8) and Example 2-2,

$$R_x = [8.0 - 3(8.7) + 2(-3.0)] \text{ m} = -24.1 \text{ m},$$

and

$$R_y = [6.0 - 3(-5.0) + 2(4.0)] \text{ m} = 29.0 \text{ m},$$

so

$$\mathbf{R} = (-24.1\mathbf{i} + 29.0\mathbf{j}) \text{ m}.$$

The magnitude and direction of **R** are

$$R = \sqrt{(-24.1)^2 + (29.0)^2} \text{ m} = 37.7 \text{ m}, \qquad (i)$$

and

$$\theta = \tan^{-1} \frac{29.0 \text{ m}}{-24.1 \text{ m}} = 130°. \qquad (ii)$$

This vector is shown in Fig. 2-19.

A warning here about vector directions. In determining θ using a calculator, we would find $-50°$, a direction *opposite* to that of **R**, displayed on the calculator. The reason is that $\tan \phi = \tan(\phi + 180°)$, and a calculator sometimes gives the inappropriate option. This problem can be avoided by using the components to determine which quadrant of the coordinate system the vector lies in. It then becomes clear whether you need to add 180° to the calculator result to obtain the true direction of the vector.

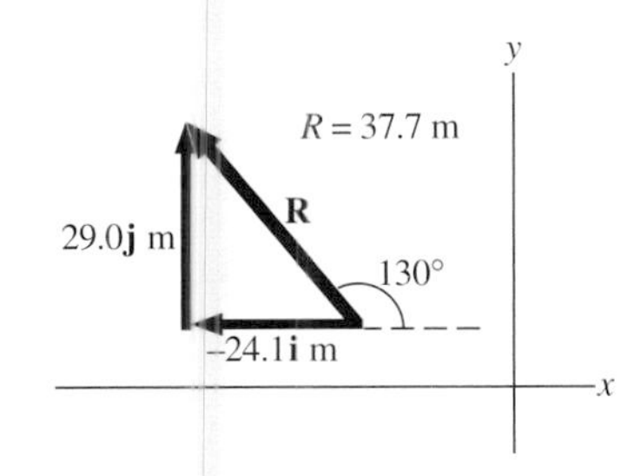

FIGURE 2-19 The resultant vector **R** of Example 2-4.

EXAMPLE 2-5 Running Up Steps

The jogger of Fig. 2-20 runs up a flight of 200 identical steps and stops at a drinking fountain. (*a*) If his displacement from the bottom of the steps (point A) to the fountain (point B) is $(-90\mathbf{i} + 30\mathbf{j})$ m, calculate the height and width of a single step in the flight. (*b*) Compare the actual distance he has run with the magnitude of his displacement. (*c*) After a rest, he jogs back down to point A. What is his overall displacement?

SOLUTION (*a*) If h and w represent the height and width of each of the steps, then the displacement in ascending the steps is given by $(-200w\mathbf{i} + 200h\mathbf{j})$. By adding this to the vector representing his horizontal displacement from the top of the steps $(-50\mathbf{i}$ m), we obtain the displacement from point A to point B:

$$(-200w\mathbf{i} + 200h\mathbf{j}) - 50\mathbf{i} \text{ m} = (-90\mathbf{i} + 30\mathbf{j}) \text{ m}.$$

Since two vectors are equal if and only if their components are equal,

$$(-200w - 50 \text{ m}) = -90 \text{ m},$$
$$(200h) = 30 \text{ m},$$

and $w = 0.20$ m, $h = 0.15$ m.

(*b*) The magnitude of the displacement from A to B is

$$\sqrt{(-90 \text{ m})^2 + (30 \text{ m})^2} = 95 \text{ m}.$$

FIGURE 2-20 A jogger runs up a flight of steps.

The *actual* distance traveled can be found by adding the distance up the flight of steps to the horizontal distance run (50 m). From part (*a*), the displacement representing the ascension of the steps is $(-40\mathbf{i} + 30\mathbf{j})$ m, a vector whose magnitude is

$$\sqrt{(-40 \text{ m})^2 + (30 \text{ m})^2} = 50 \text{ m}.$$

The jogger therefore runs 50 m + 50 m = 100 m, while the magnitude of his displacement is only 95 m.

(*c*) There is no overall displacement if the jogger returns to his starting point; the vector representing the entire trip is the null vector **0**.

DRILL PROBLEM 2-2

Solve Drill Prob. 2-1 by resolving the vectors into their rectangular components.

2-4 MULTIPLICATION OF VECTORS

Vectors may be multiplied by other vectors. There are two vector multiplication operations that are useful. They are the *scalar* (or *dot*) *product* and the *vector* (or *cross*) *product*.*

SCALAR PRODUCT

To find the scalar product of two vectors **A** and **B**, we first arrange them so that their tails are at a common point. (See Fig. 2-21.) Their dot product, which is written as $\mathbf{A}\cdot\mathbf{B}$, is by definition

$$\mathbf{A}\cdot\mathbf{B} = AB\cos\theta, \tag{2-9}$$

where θ is the smaller angle between the vectors. This product combines two vectors to form a scalar. There is no direction associated with it—only a magnitude and a unit. The units of $\mathbf{A}\cdot\mathbf{B}$ are the product of the units of **A** and the units of **B** because $\cos\theta$ is dimensionless.

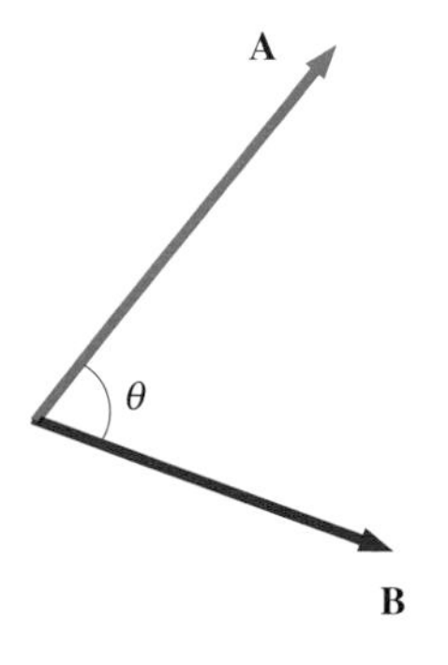

FIGURE 2-21 $\mathbf{A}\cdot\mathbf{B} = AB\cos\theta$.

*The scalar product is first applied in Chap. 7 in the calculation of mechanical work. The vector product is first applied in Chap. 9 when torque is introduced.

The dot product can also be interpreted as either the projection of **A** onto **B** multiplied by B, or as the projection of **B** onto **A** multiplied by A. This interpretation is illustrated in Fig. 2-22.

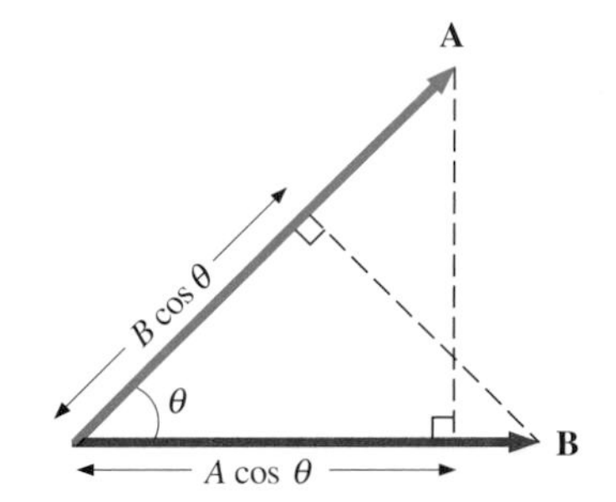

FIGURE 2-22 $\mathbf{A}\cdot\mathbf{B} = (A\cos\theta)B$ [the projection of **A** onto **B** multiplied by B] $= (B\cos\theta)A$ [the projection of **B** onto **A** multiplied by A].

The dot product obeys both the commutative and the distributive laws; that is,

$$\mathbf{A}\cdot\mathbf{B} = \mathbf{B}\cdot\mathbf{A},$$

and

$$\mathbf{A}\cdot(\mathbf{B} + \mathbf{C}) = \mathbf{A}\cdot\mathbf{B} + \mathbf{A}\cdot\mathbf{C}.$$

We often express the dot product of two vectors in terms of their rectangular components. As an example, consider the vectors $\mathbf{A} = A_x\mathbf{i} + A_y\mathbf{j} + A_z\mathbf{k}$ and $\mathbf{B} = B_x\mathbf{i} + B_y\mathbf{j} + B_z\mathbf{k}$. Using the distributive property of the dot product, we obtain

$$\mathbf{A}\cdot\mathbf{B} = A_xB_x\mathbf{i}\cdot\mathbf{i} + A_xB_y\mathbf{i}\cdot\mathbf{j} + \ldots + A_zB_z\mathbf{k}\cdot\mathbf{k}.$$

For the unit vectors the dot products are

$$\mathbf{i}\cdot\mathbf{i} = \mathbf{j}\cdot\mathbf{j} = \mathbf{k}\cdot\mathbf{k} = (1)(1)\cos 0^\circ = 1,$$

and

$$\mathbf{i}\cdot\mathbf{j} = \mathbf{i}\cdot\mathbf{k} = \mathbf{j}\cdot\mathbf{k} = (1)(1)\cos 90^\circ = 0.$$

It then follows that

$$\mathbf{A}\cdot\mathbf{B} = A_xB_x + A_yB_y + A_zB_z. \tag{2-10}$$

VECTOR PRODUCT

The vector (or cross) product is so named because two vectors are multiplied to form a third vector. Vector **C**, obtained by the cross product of **A** and **B**, is denoted by

$$\mathbf{C} = \mathbf{A}\times\mathbf{B}.$$

This product has the following two properties:

1. The magnitude of **C** is given by

$$C = AB\sin\theta, \tag{2-11}$$

where θ is the smaller of the two angles between **A** and **B** when the vectors are arranged so that their tails coincide as shown in Fig. 2-21.

2. **C** is perpendicular to the plane containing **A** and **B**. Of the two possible orientations, the one corresponding to **C** can be determined with the *right-hand rule*: You point the fingers of your right hand along **A** and turn them across θ so they align with **B** (see Fig. 2-23); your erect thumb then points along **A** × **B**.

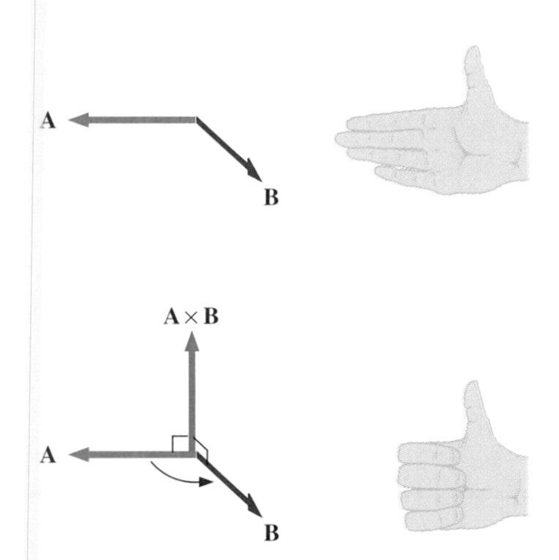

FIGURE 2-23 The direction of the vector cross product is determined by turning **A** into **B**. With the fingers of your right hand pointing along **A**, you rotate them into **B**; then your thumb points along **A** × **B**.

If you align your fingers with **B** and turn them into **A**, you have the direction of the product **B** × **A**. Since your thumb now points in the direction opposite to **A** × **B**,

$$\mathbf{A} \times \mathbf{B} = -\mathbf{B} \times \mathbf{A}.$$

Obviously, the cross product does not obey the commutative law. However, it does obey the distributive law:

$$\mathbf{A} \times (\mathbf{B} + \mathbf{C}) = \mathbf{A} \times \mathbf{B} + \mathbf{A} \times \mathbf{C}.$$

The cross products among the unit vectors are

$$\mathbf{i} \times \mathbf{i} = \mathbf{j} \times \mathbf{j} = \mathbf{k} \times \mathbf{k} = \mathbf{0};$$
$$\mathbf{i} \times \mathbf{j} = \mathbf{k}, \qquad \mathbf{j} \times \mathbf{k} = \mathbf{i}, \qquad \mathbf{k} \times \mathbf{i} = \mathbf{j}.$$

By using these relationships along with the distributive property, we obtain

$$\mathbf{A} \times \mathbf{B} = (A_y B_z - A_z B_y)\mathbf{i} + (A_z B_x - A_x B_z)\mathbf{j} + (A_x B_y - A_y B_x)\mathbf{k}. \quad (2\text{-}12)$$

EXAMPLE 2-6 Vector Products

For the vectors of Fig. 2-24, determine (*a*) **A**·**B**, (*b*) **A**·**C**, (*c*) **A** × **B**, (*d*) **A** × **C**, and (*e*) **B**·(**A** × **C**). (*f*) Is **B** × (**A**·**C**) a meaningful expression? Note that **A** is in the *xy* plane and **C** is in the *xz* plane.

SOLUTION (*a*) From Eq. (2-9),

$$\mathbf{A}\cdot\mathbf{B} = (10.0)(8.0)\cos 60° = 40.0.$$

(*b*) The dot product **A**·**C** cannot be immediately calculated from Eq. (2-9) since the angle between these two vectors is not apparent. However, we can use vector components to determine this dot product. [See Eq. (2-10).] From inspection of the figure,

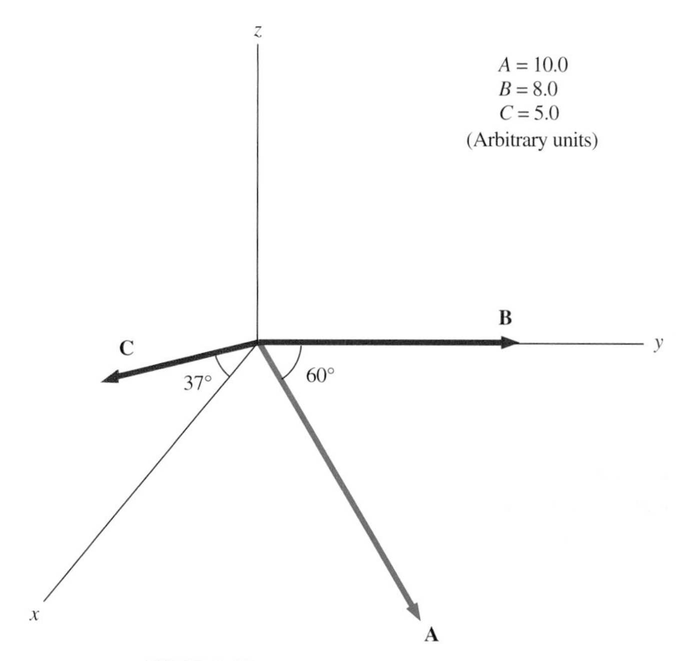

FIGURE 2-24 The vectors of Example 2-6.

$$\mathbf{A} = (10.0 \sin 60°)\mathbf{i} + (10.0 \cos 60°)\mathbf{j} = 8.7\mathbf{i} + 5.0\mathbf{j},$$
$$\mathbf{B} = 8.0\mathbf{j},$$

and

$$\mathbf{C} = (5.0 \cos 37°)\mathbf{i} + (5.0 \sin 37°)\mathbf{k} = 4.0\mathbf{i} + 3.0\mathbf{k},$$

so from Eq. (2-10),

$$\mathbf{A}\cdot\mathbf{C} = (8.7)(4.0) + (5.0)(0) + (0)(3.0) = 34.8.$$

(*c*) **A** and **B** both lie in the *xy* plane. With the right-hand rule and Eq. (2-11) furnishing the direction and magnitude, respectively, of **A** × **B**, we have

$$\mathbf{A} \times \mathbf{B} = (10.0)(8.0)(\sin 60°)\mathbf{k} = 69.3\mathbf{k}.$$

(*d*) The procedure of part (*c*) doesn't work very well for **A** × **C** because neither the angle between **A** and **C** nor the direction of **A** × **C** is obvious from simple geometric considerations. However, using Eq. (2-12), we easily find that

$$\mathbf{A} \times \mathbf{C} = 15.0\mathbf{i} - 26.1\mathbf{j} - 20.0\mathbf{k}.$$

(*e*) To determine **B**·(**A** × **C**), we use the rectangular components of **B** and **A** × **C** in Eq. (2-10):

$$\mathbf{B}\cdot(\mathbf{A} \times \mathbf{C}) = (0)(15.0) + (8.0)(-26.1) + (0)(-20.0) = -209.$$

(*f*) The dot product **A**·**C** is a scalar. Since the cross product is a multiplication operation for two vectors, the expression **B** × (**A**·**C**) is meaningless.

DRILL PROBLEM 2-3

For the vectors of Fig. 2-17, find **A**·**B**, **A**·**C**, **C**·**A**, **A** × **B**, **B** × **A**, and **A** × **C**.
ANS. 39.1; 0; 0; −92.1**k**; 92.1**k**; 50**k**.

SUMMARY

1. **Displacement**
   Vectors are quantities with both magnitude and direction. Displacement is a simple example of a vector and can be used to illustrate the behavior of vectors.
2. **Vector arithmetic**
   (*a*) Vectors are equal when both their magnitudes and directions are equal.
   (*b*) Vectors are added geometrically by laying them out tail-to-head and then drawing the resultant from the tail of the first to the head of the last.
   (*c*) **B** is subtracted from **A** by adding −**B** to **A**.
   (*d*) Vector addition and subtraction satisfy the commutative law and the associative law.
   (*e*) $k\mathbf{A}$ is a vector of magnitude $|k||\mathbf{A}|$ that is directed along **A** if $k > 0$ and opposite to **A** if $k < 0$.
3. **Components of a vector**
   (*a*) The components of a vector are the projections of that vector onto defined coordinate axes.
   (*b*) The vector **A** is written in terms of its components as
   $$\mathbf{A} = A_x\mathbf{i} + A_y\mathbf{j} + A_z\mathbf{k}.$$
   (*c*) In terms of components the sum **A** + **B** is
   $$\mathbf{A} + \mathbf{B} = (A_x + B_x)\mathbf{i} + (A_y + B_y)\mathbf{j} + (A_z + B_z)\mathbf{k}.$$
4. **Scalar and vector products**
   (*a*) $\mathbf{A}\cdot\mathbf{B} = AB\cos\theta = A_xB_x + A_yB_y + A_zB_z$.
   (*b*) $\mathbf{A}\times\mathbf{B}$ is a vector of magnitude $AB\sin\theta$ whose direction is given by the right-hand rule.

QUESTIONS

2-1. Is it possible for two vectors of different magnitude to add to zero? Is it possible for three vectors of different magnitude to add to zero?

2-2. Does the odometer in an automobile indicate a scalar or a vector?

2-3. When a miler crosses the finish line, what is his net displacement? Assume that he is running on a quarter-mile track.

2-4. If one of the two components of a vector is not zero, can the magnitude of the vector be zero?

2-5. A vector has zero magnitude. Is it necessary to specify the direction of the vector?

2-6. If two vectors have the same magnitude, do their components have to be the same?

2-7. Can the magnitude of a vector be negative?

2-8. Can the magnitude of a particle's displacement be greater than the distance traveled?

2-9. Is it possible to add a vector quantity to a scalar quantity?

2-10. If $\mathbf{V}_1 = \mathbf{V}_2$, what can you say about the components of the two vectors?

2-11. If three vectors sum to zero, what geometric condition do they satisfy?

2-12. Can the cross product of a vector with itself be nonzero? Can the dot product of a vector with itself be zero?

2-13. If $\mathbf{L}\times\mathbf{M} = 0$, what can you say about **L** and **M**?

2-14. For two arbitrary vectors $\mathbf{V}_1$ and $\mathbf{V}_2$, does $\mathbf{V}_1\cdot\mathbf{V}_2 = \mathbf{V}_2\cdot\mathbf{V}_1$? Does $\mathbf{V}_1\times\mathbf{V}_2 = \mathbf{V}_2\times\mathbf{V}_1$?

PROBLEMS

The vectors depicted in the figures at the top of the next page are to be used in this problem set whenever **A**, **B**, **C**, **D**, **E**, **F**, **G**, or **H** are referenced.

Vector Arithmetic

2-1. Find graphically (*a*) **A** + **B**; (*b*) **B** + **C**; (*c*) **D** + **E**; (*d*) **A** − **B**; (*e*) **D** − **E**; (*f*) **A** + 2**E**; (*g*) **C** − 2**D** + 3**E**; (*h*) **A** − 4**D** + 2**E**.

2-2. A deliveryman starts at P_1, drives 40 km north, then 20 km west, then 60 km northeast, and finally 50 km north to end up at P_2. What is his net displacement?

2-3. The deliveryman of Prob. 2-2 makes one more trip from P_2 to P_3, which is 50 km directly west of P_1. What is his displacement from P_2 to P_3?

2-4. A dog strays from his home, runs three blocks east, two blocks north, one block east, one block north, and two blocks west. What is his net displacement?

2-5. Using your results of Prob. 2-1, find the vector **R** such that **D** + **R** = **E**.

2-6. The magnitudes of the displacements $\mathbf{R}_1$ and $\mathbf{R}_2$ are 20 and 6 m, respectively. If the directions of $\mathbf{R}_1$ and $\mathbf{R}_2$ are arbitrary, what are the maximum and minimum values of $|\mathbf{R}_1 + \mathbf{R}_2|$?

2-7. Four vectors each have the same magnitude a. What is the maximum magnitude of the vector resulting from their summation? What is the minimum magnitude? Sketch the vector sum that gives the minimum magnitude.

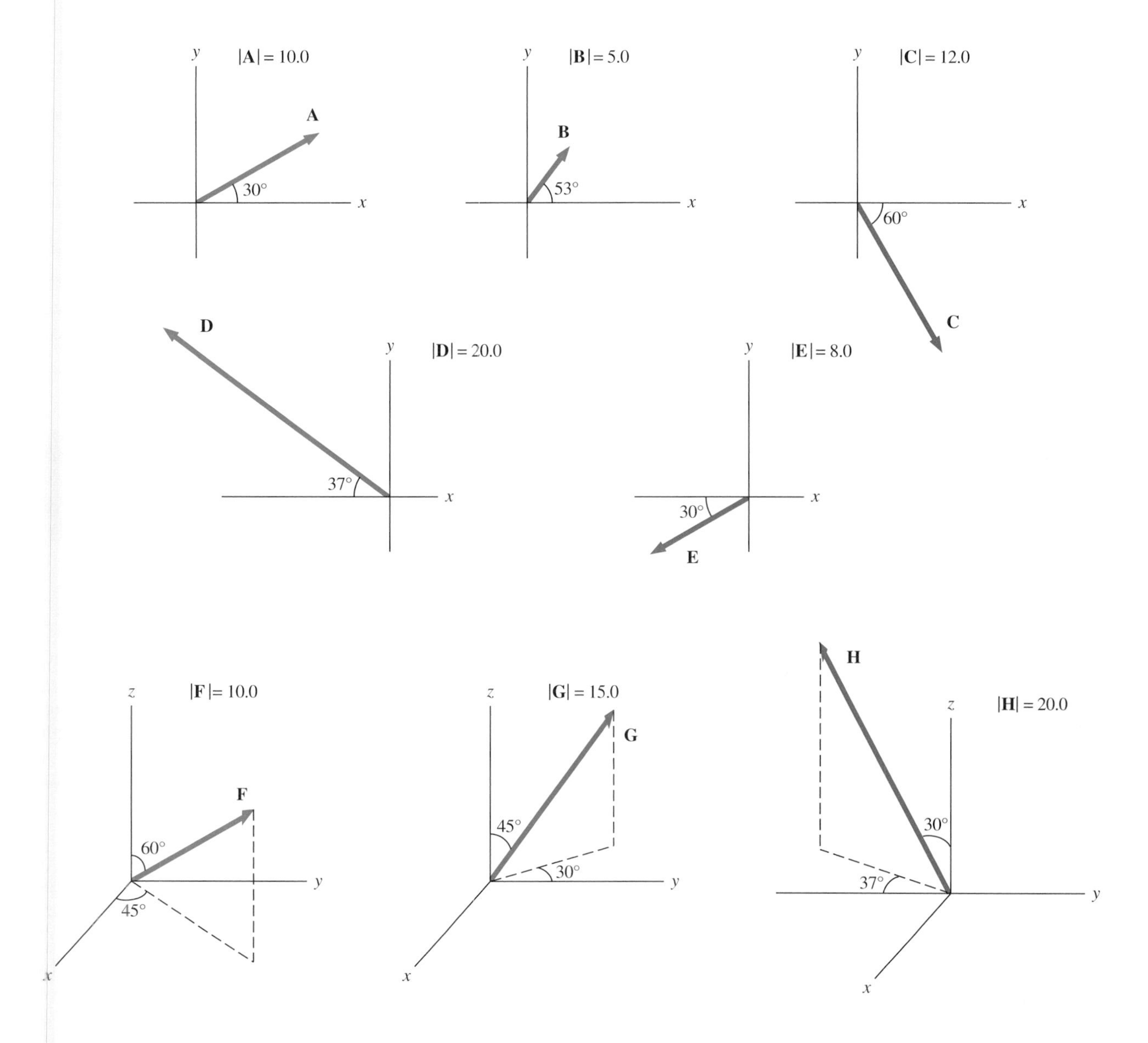

2-8. A circular portion of a roller coaster track of circumference 300 m is shown in the accompanying figure. The origin of the given coordinate system is located at P_4. What is the displacement of a cart relative to the origin when it is at P_1? at P_2? at P_3? at P_4?

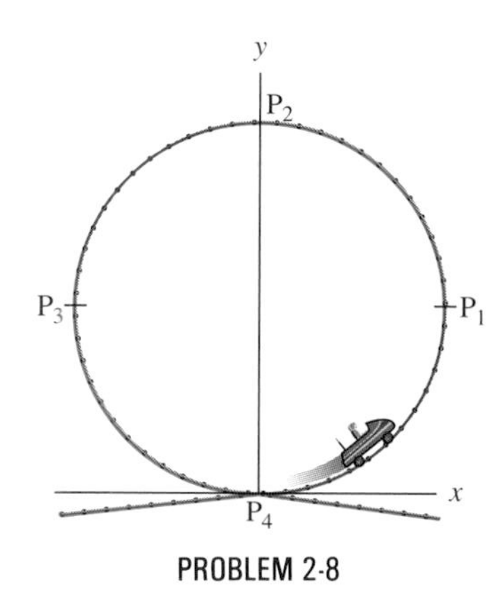

PROBLEM 2-8

Vector Components

2-9. What are the rectangular components of **A**? **D**? **F**? **H**?

2-10. Express the vectors **B**, **C**, and **G** in terms of the rectangular unit vectors.

2-11. Do the calculations of Prob. 2-1 using rectangular vector components.

2-12. An airplane in a power dive is moving at 300 m/s at an angle of 45° below the horizontal. If the sun is directly overhead, how fast is the plane's shadow moving across the ground?

2-13. A particle undergoes three consecutive displacements given by $\mathbf{d}_1 = (3.0\mathbf{i} - 4.0\mathbf{j} - 2.0\mathbf{k})$ cm, $\mathbf{d}_2 = (1.0\mathbf{i} - 7.0\mathbf{j} + 4.0\mathbf{k})$ cm, and $\mathbf{d}_3 = (-7.0\mathbf{i} + 4.0\mathbf{j} + 1.0\mathbf{k})$ cm. (*a*) Find the resultant displacement of the particle. (*b*) What is the magnitude of the resultant displacement? (*c*) If each displacement is made in a straight line, how far does the particle travel?

2-14. The displacement $\mathbf{d}_1$ is given by $\mathbf{d}_1 = (3.0\mathbf{i} - 4.0\mathbf{j})$ m. What is $\mathbf{d}_2$ such that $\mathbf{d}_1 + \mathbf{d}_2 = -4|\mathbf{d}_1|\mathbf{j}$?

2-15. The minute hand of a clock is 10 cm long. With the minute hand's position given by the vector **R**, where $\mathbf{R} = 10\mathbf{j}$ cm on the hour and $\mathbf{R} = 10\mathbf{i}$ cm at 15 min past the hour, what is **R** at (*a*) 30 min, (*b*) 45 min, (*c*) 5 min, (*d*) 20 min, and (*e*) 35 min past the hour?

2-16. Assume that the clock of the previous problem reads 12 noon. What is the displacement of the minute hand 15 min later? 30 min later? 40 min later? 1 h later?

2-17. The three vectors shown in the accompanying figure form an equilateral triangle with sides 2.0 m long. Express the following in terms of rectangular unit vectors: (*a*) $\mathbf{R}_1$, $\mathbf{R}_2$, $\mathbf{R}_3$; (*b*) $\mathbf{R}_1 + \mathbf{R}_2$; (*c*) $\mathbf{R}_2 + \mathbf{R}_3$; (*d*) $\mathbf{R}_1 - \mathbf{R}_2$; (*e*) $\mathbf{R}_1 + \mathbf{R}_2 + \mathbf{R}_3$.

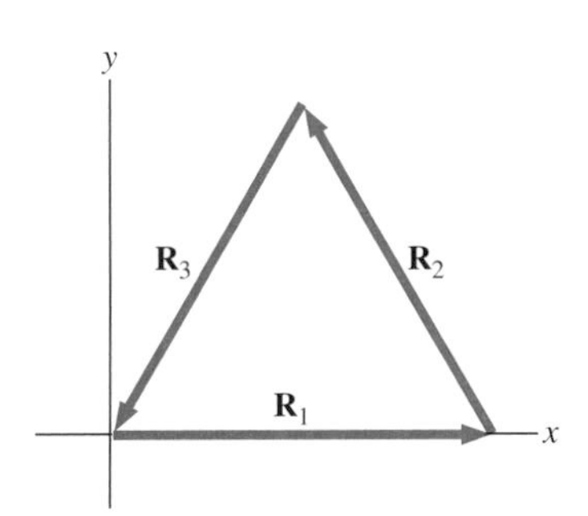

PROBLEM 2-17

2-18. A cube of side $2a$ is centered at the origin of a rectangular coordinate system. What are the vectors to the corners of the cube? What is their sum?

2-19. The four vectors shown in the accompanying figure all have a magnitude of 1.0 m. (*a*) Calculate the sum of these vectors using the component method. (*b*) Check your answer using the graphical method.

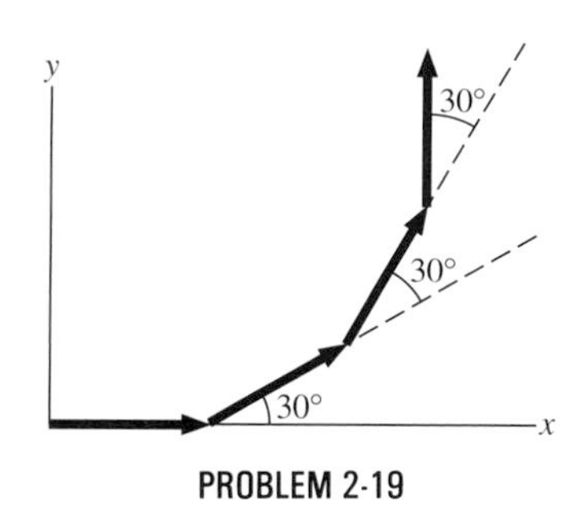

PROBLEM 2-19

Vector Multiplication

2-20. Calculate $\mathbf{A} \cdot \mathbf{B}$ and $\mathbf{B} \cdot \mathbf{C}$ using (*a*) Eq. (2-9) and (*b*) Eq. (2-10).

2-21. What is the angle between **F** and **G**?

2-22. Vector **M** is 5.0 cm long and vector **N** is 4.0 cm long. Find the angle between **M** and **N** if (*a*) $\mathbf{M} + \mathbf{N}$ is 3.0 cm long and (*b*) if $\mathbf{M} - \mathbf{N}$ is 3.0 cm long.

2-23. Find the angles that the vector $\mathbf{M} = 4.0\mathbf{i} + 3.0\mathbf{j} + 5.0\mathbf{k}$ makes with the x, y, and z axes.

2-24. Find M_x such that $\mathbf{M} = M_x\mathbf{i} + 4.0\mathbf{j}$ is perpendicular to $\mathbf{N} = 3.0\mathbf{i} + 6.0\mathbf{j}$.

2-25. Show that $\mathbf{M} = 2.0\mathbf{i} - 4.0\mathbf{j} + 1.0\mathbf{k}$ is perpendicular to $\mathbf{N} = 3.0\mathbf{i} + 4.0\mathbf{j} + 10.0\mathbf{k}$.

2-26. What is the angle between $\mathbf{M} = 1.0\mathbf{i} + 1.0\mathbf{j} + 2.0\mathbf{k}$ and $\mathbf{N} = 1.0\mathbf{i} + 3.0\mathbf{j}$?

2-27. If $\mathbf{R} = \mathbf{R}_1 + \mathbf{R}_2$, prove that $R^2 = R_1^2 + R_2^2 + 2R_1R_2\cos\theta$, where θ is the angle between $\mathbf{R}_1$ and $\mathbf{R}_2$.

2-28. What is the component of $\mathbf{M} = 2.0\mathbf{i} + 1.0\mathbf{j}$ along $\mathbf{N} = 1.0\mathbf{i} + 3.0\mathbf{k}$?

2-29. Calculate $\mathbf{A} \times \mathbf{B}$, $\mathbf{B} \times \mathbf{C}$, $\mathbf{D} \times \mathbf{G}$, and $\mathbf{F} \times \mathbf{H}$.

2-30. (*a*) If $\mathbf{A} \times \mathbf{B} = \mathbf{A} \times \mathbf{C}$, can we conclude that $\mathbf{B} = \mathbf{C}$? (*b*) If $\mathbf{A} \cdot \mathbf{B} = \mathbf{A} \cdot \mathbf{C}$, can we conclude that $\mathbf{B} = \mathbf{C}$?

2-31. **P** and **Q** are vectors in the xy plane, have the same magnitude, and are perpendicular to each other. If $\mathbf{Q} = 3.0\mathbf{i} + 4.0\mathbf{j}$, what is **P**?

2-32. Calculate $(\mathbf{A} \times \mathbf{G}) \cdot \mathbf{H}$ and $\mathbf{A} \cdot (\mathbf{G} \times \mathbf{H})$.

General Problems

2-33. A barge is pulled by two tugboats as shown in the accompanying figure. Using the fact that force is a vector, determine the net force on the barge due to the two tugboats if one pulls with a force of 4000 lb at 15° to AB and the other pulls with a force of 5000 lb at 12° to AB.

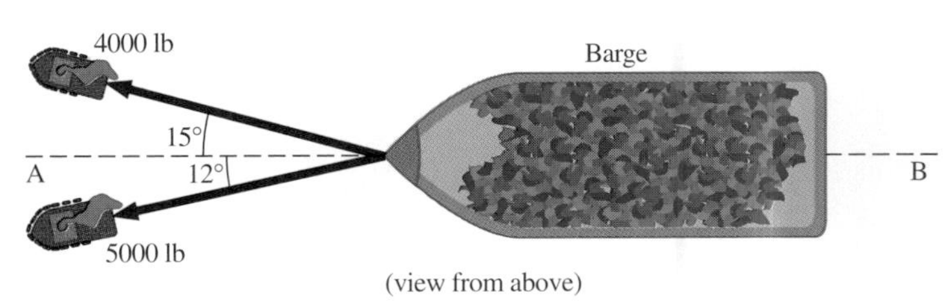

(view from above)

PROBLEM 2-33

2-34. With the control tower at the origin of his reference frame, an air-traffic controller specifies the positions of two planes as follows: A Boeing 747 is at an altitude of 2500 m, 10° above the horizontal, at 30° north of west; and a DC 10 is at an altitude of 3000 m, 5° above the horizontal and directly west. (*a*) Find the displacement of each plane relative to the control tower. (*b*) What is the distance between the planes?

2-35. Determine the equation of a plane that contains the point (x_0, y_0, z_0) and is perpendicular to the vector **M** (*a*) in vector form and (*b*) in terms of rectangular coordinates. [*Hint:* If (x, y, z) is any point in the plane, the vector **V** directed from (x_0, y_0, z_0) to (x, y, z) is perpendicular to **M**.] (*c*) What is the equation of the plane that contains (2, 2, 3) and is perpendicular to the line through the origin and the point (4, 7, 9)? Refer to the accompanying figure.

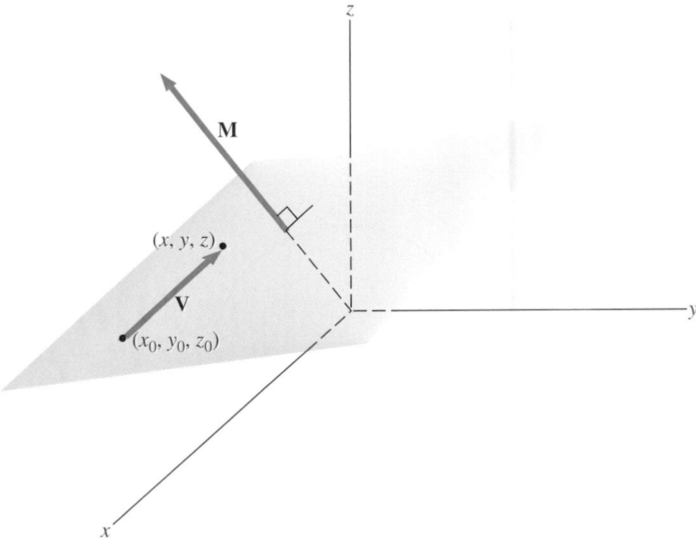

PROBLEM 2-35

2-36. Show that $\mathbf{A} \cdot (\mathbf{B} \times \mathbf{C})$ is the volume of the parallelepiped with edges formed by the three vectors. (See the accompanying figure.)

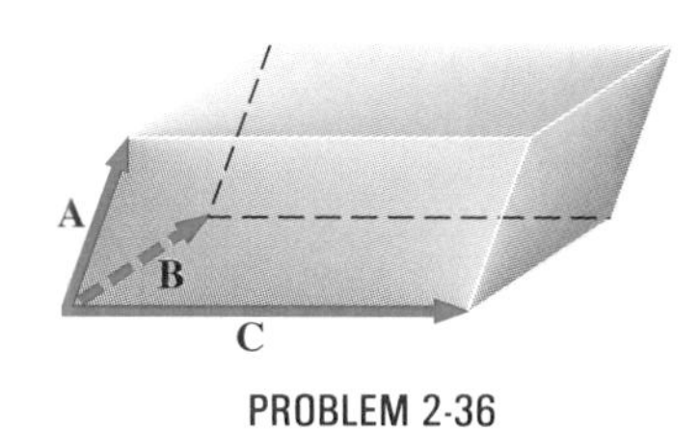

PROBLEM 2-36

2-37. The figure shows a triangle formed by the three vectors **A**, **B**, and **C**. If **C**′ is the vector drawn between the midpoints of **A** and **B**, show that $\mathbf{C}' = \mathbf{C}/2$.

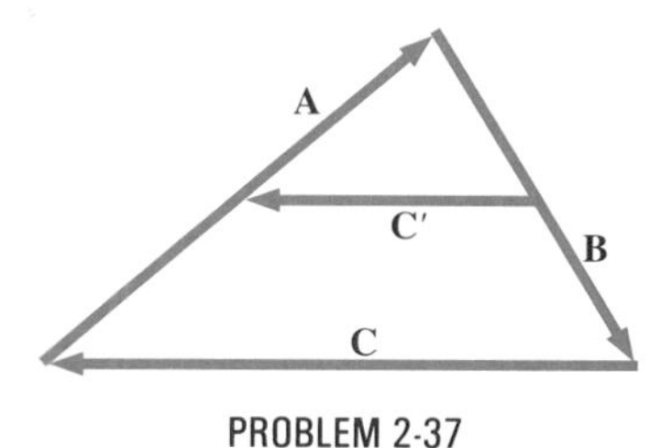

PROBLEM 2-37

2-38. An (x', y') coordinate system is rotated at an angle θ with respect to the (x, y) system as shown in the accompanying figure. (*a*) If (x_1, y_1) and (x_1', y_1') are the coordinates of the point P_1 in the two systems, show that

$$x_1' = x_1 \cos\theta + y_1 \sin\theta \qquad \text{and} \qquad y_1' = -x_1 \sin\theta + y_1 \cos\theta.$$

(*b*) Show that the formula for the distance of P_1 from the origin O is *invariant* under coordinate rotations by deriving the formula

$$\sqrt{x_1^2 + y_1^2} = \sqrt{(x_1')^2 + (y_1')^2}.$$

(*c*) Also show that the distance between the two points P_1 and P_2 is invariant under a coordinate rotation. In this case, you must show that

$$\sqrt{(x_2 - x_1)^2 + (y_2 - y_1)^2} = \sqrt{(x_2' - x_1')^2 + (y_2' - y_1')^2}.$$

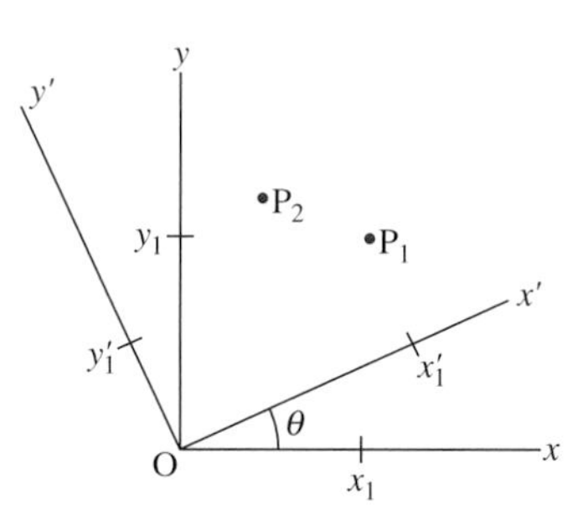

PROBLEM 2-38

Vertical launch of a space shuttle.

CHAPTER 3 ONE-DIMENSIONAL MOTION

PREVIEW

In this chapter we discuss motion along a straight line. Special emphasis is placed on the case where the acceleration is constant. The main topics to be discussed here are the following:

1. **Definitions of position, displacement, velocity, average velocity, speed, and acceleration.** These are the basic quantities used to describe motion.
2. **Graphical displays of position and velocity vs. time.** These graphs are often useful for analyzing motion.
3. **One-dimensional motion with constant acceleration.** The analysis of this motion will be made using relationships between (*a*) velocity and time (Eq. 3-10), (*b*) position and time (Eq. 3-11), and (*c*) velocity and position (Eq. 3-12). Every constant acceleration problem can be solved using one or more of these three equations.
4. **Free fall.** Free fall is a common and important example of one-dimensional motion with constant acceleration.

With this chapter we begin a comprehensive treatment of *mechanics*, which is basically the study of the motion of material objects. Systems as diverse as a rotating baseball, a vibrating guitar string, a planet orbiting the sun, and a spinning top are governed by the laws of mechanics. The study of motion is generally divided into two parts: *kinematics*, which is the description of motion without regard to its causes, and *dynamics*, which is an investigation of those causes.

For simplicity, kinematics (Chaps. 3 and 4) and dynamics (Chaps. 5 and 6) will first be discussed for bodies that can be treated as *particles*, which are point objects. This idealization is justified in the many cases where the spatial dimensions of an object can be ignored. For example, even a huge body like the earth can be assumed to be a particle when its motion around the sun is considered. This approximation is quite accurate because the earth is so much smaller than the range of its orbit. A body that moves such that all points on it follow parallel paths may also be treated as a particle. If we know the position of one point (or particle) on the body of Fig. 3-1, we can determine the position of every other point (or particle) on that body. Of course, there are many interesting situations in which the dimensions of an object simply cannot be ignored. The motion of extended bodies is a topic that will be considered in later chapters.

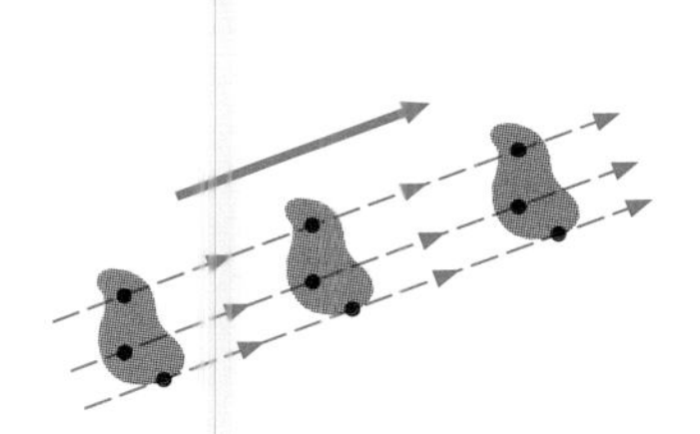

FIGURE 3-1 The motion of this object is such that all its points follow parallel paths.

Our study of kinematics begins with motion along a straight line, or *one-dimensional motion*. The basic quantities—position, displacement, velocity, and acceleration—are first introduced; then their roles in describing motion are discussed. The case of constant acceleration is emphasized, as it occurs (at least approximately) often in real life. One simple example of such motion is that of a ball tossed straight up in the air.

3-1 Position and Displacement

The **position** of a particle is a vector that is specified with respect to a chosen coordinate system. By definition, the position vector $\mathbf{r}(t) = x(t)\mathbf{i} + y(t)\mathbf{j} + z(t)\mathbf{k}$ is directed from the origin of the coordinate system to the particle at (x, y, z) as shown in Fig. 3-2. The position vector is written as a function of time t, because it changes if the particle is in motion.

Since only one-dimensional motion is presently being considered, position can be represented by

$$\mathbf{r}(t) = x(t)\mathbf{i}, \tag{3-1}$$

where it is assumed that the motion of interest is along the x axis. Consider as an example the car of Fig. 3-3*a* which is moving down a long, straight street. With the street lamp as the origin of our coordinate system (Fig. 3-3*b*), the position of the car at A is given by $+200\mathbf{i}$ m, and its later position at B is -100**i** m. Of course, any of the other coordinate axes can be used to describe one-dimensional motion. For example, you can express the position as $\mathbf{r}(t) = y(t)\mathbf{j}$ or $\mathbf{r}(t) = z(t)\mathbf{k}$.

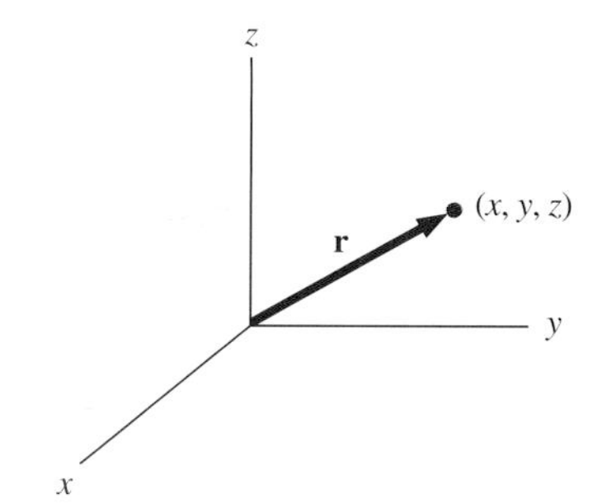

FIGURE 3-2 The vector $\mathbf{r} = x\mathbf{i} + y\mathbf{j} + z\mathbf{k}$ drawn from the origin of the coordinate system to the particle located at (x, y, z) is the position of the particle.

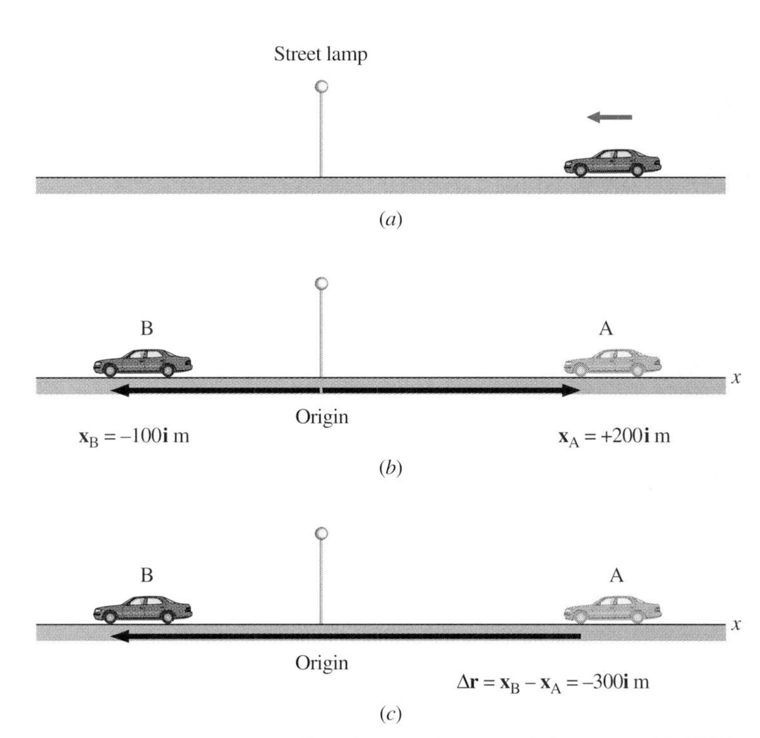

FIGURE 3-3 (*a*) A car traveling down a long, straight street. (*b*) With the street lamp as the origin, the positions of the car at A and B are $+200\mathbf{i}$ m and $-100\mathbf{i}$ m, respectively. (*c*) The displacement of the car from A to B is $-300\mathbf{i}$ m.

In Chap. 2, displacement was defined as a change in position. If the positions of a particle in one-dimensional motion at times t_1 and t_2 are given by $x(t_1)\mathbf{i}$ and $x(t_2)\mathbf{i}$, respectively, its displacement between these times is

$$\Delta\mathbf{r} = \Delta x\mathbf{i} = [x(t_2) - x(t_1)]\mathbf{i}. \tag{3-2}$$

The displacement vector $\Delta\mathbf{r}$ is drawn with its tail at the initial position and its head at the final position. Hence when the car travels from A to B as shown in Fig. 3-3*c*, its displacement is directed from A to B and equals $-300\mathbf{i}$ m.

In analyzing motion along a straight line, it is convenient to suppress the unit vector when expressing quantities like position and displacement (and, as you will see, velocity and acceleration). So the position of the car of Fig. 3-3 at point A is written as simply $x(t_1) = 200$ m, its position at B is $x(t_2) = -100$ m, and its displacement from A to B is $x = x(t_2) - x(t_1) = -300$ m. Do keep in mind, however, that position and displacement are *vectors*! A positive value for position, displacement, velocity, or acceleration means that the corresponding vector points in the positive x direction, and a negative value means that the vector points in the negative x direction.

EXAMPLE 3-1 **ONE-DIMENSIONAL DISPLACEMENT**

Between $t = 2.0$ and 4.0 s, a particle moves from $x(2.0\text{ s}) = 7.2$ m to $x(4.0\text{ s}) = -4.4$ m. What is the displacement of the particle in this time interval?

SOLUTION Since displacement is the change in position,

$$\Delta x = x(4.0\text{ s}) - x(2.0\text{ s}) = [(-4.4) - (7.2)]\text{ m} = -11.6\text{ m}.$$

The negative sign designates that the displacement vector points in the negative x direction.

DRILL PROBLEM 3-1

Suppose that the origin of the coordinate system in Example 3-1 is moved 2.0 m in the positive x direction. Find $x(2.0\text{ s})$, $x(4.0\text{ s})$, and Δx.
ANS. 5.2 m; −6.4 m; −11.6 m.

3-2 VELOCITY

In describing the motion of a particle, we must state "where it is" (its position) along with "how fast" it is moving. This latter property is given by the rate at which the position changes with respect to time. If the details of the motion at each instant are not important, the rate is usually expressed in terms of the **average velocity**. This vector quantity is simply *the net displacement between two points divided by the time taken to travel between those points*. If $x(t_1)\mathbf{i}$ and $x(t_2)\mathbf{i}$ are the positions of the particle at times t_1 and t_2, respectively, the average velocity $\bar{\mathbf{v}}$ between these two times is then

$$\bar{\mathbf{v}} = \frac{x(t_2)\mathbf{i} - x(t_1)\mathbf{i}}{t_2 - t_1}, \tag{3-3}$$

where the bar above **v** represents the fact that this quantity is an average. For example, if the car of Fig. 3-3 travels from A to B in 15 s, its average velocity is

$$\bar{v} = \frac{-300\text{ m}}{15\text{ s}} = -20\text{ m/s},$$

where the unit vector has been suppressed. Since $\bar{v}$ is negative, the average velocity vector points in the negative x direction. Average velocity is the ratio of a length to a time quantity and therefore has dimensions of length per time. Typically, its units are meters per second (m/s), centimeters per second (cm/s), feet per second (ft/s), and miles per hour (mi/h).

Notice that average velocity is defined in terms of the *net displacement* and not the distance traveled. If you drive 2 mi to the store and then return home, you have traveled a distance of 4 mi. However, your net displacement is zero. Thus your average velocity for the entire trip is also zero.

EXAMPLE 3-2 **AVERAGE VELOCITY**

The mail carrier of Fig. 3-4 starts at the corner, drives 200 m to the Smith home in 20 s, stops 10 s to deliver the family's mail, drives 300 m farther to the Perez home in 22 s, stops for 15 s to deliver that mail, then turns around and returns to the corner in 40 s. (*a*) What is the mail carrier's average velocity between the time he leaves the corner and the time he arrives at the Smiths'? (*b*) between the times he leaves the corner and arrives at the Perezes'? (*c*) between arriving at the Smiths' and leaving the Perezes'? (*d*) between leaving the Perezes' and arriving at the corner? (*e*) between leaving and returning to the corner? Use the coordinate system shown in the figure.

SOLUTION In each case, we take the displacement and divide it by the time interval:

(*a*) $$\bar{v} = \frac{200\text{ m}}{20\text{ s}} = 10\text{ m/s};$$

(*b*) $$\bar{v} = \frac{200\text{ m} + 300\text{ m}}{20\text{ s} + 10\text{ s} + 22\text{ s}} = 9.6\text{ m/s};$$

(*c*) $$\bar{v} = \frac{300\text{ m}}{10\text{ s} + 22\text{ s} + 15\text{ s}} = 6.4\text{ m/s};$$

(*d*) $$\bar{v} = \frac{-500\text{ m}}{40\text{ s}} = -12.5\text{ m/s};$$

(*e*) $$\bar{v} = \frac{0}{20\text{ s} + 10\text{ s} + 22\text{ s} + 15\text{ s} + 40\text{ s}} = 0.$$

The quantity that tells us how fast an object is moving anywhere along its path is the **instantaneous velocity** $\mathbf{v}(t)$ (usually called simply the **velocity**). It is *the average velocity between two points on the path in the limit that the time (and therefore the displacement) between the points approaches zero*. The difference between the average and instantaneous velocities is illustrated in Fig. 3-5, which depicts a particle traveling from A to B. The positions of the particle at A and B are $x(t_A)$ and $x(t_B)$, respectively, and the particle's average velocity between these two points is $\bar{v} = [x(t_B) - x(t_A)]/(t_B - t_A)$. Suppose that at some arbitrary time t during the trip, the particle is at $x(t)$ (point P on the path). To find the instantaneous velocity at P, we take position $x(t)$ at time t and position $x(t + \Delta t)$ at time $t + \Delta t$ and calculate

$$\mathbf{v}(t) = \lim_{\Delta t \to 0} \frac{x(t + \Delta t)\mathbf{i} - x(t)\mathbf{i}}{\Delta t} = \frac{dx(t)}{dt}\,\mathbf{i}. \tag{3-4a}$$

FIGURE 3-4 A mail carrier delivering to homes. The origin of the coordinate system is at the street corner.

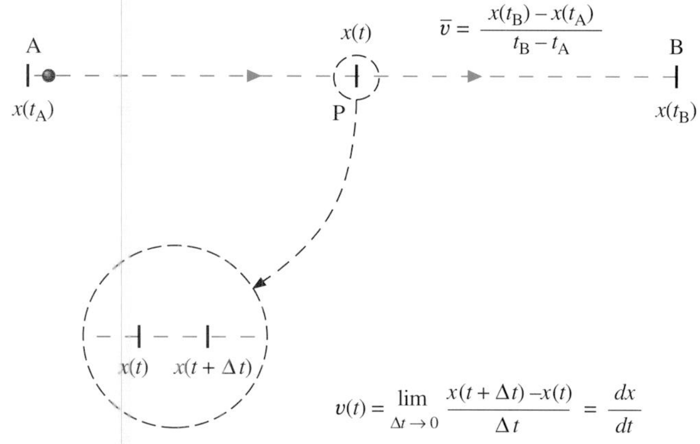

FIGURE 3-5 A particle travels from A to B where its respective positions are $x(t_A)$ and $x(t_B)$. Its average velocity over the trip is $\bar{v} = [x(t_B) - x(t_A)]/(t_B - t_A)$, while its instantaneous velocity is $v = dx/dt$ at any point.

In the notation of calculus, the limit of Eq. (3-4*a*) is called the *derivative* of x with respect to t, or $dx(t)/dt$. Like average velocity, instantaneous velocity is a vector with dimensions of length per time. If we write

$$\mathbf{v}(t) = v(t)\mathbf{i},$$

then

$$v(t) = \frac{dx(t)}{dt}. \tag{3-4b}$$

The magnitude of the instantaneous velocity is known as the instantaneous **speed** (usually referred to simply as speed); that is, instantaneous speed = $|\mathbf{v}(t)|$. A particle moving along the x axis at $+7.0$ m/s and one moving along the same axis at -7.0 m/s both have a speed of 7.0 m/s. However, their velocities are different, as the particles are traveling in opposite directions. Since speed is the magnitude of a vector, it is a scalar quantity.

At this point, we only consider positions and velocities that are polynomial functions of time. The derivatives of such functions can easily be determined by noting that for $f(t) = At^n$, where A and n are constants, the time derivative is

$$\frac{df(t)}{dt} = nAt^{n-1}. \tag{3-5}$$

This equation is discussed in detail in Supplement 3-1 at the end of this chapter.

EXAMPLE 3-3 INSTANTANEOUS VELOCITY

The position of a particle is given by

$$x(t) = (3.0t + 0.50t^3) \text{ m}.$$

(*a*) Find an approximate value for the instantaneous velocity at $t = 5.0$ s by calculating the average velocity between 4.0 and 6.0 s. (*b*) Using Eq. (3-5), find the instantaneous velocity at $t = 5.0$ s.

SOLUTION (*a*) To determine the average velocity between 4.0 and 6.0 s, we must first obtain the values of $x(4.0\text{ s})$ and $x(6.0\text{ s})$. They are

$$x(4.0\text{ s}) = [(3.0)(4.0) + (0.50)(4.0)^3]\text{ m} = 44\text{ m},$$

and

$$x(6.0\text{ s}) = [(3.0)(6.0) + (0.50)(6.0)^3]\text{ m} = 126\text{ m}.$$

Then

$$\bar{v} = \frac{126\text{ m} - 44\text{ m}}{6.0\text{ s} - 4.0\text{ s}} = 41.0\text{ m/s}.$$

This gives us the *approximate* value of the instantaneous velocity at $t = 5.0$ s.

(*b*) The instantaneous velocity is determined exactly by differentiating $x(t)$ with respect to time. From Eq. (3-5),

$$v(t) = \frac{dx(t)}{dt} = (3.0 + 1.5t^2)\text{ m/s}.$$

Upon substituting $t = 5.0$ s into this equation, we find

$$v(5.0\text{ s}) = [3.0 + (1.5)(5.0)^2]\text{ m/s} = 40.5\text{ m/s}.$$

In this example, the average velocity $\bar{v}$ between 4.0 and 6.0 s is nearly the same as the instantaneous velocity v at 5.0 s. In the limit that the time interval used to calculate $\bar{v}$ approaches zero, the value obtained for $\bar{v}$ would converge to the value of v. The connection between average and instantaneous values will be further examined in Example 3-4.

DRILL PROBLEM 3-2

The position of a particle is given by $x(t) = 3.0t^2$ m. (*a*) Find the instantaneous velocity at $t = 2.0$ s using Eq. (3-5). (*b*) What is the average velocity between 2.0 and 3.0 s?
ANS. (*a*) 12.0 m/s; (*b*) 15.0 m/s.

3-3 Acceleration

The time rate of change of the velocity of a particle at any instant is called its **acceleration**. By definition, the acceleration at time t of a particle moving in a straight line is

$$\mathbf{a}(t) = \lim_{\Delta t \to 0} \frac{v(t + \Delta t)\mathbf{i} - v(t)\mathbf{i}}{\Delta t} = \frac{dv(t)}{dt}\,\mathbf{i}. \qquad (3\text{-}6a)$$

$$a(t) = \frac{dv(t)}{dt}. \qquad (3\text{-}6b)$$

Like velocity, acceleration is a vector. Its dimensions are (length per time)/time = length per time squared. The common acceleration units are meters per second squared (m/s^2), centimeters per second squared (cm/s^2), and feet per second squared (ft/s^2). The acceleration vector is positive if it points in the positive x direction and negative if it is directed the other way. The *relative* signs of $v(t)$ and $a(t)$ at a given time are also important. If both quantities have the same sign (either positive or negative), the magnitude of the velocity is increasing and the particle is "speeding up." If $v(t)$ and $a(t)$ have opposite signs [$v(t) > 0$ and $a(t) < 0$, or vice versa], the magnitude of the velocity is decreasing and the particle is "slowing down." In this case, we sometimes say that the particle is "decelerating."

EXAMPLE 3-4 **Acceleration of a Speedboat**

An experimental speedboat is moving at a constant velocity $v(t) =$ 30 m/s when rockets in its tail are fired. The velocity then increases according to

$$v(t) = (30 + 3.0t^2)\text{ m/s}.$$

(*a*) Calculate the boat's velocity at $t = 5.000$ s and at $t + \Delta t$, where the values of Δt are 0.100, 0.010, and 0.001 s, respectively. (*b*) Estimate the boat's acceleration at $t = 5.000$ s by determining $[v(t + \Delta t) - v(t)]/\Delta t$ for the three values of Δt. (*c*) Differentiate $v(t)$ to determine $a(t)$ exactly at $t = 5.000$ s. Compare your result here with the results of part (*b*).

SOLUTION (*a*) Substituting the various times into the equation for $v(t)$, we obtain

$$v(5.000\text{ s}) = 105\text{ m/s},$$
$$v(5.100\text{ s}) = 108.03\text{ m/s},$$
$$v(5.010\text{ s}) = 105.3003\text{ m/s},$$
$$v(5.001\text{ s}) = 105.030003\text{ m/s}.$$

(*b*) For $\Delta t = 0.100$ s, we have for the approximate acceleration at $t = 5.000$ s

$$\frac{\Delta v}{\Delta t} = \frac{108.030 - 105.000}{0.100}\text{ m/s}^2 = 30.3\text{ m/s}^2.$$

Similarly, for $\Delta t = 0.010$ s, the approximate acceleration is

$$\frac{\Delta v}{\Delta t} = 30.03\text{ m/s}^2,$$

and for $\Delta t = 0.001$ s,

$$\frac{\Delta v}{\Delta t} = 30.003\text{ m/s}^2.$$

(*c*) The exact acceleration is

$$a(t) = \frac{dv(t)}{dt} = 6.0t\text{ m/s}^2,$$

which at $t = 5.000$ s is 30 m/s^2.

Notice that in accordance with the definition of the derivative, the values of $[v(t + \Delta t) - v(t)]/\Delta t$ determined in part (*b*) approach 30 m/s^2 as $\Delta t \to 0$.

DRILL PROBLEM 3-3

A flowerpot is accidentally knocked off the edge of a balcony 50 m above the ground. The position of the flowerpot relative to the ground at time t is given approximately by $50 - 5.0t^2$. Use Eq. (3-5) to find the velocity and acceleration of the flowerpot at $t = 2.0$ s.
ANS. -20 m/s; -10 m/s^2.

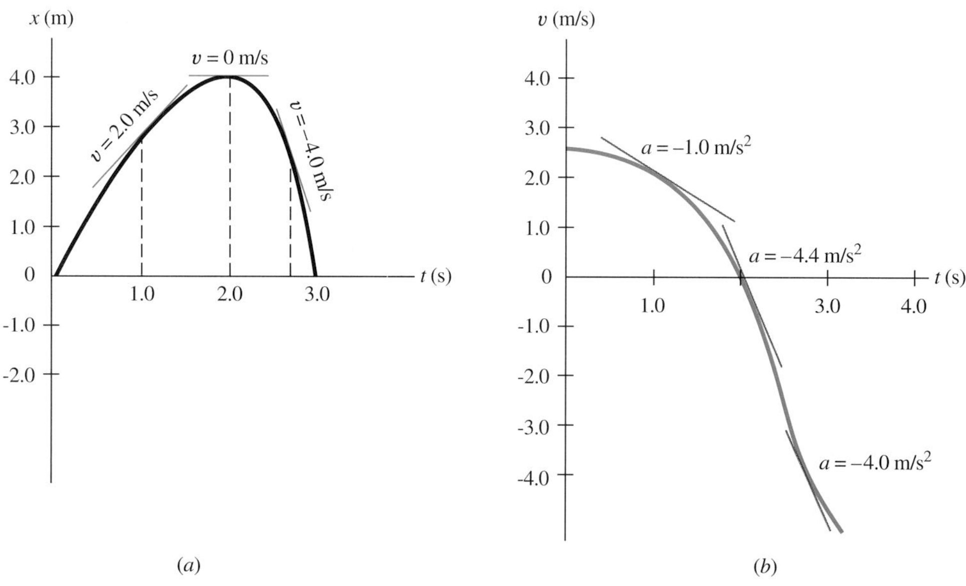

FIGURE 3-6 (*a*) The slope of the position-vs.-time graph is the velocity. (*b*) The slope of the velocity-vs.-time graph is the acceleration.

3-4 Graphical Representation of One-Dimensional Motion

Graphs of position vs. time and of velocity vs. time are very useful for studying one-dimensional motion. As discussed in Supplement 3-1, the graphical representation of the derivative of a function at a particular point is the slope of the straight line tangent at that point to the curve representing the function. Since velocity is the time derivative of position, the slope of the tangent line to the position-vs.-time curve at any point is the velocity, as illustrated in Fig. 3-6*a*. A positive slope indicates that the velocity is positive, a zero slope indicates zero velocity, and a negative slope indicates that the velocity is negative. Acceleration is the time derivative of velocity, so the slope of the tangent line to the velocity-vs.-time curve (see Fig. 3-6*b*) is the acceleration.

The velocity-vs.-time curve can also be used to determine displacement. In Fig. 3-7, the interval from t_1 to t_2 is divided into a number of small intervals of width Δt_i. The displacement during the time interval centered at Δt_i (the blue area) is

$$\Delta x_i = \bar{v}(t_i)\,\Delta t_i, \tag{3-7}$$

where $\bar{v}(t_i)$ is the average velocity over the time interval. The net displacement between t_1 and t_2 is equal to the sum of the displacements in all of the intervals between t_1 and t_2. Therefore

$$x(t_2) - x(t_1) = \sum_i \bar{v}(t_i)\,\Delta t_i. \tag{3-8}$$

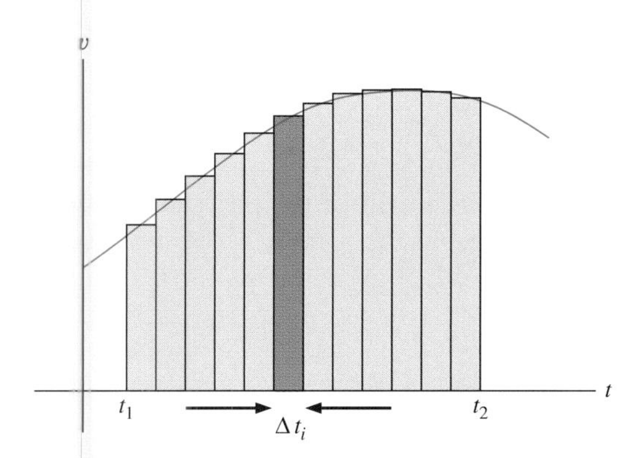

FIGURE 3-7 A plot of $v(t)$ versus t. The displacement from t_1 to t_2 is approximately the sum of the areas of the rectangles shown. In the limit as Δt_i goes to zero, this sum becomes the area under the curve.

Now $\bar{v}(t_i)$ and Δt_i are the height and width, respectively, of the ith rectangle such as the one shaded blue in Fig. 3-7. Consequently, the net displacement from t_1 to t_2 is equal to the sum of the areas of the rectangles.

If the intervals Δt_i are made smaller and smaller, the number of terms in the sum of Eq. (3-8) increases. Furthermore, $\bar{v}(t_i)$ in every interval gets closer and closer to the instantaneous velocity $v(t_i)$ in the middle of the interval. Thus in the limit as $\Delta t_i \to 0$ and the number of terms in the sum approaches infinity, the sum approaches a value equal to the area under the velocity-time curve between t_1 and t_2; that is,

$$x(t_2) - x(t_1) = \lim_{\Delta t_i \to 0} \sum_i v(t_i)\,\Delta t_i = \left\{ \begin{array}{c} \text{area under} \\ v(t) \text{ between} \\ t_1 \text{ and } t_2 \end{array} \right\}. \tag{3-9}$$

To summarize, if you know how velocity varies with time for one-dimensional motion, *you can determine the displacement between t_1 and t_2 by measuring the area under the $v(t)$ curve between t_1 and t_2.*

EXAMPLE 3-5 GRAPHICAL DETERMINATION OF VELOCITY AND ACCELERATION

A car's position varies with time as shown in Fig. 3-8. What is the velocity of the car? What is its acceleration?

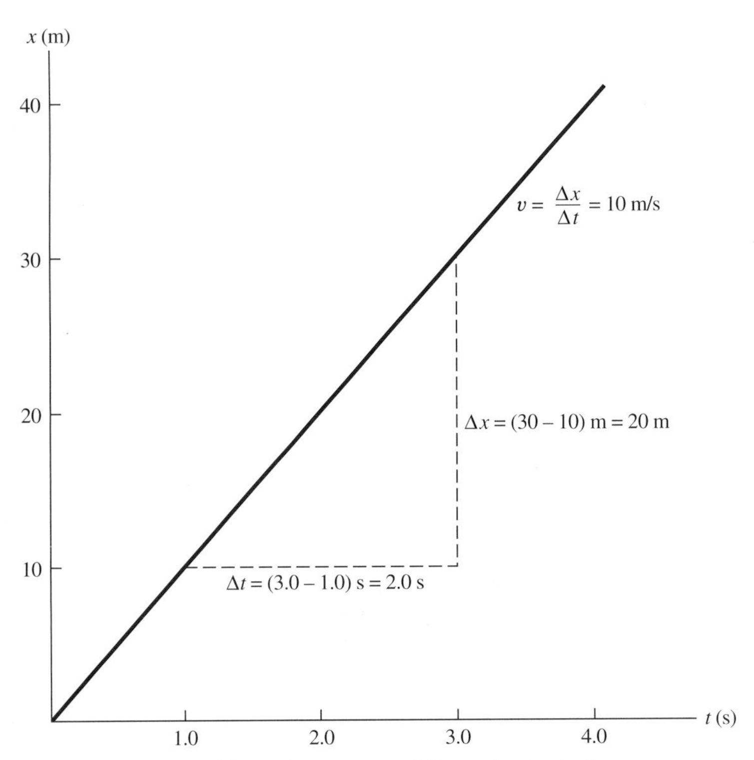

FIGURE 3-8 The position of an automobile varies with time as shown.

SOLUTION The velocity of the car is equal to the slope of the position-vs.-time curve. In this case the slope is constant, so using the interval between 1.0 and 3.0 s, we have

$$v = \frac{30 \text{ m} - 10 \text{ m}}{3.0 \text{ s} - 1.0 \text{ s}} = 10 \text{ m/s}.$$

Since the velocity is constant, its time derivative is zero, and the car is not accelerating.

EXAMPLE 3-6 GRAPHICAL DETERMINATION OF ACCELERATION AND DISPLACEMENT

The velocity of a radio-controlled toy car over the first 10.0 s of its motion is shown in Fig. 3-9. (*a*) Find the acceleration of the toy car at $t = 2.0$ s, $t = 4.0$ s, and $t = 8.0$ s. (*b*) Find the displacement of the toy car in the first 5.0 s and between $t = 3.0$ s and $t = 10.0$ s.

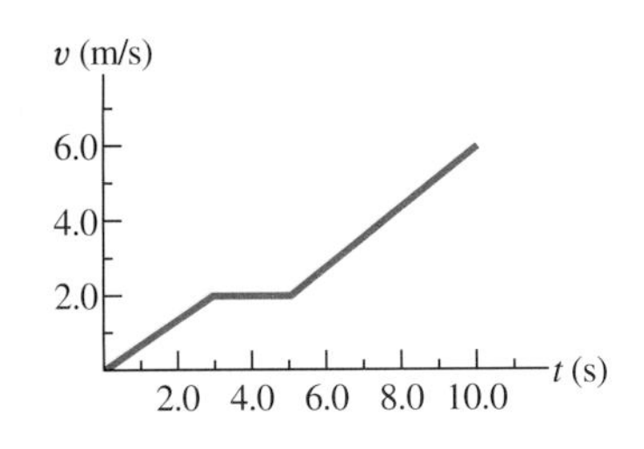

FIGURE 3-9 The velocity of a radio-controlled toy car varies with time as shown.

SOLUTION (*a*) The acceleration is the slope of the curve of Fig. 3-9, so

$$a(2.0 \text{ s}) = \frac{2.0 \text{ m/s} - 0}{3.0 \text{ s} - 0} = 0.67 \text{ m/s}^2;$$

$$a(4.0 \text{ s}) = \frac{2.0 \text{ m/s} - 2.0 \text{ m/s}}{5.0 \text{ s} - 3.0 \text{ s}} = 0;$$

$$a(8.0 \text{ s}) = \frac{6.0 \text{ m/s} - 2.0 \text{ m/s}}{10.0 \text{ s} - 5.0 \text{ s}} = 0.80 \text{ m/s}^2.$$

(*b*) The displacement is the area under the $v(t)$ curve in the appropriate time interval. Hence from $t = 0$ to $t = 5.0$ s, the displacement of the toy car is

$$x(5.0 \text{ s}) - x(0) = \tfrac{1}{2}(2.0 \text{ m/s})(3.0 \text{ s}) + (2.0 \text{ m/s})(2.0 \text{ s}) = 7.0 \text{ m},$$

and from $t = 3.0$ s to $t = 10.0$ s,

$$x(10.0 \text{ s}) - x(3.0 \text{ s}) = (2.0 \text{ m/s})(7.0 \text{ s}) + \tfrac{1}{2}(4.0 \text{ m/s})(5.0 \text{ s}) = 24.0 \text{ m}.$$

DRILL PROBLEM 3-4

The velocity of a particle varies with time as $3.0t$ m/s. Find the displacement of the particle between $t = 2.0$ s and $t = 6.0$ s.
ANS. 48 m.

DRILL PROBLEM 3-5

The velocity of a particle is given by $(2.0 - 4.0t)$ m/s. What is the particle's displacement between $t = 0$ and $t = 4.0$ s? What is the significance of the negative sign in your answer?
ANS. −24 m; the particle is displaced in the negative x direction.

3-5 MOTION IN ONE DIMENSION WITH CONSTANT ACCELERATION

If the acceleration of a particle moving in one dimension is the constant a, the graph of v vs. t must be a straight line with slope a. Since the equation of a straight line with slope m is $y = mx + b$, where b is the value of y at $x = 0$, we have analogously

$$v = v_0 + at, \tag{3-10}$$

where v_0 is the velocity at $t = 0$.* This equation is plotted in Fig. 3-10. The displacement of the particle between 0 and t is equal to the area under the curve between these two times. From simple geometry, the area is $A_1 + A_2$, where

$$A_1 = \tfrac{1}{2}(t - 0)(at) = \tfrac{1}{2}at^2$$

and

$$A_2 = v_0(t - 0) = v_0 t.$$

*In discussing the equations of motion, we will write $x(t)$ and $v(t)$ as simply x and v. Keep in mind, however, that these variables are functions of time.

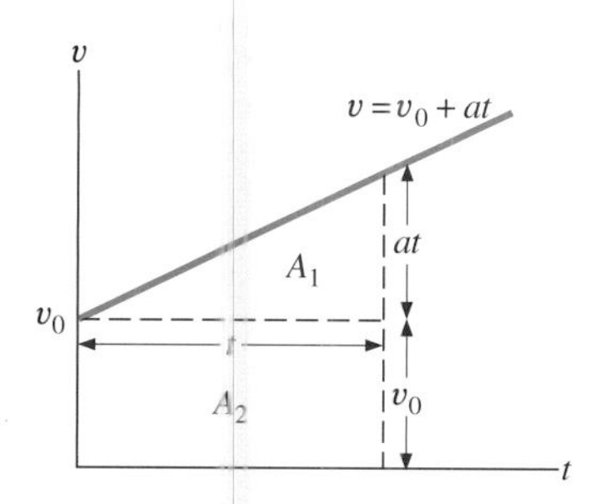

FIGURE 3-10 Velocity vs. time for constant acceleration. The area $A_1 + A_2$ is the displacement in the time interval from 0 to t.

If x_0 and x are the positions at times 0 and t, respectively, the displacement $x - x_0$ is then

$$x - x_0 = A_1 + A_2 = \tfrac{1}{2}at^2 + v_0 t,$$

so

$$x = x_0 + v_0 t + \tfrac{1}{2}at^2. \quad (3\text{-}11)$$

Often, the displacement over a given time interval is specified. In this case, it is convenient to write Eq. (3-11) as

$$x - x_0 = v_0 t + \tfrac{1}{2}at^2.$$

We now have relationships (Eqs. 3-10 and 3-11) that tell us how the velocity and position vary with time. In many applications, an equation that relates position and velocity is also very useful. To find this equation, we eliminate t between Eqs. (3-10) and (3-11). This gives, after some algebraic simplification,

$$v^2 = v_0^2 + 2a(x - x_0). \quad (3\text{-}12)$$

The three equations just found for one-dimensional motion with constant acceleration are listed in Table 3-1. For a given acceleration, all three relate two variables: Eq. (3-10), velocity and time; Eq. (3-11), position and time; and Eq. (3-12), velocity and position. You will likely find these equations easier to apply if you think of them in terms of the variables they relate rather than as formulas into which numbers are to be inserted.

EXAMPLE 3-7 Motion with Constant Velocity

A motorcycle moving with a constant velocity of 30.0 m/s travels from $x_0 = 5.0$ km to $x = 25.0$ km. How much time is required for the trip?

SOLUTION We can determine how much time is required for a specific change in position by relating position to time. With $x = 25.0$ km $= 25.0 \times 10^3$ m, $x_0 = 5.0$ km $= 5.0 \times 10^3$ m, $v_0 = 30.0$ m/s, and $a = 0$ m/s^2, we find

$$x - x_0 = v_0 t + \tfrac{1}{2}at^2$$
$$25.0 \times 10^3 \text{ m} - 5.0 \times 10^3 \text{ m} = (30.0 \text{ m/s})t + 0,$$

and

$$t = 667 \text{ s}.$$

TABLE 3-1 **The Kinematic Equations for Constant Acceleration**

Relationship	Equation
Velocity to time	$v = v_0 + at$
Position to time	$x = x_0 + v_0 t + \tfrac{1}{2}at^2$
Velocity to position	$v^2 = v_0^2 + 2a(x - x_0)$

EXAMPLE 3-8 An Accelerating Automobile

An automobile starts from rest and travels a distance of 300 m in 30 s while moving at constant acceleration. (*a*) What is the acceleration of the automobile? (*b*) What is its velocity at $t = 30$ s?

SOLUTION (*a*) The position changes by 300 m in 30 s, so from the relationship between position and time,

$$x - x_0 = v_0 t + \tfrac{1}{2}at^2$$
$$300 \text{ m} = (0)t + \tfrac{1}{2}a(30 \text{ s})^2,$$

which gives

$$a = 0.67 \text{ m/s}^2.$$

(*b*) The velocity at the time $t = 30$ s is found by relating velocity to time:

$$v = v_0 + at = 0 + (0.67 \text{ m/s}^2)(30 \text{ s}) = 20 \text{ m/s}.$$

DRILL PROBLEM 3-6

An alternate approach for Example 3-8: Use the relationship between velocity and position to find the velocity of the automobile.

EXAMPLE 3-9 Reversal of Motion Under Constant Acceleration

A particle travels with a constant acceleration of -3.00 m/s^2, as shown in Fig. 3-11. At $t = 0$, its velocity is 30.0 m/s. (*a*) How far does the particle move before turning around? (*b*) At what time does it turn around? (*c*) At what time does it return to its starting point? (*d*) What is its velocity when it returns?

SOLUTION (*a*) At the instant the particle turns around, its velocity v is zero. From the velocity-position equation,

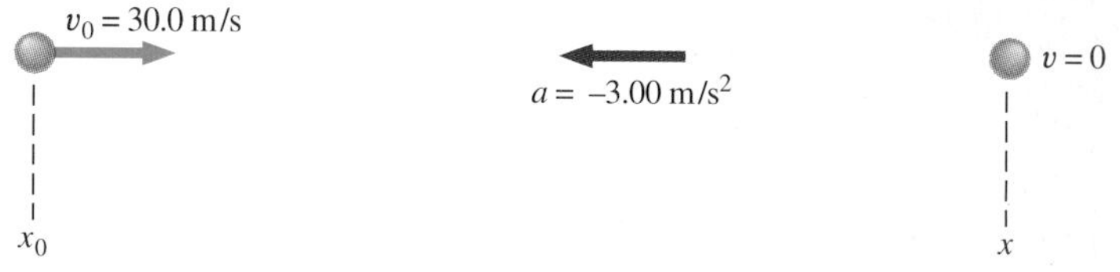

FIGURE 3-11 At $t = 0$, a particle is at x_0 moving with a velocity of 30.0 m/s. It moves to the right with a constant acceleration of -3.00 m/s^2 until it reaches x, at which point it turns around and moves to the left.

$$v^2 = v_0^2 + 2a(x - x_0)$$
$$0^2 = (30.0\ \text{m/s})^2 + 2(-3.00\ \text{m/s}^2)(x - x_0),$$

and

$$x - x_0 = 150\ \text{m}.$$

(*b*) From the velocity-time equation,

$$v = v_0 + at$$
$$0 = 30.0\ \text{m/s} + (-3.00\ \text{m/s}^2)t,$$

which yields

$$t = 10.0\ \text{s}.$$

(*c*) When the particle returns to its starting point, its displacement $x - x_0$ is zero. The time at which this occurs is found with

$$x - x_0 = v_0 t + \tfrac{1}{2}at^2,$$

so

$$0 = (30.0\ \text{m/s})t + \tfrac{1}{2}(-3.00\ \text{m/s}^2)t^2.$$

There are two values of t that satisfy this equation; they are $t = 0$ and $t = 20.0$ s. The first of these is the time when the particle departs, and $t = 20.0$ s is the time at which it returns.

(*d*) The velocity at $t = 20.0$ s is given by

$$v = v_0 + at,$$

so

$$v = 30.0\ \text{m/s} + (-3.00\ \text{m/s}^2)(20.0\ \text{s}) = -30.0\ \text{m/s}.$$

DRILL PROBLEM 3-7

An alternate approach for Example 3-9: First find the time when the particle turns around (part *b*), then use this time to determine how far the particle moves before turning around (part *a*).

DRILL PROBLEM 3-8

An automobile starts from rest, accelerates to 45 mi/h in 363 ft, then continues at this velocity. (*a*) How much time is required for the automobile to reach 45 mi/h? (*b*) How far does it travel in the first 20 s of its motion?
ANS. (*a*) 11 s; (*b*) 957 ft.

DRILL PROBLEM 3-9

Average velocity for motion at constant acceleration. For a particle moving in a straight line at constant acceleration, the velocities at t_1 and t_2 are v_1 and v_2, respectively. Show that the average velocity between t_1 and t_2 is $\bar{v} = (v_1 + v_2)/2$.

3-6 Free Fall

An interesting application of the relationships in Table 3-1 is *free fall*, which describes the motion of a body falling near the surface of the earth (and other objects of planetary size). At this point we assume that the body is falling in a straight line perpendicular to the earth's surface; its motion is then one-dimensional. "Falling" in the context of free fall does not necessarily imply a body is moving from a greater height to a lesser height. If a ball is thrown upward, the equations of free fall apply to its ascent as well as its descent.

In the absence of air resistance, all bodies, regardless of their size or composition, fall toward the earth with the same acceleration. A photograph of a popular lecture demonstration illustrating this fact is shown in Fig. 3-12. The magnitude of this acceleration, which is represented by g, is approximately 9.80 $\text{m/s}^2 = 980\ \text{cm/s}^2 = 32.2\ \text{ft/s}^2$ at the surface of the earth. The fact that all bodies fall with the same acceleration when there is no air resistance is far from obvious. Until Galileo Galilei (1564–1642) demonstrated otherwise, people believed that the heavier a body, the larger its acceleration in free fall. This belief was "substantiated" by comparing the speeds with which light objects like feathers and heavy objects like stones fell to

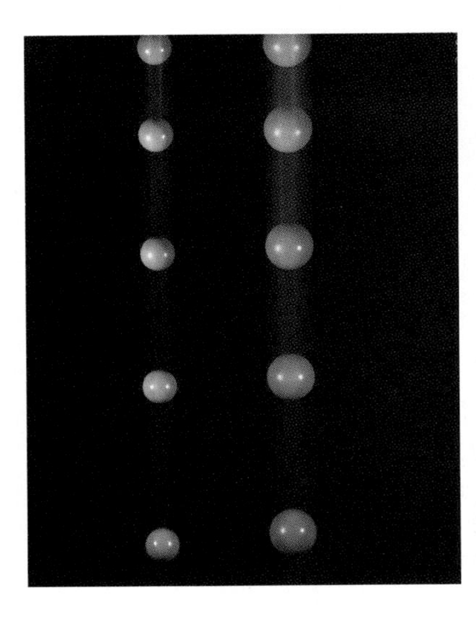

Time exposure of two falling bodies of different size. If air resistance is negligible, all bodies have the same free-fall acceleration.

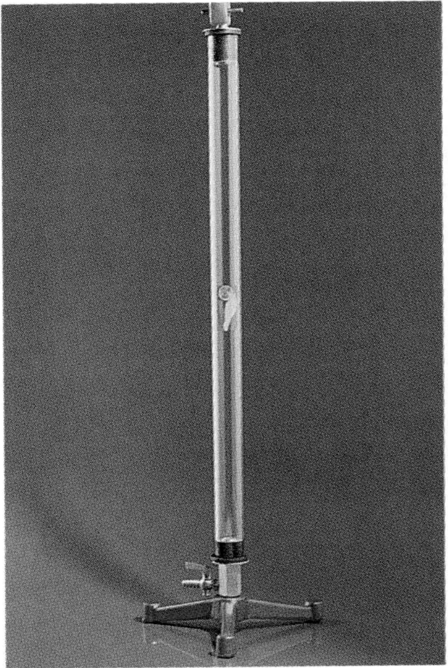

FIGURE 3-12 The long glass tube contains a small disk and a feather. When the air is pumped from the tube, the disk and feather fall at the same rate. With air in the tube, the feather floats slowly downward while the disk falls at almost the same rate as it does in the vacuum.

Courtesy Central Scientific Company.

Galileo Galilei.

the earth. People concluded—incorrectly—that light objects fall more slowly than heavy objects. We now know that they did not consider the effects of air resistance.

Why all bodies have the same free-fall acceleration will be explained in Chap. 5. Until then, this fact will be used without proof.

EXAMPLE 3-10 FREE FALL OF A BALL

Figure 3-13 shows the various positions of a ball thrown from the top of a building, which is 320 ft high, with an initial velocity of 16 ft/s downward. (*a*) How much time elapses before the ball reaches the ground? (*b*) What is its velocity when it arrives?

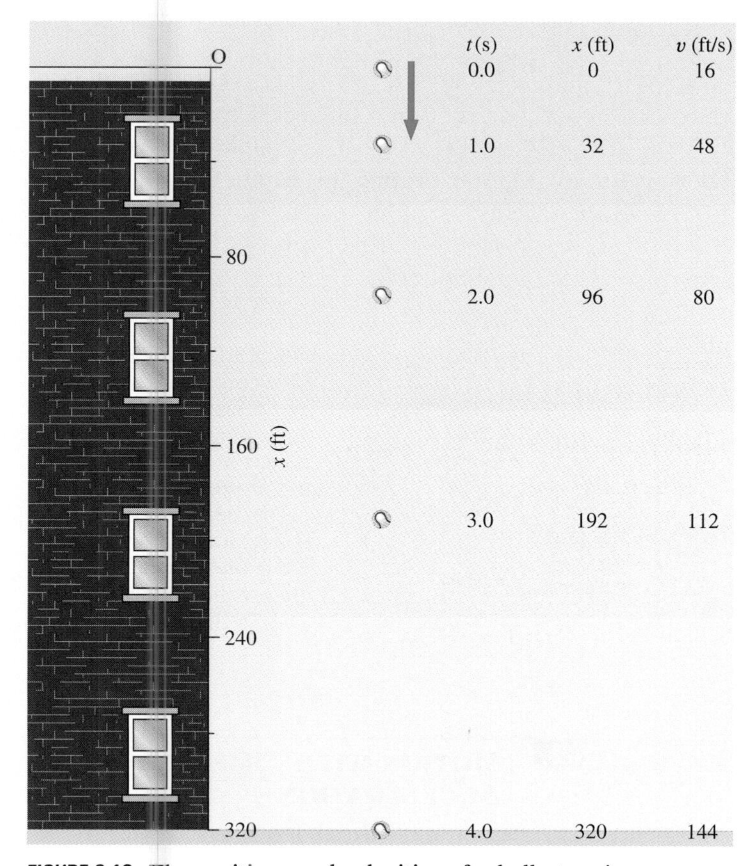

FIGURE 3-13 The positions and velocities of a ball at various instants.

SOLUTION (*a*) For a coordinate system whose origin O is at the top of the building and whose positive direction is downward, $x_0 = 0$, $v_0 = 16$ ft/s, and $a = g = 32$ ft/s^2. To find the time at which the position is 320 ft, we use

$$x = x_0 + v_0 t + \tfrac{1}{2}at^2$$
$$320 \text{ ft} = 0 + (16 \text{ ft/s})t + \tfrac{1}{2}(32 \text{ ft/s}^2)t^2.$$

This simplifies to

$$t^2 + t - 20 = 0,$$

a quadratic equation whose two roots are $t = -5.0$ s and $t = 4.0$ s. The positive root, $t = 4.0$ s, is clearly the time we are interested in. The time -5.0 s represents the fact that a ball thrown upward from the ground would have been in the air for 5.0 s when it passed by the top of the building while moving downward at 16 ft/s.

(*b*) Relating velocity to time, we have

$$v = v_0 + at$$
$$= 16 \text{ ft/s} + (32 \text{ ft/s}^2)(4.0 \text{ s}) = 144 \text{ ft/s}.$$

EXAMPLE 3-11 VERTICAL MOTION OF A BASEBALL

A batter hits a baseball staight upward at home plate, and the ball is caught by the catcher 5.0 s after it is struck. (See Fig. 3-14.) (*a*) What is the initial velocity of the ball? (*b*) What is the maximum height that the ball reaches? (*c*) How long does it take to reach this maximum height? (*d*) What is the acceleration of the ball at the top of its path? (*e*) What is the velocity of the ball when it is caught? Assume that the baseball is hit and caught at the same location.

SOLUTION (*a*) We use a coordinate system whose positive x axis points upward and whose origin is at the spot where the ball is hit and caught. Relating position to time, we have

$$x = x_0 + v_0 t + \tfrac{1}{2}at^2$$
$$0 = 0 + v_0(5.0 \text{ s}) + \tfrac{1}{2}(-9.8 \text{ m/s}^2)(5.0 \text{ s})^2,$$

and

$$v_0 = 24.5 \text{ m/s}.$$

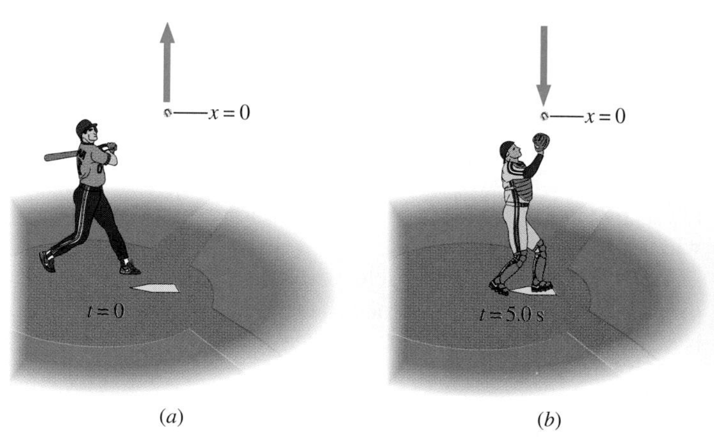

FIGURE 3-14 A baseball hit straight up is caught by the catcher 5.0 s later.

(b) At the maximum height, $v = 0$. With $v_0 = 24.5$ m/s and $a = -g = -9.8$ m/s^2, the equation relating velocity and position gives

$$v^2 = v_0^2 + 2a(x - x_0)$$
$$0^2 = (24.5 \text{ m/s})^2 + 2(-9.8 \text{ m/s}^2)(x - 0),$$

so

$$x = 30.6 \text{ m}.$$

(c) To find the time when $v = 0$, we use

$$v = v_0 + at$$
$$0 = 24.5 \text{ m/s} + (-9.8 \text{ m/s}^2)t,$$

which yields $t = 2.5$ s. Since the ball rises for 2.5 s, the time it falls is also 2.5 s. It's not difficult to show that the time of ascent is equal to the time of descent for *any* initial velocity.

(d) The acceleration is $a = -9.8$ m/s^2 *everywhere*, including the top point of the path. Even though the velocity is zero at the top, it is still changing at a rate of 9.8 m/s^2 downward.

(e) The velocity at $t = 5.0$ s can be determined with the velocity-time equation:

$$v = v_0 + at$$
$$= 24.5 \text{ m/s} + (-9.8 \text{ m/s}^2)(5.0 \text{ s}) = -24.5 \text{ m/s}.$$

Thus the ball returns with the speed it had when it left. This is a general property of free fall that holds for any initial velocity.

DRILL PROBLEM 3-10

A man throws a ball straight up in the air with an initial speed of 24 m/s. (a) How long does its take for the ball to return to him? (b) What is the maximum height of the ball above the point where it was thrown?
ANS. (a) 4.9 s; (b) 29 m.

DRILL PROBLEM 3-11

A flowerpot is observed to fall past a window 2.0 m high in 0.20 s. What is its speed at the top of the window?
ANS. 9.0 m/s.

*3-7 Kinematics Using Integral Calculus

If you have studied the techniques of integral calculus, you will see in this optional section how those techniques are used in obtaining the kinematic equations. If you have not yet learned about integration, you may wish to come back to this section when you have.

Let's begin with a particle whose acceleration $a(t)$ is a known function of time. Since the time derivative of velocity is acceleration,

$$\frac{dv(t)}{dt} = a(t),$$

which we can write as

$$dv(t) = a(t)\, dt.$$

Taking the indefinite integral of both sides, we have

$$\int dv(t) = \int a(t)\, dt + C_1,$$

where C_1 is a constant of integration. Since $\int dv(t) = v(t)$, the velocity is given by

$$v(t) = \int a(t)\, dt + C_1. \tag{3-13}$$

Now the time derivative of position is velocity $[dx(t)/dt = v(t)]$, so using the same argument as that leading to Eq. (3-13), we obtain

$$x(t) = \int dx(t) = \int v(t)\, dt + C_2, \tag{3-14}$$

where C_2 is a second constant of integration.

We can use these indefinite integrals to derive the kinematic equations for constant acceleration. With $a(t) = a$, a constant, Eq. (3-13) gives

$$v(t) = \int a\, dt + C_1 = at + C_1.$$

If the initial velocity $v(0) = v_0$, then

$$v_0 = 0 + C_1,$$

so $C_1 = v_0$, and

$$v(t) = v_0 + at,$$

which is Eq. (3-10). Substituting this expression for $v(t)$ into Eq. (3-14) then yields

$$x(t) = \int (v_0 + at)\, dt + C_2,$$

and

$$x(t) = v_0 t + \tfrac{1}{2}at^2 + C_2.$$

Finally, if $x(0) = x_0$,

$$x_0 = 0 + 0 + C_2,$$

so $C_2 = x_0$, and

$$x(t) = x_0 + v_0 t + \tfrac{1}{2}at^2,$$

which is Eq. (3-11).

EXAMPLE 3-12 **Motion with Time-Varying Acceleration**

A bullet accelerates down the barrel of a rifle according to

$$a(t) = A - Bt,$$

where $A = 3.0 \times 10^4$ m/s^2 and $B = 2.0 \times 10^6$ m/s^3. If the bullet leaves the barrel 0.010 s after it starts accelerating, what are the muzzle velocity of the bullet and the length of the barrel?

SOLUTION The bullet starts from rest at the back end of the rifle, which we designate as $x = 0$. Thus $x_0 = 0$ and $v_0 = 0$. From

$$\frac{dv(t)}{dt} = a(t) = A - Bt,$$

we have

$$\int dv(t) = \int (A - Bt)\,dt + C_1$$

and

$$v(t) = At - \tfrac{1}{2}Bt^2 + C_1.$$

At $t = 0$, $v(t) = v_0 = 0$, so $C_1 = 0$, and the velocity of the bullet as a function of time is given by

$$v(t) = At - \tfrac{1}{2}Bt^2. \qquad (i)$$

Now using

$$\frac{dx(t)}{dt} = v(t) = At - \tfrac{1}{2}Bt^2,$$

we have

$$\int dx(t) = \int (At - \tfrac{1}{2}Bt^2)\,dt + C_2,$$

and

$$x(t) = \tfrac{1}{2}At^2 - \tfrac{1}{6}Bt^3 + C_2.$$

Since $x_0 = 0$, $C_2 = 0$, and the position of the bullet as a function of time is given by

$$x(t) = \tfrac{1}{2}At^2 - \tfrac{1}{6}Bt^3. \qquad (ii)$$

The muzzle velocity v_m is the velocity at which the bullet leaves the end of the barrel. This occurs at $t = 0.010$ s, so from Eq. (i),

$$v_m = v(0.010\text{ s}) = (3.0 \times 10^4\text{ m/s}^2)(0.010\text{ s}) - \tfrac{1}{2}(2.0 \times 10^6\text{ m/s}^3)(0.010\text{ s})^2 = 200\text{ m/s}.$$

The length L of the barrel is the same as the position of the bullet at $t = 0.010$ s, which from Eq. (ii) is

$$L = x(0.010\text{ s}) = \tfrac{1}{2}(3.0 \times 10^4\text{ m/s}^2)(0.010\text{ s})^2 - \tfrac{1}{6}(2.0 \times 10^6\text{ m/s}^3)(0.010\text{ s})^3 = 1.2\text{ m}.$$

SUMMARY

1. **Definitions of position, displacement, velocity, average velocity, speed, and acceleration**
   (*a*) The position of an object in one-dimensional motion along the x axis is $x(t)\mathbf{i}$. This vector is often expressed as $x(t)$, where the sign of $x(t)$ tells us which side of the origin the particle is on.
   (*b*) If the positions of a particle at t_1 and t_2 are $x(t_1)$ and $x(t_2)$, respectively, the displacement of the particle between these two times is $x(t_2) - x(t_1)$.
   (*c*) Average velocity is the net displacement between t_1 and t_2 divided by $t_2 - t_1$.
   (*d*) Velocity $v(t) = dx(t)/dt$.
   (*e*) Speed is $|\mathbf{v}(t)|$.
   (*f*) Acceleration $a(t) = dv(t)/dt$.
2. **Graphical displays of position and velocity vs. time**
   (*a*) The slope of the position-vs.-time curve is the velocity.
   (*b*) The slope of the velocity-vs.-time curve is the acceleration.
   (*c*) The area under the velocity-vs.-time curve between t_1 and t_2 is the displacement $x(t_2) - x(t_1)$.
3. **One-dimensional motion with constant acceleration**
   (*a*) The relationship between velocity and time is
   $$v = v_0 + at.$$
   (*b*) The relationship between position and time is
   $$x = x_0 + v_0 t + \tfrac{1}{2}at^2.$$
   (*c*) The relationship between velocity and position is
   $$v^2 = v_0^2 + 2a(x - x_0).$$
4. **Free fall**
   A body in free fall near the surface of the earth has an acceleration of $g = 9.8\text{ m/s}^2$ downward.

SUPPLEMENT 3-1: THE DERIVATIVE

This supplement is devoted to two topics: the evaluation of the derivative with respect to t of the function

$$x(t) = At^n,$$

where A is a constant, and a graphical interpretation of the derivative with respect to t of an arbitrary function $x(t)$.

The Derivative with Respect to t of At^n

To evaluate the derivative of At^n, we begin with the definition of the derivative, which is

$$\frac{dx(t)}{dt} = \lim_{\Delta t \to 0} \frac{x(t + \Delta t) - x(t)}{\Delta t}.$$

For $n = 2$ in $x(t) = At^n$,

$$x(t + \Delta t) = A(t + \Delta t)^2 = A[t^2 + 2t\,\Delta t + (\Delta t)^2],$$

so

$$\frac{x(t + \Delta t) - x(t)}{\Delta t} = A(2t + \Delta t),$$

and

$$\frac{dx(t)}{dt} = \lim_{\Delta t \to 0} (2At + A\,\Delta t) = 2At.$$

Similarly, for $n = 3$,

$$x(t + \Delta t) = A(t + \Delta t)^3$$
$$= A[t^3 + 3t^2\,\Delta t + 3t(\Delta t)^2 + (\Delta t)^3],$$

so

$$\frac{x(t + \Delta t) - x(t)}{\Delta t} = A[3t^2 + 3t\,\Delta t + (\Delta t)^2],$$

and

$$\frac{dx(t)}{dt} = \lim_{\Delta t \to 0} [3At^2 + 3At\,\Delta t + A(\Delta t)^2] = 3At^2.$$

The extension to arbitrary integer powers of n now becomes apparent. We write

$$x(t + \Delta t) = At^n + Ant^{n-1}\,\Delta t + \text{terms involving } (\Delta t)^2, (\Delta t)^3, \ldots, (\Delta t)^n,$$

so

$$\frac{x(t + \Delta t) - x(t)}{\Delta t} = nAt^{n-1} + \text{terms involving } \Delta t, (\Delta t)^2, \ldots, (\Delta t)^{n-1},$$

and

$$\frac{dx(t)}{dt} = \lim_{\Delta t \to 0} \frac{x(t + \Delta t) - x(t)}{\Delta t} = nAt^{n-1}.$$

We won't prove it here, but this rule is also valid when n is negative and when n is not an integer. *For arbitrary n,*

$$\frac{d}{dt}(At^n) = nAt^{n-1}.$$

Graphical Interpretation of the Derivative

A graph of the arbitrary function $x(t)$ is shown in Fig. 3-15. For the interval Δt_1,

$$\frac{x(t + \Delta t_1) - x(t)}{\Delta t_1}$$

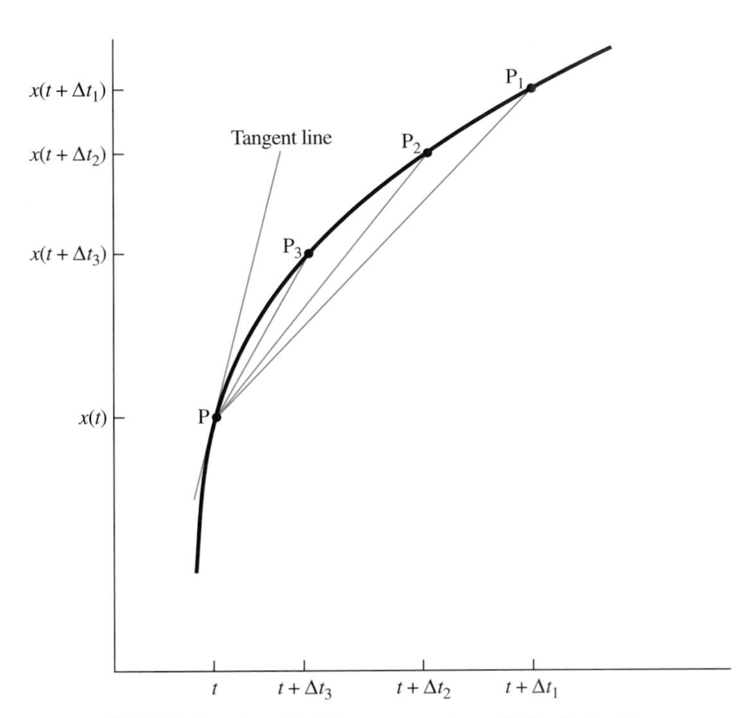

FIGURE 3-15 Graphical interpretation of the derivative.

is the slope of the straight line drawn between P and P_1. For the smaller interval Δt_2,

$$\frac{x(t + \Delta t_2) - x(t)}{\Delta t_2}$$

is the slope of the straight line drawn between P and P_2; and analogous expressions give the slopes of the straight lines drawn between P and P_3, P_4, ... for shorter and shorter intervals Δt_3, Δt_4, ...

You can see in the figure what happens in the limit as $\Delta t_i \to 0$ (or equivalently, as $P_i \to P$): The straight line between P and P_i gets closer and closer to the tangent line to the curve representing $x(t)$ at the point P. Thus *in the limit as* $\Delta t \to 0$,

$$\frac{dx(t)}{dt} = \lim_{\Delta t \to 0} \frac{x(t + \Delta t) - x(t)}{\Delta t}$$

is the slope of the line tangent at the point P to the curve representing $x(t)$.

Questions

3-1. Discuss the difference between average velocity and instantaneous velocity.

3-2. Over a given time interval, the average velocity of an object is zero. What can you conclude about its displacement over the time interval?

3-3. If a particle starts and ends at different points, can its average velocity be zero?

3-4. Compare the acceleration of a sprinter whose velocity increases from zero to 10 m/s in 1.2 s with that of an airplane whose velocity increases from 250 to 260 m/s in the same amount of time.

3-5. Can a body have a positive velocity and a negative acceleration?

3-6. An automobile is moving at a velocity of 100 km/h when it is suddenly accelerated in the direction of motion. What happens to its velocity? What happens to its velocity when it is accelerated in the opposite direction?

3-7. Does the speedometer of a car give speed or velocity?

3-8. Can a body with zero velocity be accelerating?

3-9. When a ball is thrown upward, what is its acceleration at its highest point?

3-10. A ball is thrown straight upward. Compare the acceleration vectors of the ball before and after it leaves the hand.

3-11. Compare the acceleration vectors of particles in free fall moving upward and downward.

3-12. A bolt falls off a rocket that is accelerating upward at 25 m/s^2. What is the acceleration of the bolt immediately after it breaks loose? Five seconds after it breaks loose?

3-13. A girl standing at the edge of a building throws one ball upward with an initial speed v_0 and then throws another one downward with the same speed. How do the velocities of the two balls compare when they hit the ground?

3-14. What are the conversion factors for miles per hour to feet per second, miles per hour to kilometers per second, and feet per second to meters per second?

3-15. If the slope of a particle's velocity-vs.-time curve is horizontal, what can you say about its acceleration? What if the curve has a negative slope?

3-16. As a lightning bolt passes through the air, the heat it generates causes the shock wave we know as thunder. We can estimate the distance away of the lightning by measuring the time interval between sighting the bolt and hearing the thunder. Roughly, for every 5.0-s interval the distance is 1.0 mi. Explain. (The speed of light is 1.86×10^5 mi/s, and the speed of sound is approximately 1050 ft/s.)

3-17. You have probably tried unsuccessfully to catch a dollar bill whose entire length of 15.6 cm has been dropped between your thumb and forefinger. Use this fact to estimate a minimum reaction time for humans.

3-18. An object thrown straight downward is observed by two people. One says the object's acceleration is positive, and the other says its acceleration is negative. Explain why they both might be correct.

3-19. When a bullet is shot straight upward, its muzzle speed is greater than its speed just before it hits the ground. Explain why. Would this happen on the moon?

3-20. Raindrops are moving quite slowly when they reach the ground even though they have fallen through fairly large distances. Explain why.

PROBLEMS

Position, Displacement, and Velocity

3-1. Consider a coordinate system in which the positive x axis is directed vertically upward. What are the positions of a particle (*a*) 5.0 m directly above the origin and (*b*) 3.0 m below the origin?

3-2. Repeat Prob. 3-1 with the origin moved (*a*) 2.0 m upward and (*b*) 3.0 m downward.

3-3. A car is 2.0 km west of a traffic light at $t = 0$ and 5.0 km east of the light at $t = 6.0$ min. Assume that the origin of the coordinate system is at the light and that the positive x direction is eastward. What are the car's position vectors at these two times? What is the car's displacement between 0 and 6.0 min?

3-4. The position of a particle moving along the z axis is given by $z(t) = (4.0 - 2.0t)$ m. (*a*) At what time does the particle cross the origin? (*b*) What is the displacement of the particle between $t = 3.0$ s and $t = 6.0$ s?

3-5. As a professional, Nolan Ryan could pitch a baseball at approximately 160 km/h. At that speed, how long did it take a ball thrown by Ryan to reach home plate, which is 18.4 m from the pitcher's mound?

3-6. An airplane leaves Chicago and makes the 3000-km trip to Los Angeles in 5.0 h. A second plane leaves Chicago one-half hour later and arrives in Los Angeles at the same time. Compare the average velocities of the two planes. Ignore the curvature of the earth and the difference in altitude between the two cities.

3-7. In a 100-m race, the winner is timed in 11.2 s. The second-place finisher's time is 11.6 s. How far is the runner-up behind the winner when she crosses the finish line? Assume that the velocity of each runner is constant throughout the race.

3-8. A particle moves along the y axis according to $y(t) = (3.0t^2 - 8.0t)$ m. (*a*) What is its average velocity between $t = 2.0$ s and $t = 9.0$ s? (*b*) What is its instantaneous velocity at $t = 4.0$ s? at $t = 6.0$ s? (*c*) At what time is its velocity zero?

3-9. A skydiver jumps from a plane and falls 430 m in 12 s. Then she opens her chute and falls the remaining 400 m to the ground in 150 s. Assuming the y axis is directed vertically upward, express the skydiver's average velocity (as a vector) for each part of the fall and for the entire fall.

Acceleration

3-10. An airplane, starting at rest, moves down the runway at constant acceleration for 18 s and then takes off at a speed of 60 m/s. What is the acceleration of the plane?

3-11. An object has an acceleration of $+1.2\ cm/s^2$. At $t = 4.0$ s, its velocity is -3.4 cm/s. Determine the particle's velocities at $t = 1.0$ s and at $t = 6.0$ s.

3-12. What is the acceleration of the particle of Prob. 3-8 at $t = 2.0$ s and $t = 9.0$ s?

3-13. A particle moves along the x axis according to the equation $x(t) = (2.0 - 4.0t^2)$ m. What are the velocity and acceleration at $t = 2.0$ s and $t = 5.0$ s?

3-14. A particle moving at constant acceleration has velocities of 2.0 m/s at $t = 2.0$ s and -7.6 m/s at $t = 5.2$ s. What is the acceleration of the particle?

3-15. The accelerations of automobiles are sometimes given in miles per (hour·second). (*a*) Determine the conversion factor for changing this unit to miles per hour squared. (*b*) Convert an acceleration of 15 mi/h·s to miles per second squared.

3-16. A comet moves along a straight line toward the sun with a velocity given by $v = -\sqrt{c_1 + c_2/x}$, where x is measured outward from the center of the sun. Using the chain rule for differentiation, show that the acceleration of the comet is given by $a = -c_2/2x^2$.

3-17. The position of a particle moving along the x axis varies with time according to $x(t) = (5.0t^2 - 4.0t^3)$ m. Find (*a*) the velocity and acceleration of the particle as functions of time, (*b*) the velocity and acceleration at $t = 2.0$ s, (*c*) the time at which the position is a maximum, (*d*) the time at which the velocity is zero, and (*e*) the maximum position.

Graphical Representation of One-Dimensional Motion

3-18. The position of a particle varies with time as shown in the accompanying figure. Plot corresponding graphs showing the velocity and the acceleration of the particle as functions of time.

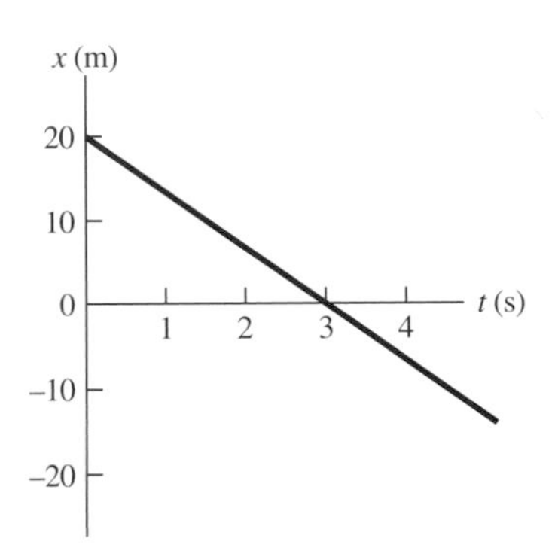

PROBLEM 3-18

3-19. The velocity of a particle moving in one dimension varies as shown in the accompanying graph. (*a*) Find the acceleration at $t = 10$ s, $t = 30$ s, and $t = 50$ s. (*b*) What is the particle's displacement between $t = 0$ and $t = 40$ s? between $t = 40$ s and $t = 60$ s?

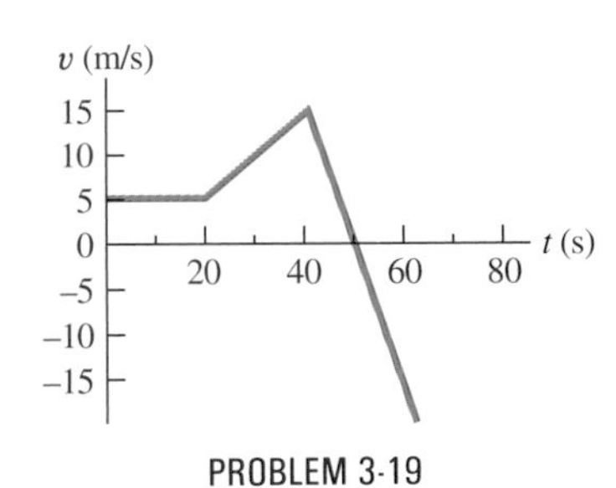

PROBLEM 3-19

3-20. The accompanying graph shows the velocity of a particle moving in one dimension plotted as a function of time. (*a*) Find the acceleration at $t = 1.0$ s, $t = 3.0$ s, $t = 5.0$ s, and $t = 8.0$ s. (*b*) What is the displacement of the particle between $t = 0$ and $t = 2.0$ s? between $t = 0$ and $t = 6.0$ s? between $t = 6.0$ s and $t = 10.0$ s? (*c*) What is the average velocity between $t = 6.0$ s and $t = 10.0$ s?

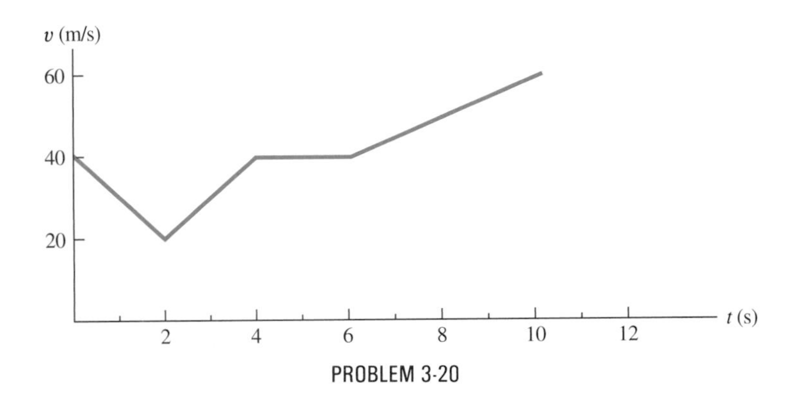

PROBLEM 3-20

3-21. The position of a particle varies with time as shown. Determine whether the velocity is positive, negative, or zero at t_1, t_2, t_3, and t_4. Do the same for the acceleration.

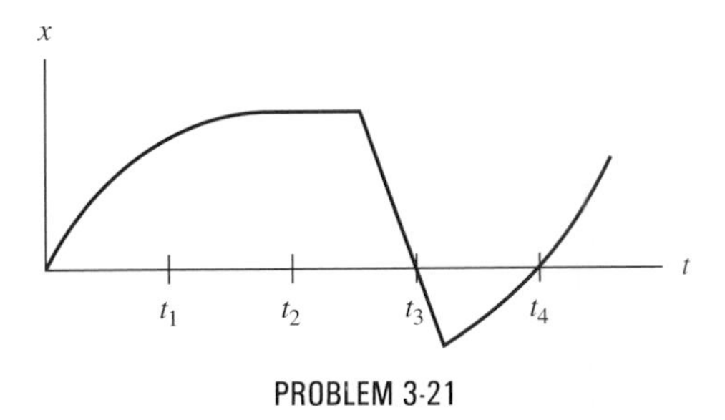

PROBLEM 3-21

3-22. The position of a particle varies with time as shown. (*a*) By measuring the slope of the graph, find the instantaneous velocity at $t = 3.0$ s. (*b*) Find the average velocity between $t = 0$ and $t = 6.0$ s.

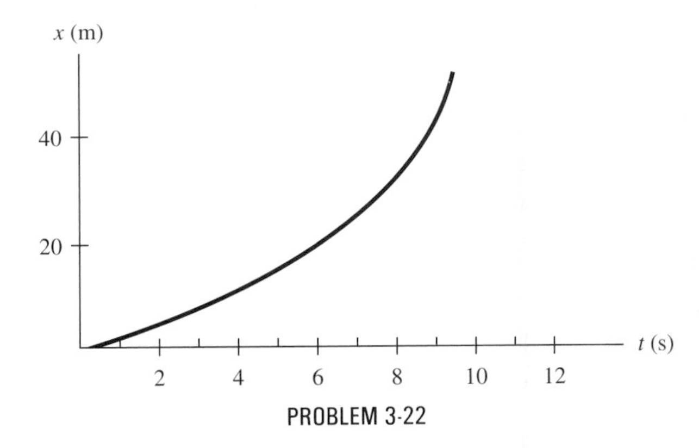

PROBLEM 3-22

Motion with Constant Acceleration

3-23. A particle moves in a straight line at a constant velocity of 30 m/s. What is its displacement between $t = 0$ and $t = 5.0$ s?

3-24. A particle moves in a straight line at a constant acceleration of 30 m/s^2. If at $t = 0$, $x = 0$ and $v = 0$, what is the particle's position at $t = 5.0$ s?

3-25. A particle moves in a straight line with an initial velocity of 30 m/s and a constant acceleration of 30 m/s^2. (*a*) What is its displacement at $t = 5.0$ s? (*b*) What is its velocity at this same time?

3-26. A particle has a constant acceleration $a = 6.0$ m/s^2. If its initial velocity is 2.0 m/s, at what time is its displacement 5.0 m? What is its velocity at that time?

3-27. The position and velocity of a particle at $t = 8.0$ s are 10 m and 2.0 m/s. If the particle is moving in a straight line with a constant acceleration of -1.0 m/s^2, what are its position and velocity at (*a*) $t = 4.0$ s and (*b*) $t = 10.0$ s? (*c*) What is the particle's displacement between $t = 8.0$ s and $t = 10.0$ s?

3-28. At $t = 10$ s, a particle is moving from left to right with a speed of 5.0 m/s. At $t = 20$ s, the particle is moving from right to left with a speed of 8.0 m/s. Assuming that the particle's acceleration is constant, determine (*a*) its acceleration, (*b*) its initial velocity, and (*c*) the instant when its velocity is zero.

3-29. A baseball pitcher accelerates a ball from rest to 90 mi/h while moving his hand 3.0 ft. Determine the ball's acceleration while in the pitcher's hand. Assume that the acceleration is constant. Compare this acceleration with g.

3-30. A freeway offramp is 200 ft long. The speed limit on the ramp is 30 mi/h. What is the minimum time taken by a law-abiding driver to travel the entire offramp and come to a complete stop at the end of it? Assume that the acceleration is constant.

3-31. What is the average velocity of the car of the previous problem?

3-32. An automobile moving at 30 km/h accelerates for 7.0 s, after which its velocity is 88 km/h. (*a*) What is its acceleration? (*b*) How far does it travel while it is accelerating?

3-33. A train is moving up a steep grade at constant velocity when its caboose breaks loose and starts rolling freely along the track. (See the accompanying figure.) If after 5.0 s, the caboose is 30 m behind the train, what is the acceleration of the caboose? Assume that the velocity of the train does not change.

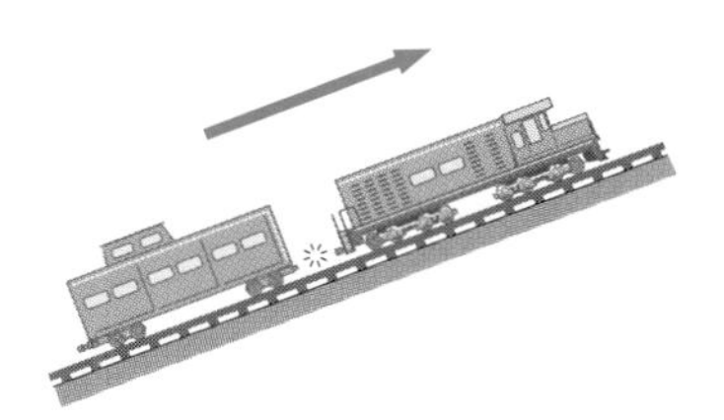

PROBLEM 3-33

3-34. Assume that when a car stops, its deceleration is the same for all initial velocities. Compare the distances required for the car to stop from speeds of 40 and 60 mi/h.

3-35. An electron is moving in a straight line with a velocity of 4.0×10^5 m/s. It enters a region 5.0 cm long where it undergoes an acceleration of 6.0×10^{12} m/s^2 along the same straight line. (*a*) What is the electron's velocity when it emerges from this region? (*b*) How long does the electron take to cross the region?

3-36. An ambulance driver is rushing a patient to the hospital. While traveling at 72 km/h, she notices that the traffic light at the upcoming intersection has turned amber. In order to reach the intersection before the light turns red, she must travel 50 m in 2.0 s. (*a*) What minimum acceleration must her ambulance have in order to reach the intersection before the light turns red? (*b*) What is the speed of the ambulance when it gets to the intersection?

3-37. A motorcycle that is uniformly slowing down covers 2 successive miles in 80 and 120 s, respectively. Calculate (*a*) the acceleration of the motorcycle and (*b*) its velocity at the beginning and end of the 2-mi trip.

3-38. A bicyclist travels from point A to point B in 10 min. During the first 2 min of her trip, she maintains a uniform acceleration of 0.30 ft/s^2. She then travels at constant velocity for the next 5 min. Then she decelerates at a constant rate so that she comes to rest at B 3 min later. (*a*) Sketch the velocity-vs.-time graph for the trip. (*b*) What is the acceleration during the last 3 min, and (*c*) how far does the bicyclist travel?

3-39. The maximum deceleration of a car on pavement is about g (32 ft/s^2). What would the police conclude about a driver who left skid marks 256 ft long?

3-40. A fly reposing on a table sees a swatter 0.50 m away accelerate from rest toward it at 2.0 m/s^2. The insect immediately takes off, accelerating to the side at 0.10 m/s^2. The swatter is a flat square of sides 0.10 m long and is aimed to hit the insect at the center of the swatter. Does the fly manage to escape the blow?

3-41. Two trains are moving at 30 m/s in opposite directions on the same track. The engineers simultaneously see that they are on a collision course and apply the brakes when they are 1000 m apart. Assuming both trains have the same acceleration, what must this acceleration be if the trains are to stop just short of colliding?

3-42. A 30-ft-long truck moving with a constant velocity of 60 mi/h passes a 10-ft-long automobile moving with a constant velocity of 50 mi/h. How much time elapses between the moment the front of the truck is even with the back of the automobile and the moment the back of the truck is even with the front of the automobile? (See the figure at bottom of page.)

3-43. A particle starts at $x = 0$ with an initial velocity v_0 and moves in a straight line with a constant acceleration $-a$. (*a*) Differentiate the position-time equation to find the time at which the position is a maximum. Interpret this result physically. (*b*) What is the maximum position?

Free Fall

3-44. A ball is dropped from the top of a 40-m building. (*a*) What is its velocity after falling for 2.0 s? (*b*) How far does it fall in 2.0 s? (*c*) What is its velocity after falling 10 m? (*d*) How much time elapses before it hits the ground? (*e*) What is its velocity when it reaches the ground?

3-45. Graph the displacement, velocity, and acceleration of the falling ball of the previous problem versus time. Use values of t such that the points you plot are about evenly spaced.

3-46. Repeat Prob. 3-44 with the ball thrown downward with an initial velocity of 5.0 m/s.

3-47. Repeat Prob. 3-44 with the ball thrown upward with an initial velocity of 8.0 m/s.

3-48. A raindrop falls from a cloud 100 m above the ground. Neglecting air resistance, what is the speed of the raindrop when it hits the ground? Is this value realistic?

3-49. A boy throws a ball straight upward and catches it 4.0 s later. (*a*) What is the initial velocity of the ball? (*b*) What is the maximum height of the ball?

3-50. Compare the time that a basketball player who jumps 1.0 m vertically is off the floor with that of a player who jumps 0.3 m vertically.

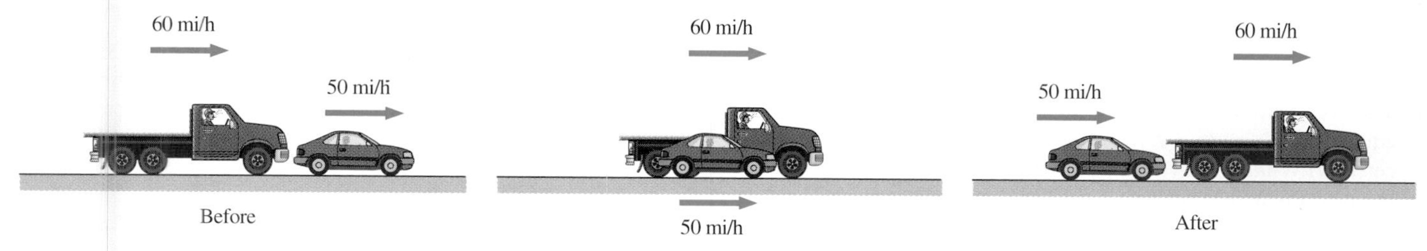

PROBLEM 3-42

3-51. It is claimed that good leapers in basketball seem to stay suspended in midair. (*a*) Explain why this is not possible. (*b*) Compare the time that a 1.0-m vertical leaper is in the air with the time that he is within 0.2 m of the top of his jump. Does this give you a possible explanation of this common misperception?

3-52. Suppose that the average person takes 0.50 s to react and move his hand to catch an object he has dropped. (*a*) How far does that object fall on the earth, where $g = 32$ ft/s^2? (*b*) How far does that object fall on the moon, where $g = 5.6$ ft/s^2?

3-53. A ball is thrown straight upward so that its speed is 19.6 m/s when it is at one half its maximum height. Find (*a*) its maximum height, (*b*) its velocity 2.0 s after it is thrown, (*c*) its position 2.0 s after it is thrown, and (*d*) its acceleration at its maximum height.

3-54. A roller coaster is 20 m above the ground and moving vertically upward at 6.0 m/s when a passenger's cap falls off. (*a*) How much higher upward does the cap rise? (*b*) How much time elapses before the cap hits the ground? (*c*) What is its velocity when it hits? Ignore air resistance.

3-55. A ball is thrown straight upward from the edge of a cliff, as shown in the accompanying figure. It rises to a height of 60 m above the cliff and then falls vertically downward, striking the ground 10 s after it is released. Assume the ball is released 1.0 m above the edge of the cliff. (*a*) What is the initial velocity of the ball? (*b*) How high is the cliff? (*c*) What is the ball's velocity when it hits the ground? (*d*) What is the acceleration of the ball at its highest point?

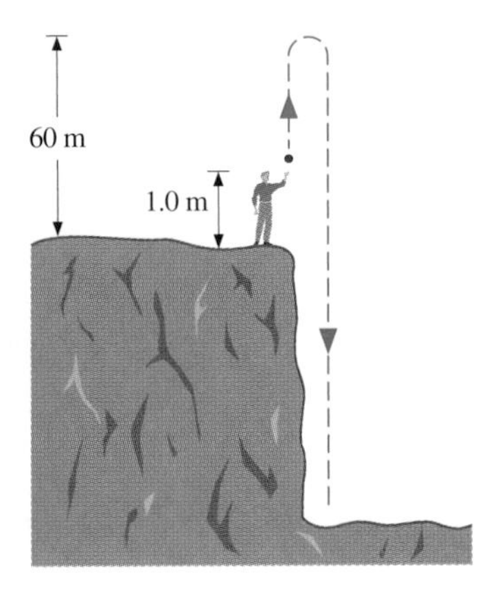

(Not to scale)

PROBLEM 3-55

3-56. An object is dropped from the roof of a tall building of height h. During the last second of its descent, it falls a distance $h/3$. Calculate the height of the building.

3-57. A hot-air balloon rises upward from the ground with a constant velocity of 3.0 m/s. One minute after liftoff, a sandbag is accidentally dropped from the balloon. Calculate (*a*) the time taken by the bag to reach the ground and (*b*) the velocity of the bag when it reaches the ground.

Kinematic Problems Involving Integration

3-58. The acceleration of a particle varies with time according to the equation $a(t) = pt^2 - qt^3$. Initially, the velocity and position are zero. (*a*) What is the velocity as a function of time? (*b*) What is the position as a function of time?

3-59. Between $t = 0$ and $t = t_0$, a rocket moves straight upward with an acceleration given by $a(t) = A - Bt^{1/2}$, where A and B are constants. (*a*) If x is in meters and t is in seconds, what are the units of A and B? (*b*) If the rocket starts at rest, how does its velocity vary with time between $t = 0$ and $t = t_0$? (*c*) If its initial position is 0, what is the rocket's position as a function of time during this same time interval?

3-60. The velocity of a particle moving along the x axis varies with time according to $v(t) = A + Bt^{-1}$, where $A = 2.0$ m/s, $B = 0.25$ m, and $1.0\text{ s} \le t \le 8.0$ s. Determine the acceleration and position of the particle at $t = 2.0$ s and $t = 5.0$ s. Assume that at $t = 1.0$ s, $x = 0$.

3-61. A particle at rest leaves the origin with its velocity increasing with time according to $v(t) = 3.2t$ m/s. At 5.0 s, the particle's velocity starts decreasing according to $v(t) = [16.0 - 1.5(t - 5.0)]$ m/s. This decrease continues until $t = 11.0$ s, after which the particle's velocity remains constant at 7.0 m/s. (*a*) What is the acceleration of the particle as a function of time? (*b*) What is the position of the particle at $t = 2.0$ s, $t = 7.0$ s, and $t = 12.0$ s?

General Problems

3-62. Suppose that during a 100-m race a sprinter accelerates uniformly at the start to a speed of 10 m/s and maintains that pace over the remainder of the race. He finishes in 11 s. Determine (*a*) the time interval in which he is accelerating, (*b*) the time interval in which he is running at a uniform speed, (*c*) the distance over which he is accelerating, and (*d*) his acceleration.

3-63. Just before the start of the Super Bowl, the national anthem is sung by a performer who is standing in the middle of the stadium, 300 ft from the speakers that reproduce the sound for the fans. If the speed of sound is 1100 ft/s, how long after the performer sings a note does she hear it coming from the speakers? If the anthem is played at 100 beats per minute, why could this time delay confuse the singer?

3-64. A police officer in a car moving at 80 km/h sees a car 60 m in front of him moving at 130 km/h. The officer then starts accelerating his car at 1.6 m/s^2 in order to overtake the speeding car. (*a*) How much time elapses before the officer catches the speeder? (*b*) What distance does the police car cover in the pursuit?

3-65. A truck and an automobile are moving in opposite directions at 60 mi/h on a collision course. When they are 300 ft apart, their drivers simultaneously slam on the brakes. The truck accelerates at -14 ft/s^2, and the automobile accelerates at -18 ft/s^2. (*a*) Do they collide? (*b*) How fast is each moving when (and if) they collide?

3-66. A truck moving at 30 m/s is rounding a curve when its driver sees a car 50 m ahead moving at a constant speed of 10 m/s. The driver slams on the brakes but, unfortunately, cannot prevent a collision. (*a*) If the truck is moving at 15 m/s when the collision occurs, what is the truck's acceleration? (*b*) How long after the truck driver slams on the brakes does the collision occur?

3-67. A ball is dropped from the edge of a building. Somewhere below, an observer watches the ball pass his 3.0-m-long window in 0.20 s. How far below the release point is the top of the window?

3-68. In order to determine the depth of a well, a physics student drops a stone in the well and waits to hear the splash; he hears it 3.13 s later. How deep is the well? Assume the speed of sound is 1100 ft/s.

3-69. When the brakes are applied, Ed's car decelerates at 6.0 m/s^2. When Ed drives straight home from work, his reaction time for applying the brakes is 0.40 s. However, if he stops at his favorite tavern before going home, this reaction time increases to 0.60 s, 0.80 s, or 1.0 s, depending on the number of drinks he has. Compare the stopping distance of Ed's car for these reaction times, assuming his initial velocity is 30 m/s.

3-70. *The Guiness Book of Records* states that the cheetah, with a maximum speed of about 27 m/s, is the fastest of all land animals over short distances (up to 550 m), while the fastest land animal over sustained distances is the pronghorn antelope, which has been observed to run at 16 m/s for 6400 m. Suppose that a hungry cheetah moving at 27 m/s starts chasing an antelope that is running at 16 m/s. The animals are moving along the same straight line and are initially 240 m apart. After 500 m, the cheetah starts slowing down with an acceleration of $a = -1.0$ m/s^2, while the antelope keeps running at the same speed. (*a*) Does the cheetah catch the antelope? (*b*) If so, how far does the cheetah have to run for his meal, and (*c*) what is his speed when he captures his prey?

3-71. The balls in the accompanying figure are tied together along a string so that the lowest one is a distance d from the floor and the others are separated by increasing distances of $3d$, $5d$, $7d$,. . . . Show that when the string is released, the balls successively hit the floor at equal time intervals. (This result was first demonstrated by Galileo.)

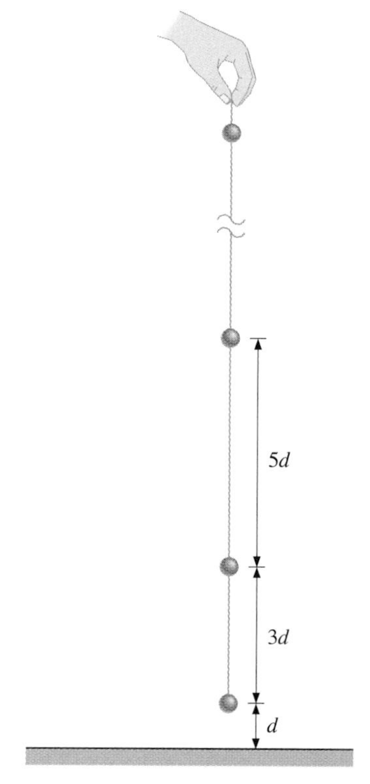

PROBLEM 3-71

An example of projectile motion.

CHAPTER 4 Motion in Two and Three Dimensions

Preview

The kinematics of two- and three-dimensional motion is studied in this chapter, with special emphasis placed on the case involving constant acceleration. The main topics to be discussed here are the following:

1. **Description of two- and three-dimensional motion.** The quantities position, displacement, velocity, average velocity, speed, and acceleration are now introduced as two- and three-dimensional vectors.
2. **Analysis of multidimensional motion.** The determination of an object's motion is studied by examining its motion along each coordinate axis.
3. **Common examples of two-dimensional motion.** Projectile motion and circular motion are discussed.

*4. **Relative velocity and relative acceleration.** The relationship between velocities (and also between accelerations) observed in two frames in relative motion is considered.

In this chapter, we'll extend our study of kinematics to motion in two and three dimensions. This is an important step to take, for the motion of nearly all objects is multidimensional. Some examples are the flight of a baseball or golf ball, the path of a bullet shot from a gun, and the movement of a satellite in its orbit. Just as in the one-dimensional case, the fundamental quantities used to describe motion in two and three dimensions are position, displacement, velocity, and acceleration. Their definitions are rather straightforward extensions of their one-dimensional counterparts. In fact, two- (or three-) dimensional motion can be treated as a combination of two (or three) one-dimensional motions.

4-1 Position and Displacement

Figure 4-1 shows a rectangular coordinate system and a particle P located at (3.0, 4.0, 5.0) m. From the previous chapter, its position is defined as the vector **r** drawn from the origin of the coordinate system to P. In terms of the unit vectors,

$$\mathbf{r} = (3.0\mathbf{i} + 4.0\mathbf{j} + 5.0\mathbf{k})\ \text{m}.$$

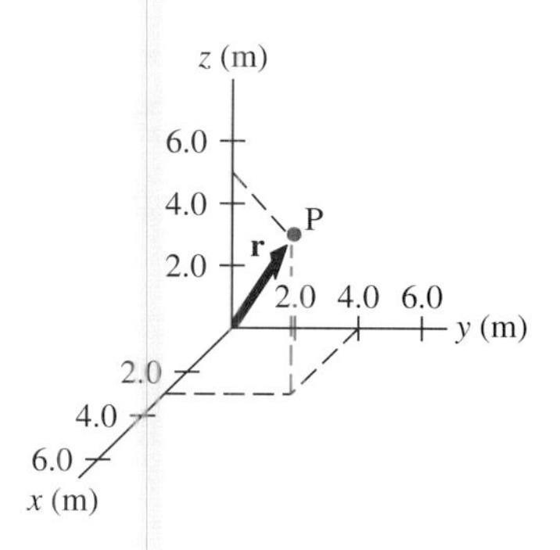

FIGURE 4-1 The position vector **r** is drawn from the origin to the particle located at (3.0, 4.0, 5.0) m.

When a particle moves, its position changes, as illustrated in Fig. 4-2. The position is therefore generally written as a function of time t; that is,

$$x = x(t) \qquad y = y(t) \qquad z = z(t), \tag{4-1a}$$

and

$$\mathbf{r}(t) = x(t)\mathbf{i} + y(t)\mathbf{j} + z(t)\mathbf{k}. \tag{4-1b}$$

The curve traced out by the head of the position vector is the **path** of the particle. Mathematically, the path is represented by Eqs. (4-1). By substituting different values of t into these equations, we obtain positions along the path.

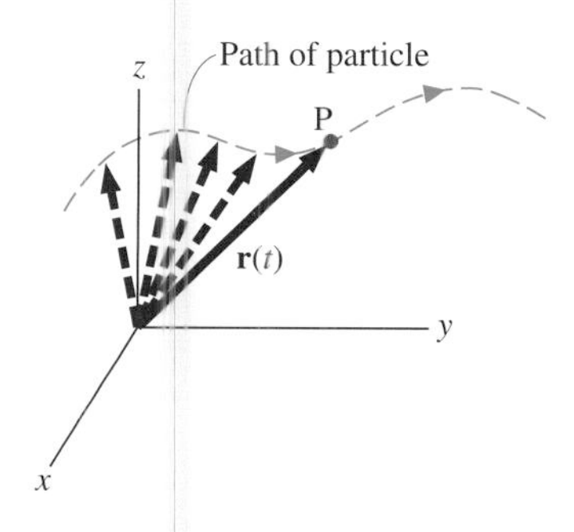

FIGURE 4-2 The vector $\mathbf{r}(t)$ drawn from the origin of the rectangular coordinate system to the particle located at P represents the position of the particle at the time t. As the particle moves through space, the head of the position vector traces out the path of the particle.

When a particle moves from A to B, its displacement (or change in position) is by definition the vector drawn from A to B. An example is shown in Fig. 4-3, where the positions of a particle at times t_1 and t_2 are $\mathbf{r}(t_1)$ and $\mathbf{r}(t_2)$, respectively. The displacement $\Delta\mathbf{r}$ from $\mathbf{r}(t_1)$ to $\mathbf{r}(t_2)$ is then

$$\Delta\mathbf{r} = \mathbf{r}(t_2) - \mathbf{r}(t_1). \tag{4-2}$$

The vector $\Delta\mathbf{r}$ is drawn from the head of $\mathbf{r}(t_1)$ to the head of $\mathbf{r}(t_2)$. As shown previously in Fig. 2-1, the displacement between two points is independent of the path actually traveled between the points.

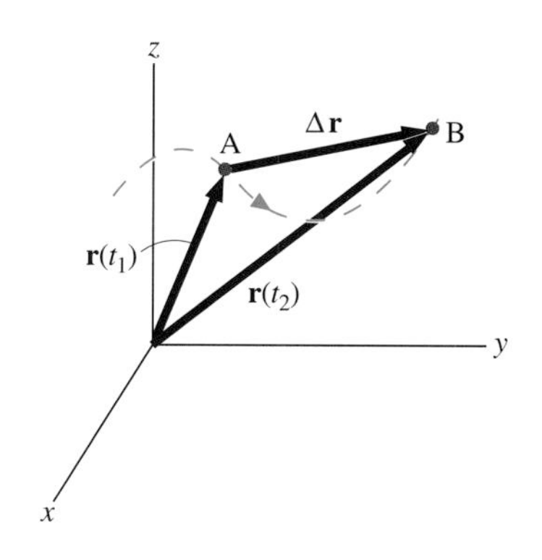

FIGURE 4-3 The particle is at A at the time t_1, and it is at B at the time t_2. The displacement of the particle from A to B is $\Delta\mathbf{r}$.

EXAMPLE 4-1 Displacements of a Billiard Ball

In one version of "eight ball," a pocket billiards game, the winner completes the game by making the eight ball enter a pocket after it has bounced off at least one rail of the table. A rather clever winning shot is depicted in Fig. 4-4. What are the displacements of the ball between A and B, between B and D, and between A and E?

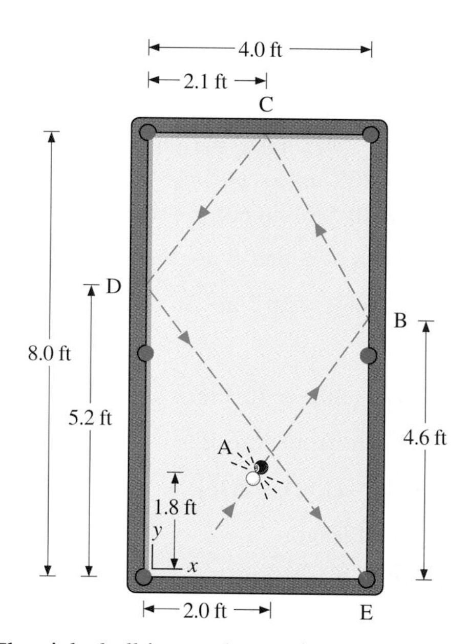

FIGURE 4-4 The eight ball is struck at point A by the cue ball. The eight ball then bounces off three rails at B, C, and D, and goes into the pocket at E. The origin of the rectangular coordinate system is at the lower left-hand corner of the table.

SOLUTION We place the origin of our rectangular coordinate system at the lower left-hand corner of the table. The displacement between A and B is then

$$\Delta\mathbf{r} = \mathbf{r}_B - \mathbf{r}_A = (4.0\mathbf{i} + 4.6\mathbf{j})\text{ ft} - (2.0\mathbf{i} + 1.8\mathbf{j})\text{ ft} = (2.0\mathbf{i} + 2.8\mathbf{j})\text{ ft}.$$

Similarly, between B and D,

$$\Delta\mathbf{r} = (0\mathbf{i} + 5.2\mathbf{j})\text{ ft} - (4.0\mathbf{i} + 4.6\mathbf{j})\text{ ft} = (-4.0\mathbf{i} + 0.6\mathbf{j})\text{ ft},$$

and between A and E,

$$\Delta\mathbf{r} = (4.0\mathbf{i} + 0\mathbf{j})\text{ ft} - (2.0\mathbf{i} + 1.8\mathbf{j})\text{ ft} = (2.0\mathbf{i} - 1.8\mathbf{j})\text{ ft}.$$

EXAMPLE 4-2

Motion of the Earth Around the Sun

Assuming the earth moves around the sun in a circular orbit of radius 1.50×10^{11} m (see Fig. 4-5), what are the positions of the earth when it is at A and when it is at B? What is the displacement of the earth when it moves from A to B?

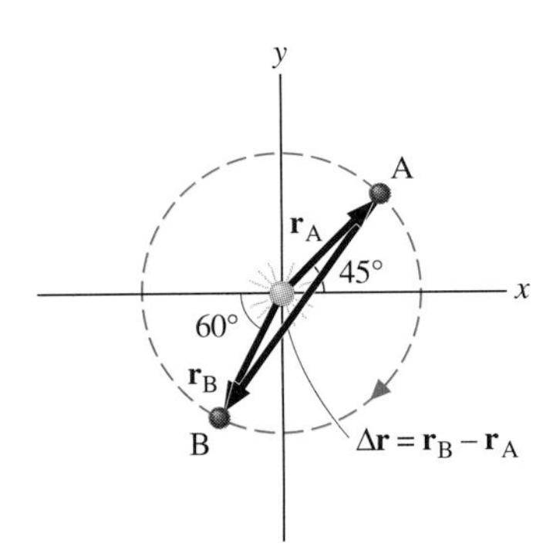

FIGURE 4-5 The earth moves from A to B along its orbit. Its displacement is $\Delta\mathbf{r}$.

SOLUTION With the sun at the origin of our coordinate system, the components of the earth's position at A and B are

$$x_A = (1.50 \times 10^{11}\text{ m})\cos 45° = 1.06 \times 10^{11}\text{ m},$$
$$y_A = (1.50 \times 10^{11}\text{ m})\sin 45° = 1.06 \times 10^{11}\text{ m},$$
$$x_B = -(1.50 \times 10^{11}\text{ m})\cos 60° = -0.75 \times 10^{11}\text{ m},$$
$$y_B = -(1.50 \times 10^{11}\text{ m})\sin 60° = -1.30 \times 10^{11}\text{ m}.$$

The position vectors at A and B are then

$$\mathbf{r}_A = (1.06\mathbf{i} + 1.06\mathbf{j}) \times 10^{11}\text{ m}$$

and

$$\mathbf{r}_B = -(0.75\mathbf{i} + 1.30\mathbf{j}) \times 10^{11}\text{ m};$$

and the displacement from A to B is

$$\Delta\mathbf{r} = \mathbf{r}_B - \mathbf{r}_A = -(1.81\mathbf{i} + 2.36\mathbf{j}) \times 10^{11}\text{ m},$$

which is a vector of length

$$|\Delta\mathbf{r}| = \sqrt{1.81^2 + 2.36^2} \times 10^{11}\text{ m} = 2.97 \times 10^{11}\text{ m}$$

directed at an angle

$$\theta = \tan^{-1}\frac{-2.36 \times 10^{11}\text{ m}}{-1.81 \times 10^{11}\text{ m}} = 233°$$

with respect to the positive x axis.

DRILL PROBLEM 4-1

The origin of the coordinate system in Example 4-2 is moved to the point where the earth's orbit intersects the negative y axis. Find $\mathbf{r}_A$, $\mathbf{r}_B$, and $\Delta\mathbf{r}$.
ANS. $(1.06\mathbf{i} + 2.56\mathbf{j}) \times 10^{11}$ m; $(-0.75\mathbf{i} + 0.20\mathbf{j}) \times 10^{11}$ m; $-(1.81\mathbf{i} + 2.36\mathbf{j}) \times 10^{11}$ m.

4-2 Velocity

A particle moves along the path shown in Fig. 4-6. At a time t the particle is at A with a position $\mathbf{r}(t)$, and at a time $t + \Delta t$ it is at B with a position $\mathbf{r}(t + \Delta t)$. Its instantaneous velocity (or just velocity) is then

$$\mathbf{v}(t) = \lim_{\Delta t \to 0} \frac{\mathbf{r}(t + \Delta t) - \mathbf{r}(t)}{\Delta t} = \frac{d\mathbf{r}(t)}{dt}. \tag{4-3}$$

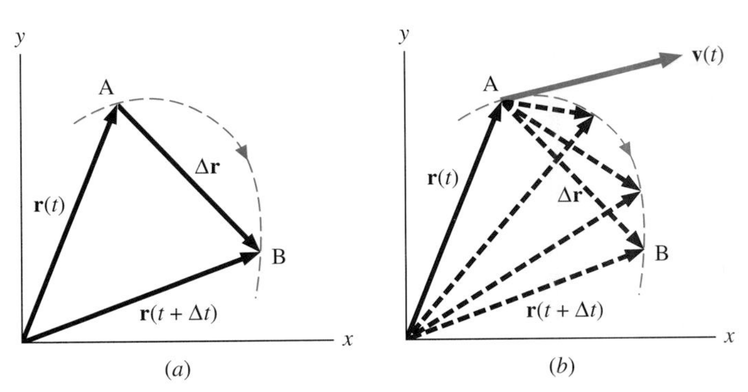

FIGURE 4-6 (*a*) A particle moves from point A to point B along the curved path. The displacement of the particle is $\Delta\mathbf{r}$. (*b*) As Δt gets smaller, $\mathbf{r}(t + \Delta t)$ gets closer to $\mathbf{r}(t)$. In the limit as $\Delta t \to 0$, the difference $\mathbf{r}(t + \Delta t) - \mathbf{r}(t)$ approaches a direction tangent to the path. Thus $\mathbf{v}(t)$ is tangent to the path.

We can determine how the velocity vector is directed relative to the path with the help of Fig. 4-6*b*. As Δt gets smaller and smaller, the displacement vector $\mathbf{r}(t + \Delta t) - \mathbf{r}(t)$ approaches a direction tangent to the path of the particle. Since the velocity is directed along the limiting displacement vector, it must be tangent to the path, as illustrated in the figure.

The unit vectors are constant in both magnitude and direction, so their time derivatives vanish. Consequently, in differentiating Eq. (4-1*b*) with respect to time, we find

$$\mathbf{v}(t) = \frac{d\mathbf{r}(t)}{dt} = \frac{dx(t)}{dt}\mathbf{i} + \frac{dy(t)}{dt}\mathbf{j} + \frac{dz(t)}{dt}\mathbf{k}, \tag{4-4a}$$

which can be written as

$$\mathbf{v}(t) = v_x(t)\mathbf{i} + v_y(t)\mathbf{j} + v_z(t)\mathbf{k}, \tag{4-4b}$$

where

$$v_x(t) = \frac{dx(t)}{dt}, \quad v_y(t) = \frac{dy(t)}{dt}, \quad v_z(t) = \frac{dz(t)}{dt} \tag{4-5}$$

are the components of the velocity vector in the x, y, and z directions, respectively.

As in the one-dimensional case, the speed of a particle moving along a two- or three-dimensional path is the magnitude of its velocity; that is, speed is the scalar quantity $|\mathbf{v}(t)|$ or v. For example, if the velocity of a particle is $(4.0\mathbf{i} + 5.0\mathbf{j})$ m/s, its speed is

$$v = \sqrt{(4.0)^2 + (5.0)^2} \text{ m/s} = 6.4 \text{ m/s}.$$

Occasionally, only the average velocity between two points is important. Its definition is a direct extension of what we used for one dimension: If the position of a particle is $\mathbf{r}(t_1)$ at t_1 and $\mathbf{r}(t_2)$ at t_2, the average velocity $\bar{\mathbf{v}}$ between these two times is

$$\bar{\mathbf{v}} = \frac{\mathbf{r}(t_2) - \mathbf{r}(t_1)}{t_2 - t_1}. \tag{4-6}$$

For example, if the eight ball of Fig. 4-4 travels from A to E in 3.0 s, its average velocity over the path is

$$\bar{\mathbf{v}} = \frac{\mathbf{r}_E - \mathbf{r}_A}{3.0 \text{ s}} = \frac{(2.0\mathbf{i} - 1.8\mathbf{j}) \text{ ft}}{3.0 \text{ s}} = (0.67\mathbf{i} - 0.60\mathbf{j}) \text{ ft/s}.$$

EXAMPLE 4-3 Velocity of a Particle in Two-Dimensional Motion

The position of a particle is given by

$$\mathbf{r}(t) = [(3.0t + 0.50t^2)\mathbf{i} + (4.0t - 0.50t^2)\mathbf{j}] \text{ m}.$$

(*a*) What are the velocity and speed of the particle at $t = 0$ and at $t = 1.0$ s? (*b*) What is the average velocity between these two times?

SOLUTION (*a*) The velocity is determined by differentiating $\mathbf{r}(t)$:

$$\mathbf{v}(t) = \frac{d\mathbf{r}(t)}{dt} = [(3.0 + 1.0t)\mathbf{i} + (4.0 - 1.0t)\mathbf{j}] \text{ m/s}.$$

Upon substituting $t = 0$ and $t = 1.0$ s into this equation, we find

$$\mathbf{v}(0) = (3.0\mathbf{i} + 4.0\mathbf{j}) \text{ m/s},$$

and

$$\mathbf{v}(1.0 \text{ s}) = (4.0\mathbf{i} + 3.0\mathbf{j}) \text{ m/s}.$$

At $t = 0$, the speed is

$$|\mathbf{v}(0)| = \sqrt{(3.0)^2 + (4.0)^2} \text{ m/s} = 5.0 \text{ m/s},$$

and at $t = 1.0$ s, it is

$$|\mathbf{v}(1.0 \text{ s})| = \sqrt{(4.0)^2 + (3.0)^2} \text{ m/s} = 5.0 \text{ m/s}.$$

Notice that the speeds, but *not* the velocities, are the same at these two times. What property of the two velocity vectors distinguishes them from each other?

(*b*) From Eq. (4-6), the average velocity is

$$\bar{\mathbf{v}} = \frac{\mathbf{r}(1.0 \text{ s}) - \mathbf{r}(0)}{1.0 \text{ s}} = \frac{(3.5\mathbf{i} + 3.5\mathbf{j}) \text{ m/s} - 0}{1.0 \text{ s}}$$
$$= (3.5\mathbf{i} + 3.5\mathbf{j}) \text{ m/s},$$

which is a vector of magnitude 4.9 m/s directed at 45° to the positive x axis.

DRILL PROBLEM 4-2

The position of a particle is $\mathbf{r}(t) = [3.0t^2\mathbf{i} + (1.0 + 1.0t^2)\mathbf{j} + 1.0t\mathbf{k}]$ m. (*a*) Find the velocity and speed at $t = 2.0$ s. (*b*) What is the average velocity between 1.0 and 3.0 s?
ANS. (*a*) $(12.0\mathbf{i} + 4.0\mathbf{j} + 1.0\mathbf{k})$ m/s, 12.7 m/s; (*b*) $(12.0\mathbf{i} + 4.0\mathbf{j} + 1.0\mathbf{k})$ m/s.

4-3 Acceleration

Acceleration is the time rate of change of velocity. If at a time t, the velocity of a particle is $\mathbf{v}(t)$, its acceleration is defined as

$$\mathbf{a}(t) = \lim_{\Delta t \to 0} \frac{\mathbf{v}(t + \Delta t) - \mathbf{v}(t)}{\Delta t} = \frac{d\mathbf{v}(t)}{dt}. \tag{4-7}$$

Since velocity is the time derivative of position, we may write

$$\mathbf{a}(t) = \frac{d^2\mathbf{r}(t)}{dt^2}. \tag{4-8}$$

The rectangular components of acceleration are therefore

$$a_x(t) = \frac{dv_x(t)}{dt} = \frac{d^2x(t)}{dt^2},$$
$$a_y(t) = \frac{dv_y(t)}{dt} = \frac{d^2y(t)}{dt^2}, \tag{4-9}$$
$$a_z(t) = \frac{dv_z(t)}{dt} = \frac{d^2z(t)}{dt^2}.$$

EXAMPLE 4-4 Two-Dimensional Acceleration

What is the acceleration of the particle of Example 4-3?

SOLUTION Since acceleration is the time derivative of velocity,

$$\mathbf{a}(t) = \frac{d}{dt}\{[(3.0 + 1.0t)\mathbf{i} + (4.0 - 1.0t)\mathbf{j}] \text{ m/s}\}$$
$$= (1.0\mathbf{i} - 1.0\mathbf{j}) \text{ m/s}^2.$$

Thus $\mathbf{a}$ is a constant vector. Its magnitude is

$$|\mathbf{a}| = \sqrt{(1.0)^2 + (1.0)^2} \text{ m/s}^2 = 1.4 \text{ m/s}^2,$$

and it is directed at an angle

$$\theta = \tan^{-1} \frac{-1.0 \text{ m/s}^2}{1.0 \text{ m/s}^2} = 315°$$

to the positive x axis.

EXAMPLE 4-5 Trajectory of a Golf Ball

Relative to the golfer, the position of a golf ball hit at $t = 0$ is given by

$$\mathbf{r}(t) = [30.0t\mathbf{i} + (30.0t - 4.9t^2)\mathbf{j}] \text{ m},$$

where air resistance has been neglected. (*a*) What is the equation of the path of the golf ball? (*b*) Find the velocity and acceleration of the ball as functions of time.

SOLUTION (*a*) The x and y positions of the golf ball are given by $x = 30.0t$ m and $y = (30.0t - 4.9t^2)$ m, respectively. The equation of the path is found by eliminating t between these two equations, thereby obtaining y as a function of x [i.e., $y = y(x)$]. From the first equation, $t = x/30.0$ s, so

$$y = 30.0\left(\frac{x}{30.0}\right) - 4.9\left(\frac{x}{30.0}\right)^2 = 1.0x - (5.4 \times 10^{-3})x^2$$

is the equation of the ball's path. Now the general equation $y = c_1x + c_2x^2$, where c_1 and c_2 are constants, represents a *parabola*. Hence the golf ball travels along a parabolic path.

(*b*) The velocity is found by differentiating $\mathbf{r}(t)$ with respect to time, and the acceleration is found by taking the second time derivative of $\mathbf{r}(t)$:

$$\mathbf{v}(t) = \frac{d\mathbf{r}(t)}{dt} = [30.0\mathbf{i} + (30.0 - 9.8t)\mathbf{j}] \text{ m/s}$$

and

$$\mathbf{a}(t) = \frac{d\mathbf{v}(t)}{dt} = -9.8\mathbf{j} \text{ m/s}^2.$$

DRILL PROBLEM 4-3

The position of a particle is given by $\mathbf{r}(t) = (1.0t^2\mathbf{i} + 4.0t^3\mathbf{j})$ m. What are the velocity and acceleration of the particle at $t = 2.0$ s?
ANS. 48.2 m/s at 85.2° to the positive x axis; 48.0 m/s^2 at 87.6° to the positive x axis.

4-4 Motion with Constant Acceleration

We can treat multidimensional motion with constant acceleration in terms of the one-dimensional equations developed in the previous chapter. For a particle moving along an arbitrary path, we have

$$v_x(t) = \frac{dx(t)}{dt}, \qquad v_y(t) = \frac{dy(t)}{dt}, \qquad v_z(t) = \frac{dz(t)}{dt},$$

and

$$a_x(t) = \frac{dv_x(t)}{dt}, \qquad a_y(t) = \frac{dv_y(t)}{dt}, \qquad a_z(t) = \frac{dv_z(t)}{dt}.$$

A comparison of these equations with their one-dimensional counterparts shows clearly that three-dimensional motion is equivalent to three one-dimensional motions. Hence by using the methods of Chap. 3 to find the position and velocity components along each coordinate axis, we can determine the position and velocity vectors.

As an example, consider a particle moving in the xy plane with the constant acceleration

$$\mathbf{a} = a_{0x}\mathbf{i} + a_{0y}\mathbf{j}.$$

We can analyze its motion along each coordinate axis using the equations of Table 3-1. With the initial positions and velocities along these axes given by (x_0, y_0) and (v_{0x}, v_{0y}), the position and velocity components are

$$x(t) = x_0 + v_{0x}t + \tfrac{1}{2}a_{0x}t^2, \qquad y(t) = y_0 + v_{0y}t + \tfrac{1}{2}a_{0y}t^2;$$

and

$$v_x(t) = v_{0x} + a_{0x}t, \qquad v_y(t) = v_{0y} + a_{0y}t.$$

Once we have determined their individual components, we can write the position and velocity vectors of the particle as

$$\mathbf{r}(t) = x(t)\mathbf{i} + y(t)\mathbf{j}$$

and

$$\mathbf{v}(t) = v_x(t)\mathbf{i} + v_y(t)\mathbf{j}.$$

EXAMPLE 4-6 **Motion of a Skier**

Figure 4-7 shows a skier at $t = 0$, moving with an acceleration of 2.1 m/s^2 down a 15° slope. With the origin of the coordinate system at the front of the ski lodge, her initial position and velocity are

$$\mathbf{r}(0) = (75.0\mathbf{i} - 50.0\mathbf{j}) \text{ m} \quad \text{and} \quad \mathbf{v}(0) = (4.1\mathbf{i} - 1.1\mathbf{j}) \text{ m/s}.$$

(*a*) What are the rectangular components of the skier's position and velocity as functions of time? (*b*) What are her position and velocity at $t = 10.0$ s?

SOLUTION (*a*) The x and y components of the skier's acceleration are

$$a_x = (2.1 \text{ m/s}^2) \cos 15° = 2.0 \text{ m/s}^2$$

FIGURE 4-7 A skier has an acceleration of 2.1 m/s^2 down a 15° slope. The origin of the coordinate system is at the ski lodge.

and

$$a_y = -(2.1 \text{ m/s}^2) \sin 15° = -0.54 \text{ m/s}^2.$$

Relating position to time in the x direction, we have

$$x(t) = x_0 + v_{0x}t + \tfrac{1}{2}a_x t^2$$
$$= 75.0 \text{ m} + (4.1 \text{ m/s})t + \tfrac{1}{2}(2.0 \text{ m/s}^2)t^2,$$

and

$$v_x(t) = v_{0x} + a_x t$$
$$= 4.1 \text{ m/s} + (2.0 \text{ m/s}^2)t.$$

Similarly, in the y direction,

$$y(t) = -50.0 \text{ m} + (-1.1 \text{ m/s})t + \tfrac{1}{2}(-0.54 \text{ m/s}^2)t^2$$

and

$$v_y(t) = -1.1 \text{ m/s} + (-0.54 \text{ m/s}^2)t.$$

(*b*) At $t = 10.0$ s,

$$x(10.0 \text{ s}) = 75.0 \text{ m} + (4.1 \text{ m/s})(10.0 \text{ s}) + \tfrac{1}{2}(2.0 \text{ m/s}^2)(10.0 \text{ s})^2$$
$$= 216 \text{ m}.$$

Similar calculations yield $v_x(10.0 \text{ s}) = 24.1$ m/s, $y(10.0 \text{ s}) = -88.0$ m, and $v_y(10.0 \text{ s}) = -6.5$ m/s. Hence the position and velocity of the skier at $t = 10.0$ s are

$$\mathbf{r}(10.0 \text{ s}) = (216\mathbf{i} - 88.0\mathbf{j}) \text{ m}$$

and

$$\mathbf{v}(10.0 \text{ s}) = (24.1\mathbf{i} - 6.5\mathbf{j}) \text{ m/s}.$$

DRILL PROBLEM 4-4

A particle moves in the xy plane with an acceleration $\mathbf{a}(t) = (2.0\mathbf{i} + 1.0\mathbf{j})$ m/s^2. At $t = 0$, $x = 0$, $y = 3.0$ m, $v_x = 4.0$ m/s, and $v_y = 6.0$ m/s. What are the velocity and position of the particle at $t = 4.0$ s?
ANS. 15.6 m/s at 39.8° to the positive x axis; 47.4 m at 47.6° to the positive x axis.

4-5 Motion of a Projectile

An object that is projected horizontally as well as vertically, such as a baseball thrown from the outfield to home plate, traverses a curved path in a plane. In the absence of air resistance, the acceleration of the *projectile* is the free-fall acceleration vector $\mathbf{g}$. As discussed in Sect. 3-6, this vector is directed toward the earth and has magnitude 9.80 m/s^2. If we use the coordinate system of Fig. 4-8, $a_x = 0$, $a_y = -g$, and the components of velocity and position vary with time according to

$$v_x = v_{0x} \qquad v_y = v_{0y} - gt,$$

and

$$x = x_0 + v_{0x}t \qquad y = y_0 + v_{0y}t - \tfrac{1}{2}gt^2.$$

We have a combination of *motion at constant velocity in the horizontal direction* and *free fall in the vertical direction*.

All golfers and baseball players who have ever played in the wind know very well that air resistance is not negligible for their particular projectiles. Nevertheless, for our examples and problems, many of which involve such objects as golf balls, baseballs, and cannon shells, air resistance will be ignored. This is sometimes not a very accurate approximation, but it does allow us to consider the motion of projectiles with simple mathematical equations.

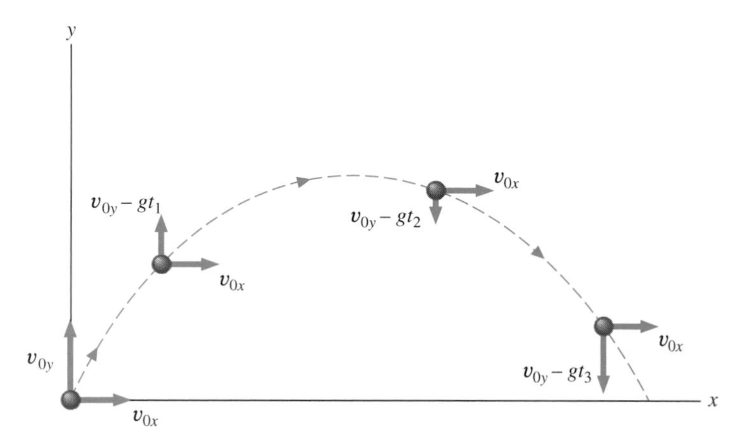

FIGURE 4-8 As a projectile falls freely in the vertical direction, it moves at constant velocity in the horizontal direction.

Courtesy Central Scientific Company.

The independence of the horizontal and vertical motions of a projectile is demonstrated with the device shown. The balls are released at the same instant, and at any time thereafter, their vertical positions are the same.

EXAMPLE 4-7 A Ball Rolling Off a Table

The ball of Fig. 4-9 rolls off the horizontal surface of a 1.5-m-high table with a speed of 2.0 m/s. Where does it land on the floor?

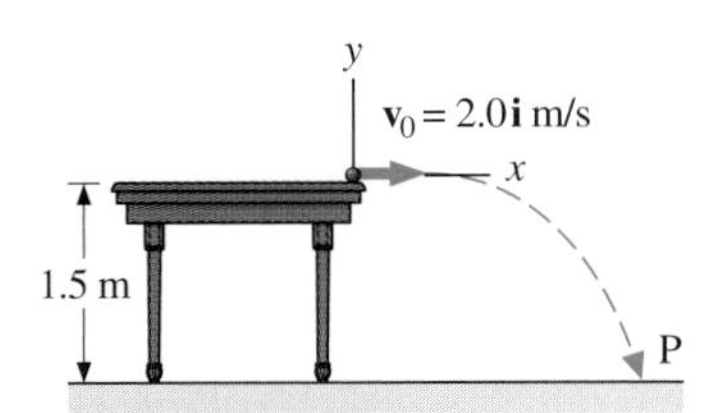

FIGURE 4-9 A ball rolls off a 1.5-m-high table with a velocity of 2.0**i** m/s and lands at point P.

SOLUTION Here we use a reference frame with the origin at the corner of the table, so $x_0 = y_0 = 0$. In order to determine where the ball lands, we need to know the time t that it is in the air. This can be found by considering the free-fall motion of the ball in the vertical direction. Since the ball's vertical displacement is -1.5 m when it lands at the time t, we have

$$y - y_0 = v_{0y}t + \tfrac{1}{2}a_{0y}t^2$$
$$-1.5 \text{ m} = 0 + \tfrac{1}{2}(-9.8 \text{ m/s}^2)t^2,$$

which yields $t = 0.55$ s.

At this time, the horizontal displacement of the ball is

$$x - x_0 = v_{0x}t + \tfrac{1}{2}a_{0x}t^2$$
$$= (2.0 \text{ m/s})(0.55 \text{ s}) + 0 = 1.1 \text{ m}.$$

Hence the ball lands on the floor at a distance of 1.1 m from the edge of the table.

DRILL PROBLEM 4-5

A bullet is shot horizontally with an initial speed of 300 m/s. If the bullet leaves the gun at shoulder height (say, 1.3 m), where does it hit the ground?
ANS. 155 m from the gun.

For the rectangular coordinate system of Fig. 4-10, the components of the projectile's initial velocity are $v_{0x} = v_0 \cos\theta$ and $v_{0y} = v_0 \sin\theta$. Since the projectile starts at the origin, the rectangular components of its position are

$$x = (v_0 \cos\theta)t$$

and

$$y = (v_0 \sin\theta)t - \tfrac{1}{2}gt^2.$$

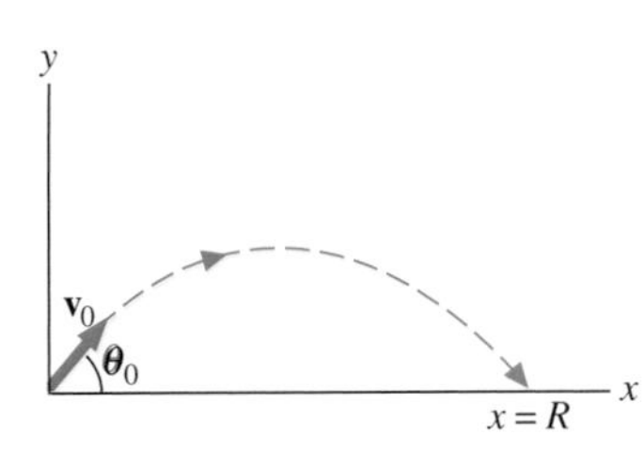

FIGURE 4-10 A projectile is shot at an angle θ to the horizontal with a speed v_0. It lands at $x = R$ and $y = 0$.

The equation of the path of the projectile, $y = y(x)$, can be obtained by eliminating t between these two equations. Solving the first for t in terms of x, then substituting this into the second, we find

$$y = (\tan\theta)x - \left(\frac{g}{2v_0^2\cos^2\theta}\right)x^2. \tag{4-10}$$

The quantities v_0, θ, and g are constants, so Eq. (4-10) has the form

$$y = ax - bx^2,$$

where a and b are constants. This is the equation of a parabola.

The projectile's horizontal displacement when it returns to the ground, assumed to be at the same elevation as the starting point, is called its **range** R. To calculate the range, we use the fact that $y = 0$ when $x = R$ in Eq. (4-10). This gives

$$0 = (\tan\theta)R - \left(\frac{g}{2v_0^2\cos^2\theta}\right)R^2,$$

which, when solved for R, yields

$$R = \frac{2v_0^2\sin\theta\cos\theta}{g} = \frac{v_0^2\sin 2\theta}{g}. \tag{4-11}$$

The value of $\sin 2\theta$ has its maximum value of 1.0 when $2\theta = 90°$, or $\theta = 45°$. Thus for a fixed speed v_0, a particle that lands at its initial elevation has a maximum horizontal range when projected at 45° to the horizontal. That maximum range is, from Eq. (4-11),

$$R_{\text{max}} = \frac{v_0^2}{g}.$$

Parabolic trajectories for fixed v_0 and varying θ are shown in Fig. 4-11. Notice how the range varies with projection angle and that R is the same for $(90° - \theta)$ and θ. This is so because $\sin 2(90° - \theta) = \sin(180° - 2\theta) = \sin 2\theta$.

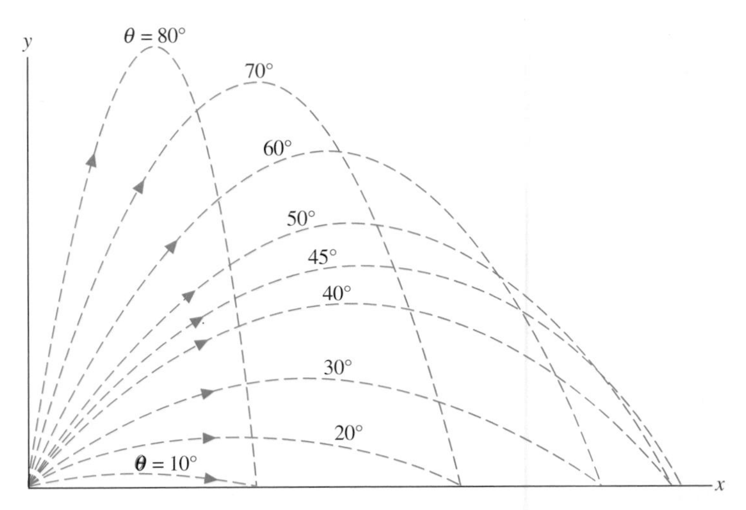

FIGURE 4-11 These parabolic trajectories have all been calculated for the same initial speed. The projection angle is varied from $\theta = 10°$ to 80°.

EXAMPLE 4-8 Projectile Motion with Starting and Landing Points at the Same Elevation

A bullet is shot with an initial velocity of 400 m/s at an angle $\theta = 37°$ to the horizontal, and it eventually returns to the same elevation at which it was fired. (*a*) Use the basic kinematic equations for constant acceleration to calculate the range of the bullet. (*b*) Use the range equation (Eq. 4-11) to calculate the range of the bullet. (*c*) What is the maximum height to which the bullet rises?

SOLUTION (*a*) For the coordinate system of Fig. 4-10, $x_0 = y_0 = 0$, $v_{0x} = (400 \text{ m/s}) \cos 37° = 320 \text{ m/s}$, $v_{0y} = (400 \text{ m/s}) \sin 37° = 240 \text{ m/s}$, $a_x = 0$, and $a_y = -9.8 \text{ m/s}^2$.

We first determine the time t that the bullet is in the air. Since $y = 0$ when it returns to the earth,

$$y = y_0 + v_{0y}t + \tfrac{1}{2}a_y t^2$$
$$0 = 0 + (240 \text{ m/s})t + \tfrac{1}{2}(-9.8 \text{ m/s}^2)t^2,$$

so $t = 49$ s. The range R is the x position when $t = 49$ s. Since $x_0 = 0$ and $a_x = 0$,

$$R = v_{0x}t = (320 \text{ m/s})(49 \text{ s}) = 1.6 \times 10^4 \text{ m}.$$

(*b*) From the range equation,

$$R = \frac{v_0^2 \sin 2\theta}{g} = \frac{(400 \text{ m/s})^2(\sin 74°)}{9.8 \text{ m/s}^2} = 1.6 \times 10^4 \text{ m},$$

which agrees with the calculation of part (*a*).

(*c*) When the bullet is at its maximum height, its vertical velocity is zero, so from the relationship between velocity and position,

$$v_y^2 = v_{0y}^2 + 2a_y(y - y_0)$$
$$(0)^2 = (240 \text{ m/s})^2 + 2(-9.8 \text{ m/s}^2)(y_{max} - 0),$$

and $y_{max} = 2.9 \times 10^3$ m.

DRILL PROBLEM 4-6

A golfer hits his tee shot along a flat fairway at 40° to the horizontal with an initial speed of 150 ft/s. (*a*) What is the range of the ball? (*b*) What is the maximum height to which the ball rises? (*c*) If the ball is hit with the same initial speed but at 60° to the horizontal, what is its range?
ANS. (*a*) 692 ft; (*b*) 145 ft; (*c*) 609 ft.

DRILL PROBLEM 4-7

Follow the steps of Example 4-8 to find the range equation directly from the kinematic equations of Table 3-1.

We now look at the case where a projectile lands at an elevation different from that of its starting point. In Fig. 4-12 a boy throws a stone that begins its motion at the elevation $y = 0$, but it lands at a point where $y = h$. Here we can calculate the time at which $y = h$ and then substitute that value into $x = (v_0 \cos\theta)t$ to get R. In the vertical direction,

$$y = (v_0 \sin\theta)t - \tfrac{1}{2}gt^2,$$

so the time t at which $y = h$ is found by solving the quadratic equation

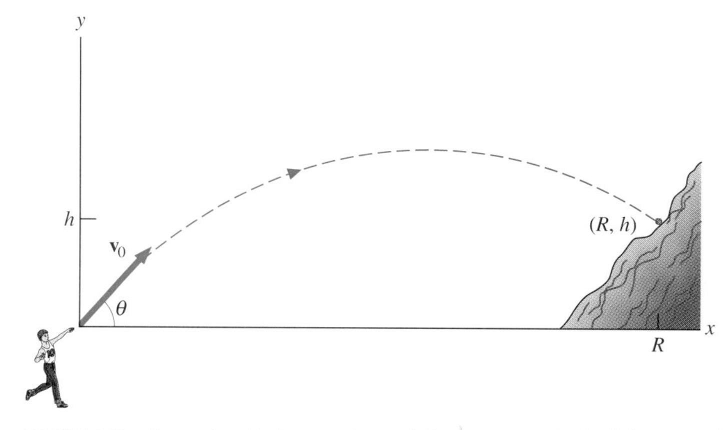

FIGURE 4-12 A projectile leaves the origin at an angle θ with a speed v_0. It lands at $x = R$, $y = h$.

$$t^2 - \left(\frac{2v_0 \sin\theta}{g}\right)t + \frac{2h}{g} = 0.$$

The two solutions for t correspond to the times when the projectile is at $y = h$, first during its ascent, and then again when it lands. (See Fig. 4-12.) As we are interested in the landing time, we use the larger of the two solutions, which is

$$t = \frac{v_0 \sin\theta}{g} + \sqrt{\frac{v_0^2 \sin^2\theta}{g^2} - \frac{2h}{g}}$$
$$= \frac{v_0 \sin\theta}{g}\left(1 + \sqrt{1 - \frac{2gh}{v_0^2 \sin^2\theta}}\right).$$

Therefore the range is

$$R = (v_0 \cos\theta)t = \frac{v_0^2 \sin 2\theta}{2g}\left(1 + \sqrt{1 - \frac{2gh}{v_0^2 \sin^2\theta}}\right), \quad (4\text{-}12)$$

where we have substituted $\sin 2\theta$ for $2 \sin\theta \cos\theta$. Notice that this reduces to our original range equation if $h = 0$.

Using $v_0 = 20$ m/s and $h = -10$ m in Eq. (4-12), we obtain $R = 49.1$ m for $\theta = 45°$ and $R = 49.5$ m for $\theta = 43°$. With different initial and final elevations, $\theta = 45°$ does not give the greatest range. To find the projection angle θ for maximum range, we take the derivative of R with respect to θ in Eq. (4-12), set that derivative equal to zero, and solve for θ.

At the low speed of this golf ball, air resistance is negligible and its path is parabolic.

EXAMPLE 4-9 Projectile Motion with Starting and Landing Points at Different Elevations

Figure 4-13 shows a golfer hitting a drive from an elevated tee to a fairway 50 ft below. The ball leaves the club at 45° to the horizontal with a speed of 160 ft/s. Use the kinematic equations of Table 3-1 to calculate the range of the ball.

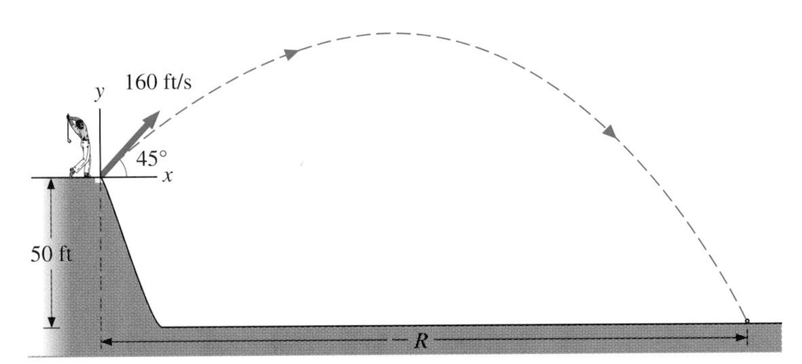

FIGURE 4-13 A golfer hits his drive from an elevated tee 50 ft above the fairway.

SOLUTION The components of the initial velocity are $v_{0x} = (160 \text{ ft/s}) \cos 45° = 113$ ft/s and $v_{0y} = (160 \text{ ft/s}) \sin 45° = 113$ ft/s. If the origin of our coordinate system is the spot where the ball is struck, then the y coordinate of the ball when it lands is -50 ft. So, from

$$y = v_{0y}t + \tfrac{1}{2}a_{0y}t^2$$

we have

$$-50 \text{ ft} = (113 \text{ ft/s})t - \tfrac{1}{2}(32 \text{ ft/s}^2)t^2,$$

which gives the following quadratic equation to be solved for t:

$$16t^2 - 113t - 50 = 0.$$

The positive solution to this equation is $t = 7.5$ s, which is the time the ball hits the fairway. During its flight the ball is traveling with a constant velocity of 113 ft/s in the horizontal direction. Thus the range is

$$R = v_{0x}t = (113 \text{ ft/s})(7.5 \text{ s}) = 848 \text{ ft}.$$

DRILL PROBLEM 4-8

Use the equation of the path of a projectile to derive Eq. (4-12).

DRILL PROBLEM 4-9

Use Eq. (4-12) to determine the range of the golf ball in Example 4-9.

4-6 Circular Motion

When a particle moves at constant speed along a straight line, neither the magnitude nor the direction of its velocity **v** changes. Consequently, $d\mathbf{v}/dt = 0$, and the particle does not accelerate. However, if a particle moves at constant speed along a curved path, $d\mathbf{v}/dt \neq 0$ because the direction of **v** changes; so in this case, there is an acceleration even when $|\mathbf{v}|$ is constant.

Motion along a particular curved path, the circle, will now be considered. Circular motion is especially important because it has so many practical applications. In addition, the descriptions of motion along a circle and along an arbitrary path are similar enough that once you understand the former, you will also have a qualitative understanding of the latter.

Circular motion at constant speed is represented by the two sketches of Fig. 4-14, which show a particle with position $\mathbf{r}(t)$ moving along a circle of radius r. At a time t, the velocity $\mathbf{v}(t)$ is tangent to the circle at P and at a time $t + \Delta t$, $\mathbf{v}(t + \Delta t)$ is tangent to the circle at P′. Since the speed is constant, $|\mathbf{v}(t)| = |\mathbf{v}(t + \Delta t)|$. The magnitude of the acceleration can be found by using the triangles of Fig. 4-14a and b to evaluate $\lim_{\Delta t \to 0}(\Delta v/\Delta t)$. The triangle with sides $\mathbf{v}(t)$ and $\mathbf{v}(t + \Delta t)$ is isosceles because the speed is constant, and the triangle with sides $\mathbf{r}(t)$ and $\mathbf{r}(t + \Delta t)$ is also isosceles because the radius is constant. Furthermore, since $\mathbf{v}(t)$ is perpendicular to $\mathbf{r}(t)$ and $\mathbf{v}(t + \Delta t)$ is perpendicular to $\mathbf{r}(t + \Delta t)$, the angle $\Delta\theta$ between the two velocity vectors is the same as that between the two position vectors. Thus the two triangles are similar, and

$$\frac{\Delta v}{v} = \frac{\Delta r}{r},$$

or

$$\Delta v = \frac{v}{r}\,\Delta r.$$

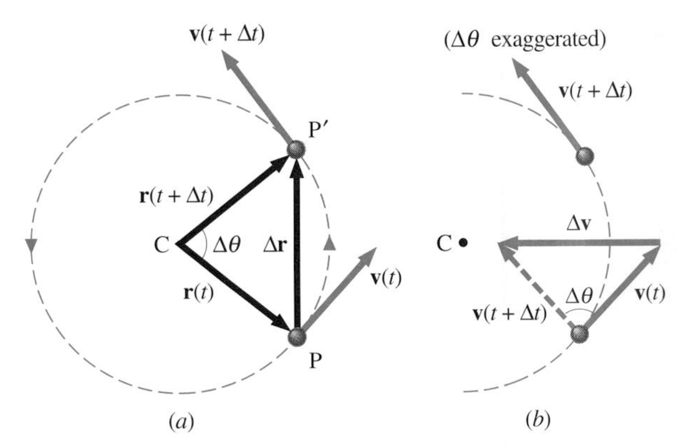

FIGURE 4-14 (a) A particle moves in a circle of radius r at constant speed. (b) The vector $\Delta\mathbf{v}$ points toward C in the limit $\Delta t \to 0$.

Dividing by Δt and taking the limit $\Delta t \to 0$, we obtain

$$a = \lim_{\Delta t \to 0}\left(\frac{\Delta v}{\Delta t}\right) = \frac{v}{r}\left(\lim_{\Delta t \to 0}\frac{\Delta r}{\Delta t}\right) = \frac{v^2}{r}.$$

The direction of the acceleration vector can be also be found with Fig. 4-14b. As Δt, and therefore $\Delta\theta$, gets smaller, the vector $\Delta\mathbf{v}$ approaches a direction perpendicular to **v**; and in the limit $\Delta t \to 0$, $\Delta\mathbf{v}$ is perpendicular to **v**. Since **v** is tangent to the circle, the acceleration $d\mathbf{v}/dt$ must point toward the center of the circle.

To summarize, for a particle moving in a circle of radius r at constant speed v:

1. The magnitude of its acceleration is given by

$$a = \frac{v^2}{r}. \tag{4-13}$$

2. The acceleration vector is directed from the particle toward the center of the circle.

The relation between **v** and **a** is shown in Fig. 4-15. Because **a** always points *toward the center* of the circle, it is called a **centripetal** (or "center-seeking") **acceleration**.

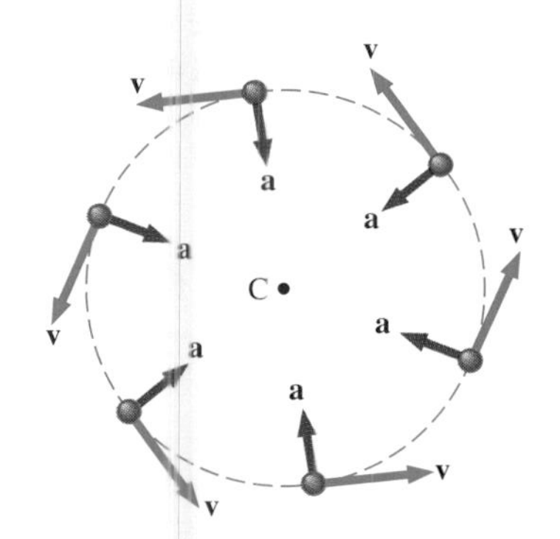

FIGURE 4-15 As a particle travels in a circle at constant speed, it accelerates toward the center C of the circle.

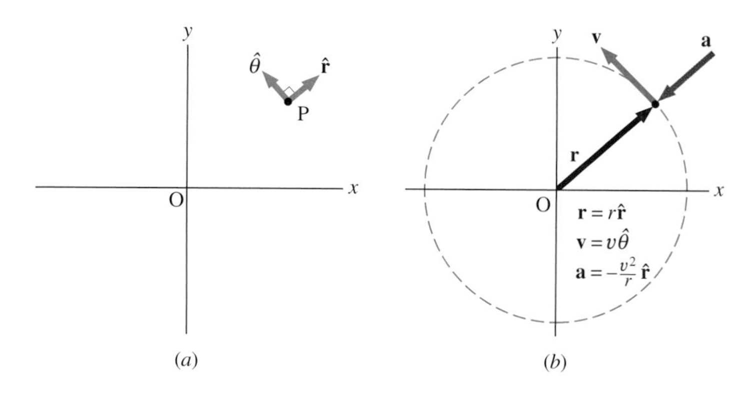

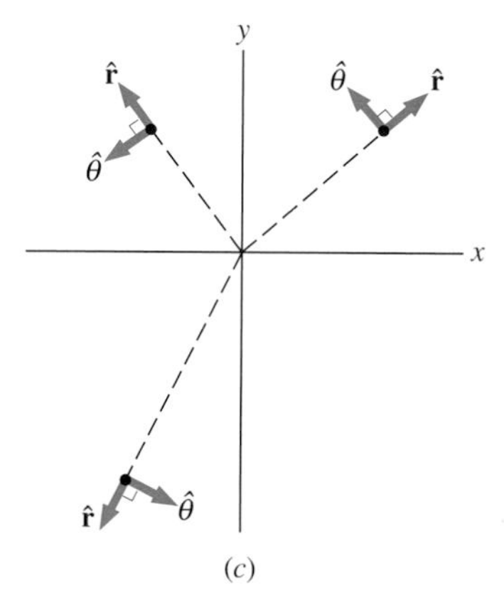

FIGURE 4-16 (*a*) The unit polar vectors $\hat{\mathbf{r}}$ and $\hat{\boldsymbol{\theta}}$. (*b*) The position, velocity, and acceleration vectors for circular motion at constant speed. (*c*) The directions of the polar vectors change with position in the plane.

The directions of vector quantities associated with circular motion are usually specified in terms of the unit polar vectors $\hat{\mathbf{r}}$ and $\hat{\boldsymbol{\theta}}$. These vectors are defined for any point P in the plane as follows (see Fig. 4-16*a*):

1. The radial unit vector $\hat{\mathbf{r}}$ is directed radially *outward* along the line connecting the origin O and the point P.
2. The tangential unit vector $\hat{\boldsymbol{\theta}}$ is directed perpendicular to $\hat{\mathbf{r}}$ in the *counterclockwise* direction.

Figure 4-16*b* shows that for a particle moving counterclockwise in a circle of radius r at a constant speed v,

$$\mathbf{r} = r\hat{\mathbf{r}}, \tag{4-14}$$

$$\mathbf{v} = v\hat{\boldsymbol{\theta}}, \tag{4-15}$$

and

$$\mathbf{a} = -\frac{v^2}{r}\hat{\mathbf{r}}.^* \tag{4-16}$$

As illustrated in Fig. 4-16*c*, the directions of the unit polar vectors are not constant, so their time derivatives, unlike those of **i**, **j**, and **k**, do not vanish. Hence the acceleration of a particle moving in a circle at a constant speed is

$$\mathbf{a} = \frac{d}{dt}(v\hat{\boldsymbol{\theta}}) = v\frac{d\hat{\boldsymbol{\theta}}}{dt} + \frac{dv}{dt}\hat{\boldsymbol{\theta}} = v\frac{d\hat{\boldsymbol{\theta}}}{dt} + 0 = v\frac{d\hat{\boldsymbol{\theta}}}{dt}.$$

Setting this equal to **a** in Eq. (4-16), we have

$$-\frac{v^2}{r}\hat{\mathbf{r}} = v\frac{d\hat{\boldsymbol{\theta}}}{dt},$$

so

$$\frac{d\hat{\boldsymbol{\theta}}}{dt} = -\frac{v}{r}\hat{\mathbf{r}}. \tag{4-17}$$

Now let's consider a particle moving in a circle at a variable speed. Once again, the acceleration is $\mathbf{a} = d\mathbf{v}/dt$, where $\mathbf{v} = v\hat{\boldsymbol{\theta}}$. However, here v is no longer constant, so $dv/dt \neq 0$, and

$$\mathbf{a} = v\frac{d\hat{\boldsymbol{\theta}}}{dt} + \frac{dv}{dt}\hat{\boldsymbol{\theta}}.$$

Now using Eq. (4-17) to substitute for $d\hat{\boldsymbol{\theta}}/dt$, we find

$$\mathbf{a} = -\frac{v^2}{r}\hat{\mathbf{r}} + \frac{dv}{dt}\hat{\boldsymbol{\theta}}. \tag{4-18}$$

Thus a particle moving in a circle at variable speed has an acceleration *toward the center* of the circle given by $a_r = -v^2/r$ and an acceleration *tangent* to the circle given by $a_\theta = dv/dt$. (See Fig. 4-17.)

*It is sometimes convenient to choose the positive $\hat{\mathbf{r}}$ direction inward rather than outward. Then the centripetal acceleration is $+v^2/r$.

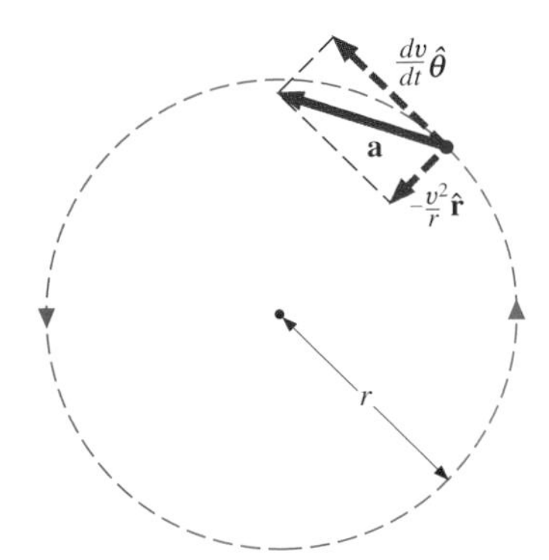

FIGURE 4-17 A particle moving in a circle at variable speed has an acceleration v^2/r toward the center of the circle and an acceleration dv/dt tangent to the circle.

For a particle moving along an arbitrary path, the acceleration can also be resolved into two components, one tangent to the path and one perpendicular to the path. The tangential component is dv/dt and the perpendicular component is v^2/ρ, where ρ is a quantity called the *radius of curvature*. In the case of circular motion, ρ is simply the radius. However, for an arbitrary path, ρ is not constant. At each point along the path, it is the radius of the circle that approximates the curve in the neighborhood of that point. Examples of locally approximate circles and their corresponding radii of curvature are shown in Fig. 4-18.

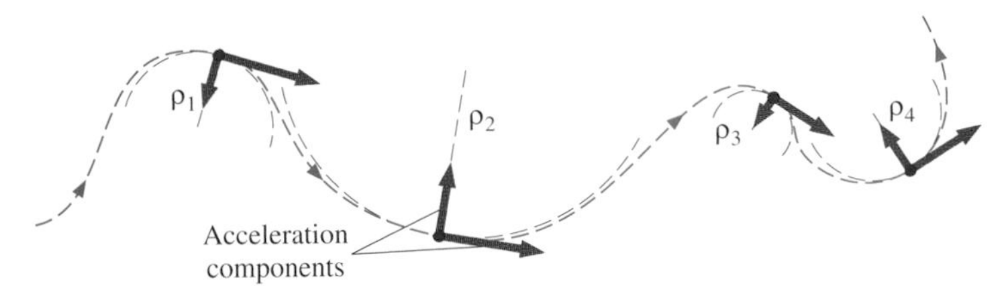

FIGURE 4-18 The curve at each point is represented by a localized circle of radius ρ. The acceleration along the curve is dv/dt, and the acceleration perpendicular to the curve is v^2/ρ.

EXAMPLE 4-10 Centripetal Acceleration of the Earth

Assume that the earth revolves around the sun at constant speed in a circular orbit of radius 1.50×10^{11} m. What is the earth's acceleration toward the sun?

SOLUTION The speed v of the earth is the distance it travels in one orbit divided by the time T it takes to complete the orbit. The value of T is 365.25 days, which to three significant figures is 3.16×10^7 s. The earth's speed is then

$$v = \frac{2\pi(1.50 \times 10^{11}\,\text{m})}{3.16 \times 10^7\,\text{s}} = 2.98 \times 10^4\ \text{m/s},$$

and the acceleration of the earth toward the sun is

$$a = \frac{v^2}{r} = \frac{(2.98 \times 10^4\ \text{m/s})^2}{1.50 \times 10^{11}\ \text{m}} = 5.93 \times 10^{-3}\ \text{m/s}^2.$$

EXAMPLE 4-11 Net Acceleration During Circular Motion

A particle moves in a circle of radius $r = 2.0$ m. During the time interval from $t = 1.0$ s to $t = 4.0$ s, its speed varies with time according to $v(t) = c_1 - c_2/t^2$, where $c_1 = 4.0$ m/s and $c_2 = 6.0$ m·s. What is the acceleration of the particle at $t = 2.0$ s?

SOLUTION At $t = 2.0$ s,

$$v = \left(4.0 - \frac{6.0}{2.0^2}\right)\ \text{m/s} = 2.5\ \text{m/s},$$

and

$$\frac{dv(t)}{dt} = \frac{2c_2}{t^3} = \frac{12.0}{2.0^3}\ \text{m/s}^2 = 1.5\ \text{m/s}^2.$$

Now from Eq. (4-18),

$$\mathbf{a} = \left(-\frac{2.5^2}{2.0}\,\hat{\mathbf{r}} + 1.5\hat{\boldsymbol{\theta}}\right)\ \text{m/s}^2 = (-3.1\hat{\mathbf{r}} + 1.5\hat{\boldsymbol{\theta}})\ \text{m/s}^2.$$

The net acceleration is then 3.4 m/s² at 64° to the $\hat{\boldsymbol{\theta}}$ direction.

DRILL PROBLEM 4-10

In the simple Bohr model of the ground state of the hydrogen atom, the electron travels in a circular orbit around a fixed proton. The radius of the orbit is 5.28×10^{-11} m, and the speed of the electron is 2.18×10^6 m/s. What is the acceleration of the electron toward the proton?
ANS. 9.00×10^{22} m/s².

DRILL PROBLEM 4-11

Assuming the moon travels around the earth in a circular orbit of radius 3.85×10^8 m with a period of 27.3 days, what is the acceleration of the moon toward the earth?
ANS. 2.73×10^{-3} m/s².

DRILL PROBLEM 4-12

The position of a particle is given by $\mathbf{r}(t) = A(\cos \omega t\,\mathbf{i} + \sin \omega t\,\mathbf{j})$, where ω is a constant. (*a*) Show that the particle moves in a circle of radius A. (*b*) Calculate $d\mathbf{r}/dt$ and then show that the speed of the particle is the constant $A\omega$. (*c*) Determine $d^2\mathbf{r}/dt^2$ and show that $\mathbf{a}$ is given by Eq. (4-16).
[*Note:* For parts (*b*) and (*c*), you will need to use $(d/dt)(\cos \omega t) = -\omega \sin \omega t$ and $(d/dt)(\sin \omega t) = \omega \cos \omega t$.]

*4-7 Relative Velocity and Relative Acceleration

When two people simultaneously measure the velocity of an object, they do not necessarily get the same result. For example, a man (A) riding in the car of Fig. 4-19 measures the velocity

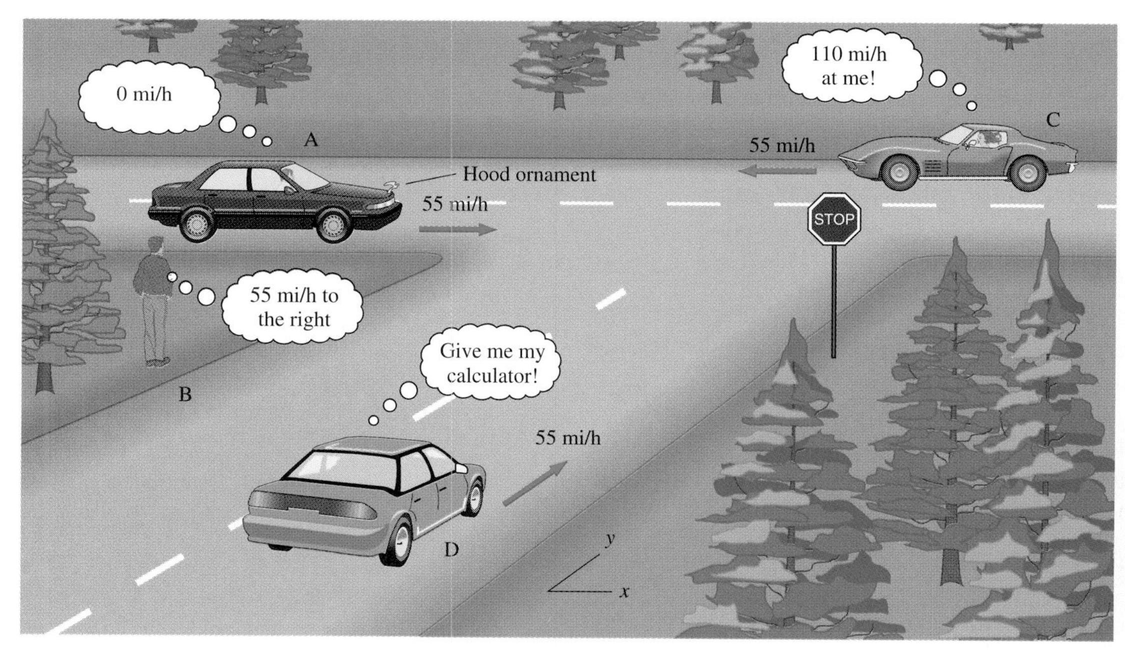

FIGURE 4-19 Different observers measure different velocities for the hood ornament.

of the hood ornament to be zero, a second observer (B) standing at the side of the road sees the ornament moving by with a velocity of $55\mathbf{i}$ mi/h, while a third observer (C) riding in the car moving from right to left at $-55\mathbf{i}$ mi/h sees the ornament moving toward her at $110\mathbf{i}$ mi/h. Finally, an observer (D) moving up the side road at $55\mathbf{j}$ mi/h sees the ornament traveling at a velocity whose magnitude and direction are not immediately obvious. To determine this observer's measurement, you must know how to calculate relative velocities.

Consider a particle P and the reference frames S and S′ of Fig. 4-20. The position of the origin of S′ as measured in S is $\mathbf{r}_{S'S}$, the position of P as measured in S′ is $\mathbf{r}_{PS'}$, and the position of P as measured in S is $\mathbf{r}_{PS}$. From inspection of the figure,

$$\mathbf{r}_{PS} = \mathbf{r}_{PS'} + \mathbf{r}_{S'S}. \tag{4-19}$$

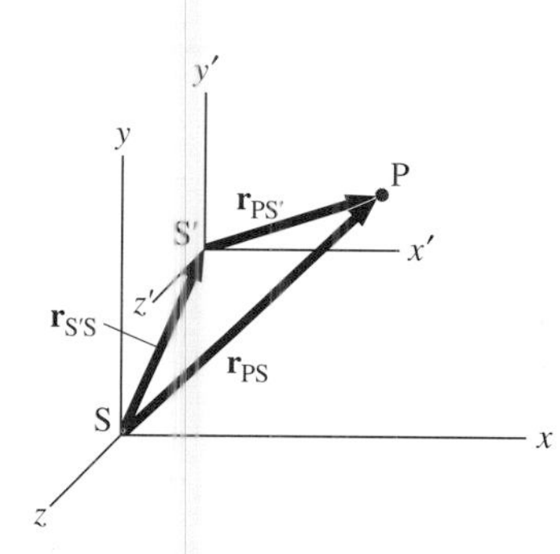

FIGURE 4-20 The positions of the particle P relative to frames S and S′ are $\mathbf{r}_{PS}$ and $\mathbf{r}_{PS'}$, respectively.

The relative velocities are just the time derivatives of the position vectors. Therefore

$$\mathbf{v}_{PS} = \mathbf{v}_{PS'} + \mathbf{v}_{S'S}. \tag{4-20}$$

The velocity of a particle relative to S is equal to its velocity relative to S′ plus the velocity of S′ relative to S.

Returning to the hood ornament O of Fig. 4-19, $\mathbf{v}_{OA}$* = 0, $\mathbf{v}_{OB} = 55\mathbf{i}$ mi/h, and $\mathbf{v}_{CB} = -55\mathbf{i}$ mi/h, so

$$\mathbf{v}_{OC} = \mathbf{v}_{OB} + \mathbf{v}_{BC} = (55\mathbf{i} + 55\mathbf{i}) \text{ mi/h} = 110\mathbf{i} \text{ mi/h}.$$

Now $\mathbf{v}_{DB} = 55\mathbf{j}$ mi/h, so the ornament's velocity relative to D is

$$\mathbf{v}_{OD} = \mathbf{v}_{OB} + \mathbf{v}_{BD} = \mathbf{v}_{OB} - \mathbf{v}_{DB} = (55\mathbf{i} - 55\mathbf{j}) \text{ mi/h}.$$

D therefore sees the ornament move with a speed

$$|\mathbf{v}_{OD}| = \sqrt{(55)^2 + (55)^2} \text{ mi/h} = 78 \text{ mi/h},$$

and at an angle (see Fig. 4-21)

$$\theta = \tan^{-1}(55/55) = 45°$$

below the positive x axis.

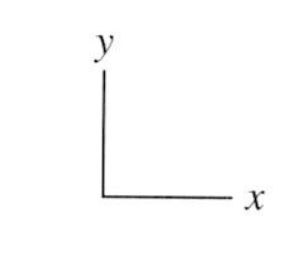

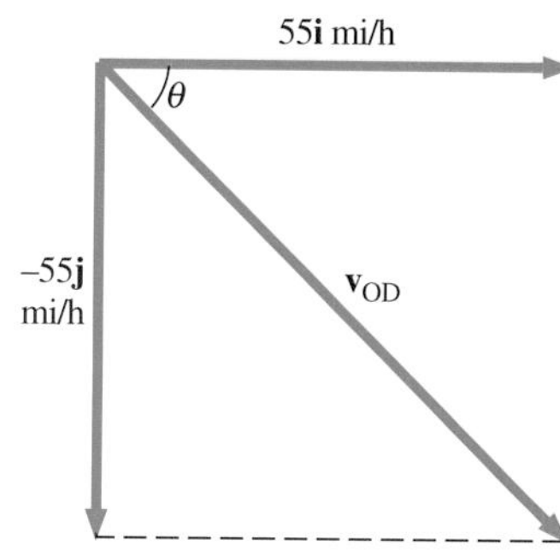

FIGURE 4-21 The velocity of the hood ornament as measured by observer D. (See Fig. 4-19.)

*$\mathbf{v}_{PA}$, the velocity of object P *with respect to* object A, is always measured in a reference system *at rest* relative to A.

The subscripts of Eq. (4-20) reveal a helpful mnemonic device you can use to remember the velocity addition rule. The subscripts on the left-hand side of the equation are the same as the two outside subscripts on the right-hand side of the equation, and the middle subscripts on the right-hand side are identical. Note the pairing in the following equation:

$$\mathbf{v}_{PB} = \mathbf{v}_{PA} + \mathbf{v}_{AB}.$$

The pairing allows us to extend the velocity addition rule to any number of reference frames. For a particle P and reference frames A, B, and C,

$$\mathbf{v}_{PC} = \mathbf{v}_{PA} + \mathbf{v}_{AB} + \mathbf{v}_{BC}.$$

How accelerations are related is determined quite easily by differentiating Eq. (4-20) with respect to time:

$$\mathbf{a}_{PS} = \mathbf{a}_{PS'} + \mathbf{a}_{S'S}. \tag{4-21}$$

You can see from this equation that if the velocity of S′ relative to S is constant (that is, if $\mathbf{a}_{S'S} = 0$), then

$$\mathbf{a}_{PS} = \mathbf{a}_{PS'}. \tag{4-22}$$

The acceleration of a particle is the same as measured by two observers moving at constant velocity relative to one another. This result, which may seem quite trivial at this point, will be very important when we consider Newton's laws in Chap. 5.

EXAMPLE 4-12 Motion of One Automobile Relative to Another

A police car is moving south on a side road at 40 mi/h, and an automobile is moving west on a highway at 55 mi/h. What is the velocity of the automobile relative to the police car?

SOLUTION We use a rectangular coordinate system attached to the earth with the x axis pointing east and the y axis pointing north. (See Fig. 4-22.) The velocity of the automobile relative to the earth is $\mathbf{v}_{AE} = -55\mathbf{i}$ mi/h, and the velocity of the police car with respect to the earth is $\mathbf{v}_{PE} = -40\mathbf{j}$ mi/h. Then

$$\mathbf{v}_{AE} = \mathbf{v}_{AP} + \mathbf{v}_{PE},$$

and

$$\mathbf{v}_{AP} = \mathbf{v}_{AE} - \mathbf{v}_{PE} = (-55\mathbf{i} + 40\mathbf{j}) \text{ mi/h}.$$

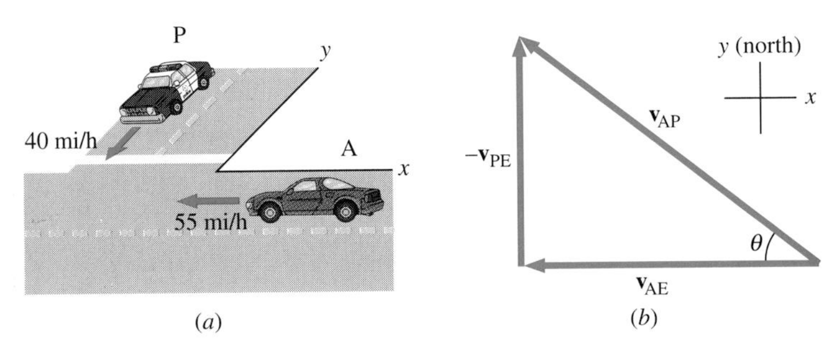

FIGURE 4-22 (*a*) A police car is traveling on a side road as an automobile moves west on a freeway. (*b*) Vector sum of the velocities.

This vector sum is shown in the figure. The police officer sees the automobile moving at $v_{AP} = 68$ mi/h in a direction $\theta = 36°$ north of west.

EXAMPLE 4-13 A Plane's Heading

A pilot must fly her plane due north to reach her destination. The plane can fly at 300 km/h in still air, and a wind is blowing out of the northeast at 90 km/h. (This is the velocity of the air with respect to the ground.) In what direction must the pilot head her plane in order to fly straight north? What is the speed of the plane relative to the ground?

SOLUTION The pilot must point her plane somewhat east of north so that $\mathbf{v}_{PG}$ is directed straight north. Using the standard notation, we have

$$\mathbf{v}_{PG} = \mathbf{v}_{PA} + \mathbf{v}_{AG},$$

an equation represented by the closed triangle of Fig. 4-23. With simple trigonometry, we find from the triangle that $v_{PG} = 230$ km/h and $\theta = 12.2°$. Hence, if the pilot heads her plane at 12.2° east of north, she will travel due north at 230 km/h relative to the ground.

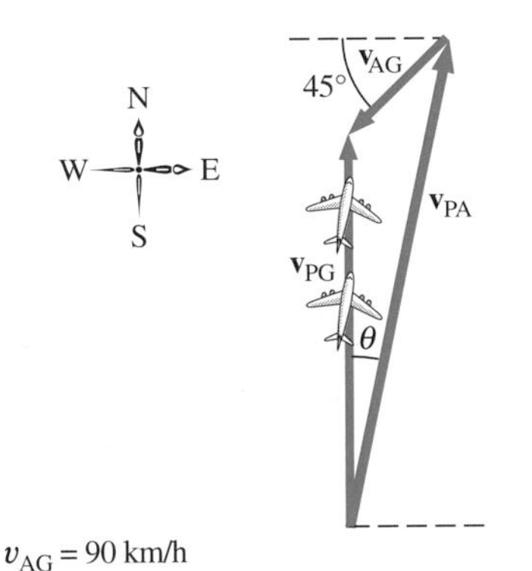

FIGURE 4-23 Vector sum of the velocities of Example 4-13.

DRILL PROBLEM 4-13

A river flows eastward with a speed of 3.0 m/s. A man who can row his boat at 5.0 m/s in still water decides to travel from the south to the north side of the river. (*a*) In what direction must he aim his boat if he is to reach a point directly north of his starting point? (*b*) If the river is 1000 m wide, how long will it take him to cross? (*c*) If he aims the boat straight north, how long will it take him to cross the river and where will he land?
ANS. (*a*) 37° west of north; (*b*) 250 s; (*c*) 200 s, 600 m downstream.

SUMMARY

1. **Description of two- and three-dimensional motion**
   (*a*) Position $\mathbf{r}(t) = x(t)\mathbf{i} + y(t)\mathbf{j} + z(t)\mathbf{k}$ is a vector drawn from the origin to the particle.
   (*b*) If the positions of a particle at t_1 and t_2 are $\mathbf{r}(t_1)$ and $\mathbf{r}(t_2)$, respectively, the displacement of the particle is $\mathbf{r}(t_2) - \mathbf{r}(t_1)$.
   (*c*) Velocity $\mathbf{v}(t) = \dfrac{d\mathbf{r}(t)}{dt} = \dfrac{dx(t)}{dt}\mathbf{i} + \dfrac{dy(t)}{dt}\mathbf{j} + \dfrac{dz(t)}{dt}\mathbf{k}$.
   (*d*) Speed is the magnitude of velocity and is represented by $|\mathbf{v}(t)|$ or $v(t)$.
   (*e*) The average velocity between t_1 and t_2 is the displacement $\mathbf{r}(t_2) - \mathbf{r}(t_1)$ in this time interval divided by $t_2 - t_1$.
   (*f*) Acceleration $\mathbf{a}(t) = \dfrac{d\mathbf{v}(t)}{dt} = \dfrac{dv_x(t)}{dt}\mathbf{i} + \dfrac{dv_y(t)}{dt}\mathbf{j} + \dfrac{dv_z(t)}{dt}\mathbf{k}$.

2. **Analysis of multidimensional motion**
   This motion can be treated as two or three one-dimensional motions. For constant acceleration, the motion can be analyzed using the methods developed in Chap. 3.

3. **Common examples of two-dimensional motion**
   (*a*) Projectile motion: The acceleration of a projectile is $\mathbf{g}$. Its motion is a combination of free fall in the vertical direction and motion at constant velocity in the horizontal direction. The path of a projectile is a parabola, and if its initial and final elevations are the same, its range is
   $$R = \frac{v_0^2 \sin 2\theta}{g}.$$
   (*b*) Circular motion: When a particle moves on a circle of radius r at a speed $v(t)$, its acceleration is
   $$\mathbf{a} = -\frac{v^2}{r}\hat{\mathbf{r}} + \frac{dv}{dt}\hat{\boldsymbol{\theta}}.$$

*4. **Relative velocity and relative acceleration**
   If frame S′ moves with respect to frame S with a velocity $\mathbf{v}_{S'S}$ and an acceleration $\mathbf{a}_{S'S}$, then a particle P moving in S′ with a velocity and an acceleration $\mathbf{v}_{PS'}$ and $\mathbf{a}_{PS'}$ has a velocity and an acceleration in S given by
   $$\mathbf{v}_{PS} = \mathbf{v}_{PS'} + \mathbf{v}_{S'S} \quad \text{and} \quad \mathbf{a}_{PS} = \mathbf{a}_{PS'} + \mathbf{a}_{S'S}.$$

QUESTIONS

4-1. "Three-dimensional motion may be analyzed as three one-dimensional motions." Discuss this statement.

4-2. Particle A starts at the origin with zero initial velocity and with the constant acceleration $\mathbf{a} = (5.0\mathbf{i} + 7.0\mathbf{j})\ \text{m/s}^2$. Particle B starts at the origin with the constant velocity $\mathbf{v} = (5.0\mathbf{i} + 7.0\mathbf{j})\ \text{m/s}$. How are the motions of the two particles similar? How do they differ?

4-3. What is the acceleration of a projectile at the top of its path?

4-4. Consider a projectile that begins and ends its motion at the same elevation. For any of the possible heights of the projectile, there are two instants during its motion when it is at that height. Which of the following quantities are identical at these two instants? (*a*) position, (*b*) velocity, (*c*) speed, (*d*) acceleration.

4-5. Two cannonballs fired simultaneously from opposing warships collide in midair. In the same situation, but without gravity, would these projectiles collide? Explain.

4-6. As a baseball player throws a ball horizontally, his cap falls off his head. Does the ball or the cap hit the ground first? Assume they start their motions at the same elevation.

4-7. Suppose you throw a ball at 33° to the horizontal and measure its horizontal range. At what other angle could you throw the ball with the same initial speed so that its range would be the same?

4-8. At what point in its path does a projectile have the smallest speed?

4-9. Is it possible for a particle to be moving such that its velocity is perpendicular to (*a*) its position, (*b*) its acceleration, and (*c*) both vectors simultaneously? For each "yes" answer, describe the motion.

4-10. A particle is moving in a circle with a constant speed. Which of the following quantities are constant? (*a*) the particle's position, (*b*) its velocity, (*c*) the magnitude of its centripetal acceleration, (*d*) its tangential acceleration, (*e*) the magnitude of its net acceleration.

4-11. Compare the unit vectors $\hat{\mathbf{r}}$ and $\hat{\boldsymbol{\theta}}$ with the unit vectors $\mathbf{i}$ and $\mathbf{j}$.

4-12. Can a car round a curve at constant velocity? at constant speed?

4-13. The moon constantly accelerates toward the earth, yet it never falls into the earth. Explain how this can happen.

4-14. If you drop a ball inside an airplane moving horizontally at constant velocity, what will you measure the acceleration of the ball to be?

4-15. Suppose that the airplane of Ques. 4-14 is moving with constant velocity, but is descending. Would you measure the acceleration of the ball to be any different than before? Explain.

4-16. If you drop an object out the window of your car while traveling west along a highway at constant velocity, where will you be relative to the object when it hits the ground?

4-17. If the earth is rotating, why doesn't an object that we drop land somewhere behind us?

4-18. A boy sitting in a train moving at constant velocity throws a ball straight upward relative to his frame. Does the ball return to him, fall behind him, or fall in front of him? What is the path of the ball as seen by an observer at rest on the ground?

4-19. A clown on a unicycle is moving backwards at constant velocity. How does he throw a ball upward so that it comes directly back to him?

4-20. When driving to the basket for a layup, a basketball player normally tosses the ball gently upward relative to himself, allowing it to fall softly off the backboard. Explain why this is done in terms of relative motion.

Problems

Two- and Three-Dimensional Motion

4-1. The coordinates of a particle in a rectangular coordinate system are (1.0, −4.0, 6.0). What is the position vector of the particle? Assume SI units.

4-2. The position of a particle changes from $\mathbf{r}_1 = (2.0\mathbf{i} + 3.0\mathbf{j})$ cm to $\mathbf{r}_2 = (-4.0\mathbf{i} + 3.0\mathbf{j})$ cm. What is the particle's displacement?

4-3. A particle initially located at $(1.5\mathbf{j} + 4.0\mathbf{k})$ m undergoes a displacement of $(2.5\mathbf{i} + 3.2\mathbf{j} - 1.2\mathbf{k})$ m. What is the final position of the particle?

4-4. During his workout, an athlete runs four complete laps at constant speed on a 440-yd oval track in 5.0 min. (*a*) What is his speed while running? (*b*) What is his net displacement for the run?

4-5. The earth takes $365\frac{1}{4}$ days to complete one orbit around the sun. Assuming that its orbit is circular, with a radius of 1.50×10^8 km, what is the orbital speed of the earth?

4-6. A bird flies straight northeast a distance of 95 km in 3.0 h. With the x axis directed from west to east and the y axis from south to north, what is the displacement (expressed as a vector) of the bird? What is its average velocity for the trip?

4-7. Suppose that the motion of the particle of Prob. 4-2 takes 0.20 s. What is the particle's average velocity during its displacement?

4-8. A particle moves around the circle of radius 8.0 m shown in the accompanying figure. What is its position when it is at A? at B? at C? at D?

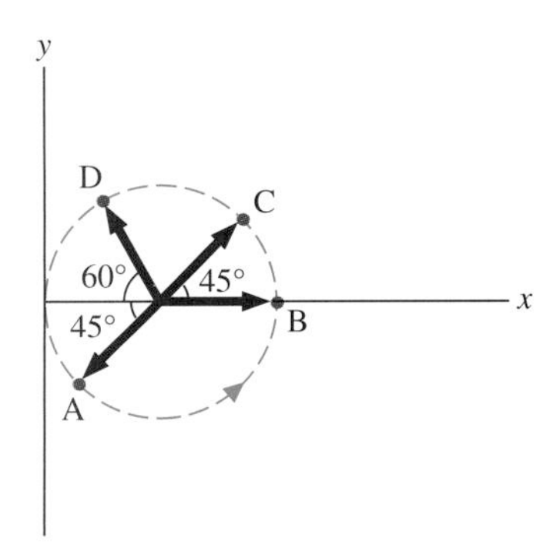

PROBLEMS 4-8, 4-9, and 4-10

4-9. What is the displacement of the particle of Prob. 4-8 when it moves from A to B? from B to C? from A to D?

4-10. Repeat Probs. 4-8 and 4-9 with the origin of the rectangular coordinate system located at B.

4-11. Calculate the average velocities for the displacements of Prob. 4-9, assuming the particle takes 10 s to move uniformly once around the circle.

4-12. The position of a particle is given for $t > 0$ by $\mathbf{r}(t) = (3.0t^2\mathbf{i} - 7.0t^3\mathbf{j} - 5.0/t^2\mathbf{k})$ cm. (*a*) What is the velocity of the particle as a function of time? (*b*) What is its acceleration as a function of time? (*c*) What is the particle's velocity at $t = 2.0$ s? (*d*) What is its speed at $t = 1.0$ s and $t = 3.0$ s? (*e*) What is the average velocity of the particle between $t = 1.0$ s and $t = 2.0$ s?

4-13. The position of a particle is given by $\mathbf{r}(t) = c_1 t\mathbf{i} - c_2 t^2\mathbf{j}$, where $c_1 = 50$ m/s and $c_2 = 4.9$ m/s^2. (*a*) What are the particle's velocity and acceleration as functions of time? (*b*) In order to produce this motion, what initial conditions would you impose on the particle, and what acceleration would you give it?

4-14. What is the average velocity of the particle of Prob. 4-13 between $t = 0$ and $t = 3.0$ s?

4-15. (*a*) Find the equation of the path of a particle whose rectangular positions are given by $x(t) = (2.0 + 3.0t)$ cm and $y(t) = (5.0 - 2.0t)$ cm. What type of curve is this? (*b*) What is the velocity of the particle at $t = 3.0$ s?

4-16. The rectangular coordinates of a particle vary with time according to $x(t) = A\cos(1.0t)$ and $y(t) = A\sin(1.0t)$. The arguments of the cosine and sine functions are in radians. (*a*) With $A =$ 5.0 cm, plot $x(t)$ and $y(t)$ and construct the path of the particle. (*b*) Eliminate t between the equations for $x(t)$ and $y(t)$ and determine the equation representing the path of the particle. (*c*) Find the particle's position at $t = 0$ and $t = 1.0$ s. (*d*) What is the particle's instantaneous velocity at $t = 1.0$ s? (See Drill Prob. 4-12.)

4-17. The positions of particles 1 and 2 are $\mathbf{r}_1 = (3.0t^2\mathbf{i} + 2.0t\mathbf{j})$ m and $\mathbf{r}_2 = (p\mathbf{i} + 4.0t^2\mathbf{j})$ m, respectively. At what time and for what value of p do the particles have the same position?

4-18. A particle's acceleration is $\mathbf{a} = (4.0\mathbf{i} + 3.0\mathbf{j})$ m/s^2. At $t = 0$, its position and velocity are zero. (*a*) What are the velocity and position of the particle as functions of time? (*b*) Find the equation representing the path of the particle.

4-19. Repeat part (*a*) of Prob. 4-18, assuming the initial position and velocity of the particle are $\mathbf{r}_0 = (5.0\mathbf{i} + 2.0\mathbf{j})$ m and $\mathbf{v}_0 = 8.0\mathbf{j}$ m/s.

4-20. The acceleration of a particle is constant. At $t = 0$, the velocity of the particle is $(10\mathbf{i} + 20\mathbf{j})$ m/s, and at $t = 4.0$ s, its velocity is $10\mathbf{j}$ m/s. (*a*) What is the particle's acceleration? (*b*) How do its position and velocity vary with time? Assume that the particle is initially at the origin.

4-21. The velocity of a particle is $\mathbf{v}(t) = (4.0t\mathbf{i} + 8.0\mathbf{j})$ m/s. Assuming that the particle is initially at the origin, determine (*a*) its acceleration, (*b*) its initial velocity, and (*c*) its position as a function of time.

4-22. The position of a particle is $\mathbf{r}(t) = (3.0t^2\mathbf{i} + 5.0\mathbf{j} - 6.0t\mathbf{k})$ m. Determine its velocity and acceleration as functions of time. What are its initial position and velocity?

4-23. If the equation of the path of a particle is $y = cx^2$, how are the velocities $v_x(t)$ and $v_y(t)$ related? (*Hint:* Use the chain rule for differentiation.)

Motion of a Projectile

4-24. A bullet is shot horizontally from shoulder height (1.5 m) with an initial speed of 200 m/s. (*a*) How much time elapses before the bullet hits the ground? (*b*) How far does the bullet travel horizontally?

4-25. A marble rolls off the edge of a horizontal tabletop 1.0 m high and hits the floor at a point 3.0 m away from the table's edge in the horizontal direction. (*a*) How long is the marble in the air? (*b*) What is the marble's speed when it leaves the table? (*c*) What is its speed when it hits the floor?

4-26. A dart is thrown horizontally, with a speed of 10 m/s, at the bull's-eye of a dartboard 2.4 m away. (See the accompanying figure.) How far below its intended target will the dart hit? What does your answer tell you about how good dart players launch their projectiles?

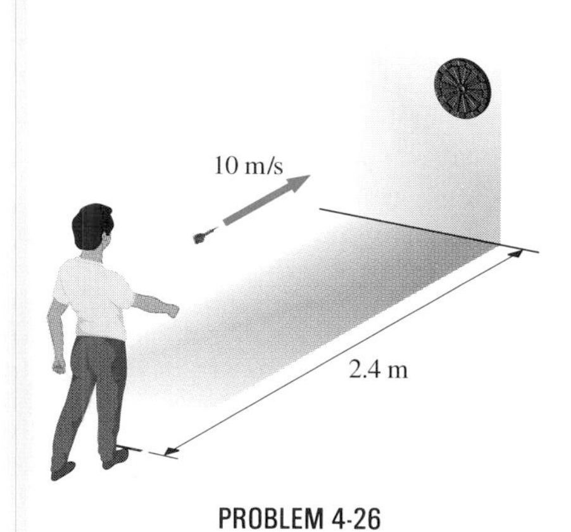

PROBLEM 4-26

4-27. A crossbow is aimed horizontally at a target 50 m away. The arrow hits 10 cm below the spot at which it was aimed. What is the initial velocity of the arrow?

4-28. In order to hit a particular spot on a target, a shooter aims the rifle horizontally at a vertical distance d above the spot. The target is 45 m away and the muzzle velocity of the bullet is 250 m/s. What is d?

4-29. An airplane flying horizontally at 500 km/h at a height of 800 m drops a crate of supplies. (See the accompanying figure.) If the parachute attached to the crate fails to open, how far in front of its release point will the crate hit the ground?

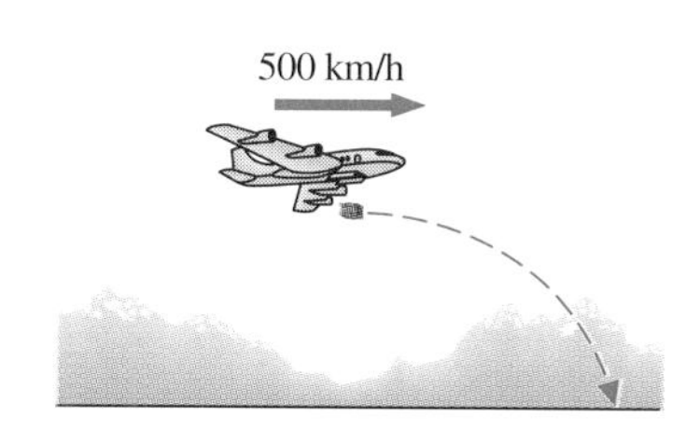

PROBLEM 4-29

4-30. Suppose that the airplane of Prob. 4-29 fires a projectile horizontally in its direction of motion at a speed of 300 m/s relative to itself. (*a*) How far in front of the release point does the projectile hit the ground? (*b*) What is its speed when it hits?

4-31. A fastball pitcher can throw the ball at speeds of about 90 mi/h. (*a*) Assuming the pitcher releases the ball at 55 ft from home plate so that it is initially moving horizontally, how long does it take the ball to reach the plate? (*b*) How far does the ball drop between the pitcher's hand and home plate? What does this imply about the "rising fastball"?

4-32. A projectile is shot at 30° to the horizontal and lands 20 s later at the same height at which it was shot. (*a*) What is the initial speed of the projectile? (*b*) What is its maximum altitude? (*c*) What is its range?

4-33. A basketball player shoots toward a basket 20 ft away and 10 ft above the floor. If the ball is released 6.0 ft above the floor at 60° above the horizontal, what must its initial speed be if it is to go through the basket?

4-34. Compare the ranges of a golf ball hit at 30° to the horizontal with an initial speed of 130 ft/s when (*a*) it is hit and lands at the same elevation; (*b*) it is hit to a fairway 40 ft below the tee; (*c*) it is hit to a green 40 ft above the tee.

4-35. At a particular instant, a hot-air balloon is 100 m in the air and descending vertically at a constant speed of 2.0 m/s. A girl riding in the balloon throws a ball horizontally (relative to herself) with an initial speed of 20 m/s. When she lands, where will she find the ball?

4-36. A man on a motorcycle traveling at a uniform speed of 10 m/s throws an empty can straight upward relative to himself; the can has an initial speed of 3.0 m/s. Find an equation representing the path of this projectile as seen by a police officer on the side of the road. Assume that $x_0 = 0$ and $y_0 = 0$ at the point where the can is thrown.

4-37. A long jumper is able to jump 8.0 m when he takes off at 45° to the horizontal. Assuming he can jump with the same initial speed for all angles, how much distance does he lose by taking off at 30°?

4-38. What is the maximum distance that the athlete of the previous problem can jump on the moon, where the gravitational acceleration is about one-sixth that of the earth?

4-39. Show that the maximum height reached by a projectile is $y_{max} = (v_0 \sin\theta)^2/2g$.

4-40. A projectile is shot at 45° to the horizontal with an initial speed of 1000 m/s. It is shot and lands at the same elevation. Determine the range of the projectile (*a*) using the basic kinematic equations for constant acceleration, (*b*) using the range formula, (*c*) using the equation of the projectile's path.

4-41. The maximum horizontal distance a young boy can throw a ball is 50 m. Assuming he can throw with the same initial speed at all angles, how high straight upward can he throw a ball?

4-42. A boy stands at the edge of a 100-m-high cliff and throws a rock at 53° above the horizontal; the rock has an initial speed of 30 m/s. (*a*) How high above the edge of the cliff does the rock rise? (*b*) How far has it moved horizontally when it is at maximum altitude? (*c*) How long after release does it hit the ground? (*d*) What is the range of the rock? (*e*) What are the horizontal and vertical positions of the rock relative to the edge of the cliff at $t = 2.0$ s? at $t = 4.0$ s? at $t = 6.0$ s?

4-43. During a physics class on the planet Xoltac, young Greels are asked to determine the gravitational acceleration by using a particle launcher that has a muzzle velocity of 10 m/s. (See the accompanying figure.) They find that the maximum horizontal range of the particles is 20 m. What is the gravitational acceleration on Xoltac? If a Greel were to aim his particle launcher at 60° above the horizontal, what would the range be?

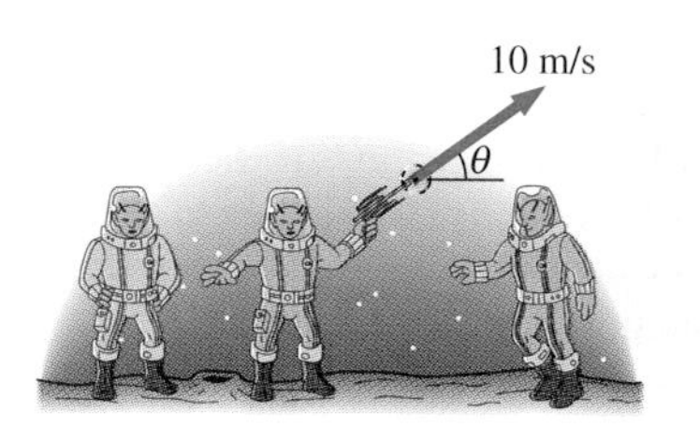

PROBLEM 4-43

4-44. A soccer player shoots the ball at the opposing goal from 15 m away. If the ball leaves his foot at 20° above the horizontal, what must its initial speed be if it is to enter the goal 2.0 m above the ground?

Circular Motion

4-45. A particle travels in a circle of radius 10 m at a constant speed of 20 m/s. What is its acceleration?

4-46. What is the acceleration of Venus toward the sun?

4-47. An experimental jet travels around the earth along its equator slightly above the earth's surface. At what speed must the jet move if its acceleration is g?

4-48. A fan is rotating at 360 rev/min. What is the acceleration of a point on one of its blades 10 cm from the axis of rotation?

4-49. A point located on the second hand of a large clock has a radial acceleration of 0.10 cm/s^2. How far is this point from the axis of rotation of the hand?

4-50. A particle travels in a circular orbit of radius 10 m. Its speed is changing at a rate of 15 m/s^2 at an instant when its speed is 40 m/s. What is the acceleration of the particle?

4-51. The driver of a car moving at 90 km/h presses down on the brake pedal as the car enters a circular curve of radius 150 m. If the speed of the car is decreasing at a rate of 9.0 km/h each second, what is the car's acceleration at the instant its speed is 60 km/h?

4-52. A body is located on the earth's surface at a latitude λ. (See the accompanying figure.) Calculate the centripetal acceleration of the body *due to the rotation of the earth around its polar axis.* Express your answer in terms of λ, the radius R_E of the earth, and the time T for one rotation of the earth. Compare your answer with g for $\lambda = 40°$.

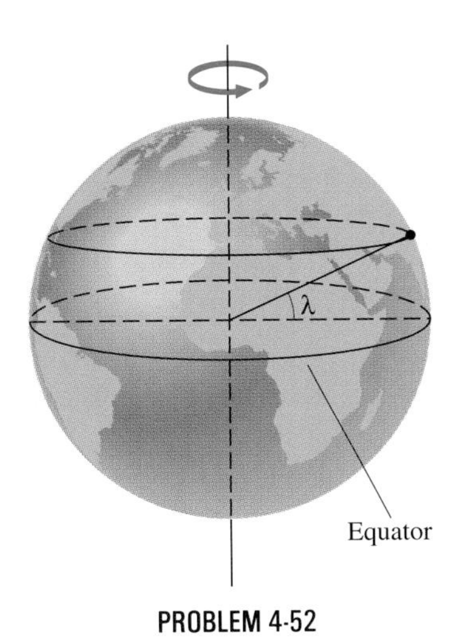

PROBLEM 4-52

Relative Motion

4-53. The coordinate axes of the reference frame S′ remain parallel to those of S, as S′ moves away from S at a constant velocity $\mathbf{v}_{S'S} = (4.0\mathbf{i} + 3.0\mathbf{j} + 5.0\mathbf{k})$ m/s. (*a*) If at time $t = 0$ the origins coincide, what is the position of the origin O′ in the S frame as a function of time? (*b*) How are a particle's positions, $\mathbf{r}(t)$ and $\mathbf{r}'(t)$, as measured in S and S′, respectively, related? (*c*) What is the relationship between the particle's velocities, $\mathbf{v}(t)$ and $\mathbf{v}'(t)$? (*d*) How are the accelerations $\mathbf{a}(t)$ and $\mathbf{a}'(t)$ related?

4-54. Raindrops fall vertically at 10 mi/h relative to the earth. What does an observer in an automobile moving at 50 mi/h in a straight line measure the velocity of the raindrops to be?

4-55. A boat can be rowed at 8.0 km/h in still water. (*a*) How much time is required to row 1.5 km downstream in a river moving at 3.0 km/h relative to the shore? (*b*) How much time is required for the return trip? (*c*) In what direction must the boat be aimed in order to row straight across the river? (*d*) Suppose the river is 0.80 km wide. What is the boat's velocity with respect to the earth and how much time is required to get to the opposite shore? (*e*) Suppose that, instead, the boat is aimed straight across the river. How much time is required to get across and how far downstream is the boat when it reaches the opposite shore?

4-56. Two speedboats are traveling at the same speed relative to the water in opposite directions in a moving river. An observer on the riverbank sees the boats moving at 4.0 and 5.5 m/s. What is the speed of the boats relative to the river? How fast is the river moving relative to the shore?

4-57. A small plane flies at 200 km/h in still air. If the wind blows directly out of the west at 50 km/h, in what direction must the pilot head her plane in order to move directly north across the land? How long does it take her to reach a point 300 km directly north of her starting point?

4-58. A truck is traveling east at 50 mi/h. At an intersection 20 mi ahead, a car is moving north at 30 mi/h. How long after this moment will the two vehicles be closest to each other? How far apart will they be at that point?

4-59. A cyclist traveling southeast along a road at 15 km/h feels a wind blowing from the southwest at 25 km/h. To a stationary observer, what are the speed and direction of the wind?

General Problems

4-60. A pebble is fired from a slingshot, at an angle of 40° below the horizontal, from the roof of a building 80 m tall. It goes through an open window 15 m above the ground of an adjacent building 3.0 s later. Calculate (*a*) the initial speed of the projectile, (*b*) the separation of the buildings, (*c*) the angle at which the pebble flies through the open window.

4-61. Trying to escape his pursuers, a secret agent skis, at a speed of 60 km/h, off a slope inclined at 30° below the horizontal. In order to survive the jump and land in the snow 100 m below, he has to clear a deep gorge 60 m wide, as shown in the accompanying figure. Does he make it? Ignore any effects due to the air.

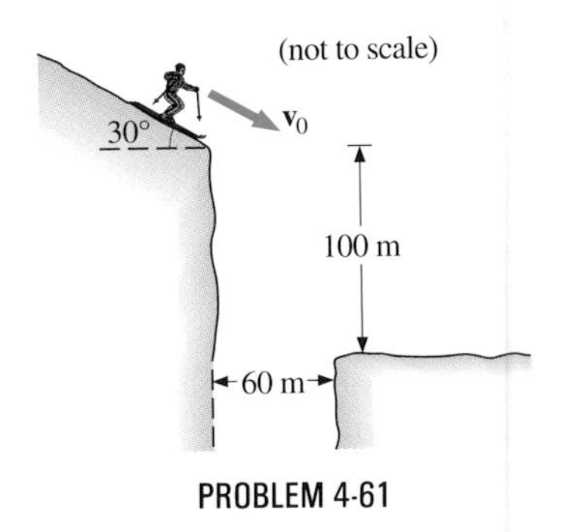

PROBLEM 4-61

4-62. A baseball player standing 300 ft from home plate sees a ball hit directly over his head. The ball is hit at 30° to the horizontal with an initial speed of 120 ft/s. If the player is to turn around, run back at constant speed, and catch the ball, how fast must he run? Assume that he takes 0.50 s to react, that the time and distance required for him to accelerate are negligible, and that he catches the ball at the same height at which it is hit.

4-63. A fawn is at A, a distance d from the riverbank, when she spots an approaching wolf. (See the accompanying figure.) She decides to head for the safety of her herd, which is located across the river at B. If she can run at a speed v_1 and swim at a speed v_2, which path (expressed in terms of θ_1 and θ_2) should she take in order to reach her herd in minimum time? Assume that the river has width w and is not moving. (*Hint:* Express the time in terms of θ_1 and θ_2, then find the minimum time by differentiating with respect to θ_1. Use the fact that l is constant to determine $d\theta_2/d\theta_1$.)

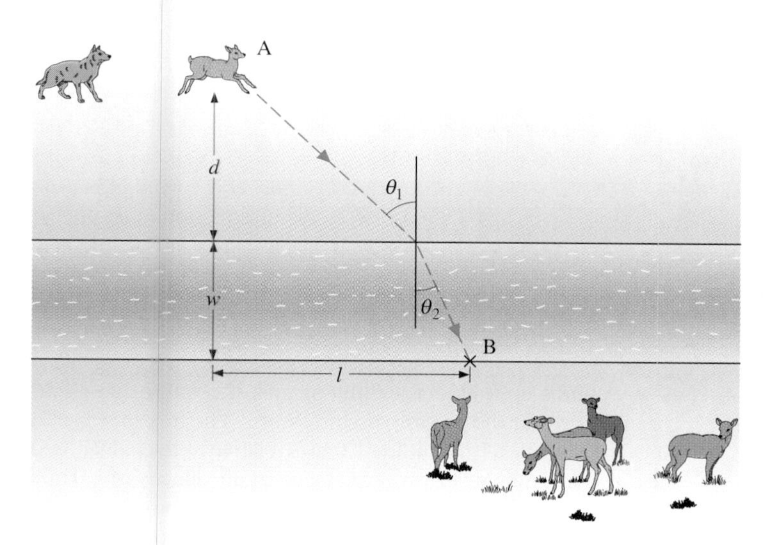

PROBLEM 4-63

4-64. During a professional football game, the Bears' receiver catches a pass and starts running upfield along the sideline at a speed v_R. At the same instant, a Packers' defensive back begins running diagonally across the field at a speed v_B to catch the receiver at the sideline before he reaches the end zone. Both players start at the same yard line, and they are initially separated by a distance x_0, as illustrated in the accompanying figure. (*a*) Express the positions of the two players as functions of time. (*b*) At what angle must the defensive back run in order to catch the receiver? (*c*) How far downfield will he catch the receiver? (*d*) If $x_0 =$ 20 yd, $v_R = 25$ ft/s, $v_B = 27$ ft/s, and the play starts at the Bears' 30-yd line, will the receiver score a touchdown?

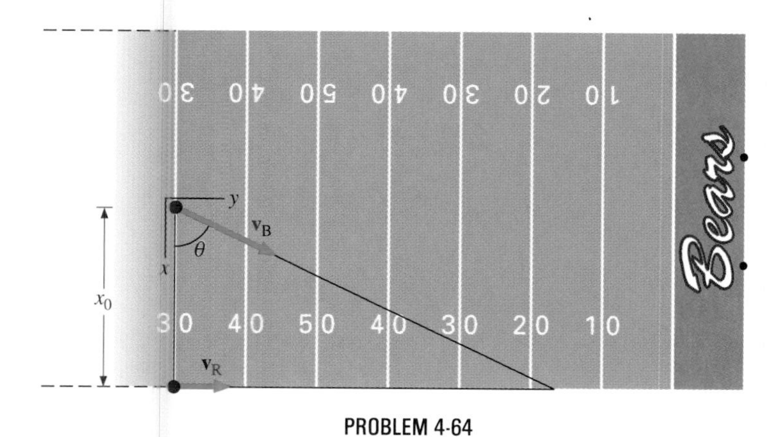

PROBLEM 4-64

4-65. Two joggers are running toward each other. Jogger A is moving at 3.0 m/s while jogger B has a speed of 5.0 m/s. Jogger A throws a pebble, which he had dislodged from his shoe, straight up relative to himself with an initial velocity of 7.0 m/s. According to jogger B, (*a*) what is the initial speed of the pebble, (*b*) what is the direction at which the pebble is thrown, and (*c*) what is the maximum height the pebble attains?

4-66. A projectile is shot at a hill whose base is 300 m away. (See the accompanying figure.) The projectile is shot at 60° above the horizontal with an initial speed of 75 m/s. The hill can be approximated by a straight line sloped at 20° to the horizontal. Relative to the coordinate system shown, the equation of this straight line is $y = [(\tan 20°)x - 109]$. Where on the hill does the projectile land?

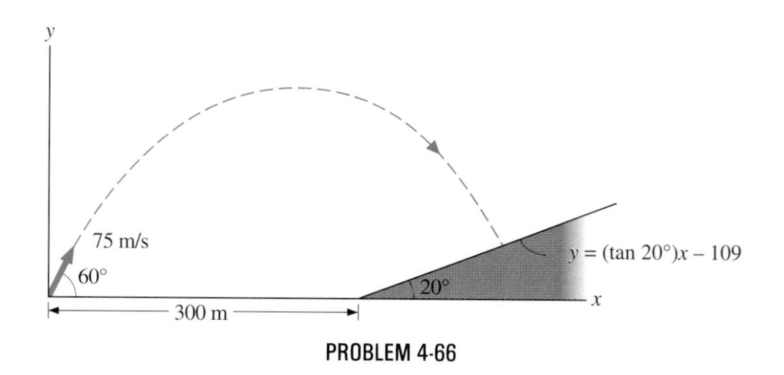

PROBLEM 4-66

4-67. At time $t = 0$, a missile in pursuit of an airplane is 200 m behind it and moving horizontally at 150 m/s with a constant acceleration of 1.0 m/s^2. Also at $t = 0$, the pilot of the plane sees on his radar screen that the missile is approaching, so he accelerates his plane to escape the missile. (*a*) If the plane is initially moving at 100 m/s and it accelerates at 2.0 m/s^2, will the missile catch the plane? (*Hint:* Analyze this problem relative to a reference frame attached to the plane.) (*b*) If instead, the plane's initial velocity is 130 m/s and its acceleration is 3.0 m/s^2, will it be caught by the missile?

4-68. Projectile A is shot with an initial speed of 100 m/s at 30° to the horizontal. Projectile B is shot with an initial speed of 200 m/s at 45° to the horizontal. (See the accompanying figure.) A coordinate system S′ moves with projectile A so that its axes are always parallel to the fixed x and y axes of system S. (*a*) What is the position of B in S′? (*b*) What are the velocity and acceleration of B in S′?

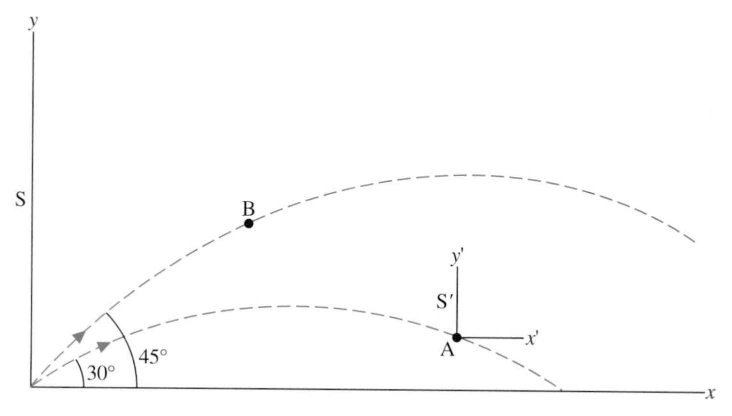

PROBLEM 4-68

4-69. When a field-goal kicker kicks a football as hard as he can at 45° to the horizontal, the ball just clears the 10-ft-high crossbar of the goalposts 50 yd away. (*a*) What is the maximum speed that the kicker can impart to the football? (*b*) In addition to clearing the crossbar, the football must be high enough in the air early in its flight to clear the reach of the onrushing defensive lineman, as illustrated in the accompanying figure. If the lineman is 5.0 yd away and has a vertical reach of 8.0 ft, will he block the 50-yd field-goal attempt? (*c*) What if the lineman is 3.0 yd away?

PROBLEM 4-69

4-70. Suppose that the kicker in the previous problem attempts a 40-yd field goal. He again kicks the ball as hard as he can but this time he gives it a 30° loft. (*a*) Show that the ball will be blocked by the lineman who is 5.0 yd away. (*b*) When the kicker reaches the sidelines, he blames his offensive linemen for the failed scoring attempt. He claims the ball would have gone over the crossbar if it hadn't been blocked. Is the kicker correct?

4-71. The quarterback stands 25 yd from the sideline when he throws a pass to a wide receiver who is 20 yd downfield, running along the sideline at a speed of 9.0 yd/s. (See the accompanying figure.) Describe how the quarterback must throw the ball if the receiver is to catch it in full stride at the sideline 40 yd downfield. What is the range of the football? Assume that the ball is thrown and caught at the same height.

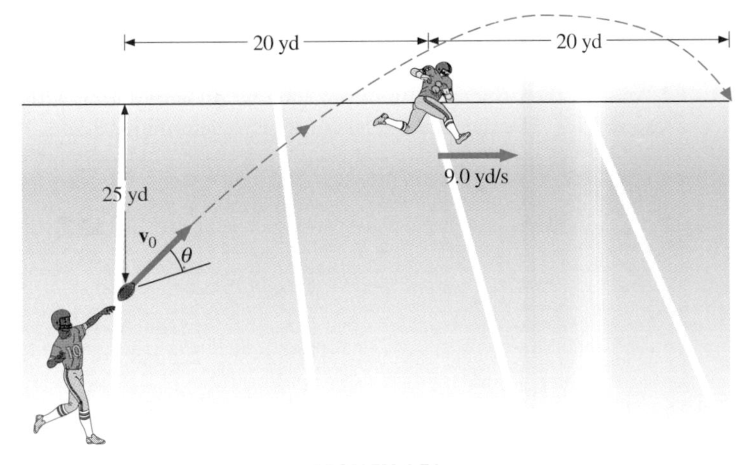

PROBLEM 4-71

4-72. A boy climbs up a ladder leaning against the wall of an elevator, as shown in the accompanying figure. The boy moves at a constant velocity of magnitude 1.0 m/s relative to the ladder, and the elevator is moving downward at a constant velocity of 2.0 m/s relative to the earth. What is the velocity of the boy as seen by a person at rest relative to the earth?

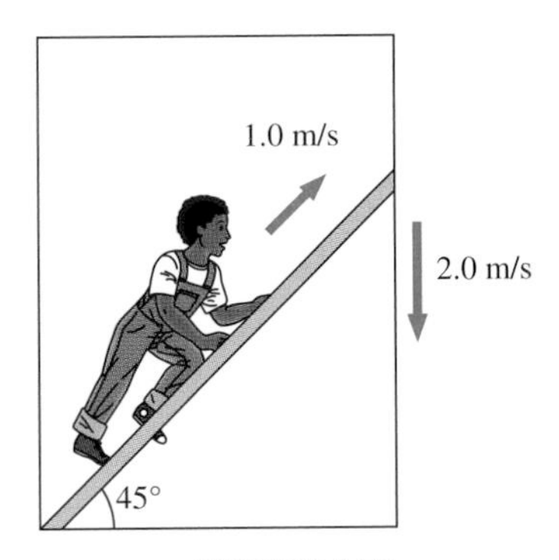

PROBLEM 4-72

Force causes a change in velocity.

CHAPTER 5 DYNAMICS I

PREVIEW

With this chapter we begin our study of dynamics, which is the investigation of the *cause* of an object's acceleration. The main topics to be discussed here are the following:

1. **Force**. The meaning of force and its relationship to motion is considered. Forces are categorized into two groups: contact forces and field forces.
2. **Newton's first law**. This law provides a method for identifying reference frames known as inertial frames. These are the frames in which Newton's second law must be applied.
3. **Newton's second law**. This law describes how a particle's acceleration is determined by the forces acting on it.
4. **Newton's third law**. This law describes the relationship between the forces two bodies exert on one another.
5. **Newton's law of universal gravitation**. Two cases are discussed: (*a*) the gravitational force between two particles and (*b*) the gravitational force of a sphere on a particle.

*6. **The gravitational field**. The concept of field is introduced. How mass creates and interacts with the gravitational field is considered.

In kinematics we use a particle's known acceleration to determine how its position and velocity vary with time. For example, if the acceleration is constant, the velocity and position are given by Eqs. (3-10) and (3-11). However, the equations of kinematics do not provide any information about the *cause* of the acceleration. This leaves us with an important and obvious question: What is responsible for the acceleration of a particle?

This question was answered in the late 1600s by Sir Isaac Newton (1642–1727).* He described how a particle's environment is responsible for its acceleration and thereby began the development of the branch of mechanics known as *dynamics*. The connection between a particle's acceleration and its environment is given by Newton's laws of motion. These laws form the cornerstone of our knowledge of classical mechanics and have an amazingly wide scope. Although they do not apply to particles moving at speeds comparable to the speed of light (3.0×10^8 m/s) or in the quantum mechanical world of atoms and nuclear particles, the motions of macroscopic systems that we encounter day in and day out—people, automobiles, planes, bridges, planets—are all governed by Newton's laws.

Isaac Newton.

5-1 Force

The environment affects a body by exerting **forces** on it. Force will be defined rigorously in Sec. 6-6, but for now, we'll forsake rigor and think of force simply as a push or a pull. When a force acts on a body, the body's velocity changes; that is, the body accelerates—and the larger the force, the larger the acceleration. If several forces act simultaneously, the resulting acceleration depends on the directions of the individual forces. A body that is pushed to the left and to the right with equal strength does not accelerate. However, if it is pushed to the right by one force and upward by another, it accelerates toward the upper-right with the exact direction determined by the relative strengths of the forces. Observations such as these indicate that the net, or resultant, force acting on a body is the vector sum of the individual forces acting on the body. *Force is therefore a vector quantity.*

In most situations, forces are grouped into two categories: *contact forces* and *field forces*. As you might guess, the former are due to direct physical contact between objects. For example, the boy of Fig. 5-1 experiences the contact forces **C**, **F**, and **T**, which are exerted by the chair on his posterior, the floor on his feet, and the table on his forearms, respectively. Field forces, however, act without the necessity of physical contact between objects. They depend on the presence of a "field" in the region of space surrounding the body under consideration. Since the boy of Fig. 5-1 is in the earth's gravitational field, he feels a gravitational force **W**.

FIGURE 5-1 The forces acting on the boy are due to the chair, the table, the floor, and the earth's gravitational field.

You can think of a field as a property of space that is detectable by the forces it exerts. Scientists believe that there are only four fundamentally different force fields in nature. These are the gravitational, electromagnetic, strong, and weak fields.* The gravitational field is responsible for the weight of a body. (For example, **W** in Fig. 5-1 is the boy's weight.) The forces of the electromagnetic field include those of static electricity and magnetism, and they are also responsible for the attraction among atoms in bulk matter. Both the strong and the weak force fields are effective only over distances roughly equal to the diameter of a nucleus (10^{-15} m). Their range is so small that neither field has an influence in the macroscopic world of Newtonian mechanics.

It's interesting to note that contact forces are fundamentally electromagnetic in nature. While the elbow of the boy in Fig. 5-1 is in contact with the tabletop, the atomic charges in his skin interact electromagnetically with the charges in the surface

*Newton's first law was actually discovered early in the seventeenth century by Galileo. However, it was Newton who first saw how this law fit into a coherent theory of motion.

*There is very strong evidence that the electromagnetic and the weak fields are manifestations of a single fundamental field called the *electroweak field*. Physicists hope that eventually all four fields can be described by a single unified field.

of the table. The net result is the force **T**. Similarly, when adhesive tape sticks to a piece of paper, the atoms of the tape are intermingled with those of the paper to cause a net electromagnetic force between the two objects. However, in the context of Newtonian mechanics, the electromagnetic origin of contact forces is not an important concern.

5-2 Newton's First Law

A stationary object does not move spontaneously. A force must be applied to move it. But what about objects that are already in motion? If you give this textbook a sharp push and cause it to slide across a table, it is quickly stopped by friction. However, if you rub a little grease on the cover of your book, you will find that it slides much farther because of a reduction in the frictional force. For the same reason, the braking distance of a car is considerably less on pavement than on ice.

Friction can be almost eliminated by supporting a body on a cushion of air. A popular demonstration of the air-supported ride is furnished by the linear air track. In Fig. 5-2 the cart is supported on a cushion of air blown through small holes in the track. With very little friction, the cart is able to glide a considerable distance before coming to rest.

Courtesy Central Scientific Company.

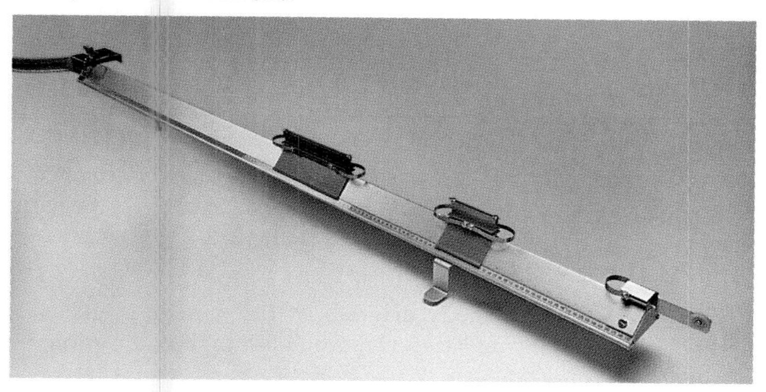

FIGURE 5-2 A linear air track.

Now let's extrapolate these observations to the limit in which friction is *completely* eliminated. The net force on a sliding body is then zero, so it should move indefinitely at a constant velocity. Newton, without the help of air tracks or skidding automobiles, came to the same conclusion, which is stated in his first law:

> Every body remains in its state of uniform motion in a straight line unless it is compelled to change that state by forces acting on it.

To Newton, "uniform motion in a straight line" meant constant velocity which, of course, includes the case of zero velocity, or rest. The first law therefore states that the velocity of an object remains constant if the net force on it is zero. A body's tendency to remain at rest or to move with a constant velocity is known as its **inertia**. For this reason, Newton's first law is sometimes referred to as the *law of inertia*.

The first law is usually considered to be a statement about reference frames. It provides a method for identifying a special type of reference frame, the *inertial reference frame*. In principle at least, we can make the net force on a body zero. If its velocity relative to a given frame is constant, then that frame is said to be inertial. So by definition, *an inertial reference frame is a reference frame in which Newton's first law is valid*.

Suppose that the reference frame S of Fig. 4-20 has been tested and found to be an inertial frame. Is S′, which is moving at constant velocity relative to S, then also inertial? From Eq. (4-22), we know that a body traveling with uniform velocity in S does the same in S′. Hence S′ is also an inertial frame. *A reference frame moving at constant velocity relative to an inertial frame is also inertial.*

However, if S′ is accelerating relative to S, a body moving at constant velocity in S is accelerating in S′. Therefore *a reference frame accelerating relative to an inertial frame is not inertial.*

Are inertial frames common in nature? It turns out that well within experimental error, a reference frame at rest relative to the most distant, or "fixed," stars is an inertial frame. All frames moving uniformly with respect to this fixed-star frame are therefore also inertial. For example, a reference frame attached to the sun is, for all practical purposes, inertial, since its velocity relative to the fixed stars does not vary by more than one part in 10^{10}. The earth accelerates relative to the fixed stars, because it rotates on its axis and revolves around the sun; hence a reference frame attached to its surface is not inertial. For most problems, however, such a frame serves as a sufficiently accurate *approximation* to an inertial frame, because the acceleration of a point on the earth's surface relative to the fixed stars is rather small ($<3.4 \times 10^{-2}$ m/s^2). So unless otherwise stated, we will consider reference frames fixed on the earth to be inertial.

Finally, no particular inertial frame is more special than any other. As far as the laws of nature are concerned, all inertial frames are equivalent. In analyzing a problem, we choose one inertial frame over another simply on the basis of convenience.

DRILL PROBLEM 5-1

Taking a frame attached to the earth as inertial, which of the following cannot have inertial frames attached to them?

a. A car moving at constant velocity
b. A car that is accelerating
c. An elevator in free fall
d. A space capsule orbiting the earth
e. An elevator descending uniformly

ANS. *b, c, d.*

5-3 Newton's Second Law

We've all seen what happens to a body when a force is applied to it. When a ball meets a bat, the velocity of the ball changes. A stationary body moves if given a push, and it eventually stops because of friction. In each case, a force causes a change in velocity, or, an acceleration. Notice that a force is required to *change* a velocity, not to keep it constant. This fact is far from obvious, and at one time, scholars thought that a net force was required to keep an object moving uniformly. It was not until the time of Galileo and Newton that the connection between force and motion was finally understood. That connection is given by Newton's second law:

> A body's acceleration, as measured relative to an inertial reference frame, is proportional to the net force acting on the body.

The net force is the vector sum of all of the forces acting on the body. With this sum written as $\sum \mathbf{F}$, Newton's second law states

$$\mathbf{a} \propto \sum \mathbf{F}.$$

The constant of proportionality is a property of the body known as its **mass** m. With mass included, Newton's second law then becomes

$$\sum \mathbf{F} = m\mathbf{a}. \tag{5-1a}$$

This is a vector equation that is equivalent to the three component equations

$$\sum F_x = ma_x, \qquad \sum F_y = ma_y, \qquad \sum F_z = ma_z. \tag{5-1b}$$

The second law is really a description of how a body responds mechanically to its environment. The influence of the environment is the net force $\sum \mathbf{F}$, the body's response is the acceleration $\mathbf{a}$, and the strength of the response is inversely proportional to the mass m. The larger the mass of an object, the smaller is its response (its acceleration) to the influence of the environment (a given net force). *A body's mass is therefore a measure of its inertia.* Mass is an intrinsic property of a body. It doesn't matter whether an object is in this room, or on the moon, or even in the Andromeda galaxy—its mass, and hence its response to an applied force, is always the same.

The phrase "relative to an inertial reference frame" is a very important part of the second law. This tells us that we have to be careful in choosing the reference frame in which we apply this law. While the second law can be used to analyze the motion of systems in frames moving at constant velocities relative to the earth, it cannot be directly applied in frames that are accelerating with respect to the earth.

The second law is stated for a *body* rather than for a *particle*. This is not a problem if the body really is a particle or if it moves in a way that all points on it are displaced parallel to one another. (See Fig. 3-1.) However, when a body rotates as it moves through space, special methods have to be used to analyze its motion with Newton's laws. These methods will be developed later, starting with Chap. 8. For now, we'll only consider particlelike motion.

The units of length, mass, and force for the three systems commonly used by scientists and engineers are given in Table 5-1. Time is measured in seconds in all three cases. Force is the product of mass and acceleration, so

$$1 \text{ N} = 1 \text{ kg}\cdot\text{m/s}^2, \qquad 1 \text{ lb} = 1 \text{ slug}\cdot\text{ft/s}^2,$$
$$1 \text{ dyne} = 1 \text{ g}\cdot\text{cm/s}^2.$$

In applications of Newton's second law, you should always draw a sketch of the body under investigation and the forces acting on it. This sketch is called a *free-body diagram*.* Usually the forces in the free-body diagram are resolved into components along defined coordinate axes.

TABLE 5-1

Physical Quantity	SI Unit	CGS Unit	British Unit
Length	Meter (m)	Centimeter (cm)	Foot (ft)
Mass	Kilogram (kg)	Gram (g)	Slug (slug)
Force	Newton (N)	Dyne (dyne)	Pound (lb)

$1.0 \text{ m} = 100 \text{ cm} = 3.281 \text{ ft}$
$1.0 \text{ kg} = 1000 \text{ g} = 6.852 \times 10^{-2} \text{ slug}$
$1.0 \text{ N} = 10^5 \text{ dyne} = 0.2248 \text{ lb}$

EXAMPLE 5-1 Single Force on a Particle

A particle of mass 2.0 kg is pushed horizontally by a constant 10-N force. (*a*) What is the acceleration of the particle? (*b*) If the particle starts at rest, how far does it travel in 4.0 s? (*c*) What is its velocity after 4.0 s?

SOLUTION The free-body diagram of the particle is shown in Fig. 5-3. (*a*) Applying Newton's second law in the x direction, we obtain

$$\sum F_x = ma_x$$
$$10 \text{ N} = (2.0 \text{ kg})a_x;$$

so

$$a_x = 5.0 \text{ m/s}^2.$$

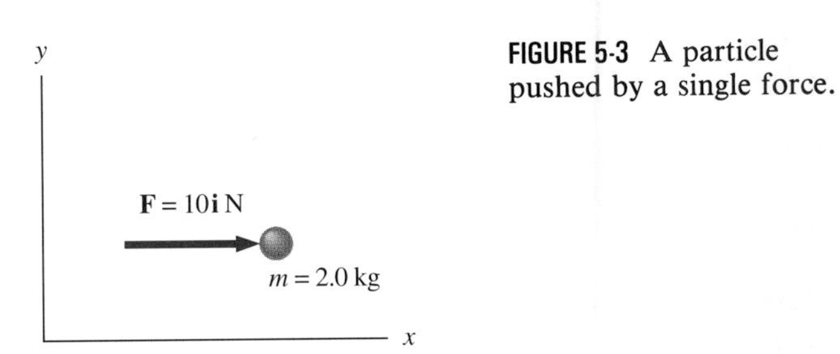

FIGURE 5-3 A particle pushed by a single force.

*Free-body diagrams will be discussed further in Chap. 6.

(*b*) Now from the kinematic equation relating position and time, we find that in 4.0 s, the particle travels

$$x - x_0 = v_{0x}t + \tfrac{1}{2}a_x t^2 = 0 + \tfrac{1}{2}(5.0\ \text{m/s}^2)(4.0\ \text{s})^2 = 40\ \text{m}.$$

(*c*) After 4.0 s, the particle's velocity is

$$v_x = v_{0x} + a_x t = 0 + (5.0\ \text{m/s}^2)(4.0\ \text{s}) = 20\ \text{m/s}.$$

EXAMPLE 5-2 Several Forces on a Particle

A particle of mass $m = 4.0$ kg is acted upon by four forces of magnitudes $F_1 = 10.0$ N, $F_2 = 40.0$ N, $F_3 = 5.0$ N, and $F_4 = 2.0$ N, as shown in the free-body diagram of Fig. 5-4. What is the acceleration of the particle?

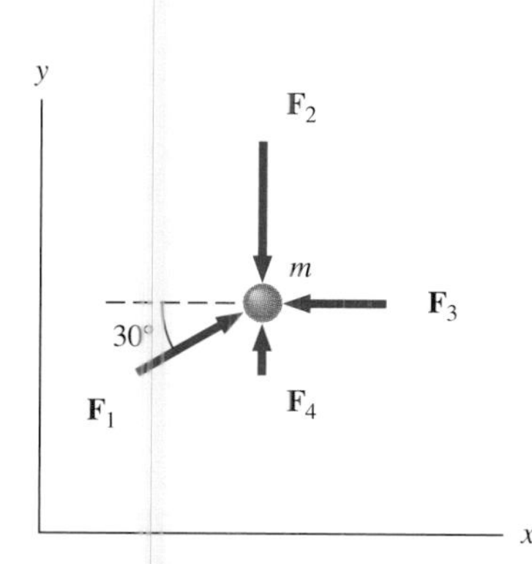

FIGURE 5-4 Four forces in the *xy* plane are applied to a 4.0-kg particle.

SOLUTION Since this is a two-dimensional problem, we must first resolve the forces along the coordinate axes shown. We can then apply the second law in each direction:

$$\sum F_x = ma_x \qquad \sum F_y = ma_y$$

$$F_1 \cos 30^\circ - F_3 = ma_x, \qquad F_1 \sin 30^\circ + F_4 - F_2 = ma_y$$

$$(10.0\ \text{N})(\cos 30^\circ) - 5.0\ \text{N} = (4.0\ \text{kg})a_x, \qquad (10.0\ \text{N})(\sin 30^\circ) + 2.0\ \text{N} - 40.0\ \text{N} = (4.0\ \text{kg})a_y,$$

$$a_x = 0.92\ \text{m/s}^2. \qquad a_y = -8.25\ \text{m/s}^2.$$

Thus the net acceleration is

$$\mathbf{a} = (0.92\mathbf{i} - 8.25\mathbf{j})\ \text{m/s}^2,$$

which is a vector of magnitude 8.30 m/s² directed at 276° to the positive *x* axis.

EXAMPLE 5-3 Forces on an Electric Train Engine

A small electric train engine of mass 0.50 kg is moving around a horizontal circular track at a constant speed of 0.80 m/s, as shown in Fig. 5-5*a*. The radius of the track is 2.0 m. (*a*) What is the horizontal component of the force of the track on the engine at points A and B? (*b*) When the engine, still traveling at 0.80 m/s, reaches point C, the boy adjusts the controls so that an additional 0.20-N force acting tangentially to the track is applied to speed up the engine. Calculate the engine's acceleration at the instant that this additional force is applied.

SOLUTION (*a*) The engine is moving with a constant speed in a circle, so it has only centripetal acceleration. The magnitude of this acceleration is

$$a = \frac{v^2}{r} = \frac{(0.80\ \text{m/s})^2}{2.0\ \text{m}} = 0.32\ \text{m/s}^2.$$

Centripetal acceleration is always directed toward the center of the circular path. The horizontal components of the force of the track on the train are shown in the overhead view of Fig. 5-5*b*. Using the

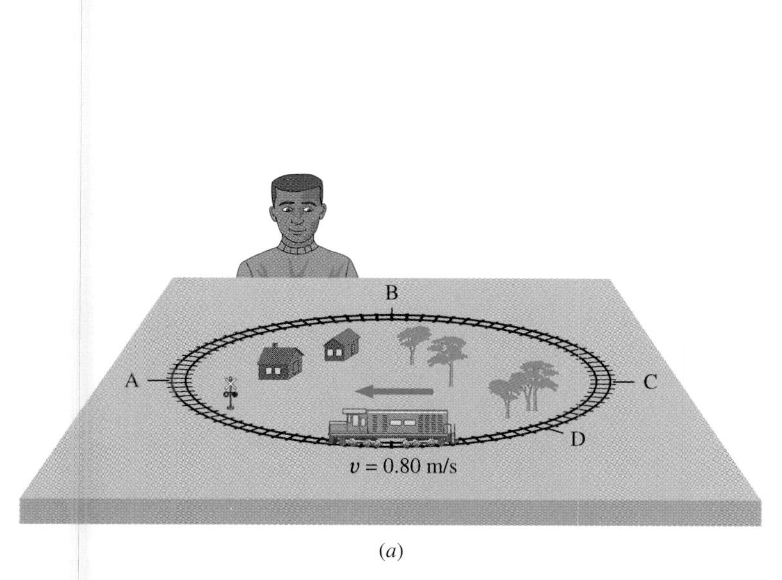

(*a*)

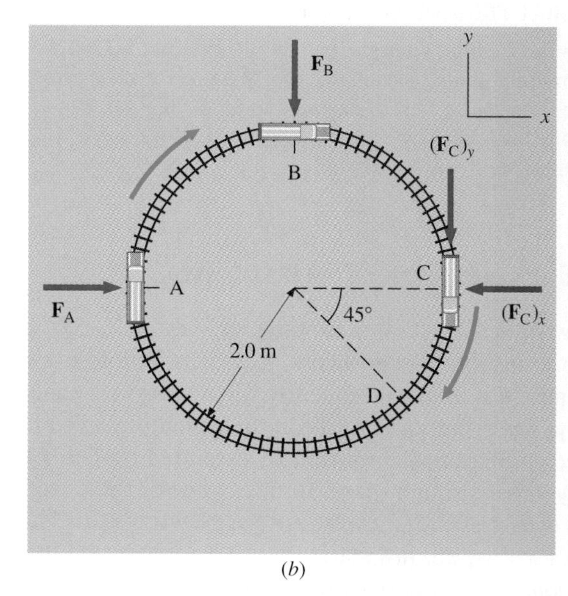

(*b*)

FIGURE 5-5 (*a*) An electric train engine is moving around a circular track. (*b*) An overhead view of the track. At point C, an additional tangential force is applied to speed up the train. Also included are free-body diagrams showing the horizontal forces on the train at A, B, and C.

coordinate system shown, this acceleration at point A is $0.32\mathbf{i}$ m/s^2, and at B it is $-0.32\mathbf{j}$ m/s^2. From Newton's second law, the horizontal components of the forces of the track on the engine at these two points are

$$\mathbf{F}_A = (0.50 \text{ kg})(0.32\mathbf{i} \text{ m/s}^2) = 0.16\mathbf{i} \text{ N},$$

and

$$\mathbf{F}_B = (0.50 \text{ kg})(-0.32\mathbf{j} \text{ m/s}^2) = -0.16\mathbf{j} \text{ N}.$$

(*b*) At point C, the acceleration of the engine immediately before the additional force is applied is the centripetal acceleration $\mathbf{a} = -0.32\mathbf{i}$ m/s^2. The applied tangential force is $-0.20\mathbf{j}$ N, which produces a tangential acceleration of $(-0.20\mathbf{j} \text{ N})/0.50 \text{ kg} = -0.40\mathbf{j}$ m/s^2. The net acceleration of the engine is the vector sum of the centripetal and tangential accelerations:

$$\mathbf{a} = (-0.32\mathbf{i} - 0.40\mathbf{j}) \text{ m/s}^2,$$

which is a vector of magnitude 0.51 m/s^2 directed at 231° to the positive x axis.

When the mountain climber pulls down on the rope, the rope pulls up on the mountain climber.

DRILL PROBLEM 5-2

By using the conversion factors for kilograms to slugs and meters to feet, find the conversion factor for newtons to pounds.
ANS. 1.00 N = 0.225 lb.

DRILL PROBLEM 5-3

When subjected to a force $\mathbf{F}_1 = -6.0\mathbf{j}$ N, a body has an acceleration of $-0.20\mathbf{j}$ m/s^2. If both $\mathbf{F}_1$ and a second force $\mathbf{F}_2 = (10.0\mathbf{i} + 2.0\mathbf{j})$ N act simultaneously on the body, what is its net acceleration?
ANS. $(0.33\mathbf{i} - 0.13\mathbf{j})$ m/s^2.

DRILL PROBLEM 5-4

Suppose that when the engine of Example 5-3 reaches point D in Fig. 5-5, its speed is 1.0 m/s. What is the net horizontal force on the engine at D? Assume that the 0.20-N tangential force applied at C by the boy has been turned off.
ANS. $(-0.18\mathbf{i} + 0.18\mathbf{j})$ N.

5-4 Newton's Third Law

Newton's third law is a statement about forces rather than about motion. Nevertheless, it is classified as a law of motion because it is used so frequently when motion is analyzed. It also plays a very important role in the development of the methods used to analyze the motion of extended bodies that rotate as they move through space. In fact, some of these methods would not even be valid if it were not for Newton's third law. In Newton's words, this law states that "To every action there is always opposed an equal reaction." "Action" and "reaction" refer to forces. Thus Newton's third law can be expressed as

> If body A exerts a force **F** on body B, then B will exert a force **−F** on A.

There are two important features of this law. First, the forces exerted (the action and reaction) are always equal in magnitude but opposite in direction. And second, these forces are acting on different bodies; A's force acts *on B* and B's force acts *on A*.

For the situation shown in Fig. 5-1, the third law indicates that since the chair is pushing upward on the boy with a force **C**, he is pushing downward on the chair with a force **−C**. Likewise, he is pushing downward with forces **−F** and **−T** on the floor and table, respectively. Finally, since the earth pulls downward on the boy with a force **W**, he pulls upward on the earth with a force **−W**.

Here are two more examples involving the third law. A student who angrily pounds his desk because of frustration with physics quickly learns the painful lesson (avoidable by studying Newton's laws!) that his desk hits back just as hard. Also, a person who is walking or running applies the third law instinctively. For example, the runner of Fig. 5-6 pushes backward on the ground so that it pushes her forward.

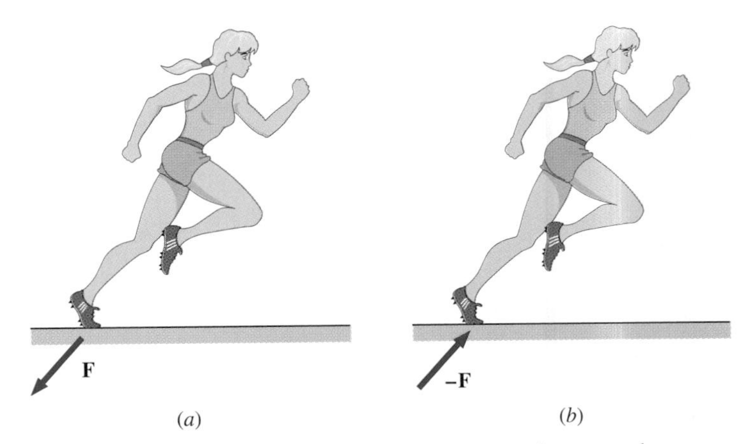

FIGURE 5-6 (*a*) The force exerted by a runner on the ground. (*b*) The reaction force of the ground on the runner pushes her forward.

EXAMPLE 5-4 **FORCES ON A STATIONARY OBJECT**

The package of Fig. 5-7 is sitting on a scale. Analyze the forces on the package and find their reaction forces.

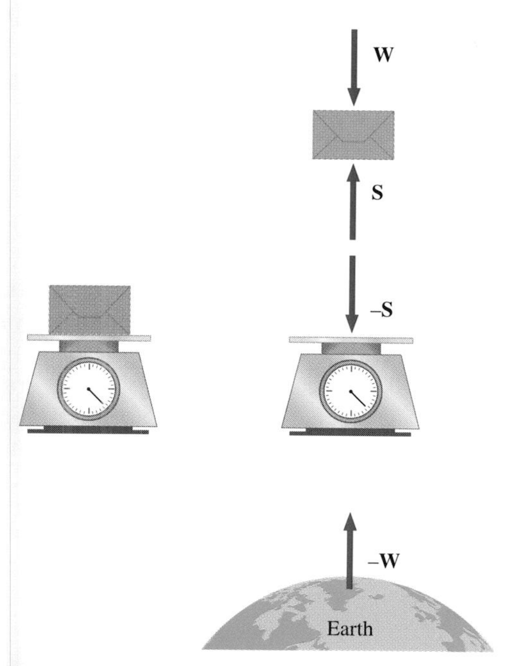

FIGURE 5-7 The forces on a package sitting on a scale along with their reaction forces. The force **W** is the weight of the package, and **S** is the force of the scale on the package.

SOLUTION The forces on the package are **S**, which is due to the scale, and **W** due to the earth's gravitational field. The reaction forces that the package exerts are $-\mathbf{S}$ on the scale and $-\mathbf{W}$ on the earth.

Since the package is not accelerating, application of the second law yields

$$\mathbf{S} + \mathbf{W} = m\mathbf{a} = 0,$$

so

$$\mathbf{S} = -\mathbf{W}.$$

Thus the scale reading gives the magnitude of the package's weight. However, the scale *does not measure* the weight of the package; it measures the force $-\mathbf{S}$ on its surface. In fact, S and W are not equal if the system is accelerating, as explained in Chap. 6.

DRILL PROBLEM 5-5

When a young girl pulls her toy wagon, the wagon pulls on her with a force equal and opposite to the force she applies to the wagon. Since the two forces sum to zero, why do the girl and wagon move? (*Hint:* Remember that the action and reaction forces act on different bodies.)

5-5 NEWTON'S LAW OF UNIVERSAL GRAVITATION

Besides the three laws of motion, Newton also formulated the *law of universal gravitation*. After comparing the acceleration of the moon in its orbit around the earth with the acceleration of a falling body at the earth's surface (see Drill Prob. 5-8), he concluded that

> Every particle in the universe attracts every other particle with a force directly proportional to the product of the masses of the two particles and inversely proportional to the square of the distance between them.

Figure 5-8 shows two particles of mass m_1 and m_2 that are separated by a distance r. The magnitude of the mutual attractive force between these particles is

$$F = G\frac{m_1 m_2}{r^2}. \tag{5-2}$$

The proportionality constant G is called the *universal gravitational constant* and is $6.6720 \times 10^{-11}\ \text{N}\cdot\text{m}^2/\text{kg}^2$ in the SI system. The adjective *universal* serves as a reminder that G has the same value everywhere in the universe.

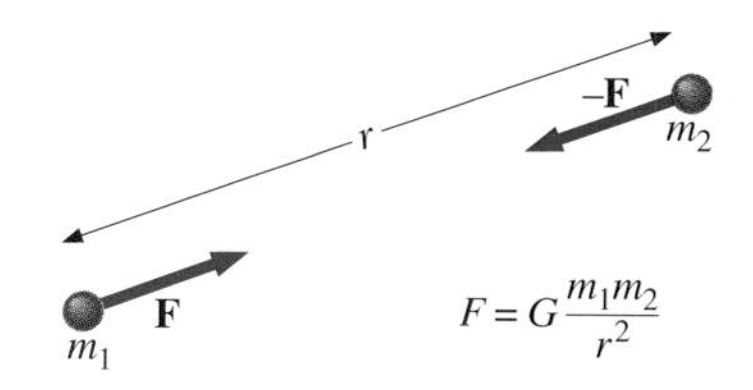

FIGURE 5-8 The gravitational force on m_1 due to m_2 is equal in magnitude and opposite in direction to the gravitational force on m_2 due to m_1.

Since *every object* attracts *every other object* with a gravitational force, why don't we feel an attraction (at least gravitationally) toward another person, or perhaps toward a building or a tree? To answer this question, let's consider two persons standing 2.0 m apart. One person has a mass of 70 kg and the other has a mass of 50 kg. A rough estimate of the attractive force they exert on each other can be obtained by treating their bodies as point masses 2.0 m apart. Then from the law of universal gravitation,

$$F = G\frac{m_1 m_2}{r^2} = (6.67 \times 10^{-11}\ \text{N}\cdot\text{m}^2/\text{kg}^2)\frac{(70\ \text{kg})(50\ \text{kg})}{(2.0\ \text{m})^2}$$
$$= 5.8 \times 10^{-8}\ \text{N}.$$

This is an extremely small force—it is only about 10^{-10} that of the weight of an average person!

While forces of this magnitude are far too small to detect by casual observation, they can be measured in the laboratory. The gravitational force between ordinary objects was first measured in 1798 by Henry Cavendish (1731–1810) with the torsion balance—a device that had been used in 1784 by Charles Augustin de Coulomb (1736–1806) to study the electrical force between charged particles. In Cavendish's experiment, two small spheres are attached to the ends of a light, rigid rod suspended by a fine, vertical fiber. Two large spheres are placed near the ends of the light rod as shown in Fig. 5-9. The gravitational force of attraction on the small spheres due to the large spheres causes the rod to rotate, thereby twisting the fiber. The twisted fiber opposes the rotation of the rod. As a result, the rod settles into a new position, which is displaced by a very small angle from its original position. This angular deflection is amplified by reflecting a beam of light from mirror R to a distant scale where it is then measured. Using the value of this deflection, knowledge of the properties of the fiber, and some simple geometry, we can calculate the attractive force between the spheres. Then G can be found using Eq. (5-2). Cavendish's original measurement was within 1 percent of the presently accepted value.

The law of gravitation gives the mutual attractive force between two *particles*. If the attracting bodies are not particles, the force must be determined using integral calculus. This is why the calculation of the gravitational force between the two people is only a rough estimate of the actual force between them. The only nonparticle gravitational force that is generally needed for our purposes is the attraction of a planet, such as the earth, for an object small enough to be considered a particle. While not entirely correct, it is reasonable to assume that a planet is a sphere whose mass density depends only on the distance from the center of the sphere. We can then represent the density as $\rho(r)$, where r is the distance from the center. In this case, the sphere's gravitational force on a particle is fairly easy to calculate. This calculation, which is done in Supplement 5-1, yields the following important result:

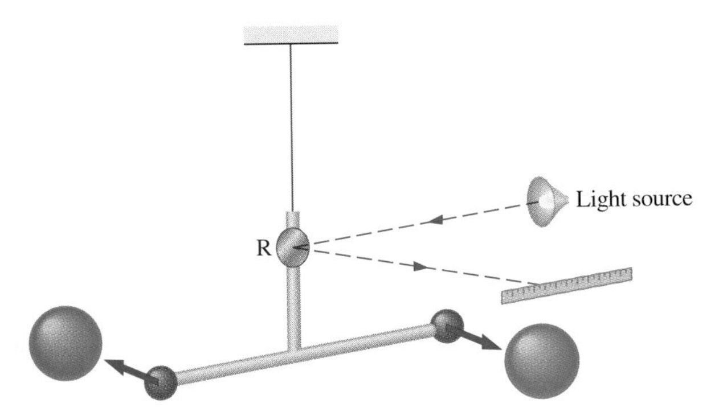

FIGURE 5-9 The Cavendish balance.

This pair of galaxies is interacting gravitationally over a distance of 35 million light-years.

> The gravitational force between a sphere of mass density $\rho(r)$ and a particle located outside the sphere is the same as if the entire mass of the sphere were located at its center.

Hence we can use Eq. (5-2) to calculate the gravitational force of a celestial body on a particle by treating the body *as a particle located at its center.*

The **weight** of an object is defined to be the gravitational force exerted on it. At or near the surface of a large celestial body, all other gravitational forces are negligible compared with that of the celestial body. Thus an object of mass m at the earth's surface has a weight given by

$$W = G\frac{mM_E}{R_E^2}, \tag{5-3}$$

where M_E and R_E represent the mass and radius of the earth, respectively. The force is, of course, directed* toward the center of the earth.

Suppose that only the gravitational force is acting on a body at the earth's surface. Then upon combining Eq. (5-3) with Newton's second law along the radial direction toward the center of the earth, we find

$$\sum F_r = G\frac{mM_E}{R_E^2} = ma_r;$$

so

$$a_r = G\frac{M_E}{R_E^2}.$$

Now a_r is the free-fall acceleration, which we have been representing by g. Thus

$$g = G\frac{M_E}{R_E^2}. \tag{5-4}$$

Notice that g is independent of the mass of the accelerating body, a property that has been used in previous chapters without proof.

Combining Eqs. (5-3) and (5-4) shows that, at the earth's surface, weight and mass are related by

$$W = mg. \tag{5-5}$$

*When weight and the acceleration due to gravity are described, their directions are frequently ignored even though they are vectors. This should not cause any confusion since we know both quantities point toward the center of the earth.

At the surface of the earth, the weight of a 5.0-kg body is (5.0 kg)(9.8 m/s^2) = 49 N, and the weight of a body of mass 3.0 slugs is (3.0 slugs)(32 ft/s^2) = 96 lb.

You should make sure that you understand the difference between weight and mass. Weight is a force that varies for a particular body as it is moved to different locations in the universe. However, mass, which measures the body's response to force, is the same everywhere. On the moon where g is 1.7 m/s^2, the 5.0-kg body has a weight of 8.5 N, while it has no weight at all at a point in space sufficiently far from any celestial object. However, the mass of the body is 5.0 kg everywhere.

EXAMPLE 5-5 Gravitational Forces on a Space Probe

Five space probes were launched during 1984 and 1985 by various nations in an effort to study Halley's comet. On March 14, 1986, the European Space Agency probe *Giotto* was at its closest approach to the comet, several hundred miles from the nucleus of the comet. Assuming the mass of *Giotto* is 1.00×10^3 kg, determine (*a*) its weight in an earth laboratory and (*b*) its weight when it reached a point 2.00×10^3 km above the earth's surface. (*c*) During its encounter with the comet, *Giotto* was 1.47×10^8 km from the center of the earth and 1.33×10^8 km from the center of the sun. Compare the gravitational attractions of the earth and the sun on the probe at this position.

SOLUTION (*a*) In the laboratory, the weight of the probe is

$$W_0 = (1.00 \times 10^3 \text{ kg})(9.80 \text{ m/s}^2) = 9.80 \times 10^3 \text{ N}.$$

(*b*) When the probe is not at the earth's surface, its weight is no longer mg. Instead, it must be determined using Eq. (5-2). For our case, $r = (6.37 \times 10^6 + 2.00 \times 10^6)$ m $= 8.37 \times 10^6$ m, the distance from the earth's center. The weight of the probe is then

$$W_1 = (6.67 \times 10^{-11} \text{ N}\cdot\text{m}^2/\text{kg}^2)\frac{(1.00 \times 10^3 \text{ kg})(5.98 \times 10^{24} \text{ kg})}{(8.37 \times 10^6 \text{ m})^2}$$
$$= 5.69 \times 10^3 \text{ N}.$$

(*c*) During the cometary encounter, the gravitational forces of the earth and the sun on *Giotto* are

$$F_E = (6.67 \times 10^{-11} \text{ N}\cdot\text{m}^2/\text{kg}^2)\frac{(1.00 \times 10^3 \text{ kg})(5.98 \times 10^{24} \text{ kg})}{(1.47 \times 10^{11} \text{ m})^2}$$
$$= 1.85 \times 10^{-5} \text{ N}$$

and

$$F_S = (6.67 \times 10^{-11} \text{ N}\cdot\text{m}^2/\text{kg}^2)\frac{(1.00 \times 10^3 \text{ kg})(1.99 \times 10^{30} \text{ kg})}{(1.33 \times 10^{11} \text{ m})^2}$$
$$= 7.50 \text{ N}.$$

Notice how small these forces are—the probe was nearly weightless during its encounter with the comet.

DRILL PROBLEM 5-6

Use Eq. (5-4) to calculate g.

DRILL PROBLEM 5-7

A body weighs 160 lb on the surface of the earth. (*a*) What is its weight on the moon's surface? (*b*) What is its mass on the moon's surface? (*c*) What is its mass on the earth's surface? (*d*) What is its weight at a point far from any celestial body? (*e*) What is its mass at this point?
ANS. (*a*) 27.8 lb; (*b*) 5.0 slugs; (*c*) 5.0 slugs; (*d*) 0; (*e*) 5.0 slugs.

DRILL PROBLEM 5-8

The moon's orbit around the earth can be approximated quite accurately by a circle of radius $R_M = 3.86 \times 10^8$ m. The period of this orbit is 27.3 days. Calculate the acceleration a_M of the moon in its orbit from $a_M = v^2/R_M$ and then compute the ratio a_M/g. Compare this ratio with $(R_E/R_M)^2$, where R_E is the radius of the earth. What does this result indicate?
ANS. The ratios are equal, which indicates that the force due to gravity varies inversely as the square of the distance from the attracting center.

*5-6 The Gravitational Field

Two bodies interact gravitationally even though they are not in contact. The transmission of a force across empty space is often called *action at a distance*. The fact that bodies are able to exert forces on one another across empty space used to bother scientists. This intuitive difficulty led to the development of the field concept, which today is an important cornerstone in the description of the fundamental forces. Rather than saying that body A exerts a gravitational force on body B, we say that A produces a gravitational field in space, and it is this field that, through contact with B, exerts a force on B. Mass plays a fundamental role in this interaction. Body A creates a gravitational field because it possesses mass, and body B interacts with the gravitational field because it has mass. In other words, *mass is responsible for both the creation of and the interaction with the gravitational field.*

Let's consider the two particles of Fig. 5-10. Particle A, assumed to be located at the origin of our coordinate system, creates a gravitational field $\mathbf{g}(\mathbf{r})$ whose spatial dependence is given by

$$\mathbf{g}(\mathbf{r}) = -G\frac{m_A}{r^2}\,\hat{\mathbf{r}}, \tag{5-6}$$

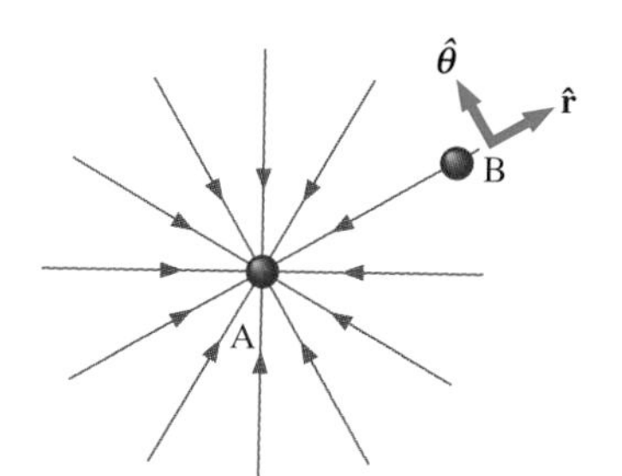

FIGURE 5-10 Particle A creates a gravitational field that points radially inward toward A.

where $\hat{\mathbf{r}}$ is a three-dimensional unit vector that at any point is directed along the straight line from A to that point. This gravitational field interacts with particle B to produce a force $m_B\mathbf{g}(\mathbf{r})$ on it:

$$\mathbf{F} = m_B\mathbf{g}(\mathbf{r}) = -G\frac{m_A m_B}{r^2}\hat{\mathbf{r}}. \qquad (5\text{-}7)$$

This is simply Newton's law of gravitation. Notice from Eq. (5-7) that $\mathbf{g}(\mathbf{r})$ has the units of force per mass, or newtons per kilogram (N/kg) in the SI system.

To find the gravitational field of a *group* of particles, we calculate the field of each particle and then compute the vector sum of the fields to get the net field. For the three particles of Fig. 5-11, the field at the point P is

$$\mathbf{g}_P = \left\{[6.67 \times 10^{-11}]\left[-\frac{3.0}{1.0^2}\mathbf{i} - \frac{4.0}{1.0^2}\mathbf{j} + \frac{2.0}{0.5^2}\mathbf{j}\right]\right\} \text{ N/kg}$$
$$= (-2.0 \times 10^{-10}\mathbf{i} + 2.7 \times 10^{-10}\mathbf{j}) \text{ N/kg}.$$

A 2.0-kg particle placed at P would then be subjected to a force

$$\mathbf{F} = m\mathbf{g}_P = (-4.0 \times 10^{-10}\mathbf{i} + 5.4 \times 10^{-10}\mathbf{j}) \text{ N}.$$

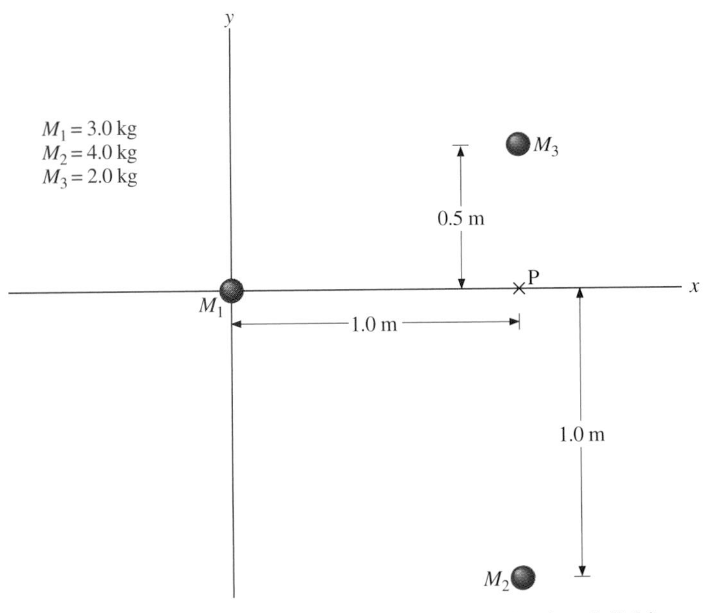

FIGURE 5-11 The three particles create a gravitational field.

Suppose a particle of mass m is in a gravitational field $\mathbf{g}(\mathbf{r})$. Since the force on the particle is $m\mathbf{g}(\mathbf{r})$, we have from Newton's second law,

$$m\mathbf{g}(\mathbf{r}) = m\mathbf{a},$$

so

$$\mathbf{a} = \mathbf{g}(\mathbf{r}).$$

Hence *the gravitational field at any point is the acceleration of a particle due to the gravitational force at that point*. Consistent with this, the units of the two quantities are the same. For example in the SI system, 1 N/kg = (1 kg·m/s²)/kg = 1 m/s².

DRILL PROBLEM 5-9

Particles of mass 2.0 kg are located at two adjacent corners of a square 1.0 m on a side, and particles of mass 1.0 kg are located at the other two corners. What is the gravitational field at the center of the square?
ANS. 1.9×10^{-10} N/kg in a direction perpendicular to the side joining the two 2.0-kg particles.

*5-7 Gravitational Field of a Continuous Mass Distribution

If the particle distribution is continuous, the gravitational field must be calculated by integration. For the continuous mass distribution of Fig. 5-12, the gravitational field at P due to the mass element Δm is given by

$$\mathbf{g} = -G\frac{\Delta m}{r^2}\hat{\mathbf{r}},$$

where r is the distance between Δm and P and $\hat{\mathbf{r}}$ is a unit vector directed from Δm toward P. The net gravitational field at P due to the entire mass distribution is then approximately

$$\mathbf{g} = -G\sum_i \frac{\Delta m_i}{r_i^2}\hat{\mathbf{r}}_i,$$

where the sum is made over all mass elements in the distribution and the index i refers to the ith element. In the limit

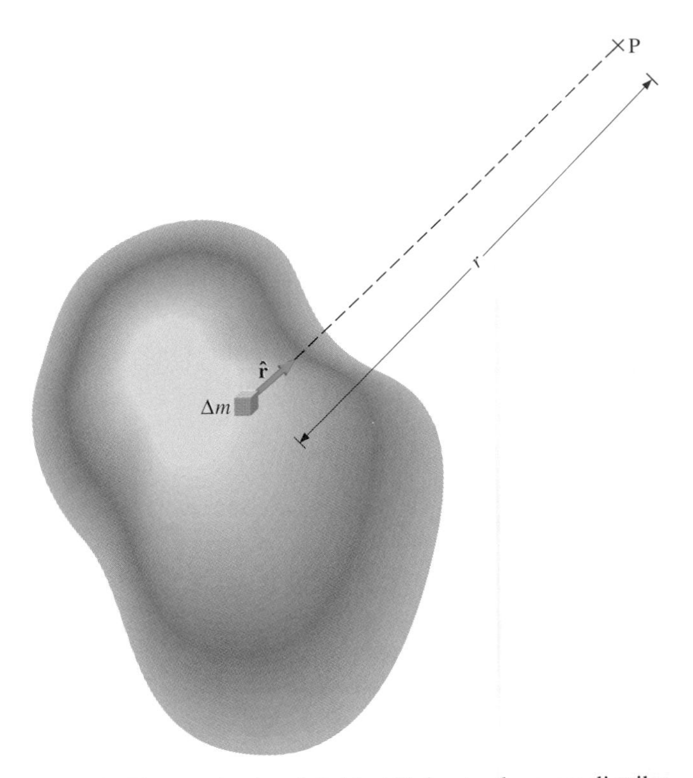

FIGURE 5-12 The gravitational field at P due to the mass distribution is calculated by integration.

$\Delta m_i \to 0$, the sum becomes an integral and the net gravitational field at P due to the entire mass distribution is then exactly

$$\mathbf{g} = -G \lim_{\Delta m_i \to 0} \sum_i \frac{\Delta m_i}{r_i^2} \hat{\mathbf{r}}_i = -G \int \frac{dm}{r^2} \hat{\mathbf{r}}.$$

It is easiest to evaluate this integral over the volume of the mass distribution. If $\rho(x, y, z)$ is the mass density of the distribution, the mass element dm can be represented by $\rho\, dv$, where dv is the volume of dm. The net gravitational field at P can then be written as

$$\mathbf{g} = -G \int \frac{\rho\, dv}{r^2} \hat{\mathbf{r}},$$

which is an integration over the volume of the distribution. If you were to evaluate this integral for the earth, you would find that at the earth's surface, $\mathbf{g} = -9.8\hat{\mathbf{r}}$ N/kg. Here the unit vector points away from the center of the earth.

SUMMARY

1. **Force**
   There are contact forces and field forces. Contact forces are due to physical contact between bodies. Field forces are exerted by a field that exists around the body under consideration. The gravitational force is a field force.
2. **Newton's first law**
   If there is no net force on a body, its velocity as measured relative to an inertial reference frame is constant.
3. **Newton's second law**
   Relative to an inertial reference frame,
   $$\sum \mathbf{F} = m\mathbf{a}.$$
4. **Newton's third law**
   If body A exerts a force $\mathbf{F}$ on body B, then B exerts a force $-\mathbf{F}$ on A.
5. **Newton's law of universal gravitation**
   (*a*) The gravitational force between two particles of mass m_1 and m_2 separated by a distance r is an attractive force along the line between the particles with a magnitude
   $$F = G\frac{m_1 m_2}{r^2}.$$
   (*b*) A sphere with mass density $\rho(r)$ attracts a particle as if all of the mass of the sphere were concentrated at its center, provided the particle is outside the sphere.

*6. **The gravitational field**
   (*a*) Mass is responsible for both the creation of and the interaction with the gravitational field. The gravitational field of a particle of mass m_A located at the origin is
   $$\mathbf{g}(\mathbf{r}) = -G\frac{m_A}{r^2}\hat{\mathbf{r}}.$$
   (*b*) The gravitational field of a group of particles is the vector sum of the fields of the individual particles.
   (*c*) The force of a gravitational field $\mathbf{g}(\mathbf{r})$ on a particle of mass m at position $\mathbf{r}$ is
   $$\mathbf{F} = m\mathbf{g}(\mathbf{r}).$$

SUPPLEMENT 5-1 GRAVITATIONAL FORCE OF A SPHERICAL MASS DISTRIBUTION ON A PARTICLE

In this supplement we will obtain an expression for the gravitational force of a sphere of mass density $\rho(r)$ on a particle located outside the surface of the sphere. Consider a particle of mass m that is a distance R from the center C of a thin-walled spherical shell, as shown in Fig. 5-13. The mass density, radius, and thickness of the shell are $\rho(r)$, r, and t, respectively. With the force on the particle due to each infinitesimal mass element dM' of the shell given by Eq. (5-2), we can find the force on the particle due to the entire shell by integrating over the mass elements.

The particle and symmetric elements dM' shown in the figure are a distance x apart. The force on m due to dM' is directed as shown, and its magnitude is

$$dF' = \frac{Gm\, dM'}{x^2}. \qquad (i)$$

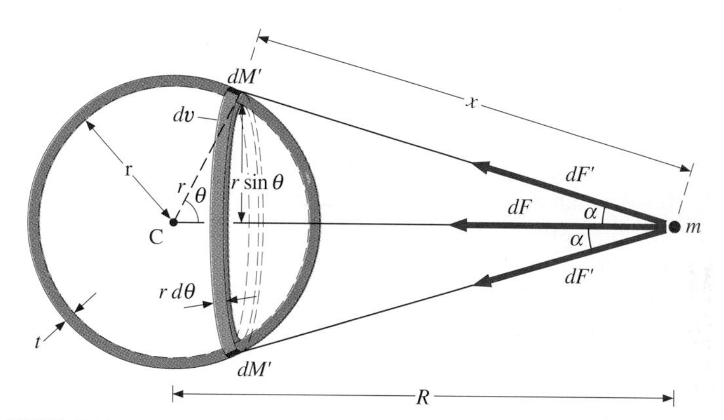

FIGURE 5-13 A particle of mass m is a distance R from the center of a thin-walled spherical shell. The gravitational attraction of a section of the shell on the particle is shown.

The magnitude of the force is the same for every dM' on the circular ring shown, because all mass elements are equidistant from the particle. The directions of those forces, however, do differ. It is clear from symmetry that for two mass elements diametrically opposite one another on the ring, the force components perpendicular to the line between C and the particle cancel, and the force components along this line add. Thus to find the net force due to all mass elements on the ring, we only have to consider the force components $dF' \cos\alpha$, which are directed toward C. Since all elements on the ring are a distance x from the particle,

$$dF = \frac{Gm\,dM}{x^2}\cos\alpha, \qquad (ii)$$

where dF is the magnitude of the force due to the entire ring and dM is the mass of the ring.

The mass dM is the volume dv of the ring multiplied by $\rho(r)$. The circumference of the base of the ring is $2\pi r\sin\theta$, its breadth is $r\,d\theta$, and its thickness is t. Therefore

$$dv = (2\pi r\sin\theta)(r\,d\theta)(t) = 2\pi t r^2 \sin\theta\,d\theta,$$
$$dM = \rho(r)\,dv = 2\pi\rho(r)tr^2\sin\theta\,d\theta,$$

and Eq. (ii) becomes

$$dF = 2\pi\rho(r)Gtmr^2\frac{\cos\alpha\sin\theta\,d\theta}{x^2}. \qquad (iii)$$

Before Eq. (iii) can be integrated over the shell, θ, α, and x must be expressed in terms of a single variable, which we choose to be x. Since x, r, and R are the three sides of a triangle,

$$x^2 = r^2 + R^2 - 2rR\cos\theta, \qquad (iv)$$

which, when differentiated, gives

$$2x\,dx = 2rR\sin\theta\,d\theta,$$

or

$$\sin\theta\,d\theta = \frac{x\,dx}{rR}. \qquad (v)$$

A relationship between α and x is found from inspection of the figure. Since

$$R = r\cos\theta + x\cos\alpha,$$
$$\cos\alpha = \frac{R - r\cos\theta}{x}. \qquad (vi)$$

With Eqs. (iv), (v), and (vi) substituted into Eq. (iii), we find

$$dF = \frac{\pi\rho(r)Gtmr}{R^2}\left(\frac{R^2 - r^2}{x^2} + 1\right)dx.$$

The force due to the entire shell is calculated by integrating this over x from its minimum value $(R - r)$ to its maximum value $(R + r)$. We then have

$$F = \frac{\pi\rho(r)Gtmr}{R^2}\int_{R-r}^{R+r}\left(\frac{R^2 - r^2}{x^2} + 1\right)dx = G\frac{\rho(r)4\pi r^2 tm}{R^2},$$

or

$$F = \frac{GMm}{R^2}, \qquad (vii)$$

where $\rho(r)(4\pi r^2 t)$ has been replaced by M, the mass of the shell. Thus the spherical shell attracts an external particle as if all of its mass were concentrated at its center.

The attraction of the solid sphere is now found by treating the sphere as an infinite number of concentric spherical shells of gradually increasing radius. Since we can place the mass of each shell at its center, we can also place the mass of the solid sphere at its center. And since the attractive force of each shell is given by Eq. (vii), the force of attraction of the solid sphere is also given by Eq. (vii). Therefore a sphere with mass density $\rho(r)$ attracts a particle as if all of the mass of the sphere were concentrated at its center, provided the particle is outside the sphere.

QUESTIONS

5-1. Only one force acts on a body. Can the body's velocity be zero? Can its acceleration be zero?

5-2. Can there be any forces on a body that remains at rest? What can you say about the net force on the body?

5-3. A car is moving down the highway at a constant velocity. What is the net force on the car?

5-4. If a rock and a feather are dropped, the rock falls quickly, while the feather flutters to the ground. Analyze the forces on each and explain why they fall so differently. What would happen on the moon?

5-5. From the objects listed here identify those that cannot have inertial reference frames attached to them: (a) an airplane moving through the air at a constant velocity; (b) an airplane taking off; (c) a stopping automobile; (d) an automobile rounding a curve at a constant speed; (e) a rotating Ferris wheel; (f) a projectile in outer space moving at a constant velocity.

5-6. Does the mass of a body depend on the inertial frame used to analyze the motion of the body?

5-7. A passenger in a car that hits a tree will be "thrown" into the windshield if he is not buckled to the seat. Explain why.

5-8. If a particle is moving to the right, is it possible for the force on it to be acting to the left? to be acting downward?

5-9. Analyze the net force on a parachutist when (a) she jumps out of the plane; (b) the parachute first opens; (c) she drifts downward at constant velocity; (d) she hits the ground.

5-10. Must a force be acting on a body that is moving in a curved path at constant speed?

5-11. How much does a 70-kg astronaut weigh in space, far from any celestial body? What is his mass at this location?

5-12. Two astronauts are floating outside their spacecraft doing repair work in deep space. Astronaut A signals to astronaut B indicating that he needs a large wrench in the latter's possession. Should B get the wrench to A by throwing it at him or by gently pushing it in his direction? Assume they are friends.

5-13. What is the direction of the net force on someone sitting in an automobile traveling in a circle? What happens if that force is eliminated?

5-14. Does the seat of a Ferris wheel push up on an occupant more when he is at the bottom or when he is at the top of his path?

5-15. Identify the action and reaction forces in the following situations: (a) the earth attracts the moon; (b) a boy kicks a football; (c) a rocket accelerates upward; (d) an automobile accelerates forward; (e) a high jumper leaps; (f) a bullet is shot from a gun.

5-16. Moe and Larry each grab an end of a rope and have a tug-of-war contest. Moe wins, but Larry claims it is a tie because he pulled on Moe as hard as Moe pulled on him. Comment.

5-17. Suppose that you are holding a cup of coffee in your hand. Identify all the forces on the cup and the reaction to each force.

5-18. When you stand on the earth, your feet push against it. Why doesn't the earth accelerate away from you?

PROBLEMS

Newton's Laws of Motion

5-1. A car of mass 1000 kg accelerates from zero to 90 km/h in 10 s. (*a*) What is its acceleration? (*b*) What is the net force on the car?

5-2. The driver in Prob. 5-1 applies the brakes when the car is moving at 90 km/h, and the car comes to rest after traveling 40 m. What is the net force on the car during its deceleration?

5-3. A particle of mass 2.0 kg is acted on by a single force $\mathbf{F}_1 = 18\mathbf{i}$ N. (*a*) What is the particle's acceleration? (*b*) If the particle starts at rest, how far does it travel in the first 5 s?

5-4. Suppose that the particle of Prob. 5-3 also experiences forces $\mathbf{F}_2 = -15.0\mathbf{i}$ N and $\mathbf{F}_3 = 6.0\mathbf{j}$ N. What is its acceleration in this case?

5-5. A 20-kg body is dropped from a height of 10 m and penetrates 0.50 m into a box of soil. What is the average retarding force that the soil exerts on the body?

5-6. Forces $\mathbf{F}_1 = 3.0\mathbf{i}$ N, $\mathbf{F}_2 = (5.0\mathbf{i} + 2.0\mathbf{j})$ N, $\mathbf{F}_3 = (-6.0\mathbf{i} + 4.0\mathbf{j})$ N, and $\mathbf{F}_4$ act on a particle of mass 4.0 kg that has an acceleration $(-2.0\mathbf{i} + 3.0\mathbf{j})$ m/s^2. What is $\mathbf{F}_4$?

5-7. Find the acceleration of the body of mass $m = 5.0$ kg shown in the accompanying figure.

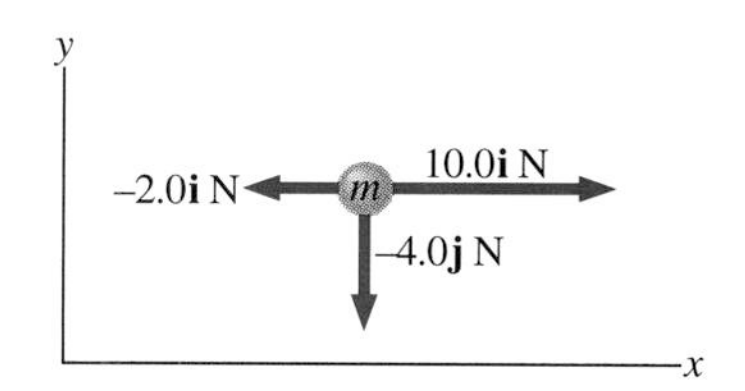

PROBLEMS 5-7 and 5-30

5-8. Find the acceleration of the body of mass $m = 10.0$ kg shown in the accompanying figure.

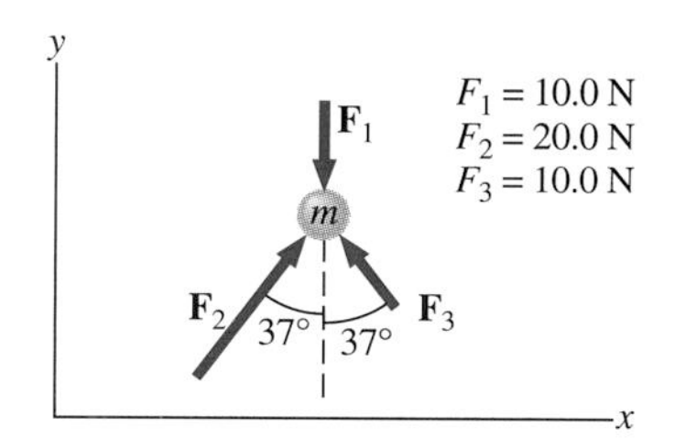

PROBLEMS 5-8 and 5-31

5-9. A particle of mass 2.5 kg is moving in a circle of radius 6.0 m. At a certain instant, $v = 3.0$ m/s and $dv/dt = 3.0$ m/s^2. What is the force on the particle?

5-10. A body of mass $m = 2.0$ kg is moving along the x axis with a speed of 3.0 m/s at the instant represented in the accompanying figure. (*a*) What is the acceleration of the body? (*b*) What is the body's velocity 10 s later? (*c*) What is its displacement after 10 s?

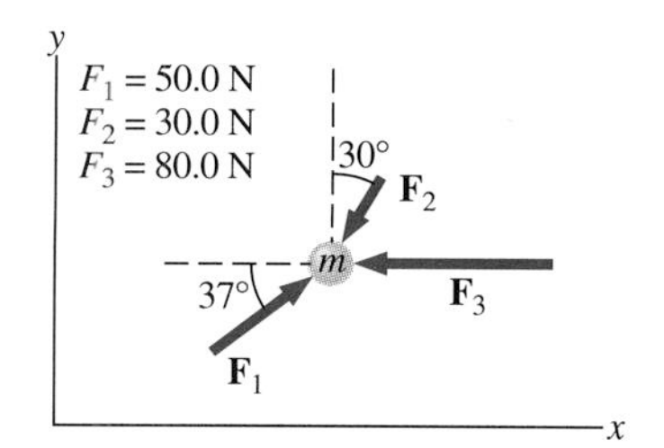

PROBLEM 5-10

5-11. Force $\mathbf{F}_B$ has twice the magnitude of force $\mathbf{F}_A$. Find the direction in which the particle shown in the accompanying figure accelerates.

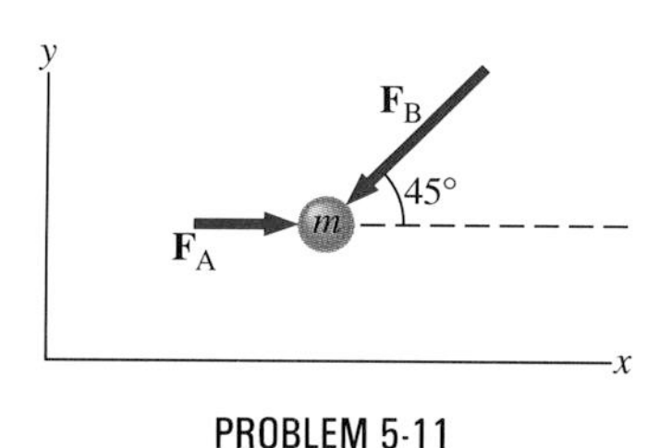

PROBLEM 5-11

5-12. The accompanying figure shows a body of mass 1.0 kg under the influence of the forces $\mathbf{F}_A$, $\mathbf{F}_B$, and $m\mathbf{g}$. If the body accelerates to the left at 0.20 m/s^2, what are $\mathbf{F}_A$ and $\mathbf{F}_B$?

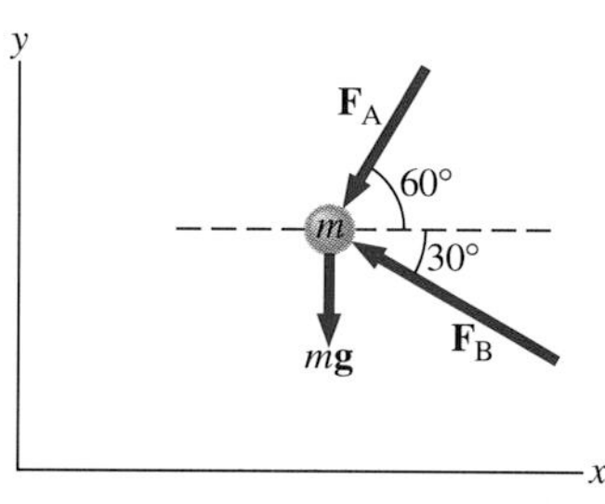

PROBLEM 5-12

5-13. The velocities of a 3.0-kg object at $t = 6.0$ s and $t = 8.0$ s are $(3.0\mathbf{i} - 6.0\mathbf{j} + 4.0\mathbf{k})$ m/s and $(-2.0\mathbf{i} + 4.0\mathbf{k})$ m/s, respectively. If the object is moving at a constant acceleration, what is the force acting on it?

5-14. A 0.50-kg particle travels from $(-2.5\mathbf{i} + 4.0\mathbf{j})$ m to $(3.0\mathbf{i} - 8.5\mathbf{j})$ m in 10 s. If its initial velocity is zero, what is the net force acting on the particle, assuming its acceleration is constant?

5-15. An automobile passenger of mass m is held firmly in his seat by a lap-and-shoulder seat belt. If the automobile is in a collision and decelerates from 100 km/h to rest in 0.30 s, what is the average force (in terms of m) that the belt exerts on the passenger? Compare this force with the weight of the passenger.

5-16. An object of mass 2.0 kg is dropped from a height of 4.9 m. As a consequence of Newton's third law, as the object falls, the earth comes up to meet it. (*a*) What is the force on the earth? (*b*) What is the acceleration of the earth? (*c*) How far does the earth move to meet the object?

5-17. A firefighter of mass 80 kg slides down a vertical pole with an average acceleration of 3.0 m/s^2. What is the average force that he exerts on the pole?

5-18. A young girl of mass 40 kg stands on a chair so she can reach a book on the top shelf of her bookcase. The mass of the chair is 10 kg. Determine (*a*) the force of the chair on the girl; (*b*) the force of the floor on the chair; (*c*) the force of the chair on the floor. (*d*) What is the reaction force to the weight of the girl? to the weight of the chair?

Newton's Law of Gravitation

5-19. The mass of a particle is 15 kg. (*a*) What is its weight on the earth? (*b*) What is its weight on the moon? (*c*) What is its mass on the moon? (*d*) What is its weight in outer space far from any celestial body? (*e*) What is its mass at this point?

5-20. What is your mass in slugs on the earth? What is your mass in slugs on the moon? What is your weight on the moon?

5-21. On a planet whose radius is 1.2×10^7 m, the acceleration due to gravity is 18 m/s^2. What is the mass of the planet?

5-22. The mean diameter of the planet Saturn is 1.2×10^8 m, and its mean mass density is 0.69 g/cm^3. Find the acceleration due to gravity at Saturn's surface.

5-23. The mean diameter of the planet Mercury is 4.88×10^6 m, and the acceleration due to gravity at its surface is 3.78 m/s^2. Estimate the mass of this planet.

5-24. The acceleration due to gravity on the surface of a planet is three times as large as it is on the surface of the earth. The mass density of the planet is known to be twice that of the earth. What is the radius of this planet in terms of the earth's radius?

5-25. A body on the surface of a planet with the same radius as the earth's weighs 10 times more than it does on the earth. What is the mass of this planet in terms of the earth's mass?

5-26. Calculate the values of g at the earth's surface for the following changes in the earth's properties: (*a*) its mass is doubled and its radius is halved; (*b*) its mass density is doubled and its radius is halved; (*c*) its mass density is halved and its mass is unchanged.

The Gravitational Field

5-27. Make a plot of the gravitational field of a particle as a function of the distance from that particle.

5-28. The masses of the two particles shown in the accompanying figure are $m_1 = 2.0$ kg and $m_2 = 4.0$ kg. What is the gravitational field at A? at B? at C?

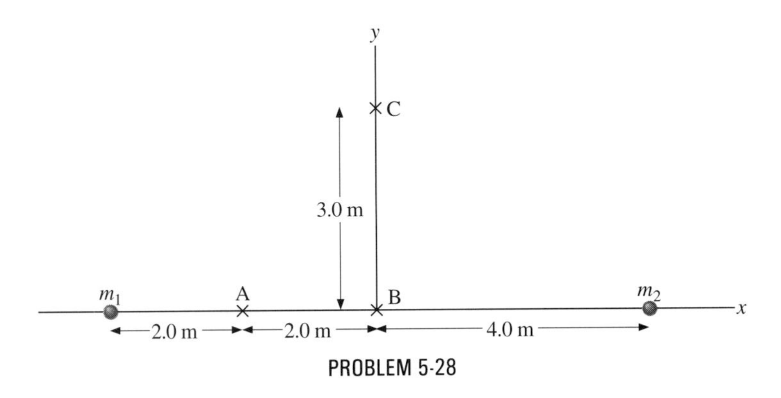

PROBLEM 5-28

5-29. What is the gravitational field at a distance R_E above the earth's surface?

5-30. Repeat your calculation of Prob. 5-7 assuming the body is in the earth's gravitational field with $\mathbf{g} = -9.8\mathbf{j}$ N/kg.

5-31. Repeat your calculation of Prob. 5-8 assuming the body is in the earth's gravitational field with $\mathbf{g} = -9.8\mathbf{j}$ N/kg.

General Problems

5-32. An electron moves through space with its position varying with time according to

$$\mathbf{r}(t) = (6.0 \times 10^6 t^2)\mathbf{i} + (4.0 \times 10^6 t^2 - 9.0 \times 10^6 t)\mathbf{j} + (4.0 \times 10^5 t^3)\mathbf{k}.$$

The mass of the electron is 9.1×10^{-31} kg, $\mathbf{r}(t)$ is in meters, and t is in seconds. What is the force on the electron at $t = 5.0$ s ?

5-33. On June 25, 1983, shot-putter Udo Beyer of East Germany threw the 7.26-kg shot 22.22 m, which at that time was a world record. (*a*) If the shot was released at a height of 2.20 m with a projection angle of 45.0°, what was its initial velocity? (*b*) If while in Beyer's hand the shot was accelerated uniformly over a distance of 1.2 m, what was the net force on it?

5-34. Suppose you can communicate with the inhabitants of a planet in another solar system. They tell you that on their planet, whose diameter and mass are 5.0×10^3 km and 3.6×10^{23} kg, respectively, the record for the high jump is 2.0 m. Given that this record is close to 2.4 m on the earth, what would you conclude about your extraterrestrial friends' jumping ability?

5-35. The accompanying figure shows a uniform rod of length l whose mass per unit length is λ. What is the gravitational force of the rod on a particle of mass m located a distance d from one end of the rod, as shown in the figure?

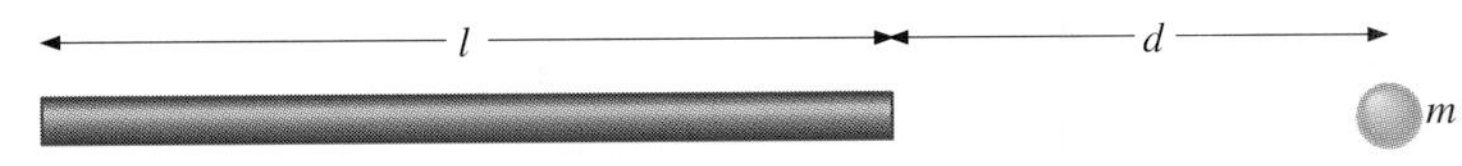

PROBLEM 5-35

5-36. Two identical uniform rods of length l and mass m are placed along the same line so that their closer ends are a distance d apart, as shown in the accompanying figure. (*a*) Show that the gravitational attraction between the rods has a magnitude of

$$F = \frac{Gm^2}{l^2} \ln \frac{(d+l)^2}{d(d+2l)}.$$

(*b*) Show that

$$\lim_{d\to\infty} F = G\frac{m^2}{d^2}.$$

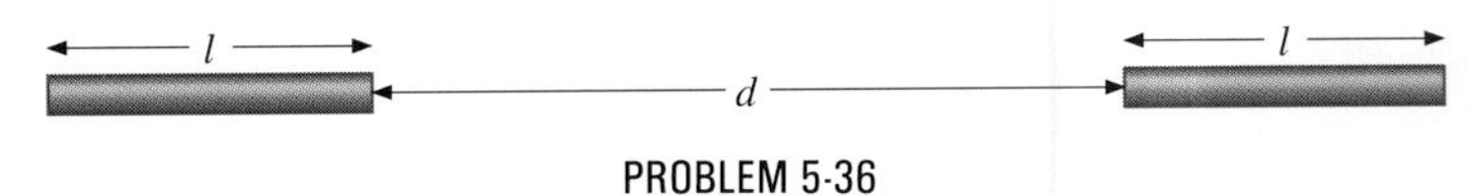

PROBLEM 5-36

5-37. For a height h above the surface of the earth such that $h/R_E \ll 1$, show that g decreases by an amount Δg such that $\Delta g = -2hg/R_E$. Use this formula to determine how much g decreases between sea level and the top of Mount Everest, which is 8848 m above sea level.

5-38. (*a*) Find the difference in gravitational fields due to the moon at points on the earth's surface nearest to and farthest away from the moon. This difference is responsible for the lunar tides on the earth. (*b*) Show that the difference in the gravitational fields of the sun at these two points is approximately 43 percent of that due to the moon.

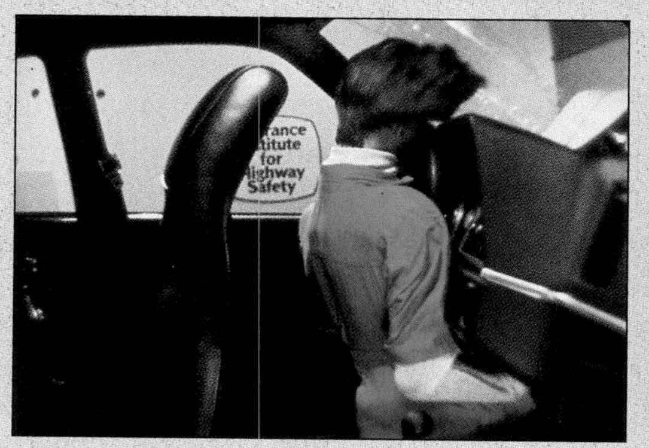

The dynamics of a crash test.

CHAPTER 6 DYNAMICS II

PREVIEW

In this chapter Newton's laws are applied to mechanical systems that are commonly encountered or observed. The primary topics to be discussed here are the following:

1. **Common forces.** These include the gravitational force, the contact force of one surface on the other, tension and compression in rigid rods, tension in strings and ropes, and the spring force.
2. **Applications of Newton's laws.** A systematic approach to analyzing problems using Newton's laws is presented. Special emphasis is placed on the importance of the free-body diagram in this approach.

*3. **Velocity-dependent frictional forces.** Motion under the influence of a velocity-dependent frictional force is studied.

*4. **Motion of planets and other satellites.** Kepler's three laws and their application to satellite motion are discussed.

*5. **Effect of the earth's rotation on *g*.**

*6. **Definitions of force and mass.** Force and mass are defined. The relationship between these definitions and Newton's second law is considered.

The primary topic of this chapter is the application of Newton's laws to realistic problems. Instead of dealing with forces of unspecified origin as we have been doing, we now consider forces that often occur in everyday life. After investigating several of these common forces, we will study how they are used in Newton's laws to calculate accelerations. These calculations are made using a rather systematic approach, which will be explained in detail in Sec. 6-2. We'll encounter a variety of interesting problems, including the effects of friction on motion, the banking of roads, the necessity of headrests in automobiles, and the orbital motion of the planets.

6-1 Common Forces

The different types of forces involved in most practical applications are surprisingly few. With just the six we now present, we are able to analyze a wide range of mechanical systems.

Force of the Gravitational Field

The gravitational force on a body of mass m is $m\mathbf{g}(\mathbf{r})$, where $\mathbf{g}(\mathbf{r})$ is the gravitational field. The gravitational field of the earth at its surface is directed toward the center of the earth and has a magnitude of 9.8 m/s^2. The gravitational field is one of the two fundamental fields that exert forces in the macroscopic world of Newtonian mechanics.

Force of the Electromagnetic Field

This is the other fundamental field force in Newtonian mechanics. We mention the electromagnetic field here only for the sake of completeness. It will be discussed thoroughly in the second half of this book. On a microscopic level, the electromagnetic force is the basis for all contact forces.

Surface Contact Forces

The force of contact between two surfaces is generally resolved into two components: one that is perpendicular to each surface and one that is parallel to each surface, as shown in Fig. 6-1. The perpendicular component is called the **normal force N** while the parallel component is known as the **frictional force f**.* For two surfaces at rest relative to each other, the frictional force is called the **force of static friction** and is represented by $\mathbf{f}_S$. The direction of the force of static friction on a body is that necessary to keep the body stationary. When two surfaces are sliding past each other, the frictional force is called the **force of sliding** (or **kinetic**) **friction**, $\mathbf{f}_K$. The force of kinetic friction always acts opposite to the direction of motion of the body under consideration.

To understand how static friction behaves, imagine that you are trying to push a stationary heavy crate across the floor by applying a force **P** parallel to the floor as shown in Fig. 6-2*a*. Other forces on the crate are its weight **W** and the contact force between the surfaces, which may be resolved into a normal force of magnitude N and a frictional force of magnitude f_S. If you push harder and the crate still doesn't move, f_S must have increased to stay equal to P. Experience shows that if you continue to push harder, the crate will eventually begin to slide. What this means is that f_S can increase to keep the block still *only to a point*. The force of static friction has a maximum value—if P is increased beyond this maximum, f_S cannot balance it.

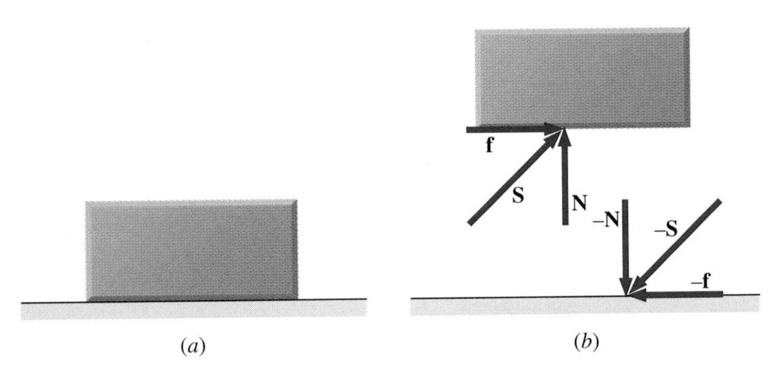

FIGURE 6-1 (*a*) A block moving on a horizontal surface. (*b*) The surface contact force **S** is resolved into a normal force **N** and a frictional force **f**. Notice that Newton's third law requires a force −**S** on the horizontal surface.

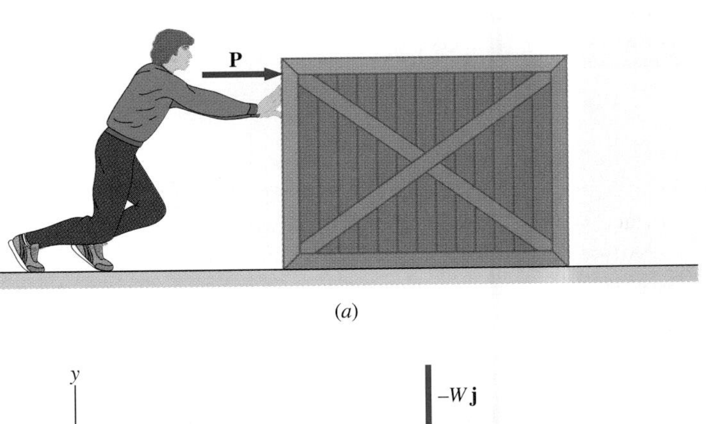

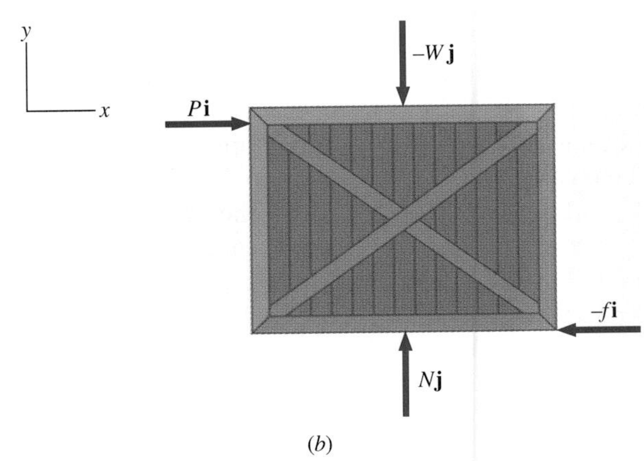

FIGURE 6-2 (*a*) A crate on a horizontal surface is pushed with a force **P**. (*b*) The forces on the crate. Here, $-f\mathbf{i}$ may represent either the static or the kinetic frictional force.

*The term *friction* is generally used to represent the contact interaction responsible for the frictional force.

For most situations, experiment shows that the *maximum* value of the static frictional force can be represented quite accurately by $f_{max} = \mu_S N$, where μ_S, the *coefficient of static friction*, is a constant that depends on the materials of the surfaces in contact. (See Table 6-1.) The force of static friction must therefore satisfy

$$f_S \leq \mu_S N. \tag{6-1}$$

TABLE 6-1 COEFFICIENTS OF STATIC AND KINETIC FRICTION

Materials	Static (μ_S)	Kinetic (μ_K)
Steel on steel	0.74	0.57
Aluminum on steel	0.61	0.47
Copper on steel	0.53	0.36
Brass on steel	0.51	0.44
Zinc on cast iron	0.85	0.21
Copper on cast iron	1.05	0.29
Glass on glass	0.94	0.4
Copper on glass	0.68	0.53
Teflon on Teflon	0.04	0.04
Teflon on steel	0.04	0.04

Once the crate of Fig. 6-2*a* is in motion, the force of kinetic friction acts. This force is also found to obey a simple formula:

$$f_K = \mu_K N, \tag{6-2}$$

where μ_K is a constant called the *coefficient of kinetic friction*. As indicated in Table 6-1, μ_K also depends on the materials in contact.

Notice that the static frictional force is variable and is *less than or equal to* $\mu_S N$, while the kinetic frictional force is always *equal to* $\mu_K N$. Both frictional forces depend on only the normal force between two given surfaces and not on the surface area in contact. Also, μ_K is generally less than μ_S. This simply means that it's harder to start an object moving than to maintain its motion.

Equations (6-1) and (6-2) are empirical laws that describe the behavior of the forces of friction. While these formulas are very useful for practical purposes, they do not have the status of mathematical statements that represent general principles (e.g., Newton's second law). In fact, there are cases for which Eqs. (6-1) and (6-2) are not even good approximations. For instance, neither formula is accurate for lubricated surfaces or for two surfaces sliding across each other at high speeds. Unless specified, we will not be concerned with these exceptions.

EXAMPLE 6-1 STATIC AND KINETIC FRICTION

A 20-kg crate is at rest on a floor as shown in Fig. 6-2*a*. The coefficient of static friction between the crate and floor is 0.70 and the coefficient of kinetic friction is 0.60. A horizontal force **P** is applied to the crate. Find the force of friction if (*a*) $P = 20$ N, (*b*) $P = 30$ N, (*c*) $P = 120$ N, and (*d*) $P = 180$ N.

SOLUTION The free-body diagram of the crate is shown in Fig. 6-2*b*. Application of Newton's second law in the horizontal and vertical directions gives

$$\sum F_x = ma_x \qquad \sum F_y = ma_y$$
$$P - f = ma_x \qquad N - W = 0.$$

Here we are using the symbol f to represent the frictional force since we haven't yet determined whether the crate is subject to static friction or kinetic friction. We will do this whenever we are unsure what type of friction is acting. Now the weight of the crate is $W = (20 \text{ kg})(9.8 \text{ m/s}^2) = 196$ N, which is also equal to N. The maximum force of static friction is therefore $(0.70)(196 \text{ N}) = 137$ N. As long as P is less than 137 N, the force of static friction keeps the crate stationary and $f_S = P$. Thus (*a*) $f_S = 20$ N, (*b*) $f_S = 30$ N, (*c*) $f_S = 120$ N.

(*d*) If $P = 180$ N, the applied force is greater than the maximum force of static friction (137 N), so the crate can no longer remain at rest. Once the crate is in motion, kinetic friction acts. Then $f_K = \mu_K N = (0.60)(196 \text{ N}) = 118$ N, and the acceleration is

$$a_x = \frac{P - f_K}{m} = \frac{180 \text{ N} - 118 \text{ N}}{20 \text{ kg}} = 3.1 \text{ m/s}^2.$$

DRILL PROBLEM 6-1

A block of mass 1.0 kg rests on a horizontal surface. The frictional coefficients for the block and surface are $\mu_S = 0.50$ and $\mu_K = 0.40$. What is the minimum horizontal force required to move the block? What is the block's acceleration when this force is applied?
ANS. 4.9 N; 1.0 m/s^2.

TENSION AND COMPRESSION FORCES IN RIGID RODS

Figure 6-3*a* shows a horizontal rigid rod fixed to a wall at one end. We apply various forces that are normal to its free end and consider the resulting forces at an arbitrary cross section AA′ within the rod. At AA′, each piece of the rod is subjected to a contact force exerted by the other piece. From Newton's third law, these contact forces are equal and opposite. We also know that since the rod is fixed, there can be no net horizontal force on either piece. Consequently, if we pull on the free end of the rod, the contact forces will be directed as shown in Fig. 6-3*b*. The two pieces then pull on each other, and the rod as a whole is being stretched. The contact force is then known as a *tensile* force, and the rod is under **tension**. If the forces at AA′ are such that the pieces are pushing each other (see Fig. 6-3*c*), the rod is under **compression**, and the normal force is a *compressive* force. Calculation of the normal force at a particular cross section determines whether the rod is under tension or compression at that point. The tangential component of the contact force, should one exist, is called a *shear* force, a force that will not be discussed in this textbook.

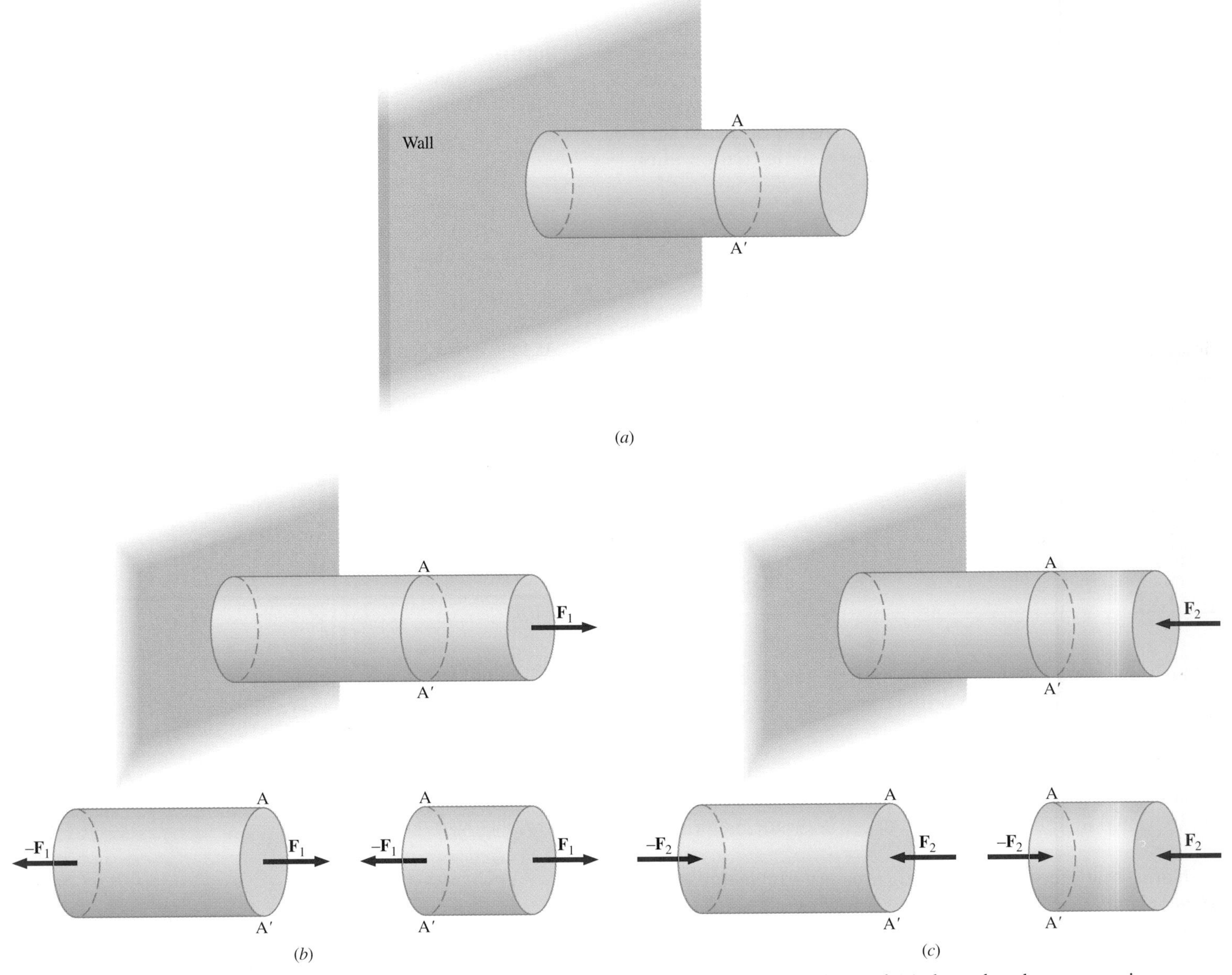

FIGURE 6-3 (*a*) A rigid cylindrical rod fixed to a wall at one end: (*b*) the rod under tension and (*c*) the rod under compression.

EXAMPLE 6-2 **TENSION AND COMPRESSION IN A ROD**

A rod of constant cross-section and density is attached to the ceiling (see Fig. 6-4). The rod weighs 120 N and is 1.50 m long. Find the normal force at the cross section 0.50 m from the bottom end. Is the rod under tension or compression?

SOLUTION The free-body diagram of the bottom 0.50-m section of the rod is shown in Fig. 6-4*b*. Since the rod has constant cross-section and density, the bottom section weighs (0.50 m/1.50 m) (120 N) = 40 N. Application of Newton's second law to this section gives

$$T - 40 \text{ N} = ma_y = 0,$$

so

$$T = 40 \text{ N}.$$

Since T is directed so that the bottom section is being pulled by the rest of the rod, the rod is under tension.

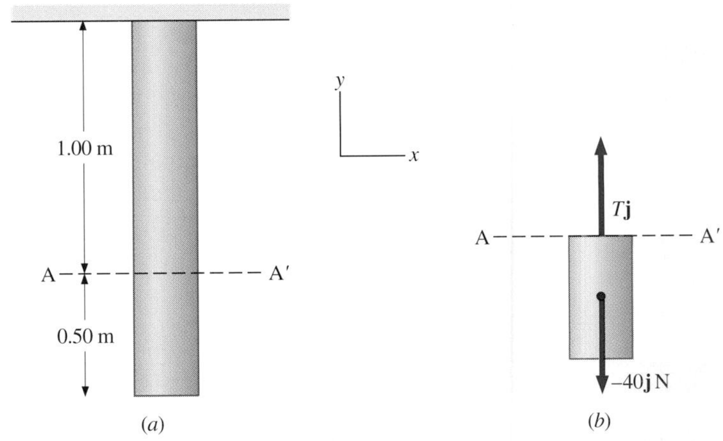

FIGURE 6-4 (*a*) A uniform rod attached to a ceiling. (*b*) The forces on the bottom 0.50-m section of the rod.

DRILL PROBLEM 6-2

A thick rope weighing 2.0 N/m is used to lift a bucket from a well at constant velocity. If the bucket and its water weigh 500 N, what is the tension in the rope at a point (*a*) 2.0 m from its bottom end? (*b*) 6.0 m from its bottom end? (*c*) 10.0 m from its bottom end?

ANS. (*a*) 504 N; (*b*) 512 N; (*c*) 520 N.

Tension in a Flexible Cord

A flexible cord such as a string or light rope can support tension but not compression. If you push on a cord, it will collapse. However, if you pull on that cord, it will expand slightly and pull back—it supports tension.

Frequently the weight (and mass) of a flexible cord is so small compared with the other forces (and masses) in the system that the cord can be assumed to be massless. As an example, consider the cord of Fig. 6-5. If the mass of the cord is negligible, the only significant forces on section AB are the tensions at the two ends. Application of the second law to this section gives

$$T_B - T_A = ma_x = (0)a_x;$$

thus

$$T_B = T_A.$$

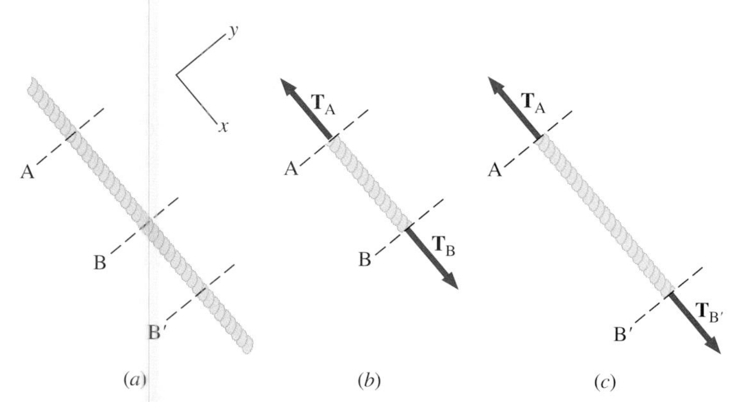

FIGURE 6-5 (*a*) Imaginary cuts are made in a straight section of massless cord, (*b*) first at A and B, (*c*) then at A and B′. Since the cord is massless, $T_A = T_B$, and $T_A = T_{B'}$; that is, the tension is the same throughout the cord.

Now A and B are arbitrary points along the cord. If we had chosen A and some other point B′, we would have found that $T_A = T_{B'}$. Hence *the tension is the same at all points along a massless cord.*

This property holds even when the cord passes over a pulley, provided both the mass of the pulley and the friction in the bearings are negligible. (See Fig. 6-6.) An explanation of this fact will be given in Chap. 9 when rotational motion is discussed.

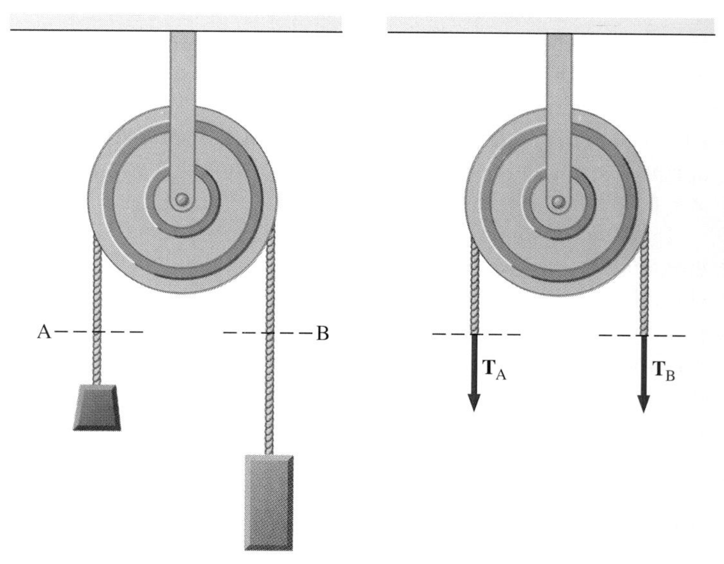

FIGURE 6-6 Imaginary cuts are made in the cord at A and B. If the cord and pulley are massless and if there is no friction in the bearings of the pulley, $T_A = T_B$.

Spring Force

Unlike a flexible cord, a spring that is loosely coiled supports both tension and compression. For small deformations, the spring will return to its relaxed length once the deforming force is removed. Figure 6-7*a* shows a relaxed spring that is attached to a wall at one end and to a small object at the other end. If we displace the object slightly so that the spring is stretched, the spring exerts a force on the object that attempts to restore it to its original position. (See Fig. 6-7*b*.) A restoring force is also produced if the object is displaced in the other direction and the spring is compressed (Fig. 6-7*c*). In either case, the force of the spring on the object acts to return the object to its original position. This **spring force** is found to be proportional to the displacement x of the end of the spring from its position when the spring is relaxed. It is represented by

$$F(x) = -kx, \qquad (6\text{-}3)$$

where k is a positive constant known as the *spring constant*. In the SI system, the units of k are newtons per meter (N/m).

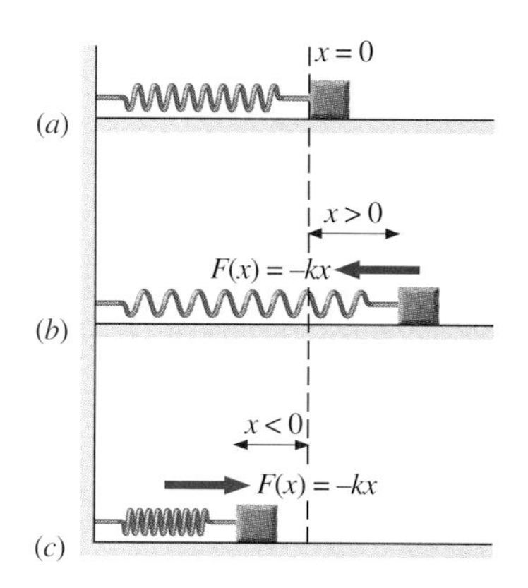

FIGURE 6-7 (*a*) A spring at its relaxed length: $x = 0$ and $F(x) = 0$. (*b*) The spring is stretched: $x > 0$ and $F(x) < 0$. (*c*) The spring is compressed: $x < 0$ and $F(x) > 0$.

Equation (6-3) is named Hooke's law after its discoverer, Robert Hooke (1635–1703). The fact that the force is directed opposite to the displacement and always tries to restore a spring to its relaxed length is represented by the minus sign. When the spring is stretched, $x > 0$ and $F(x) < 0$; when it is compressed, $x < 0$ and $F(x) > 0$.

Like the formulas describing friction, Hooke's law is a mathematically accurate approximation of a particular force and should not be taken as a statement of a fundamental physical principle. Actually, if the spring is stretched too far, the linear relationship between the force and the displacement is no longer satisfied. The spring is then said to be stretched beyond its *proportional limit*. We will generally assume that our springs obey Hooke's law. As with a massless string, *the tension (and compression) in a massless spring at any point along its length is equal in magnitude to the force it exerts at its end.*

EXAMPLE 6-3 A STRETCHED SPRING

Weights are hung at the end of a massless vertical spring (as shown in Fig. 6-8*a*), and the corresponding displacements of the end of the spring from its original position are measured. The following measurements are taken:

Weight (N)	90	135	180	225	270	315
Displacement (m)	0.100	0.150	0.200	0.250	0.310	0.390

(*a*) Plot the weight added to the end of the spring vs. the displacement and find the force constant of the spring. (*b*) What is the proportional limit of the spring?

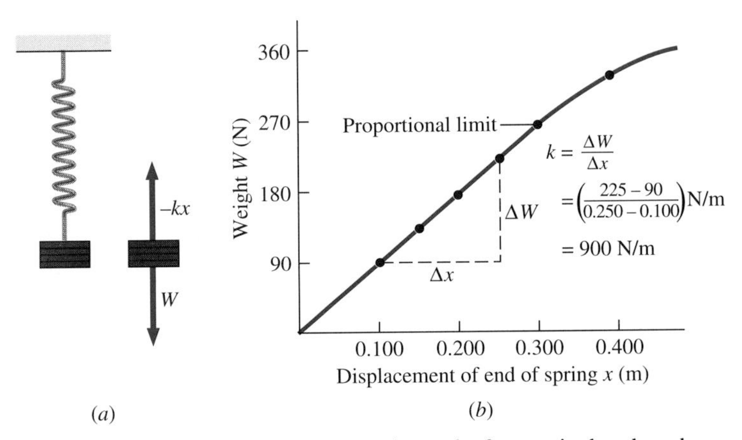

FIGURE 6-8 (*a*) Weights are hung at the end of a vertical, relaxed spring. (*b*) Weight W versus the displacement x of the end of the spring from its original position.

SOLUTION With the hanging weights at rest, the force $F(x)$ of the spring and the weight W balance each other. For displacements of the spring less than the proportional limit, we have from Newton's second law

$$\sum F = W - kx = 0,$$

so

$$W = kx,$$

where k is the force constant of the spring. In Fig. 6-8*b*, W is plotted versus x. The slope of the linear portion of the curve is k, which we find to be 900 N/m. The curve ceases to be linear at a displacement of about 0.310 m, which is therefore the proportional limit of the spring.

DRILL PROBLEM 6-3

(*a*) A horizontal massless spring is stretched 2.0 cm by a 40-N force. What is the force constant of the spring? (*b*) If this spring is hung vertically, what mass must be attached to its end to stretch it 3.0 cm?
ANS. (*a*) 2.0×10^3 N/m; (*b*) 6.1 kg.

6-2 APPLICATIONS OF NEWTON'S LAWS

Newton's laws will now be used to analyze a wide range of mechanical systems. When applying these laws, you will find the following five-step technique very useful.

1. *Clearly identify the body whose motion is to be analyzed.* Although this step may seem trivial, an inability to do this is a source of serious difficulty for many students. A system may consist of several bodies. In this case, you will frequently need to identify each body and analyze its motion individually.
2. *Pick an inertial reference frame.* The location of the frame's origin and the direction of its coordinate axes are chosen on the basis of what is convenient for solving the particular problem.
3. *Identify all forces acting on the body.* Make sure that you do not include forces acting on bodies other than the one whose motion you are analyzing.
4. *Draw a free-body diagram.* Show the reference frame and all forces on the body. The forces are usually resolved into components along the coordinate axes.
5. *Apply Newton's second law.* In this last step, you apply the second law to determine the acceleration of the body along each axis of your coordinate system.

The first three examples that follow will be discussed with explicit reference to the five steps. We will continue to use this technique thereafter, but will no longer directly refer to the steps.

EXAMPLE 6-4 THE MOTION OF A SLED

A 15-kg sled is pulled across a horizontal, snow-covered surface by the force **P** as shown in Fig. 6-9. The coefficient of kinetic friction between the sled and the snow is $\mu_K = 0.20$. (*a*) If $P = 33$ N, what

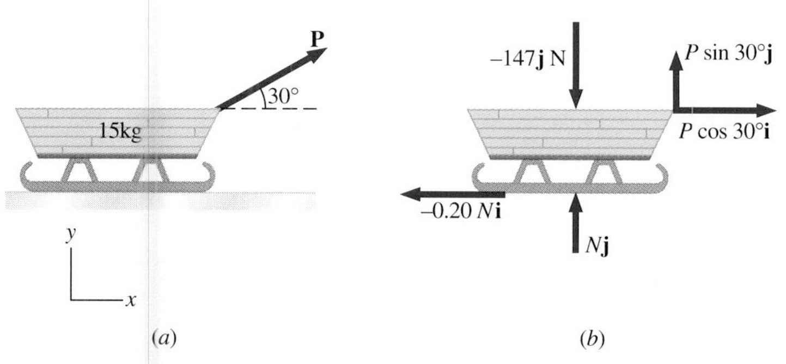

FIGURE 6-9 (*a*) A sled is pulled by the force **P**. (*b*) The free-body diagram for the sled.

is the horizontal acceleration of the sled? (*b*) What must the force be in order to pull the sled at constant velocity?

SOLUTION

Step 1: The body under consideration is the sled.

Step 2: An inertial reference frame attached to the earth is chosen with the x axis along the ground (parallel to the path of the sled) and the y axis perpendicular to the ground.

Step 3: The forces on the sled are caused by contact with the snow, by the gravitational field, and by **P**.

Step 4: The contact force with the snow is resolved into two components, a normal force $N\mathbf{j}$ and a frictional force $\mathbf{f}_K = -0.20N\mathbf{i}$. The components of **P** are $(P\cos 30°)\mathbf{i}$ and $(P\sin 30°)\mathbf{j}$. The gravitational force points in the negative y direction and is $\mathbf{W} = -(15\text{ kg})(9.8\text{ m/s}^2)\mathbf{j} = -147\mathbf{j}$ N. The free-body diagram of the sled is shown in Fig. 6-9*b*.

Step 5: Application of the second law in the x and y directions gives

$$\sum F_x = ma_x$$
$$P\cos 30° - 0.20N = (15\text{ kg})a_x$$

$$\sum F_y = ma_y$$
$$N + P\sin 30° - 147\text{ N} = (15\text{ kg})a_y.$$

(*a*) For this part of the problem, we are given that $P = 33$ N; also $a_y = 0$ because the sled does not move vertically. The substitution of these values into the equations of step 5 gives

$$(33\text{ N})\cos 30° - 0.20N = (15\text{ kg})a_x$$

and

$$N + (33\text{ N})\sin 30° - 147\text{ N} = 0.$$

We now have two equations for the two unknowns a_x and N. Solving for these unknowns, we find $a_x = 0.17\text{ m/s}^2$ and $N = 131$ N.

(*b*) If the sled moves at constant velocity, $a_x = 0$. In this case, P is unknown, so the equations of step 5 become

$$P\cos 30° - 0.20N = 0$$

and

$$N + P\sin 30° - 147\text{ N} = 0.$$

We again have two equations, this time for the unknowns P and N, which we find to be $P = 30$ N and $N = 132$ N.

DRILL PROBLEM 6-4

What happens to the sled of Fig. 6-9 if P exceeds 294 N?
ANS. It accelerates upward and leaves the horizontal surface.

DRILL PROBLEM 6-5

The sled of Example 6-4 is pushed by a force **F** directed 30° below the horizontal and moves at constant velocity. What is the magnitude of **F**?
ANS. 38 N.

EXAMPLE 6-5 A TUGBOAT PULLING TWO BARGES

The two barges of Fig. 6-10 are coupled by a cable of negligible mass. The mass of the front barge is 2.00×10^3 kg and the mass of the rear barge is 3.00×10^3 kg. A tugboat pulls the front barge with a horizontal force of magnitude 20.0×10^3 N, and the frictional forces of the water on the front and rear barges are 8.0×10^3 N and 10.0×10^3 N, respectively. Find the horizontal acceleration of the barges and the tension in the connecting cable.

SOLUTION

Step 1: In this case, each barge will be considered separately.

Step 2: An inertial reference frame attached to the earth with its x axis horizontal and its y axis vertical is used for both barges.

Step 3: There are two objects to consider, so the forces on each have to be identified separately. The forces on the

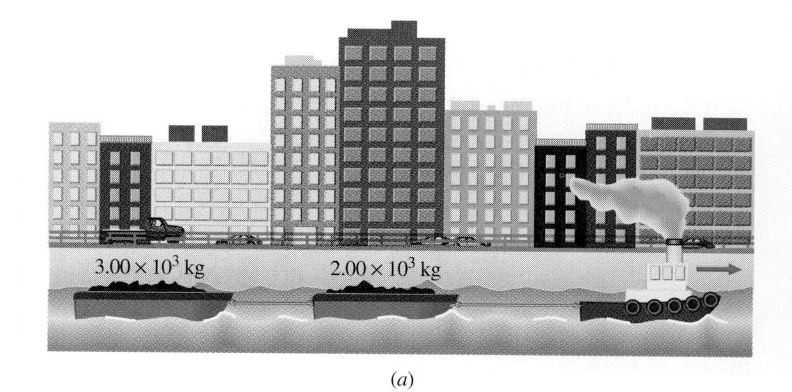

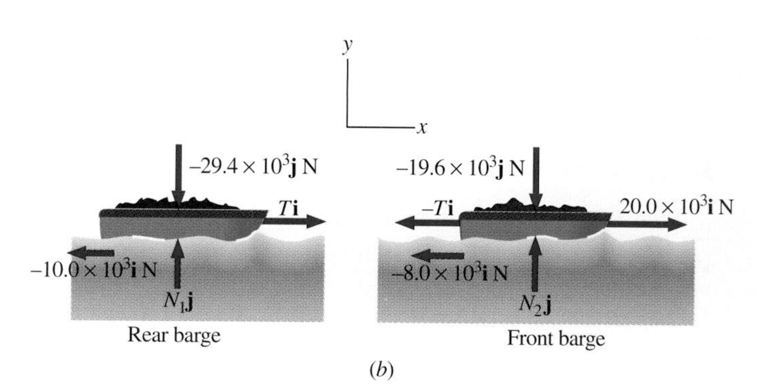

FIGURE 6-10 (*a*) Coupled barges pulled by a tugboat. (*b*) Free-body diagrams for the two barges.

rear barge are the contact force due to the water, the tension in the coupling cable, and the gravitational force. The forces on the front barge are the contact force due to the water, the tension in the cable, the gravitational force, and the 20.0×10^3-N horizontal force applied by the tugboat.

Step 4: Free-body diagrams for the barges are shown in Fig. 6-10*b*. The contact force of the water on the rear barge is resolved into a normal force $N_1\mathbf{j}$ and a frictional force $-10.0 \times 10^3\mathbf{i}$ N. The force on this barge due to the tension in the cable is $T\mathbf{i}$, and the gravitational force is $-(3.00 \times 10^3 \text{ kg})(9.80 \text{ m/s}^2)\mathbf{j} = -29.4 \times 10^3\mathbf{j}$ N. For the front barge, the contact force of the water has components $N_2\mathbf{j}$ and $-8.0 \times 10^3\mathbf{i}$ N. Since the cable's mass is negligible, the tension is constant along its length, and its force on the front barge is $-T\mathbf{i}$. The other forces on this barge are the gravitational force, $-(2.00 \times 10^3 \text{ kg})(9.80 \text{ m/s}^2)\mathbf{j} = -19.6 \times 10^3\mathbf{j}$ N, and the horizontal force $20.0 \times 10^3\mathbf{i}$ N.

Step 5: Applying Newton's second law in the horizontal direction to each barge, we have

Rear Barge

$$\sum F_x = m_1 a_x$$
$$T - 10.0 \times 10^3 \text{ N} = (3.00 \times 10^3 \text{ kg})a_x$$

Front Barge

$$\sum F_x = m_2 a_x$$
$$20.0 \times 10^3 \text{ N} - 8.0 \times 10^3 \text{ N} - T = (2.00 \times 10^3 \text{ kg})a_x,$$

where the accelerations of the barges are the same because they are coupled and thus move together. We are left with two equations for the two unknowns a_x and T. Solving these equations, we obtain $a_x = 0.40 \text{ m/s}^2$ and $T = 11.2 \times 10^3$ N.

DRILL PROBLEM 6-6

If the order of the barges of Example 6-5 is reversed so that the tugboat pulls the 3.00×10^3-kg barge with a force of 20.0×10^3 N, what are the acceleration of the barges and the tension in the coupling cable?
ANS. 0.40 m/s^2; 8.8×10^3 N.

EXAMPLE 6-6 Scale Readings on a Moving Elevator

The 160-lb man of Fig. 6-11 stands on a spring scale resting on the floor of an elevator. Find the scale reading if (*a*) the elevator accelerates upward at 8.0 ft/s² and (*b*) the elevator moves downward at constant velocity.

SOLUTION

Step 1: The scale reading depends on the motion of the man. The object whose motion we must study is therefore the man.

Step 2: The inertial reference frame is attached to the earth, with the y axis pointing vertically upward.

Step 3: There are two forces on the man, the gravitational force and the contact force due to the scale.

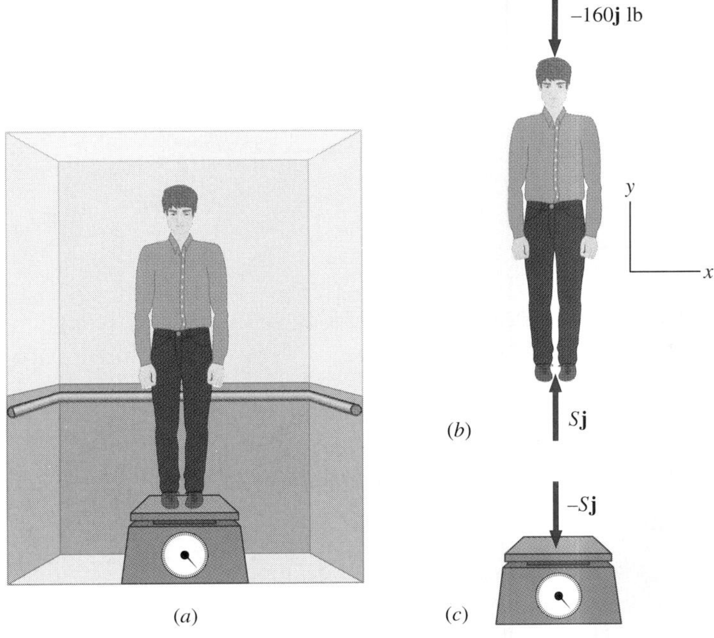

FIGURE 6-11 (*a*) A 160-lb man stands on a spring scale in an elevator. (*b*) The free-body diagram for the man. (*c*) The scale measures the force $-S\mathbf{j}$.

Step 4: Both forces are along the y axis. The man's weight is $-160\mathbf{j}$ lb and the upward force of the scale is $S\mathbf{j}$.

Step 5: Application of the second law relative to the earth-bound frame gives

$$\sum F_y = ma_y$$
$$S - 160 \text{ lb} = (5.0 \text{ slug})a_y.$$

(*a*) When $a_y = 8.0 \text{ ft/s}^2$, S is 200 lb. The scale is designed to measure the force on its surface, so its reading is 200 lb. This *apparent* increase in weight is often incorrectly attributed to an extra force (in this case, of magnitude 40 lb) pushing downward on the man. It is clear from the free-body diagram that no such force exists. This mistake is usually made when step 2 is not followed correctly. For example, you might choose the accelerating elevator, which is *noninertial*, as your reference frame. Since the man is not accelerating relative to the elevator, the only way you could explain the 200-lb reading is to invent a fictitious 40-lb force acting downward on him. In reality, S is larger than the man's weight simply because a nonzero net force is required if the man is to accelerate upward relative to an inertial frame fixed on the earth.

(*b*) If the velocity is constant, a_y is zero, so

$$S - 160 \text{ lb} = (5.0 \text{ slug})(0 \text{ m/s}^2),$$

and $S = 160$ lb. Notice that in this case the elevator frame is also inertial and can be used to analyze the man's motion. In this frame, $a_y = 0$ for the man, so $S = 160$ lb, which agrees with the result we obtained using the earth frame.

DRILL PROBLEM 6-7

What is the scale reading if the elevator of Example 6-6 is (*a*) accelerating downward at 8.0 ft/s² and (*b*) in free fall?
ANS. (*a*) 120 lb; (*b*) zero.

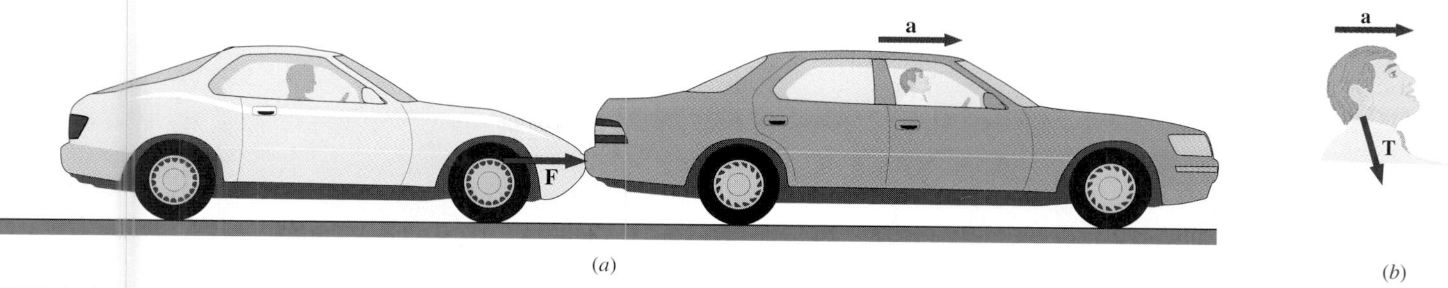

FIGURE 6-12 (a) An automobile is struck in the rear and accelerated forward. (b) The horizontal component of the neck's tension accelerates the head forward.

EXAMPLE 6-7 NECK INJURIES

When an automobile is struck in the rear, why are its passengers' necks so vulnerable to injury? Assume there are no headrests in the automobile.

SOLUTION When the automobile of Fig. 6-12 is struck in the rear, it suddenly accelerates forward. A passenger's body also accelerates forward because it is pushed by the seat, and the head, for rather obvious reasons, must follow along. However, there is nothing to push the head forward except what it is in contact with, namely, the neck. Thus tension must be created in the neck so that its horizontal component can accelerate the head. If this tension is too large, the neck is susceptible to injury.

EXAMPLE 6-8 SLIDING BLOCKS

The two blocks of Fig. 6-13 are attached to each other by a massless string that is wrapped around a frictionless pulley. When the bottom 4.0-kg block is pulled to the left by the constant force **P**, the top 2.0-kg block slides across it to the right. Find the magnitude of the force necessary to move the blocks at constant speed. Assume that the coefficient of kinetic friction between all surfaces is 0.40.

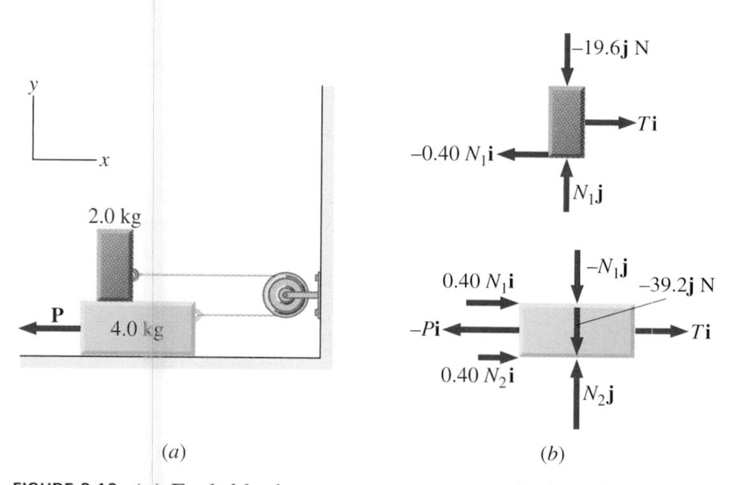

FIGURE 6-13 (a) Each block moves at constant velocity. (b) Free-body diagrams for the blocks.

SOLUTION We will analyze the motions of the two blocks separately. The top block is subjected to a contact force exerted by the bottom block. The components of this force are the normal force $N_1\mathbf{j}$ and the frictional force $-0.40N_1\mathbf{i}$. Other forces on the top block are the tension $T\mathbf{i}$ in the string and the weight of the top block itself, $-19.6\mathbf{j}$ N. The bottom block is subjected to contact forces due to the top block and due to the floor. The first contact force has components $-N_1\mathbf{j}$ and $0.40N_1\mathbf{i}$, which are simply reaction forces to the contact forces that the bottom block exerts on the top block. The components of the contact force of the floor are $N_2\mathbf{j}$ and $0.40N_2\mathbf{i}$. Other forces on this block are $-P\mathbf{i}$, the tension $T\mathbf{i}$, and the weight $-39.2\mathbf{j}$ N.

Since the top block is moving horizontally to the right at constant velocity, its acceleration is zero in both the horizontal and the vertical directions. From Newton's second law,

$$\sum F_x = m_1 a_x \qquad \sum F_y = m_1 a_y$$
$$T - 0.40N_1 = 0 \qquad N_1 - 19.6 \text{ N} = 0.$$

Solving for the two unknowns, we obtain $N_1 = 19.6$ N and $T = 0.40N_1 = 7.84$ N. The bottom block is also not accelerating, so the application of Newton's second law to this block gives

$$\sum F_x = m_2 a_x \qquad \sum F_y = m_2 a_y$$
$$T - P + 0.40N_1 + 0.40N_2 = 0 \qquad N_2 - 39.2 \text{ N} - N_1 = 0.$$

The values of N_1 and T were found with the first set of equations. When these values are substituted into the second set of equations, N_2 and P can be determined. They are $N_2 = 58.8$ N and $P = 39.2$ N.

EXAMPLE 6-9 A CRATE ON AN ACCELERATING TRUCK

A 50-kg crate rests on the bed of a truck as shown in Fig. 6-14. The coefficients of friction between the surfaces are $\mu_K = 0.30$ and $\mu_S = 0.40$. Find the frictional force on the crate when the truck is accelerating forward relative to the ground at (a) 2.0 m/s² and (b) 5.0 m/s².

SOLUTION (a) The forces on the crate are its weight and the normal and frictional forces due to contact with the truck bed. We start by *assuming* that the crate is not slipping. In this case, the static frictional force f_S acts on the crate. Furthermore, the accelerations of the crate and the truck are equal. Application of Newton's second law to the crate, using the reference frame attached to the ground, yields

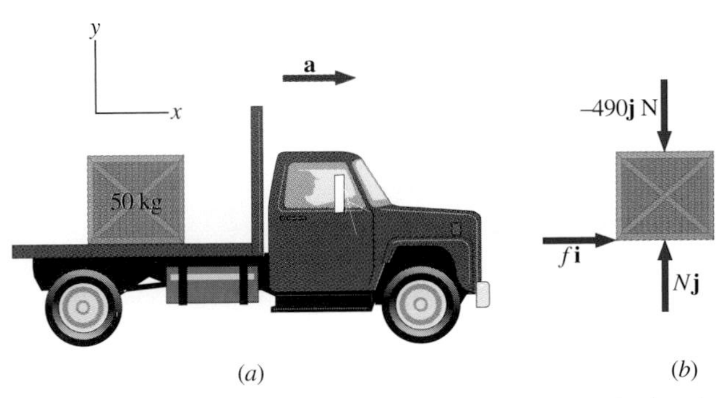

FIGURE 6-14 (*a*) A crate rests on the bed of the truck that is accelerating forward. (*b*) The free-body diagram of the crate.

$$\sum F_x = ma_x \qquad \sum F_y = ma_y$$
$$f_S = (50\ \text{kg})(2.0\ \text{m/s}^2) \qquad N - 490\ \text{N} = (50\ \text{kg})(0)$$
$$= 100\ \text{N} \qquad N = 490\ \text{N}.$$

The validity of our no-slip assumption can now be checked. The maximum value of the force of static friction is $\mu_S N = (0.40) \times (490\ \text{N}) = 196\ \text{N}$, while the *actual* force of static friction that acts when the truck is accelerating forward at $2.0\ \text{m/s}^2$ is only 100 N. Thus the assumption of no slipping is valid.

(*b*) If the crate is to move with the truck when it accelerates at $5.0\ \text{m/s}^2$, the force of static friction must be $f_S = ma_x = (50\ \text{kg}) \times (5.0\ \text{m/s}^2) = 250\ \text{N}$. Since this exceeds the maximum of 196 N, the crate must slip. The frictional force is therefore kinetic and is $f_K = \mu_K N = (0.30)(490\ \text{N}) = 147\ \text{N}$. The horizontal acceleration of the crate relative to the ground is now found from

$$\sum F_x = ma_x$$
$$147\ \text{N} = (50\ \text{kg})a_x,$$

so $a_x = 2.9\ \text{m/s}^2$.

Relative to the ground, the truck is accelerating forward at $5.0\ \text{m/s}^2$ and the crate is accelerating forward at $2.9\ \text{m/s}^2$. Hence the crate is sliding backward relative to the bed of the truck with an acceleration $2.9\ \text{m/s}^2 - 5.0\ \text{m/s}^2 = -2.1\ \text{m/s}^2$.

EXAMPLE 6-10 A Downhill Skier

The skier of Fig. 6-15 glides down a slope that is inclined at $\theta = 13°$ to the horizontal. The coefficient of kinetic friction between the skis and the snow is $\mu_K = 0.20$. What is the acceleration of the skier?

SOLUTION The forces acting on the skier are her weight and the contact force of the slope, which has a component normal to the incline and a component along the incline (force of kinetic friction). Because the skier moves along the slope, the most convenient reference frame for analyzing her motion is one with the x axis along and the y axis perpendicular to the incline. In this frame, both the normal and the frictional forces lie along coordinate axes, the components of the weight are $mg \sin\theta\,\mathbf{i}$ and $-mg\cos\theta\,\mathbf{j}$ (see Drill Prob. 6-8), and there is only an acceleration along the x axis ($a_y = 0$).

We can now apply Newton's second law to the skier:

$$\sum F_x = ma_x \qquad \sum F_y = ma_y$$
$$mg \sin\theta - \mu_K N = ma_x \qquad N - mg\cos\theta = m(0).$$

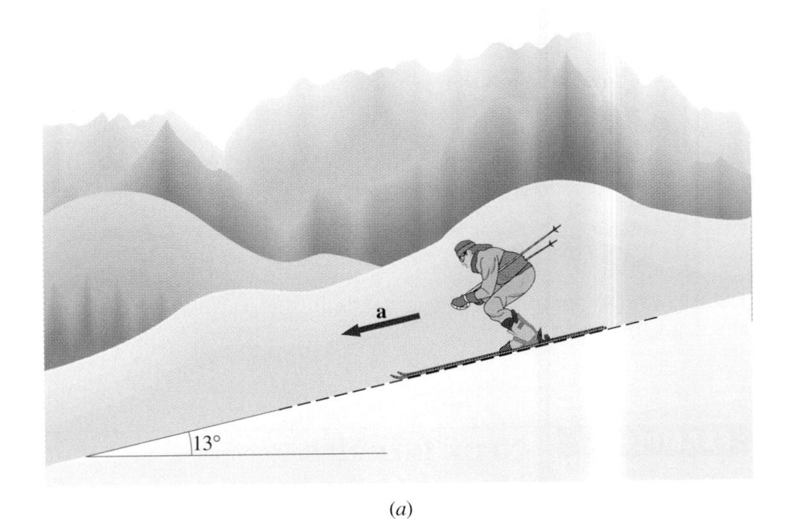

(*a*)

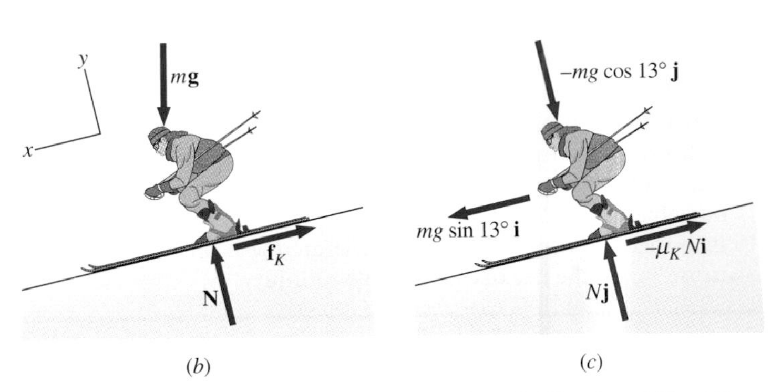

FIGURE 6-15 (*a*) A skier glides down the slope inclined at 13° to the horizontal. (*b*) The forces on the skier. (*c*) The free-body diagram of the skier.

From the second equation, $N = mg\cos\theta$. Upon substituting this into the first equation, we find

$$a_x = g(\sin\theta - \mu_K \cos\theta)$$
$$= g(\sin 13° - 0.20\cos 13°) = 0.29\ \text{m/s}^2.$$

Notice from this equation that if θ is small enough or μ_K is large enough, a_x is negative; that is, the skier slows down.

DRILL PROBLEM 6-8

A block of mass m rests on a surface inclined at an angle θ to the horizontal. Show that the block's weight can be resolved into components $mg\sin\theta$ along the incline and $mg\cos\theta$ perpendicular to the incline.

DRILL PROBLEM 6-9

What is the acceleration of the skier of Example 6-10 if she is moving down an incline of 10°?
ANS. $-0.23\ \text{m/s}^2$; the negative sign indicates she is slowing down.

EXAMPLE 6-11 TWO ATTACHED BLOCKS

Figure 6-16 shows a block of mass m_1 on a frictionless horizontal surface. It is pulled by a light string that passes over a frictionless and massless pulley. The other end of the string is connected to a block of mass m_2. Find the acceleration of the blocks and the tension in the string in terms of m_1, m_2, and g.

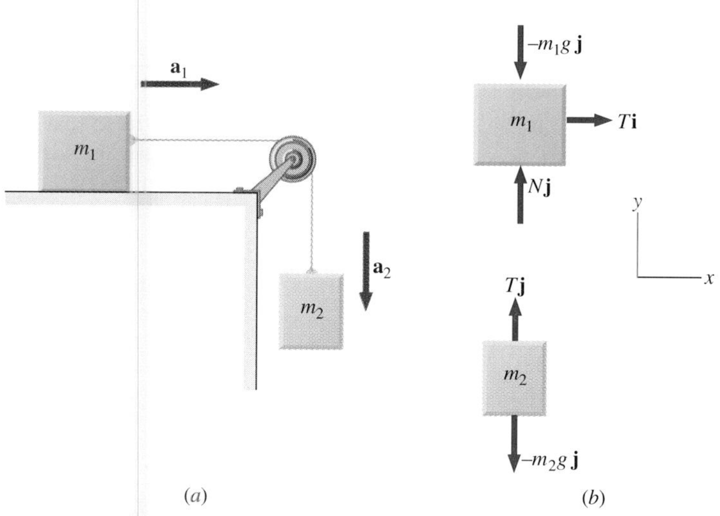

FIGURE 6-16 (*a*) Block 1 is connected by a light string to block 2. (*b*) The free-body diagrams of the blocks.

SOLUTION The forces on block 1 are the gravitational force, the contact force of the surface, and the tension in the string. Block 2 is subjected to the gravitational force and the string tension. Since the string and the pulley have negligible mass, and since there is no friction in the pulley, the tension is the same throughout the string.

Applying Newton's second law to each block, we obtain

Block 1	**Block 2**
$\sum F_x = ma_x$	$\sum F_y = ma_y$
$T = m_1 a_{1x}$	$T - m_2 g = m_2 a_{2y}.$

When block 1 moves to the right, block 2 travels an equal distance downward; thus $a_{1x} = -a_{2y}$. Writing the common acceleration of the blocks as $a = a_{1x} = -a_{2y}$, we now have

$$T = m_1 a$$

and

$$T - m_2 g = -m_2 a.$$

From these two equations, we can express a and T in terms of the masses m_1 and m_2, and g:

$$a = \frac{m_2}{m_1 + m_2} g$$

and

$$T = \frac{m_1 m_2}{m_1 + m_2} g.$$

Notice that the tension in the string is *less* than the weight of the block hanging from the end of it. A common error in problems like this is to set $T = m_2 g$. You can see from the free-body diagram of block 2 that this can't be correct if the block is accelerating.

DRILL PROBLEM 6-10

The device shown in Fig. 6-17 is called Atwood's machine. Assuming that the masses of the string and the frictionless pulley are negligible, find the acceleration of the two blocks and the tension in the string.

ANS. $a = \dfrac{m_2 - m_1}{m_1 + m_2} g$; $T = \dfrac{2m_1 m_2}{m_1 + m_2} g$.

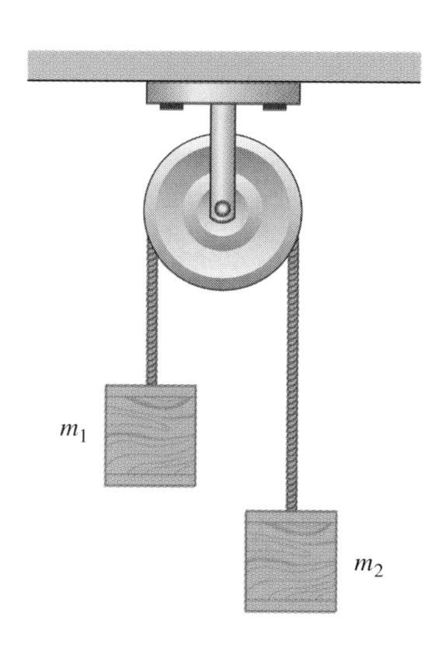

FIGURE 6-17 Atwood's machine.

EXAMPLE 6-12 CIRCULAR MOTION

A small disk of mass m is tied to the end of a light string and twirled at constant speed in a circle of radius R on a horizontal frictionless surface. (See Fig. 6-18*a*.) What is the tension in the string?

SOLUTION The free-body diagram of the disk is shown in Fig. 6-18*b*. Since there is no vertical acceleration, $mg = N$. The only horizontal force on the disk is the tension in the string, which is always

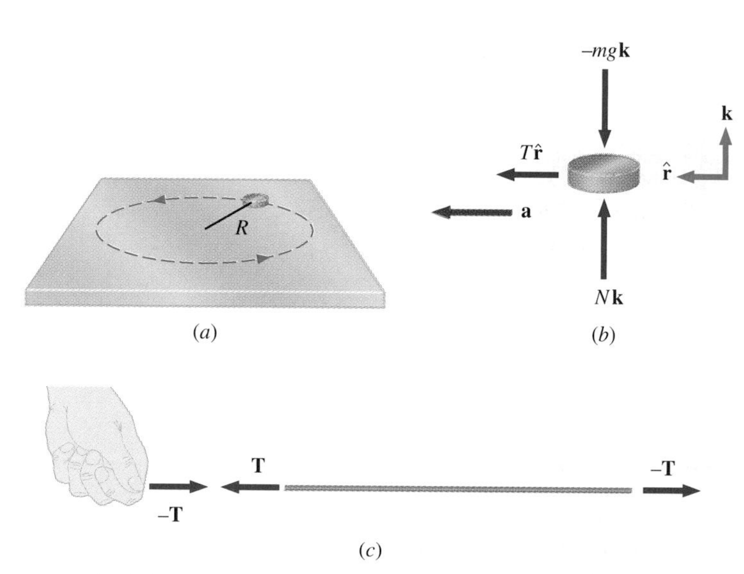

FIGURE 6-18 (*a*) A disk is tied to a light string and twirled in a horizontal circle on a frictionless surface. (*b*) The free-body diagram of the disk. (*c*) The free-body diagram of the string. Also shown is the force of the string on the hand holding it at the center of the circle.

directed toward the center of the circle. The centripetal acceleration of the disk is v^2/R, so from Newton's second law in the radial direction

$$\sum F_r = ma_r$$
$$T = m\frac{v^2}{R}.$$

Like the acceleration, the force **T** is centripetal. Centripetal forces occur frequently in nature. For example, the electron moving around the proton in the hydrogen atom is under the influence of a centripetal electrical force, and the earth in orbiting the sun is subjected to a centripetal gravitational force.

It is often incorrectly claimed that a body moves in a circle because a *centrifugal* ("center-fleeing") force keeps it from falling toward the attracting center. In order to understand why this mistake is made, let's consider the free-body diagram of Fig. 6-18*c*. Since the string pulls the disk inward, the disk pulls outward on the string. There can be no net force on the string because it is massless, so in order to balance the outward pull of the disk, the hand must exert an inward force on the string at the held end. From Newton's third law, the string must pull outward on the hand. This sensation often leads the person holding the string to surmise incorrectly that a centrifugal force is pulling the disk outward. The mistake this person has made is that he is using the force on his hand to analyze the motion of the disk.

Reprinted with special permission of King Features Syndicate.

DRILL PROBLEM 6-11

The disk of Fig. 6-18*a* has a mass of 0.10 kg and is twirled in a circle of radius 0.25 m. If the disk travels at constant speed and takes 1.0 s to complete one revolution, find (*a*) its speed, (*b*) its radial acceleration, and (*c*) the tension in the string.
ANS. (*a*) 1.6 m/s; (*b*) 9.9 m/s^2; (*c*) 0.99 N.

DRILL PROBLEM 6-12

A car moving at 60 mi/h travels around a circular curve of radius 600 ft on a flat country road. What must be the minimum coefficient of static friction to keep the automobile from slipping? (See Fig. 6-19*a*.)
ANS. 0.40.

EXAMPLE 6-13 A Banked Highway

A curve in a highway is often banked (see Fig. 6-19*b*) so that automobiles can negotiate the curve safely at normal speeds, even on days when the road is slippery. Calculate the banking angle θ necessary for an automobile to maintain a circular path of radius R at a speed v without the aid of friction.

SOLUTION Only the road and gravity exert forces on the automobile. If the road is banked as shown in Fig. 6-19*b*, the normal force

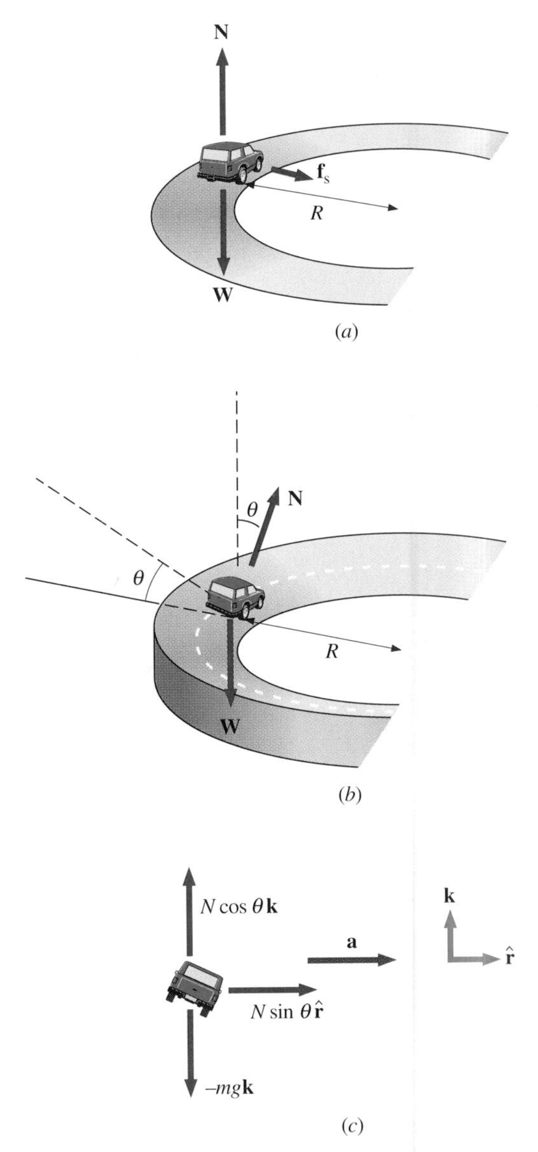

FIGURE 6-19 (*a*) The forces on the automobile as it rounds a circular curve on a level road. (*b*) The forces on the same automobile rounding a circular curve on a frictionless banked road. (*c*) The free-body diagram for the automobile on the banked road.

it exerts on the automobile has a horizontal component that provides the centripetal force necessary for circular motion. At the proper angle of banking, a frictional force is unnecessary for the automobile to maintain a circular path around the curve. This angle θ may be determined by applying Newton's second law in the vertical (z) and radial (r) directions to the free-body diagram of Fig. 6-19c:

$$\sum F_r = ma_r \qquad \sum F_z = ma_z$$
$$N \sin\theta = m\frac{v^2}{R} \qquad N\cos\theta - mg = 0.$$

Dividing the first equation by the second eliminates N, resulting in an expression for θ:

$$\frac{N\sin\theta}{N\cos\theta} = \frac{mv^2/R}{mg},$$

so

$$\tan\theta = \frac{v^2}{Rg}.$$

With knowledge of the average traffic speed and the radius of the curve, a highway designer can determine the proper angle of banking from this formula. Even under slippery conditions, an automobile of any mass can then travel around the curve without risk.

The banked track of a velodrome. The banking allows the cyclists to move around the circular track at high speeds.

EXAMPLE 6-14 A Roller Coaster Ride

A boy of mass 40 kg is in a roller coaster car that travels in a loop of radius 7.0 m. (See Fig. 6-20.) At point A the speed of the car is 10.0 m/s, and at point B the speed is 10.5 m/s. Assume the boy is not holding on and does not wear a seat belt. (a) What is the force of the car seat on the boy at these two points? (b) What minimum speed is required to keep him in his seat at A?

SOLUTION (a) The two forces on the boy are his weight and the normal force **N** of the seat. Free-body diagrams are shown for both A and B. At point A,

$$\sum F_r = ma_r$$
$$N + mg = \frac{mv^2}{R}, \qquad (i)$$

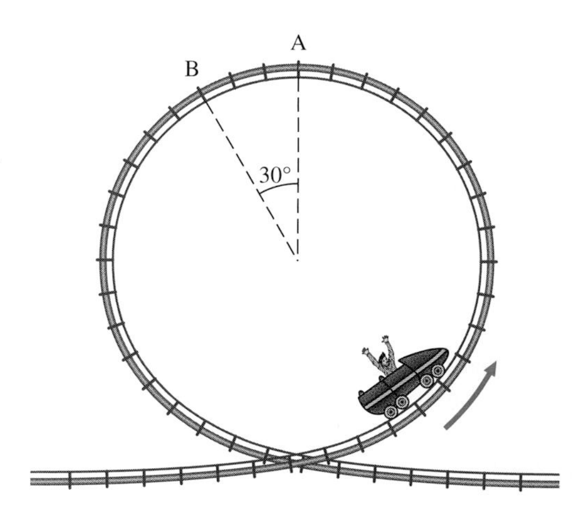

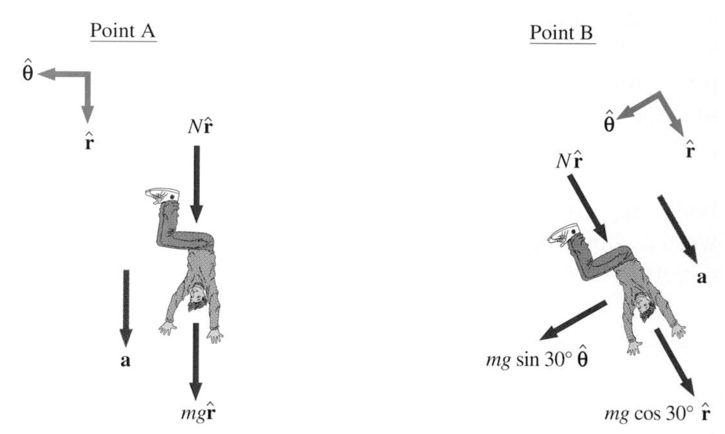

FIGURE 6-20 A roller coaster ride.

which yields

$$N = (40\text{ kg})\left[\frac{(10.0\text{ m/s})^2}{7.0\text{ m}} - 9.8\text{ m/s}^2\right] = 179\text{ N}.$$

At point B, **mg** has both a radial and a tangential component as shown. Newton's second law in the radial direction gives

$$\sum F_r = ma_r$$
$$N + mg\cos 30° = \frac{mv^2}{R}, \qquad (ii)$$

so the force that the seat exerts on the boy is

$$N = (40\text{ kg})\left[\frac{(10.5\text{ m/s})^2}{7.0\text{ m}} - (9.8\text{ m/s}^2)\cos 30°\right] = 290\text{ N}.$$

(b) Since the seat can only push on the boy, N must be greater than zero if he is to stay in his seat. From Eq. (i),

$$N = m\left(\frac{v^2}{R} - g\right) > 0;$$

and

$$v > \sqrt{Rg} = \sqrt{(7.0\text{ m})(9.8\text{ m/s}^2)} = 8.3\text{ m/s}.$$

Thus the roller coaster car has to be traveling faster than 8.3 m/s if the boy is to stay in his seat at A. At this speed would he remain in his seat at all other points on the loop?

*6-3 Velocity-Dependent Frictional Forces

When a body slides across a surface, the frictional force on it is approximately constant and given by $\mu_K N$. Unfortunately, the frictional force on a body moving through a liquid or a gas does not behave so simply. This force (also known as the **resistive force**, or **drag**) is generally a complicated function of the body's velocity. However, for a body moving in a straight line at moderate speeds through a liquid such as water, the frictional force can often be approximated by

$$f_R = -bv, \tag{6-4}$$

where b is a constant whose value depends on the dimensions and shape of the body and the properties of the liquid, and v is the velocity of the body. Two situations for which the frictional force can be represented by Eq. (6-4) are a motorboat moving through water and a small object falling slowly through a liquid.

Let's consider the object falling through a liquid. The free-body diagram of this object with the positive direction downward is shown in Fig. 6-21. Newton's second law in the vertical direction gives the differential equation

$$mg - bv = m\frac{dv}{dt},$$

where the acceleration is written as dv/dt.* As v increases, the frictional force $-bv$ increases until it matches mg. At this point, there is no acceleration and the velocity remains constant at what is known as the **terminal velocity** $\mathbf{v}_T$. From the previous equation,

$$mg - bv_T = 0,$$

so

$$v_T = \frac{mg}{b}.$$

We can find the object's velocity by integrating the differential equation for v. First, we rearrange terms in this equation to obtain

$$\frac{dv}{g - (b/m)v} = dt.$$

Assuming that $v = 0$ at $t = 0$, integration of this equation yields

$$\int_0^v \frac{dv'}{g - (b/m)v'} = \int_0^t dt',$$

or

$$-\frac{m}{b}\ln\left(g - \frac{b}{m}v'\right)\Bigg|_0^v = t'\big|_0^t,$$

where v' and t' are dummy variables of integration. With the limits given, we find

$$-\frac{m}{b}\left[\ln\left(g - \frac{b}{m}v\right) - \ln g\right] = t.$$

Since $\ln A - \ln B = \ln(A/B)$, and $\ln(A/B) = x$ implies $e^x = A/B$, we obtain

$$\frac{g - (b/m)v}{g} = e^{-bt/m},$$

and

$$v = \frac{mg}{b}(1 - e^{-bt/m}).$$

Notice that as $t \to \infty$, $v \to mg/b = v_T$, the terminal velocity.

The position at any time may be found by integrating the equation for v. With $v = dy/dt$,

$$dy = \frac{mg}{b}(1 - e^{-bt/m})\,dt.$$

Assuming $y = 0$ when $t = 0$,

$$\int_0^y dy' = \frac{mg}{b}\int_0^t (1 - e^{-bt'/m})\,dt',$$

which integrates to

$$y = \frac{mg}{b}t + \frac{m^2g}{b^2}(e^{-bt/m} - 1).$$

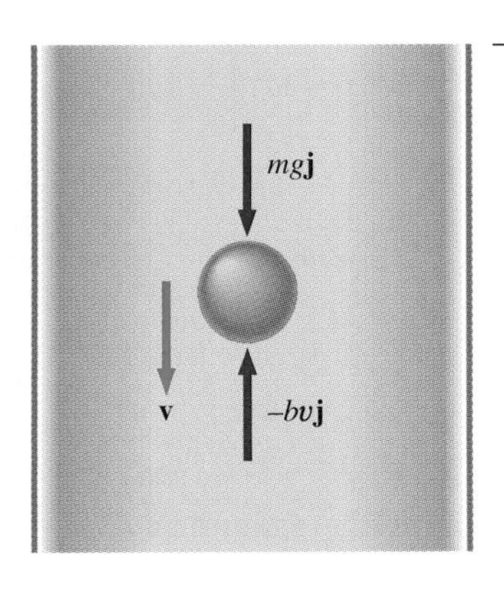

FIGURE 6-21 Free-body diagram of an object falling through a resistive medium.

*In reality, there is also a constant "buoyant force" exerted upward by the liquid on the object. We ignore the buoyant force for now but will discuss it in detail in Chap. 15.

EXAMPLE 6-15 Effect of the Resistive Force on a Motorboat

A motorboat is moving across a lake at a speed v_0 when its motor suddenly freezes up and stops. The boat then slows down under the frictional force $f_R = -bv$. (*a*) What are the velocity and position of the boat as functions of time? (*b*) If the boat slows down from 4.0 to 1.0 m/s in 10 s, how far does it travel before stopping?

SOLUTION (*a*) With the motor stopped, the only horizontal force on the boat is $f_R = -bv$, so from Newton's second law,

$$m\frac{dv}{dt} = -bv,$$

which we can write as

$$\frac{dv}{v} = -\frac{b}{m}\,dt.$$

Integrating this equation between the time zero when the velocity is v_0 and the time t when the velocity is v, we have

$$\int_{v_0}^{v} \frac{dv'}{v'} = -\frac{b}{m}\int_0^t dt',$$

so

$$\ln\frac{v}{v_0} = -\frac{b}{m}\,t,$$

which, since $\ln A = x$ implies $e^x = A$, can be written as

$$v = v_0 e^{-bt/m}. \qquad (i)$$

Now from the definition of velocity,

$$\frac{dx}{dt} = v_0 e^{-bt/m},$$

so

$$dx = v_0 e^{-bt/m}\,dt.$$

With the initial position zero, we have

$$\int_0^x dx' = v_0 \int_0^t e^{-bt'/m}\,dt',$$

and

$$x = -\frac{mv_0}{b}\,e^{-bt'/m}\Big|_0^t = \frac{mv_0}{b}\,(1 - e^{-bt/m}). \qquad (ii)$$

As time increases, $e^{-bt/m} \to 0$, and the position of the boat approaches a limiting value

$$x_{\max} = \frac{mv_0}{b}. \qquad (iii)$$

Although Eq. (*ii*) tells us that the boat takes an infinite amount of time to reach $x_{\max}$, the boat effectively stops after a reasonable time. For example, at $t = 10m/b$, Eq. (*i*) gives

$$v = v_0 e^{-10} \simeq 4.5 \times 10^{-5} v_0,$$

while from Eqs. (*ii*) and (*iii*),

$$x = x_{\max}(1 - e^{-10}) \simeq 0.99995 x_{\max},$$

so the boat's velocity and position have essentially reached their final values.

(*b*) From Eq. (*i*) with $v_0 = 4.0$ m/s and $v = 1.0$ m/s, we have

$$1.0 \text{ m/s} = (4.0 \text{ m/s})e^{-(b/m)(10 \text{ s})},$$

so

$$\ln 0.25 = -\ln 4.0 = -\frac{b}{m}\,(10 \text{ s}),$$

and

$$\frac{b}{m} = \frac{1}{10}\ln 4.0 \text{ s}^{-1} = 0.14 \text{ s}^{-1}.$$

Now from Eq. (*iii*), the boat's limiting position is

$$x_{\max} = \frac{mv_0}{b} = \frac{4.0 \text{ m/s}}{0.14 \text{ s}^{-1}} = 29 \text{ m}.$$

DRILL PROBLEM 6-13

Suppose that the resistive force of the air on a skydiver can be approximated by $f = -bv^2$. If the terminal velocity of a 100-kg skydiver is 60 m/s, what is the value of b?
ANS. 0.27 kg/m.

*6-4 Motion of Planets and Other Satellites

In addition to being an interesting topic, the motion of the planets is a very important application of Newton's laws. The study of this topic played a major role in the early development of mechanics, the law of gravitation, and calculus. The solar system (the sun and its planets) was probably the "experimental laboratory" most essential to the early development of mechanics.

When viewed from the earth, all the celestial bodies appear to orbit our planet, so it's not surprising that the first models of the universe were *geocentric*, or earth-centered. The most widely accepted model of this type was developed by the Greek astronomer Ptolemy around A.D. 140. While using somewhat arbitrary assumptions and complicated geometry based on the circle, the Ptolemaic model (and its subsequent revisions) was able to predict planetary positions with fairly good accuracy.

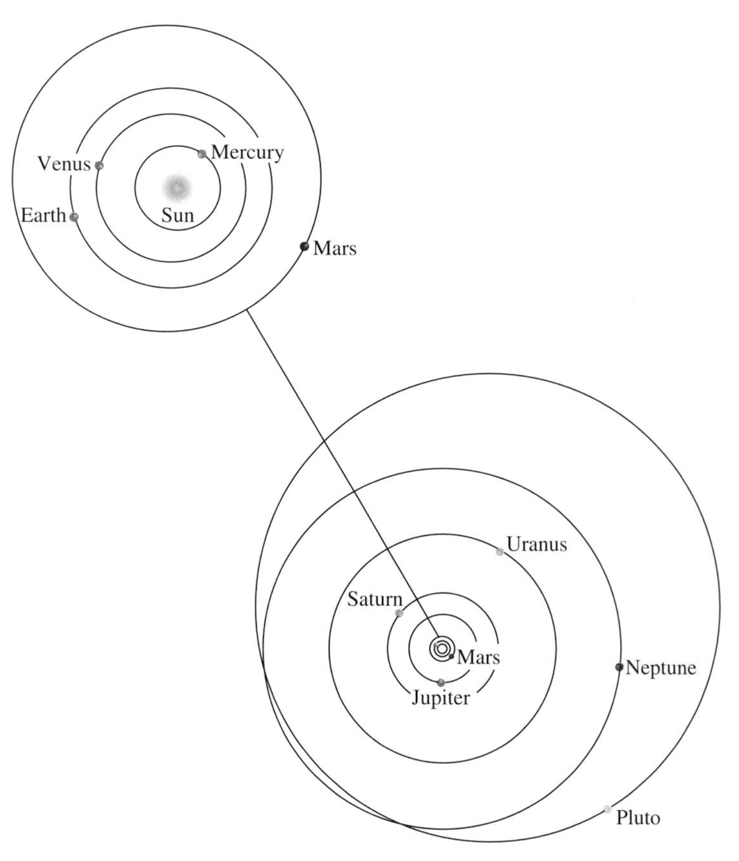

The solar system. Two different scales are necessary to show the orbits of all the planets.

It was not until the Renaissance that the idea of a sun-centered solar system began to gain acceptance. In his book, *De Revolutionibus Orbium Coelestium* (*On the Revolution of the Celestial Orbs*), Nicolaus Copernicus (1473–1543) proposed a *heliocentric* theory in which the sun was placed at the center of the universe and the earth became one of the orbiting planets. Unfortunately, his model did not predict planetary positions with greater accuracy than Ptolemy's, since both schemes were based on the same geometric principles. Nevertheless, Copernicus's work was instrumental in providing credibility for the doctrine of a sun-centered universe.

The next great breakthrough in astronomy was made nearly a century later by a German mathematician and astronomer named Johannes Kepler (1571–1630). In 1600 Kepler began his employment with Tycho Brahe, a skilled observational astronomer who acquired extensive data on the motion of the planets. When Tycho died in 1601, Kepler kept these records and began an intensive study of planetary motion based on Tycho's observations. Kepler found that the orbits of the planets were actually elliptical rather than circular, and that the speed of a planet does not remain constant during its orbit—a planet moves faster when nearer the sun and slower when farther away. These two discoveries were published in 1609 in a book called *Astronomia Nova* (*New Astronomy*). Ten years later, a third discovery, which was a mathematical relationship between the period of a planet and its distance from the sun, was included in another book, *Harmonice Mundi* (*Harmony of the Worlds*). These three discoveries are collectively known today as *Kepler's laws*:

1. Each planet moves around the sun in an elliptical orbit with the sun at a focus of the ellipse (Fig. 6-22*a*).
2. The straight line joining a planet and the sun sweeps out equal areas in equal time intervals (Fig. 6-22*b*).
3. The squares of the periods of the planets are in direct proportion to the cubes of the semimajor axes of their elliptical orbits.

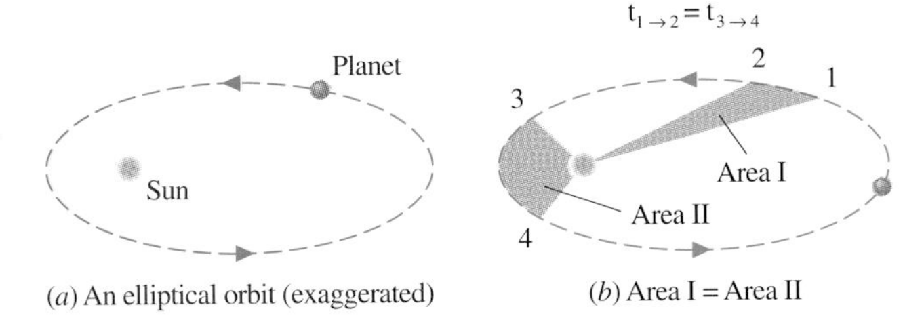

FIGURE 6-22 (*a*) The planets move around the sun in elliptical orbits with the sun at one focus of the ellipse. (*b*) If a planet travels between points 3 and 4 in the same time that it takes to move between points 1 and 2, the two shaded areas are equal.

By considering a body moving in a circular orbit around a gravitationally attracting center, we can derive Kepler's third law with a direct application of Newton's second law. In spite of its limitations, this simple circular model is instructive and does help us understand orbital motion. This is especially true for planets because most of them have orbits that are almost circular.

Astronauts orbiting the earth are subject to the earth's gravitational force and are "falling toward" the earth along with their spacecraft. They are weightless in the sense that they experience no gravitational acceleration relative to their surroundings.

Figure 6-23 depicts a body of mass m traveling in a circular orbit of radius R at a speed v. It is gravitationally attracted to the central body of mass M with a force of magnitude GmM/R^2. There is no friction, and the gravitational force is always perpendicular to the path of the orbiting body. As there is no force component along the path, there is no acceleration in that direction ($a_\theta = dv/dt = 0$), so the speed of the orbiting body is constant. Applying Newton's second law in the radial direction, we obtain

$$\sum F_r = ma_r$$
$$-\frac{GmM}{R^2} = m\left(-\frac{v^2}{R}\right),$$

and

$$v^2 = \frac{GM}{R}.$$

FIGURE 6-23 (*a*) A body of mass m moves in a circle around a gravitationally attracting center of mass M. (*b*) The free-body diagram of the orbiting body.

The period T of the orbiting body can be calculated by dividing the distance it travels in one orbit (the circumference of the circle) by its constant speed; that is,

$$T = \frac{2\pi R}{v} = \frac{2\pi R}{\sqrt{GM/R}},$$

which can be rewritten as

$$\frac{T^2}{R^3} = \frac{4\pi^2}{GM}. \qquad (6\text{-}5)$$

The right-hand side of this equation depends only on the mass of the attracting center, so all bodies orbiting the same center have the same value of T^2/R^3. For example, this quantity is the same for a satellite orbiting 200 km above the earth's surface as it is for the moon. The value of T^2/R^3 is also constant for all the planets in the solar system, because they all orbit the same attracting center. Equation (6-5) is therefore a statement of Kepler's third law.

EXAMPLE 6-16 PERIOD OF A SATELLITE

The moon's orbit around the earth is almost circular. The period and radius of the orbit are 27.3 days and 3.86×10^8 m, respectively. What is the period of a satellite traveling in a circular orbit 330 km above the earth's surface?

SOLUTION Since the radius of the earth is 6.37×10^6 m, the radius of the satellite's orbit is $(6.37 + 0.33) \times 10^6 \text{ m} = 6.70 \times 10^6$ m. Both the satellite and the moon are in orbit around the earth, so we have from Kepler's third law,

$$\frac{T^2}{(6.70 \times 10^6 \text{ m})^3} = \frac{(27.3 \text{ days})^2}{(3.86 \times 10^8 \text{ m})^3},$$

which gives for the period T of the satellite

$$T = 6.24 \times 10^{-2} \text{ days} = 89.9 \text{ min}.$$

DRILL PROBLEM 6-14

A communications satellite is placed in circular orbit above the equator so that its period is exactly 1 day. With this period, the satellite is always at the same spot above the earth and can therefore be used to send and receive radio signals constantly from fixed earth stations. At what distance from the earth's center does such a satellite orbit?
ANS. 4.26×10^7 m.

*6-5 EFFECT OF THE EARTH'S ROTATION ON g

Because the earth rotates on its axis, a coordinate system attached to its surface is accelerating relative to the fixed stars. It is therefore not an inertial frame. Up to now, we have assumed that the calculational error introduced by ignoring this fact is small. We'll see just how small this error is by comparing the readings on a spring scale used to weigh a body at the North Pole and at the equator. (See Fig. 6-24.) We use a nonrotating reference frame whose origin is fixed at the earth's center.* Suppose that the body has mass m and that the readings of the spring scale are S_N at the North Pole and S_E at the equator. At the North Pole the body is at rest relative to our nonrotating frame. The radial acceleration a_r of the body is therefore zero. From Newton's second law we find

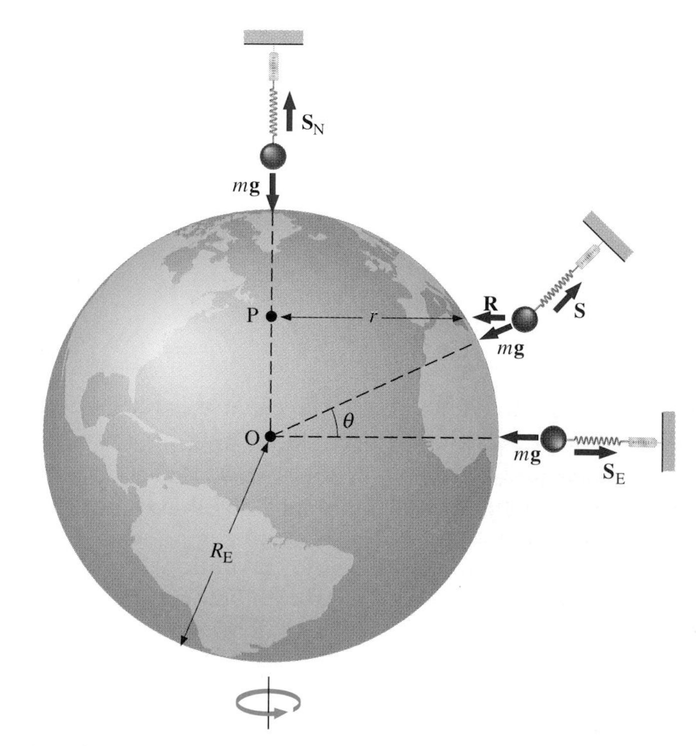

FIGURE 6-24 The resultant **R** of **S** and $m\mathbf{g}$ points in the direction of the acceleration, that is, toward the center of the circle.

$$\sum F_r = ma_r$$
$$mg - S_N = m(0),$$

so

$$S_N = mg,$$

which is the weight of the body. At the equator, the body is traveling in a circle of radius R_E and has an acceleration v_E^2/R_E toward the center of the earth, where v_E is the speed of a point on the earth's surface at the equator and R_E is the earth's radius. From Newton's second law,

$$\sum F_r = ma_r$$
$$mg - S_E = m\frac{v_E^2}{R_E},$$

which yields

$$S_E = m\left(g - \frac{v_E^2}{R_E}\right).$$

*Although this reference frame approximates an inertial frame better than one attached to the earth's surface, it is still not a perfect inertial frame because of the earth's motion around the sun.

The scale therefore reads less at the equator than at the pole by mv_E^2/R_E. This difference can be interpreted as a reduction in the effective value of g by v_E^2/R_E. Using $T = 8.64 \times 10^4$ s as the rotational period of the earth, we find

$$v_E = \frac{2\pi(6.37 \times 10^6 \text{ m})}{8.64 \times 10^4 \text{ s}} = 463 \text{ m/s},$$

and

$$\frac{v_E^2}{R_E} = \frac{(463 \text{ m/s})^2}{6.37 \times 10^6 \text{ m}} = 0.034 \text{ m/s}^2.$$

Hence the effective value of g at the equator is $(0.034/9.8) \times 100\% = 0.35\%$ less than at the North Pole because of the earth's rotation. This difference is so small that it can be ignored in most engineering applications; yet it is large enough to be easily measured.

This result does *not* imply that the gravitational force or the acceleration due to that force is reduced because the earth rotates. Rather, the acceleration in free fall is *measured* to be less at the equator than at either pole, because the measurement at the equator is made relative to a noninertial frame. In principle, motion should not be analyzed in such a frame. However, the error caused by using the earth frame is very small.

At an arbitrary latitude, the tension **S** and the weight $m\mathbf{g}$ are directed as shown in Fig. 6-24. The resultant **R** of the two forces must be directed at P to give the body an acceleration v^2/r toward that point. Consequently, **S** and the effective value of **g** decrease as we move from the pole to the equator. The magnitude of **R** has been grossly exaggerated in the figure. It is really very small because v^2/r is small, and the tension **S** and the gravitational force $m\mathbf{g}$ are always close to antiparallel.

Table 6-2 gives the values of g at various locations on the earth. Most of the variation in g is due to the rotation of the earth, but part is due to local changes in the density and shape of the earth.* Notice that the difference in the tabulated values of g at the equator and at the North Pole is 0.052 m/s^2, which is reasonably close to our simple calculation of 0.034 m/s^2.

TABLE 6-2 VALUE OF g AT VARIOUS LATITUDES (SEA LEVEL)

Latitude	g (m/s^2)	g (ft/s^2)
0°	9.7804	32.088
15°	9.7838	32.099
30°	9.7933	32.130
45°	9.8062	32.173
60°	9.8192	32.215
75°	9.8287	32.246
90°	9.8322	32.258

Source: Handbook of Chemistry and Physics, 67th Edition, 1986–87, CRC Press, Inc. (Boca Raton, FL).

*6-6 DEFINITIONS OF FORCE AND MASS

While we have covered Newton's laws thoroughly, we have not as yet presented a precise definition of force and mass. These two quantities are defined in terms of the following experimental procedure:

1. Select a standard body and assign it a mass of exactly 1 kg. This standard is located at the International Bureau of Weights and Measures at Sèvres, France.
2. Apply a single force to this standard.
3. Measure the acceleration of the standard.
4. By definition, the force is a vector in the direction of the acceleration, with a magnitude numerically equal to the acceleration's magnitude. For example, if $\mathbf{a} = 2.0\mathbf{i}$ m/s^2, $\mathbf{F} = (1.0 \text{ kg})(2.0\mathbf{i} \text{ m/s}^2) = 2.0\mathbf{i}$ N.
5. Mass can now be defined as follows: When a known force **F** gives a body an acceleration **a**, the ratio F/a is the mass of the body.

By definition, Newton's second law is satisfied for a single force. Why then is it a fundamental law rather than simply a defining equation? The answer lies in the fact that there is a significant difference between a *single* force and the *net* force acting on a body. While a single force is defined on the basis of the acceleration it imparts to the standard mass, nothing is stated about the behavior of a body subjected to multiple forces. The second law makes such a statement—it says that the *force sum* is equal to the mass times the acceleration. Newton's second law is

$$\sum \mathbf{F} = m\mathbf{a},$$

while **F** and m are defined by

$$\mathbf{F} = m\mathbf{a}$$

and the standard mass. Only the first of these two statements contains the all-important summation sign. That statement is Newton's second law.

*The earth is not a perfect sphere, as its equatorial radius is about 0.34 percent larger than its polar radius.

SUMMARY

1. **Common forces**
   (*a*) The force on a body of mass m in a gravitational field $\mathbf{g}(\mathbf{r})$ is $m\mathbf{g}(\mathbf{r})$.
   (*b*) The contact force between two surfaces is usually resolved into a component N perpendicular to the surface and a component f parallel to the surface; N is called the normal force and f the frictional force. For many materials, these two components are related by

$$f_S \leq \mu_S N \qquad \text{(force of static friction)}$$

and

$$f_K = \mu_K N \qquad \text{(force of kinetic friction).}$$

(*c*) A solid rod is under tension if the normal force at a cross section acts to stretch the rod. The rod is under compression if the force acts to compress it.
(*d*) A flexible cord supports only tension, and if its mass is negligible, the tension is the same at all cross sections of the cord.
(*e*) The spring force can be approximated by $F(x) = -kx$, where x is a small displacement of the end of the spring from its position when the spring is relaxed and k is the spring constant.

2. **Applications of Newton's laws**
A useful procedure to follow in applying Newton's laws:
(*a*) Clearly identify the body whose motion is to be analyzed.
(*b*) Pick an inertial reference frame.
(*c*) Determine what forces are acting on the body.
(*d*) Draw a free-body diagram.
(*e*) Apply Newton's second law to determine the acceleration of the body along each axis.

*3. **Velocity-dependent frictional forces**
For a body moving in a straight line at a moderate speed v through a liquid, the frictional force can often be approximated by

$$f_R = -bv,$$

where b is a constant whose value depends on the dimensions and shape of the body and the properties of the liquid.

*4. **Motion of planets and other satellites**
Orbiting planets and satellites obey Kepler's laws:
(*a*) Each planet moves around the sun in an elliptical orbit with the sun at a focus of the ellipse.
(*b*) The straight line joining a planet and the sun sweeps out equal areas in equal time intervals.
(*c*) The squares of the periods of the planets are in direct proportion to the cubes of the semimajor axes of their elliptical orbits.

*5. **Effect of the earth's rotation on g**
The effective value of **g** varies over the earth's surface, partly because a reference system at rest on the earth is not a true inertial system.

*6. **Definitions of force and mass**
(*a*) Select a standard body and assign it a mass of exactly 1.0 kg.
(*b*) Apply a single force to this standard.
(*c*) Measure the acceleration of the standard.
(*d*) By definition, the force in newtons is a vector in the direction of the acceleration **a** with a magnitude given by (1.0 kg)a, where a is in meters per second squared.
(*e*) When a known force **F** gives a body an acceleration **a**, the ratio F/a is the mass of the body.

QUESTIONS

6-1. The sum of the forces on an object is zero. Can you infer that the object is at rest?

6-2. Does a body always move in the direction of the net force on it? Give examples that verify your answer.

6-3. Our bodies "lunge forward" when we slam on the brakes in a car. Explain why.

6-4. Is it theoretically possible to stretch a rope so that it is exactly horizontal?

6-5. Why is it important to relax your legs when you land after jumping?

6-6. Can a body have a downward acceleration that is greater than **g** at the earth's surface?

6-7. How could a 96-lb crate be lowered from a building using a cord whose breaking strength is 80 lb? Would this be a practical method in general?

6-8. Could the crate of Ques. 6-7 be lifted from the ground with the cord described?

6-9. It has been said that Newtonian mechanics describes motion in an electromagnetic and/or a gravitational field. Explain.

6-10. Good wide receivers, shortstops, and basketball centers are said to have "soft" hands for catching a ball. How would you interpret this description based on what you have learned in this chapter?

6-11. What is the tension in the spring balance shown in the accompanying figure? Assume that all weights except those of the hanging blocks can be ignored.

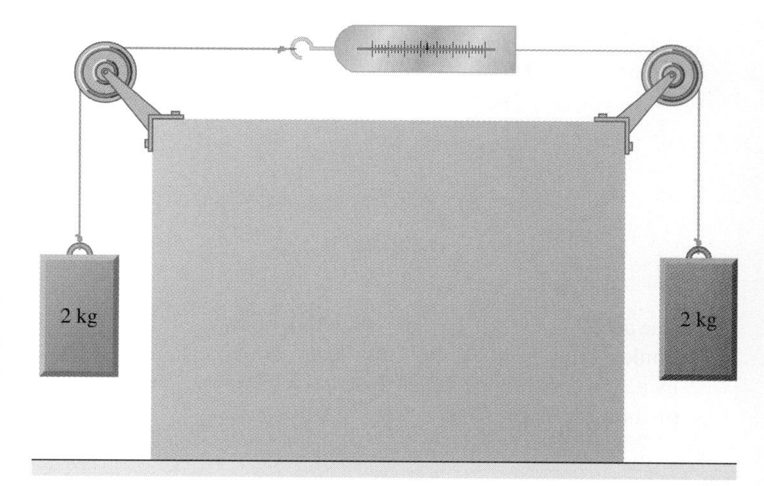

QUESTION 6-11

6-12. In a common type of home gym, weight training is done by raising a stack of heavy metal plates. This same exercise is performed in another type of machine by stretching strong, but light, elastic bands. Discuss the differences between the two ways in which this exercise is done.

6-13. Why are tire chains helpful for travel on icy roads?

6-14. Is it easier for us to move around on a rough surface or on a smooth surface? Explain.

6-15. Is it possible for the coefficients of static and kinetic friction to exceed unity?

6-16. Make a plot of how the frictional force on an object varies with increasing applied force. Include both static and kinetic friction.

6-17. People are able to jerk tablecloths out from under dishes while hardly moving the dishes. Explain why.

6-18. A person standing at the edge of a frozen pond tosses a penny so that it lands near the middle of the pond. If the frozen pond is frictionless, is it possible for the coin not to reach an edge of the pond?

6-19. Suppose you are inside a completely closed car so that you have no communication with the outside. Can you determine whether the car is moving at constant velocity? speeding up? slowing down? rounding a curve?

6-20. An automobile takes the path shown in the accompanying figure. From A to B it moves at constant speed; from B to C it slows down; from C to D it speeds up; from D to E it moves at a constant speed; and from E to F it slows down. What is the direction of the net force on the automobile in each section?

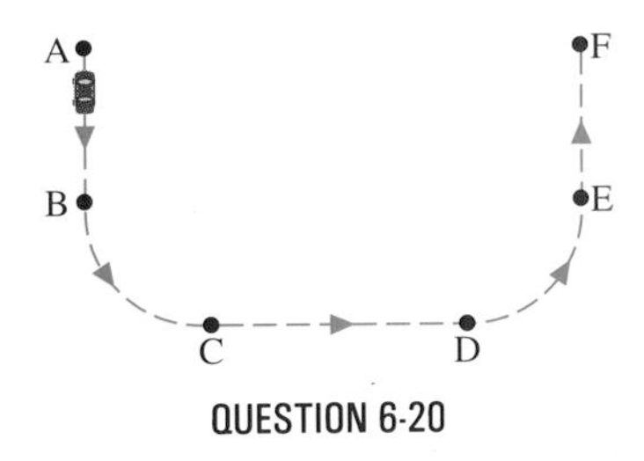

QUESTION 6-20

6-21. Why does your stomach feel strange when you are in an accelerating elevator or in a roller coaster car moving down a steep section of the track?

6-22. An astronaut in a space capsule orbiting the earth at 200 mi above its surface is said to be weightless. What is meant by weightlessness in this case? An astronaut is also said to be weightless when he or she is far from any celestial body. What is meant by weightlessness in this case?

6-23. How does a centrifuge work?

6-24. A bucket of water is twirled in a vertical circle fast enough that the water does not fall out at the top of the circle. It is claimed that the water stays in the bucket because the centrifugal force pulls up on the water to balance the downward force of the gravitational field. Is this explanation correct? How do you explain the fact that the water stays in the bucket?

6-25. Suppose that a precious-metals dealer uses a spring scale to weigh the metals she buys and sells. Would you rather buy a pound of gold from her in Cairo, Egypt, or in Anchorage, Alaska?

PROBLEMS

Common Forces

6-1. The coefficient of static friction for the block and horizontal surface shown in the accompanying figure is $\mu_S = 0.70$. What minimum force **P** applied at 30° below the horizontal is necessary to start the block moving? If $\mu_K = 0.50$, what magnitude must **P** have to keep the block moving at a constant velocity?

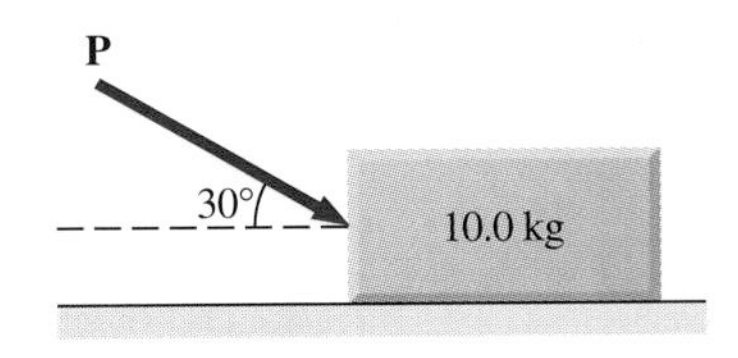

PROBLEM 6-1

6-2. The accompanying figure shows a 30-kg block resting on a frictionless ramp inclined at 60° to the horizontal. The block is held by a spring that is stretched 5.0 cm. What is the force constant of the spring?

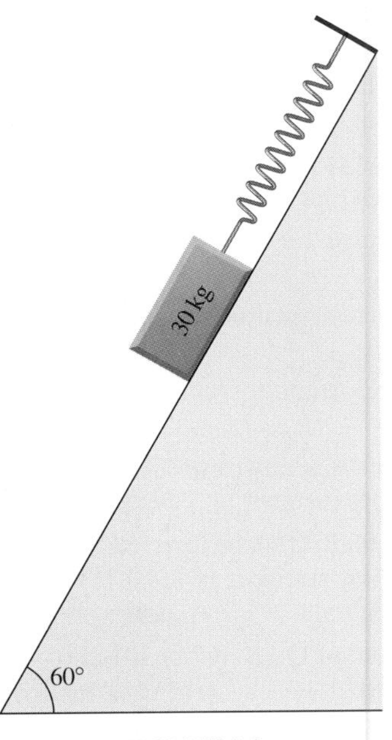

PROBLEM 6-2

6-3. The 50-kg column shown in the accompanying figure is 5.0 m long. The cross-section and density of the column are constant. If a 100-N force is applied at its end, what are the compressive forces on the column at AA′ and at BB′?

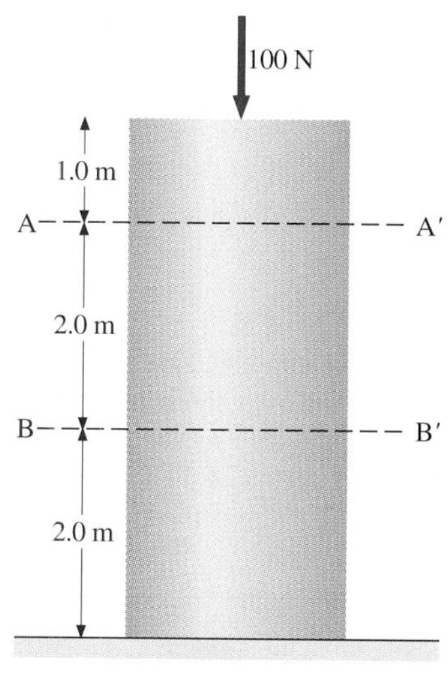

PROBLEM 6-3

6-4. A body of mass m is attached to two springs as shown in the accompanying figure. Show that the net force of the two springs is equal to that of a single spring with force constant $k = k_1 + k_2$.

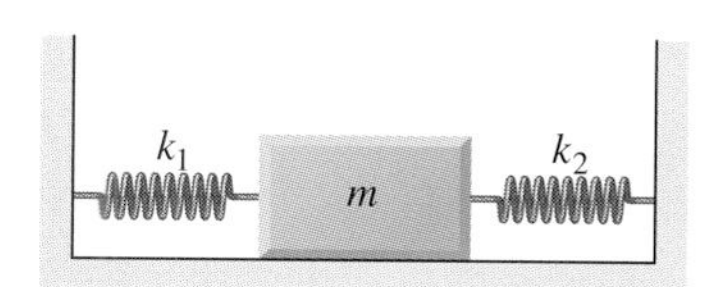

PROBLEM 6-4

Applications of Newton's Laws to Linear Motion

6-5. A crate of mass 3.0 kg is pushed across a frictionless horizontal surface by a horizontal force of magnitude 6.0 N. What is its acceleration?

6-6. A body of mass 2.0 kg is pushed straight upward by a 25-N vertical force. What is its acceleration?

6-7. A horizontal force of 10 N is applied to a 1.0-kg block sitting on a flat surface. The coefficient of static friction between the block and the surface is $\mu_S = 0.80$, and the coefficient of kinetic friction is $\mu_K = 0.70$. Plot the velocity of the block as a function of time over the first 5 s.

6-8. A constant horizontal force of 30 N acts on a cart that sits on a frictionless horizontal plane. The cart starts from rest and is observed to move 100 m in 5.0 s. (*a*) What is the mass of the cart? (*b*) If the force ceases to act at the end of 5.0 s, how far will the cart move in the next 10.0 s?

6-9. A hockey puck leaves a player's stick with a velocity of 20 m/s and slides 200 m across an ice-covered lake before coming to rest. What is the coefficient of kinetic friction between the puck and the ice?

6-10. An electron ($m = 9.1 \times 10^{-31}$ kg) leaves the cathode of a radio tube with zero initial velocity and travels in a straight line for 6.0×10^{-9} s to the anode, which is 2.0 cm away. Ignoring the force of gravity and assuming that the acceleration is constant, calculate (*a*) the acceleration of the electron; (*b*) the velocity of the electron when it reaches the anode; and (*c*) the force on the electron. (*d*) Justify that the force of gravity can be ignored.

6-11. An 80-kg passenger in an automobile traveling at 100 km/h is wearing a seat belt. The driver slams on the brakes and the automobile stops in 45 m. Estimate the force of the seat belt on the passenger.

6-12. A groundskeeper smooths the dirt in the infield of a baseball diamond by pulling a 100-kg metal plate across the field with a small tractor. The coefficient of kinetic friction between the field and the plate is $\mu_K = 0.70$. (*a*) If the force of the tractor on the plate is 600 N and directed 37° above the horizontal as shown in the accompanying figure, what is the acceleration of the plate? (*b*) What is the value of the force if the plate moves at constant velocity?

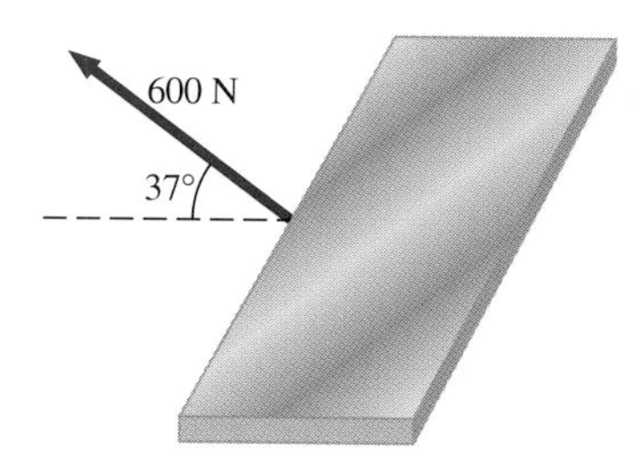

PROBLEM 6-12

6-13. A roller coaster car starts from rest at the top of a track 30 m long and inclined at 20° to the horizontal. Assume that friction can be ignored. (*a*) What is the acceleration of the car? (*b*) How much time elapses before it reaches the bottom of the track?

6-14. The accompanying figure shows a man pulling a 50-kg crate horizontally with a light rope across a surface for which $\mu_K = 0.50$. (*a*) What force must he exert on the rope in order to pull the crate at constant velocity? (*b*) What is the force of the rope on the man? (*c*) What is the force of the rope on the crate? (*d*) What is the force of the crate on the rope? The crate suddenly slides onto a surface for which $\mu_K = 0.30$. The man continues to pull with the same force. (*e*) What is the acceleration of the crate? (*f*) What is the reaction force to the man's force on the rope? (*g*) What is the force of the rope on the crate? (*h*) What is the reaction force to this force?

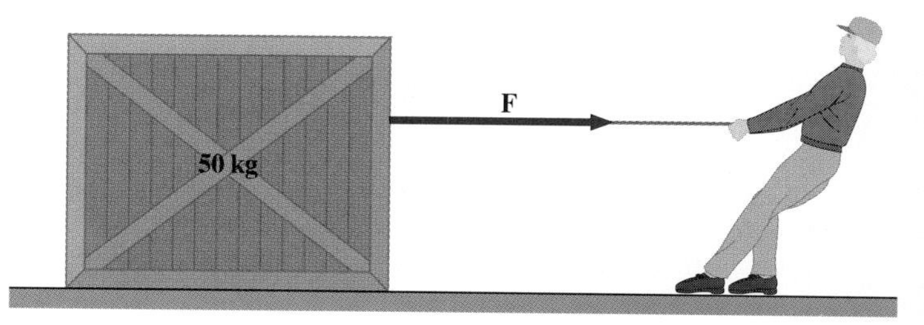

PROBLEM 6-14

6-15. The accompanying figure shows a block resting on an incline whose angle is gradually increased. The block begins to move when the angle reaches θ. What is the coefficient of static friction between the surfaces of the block and the incline?

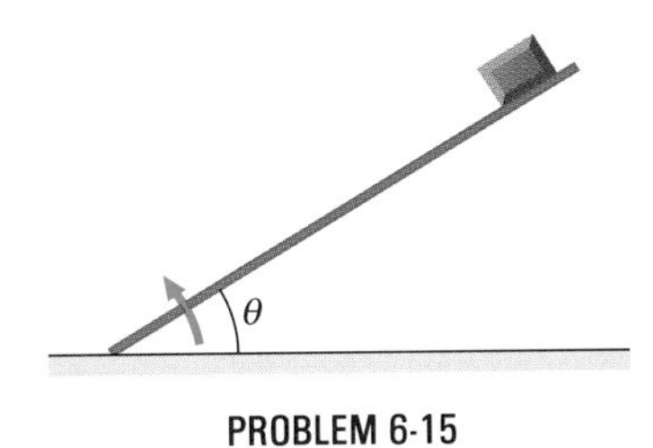

PROBLEM 6-15

6-16. A painter lifts himself and the platform up the side of a building as shown in the accompanying figure. The mass of the painter and his equipment is 80 kg, and the mass of the platform is 25 kg. In order to go up at constant velocity, what force does the painter have to exert on the rope? Compare this force with the total weight of the painter, his equipment, and the platform.

PROBLEM 6-16

6-17. Determine the acceleration of each of the identical weights of mass M seen in the accompanying figure. The masses of the pulleys and strings, along with friction in the pulleys, are negligible.

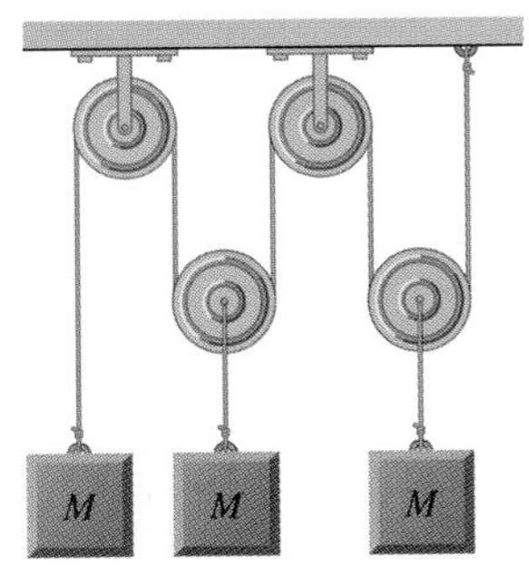

PROBLEM 6-17

6-18. A rope is looped over a frictionless pulley, and one end is attached to a bunch of bananas. The other end of the rope is held by a monkey who climbs it in order to reach the bananas, as shown in the accompanying figure. If the monkey and the bananas have the same weight and the weight of the rope is negligible, does the monkey reach the bananas?

PROBLEM 6-18

6-19. A body is hung from a spring scale attached to the ceiling of an elevator. When the elevator is accelerating downward at 3.8 m/s^2, the scale reads 60 N. (*a*) What is the mass of the body? (*b*) What does the scale read if the elevator moves upward while slowing down at a rate 3.8 m/s^2? (*c*) What does the scale read if the elevator moves upward at constant velocity? (*d*) If the cable supporting the elevator breaks loose so the elevator falls freely, what does the spring scale read?

6-20. A 0.50-kg weight hangs at the end of a massless spring attached to the ceiling of a stationary train. The train accelerates forward at 4.0 m/s^2, and the spring is then inclined at a constant angle ϕ to the vertical, as shown in the accompanying figure. The force constant of the spring is 50 N/m. (*a*) How far is the spring stretched when the train is stationary? (*b*) How far is it stretched when the train is accelerating? (*c*) Determine the angle ϕ.

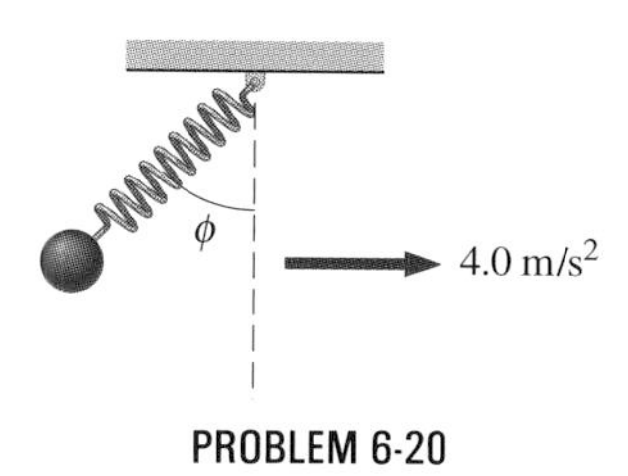

PROBLEM 6-20

6-21. A 20-g ball hangs from the roof of a freight car by a string. When the freight car begins to move, the string makes an angle of 35° with the vertical. (*a*) What is the acceleration of the freight car? (*b*) What is the tension in the string?

6-22. The accompanying figure depicts two spheres connected by a rod of mass 20.0 kg and length 50.0 cm. When a vertical force of magnitude 400 N is applied as shown, the system accelerates upward. What is the tension in the rod at (*a*) the top of the rod; (*b*) the bottom of the rod; (*c*) the middle of the rod; (*d*) a point 20.0 cm from the top of the rod? Assume that the rod has constant cross-section and density.

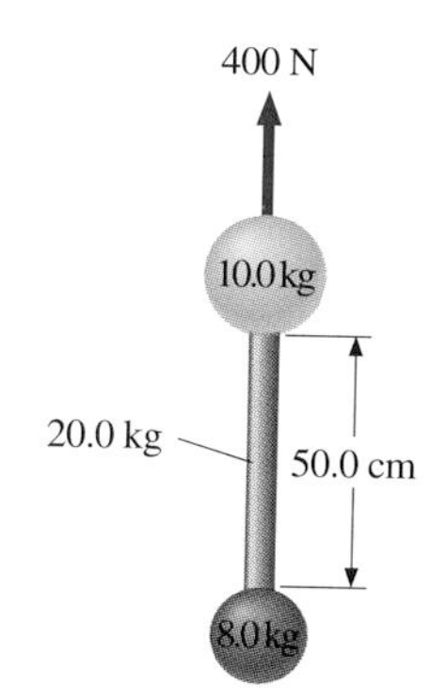

PROBLEM 6-22

6-23. A small truck pushes a car down the street by applying a 1000-N force to the left on the ground. (See the accompanying figure.) The masses of the truck and car are 1400 kg and 800 kg, respectively. Calculate the acceleration of the system and the force of contact between the truck and the car.

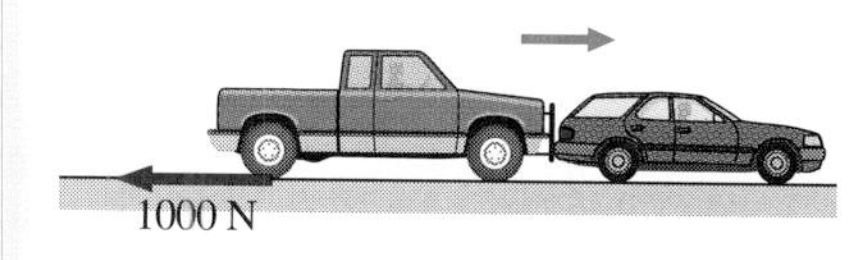

PROBLEM 6-23

6-24. A man stands in an elevator holding a 20-lb package at the end of a 25-lb-test fishing line. When the elevator starts upward, the line breaks. What is the minimum acceleration of the elevator?

6-25. A large transport truck consists of a cab and two trailers. The mass of the cab is 5000 kg and the mass of each trailer is 2000 kg. The truck leaves a traffic light with an acceleration of 0.50 m/s^2. (*a*) What is the tension in the coupling connecting the cab and trailer and the coupling connecting the two trailers? (*b*) What horizontal force must the road exert on the cab's wheels? Assume that retarding friction on the truck is negligible.

6-26. A 25-kg block is pulled by a horizontal 100-N force and slides across the floor of the elevator as shown in the accompanying figure. The coefficient of kinetic friction between the block and floor is $\mu_K = 0.40$. If the elevator is accelerating downward at 2.0 m/s^2, what is the resultant acceleration of the block?

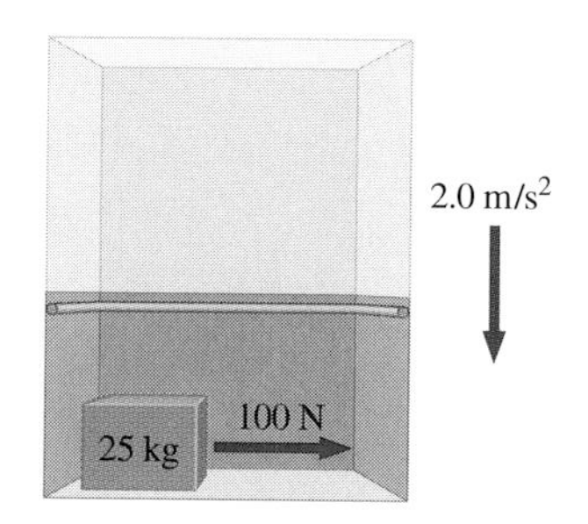

PROBLEM 6-26

6-27. The coefficients of friction between the two blocks and the floor are $\mu_S = 0.80$ and $\mu_K = 0.50$. Block A has a mass of 30 kg and block B has a mass of 45 kg. What force **P** directed as shown in the accompanying figure just starts the blocks moving? For this force, what are the acceleration of the blocks and the tension in the connecting string?

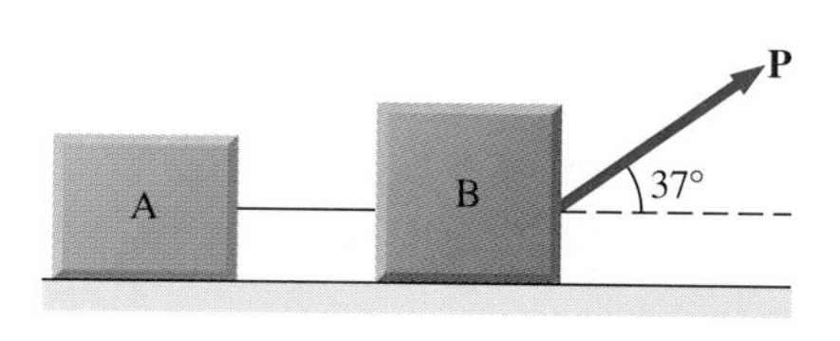

PROBLEM 6-27

6-28. The mass of each of the three freight cars shown in the accompanying figure is M, and the frictional force on each car is **f**. A locomotive exerts a horizontal force **P** on the front car. Find the acceleration of the system and the forces exerted at the coupling mechanisms between the cars.

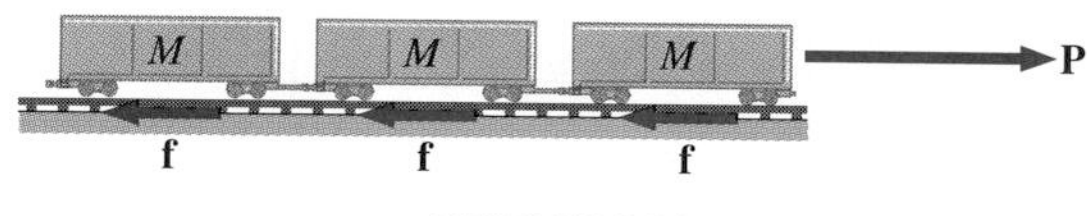

PROBLEM 6-28

6-29. What minimum acceleration must the cart shown in the accompanying figure have so that the block does not fall? The coefficient of static friction between the block and cart is 0.70.

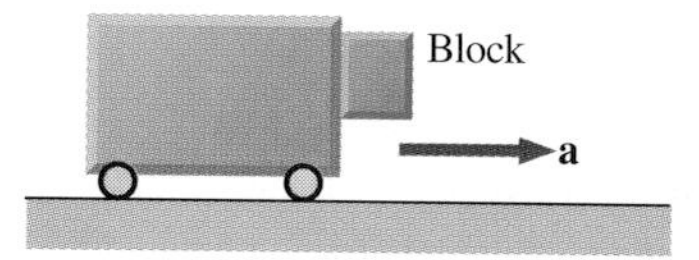

PROBLEM 6-29

6-30. For the two blocks shown in the accompanying figure, $m_1 =$ 2.0 kg and $m_2 = 4.0$ kg. They are connected by a string and slide down a 37° inclined plane. For the 2.0-kg block, $\mu_K = 0.30$ and for the 4.0-kg block, $\mu_K = 0.50$. Calculate the acceleration of the blocks and the tension in the connecting string.

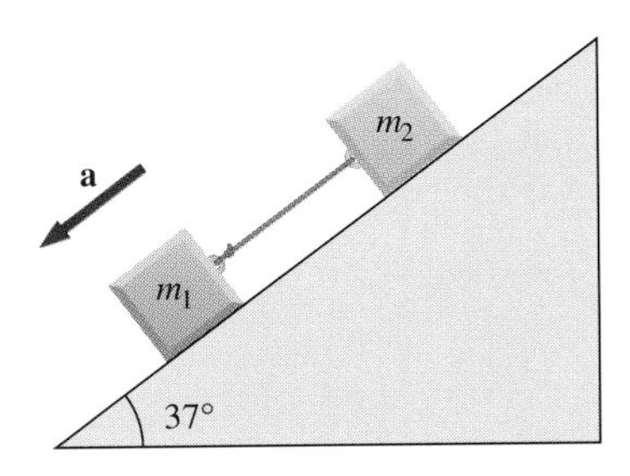

PROBLEM 6-30

6-31. The accompanying figure shows two carts connected by a cord that passes over a small frictionless pulley. Each cart rolls freely with negligible friction. Calculate the acceleration of the carts and the tension in the cord.

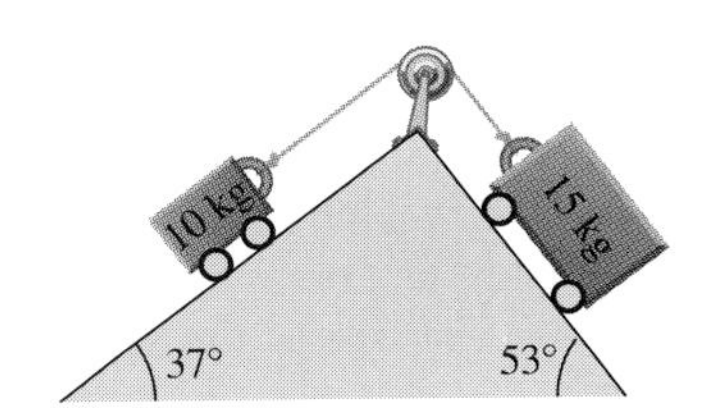

PROBLEM 6-31

6-32. In the accompanying figure, the mass of block 1 is $m_1 = 4.0$ kg, and the coefficients of friction between it and the inclined surface are $\mu_S = 0.60$ and $\mu_K = 0.40$. (*a*) What are the minimum and maximum values of m_2 that allow block 1 to remain at rest? (*b*) What are the values of m_2 that allow block 1 to slide up and down the plane at constant velocity?

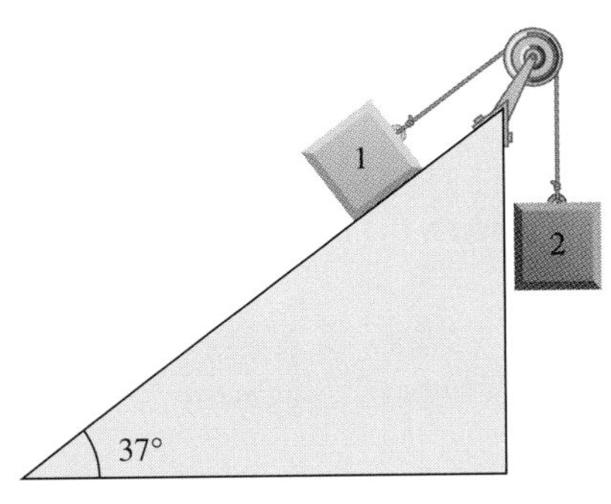

PROBLEM 6-32

6-33. The lower block in the accompanying figure is pulled to the left at constant velocity by the force **P**. The coefficient of kinetic friction between all surfaces is 0.40. What are **P** and the tension in the string?

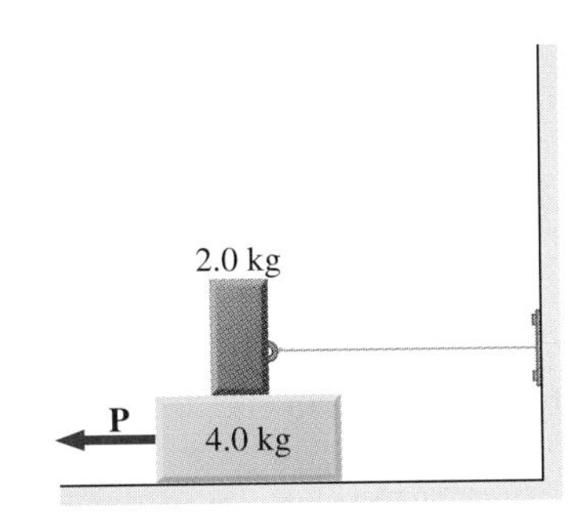

PROBLEM 6-33, 6-51, and 6-58

6-34. What is the maximum force that can be applied to the 6.0-kg cart in the accompanying figure without causing the 4.0-kg block to slip? The coefficient of static friction between the block and the cart is 0.50, and the cart rolls across the floor with negligible friction.

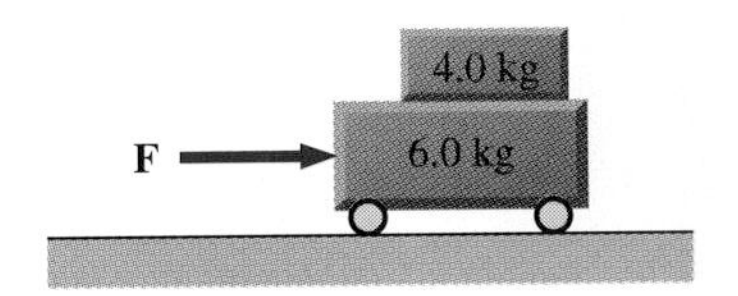

PROBLEM 6-34

6-35. A refrigerator rests on the bed of a truck moving up a 10° grade. The coefficient of static friction between the refrigerator and the bed of the truck is 0.75. (*a*) Will the refrigerator slip when the truck moves at constant velocity? (*b*) If not, at what minimum acceleration will the refrigerator slip?

6-36. (*a*) Find the ratio of the stopping distances of a truck moving with a speed v when it is on a level road to when it is on a 10° downhill slope. In both cases, $\mu_K = 0.90$. (*b*) What is this ratio if the roads are slippery and $\mu_K = 0.50$?

Applications of Newton's Laws to Circular Motion

6-37. When a body of mass 0.25 kg is attached to a vertical massless spring, it is extended 5.0 cm from its unstretched length of 4.0 cm. The body and spring are placed on a horizontal frictionless surface and rotated about the held end of the spring at 2.0 rev/s. How far is the spring stretched?

6-38. A record of diameter 30 cm rotates at $33\frac{1}{3}$ rev/min. What is the minimum coefficient of static friction needed to hold a penny at the outer edge of the record?

6-39. Railroad tracks follow a circular curve of radius 500 m and are banked at an angle of 5.0°. For trains of what speed are these tracks designed?

6-40. The CERN particle accelerator is circular with a circumference of 7.0 km. (*a*) What is the acceleration of the protons ($m = 1.67 \times 10^{-27}$ kg) that move around the accelerator at nearly the speed of light? (Use $v = 3.00 \times 10^8$ m/s.) (*b*) What is the force on the protons?

6-41. A plumb bob hangs from the roof of a railroad car. The car rounds a circular track of radius 300 m at a speed of 90 km/h. At what angle relative to the vertical does the plumb bob hang?

6-42. A small ball at the end of a string is twirled in a vertical circle of radius 1.0 m. (*a*) What is the minimum speed for which the string remains taut when the ball is at its highest point? (*b*) For this same speed, what is the tension in the string in terms of the ball's weight when the ball is at its lowest point?

6-43. A car rounds an unbanked curve of radius 65 m. If the coefficient of static friction between the road and car is 0.70, what is the maximum speed at which the car can traverse the curve without slipping?

6-44. A banked highway is designed for traffic moving at 100 km/h. The radius of the curve is 350 m. What is the angle of banking of the highway?

6-45. When a 160-lb trucker drives over the crest of a hill at 60 mi/h, his force on the seat is only 140 lb. What do you conclude about the hill?

6-46. A donut-shaped space station with an outer radius of 200 m revolves once every 50 s around its central axis. (See the accompanying figure.) What is the apparent weight of a 75-kg astronaut who is standing on the platform's outer rim?

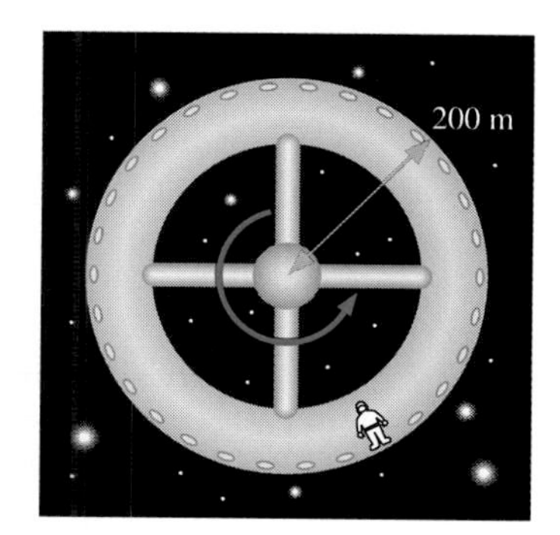

PROBLEM 6-46

Motion of Planets and Other Satellites

6-47. Io, a moon of Jupiter, orbits the planet with period 1.77 days and orbital radius 4.22×10^8 m. The orbit is very close to circular. Use this information to calculate the mass of Jupiter.

6-48. The mean radius of Saturn's orbit around the sun is 1.43×10^{12} m. What is Saturn's period?

6-49. If a planet with 1.5 times the mass of the earth was traveling in the earth's orbit, what would its period be?

6-50. Use the accompanying table of planetary data to verify Kepler's third law. What would be the period of a planet orbiting the sun at 2.8 AU?

Planet	Semimajor Axis (AU*)	Period (yr)
Mercury	0.387	0.241
Venus	0.723	0.615
Earth	1	1
Mars	1.524	1.881
Jupiter	5.203	11.862
Saturn	9.539	29.458

*One astronomical unit (AU) = 1.5×10^8 km

General Problems

6-51. What is **P** in Prob. 6-33 if the 4.0-kg block accelerates to the left at 1.0 m/s^2? What is the tension in the string?

6-52. The floor of a truck is loaded with crates. The coefficient of static friction between the floor and the crates is 0.50. If the truck is moving at 95 km/h, what is the minimum stopping distance for the truck so that the crates do not slide?

6-53. The accompanying figure shows blocks 1, 2, and 3 of mass m_1, m_2, and m_3, respectively. The coefficient of static friction between block 1 and block 2 is μ_S, and the coefficient of kinetic friction between block 2 and the horizontal surface is μ_K. What is the maximum mass of block 3 such that block 1 will not slip on block 2?

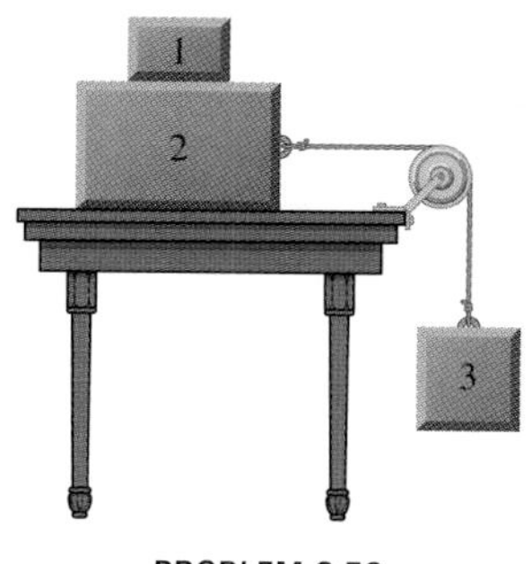

PROBLEM 6-53

6-54. The accompanying figure shows a 5.0-kg block sitting on top of a 20.0-kg block. When the 20.0-kg block is pulled by a horizontal force **F**, it accelerates at 8.0 m/s^2 and the 5.0-kg block slips and accelerates backward *relative to the bottom block* at 4.0 m/s^2. The coefficient of kinetic friction is μ_K for all surfaces. What are μ_K and **F**?

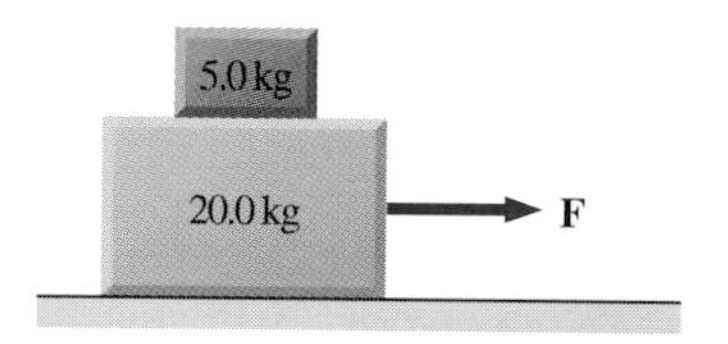

PROBLEM 6-54

6-55. The accompanying figure shows a 10-kg block being pushed by a horizontal force **F** of magnitude 200 N. The coefficient of kinetic friction between the two surfaces is 0.50. Find the acceleration of the block.

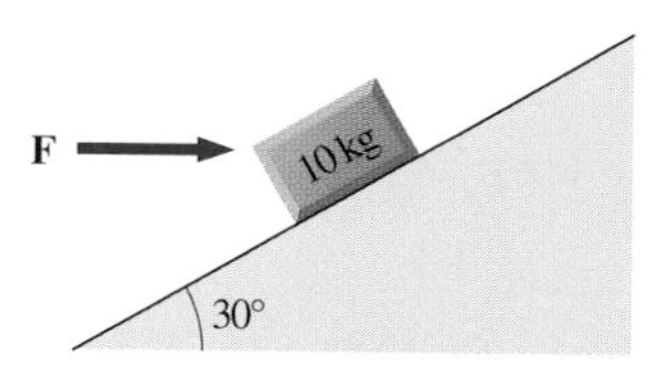

PROBLEM 6-55

6-56. For the weights A and B shown in the accompanying figure, $m_A = 10.0$ kg and $m_B = 8.0$ kg. The string is weightless, and the

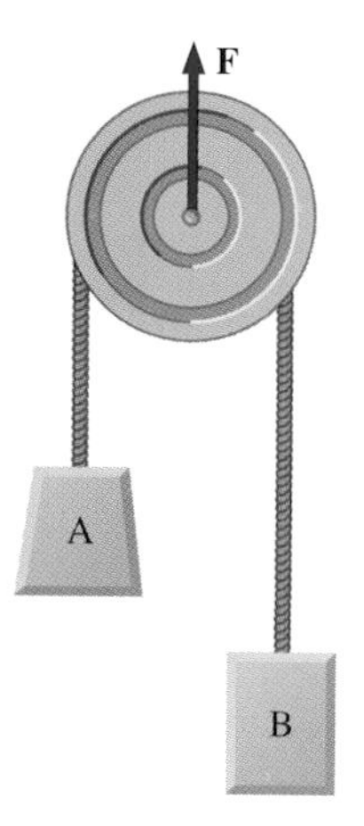

PROBLEM 6-56

pulley is both weightless and frictionless. What are the accelerations of the weights when $F = 300$ N? What is the acceleration of the pulley?

6-57. The accompanying figure shows a boxcar accelerating down the inclined track. Friction is negligible. Show that an object dropped from A will land at B regardless of the angle θ. What happens if friction is not negligible?

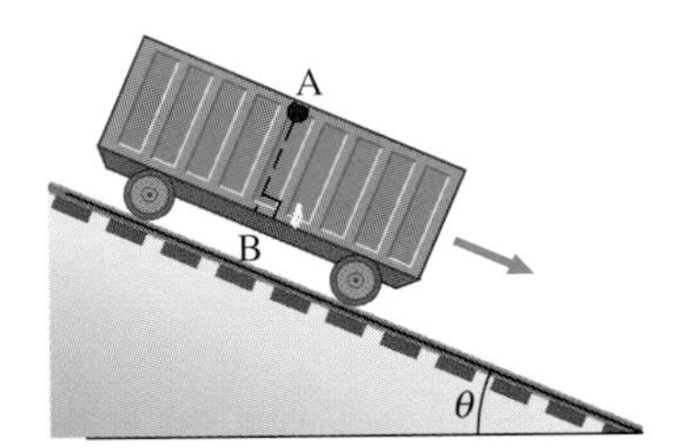

PROBLEM 6-57

6-58. Suppose that the system of Prob. 6-33 is in an elevator accelerating upward at 2.0 m/s^2. What are **P** and the tension in the string if the 4.0-kg block moves at a constant velocity in the horizontal direction?

6-59. A 120-lb girl rides on a Ferris wheel that is rotating at constant speed. When she is at the top of her path, the seat pushes on her with a force of magnitude 100 lb. What is the force of the seat on the girl when she is at the bottom of her path?

6-60. A popular amusement park ride consists of a large vertical cylinder that spins around its axis fast enough that anyone inside is held against the wall and does not fall when the floor of the ride is lowered. (See the accompanying figure.) (*a*) Assuming that the coefficient of static friction between the wall and the rider is μ_S and that the radius of the cylinder is R, show that the period of revolution T necessary to keep the rider from falling must be less than $\sqrt{4\pi^2 R\mu_S/g}$. (*b*) What is the maximum value of T if $\mu_S = 0.50$ and $R = 4.0$ m? (*c*) Must we be concerned about the weight of the rider? about the clothes he is wearing?

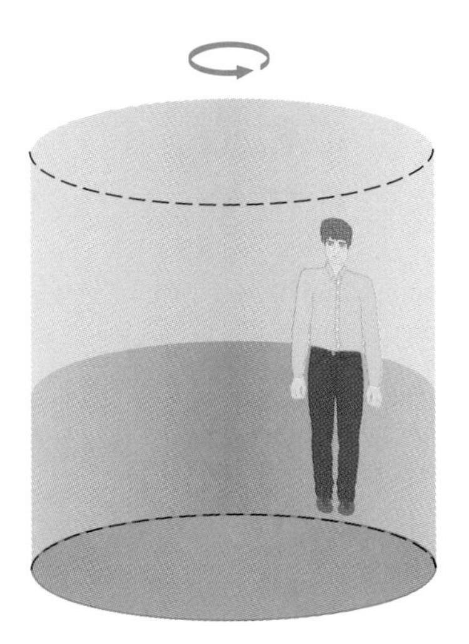

PROBLEM 6-60

6-61. A circular hoop of radius 0.20 m rotates around a vertical axis at a constant rate of 4.0 rev/s. A small bead is free to slide without friction on the hoop. (*a*) Find the angle θ at which the bead sits on the hoop. (*b*) Can the bead sit on the hoop at the same height as the center of the hoop? (*c*) What happens to the bead if the hoop rotates at 1.0 rev/s?

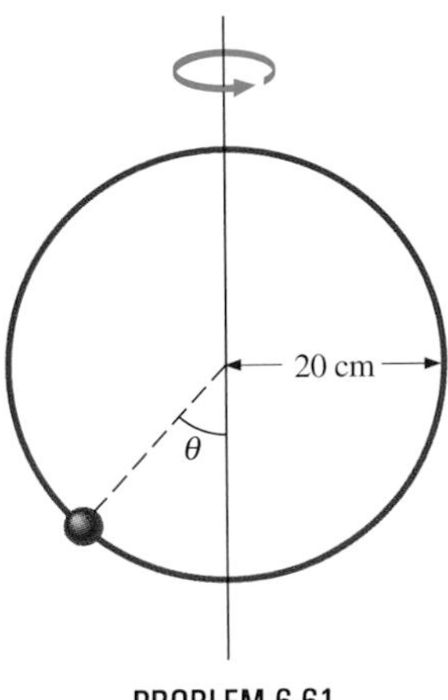

PROBLEM 6-61

6-62. Two planets in circular orbits around a star have speeds of v and $2v$. (*a*) What is the ratio of the orbital radii of the planets? (*b*) What is the ratio of their periods?

6-63. Suppose that your measured weight at the equator is one-half your measured weight at the pole on a planet whose mass and diameter are equal to those of the earth. What is the rotational period of the planet?

6-64. A body of mass 100 kg is weighed at the North Pole and at the equator with a spring scale. What is the scale reading at these two points? Assume that $g = 9.83$ m/s^2 at the pole.

6-65. A small diamond of mass 10 g drops from a swimmer's earring and falls through the water, reaching a terminal velocity of 2.0 m/s. (*a*) Assuming the frictional force on the diamond obeys $f = -bv$, what is b? (*b*) How far does the diamond fall before it reaches 90 percent of its terminal speed?

6-66. An object is subjected to one force, a frictional force $f = -Mkv^2$, where M is the mass of the object, k is a constant, and v is the velocity of the object. (*a*) Show that the velocity of the object as a function of time is given by $v = v_0/(1 + v_0 kt)$, where v_0 is the velocity of the object at $t = 0$. (*b*) Using $v = dx/dt$, show that the position of the object is given by $x = [\ln(1 + v_0 kt)]/k$, where $x = 0$ at $t = 0$. (*c*) If the object slows to 2.0 m/s in 120 s from an initial velocity of $v_0 = 10.0$ m/s, what is k? (*d*) How far does the object travel while it is slowing down to $v = 2.0$ m/s?

Ants at work.

CHAPTER 7 WORK AND MECHANICAL ENERGY

PREVIEW

In this important chapter we will analyze mechanical systems using new concepts like work and energy. The main topics to be discussed here are the following:

1. **Work done by common forces.** Work is defined for constant and variable forces. Expressions for the work done by forces commonly encountered in Newtonian mechanics are obtained.
2. **Work-energy theorem.** The kinetic energy of a particle is defined. The relationship between the change in kinetic energy and the work done by the forces on the particle is given. This is the work-energy theorem.
3. **Conservative forces and potential energy.** The significance of a conservative force is discussed. How a type of energy known as potential energy is associated with a conservative force is described.
4. **Conservation of mechanical energy.** The mechanical energy of a particle is defined. Circumstances under which this quantity is conserved are given.
5. **Mechanical energy and nonconservative forces.** The relationship between the change in the mechanical energy of a particle and the work done by the nonconservative forces acting on that particle is derived.
6. **Power.** The practical importance of power, the rate at which work is done, is discussed.

We have been analyzing mechanical systems through the direct application of Newton's laws. By calculating the net force on a body, we were able to determine its acceleration and hence its motion. In this chapter we introduce an alternative approach for investigating the motion of mechanical systems. This approach involves new concepts like *work* and *energy* and entails two important theorems: the work-energy theorem and the principle of conservation of mechanical energy. The work-energy theorem is derived directly from Newton's laws and is therefore valid for any mechanical system. It is especially useful for relating the speeds of a particle at different points along the particle's path. The principle of conservation of mechanical energy is a special application of the work-energy theorem to systems for which the work of every force involved is path-independent. In most cases, this is effectively a restriction to frictionless systems.

In the context of classical mechanics, the conservation of mechanical energy is a direct consequence of Newton's laws, since it is a special case of the work-energy theorem. However, this conservation law will be generalized in our study of thermodynamics to include other types of energy besides mechanical. This general energy-conservation principle is applicable to all known systems, including systems with friction and gaseous and electromagnetic systems.

7-1 Work Done by a Constant Force

Work is a word that has a variety of meanings in the everyday language. In solving a problem, we say we have "worked out" the answer. When we sit at our desks thinking, we claim we are working—and this type of work can be very tiring. In playing tennis we do physical work, which we also claim to do when holding a heavy object still in our hands. The colloquial uses of the word can lead to confusion. Scientists define "work" very precisely, and our everyday use of the word is often inconsistent with this precise definition. In mechanics, work is always associated with a force and a displacement. We begin by considering this relationship for a constant force.

The particle of Fig. 7-1 moves a horizontal distance d from point A to point B while subjected to a constant force $\mathbf{F}$. This force is directed at an angle θ to the displacement $\mathbf{d}$ of the particle. By definition, the work done by $\mathbf{F}$ on the particle as it travels from A to B is

$$W_{AB} = Fd\cos\theta. \tag{7-1a}$$

Notice that $F\cos\theta$ is the component of the force along the displacement of the particle. Therefore *the work done on a particle by a constant force is the product of the component of the force along the displacement of the particle and the magnitude of that displacement*. In terms of the scalar product of vectors, we may write Eq. (7-1*a*) as

$$W_{AB} = \mathbf{F}\cdot\mathbf{d}. \tag{7-1b}$$

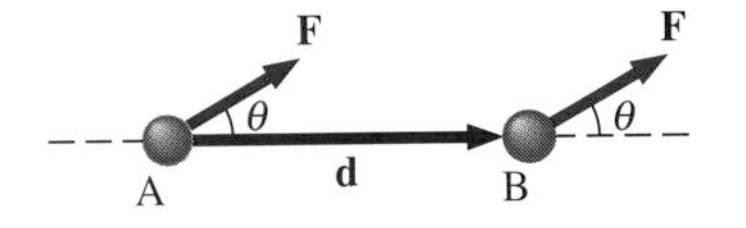

FIGURE 7-1 A particle is moved from A to B by a constant force **F** acting at an angle θ to the displacement **d**.

If the force is in the same direction as the displacement (see Fig. 7-2*a*),then $\theta = 0°$, and

$$W_{AB} = Fd\cos 0° = Fd.$$

Hence the work done by a constant force acting in the direction of motion of a particle traveling in a straight line is the product of the magnitudes of the force and the displacement.

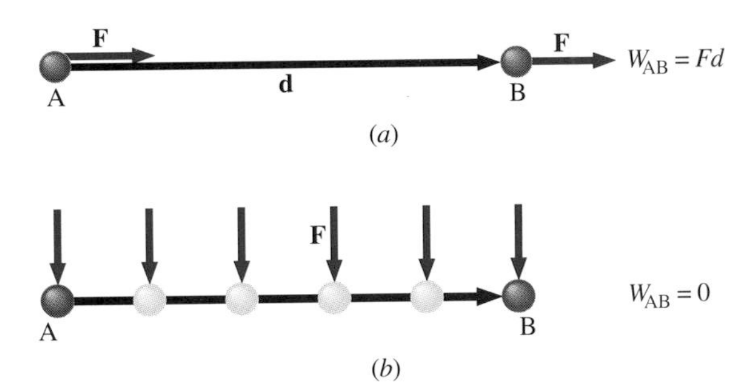

FIGURE 7-2 (*a*) The work done by a constant force in the same direction as the particle's displacement is Fd. (*b*) The work done by a force perpendicular to the displacement is 0.

Another special case is illustrated in Fig. 7-2*b*. Here the force is perpendicular to the displacement, so $\theta = 90°$. We then find

$$W_{AB} = Fd\cos 90° = 0;$$

that is, the work done by a constant force perpendicular to the direction of motion of a particle traveling in a straight line is zero.

Unlike force and displacement, work is a scalar quantity. In the SI system, the work unit is the *joule* (J), where 1 joule = 1 newton·meter (N·m). In cgs units, it is the *erg* (*erg*): 1 erg = 1 dyne·centimeter (dyne·cm). In the British system, there is no special unit; we simply express work in foot·pounds (ft·lb).

EXAMPLE 7-1 Work Done on a Toy Chest

A child is pushing a toy chest across the floor with a constant force of 140 N directed at 30° below the horizontal, as shown in Fig. 7-3*a*.

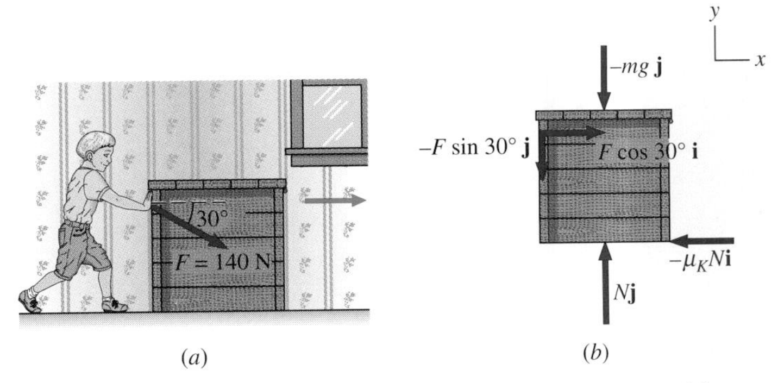

FIGURE 7-3 (*a*) A child pushes a toy chest across the floor with a constant force. (*b*) The free-body diagram of the chest.

The total mass of the chest and its contents is 8.0 kg and the coefficient of kinetic friction between the floor and the chest is 0.75. Calculate the work done by (*a*) the force of the child and (*b*) the frictional force when the chest is pushed a distance of 2.0 m.

SOLUTION (*a*) The 140-N force applied by the child is directed at an angle of 30° with respect to the displacement of the chest. The work done by this force is then

$$W = Fd\cos\theta = (140\text{ N})(2.0\text{ m})\cos 30° = 243\text{ J}.$$

(*b*) We must determine the frictional force in order to find the work that it does. Using the free-body diagram of the chest shown in Fig. 7-3*b*, we apply Newton's second law in the vertical direction to obtain

$$N - F\sin 30° - mg = 0,$$

so

$$N - (140\text{ N})\sin 30° - (8.0\text{ kg})(9.8\text{ m/s}^2) = 0,$$

and

$$N = 148\text{ N}.$$

The frictional force is therefore

$$f_K = \mu_K N = (0.75)(148\text{ N}) = 111\text{ N}.$$

This constant force is directed opposite to the chest's displacement. Its work over the 2.0 m that the chest travels is then

$$W_f = f_K d\cos\theta = (111\text{ N})(2.0\text{ m})\cos 180° = -222\text{ J}.$$

Notice that both positive and negative work are done on the chest. Can you identify a force (or forces) acting on the chest that does no work during the displacement?

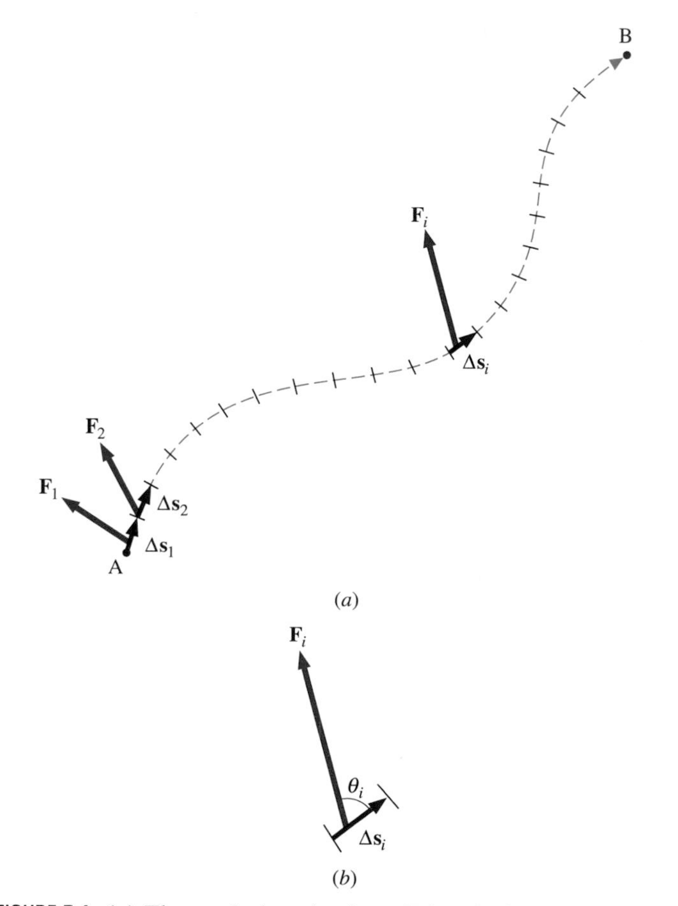

FIGURE 7-4 (*a*) The work done by force **F** is calculated by dividing the path into very small segments and summing the work done over the segments. (*b*) The *i*th segment enlarged.

DRILL PROBLEM 7-1

Find the work done by each of the following forces: (*a*) A constant 100-N force applied at an angle of 30° above the horizontal to pull a crate 3.00 m across a floor. (*b*) The tension in a vertical wire supporting a 5.0-kg flowerpot. (*c*) The tension in a string tied to a ball rotating in a circle.
ANS. (*a*) 260 J; (*b*) 0; (*c*) 0.

7-2 Work Done by a Variable Force

Suppose a particle moves from A to B along the path of Fig. 7-4*a* while under the influence of a variable force **F**. To find the work done on the particle by this force, we first divide the path into N small segments, where the displacement along the ith segment is $\Delta\mathbf{s}_i$. Since these segments are very short, the force at each segment is almost constant; so if $\mathbf{F}_i$ is the force at the ith segment, the work done over this segment is approximately

$$\Delta W_i = \mathbf{F}_i\cdot\Delta\mathbf{s}_i = F_i\,\Delta s_i\cos\theta_i,$$

where θ_i is the angle between $\mathbf{F}_i$ and $\Delta\mathbf{s}_i$. (See Fig. 7-4*b*.) Because work is a scalar quantity, we simply sum the work done over all the segments to find the net work done over the entire path. To be exact, the net work is this sum in the limit that the length of each segment goes to zero and concurrently the number of terms in the sum goes to infinity; that is, the work done on the particle when it moves over the path from A to B is

$$W_{\text{AB}} = \lim_{\Delta s_i\to 0}\sum_i \mathbf{F}_i\cdot\Delta\mathbf{s}_i = \lim_{\Delta s_i\to 0}\sum_i F_i\,\Delta s_i\cos\theta_i. \tag{7-2}$$

Suppose that the particle now returns to A from B by retracing its original path. What would the work W_{BA} done on the particle be? We can answer this question by noting that every segment $\Delta\mathbf{s}_i'$ along the path from B to A corresponds to the *negative* of some segment $\Delta\mathbf{s}_i$ along the path from A to B; that is, $\Delta\mathbf{s}_i' = -\Delta\mathbf{s}_i$. We therefore find

$$W_{\text{BA}} = \lim_{\Delta s_i'\to 0}\sum_i \mathbf{F}_i\cdot\Delta\mathbf{s}_i' = \lim_{\Delta s_i\to 0}\sum_i \mathbf{F}_i\cdot(-\Delta\mathbf{s}_i)$$
$$= -\lim_{\Delta s_i\to 0}\sum_i \mathbf{F}_i\cdot\Delta\mathbf{s}_i,$$

or

$$W_{\text{BA}} = -W_{\text{AB}}. \tag{7-3}$$

Now let's consider a particle moving along a one-dimensional path while under a variable force $F(x)$ that is directed along the path. Then $\mathbf{F}_i = F(x_i)\mathbf{i}$, $\Delta\mathbf{s}_i = \Delta x_i\mathbf{i}$, $\cos\theta_i = 1$, and from Eq. (7-2), the work done on the particle between A and B is

$$W_{\text{AB}} = \lim_{\Delta x_i\to 0}\sum_i F(x_i)\,\Delta x_i, \tag{7-4}$$

where the sum is taken from A with x coordinate x_{A} to B with x coordinate x_{B}.

In Chap. 3, we found that a one-dimensional displacement can be represented by [see (Eq. 3-9)]

$$x(t_2) - x(t_1) = \lim_{\Delta t_i \to 0} \sum_i v(t_i)\,\Delta t_i$$
$$= \{\text{area under } v(t) \text{ between } t_1 \text{ and } t_2\}.$$

Comparing this with Eq. (7-4), we see that W_{AB} can also be represented by an area. As Fig. 7-5 illustrates,

$$W_{AB} = \lim_{\Delta x_i \to 0} \sum_i F(x_i)\,\Delta x_i$$
$$= \{\text{area under } F(x) \text{ between } x_A \text{ and } x_B\}. \quad (7\text{-}5)$$

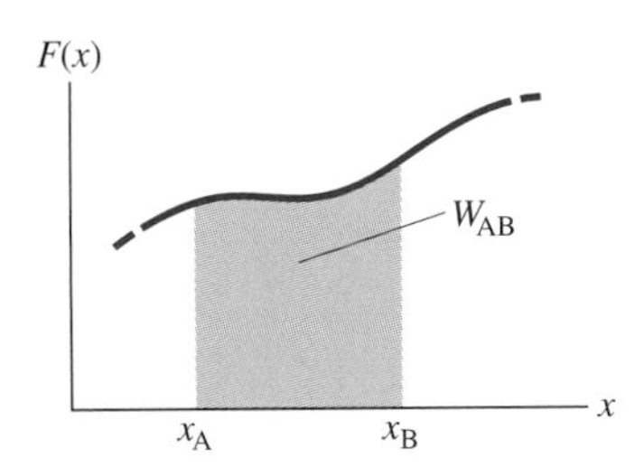

FIGURE 7-5 The work of $F(x)$ between x_A and x_B is the area under the curve representing $F(x)$ between x_A and x_B.

If you have studied integral calculus, you know that the area under $F(x)$ between x_A and x_B is the definite integral of $F(x)$ with respect to x between x_A and x_B. Thus we may write the work as

$$W_{AB} = \int_{x_A}^{x_B} F(x)\,dx. \quad (7\text{-}6)$$

For the work done over a path that is not one-dimensional (see Fig. 7-4), we have

$$W_{AB} = \int_A^B \mathbf{F}\cdot d\mathbf{s} = \int_A^B (F\cos\theta)\,ds, \quad (7\text{-}7)$$

where the integral is taken over the path from A to B. In general, the position-dependent forces that we will consider are simple functions of a single variable. Equation (7-6) is then appropriate for determining the work done by these forces.

From Eq. (7-5), the work of a position-dependent force can also be obtained by determining the area under the curve representing $F(x)$. This technique could be tedious if the force has a complicated dependence on position. Fortunately, for the two important position-dependent forces we will study, the spring force and the gravitational force, the areas are not difficult to evaluate.

7-3 Work Done by Common Forces

In Example 7-1, we calculated the work done by kinetic friction on an object traveling across a horizontal surface. We now determine the work done by some of the other forces commonly encountered in Newtonian mechanics. These forces are (1) a force that is always perpendicular to the direction of motion, (2) the spring force $-kx$, (3) the gravitational force $m\mathbf{g}$ at the earth's surface, and (4) the gravitational force $-GMm\hat{\mathbf{r}}/r^2$ between masses M and m. The second and fourth forces depend on position. In this section the work of both of these forces is calculated by integration. *If you have not yet learned about the definite integral, refer to Supplement 7-1, where the work of these two forces is calculated by evaluating the areas under the curves representing the forces as functions of position.*

Work Done by a Force that Is Always Perpendicular to the Direction of Motion

Two examples of this force are shown in Fig. 7-6. One is the normal component of the contact force between two surfaces, and the other is the tension in a string whose end is attached to a body being twirled in a circle. In either case,

$$W_{AB} = \lim_{\Delta s_i \to 0} \sum_i F_i \cos 90^\circ\,\Delta s_i = 0.$$

Note that the force does not have to be constant over the displacement—it only has to be *perpendicular* to the displacement at all points.

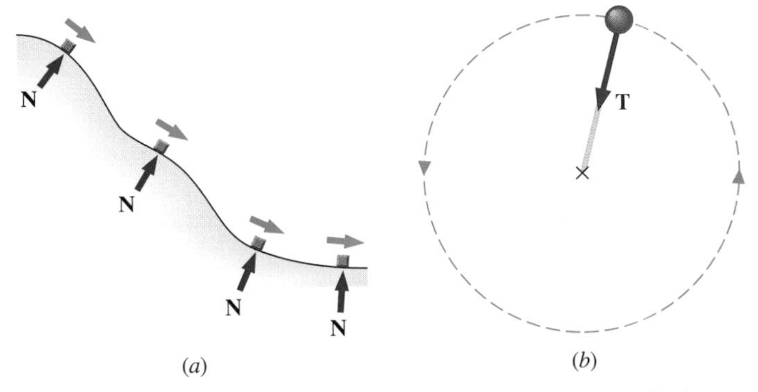

FIGURE 7-6 Two examples of forces perpendicular to the displacement. (*a*) The normal component of the contact force. (*b*) The tension in a string attached to a body being twirled in a circle.

Work Done by the Spring Force $-kx$

In Fig. 7-7, a spring is fixed at one end, and a small object is attached to its other end. We calculate *the work done by the spring force on the object as it moves from A to B.* At these points, the spring is stretched x_A and x_B, respectively, from its relaxed length.

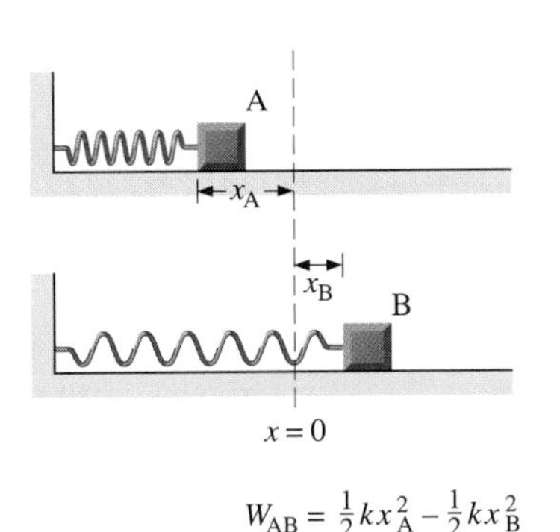

FIGURE 7-7 An object attached to a spring moves from A to B. The work of the spring force only depends on the initial and final positions.

From Eq. (7-6), the work done by the spring force is

$$W_{AB} = \int_{x_A}^{x_B} F(x)\,dx = \int_{x_A}^{x_B} -kx\,dx = \left(\frac{-kx^2}{2}\right)\Bigg|_{x_A}^{x_B},$$

so we have

$$W_{AB} = \tfrac{1}{2}kx_A^2 - \tfrac{1}{2}kx_B^2. \tag{7-8}$$

Notice that W_{AB} *depends only on the initial and final points*. The actual one-dimensional path taken from A to B is inconsequential for determining the work of the spring force. The object could have oscillated through A and B many times between the two moments under consideration, and the work done by the spring would still be given by Eq. (7-8).

Work Done by the Gravitational Force $m\mathbf{g}$

When a particle of mass m moves from point A to point B along the arbitrary path shown in Fig. 7-8a, the gravitational force $m\mathbf{g}$ on it is constant, but the direction of the force with respect to the particle's displacement varies along the path. Consequently, we must use Eq. (7-2) to find the net work done by $m\mathbf{g}$ over the entire path. This gives

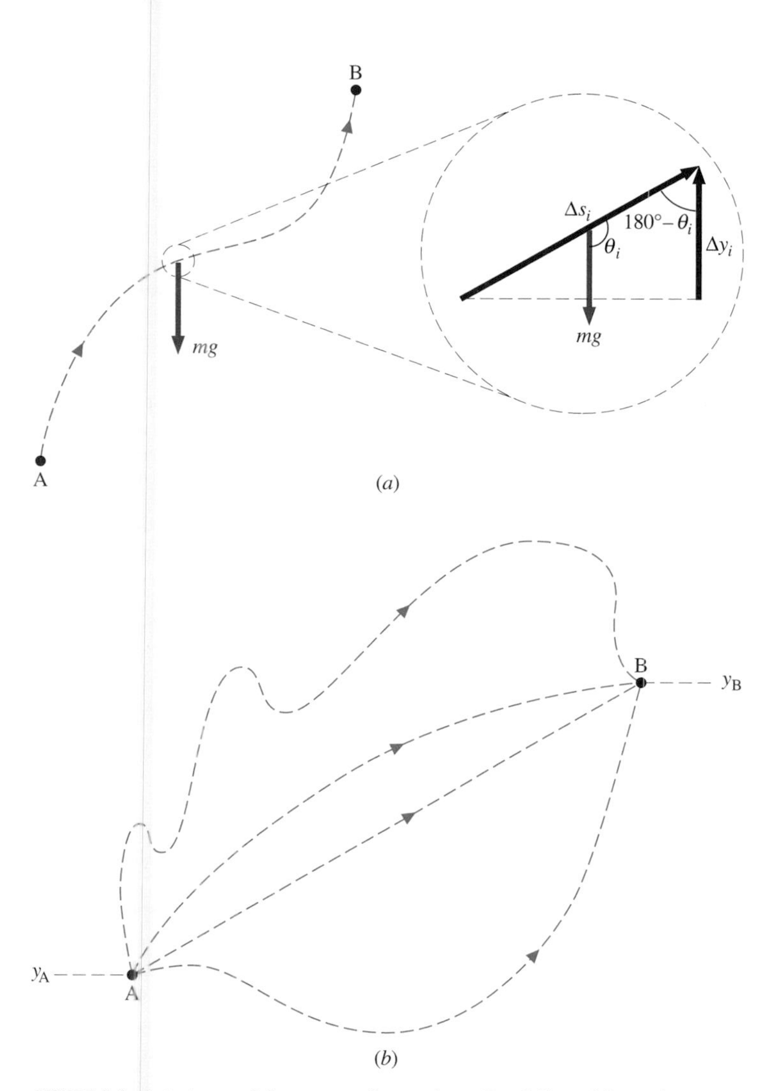

FIGURE 7-8 (a) A particle moves from A to B while subjected to the gravitational force $m\mathbf{g}$. (b) Along any of the paths shown, the work done by $m\mathbf{g}$ is $-mg(y_B - y_A)$.

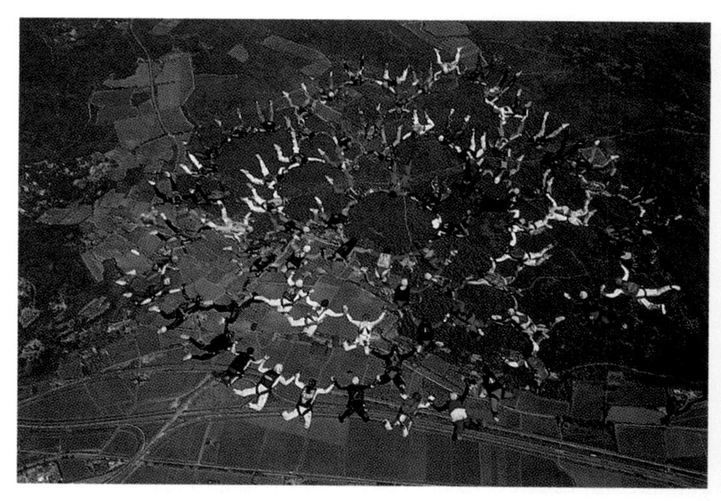

Work being done by gravity.

$$W_{AB} = \lim_{\Delta s_i \to 0} \sum_i (mg\cos\theta_i)\,\Delta s_i.$$

As Fig. 7-8a illustrates, $\Delta s_i \cos\theta_i = -\Delta s_i \cos(180° - \theta_i) = -\Delta y_i$; hence

$$W_{AB} = \lim_{\Delta y_i \to 0} \sum_i (-mg\,\Delta y_i) = -mg\left(\lim_{\Delta y_i \to 0} \sum_i \Delta y_i\right).$$

The summation of Δy_i over the path from A to B is simply $y_B - y_A$. The *work done by the force mg when a particle moves from A to B* is therefore

$$W_{AB} = -mg(y_B - y_A) = -mgh, \tag{7-9}$$

where $h = y_B - y_A$ is the change *in the vertical position* of the particle. The minus sign signifies that the work is negative if the particle rises ($y_B > y_A$), and it is positive if the particle falls ($y_B < y_A$).

Like that of the spring force, *the work of* $m\mathbf{g}$ *does not depend on the path taken by the particle* as it moves from A to B; it depends only on the change in vertical position. The work done by $m\mathbf{g}$ *along any path* between A and B (see Fig. 7-8b) is $-mg(y_B - y_A)$.

Work Done by the Gravitational Force $-GMm\hat{\mathbf{r}}/r^2$

Consider a particle of mass m moving while under the gravitational influence of a stationary body of mass M at $r = 0$. (See Fig. 7-9.) The particle is initially at point A, which is a distance r_A from M, and it moves to point B, which is r_B from M. Since the gravitational force on m due to M is $-GMm\hat{\mathbf{r}}/r^2$, the work done on m between points A and B is

$$W_{AB} = \lim_{\Delta s_i \to 0} \sum_i \mathbf{F}_i \cdot \Delta \mathbf{s}_i = \lim_{\Delta s_i \to 0} \sum_i -\left(\frac{GMm}{r_i^2}\right)\hat{\mathbf{r}}_i \cdot \Delta \mathbf{s}_i.$$

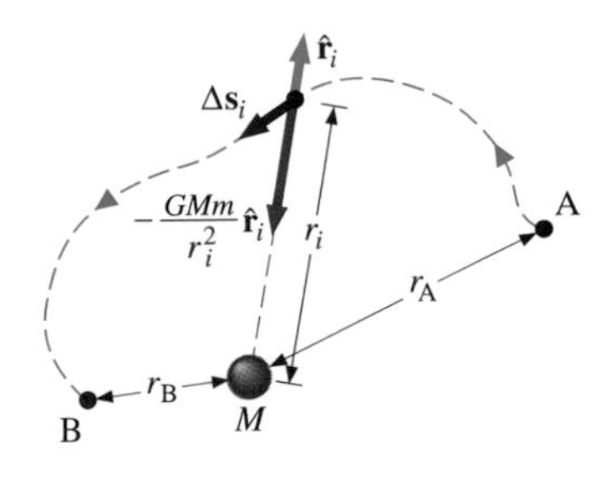

FIGURE 7-9 A particle of mass m moves from A to B while subjected to the gravitational force $-(GMm/r^2)\hat{\mathbf{r}}$ of the mass M at the origin. The work done depends on the values of r_A and r_B and not on the path.

The scalar product $\hat{\mathbf{r}}_i \cdot \Delta\mathbf{s}_i$ is equal to the projection of $\Delta\mathbf{s}_i$ onto $\hat{\mathbf{r}}_i$ times the magnitude of $\hat{\mathbf{r}}_i$. Now this projection, which is the radial component of $\Delta\mathbf{s}_i$, is Δr_i, and the magnitude of the unit vector $\hat{\mathbf{r}}_i$ is $|\hat{\mathbf{r}}_i| = 1$. We therefore find $\hat{\mathbf{r}}_i \cdot \Delta\mathbf{s}_i = \Delta r_i$. When this is substituted into the expression for W_{AB}, we have

$$W_{AB} = \lim_{\Delta r_i \to 0} \sum_i -\left(\frac{GMm}{r_i^2}\right)\Delta r_i = -\int_{r_A}^{r_B} \frac{GMm}{r^2}\,dr.$$

This integration yields

$$W_{AB} = \left.\frac{GMm}{r}\right|_{r_A}^{r_B} = \frac{GMm}{r_B} - \frac{GMm}{r_A}, \qquad (7\text{-}10)$$

which is the work done by the gravitational force $-GMm\hat{\mathbf{r}}/r^2$ on a particle moving from A to B. Once again, we have path independence, as the work done by this general gravitational force depends only on the positions of the end points A and B.

We can check to see if Eq. (7-10) reduces to Eq. (7-9) for the motion of a particle near the earth (mass M_E) as follows. Suppose that near the earth's surface, a particle moves from $r_A = R_E$ to $r_B = R_E + h$, where h, the distance above the earth's surface, is much less than R_E, the earth's radius. Then

$$W_{AB} = \frac{GM_Em}{R_E + h} - \frac{GM_Em}{R_E} = \frac{GM_Em}{R_E}\left(\frac{1}{1 + (h/R_E)} - 1\right).$$

To simplify this expression, we use the expansion

$$\frac{1}{1+\epsilon} = 1 - \epsilon + \epsilon^2 - \epsilon^3 + \ldots$$

If $\epsilon \ll 1$, then the higher-order terms in ϵ are negligible and $1/(1+\epsilon) \approx 1 - \epsilon$. Similarly, since $h \ll R_E$, we have

$$\frac{1}{1 + (h/R_E)} \approx 1 - \frac{h}{R_E},$$

so

$$W_{AB} \approx \frac{GM_Em}{R_E}\left(1 - \frac{h}{R_E} - 1\right) = -\frac{GM_Emh}{R_E^2}.$$

Upon substituting $g = GM_E/R_E^2$ (Eq. 5-4), we obtain

$$W_{AB} = -mgh,$$

which is simply Eq. (7-9).

EXAMPLE 7-2 WORK DONE ON A BLOCK MOVING ON A RAMP

A 40-kg block is attached to a spring whose other end is fixed at the top of a ramp. (See Fig. 7-10*a*.) Initially, the block is held at point A, where the spring is at its unstretched length. The block is then released and begins to slide down the ramp. Identify the forces on the block and calculate the work done by each force when the block moves to point B, which is a distance 0.10 m down the ramp. Take the spring constant to be 200 N/m, and assume that the force of kinetic friction between the block and the ramp is 40 N.

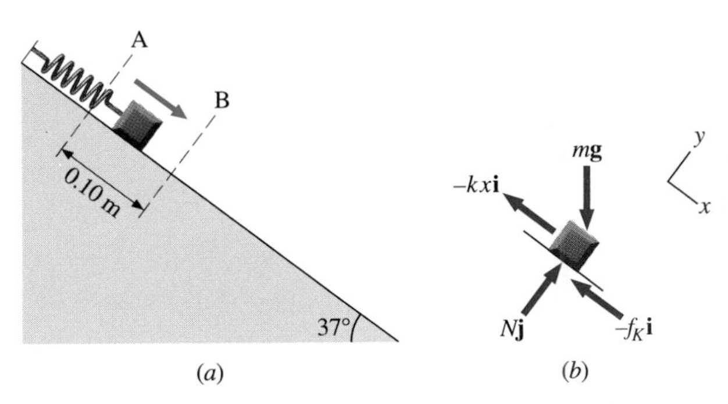

FIGURE 7-10 (*a*) A block slides from A to B. (*b*) The forces on the block.

SOLUTION We begin by identifying the forces acting on the block. (See Fig. 7-10*b*.) They are the normal force, the frictional force, the spring force, and the gravitational force. Since the normal force is always perpendicular to the motion of the block, it does no work; that is,*

$$W_n = 0.$$

The force of kinetic friction is antiparallel to the block's displacement. Its work is therefore

$$W_f = (40\text{ N})(0.10\text{ m})\cos 180° = -4.0\text{ J}.$$

When the block slides down the ramp from A to B, the spring is stretched from $x_A = 0$ to $x_B = 0.10$ m, so the work done by the spring force between these points is

$$W_s = \tfrac{1}{2}kx_A^2 - \tfrac{1}{2}kx_B^2 = \tfrac{1}{2}(200\text{ N/m})(0)^2 - \tfrac{1}{2}(200\text{ N/m})(0.10\text{ m})^2 = -1.0\text{ J}.$$

In moving down the ramp, the change in vertical position of the block is

$$h = -(0.10\text{ m})\sin 37° = -0.060\text{ m}.$$

Hence the gravitational work is

$$W_g = -mgh = -(40\text{ kg})(9.8\text{ m/s}^2)(-0.060\text{ m}) = 23.5\text{ J}.$$

*Rather than representing the beginning and end points, the subscript on the work symbol W will frequently be a letter designating the particular force doing the work.

DRILL PROBLEM 7-2

Determine the work done by each of the following forces: (*a*) The gravitational force when a 5.0-kg block slides 5.0 m up a slope inclined at 37° to the horizontal; (*b*) the force $F(x) = -3.0x$ (in newtons) acting on a particle as it moves from $x = 2.0$ m to $x = 4.0$ m; (*c*) the gravitational force when a spaceship of mass 2.0×10^4 kg moves from the surface of Mars ($M_M = 6.5 \times 10^{23}$ kg, $R_M = 3.4 \times 10^3$ km) to a point 2.6×10^3 km above the surface.

ANS. (*a*) −147 J; (*b*) −18 J; (*c*) -1.1×10^{11} J.

7-4 Work-Energy Theorem for a Particle

At this point, we can relate the work done by an arbitrary set of forces acting on a particle (or an object that moves like a particle) to a property called the **kinetic energy** of the particle. To begin, we consider the cart of Fig. 7-11 as it moves a distance $x_B - x_A = d$ across a frictionless surface from A to B. Its speed is v_A at A and v_B at B. When a force **F** is applied to the cart, it acquires an acceleration

$$a = \frac{F\cos\theta}{m}.$$

After the cart has been pulled a distance d to point B, its speed v_B may be found from

$$v_B^2 = v_A^2 + 2ad = v_A^2 + 2\left(\frac{F\cos\theta}{m}\right)d,$$

which we can rewrite as

$$\tfrac{1}{2}mv_B^2 - \tfrac{1}{2}mv_A^2 = (F\cos\theta)d.$$

The right-hand side of the equation is the work done by **F** on the cart during its motion from A to B. Since **N** and $m\mathbf{g}$ do no work (they are perpendicular to the path), $W_{AB} = (F\cos\theta)d$ is the *total* work done on the cart between the points A and B. With the symbol T used to represent it, *the kinetic energy of a particle of mass m moving with a speed v is by definition*

$$T = \tfrac{1}{2}mv^2. \tag{7-11}$$

Thus for our special case,

$$W_{AB} = T_B - T_A. \tag{7-12}$$

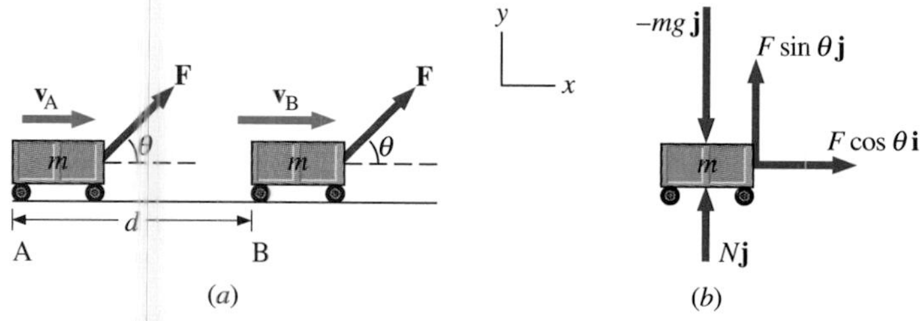

FIGURE 7-11 (*a*) A cart of mass m is pulled a distance d across a frictionless floor by a force **F**. As a result, its velocity changes from $\mathbf{v}_A$ to $\mathbf{v}_B$. (*b*) The free-body diagram of the cart.

The change in the kinetic energy of a particle is equal to the total work done on that particle. This is the *work-energy theorem.*

We now derive this theorem for the general case by considering the work done on a particle by an arbitrary set of forces. This derivation requires the use of integral calculus. From Newton's second law,

$$\sum \mathbf{F} = m\frac{d\mathbf{v}}{dt} = m\left(\frac{dv_x}{dt}\mathbf{i} + \frac{dv_y}{dt}\mathbf{j} + \frac{dv_z}{dt}\mathbf{k}\right).$$

The work done by the forces over an arbitrary infinitesimal displacement $\mathbf{ds} = dx\,\mathbf{i} + dy\,\mathbf{j} + dz\,\mathbf{k}$ is then

$$dW = \left(\sum \mathbf{F}\right)\cdot \mathbf{ds} = m\frac{dv_x}{dt}dx + m\frac{dv_y}{dt}dy + m\frac{dv_z}{dt}dz.$$

Using the chain rule for differentiation, we can reduce the first term on the right-hand side of the equation as follows:

$$m\frac{dv_x}{dt}dx = m\frac{dv_x}{dt}\frac{dx}{dt}dt = m\frac{dv_x}{dt}v_x\,dt$$
$$= mv_x\,dv_x = \frac{1}{2}md(v_x^2),$$

where in the very last step $d(v_x^2)/2$ was substituted for $v_x\,dv_x$. Substituting this and the analogous expressions for the y and z terms into the equation for dW, we get

$$dW = \tfrac{1}{2}m[d(v_x^2) + d(v_y^2) + d(v_z^2)] = \tfrac{1}{2}md(v^2),$$

where

$$v^2 = v_x^2 + v_y^2 + v_z^2.$$

Finally, if we integrate this expression between A, where the particle's speed is v_A, and B, where its speed is v_B, we find

$$\int_A^B dW = \int_A^B \frac{1}{2}md(v^2) = \frac{1}{2}mv_B^2 - \frac{1}{2}mv_A^2.$$

Thus

$$W_{AB} = T_B - T_A,$$

which is the work-energy theorem.

It's important to understand that the derivation of the work-energy theorem is based solely on Newton's laws. As a result, it is applicable to the motion of all particles that obey the laws of Newtonian mechanics.

When applying the work-energy theorem, you should consistently use the same systematic procedure. The one we will use is the following:

1. Identify the body whose motion is to be analyzed.
2. Choose an inertial reference frame.
3. Identify the path over which the work-energy theorem is to be applied.
4. Determine what forces are acting on the body.
5. Calculate the work done over the path by every force.
6. Equate the net work done on the body over the path to the change in kinetic energy of the body between the initial and final points of the path.

In the next example, each step of this procedure will be described carefully. In additional examples involving the work-energy theorem, the procedure will still be followed, but explicit reference to the steps will not be made.

EXAMPLE 7-3

Application of the Work-Energy Theorem

Find the speed of the 40-kg block of Example 7-2 when it is at point B. (See Fig. 7-10*a*.)

SOLUTION

Step 1. The body whose motion is to be analyzed is the block.

Step 2. We choose an inertial reference frame attached to the earth. Its x axis is parallel to the incline, and its y axis is perpendicular to the incline.

Step 3. The path is along the incline (the x axis) from A to B.

Step 4. The forces were identified in Example 7-2. As shown in Fig. 7-10*b*, they are the normal force, the force of kinetic friction, the spring force, and the gravitational force.

Step 5. In Example 7-2, the respective values of the work of each of these forces were found to be $W_n = 0$, $W_f = -4.0$ J, $W_s = -1.0$ J, and $W_g = 23.5$ J.

Step 6. Applying the work-energy theorem between A and B, we obtain

$$W_{AB} = T_B - T_A = \tfrac{1}{2}mv_B^2 - \tfrac{1}{2}mv_A^2$$
$$0 + (-4.0 \text{ J}) + (-1.0 \text{ J}) + (23.5 \text{ J}) = \tfrac{1}{2}(40 \text{ kg})v_B^2 - 0,$$

from which we find the speed of the block at point B to be $v_B = 0.96$ m/s.

EXAMPLE 7-4

A Sled on a Hill

Figure 7-12 shows a sled and occupant of total mass 150 kg, which starts from rest at the top of a hill and slides a distance $l = 50.0$ m to the bottom. The hill can be treated as an inclined plane with a constant slope of $\theta = 10.0°$ to the horizontal. The coefficient of kinetic friction between the sled runners and the snow is 0.070. What is the sled's speed at the bottom of the hill?

SOLUTION The body to be analyzed is the sled and its occupant as it moves along the incline from the top (point A) to the bottom (point B) of the hill. The forces acting on the body are shown in Fig. 7-12*b*; they are its weight and the normal and frictional forces of the surface on the sled. The work done on the body by each of these forces is

$$W_f = -f_K l = -\mu_K N l,$$
$$W_n = 0,$$

and

$$W_g = -mgh = -mg(-l \sin\theta) = mgl \sin\theta,$$

since the change in the vertical position of the sled in going from A to B is $-l \sin\theta$. Now, from the work-energy theorem,

$$W_g + W_n + W_f = T_B - T_A,$$

so with $v_A = 0$ and $N = mg\cos\theta$, we have

$$mgl \sin\theta - \mu_K mgl \cos\theta = \tfrac{1}{2}mv_B^2.$$

Thus

$$v_B = \sqrt{2gl(\sin\theta - \mu_K \cos\theta)}$$
$$= \sqrt{2(9.80 \text{ m/s}^2)(50.0 \text{ m})(\sin 10.0° - 0.070 \cos 10.0°)}$$
$$= 10.1 \text{ m/s}.$$

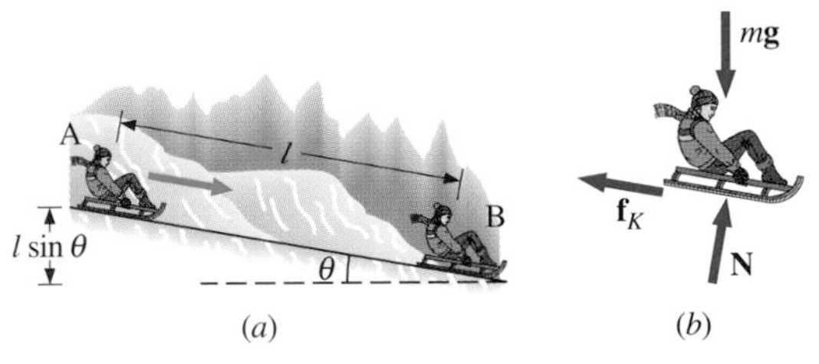

FIGURE 7-12 (*a*) Sled and occupant. (*b*) The forces on the system of the sled and its occupant.

EXAMPLE 7-5

A Particle Subjected to a Variable Force

A particle of mass $m = 0.20$ kg moves in the x direction under the influence of a single force $F(x) = c/x^2$, where $c = 3.0 \text{ N}\cdot\text{m}^2$. The particle's speed at $x_A = 3.0$ m is 7.0 m/s. What is its speed at $x_B = 6.0$ m?

SOLUTION The particle moves along the x axis from $x_A = 3.0$ m to $x_B = 6.0$ m while under the influence of the single force $F(x) = c/x^2$. From Eq. (7-10), with $c = -GMm$, the work that this force does on the particle between x_A and x_B is

$$W_{AB} = -\frac{c}{x_B} + \frac{c}{x_A} = \left(-\frac{3.0}{6.0} + \frac{3.0}{3.0}\right) \text{ J} = 0.50 \text{ J}.$$

Now from the work-energy theorem,

$$0.50 \text{ J} = \tfrac{1}{2}(0.20 \text{ kg})v_B^2 - \tfrac{1}{2}(0.20 \text{ kg})(7.0 \text{ m/s})^2,$$

and

$$v_B = 7.3 \text{ m/s}.$$

EXAMPLE 7-6

Motion of a Ball at the End of a String

A small ball is hung from the ceiling by a light string of length 1.00 m. (See Fig. 7-13.) If the ball is held at an angle of 30.0° to the vertical and then released, what is its speed when it reaches the lowest point of its arc?

SOLUTION We analyze the motion of the ball as it moves along the circular arc between the point A where it is released and the point B at the bottom of the arc. The forces on the ball are the tension **T** of the string and the gravitational force $m\mathbf{g}$. Since **T** is always perpendicular to the path, its work W_t is zero. The work W_g of mg is $-mg(y_B - y_A) = mg(y_A - y_B)$. Using the work-energy theorem, we find

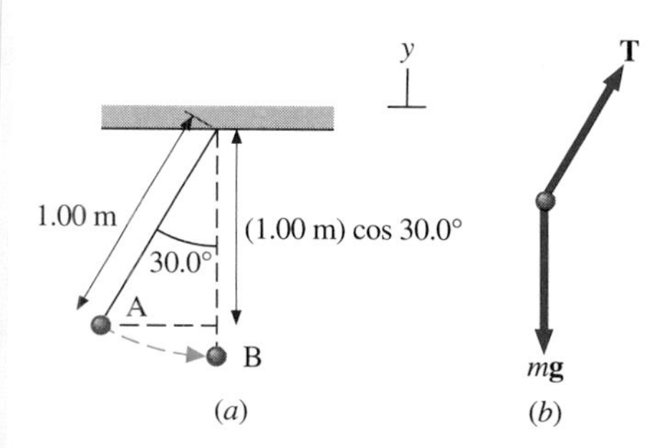

FIGURE 7-13 (*a*) A small ball is hung from the ceiling by a light string. (*b*) The forces on the ball.

$$W_{AB} = T_B - T_A$$
$$W_t + W_g = \tfrac{1}{2}mv_B^2 - \tfrac{1}{2}mv_A^2,$$

so

$$0 + mg(y_A - y_B) = \tfrac{1}{2}mv_B^2 - 0,$$

and

$$v_B = \sqrt{2g(y_A - y_B)}.$$

From the geometry of the system,

$$y_A - y_B = (1.00 \text{ m})(1 - \cos 30.0°) = 0.134 \text{ m},$$

so

$$v_B = \sqrt{2(9.80 \text{ m/s}^2)(0.134 \text{ m})} = 1.62 \text{ m/s}.$$

EXAMPLE 7-7 **Motion of a Body in a Dense Gas**

A massless spring with force constant $k = 10$ N/m hangs vertically in a container of dense gas. (See Fig. 7-14*a*.) A small body of mass $m = 0.10$ kg is attached to the end of the spring and released. The body drops a distance of 19 cm before starting back upward. Calculate the work done on the body by the resistive (frictional) force of the gas between the point A where the body is released and the point B where it stops and starts back upward.

SOLUTION Here we examine the motion of the body as it moves along its vertical path from A to B. Forces are exerted on the body by the gravitational field, the spring, and the friction of the gas, as shown in Fig. 7-14*b*. Since the body has a vertical displacement $y_B - y_A$ between A and B, the work done on it by gravity is

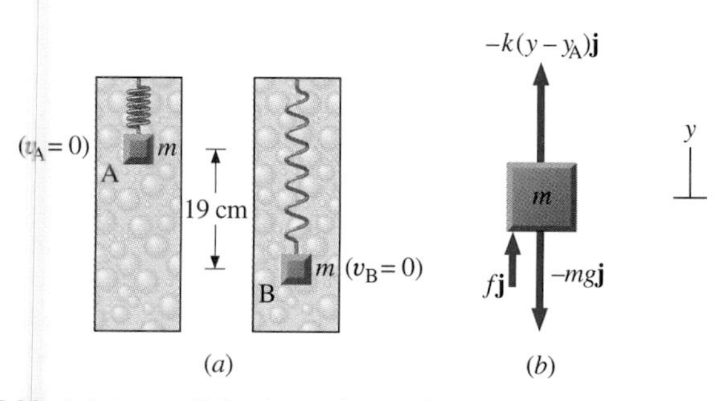

FIGURE 7-14 (*a*) A small body at the end of a spring in a dense gas. (*b*) The free-body diagram of the body.

The spring is unstretched at A and stretched a distance $y_A - y_B$ at B, so

$$W_s = 0 - \tfrac{1}{2}k(y_A - y_B)^2.$$

Writing the work of the frictional force as W_f, we have from the work-energy theorem

$$W_f + W_g + W_s = \tfrac{1}{2}mv_B^2 - \tfrac{1}{2}mv_A^2,$$

which, with $v_A = v_B = 0$, becomes

$$W_f + mg(y_A - y_B) - \tfrac{1}{2}k(y_A - y_B)^2 = 0.$$

The work of friction is then

$$\begin{aligned} W_f &= \tfrac{1}{2}k(y_A - y_B)^2 - mg(y_A - y_B) \\ &= \tfrac{1}{2}(10 \text{ N/m})(0.19 \text{ m})^2 - (0.10 \text{ kg})(9.8 \text{ m/s}^2)(0.19 \text{ m}) \\ &= -5.7 \times 10^{-3} \text{ J}. \end{aligned}$$

Notice that with the work-energy theorem we have been able to calculate the work of the frictional force even though we don't know how this force behaves.

DRILL PROBLEM 7-3

Suppose the sled of Example 7-4 starts at the top of the hill with $v_A = 3.0$ m/s. What is its speed at the bottom of the hill?
ANS. 10.6 m/s.

DRILL PROBLEM 7-4

A particle of mass 2.0 kg is constrained to move along the x axis under the influence of a spring force $F(x) = -8.0x$ (in newtons). The particle starts at the origin with a speed of 3.0 m/s. At what points is its speed zero?
ANS. $x = -1.5$ m and $x = 1.5$ m.

DRILL PROBLEM 7-5

A hockey puck of mass 0.400 kg is shot across the ice with an initial speed of 40.0 m/s. When the puck reaches the goal 20.0 m away, it is moving at a speed of 39.6 m/s. (*a*) What is the work done on the puck by the force of friction? (*b*) What is the coefficient of kinetic friction between the puck and the ice?
ANS. (*a*) −6.37 J; (*b*) 0.081.

7-5 Conservative Forces and Potential Energy

We now consider forces that are *conservative*. By definition, *a conservative force is one whose work is independent of path.* What this means is that if a particle moves between A and B while subjected to a conservative force **F**, the work done by **F** is the same for any path connecting A and B. Hence for the paths I, II, III, . . . of Fig. 7-15,

$$W_{AB}^{I} = W_{AB}^{II} = W_{AB}^{III} = \cdots$$

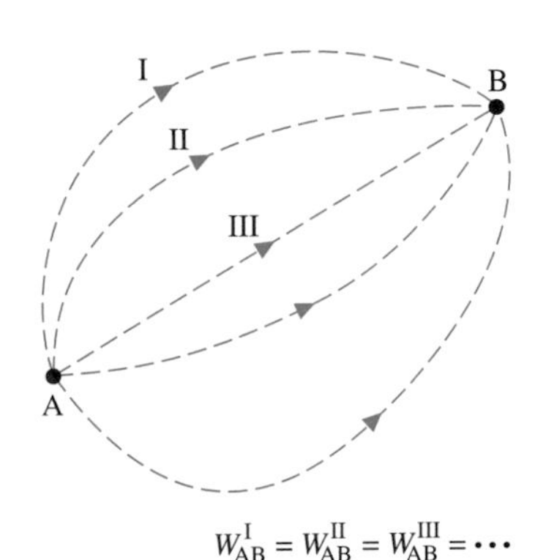

FIGURE 7-15 The work done by a conservative force is the same for any path from A to B.

For example, when a particle of mass m moves from A to B, which are distances r_A and r_B from the center of the earth, the work done on it by the gravitational force is, from Eq. (7-10),

$$W_{AB} = \frac{GM_E m}{r_B} - \frac{GM_E m}{r_A}$$

for any path between A and B.

We could just as well have defined *a conservative force as one whose work over an arbitrary closed path vanishes*. This definition is equivalent to the previous one, because each definition implies the other. To demonstrate this, let's first assume that the work is path-independent. Then, using the fact that $W_{AB} = -W_{BA}$ for any path (Eq. 7-3), we have for the closed path of Fig. 7-16

$$W_{AA} = W^{I}_{AB} + W^{II}_{BA} = W^{I}_{AB} - W^{II}_{AB};$$

which, because $W^{I}_{AB} = W^{II}_{AB}$, gives

$$W_{AA} = 0.$$

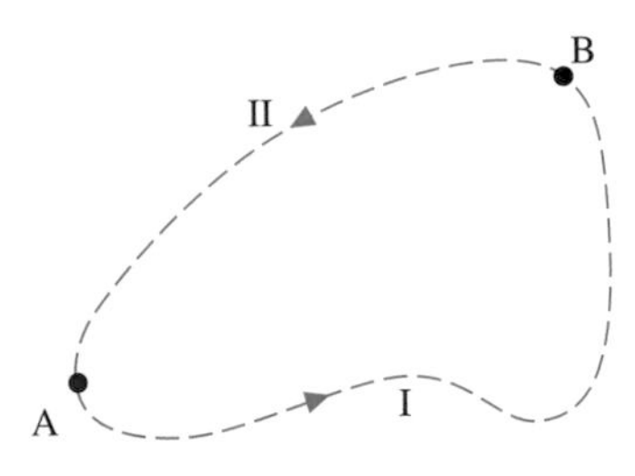

FIGURE 7-16 A particle starts at A, takes path I to B, then takes path II back to A.

Thus path independence implies zero work around a closed path. Conversely, if $W_{AA} = 0$, then

$$W_{AA} = 0 = W^{I}_{AB} + W^{II}_{BA} = W^{I}_{AB} - W^{II}_{AB},$$

and

$$W^{I}_{AB} = W^{II}_{AB}.$$

So, zero work around a closed path implies path independence.

Associated with any conservative force is a mathematical function called the **potential energy**. Actually, potential energy is defined in terms of its *change* between two points. By definition, *the difference in the potential energy at a point P and a reference point* R is the negative of the work done by the conservative force along a path from R to P*; that is,

$$U_P - U_R = -W_{RP}.$$

As shown in Supplement 7-2, the potential energy U_R at the reference point can be assigned any value without affecting the change in potential energy. For convenience, we take $U_R = 0$ so that

$$U_P = -W_{RP}. \tag{7-13}$$

This definition assigns a number, the potential energy, to every point in space. Therefore *the potential energy is a function of the spatial coordinates*. As such, it can be written as $U(x, y, z)$, where (x, y, z) are the coordinates of P.

Equation (7-13) defines the potential energy at a point in terms of the work of a conservative force. We can also relate the work of a conservative force between two points to the potential energies at those two points. For a conservative force,

$$W_{AB} = W_{AR} + W_{RB},$$

so, using $U_A = -W_{RA} = W_{AR}$ and $U_B = -W_{RB}$, we have

$$W_{AB} = U_A - U_B. \tag{7-14}$$

The work of a conservative force between A and B is equal to the difference in the potential energy $U(x, y, z)$ evaluated at A and at B.

Many forces are not conservative. For example, a time-dependent force is not conservative, because the work it does on a particle that moves from A to B depends on the time during which the particle makes the trip. A velocity-dependent force is not conservative, because the work of this type of force depends on how a particle's velocity varies along its path. The most common nonconservative force is the force of friction. If a crate is pushed along path I of Fig. 7-17, the work of the frictional force is obviously different than it would be if the crate were pushed along path II, which is longer than path I. Training in physics clearly isn't necessary in order to decide which path is easier to follow in pushing the crate from A to B!

We now determine the potential energy functions for some of the common conservative forces.

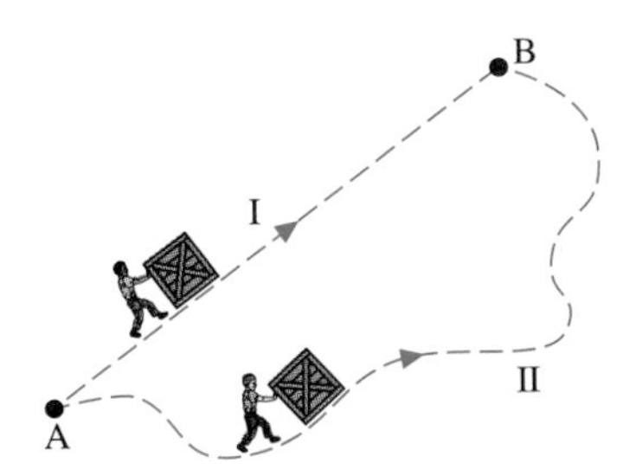

FIGURE 7-17 The work of friction, a nonconservative force, is less along path I than it is along path II.

*As you will see in Example 7-11, the choice of the reference point is arbitrary. In general, we will select the reference point that makes the potential energy function as simple as possible.

Spring Force $-kx$

From Eq. (7-8) the work that this force does on a particle as it moves from a reference point R to an arbitrary point P is

$$W_{RP} = \tfrac{1}{2}kx_R^2 - \tfrac{1}{2}kx^2,$$

where x_R is the coordinate of R and x is the coordinate of P. In order to make the potential energy function as simple as possible, we choose the reference point to be at $x_R = 0$. Then

$$U(x) = -W_{RP} = \tfrac{1}{2}kx^2. \quad (7\text{-}15)$$

Gravitational Force mg near the Earth's Surface

When a particle moves from a reference point at the elevation y_R to a point at the elevation y, the gravitational force $m\mathbf{g}$ does work $-mg(y - y_R)$. [See Eq. (7-9).] The corresponding potential energy function is therefore

$$U(y) = -[-mg(y - y_R)] = mg(y - y_R).$$

Again we make the potential-energy function as simple as possible with a judicious choice of reference point—in this case, $y_R = 0$. Then

$$U(y) = mgy. \quad (7\text{-}16)$$

This is the gravitational potential energy near the earth's surface of a particle of mass m at elevation y relative to a chosen coordinate system.

Gravitational Force $(-GMm/r^2)\hat{\mathbf{r}}$

From Eq. (7-10), the work done by this force on a particle moving from a reference point R with radial position r_R to an arbitrary point P with radial position r is

$$W_{RP} = \frac{GMm}{r} - \frac{GMm}{r_R}.$$

Here also the reference point is chosen so that the form of the potential-energy function is as simple as possible. With r_R chosen to be infinity, the potential energy at P is then

$$U(r) = -W_{\infty r} = -\frac{GMm}{r}. \quad (7\text{-}17)$$

This is the general expression for the gravitational potential energy of a particle of mass m at a distance r from a central mass M.

*One-Dimensional Variable Force $F(x)$

To determine the potential energy of a particle moving under an arbitrary force $F(x)$, we must express the work done by the force as an integral. The work done by this force on a particle moving along the x axis from a reference point R with coordinate x_R to an arbitrary point P with coordinate x is

$$W_{RP} = \int_{x_R}^{x} F(x')\,dx'.$$

This integral is a function of x, so W_{RP} is path-independent and $F(x)$ is conservative. The potential energy of the particle is then

$$U(x) = -\int_{x_R}^{x} F(x')\,dx'. \quad (7\text{-}18)$$

For example, the potential energy of the force $-Ax + Bx^2$, where A and B are constants, is

$$U(x) = -\int_0^x (-Ax' + Bx'^2)\,dx' = \frac{1}{2}Ax^2 - \frac{1}{3}Bx^3,$$

where $x_R = 0$ is chosen to be the position of the reference point.

DRILL PROBLEM 7-6

Calculate the potential energy for the force $F(x) = (ax^2 + b)$, where a and b are constants. Choose $x_R = 0$ as the position of the reference point.
ANS. $U(x) = (-ax^3/3 - bx)$.

DRILL PROBLEM 7-7

Calculate the potential energy for the force $F(x) = k/x^3$. Choose the reference point to make the potential-energy function as simple as possible.
ANS. $U(x) = k/2x^2$, with $x_R = \infty$.

7-6 Calculation of the Force from the Potential-Energy Function

When a particle makes a small displacement Δx along the x axis, the work ΔW done on it by the force $F(x)$ is approximately

$$\Delta W = F(x)\,\Delta x.$$

Now from Eq. (7-14), $\Delta W = -\Delta U$, so

$$\Delta U = -F(x)\,\Delta x.$$

Thus upon dividing both sides of this equation by Δx and taking the limit as $\Delta x \to 0$, we have

$$F(x) = -\frac{dU(x)}{dx}. \quad (7\text{-}19)$$

If you know the potential-energy function $U(x)$, you can find the force with Eq. (7-19).* For example, the potential energy associated with the spring force is $kx^2/2$, so

$$F(x) = -\frac{d}{dx}\left(\frac{1}{2}kx^2\right) = -kx.$$

DRILL PROBLEM 7-8

What is the force for the potential energy $U(x) = a \ln x$?
ANS. $F(x) = -a/x$.

7-7 Conservation of Mechanical Energy

We now come to the culmination of our study of conservative forces—the law of *conservation of mechanical energy*. For conservative systems, this law often provides insight and analysis that we simply don't find from a direct application of Newton's laws. We begin with the specific example of a projectile of mass 2.00 kg shot from the edge of a cliff; the projectile has an initial velocity of 100 m/s at an angle 37.0° above the horizontal. (See Fig. 7-18.) The speed of the projectile at various positions is calculated using the methods of Chap. 4, then the kinetic energy at these positions is found from $T = mv^2/2$. Since the only force on the projectile is $m\mathbf{g}$, its potential energy is mgy, where y is the vertical position measured relative to the origin of a chosen reference system; for this example the origin is placed at the edge of the cliff. The kinetic energy and potential energy at various positions are given in the table of Fig. 7-18. As you can see, both the kinetic energy and the potential energy vary over the path, but *their sum remains constant* at 1.00×10^4 J.

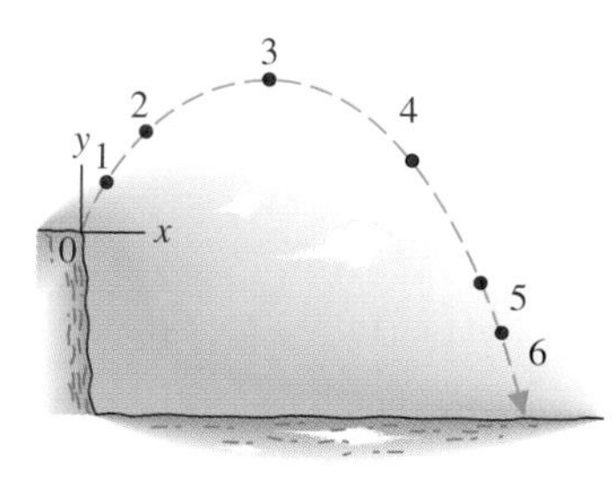

Point	y (m)	$T = \frac{1}{2}mv^2$ (J)	$U = mgy$ (J)	$T + U$ (J)
0	0.0	1.00×10^4	0.00	1.00×10^4
1	60.0	8.82×10^3	1.18×10^3	1.00×10^4
2	120.0	7.65×10^3	2.35×10^3	1.00×10^4
3	180.0	6.47×10^3	3.53×10^3	1.00×10^4
4	90.0	8.24×10^3	1.76×10^3	1.00×10^4
5	−60.0	1.118×10^4	-1.18×10^3	1.00×10^4
6	−120.0	1.235×10^4	-2.35×10^3	1.00×10^4

FIGURE 7-18 A 2.00-kg projectile is shot with a speed of 100 m/s from the edge of a cliff. Its kinetic and potential energies and the sum of these two energies for the positions shown are given in the table.

Could this sum remain constant in other situations involving conservative forces? To answer this question, let's consider a particle moving under the influence of a single conservative force $\mathbf{F}$. If the particle travels from A to B along the path represented by the solid line in Fig. 7-19, then from the work-energy theorem,

$$W_{AB} = T_B - T_A.$$

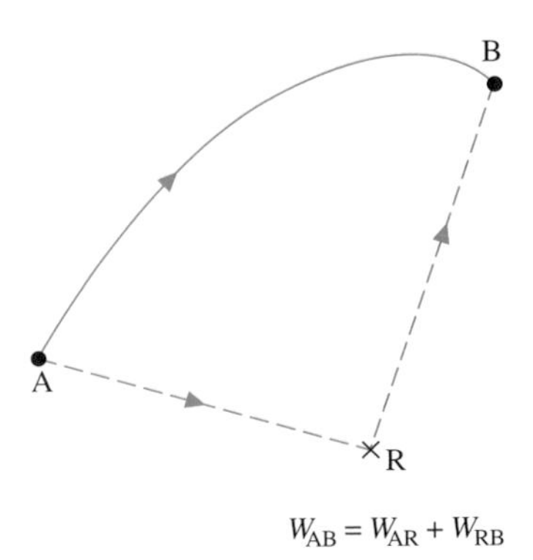

FIGURE 7-19 For a conservative force, $W_{AB} = W_{AR} + W_{RB}$. R is the reference point used in evaluating the potential energy.

Since the force is conservative, the work W_{AB} is the same along any path. We can therefore also calculate W_{AB} along the path indicated by the dashed line from A to the reference point R to B:

$$W_{AB} = W_{AR} + W_{RB} = T_B - T_A.$$

Upon reversal of the subscripts in W_{AR} [see Eq. (7-3)], this becomes

$$-W_{RA} + W_{RB} = T_B - T_A.$$

From the definition of potential energy (Eq. 7-13),

$$U_A = -W_{RA} \quad \text{and} \quad U_B = -W_{RB},$$

so

$$U_A - U_B = T_B - T_A.$$

Finally, we rearrange terms in this equation to obtain

$$T_A + U_A = T_B + U_B. \tag{7-20a}$$

Now A and B are arbitrary points. The sum of the kinetic energy and the potential energy is therefore the same *everywhere* along the path of the particle as it moves under the influence of the conservative force $\mathbf{F}$. For example, this sum is 1.00×10^4 J at any point along the parabolic path of the projectile of Fig. 7-18. The sum of the kinetic energy and the potential energy is called the **mechanical energy**, and Eq. (7-20a) is known as *the law of mechanical-energy conservation*. If E represents the mechanical energy, its conservation is expressed by

$$E = T + U = \text{constant.} \tag{7-20b}$$

*For those who have studied multivariable calculus: In the three-dimensional case, the force components are given by

$$F_x = -\frac{\partial U(x, y, z)}{\partial x}, \quad F_y = -\frac{\partial U(x, y, z)}{\partial y}, \quad F_z = -\frac{\partial U(x, y, z)}{\partial z}.$$

If a particle moves under the influence of N conservative forces, its mechanical energy is also conserved. In this case, we calculate a potential energy for each force and express the law of mechanical-energy conservation as

$$E = T + U_1 + U_2 + \ldots + U_N = \text{constant}. \qquad (7\text{-}20c)$$

The mechanical energy presently considered does *not* include terms related to other forms of energy, such as heat energy or electrical energy. Later, when we take up thermodynamics, the definition of energy will be expanded to include these other forms. You will then learn that while electrical energy can be transformed to mechanical energy, mechanical energy can be transformed to heat energy, and so on, the sum of these different types of energy still remains constant.

Finally, keep in mind that while the principle of mechanical-energy conservation only applies when conservative forces are present, the work-energy theorem is valid under any circumstances. Thus when nonconservative forces prevent you from using the law of conservation of mechanical energy, you still have the work-energy principle at your disposal.

7-8 Applications of the Law of Mechanical-Energy Conservation

You should also follow a systematic approach when you apply the principle of mechanical-energy conservation. The approach we will use is the following:

1. Identify the body or bodies to be studied. Often, in applications of the principle of mechanical-energy conservation, more than one body is investigated at the same time.
2. Identify all forces acting on the body or bodies.
3. Determine whether or not each force that does work is conservative. If a nonconservative force (e.g., friction) is doing work, then mechanical energy is not conserved. The system must then be analyzed with another principle, such as the work-energy theorem.
4. For every force that does work, choose a reference point and determine the potential-energy function for the force. The reference points for the various potential energies do not have to be at the same location.
5. Apply the principle of mechanical-energy conservation by setting the sum of the kinetic energies and potential energies equal at every point of interest.

Let's follow these steps in analyzing the motion of the roller coaster (including occupants) of Fig. 7-20 as it moves without friction along the curved track. Its mass is m, and when it passes point A, its speed is v_A. Suppose that we are interested in the speed v_B of the roller coaster when it passes point B, which is a distance d below A. We then apply the principle of conservation of mechanical energy as follows.

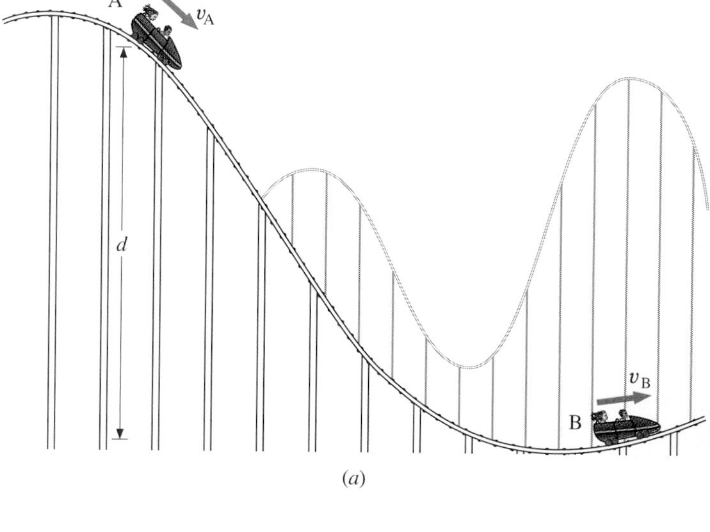

(*a*)

(*b*)

FIGURE 7-20 (*a*) A roller coaster moves without friction along a curved track. (*b*) The only forces on the roller coaster are its weight $m\mathbf{g}$ and the normal force $\mathbf{N}$ of the track.

Step 1. The body to be analyzed is the roller coaster.

Step 2. With no friction, the forces on the roller coaster are its weight $m\mathbf{g}$ and the normal force $\mathbf{N}$ due to the track.

Step 3. Since $\mathbf{N}$ is perpendicular to the path of the roller coaster everywhere, this force does no work. Thus only the gravitational force does work, and mechanical energy is conserved.

Step 4. Assume a vertical y axis and the reference $y = 0$ at the ground. The potential energy of the roller coaster is mgy.

Step 5. For the two points A and B, the conservation of mechanical energy for the roller coaster is expressed by

$$\tfrac{1}{2}mv_A^2 + mgy_A = \tfrac{1}{2}mv_B^2 + mgy_B.$$

Since B is a distance d below A, $y_A - y_B = d$, and this equation reduces to

$$\tfrac{1}{2}mv_A^2 + mgd = \tfrac{1}{2}mv_B^2.$$

Thus

$$v_B = \sqrt{v_A^2 + 2gd},$$

which gives us a relationship between the speeds at A and B.

EXAMPLE 7-8 ENERGY CONSERVATION FOR A BALL AT THE END OF A STRING

By applying the principle of mechanical-energy conservation to the small ball of Example 7-6, find its speed at the lowest point of its arc.

SOLUTION Here we analyze the motion of the ball. There are two forces acting on it, the tension **T** in the string and its weight $m\mathbf{g}$. Since the tension is always directed perpendicular to the path of the ball, it does no work on the ball. Only the conservative gravitational force $m\mathbf{g}$ does work on the ball, so mechanical energy is conserved. If the point where the string is attached to the ceiling (see Fig. 7-13) is used as the reference for potential energy, $y_A = (-1.00 \text{ m}) \times \cos 30.0° = -0.866 \text{ m}$, $y_B = -1.00$ m, and the law of mechanical-energy conservation gives

$$\tfrac{1}{2}mv_A^2 + mgy_A = \tfrac{1}{2}mv_B^2 + mgy_B$$
$$\tfrac{1}{2}m(0)^2 + m(9.80 \text{ m/s}^2)(-0.866 \text{ m}) = \tfrac{1}{2}mv_B^2 + m(9.80 \text{ m/s}^2)(-1.00 \text{ m});$$

so $v_B = 1.62$ m/s, as found in Example 7-6.

EXAMPLE 7-9 ENERGY CONSERVATION FOR A BLOCK AT THE END OF A VERTICAL SPRING

A massless spring with force constant $k = 40$ N/m hangs vertically from the ceiling. A 0.20-kg block is attached to the end of the spring and released. (See Fig. 7-21.) (*a*) What is the maximum extension of the spring? (*b*) What is the block's speed when the spring is extended 4.0 cm?

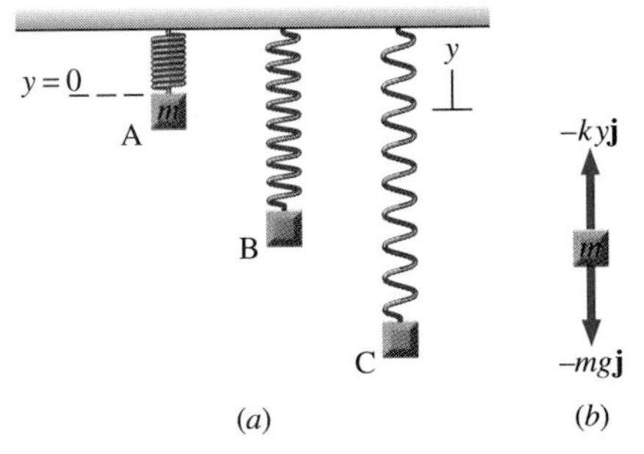

FIGURE 7-21 (*a*) A block is placed on a spring at A and falls to C before rising. Point B is 4.0 cm below point A. (*b*) The forces on the block.

SOLUTION (*a*) In this example we study the block. The forces on it are the gravitational force and the spring force. Since both of these are conservative, the mechanical energy of the block is conserved*:

$$T + U_g + U_s = \text{constant}.$$

To calculate the gravitational potential energy, we choose the reference point $y = 0$ to be at the top of the block when it is first attached to the spring, as shown in the figure. Then when the block's position is y, its gravitational potential energy is $U_g = mgy$. Since $y = 0$ is also the position of the end of the unstretched spring, the potential energy of the spring force is $ky^2/2$. The law of mechanical-energy conservation is now given by

$$T + mgy + \tfrac{1}{2}ky^2 = \text{constant}.$$

At its highest point (A) and at its lowest point (C), the speed of the block, and hence its kinetic energy, is zero; so the conservation of mechanical energy equation

$$T_A + mgy_A + \tfrac{1}{2}ky_A^2 = T_C + mgy_C + \tfrac{1}{2}ky_C^2$$

becomes

$$0 + mg(0) + \tfrac{1}{2}k(0)^2 = 0 + mgy_C + \tfrac{1}{2}ky_C^2.$$

Solving for y_C, we find

$$y_C = -\frac{2mg}{k} = -\frac{2(0.20 \text{ kg})(9.8 \text{ m/s}^2)}{(40 \text{ N/m})} = -0.098 \text{ m} = -9.8 \text{ cm}.$$

Thus the maximum extension of the spring is 9.8 cm.

(*b*) Equating the mechanical energy at A and B (where the block's position is $y_B = -4.0$ cm), we have

$$0 = T_B + mg(y_B) + \tfrac{1}{2}ky_B^2$$

so

$$0 = \tfrac{1}{2}(0.20 \text{ kg})v_B^2 + (0.20 \text{ kg})(9.8 \text{ m/s}^2)(-0.040 \text{ m}) + \tfrac{1}{2}(40 \text{ N/m})(-0.040 \text{ m})^2,$$

which yields $v_B = 0.68$ m/s.

*As we did with work, we'll often associate a subscript representing the appropriate force with a potential energy.

EXAMPLE 7-10 ESCAPE SPEED OF A SPACE CAPSULE

A space capsule of mass m is launched from the surface of the earth with a speed v_E. (*a*) If air friction can be ignored, what is the speed of the capsule as a function of height h above the earth's surface? (*b*) What minimum speed must the capsule have at the surface of the earth if it is to escape from the earth (i.e., approach $h = \infty$)?

SOLUTION (*a*) Only the conservative gravitational force acts on the capsule. Since the change in distance from the center of the earth is significant in this case, the gravitational potential energy as given by Eq. (7-17) must be used. With E representing the earth's surface and A an arbitrary point at a distance h above the surface, the law of mechanical-energy conservation gives

$$\frac{1}{2}mv_E^2 - \frac{GM_Em}{R_E} = \frac{1}{2}mv_A^2 - \frac{GM_Em}{R_E + h}, \qquad (i)$$

where R_E is the radius of the earth. We then find that the speed of the space capsule at a height h above the earth's surface is

$$v_A = \sqrt{v_E^2 - \frac{2GM_Eh}{R_E(R_E + h)}}. \qquad (ii)$$

(*b*) To find the minimum speed for escape, we rewrite Eq. (*ii*) in the form

$$v_E = \sqrt{v_A^2 + \frac{2GM_Eh}{R_E(R_E + h)}}. \qquad (iii)$$

Thus if the capsule escapes from the earth (i.e., $h \to \infty$ and $h \simeq R_E + h$), we have

$$v_E = \sqrt{v_A^2 + \frac{2GM_E}{R_E}}.$$

From this equation, the smallest value v_E can have occurs when $v_A \to 0$. The minimum escape speed is therefore

$$(v_E)_{min} = \sqrt{\frac{2GM_E}{R_E}} = \sqrt{2gR_E},$$

where Eq. (5-4) has been used to substitute g for GM_E/R_E^2. Using $g = 9.8$ m/s^2 and $R_E = 6.37 \times 10^6$ m, we find that the minimum escape speed is $(v_E)_{min} = 1.1 \times 10^4$ m/s.

EXAMPLE 7-11 MECHANICAL-ENERGY CONSERVATION AND CHANGE IN REFERENCE POINT

How a change in reference point for the potential-energy function affects the mechanical-energy-conservation equation is investigated in this example. (*a*) A particle of mass m moves under the influence of the spring force $F(x) = -kx$, where $k = 6.0$ N/m. Using $x = 0$ as a reference point, calculate the potential energy for $F(x)$, then equate the mechanical energies of the particle at two arbitrary points A and B. (*b*) Repeat part (*a*), using $x = 2.0$ m as a reference point.

SOLUTION (*a*) From Eq. (7-8), the work of the spring force between a reference point R with position x_R and an arbitrary point P with position x is

$$W_{RP} = \tfrac{1}{2}kx_R^2 - \tfrac{1}{2}kx^2,$$

so by definition the potential energy at P is

$$U(x) = -W_{RP} = \tfrac{1}{2}kx^2 - \tfrac{1}{2}kx_R^2.$$

With $x_R = 0$, the potential energy is

$$U(x) = \tfrac{1}{2}kx^2 = (3.0 \text{ N/m})x^2,$$

and mechanical-energy conservation is expressed by

$$T_A + U_A = T_B + U_B$$
$$\tfrac{1}{2}mv_A^2 + (3.0 \text{ N/m})x_A^2 = \tfrac{1}{2}mv_B^2 + (3.0 \text{ N/m})x_B^2. \qquad (i)$$

(*b*) If $x_R = 2.0$ m,

$$U(x) = \tfrac{1}{2}(6.0 \text{ N/m})x^2 - \tfrac{1}{2}(6.0 \text{ N/m})(2.0 \text{ m})^2$$
$$= (3.0 \text{ N/m})x^2 - 12 \text{ J},$$

and mechanical-energy conservation is now given by

$$\tfrac{1}{2}mv_A^2 + (3.0 \text{ N/m})x_A^2 - 12 \text{ J}$$
$$= \tfrac{1}{2}mv_B^2 + (3.0 \text{ N/m})x_B^2 - 12 \text{ J}. \qquad (ii)$$

Since the constant -12 J appears on both sides of Eq. (*ii*), it can be canceled; then Eqs. (*i*) and (*ii*) are identical. Therefore the choice of a reference point has no effect on the content of the equation that represents the law of mechanical-energy conservation. If, for example, we wish to determine v_B, both Eqs. (*i*) and (*ii*) will provide the same numerical result.

This property of the potential-energy function can be shown to be true in general. (You will do this in Prob. 7-70.) But be aware that you are only free to choose a reference point *one time* in a given situation. Once the choice is made, the reference point can no longer be changed in the course of the analysis!

DRILL PROBLEM 7-9

An artillery shell is fired at a target 200 m above the ground. When the shell is 1000 m in the air, it has a speed of 100 m/s. What is its speed when it hits its target? Neglect air friction.
ANS. 160 m/s.

DRILL PROBLEM 7-10

How far is the spring of Example 7-9 extended when the speed of the block is 0.50 m/s?
ANS. 0.015 m and 0.083 m.

DRILL PROBLEM 7-11

A single force $F(x) = -4.0x$ (in newtons) acts on a 1.0-kg body. When $x = 3.5$ m, the speed of the body is 4.0 m/s. What is its speed at $x = 2.0$ m?
ANS. 7.0 m/s.

DRILL PROBLEM 7-12

Keeping in mind that the gravitational potential energy ($-GMm/r$) goes to zero at infinity, determine the sign of the mechanical energy of (*a*) a space capsule that does not escape from the earth; (*b*) one that leaves the surface of the earth with the minimum escape speed; (*c*) one that leaves the surface of the earth with a speed greater than the minimum escape speed.
ANS. (*a*) Negative; (*b*) zero; (*c*) positive.

EXAMPLE 7-12 ENERGY CONSERVATION FOR A VARIABLE FORCE

A particle of mass 4.0 kg is constrained to move along the x axis under a single force $F(x) = -cx^3$, where $c = 8.0$ N/m^3. The particle's speed at A, where $x_A = 1.0$ m, is 6.0 m/s. What is its speed at B, where $x_B = -2.0$ m?

SOLUTION The potential energy of this force is calculated using Eq. (7-18). With the origin as the reference point, the potential energy is

$$U(x) = -\int_0^x -c(x')^3\,dx' = \frac{1}{4}cx^4,$$

and the mechanical energy of the particle is

$$E = \tfrac{1}{2}mv^2 + \tfrac{1}{4}cx^4.$$

Since E is conserved,

$$\tfrac{1}{2}mv_A^2 + \tfrac{1}{4}cx_A^4 = \tfrac{1}{2}mv_B^2 + \tfrac{1}{4}cx_B^4,$$

so the speed of the particle at point B is

$$v_B = \sqrt{v_A^2 + \frac{c}{2m}(x_A^4 - x_B^4)}$$

$$= \sqrt{(6.0\ \text{m/s})^2 + \frac{8.0\ \text{N/m}^3}{2(4.0\ \text{kg})}[(1.0\ \text{m})^4 - (-2.0\ \text{m})^4]}$$

$$= 4.6\ \text{m/s}.$$

DRILL PROBLEM 7-13

The force on a particle of mass 2.0 kg varies with position according to $F(x) = -3.0x^2$ (x in meters, $F(x)$ in newtons). The particle's velocity at $x = 2.0$ m is 5.0 m/s. Calculate the mechanical energy of the particle using (*a*) the origin as the reference point and (*b*) $x = 4.0$ m as the reference point. (*c*) Find the particle's velocity at $x = 1.0$ m. Do this part of the problem for each reference point.
ANS. (*a*) 33 J; (*b*) −31 J; (*c*) 5.7 m/s.

7-9 Mechanical Energy and the Presence of Nonconservative Forces

In using the work-energy theorem, it is often convenient to subdivide the work into two categories: W_{con}, the work done by the conservative forces, and W_{ncon}, the work done by the nonconservative forces. The work-energy theorem then becomes

$$W_{con} + W_{ncon} = T_B - T_A,$$

where A and B represent the two points between which the work is calculated. From Eq. (7-14), the work of the conservative forces between A and B may be written as

$$W_{con} = U_A - U_B,$$

where U_A and U_B are the net potential energies of all the conservative forces at A and at B. Therefore we may express the work-energy theorem as

$$U_A - U_B + W_{ncon} = T_B - T_A,$$

and

$$W_{ncon} = (T_B + U_B) - (T_A + U_A). \qquad (7\text{-}21)$$

Thus *with the mechanical energy of a nonconservative system defined to be the kinetic energy plus the potential energies of all the conservative forces, the net work done by all the nonconservative forces is equal to the change in the mechanical energy*.

This relationship between the work of the nonconservative forces and the change in mechanical energy *is simply another way of expressing the work-energy theorem*. Consequently, whenever we use the work-energy theorem, we may do so in the form given by Eq. (7-21). In fact, this approach to the work-energy theorem does give us an especially insightful way of studying the motion of an object subjected to both conservative and nonconservative forces. For example, we can think of the orbit of an artificial earth satellite as slowly decaying because its mechanical energy is continuously reduced by frictional forces.

Finally, the conservation of mechanical energy is contained in Eq. (7-21) as a special case. If $W_{ncon} = 0$,

$$T_A + U_A = T_B + U_B;$$

that is, mechanical energy is conserved.

Objects entering the earth's atmosphere are heated by friction.

EXAMPLE 7-13 A Cart at the End of a Compressed Spring

Figure 7-22 shows a small cart of mass 2.0 kg that is pressed against a massless spring with force constant $k = 2.0 \times 10^3$ N/m. The spring is compressed 0.30 m and then released. (*a*) When the spring returns to its relaxed length, what is the speed of the cart? Assume that there is no friction. (*b*) If a 40-N frictional force acts on the cart while it is in contact with the spring, what is the cart's speed when the spring returns to its relaxed length?

SOLUTION (*a*) Without friction, the only force that does work is the spring force, and mechanical energy is conserved. If the cart is at A when it is pressed against the spring and at B when the spring is relaxed (Fig. 7-22*a*), we have

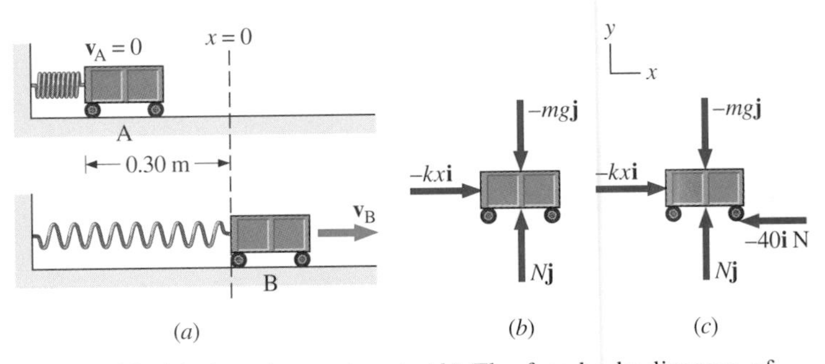

FIGURE 7-22 (*a*) A spring and cart. (*b*) The free-body diagram of the cart with no friction. (*c*) The free-body diagram of the cart in the presence of a 40-N frictional force.

$$T_A + U_A = T_B + U_B$$
$$\tfrac{1}{2}mv_A^2 + \tfrac{1}{2}kx_A^2 = \tfrac{1}{2}mv_B^2 + \tfrac{1}{2}kx_B^2$$
$$\tfrac{1}{2}(2.0\text{ kg})(0)^2 + \tfrac{1}{2}(2.0 \times 10^3\text{ N/m})(-0.30\text{ m})^2 = \tfrac{1}{2}(2.0\text{ kg})v_B^2 + \tfrac{1}{2}(2.0 \times 10^3\text{ N/m})(0)^2.$$

From this equation, we find that $v_B = 9.5$ m/s.

(*b*) With friction, mechanical energy is not conserved. However, we can determine the cart's speed when the spring is relaxed by equating the work done by friction between A and B to the change in the mechanical energy of the cart between these points; that is,

$$W_{ncon} = (T_B + U_B) - (T_A + U_A),$$
$$-(40\text{ N})(x_B - x_A) = (\tfrac{1}{2}mv_B^2 + \tfrac{1}{2}kx_B^2) - (\tfrac{1}{2}mv_A^2 + \tfrac{1}{2}kx_A^2),$$

so

$$-(40\text{ N})(0.30\text{ m}) = [\tfrac{1}{2}(2.0\text{ kg})v_B^2 + \tfrac{1}{2}(2.0 \times 10^3\text{ N/m})(0)^2] - [\tfrac{1}{2}(2.0\text{ kg})(0)^2 + \tfrac{1}{2}(2.0 \times 10^3\text{ N/m})(-0.30\text{ m})^2].$$

This yields $v_B = 8.8$ m/s. Because of the dissipative force of friction, the cart travels more slowly than it did in part (*a*).

EXAMPLE 7-14 Work of Air Friction on a Falling Object

A helicopter is hovering 1.00×10^3 m above the ground when a panel from its underside breaks loose and plummets to the earth. The mass of the panel is 15.0 kg, and just before it hits the ground, its speed is 45.0 m/s. How much work is done by the force of air friction during the descent of the panel?

SOLUTION The conservative force of gravity, with its potential energy mgy, and the nonconservative force of friction act on the panel. Setting the work of the force of friction equal to the change in mechanical energy, we have

$$W_{ncon} = (\tfrac{1}{2}mv_B^2 + mgy_B) - (\tfrac{1}{2}mv_A^2 + mgy_A).$$

With $m = 15.0$ kg, $v_A = 0$, $v_B = 45.0$ m/s, $y_A = 1.00 \times 10^3$ m, and $y_B = 0$, we find that $W_{ncon} = -1.32 \times 10^5$ J. Notice here that the initial potential energy is $mgy_A = 1.47 \times 10^5$ J, and the final kinetic energy is 0.15×10^5 J, so most of the initial mechanical energy is dissipated by friction.

DRILL PROBLEM 7-14

A bullet of mass 10.0 g is shot straight upward with a speed of 500 m/s. When it returns to the ground its speed is 250 m/s. How much work is done on the bullet by air friction?
ANS. −938 J.

7-10 Power

Time is not a consideration in the definitions of work and energy. If you have to lift four hundred 40-kg bags of cement from the ground to a truck bed 1.5 m high, you must do $(400)(40\text{ kg})(9.8\text{ m/s}^2)(1.5\text{ m}) = 2.35 \times 10^5$ J of work on the bags, no matter how long it takes to finish the job. However,

A racer with a high-horsepower engine.

the time required to complete this task is obviously important. If you were expected to get the job done in 1 h, you would probably be unsuccessful, as you would have to lift seven of these heavy bags per minute. On the other hand, if you had 4 h to finish, you would have to lift a little less than two bags per minute. This would surely be tiring, but you could get the truck loaded.

This example illustrates the importance of *the rate at which work is done*, which by definition is the **power** P:

$$P = \frac{dW}{dt}. \tag{7-22}$$

In lifting the bags onto the truck in 1 h, you do work on the bags at an average rate of $(2.35 \times 10^5\text{ J})/(3600\text{ s}) = 65$ J/s. However, by doing the job in 4 h, the average power you supply to the bags is only $(2.35 \times 10^5\text{ J})/(1.44 \times 10^4\text{ s}) = 16$ J/s.

In the SI system the power unit is the joule per second, which is called the *watt* (*W*); 1 W = 1 J/s. In the British system the power unit is a foot·pound per second. Because this unit is too small for most applications, a larger unit, the *horsepower* (*hp*) is also used; 1 hp = 550 ft·lb/s. A horsepower is presumably the rate at which a horse works, although actually, the horse would have to be indefatigable to last very long at this rate.

The power output of a particular force is conveniently expressed in terms of the velocity of the body to which the force is applied. This expression is found by considering the work done by a force **F** in an infinitesimal displacement $d\mathbf{s}$:

$$dW = \mathbf{F} \cdot d\mathbf{s}.$$

Since the power is the rate at which work is done by **F**, we have

$$P = \frac{dW}{dt} = \mathbf{F} \cdot \frac{d\mathbf{s}}{dt} = \mathbf{F} \cdot \mathbf{v}. \tag{7-23}$$

Do be careful in interpreting the equations used to calculate power. They represent the *power expended by a particular force*. They do not necessarily represent the power expended by the machine producing the force, simply because machines are not 100 percent efficient. For example, the human body has an efficiency of approximately 25 percent, which means that for every one unit of work it does on an object, it expends four units of energy. So, when you lift the bags onto the truck in 1 h, you are using energy at an average rate of about 4(65 J/s) = 260 W; when you do the job in 4 h, you expend 4(16 J/s) = 64 W.

EXAMPLE 7-15 **POWER EXPENDED IN CLIMBING STAIRS**

An 80.0-kg man walks up a flight of stairs, resulting in a change in elevation of 30.0 m. (*a*) If he climbs the stairs in 2.00 min, what is the average power used to lift his body? (*b*) Calculate the average power that the man actually generates while climbing the stairs. Assume that his efficiency is 25 percent.

SOLUTION (*a*) Since the man lifts 80.0 kg through a height of 30.0 m in 120 s, the average rate at which energy is supplied for lifting his body is

$$P = \frac{mgh}{t} = \frac{(80.0\text{ kg})(9.80\text{ m/s}^2)(30.0\text{ m})}{120\text{ s}} = 196\text{ W}.$$

(*b*) If his body is 25 percent efficient, he is producing energy at a rate 4(196 W) = 784 W while climbing the stairs.

DRILL PROBLEM 7-15

Show that 1 hp = 746 W.

DRILL PROBLEM 7-16

A 3000-lb automobile moves at a constant speed of 60 mi/h. The engine delivers 25 hp to the wheels of the automobile. What is the force of the road on the wheels? If the engine is 25 percent efficient, how much power is it developing during this time?
ANS. 156 lb; 100 hp.

DRILL PROBLEM 7-17

We are charged for electrical use in terms of a unit called the kilowatt-hour (kW·h). How much energy does this correspond to?
ANS. 3.6×10^6 J.

SUMMARY

1. **Work done by common forces**
   The following expressions are for work done between points A and B by the forces described:
   (*a*) A constant force **F** directed at an angle θ to the displacement **d**:
   $$W_{AB} = Fd\cos\theta.$$
   (*b*) A variable force $F(x)$ over a one-dimensional path:
   $$W_{AB} = \lim_{\Delta x_i \to 0} \sum_i F(x_i)\,\Delta x_i = \{\text{area under } F(x) \text{ between } x_A \text{ and } x_B\}$$
   or,
   $$W_{AB} = \int_{x_A}^{x_B} F(x)\,dx.$$
   (*c*) A variable force **F** over an arbitrary path:
   $$W_{AB} = \int_A^B \mathbf{F}\cdot d\mathbf{s} = \int_A^B (F\cos\theta)\,ds.$$
   (*d*) A force that is always perpendicular to the direction of motion:
   $$W_{AB} = 0.$$
   (*e*) The spring force $-kx$:
   $$W_{AB} = \tfrac{1}{2}kx_A^2 - \tfrac{1}{2}kx_B^2.$$
   (*f*) The gravitational force $m\mathbf{g}$:
   $$W_{AB} = -mg(y_B - y_A).$$
   (*g*) The gravitational force $-GMm\hat{\mathbf{r}}/r^2$:
   $$W_{AB} = \frac{GMm}{r_B} - \frac{GMm}{r_A}.$$
2. **Work-energy theorem**
   The total work done by the forces acting on a particle that moves from point A to point B is equal to the change in the kinetic energy of the particle between those points:
   $$W_{AB} = T_B - T_A,$$
   where the kinetic energy T of a particle of mass m moving at a velocity **v** is
   $$T = \tfrac{1}{2}mv^2.$$
3. **Conservative forces and potential energy**
   (*a*) If the work of a force **F** on a particle is independent of path, then **F** is a conservative force. Furthermore, a potential energy U at any point P can be associated with the force, where U is the negative of the work done on the particle by **F** between a reference point R and point P:
   $$U = -W_{RP}.$$
   (*b*) A conservative force can be found from its potential-energy function (for one dimension) by
   $$F(x) = -\frac{dU(x)}{dx}.$$

4. **Conservation of mechanical energy**
If only conservative forces do work on a particle, then that particle's mechanical energy E is conserved; that is,

$$E = T + U = \text{constant}.$$

5. **Mechanical energy and nonconservative forces**
The net work done by all of the nonconservative forces acting on a particle is equal to the change in the mechanical energy of the particle. For example, between points A and B, we write

$$W_{\text{ncon}} = E_B - E_A.$$

6. **Power**
Power is the rate at which work is done:

$$P = \frac{dW}{dt} = \mathbf{F}\cdot\mathbf{v}.$$

Supplement 7-1 Calculation of Work as the Area Under a Curve

In this supplement, the work done by the spring force $-kx$ and the work done by the gravitational force $-GMm\hat{\mathbf{r}}/r^2$ will be determined by calculating the area under the appropriate curve.

Work Done by the Spring Force $-kx$

In Fig. 7-7, a spring is fixed at one end, and a small object is attached to its other end. We calculate *the work done by the spring force on the object as it moves from A to B*. At these points, the spring is stretched x_A and x_B, respectively, from its relaxed length.

From Eq. (7-5), the work done by the spring force $F(x) = -kx$ between x_A and x_B is the area under the curve $-kx$ between x_A and x_B. This is simply the sum of the shaded areas A_1 and A_2 under the curve shown in Fig. 7-23. These areas lie below the x axis, so they are *negative* areas. Since A_1 is a triangle, its area is

$$A_1 = -\tfrac{1}{2}(x_B - x_A)(kx_B - kx_A),$$

while the area of the rectangle A_2 is

$$A_2 = -(x_B - x_A)(kx_A).$$

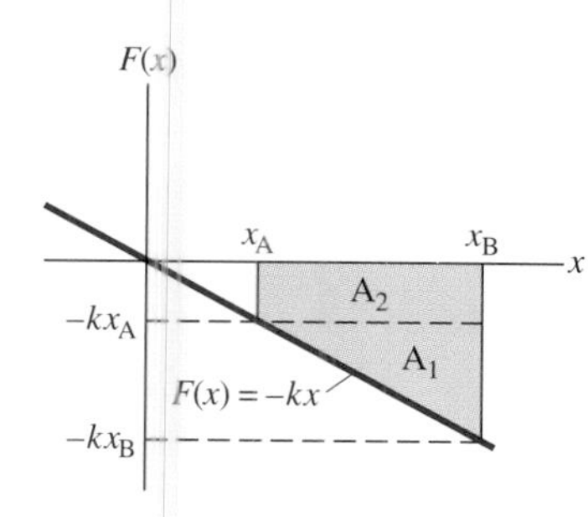

FIGURE 7-23 The work done by $F(x) = -kx$ between points x_A and x_B is $A_1 + A_2$, the area under the curve between x_A and x_B.

Adding these two areas and simplifying, we find that

$$W_{AB} = \tfrac{1}{2}kx_A^2 - \tfrac{1}{2}kx_B^2. \tag{7-8}$$

Notice that W_{AB} *depends only on the initial and final points*. The actual one-dimensional path taken from A to B is inconsequential for determining the work of the spring force. The object could have oscillated through A and B many times between the two moments under consideration; nevertheless, the work done by the spring would still be given by Eq. (7-8).

Work Done by the Gravitational Force $-GMm\hat{\mathbf{r}}/r^2$

Consider a particle of mass m that is moving while under the gravitational influence of a stationary body of mass M at $r = 0$. (See Fig. 7-9.) The particle is initially at point A, which is a distance r_A from M and moves to point B, which is r_B from M. Since the gravitational force on the particle is $-(GMm/r^2)\hat{\mathbf{r}}$, the work done on it by this force between the points A and B is

$$W_{AB} = \lim_{\Delta s_i \to 0} \sum_i \mathbf{F}_i \cdot \Delta\mathbf{s}_i = \lim_{\Delta s_i \to 0} \sum_i -\frac{GMm}{r_i^2}\,\hat{\mathbf{r}}_i \cdot \Delta\mathbf{s}_i.$$

The scalar product $\hat{\mathbf{r}}_i \cdot \Delta\mathbf{s}_i$ is equal to the projection of $\Delta\mathbf{s}_i$ onto $\hat{\mathbf{r}}_i$ times the magnitude of $\hat{\mathbf{r}}_i$. This projection, which is the radial component of $\Delta\mathbf{s}_i$, is Δr_i, and the magnitude of the unit vector $\hat{\mathbf{r}}_i$ is $|\hat{\mathbf{r}}_i| = 1$; thus $\mathbf{r}_i \cdot \Delta\mathbf{s}_i = \Delta r_i$. With this substituted into the expression for W_{AB}, we have

$$W_{AB} = \lim_{\Delta r_i \to 0} \sum_i -\left(\frac{GMm}{r_i^2}\right)\Delta r_i.$$

To calculate this sum, we change to the variable u, where

$$u = \frac{1}{r}.$$

Then

$$\Delta u = (u + \Delta u) - u = \frac{1}{r + \Delta r} - \frac{1}{r} = \frac{1}{r}\left(\frac{1}{1 + \Delta r/r}\right) - \frac{1}{r}.$$

To evaluate the right-hand side of this equation, we use the expansion

$$\frac{1}{1+\epsilon} = 1 - \epsilon + \epsilon^2 - \epsilon^3 + \ldots$$

Then

$$\Delta u = \frac{1}{r}\left[1 - \left(\frac{\Delta r}{r}\right) + \left(\frac{\Delta r}{r}\right)^2 - \left(\frac{\Delta r}{r}\right)^3 + \ldots\right] - \frac{1}{r},$$

so

$$\Delta u = -\frac{\Delta r}{r^2} + \text{terms in } (\Delta r)^2, (\Delta r)^3, \ldots$$

In the limit as $\Delta r \to 0$, the right-hand side of this equation approaches $-\Delta r/r^2$, and we can write

$$\Delta u = -\frac{\Delta r}{r^2},$$

or

$$\Delta r = -r^2\,\Delta u = -\frac{\Delta u}{u^2}.$$

Substituting $1/u_i$ for r_i and $-\Delta u_i/u_i^2$ for Δr_i in the summation representing W_{AB}, we now have

$$W_{AB} = \lim_{\Delta u_i \to 0} \sum_i -GMmu_i^2\left(-\frac{\Delta u_i}{u_i^2}\right) = \lim_{\Delta u_i \to 0} \sum_i GMm\,\Delta u_i.$$

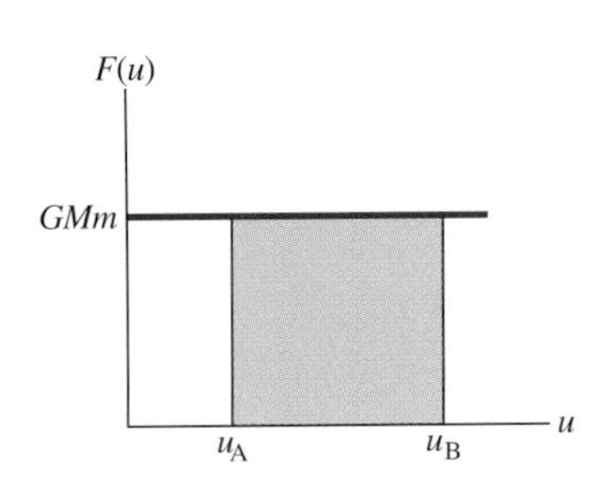

FIGURE 7-24 The area under the horizontal line $F(u) = GMm$ between u_A and u_B represents W_{AB}.

The value of this sum is the area under the horizontal line $F(u) = GMm$ between u_A and u_B. (See Fig. 7-24.) This area is simply $GMm(u_B - u_A)$; thus

$$W_{AB} = GMm(u_B - u_A).$$

Finally, returning to the variable r, we have

$$W_{AB} = \frac{GMm}{r_B} - \frac{GMm}{r_A}, \tag{7-10}$$

which is the work done by the gravitational force $-GMm\hat{\mathbf{r}}/r^2$ *on a particle moving from A to B*. Once again, we have path independence, as the work done by the gravitational force depends only on the positions of the end points A and B.

We can check to see if Eq. (7-10) reduces to Eq. (7-9) for the motion of a particle near the earth (mass M_E) as follows. Suppose that near the earth's surface, a particle moves from $r_A = R_E$ to $r_B = R_E + h$, where h, the distance above the earth's surface, is much less than R_E, the earth's radius. Then

$$W_{AB} = \frac{GM_Em}{R_E + h} - \frac{GM_Em}{R_E} = \frac{GM_Em}{R_E}\left(\frac{1}{1 + h/R_E} - 1\right).$$

To simplify this expression, we again use the expansion

$$\frac{1}{1+\epsilon} = 1 - \epsilon + \epsilon^2 - \epsilon^3 + \ldots$$

If $\epsilon \ll 1$, then the higher-order terms are negligible and $1/(1 + \epsilon) \approx 1 - \epsilon$. Similarly, since $h \ll R_E$, we have

$$\frac{1}{1 + h/R_E} \approx 1 - \frac{h}{R_E},$$

so

$$W_{AB} \approx \frac{GM_Em}{R_E}\left(1 - \frac{h}{R_E} - 1\right) = -\frac{GM_Emh}{R_E^2}.$$

Upon substituting $g = GM_E/R_E^2$ (Eq. 5-4), we obtain

$$W_{AB} = -mgh,$$

which is simply Eq. (7-9).

SUPPLEMENT 7-2 VALUE OF THE POTENTIAL ENERGY AT THE REFERENCE POINT

In this supplement we consider the value of the potential energy at the reference point. By definition, the difference in the potential energy at a point P and a reference point R is the negative of the work done by the conservative force along a path from R to P; that is,

$$U_P - U_R = -W_{RP}. \tag{i}$$

If the conservative force is represented by $\mathbf{F}$, then the work done by $\mathbf{F}$ along a path from R to P is

$$W_{RP} = \int_R^P \mathbf{F}\cdot d\mathbf{s},$$

so Eq. (i) can be rewritten as

$$U_P = -\int_R^P \mathbf{F}\cdot d\mathbf{s} + U_R. \tag{ii}$$

At a second point Q, the potential energy can similarly be written as

$$U_Q = -\int_R^Q \mathbf{F}\cdot d\mathbf{s} + U_R. \tag{iii}$$

The difference in potential energy between these two points is then

$$\begin{aligned} U_P - U_Q &= \left(-\int_R^P \mathbf{F}\cdot d\mathbf{s} + U_R\right) - \left(-\int_R^Q \mathbf{F}\cdot d\mathbf{s} + U_R\right) \\ &= -\int_R^P \mathbf{F}\cdot d\mathbf{s} + \int_R^Q \mathbf{F}\cdot d\mathbf{s} = -\int_Q^P \mathbf{F}\cdot d\mathbf{s}. \end{aligned} \tag{iv}$$

Now suppose that we assign a value U'_R to the potential energy at the reference point R. The potential energies at P and Q then become

$$U_P = -\int_R^P \mathbf{F}\cdot d\mathbf{s} + U'_R$$

and

$$U_Q = -\int_R^Q \mathbf{F}\cdot d\mathbf{s} + U'_R,$$

and the difference in potential energy is

$$\begin{aligned} U_P - U_Q &= \left(-\int_R^P \mathbf{F}\cdot d\mathbf{s} + U'_R\right) - \left(-\int_R^Q \mathbf{F}\cdot d\mathbf{s} + U'_R\right) \\ &= -\int_R^P \mathbf{F}\cdot d\mathbf{s} + \int_R^Q \mathbf{F}\cdot d\mathbf{s} = -\int_Q^P \mathbf{F}\cdot d\mathbf{s}. \end{aligned}$$

This is the same result that we found in Eq. (iv) when we assumed that the value of the potential energy at R was U_R rather than U'_R. We conclude therefore that the difference in potential energy between two points, which is the physically significant quantity, is unaffected by the value of the potential energy at the reference point. For convenience, we choose zero to be that value.

QUESTIONS

7-1. A body moves in a circle at constant speed. Does the centripetal force that accelerates the body do any work? Explain.

7-2. A particle of mass m has a velocity of $v_x\mathbf{i} + v_y\mathbf{j} + v_z\mathbf{k}$. Is its kinetic energy given by $m(v_x^2\mathbf{i} + v_y^2\mathbf{j} + v_z^2\mathbf{k})/2$? If not, what is the correct expression?

7-3. What are the units of kinetic energy in the British system?

7-4. Is it possible for (*a*) the kinetic energy or (*b*) the potential energy of a particle to be negative?

7-5. Can the normal force exerted on an object by a surface ever do work? If so, give examples.

7-6. Does friction always do negative work on a sliding object? Can friction ever do positive work?

7-7. Two marbles of masses m and $2m$ are dropped from a height h. Compare their kinetic energies when they reach the ground.

7-8. Compare the work required to accelerate a car from 30 to 40 km/h with that required for an acceleration from 50 to 60 km/h.

7-9. Suppose you are jogging at constant velocity. Are you doing any work on the environment and vice versa?

7-10. A dropped ball bounces to one-half its original height. Discuss the energy transformations that take place.

7-11. "$E = T + U =$ constant is a special case of the work-energy theorem." Discuss this statement.

7-12. Suppose you throw a ball upward and catch it when it returns at the same height. How much work does the gravitational force do on the ball over its entire trip?

7-13. A planet moves around the sun in an elliptical orbit with the sun at one focus of the ellipse. At what point(s) of the orbit is the planet's speed largest? At what point(s) is its speed smallest?

7-14. Why is it more difficult to do sit-ups while on a slant board than on a horizontal surface? (See the accompanying figure.)

QUESTION 7-14

7-15. In a common physics demonstration, a bowling ball is suspended from the ceiling by a rope. The professor pulls the ball away from its equilibrium position and holds it adjacent to his nose, as shown in the accompanying figure. He releases the ball so that it swings directly away from him. Does he get struck by the ball on its return swing? What is he trying to show in this demonstration?

QUESTION 7-15

7-16. When a body slides down a flat surface, does the work of friction depend on the body's initial speed? Answer the same question for a body sliding down a curved surface.

7-17. Does the work done in lifting an object depend on how fast it is lifted? Does the power expended depend on how fast it is lifted?

7-18. As a young man, Tarzan climbed up a vine to reach his tree house. As he got older, he decided to build and use a staircase instead. Since the work of the gravitational force mg is path-independent, what did the King of the Apes gain in using stairs?

7-19. Can the power expended by a force be negative?

7-20. How can a 50-W lightbulb use more energy than a 1000-W oven?

PROBLEMS

Work

7-1. A constant 20-N force pushes a small ball in the direction of the force over a distance of 5.0 m. What is the work done by the force?

7-2. A toy cart is pulled a distance of 6.0 m in a straight line across the floor. The force pulling the cart has a magnitude of 20 N and is directed at 37° above the horizontal. What is the work done by this force?

7-3. A 5.0-kg box rests on a horizontal surface. The coefficient of kinetic friction between the box and surface is $\mu_K = 0.50$. A horizontal force pulls the box at constant velocity for 10 cm. Find the work done by (*a*) the applied horizontal force, (*b*) the frictional force, (*c*) the net force.

7-4. The force $F(x)$ varies with position, as shown in the accompanying figure. Find the work done by this force on a particle as it moves from $x = 1.0$ m to $x = 5.0$ m.

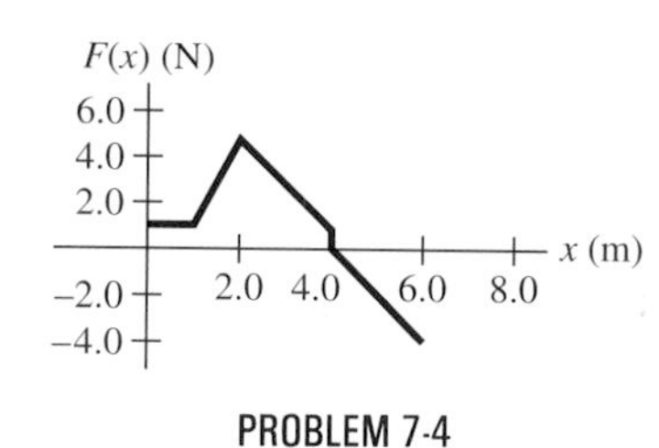

PROBLEM 7-4

7-5. A sled plus passenger with total mass 50 kg is pulled 20 m across the snow ($\mu_K = 0.20$) at constant velocity by a force directed 25° above the horizontal. Calculate (*a*) the work of the applied force, (*b*) the work of friction, (*c*) the total work.

7-6. Suppose that the sled plus passenger of Prob. 7-5 is pushed 20 m across the snow at constant velocity by a force directed 30° below the horizontal. Calculate (*a*) the work of the applied force, (*b*) the work of friction, (*c*) the total work.

*7-7. How much work does the force $F(x) = (-2.0/x)$ N do on a particle as it moves from $x = 2.0$ m to $x = 5.0$ m?

7-8. The accompanying figure shows a 40-kg crate that is pushed at constant velocity a distance 8.0 m along a 30° incline by the horizontal force **F**. The coefficient of kinetic friction between the crate and the incline is $\mu_K = 0.40$. Calculate the work done by (*a*) the applied force, (*b*) the frictional force, (*c*) the gravitational force, and (*d*) the net force.

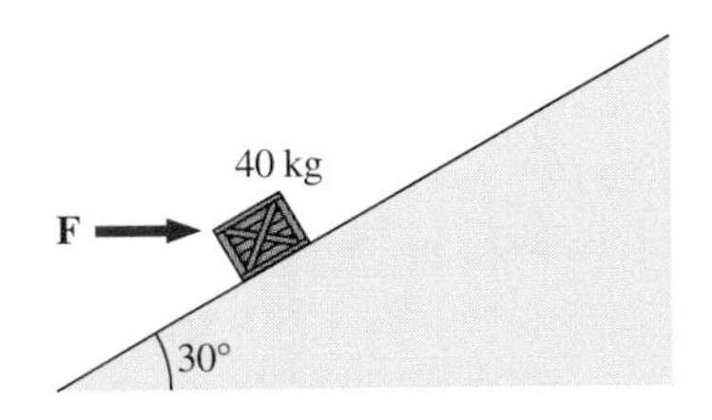

PROBLEM 7-8

7-9. How much work is done against the gravitational force on a 5.0-kg briefcase when it is carried from the ground floor to the roof of the Empire State Building, a vertical climb of 380 m?

7-10. It takes 500 J of work to compress a spring 10 cm. What is the force constant of the spring?

Work-Energy Theorem

7-11. Calculate the kinetic energies of (*a*) a 2000-kg automobile moving at 100 km/h; (*b*) an 80-kg runner sprinting at 10 m/s; and (*c*) a 9.1×10^{-31}-kg electron moving at 2.0×10^7 m/s.

7-12. A 5.0-kg body has three times the kinetic energy of an 8.0-kg body. Calculate the ratio of the speeds of these bodies.

7-13. An 8.0-g bullet has a speed of 800 m/s. (*a*) What is its kinetic energy? (*b*) What is its kinetic energy if the speed is halved?

7-14. Suppose that the box of Prob. 7-3 has an acceleration of 2.0 m/s^2 when it is pulled across the surface by a horizontal force. Find the work done over a distance of 10 cm by (*a*) the horizontal force, (*b*) the frictional force, (*c*) the net force. (*d*) What is the change in kinetic energy of the box?

7-15. A constant 10-N horizontal force is applied to a 20-kg cart at rest on a level floor. If friction is negligible, what is the speed of the cart when it has been pushed 8.0 m?

7-16. In the previous problem, the 10-N force is applied at an angle of 45° below the horizontal. What is the speed of the cart when it has been pushed 8.0 m?

7-17. Compare the work required to stop a 100-kg crate sliding at 1.0 m/s and an 8.0-g bullet traveling at 500 m/s.

7-18. A wagon with its passenger sits at the top of a hill. The wagon is given a slight push and rolls 100 m down a 10° incline to the bottom of the hill. What is the wagon's speed when it reaches the end of the incline? Assume that the retarding force of friction is negligible.

7-19. The bullet of Prob. 7-13 is shot into a wooden block and penetrates 20 cm before stopping. What is the average force of the wood on the bullet? Assume the block does not move.

7-20. The surface of Prob. 7-8 is modified so that the coefficient of kinetic friction is decreased. The same horizontal force is applied to the crate, and after being pushed 8.0 m, its speed is 5.0 m/s. How much work is now done by the force of friction? Assume that the crate starts at rest.

7-21. A 2.0-kg block starts with a speed of 10 m/s at the bottom of a plane inclined at 37° to the horizontal. The coefficient of sliding friction between the block and plane is $\mu_K = 0.30$. (*a*) Use the work-energy principle to determine how far the block slides along the plane before momentarily coming to rest. (*b*) After stopping, the block slides back down the plane. What is its speed when it reaches the bottom? (*Hint:* For the round trip, since mg is conservative, only the force of friction does work on the block.)

7-22. When a 3.0-kg block is pushed against a massless spring of force constant 4.5×10^3 N/m, the spring is compressed 8.0 cm. The block is released, and it slides 2.0 m (from the point at which it is released) across a horizontal surface before friction stops it. What is the coefficient of kinetic friction between the block and the surface?

7-23. A small block of mass 200 g starts at rest at A, slides to B where its speed is $v_B = 8.0$ m/s, then slides along the horizontal surface a distance 10 m before coming to rest at C. (See the accompanying figure.) (*a*) What is the work of friction along the curved surface? (*b*) What is the coefficient of kinetic friction along the horizontal surface?

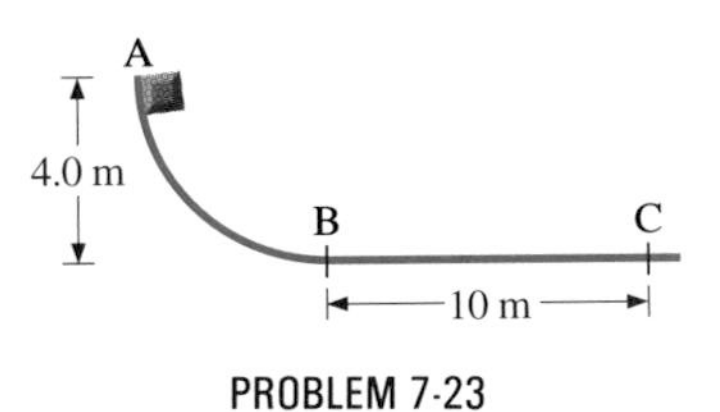

PROBLEM 7-23

7-24. A small object is placed at the top of an incline that is essentially frictionless. The object slides down the incline onto a rough horizontal surface, where it stops in 5.0 s after traveling 60 m. (*a*) What is the speed of the object at the bottom of the incline and its acceleration along the horizontal surface? (*b*) What is the height of the incline?

7-25. A typical raindrop formed in a cloud at an altitude of 600 m reaches the earth's surface with a speed of 9.0 m/s. Calculate the ratio of the mechanical energies of the raindrop before and after the trip and account for the loss in mechanical energy.

7-26. When released, a 100-g block slides down the path shown in the accompanying figure, reaching the bottom with a speed of 4.0 m/s. How much work does the force of friction do?

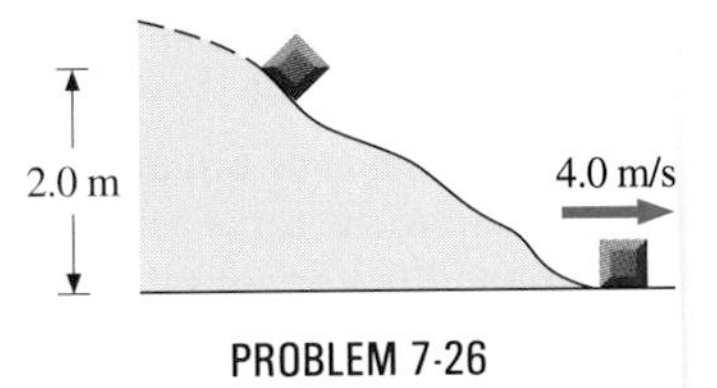

PROBLEM 7-26

Conservative Forces and Potential Energy

7-27. A force $F(x) = (3.0/x)$ N acts on a particle as it moves along the positive x axis. (*a*) How much work does the force do on the particle as it moves from $x = 2.0$ m to $x = 5.0$ m? (*b*) Pick a convenient reference point and find a potential energy for this force.

7-28. A force $F(x) = (-5.0x^2 + 7.0x)$ N acts on a particle. (*a*) How much work does the force do on the particle as it moves from $x = 2.0$ m to $x = 5.0$ m? (*b*) Pick a convenient reference point and find a potential energy for this force.

7-29. Use the gravitational potential energy to calculate the work done by the gravitational force on a 1000-kg spaceship as it moves from the surface of the earth to a height above the surface equal to the radius of the earth.

7-30. Use the gravitational potential-energy function $-(GMm/r)$ to find the corresponding gravitational force.

7-31. Find the force corresponding to the potential energy $U(x) = -a/x + b/x^2$.

7-32. The potential-energy function for either one of the two atoms in a diatomic molecule is often approximated by $U(x) = a/x^{12} - b/x^6$ where x is the distance between the atoms. (*a*) At what distance of separation is the potential energy a minimum? (*b*) What is the force on an atom at this separation? (*c*) How does the force vary with the separation distance?

Conservation of Mechanical Energy

7-33. Ignoring details associated with friction, extra forces exerted by arm and leg muscles, and other factors, we can consider a pole vault as the conversion of an athlete's running kinetic energy to gravitational potential energy. If an athlete is to lift his body 4.8 m during a vault, what speed must he have when he plants his pole?

7-34. Tarzan grabs a vine hanging vertically from a tall tree when he is running at 9.0 m/s. How high can he swing upward? Does the length of the vine affect this height?

7-35. A particle of mass 2.0 kg moves under the influence of the force $F(x) = (3/\sqrt{x})$ N. If its speed at $x = 2.0$ m is $v = 6.0$ m/s, what is its speed at $x = 7.0$ m?

7-36. A particle of mass 2.0 kg moves under the influence of the force $F(x) = (-5x^2 + 7x)$ N. If its speed at $x = -4.0$ m is $v = 20.0$ m/s, what is its speed at $x = 4.0$ m?

7-37. Assume that the force of a bow on an arrow behaves like the spring force. In aiming the arrow, an archer pulls the bow back 50 cm and holds it in position with a force of 150 N. If the mass of the arrow is 50 g and the "spring" is massless, what is the speed of the arrow immediately after it leaves the bow?

7-38. A 1.0-kg ball at the end of a 2.0-m string swings in a vertical plane. At its lowest point the ball is moving with a speed of 10 m/s. (*a*) What is its speed at the top of its path? (*b*) What is the tension in the string when the ball is at the bottom and at the top of its path?

7-39. A body is attached to a massless vertical spring and slowly lowered to its equilibrium position; this stretches the spring by an amount h. If the body is allowed to fall instead, what is the maximum extension of the spring?

7-40. A 100-kg man is skiing across level ground at a speed of 8.0 m/s when he comes to the small slope shown in the accompanying figure. (*a*) If the skier coasts up the hill, what is his speed when he reaches the top plateau? Assume friction between the snow and skis is negligible. (*b*) What is his speed when he reaches the upper level if an 80-N frictional force acts on the skis?

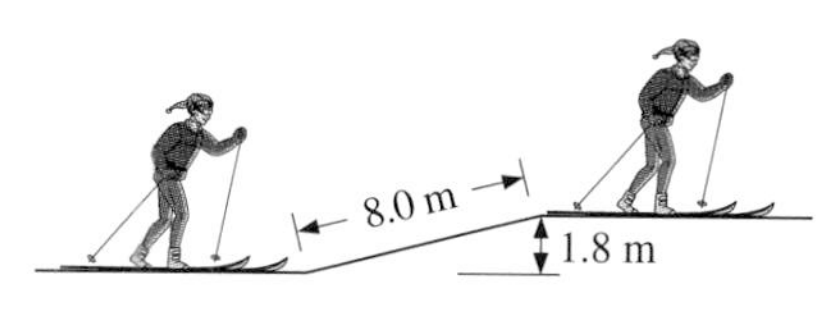

PROBLEM 7-40

7-41. A small block of mass m slides without friction around the loop-the-loop apparatus shown in the accompanying figure. (*a*) If the block starts from rest at A, what is its speed at B? (*b*) What is the force of the track on the block at B?

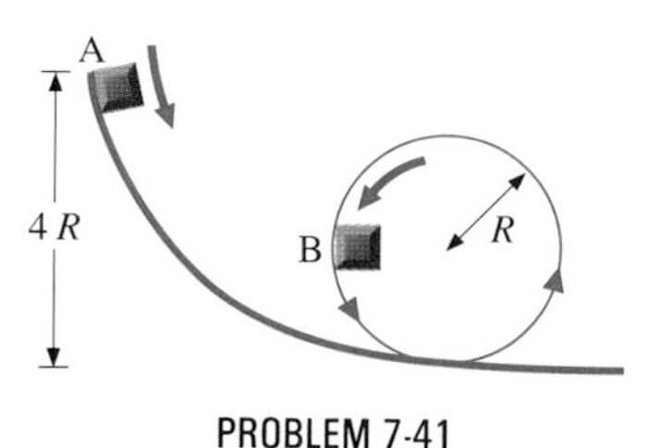

PROBLEM 7-41

7-42. The massless spring of a spring gun has a force constant $k = 12$ N/cm. When the gun is aimed vertically, a 15-g projectile is shot to a height of 5.0 m above the end of the expanded spring. (See the accompanying figure.) How much was the spring compressed initially?

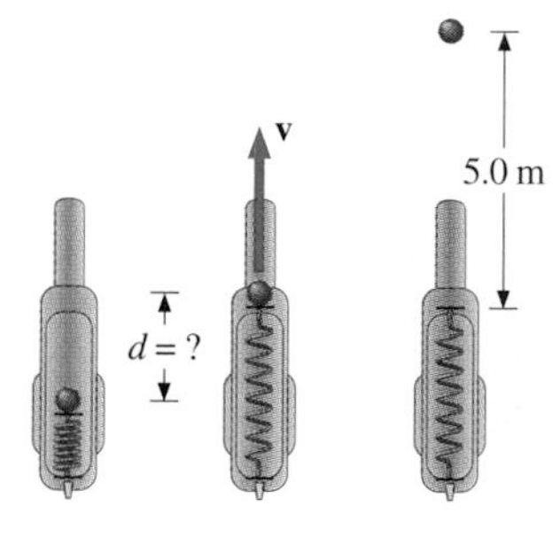

PROBLEM 7-42

7-43. A small ball is tied to a string and set rotating with negligible friction in a vertical circle. Prove that the tension in the string at the bottom of the circle exceeds that at the top of the circle by six times the weight of the ball.

7-44. How fast does a proton have to be moving at the surface of the sun in order to escape from the sun?

7-45. A projectile is launched from the earth toward the moon with the escape velocity calculated assuming an isolated earth. What is the speed of the projectile when it crashes into the moon?

7-46. Block 2 of the accompanying figure slides along a frictionless table as block 1 falls. Find the speed of the blocks after they have each moved 2.0 m. Assume that they start at rest and that the pulley has negligible mass. Use $m_1 = 2.0$ kg and $m_2 = 4.0$ kg.

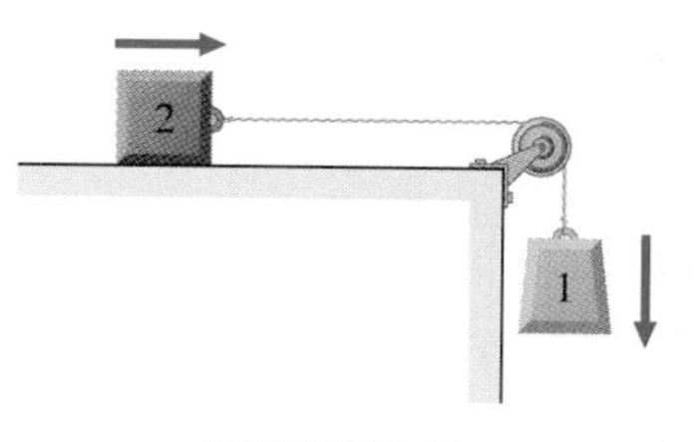

PROBLEM 7-46

7-47. A body of mass m and negligible size starts from rest and slides down the surface of a frictionless solid sphere of radius R. (See the accompanying figure.) Prove that the body leaves the sphere when $\theta = \cos^{-1}(2/3)$.

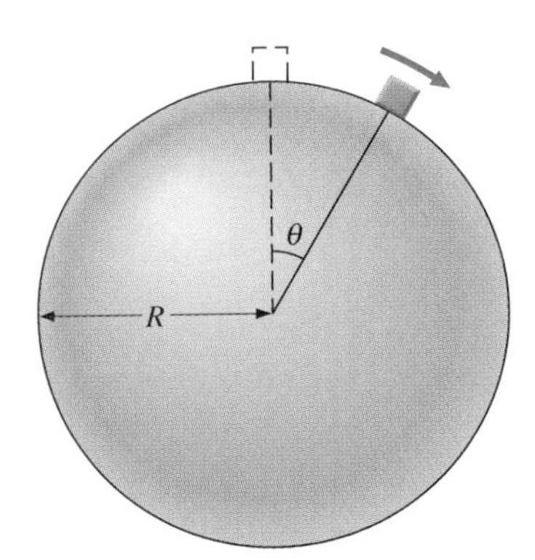

PROBLEM 7-47

7-48. With what speed does a spacecraft have to be launched from the surface of the earth if it is to escape from the solar system? Consider only the gravitational attraction of the earth and sun.

7-49. Consider a meteor moving slowly toward the earth along a straight line between the earth and sun. At a distance of 1.00×10^5 km from the center of the earth, the meteor is moving with a speed of 100 m/s. (*a*) Considering the effects of both the earth and sun, what is the speed of the meteor when it enters the earth's atmosphere at a distance of approximately 6.48×10^6 m from the center of the earth? (*b*) Is the effect of the sun important here?

Mechanical Energy and the Presence of Nonconservative Forces

7-50. A boy throws a ball of mass 0.25 kg straight upward with an initial speed of 20 m/s. When the ball returns to the boy, its speed is 17 m/s. How much work does air resistance do on the ball during its flight?

7-51. Do part (*a*) of Prob. 7-23 by relating the work of the nonconservative force to the change in total mechanical energy.

7-52. A mouse of mass 200 g falls 100 m down a vertical mine shaft and lands at the bottom with a speed of 8.0 m/s. During its fall, how much work is done on the mouse by air resistance?

7-53. Calculate the coefficient of kinetic friction in Prob. 7-22 by setting the work of friction equal to the change in total mechanical energy of the block.

7-54. Find the work of friction in Prob. 7-26 by equating the work of the nonconservative force to the change in total mechanical energy of the block.

7-55. A sled of mass 70 kg starts from rest and slides down a 10° incline 80 m long. It then travels for 20 m horizontally before starting back up an 8° incline. It travels 80 m along this incline before coming to rest. What is the net work done on the sled by friction?

7-56. A girl on a skateboard (total mass = 40 kg) is moving at a speed of 10 m/s at the bottom of a long ramp. The ramp is inclined at 20° with respect to the horizontal. If she travels 14.2 m along the ramp before stopping, what is the net frictional force on her?

7-57. A baseball of mass 0.25 kg is hit at home plate with a speed of 40 m/s. When it lands in a chair in the left-field bleachers a horizontal distance 120 m from home plate, it is moving at 30 m/s. If the ball lands 20 m above the spot where it was hit, how much work is done on it by air resistance?

7-58. A massless spring with force constant $k = 200$ N/m hangs from the ceiling. A 2.0-kg weight is attached to the free end of the spring and released. If the weight falls 17 cm before starting back upwards, how much work is done by friction during its descent?

7-59. Suppose a frictional force also acts on the particle of Prob. 7-36. If the particle's speed when it arrives at $x = 4.0$ m is 9.0 m/s, how much work is done on it by the frictional force between $x = -4.0$ m and $x = 4.0$ m?

Power

7-60. A man of mass 80 kg runs up a flight of stairs 20 m high in 10 s. (*a*) How much power is used to lift the man? (*b*) If the man's body is 25 percent efficient, how much power does he expend?

7-61. The man of the previous problem consumes approximately 1.05×10^7 J (2500 food calories) of energy per day in maintaining a constant weight. What is the average power he produces over a day? Compare this with his power production when he runs up the stairs.

7-62. The sun produces approximately 4.0×10^{33} ergs of energy each second. How long can a 100-W lightbulb stay illuminated with 1 s of the sun's energy production?

7-63. An electron in a television tube is accelerated uniformly from rest to a speed of 8.4×10^7 m/s over a distance of 2.5 cm. What is the power delivered to the electron at the instant that its displacement is 1.0 cm?

7-64. Coal is lifted out of a mine a vertical distance of 50 m by an engine that supplies 500 W to a conveyer belt. How much coal per minute can be brought to the surface? Ignore the effects of friction.

7-65. A girl pulls her 15-kg wagon along a flat sidewalk by applying a 10-N force at 37° to the horizontal. Assume that friction is negligible and that the wagon starts from rest. (*a*) How much work does the girl do on the wagon in the first 2.0 s. (*b*) How much instantaneous power does she exert at $t = 2.0$ s?

7-66. A typical automobile engine has an efficiency of 25 percent. Suppose that the engine of a 1000-kg automobile has a maximum power output of 140 hp. What is the maximum grade that the automobile can climb at 50 km/h if the frictional retarding force on it is 300 N?

7-67. When jogging at 13 km/h on a level surface, a 70-kg man uses energy at a rate of approximately 850 W. Using the fact that the "human engine" is approximately 25 percent efficient, determine the rate at which this man uses energy when jogging up a 5.0° slope at this same speed. Assume that the frictional retarding force is the same in both cases.

General Problems

7-68. Use the principle of mechanical-energy conservation to prove that the time it takes a particle of mass m to go from x_1 to x_2 is given by

$$\Delta t = \pm \int_{x_1}^{x_2} \left\{ \frac{m}{2[E - U(x)]} \right\}^{1/2} dx,$$

where the sign depends on the direction of the velocity between x_1 and x_2.

7-69. Use the result of Prob. 7-68 to calculate the time Δt it takes an object dropped from a height h to hit the ground. Compare this with the result found using the free-fall equations.

7-70. In deriving the law of mechanical-energy conservation, we used the work-energy theorem to get

$$W_{AR} + W_{RB} = T_B - T_A.$$

We next defined potential energy by

$$U_P = -W_{RP},$$

which immediately led to the mechanical-energy conservation statement

$$T_A + U_A = T_B + U_B.$$

Now suppose that instead of the reference point R we choose R′. Then

$$W_{AR'} + W_{R'B} = T_B - T_A,$$

which leads to

$$T_A + U'_A = T_B + U'_B,$$

where U'_P is defined by

$$U'_P = -W_{R'P}.$$

Use $W_{PR'} = W_{PR} + W_{RR'}$ to show that these two statements of mechanical-energy conservation are equivalent and, consequently, that the choice of reference point is arbitrary.

7-71. An object of mass 10 kg is released at point A, slides to the bottom of the 30° incline, then collides with a horizontal massless spring, compressing it a maximum distance of 0.75 m. (See the accompanying figure.) The spring constant is 500 N/m, the height of the incline is 2.0 m, and the horizontal surface is frictionless. (*a*) What is the speed of the object at the bottom of the incline? (*b*) What is the work of friction on the object while it is on the incline? (*c*) The spring recoils and sends the object back toward the incline. What is the speed of the object when it reaches the base of the incline? (*d*) What vertical distance does it move back up the incline?

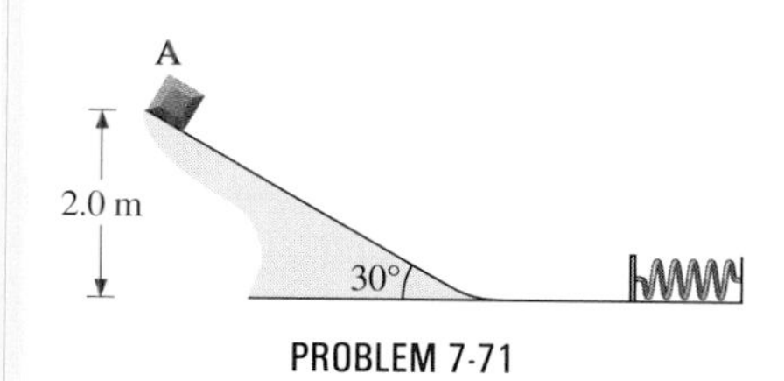

PROBLEM 7-71

7-72. The accompanying figure shows a small ball of mass m attached to a string of length a. A small peg is located a distance h below the point where the string is supported. If the ball is released when the string is horizontal, show that h must be greater than $3a/5$ if the ball is to swing completely around the peg.

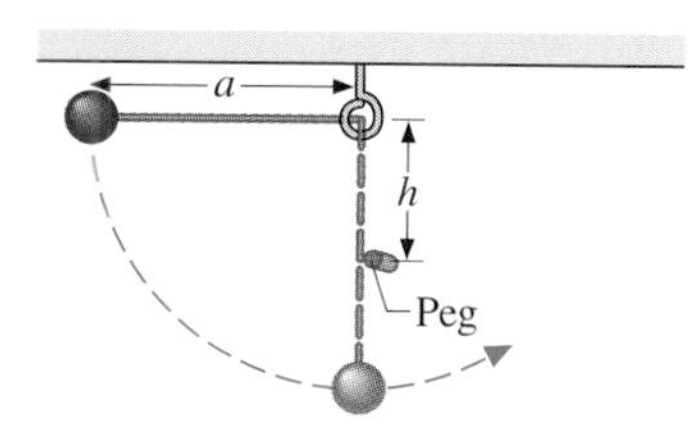

PROBLEM 7-72

7-73. (*a*) Sketch a graph of the potential-energy function $U(x) = kx^2/2 + Ae^{-\alpha x^2}$, where k, A, and α are constants. (*b*) What is the force corresponding to this potential energy? (*c*) Suppose a particle of mass m moving with this potential energy has a velocity v_a when its position is $x = a$. Show that the particle does not pass through the origin unless

$$A \le \frac{mv_a^2 + ka^2}{2(1 - e^{-\alpha a^2})}$$

Interpret this result in terms of the plot of $U(x)$.

7-74. A planet of mass m moves in a circular orbit of radius r around a star of mass M. Show that the kinetic, potential, and total mechanical energies of the planet are given, respectively, by

$$T(r) = \frac{GMm}{2r}, \qquad U(r) = -\frac{GMm}{r}, \qquad E(r) = -\frac{GMm}{2r}.$$

7-75. A crate on rollers is being pushed without frictional loss of energy across the floor of a freight car. The car is moving to the right with a constant speed v_0. If the crate starts at rest relative to the freight car, then from the work-energy theorem, $Fd = mv^2/2$, where d, the distance the crate moves, and v, the speed of the crate, are both measured relative to the freight car. (*a*) To an observer at rest beside the tracks, what distance d' is the crate pushed when it moves the distance d in the car? (*b*) What are the crate's initial and final speeds v'_0 and v' as measured by the observer beside the tracks? (*c*) Show that $Fd' = m(v')^2/2 - m(v'_0)^2/2$ and, consequently, that work is equal to the change in kinetic energy in both reference systems.

PROBLEM 7-75

7-76. A 4.0-kg particle moving along the x axis is acted upon by the force whose functional form is shown in the accompanying figure. The velocity of the particle at $x = 0$ is $v = 6.0$ m/s. Use the work-energy theorem to find the particle's speed at $x =$ (*a*) 2.0 m, (*b*) 4.0 m, (*c*) 10.0 m. (*d*) Does the particle turn around at some point and head back toward the origin? (*e*) Repeat part (*d*) if $v = 2.0$ m/s at $x = 0$.

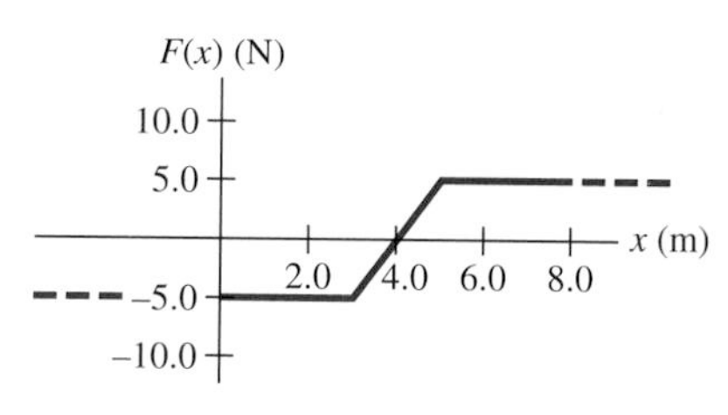

PROBLEM 7-76

7-77. A particle of mass 0.50 kg moves along the x axis with a potential energy whose dependence on x is shown in the accompanying figure. (*a*) What is the force on the particle at $x = 2.0$, 5.0, 8.0, and 12.0 m? (*b*) If the total mechanical energy E of the particle is −6.0 J, what are the minimum and maximum positions of the particle? (*c*) What are these positions if $E = 2.0$ J? (*d*) If $E = 16.0$ J, what are the speeds of the particle at the positions listed in part (*a*)?

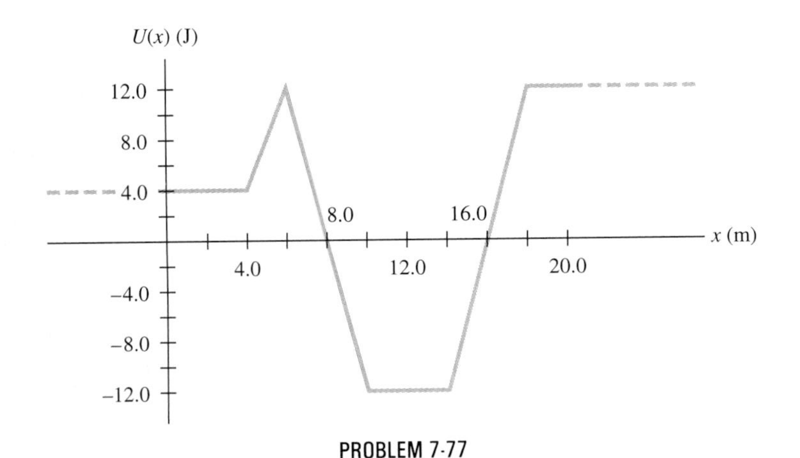

PROBLEM 7-77

7-78. A car of mass 1000 kg moves along a horizontal road at a speed of 25 m/s when its engine delivers 40 hp to the wheels. With the same power, what is the car's speed when it climbs a 5.0 percent grade? Assume that the air resistance the car encounters is independent of its speed.

7-79. A 2.0-kg block starts from rest and slides a distance d down a frictionless 37° incline before colliding with a massless spring of force constant $k = 500$ N/m. (See the accompanying figure.) After making contact with the spring, the block slides an additional 20 cm down the incline before momentarily stopping. (*a*) What is the value of d? (*b*) At what point is the speed of the block a maximum?

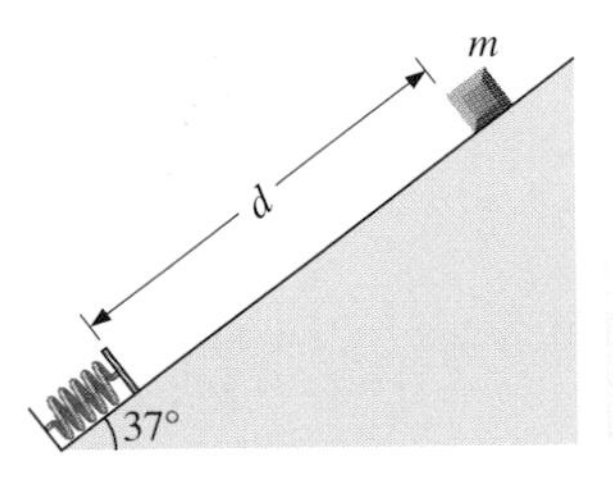

PROBLEM 7-79

7-80. Constant power P is delivered to a car of mass m by its engine. Show that if air resistance can be ignored, the distance covered in a time t by the car, starting from rest, is given by $s = (8P/9m)^{1/2}t^{3/2}$.

7-81. Suppose that the air resistance a car encounters is independent of its speed. When the car travels at 15 m/s, its engine delivers 20 hp to its wheels. (*a*) What is the power delivered to the wheels when the car travels at 30 m/s? (*b*) How much energy does the car use in covering 10 km at 15 m/s? at 30 m/s? Assume that the engine is 25 percent efficient. (*c*) Answer the same questions if the force of air resistance is proportional to the speed of the automobile. (*d*) What do these results, plus your experience with gasoline consumption, tell you about air resistance?

The motion of a complicated system of particles.

CHAPTER 8 Systems of Particles

Preview

This chapter is devoted to two very important properties of systems of particles: the center of mass and the moment of inertia. The main topics to be discussed here are the following:

1. **Center of mass.** The center of mass is defined for systems of discrete particles and for rigid bodies. Various methods used to locate the center of mass are discussed.
2. **Motion of the center of mass.** We examine the relationship between the motion of the center of mass and the motions of the individual particles of a system. How the acceleration of the center of mass is related to the net external force on the system is also considered.
3. **Moment of inertia.** The moment of inertia for a collection of discrete particles as well as for a rigid body is defined. Sample calculations of the moments of inertia of various systems are made.

Every object we have considered thus far has either been a particle or has moved like one. Most objects, however, do not move so simply. For example, all points on the rotating wheel of Fig. 8-1*a* do not follow parallel paths, while the different parts of the object of Fig. 8-1*b* don't even stay a fixed distance apart. The motion of systems such as these is obviously quite difficult to analyze, for we must describe how particles move through space along with how they move relative to one another.

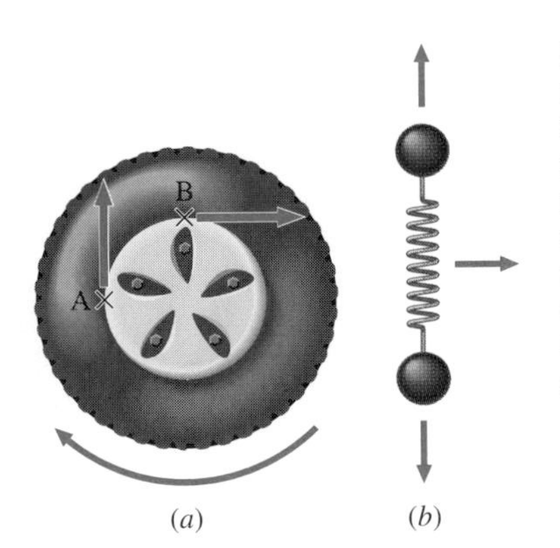

FIGURE 8-1 (*a*) A wheel is rotating clockwise around its center. At the instant represented by the figure, point A is moving upward and point B is moving to the right. (*b*) This system moves to the right as the spheres at each end of the spring oscillate back and forth.

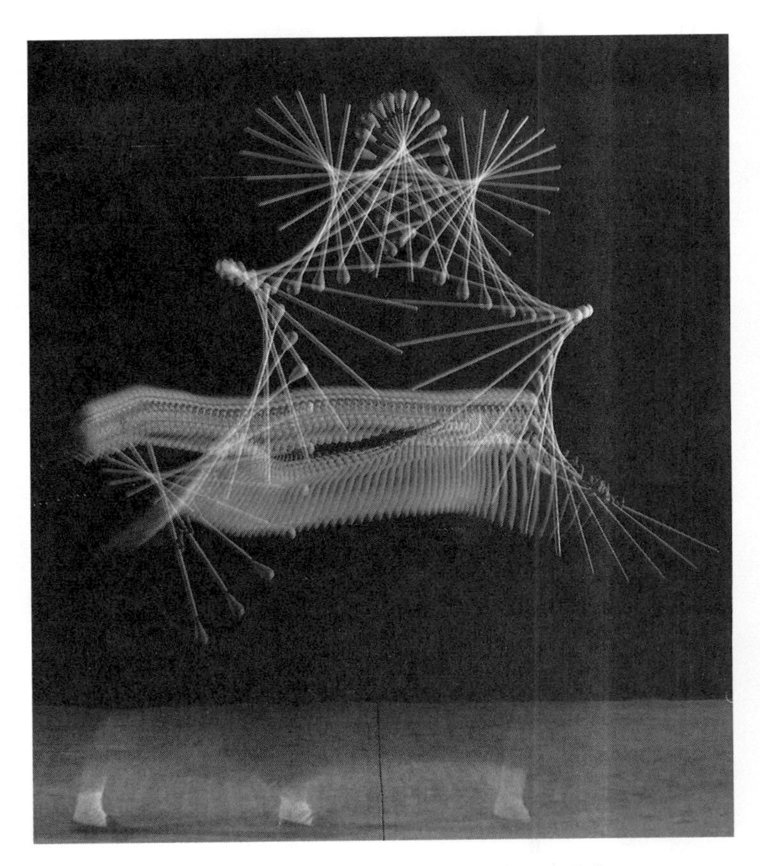

FIGURE 8-2 The center of mass of the thrown baton follows a simple parabolic path as if it were a single particle moving like a simple projectile.

A quantity that plays a very important role in the description of the motion of these complicated systems is the **center of mass**. The center of mass of a system of particles is an imaginary point whose position depends on the spatial distribution of the mass of that system. Its usefulness lies in the fact that its motion is determined only by the net external force acting on the system and is completely independent of any internal forces among the particles. Frequently, the net external force is quite simple. As a result, the motion of the center of mass is also uncomplicated. For example, the net external force on the thrown baton of Fig. 8-2 is just its weight, so its center of mass moves along the parabolic path of a projectile. It's clear, however, that the path of any other point on the spinning baton is far more difficult to determine.

There are times when the position and velocity of the center of mass are all we care to know about a body's motion. For example, when a shortstop contemplates his throw to first base, he really doesn't care exactly how the stitches on the baseball are oriented when it approaches the glove of the first baseman. However, if the shortstop spends his winters as a knife-thrower, then in this work the orientation, as well as the location, of his projectile becomes important. In this case, he must also be concerned with the rotational motion of his projectile. The **moment of inertia** is an important part of the description of this motion. The moment of inertia also depends on the distribution of the mass of the body, but unlike the center of mass, the moment of inertia is *not* a point. Instead, it is a property of a body and plays a role in rotational motion that is analogous to the role of mass in translational motion.

We will begin by defining how the center of mass is located for a system of particles. Its use in describing the motion of the system is then discussed. In the latter part of this chapter, we will define the moment of inertia and calculate it for various systems. However, the role of the moment of inertia in describing rotational motion will not be discussed until the next chapter.

8-1 Center of Mass

Let's start with a one-dimensional array of N particles such as that of Fig. 8-3*a*. If m_i and x_i are the mass and position of the ith particle, the position x_{CM} of the center of mass of the array is by definition

$$x_{CM} = \frac{m_1x_1 + m_2x_2 + \ldots + m_Nx_N}{m_1 + m_2 + \ldots + m_N}. \tag{8-1}$$

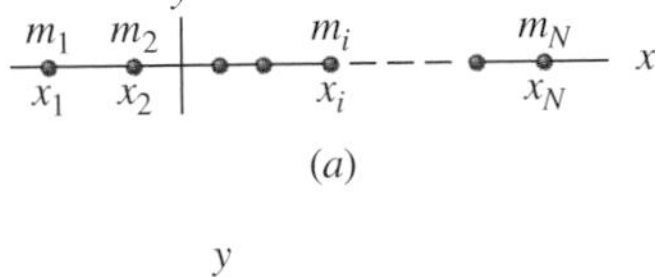

FIGURE 8-3 (*a*) A one-dimensional array of N particles. (*b*) The center of mass of this two-particle system is located at the position shown.

For example, the center of mass (CM) of the two particles of Fig. 8-3*b* is located at

$$x_{CM} = \frac{(2.0\text{ kg})(-3.0\text{ m}) + (4.0\text{ kg})(9.0\text{ m})}{2.0\text{ kg} + 4.0\text{ kg}} = 5.0\text{ m}.$$

We can easily extend this definition to an arbitrary system of particles. If the mass and position of particle i of an N-particle system (see Fig. 8-4) are m_i and (x_i, y_i, z_i), respectively, then the position (x_{CM}, y_{CM}, z_{CM}) of the center of mass is

$$x_{CM} = \frac{m_1x_1 + m_2x_2 + \ldots + m_Nx_N}{m_1 + m_2 + \ldots + m_N} = \frac{\sum_i m_ix_i}{M}$$

$$y_{CM} = \frac{m_1y_1 + m_2y_2 + \ldots + m_Ny_N}{m_1 + m_2 + \ldots + m_N} = \frac{\sum_i m_iy_i}{M}$$

$$z_{CM} = \frac{m_1z_1 + m_2z_2 + \ldots + m_Nz_N}{m_1 + m_2 + \ldots + m_N} = \frac{\sum_i m_iz_i}{M}, \qquad (8\text{-}2a)$$

where $M = \sum_i m_i$ is the total mass of the system. In terms of the position vectors $\mathbf{r}_i$ and $\mathbf{r}_{CM}$, Eq. (8-2*a*) can be written more compactly as

$$\mathbf{r}_{CM} = \frac{\sum_i m_i\mathbf{r}_i}{M}. \qquad (8\text{-}2b)$$

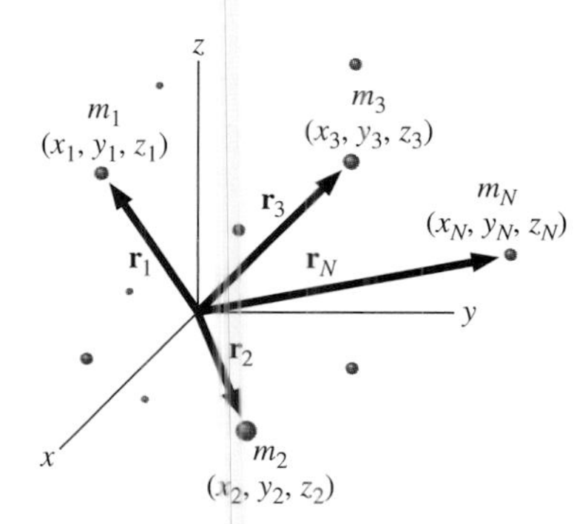

FIGURE 8-4 A system of N particles.

EXAMPLE 8-1 THE CENTER OF MASS OF THREE PARTICLES

In Fig. 8-5*a*, particle 1 has position (3.0, 2.0) m and mass m_1 = 6.0 kg; particle 2 has position (4.0, −4.0) m and mass m_2 = 2.0 kg; particle 3 has position (−2.0, 7.0) m and mass m_3 = 4.0 kg. Find the position of the center of mass of this system.

SOLUTION By definition, the coordinates of the center of mass are

$$x_{CM} = \frac{(6.0\text{ kg})(3.0\text{ m}) + (2.0\text{ kg})(4.0\text{ m}) + (4.0\text{ kg})(-2.0\text{ m})}{12.0\text{ kg}}$$
$$= 1.5\text{ m}$$

$$y_{CM} = \frac{(6.0\text{ kg})(2.0\text{ m}) + (2.0\text{ kg})(-4.0\text{ m}) + (4.0\text{ kg})(7.0\text{ m})}{12.0\text{ kg}}$$
$$= 2.7\text{ m}.$$

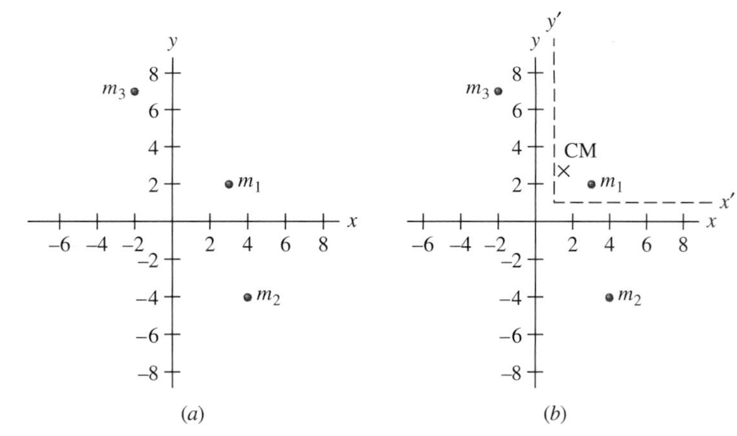

FIGURE 8-5 (*a*) A three-particle system. (*b*) A reference frame with the origin at (1.0, 1.0) m relative to the frame of part (*a*) is used. Notice that the physical location of the center of mass is unchanged.

The center of mass of the system of three particles is therefore located at (1.5, 2.7) m.

DRILL PROBLEM 8-1

(*a*) Consider the two particles shown in Fig. 8-6. Show that the center of mass is located on the line connecting them such that $d_1/d_2 = m_2/m_1$. (*b*) Find the position of the center of mass of the earth and moon. Their centers are 3.84×10^5 km apart, and their masses are $M_E = 5.98 \times 10^{24}$ kg and $M_M = 7.36 \times 10^{22}$ kg. ANS. (*b*) 4.67×10^3 km from the center of the earth, whose radius is 6.37×10^3 km.

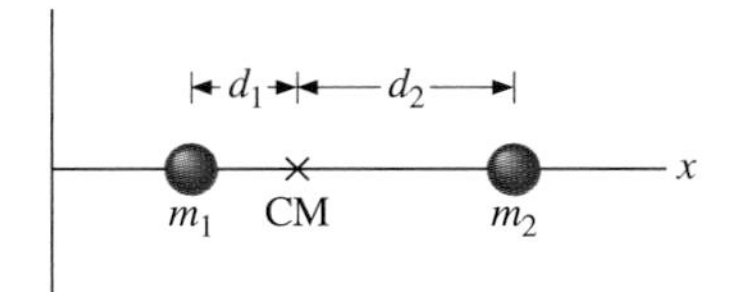

FIGURE 8-6 The position of the center of mass is between the two particles and on the line connecting them such that $d_1/d_2 = m_2/m_1$.

The physical location of the center of mass is unaffected by the choice of reference frame. As an example, consider the system described in Example 8-1. Suppose we use a new reference frame whose origin is at (1.0, 1.0) m relative to the original reference frame. (See Fig. 8-5*b*.) The coordinates of particles 1, 2, and 3 become (2.0, 1.0) m, (3.0, −5.0) m, and (−3.0, 6.0) m, and the x position of the center of mass in this frame is

$$x_{CM} = \frac{(6.0\text{ kg})(2.0\text{ m}) + (2.0\text{ kg})(3.0\text{ m}) + (4.0\text{ kg})(-3.0\text{ m})}{12.0\text{ kg}}$$
$$= 0.5\text{ m}$$

and y_{CM} = 1.7 m. As you can see in the figure, the physical location of the center of mass has not changed. The center of mass is 1.5 m to the left of and 0.7 m above particle 1 in either frame.

The proof that the physical location of the center of mass of an *arbitrary system of particles* is unaffected by the choice of reference frame is considered in Prob. 8-45. For now, we simply use this property to demonstrate how the center of mass of a system with spatial symmetry can often be found without the necessity of a formal calculation.

Suppose we have two particles of equal mass a fixed distance d apart. These particles are positioned symmetrically about the midpoint of the line joining them. If we place the origin of our coordinate system at this point, with the x axis along the line (see Fig. 8-7a), then the coordinate of one particle is $x = +d/2$ while the other is at $x = -d/2$. From Eq. (8-2a), we immediately deduce that the center of mass of the two particles must be at the origin of our coordinate system, or at the midpoint of the line joining the particles. Figure 8-7b shows six particles of equal mass distributed evenly along a circle. These particles are positioned symmetrically about the center of the circle; that is, for a particle at (x, y) relative to an origin at the center, there exists a particle at $(-x, -y)$. The center of mass therefore lies at the center of the circle.

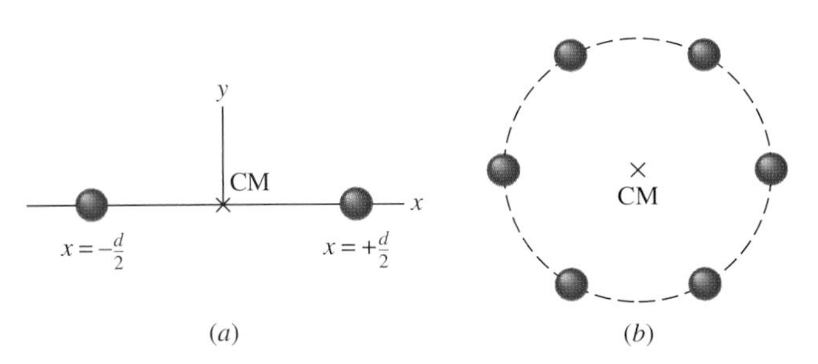

FIGURE 8-7 (a) The center of mass of two identical particles is at the midpoint of the line connecting the particles. (b) Six identical particles are distributed symmetrically along a circle. The center of mass of this system is at the center of the circle.

Our symmetry argument applies not only to systems of discrete particles, but also to rigid bodies. A *rigid body* is a system of closely spaced particles whose positions are fixed relative to one another. The particles are generally so close together that the distribution of their mass is taken to be continuous. To a good approximation, everyday objects such as tables and doors are rigid bodies.

An example of a symmetric rigid body, a solid uniform* disk of negligible thickness, is shown in Fig. 8-8a. For an infinitesimally small segment of mass Δm of the disk at the position (x, y) relative to the center of the disk, there is a corresponding segment of mass Δm at $(-x, -y)$. The center of mass must therefore be located at the center of the disk. Other examples of two-dimensional rigid bodies with symmetry are shown in the figure. Notice that the precise location of the center of mass of the isosceles triangular plate of Fig. 8-8c is not immediately obvious. The center of mass must lie, however, somewhere along the line from the apex of the triangle perpendicular to the baseline since the mass of the plate is distributed symmetrically about that line. We can locate the center of mass precisely with the help of integration techniques, a topic to be discussed in the next section. Finally, our symmetry argument can just as easily be applied to three-dimensional rigid bodies. For example,

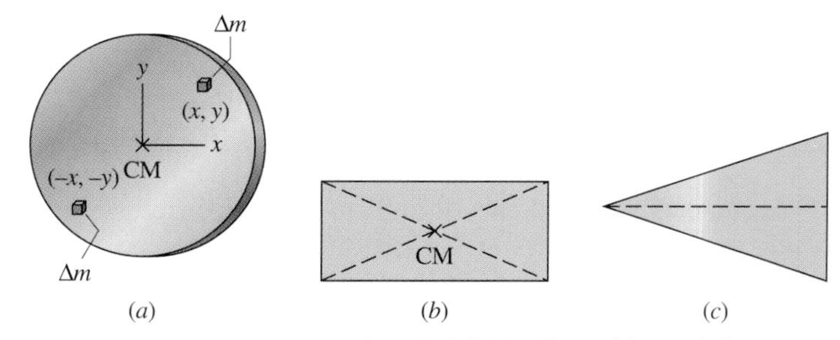

FIGURE 8-8 Three two-dimensional rigid bodies with spatial symmetry. The center of mass of the triangular plate lies along the dashed line.

the center of mass of a uniform sphere is at its geometric center, and the center of mass of a solid uniform cylinder is at the center of the circular cross section bisecting the cylinder.

It is possible to find without integration the center of mass of a complicated body if it is made up of geometrically symmetric segments. Consider the object of Fig. 8-9 as an example. It is composed of a disk of mass M_1 connected to a thin stick of mass M_2. From the definition of the center of mass, we may write

$$M_1(x_{CM})_1 = \left(\sum_i m_i x_i\right)_1$$

and

$$M_2(x_{CM})_2 = \left(\sum_i m_i x_i\right)_2,$$

where $(x_{CM})_1$ and $(x_{CM})_2$ are the x coordinates of the centers of mass of the disk and the stick, and $(\sum_i m_i x_i)_1$ and $(\sum_i m_i x_i)_2$ represent summations over all the particles of the disk and the stick. Adding these two equations, we obtain

$$M_1(x_{CM})_1 + M_2(x_{CM})_2 = \left(\sum_i m_i x_i\right)_1 + \left(\sum_i m_i x_i\right)_2 = \left(\sum_i m_i x_i\right)_{\text{all particles}}.$$

Now the x coordinate of the center of mass of the entire object is by definition

$$x_{CM} = \frac{1}{M_1 + M_2}\left(\sum_i m_i x_i\right)_{\text{all particles}}.$$

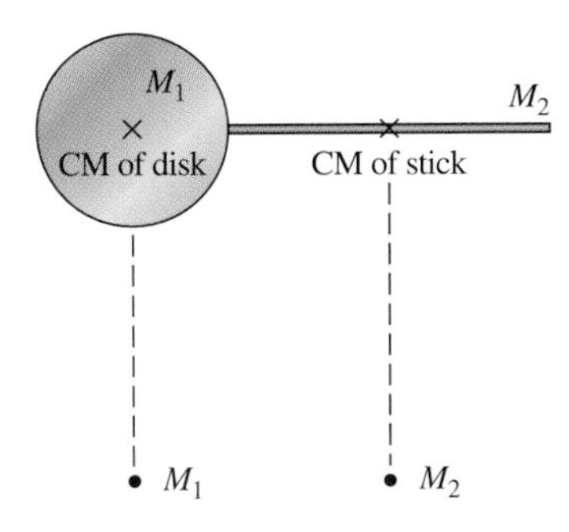

FIGURE 8-9 A rigid body made up of two segments with spatial symmetry. Each segment is replaced by a particle of the same mass located at the center of mass of that segment. The center of mass of the imaginary particles is the center of mass of the body.

*A uniform object has constant mass density.

We therefore find

$$x_{CM} = \frac{1}{M_1 + M_2}[M_1(x_{CM})_1 + M_2(x_{CM})_2].$$

What we have done in essence is to *replace each segment of the object by a particle of the same mass located at the center of mass of that segment*. (See Fig. 8-9.) We then locate the center of mass of these particles. It is at the same position as the center of mass of the entire object. For an object with an arbitrary number of segments, we write

$$x_{CM} = \frac{1}{\sum_i M_i}\left[\sum_i M_i(x_{CM})_i\right],$$

where M_i and $(x_{CM})_i$ are the mass and the x coordinate of the center of mass of the ith segment of the object. The other coordinates of the center of mass may of course be obtained in the same way.

EXAMPLE 8-2 **The Center of Mass of a System of Symmetric Objects**

A cross-sectional view of four uniform cubic blocks stacked by a small child is shown in Fig. 8-10*a*. Each block has 6.0-cm edges. Find the coordinates of the center of mass of the blocks with respect to the xy coordinate system whose origin is at point O. Treat the blocks as indicated.

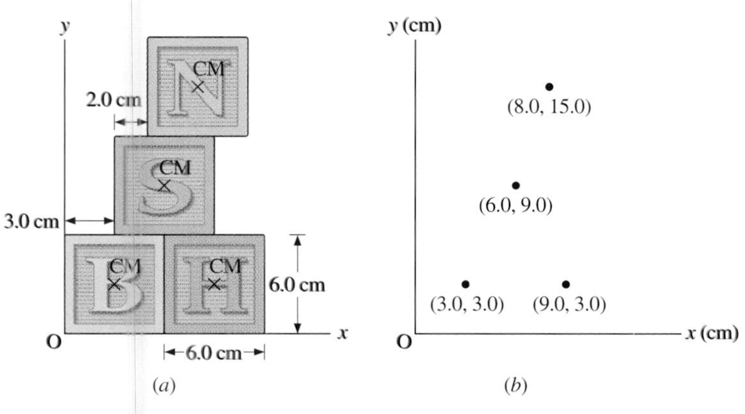

FIGURE 8-10 (*a*) A cross-sectional view of four uniform cubic blocks. (*b*) Imaginary particles at the centers of mass of the blocks are used to locate the center of mass of the system.

SOLUTION From symmetry, the center of mass of each of the blocks is at its geometric center. The coordinates of the centers of mass with respect to point O are (3.0, 3.0) cm, (9.0, 3.0) cm, (6.0, 9.0) cm, and (8.0, 15.0) cm. If m is the mass of each block, the center of mass of the system is equivalent to the center of mass of the four particles shown in Fig. 8-9*b*. The coordinates of the center of mass are then

$$x_{CM} = \frac{1}{4m}[m(3.0\text{ cm}) + m(9.0\text{ cm}) + m(6.0\text{ cm}) + m(8.0\text{ cm})] = 6.5\text{ cm}$$

and

$$y_{CM} = \frac{1}{4m}[m(3.0\text{ cm}) + m(3.0\text{ cm}) + m(9.0\text{ cm}) + m(15.0\text{ cm})] = 7.5\text{ cm}.$$

DRILL PROBLEM 8-2

Suppose the origin of the coordinate system of Fig. 8-3*b* is moved to the position of the 2.0-kg particle. What is x_{CM} now?
ANS. 8.0 m.

DRILL PROBLEM 8-3

Particles of equal mass are located at the eight corners of a cube. Find the position of the center of mass of this system.
ANS. At the center of the cube.

DRILL PROBLEM 8-4

If a fifth block is placed directly above the highest block in Fig. 8-10*a*, what is the position of the center of mass of the system with respect to point O?
ANS. (6.8, 10.2) cm.

*8-2 Locating the Center of Mass by Integration

We now see how integration is used to determine the location of the center of mass of a rigid body. This approach is especially useful when symmetry arguments cannot be applied. The position of the center of mass is determined by dividing the body into infinitesimal mass elements dm and using Eq. (8-2*a*) in integral form. The coordinates of the center of mass are then given by

$$x_{CM} = \frac{\lim_{\Delta m_i \to 0}\sum_i \Delta m_i x_i}{M} = \frac{1}{M}\int x\,dm$$

$$y_{CM} = \frac{\lim_{\Delta m_i \to 0}\sum_i \Delta m_i y_i}{M} = \frac{1}{M}\int y\,dm$$

$$z_{CM} = \frac{\lim_{\Delta m_i \to 0}\sum_i \Delta m_i z_i}{M} = \frac{1}{M}\int z\,dm. \qquad (8\text{-}3a)$$

It is easiest to evaluate these integrals if they are rewritten as integrations over the volume of the body of interest. For a body of mass density ρ, we may write

$$dm = \rho\,dV,$$

where dV is the volume of the segment of mass dm. We then have

$$x_{CM} = \frac{1}{M}\int x\rho\, dV,$$
$$y_{CM} = \frac{1}{M}\int y\rho\, dV,$$
$$z_{CM} = \frac{1}{M}\int z\rho\, dV. \quad (8\text{-}3b)$$

For one- and two-dimensional bodies, Eqs. (8-3b) become, respectively,

$$x_{CM} = \frac{1}{M}\int x\lambda\, dx,$$

and

$$x_{CM} = \frac{1}{M}\int x\sigma\, dA, \qquad y_{CM} = \frac{1}{M}\int y\sigma\, dA,$$

where λ is the linear mass density (mass per unit length) and σ is the area mass density (mass per unit area). When the mass density of a rigid body is constant, it can be factored out of the integral.

EXAMPLE 8-3 Center of Mass of a Rod

The mass per unit length of the thin, uniform, rigid rod of Fig. 8-11 is λ. Find the position of the center of mass of the rod.

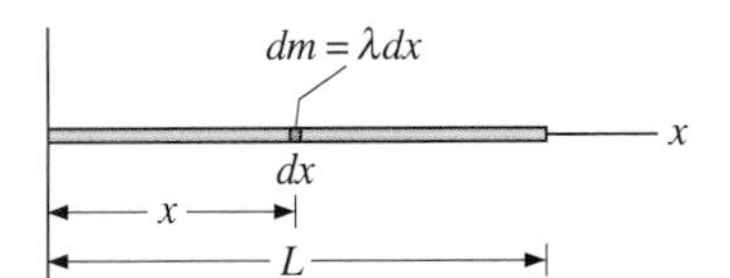

FIGURE 8-11 A thin uniform rod.

SOLUTION Since the rod is thin, it can be treated as a one-dimensional object. With $dm = \lambda\, dx$ and $M = \lambda L$, the position of the rod's center of mass is

$$x_{CM} = \frac{1}{M}\int x\, dm = \frac{1}{\lambda L}\int_0^L x\lambda\, dx = \frac{L}{2}.$$

As expected, the center of mass of the uniform rod is at its geometric center.

EXAMPLE 8-4 Center of Mass of a Triangular Plate

The mass per unit area of the uniform triangular plate of Fig. 8-12 is σ. Find the position of its center of mass.

SOLUTION Since the mass is symmetrically distributed with respect to the x axis, the center of mass must be on that axis; that is, $y_{CM} = 0$. We need, therefore, only to find x_{CM}. The area of the triangular plate is $A = (1/2)(\text{base})(\text{height}) = (1/2)(2a)(h) = ah$, so the mass of the plate is $M = \sigma ah$. The mass of the shaded infinitesimal element of height $2y$ and width dx is $dm = \sigma 2y\, dx$. We now obtain

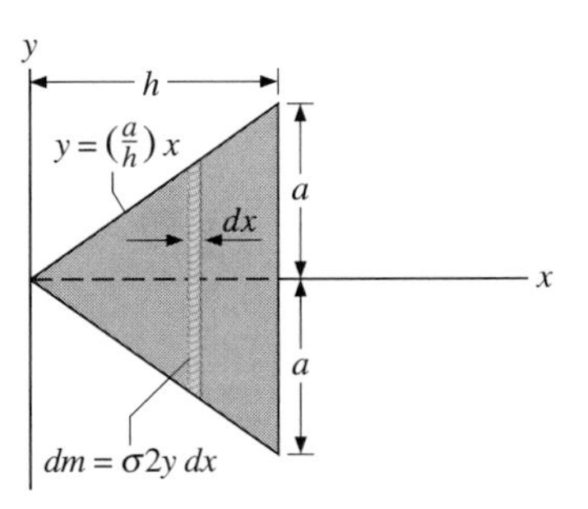

FIGURE 8-12 A uniform triangular plate.

$$x_{CM} = \frac{1}{M}\int x\, dm = \frac{1}{\sigma ah}\int_0^h x\sigma 2y\, dx = \frac{2}{ah}\int_0^h xy\, dx. \quad (i)$$

The upper edge of the plate is a straight line with slope a/h, so

$$y = \frac{a}{h}x. \quad (ii)$$

After substituting Eq. (ii) into Eq. (i), we find

$$x_{CM} = \frac{2}{ah}\int_0^h x\left(\frac{a}{h}\right)x\, dx = \frac{2}{h^2}\int_0^h x^2\, dx,$$

which, when integrated, gives

$$x_{CM} = \tfrac{2}{3}h.$$

DRILL PROBLEM 8-5

The curved side of the uniform plate of Fig. 8-13 is a parabola represented by $y = ax^2$. Find the position of the center of mass of the plate.
ANS. (0, $3h/5$).

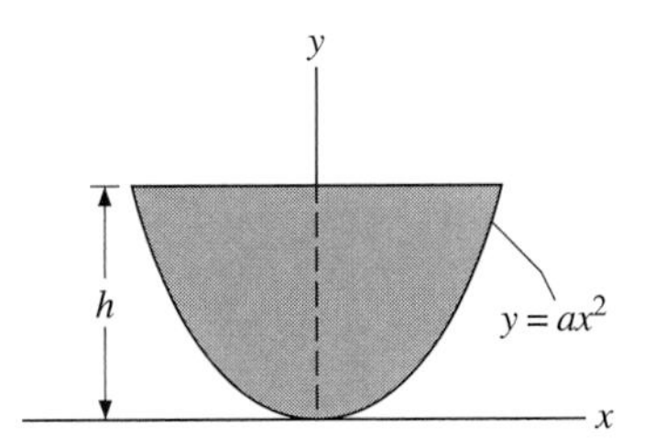

FIGURE 8-13 A uniform parabolic plate.

8-3 Motion of the Center of Mass

In this section we study the motion of the center of mass of a system. For a system of particles, the position of the center of mass is given by

$$\mathbf{r}_{CM} = \frac{1}{M}\sum_i m_i\mathbf{r}_i.$$

The velocity of the center of mass is found by differentiating this equation with respect to time. This yields

$$\frac{d\mathbf{r}_{CM}}{dt} = \frac{1}{M}\sum_i m_i\frac{d\mathbf{r}_i}{dt},$$

or

$$\mathbf{v}_{CM} = \frac{1}{M} \sum_i m_i \mathbf{v}_i. \quad (8\text{-}4a)$$

A second differentiation gives the acceleration of the center of mass:

$$\mathbf{a}_{CM} = \frac{1}{M} \sum_i m_i \mathbf{a}_i. \quad (8\text{-}5a)$$

For continuous mass distributions, these sums become integrals; that is,

$$\mathbf{v}_{CM} = \frac{1}{M} \int \mathbf{v}\, dm, \quad (8\text{-}4b)$$

and

$$\mathbf{a}_{CM} = \frac{1}{M} \int \mathbf{a}\, dm. \quad (8\text{-}5b)$$

Keep in mind that each of these represents three component equations. For example, Eq. (8-5b) can also be written as

$$(a_{CM})_x = \frac{1}{M} \int a_x\, dm; \qquad (a_{CM})_y = \frac{1}{M} \int a_y\, dm;$$

$$(a_{CM})_z = \frac{1}{M} \int a_z\, dm.$$

EXAMPLE 8-5 **MOTION OF THE CENTER OF MASS**

Four particles have the masses, positions, velocities, and accelerations given in the table. Find the position, velocity, and acceleration of the center of mass of these particles.

Particle	Mass (kg)	Position (m)	Velocity (m/s)	Acceleration (m/s^2)
1	0.60	(3.0, 2.0, −4.0)	(9.0, 0.0, 0.0)	(0.0, 0.0, 0.0)
2	0.20	(4.0, −4.0, 2.0)	(3.0, −3.0, 0.0)	(5.0, 2.0, 4.0)
3	0.40	(−2.0, 7.0, −3.0)	(1.0, 7.0, 4.0)	(0.0, 2.0, 4.0)
4	0.80	(0.0, 0.0, 0.0)	(0.0, 0.0, 0.0)	(3.0, 2.0, −7.0)

SOLUTION The x coordinate of the center of mass is

$$\begin{aligned} x_{CM} &= [(0.60 \text{ kg})(3.0 \text{ m}) + (0.20 \text{ kg})(4.0 \text{ m}) \\ &\quad + (0.40 \text{ kg})(-2.0 \text{ m}) + (0.80 \text{ kg})(0.0 \text{ m})]/(2.00 \text{ kg}) \\ &= 0.90 \text{ m}. \end{aligned}$$

Similarly, $y_{CM} = 1.6$ m and $z_{CM} = -1.6$ m.

The velocity of the center of mass is calculated with Eq. (8-4a):

$$\begin{aligned} (v_{CM})_x &= [(0.60 \text{ kg})(9.0 \text{ m/s}) + (0.20 \text{ kg})(3.0 \text{ m/s}) \\ &\quad + (0.40 \text{ kg})(1.0 \text{ m/s}) + (0.80 \text{ kg})(0.0 \text{ m/s})]/(2.00 \text{ kg}) \\ &= 3.2 \text{ m/s}, \end{aligned}$$

while $(v_{CM})_y = 1.1$ m/s and $(v_{CM})_z = 0.80$ m/s. Finally, from Eq. (8-5a),

$$\begin{aligned} (a_{CM})_x &= [(0.60 \text{ kg})(0.0 \text{ m/s}^2) + (0.20 \text{ kg})(5.0 \text{ m/s}^2) \\ &\quad + (0.40 \text{ kg})(0.0 \text{ m/s}^2) \\ &\quad + (0.80 \text{ kg})(3.0 \text{ m/s}^2)]/(2.00 \text{ kg}) \\ &= 1.7 \text{ m/s}^2, \end{aligned}$$

while $(a_{CM})_y = 1.4$ m/s^2 and $(a_{CM})_z = -1.6$ m/s^2.

Now suppose that $\mathbf{F}_i$ is the net force on the ith particle of the system. Substituting $\mathbf{F}_i$ for $m\mathbf{a}_i$ in Eq. (8-5a), we find

$$\mathbf{a}_{CM} = \frac{1}{M}(\mathbf{F}_1 + \mathbf{F}_2 + \ldots + \mathbf{F}_N). \quad (8\text{-}6)$$

The acceleration of the center of mass is therefore equal to the sum of the forces on the individual particles divided by the total mass of the system.

Each particle interacts internally with every other particle as well as with the outside environment, so for a system composed of more than a few particles, the force sum of Eq. (8-6) appears to be very formidable. Fortunately, this is not the case. Supplement 8-1 shows that, since the internal forces between the particles obey Newton's third law, they cancel each other in pairs in the sum. Thus only the sum of the *external* forces remains on the right-hand side of Eq. (8-6), so we may write

$$\sum \mathbf{F} = M\mathbf{a}_{CM}, \quad (8\text{-}7)$$

where

$$\sum \mathbf{F} = \mathbf{F}_1^E + \mathbf{F}_2^E + \ldots + \mathbf{F}_N^E$$

is the net external force on the system of particles. To find the acceleration of the center of mass of a system of particles of total mass M, we imagine that a single fictitious particle of mass M is located at the center of mass. We then use Newton's second law to determine the acceleration of that particle due to the net external force on the system. From Eq. (8-7), that acceleration is the acceleration of the center of mass of the system.

Notice that in obtaining Eq. (8-7), we did not place any restrictions on the motions of the individual particles of the system. The particles can be the molecules of a gas that is either held in a container or is completely unconfined; they can be joined together to form an amorphous blob; or they can be connected to form a rigid body. In all cases, *the center of mass acts as if it were a particle of mass M moving under the influence of just the external forces.*

The motion of the thrown baton of Fig. 8-2 is a simple example that illustrates the significance of Eq. (8-7). The only external forces on the particles of the baton are their weights, whose sum is the total weight $M\mathbf{g}$ of the baton. Since the center of mass moves under the influence of this single force, its acceleration is $\mathbf{g}$, and it follows the parabolic path of projectile motion. The exploding shell of Fig. 8-14 is another good example. In this case, the forces responsible for separating the shell

By bending his body, the high jumper is able to clear the bar with his center of mass below the bar.

FIGURE 8-14 The center of mass of an exploding artillery shell follows the parabolic path of the unexploded shell.

into its many parts are internal forces and have no effect on the motion of the center of mass. The only external force is due to the gravitational field, which makes the center of mass move along a parabolic path.

Actually, we had previously used Eq. (8-7) in Chaps. 5 and 6, when we treated nonrotating rigid bodies as if they were particles. We have now justified that procedure.

EXAMPLE 8-6 Motion of the Center of Mass due to External Forces

The three particles of Fig. 8-15 have masses $m_1 = 4.0$ kg, $m_2 = 2.0$ kg, and $m_3 = 4.0$ kg. They are subjected to forces of $4.0\mathbf{j}$ N, $-8.0\mathbf{j}$ N, and $-6.0\mathbf{i}$ N, respectively. (*a*) Find the acceleration of their center of mass. (*b*) If initially all three particles are at rest, what are the position and velocity of the center of mass at $t = 3.0$ s?

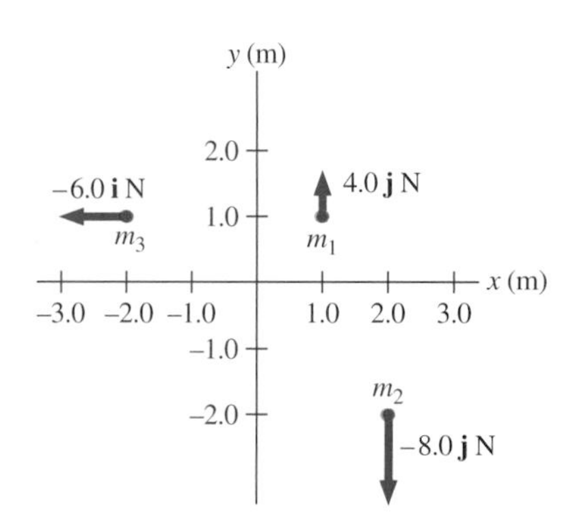

FIGURE 8-15 A three-particle system.

SOLUTION (*a*) The acceleration of the center of mass is found by dividing the net external force on the system by the total mass of the particles:

$$\mathbf{a}_{\mathrm{CM}} = \frac{\sum \mathbf{F}}{M} = \frac{[-6.0\mathbf{i} + (4.0 - 8.0)\mathbf{j}]\,\mathrm{N}}{(4.0 + 2.0 + 4.0)\,\mathrm{kg}}$$
$$= (-0.60\mathbf{i} - 0.40\mathbf{j})\ \mathrm{m/s^2}.$$

(*b*) The initial position of the center of mass is, from Eq. (8-2*b*),

$$\mathbf{r}_{\mathrm{CM}}(0) = \{[4.0(1.0\mathbf{i} + 1.0\mathbf{j}) + 2.0(2.0\mathbf{i} - 2.0\mathbf{j}) + 4.0(-2.0\mathbf{i} + 1.0\mathbf{j})]\ \mathrm{kg \cdot m}\}/(10.0\ \mathrm{kg})$$
$$= 0.40\mathbf{j}\ \mathrm{m}.$$

Since the acceleration of the center of mass is constant, its velocity and position vary with time according to

$$\mathbf{v}_{\mathrm{CM}}(t) = \mathbf{v}_{\mathrm{CM}}(0) + \mathbf{a}_{\mathrm{CM}}t,$$

and

$$\mathbf{r}_{\mathrm{CM}}(t) = \mathbf{r}_{\mathrm{CM}}(0) + \mathbf{v}_{\mathrm{CM}}(0)t + \tfrac{1}{2}\mathbf{a}_{\mathrm{CM}}t^2.$$

Substituting $\mathbf{a}_{\mathrm{CM}}$, $\mathbf{r}_{\mathrm{CM}}(0)$, and $\mathbf{v}_{\mathrm{CM}}(0) = 0$ along with $t = 3.0$ s into these equations, we find

$$\mathbf{v}_{\mathrm{CM}}(3.0\ \mathrm{s}) = (-1.8\mathbf{i} - 1.2\mathbf{j})\ \mathrm{m/s}$$

and

$$\mathbf{r}_{\mathrm{CM}}(3.0\ \mathrm{s}) = (-2.7\mathbf{i} - 1.4\mathbf{j})\ \mathrm{m}.$$

EXAMPLE 8-7 Two Colliding Asteroids

Figure 8-16 depicts two small asteroids of mass 2.0×10^{10} kg and 8.0×10^{10} kg. We assume that when the asteroids are 1.0×10^7 m apart, they are stationary relative to each other. Due to gravitational attraction, they move toward each other and eventually collide.

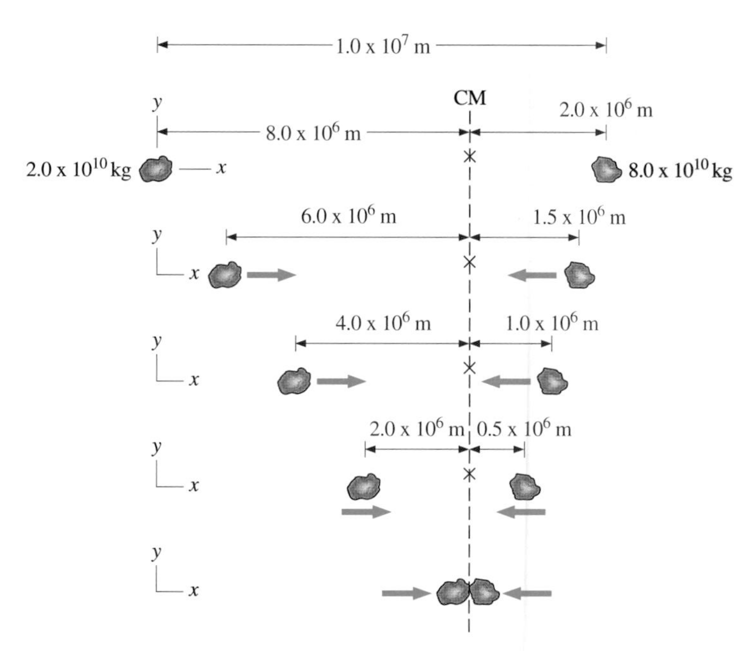

FIGURE 8-16 Two asteroids approach each other and collide such that their center of mass remains fixed.

If external forces on the system are negligible, where does the collision occur relative to the coordinate system whose origin is at the initial position of the less massive asteroid?

SOLUTION Since the net external force on the system of the two asteroids is zero, the acceleration of its center of mass is zero. Now the center of mass is initially at rest in the given coordinate system and must therefore remain stationary. The figure illustrates how the center of mass stays fixed as the asteroids approach each other and eventually collide at the center of mass whose position is

$$x_{CM} = \frac{(2.0 \times 10^{10}\ \text{kg})(0) + (8.0 \times 10^{10}\ \text{kg})(1.0 \times 10^{7}\ \text{m})}{10.0 \times 10^{10}\ \text{kg}}$$
$$= 8.0 \times 10^{6}\ \text{m}.$$

DRILL PROBLEM 8-6

What is the net external force on the system of particles of Example 8-5?
ANS. $(3.4\mathbf{i} + 2.8\mathbf{j} - 3.2\mathbf{k})$ N.

DRILL PROBLEM 8-7

A ball of radius 5.0 cm and mass 3.0 kg is attached to one end of a spring of mass 2.0 kg and force constant 1000 N/m. A ball of radius 6.0 cm and mass 5.1 kg is attached to the other end of the spring, and the combination is thrown into the air. What is the acceleration of this system's center of mass?
ANS. **g**.

8-4 Moment of Inertia

Another property of an object that is related to its mass distribution is the moment of inertia. The moment of inertia is an important quantity in the study of systems of particles that are rotating. As you will learn in the next chapter, the role of the moment of inertia in the study of rotational motion is analogous to that of mass in the study of linear motion. For now however, we will only consider the definition of the moment of inertia and how it is determined for a system of particles.

Consider the disk of Fig. 8-17*a* which is rotating about a fixed axis through its center. As illustrated in Fig. 8-17*b*, the particles of the disk all move along circular paths around that axis. The axis about which a system of particles rotates is called the *axis of rotation*. Various examples of rotating systems and their corresponding axes of rotation are shown in Fig. 8-18. By definition, the moment of inertia of a system of particles with respect to a particular axis of rotation is

$$I = \sum_i m_i r_i^2, \tag{8-8}$$

where r_i is the perpendicular distance from the axis to the ith particle, which has mass m_i. The moment of inertia is a scalar quantity with dimensions (mass)·(length)^2. Its units in the various systems are kilogram-square meter (kg·m^2), gram-square centimeter (g·cm^2), and slug-square foot (slug·ft^2). Unlike mass, the moment of inertia is not a fundamental constant of a system of particles, as its value varies with the location of the rotational axis. This fact will be illustrated in Example 8-9.

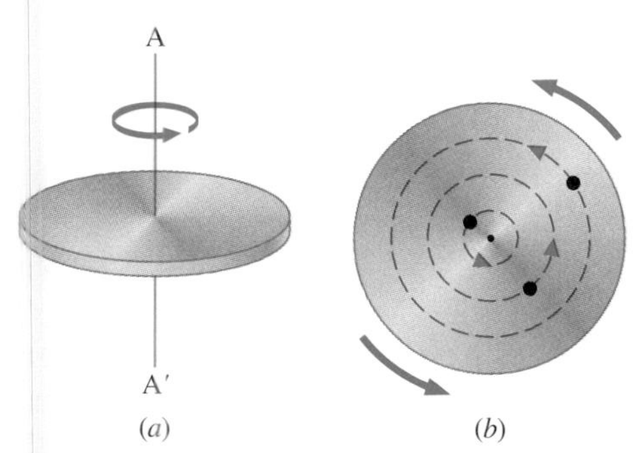

FIGURE 8-17 (*a*) A disk is rotating about a fixed axis through its center. (*b*) The rotating disk as viewed from above. The particles of the disk all move along circular paths around the axis of rotation.

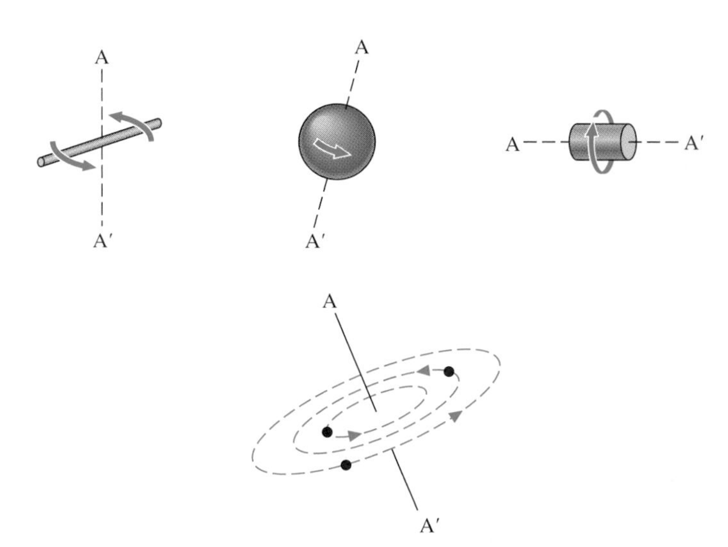

FIGURE 8-18 The axis of rotation of each system shown is designated by AA′.

EXAMPLE 8-8 Moment of Inertia of a Particle

A small ball of mass 0.10 kg is twirled at the end of a string of length 20 cm as shown in Fig. 8-19. What is the moment of inertia

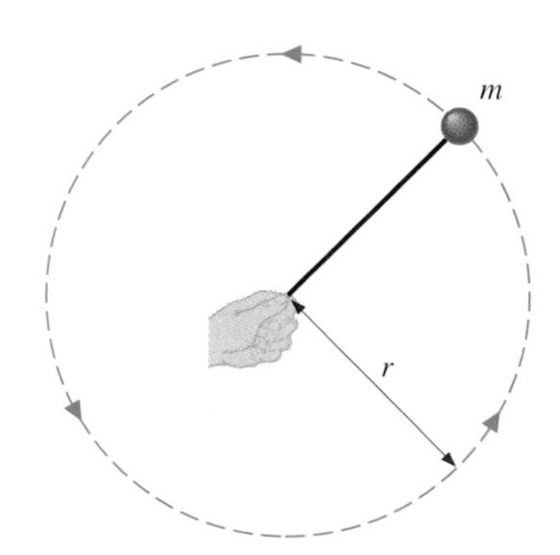

FIGURE 8-19 A small ball is twirled at the end of a string.

of the ball about the center of its circular path? Treat the ball as a particle.

SOLUTION The axis of rotation of the ball passes through the held end of the string and is perpendicular to the plane in which the ball is rotating. The distance of the ball from the axis of rotation is therefore $r = 20$ cm $= 0.20$ m. From Eq. (8-8), the moment of inertia of the ball is

$$I = mr^2 = (0.10 \text{ kg})(0.20 \text{ m})^2 = 4.0 \times 10^{-3} \text{ kg}\cdot\text{m}^2.$$

EXAMPLE 8-9 Moment of Inertia of a System of Particles

For the system of particles of Fig. 8-20, calculate the moment of inertia around (*a*) the x axis and (*b*) the axis through A that is parallel to the x axis.

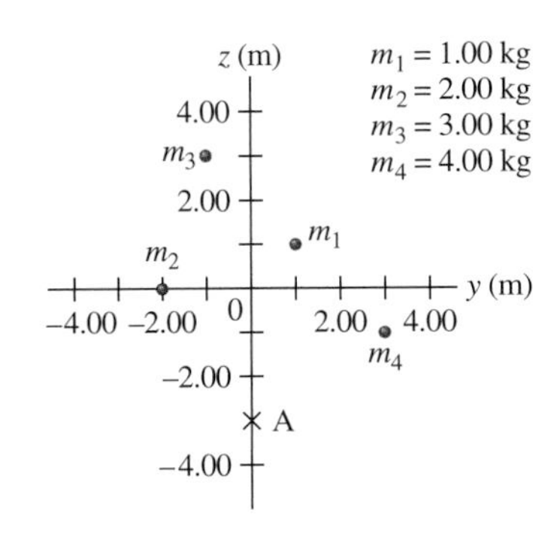

FIGURE 8-20 All four particles lie in the yz plane. The axis through A is parallel to the x axis.

SOLUTION (*a*) The respective distances of the four particles from the x axis are

$$r_1 = \sqrt{(1.00)^2 + (1.00)^2} \text{ m} = 1.41 \text{ m},$$
$$r_2 = 2.00 \text{ m},$$
$$r_3 = \sqrt{(-1.00)^2 + (3.00)^2} \text{ m} = 3.16 \text{ m},$$
$$r_4 = \sqrt{(3.00)^2 + (-1.00)^2} \text{ m} = 3.16 \text{ m}.$$

The moment of inertia about the x axis is therefore

$$I_0 = \sum_i m_i r_i^2 = (1.00 \text{ kg})(1.41 \text{ m})^2 + (2.00 \text{ kg})(2.00 \text{ m})^2 + (3.00 \text{ kg})(3.16 \text{ m})^2 + (4.00 \text{ kg})(3.16 \text{ m})^2 = 80.0 \text{ kg}\cdot\text{m}^2.$$

(*b*) For the axis through A,

$$r_1 = \sqrt{(0 - 1.00)^2 + (-3.00 - 1.00)^2} \text{ m} = 4.12 \text{ m}.$$

Similarly, $r_2 = 3.61$ m, $r_3 = 6.08$ m, and $r_4 = 3.61$ m. Thus

$$I_A = (1.00 \text{ kg})(4.12 \text{ m})^2 + (2.00 \text{ kg})(3.61 \text{ m})^2 + (3.00 \text{ kg})(6.08 \text{ m})^2 + (4.00 \text{ kg})(3.61 \text{ m})^2 = 206 \text{ kg}\cdot\text{m}^2.$$

Notice that $I_0 \neq I_A$, which demonstrates that the moment of inertia of a given system does depend on the location of the axis of rotation.

DRILL PROBLEM 8-8

Calculate the moment of inertia of the system of particles of Fig. 8-20 around the y axis and around the z axis.
ANS. 32.0 kg·m²; 48.0 kg·m².

DRILL PROBLEM 8-9

To the system of Fig. 8-20 add a fifth particle of mass $m_5 = 1.00$ kg at $y = 1.00$ m, $z = 0$. Find the moment of inertia of this system around the x axis.
ANS. 81.0 kg·m².

*8-5 Finding the Moment of Inertia by Integration

For a continuous mass distribution such as found in a rigid body, we replace the summation of Eq. (8-8) by an integral. If the system is divided into infinitesimal elements of mass dm and if r is the distance from a mass element to the axis of rotation (see Fig. 8-21), the moment of inertia is

$$I = \int r^2 \, dm,$$

where the integral is taken over the system. Denoting ρ as the mass density of the system at the volume element dV, we may write

$$dm = \rho \, dV,$$

and

$$I = \int \rho r^2 \, dV.$$

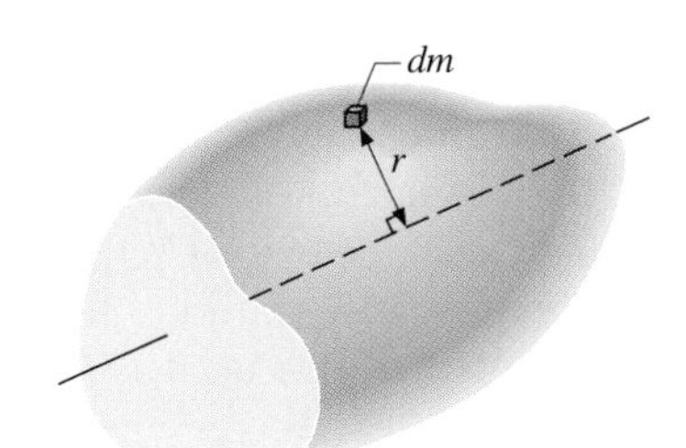

FIGURE 8-21 The moment of inertia of a continuous mass distribution is calculated from $\int r^2 \, dm$.

We will usually consider uniform bodies; in this case,

$$I = \rho \int r^2 \, dV. \tag{8-9}$$

This integral is often easy to evaluate when the axis of rotation is also an axis of symmetry. Moments of inertia of several uniform rigid bodies with symmetry are listed in Table 8-1.

TABLE 8-1 MOMENTS OF INERTIA OF SOME SIMPLY SHAPED, SYMMETRIC RIGID BODIES
In all cases but (f), the rotational axis AA′ passes through the center of mass.

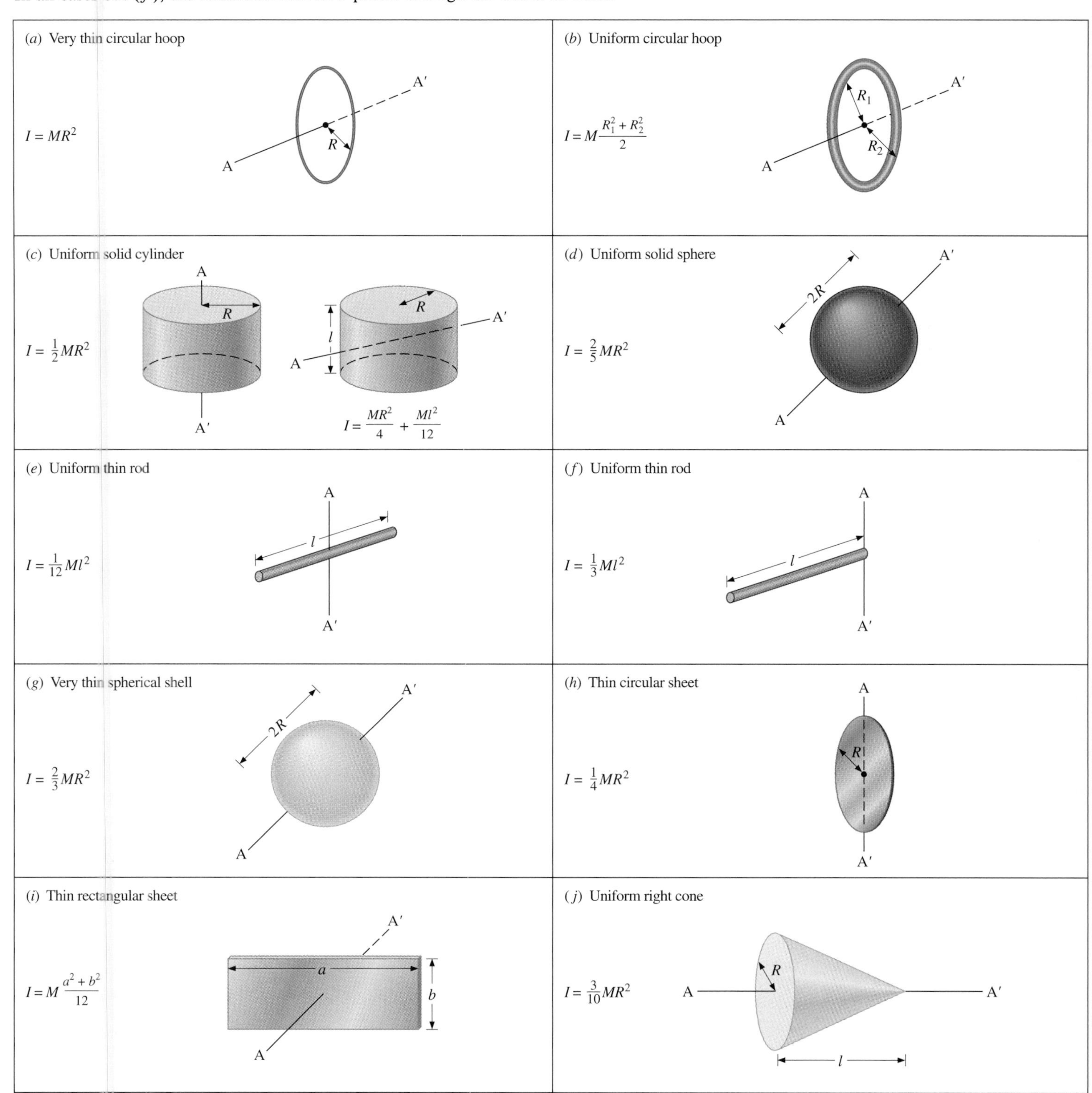

EXAMPLE 8-10 MOMENT OF INERTIA OF A UNIFORM CYLINDER

Find the moment of inertia of a uniform cylinder around the axis through its geometrical center and perpendicular to its base.

SOLUTION We first divide the cylinder into annular shells of width dr and length l, as shown in Fig. 8-22. The moment of inertia of one of these shells is

$$dI = r^2\,dm = r^2\rho\,dV = r^2\rho(2\pi lr\,dr),$$

where dV is the volume of the shell. The cylinder's moment of inertia is found by integrating this expression between 0 and R:

$$I = \int_0^R r^2(\rho 2\pi lr\,dr) = 2\pi\rho l\int_0^R r^3\,dr = \frac{\pi\rho l}{2}R^4. \qquad (i)$$

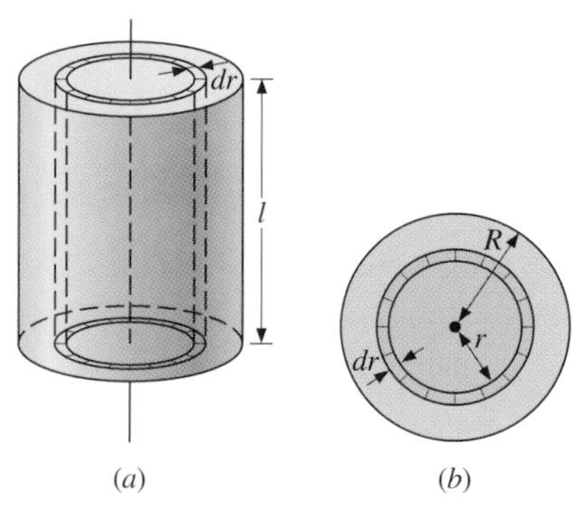

FIGURE 8-22 (*a*) A cylinder is divided into shells of length l and thickness dr. (*b*) A cross-sectional view of the cylinder.

The density ρ of the cylinder is the mass divided by the volume:

$$\rho = \frac{M}{\pi R^2 l}. \qquad (ii)$$

Upon substituting Eq. (*ii*) into Eq. (*i*), we find

$$I = \tfrac{1}{2} MR^2,$$

which is part (*c*) of Table 8-1.

DRILL PROBLEM 8-10

Derive the formula for part (*e*) in Table 8-1.

*8-6 Parallel-Axis Theorem

A very useful theorem, called the *parallel-axis theorem*, relates the moments of inertia of a rigid body around two parallel axes, one of which passes through the center of mass. Two such axes are shown in Fig. 8-23 for a body of mass M. If h is the distance between the axes and I_{CM} and I are the respective moments of inertia around them, these moments are related by

$$I = I_{CM} + Mh^2. \qquad (8\text{-}10)$$

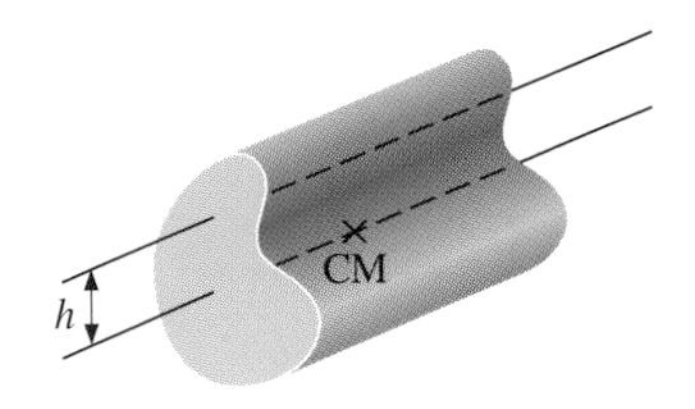

FIGURE 8-23 The parallel-axis theorem relates the moments of inertia of a rigid body around two parallel axes, one of which passes through the center of mass.

We now present a proof of Eq. 8-10, the parallel-axis theorem. A cross-sectional view of the body of Fig. 8-23 is shown in Fig. 8-24. The body is oriented so that the two axes, one through CM (the center of mass) and the other through A, are perpendicular to the plane of the page. The coordinates of these two points are $(x_{CM}, y_{CM}, 0)$ and $(x_{CM} + a, y_{CM} + b, 0)$, respectively. The distance between the two axes is $h = \sqrt{a^2 + b^2}$.

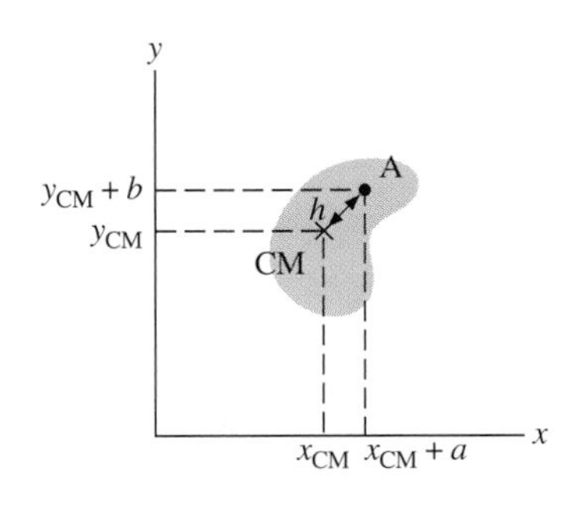

FIGURE 8-24 A cross-sectional view of the rigid body of Fig. 8-23.

The moment of inertia around the axis through CM is

$$I_{CM} = \sum_i m_i[(x_i - x_{CM})^2 + (y_i - y_{CM})^2],$$

and the moment of inertia around the parallel axis through A is

$$I_A = \sum_i m_i[(x_i - x_{CM} - a)^2 + (y_i - y_{CM} - b)^2],$$

which, after some rearrangement, can be written as

$$I_A = \sum_i m_i[(x_i - x_{CM})^2 + (y_i - y_{CM})^2] - 2a\sum_i m_i(x_i - x_{CM}) - 2b\sum_i m_i(y_i - y_{CM}) + (a^2 + b^2)\sum_i m_i.$$

The first of these terms is I_{CM}. The second and third terms are zero from the definition of the center of mass:

$$\sum_i m_i x_i = x_{CM}\sum_i m_i \quad \text{and} \quad \sum_i m_i y_i = y_{CM}\sum_i m_i.$$

Since $h^2 = a^2 + b^2$ and $M = \sum_i m_i$, the fourth term is Mh^2. Thus

$$I_A = I_{CM} + 0 + 0 + Mh^2 = I_{CM} + Mh^2,$$

which is the parallel-axis theorem.

As an example of the use of the parallel-axis theorem, consider the uniform solid sphere of Fig. 8-25. The moment of inertia around an axis through its center of mass is $I_{CM} = 2MR^2/5$, and the distance between the axes AA′ and BB′ is R; so the moment of inertia around BB′ is

$$I = \tfrac{2}{5}MR^2 + MR^2 = \tfrac{7}{5}MR^2.$$

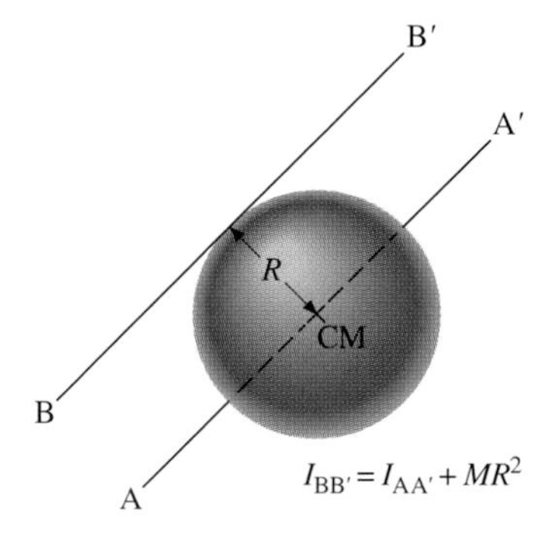

FIGURE 8-25 A sphere's moment of inertia around the BB′ axis is calculated with the parallel-axis theorem.

DRILL PROBLEM 8-11

Use the moment of inertia of part (*e*) of Table 8-1 along with the parallel-axis theorem to calculate the moment of inertia given in part (*f*) of the table.

DRILL PROBLEM 8-12

Consider the hoop of part (*a*) in Table 8-1. Find its moment of inertia around an axis that passes through the edge of the hoop and is parallel to AA′.
ANS. $2MR^2$.

SUMMARY

1. **Center of mass**
The position of the center of mass is

$$\mathbf{r}_{CM} = \frac{1}{M}\sum_i m_i \mathbf{r}_i$$

for a system of discrete particles. For a rigid body, the location of the center of mass may be found by integration.

2. **Motion of the center of mass**
(*a*) The velocity and acceleration of the center of mass of a system of particles are related to the velocities and accelerations of the individual particles of the system by

$$\mathbf{v}_{CM} = \frac{1}{M}\sum_i m_i \mathbf{v}_i \quad \text{and} \quad \mathbf{a}_{CM} = \frac{1}{M}\sum_i m_i \mathbf{a}_i.$$

(*b*) If a net external force $\sum \mathbf{F}$ acts on a system of particles of total mass M, the acceleration of the system's center of mass is

$$\mathbf{a}_{CM} = \frac{\sum \mathbf{F}}{M}.$$

3. **Moment of inertia**
(*a*) The moment of inertia of a system of discrete particles around an axis A is

$$I_A = \sum_i m_i r_i^2,$$

where r_i is the perpendicular distance from particle i to the axis. If the system is a rigid body, the moment of inertia may be found by integration.

(*b*) If I_A and I_{CM} are the moments of inertia of a rigid body around parallel axes through A and through the center of mass respectively, then

$$I_A = I_{CM} + Mh^2,$$

where M is the mass of the body and h is the distance between the axes.

SUPPLEMENT 8-1 MOTION OF THE CENTER OF MASS

We saw that the acceleration of the center of mass of a system of particles of total mass M obeys the equation

$$M\mathbf{a}_{CM} = \mathbf{F}_1 + \mathbf{F}_2 + \ldots + \mathbf{F}_N, \tag{8-6}$$

where $\mathbf{F}_i$ is the net force on the ith particle. The right-hand side of Eq. (8-6) may be simplified in the following way. We subdivide the forces on the individual particles into two types: (1) those exerted by other particles of the system and (2) those exerted by the external environment. This subdivision is illustrated in Fig. 8-26. The force on particle i due to particle j is $\mathbf{f}_{ij}$, and the net external force on particle i is $\mathbf{F}_i^E$. Now we can write Eq. (8-6) as

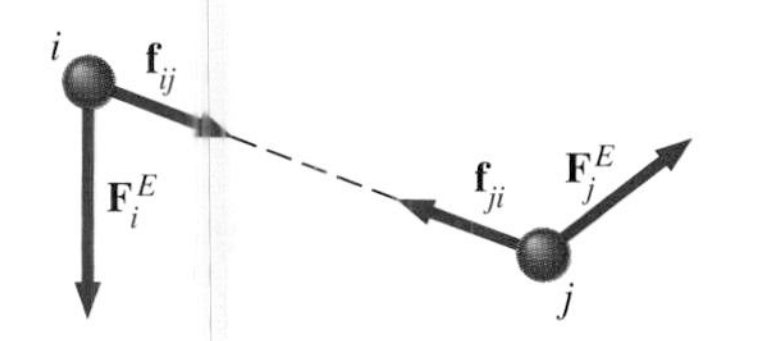

FIGURE 8-26 The force on particle i due to particle j is $\mathbf{f}_{ij}$, and the external force on particle i is $\mathbf{F}_i^E$.

$$M\mathbf{a}_{CM} = (\mathbf{F}_1^E + \mathbf{f}_{12} + \ldots + \mathbf{f}_{1N}) + (\mathbf{F}_2^E + \mathbf{f}_{21} + \ldots + \mathbf{f}_{2N}) + \ldots + (\mathbf{F}_N^E + \mathbf{f}_{N1} + \mathbf{f}_{N2} + \ldots + \mathbf{f}_{NN-1}).$$

Rearranging terms, we obtain

$$M\mathbf{a}_{CM} = (\mathbf{F}_1^E + \mathbf{F}_2^E + \ldots + \mathbf{F}_N^E) + (\mathbf{f}_{12} + \mathbf{f}_{21}) + (\mathbf{f}_{13} + \mathbf{f}_{31}) + \ldots + (\mathbf{f}_{NN-1} + \mathbf{f}_{N-1N}),$$

and a significant simplification of the equation becomes evident. By Newton's third law, $\mathbf{f}_{ij} = -\mathbf{f}_{ji}$, so the sum of every action-reaction force pair (the terms in parentheses) vanishes. Thus only the sum of the external forces remains on the right-hand side of the equation. We may therefore write

$$\sum \mathbf{F} = M\mathbf{a}_{CM}, \tag{8-7}$$

where

$$\sum \mathbf{F} = \mathbf{F}_1^E + \mathbf{F}_2^E + \ldots + \mathbf{F}_N^E$$

is the net external force on the system of particles.

Questions

8-1. Is there necessarily mass at the center of mass of a system of particles?

8-2. Most high jumpers try to clear the bar with their bodies draped over the bar. Why?

8-3. Shown in the accompanying figure are sketches of an inventor's schemes for propulsion. What is wrong with these schemes?

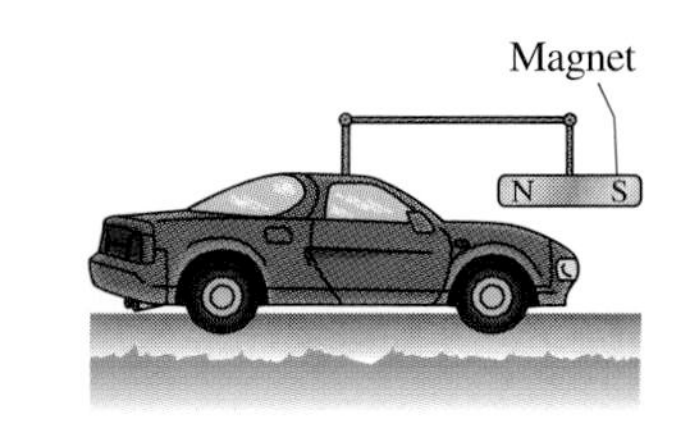

QUESTION 8-3

8-4. After the pellets from a fired shotgun leave the barrel, what is the acceleration of their center of mass?

8-5. What external force moves you when you walk or run?

8-6. What external force pushes a basketball player upward during a leap?

8-7. A young child stands at the rail of a baby crib that has casters on the legs. By repeatedly jerking on the rail, he is able to move the crib across the floor. What is the origin of the external force that pushes the bed and its occupant across the floor?

8-8. After the shell in the accompanying figure explodes, the center of mass of the pieces initially moves along the same parabolic path that the unexploded shell was following; its path then suddenly becomes quite erratic. Explain.

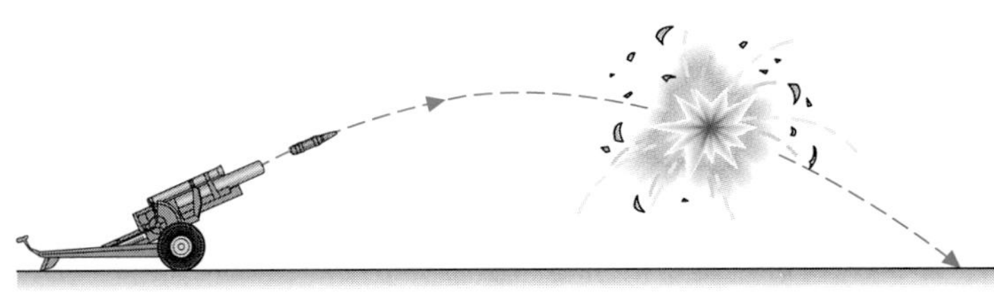

QUESTION 8-8

8-9. A man who wants to show off his strength sets up the following demonstration. He has a friend strap him to a chair, and he then attempts to lift himself (and the chair) off the floor as shown. Does his demonstration work?

QUESTION 8-9

8-10. Why do you move your torso forward and your feet backward in preparation for getting up from a chair? Push a sitting friend gently in the forehead and ask her to try to get up from her chair. What happens and why?

8-11. A sketch of the equipment used in a popular center-of-mass demonstration is shown in the accompanying figure. When the cylinder is placed on the triangular incline, it moves from A to B as you would expect. However, when the "double cone" is placed on the incline, it moves from B to A, thus appearing to move uphill. Explain why the double cone moves in this manner.

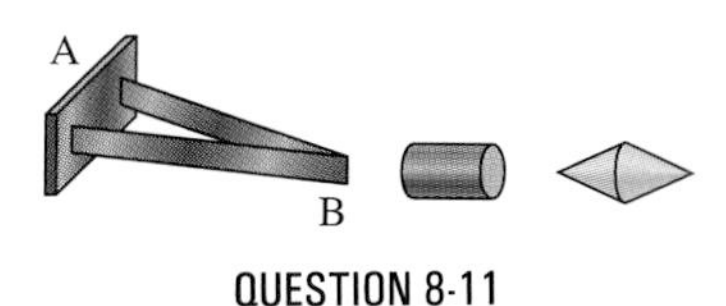

QUESTION 8-11

8-12. Two metals of different densities but the same total mass are cast into solid cylinders of the same length. Which cylinder has the larger moment of inertia around its central axis?

8-13. Explain why a hoop has a larger moment of inertia around the axis through its center than a solid cylinder with the same mass and radius as the hoop.

8-14. Suppose that a wooden bar and a copper bar of equal lengths are joined together as shown in the accompanying figure. Is the combination's moment of inertia larger around AA′ or BB′?

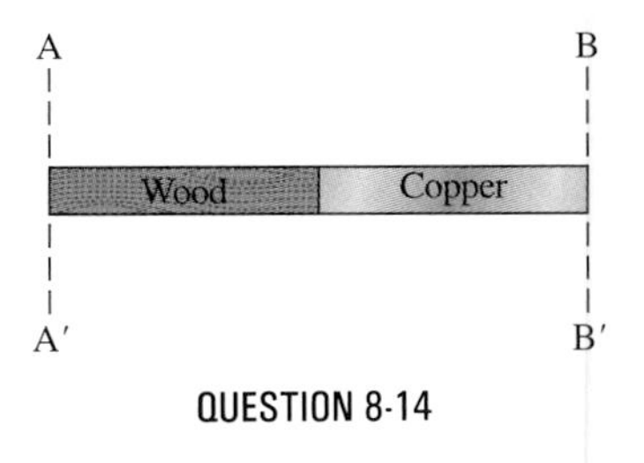

QUESTION 8-14

8-15. Around what axis is the moment of inertia of a planar body a minimum?

8-16. When calculating a body's moment of inertia, can we assume that all of its mass is located at the center of mass?

8-17. Can an object have more than one moment of inertia?

PROBLEMS

Center of Mass

8-1. Locate the center of mass of the two-particle system shown in the accompanying figure.

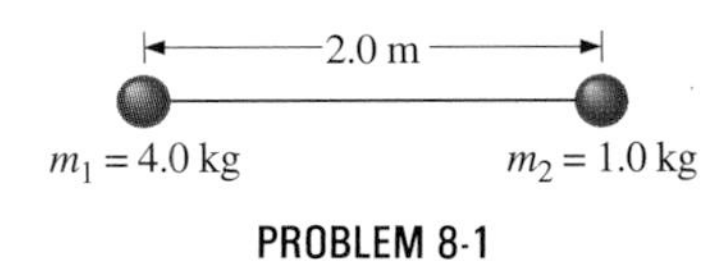

PROBLEM 8-1

8-2. The hydrogen and chlorine ions in the HCl molecule are about 1.30×10^{-10} m apart. The Cl^- ion is 35 times more massive than the H^+ ion. Where is the center of mass of the molecule?

8-3. Three particles of equal mass are located at the corners of an equilateral triangle. Where is the system's center of mass?

8-4. Where is the center of mass of the three-particle system shown in the accompanying figure?

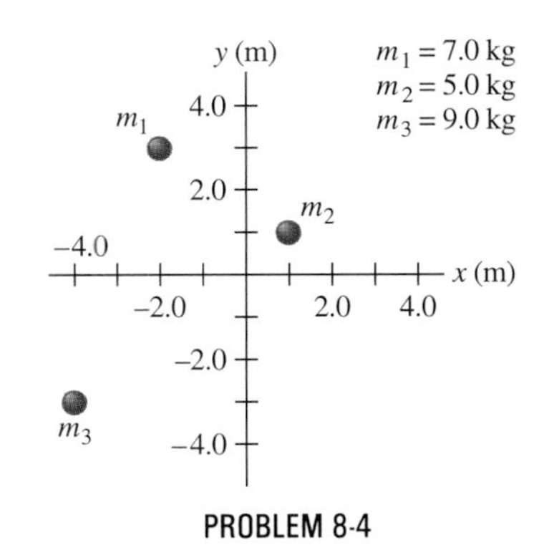

PROBLEM 8-4

8-5. If all of the planets were located on a straight line at their mean distances from the sun, how far from the sun's center would the center of mass of the solar system be located? Compare this with the radius of the sun.

8-6. Locate the center of mass of the uniform L-shaped piece shown in the accompanying figure.

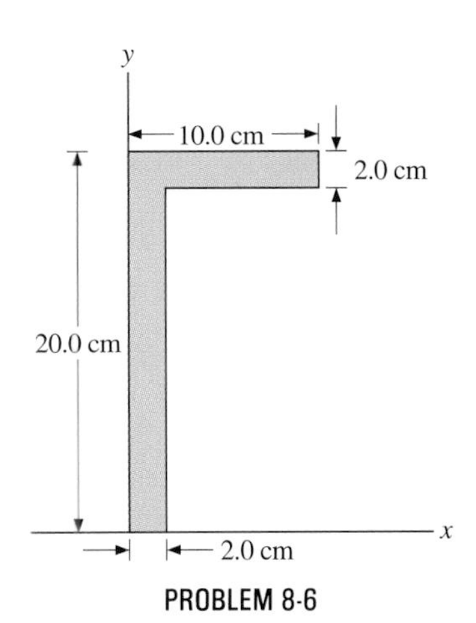

PROBLEM 8-6

8-7. A ball of radius R and mass M is attached to the end of a stick of length L and mass $2M$. The ball and stick are uniform. Where is the center of mass of this object?

8-8. Holes of radius R are cut in a uniform square plate with sides of length L, as shown in the accompanying figure. Where is the center of mass of this object?

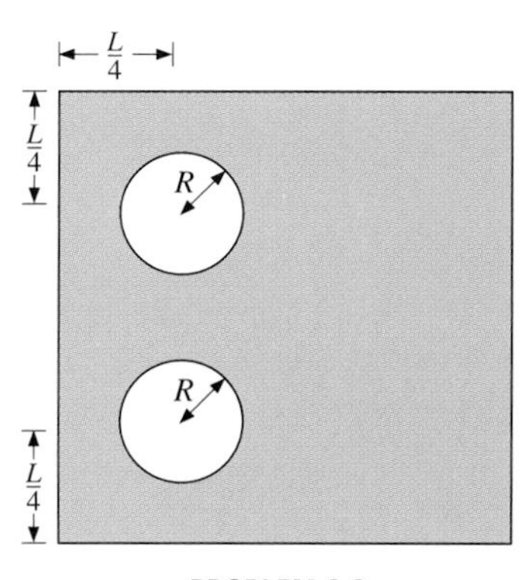

PROBLEM 8-8

8-9. The total mass of the four identical legs of the coffee table shown in the accompanying figure is equal to that of the tabletop. Assuming that the top has negligible thickness and that the table is uniform, find the position of the table's center of mass without benefit of formal calculations.

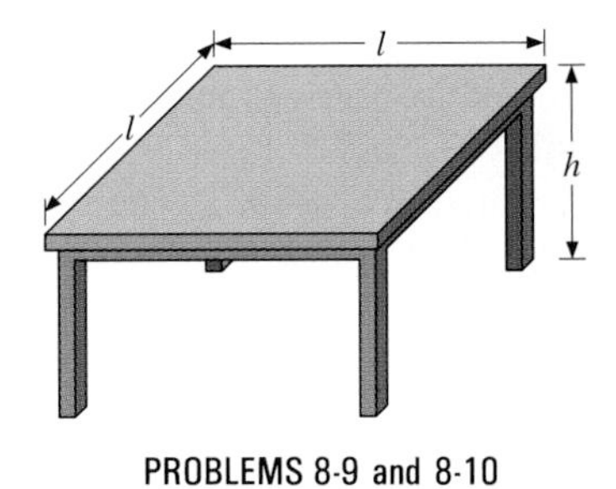

PROBLEMS 8-9 and 8-10

8-10. If two legs on opposite corners are removed from the table of Prob. 8-9, what is the position of the center of mass?

8-11. Three uniform rigid rods of length L are connected as shown in the accompanying figure. What is the position of the center of mass of the system?

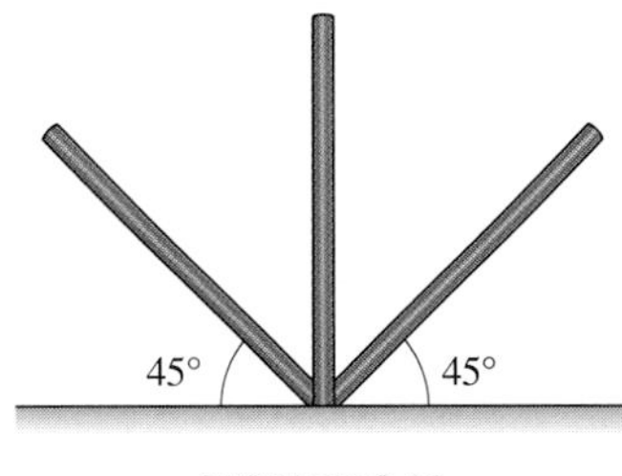

PROBLEM 8-11

8-12. The accompanying figure depicts a model that approximates the human male body. The model consists of a cylindrical trunk of mass 33.0 kg, two stick arms each of mass 4.0 kg, two stick legs each of mass 12.0 kg, and a spherical head of mass 5.0 kg. The vertical dimensions of each anatomical part are given in the figure. Find the position of the center of mass of this model body.

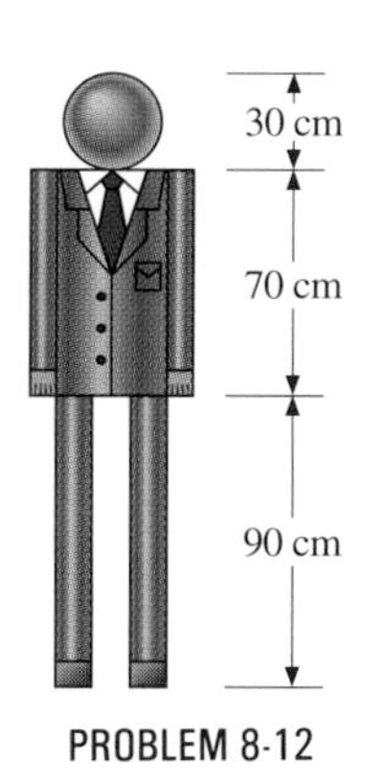

PROBLEM 8-12

8-13. A cylindrical can is 50 cm high and has a diameter of 30 cm. It is made from a sheet of material having a surface density of 10 g/cm^2. (*a*) The lid is placed on the can after it is half-filled with water (density $\rho = 1.0$ g/cm^3). Where is the center of mass of the system? (*b*) If the lid is removed, where is the center of mass? (*c*) How much water has to be added to restore the center of mass of the open can to the location found in part (*a*)?

Locating the Center of Mass by Integration

8-14. Find the position of the center of mass of a thin rod bent into a semicircle of radius R.

8-15. Find the position of the center of mass of a uniform semicircular plate of radius R.

8-16. Show that the center of mass of the uniform triangular plate represented by the shaded area in the accompanying figure is at $x = a/3$, $y = h/3$.

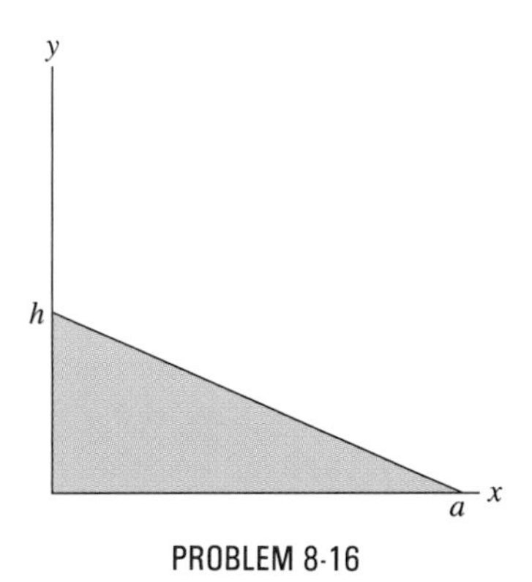

PROBLEM 8-16

8-17. Find the position of the center of mass of the uniform plate represented by the shaded area in the accompanying figure.

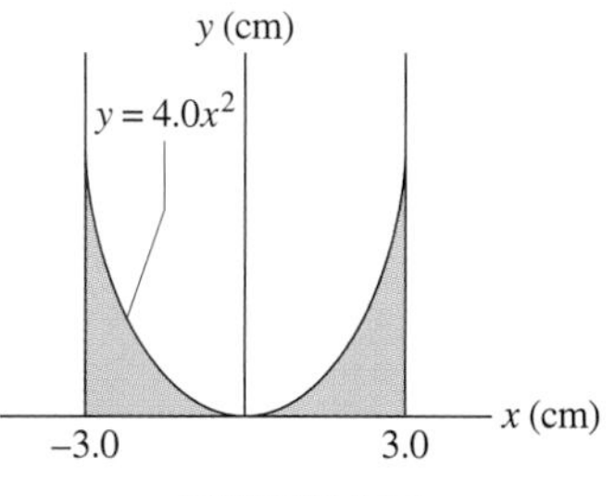

PROBLEM 8-17

8-18. Locate the center of mass of the uniform plate represented by the shaded area shown in the accompanying figure.

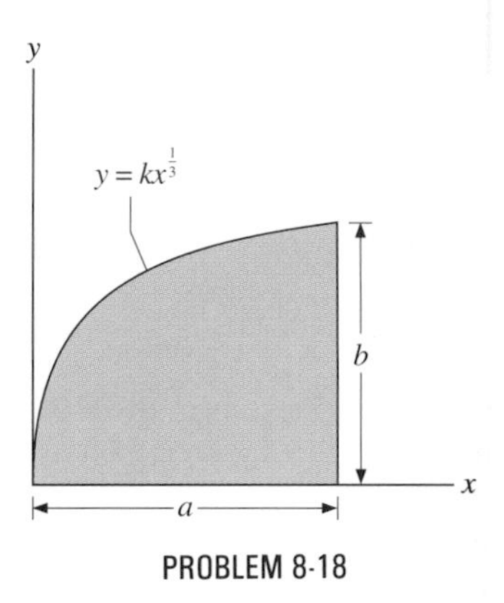

PROBLEM 8-18

8-19. Locate the center of mass of the uniform plate represented by the shaded area in the accompanying figure.

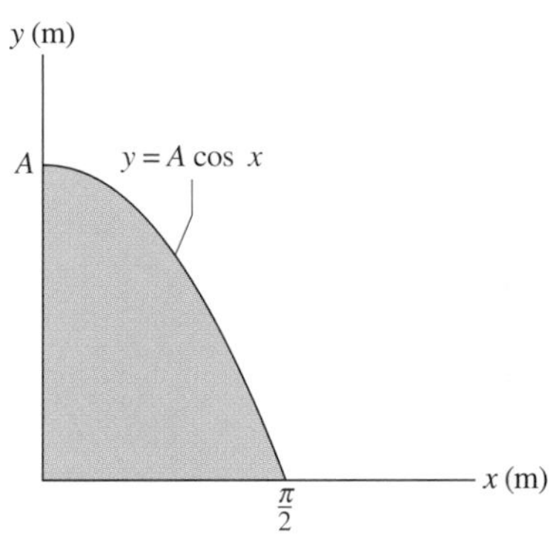

PROBLEM 8-19

Motion of the Center of Mass

8-20. Two particles of mass $m_1 = M$ and $m_2 = 3M$ are moving with constant velocities in the same direction, as shown in the accompanying figure. At $t = 0$, their separation is 2.0 m. (*a*) Relative to m_1, where is their center of mass at $t = 0$? (*b*) Where is their center of mass at $t = 10$ s?

PROBLEM 8-20

8-21. A 4.0-kg particle has a velocity of $(2.0\mathbf{i} - 7.0\mathbf{j})$ m/s, and a 2.0-kg particle has a velocity of $(5.0\mathbf{i} + 6.0\mathbf{j})$ m/s. What is the velocity of the two-particle system's center of mass?

8-22. The accelerations of the 4.0-kg and the 2.0-kg particle of the previous problem are $0.20\mathbf{i}$ m/s^2 and $(0.60\mathbf{i} - 0.50\mathbf{j})$ m/s^2, respectively. (*a*) What is the acceleration of the system's center of mass? (*b*) What is the net external force on the system?

8-23. At a particular instant, four particles have the positions, velocities, and accelerations given in the following table (cgs units). (*a*) What are the position, velocity, and acceleration of the center of mass at this instant? (*b*) What are the forces on each particle at this instant? (*c*) What is the net force on the system? (*d*) Is this net force equal to $M\mathbf{a}_{CM}$?

Mass	(x, y, z)	(v_x, v_y, v_z)	(a_x, a_y, a_z)
5.0	(−2.0, 4.0, 9.0)	(9.0, 8.0, −3.0)	(0, 0, 0)
9.0	(5.0, −7.0, 3.0)	(0, 0, 4.0)	(5.0, 2.0, −4.0)
6.0	(9.0, 6.0, 1.0)	(9.0, 0, −3.0)	(1.0, 1.0, 1.0)
5.0	(1.0, 2.0, −5.0)	(0, 4.0, 6.0)	(7.0, −3.0, 4.0)

8-24. At $t = 0$, the three particles are at rest in the positions shown in the accompanying figure. For the next 2.0 s the forces shown are applied to the particles. (*a*) What is the position of each particle at $t = 2.0$ s? (*b*) Use these positions to find the position of the system's center of mass at $t = 2.0$ s. (*c*) Use Eqs. (8-7) and (3-11) to find the position of the center of mass at $t = 2.0$ s.

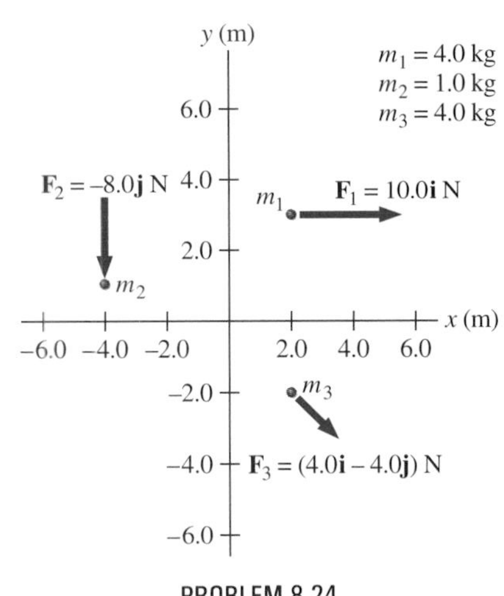

PROBLEM 8-24

8-25. At $t = 0$, the position and velocity of particle 1 are $\mathbf{r}_1 = (2.0\mathbf{i} + 3.0\mathbf{j})$ m and $\mathbf{v}_1 = (6.0\mathbf{i} - 4.0\mathbf{j})$ m/s, and the position and velocity of particle 2 are $\mathbf{r}_2 = (-6.0\mathbf{i} + 2.0\mathbf{j})$ m and $\mathbf{v}_2 = (-2.0\mathbf{i} + 1.0\mathbf{j})$ m/s. Their masses are $m_1 = 1.0$ kg and $m_2 = 3.0$ kg. (*a*) What is the position of the two-particle system's center of mass at $t = 0$? (*b*) What is the velocity of its center of mass at $t = 0$? (*c*) If particle 1 is subjected to the constant force $\mathbf{F}_1 = 2.0\mathbf{j}$ N from $t = 0$ to $t = 5.0$ s (there is no force on particle 2), what are the position and velocity of the center of mass at $t = 5.0$ s? (*d*) If the two particles attract each other with a force given by $F(r) = -kr$, where r is the distance between the particles, how does this affect the answers to parts (*a*) through (*c*)?

Moment of Inertia

8-26. Particles of mass m are located at the four corners of a square whose sides are of length d. What is the moment of inertia of the particles around (*a*) the axis through the center of the square and perpendicular to the plane of the square? (*b*) an axis through one of the corners of the square and perpendicular to the plane of the square?

8-27. Particles of mass m are located at the three corners of the equilateral triangle shown in the accompanying figure. Calculate the particles' moment of inertia around the axis that passes through P and is perpendicular to the page.

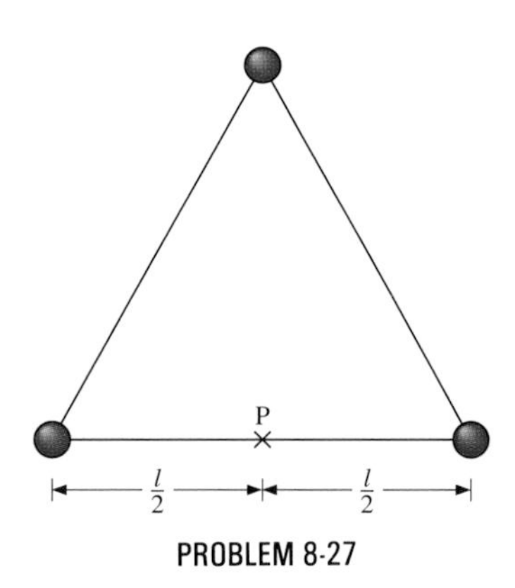

PROBLEM 8-27

8-28. Calculate the moment of inertia of the three-particle system of Prob. 8-4 around the x, y, and z axes.

8-29. Treating the sun and all planets of Prob. 8-5 as particles, calculate the system's moment of inertia around an axis through the center of the sun.

Finding the Moment of Inertia by Integration

8-30. Find the moment of inertia around the z axis of the semicircular uniform disk of mass M represented by the shaded area in the accompanying figure.

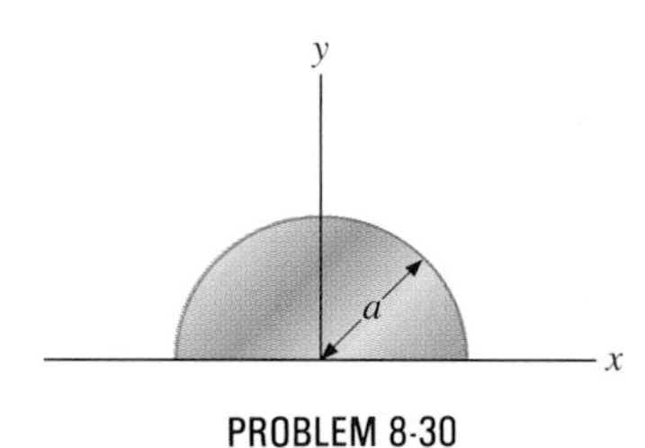

PROBLEM 8-30

8-31. The shaded area in the accompanying figure represents a uniform triangular plate of mass M. Find its moments of inertia around the x and y axes.

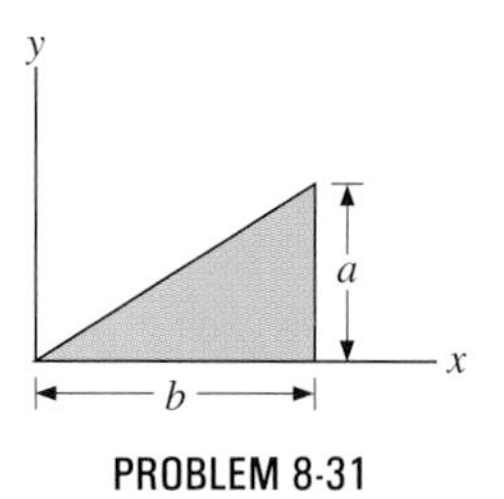

PROBLEM 8-31

8-32. A wedge-shaped section of a uniform cylinder is shown in the accompanying figure. What is this section's moment of inertia around OO′?

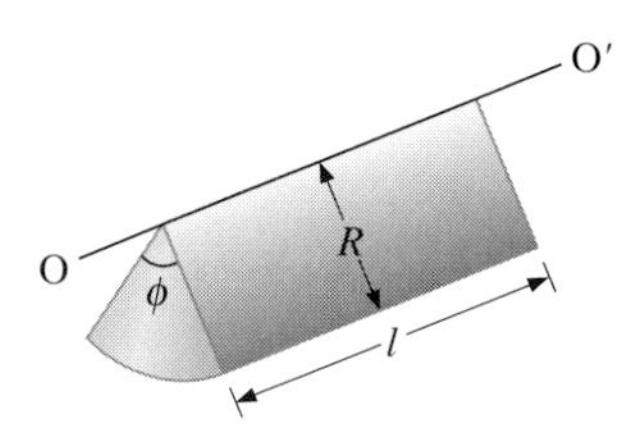

PROBLEM 8-32

8-33. The uniform disk shown in the accompanying figure has a moment of inertia of 0.60 kg$\cdot$m^2 around the axis that passes through O and is perpendicular to the plane of the page. If a segment is cut out from the disk as shown, what is the moment of inertia of the remaining piece?

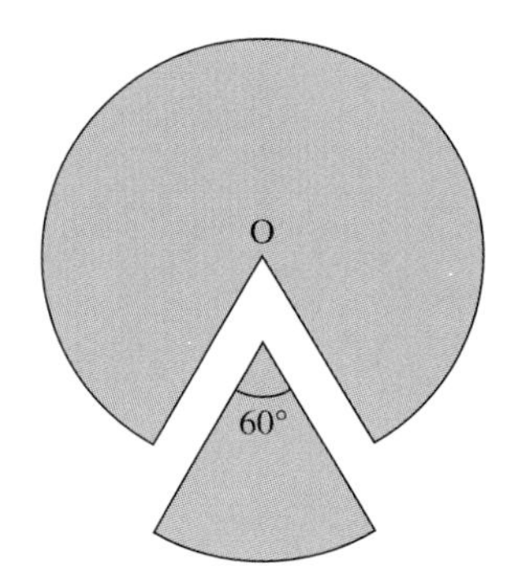

PROBLEM 8-33

8-34. The moment of inertia of a uniform rod around its end is given in Table 8-1. Use this result to calculate the moment of inertia of the plate of Prob. 8-18 around the x axis.

8-35. Using the method of Prob. 8-34, calculate the moment of inertia of the plate of Prob. 8-19 around the x axis.

Parallel-Axis Theorem

8-36. Calculate the moment of inertia of a uniform rod of length L and mass M around an axis through the rod and a distance $L/4$ from one of its ends.

8-37. Calculate the moment of inertia of a spherical shell of radius R and mass M around an axis tangent to the surface of the sphere.

8-38. The accompanying figure shows a washer of mass M with an inner radius R_1 and an outer radius R_2. Calculate its moments of inertia around the axes through A, B, and C, all of which are perpendicular to the plane of the washer.

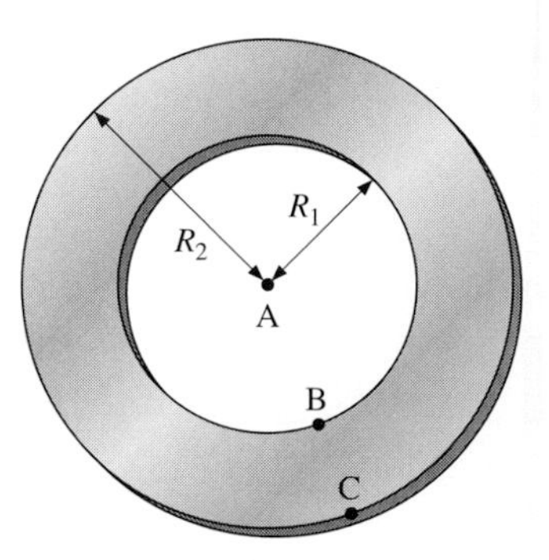

PROBLEM 8-38

8-39. Spheres are attached to each end of a thin rod of length 2.0 m. The mass of the rod is 4.0 kg, and the masses and radii of the spheres are $m_A = 1.0$ kg, $R_A = 0.10$ m and $m_B = 3.4$ kg, $R_B = 0.15$ m. (See the accompanying figure.) (*a*) Find the position of the system's center of mass. (*b*) Calculate its moment of inertia around OO′ and PP′.

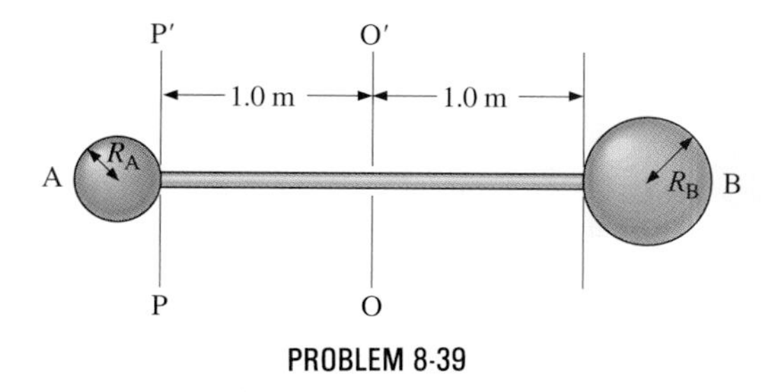

PROBLEM 8-39

General Problems

8-40. Use the result of Prob. 8-16 to locate the center of mass of the uniform plate shown in the accompanying figure.

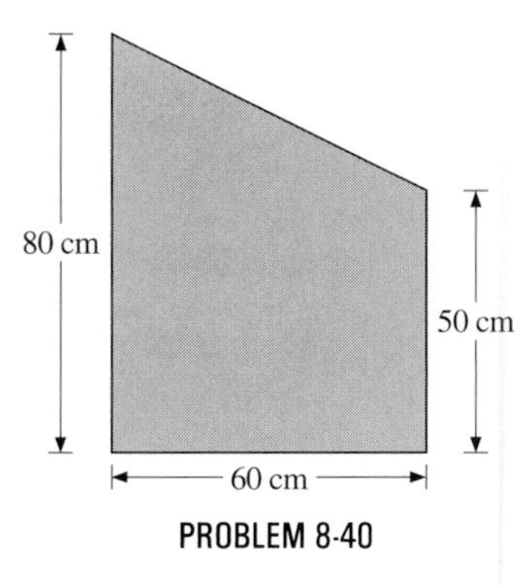

PROBLEM 8-40

8-41. A vertical cross section of a watering trough is shown in the accompanying figure. Calculate the position of the center of mass when (*a*) the trough is filled to a height of 1.5 ft and (*b*) the trough is completely full. The mass of the trough is small enough to be ignored. You will need the result of Prob. 8-16.

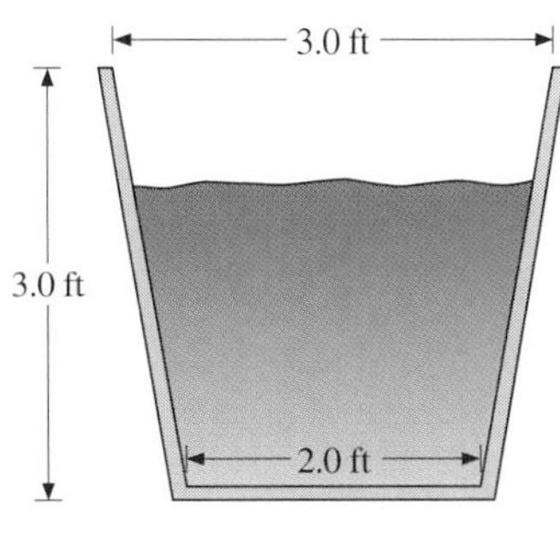

PROBLEM 8-41

8-42. A young boy of mass M sits at the back of a uniform, flat-bottomed raft, also of mass M and of length l. The front end of the raft touches the dock. The boy stands up and walks to the front end of the raft in order to jump out of the boat onto the dock. However, he finds that the raft has moved away from the dock, making it impossible for him to jump off. How far does the raft move away from the dock? Assume that there is no friction between the raft and water.

8-43. A shell of mass m is shot with an initial speed of 500 m/s at an angle of 37° to the horizontal. At the top of its path it explodes into two pieces, each of mass $m/2$. One piece starts with zero initial velocity and drops straight downward. Where does the second piece land?

8-44. The linear mass density of a thin rod of length L is given by $\lambda(x) = \lambda_0(1 + ax)$. What is the moment of inertia around an axis that passes through the end of the rod and is perpendicular to a plane containing the rod?

8-45. Particles with masses $m_1, m_2, \ldots, m_N$ have positions $\mathbf{r}_1, \mathbf{r}_2, \ldots, \mathbf{r}_N$ relative to the reference frame S. (See the accompanying figure.) Suppose the origin of a second reference system S' has the position $\mathbf{R}$ relative to S as shown. The positions of the particles relative to S' are then $\mathbf{r}'_1 = \mathbf{r}_1 - \mathbf{R}$, $\mathbf{r}'_2 = \mathbf{r}_2 - \mathbf{R}$, $\ldots$, $\mathbf{r}'_N = \mathbf{r}_N - \mathbf{R}$. Show that the position of the center of mass of the system of particles relative to S' is $\mathbf{r}'_{CM} = \mathbf{r}_{CM} - \mathbf{R}$, where $\mathbf{r}_{CM}$ is the position of the center of mass relative to S. Use this result to show that the position of the center of mass relative to particle i is the same for either reference system.

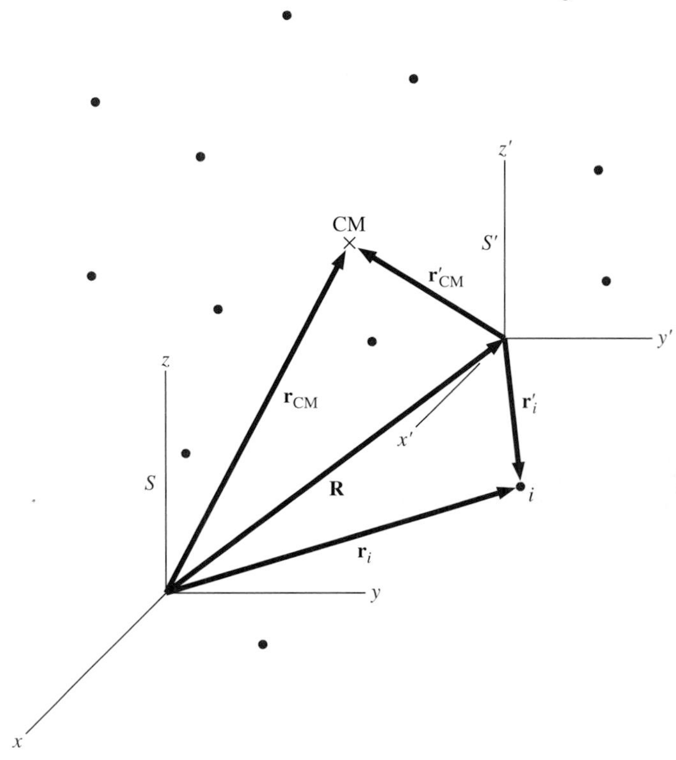

PROBLEM 8-45

A rotating system.

CHAPTER 9 ROTATIONAL MECHANICS I

PREVIEW

In this chapter we study the rotational motion of a rigid body around a fixed axis. The main topics to be discussed here are the following:

1. **Rotational kinematics**. Definitions of angular position, velocity, and acceleration are given. Special emphasis is placed on rotational motion with constant angular acceleration.
2. **Torque**. Torque is defined and interpreted in terms of its ability to rotate an object around an axis.
3. **Rotational dynamics**. The basic law used to describe the dynamics of rotational motion around a fixed axis is derived from Newton's laws of motion. Specific examples of the application of this law are investigated.
4. **Work-energy theorem for fixed-axis rotation**. The expressions for rotational kinetic energy and work are obtained. These are then used to express the work-energy theorem for fixed-axis rotational motion.
5. **Principle of conservation of mechanical energy**. How the conservation of mechanical energy is applied to fixed-axis rotational motion is considered.

In this and the following chapter, we consider the rotational motion of rigid bodies. Our study is limited to two special cases: rotation around an axis fixed in space and rolling motion in a plane. (See Fig. 9-1*a* and *b*.) This chapter is devoted to the former and Chap. 10 to the latter case.

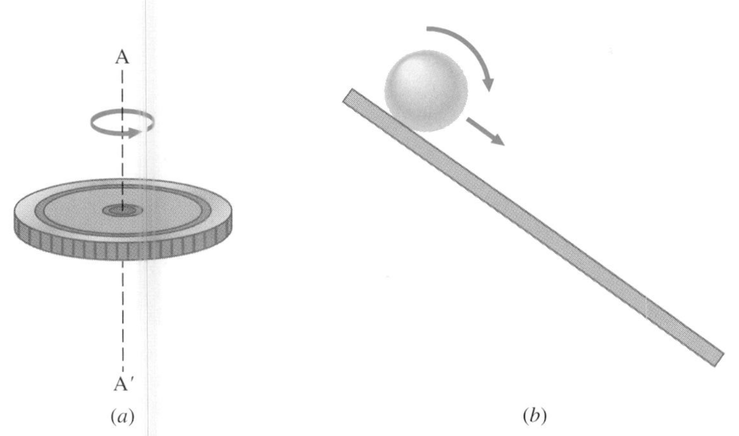

FIGURE 9-1 (*a*) A flywheel rotates around the fixed axis AA′. (*b*) A ball rolling down an incline moves translationally as it rotates.

The concept of a rigid body is actually an idealization, since bodies do distort when subjected to forces. For example, the golf club of Fig. 9-2 bends considerably while it is being swung. Fortunately, distortions are often small enough to be insignificant, as in the cases of a rotating flywheel and a rolling bowling ball. The error introduced by treating these two bodies as perfectly rigid is certainly negligible.

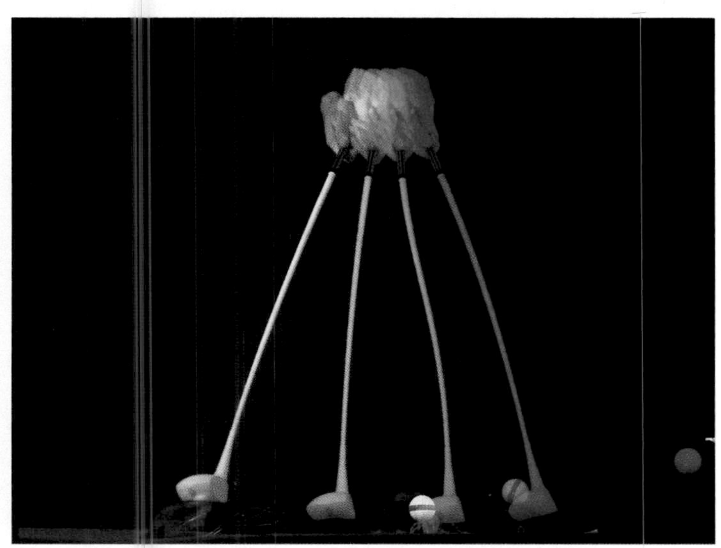

FIGURE 9-2 The shaft of the golf club is distorted while the club is swung.

Rotating rigid bodies, like all mechanical systems, are governed by Newton's laws of motion. However, Newton's laws are not applied to these bodies in the form with which you're familiar. Instead, they are used to obtain equivalent equations expressed in terms of quantities particular to rotational motion. These quantities include angular displacement, angular velocity, and angular acceleration, and also moment of inertia. The important laws of dynamics (Newton's second law, the work-energy theorem, and the principle of mechanical-energy conservation) are applied once again—only this time in their rotational forms.

9-1 ROTATIONAL KINEMATICS

When a rigid body rotates around a fixed axis, the points on that body move along circular paths centered at the axis of rotation. Radial lines between this axis and the points all sweep out the *same angle in the same time interval*, as illustrated in Fig. 9-3. A cross-sectional view of the rigid body perpendicular to the rotational axis is shown in Fig. 9-4. As the body rotates from position A to position B, the point P starts on the x axis and moves along a circular path of radius r, tracing out an arc length s while sweeping through an angle θ. From simple geometry, these three quantities are replaced by

$$\theta = \frac{s}{r}. \tag{9-1}$$

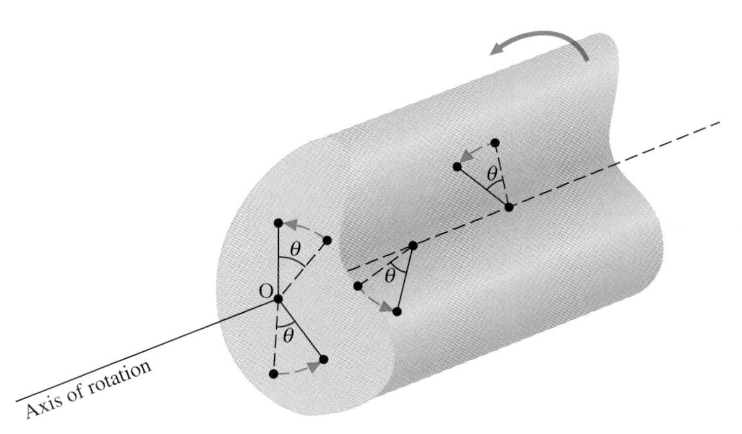

FIGURE 9-3 As a rigid body rotates around a fixed axis, every point of that body rotates through the same angle in a given time interval.

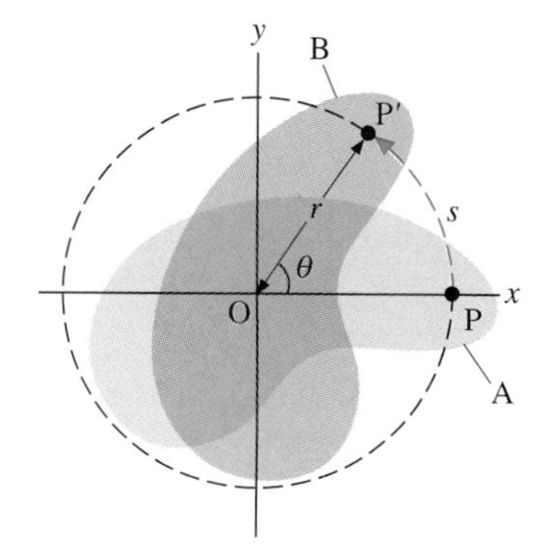

FIGURE 9-4 A cross-sectional view of the rigid body of Fig. 9-3. The rotational axis passes through O and is perpendicular to the page. As the body rotates from position A to position B, the point P on the body moves along the circular arc from P to P′.

The angle θ is known as the **angular position** of point P with respect to the x axis. Since all points on the body rotate through the same angle, we also refer to θ as the angular position of the rigid body. In Eq. (9-1), θ is expressed in *radians (rad)*. For example, when P rotates through 2π radians, the corresponding arc length is

$$s = \theta r = 2\pi r,$$

which is the circumference of the circle.

Since the radian is a ratio of length measurements, it is a dimensionless quantity. An angular measurement in radians is converted to one in degrees by using the fact that one complete rotation corresponds to either 2π rad or 360°. Thus

$$2\pi \text{ rad} = 360°,$$

or

$$1 \text{ rad} = 57.3°.$$

The **instantaneous angular velocity** ω (or just **angular velocity**) of a rigid body is, by definition,

$$\omega = \lim_{\Delta t \to 0} \frac{\Delta\theta}{\Delta t} = \frac{d\theta}{dt}. \tag{9-2}$$

For rotations around a fixed axis, every point of a rigid body has the same angular displacement $\Delta\theta$ in the time interval Δt; thus ω, like θ, describes the motion of the *entire* body rather than the motion of any specific point on the body. The dimension of angular velocity is $1/T$ (inverse time). Units commonly used to represent ω are (1) revolutions per minute (rev/min) or (2) radians per second (rad/s)—which is equivalent to 1/s because the radian is dimensionless.

If its angular velocity is not constant, a rigid body is said to have an **instantaneous angular acceleration** (or simply **angular acceleration**). This is defined as

$$\alpha = \lim_{\Delta t \to 0} \frac{\Delta\omega}{\Delta t} = \frac{d\omega}{dt}. \tag{9-3}$$

Angular acceleration is also the *same for every point* on a body rotating around a fixed axis. Its dimensions are $1/T^2$, and it is usually expressed in radians per second squared (rad/s^2), or equivalently 1/s^2.

A comparison of Eqs. (9-2) and (9-3) with Eqs. (3-4*b*) and (3-6*b*) shows that θ, ω, and α play the same roles in rotational motion that x, v, and a play in linear motion. Hence for constant α, the equations that relate the rotational quantities and time can be found by simply substituting θ for x, ω for v, and α for a in the three basic kinematic equations of Table 3-1. This yields

$$\omega = \omega_0 + \alpha t \tag{9-4}$$
$$\theta = \theta_0 + \omega_0 t + \tfrac{1}{2}\alpha t^2 \tag{9-5}$$
$$\omega^2 = \omega_0^2 + 2\alpha(\theta - \theta_0). \tag{9-6}$$

To determine how the angular and linear kinematic variables are related, let's again consider point P of Fig. 9-4. For rotation of the body around a fixed axis, P travels in a circle and its linear velocity **v** is always directed tangent to the circle. Now the arc length traversed by P is related to its angular displacement in radians by $s = r\theta$. Since the linear speed of P is $v = ds/dt$, and r is a constant, we can easily find a relationship between the linear and angular speeds:

During gear shifts, the bicycle chain is switched from one gear cog to another with a different diameter. The angular velocity of the rear wheel then changes, even though the speed of the chain remains constant.

$$v = \frac{ds}{dt} = r\frac{d\theta}{dt} = r\omega. \tag{9-7}$$

The linear acceleration of P has a tangential component, $a_\theta = dv/dt$, and a radial component (the centripetal acceleration) $a_r = v^2/r$. (See Sec. 4-6.) The tangential component is found by differentiating Eq. (9-7) with respect to time. This yields

$$a_\theta = r\alpha. \tag{9-8}$$

Finally, the radial acceleration can be written in terms of angular quantities by substituting $v = r\omega$ into $a_r = v^2/r$; we then obtain

$$a_r = r\omega^2. \tag{9-9}$$

The relationships among the linear and angular quantities are summarized in Fig. 9-5. Unlike their angular counterparts, the linear velocity and linear acceleration are *not the same for every point* on the rotating body. As Eqs. (9-7), (9-8), and (9-9) indicate, these quantities depend on r. A point that is a distance $2r$ from the axis of rotation has twice the linear velocity, but the same angular velocity, as a point a distance r from the axis. Also, since Eqs. (9-7), (9-8), and (9-9) are derived directly from $s = r\theta$, a relationship that requires θ to be in radians, these equations can only be used with the angular unit of both ω and α in radians.

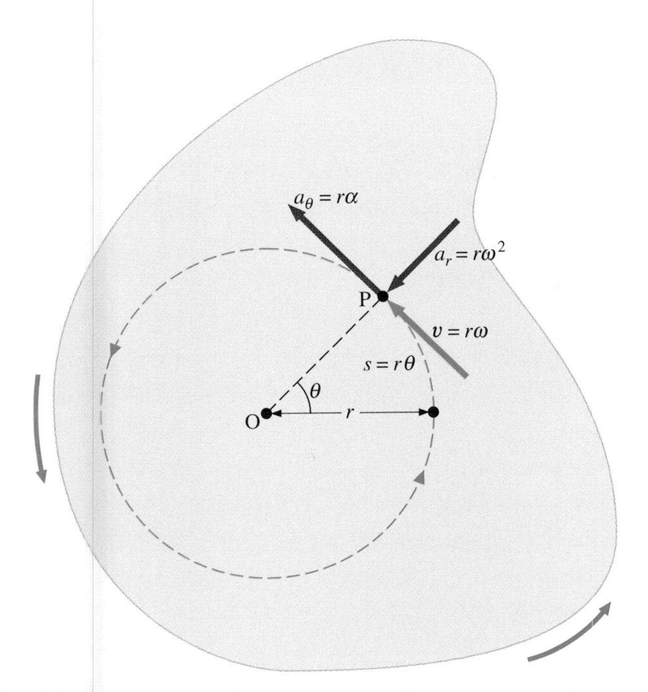

FIGURE 9-5 The relationships between the linear and angular quantities.

EXAMPLE 9-1 A SPINNING FLYWHEEL

A flywheel rotating with an angular velocity of 1.20×10^3 rev/min is brought to rest in 1.00 min. Find (*a*) the angular acceleration of the flywheel, (*b*) the angle through which the flywheel rotates while slowing to a stop, and (*c*) the linear velocity and acceleration of a point 20.0 cm from the axis of rotation at the instant when the flywheel starts to slow down.

SOLUTION The initial angular velocity ω_0 expressed in radians per second is

$$\begin{aligned}\omega_0 &= (1.20 \times 10^3 \text{ rev/min})(2\pi \text{ rad/rev})(1 \text{ min}/60 \text{ s})\\ &= 126 \text{ rad/s}.\end{aligned}$$

(*a*) The flywheel stops after 1.00 min, so upon relating angular velocity to time, we find

$$\begin{aligned}\omega &= \omega_0 + \alpha t\\ 0 &= 126 \text{ rad/s} + \alpha(60.0 \text{ s}),\end{aligned}$$

and the angular acceleration is

$$\alpha = -2.10 \text{ rad/s}^2.$$

(*b*) The angular displacement during the time that the wheel slows down is then

$$\begin{aligned}\theta - \theta_0 &= \omega_0 t + \tfrac{1}{2}\alpha t^2\\ &= (126 \text{ rad/s})(60.0 \text{ s}) + \tfrac{1}{2}(-2.10 \text{ rad/s}^2)(60.0 \text{ s})^2\\ &= 3.78 \times 10^3 \text{ rad}.\end{aligned}$$

(*c*) At the start of the deceleration, the linear velocity of a point 20.0 cm from the rotational axis is

$$v_0 = r\omega_0 = (0.200 \text{ m})(126 \text{ rad/s}) = 25.2 \text{ m/s},$$

while the initial tangential and radial accelerations of this point are

$$a_{0\theta} = r\alpha = (0.200 \text{ m})(-2.10 \text{ rad/s}^2) = -0.420 \text{ m/s}^2,$$

and

$$a_{0r} = r\omega_0^2 = (0.200 \text{ m})(126 \text{ rad/s})^2 = 3.18 \times 10^3 \text{ m/s}^2.$$

EXAMPLE 9-2 ROTATIONAL MOTION OF A PROPELLER

An airplane propeller starts from rest and rotates with a constant angular acceleration of 50 rad/s^2 for 5.0 s. (*a*) What is the propeller's angular velocity at the end of this time interval? (*b*) Through what angle does it turn during the interval? (*c*) At the end of the 5.0-s interval, what are the linear velocity and acceleration of a point on the tip of the propeller 2.00 m from the axis of rotation?

SOLUTION (*a*) The angular velocity can be determined using

$$\omega = \omega_0 + \alpha t,$$

so

$$\omega = 0 + (50 \text{ rad/s}^2)(5.0 \text{ s}) = 250 \text{ rad/s}.$$

(*b*) Relating angular displacement to time, we find that during the 5.0-s interval, the propeller rotates through an angle

$$\begin{aligned}\theta - \theta_0 &= \omega_0 t + \tfrac{1}{2}\alpha t^2\\ &= 0 + \tfrac{1}{2}(50 \text{ rad/s}^2)(5.0 \text{ s})^2 = 625 \text{ rad}.\end{aligned}$$

(*c*) After 5.0 s, the linear velocity and the tangential and radial accelerations of a point 2.00 m from the axis of rotation are

$$\begin{aligned}v &= r\omega = (2.00 \text{ m})(250 \text{ rad/s}) = 500 \text{ m/s};\\ a_\theta &= r\alpha = (2.00 \text{ m})(50 \text{ rad/s}^2) = 100 \text{ m/s}^2;\\ a_r &= r\omega^2 = (2.00 \text{ m})(250 \text{ rad/s})^2 = 1.25 \times 10^5 \text{ m/s}^2.\end{aligned}$$

DRILL PROBLEM 9-1

(*a*) What is the angular velocity of the second hand on a clock? of the minute hand? of the hour hand? (*b*) If all three hands are 10.0 cm long, what is the radial acceleration of a point on the tip of each?
ANS. (*a*) 0.105 rad/s, 1.75×10^{-3} rad/s, 1.45×10^{-4} rad/s; (*b*) 1.10×10^{-3} m/s^2, 3.06×10^{-7} m/s^2, 2.10×10^{-9} m/s^2.

DRILL PROBLEM 9-2

A grinding wheel rotating at 400 rev/min is brought to rest in 10 s. Find its angular acceleration and the angle through which it rotates while slowing down. Assume that the angular acceleration of the wheel is constant.
ANS. -4.2 rad/s^2; 210 rad.

DRILL PROBLEM 9-3

The angular position of a body varies with time according to $\theta = pt^2 - qt^3$. At what time(s) is its angular velocity zero? What is its angular acceleration then?
ANS. $t = 0$ and $t = 2p/3q$; $2p$ and $-2p$.

9-2 Torque

It's clear that a force is required to make a body rotate. For example, the child of Fig. 9-6 cannot open the door until he pushes on it. If he is smart, he quickly realizes that he shouldn't push just anywhere—the farther from the hinges he applies the force, the more easily the door is opened. However, even if he pushes at the outer edge of the door, it remains shut if his force is parallel to the surface of the door. He is most effective if he pushes perpendicular to the surface. There are many tasks that are similar to opening a door, such as turning a wrench and twisting off a bottle cap. (See Fig. 9-7.) All such activities lead to the same conclusion—that the ability of a force to rotate a rigid body around a particular axis depends on three properties: (1) the magnitude of the force, (2) the distance between the point of application of the force and the axis of rotation, and (3) the direction of the force.

(*a*)

(*b*)

FIGURE 9-6 (*a*) The farther away from the hinges that the child pushes, the more easily he can open the door. (*b*) However, he is most effective if he pushes perpendicular to the surface of the door.

All three of these properties are incorporated in the definition of a vector quantity called the **torque**. For now, we consider this definition only for situations in which all forces acting on a body lie in the plane perpendicular to the body's axis of rotation. We represent this plane by the plane of the page. Examples are shown in the sketches of Fig. 9-8, where forces are

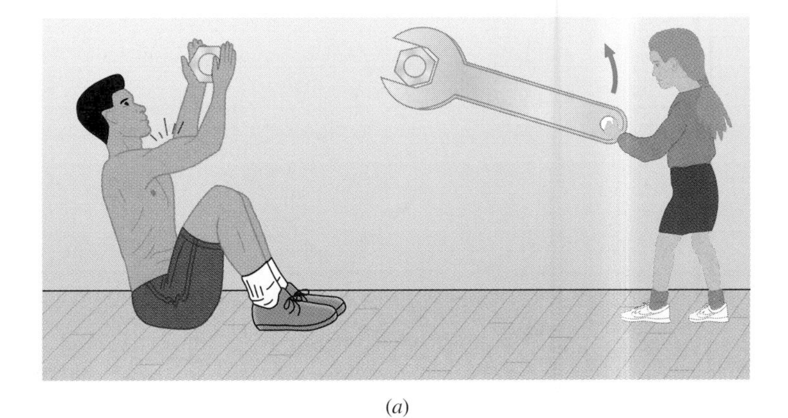

(*a*)

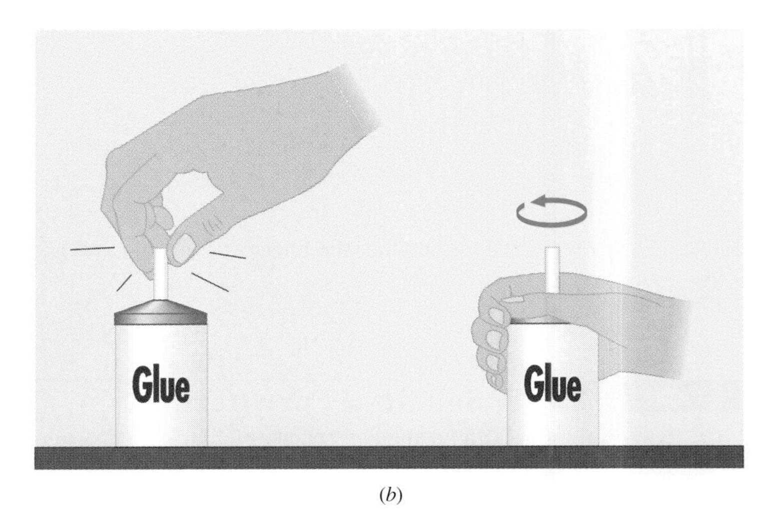

(*b*)

FIGURE 9-7 (*a*) The rusty nut cannot be removed without the wrench. (*b*) The sticky cap on the glue bottle is much easier to open if it is grasped as far from the axis of rotation as possible.

applied along various lines of action on a wrench. By definition, the magnitude of the torque $\boldsymbol{\tau}$ of a force **F** around the axis through O is

$$\tau = r_{\perp} F, \tag{9-10}$$

where $r_{\perp}$ is the **moment arm** of **F**. The moment arm is the *perpendicular distance from the axis of rotation to the line of action of the applied force*. It is important to distinguish torques according to whether they tend to rotate an object clockwise or counterclockwise. For example, if the girl of Fig. 9-7*a* rotates the wrench clockwise by pulling down on it, she tightens the bolt; however, if she rotates the wrench counterclockwise by pushing up on it, she achieves the opposite effect and loosens

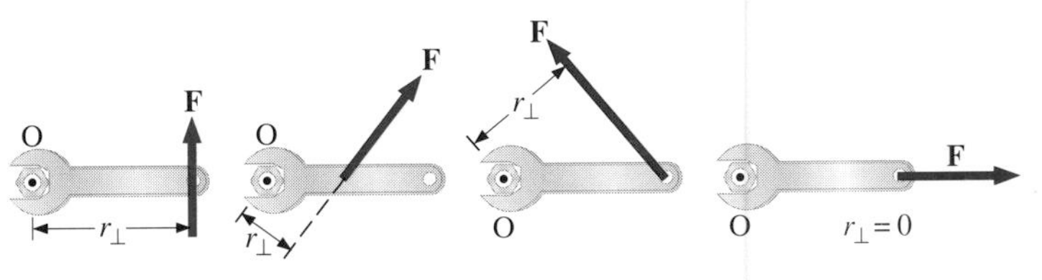

FIGURE 9-8 The torque around the axis through O (the nut) and perpendicular to the page is $r_{\perp}F$. Notice that $r_{\perp}$ depends on where and in what direction the force is applied.

Can you explain how a crowbar works in terms of applied force and moment arm?

the bolt. We usually take torques that cause counterclockwise rotations (often referred to as *counterclockwise torques*) to be *positive*, and torques that cause clockwise rotations (*clockwise torques*) to be *negative*.

DRILL PROBLEM 9-4

Explain how Eq. (9-10) incorporates the three properties that are responsible for the rotational effectiveness of a force.

EXAMPLE 9-3 **TORQUES ON A THIN ROD**

Acting on a thin rod pivoted at point O are three different forces **A**, **B**, and **C**, which lie in the plane of the page as shown in Fig. 9-9*a*, *b*, and *c*. The magnitudes of these forces are $A = 100$ N, $B = 50$ N, and $C = 80$ N. Calculate the torque of each force around the axis perpendicular to the plane of the page and passing through O.

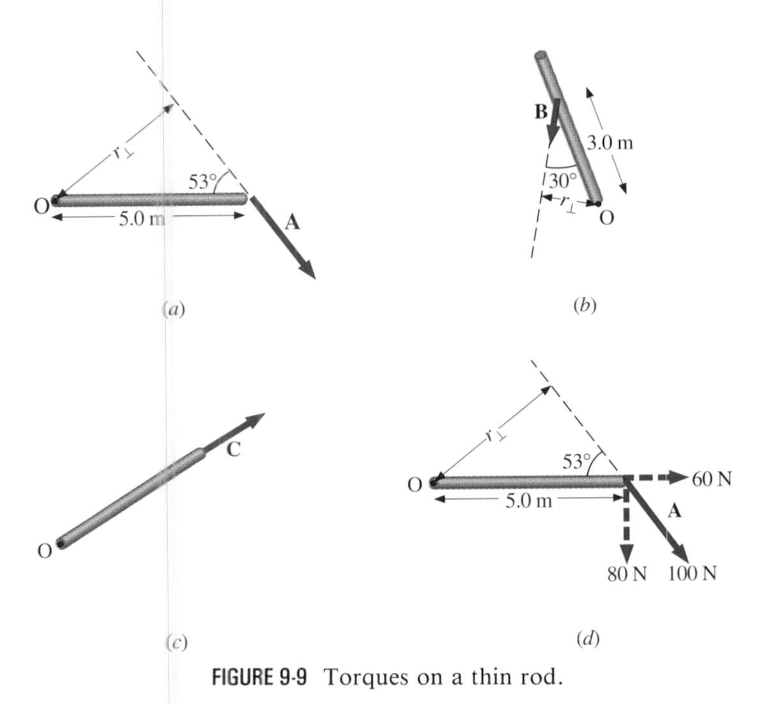

FIGURE 9-9 Torques on a thin rod.

SOLUTION To find the torques, we first calculate the moment arms, or the perpendicular distances from O to the lines of action of the forces. For **A**, **B**, and **C**, the moment arms are (5.0 m) sin 53° = 4.0 m, (3.0 m) sin 30° = 1.5 m, and 0, respectively. With counterclockwise designated as positive, the torques of **A**, **B**, and **C** are

$$\tau_A = -(4.0 \text{ m})(100 \text{ N}) = -400 \text{ N}\cdot\text{m},$$

$$\tau_B = (1.5 \text{ m})(50 \text{ N}) = 75 \text{ N}\cdot\text{m},$$

and

$$\tau_C = 0.$$

Frequently the torque of a force is calculated after first resolving the force into its rectangular components. For example, the magnitudes of the x and y components of the force **A** of the previous example are 60 and 80 N (see Fig. 9-9*d*), and the torques of these components are (0 m)(60 N) = 0 N·m and −(5.0 m)(80 N) = −400 N·m. Their sum, the net torque, is −400 N·m, which agrees with the previous calculation.

Now let's consider how torques are determined in the general three-dimensional case. When a force **F** is applied at point P, whose position is **r** relative to O (see Fig. 9-10), the torque around O is defined as

$$\boldsymbol{\tau} = \mathbf{r} \times \mathbf{F}. \tag{9-11}$$

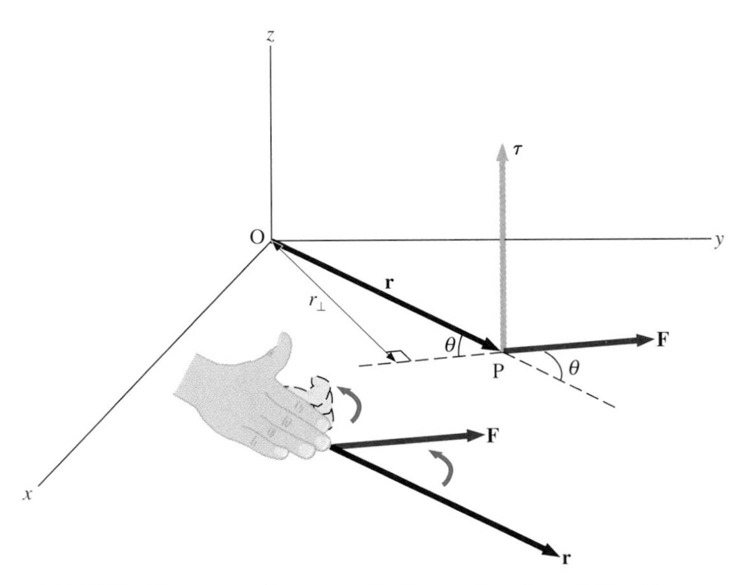

FIGURE 9-10 The torque is perpendicular to the plane defined by **r** and **F** and its direction is given by the right-hand rule.

By definition of the vector product (Sec. 2-4), $\boldsymbol{\tau}$ is perpendicular to the plane of **r** and **F** and has magnitude $rF \sin\theta$, where θ is the angle between the two vectors. To determine in which of the two perpendicular directions the torque points, you align the fingers of your right hand along **r** and rotate them across the smaller angle into **F**; your thumb then points along $\boldsymbol{\tau}$.

Equation (9-11) can, of course, be applied to planar problems such as the ones we have already discussed. In these cases, the vector product reduces to Eq. (9-10) because $r \sin\theta = r_\perp$, as illustrated in Fig. 9-10. Furthermore, the positive z direction of the torque given by the right-hand rule is consistent with the fact that the torque causes counterclockwise rotation in the xy

plane and is considered positive. If the force **F** in Fig. 9-10 were reversed in direction, $\boldsymbol{\tau} = \mathbf{r} \times \mathbf{F}$ would point in the negative z direction. This is consistent with the fact that the torque now causes clockwise rotation in the xy plane, and is therefore considered negative.

An important application of Eq. (9-11) is the calculation of the net torque on a system of particles due to the gravitational forces $m_i\mathbf{g}$ on the individual particles. For the system of Fig. 9-11, the gravitational torque about point O is

$$\boldsymbol{\tau} = \sum_i \mathbf{r}_i \times m_i\mathbf{g}.$$

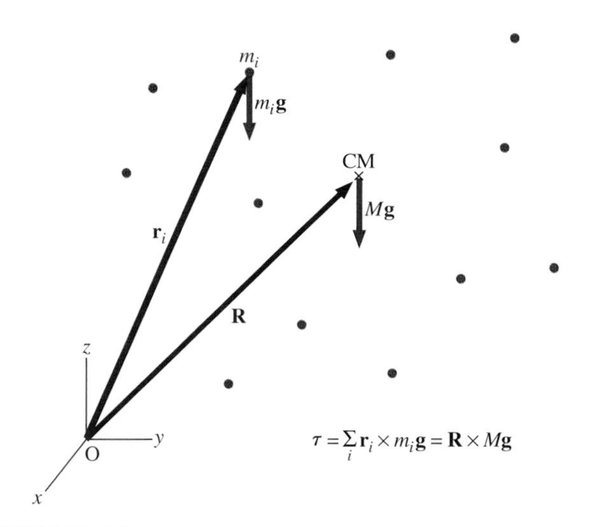

FIGURE 9-11 The net torque of the weights of the particles of a system is equal to that found by placing the net weight of the system $M\mathbf{g}$ at the center of mass (position **R**) and then calculating $\mathbf{R} \times M\mathbf{g}$.

Since **g** is constant, this can be written as

$$\boldsymbol{\tau} = \left(\sum_i m_i\mathbf{r}_i\right) \times \mathbf{g}.$$

From the definition of the center of mass, $\sum_i m_i\mathbf{r}_i = M\mathbf{R}$, where $M = \sum_i m_i$ and **R** is the position of the center of mass with respect to point O. We can therefore express the torque as

$$\boldsymbol{\tau} = M\mathbf{R} \times \mathbf{g} = \mathbf{R} \times M\mathbf{g}.$$

Hence *the net gravitational torque on a body of mass M is the same as the torque of its weight M***g** *acting at the center of mass.*

EXAMPLE 9-4 Calculations of Torque

(*a*) Using Eq. (9-10), calculate the torque of the 60-N force of Fig. 9-12 around the z axis. (*b*) Using Eq. (9-11), calculate the torque of this same force.

SOLUTION (*a*) The moment arm of the 60-N force is $r_\perp = (5.0\text{ m}) \times \sin 37° = 3.0$ m. The associated torque causes counterclockwise rotation around the z axis and is

$$\tau = (3.0\text{ m})(60\text{ N}) = 180\text{ N}\cdot\text{m}.$$

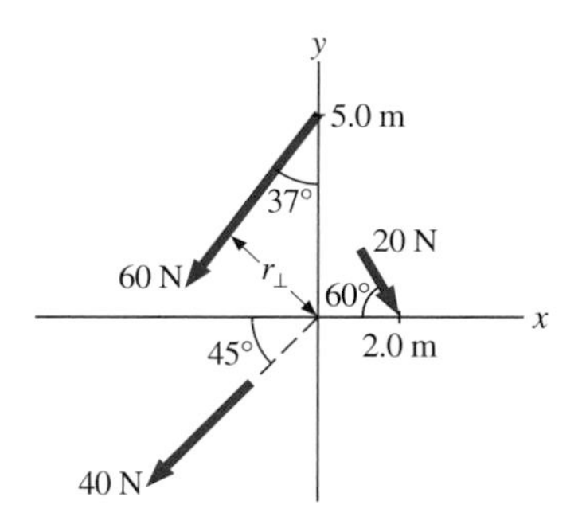

FIGURE 9-12 The forces of Example 9-4 and Drill Prob. 9-5. The z axis points out of the page.

(*b*) The force acts along a line that passes through the point $\mathbf{r} = 5.0\mathbf{j}$ m, and

$$\mathbf{F} = [(-60\sin 37°)\mathbf{i} + (-60\cos 37°)\mathbf{j}]\text{ N} = (-36\mathbf{i} - 48\mathbf{j})\text{ N}.$$

From Eq. (9-11), the torque is

$$\boldsymbol{\tau} = \mathbf{r} \times \mathbf{F} = (5.0\mathbf{j})\text{ m} \times (-36\mathbf{i} - 48\mathbf{j})\text{ N}$$
$$= (-180\mathbf{j} \times \mathbf{i} - 240\mathbf{j} \times \mathbf{j})\text{ N}\cdot\text{m} = 180\mathbf{k}\text{ N}\cdot\text{m}.$$

This agrees with the calculation of part (*a*), because a torque in the positive z direction causes counterclockwise rotation around the z axis.

EXAMPLE 9-5 Torque as a Cross Product

A 50.0-N force is applied at the point $x = 4.0$ m, $y = 3.0$ m, and $z = 0$, as shown in Fig. 9-13. The force is directed parallel to the yz plane. Calculate the torque of this force relative to the origin O.

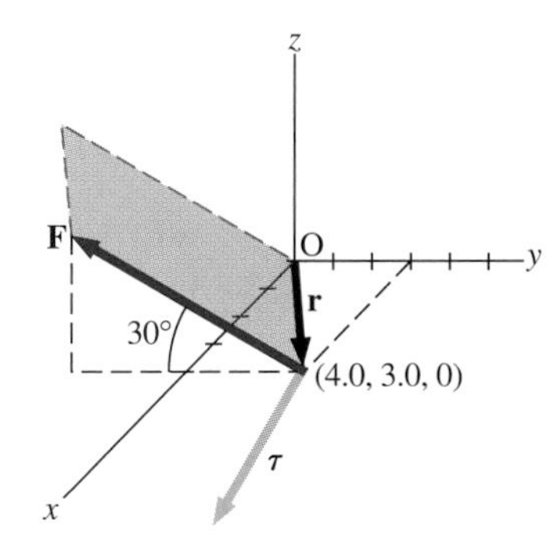

FIGURE 9-13 The force **F** is applied at the point $x = 4.0$ m, $y = 3.0$ m, $z = 0$ and is parallel to the yz plane. The torque $\boldsymbol{\tau} = \mathbf{r} \times \mathbf{F}$ is perpendicular to the plane containing **r** and **F**.

SOLUTION The rectangular components of **F** are $F_x = 0$, $F_y = -(50.0\text{ N})\cos 30° = -43.3$ N, and $F_z = (50.0\text{ N})\sin 30° = 25.0$ N, while those of **r** are $r_x = 4.0$ m, $r_y = 3.0$ m, and $r_z = 0$. From Eq. (2-12), the torque is then

$$\boldsymbol{\tau} = \{[(3.0)(25.0) - (0)(-43.3)]\mathbf{i} + [(0)(0) - (4.0)(25.0)]\mathbf{j}$$
$$+ [(4.0)(-43.3) - (3.0)(0)]\mathbf{k}\}\text{ N}\cdot\text{m}$$
$$= (75\mathbf{i} - 100\mathbf{j} - 173\mathbf{k})\text{ N}\cdot\text{m}.$$

As required by the vector product, $\boldsymbol{\tau}$ is perpendicular to the plane containing **r** and **F**, as shown in Fig. 9-13.

DRILL PROBLEM 9-5

(*a*) Use Eq. (9-10) to calculate the torques around the z axis of the 20-N and 40-N forces of Fig. 9-12. (*b*) Using Eq. (9-11), calculate the torques of these two forces.
ANS. (*a*) −35 N·m, 0; (*b*) −35**k** N·m, 0.

DRILL PROBLEM 9-6

A force $\mathbf{F} = (6.0\mathbf{i} + 4.0\mathbf{j})$ N is applied at a point whose position is $\mathbf{r} = (1.0\mathbf{i} + 2.0\mathbf{j})$ m. What is the torque of **F** around the origin of the reference frame?
ANS. −8.0**k** N·m.

9-3 DYNAMICS OF FIXED-AXIS ROTATION

In this section we investigate the law that describes the dynamics of a planar rigid body rotating around a fixed axis. A general derivation of this law is given in Supplement 9-1. For now, however, we consider the object of Fig. 9-14*a* which consists of two small spheres of masses m_1 and m_2 placed at opposite ends of a massless rigid rod. The body is free to rotate in the plane of the page around the axis through O, a distance r_1 from m_1 and r_2 from m_2. We represent the net external forces on the spheres as $\mathbf{F}_1$ and $\mathbf{F}_2$. The components of these forces along and perpendicular to the rod are shown in Fig. 9-14*b*. Applying Newton's second law in the tangential direction to each sphere, we have

$$F_{1\theta} = m_1 a_{1\theta} \quad \text{and} \quad F_{2\theta} = m_2 a_{2\theta}.$$

If α is the angular acceleration of the body, $a_{1\theta} = r_1\alpha$ and $a_{2\theta} = r_2\alpha$. Substituting these expressions for $a_{1\theta}$ and $a_{2\theta}$ into the equations for $F_{1\theta}$ and $F_{2\theta}$, then multiplying the first equation by r_1 and the second by r_2, we find

$$F_{1\theta} r_1 = m_1 r_1^2 \alpha \quad \text{and} \quad F_{2\theta} r_2 = m_2 r_2^2 \alpha.$$

The large moment of inertia of the pole protects the tightrope walker against tipping over.

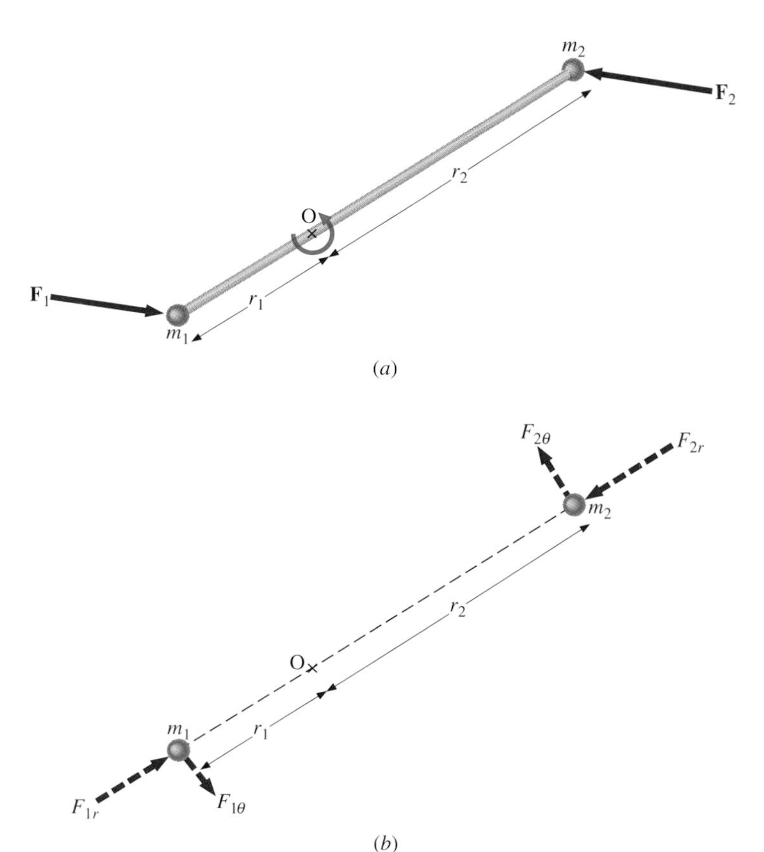

FIGURE 9-14 (*a*) Two small spheres are at opposite ends of a massless rod of length l. (*b*) The net external forces on the spheres are resolved into radial and tangential components.

If we sum these two equations, we find that the left-hand side is the net torque acting on the body around the axis through O:

$$\sum \tau = F_{1\theta} r_1 + F_{2\theta} r_2.$$

The right-hand side is

$$(m_1 r_1^2 + m_2 r_2^2)\alpha,$$

which is the moment of inertia [see Eq. (8-8)] of the body multiplied by its angular acceleration. We therefore have

$$\sum \tau = I\alpha, \tag{9-12}$$

where $\sum \tau$ is the net torque around the axis of rotation of the external forces acting on the body, I is the moment of inertia of the body around the axis, and α is the angular acceleration around the axis.

Notice the correspondence between this equation and $\sum \mathbf{F} = m\mathbf{a}$! With $\sum \tau$, I, and α as the rotational counterparts of $\sum \mathbf{F}$, m, and a, we have in Eq. (9-12) *the rotational form of Newton's second law*. $\sum \mathbf{F} = m\mathbf{a}$ tells us that when a net external force is applied to a body, it responds with an acceleration that depends on the mass (or inertia) of the body. Correspondingly, $\sum \tau = I\alpha$ tells us that when a body free to rotate around a fixed axis is subjected to a net external torque, it responds with an angular acceleration whose value depends on the body's rotational inertia.

EXAMPLE 9-6 **ANGULAR ACCELERATION OF A RIGID BODY**

At a particular instant, the 4.0-kg rigid body of Fig. 9-15 is acted on by two external forces and the force of gravity. The body is free to rotate around the axis through O, and its moment of inertia around this axis is $0.16\ \text{kg}\cdot\text{m}^2$. What is the angular acceleration of the body?

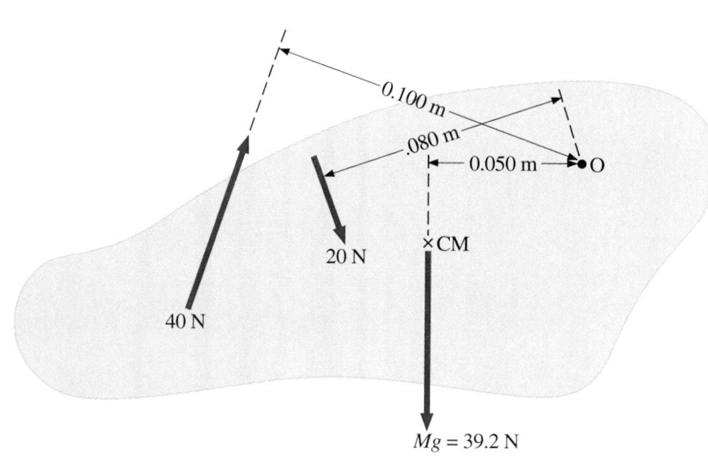

FIGURE 9-15 The forces on a rigid body.

SOLUTION The torques due to the 20-N force and the weight (which acts downward at the center of mass) turn the body counterclockwise, and their moment arms are 0.080 and 0.050 m, respectively. These torques are $(20\ \text{N})(0.080\ \text{m}) = 1.6\ \text{N}\cdot\text{m}$ and $(39.2\ \text{N}) \times (0.050\ \text{m}) = 1.96\ \text{N}\cdot\text{m}$. The 40-N force has a moment arm 0.100 m long. Its torque rotates the body clockwise and is $-(40\ \text{N}) \times (0.100\ \text{m}) = -4.0\ \text{N}\cdot\text{m}$. Now from the rotational form of the second law,

$$\sum \tau = I\alpha$$
$$(1.6 + 1.96 - 4.0)\ \text{N}\cdot\text{m} = (0.16\ \text{kg}\cdot\text{m}^2)\alpha,$$

and the angular acceleration of the body at the instant shown is

$$\alpha = -2.8\ \text{rad/s}^2.$$

DRILL PROBLEM 9-7

A 50-N·m torque is applied to a rigid body that is free to rotate around a fixed axis. The body starts from rest and rotates through 40 rad in 20 s. What is its moment of inertia?
ANS. $250\ \text{kg}\cdot\text{m}^2$.

DRILL PROBLEM 9-8

A flywheel with moment of inertia $100\ \text{kg}\cdot\text{m}^2$ starts from rest and accelerates for 10 s under the influence of a 200-N·m torque. What is the angular acceleration of the flywheel? What is the angular velocity of the flywheel after 10 s? Through what angle does it turn in this time interval?
ANS. $2.0\ \text{rad/s}^2$; 20 rad/s; 100 rad.

DRILL PROBLEM 9-9

After 10 s, the applied torque on the flywheel of Drill Prob. 9-8 is removed, and it slowly comes to rest under the influence of a 0.10 N·m frictional torque. How much time elapses before the flywheel stops? Through what angle does it turn in this time interval?
ANS. 2.0×10^4 s; 2.0×10^5 rad.

9-4 EXAMPLES IN ROTATIONAL DYNAMICS

We now have the theoretical tools needed to study the motion of a rigid body rotating around a fixed axis. The technique we will use is analogous to that used to analyze linear motion. (See Sec. 6-2.) We'll take the following steps in our analysis of rotational motion:

1. Identify the body whose rotational motion is to be analyzed. A system may consist of several bodies, some of which rotate and some of which move linearly. Any linear motion should be analyzed using the procedure outlined in Sec. 6-2.
2. Locate the body's rotational axis, which is fixed in an inertial frame.
3. Identify all forces acting on the body and draw a free-body diagram.
4. Determine the moment arms corresponding to the forces and calculate the torques on the rotating body.
5. Apply the rotational form of Newton's second law.

EXAMPLE 9-7 **ROTATION AND TRANSLATION**

A uniform disk of mass M and radius R is free to rotate around a fixed horizontal axle supported in frictionless bearings. (See Fig. 9-16.) A light cord is wrapped around the rim of the disk and then tied to a small can of mass m. Find the acceleration of the can, the angular acceleration of the disk, and the tension in the cord. Assume that the cord does not slip as it unwinds on the disk.

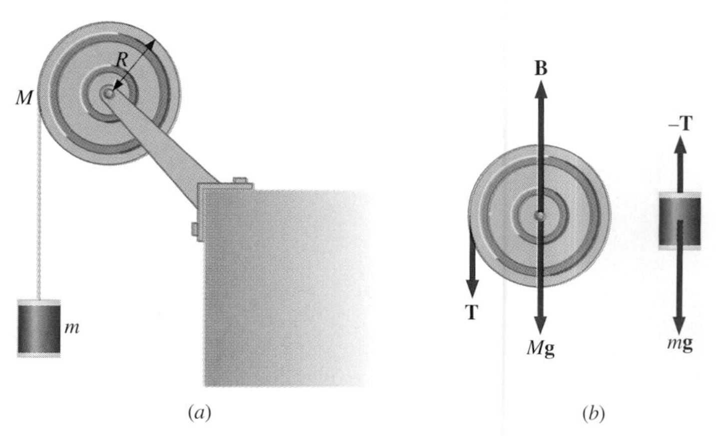

FIGURE 9-16 (*a*) A disk turns without friction as a can drops. (*b*) Free-body diagrams for the objects.

SOLUTION The system consists of two bodies: a disk rotating around a fixed horizontal axis through its center and a can moving vertically downward. As shown in the free-body diagram of Fig. 9-16*b*, the forces on the disk are the tension **T** in the cord, the force **B** of the bearings, and the weight $M\mathbf{g}$ of the disk. The forces on the can are its weight $m\mathbf{g}$ and the tension $-\mathbf{T}$ in the cord.

First we analyze the vertical motion of the can. With the positive y direction downward, Newton's second law gives

$$\sum F_y = ma_y$$
$$mg - T = ma, \qquad (i)$$

where a is the downward acceleration of the can.

The disk rotates around the axis through its center. As can be seen in the free-body diagram for the disk, only **T**, with its moment arm R, exerts a torque around the rotational axis. With $I = MR^2/2$ [part (*c*) of Table 8-1], Newton's second law for rotational motion yields

$$\sum \tau = I\alpha$$
$$TR = (\tfrac{1}{2}MR^2)\alpha,$$

so

$$\alpha = \frac{2T}{MR}. \qquad (ii)$$

There are now two equations, (*i*) and (*ii*), that relate three unknowns, T, α, and a. A third independent equation can be obtained from the information that the cord unwinds on the edge of the disk *without slipping*. In this case, the acceleration of the cord—and of the can—is the same as the tangential acceleration of the edge of the disk, so from Eq. (9-8),

$$a = R\alpha. \qquad (iii)$$

Equations (*i*), (*ii*), and (*iii*) can be easily solved for the three unknowns. We find

$$a = \frac{2m}{M + 2m}g; \qquad \alpha = \frac{2m}{M + 2m}\frac{g}{R}; \qquad T = \frac{Mm}{M + 2m}g.$$

EXAMPLE 9-8 Atwood's Machine Using a Pulley with Mass

Atwood's machine with a massless pulley was considered in Drill Prob. 6-10. Suppose now that the pulley is a solid cylinder of mass M and rotational inertia $I = MR^2/2$. What is the acceleration of the hanging weights in this case? (See Fig. 9-17.) Assume that there is no friction in the bearings and that the string does not slip on the pulley.

SOLUTION The system consists of (1) a pulley rotating around a fixed horizontal axis through its center and (2) two weights that move vertically. Free-body diagrams for the pulley and the weights are shown in Fig. 9-17. The forces on the left weight are the tension $\mathbf{T}_1$ in the string and $m_1\mathbf{g}$, and the forces on the right weight are the string's tension $\mathbf{T}_2$ and $m_2\mathbf{g}$. Notice that we have not assumed that the tensions in the string on the two sides of the pulley are equal. The reason the tensions are not equal will be given later in this solution. The forces on the pulley are the tensions $-\mathbf{T}_1$ and $-\mathbf{T}_2$, the weight $M\mathbf{g}$, and the force **B** at the axis of rotation due to the bearings.

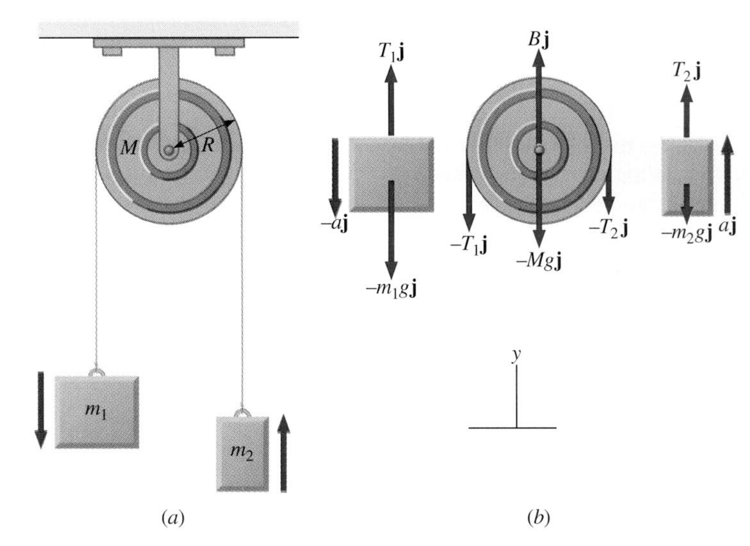

FIGURE 9-17 (*a*) Atwood's machine. The pulley's mass is not negligible. (*b*) The free-body diagrams for the objects.

Upon applying Newton's second law to each weight, we get

Weight 1:

$$\sum F_y = ma_y$$
$$T_1 - m_1 g = -m_1 a. \qquad (i)$$

Weight 2:

$$\sum F_y = ma_y$$
$$T_2 - m_2 g = m_2 a, \qquad (ii)$$

where a is the magnitude of the accelerations of the weights.

The tensions exert a counterclockwise torque of magnitude $T_1 R$ and a clockwise torque of magnitude $T_2 R$ on the pulley. Because they have zero moment arms, the forces **B** and $M\mathbf{g}$ exert no torques around the rotational axis. Now from $\sum \tau = I\alpha$, we have

$$T_1 R - T_2 R = (\tfrac{1}{2}MR^2)\alpha. \qquad (iii)$$

With a substituted for $R\alpha$ (the no-slip assumption), Eq. (*iii*) becomes

$$T_1 - T_2 = \frac{Ma}{2}. \qquad (iv)$$

The three independent equations, (*i*), (*ii*), and (*iv*), relate the three unknowns, T_1, T_2, and a. Eliminating T_1 and T_2, we find for the acceleration:

$$a = \frac{(m_1 - m_2)g}{m_1 + m_2 + M/2}.$$

You should compare this with the acceleration found in Drill Prob. 6-10. If the mass of the pulley is negligible ($M = 0$), the two equations agree.

Why the tensions in the string on the two sides of the pulley are unequal should now be clear. Very simply, they must be different if there is to be a net torque acting to rotate the pulley. [See Eq. (*iii*).] From Eq. (*iv*), the mass of the pulley must be negligible before T_1 can be set equal to T_2. Massless pulleys and the resulting equal tensions throughout massless strings were precisely the assumptions we used in the problems of Chap. 6. This example illustrates the validity and limitations of these assumptions.

EXAMPLE 9-9 A Rod Rotating About One End

Figure 9-18 shows a uniform rod of mass M and length l that is attached to a hinge so that it is free to rotate freely around one end. (*a*) What is the angular acceleration of the rod when it is inclined at an angle θ to the vertical? (*b*) What is the tangential acceleration of the free end of the rod at this angle? (*c*) What is the initial vertical acceleration of the end of the rod if it is released when it is horizontal? When the rod is released, what would happen to a coin placed on top of the free end of the rod?

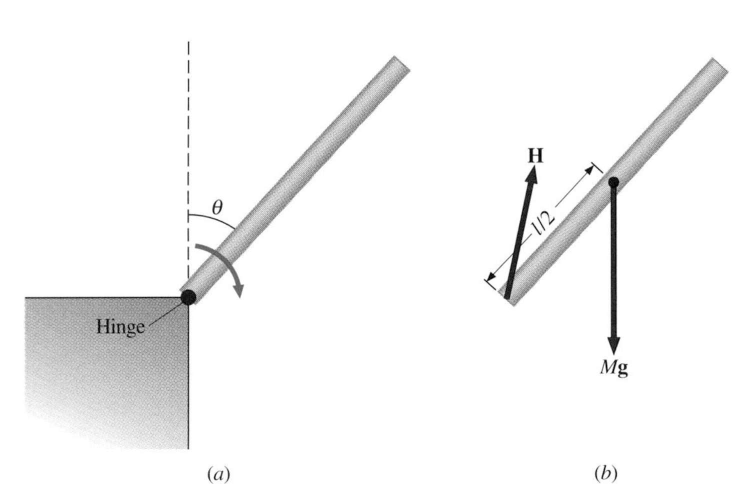

FIGURE 9-18 A rod is free to rotate about a hinge at one end. (*b*) The forces on the rod.

SOLUTION (*a*) The forces on the rod are shown in the free-body diagram of Fig. 9-18*b*. Since the force **H** at the hinge acts at the axis of rotation, it has no moment arm and therefore no torque around the axis. The weight $M\mathbf{g}$ has a moment arm $(l/2)\sin\theta$, so it exerts a clockwise torque $Mg(l/2)\sin\theta$ on the rod. Applying Newton's second law for rotational motion (taking clockwise rotation to be positive), we have

$$\sum \tau = I\alpha$$

$$\tfrac{1}{2}Mgl\sin\theta = I\alpha = \tfrac{1}{3}Ml^2\alpha,$$

where $I = (1/3)Ml^2$ is the moment of inertia of the rod. (See Table 8-1.) The angular acceleration of the rod when it is inclined at an angle θ to the vertical is therefore

$$\alpha = \frac{\tfrac{1}{2}Mgl\sin\theta}{\tfrac{1}{3}Ml^2} = \frac{3g\sin\theta}{2l}.$$

(*b*) From Eq. (9-8), the tangential acceleration of the free end of the rod, which is a distance l from the rotational axis, is

$$a_\theta = l\alpha = l\left(\frac{3g\sin\theta}{2l}\right) = \frac{3}{2}g\sin\theta.$$

(*c*) If the rod is released when it is horizontal, $\theta = \pi/2$, and the initial vertical acceleration of the end of the rod is

$$a_\theta = \frac{3}{2}g\sin\frac{\pi}{2} = \frac{3}{2}g.$$

This is greater than the acceleration of a body falling freely! Consequently, the end of the rod will fall faster than, and therefore out from under, a coin placed on it. You might verify this by using a coin and a meterstick held at one end.

DRILL PROBLEM 9-10

(*a*) If $M = 2.00$ kg, $m = 0.100$ kg, and $R = 8.00$ cm, what is the angular acceleration of the disk of Example 9-7? (*b*) If the can is removed from the cord and a tension equal to the can's weight is applied to the cord instead, what is the disk's angular acceleration?
ANS. (*a*) 11.1 rad/s^2; (*b*) 12.3 rad/s^2.

DRILL PROBLEM 9-11

Replace the uniform disk of Example 9-7 with a pulley of the same mass and radius whose moment of inertia is $I = (0.45)MR^2$. Find the acceleration of the falling body.
ANS. $a = mg/(0.45M + m)$.

9-5 Rotational Work and Kinetic Energy

In this section we investigate how the work-energy theorem is applied to a rigid body rotating around an axis fixed in space. After deriving an expression for the kinetic energy of the rotating body and determining how to calculate work done on the body, we will be able to recast the work-energy theorem into a form applicable to fixed-axis rotation. We begin by considering the rigid body of Fig. 9-19, which is rotating around the fixed axis OO′ with an angular velocity ω. From Eq. (9-7), the speed of particle i of the body is $v_i = r_i\omega$, where r_i is the distance from the rotational axis to the particle. By definition, the body's kinetic energy is the sum of the kinetic energies of all its particles:

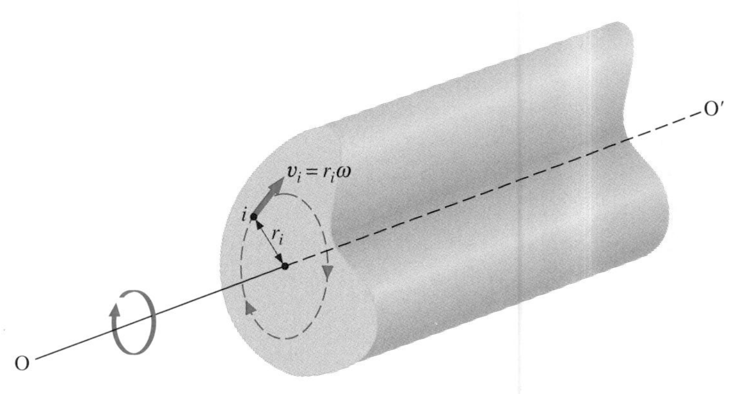

FIGURE 9-19 A rigid body is rotating around the axis OO′ with an angular velocity ω. Particle i is moving in a circle of radius r_i with a speed $r_i\omega$.

$$T = \sum_i \frac{1}{2} m_i v_i^2 = \sum_i \frac{1}{2} m_i (r_i \omega)^2.$$

The angular velocity ω is the same for all particles, and

$$I = \sum_i m_i r_i^2;$$

thus

$$T = \tfrac{1}{2}\omega^2 \sum m_i r_i^2 = \tfrac{1}{2} I\omega^2. \tag{9-13}$$

This equation represents the **rotational kinetic energy** of the body. It is analogous to the expression for the kinetic energy of a particle, with m replaced by I and v by ω. It should be noted that Eq. (9-13) is not restricted to an axis through the center of mass but holds for any axis that is fixed in an inertial frame.

DRILL PROBLEM 9-12

A flywheel with a moment of inertia of 100 kg·m^2 is rotating at an angular speed of 600 rev/min about a fixed axis. What is its rotational kinetic energy?
ANS. 1.97×10^5 J.

With the expression for rotational kinetic energy established, we now consider the work done on a rigid body, such as that shown in Fig. 9-20. An external force **F** is applied to particle (or point) P whose position is **r**, and the rigid body is constrained to rotate around the fixed axis that is perpendicular to the page and passes through O. Since the rotational axis is fixed, the tip of the vector **r** moves in a circle of radius r, and the vector $d\mathbf{r}$ is along $\hat{\boldsymbol{\theta}}$, which is perpendicular to **r**. Thus with θ in radians,

$$dr = r\, d\theta.$$

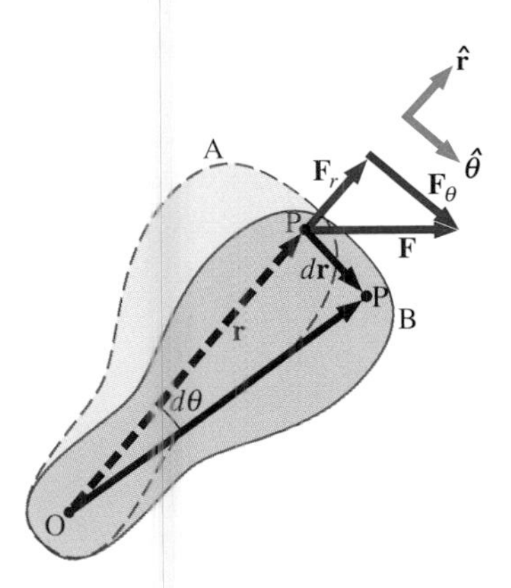

FIGURE 9-20 A rigid body rotates through an angle $d\theta$ from position A to position B while under the influence of a force **F**.

When P undergoes an infinitesimal displacement $d\mathbf{r}$ as illustrated in the figure,the work dW done by **F** is

$$dW = \mathbf{F} \cdot d\mathbf{r},$$

which can also be written as

$$dW = F_\theta\, dr = F_\theta r\, d\theta,$$

where F_θ is the component of **F** along $\hat{\boldsymbol{\theta}}$. But the torque τ of **F** around the rotational axis is

$$\tau = rF_\theta;$$

thus

$$dW = \tau\, d\theta.$$

Since all particles rotate through the same angle, the work of every external force is equal to its torque multiplied by the common angle $d\theta$. Therefore the total work done by all of the external forces on a rigid body as it rotates through $d\theta$ is

$$dW = \left(\sum \tau\right) d\theta, \tag{9-14}$$

where $\sum \tau$ represents the net torque on the body due to the external forces.

The work done by the *internal* forces is considered in Supplement 9-2. There it is shown that if the body is perfectly rigid, the total work of these forces vanishes in canceling action-reaction pairs. Consequently, *the total work done on a rigid body rotating about a fixed axis is due to only the external forces*, and it is given by Eq. (9-14).

We have calculated the total work done on a rigid body rotating around a fixed axis by summing the work done on every particle of that body—this gave us Eq. (9-14). Similarly, the kinetic energy of that body was found by summing the kinetic energy of each particle. Since the work-energy theorem is valid for every particle, it must also be valid for the sum over all particles. Therefore, for a rigid body rotating around a fixed axis from position A to position B, the work-energy theorem is

$$W_{AB} = T_B - T_A, \tag{9-15}$$

where

$$T = \tfrac{1}{2} I\omega^2,$$

and

$$W_{AB} = \int_{\theta_A}^{\theta_B} \left(\sum \tau\right) d\theta.$$

The approach we use in applying the work-energy theorem to a rigid body rotating around a fixed axis consists of the following steps:

1. Identify the forces acting on the body and draw a free-body diagram. Calculate the torque corresponding to each force.
2. Calculate the work done during the body's rotation by every torque.
3. Apply the work-energy theorem by equating the net work done on the body to the change in its rotational kinetic energy.

EXAMPLE 9-10 ROTATIONAL WORK AND ENERGY

A 12-N·m torque is applied to a flywheel that rotates about a fixed axis and has a moment of inertia of 30 kg·m^2. If the flywheel is initially at rest, what is its angular velocity after it has turned through eight revolutions?

SOLUTION The flywheel turns through 8.0 rev, which is 16π rad. Since the applied torque τ is constant, the work that it does during the flywheel's rotation is

$$W_{AB} = \tau(\theta_B - \theta_A).$$

Application of the work-energy theorem now gives

$$W_{AB} = \tau(\theta_B - \theta_A) = \tfrac{1}{2}I\omega_B^2 - \tfrac{1}{2}I\omega_A^2.$$

Using $\tau = 12$ N·m, $\theta_B - \theta_A = 16\pi$ rad, $I = 30$ kg·m^2, and $\omega_A = 0$ in this equation, we obtain

$$(12\ \text{N}\cdot\text{m})(16\pi) = \tfrac{1}{2}(30\ \text{kg}\cdot\text{m}^2)\omega_B^2 - 0,$$

so

$$\omega_B = 6.3\ \text{rad/s},$$

which is the angular velocity of the flywheel after eight revolutions.

EXAMPLE 9-11 ROTATIONAL WORK AND ENERGY

A string wrapped around the pulley of Fig. 9-21 is pulled with a constant downward force **F** of magnitude 50 N. The radius R and moment of inertia I of the pulley are 0.10 m and 2.5 × 10^{-3} kg·m^2, respectively. If the string does not slip, what is the angular velocity of the pulley after 1.0 m of string has unwound? Assume the pulley starts at rest.

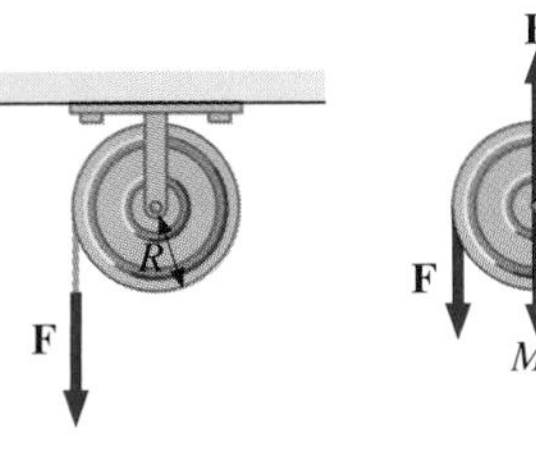

B

F

Mg

(a) (b)

FIGURE 9-21 (*a*) A string and pulley. (*b*) The free-body diagram of the pulley.

SOLUTION The forces on the pulley are shown in the free-body diagram. Neither **B**, the force of the bearings on the pulley, nor $M\mathbf{g}$, the weight of the pulley, exert a torque around the rotational axis; consequently, they also do no work on the pulley. As the pulley rotates through an angle θ, **F** acts through a distance d such that $d = R\theta$. Since the torque due to **F** has magnitude $\tau = FR$, the work of τ is

$$W = \tau\theta = (FR)\theta = Fd, \qquad (i)$$

which is just what we would have obtained had we used the force on the string to calculate W. If the force acts through a distance of 1.0 m, we have from Eq. (*i*) and the work-energy theorem

$$W_{AB} = T_B - T_A$$
$$Fd = \tfrac{1}{2}I\omega^2 - 0$$
$$(50\ \text{N})(1.0\ \text{m}) = \tfrac{1}{2}(2.5 \times 10^{-3}\ \text{kg}\cdot\text{m}^2)\omega^2.$$

Thus the angular velocity of the pulley at the instant that 1.0 m of the string has unwound is

$$\omega = 2.0 \times 10^2\ \text{rad/s}.$$

DRILL PROBLEM 9-13

Assume that the flywheel of Drill Prob. 9-12 is brought to rest by a constant torque in 20 revolutions. Use the work-energy theorem to calculate the torque.
ANS. 1.57 × 10^3 N·m.

DRILL PROBLEM 9-14

Use the work-energy theorem to calculate the speed of the can of Example 9-7 after it has fallen a distance h. Compare this result with that obtained by using the acceleration of the body found in Example 9-7 with Eq. (3-12).
ANS. $\sqrt{4mgh/(M + 2m)}$.

9-6 FIXED-AXIS ROTATION AND THE CONSERVATION OF MECHANICAL ENERGY

The principle of mechanical-energy conservation can be applied to systems containing rotating rigid bodies when only conservative forces do work on them. As seen in Chap. 7, the sum of the kinetic and potential energies remains constant. Now, however, the system may consist of some bodies that are moving linearly and some that are rotating. In this case, the total kinetic energy should contain terms like $\frac{1}{2}mv^2$ and $\frac{1}{2}I\omega^2$ to account for both types of motion. The following examples illustrate the procedure used in applying the principle of mechanical-energy conservation to fixed-axis rotation.

EXAMPLE 9-12 CONSERVATION OF MECHANICAL ENERGY FOR A PULLEY-BLOCK SYSTEM

Apply the principle of mechanical-energy conservation to the system of Fig. 9-22 to determine the speed of the hanging block of mass m after it has fallen a distance h. Assume that the block starts at rest and that the string unwinds on the pulley without slipping. The mass, radius, and moment of inertia of the pulley are M, R, and $MR^2/2$, respectively, and the mass of the string is negligible.

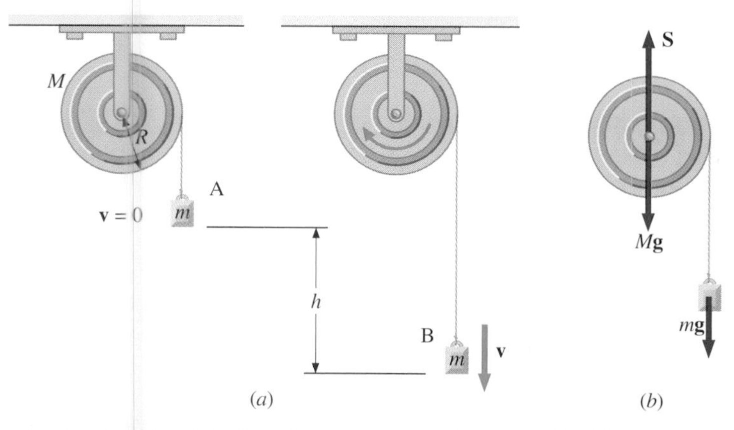

FIGURE 9-22 (*a*) A block and pulley at two instants. (*b*) The external forces on the system.

SOLUTION The system is composed of the block, the string, and the pulley. The external forces on this system are shown in the figure. Let's first consider the force **S** of the bearings supporting the pulley and the gravitational force $M\mathbf{g}$ on the pulley. Neither of these forces does any translational work because the pulley is not displaced linearly. Furthermore, these forces have zero moment arms about the axis of rotation of the pulley and their torques are zero, so no rotational work is done. The only force which does work is $m\mathbf{g}$. If the zero of potential energy is the initial position of the block, then its potential energy after falling a distance h is $-mgh$. The kinetic energy of the system is the translational kinetic energy of the falling block plus the rotational kinetic energy of the spinning pulley. Equating the mechanical energies at the point of release and at a point a distance h lower gives

$$T_A + U_A = T_B + U_B$$

$$\tfrac{1}{2}m(0)^2 + \tfrac{1}{2}I(0)^2 + mg(0) = \tfrac{1}{2}mv^2 + \tfrac{1}{2}I\omega^2 + (-mgh). \quad (i)$$

After substituting $v = R\omega$ and $I = MR^2/2$ into Eq. (*i*), we find

$$0 = \tfrac{1}{2}mv^2 + \tfrac{1}{4}Mv^2 - mgh,$$

so the speed of the block after falling a distance h is

$$v = \sqrt{\frac{4mgh}{M + 2m}}. \quad (ii)$$

EXAMPLE 9-13 Conservation of Mechanical Energy for Connected Masses

The cart and the block of Fig. 9-23 are connected by a massless string that runs over a pulley with mass M and moment of inertia I. The cart moves without friction across the horizontal surface as the block falls. What is the common speed v of the cart and block after the block has fallen a distance h? Assume that the string does not slip on the pulley and that the system is initially at rest.

SOLUTION In this case, the system we are investigating is composed of the cart, string, pulley, and block. The external forces acting on this system are shown in Fig. 9-23*b*. As in the previous example, only the conservative force $m_2\mathbf{g}$ does work. With the initial level of block 2 as the zero of potential energy, this block has a potential energy $-m_2gh$ after falling a distance h. The kinetic energy of the system is composed of the translational kinetic energies of the cart and block and the rotational kinetic energy of the pulley. Upon equating the mechanical energies at the two points, we obtain

$$T_A + U_A = T_B + U_B$$

$$\tfrac{1}{2}m_1(0)^2 + \tfrac{1}{2}m_2(0)^2 + \tfrac{1}{2}I(0)^2 + m_2g(0) = \tfrac{1}{2}m_1v^2 + \tfrac{1}{2}m_2v^2 + \tfrac{1}{2}I\omega^2 + (-m_2gh). \quad (i)$$

Since the string unwinds without slipping, $v = R\omega$. Substituting this into Eq. (*i*) and solving for the speed v, we find

$$v = \sqrt{\frac{2m_2gh}{m_1 + m_2 + I/R^2}}.$$

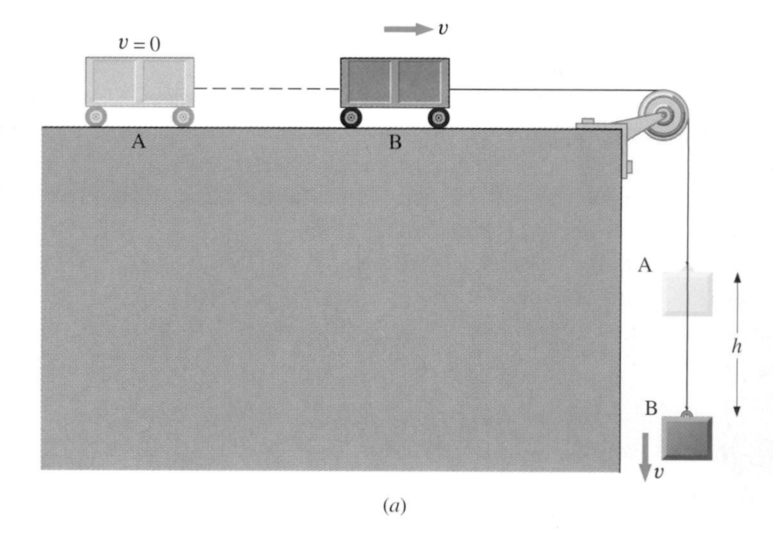

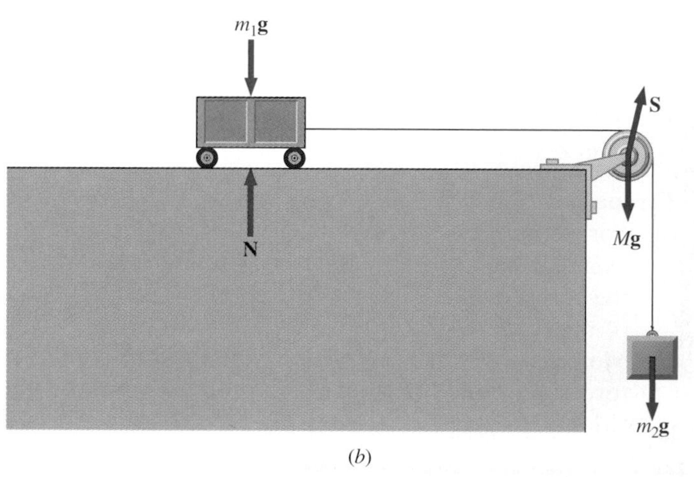

FIGURE 9-23 (*a*) A cart moves across a frictionless surface as the block falls. (*b*) The external forces on the system. **S** is the force of the support on the pulley.

DRILL PROBLEM 9-15

Use the law of conservation of mechanical energy to calculate the common speed for the two blocks of the Atwood's machine of Example 9-8 after they have each moved a distance h.
ANS. $\sqrt{4(m_1 - m_2)gh/(2m_1 + 2m_2 + M)}$.

SUMMARY

1. **Rotational kinematics**
   (*a*) Angular position is represented by the angle θ, where θ is measured with respect to a reference line. This angle is usually expressed in radians.
   (*b*) Angular velocity $\omega = d\theta/dt$.
   (*c*) Angular acceleration $\alpha = d\omega/dt$.
   (*d*) For rotational motion at constant angular acceleration α,

$$\omega = \omega_0 + \alpha t,$$
$$\theta = \theta_0 + \omega_0 t + \tfrac{1}{2}\alpha t^2,$$
$$\omega^2 = \omega_0^2 + 2\alpha(\theta - \theta_0).$$

   (*e*) If angular measurements are in radians, the linear and angular kinematic variables are related by

$$s = r\theta,$$
$$v = r\omega,$$
$$a_\theta = r\alpha,$$
$$a_r = r\omega^2.$$

2. **Torque**
   (*a*) For planar problems, the torque of a force **F** around an axis through O is $r_\perp F$, where $r_\perp$, the moment arm, is the perpendicular distance from O to the line of action of the force.
   (*b*) More generally, if a force **F** is applied at the point P, its torque $\boldsymbol{\tau}$ around the point O is $\mathbf{r} \times \mathbf{F}$, where **r** is the position vector of P relative to O.

3. **Rotational dynamics**
   For a body rotating around a fixed axis, the net external torque $\sum \tau$ and the angular acceleration α around that axis are related by

$$\sum \tau = I\alpha,$$

   where I is the body's moment of inertia around the axis.

4. **Work-energy theorem for fixed-axis rotation**
   For a rigid body rotating around a fixed axis,

$$W_{AB} = T_B - T_A,$$

   where

$$T = \tfrac{1}{2}I\omega^2,$$

   and

$$W_{AB} = \int_{\theta_A}^{\theta_B} \left(\sum \tau\right) d\theta.$$

5. **Principle of conservation of mechanical energy**
   When only conservative forces are doing work, the mechanical energy of a system containing rigid bodies rotating around a fixed axis is conserved. The net kinetic energy of the system must include the contribution $I\omega^2/2$ due to any rotating body.

SUPPLEMENT 9-1 ROTATIONAL FORM OF NEWTON'S SECOND LAW

The cross section of a rigid body constrained to rotate around the z axis of an inertial reference frame is shown in Fig. 9-24. Particle i of the body is acted on by a net external force $\mathbf{F}_i$ and a net internal force $\mathbf{f}_i$. To calculate the torques of these forces around the z axis, we only need to consider their components $F_{i\theta}$ and $f_{i\theta}$, which are perpendicular to $\mathbf{r}_i$. Since both have the same moment arm r_i, their net torque on particle i is

$$\tau_i = r_i(F_{i\theta} + f_{i\theta}).$$

By Newton's second law, $F_{i\theta} + f_{i\theta} = m_i a_{i\theta}$, and from Eq. (9-8), $a_{i\theta} = r_i\alpha$. (*Remember:* All points on the body have the same angular acceleration.) Substituting these into the equation for τ_i, we obtain

$$\tau_i = r_i(F_{i\theta} + f_{i\theta}) = m_i r_i^2 \alpha.$$

There are identical equations for every particle of the rigid body. Their addition yields

$$\sum_i r_i F_{i\theta} + \sum_i r_i f_{i\theta} = \sum_i m_i r_i^2 \alpha, \qquad (i)$$

where $\sum_i$ designates a sum over all particles.

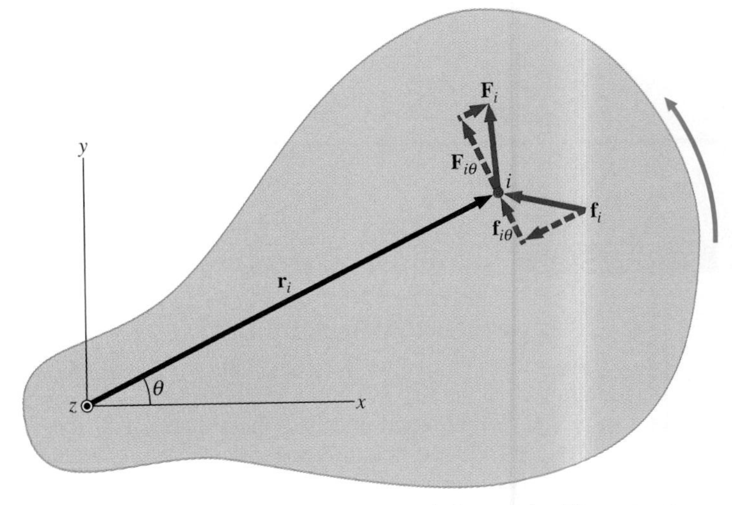

FIGURE 9-24 A rigid body rotates around the z axis. The net external force on particle i is $\mathbf{F}_i$, and the net internal force on particle i is $\mathbf{f}_i$. These forces are shown resolved into their components.

The first term, $\sum_i r_i F_{i\theta}$, is not nearly as complicated as it appears to be for two reasons: (1) With the exception of the gravitational force, the external forces act on only a small number of particles; and (2) as was shown in Sec. 9-2, the net effect of the gravitational force on all particles may be represented by a single force $M\mathbf{g}$ acting at the center of mass. Consequently, $\sum_i r_i F_{i\theta}$ can be replaced by a sum over just those particles that experience external forces and the center of mass. Hence

$$\sum_i r_i F_{i\theta} = \sum \tau,$$

where $\sum \tau$ is the net torque due to the external forces (including gravity) on the rigid body.

The second term of Eq. (*i*), $\sum_i r_i f_{i\theta}$, may be evaluated using Newton's third law. If we assume that the internal forces between each particle pair act along a line connecting the two particles, the sum over the internal torques cancels in pairs, as shown in Fig. 9-25. The second term therefore vanishes.

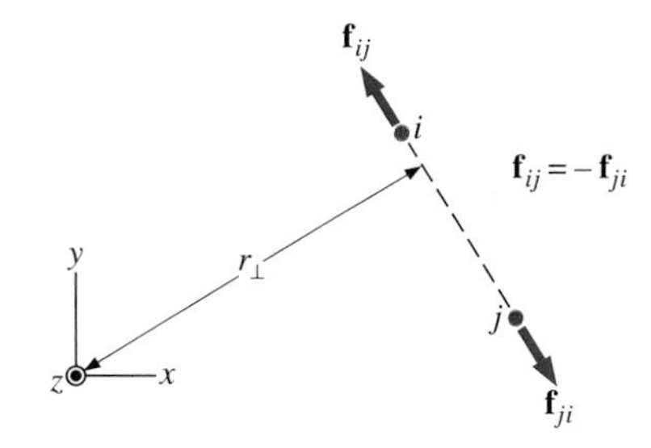

FIGURE 9-25 The internal force on i due to j is $\mathbf{f}_{ij}$, and it is equal and opposite to $\mathbf{f}_{ji}$, the internal force on j due to i. Thus the torques of these two forces around the origin cancel.

The third term of Eq. (*i*) is simply the product of the moment of inertia $I = \sum_i m_i r_i^2$ and the angular acceleration α. Therefore we are left with

$$\sum \tau = I\alpha,$$

which is the rotational form of Newton's second law.

SUPPLEMENT 9-2 WORK DONE ON A RIGID BODY BY INTERNAL FORCES

In this supplement we show that the net work done by the internal forces on a rigid body rotating around a fixed axis is zero. We begin by considering particles i and j of the rigid body of Fig. 9-26*a*. Suppose that j exerts a force $\mathbf{f}_{ij}$ on i and i exerts a force $\mathbf{f}_{ji}$ on j. This action-reaction pair is assumed to be directed along the line between the particles; that is, along $\mathbf{r}_i - \mathbf{r}_j$. The work done by this force pair in an infinitesimal displacement is

$$dW = \mathbf{f}_{ij}\cdot d\mathbf{r}_i + \mathbf{f}_{ji}\cdot d\mathbf{r}_j,$$

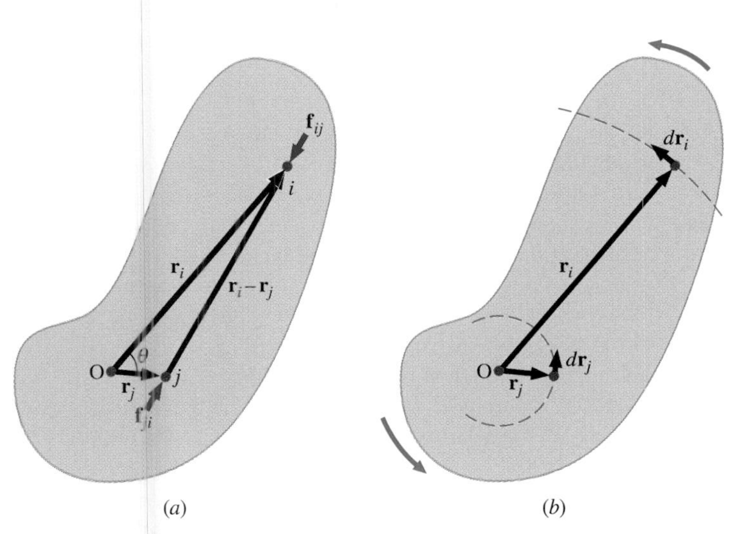

FIGURE 9-26 (*a*) A rigid body rotates around O. Forces $\mathbf{f}_{ij}$ and $\mathbf{f}_{ji}$ are the action-reaction pair between particles i and j. For a rigid body, $|\mathbf{r}_i|$, $|\mathbf{r}_j|$, and θ remain constant as the body rotates. (*b*) Since i and j move along circular paths, $d\mathbf{r}_i$ is perpendicular to $\mathbf{r}_i$ and $d\mathbf{r}_j$ is perpendicular to $\mathbf{r}_j$.

which, because $\mathbf{f}_{ij} = -\mathbf{f}_{ji}$, can be written as

$$dW = \mathbf{f}_{ij}\cdot d(\mathbf{r}_i - \mathbf{r}_j).$$

Now the vectors $\mathbf{f}_{ij}$ and $\mathbf{r}_i - \mathbf{r}_j$ lie along the same line. This means that if $d(\mathbf{r}_i - \mathbf{r}_j)$ is perpendicular to $\mathbf{r}_i - \mathbf{r}_j$, it must also be perpendicular to $\mathbf{f}_{ij}$, and

$$dW = \mathbf{f}_{ij}\cdot d(\mathbf{r}_i - \mathbf{r}_j) = 0;$$

that is, the work of the internal forces cancels in pairs.

Thus in order to prove that $dW = 0$, we need to demonstrate that $d(\mathbf{r}_i - \mathbf{r}_j)$ and $\mathbf{r}_i - \mathbf{r}_j$ are perpendicular. To accomplish this, let's consider the scalar product (denoted as P) of these two vectors:

$$P = (\mathbf{r}_i - \mathbf{r}_j)\cdot d(\mathbf{r}_i - \mathbf{r}_j),$$

which we can write as

$$P = \mathbf{r}_i\cdot d\mathbf{r}_i - \mathbf{r}_j\cdot d\mathbf{r}_i - \mathbf{r}_i\cdot d\mathbf{r}_j + \mathbf{r}_j\cdot d\mathbf{r}_j.$$

Since i and j rotate in circular paths around O (see Fig. 9-26*b*), $d\mathbf{r}_i$ is perpendicular to $\mathbf{r}_i$ and $d\mathbf{r}_j$ is perpendicular to $\mathbf{r}_j$, so $\mathbf{r}_i\cdot d\mathbf{r}_i = 0 = \mathbf{r}_j\cdot d\mathbf{r}_j$. We now have

$$P = -(\mathbf{r}_j\cdot d\mathbf{r}_i + \mathbf{r}_i\cdot d\mathbf{r}_j),$$

which, using the chain rule for differentiation, we can write as

$$P = -d(\mathbf{r}_i\cdot\mathbf{r}_j).$$

Since the body is rigid, the distances $|\mathbf{r}_i|$ from O to i and $|\mathbf{r}_j|$ from O to j along with the angle θ between $\mathbf{r}_i$ and $\mathbf{r}_j$ must remain constant as the body rotates. Thus, using the definition of the scalar product, we have

$$P = -d(|\mathbf{r}_i||\mathbf{r}_j|\cos\theta) = 0.$$

This result leads to the fact that $dW = 0$; that is, *there is no net work done on a rotating rigid body by the internal forces.*

QUESTIONS

9-1. As a baseball bat is being swung, do all points on it have the same angular velocity? the same linear velocity?

9-2. Does the tangential acceleration depend on the angular velocity or the change in angular velocity? Answer the same question for the radial acceleration.

9-3. A flywheel is spinning at constant angular velocity about an axis through its center. Which point(s) of the flywheel have the greatest and smallest (*a*) tangential velocity, (*b*) angular velocity, (*c*) tangential acceleration, (*d*) radial acceleration?

9-4. Can a point on the rim of a wheel rotating around a fixed axis through its center ever have a tangential acceleration but no radial acceleration?

9-5. Can a point on the rim of a wheel rotating around a fixed axis through its center ever have a radial velocity?

9-6. Can a single force have a zero torque?

9-7. Can a set of forces have a net torque of zero and a net force that is not zero?

9-8. Can a set of forces have a net force of zero and a net torque that is not zero?

9-9. In the expression $\mathbf{r} \times \mathbf{F}$, can $|\mathbf{r}|$ ever be less than the moment arm? Can it be equal to the moment arm?

9-10. A rod is pivoted about one end. Two forces $\mathbf{F}$ and $-\mathbf{F}$ are applied to it. Under what circumstances for these forces will the rod not rotate?

9-11. Explain how it is possible for a large force to produce a small torque. Explain how a small force can produce a large torque.

9-12. Describe some tools that are designed to produce reasonably large torques with small applied forces.

9-13. Torque and work have the same dimensions. Discuss how these quantities differ physically.

9-14. Assume that the tension applied to a magnetic tape in a cassette player is constant. Describe how this tension applies a varying torque to the reel as the tape unwinds.

9-15. Why is a doorstop most effective when placed farthest from the door's hinges?

9-16. Devise a method for measuring the moment of inertia of a body around a particular axis.

9-17. The stick shown in the accompanying figure is free to rotate around the horizontal axis through O. A penny is placed on the free end of the stick. The stick is held in the horizontal position shown and then released. Explain why the penny initially drops more slowly than the free end of the stick.

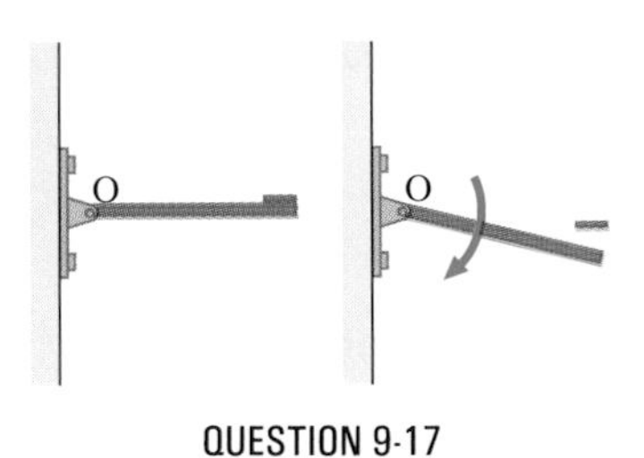

QUESTION 9-17

9-18. An electric grinder coasts for a while after its power is turned off while an electric drill stops rotating very quickly. Explain.

9-19. A solid uniform sphere and a spherical shell have the same mass and radius. Compare the amounts of work required to set the two objects into rotational motion with the same angular velocity about an axis through their centers.

9-20. What is the purpose of the long, flexible pole carried by high-wire walkers?

PROBLEMS

Rotational Kinematics

9-1. Calculate the angular velocities associated with (*a*) the rotation of the earth around its axis, and (*b*) the orbital motion of the earth around the sun.

9-2. A track star runs a 400-m race on a circular 400-m track in 45 s. (*a*) Assuming he runs at constant speed, what is his angular velocity? (*b*) What is his radial acceleration?

9-3. A wheel rotates at a constant rate of 2.00×10^3 rev/min. (*a*) What is its angular velocity in radians per second? (*b*) Through what angle does it turn in 10.0 s? Express your answer in both radians and degrees.

9-4. A particle moves 3.0 m along a circle of radius 1.5 m. (*a*) Through what angle does it rotate? Express your answer in both radians and degrees. (*b*) If the particle makes this trip in 1.0 s at constant speed, what is its angular velocity? (*c*) What is its linear acceleration?

9-5. A phonograph turntable rotating at $33\frac{1}{3}$ rev/min slows down and stops in 1.0 min. (*a*) What is the turntable's angular acceleration? (Assume that it is constant.) (*b*) How many revolutions does the turntable make while stopping?

9-6. A wheel has a constant angular acceleration of 5.0 rad/s^2. Starting from rest, it turns through 300 rad. (*a*) What then is its angular velocity? (*b*) How much time elapses while it turns through the 300 rad?

9-7. During a 6.0-s time interval, a flywheel with constant angular acceleration turns through 500 rad while acquiring an angular velocity of 100 rad/s. (*a*) What is the angular acceleration of the flywheel? (*b*) What is its angular velocity at the beginning of the 6.0-s interval?

9-8. The angular velocity of a rotating rigid body increases from 500 to 1500 rev/min in 120 s. (*a*) What is the angular acceleration of the body? (*b*) Through what angle does it turn in this 120-s interval?

9-9. A flywheel slows uniformly from 600 to 400 rev/min while rotating through 40 complete revolutions. (*a*) What is the angular acceleration of the flywheel? (*b*) How much time elapses during the 40 revolutions?

9-10. A wheel 1.0 m in diameter rotates with an angular acceleration of 4.0 rad/s^2. (*a*) If the wheel's initial angular velocity is 2.0 rad/s, what is its angular velocity after 10 s? (*b*) Through what angle does it rotate in this 10-s interval? (*c*) What are the linear velocity and acceleration of a point on the rim of the wheel at the end of this 10-s interval?

9-11. The blade of a fan rotating at 600 rev/min is 20 cm long. (*a*) What is the linear velocity of a point on the outer edge of the blade? (*b*) What is the linear acceleration of this point?

9-12. A vertical wheel with a diameter of 50 cm starts at rest and rotates with a constant angular acceleration of 5.0 rad/s^2 around a fixed axis through its center. (*a*) Where is the point that is initially at the bottom of the wheel at $t = 10$ s? (*b*) What is this point's linear acceleration at this instant?

9-13. A circular disk of radius 10 cm has a constant angular acceleration of 1.0 rad/s^2, and at $t = 0$, its angular velocity is 2.0 rad/s. Determine (*a*) the disk's angular velocity at $t = 5.0$ s, (*b*) the angle through which the disk has rotated at $t = 5.0$ s, (*c*) the linear acceleration of a point on the rim of the disk at $t = 5.0$ s. (*d*) Repeat part (*c*) for the time $t = 0.50$ s.

9-14. The angular acceleration of a rotating rigid body is given by $\alpha = at^2 - bt^3$. What are the angular velocity and angular position of the body as functions of time?

Torque

9-15. The cylindrical head bolts on a car are to be tightened by a torque of 62 N·m. If a mechanic uses a wrench of length 20 cm, what perpendicular force must he exert on the end of the wrench to tighten a bolt correctly?

9-16. Calculate the torque of the 40-N force around the axis through O and perpendicular to the plane of the page. (Refer to the accompanying figure.)

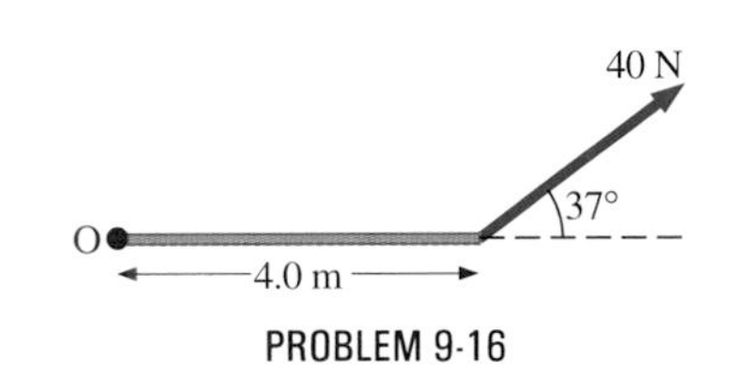

PROBLEM 9-16

9-17. In the previous problem, what vertical force must be applied 2.0 m from O so that the net torque around O is zero?

9-18. A torque around O of 5.00×10^3 N·m is required to raise the drawbridge shown in the accompanying figure. What is the tension in the rope necessary to produce this torque? Would it be easier to raise the drawbridge if θ were larger or smaller?

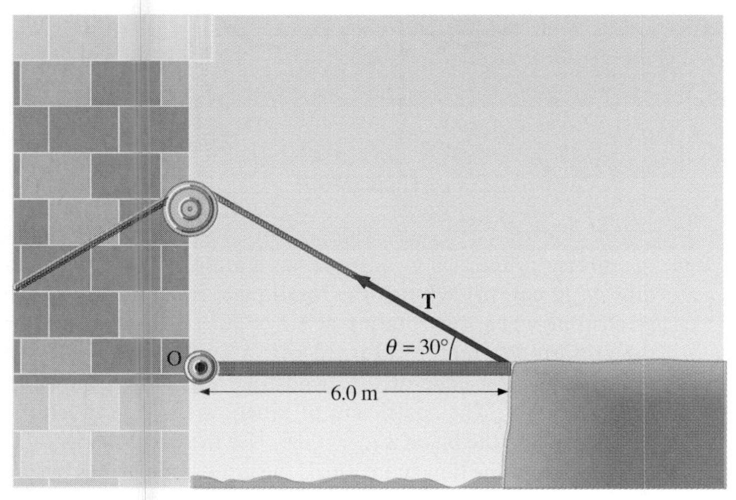

PROBLEM 9-18

9-19. The forces shown in the accompanying figure are applied in the xy plane. Calculate the net torque of these forces around the z axis (*a*) using $\tau = r_{\perp} F$ and (*b*) using $\boldsymbol{\tau} = \mathbf{r} \times \mathbf{F}$.

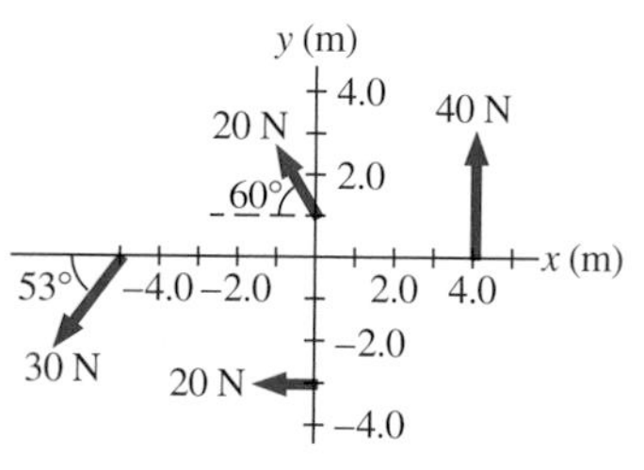

PROBLEM 9-19

9-20. The force $\mathbf{F} = 20\mathbf{j}$ N is applied at $\mathbf{r} = (4.0\mathbf{i} - 2.0\mathbf{j})$ m. What is the torque of this force around the origin?

9-21. What is the torque around the origin of the force $\mathbf{F} = (5.0\mathbf{i} - 2.0\mathbf{j} + 1.0\mathbf{k})$ N if it is applied at the point whose position is $\mathbf{r} = (-2.0\mathbf{i} + 4.0\mathbf{j})$ m?

Rotational Dynamics

9-22. A flywheel ($I = 50$ kg·m^2) starting at rest acquires an angular velocity of 200 rad/s while subjected to a constant torque from a motor for 5.0 s. (*a*) What is the angular acceleration of the flywheel? (*b*) What is the magnitude of the torque?

9-23. A constant torque is applied to a rigid body whose rotational inertia is 4.0 kg·m^2 around the axis of rotation. If the wheel starts at rest and attains an angular velocity of 20 rad/s in 10 s, what is the applied torque?

9-24. A net torque of 50 N·m is applied to a grinding wheel ($I = 20$ kg·m^2) for 20 s. (*a*) If it starts at rest, what is the angular velocity of the wheel after the torque is removed? (*b*) Through what angle does the wheel rotate while the torque is being applied?

9-25. A flywheel ($I = 100$ kg·m^2) rotating at 500 rev/min is brought to rest by friction in 2.0 min. What is the frictional torque on the flywheel?

9-26. A 50-kg grindstone, which is a solid cylinder of diameter 80 cm, is rotating with an angular velocity of 300 rev/min. The motor is turned off, and friction brings the grindstone to rest in 1 min. What is the frictional torque on the grindstone?

9-27. A uniform cylindrical grinding wheel of mass 50 kg and diameter 1.0 m is turned by an electric motor. The friction in the bearings of the grinding wheel is negligible. (*a*) What torque must be supplied by the motor to bring the wheel from rest to 120 rev/min in 20 revolutions? (*b*) A tool whose coefficient of kinetic friction with the wheel is $\mu_K = 0.60$ is pressed perpendicularly against the edge of the wheel with a force of 40 N. What torque must be supplied by the motor to keep the wheel rotating at a constant angular velocity?

9-28. Suppose that when the earth was created, it was not rotating. However, after the application of a uniform torque for 6 days, it rotated at 1 rev/day. (*a*) What was the angular acceleration of the earth during these 6 days? (*b*) What torque was applied to the earth during this period? (*c*) What force tangent to the earth's surface at the equator would produce this torque?

9-29. A pulley of moment of inertia 2.0 kg·m^2 is mounted on a wall as shown in the accompanying figure. Light strings are wrapped around the two circumferences of the pulley, and weights are tied to the free ends of the strings. What are the angular acceleration of the pulley and the linear accelerations of the weights? Assume that $r_1 = 50$ cm, $r_2 = 20$ cm, $m_1 = 1.0$ kg, and $m_2 = 2.0$ kg.

PROBLEM 9-29

9-30. The cart shown in the accompanying figure moves across the horizontal tabletop as the block falls. What is the acceleration of the cart? Friction is negligible, $m_1 = 2.0$ kg, $m_2 = 4.0$ kg, $I = 0.40$ kg·m^2, and $r = 20$ cm.

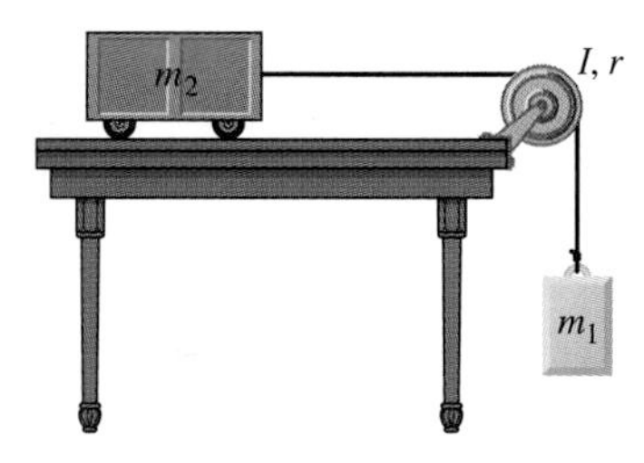

PROBLEM 9-30

9-31. A solid cylinder of mass 10 kg and radius 20 cm is rotating around an axis through its center with an angular velocity of 100 rad/s. The cylinder is pushed against a wall with a force of 10 N, as shown in the accompanying figure. If $\mu_K = 0.50$, how much time elapses before the cylinder stops spinning?

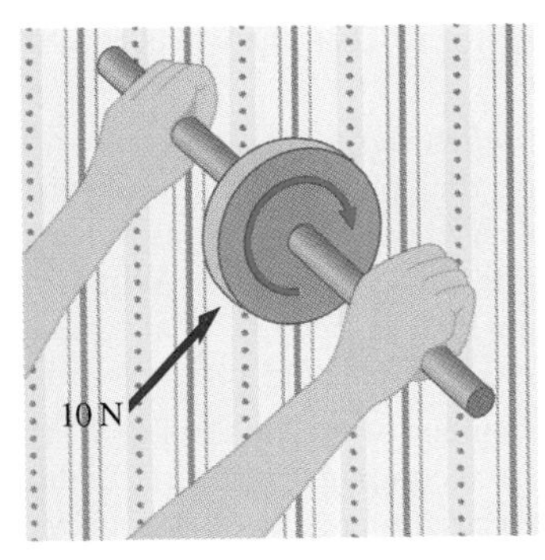

PROBLEM 9-31

Rotational Work and Mechanical Energy

9-32. A uniform stick of length L and mass M is pivoted at one end. It is pulled aside at an angle θ to the vertical and released. What is the angular velocity of the stick at the instant it swings through its vertical position?

9-33. (*a*) Consider the expression for power given by Eq. (7-23). Show that the equivalent expression for rotational motion is $P = \tau\omega$. (*b*) A boat engine operating at 8.95×10^4 W (120 hp) is running at 300 rev/min. What is the torque on the propeller shaft?

9-34. A uniform stick of length L and mass M is held vertically with one end resting on the floor as shown in the accompanying figure. When the stick is released, it rotates around its lower end until it hits the floor. Assuming the lower end of the stick does not slip, what is the linear velocity of its upper end when it hits the floor?

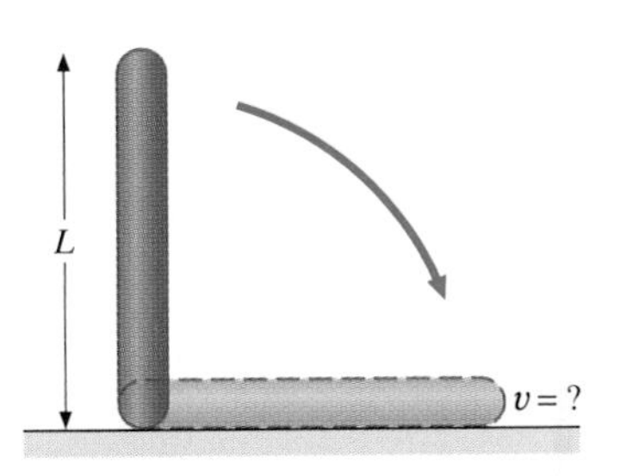

PROBLEM 9-34

9-35. A cord is wrapped around the rim of a solid cylinder of radius 0.25 m, and a constant force of 40 N is exerted on the cord. (See the accompanying figure.) The cylinder is mounted on frictionless bearings, and its moment of inertia is 6.0 kg·m^2. (*a*) Use the work-energy principle to calculate the angular velocity of the cylinder after 5.0 m of cord have been unwound. (*b*) If the 40-N force is replaced by a 40-N weight, what is the angular velocity of the cylinder after 5.0 m of cord have unwound?

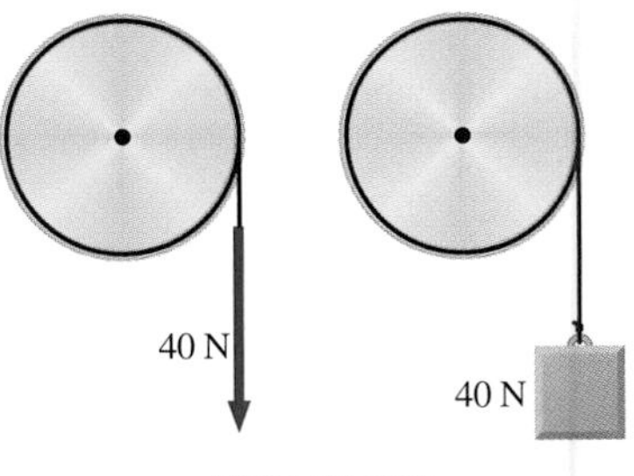

PROBLEM 9-35

9-36. A uniform cylindrical grindstone has a mass of 10 kg and a radius of 12 cm. (*a*) What is the rotational kinetic energy of the grindstone when it is rotating at 1.5×10^3 rev/min? (*b*) After the grindstone's motor is turned off, a knife blade is pressed against the outer edge of the grindstone with a perpendicular force of 5.0 N. The coefficient of kinetic friction between the grindstone and the blade is $\mu_K = 0.80$. Use the work-energy principle to determine how many turns the grindstone makes before it stops.

9-37. Small bodies of mass m_1 and m_2 are attached to opposite ends of a thin rigid rod of length L and mass M. (See the accompanying figure.) The rod is mounted so that it is free to rotate in a horizontal plane around a vertical axis. What distance d from m_1 should the rotational axis be so that a minimum amount of work is required to set the rod rotating at an angular velocity ω?

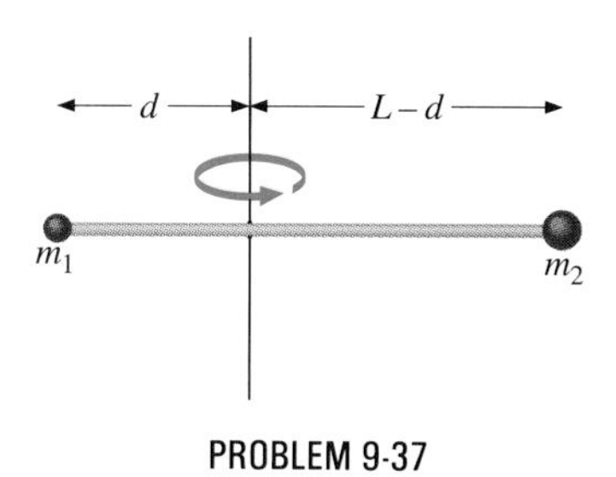

PROBLEM 9-37

9-38. A uniform disk of mass 500 kg and radius 0.25 m is mounted on frictionless bearings so that it can rotate freely around a vertical axis through its center. (See the accompanying figure.) A cord is wrapped around the rim of the disk and pulled with a constant force of 10 N. (*a*) How much work has the force done at the instant when the disk has completed three revolutions? (*b*) Determine the torque of the force, then calculate the work done by this torque at the instant when the disk has completed three revolutions. (*c*) What is the angular velocity of the disk at this instant? (*d*) What is the power output of the force at this instant?

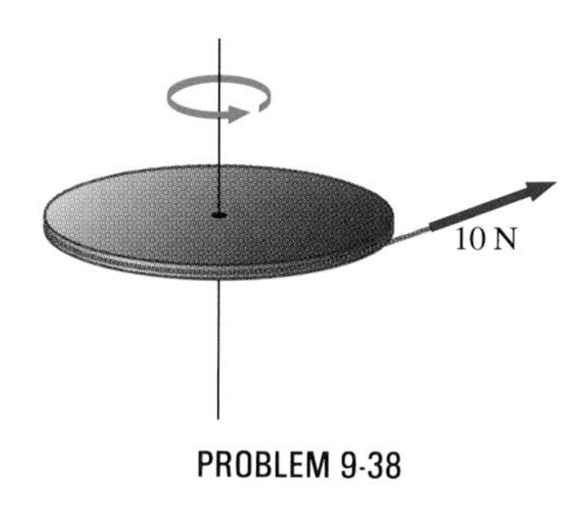

PROBLEM 9-38

General Problems

9-39. A man stands on a merry-go-round that is rotating at 2.5 rad/s. If the coefficient of static friction between the man's shoes and the merry-go-round is $\mu_S = 0.50$, how far from the axis of rotation can he stand without sliding?

9-40. One end of a string is wrapped around a cylindrical spool of diameter d, and the other end is passed over a cylindrical pulley and tied to a block of mass m_B, as shown in the accompanying figure. The spool is attached to a thin rod that passes through the center of the spool. At its upper end, the rod is attached to the middle of a thin horizontal rod of length l_R and mass m_R. Small spheres of mass m_S are attached to each end of the horizontal rod. The moment of inertia of the vertical rod plus spool is I, and the moment of inertia of the pulley is negligible. When the block is released, what is its acceleration, and what is the angular acceleration of the rotating system?

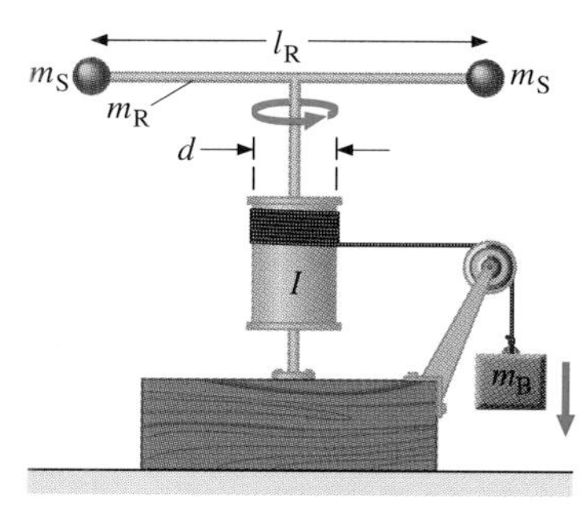

PROBLEM 9-40

9-41. A uniform rod of mass m and length l is held horizontally by two vertical strings of negligible mass, as shown in the accompanying figure. (*a*) Immediately after the right string is cut, what is the linear acceleration of the free end of the stick? (*b*) of the middle of the stick? (*c*) Apply Newton's second law to the center of mass of the stick to determine the tension in the left string immediately after the right string is cut.

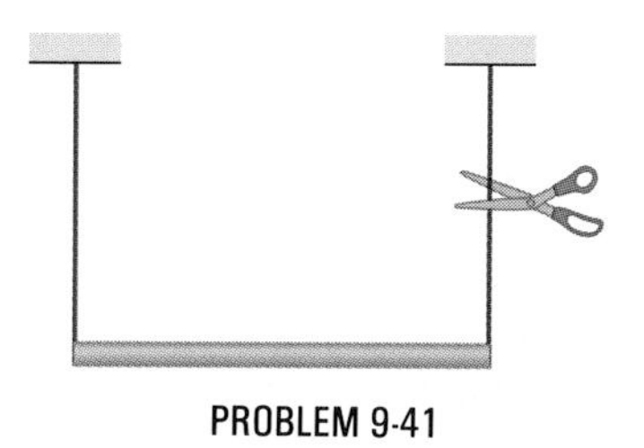

PROBLEM 9-41

9-42. In the accompanying figure, the mass of the block is 5.0 kg; the moment of inertia and radius of the pulley are 0.030 kg·m^2 and 0.10 m, respectively; the angle of the incline is 30°; and the coefficient of kinetic friction between the block and the incline is 0.40. What is the acceleration of the block down the incline?

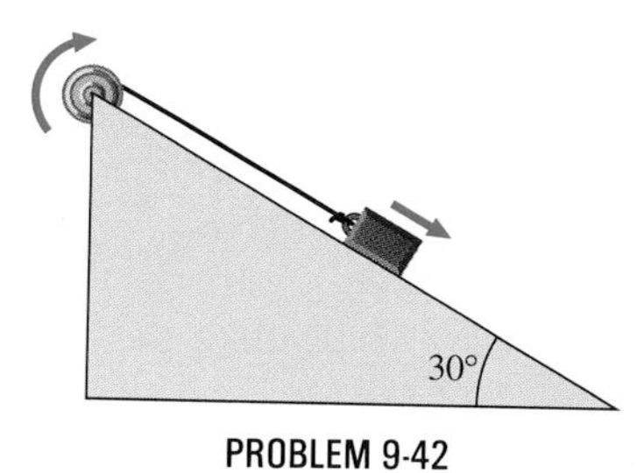

PROBLEM 9-42

9-43. Two flywheels are connected by a nonslipping belt as shown in the accompanying figure: $I_1 = 4.0$ kg·m^2, $r_1 = 0.20$ m, $I_2 =$ 20 kg·m^2, and $r_2 = 0.30$ m. If a torque of 10 N·m is applied to the smaller flywheel, what are the angular accelerations of the two flywheels?

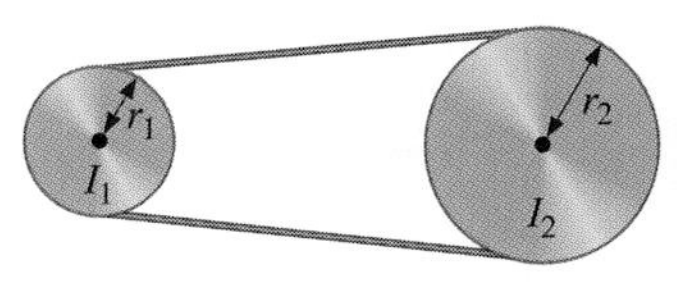

PROBLEM 9-43

9-44. The accompanying figure shows a stick of length 1.0 m and mass 6.0 kg; the stick is free to rotate around a horizontal axis through its center. Small bodies of masses 4.0 and 2.0 kg are attached to its two ends. The stick is released from the horizontal position.

(*a*) What is the angular velocity of the stick when it swings through the vertical? (*b*) What is the linear acceleration of each small body when the stick is vertical?

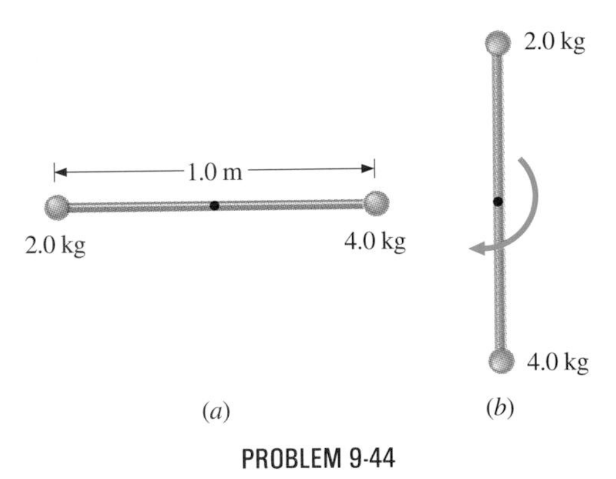

PROBLEM 9-44

9-45. A propeller is accelerated from rest to an angular velocity of 2.4×10^3 rev/min over a period of 6.0 s by a constant torque of 1.6×10^3 N·m. (*a*) What is the moment of inertia of the propeller? (*b*) What power is being provided to the propeller at 3.0 s after it starts rotating? (See Prob. 9-33.)

9-46. A thin stick of mass $m = 0.20$ kg and length $l = 0.50$ m is attached to the rim of a metal disk of mass $M = 4.0$ kg and radius $R = 0.25$ m. The stick is free to rotate around a horizontal axis through its other end, as shown in the accompanying figure. (*a*) If the combination is released with the stick horizontal, what is the speed of the center of the disk when the stick is vertical? (*b*) What is the acceleration of the center of the disk at the instant the stick is released? (*c*) at the instant the stick passes through the vertical?

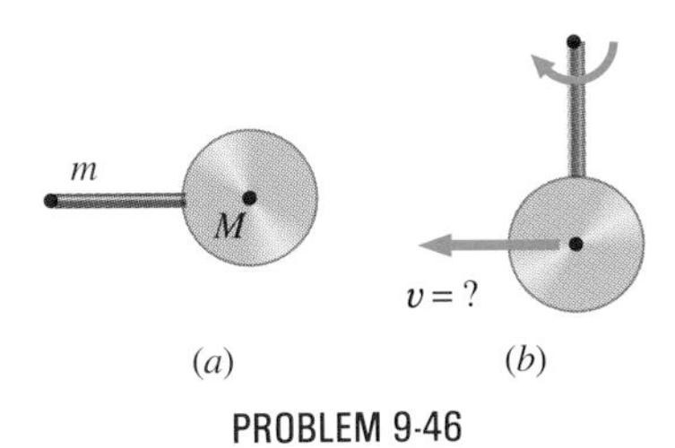

PROBLEM 9-46

9-47. The accompanying figure shows a uniform stick of mass m and length l that can rotate freely around a horizontal axis through O. The stick starts rotating from the vertical position P_1 after being given a slight push. (*a*) What is the linear velocity of the free end of the stick in the positions P_2 and P_3? (*b*) What is the acceleration of the free end of the stick in those two positions?

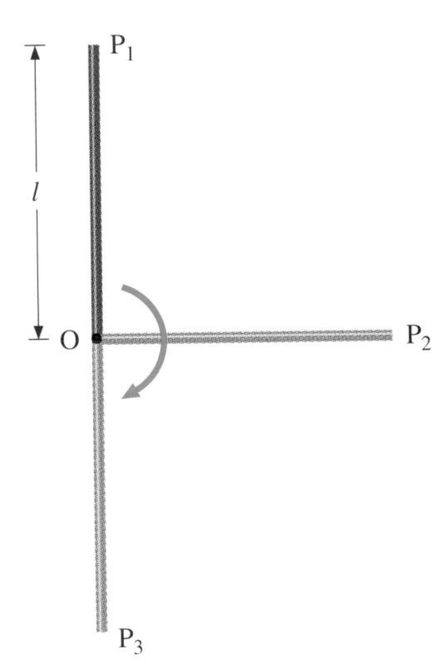

PROBLEM 9-47

9-48. The accompanying figure shows a solid sphere of radius 10 cm; this sphere is allowed to rotate freely around a horizontal axis through O. The sphere is given a sharp blow so that its center of mass starts from the position shown with a speed of 150 cm/s. What is the maximum angle θ that the diameter OO′ makes with the vertical?

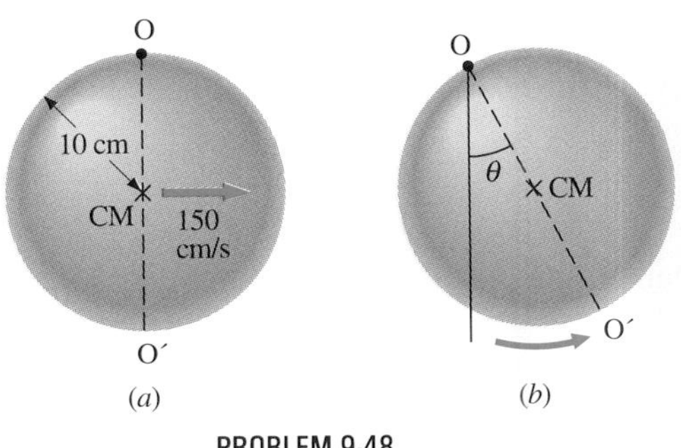

PROBLEM 9-48

9-49. Imagine an inertial reference frame whose origin is at the center of the earth. Since this is an inertial reference frame, it is not rotating with the earth. Now imagine a building of height h located at the equator, as indicated in the accompanying figure. Relative to this frame, the top of the building has an easterly velocity of $\mathbf{v}_t$, and the base of the building has an easterly velocity of $\mathbf{v}_b$. (*a*) Show that $v_t - v_b = \omega h$, where ω is the angular velocity of the earth around its polar axis and h is the height of the building. (*b*) Show that a stone dropped from the roof of the building will land east of the building's base by an amount given approximately by $d \approx \omega h\sqrt{2h/g}$. (*c*) Evaluate d for a 30-m building. (*Note:* This problem is a simple example of the effect on motion due to the earth's rotation, an effect described in general by the *Coriolis force*.)

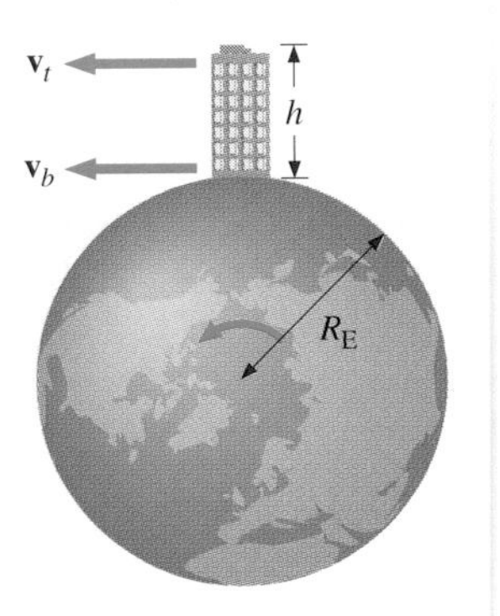

PROBLEM 9-49

Rolling and slipping.

Photo courtesy of Ringling Bros. and Barnum & Bailey Combined Shows, Inc.

CHAPTER 10 *ROTATIONAL MECHANICS II

PREVIEW

In this chapter we study the motion of bodies rolling in a plane, such as a barrel rolling down an incline or a billiard ball rolling straight across a table. The main topics to be discussed here are the following:

1. **Rolling motion**. The equations of motion for a rolling body are derived.
2. **Rolling and slipping.** Why do some objects roll without slipping while others slip? We consider the conditions under which slippage occurs during rolling motion.
3. **Conservation of mechanical energy for rolling motion**. The principle of mechanical-energy conservation is applied to the case of rolling rigid bodies.

In Chap. 9 only rigid bodies with rotational axes fixed in inertial frames were considered. We now remove that restriction and allow bodies to move translationally in a plane as they rotate around an axis perpendicular to that plane. There are many interesting applications of this type of rolling motion—a billiard ball rolling with topspin across a table and a bowling ball sliding straight down an alley are just two examples. Using the concepts introduced in this chapter, you will understand why that billiard ball continues forward after striking a second ball, and you will know how to determine whether or not the bowling ball stops sliding and starts rolling before it reaches the pins.

10-1 Rolling Motion

Since we are only interested in rigid bodies moving translationally in a plane, we can represent the bodies by their circular cross sections in that plane. As illustrated in Fig. 10-1, only three coordinates are needed to fix the position of such a rigid body. Two spatial coordinates are required to describe the planar position of a single point on the body, and one angular coordinate is needed to specify the rotational position of the entire body. These three independent coordinates are called the *three degrees of freedom* of the rigid body. The coordinates generally chosen are the horizontal and vertical positions of the center of mass and the rotational position of the body about the axis through its center of mass and perpendicular to the cross section. These are represented by (x, y) and θ in the figure.

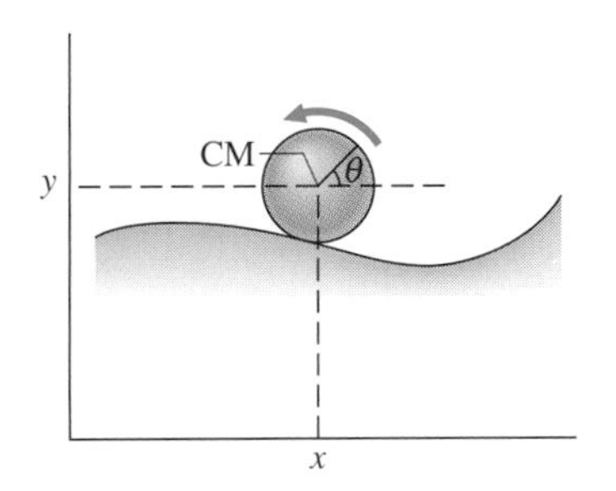

FIGURE 10-1 The coordinates x, y, and θ are normally used to describe a rigid body rolling in a plane.

The center of mass is used because its equations of motion, unlike those of any other point of a rigid body, involve only the external forces; specifically,

$$\sum F_x = M(a_{\mathrm{CM}})_x \qquad \sum F_y = M(a_{\mathrm{CM}})_y,$$

where M is the total mass of the body, $(a_{\mathrm{CM}})_x$ and $(a_{\mathrm{CM}})_y$ are the acceleration components of the center of mass, and only the external forces are included in the summations.

To find the equation of motion for the rotational angle θ, we first consider an arbitrary particle i of the rigid body shown in Fig. 10-2. The particle's position $\mathbf{r}_i$ and acceleration $\mathbf{a}_i$ are

$$\mathbf{r}_i = \mathbf{r}_i' + \mathbf{r}_{\mathrm{CM}},$$

and

$$\mathbf{a}_i = \mathbf{a}_i' + \mathbf{a}_{\mathrm{CM}},$$

where $\mathbf{r}_i'$ and $\mathbf{a}_i'$ are the position and acceleration of the particle relative to the center of mass, and $\mathbf{r}_{\mathrm{CM}}$ and $\mathbf{a}_{\mathrm{CM}}$ are the position and acceleration of the center of mass relative to the inertial frame. If $\mathbf{F}_i$ and $\mathbf{f}_i$ are the net external and internal forces, respectively, on particle i, the net torque on it around the axis through the center of mass is

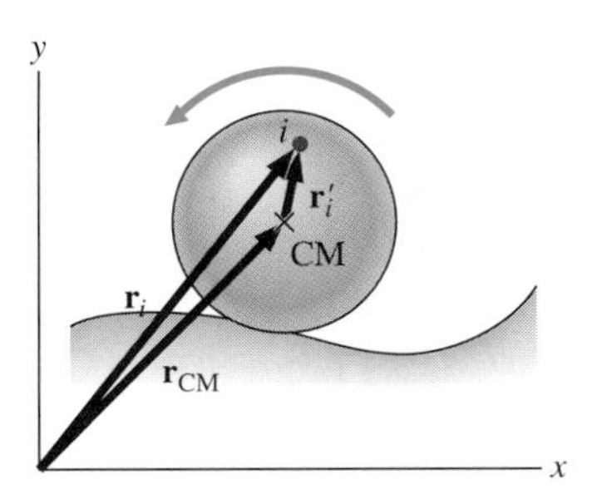

FIGURE 10-2 The position of particle i relative to the inertial frame is $\mathbf{r}_i$. Its position relative to the center of mass is $\mathbf{r}_i'$.

$$\boldsymbol{\tau}_i' = \mathbf{r}_i' \times (\mathbf{F}_i + \mathbf{f}_i).$$

From Newton's second law,

$$\mathbf{F}_i + \mathbf{f}_i = m_i\mathbf{a}_i = m_i(\mathbf{a}_i' + \mathbf{a}_{\mathrm{CM}}),$$

so upon combining these two equations, we find

$$\boldsymbol{\tau}_i' = m_i\mathbf{r}_i' \times (\mathbf{a}_i' + \mathbf{a}_{\mathrm{CM}}).$$

The net torque on the body about the axis through its center of mass is determined by summing this equation over all particles of the body. As in the fixed-axis case, the contributions of the internal forces cancel in pairs $(\sum_i \mathbf{r}_i' \times \mathbf{f}_i = 0)$, and we are left with

$$\sum_i \mathbf{r}_i' \times \mathbf{F}_i = \sum_i m_i\mathbf{r}_i' \times \mathbf{a}_i' + \sum_i m_i\mathbf{r}_i' \times \mathbf{a}_{\mathrm{CM}}. \tag{10-1}$$

Each of the terms is evaluated in Supplement 10-1, where it is shown that Eq. (10-1) reduces to

$$\sum \tau_{\mathrm{CM}} = I_{\mathrm{CM}}\alpha, \tag{10-2}$$

which we shall refer to as the *rotational form of Newton's second law for rolling motion*. Here, $\sum \tau_{\mathrm{CM}}$ is the sum of the torques around the center of mass due to the *external forces*, I_{CM} is the moment of inertia around the axis through the center of mass, and α is the angular acceleration around the axis.

Although Eqs. (10-2) and (9-12) have the same form, they represent two different situations. Equation (9-12) is valid for *an axis fixed in an inertial frame*, whether or not that axis passes through the center of mass. Equation (10-2), on the other hand, is restricted to *an axis that passes through the center of mass*. This axis is not even fixed in an inertial frame if the center of mass happens to be accelerating. There is no reason, based on Eq. (9-12), why the net torque has to be equal to $I\alpha$ for the accelerating axis. That fact has to be proven in Supplement 10-1!

10-2 Rolling and Slipping

An important question in the analysis of rolling motion is whether or not the body is slipping as it rolls. As an example, let's consider the wheel of Fig. 10-3*a* while it is being pulled across a horizontal surface by a horizontal force $\mathbf{F}$ applied at

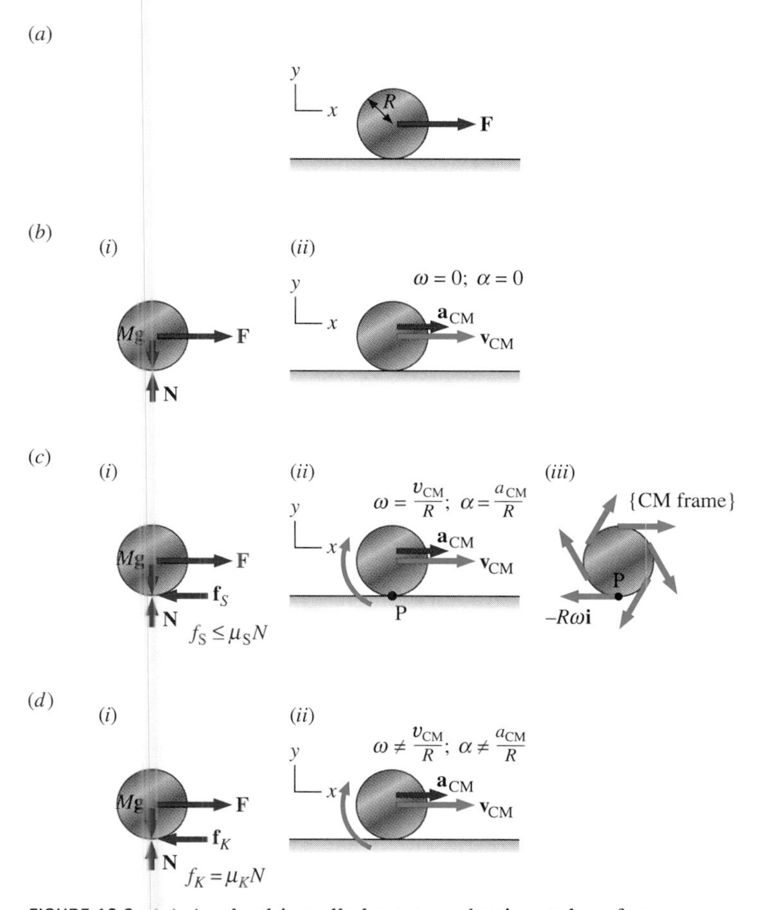

FIGURE 10-3 (*a*) A wheel is pulled across a horizontal surface. (*b*) No friction. (*c*) Friction is large enough to prevent slipping. (*d*) There is friction but the wheel slips.

its center of mass. First we'll assume that there is no friction between the wheel and surface. (See the free-body diagram of Fig. 10-3*b*.) Since all the external forces pass through the center of mass, there is no torque around the axis through that point; and by Eq. (10-2), there is no angular acceleration around that axis. Hence the wheel will not acquire an angular velocity; that is, it will not start rolling. Instead, it will slide (slip) across the surface as if it were a box or crate.

Next let's look at the other extreme, where the frictional force is so large that the wheel does not slip. (See Fig. 10-3*c*.) Then at any instant, the point P on the wheel that touches the horizontal surface is *at rest relative to that surface* (i.e., $\mathbf{v}_P = 0$), and the force of static friction $\mathbf{f}_S$ ($|\mathbf{f}_S| \leq \mu_S N$) acts between the wheel and the surface. When the wheel is rolling with an angular velocity ω, P has a linear velocity $-R\omega\mathbf{i}$ *relative to the center of mass*, as shown in sketch (*iii*) of Fig. 10-3*c*. Now, the center of mass has a velocity $v_{CM}\mathbf{i}$ relative to the horizontal surface, so the velocity of P relative to that surface is

$$\mathbf{v}_P = -R\omega\mathbf{i} + v_{CM}\mathbf{i}.$$

But by the no-slip assumption, $\mathbf{v}_P = 0$, so

$$v_{CM} = R\omega. \tag{10-3}$$

Thus if there is no slipping, the speed v_{CM} of the center of mass is equal to the product of the angular velocity ω around

The large torque causes the tires to slip as the racer takes off.

the axis through the center of mass and the radius R of the wheel.

The relationship between the acceleration of the center of mass of a nonslipping rolling body and the angular acceleration around the center of mass is obtained by differentiating Eq. (10-3). Since $\alpha = d\omega/dt$, and R is constant,

$$a_{CM} = R\alpha. \tag{10-4}$$

Also, since $dx_{CM}/dt = v_{CM}$ and $d\theta/dt = \omega$, the distance d_{CM} that the center of mass moves is related to the angle θ through which the body rotates by

$$d_{CM} = R\theta. \tag{10-5}$$

Finally, for the rolling body of Fig. 10-3*d*, the frictional force is not large enough to prevent slipping. In this case, the force of friction is kinetic ($f_K = \mu_K N$), and the contact point is no longer at rest relative to the surface. Consequently, $v_{CM} - R\omega \neq 0$, and Eqs. (10-3), (10-4), and (10-5) are not valid.

Table 10-1 presents a summary of the three situations we have considered. Notice that each case gives us one independent equation. If the wheel does not rotate at all, there is no friction and $f = 0$; for rolling without slipping, the equation is $a_{CM} = R\alpha$ (which is equivalent to either $v_{CM} = R\omega$ or $d_{CM} = R\theta$); and for rolling with slipping, the equation is $f_K = \mu_K N$.

TABLE 10-1 **Summary of Rolling Motion**

Slipping with No Rolling (Fig. 10-3*b*)	Rolling with No Slipping (Fig. 10-3*c*)	Rolling with Slipping (Fig. 10-3*d*)
1. This only occurs if there is no friction. 2. $f = 0$.	1. The point on the wheel in contact with the surface is at rest. 2. Static friction acts and $f_S \leq \mu_S N$. 3. $v_{CM} = R\omega$ $a_{CM} = R\alpha$.	1. The point in contact with the surface is moving. 2. Kinetic friction acts and $f_K = \mu_K N$.

10-3 Examples of Rolling Motion

Rolling motion is a combination of the translational motion of the center of mass and rotational motion around the center of mass. We use Newton's second law to analyze the translational motion and the rotational equation of motion to analyze the rotational motion. In addition, we need to consider whether the object is slipping as it rolls. The appropriate equation from Table 10-1 may then be applied.

EXAMPLE 10-1 Rolling and Slipping

A horizontal force **F** is applied at the center of mass of a wheel that is sitting on a flat, horizontal surface. (See Fig. 10-4*a*.) Using $I_{CM} = \frac{1}{2}MR^2$ as the moment of inertia of the wheel, do the following calculations. (*a*) If the wheel does not slip, what is its acceleration? (*b*) If $F = 6.0$ N, $\mu_S = 0.80$, $\mu_K = 0.60$, $M = 4.0$ kg, and $R = 0.10$ m, determine whether or not the no-slipping assumption is valid; then calculate the numerical value of the acceleration. (*c*) Repeat part (*b*) with the wheel pulled by a force of magnitude $F = 120$ N.

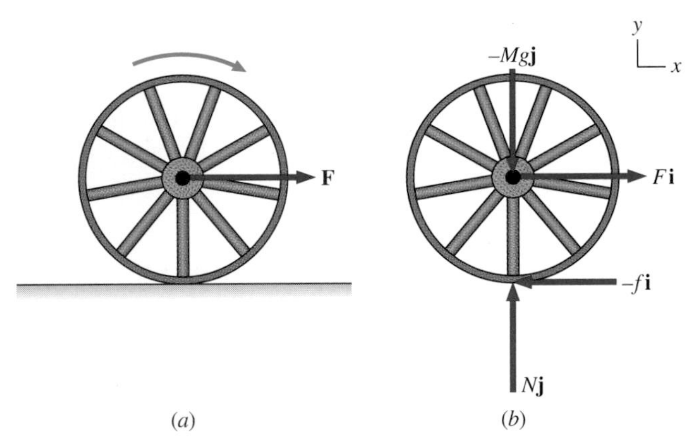

FIGURE 10-4 (*a*) A wheel is pulled across a horizontal surface by the force **F**. (*b*) The free-body diagram of the wheel. Notice that friction provides the torque that rotates the wheel.

SOLUTION (*a*) The forces on the wheel are the applied force **F**, the normal and frictional forces due to the surface, and the weight of the wheel. (See the free-body diagram of Fig. 10-4*b*.) Since we are assuming that the wheel does not slip, the frictional force is static and represented by f_S. The motion of the center of mass is analyzed using Newton's second law:

$$\sum F_x = ma_x \quad \text{and} \quad \sum F_y = ma_y$$

so

$$F - f_S = M(a_{CM})_x \tag{i}$$

and

$$N - Mg = 0 \tag{ii}$$

respectively.

To treat the rotational motion around the axis through the center of mass, we must calculate the torque of each force. The moment arms of **N**, M**g**, and **F** are zero, so they have no torques around this axis. Only the frictional force $\mathbf{f}_S$, with its moment arm R, has a torque; it is $f_S R$ and rotates the wheel clockwise. From $\sum \tau_{CM} = I_{CM}\alpha$,

$$f_S R = I_{CM}\alpha. \tag{iii}$$

If there is no slipping, $(a_{CM})_x = R\alpha$, so Eq. (*iii*) can be written as

$$f_S = I_{CM}\left[\frac{(a_{CM})_x}{R^2}\right] = \frac{1}{2}MR^2\left[\frac{(a_{CM})_x}{R^2}\right] = \frac{1}{2}M(a_{CM})_x, \tag{iv}$$

which, when substituted into Eq. (*i*), leaves

$$(a_{CM})_x = \frac{2}{3}\frac{F}{M}. \tag{v}$$

(*b*) The analysis of part (*a*) did depend on the assumption of no slipping. We now must check the validity of that assumption for our particular set of parameters. This is done by calculating the frictional force f_S and comparing it with the maximum static friction force $\mu_S N$. If $f_S \le \mu_S N$, the assumption is valid; otherwise it is not. For this example, we have

$$(a_{CM})_x = \frac{2}{3}\frac{F}{M} = \frac{2}{3}\frac{6.0 \text{ N}}{4.0 \text{ kg}} = 1.0 \text{ m/s}^2,$$
$$f_S = \tfrac{1}{2}M(a_{CM})_x = \tfrac{1}{2}(4.0 \text{ kg})(1.0 \text{ m/s}^2) = 2.0 \text{ N},$$

and

$$f_{max} = \mu_S N = \mu_S Mg = (0.80)(4.0 \text{ kg})(9.8 \text{ m/s}^2) = 31 \text{ N}.$$

Since $f_S \le f_{max}$, the no-slipping assumption is valid, and the calculated acceleration is correct.

(*c*) However, if $F = 120$ N,

$$(a_{CM})_x = \frac{2}{3}\frac{F}{M} = \frac{2}{3}\frac{120 \text{ N}}{4.0 \text{ kg}} = 20 \text{ m/s}^2,$$

and

$$f_S = \tfrac{1}{2}M(a_{CM})_x = \tfrac{1}{2}(4.0 \text{ kg})(20 \text{ m/s}^2) = 40 \text{ N}.$$

In this case, the force of static friction needed to prevent slipping is greater than $\mu_S N$, which is still 31 N. Consequently, there is slipping, and the calculated acceleration and force of friction are incorrect.

Fortunately, the problem is still solvable, because the loss of $(a_{CM})_x = R\alpha$ is compensated for by the gain of $f_K = \mu_K N$. Using $f_K = \mu_K N = \mu_K Mg$ in place of f_s in Eqs. (*i*) and (*iii*), we find

$$(a_{CM})_x = \frac{F - \mu_K Mg}{M},$$

and

$$\alpha = \frac{\mu_K MgR}{I_{CM}}.$$

Substitution of the given values of the various parameters into these two equations then yields $(a_{CM})_x = 24$ m/s^2 and $\alpha = 118$ rad/s^2.

EXAMPLE 10-2 Rolling Motion Down an Incline

A solid cylinder rolls down an incline without slipping. (See Fig. 10-5*a*.) The cylinder's mass is M and its radius is R. (*a*) What is its acceleration down the incline? (*b*) What condition must μ_S satisfy if there is no slipping?

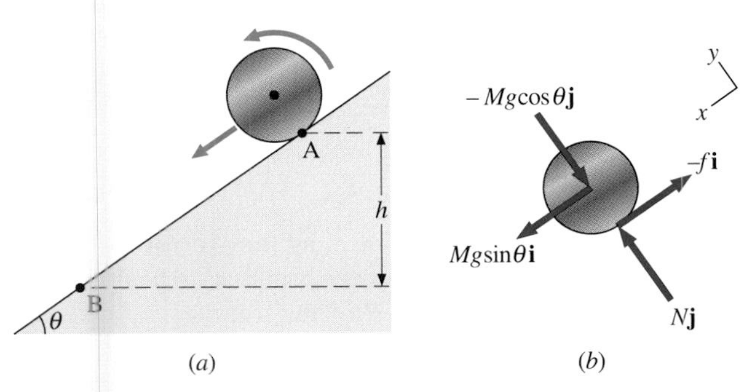

FIGURE 10-5 (*a*) A solid cylinder rolls down an incline without slipping. (*Note:* In Example 10-6, the cylinder starts from rest at A.) (*b*) The free-body diagram used in Examples 10-2 and 10-3.

SOLUTION (*a*) The free-body diagram of the cylinder is shown in Fig.10-5*b*. The forces on the cylinder are its weight and the normal and static frictional forces due to contact with the incline. The weight $M\mathbf{g}$ is resolved into components along and perpendicular to the incline. Since there is no slipping, the magnitude of the force of friction $\mathbf{f}_S$ is less than $\mu_S N$.

The translational motion of the center of mass of the cylinder is calculated from

$$\sum F_x = ma_x \quad \text{and} \quad \sum F_y = ma_y,$$

so

$$Mg\sin\theta - f_S = M(a_{CM})_x \qquad (i)$$

and

$$N - Mg\cos\theta = 0 \qquad (ii)$$

respectively. Relative to the axis through the center of mass, the moment arms of $M\mathbf{g}$ and $\mathbf{N}$ are zero, so their torques about this axis are zero. The moment arm of $\mathbf{f}_S$ is R, so its torque is $f_S R$. We therefore have

$$\sum \tau_{CM} = I_{CM}\alpha,$$
$$f_S R = I_{CM}\alpha. \qquad (iii)$$

Finally, since there is no slipping, the linear and angular accelerations are related by

$$(a_{CM})_x = R\alpha. \qquad (iv)$$

Equations (*i*), (*iii*), and (*iv*) can be solved for the three unknowns f_s, α, and $(a_{CM})_x$. The linear acceleration is found to be

$$(a_{CM})_x = \frac{Mg\sin\theta}{M + (I_{CM}/R^2)}. \qquad (v)$$

Since the body is a solid cylinder, $I_{CM} = MR^2/2$, and

$$(a_{CM})_x = \frac{Mg\sin\theta}{M + (MR^2/2R^2)} = \frac{2}{3}g\sin\theta.$$

(*b*) Because slipping does not occur, we have

$$f_S \le \mu_S N. \qquad (vi)$$

Using Eqs. (*iii*), (*iv*), and (*v*), we find that the force of friction is

$$f_S = I_{CM}\frac{\alpha}{R} = I_{CM}\frac{(a_{CM})_x}{R^2}$$
$$= \frac{I_{CM}}{R^2}\left(\frac{Mg\sin\theta}{M + (I_{CM}/R^2)}\right) = \frac{MgI_{CM}\sin\theta}{MR^2 + I_{CM}}. \qquad (vii)$$

When this equation and $N = Mg\cos\theta$ are substituted into Eq. (*vi*), we obtain

$$\frac{MgI_{CM}\sin\theta}{MR^2 + I_{CM}} \le \mu_S Mg\cos\theta,$$

so

$$\mu_S \ge \frac{\tan\theta}{1 + (MR^2/I_{CM})}.$$

For the solid cylinder,

$$\mu_S \ge \frac{\tan\theta}{1 + (2MR^2/MR^2)} = \frac{1}{3}\tan\theta.$$

DRILL PROBLEM 10-1

If they do not slip, what are the accelerations of (*a*) a sphere and (*b*) a thin-walled hoop when they roll down a slope inclined at an angle θ?
ANS. (*a*) $(5/7)g\sin\theta$; (*b*) $(1/2)g\sin\theta$.

DRILL PROBLEM 10-2

What are the minimum coefficients of static friction if (*a*) the sphere and (*b*) the thin-walled hoop are to roll down the incline without slipping?
ANS. (*a*) $(2/7)\tan\theta$; (*b*) $(1/2)\tan\theta$.

DRILL PROBLEM 10-3

Suppose that the incline is greased sufficiently so that there is negligible friction between the surface and the rolling body. What are the accelerations of the cylinder, sphere, and thin-walled hoop?
ANS. $g\sin\theta$ for all three.

The last example and three drill problems give us some rather interesting results. First they show us that the maximum acceleration occurs when there is no friction. No friction means no torque around the center of mass and therefore no angular acceleration. Hence maximum acceleration occurs when the body slides down the frictionless surface without rolling. Also, different bodies have different accelerations when there is sufficient friction to prevent slipping. If a sphere, solid cylinder, and hoop are all released together at the top of the incline, the sphere rolls to the bottom first, followed by the cylinder, with the hoop arriving last.

EXAMPLE 10-3 **A Sphere Slipping Down an Incline**

A sphere rolls down an incline steep enough that slipping occurs. What is its linear acceleration? What is its angular acceleration around the axis through the center of mass?

SOLUTION We can use Fig. 10-5*b* as the free-body diagram of the sphere. Since there is slipping, the frictional force **f** is kinetic, and

$$f_K = \mu_K N = \mu_K Mg\cos\theta.$$

Applying Newton's second law in the x direction, we find

$$\sum F_x = ma_x$$
$$Mg\sin\theta - \mu_K Mg\cos\theta = M(a_{\mathrm{CM}})_x,$$

so

$$(a_{\mathrm{CM}})_x = g(\sin\theta - \mu_K\cos\theta).$$

As in the previous example, only the frictional force $\mathbf{f}_K$ has a torque around the center-of-mass axis. The rotational equation of motion is then

$$\sum \tau_{\mathrm{CM}} = I_{\mathrm{CM}}\alpha$$
$$f_K R = I_{\mathrm{CM}}\alpha = \tfrac{2}{5}MR^2\alpha,$$

and

$$\alpha = \frac{5f_K}{2MR} = \frac{5\mu_K Mg\cos\theta}{2MR} = \frac{5\mu_K g\cos\theta}{2R}.$$

EXAMPLE 10-4 THE ACCELERATION OF A ROLLING WHEEL

A torque τ is applied around the axis through the center of mass of a wheel that is free to move across a horizontal surface. (See Fig. 10-6.) The wheel has mass M, its radius is R, and its moment of inertia around the axis through its center of mass is I. If there is sufficient friction to prevent slipping, what is the acceleration of the wheel?

SOLUTION The forces on the wheel are its weight along with the normal and static frictional forces due to contact with the surface. The translational motion of the center of mass is determined from Newton's second law:

$$\sum F_x = ma_x \quad \text{and} \quad \sum F_y = ma_y$$

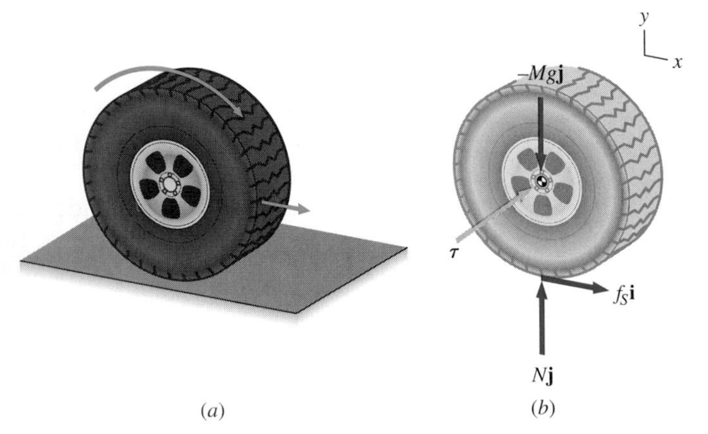

FIGURE 10-6 (*a*) A wheel accelerates to the right due to an applied torque. The wheel is assumed to roll without slipping. (*b*) The applied torque $\boldsymbol{\tau}$ is directed as shown and causes clockwise rotation of the wheel. Of the forces on the wheel, only the static frictional force has a moment arm, and hence a torque, about the center of mass. This torque is directed opposite to the applied torque $\boldsymbol{\tau}$ and causes counterclockwise rotation.

so

$$f_S = M(a_{\mathrm{CM}})_x \qquad (i)$$

and

$$N - Mg = 0 \qquad (ii)$$

respectively. The only torques around the axis through the center of mass are the applied torque τ and that due to the frictional force $\mathbf{f}_S$, which has a moment arm R. Thus from

$$\sum \tau_{\mathrm{CM}} = I_{\mathrm{CM}}\alpha,$$
$$\tau - f_S R = I\alpha. \qquad (iii)$$

When we combine this equation with Eq. (*i*) and use $(a_{\mathrm{CM}})_x = R\alpha$, we obtain

$$(a_{\mathrm{CM}})_x = \frac{\tau R}{I + MR^2}.$$

Notice that the wheel is pushed forward by the force of friction. When the wheel is rotated by the applied torque, the surface acts to prevent slipping by exerting a frictional force to the right. This frictional force then drives the wheel forward. It is this process that makes a car accelerate. When a torque is applied to the wheels by the engine, friction acts to prevent the wheels from slipping and consequently pushes the car forward.

DRILL PROBLEM 10-4

A 6.0-N·m torque is applied around the axis through the center of mass of the wheel of Fig. 10-6. The mass of the wheel is 24 kg, its radius is 0.50 m, and its moment of inertia is 3.0 kg·m^2. What is the acceleration of the wheel?
ANS. 0.33 m/s^2.

10-4 ENERGY CONSERVATION FOR ROLLING BODIES

The kinetic energy of a rigid body rolling in a plane is considered in Supplement 10-2. It is shown there that *the total kinetic energy of the body is the translational kinetic energy of the center of mass plus the kinetic energy due to the rotation around the axis through the center of mass.* Consequently, for a body whose mass is M, and whose moment of inertia around the axis through the center of mass is I_{CM} (see Fig. 10-7),

$$T = \tfrac{1}{2}Mv_{\mathrm{CM}}^2 + \tfrac{1}{2}I_{\mathrm{CM}}\omega^2, \qquad (10\text{-}6)$$

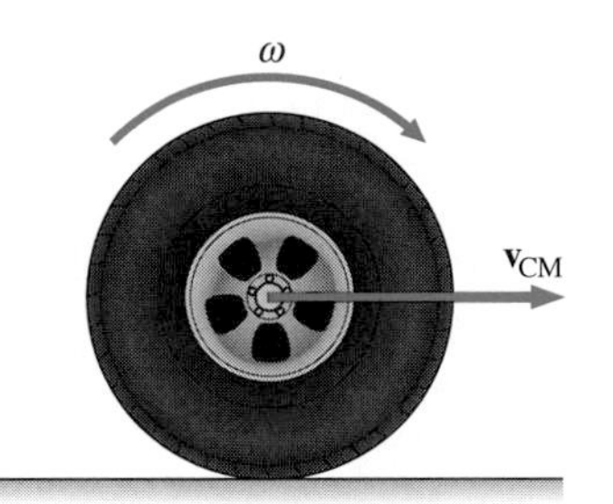

FIGURE 10-7 The kinetic energy of a rolling body is $T = (1/2)Mv_{\mathrm{CM}}^2 + (1/2)I_{\mathrm{CM}}\omega^2$.

A combination of translational and rotational kinetic energy.

where v_{CM} is the speed of the center of mass and ω is the angular velocity around the center-of-mass axis.

The equation representing mechanical-energy conservation for rolling motion is a direct extension of the equations developed for particles and for fixed-axis rotation. To find the total mechanical energy of a rolling body, we just add the kinetic energy of Eq. (10-6) to the potential energies of all of the conservative forces present.

You may be surprised that the mechanical energy of a rolling body is even conserved, since there is always a static frictional force causing the body to rotate. However, this force does no work when the body rolls without slipping. To see why, consider the point on the wheel that is instantaneously in contact with the surface. (See Fig. 10-3*c*.) As explained in Sec. 10-2, this point is at rest relative to the surface, so its displacement $d\mathbf{r} = 0$. This is also the point where the frictional force $\mathbf{f}_S$ is applied. Hence the frictional work $\mathbf{f}_S \cdot d\mathbf{r}$ vanishes because $d\mathbf{r}$ does.

But do be careful here—the no-slipping assumption is essential for energy conservation! The mechanical energy of a sliding body is *not* conserved because the kinetic frictional force does dissipative work on it. For this case, the motion must be analyzed using the dynamics methods of Sec. 10-1.

EXAMPLE 10-5

Kinetic Energy of a Rolling Hoop

A hoop of mass M and radius R rolls across the floor in a fixed direction. The speed of the hoop's center of mass is v_{CM}. If the hoop rolls without slipping, what is its kinetic energy?

SOLUTION Since the hoop rolls without slipping, $v_{CM} = R\omega$. The kinetic energy of the hoop is due to both its translational and rotational motion, which from Eq. (10-6) is

$$T = \tfrac{1}{2}Mv_{CM}^2 + \tfrac{1}{2}I_{CM}\omega^2$$
$$= \tfrac{1}{2}Mv_{CM}^2 + \tfrac{1}{2}(MR^2)\left(\frac{v_{CM}}{R}\right)^2$$
$$= Mv_{CM}^2.$$

Notice that the translational and rotational kinetic energies of a hoop that is rolling without slipping are equal. How would these two types of kinetic energy compare for a solid cylinder and a solid sphere?

DRILL PROBLEM 10-5

A solid cylinder rolls across the floor in a fixed direction without slipping. The mass of the cylinder is 50 kg, and its center of mass moves with a speed of 2.0 m/s. Calculate the kinetic energy of the cylinder. Also calculate the kinetic energy of a sphere that has the same mass, radius, and speed.
ANS. 150 J; 140 J.

EXAMPLE 10-6

Mechanical Energy Conservation for a Rolling Cylinder

What is the translational speed of the cylinder of Example 10-2 (Fig. 10-5) at point B if it is displaced a vertical distance h at that point? Assume it starts from rest at A and does not slip.

SOLUTION In this case, only the gravitational force does work, so mechanical energy is conserved. With the zero of potential energy chosen to be at point B, the potential energies of the cylinder at points A and B are $U_A = Mgh$ and $U_B = 0$. Since the cylinder starts from rest at A, the translational and rotational speeds there are zero.

From the law of mechanical-energy conservation,

$$T_A + U_A = T_B + U_B$$
$$\tfrac{1}{2}M(0)^2 + \tfrac{1}{2}I_{CM}(0)^2 + Mgh$$
$$= \tfrac{1}{2}Mv_{CM}^2 + \tfrac{1}{2}I_{CM}\omega^2 + Mg(0), \qquad (i)$$

where v_{CM} and ω represent the translational and rotational speeds at B. Since there is no slipping, $v_{CM} = R\omega$. When this, along with $I_{CM} = MR^2/2$, is substituted into Eq. (i), the translational speed at B is found to be

$$v_{CM} = \sqrt{\tfrac{4}{3}gh}.$$

EXAMPLE 10-7

A Ball Rolling Around a Track

The small ball of radius r in Fig. 10-8 starts rolling at the bottom of the vertical circular track of radius R with a translational speed v_B. What is the minimum value of v_B that allows the ball to stay on the track at its highest point? Assume that the ball does not slip and that $r \ll R$ so $R - r \approx R$.

SOLUTION The free-body diagram of the ball at the top of its path is shown in Fig. 10-8*b*. The force $\mathbf{N}$ represents the normal force of

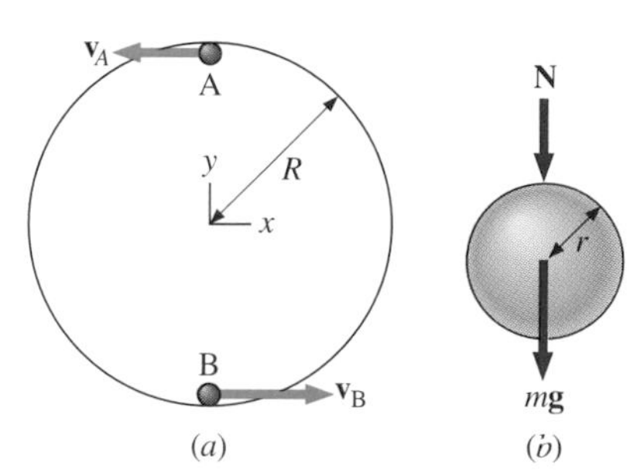

FIGURE 10-8 (*a*) A ball rolling around a track. (*b*) The free-body diagram for the ball at point A.

the track on the ball. Applying Newton's second law in the radial direction to the ball when it is at the top of the track, we find that

$$\sum F_r = ma_r$$
$$mg + N = m\frac{v_A^2}{R},$$

so

$$N = m\left(\frac{v_A^2}{R} - g\right).$$

If the ball is to remain on the track, the force N of the track on the ball has to be greater than or equal to zero. Hence

$$m\left(\frac{v_A^2}{R} - g\right) \geq 0,$$

and

$$v_A^2 \geq Rg. \qquad (i)$$

Equation (*i*) gives the minimum speed at the top of the track. The minimum speed at the bottom of the track can be found by using the law of mechanical-energy conservation to relate v_B to v_A. With the zero of potential energy at the elevation of the center of the track, we have

$$T_B + U_B = T_A + U_A$$
$$\tfrac{1}{2}mv_B^2 + \tfrac{1}{2}I_{CM}\omega_B^2 - mgR = \tfrac{1}{2}mv_A^2 + \tfrac{1}{2}I_{CM}\omega_A^2 + mgR, \qquad (ii)$$

where ω_A and ω_B are the angular velocities of the ball around its center of mass when the ball is at points A and B, respectively. Furthermore, r is ignored in the potential energy term because $r \ll R$.

If $v = r\omega$ and $I_{CM} = 2mr^2/5$ are used in Eq. (*ii*), the speeds v_A and v_B are found to be related by

$$v_A^2 = v_B^2 - \frac{20}{7}Rg. \qquad (iii)$$

Finally, with Eqs. (*iii*) and (*i*) combined,

$$v_B^2 - \frac{20}{7}Rg \geq Rg,$$

so

$$v_B \geq \sqrt{\frac{27Rg}{7}}.$$

DRILL PROBLEM 10-6

A sphere travels up a ramp inclined at 30° to the horizontal; the sphere starts with a speed of 6.0 m/s. How far along the incline does it travel before it stops? Assume that the sphere rolls without slipping.
ANS. 5.1 m.

DRILL PROBLEM 10-7

Do the calculation of Example 10-6 for (*a*) a sphere and (*b*) a hoop.
ANS. (*a*) $\sqrt{10gh/7}$; (*b*) $\sqrt{gh}$.

SUMMARY

1. **Rolling motion**
   For a rolling rigid body whose center of mass is moving in the xy plane, Newton's second law gives

   $$\sum F_x = m(a_{CM})_x \quad \text{and} \quad \sum F_y = m(a_{CM})_y,$$

   where the summations are over the external forces on the body. In addition, the rotation of the body around the axis through the center of mass and perpendicular to the xy plane is governed by

   $$\sum \tau_{CM} = I_{CM}\alpha,$$

   where this summation is over the torques applied by the external forces.
2. **Rolling and slipping**
   If a rigid body rolls without slipping, the linear acceleration of the center of mass is related to the angular acceleration around the axis through the center of mass by $a_{CM} = R\alpha$, but there is no formula to describe the frictional force. If the body slips, $a_{CM} \neq R\alpha$, but in this case the frictional force is given by $f_K = \mu_K N$.
3. **Conservation of mechanical energy for rolling motion**
   The kinetic energy of a rolling rigid body is given by

   $$T = \tfrac{1}{2}Mv_{CM}^2 + \tfrac{1}{2}I_{CM}\omega^2.$$

   If a body rolls without slipping under a net conservative force whose potential energy is U, then its mechanical energy, $E = T + U$, is conserved.

SUPPLEMENT 10-1 ROTATIONAL FORM OF NEWTON'S SECOND LAW FOR ROLLING MOTION

Here we consider each term of Eq. (10-1) and verify that this equation reduces to Eq. (10-2), the rotational form of Newton's second law for rolling motion.

The term $\sum_i \mathbf{r}_i' \times \mathbf{F}_i$ is the net torque around the axis through the center of mass due to the external forces. The weight of the body acts at the center of mass and has no torque around this axis. The only particles that contribute to the summation are those that experience an external force (other than gravity). We can therefore write

$$\text{Net external torque around the center of mass} = \sum_i \mathbf{r}_i' \times \mathbf{F}_i = \sum \tau_{CM}, \qquad (i)$$

where $\sum \boldsymbol{\tau}_{CM}$ is the sum of the torques due to just the external forces. (Notice the absence of i in this sum.)

Since the vector product is associative, the second term on the right-hand side of Eq. (10-1) can be written as

$$\left(\sum_i m_i \mathbf{r}_i'\right) \times \mathbf{a}_{CM}. \qquad (ii)$$

From Eq. (8-2*b*), the term in parentheses in Eq. (*ii*) is equal to $M\mathbf{r}_{CM}$, which in this case is zero, because the center of mass is located at $\mathbf{r}_{CM} = 0$. Hence Eq. (*ii*) vanishes and Eq. (10-1), with the substitution of Eq. (*i*), reduces to

$$\sum \boldsymbol{\tau}_{CM} = \sum_i m_i \mathbf{r}_i' \times \mathbf{a}_i'. \qquad (iii)$$

As shown in Fig. 10-9, the vectors $\mathbf{r}_i'$ and $\mathbf{a}_i'$ are both measured relative to a reference frame moving with the center of mass, and in this frame, $\mathbf{a}_i' = a_{ir}'\hat{\mathbf{r}}' + a_{i\theta}'\hat{\boldsymbol{\theta}}'$. Since $a_{ir}'\hat{\mathbf{r}}'$ is parallel to $\mathbf{r}_i'$ and $a_{i\theta}'\hat{\boldsymbol{\theta}}'$ is perpendicular to $\mathbf{r}_i'$, only $a_{i\theta}'\hat{\boldsymbol{\theta}}'$ contributes to $\mathbf{r}_i' \times \mathbf{a}_i'$. Furthermore, both $\mathbf{r}_i'$ and $\mathbf{a}_i'$ lie in the plane of the page, so their cross product is perpendicular to the page. Thus

$$|\mathbf{r}_i' \times \mathbf{a}_i'| = r_i'a_{i\theta}' \sin 90° = r_i'a_{i\theta}',$$

where, with the choice of increasing θ indicated in the figure, "out of the page" corresponds to counterclockwise rotations and "into the page" to clockwise rotations. If we keep track of torques in a clockwise-counterclockwise sense, the vector notation of Eq. (*iii*) may be suppressed, and $\mathbf{r}_i' \times \mathbf{a}_i'$ can be replaced by its magnitude $r_i'a_{i\theta}'$. With this change, Eq. (*iii*) becomes

$$\sum \tau_{CM} = \sum_i m_i r_i' a_{i\theta}'. \qquad (iv)$$

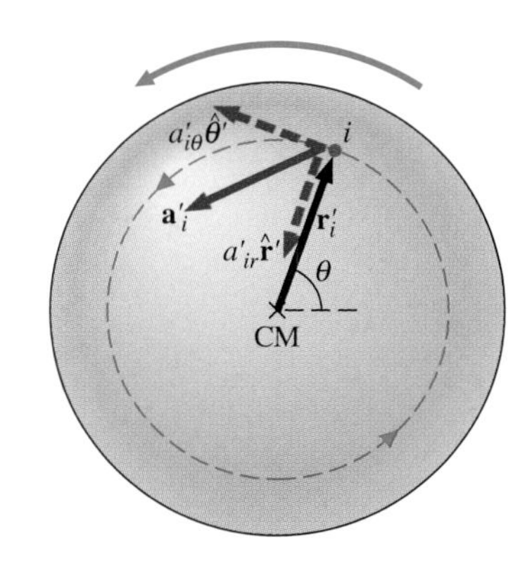

FIGURE 10-9 Relative to the center of mass (CM), particle i has position $\mathbf{r}_i'$ and is rotating in a circle of radius r_i'. The acceleration of i is $\mathbf{a}_i'$, and $\mathbf{a}_i' = a_{ir}'\hat{\mathbf{r}}' + a_{i\theta}'\hat{\boldsymbol{\theta}}'$.

From this point on, we proceed just as we did for the fixed-axis case. Since both r_i' and $a_{i\theta}'$ are measured relative to the center of mass,

$$a_{i\theta}' = r_i'\alpha, \qquad (v)$$

where α is the angular acceleration of the rigid body around the axis through the center of mass. Substituting Eq. (*v*) into Eq. (*iv*) we obtain

$$\sum \tau_{CM} = \left(\sum_i m_i r_i'^2\right)\alpha = I_{CM}\alpha.$$

This is Eq. (10-2).

SUPPLEMENT 10-2 KINETIC ENERGY OF A SYSTEM OF PARTICLES

In this supplement we calculate the kinetic energy of a system of particles in terms of the motion of the center of mass and the motion of the particles relative to the center of mass. We then apply the results of this calculation to the special case of a body rolling in a plane.

Two arbitrary particles of a system of particles are shown in Fig. 10-10. The velocity $\mathbf{v}_i$ of particle i relative to an inertial frame whose origin is at O is

$$\mathbf{v}_i = \mathbf{v}_{CM} + \mathbf{v}_i', \qquad (i)$$

where $\mathbf{v}_{CM}$ is the velocity of the center of mass relative to O, and $\mathbf{v}_i'$ is the velocity of particle i relative to the center of mass. The kinetic energy of the system of particles is by definition

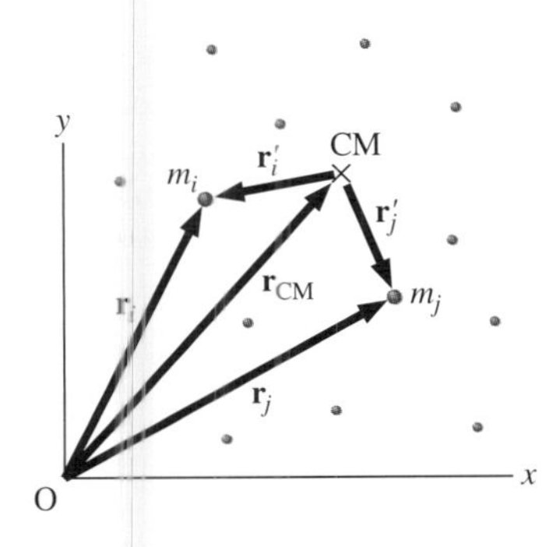

FIGURE 10-10 Two representative particles of a system of particles. The point O is the origin of an inertial reference frame, and CM is the center of mass of the system.

$$T = \frac{1}{2}\sum_i m_i v_i^2 = \frac{1}{2}\sum_i m_i \mathbf{v}_i \cdot \mathbf{v}_i.$$

Using Eq. (*i*), we can write this as

$$T = \frac{1}{2} v_{CM}^2 \sum_i m_i + \frac{1}{2}\sum_i m_i (v_i')^2 + \mathbf{v}_{CM} \cdot \sum_i m_i \mathbf{v}_i'. \qquad (ii)$$

The first of these terms is $(1/2)Mv_{CM}^2$, where $M\left(=\sum_i m_i\right)$ is the total mass of the system. Since $\mathbf{v}_i'$ is measured relative to the center of mass, we have from Eq. (8-4*a*) $\sum_i m_i \mathbf{v}_i' = 0$. With these two results substituted in Eq. (*ii*), it becomes

$$T = \frac{1}{2} Mv_{CM}^2 + \frac{1}{2}\sum_i m_i (v_i')^2. \qquad (iii)$$

Thus the kinetic energy of a system of particles is the sum of two terms: (1) the translational kinetic energy of the center of mass and (2) the total kinetic energy of the particles relative to the center of mass.

Notice that we have not yet placed any restriction on the system of particles. They could be completely independent of one another, they could be interacting with one another while in relative motion, or they could be bound together to form a rigid body. Independent of the situation, Eq. (*iii*) gives the system's kinetic energy relative to the inertial frame.

Now let's require that the particles form a rigid body that is rolling in a plane. Because the direction of the rotational axis here is fixed, $(1/2)\sum_i m_i(v_i')^2$ is evaluated just like its fixed-axis counterpart, which led to Eq. (9-13). Therefore

$$\frac{1}{2}\sum_i m_i(v_i')^2 = \frac{1}{2} I_{CM}\omega^2, \qquad (iv)$$

where $I_{CM}\left[= \sum_i m_i(r_i')^2\right]$ is the rotational inertia around the axis through the center of mass. Finally, with Eq. (*iv*) substituted in Eq. (*iii*), we have Eq. (10-6):

$$T = \tfrac{1}{2}Mv_{CM}^2 + \tfrac{1}{2}I_{CM}\omega^2.$$

The kinetic energy of a rolling rigid body is the translational kinetic energy of its center of mass plus the kinetic energy due to the body's rotation around the axis through its center of mass.

QUESTIONS

Note: Unless stated otherwise in the following Questions, objects that are "rolling" are assumed to be doing so without slipping.

10-1. Can rolling occur on a frictionless surface?

10-2. Trace the path of a point on the edge of a rolling cylinder.

10-3. Compare the velocities of the following three points on a rolling wheel: (*a*) the top point, (*b*) the center, (*c*) the bottom point.

10-4. A cylindrical can of radius R is rolling across a horizontal surface. After one complete revolution of the can, what is the distance that its center of mass has moved? Would this distance be greater or smaller if slipping occurred?

10-5. A wheel is released at the top of an incline. Is the wheel more likely to slip if the incline is steep or gently sloped?

10-6. Suppose you let a full can of soda pop roll down an incline, then drink its contents and send the empty can down the same incline. In which case does the can roll down the incline faster?

10-7. Is there a point on a ball rolling across a horizontal floor that has a vertical velocity only?

10-8. A small stick stands vertically on frictionless ice. Describe the path of its center of mass as the stick falls.

10-9. Starting from rest, two marbles roll down the inclines shown in the accompanying figure. Which marble has the greater speed at the bottom? Which marble takes longer to reach the bottom?

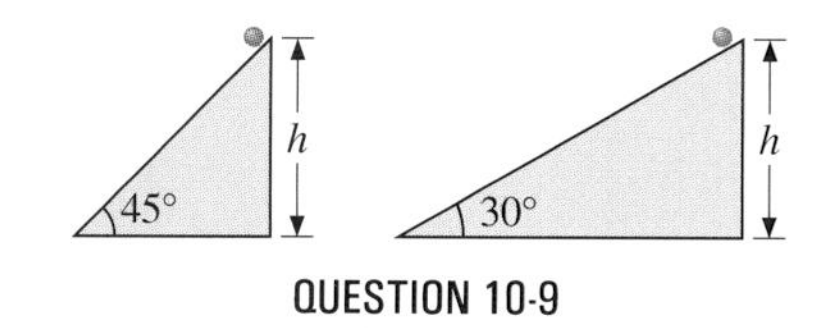

QUESTION 10-9

10-10. What is the ratio of the translational to rotational kinetic energies of the following rolling objects: (*a*) a cylinder, (*b*) a solid sphere, (*c*) a spherical shell.

10-11. A sphere and a cylinder roll up an incline with the same initial center-of-mass velocity. Which object rolls farther up the incline before stopping?

10-12. A cylinder, a sphere, and a hoop are released together at the top of an incline. Which one reaches the bottom of the incline first?

10-13. A marble and a bowling ball (without finger holes) are released together at the top of an incline. Which one reaches the bottom first?

10-14. Suppose you pull a wheel across a horizontal surface with the force **F**, as shown in the accompanying figure. In order to pull the wheel at constant velocity, you have to overcome the net retarding torque τ_f at its axle. Would this be easier to do if the wheel had a large or small radius? Assume that the mass and τ_f are the same for both cases.

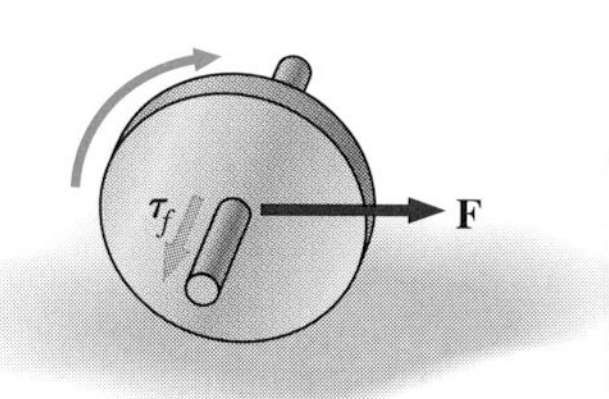

QUESTION 10-14

PROBLEMS

Note: Unless stated otherwise in the following Problems, objects that are "rolling" are assumed to be doing so without slipping.

Rolling Motion

10-1. What is the angular velocity of a 75-cm-diameter tire on an automobile traveling at 90 km/h?

10-2. A rigid body with a cylindrical cross section is released at the top of a 30° incline; it rolls 10 m to the bottom in 2.6 s. Find the moment of inertia of the body in terms of its mass M and radius R.

10-3. A marble is rolling across the floor at a speed of 7.0 m/s when it starts up a plane inclined at 30° to the horizontal. (*a*) How far along the plane does the marble travel before coming to rest? (*b*) How much time elapses while the marble moves up the plane? (*c*) After stopping, the marble rolls back down the plane. What is its speed at the bottom, and how much time does it take to reach the bottom?

10-4. A spool of inner radius r and outer radius R rolls across a horizontal floor when it is pulled by a light horizontal string wrapped around the inner cylinder. (See the accompanying figure.) What is the spool's acceleration? The mass and moment of inertia of the spool are, respectively, M and $I = MR^2/3$.

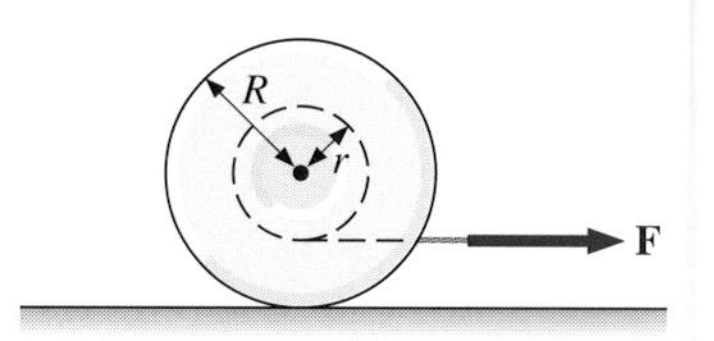

PROBLEM 10-4

10-5. Repeat Prob. 10-4 assuming the string is wrapped so that the spool is pulled as shown in the accompanying figure.

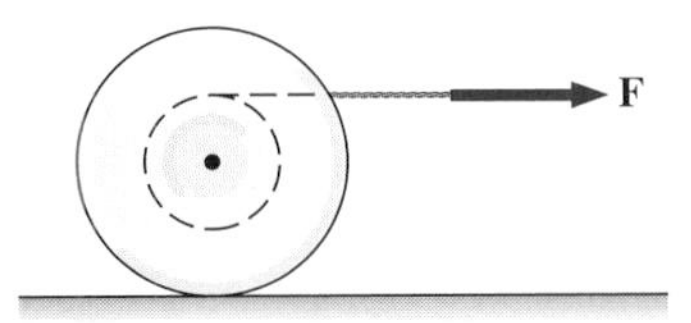

PROBLEM 10-5

10-6. A cylindrical disk of mass M and radius R has a light string wrapped around its circumference. One end of the string is held fixed in space. (See the accompanying figure.) If the disk falls as the string unwinds without slipping, what is the acceleration of the disk?

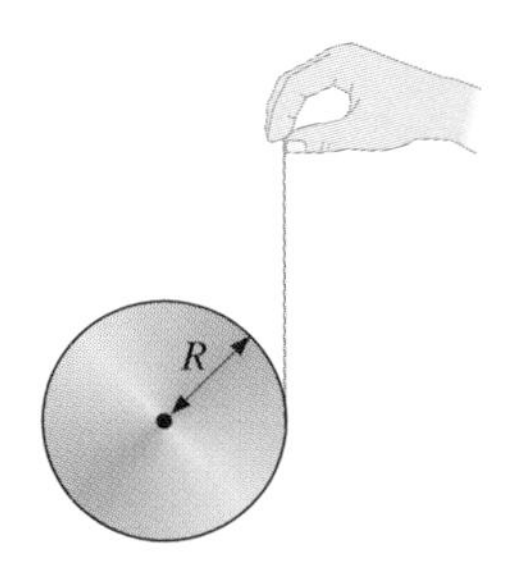

PROBLEM 10-6

10-7. A heavy roll of wrapping paper in the form of a solid cylinder is resting on a horizontal tabletop. Its mass is M and its radius is R. If a horizontal force T is applied evenly to the paper, as shown in the accompanying figure, what are the linear acceleration of the center of the roll and the angular acceleration around the center of the roll? The coefficient of kinetic friction between the paper and the table is μ_K. (*Note:* The roll of paper does slip.)

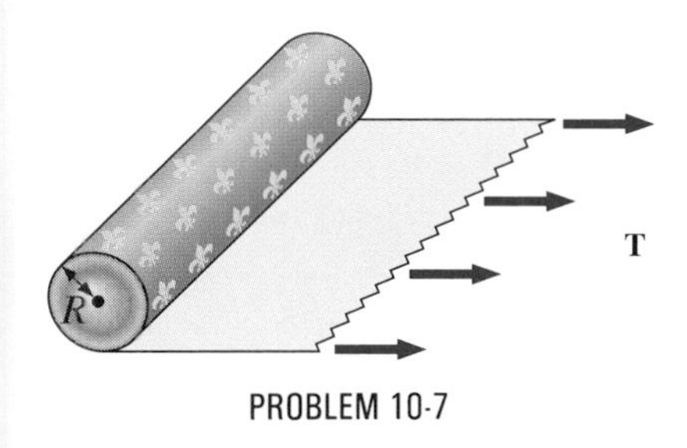

PROBLEM 10-7

10-8. A cylindrical wheel of mass M and radius R is rolling across a horizontal surface when a torque τ is applied at its rotational axis to slow it down. (*a*) What is the maximum torque that can be applied to this wheel so that it will slow down without slipping? Express your answer in terms of μ_S, M, g, and R. (*b*) If $\tau = 20\ \text{N}\cdot\text{m}$, $\mu_S = 0.60$, $M = 20$ kg, and $R = 20$ cm, what is the acceleration of the wheel?

10-9. A light string of negligible mass is wrapped around a solid cylinder of mass M and radius R. The string is pulled vertically upward with just the right tension **T** so that the center of mass of the cylinder stays fixed as the string unwinds without slipping. (See the accompanying figure.) (*a*) What is the tension in the string? (*b*) When the cylinder reaches an angular velocity ω, how much string has unwound? Express your answers in terms of ω, R, M, and g.

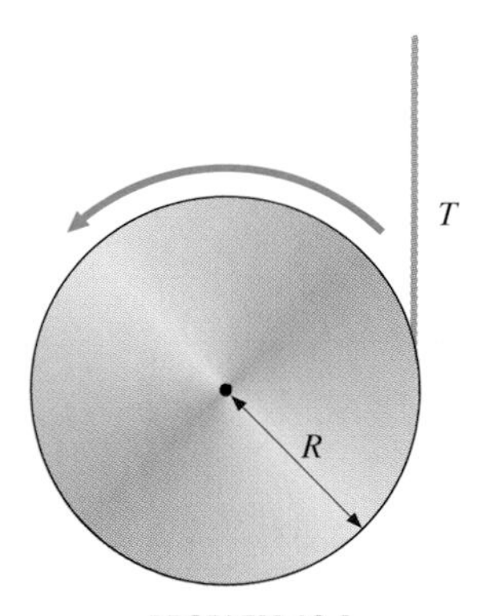

PROBLEM 10-9

10-10. A light string is wrapped around a cylindrical disk of mass $M = 8.0$ kg and radius $R = 0.20$ m. The other end of the string is pulled with a force of 20 N. The disk rolls on a 37° incline as shown. Calculate the acceleration of the disk's center of mass and the force of static friction between the incline and the disk.

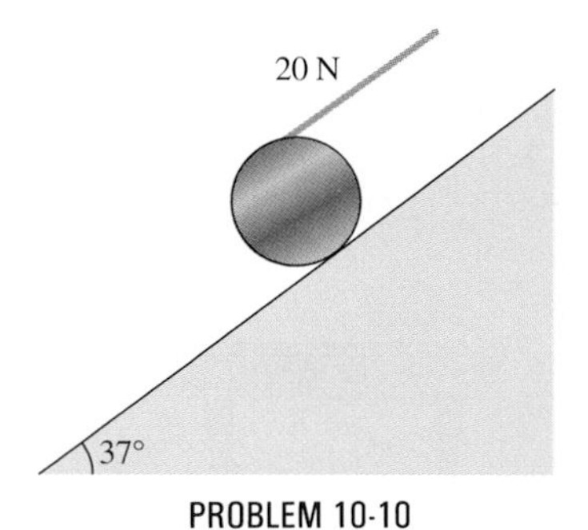

PROBLEM 10-10

10-11. A bowling ball is thrown straight down the alley. When it starts, its center of mass has a speed v_0 and it is sliding without rotating. (*a*) Determine how far the ball moves down the alley before it starts rolling without slipping. Express your answer in terms of v_0, g, and μ_K. (*b*) What is the distance if $v_0 = 12$ m/s and $\mu_K = 0.75$?

Rolling Motion and Energy

10-12. A 40-kg solid cylinder is rolling across a horizontal surface at a speed of 6.0 m/s. How much work is required to stop it?

10-13. The mass of a hoop of radius 1.0 m is 6.0 kg. It rolls across a horizontal surface with a speed of 10 m/s. (*a*) How much work is required to stop the hoop? (*b*) If the hoop starts up a surface inclined at 30° to the horizontal with a speed of 10 m/s, how far along the incline will it travel before stopping and rolling back down?

10-14. A solid sphere rolls up a plane inclined at 37° to the horizontal. If it starts at the bottom with a speed of 14 m/s, how far up the plane does it travel?

10-15. A marble starts from rest at the upper end of the track shown in the accompanying figure. It rolls until it leaves the track at the right end. If the track is horizontal at this point, determine the distance d to the right of this point at which the marble strikes the lower surface.

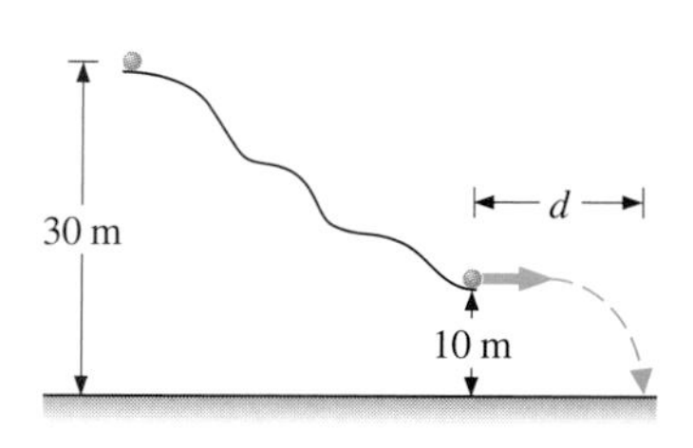

PROBLEM 10-15

10-16. A small ball of radius r starts at A and rolls down the track of Prob. 7-41. (*a*) What is the ball's speed at B? (*b*) What is the force of the track on the ball at B? Assume $R \gg r$.

10-17. In Probs. 7-41 and 10-16 it was assumed that the object on the track would make it around the top of the track. Show that this is a valid assumption.

10-18. The solid cylinder shown in the accompanying figure rolls across the table as the block falls. The pulley is frictionless, and its rotational inertia is negligible. The mass of the cylinder is 20 kg, and the mass of the block is 60 kg. (*a*) If the block starts at rest, what is its speed at the instant it has fallen 1.5 m? (*b*) What is the speed of the center of mass of the cylinder at this instant? (*Note:* The speed of the top of the cylinder is equal to the speed of the block.)

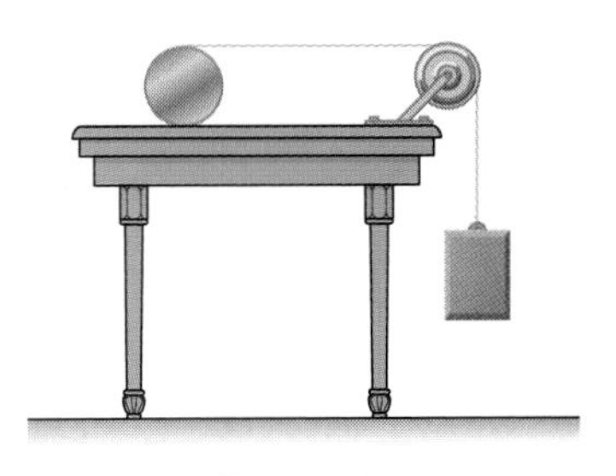
PROBLEM 10-18

General Problems

10-19. A solid cylindrical wheel of mass M and radius R is pulled by a force $\mathbf{F}$ applied to the center of the wheel at 37° to the horizontal. If the wheel is to roll without slipping, what is the maximum value of $|\mathbf{F}|$? The coefficients of static and kinetic friction are $\mu_S = 0.40$ and $\mu_K = 0.30$.

10-20. The mass of the cylinder of Prob. 10-19 is 10 kg and its radius is 10 cm. If the force $\mathbf{F}$ has magnitude Mg, what is the acceleration of the cylinder? What is its angular acceleration around the center of mass?

10-21. What are the minimum speed v and the corresponding angle θ required for the motorcyclist shown in the accompanying figure to ride on the vertical wall of the circular track? The coefficient of static friction between the wall and tires is $\mu_S = 0.70$. Assume that the center of mass of the motorcycle and rider is on the axis shown and 1.0 m above the base of the wheel. (*Hint:* The net torque *around the center of mass* must be zero.)

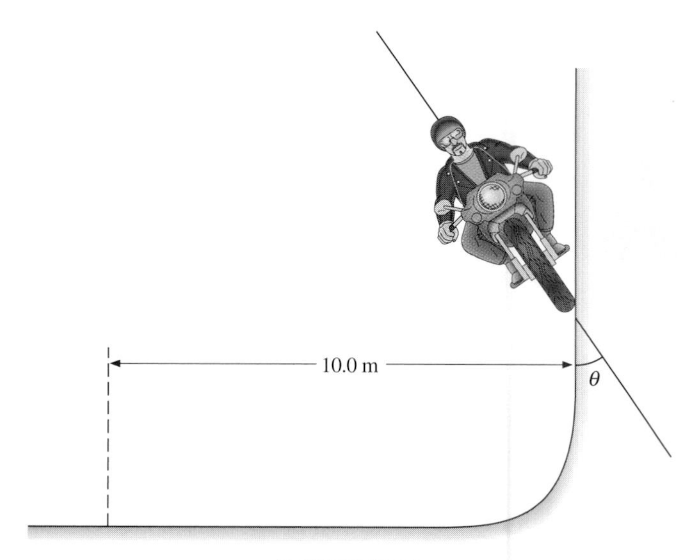

PROBLEM 10-21

10-22. A hollow, thin-walled sphere ($I_{CM} = 2MR^2/3$) of mass 20 kg is completely filled with a liquid of unknown mass. The sphere is released at the top of a plane inclined at 30° to the horizontal, and it rolls 20 m to the bottom in 3.6 s. What is the mass of the liquid?

10-23. The yo-yo shown in the accompanying figure consists of an inner cylinder of radius 0.75 cm with side caps of radius 5.0 cm. The mass of the yo-yo is 200 g, its moment of inertia is $3.0 \times 10^3\ \text{g}\cdot\text{cm}^2$, and a very light string is wrapped around the inner cylinder. A young child who is just learning to use this toy holds the end of the string between her forefinger and thumb and drops the yo-yo, which then unwinds on the string as it descends. (*a*) What are the yo-yo's acceleration and angular acceleration around the center of mass? (*b*) How far does its center of mass fall in 2.0 s?

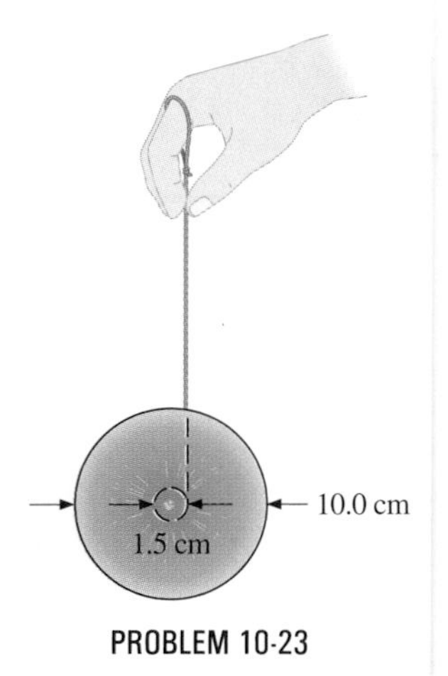

PROBLEM 10-23

10-24. A solid disk and a ring of the same mass and radius roll down a 30° incline. A string is wound around the two objects and connects them, as shown in the accompanying figure. If the mass of each object is 0.20 kg, what is the tension in the string?

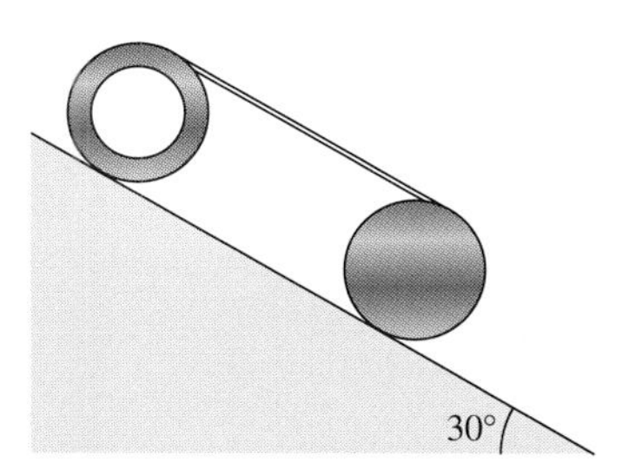

PROBLEM 10-24

10-25. The truck shown is initially at rest with a solid cylindrical roll of paper sitting on its bed. If the truck moves forward with a uniform acceleration a, what distance s does it move before the paper rolls off its back end? (*Hint:* If the roll accelerates forward at a', then it accelerates backwards relative to the truck with an acceleration of $a - a'$; in addition, $R\alpha = a - a'$.)

PROBLEM 10-25

10-26. A torque is applied at the center of mass of a solid cylinder ($M = 2.0$ kg, $R = 0.50$ m), which is free to move across a horizontal surface. The cylinder starts at rest and attains a speed of 6.0 m/s after rolling 10 m. What is the applied torque?

10-27. With the torque still being applied, the cylinder of Prob. 10-26 moves up a plane inclined at 37° to the horizontal. If the cylinder starts at the bottom with a speed of 6.0 m/s, what is its speed after it has traveled 5.0 m along the plane?

10-28. The mass of a thin-walled cylindrical drum is 20 kg. The drum is filled with dirt and given a slight push so that it rolls down an incline. The speed of the filled drum at the bottom of the incline is found to be 10 percent more than that of the empty drum when it rolls down the incline. What is the mass of the dirt in the drum?

10-29. A ball of mass m and radius r rolls along a circular path of radius R. (See the accompanying figure.) Its speed at the bottom ($\theta = 0$) of the path is v_0. Find the force of the path on the ball as a function of θ.

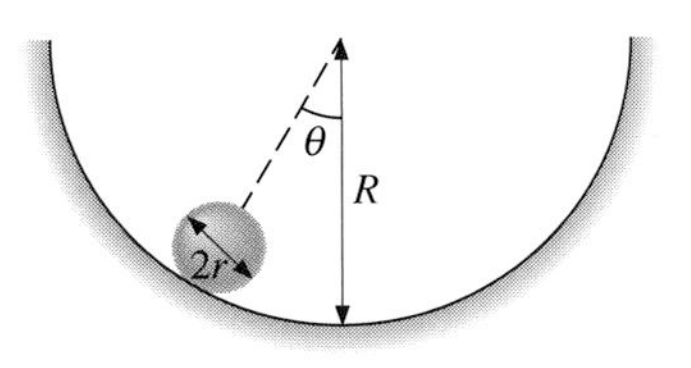

PROBLEM 10-29

10-30. A cart of mass M rolls on four wheels, each of mass m and moment of inertia $mR^2/2$. (a) Show that the kinetic energy of the cart is given by

$$T = \tfrac{1}{2}(M + 6m)v_{CM}^2,$$

where v_{CM} is the center-of-mass speed of the cart. (b) If the cart is given a slight push at the top of an incline, show that its speed at the bottom is

$$v_{CM} = \sqrt{\frac{2(M + 4m)gh}{M + 6m}},$$

where h is the height of the incline. (c) In Chap. 7 we sometimes considered such objects as rolling carts and carriages. At that point, we ignored the kinetic energy of the rotating wheels. Expand the expression for v_{CM} in powers of m/M and show that the wheels can be ignored if $m/M \ll 1$.

10-31. The wheel ($I = mR^2/2$) shown in the accompanying figure is rolling across the floor when a torque τ is applied around its center of mass to slow it down. (a) Show that the acceleration of the wheel is $a = -2\tau/3mR$. (b) Show that the maximum torque such that no slipping occurs is given by $\tau_{max} = 3\mu_S mgR/2$. Also show that $|a_{max}| = \mu_S g$. (c) Show that $a = -\mu_K g$ if $\tau > \tau_{max}$. (d) Show that the range of τ for which the absolute value of the acceleration of the wheel is greater than its slipping value is $3\mu_K mgR/2 < \tau < 3\mu_S mgR/2$. ($e$) Graph $|a|$ versus τ. (f) What does this problem tell you about how the brakes should be applied to stop a car most effectively?

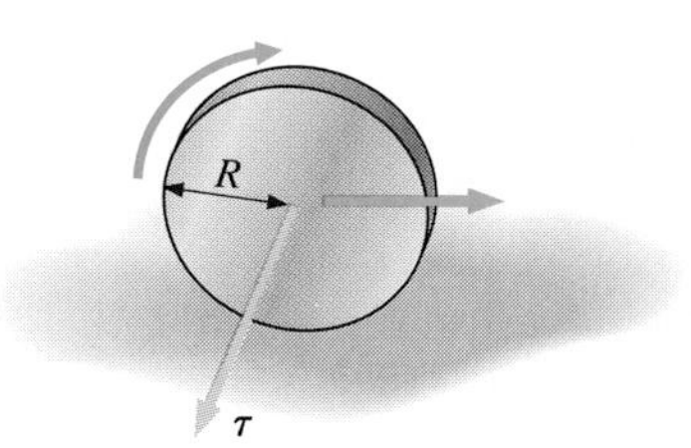

PROBLEM 10-31

Static equilibrium.

CHAPTER 11 EQUILIBRIUM OF A RIGID BODY

PREVIEW

This chapter is an introduction to the equilibrium of rigid bodies. The topics to be discussed here include the following:

1. **Equilibrium.** Equilibrium is defined, and the conditions necessary for a body to be in equilibrium are given.

*2. **Stability.** We present the concept of stability and discuss the mathematical condition necessary for the stability of a particle constrained to move along a straight line.

*3. **Stress and strain in solids.** Tensile and compressive stress and strain are defined and related to one another.

*4. **Volume compression.** We consider how a body is compressed when forces are applied uniformly over its surface.

This chapter is devoted to the mechanics of rigid bodies in equilibrium. We limit our study to cases for which the forces acting on a body all lie in the same plane. This allows us to treat equilibrium as a planar problem. A wide variety of examples will be considered, including some for which the conditions of equilibrium are insufficient for a complete specification of the forces on the body under investigation. We'll also take a brief look at the difference between stable and unstable equilibrium and at how a body is stretched or compressed under applied forces.

The concepts studied here are especially important to engineers, who often consider equilibrium problems in making design decisions: What type of cable will support this suspension bridge? Will power lines strung between these two towers snap under their own weight? What type of foundation will support this building? All of these are questions involving equilibrium.

11-1 Introduction to Equilibrium

A rigid body whose linear and angular accelerations are both zero relative to an inertial reference system is said to be in *equilibrium*. A body in equilibrium can be moving, but if it is, it must do so with constant linear and angular velocities. If a rigid body is actually at rest in our inertial frame, we say it is in *static equilibrium*. However, the distinction between rest and uniform motion is quite artificial—for example, a body may be at rest in our reference frame, yet to an observer moving at constant velocity with respect to us, the body is moving at constant velocity. What is "static" equilibrium to us is merely equilibrium to him and vice versa. Since the laws of physics are identical in all inertial reference frames, we need not make the distinction between static equilibrium and equilibrium.

11-2 Conditions of Equilibrium

The translational motion of a rigid body is governed by

$$\sum \mathbf{F} = M\mathbf{a}_{\text{CM}},$$

where the sum is over the external forces. For equilibrium, $\mathbf{a}_{\text{CM}} = 0$, so

$$\sum \mathbf{F} = 0, \qquad (11\text{-}1a)$$

or

$$\sum F_x = 0, \qquad \sum F_y = 0, \qquad \sum F_z = 0. \qquad (11\text{-}1b)$$

Equations (11-1) represent the *first equilibrium condition.*

If a rigid body is in equilibrium, its angular acceleration around any axis fixed in an inertial frame is also zero. From Eq. (9-12), this can happen if and only if there is no net external torque around any axis. For the three rectangular axes,

$$\sum \tau_x = 0, \qquad \sum \tau_y = 0, \qquad \sum \tau_z = 0, \qquad (11\text{-}2a)$$

or more succinctly,

$$\sum \boldsymbol{\tau} = 0. \qquad (11\text{-}2b)$$

Equations (11-2) represent the *second equilibrium condition.*

The two equilibrium conditions are stated for a particular reference frame. The first condition involves only forces and is therefore independent of where the origin of the reference frame is placed. But the second condition involves torques which, because of the vector $\mathbf{r}$ in $\mathbf{r} \times \mathbf{F}$, do depend on the location of the reference frame. So what happens to the second condition when the reference frame is shifted?

To answer this question, let's consider the rigid body of Fig. 11-1. It is subjected to the arbitrary set of forces $\mathbf{F}_1$, $\mathbf{F}_2$, . . . , $\mathbf{F}_N$. Since the body is assumed to be in equilibrium, the sum of these forces is zero, and the sum of their torques around the point O is also zero ($\sum \boldsymbol{\tau}_{\text{O}} = 0$). What we must do is to determine the sum of the torques when they are all calculated relative to the point O′.

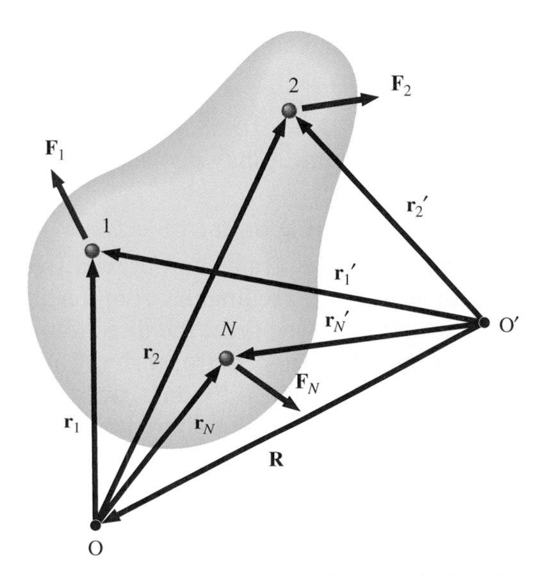

FIGURE 11-1 Three representative particles of a rigid body are shown. Relative to O, their positions are $\mathbf{r}_1$, $\mathbf{r}_2$, and $\mathbf{r}_N$. Relative to O′, their positions are $\mathbf{r}_1'$, $\mathbf{r}_2'$, and $\mathbf{r}_N'$.

The net torque about O′ is

$$\sum \boldsymbol{\tau}_{\text{O}'} = \sum_i \mathbf{r}_i' \times \mathbf{F}_i.$$

If the position of O relative to O′ is $\mathbf{R}$, then $\mathbf{r}_i' = \mathbf{r}_i + \mathbf{R}$, and

$$\sum \boldsymbol{\tau}_{\text{O}'} = \sum_i (\mathbf{r}_i + \mathbf{R}) \times \mathbf{F}_i = \sum_i \mathbf{r}_i \times \mathbf{F}_i + \sum_i \mathbf{R} \times \mathbf{F}_i$$
$$= \sum \boldsymbol{\tau}_{\text{O}} + \mathbf{R} \times \sum_i \mathbf{F}_i,$$

where $\sum_i \mathbf{r}_i \times \mathbf{F}_i = \sum \boldsymbol{\tau}_{\text{O}}$, the net torque about O, and $\sum_i \mathbf{R} \times \mathbf{F}_i = \mathbf{R} \times \sum_i \mathbf{F}_i$, because $\mathbf{R}$ is a constant factor in the sum. Finally, by assumption, $\sum \boldsymbol{\tau}_{\text{O}} = 0$ and $\sum_i \mathbf{F}_i = 0$; so,

$$\sum \boldsymbol{\tau}_{\text{O}'} = 0.$$

Thus the net torque about O′ is also zero. *When we apply the equilibrium conditions for a rigid body, we can choose any point to be the origin of our inertial reference frame.* The choice is generally made to best simplify the mathematics of the problem.

11-3 Examples Involving Systems in Equilibrium

An arbitrary rigid body has six degrees of freedom and must therefore satisfy six independent equilibrium equations (Eqs. 11-1*b* and 11-2*a*). To keep mathematical complexity at a minimum, we will consider only planar equilibrium problems. In this case, there are only three degrees of freedom and three equilibrium equations. For a rectangular reference system, they are

$$\sum F_x = 0, \qquad \sum F_y = 0, \qquad \sum \tau = 0, \tag{11-3}$$

where the net torque can be calculated around any axis perpendicular to the *xy* plane. In general, counterclockwise torques are taken to be positive and clockwise torques negative.

We will use the following approach for the analysis of planar equilibrium problems:

1. Identify the object that is to be analyzed. For some systems in equilibrium, it may be necessary to consider more than one object.
2. Identify all forces acting on the object.
3. Choose an *xy* reference frame and draw a free-body diagram indicating the rectangular components of the forces. For an unknown force, the direction of each component must be assigned arbitrarily. The correct direction is then determined by the sign of the component in the solution. A plus sign means that the assigned direction is correct; a minus sign means that the actual direction of the component is opposite to that assigned.
4. Choose a convenient axis for calculating the torques of all force components. Since this axis can be placed anywhere, you should choose it so that the torque calculations are as simple as possible.
5. Apply the two conditions of equilibrium as given by Eqs. (11-3). Equations (11-3) are simultaneous equations, which can be solved for the unknown quantities.

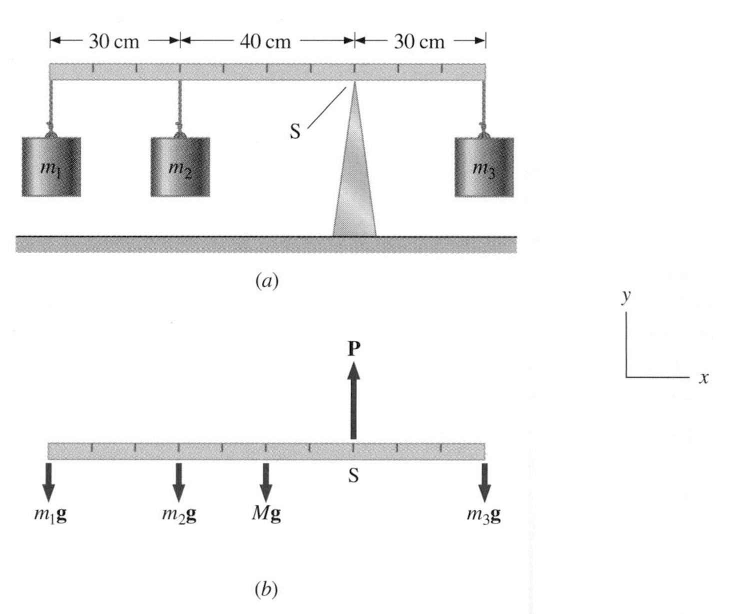

FIGURE 11-2 (*a*) Cylinders are hung from a meterstick as shown. (*b*) The free-body diagram for the stick.

EXAMPLE 11-1 Weights Hung on a Meterstick

Cylinders are tied to a uniform meterstick as shown in Fig. 11-2*a*. The masses of the two cylinders to the left of the support are $m_1 = 0.050$ kg and $m_2 = 0.075$ kg. The mass of the meterstick is 0.150 kg. Find the mass m_3 of the cylinder to the right of the support that is needed to balance the system. Also find the force of the support point on the meterstick.

SOLUTION The object to be analyzed is the meterstick. The forces on the meterstick are the tensions in the three strings holding the cylinders, the force **P** at S due to the support, and the weight $M\mathbf{g}$ of the stick. The string tensions are equal to the weights of the cylinders. As shown in the free-body diagram of Fig. 11-2*b*, all forces are vertical and are along the *y* axis of our reference system. The rotational axis is chosen to pass through S. The advantage of this axis is that the unknown force **P** has a zero moment arm and therefore is not part of the torque equilibrium equation. The first condition of equilibrium gives

$$\sum F_y = 0$$
$$P - m_1 g - m_2 g - Mg - m_3 g = 0. \tag{i}$$

With the rotational axis through S, the second condition of equilibrium gives

$$\sum \tau = 0$$
$$m_1 g(0.70 \text{ m}) + m_2 g(0.40 \text{ m}) + Mg(0.20 \text{ m}) + P(0) - m_3 g(0.30 \text{ m}) = 0. \tag{ii}$$

With our choice for the rotational axis, the moment arm of P is zero, so there is only one unknown, m_3, in Eq. (*ii*). Substituting the known masses into Eq. (*ii*), we obtain for the mass of the cylinder

$$m_3 = 0.317 \text{ kg}.$$

This value can now be substituted into Eq. (*i*) and the resulting equation solved for P. We then find that the force of the support is

$$P = 5.80 \text{ N}.$$

Notice that every term in Eq. (*ii*) is multiplied by g. As a result, the calculated mass m_3 is independent of the value of g. This meterstick balance may therefore be used to measure mass. A pan balance measures mass based on the same principle. Variations in the acceleration due to gravity at different points on the earth do not affect the measurements. This is not the case for a spring balance, since it measures the force $m\mathbf{g}$.

DRILL PROBLEM 11-1

Repeat Example 11-1 using the left end of the meterstick to calculate torques.

EXAMPLE 11-2 A Moving Object in Equilibrium

A 50-kg file cabinet is pulled across the floor as shown in Fig. 11-3. The coefficient of kinetic friction between the file cabinet and floor is $\mu_K = 0.70$. What is the magnitude of the applied force **P** if the cabinet moves at constant velocity?

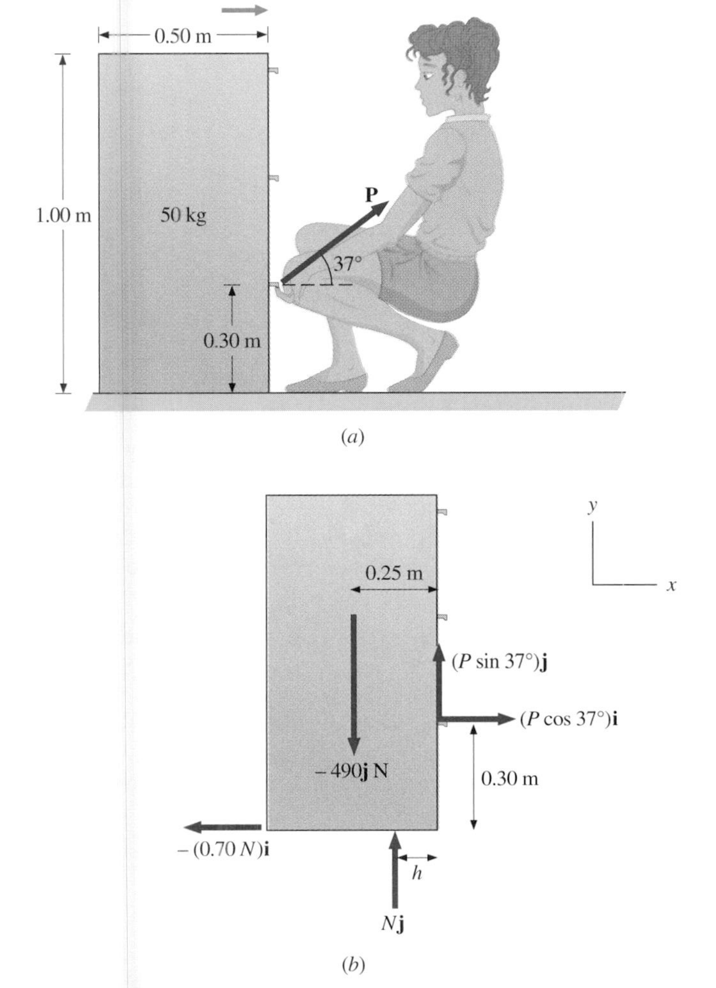

FIGURE 11-3 (*a*) A locked file cabinet is pulled across the floor at constant velocity by **P**. (*b*) The free-body diagram for the file cabinet.

SOLUTION You may recall that problems of this type were discussed in Chap. 6. However, at that point they were not identified as equilibrium problems. The body to be analyzed here is the file cabinet. The forces on the cabinet are the applied force **P**, its weight, and the normal and frictional forces of the floor. For a reference frame consisting of a horizontal *x* axis and a vertical *y* axis, the components of each force are shown in the free-body diagram of Fig. 11-3*b*. We are interested in the translational motion of the cabinet and therefore do not have to be concerned with the torques on it. With the first condition of equilibrium, we find

$$\sum F_x = 0 \qquad\qquad \sum F_y = 0$$
$$P\cos 37° - 0.70N = 0 \qquad P\sin 37° + N - 490\text{ N} = 0.$$

Eliminating N between these two equations gives

$$P\sin 37° + P\left(\frac{\cos 37°}{0.70}\right) - 490\text{ N} = 0,$$

so the magnitude of the applied force is $P = 281$ N. When this value of P is substituted into either of the original equations, we find that the normal force is $N = 321$ N.

EXAMPLE 11-3 Line of Action of a Force

Where is the line of action of the normal force **N** of the previous example? The cabinet is 50 cm wide and 100 cm tall, and **P** is applied at a point 30 cm above the floor.

SOLUTION The force **N** is assumed to act at a distance h from the lower right-hand corner of the cabinet as shown in Fig. 11-3*b*. In Example 11-2 the magnitudes of **N** and **P** were found to be 321 N and 281 N, respectively. Relative to an axis located at the lower right-hand corner of the cabinet, the second condition of equilibrium gives

$$\sum \tau = 0$$
$$-0.70N(0) - N(h) + (490\text{ N})(0.25\text{ m})$$
$$-(P\cos 37°)(0.30\text{ m}) - P\sin 37°(0) = 0.$$

With this equation and the known values for N and P, we find that $h = 0.17$ m; so the normal force **N** acts vertically upward along a line 0.17 m from the lower right-hand corner of the cabinet.

EXAMPLE 11-4 Forces in the Forearm

A weight-lifter is holding a 50-lb weight with his forearm positioned at 60° with respect to his upper arm, as shown in Fig. 11-4*a*. The forearm is supported by a contraction of the biceps muscle, which causes a torque around the elbow. What tension must the muscle exert to hold the forearm at the position shown? What is the force in the elbow joint? Assume that the forearm's weight is negligible and that the simple free-body diagram of Fig. 11-4*b* accurately represents the forces on the forearm.

SOLUTION As the free-body diagram shows, the forces on the simplified forearm are a vertical tension T_M**j** from the tendon connecting the biceps muscle to the forearm, a force F_E**j** due to contact at the elbow joint E, and a 50-lb force caused by the held weight. Notice that in our simple model the force at the elbow joint is directed vertically. This is because the forearm is in static equilibrium and the other two forces on it are also directed vertically. Around the axis through the elbow joint, the moment arm of F_E**j** is zero, the moment arm of T_M**j** is (1.5 in.) sin 60° = 1.3 in., and

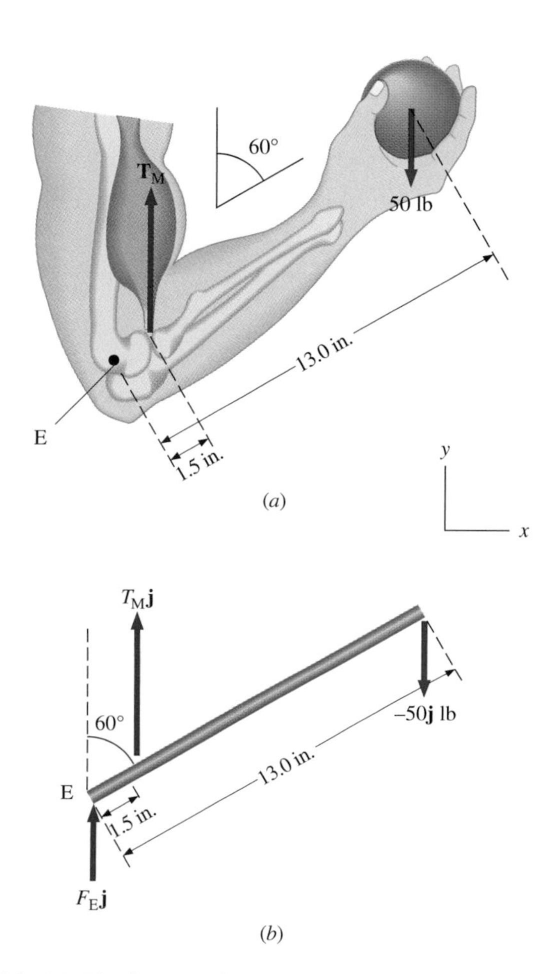

FIGURE 11-4 (*a*) The forearm is rotated around the elbow (E) by a contraction of the biceps muscle, which causes a tension $\mathbf{T}_M$. (*b*) An approximate free-body diagram for the forearm.

the moment arm of the held weight is (13.0 in.) sin 60° = 11.3 in. From the second condition of equilibrium, we have

$$\sum \tau = 0 = -(50 \text{ lb})(11.3 \text{ in.}) + T_M(1.3 \text{ in.}),$$

so the tension in the muscle is

$$T_M = 4.3 \times 10^2 \text{ lb}.$$

We can determine the force F_E at the elbow joint by summing forces in the vertical direction. This gives

$$\sum F_y = 0 = F_E + T_M - 50 \text{ lb},$$

and

$$F_E = 50 \text{ lb} - T_M = -3.8 \times 10^2 \text{ lb}.$$

Since F_E is negative, the forearm is pushed downward at the elbow joint.

We have just found that large forces in a muscle and joint are needed to support a much smaller external force. These large forces are a result of the biceps muscle's moment-arm disadvantage relative to the held weight. The arm is obviously poorly structured for holding or lifting with an extended forearm. However, its structure does allow the end of the arm (the hand) to move large distances relative to small contractions in the biceps muscle. As humans evolved, this advantage was apparently more important than the disadvantage in lifting ability.

You might consider how you rotate your torso by contracting your back muscles. In this case also, large muscle tensions with small moment arms balance much smaller forces (the weights of your torso and the objects held in your hands) with much larger moment arms. No wonder our back muscles are so susceptible to injury when we bend over to lift!

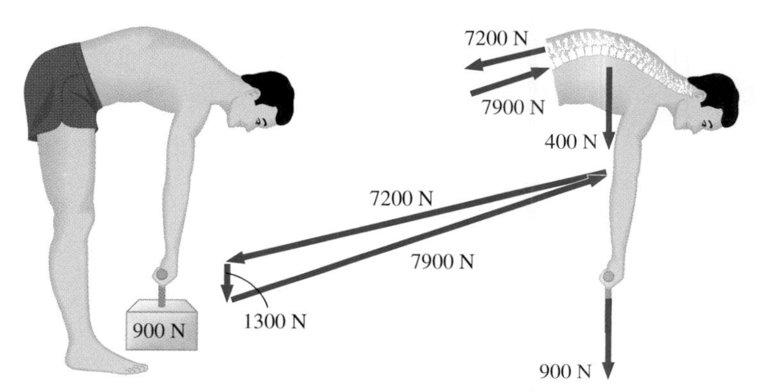

Because their moment arms are so small, the back muscles must exert large forces in order to rotate the body upright.

EXAMPLE 11-5 A Ladder Resting Against a Wall

The uniform ladder of Fig. 11-5 is 5.0 m long and weighs 400 N. The ladder rests against a slippery vertical wall that exerts a negligible frictional force on the ladder. (*a*) Find the force of the wall on the ladder and the force of the floor on the ladder. (*b*) What is the minimum coefficient of static friction needed to keep the ladder from slipping?

SOLUTION (*a*) The ladder rests against the wall and the floor so there are contact forces on it due to both surfaces. Since the wall is slippery, its contact force has just a component $-N\mathbf{i}$ normal to the wall, as shown in the free-body diagram of Fig. 11-5*b*. The contact force of the floor has a normal component $G_y\mathbf{j}$, which is vertical and a frictional component $G_x\mathbf{i}$, which is horizontal. Around

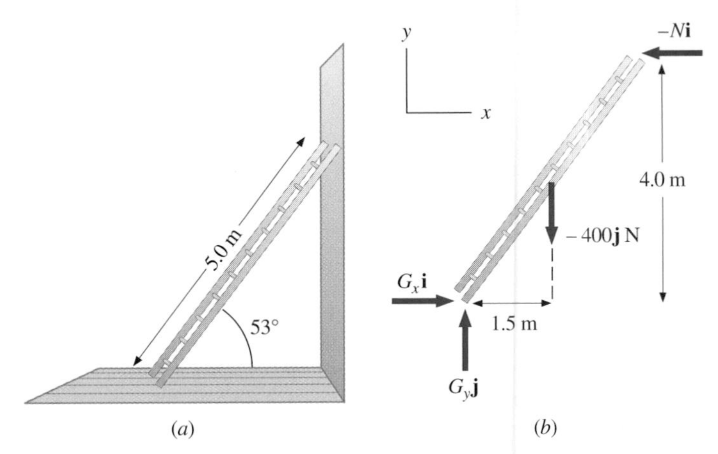

FIGURE 11-5 (*a*) A 5.0-m ladder rests against a frictionless wall. (*b*) The free-body diagram for the ladder.

the axis through the lower end of the ladder, the moment arms of G_x and G_y are zero, the moment arm of the weight of the ladder is (2.5 m) cos 53° = 1.5 m, and the moment arm of the contact force due to the wall is (5.0 m) sin 53° = 4.0 m. The conditions of equilibrium now give

$$\sum F_x = G_x - N = 0, \qquad \sum F_y = G_y - 400 \text{ N} = 0,$$

and

$$\sum \tau = -(400 \text{ N})(1.5 \text{ m}) + N(4.0 \text{ m}) = 0.$$

In solving these equations, we find that the forces on the ladder due to the floor and the wall are

$$G_x = 150 \text{ N}, \qquad G_y = 400 \text{ N}, \qquad N = 150 \text{ N}.$$

(*b*) If the ladder does not slip, the static frictional force G_x on it must be less than or equal to its maximum value $\mu_S G_y$. Thus from $G_x \leq \mu_S G_y$, we obtain

$$150 \text{ N} \leq \mu_S(400 \text{ N}),$$

and

$$\mu_S \geq 0.38.$$

EXAMPLE 11-6 Locating the Center of Mass

Describe how the position of the center of mass of a planar object can be determined.

SOLUTION (1) First we support the object at an arbitrary point O and let it hang vertically, as shown in Fig. 11-6*a*. Since $\sum \tau_O = 0$, the center of mass must lie on the vertical line AA′ that passes through the support point O. (2) Next we support the object by a second point O′ and again let it hang vertically, as shown in Fig. 11-6*b*. The center of mass must now lie on the vertical line BB′ that passes through O′. (3) The center of mass is at the intersection of AA′ and BB′.

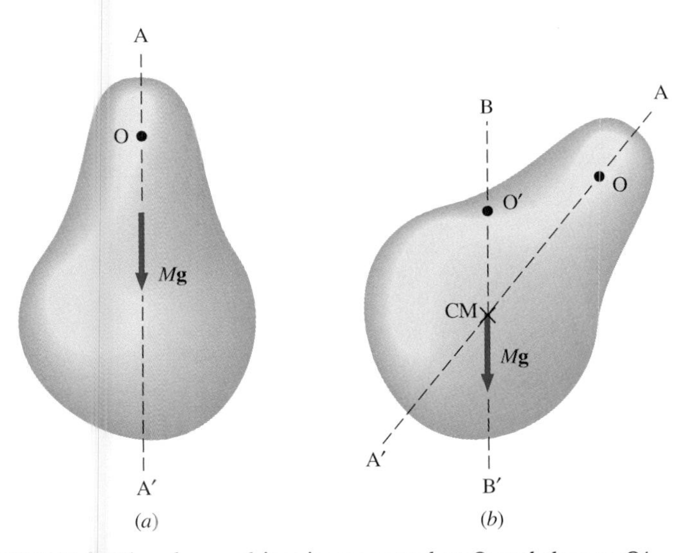

FIGURE 11-6 The planar object is supported at O and then at O′. The center of mass is at the intersection of AA′ and BB′.

EXAMPLE 11-7 Forces on a Door

A swinging door that weighs 400 N is supported by hinges at A and B. (See Fig. 11-7*a*.) Find the horizontal and vertical force components of the hinges on the door.

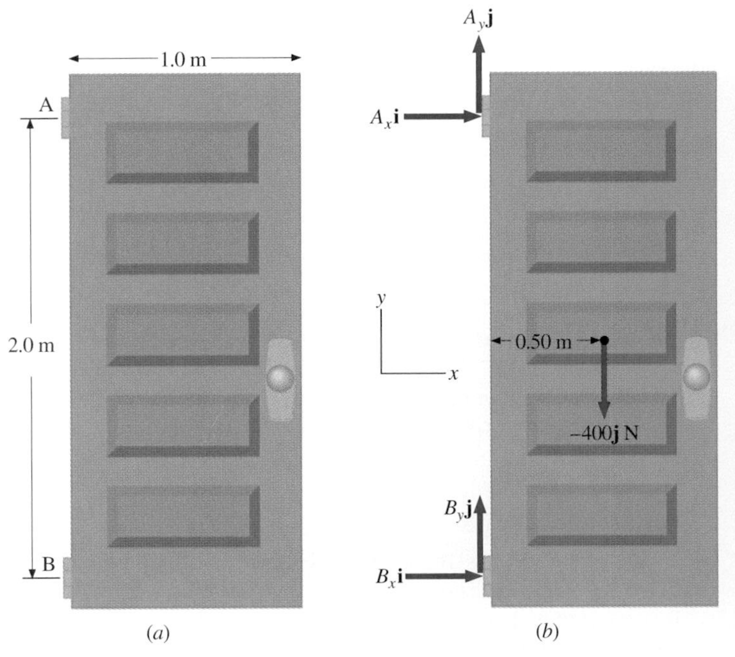

FIGURE 11-7 (*a*) A 400-N swinging door is supported by hinges at A and B. (*b*) The free-body diagram for the door.

SOLUTION The forces on the door are its weight and the contact forces at both hinges. A free-body diagram for the door is shown in Fig. 11-7*b*. There are four unknown force components, A_x, A_y, B_x, and B_y. The equilibrium conditions give only three equations that can be used to find the four unknowns. With torques calculated around the axis through A, these equations are

$$\sum F_x = A_x + B_x = 0, \qquad \sum F_y = A_y + B_y - 400 \text{ N} = 0,$$

and

$$\sum \tau = B_x(2.0 \text{ m}) - (400 \text{ N})(0.50 \text{ m}) = 0.$$

Solving these equations, we find that the horizontal components of the forces at the hinges are $B_x = 100$ N and $A_x = -100$ N. However, all we are able to determine for the vertical components A_y and B_y is that their sum is

$$A_y + B_y = 400 \text{ N}.$$

With only three equilibrium equations, there is insufficient information to determine all four of the unknowns.

The door of this example is known as a *statically indeterminate system*. It is given this name because the equilibrium conditions do not furnish sufficient information for determining all of the unknowns. Obviously, there are unique values for A_y and B_y. The fact that this problem is statically indeterminate does not mean that these two forces cannot be found. It just means that information outside the realm of equilibrium mechanics must be used in order to calculate them. Such information may be obtained from knowledge of the elastic properties of the structure.

DRILL PROBLEM 11-2

A 50-kg woman stands 1.5 m from one end of a uniform 6.0-m scaffold of mass 70 kg. (See Fig.11-8.) What are the tensions in the vertical ropes supporting the scaffold?
ANS. 710 N, 466 N.

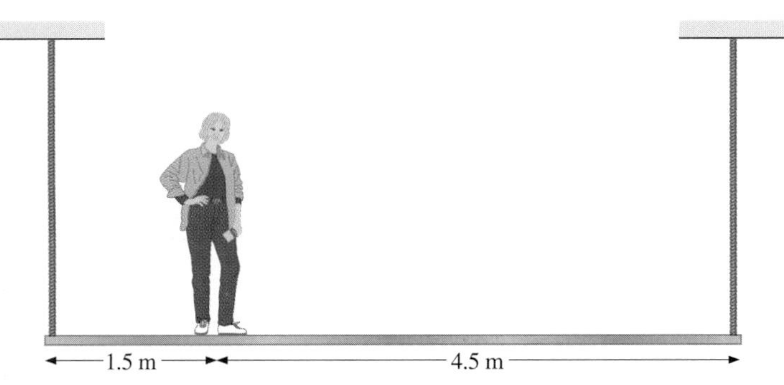

FIGURE 11-8 A 50-kg woman stands 1.5 m from the left end of a 6.0-m scaffold supported by two vertical ropes.

DRILL PROBLEM 11-3

A 400-N sign hangs from the end of the uniform strut of Fig. 11-9. The strut is 4.0 m long and weighs 600 N. The strut is supported by a hinge at the wall and by a cable whose other end is tied to the wall at a point 3.0 m above the left end of the strut. Find the tension in the supporting cable and the force of the hinge on the strut.
ANS. 1.17×10^3 N, 980 N directed upward at 18° above the horizontal.

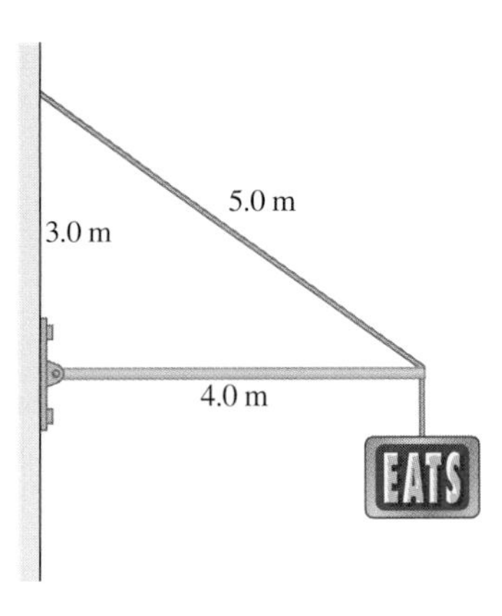

FIGURE 11-9 A 400-N sign hangs at the end of a uniform strut that weighs 600 N. The strut is supported by the cable and by the hinge at the wall.

DRILL PROBLEM 11-4

Suppose the wall exerts a frictional force on the ladder of Example 11-5. Is the system statically determinant? (*Hint:* How many unknown forces are there?)

*11-4 STABILITY

Both cones of Fig. 11-10 are in equilibrium. However, their situations are obviously quite different, for if given a slight push, cone A will return to its upright position while cone B will topple over. The first cone is said to be in *stable equilibrium*, because it moves back toward its equilibrium position after any small displacement. The second cone is in *unstable equilibrium*, since a small displacement causes it to move away from its original equilibrium position. Why these two cones behave differently can be understood with the help of Fig. 11-10. For the stable case, the torque of $M\mathbf{g}$ around the support point rotates the cone back toward its original equilibrium position. In contrast, this same torque causes the unstable cone to topple.

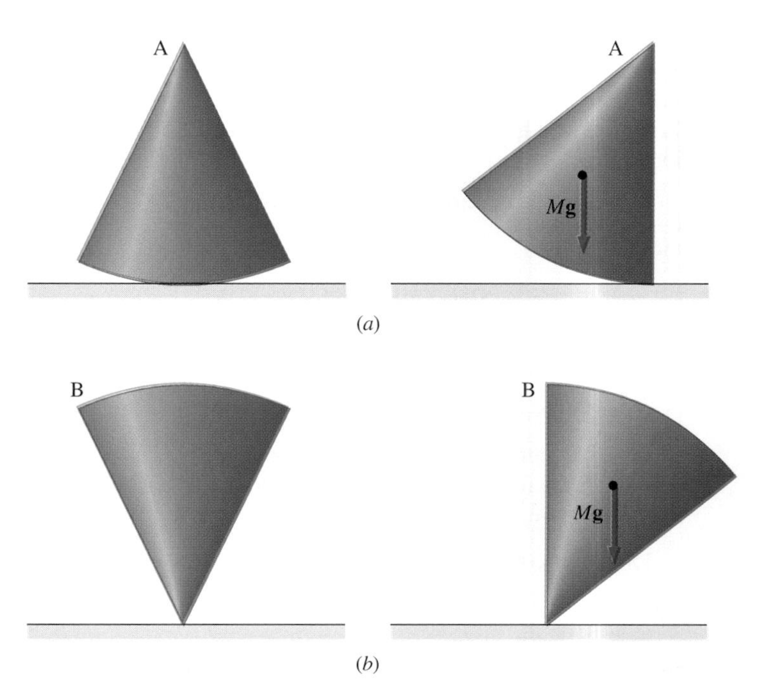

FIGURE 11-10 (*a*) Cone A is stable because the torque of $M\mathbf{g}$ rotates it upright. (*b*) Cone B is unstable because the torque of $M\mathbf{g}$ causes it to topple over, even for small displacements from equilibrium.

Unstable equilibrium.

A rigid body may be stable under small displacements from equilibrium but unstable for large displacements. For example, if the refrigerator of Fig. 11-11*a* is displaced slightly from equilibrium, it remains upright; but if it is pushed far enough, as shown in Fig. 11-11*b*, it topples. The refrigerator is stable only if the displacements do not cause the line of action of $M\mathbf{g}$ to move outside its base.

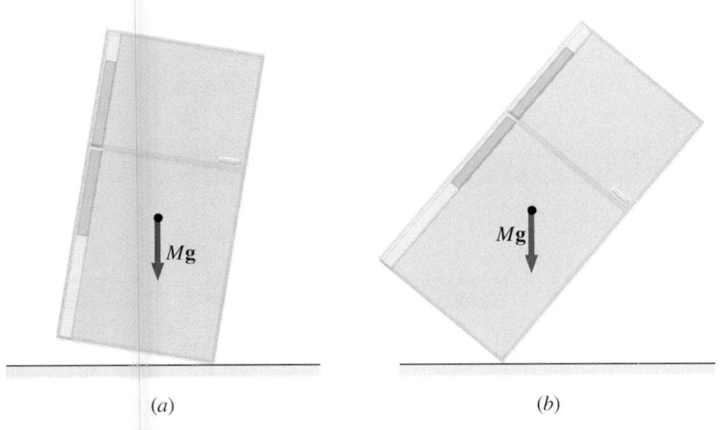

FIGURE 11-11 (*a*) The refrigerator is stable for small displacements; (*b*) it is not stable for large displacements.

A rigid body is stable under very large displacements from equilibrium if its center of mass is below its point of support. You have probably seen children's toys, such as the bird of Fig. 11-12*a*, that right themselves even when tipped quite far. The reason for this behavior can be understood with the help of Fig. 11-12*b*. As long as the center of mass is below the point of support, the torque of $M\mathbf{g}$ acts to rotate the bird upright.

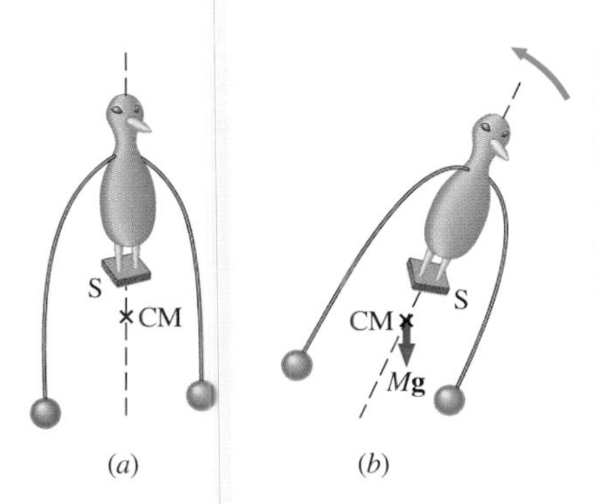

FIGURE 11-12 (*a*) A toy bird is supported at S. (*b*) Since the bird's center of mass is below S, the bird rotates back to an upright position even for large displacements from equilibrium.

Stability for small displacements can be described mathematically. Here we consider a mathematical description for the simple case of a particle constrained to move in one dimension (say, along the x axis). If it is in equilibrium at the origin under a net force $F(x)$, then from the first condition of equilibrium, $F(0) = 0$. In addition, if the particle is stable under small displacements, the force $F(x)$ must push it back toward the origin. The net force must therefore push to the left if x is positive and to the right if x is negative; that is, $F(x) < 0$ if $x > 0$, and $F(x) > 0$ if $x < 0$. (See Fig. 11-13.) In terms of the derivative of $F(x)$,

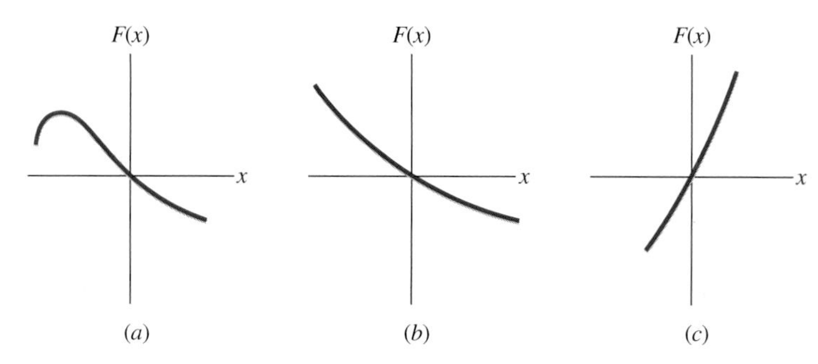

FIGURE 11-13 (*a*) and (*b*) Stable equilibrium. Since $[dF(x)/dx]_{x=0} < 0$, $F(x)$ is directed toward $x = 0$ for all small displacements. (*c*) Unstable equilibrium.

$$\left.\frac{dF(x)}{dx}\right|_{x=0} < 0 \qquad (11\text{-}4a)$$

for stable equilibrium at $x = 0$. Of course, the equilibrium position can be at x_0 rather than at the origin. Then $F(x_0) = 0$ and

$$\left.\frac{dF(x)}{dx}\right|_{x=x_0} < 0. \qquad (11\text{-}4b)$$

The condition for stability can also be expressed in terms of the potential energy corresponding to a force. Since

$$F(x) = -\frac{dU(x)}{dx},$$

x_0 is a position of stable equilibrium if

$$\left.\frac{dU(x)}{dx}\right|_{x=x_0} = 0 \quad \text{and} \quad \left.\frac{d^2U(x)}{dx^2}\right|_{x=x_0} > 0. \qquad (11\text{-}5)$$

DRILL PROBLEM 11-5

There is also a position of neutral equilibrium for which any small displacement produces no motion either toward or away from the equilibrium position. Can you think of an example? (*Hint:* The rigid body must not move in its new position.)

DRILL PROBLEM 11-6

Which of the following forces result in stable equilibrium at $x = 0$? (*a*) $F(x) = -4x$; (*b*) $F(x) = 4x^2$; (*c*) $F(x) = -4x + 7$; (*d*) $F(x) = -\sin x$.
ANS. (*a*), (*c*), and (*d*).

DRILL PROBLEM 11-7

(*a*) Sketch a potential-energy function for the region surrounding a point of stable equilibrium. (*b*) Which of the following potential energies correspond to stable equilibrium at $x = 0$? (*i*) $U(x) = U_0 \cos x$; (*ii*) $U(x) = U_0 \sin x$; (*iii*) $U(x) = kx^2/2$; (*iv*) $U(x) = U_0 \cos^2 x$; (*v*) $U(x) = U_0 \sin^2 x$.
ANS. (*b*): (*iii*) and (*v*).

*11-5 Stress and Strain in Solids

We have been treating objects as if they were perfectly rigid. Of course, this isn't exactly correct, for even a steel beam distorts slightly under an applied force—and this distortion can be an important consideration in the design of the structure that contains the beam. We will now investigate a few of the simple ideas concerning the expansion and compression of materials. We'll look at this subject from the point of view of the civil engineer who is interested in practical models that allow him or her to accurately estimate how much beams elongate, what forces objects can withstand without breaking, and other such factors. We will not attempt to relate the mechanical properties of a material to the behavior of its constituent atoms.

The body of Fig. 11-14*a* is in static equilibrium under an arbitrary set of external forces. In Fig. 11-14*b*, we see the same body with an imaginary sectional cut at CC′. Since each of the two individual parts of the body is also in static equilibrium, both internal forces and internal torques must exist at the cross section. Those on the right portion are due to the left portion, and vice versa. On the left portion, the normal and tangential components of the internal force are $\mathbf{F}_n$ and $\mathbf{F}_t$, respectively, and the net internal torque is $\boldsymbol{\tau}$. From Newton's third law, the right portion is subjected at this same cross section to force components $-\mathbf{F}_n$ and $-\mathbf{F}_t$ and the torque $-\boldsymbol{\tau}$.

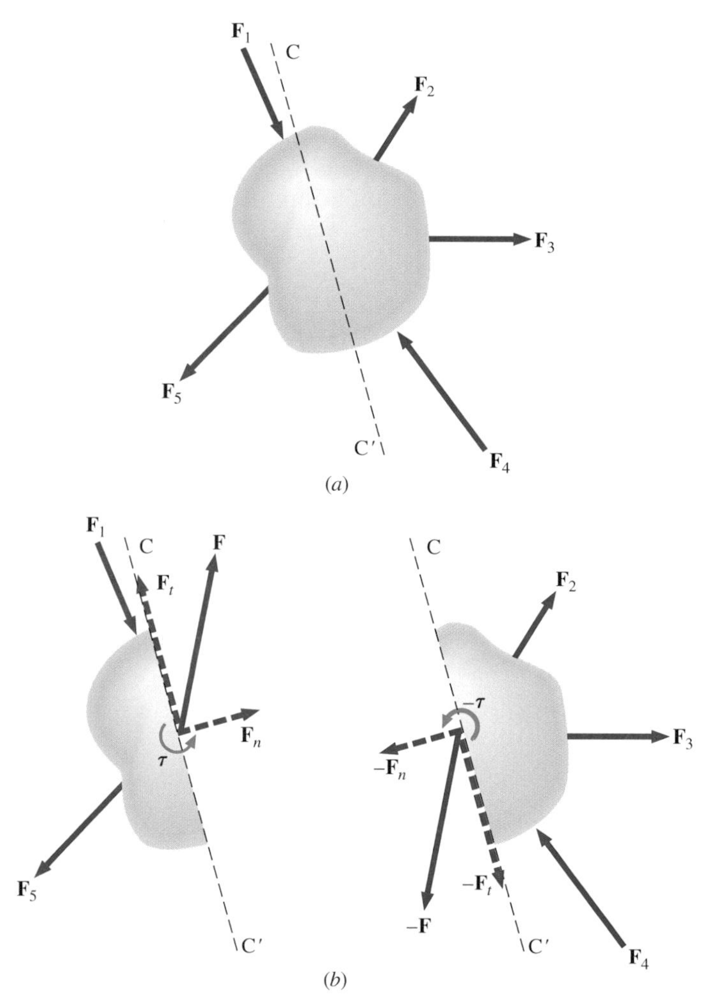

FIGURE 11-14 (*a*) A body in equilibrium under an arbitrary set of forces. (*b*) At the cross section CC′, the net force on the left piece is **F** and the net force on the right piece is −**F**. The net torques on the pieces are also equal and opposite.

We will only consider the special case for which $\mathbf{F}_t$ and $\boldsymbol{\tau}$ are negligible. This can occur when large forces are applied at the ends of a light rod. (See Fig. 11-15.) With this assumption, we need be concerned only with the normal force component $\mathbf{F}_n$ at a cross section. Recall from Sec. 6-1 that if $\mathbf{F}_n$ acts in a direction such that the two pieces pull on each other, it is known as a tensile force. If $\mathbf{F}_n$ is directed such that the pieces are pushing each other, it is a compressive force.

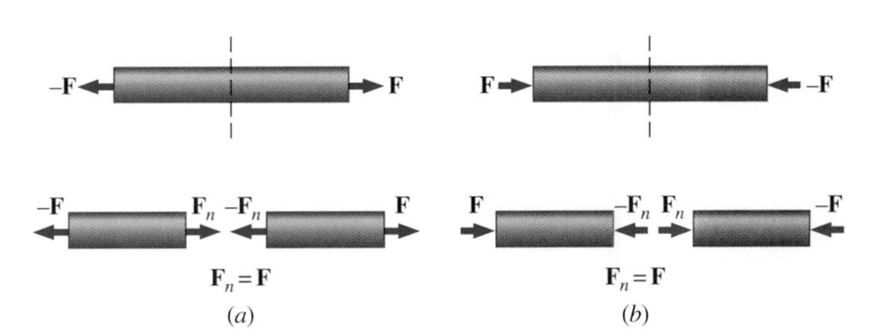

FIGURE 11-15 (*a*) A rod under tension. (*b*) A rod under compression. In each case, the force at the cross section is assumed to have only a normal component.

Associated with the normal force component at the cross section is the **normal stress** S_n, which by definition is

$$S_n = \frac{F_n}{A}, \tag{11-6}$$

where A is the cross-sectional area of the body at CC′. Since stress is a force per unit area, its units are given by newton per square meter (N/m^2), dyne per square centimeter (dyne/cm^2), or pound per square foot (lb/ft^2). If $\mathbf{F}_n$ is a tensile force (Fig. 11-15*a*), the stress is known as **tensile stress**; if $\mathbf{F}_n$ is a compressive force (Fig. 11-15*b*), the stress is **compressive stress**.

EXAMPLE 11-8 Stress in a Pillar

A stone gargoyle of weight 1.0×10^4 N is supported by the uniform pillar of Fig. 11-16*a*. The pillar's cross-sectional area is 0.20 m^2,

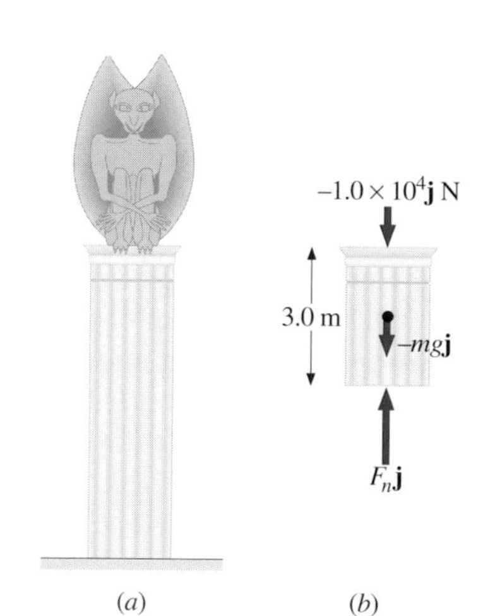

FIGURE 11-16 (*a*) A uniform pillar supporting a gargoyle. (*b*) The free-body diagram of the top 3.0-m section.

and it is made of granite of density $\rho = 2.7 \times 10^3$ kg/m^3. Find the normal stress at the cross section 3.0 m below the top of the pillar.

SOLUTION The volume of the 3.0-m piece above the cross section is

$$V = (0.20 \text{ m}^2)(3.0 \text{ m}) = 0.60 \text{ m}^3,$$

so the mass of the piece is

$$m = \rho V = (2.7 \times 10^3 \text{ kg/m}^3)(0.60 \text{ m}^3) = 1.6 \times 10^3 \text{ kg}.$$

From the free-body diagram of Fig. 11-16*b*, we find that at equilibrium,

$$F_n = 1.0 \times 10^4 \text{ N} + mg$$
$$= 1.0 \times 10^4 \text{ N} + (1.6 \times 10^3 \text{ kg})(9.8 \text{ m/s}^2) = 2.6 \times 10^4 \text{ N}.$$

The normal stress, which in this case is compressive, is then

$$S_n = \frac{F_n}{A} = \frac{2.6 \times 10^4 \text{ N}}{0.20 \text{ m}^2} = 1.3 \times 10^5 \text{ N/m}^2.$$

DRILL PROBLEM 11-8

Replace the 1.0×10^4-N force on the pillar of Example 11-8 by a downward force of magnitude 6.0×10^3 N. What is the normal stress at the horizontal cross section 2.0 m below the top surface? 3.0 m below the top surface?
ANS. 8.3×10^4 N/m^2; 1.1×10^5 N/m^2.

Steel rods under tension are often embedded in concrete structures. This enhances the ability of the concrete to support loads.

Associated with each type of stress is a corresponding strain. The **normal strain** ϵ_n is a measure of the fractional change in the length of a body due to a normal stress. When a force **F** is applied to the end of the rod, as shown in Fig. 11-17, its length changes from l to $l + \Delta l$ (or $l - \Delta l$ for $-$**F**), and the normal strain is

$$\epsilon_n = \frac{\Delta l}{l}. \tag{11-7}$$

If the rod is under tension, the normal strain is called a **tensile strain**, and if the rod is under compression, it is called a **compressive strain**. In either case, ϵ_n is a positive quantity since the *magnitude* of Δl is always used in Eq. (11-7). Finally, ϵ_n is dimensionless as it is the ratio of two length measurements.

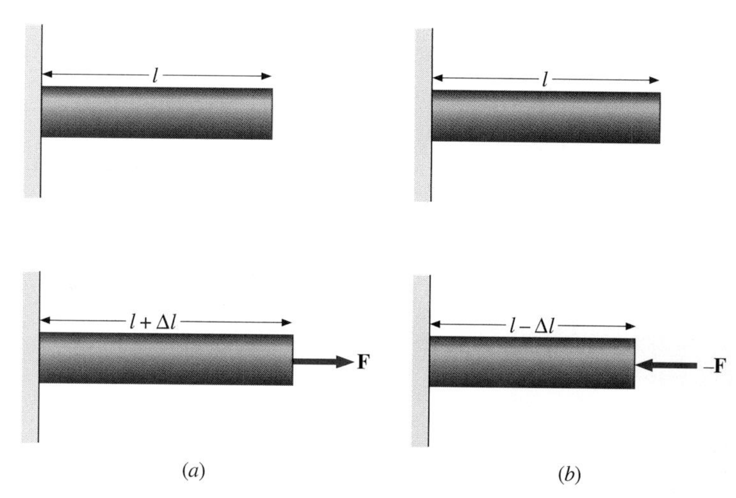

FIGURE 11-17 (*a*) Tensile stress and (*b*) compressive stress on a rod cause changes in its length.

DRILL PROBLEM 11-9

A wire 2.0 m long stretches 1.0 mm when subjected to a load. What is the normal strain of the wire?
ANS. 5.0×10^{-4}.

*11-6 Elasticity of Solids

A plot of normal stress (either tensile or compressive) vs. normal strain for a typical solid is shown in Fig. 11-18. The strain is directly proportional to the applied stress for values of stress up to S_P. In this linear region, the material returns to its original size when the stress is removed. Point P is known as the *proportional limit* of the solid. For stresses between S_P and S_E, where point E is called the *elastic limit*, the material also relaxes

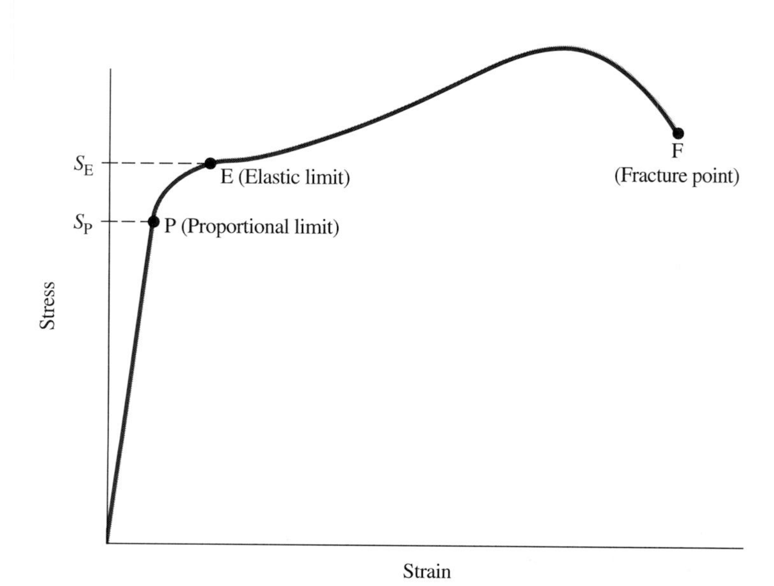

FIGURE 11-18 The stress-strain curve for a typical solid.

to its original size. However, notice that stress and strain are not proportional in this region. For deformations beyond the elastic limit, the material does not return to its original size when the stress is removed—it is permanently distorted. Finally, further stretching beyond the elastic limit leads to the eventual fracture of the solid.

A batter determining the fracture point of a baseball bat.

The proportionality constant for the linear region of the stress-vs.-strain curve is called *Young's modulus*, in honor of Thomas Young (1773–1829), an English physician and experimental physicist. With Young's modulus represented by Y_n, we have

$$S_n = Y_n \epsilon_n. \tag{11-8}$$

Since ϵ_n is dimensionless, Y_n has the same units as S_n. Values of Young's modulus for a number of common materials are listed in Table 11-1. If Y_n is different for compression and tension, both values are listed.

TABLE 11-1 TYPICAL VALUES OF YOUNG'S MODULUS FOR VARIOUS MATERIALS

Material	Young's Modulus ($\times 10^{11}$ N/m^2)
Aluminum	0.70
Brass	0.91
Copper	1.1
Human femur	
Tensile	0.16
Compressive	0.094
Iron	1.9
Lead	0.16
Nickel	2.1
Quartz	0.70
Steel	2.0
Tungsten	3.6

EXAMPLE 11-9 STRESS IN A THIN WIRE

A 4.0-kg painting is supported as shown in Fig. 11-19*a* by a thin aluminum wire of radius 0.25 mm and negligible mass. (*a*) Find the stress in the wire at C and at D. (*b*) By how much does a 50-cm section of the wire stretch?

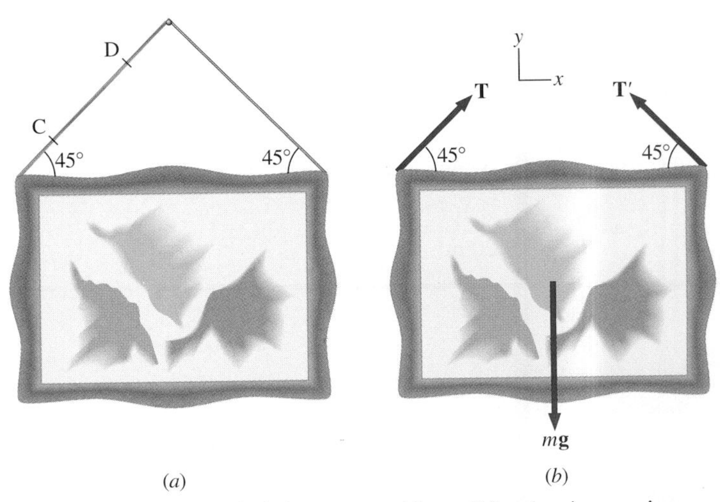

FIGURE 11-19 (*a*) A painting supported by a thin aluminum wire. (*b*) The forces on the painting.

SOLUTION (*a*) To find the stress in the wire, we need to first determine the tension in the wire. This may be done by considering the forces on the painting. As shown in Fig. 11-19*b*, these forces are its weight $m\mathbf{g}$ along with the tensions $\mathbf{T}$ and $\mathbf{T}'$ due to the sections of the wire connected at the corners of the painting. Application of the first equilibrium condition yields

$$\sum F_x = T\cos 45° - T'\cos 45° = 0$$
$$\sum F_y = T\sin 45° + T'\sin 45° - mg = 0,$$

from which we find that the tensions are

$$T = T' = \frac{mg}{2\sin 45°} = 28 \text{ N}.$$

Since the mass of the wire is negligible, the tension in it is the same throughout its length. Thus the stress at C and the stress at D are both

$$S_n = \frac{T}{A} = \frac{28 \text{ N}}{\pi(2.5 \times 10^{-4} \text{ m})^2} = 1.4 \times 10^8 \text{ N/m}^2.$$

(*b*) Combining Eqs. (11-7) and (11-8) and using the value of Young's modulus for aluminum from Table 11-1, we find that a 50-cm section of wire is stretched by an amount

$$\Delta l = l\frac{S_n}{Y_n} = (50 \text{ cm})\left(\frac{1.4 \times 10^8 \text{ N/m}^2}{7.0 \times 10^{10} \text{ N/m}^2}\right) = 0.10 \text{ cm}.$$

DRILL PROBLEM 11-10

Describe a method you could use to measure Young's modulus.

DRILL PROBLEM 11-11

A solid rod can be treated as a stiff spring for small longitudinal vibrations. Use Eq. (11-8) to determine the force constant of the rod.
ANS. $k = Y_n A/l$.

DRILL PROBLEM 11-12

A 100-kg block hangs at the end of a thin copper wire of cross-sectional area 2.0×10^{-3} cm^2 and original length 20 cm. How much is the wire stretched by the weight of the block? Ignore the weight of the wire.
ANS. 8.9 mm.

*11-7 Volume Compression

When a substance is compressed, its volume will decrease. The property of a material that is a measure of its "squeezability" is known as the **bulk modulus**. As an example, consider the solid cube of Fig. 11-20, which is subjected to compressive forces of magnitude F applied uniformly over every surface. The volume of the cube is V, and the area of each of its faces is A. Suppose that when the magnitude of the force changes by ΔF, the volume of the cube decreases from V to $V - \Delta V$. The bulk modulus B of its material is then, by definition,*

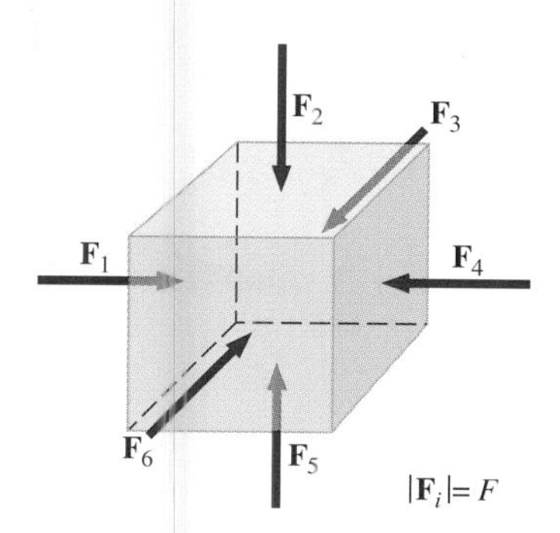

FIGURE 11-20 Compressive forces of magnitude F are applied uniformly over every surface of the cube.

*The ratio $\Delta F/A$ is the change in pressure Δp (see Chap. 15) on the surface. The bulk modulus is often written as $B = -\Delta p/(\Delta V/V)$.

$$B = -\frac{\Delta F/A}{\Delta V/V}. \tag{11-9}$$

Since an increase in the force causes a decrease in the volume, the minus sign in this equation makes B a positive quantity.

For solids and liquids, $(\Delta F/A\Delta V)$ varies only slightly; in addition, the change in volume is so small that V is almost constant. Hence B is essentially independent of the applied force for these two phases. Values of the bulk modulus for some selected solids and liquids are given in Table 11-2. When a gas is compressed, ΔV is often quite large and $(\Delta F/A\Delta V)$ varies considerably, and in this case, B varies with the applied force.

TABLE 11-2 **The Bulk Moduli of Selected Liquids and Solids**

Material	Bulk Modulus ($\times 10^{11}$ N/m^2)
Liquids	
Ethyl alcohol	0.00909
Mercury	0.270
Water	0.0218
Solids	
Aluminum	0.70
Brass	0.61
Copper	1.4
Iron	1.0
Lead	0.077
Nickel	2.6
Steel	1.6
Tungsten	2.0

DRILL PROBLEM 11-13

If the force on each face of a cubical 1.0-m^3 piece of steel changes by 1.0×10^7 N, what is its resulting change in volume?
ANS. 6.3×10^{-5} m^3.

SUMMARY

1. **Equilibrium**
   (*a*) A rigid body whose linear and angular accelerations are both zero relative to an inertial reference system is said to be in equilibrium.
   (*b*) The first condition of equilibrium: If a body is in equilibrium, the sum of the forces acting on it must be zero.
   (*c*) The second condition of equilibrium: If a body is in equilibrium, the sum of the torques acting on it must be zero.

*2. **Stability**
   (*a*) When a body displaced slightly from its equilibrium position returns to that position, it is in stable equilibrium.
   (*b*) For the one-dimensional case, if a particle under a net force $F(x)$ is in stable equilibrium at $x = x_0$, then

$$F(x_0) = 0 \quad \text{and} \quad \left.\frac{dF(x)}{dx}\right|_{x=x_0} < 0.$$

*3. **Stress and strain in solids**

(*a*) If the normal force at a cross section of a rigid body is F_n, then the normal stress at that cross section is

$$S_n = \frac{F_n}{A},$$

where A is the area of the cross section.

(*b*) If a normal stress causes a rod of length l to stretch by Δl, the normal strain of the rod is

$$\epsilon_n = \frac{\Delta l}{l}.$$

(*c*) If a rod made of a certain material is not stretched beyond the proportional limit, the normal stress and strain on it are related by

$$S_n = Y_n \epsilon_n,$$

where Y_n is Young's modulus for the material.

*4. **Volume compression**

If the volume of a body made of a certain material changes from V to $V - \Delta V$ when it is subjected to a force increase per unit area of $\Delta F/A$ applied uniformly over the surface of the body, the bulk modulus of the material is

$$B = -\frac{\Delta F/A}{\Delta V/V}.$$

QUESTIONS

Answer "true" or "false" to Ques. 11-1 through 11-4. Explain your reasoning in each case.

11-1. If there is only one external force (or torque) acting on an object, it cannot be in equilibrium.

11-2. If an object is in equilibrium, there must be an even number of forces acting on it.

11-3. Several external forces are acting on an object. If there is an odd number of forces, then the object cannot be in equilibrium.

11-4. A body moving in a circle at constant speed is in equilibrium.

11-5. A thin wire strung between two nails in the wall is used to support a large picture. Is the wire more likely to snap if it is strung tightly or if it is strung so that it sags considerably?

11-6. Is it possible to rest a ladder against a rough wall if the floor is frictionless?

11-7. For which position shown in the accompanying figure is it easiest to do push-ups: (*a*) with hands placed together, (*b*) with hands at shoulder width, or (*c*) with hands far apart?

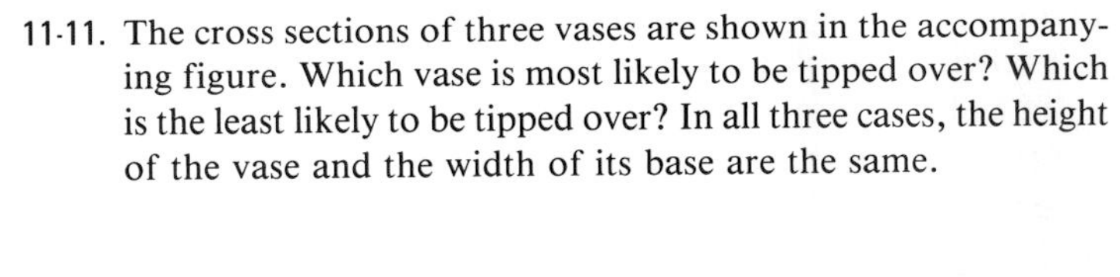

QUESTION 11-7

11-8. Show how a spring scale and a simple fulcrum can be used to weigh an object whose weight is larger than the maximum reading on the scale.

11-9. What other purpose is served by the long, flexible pole carried by high-wire walkers? (See Ques. 9-20.)

11-10. A painter climbs a ladder. Is the ladder more likely to slip when the painter is near the bottom or near the top?

11-11. The cross sections of three vases are shown in the accompanying figure. Which vase is most likely to be tipped over? Which is the least likely to be tipped over? In all three cases, the height of the vase and the width of its base are the same.

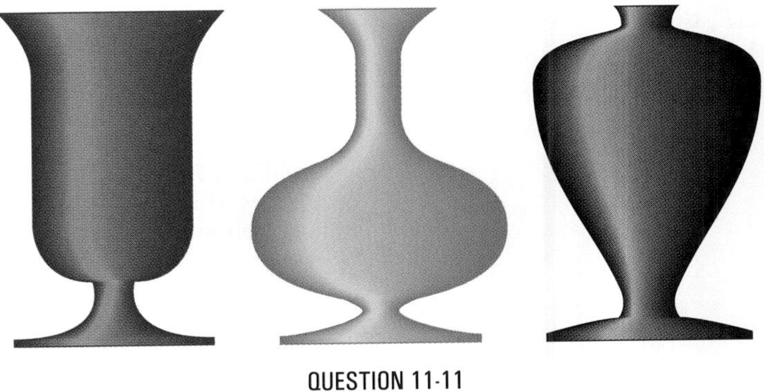

QUESTION 11-11

11-12. Review the relationship between stress and strain. Can you find any similarities between the two quantities?

11-13. What type of stress are you applying when you press on the ends of a wooden rod? When you pull on its ends?

11-14. Can compressive stress be applied to a rubber band?

11-15. At which location in the uniform pillar shown in Fig. 11-16 is the stress greatest?

11-16. What is meant when fishing line is designated as "10-lb test"?

11-17. Can Young's modulus have a negative value? What about the bulk modulus?

11-18. If a hypothetical substance has a negative bulk modulus, what happens when you squeeze a piece of it?

11-19. Steel rods are commonly placed in concrete before it sets. What is the purpose of these rods?

11-20. Discuss how you might measure the bulk modulus of a liquid.

PROBLEMS

Systems in Equilibrium

11-1. The uniform seesaw shown in the accompanying figure is balanced at its center of mass. The smaller boy on the right has a mass of 40 kg. What is the mass of his friend?

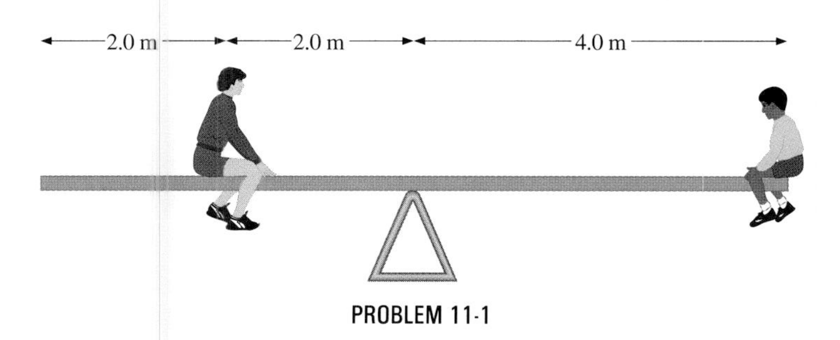

PROBLEM 11-1

11-2. The seesaw of Prob. 11-1 is moved so that the boys can sit at opposite ends of the balanced board as shown. What is the mass of the board?

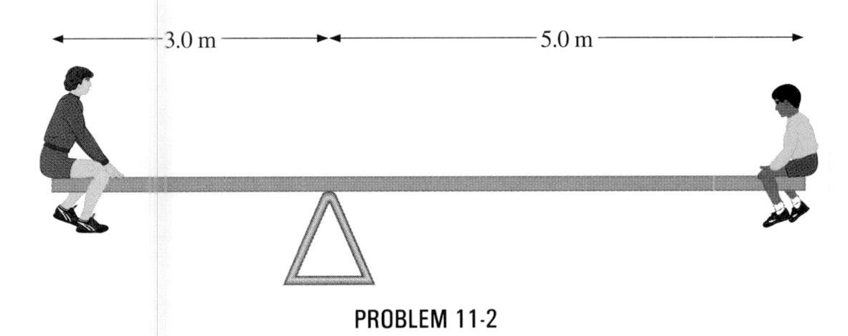

PROBLEM 11-2

11-3. A uniform plank rests on a flat surface as shown in the accompanying figure. The plank has a mass of 30 kg and is 6.0 m long. How much mass can be placed at its right end before it tips?

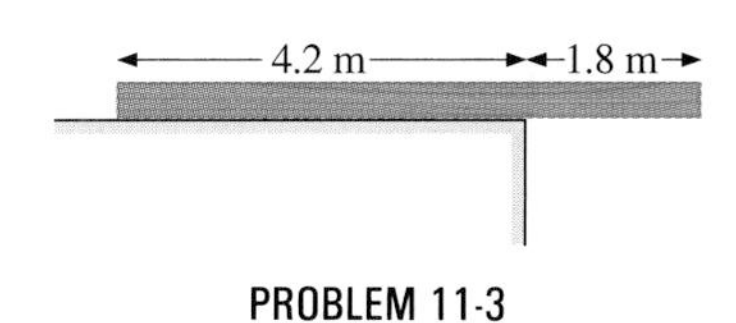

PROBLEM 11-3

11-4. In order to get his car out of the mud, a man ties one end of a rope to the front bumper and the other end to a tree 15 m away. He then pulls on the center of the rope with a force of 400 N, which causes its center to be displaced 0.30 m as shown in the figure. What is the force of the rope on the car?

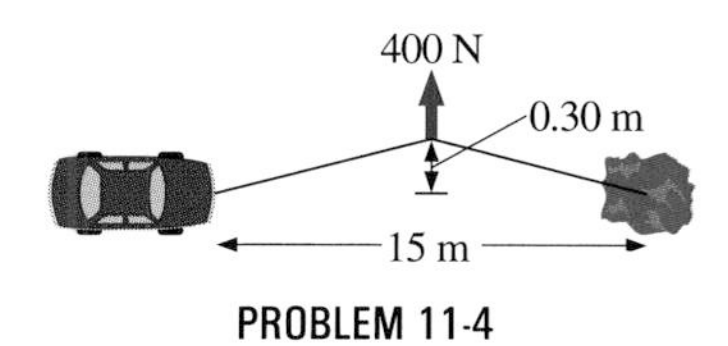

PROBLEM 11-4

11-5. A small pan of mass 22 g is supported, as shown in the accompanying figure, by two pieces of thread of lengths 5.0 and 10.0 cm. The maximum tension that the thread can support is 2.0×10^4 dyne. Mass is added slowly to the pan until one of the threads snaps. Which one is it? How much mass must be added for this to occur?

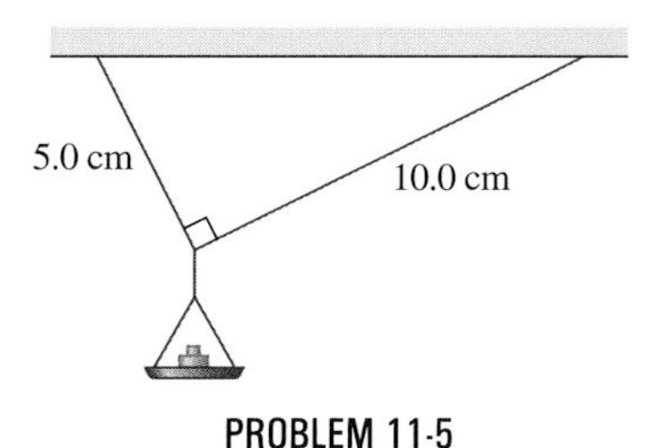

PROBLEM 11-5

11-6. A uniform scaffold whose mass and length are 40 kg and 6.0 m, respectively, is supported by two light cables, as shown in the accompanying figure. An 80-kg painter stands 1.0 m from the left end of the scaffold, and his equipment is 1.5 m from the right end. If the tension in the left cable is twice that in the right cable, what are the tensions in the cables and the mass of the equipment?

PROBLEM 11-6

11-7. The ends of a long, uniform chain are attached to objects of mass m_1 and m_2 by short strings, which are placed over two frictionless, massless pulleys, as shown in the accompanying figure. The angles that the ends of the chain make with respect to the vertical and horizontal are θ_1 and θ_2. If $m_1 = 0.10$ kg, $\theta_1 = 45°$, and $\theta_2 = 5.0°$, what are m_2 and the mass of the chain?

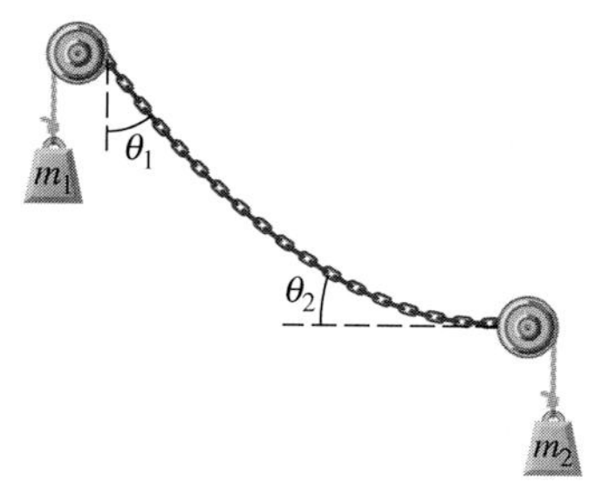

PROBLEM 11-7

11-8. What force **F** is needed to lift the 100-N weights upward at constant velocity for each of the pulley systems shown in the accompanying figure? Ignore the weights of the pulleys and ropes.

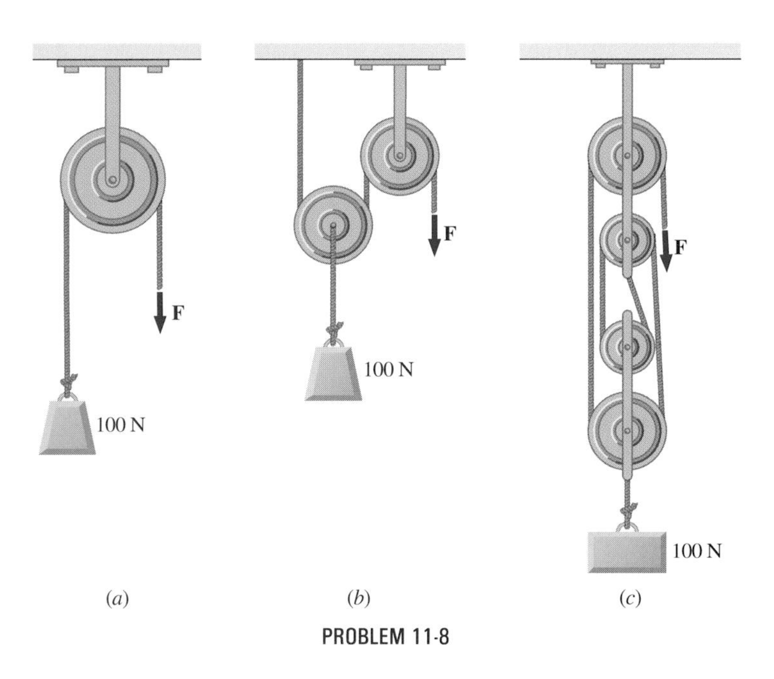

PROBLEM 11-8

11-9. The accompanying figure shows a pulley system being used to lift a 2000-kg safe up the incline. What minimum force must the men apply to move the safe up the incline? Assume that friction on the incline is negligible.

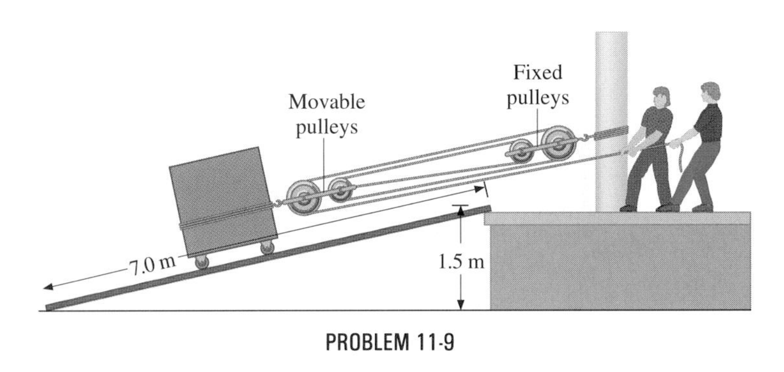

PROBLEM 11-9

11-10. A device used to lift large loads is shown in the accompanying figure. What force **F** is needed to lift the 600-kg load? Ignore the weights of the pulleys and rope.

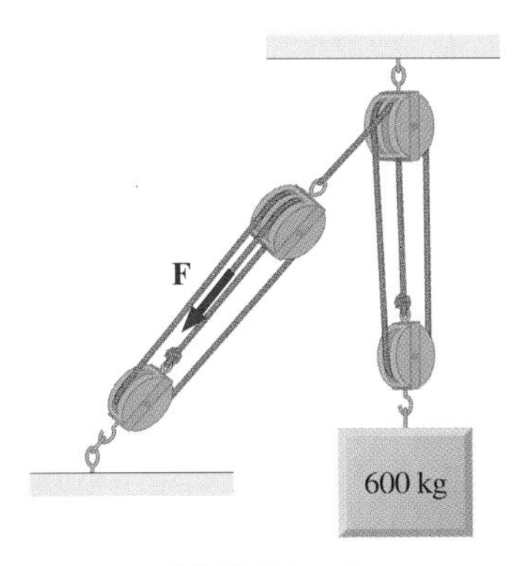

PROBLEM 11-10

11-11. What force **F** is needed to lift the 100-N weight shown in the accompanying figure at constant velocity? Ignore the weights of the pulleys.

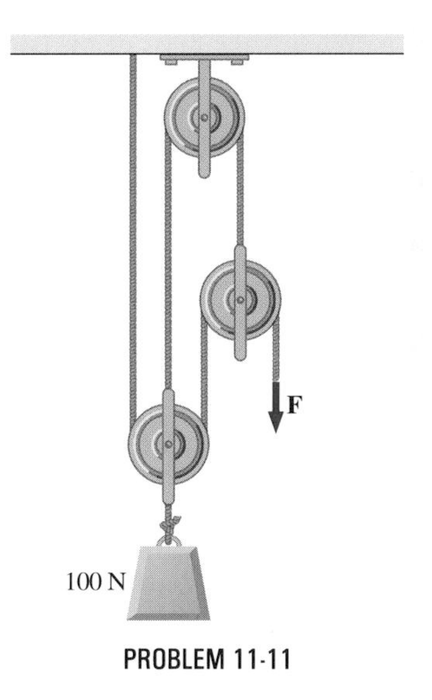

PROBLEM 11-11

11-12. What force must be applied at P to keep the structure shown in the accompanying figure in equilibrium? The weight of the structure is negligible.

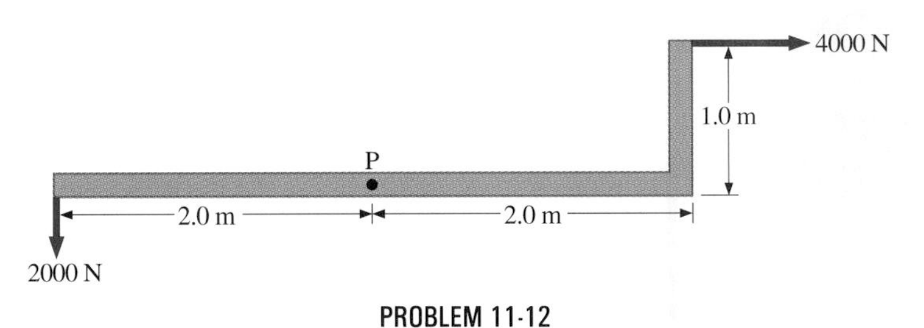

PROBLEM 11-12

11-13. Is it possible to apply a force at R to keep the structure shown in the accompanying figure in equilibrium? The weight of the structure is negligible.

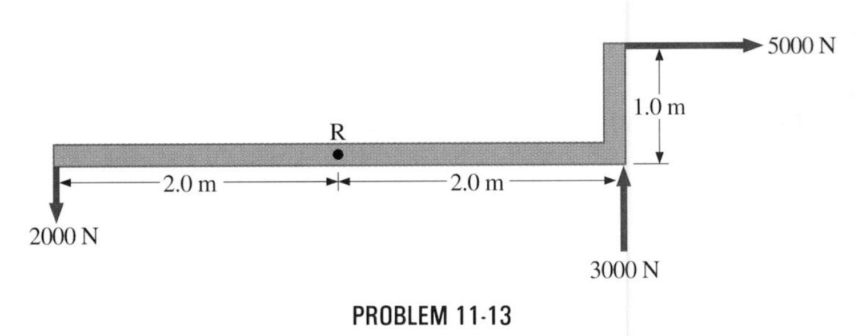

PROBLEM 11-13

11-14. The structure shown in the accompanying figure is supported at P. If it is in equilibrium, what is the magnitude of **F** and what is the force on the structure at P? The weight of the structure is negligible.

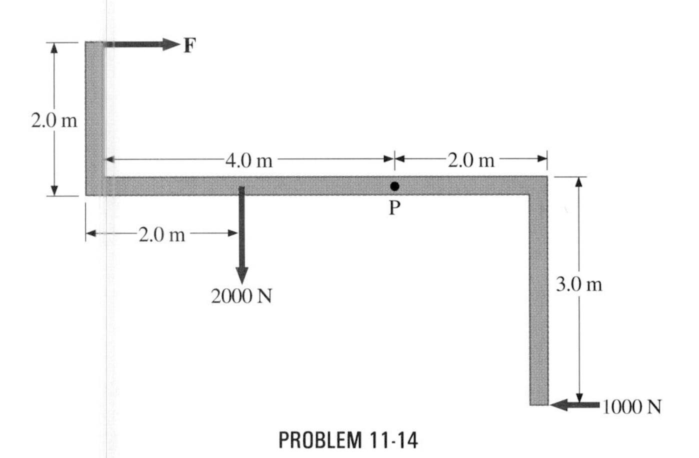

PROBLEM 11-14

11-15. In order to find her center of mass, a girl places an 8.0-ft plank between two scales, then lies on the plank, as shown in the accompanying figure. The girl is 5.0 ft 6.0 in. tall, and the plank weighs 40 lb. The girl has a friend read the two scales; the friend tells her that scale A reads 75 lb and scale B reads 85 lb. (*a*) What is the weight of the girl? (*b*) Where is her center of mass?

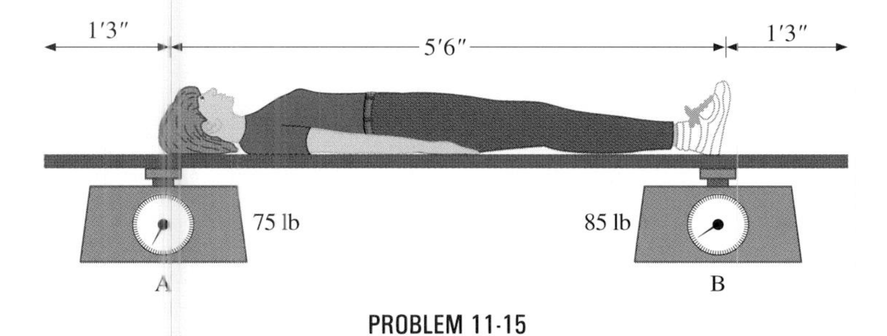

PROBLEM 11-15

11-16. A small station wagon of mass 1000 kg has a wheel base (distance between the front and rear axles) of 3.0 m. If 60 percent of the weight rests on the front wheels, how far behind the front wheels is the wagon's center of mass?

11-17. The uniform horizontal strut shown in the accompanying figure weighs 400 N. One end of the strut is attached to a hinged support at the wall, and the other end is attached to a sign that weighs 200 N. The strut is also supported by a cable attached between the end of the strut and the wall. Find the tension in the cable and the force at the hinge on the strut.

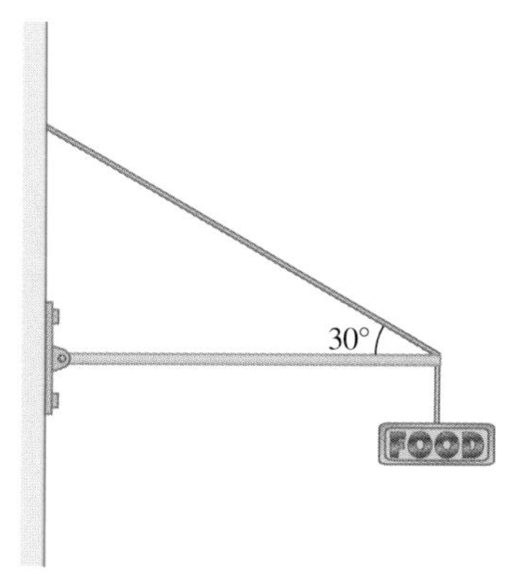

PROBLEM 11-17

11-18. Find the tension in each supporting cable shown in the accompanying figure. The weight of the suspended body is 100 N in each case, and the weights of the cables are negligible.

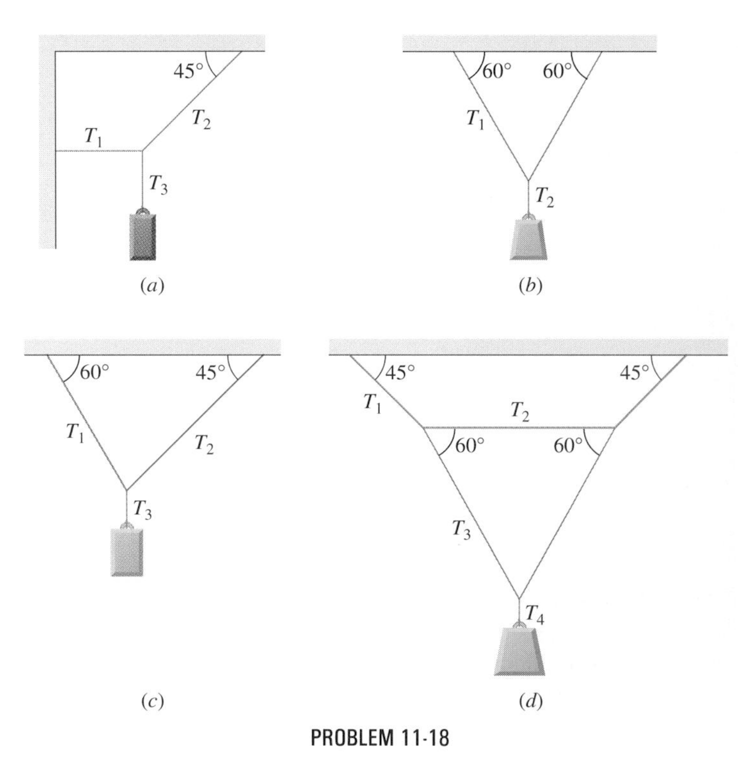

PROBLEM 11-18

11-19. Weights are gradually added to a pan until a wheel of mass M and radius R is pulled over an obstacle of height d. (See the accompanying figure.) What is the minimum mass of the weights plus the pan needed to accomplish this?

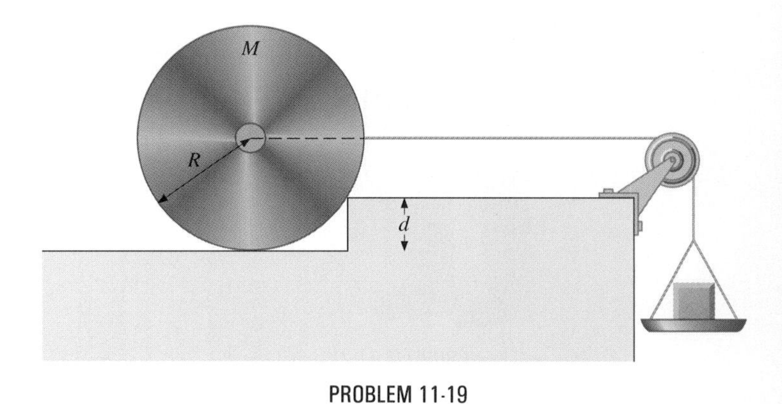

PROBLEM 11-19

11-20. The forearm seen in the accompanying figure is positioned at an angle θ with respect to the upper arm, and a 5.0-kg mass is held in the hand. The total mass of the forearm and hand is 3.0 kg, and their center of mass is 15.0 cm from the elbow.

(a) What is the force that the biceps muscle exerts on the forearm for $\theta = 60°$? (b) What is the force on the elbow joint for the same angle? (c) How do these forces depend on the angle θ?

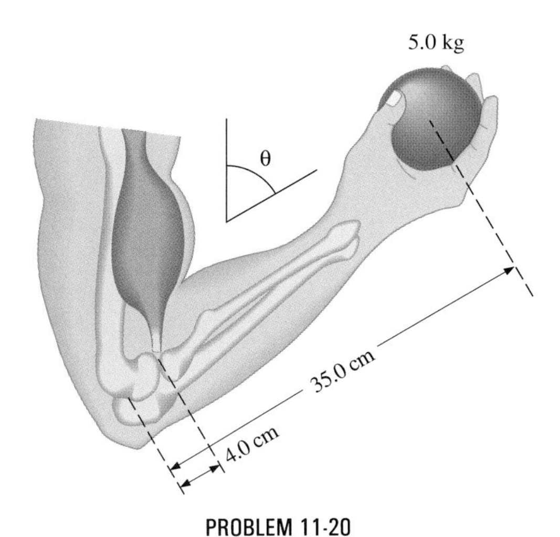

PROBLEM 11-20

11-21. In order to lift a shovelful of dirt, a gardener pushes downward on the end of the shovel and pulls upward at a distance l_2 from the end, as shown in the accompanying figure. The weight of the shovel is $m\mathbf{g}$ and acts at the point of application of $\mathbf{F}_2$. Calculate the magnitudes of the forces $\mathbf{F}_1$ and $\mathbf{F}_2$ as functions of l_1, l_2, mg, and the weight W of the load. Why don't your answers depend on θ, the angle that the shovel makes with respect to the horizontal?

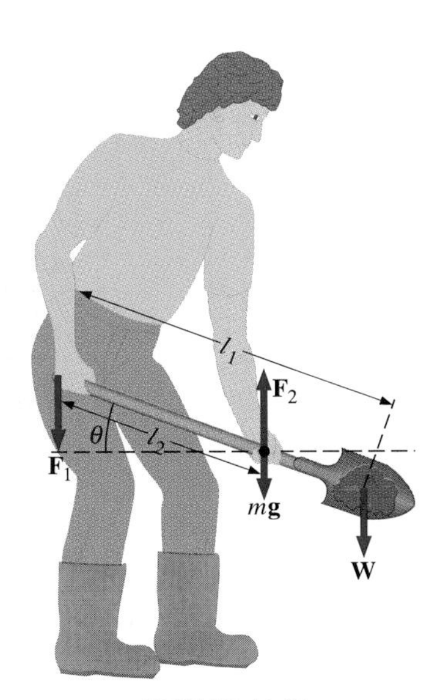

PROBLEM 11-21

11-22. The uniform boom shown in the accompanying figure weighs 3000 N. It is supported by the horizontal guy wire and by the hinged support at A. What are the forces on the boom due to the wire and the support at A? Does the force at A act along the boom?

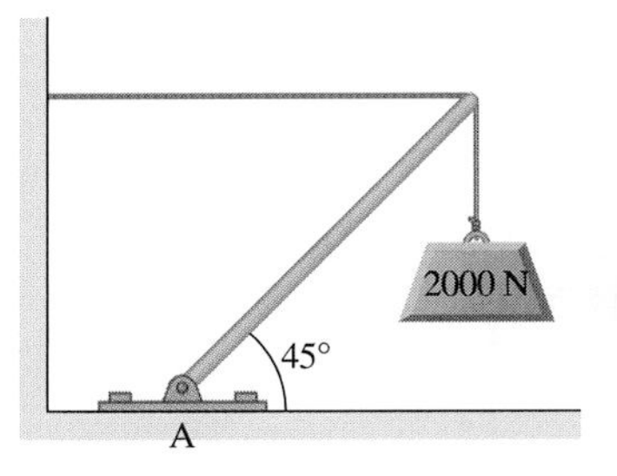

PROBLEM 11-22

11-23. The uniform boom shown in the accompanying figure weighs 700 N, and the body hanging from its right end weighs 400 N. The boom is supported by the light cable and by a hinge at the wall. Calculate the tension in the cable and the force of the hinge on the boom. Does the force at the hinge act along the boom?

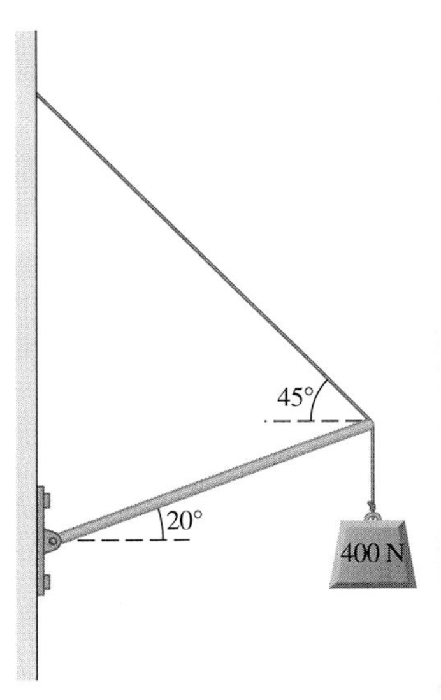

PROBLEM 11-23

11-24. The accompanying figure shows a 12-m boom AB of a crane lifting a 3000-kg load. The center of mass of the boom is at its geometric center, and the mass of the boom is 1000 kg. For the position shown, calculate the tension T in the cable and the force at the axle A.

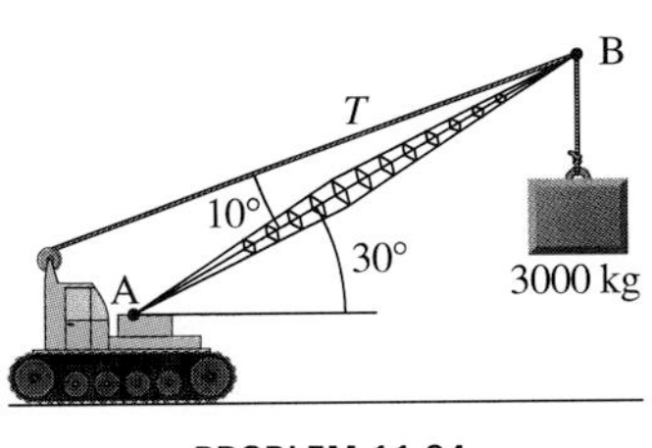

PROBLEM 11-24

11-25. The uniform trapdoor shown in the accompanying figure is 1.0 m by 1.5 m and weighs 300 N. It is supported by the single hinge H and by a light rope tied between the middle of the door and the floor. If the door is held in the position shown, what are the tension in the rope and the force at the hinge?

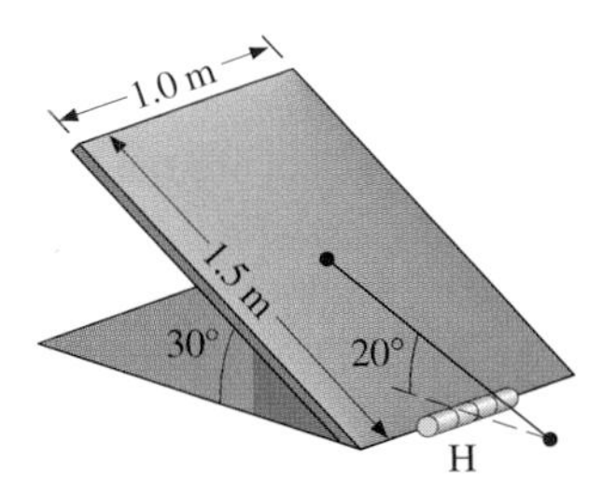

PROBLEM 11-25

11-26. A circular hoop of mass 0.40 kg and diameter 0.80 m is supported in a horizontal plane by three vertical strings attached as shown in the accompanying figure. What are the tensions in the strings?

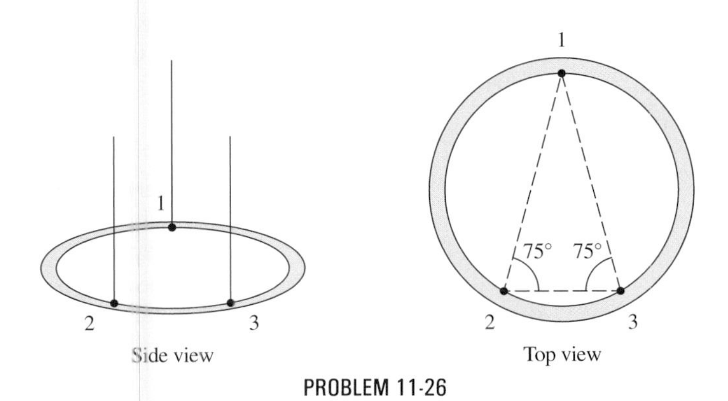

PROBLEM 11-26

11-27. A man of mass 90 kg walks across a sawhorse, as shown in the accompanying figure. The sawhorse is 2.0 m long, 1.0 m high, and its mass 25 kg. Calculate the normal force on each leg of the sawhorse due to the floor when the man is 0.5 m from the far end of the sawhorse.

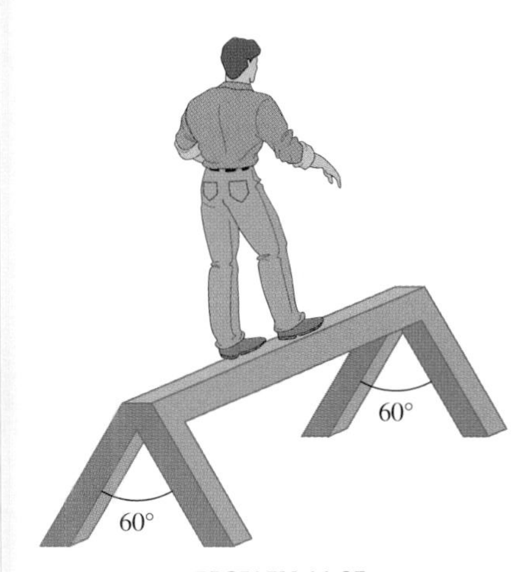

PROBLEM 11-27

11-28. The accompanying figure shows a boy of mass M standing in the middle of a uniform plank of mass m and length d. The left end of the plank is attached to the floor with a hinge. By pulling on the rope, the boy raises the plank (and himself with it). (*a*) What force must the boy apply to the rope in order to lift the plank off the floor? (*b*) Show that if m is too large, the plank will not be lifted no matter how hard the boy pulls on the rope. (*c*) What is the maximum allowable value for m if the plank is to be raised?

PROBLEM 11-28

Statically Indeterminate Systems and Stability

11-29. A uniform door weighs 400 N; it is 2.5 m high and 1.5 m wide. It is supported by two hinges, one 50 cm below the top of the door, the other 50 cm above the bottom. Calculate the net vertical force on the door at the hinges and the horizontal force on the door at each hinge. Explain why the problem is statically indeterminant.

11-30. For which of the following net forces is there stable equilibrium near $x = 0$? (*a*) $F(x) = -3.0x$; (*b*) $F(x) = -3.0x^2$; (*c*) $F(x) = 1.0 \sin x$; (*d*) $F(x) = -1.0x^3 + 2.0x^2$; (*e*) $F(x) = 1.0 \cos x - 1.0$.

11-31. The accompanying figure shows a box of height h and width w at rest on a plane whose angle of inclination can be varied. (*a*) **N** represents the normal component of the plane's force on the box. Determine how the location of **N**'s line of action (as denoted by x) varies with the angle θ. (*b*) Assuming the box does not slip, what is the maximum value θ can have before the box tips over?

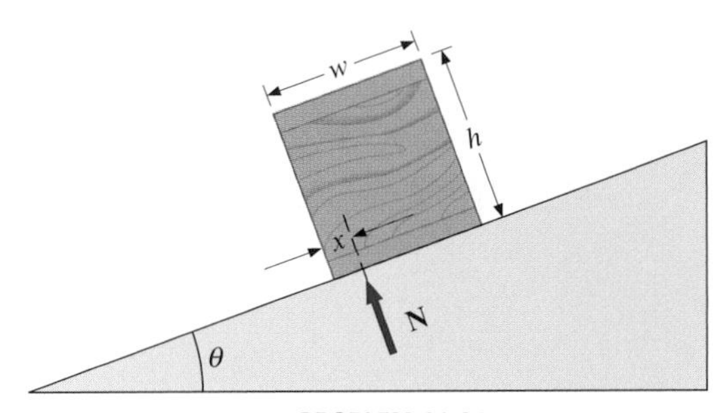

PROBLEM 11-31

Stress and Strain and the Elasticity of Solids

Note: Unless stated otherwise, the weights of the wires, rods, and other elements are assumed to be negligible.

11-32. A large, uniform cylindrical steel ($\rho = 7.8\ \text{g/cm}^3$) rod is 2.0 m long and has a diameter of 5.0 cm. The rod is fastened to a concrete floor with its long axis vertical. (*a*) What is the normal stress in the rod at a cross section 1.0 m from its lower end? (*b*) at a cross section 1.5 m from its lower end?

11-33. A uniform rope of cross-sectional area 0.50 cm^2 breaks when the tensile stress in it reaches $6.0 \times 10^6\ \text{N/m}^2$. (*a*) What is the maximum load that can be lifted slowly by the rope? (*b*) What is the maximum load that can be lifted by the rope with an acceleration of 4.0 m/s^2?

11-34. One end of a vertical metallic wire of length 2.0 m and diameter 1.0 mm is attached to the ceiling, and the other end is attached to a 5.0-N weight pan. (See the accompanying figure.) The position of the pointer before the pan is attached is 4.000 cm. Different weights are then added to the pan, and the position of the pointer is recorded in the table shown. Plot stress vs. strain for this wire, then use the resulting curve to determine Young's modulus and the proportional limit of the metal. What metal is this most likely to be?

Added load (N) (including pan)	Scale Reading (cm)
0	4.000
15	4.036
25	4.073
35	4.109
45	4.146
55	4.181
65	4.221
75	4.266
85	4.316

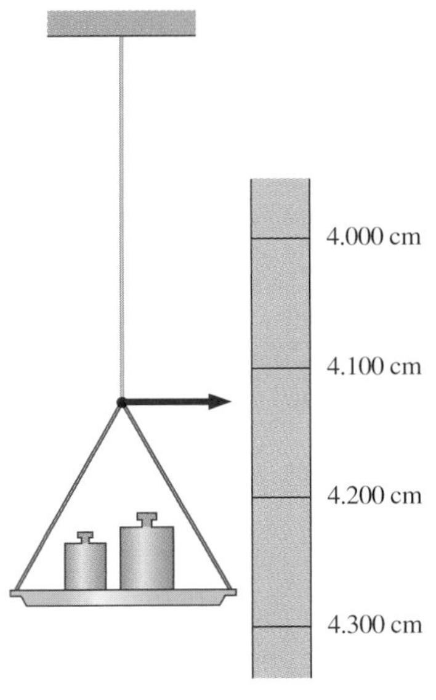

PROBLEM 11-34

11-35. What is the normal strain in the rod of Prob. 11-32 at the 1.0-m cross section?

11-36. A 90-kg mountain climber hangs from a nylon rope and stretches it by 25 cm. If the rope was originally 30 m long and its diameter is 1.0 cm, what is Young's modulus for the nylon?

11-37. A suspender rod of a suspension bridge is 25 m long. If the rod is made of steel, what must its diameter be so that it does not stretch more than 1.0 cm when a 2.5×10^4-kg truck passes by it? Assume that the rod supports all of the weight of the truck.

11-38. An aluminum ($\rho = 2.7\ \text{g/cm}^3$) wire is suspended from the ceiling and hangs vertically. How long must the wire be before the stress at its upper end reaches the proportional limit, which is $8.0 \times 10^7\ \text{N/m}^2$?

11-39. A copper wire is 1.0 m long and its diameter is 1.0 mm. If the wire hangs vertically, how much weight must be added to its free end in order to stretch it 3.0 mm?

11-40. A 100-N weight is attached to the free end of a metallic wire that hangs from the ceiling. When a second 100-N weight is added to the wire, it stretches 3.0 mm. The diameter and length of the wire are 1.0 mm and 2.0 m, respectively. What is Young's modulus of the metal used to manufacture the wire?

11-41. A copper wire of diameter 1.0 cm stretches 1.0 percent when it is used to lift a load upward with an acceleration of 2.0 m/s^2. What is the weight of the load?

Volume Compression

11-42. The bulk modulus of a material is $1.0 \times 10^{11}\ \text{N/m}^2$. What fractional change in volume does a piece of this material undergo when it is subjected to a force increase of $10^7\ \text{N/m}^2$? Assume that the force is applied uniformly over the surface.

11-43. Normal forces of magnitude 7.0×10^6 N are applied uniformly to all six faces of a cubical volume of liquid. This causes the sides of the cube to decrease from 50.000 to 49.995 cm. What is the bulk modulus of the liquid?

11-44. Normal forces are applied uniformly over the surface of a spherical volume of water whose radius is 20.0 cm. If the force per unit area over the surface is increased by $2.0 \times 10^8\ \text{N/m}^2$, by how much does the radius of the sphere decrease?

General Problems

11-45. The accompanying figure shows two blocks being pulled across one another at constant velocity by two falling weights. The coefficient of kinetic friction μ_K is the same for all surfaces. Determine μ_K and the mass M of the weight on the left.

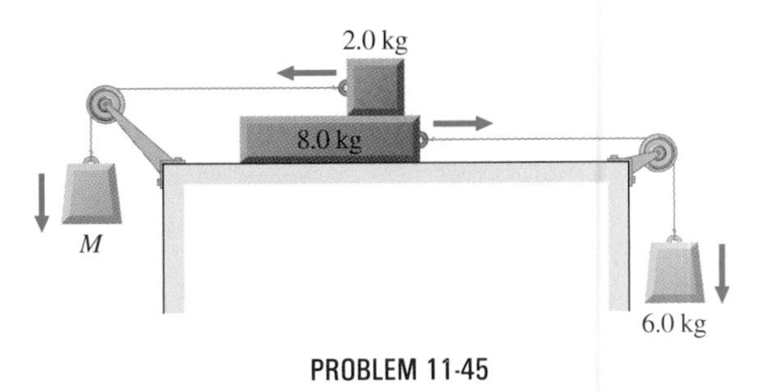

PROBLEM 11-45

11-46. The coefficient of static friction between the rubber eraser of the pencil and the tabletop is $\mu_S = 0.80$. If a force **F** is applied along the axis of the pencil, as shown in the accompanying figure, what is the minimum angle at which the pencil can stand without slipping? Ignore the weight of the pencil.

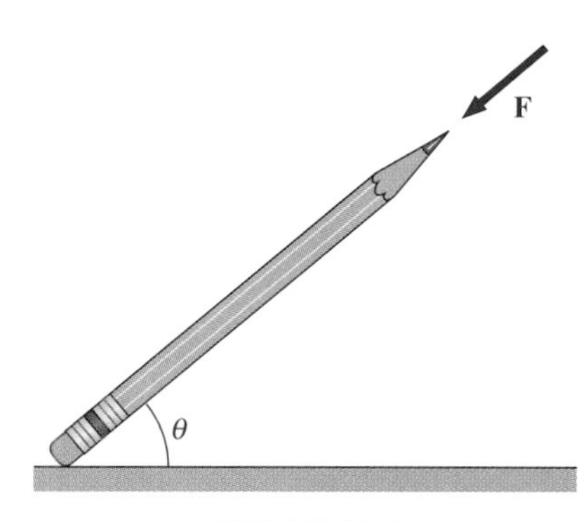

PROBLEM 11-46

11-47. Two uniform 200-N rods are joined by a pin and placed upright on a smooth horizontal surface, as shown in the accompanying figure. A chain of negligible mass connects the centers of the two rods so that the structure does not collapse. Find the tension in the chain and the force of the pin on each rod.

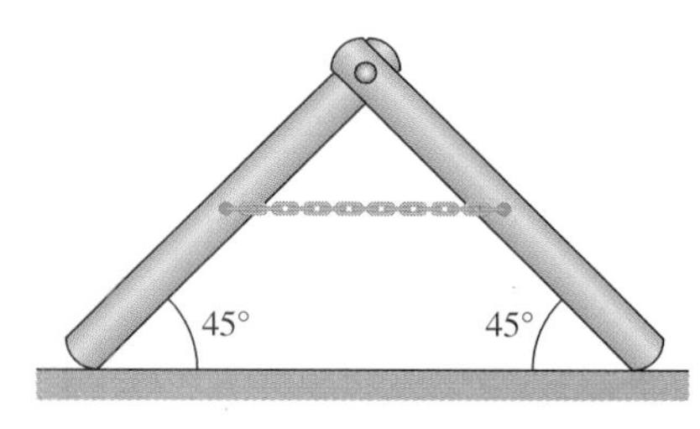

PROBLEM 11-47

11-48. A uniform 4.0-m plank weighing 200 N rests against the corner of a wall as shown in the accompanying figure. There is no frictional force at the point where the plank meets the corner. (*a*) Find the forces that the corner and the floor exert on the plank. (*b*) What is the minimum coefficient of static friction between the floor and the plank to keep the plank from slipping?

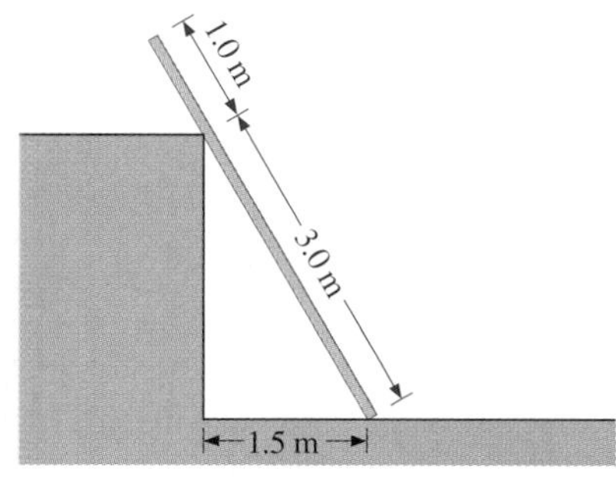

PROBLEM 11-48

11-49. A horizontal force **F** is applied to the sphere shown in the accompanying figure. What value of **F** is needed to hold the sphere in equilibrium? What is the frictional force of the incline on the sphere?

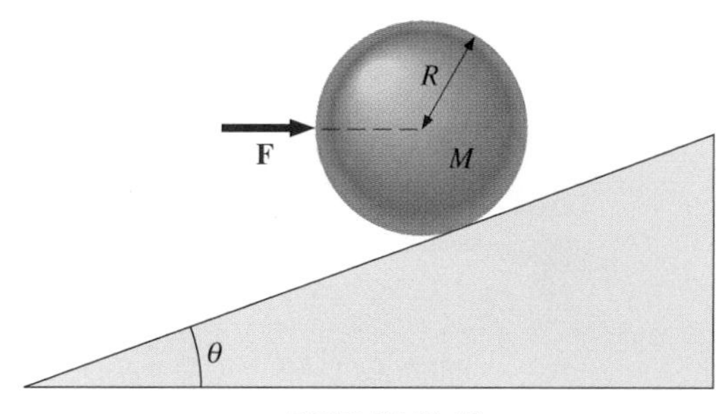

PROBLEM 11-49

11-50. The accompanying figure shows a pencil with its sharpened end resting against a smooth vertical surface and its eraser end resting on the floor. The center of mass of the pencil is 9.0 cm from the tip of the eraser and 11.0 cm from the tip of the lead. If $\mu_S = 0.80$ between the eraser and floor, what is the minimum angle θ for which the pencil does not slip?

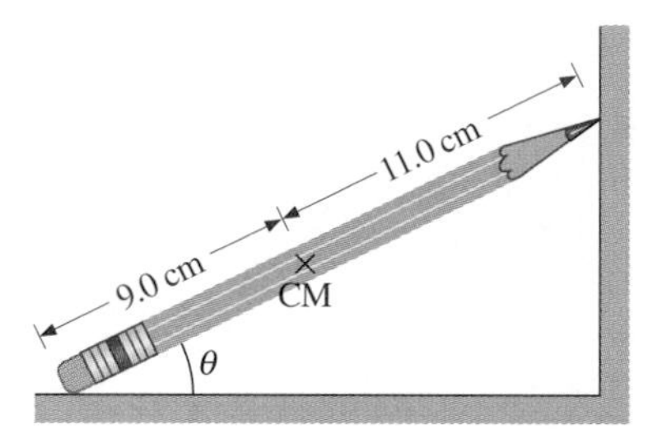

PROBLEM 11-50

11-51. When a motor is set on a pivoted mount, its weight can be used to maintain tension in the drive belt. (See the accompanying figure.) When the motor is not running, the tensions T_1 and T_2 may be assumed to be equal. The total mass of the platform and motor is 100 kg, and the diameter of the drive pulley is 16 cm. If the motor is off, what are (*a*) the tension in the belt and (*b*) the force at the hinged platform support C? Assume that the center of mass of the motor plus platform is at the center of the motor.

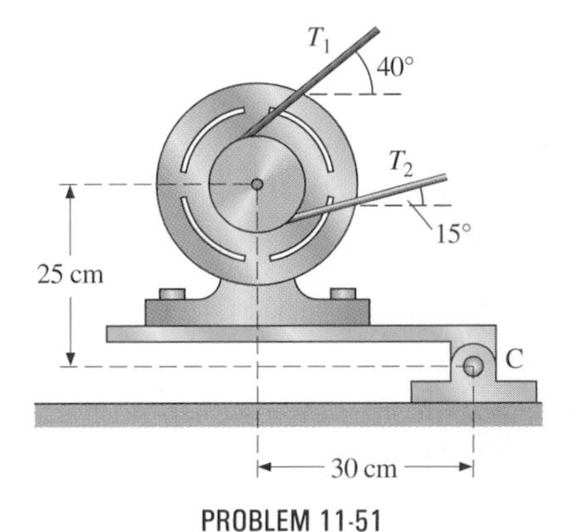

PROBLEM 11-51

11-52. Two wheels A and B of weight W and $2W$, respectively, are connected by a rod of weight $W/2$, as shown in the accompanying figure. The wheels are free to roll on the sloped surfaces. Determine the angle that the rod forms with the horizontal when the system is in equilibrium.

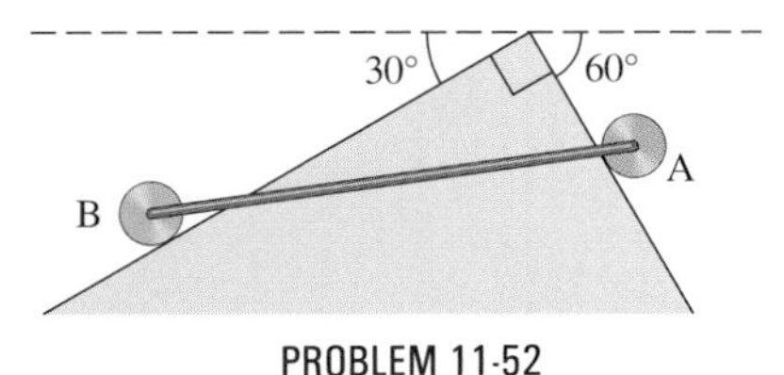

PROBLEM 11-52

11-53. A uniform rod of length $2R$ and mass M is attached to a small collar C and rests on a cylindrical surface of radius R, as shown in the accompanying figure. If the collar slides without friction along the vertical guide, what is the angle θ for equilibrium?

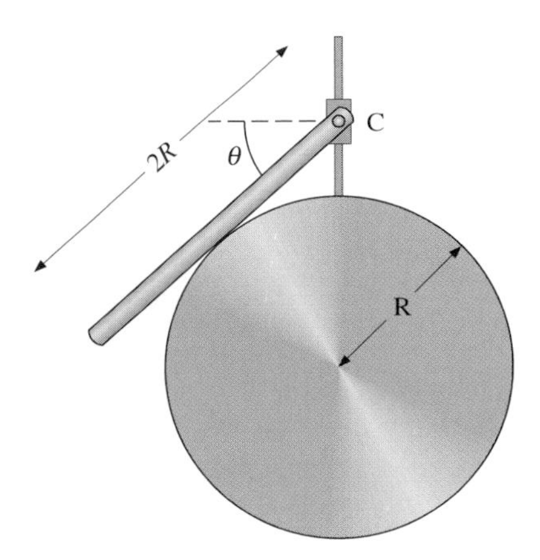

PROBLEM 11-53

11-54. When the pendulum shown in the accompanying figure is displaced slightly from equilibrium, its potential energy is given by

$$U(x) = mgy \approx mg(R - \sqrt{R^2 - x^2}).$$

Show that $x = 0$ is a point of stable equilibrium for the pendulum.

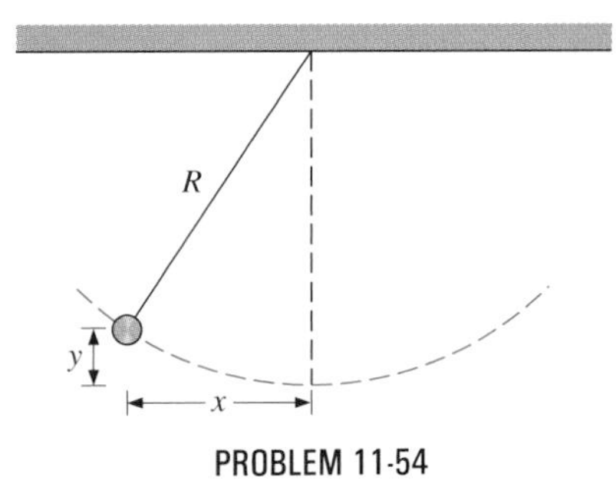

PROBLEM 11-54

11-55. The accompanying figure shows two hospital orderlies lifting a patient who weighs 700 N. Each orderly applies the same vertical force on the patient. The weight of the upper body of each orderly is 400 N and acts at the center of mass of the upper body, which rotates about P. (*a*) What is the tension T in the back muscles of each orderly when the patient is lifted as shown in part (*a*) of the figure? (*b*) Part (*b*) of the figure illustrates a trick often used to eliminate backstrain in this situation. The orderlies touch foreheads so that they exert a contact force of magnitude C on each other. What is the value of C if the tension T in the back muscles is eliminated? (*c*) What is T when $C =$ 200 N?

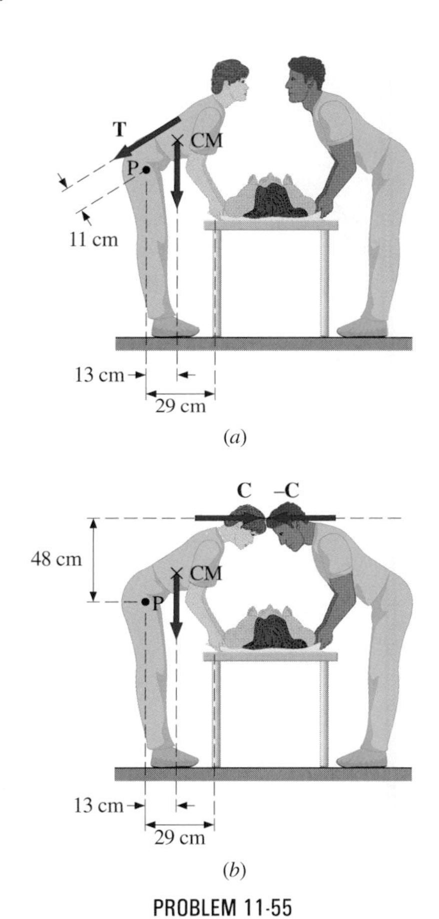

PROBLEM 11-55

11-56. A 40-kg boy jumps from a height of 3.0 m, lands on one foot, and comes to rest 0.10 s after he hits the ground. Assume that he stops with constant acceleration. If the total cross-sectional area of the bones in his leg just above the ankle is 3.0×10^{-4} m^2, what is the compressional stress in these bones? Leg bones fracture when they are subjected to stresses greater than approximately 1.7×10^8 N/m^2. Is the boy in any danger of breaking his leg?

11-57. Two thin rods, one made of steel, the other of aluminum, are joined end to end. Each rod is 2.0 m long, and the cross-sectional area of each rod is 9.1×10^{-6} m^2. If a tensile force of 1.0×10^4 N is applied at each end of the combination, (*a*) what is the stress in each rod, (*b*) what is the strain of each rod, and (*c*) what is the elongation of each rod?

11-58. Two rods, one made of copper, the other of steel, have the same dimensions. If the copper rod stretches 1.5×10^{-2} cm under a particular stress, how much does the steel rod stretch when the same stress is applied to it?

Momentum conservation.

CHAPTER 12 MOMENTUM

PREVIEW

This chapter is devoted to a quantity known as momentum. The main topics to be discussed here include the following:

1. **Momentum of a particle.** The momentum of a particle is defined, and Newton's second law is expressed in terms of momentum.
2. **Momentum of a system of particles.** Using the definition of the momentum of a particle, we obtain expressions for the momentum of a system of particles that may be discrete or may form a rigid body. We see how the system's momentum is related to the motion of its center of mass.
3. **Impulse.** Impulse and its relationship to the change in momentum of a body (the impulse-momentum theorem) are studied.
4. **Conservation of momentum.** We consider the conditions necessary for momentum to be conserved, and, using various examples, we illustrate the wide applicability of this important law.
5. **Elastic collisions.** Elastic collisions, for which both momentum and mechanical energy are conserved, are considered in both the laboratory and the center-of-mass frames.

*6. **Motion with varying mass.** Using a rocket and its fuel as a prototype, we study the motion of bodies that eject or gain mass.

There are many situations in which forces are not defined well enough for a direct application of Newton's second law. Perhaps the most common example is a *collision* between two objects. Typical collisions might involve a ball and a bat, two hockey players, or even elementary particles. A very useful quantity for studying such interactions is momentum. For example, from changes in momentum, we are able to estimate the magnitudes of sharp forces involved in many collisions, and from the constancy of momentum, we are able to determine the velocities of two bodies after they have collided. In addition, systems with variable mass, such as a rocket ejecting fuel, can be investigated in terms of momentum. We will derive all laws and equations involving momentum directly from Newton's laws. However, it is interesting to note that one of these, the law of momentum conservation, is more general than its derivation implies, for it is valid in the world of the atom and nucleus, where Newtonian mechanics is not.

12-1 Introduction to Momentum

Momentum of a Particle

If a particle of mass m has a velocity $\mathbf{v}$, its **momentum** (also known as **linear momentum**) is, by definition,

$$\mathbf{p} = m\mathbf{v}. \tag{12-1}$$

Momentum is a vector that has the same direction as the velocity. Its SI unit is simply the kilogram-meter per second (kg·m/s).

Because the mass m in Eq. (12-1) is constant,

$$\frac{d\mathbf{p}}{dt} = \frac{d}{dt}(m\mathbf{v}) = m\frac{d\mathbf{v}}{dt}.$$

Newton's second law can therefore be written in the form

$$\sum \mathbf{F} = \frac{d\mathbf{p}}{dt}, \tag{12-2}$$

where $\sum\mathbf{F}$ is the net force on the particle. *The rate of change of the momentum of a particle is equal to the net force acting on the particle.* In Newtonian mechanics, Eqs. (5-1) and (12-2) are equivalent—which of these expressions of the second law we use in a given application is strictly a matter of convenience. However, in special relativity where objects moving at speeds close to that of light are considered, only Eq. (12-2) is correct.

DRILL PROBLEM 12-1

What is the momentum of a particle of mass 5.0 kg moving with a velocity $\mathbf{v} = (2.0\mathbf{i} + 4.0t\mathbf{j})$ m/s? What is the net force acting on this particle?

ANS. $(10\mathbf{i} + 20t\mathbf{j})$ kg·m/s; $20\mathbf{j}$ N.

Momentum of a System of Particles

The momentum $\mathbf{P}$ of a system of N particles with masses $m_1, m_2, \ldots, m_N$ and velocities $\mathbf{v}_1, \mathbf{v}_2, \ldots, \mathbf{v}_N$ is defined as

$$\begin{aligned}\mathbf{P} &= m_1\mathbf{v}_1 + m_2\mathbf{v}_2 + \ldots + m_N\mathbf{v}_N \\ &= \mathbf{p}_1 + \mathbf{p}_2 + \ldots + \mathbf{p}_N;\end{aligned} \tag{12-3}$$

that is, *the momentum of a system of particles is equal to the vector sum of the momenta of the individual particles.*

From Eq. (8-4*a*), the velocity of the center of mass of a system of particles is given by $\mathbf{v}_{CM} = (\sum_i m_i\mathbf{v}_i)/M$. Upon comparing this equation to Eq. (12-3), we find that

$$\mathbf{P} = M\mathbf{v}_{CM}. \tag{12-4}$$

The momentum of a system of particles of total mass M is equal to the momentum of a single particle of mass M moving with the velocity $\mathbf{v}_{CM}$ of the center of mass.

The relationship between $\mathbf{P}$ and the net external force $\sum\mathbf{F}$ can be obtained with the help of Eq. (8-7), the relationship between the net external force on a system of particles and the acceleration of its center of mass:

$$\sum \mathbf{F} = M\mathbf{a}_{CM} = M\frac{d\mathbf{v}_{CM}}{dt} = \frac{d}{dt}(M\mathbf{v}_{CM}) = \frac{d\mathbf{P}}{dt}. \tag{12-5}$$

Notice that the internal forces have no effect on the total momentum. *The rate of change of the momentum of a system of particles is equal to the net external force on that system.*

EXAMPLE 12-1 The Momentum of a System of Particles

The three particles of Fig. 12-1 experience the constant forces shown, and at $t = 0$, they have the positions and velocities given. (*a*) What is the initial momentum of the system? (*b*) What is its momentum at $t = 3.0$ s?

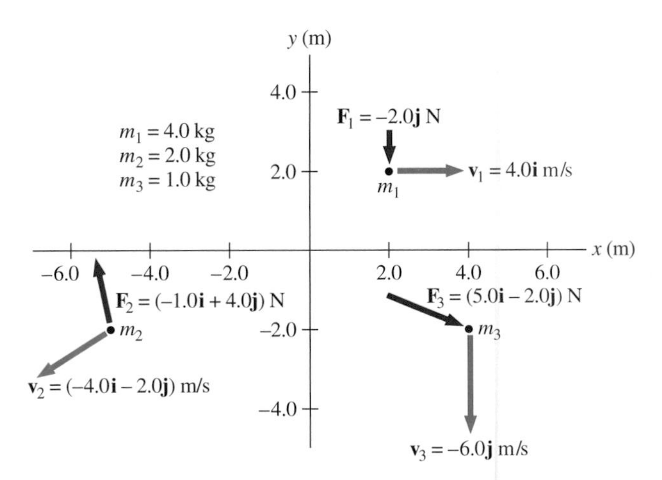

FIGURE 12-1 A system of particles.

SOLUTION (*a*) By definition, the initial momentum of the system is the sum of the initial momenta of the individual particles:

$$\begin{aligned}\mathbf{P}(0) &= m_1\mathbf{v}_1(0) + m_2\mathbf{v}_2(0) + m_3\mathbf{v}_3(0) \\ &= (4.0\text{ kg})(4.0\mathbf{i}\text{ m/s}) + (2.0\text{ kg})[(-4.0\mathbf{i} - 2.0\mathbf{j})\text{ m/s}] \\ &\quad + (1.0\text{ kg})(-6.0\mathbf{j}\text{ m/s}) \\ &= (8.0\mathbf{i} - 10.0\mathbf{j})\text{ kg}\cdot\text{m/s}.\end{aligned}$$

(*b*) We can find the momentum at $t = 3.0$ s by integrating $d\mathbf{P}/dt = \sum\mathbf{F}$ between $t = 0$ and $t = 3.0$ s. This gives

$$\mathbf{P}(3.0\text{ s}) - \mathbf{P}(0) = \int_0^{3.0\text{ s}} \left(\sum\mathbf{F}\right) dt.$$

Substituting the known expressions for the initial momentum $\mathbf{P}(0)$ and the forces, we have

$$\mathbf{P}(3.0\text{ s}) - (8.0\mathbf{i} - 10.0\mathbf{j})\text{ kg}\cdot\text{m/s}$$
$$= \int_0^{3.0\text{ s}} \{[(-1.0 + 5.0)\mathbf{i} + (-2.0 + 4.0 - 2.0)\mathbf{j}]\text{ N}\}dt$$
$$= \int_0^{3.0\text{ s}} (4.0\mathbf{i}\text{ N})\,dt = 12.0\mathbf{i}\text{ kg}\cdot\text{m/s}.$$

We then obtain

$$\mathbf{P}(3.0\text{ s}) = (20.0\mathbf{i} - 10.0\mathbf{j})\text{ kg}\cdot\text{m/s}.$$

DRILL PROBLEM 12-2

Consider the system of particles of the previous example. As an alternative approach to finding $\mathbf{P}(3.0\text{ s})$, use the accelerations of the individual particles to determine their velocities at $t = 3.0$ s, then use Eq. (12-3) to calculate the system's momentum at that instant.

12-2 IMPULSE

From Newton's second law, the change in a particle's momentum $d\mathbf{p}$ in a time interval dt is related to the net force $\sum\mathbf{F}$ acting on it by

$$d\mathbf{p} = \sum\mathbf{F}\,dt.$$

As we saw in Example 12-1, the net change in the particle's momentum over a finite time interval from t_1 to t_2 can be found by integrating this expression between t_1 and t_2:

$$\int_{t_1}^{t_2} d\mathbf{p} = \int_{t_1}^{t_2} \sum\mathbf{F}\,dt,$$

and

$$\mathbf{p}(t_2) - \mathbf{p}(t_1) = \int_{t_1}^{t_2} \sum\mathbf{F}\,dt. \qquad (12\text{-}6a)$$

The integral of the force over time is the **impulse J**:

$$\mathbf{J} = \int_{t_1}^{t_2} \sum\mathbf{F}\,dt. \qquad (12\text{-}7a)$$

Its units are newton-seconds (N·s) in the SI system. Combining Eqs. (12-6*a*) and (12-7*a*), we have

$$\Delta\mathbf{p} = \mathbf{J}, \qquad (12\text{-}6b)$$

which is the *impulse-momentum theorem*: *The change in the momentum of a particle is equal to the impulse acting on it.*

Of course, this theorem holds along any coordinate axis. Consequently, if we plot the x component, $\sum F_x(t)$, of the net force vs. time, the area under the curve between t_1 and t_2 is the impulse component J_x (see Fig. 12-2); or equivalently, it is the change in p_x between these two times.

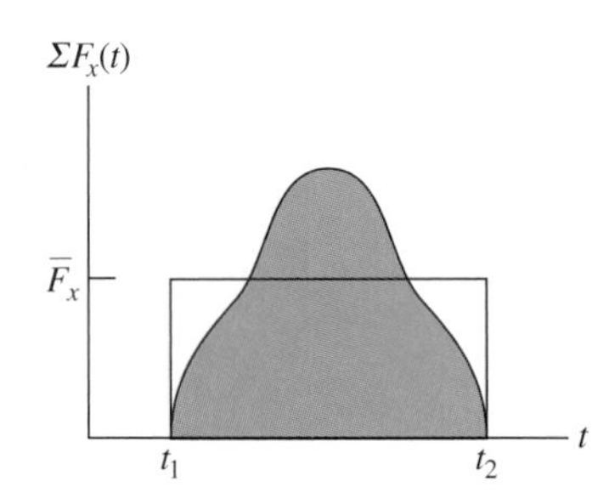

FIGURE 12-2 The impulse J_x between t_1 and t_2 is equal to the area under the curve (shaded area). The rectangle of height $\bar{F}_x$, the average force, and width $(t_2 - t_1)$ has the same area.

In many applications of the impulse-momentum theorem, we are interested in the **average force**. By definition, this is the constant force $\bar{\mathbf{F}}$ whose impulse is equal to that of the actual net force. In Fig. 12-2, the x component $\bar{F}_x$ of the average force is represented by the horizontal line drawn so that the area under it equals the shaded area under the curve representing $\sum F_x(t)$. Mathematically, this equality is written as

$$\bar{F}_x(t_2 - t_1) = \int_{t_1}^{t_2} \sum F_x(t)\,dt = J_x. \qquad (12\text{-}7b)$$

The impulse-momentum theorem is especially useful for analyzing interactions that take place over very small time intervals—for example, a bat hitting a ball or a bullet embedding itself in a target. The interactive force in such cases is usually much larger than any external forces. As a result, over the time interval of the collision, the impulses of all forces (e.g., gravity and external friction) except those between the colliding bodies can be ignored.

EXAMPLE 12-2 IMPULSE ON A BASEBALL

A baseball of mass 0.16 kg is hit while moving horizontally with a speed of 30 m/s. After leaving the bat, the ball travels in the opposite direction with a speed of 40 m/s. (See Fig. 12-3.) (*a*) Determine the impulse on the ball. (*b*) If the collision between bat and ball lasts 2.0×10^{-3} s, what is the average force on the ball due to the bat? (*c*) Show that over the time of the collision the impulse of the gravitational force on the ball is negligible.

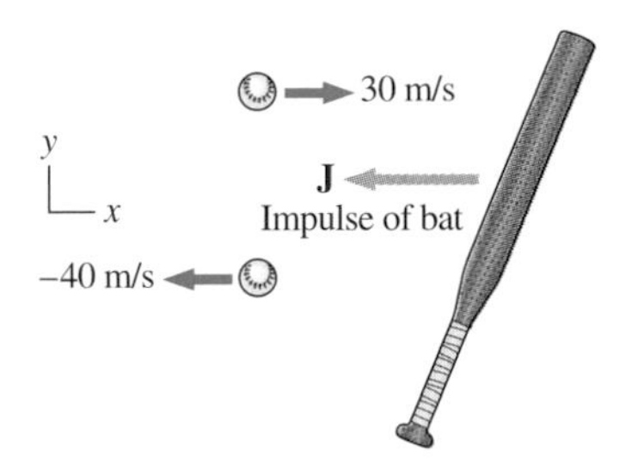

FIGURE 12-3 A baseball subjected to a horizontal impulse.

SOLUTION (*a*) Since the momenta before and after the collision are known, the impulse-momentum theorem may be used to find the impulse. If the ball is assumed to move initially in the positive x direction, we have

$$J_x = \Delta p_x = (0.16 \text{ kg})(-40 \text{ m/s}) - (0.16 \text{ kg})(30 \text{ m/s}) = -11 \text{ N}\cdot\text{s}.$$

The minus sign means that the impulse on the ball is in the negative x direction.

(*b*) We can find the average force with the help of Eq. (12-7*b*). Since

$$\bar{F}_x(t_2 - t_1) = J_x,$$
$$\bar{F}_x(2.0 \times 10^{-3} \text{ s}) = -11 \text{ N}\cdot\text{s},$$

and the average force is

$$\bar{F}_x = -5.5 \times 10^3 \text{ N}.$$

(*c*) During the collision, the vertical impulse of the gravitational force has a magnitude $mg\,\Delta t = (0.16 \text{ kg})(9.8 \text{ m/s}^2)(2.0 \times 10^{-3} \text{ s}) = 3.1 \times 10^{-3} \text{ N}\cdot\text{s}$. Since this is so much smaller than J_x, its effect is clearly negligible.

DRILL PROBLEM 12-3

A fast camera records the collision of a tennis ball and racket. One frame of the film is shown in Fig. 12-4. The collision is observed to last 3.0×10^{-2} s, and it changes the velocity of the 0.060-kg ball from 20 m/s in one direction to 25 m/s in the opposite direction. Find the impulse and the average force on the tennis ball.
ANS. 2.7 N·s; 90 N.

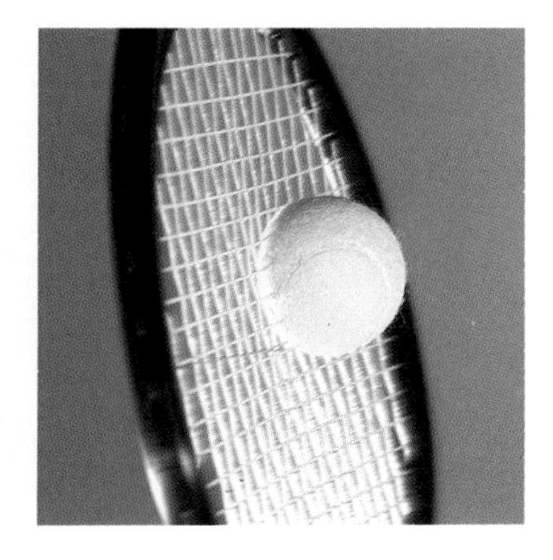

FIGURE 12-4 The collision between a tennis racket and a ball.

DRILL PROBLEM 12-4

If you are struck by a pitched baseball, you can be injured quite seriously, while a soft rubber ball with the same mass and velocity will do little harm. Explain. (*Hint:* Compare the time duration of the force in each case.)

Example 12-2 and Drill Problem 12-3 show how the impulse-momentum theorem can be used to estimate the forces in short-lived collisions. As a rule, direct measurement of these forces is not possible. However, the change in momentum can usually be determined and, with the aid of fast photography, so can the time duration of the collision. With this information, the average force can be computed.

12-3 Conservation of Momentum

Equation 12-5 tells us that the rate of change of the momentum of a system of particles is equal to the net external force $\sum\mathbf{F}$ on that system. Now suppose that $\sum\mathbf{F} = 0$. Then we have

$$\frac{d\mathbf{P}}{dt} = 0,$$

and

$$\mathbf{P} = \mathbf{p}_1 + \mathbf{p}_2 + \ldots + \mathbf{p}_N = \text{constant}. \tag{12-8a}$$

When the net external force acting on a system of particles is zero, the momentum of that system is constant. This does not mean that the momentum of each particle must remain constant! As the particles move and interact, their individual momenta change, but the *sum*, $\mathbf{p}_1 + \mathbf{p}_2 + \ldots + \mathbf{p}_N$, does not.

This is the second conservation principle we have encountered, the first being the conservation of mechanical energy. However, the conditions under which these two principles can be applied are completely different. Momentum is conserved when the net external force is zero; mechanical energy is conserved when the forces that do work are conservative. A projectile is a simple example of a system for which mechanical energy but not momentum is conserved. Since the net force $m\mathbf{g}$ on the projectile is conservative, the energy remains constant. However, since gravity is an external force, momentum is not conserved. There are also simple systems that conserve momentum but not energy. We will consider many of these in the examples and drill problems that follow.

Because momentum is a vector quantity, its conservation is represented by three equations, one for each rectangular component. Depending on the force, momentum may be conserved in one direction but not another. For example, if the net external force points in the x direction, the momentum changes in only that direction; it is conserved in the y and z directions.

We can also think of momentum conservation as a condition on the velocity of the center of mass. If $\mathbf{P}$ is constant, then from $\mathbf{P} = M\mathbf{v}_{CM}$,

$$\mathbf{v}_{CM} = \text{constant}. \tag{12-8b}$$

Equations (12-8*a*) and (12-8*b*) are simply alternative ways of describing momentum conservation. The choice of which expression to use depends only on which is more convenient in the analysis of the system under consideration.

The principle of momentum conservation is generally used to analyze the motion of a system of interacting bodies. The approach we will take in applying this principle consists of the following steps:

1. Choose an inertial reference frame. For convenience, the coordinate system is usually oriented so that one of its axes coincides with an initial velocity of one of the interacting bodies. This step is generally so straightforward that we will not mention it specifically in examples.
2. Identify the bodies that comprise the system to be analyzed. These bodies are usually interacting with one another in some way (for example, colliding).
3. Identify the net external force on the system. If there is a net external force in a particular direction, the momentum in that direction is not conserved. You must be careful here in distinguishing between external forces and internal forces on the system. Forces that the bodies of the system exert on one another are internal forces.
4. Apply the principle of momentum conservation. Equation (12-8*a*) or (12-8*b*) is applied along each coordinate axis for which the net external force is zero.

EXAMPLE 12-3 MOMENTUM CONSERVATION IN A COLLISION

Two hockey players, one of mass $m_1 = 100$ kg, the other of mass $m_2 = 75$ kg, are skating in opposite directions when they collide with one another. (See Fig. 12-5.) Just before the collision, their velocities are $v_1 = 10.0$ m/s and $v_2 = -8.0$ m/s. After the collision, the players become entangled and slide off together. (*a*) What is the velocity of this combination? (*b*) Compare the total kinetic energies before and after the collision. Assume that the frictional force of the ice on the players is negligible.

FIGURE 12-5 A collision between the two hockey players.

SOLUTION (*a*) Our system is the two hockey players. Without friction from the ice surface, there are no external horizontal forces on the system. The only forces in the horizontal direction are the *internal* action-reaction pair between the players during their collision. Momentum in this direction is therefore conserved. Setting the initial and final momentum components $(P_x)_i$ and $(P_x)_f$ equal, we have

$$(P_x)_i = (P_x)_f$$
$$(100\text{ kg})(10.0\text{ m/s}) + (75\text{ kg})(-8.0\text{ m/s}) = (100\text{ kg} + 75\text{ kg})v_x,$$

so

$$v_x = +2.3\text{ m/s}.$$

The positive sign indicates that the common velocity of the two men is to the right.

(*b*) The total kinetic energies before and after the collision are, respectively,

$$T_i = \tfrac{1}{2}(100\text{ kg})(10.0\text{ m/s})^2 + \tfrac{1}{2}(75\text{ kg})(-8.0\text{ m/s})^2 = 7.4 \times 10^3\text{ J},$$

and

$$T_f = \tfrac{1}{2}(175\text{ kg})(2.3\text{ m/s})^2 = 4.6 \times 10^2\text{ J}.$$

Thus almost 7000 J of mechanical energy are "lost" in the collision. But this shouldn't be surprising, as the bodies are distorted by the collision. The assumption of rigidity, which allows us to ignore the work of the internal forces (see Supplement 9-2), is simply not valid here. In this particular case, the internal forces are dissipative, and the net mechanical energy decreases.

It is also possible for mechanical energy to increase as a result of the internal forces. An explosion is a common example. Another is the system discussed in the next example.

EXAMPLE 12-4 MOMENTUM CONSERVATION AND A THROWN OBJECT

The 50-kg boy of Fig. 12-6 stands on a frozen pond and throws a 0.25-kg snowball horizontally at 20 m/s. (*a*) What is the velocity of the boy after he releases the snowball? Assume that friction is negligible. (*b*) Compare the mechanical energy of the system before and after the snowball is thrown.

FIGURE 12-6 As a result of throwing a snowball to the right, the boy moves to the left.

SOLUTION (*a*) Our system here is the boy and the snowball. Neglecting the friction from the ice surface, there is no net horizontal external force on the system. Thus the horizontal components of the momentum before and after the throw are the same:

$$(P_x)_i = (P_x)_f$$
$$0 = (50\text{ kg})v_x + (0.25\text{ kg})(20\text{ m/s}),$$

and

$$v_x = -0.10\text{ m/s}.$$

If the boy throws the ball to the right with a speed of 20 m/s, he will move to the left with a speed of 0.10 m/s. Once the ball is released, there is a net external vertical force on the system; it is the gravitational force on the snowball. (The weight of the boy remains balanced by the normal force of the ice surface.) Momentum in the vertical direction is then not conserved, as is apparent from the increasing vertical velocity of the falling snowball.

(*b*) There is an increase in mechanical energy of the system due to the work of the boy's muscles. Since there is no initial kinetic energy, this increase is the sum of the kinetic energies of the boy and the snowball:

$$T = \tfrac{1}{2}(0.25\text{ kg})(20\text{ m/s})^2 + \tfrac{1}{2}(50\text{ kg})(-0.10\text{ m/s})^2 = 50\text{ J}.$$

A situation similar in principle to this example is the firing of a gun—the gun recoils as the bullet is fired. However, since the rifle is so much more massive than the bullet, its recoil velocity is considerably less than the muzzle velocity of the bullet.

EXAMPLE 12-5 A Lid Closing on a Vending Cart

Figure 12-7 shows a vending cart with its lid open to the vertical position. The mass of the lid is $m/4$, and the mass of the rest of the cart is m. The lid is given a slight push and it topples over, closing on the cart. What are the horizontal position and velocity of the cart after the lid has fallen? Assume that the frictional force of the ground on the cart is negligible and that the centers of mass of both the lid and the cart are at their geometric centers. Need we be concerned with friction in the hinge?

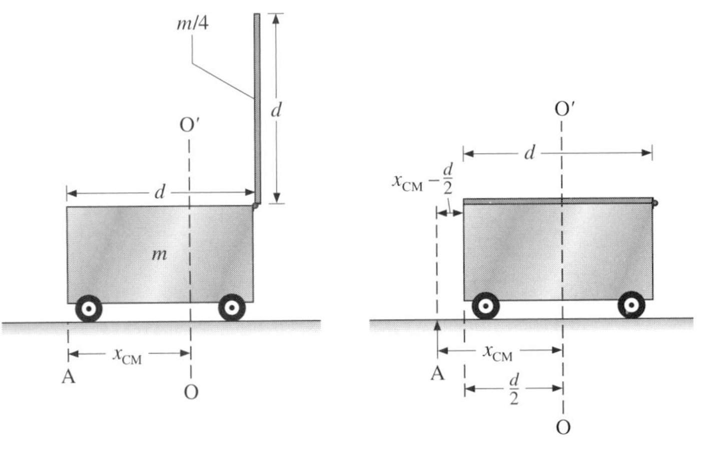

FIGURE 12-7 A vending cart with its lid open and closed. The center of mass of the system remains on the line OO′, which is fixed relative to the ground. Notice the position of the cart with respect to point A, which is also fixed relative to the ground.

SOLUTION There is no external force on the entire cart in the horizontal direction, so the momentum or, equivalently, the velocity of the center of mass of the system remains constant in this direction. Since this velocity is zero initially, it must remain zero. The center of mass therefore *does not move.* While the lid is falling, the rest of the cart moves so that the horizontal position of the pair's center of mass stays fixed; and after the lid lands, the cart no longer moves.

The position of the center of mass is calculated using Eq. (8-2*b*). With the lid upright, we have

$$\left(m + \frac{m}{4}\right)x_{CM} = (m)\frac{d}{2} + \left(\frac{m}{4}\right)d,$$

so

$$x_{CM} = \tfrac{3}{5}d.$$

When the lid lands, the center of mass is still at this position which, from symmetry, is now at the center of the cart. Hence the cart has moved a distance $3d/5 - d/2 = d/10$ to the right, as shown in the figure.

The friction in the hinge is an internal force and therefore has no effect on momentum conservation. This force affects how fast the lid falls, but the cart always ends up in the same position.

DRILL PROBLEM 12-5

Two small bodies whose masses are 4.0 and 2.0 kg are in contact with the opposite ends of a massless spring. The bodies, which are on a frictionless horizontal surface, are pushed toward one another, thereby compressing the spring. They are then released and move in opposite directions because of the spring force. Once free of the spring, the 4.0-kg body travels with a speed of 10 m/s. What is the speed of the 2.0-kg body?
ANS. 20 m/s.

DRILL PROBLEM 12-6

Two carts with masses 0.10 and 0.15 kg are connected to opposite ends of a massless spring with force constant $k = 500$ N/m. The 0.10-kg cart is given a velocity of 5.0 m/s to the right, and the 0.15-kg cart is simultaneously given a velocity of 6.0 m/s to the left. Describe the subsequent motion of the center of mass of the system. Ignore friction.
ANS. It moves at a constant velocity of 1.6 m/s to the left.

12-4 Momentum Conservation as an Approximation

Figure 12-8 shows a bullet as it approaches and then strikes a block that is resting on a rough surface. The bullet becomes embedded in the block and both objects slide across the surface as a single unit. The "interaction" between the bullet and block takes place during the time interval Δt from the instant the bullet makes contact with the block to when it stops within the block. During this time interval, the system consisting of the bullet and the block experiences a net external force due to friction between the surface and the block.

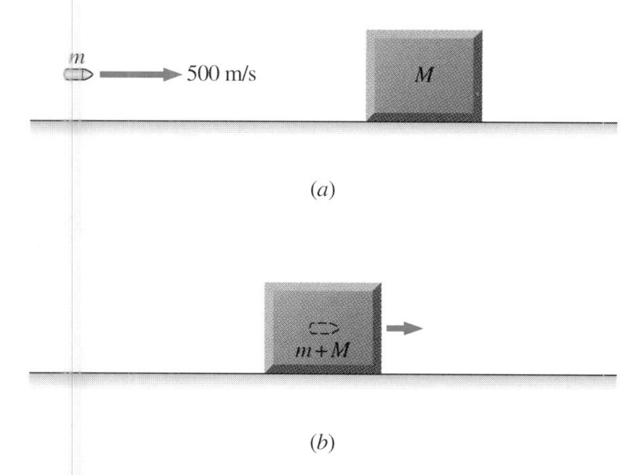

FIGURE 12-8 (a) A bullet moving horizontally strikes a stationary block. (b) The bullet and the block move off together after the short-lived interaction.

Suppose that the masses of the bullet and the block are $m =$ 10 g and $M = 1.0$ kg, respectively, and that the bullet is moving horizontally with a speed of 500 m/s just before the collision. The coefficient of kinetic friction between the block and the surface is taken to be $\mu_K = 0.60$. If the bullet penetrates 25 cm into the block and comes to rest before the block has moved a significant distance, the acceleration of the bullet (assumed constant) is

$$a = \frac{v^2 - v_0^2}{2(x - x_0)} = \frac{0^2 - (500 \text{ m/s})^2}{2(0.25 \text{ m})} = -5.0 \times 10^5 \text{ m/s}^2.$$

In addition, the time interval during which the bullet moves through the block before stopping is

$$\Delta t = \frac{v - v_0}{a} = \frac{0 - 500 \text{ m/s}}{(-5.0 \times 10^5 \text{ m/s}^2)} = 1.0 \times 10^{-3} \text{ s}.$$

The net external force on the system is the frictional force

$$\begin{aligned} f_K &= -\mu_K(m + M)g = -(0.60)(1.01 \text{ kg})(9.8 \text{ m/s}^2) \\ &= -5.9 \text{ N}, \end{aligned}$$

so the change in the system's momentum during the 1.0×10^{-3}-s interval of the interaction is

$$\begin{aligned} \Delta P_x &= f_K \, \Delta t = (-5.9 \text{ N})(1.0 \times 10^{-3} \text{ s}) \\ &= -5.9 \times 10^{-3} \text{ kg} \cdot \text{m/s}. \end{aligned}$$

Now the initial momentum of the system is

$$(P_x)_i = mv = (0.010 \text{ kg})(500 \text{ m/s}) = 5.0 \text{ kg} \cdot \text{m/s}.$$

Hence during the interaction, the momentum changes by only $(0.0059 \text{ kg} \cdot \text{m/s}/5.0 \text{ kg} \cdot \text{m/s}) \times 100\% = 0.12\%$ of its initial value, and within roughly one part in 10^3, momentum is conserved.

This collision is typical of many short-lived interactions where the net external forces are significantly less than the internal forces between the interacting bodies. In these situations, the net external force $\sum \mathbf{F}$ is small enough that over the time Δt of the interaction, $(\sum \mathbf{F}) \, \Delta t \approx 0$. Then from Eq. (12-5),

$$\Delta \mathbf{P} = (\textstyle\sum \mathbf{F}) \, \Delta t \approx 0;$$

and to a good approximation, the momenta *just before* and *just after* the interaction are equal. *The momentum of a system of particles is approximately conserved if the particles interact so quickly that the impulse of the net external force is negligible during their interaction.* For our bullet and block,

$$\begin{aligned} &(P_x)_i = (P_x)_f \\ &(0.010 \text{ kg})(500 \text{ m/s}) + (1.0 \text{ kg})(0 \text{ m/s}) = (1.01 \text{ kg})V, \end{aligned}$$

so the speed of the block immediately after the bullet stops within it is $V = 5.0$ m/s.

Fisher Scientific.

Multiple exposure of a ballistic pendulum used in an undergraduate physics laboratory.

EXAMPLE 12-6 **A Ballistic Pendulum**

The ballistic pendulum shown in Fig. 12-9 is a device that was used before the advent of modern electronics to measure the speed of a bullet. It consists of a large block suspended like a pendulum. When a bullet is shot into the block and becomes embedded, the center of mass of the combination rises some distance h. By measuring this height, the bullet's speed v can be found. What is the relationship between h and v? The masses of the bullet and block are m and M, respectively.

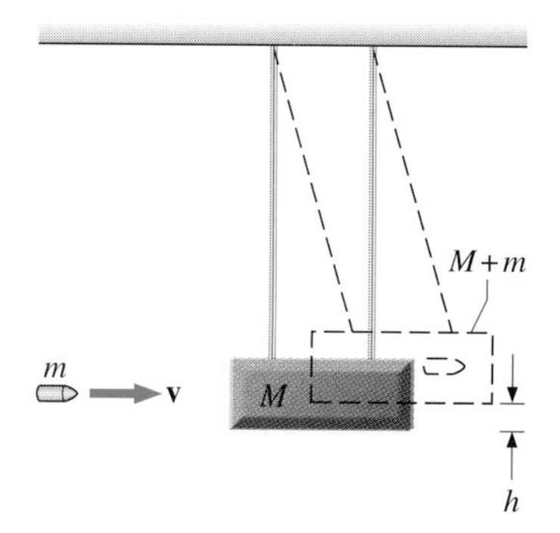

FIGURE 12-9 The ballistic pendulum.

SOLUTION Here we identify our system to be the bullet and the block. Over the short time interval that the bullet penetrates and stops within the block, the displacement of the block is negligible. Therefore, during the collision, the strings supporting the block remain essentially vertical and exert very little horizontal force on the system. There are no other horizontal forces on the system during the collision, so its momentum in the horizontal direction is, to an excellent approximation, conserved. We then have

$$mv = (m + M)V, \qquad (i)$$

where V is the speed of the combination after the interaction.

As the pendulum (and bullet) swings upward, only the gravitational force does work on it. Since this force is conservative, mechanical energy is conserved. With the zero of potential energy at the bottom of the arc, mechanical-energy conservation gives

$$\tfrac{1}{2}(m + M)V^2 = (m + M)gh. \qquad (ii)$$

By eliminating V between Eqs. (i) and (ii), we find the bullet speed v as a function of the masses and the height:

$$v = \frac{m + M}{m}\sqrt{2gh}. \qquad (iii)$$

The ratio of the final mechanical energy of the block plus bullet to the initial mechanical energy of the bullet is

$$\frac{\tfrac{1}{2}(m + M)V^2}{\tfrac{1}{2}mv^2} = \frac{(m + M)gh}{\tfrac{1}{2}[(m + M)^2/m](2gh)} = \frac{m}{m + M}, \qquad (iv)$$

which is obviously much less than 1. Although momentum is approximately conserved in the collision, a significant fraction of the mechanical energy is lost. However, after the combination starts moving, the situation is reversed—mechanical energy, but not momentum, is conserved. This energy is only a small fraction [see Eq. (iv)] of the initial kinetic energy of the bullet.

DRILL PROBLEM 12-7

A bullet of mass 8.0 g is fired into a ballistic pendulum of mass 2.0 kg. The center of mass of the pendulum rises a vertical distance of 0.10 m. What is the speed of the bullet?
ANS. 3.5×10^2 m/s.

DRILL PROBLEM 12-8

An artillery shell of mass m and horizontal velocity 400 m/s explodes into three pieces of mass $m/4$, $m/4$ and $m/2$, respectively. (See Fig. 12-10.) Immediately afterwards, the two pieces of mass $m/4$ move away with velocities 100 m/s vertically and 600 m/s horizontally. Find the speed of the third piece immediately after the explosion.
ANS. 502 m/s.

DRILL PROBLEM 12-9

Discuss the implications of friction due to the ice in Examples 12-3 and 12-4.

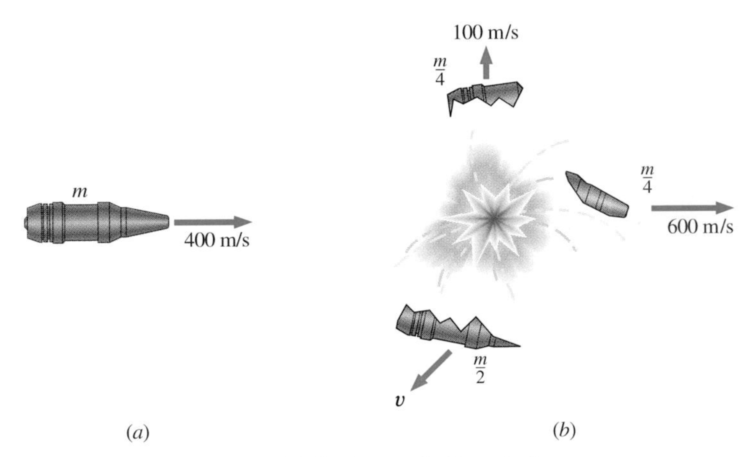

FIGURE 12-10 An exploding shell.

12-5 Elastic Collisions

In the examples involving collisions, we did not have detailed knowledge of the forces between the bodies, yet we were able to learn about the resulting motion of the bodies by using momentum conservation. Now we consider collisions in which the interactive forces are conservative and, consequently, in which mechanical energy is also conserved. Such collisions are said to be *elastic*. Information on the outcome of these collisions can be obtained through the use of both the momentum and the mechanical-energy conservation laws.

For most collisions, the interactive force acts only when the bodies are in close proximity. After separation, they move independently of one another. The potential energy is then zero (or any other constant) both before and after the collision. Mechanical-energy conservation is therefore expressed by equating just the total kinetic energies before and after an elastic collision.

Collisions between subatomic particles are frequently elastic. The bubble chamber photograph of Fig. 12-11 shows a

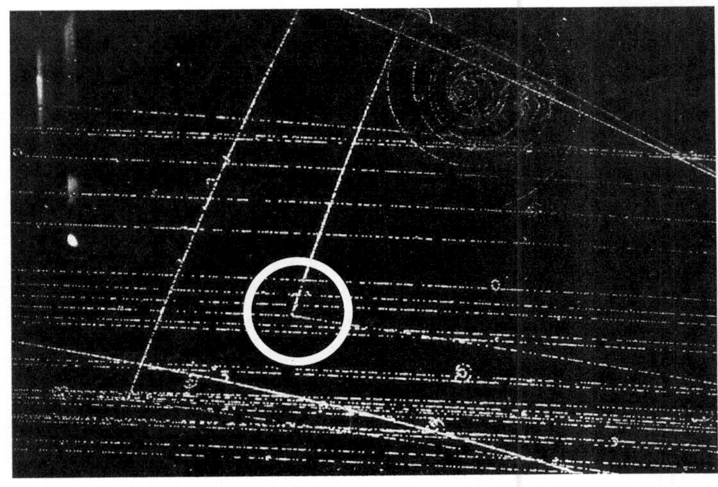

FIGURE 12-11 The circled spot in the bubble chamber designates an elastic collision between two protons.

typical example. A moving proton collides elastically with a stationary proton and the two particles move off in different directions.

A collision that does not conserve mechanical energy is called *inelastic*. Although they weren't so identified, all of the interactions considered in Secs. 12-3 and 12-4 were inelastic. For example, the system consisting of the boy and snowball in Example 12-4 gained mechanical energy from the work done by the boy's muscles, and most of the initial kinetic energy of the bullet of Example 12-6 was dissipated in the collision with the block.

In reality, all collisions in the macroscopic world are inelastic to some extent, because friction can never be completely eliminated. However, there are many cases for which the loss in kinetic energy is so small that the collision can be treated as elastic. A well-known example for which this approximation is valid is the collision between two billiard balls.

One-Dimensional Elastic Collisions

We begin our study of elastic collisions by assuming that the interacting bodies meet head-on, so that they move along the same line before and after the collision. A typical one-dimensional collision is represented by Fig. 12-12, where two bodies moving with initial velocities v_{1i} and v_{2i} collide elastically. After the bodies separate, their final velocities are v_{1f} and v_{2f}, respectively. From the law of momentum conservation,

$$m_1 v_{1i} + m_2 v_{2i} = m_1 v_{1f} + m_2 v_{2f}, \tag{12-9}$$

and from the law of mechanical-energy conservation,

$$\tfrac{1}{2} m_1 v_{1i}^2 + \tfrac{1}{2} m_2 v_{2i}^2 = \tfrac{1}{2} m_1 v_{1f}^2 + \tfrac{1}{2} m_2 v_{2f}^2. \tag{12-10}$$

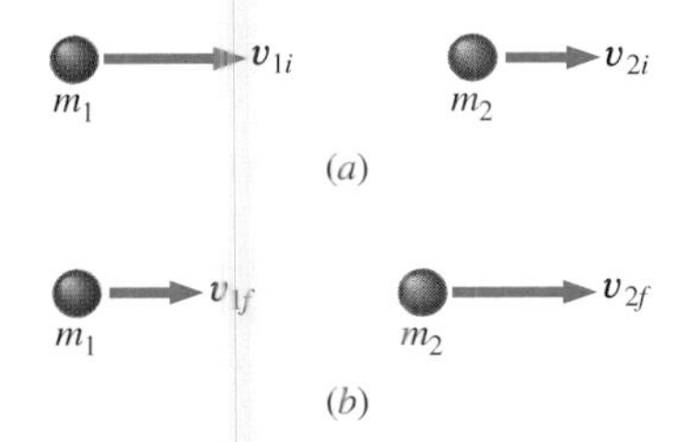

FIGURE 12-12 A one-dimensional collision between two particles: (*a*) before the collision and (*b*) after the collision.

Typically, the initial velocities v_{1i} and v_{2i} are known and the final velocities v_{1f} and v_{2f} are to be determined. We can do just that with Eqs. (12-9) and (12-10). We merely solve Eq. (12-9) for v_{2f} in terms of v_{1f} (or vice versa), substitute this into Eq. (12-10), then solve the resulting equation for v_{1f} (or v_{2f}). Notice that if the collision were inelastic, we would only have Eq. (12-9) at our disposal. It would then not be possible to obtain both final velocities only from knowledge of the initial velocities.

EXAMPLE 12-7 **A One-Dimensional Elastic Collision**

A particle of mass 2.0 kg moving at 15.0 m/s to the right collides elastically with a particle of mass 1.0 kg moving to the right at 9.0 m/s. (See Fig. 12-13.) What are the velocities of the particles after the collision?

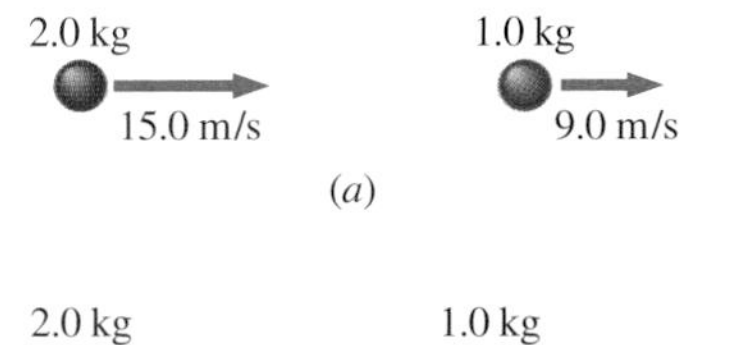

FIGURE 12-13 A one-dimensional elastic collision: (*a*) before the collision and (*b*) after the collision.

SOLUTION Since momentum is conserved,

$$m_1 v_{1i} + m_2 v_{2i} = m_1 v_{1f} + m_2 v_{2f}$$
$$(2.0\text{ kg})(15.0\text{ m/s}) + (1.0\text{ kg})(9.0\text{ m/s}) = (2.0\text{ kg})v_{1f} + (1.0\text{ kg})v_{2f},$$

and

$$39.0\text{ kg}\cdot\text{m/s} = (2.0\text{ kg})v_{1f} + (1.0\text{ kg})v_{2f}. \tag{i}$$

Since the collision is elastic, mechanical energy is also conserved:

$$\tfrac{1}{2} m_1 v_{1i}^2 + \tfrac{1}{2} m_2 v_{2i}^2 = \tfrac{1}{2} m_1 v_{1f}^2 + \tfrac{1}{2} m_2 v_{2f}^2$$
$$\tfrac{1}{2}(2.0\text{ kg})(15.0\text{ m/s})^2 + \tfrac{1}{2}(1.0\text{ kg})(9.0\text{ m/s})^2 = \tfrac{1}{2}(2.0\text{ kg})v_{1f}^2 + \tfrac{1}{2}(1.0\text{ kg})v_{2f}^2,$$

and

$$265.5\text{ J} = \tfrac{1}{2}(2.0\text{ kg})v_{1f}^2 + \tfrac{1}{2}(1.0\text{ kg})v_{2f}^2. \tag{ii}$$

Using Eq. (*i*) to find v_{2f} in terms of v_{1f} and then substituting the resulting expression into Eq. (*ii*), we obtain

$$3.0 v_{1f}^2 - 78 v_{1f} + 495 = 0.$$

This is a quadratic equation with solutions 15.0 m/s and 11.0 m/s. The first solution is just the initial velocity, which of course has to satisfy the equations. The solution of interest is the second, 11.0 m/s; this is the velocity of particle 1 after the collision. With v_{1f} known, we can now use either Eq. (*i*) or Eq. (*ii*) to obtain v_{2f}; it is 17.0 m/s.

The general formulas that give the final velocities in terms of the initial velocities are frequently useful for analyzing one-dimensional elastic collisions. They are obtained by applying the laws of momentum and mechanical-energy conservation as given by Eqs. (12-9) and (12-10) to a one-dimensional elastic collision. After rearrangement of terms, these two equations become

$$m_1(v_{1i} - v_{1f}) = m_2(v_{2f} - v_{2i}), \qquad (12\text{-}11a)$$

and

$$m_1(v_{1i}^2 - v_{1f}^2) = m_2(v_{2f}^2 - v_{2i}^2). \qquad (12\text{-}11b)$$

Upon dividing the second equation by the first (assuming $v_{1i} \neq v_{1f}$ and $v_{2i} \neq v_{2f}$), we find that the velocities are related by

$$v_{1i} + v_{1f} = v_{2f} + v_{2i}, \qquad (12\text{-}11c)$$

which can be rewritten as

$$v_{1i} - v_{2i} = -(v_{1f} - v_{2f}). \qquad (12\text{-}11d)$$

This states that the *relative velocity at which the bodies approach each other before a collision is equal and opposite to the relative velocity at which they separate after the collision.*

We now have two linear equations, Eqs. (12-11a) and (12-11d), which we can solve for v_{1f} and v_{2f} in terms of v_{1i} and v_{2i}. We find that for a *one-dimensional elastic collision*, the final velocities are given in terms of the initial velocities by

$$v_{1f} = \left(\frac{m_1 - m_2}{m_1 + m_2}\right)v_{1i} + \left(\frac{2m_2}{m_1 + m_2}\right)v_{2i} \qquad (12\text{-}12)$$

$$v_{2f} = \left(\frac{2m_1}{m_1 + m_2}\right)v_{1i} + \left(\frac{m_2 - m_1}{m_1 + m_2}\right)v_{2i}. \qquad (12\text{-}13)$$

With the help of these two equations, we can investigate some interesting special cases. One of these is a system for which $v_{2i} = 0$ and $m_2 \gg m_1$. We can approximate this situation by rolling a light plastic ball into a heavy lead ball of the same radius. Since $m_1/(m_1 + m_2) \approx 0$, $(m_2 - m_1) \approx m_2$, and $(m_2 + m_1) \approx m_2$, Eqs. (12-12) and (12-13) reduce to

$$v_{1f} \approx -v_{1i} \qquad \text{and} \qquad v_{2f} \approx 0.$$

This is just what you might expect intuitively. The heavy ball barely moves, and the light ball bounces backward almost as if it had collided with a brick wall.

If $v_{2i} = 0$ and $m_2 \ll m_1$, these same equations give

$$v_{1f} \approx v_{1i} \qquad \text{and} \qquad v_{2f} \approx 2v_{1i}.$$

The heavy incident body is hardly affected by the collision, while the light stationary body leaves with almost twice the velocity of the incident body. This result is not so obvious intuitively using a stationary reference frame. However, if you were to analyze the collision in a frame moving with body 1, the result would be quite easy to interpret. (Try it!)

Another interesting case is that for which $m_1 = m_2$. Then Eqs. (12-12) and (12-13) give

$$v_{1f} = v_{2i} \qquad \text{and} \qquad v_{2f} = v_{1i};$$

that is, the bodies exchange velocities. Anyone who has played billiards is familiar with this result. If a billiard ball moving without topspin or backspin strikes a second ball head-on, the incident ball stops, and the target ball moves off with approximately the incident ball's velocity.

The three special cases we have discussed are important considerations in the choice of an effective substance for moderating neutrons in a nuclear reactor. In a reactor, high-speed neutrons are produced. These neutrons must be slowed (or moderated) to low speeds so that a large fraction of them will react (fission) with U^{235} nuclei to produce energy and more neutrons. What substance should be used to moderate the neutrons? One with heavy nuclei or one with light nuclei? Or do electrons serve as a good moderator?

We can refer to the previous calculations for an answer. The neutron's mass is approximately equal to that of the proton, which is the nucleus of the hydrogen atom. Although lighter than any other nucleus, the proton is much heavier than an electron. If a neutron collides with a heavy nucleus, it will rebound with almost the same speed, while collisions with electrons hardly slow it down. The neutron velocity is moderated most effectively by a nucleus of almost equal mass, namely, the hydrogen proton. There are other considerations based on the properties of nuclear reactions that make hydrogen less than an ideal choice. However, its deficiencies are overcome by its moderating effectiveness and its availability and low cost as a component of water. Consequently, most nuclear reactors use water as a moderator.

DRILL PROBLEM 12-10

Use Eqs. (12-12) and (12-13) to find the final velocities of the particles of Example 12-7.

DRILL PROBLEM 12-11

A 0.60-kg cart moving along an air track at 0.50 m/s collides elastically with a 0.40-kg cart moving at 0.30 m/s in the same direction. What are the speeds of the carts after the collision?
ANS. 0.34 m/s and 0.54 m/s.

*Elastic Collisions in Two Dimensions

We now investigate elastic collisions that are not head-on and therefore not one-dimensional. For simplicity, we will only discuss the case for which one of the colliding particles is initially at rest. Much research in high-energy and nuclear physics involves studying similar collisions. Typically, a high-velocity particle produced by an accelerator collides with a stationary atomic nucleus. The colliding particles are detected and their momenta and energies measured. This information is then used to study the physical properties of the collision.

Suppose particle 1 of mass m_1 and velocity $\mathbf{v}_{1i}$ strikes particle 2, which is at rest and has mass m_2. After the collision, the velocities of the two particles are $\mathbf{v}_{1f}$ and $\mathbf{v}_{2f}$. The two velocity vectors form a plane which, since momentum is conserved, must contain $\mathbf{v}_{1i}$. The collision can therefore be represented by the planar sketch of Fig. 12-14. Along the x direction

$$m_1 v_{1i} = m_1 v_{1f}\cos\theta_1 + m_2 v_{2f}\cos\theta_2,$$

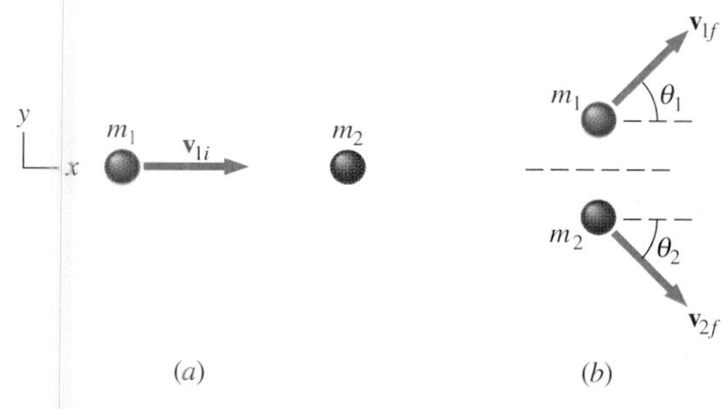

FIGURE 12-14 A two-dimensional elastic collision. Particle 2 is initially at rest. (*a*) Before collision and (*b*) after collision.

and along the *y* direction,

$$0 = m_1 v_{1f} \sin\theta_1 + m_2 v_{2f} \sin\theta_2.$$

Because the collision is elastic, mechanical energy is also conserved:

$$\tfrac{1}{2} m_1 v_{1i}^2 = \tfrac{1}{2} m_1 v_{1f}^2 + \tfrac{1}{2} m_2 v_{2f}^2.$$

If we know the masses and the initial speed v_{1i}, there are four quantities to be determined: v_{1f}, v_{2f}, θ_1, and θ_2. However, the conservation laws provide only three equations. As a result, either one of the four quantities must be known, or another mathematical condition must be found before the final state can be completely determined. One possible source of additional information is a mathematical statement about the force between the particles. (For example, the force in a head-on collision results in one-dimensional motion.) Unfortunately, we do not know how to represent the force of interaction for colliding elementary particles and are therefore unable to calculate the final velocities.

Actually, if only momentum and energy conservation were sufficient to determine the final velocities, these velocities would then be independent of the nature of the interactive force. While simplifying matters considerably, this fact would make the study of two-particle elastic collisions quite uninteresting, for then the collisions of all particles—be they billiard balls or elementary particles—would be identical.

EXAMPLE 12-8 **A TWO-DIMENSIONAL ELASTIC COLLISION**

A gas molecule of mass m and speed 400 m/s collides elastically with an identical molecule which is at rest. The incident molecule leaves at an angle of 37° to its original direction. Find the speed of each molecule after the collision and the scattering angle of the second molecule.

SOLUTION Using Fig. 12-14, we have $m_1 = m_2 = m$, $\mathbf{v}_{1i} = 400\mathbf{i}$ m/s and $\theta_1 = 37°$. The three unknowns, θ_2, v_{1f}, and v_{2f}, can be determined from the conservation equations:

$$m(400 \text{ m/s}) = m v_{1f} \cos 37° + m v_{2f} \cos\theta_2,$$
$$0 = m v_{1f} \sin 37° + m v_{2f} \sin\theta_2,$$

and

$$\tfrac{1}{2} m (400 \text{ m/s})^2 = \tfrac{1}{2} m v_{1f}^2 + \tfrac{1}{2} m v_{2f}^2.$$

After a little manipulation, these three equations become, respectively,

$$400 - 0.80 v_{1f} = v_{2f} \cos\theta_2, \qquad (i)$$
$$-0.60 v_{1f} = v_{2f} \sin\theta_2, \qquad (ii)$$

and

$$1.6 \times 10^5 = v_{1f}^2 + v_{2f}^2. \qquad (iii)$$

By squaring Eqs. (*i*) and (*ii*) and adding them, we obtain

$$(1.6 \times 10^5) - 640 v_{1f} + v_{1f}^2 = v_{2f}^2. \qquad (iv)$$

Subtracting this equation from Eq. (*iii*) gives

$$2 v_{1f}^2 = 640 v_{1f},$$

which has the nontrivial solution $v_{1f} = 320$ m/s for the final speed of molecule 1. This result can now be used in Eq. (*iii*) to determine v_{2f}, the final speed of molecule 2, and in Eq. (*ii*) to determine θ_2, the angle at which molecule 2 is directed after the collision. The results are $v_{2f} = 240$ m/s and $\theta_2 = -53°$ (53° below the horizontal line of the figure). Notice that the two molecules go off at 90° relative to each other after the collision. This result can be shown to be true for any two-dimensional elastic collision between particles of equal mass.

*COLLISIONS IN THE CENTER-OF-MASS FRAME

If the momentum of a system of particles is conserved, the velocity of that system's center of mass is constant. Thus a nonrotating coordinate system moving with the center of mass is an inertial reference frame, and the laws of mechanics, such as momentum and mechanical-energy conservation, may be employed in this frame. We now investigate how this frame can be used to greatly simplify the analysis of elastic collisions. We will only consider the one-dimensional case. The following notation (see Fig. 12-15) will be used:

1. The velocities of the colliding particles relative to the *laboratory frame* (a general term for an earth-bound reference frame) are represented by v_{1i}, v_{1f}, ...
2. The velocities of the same particles relative to their center-of-mass frame are v_{1i}^{CM}, v_{1f}^{CM}, ...
3. The velocity of the center of mass in the laboratory frame is v_{CM}. From Eq. (8-4*a*), this velocity is given by

$$v_{\text{CM}} = \frac{m_1 v_{1i} + m_2 v_{2i}}{m_1 + m_2},$$

and from Eq. (4-20), the velocities in the center of mass and the laboratory frames are related by

$$v_{1i}^{\text{CM}} = v_{1i} - v_{\text{CM}},$$
$$v_{1f}^{\text{CM}} = v_{1f} - v_{\text{CM}},$$

and so on.

Laboratory frame

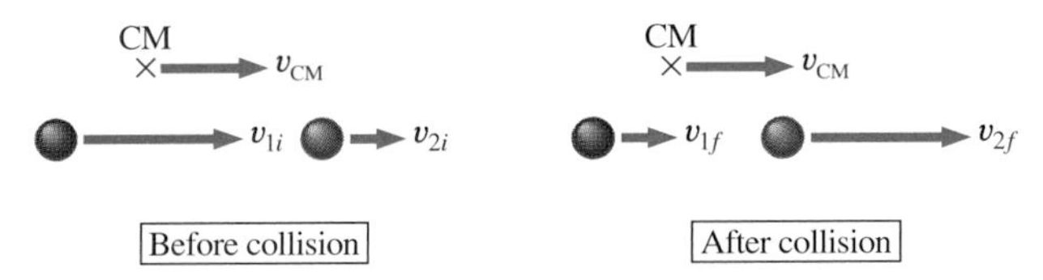

Center-of-mass frame

FIGURE 12-15 A one-dimensional elastic collision between two particles as seen in the laboratory frame and in the center-of-mass frame.

Laboratory frame

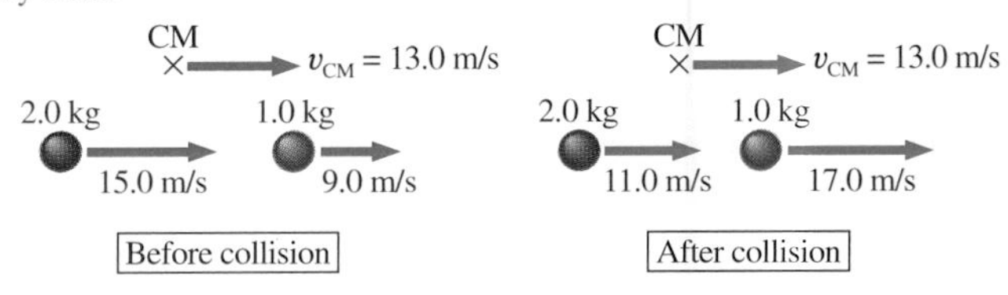

Center-of-mass frame

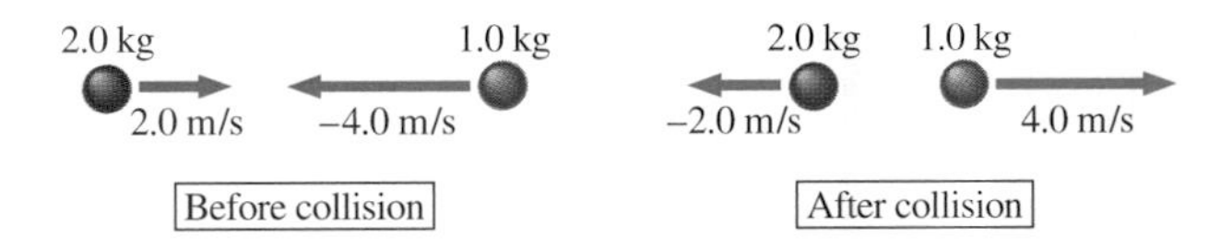

FIGURE 12-16 A collision viewed in both the laboratory and the center-of-mass frames.

The velocity of the center of mass in its own frame is, of course, zero. Therefore the net momentum in this frame is also zero, and momentum conservation is given by

$$m_1 v_{1i}^{CM} + m_2 v_{2i}^{CM} = m_1 v_{1f}^{CM} + m_2 v_{2f}^{CM} = 0.$$

For an elastic collision, the law of mechanical-energy conservation in the center-of-mass frame gives

$$\tfrac{1}{2} m_1 (v_{1i}^{CM})^2 + \tfrac{1}{2} m_2 (v_{2i}^{CM})^2 = \tfrac{1}{2} m_1 (v_{1f}^{CM})^2 + \tfrac{1}{2} m_2 (v_{2f}^{CM})^2.$$

The momentum equations can be used to find v_{2i}^{CM} in terms of v_{1i}^{CM} and v_{2f}^{CM} in terms of v_{1f}^{CM}. Substitution of these expressions into the energy equation then yields

$$(v_{1i}^{CM})^2 = (v_{1f}^{CM})^2,$$

so

$$|v_{1i}^{CM}| = |v_{1f}^{CM}|,$$

which also leads to

$$|v_{2i}^{CM}| = |v_{2f}^{CM}|.$$

Relative to the center-of-mass reference frame, a particle's speed is unchanged in an elastic collision. This fact greatly simplifies the analysis of elastic collisions, as you will see in the following example.

EXAMPLE 12-9 Analysis of a Collision Using the Center-of-Mass Frame

Determine the final velocities of the particles of Example 12-7 by analyzing the collision in the center-of-mass frame.

SOLUTION The velocity of the center of mass is

$$v_{CM} = \frac{m_1 v_{1i} + m_2 v_{2i}}{m_1 + m_2} = \frac{(2.0\text{ kg})(15.0\text{ m/s}) + (1.0\text{ kg})(9.0\text{ m/s})}{2.0\text{ kg} + 1.0\text{ kg}} = 13.0\text{ m/s},$$

so in the CM frame (as shown in Fig. 12-16), the initial velocities of the particles are

$$v_{1i}^{CM} = 15.0\text{ m/s} - 13.0\text{ m/s} = 2.0\text{ m/s}$$

and

$$v_{2i}^{CM} = 9.0\text{ m/s} - 13.0\text{ m/s} = -4.0\text{ m/s}.$$

In this frame, the speeds are unchanged by the collision. Furthermore, since this is a one-dimensional collision, the particles in their final state must move opposite to their initial directions. We therefore have

$$v_{1f}^{CM} = -2.0\text{ m/s} \quad \text{and} \quad v_{2f}^{CM} = 4.0\text{ m/s}.$$

The velocities in the laboratory frame can now be calculated. They are

$$v_{1f} = v_{1f}^{CM} + v_{CM} = -2.0\text{ m/s} + 13.0\text{ m/s} = 11.0\text{ m/s},$$
$$v_{2f} = v_{2f}^{CM} + v_{CM} = 4.0\text{ m/s} + 13.0\text{ m/s} = 17.0\text{ m/s},$$

which agree with the results of Example 12-7.

DRILL PROBLEM 12-12

Solve Drill Prob. 12-11 by analyzing the collision in the center-of-mass frame.

*12-6 Motion with Varying Mass

Up to this point, the mass of every body we have considered has remained constant. However, there are times when the body of interest either gains or loses mass. One example is a rocket plus its onboard fuel—the mass of this body obviously decreases as the fuel is ejected. Another example is a glass and the water it contains. As water enters the glass, the mass of the (glass + water contained) increases.

The analysis of such systems can be quite confusing, and mistakes are frequently made because Newton's second law is used carelessly. *It is essential that the particular set of particles under consideration at all times be identical.* Although the mass of the *body* whose motion is to be determined will be changing, the mass of the *system* to which Newton's second law is applied will not be changing.

Consider the rocket of Fig. 12-17. At a time t, the system consisting of the rocket and its unburned fuel has a mass m and velocity $\mathbf{v}$. Suppose that the fuel is being burned and ejected at a velocity $\mathbf{u}$ relative to the rocket. Then a short time Δt later, this same system consists of a rocket and its unburned fuel of mass $m - \Delta m$ moving with velocity $\mathbf{v} + \Delta\mathbf{v}$ and burned fuel of mass Δm moving with velocity $\mathbf{v} + \mathbf{u}$. The system's momenta at times t and $t + \Delta t$ are

$$\mathbf{P}(t) = m\mathbf{v},$$

and

$$\mathbf{P}(t + \Delta t) = (m - \Delta m)(\mathbf{v} + \Delta\mathbf{v}) + \Delta m(\mathbf{v} + \mathbf{u}).$$

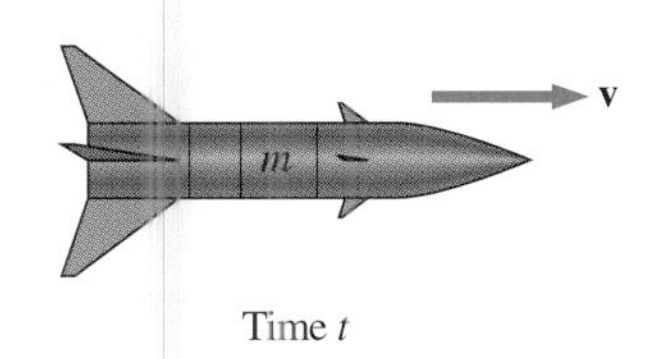

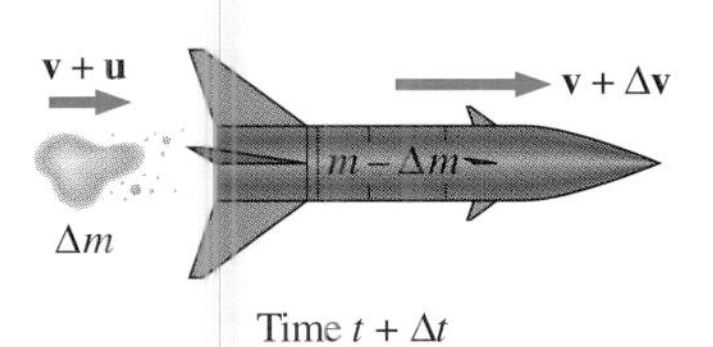

FIGURE 12-17 A rocket and its unburned fuel is a variable-mass system. In a time interval Δt, fuel of mass Δm is ejected from the rocket at a velocity $\mathbf{u}$ relative to the rocket.

Dividing the difference between these equations by Δt gives

$$\frac{\Delta\mathbf{P}}{\Delta t} = m\frac{\Delta\mathbf{v}}{\Delta t} - \Delta m\frac{\Delta\mathbf{v}}{\Delta t} + \mathbf{u}\frac{\Delta m}{\Delta t}.$$

In the limit $\Delta t \to 0$, the left-hand side of this equation is $d\mathbf{P}/dt$, which, from Newton's second law, is equal to the net external force $\sum\mathbf{F}$ on the system. Since the numerator of the second term on the right-hand side is a second-order differential, it goes to zero in this limit and we are left with

$$\sum\mathbf{F} = m\frac{d\mathbf{v}}{dt} + \mathbf{u}\frac{dm}{dt}.$$

In the derivation of this equation, dm/dt has been taken to be positive if the mass of the rocket and its unburned fuel is decreasing and negative if it is increasing. However, the opposite sign convention where the negative sign is associated with decreasing mass is more intuitive. So we'll replace dm/dt by $-dm/dt$ and write

$$\sum\mathbf{F} = m\frac{d\mathbf{v}}{dt} - \mathbf{u}\frac{dm}{dt};$$

or,

$$m\frac{d\mathbf{v}}{dt} = \sum\mathbf{F} + \mathbf{u}\frac{dm}{dt}. \qquad (12\text{-}14)$$

Newton's second law has been applied to a system consisting of a (rocket + unburned fuel) and its ejected fuel. The result is an equation of motion (Eq. (12-14) for *just* the (rocket + unburned fuel). This equation states that the mass × acceleration of the (rocket + unburned fuel) is equal to the sum of the net external force on it and the term $\mathbf{u}(dm/dt)$. This latter term, which is called the **thrust**, represents the force on the rocket due to the ejected fuel. It is a reaction to the force that the rocket engines exert on the ejected fuel.

Although Eq. (12-14) was derived for a rocket, it applies to any object that is gaining or losing mass. In general, the term $\mathbf{u}(dm/dt)$ can act as either an accelerating or decelerating force, depending on its sign.

An astronaut with a nitrogen-propelled jetpack.

EXAMPLE 12-10 **THRUST ON A SPACECRAFT**

A spacecraft is moving in gravity-free space along a straight line when its pilot decides to accelerate forward. He turns on the thrusters, and burned fuel is ejected at a constant rate of 2.0×10^2 kg/s and a relative speed of 2.5×10^2 m/s. The initial mass of the spacecraft and its unburned fuel is 2.0×10^4 kg and the thrusters are on for 30 s. (*a*) What is the thrust on the spacecraft? (*b*) What is the spacecraft's acceleration as a function of time? (*c*) What are the spacecraft's accelerations at $t = 0$, 15, 30, and 35 s?

SOLUTION We assume that the spacecraft is moving in the positive direction in this one-dimensional problem. Since fuel is ejected, $dm/dt = -2.0 \times 10^2$ kg/s, and since the fuel leaves opposite to the spacecraft's direction of motion, $u = -2.5 \times 10^2$ m/s.

(*a*) The thrust is

$$u\frac{dm}{dt} = (-2.5 \times 10^2\ \text{m/s})(-2.0 \times 10^2\ \text{kg/s}) = 5.0 \times 10^4\ \text{N}.$$

(*b*) In the absence of any external force, Eq. (12-14) becomes

$$m\frac{dv}{dt} = u\frac{dm}{dt},$$

so

$$a = \frac{dv}{dt} = \frac{u}{m}\frac{dm}{dt} = \left(\frac{1}{m}\right)5.0 \times 10^4\ \text{N}.$$

The mass of the spacecraft and its unburned fuel decreases at the constant rate of 2.0×10^2 kg/s; hence

$$m = 2.0 \times 10^4\ \text{kg} - (2.0 \times 10^2\ \text{kg/s})t,$$

and

$$a = \frac{5.0 \times 10^4\ \text{N}}{2.0 \times 10^4\ \text{kg} - (2.0 \times 10^2\ \text{kg/s})t}.$$

(*c*) At $t = 0$,

$$a = \frac{5.0 \times 10^4\ \text{N}}{2.0 \times 10^4\ \text{kg}} = 2.5\ \text{m/s}^2.$$

At $t = 15$ s,

$$a = \frac{5.0 \times 10^4\ \text{N}}{2.0 \times 10^4\ \text{kg} - (2.0 \times 10^2\ \text{kg/s})(15\ \text{s})} = 2.9\ \text{m/s}^2.$$

At $t = 30$ s,

$$a = \frac{5.0 \times 10^4\ \text{N}}{2.0 \times 10^4\ \text{kg} - (2.0 \times 10^2\ \text{kg/s})(30\ \text{s})} = 3.6\ \text{m/s}^2.$$

The thrusters are turned off after 30 s, so at $t = 35$ s, there is no acceleration.

EXAMPLE 12-11 Velocity of an Ascending Rocket

A rocket whose initial mass including fuel is m_0 is fired vertically upward from the earth's surface. Fuel is ejected out the rear at a constant relative speed u and a rate $dm/dt = -r$, where r is positive. What is the velocity of the rocket as a function of time during the period that fuel is being burned? Neglect air resistance and assume that the force of gravity is mg.

SOLUTION With the positive direction chosen to be *upward*, the relative velocity of the fuel is negative and represented by $-u$. Inserting the net external force $\sum F = -mg$ into Eq. (12-14), we find

$$m\frac{dv}{dt} = -mg + (-u)(-r),$$

or

$$\frac{dv}{dt} = -g + \frac{u}{m}r. \qquad (i)$$

The mass of the rocket and its unburned fuel decreases at a constant rate, so

$$m = m_0 - rt. \qquad (ii)$$

Substitution of Eq. (*ii*) into Eq. (*i*) then gives

$$\frac{dv}{dt} = -g + \frac{ur}{m_0 - rt}.$$

Assuming that the rocket starts at $t = 0$ with zero initial velocity, this equation can be integrated as follows:

$$\int_0^v dv' = -\int_0^t g\,dt' + ur\int_0^t \frac{dt'}{m_0 - rt'},$$

so

$$v = [-gt' - u\ln(m_0 - rt')]\Big|_0^t,$$

and

$$v = u\ln\left(\frac{m_0}{m_0 - rt}\right) - gt,$$

which gives us the velocity of the rocket as a function of time during the period that the fuel is being burned.

EXAMPLE 12-12 Snow Accumulating on a Moving Sled

A sled of mass M_0 is moving across a frictionless surface of ice at a speed v_0 when it begins to snow. (See Fig. 12-18.) The snow falls vertically with respect to the ground. (*a*) Use Eq. (12-14) to determine the velocity v_f of the sled when snow of mass m_0 has accumulated in it. (*b*) Use momentum conservation to determine v_f.

FIGURE 12-18 A variable-mass system: a sled moving across a frictionless surface and the snow that it accumulates.

SOLUTION (*a*) Suppose that at some instant the sled is moving along the positive *x* axis with velocity v relative to the ground. Since the snow is falling vertically, its horizontal velocity relative to the sled is $u = -v$. Without friction there is no external horizontal force, and Eq. (12-14) becomes

$$m\frac{dv}{dt} = -v\frac{dm}{dt}.$$

With initial values v_0 and M_0 and final values v_f and $m_0 + M_0$, integration yields

$$\int_{M_0}^{m_0+M_0} \frac{dm}{m} = -\int_{v_0}^{v_f} \frac{dv}{v},$$

so

$$\ln\frac{m_0 + M_0}{M_0} = -\ln\frac{v_f}{v_0},$$

or

$$\ln\frac{(m_0 + M_0)v_f}{M_0 v_0} = 0,$$

where we have used the relationship ln AB = ln A + ln B. Since ln 1 = 0, we have

$$\frac{(m_0 + M_0)v_f}{M_0 v_0} = 1,$$

so when snow of mass m_0 has accumulated in the sled, its velocity is

$$v_f = \frac{M_0}{m_0 + M_0} v_0.$$

(*b*) For the system composed of the sled and snow of mass m_0, there is no external force in the horizontal direction (friction at the ice surface is negligible). Consequently, the momentum of the system in this direction must remain constant. Initially, the sled has a horizontal momentum of $M_0 v_0$ and the snow has no horizontal momentum since it is falling vertically. When snow of mass m_0 has accumulated in the sled, the total horizontal momentum of the system is $(m_0 + M_0)v_f$. Thus from momentum conservation in the horizontal direction,

$$M_0 v_0 = (m_0 + M_0)v_f,$$

and

$$v_f = \frac{M_0}{m_0 + M_0} v_0,$$

in agreement with our result of part (*a*).

DRILL PROBLEM 12-13

At a particular instant, the mass of a rocket and its unburned fuel is 4.0×10^4 kg. The rocket is moving in gravity-free space and is ejecting fuel at a rate of 50 kg/s with a relative speed of 300 m/s. What is the acceleration of the rocket at this instant?
ANS. 0.38 m/s^2.

DRILL PROBLEM 12-14

Snow falls vertically and accumulates at a rate of 1.0×10^{-3} kg/s in a sled of mass 100 kg, which is moving across frictionless ice in a straight line. What net horizontal force is required to keep the sled moving at a constant velocity 0.60 m/s?
ANS. 6.0×10^{-4} N.

SUMMARY

1. **Momentum of a particle**
   (*a*) Momentum $\mathbf{p} = m\mathbf{v}$.
   (*b*) Newton's second law for a particle can be written as

   $$\sum \mathbf{F} = \frac{d\mathbf{p}}{dt},$$

   where $\sum \mathbf{F}$ is the net force acting on the particle.
2. **Momentum of a system of particles**
   (*a*) If $\mathbf{p}_1, \mathbf{p}_2, \ldots, \mathbf{p}_N$ are the momenta of the N particles of a system, the momentum of the system is

   $$\mathbf{P} = \mathbf{p}_1 + \mathbf{p}_2 + \ldots + \mathbf{p}_N.$$

   (*b*) If M and $\mathbf{v}_{CM}$ are the mass and center-of-mass velocity, respectively, of a system of particles, the momentum of the system is $\mathbf{P} = M\mathbf{v}_{CM}$.
   (*c*) If $\sum \mathbf{F}$ is the net external force acting on a system of particles, then

   $$\sum \mathbf{F} = \frac{d\mathbf{P}}{dt} = M\mathbf{a}_{CM}.$$
3. **Impulse**
   (*a*) The impulse $\mathbf{J}$ on a particle between times t_1 and t_2 is by definition

   $$\mathbf{J} = \int_{t_1}^{t_2} \sum \mathbf{F}\, dt,$$

   where $\sum \mathbf{F}$ is the net force acting on the particle.
   (*b*) The impulse-momentum theorem: The change in momentum of a particle is equal to the impulse acting on it.
4. **Conservation of momentum**
   (*a*) If the net external force on a system of particles is zero, the momentum of that system is conserved.
   (*b*) When the momentum of a system of particles is conserved, the velocity of the system's center of mass remains constant.
   (*c*) The momentum of a system of particles is approximately conserved during an interaction short enough that the impulse of the net external force is negligible.

5. **Elastic collisions**
Elastic collisions conserve both momentum and mechanical energy. Inelastic collisions conserve only momentum.

*6. **Motion with varying mass**
If a body of mass m and velocity $\mathbf{v}$ is losing or gaining mass at a rate dm/dt with a relative velocity $\mathbf{u}$, then

$$m\frac{d\mathbf{v}}{dt} = \sum \mathbf{F} + \mathbf{u}\frac{dm}{dt},$$

where $\sum \mathbf{F}$ is the net external force acting on the body.

Questions

12-1. Two bodies of unequal mass have the same kinetic energy. Does the less massive or the more massive body have the larger momentum?

12-2. Is it possible for a small force to produce a larger impulse on a given object than a large force? Explain.

12-3. Two objects of equal mass are moving with equal and opposite velocities when they collide. Can all the kinetic energy be lost in the collision?

12-4. How do you reconcile the law of conservation of momentum with the fact that an object slows down and then speeds up when it is tossed straight up in the air?

12-5. Why is a 10-m fall onto pavement far more dangerous than the same fall into water?

12-6. What external force is responsible for changing the momentum of a car moving along a horizontal road?

12-7. A piece of putty and a tennis ball with the same mass are thrown against a wall with the same velocity. For which collision is the larger impulse imparted to the wall?

12-8. A tennis ball is hit against a wall. Is momentum conserved in the collision between ball and wall? Explain.

12-9. If you are on a frictionless surface of ice covering a lake, how might you get to shore?

12-10. A ball is dropped from the top of a tall building. Define a system for which the momentum is conserved for this process. Does the momentum stay conserved after the ball hits the ground?

12-11. A 20-kg object is dropped from a height of 10 m above the earth's surface. In principle, the object and earth move toward one another until they collide. Estimate how far the earth moves in this case.

12-12. Describe a system for which mechanical energy but not momentum is conserved. Also describe a system for which momentum but not mechanical energy is conserved.

12-13. A boy points his canoe toward the shore, then starts it moving in that direction by walking from the front to the back of the canoe. If there is no friction, will the canoe keep moving after he reaches the back of the canoe and sits down?

12-14. Can momentum be conserved when dissipative forces are doing work?

12-15. Can mechanical energy be conserved when external forces are doing work?

12-16. A firecracker explodes into many pieces. Is momentum conserved in the explosion? Is mechanical energy conserved?

12-17. A girl is sitting in the back of a flat boat that is floating at rest on a calm lake. When she gets up and walks to the front end of the boat, it moves away from the shore a certain distance. Compare this distance with the distance the boat moves if the girl gets up and runs swiftly to the front end.

12-18. Suppose a rocket is maneuvering in outer space, far from any celestial body. Does the center of mass of the rocket plus its exhausted fuel accelerate?

12-19. Explain how a balloon can be used to demonstrate the principle of rocket propulsion.

12-20. A rocket can accelerate in outer space but a propeller-driven airplane cannot. Explain.

12-21. How does a spacecraft maneuver in outer space?

Problems

Momentum

12-1. Compare the momenta of a 16- and a 10-lb bowling ball if both are thrown with the same velocity.

12-2. How fast would a 1200-kg car be traveling if it had the same momentum as a 1500-kg car traveling at 100 km/h?

12-3. The center of mass of a system of particles has a velocity $(3.0\mathbf{i} - 4.0\mathbf{j})$ m/s. If the magnitude of the system's momentum is 10 N·s, what is the mass of the system?

12-4. The net momentum of a three-particle system is $(5.0\mathbf{i} - 8.0\mathbf{j} + 3.0\mathbf{k})$ N·s. Particle 1 has mass 0.50 kg and velocity $(2.0\mathbf{i} - 6.0\mathbf{k})$ m/s while particle 2 has mass 0.20 kg and velocity $(-6.0\mathbf{j} + 4.0\mathbf{k})$ m/s. If the mass of particle 3 is 1.5 kg, what is its velocity?

12-5. A 1.0-kg object is dropped from a 100-m-high building. Plot its momentum as a function of time and displacement.

Momentum Conservation

12-6. A 2.0-kg cart moving to the right with a speed of 4.0 m/s collides with and sticks to a 4.0-kg cart, which is (*a*) at rest; (*b*) moving to the right with a speed of 3.0 m/s; (*c*) moving to the left with a speed of 3.0 m/s. For each case, find the velocity of the combination after the collision, and compare the kinetic energies before and after the collision.

12-7. An 80-kg astronaut floating in gravity-free space outside his capsule throws his 0.50-kg hammer so that it moves with a speed of 20 m/s relative to the capsule. What happens to the astronaut?

12-8. A rifle of mass 4.0 kg fires a 5.0-g bullet with a speed of 800 m/s. What is the recoil speed of the rifle?

12-9. A satellite of mass m, which is at rest in gravity-free space, explodes into three pieces of mass $m/8$ and one piece of mass $5m/8$. The three pieces of equal mass leave with velocities

$(3.0\mathbf{i} - 2.0\mathbf{j}) \times 10^2$ m/s, $(5.0\mathbf{j} + 4.0\mathbf{k}) \times 10^2$ m/s, and $(-2.0\mathbf{i} + 4.0\mathbf{j} + 6.0\mathbf{k}) \times 10^2$ m/s. What is the velocity of the fourth piece after the explosion?

12-10. Two particles approach each other at right angles, as shown in the accompanying figure. Particle A has a mass $m_A = 10$ kg and a speed $v_A = 30$ m/s. Particle B has a mass $m_B = 5.0$ kg and a speed $v_B = 20$ m/s. The particles collide and stick together. (*a*) Assuming there are no outside forces, what is the velocity of the combination at any time after the collision? (*b*) What fraction of the initial kinetic energy is lost in the collision?

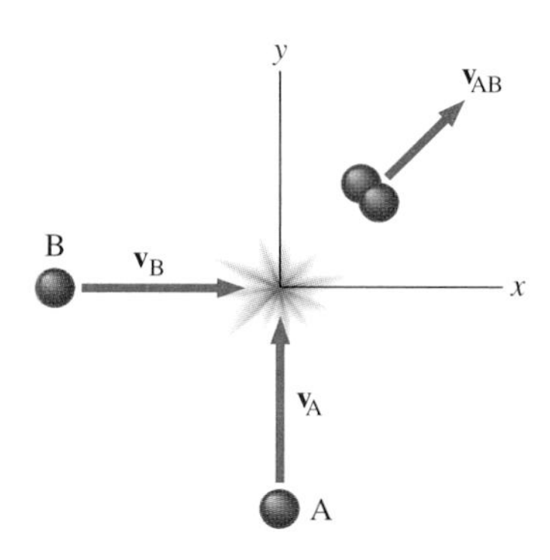

PROBLEM 12-10

12-11. A shell of mass 6.0 kg resting at the origin of a rectangular coordinate system in outer space explodes into four fragments: fragment 1 with mass $m_1 = 1.0$ kg and velocity $\mathbf{v}_1 = (2.0\mathbf{i} - 4.0\mathbf{j}) \times 10^3$ m/s, fragment 2 with mass $m_2 = 2.0$ kg and velocity $\mathbf{v}_2 = (3.0\mathbf{i} + 4.0\mathbf{j} - 2.0\mathbf{k}) \times 10^3$ m/s, fragment 3 with mass $m_3 = 1.5$ kg and velocity $\mathbf{v}_3 = -2.0 \times 10^3\,\mathbf{k}$ m/s. Find the velocity of the fourth fragment.

12-12. A 5000-kg truck moving at 15 m/s eastward collides at an intersection with a 1000-kg car moving northward at 10 m/s. If the two vehicles become completely entangled, what is the velocity of the combination immediately after the collision?

12-13. A halfback of mass 90 kg is moving with the ball from east to west at 8.0 m/s. A lineman of mass 120 kg runs northeast at 6.5 m/s and tackles the halfback. Calculate the velocity of the entangled players immediately after they meet. How much mechanical energy is lost in the collision?

12-14. A 5.0-g bullet moving horizontally at 800 m/s penetrates a 2.0-kg block at rest on a horizontal surface and emerges from the block at 400 m/s. (See the accompanying figure.) (*a*) What is the velocity of the block immediately after the bullet emerges? (*b*) What fraction of the initial kinetic energy is lost in the collision? (*c*) If the coefficient of kinetic friction between the block and surface is $\mu_K = 0.60$, how far does the block slide after the collision?

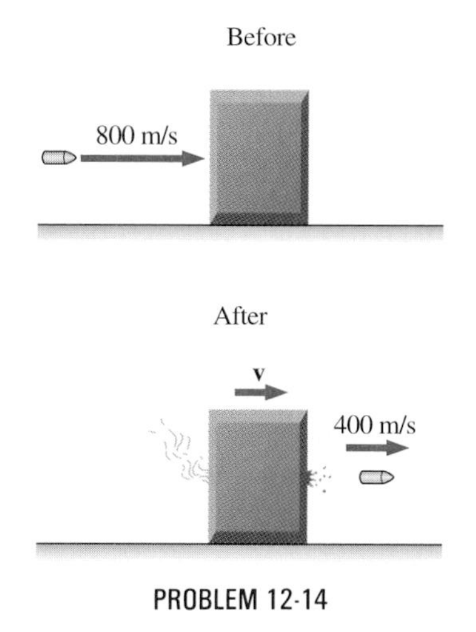

PROBLEM 12-14

12-15. A 5.0-g bullet moving with a speed of 700 m/s is stopped by a ballistic pendulum of mass 3.0 kg. Calculate the distance the pendulum rises.

12-16. A 5.0-g bullet is fired into a ballistic pendulum whose mass is 2.0 kg. The pendulum rises a vertical distance of 11 cm. What is the speed of the bullet?

12-17. A 5.0-g bullet moving with a speed of 800 m/s enters one side of a 2.0-kg ballistic pendulum and emerges from the opposite side with a speed of 200 m/s. What is the distance the pendulum moves upward?

12-18. When suspended from a massless spring, a weight pan of mass 100 g stretches the spring 5.0 cm. Find the maximum distance the pan moves downward when a lump of putty of mass 100 g is dropped from a height of 40 cm onto the pan. (See the accompanying figure.)

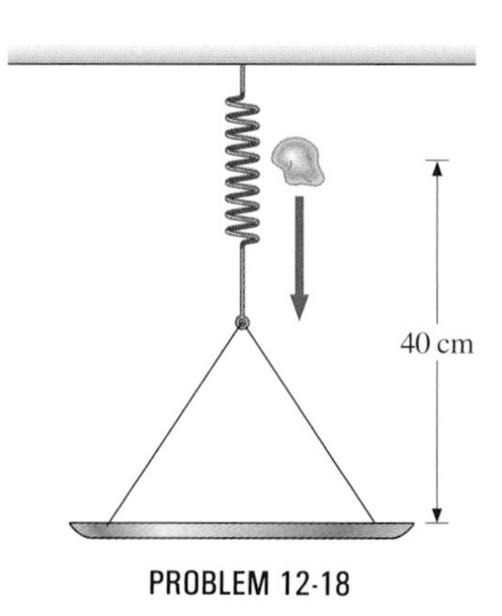

PROBLEM 12-18

12-19. Two children on ice skates, one of mass m_1, the other of mass m_2, hold onto opposite ends of a rope and pull while they are on a frozen pond. (*a*) If they start a distance *d* apart, where will they meet? (*b*) Suppose the rope is tied around the waist of the child of mass m_2, so only the child of mass m_1 pulls on the rope with her hands. Assuming they again start a distance *d* apart, where will they now meet?

12-20. The accompanying figure shows two carts pressed against opposite ends of a massless spring. The masses of the carts are 2.0 and 4.0 kg, and the force constant of the spring is $k = 4800$ N/m. When the carts are released, the spring expands, and the

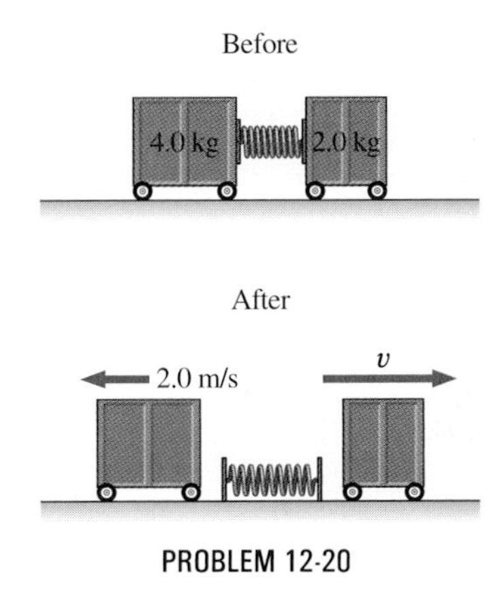

PROBLEM 12-20

4.0-kg cart moves off with a speed of 2.0 m/s. (a) What is the speed of the 2.0-kg cart? (b) How far was the spring compressed?

12-21. An 80-kg father and his 35-kg son are sitting at opposite ends of a 3.0-m-long rowboat of mass 40 kg. The father is at the front end, which is 10.0 m from the shore, and the boy is at the back end, 13.0 m from the shore. If they exchange seats, what will be the distance of the front end of the boat from the shore? Does your answer depend on whether they move simultaneously or separately? Ignore friction between the boat and the water.

12-22. A 100-kg cart moves without friction along a horizontal surface at a speed of 3.0 m/s. A young boy of mass 50 kg who is riding in the front of the cart gets up and walks to the back with a speed of 2.0 m/s relative to the cart. (a) What is the speed of the cart while the boy is walking across it? (b) What is the speed of the cart when the boy sits down again at the rear?

Impulse

12-23. A 0.20-kg ball moving horizontally with a speed of 30 m/s collides with a wall and rebounds with a horizontal velocity of 20 m/s in the opposite direction. (a) What is the impulse of the wall on the ball? (b) If the collision lasts for 1.0×10^{-3} s, what is the average force of the wall on the ball?

12-24. A pitched ball of mass 0.16 kg moving horizontally at 40 m/s is hit back along the same path at the pitcher with a speed of 50 m/s. (a) What is the impulse of the bat on the ball? (b) If the collision lasts for 1.0×10^{-3} s, what is the average force of the bat on the ball?

12-25. A 50-g rubber ball moving with a speed of 40 m/s strikes a wall at an angle of 30° to the normal and rebounds at the same angle and speed. (See the accompanying figure.) (a) What is the ball's change in momentum? (b) What is the impulse imparted to the wall by the ball?

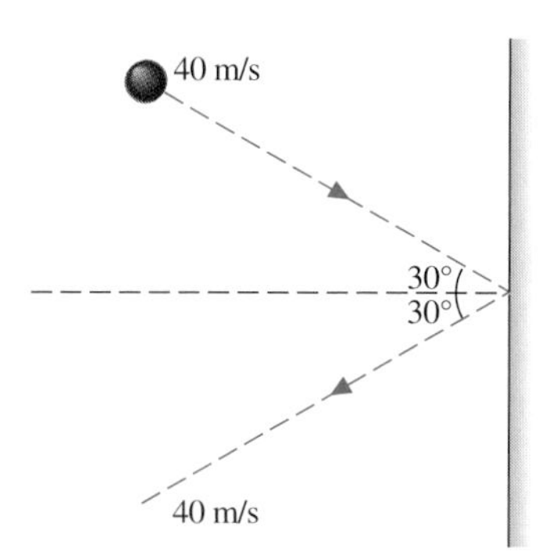

PROBLEM 12-25

12-26. A 2.0-kg particle moving along the x axis with a speed of 100 m/s at $t = 0$ is subjected to the force shown in the accompanying graph. What is the particle's velocity at the times $t = 2.0$, 4.0, 5.0, and 8.0 s?

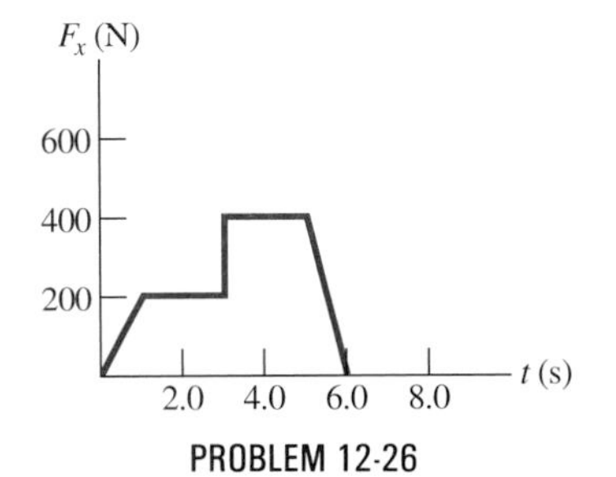

PROBLEM 12-26

12-27. What is the average value of the force of Prob. 12-26 between 0 and 3.0 s? between 0 and 6.0 s?

12-28. A 0.20-kg baseball traveling horizontally at 40 m/s is hit so that it passes back over the pitcher's head. When it leaves the bat, it is moving at 50 m/s in a direction 37° above the horizontal. (a) Find the change in momentum of the ball. Sketch a vector diagram representing this change. (b) If the bat is in contact with the ball for 2.0×10^{-3} s, what average force does it exert on the ball?

12-29. A 1.0-kg particle at rest is subjected to a force whose components vary with time as shown in the accompanying figure. What is the net impulse imparted to the particle between $t = 0$ and $t = 5.0$ s? What is the particle's momentum at $t = 2.0$ s? at $t = 5.0$ s? at $t = 7.0$ s?

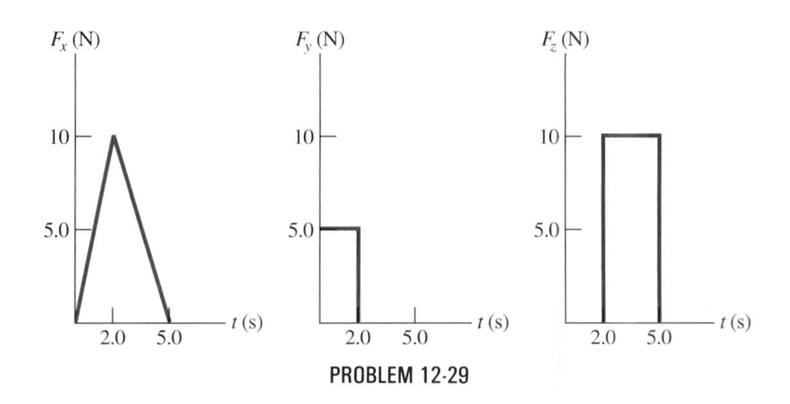

PROBLEM 12-29

12-30. A particle of mass 2.0 kg is moving with a velocity of $20\mathbf{i}$ m/s. What impulse will cause it to move with a velocity of $20\mathbf{j}$ m/s?

Elastic Collisions

12-31. A proton of mass m moving with a speed of 6.0×10^6 m/s undergoes a one-dimensional elastic collision with a helium nucleus of mass $4m$, which is initially at rest. What are the velocities of the two particles after the collision?

12-32. (a) On a frictionless air track, a 400-g sled moving to the right with a speed of 50 cm/s collides elastically with a 600-g sled moving to the left with a speed of 60 cm/s. What are the velocities of the two sleds after the collision? (b) If the two sleds stick together after the collision, what is the velocity of the combination? What fraction of the initial kinetic energy is lost?

12-33. The particles of parts (a) through (e) undergo one-dimensional elastic collisions: (a) $m_1 = 20$ g, $v_{1i} = 40$ cm/s; $m_2 = 40$ g, $v_{2i} = 20$ cm/s; (b) $m_1 = 20$ g, $v_{1i} = 40$ cm/s; $m_2 = 40$ g, $v_{2i} = -20$ cm/s; (c) $m_1 = 300$ g, $v_{1i} = 50$ cm/s; $m_2 = 200$ g, $v_{2i} = 100$ cm/s; (d) $m_1 = 3.0$ slug, $v_{1i} = 4.0$ ft/s; $m_2 = 7.0$ slug, $v_{2i} = -6.0$ ft/s; (e) $m_1 = m$, $v_{1i} = -4.0 \times 10^5$ m/s; $m_2 = 4m$, $v_{2i} = 6.0 \times 10^5$ m/s. All velocities are measured relative to the laboratory coordinate system. For each collision, calculate the velocity of the center of mass, the velocities of the two particles relative to the center of mass before and after the collision, and the velocities relative to the laboratory frame after the collision.

12-34. A proton of mass m moving with a speed of 2.0×10^5 m/s collides elastically with an alpha particle of mass $4m$ that is at rest. After the collision, the proton moves off at 30° to its original direction of motion. (See the accompanying figure.) What are the speeds of the two particles after the collision and in what direction does the alpha particle move?

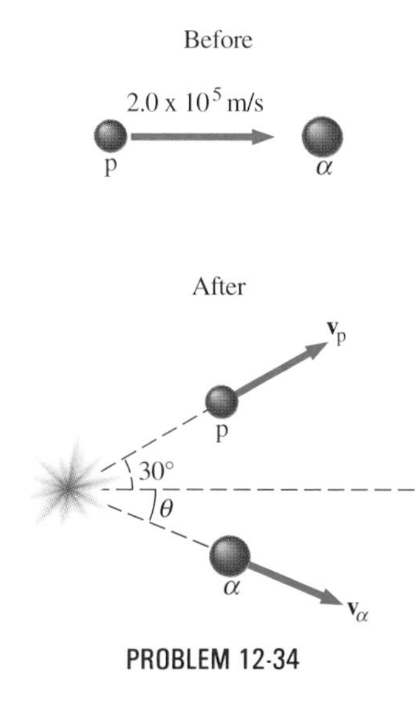

PROBLEM 12-34

12-35. In the accompanying figure, hockey puck A is moving across a smooth ice surface at 30 m/s when it collides elastically with hockey puck B, which is at rest. After the collision, B moves off with a speed of 20 m/s. Find the directions of motion of the two pucks after the collision. Specify the directions with respect to the original direction of A. Also find the speed of A after the collision.

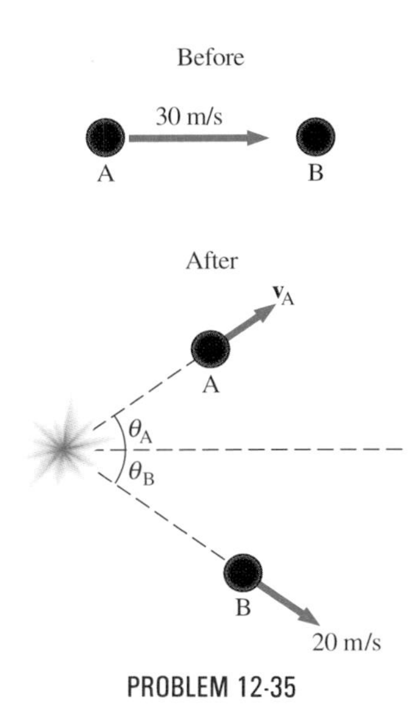

PROBLEM 12-35

Systems with Variable Mass

12-36. A rocket operating in gravity-free space burns 10 kg of fuel per second and ejects it at a relative speed of 4000 m/s. (*a*) What is the thrust of the rocket? (*b*) If at $t = 0$ the mass of the rocket plus fuel is 4000 kg and its speed is 500 m/s, what is its speed at $t = 2.0$ min? Assume the rocket moves in a straight line.

12-37. The rocket of Prob. 12-36 leaves the surface of the earth heading straight upward. What is its speed 2.0 min after launch?

12-38. A rocket whose mass with its fuel is 8000 kg is launched vertically from the surface of the earth. If the exhaust speed of the fuel is 1500 m/s, how much fuel must be ejected each second to supply the thrust necessary to (*a*) exactly balance the weight of the rocket and (*b*) give the rocket an upward acceleration of 9.8 m/s^2?

12-39. A machine gun fires 10-g bullets at a speed of 1000 m/s. If the gunner can exert an average force of 120 N on the gun while aiming and firing, what is the maximum number of bullets he can fire per minute?

12-40. A 200-g cart is pulled up a 30° incline without friction by the arrangement shown in the accompanying figure. The cart starts from rest and initially contains 1000 g of sand. If sand leaks from the cart at a rate of 40 g/s, what is the acceleration of the cart at $t = 5.0$ s? at $t = 8.0$ s?

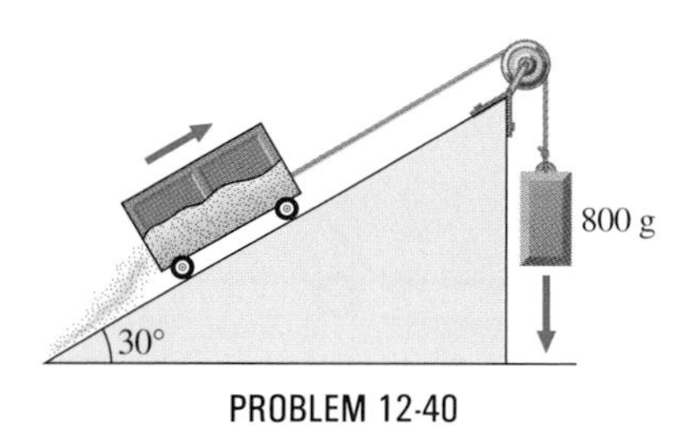

PROBLEM 12-40

12-41. A freight car that weighs 15 tons rolls along a horizontal track at 6.0 ft/s. Rain falls vertically downward with respect to the earth into the car. After the car has collected 3.0 tons of water, what is its speed? Assume that the freight car is not slowed by friction.

General Problems

12-42. A massless spring of force constant k is placed between two carts of mass m_1 and m_2. The carts are pushed toward one another until the spring is compressed a distance d. The carts are then released and the spring pushes them apart. What are the speeds of the carts after they are free of the spring?

12-43. A 2.0-kg projectile is moving in gravity-free space at a velocity of $\mathbf{v} = 5.0 \times 10^3\,\mathbf{i}$ m/s. It explodes into three pieces of equal mass when it is at the origin of a rectangular coordinate system. Three seconds later, the positions of two of the fragments are $\mathbf{r}_1 = (20\mathbf{i} - 10\mathbf{j}) \times 10^3$ m and $\mathbf{r}_2 = (-30\mathbf{i} + 25\mathbf{k}) \times 10^3$ m. What is the position of the third piece at this time?

12-44. Two children of mass m stand at the rear of a stationary cart of mass M. (*a*) If they jump off the rear of the cart together with a speed v relative to the cart, what is the subsequent speed of the cart? (*b*) If they jump off the cart one at a time with a speed v relative to the cart, what is the speed of the cart?

12-45. A freight car is 13.0 m long, 3.5 m high, and 2.5 m wide. Ice is packed solidly at the left end of the car so that it completely fills the volume of the left 1.5 m of the car. Both the ice and the freight car have masses of 12,000 kg. If the ice melts and the resulting water spreads evenly over the floor of the car, as shown in the accompanying figure, how far and in what direction does the car move? Assume the car can move without friction.

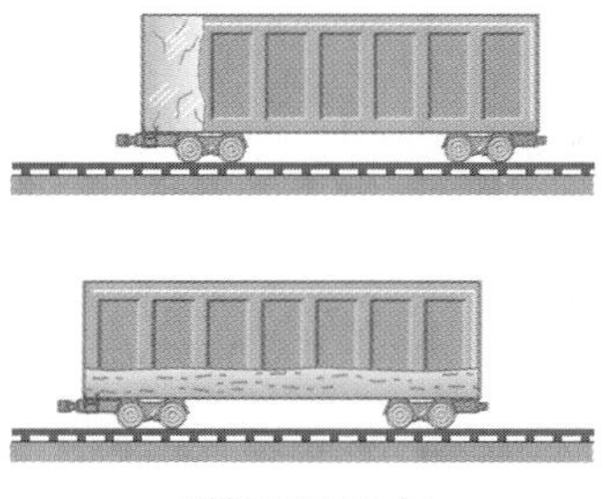

PROBLEM 12-45

12-46. A ramp of mass M is at rest on a horizontal surface. A small cart of mass m is placed at the top of the ramp and released. (See the accompanying figure.) What are the velocities of the ramp and the cart relative to the ground at the instant the cart leaves the ramp? (*Hint:* The law of mechanical-energy conservation must be used along with the law of momentum conservation.)

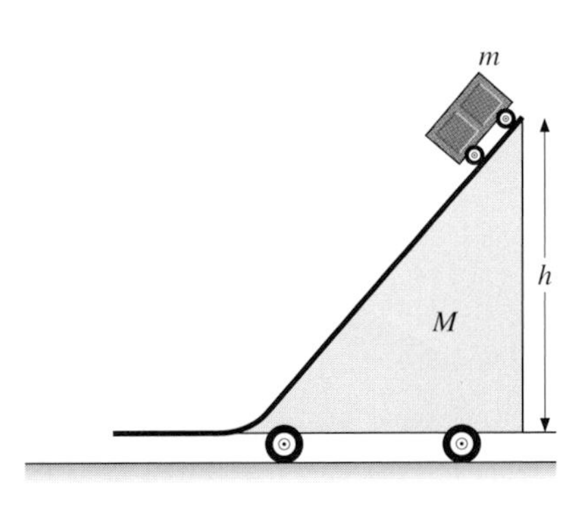

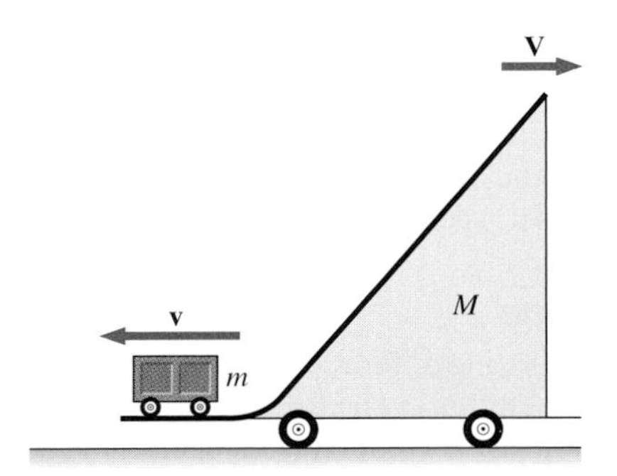

PROBLEM 12-46

12-47. A 10-g pellet shot from a gun is moving horizontally with a speed of 100 m/s when it hits and stops in the 400-g wood block shown in the accompanying figure. The block, which is sitting on a horizontal frictionless surface, is attached to a massless spring whose force constant is 100 N/m. What is the maximum compression of the spring?

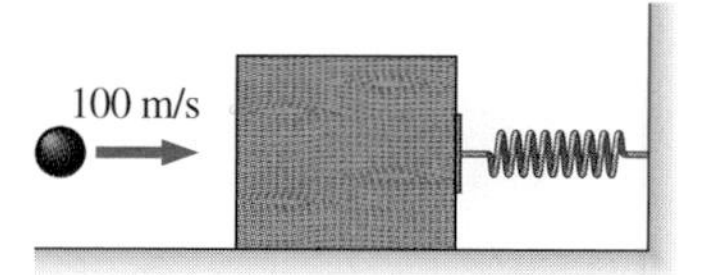

PROBLEM 12-47

12-48. A 2.0-kg block is attached to the free end of a vertical massless spring ($k = 500$ N/m) and slowly lowered to its equilibrium position. A 5.0-g pellet is then shot upward into the bottom of the block, after which the block rises upward 5.0 cm before stopping and starting back downward. What is the velocity of the pellet?

12-49. A ball of mass m_1 rolling with a speed v_1 along a flat frictionless surface collides head-on with a second ball of mass m_2, which is stationary. Assuming the one-dimensional collision is elastic, calculate (a) the kinetic energy acquired by the ball of mass m_2 and (b) the ratio m_1/m_2 such that this kinetic energy is a maximum.

12-50. Billiard ball A is moving at a speed v_0 when it collides with stationary billiard ball B. After the collision, A moves off at 30° to its original direction of motion. If the collision is elastic, what are the speeds of the two balls after the collision, and what is the direction of motion of B?

A gymnast with angular momentum.

CHAPTER 13 Angular Momentum

Preview

In this chapter we investigate angular momentum. The main topics to be discussed here include the following:

1. **Angular momentum of a particle.** The angular momentum of a particle is defined, and we derive the relationship between the rate of change of a particle's angular momentum and the torque applied to it.
2. **Angular momentum of a system of particles.** Using the definition of the angular momentum of a particle, we obtain expressions for the angular momentum of a system of particles that may be discrete or may form a rigid body. We then derive the relationship between the net external torque acting on a system of particles and the angular momentum of the system.

*3. **Angular impulse.** Angular impulse and its relationship to the change in angular momentum of a body are studied.

4. **Conservation of angular momentum.** We consider the conditions necessary for angular momentum to be conserved. The many applications of this important law are illustrated with various examples.

*5. **Precession of a top.** The mechanism responsible for the precession of a top is discussed and the rate of precession determined.

In this chapter we consider angular momentum, a quantity whose role in rotational dynamics is analogous to that of momentum in translational dynamics. Because of the obvious analogies between momentum and angular momentum, the order of topics here is quite similar to that of Chap. 12. We'll begin by defining angular momentum for three systems: a particle, a group of discrete particles, and a rigid body. Angular impulse will then be introduced and its relationship to angular momentum discussed. As with momentum, we will find that an important and widely applicable conservation law holds for angular momentum. To conclude, we will use the concept of angular momentum to analyze the intriguing motion of a rapidly rotating top.

13-1 Introduction to Angular Momentum

Angular momentum is the rotational counterpart of momentum. Like other rotational quantities, angular momentum is determined relative to the origin of a reference frame or an axis of rotation. In this section we define angular momentum for a single particle. That definition will then be used to develop expressions for the angular momentum of a system of particles, a rigid body rotating around a fixed axis, and a rigid body rolling in a plane.

Angular Momentum of a Particle

Consider the particle of Fig. 13-1. Relative to the origin O of the inertial reference frame, its position is $\mathbf{r}$ and its momentum is $\mathbf{p} = m\mathbf{v}$. By definition, the particle's **angular momentum l** with respect to the origin is

$$\mathbf{l} = \mathbf{r} \times \mathbf{p}. \tag{13-1}$$

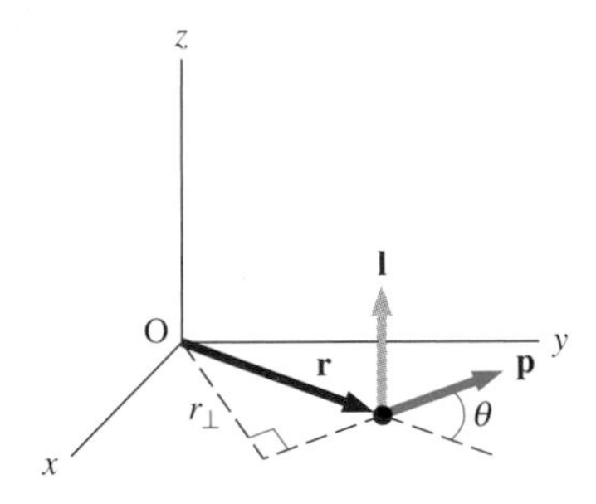

FIGURE 13-1 The angular momentum **l** of a particle with position **r** and momentum **p** is $\mathbf{l} = \mathbf{r} \times \mathbf{p}$.

From the definition of the cross product, the magnitude of the angular momentum is

$$l = rp \sin\theta = rmv \sin\theta,$$

where θ is the angle between **r** and **p**. The vector **l** is perpendicular to the plane of **r** and **p**, and its direction is given by the right-hand rule. The SI unit of angular momentum is the kilogram-square meter per second (kg·m^2/s). As with torque, we can associate a **moment arm** with angular momentum. The moment arm $r_\perp$ is the perpendicular distance from the origin to the vector **p**. As can be seen in Fig. 13-1,

$$r_\perp = r \sin\theta,$$

so

$$l = r_\perp p = r_\perp mv.$$

The length of the moment arm depends on the direction of **p** relative to the origin. If **p** lies along a line that passes through the origin ($\theta = 0$ or 180°), the moment arm is zero and the angular momentum is also zero.

The time derivative of angular momentum is found by differentiating Eq. (13-1) and using the chain rule. With $\mathbf{v} = d\mathbf{r}/dt$ and $\mathbf{p} = m\mathbf{v}$,

$$\frac{d\mathbf{l}}{dt} = \frac{d\mathbf{r}}{dt} \times \mathbf{p} + \mathbf{r} \times \frac{d\mathbf{p}}{dt} = \mathbf{v} \times m\mathbf{v} + \mathbf{r} \times \frac{d\mathbf{p}}{dt} = \mathbf{r} \times \frac{d\mathbf{p}}{dt},$$

where the term $\mathbf{v} \times m\mathbf{v}$ vanishes because it is a cross product of parallel vectors. From Newton's second law, $d\mathbf{p}/dt = \sum\mathbf{F}$, the net force acting on the particle, and by definition, $\mathbf{r} \times \sum\mathbf{F} = \sum \boldsymbol{\tau}$, the net torque on the particle. Therefore

$$\frac{d\mathbf{l}}{dt} = \sum \boldsymbol{\tau}. \tag{13-2}$$

The rate of change of a particle's angular momentum is equal to the net torque acting on the particle.

EXAMPLE 13-1 Angular Momentum and Choice of Coordinate System

Figure 13-2 shows a 2.0-kg seagull moving parallel to the surface of the ocean (the x axis of the rectangular coordinate system shown). At the instant shown, determine the seagull's angular momentum about (a) the origin and (b) an origin at (0, 4.0) m.

FIGURE 13-2 A seagull moving parallel to the surface of the ocean.

SOLUTION (a) The moment arm of the momentum is along the y axis and is 3.0 m long. By the right-hand rule, the angular momentum $\mathbf{l} = \mathbf{r} \times \mathbf{p}$ points in the negative z direction. Thus

$$\begin{aligned}\mathbf{l} &= -(r_\perp mv)\mathbf{k} = -(3.0\text{ m})(2.0\text{ kg})(4.0\text{ m/s})\mathbf{k}\\ &= -24\mathbf{k}\text{ kg}\cdot\text{m}^2/\text{s}.\end{aligned}$$

(*b*) The position of the seagull with respect to the point at (0, 4.0) m is $\mathbf{r}'$, as indicated, and the moment arm is now 1.0 m long. With the right-hand rule, we find that the angular momentum $\mathbf{l}' = \mathbf{r}' \times \mathbf{p}$ about (0, 4.0) m points in the positive z direction. Thus

$$\mathbf{l}' = (r_\perp mv)\mathbf{k} = (1.0 \text{ m})(2.0 \text{ kg})(4.0 \text{ m/s})\mathbf{k} = 8.0\mathbf{k} \text{ kg}\cdot\text{m}^2/\text{s}.$$

These results illustrate that *the angular momentum of a particle depends on the origin of the coordinate system.*

EXAMPLE 13-2 ANGULAR MOMENTUM AND TORQUE

Figure 13-3*a* shows a car of mass m moving at a speed v_0 on a freeway overpass. At $t = 0$, the driver applies the brakes, and the car decelerates under the force of kinetic friction. (*a*) Calculate the velocity of the car as a function of time and use this to determine the car's angular momentum relative to the observer sitting at point O. (*b*) Find the net torque on the car about O and show that $d\mathbf{l}/dt = \sum \boldsymbol{\tau}$ is satisfied.

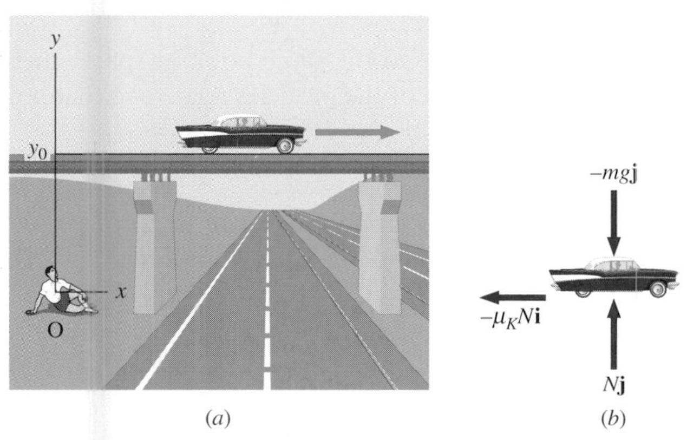

FIGURE 13-3 (*a*) A car moving on a freeway overpass. (*b*) The free-body diagram of the car.

SOLUTION (*a*) As shown in the free-body diagram of Fig. 13-3*b*, the forces on the car are its weight $-mg\mathbf{j}$ and the contact force of the road, which has a normal component $N\mathbf{j}$ and a tangential component $-\mu_K N\mathbf{i}$, where μ_K is the coefficient of kinetic friction between the car and the road. Applying Newton's second law, we have

$$\sum F_x = ma_x \qquad \sum F_y = ma_y$$
$$-\mu_K N = ma_x \qquad N - mg = 0,$$

so

$$a_x = -\frac{\mu_K N}{m} = -\frac{\mu_K mg}{m} = -\mu_K g.$$

Now from $v_x = v_{0x} + a_x t$, we obtain

$$v_x = v_0 - \mu_K gt.$$

The momentum of the car is $mv_x\mathbf{i}$, which has a moment arm y_0 about the origin O. The angular momentum of the car is therefore

$$\mathbf{l} = -y_0 mv_x\mathbf{k} = -my_0(v_0 - \mu_K gt)\mathbf{k},$$

where the direction of $\mathbf{l}$ has been determined by the right-hand rule.

(*b*) The net force on the car is $-\mu_K mg\mathbf{i}$, and the moment arm of this force is y_0. Thus the torque on the car is

$$\boldsymbol{\tau} = \mu_K mgy_0\mathbf{k},$$

where the right-hand rule has again been used to determine the direction. Differentiating $\mathbf{l}$ with respect to time, we obtain

$$\frac{d\mathbf{l}}{dt} = \frac{d}{dt}[-my_0(v_0 - \mu_K gt)\mathbf{k}] = \mu_K mgy_0\mathbf{k},$$

which is the same as the expression we found for $\boldsymbol{\tau}$. Hence $d\mathbf{l}/dt = \sum \boldsymbol{\tau}$ is satisfied.

EXAMPLE 13-3 KEPLER'S SECOND LAW

Planets travel in elliptical orbits around the sun, as illustrated in Fig. 13-4*a*. Kepler's second law states that if a planet moves from M to N in the same time that it moves from M′ to N′, then the shaded area MON is equal to the shaded area M′ON′. This is equivalent to saying that $dA/dt =$ constant, where dA is the area swept out by the planet's position vector during a time interval dt. Use Eq. (13-2) to prove Kepler's second law.

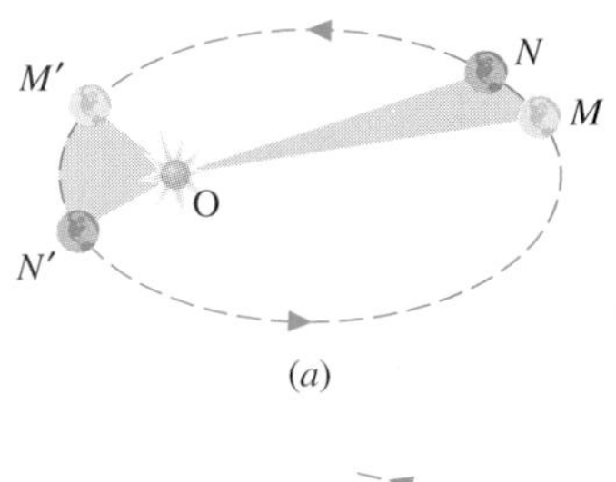

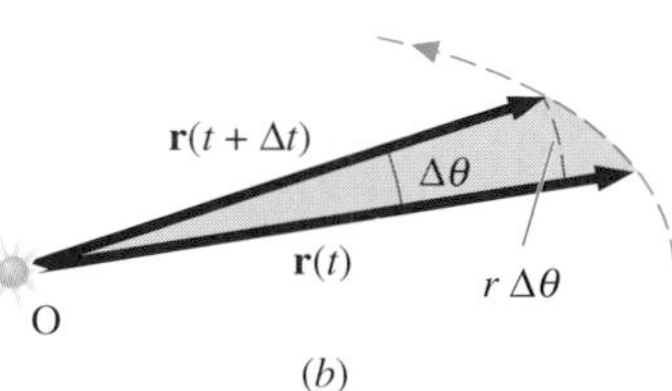

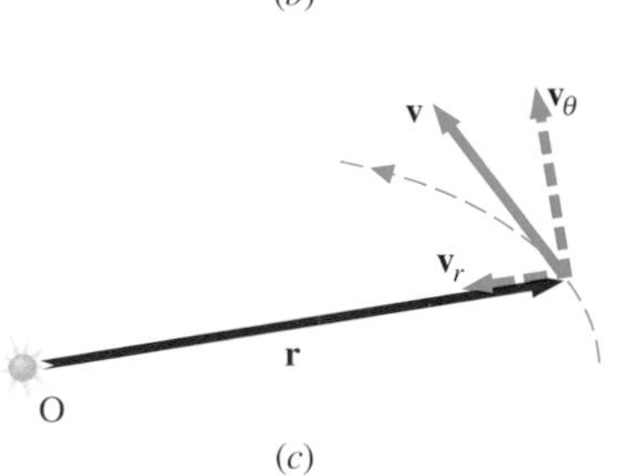

FIGURE 13-4 (*a*) The elliptical orbit (exaggerated) of a planet around the sun, which is at a focus O of the ellipse. (*b*) The planet travels from $\mathbf{r}(t)$ to $\mathbf{r}(t + \Delta t)$ and sweeps out the area shown. (*c*) Resolution of the planet's velocity into components $\mathbf{v}_r$ and $\mathbf{v}_\theta$.

SOLUTION First let's consider the position vector $\mathbf{r}$ of the planet of Fig. 13-4*b*. During a time interval Δt, the vector rotates through an angle $\Delta\theta$ and sweeps out the area ΔA. For small $\Delta\theta$, ΔA is approximately equal to the area of a triangle with base $r\,\Delta\theta$ and height r:

$$\Delta A = \tfrac{1}{2}r(r\,\Delta\theta) = \tfrac{1}{2}r^2\,\Delta\theta.$$

The rate at which the area is swept out is given by

$$\lim_{\Delta t \to 0} \frac{\Delta A}{\Delta t} = \frac{dA}{dt} = \frac{1}{2} r^2 \frac{d\theta}{dt}. \qquad (i)$$

As shown in Fig. 13-4*c*, the velocity of the planet can be divided into two components: $\mathbf{v}_\theta$, which is perpendicular to $\mathbf{r}$, and $\mathbf{v}_r$, which is along $\mathbf{r}$. Since $\mathbf{l} = \mathbf{r} \times m\mathbf{v}$, only $\mathbf{v}_\theta$ contributes to the angular momentum about O. The magnitude of $\mathbf{l}$ is therefore

$$l = mrv_\theta = mr(r\omega) = mr\left(r\frac{d\theta}{dt}\right) = mr^2 \frac{d\theta}{dt}, \qquad (ii)$$

and we find from a comparison of Eqs. (i) and (ii) that

$$\frac{dA}{dt} = \frac{l}{2m}. \qquad (iii)$$

Hence the rate at which area is swept out by the position vector of a planet is proportional to the planet's angular momentum.

The only force on the planet is the gravitational attraction of the sun. This force is directed along $\mathbf{r}$, so $\mathbf{r} \times \mathbf{F}$, the torque about the sun, is zero. We therefore conclude from Eq. (13-2) that

$$\mathbf{l} = \text{constant}.$$

When combined with Eq. (iii), this yields

$$\frac{dA}{dt} = \text{constant}.$$

This is Kepler's second law.

Although planets move in the gravitational field of the sun, the only property of planetary motion used in our derivation of Kepler's second law is the fact that the force on the planet always points toward a specific point in space (in this case, the sun). Such a force is called a *central force*. Kepler's second law is therefore not limited to the inverse-square force of the gravitational field. It is valid for *any* central force.

DRILL PROBLEM 13-1

Show that newton-meter-second (N·m·s) and joule-second (J·s) are also units of angular momentum.

DRILL PROBLEM 13-2

At a particular time, the particles of Fig. 13-5 have the positions, velocities, and applied forces shown. Find the angular momentum of and the torque applied to each particle with respect to the origin of the given coordinate system.
ANS. $-16\mathbf{k}$ kg·m²/s, $-20\mathbf{k}$ kg·m²/s, $6.0\mathbf{k}$ kg·m²/s; $6.0\mathbf{k}$ N·m, $40\mathbf{k}$ N·m, $-16\mathbf{k}$ N·m.

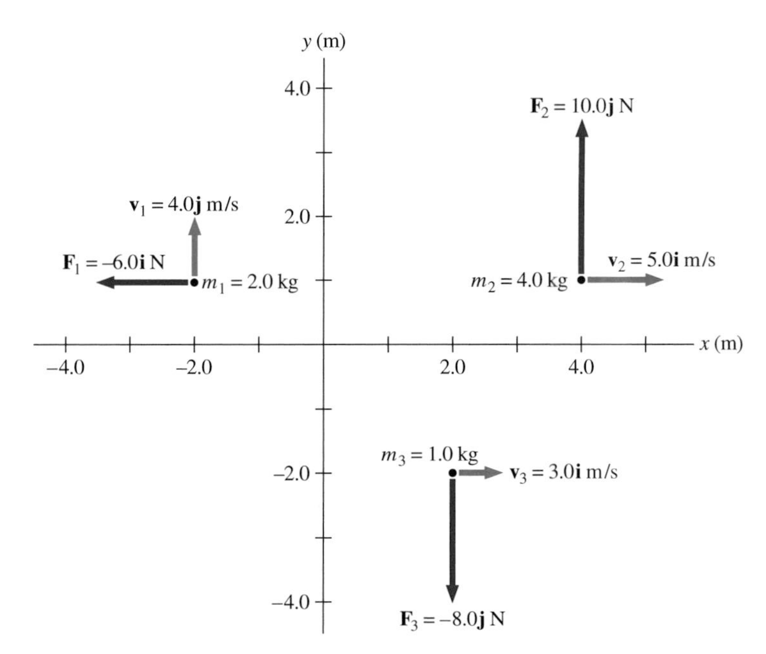

FIGURE 13-5 Particles of Drill Prob. 13-2.

ANGULAR MOMENTUM OF A SYSTEM OF PARTICLES

By definition, *the angular momentum* **L** *of a system of particles is the vector sum of the angular momenta of the individual particles*. If particles $1, 2, \ldots, N$ have angular momenta $\mathbf{l}_1, \mathbf{l}_2, \ldots, \mathbf{l}_N$, all relative to the same point in space, then the total angular momentum of the particles relative to that point is

$$\mathbf{L} = \mathbf{l}_1 + \mathbf{l}_2 + \ldots + \mathbf{l}_N. \qquad (13\text{-}3)$$

Suppose that particle i is subjected to a net torque $\boldsymbol{\tau}_i$, which is measured relative to the same point as $\mathbf{l}_i$. Since $\boldsymbol{\tau}_i = d\mathbf{l}_i/dt$, we find by differentiating Eq. (13-3) that

$$\frac{d\mathbf{L}}{dt} = \sum_i \frac{d\mathbf{l}_i}{dt} = \sum_i \boldsymbol{\tau}_i,$$

where $\sum_i \boldsymbol{\tau}_i$ is the sum of all torques (both internal and external) acting on the system. As discussed in Supplement 9-1, the internal torques of this sum cancel in pairs, leaving only the net external torques, which we designate as $\sum \boldsymbol{\tau}$. Thus

$$\sum_i \boldsymbol{\tau}_i = \sum \boldsymbol{\tau},$$

and

$$\frac{d\mathbf{L}}{dt} = \sum \boldsymbol{\tau}. \qquad (13\text{-}4)$$

This result states that *the rate of change of the angular momentum of a system of particles is equal to the net external torque acting on that system when both quantities are measured with respect to the same point in space.*

Next let's investigate the angular momentum of a system of N particles about their center of mass. By definition, this quantity is

$$\mathbf{L}_{\text{CM}} = \sum_i \mathbf{l}_i' = \sum_i \mathbf{r}_i' \times m_i \mathbf{v}_i', \qquad (13\text{-}5)$$

where $\mathbf{l}'_i$, $\mathbf{r}'_i$, and $\mathbf{v}'_i$ are the angular momentum, position, and velocity, respectively, of particle i measured relative to the origin of a reference frame moving with the center of mass. (See Fig. 13-6.) Upon differentiating Eq. (13-5), we find

$$\frac{d\mathbf{L}_{CM}}{dt} = \sum_i m_i \frac{d\mathbf{r}'_i}{dt} \times \mathbf{v}'_i + \sum_i m_i \mathbf{r}'_i \times \frac{d\mathbf{v}'_i}{dt}.$$

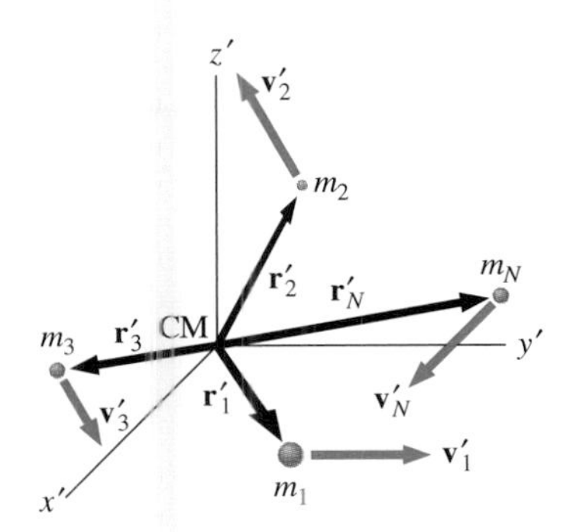

FIGURE 13-6 The system's angular momentum about its center of mass is $\mathbf{L}_{CM} = \sum_i m_i \mathbf{r}'_i \times \mathbf{v}'_i$, where $\mathbf{r}'_i$ and $\mathbf{v}'_i$ are the position and velocity of particle i relative to the center of mass.

Since $d\mathbf{r}'_i/dt = \mathbf{v}'_i$, the first term on the right-hand side of this equation vanishes, leaving

$$\frac{d\mathbf{L}_{CM}}{dt} = \sum_i m_i \mathbf{r}'_i \times \frac{d\mathbf{v}'_i}{dt} = \sum_i m_i \mathbf{r}'_i \times \mathbf{a}'_i.$$

We can substitute for the right-hand side of this equation using $\sum_i m_i \mathbf{r}'_i \times \mathbf{a}'_i = \sum \boldsymbol{\tau}_{CM}$. [You can find a proof of this in Supplement 10-1.] We then obtain

$$\frac{d\mathbf{L}_{CM}}{dt} = \sum \boldsymbol{\tau}_{CM}, \quad (13\text{-}6)$$

where $\sum \boldsymbol{\tau}_{CM}$ is the net external torque around the center of mass. *The rate of change of the angular momentum about the center of mass of a system of particles is equal to the net external torque about the center of mass acting on that system.*

Although Eqs. (13-6) and (13-4) are identical in form, they do differ in content. Equation (13-6) is stated for a reference frame moving with the center of mass, which *may or may not be an inertial reference frame*. Equation (13-4), on the other hand, is *limited to inertial frames*.

DRILL PROBLEM 13-3

Determine the angular momentum about the origin O of the three particles of Fig. 13-5 at the instant shown. What is the rate of change of the angular momentum at this instant?
ANS. $-30\mathbf{k}$ kg·m²/s; $30\mathbf{k}$ N·m.

ANGULAR MOMENTUM OF A ROTATING RIGID BODY

We now consider the case in which the particles form a rigid body rotating about a fixed axis. Figure 13-7*a* depicts a rigid body that is constrained to rotate around the z axis of a coordinate system. At the instant shown, the angular velocity of the body is ω. All mass segments of the rigid body undergo circular motion at this angular velocity around the z axis. This circular motion is illustrated in Fig. 13-7*a* for an arbitrary mass segment Δm_i whose position is $\mathbf{r}_i$ with respect to the origin. The radius of its path is R_i, and its tangential velocity is $v_i = R_i\omega$. Since the vectors $\mathbf{v}_i$ and $\mathbf{r}_i$ are perpendicular to each other, the angular momentum of this segment has the magnitude

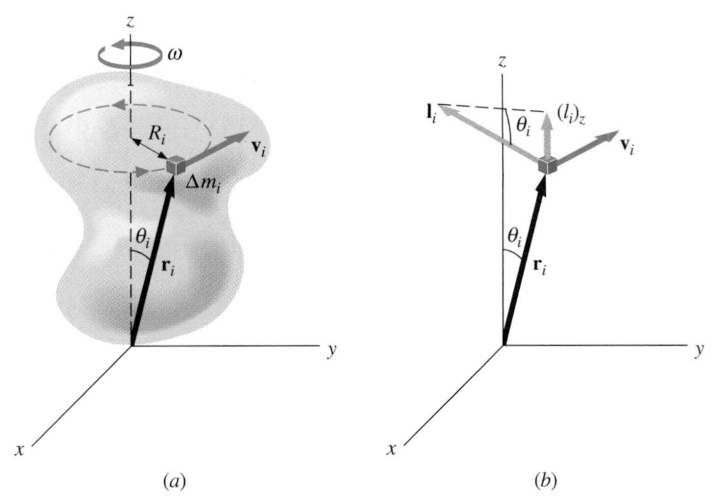

FIGURE 13-7 (*a*) A rigid body is constrained to rotate around the z axis. The circular motion of an arbitrary mass segment Δm_i is shown. (*b*) The angular momentum of the mass segment is $\mathbf{l}_i$. The component of this vector along the axis of rotation is $l_z = l_i \sin\theta_i$.

$$l_i = r_i(\Delta m_i v_i)\sin 90° = r_i\,\Delta m_i v_i.$$

From the right-hand rule, we find that $\mathbf{l}_i$ is directed as shown in Fig. 13-7*b*. The angular momentum of the body can now be found by summing vectorially the angular momenta of the mass segments. This sum yields the angular momentum components both along and perpendicular to the axis of rotation. Here we are interested in the component of the angular momentum along the axis of rotation, since this component can be easily related to the rotation of the rigid body. From Fig. 13-7*b*, the component of $\mathbf{l}_i$ along the axis of rotation is

$$\begin{aligned}(l_i)_z &= l_i \sin\theta_i = (r_i\,\Delta m_i v_i)\sin\theta_i \\ &= (r_i\sin\theta_i)(\Delta m_i v_i) = R_i\,\Delta m_i v_i.\end{aligned}$$

The net angular momentum of the rigid body along the axis of rotation is therefore

$$\begin{aligned}L &= \sum_i (l_i)_z = \sum_i R_i\,\Delta m_i v_i = \sum_i R_i\,\Delta m_i(R_i\omega) \\ &= \omega\sum_i \Delta m_i(R_i)^2.\end{aligned}$$

Now the summation $\sum_i \Delta m_i(R_i)^2$ is simply the moment of inertia I of the body around the axis of rotation. (See Sec. 8-4.) Hence the angular momentum of a rigid body rotating with an angular velocity ω around that axis is

$$L = I\omega. \quad (13\text{-}7)$$

Note that this is really the angular momentum component along the axis of rotation of the rigid body. However, Eq. 13-7 does represent *the net angular momentum around the axis of rotation* if a body has rotational symmetry about that axis. The reason is that for each mass segment on one side of the axis, there is a corresponding segment on the other side. The angular momentum components of these segments perpendicular to the axis of rotation then cancel each other, so the net angular momentum of the body lies along the axis of rotation.*

The direction of the angular momentum component along the axis of rotation of the rigid body can be determined as follows: Curl the fingers of your right hand in the same sense as the body's rotation—your thumb then points in the direction of **L**.

EXAMPLE 13-4 ANGULAR MOMENTUM OF A RIGID BODY

Two small spheres of masses $m_1 = 1.0$ kg and $m_2 = 2.0$ kg are attached to the ends of a rigid rod of length $d = 1.0$ m. The rod is rotating with an angular velocity of 8.0 rad/s around the axis that passes through its center and is perpendicular to its plane of rotation. (See Fig. 13-8.) If the mass of the rod is $M = 1.2$ kg, what is the angular momentum of the system around the axis? Assume that the spheres are small enough that they can be treated as particles.

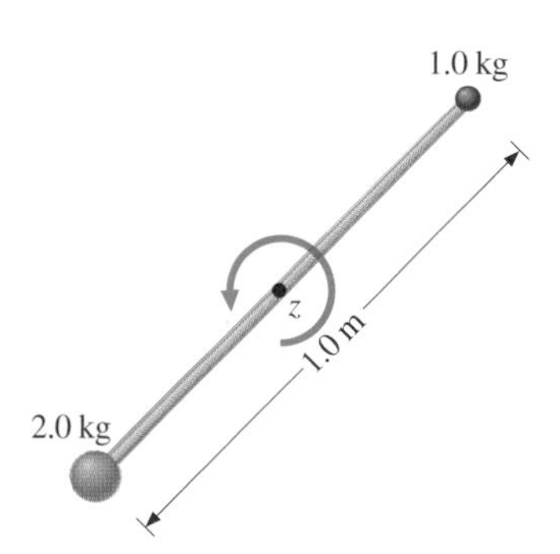

FIGURE 13-8 A rod with two attached spheres. The z axis passes through the center of the rod and is directed out of the page.

SOLUTION The moment arm for the angular momentum of each sphere is $r_\perp = 0.50$ m. For sphere 1, the magnitude of its angular momentum is

$$l_1 = r_\perp m_1 v_1 = r_\perp m_1 (r_\perp \omega) = m_1 r_\perp^2 \omega$$
$$= (1.0 \text{ kg})(0.50 \text{ m})^2(8.0 \text{ rad/s}) = 2.0 \text{ kg}\cdot\text{m}^2/\text{s}.$$

Similarly, we find for the second sphere that

$$l_2 = m_2 r_\perp^2 \omega = (2.0 \text{ kg})(0.50 \text{ m})^2(8.0 \text{ rad/s}) = 4.0 \text{ kg}\cdot\text{m}^2/\text{s}.$$

Application of the right-hand rule shows that both $\mathbf{l}_1$ and $\mathbf{l}_2$ are directed out of the page (designated as the $+z$ direction), so

$$\mathbf{l}_1 = 2.0\mathbf{k} \text{ kg}\cdot\text{m}^2/\text{s} \quad \text{and} \quad \mathbf{l}_2 = 4.0\mathbf{k} \text{ kg}\cdot\text{m}^2/\text{s}.$$

From Table 8-1, the moment of inertia of the rod that connects the two spheres is $I = Md^2/12$. The angular momentum of the rod is therefore

$$\mathbf{l} = \tfrac{1}{12} Md^2\omega\mathbf{k} = \tfrac{1}{12}(1.2 \text{ kg})(1.0 \text{ m})^2(8.0 \text{ rad/s})\mathbf{k}$$
$$= 0.8\mathbf{k} \text{ kg}\cdot\text{m}^2/\text{s},$$

where the direction is obtained using the right-hand rule. The system's angular momentum around its axis of rotation is therefore

$$\mathbf{L} = \mathbf{l}_1 + \mathbf{l}_2 + \mathbf{l} = (2.0\mathbf{k} + 4.0\mathbf{k} + 0.8\mathbf{k}) \text{ kg}\cdot\text{m}^2/\text{s}$$
$$= 6.8\mathbf{k} \text{ kg}\cdot\text{m}^2/\text{s}.$$

DRILL PROBLEM 13-4

Suppose that the rigid body of Fig. 13-7 has an angular velocity of 500 rev/min, and that its moment of inertia around the axis of rotation is 2.0 kg·m². What is its angular momentum around the axis?
ANS. 105**k** kg·m²/s.

DRILL PROBLEM 13-5

A point on the rim of a rotating disk has a tangential velocity v. What is the disk's angular momentum around the axis that passes through its center of mass and is perpendicular to its faces? The mass of the disk is M and its radius is R.
ANS. $MRv/2$.

By differentiating $L = I\omega$ [Eq. (13-7)], we can easily derive the rotational form of Newton's second law. Since $\sum \boldsymbol{\tau} = d\mathbf{L}/dt$, we have (vector notation suppressed)

$$\sum \tau = \frac{d}{dt}(I\omega) = I\frac{d\omega}{dt},$$

which, with α substituted for $d\omega/dt$, becomes

$$\sum \tau = I\alpha. \tag{13-8}$$

This is Eq. (9-12) for fixed-axis rotation and Eq. (10-2) for rolling motion.

A comparison of the linear and angular equations involving momentum gives us some interesting analogies. For linear motion we have $\sum\mathbf{F} = d\mathbf{P}/dt$. With $\sum\mathbf{F}$ and $\mathbf{P}$ replaced by $\sum\boldsymbol{\tau}$ and $\mathbf{L}$, respectively, we obtain the linear equation's rotational counterpart, $\sum\boldsymbol{\tau} = d\mathbf{L}/dt$. The relationship between external torque and angular momentum is analogous to that between external force and momentum. In addition, by replacing mass M with moment of inertia I, and velocity $\mathbf{v}_{CM}$ with angular velocity ω, we obtain $L = I\omega$ from $\mathbf{P} = M\mathbf{v}_{CM}$.

*13-2 ANGULAR IMPULSE

Angular impulse is the rotational analogue of impulse. By integrating $d\mathbf{L}/dt = \sum\boldsymbol{\tau}$ between the times t_1 and t_2, we get

*If we apply a similar argument to a rigid body rolling in a plane, we find that the body's angular momentum also satisfies Eq. (13-7). However, in this case *the axis of rotation must pass through the center of mass, and it must be perpendicular to the plane in which the body is rolling.*

$$\mathbf{L}(t_2) - \mathbf{L}(t_1) = \int_{t_1}^{t_2} \sum \boldsymbol{\tau}\, dt. \qquad (13\text{-}9a)$$

This equation is generally written in the form

$$\Delta \mathbf{L} = \mathbf{A}, \qquad (13\text{-}9b)$$

where

$$\mathbf{A} = \int_{t_1}^{t_2} \sum \boldsymbol{\tau}\, dt \qquad (13\text{-}10)$$

is the **angular impulse**. Thus *the angular impulse on a body is equal to the change in its angular momentum.*

Like its linear counterpart impulse, angular impulse is especially useful for studying the motion of a system that experiences a sharp, short-lived force. It is, of course, used to analyze the rotational part of the motion, as the following examples illustrate.

EXAMPLE 13-5 The Center of Percussion

The rigid body of Fig. 13-9 is free to rotate around the fixed axis that passes through O and is perpendicular to the plane of the page. The body's moment of inertia around this axis is I, and its center of mass is a distance a from point O. A sharp horizontal force **F** is applied to the body a distance h from the axis. At what point must **F** be applied so that there is no horizontal impulse ($S_x = 0$) at O? This point is called *the center of percussion.*

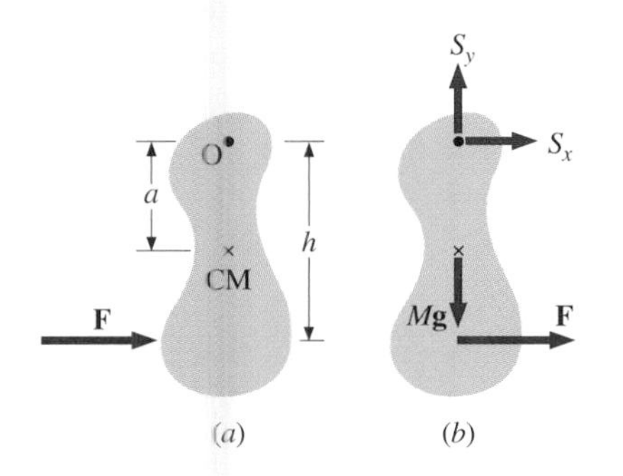

FIGURE 13-9 (*a*) A rigid body is free to rotate about the axis through O. (*b*) The free-body diagram of the body.

SOLUTION The force is assumed to be sharp enough that the rigid body moves only a negligible distance during the time it is applied. A free-body diagram of the body is shown in Fig. 13-9*b*. The horizontal forces on it are the applied force F and S_x, which is due to the support point at O. The horizontal impulse is

$$J = \int_{t_1}^{t_2} F\, dt + \int_{t_1}^{t_2} S_x\, dt = J_F + J_S,$$

which, from the impulse-momentum theorem, is equal to the horizontal change in the momentum of the system:

$$J_F + J_S = \Delta P = Mv_{CM} - 0 = Mv_{CM}, \qquad (i)$$

where v_{CM} is the initial velocity (in the x direction) of the center of mass. The angular impulse around the axis through O is

$$A = \int_{t_1}^{t_2} (Fh)\, dt = h \int_{t_1}^{t_2} F\, dt = hJ_F.$$

From Eq. (13-9*b*), it is equal to the change in angular momentum around the axis. Thus

$$hJ_F = I\omega - 0 = I\omega, \qquad (ii)$$

where ω is the initial angular velocity of the rigid body.

By combining Eqs. (*i*) and (*ii*) and using $v_{CM} = a\omega$, we obtain

$$J_S = J_F\left(\frac{Mha}{I} - 1\right). \qquad (iii)$$

The point at which **F** must be applied for zero impulse at O is now found by setting J_S in Eq. (*iii*) equal to zero. Then

$$\frac{Mha}{I} - 1 = 0,$$

and the center of percussion is a distance

$$h = \frac{I}{Ma} \qquad (13\text{-}11)$$

from the point O.

The center of percussion is an important point for a hammer, a baseball bat, a tennis racket, or any other device used for striking other objects. People using such devices naturally attempt to have them make contact at the center of percussion; for if they do, there is no force transmitted to the rotational axis, which is frequently through the wrists. In sports jargon, this point is called the "sweet spot." Anyone who has played baseball has experienced the sting in his or her wrists when hitting the ball near the end or the handle of the bat, both of which are far from the center of percussion. However, when the ball is struck at the bat's center of percussion, batters feel almost as if they haven't hit the ball at all, since there is now no force at the axis of rotation through the wrists. Striking the ball at the center of percussion also allows batters to follow through more effectively, because they do not have to contend with the retarding forces at their wrists.

EXAMPLE 13-6 Impulse on a Billiard Ball

The stationary billiard ball of Fig. 13-10*a* is struck horizontally at a distance h above its center. What must h be if the ball is to start rolling without slipping? Assume that while the ball is being struck,

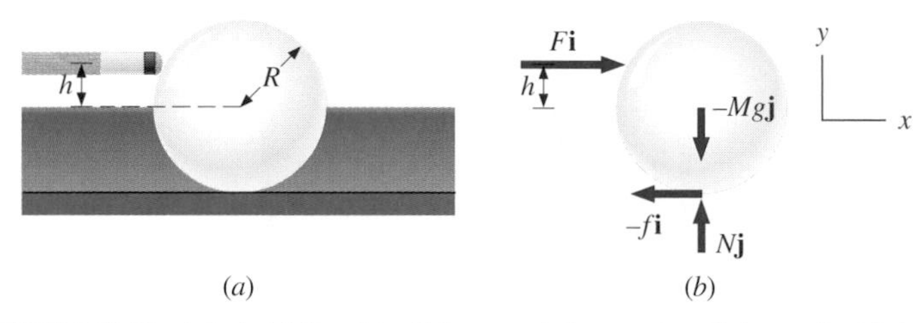

FIGURE 13-10 (*a*) A billiard ball is struck by a cue stick so that it rolls without slipping. (*b*) The free-body diagram of the ball.

the frictional force exerted by the table is negligibly small compared with the force of the cue stick.

SOLUTION As shown in the free-body diagram of Fig. 13-10b, the forces on the ball are its weight $M\mathbf{g}$, the force $\mathbf{F}$ of the cue stick, and the contact force of the table, which is resolved into a normal force $N\mathbf{i}$ and a frictional force $-f\mathbf{i}$. If the frictional force $-f\mathbf{i}$ is negligible, the only force that causes a horizontal impulse is $\mathbf{F}$. It is also the only force that creates a torque around the center of mass. Since the change in the momentum is equal to the impulse on the ball,

$$Mv_{\text{CM}} - 0 = \int_{t_1}^{t_2} F\,dt. \qquad (i)$$

The torque of $\mathbf{F}$ around the center of mass of the ball is Fh. From Eq. (13-9a), the change in the angular momentum is equal to the angular impulse, so

$$I_{\text{CM}}\omega - 0 = \int_{t_1}^{t_2} Fh\,dt = h\int_{t_1}^{t_2} F\,dt. \qquad (ii)$$

Dividing Eq. (i) by Eq. (ii), we obtain the ratio of the speed to the angular speed:

$$\frac{v_{\text{CM}}}{\omega} = \frac{I_{\text{CM}}}{Mh}. \qquad (iii)$$

We are assuming that the ball starts rolling without slipping. This means that v_{CM} and ω are related by $v_{\text{CM}} = R\omega$. When this relationship and $I_{\text{CM}} = 2MR^2/5$ are substituted into Eq. (iii), we find that the ball must be struck a distance

$$h = \tfrac{2}{5}R$$

above its center if it is to start rolling without slipping.

DRILL PROBLEM 13-6

Where is the center of percussion of a uniform meterstick that is free to rotate around an axis through one end?
ANS. 67 cm from the rotational axis.

DRILL PROBLEM 13-7

Gently hold a meterstick at one end; then let it rotate and strike the edge of a table, as shown in Fig. 13-11. Observe what happens when the meterstick strikes far from its center of percussion and at its center of percussion.

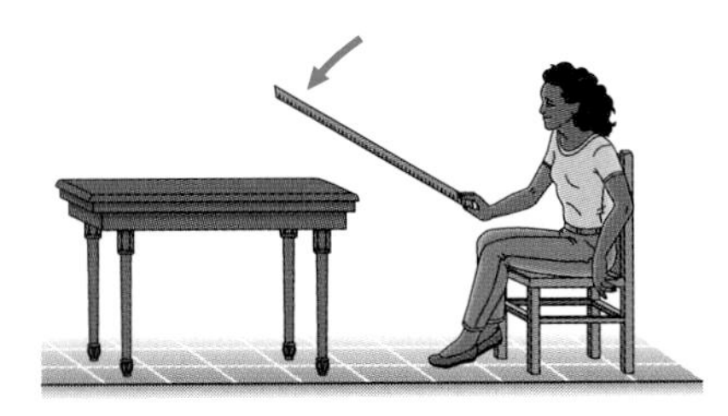

FIGURE 13-11 A meterstick is held gently between two fingers and then allowed to rotate so that it hits the edge of a table.

13-3 CONSERVATION OF ANGULAR MOMENTUM

We have seen that the rate of change of the angular momentum $\mathbf{L}$ of a system of particles is equal to the net external torque $\sum \boldsymbol{\tau}$ acting on that system. Now suppose that $\sum \boldsymbol{\tau} = 0$. Equations (13-4) and (13-6) then reduce to

$$\frac{d\mathbf{L}}{dt} = 0,$$

so

$$\mathbf{L} = \mathbf{l}_1 + \mathbf{l}_2 + \ldots + \mathbf{l}_N = \text{constant}. \qquad (13\text{-}12)$$

The angular momentum of a system of particles around a point fixed in an inertial frame (or around a moving center of mass) is conserved if there is no net external torque around that point. Notice that it is the *total* angular momentum $\mathbf{L}$ that is conserved. Any of the individual angular momenta $\mathbf{l}_i$ can change; however, their sum remains constant.

The angular momentum of a rigid body around a rotational axis fixed in an inertial frame (or moving with the center of mass) is $I\omega$. Consequently, if this angular momentum is conserved, the angular velocity must increase when the moment of inertia decreases, and vice versa. As an example, suppose that $I = 2.0\ \text{kg}\cdot\text{m}^2$ and $\omega = 4.0$ rad/s. Then if I changes to $1.0\ \text{kg}\cdot\text{m}^2$, ω must correspondingly increase to 8.0 rad/s so that the product $I\omega$ stays at $8.0\ \text{kg}\cdot\text{m}^2/\text{s}$. This principle is often used by gymnasts, divers, ice-skaters, and others who must vary the speed at which they rotate. An ice-skater pulls his arms in as close as possible to the axis of rotation in order to decrease his moment of inertia, thereby increasing his angular velocity. (See Fig. 13-12a.) The diver of Fig. 13-12b is performing a dive that

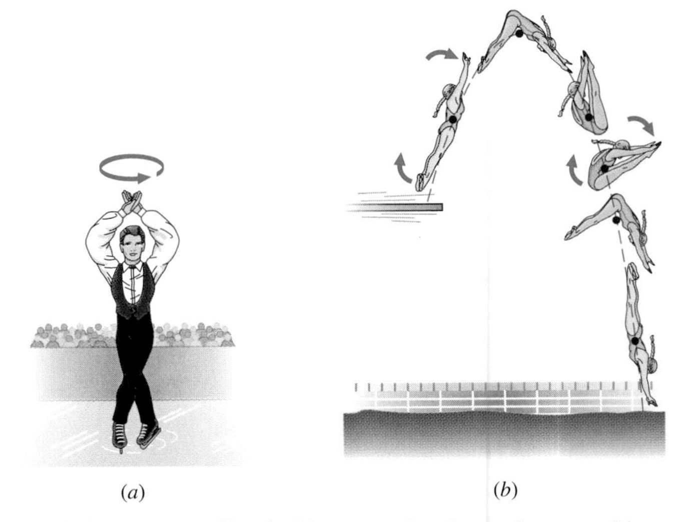

FIGURE 13-12 (a) By pulling in his arms, the skater decreases his moment of inertia and increases his angular velocity. (b) The diver leaves the board with her arms and legs outstretched. When she pulls her limbs inward, she decreases her moment of inertia and increases her angular velocity.

requires considerable body rotation. She leaves the board at a small rotational velocity with her arms and legs outstretched. By pulling her limbs in close to the rotational axis through her center of mass, she decreases her moment of inertia around this axis. Since there is no external torque ($M\mathbf{g}$ acts at the center of mass), her angular velocity must increase to keep $I\omega$ constant.

Although their angular momenta remain constant, the kinetic energies of these athletes do not. If, for example, the diver leaves the board with an angular velocity ω_i and a moment of inertia I_i, her initial angular momentum and kinetic energy are, respectively,

$$L_i = I_i\omega_i \qquad \text{and} \qquad T_i = \tfrac{1}{2}I_i\omega_i^2.$$

When she decreases her moment of inertia to I_f, her angular velocity increases to ω_f, where, by the law of angular-momentum conservation,

$$I_i\omega_i = I_f\omega_f.$$

Her new kinetic energy T_f is therefore

$$T_f = \frac{1}{2}I_f\omega_f^2 = \frac{1}{2}I_f\left(\frac{I_i}{I_f}\omega_i\right)^2 = \frac{1}{2}I_i\omega_i^2\left(\frac{I_i}{I_f}\right) = \frac{I_i}{I_f}T_i,$$

which, since $I_i > I_f$, is greater than her initial kinetic energy.

What is the origin of the work that causes this increase in kinetic energy? The answer to this question is fairly clear to the people who perform such maneuvers. They will tell you that their muscles have to do work in order to pull their limbs inward. This work causes an increase in their rotational kinetic energy.

An instructive classroom demonstration that illustrates the principle of angular-momentum conservation is depicted in Fig. 13-13. A student is standing at the center of a horizontal platform mounted on frictionless bearings. He holds the axle of a bicycle wheel of moment of inertia I_0. The wheel is spinning with an angular velocity ω_0 around the vertical axis, and the platform is stationary. If the system consists of the student, the platform, and the wheel, then its overall initial angular momentum around the vertical axis is $I_0\omega_0$.

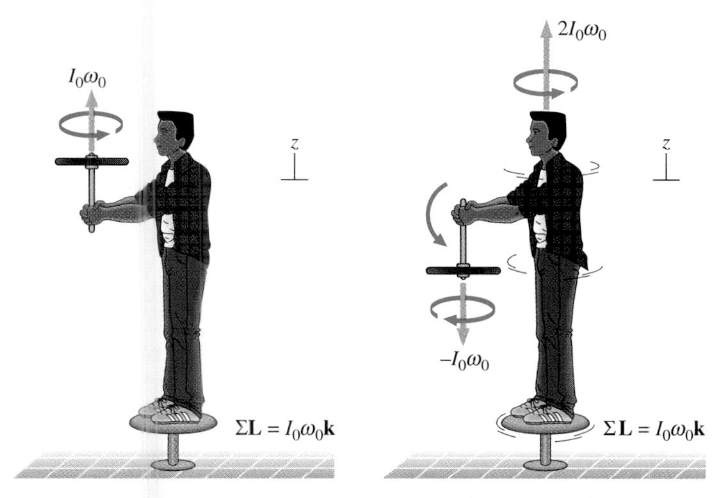

FIGURE 13-13 When the student inverts the bicycle wheel, he starts rotating because the system's angular momentum must remain constant.

Suppose the student suddenly turns the axle through 180° so that the angular momentum of the wheel becomes $-I_0\omega_0$, where the negative sign indicates the downward direction. Since there is no net external torque around the rotational axis (the moment arm of the weight $M\mathbf{g}$ is zero), the system's angular momentum must remain equal to its initial value, $I_0\omega_0$. Thus the student and platform must now rotate with an angular momentum $2I_0\omega_0$ so that the total angular momentum, which is $2I_0\omega_0 - I_0\omega_0$, remains at $I_0\omega_0$. A person watching the student would then see him rotate once he inverts the wheel.

The helicopter has front and rear blades that rotate in opposite directions. Without the rear blade, angular-momentum conservation would require that the helicopter body rotate opposite to the front blade.

EXAMPLE 13-7 COUPLED FLYWHEELS

A flywheel rotates without friction at an angular velocity $\omega_0 =$ 600 rev/min on a frictionless vertical shaft of negligible rotational inertia. A second flywheel, which is at rest and has a moment of inertia three times that of the rotating flywheel, is dropped onto it. (See Fig. 13-14.) Because of friction between their surfaces, the flywheels very quickly reach the same rotational velocity, after

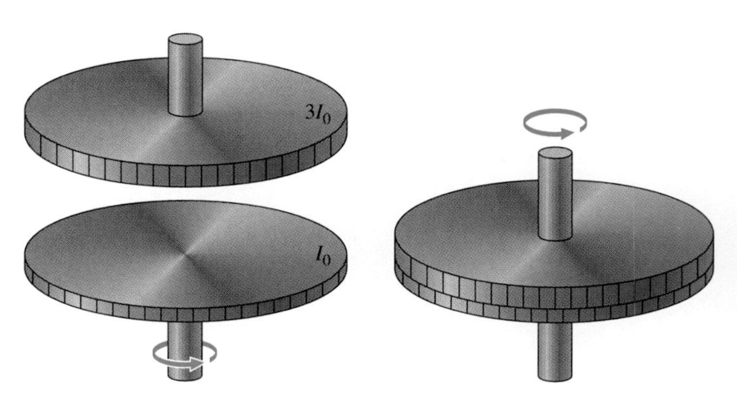

FIGURE 13-14 Two flywheels are coupled and rotate together.

which they spin together. (*a*) Use the law of conservation of angular momentum to determine the angular velocity ω of the combination. (*b*) What fraction of the original kinetic energy is lost in the coupling of the flywheels?

SOLUTION (*a*) Our system is the two flywheels. The frictional force of each wheel on the other is an internal force; hence the torque due to friction is an internal torque and does not affect the angular momentum of the system. There are no external torques on the system, so the angular momentum of our system along the axis of the shaft is conserved. If we represent the moments of inertia of the two flywheels by I_0 and $3I_0$, the principle of conservation of angular momentum yields

$$I_0\omega_0 = (I_0 + 3I_0)\omega,$$

and the common angular velocity of the two flywheels is

$$\omega = \tfrac{1}{4}\omega_0 = 150 \text{ rev/min} = 15.7 \text{ rad/s}.$$

(*b*) Since only the flywheel with the moment of inertia I_0 is rotating before contact, the initial kinetic energy of the system is $(1/2)I_0\omega_0^2$. The final kinetic energy is $(1/2)(4I_0)\omega^2 = (1/2)(4I_0) \times (\omega_0/4)^2 = (1/8)I_0\omega_0^2$. The ratio of the final kinetic energy to the initial kinetic energy is therefore

$$\frac{\frac{1}{8}I_0\omega_0^2}{\frac{1}{2}I_0\omega_0} = \frac{1}{4},$$

so three-fourths of the original kinetic energy is lost in the coupling of the flywheels.

EXAMPLE 13-8 GIRL ON A SPINNING PLATFORM

The girl of Fig. 13-15 stands on a horizontal platform that rotates without friction around the vertical axis at an angular velocity of 1.0 rad/s. Her arms are outstretched and she holds a weight in each hand. In this position, the moment of inertia of the system consisting of the girl, weights, and platform is 8.0 kg·m^2. When the girl pulls the weights in close to her body and decreases the system's moment of inertia to 3.0 kg·m^2, what happens to the angular velocity and the kinetic energy of the system?

FIGURE 13-15 (*a*) A girl stands on a rotating platform with her arms outstretched. (*b*) When she pulls in her arms, her angular velocity increases.

SOLUTION Our system consists of the girl, the weights, and the platform. Since there are no external torques on the system around the axis of rotation, angular momentum is conserved, and the angular momenta of the system before and after the girl pulls the weights inward are equal:

$$I_i\omega_i = I_f\omega_f,$$

$$(8.0 \text{ kg}\cdot\text{m}^2)(1.0 \text{ rad/s}) = (3.0 \text{ kg}\cdot\text{m}^2)\omega_f,$$

from which we find that the angular velocity increases to $\omega_f =$ 2.67 rad/s after the girl draws in the weights.

The kinetic energy increases by the factor

$$\frac{T_f}{T_i} = \frac{\frac{1}{2}(3.0 \text{ kg}\cdot\text{m}^2)(2.67 \text{ rad/s})^2}{\frac{1}{2}(8.0 \text{ kg}\cdot\text{m}^2)(1.0 \text{ rad/s})^2} = \frac{8}{3}.$$

Each weight is rotating with a decreasing radius as it is being pulled inward by the centripetal force supplied by the girl. Unlike that corresponding to circular motion, this centripetal force does work because the weight is displaced parallel to the force. The work of the centripetal force on the weights is responsible for the increase in kinetic energy.

EXAMPLE 13-9 CONSERVATION OF ANGULAR MOMENTUM IN A COLLISION

A bullet of mass $m = 2.0$ g is moving horizontally with a speed $v = 500$ m/s. The bullet strikes and becomes embedded in the rim of a solid cylinder of mass $M = 3.2$ kg and radius $R = 0.50$ m. The cylinder is free to rotate around a vertical axis through its center, and it is initially at rest. (See Fig. 13-16.) What is the angular velocity of the cylinder immediately after the bullet is embedded?

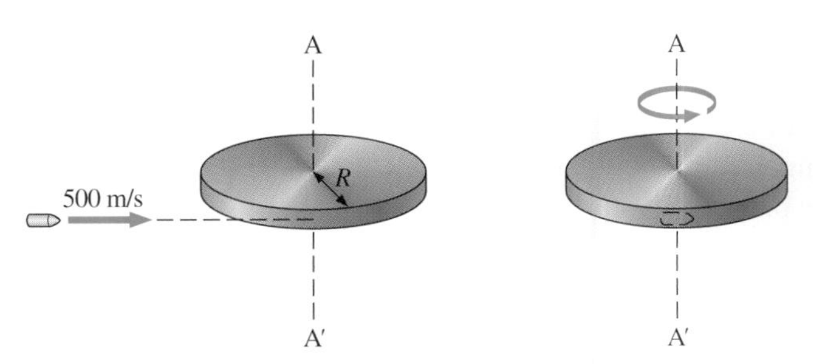

FIGURE 13-16 A bullet is fired into the edge of a cylinder that is free to rotate around the vertical axis AA′.

SOLUTION For the system consisting of the bullet and the cylinder, there is no net external torque along the vertical axis through the center of the cylinder; hence angular momentum along this axis is conserved. The initial angular momenta of the bullet and the cylinder are mvR and zero, respectively, so the net initial angular momentum of the system is mvR. After the bullet is embedded in the cylinder, the moment of inertia of the bullet plus cylinder around the rotational axis is

$$I = mR^2 + \frac{1}{2}MR^2 = \left(m + \frac{M}{2}\right)R^2.$$

The angular momentum is then

$$I\omega_f = \left(m + \frac{M}{2}\right)R^2\omega_f,$$

where ω_f is the final angular velocity of the cylinder. From the law of conservation of angular momentum,

$$L_i = L_f$$

$$mvR = \left(m + \frac{M}{2}\right)R^2\omega_f,$$

and the final angular velocity is

$$\omega_f = \frac{mvR}{(m + M/2)R^2} = \frac{(2.0 \times 10^{-3}\text{ kg})(500\text{ m/s})(0.50\text{ m})}{(2.0 \times 10^{-3}\text{ kg} + 1.6\text{ kg})(0.50\text{ m})^2}$$
$$= 1.2\text{ rad/s}.$$

DRILL PROBLEM 13-8

If the girl of Fig. 13-15*b* pushes the weights back out to their original positions, what happens to the rotational velocity and the kinetic energy of the system? Account for the change in the kinetic energy.

ANS. Rotational velocity and kinetic energy return to their original values.

DRILL PROBLEM 13-9

Suppose that the bullet of Example 13-9 is moving such that its velocity is directed at the center of the cylinder. What is the angular velocity of the cylinder after the bullet becomes embedded?

ANS. The cylinder will not rotate.

DRILL PROBLEM 13-10

The sun's mass is 2.0×10^{30} kg, its radius is 7.0×10^5 km, and it has a rotational period of approximately 28 days. If the sun should collapse to a white dwarf of radius 3.5×10^3 km, what would its period be? Assume that no mass is ejected and that the sun is a sphere of uniform density both before and after its collapse.

ANS. 1.0 min.

DRILL PROBLEM 13-11

A cylinder with rotational inertia $I_1 = 2.0\text{ kg}\cdot\text{m}^2$ rotates clockwise around a vertical axis through its center with an angular speed $\omega_1 = 5.0$ rad/s. A second cylinder with rotational inertia $I_2 = 1.0\text{ kg}\cdot\text{m}^2$ is rotating counterclockwise around the same vertical axis with an angular speed $\omega_2 = 8.0$ rad/s. If the cylinders are coupled so that their rotational axes coincide, what is the angular velocity of the combination? What percentage of the original kinetic energy is lost to friction?

ANS. 0.67 rad/s clockwise; 99%.

*13-4 Precession of a Top

We now conclude our study of angular momentum with the spinning symmetric top. While a detailed analysis of the top's motion is quite complicated, we can make a realistic simplifying assumption that allows us to investigate this motion at an appropriate mathematical level.

Figure 13-17 shows a top symmetric with respect to the axis OO′, which lies in the *yz* plane. The top is attached to a frictionless bearing at the fixed point O. First let's suppose that the top is not spinning. When it is released, the only torque on it with respect to O is due to the gravitational force $M\mathbf{g}$; this torque $\boldsymbol{\tau} = \mathbf{r} \times M\mathbf{g}$ points horizontally as shown in Fig. 13-17*a*. Since $d\mathbf{L}/dt = \boldsymbol{\tau}$,

$$d\mathbf{L} = \boldsymbol{\tau}\,dt,$$

and $d\mathbf{L}$ is directed along $\boldsymbol{\tau}$. Because the top is released from rest, its initial angular momentum is zero. The top therefore acquires angular momentum simply by rotating around the *x* axis as it topples over.

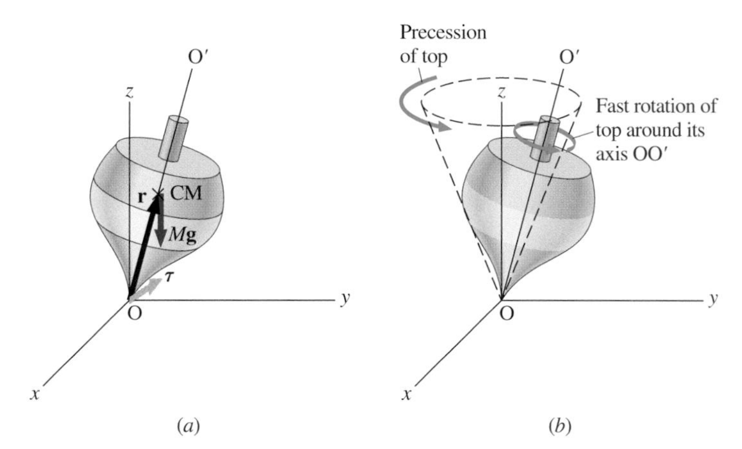

FIGURE 13-17 (*a*) The top is symmetric with respect to OO′, which lies in the *yz* plane. If the top is not spinning, the torque $\mathbf{r} \times M\mathbf{g}$ causes it to topple over by rotating around the *x* axis. (*b*) If the top is spinning very fast around its symmetry axis when released, it precesses around the *z* axis with OO′ tracing out a cone as shown.

Next suppose that before it is released, the top is given a very large angular velocity ω around its axis of symmetry. It is then observed that the top does not simply topple over in response to gravity. Instead, its axis of symmetry OO′ sweeps out a cone about the *z* axis, as shown in Fig. 13-17*b*. This motion is known as the *precession* of the top. The initial angular momentum is directed along OO′ and has magnitude $L = I\omega$, where I is its moment of inertia around the axis. Upon release, there is also angular momentum associated with the top's precessional motion; however, the angular velocity of precession is generally so slow compared to ω that we can ignore this precessional angular momentum. With this assumption, the angular momentum of the top continues to be only that due to the rotation around the symmetry axis.

The top at some instant after being released is shown in Fig. 13-18. By the right-hand rule, the torque

$$\boldsymbol{\tau} = \mathbf{r} \times M\mathbf{g}$$

is parallel to the xy plane, and since it is perpendicular to $\mathbf{r}$, it is also perpendicular to $\mathbf{L}$. Hence the change in $\mathbf{L}$,

$$d\mathbf{L} = \boldsymbol{\tau}\, dt,$$

is also perpendicular to $\mathbf{L}$. With $\mathbf{L}$ and $d\mathbf{L}$ perpendicular to each other, the magnitude of $\mathbf{L}$ must remain constant while its direction changes. This can only happen if the tip of $\mathbf{L}$ traces out a circle in a horizontal plane as it "chases" $d\mathbf{L}$ around that circle, as illustrated in Fig. 13-18.

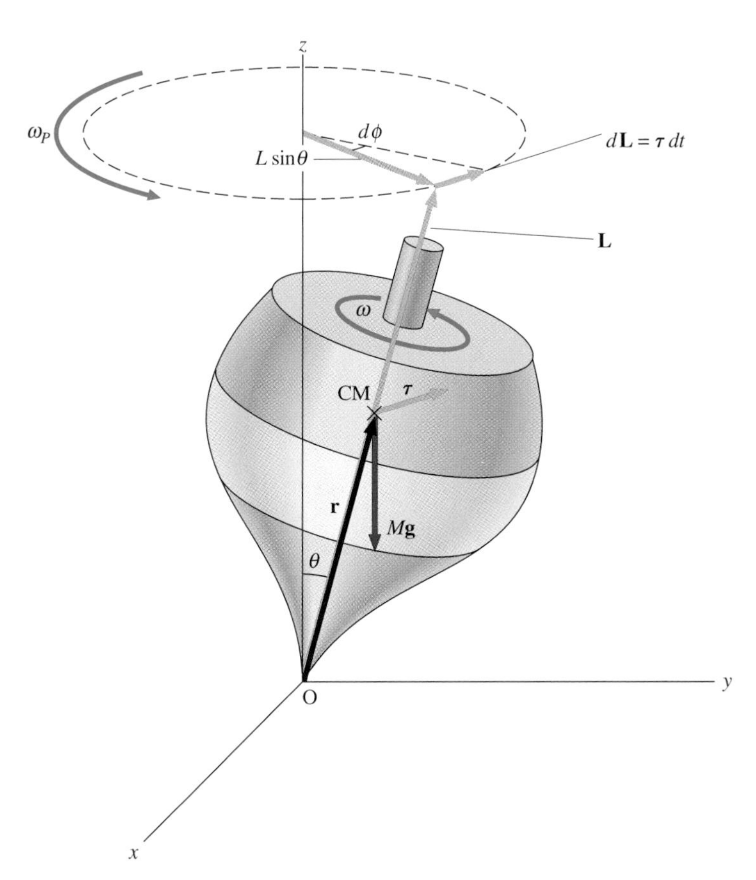

FIGURE 13-18 During a time dt, the top precesses through an angle $d\phi = dL/(L \sin\theta)$.

We can calculate the rate at which the top precesses as follows. If the center of mass of the top is a distance r from the pivot point O, then

$$\tau = rMg \sin\theta,$$

and from $dL = \tau\, dt$,

$$dL = rMg \sin\theta\, dt.$$

Figure 13-18 shows that the angle $d\phi$ through which the top precesses in a time dt is

$$d\phi = \frac{dL}{L \sin\theta} = \frac{rMg \sin\theta}{L \sin\theta}\, dt = \frac{rMg}{L}\, dt,$$

where we have substituted the expression found for dL. The precessional angular velocity is $\omega_P = d\phi/dt$, so we obtain

$$\omega_P = \frac{rMg}{L}.$$

Since $L = I\omega$ with our approximation, the precessional angular velocity can be written as

$$\omega_P = \frac{rMg}{I\omega}. \qquad (13\text{-}13)$$

In calculating ω_p, we have made the assumption that the angular momentum vector lies exactly along the axis of symmetry. This assumption does introduce a slight error into our analysis because we are ignoring the angular momentum component along the vertical axis associated with the precession of the top. In reality, the angular momentum does not point exactly along the symmetry axis of the top. As a result, the change in angular momentum due to the gravitational torque actually causes the top to bob up and down slightly as it precesses. The precise calculation of the top's motion does explain this bobbing.

Courtesy Central Scientific Company.

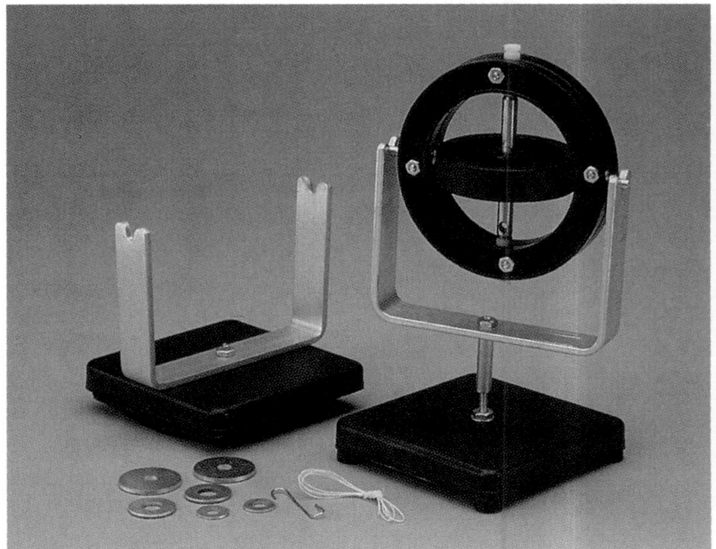

A student gyroscope is supported so that there is no torque on it; hence its angular momentum is constant. Once spinning, the orientation of the gyroscope's spin axis doesn't change, even when it is moved. As a result, the gyroscope can be used to fix a direction in space.

SUMMARY

1. **Angular momentum of a particle**
   (*a*) Angular momentum $\mathbf{l} = \mathbf{r} \times \mathbf{p}$.
   (*b*) If $\sum \boldsymbol{\tau}$ is the net torque on a particle with angular momentum $\mathbf{l}$,

$$\sum \tau = \frac{d\mathbf{l}}{dt}.$$

2. **Angular momentum of a system of particles**
   (*a*) If $\mathbf{l}_1, \mathbf{l}_2, \ldots, \mathbf{l}_N$ are the angular momenta of the N particles of a system, the angular momentum of the system is

$$\mathbf{L} = \mathbf{l}_1 + \mathbf{l}_2 + \ldots + \mathbf{l}_N.$$

   (*b*) If a rigid body is rotating around a fixed axis with an angular velocity ω, its angular momentum around that axis is

$$L = I\omega,$$

   where I is the body's moment of inertia about the axis.
   (*c*) If a net external torque $\sum \boldsymbol{\tau}$ is acting on a system of particles of angular momentum $\mathbf{L}$, then

$$\sum \tau = \frac{d\mathbf{L}}{dt}.$$

   For a rotating rigid body, $\sum \boldsymbol{\tau} = d\mathbf{L}/dt$ can be expressed as

$$\sum \tau = I\alpha.$$

*3. **Angular impulse**
   (*a*) The angular impulse on a body between times t_1 and t_2 is by definition

$$\mathbf{A} = \int_{t_1}^{t_2} \sum \boldsymbol{\tau}\, dt,$$

   where $\sum \boldsymbol{\tau}$ is the net external torque acting on the body.
   (*b*) The angular impulse acting on a body between times t_1 and t_2 is equal to the change in angular momentum of the body between these two times.

4. **Conservation of angular momentum**
   If the net external torque on a system of particles is zero, the angular momentum of that system is conserved.

*5. **Precession of a top**
   A top spinning rapidly around its axis of symmetry with an angular velocity ω precesses around the vertical axis through its fixed base at a rate

$$\omega_P = \frac{rMg}{I\omega}.$$

QUESTIONS

13-1. Can you assign an angular momentum to a particle without first defining a reference point?

13-2. For a particle traveling in a straight line, are there any points about which its angular momentum is zero?

13-3. Can a particle have angular momentum but not momentum?

13-4. Can a particle have momentum but not angular momentum?

13-5. What is the direction of the earth's angular momentum vector associated with its motion around the sun?

13-6. The torque on a particle about O is zero. What can you say about the angular momentum of the particle about O? about another point?

13-7. A particle is moving with constant velocity. Can its angular momentum about any axis be changing with time?

13-8. If you know the velocity of a particle, can you say anything about the particle's angular momentum?

13-9. Compare how Newton's second law for a particle is written in terms of momentum with how it is written in terms of angular momentum.

13-10. What is the purpose of the small propeller at the back of a helicopter that rotates in a plane perpendicular to that of the large propeller?

13-11. A spinning rubber ball falls vertically on a rough, flat surface, as shown in the accompanying figure. Does the ball bounce straight up, to the left, or to the right? Explain.

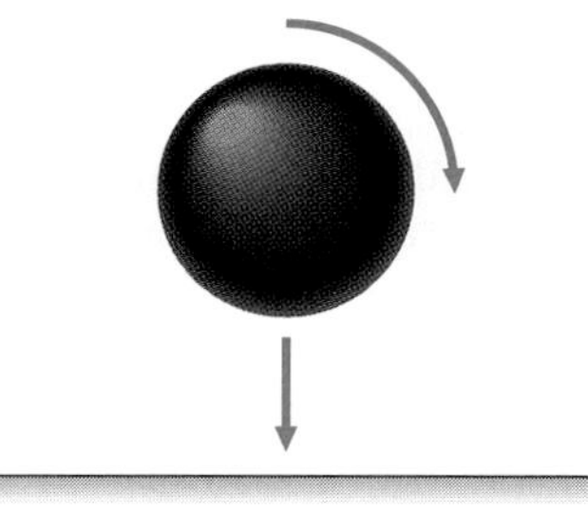

QUESTION 13-11

13-12. As the rope of a tetherball winds around the pole, what happens to the angular velocity of the ball?

13-13. Suppose all the ice from the polar caps was transported to equatorial regions in order to irrigate the deserts. What effect would this have on the angular velocity of the earth around its polar axis?

13-14. When stars collapse, they spin faster. Can you explain why?

13-15. How can you distinguish unboiled eggs from hard-boiled eggs by spinning them on a table?

13-16. A flywheel is rotating around a horizontal axis through its center when a large piece breaks off its edge and flies away. What happens to the angular velocity of the remaining flywheel?

13-17. In order to get more weight in the head of his driver, a golfer adds lead tape to the back side of the club's head. What problem might he encounter?

13-18. Why is a moving bicycle easier to balance than a stationary one?

13-19. Baseball players sometime "choke up" (grip the bat up from the end of the handle) when they are batting against a good fastball pitcher. What is the advantage of this strategy?

PROBLEMS

Angular Momentum

13-1. A 0.20-kg particle is traveling along the line $y = 2.0$ m with a velocity of $5.0\mathbf{i}$ m/s. What is the angular momentum of the particle about the origin?

13-2. What is the angular momentum of the particle of the previous problem about the point (2.0, 0) m? about the point (5.0, 0) m?

13-3. Use the right-hand rule to determine the directions of the angular momenta about the origin of the particles in the accompanying figure.

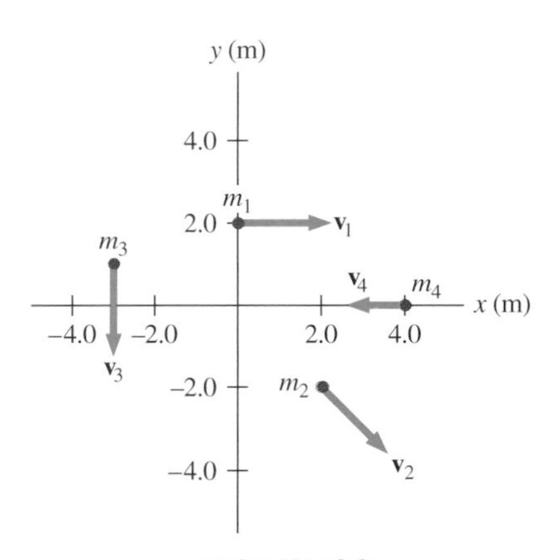

PROBLEM 13-3

13-4. Suppose that the particles in the figure for the previous problem have masses $m_1 = 0.10$ kg, $m_2 = 0.20$ kg, $m_3 = 0.30$ kg, and $m_4 = 0.40$ kg. The velocities of the particles are $\mathbf{v}_1 = 2.0\mathbf{i}$ m/s, $\mathbf{v}_2 = (3.0\mathbf{i} - 3.0\mathbf{j})$ m/s, $\mathbf{v}_3 = -1.5\mathbf{j}$ m/s, and $\mathbf{v}_4 = -4.0\mathbf{i}$ m/s. (*a*) Calculate each particle's angular momentum around the origin. (*b*) What is the angular momentum of the four-particle system around the origin?

13-5. Two particles of equal mass travel in opposite directions with the same speed along parallel lines separated by a distance *d*. Show that the angular momentum of this two-particle system is the same no matter what point is used as the reference for calculating the angular momentum.

13-6. A 4.0-kg particle moves in a circle of radius 2.0 m. The angular momentum of the particle varies with time according to $l = 5.0t^2$, where SI units are used. At $t = 3.4$ s, (*a*) What is the torque on the particle about the center of the circle? and (*b*) What is the angular velocity of the particle?

13-7. Show that a particle's angular momentum about an arbitrary point is constant if the velocity of the particle is constant.

13-8. An airplane of mass 1.40×10^4 kg flies horizontally at an altitude of 10 km with a constant speed of 250 m/s relative to the earth. (*a*) What is the magnitude of the airplane's angular momentum relative to a ground observer who is directly below the plane? (*b*) Does this angular momentum change as the airplane moves along its horizontal path?

13-9. At a particular instant, a 1.0-kg particle's position is $\mathbf{r} = (2.0\mathbf{i} - 4.0\mathbf{j} + 6.0\mathbf{k})$ m, its velocity is $\mathbf{v} = (-1.0\mathbf{i} + 4.0\mathbf{j} + 1.0\mathbf{k})$ m/s, and the force on it is $\mathbf{F} = (10\mathbf{i} + 15\mathbf{j})$ N. (*a*) What is the angular momentum of the particle about the origin? (*b*) What is the torque on the particle about the origin? (*c*) What is the time rate of change of the angular momentum at this instant?

13-10. A particle of mass *m* is dropped at the point $(-d, 0)$, and it falls vertically in the earth's gravitational field $-g\mathbf{j}$. (*a*) Find the expression for the particle's angular momentum around the *z* axis, which points out of the page in the accompanying figure. (*b*) Calculate the torque on the particle around this same axis. (*c*) Is the torque equal to the time rate of change of the angular momentum?

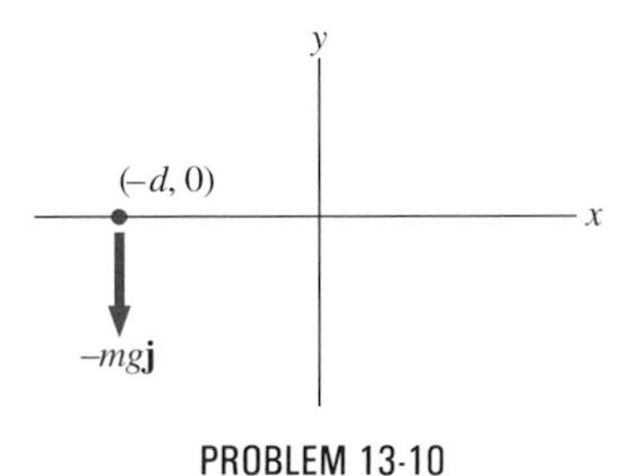

PROBLEM 13-10

13-11. A thin meterstick of mass 150 g rotates around an axis perpendicular to the stick's long axis at an angular velocity of 120 rev/min. What is the angular momentum of the stick if the rotational axis (*a*) passes through the center of the stick; (*b*) passes through one end of the stick? (*c*) What is the rotational kinetic energy for these two cases?

13-12. (*a*) What is the angular momentum of the earth around the axis through the two poles? (*b*) What is its rotational kinetic energy around this axis? Assume the earth is a uniform sphere.

13-13. Compare your result from Prob. 13-12 with the angular momentum of the earth in its orbit around the sun. Treat the orbit as circular and the earth as a particle.

13-14. At $t = 0$, a 2.0-kg particle has the position $\mathbf{r}(0) = (4.0\mathbf{i} + 2.0\mathbf{j})$ m and the velocity $\mathbf{v}(0) = (5.0\mathbf{i} - 2.0\mathbf{j})$ m/s. The particle is acted on by the constant force $\mathbf{F} = (-6.0\mathbf{i} + 2.0\mathbf{j})$ N. (*a*) What is the particle's angular momentum about the *z* axis at $t = 0$? (*b*) What is the torque around the *z* axis on the particle at this instant? (*c*) What is the time rate of change of the angular momentum at this instant? (*d*) Determine how the position and velocity vary with time, and use these to determine the time dependence of the particle's angular momentum. (*e*) What is the torque on the particle as a function of time? (*f*) Compare the time rate of change of the particle's angular momentum with this torque.

13-15. The moon orbits the earth so that the same side of the moon always faces the earth. What does this observation tell you about the rotational and orbital angular velocities of the moon? Treating the moon as a uniform sphere orbiting in a circle around a stationary earth, determine the ratio of the moon's rotational angular momentum around its polar axis to its orbital angular momentum around the earth.

13-16. Particle A of mass 1.0 kg moves along a straight line parallel to and 1.0 m from the x axis. There is a constant force $2.5\mathbf{i}$ N on this particle, and its velocity as it crosses the y axis at $t = 0$ is $4.0\mathbf{i}$ m/s. Particle B of mass 3.0 kg moves along a straight line parallel to and 2.0 m from the y axis. There is a constant force $1.5\mathbf{j}$ N on this particle, and its velocity as it crosses the x axis at $t = 0$ is $2.0\mathbf{j}$ m/s. The motion of the two particles is illustrated in the accompanying figure. (a) Calculate the position and velocity of each particle as functions of time. (b) Calculate the angular momentum of the two-particle system about the origin as a function of time. (c) Calculate the net torque on the two-particle system. (d) Is the time derivative of the angular momentum equal to the torque?

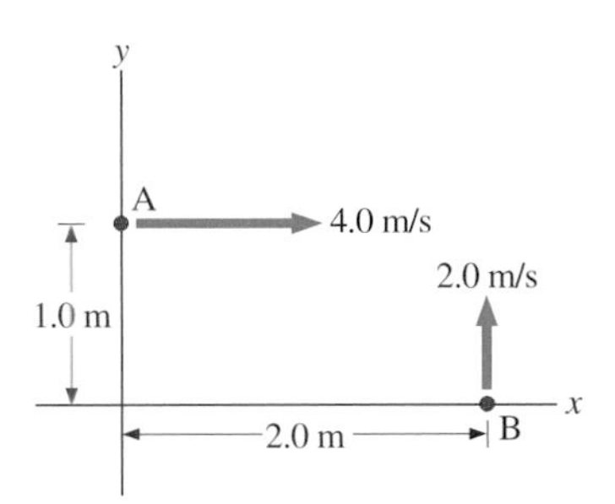

PROBLEM 13-16

13-17. A small ball of mass 0.50 kg is attached by a massless string to a thin vertical rod that is spinning as shown in the accompanying figure. When the rod has an angular velocity of 6.0 rad/s, the string makes an angle of 30° with respect to the vertical. (a) If the rotational velocity is increased to 10.0 rad/s, what is the new angle of the string? (b) Calculate the initial and final angular momenta of the ball. Is the law of conservation of angular momentum violated? (c) Can the rod spin fast enough so that the string is horizontal?

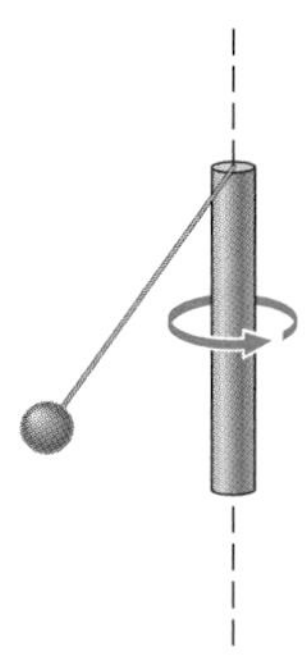

PROBLEM 13-17

Angular Impulse

13-18. Calculate the angular impulse required to change the angular momentum of a particle from $4.0\mathbf{k}$ kg·m²/s to $-4.0\mathbf{k}$ kg·m²/s.

13-19. Immediately after an angular impulse is applied, the angular momentum of a particle is $0.80\mathbf{j}$ kg·m²/s. If the angular impulse is $2.0\mathbf{j}$ kg·m²/s, what was the initial angular momentum of the particle?

13-20. A uniform stick of length d and mass m hangs vertically. It is free to rotate around a horizontal axis that passes through its end and is perpendicular to the plane of the page. The stick is struck sharply at its center by a force $\mathbf{F}$, as indicated in the accompanying figure, and it starts rotating with an angular velocity ω. (a) What is the angular impulse on the stick? (b) If the force is applied for a short time Δt, what is the average value of the force? Assume that $m = 150$ g, $d = 1.0$ m, $\omega = 4.0$ rad/s, and $\Delta t = 2.0 \times 10^{-3}$ s.

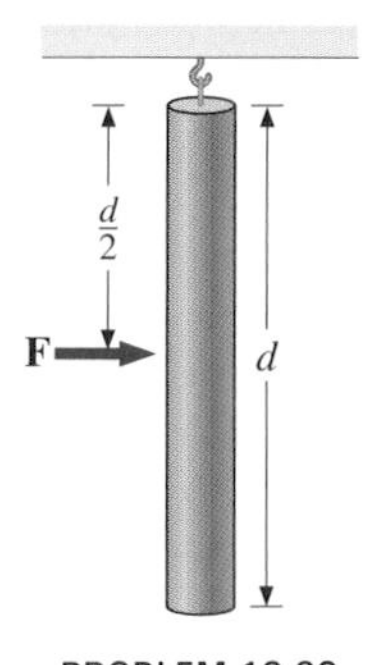

PROBLEM 13-20

13-21. A uniform thin meterstick of mass 200 g lies on a frictionless marble floor. The stick is struck by a force applied perpendicular to its length at a point 20 cm from one end. If the impulse of the force is 1.0 N·s, what is the resulting motion of the meterstick?

13-22. Where can a spherical, solid Christmas tree ornament be struck without causing it to fall off the tree? (See the accompanying figure.) Ignore the moment of inertia and the length of the hook.

PROBLEM 13-22

Conservation of Angular Momentum

13-23. A solid cylinder of mass 2.0 kg and radius 20 cm is rotating counterclockwise around a vertical axis through its center at 600 rev/min. A second identical cylinder is rotating clockwise around the same vertical axis at 900 rev/min. If the cylinders are coupled so that their rotational axes coincide, what is the angular velocity of the combination?

13-24. Tom is standing at the center of a solid rectangular raft when he turns around (rotates 180° relative to the earth). The mass of the raft is 100 kg and its dimensions are 2.0 by 3.0 m. Tom's moment of inertia about the vertical axis through the center of his body is $1.1 \text{ kg}\cdot\text{m}^2$. Through what angle does the raft rotate when Tom turns around?

13-25. A boy stands on a platform that is rotating without friction at an angular velocity of 1.0 rev/s. The boy holds weights as far from his body as possible. In this position the total moment of inertia of the boy, platform, and weights is $5.0 \text{ kg}\cdot\text{m}^2$. The boy draws the weights in close to his body, thereby decreasing the total rotational inertia to $1.5 \text{ kg}\cdot\text{m}^2$. (*a*) What is the resulting angular velocity of the platform? (*b*) By how much does the rotational kinetic energy increase?

13-26. When a satellite moves in an elliptical orbit around a fixed gravitationally attracting body (mass M), the body is at one focus of the ellipse, as illustrated in the accompanying figure. The satellite's point of closest approach to the attracting body is called the *perigee* (P), and the point at which it is the farthest from the body is called the *apogee* (A). (*a*) Use the law of angular-momentum conservation to show that $r_A/r_P = v_P/v_A$. (*b*) Use the law of mechanical-energy conservation to show that

$$2GM\left(\frac{1}{r_A} - \frac{1}{r_P}\right) = v_A^2 - v_P^2.$$

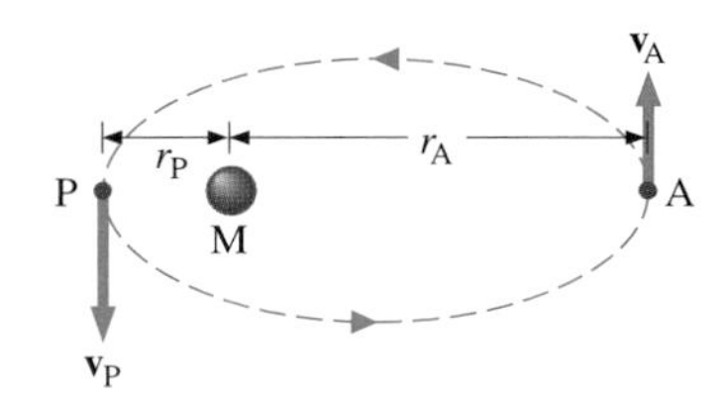

PROBLEM 13-26

13-27. Consider an earth satellite whose minimum and maximum distances from the surface of the earth are 500 and 2500 km, respectively. Use the results of Prob. 13-26 to find the speeds of the satellite at perigee and apogee.

13-28. An earth satellite moves in an elliptical orbit. At perigee it is 400 km above the earth's surface and moving with a speed of 10 km/s. What are its speed and its distance above the earth's surface at apogee? (See Prob. 13-26.)

13-29. Eight children, each of mass 40 kg, climb on a small merry-go-round, position themselves evenly around its outer edge, and join hands. The merry-go-round has a radius of 4.0 m and a moment of inertia of $1000 \text{ kg}\cdot\text{m}^2$. After the merry-go-round is given an angular velocity of 6.0 rev/min, the children walk inward and stop when they are 0.75 m from the axis of rotation. What is the new angular velocity of the merry-go-round? Assume that the frictional torque on the structure is negligible.

13-30. The accompanying figure shows a small particle of mass 20 g that is moving with a speed of 10 m/s when it collides and sticks to the edge of a uniform solid cylinder. The cylinder is free to rotate around the axis that passes through its center and is perpendicular to the plane of the page. The cylinder has a mass of 0.50 kg and a radius of 10 cm, and it is initially at rest. (*a*) Find the angular velocity of the system after the collision. (*b*) How much kinetic energy is lost in the collision?

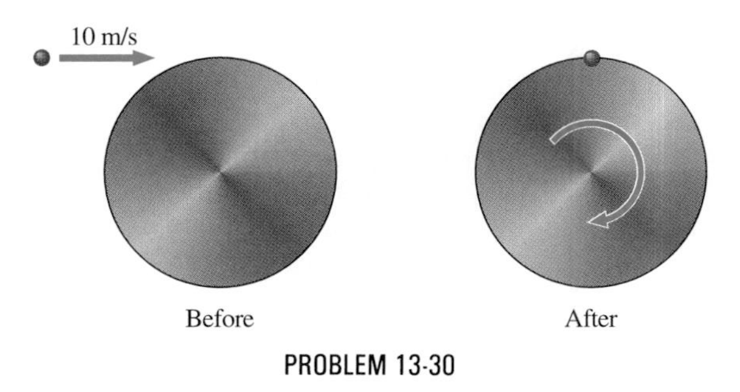

PROBLEM 13-30

13-31. A bug of mass 0.020 kg is at rest on the edge of a solid cylindrical disk ($M = 0.10$ kg, $R = 0.10$ m) rotating in a horizontal plane around the vertical axis through its center. The disk is rotating at 10 rad/s. The bug crawls to the center of the disk. (*a*) What is the new angular velocity of the disk? (*b*) What is the change in the kinetic energy of the system? (*c*) If the bug crawls back to the outer edge of the disk, what is the angular velocity of the disk then? (*d*) What is the new kinetic energy of the system? (*e*) What is the cause of the increase and decrease of kinetic energy?

13-32. A merry-go-round has a radius of 2.0 m and a moment of inertia $300 \text{ kg}\cdot\text{m}^2$. A boy of mass 50 kg runs tangent to the rim at a speed of 4.0 m/s and jumps on. If the merry-go-round is initially at rest, what is its angular velocity after the boy becomes a passenger?

13-33. A uniform rod of mass 200 g and length 100 cm is free to rotate in a horizontal plane around a fixed vertical axis through its center. Two small beads, each of mass 20 g, are mounted in grooves along the rod. Initially, the two beads are held by catches on opposite sides of the rod's center, 10 cm from the axis of rotation. With the beads in this position, the rod is rotating with an angular velocity of 10 rad/s. When the catches are released, the beads slide outward along the rod. (*a*) What is the rod's angular velocity when the beads reach the ends of the rod? (*b*) What is the rod's angular velocity if the beads fly off the rod?

Precession of a Top

13-34. A top spins at 25 rev/s around an axis making an angle of 30° with the vertical. Its mass is 0.50 kg, it has a moment of inertia of $4.5 \times 10^{-4} \text{ kg}\cdot\text{m}^2$ around its rotational axis, and its center of mass is 5.0 cm from the pivot point. If the spin of the top is clockwise as seen from above, what are the magnitude and direction of the precessional angular velocity?

13-35. The top of the previous problem is seen to precess at a rate of 0.40 rad/s at an angle of 45°. What is the top's angular velocity around its rotational axis?

General Problems

13-36. A bullet of mass 10 g is shot through a swinging door that is initially at rest. The door is a uniform rectangular slab of mass 10 kg, it is 1.0 m wide and 1.5 m tall, and it is hinged along one of the 1.5-m sides. The bullet is fired perpendicular to the door with an initial speed of 500 m/s. The bullet hits exactly in the center of the door and passes through, after which the door rotates with an angular velocity of 0.25 rad/s. (*a*) What is the moment of inertia of the door around the axis through its hinges? (*b*) What is the speed of the bullet when it emerges from the door? (*c*) What fraction of the initial kinetic energy is lost in the collision of the bullet and door?

13-37. A bug flying horizontally at 1.0 m/s collides and sticks to the end of a uniform stick hanging vertically. After the impact, the stick swings out to a maximum angle of 5.0° from the vertical before rotating back. If the mass of the stick is 10 times that of the bug, calculate the length of the stick.

13-38. A particle is subject to an attractive force from a fixed force center. Show that unless the particle starts out so that its velocity is along a line that passes through the force center, it will never fall into the force center.

13-39. A meterstick of mass 150 g is free to rotate around a horizontal axis through its center. When the stick is at rest in the position shown in the accompanying figure, a piece of putty of mass 25 g moving horizontally at 500 cm/s strikes and adheres to it at the 80-cm mark. Use the law of angular-momentum conservation to find the resulting angular velocity of the stick. Is momentum conserved in the collision?

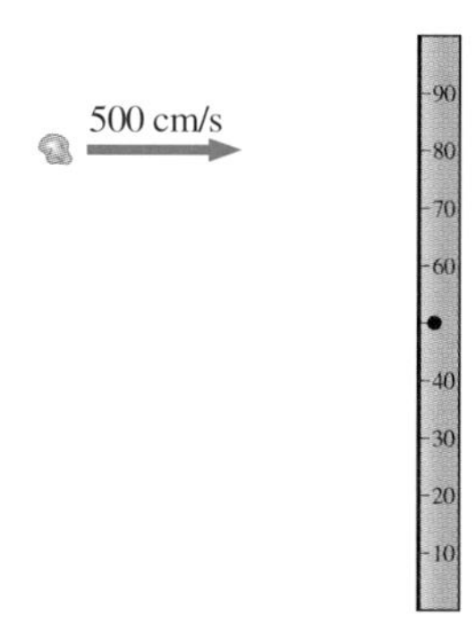

PROBLEM 13-39

13-40. An ice cube of mass M and with sides of length a is sliding without friction across a countertop with a speed v_0 when it hits a ridge E at the edge of the counter. (See the accompanying figure.) This collision causes the cube to tilt as shown. Use the principle of angular-momentum conservation around the axis through E to show that the minimum value of v_0 needed for the cube to fall off the table is given by

$$v_0 = \sqrt{1.1ag}.$$

(*Note:* The moment of inertia of a cube around any of its edges is $2Ma^2/3$.)

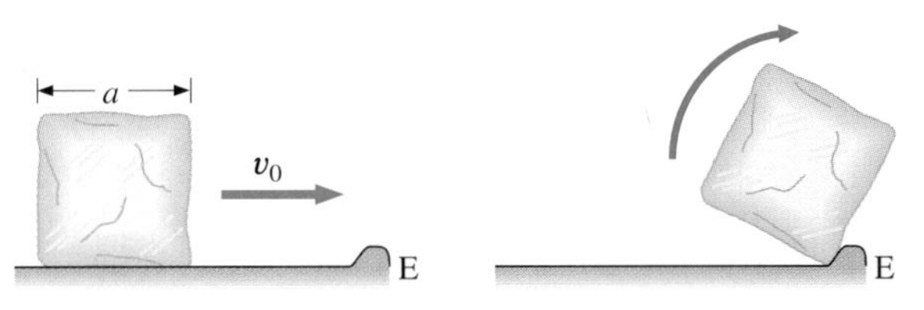

PROBLEM 13-40

13-41. Consider the model of the human male given in Prob. 8-12. Now assume that the arms are extended straight upward as shown in part (*a*) of the figure. (*a*) Where is the center of mass (CM) for this position? (*b*) What is the moment of inertia around the axis AA′ that passes through the center of mass? (*c*) The "man" jumps off a diving board with his arms extended as just described while rotating at 0.50 rev/s. He then pulls his arms and legs in and tucks his head, thereby forming an approximate cylinder of diameter 80 cm with axis AA′ now perpendicular to the page as shown in part (*b*) of the figure. What is his new angular velocity? (*d*) If he jumps from a 10-m board, approximately how many revolutions does he make before hitting the water?

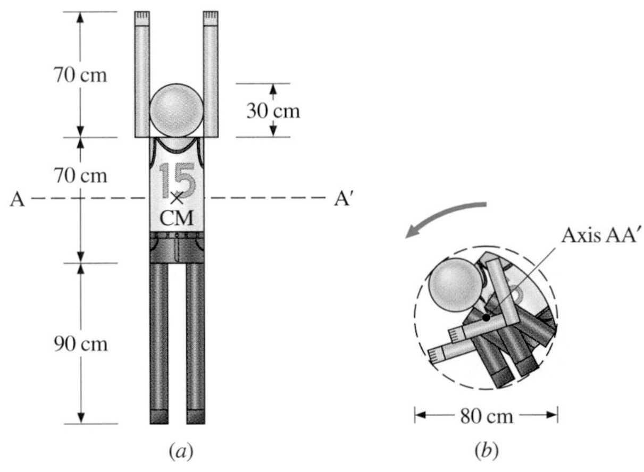

PROBLEM 13-41

The seismometer records vibrations in the earth's crust during an earthquake.

CHAPTER 14 Oscillatory Motion

Preview

The oscillatory motion of simple mechanical systems is considered, with special emphasis placed on harmonic motion. The main topics to be discussed here include the following:

1. **Simple harmonic motion.** Using a body oscillating without friction at the end of a massless spring as our prototype, we investigate simple harmonic motion. The motion is analyzed in terms of the conservation of mechanical energy, and we see how position, velocity, and acceleration vary with time.
2. **Parameters of simple harmonic motion.** Phase, phase constant, amplitude, period, frequency, and angular frequency are defined, and the formulas used to calculate them for the mass-spring system are given.

*3. **Circular motion and simple harmonic motion.** The relationship between simple harmonic motion and the projection of uniform circular motion onto a line is discussed.

4. **Additional examples of simple harmonic motion.** The simple pendulum, the compound pendulum, and the torsional pendulum are considered. The motion of each of these is approximately simple harmonic for small angular displacements.

*5. **Damped oscillations.** We look at how the simple harmonic oscillator is affected when its motion is damped by friction.

*6. **Forced oscillations and resonance.** We examine the effect of an externally applied sinusoidally varying force on the motion of a damped harmonic oscillator.

If a body moves back and forth at least approximately over the same path, its motion is said to be oscillatory. Nature abounds in motion of this type. A pendulum, a plucked guitar string, the air column of an organ pipe, and atoms in a solid can all oscillate. Less obvious, but equally important, are the oscillations of nonmechanical phenomena, such as the alternating currents in household electric circuits and the electromagnetic waves that make up radio and television signals.

In this chapter we consider the oscillations of several simple mechanical systems. Our primary focus will be on systems undergoing simple harmonic motion. The most familiar example of such systems is a body oscillating without friction at the end of a spring. The effects of friction and sinusoidally varying forces on oscillating systems will also be investigated, but not in detail. Friction causes the oscillations to cease eventually while applied forces are responsible for many of the unwanted vibrations found in mechanical devices like automobiles and machinery.

14-1 Simple Harmonic Motion

By definition, the motion of a particle is simple harmonic when the net force acting on it is proportional to the negative of its displacement from its equilibrium position. If x and $F(x)$ represent the displacement and net force, respectively, then

$$F(x) = -kx, \tag{14-1}$$

where the proportionality constant k is the **force constant**. The two- or three-dimensional motion of a particle is simple harmonic if the force component along each axis obeys Eq. (14-1).

A familiar system that undergoes simple harmonic motion is shown in Fig. 14-1, where we see a cart attached to a massless spring. If there is no friction and the end of the relaxed spring is at $x = 0$, the force on the cart is $F(x_1) = |kx_1|$ to the right when the spring is compressed a distance x_1 and $F(x_2) = |kx_2|$ to the left when the spring is stretched a distance x_2. The force, which is always opposite to the displacement, can therefore be represented by Eq. (14-1)—and the motion of the cart is simple harmonic.

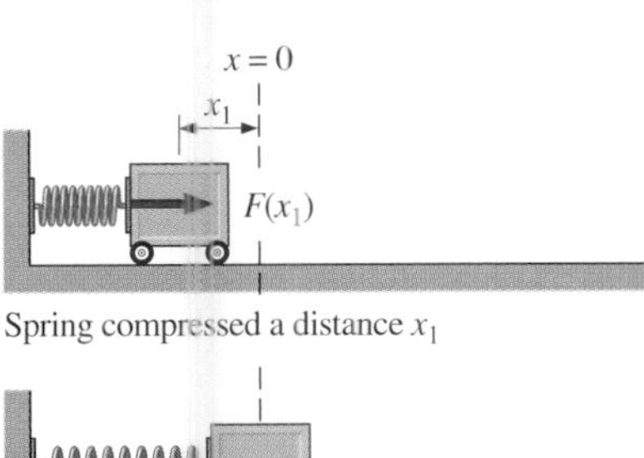

FIGURE 14-1 The cart at the end of the spring is undergoing simple harmonic motion. $F(x) = -kx$ is the restoring force.

Suppose that a force $F(x)$ acts on a body moving on the x axis around a point of stable equilibrium at $x = 0$. It was shown in Sec. 11-4 that $F(x)$ must satisfy: (1) $F(0) = 0$ and (2) $[dF(x)/dx]_{x=0} < 0$. For small displacements from $x = 0$, $F(x)$ can be expanded in the power series given in Supplement 14-1:

$$F(x) = F(0) + \left[\frac{dF(x)}{dx}\right]_{x=0} x + \frac{1}{2}\left[\frac{d^2F(x)}{dx^2}\right]_{x=0} x^2 + \cdots$$

If the maximum displacement is so small that the series can be terminated with the term linear in x, then

$$F(x) = \left[\frac{dF(x)}{dx}\right]_{x=0} x. \tag{14-2}$$

Defining the positive number k as

$$k = -\left[\frac{dF(x)}{dx}\right]_{x=0}, \tag{14-3}$$

we have

$$F(x) = -kx,$$

which is the force function for simple harmonic motion. Therefore, if the maximum displacement of a body from a point of stable equilibrium is small enough that the net force on it can be accurately represented by Eq. (14-2), the motion of that body is simple harmonic. Many of the small vibrations of objects as diverse as bridges, airplane wings, and atoms are approximately simple harmonic. The mass at the end of a spring is indeed an important problem!

EXAMPLE 14-1 A Body at the End of a Vertical Spring

The massless spring of Fig. 14-2 hangs vertically from the ceiling. A body of mass m is attached to the free end of the spring and then lowered slowly until the spring is stretched a distance l. In this position the body is in equilibrium under the forces of the spring and gravity. Show that the motion of the body about its equilibrium position is simple harmonic.

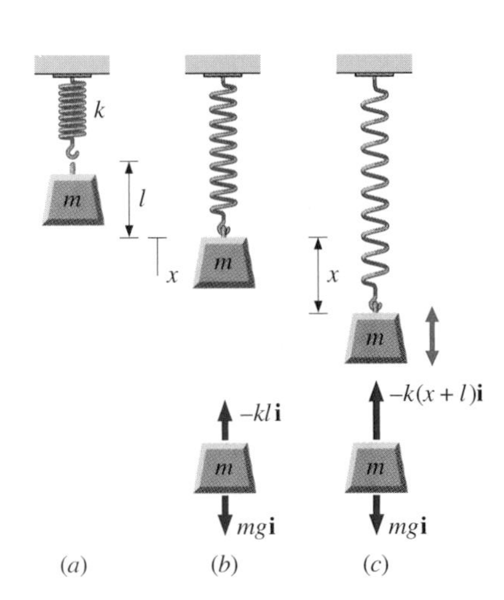

FIGURE 14-2 (*a*) A mass-spring system. (*b*) When the spring is stretched a distance l, the block is in equilibrium under the influence of the spring force and the gravitational force. (*c*) When the block is displaced from its equilibrium position, it undergoes simple harmonic motion.

SOLUTION When the body is in equilibrium (see Fig. 14-2*b*), the net force on it must vanish, so

$$-kl + mg = 0.$$

Figure 14-2*c* shows the body when it is displaced a distance x from equilibrium. In this case,

$$F(x) = -k(x + l) + mg,$$

which, since $kl = mg$, reduces to

$$F(x) = -kx.$$

Thus the motion of the body is simple harmonic.

DRILL PROBLEM 14-1

A particle moves under the force $F(x) = (1.0x^2 - 6.0x)$ N, where x is in meters. For small displacements from the origin, what is the force constant in the simple harmonic motion approximation?
ANS. $k = 6.0$ N/m.

14-2 Equations of Simple Harmonic Motion

For a body undergoing simple harmonic motion, Newton's second law gives

$$-kx = ma, \quad (14\text{-}4a)$$

which, since $a = d^2x/dt^2$, can be written in the form

$$\frac{d^2x}{dt^2} + \frac{k}{m}x = 0. \quad (14\text{-}4b)$$

This is a linear, homogeneous, second-order differential equation that we want to solve for x as a function of t. Solving this equation using the techniques of differential equations is inappropriate for the level of this textbook. However, there are other methods we can use to find the equation's solution. One method is based on the fact that mechanical energy is conserved. Because the potential energy is $kx^2/2$ for $F(x) = -kx$, the total mechanical energy is

$$E = \tfrac{1}{2}mv^2 + \tfrac{1}{2}kx^2. \quad (14\text{-}5a)$$

It's clear from this equation and the principle of mechanical-energy conservation that the position x is a maximum when the velocity v is zero. Designating this maximum position by A, we have

$$E = \tfrac{1}{2}m(0)^2 + \tfrac{1}{2}kA^2 = \tfrac{1}{2}kA^2, \quad (14\text{-}5b)$$

so for any $x \leq A$,

$$\tfrac{1}{2}mv^2 + \tfrac{1}{2}kx^2 = \tfrac{1}{2}kA^2, \quad (14\text{-}5c)$$

and

$$v = \pm\sqrt{\frac{k}{m}}\sqrt{A^2 - x^2}. \quad (14\text{-}6)$$

By substituting dx/dt for v and using the positive root, we find

$$\int \frac{dx}{\sqrt{A^2 - x^2}} = \sqrt{\frac{k}{m}}\int dt + \phi,$$

where ϕ is a constant of integration. (The only effect of using the negative root is to change the value of ϕ.) Finally, we integrate to obtain

$$\sin^{-1}\frac{x}{A} = \sqrt{\frac{k}{m}}\,t + \phi,$$

which can be written as

$$x(t) = A\sin\left(\sqrt{\frac{k}{m}}\,t + \phi\right). \quad (14\text{-}7a)$$

Hence *the position of a body in simple harmonic motion varies sinusoidally with time*. As illustrated in Fig. 14-3, the body oscillates back and forth between the extreme positions of $x = -A$ and $x = A$.

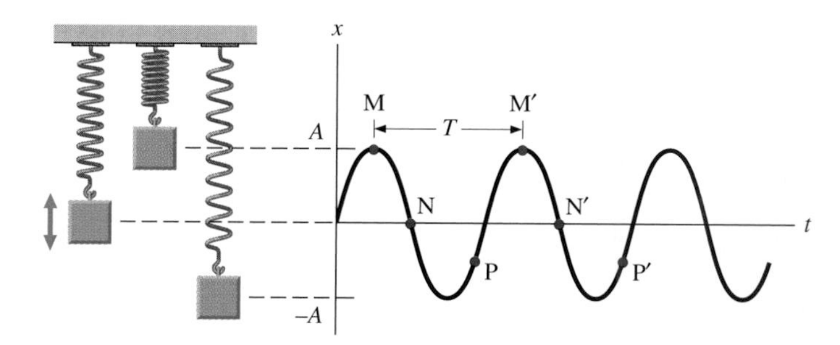

FIGURE 14-3 The body at the end of the spring oscillates back and forth sinusoidally between $x = +A$ and $x = -A$. The time interval T represents the period of the oscillations (discussed in Sec. 14-3).

Since ϕ is a constant of integration, we can represent it by $\phi = \phi' + \pi/2$, where ϕ' is another constant. Substituting this into Eq. (14-7*a*), we have

$$x(t) = A\sin\left(\sqrt{\frac{k}{m}}\,t + \phi' + \frac{\pi}{2}\right),$$

which, using $\sin(\theta + \pi/2) = \cos\theta$, we can write as

$$x(t) = A\cos\left(\sqrt{\frac{k}{m}}\,t + \phi'\right). \quad (14\text{-}7b)$$

Equation (14-7*b*) also gives the variation with time of the position of a body oscillating in simple harmonic motion. Whether you use Eq. (14-7*a*) or (14-7*b*) to represent simple harmonic motion is strictly a matter of convenience. In either case, the position is a simple trigonometric function of time involving two constants of integration. As you'll soon see, these two constants are determined by the initial position and velocity of the oscillating body. These constants will be represented from now on simply as A and ϕ regardless of whether the sine or cosine function is chosen to represent $x(t)$.

The potential energy $U(x) = kx^2/2$ gives us an insightful look at simple harmonic motion. This function, which is graphed in Fig. 14-4, is parabolic. The horizontal line at $U(x) = E$ represents the total mechanical energy, and the kinetic energy, $E - U(x)$, is given by the vertical distance from the parabola to the line. Since the kinetic energy must be positive, motion is restricted to values of x for which $E \geq U(x)$. The end points are at $x = \pm A$, the two positions where the kinetic energy is zero. A particle undergoing simple harmonic motion can be pictured as being trapped inside its potential-energy parabola as it bounces back and forth between $x = -A$ and $x = +A$. The parabola is often called a "potential well." Can you picture how the speed of the particle varies as it moves inside the well?

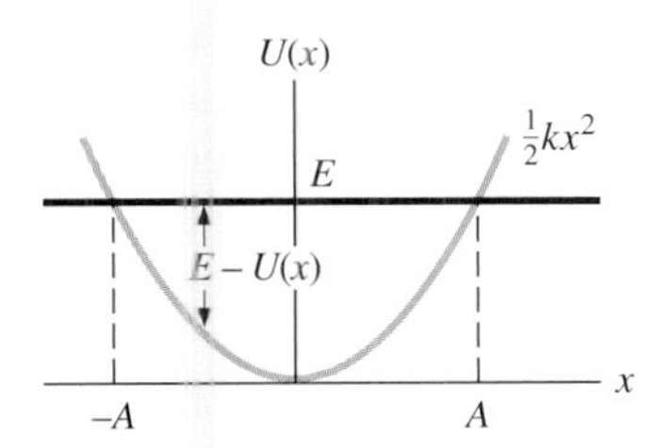

FIGURE 14-4 The potential energy $U(x) = \frac{1}{2}kx^2$ is plotted versus x. The mechanical energy is represented by the horizontal line $U(x) = E$. The kinetic energy is $E - U(x)$.

EXAMPLE 14-2 Energy in Simple Harmonic Motion

A small object of mass 0.10 kg is attached to a spring of force constant $k = 200$ N/m, as shown in Fig. 14-1. If the object oscillates between the points $x = -5.0$ cm and $x = +5.0$ cm, what are (a) the total mechanical energy of the system and (b) the maximum speed of the object? (c) Write an expression for $x(t)$, the position of the object, assuming that $\phi = 0$. Ignore any effects of friction.

SOLUTION (a) The total mechanical energy of the system is

$$E = \tfrac{1}{2}mv^2 + \tfrac{1}{2}kx^2 = \tfrac{1}{2}kA^2.$$

Since $k = 200$ N/m and $A = 0.050$ m, we obtain

$$E = \tfrac{1}{2}(200\ \text{N/m})(0.050\ \text{m})^2 = 0.25\ \text{J}.$$

(b) Since the mechanical energy is constant, the kinetic energy (and hence speed) of the object is greatest when its potential energy is smallest. This occurs at $x = 0$, where

$$\tfrac{1}{2}(0.10\ \text{kg})v_{\text{max}}^2 + \tfrac{1}{2}(200\ \text{N/m})(0)^2 = 0.25\ \text{J},$$

so the maximum speed is

$$v_{\text{max}} = 2.2\ \text{m/s}.$$

(c) Assuming $\phi = 0$, we can use Eq. (14-7a) to express the position as

$$x(t) = (0.050\ \text{m}) \sin\left(\sqrt{\frac{200\ \text{N/m}}{0.10\ \text{kg}}}\, t\right) = (0.050\ \text{m}) \sin (45\ \text{rad/s})t.$$

DRILL PROBLEM 14-2

Suppose that in Fig. 14-1, $m = 2.0$ kg and $k = 500$ N/m. The object is set oscillating so that its maximum displacement is 10 cm. Find the total mechanical energy and the speed of the object when its position is 5.0 cm.
ANS. 2.5 J; 1.4 m/s.

DRILL PROBLEM 14-3

If $x = A/2$, what fraction of the mechanical energy is potential? What fraction is kinetic?
ANS. 1/4; 3/4.

14-3 Parameters of Simple Harmonic Motion

The **angular frequency** ω for the mass-spring system is defined as

$$\omega = \sqrt{\frac{k}{m}}, \tag{14-8}$$

and is expressed in radians per second (rad/s). With ω substituted for $\sqrt{k/m}$, Eqs. (14-7a) and (14-7b) become

$$x(t) = A \sin (\omega t + \phi) \tag{14-9a}$$

and

$$x(t) = A \cos (\omega t + \phi). \tag{14-9b}$$

Notice that explicit reference to the parameters of the mass-spring system has now been eliminated, so the variable x in Eqs. (14-9) can represent the position, whether linear or angular, for any type of simple harmonic motion.

The **phase** is the quantity $\omega t + \phi$, and the **phase constant** is ϕ. Both of these are angular quantities and are usually expressed in radians. The value of the phase constant ϕ is related to the initial position and velocity of the simple harmonic oscillator.

The **amplitude** A in Eqs. (14-9) indicates the extremes of the position, which are $+A$ and $-A$. The mechanical energy is related to the amplitude by $E = kA^2/2$.

The **period** T is the time required for one complete oscillation. In Fig. 14-3, it is the time interval between M and M′, or N and N′, or P and P′. When a particle moves through one complete oscillation, the phase changes by 2π rad as the time advances from t to $t + T$. Thus

$$\omega(t + T) + \phi = (\omega t + \phi) + 2\pi,$$

which, when solved for T, gives

$$T = \frac{2\pi}{\omega}. \tag{14-10}$$

For the mass-spring system,

$$T = 2\pi\sqrt{\frac{m}{k}}. \tag{14-11}$$

The period is independent of the amplitude. Consequently, an oscillating spring can be used as a clock. As the amplitude decreases due to the small (but negligible over one cycle) effect of friction, the time for each oscillation does not change. Time can therefore be measured in terms of the number of oscillations of the spring. The accuracy of a similar clock will be investigated when we consider the simple pendulum.

The **frequency** f is the number of complete oscillations that occur per unit time. It is given in terms of the period by

$$f = \frac{1}{T}, \tag{14-12a}$$

or in terms of the angular frequency by

$$f = \frac{\omega}{2\pi}. \tag{14-12b}$$

For the mass-spring system,

$$f = \frac{1}{2\pi}\sqrt{\frac{k}{m}}. \tag{14-13}$$

The SI frequency unit is a 1/second (1/s). It is called a *hertz* (*Hz*) in honor of Heinrich Hertz (1857–1894), whose experimental research confirmed the existence of electromagnetic radiation.

The velocity and acceleration of a particle undergoing simple harmonic motion are found by differentiating either Eq. (14-9*a*) or (14-9*b*) with respect to time. The first derivative gives the velocity, and the second derivative gives the acceleration. With Eq. (14-9*a*), we have

$$v(t) = \frac{dx}{dt} = A\omega \cos(\omega t + \phi) \tag{14-14a}$$

and

$$a(t) = \frac{d^2x}{dt^2} = -A\omega^2 \sin(\omega t + \phi). \tag{14-15a}$$

Similarly, the derivatives of Eq. (14-9*b*) yield

$$v(t) = -A\omega \sin(\omega t + \phi) \tag{14-14b}$$

and

$$a(t) = -A\omega^2 \cos(\omega t + \phi). \tag{14-15b}$$

The position, velocity, and acceleration are plotted as functions of time in Fig. 14-5 [assuming $x(t) = A \sin(\omega t + \phi)$]. For the case shown, $x(0) = A/2 = A \sin(0 + \phi)$, so the phase constant is $\phi = \pi/6$ rad. These plots demonstrate the phase relationships between the three kinematic variables. For example, when the position is a maximum in either direction, the acceleration is a maximum in the opposite direction, and the velocity is zero; and when the velocity is a maximum, the position and acceleration are zero. The velocity oscillates one-quarter of a cycle and the acceleration one-half of a cycle ahead of the position. In terms of the phase, we say that the velocity leads the position by $\pi/2$ rad, the acceleration leads the position by π rad, and the acceleration leads the velocity by $\pi/2$ rad.

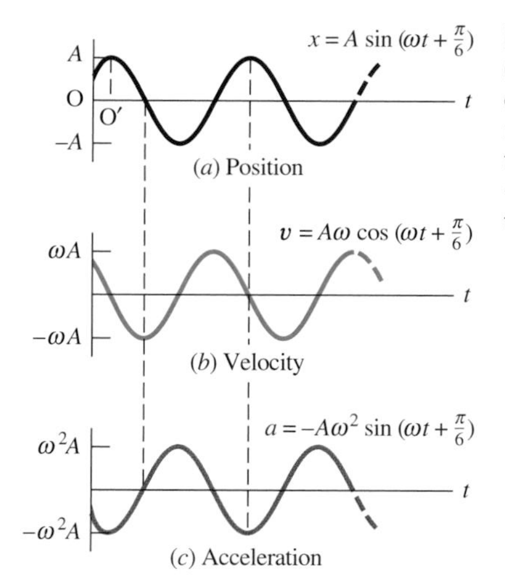

FIGURE 14-5 (*a*) Position, (*b*) velocity, and (*c*) acceleration as functions of time for simple harmonic motion. The phase constant is $\pi/6$ rad. Point O′ pertains to Drill Prob. 14-4.

EXAMPLE 14-3 Initial Conditions of Simple Harmonic Motion

A particle in simple harmonic motion with angular frequency ω starts at a position $x(0)$ with a velocity $v(0)$. (*a*) What is the phase constant of the motion? (*b*) What is the amplitude of the motion?

SOLUTION (*a*) With $x(0)$, $v(0)$, and $t = 0$ substituted into Eqs. (14-9*a*) and (14-14*a*), we find

$$x(0) = A \sin\phi \tag{i}$$

and

$$v(0) = A\omega \cos\phi. \tag{ii}$$

Dividing Eq. (*i*) by Eq. (*ii*) then gives

$$\frac{x(0)}{v(0)} = \frac{1}{\omega}\frac{\sin\phi}{\cos\phi}$$

so the phase constant is

$$\phi = \tan^{-1}\frac{\omega x(0)}{v(0)}. \tag{iii}$$

Notice from this equation that the phase constant is determined from the initial conditions—that is, $x(0)$, $v(0)$—of the motion of a given simple harmonic oscillator.

(*b*) The amplitude is also determined from the initial conditions. From Eq. (*i*),

$$A = \frac{x(0)}{\sin\phi}.$$

In Eq. (*iii*) ϕ is given in terms of $x(0)$ and $v(0)$. Hence the amplitude A can also be expressed in terms of $x(0)$ and $v(0)$, the initial conditions.

EXAMPLE 14-4 Parameters of Simple Harmonic Motion

A body of mass $m = 4.00$ kg is attached to a horizontal spring with force constant $k = 100$ N/m. The body is displaced 10.0 cm from its equilibrium position and released. For the resulting simple harmonic motion, find (*a*) the amplitude, (*b*) the period, (*c*) the frequency, (*d*) the angular frequency, (*e*) the mechanical energy, (*f*) the maximum velocity, and (*g*) the maximum acceleration. Assuming the displacement is represented by Eq. (14-9*a*), (*h*) determine the phase constant, (*i*) calculate the velocity and acceleration when the position is $x = 5.0$ cm, and (*j*) write expressions for the position, velocity, and acceleration as functions of time. (*k*) If the displacement is represented by Eq. (14-9*b*), what is the phase constant?

SOLUTION (*a*) At $x = 10.0 \text{ cm} = 0.100$ m, the velocity is zero so the position is a maximum. The amplitude of the motion is therefore $A = 0.100$ m.

(*b*) The period is calculated with Eq. (14-11):

$$T = 2\pi\sqrt{\frac{m}{k}} = 2\pi\sqrt{\frac{4.00 \text{ kg}}{100 \text{ N/m}}} = 1.25 \text{ s}.$$

(*c*) The frequency is the inverse of the period:

$$f = \frac{1}{T} = \frac{1}{1.25 \text{ s}} = 0.800 \text{ Hz}.$$

(*d*) The angular frequency is found with Eq. (14-12*b*):

$$\omega = 2\pi f = 2\pi(0.800 \text{ Hz}) = 5.00 \text{ rad/s}.$$

(*e*) From Eq. (14-5*b*), the mechanical energy is

$$E = \tfrac{1}{2}kA^2 = \tfrac{1}{2}(100 \text{ N/m})(0.100 \text{ m})^2 = 0.500 \text{ J}.$$

(*f*) Since the cosine function oscillates between -1 and $+1$, the maximum velocity is the multiplier of the cosine function in Eq. (14-14*a*):

$$v_{max} = A\omega = (0.100 \text{ m})(5.00 \text{ rad/s}) = 0.500 \text{ m/s}.$$

(*g*) Similarly, the maximum acceleration is the multiplier of the sine function in Eq. (14-15*a*):

$$a_{max} = A\omega^2 = (0.100 \text{ m})(5.00 \text{ rad/s})^2 = 2.50 \text{ m/s}^2.$$

(*h*) At $t = 0$, the position is $x(0) = 0.100$ m, so from Eq. (14-9*a*),

$$x(0) = A \sin(0 + \phi) = A \sin\phi,$$

and the phase constant is

$$\phi = \sin^{-1}\frac{x(0)}{A} = \sin^{-1}\frac{0.100 \text{ m}}{0.100 \text{ m}} = \frac{\pi}{2} \text{ rad}.$$

(*i*) From Eq. (14-6), the velocity when $x = 5.0$ cm is

$$v = \pm\sqrt{\frac{k}{m}}\sqrt{A^2 - x^2}$$

$$= \pm\sqrt{\frac{100 \text{ N/m}}{4.00 \text{ kg}}}\sqrt{(0.100 \text{ m})^2 - (0.050 \text{ m})^2} = \pm 0.433 \text{ m/s}.$$

Both signs are meaningful because the body can be moving either toward or away from the equilibrium position.

Combining Eqs. (14-4*a*) and (14-8), we find for the acceleration

$$a = -x\omega^2 = -(0.050 \text{ m})(5.00 \text{ rad/s})^2 = -1.25 \text{ m/s}^2.$$

(*j*) Substituting the values of the various parameters into Eqs. (14-9*a*), (14-14*a*), and (14-15*a*), we have

$$x(t) = (0.100 \text{ m}) \sin\left[(5.00 \text{ rad/s})t + \frac{\pi}{2} \text{ rad}\right],$$

$$v(t) = (0.500 \text{ m/s}) \cos\left[(5.00 \text{ rad/s})t + \frac{\pi}{2} \text{ rad}\right],$$

and

$$a(t) = -(2.50 \text{ m/s}^2) \sin\left[(5.00 \text{ rad/s})t + \frac{\pi}{2} \text{ rad}\right].$$

(*k*) From Eq. (14-9*b*), the initial position is

$$x(0) = A \cos(0 + \phi) = A \cos\phi,$$

so

$$\phi = \cos^{-1}\frac{x(0)}{A} = \cos^{-1}\frac{0.100 \text{ m}}{0.100 \text{ m}} = 0.$$

DRILL PROBLEM 14-4

Suppose that the origin of Fig. 14-5*a* is shifted to O′. Now what are the initial position and velocity of the simple harmonic oscillator? What is the phase constant that represents these initial conditions? Use Eq. (14-9*a*) to represent $x(t)$.

ANS. $x(0) = A$; $v(0) = 0$; $\phi = \pi/2$ rad.

DRILL PROBLEM 14-5

The initial position and velocity of a body moving in simple harmonic motion with period $T = 0.25$ s are $x(0) = 5.0$ cm and $v(0) = 218$ cm/s. What are the amplitude and phase constant of the motion? Use Eq. (14-9*a*) to represent $x(t)$.

ANS. 10.0 cm; $\pi/6$ rad.

DRILL PROBLEM 14-6

A cart of mass 2.00 kg is attached to the end of a horizontal spring with force constant $k = 150$ N/m. The cart is displaced 15.0 cm from its equilibrium position and released. What are (*a*) the amplitude, (*b*) the period, (*c*) the frequency, (*d*) the mechanical energy, and (*e*) the maximum velocity of the motion of the cart? Assume there is no friction.

ANS. (*a*) 15.0 cm; (*b*) 0.726 s; (*c*) 1.38 Hz; (*d*) 1.69 J; (*e*) 1.30 m/s.

DRILL PROBLEM 14-7

A 100-g block is attached to a vertical spring and slowly lowered to its equilibrium position; this stretches the spring 5.0 cm. If the block is then displaced from its equilibrium position and released, what is the period of the motion?

ANS. 0.45 s.

*14-4 Circular Motion and Simple Harmonic Motion

It is sometimes helpful to investigate simple harmonic motion in terms of the projection of uniform circular motion onto a line. The shadow of the small ball undergoing circular motion in Fig. 14-6 is an example of that projection. We'll examine the connection between the two motions with the help of Fig. 14-7. It shows a ball that is moving on a circular path of radius A at a constant angular velocity ω. The ball's velocity is $A\omega$ in a direction tangent to the circle, its acceleration is $A\omega^2$ directed toward the center of the circle, and its angular position is $\theta = \omega t + \phi$. As the ball moves from P to P′, its projection moves on the x axis from Q to Q′. From simple trigonometry, the projected position, velocity, and acceleration are

$$x(t) = A \sin \theta = A \sin (\omega t + \phi),$$
$$v(t) = A\omega \cos \theta = A\omega \cos (\omega t + \phi),$$

and

$$a(t) = -A\omega^2 \sin \theta = -A\omega^2 \sin (\omega t + \phi).$$

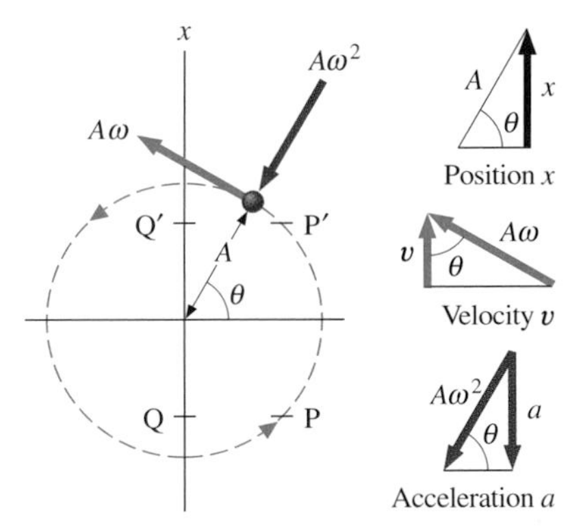

FIGURE 14-7 The projection of circular motion at constant angular velocity onto a coordinate axis is simple harmonic motion.

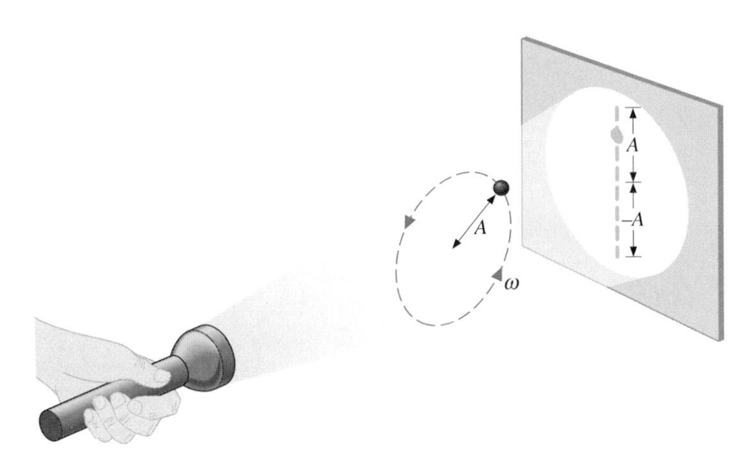

FIGURE 14-6 A small ball moves in a circle of radius A at constant angular velocity ω. The motion of its shadow is simple harmonic.

These three equations are identical to Eqs. (14-9a), (14-14a), and (14-15a); therefore the ball's projection is undergoing simple harmonic motion. The radius, angular velocity, and initial angular position for the circular motion are the amplitude, angular frequency, and phase constant, respectively, of the projection.

Although the vertical axis is used in this discussion, the motion of the projection along any axis through the center of the circle is simple harmonic. For example, if you make your projection onto the horizontal axis, you will find that the motion is represented by Eqs. (14-9b), (14-14b), and (14-15b).

EXAMPLE 14-5 Circular Motion and Simple Harmonic Motion

How much time is required for the body of Example 14-4 to travel from $x = -5.0$ cm to $x = 5.0$ cm?

SOLUTION We can analyze this problem by using the relationship between circular motion and simple harmonic motion. Suppose that we treat the simple harmonic motion of the body of Example 14-4 as the projection of the uniform circular motion of a small imaginary object. As this imaginary object travels from P to P′ in Fig. 14-8 at a constant angular velocity ω, its projection moves from $x = -5.0$ cm to $x = 5.0$ cm. From the geometry shown, the angle covered in going from P to P′ is

$$\theta = \pi - 2\alpha = \pi - 2 \cos^{-1} \frac{5.0 \text{ cm}}{10.0 \text{ cm}} = \frac{\pi}{3} \text{ rad}.$$

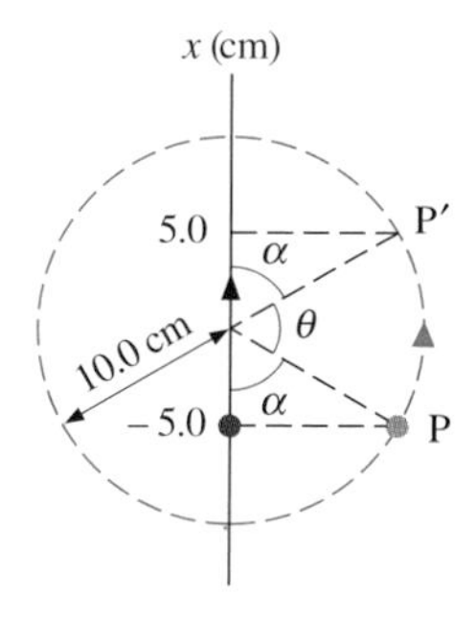

FIGURE 14-8 As the body moves from $x = -5.0$ cm to $x = +5.0$ cm, the imaginary object on the circle moves from P to P′.

The time the imaginary object takes to traverse this angle may be found from $\theta = \omega t$. Thus

$$\frac{\pi}{3} \text{ rad} = (5.0 \text{ rad/s})t,$$

and

$$t = 0.21 \text{ s},$$

which is also the time taken by the body to travel from $x = -5.0$ cm to $x = 5.0$ cm.

DRILL PROBLEM 14-8

How much time is required for the body of Example 14-4 to move from $x = -8.0$ cm to $x = 4.0$ cm?
ANS. 0.27 s.

DRILL PROBLEM 14-9

A body of mass 0.10 kg is attached to a vertical massless spring with force constant 4.0×10^3 N/m. The body is displaced 10.0 cm from its equilibrium position and released. How much time elapses as the body moves from a point 8.0 cm on one side of the equilibrium position to a point 6.0 cm on the same side of the equilibrium position?
ANS. 1.4×10^{-3} s.

14-5 Additional Examples of Simple Harmonic Motion

The body at the end of a spring is just one of many systems whose motion is simple harmonic. We now look at some others.

A Simple Pendulum

A simple pendulum is an idealized system consisting of a point mass attached to the end of a massless, inextensible cord moving without friction in a vertical plane. Figure 14-9 shows a simple pendulum of mass m and length l. Since the mass of the pendulum is concentrated in a point at the lower end of the string, its moment of inertia around the horizontal axis through O is ml^2. There is no torque around this axis due to the tension **T**, because **T** is directed toward O. The moment arm of the gravitational force is $l \sin\theta$, and the torque due to this force is $-mgl\sin\theta$. Since $\sum \tau = I\alpha$,

$$-mgl\sin\theta = ml^2\frac{d^2\theta}{dt^2},$$

which simplifies to

$$\frac{d^2\theta}{dt^2} + \frac{g}{l}\sin\theta = 0. \tag{14-16}$$

For small oscillations, $\sin\theta \approx \theta$ (see Supplement 14-1), where θ is expressed in radians. Equation (14-16) then becomes

$$\frac{d^2\theta}{dt^2} + \frac{g}{l}\theta = 0. \tag{14-17}$$

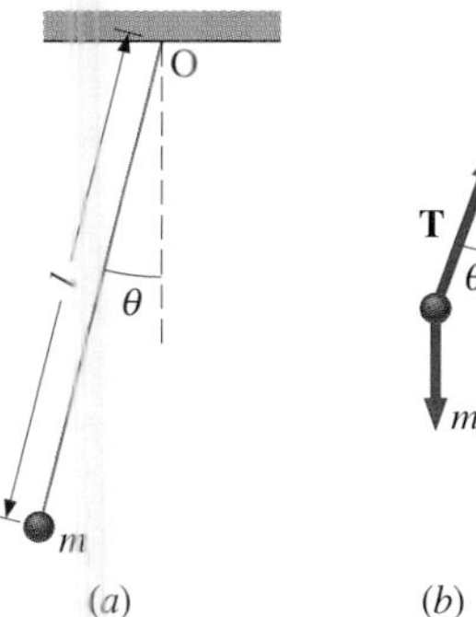

FIGURE 14-9 (*a*) A simple pendulum. (*b*) The free-body diagram for the pendulum bob. The force $m\mathbf{g}$ inserts a torque $-mgl\sin\theta$ around the axis through O.

You can see just how good this approximation is by comparing $\theta = 10° = 0.1745$ rad to $\sin 10° = 0.1736$: They differ by only 0.52 percent.

Now compare Eqs. (14-17) and (14-4*b*). Since the latter represents simple harmonic motion, so does the former. *The small oscillations of a simple pendulum are therefore simple harmonic.* Identifying θ with x and g/l with m/k, we have

$$\theta = A\sin(\omega t + \phi),$$

where A is the amplitude, ϕ is the phase constant, and ω, the angular frequency, is $\sqrt{g/l}$. In addition, the period of the simple pendulum is

$$T = \frac{2\pi}{\omega} = 2\pi\sqrt{\frac{l}{g}}. \tag{14-18}$$

Analysis of the simple pendulum's motion for amplitudes that are not small requires advanced mathematical techniques. The period is found to be

$$T = 2\pi\sqrt{\frac{l}{g}}\left(1 + \frac{1}{2^2}\sin^2\frac{A}{2} + \frac{1\cdot 3^2}{2^2\cdot 4^2}\sin^4\frac{A}{2} + \cdots\right). \tag{14-19}$$

The period may be calculated to any number of significant figures simply by using a sufficient number of terms in the infinite series. For $A = 10°$, Eq. (14-19) yields $T = 6.2952\sqrt{l/g}$. With the small-angle approximation, $T = 2\pi\sqrt{l/g} = 6.2832\sqrt{l/g}$. These two values differ by only 0.19 percent.

The pendulum may be used as a fairly accurate timekeeper because its period is almost independent of its amplitude. As the pendulum swings back and forth, the damping force of friction causes a gradual decrease in its amplitude. However, as the amplitude decreases, the change in the period is very small—just how small is discussed in the following example.

The oscillations of the pendulum keep time in the grandfather clock.

EXAMPLE 14-6 Time Measured by a Pendulum Clock

A pendulum clock is calibrated to keep accurate time when its amplitude is $A = 10°$. At the start of the workday (9:00 A.M. on the clock), its amplitude is 10°, but by the end of the workday (5:00 P.M. on the clock), the amplitude has decreased to 5°. How much time has actually elapsed corresponding to an advancement of 8 h on the clock? Assume that the average period of the pendulum can be estimated by averaging the periods calculated from Eq. (14-19) with $A = 10°$ and $A = 5°$.

SOLUTION The periods for $A = 10°$ and $A = 5°$ can be found from Eq. (14-19). They are, respectively,

$$T_{10} = 6.2952\sqrt{\frac{l}{g}} \quad \text{and} \quad T_5 = 6.2860\sqrt{\frac{l}{g}}. \qquad (i)$$

The average of these two periods is

$$\bar{T} = 6.2906\sqrt{\frac{l}{g}}. \qquad (ii)$$

The clock is calibrated so that in N oscillations at 10°, 8 h (or 2.88×10^4 s) elapse *on the clock*; therefore

$$NT_{10} = 2.88 \times 10^4 \text{ s}. \qquad (iii)$$

During the N oscillations, the average period is less than T_{10}, so rather than 8 h, which the clock reads, an *actual* time

$$t = N\bar{T} \qquad (iv)$$

elapses. Finally, substituting N from Eq. (iii), $\bar{T}$ from Eq. (ii), and T_{10} from Eq. (i) into Eq. (iv), we find

$$t = \left(\frac{6.2906}{6.2952}\right)(2.88 \times 10^4 \text{ s}) = 28779 \text{ s},$$

which is 21 s short of the 8 h that elapse on the clock. The pendulum clock is obviously not suitable for precise scientific measurements; nevertheless, it does remain reasonably accurate over the interval from 9:00 A.M. to 5:00 P.M. For all practical purposes, the employees are going to work their eight-hour day!

The simple pendulum can be used to make fairly precise measurements of the acceleration due to gravity. If l is known and T is measured, then g can be determined with Eq. (14-18). Geologists actually use more complicated and more accurate pendulums to look for small local variations in g. These variations are often evidence of deposits of valuable ores or oil.

A Physical Pendulum

The physical pendulum is a rigid body that is free to rotate around a fixed axis. An example is shown in Fig. 14-10. The distance between that pendulum's center of mass CM and O, the point through which its rotational axis passes, is d. The mass of the pendulum is m, and its moment of inertia around the rotational axis is I. The pendulum oscillates because the gravitational force $m\mathbf{g}$ exerts a torque

$$\tau = -mgd \sin\theta$$

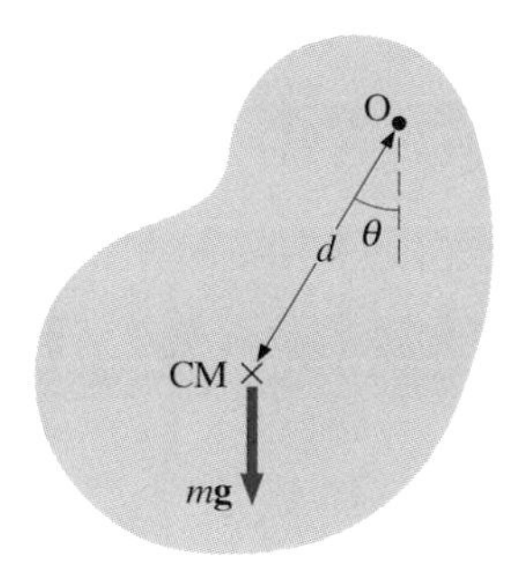

FIGURE 14-10 A physical pendulum. The force $m\mathbf{g}$ exerts a torque $-mgd \sin\theta$ around the axis through O.

around the rotational axis. From $\sum \tau = I\alpha$,

$$-mgd \sin\theta = I\frac{d^2\theta}{dt^2}$$

and

$$\frac{d^2\theta}{dt^2} + \frac{mgd}{I}\sin\theta = 0.$$

This differential equation, like that for the simple pendulum, can be easily solved if the oscillations are small enough that $\sin\theta \approx \theta$. Then by analogy with either the spring or the simple pendulum, we find that the period of the physical pendulum is

$$T = 2\pi\sqrt{\frac{I}{mgd}}. \qquad (14\text{-}20)$$

Of some interest is the length of a simple pendulum whose period is equal to that of a physical pendulum of known dimensions and mass. This length is easily found by equating Eq. (14-18) and Eq. (14-20). We then have

$$2\pi\sqrt{\frac{l}{g}} = 2\pi\sqrt{\frac{I}{mgd}},$$

and

$$l = \frac{I}{md}. \qquad (14\text{-}21)$$

This equation tells us that a simple pendulum of length l has the same period as a physical pendulum with moment of inertia I and mass m whose center of mass is located a distance d from the axis of rotation. As far as the period is concerned, the entire mass of the physical pendulum could be concentrated at a point that is a distance l from the axis of rotation. This point is called the *center of oscillation*.

Notice that Eq. (14-21) is identical in form to Eq. (13-11), which was used to determine the center of percussion of a rigid body. The center of percussion and the center of oscillation are therefore located at the same point. If you never seem to be able to find the sweet spot of a golf club or tennis racket by striking a ball, do not despair. You can at least find it by measuring a period!

EXAMPLE 14-7 An Oscillating Sphere

A uniform sphere of radius R is attached to the ceiling (see Fig. 14-11) so that it is free to oscillate around a horizontal axis through the point of attachment O. Find its period of oscillation.

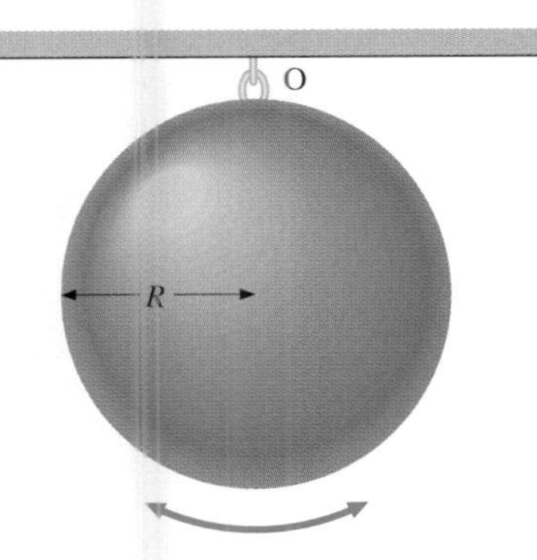

FIGURE 14-11 A uniform sphere is free to oscillate around a horizontal axis through O.

SOLUTION The center of mass of the sphere is at its geometric center, which is a distance R from the axis of rotation at O. By the parallel-axis theorem, the moment of inertia about the axis through O is

$$I = I_{CM} + mh^2 = \tfrac{2}{5}mR^2 + mR^2 = \tfrac{7}{5}mR^2.$$

Now from Eq. (14-20), the period of oscillation of this physical pendulum is

$$T = 2\pi\sqrt{\frac{I}{mgh}} = 2\pi\sqrt{\frac{7mR^2/5}{mgR}} = 2\pi\sqrt{\frac{7R}{5g}}.$$

DRILL PROBLEM 14-10

A hoop is free to rotate around an axis through its rim. Find its period for small oscillations.
ANS. $T = 2\pi\sqrt{2R/g}$.

DRILL PROBLEM 14-11

Find the centers of oscillation for the sphere of Example 14-7 and for the hoop of Drill Prob. 14-10.
ANS. $7R/5$, $2R$.

DRILL PROBLEM 14-12

The torsional pendulum illustrated in Fig. 14-12 is a uniform disk suspended by a vertical wire that is attached to the center of the disk. When the disk is twisted through a small angle θ, the wire exerts a restoring torque given by $-\kappa\theta$, where κ is a constant that depends on the properties of the wire. What is the period of the pendulum?
ANS. $T = 2\pi\sqrt{I/\kappa}$, where I is the disk's moment of inertia around its center.

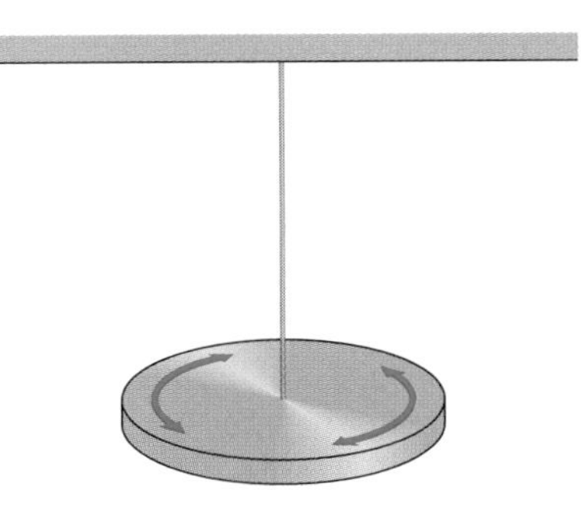

FIGURE 14-12 A torsional pendulum. The disk is free to oscillate around an axis along the suspension wire.

*14-6 Damped Oscillations

Because we have ignored friction, all of the systems we have studied thus far conserve mechanical energy and oscillate indefinitely. This situation is obviously an idealization. Since friction is a dissipative force, mechanical energy and the amplitude of oscillation must decrease with time.

The frictional force is often caused by the medium in which the oscillating body is immersed. As we discussed in Sec. 6-3, if the medium is a liquid (see Fig. 14-13), the frictional force may often be represented by

$$f_R = -bv = -b\,\frac{dx}{dt}, \tag{14-22}$$

where b, the **damping constant**, depends on the properties of both the body and the medium. With this added to the restoring force, Newton's second law applied to the spring-mass system gives

$$\sum F_x = ma_x$$
$$-kx - b\,\frac{dx}{dt} = m\,\frac{d^2x}{dt^2},$$

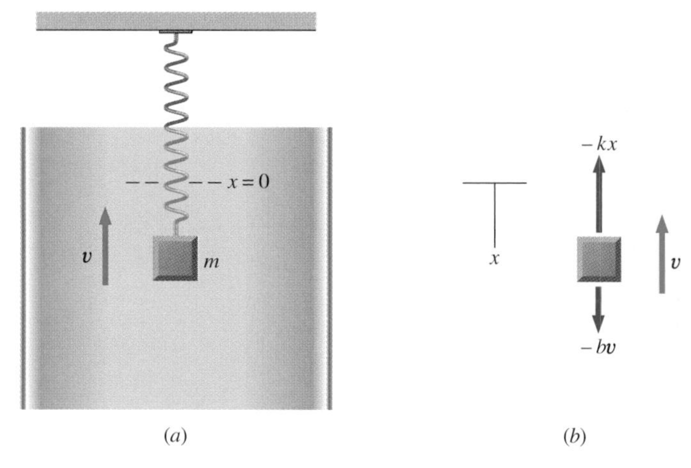

FIGURE 14-13 (*a*) A body is below its equilibrium position and is moving upward in a liquid medium. (*b*) The free-body diagram of the body. (*Remember*: The force of gravity is not included if x is measured from the equilibrium position.)

so the position satisfies the differential equation

$$m\frac{d^2x}{dt^2} + b\frac{dx}{dt} + kx = 0. \tag{14-23}$$

Systems that obey this equation are said to execute *damped harmonic motion*.

It can be shown that the solution to this differential equation takes three forms, depending on whether the angular frequency of the undamped spring $\omega_0 = \sqrt{k/m}$ is greater than, equal to, or less than $b/2m$. Graphs of the three solutions are given in Fig. 14-14. The solutions represented in the plots of parts (*a*), (*b*), and (*c*) represent motions known as *underdamped*, *critically damped*, and *overdamped*, respectively. They do behave differently, but in each case, the position x decreases with time and eventually converges to zero.

Note that there are oscillations only for the underdamped case and that the time for the return to zero displacement is shortest for critical damping. A practical application of the concept of critical damping is the design of shock absorbers for an automobile. Shock absorbers are connected between the frame and the body of an automobile to dampen the oscillations of the body. For the most comfortable ride, the damping constant of the shock absorbers should produce critical damping. (See Fig. 14-14*b*.) You may have noticed the "bouncy" ride of an automobile with worn-out shock absorbers. In this case, b has decreased sufficiently that the oscillations of the body are underdamped. (See Fig. 14-14*a*.)

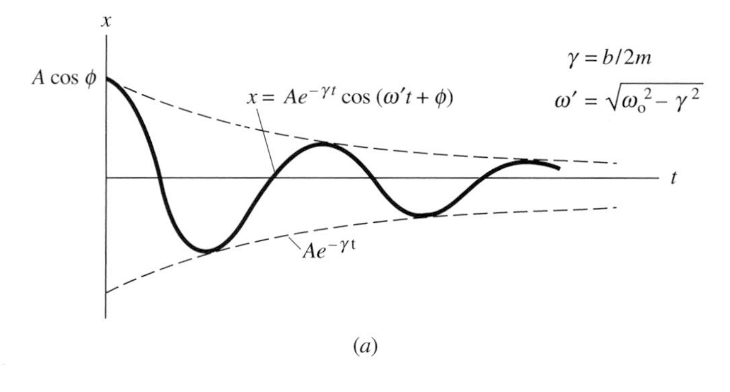

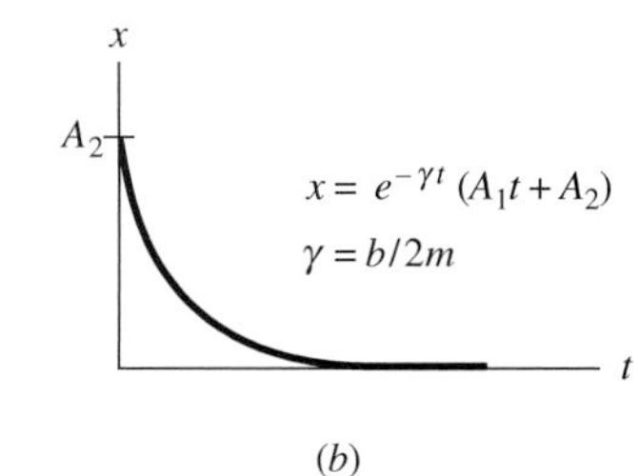

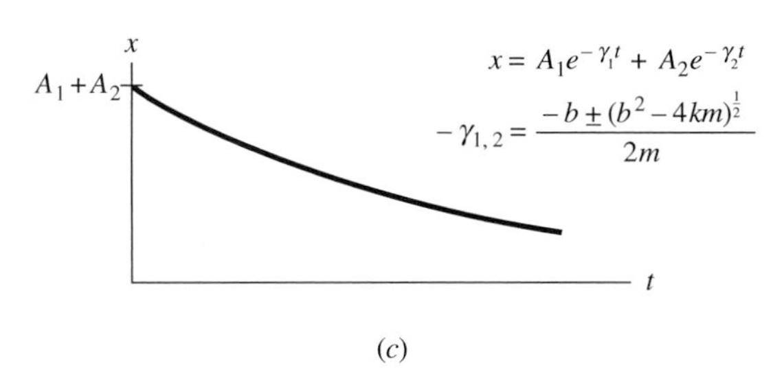

FIGURE 14-14 (*a*) Underdamped harmonic motion; (*b*) critically damped harmonic motion; (*c*) overdamped harmonic motion.

The shock absorbers quickly damp the vibrations of the car on its springs.

*14-7 Forced Oscillations and Resonance

We now discuss what happens to the damped oscillator when an external harmonic driving force, $F_E \sin \omega_E t$, is applied to it. The forces on the oscillator are then the restoring force, the frictional force, and the driving force. From Newton's second law,

$$\sum F_x = ma_x$$

$$-kx - b\frac{dx}{dt} + F_E \sin \omega_E t = m\frac{d^2x}{dt^2},$$

so the position x obeys the differential equation

$$m\frac{d^2x}{dt^2} + b\frac{dx}{dt} + kx = F_E \sin \omega_E t. \tag{14-24}$$

With the methods used for solving differential equations, the solution to this equation is found to be the sum of two terms. One term, called the *transient*, is the solution to the damped harmonic oscillator as given in Fig. 14-14. It decays with time and eventually becomes insignificant. The second term, which represents the behavior of x *after* the transient has effectively decayed, is called the *steady-state solution*. It is

$$x = A\cos(\omega_E t + \phi), \tag{14-25}$$

where

$$A = \frac{F_E/m}{\sqrt{(\omega_0^2 - \omega_E^2)^2 + b^2\omega_E^2/m^2}} \tag{14-26}$$

and

$$\omega_0 = \sqrt{\frac{k}{m}}. \tag{14-27}$$

Notice that ω_0 is the angular frequency of the undamped, unforced oscillator. It is called the **natural angular frequency**.

The steady-state response is harmonic with the angular frequency ω_E of the driving force, and its amplitude varies considerably with ω_E. As Fig. 14-15 shows, the amplitude is small for small values of ω_E, rises to a maximum, and falls to zero as ω_E goes to infinity. The quantity ω_R is the angular frequency at which the amplitude is a maximum. It is called the **resonant angular frequency** and is given by

$$\omega_R = \sqrt{\omega_0^2 - \frac{b^2}{2m^2}}. \tag{14-28}$$

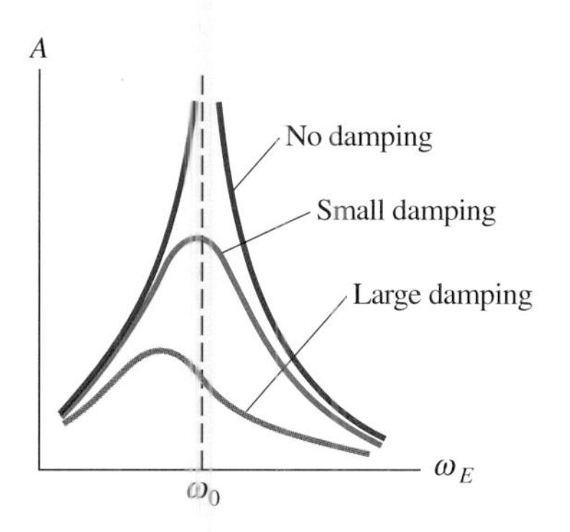

FIGURE 14-15 The amplitude of a driven, damped harmonic oscillator is plotted vs. the frequency of the driving force.

Frequently, the damping term, $b^2/2m^2$, is small compared with ω_0^2 and can be ignored in Eq. (14-28). In this case,

$$\omega_R \simeq \omega_0 \approx \sqrt{\frac{k}{m}},$$

as shown in Fig. 14-15. The resonance property of forced oscillations has many applications. One that is quite interesting is the microwave oven. The microwaves produced by the oven have a frequency that is equal to a natural vibrational frequency of the water molecule. As a result, the water in food absorbs large amounts of microwave energy, gets hot, and then warms the food it is part of. A mechanical example of resonance is the simple act of pumping a swing. A child quickly learns that by pumping at the natural frequency of the swing, she is able to build her oscillation amplitude very quickly. As an experiment, you might try oscillating a vertical mass-spring system at various frequencies. When you move the spring at a frequency far from resonance, the oscillations do not get very large. However, if you force the system close to its natural frequency, the oscillations quickly build to unmanageable amplitudes.

The large amplitudes of resonant vibrations are not always beneficial. Most of us have experienced irritating vibrations in the dashboards of our automobiles caused by a particular frequency coming from the radio speakers. Vibrations in the steering column of a car due to resonance between the wheels and the suspension system are also quite common. They become particularly bad if the front tires are poorly balanced. Engineers have to be especially careful to avoid resonant frequencies when designing machinery. For example, care must be taken to prevent a machine mount from vibrating excessively because it has a natural frequency equal to the rotational frequency of the motor shaft. Finally, there's the infamous Tacoma Narrows Bridge (see Fig. 14-16), formerly of Puget Sound, Washington. In 1940, just four months after the bridge was opened, strong winds set it vibrating with oscillations so large that its main span broke loose and fell into the water within a few hours after the vibrations started. It was later determined that the wind produced a fluctuating force that resonated with one of the natural frequencies of the bridge.

FIGURE 14-16 The Tacoma Narrows Bridge before and after the collapse of the main span.

SUMMARY

1. **Simple harmonic motion**
When a body of mass m moves in one dimension under the influence of the force $F(x) = -kx$, its position, velocity, and acceleration vary with time according to

$$x(t) = A \sin(\omega t + \phi),$$
$$v(t) = A\omega \cos(\omega t + \phi),$$
$$a(t) = -A\omega^2 \sin(\omega t + \phi),$$

or equivalently,

$$x(t) = A \cos(\omega t + \phi),$$
$$v(t) = -A\omega \sin(\omega t + \phi),$$
$$a(t) = -A\omega^2 \cos(\omega t + \phi),$$

where A is the amplitude, ω is the angular frequency, and ϕ is the phase constant. For the spring-mass system, $\omega = \sqrt{k/m}$.

2. **Parameters of simple harmonic motion**
The angular frequency of a simple harmonic oscillator is ω.
The phase is $\omega t + \phi$.
The phase constant is ϕ.
The amplitude A is the maximum displacement.
The period T $(= 2\pi/\omega)$ is the time for one complete oscillation.
The frequency f $(= 1/T)$ is the number of oscillations per unit time.

*3. **Circular motion and simple harmonic motion**
When an object moves around a circle of radius A at a constant angular velocity ω, the projection of that motion onto a line passing through the center of the circle is simple harmonic motion with an amplitude A and an angular frequency ω.

4. **Additional examples of simple harmonic motion**
For small vibrations, the simple, compound, and torsional pendulums execute simple harmonic motion with periods given by $2\pi\sqrt{l/g}$, $2\pi\sqrt{I/mgh}$, and $2\pi\sqrt{I/\kappa}$, respectively.

*5. **Damped oscillations**
When a frictional force $-b(dx/dt)$ is applied to a simple harmonic oscillator with angular frequency $\sqrt{k/m}$, the motion becomes underdamped, critically damped, or overdamped depending on whether $\sqrt{k/m}$ is, respectively, greater than, equal to, or less than $b/2m$.

*6. **Forced oscillations and resonance**
When a damped simple harmonic oscillator is subjected to a driving force $F_E \sin \omega_E t$, its steady-state oscillations are simple harmonic with an angular frequency ω_E. The amplitude of the oscillations varies with the frequency of the driving force and is a maximum at the resonant frequency

$$\omega_R = \sqrt{\omega_0^2 - \frac{b^2}{2m^2}}.$$

SUPPLEMENT 14-1 TAYLOR SERIES EXPANSIONS

If a function $f(x)$ has continuous derivatives of all orders, it can be expanded about $x = a$ in a Taylor series. This series is given by

$$f(x) = f(a) + \left(\frac{df}{dx}\right)_{x=a}(x-a) + \frac{1}{2!}\left(\frac{d^2f}{dx^2}\right)_{x=a}(x-a)^2 + \cdots + \frac{1}{n!}\left(\frac{d^nf}{dx^n}\right)_{x=a}(x-a)^n + \cdots, \qquad (i)$$

where $(d^nf/dx^n)_{x=a}$ represents the nth derivative of $f(x)$ evaluated at $x = a$. The Taylor series of some selected functions follow.

1. e^x expanded about $x = 0$: Since $de^x/dx = e^x$, $(d^nf/dx^n)_{x=0} = 1$. Then from Eq. (i),

$$e^x = 1 + x + \tfrac{1}{2}x^2 + \cdots \qquad (ii)$$

2. $\sin(x)$ expanded about $x = 0$: Since $d \sin(x)/dx = \cos(x)$ and $d \cos(x)/dx = -\sin(x)$, $f(0) = \sin(0) = 0$, $(df/dx)_{x=0} = \cos(0) = 1$, $(d^2f/dx^2)_{x=0} = -\sin(0) = 0$, $(d^3f/dx^3)_{x=0} = -\cos(0) = -1$, etc. From Eq. (i),

$$\sin(x) = x - \frac{x^3}{3!} + \frac{x^5}{5!} - \cdots$$

3. $(1-x)^{-1}$ expanded about $x = 0$: For this function, $d(1-x)^{-1}/dx = 1(1-x)^{-2}$, $d^2(1-x)^{-1}/dx^2 = 1\cdot 2(1-x)^{-3}$, $d^3(1-x)^{-1}/dx^3 = 1\cdot 2\cdot 3(1-x)^{-4}$, etc., so

$$(1-x)^{-1} = 1 + x + x^2 + x^3 + x^4 + \cdots$$

4. Some other expansions, all about $x = 0$, are

$$(1+x)^{-1} = 1 - x + x^2 - x^3 + x^4 - \cdots$$
$$(1 \pm x)^n = 1 \pm nx + \frac{n(n-1)}{2!}x^2 \pm \frac{n(n-1)(n-2)}{3!}x^3 + \cdots$$
$$\cos(x) = 1 - \frac{x^2}{2!} + \frac{x^4}{4!} - \cdots$$
$$\ln(1+x) = x - \tfrac{1}{2}x^2 + \tfrac{1}{3}x^3 - \cdots$$

QUESTIONS

Note: Unless stated otherwise in the following questions, all springs are assumed to have negligible mass.

14-1. What happens to the frequency of oscillation when the mass at the end of a spring is doubled?

14-2. How long (in terms of the period T) does it take a simple harmonic oscillator to travel a distance equal to its amplitude A?

14-3. A spring-mass system is oscillating with a frequency f and an amplitude A. (*a*) What happens to the frequency if the amplitude is doubled? (*b*) What happens to the total mechanical energy if the amplitude is doubled?

14-4. Atoms in a crystal lattice are sometimes viewed as particles attached to their neighbors by springs. What does this tell you about the approximate potential-energy function of the atoms? What happens to one of these atoms when given a slight displacement from its equilibrium position?

14-5. You are given two identical springs. One is attached to a body of mass M, the other to a body of mass $2M$. How could you ensure (*a*) that the total mechanical energies of the two simple harmonic oscillators are equal? (*b*) that the total mechanical energy of one is twice as great as the other?

14-6. What happens to the period of a simple pendulum if it is in a satellite orbiting the earth?

14-7. Make a table for a spring-mass system as follows: Label the columns for amplitudes "$-A$," "0," and "$+A$" and the rows "Displacement," "Speed," "Acceleration," "Kinetic energy," "Potential energy," and "Total energy." Fill the table. Wherever appropriate, mark the entries as "Maximum" or "Minimum."

14-8. In a stationary elevator, the period of a pendulum suspended from the ceiling is determined to be T. What happens to this value when the elevator (*a*) accelerates upward, (*b*) accelerates downward, (*c*) moves with a constant velocity?

14-9. Answer the previous question for (*a*) a spring-mass system undergoing oscillations on a frictionless tabletop in the elevator; (*b*) a rod suspended from the ceiling; (*c*) a torsional pendulum suspended from the ceiling.

14-10. Is it possible to make an ideal simple pendulum?

14-11. Pendulum A is a small body of mass m hanging at the end of a thread of length l. Pendulum B is a small body of mass m hanging at the end of a cord of length l and mass m. Which pendulum would you expect to have the greater frequency?

14-12. Suppose you have a spring of unknown force constant and a small body of unknown mass. Show how you can calculate the period of the spring-mass system's oscillations simply by measuring how far the body stretches the spring.

14-13. Explain qualitatively what effect the mass of the spring has on the period of oscillation.

PROBLEMS

Note: Unless stated otherwise in the following problems, all springs are assumed to have negligible mass.

Simple Harmonic Motion

14-1. Two passengers with a combined mass of 150 kg are observed to compress the springs of a 1200-kg car 2.0 cm. What is the vibrational period of the car and its passengers?

14-2. A body of mass 4.0 kg is attached to a vertical spring, and the system oscillates with a frequency of 4.0 Hz. How much will the spring shorten when the body is removed?

14-3. A 20-N force stretches a spring 10 cm. A body is attached to the spring and the system oscillates with a period of 0.393 s. What is the mass of the body?

14-4. A 400-g body hanging from a spring is observed to go through 10 complete oscillations in 5.0 s. What is the force constant of the spring?

14-5. A body of mass 200 g is in equilibrium at $x = 0$ under the influence of the force $F(x) = (-100x + 10x^2)$ N. (*a*) If the body is displaced a small distance from equilibrium, what is the period of its oscillations? (*b*) If the amplitude is 4.0 cm, by how much do we err in assuming that $F(x) = -kx$ at the end points of the motion?

14-6. A 100-g body oscillates at the end of a spring whose force constant is $k = 19.6$ N/m. Calculate the period, frequency, and angular frequency of the motion.

14-7. What is the mechanical energy of a 2.0-kg particle executing simple harmonic motion with a period of 0.25 s and an amplitude of 10 cm?

14-8. A 0.20-kg body is attached to a vertical spring and slowly lowered 5.0 cm to its equilibrium position. The body is then pulled down 6.0 cm from its equilibrium position and released with an upward velocity of magnitude 6.0 m/s. What is the amplitude of the motion?

14-9. A 0.50-kg body is attached to the end of a vertical spring and slowly lowered to its equilibrium position; this stretches the spring 10.0 cm. The body is then pulled down 5.0 cm more and released. Calculate (*a*) the spring's force constant and the resulting (*b*) amplitude, (*c*) period, (*d*) frequency, (*e*) angular frequency, (*f*) phase constant [assuming Eq. (14-9*a*)]. Also find the body's (*g*) maximum speed, (*h*) maximum acceleration, and (*i*) the total mechanical energy.

14-10. Suppose the 0.50-kg body of Prob. 14-9 is released rather than slowly lowered to its equilibrium position. (*a*) What is the maximum extension of the spring? (*b*) What are the frequency and amplitude of the resulting oscillations?

14-11. The position of a body moving in simple harmonic motion is given by $x = 0.25 \sin(20t + \pi/4)$ (in SI units). For this motion, what are the (*a*) amplitude, (*b*) period, (*c*) phase constant, (*d*) frequency, and (*e*) angular frequency? Determine the body's (*f*) maximum speed and (*g*) maximum acceleration.

14-12. The position of a body moving in simple harmonic motion is given by $x = 5.0 \cos 40t$ (in cgs units). For this motion, what are the (*a*) amplitude, (*b*) period, (*c*) phase constant, (*d*) frequency, and (*e*) angular frequency? What are the body's (*f*) maximum speed and (*g*) maximum acceleration?

14-13. A small body of mass 150 g is undergoing simple harmonic motion with an amplitude of 5.0 cm and a period of 0.25 s. (*a*) What are the maximum speed and acceleration of the body? (*b*) What are the velocity and acceleration of the body when its displacement is 3.0 cm? (*c*) If the oscillations are produced by a spring, what is its force constant? (*d*) What is the mechanical energy of the system?

Simple Harmonic Motion and Circular Motion

14-14. A body undergoes simple harmonic motion with a period of 0.20 s and an amplitude of 10.0 cm. What is the shortest time required for the particle to move from (*a*) −10.0 cm to +10.0 cm, (*b*) −5.0 cm to + 5.0 cm, (*c*) −6.0 cm to + 8.0 cm, (*d*) 5.0 cm to −6.0 cm?

14-15. A 2.0-kg particle undergoes simple harmonic motion according to

$$x = 1.5 \sin(\pi t/4 + \pi/6) \qquad \text{(in SI units)}.$$

(*a*) What is the total mechanical energy of the particle? (*b*) What is the shortest time required for the particle to move from $x = 0.50$ m to $x = -0.75$ m?

14-16. A body is oscillating with simple harmonic motion of amplitude 10.0 cm and frequency 8.0 Hz. Find (*a*) the body's maximum speed and acceleration, (*b*) the body's velocity and acceleration when its position is 6.0 cm, and (*c*) the minimum time required for it to move from $x = -6.0$ cm to $x = 8.0$ cm.

Applications of Simple Harmonic Motion

14-17. At a particular location on the earth, a simple pendulum of length 0.250 m is observed to oscillate 100 times in 100.46 s. What is the acceleration due to gravity at that location?

14-18. A simple pendulum has a period of 1.00 s at a point on the earth where $g = 9.80\ \text{m/s}^2$. What is the period of the pendulum at a point on the moon where $g = 1.70\ \text{m/s}^2$?

14-19. A simple pendulum is 2.0 m long and swings through an arc of 20° (−10° to +10°). (*a*) What is the angular velocity of the pendulum when it swings through its lowest point? (*b*) What is the linear speed of the pendulum bob at this point? (*c*) If the mass of the bob is 200 g, what is the total mechanical energy of the pendulum? (*d*) Suppose the mass of the bob is increased to 400 g. Answer parts (*a*), (*b*), and (*c*) for this case.

14-20. A simple pendulum that is used to keep time has a period of 2.0 s. When placed in a uniformly accelerating elevator, its period increases to 2.1 s. What are the magnitude and direction of the acceleration of the elevator?

14-21. A meterstick is pivoted at one end and set oscillating in a vertical plane. (*a*) What is its period? (*b*) If the axis of rotation is moved to the 30-cm mark on the meterstick, what does its period become?

14-22. A thin hoop of radius 40 cm is placed on a wedge and set oscillating. (See the accompanying figure.) (*a*) What is the period of its simple harmonic motion? (*b*) What is the length of a simple pendulum with the same period?

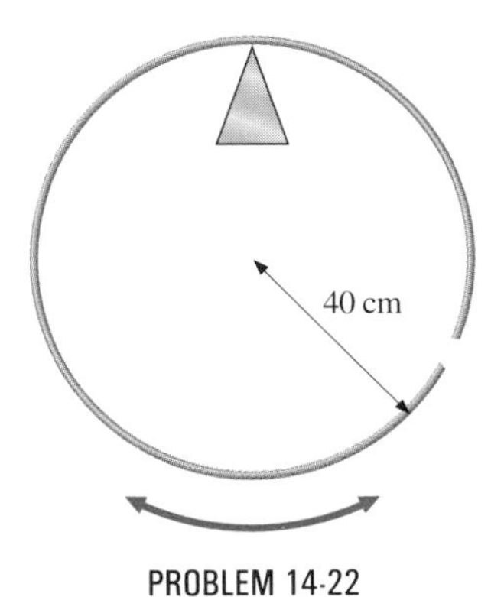

PROBLEM 14-22

14-23. A long, thin wire is attached to the center of one end of a solid cylinder of mass 1.0 kg and radius 0.25 m. The other end of the wire is fixed to the ceiling, so the cylinder hangs such that its circular faces are in horizontal planes. A force of 10 N applied tangent to the rim of the cylinder causes it to rotate through 3.0° before stopping. If the cylinder is set oscillating, what is its period?

14-24. A uniform stick of length 2.0 m is oscillating in a vertical plane. If the period of its oscillations is measured to be 2.5 s, what are the points on the stick through which the axis of rotation may pass?

14-25. A baseball bat is set oscillating around an axis through the end of its handle. The bat has a period of 1.6 s. Where is the center of percussion of the bat?

Damped and Forced Harmonic Motion

14-26. A body of mass $m = 200$ g is attached to a spring with force constant $k = 6.0$ N/m. The body is suspended in a viscous medium with a damping constant $b = 0.50$ kg/s and set oscillating. (*a*) Is the motion underdamped, critically damped, or overdamped? (*b*) What is the period of the motion? (*c*) How much time elapses before the amplitude drops to half its initial value? (*d*) What value must b have if the motion is to be critically damped?

14-27. If the oscillator of Prob. 14-26 is driven by the force (in newtons) $F(t) = 0.20 \sin(2.2t)$, what is the amplitude of its steady-state oscillations?

14-28. What are the resonant frequencies of the two oscillators shown in the accompanying figure? Assume damping is negligible.

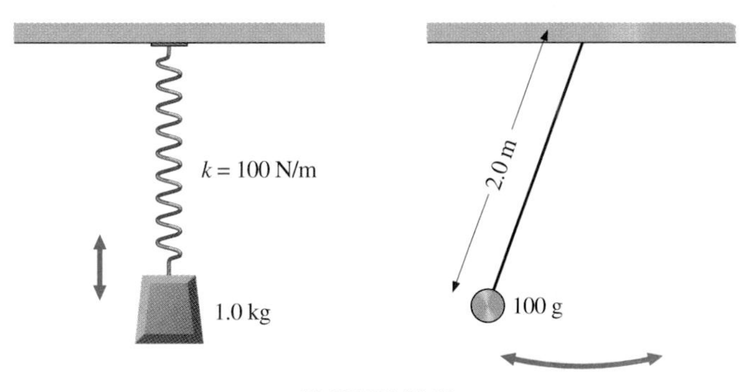

PROBLEM 14-28

General Problems

14-29. (*a*) Two identical springs are joined, and the lower spring is connected to a block of mass m, as shown in part (*a*) of the accompanying figure. Find the period of oscillation of the system. (*b*) A spring has a force constant k. If the spring is cut into two

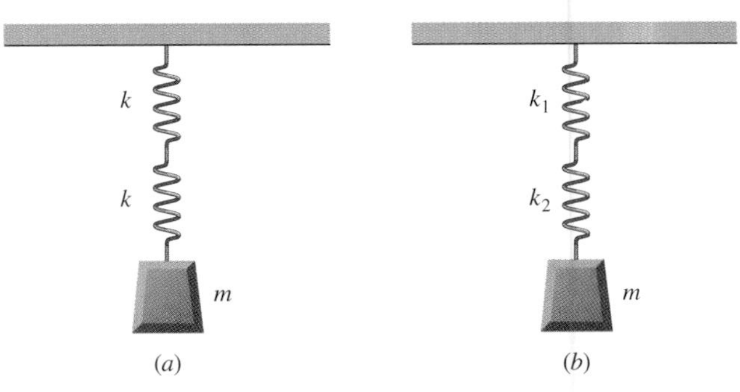

PROBLEM 14-29

pieces, show that the force constants k_1 and k_2 of the two pieces are related to the force constant of the original spring by (see part *b* of the accompanying figure)

$$\frac{1}{k} = \frac{1}{k_1} + \frac{1}{k_2}.$$

14-30. Consider the spring-mass system of Prob. 14-9. (*a*) What is the tension in the spring when the 0.50-kg body is 3.0 cm below its equilibrium position? (*b*) Subtract the weight of the 0.50-kg body from the tension. Compare this with the restoring force k (0.030 m). What does the equality of the values imply?

14-31. A body in simple harmonic motion has a speed of 5.0 m/s when it is displaced 1.0 m from its equilibrium position, and its speed is 3.0 m/s when it is at 1.5 m. Find the period and amplitude of the motion.

14-32. A 2.0-kg block that rests on a frictionless surface is attached to a horizontal spring with force constant $k = 600$ N/m, as shown in the accompanying figure. A 10-g bullet traveling horizontally at 600 m/s penetrates and stops in the block. (*a*) What is the period of the resulting simple harmonic motion? (*b*) What is the amplitude of the motion?

PROBLEM 14-32

14-33. A solid cylinder of mass m is attached to a horizontal spring with force constant k. The cylinder can roll without slipping along the horizontal plane. (See the accompanying figure.) Show that the center of mass of the cylinder executes simple harmonic motion with a period

$$T = 2\pi\sqrt{\frac{3m}{2k}}$$

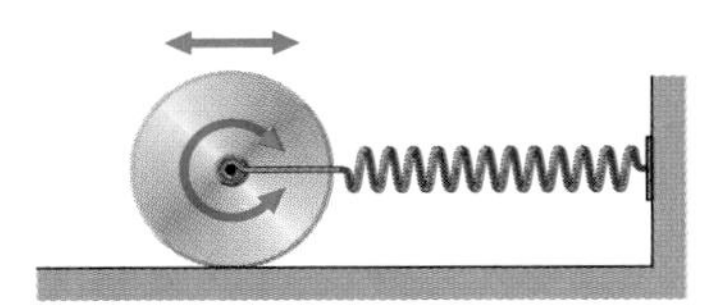

PROBLEM 14-33

14-34. The cylinder of the previous problem has a speed v_0 when it passes through equilibrium. Calculate the amplitude of the motion and the maximum value of the force of static friction between the cylinder and the floor.

14-35. Particles of dust are resting on the horizontal diaphragm of a loudspeaker. When the loudspeaker is turned on, the diaphragm moves in simple harmonic motion in the vertical direction with a fixed amplitude of 1.0×10^{-4} m. What is the highest frequency at which the diaphragm can oscillate so that the dust particles are always able to remain on the surface?

14-36. A small marble is set oscillating with small angular amplitude inside a hollow cylinder, as shown in the accompanying figure. (*a*) Determine the acceleration of the marble along the surface in terms of R, θ, and g. Assume the marble rolls without slipping and that its radius r is much less than the cylinder's radius R, so $R - r \approx R$. (*b*) Approximate $\sin\theta$ by θ, use $a = R(d^2\theta/dt^2)$, and determine the period of oscillation of the marble.

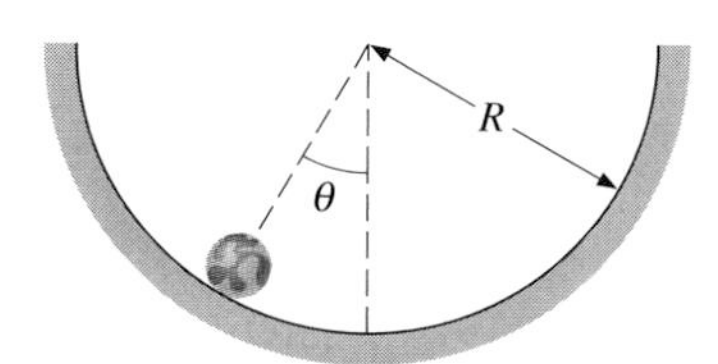

PROBLEM 14-36

14-37. A cord is attached between a 0.50-kg body and a spring with force constant $k = 20$ N/m. The other end of the spring is attached to the wall, and the cord is placed over a pulley ($I = 0.60MR^2$) of mass 5.0 kg and radius 0.50 m. (See the accompanying figure.) Assuming no slipping occurs, what is the frequency of the oscillations when the body is set into motion?

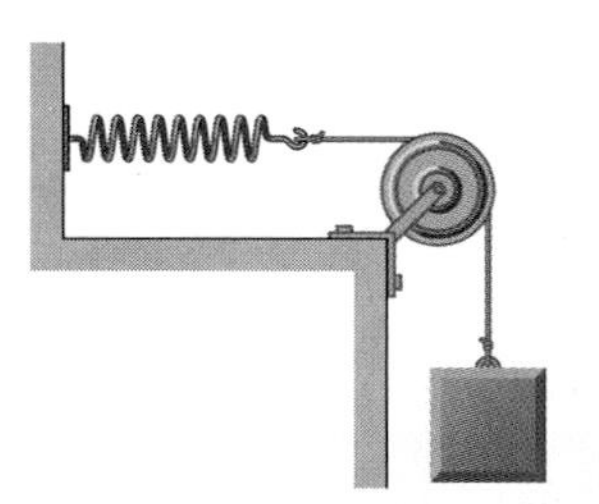

PROBLEM 14-37

14-38. Treat the simple pendulum as a compound pendulum consisting of a spherical solid ball of radius R and mass M at the end of a string of length l and mass m. (See the accompanying figure.) (*a*) What is its period? (*b*) By how much is the simple-pendulum formula, $T = 2\pi\sqrt{l/g}$, in error if $l = 1.0$ m, $R = 2.0$ cm, $m = 1.0$ g, and $M = 320$ g?

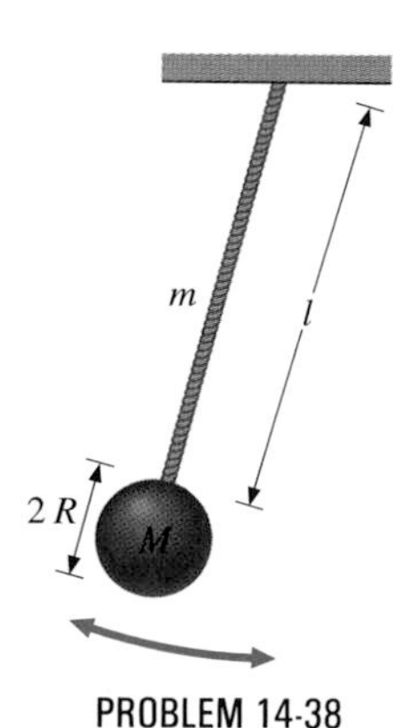

PROBLEM 14-38

14-39. A uniform rod of length l and mass m is suspended from the ceiling by a frictionless pin. A bullet of mass m that is traveling horizontally strikes and embeds itself in the bottom of the rod. The rod then starts oscillating in simple harmonic motion. What is the frequency of the oscillations?

An application of Archimedes' principle.

CHAPTER 15 FLUID MECHANICS

PREVIEW

This chapter is devoted to the mechanics of fluids. Both fluids at rest and fluids in motion will be considered. The main topics to be discussed here include the following:

1. **States of matter.** The three states of matter are discussed briefly, and the definition of a fluid is presented.
2. **Density and pressure.** The concept of density of a substance is reviewed. Pressure is defined and its relationship to the force of a fluid on a surface is described.
3. **Pressure variation with depth.** The first condition of static equilibrium is used to determine how pressure varies with depth in a fluid. Applications considered are pressure measurements and units, the barometer, the force on the vertical face of a dam, and hydraulic lifts.
4. **Archimedes' principle.** We consider the principle that describes the buoyant force that a fluid exerts on an object immersed in it.
5. **Fluids in motion.** The quantities that characterize a moving fluid are described, and various types of fluid flow are defined.
6. **The equation of continuity.** We discuss how the speed of a fluid varies with the cross-sectional area of the pipe through which it is flowing.
7. **Bernoulli's equation.** Bernoulli's equation, which describes the behavior of an ideal fluid (a fluid flowing without energy loss to friction), is derived. Applications covered include flow through pipes, forces on an airplane wing and a spinning ball, and the Venturi meter.

*8. **Laminar flow in a cylindrical pipe.** The effect of friction on the flow of a fluid through a horizontal pipe with a cylindrical cross section is investigated. One interesting application to be considered is the flow of blood through the circulatory system.

We generally think of a fluid as a substance that can flow. Hence liquids and gases are fluids. Like a rigid body, a fluid is composed of a very large number of particles, each of which is governed by the laws of Newtonian mechanics. However, unlike those of a rigid body, the particles of a fluid do not maintain fixed separations. Some may be stationary at the same time that others are moving in straight lines or even circles. As you might imagine, the general motion of a fluid is very complex.

Fortunately, many real-life situations involving fluids can be approximated very well by relatively simple mathematical models. We'll consider such models for the following three cases:

1. Fluids at rest
2. Friction-free fluid flow
3. Fluid flow with friction in a horizontal pipe

The branch of fluid mechanics that deals with fluids at rest is known as *hydrostatics*, while the study of fluids in motion is called *hydrodynamics*. Using the theoretical framework to be developed for these three cases, we will be able to investigate a number of interesting phenomena, including the variation of pressure with depth, buoyancy, the lift on an airplane wing and a golf ball, the "curve ball," and the flow of blood in the circulatory system.

15-1 States of Matter

Except at very high temperatures, matter exists in one of three states: *solid*, *liquid*, or *gas*.* Which state a particular material is in depends on its temperature and pressure. For example, water is a liquid at room temperature and atmospheric pressure. At atmospheric pressure, it does not solidify until its temperature drops below 0°C, and it does not become a gas until its temperature reaches 100°C. Iron, on the other hand, is a solid at normal pressures and temperatures. In order to melt iron at atmospheric pressure, we must heat it to a temperature above 1530°C. Another familiar substance, oxygen, is a gas at normal temperatures and pressures. At atmospheric pressure, oxygen does not become a liquid until it is cooled below −183°C, and it does not freeze until its temperature is reduced to −219°C.

The three states of a substance can be distinguished by its behavior when placed in a container. In the solid state the shape of the substance depends only slightly on its surroundings. A block of wood remains a block of wood when it is moved from a tabletop to the interior of a container. A liquid behaves quite differently when placed in a container. The liquid does have a top surface, but it flows outward to the sides of the container. If the shape of the container changes, the shape of the liquid also changes—but its volume remains the same. Finally, in the gaseous state a material also expands to the sides of the container. However, a gas is different from a liquid in that it does not have a well-defined upper surface, unless a lid is placed on the container. Without the lid, the gas escapes through the container opening.

The molecules in a solid are closely packed and the interatomic forces are strong. Consequently, it is quite difficult to distort a solid, because in doing so the relative positions of its molecules must be altered. A person does not crush a block of steel merely by squeezing it in his or her hand.

In the gaseous state the molecules are almost free from one another. They wander about in a container experiencing almost no forces due to the other molecules, except during collisions. When the lid is removed from a container of gas, the intermolecular forces are not even strong enough to confine the molecules to the container. These forces are also not much of a deterrent to the compression of a gas. Its volume can be decreased quite easily by compressing its container.

The forces between the molecules of a liquid are intermediate in strength between those of a solid and those of a gas. The intermolecular forces are strong enough to prevent the easy compression of a liquid, but they cannot keep the liquid from flowing outward to the sides of its container.

Liquids and gases are known collectively as *fluids*. When a tangential force acts on a fluid surface, the surface yields and the fluid flows. For example, water readily flows around the hull of a moving sailboat. Since a fluid flows in response to tangential surface forces, *the net force on any surface of a fluid at rest must be directed perpendicular to the surface.*

In treating fluids, we ignore individual molecules and assume that fluid matter is distributed *continuously* throughout the region that it occupies. We can then take any section, be it finite or infinitesimal, of that fluid and consider it to be a body or a particle, respectively, that obeys the laws of Newtonian mechanics. For example, we can ascribe the usual dynamic variables such as energy and momentum to an infinitesimal volume element of a fluid and then analyze the motion of the element as if it were an ordinary particle.

15-2 Density and Pressure

The laws of fluid mechanics are generally formulated in terms of mass density and pressure. By definition, mass density (or just density) is mass per unit volume. If a homogeneous substance has mass m and volume V, its density is

$$\rho = \frac{m}{V}. \qquad (15\text{-}1)$$

Of course, density varies with location in an inhomogeneous fluid. The densities of some of the commonly occurring substances are listed in Table 15-1. Notice that, as expected, the densities of gases are much smaller than those of solids and liquids. In the chapters on thermodynamics, you'll see that the volume of a sample of material is temperature-dependent. Since the mass of the sample is constant, this means that the density of the material must vary with temperature. This variation is small for solids and liquids but can be quite pronounced for gases.

*At very high temperatures (above approximately 10^5°C) the atoms of matter ionize and form a collection of negatively charged electrons and positively charged ions. This collection is a fourth state of matter called the *plasma state*. It will not be considered in this textbook.

TABLE 15-1 **DENSITIES OF SELECTED SUBSTANCES** (Unless stated otherwise, the values given are for 20°C and atmospheric pressure. To convert the units to kg/m^3, multiply by 10^3.)

Substance	Density (g/cm^3)
Air (0°C)	1.29×10^{-3}
Air (20°C)	1.20×10^{-3}
Aluminum	2.70
Benzene	0.90
Brass	8.6
Cement (set)	2.7–3.0
Copper (cast)	8.30–8.95
Ethyl alcohol	0.789
Gold (cast)	19.3
Helium (0°C)	1.79×10^{-1}
Ice (0°C)	0.922
Iron (pure)	7.85–7.88
Lead	11.3
Mercury	13.6
Oak	0.60–0.90
Seawater (15°C)	1.025
Steel	7.60–7.80
Water (0°C)	0.999841
Water (3.98°C)	1.000000
Water (20°C)	0.998203

The **pressure** at any point O within a fluid at rest is defined in terms of the force of the fluid on one side of an arbitrary surface of infinitesimal area ΔA containing the point. (See Fig. 15-1.) Since the fluid is at rest, this force must *always be normal* to that surface, as illustrated in Fig. 15-2. If the normal force on the surface of area ΔA in Fig. 15-1 is ΔF_n, then by definition, the pressure p at O is

$$p = \lim_{\Delta A \to 0} \frac{\Delta F_n}{\Delta A}, \tag{15-2}$$

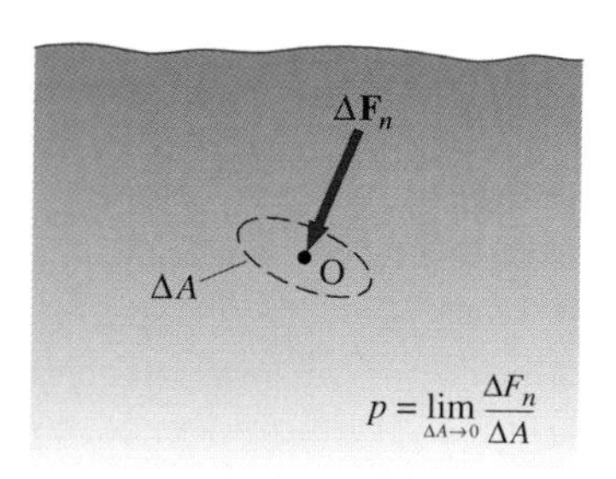

FIGURE 15-1 The pressure at any point O within a fluid is defined in terms of the normal force of the fluid on one side of an infinitesimal area ΔA that contains the point.

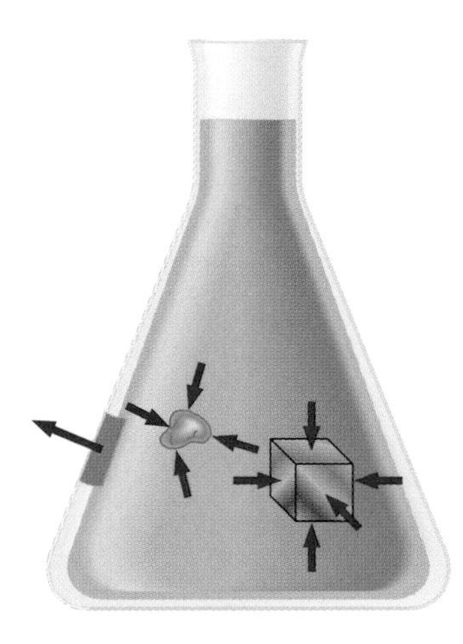

FIGURE 15-2 The force on any arbitrary surface within a fluid is always normal to that surface.

where the limit is written as a reminder that ΔA is an infinitesimal area.

Although pressure is defined in terms of a particular infinitesimal surface surrounding a point, it is actually independent of that surface. For a fluid at rest, the pressure only depends on the *location of the point*. This can be shown with the help of the very small wedge-shaped fluid section of Fig. 15-3. This section surrounds a point O within a fluid of density ρ, and it is in static equilibrium. Because the net vertical force on the section must vanish,

$$\Delta F_3 \cos\theta - \Delta F_1 - (\Delta m)g = 0,$$

where Δm is the mass of the fluid within the wedge. Now $\Delta m = \rho\Delta V = \rho(\Delta l_1\, \Delta l_2\, \Delta l_4/2)$, so, from Eq. (15-2),

$$(p_3\, \Delta l_3\, \Delta l_4)\cos\theta - (p_1\, \Delta l_1\, \Delta l_4) - \rho(\tfrac{1}{2}\, \Delta l_1\, \Delta l_2\, \Delta l_4)g = 0.$$

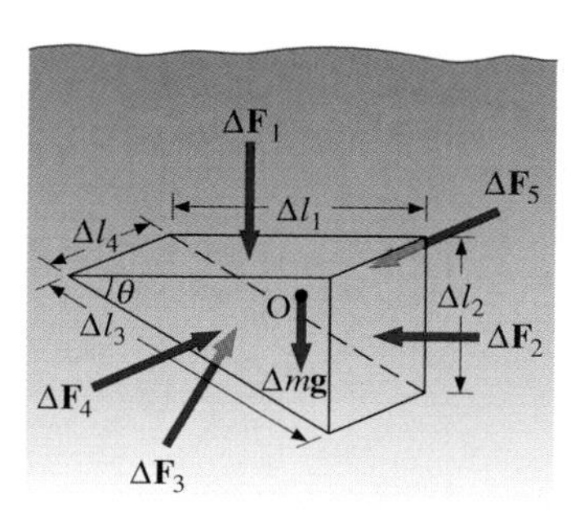

FIGURE 15-3 The forces on a wedge-shaped volume (assumed to be infinitesimally small) in a fluid.

In the limit as the volume shrinks to zero, the last term in this equation, a third-order infinitesimal, is negligible compared with the first two terms, which are second-order infinitesimals. Then with Δl_1 substituted for $\Delta l_3 \cos\theta$, we find in this limit that

$$p_3(\Delta l_1\, \Delta l_4) - p_1(\Delta l_1\, \Delta l_4) = 0;$$

and

$$p_3 = p_1.$$

Notice that this result is independent of the angle chosen for the slanted side of the wedge. Consequently, the pressure is the same on *any* infinitesimal surface surrounding the point O. Furthermore, since there is no direction associated with it, *pressure is a scalar quantity*.

Pressure is a force per unit area and is commonly expressed in newtons per square meter (N/m^2) and pounds per square foot (lb/ft^2). Other pressure units (to be discussed shortly) include the atmosphere, the bar, and centimeters of mercury.

EXAMPLE 15-1 PRESSURE ON THE FLOOR

A girl of mass 60 kg stands on a level floor. The bottoms of her shoes can be assumed to be rectangles with dimensions 8.0 cm by 25.0 cm. What pressure is exerted on the floor by her shoes?

SOLUTION The girl is in static equilibrium, so the net upward force exerted on her shoes by the floor is equal to her weight $mg = (60\text{ kg})(9.8\text{ m/s}^2) = 588\text{ N}$. By Newton's third law, the net force that her shoes exert on the floor is 588 N directed downward. Assuming this force is distributed evenly over the bottoms of her two shoes, the pressure exerted on the floor by her shoes is then

$$p = \frac{F}{A} = \frac{588\text{ N}}{2(0.080\text{ m})(0.25\text{ m})} = 1.5 \times 10^4\text{ N/m}^2.$$

EXAMPLE 15-2 Force Due to a Difference in Air Pressure

Assume that the air pressure at a height of 10,000 m above sea level is $2.7 \times 10^4\text{ N/m}^2$. If the inside of an airplane flying at this altitude is pressurized to $1.0 \times 10^5\text{ N/m}^2$ (atmospheric pressure), what is the net force due to the pressure difference on a cabin door whose dimensions are 2.0 m by 1.5 m? (See Fig. 15-4.)

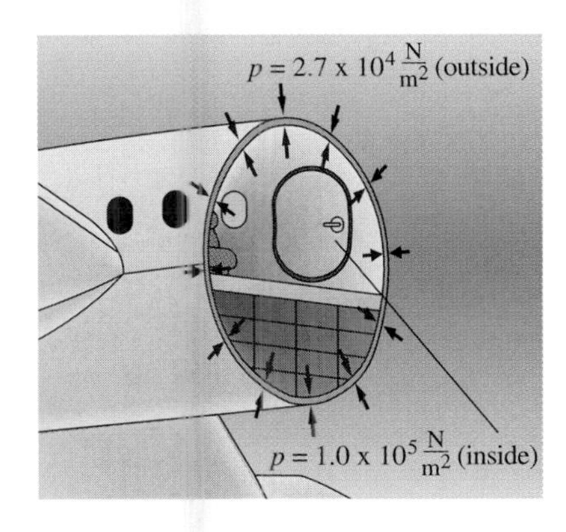

FIGURE 15-4 The cabin door of the airplane is subjected to a large net force directed outward from the airplane.

SOLUTION The area of the door is 3.0 m^2. The force on the inside surface of the door is $(1.0 \times 10^5\text{ N/m}^2)(3.0\text{ m}^2) = 3.0 \times 10^5\text{ N}$, and the force on the outside surface is $(2.7 \times 10^4\text{ N/m}^2)(3.0\text{ m}^2) = 8.1 \times 10^4\text{ N}$. The net force is their difference, 2.2×10^5 N. Notice how large this net force is—the weight of a 100-kg door is only 0.45 percent of this force!

DRILL PROBLEM 15-1

The footprint of an elephant is shallower than that of a human. Can you explain why?

DRILL PROBLEM 15-2

(*a*) A cylindrical piston made of aluminum is 5.0 cm thick and has a diameter of 20.0 cm. It rests in a vertical combustion chamber and is kept from falling by a pressure difference between its two faces. What is that pressure difference? (*b*) If the pressure inside the chamber is suddenly increased by 200 N/m², what is the instantaneous acceleration of the piston?
ANS. (*a*) $1.3 \times 10^3\text{ N/m}^2$; (*b*) 1.5 m/s^2.

DRILL PROBLEM 15-3

What is the force of the fluid on the other side of the infinitesimal area ΔA shown in Fig. 15-1?
ANS. $-\Delta \mathbf{F}_n$.

15-3 Pressure Variation with Depth

If a fluid is in static equilibrium, the sum of the forces on any portion of it must vanish. To investigate what this tells us about how pressure varies within the fluid, let's consider the thin, vertical cylindrical fluid section of Fig. 15-5. With positive y upward, this section's cross-sectional area and height are A and dy, respectively, and its weight is $\rho(\text{volume})g = \rho Ag\,dy$, where

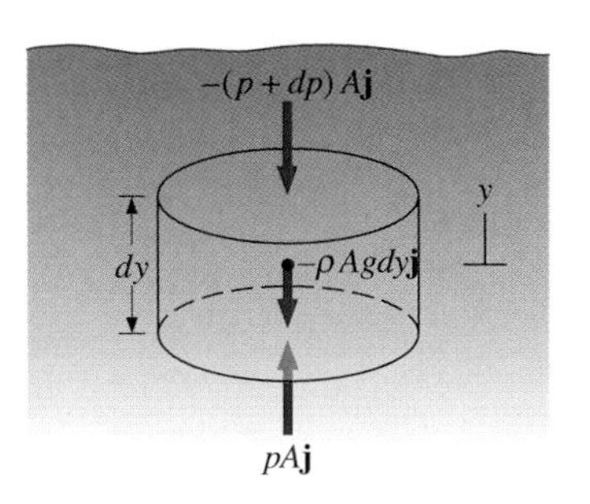

FIGURE 15-5 A thin cylindrical section of fluid is in equilibrium under the forces shown.

ρ is the density of the fluid. Since the section is in static equilibrium, there is no net force on it. So, in the horizontal direction, the normal forces on the cylindrical sides sum to zero, while in the vertical direction, the difference in the normal forces on the two end caps balances the weight of the fluid section. The force on the top cap is $(p + dp)A$ downward, and the force on the bottom cap is pA upward. Summing forces in the vertical direction, we find

$$pA - (p + dp)A - \rho Ag\,dy = 0,$$

so

$$\frac{dp}{dy} = -\rho g, \tag{15-3}$$

where the negative sign indicates that $dp > 0$ when $dy < 0$; that is, the pressure increases as one descends within a fluid. If the fluid is a *liquid*, it is not compressed much even for large increases in the pressure. In this case, we can take ρ to be a constant in Eq. (15-3) and then easily integrate the equation. If the pressure is p_1 at the vertical position y_1 and p_2 at the vertical position y_2, we have

$$p_2 - p_1 = \int_{p_1}^{p_2} dp = -\int_{y_1}^{y_2} \rho g\,dy = -\rho g \int_{y_1}^{y_2} dy = -\rho g(y_2 - y_1),$$

and the pressures at two vertical positions are related by

$$p_2 = p_1 + \rho g(y_1 - y_2). \tag{15-4a}$$

Thus the pressure difference between two points in a liquid of constant density is proportional to the difference in the *vertical* positions of the points. Frequently, the difference in vertical position is called the **depth** that point 2 is below point 1. Then when point 2 is a depth $h = y_1 - y_2$ below point 1, the pressures at these two points are related by

$$p_2 = p_1 + \rho g h. \tag{15-4b}$$

Notice that Eqs. (15-4) contain no reference to the amount of fluid involved in a particular situation—only the depth of a given point is significant. For example, the pressure is the same at a depth h below each of the surfaces of Fig. 15-6. This rather surprising result was demonstrated by Blaise Pascal (1623–1662) in the simple experiment represented by Fig. 15-7. A stout oak barrel was first filled with water. Its lid, which had a long, thin tube inserted into a hole at its center, was then set in place and water was poured into the tube. It was found that when the column of water reached a height of about 12 m, the pressure became great enough to burst the barrel. A narrow tube was purposely used by Pascal so that the actual amount of water poured into the tube would be small. This demonstrated that what was important in determining the pressure on the barrel was the *height* of the water column. In doing Drill Prob. 15-5, you will calculate the pressure and force involved in this experiment.

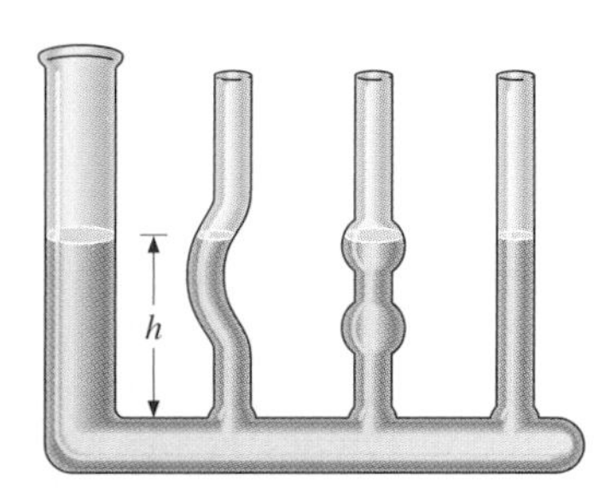

FIGURE 15-6 The pressure is the same at the bottom of each of the containers.

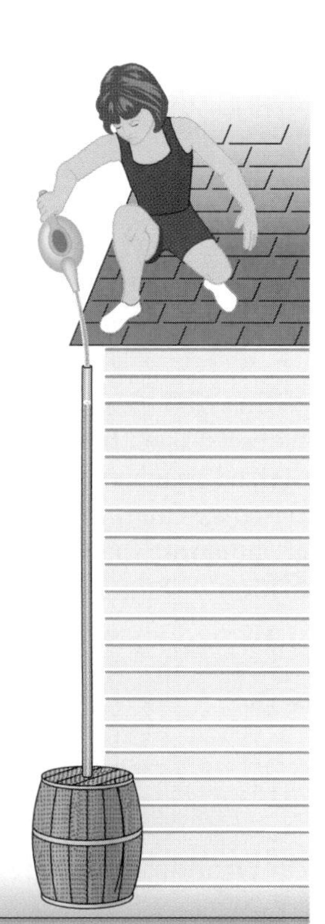

FIGURE 15-7 A representation of Pascal's experiment. The barrel burst when the column of water reached a height of about 12 m.

Equations (15-4) describe pressure variations in liquids, because they were derived using the fact that liquid densities are essentially independent of pressure. How then does pressure change within a gas? Unlike a liquid, a gas is easily compressed, so its density varies considerably with pressure. To integrate Eq. (15-3), we must know how the pressure and density are related. This calculation is considered in Prob. 15-65 for the atmosphere. Interestingly enough, Eqs. (15-4) turn out to be reasonably accurate for calculating air pressure at low altitudes. For heights less than 3 km above sea level, the error introduced with these equations is less than 4 percent.

If the pressure p_1 at point 1 of a liquid in a container is increased by Δp, then from Eq. (15-4*b*), the pressure p_2 at point 2, which is a depth h below point 1, is also increased by Δp. This was first recognized by Pascal and is called *Pascal's principle*:

> When the pressure at a particular point in a confined fluid is changed, the pressure at every other point in the fluid changes by the same amount.

One familiar application of Pascal's principle is the hydraulic lift. A simple schematic representation of the device is shown in Fig. 15-8. A fluid, usually oil, fills the two connected cylindrical chambers. At the top of each chamber is a movable piston. Any motion of the pistons takes place slowly enough that the confined fluid is effectively at rest and therefore obeys Pascal's principle. We assume that friction between the pistons and the walls of the chamber is negligible. Then when a force F_1 is applied to the smaller piston of area A_1, the pressure on the fluid at this piston is increased by $\Delta p = F_1/A_1$. By Pascal's principle, the fluid pressure at the larger piston (of area A_2) must also increase by Δp, which causes an increase in the upward force on the larger piston by an amount $\Delta p A_2 = F_1 A_2/A_1$. Since $A_2 > A_1$, the application of F_1 to the smaller piston causes a larger force F_2 to be exerted on the larger piston. For example, if $A_1 = 4.0 \times 10^{-2}\ \text{m}^2$, $A_2 = 2.0\ \text{m}^2$, and $F_1 = 1000$ N, then

$$F_2 = (1000\ \text{N})(2.0\ \text{m}^2)/(4.0 \times 10^{-2}\ \text{m}^2) = 50{,}000\ \text{N}.$$

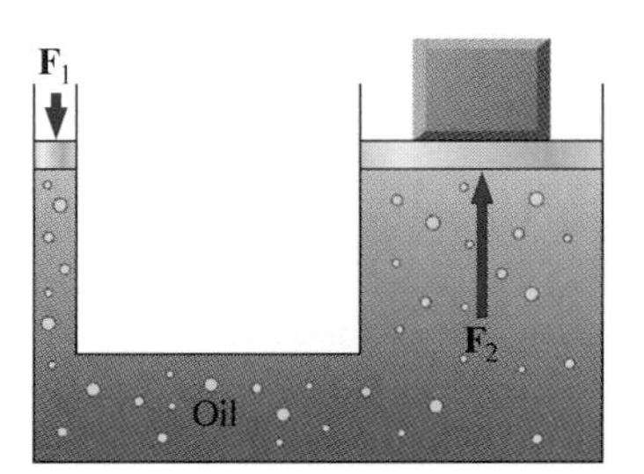

FIGURE 15-8 A representation of the hydraulic lift.

When the smaller piston is pushed downward a distance d_1, a volume of oil $A_1 d_1$ must move from the smaller to the larger chamber. This causes the larger piston to move upward a distance d_2 such that $A_2 d_2 = A_1 d_1$. Now the work done on the smaller piston is $W_1 = F_1 d_1$, and the work done by the larger piston is $W_2 = F_2 d_2$; so, since $F_1/A_1 = F_2/A_2$,

$$W_2 = F_2 d_2 = \left(F_1 \frac{A_2}{A_1}\right)\left(\frac{A_1 d_1}{A_2}\right) = F_1 d_1 = W_1.$$

The work input to the smaller piston is equal to the work output of the larger piston.

Devices that use Pascal's principle to transmit forces are quite common. Examples are automobile lifts in repair shops, chair lifts in barber shops, and automobile brakes. In all cases, a force is applied to a small piston, resulting in a larger force exerted on a larger piston. However, we are not "getting something for nothing" because the input and output work must be balanced. This means that the large piston moves a load a smaller distance than the small piston is displaced. While getting your hair cut you may have noticed how many times a barber pumps the handle in order to move the chair a slight distance upward.

Starting with a small input force, this device uses hydraulic fluid to produce a large output force.

EXAMPLE 15-3 **PRESSURE UNDER WATER**

If atmospheric pressure is 1.01×10^5 N/m^2, at what depth below the surface of a freshwater lake is the pressure doubled?

SOLUTION Suppose that point 1 is at the surface of the lake and point 2 is at the depth where the pressure has doubled. From Eq. (15-4*b*) we have

$$p_2 = p_1 + \rho g h$$
$$2(1.01 \times 10^5 \text{ N/m}^2) = 1.01 \times 10^5 \text{ N/m}^2 + (1.00 \times 10^3 \text{ kg/m}^3)(9.80 \text{ m/s}^2)h.$$

We then find that the pressure doubles at a depth of $h = 10.3$ m below the surface of the lake.

EXAMPLE 15-4 **FORCE DUE TO A DIFFERENCE IN WATER PRESSURE**

A research submarine is 50.0 m below the surface of the ocean. What is the net force on a circular submarine window of diameter 40.0 cm due to the pressure difference between the two sides of the window? Assume that the window is also 50.0 m below the surface, and that the pressure inside the submarine is 0.90×10^5 N/m^2. Atmospheric pressure is 1.01×10^5 N/m^2, and the density of seawater is 1.03×10^3 kg/m^3.

SOLUTION At a depth of 50.0 m, the pressure on the outside of the window is

$$\begin{aligned} p_2 &= p_1 + \rho g h \\ &= 1.01 \times 10^5 \text{ N/m}^2 + (1.03 \times 10^3 \text{ kg/m}^3) \\ &\quad \times (9.80 \text{ m/s}^2)(50.0 \text{ m}) \\ &= 6.06 \times 10^5 \text{ N/m}^2. \end{aligned}$$

The pressure difference between the the two sides of the window is

$$\Delta p = (6.06 \times 10^5 - 0.90 \times 10^5) \text{ N/m}^2 = 5.16 \times 10^5 \text{ N/m}^2,$$

so the force on the window is inward with a magnitude of

$$\begin{aligned} F &= \Delta p A = \Delta p\left(\pi \frac{d^2}{4}\right) \\ &= (5.16 \times 10^5 \text{ N/m}^2)\left[\pi \frac{(0.400 \text{ m})^2}{4}\right] = 6.48 \times 10^4 \text{ N}. \end{aligned}$$

DRILL PROBLEM 15-4

A diver is 100 m below the surface of the ocean. What pressure is she subjected to at this depth? The density of seawater is 1.03×10^3 kg/m^3, and atmospheric pressure is 0.99×10^5 N/m^2.
ANS. 1.11×10^6 N/m^2.

DRILL PROBLEM 15-5

Suppose the cross-sectional area of the long, thin tube connected to the oak barrel of Fig. 15-7 is 5.0×10^{-5} m^2, the area of the lid is $A_2 = 0.20$ m^2, and the barrel bursts when the water column reaches a height of 12 m. (*a*) What is the weight of the water in the tube? (*b*) What is the pressure of the water on the lid of the barrel? (*c*) What is the net force on the lid due to the pressure difference?
ANS. (*a*) 5.9 N; (*b*) 2.2×10^5 N/m^2; (*c*) 2.4×10^4 N.

DRILL PROBLEM 15-6

Some people feel their ears "pop" when they travel upward quickly. This occurs when the pressure behind the eardrum does not change during the ascent. The decrease in air pressure with altitude then results in a net outward force on the eardrum, which causes the "pop." (*a*) Use Eq. (15-4*b*) to determine the decrease in air pressure a hiker experiences when she climbs a vertical distance of 200 m up a mountain. Assume that the density of air is 1.20 kg/m^3. (*b*) If the pressure behind the eardrum does not decrease during the upward trip, what is the net force on it if its area is 0.60 cm^2?
ANS. (*a*) 2.35×10^3 N/m^2; (*b*) 0.14 N.

15-4 Measurement of Pressure

Pressure measurements are often simple applications of Eqs. (15-4). One such application is the measurement of atmospheric pressure with the mercury barometer of Fig. 15-9. The barometer is a glass tube, usually about 80 cm long, which is filled with mercury and then inverted in a container of mercury. Since the tube is filled before it is inverted, there is no air above the mercury column. The only gas in this region is mercury vapor, whose pressure at room temperature is so small that it can be ignored. The lower end of the mercury column is open to the atmosphere. With 1 and 2 representing the top and bottom of the tube, respectively, $p_1 = 0$ and $p_2 = p_{atm}$, and from Eq. (15-4*b*),

$$p_{atm} = p_1 + \rho gh = 0 + \rho_{Hg} gh;$$

so

$$p_{atm} = \rho_{Hg} gh.$$

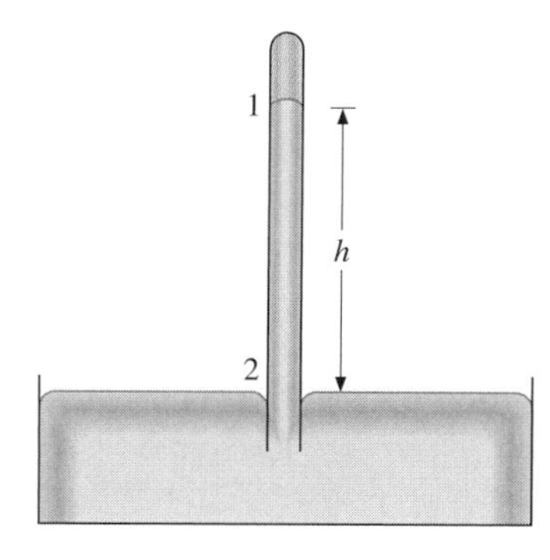

FIGURE 15-9 A simple barometer. The mercury rises to a height proportional to the atmospheric pressure.

Atmospheric pressure is proportional to the height of the mercury column. For example, when the column is 76.0 cm high, atmospheric pressure is

$$p_{atm} = (1.36 \times 10^4 \text{ kg/m}^3)(9.80 \text{ m/s}^2)(0.760 \text{ m})$$
$$= 1.01 \times 10^5 \text{ N/m}^2.$$

Atmospheric pressure is often expressed in centimeters of mercury (cmHg) or inches of mercury (inHg). As was just illustrated for the 76.0-cm column of mercury, the conversion from the height of a mercury column to a force per unit area is easy to make.

Other pressure units frequently used are the *atmosphere* (1 atm = 1.013×10^5 N/m^2); the *pascal* (1 Pa = 1 N/m^2), named in honor of Blaise Pascal; the *bar* (1 bar = 10^5 N/m^2), the unit used by the National Weather Service; and the *torr* (1 torr = 1 mmHg), named after Evangelista Torricelli (1608–1647), the inventor of the mercury barometer.

Pressure measurements relative to atmospheric pressure are often made with an *open-tube manometer* such as that of Fig. 15-10. One end of the tube is connected to the vessel that contains the gas whose pressure is to be measured, while the other end is open to the atmosphere. With mercury in the tube, the difference in the heights of the two columns (h in the figure) gives the pressure of the gas in centimeters of mercury (cmHg) relative to atmospheric pressure. The pressure of the gas is then found by adding the column height to the atmospheric pressure.

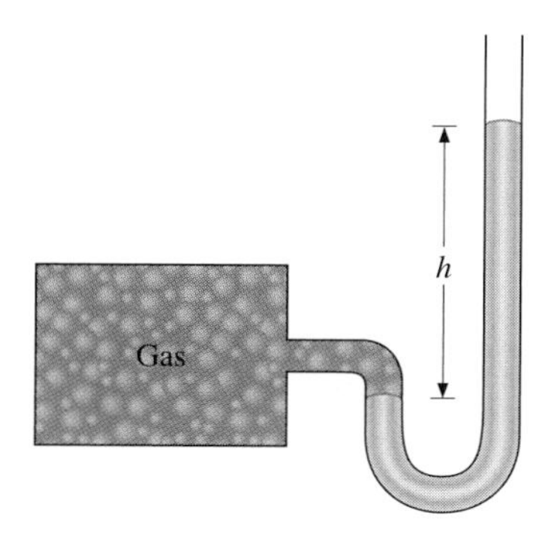

FIGURE 15-10 The height of the mercury column gives the gauge pressure of the gas in the container.

For example, if h = 20 cm and p_{atm} = 75 cmHg, the pressure of the gas is 75 cmHg + 20 cmHg = 95 cmHg; and if h = −10 cm, the gas pressure is 75 cmHg − 10 cmHg = 65 cmHg.

Like the open-tube manometer, most pressure gauges have one side open to atmospheric pressure. Quantities such as the height of a mercury column and the tension in a spring are therefore proportional to the *difference between the pressure being measured and atmospheric pressure.* This difference is called the **gauge pressure**. Since it may be greater or less than atmospheric pressure, the gauge pressure is designated as either positive or negative. For example, the gauge pressures of the gases just considered are +20 cmHg and −10 cmHg.

To distinguish it from gauge pressure, the actual pressure is called the **absolute pressure**. These pressures are related by

$$p_{gauge} = p_{abs} - p_{atm}.$$

In situations where there is no chance for confusion, the adjectives *absolute* and *gauge* are usually omitted.

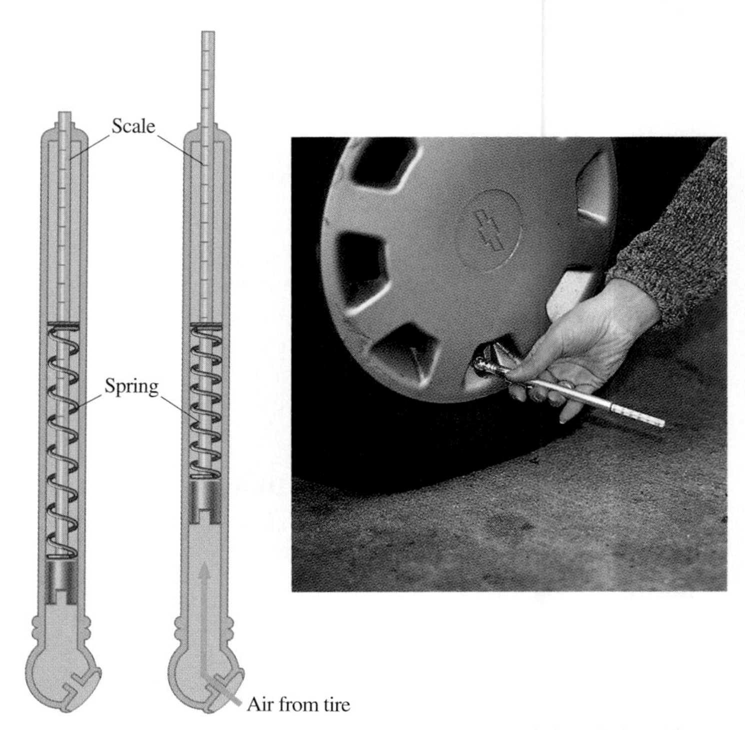

A pressure gauge measures the gauge pressure of the air in a tire. The compression of the spring is proportional to the force on it, which is, in turn, proportional to the air pressure in the tire. When the gauge is removed from the tire, the spring expands, but the scale, which is not attached to the spring, is held in place by friction.

EXAMPLE 15-5 PRESSURE MEASUREMENTS

On a day when a mercury barometer reads 76.2 cmHg, an open-tube manometer connected to a vessel of gas measures a gauge pressure of 25.7 cmHg. (*a*) What is the absolute pressure of the gas in pascals? (*b*) If an impending thunderstorm causes atmospheric pressure to drop to 74.2 cmHg, what is the gauge pressure of the gas?

SOLUTION (*a*) The absolute pressure of the gas in centimeters of mercury is (76.2 + 25.7) cmHg = 101.9 cmHg. Since 76.0 cmHg = 1.01×10^5 Pa, the absolute pressure is

$$101.9 \text{ cmHg} = \frac{101.9 \text{ cmHg}}{76.0 \text{ cmHg}}(1.01 \times 10^5 \text{ Pa}) = 1.35 \times 10^5 \text{ Pa}.$$

(*b*) The absolute pressure of the gas is still 101.9 cmHg. Its gauge pressure is now

$$p = 101.9 \text{ cmHg} - 74.2 \text{ cmHg} = 27.7 \text{ cmHg} = 3.68 \times 10^4 \text{ Pa}.$$

EXAMPLE 15-6 FORCE ON A DAM

Water stands at a height h behind the dam of Fig. 15-11. If the face of the dam is rectangular, what is the total force on it due to the water?

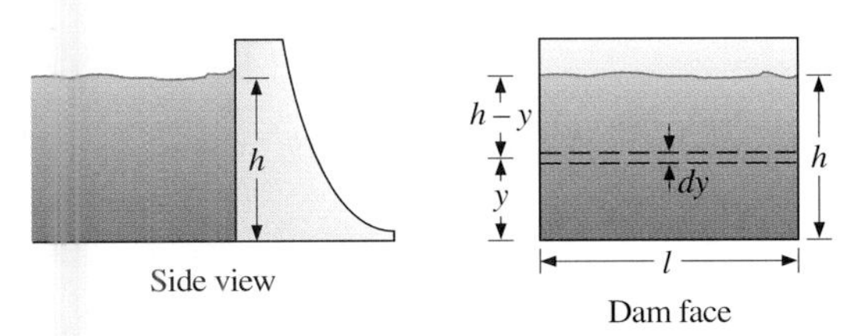

FIGURE 15-11 Water stands at a height h against the rectangular face of a dam.

SOLUTION The pressure on the left side of the dam is that of the atmosphere plus the water, but only atmospheric pressure acts on the right side. The net pressure at a height y above the bottom of the dam is therefore the gauge pressure due to the water and given by $\rho g(h - y)$. The infinitesimal force on a section of area $l\,dy$ is then

$$dF = p\,dA = \rho g(h - y)l\,dy.$$

The total force on the side of the dam is found by integrating the infinitesimal force from $y = 0$ to $y = h$. This gives

$$F = \int_0^h \rho g(h - y)l\,dy = \frac{1}{2}\rho g l h^2.$$

DRILL PROBLEM 15-7

In the eye of a hurricane, atmospheric pressure can drop to as low as 67 cmHg. What is this pressure in newtons per square meter?
ANS. 8.9×10^4 N/m^2.

DRILL PROBLEM 15-8

Why isn't water used instead of mercury in barometers?

DRILL PROBLEM 15-9

Find the gauge pressure and the absolute pressure of the gas in the tank of Fig. 15-12. The right-hand column of mercury is open to the atmosphere, whose pressure is 75 cmHg.
ANS. −15 cmHg; 60 cmHg.

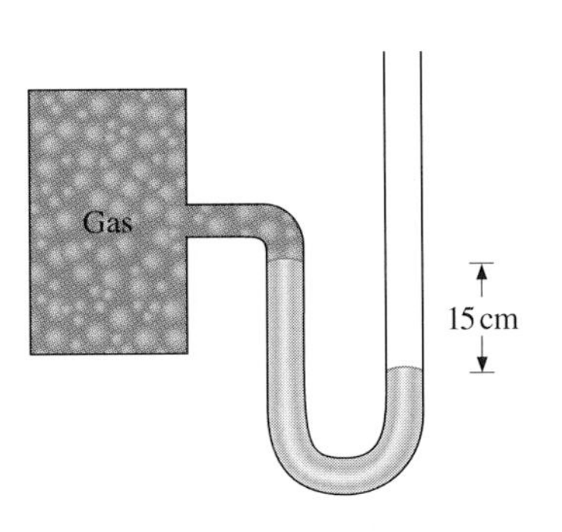

FIGURE 15-12 A pressure gauge.

15-5 ARCHIMEDES' PRINCIPLE

You have probably experienced in one way or another the **buoyant force** exerted on a body by the fluid surrounding it. This force is responsible for making wood float and a helium-filled balloon rise. *The buoyant force on a body is the resultant of the normal forces due to the fluid pressure at the surface of the body.* For example, when the rectangular slab of Fig. 15-13 is completely submerged in a fluid of density ρ, it experiences an upward force $p_2 A$ at its lower surface and a downward force $p_1 A$ at its upper surface. The buoyant force $\mathbf{F}_B$ is then

$$\mathbf{F}_B = (p_2 A - p_1 A)\mathbf{j},$$

which, since $p_2 = p_1 + \rho g h$, reduces to

$$\mathbf{F}_B = \rho A h g \mathbf{j}.$$

Notice that there is an upward buoyant force because the pressure increases with depth. If, for example, the pressure were independent of depth, $p_2 = p_1$, and $\mathbf{F}_B = 0$.

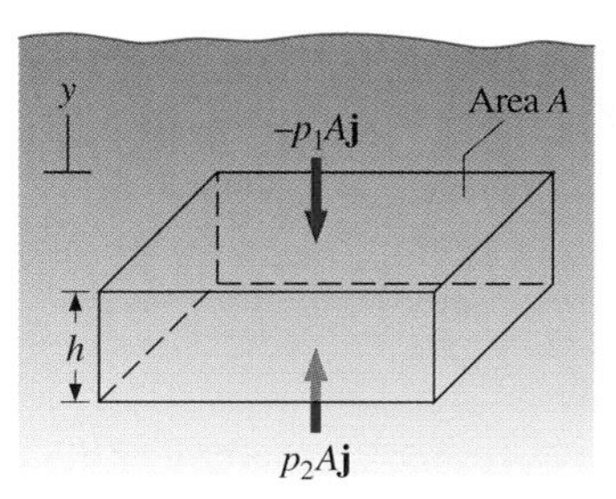

FIGURE 15-13 The net upward force on the rectangular slab due to the surrounding fluid is $(p_2 A - p_1 A)\mathbf{j}$.

For more complicated geometries, an analogous calculation of $\mathbf{F}_B$ requires that we integrate the vertical component of $p\,dA$ over the surface of a body immersed in the fluid (see Fig. 15-14*a*). Fortunately, the buoyant force is much easier to determine with *Archimedes' principle*, named after its discoverer, a Greek scientist who lived in the third century, B.C. Archimedes supposedly made his discovery in response to King Hieron's request that he determine whether the royal crown was pure gold or a gold alloy. Archimedes' principle states:

> When a body is immersed in a fluid at rest, the fluid exerts a net upward force on that body which is equal in magnitude to the weight of the displaced fluid.

For example, the rectangular slab of Fig. 15-13 displaces fluid of volume Ah. This volume of fluid has a weight of ρAhg, which is the magnitude of the buoyant force on the slab.

Archimedes' principle is a necessary consequence of the first condition of static equilibrium ($\sum\mathbf{F} = 0$). To derive Archimedes' principle, imagine that the body of Fig. 15-14*a* is replaced by a fluid section of the same size and shape, as illustrated in Fig. 15-14*b*. Since the fluid is at rest, this section must be in static equilibrium. The forces on it are its weight $m\mathbf{g}$ and the net force of the surrounding fluid, which, as in the case of the body, is the resultant of the normal forces of magnitude $p\,dA$ acting over the surface. The pressures at all points on the surface of the fluid section are identical to those on the body, so the buoyant forces on the fluid section and on the body must be *identical*. Applying the first condition of static equilibrium to the fluid section of Fig. 15-14*c*, we obtain

$$\mathbf{F}_B + m\mathbf{g} = 0,$$

and

$$\mathbf{F}_B = -m\mathbf{g}. \tag{15-5}$$

Therefore *the buoyant force on a body acts upward and is equal in magnitude to the weight of the fluid displaced*. This is Archimedes' principle.

Whether a body sinks or floats in a fluid can be easily determined with Archimedes' principle. If the volume of the body of Fig. 15-15*a* is V, the vertical forces on it are the buoyant force $\rho_f Vg$ and its weight $-\rho_b Vg$, where ρ_f and ρ_b are the densities of the fluid and the body, respectively. The net force upward is then $(\rho_f - \rho_b)Vg$, which is positive if $\rho_f > \rho_b$ and negative if $\rho_f < \rho_b$. Thus a body rises if its density is less than that of the surrounding fluid, and it sinks if its density is greater.

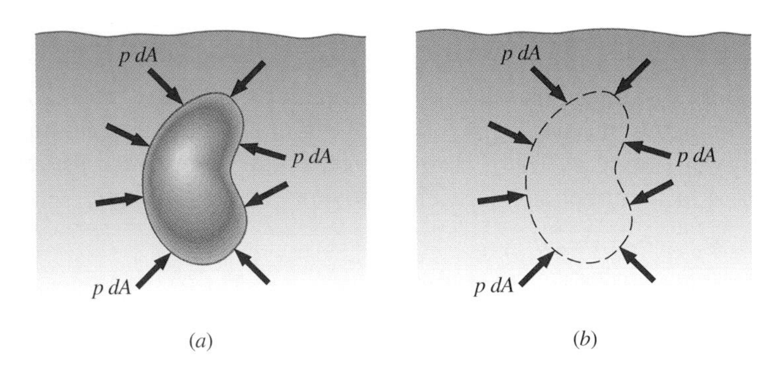

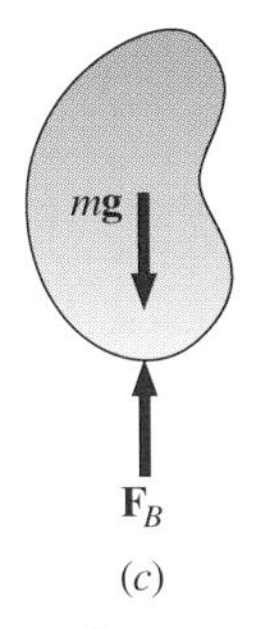

FIGURE 15-14 The force due to the surrounding fluid on both (*a*) the body and (*b*) the fluid section is $\mathbf{F}_B$. (*c*) The free-body diagram of the fluid section.

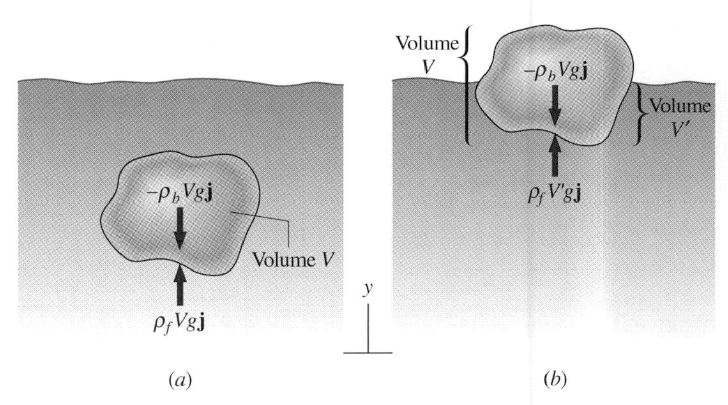

FIGURE 15-15 (*a*) The body rises if $\rho_b < \rho_f$. (*b*) The fraction of volume above the surface depends on the ratio ρ_b/ρ_f.

If there is an upper surface to the fluid, a body less dense than the fluid will rise to the top and come to rest partially submerged. (See Fig. 15-15*b*.) Since the floating body is in static equilibrium, the downward gravitational force is balanced by the upward buoyant force:

$$\rho_f V'g = \rho_b Vg,$$

where V' is the volume submerged. Hence

$$\frac{V'}{V} = \frac{\rho_b}{\rho_f},$$

and the fraction of the volume submerged is equal to the ratio of the densities of the body and the fluid.

EXAMPLE 15-7 A Helium Balloon

A girl holds a thin string tied to a helium-filled rubber balloon of volume $V = 0.340\ \text{m}^3$, as shown in Fig. 15-16*a*. The mass of the rubber is m = 15.2 g and the mass of the string is negligible. Taking the densities of helium and air to be $\rho_{\text{He}} = 0.18\ \text{kg/m}^3$ and $\rho_{\text{air}} = 1.25\ \text{kg/m}^3$ respectively, determine the tension in the string.

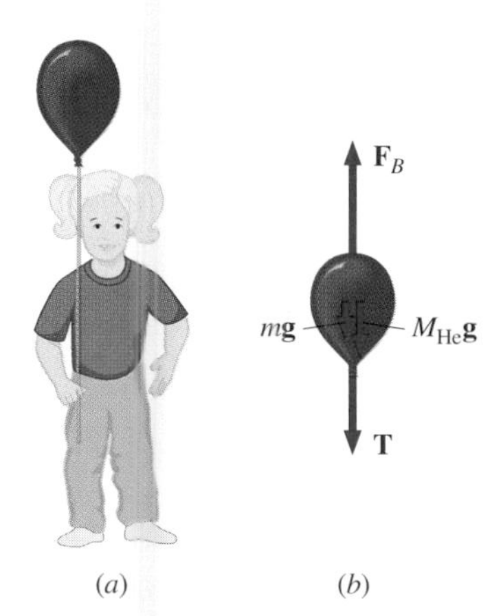

FIGURE 15-16 (*a*) A girl and her helium-filled balloon. (*b*) The free-body diagram for the balloon.

SOLUTION Ignoring the thickness of the rubber, the volume of the air displaced is equal to the inflated volume of the balloon. As shown in the free-body diagram in Fig. 15-16*b*, the forces on the balloon are the buoyant force $\mathbf{F}_B$ of the air, the weight $m\mathbf{g}$ of the rubber, the weight $M_{\text{He}}\mathbf{g}$ of the helium, and the tension $\mathbf{T}$ in the string. Since the balloon is in static equilibrium,

$$\sum F_y = F_B - T - M_{\text{He}}g - mg = 0.$$

From $M_{\text{He}} = \rho_{\text{He}}V$ and $F_B = \rho_{\text{air}}Vg$, the weight of the displaced air, we obtain

$$T = F_B - M_{\text{He}}g - mg = [(\rho_{\text{air}} - \rho_{\text{He}})V - m]\,g.$$

Substituting the known values into this equation, we find that the tension in the string is $T = 3.42$ N.

EXAMPLE 15-8 Acceleration Under Water

A piece of wood of density $\rho_w = 600\ \text{kg/m}^3$ is held beneath the surface of a pool of water. When it is released, what is its acceleration?

SOLUTION There are two vertical forces on the wood, the water's buoyancy $\rho_{\text{H}_2\text{O}}Vg$ and the wood's weight $-\rho_w Vg$. With Fig. 15-15*a* representing the free-body diagram of the piece of wood, application of Newton's second law in the vertical direction gives

$$\sum F_y = ma_y$$

$$\rho_{\text{H}_2\text{O}}Vg - \rho_w Vg = \rho_w V a_y,$$

so

$$a_y = \left(\frac{\rho_{\text{H}_2\text{O}} - \rho_w}{\rho_w}\right)g = \left(\frac{1000\ \text{kg/m}^3 - 600\ \text{kg/m}^3}{600\ \text{kg/m}^3}\right)g = \frac{2}{3}\,g.$$

This is the acceleration at the instant the block is released. After the block starts moving upward, the frictional force of the water quickly brings the block to a slow upward terminal velocity.

DRILL PROBLEM 15-10

A brass figurine hangs from a calibrated vertical scale as shown in Fig. 15-17. When the figurine is suspended in air, the scale reads 4.90 N. What is the scale reading W when the figurine is completely submerged in water?
ANS. 4.33 N.

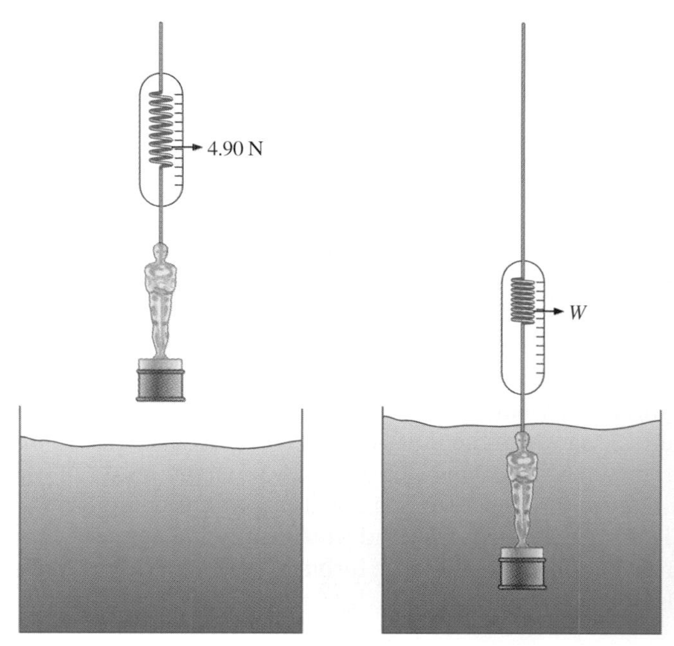

FIGURE 15-17 A brass figurine and scale.

DRILL PROBLEM 15-11

A child drops a wooden block of density $0.40\ \text{g/cm}^3$ into a swimming pool. What percentage of the block's volume is submerged?
ANS. 40%.

The scale is read when the person is in air and when he is immersed in water. With these two readings, the person's density can be determined. The density can then be used to estimate the percentage of fat on his body.

15-6 Fluids in Motion

As first suggested by Leonard Euler (1707–1783), a moving fluid can be characterized at each point in the fluid and at each instant in time by its density, its pressure, and a **fluid velocity** $\mathbf{v}(x, y, z, t)$. The function $\mathbf{v}(x, y, z, t)$ represents the velocity of the infinitesimal volume element of fluid that at a time t is moving through the point (x, y, z). When t changes to t', a different infinitesimal volume element is at (x, y, z); $\mathbf{v}(x, y, z, t')$ then represents the velocity of that element. We now discuss the common terms used to describe fluid flow.

A color-enhanced photograph of the turbulent flow of gases in the atmosphere of Jupiter.

Steady and Nonsteady Flow

In *steady flow*, the fluid velocity at every point is independent of time. If this is not the case, the flow is said to be *nonsteady* or *turbulent*.

Figure 15-18 depicts a curve consisting of points at which the fluid velocity is tangent to the curve. In steady flow, every infinitesimal fluid volume element passing through a particular point has the same velocity, so this curve represents an actual path along which the fluid moves. It is known as a *streamline*. Streamlines can often be observed by inserting smoke or paper streamers into a moving gas or dye into a moving liquid. In Fig. 15-19, the streamlines around a spinning baseball are delineated by smoke trails. The turbulence behind the baseball can also be clearly seen.

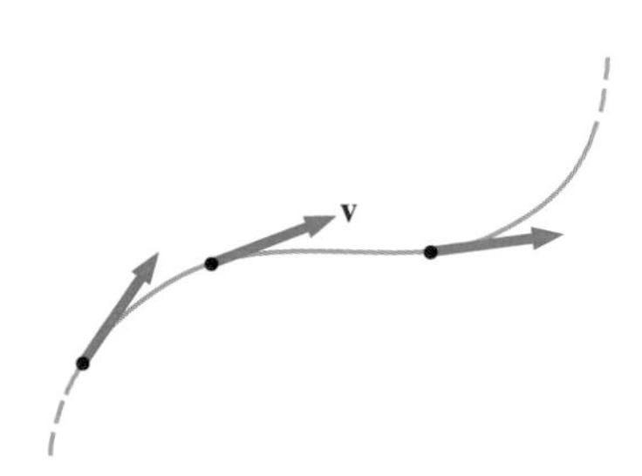

FIGURE 15-18 A streamline. The velocity of the fluid is tangent to this curve at any point on it.

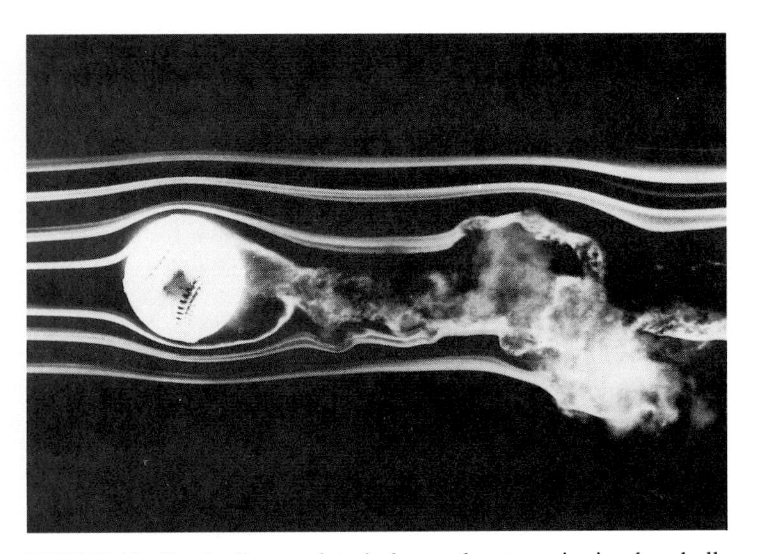

FIGURE 15-19 Steady flow and turbulence about a spinning baseball are depicted by smoke trails.

A *flow tube* is shown in Fig. 15-20. It is a volume whose side boundaries are streamlines and whose end boundaries are cross-sectional areas perpendicular to the streamlines. Since the fluid velocity is directed along the streamlines, fluid does not cross the side boundaries of a flow tube; it enters or leaves the tube through the cross-sectional areas at each end. You can think of a flow tube as *an imaginary pipe that confines the fluid to flow through its interior*. Of course, the sides of a real pipe are always boundaries of a flow tube.

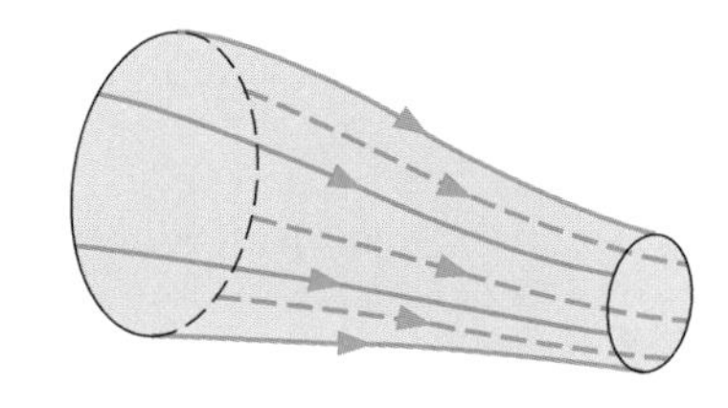

FIGURE 15-20 A flow tube. Fluid does not cross the side boundaries since they are composed of streamlines.

Compressible and Incompressible Flow

The flow of a fluid whose density is independent of both position and time is said to be *incompressible*. The assumption of incompressibility is very accurate for liquids, whose volumes don't decrease much for large increases in pressure. Surprisingly, even highly compressible gases often flow with very little variation in density. Examples include the air when it moves across the earth's surface (the wind) and the air moving by an airplane wing in subsonic flight.

Rotational and Irrotational Flow

Imagine that you are standing at the edge of a stream watching a flower slowly float by. If its petals do not rotate as the flower moves, then the flow of water at the surface is said to be *irrotational*. If the petals are spinning, then the flow is *rotational*. In rotational flow, the fluid element at any point has a nonzero angular velocity. The motion of whirlpools and tornadoes are two common examples of this type of flow.

VISCOUS AND NONVISCOUS FLOW

If the frictional forces in a moving fluid are not negligible, the flow is said to be *viscous*, or to exhibit *viscosity*. Frictionless or nonviscous fluid flow is often described as *ideal*. Since an ideal fluid flows without friction, its mechanical energy is conserved.

Our study of fluids in motion will be restricted to flow that is *steady*, *incompressible*, and *irrotational*. We'll also be studying only *ideal* fluids until the last section of this chapter, at which point viscous flow in a horizontal pipe will be investigated.

15-7 THE EQUATION OF CONTINUITY

A flow tube for an ideal fluid is shown in Fig. 15-21. At points 1 and 2, the cross-sectional areas of the tube are A_1 and A_2, respectively. The tube is assumed to be so narrow at these points that any variation in fluid velocity over either A_1 or A_2 is insignificant. We can therefore associate fluid velocities $\mathbf{v}_1$ with A_1 and $\mathbf{v}_2$ with A_2. During a time Δt, the cylindrical volume with base area A_1 and thickness $v_1 \Delta t$ enters the flow tube at 1, while the cylindrical volume with base area A_2 and thickness $v_2 \Delta t$ leaves the tube at 2. Since the fluid is incompressible, the volume entering the tube at point 1 in the time interval Δt must equal the volume leaving the tube at point 2 during the same time interval. Thus

$$A_1 v_1 \Delta t = A_2 v_2 \Delta t,$$

and

$$A_1 v_1 = A_2 v_2; \tag{15-6a}$$

or equivalently,

$$Av = \text{constant}. \tag{15-6b}$$

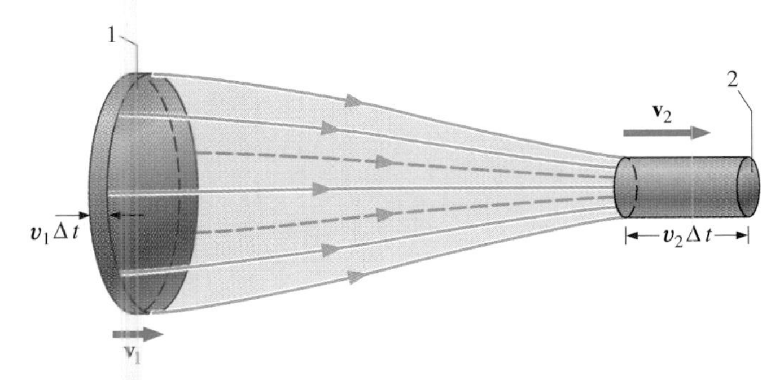

FIGURE 15-21 During a time interval Δt, the cylindrical volume at point 1 enters the flow tube, and the cylindrical volume at point 2 leaves the tube. These two volumes are equal.

Equation (15-6), in either form, is called *the equation of continuity*. It tells us that *the speed of an incompressible fluid is largest where the flow tube is narrowest and smallest where the flow tube is widest*. If you've washed a car or watered a garden, you have probably experienced firsthand how the equation of continuity works. In order to increase the velocity of the water from the hose you are using, you simply place a finger over its end to reduce the opening. As predicted by the equation of continuity, the water then leaves the end of the hose at a higher speed.

The product Av represents the fluid volume that crosses the area A per unit time and is called the **flow rate**. Representing it by Q, we have

$$Q = Av. \tag{15-7}$$

Typical units for Q are cubic meters per second (m^3/s) and cubic feet per second (ft^3/s).

EXAMPLE 15-9 A NOZZLE ON THE END OF A HOSE

Water flows through a hose of cross-sectional area $A_1 = A$ with a speed of $v_1 = 3.0$ m/s. A nozzle whose cross-sectional area is $A_2 = A/4$ is placed over the end of the hose, as illustrated in Fig. 15-22. (*a*) What is the speed v_2 of the water in the nozzle? (*b*) What volume of water leaves the nozzle in 5 min if $A_2 = 5.00$ cm^2?

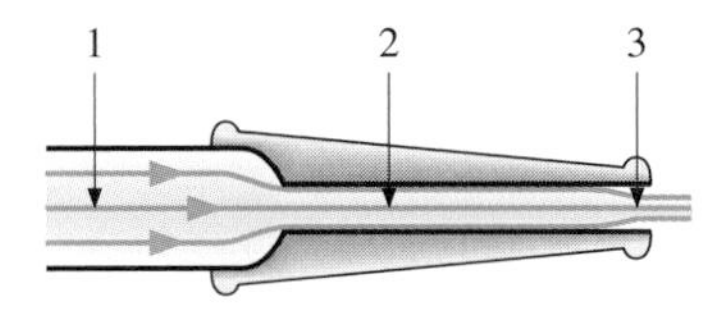

FIGURE 15-22 A nozzle is placed on the end of a hose. (Point 3 is considered in Example 15-11.)

SOLUTION (*a*) If the flow rate is not changed after adding the nozzle, we have from the equation of continuity,

$$A_1 v_1 = A_2 v_2,$$

so the speed of the water in the nozzle is

$$v_2 = \frac{A_1}{A_2} v_1 = 4v_1 = 12.0 \text{ m/s}.$$

(*b*) The flow rate is Av, so the amount of water leaving the nozzle in 5 min is

$$A_2 v_2 t = (5.00 \times 10^{-4} \text{ m}^2)(12.0 \text{ m/s})(300 \text{ s}) = 1.80 \text{ m}^3.$$

DRILL PROBLEM 15-12

At point 1 in a pipeline, water flows with a speed of 10.0 m/s. (*a*) What is its speed at point 2 where the cross-sectional area of the pipe is twice as large as that at point 1? (*b*) What is the water's speed at point 3 where the pipe's diameter is one-half as large as that at point 1?
ANS. (*a*) 5.0 m/s; (*b*) 40.0 m/s.

15-8 BERNOULLI'S EQUATION

The forces on a section of an ideal fluid are its weight and the forces on its surface due to the pressure of the surrounding fluid. For a fluid at rest, the net force on the section is zero, and this leads to Eqs. (15-4), the relationship between pressure and depth. However, if a fluid is moving, it is not necessarily in equilibrium. The work done by the forces on a fluid section may then result in a change in the kinetic energy of that section.

The mechanics of a moving fluid were first studied successfully by Daniel Bernoulli (1700–1782). By applying the work-energy theorem to a section of an ideal fluid, he derived an equation that relates the pressure, speed, and vertical position at arbitrary points in the fluid. We will derive this equation for the section of fluid between points 1 and 2 in the flow tube of Fig. 15-23. In a time interval Δt, this section moves to the region between 1′ and 2′. The speed and pressure of the fluid at the lower face are v_1 and p_1, respectively, while v_2 and p_2 are the corresponding quantities at the upper face. The cross-sectional areas of the faces are A_1 and A_2, the vertical positions of their midpoints are y_1 and y_2, and the density of the fluid is ρ. We'll assume that the tube is narrow enough at points 1 and 2 that neither the speed nor pressure varies significantly over the faces.

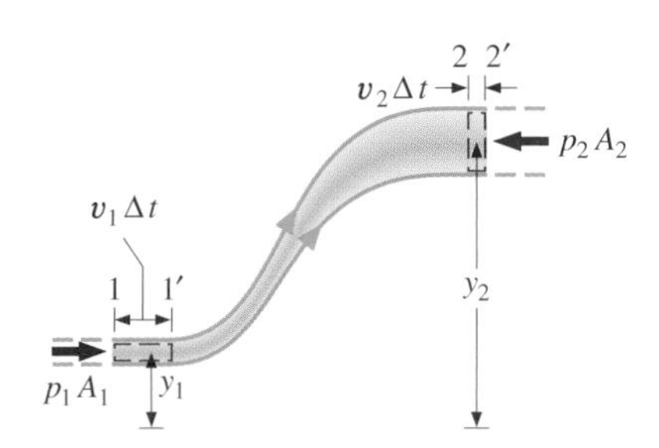

FIGURE 15-23 In a time Δt, the fluid section between points 1 and 2 moves to the region between 1′ and 2′.

Since there is no viscosity, only the forces p_1A_1 and p_2A_2, along with the force of gravity, do work on the fluid section as it moves. From the work-energy theorem, the net work of these forces can be set equal to the change in kinetic energy of the section. The details of that calculation are given in Supplement 15-1. The fluid is assumed to be incompressible, and its flow is assumed to be steady, nonviscous, and irrotational. The result is *Bernoulli's equation*:

$$p_1 + \tfrac{1}{2}\rho v_1^2 + \rho g y_1 = p_2 + \tfrac{1}{2}\rho v_2^2 + \rho g y_2, \qquad (15\text{-}8a)$$

which, because 1 and 2 are arbitrary points, is equivalent to

$$p + \tfrac{1}{2}\rho v^2 + \rho g y = \text{constant}. \qquad (15\text{-}8b)$$

Each term in Bernoulli's equation is an energy density. The first term is a force per unit area = (force)(length)/[(area) × (length)] = work per unit volume, the second is a kinetic energy per unit volume, and the third is a work per unit volume. This dimensional observation provides an instructive view of Bernoulli's equation. As shown in the supplement, when the fluid section of Fig. 15-23 moves from the region between points 1 and 2 to the region between points 1′ and 2′, $p_1 - p_2$ is the work per unit volume done on it by the pressure difference, and $\rho g(y_1 - y_2)$ is the work per unit volume done on it by gravity. The sum of these two terms is equal to its change in kinetic energy per unit volume, which is $\rho(v_2^2 - v_1^2)/2$.

A simple but important application of Bernoulli's equation is a determination of how pressure varies in a fluid that flows through a horizontal pipe. Then $y_1 = y_2$, and we have from Bernoulli's equation

$$p_1 + \tfrac{1}{2}\rho v_1^2 = p_2 + \tfrac{1}{2}\rho v_2^2.$$

This equation tells us that the pressure is highest where the speed is lowest. Now from the equation of continuity, the speed varies inversely with the cross-sectional area of the pipe. Consequently, when an ideal fluid flows through a horizontal pipe, the pressure is highest where the cross-sectional area of the pipe is largest. This result may surprise you, for most people would guess (incorrectly) that the pressure is greatest where the fluid flows through the smallest opening.

EXAMPLE 15-10 WATER FLOWING THROUGH A PIPE

An unusual water pipe is shown in Fig. 15-24. The cross-sectional areas of the pipe at points 1 and 2 are $A_1 = 0.060\ \text{m}^2$ and $A_2 = 0.020\ \text{m}^2$, and their midpoints are at elevations $y_1 = 1.0$ m and $y_2 = 3.0$ m, respectively. If water moves steadily without friction through the pipe at a rate of $0.040\ \text{m}^3/\text{s}$, what is the speed of the water at point 1 and at point 2? What is the pressure difference between these two points?

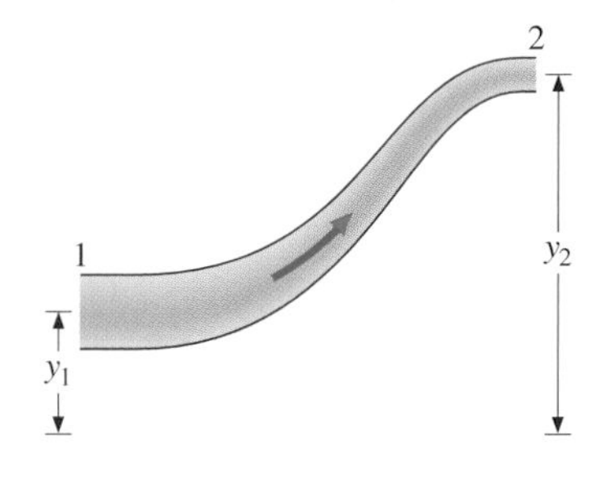

FIGURE 15-24 A water pipe.

SOLUTION Since water is incompressible and the flow is steady, the flow rate is the same at all points in the pipe. Thus at point 1,

$$Q = 0.040\ \text{m}^3/\text{s} = A_1 v_1 = (0.060\ \text{m}^2)v_1,$$

so

$$v_1 = 0.67\ \text{m/s};$$

and at point 2,

$$Q = 0.040\ \text{m}^3/\text{s} = A_2 v_2 = (0.020\ \text{m}^2)v_2,$$

so

$$v_2 = 2.0\ \text{m/s}.$$

The pressure difference is determined with Bernoulli's equation, which can be written as

$$p_1 - p_2 = \tfrac{1}{2}\rho(v_2^2 - v_1^2) + \rho g(y_2 - y_1).$$

With the known values substituted into the right-hand side, we find that the pressure difference between points 1 and 2 is

$$p_1 - p_2 = 2.1 \times 10^4 \text{ N/m}^2.$$

EXAMPLE 15-11 Water Leaving a Hose

If the gauge pressure in the hose of Example 15-9 (and Fig. 15-22) is 3.0×10^5 N/m², (*a*) what is the gauge pressure in the nozzle? (*b*) What is the speed of the water just as it leaves the nozzle (point 3 in the figure)? (*c*) What happens to the cross-sectional area of the water stream just as it leaves the nozzle?

SOLUTION (*a*) Using gauge pressure in Bernoulli's equation (why can we do this?), we find that in the nozzle the gauge pressure is

$$\begin{aligned} p_2 &= p_1 + \tfrac{1}{2}\rho(v_1^2 - v_2^2) \\ &= 3.0 \times 10^5 \text{ N/m}^2 + \tfrac{1}{2}(1000 \text{ kg/m}^3) \\ &\quad \times [(3.0 \text{ m/s}^2)^2 - (12.0 \text{ m/s}^2)^2] \\ &= 2.3 \times 10^5 \text{ N/m}^2. \end{aligned}$$

Notice we have used $y_1 = y_2$ in Bernoulli's equation because the hose is horizontal.

(*b*) Outside the nozzle (point 3) the gauge pressure is zero, so from Bernoulli's equation,

$$\begin{aligned} v_3^2 &= v_2^2 + \frac{2}{\rho}(p_2 - p_3) \\ &= (12.0 \text{ m/s})^2 + \frac{2(2.3 \times 10^5 \text{ N/m}^2 - 0)}{(1000 \text{ kg/m}^3)}, \end{aligned}$$

and the speed of the water just as it leaves the nozzle is $v_3 = 25$ m/s.

(*c*) From the equation of continuity, the cross-sectional area of the stream of water immediately outside the nozzle is

$$A_3 = A_2\frac{v_2}{v_3} = A_2\frac{12 \text{ m/s}}{25 \text{ m/s}} = 0.48A_2.$$

Surprisingly, this is *smaller* than the cross-sectional area of the nozzle opening.

EXAMPLE 15-12 Venturi Meter

A device known as a Venturi meter is illustrated in Fig. 15-25. It is used to measure the speed of a fluid flowing through a pipe. The meter consists of a manometer and a constriction in the pipe (point 2 in the figure) with a cross-sectional area A_2 that is smaller than the cross-sectional area A_1 of the pipe itself (point 1 in the figure). Relate the difference h in the heights of the two mercury columns in the manometer to the speed v_1 of the fluid in the pipe.

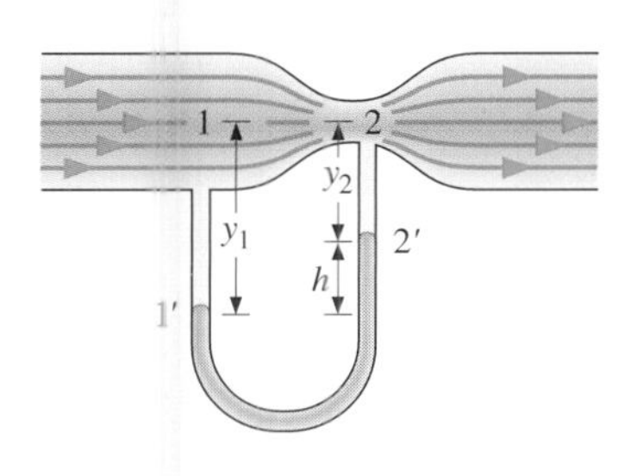

FIGURE 15-25 A Venturi meter.

SOLUTION With ρ_f representing the density of the fluid, Bernoulli's equation relates the fluid pressures and speeds at points 1 and 2 by

$$p_1 - p_2 = \tfrac{1}{2}\rho_f(v_2^2 - v_1^2). \qquad (i)$$

From the equation of continuity, the speeds of the fluid at these two points are related by

$$v_2 = (A_1/A_2)v_1,$$

which, when substituted into Eq. (*i*), yields

$$p_1 - p_2 = \tfrac{1}{2}\rho_f[(A_1/A_2)^2 - 1]v_1^2. \qquad (ii)$$

We can relate $p_1 - p_2$ to the difference h in the heights of the two mercury columns in the manometer by repeated applications of Eq. (15-4*b*). As shown in Fig. 15-25, points 1′ and 2′ represent the interfaces between the fluid and the left and right mercury columns, respectively, of the manometer. The pressures at points 1′ and 2′ are given by

$$\begin{aligned} p_{1'} &= p_1 + \rho_f g y_1 \\ p_{2'} &= p_2 + \rho_f g y_2, \end{aligned}$$

where y_1 is the difference in the vertical positions of 1 and 1′ and y_2 is the difference in the vertical positions of 2 and 2′. Furthermore, the pressures at 1′ and 2′ are related by

$$p_{1'} = p_{2'} + \rho_{\text{Hg}} g h,$$

where ρ_{Hg} is the density of mercury.

When these three equations are combined appropriately and h is set equal to $y_1 - y_2$, the pressure difference between points 1 and 2 is found to be given in terms of the difference h in the heights of the two mercury columns by

$$p_1 - p_2 = (\rho_{\text{Hg}} - \rho_f)gh. \qquad (iii)$$

Finally, by equating Eqs. (*ii*) and (*iii*) and solving the resulting equation for v_1, we find that the speed of the fluid in the pipe is related to h by

$$v_1 = A_2\sqrt{\frac{2(\rho_{\text{Hg}} - \rho_f)gh}{\rho_f(A_1^2 - A_2^2)}}.$$

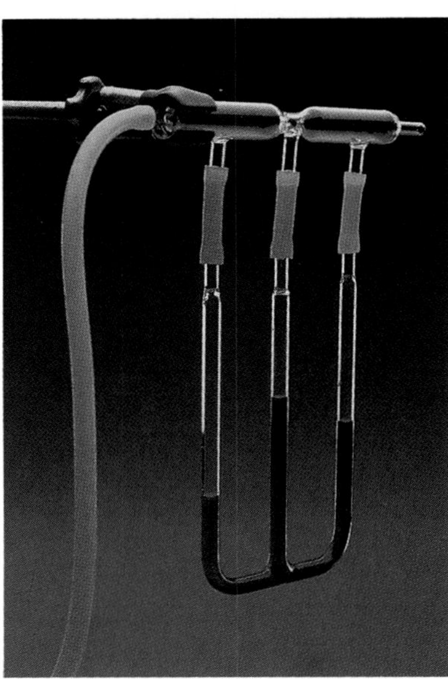

A student Venturi meter. Note the difference in pressure between wide parts of the tube and the constriction. The rightmost column is added to show the reduction in fluid pressure due to viscosity.

Courtesy Central Scientific Company.

DRILL PROBLEM 15-13

Assume that all three points of the pipe of Drill Prob. 15-12 are at the same height. What is the pressure difference between points 1 and 2? between points 1 and 3?
ANS. -3.8×10^4 N/m^2; 7.5×10^5 N/m^2.

DRILL PROBLEM 15-14

Water is moving at a speed of 6.0 m/s through a pipe whose cross-sectional area is 5.0 cm^2. The pipe gradually descends 10 m as its cross-sectional area increases to 12.0 cm^2. (*a*) What is the flow speed at the lower level? (*b*) What is the difference in pressure between the two levels?
ANS. (*a*) 2.5 m/s; (*b*) 1.1×10^5 N/m^2.

15-9 Moving Air

There are many interesting applications of Bernoulli's equation that involve the flow of air around objects. In this section we will consider three of these applications.

Buildings in High Winds

Figure 15-26 depicts a high wind blowing over a flat roof. With 1 and 2 representing points below and above the roof, respectively, we have from Bernoulli's equation

$$p_1 + \tfrac{1}{2}\rho v_1^2 = p_2 + \tfrac{1}{2}\rho v_2^2,$$

where v_1 and v_2 are the wind speeds at the two locations. Since point 1 is inside the building, $v_1 = 0$, and there is a pressure difference

$$p_1 - p_2 = \tfrac{1}{2}\rho v_2^2$$

pushing upward on the roof. If v_2 is large enough, this pressure difference can actually "blow" the roof off the building. Windows are also susceptible to this same effect, with the excess pressure of the internal stagnant air causing the windows to break outward.

The Hancock Building in Boston, Massachusetts, 1973. Due to a design flaw, the pressure difference between the inside and outside of the building during a strong wind was partially responsible for the windowpanes shattering. In this picture, the broken windows are covered with plywood.

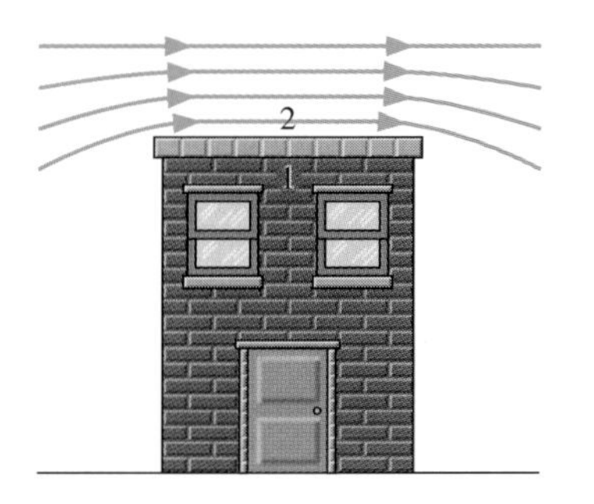

FIGURE 15-26 When a strong wind blows over the flat roof, there is a substantial pressure difference between points 1 and 2.

Atomizers

Figure 15-27 shows a sketch of a perfume atomizer. Devices very similar to this are used to spray paint and insecticides, and a comparable mechanism is used to transport gasoline in the carburetor of an automobile. In all cases, a stream of air is passed over one end of an open tube whose other end is inserted into the vessel containing the liquid to be sprayed. The pressure of the moving air is less than that of stagnant air inside the vessel, so there is a pressure difference between the two ends of the tube. This pressure difference is large enough to push the liquid up the tube and into the moving air. The liquid is then dispersed in a fine spray.

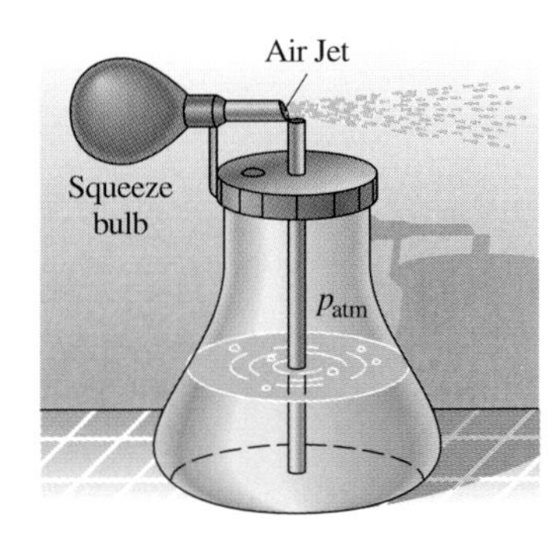

FIGURE 15-27 The pressure difference between the ends of the tube in an atomizer causes liquid to be pushed up the tube. The liquid is then dispersed by the moving air.

Airplane Wings

The cross section of a wing as seen by an observer at rest relative to the wing is shown in Fig. 15-28. If the plane is moving from right to left, then, relative to the wing, air is moving by it from left to right as indicated. The wing is shaped so that the air has to travel farther (and faster) along its upper surface than its lower surface. By Bernoulli's principle, there is then a higher pressure below than above the wing. This pressure difference is partially responsible for the lift that the air exerts on an airplane.

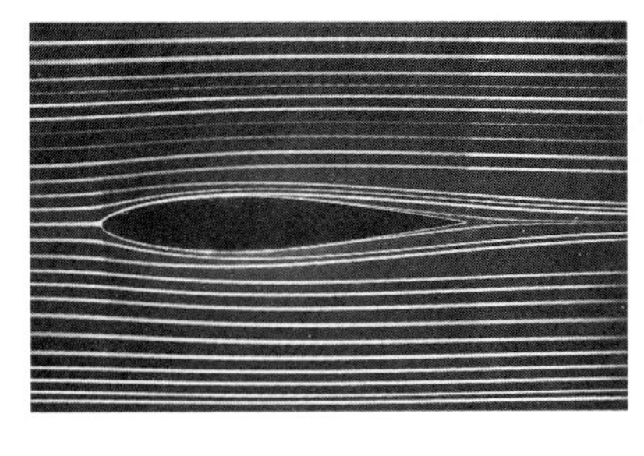

FIGURE 15-28 The streamlines around an airplane wing moving from right to left.

This is an extremely simplified model of lift, for both viscosity and turbulence are an important part of air flow around a wing. Since Bernoulli's equation is only valid for an ideal fluid, calculations based on this model must be considered as rough approximations. Also, airplane wings are actually tilted upward, which causes the air hitting the underside of the wing to be deflected downward. By Newton's third law, the air must then push the wing upward. This action is also responsible for some of the lift on an airplane wing.

EXAMPLE 15-13 **LIFT ON AN AIRPLANE WING**

The air speeds across the top and bottom of the wings of an airplane are 200 and 175 m/s, respectively. (*a*) If the density of the air is 0.75 kg/m^3, what is the difference in air pressure between the bottom and top of the wings? (*b*) If the total area of the bottoms of the wings is 100 m^2 and if the plane is traveling horizontally, what is the mass of the plane? In answering these questions, assume that the lift on the wings is given by Bernoulli's equation.

SOLUTION (*a*) The difference in height between the top and bottom of each wing is small enough that the gravitational term in Bernoulli's equation can be ignored. Then the pressure difference between the bottom and top of each wing is

$$\begin{aligned} p_1 - p_2 &= \tfrac{1}{2}\rho(v_2^2 - v_1^2) \\ &= \tfrac{1}{2}(0.75\ \text{kg/m}^3)[(200\ \text{m/s})^2 - (175\ \text{m/s})^2] \\ &= 3.5 \times 10^3\ \text{N/m}^2. \end{aligned}$$

(*b*) The force due to this pressure difference must balance the weight of the airplane, so

$$(p_1 - p_2)A = mg,$$

and the mass of the airplane is

$$m = (3.5 \times 10^3\ \text{N/m}^2)(100\ \text{m}^2)/(9.8\ \text{m/s}^2) = 3.6 \times 10^4\ \text{kg}.$$

DRILL PROBLEM 15-15

Air moves past the lower surface of a wing at a speed of 100 m/s. What flow speed over the wing's upper surface results in a pressure difference of 1.2×10^3 N/m^2 that acts to lift the wing? Use $\rho = 1.1$ kg/m^3 for the density of air.
ANS. 110 m/s.

DRILL PROBLEM 15-16

(*a*) An overhead view of a spinning baseball moving from left to right is shown in Fig. 15-29. In the ball's rest frame, the air moves from right to left. As any pitcher knows, this ball will curve as indicated when seen from overhead in the earth's frame. What does this imply about the velocity distribution of the air around the baseball? (*b*) What kind of spin must a skillful golfer impart to a golf ball in order to give it lift, thereby keeping it in the air longer so it travels farther?
ANS. (*a*) $v_2 < v_1$; (*b*) backspin (backward rotation of the ball).

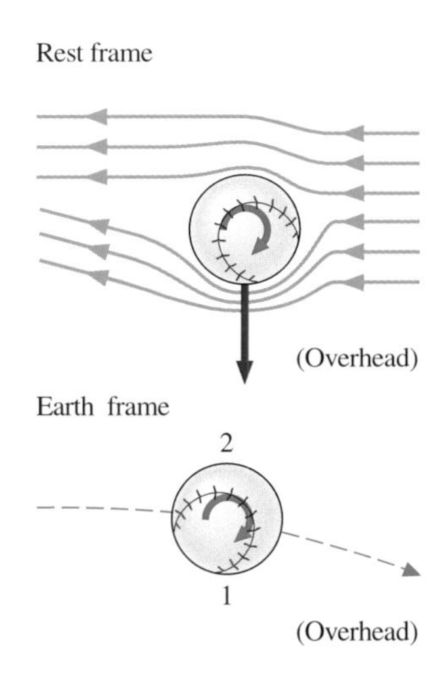

FIGURE 15-29 The flow of air around a spinning baseball causes it to curve.

*15-10 LAMINAR FLOW IN A CYLINDRICAL PIPE

We end this chapter by looking at the effects of viscosity on fluid flow. We know from experiment that viscous fluids often move in layers. These layers do not all have the same velocity, and even adjacent layers move relative to one another. This is known as *laminar flow*. Here we will consider just one special case: laminar fluid flow along a horizontal cylindrical pipe.

In a cylindrical pipe the fluid layers are concentric cylinders whose axes are along the central axis of the pipe. The velocities of the layers increase from zero at the wall of the pipe to a maximum at the pipe's central axis, as depicted in Fig. 15-30. Since adjacent layers move over one another, frictional forces are generated between them. These forces dissipate mechanical energy, so a moving fluid in the absence of a driving force must eventually come to rest.

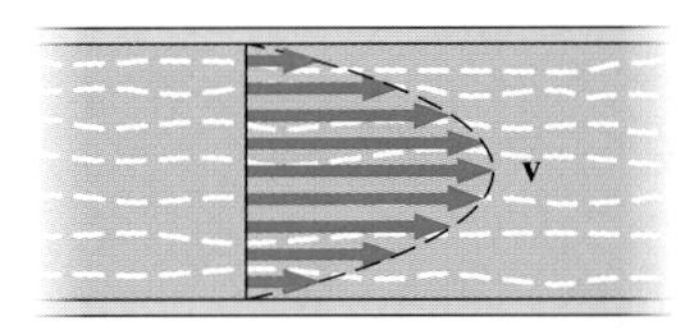

FIGURE 15-30 In laminar flow down a cylindrical pipe, cylindrical fluid layers slide over one another with velocities that increase from zero at the wall to a maximum at the center of the pipe. The heads of the velocity vectors form a parabolic surface.

Laminar flow through a horizontal cylindrical pipe of radius R and length L is illustrated in Fig. 15-31. The horizontal forces on the cylindrical fluid section of radius r are $F_p(r)$ due to the pressure difference between the ends of the section and the frictional force $F_f(r)$ at the cylindrical boundary. The force $F_p(r)$ is just the pressure difference multiplied by the cross-sectional area of the cylinder:

$$F_p(r) = (p_1 - p_2)\pi r^2.$$

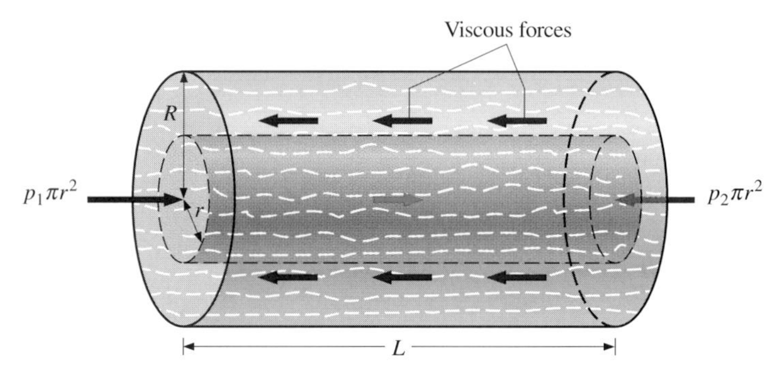

FIGURE 15-31 The horizontal forces on a cylindrical section of fluid in viscous flow down a cylindrical pipe.

The functional form of the viscous force is known from experiment. You should consult a book on fluid mechanics if you're interested in a complete explanation of this force.* Here we will simply state that the force is proportional to the product of the cylinder's surface area $2\pi rL$ and the change of the fluid velocity with radius dv/dr. Mathematically,

$$F_f(r) = \eta 2\pi rL \frac{dv}{dr}, \tag{15-9}$$

where η, the *coefficient of viscosity*, is the proportionality constant. Since dv/dr is negative (the velocity decreases with increasing radius), $F_f(r)$ is negative and acts in a direction opposite to that of the velocity vector. Values of η for some representative fluids are given in Table 15-2. Notice that η decreases with temperature for liquids, but increases with temperature for gases.

In steady flow, the velocity of this cylindrical section is constant, so the net force on it must vanish. Hence

$$F_p(r) + F_f(r) = (p_1 - p_2)\pi r^2 + \eta 2\pi rL \frac{dv}{dr} = 0,$$

and

$$\frac{dv}{dr} = -\frac{1}{2\eta}\left(\frac{p_1 - p_2}{L}\right)r.$$

The dependence of the velocity v on the radius r can be found by integrating this equation between r and the outer radius R, where $v(R) = 0$. We then have

$$\int_v^0 dv' = -\frac{1}{2\eta}\left(\frac{p_1 - p_2}{L}\right)\int_r^R r'\,dr'$$

so

$$v(r) = \frac{1}{4\eta}\left(\frac{p_1 - p_2}{L}\right)(R^2 - r^2). \tag{15-10}$$

This is the equation of the parabola drawn through the tips of the velocity vectors in Fig. 15-30.

*See, for example, Chap. 8 of W. M. Swanson, *Fluid Mechanics*, Holt, Rinehart and Winston, New York, 1970.

TABLE 15-2 **Coefficients of Viscosity of Selected Fluids**

Gases	η (N·s/m²) $T = 0°C$	$T = 20°C$	$T = 40°C$
Air	1.71×10^{-5}	1.81×10^{-5}	1.90×10^{-5}
Carbon dioxide	1.39×10^{-5}	1.48×10^{-5}	1.57×10^{-5}
Hydrogen	0.84×10^{-5}	0.89×10^{-5}	
Liquids	**η (N·s/m²) $T = 0°C$**	**$T = 20°C$**	**$T = 40°C$**
Acetone	0.395×10^{-3}	0.303×10^{-3}	0.280×10^{-3}
Benzene	0.906×10^{-3}	0.654×10^{-3}	0.498×10^{-3}
Blood	—	—	4×10^{-3}
Castor oil	53.0×10^{-3}	9.86×10^{-3}	2.31×10^{-3}
Ether	0.284×10^{-3}	0.233×10^{-3}	0.197×10^{-3}
Ethyl alcohol	1.79×10^{-3}	1.72×10^{-3}	1.65×10^{-3}
Water	1.79×10^{-3}	1.01×10^{-3}	0.656×10^{-3}

A quantity of more practical interest than the velocity is the flow rate. From Eq. (15-7), the volume of fluid passing through an area dA per second is $dQ = v\,dA$. The cross section of the cylindrical pipe can be divided into circular disks of radius r and thickness dr. Then $dA = 2\pi r\,dr$ and

$$dQ = v(2\pi r\,dr) = \frac{1}{4\eta}\left(\frac{p_1 - p_2}{L}\right)(R^2 - r^2)(2\pi r\,dr).$$

The flow rate Q through the pipe can now be found by integration:

$$Q = \int_0^Q dQ' = \frac{2\pi}{4\eta}\left(\frac{p_1 - p_2}{L}\right)\int_0^R [R^2 - (r')^2]r'\,dr'$$

so

$$Q = \frac{\pi R^4}{8\eta}\left(\frac{p_1 - p_2}{L}\right). \tag{15-11}$$

This equation is known as *Poiseuille's law* after its discoverer Jean Louis Marie Poiseuille (1799–1869). It states that *for a fixed pressure difference between the ends of a pipe, the volume flow rate depends on the fourth power of the radius.* Double the radius and 16 times as much fluid flows; halve the radius and one-sixteenth as much fluid flows.

Poiseuille's law is frequently used to describe blood flow through the vessels of our circulatory system, although its application here is at best a rough approximation for many reasons, including the following: (1) blood flow is not completely laminar; (2) many capillaries and veins are so short that Eq. (15-9) does not accurately describe the frictional force even when the flow is laminar; (3) blood vessels are not rigid, as they expand and contract; and (4) the flow is not necessarily horizontal. Despite these shortcomings, Poiseuille's law provides us with valuable insight about the circulatory system, as demonstrated in some of the following examples and drill problems.

EXAMPLE 15-14 Pressure Difference in Viscous Flow

A horizontal pipe is 5.0 m long and has an inside diameter of 4.0 cm. What pressure difference between the ends of the pipe is required to get 3.0 m^3/s of water at 20°C flowing through it?

SOLUTION With η from Table 15-2, we find from Poiseuille's law that the pressure difference between the ends of the pipe required for the specified flow rate is

$$\Delta p = \frac{8\eta QL}{\pi R^4} = \frac{8(1.0 \times 10^{-3}\ \text{N}\cdot\text{s/m}^2)(3.0\ \text{m}^3/\text{s})(5.0\ \text{m})}{\pi(2.0 \times 10^{-2}\ \text{m})^4}$$
$$= 2.4 \times 10^5\ \text{N/m}^2.$$

EXAMPLE 15-15 Blood Flow in a Capillary

A typical capillary of the circulatory system is about 1.0 mm long and has a diameter of 4.0×10^{-4} cm. The average speed of the blood flowing through it is approximately 3.3×10^{-2} cm/s. (*a*) What is the blood's flow rate through the capillary? (*b*) What pressure difference is required to maintain this flow rate?

SOLUTION (*a*) The blood's flow rate is, from Eq. (15-7),

$$Q = Av = \pi r^2 v = \pi(2.0 \times 10^{-6}\ \text{m})^2(3.3 \times 10^{-4}\ \text{m/s})$$
$$= 4.1 \times 10^{-15}\ \text{m}^3/\text{s}.$$

(*b*) Using the coefficient of viscosity of blood from Table 15-2, we can find the required pressure difference to maintain the flow rate of part (*a*) with Poiseuille's law:

$$\Delta p = \frac{8\eta QL}{\pi R^4}$$
$$= \frac{8(4.0 \times 10^{-3}\ \text{N}\cdot\text{s/m}^3)(4.1 \times 10^{-15}\ \text{m}^3/\text{s})(1.0 \times 10^{-3}\ \text{m})}{\pi(2.0 \times 10^{-6}\ \text{m})^4}$$
$$= 2.6 \times 10^3\ \text{N/m}^2 = 20\ \text{mmHg}.$$

EXAMPLE 15-16 Power Output of the Heart

In an adult, the heart pumps blood at an approximate rate of $Q = 1.0 \times 10^{-4}\ \text{m}^3/\text{s}$ through the circulatory system. This is done by maintaining an average pressure difference between the arterial and venous systems of about $\Delta p = 1.2 \times 10^4\ \text{N/m}^2$. Determine the average power output of the heart.

SOLUTION The heart circulates blood by pumping it into a large artery known as the aorta. Assuming that the cross-sectional area of the aorta is A, the force applied to the blood by the heart is $F = (\Delta p)A$. The power output P of the heart is the force times the velocity of the blood, so

$$P = Fv = (\Delta p)Av = (\Delta p)Q,$$

where $Q = Av$. Substituting the given values into this equation, we obtain

$$P = (1.2 \times 10^4\ \text{N/m}^2)(1.0 \times 10^{-4}\ \text{m}^3/\text{s}) = 1.2\ \text{W}$$

as the power output of the heart.

From Poiseuille's law, $Q \propto \Delta p R^4$. Consequently, partial obstructions in the arterial system make the heart maintain a higher blood pressure in order to produce the same flow rate Q. Since the power P is proportional to Δp, the obstructed arteries make the heart pump with a greater power output. This extra burden may eventually damage the heart.

DRILL PROBLEM 15-17

If the diameter of the pipe of Example 15-14 is doubled, how much water per second flows through the pipe when the pressure difference between its ends is $2.4 \times 10^5\ \text{N/m}^2$?
ANS. 48 m^3/s.

DRILL PROBLEM 15-18

If the capillary of Example 15-15 becomes partially clogged due to cholesterol buildup so that its radius decreases by 10 percent, what pressure difference is required to keep the flow rate at $4.1 \times 10^{-15}\ \text{m}^3/\text{s}$?
ANS. 30 mmHg, or an increase of 50% over the unclogged artery.

DRILL PROBLEM 15-19

When a person exercises heavily, the flow of blood to the muscles increases. Speculate on how this is accomplished physiologically.
ANS. The blood pressure increases and the blood vessels expand.

SUMMARY

1. **States of matter**
   (*a*) Except at very high temperatures, matter exists in one of three states: solid, liquid, or gas. Which state a particular material is in depends on its temperature and pressure.
   (*b*) Liquids and gases are known collectively as fluids. A fluid at rest cannot support a force applied tangent to any surface of the fluid.
2. **Density and pressure**
   (*a*) The density of a substance is its mass per unit volume.
   (*b*) The pressure at any point in a fluid is the normal force per unit area exerted by the fluid on an infinitesimal surface located at that point.

3. **Pressure variation with depth**
   (*a*) In a static fluid, the pressure variation with depth is given by

   $$p_2 = p_1 + \rho g h.$$

   (*b*) Pascal's principle states that a pressure applied to the surface of an enclosed static fluid is transmitted undiminished to every point in the fluid.
   (*c*) Atmospheric pressure is proportional to the height of the mercury column in a barometer. A height of 76.0 cm corresponds to a pressure of 1.01×10^5 N/m^2. Gauge pressure is the difference between the pressure being measured and atmospheric pressure.
4. **Archimedes' principle**
   When a body is immersed in a fluid at rest, the fluid exerts a net upward force (the buoyant force) on that body which is equal in magnitude to the weight of the fluid displaced.
5. **Fluids in motion**
   (*a*) At any instant, a moving fluid can be characterized by its density, pressure, and velocity at every point in the fluid.
   (*b*) Fluid flow is characterized as (*i*) steady or nonsteady, (*ii*) compressible or incompressible, (*iii*) rotational or irrotational, and (*iv*) viscous or nonviscous.
   (*c*) For steady flow, the streamline passing through a point represents the direction of the fluid velocity at that point.
6. **The equation of continuity**
   If A is the cross-sectional area of a flow tube at a point where the speed of the fluid is v, then

   $$Av = \text{constant}.$$

7. **Bernoulli's equation**
   For an ideal fluid,

   $$p + \tfrac{1}{2}\rho v^2 + \rho g h = \text{constant}.$$

*8. **Laminar flow in a cylindrical pipe**
   Laminar flow in a horizontal cylindrical pipe obeys Poiseuille's law:

   $$Q = \frac{\pi R^4}{8\eta}\left(\frac{p_1 - p_2}{L}\right).$$

SUPPLEMENT 15-1 BERNOULLI'S EQUATION

Here we'll derive Bernoulli's equation using the work-energy theorem. We start by considering the fluid section shown in Fig. 15-23. We assume that the fluid is incompressible and that its flow is steady, nonviscous, and irrotational. With no viscosity, only the forces due to gravity and the surrounding fluid do work on this section as it moves along the flow tube. During a short time interval Δt, the work done by the surrounding fluid at A_1 is

$$W_1 = F_1\,\Delta x_1 = (p_1 A_1)(v_1\,\Delta t) = p_1 A_1 v_1\,\Delta t,$$

and at A_2 it is

$$W_2 = -p_2 A_2 v_2\,\Delta t.$$

The quantity W_2 is negative because the force acts opposite to the direction of motion. Now the fluid is incompressible, so its density is constant. Also, because the flow is steady and irrotational, the motion of the entire section over a time interval Δt is equivalent to the transport of the dotted cylindrical region of thickness $v_1\,\Delta t$ at the lower end of the flow tube to the dotted cylindrical region of thickness $v_2\,\Delta t$ at the upper end; that is, it is equivalent to a fluid section of mass $\rho A_1 v_1\,\Delta t$ (which equals $\rho A_2 v_2\,\Delta t$) being raised through a vertical height $y_2 - y_1$. The work of gravity is therefore

$$W_g = -\rho g A_1 v_1\,\Delta t(y_2 - y_1).$$

The total work W done on the fluid section during Δt is the sum

$$\begin{aligned} W &= W_1 + W_2 + W_g \\ &= p_1 A_1 v_1\,\Delta t - p_2 A_2 v_2\,\Delta t - \rho g A_1 v_1\,\Delta t(y_2 - y_1) \\ &= A_1 v_1\,\Delta t[(p_1 - p_2) - \rho g(y_2 - y_1)], \end{aligned}$$

where in the last line, $A_1 v_1$ has been substituted for $A_2 v_2$.

To determine the change in the kinetic energy of the fluid, we again use the fact that the motion of the entire section is equivalent to the transport of the cylindrical region of mass $\rho A_1 v_1\,\Delta t$ at the lower end of the flow tube to the cylindrical region of equal mass $\rho A_2 v_2\,\Delta t$ at the upper end. The overall change in the kinetic energy of the fluid is therefore equal to the difference in the kinetic energies of the cylindrical regions:

$$\begin{aligned} \Delta T &= \tfrac{1}{2}(\rho A_2 v_2\,\Delta t)v_2^2 - \tfrac{1}{2}(\rho A_1 v_1\,\Delta t)v_1^2 \\ &= \tfrac{1}{2}\rho A_1 v_1\,\Delta t(v_2^2 - v_1^2), \end{aligned}$$

where again we have used $A_1 v_1 = A_2 v_2$.

Since the total work done on the fluid section is equal to the change in its kinetic energy,

$$A_1 v_1\,\Delta t[(p_1 - p_2) - \rho g(y_2 - y_1)] = \tfrac{1}{2}\rho A_1 v_1\,\Delta t(v_2^2 - v_1^2),$$

which simplifies to

$$p_1 + \tfrac{1}{2}\rho v_1^2 + \rho g y_1 = p_2 + \tfrac{1}{2}\rho v_2^2 + \rho g y_2.$$

This is Bernoulli's equation.

QUESTIONS

15-1. Which occupies a greater volume—1 kg of water or 1 kg of iron?

15-2. A fluid conforms to the shape of its container. Does this imply that the density of a given mass of fluid depends on the container into which it is poured?

15-3. Old windowpanes are often found to be slightly thicker at the bottom than at the top. Why is this?

15-4. The air pressure on you is approximately 1.0×10^5 N/m^2. Why doesn't this pressure prevent you from jumping upward?

15-5. Why is one end of a nail flat while the other end is pointed?

15-6. Explain how a suction cup works.

15-7. Why does the increase in pressure with depth serve to push an object upward rather than downward?

15-8. Contrary to the simplified description of Example 15-6, dams are actually built so that they are convex when viewed from the water side. They are also made thicker at the bottom than at the top. (See the accompanying figure.) Explain why dams are built this way.

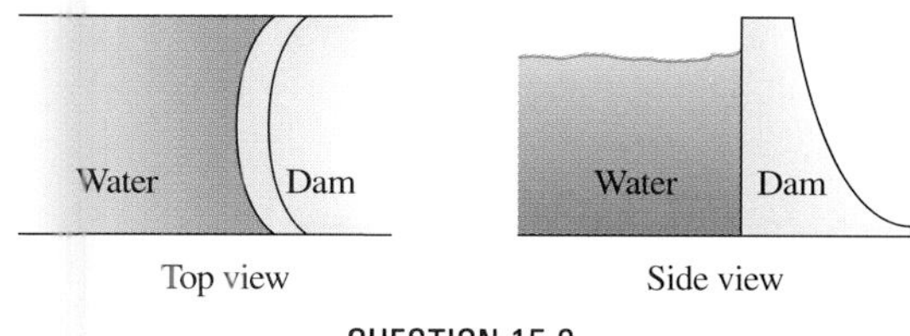

QUESTION 15-8

15-9. What happens to the water level in a glass of ice water when the ice cubes melt? Explain.

15-10. The bottom of a glass is packed with several ice cubes. When water is first added, the cubes remain at the bottom because they are jammed up against the walls of the glass. After a few minutes, however, the cubes rise to the top. What happens to the water level?

15-11. Does whether an object float in a liquid depend on the value of g? Explain.

15-12. Why do some people float in water better than others? What body type is most adaptable to floating?

15-13. Will a solid steel ball float in a pool of mercury?

15-14. People frequently say that objects heavier than water sink and those lighter than water float. What is wrong with this statement?

15-15. Most people tread water more easily in the ocean than in a swimming pool. Can you explain why?

15-16. Explain why a ship that is filling with water frequently turns over before sinking.

15-17. A rather large toy boat that contains lead shot is floating in a bathtub. If the shot is removed from the boat and placed on the bottom of the tub, what happens to the water level in the tub?

15-18. Air is heated at the bottom of a hot-air balloon with a gas burner. The air then rises into the inside of the balloon, thereby causing the balloon to float upward. Explain why the balloon ascends.

15-19. Describe how Archimedes' principle can be used to distinguish a gold-plated lead crown from a solid-gold crown.

15-20. A bucket of water containing a piece of ice is placed in an elevator. Does the fraction of ice floating above the surface of the water depend on the vertical acceleration of the elevator?

15-21. What important assumptions make the equation of continuity and Bernoulli's equation only approximately valid for the flow of air?

15-22. Why does the height of water waves increase when the wind speed increases?

15-23. Why is it easier to make a tennis ball curve than to make a baseball curve?

15-24. Explain why the lift on an airplane wing depends on the altitude of the plane.

15-25. If possible, airplanes take off and land heading into the wind. Why?

15-26. Explain why when you pass a truck on the highway, your car is pushed toward the truck.

PROBLEMS

Density and Pressure

15-1. A rectangular block of aluminum measures 10.0 cm by 8.0 cm by 30.0 cm. What is the mass of this block?

15-2. A block of unknown metal has the same dimensions as the aluminum block of the previous problem but is found to weigh 50 percent more. What is the density of the unknown metal?

15-3. Compare the volume of 500 kg of freshwater to that of seawater.

15-4. White dwarfs are stars that have collapsed to the size of a planet and are slowly cooling. Suppose that a white dwarf has the radius of the earth and the mass of the sun. What is the average density of this star? Compare your answer to the densities of ordinary substances such as metals.

15-5. Stars more massive than the sun often supernova and leave behind cores known as neutron stars, which are much smaller but more massive than white dwarfs. Consider a neutron star that has a radius of 10 km but is twice the mass of the sun. What is the average density of this star? Compare your answer with that of the previous problem.

15-6. The head of a nail has a diameter of 1/8 in. It is struck by a hammer with a force of 40 lb. (*a*) What is the pressure on the head of the nail? (*b*) If the sharp end of the nail has a diameter of 1/64 in., what is the pressure on this end?

15-7. A 5.0-N force is applied to the 3.0-cm^2 plunger of the hypodermic needle shown in the accompanying figure. (*a*) What is the gauge pressure of the fluid in the chamber of the syringe? (*b*) If the cross-sectional area of the needle is 6.0×10^{-3} cm^2, what force would the fluid exert on a solid obstruction placed against the exit end of the needle? (*c*) What minimum force must be applied to the plunger to inject fluid into a vein where the blood pressure is 100 mmHg? (*Note:* The variation of pressure with height should be ignored in this problem. Justify this assumption.)

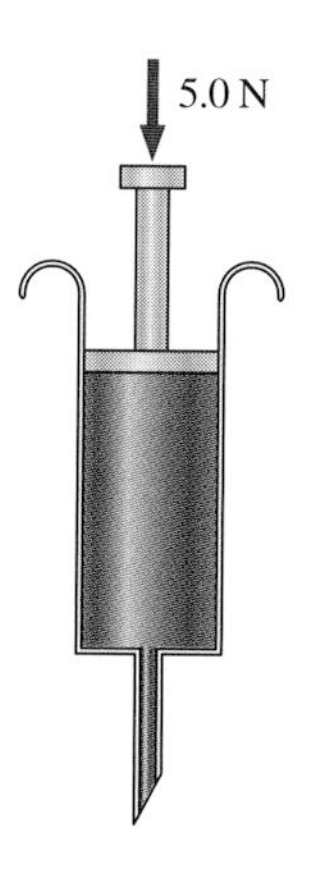

PROBLEM 15-7

15-8. The pressure p is constant over the hemispherical surface of radius r shown in the accompanying figure. Prove that the net force *on the outer spherical surface* of the hemisphere due to the pressure is $p\pi r^2$ and acts perpendicular to the imaginary flat surface that encloses the open end of the hemisphere. (*Hint:* The force can be calculated either by integrating $p\,dA$ over the hemisphere or by adding a flat surface to the open end of the hemisphere and then applying the first condition of equilibrium to the enclosed hemisphere.)

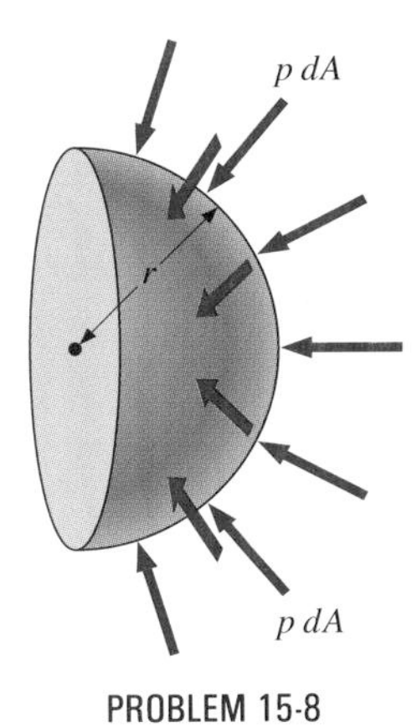

PROBLEM 15-8

Pressure Variation with Depth

15-9. Express a pressure of 2.3×10^4 N/m^2 (*a*) in atmospheres and (*b*) in centimeters of mercury.

15-10. What is the pressure 300 m below the surface of the ocean?

15-11. The pressure on the hull of a submarine is 5.0×10^6 N/m^2. How far is the submarine below the ocean surface?

15-12. The *Guinness Book of World Records* states that the greatest recorded depth to which a whale has dived is 1134 m. (*a*) What is the pressure in newtons per square meter on the whale at this depth? (*b*) What is the pressure in atmospheres? (*c*) in centimeters of mercury?

15-13. The record ocean descent by a human was achieved by Dr. Jacques Picard and Lt. Donald Walsh in the bathyscaphe *Trieste*. They reached a depth of 10,918 m on January 23, 1960. (*a*) What is the pressure in atmospheres at this depth? (*b*) What is the force due to the water pressure on a window of area 2.0 m^2? Compare this force with the weight of an 80-kg diver.

15-14. Assuming that the masses of the two pistons in the accompanying figure can be ignored, (*a*) what force must be applied to piston 1 to lift piston 2 slowly? (*b*) If piston 1 is depressed a distance Δl_1, through what vertical height Δl_2 is piston 2 lifted? (*c*) If $\Delta l_1 = 2.0$ cm, how much work is done on piston 1 by the applied force? How much work is done on the 10,000-N weight by piston 2?

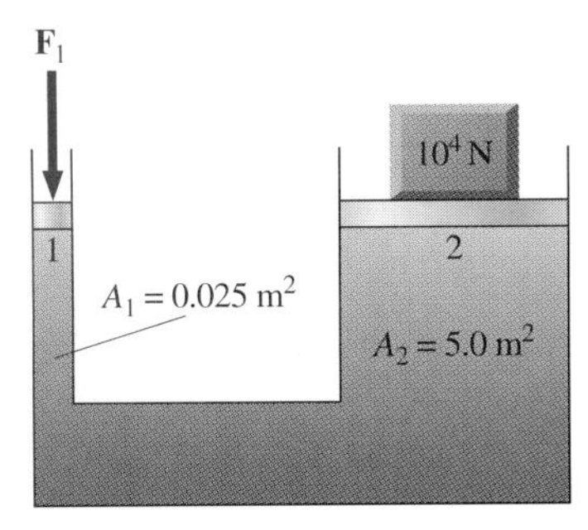

PROBLEM 15-14

Measurement of Pressure

15-15. A change of one floor in a building corresponds to a vertical height of approximately 4 m. Can you use an ordinary barometer to measure the difference in air pressure between the ground level and the roof of a three-story building?

15-16. How tall is a column of ethyl alcohol that corresponds to a pressure of 1 atmosphere?

15-17. For each case shown in the accompanying figure, calculate the absolute and the gauge pressures of the confined gas in newtons per square meter. Now calculate the absolute pressures in centimeters of mercury, centimeters of water, torrs, pascals, atmospheres, and bars.

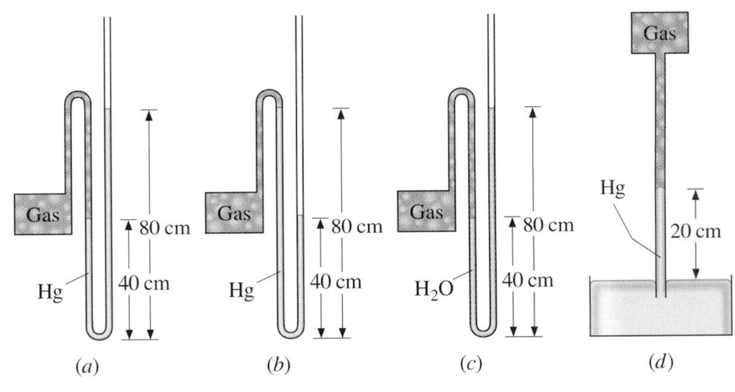

PROBLEM 15-17

15-18. An air-filled cylinder of cross-sectional area 7.5×10^{-3} m^2 is connected to one side of a manometer as shown in the accompanying figure. By how much does the gauge pressure change when a 2.0-kg block is set on the piston of the cylinder?

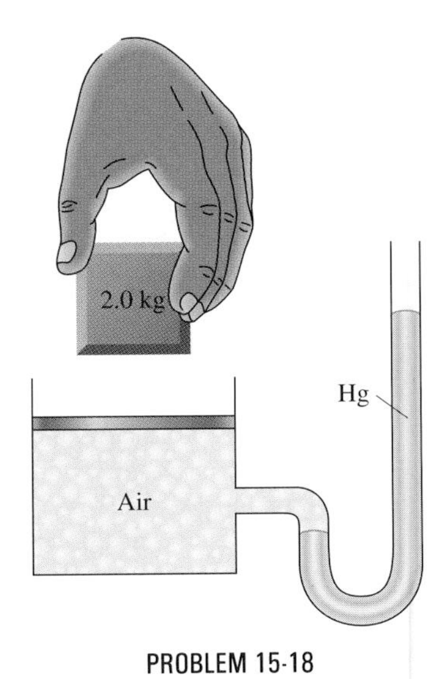

PROBLEM 15-18

Archimedes' Principle

15-19. How much force does a child have to apply to keep his beach ball (negligible mass) of 25-cm radius submerged below the ocean surface?

15-20. A cube with sides of length l is completely immersed in water. (*a*) What is the pressure difference between the bottom and top of the cube? (*b*) What is the net vertical force on the cube due to this pressure difference? (*c*) What is the weight of a section of water whose volume is l^3? (*d*) Relate these calculations to Archimedes' principle.

15-21. What fraction of an iceberg floats above the surface in the ocean?

15-22. A hot-air balloon of volume 3.0×10^3 m^3 is filled with air of density 1.0 kg/m^3. What maximum load (this includes the weight of the balloon) can the balloon lift when it is in air of density 1.3 kg/m^3?

15-23. In the accompanying figure, a block of volume 0.25 m^3 is held submerged in (*a*) water and (*b*) an unknown liquid. What is the density of the unknown liquid?

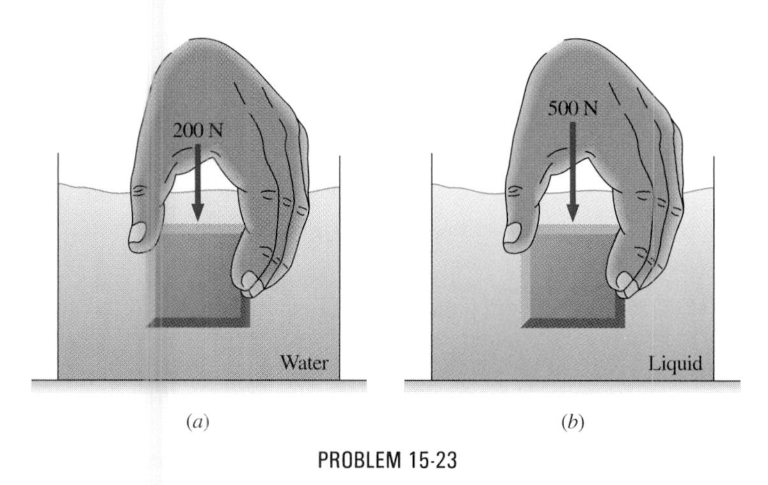

PROBLEM 15-23

15-24. A slab of ice floats on a freshwater lake. What is the minimum volume of the slab such that a 100-kg man can stand on it without getting his feet wet?

15-25. What is the minimum volume of lead that must be tied to a wood block of density 0.60 g/cm^3 and volume 8.0×10^3 cm^3 in order to sink the block in freshwater?

15-26. The weight of the pan and water shown in the accompanying figure is 100 N. (*a*) When a copper block of volume 1.00×10^{-3} m^3 is placed in the water, what is the reading on the scale? (*b*) If the string supporting the block is cut, what is the reading on the scale after the block settles on the bottom of the pan? (*c*) What is the acceleration of the block immediately after the string is cut?

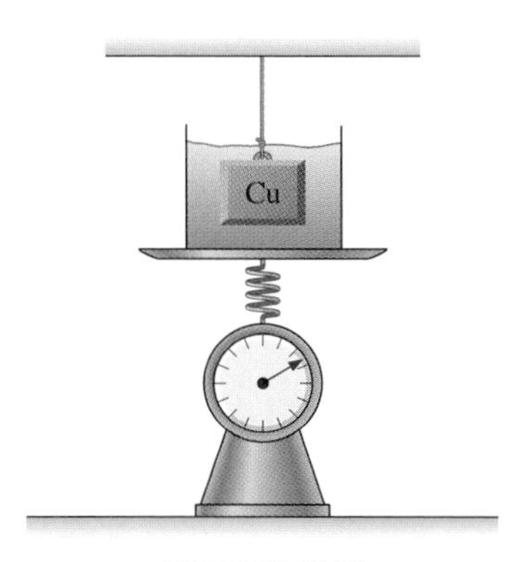

PROBLEM 15-26

15-27. A hollow cubical box is made out of a thin metal sheet whose surface mass density is 2.0 g/cm^2. (*a*) If the box is 50 cm on an edge, will it sink or float in freshwater? (*b*) If the edges are 10 cm long, will the box sink or float? (*c*) What is the smallest box that will float? (*d*) Use these calculations to explain why a metal-hulled ocean liner floats while a canoe made of the same material would sink.

15-28. The box of part (*a*), Prob. 15-27, is taken to the bottom of a swimming pool and released. What is its acceleration at the instant it is released?

The Equation of Continuity

15-29. Water in steady flow moves through a horizontal pipe of variable diameter. If the water's speed is 10 m/s at a point where the pipe's diameter is 1.0 m, what is the water's speed at a point where the diameter is 2.0 m? where the diameter is 0.50 m?

15-30. Water is flowing steadily through a section of a conduit whose circular cross section is gradually decreasing, as shown in the accompanying figure. Make a plot of water velocity vs. horizontal position in the section. The speed of the water at the left end of the section is 20 cm/s.

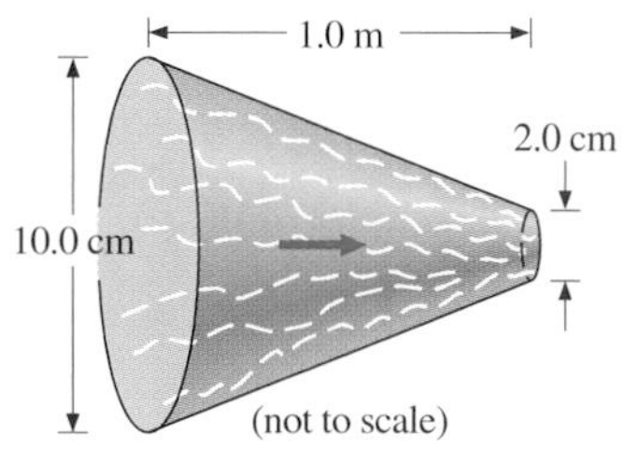

PROBLEM 15-30

15-31. Water flows through a pipe of diameter 10.0 cm with a speed of 5.0 m/s. At a junction, the water divides and flows into two smaller pipes, each of diameter 6.0 cm. What is the speed of the water in the smaller pipes? Ignore turbulence at the junction.

15-32. Blood is pumped from the heart into the aorta at a rate of approximately 80 cm^3/s. The blood subsequently travels through roughly 6×10^9 capillaries whose radii average around 8×10^{-4} cm. What is the average speed of the blood through these capillaries?

15-33. Water flows through a hose at a rate of 0.85 m^3/min. If the radius of the hose is 2.5 cm, at what speed is the water moving?

15-34. A nozzle of radius 1.5 cm is placed on the end of the hose of Prob. 15-33. What is the speed of the water in the nozzle?

15-35. A water main must be able to deliver a maximum of 2.0 gal/min to each of 40 identical apartments. If the diameter of the water main is 3.0 in., at what rate must water move through it during maximum flow? (*Note:* 1.0 gal = 231 in.^{3})

15-36. Ten sewer pipes, each of diameter 10.0 cm, drain into a single pipe of diameter 25.0 cm. If the average speed of the water in each 10.0-cm pipe is 3.0 m/s, what is its average speed in the 25.0-cm pipe?

Bernoulli's Equation

15-37. A horizontal pipe tapers from a diameter of 10.0 cm to one of 5.0 cm. At the larger end, water flows with a speed of 2.0 m/s and a pressure gauge reads 0.75 atm. What is the speed of the water at the smaller end? What is the gauge pressure at this end?

15-38. Water moves through a pipe of constant cross section. What is the pressure difference between two points that are separated by a height of 5.0 m?

15-39. Water moves by point 1 in a uniform pipeline with a speed of 10 m/s; the gauge pressure there is 2.0×10^5 N/m^2. At point 2, which is 3.0 m below point 1, the speed of the water is 20 m/s. What is the gauge pressure at point 2?

15-40. Water flows through the pipeline shown in the accompanying figure at a rate of 0.040 m^3/s. (*a*) What is the speed of the water at point 1 and at point 2? (*b*) What is the difference in pressure between points 1 and 2?

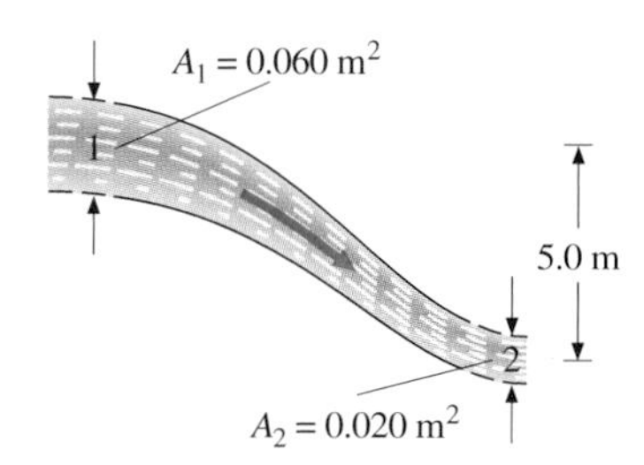

PROBLEM 15-40

15-41. A small hole is cut in the side of a water tower at a distance h below the surface of the water. (See the accompanying figure.) At what speed does water leave the hole?

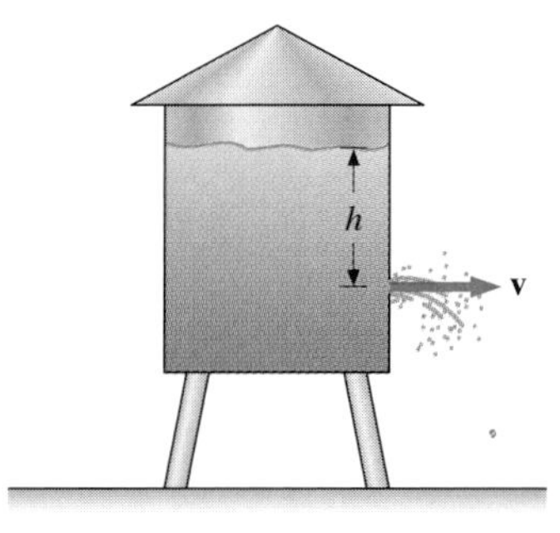

PROBLEM 15-41

15-42. A small hole of diameter 1.0 cm is cut in the side of a cylindrical water tank of diameter 10.0 m. (*a*) When the hole is 5.0 m below the the surface of the water, what is the speed at which water leaves the hole? (*b*) What is the flow rate of the water in cubic meters per second? (*c*) What is the rate at which the depth of the water is decreasing?

15-43. Water flows out of a 1.0-cm diameter circular hole in the side of a water tank at a rate of 5.0×10^{-2} m^3/min. How far is the hole below the surface of the water?

15-44. The accompanying figure shows a pump P that takes standing water from the tank and delivers it to the atmosphere at point A with a speed of 30 m/s. The water leaves the pump through a pipe of diameter 4.0 cm, it enters the atmosphere through a pipe of diameter 2.0 cm, and A is 3.0 m above the pump outlet. (*a*) What is the gauge pressure at the pump? (*b*) What is the power output of the pump?

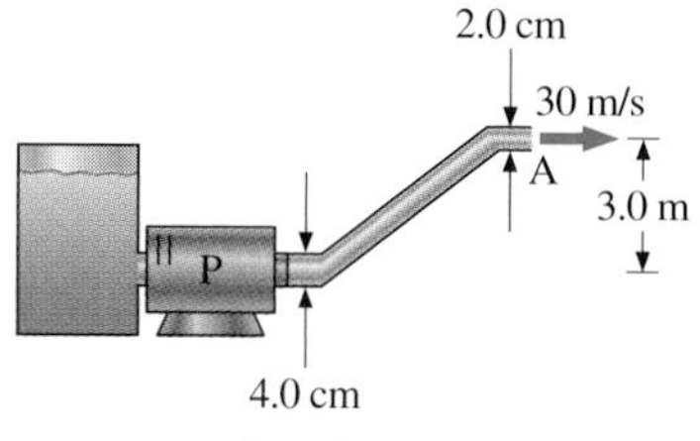

PROBLEM 15-44

15-45. A drain pipe is attached to a water tower as shown in the accompanying figure. (*a*) What is the pressure at V when the valve is closed? (*b*) What is the speed of the water at this same point when the valve is open? (*c*) How much water passes through the valve per second?

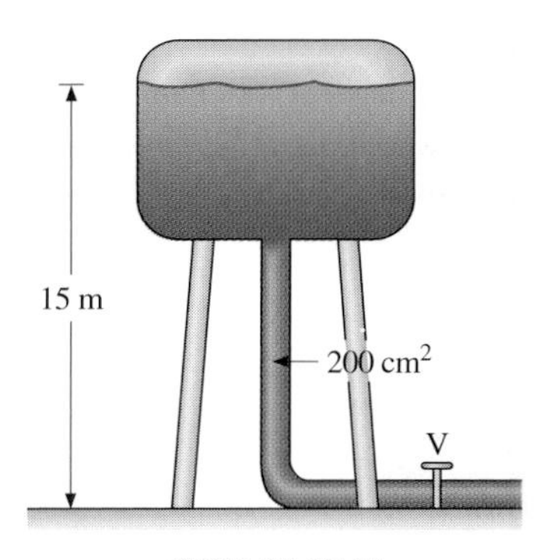

PROBLEM 15-45

15-46. One end of a horizontal hose of cross-sectional area 100 cm^2 is connected to a pump, and the other end is attached to a nozzle whose opening has an area of 20 cm^2. What is the gauge pressure developed by the pump when water leaves the nozzle at a rate of 5.0×10^{-2} m^3/s?

15-47. The pressure difference between the main pipeline and the throat of a Venturi meter is 2.0×10^5 Pa. The areas of the pipe and the constriction are 200 and 10 cm^2, respectively. What is the flow rate of the water in the pipe?

Moving Air

15-48. A 30-m/s wind blows across the top of a flat roof whose area is 100 m^2. If the pressure inside the house is 1.0 atm, what is the net force on the roof due to the difference in air pressure? Assume that the density of air is 1.3 kg/m^3.

15-49. When air of density 1.3 kg/m^3 flows across the top of the tube shown in the accompanying figure, water rises in the tube to a height of 1.0 cm. What is the speed of the air?

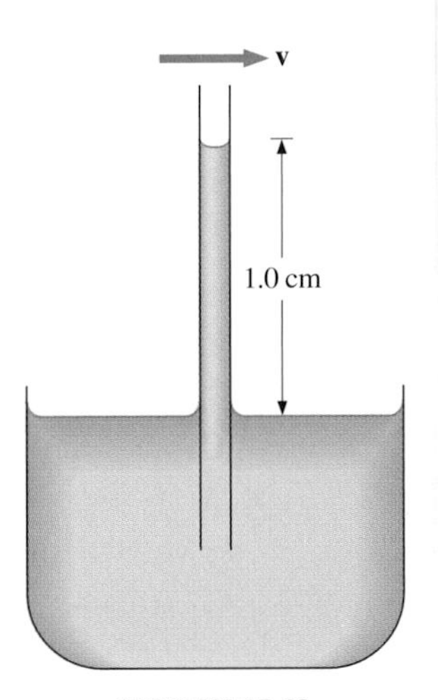

PROBLEM 15-49

15-50. Air of density 1.2 kg/m^3 streams past the upper and lower surfaces of the wings of a small plane with speeds of 50 and 40 m/s, respectively. The plane is moving horizontally. (*a*) If the total lower surface area of the two wings is 8.0 m^2, what is the net force on the airplane due to the pressure difference calculated from Bernoulli's equation? (*b*) What is the mass of the airplane?

15-51. A person blows air across the top of one side of a U-tube containing water. If the difference in water levels between the two sides of the tube is 1.5 cm, what is the speed of the air? Assume that the density of air is 1.3 kg/m^3.

15-52. If air of density 1.1 kg/m^3 moves past the lower surface of a wing at 90 m/s, what is its flow speed over the upper surface of the wing if the pressure difference between the lower and upper surfaces is 1.1×10^3 N/m^2? Assume that the pressure difference can be calculated with Bernoulli's equation.

Laminar Flow in a Cylindrical Pipe

15-53. Suppose a fluid flows through a small horizontal capillary of radius R and length L at a rate Q. (*a*) If both the radius and length of the capillary are doubled, at what rate does the fluid flow through it? (*b*) Answer the same question if both the radius and length are halved.

15-54. A certain volume of air at 20°C is observed to flow through a horizontal tube in 120 s. The same volume of a second gas at 20°C flows through the same tube under the same pressure difference in 180 s. What is the viscosity of this second gas?

15-55. (*a*) If the temperature is 20°C, what pressure difference is needed to pump water at a rate of 0.10 m^3/s along a 1000-m section of pipe whose diameter is 10.0 cm? (*b*) What pressure difference is needed if the temperature is 0°C?

15-56. Water at 20°C flows along a horizontal pipe of length 10.0 m and radius 5.0 cm with an average speed of 10.0 m/s. (*a*) What is the pressure difference between the ends of this section of pipe? (*b*) What is the pressure difference if the radius of the pipe is 25.0 cm? (*c*) Suppose the pressure at the inlet to the 10.0-m section is 1.0 atm. For part (*a*), is the pressure difference significant? Is it significant for part (*b*)? Does Bernoulli's equation accurately describe the flow of part (*a*)? of part (*b*)?

15-57. A blood vessel is 0.080 m long and has a radius of 1.3×10^{-3} m. Blood flows through the vessel at a rate of 8.0×10^{-8} m^3/s. What is the difference in pressure between the two ends of the vessel?

General Problems

15-58. The atmospheric pressure on the surface of the planet Venus is about 90 times that of the earth. How far below the ocean would a person have to be to experience this same pressure on the earth?

15-59. Both sides of a U-tube that contains mercury are open to the atmosphere. After 45 cm of water is poured into the left side of the tube, how far does the mercury in the right side rise above its original level?

15-60. A demonstration of the large forces that can be produced by pressure differences was devised by Otto von Guericke in 1654. He split a spherical shell into hemispheres, placed the hemispheres back together, and then used a pump (which he invented) to extract air from the interior of the spherical shell. Because of the pressure difference between the outside and inside of the sphere, two teams of eight horses were unable to separate the hemispheres by pulling in opposite directions. (See the accompanying figure.) Use the result of Prob. 15-8 to calculate the force each set of horses would have had to apply to separate the hemispheres. Assume that the radius of the sphere is 0.50 m and that the gauge pressure inside the sphere is −0.97 atm.

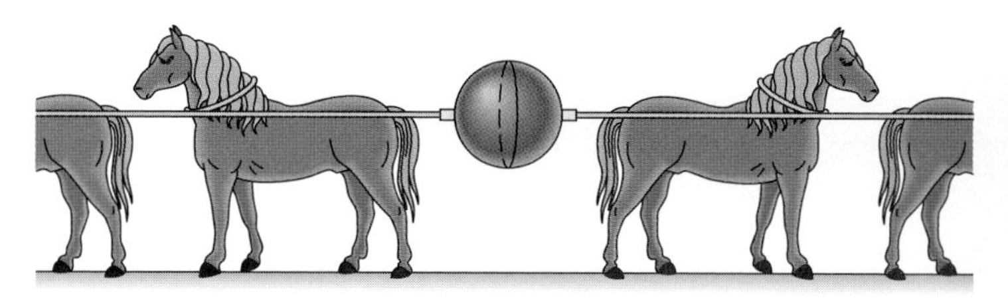

PROBLEM 15-60

15-61. Both sides of a U-tube are open to the atmosphere. Show that in the absence of friction, when the liquid on one side of the tube is depressed and then released, the liquid level on either side oscillates in simple harmonic motion with a period given by $\pi\sqrt{2L/g}$, where L is the total length of the liquid column. (*Hint:* Determine the total mechanical energy of the oscillating liquid and compare it to the total mechanical energy of the simple harmonic oscillator.)

15-62. A window in the shape of an equilateral triangle with sides of length l is inserted into the side wall of a swimming pool as shown in the accompanying figure. (*a*) What is the total force on the window due to the water? (*b*) What force must the wall exert on the window to keep it in place?

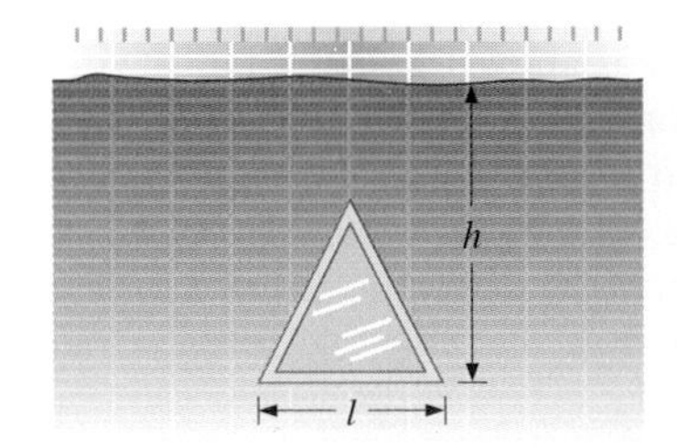

PROBLEM 15-62

15-63. When we expand our lungs, the pressure inside them drops, and air is pushed into them from the outside. If a scuba diver is able to reduce the pressure in his lungs to 70 mmHg below the pressure outside his body, what is the maximum depth d below the surface of a lake at which he can use a snorkel to breathe? (See the accompanying figure.)

PROBLEM 15-63

15-64. The trough seen in the accompanying figure has a semicircular cross section. The trough is filled to the top with water. What is the net force on an end of the trough due to the water?

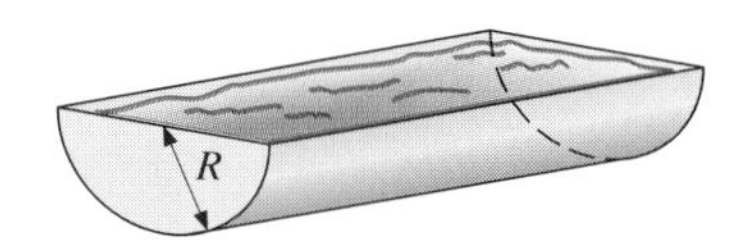

PROBLEM 15-64

15-65. If the temperature of the air does not vary, the ratio of its pressure to its density is almost constant. In this case,

$$\frac{\rho}{\rho_0} = \frac{p}{p_0},$$

where p_0 and ρ_0 are the pressure and density of air, respectively, at sea level. (*a*) Show that with this approximation, the variation of air pressure with altitude y is given by

$$p = p_0 e^{-ay},$$

where $a = g\rho_0/p_0$. Show that if $p_0 = 1.01 \times 10^5$ N/m^2 and $\rho_0 = 1.30$ kg/m^3, then $a = 1.26 \times 10^{-4}$ m^{-1}. (*Note:* With this value of a, the equation is accurate within 5 percent or less for altitudes less than 4 km.) (*b*) Show that if $ay \ll 1$, this equation reduces to the form of Eq. (15-4*b*). (*c*) Compare calculations using these two equations for altitudes of 50, 100, 500, and 1000 m.

15-66. A cubical block of copper floats in mercury. (*a*) What fraction of the block is above the surface of the mercury? (*b*) If water is poured on the mercury until the block is just covered by water, what fraction of the block is in water?

15-67. When in seawater, the submerged depth (called the *draft*) of a tanker is 30 m. What is the draft of the tanker when it is in a freshwater river? Assume that the sides of the tanker are vertical.

15-68. Two rectangular beams with the same dimensions are bonded together as shown in the accompanying figure. One beam is cast from aluminum and the other from steel (use $\rho_{\text{steel}} = 7600$ kg/m^3). The composite beam is suspended horizontally from two cables, and it is submerged in water. If each beam is 2.0 m long and has a cross-sectional area of 0.15 m^2, what are the tensions in the two cables?

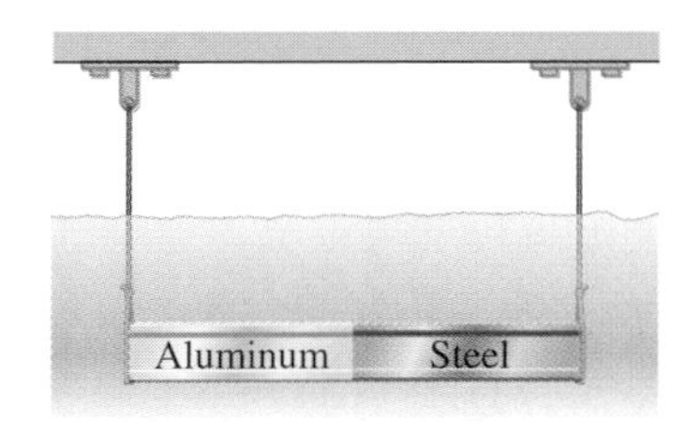

PROBLEM 15-68

15-69. When precise measurements of mass are made with a pan balance (see the accompanying figure), corrections for the buoyancy of air must be made, especially if the density of the object whose mass is being determined is quite different than that of the standard masses. (*a*) Suppose that the densities of the object O being measured, the standard masses S, and the air are ρ_O, ρ_S, and ρ_A, respectively, and that the volumes of the object and the standard masses are V_O and V_S, respectively. Show that the actual mass m_O of the object is given by

$$m_O = m_S + \rho_A(V_O - V_S),$$

where m_S is the total mass of the standards. (*b*) If a piece of wood of density 500 kg/m^3 is balanced by exactly 0.120 kg of a brass standard ($\rho = 8600$ kg/m^3), what is the mass of the piece of wood? (*Note:* V_O can be approximated by $V_O = (0.120 \text{ kg})/(500 \text{ kg/m}^3) = 2.40 \times 10^{-4}$ m^3.)

PROBLEM 15-69

15-70. A small hole of area 0.50 cm^2 is drilled in the bottom of a cylindrical container whose radius is 25 cm. If water is poured into the top of the container at a rate of 700 cm^3/s, what is the height of the water when it stops rising in the container?

15-71. A thin, square plate of area 4.0 cm^2 and mass 250 g is hinged so that it can rotate around a horizontal axis through its attached side. (See the accompanying figure.) If air of density 1.3 kg/m^3 is blown over the plate's top surface, what must be the speed of the air if the equilibrium position of the plate is horizontal?

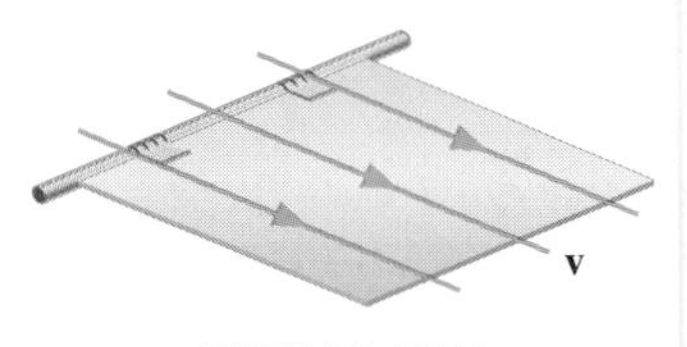

PROBLEM 15-71

PART II

MECHANICAL WAVES

A surfer riding on an ocean wave.

CHAPTER 16 WAVES

PREVIEW

In this chapter we examine some of the general properties of waves. We will focus on waves in strings and springs as well as sound waves. The main topics to be discussed here include the following:

1. **Types of waves.** Different types of waves are identified, and longitudinal and transverse waves are described.
2. **Traveling waves.** The mathematical representation of a traveling wave is considered.
3. **Wave speed.** Equations representing the speed of a wave on a string and the speed of sound in a fluid are derived.
4. **Harmonic waves.** The harmonic wave is a particularly useful waveform whose properties are discussed in detail.
5. **Energy transport in harmonic waves.** We explore the relationship between the power transmitted by a harmonic wave and the amplitude and frequency of the wave.
6. **Circular and spherical waves.** We study two-dimensional (circular) and three-dimensional (spherical) waves.
7. **Plane-wave approximation.** We see how a spherical wave can be approximated by a plane wave at points located far from the source of the wave.

*8. **Sound intensity in decibels.** A logarithmic scale for sound intensity is described.

*9. **Doppler effect.** We describe the effect on the frequency of detected waves due to the relative motion of the source and observer.

If asked to describe what a wave is, you would probably respond on the basis of your observations of water waves—that they oscillate and that they transport energy and momentum. While our notion of what a wave "looks like" or "acts like" is commonly based on those in water, other types of waves are encountered far more often in everyday life. We hear because sound waves fall on our ears; we see because light waves are reflected into our eyes; and we feel heat from the infrared radiation (electromagnetic waves) produced by warm objects.

Although sound, light, and water waves may seem very different, they have many properties that are common to all waves. For example, wave "frequency" doesn't just tell us how often ocean waves crash against a cliff; it also specifies the pitch of a musical note or the color of a light wave. And "amplitude" is not simply a measure of how high those ocean waves are; it also allows us to calculate the loudness of a sound wave and the brightness of a light wave.

There are also fundamental differences between waves. For example, light waves travel through a vacuum, but a medium is required for the propagation of sound and water waves. In addition, some waves vibrate perpendicular and some parallel to their direction of propagation.

In this chapter we will examine some of the general characteristics of waves, using waves on strings and springs along with sound waves as our models. We'll focus on harmonic waves, which are especially important because more complicated waves can be treated as superpositions of these simple waves. Among the wave properties to be considered are wavelength, frequency, and amplitude. We will also investigate the Doppler effect, which is responsible for the change in pitch we hear from a moving sound source.

16-1 Types of Waves

A wave that propagates by disturbing the particles of a medium is called a *mechanical wave*. All mechanical waves are governed by the laws of Newtonian mechanics, because the particles of the medium are so governed. Examples of mechanical waves are sound waves, water waves, and waves on strings and springs. In contrast, *electromagnetic waves* do not require a medium for propagation. We are able to see a star because its light can traverse the vast emptiness of space to reach the earth. Visible light is only one type of electromagnetic wave; others include x-rays, microwaves, and radio waves. The electromagnetic wave and its many interesting and important properties will be discussed in Chap. 35. Elementary particles such as electrons and protons also have wave properties. The behavior of such *matter waves* is determined by the laws of quantum mechanics. This topic will be considered briefly near the end of this textbook.

Although a mechanical wave may travel a long distance, the individual particles of the medium are generally not displaced far. When the end of a string is wiggled once, as shown in Fig. 16-1*a*, a disturbance known as a *wave pulse* is produced, which moves along the string. While the pulse moves horizontally from left to right, the particles of the string are displaced vertically, or perpendicular to the direction in which the pulse is moving. As the pulse travels down the string, its propagation is seen through the sequential vertical displacement of the particles of the string. Because this wave pulse displaces its medium (the string) perpendicular to its direction of travel, it is called a *transverse* wave.

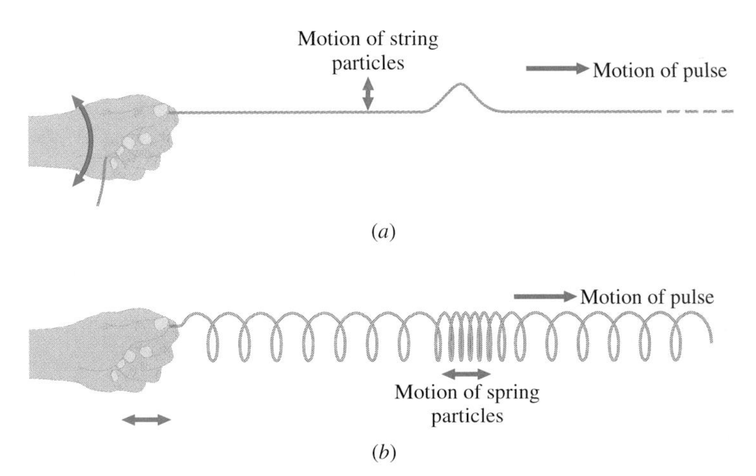

FIGURE 16-1 (*a*) A transverse wave pulse traveling along a string. (*b*) A longitudinal wave pulse traveling along a spring.

In Fig. 16-1*b*, a mechanical wave pulse is shown traveling along a horizontal spring. This pulse is produced by giving the held end a single push and pull and appears as a compression moving down the spring. Since the particles of the spring are displaced parallel to the direction of propagation of the pulse, the wave is called a *longitudinal* wave. Sound is the best-known longitudinal wave. Its vibrations are the longitudinal displacements of the particles of the medium through which the sound is traveling, as illustrated in Fig. 16-2*a*.

A mechanical wave doesn't have to be purely transverse or longitudinal. For example, when a wave moves through water, the particles of the medium move in circles. They are therefore displaced both parallel and perpendicular to the direction in which the wave is moving. (See Fig. 16-2*b*.)

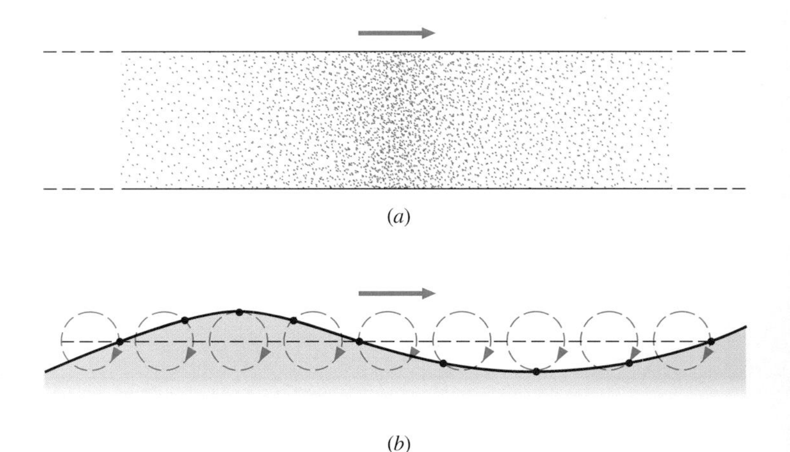

FIGURE 16-2 (*a*) A longitudinal sound pulse traveling in the elastic medium filling a long tube. The concentration of dots is proportional to the density of molecules of the medium. (*b*) Particles of water are displaced both parallel and perpendicular to the wave's direction of propagation.

16-2 Traveling Waves

How is the motion of a wave represented mathematically? We'll answer this question by studying a wave pulse traveling along a string. We will assume that the shape of the pulse does not change with time and that at $t = 0$ it is represented by the function $y = F(x)$. For simplicity, the coordinate system of Fig. 16-3*a* will be used. Then at $t = 0$, the center of the pulse is at the origin. If the pulse moves in the positive x direction at a constant speed v, it is displaced a horizontal distance $x = vt$ after a time t. (See Fig. 16-3*b*.) That is, after a time t, *every point on the pulse* is a distance vt farther along the x axis, as the figure illustrates. Thus $F(x - vt)$ centered at $x = vt$ has exactly the same shape as $F(x)$ centered at $x = 0$. This means that the time-dependent traveling wave pulse can be represented by

$$y(x, t) = F(x - vt). \tag{16-1a}$$

The representation of a pulse traveling in the negative x direction with a speed v is similar. We simply replace v by $-v$ in Eq. (16-1*a*) and obtain

$$y(x, t) = F(x + vt). \tag{16-1b}$$

Notice that the mathematical representation of a pulse, as given by Eqs. (16-1), is a function of two variables: position x and time t. For a fixed time t_0, $y(x, t_0)$ is a function of the single variable x. You can think of $y(x, t_0)$ as representing a "snapshot" of the pulse at the time t_0. An example is a photograph of the ripples on the surface of a lake. On the other hand, the function $y(x_0, t)$ represents the motion of a particle of the medium at x_0 as the pulse passes by.

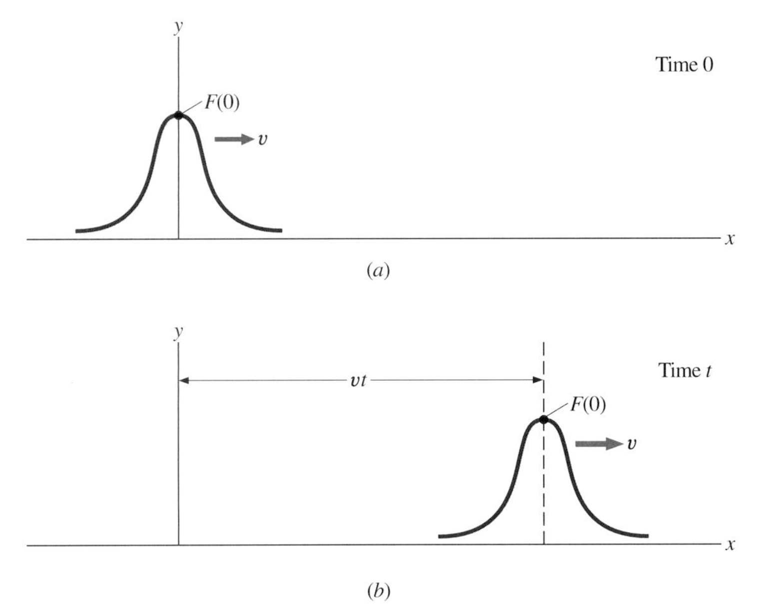

FIGURE 16-3 A pulse traveling in the $+x$ direction with a constant speed v has the form $y(x, t) = F(x - vt)$. The shape of the pulse does not change with time. (*a*) At $t = 0$, the center of the pulse is at the origin. (*b*) At time t, this point has moved to $x = vt$.

EXAMPLE 16-1 Wave Pulse on a String

Of the following functions, which ones represent a wave pulse traveling along a horizontal string? What is the velocity of the pulse in each case? Assume SI units.

(*a*) $y(x, t) = 2/[3 + (x - t)^2]$; (*b*) $y(x, t) = (x - 2)/(t + 4)$; (*c*) $y(x, t) = 0.5e^{-(x+2t)-3}$; (*d*) $y(x, t) = (x - 3t)/4x$.

SOLUTION Only the functions of parts (*a*) and (*c*) have the form $F(x \pm vt)$ required to represent a traveling wave pulse. By inspection, $v = +1.0$ m/s for part (*a*). Since the function of part (*c*) can be written as $(0.5e^{-3})e^{-(x+2t)}$, the pulse it describes is moving with a velocity of -2.0 m/s.

EXAMPLE 16-2 Plotting a Wave Pulse

A wave pulse on a horizontal string is represented by the function

$$y(x, t) = \frac{5.0}{1.0 + (x - 2.0t)^2} \quad \text{(cgs units)}.$$

Plot this function at $t = 0$, 2.5, and 5.0 s.

SOLUTION At the given times, the function representing the wave pulse is

$$y(x, 0) = \frac{5.0}{1.0 + x^2};$$

$$y(x, 2.5\text{ s}) = \frac{5.0}{1.0 + (x - 5.0)^2};$$

$$y(x, 5.0\text{ s}) = \frac{5.0}{1.0 + (x - 10.0)^2}.$$

The maximum of $y(x, 0)$ is 5.0 cm; it is located at $x = 0$. Notice in the plot of Fig. 16-4 that the pulse is also centered at this position. At $t = 2.5$ and 5.0 s, the center of the pulse has moved to $x = 5.0$ and 10.0 cm, respectively. So, in each 2.5-s time interval, the pulse moves 5.0 cm in the positive x direction. Its velocity is therefore $+2.0$ cm/s, a value that is also evident from the given function.

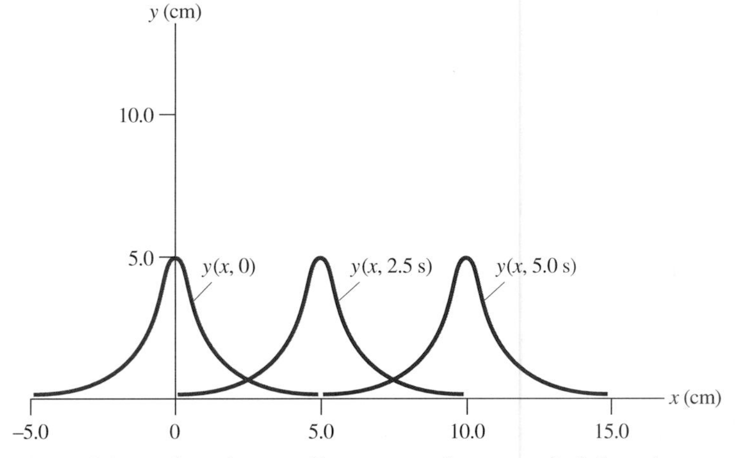

FIGURE 16-4 A plot of a traveling wave pulse at $t = 0$, 2.5, and 5.0 s.

DRILL PROBLEM 16-1

Is the function

$$y(x, t) = 2.0e^{-(3.0x-2.0t)^2/4.0} \quad \text{(SI units)}$$

appropriate for describing a moving pulse? If so, what is the velocity of the pulse?

ANS. Yes; +0.67 m/s.

DRILL PROBLEM 16-2

Plot the function of Drill Prob. 16-1 at $t = 0$, 3.0, and 6.0 s.

16-3 WAVE SPEED

Since a mechanical wave propagates by displacing particles in a medium, its speed must depend on the properties of that medium. For example, sound travels at about 340 m/s in air, but at nearly 1500 m/s in freshwater. (See Table 16-1 for values of the speed of sound in various media.) We now investigate the relationship between wave speed and medium for two important cases: a transverse wave on a stretched string and a longitudinal wave in a fluid.

TABLE 16-1 THE SPEED OF SOUND IN VARIOUS MEDIA

Substance	Temperature (°C)	Speed (m/s)
Gases		
Air	0	331
Air	20	343
Hydrogen	0	1286
Helium	0	972
Oxygen	0	317
Carbon dioxide	0	259
Liquids		
Chloroform	25	1004
Methyl alcohol	25	1143
Water (fresh)	25	1493
Water (sea)	25	1533
Solids (in thin rods)		
Aluminum	–	5100
Copper	–	3560
Glass (pyrex)	–	5170
Lead	–	1322
Steel	–	5200

SPEED OF A TRANSVERSE WAVE PULSE ON A STRING

Suppose a horizontal string of linear mass density μ is stretched to a tension F. If a pulse is moving from left to right with a speed v in the laboratory frame, then in the reference frame of the pulse, the string is moving from right to left with the same speed. (See Fig. 16-5*a*.) The pulse frame is inertial because it is moving at constant velocity relative to the laboratory. Figure 16-5*b* shows a small section of the string in this frame. The top of the string is approximated by a circular arc subtending

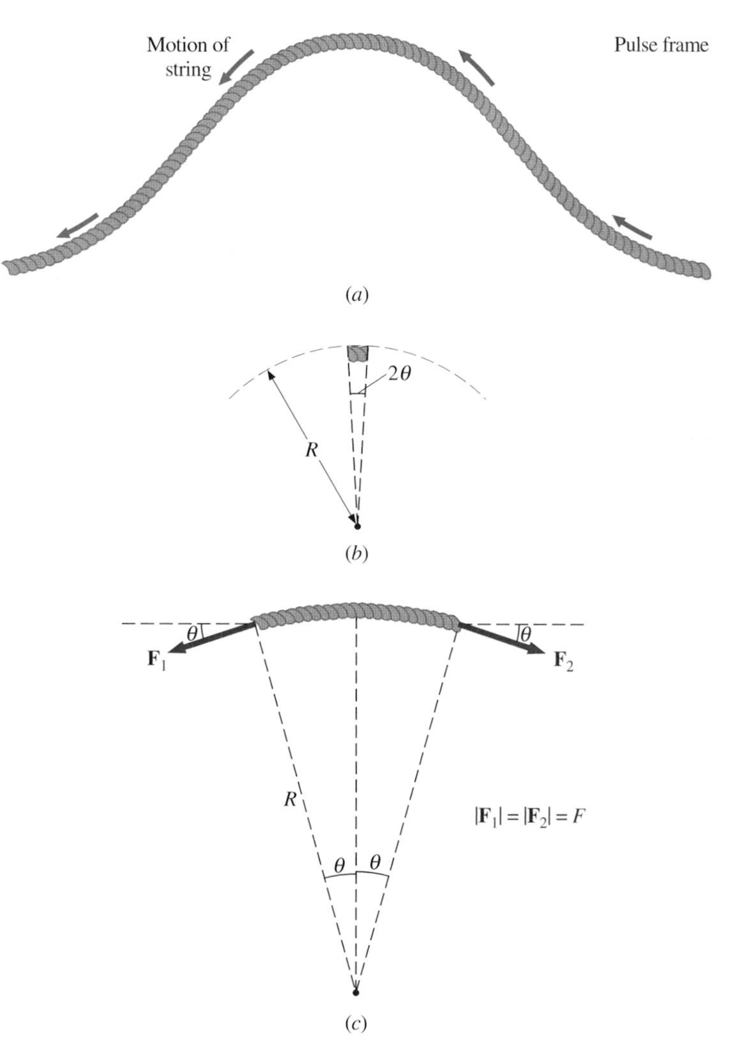

FIGURE 16-5 (*a*) A string is moving right to left with a speed v in the reference frame of the pulse. (*b*) A small section at the top of the pulse can be represented as an arc of a circle. (*c*) The free-body diagram of the section (expanded for clarity).

an angle 2θ at the center of a circle of radius R. The length of the section is therefore $2R\theta$ and its mass is $2\mu R\theta$. The forces acting on it are shown in the free-body diagram of Fig. 16-5*c*. The string is assumed to be light enough that the weight of the section is negligible. Since the pulse moves at constant velocity in the horizontal direction, there is no net force in this direction on the section of string. The tensions at the ends of the section must therefore be directed at the same angle θ with the horizontal.

In the vertical direction, the net force on the section is $2F \sin\theta \approx 2F\theta$ pointing toward the center of the circle. This is the centripetal force on the section, so from

$$\sum F_r = ma_r,$$

$$2F\theta = (2\mu R\theta)\frac{v^2}{R},$$

which yields for the speed of the pulse

$$v = \sqrt{\frac{F}{\mu}}. \tag{16-2}$$

Notice that we never specified the shape of the pulse in this simplified derivation. A small section at the top of the pulse was merely approximated as a circular arc. Hence $v = \sqrt{F/\mu}$ should give the speed of a pulse of any shape on the string. It turns out, however, that *this expression is correct only for small pulses.* If the pulse height were comparable to the length of the string, then neither the approximation $\sin\theta \approx \theta$ nor the assumption that the tension in the string is unaffected by the pulse would be accurate, and our derivation would not be valid.

EXAMPLE 16-3 **WAVE SPEED ON A PIANO WIRE**

A steel piano wire of linear mass density 0.020 kg/m is stretched with a tension of 8.0×10^2 N. (*a*) What is the speed of a wave on the wire? (*b*) What is the tension if the speed is half this value?

SOLUTION (*a*) From $v = \sqrt{F/\mu}$, the wave speed for a tension of 8.0×10^2 N in the piano wire is

$$v = \sqrt{\frac{8.0 \times 10^2 \text{ N}}{0.020 \text{ kg/m}}} = 2.0 \times 10^2 \text{ m/s}.$$

(*b*) Suppose F' is the tension for a wave speed of 1.0×10^2 m/s. Since the wave speed is proportional to the square root of the tension, we have

$$\frac{2.0 \times 10^2 \text{ m/s}}{1.0 \times 10^2 \text{ m/s}} = \sqrt{\frac{8.0 \times 10^2 \text{ N}}{F'}},$$

and

$$F' = 2.0 \times 10^2 \text{ N}.$$

DRILL PROBLEM 16-3

Show that $v = \sqrt{F/\mu}$ is dimensionally correct.

DRILL PROBLEM 16-4

A wave pulse propagates with a speed of 50 m/s along a taut string whose linear mass density is 2.0×10^{-3} kg/m. What is the tension in the string?
ANS. 5.0 N.

SPEED OF A LONGITUDINAL WAVE PULSE IN A FLUID

The long cylinder of Fig. 16-6*a* is filled with a fluid of uniform density ρ. The left end of the cylinder is a movable piston of cross-sectional area A. At $t = 0$, the fluid is in equilibrium with its surroundings and under a pressure p. Suppose we produce a traveling wave pulse by pushing the piston against the fluid with a constant speed v_{pn}. After a time interval Δt, the piston is displaced a distance $v_{pn}\,\Delta t$. (See Fig. 16-6*b*.) If the wave speed (that is, the speed of sound) in the fluid is v, then the "leading edge" OO′ of the compressed region in the fluid has

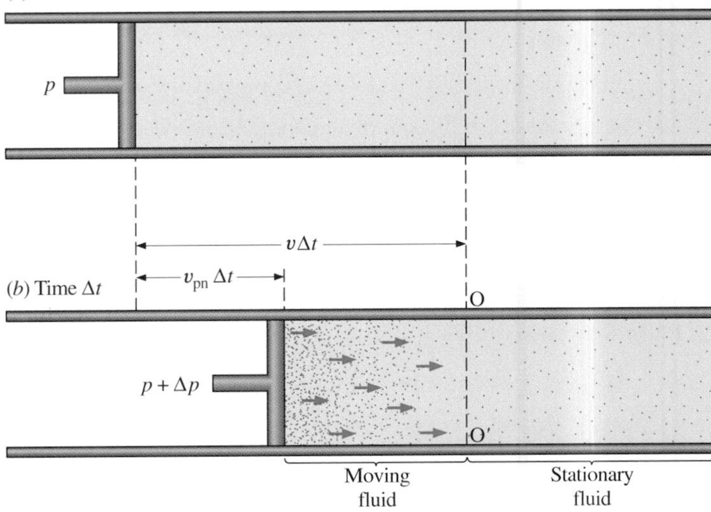

FIGURE 16-6 (*a*) At $t = 0$, the fluid in the tube is in equilibrium at a pressure p. (*b*) When an additional pressure Δp is applied to the piston, a longitudinal pulse is produced that travels from left to right through the fluid.

moved a distance $v\,\Delta t$. The particles to the right of OO′ are as yet undisturbed.

To create this pulse, an additional pressure Δp had to be applied to the piston in order to move it. The net force on the fluid is therefore $F = (\Delta p)A$, and the corresponding impulse on it during the time Δt is

$$J = F\,\Delta t = (\Delta p)A\,\Delta t.$$

As Fig. 16-6*b* shows, the fluid set in motion during the time Δt has an initial volume of $Av\,\Delta t$. Thus during Δt, the mass of fluid set in motion by the piston is $\rho(Av\,\Delta t)$. Assuming that the fluid particles have the same speed v_{pn} as the piston, the momentum of the moving fluid after Δt is $\rho Av\,\Delta t\,v_{pn}$. This is also the *change* in momentum as the fluid was initially at rest. Since the impulse on a system is equal to its change in momentum,

$$J = (\Delta p)A\,\Delta t = \rho Av\,\Delta t\,v_{pn},$$

and the additional pressure applied to move the piston is

$$\Delta p = \rho v v_{pn}. \quad (16\text{-}3)$$

A comparison of Fig. 16-6*a* and *b* shows that the pressure increase Δp has compressed the fluid section by $\Delta V = -Av_{pn}\,\Delta t$ from its initial volume $V = Av\,\Delta t$. From Sec. 11-7, the bulk modulus B of a fluid is the ratio of the pressure change Δp to the resulting fractional change in volume; that is, $B = -(\Delta p)/(\Delta V/V)$, so

$$B = -\frac{\Delta p}{(-Av_{pn}\,\Delta t)/(Av\,\Delta t)} = \Delta p\left(\frac{v}{v_{pn}}\right),$$

and

$$\Delta p = B\left(\frac{v_{pn}}{v}\right).$$

Substituting this last expression into Eq. (16-3), we find

$$B\left(\frac{v_{pn}}{v}\right) = \rho v v_{pn},$$

so

$$v = \sqrt{\frac{B}{\rho}}, \quad (16\text{-}4a)$$

which is the relationship between longitudinal wave speed and the properties of the fluid. Although Eq. (16-4*a*) was derived for a fluid, it also represents the speed of longitudinal waves in bulk solids. However, in calculating wave speeds in solid rods or bars, Young's modulus Y_n* must be used instead of the bulk modulus B. We then have

$$v = \sqrt{\frac{Y_n}{\rho}}. \quad (16\text{-}4b)$$

Notice the similarity between Eqs. (16-2) and (16-4). In each case, the density of the medium is the denominator and a term directly related to the force of interaction among the molecules of the medium is the numerator. *The expressions for the speeds of all mechanical waves follow this general form—the square root of an elastic property divided by an inertial property of the medium.*

EXAMPLE 16-4 **Bulk Modulus and the Speed of Sound**

Sound travels approximately 15 times faster in bulk aluminum than in air at a temperature of 0°C and a pressure of 1.0 atm. The density of aluminum is about 2000 times greater than that of air. Find the ratio of the bulk moduli of the two media.

SOLUTION From $v = \sqrt{B/\rho}$, we have

$$\sqrt{\frac{B_{Al}}{\rho_{Al}}} = 15\sqrt{\frac{B_{air}}{\rho_{air}}},$$

so

$$\sqrt{\frac{B_{Al}}{B_{air}}} = 15\sqrt{\frac{\rho_{Al}}{\rho_{air}}} = 15\sqrt{2000},$$

and the ratio of the bulk moduli of the two media is

$$\frac{B_{Al}}{B_{air}} = 4.5 \times 10^5.$$

This result is a reflection of a well-known fact—it's far more difficult to "squeeze" a solid than a gas.

*As discussed in Sec. 11-5, Young's modulus is the ratio of the normal force applied at the cross section of a rod to the resulting fractional change in the length of the rod.

DRILL PROBLEM 16-5

At 0°C, air has a density of 1.29 kg/m^3, and sound travels through it with a speed of 331 m/s. What is the bulk modulus of air at this temperature?
ANS. 1.41×10^5 N/m^2.

16-4 Harmonic Waves

When the end of the string of Fig. 16-1*a* was given a single wiggle, a traveling wave pulse was formed. Suppose that this end is now displaced up and down continuously so that it is in simple harmonic motion. The wave that is then created on the string is known as a *harmonic wave*. Another common example of a harmonic wave is the sound produced by a tuning fork, which vibrates at essentially a single frequency.

The harmonic wave is especially important, because it serves as a basic building block for complicated waveforms. The Fourier theorem, developed by Jean Baptiste Joseph Fourier (1786–1830), states that *all waveforms can be written as linear combinations of harmonic waves*. This theorem is used extensively in almost every area of physics and engineering.

A transverse harmonic wave on a string that is produced by a source in simple harmonic motion is shown in Fig. 16-7*a*. The shape of the wave is sinusoidal. We assume that at the instant we designate as $t = 0$, the position of the held end of the string is $y = 0$. The harmonic wave at this time can then be represented mathematically by (see Fig. 16-7*b*)

$$y(x) = A \sin \frac{2\pi}{\lambda} x, \quad (16\text{-}5)$$

where A is the **amplitude** of the wave and λ is its **wavelength**. As Fig. 16-7*a* indicates, A is the maximum transverse displacement of a particle on the string, and λ is the distance between any two corresponding points at the same position on successive repeating wave shapes. Points on the string that are separated by an integral number of wavelengths have the same

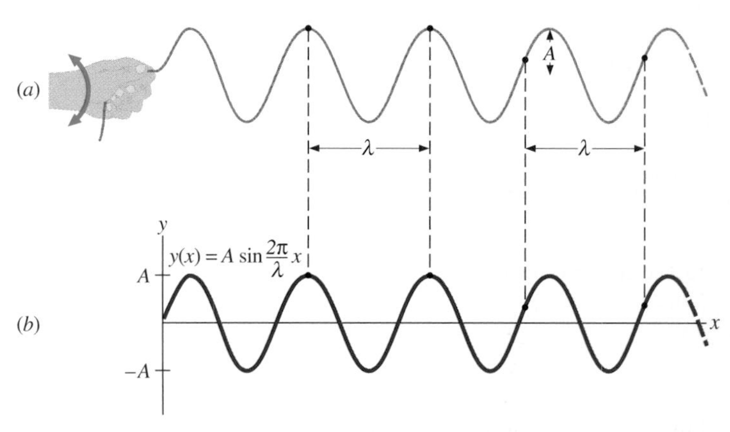

FIGURE 16-7 (*a*) A transverse harmonic wave on a string at the instant designated as $t = 0$. (*b*) The mathematical representation of this harmonic wave.

transverse displacement; that is, $y(x) = y(x + n\lambda)$. *Wavelength therefore describes the spatial periodicity of a harmonic wave.*

Equation (16-5) only represents the traveling harmonic wave at the instant $t = 0$. Since a wave propagating with a speed v must have the form $F(x - vt)$, the time-dependent expression for the harmonic wave is

$$y(x, t) = A \sin \frac{2\pi}{\lambda}(x - vt). \tag{16-6a}$$

The time periodicity or **period** T of a harmonic wave is defined as the *time it takes the wave to travel a distance of one wavelength*. (See Fig. 16-8.) Since the wave travels at constant speed,

$$v = \frac{\lambda}{T}. \tag{16-7a}$$

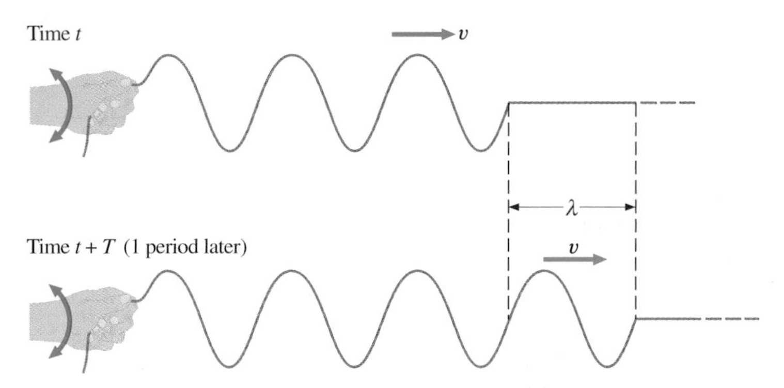

FIGURE 16-8 During a period T, the harmonic wave travels a distance equal to its wavelength.

The **frequency** f of the harmonic wave is *the number of cycles of the wave that cross an imaginary plane perpendicular to the direction of propagation per unit time*. Its SI unit is the hertz (Hz). Since one cycle crosses the plane in a time T, the frequency is the inverse of the period:

$$f = \frac{1}{T}. \tag{16-8}$$

Furthermore, from $v = \lambda/T$,

$$v = f\lambda. \tag{16-7b}$$

By substituting $T = \lambda/v$ into Eq. (16-6a) we obtain an alternative representation of the harmonic wave:

$$y(x, t) = A \sin 2\pi\left(\frac{x}{\lambda} - \frac{t}{T}\right). \tag{16-6b}$$

This expression highlights both the space and time periodicities of the wave. If x changes by $n\lambda$ (n an integer) or if t changes by nT, then the argument of the sine function changes by $2\pi n$ and $y(x, t)$ remains the same.

There's yet a third (and more compact) way of describing the harmonic wave. This uses the **wave number** k and the **angular frequency** ω, where

$$k = \frac{2\pi}{\lambda} \tag{16-9}$$

and

$$\omega = \frac{2\pi}{T} = 2\pi f. \tag{16-10}$$

The dimensions of k and ω are (1/length) and (1/time), respectively, and ω is expressed in radians per second (rad/s). How the wave speed is related to these two quantities is determined by dividing ω by k:

$$\frac{\omega}{k} = \frac{2\pi f}{2\pi/\lambda} = f\lambda = v. \tag{16-11}$$

Substitution of k and ω into Eq. (16-6b) then gives

$$y(x, t) = A \sin (kx - \omega t). \tag{16-6c}$$

We still don't have the most general expression for $y(x, t)$, because our equations only give $y = 0$ when both $x = 0$ and $t = 0$. As Fig. 16-9 illustrates, *any* initial value of y at $x = 0$ is

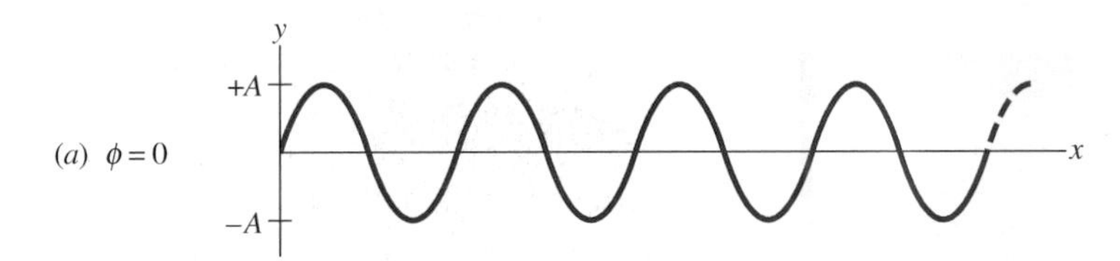

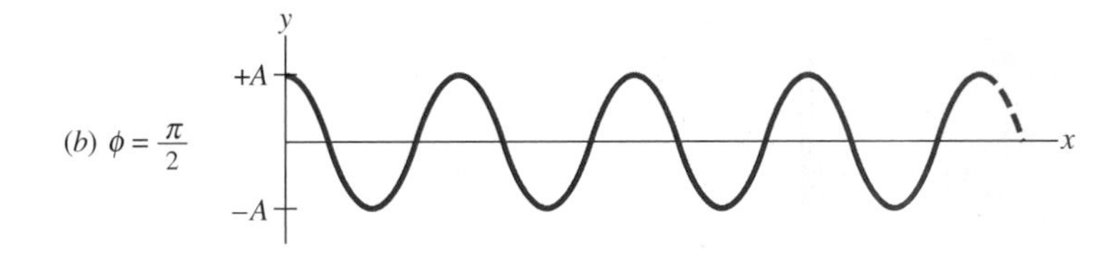

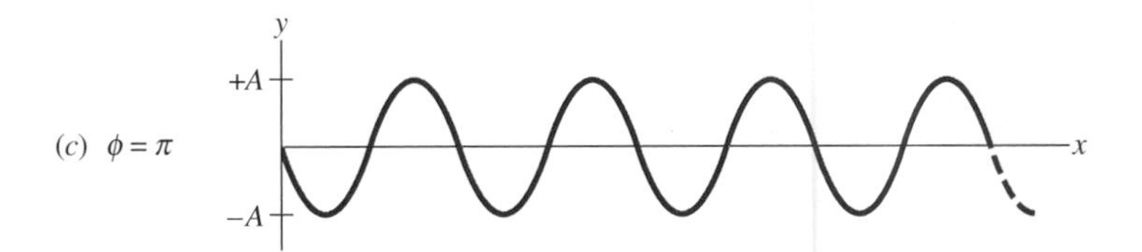

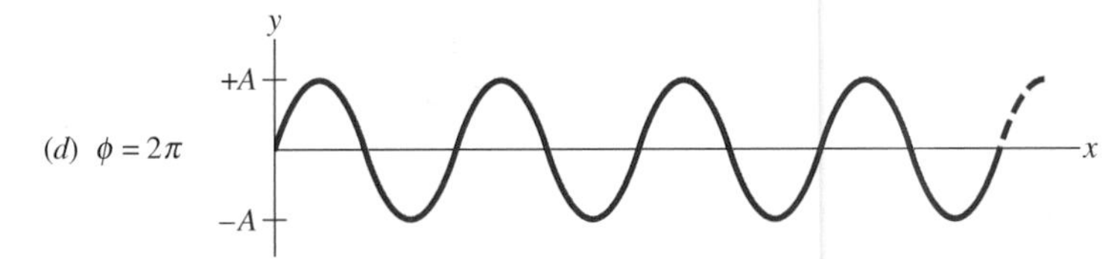

FIGURE 16-9 The harmonic wave $y(x, t) = A \sin (kx - \omega t + \phi)$ at $t = 0$ for different values of the phase constant ϕ.

allowed if we simply add a phase constant ϕ to the argument of the sine function. We then have

$$y(x, t) = A \sin (kx - \omega t + \phi), \qquad (16\text{-}12)$$

where $(kx - \omega t + \phi)$ is the phase of the wave. Now $y(0, 0) = A \sin \phi$, so with the appropriate choice of ϕ, we can set $y(0, 0)$ to any value between $-A$ and A. When $\phi = \pi/2$ rad, $y(x, t) = A \sin (kx - \omega t + \pi/2) = A \cos (kx - \omega t)$. Since it is now a cosine function, $y(x, t)$ has a displacement A at $x = 0$ and $t = 0$, as illustrated in Fig. 16-9*b*. We can represent any harmonic wave by either a sine or a cosine function, as long as we use the appropriate phase constant.

The spatial dependence of a transverse harmonic wave at a single point in time was shown in Fig. 16-7. Now let's examine the wave from a different perspective by looking at the motion of a single particle at x_0 of the medium as the wave passes by. (See Fig. 16-10*a* through *e*.) From Eq. (16-6*c*), the displacement of the particle is

$$y(x_0, t) = A \sin (kx_0 - \omega t) = A \sin (\omega t - kx_0 + \pi)$$
$$= A \sin (\omega t + \phi'),$$

where the phase constant ϕ' is $(-kx_0 + \pi)$. Hence at a fixed position, y varies with time just like a body at the end of a spring (see Fig. 16-10*f*); that is, the motion of the medium at x_0 is simple harmonic. This gives us another interpretation of the period T—it is also *the time that a given particle of the medium takes to oscillate through one complete cycle.*

The mathematical description of a longitudinal harmonic wave is essentially the same as that for a transverse wave. However, the particles of the medium are displaced along the direction of propagation of the wave (in our case, the x axis), so we recast Eq. (16-12) in the form

$$s(x, t) = s_0 \sin (kx - \omega t + \phi). \qquad (16\text{-}13a)$$

Here, $s(x, t)$ represents the longitudinal displacement at a time t of the particle whose position is x when there is no disturbance passing through the medium.

A longitudinal harmonic sound wave can be produced by making a piston vibrate in simple harmonic motion against an air column in a long narrow tube. (See Fig. 16-11*a*.) When the piston is pushed in, the nearby air is compressed. As time progresses, this compression is transmitted down the air column by means of collisions between adjacent layers of air molecules. When the piston is pulled outward, the nearby air expands. This expansion is also transmitted down the tube through intermolecular collisions. With the piston in continuous simple harmonic motion, alternate compressions and expansions of the air next to the piston result in the transmission of a harmonic displacement wave down the tube, as illustrated in Fig. 16-11*b*.

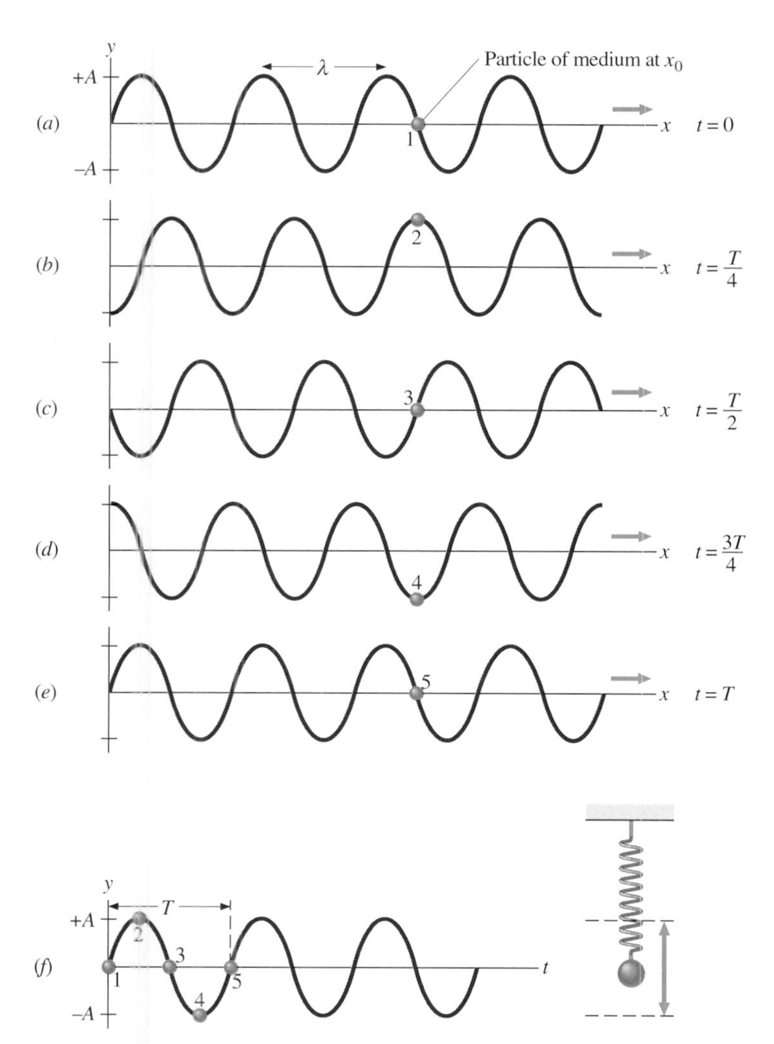

FIGURE 16-10 (*a*) through (*e*). The displacements at various times of a particle of a medium through which a harmonic wave is traveling. (*f*) A plot of the particle's displacement vs. time shows that its motion is simple harmonic with the same period T as the wave. The particle's motion is analogous to that of a mass at the end of a spring.

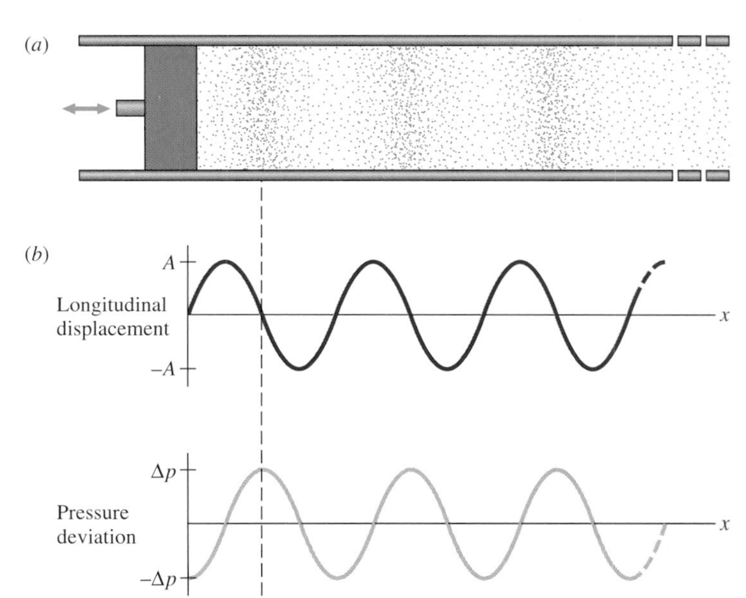

FIGURE 16-11 (*a*) While the piston vibrates in simple harmonic motion, alternate compressions and expansions are transmitted through the air in the tube. (*b*) The result is a harmonic displacement wave traveling through the air. The curves represent the displacement and pressure variation of the air at one particular instant as a function of position along the tube.

The frequency of the displacement wave is the same as that of the oscillations of the piston; and if v is the speed of sound in the air column, the wave number of the harmonic wave is $k = \omega/v$.

In Fig. 16-12*a*, the harmonic displacement wave $s(x, t)$ is plotted as a function of x for one particular time t. Let's first consider the point D_1, where the displacement of the medium is zero. As Fig. 16-12*b* shows, air molecules on either side of D_1 are moving away from this point, because $s(x, t)$ is negative to the left of it and positive to its right. Since air molecules are leaving D_1, the density of air at this point is a minimum—and since the pressure is proportional to the density, the air pressure there is also a minimum. Conversely, at the point of zero displacement D_3, which is a distance $\lambda/2$ away from D_1, air molecules on either side of the point are moving toward it. Thus the density and pressure there are a maximum. Finally, at D_2 and its surrounding region, $s(x, t)$ is positive, so all air molecules in this region are moving to the right, as illustrated in Fig. 16-12*b*. Therefore the density and pressure at D_2 at the instant represented by the figures are the same as their ambient values.

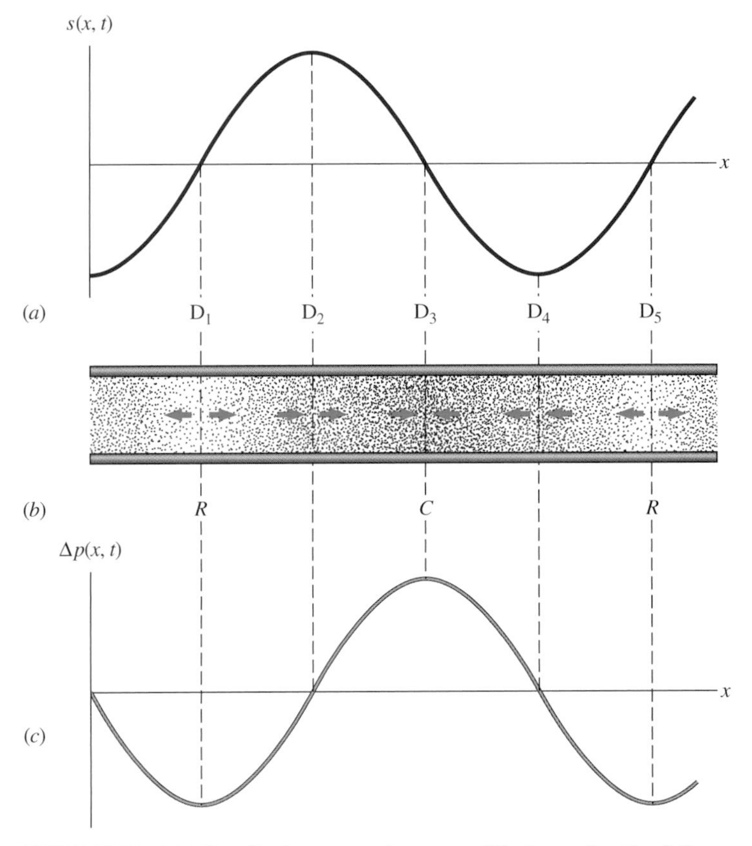

FIGURE 16-12 (*a*) The displacement from equilibrium $s(x, t)$ of the air molecules at one instant of time t in a tube through which a harmonic sound wave is traveling. (*b*) The distribution and motion of the air molecules at this same time t near each of the five points $D_1, D_2, \ldots, D_5$. The arrows represent the direction of motion of the molecules at this time. (*c*) The change in air pressure $\Delta p(x, t)$ from its ambient value along the tube at the time t. Notice that the pressure wave is 90° out of phase with the displacement wave.

We can now construct the pressure (or density) wave corresponding to the displacement wave. The pressure wave $\Delta p(x, t)$, which represents the variation in pressure from its ambient value, is shown in Fig. 16-12*c*. It is out of phase with the displacement wave by $\pi/2$ rad, because $\Delta p(x, t)$ is a minimum or maximum where $s(x, t)$ is zero and $\Delta p(x, t)$ is zero where $s(x, t)$ is a minimum or maximum. With $s(x, t)$ given by Eq. (16-13*a*), the deviation in pressure from its ambient value is

$$\Delta p(x, t) = -\Delta p_0 \cos(kx - \omega t + \phi), \tag{16-13b}$$

where Δp_0 is the amplitude of the pressure wave. This equation is derived in Supplement 16-1 in a way that also gives the relationship between the amplitudes of Eqs. (16-13*a*) and (16-13*b*). That relationship is

$$\Delta p_0 = \rho k v^2 s_0, \tag{16-14}$$

where ρ is the density of the undisturbed air.

The regions of maximum pressure ($\Delta p = \Delta p_0$) in the air column are called *condensations*, and the regions of minimum pressure ($\Delta p = -\Delta p_0$) are called *rarefactions*. These regions are identified by C and R, respectively, in the figure. The distance between successive condensations (or rarefactions) is the wavelength of the sound wave.

The pressure fluctuations associated with sound waves are usually very small. For example, normal conversation has a pressure amplitude of about 0.03 N/m^2, which is much less than the ambient atmospheric pressure of approximately 1.0×10^5 N/m^2.

EXAMPLE 16-5 A Harmonic Wave

A harmonic wave on a string is given by

$$y(x, t) = (1.0 \times 10^{-3}) \sin(20x - 600t) \qquad \text{(SI units)}.$$

From this expression, deduce the amplitude, wavelength, period, frequency, and speed of the wave.

SOLUTION We can obtain these properties of the wave by comparing the given expression with Eq. (16-6*c*). By inspection, $A = 1.0 \times 10^{-3}$ m, $k = 20$ m^{-1}, and $\omega = 600$ rad/s. Since $k = 2\pi/\lambda$, we find for the wavelength

$$\lambda = \frac{2\pi}{k} = \frac{2\pi}{20\ \text{m}^{-1}} = 0.31\ \text{m}.$$

Also, $\omega = 2\pi/T$ and $f = 1/T$, so

$$T = \frac{2\pi}{\omega} = 1.05 \times 10^{-2}\ \text{s} \qquad \text{and} \qquad f = \frac{1}{T} = 95\ \text{Hz}.$$

Finally, from $v = f\lambda$, the speed of the wave is

$$v = (95\ \text{Hz})(0.31\ \text{m}) = 30\ \text{m/s}.$$

EXAMPLE 16-6 **ANOTHER HARMONIC WAVE**

A harmonic wave on a string has an amplitude of 2.0 cm, a wavelength of 25 cm, and a frequency of 50 Hz. If the wave is moving in the negative x direction, what is the transverse displacement of the string as a function of x and t? Assume that at $t = 0$, the string particle at the origin is not displaced and has a transverse velocity in the positive y direction.

SOLUTION The general expression for a harmonic wave propagating in the negative x direction is

$$y(x, t) = A \sin(kx + \omega t + \phi) = A \sin\left(\frac{2\pi}{\lambda}x + 2\pi ft + \phi\right). \qquad (i)$$

Using SI units for this wave, we have $A = 0.020$ m, $\lambda = 0.25$ m, and $f = 50$ Hz. Substituting these values into $y(x, t)$, we find

$$y(x, t) = 0.020 \sin[2\pi(4.0x + 50t) + \phi]. \qquad (ii)$$

Since

$$y(0, 0) = 0.020 \sin \phi = 0,$$

the value of the phase constant ϕ is either 0 or π rad. To determine which one is correct, we use the fact that the particle at the origin is moving in the positive y direction at $t = 0$. The transverse velocity of the string particle at the position x and time t may be found by simply taking the partial* time derivative of Eq. (ii):

$$v(x, t) = \frac{\partial y}{\partial t} = 2.0\pi \cos[2\pi(4.0x + 50t) + \phi]. \qquad (iii)$$

Note that $v(x, t)$ is *not* the velocity of the wave on the string. It is the velocity in the y direction at a time t of the string particle located at the position x. For $x = 0$ and $t = 0$, this expression becomes

$$v(0, 0) = 2.0\pi \cos \phi,$$

so $v(0, 0) = 2.0\pi$ m/s for $\phi = 0$ and $v(0, 0) = -2.0\pi$ m/s for $\phi = \pi$ rad. Now we are given that $v(0, 0) > 0$; thus $\phi = 0$, and the appropriate expression for the wave is (in SI units)

$$y(x, t) = 0.020 \sin[2\pi(4.0x + 50t)].$$

EXAMPLE 16-7 **A LONGITUDINAL WAVE ON A SPRING**

As a harmonic wave moves down a spring such as a Slinky, the motion of the coils is analogous to the motion of the molecules in an air-filled tube as a harmonic sound wave moves through the tube. (See Fig. 16-11a.) Figure 16-13 shows a longitudinal harmonic wave traveling down a long Slinky. In this case, the condensations C are at the points where the coils are closest together and the rarefactions R occur where the coils are farthest apart. As with the air molecules, the displacement of the coils of the Slinky is zero at any condensation or rarefaction. The frequency of the wave is 30 Hz, and the distance between adjacent condensations and rarefactions is 0.10 m. What are (a) the wavelength, (b) the wave number, (c) the angular frequency, and (d) the speed of the wave?

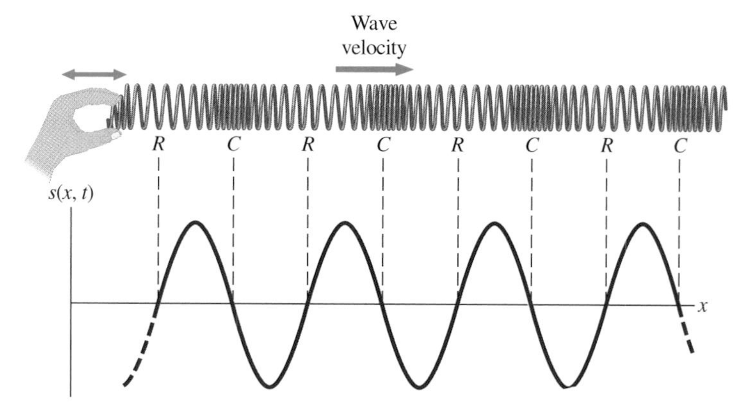

FIGURE 16-13 A harmonic displacement wave travels down a long Slinky. The longitudinal displacement of the coils at any instant t is zero at the condensations C and the rarefactions R.

SOLUTION (a) The wavelength is the distance between successive condensations (or rarefactions) of the spring, which is twice the distance between a condensation and the adjacent rarefaction. Hence $\lambda = 0.20$ m.

(b) The wave number is related to the wavelength by $k = 2\pi/\lambda$, so

$$k = \frac{2\pi}{0.20 \text{ m}} = 31 \text{ m}^{-1}.$$

(c) The angular frequency ω is

$$\omega = 2\pi f = 2\pi(30 \text{ Hz}) = 190 \text{ rad/s}.$$

(d) From $v = f\lambda$, we find for the speed of the wave

$$v = (30 \text{ Hz})(0.20 \text{ m}) = 6.0 \text{ m/s}.$$

DRILL PROBLEM 16-6

The frequency of the vibrating source of Example 16-7 is increased to 60 Hz. (a) Is the speed of the wave affected? Determine the (b) wavelength, (c) period, (d) wave number, and (e) angular frequency of the wave.

ANS. (a) No; (b) 0.10 m; (c) 0.017 s; (d) 63 m^{-1}; (e) 380 rad/s.

DRILL PROBLEM 16-7

Write an expression for the displacement wave described in Example 16-7. Assume that the amplitude is 2.0×10^{-3} m and that at $t = 0$ and $x = 0$, the displacement of the Slinky is 2.0×10^{-3} m.

ANS. $s(x, t) = (2.0 \times 10^{-3}) \cos(31x - 190t)$ (SI units).

*If you haven't yet done so, you will soon learn in calculus about partial derivatives such as $\partial y(x, t)/\partial x$ and $\partial y(x, t)/\partial t$. For our purposes, you just need to know that $\partial y(x, t)/\partial t$ is calculated by assuming x is constant and then differentiating in the normal manner with respect to t. Examples: $\partial(x^2 t)/\partial t = x^2$ and $\partial(x^2 t)/\partial x = 2xt$.

16-5 Energy Transport in Harmonic Waves

Mechanical harmonic waves transport energy by setting the particles of the medium into oscillatory motion. To determine how this energy transport is related to the properties of a wave, let's first consider the harmonic wave $y(x, t) = A \sin(kx - \omega t)$ propagating along a string of linear mass density μ. Figure 16-14 shows the string at two instants separated in time by the period T of the wave. As indicated in the figure, the vibrating source produces one additional complete wave during this time interval.

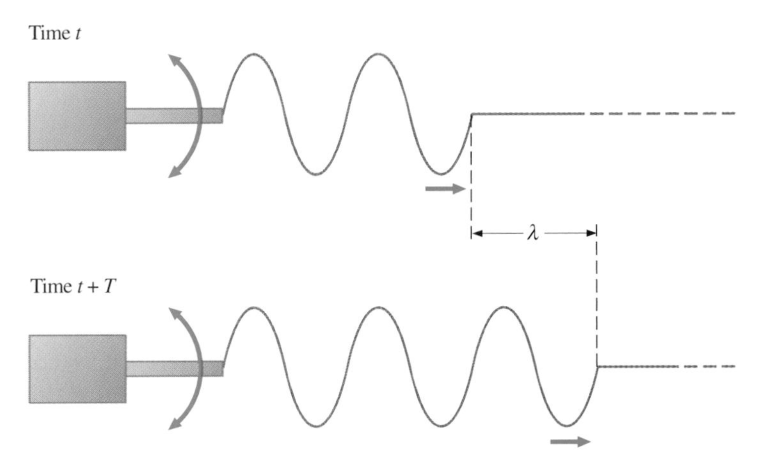

FIGURE 16-14 In the time T, one additional complete cycle is produced in the string. The energy added by the source is the energy of oscillation contained in a length λ of the string.

Each particle on the string is in simple harmonic motion with an amplitude A. From Chap. 14, the maximum speed of the particle is ωA and the mechanical energy dE of a section of string of length dx and mass $dm = \mu\, dx$ is

$$dE = \tfrac{1}{2}dm(\omega A)^2 = \tfrac{1}{2}(\mu\, dx)\omega^2 A^2 = \tfrac{1}{2}\mu\omega^2 A^2\, dx.$$

The energy E added to the string by the source during one oscillation is the energy contained in a length of string equal to one wavelength. It is calculated by integrating dE over one wavelength of the string:

$$E = \int_0^\lambda \frac{1}{2}\,\mu\omega^2 A^2\, dx = \frac{1}{2}\,\mu\lambda\omega^2 A^2.$$

By definition, the average power $\bar{P}$ of the wave is the average rate at which energy is transported past an arbitrary point on the string. This is equal to the energy we just calculated divided by the period T:

$$\bar{P} = \frac{E}{T} = \frac{1}{2}\,\mu v\omega^2 A^2 \qquad (16\text{-}15a)$$

where v, the speed of the wave, has been substituted for λ/T. Recall that, in SI units, power is expressed in watts (W).

A similar argument gives the energy transported by a harmonic sound wave. As the piston of Fig. 16-11*a* oscillates at an angular frequency ω, it transmits energy to the air in the cylinder, and a longitudinal harmonic wave, also of angular frequency ω, propagates through the air. Analogously, the mechanical energy dE of a section of air of mass dm is

$$dE = \tfrac{1}{2}(dm)(\omega s_0)^2,$$

where s_0 is the displacement amplitude of the harmonic wave. Assuming a tube of cross-sectional area $\mathcal{A}$ and air of density ρ, the mass of a section of air of length dx is $dm = \rho\mathcal{A}\, dx$, so

$$dE = \tfrac{1}{2}\rho\mathcal{A}(\omega s_0)^2\, dx;$$

and the total energy in a section of length λ is

$$E = \int_0^\lambda dE = \frac{1}{2}\,\rho\mathcal{A}(\omega s_0)^2\lambda.$$

As with the wave on a string, if we divide by the period T and use $v = \lambda/T$, we find that the average power $\bar{P}$ transmitted through the air by the wave is

$$\bar{P} = \tfrac{1}{2}\rho\mathcal{A}v\omega^2 s_0^2. \qquad (16\text{-}15b)$$

With the help of Eq. (16-14) and $v = \omega/k$, this is given in terms of the pressure amplitude Δp_0 by

$$\bar{P} = \frac{\mathcal{A}\,\Delta p_0^2}{2\rho v}. \qquad (16\text{-}15c)$$

Although obtained for two specific cases, the general form of Eqs. (16-15*a*) and (16-15*b*) holds for any harmonic mechanical wave, be it longitudinal or transverse, propagating in one, two, or three dimensions. That is, *the average power of a harmonic mechanical wave is proportional to the product of the square of the displacement amplitude and the square of the frequency.*

The glass absorbs so much energy from a sound wave that it shatters.

DRILL PROBLEM 16-8

A vibrator is used to produce a harmonic wave on a string. How is the power of the wave affected by (*a*) doubling the frequency of the vibrator? (*b*) halving the amplitude of the oscillations of the vibrator arm? (*c*) doing parts (*a*) and (*b*) simultaneously?
ANS. (*a*) Four times greater; (*b*) four times smaller; (*c*) unchanged.

16-6 Circular and Spherical Waves

Most waves do not travel in a one-dimensional medium such as a string. If a pebble is dropped into the still water of a pond, circular ripples will emanate from the point where the pebble broke the surface. This is an example of a two-dimensional wave since the ripples propagate in the plane of the water surface. Three-dimensional waves also emanate from point sources. For example, a small lightbulb in the middle of a room illuminates its surroundings because its spherical light waves spread radially in all directions.

A circular wave.

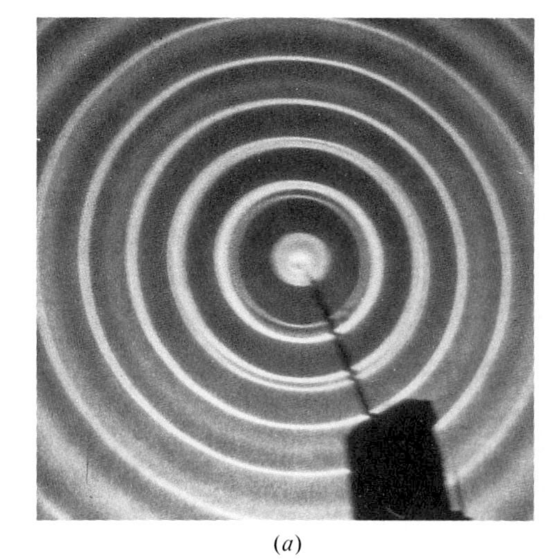

(a)

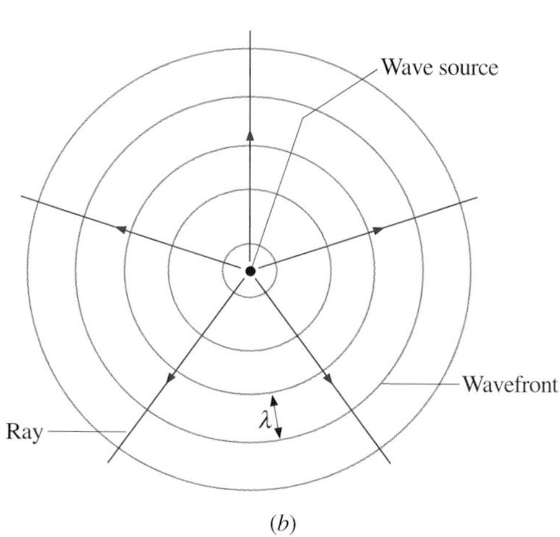

(b)

FIGURE 16-15 (a) Circular waves on a water surface and (b) their schematic representation.

Figure 16-15 shows a photograph and a corresponding schematic representation of circular waves on a water surface. The point where the waves originate is called the *wave source*, and the concentric circles representing the waves are called *wavefronts. All points on a wavefront have the same phase, and the wavefronts are usually drawn with a spacing of* λ. The radial lines that are normal to the wavefront are called *rays*. They are imaginary lines that indicate the direction of propagation of a *circular wave*.

A three-dimensional wave that radiates isotropically from a point source is called a *spherical wave*. In this case, the wavefronts are spherical surfaces centered at the source, and the rays are radial lines originating from the source. We usually describe energy transport by a spherical wave in terms of **intensity** rather than power. By definition, *the intensity I is the average power transmitted by the wave across a unit area normal to the direction of propagation of the wave.* Intensity represents the brightness of a light wave and the loudness of a sound wave. In the SI system, I is expressed in watts per square meter (W/m^2).

Suppose that the average power output of a source is $\bar{P}$. If no energy is dissipated in the medium, the power crossing any imaginary surface enclosing the source must be the same. We'll assume that the radiation is isotropic; that is, the intensity depends only on the distance r from the source. Then for spherical surfaces of radii r_1 and r_2 (see Fig. 16-16),

$$\bar{P} = I(r_1)4\pi r_1^2 = I(r_2)4\pi r_2^2,$$

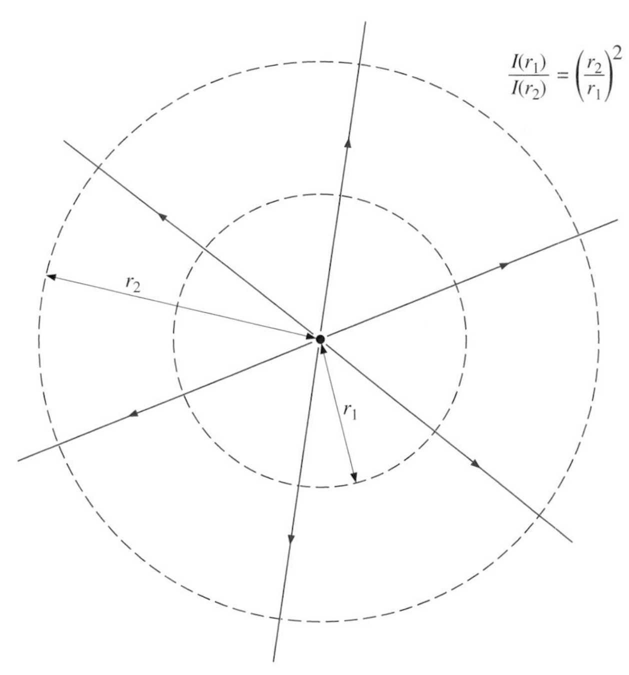

FIGURE 16-16 Isotropic radiation from a point source. If no energy is absorbed by the medium, the intensity of the radiation varies as $1/r^2$.

and

$$\frac{I(r_1)}{I(r_2)} = \frac{r_2^2}{r_1^2}.$$

Therefore the intensity decreases with the square of the distance from the source; that is,

$$I(r) \propto \frac{1}{r^2}. \qquad (16\text{-}16)$$

For this reason, light is dimmer and sound is not as loud the farther you recede from their respective point sources.

Like power, the intensity of a wave is proportional to the square of the amplitude. Thus from Eq. (16-16), the amplitude of a spherical wave is inversely proportional to the distance r from the source. A spherical harmonic wave emanating from a point source can therefore be written as

$$y(r, t) = \frac{A}{r} \sin (kr - \omega t), \qquad (16\text{-}17)$$

where A is a constant.

EXAMPLE 16-8 Power from the Sun and Alpha Centauri Compared

The sun produces energy at the rate of about 4.0×10^{26} W, and it is approximately 1.5×10^8 km from the earth. The next nearest star, Alpha Centauri, is about 4 light-years away. (One light-year is the distance traveled in one year by light moving through a vacuum at a speed of 3.0×10^5 km/s.) The earth has a diameter of 1.3×10^4 km. Calculate the power the earth receives from (*a*) the sun and (*b*) Alpha Centauri. Assume that the rate of energy production by the two stars is the same.

SOLUTION Imagine a spherical object of diameter D located a distance R from a small light source. (See Fig. 16-17.) At this distance, the power from the source is distributed uniformly over a sphere of surface area $4\pi R^2$. The object, however, only intercepts the power that is incident on its cross-sectional area $\pi(D/2)^2$. Therefore the fraction of the power produced by the source that is received by the object is given by

$$(\pi D^2/4)/(4\pi R^2) = (D/4R)^2.$$

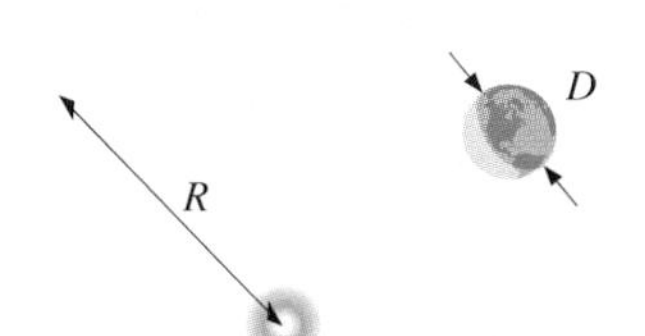

FIGURE 16-17 A spherical object of diameter D located a distance R from a light source.

(*a*) With the sun as the source and the earth as the object, $R = 1.5 \times 10^8$ km and $D = 1.3 \times 10^4$ km, so the solar power received by the earth is

$$(4.0 \times 10^{26}\ \text{W}) \times \left[\frac{1.3 \times 10^4\ \text{km}}{4(1.5 \times 10^8\ \text{km})}\right]^2 = 1.9 \times 10^{17}\ \text{W}.$$

(*b*) With Alpha Centauri as the source, $R = 4$ light-years and D remains at 1.3×10^4 km. The distance corresponding to 1 light-year is the product of the number of seconds in 1 year and the speed of light; this is 9.5×10^{12} km. The power received from the star is therefore

$$(4.0 \times 10^{26}\ \text{W}) \times \left[\frac{1.3 \times 10^4\ \text{km}}{4(4 \times 9.5 \times 10^{12}\ \text{km})}\right]^2 = 2.9 \times 10^6\ \text{W},$$

which, as expected, is negligible compared with what we get from the sun.

DRILL PROBLEM 16-9

A small lightbulb emits 10 W of radiant power. What is the intensity of the light at a distance of 20 m from the bulb? How does the intensity change if this distance is halved? doubled?
ANS. 2.0×10^{-3} W/m^2; 8.0×10^{-3} W/m^2; 5.0×10^{-4} W/m^2.

16-7 Plane-Wave Approximation

Far from a source of spherical waves, small sections on the surfaces of the spherical wavefronts have little curvature and can be approximated very well by parallel planes. (See Fig. 16-18*a*.) For example, the stars are so far away that the wavefronts of light we receive from them are effectively parallel. A wave with such wavefronts is called a *plane wave*. As shown in Fig. 16-18*b*, the rays of a plane wave are parallel and denote the single direction in which the wave is traveling.

Plane waves from the sun cause the shadows to have the same size and orientation.

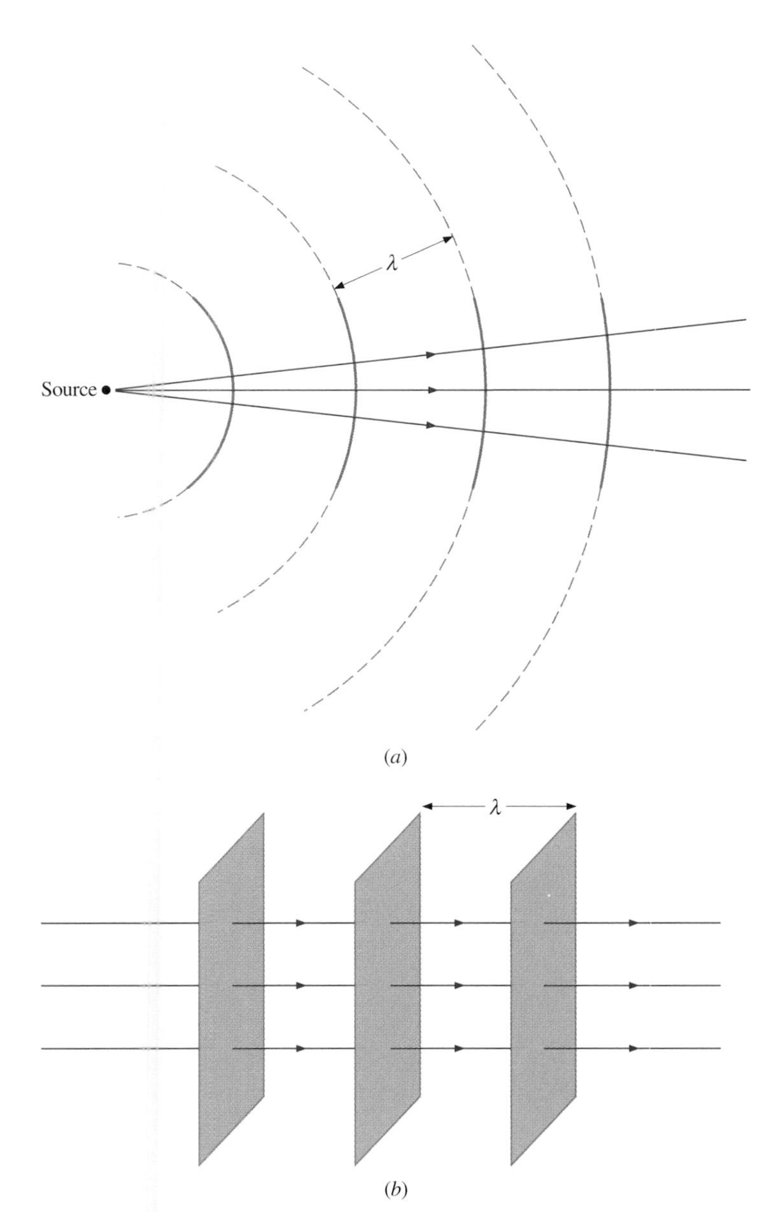

FIGURE 16-18 (*a*) A section of a spherical wavefront has less curvature when it is farther away from the source. This leads to the plane-wave approximation. (*b*) Wavefronts and rays of a plane wave.

EXAMPLE 16-9 Example of the Plane-Wave Approximation

The star Alpha Centauri and the earth at points A and B are represented in the sketch of Fig. 16-19*a*. The distance AB is the diameter of the earth's orbit. Determine the angle θ in the figure.

SOLUTION If the light that reaches the earth from Alpha Centauri were truly composed of plane waves, as depicted in Fig. 16-19*b*, the rays would be parallel and θ would be exactly 0°. Of course, the star actually produces spherical waves, so to check the validity of the plane-wave approximation, we have to determine how small θ really is. The diameter of the earth's orbit is AB = 3.0×10^8 km, and the distance of the star from the earth-sun system is 4 light-years = 3.8×10^{13} km. The angle θ is then

$$\theta = \frac{3.0 \times 10^8 \text{ km}}{3.8 \times 10^{13} \text{ km}} = 7.9 \times 10^{-6} \text{ rad} = 4.5 \times 10^{-4} \text{ degrees},$$

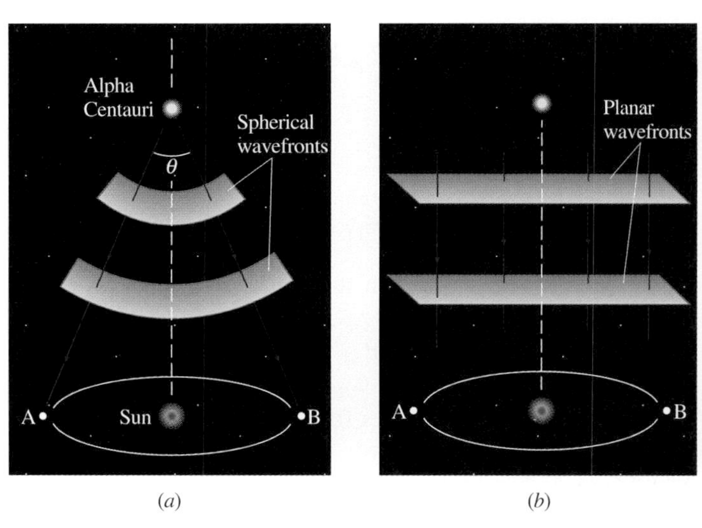

FIGURE 16-19 (*a*) The sun, Alpha Centauri, and the earth in two different positions. (*b*) If plane waves reached the earth, θ would be zero.

which is indeed very close to 0°! What does this tell us about the relative positions of stars as seen from the earth at different times of the year?

DRILL PROBLEM 16-10

Two people on opposite sides of the earth are viewing the sun. (See Fig. 16-20.) Assuming that the sun is a point source of light, calculate the angle θ between the rays reaching the observers. Is it valid to say that light rays that reach any two points on the earth are approximately parallel? Take the diameter of the earth to be 1.3×10^7 m.
ANS. 8.7×10^{-5} rad; yes.

FIGURE 16-20 Two observers viewing the sun from opposite sides of the earth. The sun is assumed to be a point source.

*16-8 Sound Intensity in Decibels

The physiological sensation of loudness is closely related to the intensity of the wave producing the sound. At a frequency of 1 kHz, people are able to detect sounds with intensities as low as 10^{-12} W/m^2. At the other extreme, an intensity of 1 W/m^2 can cause pain, and prolonged exposure to sound at this level will damage a person's ears. Because the range in intensities over which people hear is so large, it is convenient to use a logarithmic scale to specify intensities. This scale is defined as follows:

If the intensity of sound in watts per square meter is I, then the **intensity level** β in decibels (dB) is given by

$$\beta = 10 \log \frac{I}{I_0}, \tag{16-18}$$

where the base of the logarithm is 10, and $I_0 = 10^{-12}$ W/m^2 (roughly the minimum intensity that can be heard).

On the decibel scale, the pain threshold of 1 W/m^2 is then

$$\beta = 10 \log \frac{1}{10^{-12}} = 120 \text{ dB}.$$

Table 16-2 gives typical values for the intensity levels of some of the common sounds.

TABLE 16-2 SOUND INTENSITY LEVELS IN DECIBELS (Threshold of hearing = 0 dB; threshold of pain = 120 dB.)

Source of Sound	dB
Rustling leaves	10
Whisper	20
Quiet room	30
Normal level of speech (inside)	65
Street traffic (inside car)	80
Riveting tool	100
Thunder	110
Indoor rock concert	120

Ear protectors are worn by these servicemen to avoid hearing loss caused by intense noise.

EXAMPLE 16-10 THE INTENSITIES OF ROCK MUSIC AND CONVERSATION COMPARED

Use Table 16-2 to compare the sound intensities in watts per square meter of the music at a rock concert and of ordinary conversation.

SOLUTION From the definition of the decibel, the sound intensity in watts per square meter is related to the intensity level in decibels by

$$I = I_0 10^{\beta/10}.$$

For rock music,

$$I_r = I_0 10^{120/10} = 10^{12} I_0,$$

and for ordinary conversation,

$$I_c = I_0 10^{65/10} = 10^{6.5} I_0.$$

The ratio of these two intensities is

$$\frac{I_r}{I_c} = \frac{10^{12} I_0}{10^{6.5} I_0} = 10^{5.5} = 3.2 \times 10^5.$$

No wonder there isn't much conversation while the band is playing!

DRILL PROBLEM 16-11

If the sound intensity increases by 1.0 dB, what is the corresponding increase in intensity in watts per square meter?
ANS. It increases by a factor of 1.26.

*16-9 DOPPLER EFFECT

The *Doppler effect* is a wave phenomenon discovered by Christian Johann Doppler (1803–1853) for light. However, this effect is more easily detected, and therefore better known, for sound. *The Doppler effect is the change in detected frequency of the signal emanating from a source due to the relative motion of the source and observer.* For example, if you are driving on a two-lane road and a car passes you going in the opposite direction, you can hear the change in the pitch (or frequency) of the whine of that car's engine as it goes by. When the source and observer are *approaching* each other, the frequency detected by the observer is *increased*, and when they are *receding* from each other, the detected frequency is *reduced*.

We study the Doppler effect assuming a source S produces mechanical waves that travel through a medium to an observer O. The waves from S are found by O to have a frequency f_S and wavelength λ_S when both S and O are stationary relative to the medium. We'll determine what frequency and wavelength O detects when either S or O is in motion relative to the medium. The speed of the waves *relative to the medium* will be represented by v. Where applicable, the speeds of the source and the observer relative to the medium are represented by v_S and v_O, respectively. It should be noted that v is unaffected by any motion of either the source or observer, because it only depends on the properties of the medium. Finally, it should be pointed out that the following analysis does not apply to electromagnetic waves. Their Doppler effect must be described using Einstein's special theory of relativity.

STATIONARY SOURCE AND MOVING OBSERVER

With O moving toward S (Fig. 16-21*a*), the relative velocity of a wave *with respect to O* is $v + v_O$. The number of wavefronts detected by O during a time interval Δt is therefore $(v + v_O)\,\Delta t/\lambda_S$. This quantity divided by Δt is the wave frequency f_O measured by the moving observer:

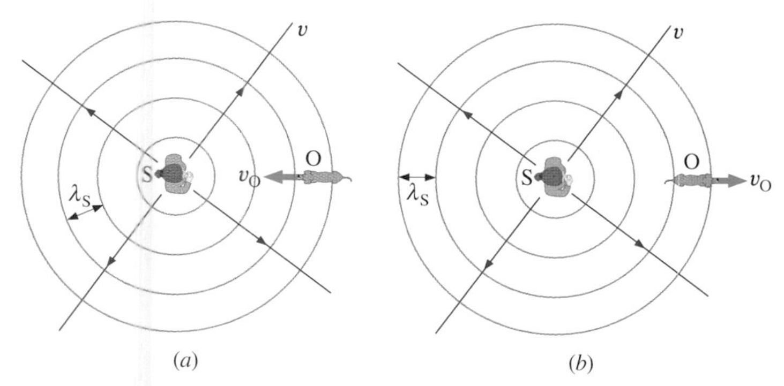

FIGURE 16-21 Stationary source S and moving observer O. (*a*) O moving toward S; (*b*) O moving away from S.

$$f_O = \frac{v + v_O}{\lambda_S} = \left(\frac{v + v_O}{v}\right)\left(\frac{v}{\lambda_S}\right) = \left(1 + \frac{v_O}{v}\right)f_S. \tag{16-19a}$$

Thus when the observer is moving *toward* a stationary source, the frequency he records is *higher* by $(v_O/v)f_S$ than when he is at rest.

Figure 16-21*b* shows the observer receding from the source. If his speed is v_O, then the relative velocity of the waves approaching him is $v - v_O$. Using the same reasoning as before, we find that the frequency he measures is

$$f_O = \frac{v - v_O}{\lambda_S} = \left(1 - \frac{v_O}{v}\right)f_S. \tag{16-19b}$$

So the frequency measured by an observer moving *away from* a wave source is *lower* than when he is at rest by the amount $(v_O/v)f_S$. Equations (16-19*a*) and (16-19*b*) show that the change in frequency is directly proportional to the velocity of the observer. The faster an observer approaches/recedes from a wave source, the greater is the increase/decrease in the measured frequency. These two equations can be combined into the single equation

$$f_O = \left(1 \pm \frac{v_O}{v}\right)f_S, \tag{16-19c}$$

where the positive sign corresponds to the observer approaching and the negative sign to the observer receding from the source.

EXAMPLE 16-11 CHANGE IN A WHISTLE'S PITCH FOR A MOVING OBSERVER

To summon her dog, Virginia blows a whistle that has a frequency of 500 Hz. (*a*) What is the frequency heard by the dog if he is running toward Virginia at 10 m/s? (*b*) If the dog runs past Virginia without slowing down, what is the change in the pitch of the whistle that he hears? Assume that the speed of sound is 330 m/s.

SOLUTION (*a*) Since the dog is approaching the wave source, he hears a higher frequency, as given by Eq. (16-19*a*):

$$f_O = \left(1 + \frac{10\text{ m/s}}{330\text{ m/s}}\right)(500\text{ Hz}) = 515\text{ Hz}.$$

(*b*) Here the dog is receding from the source and hears the lower frequency

$$f_O = \left(1 - \frac{10\text{ m/s}}{330\text{ m/s}}\right)(500\text{ Hz}) = 485\text{ Hz}.$$

The shift in frequency that the dog hears in passing Virginia is therefore 30 Hz.

DRILL PROBLEM 16-12

Answer the questions of the previous example for a whistle of frequency 1000 Hz and a dog running at 8.0 m/s.
ANS. 1024 Hz; 976 Hz.

MOVING SOURCE AND STATIONARY OBSERVER

As the source in the photograph of Fig. 16-22*a* moves from left to right across the ripple tank, the wavefronts ahead are bunched together so the wavelength, or distance between the fronts, is smaller than the distance between wavefronts when the source is stationary. Behind the source, the opposite effect occurs and the wavelength is larger. In Fig. 16-22*b*, five positions of the source in its left-to-right motion are shown and numbered 1 through 5. The wavefronts produced by the source at these positions are also depicted. Now a wavefront generated at an earlier time will have propagated farther than one produced later. Wavefront 1 therefore has the largest radius and wavefront 5 the smallest. Finally, because the wavefronts are all circular but centered at different points, they are grouped more closely together in front of the moving source and farther apart behind it.

When waves from a radar gun reflect off an approaching automobile, their frequency is shifted because of the Doppler effect. This shift in frequency is used to determine the speed of the automobile.

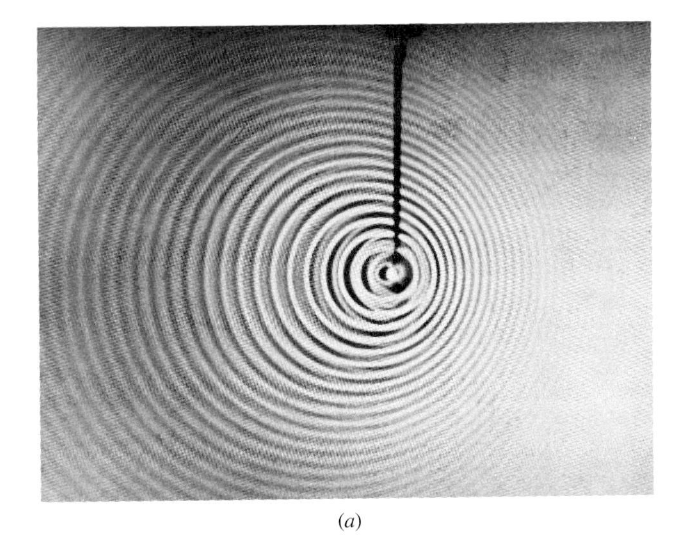

(a)

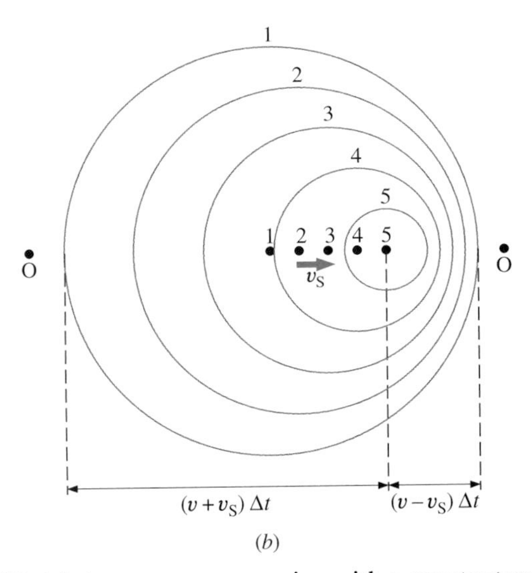

(b)

FIGURE 16-22 (*a*) A wave source moving with a constant velocity in a ripple tank. (*b*) Schematic representation of part (*a*). Each numbered wavefront was emitted when the source was at the correspondingly numbered position.

In a time interval Δt, the source produces $N = f_S\,\Delta t$ wavefronts. For example, if Δt represents the time interval between positions 1 and 5 of a moving source, then $N = 5$. For any value of Δt, the first wavefront will travel a distance $v\,\Delta t$ as the source moves $v_S\,\Delta t$ to the right. This means that directly in front of the source, the N wavefronts are clustered within a distance $(v - v_S)\,\Delta t$, while behind the source, they are clustered within $(v + v_S)\,\Delta t$. The wavelength λ_O measured by the observer is simply the spacing between successive wavefronts that pass him; that is,

$$\lambda_O = \frac{(v \mp v_S)\,\Delta t}{N},$$

where the negative and positive signs correspond to S approaching and S receding, respectively, from O. Substituting for N in this equation, we obtain

$$\lambda_O = \frac{(v \mp v_S)\,\Delta t}{f_S\,\Delta t} = \frac{v \mp v_S}{f_S}.$$

The observer is stationary in the medium and therefore measures the speed of the waves to be v; so, since $v = f_O \lambda_O$,

$$f_O = v\left(\frac{f_S}{v \mp v_S}\right) = \left(\frac{1}{1 \mp v_S/v}\right) f_S. \tag{16-20}$$

When the source is approaching the observer, the negative sign in Eq. (16-20) applies, and $f_O > f_S$. The source moving away from the observer corresponds to the positive sign in the equation, in which case $f_O < f_S$.

EXAMPLE 16-12 Change in Pitch for a Moving Siren

A siren on a police car has a frequency of 1000 Hz. (*a*) What is the frequency heard by a felon (assumed stationary) when the car is approaching him at 30 m/s? (*b*) When the car stops, the felon runs away at a speed of 8.0 m/s. What siren frequency does he hear? Use 330 m/s for the speed of sound.

SOLUTION (*a*) From Eq. (16-20) with the negative sign, the frequency heard by the felon is

$$f_O = \left[\frac{1}{1 - (30\text{ m/s}/330\text{ m/s})}\right](1000\text{ Hz}) = 1100\text{ Hz}.$$

(*b*) Here the observer is receding from a stationary source, so from Eq. (16-19*c*) (using the negative sign),

$$f_O = \left(1 - \frac{8.0\text{ m/s}}{330\text{ m/s}}\right)(1000\text{ Hz}) = 976\text{ Hz}.$$

DRILL PROBLEM 16-13

If the police car's speed in part (*a*) of Example 16-12 is reduced to 15 m/s, what is the observed frequency?
ANS. 1050 Hz.

Moving Source and Moving Observer

Finally, we determine the frequency shift when both the source and observer are moving. If the observer is at rest, he detects the frequency given by Eq. (16-20), so if he is moving, this frequency must be modified by the factor multiplying f_S in Eq. (16-19*c*). Hence the Doppler shift is determined by substituting f_O from Eq. (16-20) for f_S in Eq. (16-19*c*). This gives

$$f_O = \left(1 \pm \frac{v_O}{v}\right)\left(\frac{1}{1 \mp v_S/v}\right) f_S,$$

which simplifies to

$$f_O = \left(\frac{v \pm v_O}{v \mp v_S}\right) f_S. \tag{16-21}$$

Notice that this equation reduces to Eq. (16-19*c*) if $v_S = 0$ and to Eq. (16-20) if $v_O = 0$.

DRILL PROBLEM 16-14

Suppose that in Example 16-11, Virginia is blowing her whistle as she chases after her dog at 4.0 m/s. If the dog is running away from her at 10.0 m/s, what is the frequency of the whistle heard by the dog?
ANS. 491 Hz.

Shock Waves

A common bond among the derivations leading to the Doppler-shift equations is the assumption that the source has a smaller speed than the waves ($v_S < v$). If this condition is not met, there can be no waves in front of the source because it is moving faster than the waves themselves. This situation is illustrated in Fig. 16-23*a*, where a source is shown moving across a ripple tank with a speed larger than that of the water waves. The wavefronts form a cone with the source at its apex. (See Fig. 16-23*b*.) Probably the most common manifestation of this case is the "sonic boom" caused by the shock waves produced when a jet plane exceeds the speed of sound. As shown in Fig. 16-24, the plane produces not one but two cones, with the first emanating from the nose of the plane and the second from the tail. The sonic boom is due to pressure variations in the region between the cones, and a person within that region (for example, Ed in the figure) hears that variation. The shock waves are produced continuously by the plane while it is traveling at supersonic speeds. However, only a person located in the appropriate region will hear the boom. Since the plane is traveling from left to right in the figure, Ralph has already experienced the shock waves, while Alice and Trixie await their effect.

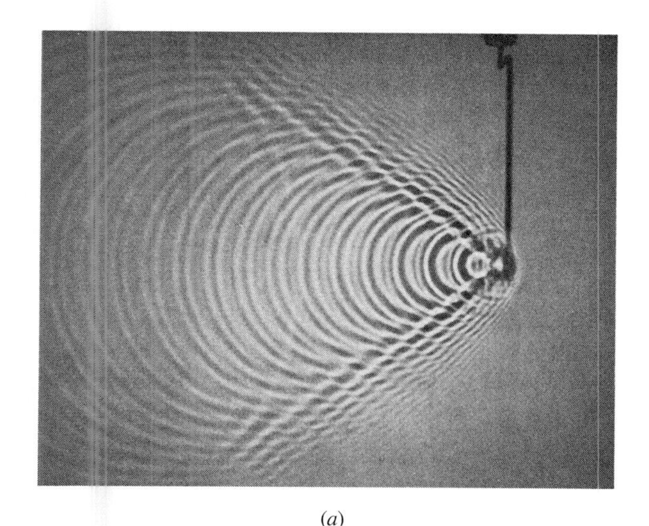

(*a*)

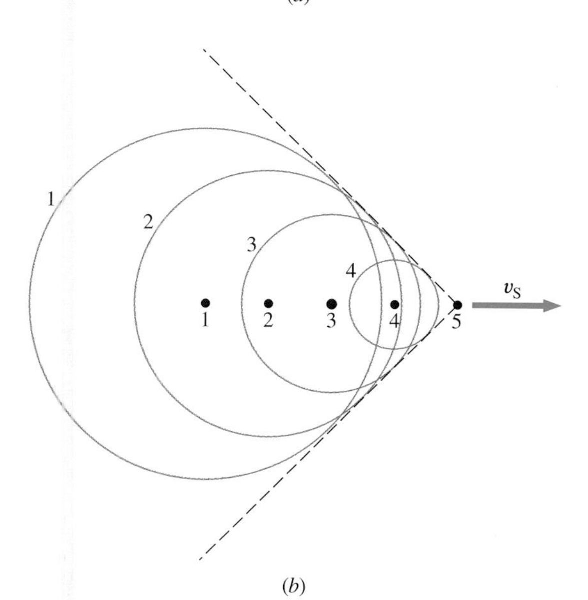

(*b*)

FIGURE 16-23 (*a*) A wave source that is moving faster than the speed of the waves in the medium. A cone with the source at the apex is formed by the wavefronts. (*b*) Notice that the wavefronts are all behind the source.

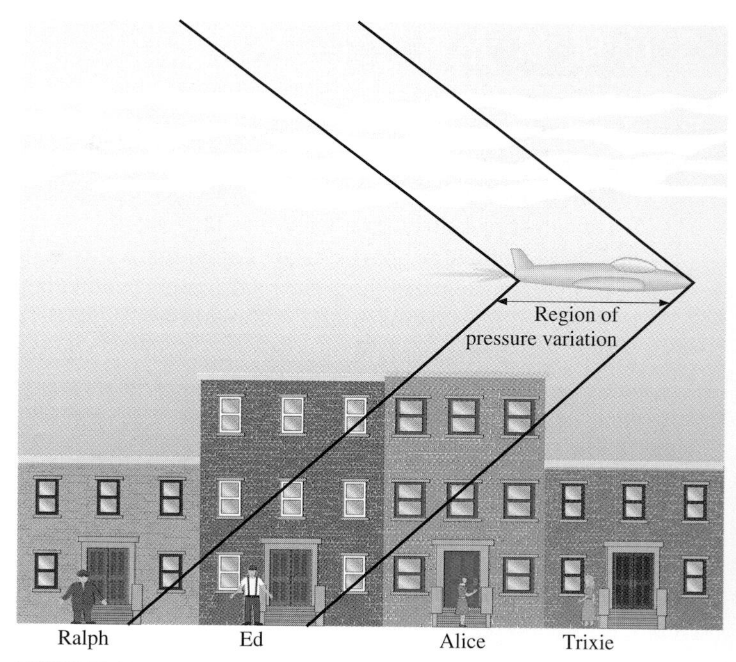

FIGURE 16-24 The sonic boom is heard by a listener who is within the region of pressure variation, which is located between the cones.

Shock wave produced by T-38 Talon flying at 1,200 km/h, a little faster than the speed of sound.

SUMMARY

1. **Types of waves**
   (*a*) Mechanical waves travel through a medium by disturbing the particles of the medium. An electromagnetic wave can travel through a vacuum. The waves associated with elementary particles are called matter waves.
   (*b*) A transverse wave displaces the medium perpendicular to its direction of propagation. A longitudinal wave displaces the medium parallel to its direction of propagation.
2. **Traveling waves**
   A one-dimensional traveling wave can be represented by
   $$y(x, t) = F(x \pm vt),$$
   where v is the speed of the wave.
3. **Wave speed**
   (*a*) The speed of a wave on a string is
   $$v = \sqrt{\frac{F}{\mu}},$$
   where F is the tension in the string and μ is the linear mass density of the string.
   (*b*) The speed of a sound wave in a bulk medium is
   $$v = \sqrt{\frac{B}{\rho}},$$
   where B is the bulk modulus of the medium through which the sound wave is moving and ρ is the density of the medium.
4. **Harmonic waves**
   (*a*) The harmonic wave can be represented by
   $$y(x, t) = A \sin (kx \pm \omega t + \phi),$$
   where A is the amplitude, k the wave number, ω the angular frequency, and ϕ the phase constant of the wave.
   (*b*) The wave number and angular frequency are related to the wavelength λ and frequency f by
   $$k = \frac{2\pi}{\lambda} \quad \text{and} \quad \omega = 2\pi f.$$
   (*c*) The speed of the harmonic wave is
   $$v = f\lambda = \frac{\omega}{k}.$$
5. **Energy transport in harmonic waves**
   The average power transmitted by a mechanical harmonic wave is proportional to the product of the square of the amplitude and the square of the frequency.
6. **Circular and spherical waves**
   (*a*) Two-dimensional waves emanating from point sources are represented by circular wavefronts spaced a distance λ apart. Three-dimensional waves from a point source are represented by spherical wavefronts spaced a distance λ apart.
   (*b*) The intensity of a spherical wave is the average power transmitted by the wave across a unit area normal to the direction of propagation of the wave. The intensity of a spherical wave decreases with the inverse square of the distance from the source.
7. **Plane-wave approximation**
   Far from a source of spherical waves, small sections on a spherical wavefront can be approximated by planes.

*8. **Sound intensity in decibels**
   If the intensity of sound in watts per square meter is I, then the intensity level β (in decibels) is given by
   $$\beta = 10 \log \left(\frac{I}{I_0}\right),$$
   where the base of the logarithm is 10 and $I_0 = 10^{-12}\ \text{W/m}^2$.

*9. **Doppler effect**
   (*a*) When an observer (O) is moving toward or away from a stationary source (S) of waves of frequency f_S, the frequency detected by the observer is
   $$f_O = \left(1 \pm \frac{v_O}{v}\right) f_S,$$
   where the positive sign corresponds to the observer approaching and the negative sign to the observer receding from the source.
   (*b*) If the observer is stationary and the source is moving,
   $$f_O = \left(\frac{1}{1 \mp v_S/v}\right) f_S,$$
   where the positive sign corresponds to the source moving away from the observer and the negative sign to the source approaching the observer.

Supplement 16-1 Relationship Between the Harmonic Displacement Wave and the Harmonic Pressure Wave for Sound

In this supplement we will derive the relationship between the harmonic displacement wave and the harmonic pressure wave for sound. Figure 16-25 shows a harmonic displacement wave moving through air contained in a long tube of cross-sectional area $\mathcal{A}$. The mathematical representation of this wave is given by Eq. (16-13*a*). At a time t, the displacement of the air at $x + \Delta x$ is (assuming $\phi = 0$)

$$s(x + \Delta x, t) = s_0 \sin [k(x + \Delta x) - \omega t],$$

while its displacement at x is

$$s(x, t) = s_0 \sin (kx - \omega t).$$

Before the wave moved down the tube, the volume of the air between x and $x + \Delta x$ was $V = \mathcal{A}\,\Delta x$. At the time t, the displacement wave causes the volume of this same air to change by an amount

$$\Delta V = \mathcal{A}[s(x + \Delta x, t) - s(x, t)]. \qquad (i)$$

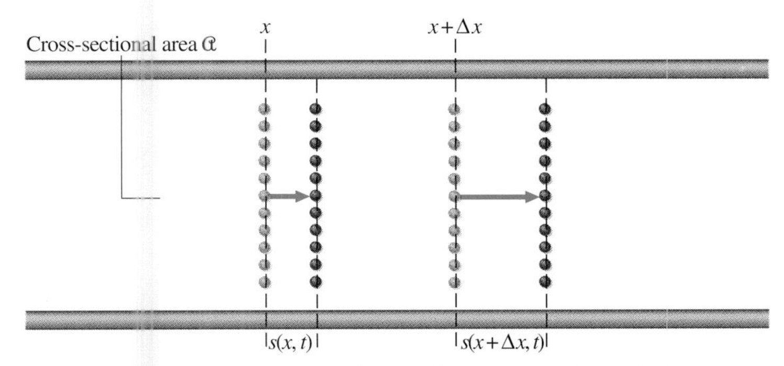

FIGURE 16-25 The change in volume of a given section of air as a harmonic displacement wave passes through the air. At a time t, the wave has displaced the molecules at x an amount $s(x, t)$ and the molecules at $x + \Delta x$ an amount $s(x + \Delta x, t)$.

Using the trigonometric identity $\sin(\alpha + \beta) = \sin\alpha \cos\beta + \cos\alpha \sin\beta$, we can write

$$s(x + \Delta x, t) = s_0 [\sin (kx - \omega t) \cos (k\,\Delta x) + \cos (kx - \omega t) \sin (k\,\Delta x)].$$

But Δx, and therefore $k\,\Delta x$, is small, so $\cos (k\,\Delta x) \simeq 1$ and $\sin (k\,\Delta x) \simeq k\,\Delta x$; thus

$$s(x + \Delta x, t) = s_0 [\sin (kx - \omega t) + (k\,\Delta x) \cos (kx - \omega t)].$$

Now from Eq. (*i*),

$$\Delta V = \mathcal{A} s_0 [\sin (kx - \omega t) + (k\,\Delta x) \cos (kx - \omega t) - \sin (kx - \omega t)],$$

which simplifies to

$$\Delta V = s_0 k \mathcal{A}\,\Delta x \cos (kx - \omega t). \qquad (ii)$$

From the definition of the bulk modulus (Eq. 11-9),

$$\Delta V = -\frac{\Delta p V}{B} = -\frac{\Delta p \mathcal{A}\,\Delta x}{B}.$$

Combining this with Eq. (*ii*) gives

$$s_0 k \mathcal{A}\,\Delta x \cos (kx - \omega t) = -\frac{\Delta p \mathcal{A}\,\Delta x}{B},$$

which when solved for Δp leaves

$$\Delta p = -s_0 k B \cos (kx - \omega t).$$

Finally, using $v = \sqrt{B/\rho}$, we have the expression for the pressure wave:

$$\Delta p = -\rho k v^2 s_0 \cos (kx - \omega t).$$

Notice that the amplitude of this wave is $\Delta p_0 = \rho k v^2 s_0$, which is in agreement with Eq. (16-14).

Questions

16-1. Discuss the differences and similarities between a wave on a string and a sound wave propagating in a fluid in a long tube.

16-2. If $F(x - vt)$ describes a pulse traveling in the positive x direction, what does $-F(x - vt)$ describe?

16-3. Does the speed of a pulse traveling along a wire depend on the wire's diameter?

16-4. What happens to the speed of a wave pulse on a string if we (*a*) increase the length of the string? (*b*) decrease the tension? (*c*) wrap tape around the string?

16-5. What happens to the speed of a wave pulse in a long tube of air if we (*a*) pump more air into the tube? (*b*) replace some of the air with helium? (*c*) increase the diameter of the tube while keeping the air density constant?

16-6. Would air in a narrow tube better approximate a one-dimensional medium for high- or low-frequency sound waves? Explain.

16-7. A popular lecture demonstration is performed by placing a ringing bell inside a jar attached to a vacuum pump. When the air is pumped from the jar, the bell can no longer be heard. Why? If the electrical power operating the bell is turned off, will the bell keep vibrating longer when there is air in the jar or when the air is evacuated?

16-8. The densities of most solids are more than 1000 times that of air; yet the speed of sound in a solid is usually greater than its speed in air. What does this tell you about the bulk moduli of solids as compared with that of air?

16-9. Is it possible to have a transverse or longitudinal wave if the vibrational motion of its source is not simple harmonic?

16-10. A string is attached to a vibrator of constant frequency. How would the wavelength of the harmonic wave produced be affected if the string were twice as thick?

16-11. Suppose that the vibrator of the previous question is set to a new frequency while oscillating at the same amplitude. Which of the following properties of the harmonic wave it produces would be affected? (*a*) frequency; (*b*) wavelength; (*c*) speed; (*d*) wave number; (*e*) amplitude; (*f*) period.

16-12. For the following properties of a harmonic wave, group those whose values depend on each other. Are there any that are independent of all the other properties? Properties: f, T, v, λ, k, ω, A, and ϕ.

16-13. At which points in its motion does a particle on a string through which a harmonic wave passes have zero velocity?

16-14. At $t = 0$, a harmonic wave is given by $y(x) = A \sin (2\pi/\lambda)x$. Which properties of the wave can you deduce from this expression? Which properties cannot be deduced?

16-15. Are $y(x) = A \cos (2\pi/\lambda)x$ and $y(x) = A \sin (2\pi/\lambda)x$ valid representations of a harmonic wave at $t = 0$? Why are $A \sin^2 (2\pi/\lambda)x$ and $A \tan (2\pi/\lambda)x$ inappropriate for describing a harmonic wave at a particular instant?

16-16. At a given distance from a point source, would the plane-wave approximation work better for high-frequency or low-frequency waves? Or is this approximation even affected by the frequency?

16-17. Sound waves of constant amplitude and frequency are emitted by a loudspeaker. What happens to the amplitude, frequency, and intensity of these waves as detected by a listener when he moves twice as far away from the loudspeaker?

16-18. A tennis player finds that the sun is shining directly in her eyes when she serves. She therefore moves to the equivalent position on an adjacent court to serve. Has she solved her problem? Explain.

16-19. Do the wavelengths of sound waves increase or decrease as their source moves toward the listener? away from the listener? How are these wavelengths affected if the listener is in motion, either directly toward or away from the source?

16-20. Ripples on the surface of a large tank of water are produced by the vibrating vertical rod shown in the accompanying figure. If the rod moves slowly eastward while continuing to vibrate at its original frequency, discuss how the spacing between the ripples is changed from the case where the rod is stationary. Consider points to the north, south, east, and west of the moving rod.

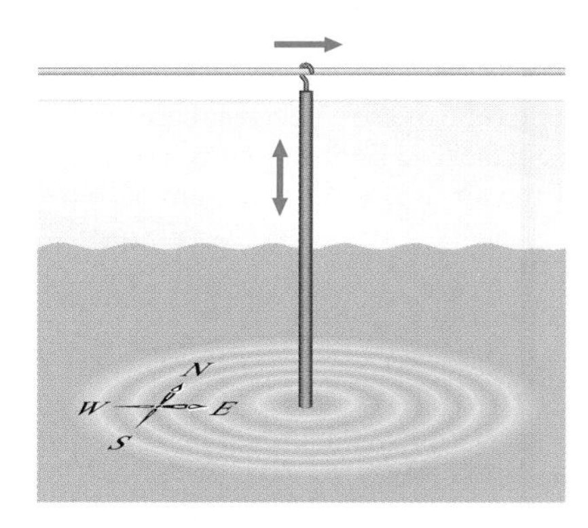

QUESTION 16-20

16-21. Can an observer who moves with the same velocity as the wave source detect a Doppler shift due to the motion of the source? due to his own motion?

16-22. A figure skater decides to perform with stereophonic music produced by speakers at opposite ends of the rink. What difficulty might she encounter?

16-23. A source and receiver of sound waves are stationary relative to each other, but a strong wind is blowing away from the receiver and toward the source. Does a Doppler shift occur in this case?

PROBLEMS

Traveling Waves

16-1. What are the velocities of the following wave pulses (all in SI units)? $F_1(x - 3.0t)$; $F_2(x + 2.0t)$; $F_3(y + 5.0t)$.

16-2. A wave pulse travels the length of an 8.0-m-long horizontal string in 1.6 s. Write the functional form of this pulse, assuming it is moving in the negative x direction.

16-3. Consider the function $y(x, t) = A(x - vt)^2$ where A and v are constants. Is it appropriate to use this function to represent an actual wave pulse? [*Hint:* Sketch the shape of $F(x, t)$ at $t = 0$.]

16-4. A traveling wave pulse is represented by

$$y(x, t) = \frac{1.0}{2.0 + (x + 0.50t)^2} \quad \text{(cgs units).}$$

What is the velocity of the pulse? Verify your answer by plotting the shape of the pulse at $t = 0.0$, 1.0, and 2.0 s.

16-5. A wave pulse on a string is given by

$$y(x, t) = 2.0e^{-(x-1.5t)^2/2.0} \quad \text{(cgs units).}$$

(*a*) What is the velocity of the pulse? (*b*) Plot the displacements of the string particles located at $x = -2.0$, 0.0, and $+2.0$ cm versus time.

Wave Speed

16-6. Calculate the speed of a wave along a string that has a linear mass density of 0.010 kg/m and is under a tension of 4.0 N.

16-7. What is the tension in the string of the previous problem if the wave speed is 15 m/s?

16-8. A horizontal wire of length 10.0 m is stretched between two supports; the tension in the wire is 800 N. A wave pulse travels with a speed of 120 m/s along this wire. What is the mass of the wire?

16-9. A steel wire is attached to a body of mass 10 kg at one end and fixed to a wall at the other, as shown in the accompanying figure. A transverse pulse takes 8.0×10^{-2} s to travel 5.0 m along the wire. What is the linear mass density of the wire?

PROBLEM 16-9

16-10. A 5.0-m cord has a mass of 0.10 kg. What tension is required to produce a wave pulse traveling with a speed of 20 m/s in the cord?

16-11. Suppose that the cord of the previous problem is wrapped with tape so that its total mass is doubled. What tension would now be required to produce a wave with a speed of 20 m/s?

16-12. Two horizontal wires A and B are held under the same tension. If the linear mass density of B is 1.5 times that of A, what is the ratio of the wave speeds on the two wires?

16-13. Two steel wires are held under the same tension. If the diameter of one wire is 1.2 times that of the other wire, what is the ratio of the speeds of the waves traveling in the wires?

16-14. A wave pulse traverses the length of a taut horizontal string in 1.2×10^{-2} s. The string has a mass of 20 g and the tension in it is 2.0×10^2 N. What is the length of the string?

16-15. The following data are from measurements of the speed of a wave pulse on a wire whose tension is being varied.

Tension T (N)	Speed v (m/s)
100	98
300	175
500	224
700	267
900	297

(*a*) By plotting v^2 versus T, determine the pulse speed when $T = 200$ and 350 N. (*b*) From the graph, determine the linear mass density of the wire.

16-16. The density of copper is 8.9×10^3 kg/m^3, and the density of gold is 1.9×10^4 kg/m^3. When two wires of these metals are held under the same tension, the wave speed in the gold wire is found to be half that in the copper wire. What is the ratio of the diameters of the two wires?

16-17. Wave pulses travel with the same speed down two wires made of different metals. The wires are stretched between the same two supports. If the tension in wire A is twice that in wire B, and if the radius of wire A is three times that of wire B, what is the ratio of the densities of the metals in the two wires?

16-18. Copper has a density of 8.9×10^3 kg/m^3 and a bulk modulus of 1.4×10^{11} N/m^2. Calculate the speed of sound in bulk copper.

16-19. The speed of sound in water is 1.45×10^3 m/s. Find the bulk modulus of water.

16-20. Sound waves travel in lead rods with a speed of about 1.2×10^3 m/s. Given that Young's modulus for lead is 1.6×10^{10} N/m^2, what is the density of lead?

16-21. Use Eq. (16-4*a*) to show that bulk modulus and pressure have the same dimensions.

16-22. In the old Western movies, the train robbers detected an approaching train by placing their ears near the steel tracks rather than by listening for the sound of the whistle coming through the air. If the train is 5.0 km away, how much sooner will the sound from it arrive through the rails as opposed to through the air? Assume that the speed of sound in air is 340 m/s and that Young's modulus and the density of steel are 2.0×10^{11} N/m^2 and 7.8×10^3 kg/m^3, respectively.

Harmonic Waves

16-23. At $t = 0$ a harmonic wave on a string is given by

$$y(x) = 0.50 \sin (2.0x) \qquad \text{(SI units)}.$$

What are the wavelength and amplitude of this wave? Plot $y(x)$ versus x in the range $0.0 \text{ m} \leq x \leq 4.0$ m.

16-24. Repeat the previous problem for a harmonic wave which at $t = 0$ is given by

$$y(x) = 0.20 \cos (4.0x + \pi/4) \qquad \text{(SI units)}.$$

16-25. A photograph of a transverse harmonic wave traveling along a wire shows its amplitude to be 2.0 cm and its wavelength to be 4.0 cm. Write an expression that represents the wave at $t = 0$. Can you write an expression representing the wave at an arbitrary time t?

16-26. The period of a longitudinal harmonic wave is 0.020 s and its wavelength is 8.0 m. Calculate the wave's speed, frequency, angular frequency, and wave number.

16-27. The normal human ear can hear frequencies ranging typically from 20 to 2.0×10^4 Hz. What range in wavelength does this correspond to? Assume the speed of sound in air is 340 m/s.

16-28. The note A above middle C on the piano has a frequency of 440 Hz. (*a*) What is the wavelength of the corresponding sound wave? Use 340 m/s for the speed of sound in air. (*b*) What is the wavelength of this sound in water where its speed is 1.46×10^3 m/s?

16-29. Two traveling harmonic waves propagate with the same speed, but the frequency of one wave is twice that of the other. Calculate the ratios of their periods, wavelengths, and wave numbers.

16-30. The wave number of a longitudinal harmonic wave traveling along a spring at 50 m/s is 1.0 m^{-1}. What are the period, wavelength, and frequency of the wave?

16-31. A transverse harmonic wave is represented by

$$y(x, t) = 0.50 \sin (1.2x - 100t) \qquad \text{(cgs units)}.$$

Find the amplitude, frequency, wavelength, period, and velocity of the wave.

16-32. A body of mass M is suspended by a wire of linear mass density 8.0×10^{-3} kg/m, as shown in the accompanying figure. A harmonic wave produced by a source vibrating at 45 Hz travels along the wire with a wavelength of 1.5 m. What is the value of M?

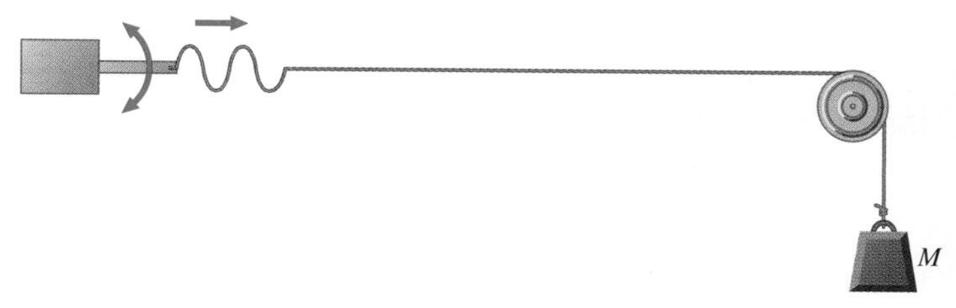

PROBLEM 16-32

16-33. A vibrating arm fixed to one end of a taut string produces a harmonic wave, as shown in the accompanying figure. Determine how the wavelength, frequency, speed, and amplitude of the

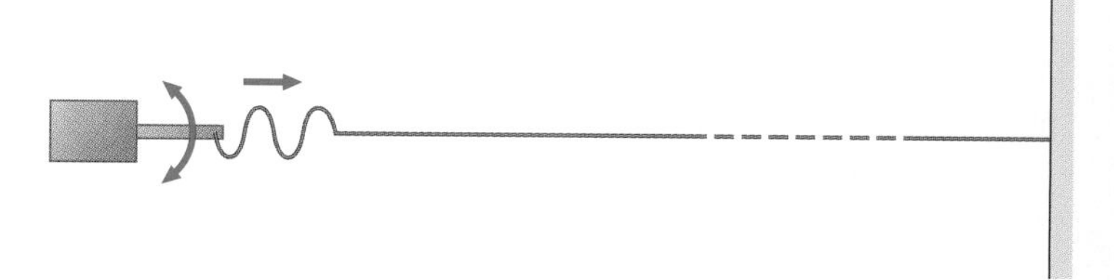

PROBLEM 16-33

wave are affected if (*a*) the frequency of vibration of the arm is doubled, (*b*) the arm vibrates with twice its original amplitude, and (*c*) the tension in the string is reduced to 0.8 of its original value.

16-34. Write a mathematical representation for a transverse harmonic wave traveling along a string in the negative x direction with a speed of 10 m/s and a wavelength of 0.020 m. The maximum transverse displacement of a string particle is 0.010 m.

16-35. What are the possible phase constants of the harmonic wave of Prob. 16-34 if at $t = 0$ the particle on the string located at the origin of the chosen coordinate system has a transverse displacement of 2.5×10^{-3} m?

16-36. Repeat the calculation of Prob. 16-35 if the string particle at $x = -0.50$ m has a transverse displacement of 1.5×10^{-3} m at $t = 5.0$ s.

16-37. The equation for a wave on a cord is

$$y(x, t) = 0.020 \cos 2\pi\left(\frac{x}{0.40} - 25t + \frac{1}{6}\right) \quad \text{(SI units)}.$$

(*a*) Determine the wavelength, frequency, and velocity of the wave. (*b*) Determine the maximum transverse velocity and acceleration of any point on the cord.

16-38. A transverse harmonic wave travels down a string such that at $t = 0$ and $x = 0$ its displacement is a maximum. The amplitude of the wave is 0.25 cm, its wavelength is 5.0 cm, and its frequency is 100 Hz. Write an equation that represents how the wave depends on time and position.

16-39. Write a mathematical representation for a longitudinal harmonic wave traveling in the positive x direction along a spring with a frequency of 20 Hz and a speed of 15 m/s. The amplitude of the wave is 3.0×10^{-2} m, and the spring particle at the origin has a displacement of 1.2×10^{-2} m when $t = 0$.

16-40. The expression for a harmonic wave on a string is given by

$$y(x, t) = 0.10 \sin\left(0.20x + 40t + \frac{\pi}{3}\right) \quad \text{(cgs units)}.$$

What are the displacement and velocity of the string particle (*a*) at the origin of the coordinate system at $t = 0$? (*b*) at $x = -15$ cm at $t = 10$ s?

16-41. A harmonic sound wave with a displacement amplitude of 2.0×10^{-8} m moves through air at a speed of 340 m/s. What is the pressure amplitude of the wave at a frequency of 10 kHz? Assume that the density of air is 1.3 kg/m^3.

16-42. A harmonic sound wave of frequency 5.0×10^3 Hz moves through air of density 1.2 kg/m^3 at a speed of 330 m/s. If the pressure amplitude of the wave is 2.0×10^{-3} N/m^2, what is its displacement amplitude?

16-43. A harmonic displacement sound wave of frequency 3.00 kHz moves through air of density 1.24 kg/m^3 at a speed of 340 m/s. The amplitude and phase constant of the wave are 4.0×10^{-9} m and 0.50π rad, respectively. (*a*) Write an equation that represents the displacement wave. (*b*) Write the corresponding equation for the pressure wave.

16-44. A harmonic displacement wave for sound in a gas of density 2.5 kg/m^3 is given by

$$s = (2.0 \times 10^{-8}) \sin\left[30x - (2.0 \times 10^4)t\right] \quad \text{(SI units)}.$$

(*a*) What is the velocity of the wave? (*b*) What is the amplitude of the corresponding harmonic pressure wave? (*c*) Write an equation representing the pressure wave.

Energy Transport in Harmonic Waves

16-45. The harmonic wave of Prob. 16-31 moves down a string whose mass per unit length is 10 g/m. Calculate the power transmitted down the string.

16-46. A wire of mass 0.25 kg is 50 m long and under a tension of 70 N. A vibrating source operating at an angular frequency of 350 rad/s produces a harmonic wave in the wire. If the source can supply energy to the wire at a maximum rate of 30 J/s, what is the maximum amplitude of the waves it generates?

16-47. A vibrating source produces harmonic waves on a taut string. By what factor will the power transmitted down the string be changed if (*a*) the amplitude of the waves is halved; (*b*) the frequency of the vibrations is doubled; (*c*) both the wavelength and amplitude of the vibrations are halved; (*d*) the frequency of the vibrations is halved and their amplitude is doubled?

16-48. An average power of 30 W is transmitted down a string by the harmonic wave of Prob. 16-37. What is the tension in the string?

16-49. The mass of a string 3.1 m long is 0.25 kg. The string is under a tension of 40 N. What must be the frequency of a harmonic wave of amplitude 8.2 mm in order that the power transmitted down the string be 80 W?

Circular and Spherical Waves/Plane-Wave Approximation

16-50. For a given point source with a constant power output, calculate the ratio of the intensities received by an observer at distances of 1.0, 2.0, and 4.0 m from the source.

16-51. A point source has a constant power output of 1.5×10^3 W. Make a plot of intensity vs. distance from the source between 10 and 20 m.

16-52. At what distance from a lightbulb producing 60 W of radiant power must an observer be in order to detect a light intensity of 4.0×10^{-3} W/m^2? If he approaches the source to one-third his original distance, what is the light intensity detected?

16-53. What is the power output of a source if the intensity received 15 m away is 9.0×10^{-2} W/m^2?

16-54. By what fraction of her original distance from a source of constant power must a person move so as to increase the intensity she measures by 50 percent?

16-55. Using the values given in Example 16-8, calculate the intensity of sunlight received on the earth above the atmosphere.

16-56. A circular light detector of diameter of 1.0 cm is placed 3.0 m from a point source whose power output is 100 W. (*a*) How much power is absorbed by the detector? (*b*) How much energy does it absorb in 5.0 s?

16-57. Compare the amount of radiant energy and the intensity of sunlight received by the planets listed in the table with that received by the earth.

Planet	Distance from Sun (Earth = 1.0)	Diameter (Earth = 1.0)
Venus	0.72	0.95
Mars	1.5	0.53
Saturn	9.5	9.5
Pluto	39	0.24

16-58. The radiant energy intensity of sunlight at the earth's surface is about 700 W/m^2. The faintest light that the normal human eye can detect comes from a star whose radiant energy intensity on the earth is about e^{-33} that of the sun. Calculate the energy entering the eye per second from this star. Assume that the diameter of the eye's pupil is 5.0 mm.

16-59. Are the differences in the light intensity received from Alpha Centauri by the planets listed in Prob. 16-58 completely negligible? What about the differences in the radiant energy?

Sound Intensity in Decibels

16-60. The intensity level of a sound wave is 70 dB. Calculate the intensity of the sound in watts per square meter.

16-61. The intensity of a sound wave is 2.5×10^{-6} W/m^2. What is the intensity level in decibels of this wave?

16-62. A siren emits sound of total power 10 W uniformly in all directions. At what distance from the source is the intensity level (*a*) 80 dB and (*b*) 50 dB?

16-63. The sound intensity level at a window of area 2.0 m^2 is 50 dB. What is the total energy per second that enters the window?

16-64. When one of two speakers is operating, the intensity level at a particular point is 70 dB. When that speaker is turned off and the other one is turned on, the intensity level at the same point is 80 dB. What is the intensity level at this point when both speakers are operating?

Doppler Effect

Note: In all of these problems, assume that the speed of sound in air is 340 m/s.

16-65. How fast must a train be moving toward a listener in order for the frequency of the sound from the train's whistle to increase 10 percent above its stationary value?

16-66. If the train of the previous problem has passed and is moving away from the listener at 20 m/s, what is the fractional change in the frequency of the whistle's sound from its stationary value?

16-67. A passenger on a train approaching a station detects sound waves of frequency 450 Hz from a ringing bell at the station. What frequency is heard if the listener is motionless? Assume that the train is moving at 140 km/h.

16-68. The driver of a car traveling at 60 km/h sees a distant car traveling directly toward him. He sounds his horn, which has a frequency of 500 Hz. The driver of the other car hears a frequency of 560 Hz. At what speed is this car traveling?

16-69. A tuning fork of frequency 4.0×10^2 Hz is fixed to the edge of a disk that is spinning at the rate of 40 rev/min. (See the accompanying figure.) If the radius of the disk is 1.0 m, what are the maximum and minimum frequencies heard by the listener shown? To which positions of the tuning fork do these frequencies correspond?

16-70. On a distant planet, the occupant of a small, low-flying vehicle hears a frequency of 1.5×10^3 Hz emitted from a beeper on a stationary space buoy she is approaching directly at 80 m/s. She comes to a halt and now detects a frequency of 1.4×10^3 Hz. What is the speed of sound on this planet?

16-71. The frequency of the horn of a sports car is 400 Hz. Suppose this car is traveling at 130 km/h on the highway. What is the frequency of the horn heard by the driver of a second vehicle moving in the same direction at 80 km/h (*a*) before the sports car passes him? (*b*) after the sports car passes him?

General Problems

16-72. Using the equation for wave velocity on a string ($v = \sqrt{T/\mu}$), show that (*a*) for constant μ, $dv/v = dT/2T$ and (*b*) for constant T, $dv/v = -d\mu/2\mu$.

16-73. If the tension of a string is decreased by 2 percent of its original value, what is the corresponding change in the wave speed? Use the result of Prob. 16-72. If the tension is decreased by 50 percent, what is the corresponding change in speed according to Prob. 16-72? Repeat your calculation using Eq. (16-2). Why is there a discrepancy in the results? Which is the correct result?

16-74. What is the maximum transverse speed that a string particle can have when the harmonic wave represented by $y(x, t) = A \sin(kx - \omega t)$ moves down the string? What is the transverse displacement of the particle when it is at its maximum speed?

16-75. A vibrating tuning fork sends waves down a string with an amplitude of 2.0×10^{-3} m and a wavelength of 0.20 m. What is the ratio of the wave speed to the maximum transverse speed of a string particle?

16-76. The average period for the rotation of the earth about its polar axis can be measured either with respect to the sun or with respect to the distant stars. The former period is 24 h and is known as a *solar day*; the latter period is called a *sidereal day*. The difference between these two periods is illustrated in the accompanying figure. The solar day is the interval between successive times when the pointer on the earth is directed toward the sun. The sidereal day is the corresponding time interval for the distant stars rather than the sun. Is the sidereal day longer or shorter than the solar day? By how much?

PROBLEM 16-69

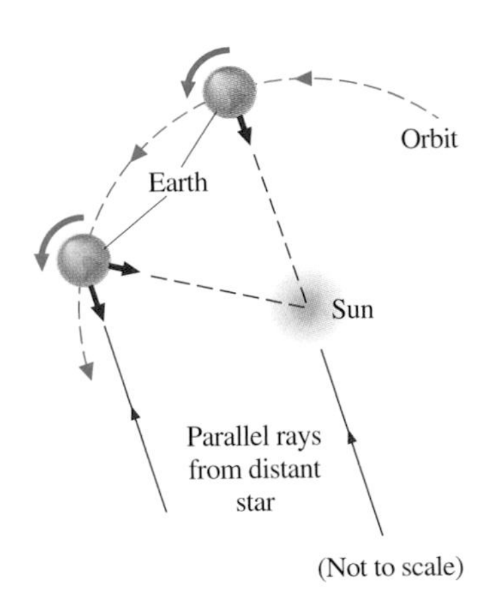

PROBLEM 16-76

16-77. Imagine a spherical radio wave falling on an antenna of length $2a$, as shown in the accompanying figure. (*a*) If the wave is produced a distance $R \gg a$ from the antenna, show that the spherical wavefront arrives at the ends of the antenna at a time $t = a^2/2cR$ (c = speed of light) later than it arrives at the antenna's center. In deriving this formula, use $\sqrt{R^2 + a^2} = R\sqrt{1 + a^2/R^2} \approx R(1 + a^2/2R^2)$. (*b*) If $R = 20$ km and $a = 10$ m, how does this time compare with the period of a 100-kHz signal from an AM radio station? Is the plane-wave approximation accurate for this case?

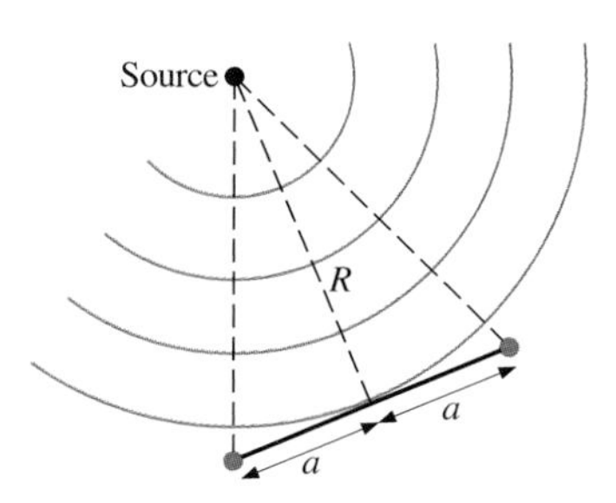

PROBLEM 16-77

16-78. The frequency of a siren on an ambulance is 700 Hz. When the ambulance is speeding away from a stationary pedestrian with a speed v_S, he hears a frequency f_O. When the vehicle increases its speed by 50 percent, the frequency heard by the pedestrian is reduced by 2.5 percent. Determine v_S and f_O. Assume that the speed of sound in air is 340 m/s.

16-79. (*a*) If there is a 50-km/h wind blowing from the station toward the moving train of Prob. 16-67, what frequency does the passenger detect? (*b*) What frequency is heard if the wind blows in the opposite direction? Assume that the speed of sound relative to still air is 340 m/s. (*Hint:* Work in the reference frame of the medium.)

16-80. What is the horn frequency that the driver of the other car in Prob. 16-68 would hear if a wind of 30 km/h were blowing from the honking car toward him?

Music is the superposition of many harmonic waves.

CHAPTER 17 Superposition of Waves

Preview

In this chapter we investigate what happens when there is more than one wave in a particular region of space. The main topics to be discussed here include the following:

1. **Principle of superposition.** This principle describes how waves are added.
2. **Superposition of two harmonic waves of the same frequency.** The superposition principle is used to find the sum of two harmonic waves of the same frequency. Constructive and destructive interference are defined.
3. **Beats.** The superposition principle is applied to two harmonic waves of different frequency. The concept of beats is introduced.
4. **Boundary conditions.** When a wave encounters the boundary between two media, it is partially reflected and partially transmitted. How the reflected and transmitted waves are related to the incident wave is determined.
5. **Standing waves.** We see how two waves of the same frequency that are moving in opposite directions combine to form a standing wave. Properties of the standing wave are considered.
6. **Standing waves on a string.** We determine the conditions necessary for the production of a standing wave on a string fixed at both ends.
7. **Standing waves in a tube.** We determine the conditions necessary for the production of standing waves in open and closed tubes.

*8. **Music and musical instruments.** We see how standing waves on strings and in tubes are related to the production of musical notes.

It is very common for many waves to propagate in the same region of space. At this moment, your book is being illuminated by light waves, some coming to you directly from sources and others from reflections off the walls and various objects in the room. You are also hearing many sound waves that are falling simultaneously on your ears. Like the light waves, these sound waves arrive both directly from sources and from reflecting bodies. Multiwave phenomena are, of course, not restricted to light or sound but occur for all types of waves. For example, a television signal is produced by broadcasting electromagnetic waves over a frequency band approximately 4 million hertz wide.

In this chapter we examine the effects of combining traveling waves. We begin by considering transverse waves on a stretched string, because it is easy to visualize what happens when their paths cross. The results will then be generalized to waves in other media. We will also study the superposition of longitudinal waves, an interesting application of which is the production of sound in musical instruments.

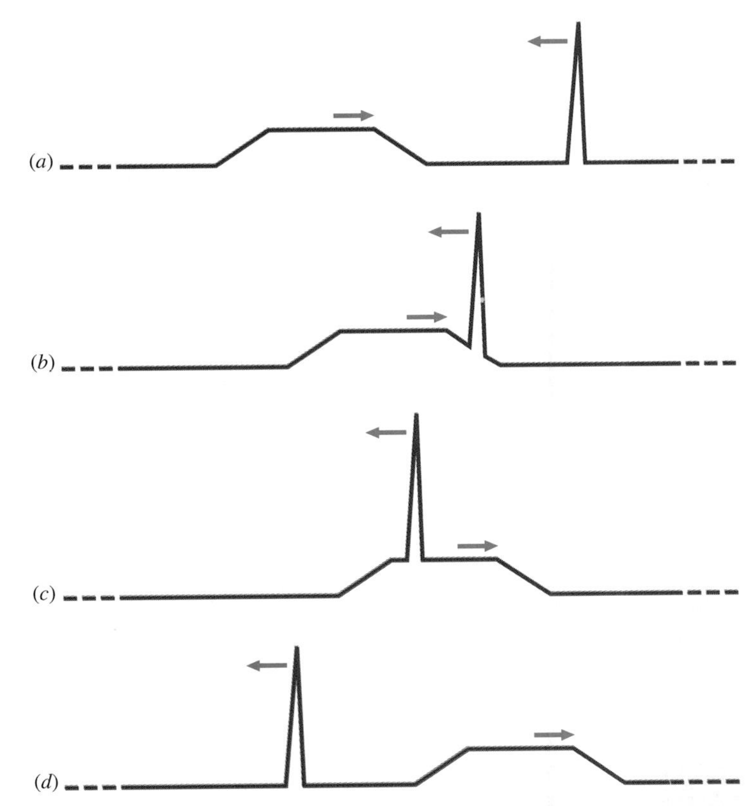

FIGURE 17-1 The superposition of two wave pulses. At any instant, the overall displacement of the medium is the sum of the displacements due to the individual pulses.

17-1 Principle of Superposition

The principle of superposition is used throughout physics. We applied it in Newtonian mechanics when we added individual forces to obtain the resultant force and when we determined the net gravitational field of a number of masses by summing their individual fields. Waves are superposed when they occupy the same region of a medium such as a given section of a string. The principle of superposition for waves states that

> The resultant wave due to two or more waves is the sum of the individual waves.

Implicit in the superposition principle is the assumption that the individual waves do not affect one another. When many waves are passing through the same region, each behaves as if it were the only wave there. The superposition principle is illustrated in Fig. 17-1 for two pulses traveling in opposite directions along a string. Notice how the overall shape of the resultant wave changes when the two pulses "move through" each other in Fig. 17-1*b*, *c*, and *d*. Mathematically, the function representing the resultant wave is found by simply adding those functions describing the individual waves. For example, if the pulse traveling from left to right is given by $y_1(x - v_1 t)$ and the one traveling in the opposite direction by $y_2(x + v_2 t)$, then the resultant wave at x and t is

$$y(x, t) = y_1(x - v_1 t) + y_2(x + v_2 t).$$

For mechanical waves of small amplitudes, the assumption that the individual waves do not affect each other holds very well, and the superposition principle is valid. However, for high-amplitude waves such as the "shock waves" produced by a strong explosion, superposition isn't applicable because these waves significantly alter the medium through which they propagate. For example, a strong shock wave may heat the air as it passes through. The passage of such a wave therefore affects the propagation of a second wave due to its effect on the medium. We will not treat these severe conditions; we will only consider waves that obey the superposition principle.

17-2 Superposition of Two Harmonic Waves of the Same Frequency

Let's consider two harmonic waves traveling through a medium in the same direction. The waves have the same amplitude, wavelength, and frequency, but different phase constants. We can represent them by $y_1(x, t) = A \sin(kx - \omega t)$ and $y_2(x, t) = A \sin(kx - \omega t + \phi)$. The difference between the phases, which in this case is just ϕ, is the **phase difference** of the two waves. From the principle of superposition, the resultant wave is

$$y(x, t) = y_1 + y_2 = A[\sin(kx - \omega t) + \sin(kx - \omega t + \phi)].$$

This expression can be simplified by using the trigonometric identity

$$\sin L + \sin M = 2 \sin \tfrac{1}{2}(L + M) \cos \tfrac{1}{2}(L - M) \qquad (17\text{-}1)$$

and substituting $kx - \omega t$ for L and $kx - \omega t + \phi$ for M. We then find that

$$y(x, t) = \left(2A \cos \frac{\phi}{2}\right) \sin\left(kx - \omega t + \frac{\phi}{2}\right). \qquad (17\text{-}2)$$

The superposition of two circular water waves.

The resultant wave therefore has the *same* frequency as y_1 and y_2, and it has an amplitude of $2A \cos(\phi/2)$. This amplitude depends on the phase difference ϕ between the individual waves and can vary from a minimum of 0 to a maximum of $2A$.

It is instructive to think of the phase difference in terms of how much one wave is "shifted" with respect to the other. For example, if $\phi = 0$ (no shift), y_1 and y_2 exactly coincide, as shown in Fig. 17-2*a*. The resultant wave then has an amplitude of $2A$, or twice the amplitude of the individual wave. This is known as *constructive interference*. If $\phi = \pi$ rad (a shift of half the wavelength), the crests of one wave match with the troughs of the other (Fig. 17-2*b*) and the two waves completely cancel. This is *destructive interference*. If $\phi = 2\pi$ rad, the waves again coincide (Fig. 17-2*c*) because they are now shifted from each other by exactly one wavelength, and constructive interference occurs once again. For values of the phase difference other than 0 or $n\pi$ (n an integer), y_1 and y_2 neither completely add nor cancel, but the shape of the resultant wave is still known from Eq. (17-2). For example, if $\phi = \pi/3$ rad, the resultant wave is that shown in Fig. 17-2*d*.

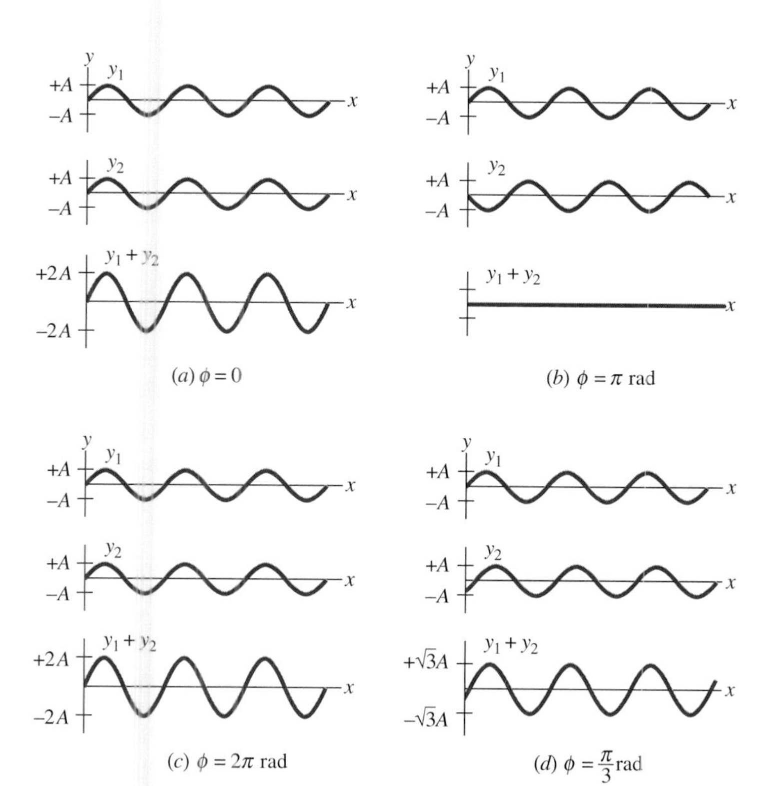

FIGURE 17-2 The addition of two harmonic waves of the same frequency at a fixed time t. The amplitude of the resultant wave depends on the phase difference ϕ between the waves.

EXAMPLE 17-1 The Addition of Two Harmonic Waves

Two harmonic waves are represented in SI units by $y_1(x, t) = 0.20 \sin(x - 3.0t)$ and $y_2(x, t) = 0.20 \sin(x - 3.0t + \phi)$. (*a*) Write the expression for the sum $y_1 + y_2$ for $\phi = \pi/2$ rad. (*b*) Suppose the phase difference ϕ between the waves is unknown. If the amplitude of their sum is 0.32 m, what is ϕ?

SOLUTION (*a*) From Eq. (17-2), the resultant wave can be written as

$$y(x, t) = 2(0.20)\left(\cos\frac{\pi}{4}\right)\sin\left(x - 3.0t + \frac{\pi}{4}\right)$$
$$= 0.28 \sin\left(x - 3.0t + \frac{\pi}{4}\right).$$

(*b*) Since the amplitude of the resultant wave is 0.32 m and $A = 0.20$ m, we have from Eq. (17-2)

$$2(0.20\text{ m})\cos\frac{\phi}{2} = 0.32\text{ m},$$

so the phase difference between the two given harmonic waves is

$$\phi = 2\cos^{-1}(0.80) = \pm 1.29\text{ rad}.$$

DRILL PROBLEM 17-1

Verify the calculations of part (*a*) of Example 17-1 by plotting the individual waves at $t = 0$ for values of x ranging from 0 to 2.0 m. Determine the shape of the resultant wave graphically by adding the values of the individual waves at a sufficient number of points along the x axis in the given range.

DRILL PROBLEM 17-2

What is the resultant wave obtained when $y_1(x, t) = A \sin(kx - \omega t + \phi_1)$ and $y_2(x, t) = A \sin(kx - \omega t + \phi_2)$ are added? Use $\phi_1 = \pi/6$ rad and $\phi_2 = \pi/2$ rad.
ANS. $y(x, t) = \sqrt{3}A \sin(kx - \omega t + \pi/3)$.

17-3 Beats

We have just seen the effects of superposing two harmonic waves of the same frequency. What happens if the wave frequencies are different? To answer this question, let's consider the harmonic waves

$$y_1 = A \sin(k_1 x - \omega_1 t)$$

and

$$y_2 = A \sin(k_2 x - \omega_2 t),$$

which, for simplicity, we assume have the same amplitude and zero phase constant. By applying the trigonometric identity of Eq. (17-1), we find that their sum

$$y = A[\sin(k_1 x - \omega_1 t) + \sin(k_2 x - \omega_2 t)]$$

can be written as

$$y = 2A \sin \tfrac{1}{2}[(k_1 + k_2)x - (\omega_1 + \omega_2)t] \times \cos \tfrac{1}{2}[(k_1 - k_2)x - (\omega_1 - \omega_2)t]. \quad (17\text{-}3)$$

Now imagine that we are detecting this signal (e.g., hearing it if it is sound) at a given position x_0. The terms $(k_1 + k_2)x_0$ and $(k_1 - k_2)x_0$ are then constants that can be expressed in terms of two phase constants ϕ_a and ϕ_b, where

$$(k_1 + k_2)x_0 = \pi - 2\phi_a,$$

and

$$(k_1 - k_2)x_0 = -2\phi_b.$$

The resultant wave then becomes

$$y(t) = 2A \cos\left[2\pi\left(\frac{f_1 + f_2}{2}\right)t + \phi_a\right] \times \cos\left[2\pi\left(\frac{f_1 - f_2}{2}\right)t + \phi_b\right], \quad (17\text{-}4)$$

where we have used $\omega = 2\pi f$ wherever appropriate.

Figure 17-3 is a typical plot of $y(t)$ at a given point in space for two superposed waves of nearly equal frequencies. The resultant waveform consists of fast oscillations at a frequency $(f_1 + f_2)/2$ modulated by an amplitude $A_R(t)$ that varies much more slowly with time according to

$$A_R(t) = 2A \cos\left[2\pi\left(\frac{f_1 - f_2}{2}\right)t + \phi_b\right]. \quad (17\text{-}5)$$

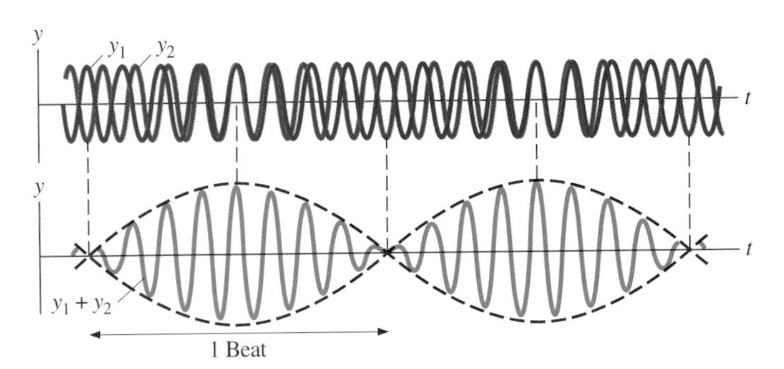

FIGURE 17-3 The addition of two waves of slightly different frequencies at a given point in space. The amplitude (represented by the dashed envelope) of the resultant wave varies periodically with time.

Notice that A_R reaches its extreme values of $\pm 2A$ whenever $\cos\{2\pi[(f_1 - f_2)/2]t + \phi_b\} = \pm 1$. This occurs *twice* during each cycle of the cosine function, or at a frequency of $2[(f_1 - f_2)/2] = f_1 - f_2$. The intensity of the signal, which is proportional to the amplitude squared, must therefore also oscillate at this frequency. This variation in intensity is known as *beating*, and $f_1 - f_2$ is called the **beat frequency**. As a practical matter, beating is only observed when f_1 and f_2 are nearly equal. By comparing Fig. 17-3 with Fig. 17-4, for which $f_1 = 9.0f_2$, you can see that beats are just not evident in the latter case.

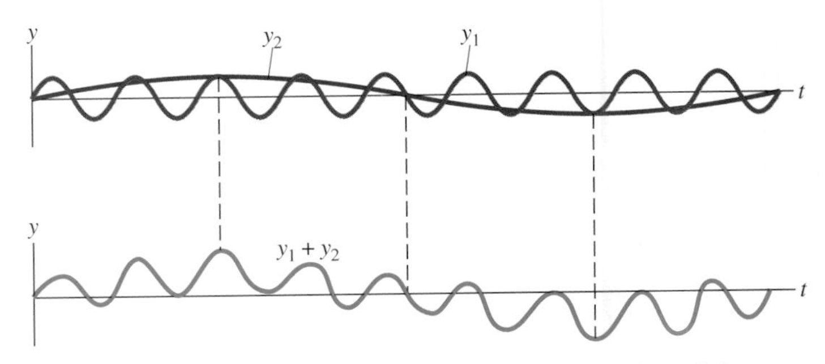

FIGURE 17-4 Beats are not evident when the frequencies of the added waves are very different. Here, $f_1 = 9f_2$.

Suppose that y_1 and y_2 represent harmonic sound waves with slightly different frequencies, produced perhaps by musical instruments or tuning forks. The listener then hears the resultant $y_1 + y_2$ as a note of frequency $(f_1 + f_2)/2$, which varies in loudness at the beat frequency $f_1 - f_2$. If the beat frequency is low enough (fewer than about 10 beats per second), the listener is able to detect the variations in loudness. At higher beat frequencies, the two-note combination may sound either harmonious or dissonant, depending on the relative values of the frequencies. However, the intensity seems to remain constant, because its variation is too rapid to be detected by the listener.

One practical application of beats is the tuning of musical instruments. For example, the note from a piano string to be tuned might be compared with a note from a tuning fork of the desired frequency. If beats are heard, the string tension is adjusted until they disappear. At that point, the note from the piano string will have the same frequency as that from the tuning fork.

EXAMPLE 17-2 **TUNING A VIOLIN**

When two music students simultaneously bow the A strings of their violins without fingering, they hear two beats per second. (*a*) One of the students then decides to tune her instrument by playing the A string simultaneously with a tuning fork of frequency 440 Hz. She hears one beat per second. What are the possible frequencies of the note produced by the string? (*b*) What are the possible frequencies of the note produced by the A string of the other violin?

SOLUTION (*a*) The beat frequency is equal to the difference between the frequencies of the two individual notes. The frequency of the note produced by the violin string must therefore differ from that of the tuning fork by 1 Hz. Its possible values are 439 and 441 Hz.

(*b*) If the actual frequency of the note of the violin of part (*a*) is 439 Hz, then the A string of the other violin would produce notes of either 437 or 441 Hz, since two beats per second are heard when the strings are played simultaneously. However, if the actual frequency of the note of the violin of part (*a*) is 441 Hz, then the other violin would produce notes of either 439 or 443 Hz. Based on the information given, we can only conclude that the A string of the other violin produces a note of one of four possible frequencies.

17-4 Boundary Conditions

When a traveling wave encounters the boundary of another medium, it is partially reflected and partially transmitted. In this section we will investigate the division of the incident wave into reflected and transmitted components, as specified by the *boundary conditions*. We'll do so with the help of our old friend, the string. Once again, the rules obtained with waves on a string can be easily generalized to the reflection and transmission of other types of waves.

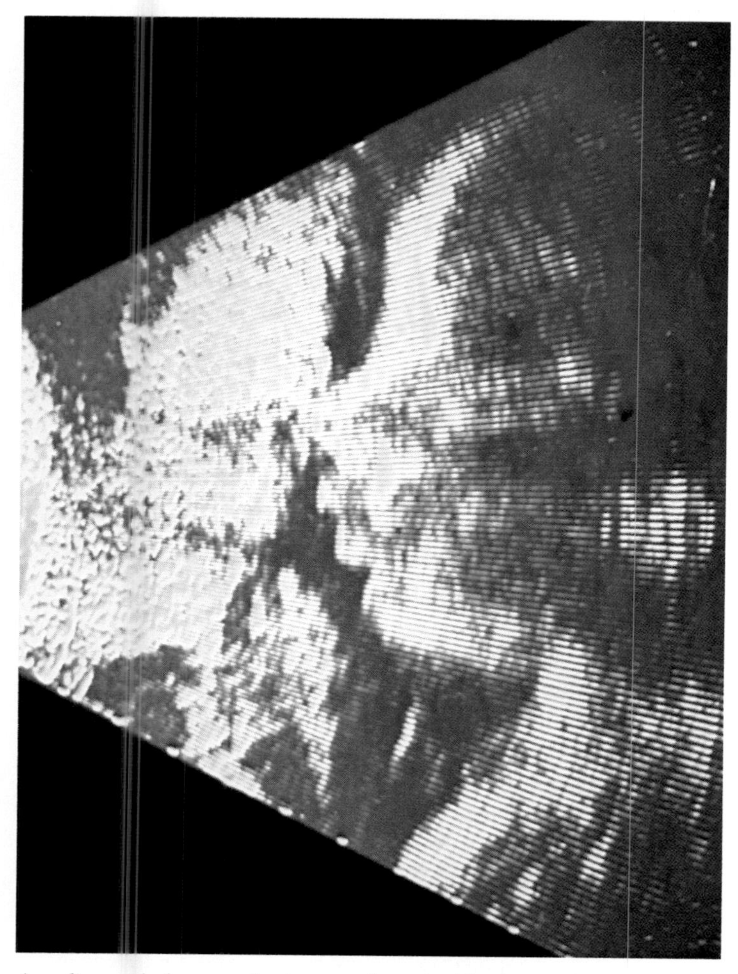

An ultrasonic image of a human fetus inside the mother's body. The image is constructed from information contained in the amplitude and phase of reflected ultrasonic waves (sound waves whose frequencies are higher than the upper limit heard by human beings).

Our boundary is the junction between the strings of Fig. 17-5. Both strings are horizontal and under the same tension F, and the linear mass densities of the left and right strings are μ_1 and μ_2, respectively. The incident pulse at the boundary is represented by $y_{\text{in}}(x - v_1 t)$, the reflected pulse by $y_{\text{re}}(x + v_1 t)$, and the transmitted pulse by $y_{\text{tr}}(x - v_2 t)$. Both the incident and the reflected pulses travel on the left string at a speed $v_1 = \sqrt{F/\mu_1}$, while the transmitted pulse propagates down the right string at a speed $v_2 = \sqrt{F/\mu_2}$. By the superposition principle, the net vertical displacement of the left string is $y_{\text{in}}(x - v_1 t) + y_{\text{re}}(x + v_1 t)$.

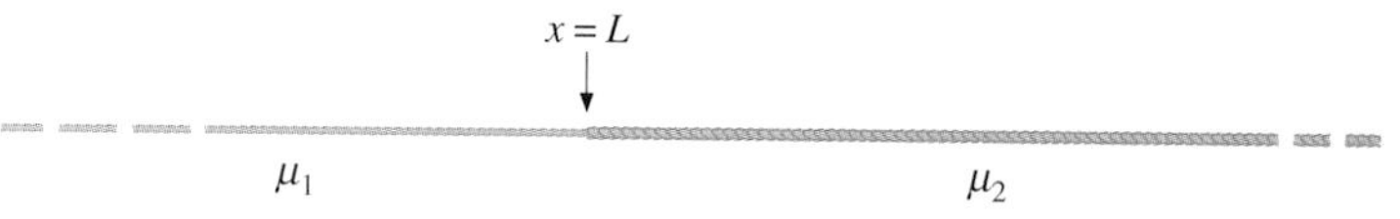

FIGURE 17-5 A junction between dissimilar strings. The tensions in the strings are the same, but the strings have different linear mass densities.

There are two independent conditions that the pulses must satisfy at the boundary. First, as Fig. 17-6*a* shows, the net vertical displacement on both sides of the boundary must be the same at all times. This is expressed mathematically by

$$[y_{\text{in}}(L - v_1 t) + y_{\text{re}}(L + v_1 t)] = y_{\text{tr}}(L - v_2 t),$$

where L, the position of the boundary, has been substituted for x. The second condition is a consequence of the fact that the two strings must exert equal and opposite forces on each other at the junction. As Fig. 17-6*b* illustrates, $\mathbf{F}_1 = -\mathbf{F}_2$ only holds if the slopes of the two strings are the same at the junction. We therefore have

$$\left\{\frac{\partial}{\partial x}[y_{\text{in}}(x - v_1 t) + y_{\text{re}}(x + v_1 t)]\right\}_{x=L} = \left\{\frac{\partial}{\partial x}[y_{\text{tr}}(x - v_2 t)]\right\}_{x=L}.$$

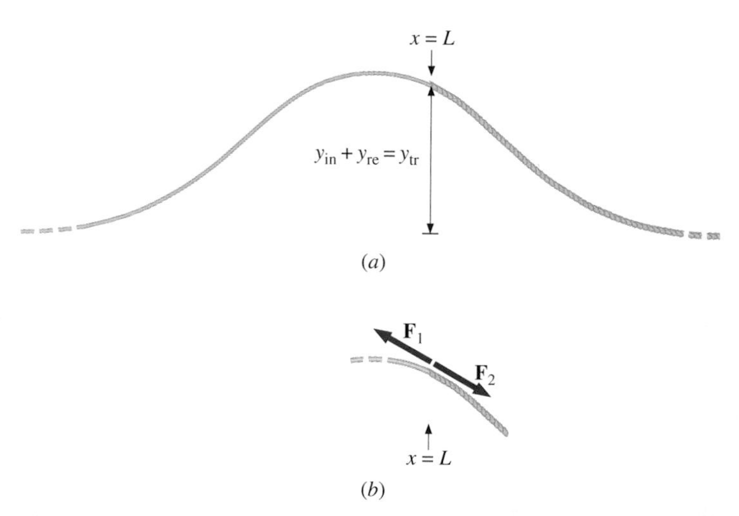

FIGURE 17-6 Boundary conditions: (*a*) The vertical displacements of the strings must be the same; (*b*) The slopes of the strings must be the same if they are to exert equal and opposite forces on each other.

Supplement 17-1 shows how these two conditions give the transmitted and the reflected pulses *at* $x = L$ in terms of the incident pulse. We find there that

$$y_{\text{tr}}(L - v_2 t) = \left(\frac{2v_2}{v_1 + v_2}\right) y_{\text{in}}(L - v_1 t) \qquad (17\text{-}6)$$

and

$$y_{re}(L + v_1 t) = \left(\frac{v_2 - v_1}{v_1 + v_2}\right) y_{in}(L - v_1 t). \quad (17\text{-}7)$$

The situations for $v_1 > v_2$ and for $v_1 < v_2$ are depicted in Fig. 17-7*a* and *b*, respectively. In both cases, the transmitted pulse has the same orientation as the incident pulse. The reflected pulse, however, has the same orientation as the incident pulse only if $v_1 < v_2$; it becomes inverted relative to the incident pulse if $v_1 > v_2$.

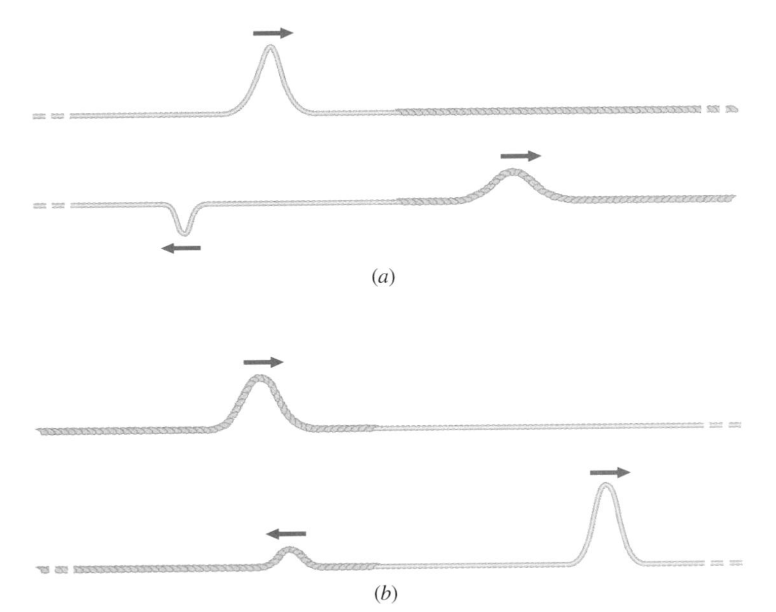

FIGURE 17-7 The reflection and transmission of a wave pulse at a boundary between two strings. (*a*) Incident pulse in the lighter string. (*b*) Incident pulse in the heavier string.

Now let's take a look at what happens under certain limiting conditions for the right string. One limit is that this string becomes infinitely dense and cannot move. It then acts like a rigid structure, and we are left with a single string with a fixed end. The other limiting condition occurs when the density of the right string is so small that it has no effect on the left string. We then have a single string whose end is completely free. A discussion of these two cases follows.

Fixed End of a String

The pulse of Fig. 17-8 is produced by exerting a force, first directed up, then down, on the held end of the string. When the pulse reaches the opposite end of the string, which is fixed, the string exerts the identical up-down force on the support. From Newton's third law, the support will exert an equal and opposite force on the string at all instants. This force is therefore directed first down and then up. It produces a pulse that is inverted but otherwise identical to the original. This reflected pulse travels with the same speed as the incident pulse as it moves back along the same string.

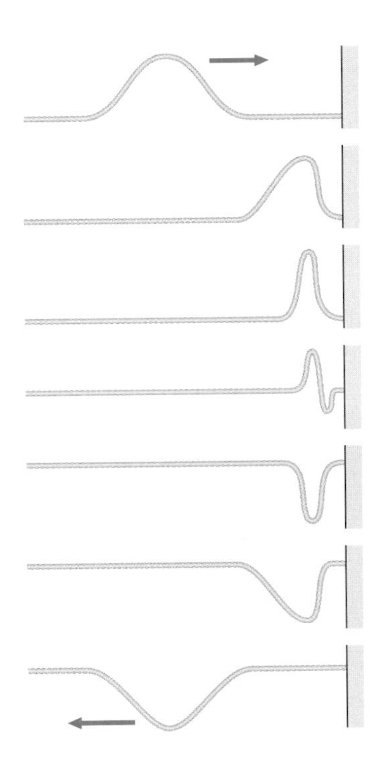

FIGURE 17-8 Reflection of a wave pulse by the fixed end of a string. The reflected pulse is always inverted.

Equation (17-7) gives us a mathematical description of the reflected pulse. Since the rigid support is equivalent to a string of infinite density, $v_2 = \sqrt{F/\mu_2} = 0$, and we obtain

$$y_{re}(L + v_1 t) = \left(\frac{0 - v_1}{v_1 + 0}\right) y_{in}(L - v_1 t) = -y_{in}(L - v_1 t).$$

Thus at the boundary, the reflected pulse is an inverted image of the incident pulse.

A helpful way to visualize the reflection of a pulse at a fixed boundary is illustrated in Fig. 17-9*a*. Think of the string as extending indefinitely beyond the support point and imagine that the pulse passes into this other region as though the boundary weren't there. At the same time that the pulse approaches and crosses the boundary, an identical but inverted pulse approaches and crosses the boundary from the other side. The actual displacement of the string is then the sum of these two pulses, as indicated in the figure.

Free End of a String

To make the end of the string completely free in the transverse direction, we attach it to a ring of negligible mass that slides without friction along a vertical rod. (See Fig. 17-10.) When the pulse reaches the end, an upward force is exerted on the attached ring. Since its inertia is negligible, the ring accelerates upward very quickly only to be limited by the inertia of the attached string. The ring reaches a maximum height above the crest of the original pulse and is then pulled back down by the latter half of the pulse. The result is a reflected pulse that is upright like the incident pulse.

We again turn to Eq. (17-7) for a mathematical description of this process. In this case, $\mu_2 \to 0$, so $v_2 = \sqrt{F/\mu_2} \to \infty$, and at the boundary,

$$y_{re}(L + v_1 t) = \left(\frac{\infty - v_1}{v_1 + \infty}\right) y_{in}(L - v_1 t) = y_{in}(L - v_1 t).$$

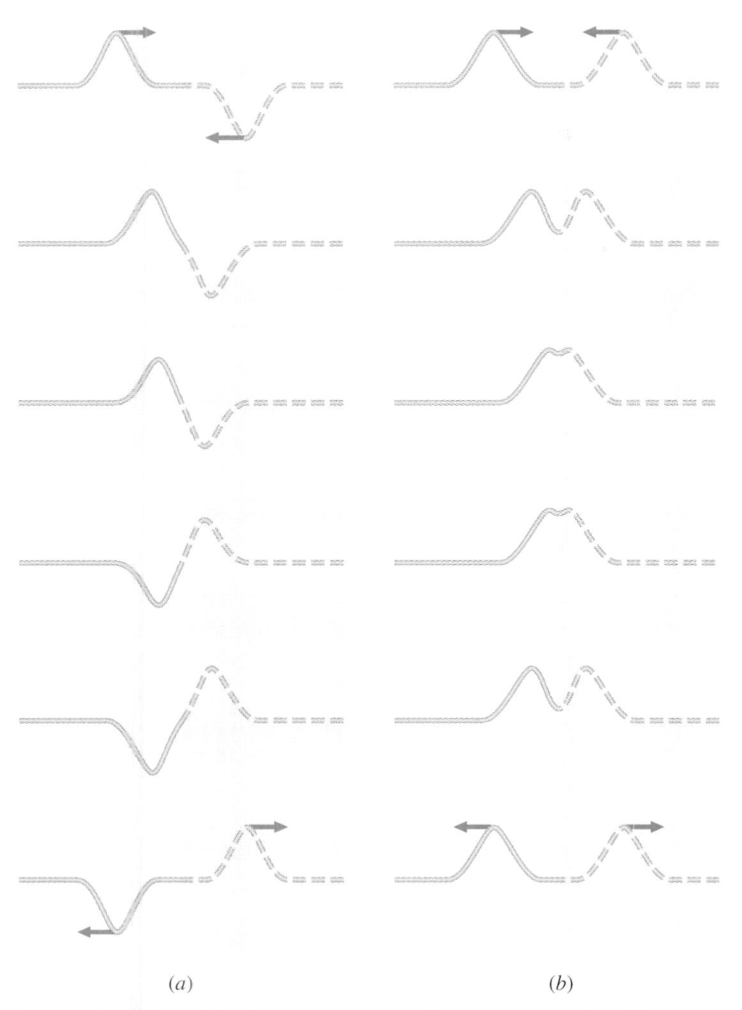

FIGURE 17-9 The reflection of a wave pulse can be visualized by considering an imaginary pulse approaching the boundary from the other side. The actual displacement of the string is then the sum of the pulses as they continue through the boundary. (*a*) Reflection at a fixed end; (*b*) reflection at a free end.

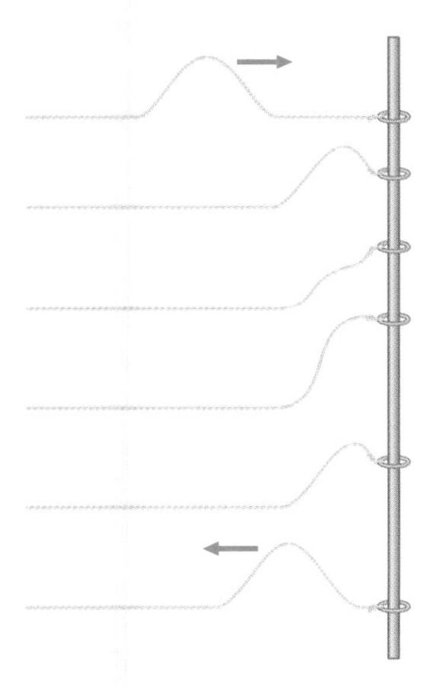

FIGURE 17-10 Reflection of a wave pulse by the free end of a string. The reflected pulse is always upright.

Except for its direction of motion, the reflected pulse is identical to the incident pulse.

We can also treat reflection at a free end as the superposition of the actual pulse and an imaginary pulse moving in opposite directions toward and through the boundary. (See Fig. 17-9*b*.) However, in this case the imaginary pulse is not inverted.

EXAMPLE 17-3 THE SPEED, FREQUENCY, AND WAVELENGTH OF REFLECTED AND TRANSMITTED HARMONIC WAVES

A harmonic wave is traveling along a string to which is attached a second string of greater linear mass density. Compare (*a*) the speed, (*b*) the frequency, and (*c*) the wavelength of the transmitted and reflected harmonic waves with those of the incident wave.

SOLUTION (*a*) The speed of a wave on a string is $\sqrt{F/\mu}$. Since the strings are under the same tension, a wave will propagate faster along the lighter string. The incident and reflected waves, which have the same speed, therefore travel faster than the transmitted wave.

(*b*) If the wave frequencies in the two strings were different, the strings on either side of the junction would oscillate at different rates and separate. Since this can't happen, the frequencies of the harmonic waves are the same in both strings.

(*c*) The speed v of the transmitted wave decreases and its frequency f is unchanged. From the relationship $\lambda = v/f$, we see that the wavelength of the transmitted wave is less than that of the incident wave. Finally, because v and f are the same for the reflected and incident waves, their wavelengths are equal.

EXAMPLE 17-4 TRANSMITTED AND REFLECTED HARMONIC WAVES

(*a*) Suppose that the linear mass density of the second string of Example 17-3 is four times that of the first string, and that the boundary between the two strings is at $x = 0$. If the expression for the incident wave is

$$y_{\text{in}}(x, t) = A_{\text{in}} \cos(k_1 x - \omega_1 t),$$

what are the expressions for the transmitted and the reflected waves in terms of A_{in}, k_1, and ω_1? (*b*) Show that the average power carried by the incident wave is equal to the sum of the average power carried by the transmitted and reflected waves.

SOLUTION (*a*) We begin by considering the wave speeds in the two strings. Since $v = \sqrt{F/\mu}$, $F_2 = F_1$, and $\mu_2 = 4\mu_1$, we have

$$v_2 = \sqrt{\frac{F_2}{\mu_2}} = \sqrt{\frac{F_1}{4\mu_1}} = \frac{1}{2}\sqrt{\frac{F_1}{\mu_1}} = \frac{1}{2} v_1. \qquad (i)$$

From Example 17-3, we know that the frequencies of the incident, reflected, and transmitted waves are the same, that is,

$$f_1 = f_2,$$

so

$$\omega_1 = \omega_2. \qquad (ii)$$

Also, because $k = \omega/v$, the wave numbers of the harmonic waves in the two strings are related by

$$k_2 = \frac{\omega_2}{v_2} = \frac{\omega_1}{(v_1/2)} = 2\frac{\omega_1}{v_1} = 2k_1. \qquad (iii)$$

Now we can compare the amplitudes of the waves. We assume that the transmitted and reflected waves are given by

$$y_{\text{tr}}(x, t) = A_{\text{tr}} \cos (k_2 x - \omega_2 t)$$

and

$$y_{\text{re}}(x, t) = A_{\text{re}} \cos (k_1 x + \omega_1 t),$$

respectively. Then, using Eqs. (17-6), (17-7), (*i*), and (*ii*), along with $\cos(-\omega_1 t) = \cos(\omega_1 t)$, we have

$$A_{\text{tr}} = \left(\frac{2v_2}{v_1 + v_2}\right)A_{\text{in}} = \left[\frac{2(v_1/2)}{v_1 + (v_1/2)}\right]A_{\text{in}} = \frac{2}{3}A_{\text{in}}, \qquad (iv)$$

and

$$A_{\text{re}} = \left(\frac{v_2 - v_1}{v_1 + v_2}\right)A_{\text{in}} = \left[\frac{(v_1/2) - v_1}{v_1 + (v_1/2)}\right]A_{\text{in}} = -\frac{1}{3}A_{\text{in}}. \qquad (v)$$

Now with Eqs. (*ii*), (*iii*), and (*iv*), the transmitted wave can be written in terms of A_{in}, k_1, and ω_1 as

$$y_{\text{tr}}(x, t) = \tfrac{2}{3}A_{\text{in}} \cos (2k_1 x - \omega_1 t).$$

Similarly, the reflected wave can be expressed as

$$y_{\text{re}}(x, t) = -\tfrac{1}{3}A_{\text{in}} \cos (k_1 x + \omega_1 t)$$
$$= \tfrac{1}{3}A_{\text{in}} \cos (k_1 x + \omega_1 t + \pi).$$

(*b*) The average power of a harmonic wave on a string is given by Eq. (16-15*a*). The average power of the incident wave is then

$$\bar{P}_{\text{in}} = \tfrac{1}{2}\omega_1^2 A_{\text{in}}^2 \mu_1 v_1;$$

and the average transmitted and reflected powers are

$$\bar{P}_{\text{tr}} = \tfrac{1}{2}\omega_1^2(\tfrac{2}{3}A_{\text{in}})^2(4\mu_1)(\tfrac{1}{2}v_1) = \tfrac{4}{9}\omega_1^2 A_{\text{in}}^2 \mu_1 v_1$$

and

$$\bar{P}_{\text{re}} = \tfrac{1}{2}\omega_1^2(-\tfrac{1}{3}A_{\text{in}})^2 \mu_1 v_1 = \tfrac{1}{18}\omega_1^2 A_{\text{in}}^2 \mu_1 v_1.$$

Now,

$$\bar{P}_{\text{tr}} + \bar{P}_{\text{re}} = (\tfrac{4}{9} + \tfrac{1}{18})\omega_1^2 A_{\text{in}}^2 \mu_1 v_1 = \tfrac{1}{2}\omega_1^2 A_{\text{in}}^2 \mu_1 v_1 = \bar{P}_{\text{in}}.$$

Thus the power leaving the junction between the two strings ($\bar{P}_{\text{tr}} + \bar{P}_{\text{re}}$) is equal to the power entering the junction ($\bar{P}_{\text{in}}$), and energy is conserved.

DRILL PROBLEM 17-3

Assume that the incident pulse of Fig. 17-7*b* has a speed of 10 cm/s. (*a*) If the linear mass density of the right string is 0.25 that of the left string, at what speed does the transmitted pulse travel? (*b*) Compare the heights of the transmitted pulse and the reflected pulse to that of the incident pulse.
ANS. (*a*) 20 cm/s; (*b*) $y_{\text{tr}}(L, t)/y_{\text{in}}(L, t) = 4/3$ and $y_{\text{re}}(L, t)/y_{\text{in}}(L, t) = 1/3$.

17-5 Standing Waves

The expressions

$$y_1(x, t) = A \sin (kx - \omega t) \qquad (17\text{-}8a)$$

and

$$y_2(x, t) = A \sin (kx + \omega t) \qquad (17\text{-}8b)$$

represent two harmonic waves of the same amplitude and frequency moving in opposite directions with the same speed. By the superposition principle, the resultant of the two waves is

$$y(x,t) = y_1 + y_2 = A[\sin (kx - \omega t) + \sin (kx + \omega t)],$$

which can be simplified with the trigonometric identity of Eq. (17-1) to

$$y(x, t) = 2A \sin kx \cos \omega t. \qquad (17\text{-}9)$$

This expression is different from the wave representations that we have encountered up to now. It doesn't have the form $F(x \pm vt)$ and therefore does *not* describe a traveling wave. Instead, Eq. (17-9) represents what is known as a *standing wave*.

The two traveling waves and their resultant standing wave are sketched in Fig. 17-11 for four different times over one period T of the traveling waves: $t = 0$, $T/4$, $T/2$, and $3T/4$. At $t = 0$, the two traveling waves have the same displacement everywhere and add to produce the standing wave shown. At $t = T/4$, the waves have each moved a distance of $\lambda/4$ in opposite directions, so they differ in phase by π rad and completely

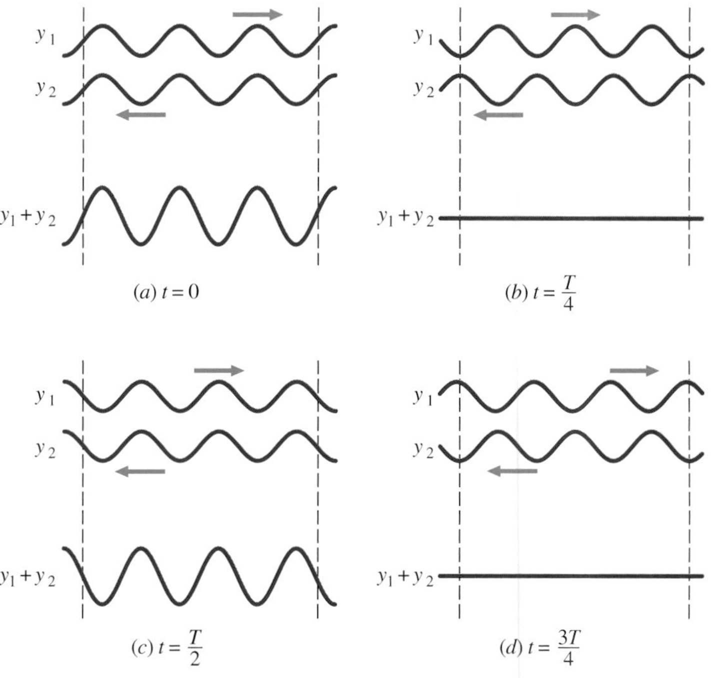

FIGURE 17-11 The superposition of two harmonic waves with equal amplitudes traveling in opposite directions produces a standing wave. The sum is shown for the times $t = 0$, $T/4$, $T/2$, and $3T/4$, where T is the period of the traveling waves.

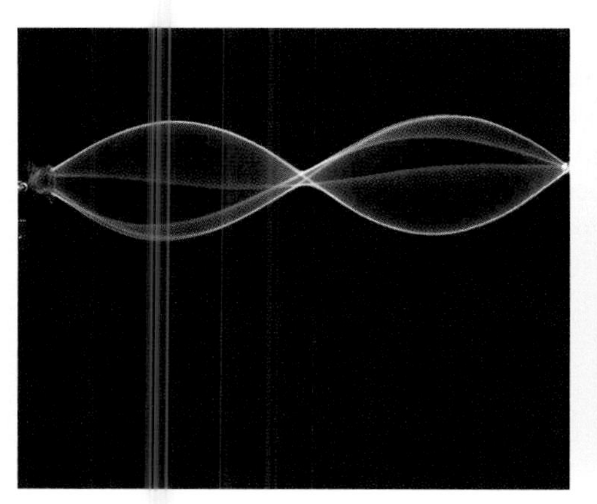
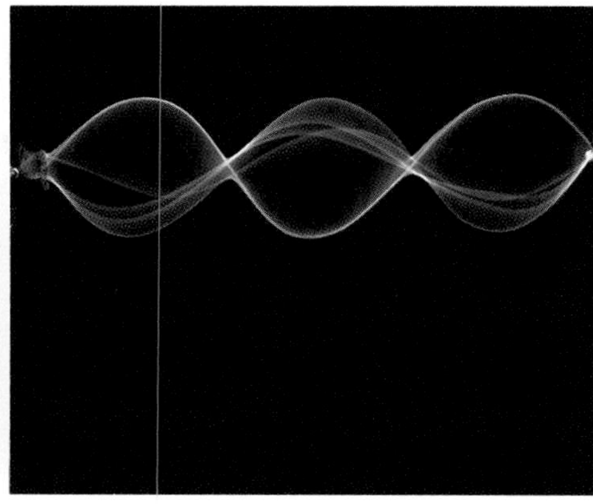
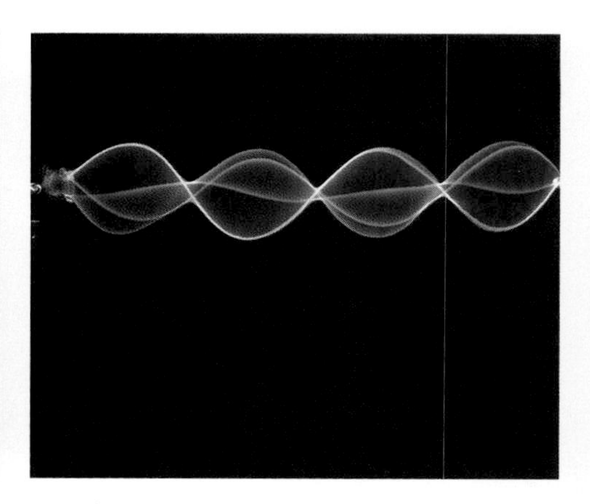

FIGURE 17-12 Time exposures of standing waves on a stretched string.

cancel. At $t = T/2$, they again align spatially. However, by comparing Fig. 17-11a and c, we see that the positions of the peaks and troughs of the standing waves at $t = 0$ and $T/2$ have been interchanged. At $t = 3T/4$, the traveling waves completely cancel once more. Finally at $t = T$, or after one period, the standing wave reassumes the shape it had at $t = 0$.

Time exposures of standing waves on a string over many periods are shown in Fig. 17-12. There are points on the string that are never displaced—these are known as the *nodes* of the standing wave. Between the nodes are points of maximum displacement called the *antinodes*. They are located at the centers of the envelopes of the standing wave pattern. While all points on the string are in simple harmonic motion with a period T, the amplitude of the motion, $2A \sin kx$, varies from point to point, ranging from zero at the nodes to $2A$ at the antinodes.

We can find where the nodes are by using the condition that $y(x, t) = 0$ at a node for any t. Then from Eq. (17-9),

$$2A \sin kx = 0.$$

This equation is satisfied by those values of x such that $kx = n\pi$, where $n = 0, 1, 2, \ldots$. Because $k = 2\pi/\lambda$, the positions of the nodes are

$$x = 0, \frac{\lambda}{2}, \lambda, \frac{3\lambda}{2}, \ldots . \qquad (17\text{-}10)$$

The separation of the nodes is therefore $\lambda/2$, where λ is the wavelength of the traveling waves. To locate the antinodes, we simply find those values of x for which the displacement is a maximum. They correspond to $\sin kx = \pm 1$, so $kx = (n + 1/2)\pi$ and the positions of the antinodes are

$$x = \frac{\lambda}{4}, \frac{3\lambda}{4}, \frac{5\lambda}{4}, \frac{7\lambda}{4}, \ldots . \qquad (17\text{-}11)$$

As expected, the separation of the antinodes is also $\lambda/2$.

Since the nodes of the string are always at rest, energy cannot flow past them. Consequently, energy is not transported along a string vibrating in a standing-wave pattern. The energy remains "standing" in the string as each string particle between the nodes oscillates in simple harmonic motion.

17-6 Standing Waves on a String

In Fig. 17-12 the standing waves are produced by fixing the string at one end and attaching the other end to a vibrator oscillating at the appropriate frequency. The harmonic waves created by the vibrations travel down the string, are reflected when they reach the fixed end, and then move back toward the vibrator. If the string has the appropriate length, the waves traveling in opposite directions add to produce standing waves. Since the oscillations of a vibrator are generally very small, we can treat the string as fixed at both ends. These ends must therefore be two of the nodes of any standing wave that is formed. So from Eq. (17-10), the condition that must be satisfied by a standing wave on the string is

$$L = n\left(\frac{\lambda_n}{2}\right), \qquad (17\text{-}12)$$

where L is the length of the string, $n = 1, 2, 3, \ldots$, and λ_n are the possible wavelengths of the traveling waves. Since the wave speed on the string is $v = \sqrt{F/\mu}$, the corresponding frequencies at which particles of the string oscillate are

$$f_n = \frac{v}{\lambda_n} = n\left(\frac{v}{2L}\right) = \frac{n}{2L}\sqrt{\frac{F}{\mu}} \qquad (n = 1, 2, 3, \ldots). \qquad (17\text{-}13)$$

Several examples of standing-wave patterns are shown in Fig. 17-13. These patterns are called the **normal modes** of the string. The wave pattern with the lowest frequency, $f_1 = v/2L$, is shown in Fig. 17-13a. It is the **fundamental mode**, or **first harmonic**, of the string, and f_1 is the string's **fundamental frequency**. From Eq. (17-12), the wavelength λ_n and the length L of the string are related by $\lambda_n = 2L/n$. Hence, for the first harmonic, $\lambda_1 = 2L$. Figure 17-13b shows the **first overtone**, or the second harmonic, of the string, whose oscillation frequency is f_2. When $n = 3$, the frequency is f_3 and the string is vibrating in the second overtone, or third harmonic, and so on.

Notice that all standing-wave frequencies are related to the first harmonic by

$$f_n = nf_1 \qquad (n = 1, 2, 3, \ldots). \qquad (17\text{-}14)$$

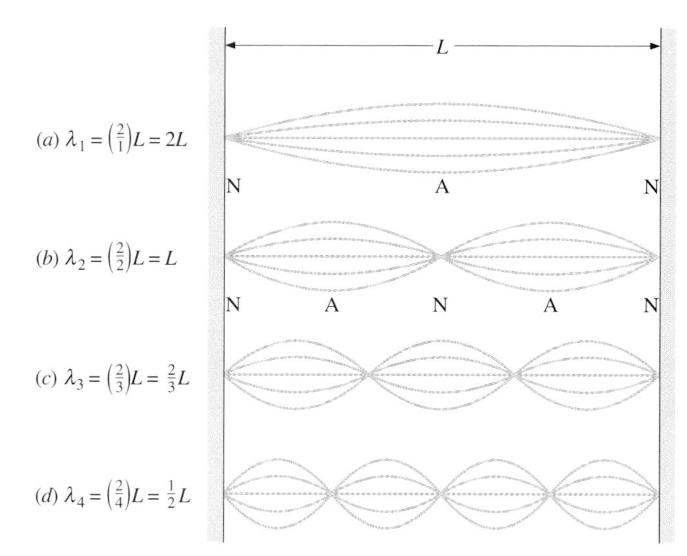

FIGURE 17-13 Standing-wave patterns on a string. The envelopes represent the shapes of the string at different instants. Cases (*a*) through (*d*) correspond to the first four harmonics, respectively. The nodes (N) and the antinodes (A) are denoted for (*a*) and (*b*).

This set of frequencies represents the **resonant frequencies**, or **natural frequencies**, of oscillation of the string. At resonance, energy is easily transferred from the vibrator to the string while essentially no transfer in the opposite direction occurs. As a result, the oscillations quickly grow in amplitude until the string loses energy to friction as fast as it gains energy from the vibrator.

If the frequency of the vibrator does not match one of the resonant frequencies of the string, the reflected wave will be out of phase with the vibrator when it returns to the vibrator. In this case, the string can transmit energy to the vibrator as well as receive energy from it. As a result, energy transfer to the string is inefficient, so the vibrations are small and standing-wave patterns are not produced.

Here's an instructive way to look at the connection between a standing wave and the traveling waves that produce it. Consider what happens to a particular crest of the traveling wave between two instants when it has exactly the same position and velocity. During that time interval, it travels to one end of the string, is reflected, travels to the other end, is reflected again, and then returns to its original position. In this process, the crest has traveled a distance $2L$ and undergone two reflections. Since it is inverted each time it is reflected, it is erect when it gets back to its original position. If $2L$ is equal to an integer number of wavelengths, then the crest will exactly coincide with another crest produced at a later instant. Through this process, a standing wave is produced with an amplitude much greater than that of its constituent traveling waves. This situation can only occur if $2L = n\lambda_n$, which is the standing-wave condition of Eq. (17-12).

The guitarist varies the resonant frequencies of the strings by pressing them down at different points along the fingerboard.

EXAMPLE 17-5 A Violin String

The length of the G string on a violin is $L = 32.0$ cm, and its mass is $m = 4.50 \times 10^{-4}$ kg. (*a*) What tension F must the string be subjected to so that its fundamental frequency will be $f_1 = 196$ Hz? (*b*) What are the frequencies of the second and third harmonics of the string?

SOLUTION (*a*) From Eq. (17-13), the fundamental frequency of the string is

$$f_1 = \frac{1}{2L}\sqrt{\frac{F_1}{\mu}},$$

which when solved for the tension gives $F_1 = 4f_1^2L^2\mu$. Since the linear mass density of the string is $\mu = m/L$, the tension is also given by $F = 4f_1^2Lm$, so the tension in the G string is

$$F = 4(196\text{ Hz})^2(0.320\text{ m})(4.50 \times 10^{-4}\text{ kg}) = 22.1\text{ N}.$$

(*b*) Since the frequencies of the higher harmonics are integer multiples of the fundamental frequency, the frequencies of the second and third harmonics are

$$f_2 = 2f_1 = 392\text{ Hz} \quad \text{and} \quad f_3 = 3f_1 = 588\text{ Hz}.$$

EXAMPLE 17-6 Fundamental Frequency of a Vibrating Cord

A cord of length 1.00 m and mass 20.0 g is stretched to a tension of 500 N and clamped at both ends. (*a*) What is its fundamental frequency? (*b*) Suppose that the breaking tension of the cord is 1200 N. What is the maximum fundamental frequency that can be produced in the cord?

SOLUTION (*a*) The linear mass density of the cord is $\mu = (2.00 \times 10^{-2}\text{ kg})/(1.00\text{ m}) = 2.00 \times 10^{-2}$ kg/m. The wave speed is therefore

$$v = \sqrt{\frac{F}{\mu}} = \sqrt{\frac{500\text{ N}}{2.00 \times 10^{-2}\text{ kg/m}}} = 160\text{ m/s},$$

and the fundamental frequency is

$$f_1 = \frac{v}{2L} = \frac{160\text{ m/s}}{2(1.00\text{ m})} = 80\text{ Hz}.$$

(*b*) Increasing the tension produces an increase in the fundamental frequency. For the maximum tension the cord can support, we find from Eq. (17-13) that the corresponding fundamental frequency is

$$f_1 = \frac{1}{2L}\sqrt{\frac{F}{\mu}} = \frac{1}{2(1.00\text{ m})}\sqrt{\frac{1200\text{ N}}{2.00 \times 10^{-2}\text{ kg/m}}} = 122\text{ Hz}.$$

DRILL PROBLEM 17-4

A string fixed at both ends has a fundamental frequency of 100 Hz. The wave speed on the string is 500 m/s. What is the length of the string?
ANS. 2.50 m.

DRILL PROBLEM 17-5

Suppose you wanted to triple the fundamental frequency of the cord of Example 17-6 by adjusting its tension. (*a*) Would the cord break during this process? (*b*) If the cord were not to break, what would be its new tension? (*c*) How would the higher harmonics be affected?
ANS. (*a*) Yes; (*b*) 4500 N; (*c*) their frequencies would be tripled.

17-7 Standing Waves in a Tube

Standing waves can also be produced by longitudinal waves traveling in an elastic medium. The sounds from musical instruments such as horns and organ pipes are produced by standing waves in tubes filled with air. A simple method that can be used to produce standing waves in a tube is illustrated in Fig. 17-14. The tuning fork produces harmonic sound waves that travel down the tube, are reflected at the far end, and move back toward the tuning fork. If the tube has the right length, the waves traveling in opposite directions interfere to produce standing waves.

FIGURE 17-14 Standing waves can be produced in an air-filled tube of the appropriate length.

The boundary conditions for reflections at a closed end and at an open end of a tube are different, as illustrated in Fig. 17-15*a* and *b*. Since there is no displacement of the air at a closed end, *the closed end of a tube is a displacement node* ($s = 0$). We saw in Sec. 16-4 that the amplitudes of the pressure oscillations are largest where those of the displacement oscillations are smallest and vice versa. Consequently, *the closed end of a tube is a pressure antinode.* At the open end of a tube, the air pressure is always equal to that of the surrounding atmosphere, so $\Delta p = 0$ there. Thus *the open end of a tube is a pressure node and a displacement antinode.**

*Actually, the pressure node is located slightly beyond the end. For our purposes, however, we can neglect this small correction if the length of the tube is much larger than its diameter.

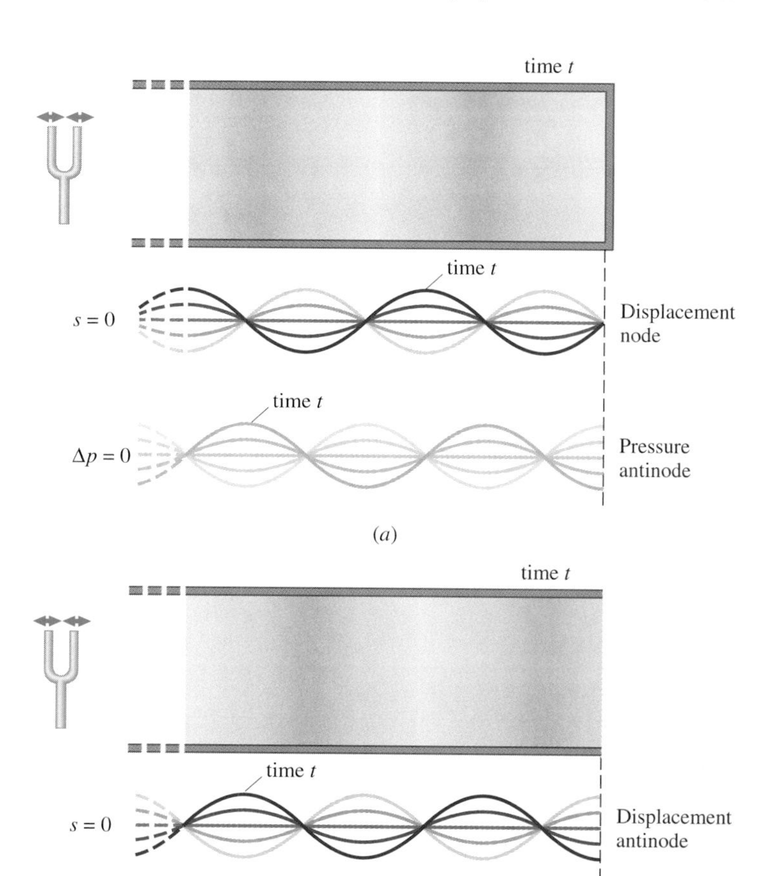

FIGURE 17-15 Boundary conditions at (*a*) the closed end and (*b*) the open end of an air-filled tube. The standing waves shown represent the displacement and the pressure deviation at various instants.

We will first investigate the standing waves formed in a tube open at both ends. Since both ends of the tube are open, there are pressure nodes (or displacement antinodes) at both ends. Figure 17-16 shows the resulting standing waves for the four lowest resonant frequencies. Since the distance between pressure nodes is $\lambda/2$, the resonance condition is $L = n(\lambda_n/2)$, where $n = 1, 2, 3, \ldots$ and L is the length of the tube. The resonant frequencies for a tube open at both ends are then

$$f_n = \frac{v}{\lambda_n} = n\,\frac{v}{2L} = nf_1 \qquad (n = 1, 2, 3, \ldots), \tag{17-15}$$

where v is the speed of sound in the tube and f_1 is the fundamental frequency $v/2L$.

If the tube is closed at one end and open at the other, resonance corresponds to a pressure antinode at the closed end and a pressure node at the open end. The standing-wave patterns

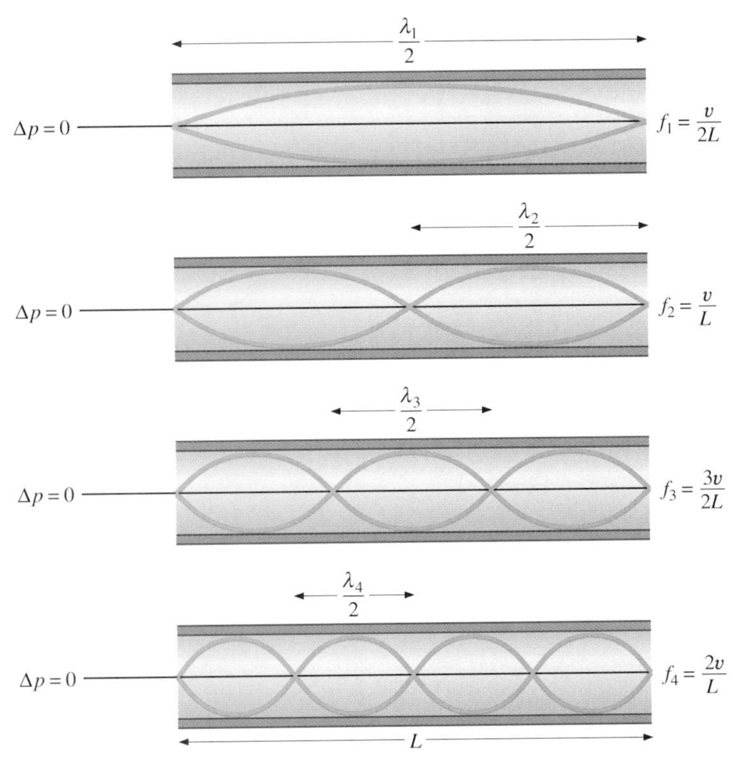

FIGURE 17-16 Standing pressure waves for the four lowest resonant frequencies in a tube with both ends open. The envelopes correspond to the extremes of the pressure variation.

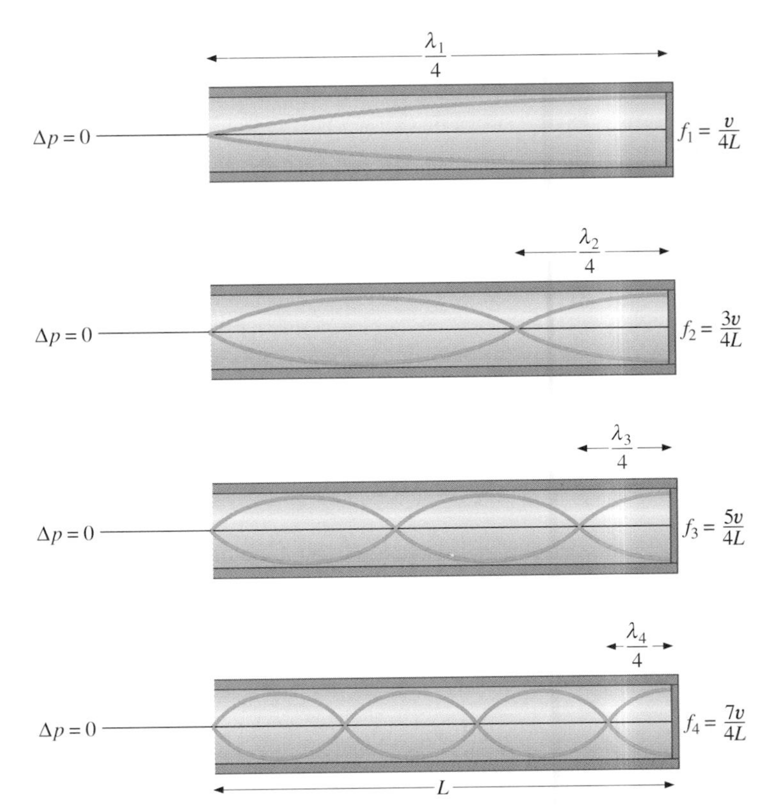

FIGURE 17-17 Standing pressure waves for the four lowest resonant frequencies in a tube with one end open and one end closed. The envelopes correspond to the extremes of the pressure variation.

for the four lowest harmonics in this situation are shown in Fig. 17-17. Since the node-antinode separation is $\lambda/4$, the resonance condition for the first harmonic is $L = \lambda_1/4$. Similarly, the resonance conditions for the higher harmonics are $L = 3\lambda_2/4$, $5\lambda_3/4$, $7\lambda_4/4, \ldots$. The natural frequencies of oscillation of the air in the tube closed at one end and open at the other are therefore

$$f_n = \frac{v}{\lambda_n} = n\frac{v}{4L} = nf_1 \qquad (n = 1, 3, 5, \ldots), \tag{17-16}$$

where the fundamental frequency is now $f_1 = v/4L$.

A pipe organ. The lengths of the pipes determine the frequencies of the notes.

Notice the difference between the resonant frequencies for the two types of tubes. For the tube open at both ends, the fundamental frequency is $v/2L$ and *all* higher harmonics exist. If one end of the tube is closed, the fundamental frequency is $v/4L$ and only the *odd* harmonics exist. The air in this tube resonates only at odd multiples of the fundamental frequency.

EXAMPLE 17-7 Harmonics of Air Vibrating in a Tube

A thin uniform tube filled with air is 0.850 m long. Calculate the frequencies of the first three harmonics if (*a*) both ends of the tube are open and (*b*) one end is open. Assume that the speed of sound in air is 340 m/s.

SOLUTION (*a*) In a tube with both ends open, the fundamental frequency is

$$f_1 = \frac{v}{2L} = \frac{340 \text{ m/s}}{2(0.850 \text{ m})} = 200 \text{ Hz}.$$

The frequencies of the second and third harmonics are therefore

$$f_2 = 2f_1 = 400 \text{ Hz} \qquad \text{and} \qquad f_3 = 3f_1 = 600 \text{ Hz}.$$

(*b*) In a tube with one end open, the fundamental frequency is

$$f_1 = \frac{v}{4L} = \frac{340 \text{ m/s}}{4(0.850 \text{ m})} = 100 \text{ Hz}.$$

From Eq. (17-16), the second harmonic does not exist, and the frequency of the third harmonic is

$$f_3 = 3f_1 = 300 \text{ Hz}.$$

DRILL PROBLEM 17-6

Two tubes filled with air are found to have the same fundamental frequency. One tube has both ends open while the other has one open end. What is the ratio of the lengths of the tubes? Compare the frequencies of their third harmonics.
ANS. 2:1 with the tube open at both ends longer; they are the same.

Four of the many vibrational modes of a kettledrum. The powder on the drumhead is distributed along the nodes.

*17-8 Music and Musical Instruments

It's easy to distinguish between the sounds that a violin, a piano, and a trumpet make. Each instrument produces musical notes that are particular to that instrument. Yet the basis for the production of the notes is the same—some component of the instrument is made to vibrate at its natural frequencies. In a string instrument such as a violin or a piano, it is clearly the strings that are set into oscillatory motion. In wind instruments such as a clarinet or an organ, standing waves are formed in a tube filled with air. In a drum, a stretched membrane vibrates at its resonant frequencies. Finally, rods and plates are also capable of resonance, and they are used in musical instruments such as the xylophone.

Let's consider how a violin is designed for "making music." Its four strings are all fixed at both ends and have the same length; however, they differ in linear mass density and therefore in their fundamental frequencies. The thickest string has the lowest fundamental frequency because $f_1 = v/2L = (1/2L) \times \sqrt{F/\mu}$. A wide range of musical notes can be produced by pressing the various strings against the neck of the violin and, in essence, changing their lengths. When plucked or bowed, the strings are set into oscillations that are significant only at their harmonics. A musical note produced by a violin string is composed of several harmonics, with the lowest harmonic, or fundamental, determining the frequency of the note. The body of a violin is hollow and is set into vibration by the strings. Since the body has a much larger surface area in contact with the air than the strings have, it can produce a much more intense sound wave; thus the body acts to amplify the sound.

This Stradivarius violin is over 300 years old and is a highly prized instrument because of its beautiful tone. Copyright © 1979 by The Metropolitan Museum of Art.

Other musical instruments follow basically the same principles. Instead of plucking or bowing, we set a piano string into oscillatory motion by tapping one of its keys. This causes a felt-covered hammer to strike the corresponding string within the piano. The vibrating string sets a sounding board oscillating, which then transmits the note to the air. While an organ contains long tubes of air as counterparts to the piano strings, a woodwind such as an oboe or a clarinet has only a single tube (the bore, as shown in Fig. 17-18). Its harmonics, which are determined by its dimensions, are modified by side holes, which may be opened or closed. The device that is used to vibrate the

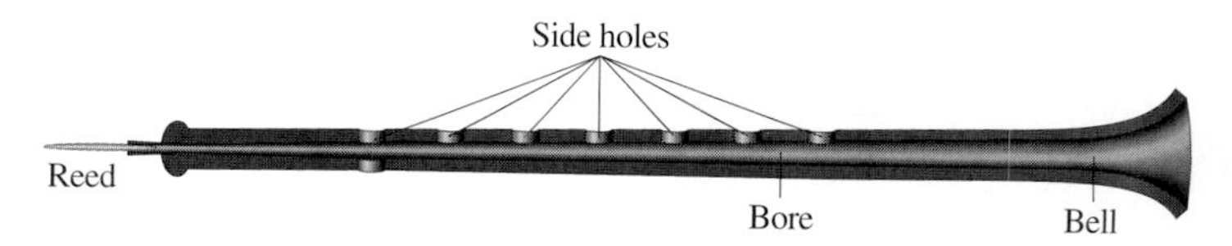

FIGURE 17-18 Cross-sectional representation of an oboe.

air in the bore is known as a reed. It is usually a blade of cane that vibrates simultaneously over a range of frequencies. However, the only frequencies for which the standing sound waves can build up to audible intensities are those corresponding to the resonant frequencies of the bore.

As noted earlier, the sounds produced by different instruments playing the same note are quite distinct. Even the untrained ear can easily distinguish a particular note produced by a trumpet from that produced by a clarinet or by a violin. The characteristic quality of the sound of an instrument is often referred to as the *timbre* of the instrument. The timbre depends on both the particular overtones in a note produced by an instrument and the relative amplitudes of these overtones. Figure 17-19 shows the overtones and their relative amplitudes for different instruments that are playing the notes A and B-flat just below middle C. You can see that the mixture of harmonics varies considerably among the instruments. Notice that the first harmonic does not necessarily have the largest amplitude.

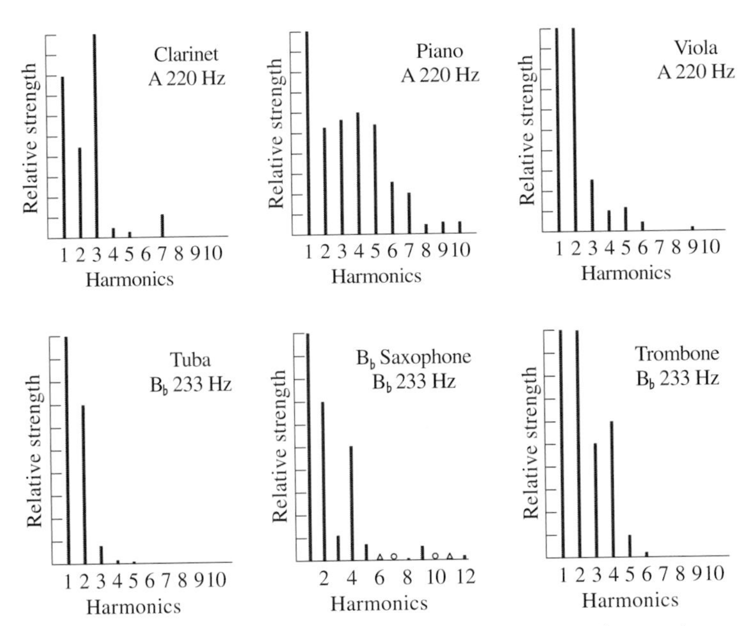

FIGURE 17-19 The relative strengths of the harmonics when various instruments play A below middle C and B-flat below middle C. (*From Charles A. Culver,* Musical Acoustics, *4th ed., McGraw Hill, New York, 1956.*)

The timbre of an instrument is also affected by the way it is played. For example, a note produced by a plucked violin string sounds quite different from the same note produced by pulling a bow across the string (although both notes have the same fundamental frequency). The timbre can also change while a string or air column is vibrating because different harmonics dissipate at different rates. Try listening to a sustained note produced by a piano or guitar. The change in timbre with time should be quite easy to detect.

Most vibrating objects do not produce what we normally interpret as "music." For example, the sound made by a flag flapping in the wind is not considered melodious. Nonmusical sounds are usually composed of a continuous (or almost continuous) range of frequencies, without fundamental modes or harmonics. We commonly refer to such sounds as "noise."

EXAMPLE 17-8 PLAYING A VIOLIN

The strings of a violin are each 30.0 cm long. When the A string is played without fingering, it vibrates at its fundamental frequency of 440 Hz (the musical note "middle A"). The next two higher notes on the C scale are B (494 Hz) and C (523 Hz). How far from the end of the string must one press to play these notes?

SOLUTION The string cannot move at the point where it is pressed, so a node exists there and the string is effectively shortened. The wave speed can be determined by applying Eq. (17-13) to the unfingered string of length $L_A = 0.300$ m:

$$f_1 = \frac{v}{2L_A}$$

$$440\text{ Hz} = \frac{v}{2(0.300\text{ m})},$$

so

$$v = 264\text{ m/s}.$$

If L_B is the length of the string when it is vibrating at 494 Hz (the B note), then

$$494\text{ Hz} = \frac{v}{2L_B} = \frac{264\text{ m/s}}{2L_B},$$

and

$$L_B = 0.267\text{ m} = 26.7\text{ cm}.$$

So in order to play this note, a violinist must press $(30.0 - 26.7)$ cm = 3.3 cm from the end of the A string.

Similarly, if L_C is the length of the string when it is vibrating at 523 Hz (the C note), then

$$523\text{ Hz} = \frac{v}{2L_C} = \frac{264\text{ m/s}}{2L_C},$$

and

$$L_C = 0.252\text{ m} = 25.2\text{ cm}.$$

This corresponds to a point 4.8 cm from the end of the A string.

DRILL PROBLEM 17-7

How far from the end of the A string on the violin of Example 17-8 must one press in order to produce an E note of frequency 659 Hz when the string is plucked?
ANS. 10.0 cm.

SUMMARY

1. **Principle of superposition**
The resultant of two or more waves is the sum of the individual waves.

2. **Superposition of two harmonic waves of the same frequency**
If $y_1(x, t) = A \sin(kx - \omega t)$ and $y_2(x, t) = A \sin(kx - \omega t + \phi)$, then the sum $y(x, t) = y_1(x, t) + y_2(x, t)$ is

$$y(x, t) = \left(2A \cos\frac{\phi}{2}\right) \sin\left(kx - \omega t + \frac{\phi}{2}\right).$$

When $\phi = 2n\pi$ ($n = 0, 1, 2, \ldots$), the waves interfere constructively, and when $\phi = (2n+1)\pi$ ($n = 0, 1, 2, \ldots$), the waves interfere destructively.

3. **Beats**
When two waves with slightly different frequencies f_1 and f_2 are superimposed, the resulting wave has a frequency of $(f_1 + f_2)/2$ and an amplitude that varies at a beat frequency of $f_1 - f_2$.

4. **Boundary conditions**
At a junction between dissimilar strings, the reflected pulse $y_{re}(x + v_1 t)$ and the transmitted pulse $y_{tr}(x - v_2 t)$ are related to the incident pulse $y_{in}(x - v_1 t)$ by

$$y_{re}(L + v_1 t) = \left(\frac{v_2 - v_1}{v_1 + v_2}\right) y_{in}(L - v_1 t)$$

and

$$y_{tr}(L - v_2 t) = \left(\frac{2v_2}{v_1 + v_2}\right) y_{in}(L - v_1 t),$$

where the junction is at $x = L$, and v_1 and v_2 are the speeds of the pulses on the two strings. If the junction is rigid, $v_2 = 0$, and

$$y_{re}(L + v_1 t) = -y_{in}(L - v_1 t).$$

If the junction is free, $v_2 \to \infty$, and

$$y_{re}(L + v_1 t) = y_{in}(L - v_1 t).$$

5. **Standing waves**
With a standing wave represented by

$$y(x, t) = 2A \sin kx \cos \omega t,$$

the nodes are located at $x = 0, \lambda/2, \lambda, 3\lambda/2, \ldots$, and the antinodes at $x = \lambda/4, 3\lambda/4, 5\lambda/4, \ldots$.

6. **Standing waves on a string**
For a string fixed at both ends, the frequencies of its standing waves are

$$f_n = \frac{n}{2L}\sqrt{\frac{F}{\mu}} \qquad (n = 1, 2, 3, \ldots),$$

where F is the tension in the string and μ is the linear mass density of the string.

7. **Standing waves in a tube**
The frequencies of the standing waves in a tube open at both ends are

$$f_n = n\frac{v}{2L} \qquad (n = 1, 2, 3, \ldots),$$

where v is the speed of sound in the gas filling the tube and L is the length of the tube. If the tube is closed at one end and open at the other, the frequencies are

$$f_n = n\frac{v}{4L} \qquad (n = 1, 3, 5, \ldots).$$

*8. **Music and musical instruments**
The notes played by strings and tubes correspond to the frequencies of their first harmonics. The overtones of these notes correspond to the frequencies of the higher harmonics.

SUPPLEMENT 17-1 BOUNDARY CONDITIONS

In this supplement we will obtain expressions for the transmitted and reflected pulses at a boundary between two strings in terms of the incident pulse. The conditions at the boundary, where $x = L$, are

$$y_{in}(L - v_1 t) + y_{re}(L + v_1 t) = y_{tr}(L - v_2 t) \qquad (i)$$

and

$$\left\{\frac{\partial}{\partial x}[y_{in}(x - v_1 t) + y_{re}(x + v_1 t)]\right\}_{x=L} = \left\{\frac{\partial}{\partial x}[y_{tr}(x - v_2 t)]\right\}_{x=L}. \qquad (ii)$$

We'll start with Eq. (*ii*). With $\xi_{in} = x - v_1 t$, $\xi_{re} = x + v_1 t$, $\xi_{tr} = x - v_2 t$, Eq. (*ii*) becomes

$$\left[\frac{dy_{in}(\xi_{in})}{d\xi_{in}}\frac{\partial \xi_{in}}{\partial x} + \frac{dy_{re}(\xi_{re})}{d\xi_{re}}\frac{\partial \xi_{re}}{\partial x}\right]_{x=L} = \left[\frac{dy_{tr}(\xi_{tr})}{d\xi_{tr}}\frac{\partial \xi_{tr}}{\partial x}\right]_{x=L},$$

which, since the partial derivatives are all equal to 1, reduces to

$$\left[\frac{dy_{in}(\xi_{in})}{d\xi_{in}} + \frac{dy_{re}(\xi_{re})}{d\xi_{re}}\right]_{x=L} = \left[\frac{dy_{tr}(\xi_{tr})}{d\xi_{tr}}\right]_{x=L}. \qquad (iii)$$

So that we are working with derivatives for both boundary conditions, we differentiate both sides of Eq. (*i*) with respect to time and follow the same procedure that led to Eq. (*iii*). This yields

$$\left[-v_1\frac{dy_{in}(\xi_{in})}{d\xi_{in}} + v_1\frac{dy_{re}(\xi_{re})}{d\xi_{re}}\right]_{x=L} = \left[-v_2\frac{dy_{tr}(\xi_{tr})}{d\xi_{tr}}\right]_{x=L}. \qquad (iv)$$

Next, with Eqs. (*iii*) and (*iv*) solved for $dy_{tr}/d\xi_{tr}$ and $dy_{re}/d\xi_{re}$ in terms of $dy_{in}/d\xi_{in}$, we find (from here on we'll suppress the subscript $x = L$)

$$\frac{dy_{tr}(\xi_{tr})}{d\xi_{tr}} = \frac{2v_1}{v_1 + v_2}\frac{dy_{in}(\xi_{in})}{d\xi_{in}} \qquad (v)$$

and

$$\frac{dy_{re}(\xi_{re})}{d\xi_{re}} = \frac{v_1 - v_2}{v_1 + v_2}\frac{dy_{in}(\xi_{in})}{d\xi_{in}}. \qquad (vi)$$

At $x = L$,

$$d\xi_{in} = -v_1\,dt, \qquad d\xi_{re} = v_1\,dt, \qquad d\xi_{tr} = -v_2\,dt.$$

When dt is eliminated between pairs of these equations, we find

$$d\xi_{tr} = \frac{v_2}{v_1}\,d\xi_{in} \qquad (vii)$$

and

$$d\xi_{re} = -d\xi_{in}. \qquad (viii)$$

Finally, we multiply Eq. (v) by $d\xi_{tr}$ and use Eq. (vii) to obtain

$$dy_{tr} = \frac{2v_2}{v_1 + v_2}\,dy_{in},$$

which integrates to

$$y_{tr}(L - v_2 t) = \left(\frac{2v_2}{v_1 + v_2}\right) y_{in}(L - v_1 t).$$

This is Eq. (17-6).

Equation (17-7) is found in the same way. By multiplying Eq. (vi) by $d\xi_{re}$, substituting Eq. ($viii$), and then integrating, we get

$$y_{re}(L + v_1 t) = \left(\frac{v_2 - v_1}{v_1 + v_2}\right) y_{in}(L - v_1 t),$$

which is Eq. (17-7).

QUESTIONS

17-1. What is the phase difference between the waves $y_1(x, t) = A \sin(kx + \omega t)$ and $y_2(x, t) = A \cos(kx + \omega t)$?

17-2. When two waves interfere, how are their motions affected?

17-3. Does constructive interference imply a gain of energy and destructive interference a loss of energy?

17-4. Can wave pulses interfere constructively or destructively?

17-5. Is it possible to add two waves and obtain a resultant wave that has an amplitude greater than the sum of the individual amplitudes?

17-6. Can two harmonic waves with different amplitudes produce total destructive interference? Is it possible for three such waves with different amplitudes?

17-7. As a person moves toward a pair of loudspeakers producing harmonic signals of equal wavelength and intensity, he hears a decrease in sound intensity. Explain how this can happen.

17-8. Standing waves are an example of wave interference in space, whereas beats are an example of wave interference in time. Discuss this statement.

17-9. Compare the beats produced by adding two waves of frequency 100 and 101 Hz with those produced by waves of frequency 1000 and 1001 Hz.

17-10. Two notes an octave apart have a frequency ratio of 2:1. If one of two such notes on a piano is slightly out of tune, beats are heard when the notes are played simultaneously. How are the beats produced?

17-11. Upon reflection at a boundary, a wave pulse on a string is inverted. There is no transmitted pulse. How is the string attached at that boundary?

17-12. Compare the frequencies, wavelengths, and speeds of the incident, reflected, and transmitted harmonic waves when (a) the first string has a smaller linear mass density than the second string and (b) the first string has a greater linear mass density than the second string.

17-13. What happens to the reflected pulse if the linear mass densities of two strings tied together are almost equal?

17-14. Discuss how the reflected and transmitted pulses are affected if the tension in two strings that are tied together is slowly increased.

17-15. Can a reflected or a transmitted harmonic wave have a larger amplitude than the incident wave?

17-16. If the wavelengths of the incident and transmitted harmonic waves at a boundary are the same, what can be said about the linear mass densities of the strings?

17-17. The sum of the amplitudes of the reflected and transmitted harmonic waves at a boundary equals the amplitude of the incident harmonic wave. True or false?

17-18. Can a standing wave be formed from the addition of two traveling harmonic waves that have different amplitudes?

17-19. If a string is vibrating in a standing-wave pattern, is energy transmitted down the string? Explain your answer.

17-20. If you play a song by raising all notes one octave, the melody sounds the same. For example, suppose three consecutive notes are played: C (264 Hz), G (396 Hz), E (330 Hz). If the three notes are then played in the next octave as C (528 Hz), G (792 Hz), E (660 Hz), the melody sounds the same. What does this tell you about how we distinguish the melody when it is played in different octaves?

17-21. Discuss the differences between the ear's and the eye's ability to distinguish the frequency components of the signals that each one processes.

17-22. Which of the quantities describing a sinusoidal wave is most closely related to musical pitch? to loudness?

17-23. In an organ, are the longest or shortest pipes used to produce the highest notes?

17-24. A tube open at one end and closed at the other and a tube open at both ends are adjusted in length so that they both have the same fundamental vibrational frequency. How will the sound heard from these two tubes differ?

17-25. Compare and discuss the effects of fingering a string of a violin and covering a side hole of a flute.

17-26. What generalization can be made about the relationship between the size of a wind instrument and the frequency range of the notes it produces?

17-27. Is the tuning of a stringed instrument affected by the air temperature? What about the tuning of a horn?

17-28. In a speaker, the tweeters are used to produce the high frequencies, while the woofers produce the low frequencies. Which of these two structures would you expect to be larger? Explain.

17-29. Many men sound better when they sing in a shower stall. Can you guess why? Do most women gain the same acoustical benefit from the shower?

17-30. The greater the speed of sound in the medium enclosed by a tube, the higher the fundamental frequency of the tube. Is it feasible, therefore, to fill an instrument such as a flute with a solid in order to obtain very high frequency notes?

17-31. At 300 K, the speeds of sound in air, helium, and carbon dioxide are 340, 1020,and 270 m/s, respectively. Can you guess what happens to your voice when you inhale helium? carbon dioxide?

17-32. The standing waves on a circular drum are represented by mathematical functions called Bessel functions. These functions oscillate back and forth through zero as their dependent variable increases. In the case of the circular drum, the dependent variable is the radial distance from the center of the drum. In light of the fact that the sounds of a drum are not musical like those from a horn or stringed instrument, what does this indicate about the zeros of the Bessel functions?

PROBLEMS

Note: In the following problems, take the speed of sound in air to be 340 m/s whenever applicable.

Principle of Superposition

17-1. Two pulses traveling in opposite directions along a string are shown for $t = 0$ in the accompanying figure. Plot the shape of the string at $t = 1.0$, 2.0, 3.0, 4.0, and 5.0 s, respectively.

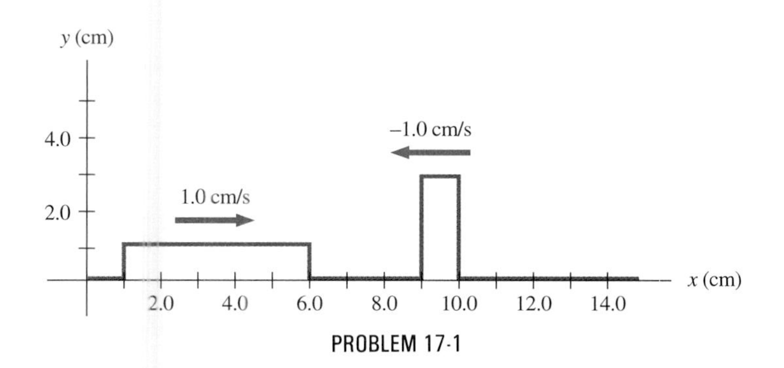

PROBLEM 17-1

17-2. Repeat the calculation of the previous problem for the pulses shown in the accompanying figure.

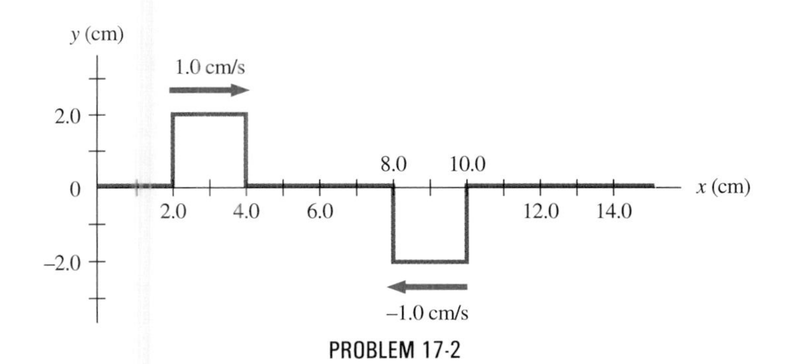

PROBLEM 17-2

17-3. Two rectangular pulses of heights 3.0 and 2.0 cm and the same width travel in opposite directions toward each other in a one-dimensional medium. When these pulses coincide, what is the maximum possible displacement of a particle of the medium? Consider the two cases in which the pulses have the same and the opposite orientations.

17-4. Suppose that the width of each of the pulses of Prob. 17-3 is 6.0 cm. If the velocity of the 3.0-cm-high pulse is +1.5 cm/s and the velocity of the 2.0-cm pulse is −0.5 cm/s, over what time interval are the pulses merged together?

17-5. When transmitter 1 is turned on, it sends out a plane-wave signal such that the energy per unit time received at a detector is I. When transmitter 2 is turned on and transmitter 1 is off, the energy per unit time received by the detector is also I. What are the minimum and maximum values of the energy per unit time that can be received by the detector when both transmitters are turned on? Assume that both transmit at the same frequency.

Superposition of Two Harmonic Waves of the Same Frequency

17-6. When two harmonic waves with the same amplitude (0.50 cm), wavelength, and frequency are superposed, the resultant wave has an amplitude of 0.80 cm. What is the phase difference between the waves?

17-7. Two longitudinal harmonic waves have a phase difference of 35° but are otherwise identical. If the amplitude of each wave is 2.0 cm, what is the amplitude of the resultant wave when they are added?

17-8. The phase difference between two otherwise identical harmonic waves is $\pi/4$ rad. If the resultant harmonic wave has an amplitude of 0.20 m, what is the amplitude of the individual waves?

17-9. Find the expression for the sum of $y_1(x, t) = 1.5 \sin(x + 0.50t)$ and $y_2(x, t) = 1.5 \sin(x + 0.50t + \pi/2)$. Assume SI units.

17-10. What is the amplitude of the resultant wave when $y_1(x, t) = 0.40 \sin(x - 0.20t)$ and $y_2(x, t) = 0.40 \cos(x - 0.20t)$ are added? Write an expression for this wave. Assume SI units.

17-11. Find the sum of the following waves:

$$y_1(x, t) = 3.0 \sin\left(2.0x + 40t + \frac{\pi}{4}\right)$$

and

$$y_2(x, t) = 3.0 \sin\left(2.0x + 40t + \frac{2\pi}{3}\right) \qquad \text{(SI units).}$$

17-12. When two harmonic waves $y_1(x, t)$ and $y_2(x, t)$, which differ only in phase, are summed, the resultant wave is given by

$$y(x, t) = 6.0 \sin\left(2.5x - 600t + \frac{\pi}{3}\right) \qquad \text{(cgs units).}$$

(*a*) If $y_1(x, t) = A \sin(2.5x - 600t)$, what is the phase of the other wave? (*b*) What is the amplitude A of the individual waves?

17-13. The sum of two harmonic waves $y_1(x, t)$ and $y_2(x, t)$, which differ only in phase, is

$$y(x, t) = 3.5 \sin\left(0.50x + 400t + \frac{\phi}{2}\right) \qquad \text{(cgs units)}.$$

(*a*) What is the value of ϕ if each individual wave has an amplitude of 2.0 cm? (*b*) Write expressions for $y_1(x, t)$ and $y_2(x, t)$.

17-14. Identical points on two harmonic waves with the same wavelength (0.50 m) and frequency are separated by a distance of 0.15 m. What is the phase difference between the waves?

17-15. Two otherwise identical harmonic waves have a phase difference of $\pi/6$ rad and are shifted relative to one another by 30 cm. What is their wavelength?

17-16. The two identical speakers shown in the accompanying figure are driven by the same oscillator and emit harmonic sound waves of frequency 680 Hz. A listener is at a distance d_1 from one speaker and a distance d_2 from the other, as shown. (*a*) For what values of $|d_2 - d_1|$ would the listener detect minima in the sound intensity from the speakers? (*b*) For what values of $|d_2 - d_1|$ would he detect maxima? (*c*) Would he ever detect a minimum if he walked along the center line between the speakers?

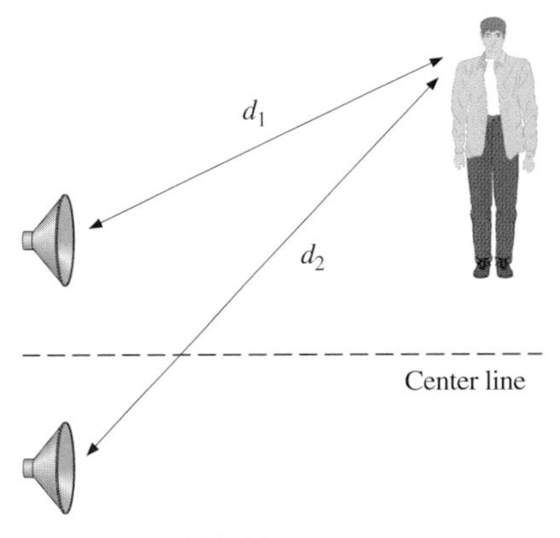

PROBLEM 17-16

Beats

17-17. What is the beat frequency detected when two loudspeakers vibrate simultaneously with frequencies of 500 and 508 Hz?

17-18. A tuning fork of frequency 196 Hz is used to tune a violin string. When the string and tuning fork are vibrated simultaneously, two beats per second are heard. What are the possible frequencies of vibration of the violin string?

17-19. What is the beat frequency when the following waves are superposed:

$$y_1(x, t) = A \sin \pi(x - 800t)$$
$$y_2(x, t) = A \sin \pi(x - 802t) \qquad \text{(SI units)}.$$

17-20. Two tuning forks have identical frequencies of 440 Hz. If one is held by an observer while the other moves away from her at a speed of 4.7 m/s, what does the observer hear? [*Hint*: See Sec. 16-9].

Boundary Conditions

17-21. Draw the approximate shapes of the reflected waves for the situations shown in the accompanying figure.

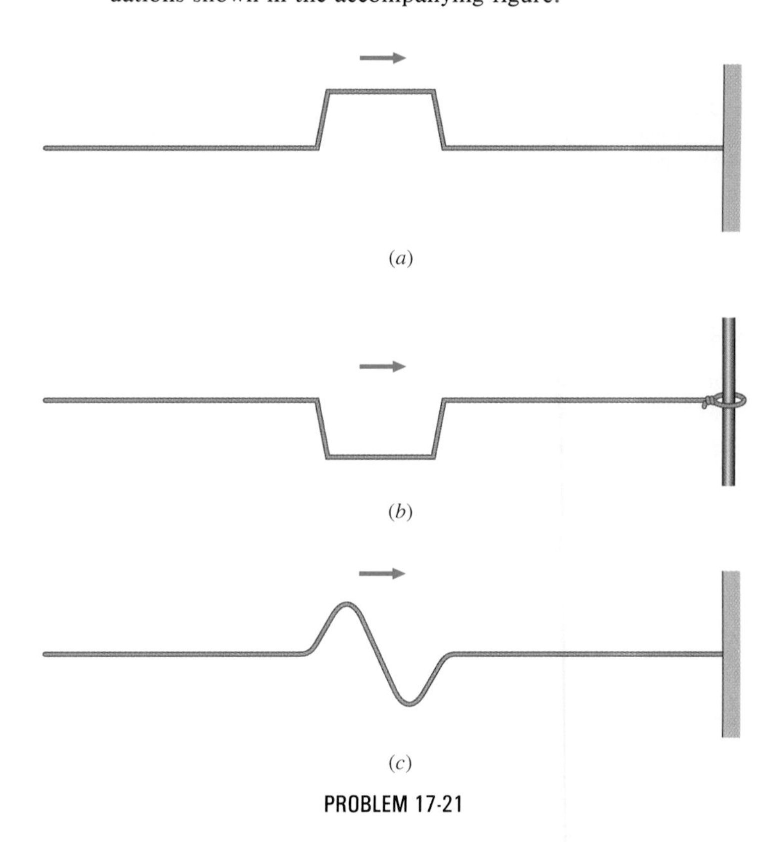

PROBLEM 17-21

17-22. A wave pulse $F(x - vt)$ is traveling along a string. Write expressions for the reflected wave if (*a*) the string has a fixed end and (*b*) the string has a free end.

17-23. A wave $y_{in}(x, t) = 0.30 \cos(2.0x - 40t)$ is traveling along a string toward a boundary at $x = 0$. Write expressions for the reflected waves if (*a*) the string has a fixed end at $x = 0$ and (*b*) the string has a free end at $x = 0$. Assume SI units.

17-24. A harmonic wave is reflected at the free end of a string located at $x = 0$. The reflected wave is represented by $y_{re}(x, t) = 1.5 \cos \pi(0.20x + 60t)$. Write an expression for the incident wave. Assume cgs units.

17-25. The harmonic wave $y_{in}(x, t) = (2.0 \times 10^{-3}) \cos \pi(2.0x - 50t)$ travels along a string toward a boundary at $x = 0$ with a second string. The wave speed on the second string is 50 m/s. Write expressions for the reflected and transmitted waves. Assume SI units.

17-26. A string that is 5.0 m long is fixed at both ends. At $t = 0$, a pulse traveling from left to right at 10.0 m/s is 2.0 m from the right end, as shown in the accompanying figure. Determine the next five times when the pulse will be at that point again. State in each case whether the pulse is upright or inverted.

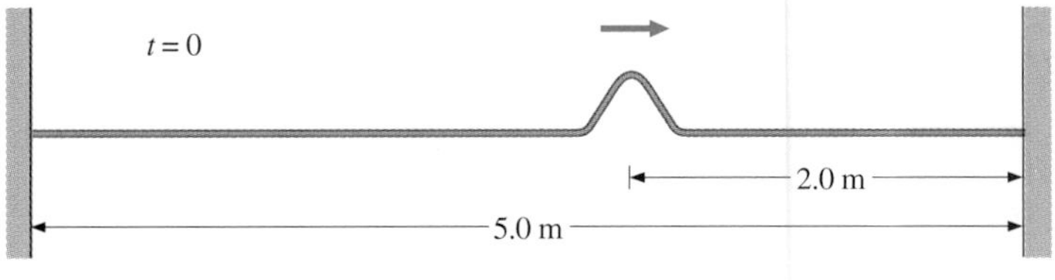

PROBLEM 17-26

17-27. Repeat the calculation of Prob. 17-26 if the left end of the string is completely free to move in the vertical direction and the right end of the string is fixed.

17-28. The ratio of the speeds of the transmitted harmonic wave and the incident harmonic wave in two attached strings is 2:1. (*a*) Calculate the relative amplitudes of these waves. (*b*) Is the reflected wave upright or inverted? (*c*) Determine its amplitude relative to that of the incident wave.

17-29. At the boundary between two strings, it is found that the reflected harmonic wave is inverted and has an amplitude that is 1.2 times that of the transmitted harmonic wave. (*a*) What is the ratio of the wave speeds in the two strings? (*b*) What are the amplitudes of the transmitted and reflected harmonic waves relative to that of the incident wave?

17-30. The harmonic wave transmitted through the boundary between two strings has a wavelength that is 1.3 times that of the incident harmonic wave. Determine the ratio of the linear mass densities of the strings.

17-31. For the transmitted and incident waves of Prob. 17-30, find the ratios of (*a*) their speeds, (*b*) their frequencies, and (*c*) their amplitudes.

17-32. A harmonic wave represented by $y_{\text{in}}(x, t) = 0.020 \cos(10x - 100t)$ propagates along a string that is tied to a second string at $x = 0$. The linear mass density of the second string is 0.80 times that of the first. Write expressions for the reflected and transmitted waves. Assume SI units.

17-33. The wave reflected at a boundary located at $x = 0$ between two strings is given by

$$y_{\text{re}}(x, t) = 0.040 \cos(4.0x + 200t) \qquad \text{(SI units)}.$$

If this wave travels at 0.60 times the speed of the transmitted wave, write expressions for the incident and transmitted waves.

Standing Waves

17-34. Write an expression for a standing wave formed by the addition of the waves $2.5 \sin(\pi x - 30t)$ and $2.5 \sin(\pi x + 30t)$. Assume cgs units.

17-35. A standing wave is represented by $4.0 \sin 2\pi x \cos \pi t$. What are the amplitude, speed, frequency, and wavelength of the two traveling harmonic waves that are superposed to form this standing wave? Assume cgs units.

17-36. Consider the standing wave of Prob. 17-35. Plot the shape of this wave from $x = 0$ cm to $x = 1.0$ cm for values of t ranging from 0 to 2.0 s at 0.25-s intervals.

17-37. The nodes of a standing wave are 0.40 m apart. What is the wavelength of the traveling waves that produce this standing wave?

17-38. A series of photographs of the standing wave of Prob. 17-37 shows that it has a period of 0.050 s and a maximum displacement of 3.0×10^{-2} m. Write expressions for the two traveling waves whose sum produced this standing wave.

17-39. A standing wave is given by

$$y(x, t) = 2.0 \sin \frac{\pi x}{2.0} \cos 25\pi t,$$

where cgs units are used. What is the distance between a node and an adjacent antinode? Plot $y(x, t)$ at $x = 1.0$ cm and $x = 2.0$ cm as a function of time over one period.

Standing Waves on a String

17-40. The wave speed along a wire is 200 m/s. If the wire is 2.0 m long, what is its fundamental frequency?

17-41. A wave pulse travels along a string with a speed of 60 m/s. If the string is 2.0 m long, what are the frequencies of its four lowest harmonics?

17-42. Determine the length of a string whose third harmonic has a frequency of 24 Hz and along which waves travel at 50 m/s.

17-43. The frequency of the second harmonic of a string is 80 Hz. If the string is 2.0 m long, what is the speed of a wave propagating along it?

17-44. A wire that is 3.0 m long resonates in its third harmonic with a frequency of 90 Hz. If the tension in the string is 20 N, what is the mass of the string?

17-45. What tension is needed to produce a third harmonic of frequency 120 Hz in the string of Prob. 17-44?

17-46. By what fraction of its original value does the frequency of the first harmonic of a string change if the tension in the string is doubled? Repeat the calculation for the next two higher harmonics.

17-47. If the frequencies of the second and fifth harmonics of a string differ by 54 Hz, what is the fundamental frequency of the string?

17-48. The mass of a 2.0-m wire is 0.040 kg. If the difference in the frequencies of its second and third harmonics is 8.0 Hz, what is the tension in the wire?

17-49. A wire is attached to a pan of mass 200 g that contains a 2.0-kg mass, as shown in the accompanying figure. When plucked, the wire vibrates at a fundamental frequency of 220 Hz. An additional unknown mass M is then added to the pan and a fundamental frequency of 260 Hz is detected. What is the value of M?

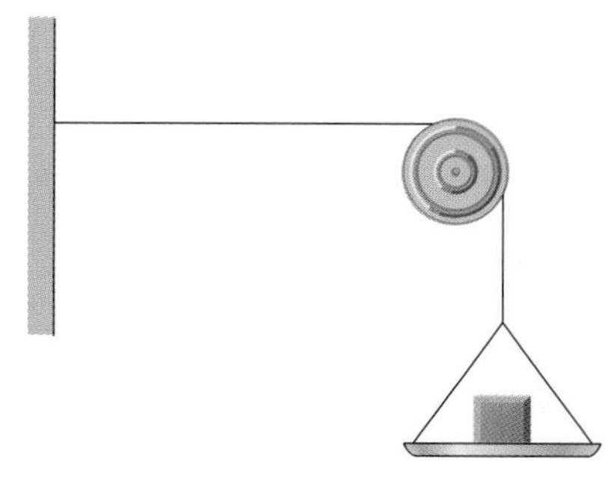

PROBLEM 17-49

17-50. A 100-Hz vibrator forms a standing-wave pattern with three envelopes on the string, as shown in the accompanying figure. (*a*) If the string is 2.0 m long and has a mass of 15 g, under what tension is it held? (*b*) For the same tension, what must be the frequency of the vibrator if five envelopes are to be formed in the standing-wave pattern?

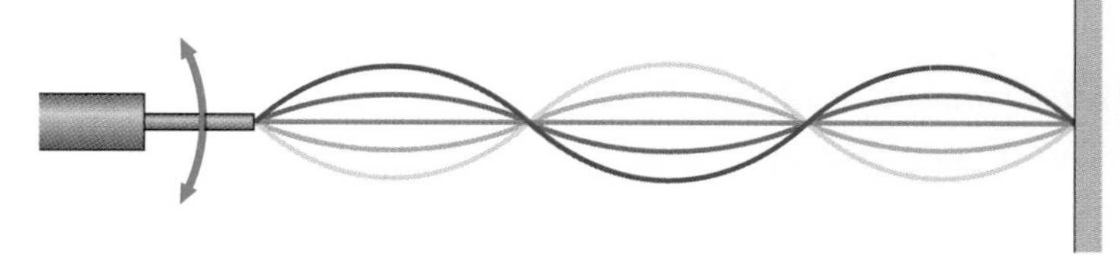

PROBLEM 17-50

17-51. A wire fixed at both ends is 1.0 m long and has a mass of 36 g. One of its resonant frequencies is 250 Hz and the next-higher one is 300 Hz. (*a*) Which harmonics do these frequencies represent? (*b*) What is the tension in the wire?

17-52. A string resonating in its third harmonic exhibits a standing-wave pattern in which the separation between the nodes is 0.30 m. What will this separation be when the string vibrates at its fundamental frequency?

17-53. The tension in a wire of mass 4.0 g is 200 N, and the frequency of the second harmonic of the wire is found to be 240 Hz. (*a*) What is the length of the wire? (*b*) What is the separation between a node and its neighboring antinode when the wire is vibrating in the second harmonic?

Standing Waves in a Tube

17-54. What is the length of an air-filled tube closed at both ends and having a fundamental frequency of 360 Hz?

17-55. If one of the ends of the tube of Prob. 17-54 were open, what would be its fundamental frequency?

17-56. The human ear canal is approximately 2.7 cm long, open at one end and closed at the other. The ear is most sensitive to sound at a frequency of about 3000 Hz. Compare this with the resonant frequencies of the ear canal.

17-57. A tube closed at one end and filled with a gas has a third harmonic of 360 Hz. If the length of the tube is 0.50 m, what is the speed of sound in the gas?

17-58. The distance between the displacement nodes of the air in a long tube closed at one end and vibrating at its third harmonic is 1.0 m. (*a*) What is the length of the tube? (*b*) What is its fundamental frequency?

17-59. The air in the tube shown in the accompanying figure is set vibrating at its fundamental frequency of 100 Hz. If the tube is filled with helium, what is its new fundamental frequency? Take the speed of sound in helium to be 1020 m/s.

PROBLEM 17-59

17-60. The fundamental frequency of an air-filled tube that is open at both ends is three times greater than that of a tube that is open at one end. Which is the longer tube? Calculate the ratio of the lengths of the tubes.

17-61. A resonant frequency of a tube closed at one end and open at the other is 300 Hz. The next-higher resonant frequency is found to be 500 Hz.To which harmonics do these frequencies correspond? What is the length of the tube?

Music and Musical Instruments

17-62. If a guitar string is tuned to a certain frequency, by what factor must its tension be increased if the string is to produce a note one octave higher (i.e., double the initial frequency)?

17-63. The C below middle C has a frequency of 130 Hz. If this note is to be played on a piano wire of length 2.1 m and linear mass density 5.1 g/m, what tension must be placed on the wire?

17-64. A violin string is 29.8 cm long and has a mass of 1.98 g. If this string is tuned to play the A note of frequency 440 Hz, (*a*) what is the tension in the string? (*b*) Where must a violinist place her finger to play a C note of frequency 528 Hz?

17-65. The steel B ($f = 247.5$ Hz) string of an acoustic guitar is 64.1 cm long and is made of 16-gauge wire (diameter = 0.406 mm). What is the tension in the string?

17-66. (*a*) How long must an organ pipe open at both ends be if it is to have a fundamental frequency of 294 Hz (D above middle C)? (*b*) If a pipe closed at one end plays the same note, what is its length? (*c*) Would the notes played by the two organ pipes make the same sound?

17-67. The four strings of a violin are the same length and are tuned to frequencies 196, 294, 440, and 659 Hz, respectively. If all four strings are under the same tension, what must be the linear mass density of each string relative to that of the lowest string?

17-68. A piccolo and a flute are similar, except that the piccolo is considerably shorter than the flute. When all holes are closed on either one, the instrument can be treated as a cylindrical tube open at both ends, with the length L being the distance from the mouthpiece to its far end, as illustrated in the accompanying figure. The lowest fundamental frequency of the piccolo is 587 Hz, and that of the flute is 262 Hz. What is the length of the piccolo compared with that of the flute?

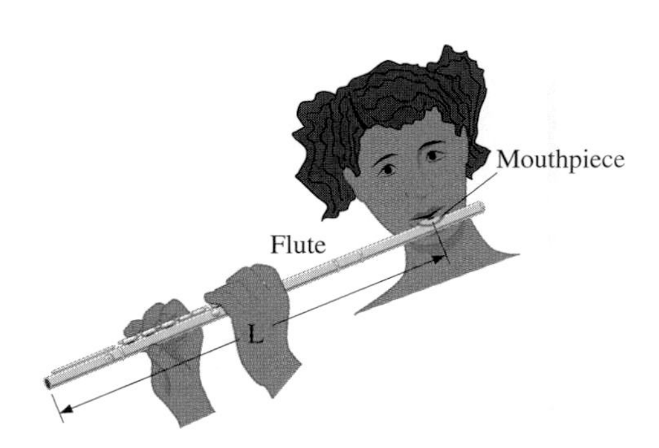

PROBLEM 17-68

17-69. The finger holes on a piccolo or a flute (see the previous problem) effectively move the far end of the instrument to the locations of the holes. (*a*) How far from the mouthpiece of the piccolo is the hole that must be opened to play the A note of frequency 880 Hz? (*b*) How far from the mouthpiece of the flute is the hole that must be opened for the A note of frequency 440 Hz?

General Problems

17-70. The sum of two harmonic waves $y_1(x, t)$ and $y_2(x, t)$ differing only in phase is given by

$$y(x, t) = 4.2 \sin\left(1.5x - 200t + \frac{\pi}{4}\right).$$

(*a*) If $y_1(x, t) = A \sin(1.5x - 200t + \pi/3)$, what is the phase of the other wave? (*b*) What is the amplitude A of the individual waves? Assume cgs units.

17-71. Two tuning forks A and B, which are 0.50 m apart, oscillate in phase at a frequency of 200 Hz. A listener aligns himself with the tuning forks as shown in the accompanying figure. (*a*) What is the phase difference between the harmonic waves that the listener detects? (*b*) The separation between the tuning forks is now varied by moving A to the left. What must be the new separation in order for the waves to interfere constructively at the listener's position? (*c*) for them to interfere destructively at that point?

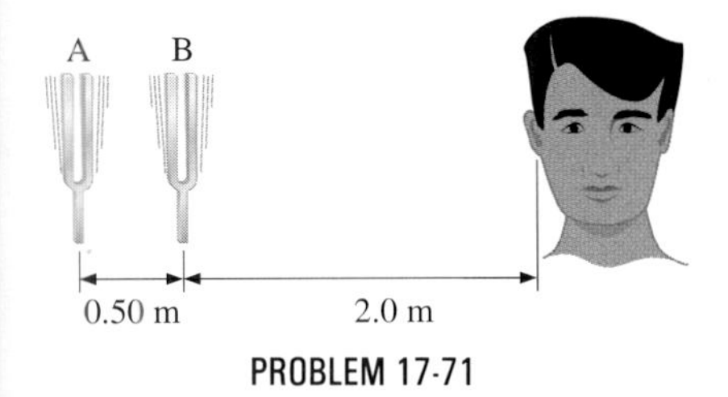

PROBLEM 17-71

17-72. Refer again to the figure of Prob. 17-71. Tuning fork B is now free to be moved to the right while A is fixed. (*a*) How far must B be moved for constructive interference to occur at the listener's position? (*b*) for destructive interference to occur there?

17-73. A harmonic wave of frequency 200 Hz and wavelength 0.50 m is traveling along a string toward an end attached to a second string whose linear mass density is twice as great as that of the first string. (*a*) Calculate the speeds, frequencies, and wavelengths of the reflected and transmitted waves. (*b*) Write expressions for the incident, reflected, and transmitted harmonic waves. Assume that the amplitude of the incident wave is 0.10 m and that the junction of the two strings is at $x = 0$.

17-74. Two strings, one with half the linear mass density of the other, are connected to make a single taut line. A harmonic wave of amplitude 1.2 cm travels along the heavier string toward the junction. (*a*) After it reaches the junction, what are the amplitude and phase of the transmitted wave with respect to the incident wave? (*b*) Answer the same question for the reflected wave. (*c*) If the average power in the incident wave is 2.0 W, what are the average powers in the transmitted and the reflected waves?

17-75. A short string of linear mass density 3μ is tied at both ends to long strings of linear mass density μ, as shown in the accompanying figure. A harmonic wave of amplitude A starts in the left string and travels from left to right. Determine the amplitude of the harmonic wave that is transmitted down the right string.

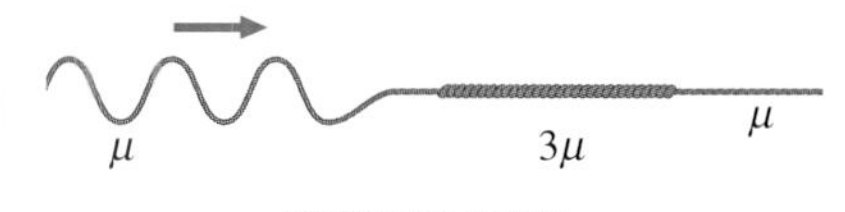

PROBLEM 17-75

17-76. A string of unknown length has a mass of 3.0×10^{-3} kg and is stretched to a tension of 50 N. A wave pulse is found to traverse its entire length in 1.2×10^{-2} s. What is the fundamental frequency of this string?

17-77. The separation between nodes on a string vibrating in its nth harmonic is 0.24 m. In the $(n + 1)$st harmonic, this separation is reduced to 0.20 m. (*a*) What is the value of n? (*b*) What is the separation between nodes when the string vibrates at its fundamental frequency? (*c*) What is the length of the string?

17-78. A string of linear mass density 5.0×10^{-3} kg/m is stretched under a tension of 65 N between two rigid supports 60 cm apart. (*a*) If the string is vibrating in its second overtone so that the amplitude at one of its antinodes is 0.25 cm, what are the maximum transverse speed and acceleration of the string at the antinodes? (*b*) What are these quantities at a distance 5.0 cm from an antinode? (*c*) What are they at a node?

17-79. An air-filled tube whose ends are closed has a fundamental frequency of 80 Hz. When it contains a gas of density 1.0×10^{-3} g/cm^3, its third harmonic is found to be 330 Hz. What is the bulk modulus of the gas?

17-80. Suppose the air temperature drops, causing the speed of sound to change from 340 to 335 m/s. An organ pipe open at both ends is tuned to middle C (261 Hz) at the higher temperature. What is the fundamental frequency of the pipe after the temperature drops? Neglect the variation in the length of the pipe with temperature.

17-81. A tuning fork whose natural frequency is 440 Hz is placed just above the open end of a tube that contains water. The water is slowly drained from the tube while the tuning fork remains in place and is kept vibrating. The sound is found to be enhanced when the air column is 60 cm long and when it is 100 cm long. Use this information to determine the speed of sound in the air.

17-82. Two identical A strings on different violins have a fundamental frequency of 440 Hz. What fractional change in the tension of one of the strings will produce six beats per second when both strings are played simultaneously?

PART III

THERMODYNAMICS

The temperature of molten lava is about 1150°C.

CHAPTER 18 TEMPERATURE AND THE PROPERTIES OF GASES

PREVIEW

This chapter is devoted to temperature, the relationship of temperature to thermal equilibrium, and the properties of gases. The main topics to be discussed here are the following:

1. **Temperature and thermal equilibrium.** The concept of temperature is introduced and its relationship to thermal equilibrium is explained.
2. **Thermometers and temperature scales.** Different types of thermometers are described, and the Celsius, Fahrenheit, and Kelvin temperature scales are defined and related.
3. **Ideal-gas law.** The ideal-gas law is discussed, and examples of its application are given.
4. **A statistical interpretation of an ideal gas.** An ideal gas is described in terms of a simple statistical model. Pressure is interpreted in terms of the average force of the molecules on the walls of their container. We derive the relationship between the average translational kinetic energy of the molecules and the temperature of the ideal gas.

*5. **Phase diagrams of nonideal gases.** Pressure-temperature and pressure-volume phase diagrams are described. Phase transitions are explained in terms of these diagrams.

*6. **The van der Waals equation.** We investigate an equation that provides a rough description of the behavior of a nonideal gas.

We now begin a study of thermal physics. Familiar quantities like *heat* and *temperature* play fundamental roles in this branch of physics. We will investigate the transfer of heat between systems and the conditions and effects of that transfer. One well-known effect is a change in the temperature of a system. Examples of common thermal processes include the melting of snow in the spring, the warming of the earth by the sun, and the cooling of food in a refrigerator.

There are two approaches used in thermal physics. (1) The *macroscopic* approach is based on experimental observations of how the bulk properties of a system (such as its temperature, pressure, and volume) behave under different thermal conditions. The laws that result from these observations form the basis for *thermodynamics*. (2) In the *microscopic* approach Newtonian or quantum mechanics is used to determine the average behavior of the molecules of a system. This behavior in turn provides information about the system's bulk properties. For example, knowing the average speed of the molecules of a dilute (low-density) gas allows us to find the temperature of the gas, while knowing the average force exerted on the walls of the container by the gas molecules allows us to determine the pressure of the gas. The microscopic approach forms the basis for *statistical mechanics*.

A major portion of this chapter is devoted to thermodynamic equilibrium and temperature. These two concepts are closely related, for two systems in thermal equilibrium have the same temperature. The rest of the chapter is concerned with the properties of gases. Among the topics we consider there are condensation, sublimation, and equations of state.

18-1 Temperature and Thermal Equilibrium

Temperature is an important quantity in thermodynamics. It is used to specify how hot or cold an object feels. A hot object is said to have a high temperature, while a cold object is said to have a low temperature. However, the concept of "hotness" is rather subjective, for a visitor from the Arctic will likely describe a late-fall day in the northern continental United States as warm, while the same day will probably seem quite cold to a visitor from an equatorial country. Clearly, something more precise than touch or feel must be used to specify relative temperatures.

The surface of Venus produced from digitally combined radar images. The surface temperature of this planet is about 750 K, much hotter than the average household oven.

Our ability to associate a temperature with an object is a direct consequence of the *zeroth law of thermodynamics*.* This law is usually stated in terms of the following imaginary experiment. We first bring objects A and B together (see Fig. 18-1*a*), place them in contact, isolate them from any external thermal influences, then monitor them. Certain properties of the objects will begin to change—for example, if one of the objects is a column of mercury, the length of the column will change. Eventually however, every property of both objects will cease changing and thereafter remain constant with time. Objects A and B are then by definition in *thermal equilibrium*.

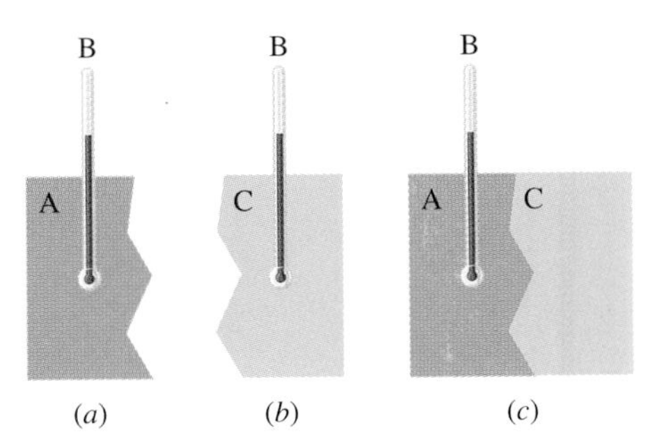

FIGURE 18-1 (*a*) Bodies A and B are placed in contact and allowed to reach thermal equilibrium. (*b*) Then B is placed in thermal contact with C, and no property of either body changes. Consequently, B and C are also in thermal equilibrium. (*c*) Now when A and C are brought together, these two bodies are also found to be in thermal equilibrium.

Next, we remove B from A and place it in thermal contact with C, as illustrated in Fig. 18-1*b*. Suppose that in so doing, we detect no change in the properties of B; that is, we find that B is in thermal equilibrium with C. If this is the case, when we move C over to A, giving the arrangement of Fig. 18-1*c*, we will find that C and A are also in thermal equilibrium. We conclude therefore that

> If objects A and C are separately in thermal equilibrium with object B, then A and C are in thermal equilibrium with each other.

This is the zeroth law of thermodynamics.

With the zeroth law, we can now quantify temperature. This is accomplished by choosing an object as a temperature-measuring device, or as it is commonly called, a *thermometer*. To keep the discussion simple, let's assume that the thermometer is a column of mercury. In this case, a *single* property, the length of the column, changes as the thermometer is placed in thermal equilibrium with different objects. We can therefore

*The zeroth law was developed in the 1930s, long after the first and second laws of thermodynamics had been discovered and numbered. It was labeled "zeroth" because it defines temperature, a quantity used in the first and second laws.

associate a particular temperature with every length of the column. From the zeroth law, all systems that make the mercury column reach thermal equilibrium at the same length are in equilibrium with one another—and all therefore have the same temperature. All bodies in thermal equilibrium with one another have the same **temperature**.

There are many different types of thermometers. Each has a measurable property that changes with temperature. This property is calibrated in temperature units. Of course, different thermometers must agree when measuring the same temperature. A description of some of the more common and useful thermometers follows.

18-2 Thermometers and Temperature Scales

If a measurable property of a system varies reproducibly with temperature, that system can be used as a thermometer. To calibrate the thermometer, reproducible reference points must be selected and temperatures assigned to them. Two reference points normally used are the freezing point and the boiling point of water at a pressure of 1 atm.

In everyday applications, temperature is often expressed using the *Celsius scale* (formerly the centigrade scale) of Anders Celsius (1701–1744). Its unit is the degree Celsius (°C). The freezing temperature of water at 1 atm is chosen to be the zero point of the Celsius scale (0°C), and the boiling temperature of water at 1 atm is assigned the temperature 100°C.

A scale used widely in the United States is the *Fahrenheit scale*, named after Gabriel Fahrenheit (1686–1736). There are 180 degrees Fahrenheit (°F) between the freezing and boiling points of water at 1 atm, which are chosen to be 32°F and 212°F, respectively. One Fahrenheit degree is therefore exactly five-ninths (5/9) as large as one Celsius degree. If T_F and T_C represent a particular temperature measured on the Fahrenheit and the Celsius scales, then

$$T_F = \tfrac{9}{5} T_C + 32°\text{F} \qquad (18\text{-}1)$$

"LET'S GO OVER TO CELSIUS'S PLACE. I HEAR IT'S ONLY 36° OVER THERE."

The most common type of thermometer is constructed simply by pouring mercury into a thin glass tube. As the mercury is heated, it expands and rises in the tube. With graduations marked on the tube as shown in Fig. 18-2*a*, the height of the mercury column gives the temperature measurement. Over the range of temperatures that the thermometer is normally used, the expansion of mercury is almost linear, so the scale is for all practical purposes linear. The mercury thermometer is portable and easy to use but not very precise. It serves quite adequately for day-to-day measurements of body and air temperatures, but its limited accuracy makes it unsuitable when precise scientific determinations of temperatures are needed. Also, since mercury freezes at −39°C, it cannot be used to measure low temperatures.

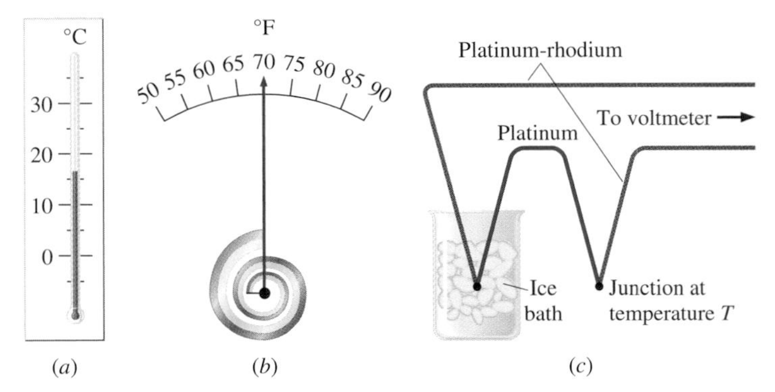

FIGURE 18-2 Various types of thermometers. (*a*) A common mercury thermometer. (*b*) A bimetallic strip. (*c*) A thermocouple.

Another common type of thermometer is made from a bimetallic strip. (See Fig. 18-2*b*.) It is constructed by placing two metals together to form a double-layered strip. As they are heated, the metals expand differently, causing the strip to bend and move a pointer across a calibrated scale. This thermometer is also limited in accuracy, but it does give sufficiently precise readings for use in thermostats and other temperature-controlling devices.

Very precise measurements (typically, ±0.1°C) of temperature can be made with resistance thermometers. As the temperature of a conductor changes, its electrical resistance* also changes. Since resistance can be determined very accurately, correspondingly accurate measurements of temperature result. The resistance thermometer is useful because it can be employed over a wide temperature range. Different resistance thermometers cover from about −270 to 2300°C. The standard platinum wire by itself works very well from −200 to 1200°C.

Among the most popular instruments used in laboratories for accurate temperature measurements is the thermocouple. (See Fig. 18-2*c*.) The thermocouple consists of two types of metallic wires connected to form two junctions. When the junctions are maintained at different temperatures, a voltage (also to be discussed in Chap. 26) is generated between them. The

*Resistance is a measure of how well or poorly a piece of material such as a wire allows electric current to flow. You will study this concept in Chap. 26.

value of the voltage is related to the temperature difference between the junctions. If we maintain one junction at a fixed reference temperature (e.g., by immersing it in an ice-water mixture), we obtain the temperature of the other junction relative to this reference. Because the mass of a thermocouple is so small, it quickly reaches thermal equilibrium with the substance whose temperature is being measured. Thermocouples are easy to construct and to use, and they have temperature ranges approximately equal to those of resistance thermometers.

Finally, there is the constant-volume gas thermometer of Fig. 18-3*a*. The bulb, which contains a dilute gas, is immersed in the liquid bath whose temperature is to be determined. A constant gas volume is maintained by raising or lowering the reservoir so that the level of the mercury remains at L. The gauge pressure of the gas is given by the difference in the mercury levels. The absolute pressure p (hereafter referred to simply as **pressure**) of the gas can then be determined by adding atmospheric pressure to the gauge pressure. To a good approximation, the temperature and pressure of the gas in this device are proportional to each other; that is, $T \propto p$. Consequently, the pressure of the gas can be used to measure temperature. This temperature is on the *Kelvin scale*,* and it is expressed in kelvins. A temperature of, for example, 200 kelvins is written as 200 K. The Kelvin scale is named in honor of Lord Kelvin (1824–1907), an English physicist who was an important contributor to the development of the theory of thermodynamics.

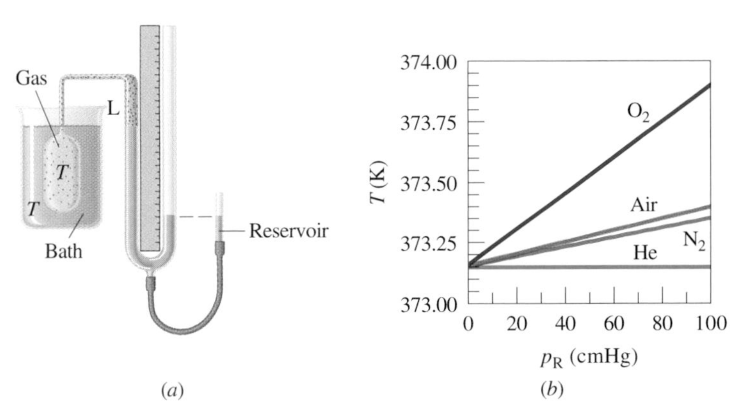

FIGURE 18-3 (*a*) A constant-volume gas thermometer. By adjusting the position of the reservoir, the level of the gas is kept at point L. Hence its volume is kept constant. The gauge pressure of the gas is proportional to the difference in the heights of the two mercury columns. (*b*) As the different gases become more dilute, their temperature measurements converge to the ideal-gas temperature. In this case, the temperature of boiling water is being measured.

To calibrate the constant-volume gas thermometer, a reproducible reference point must be selected. Then the temperature T of the gas at a pressure p is calculated from

$$\frac{T}{T_R} = \frac{p}{p_R},$$

where T_R and p_R are the gas temperature and pressure at the reference point. In 1954 the International Committee on Weights and Measures selected the *triple point* of water as the reference point because it is easily reproduced. At the triple point, ice, water, and water vapor coexist in thermal equilibrium at a single temperature and pressure. This temperature is designated to be exactly 273.16 K. The previous equation then becomes

$$T = 273.16\text{ K}\,\frac{p}{p_R}, \tag{18-2a}$$

where p_R is the pressure of the gas in the thermometer at the triple-point temperature, p is the gas pressure when the thermometer is in thermal equilibrium with the system whose temperature is to be determined, and T is the temperature of the system in kelvins.

Now let's consider what happens as the amount of gas in the constant-volume gas thermometer is reduced. Suppose that we are measuring the temperature of water boiling at a pressure of 1 atm. With a given amount of gas in the bulb of the thermometer, p is measured when the thermometer is in thermal equilibrium with the boiling water, and p_R is measured when it is in thermal equilibrium with water at the triple point. The temperature of the boiling water is then calculated with Eq. (18-2*a*). Next, some gas is removed from the thermometer so that p_R is reduced. The pressures of the gas in the thermometer at the boiling point and the triple point of water are measured again and T is recalculated with Eq. (18-2*a*). This process is repeated over and over, each time with less gas, and the calculated temperatures are plotted against p_R. The graph of Fig. 18-3*b* represents this procedure for different gases. You can see that for very low p_R, the measured temperatures of the boiling point of water are almost the same for every gas used in the thermometer. Furthermore, in the limit as $p_R \to 0$, the temperature determinations converge to a common value T, which is found by extrapolating the various graphs to the point of zero pressure; that is,

$$T = \lim_{p_R \to 0} (273.16\text{ K})\left(\frac{p}{p_R}\right). \tag{18-2b}$$

As shown in Fig. 18-3*b*, this limiting temperature is 373.15 K for boiling water at a pressure of 1 atm. With this same procedure, the freezing temperature of water at 1 atm is found to be 273.15 K.

A temperature measured in the limit described by Eq. (18-2*b*) is called an **ideal-gas temperature**. The constant-volume gas thermometer used in this limit is known as an *ideal-gas thermometer*. This thermometer is not practical for common usage, as it is slow in reaching thermal equilibrium and is cumbersome and time-consuming to use. Also, because gases liquefy when cold, very low temperatures cannot be measured with this thermometer. The lowest temperature it can measure is about 1 K. In this case, helium, which becomes a liquid at a lower temperature than any other gas, must be employed. For these reasons, the ideal-gas thermometer is primarily restricted to accurate research and scientific measurements.

*You will see in Chap. 21 that this scale is actually defined in terms of a cyclic process for an ideal heat engine.

The International Committee on Weights and Measures chose the triple-point temperature to be 273.16 K so that the freezing and boiling temperatures of water would differ by *exactly* 100 K. This choice also made the size of the degree the same on both the Celsius and Kelvin scales. With T_C and T representing temperature measurements on the respective scales,

$$T_C = T - 273.15°C. \quad (18\text{-}3)$$

The temperatures associated with selected systems and transitions are given in Table 18-1.

The ideal-gas temperature scale has a lower limit of $T = 0$ K, because pressure is a positive quantity. This temperature is often called *absolute zero*. Zero kelvin is now generally accepted as the lower limit on temperature, because it is found experimentally that the closer the temperature of a body is to absolute zero, the more difficult it becomes to reduce the body's temperature further.* However, temperatures as low as 10^{-3} K are not uncommon in low-temperature research laboratories.

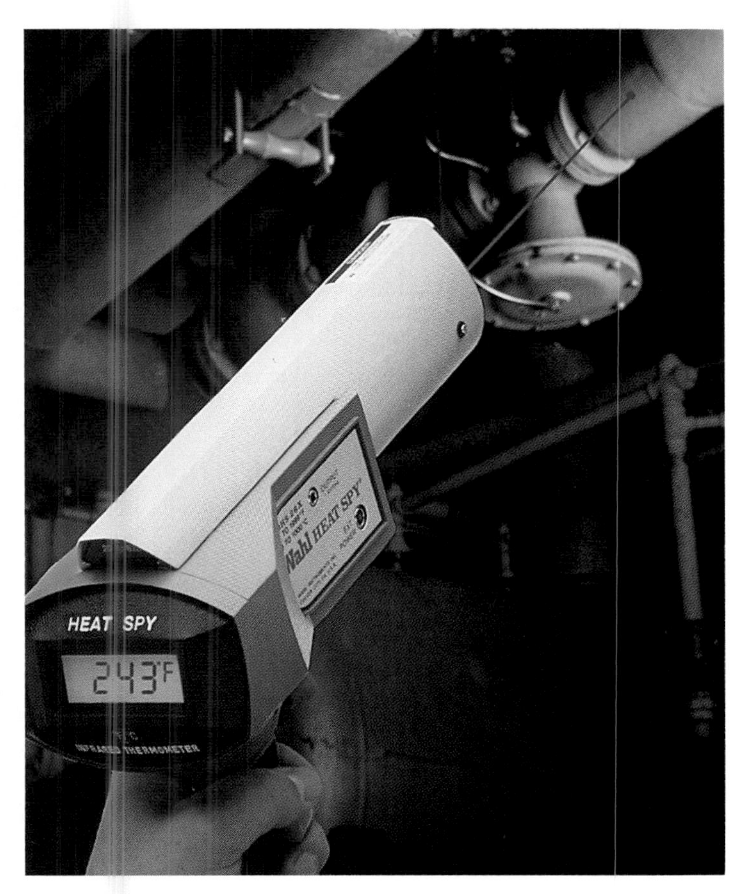

An infrared thermometer. When pointed at a surface, the instrument detects the infrared radiation emitted by that surface. The frequency for which the radiation intensity is a maximum is related to the temperature of the surface.

*The inability to reach absolute zero is now given the status of a basic principle – the *third law of thermodynamics*. This law will not be discussed in this textbook.

TABLE 18-1 **Temperatures Associated with Selected Systems and Transitions**

System or Transition	Temperature (K)
Core of sun	1.5×10^7
Core of earth	1.6×10^4
Surface of sun	6.0×10^3
Gas flame from stove	1.9×10^3
Melting of iron	1.8×10^3
Melting of gold	1.3×10^3
Fireplace fire	1.1×10^3
Boiling of water	373.15
Human body	310
Hybernating squirrel	275
Freezing of water	273.15
Liquefaction of nitrogen	77
Liquefaction of hydrogen	20
Cosmic background radiation	2.7
Lowest temperature attained	$\sim 5 \times 10^{-8}$

EXAMPLE 18-1 **Conversions Between Temperature Scales**

What is 0 K on (*a*) the Celsius scale and (*b*) the Fahrenheit scale? (*c*) What is a room temperature of 72°F on the Celsius scale?

SOLUTION (*a*) With $T = 0$ K, we have from Eq. (18-3) that the equivalent temperature in degrees Celsius is

$$T_C = 0 - 273.15°C = -273.15°C.$$

(*b*) Equation (18-1) gives us the equivalent temperature in degrees Fahrenheit:

$$T_F = [\tfrac{9}{5}(-273.15) + 32]°F = -459.67°F.$$

(*c*) Again from Eq. (18-1), the temperature on the Celsius scale that is equivalent to 72°F is

$$T_C = \tfrac{5}{9}(72 - 32)°C = 22°C.$$

DRILL PROBLEM 18-1

(*a*) The normal temperature of the human body is 98.6°F. What is it on the Celsius scale? on the Kelvin scale? (*b*) If the air temperature is −15°C, what is it in degrees Fahrenheit? (*c*) What is a temperature change of 20.0°C expressed in kelvins? in degrees Fahrenheit?
ANS. (*a*) 37.0°C, 310.2 K; (*b*) 5.0°F; (*c*) 20.0 K, 36.0°F.

18-3 Ideal-Gas Law

Figure 18-3*b* illustrates the similar behavior of different gases at low densities. The plots of temperature vs. pressure are linear for every gas. Furthermore, as p_R (or the gas density) approaches zero, all plots intersect at the same point on the temperature axis. This zero-density limit corresponds to an *ideal gas*. Most gases behave almost like an ideal gas when their temperatures are high enough and their pressures low enough that they are not close to the conditions under which they condense (become liquids). Such is the case for most gases at room temperature and atmospheric pressure.

In general, a gas law relates the Kelvin temperature T, the pressure p (recall from Sec. 18-2 that pressure represents absolute pressure), the volume V, and the mass m of a gas. The mass is usually specified in terms of the number (n) of moles of the gas. By definition, one *mole* (*mol*) of a substance is the amount of that substance containing Avogadro's number N_A (to four significant figures, $N_A = 6.022 \times 10^{23}$ molecules/mol) of molecules. The *molecular mass* M of a substance is the mass of one mole. The number of moles of a substance is related to its mass m by

$$n = \frac{m}{M}.$$

For example, the molecular mass of oxygen is 32.00 g/mol, so 64.00 g of the gas corresponds to 2.000 mol and 12.044×10^{23} molecules.

The ideal-gas law is based on experiments performed by Robert Boyle (1627–1691) and Jacques Charles (1746–1823). Boyle found that when a fixed quantity of a dilute gas is kept at constant temperature, its pressure is inversely proportional to its volume. Charles discovered that when the pressure of a dilute gas is kept constant, its volume is directly proportional to its temperature *on the Kelvin scale*. These experimental observations are combined in the *ideal-gas law*:

$$pV = nRT, \qquad (18\text{-}4a)$$

where R is a constant called the *universal gas constant*. As its name implies, R is the same for all gases that behave as ideal gases. In the SI system of units,

$$R = 8.314 \text{ J/mol}\cdot\text{K}.$$

When the pressure is expressed in atmospheres and the volume in liters ($1 \text{ L} = 10^{-3} \text{ m}^3$),

$$R = 0.08207 \text{ L}\cdot\text{atm/mol}\cdot\text{K}.$$

The universal gas constant can be determined experimentally. For example, various values of p, V, and T can be measured for a fixed quantity of gas (n moles). A plot of pV versus T will then be a straight line with slope nR. With the slope determined from the graph, R can be calculated.

The ideal-gas law can also be expressed in terms of the number of gas molecules. If there are n moles of a gas, then the number N of molecules in the gas is

$$N = nN_A,$$

where N_A is Avogadro's number. Substituting N/N_A for n in Eq. (18-4a), we have

$$pV = N\left(\frac{R}{N_A}\right)T,$$

or

$$pV = NkT \qquad (18\text{-}4b)$$

where

$$k = \frac{R}{N_A} = \frac{8.314 \text{ J/mol}\cdot\text{K}}{6.022 \times 10^{23}/\text{mol}} = 1.381 \times 10^{-23} \text{ J/K}.$$

The quantity k is called *Boltzmann's constant* in honor of Ludwig Boltzmann (1844–1906), one of the pioneers in the development of the statistical description of gases.

As the temperature of the gas in the balloon increases, its volume also increases.

EXAMPLE 18-2 NITROGEN AS AN IDEAL GAS

A gas of pure N_2 is confined to a cylinder with a movable piston at one end. The initial pressure, volume, and temperature of the gas are $2.00 \times 10^5 \text{ N/m}^2$, $1.00 \times 10^{-3} \text{ m}^3$, and 30°C, respectively. (a) How many moles of the gas are in the container? (b) If the piston is compressed so that the pressure is increased to $4.20 \times 10^5 \text{ N/m}^2$ and the volume is decreased to $5.10 \times 10^{-4} \text{ m}^3$, what is the new temperature of the gas?

SOLUTION (a) From the ideal-gas law, the number of moles is given by

$$n = \frac{pV}{RT} = \frac{(2.00 \times 10^5 \text{ N/m}^2)(1.00 \times 10^{-3} \text{ m}^3)}{(8.314 \text{ J/mol}\cdot\text{K})(303 \text{ K})} = 0.0794 \text{ mol}.$$

(b) Since we have a fixed amount of gas, the number of moles is constant and

$$\frac{p_1V_1}{T_1} = \frac{p_2V_2}{T_2}.$$

With the appropriate values substituted into this equation, we have

$$T_2 = \left(\frac{p_2V_2}{p_1V_1}\right)T_1 = \left[\frac{(4.20 \times 10^5 \text{ N/m}^2)(5.10 \times 10^{-4} \text{ m}^3)}{(2.00 \times 10^5 \text{ N/m}^2)(1.00 \times 10^{-3} \text{ m}^3)}\right] 303 \text{ K},$$

so the new gas temperature is $T_2 = 325$ K, or 52°C.

EXAMPLE 18-3 **STANDARD TEMPERATURE AND PRESSURE**

(*a*) What volume does 1.00 mol of an ideal gas occupy at a pressure of 1 atm and a temperature of 0°C (called *standard temperature and pressure*, or *STP*)? (*b*) If this ideal gas is air, what is its density?

SOLUTION (*a*) With $T = 273$ K, we have from the ideal-gas law,

$$V = \frac{nRT}{p} = \frac{(1.00\ \text{mol})(8.31\ \text{J/mol}\cdot\text{K})(273\ \text{K})}{1.01 \times 10^5\ \text{N/m}^2}$$
$$= 22.4 \times 10^{-3}\ \text{m}^3 = 22.4\ \text{L}.$$

This is the approximate volume occupied by 1 mol of *any dilute gas* at STP.

(*b*) The average molecular mass of air is 29.0 g/mol, so the mass of 1.00 mol of air is $M = 29.0\ \text{g} = 0.0290$ kg. The density of the air at STP is then

$$\rho = \frac{m}{V} = \frac{0.0290\ \text{kg}}{22.4 \times 10^{-3}\ \text{m}^3} = 1.29\ \text{kg/m}^3.$$

EXAMPLE 18-4 **OXYGEN AS AN IDEAL GAS**

A dilute O_2 gas is confined to a rigid container of volume 1.50 m^3 at a pressure of 0.400 atm and a temperature of 40°C. (*a*) If the temperature is raised to 100°C, what is the pressure of the gas? (*b*) What is the mass of the gas?

SOLUTION (*a*) Since both the number of moles and the volume of the gas are constant, we have from the ideal-gas law

$$\frac{p_1}{T_1} = \frac{p_2}{T_2} = \text{constant},$$

so

$$p_2 = p_1 \frac{T_2}{T_1}.$$

Here the initial pressure and temperature are $p_1 = 0.400$ atm and $T_1 = (40 + 273)\ \text{K} = 313$ K, and the final temperature is $T_2 = (100 + 273)\ \text{K} = 373$ K; thus the final pressure is

$$p_2 = p_1 \frac{T_2}{T_1} = (0.400\ \text{atm})\left(\frac{373\ \text{K}}{313\ \text{K}}\right) = 0.477\ \text{atm}.$$

(*b*) The number of moles of the gas can be calculated from the ideal-gas law. With $p_1 = 0.400\ \text{atm} = 4.04 \times 10^4\ \text{N/m}^2$, we have

$$n = \frac{p_1 V_1}{RT_1} = \frac{(4.04 \times 10^4\ \text{N/m}^2)(1.50\ \text{m}^3)}{(8.31\ \text{J/mol}\cdot\text{K})(313\ \text{K})} = 23.3\ \text{mol}.$$

The molecular mass of oxygen is 0.0320 kg/mol, so the mass of the gas in the container is

$$(0.0320\ \text{kg/mol})(23.3\ \text{mol}) = 0.746\ \text{kg}.$$

DRILL PROBLEM 18-2

If dilute nitrogen gas at 1.00 atm pressure is at the same temperature as boiling water at 1.00 atm pressure, what is the density of the gas?
ANS. 0.912 kg/m^3.

DRILL PROBLEM 18-3

An ideal gas is confined to a volume of 4.0 L with a pressure of 5.0 atm. If the gas expands to a volume of 10.0 L without any change in its temperature, what is its new pressure?
ANS. 2.0 atm.

DRILL PROBLEM 18-4

The stopcock on a container is opened and a portion of the ideal gas it contains is released. When the remaining gas returns to its original temperature, its pressure is found to have dropped by a factor of 3. What fraction of the gas is released?
ANS. 2/3.

18-4 A STATISTICAL INTERPRETATION OF AN IDEAL GAS

The properties of an ideal gas can be understood in terms of a simple statistical model. With this model, we can determine averages of certain mechanical properties of the molecules and then relate these averages to measurable macroscopic properties of the gas. The model is based on the following assumptions:

1. The gas is in thermal equilibrium with its environment.
2. The molecules of the gas move randomly and obey Newton's laws. The gas can be pictured as a very large number of point masses that collide elastically as they move randomly throughout the container. (See Fig. 18-4.)

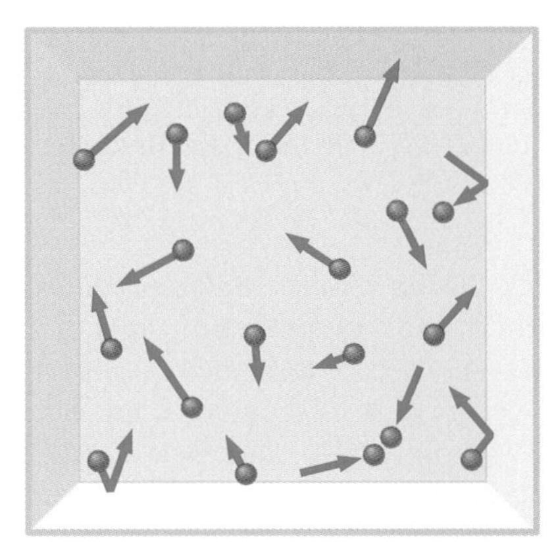

FIGURE 18-4 An ideal gas can be thought of as a large number of point masses moving randomly throughout the container and colliding elastically with each other and the walls.

3. The total number of molecules is large enough that averages of molecular dynamic variables such as speed are meaningful physically.
4. The total volume of the molecules is negligible compared with the volume that the gas occupies.
5. Intermolecular collisions and collisions between molecules and the walls of the container are elastic and occur over time

intervals that are negligible compared with the average time interval between collisions.

6. The molecules do not interact except during collisions.

We'll first calculate the pressure of the gas. For simplicity, we assume that the gas is composed of a single type of molecule and is confined to a cubical box whose sides are of length l. The potential energy associated with the force of gravity may be ignored, as its effect is negligible for a small container. (This is illustrated in Prob. 18-31.) We begin with the single molecule of Fig. 18-5*a*. It bounces back and forth between the left and right walls of the container at a constant speed. Actually, no molecule of the gas behaves so simply, as all molecules collide many times with one another while traversing the container. However, because there are so many collisions, all of which are elastic, the number of molecules with one particular speed remains essentially constant. The concept of a single representative molecule moving as described is therefore meaningful.

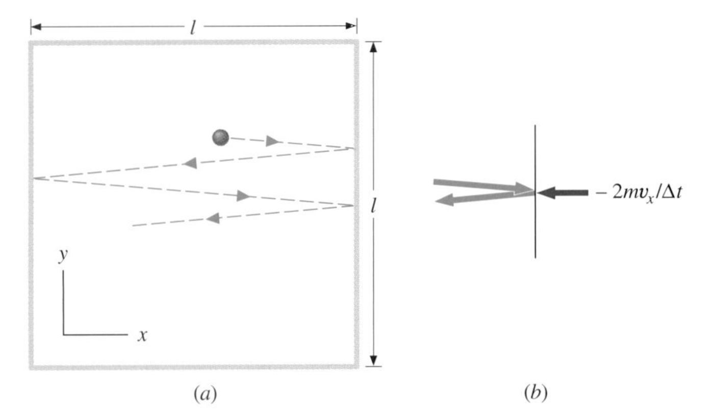

FIGURE 18-5 (*a*) An ideal-gas molecule moving with a constant speed collides elastically with the walls of the container. The molecule moves from the right wall to the left wall and back to the right wall in a time $2l/v_x$. (*b*) During the time of collision Δt, the wall exerts a force $-2mv_x/\Delta t$ on the molecule.

In the elastic collision with the wall represented in Fig. 18-5*b*, the x component of the momentum of the molecule is changed from mv_x to $m(-v_x)$ where v_x is the magnitude of the x component of the velocity. If the time of contact is Δt, the force exerted by the wall on the molecule is $-2mv_x/\Delta t$, so by Newton's third law, the force that the molecule exerts on the wall during this time is $2mv_x/\Delta t$. After the collision, the molecule travels to the opposite side of the container, collides there, and returns to the original wall in a time interval $2l/v_x$. Throughout this period, there is no interaction between the molecule and the original wall because there is no contact between them. This sequence of events recurs indefinitely because no energy is dissipated in the collisions.

In one round-trip, the fraction of time that the molecule is in contact with the wall is $\Delta t/(2l/v_x)$. Consequently, of the n molecules moving at v_x, $n\,\Delta t/(2l/v_x)$ are in contact with the wall at *any* instant. To determine the net pressure of the gas, we need to consider all of its constituent molecules. Suppose there are n_1 gas molecules with a velocity component in the x direction of magnitude v_{x1}, n_2 molecules with v_{x2}, and so on. At any instant, the number of molecules with v_{x1} that are in contact with the wall is given by $n_1\,\Delta t/(2l/v_{x1})$. Multiplying this by the force per molecule yields the total force exerted by these molecules on the wall:

$$F_{x1} = \left(\frac{n_1\,\Delta t}{2l/v_{x1}}\right)\left(\frac{2mv_{x1}}{\Delta t}\right) = n_1\left(\frac{mv_{x1}^2}{l}\right).$$

Similarly, the forces due to the molecules with $v_{x2}, v_{x3}, \ldots$ are $n_2(mv_{x2}^2/l)$, $n_3(mv_{x3}^2/l)$, etc. Thus at any instant, the net force exerted on the wall by the gas is

$$F_x = n_1\left(\frac{mv_{x1}^2}{l}\right) + n_2\left(\frac{mv_{x2}^2}{l}\right) + \cdots$$
$$= \frac{m}{l}\,(n_1 v_{x1}^2 + n_2 v_{x2}^2 + \cdots).$$

By definition, the average value of v_x^2 (denoted by $\overline{v_x^2}$) is

$$\overline{v_x^2} = \frac{n_1 v_{x1}^2 + n_2 v_{x2}^2 + \cdots}{N},$$

where $N = n_1 + n_2 + n_3 + \cdots$ is the total number of molecules of the gas. The expression for F_x can then be written as

$$F_x = N\frac{m}{l}\,\overline{v_x^2}.$$

Since there is no preferred direction in space, the averages of all three components of the velocity must be the same; that is,

$$\overline{v_x^2} = \overline{v_y^2} = \overline{v_z^2}.$$

Consequently,

$$\overline{v^2} = \overline{v_x^2} + \overline{v_y^2} + \overline{v_z^2} = 3\overline{v_x^2},$$

and

$$F_x = \frac{Nm\overline{v^2}}{3l}.$$

By dividing F_x by the area l^2 of the wall, we obtain the pressure (or force per unit area) p of the gas on the wall. It is

$$p = \frac{F_x}{l^2} = \frac{Nm}{l^3}\,\frac{\overline{v^2}}{3},$$

which we can also write as

$$pV = \tfrac{1}{3}Nm\overline{v^2} = \tfrac{2}{3}[N(\tfrac{1}{2}m\overline{v^2})], \tag{18-5}$$

where $V = l^3$ is the volume of the container and $N(m\overline{v^2}/2)$ is the total translational kinetic energy of the gas molecules. Equation (18-5) is a relationship between the macroscopic properties of the gas and the microscopic properties of its molecules. It tells us that *the product of the pressure and the volume of an ideal gas is proportional to the total translational kinetic energy of its molecules.*

The total translational kinetic energy of the molecules can be related to the temperature of the gas simply by combining the ideal-gas law and Eq. (18-5). We then obtain

$$nRT = \tfrac{2}{3}[N(\tfrac{1}{2}m\overline{v^2})],$$

which, since $n = N/N_A$, becomes

$$\frac{1}{2} m\overline{v^2} = \frac{3}{2}\left(\frac{R}{N_A}\right)T,$$

where T is the temperature expressed in kelvins. Now the ratio R/N_A is Boltzmann's constant k. (See Sec. 18-3.) Thus the average translational kinetic energy of an ideal-gas molecule is

$$\tfrac{1}{2}m\overline{v^2} = \tfrac{3}{2}kT. \tag{18-6}$$

The **root mean square (rms) speed** of the molecules of a gas is by definition

$$v_{rms} = \sqrt{\overline{v^2}}.$$

For the ideal gas, we have from Eq. (18-6)

$$v_{rms} = \sqrt{\frac{3kT}{m}}, \tag{18-7a}$$

or in terms of the universal gas constant R $(= N_A k)$ and the molecular mass M $(= N_A m)$,

$$v_{rms} = \sqrt{\frac{3RT}{M}}.^{*} \tag{18-7b}$$

For example, at a room temperature of 293 K, the rms speed of O_2 molecules is

$$v_{rms} = \sqrt{\frac{3(8.31\ \text{J/mol}\cdot\text{K})(293\ \text{K})}{0.0320\ \text{kg/mol}}} = 478\ \text{m/s}.$$

The rms speed represents how fast the typical molecule of a particular gas is moving between collisions. Notice that at a given temperature, lighter molecules move faster, on the average, than heavier molecules. For example, molecules of N_2, whose molecular mass is 28 g/mol, move slightly faster through the air than molecules of O_2, whose molecular mass is 32 g/mol. From Eq. (18-7b), the ratio of their rms speeds is $\sqrt{32/28} = 1.1$.

*18-5 Phase Diagrams of Nonideal Gases

Pressure vs. Temperature

According to the ideal-gas law, pressure varies linearly with temperature for a given quantity of ideal gas occupying a fixed volume. This behavior is represented by the plot of pressure vs. temperature for various volumes of an ideal gas, as shown in Fig. 18-6. However, real gases do not behave so simply. For example, reducing the temperature of water vapor changes it into

*According to Eqs. (18-6) and (18-7), the molecules have no translational motion at $T = 0$ K. This is not strictly correct, for at this temperature molecules vibrate in what is known as *zero-point motion*, an effect that can only be explained with quantum mechanics.

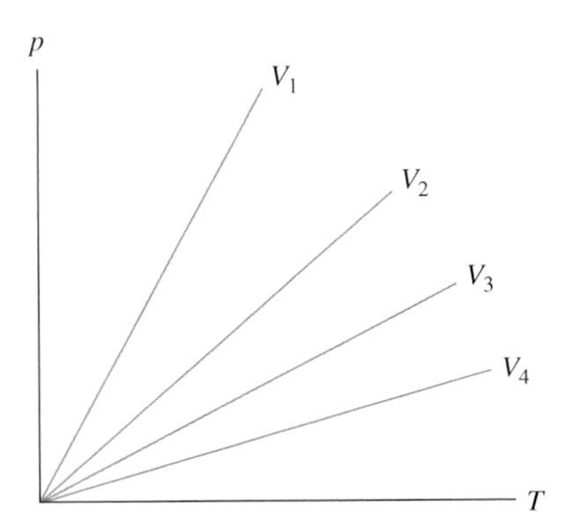

FIGURE 18-6 A plot of pressure vs. temperature for various volumes of an ideal gas.

a liquid, and further cooling produces ice. These different states are known as *phases*. Water vapor* (or steam) exemplifies the *gas phase*, water the *liquid phase*, and ice the *solid phase*. Very often, we become accustomed to seeing a substance in a particular phase—for example, it's hard to imagine that oxygen can be anything but a gas. However, if oxygen is cooled at 1 atm to −183°C, it liquefies; and at −219°C, it freezes into a solid.

Matter exists in three phases. Here the liquid phase is the water and the solid phase is the ice. The third phase is the water vapor in the air.

The three phases of a real substance are commonly represented on a *phase diagram* such as that shown in Fig. 18-7a. This plot is also known as a *pT diagram*. The various phases are delineated by lines known as *phase-equilibrium lines*. You can immediately determine what phase a substance is in by locating on the diagram the point corresponding to the pressure and temperature of the substance. The points on a phase-equilibrium line represent the pressures and temperatures for which two phases coexist in equilibrium. For example, in Fig. 18-7a, the pressures and temperatures for solid-gas phase equilibrium are on line OA. Similarly, the other two lines represent equilibrium between the solid and liquid phases and the

*The gas that is in equilibrium with the liquid is commonly called *vapor*.

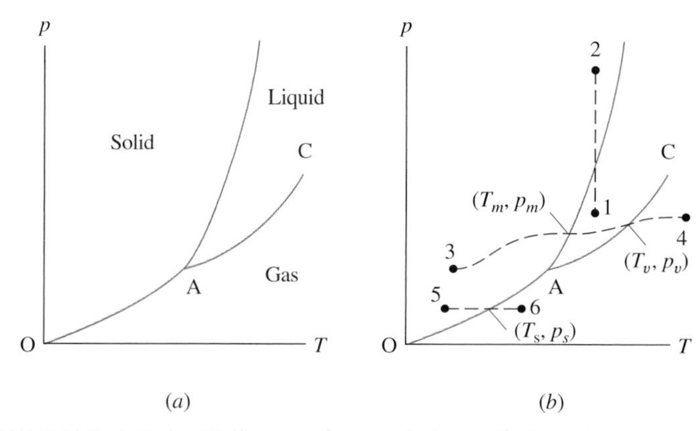

FIGURE 18-7 (*a*) A pT diagram for a substance that contracts upon freezing. Most substances have pT diagrams similar to the one shown here. (*b*) Various phase transitions are represented by the dashed lines.

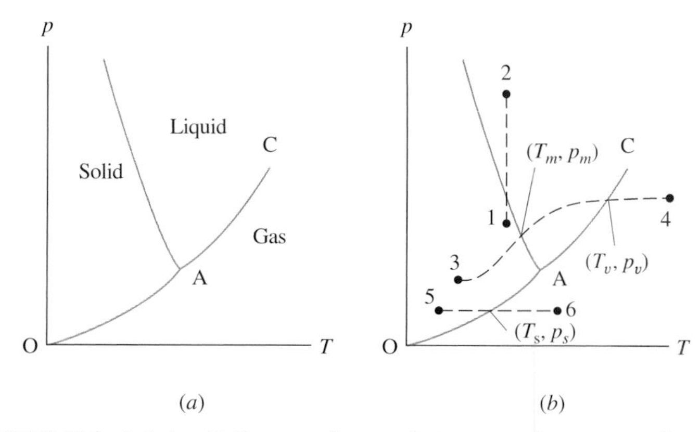

FIGURE 18-8 (*a*) A pT diagram for a substance such as water that expands upon freezing. (*b*) Various phase transitions are represented by the dashed lines.

liquid and gas phases. The intersection of the three phase-equilibrium lines is known as the *triple point* (A). At the pressure and temperature of this point, all three phases of the substance coexist in equilibrium. As stated in Sec. 18-2, the triple point of water has been chosen as a reference state in thermometry and assigned a temperature of 273.16 K. Another interesting feature of the pT diagram is the *critical point* C where the liquid-gas equilibrium line terminates. Beyond this point, the liquid and gas are indistinguishable and there is no phase separation. Table 18-2 gives the temperatures and pressures of the triple points and the critical points of some common substances.

TABLE 18-2 **Triple-Point and Critical-Point Data**

	Triple Point		Critical Point	
Substance	***T* (K)**	***p* (atm)**	***T* (K)**	***p* (atm)**
CO_2	216.55	5.105	304	72.9
H_2	13.84	0.0695	33.2	12.8
H_2O	273.16	0.00603	647.2	218.3
N_2	63.18	0.12	126	33.5
NH_3	195.40	0.05996	405.6	112.5
O_2	54.35	0.00150	154.6	50.1
SO_2	197.68	0.001653	430.9	77.7

The slope of the solid-liquid equilibrium line determines whether an increase in pressure at a fixed temperature causes a liquid to change to a solid or a solid to change to a liquid. For most substances, the slope is positive, as shown in Fig. 18-7, indicating that increased pressure ($1 \rightarrow 2$) causes the liquid-to-solid transition. However, a small group of substances have a pT diagram like that of Fig. 18-8. For these substances, an increase in pressure at constant temperature ($1 \rightarrow 2$) produces a transition from the solid to the liquid phase. Water is an example from this latter group, as is evident from the fact that ice melts when pressure is applied to it. It turns out that the slope of the pT diagram also determines whether a substance expands or contracts when it makes the transition from a solid to a liquid at a fixed temperature. Figure 18-7 corresponds to expansion and Fig. 18-8 to contraction.

Each time a phase-equilibrium line is crossed because of a change in the pressure or temperature, a phase transformation occurs. When a solid is converted into liquid, the process is called *melting*; when a liquid becomes a gas, it undergoes *vaporization*. On the pT diagrams of Figs. 18-7 and 18-8, melting occurs for the transition $3 \rightarrow 4$ at the temperature T_m and pressure p_m, which is the point where the dashed line between 3 and 4 crosses the liquid-solid phase-equilibrium line. For this same transition, vaporization occurs at the temperature T_v and pressure p_v, or the point where the dashed line and the liquid-vapor phase-equilibrium line intersect. Finally, when the pressure is less than the triple-point pressure, a solid can go directly to the gas phase and bypass the liquid phase. This transformation is known as *sublimation*. For example, sublimation occurs for the transition $5 \rightarrow 6$ at the temperature T_s and pressure p_s. A well-known substance that sublimes at atmospheric pressure is carbon dioxide. Solid CO_2 (known as dry ice), when warmed, goes into the gas phase without passing through its liquid state.

A change of phase.

This phenomenon has been used in the movie industry—for example, when an eerie ground-covering fog is needed for nocturnal cemetery scenes.

Not all substances make distinct phase changes. For example, when tar is heated, it gets softer and softer, eventually becoming a liquid. However, there is no single fixed temperature at which it makes the solid-liquid transition. Substances that behave in this manner do not have a well-defined crystalline structure in the solid state; they are generally called *amorphous* materials.

EXAMPLE 18-5 **PHASE CHANGES OF CO_2**

The pT diagram of carbon dioxide is shown in Fig. 18-9. (*a*) Assuming the pressure is constant, how must the temperature be aried so that carbon dioxide goes from point 1 in its solid phase to its liquid phase? to its gas phase? (*b*) If the temperature is constant, must the pressure be increased or decreased in order for the CO_2 to go from point 2 in the liquid phase to the gas phase? to the solid phase? (*c*) What is the maximum pressure under which sublimation can occur?

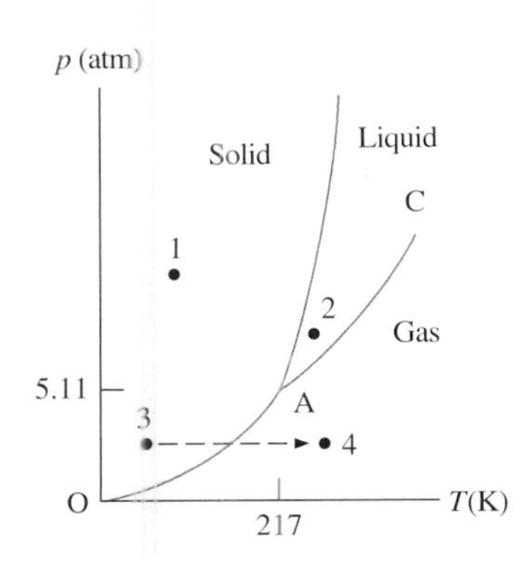

FIGURE 18-9 The pT diagram of carbon dioxide.

SOLUTION (*a*) If the temperature is increased at constant pressure, the path taken on the pT diagram is horizontal and from left to right. Starting from point 1, an increase in temperature first takes solid CO_2 into its liquid phase. A further increase, if sufficiently large, will transform the liquid into a gas.

(*b*) At a fixed temperature, a pressure increase corresponds to a vertical path upward and a pressure decrease to a vertical path downward. By inspection, if the pressure on the liquid CO_2 at point 2 is decreased, a liquid-gas transformation will occur. Alternatively, a pressure increase denotes a path crossing the liquid-solid equilibrium line, and liquid CO_2 is transformed into a solid.

(*c*) The figure indicates that if the pressure is less than the triple-point pressure of 5.11 atm, solid CO_2 goes directly into its gas phase with increasing temperature. (See path $3 \rightarrow 4$.) Therefore, at normal pressures of about 1.0 atm, solid CO_2 sublimes when warmed. If the pressure is greater than the triple-point value, any horizontal path that originates from a point in the solid phase must first cross the liquid phase, so sublimation cannot occur because of a simple increase in the temperature.

DRILL PROBLEM 18-5

(*a*) An unknown substance in the liquid state completely fills a sealed container. When it freezes, there is empty space in the container. Is this substance's pT diagram like that of Fig. 18-7 or that of Fig. 18-8? (*b*) Another substance in the liquid state completely fills a sealed container. When this substance freezes, the container cracks. Answer the same question for this substance.

ANS. (*a*) Fig. 18-7; (*b*) Fig. 18-8.

PRESSURE VS. VOLUME

At a constant temperature the volume of a given quantity of ideal gas varies inversely with the pressure. A series of constant-temperature curves (*isotherms*) depicting this relationship is shown in the diagram of Fig. 18-10.

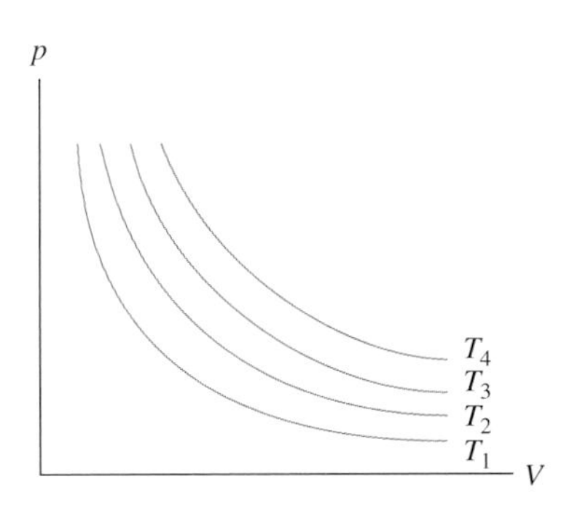

FIGURE 18-10 The pressure-vs.-volume diagram of an ideal gas. The curves correspond to different fixed temperatures T_1, T_2, ...

The three phases of a real substance can also be distinguished on a pressure vs. volume (pV) diagram. Figure 18-11 is the pV diagram for CO_2. It resembles an ideal-gas diagram only at high temperatures. At the lower temperatures, the behavior is quite interesting. We analyze this behavior starting at point O

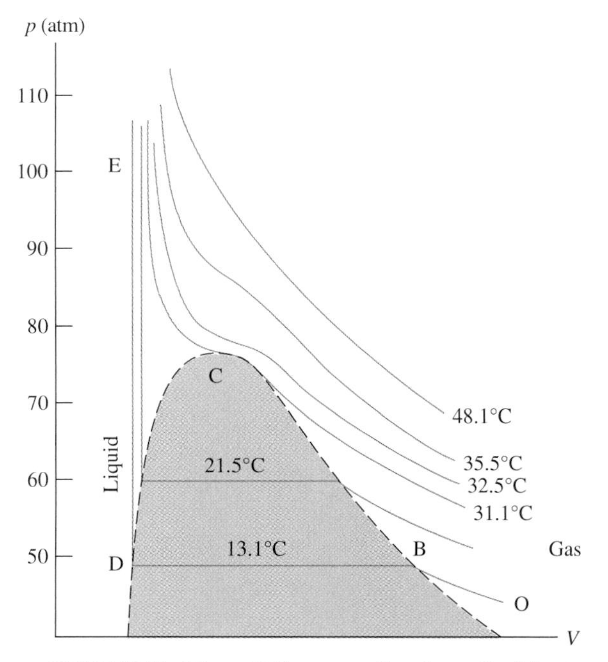

FIGURE 18-11 The pV diagram of carbon dioxide.

on the curve corresponding to 13.1°C, where CO_2 is in its gaseous state. Increasing the pressure reduces its volume until it is at B, where it begins to condense into a liquid. This condensation (at a constant pressure of about 49 atm) is represented by the horizontal section of the curve along which the gas and liquid coexist in equilibrium. As the path is traversed from B to D, more and more of the gas condenses into liquid until at point D the process is complete and the CO_2 is in its liquid state. Since it is much more difficult to compress a liquid than a gas, a further increase in pressure produces only a small reduction in volume, as represented by section DE of the curve.

At the next higher temperature shown (21.5°C), the general shape of the curve is unchanged. However, a greater pressure is required before the gas begins to condense, and the horizontal section along which the gas and liquid phases are in equilibrium is shorter. This trend continues with increasing temperature until the length of the horizontal section is zero (point C). The temperature at which this occurs is the critical temperature, which is about 31.1°C for CO_2. Above the critical temperature of a gas, no amount of applied pressure can produce a well-defined phase transformation. Point C, the critical point, was represented earlier in the pT diagram of CO_2 (Fig. 18-9).

*18-6 The van der Waals Equation

A Dutch scientist named J. D. van der Waals (1837–1923) made one of the earliest successful attempts to modify the ideal-gas law in order to describe the behavior of nonideal gases. His result is known as the van der Waals equation which, for one mole of gas, is

$$\left(p + \frac{a}{v^2}\right)(v - b) = RT, \tag{18-8}$$

where a and b are constants that depend on the gas under consideration and v is the molar volume.

By comparing the van der Waals equation to the ideal-gas law, we see that two modifications to the latter have been made. The first is the subtraction of b from the volume v. Ideal-gas molecules are assumed to be point masses that occupy no volume, so the entire volume of the container is available to each molecule. In a real gas, however, the finite size of the molecules reduces the effective volume of the space within the container available to the molecules. This reduction is represented by $-b$.

The pressure-correction term a/v^2 is due to the attractive forces among the molecules of a nonideal gas. A molecule approaching the wall of a container is subjected to a net inward force due to the inner gas molecules. This results in a reduction in the momentum of the molecule striking the wall and, consequently, a reduction in the effective pressure of the gas. The ideal-gas pressure is therefore greater than the actual-gas pressure because of the attractive intermolecular forces. The difference between these two pressures must be proportional to the inward force exerted by the bulk of the molecules on those near the wall. Now this force depends on two factors: the number of molecules within the gas and the number of molecules striking the wall per unit time. Since both of these are proportional to the gas density, the pressure difference is inversely proportional to the square of the volume occupied by the gas; thus the pressure-correction term in Eq. (18-8) may be written as a/v^2.

A pV diagram of a gas that obeys the van der Waals equation is shown in Fig. 18-12. By comparing it with the experimental curve for carbon dioxide in Fig. 18-11, you can see that the van der Waals equation clearly represents the nonideal gas better than does the ideal-gas law. While the horizontal sections of the isotherms in the transition region are not reproduced, the van der Waals model does shows a clear distinction between the gas and the liquid phases.

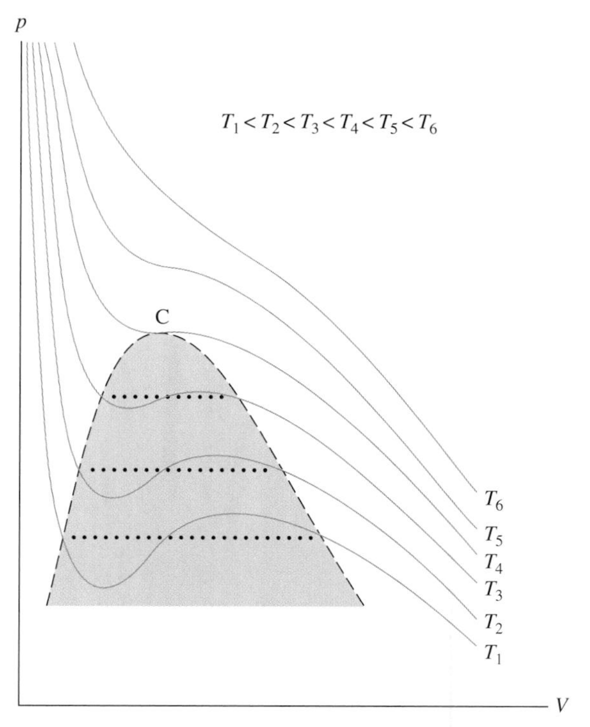

FIGURE 18-12 The pV diagram for one mole of a van der Waals gas. The solid lines are plots of Eq. (18-8) for temperatures T_1, T_2, . . . , T_6. The dotted horizontal lines represent liquid-gas transitions for a real substance and are shown for comparison purposes.

DRILL PROBLEM 18-6

What happens to the correction terms in the van der Waals equation in the limit that the gas density goes to zero? What is the appropriate equation that describes the behavior of this gas?
ANS. $pv = RT$.

SUMMARY

1. **Temperature and thermal equilibrium**
   (*a*) If bodies A and B are separately in thermal equilibrium with body C, then A and B are in thermal equilibrium with each other.
   (*b*) All bodies in thermal equilibrium with one another have the same temperature.

2. **Thermometers and temperature scales**
   (*a*) The freezing temperature of water at a pressure of 1 atm is 0°C or 32°F. The boiling temperature of water at 1 atm is 100°C or 212°F.
   (*b*) The Fahrenheit and Celsius scales are related by

   $$T_F = \tfrac{9}{5}T_C + 32°F.$$

   (*c*) The size of the degree on the Celsius scale is the same as that on the Kelvin scale. Temperatures on the two scales are related by

   $$T_C = T - 273.15°C.$$

3. **Ideal-gas law**
   To a good approximation, dilute gases satisfy

   $$pV = nRT,$$

   where p is the pressure, V is the volume, n is the number of moles, R is the universal gas constant, and T is the Kelvin temperature.

4. **A statistical interpretation of an ideal gas**
   (*a*) The average translational kinetic energy of an ideal-gas molecule is

   $$\overline{\tfrac{1}{2}mv^2} = \tfrac{3}{2}kT,$$

   where k, Boltzmann's constant, is 1.381×10^{-23} J/K and T is the Kelvin temperature.
   (*b*) The rms speed of an ideal-gas molecule of mass m is

   $$v_{rms} = \sqrt{\frac{3kT}{m}},$$

   where T is the Kelvin temperature of the gas.

*5. **Phase diagrams of nonideal gases**
   With phase diagrams, we can determine the phase of a substance when we know its pressure, temperature, and volume.

*6. **The van der Waals equation**
   One modification of the ideal-gas law is the van der Waals equation. For 1 mol of gas, this equation is

   $$\left(p + \frac{a}{v^2}\right)(v - b) = RT,$$

   where a and b are constants whose values depend on the gas under consideration. This equation generally describes the behavior of a nonideal gas more accurately than does the ideal-gas law.

QUESTIONS

18-1. Discuss the factors that must be considered before choosing a gas for a constant-volume gas thermometer.

18-2. (*a*) How is a temperature assigned to a region in outer space? (*b*) The temperature at the outer edge of the earth's atmosphere is about 1000 K. What does this mean? (*c*) If the temperature is so high in this region, why do the space-walking astronauts have to be thermally insulated against freezing?

18-3. Suppose someone tells you that a thermometer is useless because it measures its own temperature rather than that of the substance it is immersed in. How would you respond?

18-4. Why isn't water used rather than mercury in thermometers that measure air temperature?

18-5. Why does the air pressure in a tire increase when a car is moving?

18-6. The temperature of an ideal gas is changed from 300 to 100 K. Has the average translational kinetic energy of the gas molecules increased or decreased? By what fraction of its original value?

18-7. (1) The properties of an ideal gas were derived under the assumption that it contained a large number of molecules. (2) Real gases behave like an ideal gas at low densities. Relate these two statements.

18-8. What is the average molecular *velocity* of an ideal gas in a stationary container?

18-9. Using a pT diagram, explain why ice melts when pressure is applied to it.

18-10. Why does dew form when the temperature drops during the night? When is that dew frost?

18-11. Why does the outside of a glass containing a cold drink get wet?

18-12. Why does the bathroom get "steamy" when you take a long, hot shower? Is this effect more pronounced in high or low humidity?

18-13. Explain why you can often see your breath on a cold winter day.

18-14. Explain why ice-skating is difficult during very, very cold weather.

PROBLEMS

Thermometers and Temperature Scales

18-1. How many divisions on (*a*) the Celsius scale and (*b*) the Kelvin scale correspond to 100 divisions on the Fahrenheit scale?

18-2. What temperature expressed in degrees Celsius has the same numerical value in degrees Fahrenheit?

18-3. What temperature expressed in degrees Fahrenheit has the same numerical value in kelvins?

18-4. The surface temperature of the sun is approximately 6.0×10^3 K. What is this temperature in degrees Fahrenheit? in degrees Celsius?

18-5. What temperatures on the Fahrenheit and Kelvin scales correspond to 27°C?

18-6. Suppose we define a hypothetical scale in which temperatures are measured in degrees H. If the freezing and boiling points of water at 1 atm are 800 and 1200°H, respectively, what does 1450°H correspond to on (*a*) the Celsius, (*b*) the Fahrenheit, and (*c*) the Kelvin scales?

18-7. Consider again the hypothetical temperature scale of Prob. 18-6. Express the following temperature measurements in degrees H: (*a*) 74°C, (*b*) −40°F, (*c*) 0 K.

18-8. If measurements with a constant-volume ideal-gas thermometer of the boiling-point temperature of water converge to 373.15 K, what is the limiting value of the ratio of the pressure of the ideal gas at the boiling point and its pressure at the triple point of water?

Ideal-Gas Law

18-9. Use the ideal-gas law to deduce Boyle's observations and Charles' observations.

18-10. A cubic box of side 0.80 m contains 5.0×10^{20} atoms of an ideal gas at 30°C. What is the pressure of the gas?

18-11. What is the average separation between ideal-gas atoms at standard temperature and pressure?

18-12. The best vacuum that can be easily produced in the laboratory is about 10^{-9} mmHg. How many molecules are there per cubic centimeter of gas for such a vacuum? Assume the temperature is 20°C.

18-13. Verify that the values of the universal gas constant R expressed in the two different sets of units given in Sec. 18-3 are equivalent.

18-14. How many moles are there in (*a*) 0.050 g of N_2 gas? (*b*) 10 g of CO_2 gas? (*c*) How many molecules are present in each case?

18-15. Complete the following tables for a given quantity of ideal gas:

(*a*) At constant volume

Pressure (mmHg)	150	?	80	?	320	?
Temperature (K)	60	100	?	170	?	300

(*b*) At constant pressure

Volume (L)	0.50	0.80	?	0.10	?	1.50
Temperature (K)	300	?	230	?	700	?

(*c*) At constant temperature

Pressure (mmHg)	800	?	400	?	150	?
Volume (L)	1.20	3.40	?	0.90	?	0.50

Plot three graphs corresponding to the values found in tables (*a*), (*b*), and (*c*), respectively.

18-16. The product of the pressure and volume of a certain mass of hydrogen gas at 0°C is 80 J. (*a*) How many moles of the gas are present? (*b*) What is the average translational kinetic energy of the gas molecules? (*c*) What is the value of the product of pressure and volume at 200°C?

18-17. Carbon dioxide has a density of 1.976 g/L at 0°C and a pressure of 1.00 atm. Assuming that CO_2 is an ideal gas, determine the value of the universal gas constant R.

18-18. A cubic container of volume 2.0 L holds 0.10 mol of nitrogen gas at a temperature of 20°C. What is the net force due to the gas on a wall of the container?

18-19. Suppose a mixture of several ideal gases is stored in a container. *Dalton's law of partial pressures* states that the net pressure of the mixture is the sum of the pressures that each constituent gas would exert if it alone filled the container. Using Dalton's law, calculate the net pressure of a mixture of 0.050 mol of H_2 and 0.030 mol of He in a container of volume 10 L at a temperature of 20°C.

18-20. Air is composed of about 78 percent N_2 molecules, 21 percent O_2 molecules, and 1 percent molecules of other gases. Ignoring the 1 percent of other gases, use Dalton's law of partial pressures (see Prob. 18-19) to find the mass of a cubic meter of air at 27°C and atmospheric pressure.

18-21. The partial pressure of CO_2 in the lungs is about 35 mmHg. (See Prob. 18-19.) What molecular percentage of the air in the lungs is CO_2? Compare your result with the molecular percentage of CO_2 in the atmosphere (~0.033 percent). Assume that the total pressure in the lungs is 760 mmHg.

18-22. A gas company advertises that it delivers helium at a gauge pressure of 1.72×10^7 N/m^2 (at 21°C) in a cylinder of volume 43.8 L. How many balloons of volume 4.0 L can be inflated with this amount of helium? Assume that the temperature and pressure of the helium in the balloons are 25°C and 76 cmHg, respectively.

18-23. (*a*) When the temperature is 0°C, what is the total mass of the air (molecular mass = 29 g/mol) inside a house of volume 500 m^3? (*b*) If the temperature rises to 30°C, what is the mass of the air that leaves the house? Assume that atmospheric pressure is 76 cmHg at both temperatures.

18-24. An automobile tire is filled to a gauge pressure of 32 $lb/in.^2$ when the temperature is 20°C. After a drive, the gauge pressure is measured to be 35 $lb/in.^2$. What is the temperature of the air in the tire now, assuming that the volume of the tire remains the same? If the tire volume had increased after the drive, would the temperature be greater or smaller than your calculated value?

18-25. At 300 K, a flask is open to the atmosphere through a thin tube at its top, as shown in the accompanying figure. If the flask is cooled to 200 K, does air enter or leave it? What is the fractional change of the mass of the air inside the flask?

PROBLEM 18-25

18-26. When an air bubble rises from the bottom to the top of a freshwater lake, its volume increases by 80 percent. If the temperatures at the bottom and the top of the lake are 4 and 10°C, respectively, what is the depth of the lake?

A Statistical Interpretation of the Ideal Gas

18-27. A cubic box of side 0.10 m contains 10^{15} atoms of an ideal gas. If the gas pressure is 0.50 atm, what is the average translational kinetic energy of the atoms?

18-28. The average translational kinetic energy of the molecules of an ideal gas in a container is doubled. How is the pressure of the gas affected?

18-29. An ideal gas in a container is heated until its pressure increases by 50 percent. How has the average translational kinetic energy of the gas molecules changed?

18-30. Find the total translational kinetic energy of 1 mol of an ideal gas at 100°C.

18-31. A cubic box of side 1.0 m contains dilute H_2 gas at a temperature of 300 K. What is the average translational kinetic energy of a gas molecule? What is the difference in a molecule's potential energy between the top and bottom of the box? Use these results to explain why gravitational effects could be neglected in our calculations of Sec. 18-4.

18-32. Determine the root mean square speed of an atom of a helium gas at 300 K. How much must the temperature be reduced in order to halve this speed?

18-33. Three separate containers hold dilute quantities of H_2, N_2, and CO_2 gases. If the gases are at the same temperature, what are the ratios of (*a*) the average translational kinetic energies and (*b*) the rms speeds of the molecules?

18-34. Two moles of an ideal gas are confined to a container of volume 2.0 m^3. If the pressure of the gas is 2.5×10^5 N/m^2, what is the average translational kinetic energy of the gas molecules?

18-35. The speeds of five different gas molecules are 500, 480, 540, 620, and 420 m/s, respectively. What is the rms speed of the molecules?

18-36. At what temperature do the atoms of a helium gas have the same rms speed as the molecules of a hydrogen gas at 50°C?

18-37. One hundred pellets, each of mass 1.0 g, strike at random positions over a 2.0-cm^2 surface of a metal plate. The pellets are all moving at a speed of 100 m/s and strike the plate at an angle of 60° to the normal to its surface. Assume that the collisions are elastic. If the pellets strike the plate at random times over a time interval of 10^{-3} s, what are the average force and the pressure exerted on the surface section?

Phase Diagrams of Nonideal Gases and the Van Der Waals Equation

18-38. The critical point of a substance is at a pressure of 20.0 atm and a temperature of 800 K, and its triple point is at 8.0 atm and 300 K. When $p = 16.0$ atm and $T = 100$ K, the solid and liquid states of the substance are found to be in equilibrium. From the information given, plot approximately the pT diagram of the substance. Does it expand or contract upon melting?

18-39. Consider an imaginary substance whose solid-liquid and liquid-gas phase-equilibrium lines *both* have negative slopes. Discuss the behavior of such a substance and its possible applications.

18-40. The following table lists the values of the van der Waals constants a and b for several gases. From these values, which gas would you expect to be the best approximation of an ideal gas and which the worst? If equal quantities of the different gases are condensed, which liquid would have the smallest volume? the greatest volume? Explain.

Gas	a ($L^2 \cdot atm/mol^2$)	b (L/mol)
H_2	0.2444	0.02661
He	0.03412	0.02370
N_2	1.390	0.03913
O_2	1.360	0.03183
CO_2	3.592	0.04267

18-41. Show that to a first approximation the deviation of van der Waals equation from the ideal-gas law goes as $1/v$. Use the approximation

$$\frac{v}{v-b} \approx 1 + \frac{b}{v}.$$

General Problems

18-42. A 5.0-L container holds 0.20 mol of an ideal gas at a pressure of 1.0 atm. (*a*) What is the temperature of the gas? (*b*) The container is partially evacuated using a vacuum pump and then sealed off again. The gas pressure is measured to be 10 mmHg, while the temperature is unchanged. How many moles of the gas remain in the container? (*c*) On the average, how many gas molecules are there per cubic centimeter after the partial evacuation? (*d*) The temperature of the gas is now changed to 77 K. What is the new gas pressure? (*e*) How is the density of the gas affected by this temperature change? (*f*) How is the average molecular kinetic energy affected?

18-43. The top of a vertical cylindrical tank is a tight-fitting, frictionless piston of diameter 10 cm. When the tank is filled with hydrogen at 300 K, the piston is in the position shown in the accompanying figure. The tank is now warmed and the piston rises a distance of 15 cm. What is the new temperature of the gas?

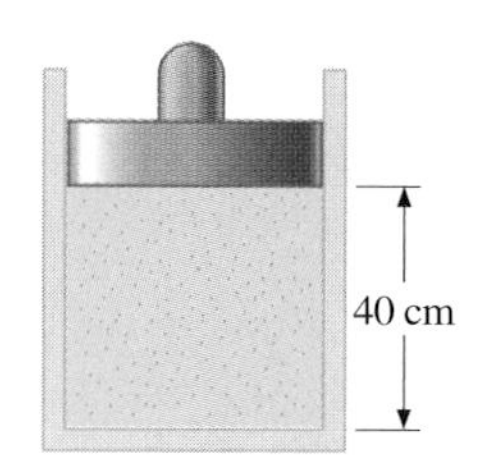

PROBLEM 18-43

18-44. The accompanying figure shows a flask containing an ideal gas at a pressure of 400 mmHg. A portion of the gas is removed and stored in a capped test tube of volume 20 cm^3. The gas pressure in the test tube is found to be 200 mmHg, and the pressure of

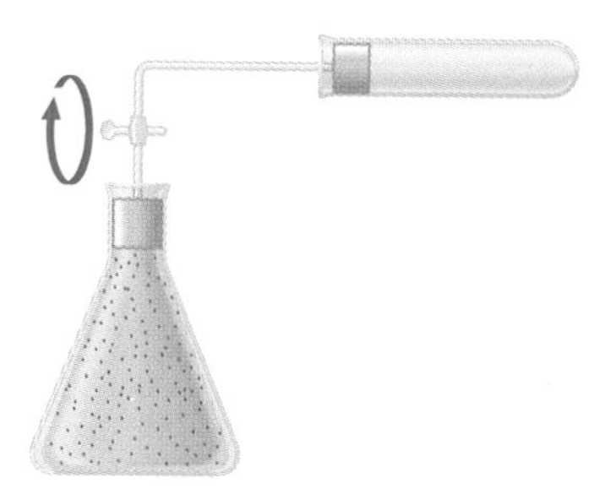

PROBLEM 18-44

the gas that remains in the flask is 350 mmHg. Assuming that all measurements are made at the same temperature and that 2.0×10^{-4} mol are transferred to the tube, determine (*a*) the volume of the flask, (*b*) the temperature, (*c*) the total number of moles of the gas, and (*d*) the density (in molecules per cubic centimeter) of the gas remaining in the flask.

18-45. The container shown at the left in the accompanying figure is kept at 100°C and holds an ideal gas at 1.0 atm. When the stopcock is opened, some of the gas leaks into the empty container on the right, which is immersed in ice at 0°C. Each container has a volume of 4.0 L. (*a*) How many moles of ideal gas are there? (*b*) After a sufficiently long time, the gas pressures in the containers are equalized. Determine the number of moles that are now in each container.

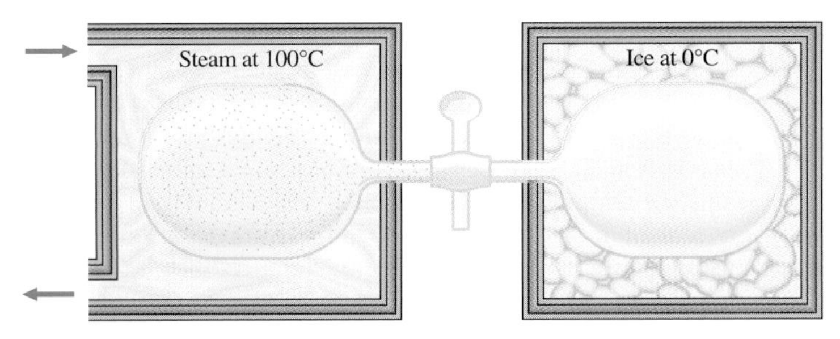

PROBLEM 18-45

18-46. A cylindrical tube of length 1.00 m is closed at one end and open at the other. It is pushed vertically into the ocean ($\rho = 1030\ kg/m^3$), with the closed end up. How far is the closed end from the surface of the ocean when the water has risen 0.30 m inside the tube? (*Hint:* The water level is lower inside than outside the tube.)

18-47. Consider once again the tube of the previous problem. When its open end is first in contact with the water surface, it is partially evacuated so that water rises half-way up its length. (See the accompanying figure.) (*a*) What is the air pressure inside the tube? (*b*) If the tube is then pushed downward into the water, how far must the closed end be from the surface in order that the water rise another 0.10 m in the tube?

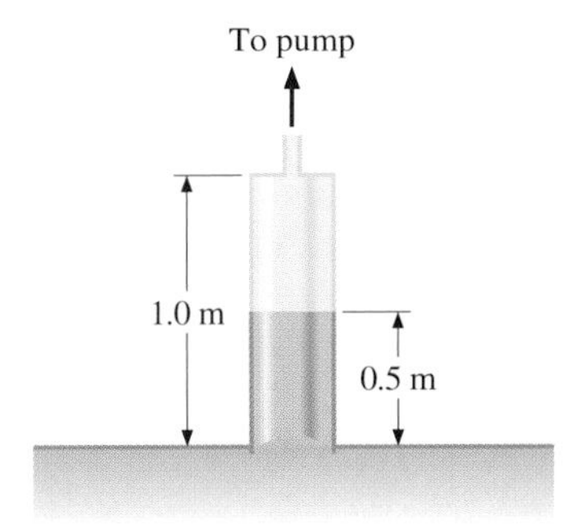

PROBLEM 18-47

18-48. When a substance's liquid and vapor phases are in equilibrium at a given temperature, the pressure of the vapor is known as the *vapor pressure* of the substance at that temperature. For example, in Fig. 18-11, the pressures corresponding to the horizontal sections of the isotherms in the region where the liquid and vapor phases are in equilibrium are the vapor pressures at the temperatures of the isotherms. At any temperature, the partial pressure (see Prob. 18-19) of the water in the air cannot exceed the vapor pressure of water at that temperature, for whenever the partial pressure reaches the vapor pressure, water condenses out of the air. The morning dew is an example of this condensation. The temperature at which condensation occurs for a particular sample of air is called the *dew point*. The dew point is easily determined by slowly cooling a metal ball and observing at what temperature condensation first appears on it.

The vapor pressures and corresponding densities of water vapor at some selected temperatures are given in the following table.

T (°C)	Vapor Pressure (mmHg)	Density (kg/m^3)
0	4.579	0.00485
3	5.685	0.00595
5	6.543	0.00680
8	8.045	0.00827
10	9.209	0.00941
13	11.231	0.01135
15	12.788	0.01283
18	15.477	0.01536
20	17.535	0.01730
23	21.068	0.02058
25	23.756	0.02304
30	31.824	0.03035
35	42.175	0.03960
40	55.324	0.05110

The *relative humidity* (R.H.) at temperature T is by definition

$$\text{R.H.} = \frac{\text{Partial pressure of water vapor at } T}{\text{Vapor pressure of water vapor at } T} \times 100\%.$$

This is easily determined by using the table to find the vapor pressures at dew point and at air temperature. For example, if the dew point is 15°C at an air temperature of 25°C, the relative humidity is

$$\text{R.H.} = \frac{12.788\ \text{mmHg}}{23.756\ \text{mmHg}} \times 100\% = 53.8\%.$$

Fog forms when the air temperature drops below the dew point.

(*a*) If the partial pressure of water vapor is 8.045 mmHg, what is the dew point? (*b*) If the dew point is 15°C when the air temperature is 23°C, what are the partial pressure of the water in the air and the relative humidity? (*c*) On a warm day when the air temperature is 35°C, the dew point is 25°C. What is the relative humidity?

18-49. On a winter day when the air temperature is 0°C, the relative humidity is 50 percent. (See Prob. 18-48.) Outside air is brought inside and heated to a room temperature of 20°C. What is the relative humidity of the air in the room? (*Note:* This problem shows why inside air is so dry in the winter.)

18-50. On a warm day when the air temperature is 30°C, a metal can is slowly cooled by adding ice water to it. Condensation first appears when the can reaches 15°C. What is the relative humidity of the air? (See Prob. 18-48.)

18-51. (*a*) People often think of humid air as "heavy." Evaluate this perception by comparing the densities of dry air (0 percent relative humidity) and moist air (100 percent relative humidity) when both are at 76 cmHg and 30°C. (See Prob. 18-48.) Assume that the dry air is an ideal gas composed of molecules whose molecular mass is 29 g/mol and that the moist air is an ideal gas of molecular mass 29 g/mol mixed with water vapor. (*b*) The air resistance of "high-speed" projectiles such as baseballs and golf balls is approximated by

$$f = K\rho v^2,$$

where ρ is the mass density of the air, v is the speed of the projectile, and K is a constant for a particular projectile. *For a fixed air pressure*, describe qualitatively how the range of a projectile changes with the relative humidity. (*c*) An impending thunderstorm is usually associated with high humidity and low atmospheric pressure. If these conditions occur during a baseball game, would the home-run hitters have an advantage?

18-52. Air is trapped above the mercury in the carelessly constructed barometer shown in the accompanying figure. When atmospheric pressure is 76 cmHg, the height of the mercury column is 73 cm and the air space above the column is 10 cm long. What is the atmospheric pressure when the barometer reads 69 cmHg? Assume that the temperature is the same for both cases.

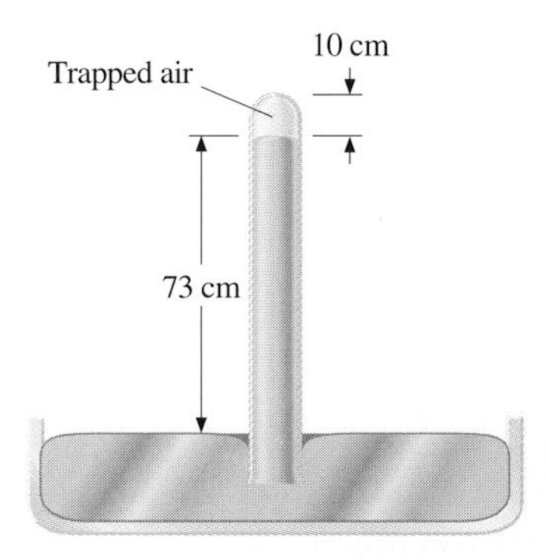

PROBLEM 18-52

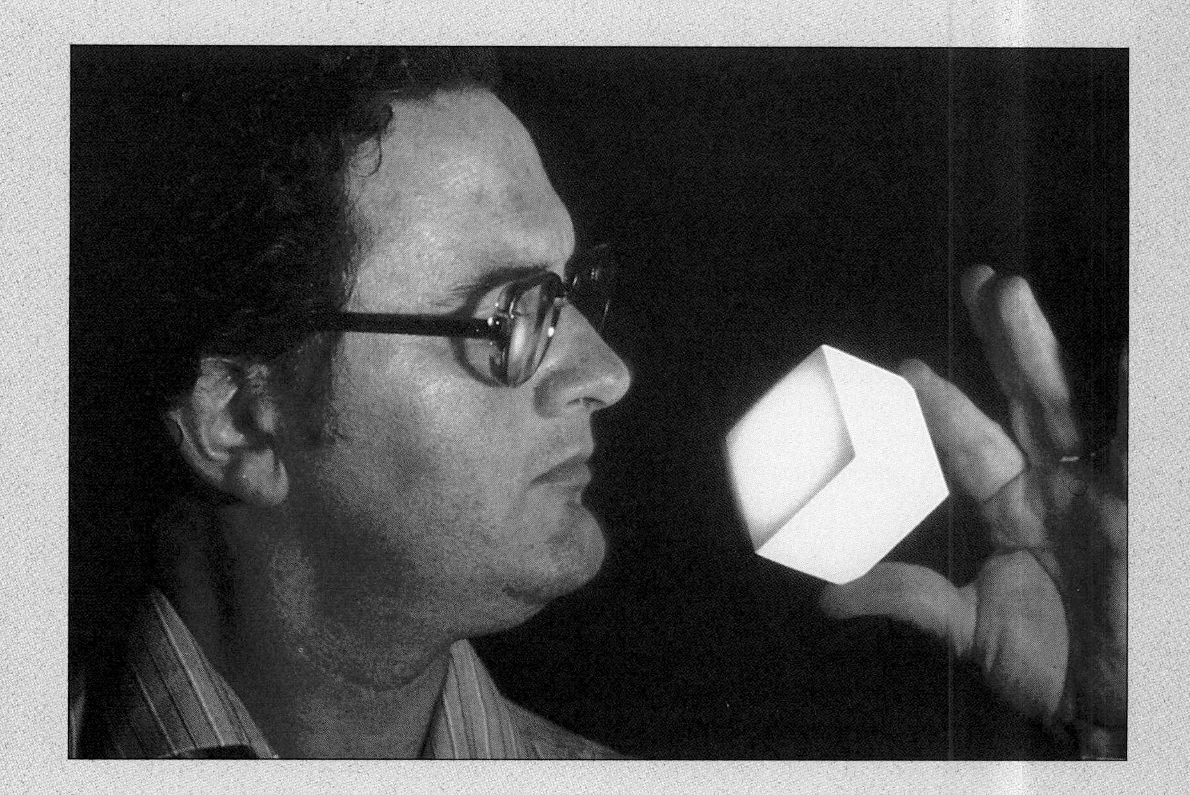

This cube of the lightweight silicate-based material used in a space shuttle's thermal tile insulation is at a temperature of 2300°F! It can be held at the corners because of its extremely poor thermal conductivity and low heat capacity.

CHAPTER 19 Thermal Properties of Matter

Preview

In this chapter we study the thermal properties of matter. The main topics to be discussed here are the following:

1. **Internal energy and heat.** These quantities are defined and their relationship to temperature is discussed.
2. **Specific heat.** Specific heat, a property of a substance that relates temperature change and heat exchange, is considered.
3. **Change of phase.** We investigate the heat exchange that occurs during a change of phase.
4. **Heat exchange.** We study the transfer of heat among bodies in thermal contact.
5. **Conduction, convection, and radiation.** We look at the three processes responsible for the transfer of heat.
6. **Thermal expansion.** How the size of a body changes with temperature is considered.

We have seen that when two bodies are placed in contact, they eventually reach thermal equilibrium and are then at the same temperature. During the approach to thermal equilibrium, the temperatures of the two bodies change because heat is transferred from the hotter body to the colder body. However, heat does not always produce a temperature change—it can also cause the phase of a substance to change. For example, when heat is added to ice at 0°C and atmospheric pressure, the ice melts to liquid water without changing temperature.

This chapter is primarily devoted to a study of heat exchange. We will investigate how the temperature and the phase of a substance are affected by the heat it exchanges with its surroundings and will briefly discuss the different mechanisms involved in the transfer of heat. We will also investigate how the dimensions of solids and liquids vary with temperature.

19-1 INTERNAL ENERGY AND HEAT

The **internal energy** U of a system of molecules is by definition the sum of the mechanical energies of the molecules. If the kinetic and potential energies of molecule i are $\mathfrak{T}_i$ and $\mathfrak{U}_i$, respectively, then

$$U = \sum_i (\mathfrak{T}_i + \mathfrak{U}_i), \tag{19-1}$$

where the summation is over all the molecules of the system. The kinetic energy $\mathfrak{T}_i$ of an individual molecule includes contributions due to its rotation and vibration, as well as its translational energy $m_i v_i^2/2$, where v_i is the molecule's speed measured relative to the center of mass of the system. The potential energy $\mathfrak{U}_i$ is associated only with the interactions between molecule i and the other molecules of the system. In effect, neither the system's location nor its motion is of any consequence as far as internal energy is concerned. The internal energy of the system is not affected by moving it from the basement to the roof of a hundred-story building or by placing it on a moving train.

In an ideal monatomic gas, each molecule is a single atom. Consequently, there is no rotational or vibrational kinetic energy, and $\mathfrak{T}_i = m_i v_i^2/2$. Furthermore, there is no interatomic interaction (collisions notwithstanding) so $\mathfrak{U}_i = 0$. The internal energy is therefore due to translational kinetic energy only, and

$$U = \sum_i \mathfrak{T}_i = \sum_i \frac{1}{2} m_i v_i^2.$$

From Eq. (18-6), the average kinetic energy of a molecule in an ideal monatomic gas is

$$\overline{\tfrac{1}{2} m_i v_i^2} = \tfrac{3}{2} kT,$$

where T is the Kelvin temperature of the gas. Consequently, the average mechanical energy per molecule of an ideal monatomic gas is also $3kT/2$; that is,

$$\overline{\mathfrak{T}_i + \mathfrak{U}_i} = \overline{\mathfrak{T}_i} = \tfrac{3}{2} kT.$$

The internal energy is just the number of molecules multiplied by the average mechanical energy per molecule. Thus for n moles of an ideal monatomic gas,

$$U = nN_A(\tfrac{3}{2}kT) = \tfrac{3}{2}nRT. \tag{19-2}$$

Notice that the internal energy of a given quantity of an ideal monatomic gas depends on just the temperature and is completely independent of the pressure and volume of the gas. For other systems, the internal energy cannot be expressed so simply. However, an increase in internal energy can often be associated with an increase in temperature.

We know from the zeroth law of thermodynamics that when two systems are placed in thermal contact, they eventually reach thermal equilibrium, at which point they are at the same temperature. As an example, suppose we mix two monatomic ideal gases. Now the energy per molecule of an ideal monatomic gas is proportional to its temperature. Thus when the two gases are mixed, the molecules of the hotter one must lose energy and the molecules of the colder one must gain energy. This continues until thermal equilibrium is reached, at which point the temperature, and therefore the average translational kinetic energy per molecule, is the same for both gases. The approach to equilibrium for real systems is somewhat more complicated than for an ideal monatomic gas. Nevertheless, we can still say that energy is exchanged between the systems until their temperatures are the same.

The energy exchanged due to a temperature difference is called **heat** (represented by the symbol Q). For example, when a piece of hot metal is placed in a vat of cold water, the metal cools off by exchanging heat with the water. If there is a decrease of 10 J in the internal energy of the metal, there is an increase of 10 J in the water's internal energy; that is, 10 J of heat have been transferred from the metal to the water.

In this example, we have assumed that the system composed of the metal and the water is thermally isolated (or *insulated*) from its surroundings. This means that the system *does not exchange heat with its surroundings*. For practical purposes, we can come close to insulating a system by surrounding it with a substance such as Styrofoam that conducts heat very poorly. Inexpensive picnic coolers are often made from Styrofoam.

Two centuries ago, heat was believed to be an invisible fluid called *caloric* that existed in all bodies. In a thermal interaction, caloric was assumed to be conserved as it flowed from one body to another. Since caloric was supposedly a substance contained within a body, it was reasoned that there must be a finite amount of it in any given object. Consequently, an object giving up caloric would eventually have its supply exhausted. The caloric theory explained simple heat exchange processes such as that involving the metal and water. However, it eventually fell into disfavor because of the work of Benjamin Thompson (1753–1814), an American who emigrated to Europe after the Revolutionary War because he sided politically with the Tories.

Thompson (by this time Count Rumford of Bavaria) was responsible for supervising the boring of cannons. He noticed that during the time a cannon was being bored, water had to be continually added to cool it off. No matter how long the boring continued, heat continued to flow from the cannon to the water; the source of caloric appeared to be inexhaustible. The caloric also seemed to be weightless, for as it was being freed from the cannon, the weight of the cannon was not changing. The concept of an inexhaustible, weightless fluid was troubling and led to an alternative idea. Thompson suggested, correctly, that the idea of the caloric was a fallacious concept, and that the heat flowing from the cannon to the water was energy originally produced by the work of the frictional force between the bore and the cannon.*

Since heat is a form of energy, its units are joules, ergs, etc. There is also a unit normally used only for heat: the *calorie* (*cal*), which was once defined as the amount of heat required to raise the temperature of one gram of water at atmospheric pressure from 14.5 to 15.5°C. The relationship between the calorie and the joule was determined experimentally by doing work on water, measuring its temperature rise, then comparing this work with the amount of heat needed to get the same temperature increase. The first accurate experiments relating the calorie and the joule (see Fig. 19-1) were performed by James Joule (1818–1889), after whom the SI unit of energy is named. In 1948 the Ninth General Conference on Weights and Measures adopted the joule as the official unit of heat and then defined the calorie as

$$1 \text{ cal} = 4.186 \text{ J},$$

which, of course, is in agreement with the conversion factor determined experimentally prior to 1948.

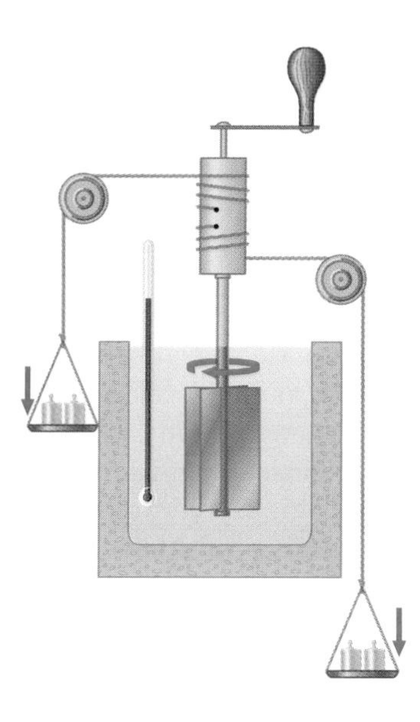

FIGURE 19-1 Joule measured the relationship between the joule and the calorie with a device like the one shown. When the weights fall, the paddles turn and do work on the water, and the temperature of the water rises.

Nutritionists also use a unit called the *food Calorie*, which is actually a kilocalorie (10^3 calories). After its water is removed, a 200-food-Calorie piece of chocolate candy will produce $(200 \times 10^3 \text{ cal})(4.186 \text{ J/cal}) = 8.372 \times 10^5$ J of energy when it is burned in an atmosphere of pure oxygen.

*The connection between work and heat will be explained when the first law of thermodynamics is covered in the next chapter.

A word of caution is needed at this point. While the words heat and temperature are often used interchangeably in everyday conversation, you should keep in mind that the two quantities are quite different. Heat is energy exchanged by two bodies, while temperature is a determinant of thermal equilibrium.

DRILL PROBLEM 19-1

An average person dissipates to the environment about 2500 food Calories of heat per day. What is the average rate in watts at which heat is dissipated?
ANS. 121 W. (*Note:* It is interesting to compare this number with the power output of a lightbulb. It then becomes clear why a room full of people gets warm.)

19-2 SPECIFIC HEAT

The amount of heat required to produce the same temperature increase for a given amount of substance varies from one substance to another. For example, at room temperature 4.18×10^3 J of heat are required to raise the temperature of 1 kg of water 1°C, while only 130 J are needed to increase the temperature of the same mass of lead 1°C. The relationship between heat exchanged and the corresponding temperature change is characterized by the **specific heat** c of a substance. If the temperature of a substance of mass m changes from T to $T + dT$ when it exchanges an amount of heat dQ with its surroundings, then its specific heat is

$$c = \frac{1}{m}\frac{dQ}{dT}. \tag{19-3}$$

In the SI system, specific heat is given in joules per kilogram-degree Celsius (J/kg·°C). Because heat is so frequently measured in calories, the unit calorie per gram-degrees Celsius (cal/g·°C) is also used quite often.

A closely related quantity is the **molar heat capacity** C. It is defined as

$$C = \frac{1}{n}\frac{dQ}{dT}, \tag{19-4a}$$

where n is the number of moles of the substance. If M is the molecular mass of a substance of mass m, then $n = m/M$ and

$$C = \frac{M}{m}\frac{dQ}{dT} = Mc. \tag{19-4b}$$

The specific heat depends on the pressure, volume, and temperature of the substance. For liquids and solids, specific-heat measurements are most often made at constant pressure as functions of temperature, because constant pressure is quite easy to produce experimentally. In addition, the specific heat at constant volume can be related theoretically to the specific heat at constant pressure. These two specific heats are generally almost equal for solids and liquids, because the volumes of these two phases change so little with temperature. Gases, however, expand much more readily, and their two specific heats differ considerably.

The temperature dependence of the specific heat of water at 1 atm is shown in Fig. 19-2. Its variation is less than 1 percent over the interval from 0 to 100°C. Such a small variation is typical for most solids and liquids, so their specific heats can generally be taken to be constant over fairly large temperature ranges. Instead of the differential form of Eq. (19-3), we can then write

$$Q = mc(T_2 - T_1), \tag{19-5}$$

where Q is the amount of heat required to change a body's temperature from T_1 to T_2. The specific heats of some common solids and liquids are listed in Table 19-1. You can assume that they do not change with temperature whenever you use them to do the problems in this textbook.

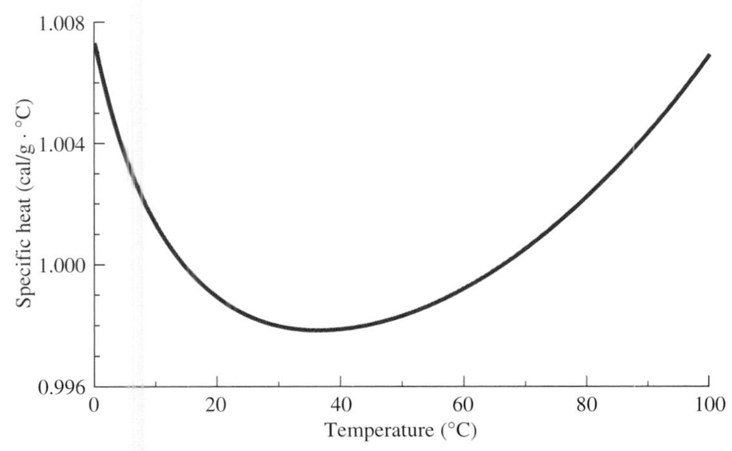

FIGURE 19-2 The temperature dependence of the specific heat of water at 1 atm.

TABLE 19-1 **Specific Heats of Various Substances** Unless stated otherwise, the specific heats are for 25°C and 1.0 atm.

Substance	Specific Heat (J/kg · °C)	Specific Heat (cal/g · °C)	Molar Heat Capacity (J/mol · °C)
Solids			
Aluminum	900	0.215	24.1
Brass	380	0.091	
Copper	390	0.093	24.5
Glass	840	0.201	
Gold	130	0.031	25.4
Ice (−10°C)	2220	0.530	39.6
Iron or steel	450	0.108	25.1
Lead	130	0.031	27.2
Silver	235	0.056	25.5
Sodium chloride	880	0.210	51.4
Wood	1700	0.406	
Liquids			
Ethyl alcohol	2400	0.573	87.8
Mercury	140	0.033	28.0
Water	4180	0.999	75.3
Water (15°C)	4186	1.000	75.4

Notice that the specific heat of water is much larger than that of most other substances. Consequently, for the same amount of added heat, the temperature change of a given mass of water is generally less than that for the same mass of another substance. For this reason, a large body of water moderates the climate of nearby land. In the winter, the water cools off more slowly than the surrounding land and tends to warm the land. In the summer, the opposite effect occurs, as the water heats up more slowly than the land. This tempering effect is especially pronounced where the prevailing winds blow from the water to the land—for example, on the west coast of the United States.

EXAMPLE 19-1 Comparing the Heating of Water and Lead

(*a*) How much heat is added to a 200-g piece of lead whose temperature increases from 10 to 70°C? (*b*) How much heat is needed for the same temperature change in 200 g of water?

SOLUTION (*a*) The heat required for this temperature change in the lead piece is easily found with Eq. (19-5):

$$Q = mc\,\Delta T = (200 \text{ g})(0.130 \text{ J/g}\cdot°\text{C})(60°\text{C}) = 1.56 \times 10^3 \text{ J}.$$

(*b*) For water, the heat required for the same temperature change is

$$Q = mc\,\Delta T = (200 \text{ g})(4.18 \text{ J/g}\cdot°\text{C})(60°\text{C}) = 5.02 \times 10^4 \text{ J}.$$

Notice the effect of the relatively high specific heat of water. The masses of both substances are the same; yet for the same temperature increase, Q is more than 32 times greater for water than for lead.

DRILL PROBLEM 19-2

When 400 J of heat are added to a 0.10-kg sample of metal, its temperature increases by 20°C. What is the specific heat of the metal?
ANS. 200 J/kg · °C.

DRILL PROBLEM 19-3

When 860 cal of heat are added to a 400-g metal sample, its temperature increases from 20 to 30°C. Use Table 19-1 to determine what the metal is.
ANS. Aluminum.

19-3 Change of Phase

Suppose that we slowly heat a cube of ice whose temperature is below 0°C at atmospheric pressure. What changes do we observe in the ice? Initially, we find that its temperature increases according to Eq. (19-5). Once 0°C is reached, however, additional heat no longer changes the temperature of the ice. Instead, the ice melts until just water at 0°C remains. The temperature of the water then starts to rise and eventually reaches 100°C, whereupon the water vaporizes into steam at this same temperature. Not until all the water has evaporated does the temperature of the steam increase, provided the heat is added slowly enough.

During phase transitions such as those just described for water, the added heat causes a change in the positions of the molecules relative to one another *without affecting the temperature*. In the solid-liquid transition, the molecules are moved from well-defined positions in a crystal to the semifree states of a liquid. When a liquid becomes a gas, the molecules are further liberated, ending up in a state in which they are almost completely independent of one another.

The heat necessary to change a unit mass of a substance from one phase to another is called the **latent heat**. For the solid-liquid transition it is known as the **latent heat of fusion** (L_f); for the liquid-gas transition, it is known as the **latent heat of vaporization** (L_v). The amounts of heat required for melting and vaporizing a substance of mass m are given by

$$Q_f = mL_f \quad (19\text{-}6a)$$

and

$$Q_v = mL_v. \quad (19\text{-}6b)$$

For water at 1 atm, $L_f = 80.0$ cal/g $= 3.35 \times 10^5$ J/kg. This really has two meanings: First, 80.0 cal of heat are required to melt 1.00 g of ice; second, 1.00 g of water will liberate 80.0 cal of heat upon freezing. Similarly, when 1.00 g of water is vaporized/condensed, it absorbs/liberates 539 cal of heat. The latent heats of some selected substances are given in Table 19-2.

TABLE 19-2 **Latent Heats of Fusion and Vaporization of Various Substances at Atmospheric Pressure**

Heats of Fusion			
Substance	**Melting Point (°C)**	**J/kg**	**cal/g**
Aluminum	660	3.97×10^5	95.3
Copper	1083	2.05×10^5	48.9
Gold	1063	6.45×10^4	15.4
Iron	1535	2.75×10^5	65.6
Lead	328	2.45×10^4	5.85
Mercury	−39	1.18×10^4	2.82
Nitrogen	−210	2.55×10^4	6.10
Oxygen	−219	1.38×10^4	3.30
Water	0	3.35×10^5	80.0
Heats of Vaporization			
Substance	**Boiling Point (°C)**	**J/kg**	**cal/g**
Aluminum	2467	1.23×10^7	2940
Copper	2567	4.80×10^6	1147
Gold	2660	1.58×10^6	377
Iron	2750	6.29×10^6	1503
Lead	1740	8.70×10^5	208
Mercury	357	2.90×10^5	69.3
Nitrogen	−196	2.00×10^5	47.8
Oxygen	−183	2.10×10^5	50.2
Water	100	2.26×10^6	539

EXAMPLE 19-2 Changing Ice to Steam

At atmospheric pressure, how much heat is required to change 10 g of ice at −20°C into steam at 100°C?

SOLUTION The steps required to make this change are (*i*) heat the ice from −20 to 0°C, (*ii*) melt the ice, (*iii*) heat the water from 0 to 100°C, (*iv*) vaporize the water. The quantities of heat corresponding to each of these steps are:

$$(i)\ Q_i = mc_{\text{ice}}[T - (-20°\text{C})],$$

which, with $T = 0°$C, gives

$$Q_i = (10 \text{ g})(0.530 \text{ cal/g}\cdot°\text{C})(20°\text{C}) = 106 \text{ cal}.$$

$$(ii)\ Q_{ii} = mL_f = (10 \text{ g})(80.0 \text{ cal/g}) = 800 \text{ cal}.$$

$$(iii)\ Q_{iii} = mc_{H_2O}(T - 0°\text{C}),$$

which, with $T = 100°$C, gives

$$Q_{iii} = (10 \text{ g})(1.00 \text{ cal/g}\cdot°\text{C})(100°\text{C}) = 1.00 \times 10^3 \text{ cal}.$$

$$(iv)\ Q_{iv} = mL_v = (10 \text{ g})(539 \text{ cal/g}) = 5.39 \times 10^3 \text{ cal}.$$

The net heat required for this process is then

$$Q_i + Q_{ii} + Q_{iii} + Q_{iv} = 7.30 \times 10^3 \text{ cal}.$$

The heat added as a function of temperature is plotted in Fig. 19-3.

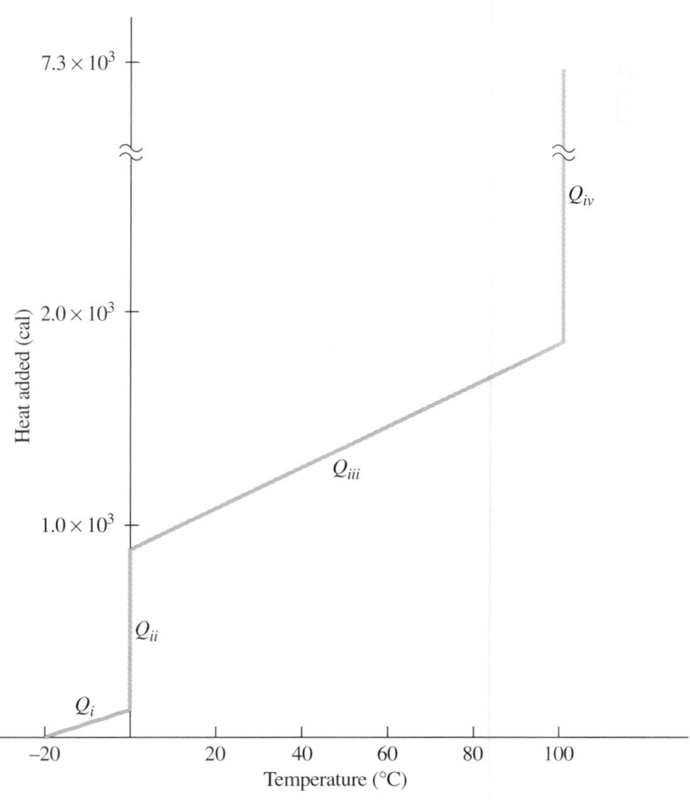

FIGURE 19-3 Heat is added to ice at −20°C to convert it to steam at 100°C.

DRILL PROBLEM 19-4

How much heat is liberated when 20.0 g of steam condense at atmospheric pressure?
ANS. 1.08×10^4 cal.

19-4 Heat Exchange

When bodies at different temperatures are placed in contact with one another, they exchange heat until thermal equilibrium is attained. The final temperature reached during these heat-exchange processes can be found by simply equating the total heat lost by some bodies to the total heat gained by the others. Examples of such calculations follow.

Fisher Scientific.

A student calorimeter. The cup is held in the center of the can by an insulating ring. A thermometer and stirrer can be placed into the cup through holes in the lid. A heating element is also attached to the lid.

EXAMPLE 19-3 A Cup of Coffee

An insulated aluminum cup at 20°C with a mass of 200 g is filled with 400 g of coffee at 90°C. What is the equilibrium temperature of the cup and coffee? Assume that the specific heat of coffee is the same as that of water.

SOLUTION If the equilibrium temperature of the cup and coffee is T, the heat gained by the cup is

$$Q_+ = (200\ \text{g})(0.900\ \text{J/g}\cdot{}^\circ\text{C})(T - 20^\circ\text{C}),$$

and the heat lost by the coffee is

$$Q_- = (400\ \text{g})(4.18\ \text{J/g}\cdot{}^\circ\text{C})(90^\circ\text{C} - T).$$

Since the heat gained equals the heat lost,

$$Q_+ = Q_-,$$
$$(200\ \text{g})(0.900\ \text{J/g}\cdot{}^\circ\text{C})(T - 20^\circ\text{C}) = (400\ \text{g})(4.18\ \text{J/g}\cdot{}^\circ\text{C})(90^\circ\text{C} - T),$$

and

$$T = 83^\circ\text{C}.$$

The cup doesn't cool the coffee very much because the specific heat of water is so much larger than that of aluminum.

EXAMPLE 19-4 Cooling Water with Ice

A 5.0-g piece of ice at 0°C is mixed with 100 g of water at 90°C. What is the equilibrium temperature of the mixture?

SOLUTION The total heat gained by the ice is made up of two parts: the amount required to melt it at 0°C and the amount it then gains in reaching its equilibrium temperature T:

$$Q_+ = (5.0\ \text{g})(335\ \text{J/g}) + (5.0\ \text{g})(4.18\ \text{J/g}\cdot{}^\circ\text{C})(T - 0^\circ\text{C}).$$

The heat lost by the 100 g of water initially at 90°C is

$$Q_- = (100\ \text{g})(4.18\ \text{J/g}\cdot{}^\circ\text{C})(90^\circ\text{C} - T).$$

Equating Q_+ and Q_-, we find T to be 82°C.

DRILL PROBLEM 19-5

What is the maximum amount of ice at 0°C that the water in Example 19-4 can melt? What is the equilibrium temperature of the mixture?
ANS. 112 g; 0°C.

19-5 Conduction, Convection, and Radiation

There are three ways in which heat can be transferred from one location to another: *conduction*, *convection*, and *radiation*. In the first two processes, a medium is necessary for the heat transfer. Radiation, however, does not have this restriction.

Conduction is the most effective means of heat transport in solids like the hard plastic handle on the pan of Fig. 19-4. When the pan is heated, the end of the handle attached to the pan quickly gets hot. The oscillations of the molecules at this end of the handle then increase, and in dominolike fashion, successive groups of molecules along the handle acquire some of this vibrational energy through a series of collisions. As a result, the other end of the handle becomes warmer.

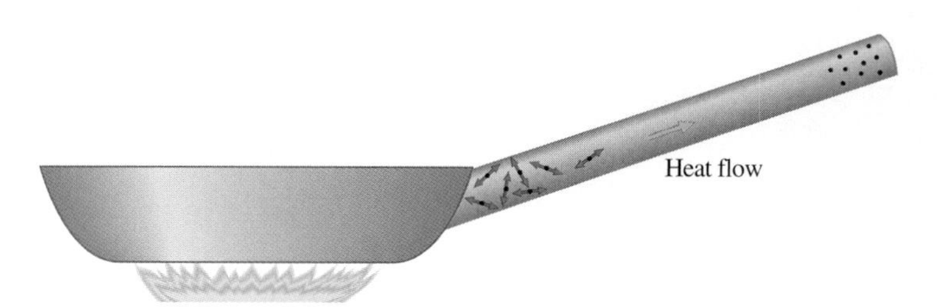

FIGURE 19-4 A representation of heat conduction along the handle of a frying pan.

We know from everyday experience that the ability to conduct heat varies from one material to another. For example, if the handle on the pan of Fig. 19-4 were made of metal, we would probably burn our fingers when grabbing its free end. Air, which is a poor thermal conductor, plays an important part in keeping us warm. The material used in winter clothing and for home insulation traps air in its fibers. This layer of air allows us to retain heat better because it is a poor conductor.

How well or how poorly a substance conducts heat is specified by its **thermal conductivity** κ. Figure 19-5*a* shows a rod of uniform cross section A and length l. The left end of the rod is kept at a constant temperature T_2 while its right end is kept at a constant lower temperature T_1. Eventually, thermal "steady state" is reached, after which the temperature decreases linearly from T_2 to T_1 along the rod (see Fig. 19-5*b*), and the temperature at any point along the rod remains constant. For steady state, experiment shows that the time rate of heat flow dQ/dt through the rod is related to the dimensions of the rod as well as to the temperature difference between its ends in the following manner:

$$\frac{dQ}{dt} = \kappa(T_2 - T_1)\frac{A}{l}, \tag{19-7}$$

where the constant of proportionality κ is the thermal conductivity of the material of the rod. From this equation, the SI unit of κ is joule per second-meter-degree Celsius (J/s·m·°C).

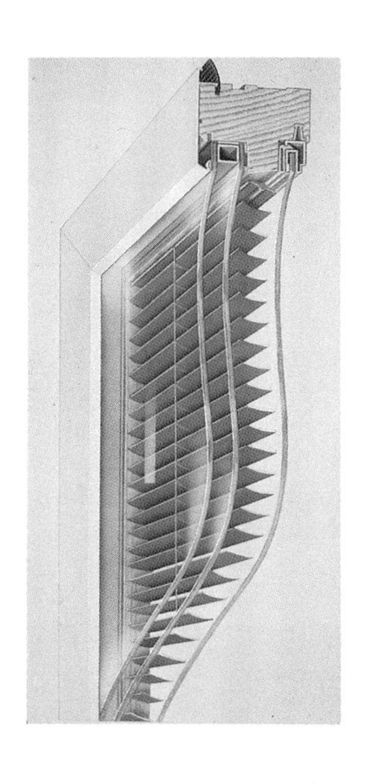

A double-paned window. The air between the panes serves as an insulator and reduces heat loss through the window to approximately 10 percent the loss through a single-sheeted pane.

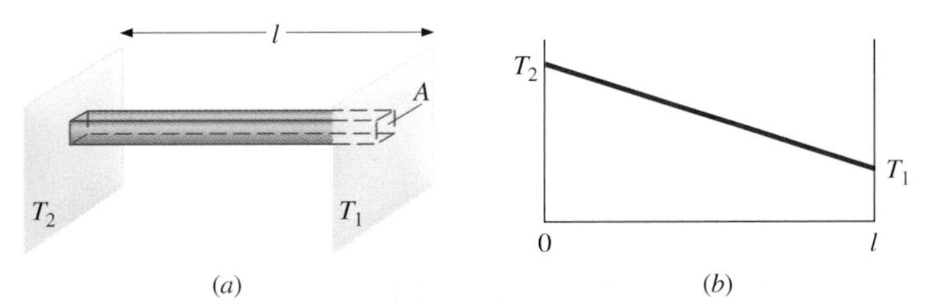

FIGURE 19-5 (*a*) A uniform rod of cross-sectional area A and length l whose ends are at temperatures T_1 and T_2. (*b*) The steady-state temperature distribution along the rod.

The thermal conductivities of various materials at room temperature are listed in Table 19-3. The greatest values belong to those metals that are also characterized by high electrical conductivities. This is because conduction electrons in a metal are not only involved in the flow of an electric current but also contribute significantly to the transport of heat.

TABLE 19-3 **Thermal Conductivities of Various Substances Around Room Temperature**

Substance	Thermal Conductivity	
	W/m·°C	cal/s·cm·°C
Aluminum	237	0.566
Air	0.024	5.7×10^{-5}
Iron	79.5	0.190
Copper	385	0.920
Helium	0.14	3.3×10^{-4}
Lead	34.7	0.083
Silver	406	0.970
Steel	50.2	0.120
Representative Values for		
Concrete	0.8	0.002
Glass	0.8	0.002
Ice	1.6	0.0038
Insulating brick	0.15	3.6×10^{-4}
Red brick	0.60	0.0015
Styrofoam	0.01	2×10^{-5}
Wood	0.08	2×10^{-4}

Heat conduction for non-steady-state conditions and different geometries can be described by a generalization of Eq. (19-7). At a point in the material where the thermal conductivity is κ, we let dQ/dt represent the time rate of heat flow through an infinitesimally thin slab of cross-sectional area A and thickness dx. The direction of the heat flow is along x, which is perpendicular to the area element. (See Fig. 19-6.) If at the cross section the temperature variation with x is dT/dx (called the **temperature gradient**), then

$$\frac{dQ}{dt} = -\kappa A\frac{dT}{dx}. \tag{19-8}$$

Notice that when dT/dx is negative, dQ/dt is positive. This simply corresponds to the fact that heat flows in the direction of decreasing temperature.

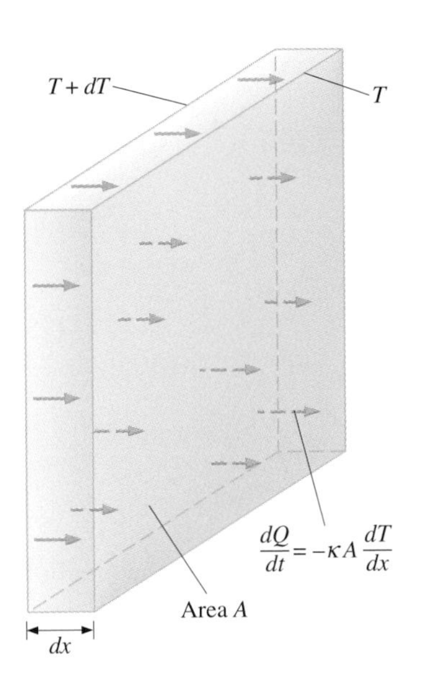

FIGURE 19-6 Heat transfer through an infinitesimal slab of cross-sectional area A and thickness dx.

EXAMPLE 19-5 **Heat Conduction Along a Copper Rod**

Heat is conducted through a uniform copper rod that connects a boiler holding pressurized steam at 200°C and a large insulated container of ice water at 0°C. The rod's length is $l = 20.0$ cm, and its cross-sectional area is $A = 4.00\ \text{cm}^2$. Once steady state is reached, how much time elapses while enough heat is conducted through the rod to melt 500 g of ice?

SOLUTION At steady state, the rate at which heat is transferred is given by Eq. (19-7). With $\kappa = 385\ \text{W/m}\cdot{}^\circ\text{C}$ for copper, we have

$$\frac{dQ}{dt} = \kappa(T_2 - T_1)\frac{A}{l}$$
$$= (385\ \text{W/m}\cdot{}^\circ\text{C})(200^\circ\text{C})\left(\frac{4.00 \times 10^{-4}\ \text{m}^2}{0.200\ \text{m}}\right) = 154\ \text{J/s}.$$

Since $Q = mL_f = (0.500\ \text{kg})\ (3.35 \times 10^5\ \text{J/kg}) = 1.68 \times 10^5$ J are required to melt 500 g of ice, the time that elapses during this process is

$$t = \frac{1.68 \times 10^5\ \text{J}}{154\ \text{J/s}} = 1.09 \times 10^3\ \text{s}.$$

Although conduction does occur in liquids and gases, heat is transported in these media mostly by convection. In this process, the actual motion of the material is responsible for the heat transfer. Because of the ordered crystalline structure of a solid, the motion of an atom is limited to small vibrations about a site fixed in the crystal. However, the molecules of liquids and gases are much freer to move, so convection currents can be produced.

The convection process is illustrated in Fig. 19-7. The warm liquid at the bottom of the container expands slightly (see Sec. 19-6) and rises because of its reduced density. It is replaced by cooler, denser liquid, which then becomes heated and consequently rises as well. Upon reaching the top, it cools and therefore sinks back to the bottom. These convection currents flow continuously through the liquid, as depicted in the figure.

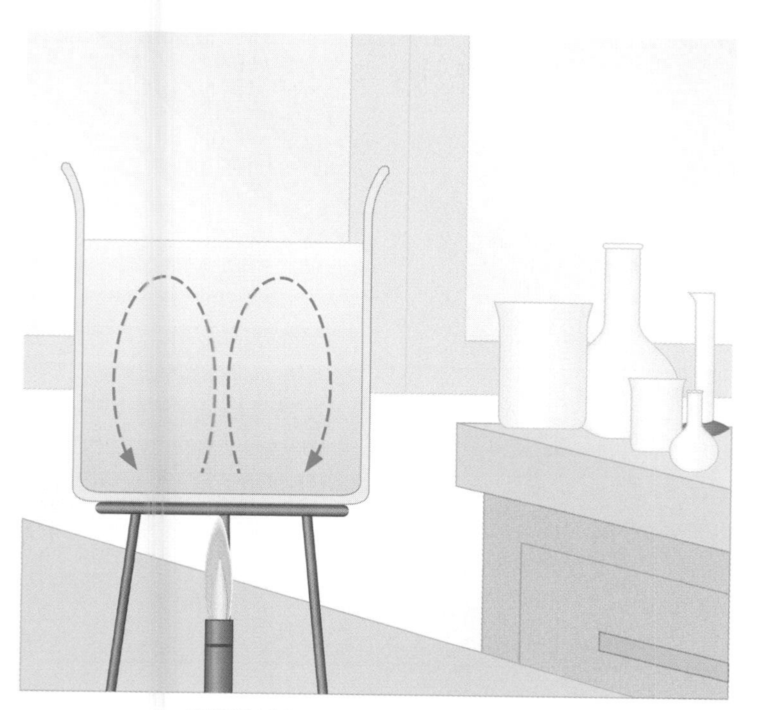

FIGURE 19-7 Convection in a liquid.

In Los Angeles, the air temperature often increases with altitude. As a result, there are no rising convection air currents, and an excessive concentration of pollutants develops at ground level.

A similar phenomenon occurs on the outer surface of the sun. When viewed without magnification from the earth, the sun appears smooth, but in fact, closer inspection reveals that it is granulated, as shown in Fig. 19-8*a*. These granules are actually rising columns of hydrogen gas heated by the solar surface whose temperature is about 6000 K. The rising gas eventually cools and falls back toward the surface to be heated again. (See Fig. 19-8*b*.) The dark boundaries of the granules are formed by this cooler, descending gas.

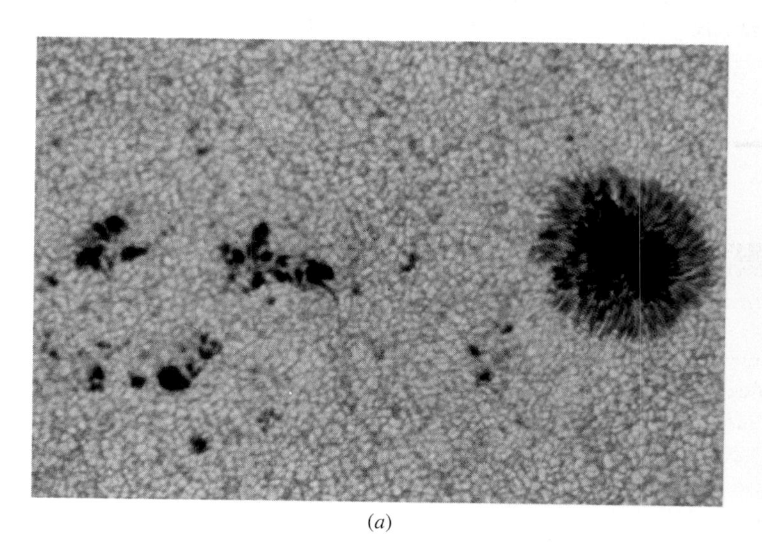

(*a*)

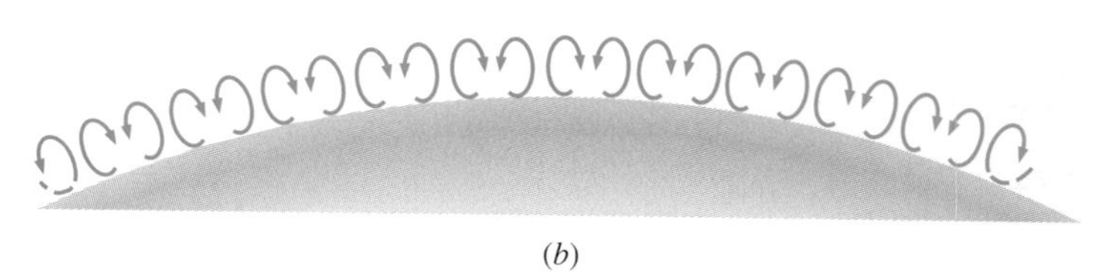

(*b*)

FIGURE 19-8 (*a*) Granular appearance of the sun's surface. The dark patches are sunspots. (*b*) Convection currents on the sun.

Radiation is a form of heat transfer that does not require a medium. Probably the best-known example of this process is the radiation from the sun. This energy comes to us across a vast expanse of space, which contains less matter per unit volume than any vacuum we can create on the earth. Radiation is composed of time-varying electric and magnetic fields, called electromagnetic waves, propagating through space. One particular class of these waves is visible light. Other examples of electromagnetic waves are x-rays, microwaves, radio waves, and infrared and ultraviolet light. All objects emit and absorb electromagnetic radiation. The higher the temperature of an object, the more it radiates. Quantum theory tells us (see Chap. 40) that the energy radiated by a body is proportional to the fourth power of its Kelvin temperature.

The infrared radiation emitted by these soldiers is used to produce this nighttime image.

DRILL PROBLEM 19-6

A carpeted and tile floor are both at a room temperature of 21°C. Which surface would feel colder to your bare feet? Explain. (*Hint:* Consider the thermal conductivities of the materials.)

DRILL PROBLEM 19-7

Suppose the liquid of Fig. 19-7 is heated at the top rather than at the bottom. What is the main process by which the rest of the liquid becomes hot?
ANS. Conduction.

19-6 Thermal Expansion

Practically all substances expand when heated. The average kinetic energy of the molecules of a substance increases with temperature, and, in general, this results in a greater average molecular separation. Consequently, the dimensions of the entire substance also increase with temperature.

Suppose that the temperature of a thin rod of length l is changed from T to $T + \Delta T$. It is found experimentally that, if ΔT is not too large, the corresponding change in length Δl of the rod is related to l and ΔT according to

$$\Delta l = \alpha l \, \Delta T. \tag{19-9}$$

The constant α is called the **coefficient of linear expansion** of the material of the rod and its SI unit is inverse degree Celsius $[(°C)^{-1}]$. Actually, α does depend slightly on the temperature, but its variation is usually small enough to be negligible, even over a temperature range of 100°C. We will always assume that α is constant.

Table 19-4 lists the coefficients of linear expansion for several materials. Typically, α is of the order $10^{-5}\ (°C)^{-1}$, so a 1.0-m rod whose temperature is increased by 1.0°C expands by only 0.010 mm. While this may seem insignificant, it is a concern to the engineer who is designing a large structure such as a bridge. For example, a 1000-m bridge whose temperature varies over a range of 50°C expands by about 0.50 m! The bridge must therefore be constructed so that it can withstand this change without structural damage. (See Fig. 19-9.)

Since length changes with temperature, so does volume. Let's consider a cube whose sides are a length l. For a temperature increase ΔT, each side expands to a length $l + \Delta l = l(1 + \alpha \Delta T)$, resulting in a volume change of

$$\begin{aligned}\Delta V &= [l(1 + \alpha \Delta T)]^3 - l^3 \\ &= 3l^3\alpha \Delta T + 3l^3(\alpha \Delta T)^2 + l^3(\alpha \Delta T)^3.\end{aligned}$$

TABLE 19-4 Coefficients of Expansion for Various Materials near Room Temperature

Material	Coefficient of Linear Expansion $(°C)^{-1}$
Aluminum	24×10^{-6}
Brass	20×10^{-6}
Celluloid	110×10^{-6}
Concrete	12×10^{-6}
Copper	17×10^{-6}
Glass (most types)	9×10^{-6}
Glass (Pyrex)	3.2×10^{-6}
Ice	51×10^{-6}
Lead	29×10^{-6}
Steel	12×10^{-6}
Oak wood (along grain)	5×10^{-6}
Oak wood (cross grain)	54×10^{-6}
Material	**Coefficient of Volume Expansion $(°C)^{-1}$**
Air	36.7×10^{-4}
Alcohol (ethyl)	11.2×10^{-4}
Alcohol (methyl)	12.2×10^{-4}
Gasoline	9.5×10^{-4}
Mercury	1.8×10^{-4}
Water	2.1×10^{-4}

FIGURE 19-9 The expansion joints in a bridge. The space in the joints allows the sections of the bridge to expand without buckling.

If $\alpha\,\Delta T$ is sufficiently small, the higher-order terms can be neglected, and

$$\Delta V = 3l^3\alpha\,\Delta T = 3\alpha V\,\Delta T = \beta V\,\Delta T, \qquad (19\text{-}10a)$$

where

$$\beta = 3\alpha \qquad (19\text{-}10b)$$

is the **coefficient of volume expansion**. Its SI unit is also inverse degree Celsius $[(°C)^{-1}]$.

What happens to the volume of a cavity in a solid that is heated? If the temperature of the solid increases by ΔT, then the volume of the cavity also increases according to Eq. (19-10*a*). You might convince yourself of this by imagining that the cavity is first filled with the same material and then the entire solid is heated.

Most liquids also expand when their temperatures increase. Their expansion can also be described by Eq. (19-10*a*). As Table 19-4 indicates, the volume expansion coefficients for liquids are about 100 times larger than those for solids.

Some substances contract when heated over a certain temperature range. The most common example is water. Figure 19-10 shows how the volume of 1 g of water varies with temperature at atmospheric pressure. Notice that the volume decreases as the temperature is raised from 0°C to about 4°C, at which point the volume is a minimum and the density a maximum. Above 4°C, water expands with increasing temperature like most substances.

This anomalous behavior of water causes ice to form first at the surface of a lake in cold weather. As winter approaches, the water temperature decreases initially at the surface. The water there sinks because of its increased density and is replaced by warmer liquid from the lake bottom. This liquid in turn becomes cold and sinks, and the convection currents cool the entire body of water until a uniform temperature of about 4°C

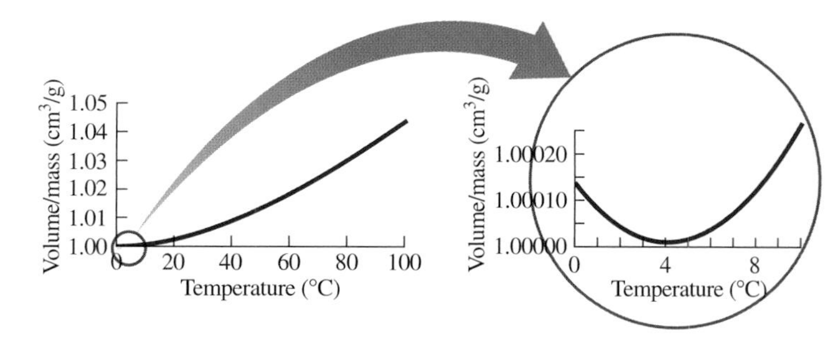

FIGURE 19-10 The anomalous expansion of water.

is reached. Any further drop in temperature at the surface only decreases the density of the water there so it does not circulate. The body of the lake can now only be cooled by conduction from its surface. Consequently, the surface reaches 0°C first, and the lake becomes covered with ice. Aquatic life is able to survive the cold winter as the lake bottom remains unfrozen at a temperature of about 4°C.

EXAMPLE 19-6 Expansion and Contraction

A steel ruler exactly 20 cm long is graduated to give correct measurements at 20°C. (*a*) Will it give readings that are too long or too short at lower temperatures? (*b*) What will the actual length of the ruler be when it is used in the desert at a temperature of 40°C?

SOLUTION (*a*) If the temperature decreases, the length of the ruler also decreases through thermal contraction. Below 20°C, each centimeter division is actually somewhat shorter than 1.0 cm so the steel ruler gives readings that are too long.

(*b*) From Table 19-4, the coefficient of thermal expansion of steel is $1.2 \times 10^{-5}\,(°C)^{-1}$. At 40°C, the increase in length of the ruler is $[1.2 \times 10^{-5}\,(°C)^{-1}]\,(20\text{ cm})\,(20°C) = 0.48 \times 10^{-2}$ cm. The actual length of the ruler is then 20.0048 cm.

DRILL PROBLEM 19-8

Consider a thin, square sheet of material of side l. By repeating the steps leading to Eqs. (19-10), calculate the coefficient of area expansion γ of the material in terms of α. If the sheet is made of aluminum, and $l = 1.0$ m, by how much does its area increase for a temperature rise of 100°C?
ANS. $\gamma = 2\alpha$; $4.8 \times 10^{-3}\text{ m}^2$.

DRILL PROBLEM 19-9

Compare the values of $3\alpha\,\Delta T$, $3(\alpha\,\Delta T)^2$, and $(\alpha\,\Delta T)^3$ to see if the last two terms are negligible relative to the first term. Do this for $\Delta T = 100°C$ for several different materials listed in Table 19-4.

SUMMARY

1. **Internal energy and heat**
   (*a*) The internal energy of a system of molecules is the sum of the mechanical energies of the molecules.
   (*b*) The internal energy of n moles of an ideal monatomic gas at a Kelvin temperature T is
   $$U = \tfrac{3}{2}nRT.$$
   (*c*) The energy exchanged due to a temperature difference is called heat.
   (*d*) A unit frequently used to specify heat is the calorie (cal), where 1 cal = 4.186 J.
2. **Specific heat**
   (*a*) If the temperature of a substance of mass m changes from T to $T + dT$ when it exchanges an amount of heat dQ with its surroundings, then the specific heat of the substance is
   $$c = \frac{1}{m}\frac{dQ}{dT}.$$
   (*b*) The molar heat capacity of a substance is
   $$C = \frac{1}{n}\frac{dQ}{dT},$$
   where n is the number of moles of the substance.
   (*c*) The specific heat and molar heat capacity of a substance are related by
   $$C = Mc,$$
   where M is the molecular mass of the substance.
3. **Change of phase**
   (*a*) The heat exchanged when a unit mass of a substance makes the solid-liquid transition is called the latent heat of fusion.
   (*b*) The heat exchanged when a unit mass of a substance makes the liquid-gas transition is called the latent heat of vaporization.
4. **Heat exchange**
   When bodies at different temperatures are placed in contact, they exchange heat until thermal equilibrium is attained. The total heat lost by some bodies is equal to the total heat gained by the others.
5. **Conduction, convection, and radiation**
   (*a*) Heat is transferred from one body to another by means of one or a combination of three processes: conduction, convection, and radiation.
   (*b*) For the steady-state conduction of heat down a rod,
   $$\frac{dQ}{dt} = \kappa(T_2 - T_1)\frac{A}{l},$$
   where κ is the thermal conductivity of the material of the rod.
6. **Thermal expansion**
   (*a*) When the temperature of a rod of length l changes from T to $T + \Delta T$, the corresponding change in length Δl of the rod is given by
   $$\Delta l = \alpha l\,\Delta T,$$
   where α is the coefficient of linear expansion of the material of the rod.
   (*b*) The corresponding change in volume from V to $V + \Delta V$ for this temperature change is
   $$\Delta V = \beta V\,\Delta T,$$
   where $\beta = 3\alpha$ is the coefficient of volume expansion of the material of the rod. This equation also holds for liquids.

QUESTIONS

19-1. When you hold a lit match, why doesn't the heat burn your fingers?

19-2. In cold weather, birds often fluff their feathers to stay warm. Explain.

19-3. Simply by touch, a mother can tell when her baby's body temperature is as little as 1 K above normal. Given that the normal body temperature is 310 K, how can she detect such a small percentage change in body temperature?

19-4. When a beverage in an aluminum can is placed in a refrigerator, it cools faster than one in a glass bottle. This fact points out both an advantage and a disadvantage of aluminum cans over glass bottles. What are the advantage and the disadvantage?

19-5. On a hot day, the land is heated to a higher temperature than the ocean because it has a lower specific heat. Does this result in on-shore or off-shore ocean breezes? What happens during a cold night?

19-6. Is it possible to add heat to a substance without changing its temperature? If so, give an example.

19-7. Which causes the more severe burn—steam at 100°C or boiling water? Explain.

19-8. Perspiring is a mechanism our bodies use to prevent overheating. Explain.

19-9. How do people lose body heat when the air temperature gets above 37°C?

19-10. Cold drinking glasses often crack when very hot water is poured into them. Explain why.

19-11. When a mercury thermometer is heated, the mercury level actually drops slightly before it starts to rise. Why?

19-12. Suppose two drinking glasses stacked as shown in the accompanying figure become stuck together. Describe how you can separate them.

QUESTION 19-12

19-13. The horizontal bimetallic strip shown in the accompanying figure is made of copper and stainless steel riveted together. As the strip is heated, does it bend such that its free end rises or falls?

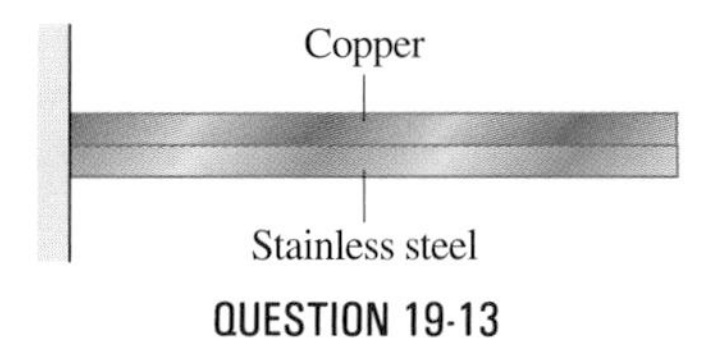

QUESTION 19-13

19-14. A hole is drilled in a sheet of aluminum. As the aluminum is heated, does the diameter of the hole increase or decrease?

PROBLEMS

Internal Energy and Heat

19-1. What is the average mechanical energy of the atoms of an ideal monatomic gas at 300 K?

19-2. What is the internal energy of 6.0 mol of an ideal monatomic gas at 200°C?

19-3. Calculate the internal energy of 15 mg of helium at a temperature of 0°C.

19-4. Two monatomic ideal gases A and B are at the same temperature. If 1.0 g of gas A has the same internal energy as 0.10 g of gas B, what are (*a*) the ratio of the number of moles of each gas and (*b*) the ratio of the atomic masses of the two gases?

19-5. Express 450 J in calories.

19-6. A chocolate bar has about 200 food calories. How many joules is this equivalent to?

19-7. When a young man is resting, his body is consuming internal energy at a rate of approximately 1.2 watts per kilogram body mass. This rate of energy consumption is called the *basal metabolism rate*. Some of this energy is used to drive basic metabolic activities such as breathing and digestion, but most of it is converted directly into heat. (*a*) How much internal energy is used in 1 h by a 70-kg young man at rest? (*b*) Compare this with the energy content of a 150-food-Calorie doughnut.

19-8. Answer the questions of the previous problem for a young man who is using internal energy at a rate of 7.5 W/kg while he is climbing stairs. Is exercise by itself a feasible way to lose weight?

Heat Exchange

19-9. What is the increase in temperature of a 500-g piece of lead to which 900 J of heat are added?

19-10. When 500 cal are added to a 700-g metal block, its temperature increases by 10°C. What is the specific heat of the metal?

19-11. The temperature of an iron pellet increases from 20 to 45°C when 75 J of heat are added. What is the mass of the pellet?

19-12. In the Joule experiment shown in Fig. 19-1, the temperature of 0.10 kg of water is raised 0.10°C after the two equal masses have each dropped a distance of 1.0 m. What is the total amount of mass that is falling?

19-13. Using Table 19-1 and the periodic table, calculate the molar heat capacities of (*a*) copper, (*b*) aluminum, and (*c*) sodium chloride.

19-14. How much heat is necessary to melt 15 g of gold at 1063°C?

19-15. A thermos flask is filled with 200 g of ice at a temperature of −10°C. After 24 h, it is found that the ice has completely melted into water at 5.0°C. What is the rate at which heat flows into the flask?

19-16. A 0.500-kg aluminum electric kettle is rated at 1500 W. It takes 5.0 min to raise the temperature of 0.50 L of water from 25°C to its boiling point. How much heat is lost in this process? What are the different ways by which this heat may be lost?

19-17. How much heat is released when ice at −10°C is formed from 50 g of steam at 100°C?

19-18. A copper ball at a temperature of 150°C is immersed in 2.0 kg of water at 20°C. The equilibrium temperature reached is 25°C. What is the mass of the ball?

19-19. Ten grams of water vapor at 100°C are added to 50 g of ice at 0°C. Describe the final state of the mixture.

19-20. Two solid spheres A and B, which are made of the same material, are at temperatures of 0 and 100°C, respectively. The spheres are placed in thermal contact and a common temperature of 20°C is reached. Which is the larger sphere? What is the ratio of their diameters?

19-21. An 800-g iron cylinder at a temperature of 1000°C is dropped into an insulated chest of ice. How much ice melts?

19-22. Object A has a mass of 200 g and is at a temperature of 250°C. Object B has a mass of 45 g and is at 90°C. When A and B are placed in thermal contact, they reach an equilibrium temperature of 130°C. What is the ratio of the specific heats of the objects?

19-23. An insulated calorimeter of negligible specific heat contains a mixture of 100 g of ice and 1000 g of water at 0°C. Steam at 100°C and atmospheric pressure is added to the mixture through a tube until the mixture's temperature reaches 30°C. How much steam is added?

19-24. The following table is a summary of temperature measurements of a 50-g solid originally at −10°C as it is being heated at the constant rate of 50.0 cal/min.

Time (min)	0.0	1.0	2.0	3.0	4.0	5.0	6.0	7.0	8.0	9.0	10.0
Temperature (°C)	−10	−4.0	2.0	8.0	8.0	8.0	8.0	17.0	26.0	35.0	46.0

Plot these measurements on a graph of "Heat added to the solid" vs. "Temperature" and answer the following questions: (*a*) What is the specific heat of the substance at 0°C? (*b*) What is the specific heat of the substance at 30.0°C? (*c*) What is happening to the substance at 8.0°C? (*d*) What is the latent heat of fusion of the substance?

Conduction, Convection, and Radiation

19-25. The rate of heat flow across a slab of cross-sectional area 0.10 m^2 and thickness 0.020 m is measured to be 30 W. What is the thermal conductivity of the slab material, given that the temperature difference between the walls of the slab is 10°C?

19-26. Consider the rod of Fig. 19-5*a*. Let $T_1 = 20°C$, $T_2 = 90°C$, $A = 0.010\ m^2$, and $l = 1.50$ m. (*a*) What is the temperature of a point on the rod at a distance of 0.50 m from the hot end? (*b*) What is the location of the point on the rod which is at a temperature of 75°C? (*c*) If the rod is made of lead, how much heat is conducted down the rod in 1.0 s? in 5.0 min?

19-27. It is found that 3.2×10^2 J of heat are lost each second through a wooden wall of cross-sectional area 16 m^2 and thickness 0.040 m. What is the temperature difference between the two sides of the wall?

19-28. A glass windowpane has an area of 1.5 m^2, and it is 0.50 cm thick. If the two sides of the window are at 21 and 0°C, how much heat flows through the pane in 1 h?

19-29. A homeowner finds that if she turns down her thermostat by 3°C, she reduces the rate at which heat is lost to the external environment by 10 percent. If the outside temperature is −5°C, what are the thermostat readings before and after the adjustment?

19-30. A solid copper rod of diameter 2.0 cm and length 200 m is placed between a large container of boiling water at 100°C and a large container of crushed ice at 0°C. How much ice is changed to water every second by the heat conducted through the rod?

Thermal Expansion

19-31. A metal rod is heated from 0 to 100°C. At the higher temperature, its length is measured to be 2.00 m. If the rod is made of brass, determine its length at the lower temperature.

19-32. The following table is a summary of measurements of the length of a rod of unknown material as a function of temperature. Make a plot of the data and determine from your graph the coefficient of linear expansion of the material.

Length (m)	10.40	10.46	10.50	10.55	10.61	10.66	10.70
Temperature (°C)	−50	0	50	100	150	200	300

19-33. A copper sheet is heated until its area has increased by 0.40 percent. (See Drill Prob. 19-8.) What is the temperature increase corresponding to this change?

19-34. A metal rod that is 1.020 m long at a temperature of 100°C is cooled to 41°C. Its length at this lower temperature is found to be 1.019 m. With the help of Table 19-4, determine the metal of the rod.

19-35. When the temperature of a metal plate is raised from 0 to 100°C, its surface area increases by 0.24 percent. (See Drill Prob. 19-8.) (*a*) Find the metal's coefficient of linear expansion. (*b*) What is the metal?

19-36. Show that when the temperature of a solid object changes by ΔT, its moment of inertia I changes by $\Delta I = 2\alpha I\,\Delta T$.

19-37. Use the result of the previous problem to show that the period τ of a physical pendulum changes by approximately $\Delta\tau = \frac{1}{2}\alpha\tau\,\Delta T$ when the temperature is changed from T to $T + \Delta T$.

19-38. A solid copper cube is cooled from 80 to 25°C. What is the fractional change in the density of the copper?

19-39. An engineer wishes to design a structure for which the difference in lengths between a steel and an aluminum beam remains at 0.50 m for all ordinary temperatures. What must be the lengths of the beams?

19-40. At 20°C a copper ring has an inner diameter of 9.90 cm, and a steel cylinder has a diameter of 9.91 cm. At what common temperature will the ring just fit over the cylinder?

19-41. An aluminum disk just fits in a circular hole in a brass plate when both objects are at a temperature of 120°C. When they are cooled to 25°C, the diameter of the disk shrinks to 10.0 cm. What is the area of the aperture between the disk and the hole at the latter temperature?

19-42. The bulk modulus of aluminum is $7.0 \times 10^{10}\ N/m^2$. What hydrostatic pressure is necessary to prevent an aluminum block from expanding when its temperature is increased from 20 to 40°C?

General Problems

19-43. A wall of a red brick house measures 8.0 m by 4.0 m and is 10 cm thick. Inset in this wall is a window 2.0 m by 1.0 m made of glass of thickness 5.0 mm. What is the ratio of the amounts of heat lost to the exterior by conduction through the glass and through the brick?

19-44. A composite slab of cross-sectional area A is made up of two materials I and II of thicknesses d_I and d_{II} and thermal conductivities κ_I and κ_{II}, respectively. (See the accompanying figure.) At steady state, the temperatures of the hot and cold sides are T_1 and T_3, while the interface is at an intermediate temperature of T_2. Show that the rate of heat flow through the slab is given by

$$\frac{dQ}{dt} = \frac{(T_1 - T_3)A}{(d_I/\kappa_I) + (d_{II}/\kappa_{II})}.$$

(*Hint:* In steady state, what is the relationship between the quantities of heat conducted across the two materials?)

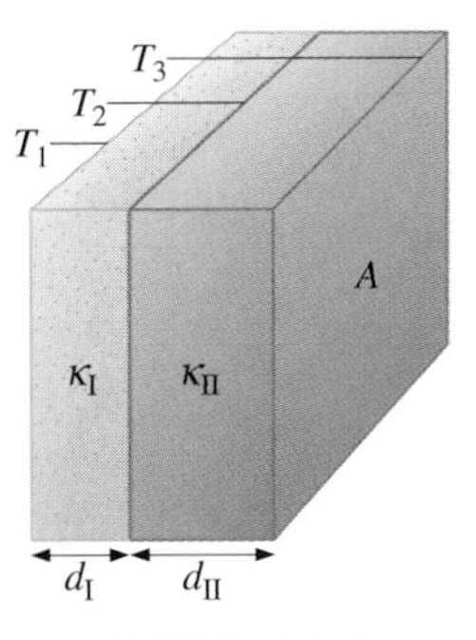

PROBLEM 19-44

19-45. A brick wall of a house measures 8.0 m by 4.0 m and is 10 cm thick. The owner of the house decides to insulate this wall by adding a layer of insulating material 2.0 cm thick with a thermal conductivity of 8.0×10^{-2} J/m·s·°C. By what fraction does he reduce energy loss through the wall? (*Hint:* Use the result of Prob. 19-44.)

19-46. A clock controlled by a brass pendulum keeps correct time at 10°C. If the room temperature is 30°C, does the clock run faster or slower? What is its error in seconds per day? (See Prob. 19-37.)

19-47. The inside and outside walls of the hollow cylinder shown in the accompanying figure are maintained at temperatures T_1 and T_2, respectively, with $T_1 > T_2$. The thermal conductivity of the material between the walls is κ. Show that if end effects can be ignored, the rate of heat flow in the radial direction from the inner to the outer wall is given by

$$\frac{dQ}{dt} = 2\pi L\kappa\left(\frac{T_1 - T_2}{\ln R_2/R_1}\right).$$

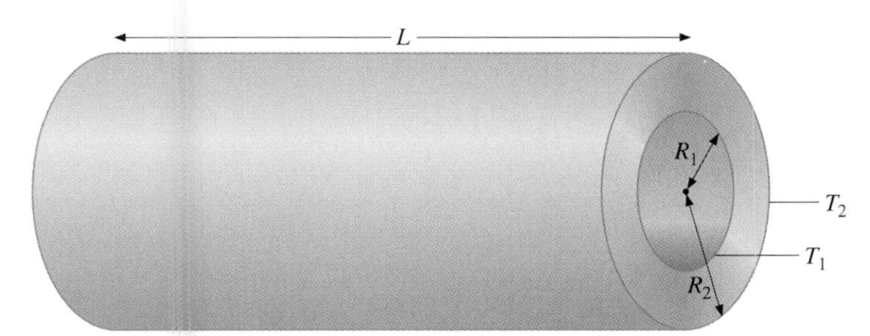

PROBLEM 19-47

19-48. Assume that a thermos bottle can be approximated by the cylinder of Prob. 19-47, with $R_1 = 0.045$ m, $R_2 = 0.050$ m, $L = 0.40$ m, and $\kappa = 1.0 \times 10^{-2}$ J/s·m·°C. If 1.0 kg of soup at 95°C is poured into the thermos at 7:00 A.M., what is its temperature at lunchtime (noon)? Assume that the air temperature remains constant at 20°C and that the soup has the specific heat of water.

19-49. A 1-L glass bottle is completely filled with water at 20°C. The bottle and water are then heated to 30°C. How much water spills out of the bottle? Take β to be 1.2×10^{-5} $(°C)^{-1}$ for glass and 25×10^{-5} $(°C)^{-1}$ for water.

19-50. The inner and outer surfaces of the hollow sphere shown in the accompanying figure are maintained at temperatures T_1 and T_2, respectively, with $T_1 > T_2$. The thermal conductivity of the material between the two surfaces is κ. Show that the rate of heat flow from the inner to the outer surface is given by

$$\frac{dQ}{dt} = 4\pi\kappa R_1 R_2\left(\frac{T_1 - T_2}{R_2 - R_1}\right).$$

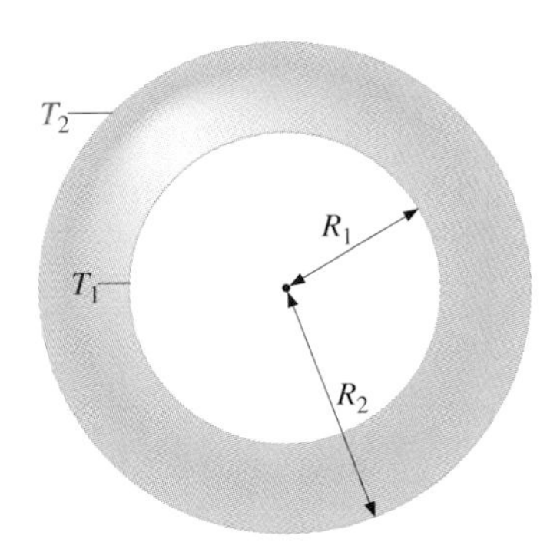

PROBLEM 19-50

19-51. The 1.0-m-long aluminum pipe shown in the accompanying figure has an outer diameter D of 0.100 m and an inner diameter d of 0.080 m. If the temperature of the pipe is increased from 20 to 70°C, calculate the changes in (*a*) the length, (*b*) the outer diameter, (*c*) the inner diameter, (*d*) the cross-sectional area, and (*e*) the volume of the aluminum.

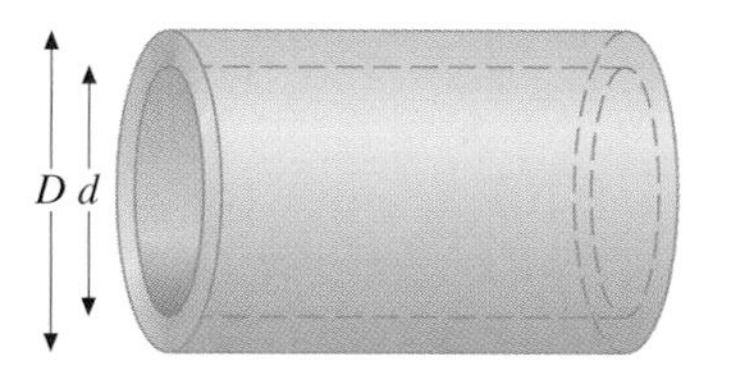

PROBLEM 19-51

19-52. Use Eq. (19-10*a*) to show that the volume coefficient of expansion for an ideal gas at constant pressure is given by $\beta = 1/T$.

19-53. A 200-g aluminum cylinder at 100°C is placed in an insulated cup of negligible heat capacity. The cup contains a mixture of 80 g of ice and 20 g of water, both at 0°C. What is the temperature of the aluminum cylinder when it reaches thermal equilibrium with the water?

19-54. Three objects A, B, and C have masses m, $2m$, and $3m$, and they are at temperatures of 10, 16, and 20°C. First A and B are placed in thermal contact and reach an equilibrium temperature of 14°C. Next A and C are placed in thermal contact and reach an equilibrium temperature of 17°C. What then is the equilibrium temperature when B and C are placed in thermal contact?

19-55. The molar heat capacity of NaCl at some representative low temperatures is given in the following table. Plot these data and estimate by a rough numerical integration how much heat is required to raise the temperature of 2 mol of NaCl from 30 to 80 K.

T (K)	10	20	30	40	50	60	70	80
C (J/mol·K)	0.151	1.30	4.76	9.98	15.7	21.0	25.5	29.3

19-56. For "low" temperatures, the molar heat capacity of a solid can be represented by the Debye T^3 law,

$$C = 234R\left(\frac{T}{\Theta}\right)^3,$$

where Θ, the Debye temperature, has a particular value for each material. This formula works quite well as long as $T < 0.04\Theta$. For germanium, $\Theta = 363$ K; for NaCl, $\Theta = 321$ K; for graphite, $\Theta = 420$ K. How much heat is required to raise the temperature of 2 mol of graphite from 5 to 15 K? of 5 mol of germanium from 8 to 10 K?

This infrared image of Jupiter shows two impact sites of Comet Shoemaker/Levy-9. The sites glow brightly due to the heat of the impact.

CHAPTER 20 FIRST LAW OF THERMODYNAMICS

PREVIEW

The first law of thermodynamics is introduced in this chapter. The main topics to be discussed here are the following:

1. **Thermodynamic processes.** We introduce the terminology used to describe the thermal behavior of a system and the processes that it undergoes.
2. **Thermodynamic work in volume changes.** Thermodynamic work is defined, and examples showing how it is calculated are presented.
3. **First law of thermodynamics.** The first law of thermodynamics is explained and then applied to specific systems.
4. **Molar heat capacities of an ideal gas.** The molar heat capacities at constant volume and at constant pressure are determined for an ideal gas.
5. **Adiabatic expansion of an ideal gas.** We examine the behavior of an ideal gas when it expands without exchanging heat with the environment.

By definition, heat is the energy exchanged by two objects because of a temperature difference between the objects. Energy can also be transferred when one object does mechanical work on another. These two types of energy transfer are combined in the first law of thermodynamics, which is the main topic of this chapter. This law states that when a system exchanges heat and does work, the net energy exchanged by the system is exactly equal to the change in the total mechanical energy of its atoms and molecules. The first law is also a law of energy conservation. Different parts of an *isolated system* may exchange energy by transferring heat and doing work on one another—but the total energy of the system must remain constant.

20-1 Thermodynamic Processes

In solving mechanics problems, we isolate the body under consideration, analyze the external forces on the body, then use Newton's laws to predict its behavior. In thermodynamics, a similar approach is taken. We start by identifying the part of the universe we wish to study; it is known as our *system* (a term used freely, but without definition, in the last two chapters). A system is anything whose properties are of interest to us; it can be a single atom or the entire earth. Once our system is selected, we determine how the *environment*, or surroundings, interacts with the system. Finally, with the interaction understood, we study the thermal behavior of the system with the help of the laws of thermodynamics.

The thermal behavior of a system is described in terms of *thermodynamic variables*. For an ideal gas these variables are pressure, volume, temperature, and the number of molecules or moles of the gas. Different types of systems are generally characterized by different sets of variables. For example, the thermodynamic variables for a stretched rubber band are tension, length, temperature, and mass.

When a system is in thermodynamic equilibrium with its environment, its thermodynamic variables are related by an *equation of state*. We have discussed two of these equations: Eqs. (18-4) for an ideal gas and Eq. (18-8) for a van der Waals gas. However, if a system is not in thermal equilibrium, there is no equation of state. In fact, there may not even be a well-defined temperature, pressure, etc., that we can specify for the system.

In changing from one equilibrium state to another, a system usually passes through a series of nonequilibrium states. Such is the case during the rapid expansion of a gas. Since there are no equations of state for systems not in equilibrium, the mathematical analysis of these transitions is severely limited. Suppose, however, that a system makes the transition so slowly that it is always essentially in a state of equilibrium. For example, the confined gas of Fig. 20-1 is placed in thermal contact with a very large heat source at temperature T_0. This source (or *heat reservoir*) is so large that any change in its temperature due to heat exchange with the gas is insignificant. Consequently, the temperature of the gas when it is in equilibrium with the reservoir is T_0. The gas is then allowed to expand by removing one grain of sand at a time from the top of the piston. As each grain

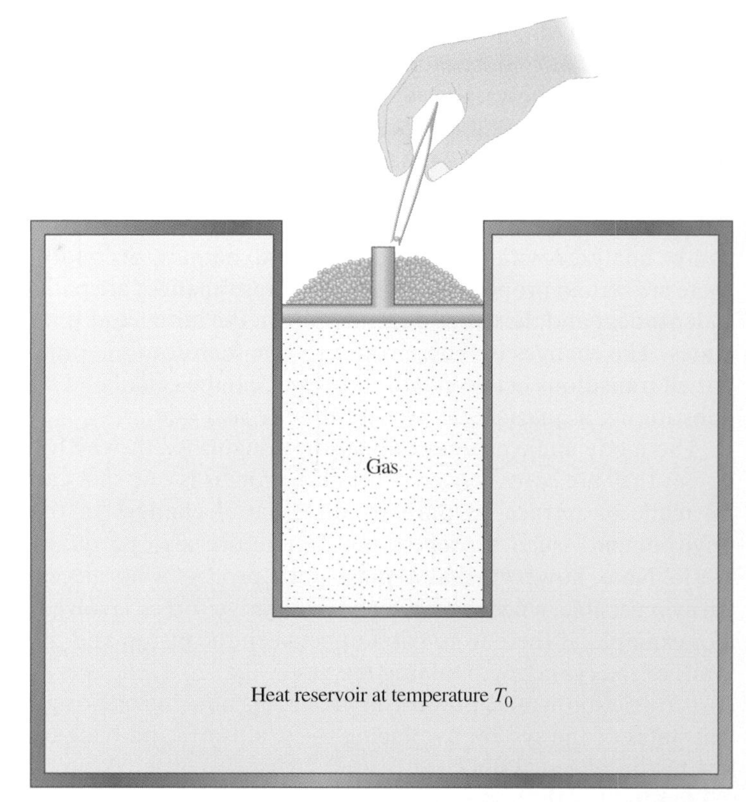

FIGURE 20-1 When sand is removed one grain at a time from the top of the piston, the gas expands through a series of equilibrium states while remaining at the temperature T_0.

is removed, the resulting decrease in pressure is so slight that for all practical purposes, the gas passes to a new equilibrium state with a slightly lower pressure. With the removal of more grains, the gas continues to expand through a series of equilibrium states at the constant temperature T_0.

Figure 20-2 depicts another transition involving only equilibrium states. Here the gas is subjected to the constant pressure of the outside atmosphere. It is also insulated from the atmosphere so that no heat exchange between the gas and the atmosphere can occur. The gas is allowed to expand at constant pressure as the temperature is increased in small, discrete steps. This is accomplished (theoretically, not practically) by moving the gas container from one heat reservoir to another as shown. The temperature changes are all so slight that the gas can be assumed to always be in an equilibrium state.

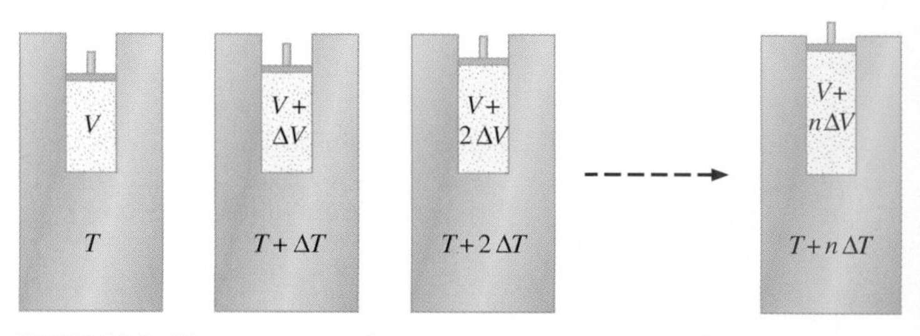

FIGURE 20-2 The gas expands at constant pressure as its temperature is increased in small steps through the use of a series of heat reservoirs. This transition, like that shown in Fig. 20-1, involves only equilibrium states.

Thermodynamic processes like the two just described are called *quasi-static*. In such processes, the changes are so slow that the system moves through a series of equilibrium states. Even though a quasi-static process *may not* represent how a system *actually* passes from one state to another, it is nevertheless important for two reasons. First, the actual transition can be at least approximately represented by a quasi-static one, which can be analyzed with the laws of thermodynamics. Secondly, there are certain properties of a system whose changes are path-independent and therefore depend only on the initial and final states. The changes in these properties are therefore the same for all transitions between two states and can be calculated by substituting a quasi-static process for the real one.

Thermodynamic processes are also distinguished by whether or not they are *reversible*. A reversible process is one that can be made to retrace its path by differential changes in the environment—such a process must therefore also be quasi-static. Note, however, that a quasi-static process is not necessarily reversible, since there may be dissipative forces involved. For example, if there were friction between the piston and the walls of the cylinder containing the gas in Fig. 20-1, the energy lost to friction would prevent us from reproducing the original states of the system by placing the grains of sand back on the piston one at a time.

Among the thermodynamic processes we will consider are the following:

1. An *isothermal* process during which the system's temperature remains constant
2. An *adiabatic* process during which no heat is transferred to or from the system
3. An *isobaric* process during which the pressure of the system is constant
4. An *isochoric* process during which the system's volume does not change

There are, of course, many other processes that do not fit into any of these four categories.

20-2 Thermodynamic Work in Volume Changes

Figure 20-3 shows a gas confined to a cylinder that has a movable piston at one end. If the gas expands against the piston, it exerts a force through a distance and does work on the piston. If the piston compresses the gas as it is moved inward, work is also done—in this case, on the gas. The work associated with such volume changes can be determined as follows. Let the gas pressure on the piston face be p. Then the force on the piston due to the gas is pA, where A is the area of the face. When the piston is pushed outward an infinitesimal distance dx, the work done by the gas is

$$dW = F\,dx = pA\,dx,$$

which, since the change in volume of the gas is $dV = A\,dx$, becomes

$$dW = p\,dV. \tag{20-1a}$$

For a finite change in volume from V_1 to V_2, this equation is then integrated between V_1 and V_2 to find the net work:

$$W = \int_{V_1}^{V_2} p\,dV. \tag{20-1b}$$

This integral is *only meaningful for a quasi-static process*, for only then is there a well-defined mathematical relationship (the equation of state) between the pressure and volume that can be represented on a pV diagram. The integral is interpreted graphically as the *area under the pressure-vs.-volume curve* (the shaded area of Fig. 20-4). Work done by the gas is positive for an expansion and negative for a compression.

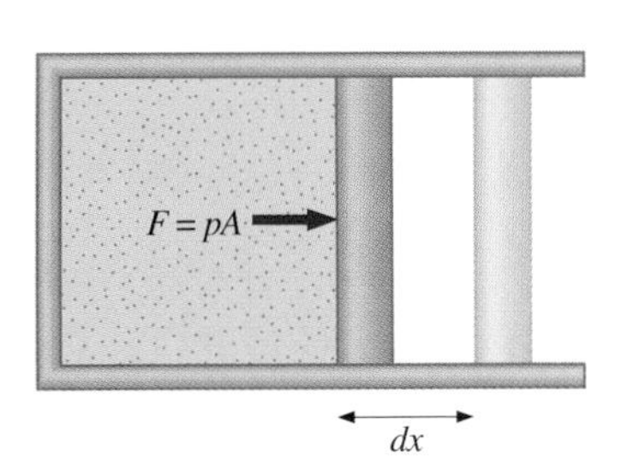

FIGURE 20-3 The work done by the confined gas in moving the piston a distance dx is given by $dW = F\,dx = p\,dV$.

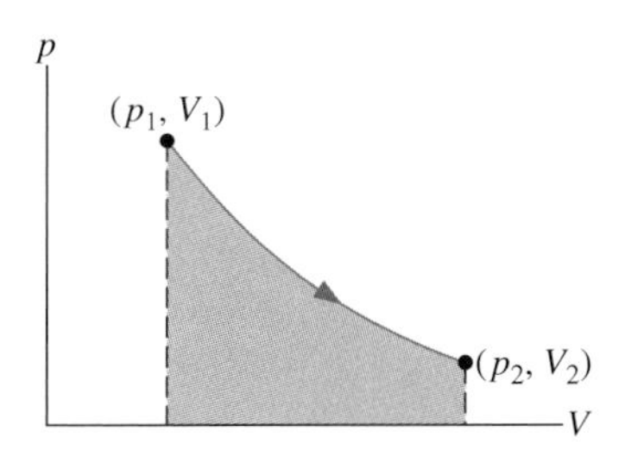

FIGURE 20-4 When a gas expands quasi-statically from V_1 to V_2, the work that it does is represented by the shaded area under the pV curve.

Consider the two quasi-static processes involving an ideal gas that are represented by paths AC and ABC in Fig. 20-5. The first process is an isothermal expansion, with the volume of the gas changing from V_1 to V_2. The expansion is represented by the curve between A and C. The gas is kept at a constant temperature T by keeping it in thermal equilibrium with a heat reservoir at that temperature. From Eq. (20-1b) and the ideal-gas law,

$$W = \int_{V_1}^{V_2} p\,dV = \int_{V_1}^{V_2} \left(\frac{nRT}{V}\right) dV.$$

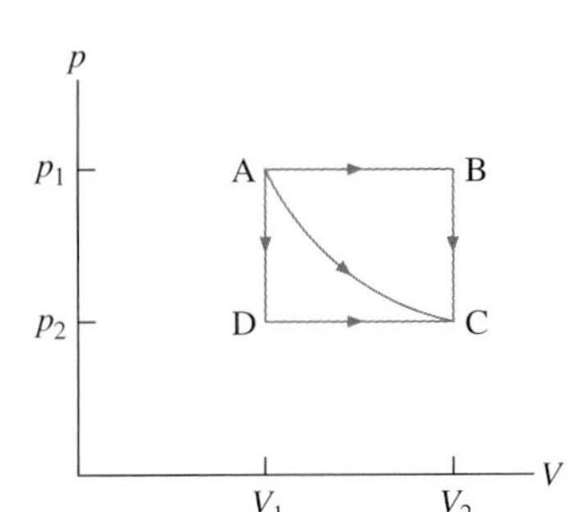

FIGURE 20-5 ABC, AC, and ADC represent three different quasi-static transitions between the equilibrium states A and C.

The expansion is isothermal, so T is constant. Since n and R are also constant, the only variable in the integrand is V, and

$$W = nRT \int_{V_1}^{V_2} \frac{dV}{V} = nRT \ln \frac{V_2}{V_1} \qquad (20\text{-}2)$$

is the work done by an ideal gas in an isothermal process. Notice that if $V_2 < V_1$ (compression), W is negative, and if $V_2 > V_1$ (expansion), W is positive, as expected.

The straight lines from A to B and then from B to C represent the other process. Here a gas at a pressure p_1 first expands isobarically and quasi-statically from V_1 to V_2, after which it cools quasi-statically at the constant volume V_2 until its pressure drops to p_2. From A to B, the pressure is constant at p_1, so the work over this part of the path is

$$W = \int_{V_1}^{V_2} p\, dV = p_1 \int_{V_1}^{V_2} dV = p_1(V_2 - V_1).$$

From B to C, there is no change in volume and therefore no work is done. The net work over the path ABC is then

$$W = p_1(V_2 - V_1) + 0 = p_1(V_2 - V_1).$$

A comparison of the expressions for the work done by the gas in the two processes of Fig. 20-5 shows that they are quite different.* This illustrates a very important property of thermodynamic work: It is *path-dependent*. We cannot determine the work done by a system as it goes from one equilibrium state to another unless we know its thermodynamic path. *Different values of the work are associated with different paths.*

EXAMPLE 20-1 **Isothermal Expansion of a van der Waals Gas**

One mole of a van der Waals gas expands isothermally and quasi-statically from a volume v_1 to a volume v_2. How much work is done by the gas during the expansion?

SOLUTION From Eq. (18-8), the equation of state for 1 mol of the gas is

$$\left(p + \frac{a}{v^2}\right)(v - b) = RT.$$

The work is calculated from

$$W = \int_{v_1}^{v_2} p\, dv.$$

To evaluate this integral, we must express p as a function of v. From Eq. (i), the gas pressure is

$$p = \frac{RT}{v - b} - \frac{a}{v^2},$$

so with T constant, the work done by 1 mol of a van der Waals gas in expanding from a volume v_1 to a volume v_2 is

$$W = \int_{v_1}^{v_2} \left(\frac{RT}{v - b} - \frac{a}{v^2}\right) dv = RT \ln (v - b) + \frac{a}{v}\Bigg|_{v_1}^{v_2}$$

$$= RT \ln \left(\frac{v_2 - b}{v_1 - b}\right) + a\left(\frac{1}{v_2} - \frac{1}{v_1}\right).$$

*Since work is the area under the path in the pV diagram, you can see from Fig. 20-5 that more work is done by the gas in going along ABC than in the isothermal process AC.

DRILL PROBLEM 20-1

How much work is done by the gas of Fig. 20-5 when it expands quasi-statically along the path ADC?
ANS. $p_2(V_2 - V_1)$.

20-3 First Law of Thermodynamics

The internal energy of a system can change either through exchange of heat with its environment or through work done. The first law of thermodynamics is in part a consolidation of these two processes. Suppose Q represents the heat exchanged between the system and the environment and W is the work done by or on the system. The first law states that the change in internal energy of that system is $Q - W$. Since added heat increases the internal energy of a system, *Q is positive when it is added to the system and negative when it is removed from the system.* When a gas expands it does work and its internal energy decreases. Thus *W is positive when work is done by the system and negative when work is done on the system.* This sign convention is summarized in Table 20-1. The first law of thermodynamics is stated as follows:

Associated with every equilibrium state of a system is its internal energy U. The change in U for any transition between two equilibrium states is

$$\Delta U = Q - W, \qquad (20\text{-}3a)$$

where Q and W represent, respectively, the heat exchanged by the system and the work done by or on the system.

James Joule's work led to the concept of the equivalence of heat and work.

TABLE 20-1 **Thermodynamic Sign Conventions for Heat and Work**

Process	Convention
Heat added to system	$Q > 0$
Heat removed from system	$Q < 0$
Work done by system	$W > 0$
Work done on system	$W < 0$

The first law is a statement of energy conservation. It tells us that a system can exchange energy with its surroundings by the transmission of heat and by the performance of work. The net energy exchanged is then equal to the change in the total mechanical energy of the molecules of the system (i.e., the system's internal energy). Thus if a system is isolated, its internal energy must remain constant.

While Q and W both depend on the thermodynamic path taken between two equilibrium states, their difference $Q - W$ does not. Figure 20-6 is the pV diagram of a system that is making the transition from A to B repeatedly along different thermodynamic paths. Along path 1, the system absorbs heat Q_1 and does work W_1; along path 2, it absorbs heat Q_2 and does work W_2, and so on. While the values of Q_i and W_i may vary from path to path, we have

$$Q_1 - W_1 = Q_2 - W_2 = \cdots = Q_i - W_i = \cdots,$$

or

$$\Delta U_1 = \Delta U_2 = \cdots = \Delta U_i = \cdots;$$

that is, *the change in the internal energy of the system between A and B is path-independent*. In Chap. 7, we encountered another path-independent quantity, the change in potential energy between two arbitrary points in space. This change represents the negative of the work done by a conservative force between the two points. [See Eq. (7-14).] While the potential energy is a function of the spatial coordinates, the internal energy is a function of the thermodynamic variables. For example, we might write $U(T, p)$ for the internal energy. Functions like internal energy and potential energy are known as *state functions* because their values depend solely on the state of the system.

Often the first law must be used in its differential form, which is

$$dU = đQ - đW. \qquad (20\text{-}3b)$$

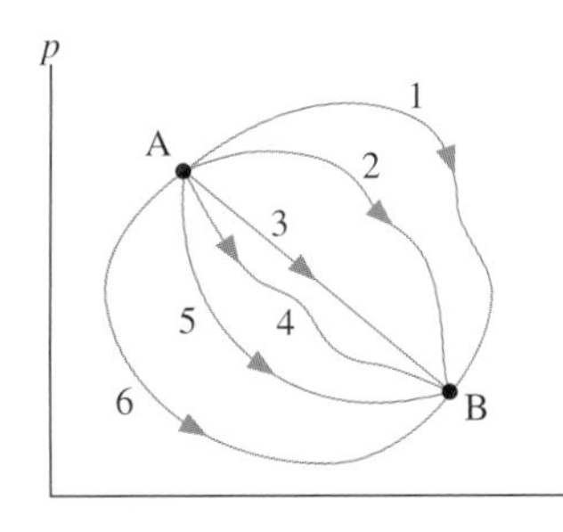

FIGURE 20-6 Different thermodynamic paths taken by a system in going from state A to state B. For all transitions, the change in the internal energy of the system, $\Delta U = Q - W$, is the same.

Here, dU is an infinitesimal change in internal energy when an infinitesimal amount of heat $đQ$ is exchanged with the system and an infinitesimal amount of work $đW$ is done by or on the system. Notice the bars on the differentials ($đ$) for Q and W. They serve to remind us that heat and work are path-dependent.

EXAMPLE 20-2 **Changes of State and the First Law**

When a system goes from state A to state B, it is supplied with 400 J of heat and does 100 J of work. (*a*) For this transition, what is the system's change in internal energy? (*b*) If the system moves from B to A, what is its change in internal energy? (*c*) If in moving from A to B along a different path, $W'_{AB} = 400$ J of work is done on the system, how much heat does it absorb?

SOLUTION (*a*) From the first law, the change in the system's internal energy is

$$\Delta U_{AB} = Q_{AB} - W_{AB} = 400 \text{ J} - 100 \text{ J} = 300 \text{ J}.$$

(*b*) Consider a closed path that passes through the states A and B. Internal energy is a state function so ΔU is zero for a closed path. Thus

$$\Delta U = \Delta U_{AB} + \Delta U_{BA} = 0,$$

and

$$\Delta U_{AB} = -\Delta U_{BA},$$

which yields

$$\Delta U_{BA} = -300 \text{ J}.$$

(*c*) The change in internal energy is the same for any path, so

$$\Delta U_{AB} = \Delta U'_{AB} = Q'_{AB} - W'_{AB}$$
$$300 \text{ J} = Q'_{AB} - (-400 \text{ J}),$$

and the heat exchanged is

$$Q'_{AB} = -100 \text{ J}.$$

The negative sign indicates that the system loses heat in this transition.

Notice that in the last example we did not assume that the transitions were quasi-static. This is because the first law is not subject to such a restriction! It describes transitions between equilibrium states but is not concerned with the intermediate states. The system does not have to pass through only equilibrium states. For example, if a gas in a steel container at a well-defined temperature and pressure is made to explode by means of a spark, some of the gas may condense, different gas molecules may combine to form new compounds, and there may be all sorts of turbulence in the container—but eventually the system will settle down to a new equilibrium state. This system is clearly not in equilibrium during its transition; however, its behavior is still governed by the first law since the process starts and ends with the system in equilibrium states.

EXAMPLE 20-3 **POLISHING A FITTING**

A machinist polishes a 0.50-kg copper fitting with a piece of emery cloth for 2.0 min. He moves the cloth across the fitting at a constant speed of 1.0 m/s by applying a force of 20 N tangent to the surface of the fitting. (*a*) What is the total work done on the fitting by the machinist? (*b*) What is the increase in the internal energy of the fitting? Assume that the change in the internal energy of the cloth is negligible and that no heat is exchanged between the fitting and its environment. (*c*) What is the increase in the temperature of the fitting?

SOLUTION (*a*) From Eq. (7-23), the rate at which the machinist does frictional work on the fitting is $\mathbf{F}\cdot\mathbf{v} = -Fv$. Thus in a time Δt (2.0 min) the work done on the fitting is

$$W = -Fv\,\Delta t = -(20\text{ N})(1.0\text{ m/s})(1.2\times 10^2\text{ s}) = -2.4\times 10^3\text{ J}.$$

(*b*) By assumption, no heat is exchanged between the fitting and its environment, so the first law gives for the change in the internal energy of the fitting

$$\Delta U = -W = 2.4\times 10^3\text{ J}.$$

(*c*) Since ΔU is path-independent, the effect of the 2.4×10^3 J of work is the same as if it were supplied at atmospheric pressure by a transfer of heat. Thus from Eq. (19-5),

$$2.4\times 10^3\text{ J} = mc\,\Delta T = (0.50\text{ kg})(3.9\times 10^2\text{ J/kg}\cdot{}^\circ\text{C})\,\Delta T,$$

and the increase in the temperature of the fitting is

$$\Delta T = 12^\circ\text{C},$$

where we have used the value for the specific heat of copper given in Table 19-1.

DRILL PROBLEM 20-2

The quantities in the following table represent four different transitions between the same initial and final state. Fill in the blanks.

Q (J)	*W* (J)	ΔU (J)
−80	−120	
90		
	40	
	−40	

EXAMPLE 20-4 **AN IDEAL GAS MAKING TRANSITIONS BETWEEN TWO STATES**

Consider the quasi-static expansions of an ideal gas between the equilibrium states A and C of Fig. 20-5. If 515 J of heat are added to the gas as it traverses the path ABC, how much heat is required for the transition along ADC? Assume that $p_1 = 2.10\times 10^5\text{ N/m}^2$, $p_2 = 1.05\times 10^5\text{ N/m}^2$, $V_1 = 2.25\times 10^{-3}\text{ m}^3$, and $V_2 = 4.50\times 10^{-3}\text{ m}^3$.

SOLUTION For path ABC, the heat added is $Q_{ABC} = 515$ J, and the work done by the gas is the area under the path on the pV diagram, which is

$$W_{ABC} = p_1(V_2 - V_1) = 473\text{ J}.$$

Therefore

$$\Delta U_{ABC} = Q_{ABC} - W_{ABC} = 42\text{ J}.$$

The change in internal energy between two equilibrium states is path-independent, so for both transitions ABC and ADC, $\Delta U = 42$ J. Along ADC, the work done by the gas is again the area under the path:

$$W_{ADC} = p_2(V_2 - V_1) = 236\text{ J}.$$

The first law now yields

$$42\text{ J} = Q_{ADC} - 236\text{ J},$$

and

$$Q_{ADC} = 278\text{ J}.$$

EXAMPLE 20-5 **ISOTHERMAL EXPANSION OF AN IDEAL GAS**

Heat is added to 1 mol of an ideal monatomic gas confined to a cylinder with a movable piston at one end. The gas expands quasi-statically at a constant temperature of 300 K until its volume increases from V to $3V$. (*a*) What is the change in internal energy of the gas? (*b*) How much work does the gas do? (*c*) How much heat is added to the gas?

SOLUTION (*a*) We saw in Sec. 19-1 that the internal energy of an ideal monatomic gas is a function only of temperature. Since $\Delta T = 0$ for this process, $\Delta U = 0$.

(*b*) The quasi-static isothermal expansion of an ideal gas was considered in Sec. 20-2. The work done by the gas during this expansion is

$$W = nRT\ln\frac{V_2}{V_1} = nRT\ln\frac{3V}{V} = (1.00\text{ mol})(8.314\text{ J/K}\cdot\text{mol})(300\text{ K})(\ln 3) = 2.74\times 10^3\text{ J}.$$

(*c*) With the results of parts (*a*) and (*b*), we can use the first law to determine the heat added:

$$\Delta U = Q - W$$
$$0 = Q - 2.74\times 10^3\text{ J},$$

so

$$Q = 2.74\times 10^3\text{ J}.$$

DRILL PROBLEM 20-3

Why was it necessary to state that the process of Example 20-5 is quasi-static?

EXAMPLE 20-6 **Vaporizing Water**

When 1.00 g of water at 100°C changes from the liquid to the gas phase at atmospheric pressure, its volume change is 1.67×10^{-3} m^3. (*a*) How much heat must be added to vaporize the water? (*b*) How much work is done by the water against the atmosphere in its expansion? (*c*) What is the change in the internal energy of the water?

SOLUTION (*a*) With L_v representing the latent heat of vaporization, the heat required to vaporize the water is

$$Q = mL_v = (1.00\text{ g})(2.26 \times 10^3\text{ J/g}) = 2.26 \times 10^3\text{ J}.$$

(*b*) Since the pressure on the system is constant at 1.00 atm = 1.01×10^5 N/m^2, the work done by the water as it is vaporized is

$$W = p\,\Delta V = (1.01 \times 10^5\text{ N/m}^2)(1.67 \times 10^{-3}\text{ m}^3) = 169\text{ J}.$$

(*c*) From the first law, the internal energy of the water during its vaporization changes by

$$\Delta U = Q - W = 2.26 \times 10^3\text{ J} - 169\text{ J} = 2.09 \times 10^3\text{ J}.$$

DRILL PROBLEM 20-4

When 1.00 g of ammonia boils at atmospheric pressure and −33.0°C, its volume changes from 1.47 to 1130 cm^3. Its heat of vaporization at this pressure is 1.37×10^6 J/kg. What is the change in the internal energy of the ammonia when it vaporizes?
ANS. 1.26×10^3 J.

20-4 Molar Heat Capacities of an Ideal Gas

Figure 20-7 shows two vessels A and B, each containing 1 mol of the same type of ideal gas at a temperature T and a volume V. The only difference between the two vessels is that the piston at the top of A is fixed while the one at the top of B is free to move against a constant external pressure p. We now consider what happens when the temperature of the gas in each vessel is slowly increased to $T + dT$ with the addition of heat.

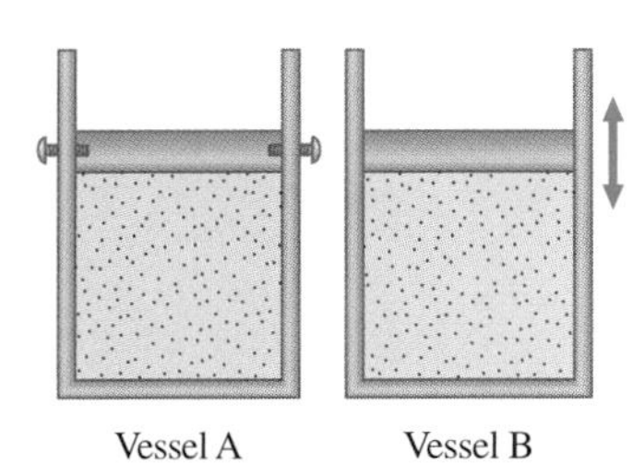

FIGURE 20-7 The two vessels are identical except that the piston at the top of A is fixed while that atop B is free to move against a constant external pressure p.

Since the piston of vessel A is fixed, the volume of the enclosed gas does not change. Consequently, the gas does no work and we have from the first law

$$dU = đQ - đW = đQ.$$

We represent the fact that the heat is exchanged at constant volume by writing

$$đQ = C_v\,dT,$$

where C_v is the **molar heat capacity at constant volume** of the gas. In addition, since $dU = đQ$ for this particular process,

$$dU = C_v\,dT. \tag{20-4}$$

Now this equation was obtained assuming the volume of the gas was fixed. However, internal energy is a state function that depends on only the temperature of an ideal gas. Therefore *$dU = C_v\,dT$ gives the change in internal energy of an ideal gas for any process involving a temperature change dT.*

When the gas in vessel B is heated, it expands against the movable piston and does work $đW = p\,dV$. In this case, the heat is added at constant pressure, and we write

$$đQ = C_p\,dT,$$

where C_p is the **molar heat capacity at constant pressure** of the gas. Furthermore, since the ideal gas expands against a constant pressure,

$$d\,(pV) = d\,(RT)$$

becomes

$$p\,dV = R\,dT.$$

Finally, inserting the expressions for $đQ$ and $p\,dV$ in the first law, we obtain

$$dU = đQ - p\,dV = (C_p - R)\,dT.$$

We have found dU for both an isochoric and an isobaric process. Because the internal energy of an ideal gas depends on only the temperature, dU must be the same for both processes. Thus

$$C_v\,dT = (C_p - R)\,dT,$$

and

$$C_p = C_v + R. \tag{20-5}$$

The derivation of Eq. (20-5) was based only on the ideal-gas law. Consequently, this relationship is approximately valid for all dilute gases, be they monatomic like He, diatomic like O_2, or polyatomic like CO_2 or NH_3. (See Table 20-2.) However, the actual values of U, C_p, and C_v do depend on the number of atoms in each molecule of the gas. Let's first consider these quantities for a monatomic ideal gas. In Sec. 18-4 we found that the average energy of a monatomic ideal-gas molecule is

$$\overline{\tfrac{1}{2}mv^2} = \tfrac{3}{2}kT,$$

where k is Boltzmann's constant and T is the temperature of the gas. The internal energy of 1 mol of an ideal monatomic gas is then

$$U = \tfrac{3}{2}N_A kT = \tfrac{3}{2}RT.$$

Now from $dU = C_v\,dT$, the molar heat capacity at constant volume is

$$C_v = \tfrac{3}{2}R, \tag{20-6a}$$

TABLE 20-2 **Molar Heat Capacities of Various Dilute Gases at Room Temperature**

Type of Molecule	Gas	C_p (J/mol · K)	C_v (J/mol · K)	$C_p - C_v$ (J/mol · K)
Monatomic	Ideal	$\frac{5}{2}R = 20.79$	$\frac{3}{2}R = 12.47$	$R = 8.31$
	He	20.8	12.5	8.3
	Ar	20.8	12.5	8.3
	Kr	20.9	12.3	8.6
	Xe	21.0	12.6	8.4
Diatomic	Ideal	$\frac{7}{2}R = 29.10$	$\frac{5}{2}R = 20.79$	$R = 8.31$
	H_2	28.8	20.4	8.4
	N_2	29.1	20.8	8.3
	O_2	29.4	21.1	8.3
	Cl_2	34.1	25.1	9.0
Polyatomic	Ideal	$4R = 33.26$	$3R = 24.94$	$R = 8.31$
	CO_2	36.6	28.3	8.3
	NH_3	37.3	28.5	8.8
	SO_2	39.4	31.1	8.3
	C_2H_6	51.7	43.3	8.4

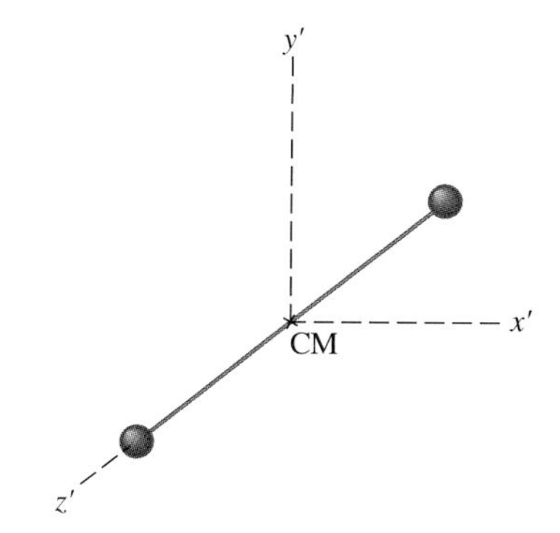

FIGURE 20-8 A diatomic molecule with orthogonal axes x', y', and z' attached to its center of mass. Since the atoms lie along the z' axis and are assumed to be point masses, $I_{z'} = 0$, and there is no contribution to the rotational kinetic energy corresponding to motion around this axis.

which, when substituted into $C_p = C_v + R$, gives for the molar heat capacity at constant pressure

$$C_p = \tfrac{5}{2}R. \tag{20-6b}$$

In deriving $\overline{mv^2/2} = 3kT/2$, we used the fact that all three spatial dimensions are equivalent, a consequence of which is

$$\overline{v_x^2} = \overline{v_y^2} = \overline{v_z^2}.$$

This allows us to write

$$\overline{\tfrac{1}{2}mv^2} = \overline{\tfrac{1}{2}mv_x^2} + \overline{\tfrac{1}{2}mv_y^2} + \overline{\tfrac{1}{2}mv_z^2} = 3(\overline{\tfrac{1}{2}mv_x^2}),$$

which, coupled with $\overline{mv^2/2} = 3kT/2$, gives

$$\overline{\tfrac{1}{2}mv_x^2} = \overline{\tfrac{1}{2}mv_y^2} = \overline{\tfrac{1}{2}mv_z^2} = \tfrac{1}{2}kT.$$

We interpret this equation as a statement of the fact that each *degree of freedom* of the molecule contributes $kT/2$ to the average energy per molecule of the ideal gas. For our purposes, we can say that there is one degree of freedom for each independent motion contributing to the energy of the molecule. For example, a monatomic molecule with energy $mv_x^2/2 + mv_y^2/2 + mv_z^2/2$ has *three* degrees of freedom, with each degree corresponding to motion along a coordinate axis.

Statistical mechanics provides us with a generalization of this result, known as the *equipartition theorem*. This theorem states that each degree of freedom of a molecule contributes $kT/2$ to the average energy per molecule of an ideal gas. Now a diatomic molecule also has energy due to its rotational motion. Figure 20-8 is a representation of a diatomic molecule with the orthogonal axes x', y', and z' attached to its center of mass. We can write the rotational kinetic energy of this molecule as $I_{x'}\omega_{x'}^2/2 + I_{y'}\omega_{y'}^2/2$, where $I_{x'}$ and $I_{y'}$ are the moments of inertia of the molecule around the x' and y' axes and $\omega_{x'}$ and $\omega_{y'}$ are its angular velocities around those axes. Since the atoms (assumed to be point masses) lie along the z' axis, $I_{z'} = 0$, and there is no contribution to the rotational kinetic energy corresponding to motion around the z' axis. The energy of a diatomic molecule is then $mv_x^2/2 + mv_y^2/2 + mv_z^2/2 + I_{x'}\omega_{x'}^2/2 + I_{y'}\omega_{y'}^2/2$. As there are now five degrees of freedom, we obtain from the equipartition theorem that the average internal energy of the diatomic ideal-gas molecule is $5(kT/2) = 5kT/2$. Thus the internal energy of 1 mol of a diatomic ideal gas is

$$U = \tfrac{5}{2}N_A kT = \tfrac{5}{2}RT.$$

Now from $dU = C_v\,dT$ and $C_p = C_v + R$,

$$C_v = \tfrac{5}{2}R \tag{20-7a}$$

and

$$C_p = \tfrac{7}{2}R \tag{20-7b}$$

for a diatomic ideal gas.

As you can see in Table 20-2, these results agree quite well with the measured heat capacities of diatomic molecules at room temperature. However, this is not the case at temperatures far below and above room temperature. This is illustrated in Fig. 20-9 where C_v for hydrogen is plotted vs. temperature. At low temperatures, $C_v = 3R/2$, and at high temperatures $C_v = 7R/2$. Only in the region around room temperature is $C_v = 5R/2$.

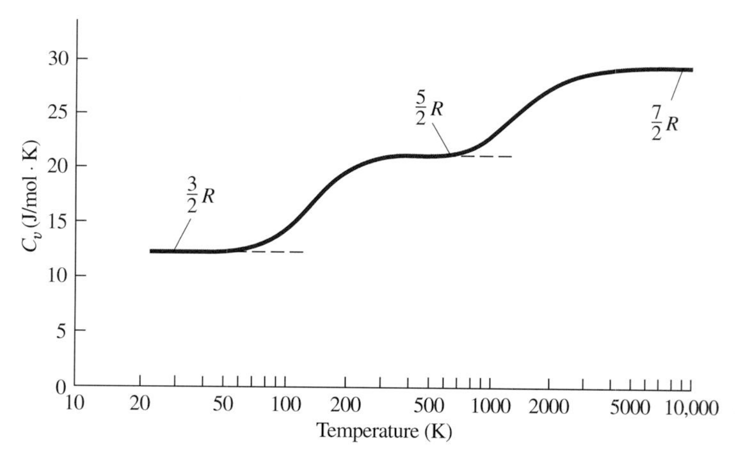

FIGURE 20-9 The molar heat capacity C_v of hydrogen as a function of temperature. The "steps" in C_v are due to changes in the number of degrees of freedom that contribute to the energy of the molecule.

The high-temperature limit suggests that we should also consider vibrations of the diatomic molecule. This results from the fact that if we consider the molecule to be two atoms joined by an imaginary spring, we then obtain two more degrees of freedom, one for the vibrational kinetic energy and one for the spring potential energy. Each of these contributes an additional $kT/2$ to the average energy per molecule and an additional $R/2$ to C_v. As a result, $C_v = 7R/2$, which agrees with the experimental high-temperature limit. At low temperatures, both the rotational and the vibrational motions of the molecule become insignificant compared with its translational motion; as a result, there are effectively only three degrees of freedom and $C_v = 3R/2$. The variation of C_v with temperature can be explained correctly only with quantum mechanics, which is outside the realm of this textbook.

The number of nonvibrational degrees of freedom for a polyatomic molecule is six, or an increase of one over the number for diatomic molecules. This extra degree of freedom is due to the fact that the atoms of polyatomic molecules do not generally lie along a straight line, so there is no axis such as the z' axis of Fig. 20-8 around which the moment of inertia is zero. Hence the energy of a polyatomic molecule is $mv_x^2/2 + mv_y^2/2 + mv_z^2/2 + I_{x'}\omega_{x'}^2/2 + I_{y'}\omega_{y'}^2/2 + I_{z'}\omega_{z'}^2/2$, which is equivalent to $6(kT/2) = 3kT$. The molar heat capacities of a polyatomic molecule are then

$$C_v = 3R \tag{20-8a}$$

and

$$C_p = 4R \tag{20-8b}$$

Table 20-2 shows that molar heat capacities at room temperature are somewhat higher than Eqs. (20-8) predict. This indicates that vibrational motion in polyatomic molecules is significant even at room temperature. Nevertheless, the difference in the molar heat capacities, $C_p - C_v$, is very close to R, even for the polyatomic gases.

20-5 Adiabatic Expansion of an Ideal Gas

When an ideal gas is compressed adiabatically ($Q = 0$), work is done on it and its temperature increases; in an adiabatic expansion, the gas does work and its temperature drops. Adiabatic compressions actually occur in the cylinders of a car, where the compression of the gas-air mixture takes place so quickly that there isn't time for the mixture to exchange heat with its environment. Nevertheless, because work is done on the mixture during the compression, its temperature does rise significantly. In fact, the temperature increase can be so large that the mixture can actually explode without the addition of a spark. Such explosions, since they are not timed, make a car run poorly—it usually "knocks." Because ignition temperature rises with the octane of gasoline, one way to overcome this problem is to use a higher-octane gasoline.

Another interesting adiabatic process is the free expansion of a gas. Figure 20-10 shows a gas confined by a membrane to one side of a two-compartment, thermally insulated container. When the membrane is punctured, gas rushes into the empty side of the container, thereby expanding freely. Because the gas expands "against a vacuum" ($p = 0$), it does no work; and because the vessel is thermally insulated, the expansion is adiabatic. With $Q = 0$ and $W = 0$ in the first law, $\Delta U = 0$, so $U_i = U_f$ for the free expansion.

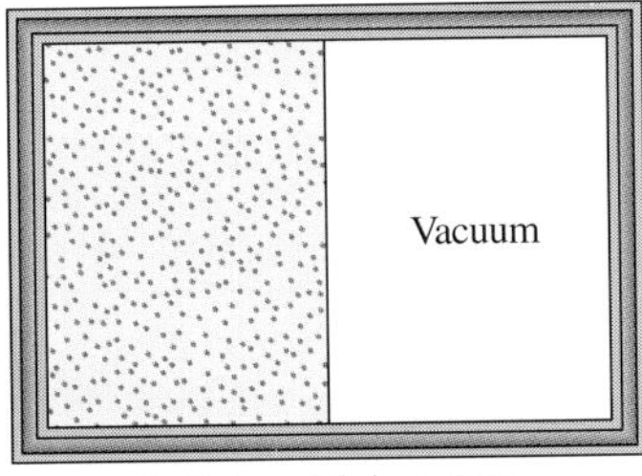

Initial equilibrium state

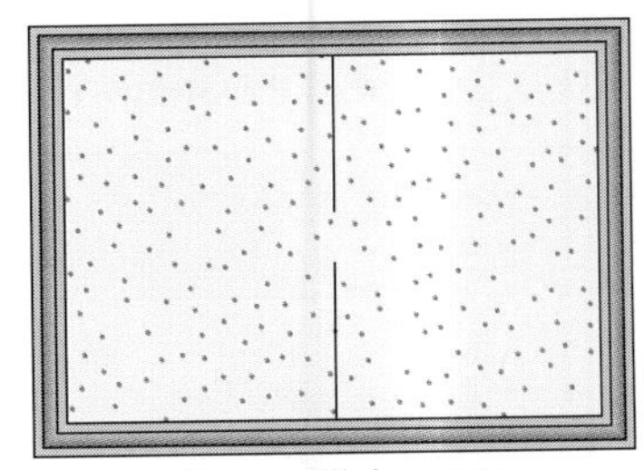
Final equilibrium state

FIGURE 20-10 The gas in the left chamber expands freely into the right chamber when the membrane is punctured.

If the gas is ideal, the internal energy depends on only the temperature. Therefore *when an ideal gas expands freely, its temperature does not change.*

A quasi-static, adiabatic expansion of an ideal gas is represented in Fig. 20-11, which shows an insulated cylinder that contains 1 mol of an ideal gas. The gas is made to expand quasi-statically by removing one grain of sand at a time from the top of the piston. When the gas expands by dV, the change in its temperature is dT. The work done by the gas in the expansion is $đW = p\,dV$; $đQ = 0$ because the cylinder is insulated; and the change in the internal energy of the gas is, from Eq. (20-4), $dU = C_v\,dT$. From the first law,

$$C_v\,dT = 0 - p\,dV = -p\,dV,$$

so

$$dT = -\frac{p\,dV}{C_v}.$$

Also, for 1 mol of an ideal gas,

$$d(pV) = d(RT),$$

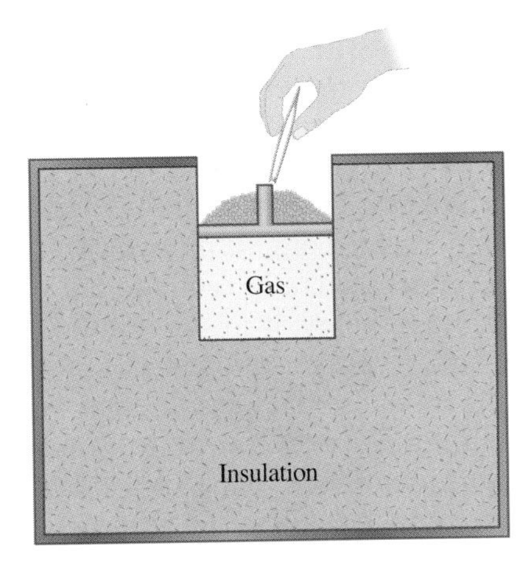

FIGURE 20-11 When sand is removed from the piston one grain at a time, the gas expands adiabatically and quasi-statically in the insulated vessel.

so

$$p\,dV + V\,dp = R\,dT,$$

and

$$dT = \frac{p\,dV + V\,dp}{R}.$$

We now have two equations for dT. Upon equating them, we find that

$$C_v V\,dp + (C_v + R)p\,dV = 0.$$

Next we divide this equation by pV and use $C_p = C_v + R$. We are then left with

$$C_v \frac{dp}{p} + C_p \frac{dV}{V} = 0,$$

which becomes

$$\frac{dp}{p} + \gamma \frac{dV}{V} = 0,$$

where we define γ as the ratio of the molar heat capacities:

$$\gamma = \frac{C_p}{C_v}. \tag{20-9}$$

Thus

$$\int \frac{dp}{p} + \gamma \int \frac{dV}{V} = 0,$$

and

$$\ln p + \gamma \ln V = \text{constant}.$$

Finally, using $\ln(A^x) = x \ln A$ and $\ln AB = \ln A + \ln B$, we can write this in the form

$$pV^\gamma = \text{constant}. \tag{20-10a}$$

This equation is the *condition that must be obeyed by an ideal gas in a quasi-static adiabatic process*. For example, if an ideal gas makes a quasi-static adiabatic transition from a state with pressure and volume p_1 and V_1 to a state with p_2 and V_2, then it must be true that $p_1V_1^\gamma = p_2V_2^\gamma$.

The adiabatic condition of Eq. (20-10a) can be written in terms of other pairs of thermodynamic variables by combining it with the ideal-gas law. In doing this, you will find that

$$p^{1-\gamma}T^\gamma = \text{constant} \tag{20-10b}$$

and

$$TV^{\gamma-1} = \text{constant}. \tag{20-10c}$$

A reversible adiabatic expansion of an ideal gas is represented on the pV diagram of Fig. 20-12. The slope of the curve at any point is

$$\frac{dp}{dV} = \frac{d}{dV}\left(\frac{\text{constant}}{V^\gamma}\right) = -\gamma\,\frac{p}{V}.$$

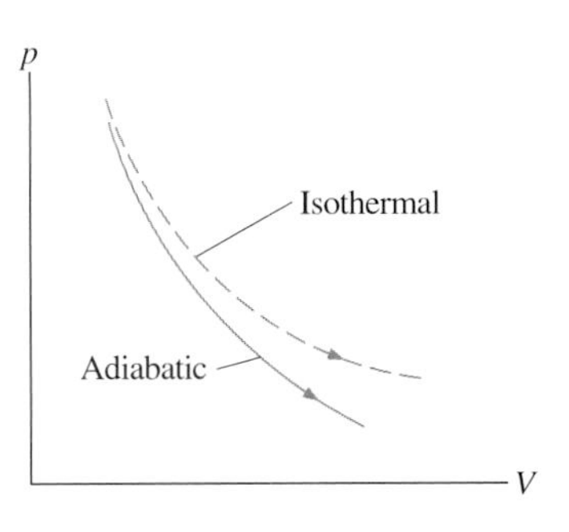

FIGURE 20-12 Quasi-static adiabatic and isothermal expansions of an ideal gas.

The dashed curve shown on this pV diagram represents an isothermal expansion where T (and therefore pV) is constant. The slope of this curve is useful when we consider the second law of thermodynamics in the next chapter. This slope is

$$\frac{dp}{dV} = \frac{d}{dV}\,\frac{nRT}{V} = -\frac{p}{V}.$$

Because $\gamma > 1$, the isothermal curve is not as steep as that for the adiabatic expansion.

EXAMPLE 20-7 Compression of an Ideal Gas in an Automobile Engine

Gasoline vapor is injected into the cylinder of an automobile engine when the piston is in its expanded position. The temperature, pressure, and volume of the resulting gas-air mixture are 20°C, 1.00×10^5 N/m², and 240 cm³, respectively. The mixture is then compressed adiabatically* to a volume of 40 cm³. For the gas-air mixture, γ can be assumed to be 1.40. (a) What are the pressure and temperature of the mixture after the compression? (b) How much work is done by the mixture during the compression?

SOLUTION (a) For an adiabatic compression we have

$$p_2 = p_1\left(\frac{V_1}{V_2}\right)^\gamma,$$

so after the compression, the pressure of the mixture is

$$p_2 = (1.00 \times 10^5\ \text{N/m}^2)\left(\frac{240 \times 10^{-6}\ \text{m}^3}{40 \times 10^{-6}\ \text{m}^3}\right)^{1.40}$$
$$= 1.23 \times 10^6\ \text{N/m}^2.$$

From the ideal-gas law, the temperature of the mixture after the compression is

$$T_2 = \left(\frac{p_2V_2}{p_1V_1}\right)T_1 = \frac{(1.23 \times 10^6\ \text{N/m}^2)(40 \times 10^{-6}\ \text{m}^3)}{(1.00 \times 10^5\ \text{N/m}^2)(240 \times 10^{-6}\ \text{m}^3)}\ 293\ \text{K}$$
$$= 601\ \text{K} = 328°\text{C}.$$

(b) The work done by the mixture during the compression is

$$W = \int_{V_1}^{V_2} p\,dV.$$

*In the actual operation of an automobile engine, the compression is not quasi-static, although we are making that assumption here.

With the adiabatic condition of Eq. (20-10a), we may write p as K/V^γ, where $K = p_1V_1^\gamma = p_2V_2^\gamma$. The work is therefore

$$W = \int_{V_1}^{V_2} \frac{K}{V^\gamma}\,dV = \frac{K}{1-\gamma}\left(\frac{1}{V_2^{\gamma-1}} - \frac{1}{V_1^{\gamma-1}}\right)$$
$$= \frac{1}{1-\gamma}\left(\frac{p_2V_2^\gamma}{V_2^{\gamma-1}} - \frac{p_1V_1^\gamma}{V_1^{\gamma-1}}\right) = \frac{1}{1-\gamma}(p_2V_2 - p_1V_1)$$
$$= \frac{1}{1-1.40}[(1.23\times10^6\ \text{N/m}^2)(40\times10^{-6}\ \text{m}^3)$$
$$-(1.00\times10^5\ \text{N/m}^2)(240\times10^{-6}\ \text{m}^3)] = -63\ \text{J}.$$

The negative sign indicates that the piston does work on the gas-air mixture.

DRILL PROBLEM 20-5

Helium, a monatomic gas assumed here to be ideal, has a pressure and volume of 2.0 atm and 4.0 L, respectively. If the gas expands adiabatically and quasi-statically to a volume of 8.0 L, what is the new pressure of the gas?
ANS. 0.63 atm.

SUMMARY

1. **Thermodynamic processes**
   (*a*) The thermal behavior of a system is described in terms of thermodynamic variables. For an ideal gas, these variables are pressure, volume, temperature, and number of molecules or moles of the gas.
   (*b*) For systems in thermodynamic equilibrium, the thermodynamic variables are related by an equation of state.
   (*c*) A heat reservoir is so large that when it exchanges heat with other systems its temperature does not change.
   (*d*) A quasi-static process takes place so slowly that the system involved is always in thermodynamic equilibrium.
   (*e*) A reversible process is one that can be made to retrace its path.
   (*f*) There are several types of thermodynamic processes, including the following:

   Isothermal, where the system's temperature is constant
   Adiabatic, where no heat is exchanged by the system
   Isobaric, where the system's pressure is constant
   Isochoric, where the system's volume is constant.

2. **Thermodynamic work in volume changes**
   (*a*) When a gas expands quasi-statically, with its volume changing from V_1 to V_2, the work it does is

$$W = \int_{V_1}^{V_2} p\,dV.$$

   (*b*) If the expansion is isothermal,

$$W = nRT\ln\frac{V_2}{V_1}.$$

3. **First law of thermodynamics**
   Associated with every equilibrium state of a system is its internal energy U. The change in U for any transition between two equilibrium states is

$$\Delta U = Q - W,$$

   where Q and W represent, respectively, the heat the system exchanges with its environment and the work done by or on the system.

4. **Molar heat capacities of an ideal gas**
   For an ideal gas, the molar heat capacity at constant volume, C_v, and the molar heat capacity at constant pressure, C_p, are related by

$$C_p = C_v + R.$$

5. **Adiabatic expansion of an ideal gas**
   When an ideal gas expands adiabatically and quasi-statically,

$$pV^\gamma = \text{constant},$$

   where $\gamma = C_p/C_v$.

QUESTIONS

20-1. When a gas expands isothermally, it does work. What is the source of energy needed to do this work?

20-2. What does the first law of thermodynamics tell us about the energy of the universe?

20-3. Is it possible to determine whether a change in internal energy is caused by heat transferred, by work performed, or by a combination of the two?

20-4. When a liquid is vaporized, its change in internal energy is not equal to the heat added. Why?

20-5. When a wet towel is rapidly twirled in a circle, it cools off. Explain why.

20-6. Is it possible for γ to be smaller than unity?

20-7. Would you expect γ to be larger for a gas or a solid? Explain.

20-8. In cool, foggy weather, ice sometimes forms in the throat of the carburetor of a moving automobile, even though the air temperature is above freezing. Why?

20-9. Why is the heat of fusion smaller than the heat of vaporization?

20-10. Is it possible for the temperature of a system to remain constant when heat flows into or out of it? If so, give examples.

20-11. There is no change in the internal energy of an ideal gas undergoing an isothermal process since the internal energy depends only on the temperature. Is it therefore correct to say that an isothermal process is the same as an adiabatic process for an ideal gas? Explain your answer.

PROBLEMS

Thermodynamic Work in Volume Changes

20-1. A gas at a pressure of 2.0 atm undergoes a quasi-static isobaric expansion from 3.0 to 5.0 L. How much work is done by the gas?

20-2. It takes 500 J of work to compress quasi-statically 0.50 mol of an ideal gas to one-fifth its original volume. Calculate the temperature of the gas, assuming it remains constant during the compression.

20-3. It is found that, when a dilute gas expands quasi-statically from 0.50 to 4.0 L, it does 250 J of work. Assuming that the gas temperature remains constant at 300 K, how many moles of gas are present?

20-4. In a quasi-static isobaric expansion, 500 J of work are done by the gas. If the gas pressure is 0.80 atm, what is the fractional increase in the volume of the gas, assuming it was originally at 20.0 L?

20-5. When a gas undergoes a quasi-static isobaric change in volume from 10.0 to 2.0 L, 15 J of work from an external source are required. What is the pressure of the gas?

20-6. An ideal gas expands quasi-statically and isothermally from a state with pressure p and volume V to a state with volume $4V$. Show that the work done by the gas in the expansion is $pV(\ln 4)$.

20-7. Calculate the work done by the gas in the quasi-static processes represented by the paths AB, ADB, ACB, and ADCB in the accompanying figure.

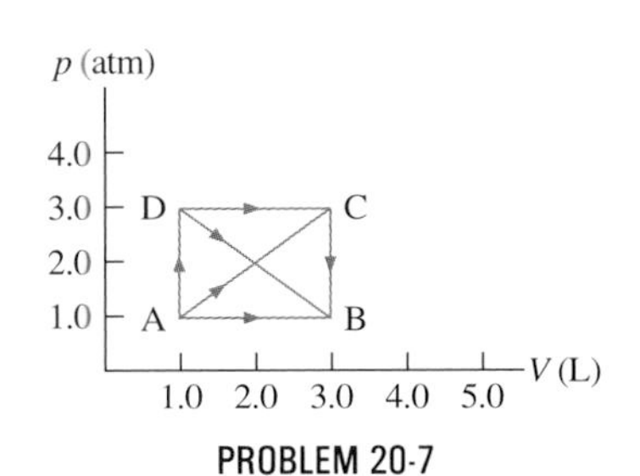

PROBLEM 20-7

20-8. Calculate the work done by the gas along the reversible closed path shown in the accompanying figure. The curved section between R and S is semicircular. If the process is carried out in the opposite direction, what is the work done by the gas?

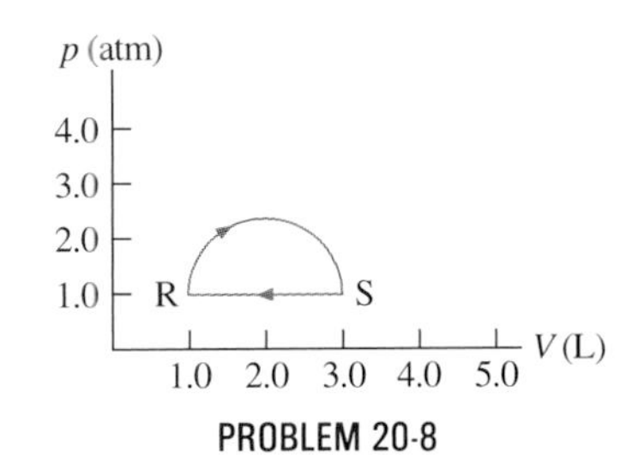

PROBLEM 20-8

20-9. An ideal gas expands quasi-statically to three times its original volume. Which process requires more work from the gas, an isothermal process or an isobaric one? Determine the ratio of the work done in these processes.

20-10. A dilute gas at a pressure of 2.0 atm and a volume of 4.0 L is taken through the following quasi-static steps: (*a*) an isobaric expansion to a volume of 10.0 L, (*b*) an isochoric change to a pressure of 0.50 atm, (*c*) an isobaric compression to a volume of 4.0 L, and (*d*) an isochoric change to a pressure of 2.0 atm. Show these steps on a pV diagram and determine from your graph the net work done by the gas.

First Law of Thermodynamics

20-11. Consider the quasi-static isothermal expansion of Prob. 20-3. What is the change in the internal energy of the gas? How much heat is absorbed by the gas in this process?

20-12. If the internal energy of the gas increases by 80 J in the expansion of Prob. 20-4, how much heat does the gas absorb?

20-13. How much heat is added to the expanding gas of Prob. 20-6?

20-14. Consider the processes of Prob. 20-7. If the heat absorbed by the gas along AB is 400 J, determine the quantities of heat absorbed along ADB, ACB, and ADCB.

20-15. During the isobaric expansion from A to B represented in the accompanying figure, 130 J of heat are removed from the gas. What is the change in its internal energy?

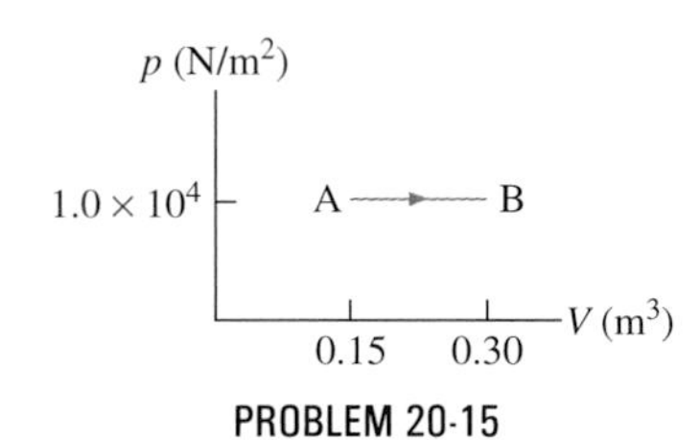

PROBLEM 20-15

20-16. In each of the following cases, determine the change in internal energy of the system: (*a*) The system absorbs 400 cal of heat while doing 1000 J of work; (*b*) the system ejects 500 cal of heat while 400 J of work are done on it; (*c*) the system absorbs 1000 J of heat at constant volume.

20-17. What is the change in internal energy for the process represented by the closed path of Prob. 20-8? How much heat is exchanged? If the path is traversed in the opposite direction, how much heat is exchanged?

20-18. When a gas expands along path AC of the accompanying figure, it does 400 J of work and absorbs 300 J of heat. (*a*) Suppose you are told that along path ABC, the gas absorbs either 200 or 400 J of heat. Which of these values is correct? (*b*) Given the correct answer from part (*a*), how much work is done by the gas along ABC? (*c*) Along CD, the internal energy of the gas decreases by 50 J. How much heat is exchanged by the gas along this path?

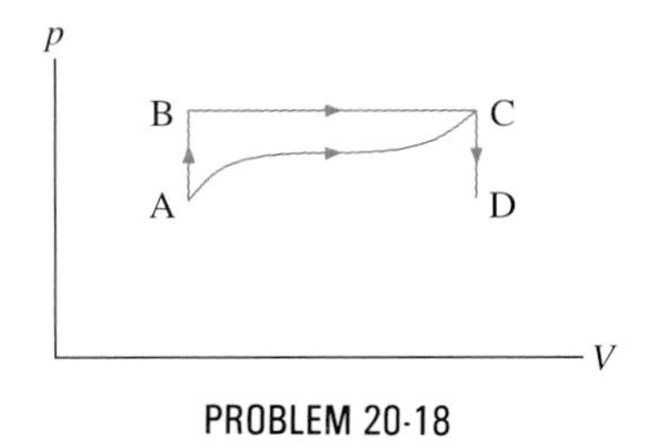

PROBLEM 20-18

20-19. When a gas expands along AB (see the accompanying figure), it does 500 J of work and absorbs 250 J of heat. When the gas expands along AC, it does 700 J of work and absorbs 300 J of heat. (*a*) How much heat does the gas exchange along BC? (*b*) When the gas makes the transition from C to A along CDA, 800 J of work are done on it from C to D. How much heat does it exchange along CDA?

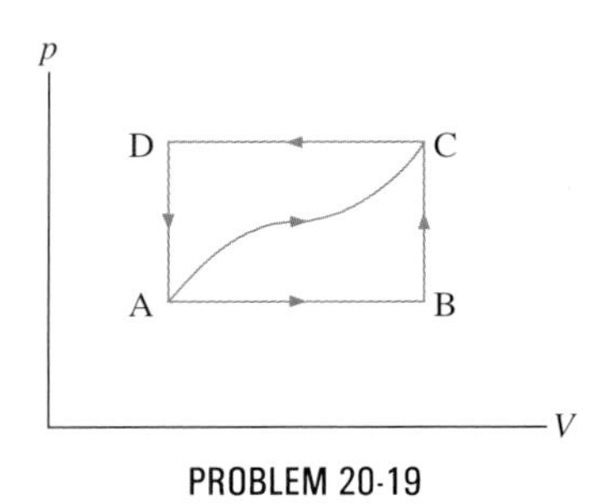

PROBLEM 20-19

20-20. An ideal gas expands isothermally along AB and does 700 J of work. (See the accompanying figure.) (*a*) How much heat does the gas exchange along AB? (*b*) The gas then expands adiabatically along BC and does 400 J of work. When the gas returns to A along CA, it exhausts 100 J of heat to its surroundings. How much work is done on the gas along this path?

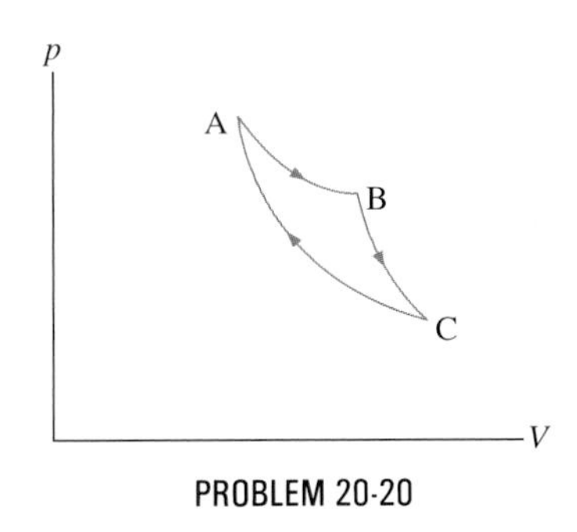

PROBLEM 20-20

20-21. A dilute gas is stored in the left chamber of a container whose walls are perfectly insulating (see the accompanying figure), and the right chamber is evacuated. When the partition is removed, the gas expands and fills the entire container. Calculate the work done by the gas. Does the internal energy of the gas change in this process?

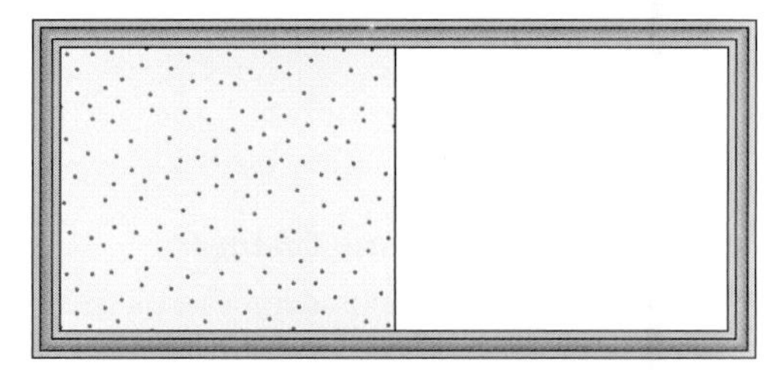

PROBLEM 20-21

20-22. Ideal gases A and B are stored in the left and right chambers of an insulated container, as shown in the accompanying figure. The partition is removed and the gases mix. Is any work done in this process? If the temperatures of A and B are initially equal, what happens to their common temperature after they are mixed?

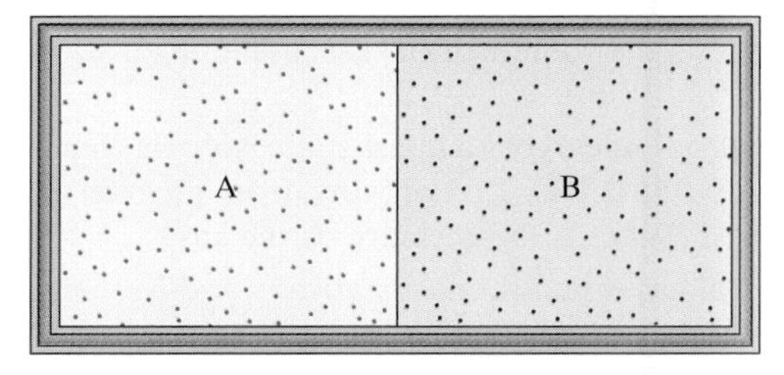

PROBLEM 20-22

20-23. Refer to the accompanying figure. (*a*) What is the change in internal energy of the two blocks due to an inelastic collision? (*b*) Both blocks are initially at the same temperature and are made from the same material, whose specific heat is $c = 300\ \text{J/kg}\cdot°\text{C}$. What is the increase in temperature of the blocks?

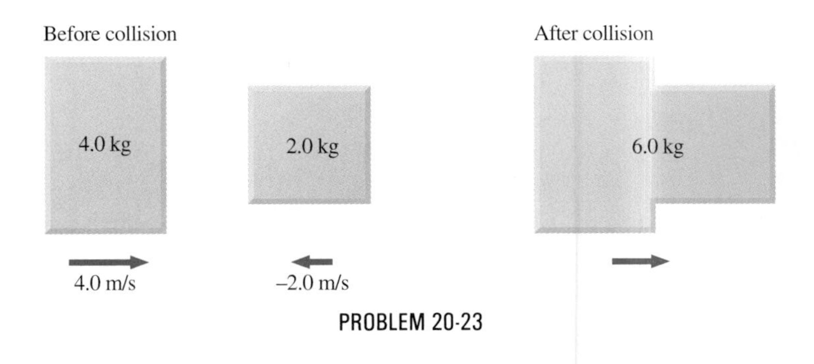

PROBLEM 20-23

20-24. A block of wood of mass 0.50 kg slides along a horizontal surface with an initial speed of 5.0 m/s. It comes to rest after sliding a distance of 10 m. (*a*) What is the work done on the block? (*b*) Assuming that 80 percent of the work done on the block shows up as an increase in its internal energy, what is the change in temperature of the block? Use $c = 0.42\ \text{cal/g}\cdot°\text{C}$ for the wood.

20-25. A 1.0-kg object is dropped through a vertical distance z into a large box of ice shavings. It is found that 2.0×10^{-4} kg of ice is melted in the inelastic collision. Assuming that all of the work done on the object during the collision goes into melting the ice, determine the value of z.

20-26. An ideal monatomic gas at a pressure of 2.0×10^5 N/m^2 and a temperature of 300 K undergoes a quasi-static isobaric expansion from 2.0×10^3 to 4.0×10^3 cm^3. (*a*) What is the work done by the gas? (*b*) What is the temperature of the gas after the expansion? (*c*) How many moles of gas are there? (*d*) What is the change in internal energy of the gas? (*e*) How much heat is added to the gas?

20-27. A spring with force constant $k = 1.0 \times 10^5$ N/m is compressed 10 cm and placed in 2.0 kg of water. The spring then expands, causing the water temperature to rise. If the heat absorbed by the spring and the water container is negligible, what is the temperature increase of the water?

Molar Heat Capacities of an Ideal Gas

20-28. The temperature of an ideal monatomic gas rises by 8.0 K. What is the change in the internal energy of 1 mol of the gas?

20-29. For a temperature increase of 10°C at constant volume, what is the heat absorbed by (*a*) 3.0 mol of a dilute monatomic gas, (*b*) 0.50 mol of a dilute diatomic gas, and (*c*) 15 mol of a dilute polyatomic gas?

20-30. If the gases of Prob. 20-29 are initially at 300 K, what are their internal energies after they absorb the heat?

20-31. Consider 0.40 mol of dilute carbon dioxide at a pressure of 0.50 atm and a volume of 50 L. What is the internal energy of the gas?

20-32. When 400 J of heat are slowly added to 10 mol of an ideal monatomic gas, its temperature rises by 10°C. What is the work done by the gas?

20-33. One mole of a dilute diatomic gas occupying a volume of 10 L expands against a constant pressure of 2.0 atm when it is slowly heated. If the temperature of the gas rises by 10 K and 400 J of heat are added in the process, what is its final volume?

Adiabatic Expansion of an Ideal Gas

20-34. A monatomic ideal gas undergoes a quasi-static adiabatic expansion in which its volume is doubled. How is the pressure of the gas changed?

20-35. An ideal gas has a pressure of 0.50 atm and a volume of 10 L. It is compressed adiabatically and quasi-statically until its pressure is 3.0 atm and its volume is 2.8 L. Is the gas monatomic, diatomic, or polyatomic?

20-36. The following table lists pressure and volume measurements of a dilute gas undergoing a quasi-static adiabatic expansion. Plot $\ln p$ vs. $\ln V$ and determine γ for this gas from your graph.

p (atm)	20.0	17.0	14.0	11.0	8.0	5.0	2.0	1.0
V (L)	1.0	1.1	1.3	1.5	2.0	2.6	5.2	8.4

20-37. An ideal monatomic gas at 300 K expands adiabatically and reversibly to twice its volume. What is its final temperature?

20-38. An ideal diatomic gas at 80 K is slowly compressed adiabatically to one-third its original volume. What is its final temperature?

20-39. Compare the change in internal energy of an ideal gas for a quasi-static adiabatic expansion with that for a quasi-static isothermal expansion. What happens to the temperature of an ideal gas in an adiabatic expansion?

20-40. The temperature of n moles of an ideal gas changes from T_1 to T_2 in a quasi-static adiabatic transition. Show that the work done by the gas is given by

$$W = \frac{nR}{\gamma - 1}(T_1 - T_2).$$

20-41. A dilute gas expands quasi-statically to three times its initial volume. Is the final gas pressure greater for an isothermal or an adiabatic expansion? Does your answer depend on whether the gas is monatomic, diatomic, or polyatomic?

General Problems

20-42. An aluminum block of mass 0.50 kg slides from rest down a plane inclined at 30° to the horizontal. At the bottom of the incline, it is moving at 8.0 m/s and its temperature is found to have increased by 0.01°C. What is the distance that the block has moved along the incline? Assume that the plane does not absorb any of the heat generated by friction.

20-43. One mole of an ideal monatomic gas occupies a volume of 1.0×10^{-2} m^3 at a pressure of 2.0×10^5 N/m^2. (*a*) What is the temperature of the gas? (*b*) The gas undergoes a quasi-static adiabatic compression until its volume is decreased to 5.0×10^{-3} m^3. What is the new gas temperature? (*c*) How much work is done on the gas during the compression? (*d*) What is the change in the internal energy of the gas?

20-44. One mole of an ideal gas is initially in a chamber of volume 1.0×10^{-2} m^3 and at a temperature of 27°C. (*a*) How much heat is absorbed by the gas when it slowly expands isothermally to twice its initial volume? (*b*) Suppose the gas is slowly transformed to the same final state by first decreasing the pressure at constant volume and then expanding it isobarically. What is the heat transferred for this case? (*c*) Calculate the heat transferred when the gas is transformed quasi-statically to the same final state by expanding it isobarically, then decreasing its pressure at constant volume.

20-45. A bullet of mass 10 g is traveling horizontally at 200 m/s when it strikes and embeds in a pendulum bob of mass 2.0 kg. (*a*) How much mechanical energy is dissipated in the collision? (*b*) Assuming that C_v for the bob plus bullet is $3R$, calculate the temperature increase of the system due to the collision. Take the molecular mass of the system to be 200 g/mol.

20-46. The insulated cylinder shown in the accompanying figure is closed at both ends and contains an insulating piston that is free to move on frictionless bearings. The piston divides the chamber into two compartments containing gases A and B. Originally, each compartment has a volume of 5.0×10^{-2} m^3 and contains a monatomic ideal gas at a temperature of 0°C and a pressure of 1.0 atm. (*a*) How many moles of gas are in each compartment? (*b*) Heat Q is slowly added to A so that it expands and B is compressed until the pressure of both gases is 3.0 atm. Use the fact that the compression of B is adiabatic (why?) to determine the final volume of both gases. (*c*) What are their final temperatures? (*d*) What is the value of Q?

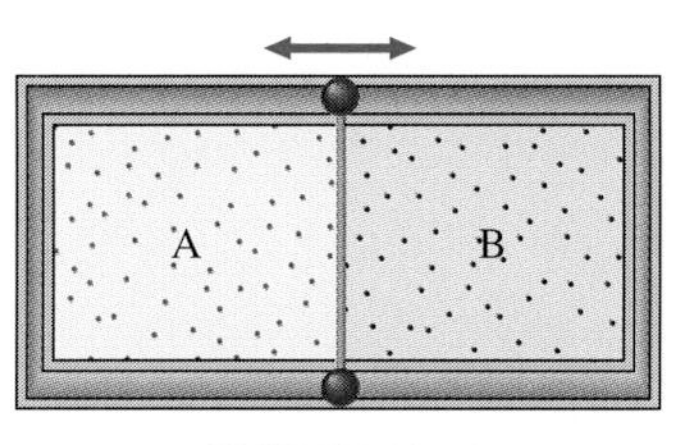

PROBLEM 20-46

20-47. A cast-iron ($\rho = 7.8 \times 10^3$ kg/m^3, $C_v = 0.47$ J/g·°C) flywheel of diameter 1.0 m and thickness 10.0 cm is spinning clockwise around a vertical axis at 2000 rev/min when it is dropped onto an identical flywheel rotating counterclockwise around the same axis at 2000 rev/min. What is the final angular velocity of the flywheels? What is their change in temperature when they reach this angular velocity? Assume that $I = MR^2/2$ for the flywheels and that they are at the same temperature. Ignore any heat exchanged with the environment during the coupling.

20-48. In a diesel engine, the fuel is ignited without a spark plug. Instead, air in a cylinder is compressed adiabatically to a temperature above the ignition temperature of the fuel; then at the point of maximum compression, the fuel is injected into the cylinder. Suppose that air at 20°C is taken into the cylinder at a volume V_1 and then compressed adiabatically and quasi-statically to a temperature of 600°C and a volume V_2. If $\gamma = 1.4$, what is the ratio V_1/V_2? (*Note:* In an operating diesel engine, the compression is *not* quasi-static.)

A heat engine.

CHAPTER 21 SECOND LAW OF THERMODYNAMICS

PREVIEW

This chapter is devoted to the second law of thermodynamics. The main topics to be discussed here are the following:

1. **Heat engines.** We discuss the properties of heat engines and refrigerators. The cyclic sequence of steps involved in the operation of a Carnot heat engine (and refrigerator) is described in detail.
2. **Second law of thermodynamics.** The Kelvin and Clausius formulations of the second law of thermodynamics are presented and explained.
3. **Second law of thermodynamics and the reversible heat engine.** The reversible engine is shown to be the most efficient engine operating between two given heat reservoirs.
4. **Absolute temperature scale.** We define the absolute temperature scale in terms of the Carnot cycle.
5. **Entropy.** The change in entropy for a system undergoing a reversible transition is defined mathematically.
6. **Entropy and the second law of thermodynamics.** The connection between entropy and the second law of thermodynamics is described. We then calculate the changes in entropy for systems that make irreversible transitions.

*7. **Entropy and disorder.** The physical significance of entropy is seen through its relationship to the disorder of a system.

According to the first law of thermodynamics, the only processes that can occur are those for which energy is conserved. But this is obviously not the only restriction imposed by nature, because there are many thermodynamic processes which, while they would conserve energy, do not occur. For example, when two bodies are in thermal contact, heat never flows from the cooler to the warmer body. The absence of such processes leads to the inevitable conclusion that there must be another thermodynamic law controlling the behavior of physical systems. This law is the *second law of thermodynamics*. The second law is basically a restriction on the usage of energy. Nature does not allow energy to be distributed among systems in an arbitrary manner. For example, we cannot transfer heat from a cold object to a hot object without doing work, nor can we unmix our cream and coffee without the help of a sophisticated chemical process. Also, we cannot use the internal energy stored in the air to propel a car, nor can we directly use the energy of ocean water to run a ship.

21-1 Heat Engines and the Carnot Cycle

The second law can be formally stated in several ways. One statement is based on the *heat engine*, a device that converts heat to work. *Whenever we consider heat engines (and refrigerators) in this chapter, we will not use the normal sign convention for heat and work*. For convenience, *we will assume that symbols such as Q_h, Q_c, and W represent the magnitudes of heat and work*. Whether energy is entering or leaving a system will be indicated by signs in front of the symbols and by the directions of arrows in diagrams. The heat engine is represented schematically in Fig. 21-1*a*. It absorbs heat Q_h from a high-temperature reservoir of Kelvin temperature T_h, uses some of that energy to produce useful work W, and discards the remaining energy as heat Q_c to a low-temperature reservoir of Kelvin temperature T_c. Power plants and internal combustion engines (such as those in cars) are examples of heat engines. Power plants use steam produced at high temperatures to drive electric generators while exhausting heat to the atmosphere or a nearby body of water. In an internal combustion engine, a hot gas-air mixture is used to push a piston, and heat is exhausted to the atmosphere.

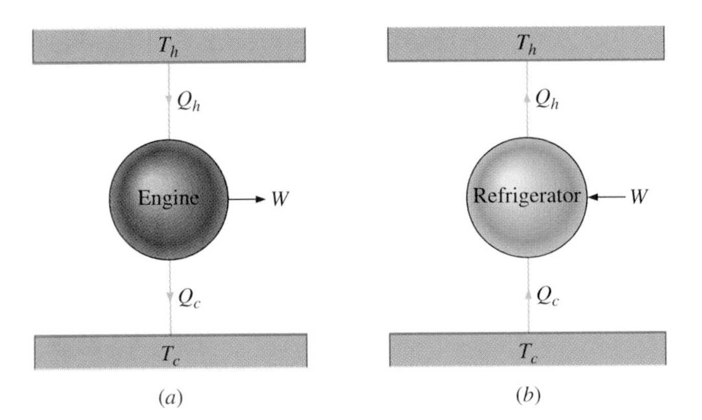

FIGURE 21-1 (*a*) Schematic representation of a reversible heat engine. (*b*) Schematic representation of a reversible refrigerator.

The heat exhausted from the nuclear power plant goes to the cooling tower, where it is released to the atmosphere.

Heat engines operate by carrying a *working substance* through a cycle. In a steam power plant, the working substance is water, which starts as a liquid, becomes vaporized, is then used to drive a turbine, and is finally condensed back to the liquid state. As is the case for all working substances in cyclic processes, once the water returns to its initial state, it repeats the same sequence.

For now, we assume that the cycles of heat engines are reversible. Then there is no energy loss to friction or other irreversible effects. Suppose that the engine of Fig. 21-1*a* goes through one complete cycle and that Q_h, Q_c, and W represent the heats exchanged and the work done for that cycle. Since the initial and final states of the system are the same, $\Delta U = 0$ for the cycle. We therefore have from the first law

$$W = Q - \Delta U = (Q_h - Q_c) - 0,$$

so

$$W = Q_h - Q_c. \tag{21-1}$$

A convenient parameter usually used to describe heat engines is the **thermal efficiency** ϵ. It is simply "what we get out" divided by "what we put in" during each cycle, as defined by

$$\epsilon = \frac{W_{\text{OUT}}}{Q_{\text{IN}}}.$$

With this engine, we get out W and put in Q_h, so

$$\epsilon = \frac{W}{Q_h}. \tag{21-2a}$$

We can write the efficiency in terms of just the exchanged heats by substituting Eq. (21-1) into Eq. (21-2a). We then find

$$\epsilon = 1 - \frac{Q_c}{Q_h}. \tag{21-2b}$$

Since the cycle is reversible, its sequence of steps can just as easily be performed in the opposite direction. In this case, the engine is known as a *refrigerator*. As illustrated in Fig. 21-1b, a refrigerator absorbs heat Q_c from a low-temperature reservoir of Kelvin temperature T_c and discards heat Q_h to a high-temperature reservoir of Kelvin temperature T_h, while work W is done on the working substance. A household refrigerator removes heat from the food within it while exhausting heat to the surrounding air. The required work, which we pay for in our electricity bill, is performed by the motor that moves a coolant through the coils.

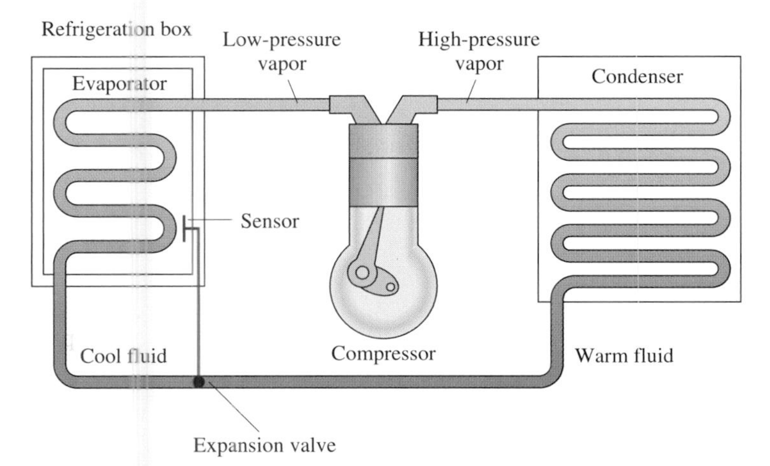

A schematic diagram of a refrigerator. A coolant with a boiling temperature below the freezing point of water is sent through the cycle shown. The coolant extracts heat from the refrigerator at the evaporator, causing coolant to vaporize. It is then compressed and sent through the condenser, where it exhausts heat to the outside.

The effectiveness or **coefficient of performance** κ of a refrigerator is measured by the heat removed from the low-temperature reservoir divided by the work done on the working substance per cycle:

$$\kappa = \frac{Q_c}{W}, \tag{21-3a}$$

which, from the first law, can be written as

$$\kappa = \frac{Q_c}{Q_h - Q_c}. \tag{21-3b}$$

There are many types of cyclic processes that heat engines can undergo. Two very important practical examples are the cycles for the automobile engine (Prob. 21-57) and for the diesel engine (Prob. 21-58). For now, however, we consider the two simple cycles depicted in Fig. 21-2. In each case, the working substance is assumed to be n moles of an ideal gas. The molar heat capacity at constant volume of the gas is C_v.

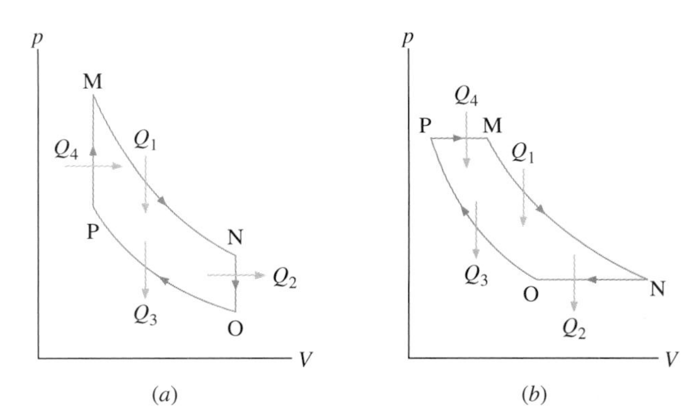

FIGURE 21-2 (a) A reversible cycle (MNOPM) consisting of an isothermal expansion, a decrease in pressure at constant volume, an isothermal compression, and an increase in pressure at constant volume. (b) A reversible cycle (MNOPM) consisting of an isothermal expansion, an isobaric compression, an isothermal compression, and an isobaric expansion. In each case, the work done per cycle is equal to the area enclosed by the path.

The cycle of Fig. 21-2a is composed of the following steps: (1) an isothermal expansion of the gas from M to N during which heat Q_1 is absorbed from a heat reservoir of temperature T_h; (2) a decrease in pressure from N to O at the constant volume $V_N = V_O$; (3) an isothermal compression from O to P during which heat Q_3 is exhausted to a heat reservoir of temperature T_c; (4) an increase in pressure from P to M at the constant volume $V_P = V_M$. For the two isothermal processes (steps 1 and 3), $\Delta U = 0$, so $Q = W$ and the heat exchanged by the ideal gas is given by Eq. (20-2). We therefore have

$$Q_1 = nRT_h \ln \frac{V_N}{V_M} \quad \text{and} \quad Q_3 = nRT_c \ln \frac{V_O}{V_P}.$$

For the reversible decrease in pressure at the constant volume V_N (step 2), the ideal gas must be placed in contact with a series of heat reservoirs whose temperatures decrease in order from T_h to T_c. During this process, the gas gives up heat

$$Q_2 = nC_v(T_h - T_c).$$

Similarly, during the reversible increase in pressure (step 4) the ideal gas absorbs heat

$$Q_4 = nC_v(T_h - T_c)$$

from the same series of reservoirs.

Since $Q_2 = Q_4$, there is no *net* heat absorbed during the two constant-volume transitions of the cycle. Thus this system is equivalent to the two-reservoir system of Fig. 21-1a, with $Q_1 = Q_h$ and $Q_3 = Q_c$, and its efficiency is, from Eq. (21-2b),

$$\epsilon = 1 - \frac{Q_3}{Q_1}.$$

Now $V_N = V_O$ and $V_M = V_P$, so

$$\frac{Q_3}{Q_1} = \frac{nRT_c \ln V_O/V_P}{nRT_h \ln V_N/V_M} = \frac{T_c}{T_h},$$

and the efficiency of this reversible engine is given in terms of the temperatures of the hot and cold reservoirs by

$$\epsilon = 1 - \frac{T_c}{T_h}.$$

The cycle of Fig. 21-2*b* is composed of two isothermal processes and two isobaric changes in volume between the temperatures T_h and T_c. The isothermal processes are the same as those of Fig. 21-2*a*, while for the two isobaric transitions,

$$Q_2 = nC_p(T_h - T_c) \quad \text{and} \quad Q_4 = nC_p(T_h - T_c),$$

where C_p is the molar heat capacity at constant pressure of the gas. As in the previous case, $Q_2 = Q_4$, so this cycle is also equivalent to that of Fig. 21-1*a*, and

$$\epsilon = 1 - \frac{Q_3}{Q_1}.$$

Since the transitions from M to N and from O to P are isothermal,

$$p_M V_M = p_N V_N \quad \text{and} \quad p_O V_O = p_P V_P,$$

so

$$Q_1 = nRT_h \ln \frac{V_N}{V_M} = nRT_h \ln \frac{p_M}{p_N}$$

and

$$Q_3 = nRT_c \ln \frac{V_O}{V_P} = nRT_c \ln \frac{p_P}{p_O}.$$

Furthermore, the transitions from N to O and from P to M are isobaric, so $p_O = p_N$ and $p_M = p_P$. Using these relationships in the expressions for Q_1 and Q_3, we find that $Q_3/Q_1 = T_c/T_h$, and again the efficiency is given by

$$\epsilon = 1 - \frac{T_c}{T_h}.$$

In the early 1820s, Sadi Carnot (1786–1832), a French engineer, became interested in improving the efficiencies of practical heat engines. His studies led him to propose in 1824 a hypothetical working cycle called the *Carnot cycle*. An engine operating in this cycle is called a *Carnot engine*. The Carnot cycle is of special importance for a variety of reasons. At a practical level, this cycle represents a reversible model for both the steam power plant and the refrigerator. Yet it is also very important theoretically, for it plays a major role in the development of the entropy statement of the second law of thermodynamics. Finally, because only two reservoirs are involved in its operation, it can be used along with the second law of thermodynamics to define an absolute temperature scale that is truly independent of any substance used in the temperature measurement. With an ideal gas as the working substance, the steps of the Carnot cycle, as represented by the sketches of Fig. 21-3, are the following:

1. *Isothermal expansion*. The gas is placed in thermal contact with a heat reservoir at a temperature T_h. It absorbs heat Q_h from the reservoir and is allowed to expand isothermally,

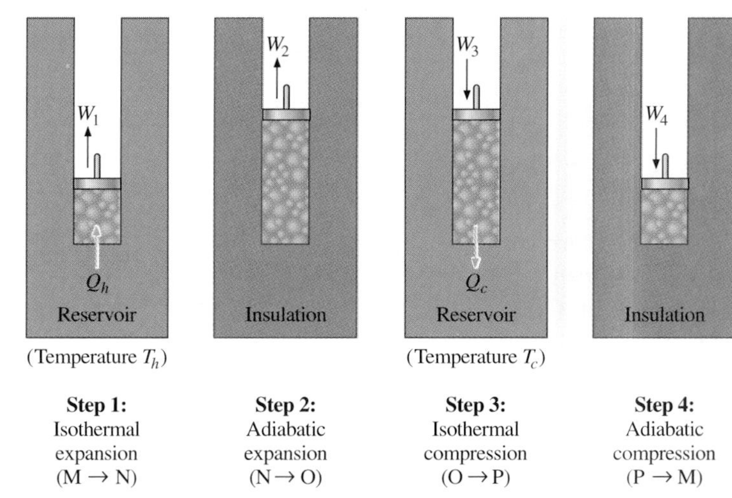

FIGURE 21-3 The four steps of the Carnot cycle. The working substance is an ideal gas whose thermodynamic path is represented in Fig. 21-4.

doing work W_1. Since the internal energy U of an ideal gas is a function only of temperature, $\Delta U = 0$ in this step. With the first law and Eq. (20-2), we find that the gas absorbs heat given by

$$Q_h = W_1 = nRT_h \ln \frac{V_N}{V_M}.$$

2. *Adiabatic expansion*. The gas is thermally isolated and allowed to expand further, doing work W_2. Since this expansion is adiabatic, the temperature of the gas falls—in this case, from T_h to T_c. From Eq. (20-10*c*), the temperature and volume of the gas are related during this step by

$$TV^{\gamma-1} = \text{constant},$$

so

$$T_h V_N^{\gamma-1} = T_c V_O^{\gamma-1}.$$

3. *Isothermal compression*. The gas is placed in thermal contact with a heat reservoir at a temperature T_c and compressed isothermally. During this process, work W_3 is done on the gas and it gives up heat Q_c to the reservoir. The reasoning used in step 1 now yields

$$Q_c = nRT_c \ln \frac{V_O}{V_P},$$

where Q_c is the heat rejected to the reservoir.

4. *Adiabatic compression*. The gas is thermally isolated and returned to its initial state by compression. In this process, work W_4 is done on the gas. Since the compression is adiabatic, the temperature of the gas rises—from T_c to T_h in this particular case. The reasoning of step 2 now gives

$$T_c V_P^{\gamma-1} = T_h V_M^{\gamma-1}.$$

The total work done by the gas in the reversible cycle is

$$W = W_1 + W_2 - W_3 - W_4.$$

This work is equal to the area enclosed by the path in the pV diagram of Fig. 21-4. Since the initial and final states of the system are the same, $\Delta U = 0$ for the cycle, and the first law of thermodynamics gives

$$W = Q - \Delta U = (Q_h - Q_c) - 0,$$

and

$$W = Q_h - Q_c.$$

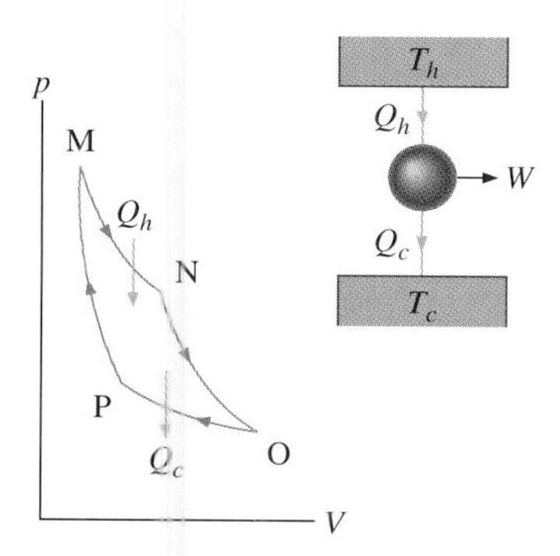

FIGURE 21-4 The work done by the gas in one Carnot cycle is given by the area enclosed by the path MNOPM.

To find the efficiency of this engine, we first divide Q_c by Q_h:

$$\frac{Q_c}{Q_h} = \frac{T_c}{T_h}\frac{\ln V_O/V_P}{\ln V_N/V_M}.$$

When the equation of step 2 is divided by that of step 4, we find

$$\frac{V_O}{V_P} = \frac{V_N}{V_M}.$$

Substituting this into the equation for Q_c/Q_h, we obtain

$$\frac{Q_c}{Q_h} = \frac{T_c}{T_h}. \tag{21-4}$$

Finally, with Eq. (21-2b) we find that the efficiency of this ideal-gas Carnot engine is also given by

$$\epsilon = 1 - \frac{T_c}{T_h}. \tag{21-5}$$

We have just analyzed three reversible engines, all of which have an efficiency given by Eq. (21-5). Also, the *net* effect of all three of these engines is the absorption of heat from a high-temperature reservoir, the production of work, and the discarding of heat to a low-temperature reservoir. This leads us to ask: Do all reversible cycles operating between the same two reservoirs have the same efficiency? The answer to this question comes from the second law of thermodynamics. You'll soon see that *all reversible cycles do indeed have the same efficiency.* Also, as you might expect, all real engines operating between two reservoirs are less efficient than reversible engines operating between the same two reservoirs. This too is a consequence of the second law of thermodynamics.

The cycle of an ideal-gas Carnot refrigerator is represented by the pV diagram of Fig. 21-5. It is a Carnot engine operating in reverse. The refrigerator extracts heat Q_c from a cold-temperature reservoir at T_c when the ideal gas expands isothermally. The gas is then compressed adiabatically until its temperature reaches T_h, after which an isothermal compression of the gas results in heat Q_h being discarded to a high-temperature reservoir at T_h. Finally, the cycle is completed by an adiabatic expansion of the gas, causing its temperature to drop to T_c.

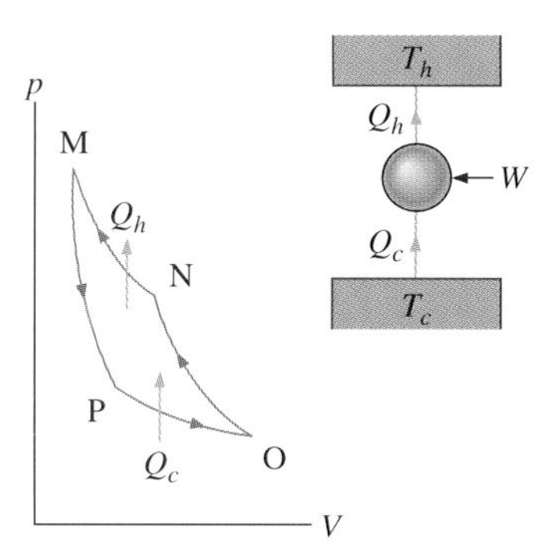

FIGURE 21-5 The work done on the gas in one cycle of the Carnot refrigerator is given by the area enclosed by the path MPONM.

The work done on the ideal gas is equal to the area enclosed by the path of the pV diagram. From the first law, this work is given by

$$W = Q_h - Q_c.$$

An analysis just like the analysis done for the Carnot engine gives

$$\frac{Q_c}{T_c} = \frac{Q_h}{T_h},$$

which, when combined with Eq. (21-3b), yields

$$\kappa = \frac{T_c}{T_h - T_c} \tag{21-6}$$

for the coefficient of performance of the ideal-gas Carnot refrigerator.

Equations representing the efficiency of a Carnot engine and the coefficient of performance of a Carnot refrigerator have just been found assuming an ideal gas for the working substance in both devices. However, these equations are more general than their derivations imply. It will be shown shortly that they are both valid no matter what the working substance is.

EXAMPLE 21-1 THE CARNOT ENGINE

A Carnot engine has an efficiency of 0.60, and the temperature of its cold reservoir is 300 K. (*a*) What is the temperature of the hot reservoir? (*b*) If the engine does 300 J of work per cycle, how much heat is removed from the high-temperature reservoir per cycle? (*c*) How much heat is exhausted to the low-temperature reservoir per cycle?

SOLUTION (*a*) From Eq. (21-5) we have

$$0.60 = 1 - \frac{300\text{ K}}{T_h},$$

so the temperature of the hot reservoir is

$$T_h = 750\text{ K}.$$

(*b*) By definition, the efficiency of the engine is $\epsilon = W/Q_h$, so the heat removed from the high-temperature reservoir per cycle is

$$Q_h = \frac{W}{\epsilon} = \frac{300 \text{ J}}{0.60} = 500 \text{ J}.$$

(*c*) Now from the first law, the heat exhausted to the low-temperature reservoir per cycle by the engine is

$$Q_c = Q_h - W = 500 \text{ J} - 300 \text{ J} = 200 \text{ J}.$$

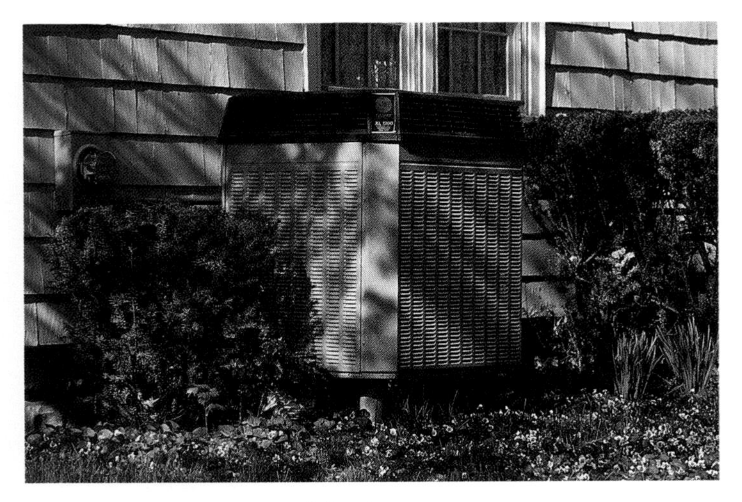

A heat pump.

EXAMPLE 21-2 A Heat Pump

A device that extracts heat from the ground outside a house and then delivers heat to the inside of the house is called a *heat pump*. This transfer of heat is accomplished by circulating a fluid (the working substance) between the exterior and the interior of the house. If the heat pump is a Carnot refrigerator operating between an outside temperature of 0°C and an inside temperature of 20°C, what is the ratio of the heat Q_h delivered to the inside of the house to the work W done by the pump per cycle?

SOLUTION We use Fig. 21-5 to represent our Carnot refrigerator. Applying the first law of thermodynamics to one cycle of the refrigerator, we obtain

$$\Delta U = 0 = Q_c - Q_h + W. \qquad (i)$$

Since the refrigerator is operating in a Carnot cycle, we have, from Eq. (21-4),

$$\frac{Q_h}{T_h} = \frac{Q_c}{T_c}. \qquad (ii)$$

We can compare Q_h to W by solving Eq. (*ii*) for Q_c in terms of Q_h and then substituting this into Eq. (*i*). This leaves

$$0 = \left(\frac{T_c}{T_h} - 1\right) Q_h + W,$$

which simplifies to

$$\frac{Q_h}{W} = \frac{T_h}{T_h - T_c}. \qquad (iii)$$

For $T_c = 273$ K and $T_h = 293$ K, this ratio is 14.7; that is, for every unit of work done on the circulating fluid, 14.7 units of heat are delivered to the interior of the house. At these temperatures, a typical commercial heat pump used for heating buildings does not do nearly as well as this Carnot engine; it furnishes only about four units of heat for every unit of work. The work represents the electric energy we must buy in order to operate the heat pump.

In terms of energy costs, the heat pump is a very economical means for heating buildings. Contrast this method with turning electric energy directly into heat with resistive heating elements. In this case, one unit of electric energy furnishes at most only one unit of heat. Unfortunately, there are problems with heat pumps that do limit their usefulness. They are quite expensive, and, as Eq. (*iii*) shows for a Carnot heat pump, they become less effective as the outside temperature decreases. In fact, below about −10°C the heat they furnish is less than the energy used to operate them.

DRILL PROBLEM 21-1

A Carnot engine operates between reservoirs at 400 and 30°C. (*a*) What is its efficiency? (*b*) If the engine does 5.0 J of work per cycle, how much heat per cycle does it absorb from the high-temperature reservoir? (*c*) How much heat per cycle does it exhaust to the cold-temperature reservoir?
ANS. (*a*) 0.55; (*b*) 9.1 J; (*c*) 4.1 J.

DRILL PROBLEM 21-2

A Carnot refrigerator operates between two heat reservoirs whose temperatures are 0 and 25°C. (*a*) What is the coefficient of performance of the refrigerator? (*b*) If 200 J of work are done on the working substance per cycle, how much heat per cycle is extracted from the cold reservoir? (*c*) How much heat per cycle is discarded to the hot reservoir?
ANS. (*a*) 10.9; (*b*) 2.18×10^3 J; (*c*) 2.38×10^3 J.

21-2 Second Law of Thermodynamics

In terms of heat engines, the second law of thermodynamics may be stated as follows:

> It is impossible to construct a heat engine that, when operating in a cycle, completely converts heat into work.

This is known as *Kelvin's formulation of the second law*. Such an unattainable "perfect engine" is represented schematically in Fig. 21-6*a*. Now an engine can absorb heat and turn it all into work, *but not if it completes a cycle*. For example, a chamber of gas could absorb heat from a heat reservoir and do work isothermally against a piston as it expanded. However, if the gas were to be returned to its initial state (that is, made to complete a cycle), it would have to be compressed and heat would have to be extracted from it.

Kelvin's statement is a manifestation of a well-known engineering problem. Despite advancing technology, we are not able to build a heat engine that is 100 percent efficient. Even the reversible engines discussed in the previous section exhaust heat. The first law does not exclude the possibility of constructing a perfect engine; it is the second law that forbids it.

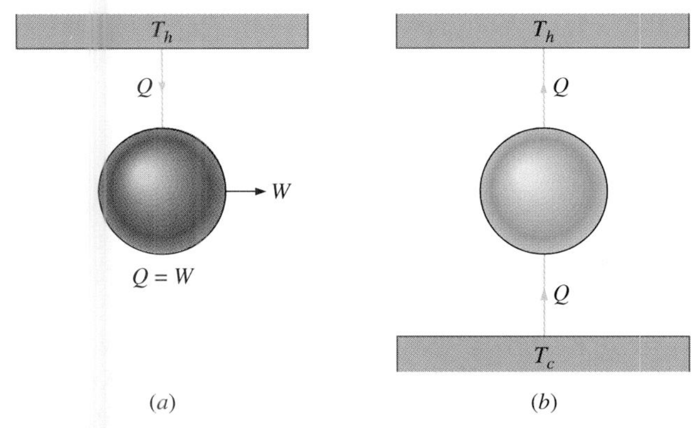

FIGURE 21-6 (*a*) A "perfect heat engine." (*b*) A "perfect refrigerator." Neither of these engines is achievable.

A statement equivalent to Kelvin's was made by Clausius:

> It is impossible to construct a refrigerator that transfers heat from a cold reservoir to a hot reservoir without aid from an external source.

Clausius' formulation is related to the everyday observation that heat never flows spontaneously from a cold object to a hot object. *Heat transfer in the direction of increasing temperature always requires some energy input.* A "perfect refrigerator" (Fig. 21-6*b*), which works without such external aid, is impossible to construct.

To prove the equivalence of the Kelvin and Clausius statements, we show that if one statement is false, it necessarily follows that the other statement is also false. Let's first assume that the Clausius formulation is false, so that the perfect refrigerator of Fig. 21-6*b* does exist. This refrigerator removes heat Q from a cold reservoir at a temperature T_c and transfers all of it to a hot reservoir at a temperature T_h. Now consider a real heat engine working in the same temperature range. It extracts heat $Q + \Delta Q$ from the hot reservoir, does work W, and discards heat Q to the cold reservoir. From the first law, these quantities are related by

$$W = (Q + \Delta Q) - Q = \Delta Q.$$

Suppose these two engines are combined as shown in Fig. 21-7. The net heat removed from the hot reservoir is ΔQ, there is no net heat transfer to or from the cold reservoir, and work W is done on some external body. Since $W = \Delta Q$, the combination of a perfect refrigerator and a real heat engine is itself a perfect heat engine, thereby contradicting Kelvin's statement. Thus if the Clausius statement is false, the Kelvin statement must also be false. To complete the proof of the equivalence of the two statements, we must show that the reverse holds as well; that is, if the Kelvin statement is false, the Clausius statement must also be false. This proof is an exercise for you in the following Drill Problem.

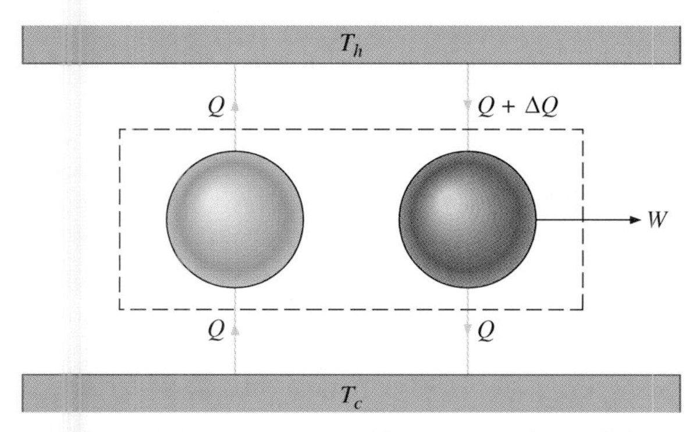

FIGURE 21-7 Combining a perfect refrigerator and a real heat engine yields a perfect heat engine since $W = \Delta Q$.

DRILL PROBLEM 21-3

Show that if the Kelvin statement is false, it follows that the Clausius statement is false.

21-3 Second Law of Thermodynamics and the Reversible Heat Engine

Using the second law of thermodynamics, we now prove two important properties of heat engines operating between two heat reservoirs. The first property is that *any reversible engine operating between two reservoirs has a greater efficiency than any irreversible engine operating between the same two reservoirs.* Other reservoirs can be involved in the reversible cycles, as they were for the two cycles of Fig. 21-2 in which a series of heat reservoirs were used. However, there must be no *net* heat exchanged between these other reservoirs and the engine. Only the high-temperature and the low-temperature reservoirs are allowed a net exchange of heat with the reversible engine. The second property to be demonstrated is that *all reversible engines operating between the same two reservoirs have the same efficiency.*

We start with the two engines D and E of Fig. 21-8*a*, which are operating between two common heat reservoirs at temperatures T_h and T_c. First we assume that D is a reversible engine and that E is a hypothetical irreversible engine that has a greater efficiency than D. If both engines perform the same amount of work W per cycle, it follows from Eq. (21-2*a*) that $Q_h > Q_h'$. It then follows from the first law that $Q_c > Q_c'$.

Suppose the cycle of D is reversed so that it operates as a refrigerator, and the two engines are coupled such that the work output of E is used to drive D. (See Fig. 21-8*b*.) Since $Q_h > Q_h'$ and $Q_c > Q_c'$, the net result of each cycle is equivalent to a

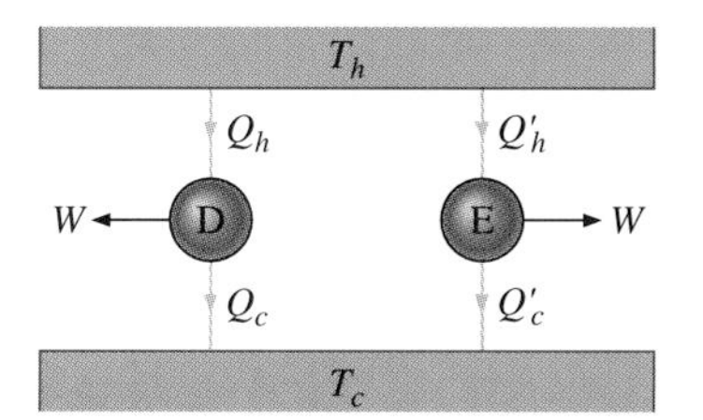

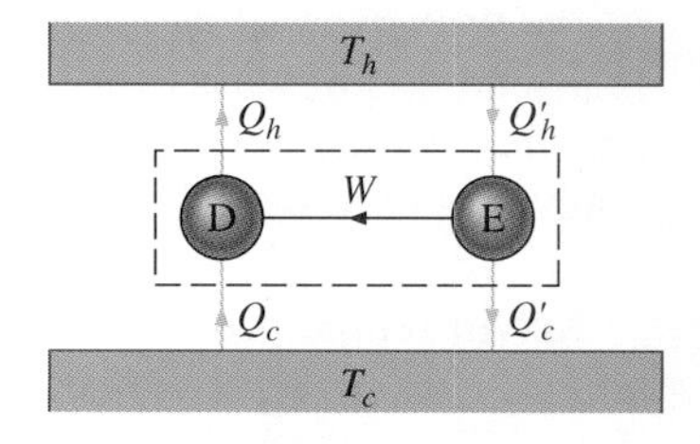

FIGURE 21-8 (*a*) Two uncoupled engines D and E working between the same reservoirs. (*b*) The coupled engines, with D working in reverse.

spontaneous transfer of heat from the cold reservoir to the hot reservoir, a process the second law does not allow. The original assumption must therefore be wrong, and it *is* impossible to construct an irreversible engine such as E that is more efficient than the reversible engine D.

Now it's quite easy to demonstrate that the efficiencies of all reversible engines operating between the same reservoirs are equal. Suppose that D and E are both reversible engines. If they are coupled as shown in Fig. 21-8*b*, the efficiency of E cannot be greater than the efficiency of D in order that the second law not be violated. If both engines are then reversed, the same reasoning implies that the efficiency of D cannot be greater than the efficiency of E. Combining these results leads to the conclusion that all reversible engines working between the same two reservoirs have the same efficiency. This includes the three reversible engines whose efficiencies were calculated in Sec. 21-1. Thus all reversible engines operating between two reservoirs at temperatures T_h and T_c have an efficiency given by Eq. (21-5). In addition, all such reversible engines satisfy Eq. (21-4). This important property can be easily proved by combining Eqs. (21-2*b*) and (21-5).

DRILL PROBLEM 21-4

What is the efficiency of a perfect heat engine? What is the coefficient of performance of a perfect refrigerator?
ANS. 1.0; ∞.

DRILL PROBLEM 21-5

Show that $Q_h - Q_h' = Q_c - Q_c'$ for the hypothetical engine of Fig. 21-8*b*.

21-4 Absolute Temperature Scale

All reversible engines operating between the same two heat reservoirs have the same efficiency $\epsilon = 1 - (Q_c/Q_h)$, where Q_c and Q_h are the magnitudes of the heat exchanged with the cold and the hot reservoirs, respectively. If we had a number of Carnot engines operating with different working substances between the same two reservoirs, the efficiencies of all these engines would be the same; that is, the efficiency of a Carnot engine is independent of the working substance. A temperature scale based on the Carnot engine is therefore *independent of the properties of any particular substance*. For this reason, this scale is known as the *absolute temperature scale* or the *thermodynamic temperature scale*. Because this scale was originally proposed by Kelvin, it is also called the *Kelvin scale*. (See Chap. 18.)

Kelvin temperatures are defined in terms of the heats Q_c and Q_h exchanged at the heat reservoirs of a Carnot engine. If T_c and T_h are the Kelvin temperatures of the cold and hot reservoirs, then according to Kelvin, they must satisfy

$$\frac{T_c}{T_h} = \frac{Q_c}{Q_h}.$$

By substituting this equation into $\epsilon = 1 - (Q_c/Q_h)$, we obtain for the Carnot engine efficiency

$$\epsilon = 1 - \frac{T_c}{T_h}.$$

This equation is identical to Eq. (21-5)—with one important difference. The reservoir temperatures here are expressed on the Kelvin scale, while in Eq. (21-5) they are expressed on the ideal-gas scale. To complete the definition of the Kelvin scale, the triple point of water is taken to be exactly 273.16 K, which is the same as its value on the ideal-gas scale. *Absolute (Kelvin) temperature and ideal-gas temperature are therefore identical.* Throughout the last three chapters, we've used these temperature scales interchangeably—we now have our justification.

21-5 Entropy

The second law is often expressed in terms of a *change* in the thermodynamic variable known as **entropy**, which is represented by the symbol S. Entropy, like internal energy, is a state function. This means that when a system makes a transition from one state to another, the change in entropy ΔS is independent of path and depends only on the thermodynamic variables of the two states.

We first consider ΔS for a system undergoing a *reversible process at a constant temperature*. In this case, the change in entropy of the system is

$$\Delta S = \frac{Q}{T}, \qquad (21\text{-}7)$$

where Q is the heat exchanged by the system kept at a temperature T (in kelvins). If the system absorbs heat ($Q > 0$*), its entropy increases; if heat is lost during the process ($Q < 0$), the entropy of the system decreases. As an example, suppose a gas is kept at a constant temperature of 300 K while it absorbs 10 J of heat in a reversible process. Then, from $\Delta S = Q/T$, the entropy change of the gas is

$$\Delta S = \frac{10 \text{ J}}{300 \text{ K}} = 0.033 \text{ J/K}.$$

Similarly, if the gas loses 5.0 J of heat ($Q = -5.0$ J) at $T =$ 200 K,

$$\Delta S = \frac{-5.0 \text{ J}}{200 \text{ K}} = -0.025 \text{ J/K}.$$

EXAMPLE 21-3 Entropy Change of Melting Ice

Heat is slowly added to a 50-g chunk of ice at 0°C until it completely melts to water at the same temperature. What is the entropy change of the ice?

*Since we are no longer discussing heat engines, we return to the normal sign convention of Sec. 20-3.

SOLUTION The ice is melted by adding an amount of heat

$$Q = mL_f = (50\text{ g})(335\text{ J/g}) = 1.68 \times 10^4\text{ J}.$$

In this reversible process, the temperature of the ice-water mixture is 273 K. Now from $\Delta S = Q/T$, the entropy change of the ice is

$$\Delta S = \frac{1.68 \times 10^4\text{ J}}{273\text{ K}} = 61.5\text{ J/K}$$

when it melts to water at 0°C.

The change in entropy for an arbitrary reversible transition for which the temperature is not necessarily constant is defined by modifying $\Delta S = Q/T$. Imagine a system making a transition from state A to state B in small, discrete steps. The temperatures associated with these states are T_A and T_B, respectively. During each step of the transition, the system exchanges heat ΔQ_i reversibly at a temperature T_i. This can be accomplished experimentally by placing the system in thermal contact with a large number of heat reservoirs of varying temperature T_i, as illustrated in Fig. 20-2. The change in entropy for each step is $\Delta S_i = \Delta Q_i/T_i$. The net change in entropy for the transition is then

$$S_B - S_A = \sum_i \Delta S_i = \sum_i \frac{\Delta Q_i}{T_i}.$$

We now take the limit as $\Delta Q_i \to 0$ and the number of steps approaches infinity. Then, replacing the summation by an integral, we obtain

$$S_B - S_A = \int_A^B \frac{đQ}{T}, \tag{21-8}$$

where the integral is taken between the initial state A and the final state B. This equation is *valid only if the transition from A to B is reversible.*

As an example, let's determine the net entropy change of a reversible engine while it undergoes a single Carnot cycle. In the adiabatic steps 2 and 4 of the cycle shown in Fig. 21-3, no heat exchange takes place, so $\Delta S_2 = \Delta S_4 = \int đQ/T = 0$. In step 1, the engine absorbs heat Q_h at a temperature T_h, so its entropy change is $\Delta S_1 = Q_h/T_h$. Similarly, in step 3, $\Delta S_3 = -Q_c/T_c$. The net entropy change of the engine in one cycle of operation is then

$$\Delta S_E = \Delta S_1 + \Delta S_2 + \Delta S_3 + \Delta S_4 = \frac{Q_h}{T_h} - \frac{Q_c}{T_c}.$$

However, from Eq. (21-4),

$$\frac{Q_h}{T_h} = \frac{Q_c}{T_c},$$

so

$$\Delta S_E = 0.$$

There is no net change in the entropy of the Carnot engine over a complete cycle. Although this result was obtained for a particular case, its validity can be shown to be far more general: *There is no net change in the entropy of a system undergoing any complete reversible cyclic process.* Mathematically, we write this statement as

$$\oint dS = \oint \frac{đQ}{T} = 0, \tag{21-9}$$

where $\oint$ represents the integral over a *closed reversible path.*

We can use Eq. (21-9) to show that the entropy change of a system undergoing a reversible process between two given states is path-independent. An arbitrary closed path representing a reversible process that passes through the states A and B is shown in Fig. 21-9. From Eq. (21-9), $\oint dS = 0$ for this path. We may split this integral into two segments, one along I, which leads from A to B, the other along II, which leads from B to A; then

$$\left[\int_A^B dS\right]_I + \left[\int_B^A dS\right]_{II} = 0.$$

Since the process is reversible, $[\int_B^A dS]_{II} = -[\int_A^B dS]_{II}$, so

$$\left[\int_A^B dS\right]_I = \left[\int_A^B dS\right]_{II}.$$

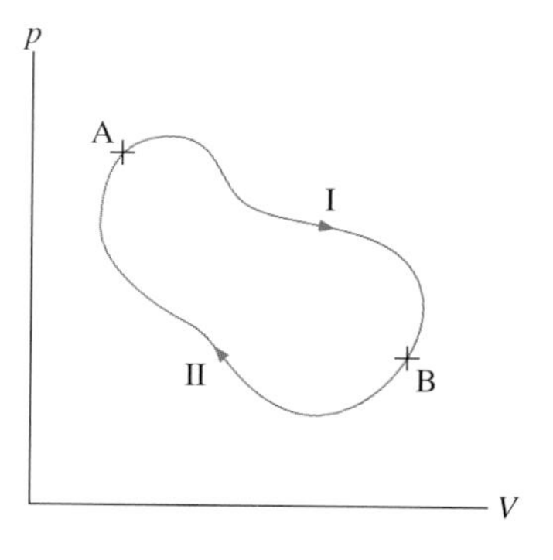

FIGURE 21-9 The closed path passing through states A and B represents a reversible process.

Hence the entropy change in going from A to B is the same for paths I and II. Since I and II are arbitrary reversible paths, *the entropy change in a transition between two equilibrium states is the same for all reversible processes joining these states. Entropy, like internal energy, is therefore a state function.*

21-6 Entropy and the Second Law of Thermodynamics

Most of the processes that occur naturally are irreversible. Examples we have considered include the free expansion of a gas and the conduction of heat along a metal rod. During these processes, the values of thermodynamic variables such as temperature, pressure, and volume may not even be well defined.

To determine the change in entropy for a system that undergoes an irreversible process (assuming it starts and ends in an equilibrium state), we use the fact that entropy is a state function. This means that the entropy of a system in a particular state depends on the thermodynamic variables that define that state. Consequently, it doesn't matter whether the system makes a transition between two equilibrium states by a reversible or an irreversible process—its entropy change will always be the same. So, in order to calculate the change in entropy of a system when it makes an irreversible transition between equilibrium states A and B, we simply *use a reversible path* that also takes the system from A to B. The only necessary similarity between the path representing the actual process and the substituted reversible path is that they begin and end at the same equilibrium states. Other than that, they can be (and most often are) entirely different processes. The following examples illustrate this point.

EXAMPLE 21-4 ENTROPY CHANGE FOR AN ISOBARIC PROCESS

Determine the entropy change of an object of mass m and specific heat c that is cooled rapidly (and irreversibly) at constant pressure from T_h to T_c.

SOLUTION To replace this rapid cooling with a process that proceeds reversibly, we imagine that the hot object is put in thermal contact with successively cooler heat reservoirs whose temperatures range from T_h to T_c. Throughout the substitute transition, the object loses heat reversibly, so Eq. (21-8) can be used to calculate its entropy change:

$$\Delta S = \int_{T_h}^{T_c} \frac{đQ}{T}. \qquad (i)$$

From Eq. (19-3), an infinitesimal heat exchange $đQ$ is related to the temperature change dT of the object by

$$đQ = mc\,dT. \qquad (ii)$$

Substituting Eq. (*ii*) into Eq. (*i*), we obtain for the entropy change of the object as it is cooled at constant pressure from T_h to T_c:

$$\Delta S = \int_{T_h}^{T_c} \frac{mc\,dT}{T} = mc \ln \frac{T_c}{T_h}. \qquad (21\text{-}10)$$

EXAMPLE 21-5 ENTROPY AND A FALLING ROCK

An 80-kg rock falls from a cliff into a lake 40 m below. The lake is 10 m deep, and both the rock and lake are at 20°C. What is the change in entropy of the rock-lake system once thermal equilibrium has been reached with the rock at the bottom of the lake? Assume that the velocity of the rock is negligible when it reaches the bottom of the lake.

SOLUTION From Eq. (7-21), the work done by the nonconservative forces on an object moving between two points equals the change in the sum of the kinetic energy $\mathcal{K}$ and the potential energy $\mathcal{U}$ of the object; that is,

$$W_{\text{ncon}} = (\mathcal{K}_B + \mathcal{U}_B) - (\mathcal{K}_A + \mathcal{U}_A). \qquad (i)$$

In this example, W_{ncon} is the work done on the rock by the lake, $\mathcal{K}_A = \mathcal{K}_B = 0$ since the rock starts and ends with zero velocity, and finally, $\mathcal{U}_B - \mathcal{U}_A = -mgh = -(80\text{ kg})(9.8\text{ m/s}^2)(50\text{ m}) = -3.92 \times 10^4$ J. We then find from Eq. (*i*) that

$$W_{\text{ncon}} = -mgh = -3.92 \times 10^4\text{ J}.$$

This work results in an increase of 3.92×10^4 J in the rock-lake system's internal energy. Since the lake is essentially a heat reservoir, this energy change has negligible effect on the temperature of both the rock or lake once thermal equilibrium has been reached. To calculate this system's change in entropy, we have to replace the fall of the rock by a reversible process that deposits 3.92×10^4 J of heat in the system and gets the rock to the bottom of the lake. One such process is a slow lowering of the rock to the lake bottom, followed by the transfer of 3.92×10^4 J of heat from another heat reservoir at 20°C to the lake (plus its rock) at 20°C. The entropy change of the system is then

$$\Delta S = \frac{Q}{T} = \frac{3.92 \times 10^4\text{ J}}{293\text{ K}} = 134\text{ J/K}.$$

EXAMPLE 21-6 ENTROPY CHANGE IN THE ADIABATIC FREE EXPANSION OF AN IDEAL GAS

An ideal gas occupies a partitioned volume V_1 inside a box whose walls are thermally insulating. (See Fig. 21-10*a*.) When the partition is removed, the gas expands and fills the entire volume V_2 of the box, as shown in Fig. 21-10*b*. What is the entropy change of the *universe* (the *system plus its environment*)?

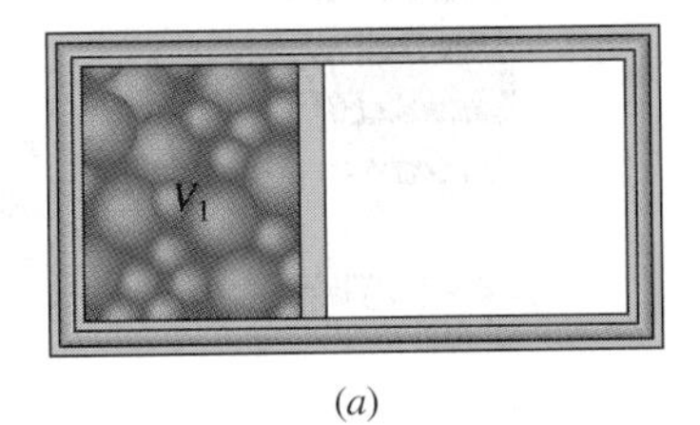

(*a*)

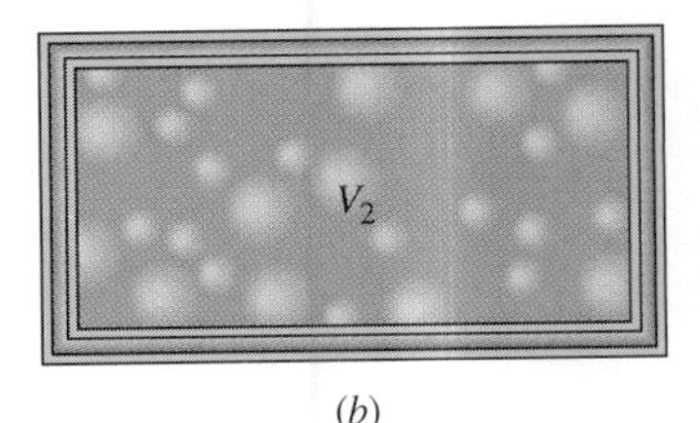

(*b*)

FIGURE 21-10 The adiabatic free expansion of an ideal gas from volume V_1 to volume V_2.

SOLUTION The adiabatic free expansion of an ideal gas is an irreversible process. In Sec. 20-5, we saw that there is no change in the internal energy (and hence temperature) of the gas in such an expansion. Thus a convenient reversible path connecting the same two equilibrium states is a slow, isothermal expansion from V_1 to V_2. In this process, the gas could be expanding against a piston while in thermal contact with a heat reservoir, as in step 1 of the Carnot cycle. Since the temperature is constant, the entropy change is given by $\Delta S = Q/T$, where

$$Q = W = \int_{V_1}^{V_2} p\,dV$$

because $\Delta U = 0$. Now, with the help of the ideal-gas law, we have

$$Q = nRT \int_{V_1}^{V_2} \frac{dV}{V} = nRT \ln \frac{V_2}{V_1},$$

so the change in entropy of the gas is

$$\Delta S = \frac{Q}{T} = nR \ln \frac{V_2}{V_1}.$$

Because $V_2 > V_1$, ΔS is positive and the entropy of the gas increases.

What about the environment? The walls of the container are thermally insulating, and no heat exchange takes place between the gas and its surroundings. The entropy of the environment is therefore constant during the expansion. The net entropy change of the universe is then simply the entropy change of the gas. Since this is positive, the entropy of the universe increases in the free expansion of the gas.

EXAMPLE 21-7 Entropy Change During a Heat Transfer

Heat flows from a steel object of mass 4.00 kg whose temperature is 400 K to an identical object at 300 K. Assuming that the objects are thermally isolated from the environment, what is the net entropy change of the universe after thermal equilibrium has been reached?

SOLUTION Since the objects are identical, their common temperature at equilibrium is 350 K. To calculate the entropy changes associated with their transitions, we substitute the reversible process of Example 21-4 for the actual process. Using $c = 450\ \text{J/kg}\cdot\text{K}$ for the specific heat of steel, we have for the hotter object

$$\Delta S_h = \int_{T_1}^{T_2} \frac{mc\,dT}{T} = mc \ln \frac{T_2}{T_1}$$

$$= (4.00\ \text{kg})(450\ \text{J/kg}\cdot\text{K}) \ln \frac{350\ \text{K}}{400\ \text{K}} = -240\ \text{J/K}.$$

Similarly, the entropy change of the cooler object is

$$\Delta S_c = (4.00\ \text{kg})(450\ \text{J/kg}\cdot\text{K}) \ln \frac{350\ \text{K}}{300\ \text{K}} = 277\ \text{J/K}.$$

The net entropy change of the two objects during the heat transfer is then

$$\Delta S_h + \Delta S_c = 37\ \text{J/K}.$$

The objects are thermally isolated from the environment, so its entropy must remain constant. Thus the entropy of the universe also increases by 37 J/K.

The irreversible processes described in the last three examples—the falling of a rock into a lake, the free expansion of a gas, and the flow of heat from a hot to a cold object—occur spontaneously in nature. In each case there is an increase in the entropy of the universe. The second law, stated in terms of entropy, is a generalization of such observations. It may be stated as follows:

> The entropy of the universe always increases in an irreversible process and remains constant in a reversible process.

Like the formulations of Kelvin and Clausius, the entropy version of the second law is a way of saying that certain processes, although energetically possible, never occur. The reason is that such processes would result in a decrease in the entropy of the universe.

The efficiency of a real engine is always less than that of a Carnot engine because of the presence of irreversible effects, the most common of which is friction. As a result, the entropy change of the universe is positive during one cycle of a real engine, whereas this change is zero during the operation of a Carnot engine.

According to the second law, all natural processes proceed so that the entropy of the universe either increases or remains the same. Eventually, the entropy of the universe will reach a maximum, at which point all objects will have the same temperature. Now in order to convert energy into work, we need two heat reservoirs at different temperatures. This condition cannot be met if the temperature of the universe is uniform and useful work can no longer be produced. The second law therefore implies that the universe is invariably heading toward a state in which energy can no longer be converted into useful work and all life must consequently cease.

It is important to understand that in reaching this final equilibrium state, the net energy of the universe has not decreased—this is forbidden by the first law. Instead, it is the ability to convert heat to work that has been lost!

EXAMPLE 21-8 The Second Law and the Perfect Refrigerator

Consider the perfect refrigerator of Fig. 21-6*b*. Show that its operation violates the second law of thermodynamics.

SOLUTION In one cycle, a quantity of heat Q is transferred reversibly by the refrigerator from a cold reservoir at temperature T_c to a hot reservoir at temperature T_h. The entropy changes of the hot and cold reservoirs are Q/T_h and $-Q/T_c$, respectively. Since the refrigerator returns to its original state after one complete cycle, its entropy remains the same, so the net entropy change of the system (the refrigerator and the reservoirs) is

$$\Delta S = \frac{Q}{T_h} - \frac{Q}{T_c}.$$

No other part of the universe is involved in this process, so the net entropy change of the universe must also be $Q/T_h - Q/T_c$, which, since $T_h > T_c$, is negative. The operation of the perfect refrigerator is therefore a violation of the second law of thermodynamics. This is equivalent to the Clausius statement discussed in Sec. 21-2.

DRILL PROBLEM 21-6

A quantity of heat Q is absorbed from a reservoir at a temperature T_h by a cooler reservoir at a temperature T_c. What is the entropy change of the hot reservoir? the cold reservoir? the universe?
ANS. $-Q/T_h$; Q/T_c; $Q(1/T_c - 1/T_h)$.

DRILL PROBLEM 21-7

Repeat Example 21-8 for a perfect heat engine.

DRILL PROBLEM 21-8

Match the thermodynamic variables of the first column with the laws of thermodynamics listed in the second column. Explain your choices.

Entropy	Zeroth law
Temperature	First law
Internal energy	Second law

DRILL PROBLEM 21-9

A 50-g copper piece at a temperature of 20°C is placed into a large insulated vat of water at 100°C. (*a*) What is the entropy change of the copper piece when it reaches thermal equilibrium with the water? (*b*) What is the entropy change of the water? (*c*) What is the entropy change of the universe?
ANS. (*a*) +4.71 J/K; (*b*) −4.18 J/K; (*c*) +0.53 J/K.

*21-7 Entropy and Disorder

We have seen how entropy is related to heat exchange at a particular temperature. In this section, we consider entropy from a statistical viewpoint. Although the details of the argument are not appropriate for this textbook, it turns out that entropy can be related to how *disordered* or *randomized* a system is—the more it is disordered, the higher is its entropy. For example, a new deck of cards is very ordered, as the cards are arranged numerically by suit. In shuffling this new deck, we randomize the arrangement of the cards and therefore increase its entropy.

The entropy of the new deck of cards is increased by the dealer.

Now the second law of thermodynamics requires that the entropy of the universe increase in any irreversible process. Thus in terms of order, the second law may be stated as follows:

> In any irreversible process, the universe becomes more disordered.

For example, the irreversible free expansion of the ideal gas of Example 21-6 results in a larger volume for the gas molecules to occupy. Since a larger volume means more possible arrangements for the same number of atoms, disorder is also increased. As a result, the entropy of the universe increases.

Changes in phase also illustrate the connection between entropy and disorder. Suppose we place 50 g of ice at 0°C in contact with a heat reservoir at 20°C. Heat spontaneously flows from the reservoir to the ice, which melts and eventually reaches a temperature of 20°C. During this transition, the ice gains an amount of heat from the reservoir equal to

$$\begin{aligned} Q &= mL_f + mc\,\Delta T \\ &= (50\text{ g})(335\text{ J/g}) + (50\text{ g})(4.19\text{ J/g}\cdot\text{K})(20\text{ K}) \\ &= 1.68 \times 10^4\text{ J} + 4.19 \times 10^3\text{ J} \\ &= 2.10 \times 10^4\text{ J}. \end{aligned}$$

From Eqs. (21-7) and (21-10), the increase in entropy of the ice is

$$\Delta S_I = \frac{1.68 \times 10^4\text{ J}}{273\text{ K}} + (50\text{ g})(4.19\text{ J/g}\cdot\text{K})\ln\frac{293\text{ K}}{273\text{ K}} = 76.3\text{ J/K},$$

while the decrease in entropy of the reservoir is

$$\Delta S_R = \frac{-2.10 \times 10^4\text{ J}}{293\text{ K}} = -71.7\text{ J/K}.$$

The increase in entropy of the universe is therefore

$$\Delta S_U = 76.3\text{ J/K} - 71.7\text{ J/K} = 4.6\text{ J/K}.$$

This process also results in a more disordered universe. The ice changes from a solid with molecules located at specific sites to a liquid whose molecules are much freer to roam. The molecular arrangement has therefore become more randomized. Although the change in average kinetic energy of the molecules of the heat reservoir is negligible, there is nevertheless a significant decrease in the entropy of the reservoir because it has so many molecules. However, its decrease in entropy is still not as large as the increase in entropy of the ice. The increased disorder of the ice more than compensates for the increased order of the reservoir, and the entropy of the universe increases by 4.6 J/K.

One might suspect that the embryonic development of different forms of life might be a net ordering process and therefore a violation of the second law. After all, a single cell gathers molecules and eventually becomes a highly structured organism such as a human being. However, this ordering process is more than compensated for by the disordering of the rest of the universe. In its development, the embryo sheds waste products and uses energy that the sun furnishes. The net result is an increase in entropy and an increase in disorder of the universe.

DRILL PROBLEM 21-10

In Example 21-7, the spontaneous flow of heat from a hot object to a cold object results in a net increase in entropy of the universe. Discuss how this result can be related to an increase in disorder of the system.

SUMMARY

1. **Heat engines**
   (*a*) If an engine absorbs heat Q_h from a heat reservoir and does work W, its efficiency is
   $$\epsilon = \frac{W}{Q_h}.$$
   (*b*) If a refrigerator removes heat Q_c from a heat reservoir by doing work W, its coefficient of performance is
   $$\kappa = \frac{Q_c}{W}.$$
   (*c*) The efficiency of a reversible engine operating between hot and cold reservoirs at temperatures T_h and T_c, respectively, is
   $$\epsilon = 1 - \frac{T_c}{T_h}.$$
   (*d*) The coefficient of performance of a reversible refrigerator operating between a cold reservoir at a temperature T_c and a hot reservoir at a temperature T_h is
   $$\kappa = \frac{T_c}{T_h - T_c}.$$
2. **Second law of thermodynamics**
   (*a*) The Kelvin formulation: It is impossible to construct a heat engine that, when operating in a cycle, completely converts heat into work.
   (*b*) The Clausius formulation: It is impossible to construct a refrigerator that transfers heat from a cold reservoir to a hot reservoir without aid from an external source.
3. **Second law of thermodynamics and the reversible heat engine**
   The reversible engine is the most efficient engine operating between two heat reservoirs.
4. **Absolute temperature scale**
   (*a*) If a Carnot engine extracts heat Q_h from the hot reservoir and exhausts heat Q_c to the cold reservoir, then the absolute, or Kelvin, temperatures of the two reservoirs satisfy
   $$\frac{T_c}{T_h} = \frac{Q_c}{Q_h}.$$
   (*b*) Absolute and ideal-gas temperatures are identical.
5. **Entropy**
   (*a*) The change in entropy of a system that undergoes a reversible process at a constant temperature T is
   $$\Delta S = \frac{Q}{T},$$
   where Q is the heat exchanged by the system during the process.
   (*b*) When a system undergoes a reversible transition between states A and B of different temperatures, its change in entropy is
   $$\Delta S = \int_{\mathrm{A}}^{\mathrm{B}} \frac{đQ}{T}.$$
   (*c*) A system's change in entropy between two states is independent of the reversible thermodynamic path taken by the system when it makes a transition between the states.
6. **Entropy and the second law of thermodynamics**
   (*a*) In order to calculate the change in entropy of a system during an irreversible transition between equilibrium states A and B, we use a reversible path that also takes the system from A to B.
   (*b*) The entropy of the universe always increases in an irreversible process and remains constant in a reversible process.

*7. **Entropy and disorder**
   Entropy can be related to how disordered a system is—the more it is disordered, the higher is its entropy. In any irreversible process, the universe becomes more disordered.

Questions

21-1. In order to increase the efficiency of a reversible engine, should the temperature of the hot reservoir be raised or lowered? What about the cold reservoir?

21-2. If the refrigerator door is left open, what happens to the temperature of the kitchen?

21-3. Is it possible for the efficiency of a reversible engine to be greater than 1.0? Is it possible for the coefficient of performance of a reversible refrigerator to be less than 1.0?

21-4. Why don't we operate ocean liners by extracting heat from the ocean or operate airplanes by extracting heat from the atmosphere?

21-5. Discuss the practical advantages and disadvantages of heat pumps, furnaces, and electric heating.

21-6. The energy output of a heat pump is greater than the energy used to operate the pump. Why doesn't this statement violate the first law of thermodynamics?

21-7. Speculate as to why nuclear power plants are less efficient than fossil-fuel plants.

21-8. Explain in practical terms why efficiency is defined as W/Q_h and the coefficient of performance as Q_c/W.

21-9. Is it possible for a system to have an entropy change if it neither absorbs nor emits heat during a reversible transition? during an irreversible transition?

21-10. Are the entropy changes of the *systems* in the following processes positive or negative? (*a*) *water vapor* that condenses on a cold surface, (*b*) *gas* in a container that leaks into the surrounding atmosphere; (*c*) an *ice cube* that melts in a glass of lukewarm water; (*d*) the *lukewarm water* of part (*c*); (*e*) a *real heat engine* performing a cycle; (*f*) *food* cooled in a refrigerator.

21-11. An ideal gas goes from state (p_i, V_i) to state (p_f, V_f) when it is allowed to expand freely. Is it possible to represent the actual process on a pV diagram? Explain.

21-12. Discuss the entropy changes in the systems of Ques. 21-10 in terms of disorder.

Problems

Heat Engines and the Second Law of Thermodynamics

21-1. A reversible engine is found to have an efficiency of 0.40. If it does 200 J of work per cycle, what are the corresponding quantities of heat absorbed and rejected?

21-2. In performing 100 J of work, a reversible engine rejects 50 J of heat. What is the efficiency of the engine?

21-3. A reversible engine with an efficiency of 0.30 absorbs 500 J of heat per cycle. (*a*) How much work does it perform per cycle? (*b*) How much heat does it reject per cycle?

21-4. It is found that a reversible engine rejects 100 J while absorbing 125 J each cycle of operation. (*a*) What is the efficiency of the engine? (*b*) How much work does it perform per cycle?

21-5. A reversible refrigerator has a coefficient of performance of 3.0. (*a*) If it requires 200 J of work per cycle, how much heat per cycle does it remove from the cold reservoir? (*b*) How much heat per cycle is discarded to the hot reservoir?

21-6. During one cycle, a reversible refrigerator removes 500 J from a cold reservoir and rejects 800 J to its hot reservoir. (*a*) What is its coefficient of performance? (*b*) How much work per cycle does it require to operate?

21-7. If a reversible refrigerator discards 80 J of heat per cycle and its coefficient of performance is 6.0, what are (*a*) the quantity of heat it removes per cycle from a cold reservoir and (*b*) the amount of work per cycle required for its operation?

21-8. If the temperature of the cold reservoir of the reversible engine in Prob. 21-3 is 300 K, what is the temperature of the hot reservoir?

21-9. The Kelvin temperature of the hot reservoir of a reversible engine is twice that of the cold reservoir, and the work done by the engine per cycle is 50 J. Calculate (*a*) the efficiency of the engine, (*b*) the heat absorbed per cycle, and (*c*) the heat rejected per cycle.

21-10. What is the ratio of the temperatures of the hot and cold reservoirs of the reversible refrigerator of Prob. 21-5?

21-11. The temperatures of the cold and hot reservoirs between which a Carnot refrigerator operates are −73 and 27°C, respectively. What is its coefficient of performance?

21-12. Suppose a Carnot refrigerator operates between T_c and T_h. Calculate the amount of work required to extract 1.0 J of heat from the cold reservoir if (*a*) $T_c = 7°C$, $T_h = 27°C$; (*b*) $T_c = -73°C$, $T_h = 27°C$; (*c*) $T_c = -173°C$, $T_h = 27°C$; (*d*) $T_c = -270°C$, $T_h = 27°C$.

21-13. A Carnot engine operates between reservoirs at 600 and 300 K. If the engine absorbs 100 J per cycle at the hot reservoir, what is its work output per cycle?

21-14. A 500-W motor operates a Carnot refrigerator between −5 and 30°C. (*a*) What is the amount of heat per second extracted from the inside of the refrigerator? (*b*) How much heat is exhausted to the outside air per second?

21-15. Sketch a Carnot cycle on (*a*) a temperature-volume diagram and (*b*) a temperature-entropy diagram.

21-16. A Carnot heat pump operates between 0 and 20°C. How much heat is exhausted into the interior of a house for every 1.0 J of work done by the pump?

21-17. Suppose a Carnot motor can be operated between two reservoirs as either a heat engine or a refrigerator. How is the coefficient of performance of the refrigerator related to the efficiency of the heat engine?

21-18. An engine operating between heat reservoirs at 20 and 200°C extracts 1000 J per cycle from the hot reservoir. (*a*) What is the maximum possible work that this engine can do per cycle? (*b*) For this maximum work, how much heat is exhausted to the cold reservoir per cycle?

21-19. An inventor claims to have developed a heat engine that operates between heat reservoirs at temperatures 200 and 20°C with an efficiency of 45 percent. Would you invest money in her project? What if she claims an efficiency of 35 percent?

21-20. What is the minimum work required of a refrigerator if it is to extract 50 J per cycle from the inside of a freezer at −10°C and exhaust heat to the air at 25°C?

21-21. A 300-W heat pump operates between the ground whose temperature is 0°C and the interior of a house at 22°C. What is the maximum amount of heat per hour that the heat pump can supply to the house?

21-22. An engineer must design a refrigerator that does 300 J of work per cycle to extract 2100 J of heat per cycle from a freezer whose temperature is −10°C. What is the maximum air temperature for which this condition can be met? Is this a reasonable condition to impose on the design?

21-23. A Carnot engine is used to measure the temperature of a heat reservoir. The engine operates between the heat reservoir and a reservoir consisting of water at its triple point. (*a*) If 400 J per cycle are removed from the heat reservoir while 200 J per cycle are deposited in the triple-point reservoir, what is the temperature of the heat reservoir? (*b*) If 400 J per cycle are removed from the triple-point reservoir while 200 J per cycle are deposited in the heat reservoir, what is the temperature of the heat reservoir?

Entropy and the Second Law of Thermodynamics

21-24. Two hundred joules of heat are removed from a heat reservoir at a temperature of 200 K. What is the entropy change of the reservoir?

21-25. In an isothermal reversible expansion at 27°C, an ideal gas does 20 J of work. What is the entropy change of the gas?

21-26. An ideal gas at 300 K is compressed isothermally to one-fifth its original volume. Determine the entropy change per mole of the gas.

21-27. What is the entropy change of 10 g of steam at 100°C when it condenses to water at the same temperature?

21-28. A metal rod is used to conduct heat between two reservoirs at temperatures T_h and T_c, respectively. When an amount of heat Q flows through the rod from the hot to the cold reservoir, what is the net entropy change of the rod? of the hot reservoir? of the cold reservoir? of the universe?

21-29. For the Carnot cycle of Fig. 21-4, what is the entropy change of the hot reservoir? the cold reservoir? the universe?

21-30. A 5.0-kg piece of lead at a temperature of 600°C is placed in a lake whose temperature is 15°C. Determine the entropy change of (*a*) the lead piece, (*b*) the lake, and (*c*) the universe.

21-31. One mole of an ideal gas doubles its volume in a reversible isothermal expansion. (*a*) What is the change in entropy of the gas? (*b*) If 1500 J of heat are added in this process, what is the temperature of the gas?

21-32. One mole of an ideal monatomic gas is confined to a rigid container. When heat is added reversibly to the gas, its temperature changes from T_1 to T_2. (*a*) How much heat is added? (*b*) What is the change in entropy of the gas?

21-33. (*a*) A 5.0-kg rock at a temperature of 20°C is dropped into a shallow lake also at 20°C from a height of 1.0×10^3 m. What is the resulting change in entropy of the universe? (*b*) If the temperature of the rock is 100°C when it is dropped, what is the change of entropy of the universe? Assume that air friction is negligible (not a good assumption) and that $c = 860$ J/kg·°C is the specific heat of the rock.

21-34. A 5.0-kg wood block starts with an initial speed of 8.0 m/s and slides across the floor until friction stops it. Estimate the resulting change in entropy of the universe. Assume that everything stays at a room temperature of 20°C.

21-35. A copper rod of cross-sectional area 5.0 cm² and length 5.0 m conducts heat from a heat reservoir at 373 K to one at 273 K. What is the time rate of change of the universe's entropy for this process?

21-36. Fifty grams of water at 20°C is heated until it becomes vapor at 100°C. Calculate the change in entropy of the water in this process.

21-37. Fifty grams of frozen water at 0°C is changed into vapor at 100°C. What is the change in entropy of the water in this process?

21-38. In an isochoric process, heat is added to 10 mol of a monatomic ideal gas whose temperature increases from 273 to 373 K. What is the entropy change of the gas?

21-39. Two hundred grams of water at 0°C is brought into contact with a heat reservoir at 80°C. After thermal equilibrium is reached, what is the temperature of the water? of the reservoir? How much heat has been transferred in the process? What is the entropy change of the water? of the reservoir? What is the entropy change of the universe?

21-40. Suppose that the temperature of the water in the previous problem is raised by first bringing it to thermal equilibrium with a reservoir at a temperature of 40°C and then with a reservoir at 80°C. Calculate the entropy changes of each reservoir, of the water, and of the universe.

21-41. Repeat the calculations of Prob. 21-39 for a four-step process in which the water is brought into thermal equilibrium successively with reservoirs at 20, 40, 60, and 80°C. Compare the entropy change of the universe in this process with those of Probs. 21-39 and 21-40. What is the entropy change of the universe if the water is brought from 0 to 80°C by placing it in thermal contact with an infinite number of reservoirs between these temperatures? What kind of process is this?

21-42. (*a*) Ten grams of H_2O starts as ice at 0°C. The ice absorbs heat from the air (just above 0°C) until all of it melts. Calculate the entropy change of the H_2O, of the air, and of the universe. (*b*) Suppose that the air in part (*a*) is at 20°C rather than 0°C and that the ice absorbs heat until it becomes water at 20°C. Calculate the entropy change of the H_2O, of the air, and of the universe. (*c*) Is either of these processes reversible?

21-43. The Carnot cycle is represented by the temperature-entropy diagram shown in the accompanying figure. (*a*) How much heat is absorbed per cycle at the high-temperature reservoir? (*b*) How much heat is exhausted per cycle at the low-temperature reservoir? (*c*) How much work is done per cycle by the engine? (*d*) What is the efficiency of the engine?

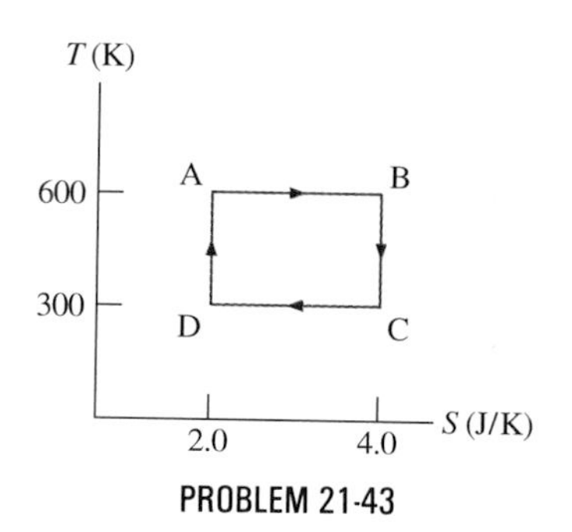

PROBLEM 21-43

21-44. A Carnot engine operating between heat reservoirs at 500 and 300 K absorbs 1500 J per cycle at the high-temperature reservoir. (*a*) Represent the engine's cycle on a temperature-entropy diagram like that of Prob. 21-43. (*b*) How much work per cycle is done by the engine?

General Problems

21-45. A Carnot engine has an efficiency of 0.60. When the temperature of its cold reservoir changes, the efficiency drops to 0.55. If initially $T_c = 27°\text{C}$, determine (*a*) the constant value of T_h and (*b*) the final value of T_c.

21-46. A Carnot engine performs 100 J of work while rejecting 200 J of heat each cycle. After the temperature of the hot reservoir only is adjusted, it is found that the engine now does 130 J of work while discarding the same quantity of heat. (*a*) What are the initial and final efficiencies of the engine? (*b*) What is the fractional change in the temperature of the hot reservoir?

21-47. A Carnot refrigerator exhausts heat to the air, which is at a temperature of 25°C. How much power is used by the refrigerator if it freezes 1.5 g of water per second? Assume the water is at 0°C.

21-48. An ideal gas at temperature T is stored in the left half of an insulating container of volume V using a partition of negligible volume. (See the accompanying figure.) What is the entropy change per mole of the gas in each of the following cases? (*a*) The partition is suddenly removed and the gas quickly fills the entire container. (*b*) A tiny hole is punctured in the partition and after a long period, the gas reaches an equilibrium state such that there is no net flow through the hole. (*c*) The partition is moved very slowly and adiabatically all the way to the right wall so that the gas finally fills the entire container.

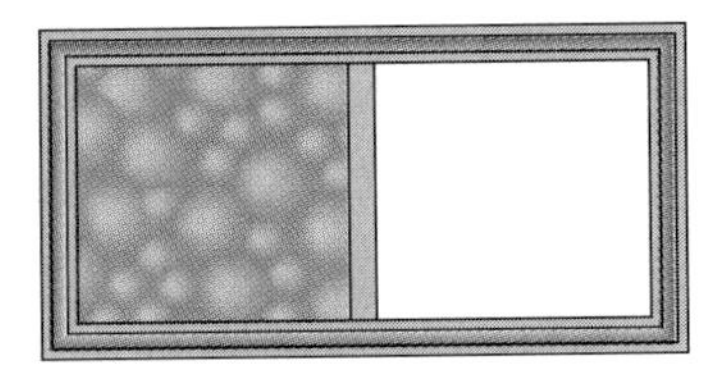

PROBLEM 21-48

21-49. (*a*) An infinitesimal amount of heat is added reversibly to a system. By combining the first and second laws, show that $dU = TdS - đW$. (*b*) When heat is added to an ideal gas, its temperature and volume change from T_1 and V_1 to T_2 and V_2. Show that the entropy change of n moles of the gas is given by

$$\Delta S = nC_v \ln \frac{T_2}{T_1} + nR \ln \frac{V_2}{V_1}.$$

21-50. Using the result of the previous problem, show that for an ideal gas undergoing an adiabatic process, $TV^{\gamma-1}$ is constant.

21-51. With the help of Prob. 21-49, show that ΔS between states 1 and 2 of n moles an ideal gas is given by

$$\Delta S = nC_p \ln \frac{T_2}{T_1} - nR \ln \frac{p_2}{p_1}.$$

21-52. A diatomic ideal gas is brought from an initial equilibrium state at $p_1 = 0.50$ atm and $T_1 = 300$ K to a final state with $p_2 = 0.20$ atm and $T_2 = 500$ K. Use the results of the previous problem to determine the entropy change per mole of the gas.

21-53. A 0.50-kg piece of aluminum at 250°C is dropped into 1.0 kg of water at 20°C. After equilibrium is reached, what is the net entropy change of the system?

21-54. Suppose 20 g of ice at 0°C is added to 300 g of water at 60°C. What is the total change in entropy of the mixture after it reaches thermal equilibrium?

21-55. A cylinder contains 500 g of helium at 120 atm and 20°C. The valve is leaky, and all the gas slowly escapes isothermally into the atmosphere. Use the results of Prob. 21-51 to determine the resulting change in entropy of the universe.

21-56. Consider the system shown in the accompanying figure. An ideal gas is constrained by a partition to be on the left side of the container. It has pressure p_0, volume V_0, internal energy U_0, entropy S_0, and temperature T_0. (*a*) The partition is removed so that the gas now fills a volume V_i, and its pressure, temperature, entropy, and internal energy are p_i, T_i, S_i, U_i, respectively. Describe and explain what happens to the temperature, entropy, and internal energy in this process (increase, decrease, stay the same). (*b*) Now the gas expands adiabatically and reversibly against the piston so that $V_i \to V_f$, and $U_i, S_i, T_i, p_i \to U_f, S_f, T_f, p_f$. Describe and explain what happens to its temperature, entropy, and internal energy in this process. (*c*) Instead of the process described in part (*b*), heat is added to the gas so that p_i remains constant in the reversible expansion to V_f. Again explain what happens to the temperature, entropy, and internal energy of the gas.

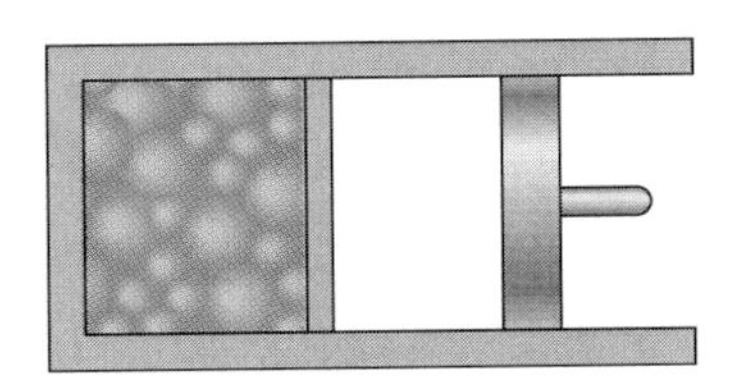

PROBLEM 21-56

21-57. The gasoline internal combustion engine operates in a cycle consisting of six parts. Four of these parts involve motion of the piston and are called *strokes*. Since the cycle involves, among other things, friction, heat exchange through finite temperature differences, and accelerations of the piston, it is irreversible. Nevertheless, it is represented by the ideal reversible *Otto cycle*, which is illustrated in the accompanying figure. The working substance of the cycle is assumed to be air. The six steps of the Otto cycle are the following:

1. Isobaric intake stroke (OA). A mixture of gasoline and air is drawn into the combustion chamber at atmospheric pressure p_0 as the piston expands, increasing the volume of the cylinder from zero to V_A.

2. Adiabatic compression stroke (AB). The temperature of the mixture rises as the piston compresses it adiabatically from a volume V_A to V_B.

3. Ignition at constant volume (BC). The mixture is ignited by a spark. The combustion happens so fast that there is essentially no motion of the piston. During this process, the added heat Q_1 causes the pressure to increase from p_B to p_C at the constant volume V_B (= V_C).

4. Adiabatic expansion (CD). The heated mixture of gasoline and air expands against the piston, increasing the volume from V_C to V_D. This is called the *power stroke*, as it is the part of the cycle that delivers most of the power to the crankshaft.

5. Constant-volume exhaust (DA). When the exhaust valve opens, some of the combustion products escape. There is almost no movement of the piston during this part of the cycle, so the volume remains constant at V_A $(= V_D)$. Most of the available energy is lost here, as represented by the heat exhaust Q_2.

6. Isobaric compression (AO). The exhaust valve remains open, and the compression from V_A to zero drives out the remaining combustion products.

(*a*) Using (*i*) $\epsilon = W/Q_1$; (*ii*) $W = Q_1 - Q_2$; and (*iii*) $Q_1 = nC_v(T_C - T_B)$, $Q_2 = nC_v(T_D - T_A)$, show that

$$\epsilon = 1 - \frac{T_D - T_A}{T_C - T_B}.$$

(*b*) Use the fact that steps 2 and 4 are adiabatic to show that

$$\epsilon = 1 - \frac{1}{r^{\gamma-1}},$$

where $r = V_A/V_B$. The quantity r is called the *compression ratio* of the engine.

(*c*) In practice, r is kept less than around 7. For larger values, the gasoline-air mixture is compressed to temperatures so high that it explodes before the finely timed spark is delivered. This *preignition* causes engine knock and loss of power. Show that for $r = 6$ and $\gamma = 1.4$ (the value for air), $\epsilon = 0.51$, or an efficiency of 51 percent. Because of the many irreversible processes, an actual internal combustion engine has an efficiency much less than this ideal value. A typical efficiency for a tuned engine is about 25 to 30 percent.

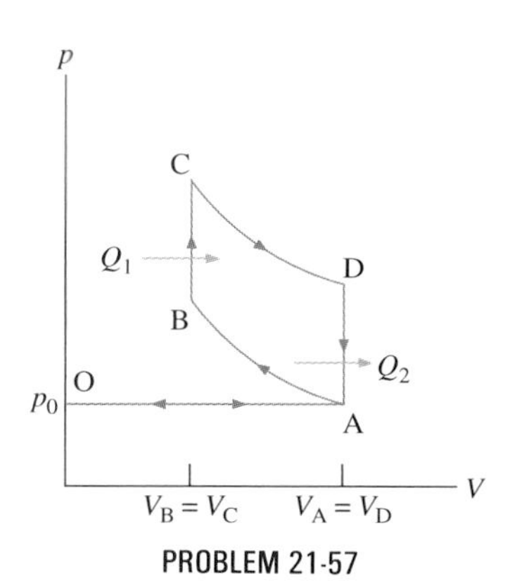

PROBLEM 21-57

21-58. An ideal *diesel cycle* is shown in the accompanying figure. This cycle consists of five strokes. In this case, only air is drawn into the chamber during the the intake stroke OA. The air is then compressed adiabatically from state A to state B raising its temperature high enough so that when fuel is added during the power stroke BC, it ignites. After ignition ends at C, there is a further adiabatic power stroke CD. Finally, there is an exhaust at constant volume as the pressure drops from p_D to p_A, followed by a further exhaust when the piston compresses the chamber volume to zero. (*a*) Use $W = Q_1 - Q_2$, $Q_1 = nC_p(T_C - T_B)$, and $Q_2 = nC_v(T_D - T_A)$ to show that

$$\epsilon = \frac{W}{Q_1} = 1 - \frac{T_D - T_A}{\gamma(T_C - T_B)}.$$

(*b*) Use the fact that $A \to B$ and $C \to D$ are adiabatic to show that

$$\epsilon = 1 - \frac{1}{\gamma}\frac{\left(\frac{V_C}{V_D}\right)^{\gamma} - \left(\frac{V_B}{V_A}\right)^{\gamma}}{\left(\frac{V_C}{V_D}\right) - \left(\frac{V_B}{V_A}\right)}.$$

(*c*) Since there is no preignition (remember, the chamber doesn't contain any fuel during the compression), the compression ratio can be larger than that for a gasoline engine. Typically, $V_A/V_B = 15$ and $V_D/V_C = 5$. For these values and $\gamma = 1.4$, show that $\epsilon = 0.56$, or an efficiency of 56 percent. Diesel engines actually operate at an efficiency of about 30 to 35 percent compared with 25 to 30 percent for gasoline engines.

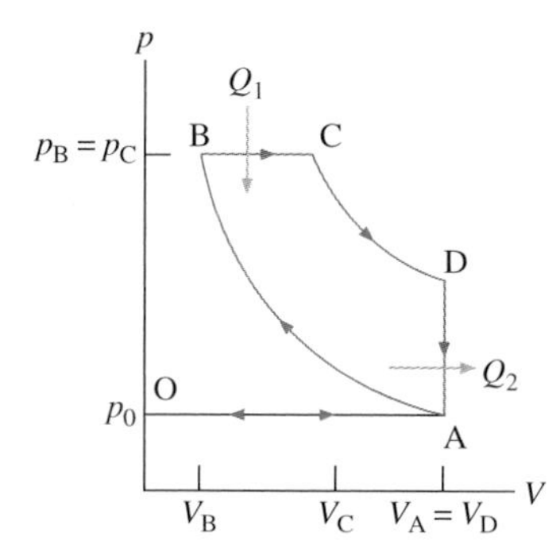

PROBLEM 21-58

PART IV

ELECTRICITY AND MAGNETISM

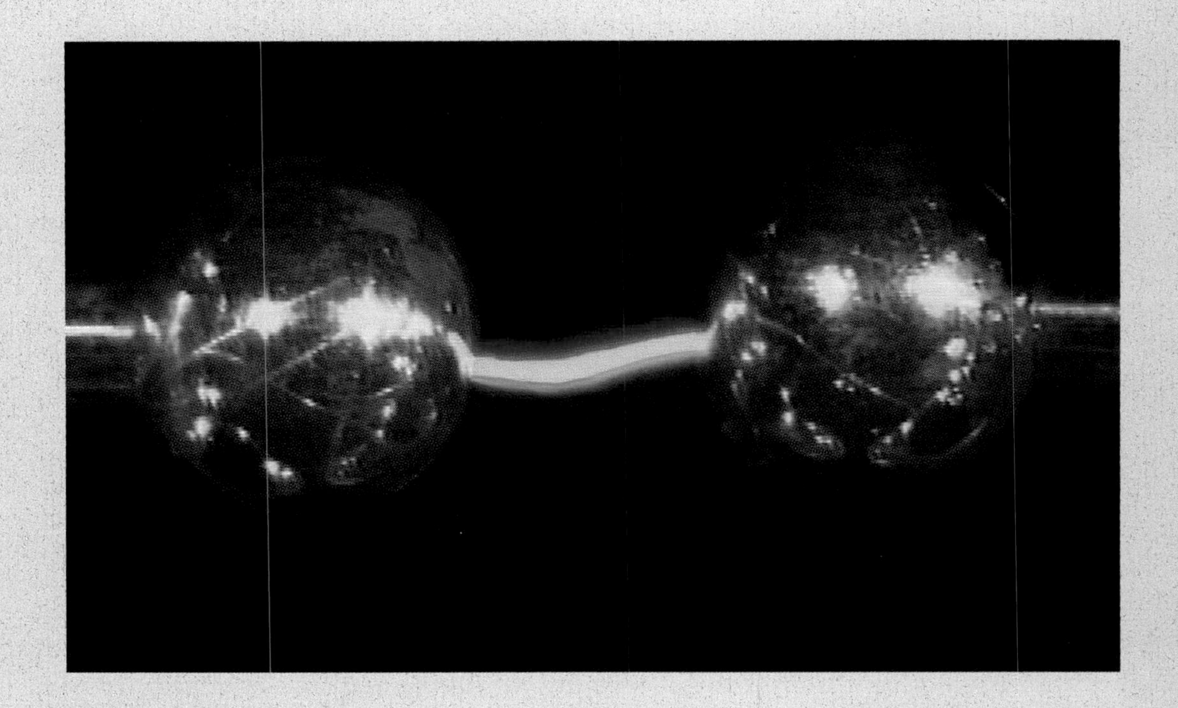

Electric discharge between oppositely charged conductors.

CHAPTER 22

ELECTRIC CHARGE AND COULOMB'S LAW

PREVIEW

In this chapter we investigate the nature of electric charge and the law that describes how electric charges interact. The main topics to be discussed here are the following:

1. **Electric charge.** The nature of electric charge is discussed, and how an object acquires a net positive or negative charge is explained. Charge conservation and charge quantization are considered.
2. **Conductors and insulators.** Very simple definitions of these two types of materials are presented. Charging by conduction and by induction are explained, and the grounding of a conductor is described.
3. **Coulomb's law.** We discuss this law, which describes how pairs of point charges interact.

With this chapter we begin a detailed investigation of the branch of physics known as *electromagnetism*. Many phenomena are electromagnetic in nature, including electric fields and magnetic fields, light waves, radio signals and even contact forces. (See Sec. 5-1.) There are also many important and interesting applications of electromagnetic theory. Some familiar examples are electronic circuits, electric power generation, electric motors, and particle accelerators.

The first recorded observation of an electrical effect was made about 600 B.C. by the Greek mathematician and astronomer Thales. He found that when amber is rubbed, it attracts light objects such as straw and feathers. We now know that this force is caused by the electrical attraction between charged particles. The permanent magnet was also discovered in ancient Greece. In fact, the word *magnetism* is derived from Magnesia, the name of a Greek city-state near which magnetic iron oxide was found.

After the discoveries of the Greeks, very little was done to expand human understanding of electromagnetism for the next 2000 years. This long period was marked by just two advances: the invention of the magnetic compass in China in the eleventh century and the discovery of magnetic poles—and the conditions under which they attract and repel—by Pierre de Maricourt in the thirteenth century.

The first systematic investigations of electromagnetic phenomena are attributed to William Gilbert (1540–1603), an English physician. Gilbert discovered that amber is just one of many materials which when rubbed attract light objects. He was also the first to suggest that the earth affects a compass needle because the earth itself is a large magnet.

Gilbert's work was certainly not followed by a flurry of scientific activity, especially by today's standards. The next significant discovery did not occur until 1730, when Stephen Gray found that the ability to attract light objects could be transferred from one piece of amber to another by connecting a metal wire between the amber pieces; whatever was transferred was then called *electricity*. Soon after Gray's experiments, Charles Du Fay (1698–1739) discovered that electrified bodies could repel as well as attract. This led him to suggest that there must be two types of electricity.

In 1747 Benjamin Franklin (1706–1790) proposed that only one of these two types of electricity is actually transferred. He labeled the two types "plus" and "minus" electricity and assigned plus to the transferable electricity. In Franklin's model, a body becomes minus by losing plus electricity, and a body becomes plus by gaining plus electricity. Implicit in Franklin's model is the concept of the conservation of electricity. Electricity is neither created nor destroyed; it is just transferred between bodies.

We now know that Franklin was essentially correct. Electricity (we now call it **charge**) is actually transferred by the exchange of electrons. However, contrary to Franklin's model, the accepted convention today assigns minus electricity to electrons.

The mathematical laws that describe the electric and magnetic interactions were discovered in the late eighteenth and the nineteenth centuries. Among the many talented scientists who worked in this period were Charles Augustin de Coulomb (1736–1806), who discovered the law that describes the electric force between charged particles; Hans Oersted (1777–1851), who found that an electric current can exert a force on a magnet; André Ampère (1775–1836), who studied the magnetic interaction between current-carrying wires; Michael Faraday (1791–1867), who discovered the connection between changing magnetic fields and electric fields; Georg Ohm (1787–1854), whose law is so important in electric circuit theory; and finally, James Clerk Maxwell (1831–1879), who proposed in 1873 that all of electricity and magnetism could be explained in terms of just four equations, which are now known as Maxwell's equations. Time has shown Maxwell to be correct. No experiment or observation in the macroscopic world has ever contradicted his hypothesis.

In this chapter we will consider electric charge and how it interacts when at rest (i.e., its *electrostatic* interaction). Topics include the relationship between charge and the atomic nature of matter and the transfer of charge between bodies. The basic properties of the electrostatic force between charges will be described, and we'll conclude the chapter by investigating Coulomb's law, which is the fundamental force law between stationary charges.

22-1 Electric Charge

The property of matter known as mass plays a fundamental role in the description of the gravitational force. A particle exerts a gravitational force because it possesses mass, and a particle experiences a gravitational force also because it possesses mass. The property of matter that plays an analogous role for the electric force is charge. A particle can neither exert nor experience an electric force unless it possesses charge.

A typical experiment designed to determine the nature of electric charge is illustrated in Fig. 22-1. Two Lucite (a hard plastic) rods are rubbed with fur, and two glass rods are rubbed with silk. When pairs of these rods are placed close together, they are observed to exert forces on one another. Lucite rods repel each other, as do glass rods, while a Lucite rod and a glass rod attract each other. Similar experiments demonstrate that many substances when rubbed with the appropriate material behave like either Lucite or glass; that is, they are either repelled by Lucite and attracted by glass, or vice versa.

We explain these observations by asserting that when substances are rubbed, they acquire a charge. This charge is one of two types. That acquired by the Lucite rod rubbed with fur is *negative*; that acquired by the glass rod rubbed with silk is

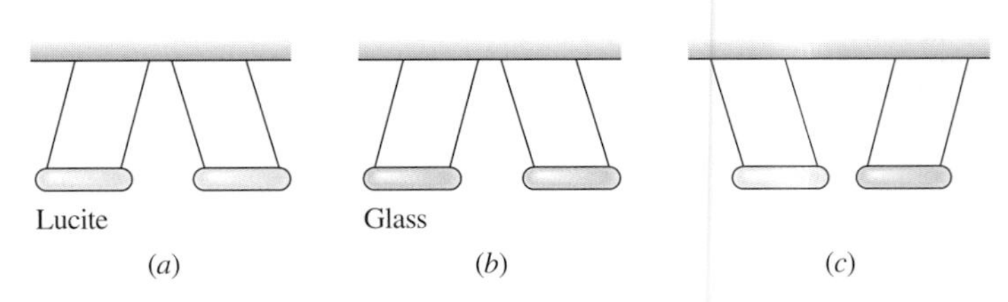

FIGURE 22-1 (*a*) Two Lucite rods, each rubbed with fur, repel, as do (*b*) two glass rods, each rubbed with silk. (*c*) However, the Lucite rod rubbed with fur attracts the glass rod rubbed with silk.

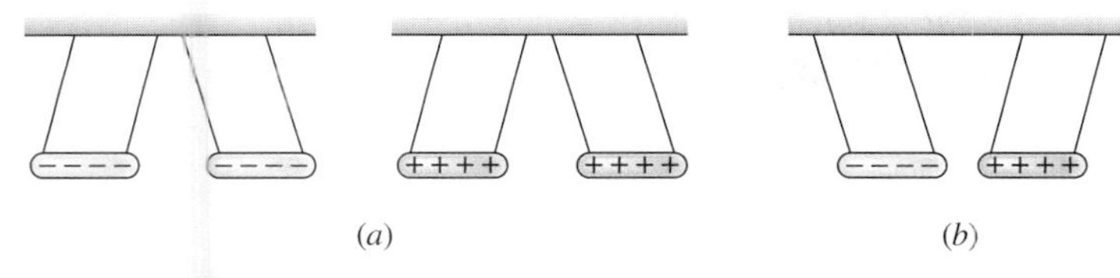

FIGURE 22-2 (*a*) Like charges of either sign repel, while (*b*) unlike charges attract.

positive. *Charges exert forces on one another*. The force between *like charges* is *repulsive*, while the force between *unlike charges* is *attractive*. (See Fig. 22-2.)

It's easy to understand how a body acquires a charge in terms of the atomic model of matter. Matter is composed of atoms that have positively charged nuclei surrounded by negatively charged electrons. Normally, matter is electrically neutral, since the total negative charge of the electrons is exactly balanced by the total positive charge of the atomic nuclei. However, when two bodies are rubbed together, this balance can be disturbed as one body can give up electrons to the other. As a result, the body from which the electrons have been transferred becomes positively charged, while the body receiving the electrons becomes negatively charged.

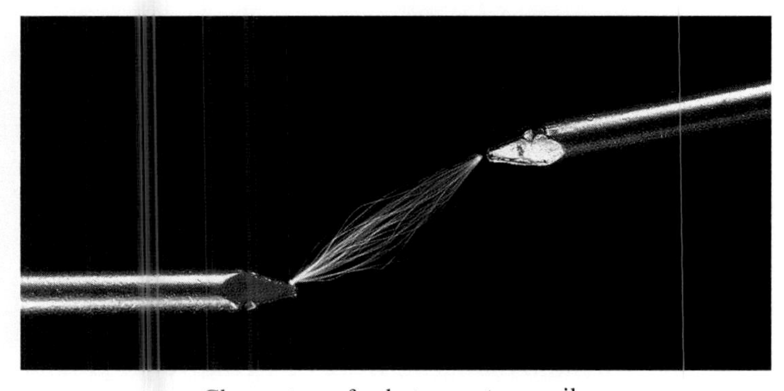

Charge transfer between two nails.

Because charging occurs by a transfer of electrons, the charge on any body is an integer multiple of the electron's charge. If $-e$ is the charge of an electron, where e is a positive number, then the charge on any body is $\pm ne$. Charge is said to be *quantized* in units of e. The charge of every observed elementary particle is also an integer multiple of e.* The charge of the proton is $+e$, the charge of the negative pion is $-e$, the charge of the alpha particle (a helium nucleus) is $+2e$, etc. Table 22-1 shows the electric charges of various elementary particles.

The number of electrons involved in the charging of a body is typically a very small fraction of the total number of electrons in the body. In Example 22-1, we will illustrate this by considering the maximum amount of charge that can be placed on an

*There is considerable evidence that most subatomic particles are actually combinations of even more elementary particles called *quarks*, which carry charges of $-(1/3)e$ or $+(2/3)e$. However, no free quark has ever been detected. The charge of every *observed* particle is an integer multiple of e.

TABLE 22-1 **Charges of Various Elementary Particles**

Particle	Charge	Particle	Charge
Proton (p)	$+e$	π mesons (π^+)	$+e$
Neutron (n)	0	(π^0)	0
Electron (e^-)	$-e$	(π^-)	$-e$
Positron (e^+)	$+e$	Muons (μ^+)	$+e$
K mesons (K^+)	$+e$	(μ^-)	$-e$
(K^0)	0	Neutrino (ν)	0
(K^-)	$-e$	Omegas (Ω^-)	$-e$
		(Ω^+)	$+e$

iron ball such as that used in the shot put event of track and field competition. We find there that the excess charge on the ball corresponds to only about 3×10^{-13} percent of the neutral ball's electrons!

When charge is transferred from one body to another, the total charge of the two bodies is unchanged; that is, charge is *conserved*. Charge conservation was first hypothesized by Benjamin Franklin. Interestingly, Franklin suggested this conservation law long before the atomic model of matter was known. What might seem rather apparent to us (at least for charging phenomena) was not so obvious in Franklin's time!

The conservation of charge is a universal principle that applies to all physical processes, not just those involving charge transfer in bulk matter. One example is the elementary particle interaction

$$\pi^+ + p \rightarrow \pi^- + \pi^+ + \pi^+ + p.$$

Here a positive pion π^+ (charge $+e$) collides with a proton p (charge $+e$) to form a negative pion π^- (charge $-e$), two positive pions (each with charge $+e$), and a proton (charge $+e$). The total charge before the interaction is $+2e$; the total charge after the interaction is also $+2e$.

DRILL PROBLEM 22-1

When a Lucite rod is rubbed with fur, the rod acquires a negative charge while the fur becomes positively charged. Explain this effect in terms of the transfer of electric charge.

22-2 Conductors and Insulators

For present purposes, we can identify all materials as either *conductors* or *insulators*. The common metals such as copper, aluminum, and silver are conductors. Glass, Lucite, and wood are examples of insulators. In a conductor, the outer electrons of the atoms can move throughout its volume and are said to be "free." Any excess charge placed in a small region of a conductor almost instantaneously becomes distributed over its entire *surface*.* An insulator, on the other hand, has very few free electrons. Excess charge placed at a particular location on an

*It will be shown in Sec. 25-7 that the excess charge on a conductor resides on its surface.

insulator tends to remain there. Very simply, we can think of conductors as materials in which charge moves freely and insulators as materials through which charge does not move.

Conductors can be charged both positively and negatively by a process known as *conduction*. Figure 22-3*a* shows two metal spheres that sit on insulating stands. When the spheres are placed in contact, electrons move from the surface of the negatively charged conductor to the uncharged conductor, causing both to become negatively charged. On the other hand, when the metal spheres of Fig. 22-3*b* touch, electrons from the uncharged body move onto the positively charged body. The uncharged sphere loses electrons and becomes positively charged, while the other sphere gains electrons, causing its net positive charge to decrease.

Suppose that the spheres of Fig. 22-3 have radii r_1 and r_2. You will learn in Example 25-16 that when they are in contact, charge will flow between them until the ratio of the excess charge on each sphere is equal to the ratio of their radii:

$$\frac{q_1}{q_2} = \frac{r_1}{r_2}.$$

Of course, not all conductors are spheres. Nevertheless, this equation does indicate that if two charged conductors of vastly different sizes are placed in contact, the excess charge moves almost entirely to the larger conductor; the smaller one becomes essentially discharged.

Imagine now that the initially uncharged sphere of Fig. 22-3*a* is the earth, which is a relatively good conductor because of the metals in the soil and the dissolved salts in the oceans. Quite obviously, it is also a rather large sphere. A charged conductor connected electrically to the earth therefore becomes discharged. It is said to be *grounded*. The grounding of a conductor is represented by the symbol shown in Fig. 22-4.

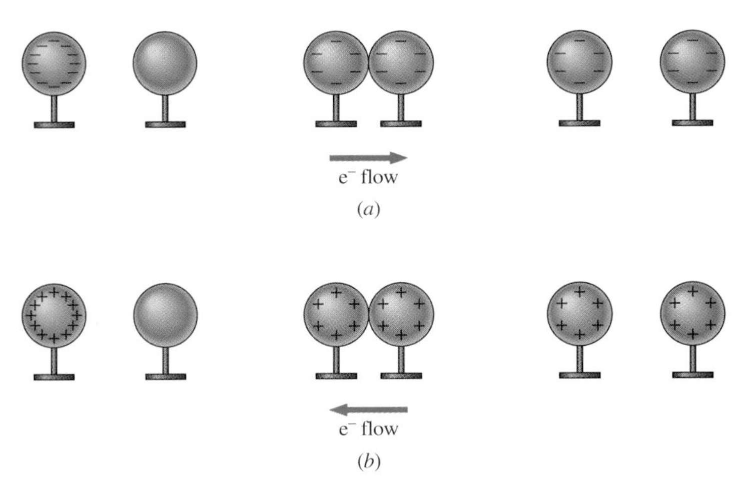

FIGURE 22-3 Charging by conduction. Notice that in both (*a*) and (*b*) the charges that flow between the spheres when they are in contact are electrons.

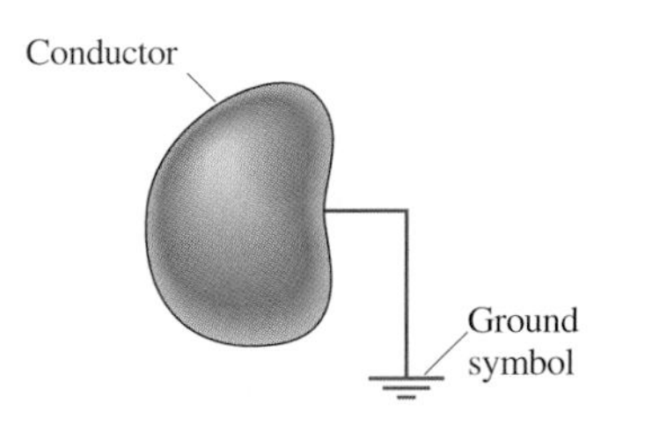

FIGURE 22-4 Representation of grounding (connecting electrically to the earth).

A conductor can also be charged by *induction*. This process is illustrated by the sketches of Fig. 22-5. When the negatively charged rod is brought near the uncharged sphere, it repels the negative charge on the sphere, resulting in a charge distribution like that of Fig. 22-5*a*. Now suppose that we ground the sphere. As shown in Fig. 22-5*b*, the negative charge on the sphere can move even farther away from the rod by flowing through the grounding wire into the earth. If we now remove the grounding wire and then the rod in that order, the sphere is left with an excess of positive charge (Fig. 22-5*c*).

Notice that there is an important difference between the two charging processes. In the conduction process, a body with one

In 1752 Benjamin Franklin flew a kite into a thundercloud and demonstrated that lightning is an electrical phenomenon. He was fortunate that he wasn't killed doing this dangerous experiment.

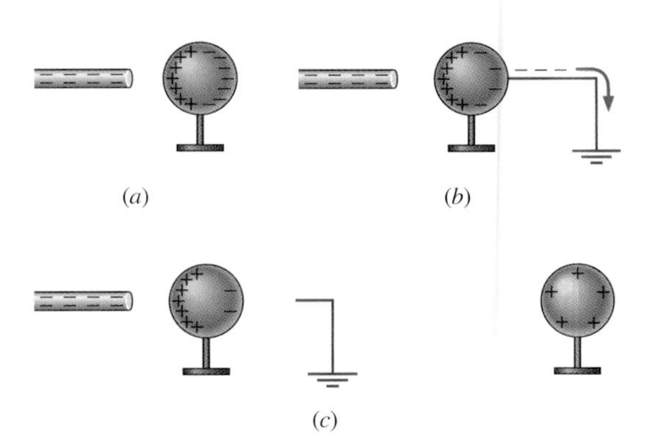

FIGURE 22-5 Charging by induction. (*a*) The negatively charged rod causes a redistribution of the charges on the uncharged sphere. (*b*) The grounding of the sphere results in the flow of some electrons into the earth. (*c*) With the removal of the grounding wire and then the rod, the sphere is left with a net positive charge.

type of charge produces the *same* type of charge on a conductor, while in the induction process, the *opposite* type of charge is produced.

Insulators are also affected by nearby charge. For example, consider a negatively charged rod held near an uncharged insulator. Since charges are unable to move freely within an insulator, no significant redistribution of charge occurs. Instead, the electric force of the charged rod causes a slight shift in the positions of the positive and negative charges within the molecules of the insulator. The positive and negative charge centers of a molecule are then displaced relative to each other. As a result, an excess of positive charge is produced at the surface of the insulator near the negatively charged rod and an excess of negative charge is produced at the far surface. (See Fig. 22-6.) This process is known as *polarization*. Notice that the insulator does not acquire any excess charge and remains uncharged overall. We will discuss polarization in greater detail in Sec. 32-7.

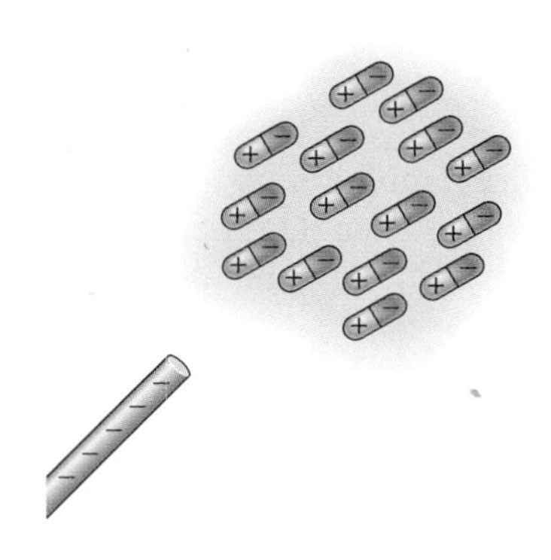

FIGURE 22-6 Polarization of charge in an insulator.

DRILL PROBLEM 22-2

Describe what happens to the excess charge on a metal sphere when you place your hand on its surface. (*Hint:* The human body is a good conductor.)

DRILL PROBLEM 22-3

What happens if (*a*) the negatively charged rod of Fig. 22-5*b* is removed before the grounding wire is connected to the sphere? (*b*) the negatively charged rod is removed before the grounding wire is disconnected?

DRILL PROBLEM 22-4

You are given two uncharged metal spheres and a negatively charged rod. Describe how you can produce equal but opposite charges on the spheres. (*Hint:* Begin with the two spheres in contact with each other.)

22-3 Coulomb's Law

The law that describes how charges interact with one another was discovered by Charles Augustin de Coulomb in 1785. With

Charles Augustin de Coulomb, a French scientist and engineer.

a sensitive torsion balance similar to the one Cavendish used in 1798 to study the gravitational force between lead spheres (see Sec. 5-5), Coulomb measured the electric force between charged spheres. After determining the force as a function of charge and distance between the spheres, he concluded that

1. The magnitude of the electric force $\mathbf{F}_e$ between two charged particles (or *point charges*) varies inversely with the square of the distance r between them:

$$F_e \propto \frac{1}{r^2}.$$

2. The electric force on each particle is directed along the line joining the particles. Like charges repel, unlike charges attract, and the forces obey Newton's third law.

3. The magnitude of the electric force between two charged particles is proportional to the product of the charges of the particles. If the charges of the two particles are q_1 and q_2,

$$F_e \propto \frac{|q_1||q_2|}{r^2};$$

that is,

$$F_e = k\frac{|q_1||q_2|}{r^2}. \tag{22-1a}$$

Coulomb's experimental conclusions are summarized in Fig. 22-7*a* and *b*, which show two charges q_1 and q_2 separated by a distance r. In Fig. 22-7*a*, q_1 and q_2 are unlike charges and therefore attract each other. The force on q_1 by q_2 is then

$$\mathbf{F}_{12} = k\frac{|q_1||q_2|}{r^2}\,\hat{\mathbf{r}},$$

while the force on q_2 by q_1 is

$$\mathbf{F}_{21} = -k\frac{|q_1||q_2|}{r^2}\,\hat{\mathbf{r}},$$

where $\hat{\mathbf{r}}$ is a unit vector directed from q_1 to q_2. In Fig. 22-7*b*, we assume that q_1 and q_2 are like charges that repel each other. The forces on the charges then become

$$\mathbf{F}_{12} = -k\frac{|q_1||q_2|}{r^2}\,\hat{\mathbf{r}},$$

and

$$\mathbf{F}_{21} = k\frac{|q_1||q_2|}{r^2}\,\hat{\mathbf{r}}.$$

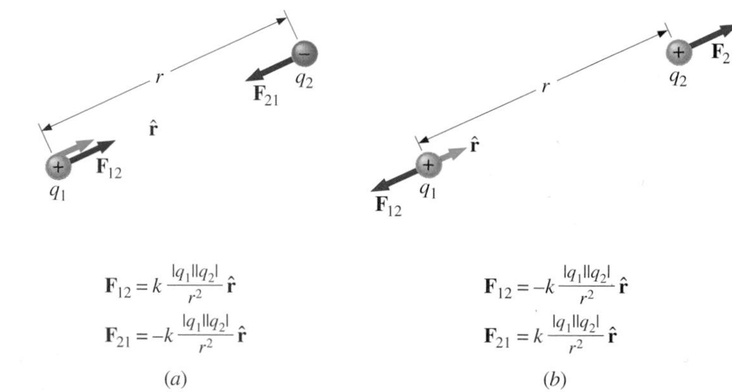

FIGURE 22-7 The electric force between two charges. (*a*) Here q_1 and q_2 are unlike charges and attract each other; (*b*) q_1 and q_2 are like charges and repel each other.

The constant of proportionality in Coulomb's law (Eq. 22-1*a*) depends on the unit used to represent charge. In the SI system that unit is the *coulomb (C)**; the constant is then

$$k = 8.98742 \times 10^9 \text{ N}\cdot\text{m}^2/\text{C}^2,$$

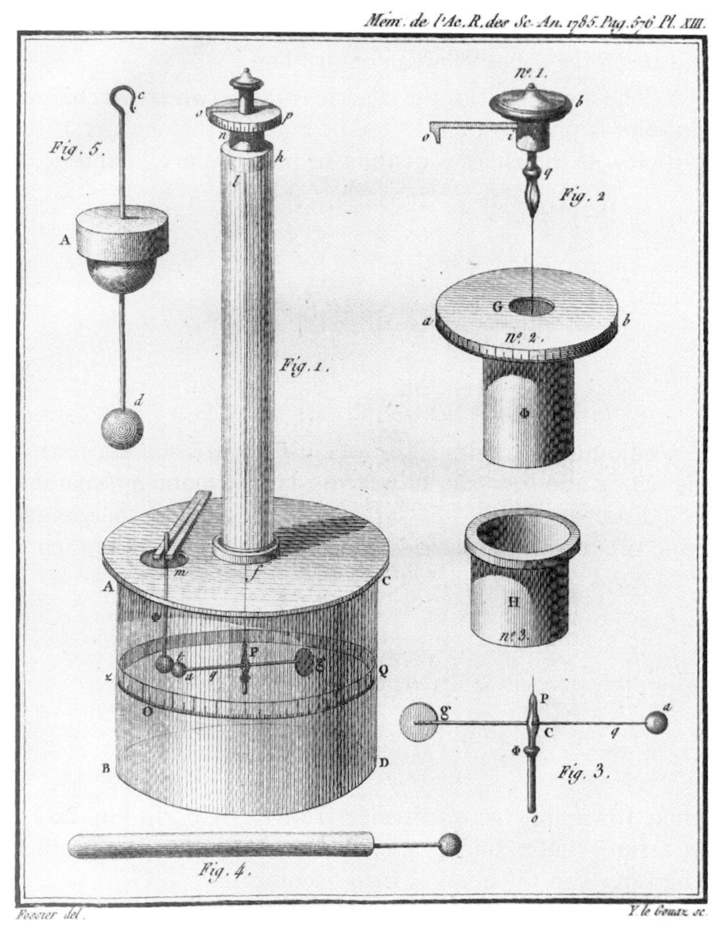

The torsion balance that Coulomb used to investigate the force between electric charges.

*The coulomb is defined in terms of the magnetic force between current-carrying wires (Sec. 29-2).

which is usually rounded off to $9.0 \times 10^9 \text{ N}\cdot\text{m}^2/\text{C}^2$.*

One coulomb is an enormous amount of charge! For example, if two 1.0-C point charges are placed 1.0 m apart, the force between them is 9.0×10^9 N, which is much larger than any electric force ever encountered in the laboratory. In electrostatic experiments, objects typically acquire charges on the order of 10^{-6} C (or 1 μC). As you might expect, the charges of elementary particles have extremely small values. The electron, the smallest "unit" of charge, possesses a charge of

$$-e = -1.6021917 \times 10^{-19} \text{ C}.$$

Many of the equations used to describe the electric interaction are simpler in form if k is replaced by $1/4\pi\epsilon_0$. The constant ϵ_0 is called the *permittivity of free space*. Its numerical value is

$$\epsilon_0 = \frac{1}{4\pi k} = 8.85418 \times 10^{-12} \text{ C}^2/\text{N}\cdot\text{m}^2.$$

With $1/4\pi\epsilon_0$ substituted for k, Coulomb's law becomes

$$F_e = \frac{1}{4\pi\epsilon_0} \frac{|q_1||q_2|}{r^2}. \qquad (22\text{-}1b)$$

The equations representing Coulomb's law and Newton's law of gravitation (Eq. 5-2) are similar in form. One depends on the product of the interacting charges, the other on the product of the interacting masses; and both vary inversely with the square of the distance separating the interacting particles. However, there is one important difference in the two forces: The gravitational force is *always attractive*, while the electric force can be *either attractive or repulsive*. This difference is related to the fact that there is one type of mass but two types of charge.

When using Coulomb's law, you must remember to *treat the electric force as a vector*. When there are more than two point charges present, the net force on any one of the charges is obtained by calculating the vector sum of the electric forces on it due to the other charges. For example, the net force on q of Fig. 22-8 due to q_1, q_2, and q_3 is the vector sum shown.

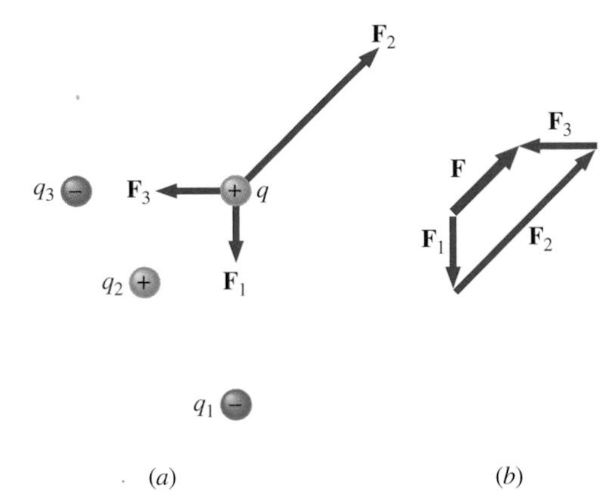

FIGURE 22-8 (*a*) The electric forces on q due to q_1, q_2, and q_3 are represented by the vectors $\mathbf{F}_1$, $\mathbf{F}_2$, and $\mathbf{F}_3$, respectively. (*b*) The net force on q is the vector sum $\mathbf{F} = \mathbf{F}_1 + \mathbf{F}_2 + \mathbf{F}_3$.

*This value of k is actually for point charges in a vacuum. It is slightly different for interactions in air. However, the values of k in a vacuum and in air differ by only 0.06 percent, which is smaller than the round-off error made in setting k equal to $9.0 \times 10^9 \text{ N}\cdot\text{m}^2/\text{C}^2$.

EXAMPLE 22-1 A Charged Iron Ball

Suppose that a 7.256-kg iron ball used in the shot put event of track and field competition is given an excess negative charge of magnitude 1.00×10^{-6} C. This is nearly the maximum amount of charge that can be placed on a ball of this size. (You will learn how to calculate the maximum amount of charge a conductor can hold in Chap. 25.) How many excess electrons are on the ball? What percentage of the total number of electrons in the neutral ball does this correspond to?

SOLUTION The number of electrons that corresponds to a charge of magnitude 1.00×10^{-6} C is

$$N = \frac{1.00 \times 10^{-6}\ \text{C}}{1.60 \times 10^{-19}\ \text{C/electron}} = 6.25 \times 10^{12}\ \text{electrons}.$$

Now iron has a molecular mass of 55.85 g/mol, so the ball contains

$$(7256\ \text{g})\left(\frac{1\ \text{mol}}{55.85\ \text{g}}\right) = 130\ \text{mol}$$

of iron. Since each neutral iron atom contains 26 electrons, there are

$$\begin{aligned} &26(130\ \text{mol})(6.02 \times 10^{23}\ \text{electrons/mol}) \\ &\quad = 2.03 \times 10^{27}\ \text{electrons} \end{aligned}$$

in an uncharged ball. Thus the excess charge on the ball corresponds to only

$$\frac{6.25 \times 10^{12}}{2.03 \times 10^{27}}(100\%) = 3.08 \times 10^{-13}\%$$

of the neutral ball's electrons.

EXAMPLE 22-2 Comparison of Gravitational and Electric Forces in the Bohr Model

In the Bohr model of the hydrogen-atom ground state, the electron moves around the proton in a circular orbit of radius 5.3×10^{-11} m. Compare the magnitudes of the gravitational and electric forces between the proton and the electron.

SOLUTION The mutual gravitational force between the proton and the electron is

$$\begin{aligned} F_g &= G\frac{m_1 m_2}{r^2} = (6.67 \times 10^{-11}\ \text{N}\cdot\text{m}^2/\text{kg}^2) \\ &\times \frac{(9.1 \times 10^{-31}\ \text{kg})(1.67 \times 10^{-27}\ \text{kg})}{(5.3 \times 10^{-11}\ \text{m})^2} = 3.6 \times 10^{-47}\ \text{N}, \end{aligned}$$

and the mutual electric force is

$$\begin{aligned} F_e &= k\frac{|q_1||q_2|}{r^2} = (9.0 \times 10^9\ \text{N}\cdot\text{m}^2/\text{C}^2) \\ &\times \frac{(1.6 \times 10^{-19}\ \text{C})(1.6 \times 10^{-19}\ \text{C})}{(5.3 \times 10^{-11}\ \text{m})^2} = 8.2 \times 10^{-8}\ \text{N}. \end{aligned}$$

A comparison of these two forces gives us some interesting information. First, the electric force is approximately 10^{39} times larger than the gravitational force. Second, since both F_g and F_e are proportional to $1/r^2$, $F_e/F_g \approx 10^{39}$ everywhere. Consequently, the gravitational forces between elementary charged particles are always insignificant compared with the electric forces they exert on one another.

EXAMPLE 22-3 Net Electric Force

Three particles with charges $q_1 = -4.0 \times 10^{-6}$ C, $q_2 = 6.0 \times 10^{-6}$ C, and $q_3 = 2.0 \times 10^{-6}$ C are placed at the corners of an equilateral triangle with sides of length 10 cm. (See Fig. 22-9.) What is the force on q_3?

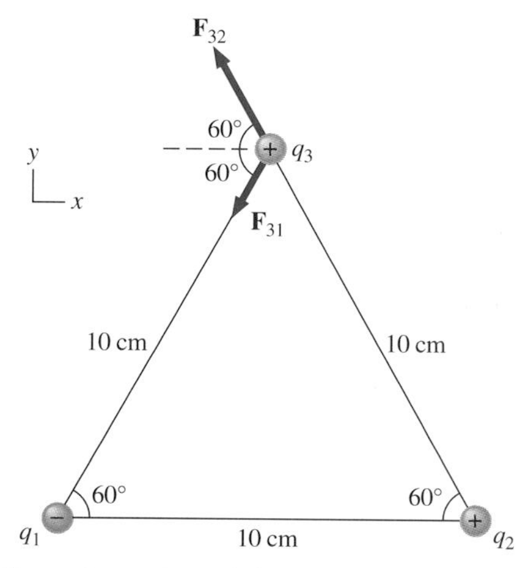

FIGURE 22-9 Three charged particles at the corners of an equilateral triangle.

SOLUTION To solve force problems involving several charges, we must first obtain the electric forces between individual pairs of the charges. You will find that the best way to determine the force vector between two charges is to first calculate its magnitude using Eq. (22-1) and then deduce its direction based on the signs of the charges. For this example, the magnitudes of the forces on q_3 due to q_1 and q_2 are

$$\begin{aligned} F_{31} &= k\frac{|q_1||q_3|}{r^2} = (9.0 \times 10^9\ \text{N}\cdot\text{m}^2/\text{C}^2) \\ &\times \frac{(4.0 \times 10^{-6}\ \text{C})(2.0 \times 10^{-6}\ \text{C})}{(0.10\ \text{m})^2} = 7.2\ \text{N}, \end{aligned}$$

and

$$\begin{aligned} F_{32} &= k\frac{|q_2||q_3|}{r^2} = (9.0 \times 10^9\ \text{N}\cdot\text{m}^2/\text{C}^2) \\ &\times \frac{(6.0 \times 10^{-6}\ \text{C})(2.0 \times 10^{-6}\ \text{C})}{(0.10\ \text{m})^2} = 10.8\ \text{N}. \end{aligned}$$

Since q_3 is positive, it is attracted by q_1 and repelled by q_2, as shown in the figure. The forces on q_3 can then be written as

$$\begin{aligned} \mathbf{F}_{31} &= -(7.2\ \text{N})\cos 60°\mathbf{i} - (7.2\ \text{N})\sin 60°\mathbf{j} \\ &= (-3.6\mathbf{i} - 6.2\mathbf{j})\ \text{N} \end{aligned}$$

and

$$\mathbf{F}_{32} = -(10.8\ \text{N})\cos 60°\mathbf{i} + (10.8\ \text{N})\sin 60°\mathbf{j}$$
$$= (-5.4\mathbf{i} + 9.3\mathbf{j})\ \text{N}.$$

The net force on q_3 is the vector sum of $\mathbf{F}_{31}$ and $\mathbf{F}_{32}$:

$$\mathbf{F} = \mathbf{F}_{31} + \mathbf{F}_{32} = (-9.0\mathbf{i} + 3.1\mathbf{j})\ \text{N},$$

which is a vector of magnitude 9.5 N directed at 161° with respect to the positive x axis.

EXAMPLE 22-4 Two Charged Balls in Equilibrium

Two identical balls each have an excess charge of 5.0×10^{-7} C, and they are attached to the ceiling by light strings 0.80 m long, as shown in Fig. 22-10a. If the angle that the threads make with respect to each other is 30°, determine the mass of each ball. Assume that the balls are small enough that they can be treated as particles.

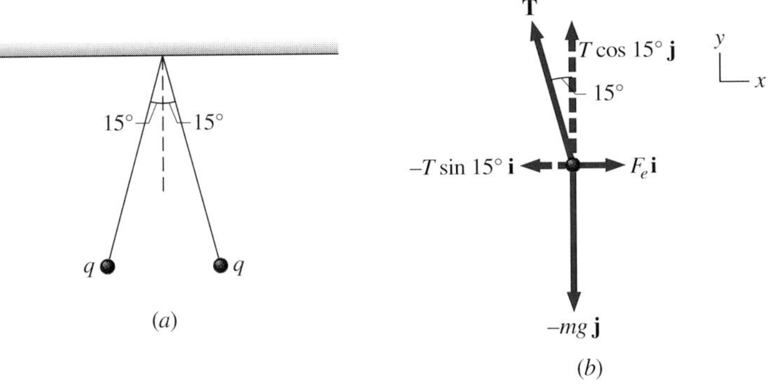

FIGURE 22-10 (a) Charged balls are in equilibrium as shown. (b) The free-body diagram of the ball on the right.

SOLUTION This is an equilibrium problem that can be solved by considering the forces on either ball. The free-body diagram for the ball on the right is shown in Fig. 22-10b. The forces that act on it are the string tension $\mathbf{T}$, the gravitational force $m\mathbf{g}$ and the electric force $\mathbf{F}_e$ of the other ball. Since the balls have the same charge, $\mathbf{F}_e$ must be directed as indicated. From Newton's second law,

$$\sum F_x = 0 \qquad \sum F_y = 0$$
$$F_e - T\sin 15° = 0 \qquad T\cos 15° - mg = 0.$$

These two equations can be combined to eliminate T. We then find

$$F_e = mg\tan 15°. \qquad (i)$$

The magnitude of the electric force can be obtained from Coulomb's law:

$$F_e = k\frac{q^2}{r^2} = (9.0 \times 10^9\ \text{N}\cdot\text{m}^2/\text{C}^2)\frac{(5.0 \times 10^{-7}\ \text{C})^2}{[2(0.80\ \text{m})\sin 15°]^2}$$
$$= 0.013\ \text{N}.$$

Finally, from Eq. (i), the mass of the ball is

$$m = \frac{F_e}{g\tan 15°} = \frac{0.013\ \text{N}}{(9.8\ \text{m/s}^2)\tan 15°} = 5.0 \times 10^{-3}\ \text{kg}.$$

DRILL PROBLEM 22-5

How many electrons must be removed from a conductor in order to give it a charge of 2.30×10^{-10} C?
ANS. 1.44×10^9 electrons.

DRILL PROBLEM 22-6

Suppose that three point charges $q_1 = 4.0 \times 10^{-6}$ C, $q_2 = -2.0 \times 10^{-6}$ C, and $q_3 = 5.0 \times 10^{-6}$ C are placed along a straight line as shown in Fig. 22-11. What is the electric force on each charge?
ANS. $13.0 \times 10^{-3}\mathbf{i}$ N, $-12.4 \times 10^{-3}\mathbf{i}$ N, $-0.63 \times 10^{-3}\mathbf{i}$ N.

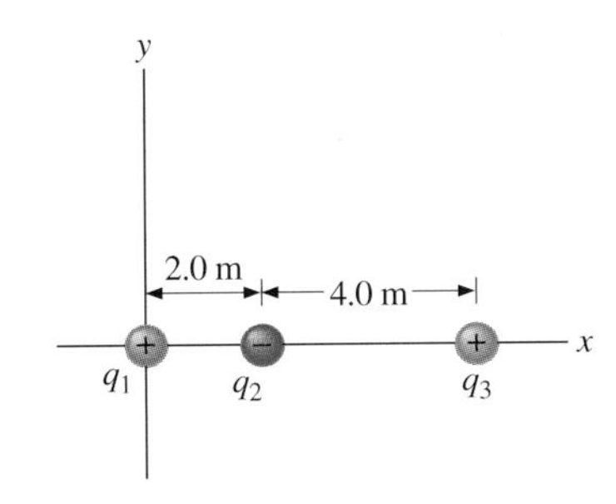

FIGURE 22-11 The three charges of Drill Prob. 22-6.

SUMMARY

1. **Electric charge**
   (a) An uncharged body becomes negatively charged when it acquires electrons and positively charged when it loses electrons.
   (b) Charge quantization: The charge on a body is an integer multiple of the electron's charge, -1.602×10^{-19} C.
   (c) Charge conservation: When bodies exchange charge, the net charge of the bodies does not change.

2. **Conductors and insulators**
   (a) The conduction electrons are free to move throughout a conductor. There are very few free electrons in an insulator.
   (b) In the conduction process, a conductor with one type of charge is used to produce the same type of charge on a second conductor through transfer of electrons. In the induction process, a conductor is charged without ben-

efit of contact with a charged object. The charge induced on the conductor is opposite to that of the charged object. Polarization occurs when the positions of the charges of the molecules of an insulator are shifted by an external electric force.

3. **Coulomb's law**

Two point charges q_1 and q_2 exert equal and opposite forces on each another of magnitude

$$F = k\frac{|q_1||q_2|}{r^2},$$

where $k = 9.0 \times 10^9\ \text{N}\cdot\text{m}^2/\text{C}^2$. The force is directed along the line between the two charges. It is attractive if q_1 and q_2 have opposite signs, and it is repulsive if q_1 and q_2 have the same sign. The force on a point charge due to a group of charges must be calculated by adding vectorially the forces that each charge exerts on that point charge.

QUESTIONS

22-1. A positively charged rod attracts a small piece of cork. (*a*) Can we conclude that the cork is negatively charged? (*b*) The rod repels another small piece of cork. Can we conclude that this piece is positively charged?

22-2. Two bodies attract each other electrically. Do they both have to be charged? Answer the same question if the bodies repel one another.

22-3. How would you determine whether the charge on a particular rod is positive or negative?

22-4. Does the uncharged conductor shown in the accompanying figure experience a net electric force?

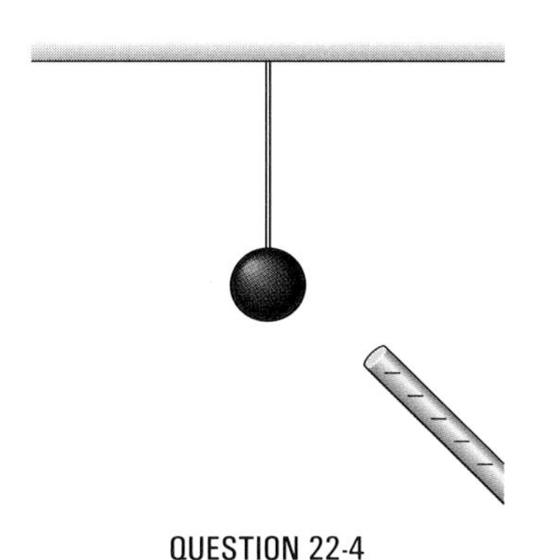

QUESTION 22-4

22-5. While walking across a rug, a person frequently becomes charged because of the rubbing between his shoes and the rug. This charge then causes a spark and a slight shock when the person gets close to a metal object. Why are these shocks so much more common on a dry day?

22-6. Compare charging by conduction to charging by induction.

22-7. Small pieces of tissue are attracted to a charged comb. Soon after sticking to the comb, the pieces of tissue are repelled from it. Explain.

22-8. Trucks that carry gasoline often have chains dangling from their undercarriages and brushing the ground. Why?

22-9. Why do electrostatic experiments work so poorly in humid weather?

22-10. Why do some clothes cling together after being removed from the clothes dryer? Does this happen if they're still damp?

22-11. Would defining the charge on an electron to be positive have any effect on Coulomb's law?

22-12. An atomic nucleus contains positively charged protons and uncharged neutrons. Since nuclei do stay together, what must we conclude about the forces between these nuclear particles?

22-13. Is the force between two fixed charges influenced by the presence of other charges?

22-14. Can induction be used to produce charge on an insulator?

22-15. Suppose someone tells you that rubbing quartz with cotton cloth produces a third kind of charge on the quartz. Describe what you might do to test this claim.

22-16. A handheld copper rod will not acquire a charge when you rub it with a cloth. Explain why.

PROBLEMS

Electric Charge

22-1. What is the net charge of all of the electrons in 1.0 mol of hydrogen gas (H_2)?

22-2. A 2.5-g copper penny is given a charge of -2.0×10^{-9} C. (*a*) How many excess electrons are on the penny? (*b*) By what percent do the excess electrons change the mass of the penny?

22-3. A 2.5-g copper penny is given a charge of 4.0×10^{-9} C. (*a*) How many electrons are removed from the penny? (*b*) If no more than one electron is removed from an atom, what percent of the atoms are ionized by this charging process?

Coulomb's Law

22-4. In a salt crystal, the distance between adjacent sodium and chlorine ions is 2.82×10^{-10} m. What is the force of attraction between the two singly charged ions?

22-5. Estimate how many electrons must be added to each of the lead spheres of the Cavendish experiment (see Sec. 5-5) to produce an electric force one-tenth as large as the gravitational force. Assume that each sphere has a mass of 5.0 kg and that the centers of the spheres are 0.25 m apart.

22-6. Protons in an atomic nucleus are typically 10^{-15} m apart. What is the electric force of repulsion between nuclear protons?

22-7. Suppose you could separate all of the electrons from the hydrogen and oxygen nuclei in a cup (8 oz) of water. If you could then place all of the electrons on a small sphere and all of the nuclei on a second small sphere, what would be the force of attraction in pounds between the spheres when they are 6.0 ft apart?

22-8. Suppose the earth and the moon each carried a net negative charge $-Q$. (*a*) What value of Q is required to balance the gravitational attraction between the two bodies? (*b*) Does the distance between the earth and moon affect your answer? (*c*) How many electrons would be needed to produce this charge? (*d*) Calculate the minimum number of moles of hydrogen needed to produce this number of electrons.

22-9. Point charges $q_1 = 50\ \mu C$ and $q_2 = -25\ \mu C$ are placed 1.0 m apart. What is the force on a third charge $q_3 = 20\ \mu C$ placed midway between q_1 and q_2?

22-10. Where must q_3 of the previous problem be placed so that the net force on it is zero?

22-11. A charge $q = 2.0\ \mu C$ is placed at the point P shown in the accompanying figure. What is the force on q?

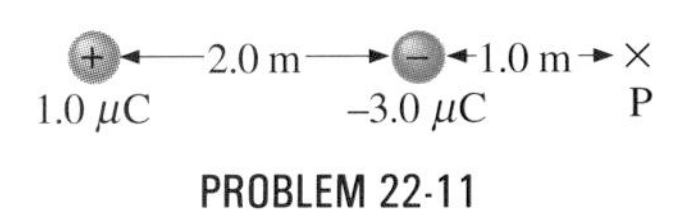

PROBLEM 22-11

22-12. Two small balls, each of mass 25 g, are 1.0 m apart. (*a*) How much negative charge must be placed on each so that the electric and the gravitational forces are equal in magnitude? (*b*) How many electrons is this charge equivalent to?

22-13. Two small balls, each of mass 5.0 g, are attached to silk threads 50 cm long, which are in turn tied to the same point on the ceiling, as shown in the accompanying figure. When the balls are given the same charge q, the threads hang at 5.0° to the vertical. What is the magnitude of q? Can you determine the sign of q?

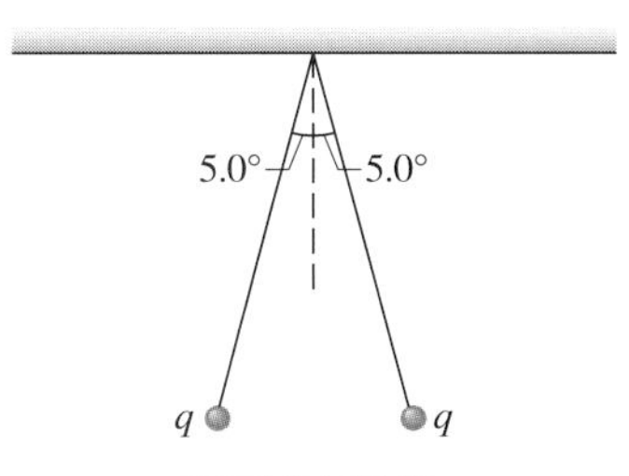

PROBLEM 22-13

22-14. What is the net electric force on the charge located at the lower-right-hand corner of the triangle shown in the accompanying figure?

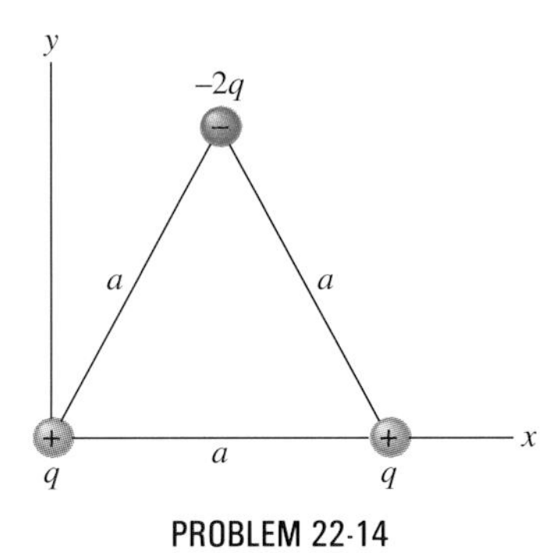

PROBLEM 22-14

22-15. Two fixed particles, each of charge 5.0×10^{-6} C, are 24 cm apart. What force do they exert on a third particle of charge -2.5×10^{-6} C that is 13 cm from each of them?

22-16. The charges $q_1 = 2.0 \times 10^{-7}$ C, $q_2 = -4.0 \times 10^{-7}$ C, and $q_3 = -1.0 \times 10^{-7}$ C are placed at the corners of the triangle shown in the accompanying figure. What is the force on q_1?

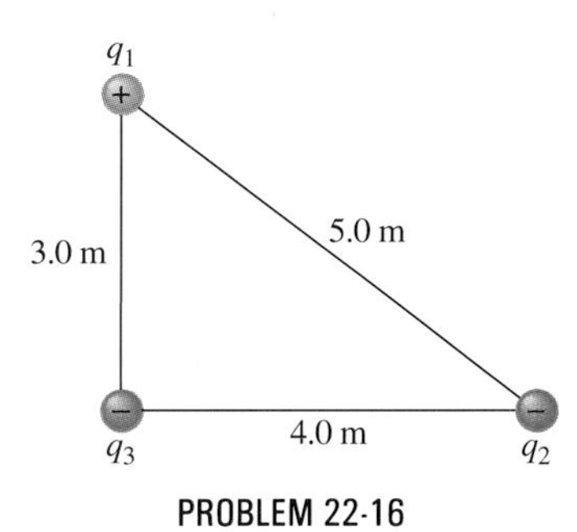

PROBLEM 22-16

22-17. What is the force on the charge q at the lower-right-hand corner of the square shown in the accompanying figure?

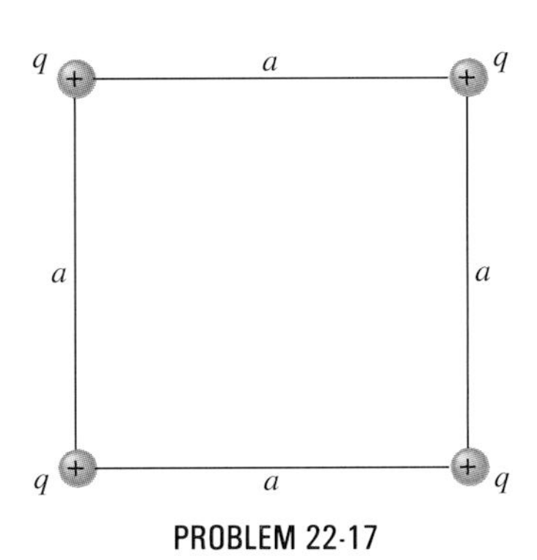

PROBLEM 22-17

22-18. Point charges $q_1 = 10\ \mu C$ and $q_2 = -30\ \mu C$ are fixed at $\mathbf{r}_1 = (3.0\mathbf{i} - 4.0\mathbf{j})$ m and $\mathbf{r}_2 = (9.0\mathbf{i} + 6.0\mathbf{j})$ m. What is the force of q_2 on q_1?

22-19. Point charges $q_1 = 2.0\ \mu C$ and $q_2 = 4.0\ \mu C$ are located at $\mathbf{r}_1 = (4.0\mathbf{i} - 2.0\mathbf{j} + 5.0\mathbf{k})$ m and $\mathbf{r}_2 = (8.0\mathbf{i} + 5.0\mathbf{j} - 9.0\mathbf{k})$ m. What is the force of q_2 on q_1?

22-20. What is the force on the 5.0-μC charge shown in the accompanying figure?

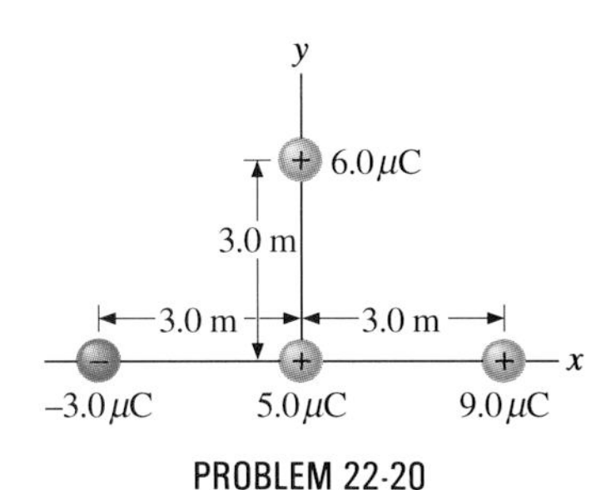

PROBLEM 22-20

22-21. What is the force on the 2.0-μC charge placed at the center of the square shown in the accompanying figure?

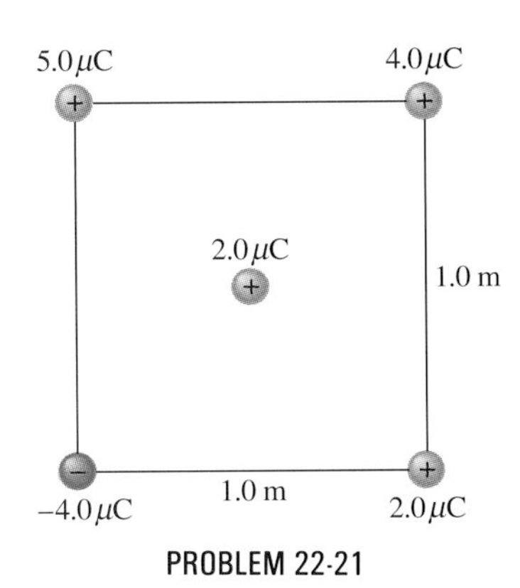

PROBLEM 22-21

22-22. Four charged particles are positioned at the corners of a parallelogram as shown in the accompanying figure. If $q = 5.0\ \mu$C and $Q = 8.0\ \mu$C, what is the net force on q?

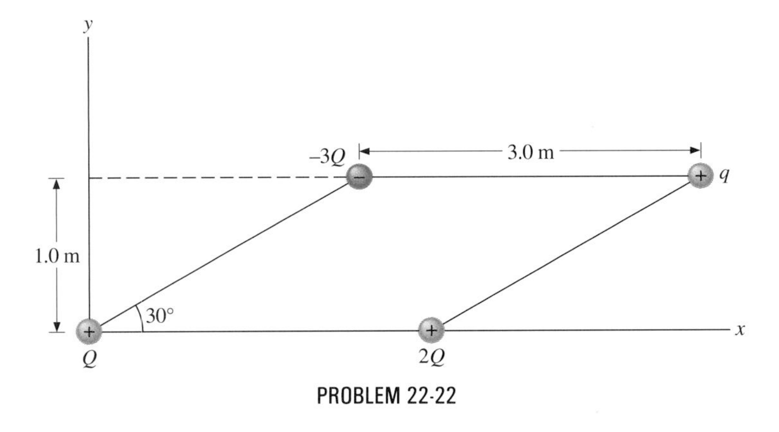

PROBLEM 22-22

General Problems

22-23. The net excess charge on two small spheres is Q. Show that the force of repulsion between the spheres is greatest when each sphere has an excess charge $Q/2$. Assume that the distance between the spheres is so large compared with their radii that the spheres can be treated as point charges.

22-24. Two small, identical conducting spheres repel each other with a force of 0.050 N when they are 0.25 m apart. After a conducting wire is connected between the spheres and then removed, they repel each other with a force of 0.060 N. What is the original charge on each sphere?

22-25. A charge Q is fixed at the origin and a second charge q moves along the x axis, as shown in the accompanying figure. How much work is done on q by the electric force when q moves from x_1 to x_2?

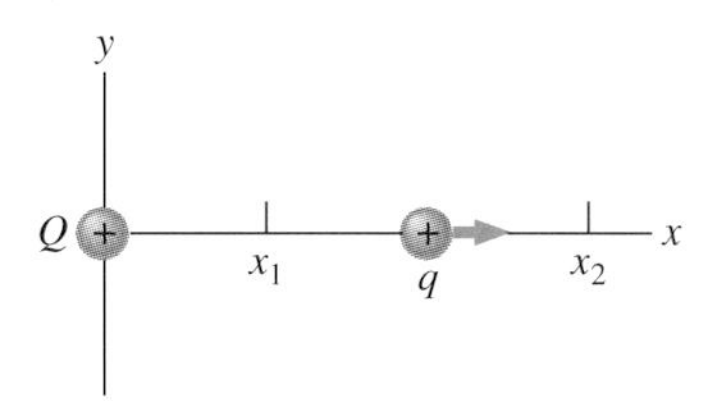

PROBLEM 22-25

22-26. A charge $q = -2.0\ \mu$C is released from rest when it is 2.0 m from a fixed charge $Q = 6.0\ \mu$C. What is the kinetic energy of q when it is 1.0 m from Q?

22-27. The charge per unit length on the thin rod of the accompanying figure is λ. What is the electric force on the point charge q? Solve this problem by first considering the electric force $d\mathbf{F}$ on q due to a small segment dx of the rod, which contains charge $\lambda\ dx$. Then find the net force by integrating $d\mathbf{F}$ over the length of the rod.

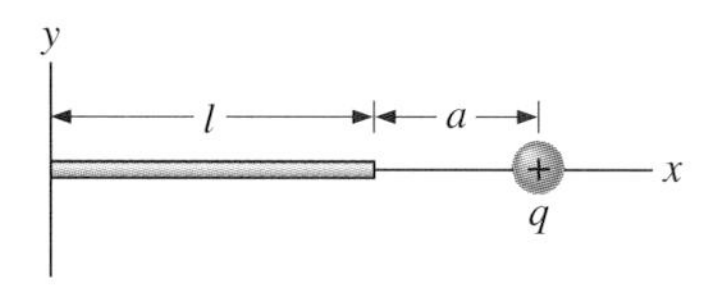

PROBLEM 22-27

22-28. The charge per unit length on the thin rod shown in the accompanying figure is λ. What is the electric force on the point charge q? (See Prob. 22-27.)

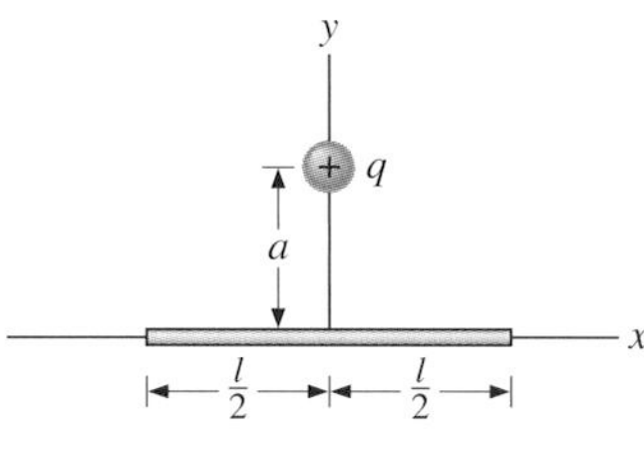

PROBLEM 22-28

22-29. The charge per unit length on the thin semicircular wire shown in the accompanying figure is λ. What is the electric force on the point charge q? (See Prob. 22-27.)

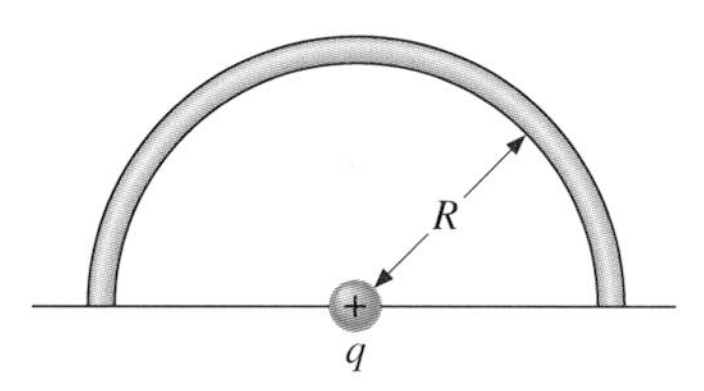

PROBLEM 22-29

22-30. Charge Q is distributed uniformly over a spherical volume of radius r. What is the force on a point charge q at a distance R ($> r$) from the center of the distribution? (*Hint:* See Suppl. 5-1.)

This brave volunteer is showing her friends an effect of the electric field.

CHAPTER 23 THE ELECTRIC FIELD

PREVIEW

In this chapter we define the electric field and describe how the electric fields of fixed charge distributions are calculated. The main topics to be discussed here are the following:

1. **Definition of the electric field.** The electric field is defined in terms of the force it exerts on a test charge.
2. **Electric fields of charge distributions.** The electric field of an isolated point charge is determined, and the method used to calculate the electric fields of arbitrary charge distributions is introduced.
3. **Field lines.** We consider this means of visualizing the electric field.
4. **Calculating the electric field.** The electric fields of various charge distributions are calculated.
5. **Motion of a charged particle in a uniform electric field.** This motion is analyzed by determining the acceleration of a charged particle due to the electric force and then applying the kinematic equations for constant acceleration.

In this chapter we discuss the interaction of electric charges in terms of an electric field. The electric field is introduced and defined through a comparison with the gravitational field, with which it is remarkably similar in form. The electric fields of various charge distributions are calculated and the visualization of the electric field through the use of field lines is discussed. We end this chapter by investigating the motion of a charged particle in a uniform electric field, an important example of which is Millikan's measurement of the charge of the electron.

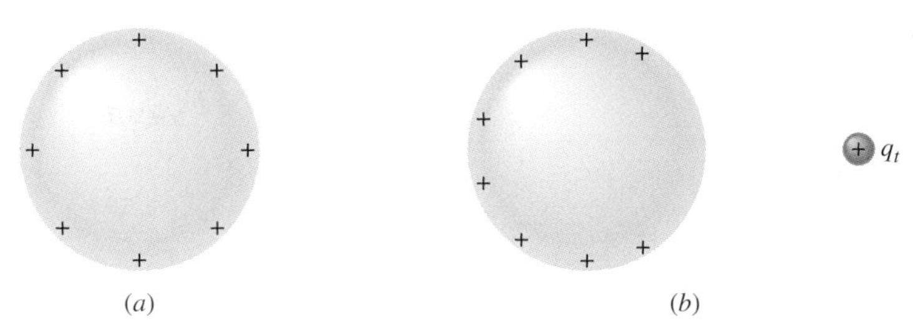

FIGURE 23-1 (*a*) The charged isolated sphere creates a spherically symmetric electric field. (*b*) The test charge q_t alters the charge distribution on the sphere and consequently its electric field as well.

23-1 Definition of the Electric Field

In Sec. 5-6 we learned that two masses exert a mutual gravitational force on each other across empty space. We overcame the difficulty of conceptualizing this action-at-a-distance by introducing the idea of the gravitational field. We discarded the notion that m_1 directly exerts a gravitational force on m_2 and instead used the idea that m_1 first creates a gravitational field, which then exerts the force on m_2 (or vice versa). The electric interaction between two charges is likewise described using the concept of the **electric field**, one of the most significant and useful concepts in electromagnetism. We say that charge q_1 creates an electric field which then exerts a force on q_2.

A gravitational field can be determined experimentally at any location by placing a small test mass m_t there and measuring the gravitational force $\mathbf{F}_g$ on it. The field is then

$$\mathbf{g} = \frac{\mathbf{F}_g}{m_t}.$$

Like its gravitational counterpart, the electric field is determined with a test particle, in this case a positive test charge q_t. If at a particular point the electric force on q_t is $\mathbf{F}_e$, the electric field there is by definition

$$\mathbf{E} = \frac{\mathbf{F}_e}{q_t}. \tag{23-1}$$

In measuring $\mathbf{E}$, we must make sure that q_t does not disturb the charge distribution that creates the field. For example, if q_t is large enough that it has the effect shown in Fig. 23-1, it will distort the original electric field, the one being measured. This experimental restriction is included in the definition of $\mathbf{E}$ by making q_t vanishingly small:

$$\mathbf{E} = \lim_{q_t \to 0} \frac{\mathbf{F}_e}{q_t}.$$

You can see from the defining equation for $\mathbf{E}$ that it is a vector whose SI unit is the newton per coulomb (N/C). The direction of $\mathbf{E}$ is the same as that of the force on the positive test charge, which, of course, is opposite to the force on a negative test charge. Also notice that q_t is not part of the charge distribution that produces $\mathbf{E}$. *The field remains when the test charge is removed.*

In the next two chapters, we'll limit our investigation to *electrostatic fields*, which are time-independent electric fields produced by fixed charge distributions. This is an important restriction because fields that vary with time behave differently from electrostatic fields. Time-varying fields will be studied in later chapters.

EXAMPLE 23-1 Forces of the Gravitational and Electric Fields

The magnitude of a typical electric field produced in the laboratory is 1.0×10^5 N/C. Compare the weight of an electron with the force this electric field exerts on it.

SOLUTION From the defining equations for the electric and gravitational fields, the magnitudes of the electric force $\mathbf{F}_e$ and the gravitational force $\mathbf{F}_g$ on the electron are

$$F_e = eE = (1.6 \times 10^{-19}\ \text{C})(1.0 \times 10^5\ \text{N/C}) = 1.6 \times 10^{-14}\ \text{N},$$

and

$$F_g = mg = (9.1 \times 10^{-31}\ \text{kg})(9.8\ \text{m/s}^2) = 8.9 \times 10^{-30}\ \text{N},$$

so $F_e/F_g = 1.8 \times 10^{15}$. This example illustrates what is always the case in the laboratory: The gravitational forces on elementary charged particles are negligible compared with the electric forces they experience.

EXAMPLE 23-2 A Charged Sphere in Static Equilibrium

A small sphere of mass 20 g that carries an excess charge of 2.0×10^{-6} C is suspended at the end of a light string. When a uniform horizontal electric field $\mathbf{E}$ is applied as shown in Fig. 23-2*a*, the sphere reaches static equilibrium with the string under tension $\mathbf{T}$ and making an angle of $\theta = 5.0°$ to the vertical. What is the magnitude of the applied field?

SOLUTION Since the sphere is in static equilibrium, we have

$$\mathbf{T} + m\mathbf{g} + q\mathbf{E} = 0.$$

Hence in the x and y directions (see Fig. 23-2*b*),

$$T \sin\theta = qE, \tag{i}$$

$$T \cos\theta = mg. \tag{ii}$$

Dividing (*i*) by (*ii*), we obtain

$$\tan\theta = \frac{qE}{mg},$$

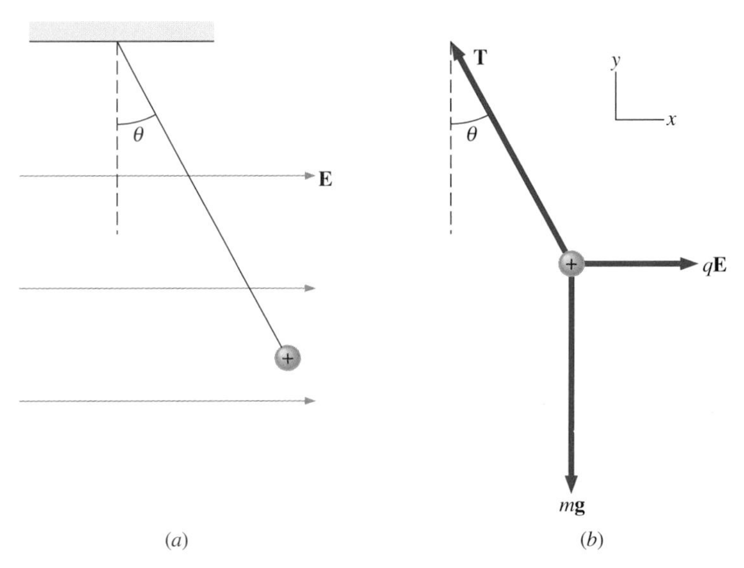

FIGURE 23-2 A charged sphere in static equilibrium.

and the magnitude of the electric field is

$$E = \frac{mg \tan \theta}{q} = \frac{(0.020 \text{ kg})(9.8 \text{ m/s}^2)(\tan 5.0°)}{2.0 \times 10^{-6} \text{ C}}$$
$$= 8.6 \times 10^3 \text{ N/C}.$$

Notice that in this case the electric and gravitational forces on the sphere are comparable: Here, in contrast to Example 23-1, we are not dealing with an elementary particle but with an object whose mass-to-charge ratio is much greater.

DRILL PROBLEM 23-1

Describe some common situations for which the electric force and the gravitational force on a body are comparable.

23-2 THE ELECTRIC FIELD OF A CHARGE DISTRIBUTION

To determine the electric field of an isolated point charge q, let's consider a test charge q_t at an arbitrary point P, which is a distance r from q. (See Fig. 23-3.) From Coulomb's law, the magnitude of the force of q on q_t is

$$F_e = \frac{1}{4\pi\epsilon_0} \frac{|q||q_t|}{r^2},$$

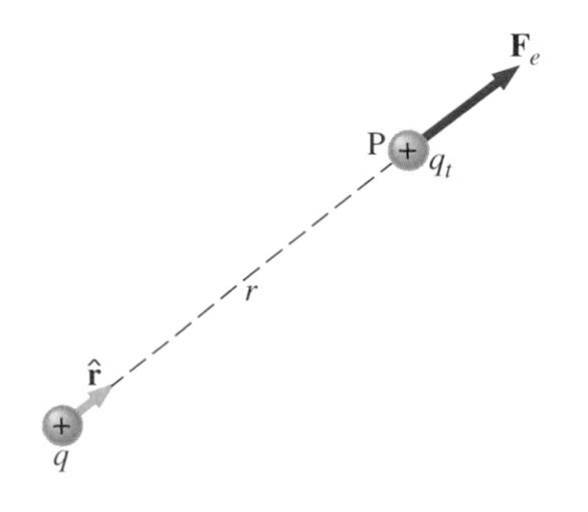

FIGURE 23-3 The electric field at point P is determined by measuring the force on a test charge q_t placed there.

so, from Eq. (23-1), the magnitude of the electric field at P due to q is

$$E = \frac{F_e}{|q_t|} = \frac{1}{4\pi\epsilon_0} \frac{|q|}{r^2}.$$

To specify the direction of the electric field, we let $\hat{\mathbf{r}}$ represent a unit vector pointing from q to q_t. Then the electric field at P due to q is

$$\mathbf{E} = \frac{1}{4\pi\epsilon_0} \frac{q}{r^2} \hat{\mathbf{r}}. \tag{23-2}$$

From this expression, we see that *the electric field of a positive charge* ($q > 0$) *is directed radially away from the charge, while the electric field of a negative charge* ($q < 0$) *is directly radially toward that charge.* Furthermore, since E depends only on distance from the charge, the electric field of a point charge is isotropic. These two properties are illustrated in Fig. 23-4, which shows the electric field vectors at four points P_1, P_2, P_3, and P_4 due to an isolated positive charge (Fig. 23-4*a*) and an isolated negative charge (Fig. 23-4*b*). At point P_1, which is closest to the charge, the magnitude of the electric field is the greatest, while at P_4, the farthest point, the magnitude of the field is smallest. Since P_2 and P_3 are at the same distance from the charge, the electric fields at these two points have the same magnitude.

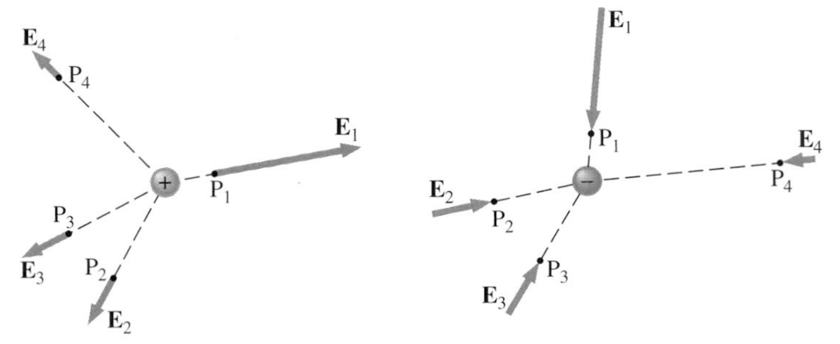

FIGURE 23-4 The electric field of a point charge is directed radially *away from* the charge if it is positive and *toward* the charge if it is negative. The magnitude of the electric field decreases as the square of the distance from the charge.

The electric field of a discrete charge distribution at an arbitrary point is the vector sum (or *superposition*) of the electric fields of the individual point charges; that is,

$$\mathbf{E} = \frac{1}{4\pi\epsilon_0} \sum_i \frac{q_i}{r_i^2} \hat{\mathbf{r}}_i, \tag{23-3}$$

where $\sum_i$ represents a sum over all charges of the distribution, r_i is the distance from a charge q_i to the point where the net electric field is to be determined, and $\hat{\mathbf{r}}_i$ is the radial unit vector directed from charge q_i to that point. Figure 23-5*a* shows the electric fields at point P of three charges. The net electric field at P is found from Eq. (23-3), that is, by adding vectorially the fields of the individual charges, as shown in Fig. 23-5*b*.

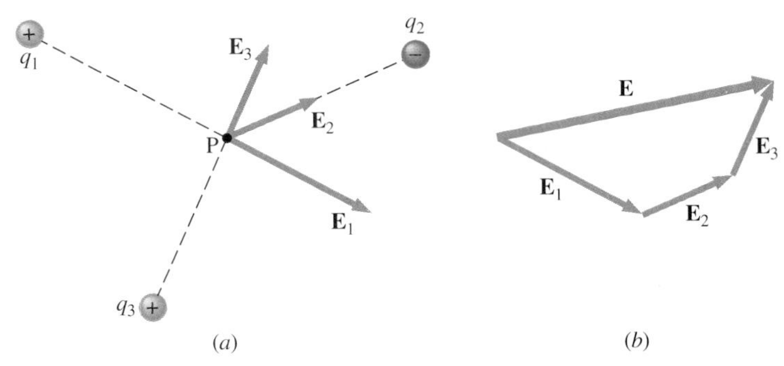

FIGURE 23-5 The net electric field **E** at P can be found by adding vectorially the fields of the individual charges.

Often the number of point charges involved is so large that their resulting charge distribution can be treated as continuous. The field of a continuous charge distribution is determined by dividing it into infinitesimal elements dq and evaluating

$$\mathbf{E} = \frac{1}{4\pi\epsilon_0}\int \frac{dq}{r^2}\,\hat{\mathbf{r}}. \tag{23-4a}$$

This integration, like the summation of Eq. (23-3), is a vector operation. Equation (23-4a) can best be evaluated by integrating over the spatial dimensions of the charge distribution. In so doing, we need to know the **charge density** of the distribution. Consider a spherical volume of radius $r = 10$ cm that contains a charge $Q = 1.0 \times 10^{-6}$ C uniformly distributed throughout; the *volume charge density* (or charge per unit volume) ρ within the sphere is

$$\rho = \frac{Q}{\frac{4}{3}\pi r^3} = \frac{1.0 \times 10^{-6}\ \text{C}}{\frac{4}{3}\pi(0.10\ \text{m})^3} = 2.4 \times 10^{-4}\ \text{C/m}^3.$$

The volume charge density is not necessarily constant throughout a charge distribution; nevertheless, if we consider an infinitesimal volume element dv of the distribution, we may always write for the amount of charge dq within that element

$$dq = \rho\, dv,$$

and Eq. (23-4a) can then be expressed as

$$\mathbf{E} = \frac{1}{4\pi\epsilon_0}\int \frac{\rho\, dv}{r^2}\,\hat{\mathbf{r}}, \tag{23-4b}$$

which is simply an integration over the volume of the charge distribution.

Similarly, if charge is distributed over a surface with *surface charge density* (or charge per unit area) σ, then the amount of charge dq in an infinitesimal area element dA is

$$dq = \sigma\, dA,$$

and Eq. (23-4a) becomes

$$\mathbf{E} = \frac{1}{4\pi\epsilon_0}\int \frac{\sigma\, dA}{r^2}\,\hat{\mathbf{r}}. \tag{23-4c}$$

Finally, for charge distributed along a line with *linear charge density* (charge per unit length) λ, we have

$$dq = \lambda\, dl,$$

$$\mathbf{E} = \frac{1}{4\pi\epsilon_0}\int \frac{\lambda\, dl}{r^2}\,\hat{\mathbf{r}}. \tag{23-4d}$$

23-3 Field Lines

Field lines, a concept introduced by Michael Faraday, provide us with an easy way to visualize the electric field. The lines are drawn such that

1. They originate at positive charges and terminate at negative charges.
2. The number of field lines originating or terminating at a charge is proportional to the magnitude of that charge.
3. At each point in space, a field line is tangent to the electric field passing through that point.
4. The number of lines per unit area passing through a small cross section perpendicular to the electric field is proportional to the magnitude of the field at the cross section. If the lines are close together, the field is large; if they are far apart, the field is small.

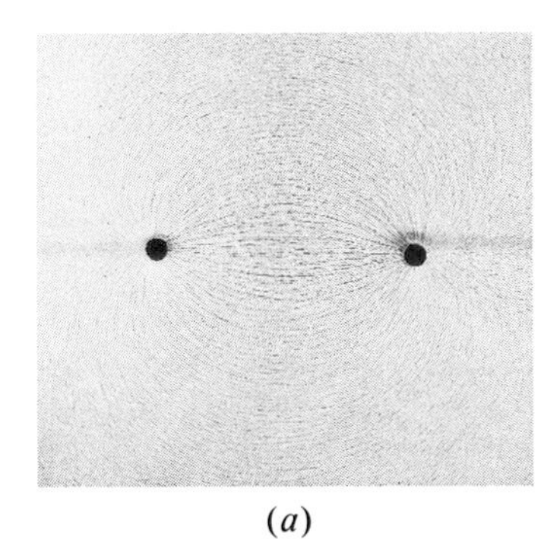

(a)

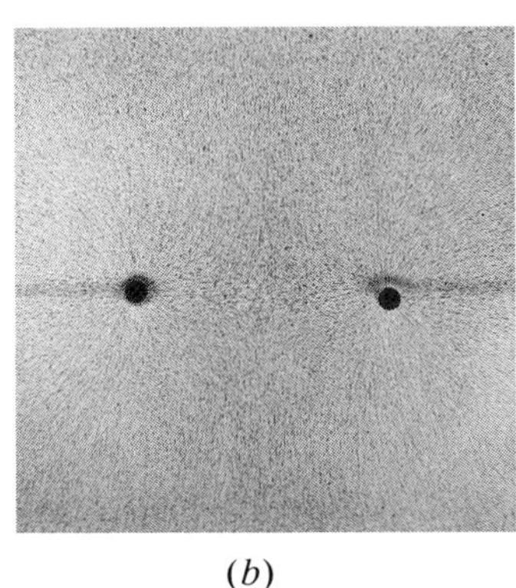

(b)

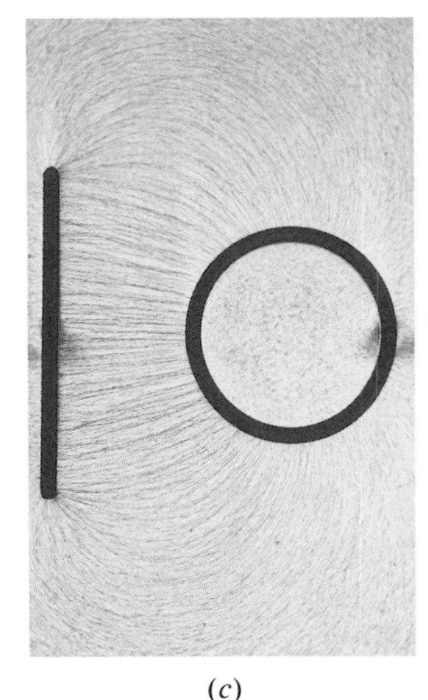

(c)

The electric field lines for (a) two small, oppositely charged conductors, (b) two conductors with like charge, and (c) a charged conducting plate near a charged conducting cylinder. The lines are delineated by small pieces of thread suspended in oil, which align with the electric field.

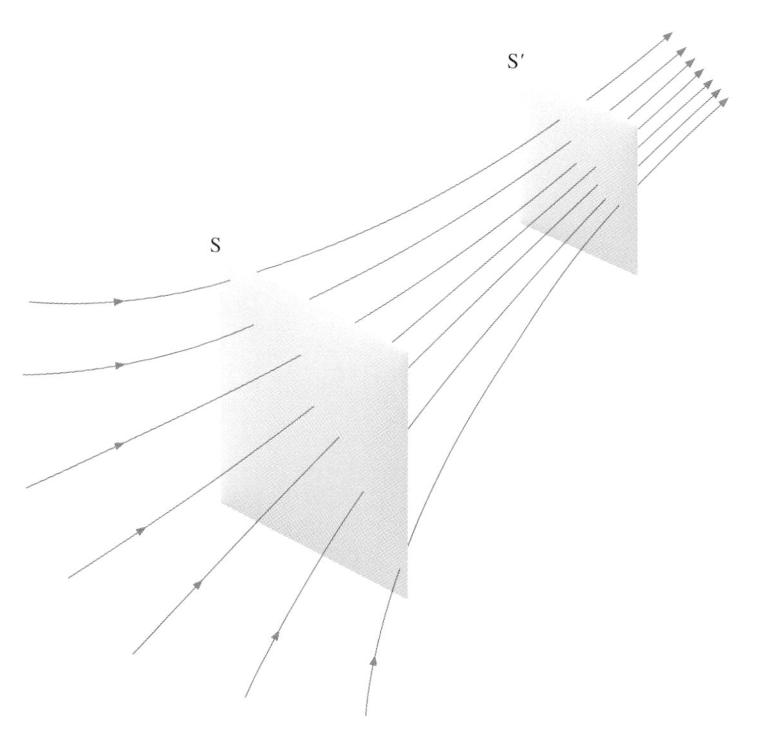

FIGURE 23-6 The density of the electric field lines is an indication of the strength of the field. Here the field is stronger at S′ than at S.

A set of field lines is shown in Fig. 23-6. The same number of lines pass through S and the smaller area S′, so the electric field is stronger at S′ than at S. The electric field of an isolated positive charge q is represented by uniformly distributed, continuous field lines emanating radially from the charge, as shown in Fig. 23-7*a*. The field lines of an isolated negative charge $-q$ are depicted in Fig. 23-7*b*. They behave just like those of a positive charge, except they are directed radially inward to indicate the direction of **E**.

In order to show that these field lines are consistent with the expression for the electric field of an isolated point charge (Eq. 23-2), let's consider a series of imaginary concentric spheres centered on the charge. (See Fig. 23-7*b*.) The same number of field lines cross the surface of each sphere. Since the surface areas of these concentric spheres vary as r^2, the density of the field lines (and hence the strength of the field) decreases with distance r from the charge as $1/r^2$, in accordance with Eq. (23-2).

Always keep in mind that field lines serve only as a convenient way to visualize the electric field; they are *not* physical entities. Although the direction and relative intensity of the electric field can be deduced from a set of field lines, the lines can also be misleading. For example, the field lines drawn to represent the electric field in a region must, by necessity, be discrete. However, the electric field in that region exists at every point.

Field lines for three groups of discrete charges are shown in Fig. 23-8. Since the charges of Fig. 23-8*a* have the same magnitude, the same number of field lines are shown leaving or entering each charge. This is also the case for Fig. 23-8*b*. In Fig. 23-8*c* we draw three times as many field lines leaving the $+3q$ charge as entering the $-q$ charge. The lines that do not terminate at $-q$ emanate outward from the charge configuration.

It's clear from the figure that in the region immediately surrounding any single charge within a group of charges the field lines are very nearly radial. How do the field lines appear at distances that are much larger than the separation between the charges? We can get a qualitative answer to this question by comparing the charge configurations of Fig. 23-8*b* and 23-8*c*. Suppose that for a charge of magnitude q, there are N field lines entering or leaving the charge. In Fig. 23-8*b*, none of the lines

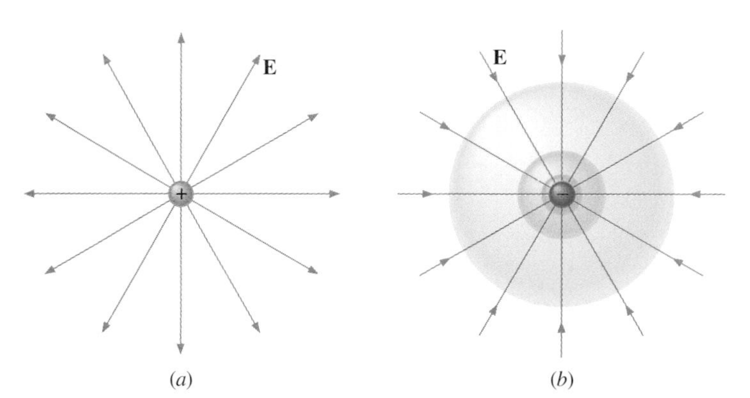

FIGURE 23-7 Electric field lines of (*a*) a positive point charge and (*b*) a negative point charge. The magnitude of the electric field (or the density of the field lines) of a point charge decreases as $1/r^2$.

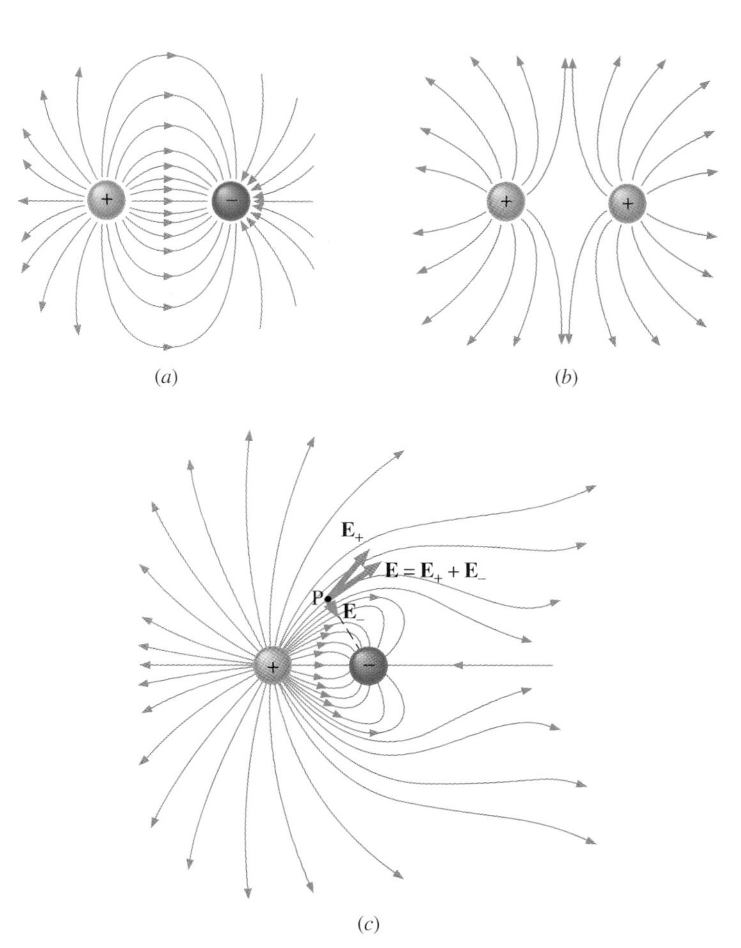

FIGURE 23-8 The field lines for (*a*) $+q$ and $-q$, (*b*) $+q$ and $+q$, and (*c*) $+3q$ and $-q$. At any point P in part (*c*), the net field is $\mathbf{E}_+ + \mathbf{E}_-$, and it is tangent to the field line.

terminates at a charge; hence there are $2N$ lines emanating from the two charges. At very large distances, these lines appear to be directed radially from the charge configuration. In other words, it appears as though the $2N$ field lines were coming from a single charge $+2q$. In Fig. 23-8*c* $3N$ lines emerge from the $+3q$ charge, while N lines terminate at the $-q$ charge. The remaining number, $2N$, emanate outward. Once again, at large distances from the charges, the field lines resemble those from a single $+2q$ charge. Hence we see that far away from a group of charges, the field lines (and thus the electric field) appear as though they are due to a single charge that is the sum of all the charges in the group. Try applying this result to the two equal and opposite charges of Fig. 23-8*a*. Can you picture what happens to the field lines (and the electric field) at progressively greater distances away from the charges?

DRILL PROBLEM 23-2

Do field lines ever cross? (*Hint:* Consider the definition of a field line.)

23-4 Calculating the Electric Field

The following examples illustrate how the electric fields of various charge distributions are determined. In calculating **E**, you should remember that (1) the electric field of a positive charge is directed away from that charge, while the electric field of a negative charge is directed toward the charge, and (2) electric fields are vectors and must be added accordingly. For a group of discrete charges, the net field is simply the vector sum of the fields of the individual charges (Eq. 23-3). For a continuous charge distribution, you must perform an integration (Eqs. 23-4) to find the net field. The continuous charge distributions we consider generally have some spatial symmetry, and as a result, the integral is simplified. Such continuous distributions are analyzed in Examples 23-5 through 23-9.

EXAMPLE 23-3 The Electric Field of Two Identical Point Charges

Two positive point charges q are separated by a distance $2a$. (See Fig. 23-9.) Find the electric field due to the two charges at P and at P′.

SOLUTION (*a*) At P, the electric field of the left charge points in the positive x direction, the field of the right charge points in the negative x direction, and the magnitude of each field equals $q/4\pi\epsilon_0 a^2$. The resultant electric field at P is therefore zero.

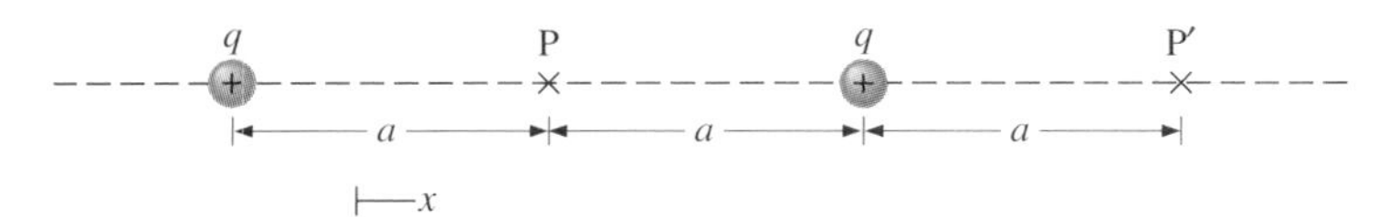

FIGURE 23-9 Two point charges.

(*b*) At P′, the electric fields due to the left and right charges are $[q/4\pi\epsilon_0(3a)^2]\mathbf{i}$ and $(q/4\pi\epsilon_0 a^2)\mathbf{i}$, respectively. The resultant electric field at P′ is then

$$\mathbf{E} = \frac{1}{4\pi\epsilon_0}\left[\frac{q}{(3a)^2}\mathbf{i} + \frac{q}{a^2}\mathbf{i}\right] = \frac{5}{18\pi\epsilon_0}\frac{q}{a^2}\mathbf{i}.$$

DRILL PROBLEM 23-3

Suppose the right particle of Fig. 23-9 is replaced by a particle of charge $-q$. Now what are the electric fields at P and at P′?
ANS. $(q/2\pi\epsilon_0 a^2)\mathbf{i}$; $(-2q/9\pi\epsilon_0 a^2)\mathbf{i}$.

EXAMPLE 23-4 The Electric Field of Two Point Charges

(*a*) Point charges $q_1 = 6.0 \times 10^{-9}$ C and $q_2 = -4.0 \times 10^{-9}$ C are placed 10.0 cm apart, as shown in Fig. 23-10*a*. Calculate the electric field due to these two charges at P and at P′. Point P is between the charges, and P′ and the two charges are at the corners of an equilateral triangle. (*b*) A third point charge $q_3 = 3.0 \times 10^{-9}$ C is placed on the positive y axis 4.0 cm above q_1. (See Fig. 23-10*b*.) Now what is the electric field at P′?

SOLUTION (*a*) At P, the field due to q_1 points in the positive x direction and has magnitude

$$E_1 = (9.0 \times 10^9\ \mathrm{N \cdot m^2/C^2})\frac{6.0 \times 10^{-9}\ \mathrm{C}}{(0.040\ \mathrm{m})^2} = 3.4 \times 10^4\ \mathrm{N/C}.$$

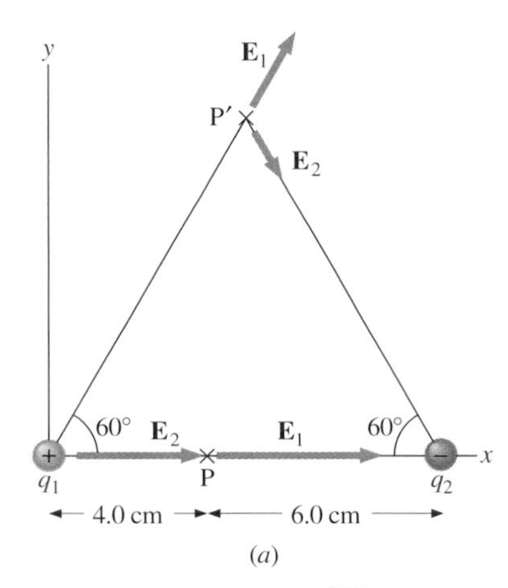

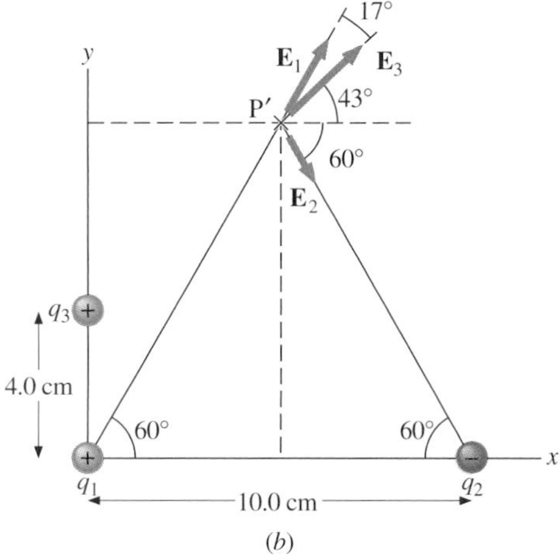

FIGURE 23-10 Simple charge distributions.

The field at P due to q_2 also points in the positive x direction; its magnitude is

$$E_2 = (9.0 \times 10^9\ \mathrm{N \cdot m^2/C^2})\,\frac{4.0 \times 10^{-9}\ \mathrm{C}}{(0.060\ \mathrm{m})^2} = 1.0 \times 10^4\ \mathrm{N/C}.$$

The resultant field at P is the vector sum of $\mathbf{E}_1$ and $\mathbf{E}_2$:

$$\mathbf{E} = (3.4 \times 10^4\mathbf{i} + 1.0 \times 10^4\mathbf{i})\ \mathrm{N/C} = 4.4 \times 10^4\mathbf{i}\ \mathrm{N/C}.$$

At P′, the fields are directed as shown in the figure. Their magnitudes are

$$E_1 = (9.0 \times 10^9\ \mathrm{N \cdot m^2/C^2})\,\frac{6.0 \times 10^{-9}\ \mathrm{C}}{(0.10\ \mathrm{m})^2} = 0.54 \times 10^4\ \mathrm{N/C},$$

and

$$E_2 = (9.0 \times 10^9\ \mathrm{N \cdot m^2/C^2})\,\frac{4.0 \times 10^{-9}\ \mathrm{C}}{(0.10\ \mathrm{m})^2} = 0.36 \times 10^4\ \mathrm{N/C}.$$

The x component of the resultant field is

$$E_x = (0.54 \times 10^4\ \mathrm{N/C})\cos 60° + (0.36 \times 10^4\ \mathrm{N/C})\cos 60°$$
$$= 0.45 \times 10^4\ \mathrm{N/C},$$

and the y component is

$$E_y = (0.54 \times 10^4\ \mathrm{N/C})\sin 60° - (0.36 \times 10^4\ \mathrm{N/C})\sin 60°$$
$$= 0.16 \times 10^4\ \mathrm{N/C}.$$

The electric field at P′ is therefore

$$\mathbf{E} = (0.45 \times 10^4\mathbf{i} + 0.16 \times 10^4\mathbf{j})\ \mathrm{N/C},$$

which is a vector of magnitude 0. 48 × 10^4 N/C directed 20° above the x axis.

(*b*) The point P′ is a distance (0.10 m) sin 60° = 0.087 m above the x axis. Consequently, the distance between P′ and q_3 is

$$r_3 = \sqrt{(0.050\ \mathrm{m})^2 + (0.087\ \mathrm{m} - 0.040\ \mathrm{m})^2} = 0.068\ \mathrm{m}.$$

At P′ the electric field $\mathbf{E}_3$ due to q_3 is directed at an angle

$$\theta = \tan^{-1}\frac{0.087\ \mathrm{m} - 0.040\ \mathrm{m}}{0.050\ \mathrm{m}} = 43°$$

above the x axis, as illustrated in the figure; and the magnitude of $\mathbf{E}_3$ is

$$E_3 = (9.0 \times 10^9\ \mathrm{N \cdot m^2/C^2})\,\frac{3.0 \times 10^{-9}\ \mathrm{C}}{(0.068\ \mathrm{m})^2} = 0.58 \times 10^4\ \mathrm{N/C}.$$

Using the results from part (*a*), we have for the resultant field at P′ due to the three charges

$$E_x = 0.45 \times 10^4\ \mathrm{N/C} + (0.58 \times 10^4\ \mathrm{N/C})\cos 43°$$
$$= 0.87 \times 10^4\ \mathrm{N/C},$$

and

$$E_y = 0.16 \times 10^4\ \mathrm{N/C} + (0.58 \times 10^4\ \mathrm{N/C})\sin 43°$$
$$= 0.56 \times 10^4\ \mathrm{N/C},$$

so the electric field at P′ is now

$$\mathbf{E} = (0.87 \times 10^4\mathbf{i} + 0.56 \times 10^4\mathbf{j})\ \mathrm{N/C},$$

which is a vector of magnitude 1.03 × 10^4 N/C directed 33° above the x axis.

DRILL PROBLEM 23-4

(*a*) Suppose an electron is placed at point P of Fig. 23-10*a*. What is the force on the electron? (*b*) What is the force on an electron placed at P′?
ANS. (*a*) $-7.0 \times 10^{-15}\mathbf{i}$ N; (*b*) $-(0.72 \times 10^{-15}\mathbf{i} + 0.26 \times 10^{-15}\mathbf{j})$ N.

EXAMPLE 23-5

THE ELECTRIC FIELD OF AN INFINITE LINE OF CHARGE

What is the electric field at a distance r from a thin, uniformly charged wire of infinite length? (See Fig. 23-11.) The charge per unit length on the wire is λ.

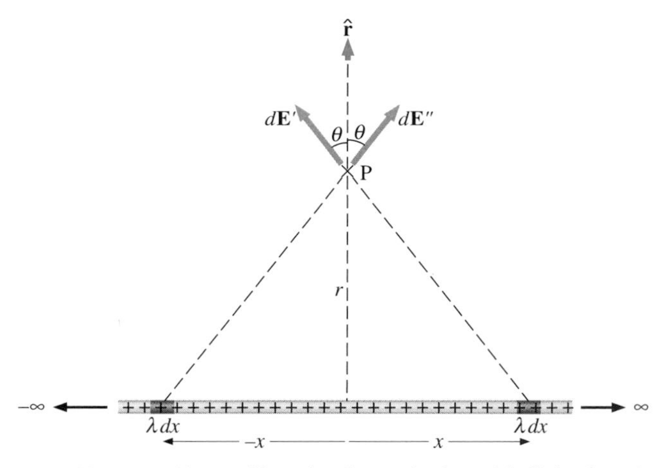

FIGURE 23-11 A thin, uniformly charged wire of infinite length.

SOLUTION At the point P, which is a distance r from the wire, the magnitude of the electric field $d\mathbf{E}'$ produced by the charge $\lambda\,dx$ in the element between x and $x + dx$ is

$$dE' = \frac{1}{4\pi\epsilon_0}\frac{\lambda\,dx}{x^2 + r^2}. \qquad (i)$$

The field is directed along the line from x to P. Because the wire is infinite, there are symmetric charge elements $\lambda\,dx$ at x and $-x$. The electric field $d\mathbf{E}''$ due to the element at $-x$ is directed along the line from $-x$ to P, and its magnitude is also given by the right-hand side of Eq. (*i*). You can see by inspecting the figure that the components of $d\mathbf{E}'$ and $d\mathbf{E}''$ cancel in the x direction and add in the radial ($\hat{\mathbf{r}}$) direction. Consequently, the net field $\mathbf{E}$ at P is found by integrating the *radial component* dE_r' of $d\mathbf{E}'$ from $x = -\infty$ to $x = \infty$:

$$\mathbf{E} = \hat{\mathbf{r}}\int dE_r' = \hat{\mathbf{r}}\int dE'\cos\theta = \hat{\mathbf{r}}\int_{-\infty}^{\infty}\frac{\lambda\cos\theta\,dx}{4\pi\epsilon_0(x^2 + r^2)}. \qquad (ii)$$

To evaluate this integral, we express $\cos\theta$ as a function of x. From the figure,

$$\cos\theta = \frac{r}{(x^2 + r^2)^{1/2}},$$

which, when substituted into Eq. (*ii*), gives

$$\mathbf{E} = \frac{\lambda r\hat{\mathbf{r}}}{4\pi\epsilon_0}\int_{-\infty}^{\infty}\frac{dx}{(x^2+r^2)^{3/2}}$$
$$= \frac{\lambda r\hat{\mathbf{r}}}{4\pi\epsilon_0}\left(\frac{x}{r^2\sqrt{x^2+r^2}}\right)\Bigg|_{-\infty}^{\infty} = \frac{\lambda}{2\pi\epsilon_0 r}\hat{\mathbf{r}}$$

for the electric field of an infinite line of charge. This result is consistent with what we should expect from the cylindrical symmetry of the system. The field lines are directed radially (outward if λ is positive and inward if λ is negative), and **E** is constant in magnitude over any infinitely long cylinder that is concentric with the wire.

EXAMPLE 23-6 The Electric Field of a Ring of Charge

Charge q is distributed uniformly around a thin circular ring of radius a. (See Fig. 23-12.) Calculate the electric field due to this charge at a point P on the axis of the ring and located a distance x from the center of the ring.

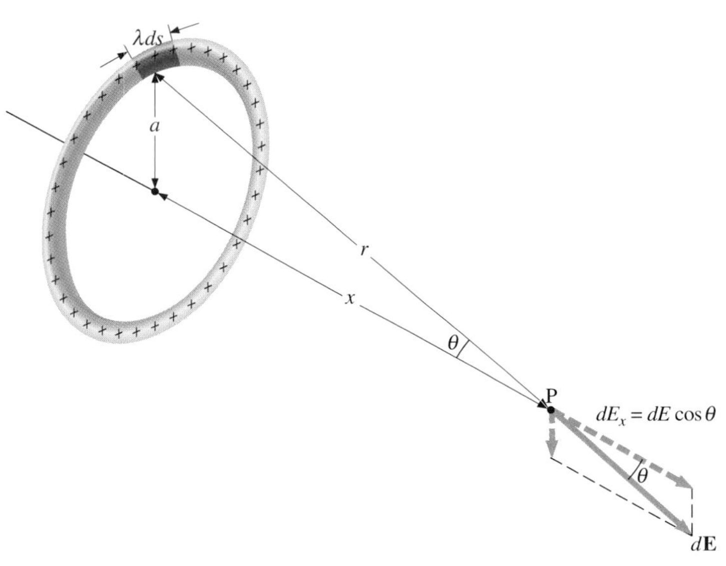

FIGURE 23-12 A ring of uniform charge.

SOLUTION Since the ring is thin, we assume that the charge is distributed uniformly around a circle of radius a with a linear charge density λ, where

$$\lambda = \frac{q}{2\pi a}. \qquad (i)$$

The charge dq in an element of length ds on the ring is

$$dq = \lambda\, ds,$$

and the magnitude of the field at P due to this element of charge is

$$dE = \frac{1}{4\pi\epsilon_0}\frac{dq}{r^2} = \frac{1}{4\pi\epsilon_0}\frac{\lambda\, ds}{r^2}. \qquad (ii)$$

To find the resultant electric field at P, we add vectorially the fields due to all charge elements. It is clear from the geometry that contributions due to elements directly opposite one another on the ring add along the axis of the ring (the x axis) and cancel along the two coordinate axes parallel to the plane of the ring. The field at P due to the entire ring of charge must therefore point along the x axis. Its magnitude is

$$E = \int_{\text{RING}} dE_x = \int_{\text{RING}} dE\cos\theta = \int_{\text{RING}}\frac{\lambda\, ds}{4\pi\epsilon_0 r^2}\cos\theta,$$

where

$$\cos\theta = \frac{x}{r}. \qquad (iii)$$

All of the quantities in the integrand are constant over the ring, so

$$E = \frac{\lambda\cos\theta}{4\pi\epsilon_0 r^2}\int_{\text{RING}} ds = \frac{\lambda\cos\theta}{4\pi\epsilon_0 r^2}(2\pi a),$$

which, when Eqs. (*i*) and (*iii*) are used to replace λ and $\cos\theta$, becomes

$$E = \frac{qx}{4\pi\epsilon_0 r^3} = \frac{qx}{4\pi\epsilon_0(x^2+a^2)^{3/2}}.$$

Thus at point P, the electric field of the ring of charge is

$$\mathbf{E} = \frac{qx}{4\pi\epsilon_0(x^2+a^2)^{3/2}}\,\mathbf{i}.$$

Notice that when $x \gg a$,

$$\mathbf{E} \approx \frac{1}{4\pi\epsilon_0}\frac{q}{x^2}\,\mathbf{i},$$

which is just what we'd expect, because far out on the x axis the ring of charge looks much like a point charge located at the origin. In the other extreme, when $x \ll a$, $\mathbf{E} \approx 0$. Can you show that at $x = a/\sqrt{2}$, the axial field is a maximum?

EXAMPLE 23-7 The Electric Field of a Sheet of Charge

An infinite plane sheet of charge is shown in Fig. 23-13*a*. If the charge per unit area σ is constant over the sheet, what is the electric field at the point P, which is a distance x in front of the sheet?

SOLUTION Let's first consider the electric field at P due to a ring of radius r and thickness dr that is on the sheet (see the figure). The area of the ring is $2\pi r\, dr$, so the charge dq in this ring is

$$dq = \sigma 2\pi r\, dr. \qquad (i)$$

From Example 23-6, we know that the electric field due to the ring is perpendicular to the sheet and has magnitude

$$dE = \frac{x\, dq}{4\pi\epsilon_0(x^2+r^2)^{3/2}} = \frac{\sigma x r\, dr}{2\epsilon_0(x^2+r^2)^{3/2}}. \qquad (ii)$$

To find the net field at P, we sum the contributions of all rings from $r = 0$ to $r = \infty$. Since the electric field of every ring points in the same direction, the net field can be determined simply by integrating Eq. (*ii*) from $r = 0$ to $r = \infty$ (x is a constant in this case). Thus

$$E = \frac{\sigma x}{2\epsilon_0}\int_0^{\infty}\frac{r\, dr}{(x^2+r^2)^{3/2}}, \qquad (iii)$$

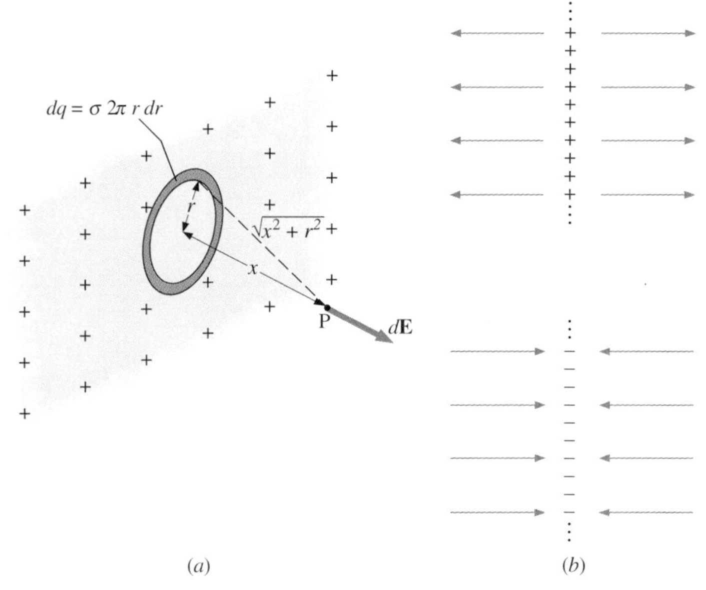

FIGURE 23-13 (*a*) Calculation of the field due to an infinite plane sheet of charge. (*b*) The electric field lines of a positive sheet and a negative sheet of charge.

which gives for the electric field of an infinite sheet of charge

$$E = \frac{\sigma}{2\epsilon_0}. \tag{23-5}$$

The electric field is perpendicular to the sheet and is directed away from or toward the sheet, depending on whether the charge is positive or negative. (See Fig. 23-13*b*.) Also, since E is independent of x, *the magnitude of the electric field is the same at any distance from the sheet.*

DRILL PROBLEM 23-5

Assume that the infinite sheet of charge of Example 23-7 is replaced by a circular sheet of radius R containing charge of uniform surface density σ. Point P is a distance x directly above the center of the sheet, as shown in Fig. 23-14. (*a*) Replace the upper limit of Eq. (*iii*) in Example 23-7 by R and then determine the electric field at point P. (*b*) What condition must be satisfied if the circular sheet is to be treated as an infinite sheet?
ANS. (*a*) $E = (\sigma/2\epsilon_0)(1 - x/\sqrt{x^2 + R^2})$; (*b*) $x/\sqrt{x^2 + R^2} \approx x/R \ll 1$.

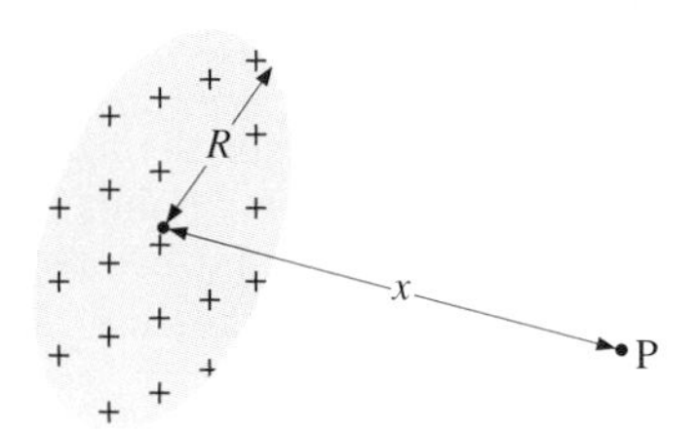

FIGURE 23-14 A circular sheet of uniform charge density.

EXAMPLE 23-8 THE ELECTRIC FIELD OF A CONDUCTING PLATE

Consider the infinite, uniformly charged conducting plate whose cross section is shown in Fig. 23-15*a*. (*a*) What is the electric field at any point outside the plate? (*b*) What is the field at any point inside the metal plate?

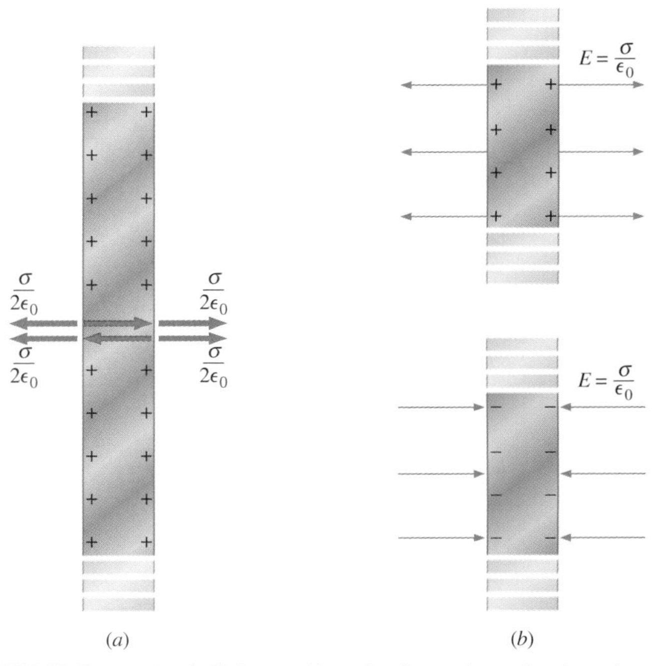

FIGURE 23-15 (*a*) An infinite, uniformly charged conducting plate. The plate is equivalent to two parallel sheets of charge. (*b*) The field lines for a positively charged plate and a negatively charged plate.

SOLUTION Electric charges are free to move within a conductor. This means that when excess charges are placed on a conductor, they redistribute until a state of equilibrium is reached where there is no net force on any charge. The distribution of charge in a conductor is an important topic that will be discussed in Sec. 25-7. There we show that *any excess charge on a conductor in electrostatic equilibrium resides on the surface of that conductor.* For now we will simply use this result without proof.

(*a*) Since excess charge resides on the surface of a conductor, we can consider the charged plate to be two parallel sheets of charge like the one of Example 23-7. The net electric field due to the entire charge is the vector sum of the electric fields due to the two sheets. For positive charge, each sheet produces an electric field that points away from the plate and has magnitude $\sigma/2\epsilon_0$. The magnitude of the net electric field at *any point outside the conducting plate* is therefore given by

$$E = \frac{\sigma}{\epsilon_0}. \tag{23-6}$$

The field is directed away from the plate if σ is positive and toward the plate if σ is negative. The field lines for both cases are shown in Fig. 23-15*b*.

(*b*) Inside the plate, the electric fields of the two charged sheets point in opposite directions and cancel, as shown in Fig. 23-15*a*. Hence there is *no electric field inside the conducting plate.*

EXAMPLE 23-9 THE ELECTRIC FIELD OF TWO OPPOSITELY CHARGED PLATES

The two large, oppositely charged conducting plates of Fig. 23-16 are placed close together and parallel to one another. Because opposite charges attract, all excess charge resides on the inside surfaces of the plates. The surface charge density is σ on one plate and $-\sigma$ on the other. Assuming edge effects can be ignored (i.e., assuming the plates are infinite), calculate the electric field in the region between the plates.

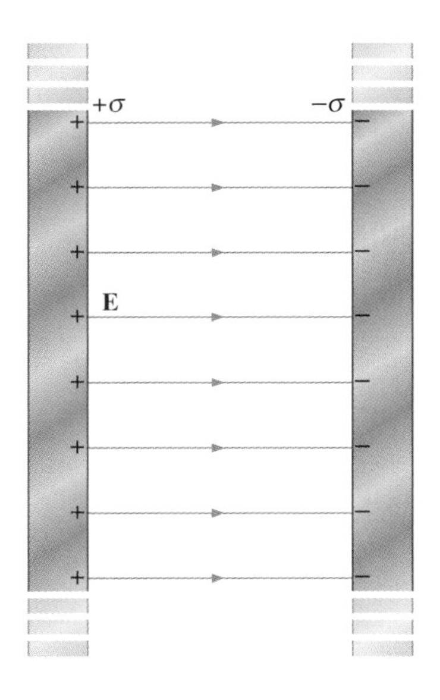

FIGURE 23-16 Two parallel, oppositely charged plates.

SOLUTION We know from Eq. (23-5) that the positive and negative surface charges each produce an electric field of magnitude $\sigma/2\epsilon_0$. Both fields are directed from the positive plate to the negative plate. The net electric field *everywhere between the two oppositely charged conducting plates* is therefore a vector of magnitude

$$E = \frac{\sigma}{\epsilon_0}, \tag{23-7}$$

which is directed from the positive to the negative plate as shown in Fig. 23-16.

DRILL PROBLEM 23-6

The charge densities on the positive and negative plates of Example 23-9 are 1.0 μC/m^2 and -1.0 μC/m^2, respectively. (*a*) What is the electric field in the region between the plates? (*b*) What force does an electron in this region experience? (*c*) What is the acceleration of the electron?
ANS. (*a*) 1.1×10^5 N/C; (*b*) 1.8×10^{-14} N; (*c*) 2.0×10^{16} m/s^2.

DRILL PROBLEM 23-7

Consider the parallel conducting plates of Example 23-9. (*a*) What is the electric field at any point inside the material of the plates? (*b*) What is the electric field at any other point not between the plates?
ANS. (*a*) $\mathbf{E} = 0$; (*b*) $\mathbf{E} = 0$.

23-5 MOTION OF A CHARGED PARTICLE IN A UNIFORM ELECTRIC FIELD

When a particle of charge q and mass m enters a uniform electric field $\mathbf{E}$, the electric force on it is $q\mathbf{E}$. If this is the only force on the particle, then from Newton's second law, its acceleration is

$$\mathbf{a} = \frac{q\mathbf{E}}{m}. \tag{23-8}$$

To analyze the particle's motion, we can use the kinematic equations of Chap. 3 (Eqs. 3-10, 3-11, and 3-12). For example, suppose the particle is moving in the constant electric field $-E_0\mathbf{j}$ produced by the oppositely charged parallel plates of Fig. 23-17. Then $a_x = 0$ and $a_y = -qE_0/m$. You may recall learning about such motion before—it's simply the motion of a projectile in Newtonian mechanics (with qE_0/m equivalent to g). The motion of a charged particle in a uniform electric field and the motion of a particle in the earth's gravitational field are therefore completely analogous, with one important exception: there is no "negative mass" that moves against the gravitational field!

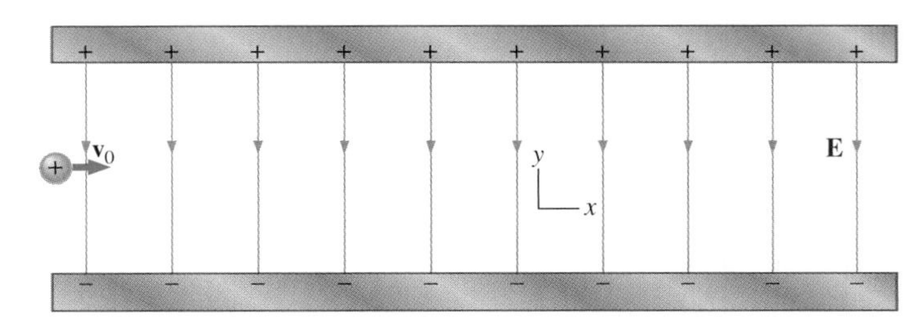

FIGURE 23-17 A particle of mass m and charge q moving between two parallel, oppositely charged plates.

If the charged particle enters the electric field moving horizontally with a speed v_0, then after a time t, its horizontal displacement is $x - x_0 = v_0 t$, while its vertical displacement and velocity are given respectively by

$$y - y_0 = -\frac{1}{2}\left(\frac{qE_0}{m}\right)t^2$$

and

$$v_y = -\left(\frac{qE_0}{m}\right)t.$$

The work-energy theorem is also useful for analyzing the motion of a charged particle in an electric field. As an example, suppose the particle is released from rest at the positive plate of Fig. 23-17 and travels a vertical distance d to the negative plate. The work done on it by the electric force is then

$$W = qE_0 d.$$

From the work-energy theorem, $qE_0 d = (1/2)mv^2$, so the speed of the charged particle when it reaches the negative plate is

$$v = \sqrt{\frac{2qE_0 d}{m}}.$$

Fisher Scientific.

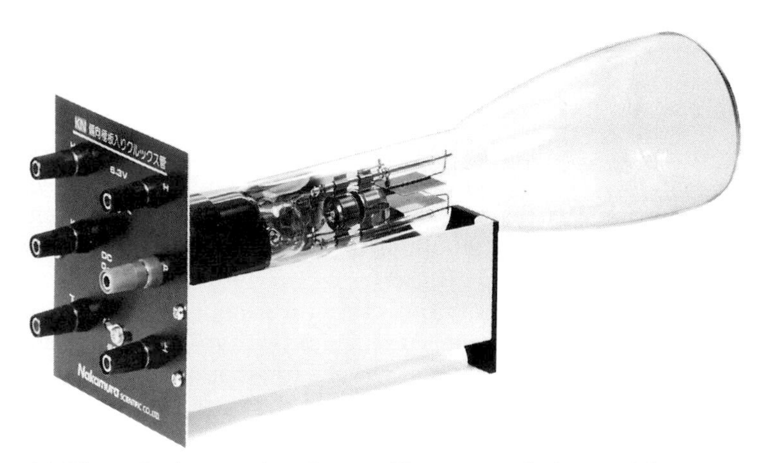

(a) The cathode ray tube of an oscilloscope, a device used for measuring time-varying electrical signals.

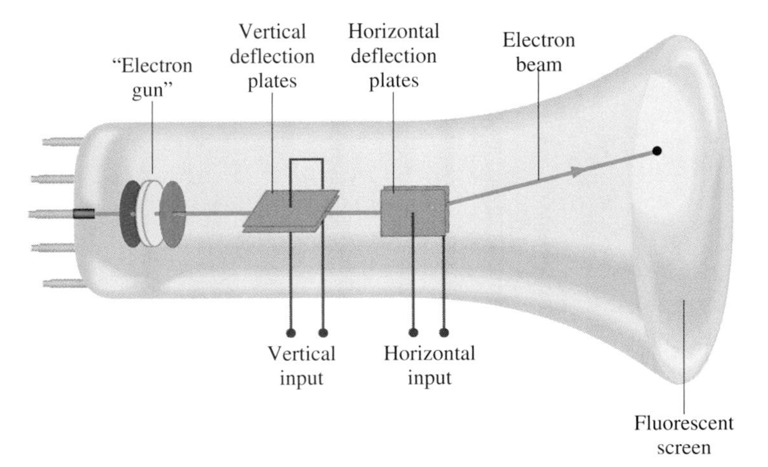

(b) A schematic representation of the cathode ray tube. The deflection plates are charged by the input signals. Their resulting electric fields then deflect the electron beam, whose position on the fluorescent screen is a measure of the input signals.

EXAMPLE 23-10 **AN ELECTRON MOVING BETWEEN TWO CHARGED PLATES**

An electron enters the oppositely charged parallel plates of Fig. 23-18 while moving horizontally with a speed v_0. The field between the plates is directed downward and has magnitude E_0, and the length of the plates is l. What distance h above its starting point will it be when it emerges at the other end of the plates? Ignore any variation in the electric field near the edges of the plates.

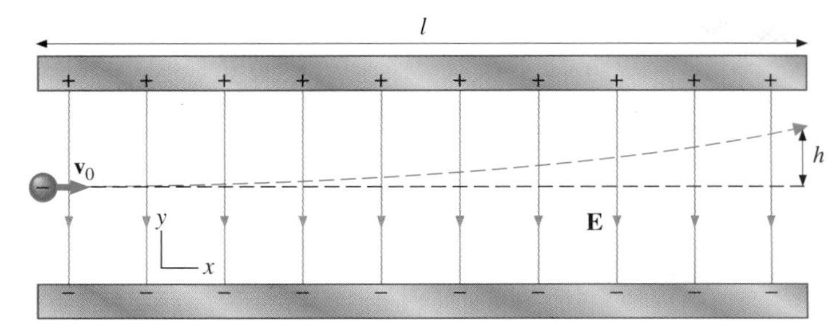

FIGURE 23-18 An electron enters between the parallel plates with a horizontal initial velocity.

SOLUTION The electric field points downward so the electron is deflected upward with an acceleration of magnitude $a_y = eE_0/m$. Since there is no horizontal force, the velocity v_0 in this direction is constant, and the electron moves between the plates in a time $t = l/v_0$. During this time, the electron is deflected upward a distance

$$h = \frac{1}{2}a_y t^2 = \frac{1}{2}\left(\frac{eE_0}{m}\right)\left(\frac{l}{v_0}\right)^2 = \frac{eE_0 l^2}{2mv_0^2}.$$

DRILL PROBLEM 23-8

The electron of Example 23-10 enters the parallel plates with a horizontal velocity of 1.0×10^7 m/s. The plates are 5.0 cm long, and the electric field between them is 1.0×10^3 N/C. What is the vertical deflection of the electron?
ANS. 2.2 mm.

DRILL PROBLEM 23-9

The electron of Drill Prob. 23-8 also has an initial vertical velocity of 1.0×10^5 m/s. What is the vertical deflection of the electron?
ANS. 2.7 mm

*23-6 THE MILLIKAN OIL-DROP EXPERIMENT

The charge of the electron was first measured in 1909 by R. A. Millikan (1868–1953) and his colleagues. A simple representation of the apparatus they used is shown in Fig. 23-19a. First, tiny oil drops are sprayed into the region between the plates. With the switch S open, the plates are uncharged so there is no electric field between them. A single drop of oil is observed through the telescope as it descends under the influence of the gravitational force $-mg\mathbf{j}$ and a resistive force $bv\mathbf{j}$. The speed of the drop increases until these two forces balance,

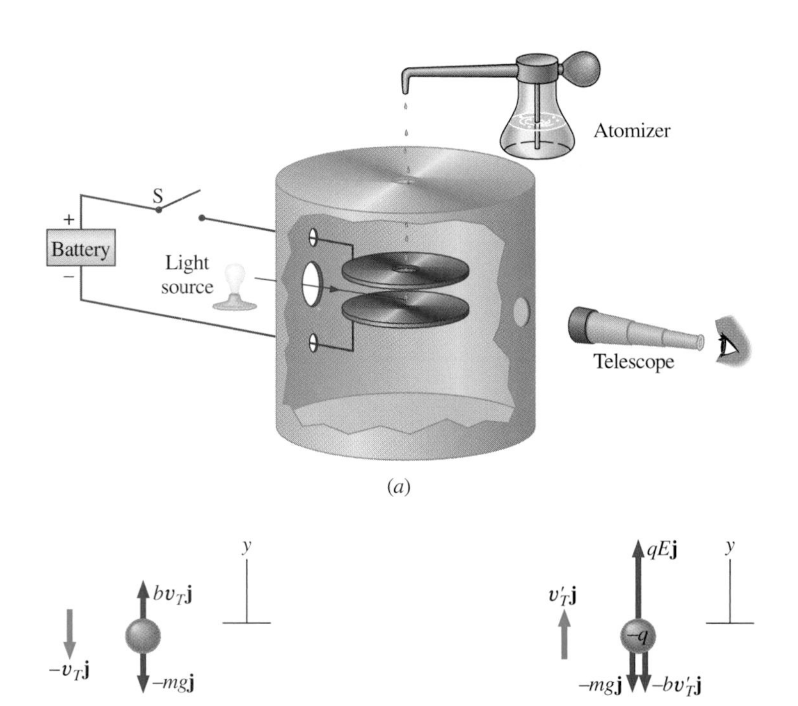

(b) (c)

FIGURE 23-19 The Millikan oil-drop experiment. (*a*) The atomizer sprays oil droplets between the plates, where they are observed with the telescope. (*b*) A neutral droplet falls at a constant velocity under the influence of the two forces. (*c*) A charged droplet rises at constant velocity under the influence of the three forces.

after which it falls at a constant terminal velocity* given by (see Fig. 23-19*b*)

$$v_T = \frac{mg}{b}.$$

Since the drop is very small, it cannot be weighed directly. However, v_T can be measured by observing the descent of the drop with the telescope, and b can be calculated with a formula we will not discuss here. The mass can then be determined from the equation for v_T.

After the drop reaches its terminal velocity, the region between the plates is exposed to a short burst of ionizing x-rays. The x-rays strip electrons from air molecules, and a few of the electrons attach to the descending oil drop, which acquires a net charge $-q$. The switch is then closed, and the battery charges the plates, producing an electric field $\mathbf{E}$ between them. This field exerts an upward force on the negatively charged oil drop. It then rises and quickly reaches a terminal velocity v'_T as shown in Fig. 23-19*c*, this time while under the influence of the forces $-mg\mathbf{j}$, $-bv'_T\mathbf{j}$, and $qE\mathbf{j}$. Thus

$$mg + bv'_T = qE,$$

and the *magnitude* of the charge on the oil drop is given by

$$q = \frac{mg + bv'_T}{E}.$$

The mass m and the resistive constant b are known, v'_T is measured by watching the drop through the telescope, E is known from the rating of the battery, so q can be determined from this equation.

Courtesy Central Scientific Company.

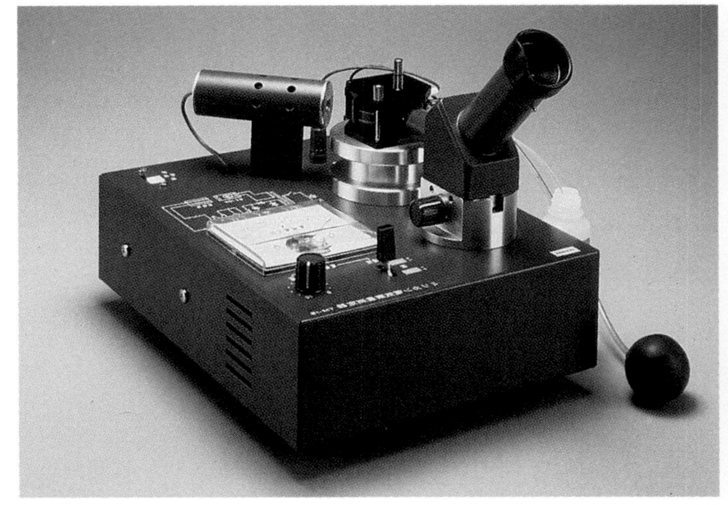

A student oil-drop apparatus.

Millikan's group measured the charge on thousands of drops. They found that within experimental error, q was always a small integer multiple of a basic charge e; that is,

$$q = ne, \tag{23-9}$$

where $n = 1,2,3,\ldots$. The value of the basic charge was found to be $e = 1.60 \times 10^{-19}$ C, which we now know is the magnitude of the charge of an electron.

For precise measurements of the electron charge, the analysis presented here has to be slightly modified. First, a correction must be made for the buoyancy of the air through which the oil drop falls. Secondly, air is not a continuous fluid, but instead a gas composed of molecules separated by distances roughly the same size as the radius of the drops. This requires a modification to the formula $\mathbf{f} = -b\mathbf{v}$ for the resistive force.

EXAMPLE 23-11 The Millikan Oil-Drop Experiment

(*a*) Suppose that an oil drop in Millikan's experiment is observed to fall with a terminal velocity of 0.250 mm/s. If $b = 1.60 \times 10^{-10}$ N·s/m, what is the mass of the drop? (*b*) In an electric field of magnitude 1.00×10^5 N/C, the drop is observed to rise with a terminal velocity of 5.00×10^{-2} mm/s. What is the magnitude of the charge on the drop? How many excess electrons are on the drop? (*c*) Other drops are found to have excess charge of magnitudes 4.81×10^{-19} C, 3.18×10^{-19} C, 9.55×10^{-19} C, and 8.07×10^{-19} C. Confirm that all four of these satisfy Eq. (23-9).

SOLUTION (*a*) The mass of the oil drop is

$$m = \frac{bv_T}{g} = \frac{(1.60 \times 10^{-10}\ \text{N·s/m})(2.50 \times 10^{-4}\ \text{m/s})}{9.80\ \text{m/s}^2}$$
$$= 4.08 \times 10^{-15}\ \text{kg}.$$

*See Sec. 6-3 for a discussion of resistive force and terminal velocity.

(*b*) We can now find q from its relationship to the external forces:

$$q = \frac{mg + bv_T'}{E}$$
$$= [(4.08 \times 10^{-15}\ \text{kg})(9.80\ \text{m/s}^2) + (1.60 \times 10^{-10}\ \text{N}\cdot\text{s/m})(5.00 \times 10^{-5}\ \text{m/s})]/(1.00 \times 10^5\ \text{N/C})$$
$$= 4.80 \times 10^{-19}\ \text{C}.$$

Since $q/e = (4.80 \times 10^{-19}\ \text{C})/(1.60 \times 10^{-19}\ \text{C}) = 3.00$, there are three excess electrons on the drop.

(*c*) For these oil drops, dividing the excess charge by the electron charge gives

$$\frac{4.81 \times 10^{-19}\ \text{C}}{1.60 \times 10^{-19}\ \text{C}} = 3.01, \qquad \frac{3.18 \times 10^{-19}\ \text{C}}{1.60 \times 10^{-19}\ \text{C}} = 1.99,$$
$$\frac{9.55 \times 10^{-19}\ \text{C}}{1.60 \times 10^{-19}\ \text{C}} = 5.97, \qquad \frac{8.07 \times 10^{-19}\ \text{C}}{1.60 \times 10^{-19}\ \text{C}} = 5.04.$$

Within experimental error, each of these ratios is an integer, in agreement with Eq. (23-9).

SUMMARY

1. **Definition of the electric field**
   If the electric force on a positive test charge q_t at a given point is $\mathbf{F}_e$, the electric field $\mathbf{E}$ at that point is
   $$\mathbf{E} = \lim_{q_t \to 0} \frac{\mathbf{F}_e}{q_t},$$
   where q_t is chosen to be as small as possible to minimize its effect on the measured field.
2. **Electric field of a charge distribution**
   (*a*) The electric field of an isolated point charge is given by
   $$\mathbf{E} = \frac{1}{4\pi\epsilon_0} \frac{q}{r^2} \hat{\mathbf{r}}.$$
   (*b*) The electric field of a discrete charge distribution is the vector sum of the electric fields of the individual charges:
   $$\mathbf{E} = \frac{1}{4\pi\epsilon_0} \sum_i \frac{q_i}{r_i^2} \hat{\mathbf{r}}_i.$$
   (*c*) If the charge distribution can be treated as continuous, its net field is
   $$\mathbf{E} = \frac{1}{4\pi\epsilon_0} \int \frac{dq}{r^2} \hat{\mathbf{r}}.$$
3. **Field lines**
   Field lines are drawn such that (*a*) they originate at positive charges and terminate at negative charges; (*b*) the number of lines originating or terminating at a charge is proportional to the magnitude of that charge; (*c*) the electric field at a particular point in space is tangent to the field line passing through that point; and (*d*) the number of field lines per unit area passing through a small cross-sectional area perpendicular to the field is proportional to the magnitude of the field at the cross section.
4. **Calculating the electric field**
   The electric fields of discrete and continuous charge distribution are calculated using the vector sums given above in 2(*b*) and 2(*c*), respectively.
5. **Motion of a charged particle in a uniform electric field**
   A charge q moves in a uniform electric field just like a particle moves in a uniform gravitational field. The acceleration of the charge is
   $$\mathbf{a} = \frac{q\mathbf{E}}{m}.$$

QUESTIONS

23-1. When measuring an electric field, could we use a negative rather than a positive test charge?

23-2. During fair weather, the electric field due to the net charge on the earth points downward. Is the earth charged positively or negatively?

23-3. If a point charge is released from rest in a uniform electric field, will it follow a field line? Will it do so if the electric field is not uniform?

23-4. Under what conditions, if any, will the trajectory of a charged particle not follow a field line?

23-5. How would you experimentally distinguish an electric field from a gravitational field?

23-6. A representation of an electric field shows 10 field lines perpendicular to a square plate. How many field lines should pass perpendicularly through the plate to depict a field with twice the magnitude?

23-7. What is the ratio of the number of electric field lines leaving a charge $10q$ and a charge q?

23-8. Give a plausibility argument as to why the electric field outside an infinite charged sheet is constant.

23-9. Compare the electric fields of an infinite sheet of charge, an infinite, charged conducting plate, and infinite, oppositely charged parallel plates.

23-10. Describe the electric fields of an infinite charged plate and of two infinite, charged parallel plates in terms of the electric field of an infinite sheet of charge.

23-11. Suppose you place a charge q near a large metal plate. (*a*) If q is attracted to the plate, is the plate necessarily charged? (*b*) If q is repelled by the plate, is the plate necessarily charged?

23-12. A negative charge is placed at the center of a ring of uniform positive charge. What is the motion (if any) of the charge? What if the charge were placed at a point on the axis of the ring other than the center?

23-13. If the electric field at a point on the line between two charges is zero, what do you know about the charges?

23-14. Two charges lie along the x axis. Is it true that the net electric field always vanishes at some point (other than infinity) along the x axis?

PROBLEMS

Electric Field

23-1. A particle of charge 2.0×10^{-8} C experiences an upward force of magnitude 4.0×10^{-6} N when it is placed at a particular point in an electric field. (*a*) What is the electric field at that point? (*b*) If a charge $q = -1.0 \times 10^{-8}$ C is placed there, what will be the force on it?

23-2. What is the electric field at a point where the force on a -2.0×10^{-6}-C charge is $(4.0\mathbf{i} - 6.0\mathbf{j}) \times 10^{-6}$ N?

23-3. A proton is suspended in the air by an electric field at the surface of the earth. What is the strength of this electric field?

23-4. On a typical clear day, the atmospheric electric field points downward and has a magnitude of approximately 100 N/C. Compare the gravitational and electric forces on a small dust particle of mass 2.0×10^{-15} g that carries a single electron charge. What is the acceleration (both magnitude and direction) of the dust particle?

23-5. The electric field in a particular thundercloud is 2.0×10^5 N/C. What is the acceleration of an electron in this field?

23-6. A small piece of cork whose mass is 2.0 g is given a charge of 5.0×10^{-7} C. What electric field is needed to place the cork in equilibrium under the combined electric and gravitational forces?

Electric Fields of Discrete Charge Distributions

23-7. If the electric field is 100 N/C at a distance of 50 cm from a point charge q, what is the value of q?

23-8. What is the electric field of a proton at the first Bohr orbit for hydrogen ($r = 5.29 \times 10^{-11}$ m)? What is the force on the electron in that orbit?

23-9. (*a*) What is the electric field of an oxygen nucleus at a point that is 10^{-10} m from the nucleus? (*b*) What is the force this electric field exerts on a second oxygen nucleus placed at that point?

23-10. Consider an electron that is 10^{-10} m from an alpha particle ($q = +3.2 \times 10^{-19}$ C). (*a*) What is the electric field due to the alpha particle at the location of the electron? (*b*) What is the electric field due to the electron at the location of the alpha particle? (*c*) What is the electric force on the alpha particle? on the electron?

23-11. Suppose the electric field of an isolated point charge decreased with distance as $1/r^{2+\delta}$ rather than as $1/r^2$. Show that it is then impossible to draw continuous field lines so that their number per unit area is proportional to E.

23-12. Each of the balls shown in the accompanying figure carries a charge q and has a mass m. The length of each thread is l, and at equilibrium, the balls are separated by an angle 2θ. Show that θ satisfies

$$\tan\theta \sin^2\theta = \frac{q^2}{16\pi\epsilon_0 mgl^2}.$$

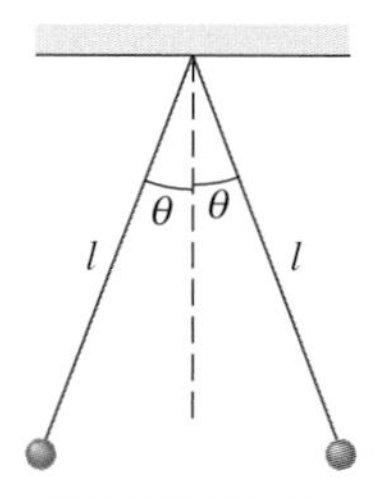

PROBLEM 23-12

23-13. Two point charges, $q_1 = 2.0 \times 10^{-7}$ C and $q_2 = -6.0 \times 10^{-8}$ C, are held 25.0 cm apart. (*a*) What is the electric field at a point 5.0 cm from the negative charge and along the line between the two charges? (*b*) What is the force on an electron placed at that point?

23-14. Point charges $q_1 = 50\ \mu$C and $q_2 = -25\ \mu$C are placed 1.0 m apart. (*a*) What is the electric field at a point midway between them? (*b*) What is the force on a charge $q_3 = 20\ \mu$C situated there?

23-15. Can you arrange the two point charges $q_1 = -2.0 \times 10^{-6}$ C and $q_2 = 4.0 \times 10^{-6}$ C along the x axis so that $\mathbf{E} = 0$ at the origin?

23-16. Point charges $q_1 = q_2 = 4.0 \times 10^{-6}$ C are fixed on the x axis at $x = -3.0$ m and $x = 3.0$ m. What charge q must be placed at the origin so that the electric field vanishes at $x = 0$, $y = 3.0$ m?

23-17. What is the electric field at the midpoint M of the hypotenuse of the triangle shown in the accompanying figure?

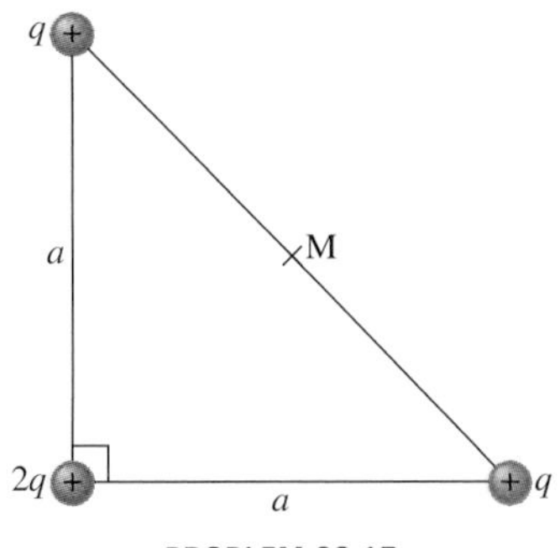

PROBLEM 23-17

23-18. Find the electric field at P for the charge configurations shown in the accompanying figures.

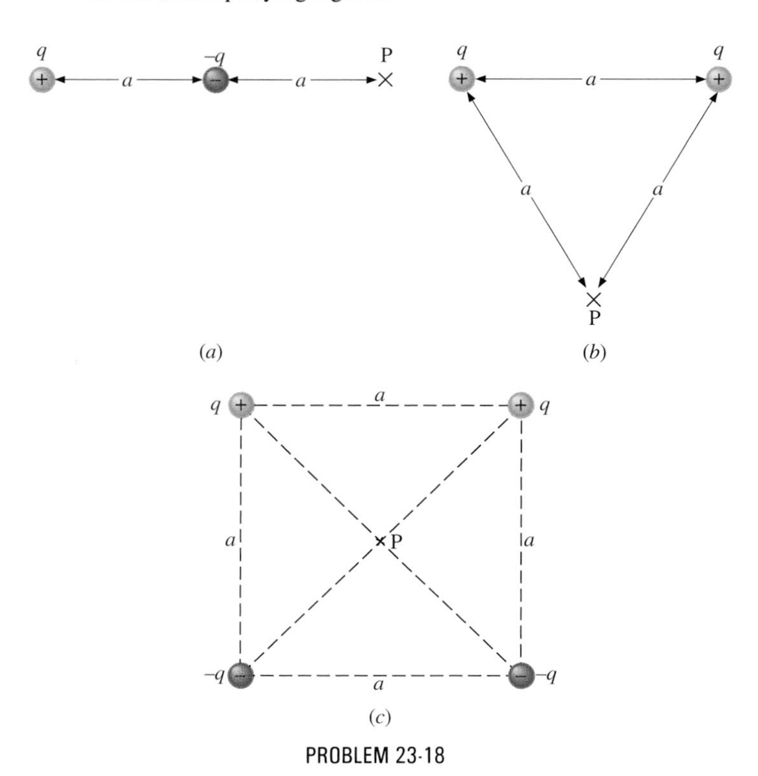

PROBLEM 23-18

23-19. (*a*) What is the electric field at the lower-right-hand corner of the square shown in the accompanying figure? (*b*) What is the force on a charge q placed at that point?

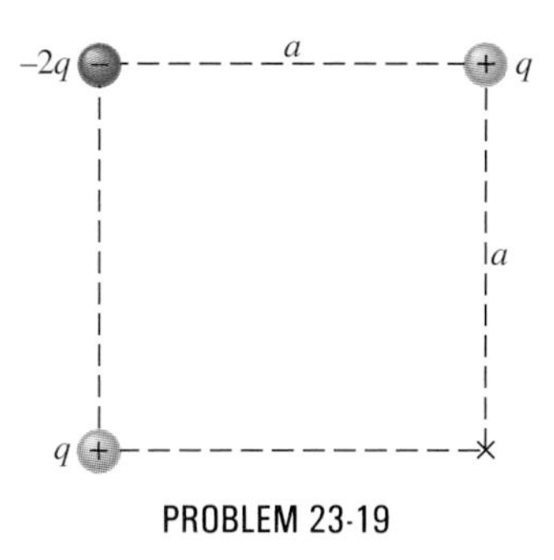

PROBLEM 23-19

23-20. Point charges are placed at the four corners of a rectangle as shown in the accompanying figure: $q_1 = 2.0 \times 10^{-6}$ C, $q_2 = -2.0 \times 10^{-6}$ C, $q_3 = 4.0 \times 10^{-6}$ C, and $q_4 = 1.0 \times 10^{-6}$ C. What is the electric field at P?

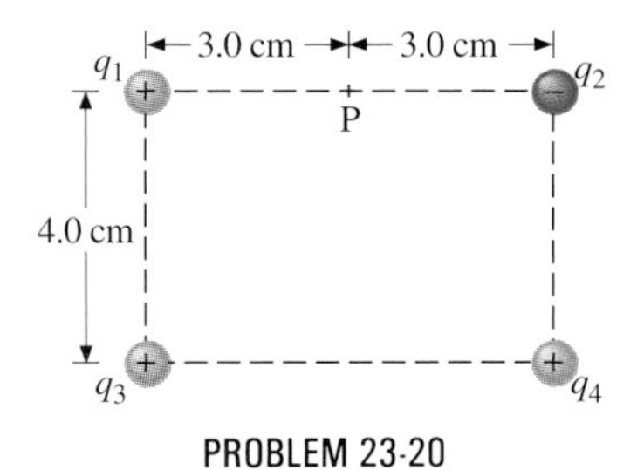

PROBLEM 23-20

Electric Fields of Continuous Charge Distributions

23-21. A thin conducting plate 1.0 m on a side is given a charge of -2.0×10^{-6} C. An electron is placed 1.0 cm above the center of the plate. What is the acceleration of the electron?

23-22. Calculate the magnitude and direction of the electric field 2.0 m from a long wire that is charged uniformly at $\lambda = 4.0\ \mu$C/m.

23-23. Two thin conducting plates, each 25.0 cm on a side, are situated parallel to one another and 5.0 mm apart. If 10^{11} electrons are moved from one plate to the other, what is the electric field between the plates?

23-24. (*a*) A total charge of 8.0 μC is spread evenly over a plane square *surface* 2.0 m on a side. What is the charge density of the surface? What is the electric field at a distance 1.0 cm above the center of the surface? (*b*) A total charge of 8.0 μC is spread evenly over the entire surface of a thin, square conducting *plate* that is 2.0 m on a side. What is the charge density on the plate? What is the electric field at a distance 1.0 cm above the center of the plate's surface?

23-25. Two thin parallel conducting plates are placed 2.0 cm apart. Each plate is 2.0 m on a side; one plate carries a net charge of 8.0 μC, and the other plate carries a net charge of $-8.0\ \mu$C. What is the charge density on the inside surface of each plate? What is the electric field between the plates?

23-26. A thin conducting plate 2.0 m on a side is given a total charge of $-10.0\ \mu$C. (*a*) What is the electric field 1.0 cm above the plate? (*b*) What is the force on an electron at this point? (*c*) Repeat these calculations for a point 2.0 cm above the plate. (*d*) When the electron moves from 1.0 to 2.0 cm above the plate, how much work is done on it by the electric field?

23-27. A total charge q is distributed uniformly along a thin, straight rod of length L. (See the accompanying figure.) What is the electric field at P_1? at P_2?

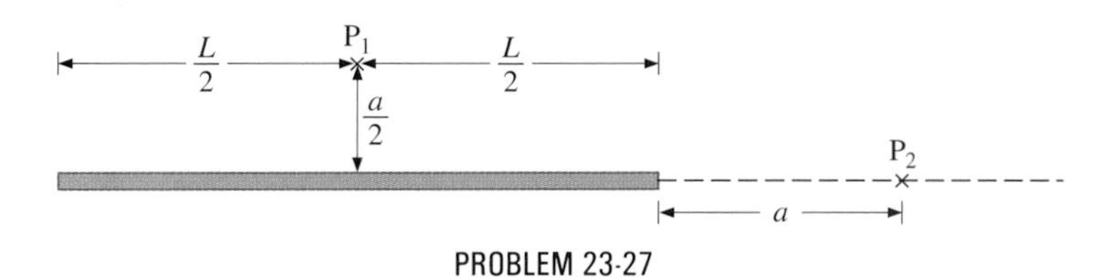

PROBLEM 23-27

23-28. Charge is distributed along the entire x axis with uniform density λ. How much work does the electric field of this charge distribution do on an electron that moves along the y axis from $y = a$ to $y = b$?

23-29. Charge is distributed along the entire x axis with uniform density λ_x and along the entire y axis with uniform density λ_y. Calculate the resulting electric field at (*a*) $\mathbf{r} = a\mathbf{i} + b\mathbf{j}$ and (*b*) $\mathbf{r} = c\mathbf{k}$.

23-30. A total charge q is distributed uniformly around a semicircular ring of radius r. (See the accompanying figure.) What is the electric field at P?

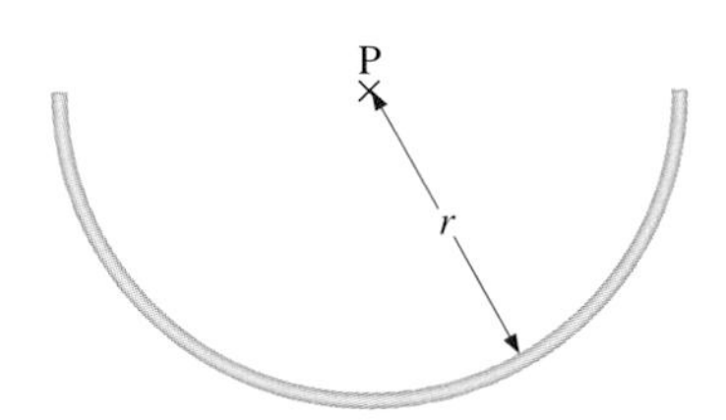

PROBLEM 23-30

23-31. A rod bent into the arc of a circle subtends an angle 2θ at the center P of the circle. (See the accompanying figure.) If the rod is charged uniformly with a total charge Q, what is the electric field at P?

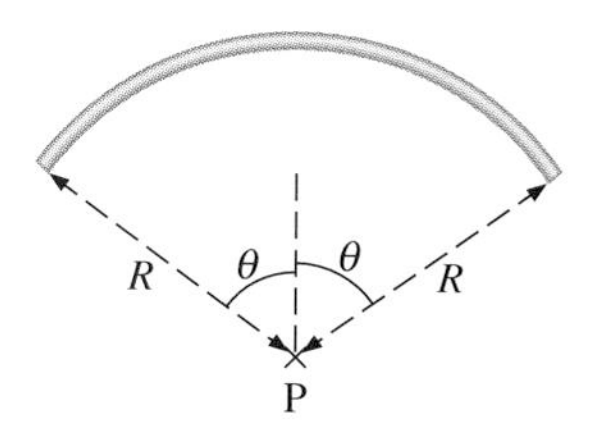

PROBLEM 23-31

Motion of a Charged Particle in a Uniform Electric Field

23-32. A proton moves in the electric field $\mathbf{E} = 200\mathbf{i}$ N/C. (*a*) What are the force on and the acceleration of the proton? (*b*) Do the same calculation for an electron moving in this field.

23-33. An electron is moving with a velocity $\mathbf{v} = 2.0 \times 10^7\mathbf{i}$ m/s when it enters an electric field $\mathbf{E} = 8.0 \times 10^3\mathbf{i}$ N/C. (*a*) How far will the electron travel before turning around? (*b*) How much time will elapse before it returns to its starting point?

23-34. An electron and a proton, each starting from rest, are accelerated by the same uniform electric field of 200 N/C. Determine the distance and time for each particle to acquire a kinetic energy of 3.2×10^{-16} J.

23-35. A spherical water droplet of radius 25 μm carries an excess 250 electrons. What vertical electric field is needed to balance the gravitational force on the droplet at the surface of the earth?

23-36. An electron is released with zero initial velocity at a point 5.0 cm above a large conducting plate. The plate is charged positively and produces a uniform electric field of magnitude 1.0×10^2 N/C. (*a*) How much time elapses before the electron strikes the plate? (*b*) What is the speed of the electron when it strikes the plate?

23-37. The following is an alternative method of performing Millikan's experiment; it is based on the *assumption that the radius of an oil drop can be accurately determined.* After finding the radius of a particular drop, the drop is charged as described in Sec. 23-6 by the ionizing x-rays. Suppose that the electric field is then turned on and adjusted to balance the gravitational force so that the drop is suspended in the air. (*a*) If the radius of the drop is r, the mass density of the oil is ρ, and the magnitude of the electric field at balance is E, show that when the drop is suspended, the magnitude of the excess charge on it is given by

$$q = \frac{4\pi r^3 \rho g}{3E}.$$

(*b*) Suppose that measurements on three different drops of oil ($\rho = 0.851$ g/cm^3) yield the following data:

E (N/C)	*r* (m)
2.55×10^5	1.52×10^{-6}
2.82×10^5	1.73×10^{-6}
2.24×10^5	1.27×10^{-6}

What is the charge on each drop? (*c*) Divide these charges by 1.60×10^{-19} C. What do you conclude about the number of electrons on each drop?

23-38. An electron has an initial velocity $\mathbf{v}_0 = 2.0 \times 10^6\mathbf{i}$ m/s when it enters a uniform electric field $\mathbf{E} = 500\mathbf{j}$ N/C. (*a*) What is the acceleration of the electron? (*b*) When the electron has traveled 10 cm in the x direction, what is its deflection in the y direction?

23-39. A proton traveling with an initial velocity of $4.0 \times 10^6\mathbf{i}$ m/s enters a region where there is a uniform electric field $\mathbf{E} = 2.0 \times 10^4\mathbf{i}$ N/C. (*a*) What is the force on the proton? (*b*) What is the acceleration of the proton? (*c*) What are the velocity and the displacement of the proton 2.0×10^{-6} s after it enters the field?

23-40. A proton enters the uniform electric field produced by the two charged plates shown in the accompanying figure. The magnitude of the electric field is 4.0×10^5 N/C, and the speed of the proton when it enters is 1.5×10^7 m/s. What distance d has the proton been deflected downward when it leaves the plates?

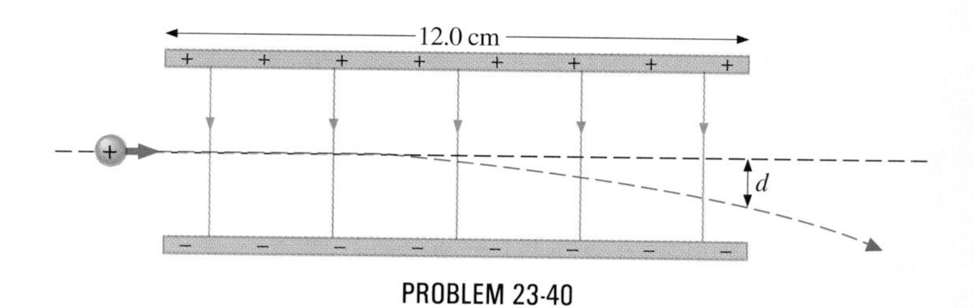

PROBLEM 23-40

23-41. An electron is projected at 45° above the horizontal with an initial speed of 5.0×10^5 m/s. If the electron is in a uniform vertical electric field $\mathbf{E} = 200\mathbf{j}$ N/C, find (*a*) the time that elapses before the electron returns to its initial height, (*b*) the maximum height reached by the electron, and (*c*) the horizontal displacement of the electron when it returns to its initial height.

23-42. An electron enters midway between two parallel plates with a velocity that is directed parallel to the plates and has magnitude 5.0×10^7 m/s. (See the accompanying figure.) The plates are 10 cm on a side and are separated by 1.0 cm. The strength of the electric field between the plates is 1.3×10^4 N/C. (*a*) What is the vertical deflection d of the electron when it leaves the plates? (*b*) Through what angle θ has the electron been deflected at the instant it leaves the plates?

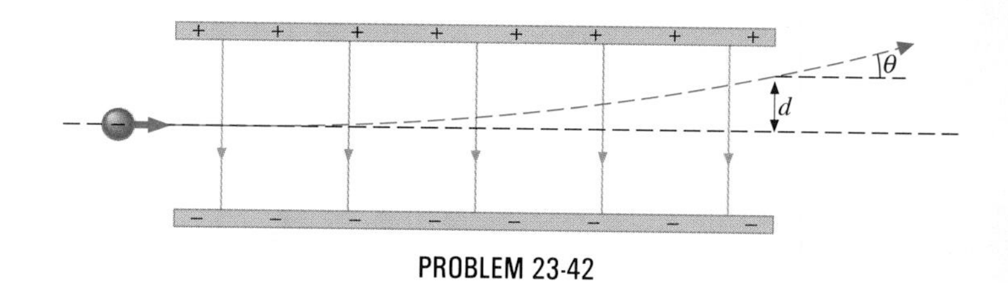

PROBLEM 23-42

General Problems

23-43. Three charges are positioned at the corners of a parallelogram as shown in the accompanying figure. (*a*) If $Q = 8.0\ \mu$C, what is the electric field at the unoccupied corner? (*b*) What is the force on a 5.0-μC charge placed at this corner?

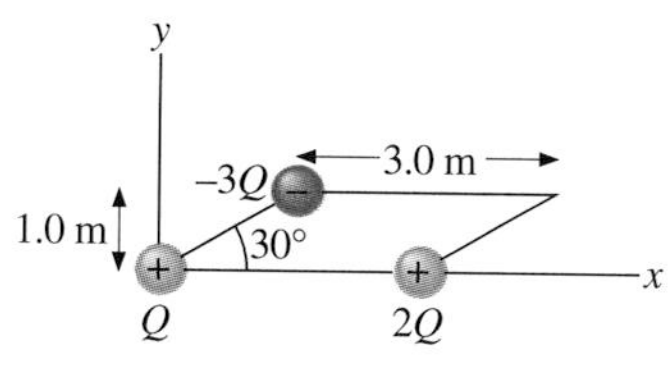

PROBLEM 23-43

23-44. Consider the equal and opposite charges shown in the accompanying figure. (*a*) Show that at all points on the x axis for which $|x| \gg a$, $E \approx Qa/2\pi\epsilon_0 x^3$. (*b*) Show that at all points on the y axis for which $|y| \gg a$, $E \approx Qa/\pi\epsilon_0 y^3$. (*Note:* This charge configuration is known as an *electric dipole* and will be discussed thoroughly in Chap. 25.)

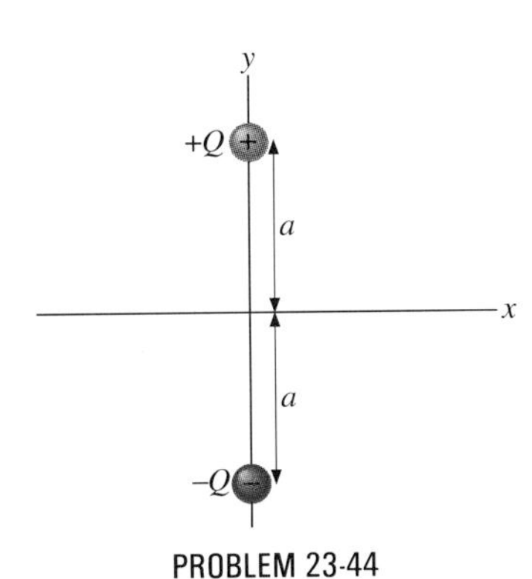

PROBLEM 23-44

23-45. The accompanying figure shows a small sphere of mass 0.25 g that carries a charge of 9.0×10^{-10} C. The sphere is attached to one end of a very thin silk string 5.0 cm long. The other end of the string is attached to a large vertical conducting plate that has a charge density of 30×10^{-6} C/m². What is the angle that the string makes with the vertical?

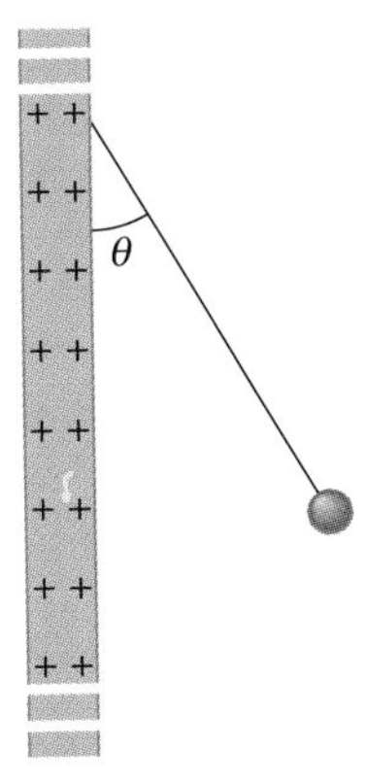

PROBLEM 23-45

23-46. Two infinite rods, each carrying a uniform charge density λ, are parallel to one another and perpendicular to the plane of the page. (See the accompanying figure.) What is the electric field at P_1? at P_2?

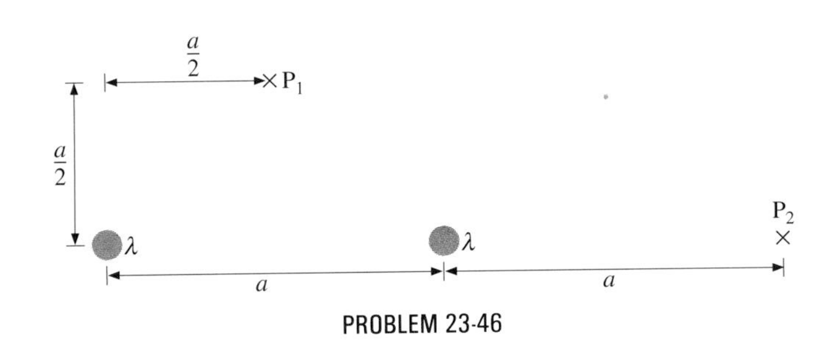

PROBLEM 23-46

23-47. Positive charge is distributed with a uniform density λ along the positive x axis from r to ∞, along the positive y axis from r to ∞, and along a 90° arc of a circle of radius r, as shown in the accompanying figure. What is the electric field at O?

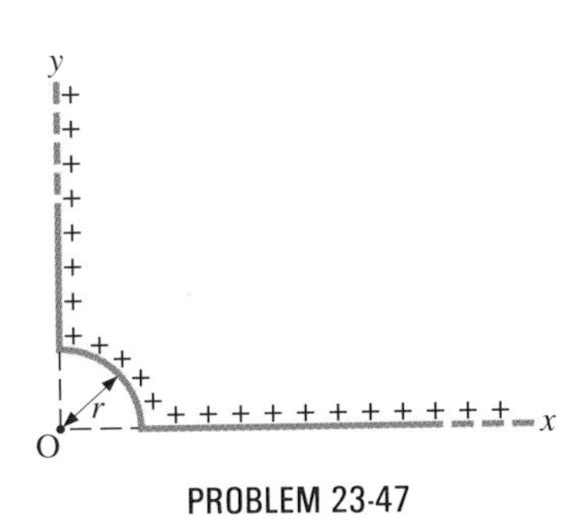

PROBLEM 23-47

23-48. From a distance of 10 cm, a proton is projected with a speed of $v = 4.0 \times 10^6$ m/s directly at a large, positively charged plate whose charge density is $\sigma = 2.0 \times 10^{-5}$ C/m². (See the accompanying figure.) (*a*) Does the proton reach the plate? (*b*) If not, how far from the plate does it turn around?

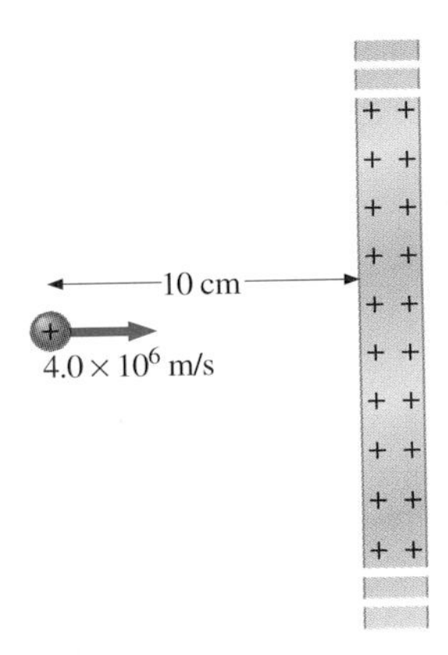

PROBLEM 23-48

23-49. At $t = 0$, the velocity of an electron is $\mathbf{v} = (1.0\mathbf{i} - 3.0\mathbf{j} + 2.0\mathbf{k}) \times 10^6$ m/s. The electron is in an electric field $\mathbf{E} = (2.0\mathbf{i} - 4.0\mathbf{j}) \times 10^3$ N/C. Find the velocity and displacement of the electron at $t = 2.0 \times 10^{-8}$ s.

23-50. Consider the electron beam shown in the accompanying figure. Prove that the angle θ through which it is deflected as it leaves the plates is given by $\theta = \tan^{-1}(eEl/m_e v_0^2)$.

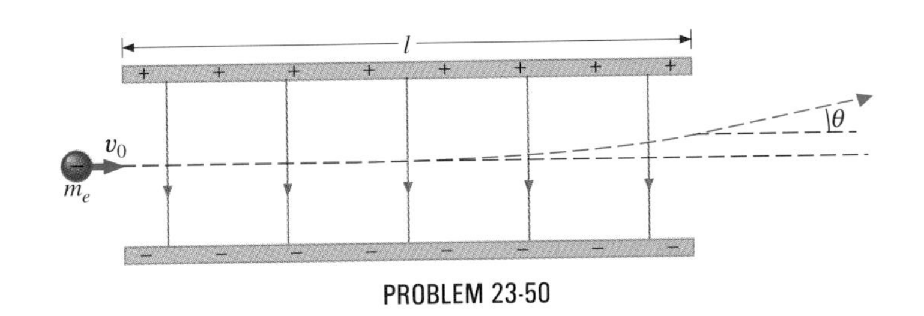

PROBLEM 23-50

23-51. Solve Prob. 23-39, given the initial velocity of the proton as $(4.0\mathbf{i} + 3.0\mathbf{j}) \times 10^6$ m/s.

23-52. A particle of mass m and charge $-q$ moves along a straight line away from a fixed particle of charge Q. When the distance between the two particles is r_0, $-q$ is moving with a speed v_0. (*a*) Use the work-energy theorem to calculate the maximum separation of the charges. (*b*) What do you have to assume about v_0 to make this calculation? (*c*) What is the minimum value of v_0 such that $-q$ escapes from Q?

23-53. A positive charge q is released from rest at the origin of a rectangular coordinate system and moves under the influence of the electric field $\mathbf{E} = E_0(1 + x/a)\mathbf{i}$. What is the kinetic energy of q when it passes through $x = 3a$?

23-54. A particle of charge $-q$ and mass m is placed at the center of a uniformly charged ring of total charge Q and radius R. The particle is displaced a small distance along the axis perpendicular to the plane of the ring and released. Assuming that the particle is constrained to move along the axis, show that the particle oscillates in simple harmonic motion with a frequency

$$f = \frac{1}{2\pi}\sqrt{\frac{qQ}{4\pi\epsilon_0 mR^3}}.$$

23-55. Charge is distributed uniformly along the entire y axis with a density λ_y and along the positive x axis from $x = a$ to $x = b$ with a density λ_x. What is the force between the two distributions?

23-56. The circular arc shown in the accompanying figure carries a charge per unit length $\lambda = \lambda_0 \cos\theta$, where θ is measured from the x axis. What is the electric field at the origin?

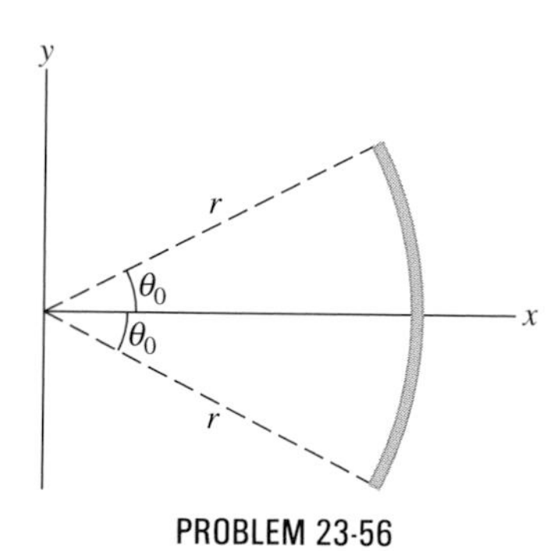

PROBLEM 23-56

23-57. Solve Prob. 23-56, given $\lambda = \lambda_0 e^{-\alpha|\theta|}$.

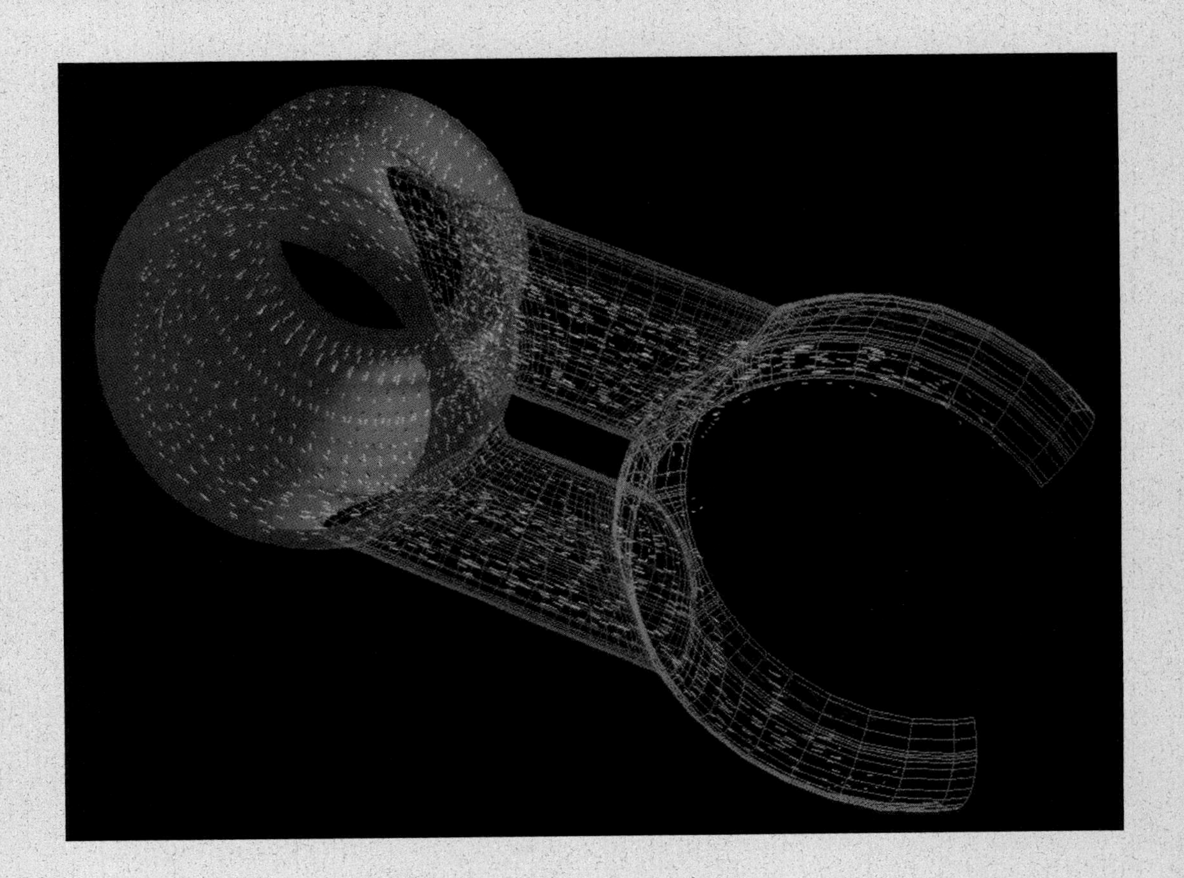

A computer-generated picture of an unusual surface, certainly not one we would use in Gauss' law!

CHAPTER 24 GAUSS' LAW

PREVIEW

In this chapter we introduce the first of Maxwell's four equations—Gauss' law. This law is particularly useful in finding the electric fields of charge distributions exhibiting spatial symmetry. The main topics to be discussed here are the following:

1. **Electric flux.** Electric flux is defined for both open and closed surfaces.
2. **Gauss' law.** Gauss' law is derived for an arbitrary charge distribution. The role of electric flux in Gauss' law is considered.
3. **Calculating electric fields with Gauss' law.** The use of Gauss' law in finding the electric fields of spatially symmetric charge distributions is explained. The importance of the choice of Gaussian surface is discussed, and examples involving the applications of Gauss' law are given.

So far, we have found that the electrostatic field begins and ends at point charges and that the field of a point charge varies inversely with the square of the distance from that charge. These characteristics of the electrostatic field lead to an important mathematical relationship known as Gauss' law (Eq. 24-4), the first of Maxwell's four equations that we will study. This law is named in honor of the extraordinary German mathematician and scientist, Karl Friedrich Gauss (1777–1855). Gauss' law relates the electric field, whether it be static or time-dependent, over a closed surface to the net charge within that surface. In electrostatics, Gauss' law gives us an elegantly simple way of finding the electric field. As you will see in the examples to follow, it is much easier to use than the integration method described in Chap. 23. However, there is a catch! Gauss' law has a limitation in that it can be readily applied only for charge distributions with certain symmetries.

24-1 Electric Flux

Figure 24-1a shows a planar surface S_1 of area A_1 that is perpendicular to the constant electric field $\mathbf{E} = E\mathbf{j}$. If N field lines pass through S_1, then we know from Sec. 23-3 that $N/A_1 \propto E$, or

$$N \propto EA_1. \qquad (24\text{-}1a)$$

The quantity EA_1 is the **electric flux** through S_1. We will represent the electric flux through an open surface like S_1 by the symbol ϕ_e. It is a scalar quantity and has an SI unit of newton-square meter per coulomb ($\mathrm{N \cdot m^2/C}$). Since Eq. (24-1a) may also be written as $N \propto \phi_e$, *electric flux is a measure of the number of field lines crossing a surface.*

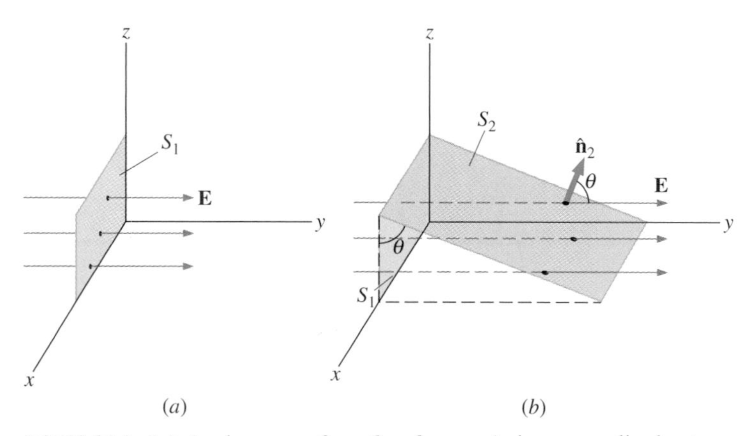

FIGURE 24-1 (a) A planar surface S_1 of area A_1 is perpendicular to the electric field $E\mathbf{j}$. N field lines cross surface S_1. (b) A surface S_2 of area A_2 whose projection onto the xz plane is S_1. The same number of field lines cross each surface.

Now let's consider a planar surface that is not perpendicular to the field. How would the electric flux then be represented? Figure 24-1b shows a surface S_2 of area A_2 that is inclined at an angle θ to the xz plane and whose projection in that plane is S_1 (area A_1). The areas are related by

$$A_2 \cos\theta = A_1.$$

Because the same number of field lines cross both S_1 and S_2, the fluxes through both surfaces must be the same. The flux through S_2 is therefore $\phi_e = EA_1 = EA_2\cos\theta$. Designating $\hat{\mathbf{n}}_2$ as a unit vector normal to S_2 (see Fig. 24-1b), we obtain

$$\phi_e = \mathbf{E}\cdot\hat{\mathbf{n}}_2 A_2. \qquad (24\text{-}1b)$$

This result can be easily generalized to the case of an arbitrary electric field $\mathbf{E}$ varying over an arbitrary surface S. As illustrated in Fig. 24-2, we first divide S into infinitesimal elements with areas da and unit normals $\hat{\mathbf{n}}$. Since the elements are infinitesimal, they may be assumed to be planar and $\mathbf{E}$ may be taken as constant over any element. The flux $d\phi_e$ through an area da is given by

$$d\phi_e = \mathbf{E}\cdot\hat{\mathbf{n}}\, da.$$

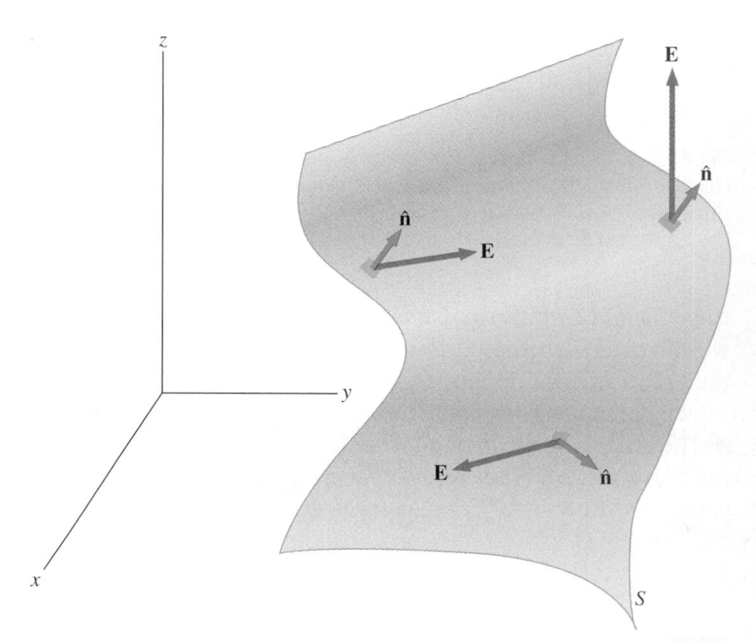

FIGURE 24-2 The flux through an open surface S is given by the integral $\phi_e = \int_S \mathbf{E}\cdot\hat{\mathbf{n}}\, da$.

It is positive when the angle between $\mathbf{E}$ and $\hat{\mathbf{n}}$ is less than 90° and negative when the angle is greater than 90°. The net flux is the sum of the infinitesimal flux elements over the entire surface. With $\int_S$ representing the integral over S,

$$\phi_e = \int_S \mathbf{E}\cdot\hat{\mathbf{n}}\, da \quad \text{(open surface).} \qquad (24\text{-}2a)$$

To distinguish between the flux through an open surface like that of Fig. 24-2 and the flux through a closed surface (one which completely bounds some volume), we represent flux for the latter case by

$$\Phi_e = \oint_S \mathbf{E}\cdot\hat{\mathbf{n}}\, da \quad \text{(closed surface).} \qquad (24\text{-}2b)$$

Since $\hat{\mathbf{n}}$ is a unit normal to a surface, it has two possible directions at every point on that surface. (See Fig. 24-3a.) On a *closed surface* such as that of Fig. 24-3b, $\hat{\mathbf{n}}$ is usually chosen

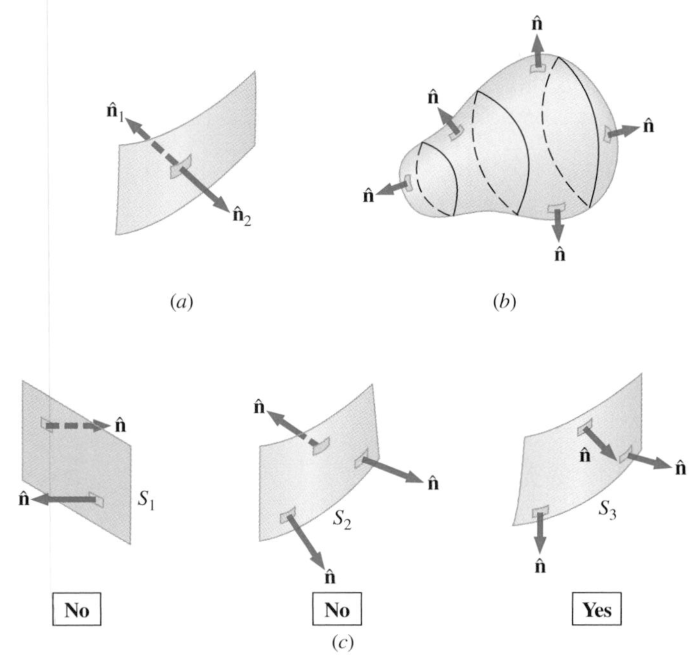

FIGURE 24-3 (*a*) There are two normals at every point on a surface. (*b*) The outward normal is generally used to calculate the flux through a closed surface. (*c*) The flux through the surfaces S_1 and S_2 cannot be determined by using the sets of normals shown.

to be the *outward normal* at every point. For an open surface, we can use either direction, as long as we are consistent over the entire surface. Various cases are depicted in Fig. 24-3*c*.

It is sometimes helpful to visualize electric flux as the "flow" of the electric field through a surface. When field lines leave (or "flow out of") a closed surface, Φ_e is positive; when they enter (or "flow into") the surface, Φ_e is negative. But do be careful with this interpretation! Field lines are just a visual aid. The electric field does not actually "flow."

EXAMPLE 24-1 Electric Flux Through a Plane

A uniform electric field **E** of magnitude 10 N/C is directed parallel to the *yz* plane at 30° above the *xy* plane, as shown in Fig. 24-4. What is the electric flux through the planar surface of area 6.0 m^2 located in the *xz* plane? Assume that $\hat{\mathbf{n}}$ points in the positive *y* direction.

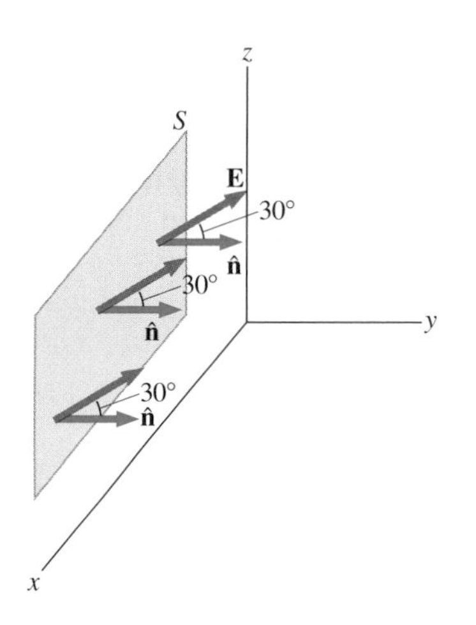

FIGURE 24-4 The electric field produces a net electric flux through the surface *S*.

SOLUTION The angle between the uniform electric field **E** and the unit normal $\hat{\mathbf{n}}$ to the planar surface is 30°. Using Eq. (24-2*a*), we find that the electric flux through the surface is

$$\phi_e = \int_S \mathbf{E}\cdot\hat{\mathbf{n}}\,da = EA\cos\theta$$
$$= (10\text{ N/C})(6.0\text{ m}^2)(\cos 30°) = 52\text{ N}\cdot\text{m}^2/\text{C}.$$

EXAMPLE 24-2 Electric Flux Through a Rectangular Surface

What is the total flux of the electric field $\mathbf{E} = cy^2\mathbf{k}$ through the rectangular surface shown in Fig. 24-5?

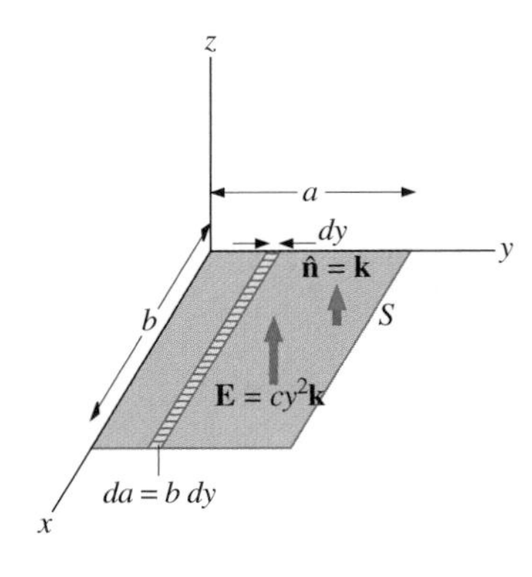

FIGURE 24-5 Since the electric field is not constant over the surface, an integration is necessary to determine the flux.

SOLUTION We assume that the unit normal $\hat{\mathbf{n}}$ to the given surface points in the positive *z* direction, so $\hat{\mathbf{n}} = \mathbf{k}$. Since the electric field is not uniform over the surface, it is first necessary to divide the surface into infinitesimal strips along which **E** is essentially constant. As shown in Fig. 24-5, these strips are parallel to the *x* axis and each strip has an area $da = b\,dy$. Now from Eq. (24-2*a*), we find that the net flux through the rectangular surface is

$$\phi_e = \int_S \mathbf{E}\cdot\hat{\mathbf{n}}\,da = \int_0^a (cy^2\mathbf{k})\cdot\mathbf{k}(b\,dy)$$
$$= cb\int_0^a y^2\,dy = \frac{1}{3}a^3bc.$$

EXAMPLE 24-3 Electric Flux Through a Spherical Surface Surrounding a Point Charge

Find the electric flux through a closed spherical surface of radius *R* due to the electric field of a point charge *q* positioned at the center of the sphere. (See Fig. 24-6.)

SOLUTION The electric field **E** at a distance *r* from *q* is given by

$$\mathbf{E} = \frac{1}{4\pi\epsilon_0}\frac{q}{r^2}\,\hat{\mathbf{r}},$$

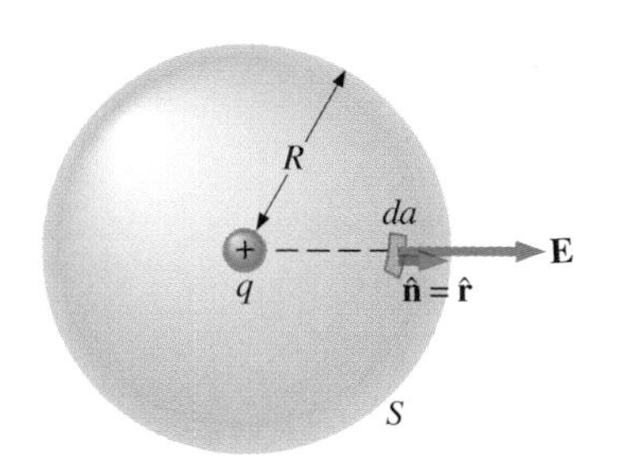

FIGURE 24-6 A closed spherical surface surrounding a point charge q.

Karl Friedrich Gauss, a legendary mathematician of the nineteenth century. Although his major contributions were to the field of mathematics, he also did important work in physics and astronomy.

where $\hat{\mathbf{r}}$ is the unit vector directed radially away from q. On the sphere, $\hat{\mathbf{n}} = \hat{\mathbf{r}}$ and $r = R$, so for an infinitesimal area da,

$$d\Phi_e = \mathbf{E}\cdot\hat{\mathbf{n}}\, da = \frac{1}{4\pi\epsilon_0}\frac{q}{R^2}\,\hat{\mathbf{r}}\cdot\hat{\mathbf{r}}\, da = \frac{1}{4\pi\epsilon_0}\frac{q}{R^2}\, da. \qquad (i)$$

We now find the net flux by integrating Eq. (i) over the surface of the sphere:

$$\Phi_e = \frac{1}{4\pi\epsilon_0}\frac{q}{R^2}\oint_S da = \frac{1}{4\pi\epsilon_0}\frac{q}{R^2}(4\pi R^2) = \frac{q}{\epsilon_0}. \qquad (ii)$$

Notice that Φ_e is *independent of the radius of the sphere*. The flux through *any sphere* with a single point charge q at its center is given by Eq. (ii). *This result is obtained only because* $E \propto 1/r^2$ *for a point charge*. If the electric field varied with distance in some other way, the flux through a sphere surrounding the charge would depend on the radius of the sphere. For example, for a point-charge field with the fictitious form $\mathbf{E} = (q/4\pi\epsilon_0 r^3)\hat{\mathbf{r}}$, the flux through a sphere of radius R would be $\Phi_e = \oint_S (q/4\pi\epsilon_0 r^3)\hat{\mathbf{r}}\cdot\hat{\mathbf{r}}\, da = (q/4\pi\epsilon_0 R^3)\oint_S da = (q/4\pi\epsilon_0 R^3)(4\pi R^2) = q/\epsilon_0 R$, which is a function of the sphere's radius.

DRILL PROBLEM 24-1

How would you reposition the surface shown in Fig. 24-1a so that no electric flux passes through it?
ANS. Place it so that its unit normal is perpendicular to **E**.

DRILL PROBLEM 24-2

If the electric field in Example 24-2 is $\mathbf{E} = mx\mathbf{k}$, what is the flux through the rectangular area?
ANS. $mab^2/2$.

DRILL PROBLEM 24-3

The infinite charged wire of Example 23-5 lies along the central axis of a cylindrical surface of radius r and length l. What is the flux through the surface due to the electric field of the charged wire?
ANS. $\lambda l/\epsilon_0$.

24-2 Gauss' Law

We can now determine the electric flux through an arbitrary closed surface due to an arbitrary charge distribution. This will give us Gauss' law. To begin, we consider the positive point charge of Fig. 24-7a. It is at the center of the spherical surface S', which is completely enclosed by the arbitrary surface S. Every field line that originates at q passes through both S and S', so the flux is the same through both surfaces. We found in Example 24-3 that the flux through S' is q/ϵ_0; consequently, the flux through S is also q/ϵ_0. We therefore conclude that *the net flux through any closed surface due to the electric field of an enclosed point charge is*

$$\Phi_e = \frac{q}{\epsilon_0}. \qquad (24\text{-}3)$$

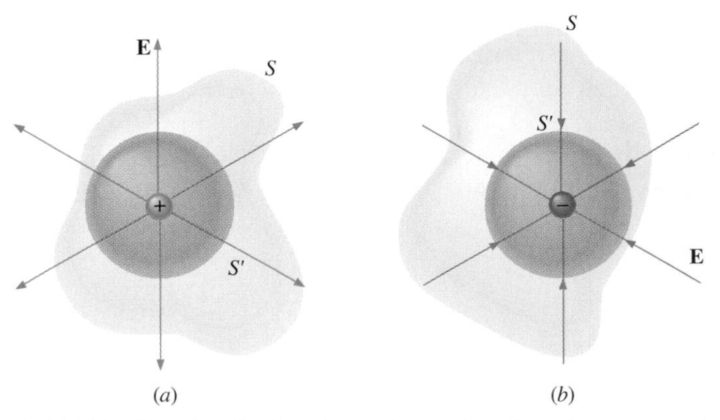

FIGURE 24-7 The electric flux through *any* closed surface surrounding a point charge q is given by $\Phi_e = q/\epsilon_0$.

This equation holds for *charges of either sign*. If the enclosed charge is negative (see Fig. 24-7b), then the flux through either S or S' as given by Eq. (24-3) is negative.

In Fig. 24-8, the point charge q is outside the closed surface. A typical field line enters the surface at da_1 and leaves at da_2. Every line that enters the surface must also leave that surface. Hence the net "flow" of the field lines into or out of the surface is zero. In other words, *the total electric flux through a closed surface due to the electric field of a charge that is outside that surface is zero*.

In addition to the information it provides about flux, Eq. (24-3) also serves as a construction rule for field lines. Since

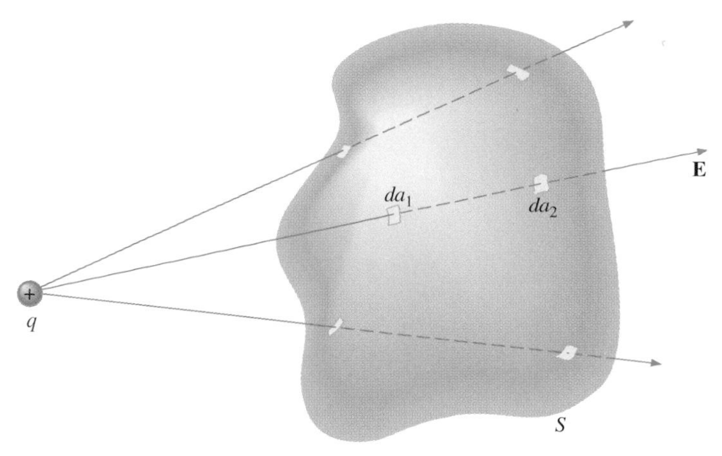

FIGURE 24-8 The electric flux through a closed surface due to a charge outside that surface is zero.

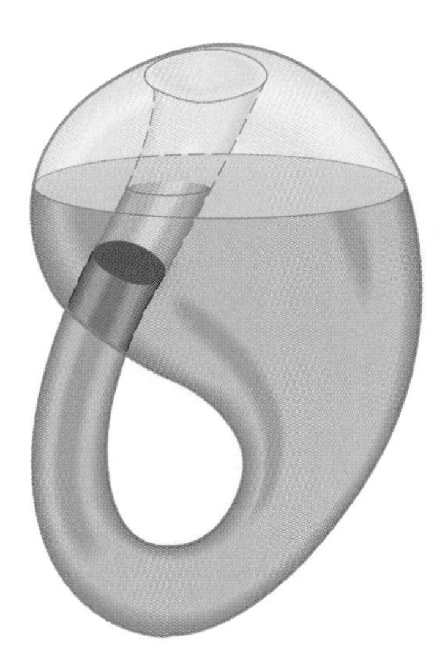

A *Klein bottle* partially filled with a liquid. Could the Klein bottle be used as a Gaussian surface?

$N \propto \Phi_e$ and $\Phi_e = q/\epsilon_0$, the number of lines originating on a positive charge (or terminating on a negative charge) must be proportional to the magnitude of the charge.* For example, if you decide to represent the field of a 1-μC charge with 8 field lines emanating from the charge, then you should draw 16 lines originating from a 2-μC charge and 12 lines terminating at a -1.5-μC charge if these charges are also in the figure.

Now let's consider the electric flux through an arbitrary closed surface due to any given charge distribution. Figure 24-9 shows a charge distribution and a closed surface S. From now on, we will refer to closed surfaces used in flux calculations as *Gaussian surfaces*. If $\mathbf{E}_1$ is the electric field of q_1, $\mathbf{E}_2$ is the electric field of q_2, etc., then the net field $\mathbf{E}$ is $\mathbf{E} = \mathbf{E}_1 + \mathbf{E}_2 + \ldots + \mathbf{E}_n$, and the total flux through S is

$$\Phi_e = \oint_S \mathbf{E}\cdot\hat{\mathbf{n}}\,da = \oint_S \mathbf{E}_1\cdot\hat{\mathbf{n}}\,da + \oint_S \mathbf{E}_2\cdot\hat{\mathbf{n}}\,da + \ldots + \oint_S \mathbf{E}_n\cdot\hat{\mathbf{n}}\,da.$$

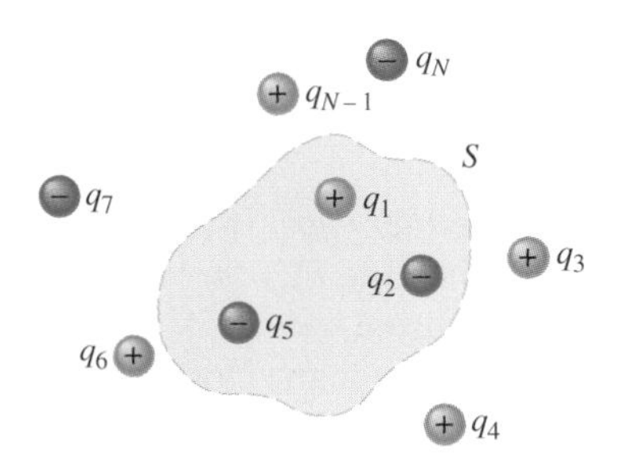

FIGURE 24-9 The flux through the Gaussian surface shown due to the charge distribution is $\Phi_e = (q_1 + q_2 + q_5)/\epsilon_0$.

For each point charge, $\oint_S \mathbf{E}_i\cdot\hat{\mathbf{n}}\,da$ equals q_i/ϵ_0 or zero, depending on whether q_i is inside or outside S. Consequently, each term on the right-hand side of the last equation can be replaced by either q_i/ϵ_0 or zero, leaving

$$\Phi_e = \oint_S \mathbf{E}\cdot\hat{\mathbf{n}}\,da = \frac{Q_{\text{in}}}{\epsilon_0}, \qquad (24\text{-}4)$$

where Q_{in} is the *total charge enclosed by the Gaussian surface S*. For example, the flux through the Gaussian surface S of Fig. 24-9 is $\Phi_e = (q_1 + q_2 + q_5)/\epsilon_0$.

Equation (24-4) is *Gauss' law*. To use it effectively, you must have a clear understanding of what each term in the equation represents. The field $\mathbf{E}$ is the *total electric field* at every point on the Gaussian surface. This total field includes contributions from charges both *inside* and *outside* the Gaussian surface. However, Q_{in} is just the charge *inside* the Gaussian surface. Finally, the Gaussian surface is *any closed surface* in space. That surface can coincide with the actual surface of a conductor or it can be an imaginary geometric surface. The only requirement imposed on a Gaussian surface is that it be closed.

EXAMPLE 24-4 Electric Flux Through Gaussian Surfaces

Calculate the electric flux through each Gaussian surface shown in Fig. 24-10.

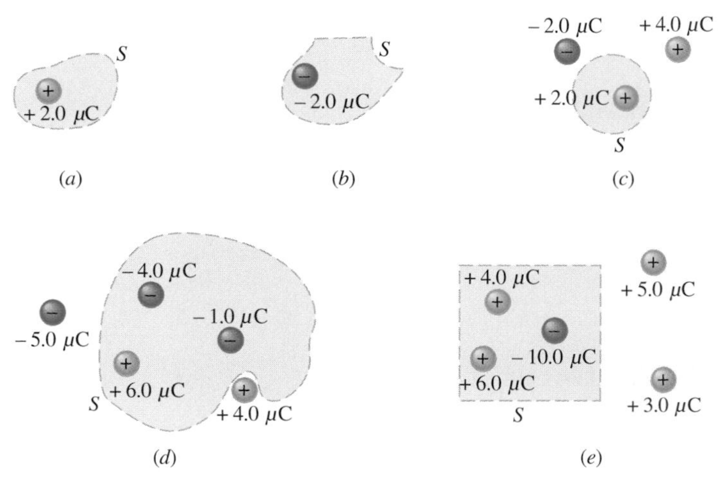

FIGURE 24-10 Various Gaussian surfaces and charges.

*This is rule 2 for drawing field lines, as discussed in Sec. 23-3.

SOLUTION From Gauss' law, the flux through each surface is given by Q_{in}/ϵ_0, where Q_{in} is the charge *enclosed* by that surface. Hence, for the surfaces and charges shown, we find

$$(a)\ \Phi_e = \frac{2.0\ \mu\text{C}}{\epsilon_0} = 2.26 \times 10^5\ \text{N}\cdot\text{m}^2/\text{C}.$$

$$(b)\ \Phi_e = \frac{-2.0\ \mu\text{C}}{\epsilon_0} = -2.26 \times 10^5\ \text{N}\cdot\text{m}^2/\text{C}.$$

$$(c)\ \Phi_e = \frac{2.0\ \mu\text{C}}{\epsilon_0} = 2.26 \times 10^5\ \text{N}\cdot\text{m}^2/\text{C}.$$

$$(d)\ \Phi_e = \frac{-4.0\ \mu\text{C} + 6.0\ \mu\text{C} - 1.0\ \mu\text{C}}{\epsilon_0} = 1.13 \times 10^5\ \text{N}\cdot\text{m}^2/\text{C}.$$

$$(e)\ \Phi_e = \frac{4.0\ \mu\text{C} + 6.0\ \mu\text{C} - 10.0\ \mu\text{C}}{\epsilon_0} = 0.$$

DRILL PROBLEM 24-4

Calculate the electric flux through the closed cubical surface for each charge distribution shown in Fig. 24-11.
ANS. (*a*) 3.4×10^5 N·m²/C; (*b*) -3.4×10^5 N·m²/C; (*c*) 3.4×10^5 N·m²/C; (*d*) 0.

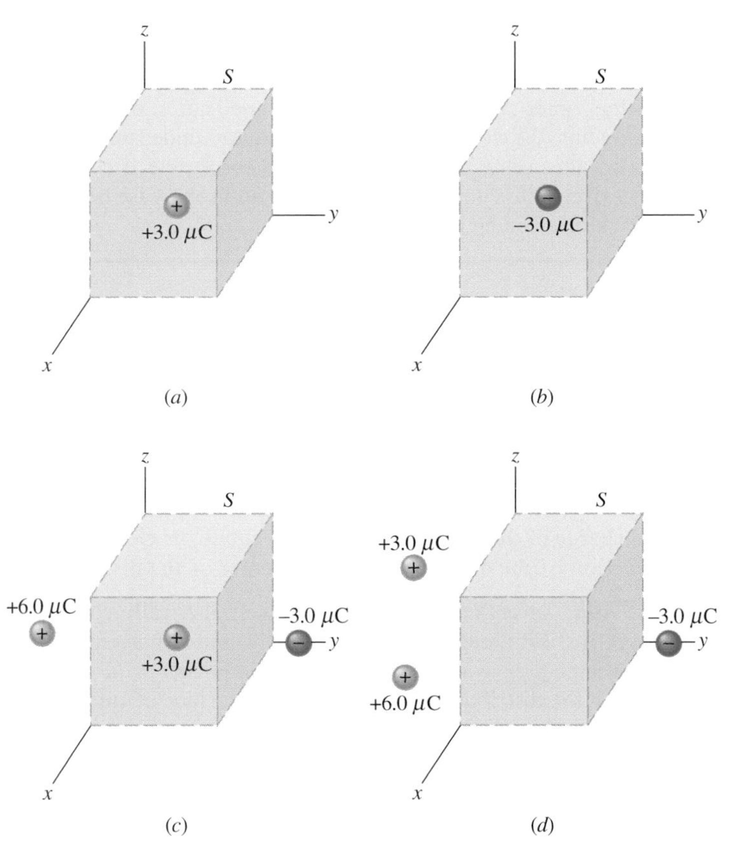

FIGURE 24-11 A cubical Gaussian surface with various charge distributions.

24-3 Calculating Electric Fields with Gauss' Law

In addition to making an insightful statement about electric flux, Gauss' law also serves as a valuable tool for calculating the electric fields of certain spatially symmetric charge distributions. When a charge distribution possesses spatial symmetry, we can often find a Gaussian surface S over which the electric field has constant magnitude. Furthermore, if **E** is parallel to $\hat{\mathbf{n}}$ everywhere on the surface, then $\mathbf{E}\cdot\hat{\mathbf{n}} = E$.* The surface integral in Gauss' law then simplifies to

$$\oint_S \mathbf{E}\cdot\hat{\mathbf{n}}\, da = E\oint da = EA,$$

where A is the area of the surface. With the integral simplified, Gauss' law can be used to determine **E**. Here is a summary of the steps we will follow:

1. *Identify the spatial symmetry of the charge distribution.* This is an important first step that will allow us to choose the appropriate Gaussian surface. As examples, an isolated point charge has spherical symmetry and an infinite line of charge has cylindrical symmetry.

2. *Choose a Gaussian surface with the same symmetry as the charge distribution.* With such a choice, $\mathbf{E}\cdot\hat{\mathbf{n}}$ is easily determined over the Gaussian surface.

3. *Evaluate the integral $\oint_S \mathbf{E}\cdot\hat{\mathbf{n}}\, da$ over the Gaussian surface.* The symmetry of the Gaussian surface allows us to factor $\mathbf{E}\cdot\hat{\mathbf{n}}$ outside the integral.

4. *Determine the amount of charge enclosed by the Gaussian surface.* This is in essence an evaluation of the right-hand side of the equation representing Gauss' law. It may be necessary to perform an integration to obtain the net enclosed charge.

5. *Evaluate the electric field of the charge distribution.* The field may now be found using the results of steps 3 and 4.

EXAMPLE 24-5 Electric Field of a Point Charge

Use Gauss' law to find the electric field of an isolated point charge.

SOLUTION By symmetry, the electric field of an isolated point charge q is directed radially toward or away from the charge. Furthermore, the magnitude of the field at any point depends only on the distance of that point from the charge. To use the spherical symmetry of the field, we define a spherical Gaussian surface of radius r centered at the charge. (See Fig. 24-12.) At the surface, the field is $\mathbf{E} = E(r)\hat{\mathbf{r}}$, where $E(r)$ is constant. Since $\hat{\mathbf{n}} = \hat{\mathbf{r}}$ on the Gaussian surface,

$$\oint_S \mathbf{E}\cdot\hat{\mathbf{n}}\, da = \oint_S E(r)\, da = E(r)\oint_S da = E(r)4\pi r^2;$$

*If **E** and $\hat{\mathbf{n}}$ are antiparallel everywhere on the surface, then $\mathbf{E}\cdot\hat{\mathbf{n}} = -E$.

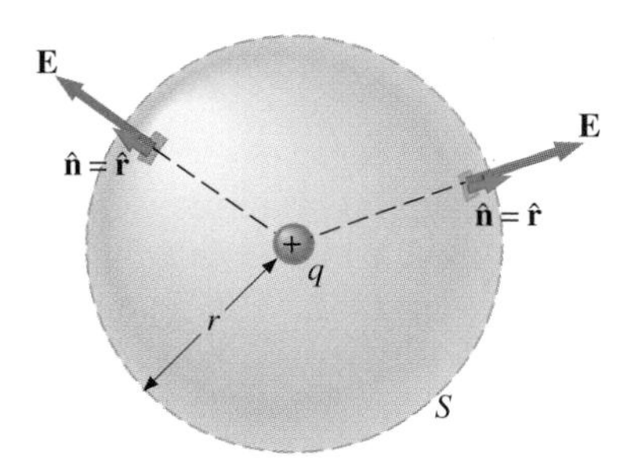

FIGURE 24-12 A spherical Gaussian surface centered on a point charge can be used to determine the electric field of that charge.

so from Gauss' law,

$$E(r)4\pi r^2 = \frac{q}{\epsilon_0},$$

and

$$E(r) = \frac{1}{4\pi\epsilon_0}\frac{q}{r^2}.$$

Thus

$$\mathbf{E} = \frac{1}{4\pi\epsilon_0}\frac{q}{r^2}\hat{\mathbf{r}},$$

which is the electric field of a point charge.

Gauss' law has just been used to obtain the electric field **E** of a point charge. If we now place a test charge q_t in this field, the force $q_t\mathbf{E}$ on the test charge gives us Coulomb's law. In the previous two sections, we used the electric field of a point charge (as found with Coulomb's law) to derive Gauss' law. The two laws are therefore equivalent ways of describing the electrostatic field, and the validity of Coulomb's law can be tested by searching for experimental disagreement with the predictions of Gauss' law. One very precise experiment will be discussed in Sec. 25-7. You will see there that Coulomb's law has been verified with amazing precision.

EXAMPLE 24-6 **ELECTRIC FIELD OF A LONG, CHARGED WIRE**

Figure 24-13*a* shows an infinitely long, thin wire with a uniform linear charge density λ. What is the electric field at a distance r from the wire? (*Note:* This situation can be realized experimentally by making r much less than the length of the wire.)

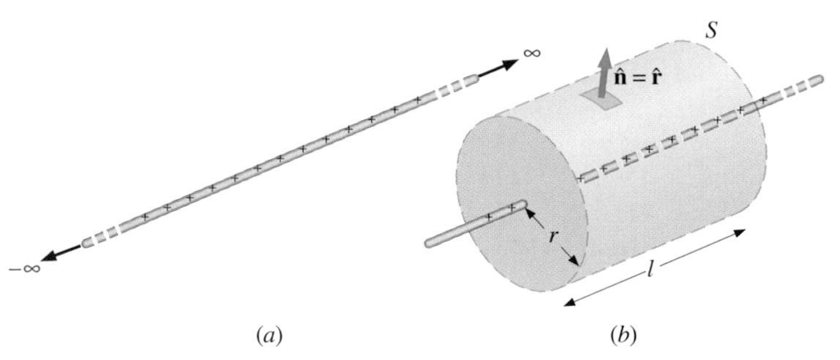

FIGURE 24-13 (*a*) An infinitely long, thin wire of linear charge density λ. (*b*) A cylindrical Gaussian surface concentric with the wire is used to determine the electric field.

SOLUTION Since the uniformly charged wire is both infinitely long and straight, the electric field is cylindrically symmetric and directed radially. To calculate **E**, we choose a cylindrical Gaussian surface of radius r and length l that is concentric with the wire and closed at each end by plane circular caps. (See Fig. 24-13*b*.) The electric field is perpendicular to the cylindrical side and parallel to the planar end caps of the surface. The flux through the cylindrical part is

$$\int_S \mathbf{E}\cdot\hat{\mathbf{n}}\,da = E\int_S da = E(2\pi rl),$$

while the flux through the end caps is zero since $\mathbf{E}\cdot\hat{\mathbf{n}} = 0$ there. Thus

$$\oint_S \mathbf{E}\cdot\hat{\mathbf{n}}\,da = E(2\pi rl) + 0 + 0 = 2\pi rlE.$$

Using $Q_{\text{in}} = \lambda l$, we now have from Gauss' law

$$E2\pi rl = \frac{\lambda l}{\epsilon_0},$$

and the electric field of a long, charged wire is

$$\mathbf{E} = \frac{\lambda}{2\pi\epsilon_0 r}\hat{\mathbf{r}},$$

where $\hat{\mathbf{r}}$ is the unit vector directed radially away from the wire. This agrees with the calculation of Example 23-5, where the electric field was found by integrating over the charged wire. Notice how much simpler the calculation of this electric field is with Gauss' law!

What happens if the wire is finite? The translational symmetry now disappears, and we can no longer assume that **E** is along $\hat{\mathbf{r}}$ or that E is constant over the cylindrical surface. Without these assumptions, the integral $\oint_S \mathbf{E}\cdot\hat{\mathbf{n}}\,da$ cannot be factored into an electric field times an area, precluding the application of Gauss' law to determine the electric field. It's important to understand that this does not mean that Gauss' law is invalid for the electric field of a finite wire; Gauss' law is simply not useful in this case because of the lack of symmetry.

EXAMPLE 24-7 **ELECTRIC FIELD OF A SPHERICAL CHARGE DISTRIBUTION**

A spherical uniform charge distribution* is represented by the shaded region in Fig. 24-14. The radius of the region is R and the total charge of the distribution is Q. Calculate the electric field as a function of the distance r from the center of the distribution.

SOLUTION Since the charge distribution is spherically symmetric, the electric field **E** must be directed radially, and its magnitude must be the same everywhere on a spherical Gaussian surface concentric with the distribution. For a spherical surface of radius r,

$$\oint_S \mathbf{E}\cdot\hat{\mathbf{n}}\,da = \oint_S E\,da = E\oint_S da = E4\pi r^2,$$

*A volume charge distribution can be produced in an insulator but not a conductor, where all excess charge resides on the surface. However, when excess charge is placed on an insulator, its molecules become polarized and produce their own electric field. For simplicity, we will always ignore the field of the polarized molecules.

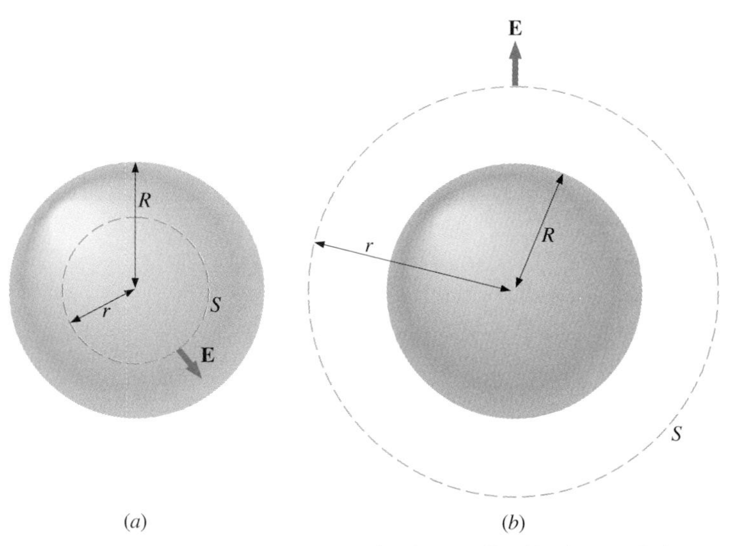

FIGURE 24-14 A spherically symmetric charge distribution and the Gaussian surfaces used for finding the field (a) inside and (b) outside the distribution.

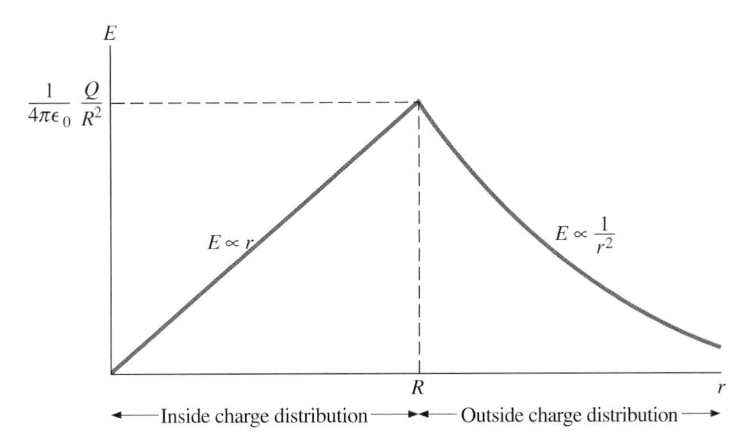

FIGURE 24-15 The electric field due to a spherical charge distribution of uniform charge density and total charge Q as a function of distance from the center of the distribution.

so from Gauss' law,

$$E4\pi r^2 = \frac{Q_{\text{in}}}{\epsilon_0},$$

and

$$\mathbf{E} = \frac{1}{4\pi\epsilon_0}\frac{Q_{\text{in}}}{r^2}\,\hat{\mathbf{r}}, \qquad (i)$$

where Q_{in} is the charge enclosed by the Gaussian surface.

The electric field within the charge distribution is found by using a spherical Gaussian surface that is within the distribution ($r < R$), as shown in Fig. 24-14a. The charge enclosed by the Gaussian surface is given by

$$Q_{\text{in}} = \int \rho_0\, dv = \int_0^r \rho_0 4\pi r'^2\, dr' = \rho_0\left(\frac{4}{3}\pi r^3\right),$$

where we have represented the uniform charge density by ρ_0. Since Q is the total charge of the distribution, the charge density is $\rho_0 = Q/(\frac{4}{3}\pi R^3)$, and the charge within our Gaussian surface can be expressed as

$$Q_{\text{in}} = \frac{Q}{(\frac{4}{3}\pi R^3)}\left(\frac{4}{3}\pi r^3\right) = Q\left(\frac{r}{R}\right)^3.$$

From Eq. (i), the field *inside* the charge distribution is therefore

$$\mathbf{E} = \frac{1}{4\pi\epsilon_0}\frac{Q_{\text{in}}}{r^2}\,\hat{\mathbf{r}} = \frac{Qr}{4\pi\epsilon_0 R^3}\,\hat{\mathbf{r}}. \qquad (ii)$$

The electric field outside the distribution is found by using a Gaussian surface where $r > R$. (See Fig. 24-14b.) This surface completely encloses the charge distribution, so the charge within the Gaussian surface is $Q_{\text{in}} = Q$, the total charge of the distribution. Now from Eq. (i), the electric field *outside* the distribution is

$$\mathbf{E} = \frac{1}{4\pi\epsilon_0}\frac{Q}{r^2}\,\hat{\mathbf{r}}. \qquad (iii)$$

The results of these calculations are summarized in Fig. 24-15.

Notice that Eq. (i) has the same form as the equation of the electric field of an isolated point charge. In determining the electric field of a uniform spherical charge distribution, we can therefore assume that all of the charge inside the appropriate spherical Gaussian surface [Q_{in} in Eq. (ii) or Q in Eq. (iii)] is located at the center of the distribution. Also notice the similarities between the gravitational field of a uniform spherical mass distribution (see Supplement 5-1) and the electric field of a uniform spherical charge distribution. This is not surprising, for both fields are vector sums of inverse-square fields created by uniformly distributed point sources. However, the calculation of this example is much simpler than that of Supplement 5-1. Rather than evaluating a complicated integral to determine **E**—which we would have had to do if we had used Eqs. (23-4)—we use Gauss' law and symmetry to effect a very simple calculation of **E**. Think of the mathematical effort we could have saved in Chap. 5 if we had developed a Gauss' law for the gravitational field!

EXAMPLE 24-8 ELECTRIC FIELD OF A SHEET OF CHARGE

An infinite plane sheet of charge of surface charge density σ is shown in Fig. 24-16. What is the electric field at a distance x from the sheet? Compare the result of this calculation with that of Example 23-7.

SOLUTION The symmetry of the system forces **E** to be perpendicular to the sheet and constant over any plane parallel to the sheet. To calculate the electric field, we choose the cylindrical Gaussian surface shown in the figure. The cross-sectional area and the height of the cylinder are A and $2x$, respectively, and the cylinder is positioned so that it is bisected by the plane sheet. Since **E** is perpendicular to each end and parallel to the side of the cylinder, we have

$$\oint_S \mathbf{E}\cdot\hat{\mathbf{n}}\, da = EA + EA + 0 = 2EA,$$

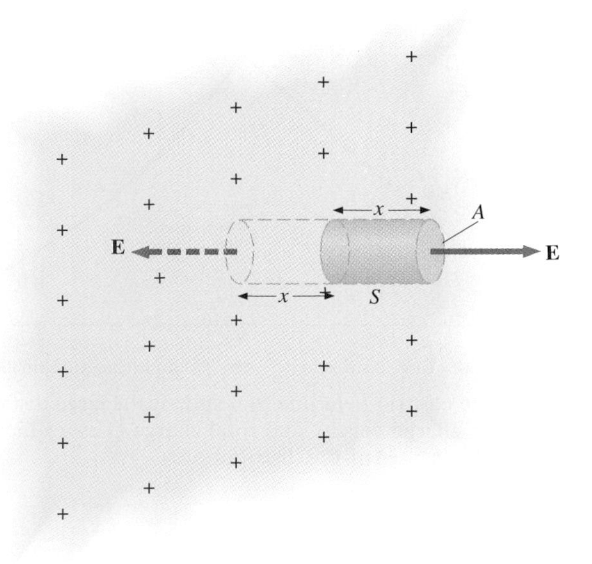

FIGURE 24-16 An infinite plane sheet of charge. To use the symmetry of the electric field, we choose the cylindrical Gaussian surface shown.

where EA is the flux through each end and there is no flux through the side where $\mathbf{E}\cdot\hat{\mathbf{n}} = 0$. The charge enclosed by the cylinder is σA, so from Gauss' law,

$$2EA = \frac{\sigma A}{\epsilon_0},$$

and the electric field of an infinite sheet of charge is

$$E = \frac{\sigma}{2\epsilon_0},$$

in agreement with the calculation of Example 23-7.

DRILL PROBLEM 24-5

What is Gauss' law for the gravitational field?
ANS. $\oint_S \mathbf{g}\cdot\hat{\mathbf{n}}\, da = 4\pi GM$, where M is the total mass inside S.

DRILL PROBLEM 24-6

Suppose that the charge density of the spherical charge distribution shown in Fig. 24-14 is $\rho(r) = \rho_0 r/R$ for $r \leq R$ and zero for $r > R$. Obtain expressions for the electric field both inside and outside the distribution.
ANS. $\mathbf{E} = (\rho_0 r^2/4\epsilon_0 R)\hat{\mathbf{r}}$ for $r \leq R$; $\mathbf{E} = (\rho_0 R^3/4\epsilon_0 r^2)\hat{\mathbf{r}}$ for $r \geq R$.

DRILL PROBLEM 24-7

Discuss the restrictions on the Gaussian surface used in Example 24-8. For example, is its length important? Does the cross section have to be circular? Must the end faces be on opposite sides of the sheet?

SUMMARY

1. **Electric flux**
   The electric flux through a surface is proportional to the number of field lines crossing that surface. The electric flux is obtained by evaluating the surface integral

$$\Phi_e = \oint_S \mathbf{E}\cdot\hat{\mathbf{n}}\, da,$$

   where the notation used here is for a closed surface S.

2. **Gauss' law**
   Gauss' law is

$$\oint_S \mathbf{E}\cdot\hat{\mathbf{n}}\, da = \frac{Q_{in}}{\epsilon_0},$$

   where Q_{in} is the total charge inside the Gaussian surface S.

3. **Calculating electric fields with Gauss' law**
   For a charge distribution with spatial symmetry, we can often find a Gaussian surface over which $\mathbf{E}\cdot\hat{\mathbf{n}} = E$, where E is constant over the surface. The electric field is then easily determined with Gauss' law.

Questions

24-1. Discuss how you would orient a planar surface of area A in a uniform electric field of magnitude E_0 to obtain (*a*) the maximum flux and (*b*) the minimum flux through the area.

24-2. What are the maximum and minimum values of the flux in Ques. 24-1?

24-3. The net electric flux crossing a closed surface is always zero. True or false?

24-4. The net electric flux crossing an open surface is never zero. True or false?

24-5. (*a*) If the electric flux through a closed surface is zero, is the electric field necessarily zero at all points on the surface? (*b*) What is the net charge inside the surface?

24-6. Two concentric spherical surfaces enclose a point charge q. The radius of the outer sphere is twice that of the inner one. Compare the electric fluxes crossing the two surfaces.

24-7. Compare the electric flux through the surface of a cube of side length a that has a charge q at its center to the flux through a spherical surface of radius a with a charge q at its center.

24-8. Discuss how Gauss' law would be affected if the electric field of a point charge did not vary as $1/r^2$.

24-9. Discuss the similarities and differences between the gravitational field of a point mass m and the electric field of a point charge q.

24-10. Is the term **E** in Gauss' law the electric field produced by just the charge inside the Gaussian surface?

24-11. Reformulate Gauss' law by choosing the unit normal of the Gaussian surface to be the one directed inward.

24-12. Would Gauss' law be helpful for determining the electric field of two equal but opposite charges a fixed distance apart?

24-13. Discuss the role that symmetry plays in the application of Gauss' law. Give examples of continuous charge distributions whose fields can and cannot be determined with Gauss' law.

Problems

Electric Flux

24-1. A uniform electric field of magnitude 1.1×10^4 N/C is perpendicular to a square sheet with sides 2.0 m long. What is the electric flux through the sheet?

24-2. Calculate the flux through the sheet of the previous problem if the plane of the sheet is at an angle of 60° to the field. Find the flux for both directions of the unit normal to the sheet.

24-3. Consider the uniform electric field $\mathbf{E} = (4.0\mathbf{j} + 3.0\mathbf{k}) \times 10^3$ N/C. What is its electric flux through a circular area of radius 2.0 m that lies in the xy plane?

24-4. Repeat the previous problem, given that the circular area is (*a*) in the yz plane and (*b*) 45° above the xy plane.

24-5. A vector **F** is given by $\mathbf{F} = 3x^2\mathbf{k}$. Calculate $\int_S \mathbf{F}\cdot\hat{\mathbf{n}}\,da$, where S is the area shown in the accompanying figure. Assume that $\hat{\mathbf{n}} = \mathbf{k}$.

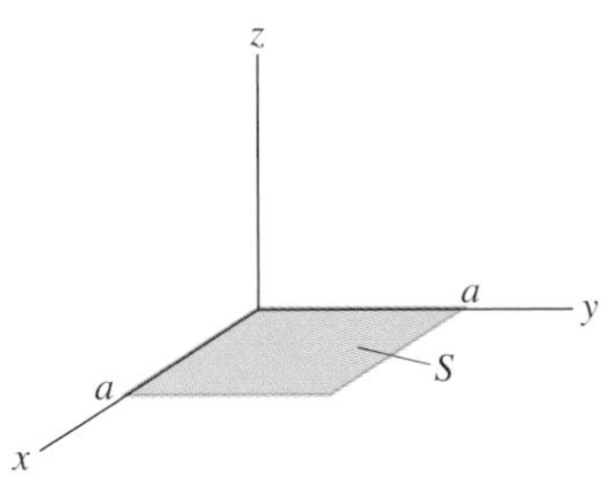

PROBLEM 24-5

24-6. Do Prob. 24-5, with $\mathbf{F} = 2x\mathbf{i} + 3x^2\mathbf{k}$.

Gauss' Law

24-7. Determine the electric flux through each surface whose cross section is shown in the accompanying figure.

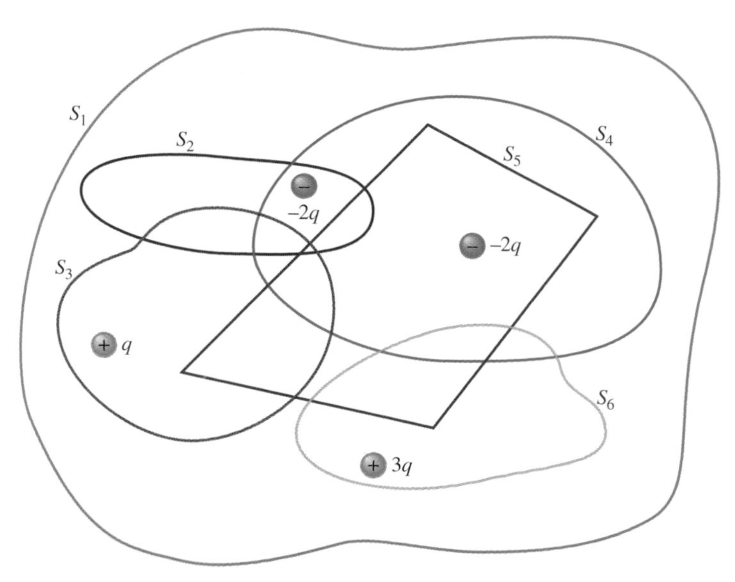

PROBLEM 24-7

24-8. Find the electric flux through the closed surfaces whose cross sections are shown in the accompanying figure.

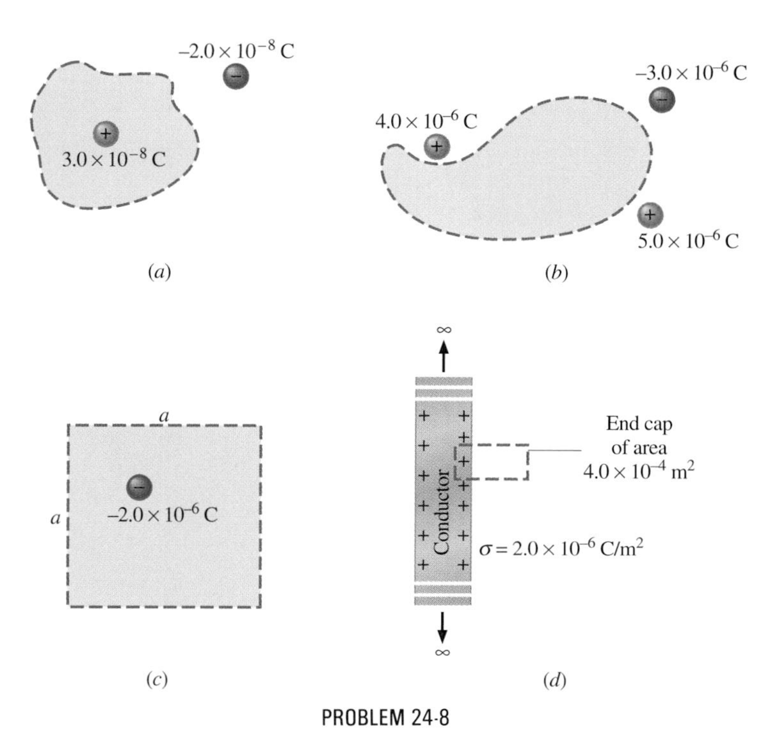

PROBLEM 24-8

24-9. A point charge q is located at the center of a cube whose sides are of length a. If there are no other charges inside the cube, what is the electric flux through its surface?

24-10. The electric flux through a cubical box 8.0 cm on a side is 1.2×10^3 N·m²/C. What is the total charge enclosed by the box?

24-11. The electric flux through a spherical surface is 4.0×10^4 N·m²/C. What is the net charge enclosed by the surface?

24-12. A cube whose sides are of length d is placed in a uniform electric field of magnitude $E = 4.0 \times 10^3$ N/C so that the field is perpendicular to two opposite faces of the cube. What is the net flux through the cube?

24-13. Repeat Prob. 24-12, assuming that the electric field is directed along a body diagonal of the cube.

24-14. A point charge q is located at the center of a cube of side l. What is the electric flux through one face of the cube?

24-15. A total charge of 5.0×10^{-6} C is distributed uniformly throughout a cubical volume whose edges are 8.0 cm long. (*a*) What is the charge density in the cube? (*b*) What is the electric flux through a cube with 12.0-cm edges that is concentric with the charge distribution? (*c*) Do the same calculation for cubes whose edges are 10.0 cm long and 5.0 cm long. (*d*) What is the electric flux through a spherical surface of radius 3.0 cm that is also concentric with the charge distribution?

24-16. Suppose that the electric field of an isolated point charge were proportional to $1/r^{2+\delta}$ rather than $1/r^2$. Determine the flux that passes through the surface of a sphere of radius R centered at the charge. Would Gauss' law remain valid?

Calculating Electric Fields with Gauss' Law

24-17. A very long, thin wire has a uniform linear charge density of 50 μC/m. What is the electric field at a distance 2.0 cm from the wire?

24-18. Charge of uniform density 3.0×10^{-2} C/m³ is spread throughout a spherical volume of radius 25 cm. Determine the electric field due to this charge distribution at distances of 10, 25, and 50 cm from the center of the distribution.

24-19. A charge of −30 μC is distributed uniformly throughout a spherical volume of radius 10.0 cm. Determine the electric field due to this charge at a distance of (*a*) 2.0 cm, (*b*) 5.0 cm, (*c*) 20.0 cm from the center of the sphere.

24-20. Repeat your calculations for the previous problem, given that the charge is distributed uniformly over the surface of a spherical conductor of radius 10.0 cm.

24-21. A total charge Q is distributed uniformly throughout a spherical shell of inner and outer radii r_1 and r_2, respectively. Show that the electric field due to the charge is

$$\mathbf{E} = 0 \qquad (r \le r_1);$$

$$\mathbf{E} = \frac{Q}{4\pi\epsilon_0 r^2}\left(\frac{r^3 - r_1^3}{r_2^3 - r_1^3}\right)\hat{\mathbf{r}} \qquad (r_1 \le r \le r_2);$$

$$\mathbf{E} = \frac{Q}{4\pi\epsilon_0 r^2}\,\hat{\mathbf{r}} \qquad (r \ge r_2).$$

24-22. A large sheet of charge has a uniform charge density of 10 μC/m². What is the electric field due to this charge at a point just above the surface of the sheet?

24-23. Charge is spread uniformly over the surface of a long cylindrical wire of radius R. The surface density of the charge is σ. Assume that the wire is infinitely long and calculate the electric field due to the charge for $r \le R$ and $r > R$.

24-24. Charge is distributed uniformly with a density ρ throughout an infinitely long cylindrical volume of radius R. Show that the field of this charge distribution is directed radially with respect to the cylinder and that

$$E = \frac{\rho r}{2\epsilon_0} \qquad (r \le R);$$

$$E = \frac{\rho R^2}{2\epsilon_0 r} \qquad (r \ge R).$$

23-25. Charge is distributed throughout a very long cylindrical volume of radius R such that the charge density increases with the distance r from the central axis of the cylinder according to $\rho = \alpha r$, where α is a constant. Show that the field of this charge distribution is directed radially with respect to the cylinder and that

$$E = \frac{\alpha r^2}{3\epsilon_0} \qquad (r \le R);$$

$$E = \frac{\alpha R^3}{3\epsilon_0 r} \qquad (r \ge R).$$

24-26. The electric field 10.0 cm from the surface of a copper ball of radius 5.0 cm is directed toward the ball's center and has magnitude 4.0×10^2 N/C. How much charge is on the surface of the ball?

24-27. The electric field at the surface of a spherical conductor of radius 5.0 cm is directed radially inward, and its magnitude is 200 N/C. What is the density of charge on the surface of the sphere?

General Problems

24-28. A circular surface S is concentric with the origin, has radius a, and lies in the yz plane. Calculate $\int_S \mathbf{F}\cdot\hat{\mathbf{n}}\,da$ for $\mathbf{F} = 3z^2\mathbf{i}$.

24-29. (*a*) Calculate the electric flux through the open hemispherical surface due to the electric field $\mathbf{E} = E_0\mathbf{k}$. (See the accompanying figure.) (*b*) If the hemisphere is rotated by 90° around the x axis, what is the flux through it?

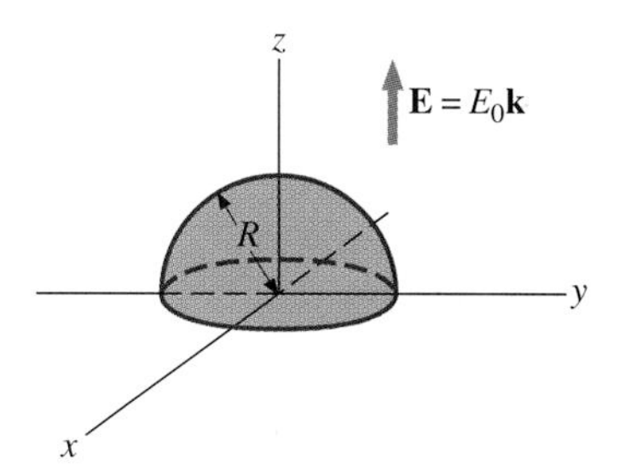

PROBLEM 24-29

24-30. A spherical rubber balloon carries a total charge Q distributed uniformly over its surface. At $t = 0$, the radius of the balloon is R. The balloon is then slowly inflated until its radius reaches $2R$ at the time t_0. Determine the electric field due to this charge as a function of time (*a*) at the surface of the balloon, (*b*) at the surface of radius R, and (*c*) at the surface of radius $2R$. Ignore any effect on the electric field due to the material of the balloon and assume that the radius increases uniformly with time.

24-31. Two equal and opposite charges of magnitude Q are located on the x axis at the points $+a$ and $-a$, as shown in the accompanying figure. What is the net flux due to these charges through a square surface of side $2a$ that lies in the yz plane and is centered at the origin? (*Hint:* Determine the flux due to each charge separately, then use the principle of superposition.)

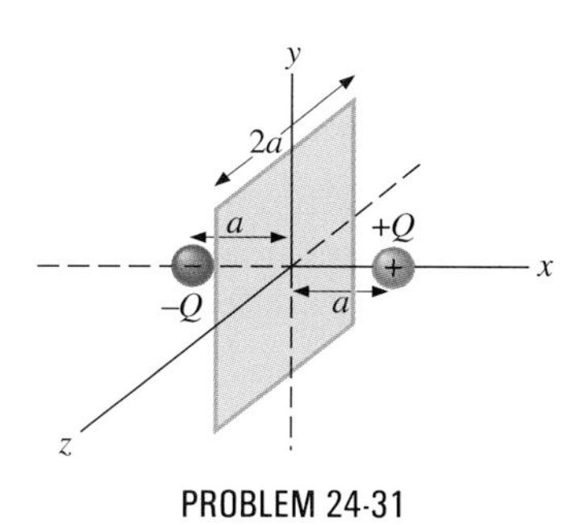

PROBLEM 24-31

24-32. Charge is distributed throughout a spherical shell of inner radius r_1 and outer radius r_2 with a volume density given by $\rho = \rho_0 r_1/r$, where ρ_0 is a constant. Determine the electric field due to this charge as a function of r, the distance from the center of the shell.

24-33. Charge is distributed throughout a spherical volume of radius R with a density $\rho = \alpha r^2$, where α is a constant. Determine the electric field due to the charge at points both inside and outside the sphere.

24-34. The electric field in a region is given by $\mathbf{E} = a\mathbf{i}/(b + cx)$, where $a = 200$ N·m/C, $b = 2.0$ m, and $c = 2.0$. What is the net charge enclosed by the shaded volume shown in the accompanying figure?

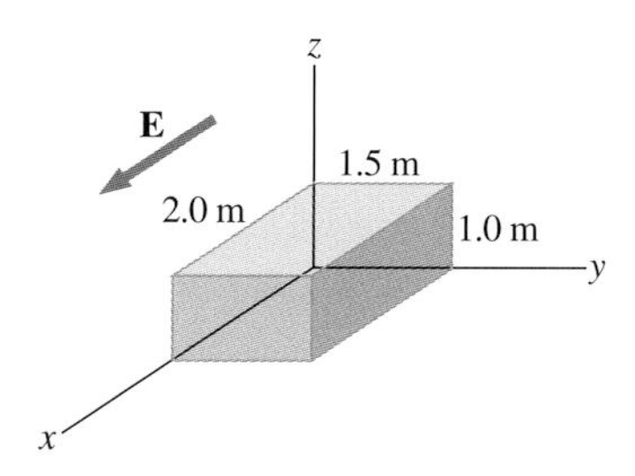

PROBLEM 24-34

24-35. Consider a uranium nucleus to be a sphere of radius $R = 7.4 \times 10^{-15}$ m with a charge of $92e$ distributed uniformly throughout its volume. (*a*) What is the electric force exerted on an electron when it is 3.0×10^{-15} m from the center of the nucleus? (*b*) What is the acceleration of the electron at this point?

24-36. The volume charge density of a spherical charge distribution is given by $\rho(r) = \rho_0 e^{-\alpha r}$, where ρ_0 and α are constants. What is the electric field produced by this charge distribution?

24-37. The infinite slab between the planes defined by $z = -a/2$ and $z = a/2$ contains a uniform volume charge density ρ. (See the accompanying figure.) What is the electric field produced by this charge distribution?

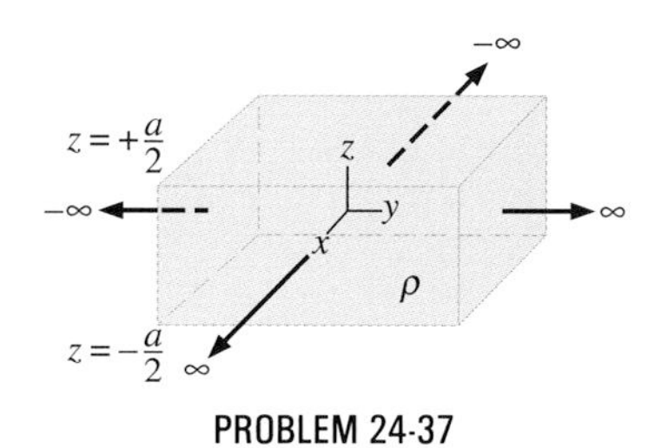

PROBLEM 24-37

24-38. A total charge Q is distributed uniformly throughout a spherical volume that is centered at O_1 and has a radius R. Without disturbing the charge remaining, charge is removed from the spherical volume that is centered at O_2. (See the accompanying figure.) Show that the electric field everywhere in the empty region is given by

$$\mathbf{E} = \frac{Q\mathbf{r}}{4\pi\epsilon_0 R^3},$$

where $\mathbf{r}$ is the displacement vector directed from O_1 to O_2.

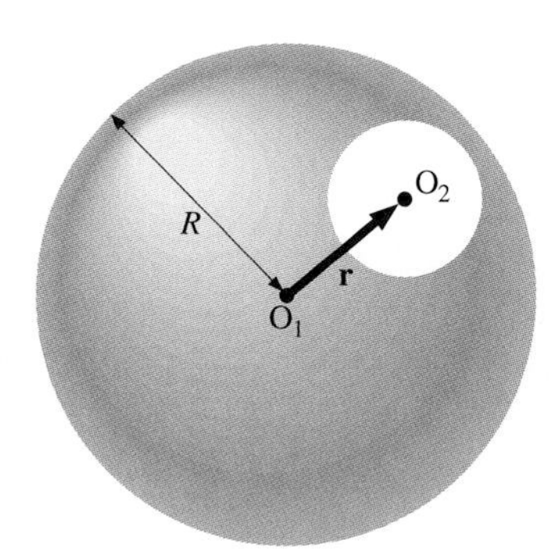

PROBLEM 24-38

A spectacular lightning storm in the desert.

CHAPTER 25 ELECTRIC POTENTIAL

PREVIEW

An important property of the electrostatic field is that the work it does on charged particles is path-independent. In this chapter we consider the consequences of this property. The main topics to be discussed here are the following:

1. **Electric potential of point charges.** The electric potential is defined in terms of the electric potential energy of a test charge in an electrostatic field. The expressions for the electric potentials of an isolated point charge and an arbitrary discrete charge distribution are given.
2. **Calculating the electric potential.** We calculate the electric potentials of various charge distributions, both discrete and continuous.
3. **Electric potential difference.** Electric potential difference is defined, and examples illustrating how it is calculated are given.
4. **Calculating the electric field from the electric potential.** We see how the components of the electric field are determined from the expression for the electric potential.
5. **Equipotential surfaces.** Equipotential surfaces are introduced, and examples of these surfaces are given.

*6. **Electric dipole.** The electric dipole is defined and its potential and electric field are determined. We examine the interaction of a dipole with an electric field.

7. **Properties of conductors in electrostatic equilibrium.** Using the concept of electric potential, we investigate the behavior of charged conductors in electrostatic equilibrium.

In Chap. 7 we learned that the gravitational force is conservative. From that fact we were able to define a potential-energy function for a mass in a gravitational field. Now the electrostatic force has the same spatial dependence ($1/r^2$) as the gravitational force; hence it too is conservative and an electrostatic potential-energy function can be found for a charge in an electric field. In the study of electric fields, it is useful to define a function known as the electric potential that is simply the electrostatic potential energy per unit charge. This important function will be considered in this chapter.

25-1 Electric Potential of Point Charges

The similarity in the formulations of the gravitational force and the electrostatic force is striking. The mutual gravitational force between two masses M and m separated by a distance r is GMm/r^2, while the mutual electrostatic force between two charges q_1 and q_2 with that same separation is kq_1q_2/r^2 (see Fig. 25-1). Because of this similarity, our discussion of Chap. 7 on the conservative nature of the gravitational force is valid also for the electrostatic force. For example, by substituting kq_1q_2 for $-GMm$ in the expression for gravitational potential energy $-GMm/r$, we obtain the electrostatic potential energy of two charges, q_1 and q_2. By analogy then, we have the following:

1. The electrostatic force is conservative. Mathematically, this property is written as

$$\oint \mathbf{F}_e \cdot d\mathbf{l} = q_t \oint \mathbf{E} \cdot d\mathbf{l} = 0, \qquad (i)$$

where q_t is a test charge in an electric field $\mathbf{E}$.

2. Associated with any conservative force is a potential energy. Specifically, if W_{RP} is the work done on a test charge q_t by the electric force $q_t\mathbf{E}$ between a reference point R, where

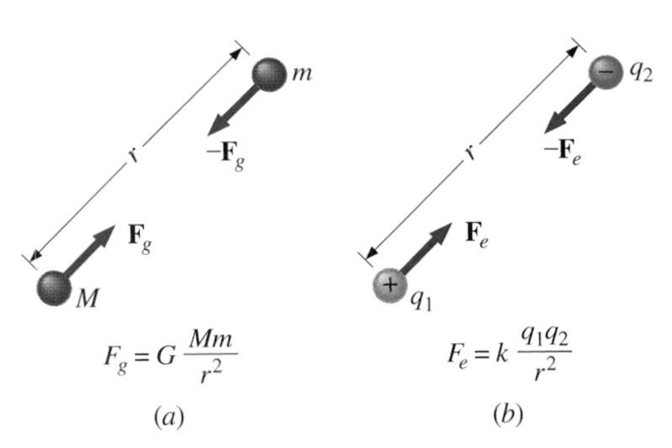

FIGURE 25-1 A comparison of the gravitational force and the electrostatic force.

the potential energy is taken to be zero (see Supplement 7-2), and some point P, the electric potential energy of q_t at P is

$$U_P = -W_{RP} = -q_t \int_R^P \mathbf{E} \cdot d\mathbf{l}. \qquad (ii)$$

3. If q_t is in the field of a point charge q that is fixed at $\mathbf{r} = 0$, then by analogy with $-GMm/r$, its potential energy is

$$U_P = \frac{1}{4\pi\epsilon_0}\frac{q_t q}{r}, \qquad (iii)$$

where r is the distance between q and q_t and the reference point is at infinity.

4. Potential energies are additive. Hence for a charge q_t located at the point P in the net field of $q_1, q_2, \ldots, q_N$, the potential energy is given by the *scalar* sum

$$U_P = \frac{q_t}{4\pi\epsilon_0}\sum_i \frac{q_i}{r_i}, \qquad (iv)$$

where r_i is the distance from q_i to P.

Just as we used the force on the test charge q_t to define the electric field $\mathbf{E}$ ($= \mathbf{F}_e/q_t$), we can use potential energy to define a scalar quantity called the **electric potential** (often referred to simply as **potential**). This quantity is very useful in calculating energies of charges as well as electric fields. If, at a particular point, the potential energy of q_t is $U(x, y, z)$, the electric potential there is, by definition,

$$V(x, y, z) = \frac{U(x, y, z)}{q_t},$$

or

$$V_P = \frac{U_P}{q_t}, \qquad (25\text{-}1)$$

where P represents the point (x, y, z).

Now let's consider the concepts represented by Eqs. (*i*) through (*iv*) in terms of electric field and potential rather than electric force and potential energy. We can do so by dividing those four equations by the charge q_t. We then obtain:

1′. The electric field is conservative. This is represented mathematically by

$$\oint \mathbf{E} \cdot d\mathbf{l} = 0. \qquad (25\text{-}2)$$

2′. Associated with an electric field is the electric potential. At an arbitrary point P, we can find the potential from the electric field by

$$V_P = -\int_R^P \mathbf{E} \cdot d\mathbf{l}, \qquad (25\text{-}3)$$

where R is a reference point chosen such that $V_R = 0$. Since **E** is a force per unit charge, Eq. (25-3) tells us that we can think of V_P as representing the *negative of the work per unit charge done by* **E** *on charges that move from R to P.*

3′. The electric potential at a point P due to an isolated point charge q is

$$V_P = \frac{1}{4\pi\epsilon_0}\frac{q}{r}, \tag{25-4}$$

where r is the distance from q to P. The sign of the point charge must be included when you use this equation. Hence the potential of an isolated negative charge is negative. Equation (25-4) can be obtained directly from Eq. (25-3) by taking the electric field to be that of a point charge at $r = 0$ and the reference point at infinity.

4′. The electric potential at P due to a number of fixed point charges $q_1, q_2, \ldots, q_N$ is the scalar sum

$$V_P = \frac{1}{4\pi\epsilon_0}\sum_i \frac{q_i}{r_i}, \tag{25-5}$$

where r_i is the distance from q_i to P.

From Eq. (25-1), the SI unit for electric potential is the joule per coulomb (J/C). This unit is given the special name *volt* (*V*); 1 V = 1 J/C. The volt is so named to honor Alessandro Volta (1745–1827), an Italian physicist who constructed the first chemical cell. The SI units combine in Eq. (25-3) to give 1 V = (1 N/C) (1 m), so the electric field may be expressed in volts per meter (V/m) as well as newtons per coulomb (N/C).

EXAMPLE 25-1 Work and Electric Potential Energy

With the reference point at ∞, the electric potential at point A is $V_A = 60$ V. (*a*) What is the electric potential energy of a proton placed at A? (*b*) How much work is done by the electric field on a proton that is brought from ∞ to A?

SOLUTION (*a*) Since electric potential is electric potential energy per unit charge, the electric potential energy of a proton at point A is

$$U_A = qV_A = (1.6 \times 10^{-19}\text{ C})(60\text{ V}) = 9.6 \times 10^{-18}\text{ J}.$$

(*b*) By definition, V_A is the negative work per unit charge done by the electric field on a charge moving from ∞ to A. Thus the work done by the field on the proton is

$$(W_{\infty A})_E = -qV_A = -(1.6 \times 10^{-19}\text{ C})(60\text{ V})$$
$$= -9.6 \times 10^{-18}\text{ J}.$$

The negative value reflects the fact that overall the force of the electric field is opposite to the displacement of the proton as it is brought from ∞ to A.

Note that in moving the proton from ∞ to A, you must apply a force opposite to the force of the electric field. Hence the work you do is positive and is given by

$$(W_{\infty A})_{\text{YOU}} = -(W_{\infty A})_E = 9.6 \times 10^{-18}\text{ J}.$$

EXAMPLE 25-2 The Bohr Model for Hydrogen

In the simple Bohr model of the hydrogen atom, an electron moves in a circular orbit of radius r around a fixed proton. (*a*) What is the potential energy of the electron? (*b*) What is the kinetic energy of the electron? (*c*) Calculate the total energy of the electron when it is in its ground (lowest-energy) state with an orbital radius of $r = 5.30 \times 10^{-11}$ m. (*d*) How much energy has to be added to a hydrogen atom in its ground state to ionize it?

SOLUTION (*a*) At a distance r from the proton (charge $+e$), the electric potential is

$$V = \frac{1}{4\pi\epsilon_0}\frac{e}{r},$$

so the potential energy U of the electron (charge $-e$) is

$$U = qV = -e\left(\frac{1}{4\pi\epsilon_0}\frac{e}{r}\right) = -\frac{1}{4\pi\epsilon_0}\frac{e^2}{r}. \tag{i}$$

(*b*) The electron moves in a circular orbit while under the influence of the Coulomb force. From Newton's second law,

$$\sum F_r = ma_r$$
$$\frac{1}{4\pi\epsilon_0}\frac{e^2}{r^2} = m_e\frac{v^2}{r}$$

where m_e is the mass of the electron. From this equation, the kinetic energy of the electron is given by

$$T = \frac{1}{2}m_e v^2 = \frac{1}{8\pi\epsilon_0}\frac{e^2}{r}. \tag{ii}$$

(*c*) The total energy E of the electron when it is a distance r from the proton is the sum of Eqs. (*i*) and (*ii*):

$$E = T + U = -\frac{1}{8\pi\epsilon_0}\frac{e^2}{r}, \tag{iii}$$

which, with the accepted values for e and ϵ_0 and with $r = 5.30 \times 10^{-11}$ m, gives $E = -2.17 \times 10^{-18}$ J for the ground-state energy.

(*d*) The atom is ionized when the electron and proton are completely separated (an infinite distance apart). If the kinetic energies of the two particles are zero after they are completely separated, the total energy of the system is zero. Therefore, *at least* 2.17×10^{-18} J must be added to the ground-state hydrogen atom to ionize it.

DRILL PROBLEM 25-1

(*a*) If in Fig. 25-2*a*, $q_1 = 3.0\ \mu C$, $q_2 = -1.0\ \mu C$, and $a = 50$ cm, what is the electric potential energy of a charge $q_t = -2.0\ \mu C$ when it is at P? (*b*) How much work is done by the net electric field of q_1 and q_2 when q_t is brought from ∞ to P? (*c*) How much work must you do to move q_t from ∞ to P?
ANS. (*a*) −0.072 J; (*b*) 0.072 J; (*c*) −0.072 J.

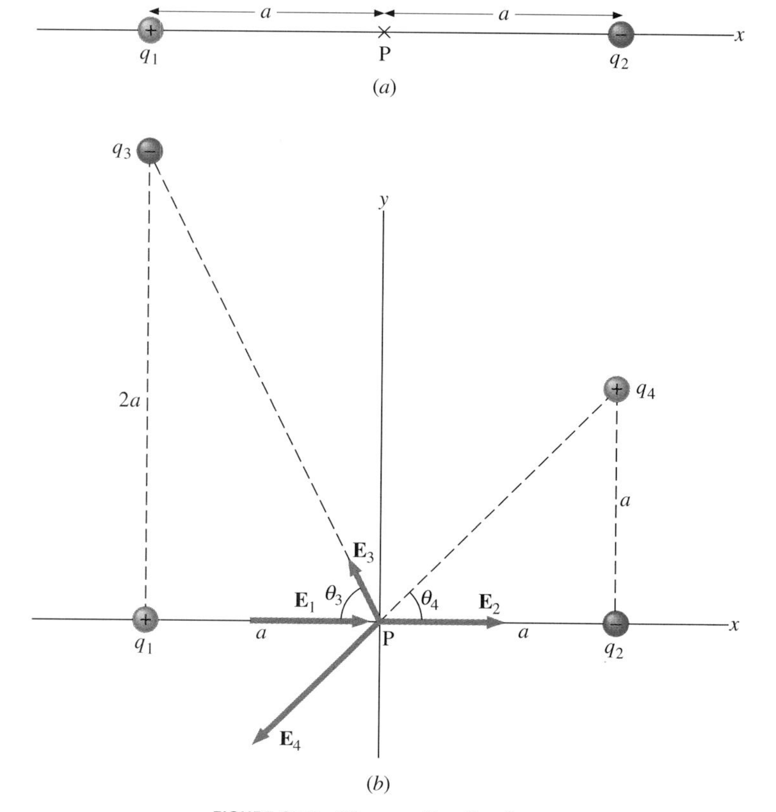

FIGURE 25-2 Charge distributions.

25-2 Calculating the Electric Potential

The electric potential due to a group of fixed discrete charges is the scalar sum of the electric potentials due to each of the charges (Eq. 25-5). If the charge distribution is continuous, the sum must be replaced by an integral:

$$V_P = \frac{1}{4\pi\epsilon_0}\int \frac{dq}{r}, \tag{25-6a}$$

where r is the distance from dq to P. For a distribution with a volume charge density ρ, we may write $dq = \rho\, dv$, and

$$V_P = \frac{1}{4\pi\epsilon_0}\int \frac{\rho\, dv}{r}. \tag{25-6b}$$

Similarly, for a surface charge distribution (say, on a conductor) of density σ and for a linear charge distribution of density λ, the electric potentials are given by, respectively,

$$V_P = \frac{1}{4\pi\epsilon_0}\int \frac{\sigma\, da}{r} \tag{25-6c}$$

and

$$V_P = \frac{1}{4\pi\epsilon_0}\int \frac{\lambda\, dl}{r}. \tag{25-6d}$$

EXAMPLE 25-3 Electric Field and Electric Potential

Find the electric field and the electric potential at point P of Fig. 25-2. (*a*) Suppose the charges of Fig. 25-2*a* are both positive, with $q_1 = q_2 = q$. (*b*) Repeat your calculations for $q_1 = q$, $q_2 = -q$. (*c*) Suppose two additional point charges $q_3 = -2q$ and $q_4 = 4q$ are placed as shown in Fig. 25-2*b*. With $q_1 = q$ and $q_2 = -q$, what are the electric field and electric potential at P? For all parts, assume that $q = 6.00\ \mu C$ and $a = 2.00$ m.

SOLUTION (*a*) The electric field at P due to the two charges is the vector sum of the fields of the individual charges:

$$\mathbf{E}_P = \frac{1}{4\pi\epsilon_0}\left(\frac{q}{a^2}\mathbf{i} - \frac{q}{a^2}\mathbf{i}\right) = 0;$$

and from Eq. (25-5), the electric potential at P is

$$V_P = \frac{1}{4\pi\epsilon_0}\left(\frac{q}{a} + \frac{q}{a}\right) = \frac{q}{2\pi\epsilon_0 a} = 5.39 \times 10^4 \text{ V},$$

where we have substituted $q = 6.00 \times 10^{-6}$ C and $a = 2.00$ m.
(*b*) For this case the electric field at P is

$$\mathbf{E}_P = \frac{1}{4\pi\epsilon_0}\left(\frac{q}{a^2}\mathbf{i} + \frac{q}{a^2}\mathbf{i}\right) = \frac{q}{2\pi\epsilon_0 a^2}\mathbf{i}$$
$$= 2.70 \times 10^4\mathbf{i} \text{ N/C};$$

while the electric potential there is

$$V_P = \frac{1}{4\pi\epsilon_0}\left(\frac{q}{a} - \frac{q}{a}\right) = 0.$$

Notice that the electric field may vanish when the electric potential does not and vice versa.
(*c*) The charge q_3 is a distance $\sqrt{a^2 + (2a)^2} = 2.236a$ from P, and q_4 is $\sqrt{a^2 + a^2} = 1.414a$ from P, so the magnitudes of the electric fields of q_3 and q_4 at P are

$$E_3 = \frac{1}{4\pi\epsilon_0}\frac{2q}{5a^2} = \frac{q}{10\pi\epsilon_0 a^2}$$

and

$$E_4 = \frac{1}{4\pi\epsilon_0}\frac{4q}{2a^2} = \frac{q}{2\pi\epsilon_0 a^2}.$$

From Fig. 25-2*b*, the angles θ_3 and θ_4 are

$$\theta_3 = \tan^{-1} 2a/a = \tan^{-1} 2 = 63.4°$$

and

$$\theta_4 = \tan^{-1} a/a = \tan^{-1} 1 = 45.0°.$$

The components of $\mathbf{E}_3$ and $\mathbf{E}_4$ are then

$$E_{3x} = -E_3 \cos\theta_3 = -\frac{q}{10\pi\epsilon_0 a^2}\cos 63.4° = -0.0447\frac{q}{\pi\epsilon_0 a^2},$$

$$E_{3y} = E_3 \sin\theta_3 = \frac{q}{10\pi\epsilon_0 a^2}\sin 63.4° = 0.0894\frac{q}{\pi\epsilon_0 a^2},$$

$$E_{4x} = -E_4 \cos\theta_4 = -\frac{q}{2\pi\epsilon_0 a^2}\cos 45.0° = -0.354\frac{q}{\pi\epsilon_0 a^2},$$

$$E_{4y} = -E_4 \sin\theta_4 = -\frac{q}{2\pi\epsilon_0 a^2}\sin 45.0° = -0.354\frac{q}{\pi\epsilon_0 a^2}.$$

Now with the help of part (b), the components of the resultant field at P are

$$E_x = (0.500 - 0.0447 - 0.354)\frac{q}{\pi\epsilon_0 a^2} = 0.101\frac{q}{\pi\epsilon_0 a^2},$$

$$E_y = (0.000 + 0.0894 - 0.354)\frac{q}{\pi\epsilon_0 a^2} = -0.265\frac{q}{\pi\epsilon_0 a^2}.$$

Substituting $q = 6.00 \times 10^{-6}$ C and $a = 2.00$ m, we find that the electric field at P is

$$\mathbf{E}_P = (5.45 \times 10^3\,\mathbf{i} - 1.43 \times 10^4\,\mathbf{j})\ \text{N/C}.$$

This is a vector of magnitude 1.53×10^4 N/C directed 69.1° below the x axis.

The electric potential at P is much easier to calculate because it is a scalar quantity. From Eq. (25-5),

$$V_P = \frac{1}{4\pi\epsilon_0}\left(\frac{q}{a} + \frac{-q}{a} + \frac{-2q}{2.236a} + \frac{4q}{1.414a}\right) = 5.22 \times 10^4\ \text{V}.$$

An electron micrograph of anhydrous caffeine crystals.

EXAMPLE 25-4 A One-Dimensional Ionic Crystal

Solids made up of positive and negative ions are called *ionic crystals* (for example, Na^+Cl^-). A simplified one-dimensional representation of such a crystal is shown in Fig. 25-3. This infinite chain consists of alternating charges $\pm e$ spaced a distance $d = 1.0 \times 10^{-10}$ m apart. Determine the potential energy of an ion in the chain.

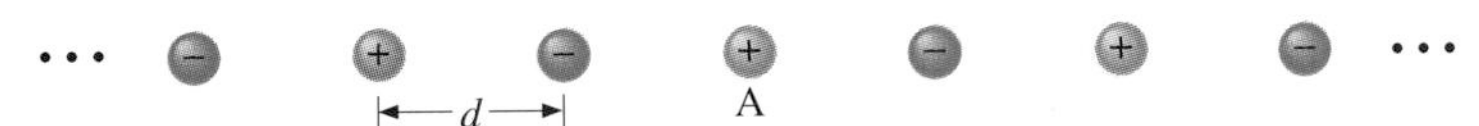

FIGURE 25-3 A simplified one-dimensional representation of an ionic crystal.

SOLUTION We'll calculate the potential energy of ion A shown in the figure. The net electric potential at A due to all the other ions is

$$V_A = \frac{1}{4\pi\epsilon_0}\left[\frac{(-e)}{d} + \frac{(-e)}{d} + \frac{(+e)}{2d} + \frac{(+e)}{2d} + \frac{(-e)}{3d} + \frac{(-e)}{3d} + \dots\right]$$

$$= -\frac{e}{2\pi\epsilon_0 d}\left(1 - \frac{1}{2} + \frac{1}{3} - \frac{1}{4} + \dots\right). \qquad (i)$$

From Supplement 14-1,

$$\ln(1 + x) = x - \frac{x^2}{2} + \frac{x^3}{3} - \frac{x^4}{4} + \dots,$$

so

$$(1 - \tfrac{1}{2} + \tfrac{1}{3} - \tfrac{1}{4} + \dots) = \ln(1 + 1) = \ln 2.$$

We can therefore write Eq. (i) as

$$V_A = -\frac{e\ln 2}{2\pi\epsilon_0 d},$$

and the potential energy of ion A becomes

$$U_A = eV_A = -\frac{e^2\ln 2}{2\pi\epsilon_0 d}.$$

Substituting the given values of e and d into this equation, we find $U_A = -3.2 \times 10^{-18}$ J. The *negative* electrostatic potential energy is primarily responsible for the stability of this ionic structure. In a more complete calculation of the energy per ion, a potential energy associated with the repulsion of the inner-shell electrons of adjacent ions and a kinetic energy representing the vibrational motion of the ions must be included. However, when these two energies are added to the electrostatic potential energy, the total energy is still negative, indicating that the ionic structure is stable.

EXAMPLE 25-5 Electric Potential of a Charged Ring

Find the electric potential at the point P on the axis of the uniformly charged circular ring of Fig. 23-12 (Example 23-6). The ring has a net charge q and radius a, and point P is a distance x from the center of the ring.

SOLUTION The ring has negligible thickness, so the potential can be found with Eq. (25-6d), which gives the potential of a line of charge. Since the ring is uniformly charged, λ is constant, and the electric potential at P is

$$V_P = \frac{1}{4\pi\epsilon_0}\int_{\text{RING}}\frac{\lambda\,dl}{r} = \frac{\lambda}{4\pi\epsilon_0}\int_{\text{RING}}\frac{dl}{r}.$$

Over the ring, the distance r between P and any charge element $\lambda\, dl$ of the ring is constant. The term $1/r$ can therefore be factored out of the integral, and

$$V_P = \frac{\lambda}{4\pi\epsilon_0 r}\int_{\text{RING}} dl = \frac{\lambda}{4\pi\epsilon_0 r}(2\pi a) = \frac{1}{4\pi\epsilon_0}\frac{q}{\sqrt{x^2+a^2}},$$

where we have used $q = \lambda(2\pi a)$ and $r = \sqrt{x^2+a^2}$.

DRILL PROBLEM 25-2

Four particles, each of charge q, are positioned at the corners of a square whose sides are of length a. Determine the electric field and the electric potential at the center of the square.
ANS. $\mathbf{E} = 0$; $V = \sqrt{2}q/\pi\epsilon_0 a$.

DRILL PROBLEM 25-3

In doing Example 25-4, we determined the potential energy of a positive ion in the chain. Would the potential energy of a negative ion be different? Verify your answer by repeating the calculation of Example 25-4 for a negative ion in the chain.
ANS. The potential energy is the same for charges of either sign.

DRILL PROBLEM 25-4

Use the electric field calculated in Example 23-6 along with Eq. (25-3) to determine the potential at any point on the axis of a charged circular ring. Compare your result with that found in Example 25-5.

25-3 Electric Potential Difference

If V_A and V_B are the electric potentials at points A and B, then the **electric potential difference** (or simply **potential difference**) between A and B is $V_B - V_A$. This quantity is especially useful in the analysis of electric circuits. Like electric potential, potential difference is a scalar and is expressed in volts.

Since the electrostatic field is conservative, the potential difference between any two points is path-independent. For example, if $V_B - V_A = 100$ V for path I of Fig. 25-4, then $V_B - V_A = 100$ V for path II, or for path III, or for any other path connecting A and B. By using Eq. (25-3) to represent V_A and V_B, we find

$$V_B - V_A = -\int_R^B \mathbf{E}\cdot d\mathbf{l} + \int_R^A \mathbf{E}\cdot d\mathbf{l},$$

which reduces to

$$V_B - V_A = -\int_A^B \mathbf{E}\cdot d\mathbf{l}. \tag{25-7}$$

Thus $V_B - V_A$ is *the negative of the work done by the electric field on a unit charge that moves from A to B.*

A unit often used to represent the energy of atomic and nuclear particles is the *electron-volt* (*eV*). It is defined in terms of potential difference as follows: *One electron-volt (1 eV) is the kinetic energy acquired by an electron that has been accelerated across a potential difference of magnitude one volt (1 V).* We can find the conversion factor between the electron-volt and the joule by considering an electron that is accelerated across a potential difference $\Delta V = 1$ V. The work done on it is $e\,\Delta V$ which, by the work-energy theorem, is equal to the change in its kinetic energy ΔT:

$$\Delta T = e\,\Delta V = (1.6\times10^{-19}\text{ C})(1\text{ J/C}) = 1.6\times10^{-19}\text{ J},$$

so

$$1\text{ eV} = 1.6\times10^{-19}\text{ J}.$$

Nuclear and particle physicists frequently use multiples of this unit such as the keV (10^3 eV), the MeV (10^6 eV), the GeV (10^9 eV), and the TeV (10^{12} eV). A smaller unit, commonly used in atomic and solid state physics, is the meV (10^{-3} eV).

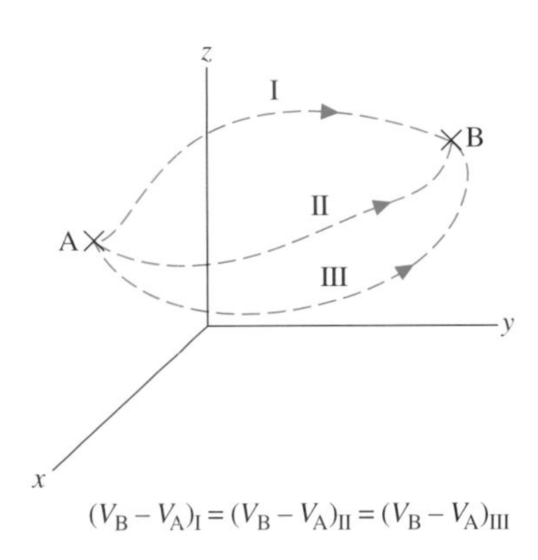

FIGURE 25-4 The potential difference between points A and B is independent of the path connecting them.

EXAMPLE 25-6 POTENTIAL DIFFERENCE

(*a*) Points A and B are, respectively, 3.0 and 6.0 m away from an isolated 9.0×10^{-9}-C point charge. What is the electric potential difference $V_B - V_A$ between A and B? (*b*) How much work is done by the electric field on a -4.0×10^{-9}-C charge that is moved from A to B?

SOLUTION (*a*) From Eq. (25-4), the electric potentials at A and B due to the point charge are

$$V_A = \frac{1}{4\pi\epsilon_0}\frac{q}{r_A} = (9.0\times10^9\text{ N}\cdot\text{m}^2/\text{C}^2)\frac{9.0\times10^{-9}\text{ C}}{3.0\text{ m}}$$
$$= 27.0\text{ V},$$

and

$$V_B = \frac{1}{4\pi\epsilon_0}\frac{q}{r_B} = (9.0\times10^9\text{ N}\cdot\text{m}^2/\text{C}^2)\frac{9.0\times10^{-9}\text{ C}}{6.0\text{ m}}$$
$$= 13.5\text{ V},$$

so the potential difference between A and B is

$$V_B - V_A = 13.5\text{ V} - 27.0\text{ V} = -13.5\text{ V}.$$

(*b*) Since $V_B - V_A$ is the negative of the work done by the electric field on a unit charge that moves from A to B, the work done on the -4.0×10^{-9}-C charge by the electric field is

$$W_{AB} = q[-(V_B - V_A)] = (-4.0 \times 10^{-9}\ \text{C})(13.5\ \text{V}) = -5.4 \times 10^{-8}\ \text{J}.$$

What do you think is the significance of the minus sign?

EXAMPLE 25-7 Conservative Nature of the Electric Field

An electric field has magnitude 200 V/m and points in the negative x direction of the coordinate system shown in Fig. 25.5. (*a*) Determine the potential difference between the origin O and the point B, which is at (6.00, 3.00) m, using the direct path between these points. (*b*) Determine the potential difference between O and B using the path OAB shown.

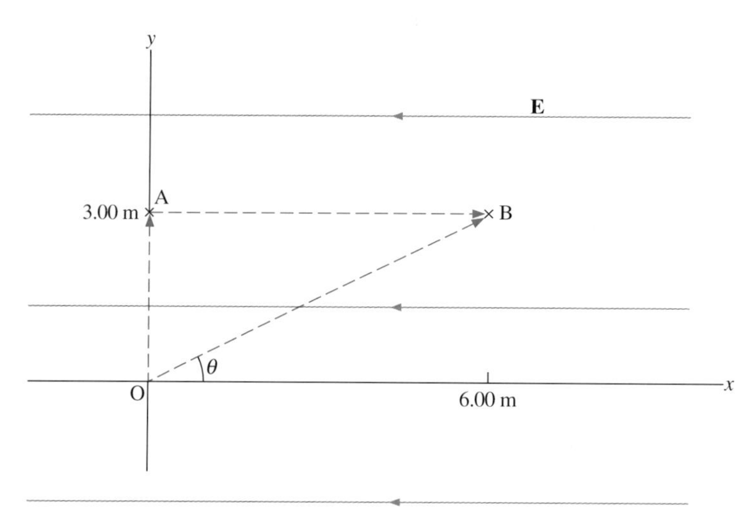

FIGURE 25-5 The potential difference between O and B is calculated using paths OB and OAB.

SOLUTION (*a*) The displacement vector from O to B makes an angle $\theta = \tan^{-1} 3.00/6.00 = 26.6°$ with respect to the x axis. The angle between the electric field and this displacement vector is therefore $180° - 26.6° = 153°$. The potential difference between O and B calculated using the direct path is then

$$V_B - V_O = -\int_O^B \mathbf{E}\cdot d\mathbf{l} = -\int_O^B E \cos 153°\, dl$$
$$= -(200\ \text{V/m}) \cos 153° \int_O^B dl.$$

The length of the path from O to B is

$$\int_O^B dl = \sqrt{(6.00\ \text{m})^2 + (3.00\ \text{m})^2} = 6.71\ \text{m},$$

so the potential difference between O and B is $V_B - V_O = 1.20 \times 10^3$ V.

(*b*) Along the segment OA of path OAB, the electric field is perpendicular to the displacement and $\mathbf{E}\cdot d\mathbf{l} = E\,dl \cos 90° = 0$; that is, points O and A are at the same potential. Along segment AB, the electric field and the displacement are antiparallel, so $\mathbf{E}\cdot d\mathbf{l} = E\,dl \cos 180° = -E\,dl$. The potential difference between O and B using path OAB is therefore

$$V_B - V_O = -\int_O^A \mathbf{E}\cdot d\mathbf{l} - \int_A^B \mathbf{E}\cdot d\mathbf{l} = 0 + E\int_A^B dl$$
$$= (200\ \text{V/m})(6.00\ \text{m}) = 1.20 \times 10^3\ \text{V}.$$

As expected, this result is the same as that obtained in part (*a*) since the electric force (and field) is conservative so the electric potential energy (and potential) is path-independent. In calculating potential differences, it is often useful to substitute an alternative path for the direct path between two points so that $-\int \mathbf{E}\cdot d\mathbf{l}$ is more easily determined. This alternative path generally has a segment parallel to the electric field and a segment perpendicular to the field, such as OAB in Fig. 25-5.

DRILL PROBLEM 25-5

In a region of space there exists a constant electric field $\mathbf{E} = (5.00\mathbf{i} + 2.50\mathbf{j}) \times 10^5$ V/m. What is the potential difference between point A at (3.50, 4.00) m and point B at (2.00, 4.00) m?
ANS. $V_B - V_A = 7.50 \times 10^5$ V.

DRILL PROBLEM 25-6

What are the energies in electron-volts of the electron in Example 25-2 and the ion in Example 25-4?
ANS. −13.6 eV; −20.0 eV.

25-4 Calculating the Electric Field from the Electric Potential

In Fig. 25-6, the electric potentials at A and B are V and $V + \Delta V$, respectively, so the potential difference between the two points is

$$V_B - V_A = (V + \Delta V) - V = \Delta V.$$

If the displacement $\Delta\mathbf{l}$ from A to B is small enough that the electric field is essentially constant over it, we may write

$$V_B - V_A = \Delta V = -\mathbf{E}\cdot\Delta\mathbf{l} = -E\,\Delta l \cos\theta,$$

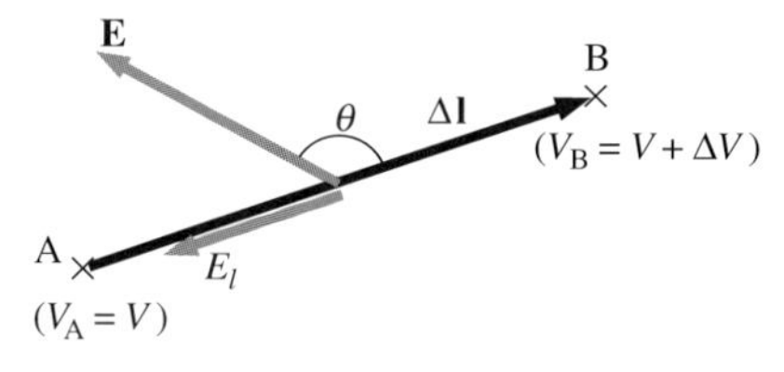

FIGURE 25-6 The electric field component along the displacement $\Delta\mathbf{l}$ is given by $E_l = -\Delta V/\Delta l$. Note that A and B are assumed to be so close together that the field is constant along $\Delta\mathbf{l}$.

where θ is the angle between $\mathbf{E}$ and $\Delta\mathbf{l}$. Designating E_l as the component of $\mathbf{E}$ along $\Delta\mathbf{l}$ ($E_l = E\cos\theta$), we have

$$\Delta V = -E_l\,\Delta l,$$

so

$$E_l = -\frac{\Delta V}{\Delta l}.$$

In the limit $\Delta l \to 0$, this equation becomes

$$E_l = -\frac{dV}{dl}. \qquad (25\text{-}8a)$$

At a given point, the component of the electric field in a particular direction is the negative rate of change of the electric potential with respect to displacement in that direction.

If the direction of $\Delta\mathbf{l}$ is taken in turn to be along the x, y, and z axes, Eq. (25-8a) will give the electric field components along those coordinate axes; that is,

$$E_x = -\frac{\partial V}{\partial x}; \qquad E_y = -\frac{\partial V}{\partial y}; \qquad E_z = -\frac{\partial V}{\partial z}. \qquad (25\text{-}8b)$$

We now have an alternative method for determining the electric field of a static charge distribution. Rather than summing electric fields vectorially to determine a net field, we can first calculate the scalar electric potential, then differentiate this potential to find the electric field. Because they involve scalar quantities, these two steps are frequently easier to perform than the vector sum of Eq. (23-3).

EXAMPLE 25-8 Obtaining the Electric Field from the Electric Potential

Use the electric potential of an isolated point charge q to find the electric field of that charge.

SOLUTION The radial electric field is calculated by differentiating Eq. (25-4) with respect to r:

$$E_r = -\frac{dV}{dr} = -\frac{q}{4\pi\epsilon_0}\frac{d}{dr}\left(\frac{1}{r}\right) = \frac{1}{4\pi\epsilon_0}\frac{q}{r^2}.$$

From symmetry, the field has only a radial component. Thus

$$\mathbf{E} = \frac{1}{4\pi\epsilon_0}\frac{q}{r^2}\,\hat{\mathbf{r}},$$

which is the expected result.

DRILL PROBLEM 25-7

Use the potential function of Example 25-5 to calculate the electric field along the axis of a ring of charge. Compare this result with that found in Example 23-6.

25-5 Equipotential Surfaces

The electric potential of an arbitrary charge distribution is a scalar function of the spatial coordinates. The locus of points for which $V(x, y, z)$ is a constant—say, V_1—is defined by

$$V(x, y, z) = V_1.$$

This equation represents a surface in space called an *equipotential surface*. By setting $V(x, y, z)$ equal to different constants, $V_1, V_2, V_3, \ldots, V_N$, we obtain a family of equipotential surfaces. These surfaces are very useful for visualizing how the electric potential and the electric field vary in space.

In the previous section, we saw that the component of the field in a particular direction depends on how the potential is changing with position in that direction. This means that in a region where equipotential surfaces are close together, $-\Delta V/\Delta l$ and consequently $\mathbf{E}$ are large; where the equipotential surfaces are far apart, $-\Delta V/\Delta l$ and $\mathbf{E}$ are small. Additionally, since $\Delta V = 0$ along an equipotential surface, the electric field component tangent to the surface is zero. Hence *electric field lines are everywhere perpendicular to equipotential surfaces*, and no work is done by the electric field on a charge that moves along an equipotential surface. Two examples of equipotential surfaces and their corresponding field lines are shown in Fig. 25-7a and b.

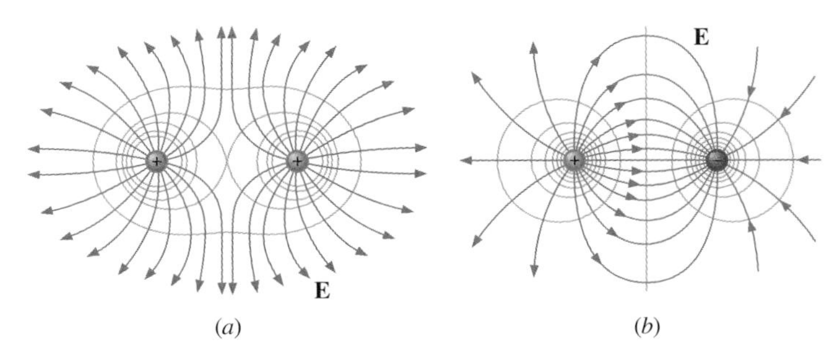

FIGURE 25-7 Equipotential surfaces and field lines of (a) two equal positive charges and (b) two equal but opposite charges.

EXAMPLE 25-9 Equipotential Surfaces of Parallel Plates

Describe the equipotential surfaces between two parallel plates that carry equal and opposite charge of surface densities $\pm\sigma$. Ignore edge effects.

SOLUTION From Example 23-9, the electric field between the plates has magnitude σ/ϵ_0 and is directed from the positive plate to the negative plate. For the geometry of Fig. 25-8, $\mathbf{E} = (\sigma/\epsilon_0)\mathbf{i}$. Since electric field lines are everywhere perpendicular to equipotential surfaces, these surfaces must be parallel planes corresponding to constant values of x. Some representative equipotential surfaces are shown in the figure.

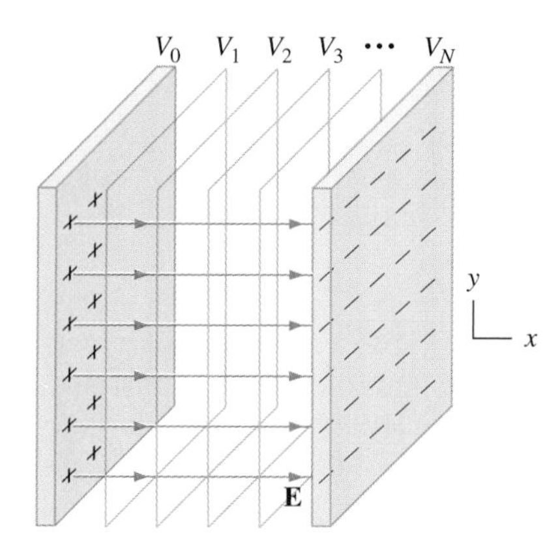

FIGURE 25-8 Equipotential surfaces and field lines between parallel charged plates.

EXAMPLE 25-10 Equipotential Surfaces of a Point Charge

What are the equipotential surfaces of an isolated point charge?

SOLUTION The electric potential of an isolated point charge is

$$V = \frac{1}{4\pi\epsilon_0}\frac{q}{r}.$$

Since q and $4\pi\epsilon_0$ are constants, the equipotential surfaces are given by

$$r = \text{constant},$$

which represents concentric spheres centered at the charge, as shown in Fig. 25-9.

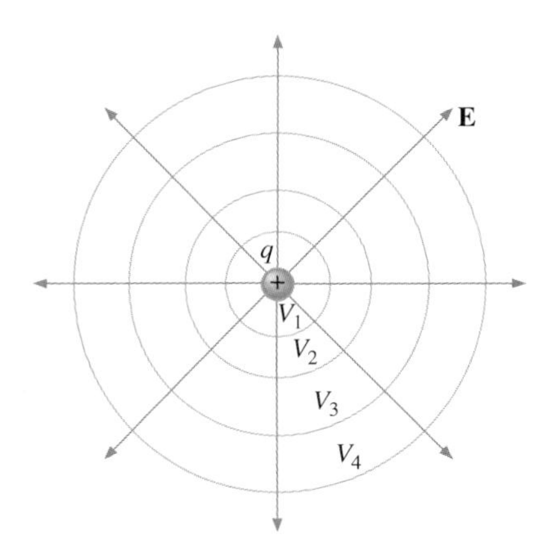

FIGURE 25-9 Equipotential surfaces and field lines of an isolated positive charge.

EXAMPLE 25-11 Infinite Line of Charge

In this and the previous chapter, we have discussed a variety of different yet related concepts. As a summary, we will now examine these concepts for a specific example, the infinite line of charge with uniform charge density λ. The electric field of this configuration was calculated in Example 24-6. (*a*) Determine the *electric potential* at an arbitrary radial distance r from the wire. Designate r_0 as the radial distance of the reference point from the wire. (*b*) What is the *electric potential energy* of a charge q that is a distance r away from the wire? (*c*) What is the *potential difference* between two points B and A, which are distances r_B and r_A, respectively, from the wire? (*d*) How much *work is done by the field* on a charge q that moves from A to B? (*e*) Use the electric potential to determine the *electric field* of the line of charge. (*f*) Describe the *equipotential surfaces* of this electric field.

SOLUTION (*a*) From Eq. (25-3), the potential at an arbitrary point a distance r from the wire, using r_0 as the distance of the reference point from the wire, is

$$V(r) = -\int_{r_0}^{r} \mathbf{E}\cdot d\mathbf{r} = -\frac{\lambda}{2\pi\epsilon_0}\int_{r_0}^{r}\frac{dr'}{r'} = -\frac{\lambda}{2\pi\epsilon_0}\ln\frac{r}{r_0}.$$

(*b*) The potential energy of a charge q at a distance r from the wire is

$$U(r) = qV(r) = -\frac{\lambda q}{2\pi\epsilon_0}\ln\frac{r}{r_0}.$$

(*c*) The potential difference between B and A is

$$V(r_B) - V(r_A) = -\frac{\lambda}{2\pi\epsilon_0}\left(\ln\frac{r_B}{r_0} - \ln\frac{r_A}{r_0}\right) = -\frac{\lambda}{2\pi\epsilon_0}\ln\frac{r_B}{r_A}.$$

(*d*) The work done by the field of the wire when a charge q moves from A to B is

$$W_{AB} = -q[V(r_B) - V(r_A)] = \frac{\lambda q}{2\pi\epsilon_0}\ln\frac{r_B}{r_A}.$$

(*e*) From Eq. (25-8*a*), the electric field is determined from the electric potential by

$$E_r = -\frac{dV}{dr} = \frac{\lambda}{2\pi\epsilon_0}\frac{d}{dr}\left(\ln\frac{r}{r_0}\right) = \frac{\lambda}{2\pi\epsilon_0}\frac{1}{r}.$$

(*f*) Equipotential surfaces are given by

$$V(r) = -\frac{\lambda}{2\pi\epsilon_0}\ln\frac{r}{r_0} = \text{constant},$$

which is equivalent to $r = \text{constant}$. The equipotential surfaces are therefore infinite cylinders whose common central axis is along the line of charge.

*25-6 Electric Dipole

A pair of equal and opposite point charges $\pm q$ that are separated by a fixed distance is known as an *electric dipole*. This charge configuration is characterized by its **electric dipole moment**, which is a vector **p** directed from the negative to the positive charge. The magnitude of the dipole moment is $p = ql$, with l being the distance between the two charges. For example, the electric dipole shown in Fig. 25-10 has an electric dipole moment

$$\mathbf{p} = 2aq\mathbf{j}. \tag{25-9}$$

The electric potential due to this dipole at the point P shown is simply the sum of the potentials due to the two charges:

$$V = \frac{1}{4\pi\epsilon_0}\left[\frac{q}{\sqrt{x^2+(y-a)^2+z^2}} - \frac{q}{\sqrt{x^2+(y+a)^2+z^2}}\right].$$

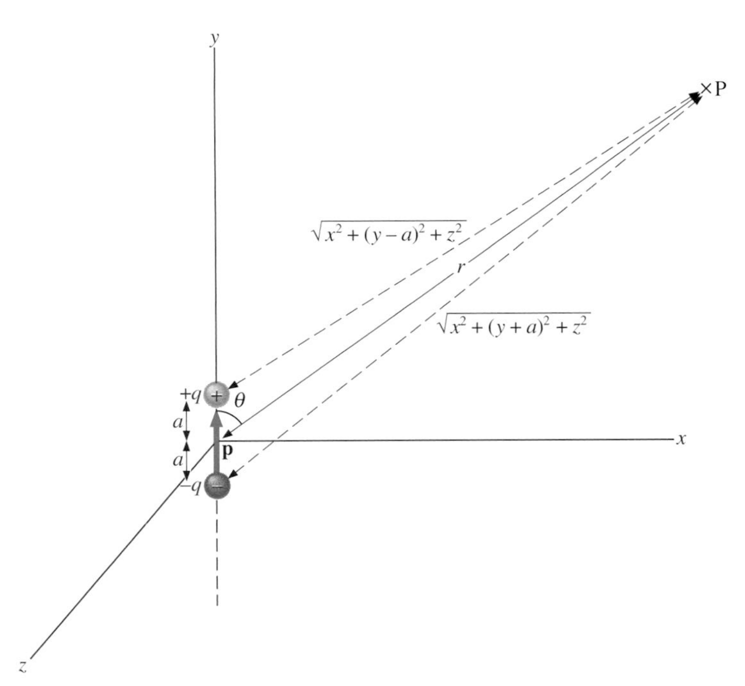

FIGURE 25-10 An electric dipole of electric dipole moment $\mathbf{p} = 2aq\mathbf{j}$. If $r \gg a$, the dipole is known as a point dipole.

By differentiating this function, we obtain for the electric field of the dipole:

$$E_x = -\frac{\partial V}{\partial x} = \frac{q}{4\pi\epsilon_0}\left\{\frac{x}{[x^2+(y-a)^2+z^2]^{3/2}} - \frac{x}{[x^2+(y+a)^2+z^2]^{3/2}}\right\},$$

$$E_y = -\frac{\partial V}{\partial y} = \frac{q}{4\pi\epsilon_0}\left\{\frac{y-a}{[x^2+(y-a)^2+z^2]^{3/2}} - \frac{y+a}{[x^2+(y+a)^2+z^2]^{3/2}}\right\},$$

$$E_z = -\frac{\partial V}{\partial z} = \frac{q}{4\pi\epsilon_0}\left\{\frac{z}{[x^2+(y-a)^2+z^2]^{3/2}} - \frac{z}{[x^2+(y+a)^2+z^2]^{3/2}}\right\}.$$

The electric dipole is especially useful in calculations involving the interactions of, or the electromagnetic radiation from, atoms and molecules. In such situations, the distance between the two opposite charges is generally very small compared with the distance from the center of the dipole to the point of interest. This allows us to approximate the equations for V and $\mathbf{E}$ that we just obtained by expansions to first order in $2a/r$, where $r\ (=\sqrt{x^2+y^2+z^2})$ is the distance from the center of the dipole. In this case, the electric dipole is commonly referred to as a *point electric dipole*. You will find the details of this approximation in Supplement 25-1. The electric potential, written in terms of the dipole moment $\mathbf{p}$, is

$$V = \frac{py}{4\pi\epsilon_0 r^3} = \frac{p\cos\theta}{4\pi\epsilon_0 r^2} = \frac{\mathbf{p}\cdot\hat{\mathbf{r}}}{4\pi\epsilon_0 r^2}, \tag{25-10}$$

where $\hat{\mathbf{r}}$ is the unit vector directed from the point dipole to P and θ is the angle between $\mathbf{p}$ and $\hat{\mathbf{r}}$. And the electric field components are

$$E_x = \frac{3pxy}{4\pi\epsilon_0 r^5},\quad E_y = \frac{p(2y^2-x^2-z^2)}{4\pi\epsilon_0 r^5},\quad E_z = \frac{3pzy}{4\pi\epsilon_0 r^5}. \tag{25-11}$$

The equipotential surfaces and electric field lines corresponding to these equations are shown in Fig. 25-11.

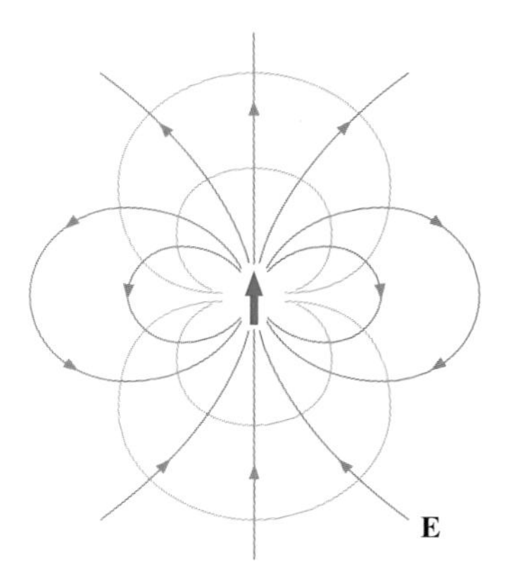

FIGURE 25-11 Equipotential surfaces and field lines of a point electric dipole.

Since an electric dipole consists of charges, it is influenced by an applied electric field. To study this interaction, let's consider the dipole of Fig. 25-12, which has charges $-q$ at y and $+q$ at $y + \Delta y$. The force exerted on it by a nonuniform electric field $E(y)$ is

$$F = qE(y+\Delta y) - qE(y).$$

Because Δy is small, $E(y+\Delta y) \approx E(y) + [dE(y)/dy]\Delta y$, and

$$F = q\left[E(y) + \frac{dE(y)}{dy}\Delta y - E(y)\right] = p\frac{dE(y)}{dy}, \tag{25-12}$$

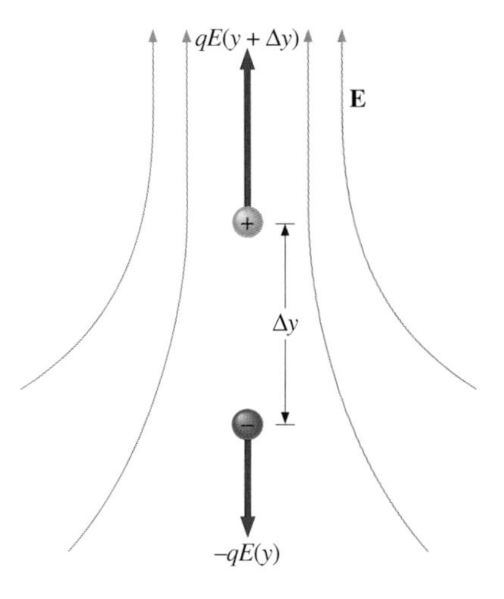

FIGURE 25-12 An electric dipole in a nonuniform electric field.

where p $(= q\Delta y)$ is the magnitude of the electric dipole moment. Notice that the electric field must vary with position if it is to exert a force on an electric dipole: *There is no net force if E is constant.*

However, a uniform field does exert a torque on an electric dipole. For example, the dipole of Fig. 25-13 is oriented at an angle θ with respect to the field, so it experiences a counterclockwise torque τ around C given by

$$\tau = qE(a\sin\theta) + qE(a\sin\theta) = (2aq)E\sin\theta = pE\sin\theta. \tag{25-13a}$$

This torque tends to align the dipole with the electric field. To express the torque as a vector, we use the fact that **p** is a vector of magnitude $2aq$ that points from $-q$ to $+q$. Then Eq. (25-13*a*) becomes

$$\boldsymbol{\tau} = \mathbf{p} \times \mathbf{E}. \tag{25-13b}$$

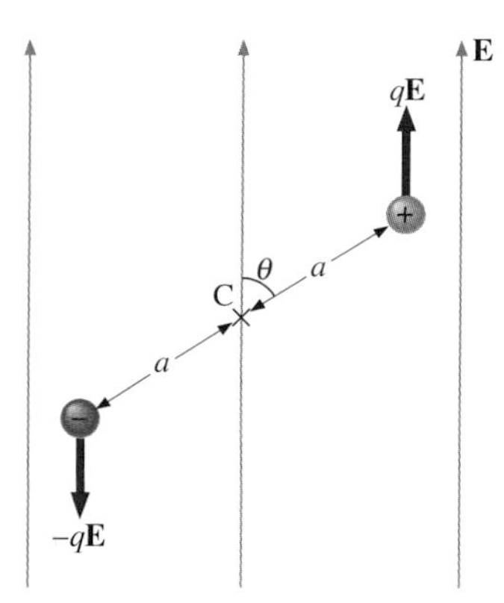

FIGURE 25-13 An electric dipole in a uniform electric field.

Since work must be done to rotate an electric dipole in an electric field, we can assign a potential energy to its rotational position. If the dipole of Fig. 25-13 is rotated through an angle $d\theta$, the work dW done on it by the electric field **E** is, from Eqs. (9-14) and (25-13*a*),

$$dW = -\tau\, d\theta = -pE\sin\theta\, d\theta,$$

where the negative sign signifies that a positive torque tends to rotate the dipole toward decreasing θ. By definition, the potential energy U of the electric dipole oriented at an angle θ to the electric field is the negative of the work done by the field when the dipole rotates from a reference angle θ_R to θ:

$$U = -\left[\int_{\theta_R}^{\theta} -\tau\, d\theta'\right] = \int_{\theta_R}^{\theta} pE\sin\theta'\, d\theta'$$
$$= -pE\cos\theta + pE\cos\theta_R.$$

The simplest form for U is obtained by choosing $\theta_R = \pi/2$; then the potential energy of the dipole is

$$U = -pE\cos\theta = -\mathbf{p}\cdot\mathbf{E}. \tag{25-14}$$

EXAMPLE 25-12 AN ELECTRIC DIPOLE

(*a*) Equal and opposite charges of magnitude 4.0×10^{-12} C are separated by a distance of 4.0×10^{-5} m. What is the magnitude of their electric dipole moment? (*b*) What is the maximum torque exerted on the dipole when it is placed in a uniform electric field of magnitude 3.0×10^5 V/m? (*c*) What is the potential energy of the dipole when it is oriented at $\theta = 30°$ to the electric field? (*d*) How much work does the field do when the dipole rotates from $\theta = 120°$ to $\theta = 30°$?

SOLUTION (*a*) From Eq. (25-9), the magnitude of the electric dipole moment is

$$p = 2aq = (4.0 \times 10^{-5}\ \text{m})(4.0 \times 10^{-12}\ \text{C})$$
$$= 1.6 \times 10^{-16}\ \text{C}\cdot\text{m}.$$

(*b*) The maximum torque is exerted on the dipole when its dipole moment is perpendicular to the applied electric field, that is, when $\theta = 90°$:

$$\tau_{max} = pE\sin 90° = (1.6 \times 10^{-16}\ \text{C}\cdot\text{m})(3.0 \times 10^5\ \text{V/m})$$
$$= 4.8 \times 10^{-11}\ \text{N}\cdot\text{m}.$$

(*c*) From Eq. (25-14), the potential energy of the dipole at 30° to the field is

$$U_{30} = -pE\cos 30°$$
$$= -(1.6 \times 10^{-16}\ \text{C}\cdot\text{m})(3.0 \times 10^5\ \text{V}\cdot\text{m})\cos 30°$$
$$= -4.2 \times 10^{-11}\ \text{J}.$$

(*d*) The work W done by the field when the dipole rotates between two orientations is the negative of the change in the dipole's potential energy:

$$W = -(U_{30} - U_{120}) = 4.2 \times 10^{-11}\ \text{J} - (-2.4 \times 10^{-11}\ \text{J})$$
$$= 6.6 \times 10^{-11}\ \text{J}.$$

DRILL PROBLEM 25-8

(*a*) What is the torque on the electric dipole of Example 25-12 when it is oriented at $\theta = 30°$ to the electric field? (*b*) How much work is done by the field when the dipole is rotated from $\theta = 0°$ to $\theta = 30°$?
ANS. (*a*) 2.4×10^{-11} N·m; (*b*) -6.5×10^{-12} J.

25-7 PROPERTIES OF CONDUCTORS IN ELECTROSTATIC EQUILIBRIUM

In the presence of an external electric field, the free charge in a conductor redistributes and very quickly reaches electrostatic equilibrium. The resulting charge distribution and its electric field have many interesting properties which we will now investigate with the help of Gauss' law and the concept of electric potential.

1. *The electric field inside a conductor vanishes.* If there were an electric field inside a conductor, it would exert forces on the free electrons, which would then be in motion. However, moving charges imply nonstatic conditions, contrary to our assumption. Therefore, when electrostatic equilibrium is reached, the charge is distributed in such a way that the electric field inside the conductor vanishes.

There is no electric field inside a conductor.

2. *Any excess charge placed on a conductor resides entirely on the surface of the conductor.* The Gaussian surface of Fig. 25-14 (the dashed line) follows the contour of the actual surface of the conductor and is located an infinitesimal distance *within* it. Since $\mathbf{E} = 0$ everywhere inside a conductor,

$$\oint_S \mathbf{E}\cdot\hat{\mathbf{n}}\, da = 0,$$

so from Gauss' law, there is no net charge inside the Gaussian surface. But the Gaussian surface lies just below the actual surface of the conductor; consequently, there is no net charge inside the conductor. Any excess charge must lie on its surface.

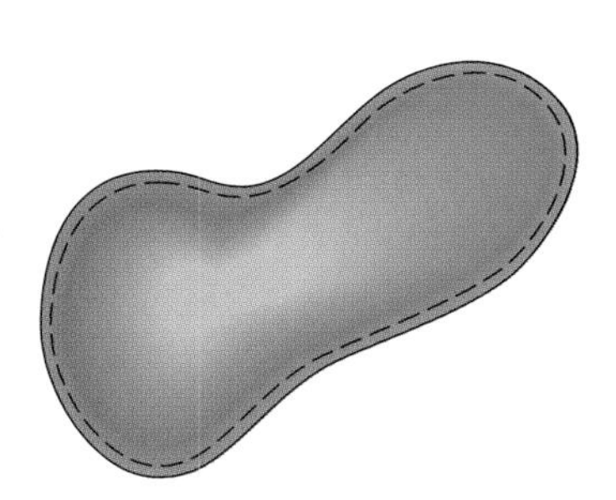

FIGURE 25-14 The dashed line represents a Gaussian surface that is just beneath the actual surface of the conductor.

This particular property of conductors is the basis for an extremely accurate method developed by Plimpton and Lawton in 1936 to verify Gauss' law and, correspondingly, Coulomb's law. A sketch of their apparatus is shown in Fig. 25-15. The two spherical shells are connected to one another through an electrometer E, a device that can detect a very slight amount of charge flowing from one shell to the other. When switch S is thrown to the left, charge is placed on the outer shell by the battery B. None of that charge should flow through the electrometer to the inner shell since that would mean a violation of

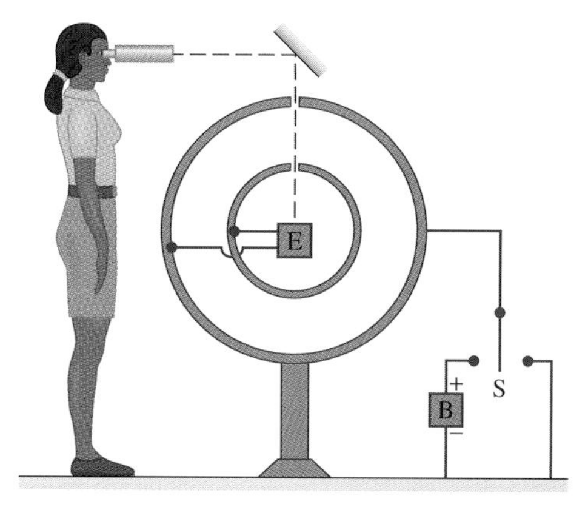

FIGURE 25-15 A representation of the apparatus of Plimpton and Lawton. Any transfer of charge between the spheres is detected by the electrometer E.

Gauss' law. Plimpton and Lawton did not detect any flow and, knowing the sensitivity of their electrometer, concluded that if the radial dependence in Coulomb's law were $1/r^{2+\delta}$, δ would be less than 2×10^{-9}! More recent measurements place δ at less than 3×10^{-16}, a number so small that the validity of Coulomb's law seems indisputable.

3. *The electric field is perpendicular to the surface of a conductor everywhere on that surface.* If the electric field had a component parallel to the surface, free charges on the surface would move, a situation contrary to the assumption of electrostatic equilibrium.

4. *The surface of a conductor is an equipotential surface.* This is an immediate consequence of property 3 and the fact that equipotential surfaces and field lines are perpendicular.

5. *The interior and the surface of a conductor are at the same potential.* The potential difference between any point S on the surface and an internal point I is given by

$$V_S - V_I = -\int_I^S \mathbf{E}\cdot d\mathbf{l}.$$

Since $\mathbf{E} = 0$ everywhere inside a conductor, the integral vanishes, leaving

$$V_S = V_I.$$

A direct consequence of properties 4 and 5 is that the difference in electric potential between any arbitrary point of one conductor and any arbitrary point of a second conductor is the same. Hence we usually speak of a potential difference between two conductors rather than a potential difference between specific points of the conductors.

6. *At any point just above the surface of a conductor, the surface charge density σ and the magnitude of the electric field E are related by*

$$E = \frac{\sigma}{\epsilon_0}. \qquad (25\text{-}15)$$

The infinitesimal Gaussian cylinder that surrounds the point P on the surface of the conductor of Fig. 25-16 has one end face inside and one end face outside the surface. The height and cross-sectional area of the cylinder are δ and ΔA, respectively. The cylinder's sides are perpendicular to the surface of the conductor and its end faces are parallel to the surface. Because the cylinder is infinitesimally small, the charge density σ is essentially constant over the surface enclosed, so the total charge inside the Gaussian cylinder is $\sigma\,\Delta A$. Now **E** is perpendicular to the surface of the conductor outside the conductor and vanishes within it. Electric flux therefore crosses only the outer end face of the Gaussian surface and may be written as $E\,\Delta A$, since ΔA is assumed to be small enough that **E** is approximately constant over that area. From Gauss' law,

$$E\,\Delta A = \frac{\sigma\,\Delta A}{\epsilon_0}.$$

Thus

$$E = \frac{\sigma}{\epsilon_0},$$

which is Eq. (25-15).

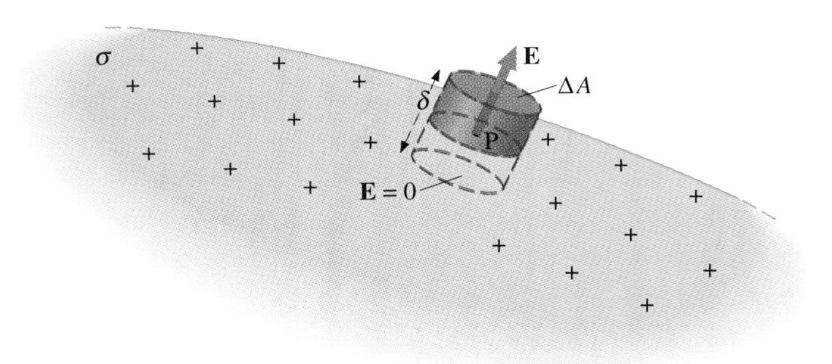

FIGURE 25-16 An infinitesimal cylindrical Gaussian surface surrounds the point P, which is on the surface of the conductor. The field **E** is perpendicular to the surface of the conductor outside the conductor and vanishes within it.

EXAMPLE 25-13 THE ELECTRIC FIELD OF A CONDUCTING PLATE

The infinite conducting plate shown in Fig. 25-17 has a uniform surface charge density σ. Use Gauss' law to find the electric field outside the plate. Compare this result with that of Example 23-8.

SOLUTION For this case, we use a cylindrical Gaussian surface, a side view of which is shown. The flux calculation is similar to that for an infinite sheet of charge (Example 24-8) with one major exception: The left face of the Gaussian surface is inside the conductor where $\mathbf{E} = 0$, so the total flux through the Gaussian surface is EA rather that $2EA$. Now from Gauss' law,

$$EA = \frac{\sigma A}{\epsilon_0}$$

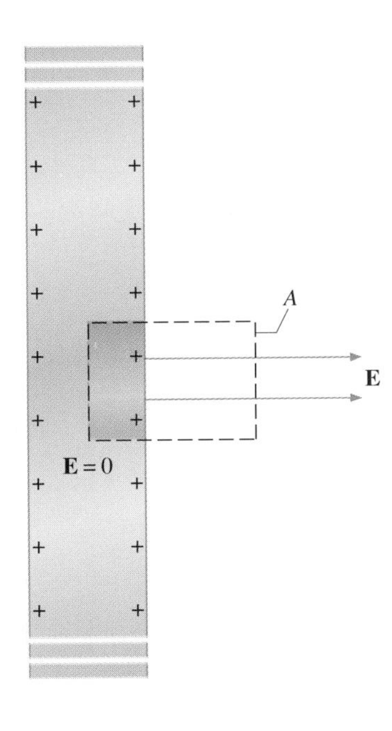

FIGURE 25-17 A side view of an infinite conducting plate and Gaussian cylinder with cross-sectional area A.

and the electric field outside the plate is

$$E = \frac{\sigma}{\epsilon_0},$$

in agreement with the result of Example 23-8.

EXAMPLE 25-14 POTENTIAL DIFFERENCE BETWEEN OPPOSITELY CHARGED PARALLEL PLATES

Two large conducting plates carry equal and opposite charges, with a surface charge density σ of magnitude 6.81×10^{-7} C/m^2, as shown in Fig. 25-18. The separation between the plates is $l =$ 6.50 mm. (*a*) What is the electric field between the plates? (*b*) What is the potential difference between the plates? (*c*) A proton is released at the positive plate and is accelerated by the electric field toward the negative plate. What is the kinetic energy of the proton in electron-volts when it arrives at the negative plate? What is the speed of the proton when it arrives?

SOLUTION (*a*) The electric field is directed from the positive to the negative plate as shown in the figure, and its magnitude is given by

$$E = \frac{\sigma}{\epsilon_0} = \frac{6.81 \times 10^{-7}\ \text{C/m}^2}{8.85 \times 10^{-12}\ \text{C}^2/\text{N}\cdot\text{m}^2} = 7.69 \times 10^4\ \text{V/m}.$$

(*b*) To find the potential difference ΔV between the plates, we use a path from the negative to the positive plate that is directed against the field. The displacement vector $d\mathbf{l}$ and the electric field **E** are antiparallel so $\mathbf{E}\cdot d\mathbf{l} = -E\,dl$. The potential difference between the positive plate and the negative plate is then

$$\Delta V = -\int \mathbf{E}\cdot d\mathbf{l} = E\int dl = El$$
$$= (7.69 \times 10^4\ \text{V/m})(6.50 \times 10^{-3}\ \text{m}) = 500\ \text{V}.$$

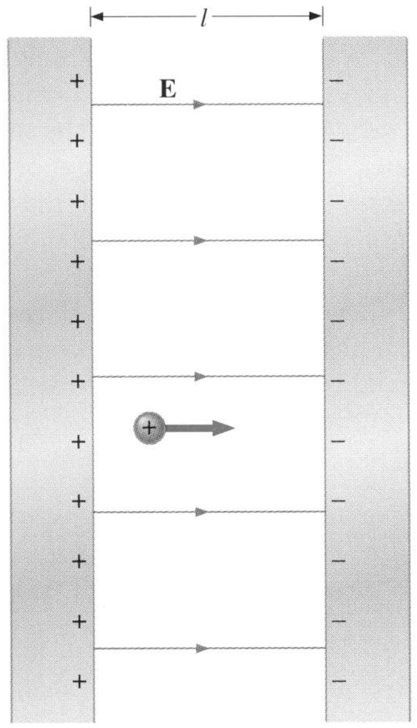

FIGURE 25-18 The electric field between oppositely charged parallel plates. A proton is released at the positive plate.

(*c*) The proton is accelerated across a potential difference of 500 V. Since the charge of the proton is e, the work done on it is 500 eV. The proton starts at rest, so from the work-energy theorem its kinetic energy when it arrives at the negative plate is also 500 eV, which in joules is $T = (1.60 \times 10^{-19}\ \text{J/eV})(500\ \text{eV}) = 8.00 \times 10^{-17}\ \text{J}$. Finally, from the definition of kinetic energy, the speed of the proton at the negative plate is

$$v = \sqrt{\frac{2T}{m}} = \sqrt{\frac{2(8.00 \times 10^{-17}\ \text{J})}{1.67 \times 10^{-27}\ \text{kg}}} = 3.10 \times 10^{5}\ \text{m/s}.$$

EXAMPLE 25-15 A CONDUCTING SPHERE

The isolated conducting sphere of Fig. 25-19 has a radius R and an excess charge q. (*a*) What is the electric field both inside and outside the sphere? (*b*) What is the electric potential inside and outside the sphere?

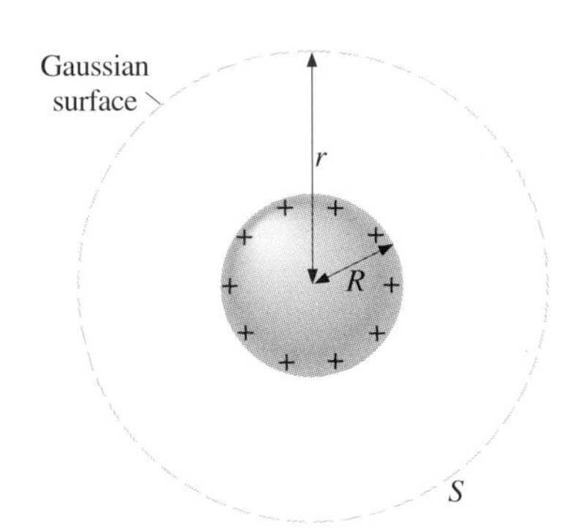

FIGURE 25-19 An isolated conducting sphere.

SOLUTION (*a*) The sphere is isolated, so its surface charge distribution and the electric field of that distribution are spherically symmetric. We can therefore represent the field as $\mathbf{E} = E(r)\hat{\mathbf{r}}$. To calculate $E(r)$, we apply Gauss' law over a closed spherical surface S of radius r that is concentric with the conducting sphere. Since r is constant and $\hat{\mathbf{n}} = \hat{\mathbf{r}}$ on the sphere,

$$\oint_S \mathbf{E} \cdot \hat{\mathbf{n}}\, da = E(r) \oint da = E(r) 4\pi r^2.$$

For $r < R$, S is within the conductor so $Q_{in} = 0$, and Gauss' law gives

$$E(r) = 0,$$

as expected inside a conductor. If $r > R$, S encloses the conductor so $Q_{in} = q$. From Gauss' law,

$$E(r) 4\pi r^2 = \frac{q}{\epsilon_0}.$$

The electric field of the sphere may therefore be written as

$$\mathbf{E} = 0 \qquad (r < R),$$

$$\mathbf{E} = \frac{1}{4\pi\epsilon_0} \frac{q}{r^2} \hat{\mathbf{r}} \qquad (r \geq R).$$

Notice that in the region $r \geq R$, the electric field due to a charge q placed on an isolated conducting sphere of radius R is identical to the electric field of a point charge q located at the center of the sphere.

(*b*) For $r \geq R$, the potential is the same as that of an isolated point charge q located at $r = 0$:

$$V(r) = \frac{1}{4\pi\epsilon_0} \frac{q}{r} \qquad (r \geq R).$$

For $r < R$, $\mathbf{E} = 0$, so $V(r)$ is constant in this region. Since $V(R) = q/4\pi\epsilon_0 R$,

$$V(r) = \frac{1}{4\pi\epsilon_0} \frac{q}{R} \qquad (r < R).$$

EXAMPLE 25-16 TWO CONNECTED CONDUCTING SPHERES

Two conducting spheres of radii R_1 and R_2, respectively, are connected by a thin wire, as shown in Fig. 25-20. The spheres are sufficiently separated so that each can be treated as if it were isolated. Show that

$$\sigma_1 R_1 = \sigma_2 R_2,$$

where σ_1 and σ_2 are the surface charge densities of the spheres.

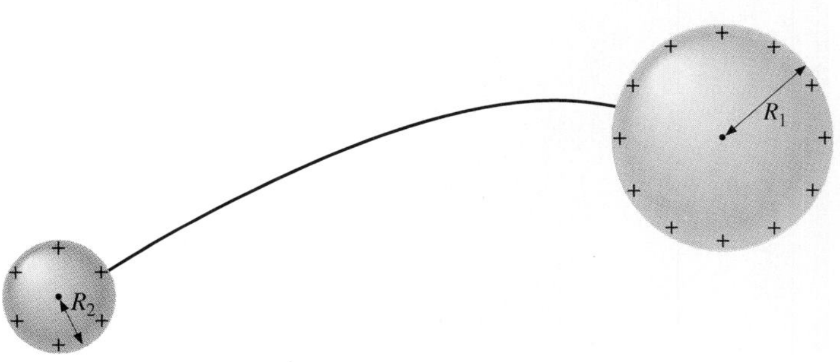

FIGURE 25-20 Two conducting spheres are connected by a thin conducting wire.

SOLUTION We have just seen that the electric potential at the surface of an isolated, charged conducting sphere of radius R is

$$V = \frac{1}{4\pi\epsilon_0}\frac{q}{R}.$$

In this example, the spheres are connected and are therefore at the same potential; hence

$$\frac{1}{4\pi\epsilon_0}\frac{q_1}{R_1} = \frac{1}{4\pi\epsilon_0}\frac{q_2}{R_2},$$

and

$$\frac{q_1}{R_1} = \frac{q_2}{R_2}. \qquad (i)$$

The net charge on a conducting sphere and its surface charge density are related by $q = \sigma(4\pi R^2)$. Substituting this equation into Eq. (i), we find

$$\sigma_1 R_1 = \sigma_2 R_2. \qquad (25\text{-}16)$$

Obviously, two spheres connected by a thin wire do not constitute a typical conductor with a variable radius of curvature. Nevertheless, Eq. (25-16) does at least provide a qualitative idea of how charge density varies over the surface of a conductor. The equation indicates that where the radius of curvature is small (e.g., points A and C of Fig. 25-21), σ, and therefore E, is large. Where the radius of curvature is large (points B and D), σ and E are small.

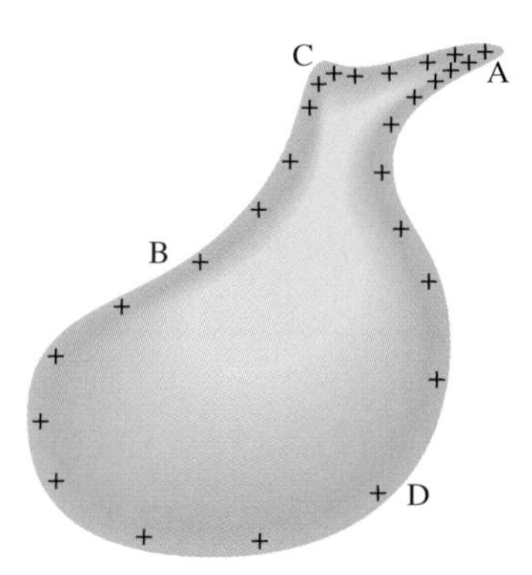

FIGURE 25-21 The surface charge density and the electric field of a conductor are greater at regions with smaller radii of curvature.

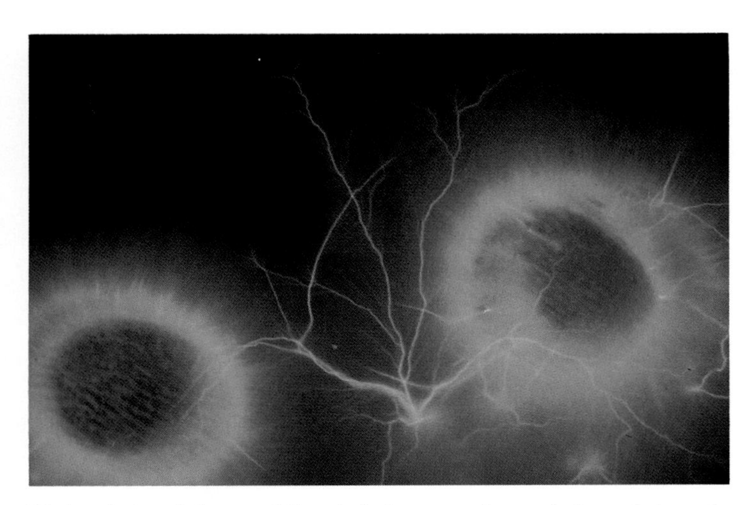

Dielectric breakdown of the air between a charged glass plate and two nearby fingertips.

A practical application of this phenomenon is the lightning rod, which is simply a grounded metal rod with a sharp end pointing upward. As positive charge accumulates in the ground due to a negatively charged cloud overhead, the electric field around the sharp point gets very large. When the field reaches a value of approximately 3.0×10^6 V/m (the *dielectric strength* of the air), the free ions in the air are accelerated to such high energies that their collisions with air molecules actually ionize the molecules. The resulting free electrons in the air then flow through the rod to the earth thereby neutralizing some of the positive charge. This keeps the electric field between the cloud and the ground from getting large enough to produce a lightning bolt in the region around the rod.

Damage done by a lightning bolt.

DRILL PROBLEM 25-9

An uncharged conductor with an internal cavity is shown in Fig. 25-22. Use the closed surface S along with Gauss' law to show that when a charge q is placed in the cavity, a total charge $-q$ is induced on the inner surface of the conductor. What is the charge on the outer surface of the conductor?

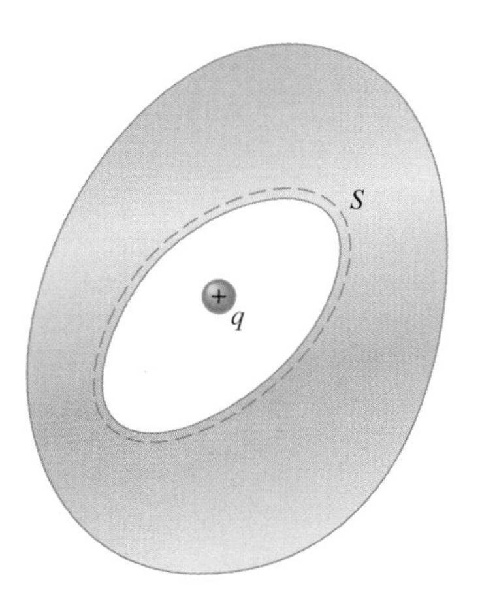

FIGURE 25-22 A hollow conductor with a charge q placed within it.

DRILL PROBLEM 25-10

(*a*) Use Gauss' law to calculate the electric field between the two large, oppositely charged parallel plates of Fig. 23-16. Compare this result with that of Example 23-9. (*b*) What is the potential difference between the plates if they are separated by a distance l?
ANS. (*a*) σ/ϵ_0; (*b*) $\sigma l/\epsilon_0$.

DRILL PROBLEM 25-11

With Example 23-8 as a guide, speculate as to how the electric fields due to the charge elements distributed over the surface of a conductor add to give $E = \sigma/\epsilon_0$ immediately outside and $E = 0$ everywhere inside the conductor.

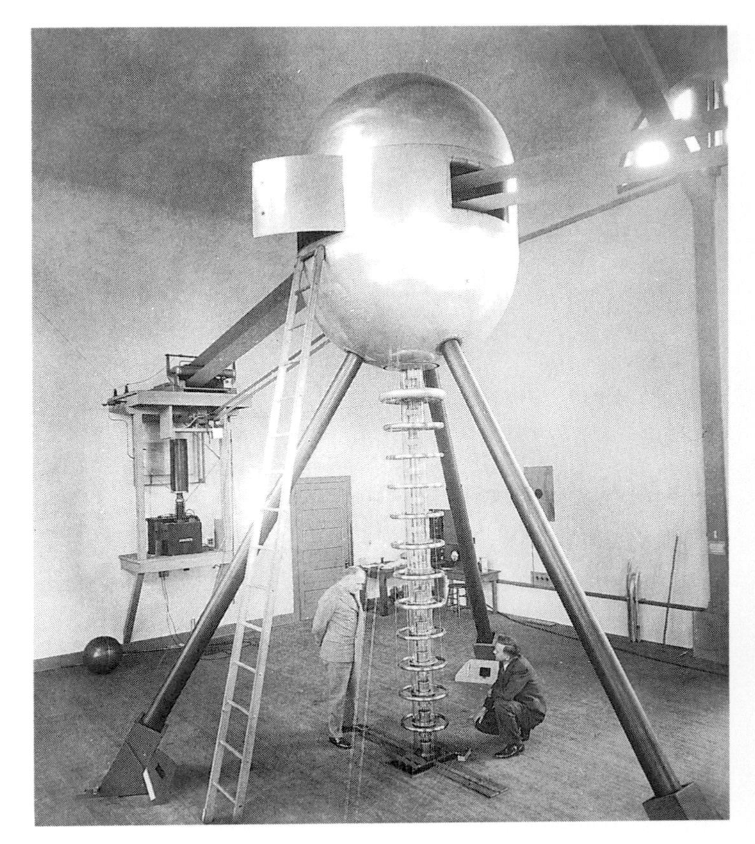

A research Van de Graaff generator.

*25-8 The Van de Graaff Generator

A simplified sketch of the electrostatic generator developed in the early 1930s by Robert Van de Graaff is shown in Fig. 25-23. Basically, the *Van de Graaff generator* consists of a hollow metal sphere that has charge delivered to it by a moving belt. A very large potential difference between the sphere and ground can then be produced. Electrostatic generators range from the simple models used to produce electric fields for classroom demonstrations to sophisticated research instruments that can accelerate charged particles through potential differences up to 20 million volts for nuclear physics experiments.

For the device shown in Fig. 25-23, the rotating drums move the belt between the high-voltage electrode and the inside of the sphere. At the electrode, discharge between metal wires and a grounded grid G delivers charge to the belt. Inside the sphere, the charge comes into contact with metal wires W, which brush across the belt and are in electric contact with the sphere. Since excess charge always resides on the surface of a conductor, the charge that collects on the wires moves to the surface of the sphere. The amount of charge that can accumulate on the sphere is limited by the fact that the field at the surface of the sphere cannot exceed the dielectric strength of the surrounding air ($\sim 3.0 \times 10^6$ V/m). If this occurs, the air is ionized and becomes a conducting medium that dissipates the charge on the sphere.

The charge on the belt experiences a repulsive force due to the like charge on the sphere. Consequently, work must be done against the electric field to move charge from the point where it is sprayed onto the belt to the point where it is removed. That work is supplied by the motors that turn the drums.

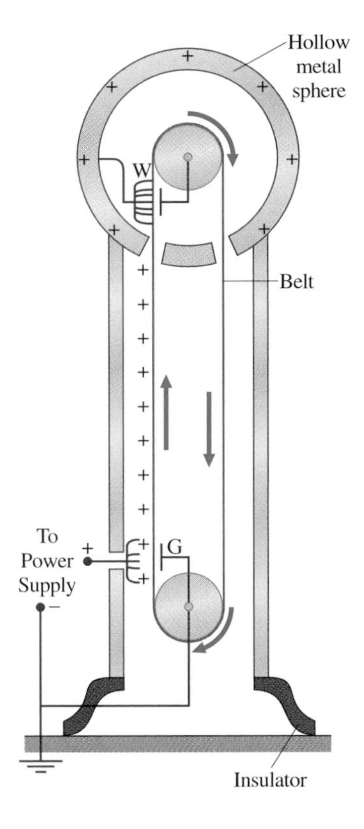

FIGURE 25-23 A schematic representation of the Van de Graaff generator.

EXAMPLE 25-17 THE VAN DE GRAAFF GENERATOR

(*a*) If the radius R of the sphere of a Van de Graaff generator is 1.0 m, how much charge must be placed on it to raise its potential to 1.0×10^6 V? (*b*) If the sphere is in air, what is the maximum charge that can be placed on it? (*c*) The work required to transfer a charge dq from the lower electrode to the sphere, which is at a potential V, is $V\,dq$. Integrate this expression to determine the work required to charge the sphere to a potential $V = 1.0 \times 10^6$ V.

SOLUTION (*a*) At the surface of the sphere, the electric potential is $V = q/4\pi\epsilon_0 R$, so for a potential 1.0×10^6 V,

$$1.0 \times 10^6 \text{ V} = (9.0 \times 10^9 \text{ N}\cdot\text{m}^2/\text{C}^2)\frac{q}{1.0 \text{ m}},$$

and the charge needed to provide this potential is

$$q = 1.1 \times 10^{-4} \text{ C}.$$

(*b*) Since the electric field is largest just outside the surface of the sphere, the air is most likely to be ionized at this location. Using 3.0×10^6 V/m as the dielectric strength of air, we have, just outside the surface,

$$3.0 \times 10^6 \text{ V/m} = \frac{1}{4\pi\epsilon_0}\frac{q_{max}}{R^2}$$
$$= (9.0 \times 10^9 \text{ N}\cdot\text{m}^2/\text{C}^2)\frac{q_{max}}{(1.0 \text{ m})^2},$$

so the maximum charge that can be placed on the sphere is

$$q_{max} = 3.3 \times 10^{-4} \text{ C}.$$

(*c*) The work required to charge the sphere is determined by integrating $dW = V\,dq$ between $V = 0$ and $V = 1.0 \times 10^6$ V. Since $V = q/4\pi\epsilon_0 R$,

$$dq = 4\pi\epsilon_0 R\,dV,$$

and

$$W = \int_0^Q V\,dq = \int_0^{1.0\times10^6\text{ V}} V(4\pi\epsilon_0 R\,dV)$$
$$= 4\pi\epsilon_0 R\int_0^{1.0\times10^6\text{ V}} V\,dV = 2\pi\epsilon_0 R V^2\Big|_0^{1.0\times10^6\text{ V}} = 55.6 \text{ J}.$$

25-9 ELECTROSTATIC FIELD: A SUMMARY

We have now completed our study of the electrostatic field. The properties of this field are outlined below:

1. The fundamental source of the electrostatic field is the point charge q.
2. This source creates an electrostatic field that decreases as the square of the distance from the source:

$$\mathbf{E} = \frac{1}{4\pi\epsilon_0}\frac{q}{r^2}\,\hat{\mathbf{r}}. \tag{23-2}$$

3. The electric field $\mathbf{E}$ exerts a force $\mathbf{F}_e$ on a test charge q_t, which is given by

$$\mathbf{F}_e = q_t\mathbf{E}. \tag{23-1}$$

4. The flux of the electrostatic field through an arbitrary Gaussian surface satisfies

$$\oint_S \mathbf{E}\cdot\hat{\mathbf{n}}\,da = \frac{Q_{in}}{\epsilon_0}, \tag{24-4}$$

where Q_{in} is the total charge enclosed by the surface. This equation is called Gauss' law.

5. The line integral of the electrostatic field around an arbitrary closed path satisfies

$$\oint \mathbf{E}\cdot d\mathbf{l} = 0. \tag{25-2}$$

This equation represents the fact that the electric field of a fixed charge distribution is conservative and allows an electric potential at the point P to be defined as

$$V_P = -\int_R^P \mathbf{E}\cdot d\mathbf{l}, \tag{25-3}$$

where R is the reference point.

SUMMARY

1. **Electric potential of point charges**
   (*a*) The electric potential at the point P is the negative of the work per unit charge done on a positive test charge by the electric field between a reference point R and the point P:

$$V_P = -\int_R^P \mathbf{E}\cdot d\mathbf{l}.$$

   (*b*) The electric potential of an isolated point charge is

$$V = \frac{1}{4\pi\epsilon_0}\frac{q}{r}.$$

   (*c*) The electric potential of a discrete charge distribution is the scalar sum

$$V = \frac{1}{4\pi\epsilon_0}\sum_i \frac{q_i}{r_i}.$$

2. **Calculating the electric potential**
The electric potential of a charge distribution is determined with the scalar sum given in part 1(*c*) above. For a continuous charge distribution, this sum is replaced by the integral

$$V = \frac{1}{4\pi\epsilon_0}\int \frac{dq}{r}.$$

3. **Electric potential difference**
If V_A and V_B are the electric potentials at A and B, then the potential difference between A and B is

$$V_B - V_A = -\int_A^B \mathbf{E}\cdot d\mathbf{l}.$$

4. **Calculating the electric field from the electric potential**
If the electric potential is $V(x, y, z)$ at a particular point, then the components of the electric field at that point are

$$E_x = -\frac{\partial V}{\partial x}, \qquad E_y = -\frac{\partial V}{\partial y}, \qquad E_z = -\frac{\partial V}{\partial z}.$$

5. **Equipotential surfaces**
The locus of points for which $V(x, y, z)$ is a constant is an equipotential surface.

*6. **Electric dipole**
An electric dipole is made up of two equal and opposite point charges separated by a fixed distance.
(*a*) If $\pm q$ are separated by a distance $2a$, the electric dipole moment is a vector directed from $-q$ to $+q$ with magnitude $2aq$.
(*b*) If an electric dipole in an electric field $E(y)\mathbf{j}$ points in the positive y direction, the force on it in this direction is

$$F(y) = p\,\frac{dE(y)}{dy}.$$

(*c*) The torque on an electric dipole in an electric field **E** is

$$\boldsymbol{\tau} = \mathbf{p} \times \mathbf{E}.$$

(*d*) The potential energy of an electric dipole in an electric field **E** is

$$U = -\mathbf{p}\cdot\mathbf{E}.$$

7. **Properties of conductors in electrostatic equilibrium**
(*a*) The electric field inside a conductor vanishes.
(*b*) Any excess charge placed on an isolated conductor resides entirely on the outer surface of the conductor.
(*c*) The electric field is perpendicular to the surface of a conductor everywhere on that surface.
(*d*) The surface of a conductor is an equipotential surface.
(*e*) The interior and surface of a conductor are at the same potential.
(*f*) At any point just above the surface of a conductor,

$$E = \frac{\sigma}{\epsilon_0}.$$

SUPPLEMENT 25-1 THE ELECTRIC POTENTIAL AND FIELD OF A POINT ELECTRIC DIPOLE

Consider the electric dipole of moment $\mathbf{p} = 2aq\mathbf{j}$ shown in Fig. 25-10. At point P, the electric potential is given by

$$V = \frac{1}{4\pi\epsilon_0}\left[\frac{q}{\sqrt{(x^2 + (y-a)^2 + z^2}} - \frac{q}{\sqrt{x^2 + (y+a)^2 + z^2}}\right]. \quad (i)$$

If $r \gg a$, where $r = \sqrt{x^2 + y^2 + z^2}$, the electric dipole is known as a point electric dipole, and V can be approximated by a much simpler expression than Eq. (*i*). We'll begin the approximation with the square roots on the right-hand side of Eq. (*i*). They can be written as

$$\sqrt{x^2 + (y-a)^2 + z^2} = \sqrt{r^2 + a^2 - 2ay} = r\sqrt{1 + \frac{a^2 - 2ay}{r^2}}, \quad (ii)$$

and

$$\sqrt{x^2 + (y+a)^2 + z^2} = \sqrt{r^2 + a^2 + 2ay} = r\sqrt{1 + \frac{a^2 + 2ay}{r^2}}. \quad (iii)$$

Hence

$$V = \frac{q}{4\pi\epsilon_0 r}\left\{\frac{1}{\sqrt{1 + [(a^2 - 2ay)/r^2]}} - \frac{1}{\sqrt{1 + [(a^2 + 2ay)/r^2]}}\right\}. \quad (iv)$$

Since $r \gg a$, we have

$$\frac{a^2 - 2ay}{r^2} \ll 1 \quad \text{and} \quad \frac{a^2 + 2ay}{r^2} \ll 1.$$

Now using the approximation

$$\frac{1}{\sqrt{1+\alpha}} \approx 1 - \frac{\alpha}{2}$$

for $\alpha \ll 1$, we obtain

$$V \approx \frac{q}{4\pi\epsilon_0 r}\left[1 - \frac{1}{2}\left(\frac{a^2 - 2ay}{r^2}\right) - 1 + \frac{1}{2}\left(\frac{a^2 + 2ay}{r^2}\right)\right],$$

which simplifies to

$$V \approx \frac{y}{4\pi\epsilon_0 r^3}(2aq). \qquad (v)$$

Finally, with $p = 2aq$ and $\cos\theta = y/r$ (see Fig. 25-10), Eq. (v) can be written as

$$V \approx \frac{py}{4\pi\epsilon_0 r^3} = \frac{p\cos\theta}{4\pi\epsilon_0 r^2} = \frac{\mathbf{p}\cdot\hat{\mathbf{r}}}{4\pi\epsilon_0 r^2},$$

which is Eq. (25-10).

If the same approximation ($r \gg a$) is used on the equations for the components of **E**, it is found after considerable algebraic manipulation that

$$E_x \approx \frac{3xy}{4\pi\epsilon_0 r^5}(2aq)$$

$$E_y \approx \frac{2y^2 - x^2 - z^2}{4\pi\epsilon_0 r^5}(2aq)$$

and

$$E_z \approx \frac{3zy}{4\pi\epsilon_0 r^5}(2aq).$$

Once again with $p = 2aq$, these components can be written as

$$E_x \approx \frac{3pxy}{4\pi\epsilon_0 r^5},$$

$$E_y \approx \frac{p(2y^2 - x^2 - z^2)}{4\pi\epsilon_0 r^5},$$

and

$$E_z \approx \frac{3pzy}{4\pi\epsilon_0 r^5},$$

which are Eqs. (25-11).

QUESTIONS

25-1. Would electric potential be meaningful if the electric field were not conservative?

25-2. Can a particle move in a direction of increasing electric potential, yet have its electric potential energy decrease? Explain.

25-3. If points A and B are at the same potential, must you do net work to move a charge between the two points? Would you have to exert a force to move the charge?

25-4. The potential difference between the terminals of a battery is 12 V. How much work does the battery do on 1.0 C of charge that moves through a wire from the plus to the minus terminal?

25-5. If a proton is released from rest in an electric field, will it move in the direction of increasing or decreasing potential? Also answer this question for an electron and a neutron.

25-6. What is the electric field in a region where the electric potential is constant?

25-7. If the electric field is zero throughout a region, must the electric potential also be zero in that region?

25-8. Explain why knowledge of **E** at one point is not sufficient to determine V at that point. What about the other way around?

25-9. Can equipotential surfaces intersect?

25-10. If two points are at the same potential, are there any electric field lines connecting them?

25-11. Suppose you have a map of equipotential surfaces spaced 1.0 V apart. What do the distances between surfaces in a particular region tell you about the strength of **E** in that region?

25-12. Compare the electric dipole moments of charges $\pm Q$ separated by a distance d and charges $\pm Q/2$ separated by a distance $d/2$.

25-13. Would Gauss' law be helpful for determining the electric field of a dipole?

25-14. Is the electric potential necessarily constant over the surface of a conductor?

25-15. Under electrostatic conditions, the excess charge on a conductor resides on its surface. Does this mean that all of the conduction electrons in a conductor are on the surface?

25-16. Can a positively charged conductor be at a negative potential? Explain.

25-17. A charge q is placed in the cavity of the conductor as shown. Will charge Q experience an electric field due to the presence of q?

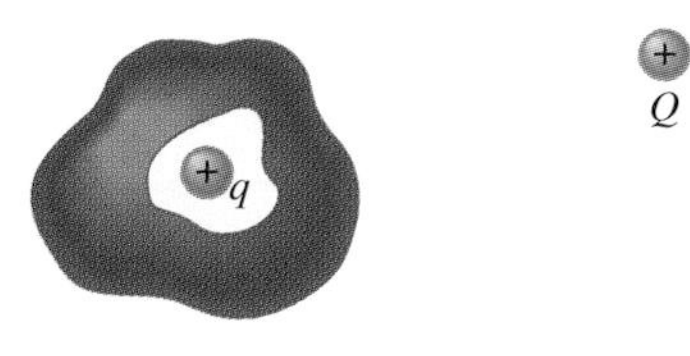

QUESTION 25-17

25-18. The metal block shown in the accompanying figure has an excess charge of $-5.0\ \mu$C. If a 2.0-μC point charge is placed in the cavity, what is the net charge on the surface of the cavity and on the outer surface of the block?

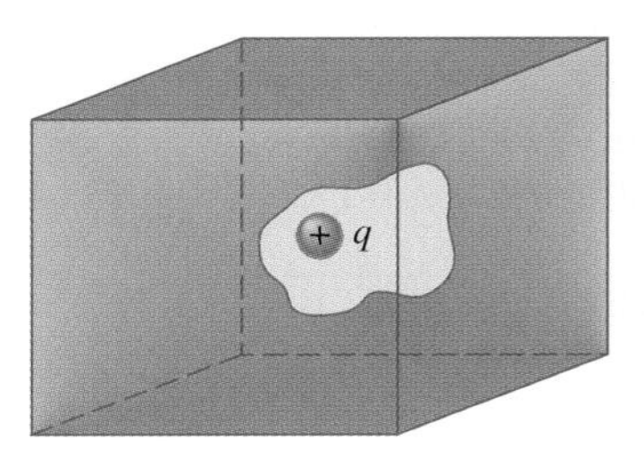

QUESTION 25-18

25-19. A solid metal cube has a spherical cavity at its center. A point charge $-4q$ is placed at the center of the cavity. The metal cube itself carries a charge $-2q$. What is the total charge on the surface of the spherical cavity and on the outer surface of the cube?

25-20. Why are the metal support rods for television antennae generally grounded?

25-21. Why are fish reasonably safe in an electrical storm?

25-22. If the electric field **E** is constant, the potential difference ΔV between two points a distance d apart on a field line is given by $\Delta V = Ed$. For an isolated point charge, the magnitudes of the electric field and the electric potential at a distance r from the charge are given by

$$E = \frac{1}{4\pi\epsilon_0}\frac{q}{r^2} \quad \text{and} \quad V = \frac{1}{4\pi\epsilon_0}\frac{q}{r},$$

respectively. Can we conclude that for a distribution of point charges, V and E are related by $V = Er$?

25-23. Consider the figure for Ques. 25-17. Does the presence of Q have any effect on q? Are your answers for this question and Ques. 25-17 consistent with Newton's third law?

Problems

Electric Potential

25-1. What is the electric potential at a point 5.0×10^{-10} m from a proton? from an electron?

25-2. At a distance r from a point charge q, the magnitude of the electric field is 400 N/C and the electric potential is 1200 V. Determine the values of q and r.

25-3. (*a*) What is the electric potential at a point that is 1.0 cm from a fixed particle of charge $-5.0\ \mu$C? (*b*) How much work must we do in order to move a particle of charge 2.0 μC from infinity to that point?

25-4. (*a*) At what distance from a 5.0-μC point charge is the electric potential 2.5×10^4 V? (*b*) At what distance from the 5.0-μC point charge is the electric field 2.5×10^4 V/m? (*c*) How much work does the field of the 5.0-μC charge do on a 1.0-μC point charge that moves from infinity to a distance of 2.5 cm from the 5.0-μC charge?

25-5. What is the potential energy of the 1.0-μC charge in part (*c*) of Prob. 25-4?

25-6. Three 8.0-μC point charges are at the corners of an equilateral triangle whose sides are 10 cm long. What is the potential energy of any one of the point charges of this system?

25-7. Replace one of the charges of Prob. 25-6 by a -4.0-μC point charge. What is the potential energy of this charge?

25-8. How much work must you do to bring an electron from a great distance to a point 1.0×10^{-10} m from a fixed proton?

25-9. An electron is released from rest when it is very far from a fixed particle of charge 4.0×10^{-6} C. What is the speed of the electron when it passes through a point 1.0 m from the fixed particle?

25-10. Seven of the eight corners of the cube shown in the accompanying figure are occupied by point charges $q = 4.0\ \mu$C. What is the electric potential at the unoccupied corner?

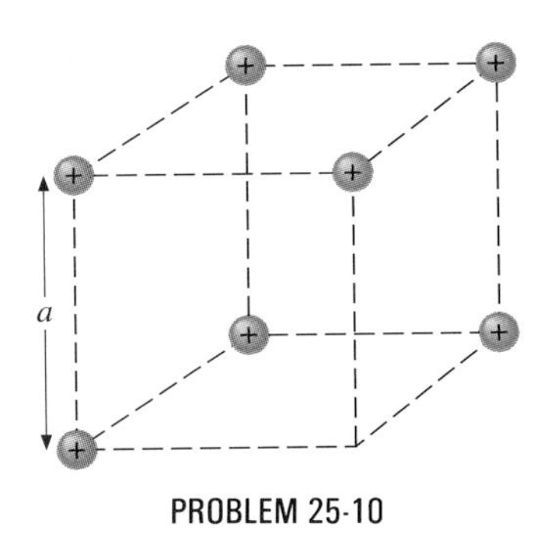

PROBLEM 25-10

25-11. Point charges are located as shown in the accompanying figures. For each case, find the electric field and the electric potential at the origin.

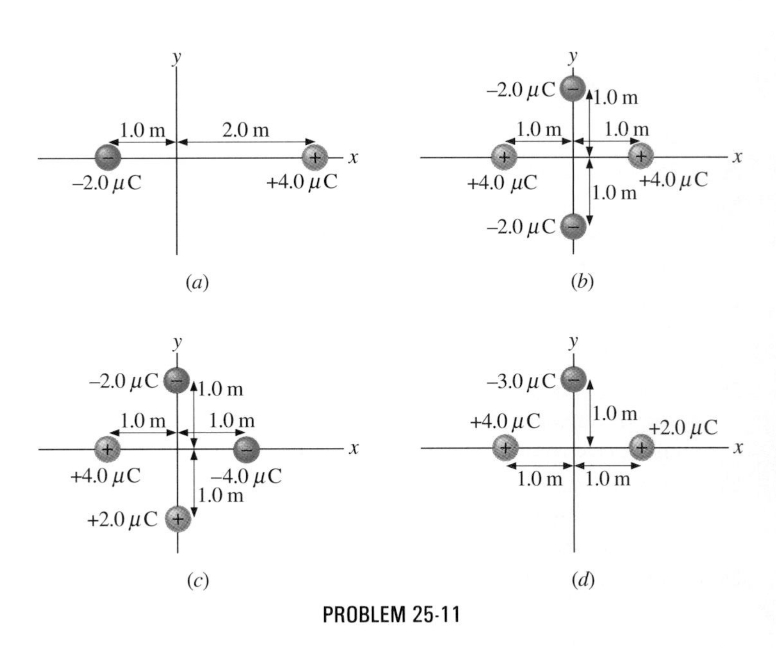

PROBLEM 25-11

25-12. (*a*) What is the net work needed to move an electron from infinity to the origin in the presence of the electric field of each charge configuration of Prob. 25-11? (*b*) What is the potential energy of the electron when it is at the origin?

25-13. Consider the charge distribution of Prob. 23-17. Calculate the electric potential at the point designated in that problem.

25-14. Consider the charge distributions of Prob. 23-18. Calculate the electric potentials at the points designated.

25-15. Consider the charge distribution of Prob. 23-19. Calculate the electric potential at the point designated.

25-16. Consider the charge distributions of Prob. 23-20. Calculate the electric potential at the point designated.

25-17. What is the electric potential at the point P on the axis of the annulus shown in the accompanying figure? The annulus has a uniform surface charge density σ.

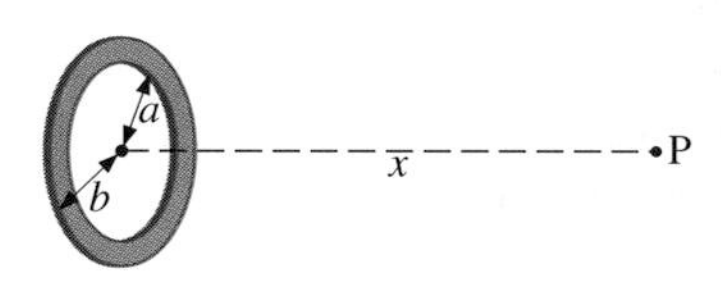

PROBLEM 25-17

25-18. Find the electric potential at the point P of the accompanying figure. The rod has a uniform linear charge density λ.

PROBLEM 25-18

Electric Potential Difference

25-19. A positive point charge $q = 2.0 \times 10^{-6}$ C is fixed at $r = 0$. (*a*) What is the potential difference between points at $r = 2.0$ m and $r = 4.0$ m? (*b*) How much work does the electric field of q do on a second point charge $q' = -4.0 \times 10^{-6}$ C that moves from $r = 2.0$ m to $r = 4.0$ m?

25-20. The electric potential at A is 50 V and at B it is 125 V. (*a*) How much work is done by the electric field on a unit charge that moves from the reference point R to A and from R to B? (*b*) How much work is done by the electric field on a unit charge that moves from A to B? (*c*) How much work must we do in order to move a unit charge from A to B?

25-21. Do Prob. 25-20 for an electron instead of a unit charge.

25-22. An electron accelerates between two points whose potential difference is 100 V. What is the change in kinetic energy of the electron? Express your answer in electron-volts.

25-23. A proton starting from rest acquires a kinetic energy of 30 MeV. Through what potential difference has the proton been accelerated?

25-24. The electron beam in a television picture tube is accelerated from rest through a potential difference of 20 kV. What is the speed of an electron at the end of this acceleration?

25-25. A particle of charge 5.0 μC is moved between two points whose potential difference is 50 V. What is the change in its potential energy? Express your answer in joules and in electron-volts.

25-26. A proton starting from rest in a uniform electric field acquires a speed of 4.0×10^4 m/s after having traveled a distance of 2.0 mm. (*a*) Through what potential difference has the proton moved? (*b*) What is the strength of the electric field?

25-27. Two stationary point charges $q_1 = 2.0 \times 10^{-10}$ C and $q_2 = -1.5 \times 10^{-10}$ C are 6.0 cm apart. Points A and B are on the line between the two charges, with A 1.0 cm from q_1 and B 1.0 cm from q_2. (*a*) What is the potential difference between A and B? (*b*) An electron is released from rest at B and accelerates toward the positive charge. What is the speed of the electron when it passes through A?

25-28. What is the average translational kinetic energy in electron-volts of an oxygen molecule in the atmosphere at 0°C?

25-29. A charge is placed in a uniform electric field $\mathbf{E} = 5.0 \times 10^4\mathbf{i}$ N/C. Through what potential difference does the charge move when it travels (*a*) parallel to the x axis from $x = 2.0$ m to $x = 4.0$ m? (*b*) parallel to the y axis from $y = 2.0$ m to $y = 4.0$ m? (*c*) along a line directed at 45° to the x axis for a distance of 5.0 m? (*d*) completely around a circle of radius 5.0 m?

25-30. In a region where the electric field is $\mathbf{E} = (1.0\mathbf{i} + 4.0\mathbf{j})$ V/m, what is the potential difference between the points whose positions are $\mathbf{r}_1 = (-2.0\mathbf{i} - 4.0\mathbf{j})$ m and $\mathbf{r}_2 = (3.0\mathbf{i} + 2.0\mathbf{j})$ m?

25-31. A lightning bolt transfers 25 C of charge to the earth through an average potential difference of 4.0×10^7 V. (*a*) How much energy is dissipated in the bolt? (*b*) How much water at 20°C could be turned to steam at 100°C with this energy?

25-32. The accompanying figure shows a particle of charge q near a large sheet of uniform surface charge density σ. (*a*) What is the potential difference between A and B? (*b*) What work must you do to move the charge from A to B. (*c*) What is the work done by the electric field when the charge moves from A to B?

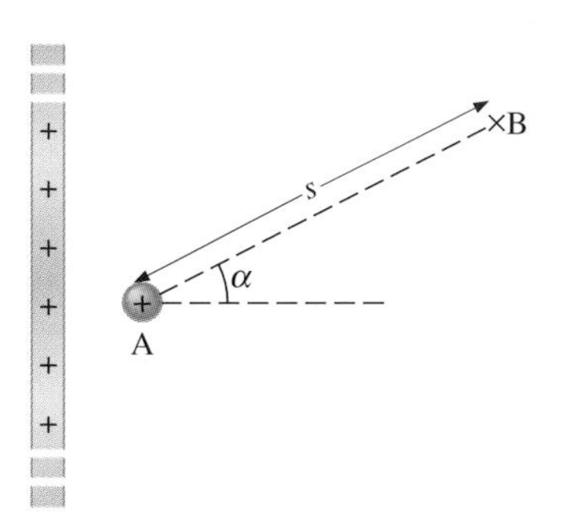

PROBLEM 25-32

Calculating the Electric Field from the Electric Potential

25-33. Throughout a region, equipotential surfaces are given by z = constant. The surfaces are equally spaced with $V = 100$ V for $z = 0.00$ m, $V = 200$ V for $z = 0.50$ m, and $V = 300$ V for $z = 1.00$ m. What is the electric field in this region?

25-34. Use the calculated potential of Prob. 25-17 to find the electric field at P.

25-35. Use the calculated potential of Prob. 25-18 to find the electric field at P. Find the electric field in the limit that $a \to 0$ and $b \to \infty$. Interpret this result.

25-36. In a particular region, the electric potential is given by $V = -xy^2z + 4xy$. What is the electric field in this region?

Electric Dipole

25-37. A stationary electron and proton are 5.0×10^{-11} m apart. What is their electric dipole moment?

25-38. The electric dipole moment due to two equal and opposite charges separated by 1.0 cm is 3.0×10^{-8} C·m. What is the magnitude of each charge?

25-39. An electric dipole is located at the origin and its electric dipole moment is $\mathbf{p} = (5.0 \times 10^{-30}\mathbf{j})$ C·m. What are the electric potential and the electric field due to the dipole (*a*) on the positive y axis at a distance 2.5×10^{-9} m away? (*b*) in the xy plane, 45° from the axis of the dipole and 2.5×10^{-9} m away? (*c*) on the x axis 2.5×10^{-9} m away?

25-40. Two point charges $q_1 = 2.0 \times 10^{-12}$ C and $q_2 = -2.0 \times 10^{-12}$ C are separated by 5.0×10^{-6} m. (*a*) What is their electric dipole moment? (*b*) If this electric dipole is oriented at 30° with respect to an electric field of magnitude 1.0×10^4 V/m, what is the torque on it? (*c*) What is the potential energy of the dipole in this position? (*d*) How much work must you do to rotate the dipole so that it is oriented parallel to the electric field?

25-41. An electric dipole with dipole moment 1.6×10^{-29} C·m is in a uniform electric field of strength 5.0×10^4 N/C. What is the torque on the dipole when it (*a*) is parallel to the field? (*b*) is perpendicular to the field? (*c*) makes an angle of 45° with respect to the field?

Conductors in Electrostatic Equilibrium

25-42. In fair weather the electric field of the earth is about 100 N/C. (*a*) Assuming a region on the surface of the earth can be treated as a large, flat conductor, what is the density of charge at the surface of the earth? (*b*) Do the same calculation treating the earth as a conducting sphere. (*c*) Explain why these two calculations yield the same result.

25-43. The electric field between parallel plates connected to a 100-V power supply is 5.00 kV/m. How far apart are the plates?

25-44. How many electrons must be removed from an uncharged spherical conductor of radius 25 cm in order to produce a potential of 5.0 kV at its surface?

25-45. At the surface of *any* conductor in electrostatic equilibrium, $E = \sigma/\epsilon_0$. Show that this equation is consistent with the fact that $E = kq/r^2$ at the surface of a spherical conductor.

25-46. Two parallel plates 10 cm on a side are given equal and opposite charges of magnitude 5.0×10^{-9} C. The plates are 1.5 mm apart. (*a*) What is the electric field in the region between the plates? (*b*) What is the potential difference between the plates?

25-47. Suppose there is a potential difference of 100 V between two parallel plates that are 2.0 mm apart. (*a*) What is the electric field in the region between the plates? (*b*) What is the charge density on the positive plate? on the negative plate?

25-48. Two large, parallel conducting plates that are 2.0 mm apart are given equal and opposite charge densities of magnitude 1.0×10^{-6} C/m^2. (*a*) What is the electric field between the plates? (*b*) What is the potential difference between the positive and the negative plate? (*c*) What is the potential difference between the positive plate and a point between the plates that is 1.25 mm from the positive plate?

25-49. Two parallel conducting plates, each of cross-sectional area 400 cm^2, are 2.0 cm apart and uncharged. If 1.0×10^{12} electrons are transferred from one plate to the other, what are (*a*) the charge density on each plate? (*b*) the electric field between the plates? (*c*) the potential difference between the plates?

25-50. Two parallel conducting plates, each 8.0 cm by 8.0 cm, are separated by 5.0 mm. (*a*) What charge must be placed on the plates to produce a potential difference of 100 V between them? (*b*) What is the electric field between the plates?

25-51. The surface charge density on a long, straight metallic pipe is σ. (See the accompanying figure.) What are the electric field and electric potential outside the pipe? inside the pipe? Assume that the diameter of the pipe is $2a$.

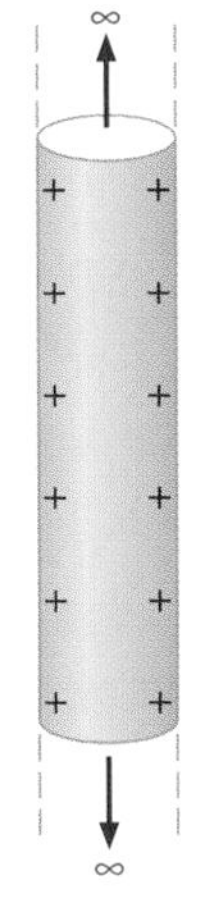

PROBLEM 25-51

25-52. A potential difference of 5.0 kV is established between parallel plates in air. If the air becomes electrically conducting when the electric field exceeds its dielectric strength, what is the minimum separation of the plates?

25-53. What is the maximum amount of charge that can be placed on a conducting sphere of radius 25 cm without its field exceeding the dielectric strength of air?

25-54. What is the minimum radius that the conducting sphere of an electrostatic generator can have if it is to be charged to 100 kV without its field exceeding the dielectric strength of air? How much charge will it carry?

25-55. A point charge $q = -5.0 \times 10^{-12}$ C is placed at the center of a spherical conducting shell of inner radius 3.5 cm and outer radius 4.0 cm. The electric field just above the surface of the conductor is directed radially outward and has magnitude 8.0 N/C. (*a*) What is the charge density on the inner surface of the shell? (*b*) What is the charge density on the outer surface of the shell? (*c*) What is the net charge on the conductor?

25-56. A hollow metal sphere of radius 12.0 cm is given a charge of 4.0 μC. What are the electric field and electric potential at the center of the sphere?

25-57. Concentric conducting spherical shells carry charges Q and $-Q$, respectively. The inner shell has negligible thickness. (See the accompanying figure.) Determine the electric field for (*a*) $r < a$, (*b*) $a < r < b$, (*c*) $b < r < c$, and (*d*) $r > c$. (*e*) What is the potential difference between the shells?

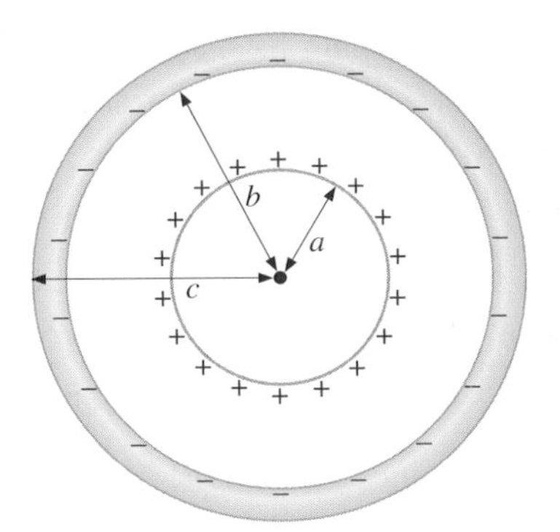

PROBLEM 25-57

25-58. The accompanying figure shows two concentric spherical shells of negligible thicknesses and radii R_1 and R_2. The inner and outer shells carry net charges q_1 and q_2, respectively, where both q_1 and q_2 are positive. What is the electric field for (*a*) $r < R_1$, (*b*) $R_1 < r < R_2$, and (*c*) $r > R_2$? (*d*) What is the net charge on the inner surface of the inner shell, the outer surface of the

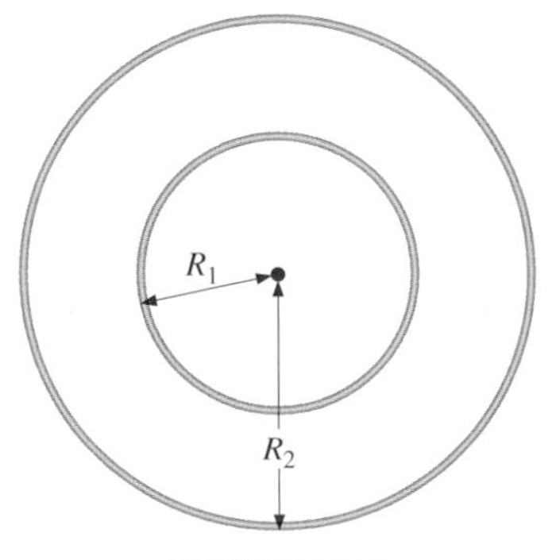

PROBLEM 25-58

inner shell, the inner surface of the outer shell, and the outer surface of the outer shell? (*e*) Determine the electric potential in the regions described in parts (*a*) through (*c*).

25-59. A point charge $q = 5.0 \times 10^{-8}$ C is placed at the center of an uncharged spherical conducting shell of inner radius 6.0 cm and outer radius 9.0 cm. Find the electric field at (*a*) $r = 4.0$ cm, (*b*) $r = 8.0$ cm, (*c*) $r = 12.0$ cm. (*d*) What is the electric potential at these three radii? (*e*) What are the charges induced on the inner and outer surfaces of the shell?

25-60. A solid cylindrical conductor of radius a is surrounded by a concentric cylindrical shell of inner radius b. The solid cylinder and the shell carry charges Q and $-Q$, respectively. Assuming that the length L of both conductors is much greater than a or b, determine the electric field as a function of r, the distance from the common central axis of the cylinders, for (*a*) $r < a$, (*b*) $a < r < b$, and (*c*) $r > b$. (*d*) What is the potential difference between the two conductors?

25-61. A charged conductor is shown in the accompanying figure. Sketch the charge distribution at the surface of the conductor and the equipotential surfaces that surround it.

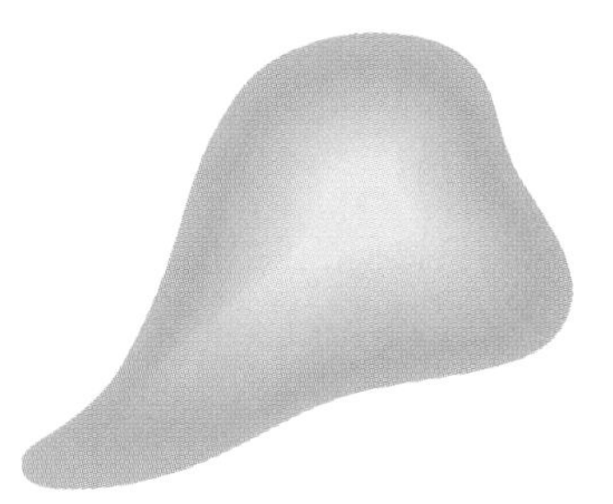

PROBLEM 25-61

25-62. A large metal plate is charged uniformly to a density $\sigma = 2.0 \times 10^{-9}$ C/m^2. How far apart are equipotential surfaces that represent a potential difference of 25 V?

General Problems

25-63. A spherical conductor of radius R is given an excess negative charge $-Q$. An electron is fired, with an initial speed v_0, from a large distance away directly at the conductor. What is the minimum value of v_0 necessary for the electron to just reach the surface of the conductor? (*Hint:* Use the principle of mechanical-energy conservation, with the speed of the electron at the surface of the conductor set equal to zero.)

25-64. A proton moves along a straight line directly at a fixed uranium nucleus ($q = 1.47 \times 10^{-17}$ C). If the proton gets to a distance of 2.0×10^{-10} m from the nucleus before turning around, what is its speed when it is very far from the nucleus?

25-65. How much work must we do to assemble the set of charges illustrated in the accompanying figure?

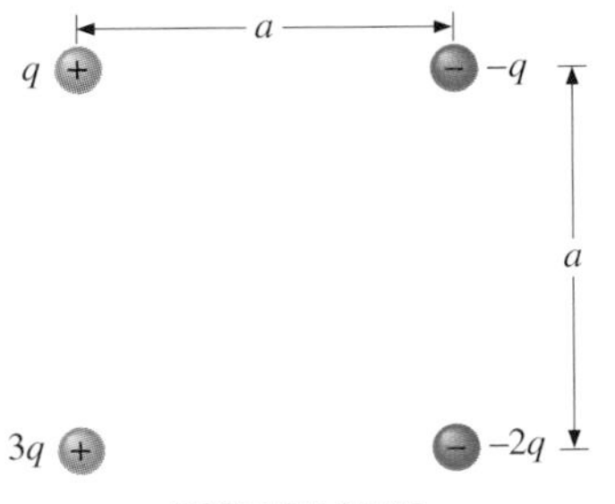

PROBLEM 25-65

25-66. In one of the nuclear reactions responsible for energy production in the sun (also, for the thermonuclear bomb and perhaps eventually for controlled-fusion reactors), two deuterium nuclei collide to produce an isotope of helium, a neutron, and energy. Before this reaction can occur, the deuterium nuclei must be moving fast enough so that they can overcome their mutual electrical repulsion and approach within at least 5.0×10^{-15} m of one another. Assuming the two nuclei are moving along a straight line toward one another at the same speed v, what is the minimum value of v such that they do get this close together? The mass and charge of a deuterium nucleus are 3.34×10^{-27} kg and 1.60×10^{-19} C, respectively. (*Hint:* Use the principle of mechanical-energy conservation, with the nuclei at rest when they are 5.0×10^{-15} m apart.)

25-67. (*a*) Charge is distributed uniformly with a surface density σ over a circular sheet of radius R. Calculate the electric potential due to this charge at a point on the axis of the circular sheet at a distance z away. (See the accompanying figure.) (*b*) Use this potential to calculate the electric field along the axis. (*c*) Compare your result with that of Drill Prob. 23-5.

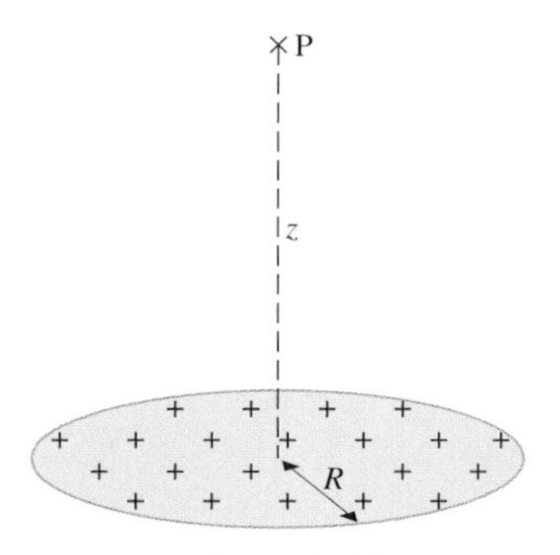

PROBLEM 25-67

25-68. Repeat parts (*a*) and (*b*) of the previous problem for a surface charge density $\sigma = \alpha r^2$, where r is the distance from the center of the distribution.

25-69. A long, straight wire of radius 1.0 mm is charged uniformly at -5.0×10^{-8} C/m. The wire is surrounded by a cylindrical conducting shell of inner radius 5.0 mm. (See the accompanying figure for a cross-sectional view of the wire and shell.) If an electron is released at the wire, what is its speed when it reaches the shell?

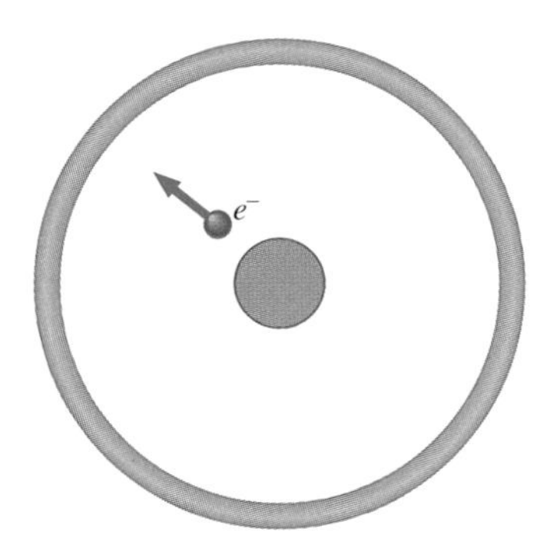

PROBLEM 25-69

25-70. An electric dipole with dipole moment $\mathbf{p}$ is oriented along a radial line emanating from a point charge q, as shown in the accompanying figure. The distance between $\mathbf{p}$ and q is r. (*a*) Show that the charge attracts the dipole with a force of magnitude $qp/2\pi\epsilon_0 r^3$. (*b*) By calculating the force of the dipole on the point charge, show that it is equal and opposite to the force calculated in part (*a*).

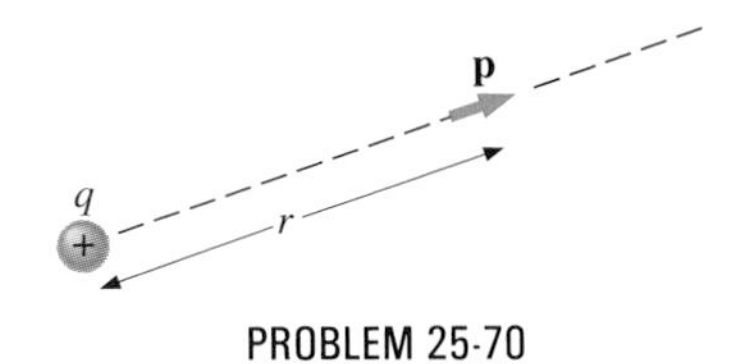

PROBLEM 25-70

25-71. Two electric dipoles with dipole moments p_1 and p_2 are aligned as shown in the accompanying figure. Show that the potential energy of one dipole in the electric field of the other is

$$U = -\frac{1}{2\pi\epsilon_0}\frac{p_1 p_2}{r^3},$$

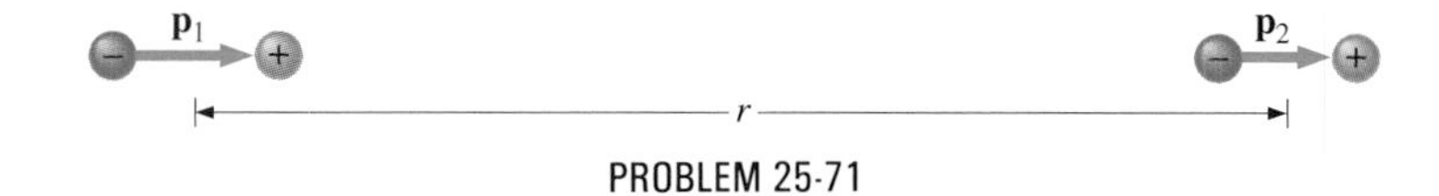

PROBLEM 25-71

where r is the distance between the dipoles by (*a*) using the equation $U = -\mathbf{p}\cdot\mathbf{E}$ and (*b*) calculating the potential energies of two charges q and $-q$ which are at distances $r - \Delta r$ and $r + \Delta r$, respectively, from p_1.

25-72. An electric dipole is free to swing about the axis through its center and perpendicular to the page. (See the accompanying figure.) If the dipole is aligned parallel to a uniform electric field $\mathbf{E}$ and then displaced slightly from this position, it oscillates. Show that these oscillations are simple harmonic with a frequency f given by

$$f = \frac{1}{2\pi}\sqrt{\frac{pE}{I}},$$

where I is the moment of inertia of the dipole and p is its electric dipole moment.

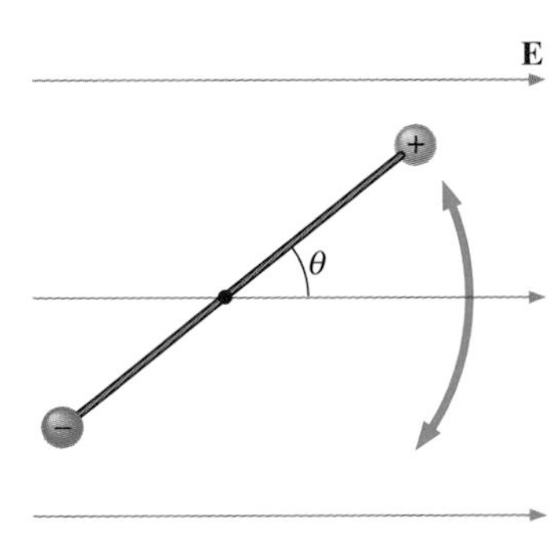

PROBLEM 25-72

Electric lights along Fremont Street in Las Vegas.

CHAPTER 26 Electric Current and Electromotive Force

Preview

This chapter is devoted to the quantities used to describe electric circuits. The main topics to be discussed here are the following:

1. **Electric current.** Electric current is defined and related to the motion of charged particles through a medium.
2. **Resistance and Ohm's law.** The relationship between the potential difference across and the current through a conducting material is investigated.
3. **Temperature dependence of resistivity.** We describe how resistance and resistivity depend on temperature.
4. **The battery and electromotive force.** Electromotive force, an important property of energy-producing elements in electric circuits, is discussed using the example of the common battery.
5. **Single-loop circuits.** We present the procedure used to analyze these simple circuits.
6. **Power in electric circuits.** The power produced or dissipated by various circuit elements is formulated in terms of current, resistance, and voltage.

Thus far our study of electricity has been limited to phenomena involving stationary charges. We now remove that restriction and consider the movement of charge—in particular, that which occurs in conducting wires. This will naturally lead us to electric circuits and their analysis. The practical applications involving electric circuits are countless. Even a list of the number of devices in a typical household that run on electricity would be formidable!

In this chapter we consider three of the basic quantities used in the analysis of electric circuits. These quantities are electric current, resistance, and electromotive force. Current represents the rate at which charge travels along a conductor. Resistance is related to how much that charge movement is inhibited by the atoms in the conductor. Electromotive force is an indication of the work done by energy-producing devices such as batteries and electric generators in transporting current around a circuit. Ohm's law, an important relationship among these quantities, is introduced and used to analyze simple circuits.

26-1 Electric Current

At room temperature, the free electrons in a conductor (see Sec. 22-2) such as the wire of Fig. 26-1 move randomly with speeds on the order of 10^5 m/s. Since the motion of the electrons is random, there is no net charge flow in any direction. For any imaginary plane passing through the conductor, the number of electrons crossing that plane in one direction is equal to the number crossing it in the other direction.

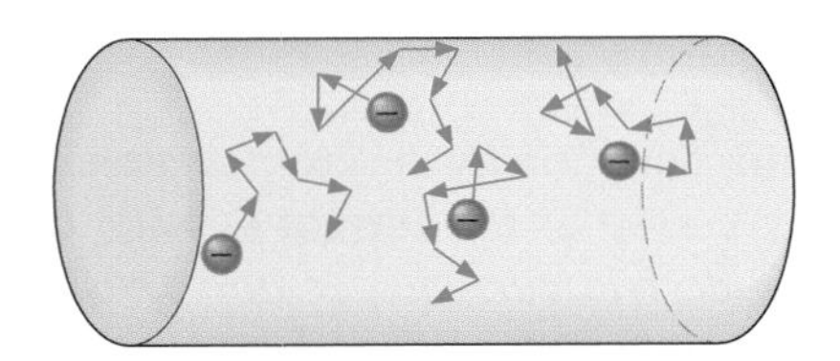

FIGURE 26-1 Random motion of free electrons in a conductor.

When a constant potential difference V is applied between the ends of the wire (see Fig. 26-2), an electric field **E** is produced inside the wire. The conduction electrons within the wire are then subjected to a force $-e\mathbf{E}$ and move overall in the direction of increasing potential. However, this force does not cause the electrons to move faster and faster as they travel from one end of the wire to the other. Instead, a typical conduction electron accelerates through a very small distance (about 5×10^{-8} m) and then collides with one of the atoms of the wire. Figure 26-2*a* depicts the paths of several electrons as they engage in a series of these collisions. Each collision transfers some of the electron's kinetic energy to the atomic lattice, resulting in an increase in the vibrational energy (and therefore in the temperature) of the lattice. Because of the competing electric and collision forces, a typical conduction electron moves slowly along the wire as it is bounced back and forth somewhat randomly by collisions with the ions.

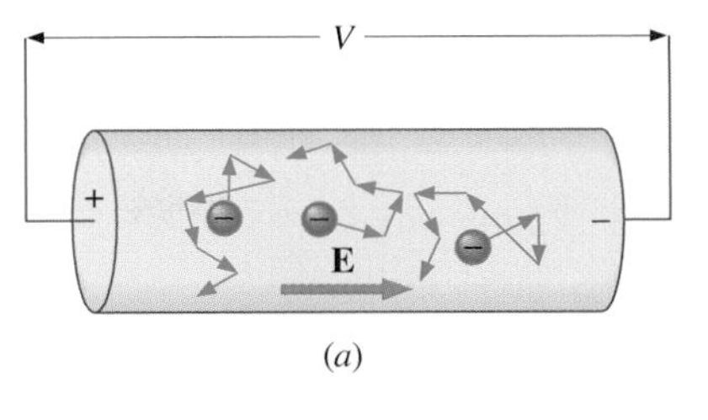

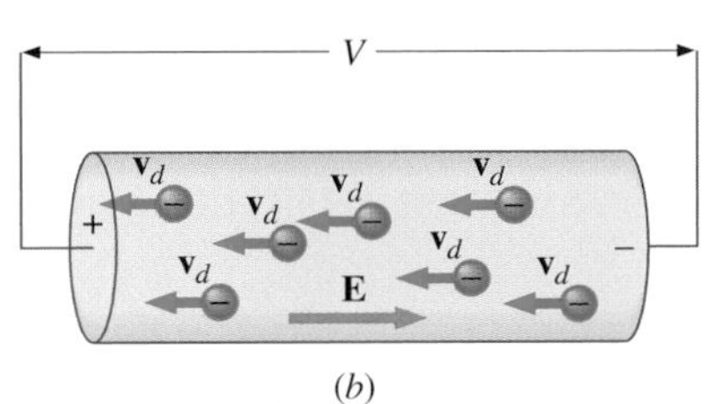

FIGURE 26-2 (*a*) Electrons under the influence of the applied electric field move down the wire and engage in a series of collisions with the atomic lattice. (*b*) The chaotic motion of part (*a*) is equivalent to motion down the wire at a slow, constant drift velocity $\mathbf{v}_d$.

This chaotic motion of the electrons can be represented quite well by a simple model. Let's imagine that the conduction electrons form an "electron gas" whose particles are moving randomly, while the gas as a whole is moving slowly down the wire with a **drift velocity** $\mathbf{v}_d$ in the direction opposite to **E**. When averaged over time, the random motion of an electron due to its collisions does not result in any net displacement. Hence we can ignore this random motion and treat the electrons as though they were all moving down the wire at the drift velocity $\mathbf{v}_d$. (See Fig. 26-2*b*.) It's interesting to note that the magnitude of the drift velocity is on the order of 10^{-4} m/s, or about 10^9 times smaller than the average speed of the electrons between collisions! Any effect related to charge movement in a wire that we investigate can be described in terms of this simple model.

The movement of conduction electrons through a wire is similar to the slow descent of a feather toward the earth. Just as the feather is pulled in one direction by the force of gravity, the electrons are pushed in one direction by an electric force. And just as the feather is continually decelerated by collisions with air molecules, the electrons are continually decelerated by collisions with lattice ions. The net result in both cases is the acquisition of a slow drift velocity in the direction of the driving force—gravitational for the feather, electric for the electrons.

Although we are primarily concerned with electron flow in a wire, there are other types of charge transfer. For example, positive and negative ions as well as electrons move through electrolytes and gaseous conductors.

The net movement of charge through a conductor is represented by **electric current** (or just **current**) I. Current is defined quantitatively in terms of the rate at which net charge passes through a cross-sectional area of the conductor. In Fig. 26-3, the current I flowing through the straight wire is the net charge dq passing through the area shown divided by the time dt it takes to pass:

$$I = \frac{dq}{dt}. \tag{26-1}$$

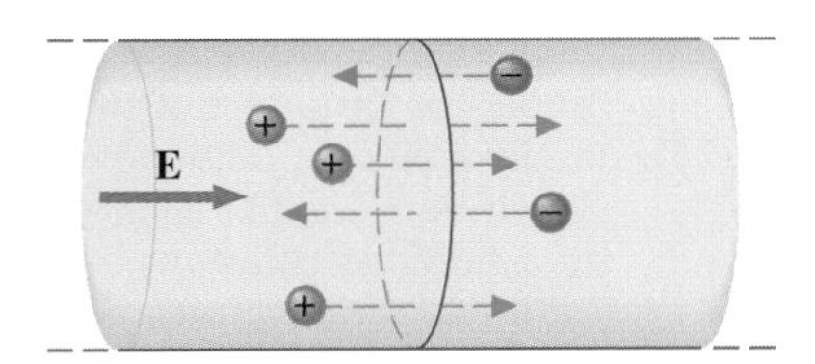

FIGURE 26-3 Electric current is the rate dq/dt at which net charge passes through the cross-sectional area of the wire.

By convention, the direction of the current is assumed to be that in which positive charge moves. For example, if +3.0 C/s is moving across the area shown in Fig. 26-4 from left to right, and −2.0 C/s is moving across it from right to left, the current is +5.0 C/s from left to right.

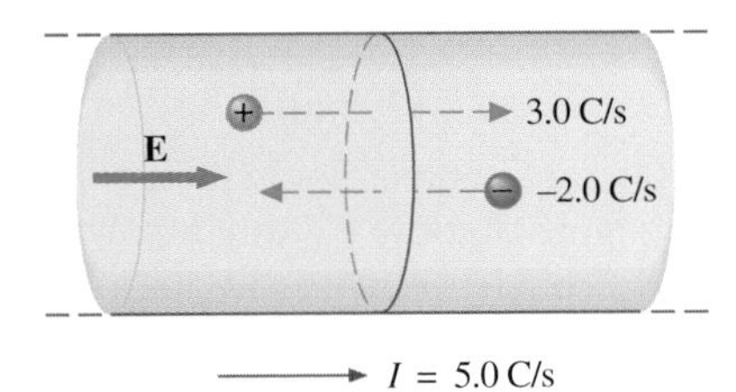

FIGURE 26-4 Calculation of the current when both positive and negative charges are in motion.

In a metal, the actual charge carriers are electrons, which are, of course, negatively charged. However, we can still describe their flow using the convention of a positive current (see Fig. 26-5), since in nearly all situations,* the flow of negative charge in a given direction is equivalent to the flow of positive charge in the opposite direction.

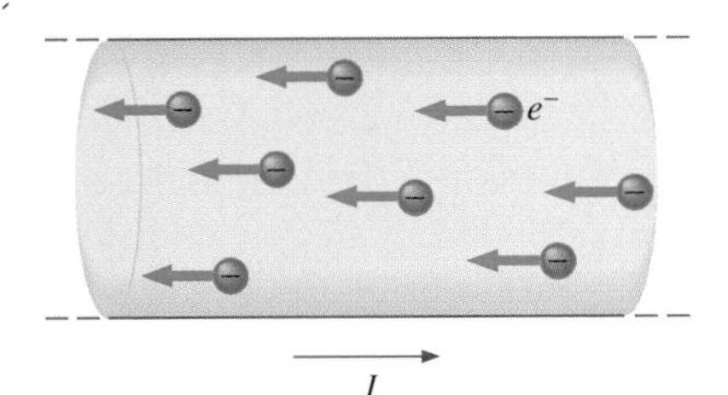

FIGURE 26-5 Current is treated as though it were due to the flow of positive charge. Hence, in a metal, the direction of the current is opposite to the motion of the electrons.

In the SI system, the unit of current is the *ampere (A)*, named in honor of André Ampère, whose contributions to our understanding of magnetism will be discussed shortly. From its definition, current has the unit of charge per time, so

$$1\ \text{A} = 1\ \text{C/s}.^{\dagger}$$

Typical household currents are of the order of a few amperes. Smaller currents are usually found in electronic devices, and they are commonly expressed in milliamperes (1 mA = 10^{-3} A) or microamperes (1 μA = 10^{-6} A).

*A phenomenon that does depend on the sign of the charge carriers is the Hall effect, to be discussed in Chap. 28.

†The formal definitions of the coulomb and the ampere are given in Sec. 29-2. It turns out that the ampere is defined on the basis of a fundamental measurement and the coulomb is then defined in terms of the ampere.

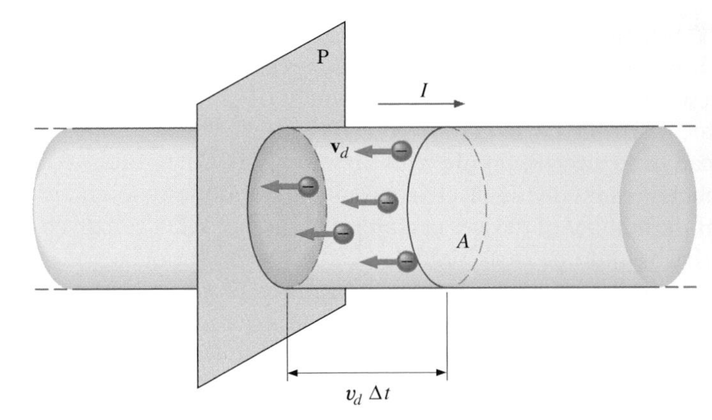

FIGURE 26-6 In a time interval Δt, the charge carriers in the volume of cross-sectional area A and length $v_d\,\Delta t$ pass through the plane P.

Current can be related to the properties of the charge carriers in the following way. Suppose there are n charge carriers per unit volume moving down the thin conducting wire of Fig. 26-6 with an average drift speed v_d. In a time Δt, all carriers in the cylinder of length $v_d\,\Delta t$ and base area A pass through the plane P. The volume of the cylinder is $Av_d\,\Delta t$, and the number of charge carriers within it is $nAv_d\,\Delta t$. If the magnitude of the charge of each carrier is e, the total charge Δq crossing P in the time Δt is

$$\Delta q = e(nAv_d\,\Delta t),$$

which yields for the current

$$I = \frac{\Delta q}{\Delta t} = neAv_d. \tag{26-2}$$

By dividing both sides of this equation by the cross-sectional area A, we obtain a quantity known as the **current density** J, which, by definition, is *the electric current per unit cross-sectional area* of the thin wire:

$$J = \frac{I}{A} = nev_d. \tag{26-3}$$

For geometries more complicated than the thin wire of Fig. 26-6, the vector nature of the current density becomes important. We then write

$$\mathbf{J} = ne\mathbf{v}_d, \tag{26-4}$$

and the current I across a surface S is related to $\mathbf{J}$ by

$$I = \int_S \mathbf{J}\cdot\hat{\mathbf{n}}\,da. \tag{26-5}$$

For a thin wire, we assume that $\mathbf{J}$ is essentially constant over a cross-sectional surface S, so that Eq. (26-5) reduces to $I = JA$.

The distinctions between current and current density show up clearly in Eq. (26-5). First, current density is a vector while current is a scalar. The current through a surface is the flux of the current density over the surface. Secondly, current density is a function of position and can therefore vary from point to point within a conductor. Current, on the other hand, is not a position-dependent quantity. Instead, it represents charge flow across an entire surface.

EXAMPLE 26-1 Drift Speed in a Copper Wire

Copper has one conduction electron per atom. Its density is 8.89 g/cm^3 and its atomic mass is 63.54 g/mol. If a copper wire of diameter 1.0 mm carries a current of 2.0 A, what is the drift speed of the electrons in the wire?

SOLUTION The number of moles of copper per cubic centimeter is

$$\frac{8.89\ g/cm^3}{63.54\ g/mol} = 0.140\ mol/cm^3,$$

so there are $(0.140\ mol/cm^3)(6.02 \times 10^{23}\ atoms/mol) = 8.42 \times 10^{22}\ atoms/cm^3$. Since there is one conduction electron per atom,

$$n = 8.42 \times 10^{22}\ electrons/cm^3 = 8.42 \times 10^{28}\ electrons/m^3.$$

Now from $I = neAv_d$, we have

$$2.0\ C/s = (8.42 \times 10^{28}\ electrons/m^3)(1.6 \times 10^{-19}\ C/electron) \times \pi(5.0 \times 10^{-4}\ m)^2 v_d,$$

and the drift speed is

$$v_d = 1.9 \times 10^{-4}\ m/s.$$

This example illustrates how slowly electrons move down a wire. It takes the average electron almost 1.5 h to move 1 m. However, this certainly does *not* mean that the electric signal also takes 1.5 h to travel this distance. These signals travel down wires at nearly the speed of light (3×10^8 m/s). So, although the flow of electrons along the wire is slow, the signal that tells them to start moving travels 1 m in approximately $(1\ m)/(3 \times 10^8\ m/s) \approx 3 \times 10^{-7}$ s.

Charge flow in a wire is, in many ways, analogous to water flow in a pipe. When water enters one end of a filled pipe, water flows out the other end almost immediately, yet a particular water molecule takes much longer to go through the pipe. Similarly, when a switch is closed to complete a circuit, electrons throughout the wire start to move almost instantaneously; yet, as we have just seen, a particular electron moves quite slowly in the wire.

EXAMPLE 26-2 Comparisons of Current Densities and Drift Speeds

Wire C has twice the cross-sectional area of wire D. Both wires are made of silver and the current in wire C is four times the current in wire D. Compare (*a*) the current densities and (*b*) the drift speeds of the electrons in the wires.

SOLUTION (*a*) The current density is the current per unit cross-sectional area of the conductor. Since $I_C = 4I_D$, and $A_C = 2A_D$, the current densities are related by

$$J_C = \frac{I_C}{A_C} = \frac{4I_D}{2A_D} = 2\left(\frac{I_D}{A_D}\right) = 2J_D.$$

(*b*) Both wires are made of silver, so the densities of the charge carriers (the free electrons) are the same. Using $J = nev_d$, we find that the ratio of the drift speeds is

$$\frac{v_d^C}{v_d^D} = \frac{J_C}{J_D} = 2.$$

DRILL PROBLEM 26-1

In an electrolytic solution, positive and negative ions move in opposite directions at rates of 4.0 C/s and −3.0 C/s, respectively. What is the current in the solution?
ANS. 7.0 C/s.

DRILL PROBLEM 26-2

The current density in a copper wire of cross-sectional radius 0.20 cm is 40 A/cm^2. Calculate (*a*) the current in the wire and (*b*) the drift speed of the conduction electrons.
ANS. (*a*) 5.0 A; (*b*) 3.0×10^{-3} cm/s.

DRILL PROBLEM 26-3

Two wires M and N have the same cross-sectional area but are made from different metals. If $v_d^M/v_d^N = 1.6$ and $I_M/I_N = 0.40$, what is the ratio of the charge-carrier densities?
ANS. $n_M/n_N = 0.25$.

A cable containing many separate wires.

26-2 Resistance and Ohm's Law

If a current I flows through a conductor when the potential difference between its ends is V, then by definition, the **resistance** R of the conductor is

$$R = \frac{V}{I}. \tag{26-6}$$

The SI unit for resistance is the *ohm*, which is abbreviated by the capital Greek letter omega (Ω). Since $R = V/I$, 1 Ω = 1 V/A. Other units commonly used to represent resistance are the kilohm (1 kΩ = 10^3 Ω), the megohm (1 MΩ = 10^6 Ω), and the milliohm (1 mΩ = 10^{-3} Ω).

The resistances of the connecting wires in most circuits are negligible, and devices known as *resistors* are generally inserted to provide essentially all the desired resistance in a typical circuit. Their resistances typically range from a fraction of an ohm to hundreds of thousands of ohms. There are many types of resistors, a few of which are shown in the photograph of Fig. 26-7. The most common type of resistor is made from a small slab of carbon, which provides the resistance, coated with a layer of enamel for protection. Heating elements in stoves and the filaments in lightbulbs are also resistors.

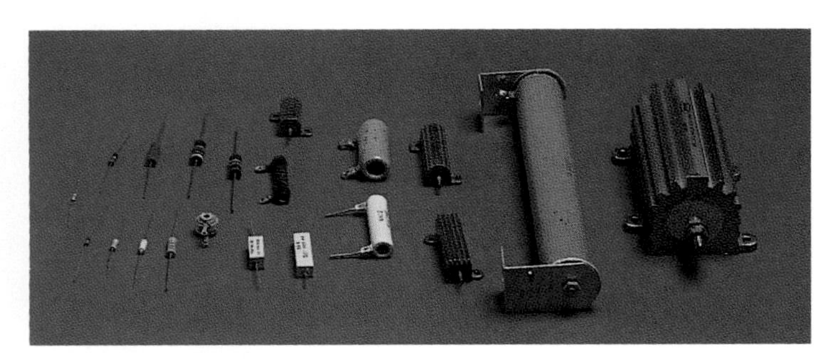

FIGURE 26-7 Various types of resistors.

In a circuit diagram, a fixed resistance is represented by the symbol shown in Fig. 26-8*a*. It is sometimes necessary to use a resistance whose value can be adjusted. The circuit representation of such a variable resistance is shown in Fig. 26-8*b*.

(*a*) (*b*)

FIGURE 26-8 Circuit representation for (*a*) fixed resistance and (*b*) variable resistance.

To determine the resistance of a particular sample, we apply a potential difference V across its ends and measure the current I flowing through it. The ratio V/I then gives us the resistance R. Now suppose we perform these steps for a number of different values of the applied potential difference V. The results of these measurements on two different samples are shown in Figs. 26-9 and 26-10. In each case, the experiments were done at a constant temperature, because resistance is temperature-dependent. In Fig. 26-9, the relationship between V and I is linear, so R (represented by the slope) is constant and we can write

$$V = IR. \qquad (R \text{ constant}) \tag{26-7}$$

Equation (26-7) is known as *Ohm's law*, in honor of its discoverer, Georg S. Ohm. Materials that obey this law are called

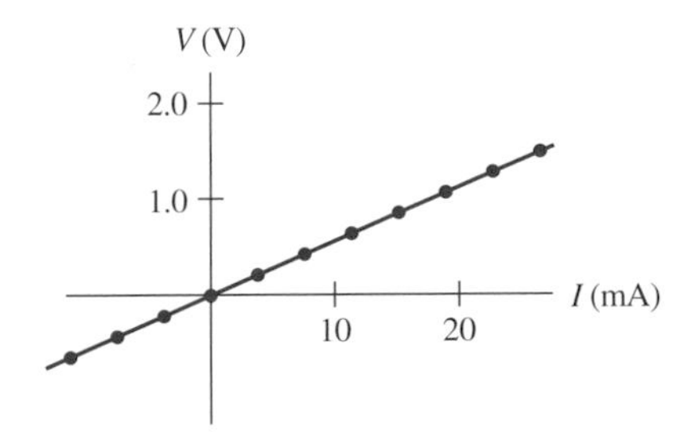

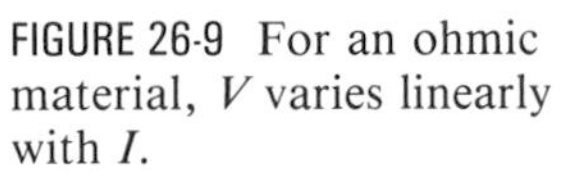

FIGURE 26-9 For an ohmic material, V varies linearly with I.

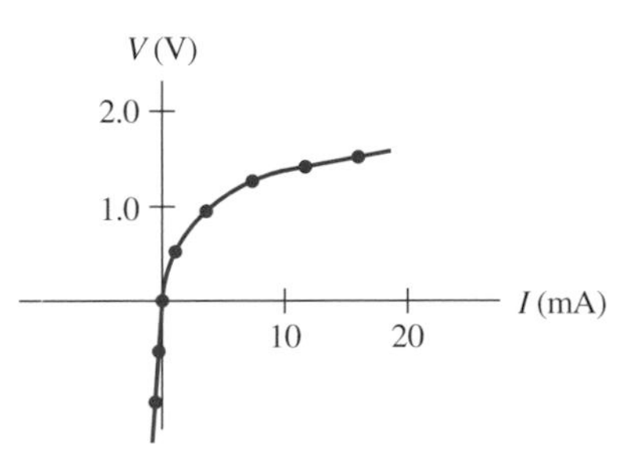

FIGURE 26-10 A typical V-I curve for a silicon diode, a nonohmic device.

ohmic. The resistance of an ohmic sample is independent of the applied potential difference across its ends (or current flowing through it). The doped silicon diode* whose V-I curve is shown in Fig. 26-10 is a good example of a *nonohmic* device. Since V varies nonlinearly with I, the resistance of the diode does not remain constant. In fact, the resistance is different by orders of magnitude for the two directions of current.

Ohm's law can also be written in terms of the electric field and the current density. Let's consider the cylindrical conductor of length l and cross-sectional area A shown in Fig. 26-11. Substituting $V = El$ and $I = JA$ into Ohm's law, we find

$$El = (JA)R,$$

so

$$E = \frac{RA}{l}J.$$

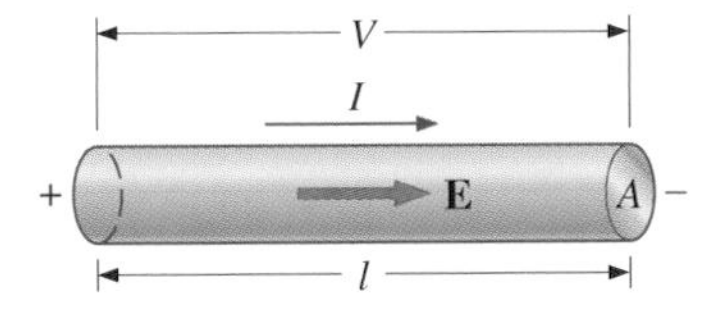

FIGURE 26-11 An alternative form of Ohm's law is $E = \rho J$.

With the **resistivity** ρ defined for ohmic materials to be

$$\rho = \frac{RA}{l}, \tag{26-8}$$

we can write Ohm's law as

$$E = \rho J, \tag{26-9a}$$

*The silicon diode will be discussed further in Chap. 34.

or, using the fact that the electric field and the current density are vectors,

$$\mathbf{E} = \rho\mathbf{J}. \tag{26-9b}$$

In this form, Ohm's law relates two vector quantities at an *arbitrary position* in the sample. The other form, $V = IR$, relates two scalar quantities associated with the *entire* sample.

Resistivity is a property of the material of a sample and, unlike resistance, does not depend on the size or shape of the sample.* Its SI unit is the ohm-meter ($\Omega \cdot$m). The resistivities of a number of common materials are listed in Table 26-1. Notice the strikingly wide range in the values. For the materials listed, the best conductor, silver, has a resistivity about 10^{24} times smaller than that of the worst conductor, quartz!

Frequently, how well or how poorly a material conducts electricity is expressed in terms of its **conductivity** rather than its resistivity. By definition, the conductivity σ is the reciprocal of the resistivity:

$$\sigma = \frac{1}{\rho}. \tag{26-10}$$

Thus a good conductor has low resistivity and high conductivity. In terms of σ, Ohm's law (Eq. 26-9*b*) can be written as

$$\mathbf{J} = \sigma\mathbf{E}. \tag{26-9c}$$

It must be emphasized that Ohm's law is not a fundamental equation of electromagnetism in the same sense as Gauss' law. Instead, it is a very useful and accurate representation of how potential difference and current are related in many materials. You can think of Ohm's law as the electromagnetic counterpart of the frictional laws (Eqs. 6-1 and 6-2) of mechanics.

TABLE 26-1 RESISTIVITIES AND TEMPERATURE COEFFICIENTS OF RESISTIVITY FOR VARIOUS MATERIALS Unless stated otherwise, the values given are for 20°C.

Material	Resistivity ($\Omega \cdot$m)	Temperature Coefficient [(°C)$^{-1}$]
Silver	1.59×10^{-8}	3.8×10^{-3}
Copper	1.67×10^{-8}	6.8×10^{-3}
Gold	2.35×10^{-8}	4.0×10^{-3}
Aluminum	2.65×10^{-8}	4.29×10^{-3}
Tungsten (27°C)	5.65×10^{-8}	4.25×10^{-3}
Nickel	6.84×10^{-8}	6.9×10^{-3}
Iron	9.71×10^{-8}	6.51×10^{-3}
Platinum	10.6×10^{-8}	3.93×10^{-3}
Nichrome	1.00×10^{-6}	4×10^{-4}
Carbon (0°C)	3.5×10^{-5}	-5×10^{-4}
Germanium	4.6×10^{-2}	-4.8×10^{-2}
Yellow brass	7.0×10^{-8}	2×10^{-3}
Silicon	2.5×10^{3}	-7.0×10^{-2}
Wood	10^{8}–10^{11}	
Glass	10^{10}–10^{14}	
Quartz (fused)	10^{16}	

*Consider this helpful analogy: mass density to mass and resistivity to resistance.

EXAMPLE 26-3 RESISTANCE FROM V-I PLOTS

The results of potential difference-current measurements for two different pieces of copper are given in the following tables. Plot these variables and determine the resistance of each sample.

Sample A		Sample B	
V (mV)	*I* (A)	*V* (mV)	*I* (A)
1.00	5.81	1.00	0.73
2.00	11.65	2.00	1.44
3.00	17.41	3.00	2.20
4.00	23.24	4.00	2.94
5.00	29.00	5.00	3.60
6.00	34.80	6.00	4.39
7.00	40.67	7.00	5.10

SOLUTION The *V-I* plots for both samples are shown in Fig. 26-12. From Ohm's law, the slopes of the lines give the resistances of the samples:

$$R_A = \frac{4.00 \times 10^{-3}\ \text{V}}{23.26\ \text{A}} = 1.72 \times 10^{-4}\ \Omega$$

and

$$R_B = \frac{4.50 \times 10^{-3}\ \text{V}}{3.27\ \text{A}} = 13.8 \times 10^{-4}\ \Omega.$$

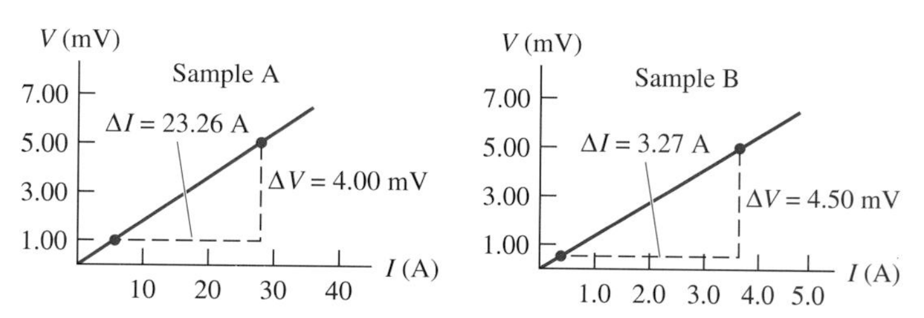

FIGURE 26-12 *V-I* plots for the two samples.

EXAMPLE 26-4 RESISTIVITY CALCULATIONS

The cross-sectional area and the length of sample A in Example 26-3 are 1.00×10^{-4} m^2 and 1.00 m; for sample B these values are 2.50×10^{-5} m^2 and 2.00 m. Determine the resistivities of the two copper pieces.

SOLUTION From $\rho = RA/l$, the resistivities of the copper pieces are

$$\rho_A = \frac{(1.72 \times 10^{-4}\ \Omega)(1.00 \times 10^{-4}\ \text{m}^2)}{1.00\ \text{m}} = 1.72 \times 10^{-8}\ \Omega \cdot \text{m},$$

and

$$\rho_B = \frac{(13.8 \times 10^{-4}\ \Omega)(2.50 \times 10^{-5}\ \text{m}^2)}{2.00\ \text{m}} = 1.72 \times 10^{-8}\ \Omega \cdot \text{m}.$$

As expected, ρ_A and ρ_B are equal since both samples are made from the same material.

EXAMPLE 26-5 RESISTANCE AND OHM'S LAW

(*a*) What is the resistance of an aluminum wire 10.0 m long with a cross-sectional area of 5.00×10^{-7} m^2? (*b*) If there is a potential difference of 0.200 V between the ends of the wire, what is the current flowing through the wire?

SOLUTION (*a*) From Table 26-1, the resistivity of aluminum is $\rho = 2.65 \times 10^{-8}\ \Omega\cdot\text{m}$. The resistance of the wire is then

$$R = \frac{\rho l}{A} = (2.65 \times 10^{-8}\ \Omega\cdot\text{m})\left(\frac{10.0\ \text{m}}{5.00 \times 10^{-7}\ \text{m}^2}\right) = 0.530\ \Omega.$$

(*b*) From Ohm's law,

$$0.200\ \text{V} = I(0.530\ \Omega),$$

so the current through the wire is

$$I = 0.377\ \text{A}.$$

EXAMPLE 26-6 RESISTANCE AND RESISTIVITY

The electric field inside a wire of cross-sectional area 1.00×10^{-7} m^2 is 5.30×10^{-3} V/m when a current of 2.00×10^{-2} A passes through it. What is the resistance of 1.00 m of the wire? What material is the wire made from?

SOLUTION The potential difference between two points 1.00 m apart on the wire is

$$V = El = (5.30 \times 10^{-3}\ \text{V/m})(1.00\ \text{m}) = 5.30 \times 10^{-3}\ \text{V}.$$

From Ohm's law, the resistance of 1.00 m of the wire is

$$R = \frac{V}{I} = \frac{5.30 \times 10^{-3}\ \text{V}}{2.00 \times 10^{-2}\ \text{A}} = 0.265\ \Omega.$$

With R known, we can use Eq. (26-8) to determine ρ:

$$\rho = R\frac{A}{l} = (0.265\ \Omega)\left(\frac{1.00 \times 10^{-7}\ \text{m}^2}{1.00\ \text{m}}\right) = 2.65 \times 10^{-8}\ \Omega\cdot\text{m}.$$

Comparing this value with the resistivities listed in Table 26-1, we conclude that the wire is made of aluminum.

DRILL PROBLEM 26-4

The ends of an iron wire are kept at a potential difference of 0.500 V. The length of the wire is 4.00 m, and its resistance is 0.200 Ω. Determine (*a*) the current flowing through the wire, (*b*) the electric field inside the wire, (*c*) the current density, and (*d*) the cross-sectional area of the wire.
ANS. (*a*) 2.50 A; (*b*) 0.125 V/m; (*c*) 1.29×10^6 A/m^2; (*d*) 1.94×10^{-6} m^2.

DRILL PROBLEM 26-5

A silver and an aluminum wire have the same length and resistance. Which is the thicker wire? What is the ratio of the cross-sectional areas of the wires?
ANS. Aluminum; $A_{al}/A_{ag} = 1.67$.

26-3 TEMPERATURE DEPENDENCE OF RESISTIVITY AND RESISTANCE

Up to this point, we've ignored the fact that resistivity and resistance are both temperature-dependent. For most applications, the resistivity ρ of a material and the temperature T can be related with the empirical equation

$$\rho = \rho_0[1 + \alpha(T - T_0)], \tag{26-11a}$$

where T_0 is a selected reference temperature, ρ_0 is the resistivity at that temperature, and α is a constant called the **temperature coefficient of resistivity**. If $\alpha > 0$, the resistivity increases with temperature; if $\alpha < 0$, the resistivity decreases with temperature. Values of α for different materials are listed in Table 26-1. For metals and most substances, the resistivity goes up with temperature. Notice however that for carbon, silicon, and germanium, there is a decrease in the resistivity as the temperature is increased. These three materials belong to the class of substances known as *semiconductors*, whose conducting properties are intermediate between those of conductors and insulators. Semiconductors have found widespread use in the fabrication of transistors and integrated circuits. The semiconductors in electronic devices are "doped" semiconductors in which "impurity atoms" are added to a pure semiconductor such as silicon or germanium. These impurity atoms are generally atoms of another semiconducting material with a different number of valence electrons than the atoms of the pure semiconductor. The resistivity of the doped semiconductor can be controlled by the amount of doping.

The difference in the sign of α for metals and the pure semiconductors of Table 26-1 is because there are two competing effects responsible for the temperature dependence of the resistivity. One is an increase in the number of charge carriers with temperature. This increase is very small for metals but is significant for semiconductors. The other effect is an increase in the average amplitude of the lattice vibrations with temperature. This results in a greater number of collisions between conduction electrons and lattice ions, which is in turn the primary cause of the increase in a metal's resistivity with temperature.

Since resistance is proportional to resistivity, we can also write

$$R = R_0[1 + \alpha(T - T_0)], \tag{26-11b}$$

where R_0 is the resistance at T_0. Resistance-temperature curves are typically so reproducible [even when they don't satisfy Eq. (26-11*b*)] that resistance measurements of specially designed probes are often used to monitor temperatures.

There are many materials whose resistance vanishes completely below what is called the *critical temperature*. These materials are called *superconductors*. They are so resistance-free that steady currents have been observed to persist undiminished for years in superconducting rings without the help of an applied voltage! Superconductivity was observed first in mercury by the Dutch physicist Kammerlingh Onnes (1853–1926) in

1911. The resistance of mercury is plotted as a function of temperature near absolute zero in Fig. 26-13. You can see in the plot that its critical temperature is very close to 4.2 K. The critical temperatures of a number of substances are listed in Table 26-2.

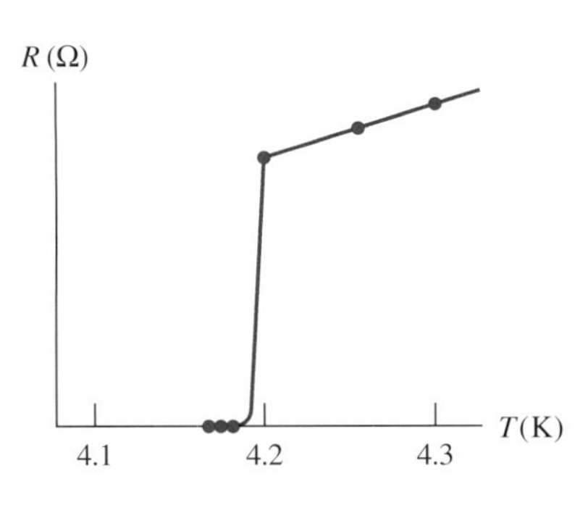

FIGURE 26-13 The resistance of a mercury sample as a function of temperature in the vicinity of 4 K.

TABLE 26-2 **Critical Temperatures of Various Materials**

Material	Critical Temperature (K)
Thallium (Th)	2.4
Mercury (Hg)	4.2
Lead (Pb)	7.2
Niobium (Nb)	9.2
Nb_3Al	17.5
Nb_3Sn	18
Nb_3Ge	23.2
$(La,Ba)_2CuO_4$	35
$YBa_2Cu_3O_7$	92
$Bi_2Sr_2Ca_2Cu_3O_{10}$	110
$Tl_2Ba_2Ca_2Cu_3O_{10}$	125

Until recently, the highest known critical temperature was about 23 K. The most common way of keeping a superconductor below its critical temperature is immersion in liquid helium, which has a temperature of 4.2 K at atmospheric pressure. However, liquid helium is both expensive and somewhat difficult to use, as it has to be frequently replenished because of evaporation. Hence the practical applications of superconductivity have been limited to specialized high-current magnets such as those used in magnetic resonance imaging apparatus and particle accelerators.

The first step in overcoming this practical limitation may have occurred in 1986, when K. Alex Müller and J. Georg Bednorz discovered a copper oxide ceramic whose critical temperature is 35 K. Their work inspired scientists to search for other ceramics with still higher critical temperatures—and success followed quickly. Only six months after the work of Müller and Bednorz was reported, a group led by Ching-Wu Chu of the University of Houston found a material whose critical temperature is 92 K. This temperature is extremely important, because it can be reached with liquid nitrogen, whose boiling temperature is 77 K. Compared with liquid helium, liquid nitrogen is inexpensive and easy to use. As this textbook is being written, the maximum confirmed critical temperature is around 125 K, and there have been hints of still higher values.

A small magnet is floating in air above a disk of superconducting material in a bath of liquid nitrogen at 77 K.

But we're a long way yet from developing practical applications for these recently discovered superconductors. They cannot carry very high currents, and they also are not flexible enough to be used as conducting wires. Researchers worldwide have been working to overcome these difficulties. Their success could lead to a major technological revolution, with possible applications including lossless power lines, miniature motors and supercomputers, and magnetically levitated trains.

26-4 The Battery and Electromotive Force

A steady current flows through a wire when a constant potential difference is maintained across the ends of the wire. Certainly the best known device capable of producing such a potential difference is the battery, common examples of which include the 12-V car battery and the 1.5-V flashlight battery.

A single-cell lead-acid battery is depicted in Fig. 26-14*a*. One plate of the battery is coated with lead (Pb), the other with lead dioxide (PbO_2), and both plates are immersed in a dilute ionic

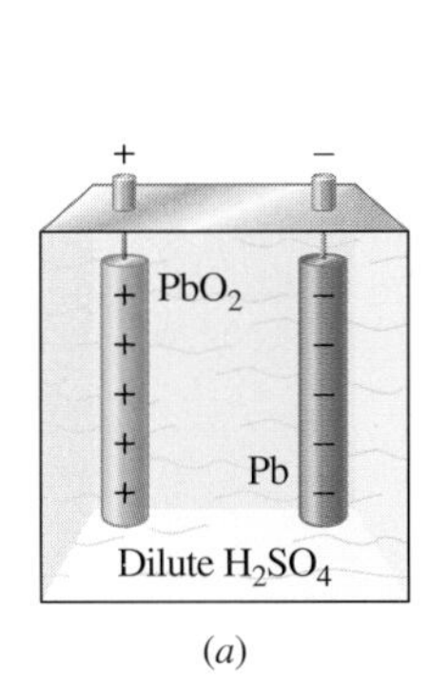

(*a*)

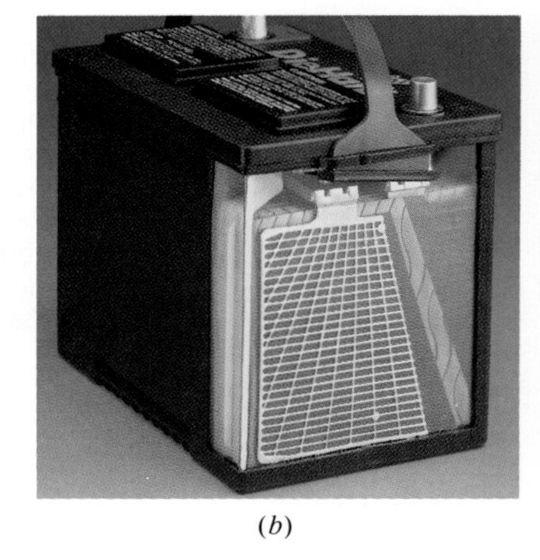

(*b*)

FIGURE 26-14 (*a*) A single-cell lead-acid battery. Electrons are chemically deposited at the Pb plate, which becomes negatively charged. Electrons are chemically removed from the PbO_2 plate, which becomes positively charged. (*b*) A car battery has six cells.

(H^+ and SO_4^{2-}) solution of sulfuric acid. At the Pb plate, the chemical reaction

$$Pb + SO_4^{2-} \rightarrow PbSO_4 + 2e^-,$$

deposits electrons on the plate, thereby charging it negatively. At the PbO_2 plate, the reaction

$$PbO_2 + 4H^+ + SO_4^{2-} + 2e^- \rightarrow PbSO_4 + 2H_2O$$

occurs. Here two electrons are removed from the plate, leaving it with a net positive charge. These two reactions proceed until the electrostatic field produced by the oppositely charged plates is sufficient to prevent any further transfer of electrons to or from them. For our combination of Pb and PbO_2, this occurs when the potential difference between the plates is very close to 2.0 V. In a car battery, six of these cells are connected together as shown in Fig. 26-14*b* to produce a potential difference of 12 V.

A battery's plates are usually connected to terminals that protrude from its case. The positive and negative terminals are designated by (+) and (−), as illustrated in Fig. 26-14*a*. In a circuit diagram, a battery is represented by the symbol shown in Fig. 26-15, with the longer and thinner vertical line corresponding to the positive terminal.

FIGURE 26-15 Circuit representation of a battery.

When the terminals of the battery are connected externally, as shown in Fig. 26-16, electrons are driven through the connecting wire by the battery from the negative to the positive terminal. However, this charge transfer does not deplete the supply of electrons on the Pb plate. As electrons leave the plate, they are replaced by others chemically deposited from the ionic solution, resulting in a constant negative charge on the plate. Likewise, the positive charge on the PbO_2 plate remains constant.

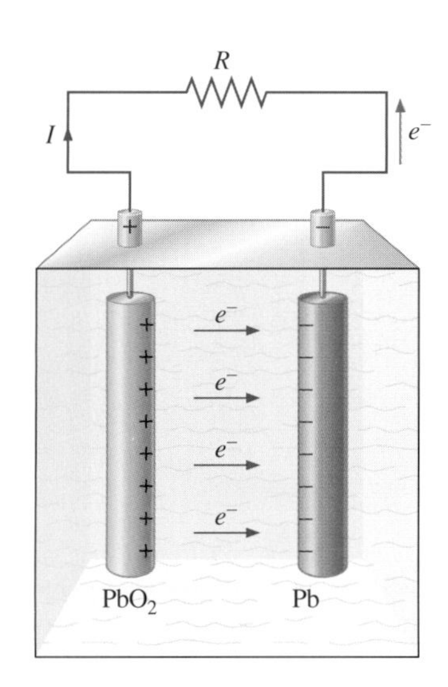

FIGURE 26-16 A battery in a circuit. The internal chemical reactions are equivalent to an electron flow from the PbO_2 plate to the Pb plate. Externally, electrons flow through the wire from the Pb plate to the PbO_2 plate.

Notice that the chemical reactions at both plates result in a depletion of the sulfuric acid ions H^+ and SO_4^{2-}. Eventually, the supply of these ions is exhausted and the battery is *discharged* (incapable of producing additional current). The battery can be *charged* by using another source to send current through it in the backward direction. This reverses the reactions at the plates and restores the supply of sulfuric acid.

The battery is our first example of a *source of electromotive force* which, by definition, is a *device in which nonelectric energy is converted to electric energy*. In a battery, the nonelectric energy comes from the chemical reaction. It acts as a "charge pump" that lifts electrons against a retarding potential difference. Other familiar sources of emf are the photocell, which converts light energy into electric energy, and the electric generator, which uses the mechanical energy of steam or falling water to produce electric energy.

A mechanical analogue to the chemical charge pump is illustrated in Fig. 26-17. Here the electrons are transported by a gremlin who resides inside the battery. He picks an electron off the positive terminal, does work by pushing the electron against the electric force as he moves toward the negative terminal, deposits the electron when he reaches that point, then returns to the positive terminal to get another electron. The work done by our diminutive laborer, which is equivalent to the work done by a chemical reaction in a real battery, is stored in the increased potential energy of the transported electron. This potential energy is then dissipated when the electron moves from the negative to the positive terminal through an external circuit.

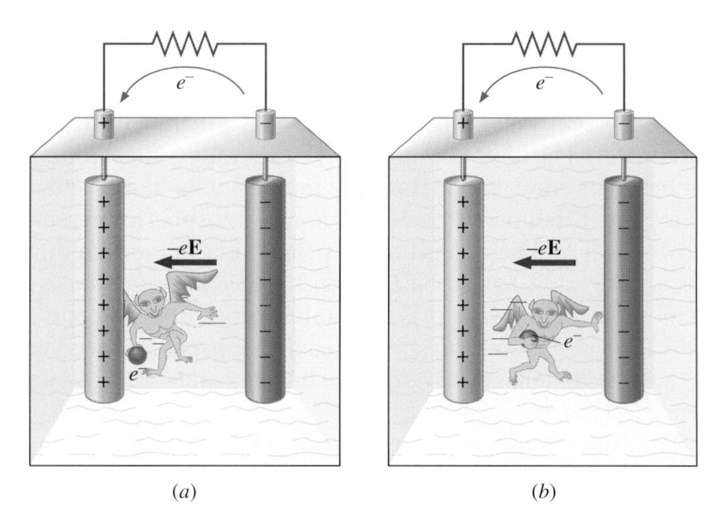

FIGURE 26-17 Mechanical analogue to the chemical charge pump in a battery: (*a*) A gremlin pulls an electron off the positive plate and (*b*) does work in transporting the electron against the electric field to deposit it on the negative plate.

The term **electromotive force** originated a long time ago when the concepts of force and energy were not well understood. It is actually a misnomer, for electromotive force is not a force; it is *the work per unit charge done by the source* in transporting electrons internally from its positive terminal to its negative terminal. Equivalently, electromotive force is the *electric potential energy per unit charge* gained by the electrons in being transported internally across the source. Unfortunately, this term, like the concept of positive charge carriers in wires, has survived. To avoid any possible confusion, we will simply refer to electromotive force as **emf**, represented by the symbol $\mathcal{E}$. Since emf is the work per unit charge, its SI unit is the volt. A "12-V battery" is therefore a battery whose emf is 12 V.

So far, we've ignored the internal resistance of a source of emf. In batteries, the internal resistance is basically caused by forces in the electrolytic solution that retard the flow of electrons. A battery with an internal resistance r may be represented by either of the symbols shown in Fig. 26-18, where the box on the right depicts the battery case. It's important to remember that the physical processes responsible for $\mathcal{E}$ and r occur *inside* the battery. You cannot remove one and leave the other. The internal resistances of sources are often so small (around 0.005 Ω for a fresh automobile battery) compared with other resistances in a circuit that they can be ignored.

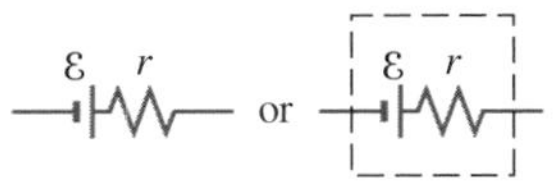

FIGURE 26-18 Circuit representations for a battery with internal resistance r.

26-5 Single-Loop Circuits

In this section we consider the simplest type of circuit—one whose components are connected together by resistanceless wires to form a single complete loop. Such a circuit is shown in Fig. 26-19, where a resistor R is connected across a battery with emf $\mathcal{E}$ and internal resistance r. Of primary importance in the analysis of this or any other circuit is that electric potential difference is a *path-independent quantity*. Consequently, if we start at the arbitrary point P, then traverse the circuit, adding all potential changes in the various elements, the sum of these changes must be zero when we return to P. Let's make that excursion clockwise around the circuit as shown, assuming that the current I also circulates clockwise. As we go through the battery from the negative to the positive terminal, we encounter an increase in potential equal to the battery emf* $\mathcal{E}$ and, simultaneously, a potential change due to its internal resistance r. Since we are crossing r in the direction of I, this potential change is $-Ir$, where the minus sign represents the fact that positive charge always moves through a resistor from a higher to a lower potential. The net potential change in crossing the battery is therefore $\mathcal{E} - Ir$. The change in potential across the resistor R is $-IR$ as we are moving in the direction of I. The sum of the changes in potential over the closed loop is then

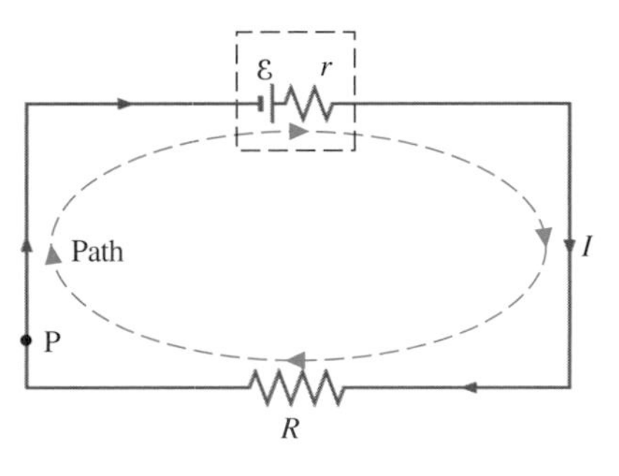

FIGURE 26-19 A circuit with a resistor of resistance R and a battery of emf $\mathcal{E}$ and internal resistance r.

$$\mathcal{E} - Ir - IR = 0,$$

so the current flowing in this circuit is

$$I = \frac{\mathcal{E}}{r + R}.$$

A slightly more complicated circuit is shown in Fig. 26-20. Following a clockwise path, which is also in the assumed direction of current flow, we find by summing potential differences that

$$\mathcal{E}_1 - Ir_1 - IR - Ir_2 - \mathcal{E}_2 = 0,$$

so

$$I = \frac{\mathcal{E}_1 - \mathcal{E}_2}{r_1 + r_2 + R} = \frac{12.0\ \text{V} - 6.0\ \text{V}}{1.0\ \Omega + 0.5\ \Omega + 4.5\ \Omega} = 1.0\ \text{A}.$$

Notice that the current flows through the 12.0-V battery from its negative terminal to its positive terminal, but through the 6.0-V battery from the positive to the negative terminal. Hence the 12.0-V battery is discharging while the 6.0-V is being charged.

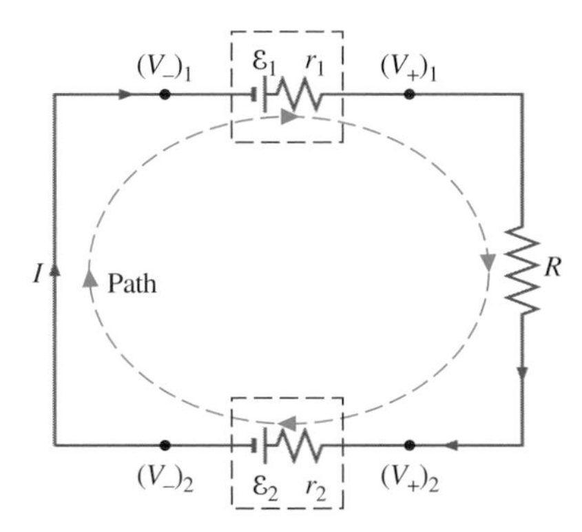

FIGURE 26-20 A single-loop circuit consisting of two batteries and a resistor. The values of the emf's and the resistances are as follows: $\mathcal{E}_1 = 12.0$ V, $r_1 = 1.0\ \Omega$, $\mathcal{E}_2 = 6.0$ V, $r_2 = 0.5\ \Omega$, and $R = 4.5\ \Omega$.

The potential difference between the terminals of a battery is known as its **terminal voltage**. We can determine how the terminal voltage is related to the emf with the help of the two batteries in Fig. 26-20. Let the potentials at the positive and

*By definition, the emf is the potential energy per unit charge gained by a charge carrier in crossing the source. Hence $\mathcal{E}$ is equivalent to the potential difference across the source.

negative terminals of battery 1 be $(V_+)_1$ and $(V_-)_1$, respectively. The terminal voltage is then $(V_+)_1 - (V_-)_1$. In traversing battery 1 from its positive terminal to its negative terminal, we have

$$(V_+)_1 + Ir_1 - \mathcal{E}_1 = (V_-)_1,$$

so

$$\begin{aligned}(V_+)_1 - (V_-)_1 &= -Ir_1 + \mathcal{E}_1 \\ &= -(1.0\ \text{A})(1.0\ \Omega) + 12.0\ \text{V} = 11.0\ \text{V}.\end{aligned}$$

The terminal voltage of battery 1 is therefore 11.0 V, which is less than its emf of 12.0 V. The "missing" 1.0 V is simply the potential difference across the internal resistance of the battery.

For battery 2, which is being charged, we find

$$(V_+)_2 - Ir_2 - \mathcal{E}_2 = (V_-)_2$$

so

$$\begin{aligned}(V_+)_2 - (V_-)_2 &= Ir_2 + \mathcal{E}_2 \\ &= (1.0\ \text{A})(0.5\ \Omega) + 6.0\ \text{V} = 6.5\ \text{V}.\end{aligned}$$

The terminal voltage of 6.5 V is *greater than* the 6.0-V emf of the battery. In this case, the potential difference across the battery's internal resistance furnishes the "extra" 0.5 V.

Finally, notice that if no current is flowing through a battery, then

$$V_+ - V_- = \mathcal{E} \pm (0)r = \mathcal{E};$$

that is, *the terminal voltage of a battery is equal to its emf if no current flows through the battery.*

EXAMPLE 26-7 A Single-Loop Circuit

(*a*) What is the current in the circuit of Fig. 26-21 when the switch S is closed? (*b*) What is the terminal voltage of the battery? (*c*) What is the potential difference across the resistor? (*d*) If the switch is open, what is the current? Now what are the potential differences across the battery, the resistor, and the switch?

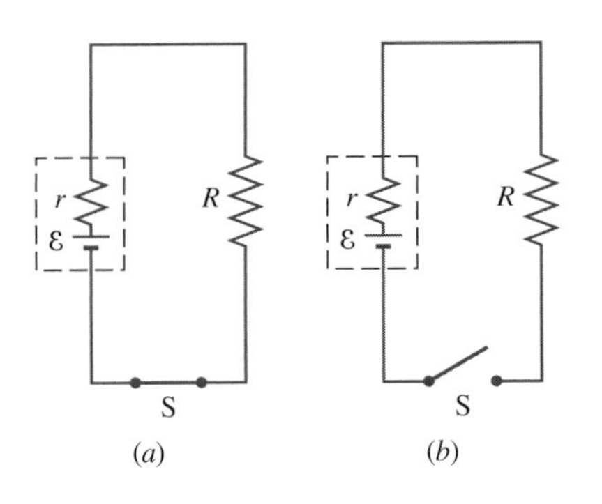

FIGURE 26-21 A circuit with (*a*) the switch S closed and (*b*) the switch S open. Here, $\mathcal{E} = 6.0$ V, $r = 1.0\ \Omega$, and $R = 2.0\ \Omega$.

SOLUTION (*a*) The sum of the changes in potential for the clockwise path around the circuit is

$$\mathcal{E} - Ir - IR = 0,$$

so

$$I = \frac{\mathcal{E}}{r + R} = \frac{6.0\ \text{V}}{1.0\ \Omega + 2.0\ \Omega} = 2.0\ \text{A}.$$

(*b*) We designate the potentials at the positive and negative terminals of the battery by V_+ and V_-, respectively. Traversing the battery from its negative terminal to its positive terminal, we obtain

$$V_- + 6.0\ \text{V} - (2.0\ \text{A})(1.0\ \Omega) = V_+,$$

which gives $V_+ - V_- = 4.0$ V. The terminal voltage of the battery is therefore 4.0 V, or 2.0 V less than its emf.

(*c*) Since the wires are assumed to have zero resistance, the two terminals of the battery are electrically equivalent to the corresponding ends of the resistor, so the potential difference across the resistor is also 4.0 V.

(*d*) Electrically, the open switch is equivalent to an infinite resistance; thus no current flows through the circuit of Fig. 26-21*b*. The terminal voltage is then

$$V_+ - V_- = \mathcal{E} - (0)r = 6.0\ \text{V},$$

which is the same as the emf of the battery.

To find the potential difference V_S across the switch, we sum the changes in potential around the circuit of part (*b*):

$$\mathcal{E} - Ir - V_S - IR = 0.$$

With $I = 0$,

$$V_S = \mathcal{E} = 6.0\ \text{V}.$$

DRILL PROBLEM 26-6

When the switch in the circuit of Fig. 26-22 is open, the terminal voltage of the battery is 12.0 V; when it is closed, the terminal voltage is 10.0 V. What is the emf of the battery? What is the current in the circuit when the switch is closed? What is the internal resistance of the battery?
ANS. 12.0 V; 1.0 A; 2.0 Ω.

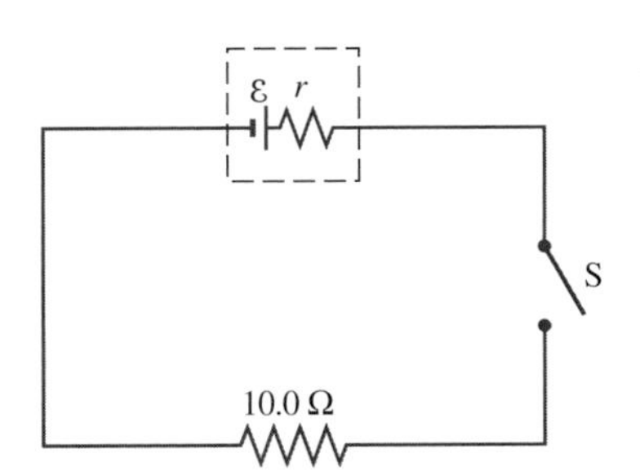

FIGURE 26-22 The circuit of Drill Prob. 26-6.

26-6 Power in Electric Circuits

When charges are transported across a source of emf, their potential energy changes. If a net charge Δq moves through a potential difference $\mathcal{E}$ in a time Δt, the change in electric potential energy of the charge is $\mathcal{E}\,\Delta q$. Thus the source of emf does work

$$\Delta W = \mathcal{E}\,\Delta q$$

on Δq during the time interval Δt. Dividing both sides of this equation by Δt, then taking the limit as $\Delta t \to 0$, we find

$$\frac{dW}{dt} = \mathcal{E}\,\frac{dq}{dt}.$$

By definition, $dq/dt = I$, the current through the battery, and $dW/dt = P$, the power output of (or input to) the battery. Hence

$$P = \mathcal{E}I. \tag{26-12}$$

The quantity P represents the rate at which energy is transferred *from* a discharging battery or *to* a charging battery. Power transfer in a source of emf is summarized in Fig. 26-23. From $P = \mathcal{E}I$, the SI unit of electric power is (1 A) (1 V) = (1 C/s) × (1 J/C) = 1 J/s = 1 W.

FIGURE 26-23 (*a*) Energy is transferred *from* the source of emf at a rate $\mathcal{E}I$. (*b*) Energy is transferred *to* the source at a rate $\mathcal{E}I$.

Now let's consider the power dissipated in a conducting element such as the one shown in Fig. 26-24. It has a resistance R and the potential difference between its ends is V. In moving through the element from higher to lower potential, a positive charge Δq loses energy $\Delta U = V\Delta q$. This electric energy is absorbed by the conductor through collisions between its atomic lattice and the charge carriers, causing its temperature to rise. This effect is commonly called *Joule heating*. Since power is the rate at which energy is transferred, we have

$$P = \frac{\Delta U}{\Delta t} = V\frac{\Delta q}{\Delta t} = VI, \tag{26-13a}$$

which, with the help of Ohm's law, can also be written in the forms

$$P = I^2R \tag{26-13b}$$

and

$$P = \frac{V^2}{R}. \tag{26-13c}$$

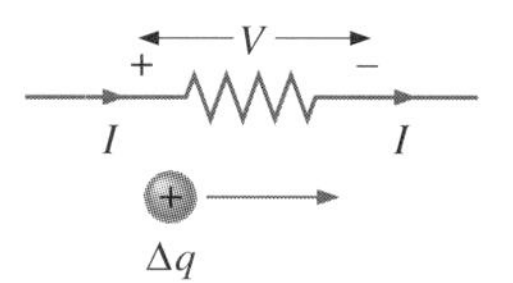

FIGURE 26-24 Power is always dissipated in a resistance as a current passes through it. Here the charge Δq loses energy $V\Delta q$ as it transverses the resistance.

Joule heating occurs whenever a current passes through an element that has resistance. To prevent the overheating of delicate electronic components, many electric devices like videocassette recorders, televisions, and computer monitors have grilles in their chassis to allow some of the heat produced to escape. If you place your hand above the grille of one of these devices, you can directly experience the effect of Joule heating.

The power I^2R dissipated in the lightbulb filament causes it to glow brightly.

EXAMPLE 26-8 POWER IN A CIRCUIT

Consider the circuit of Fig. 26-20. (*a*) Determine the power produced by the source of emf $\mathcal{E}_1$ in the 12.0-V battery. Determine the power absorbed by the source of emf $\mathcal{E}_2$ in the 6.0-V battery. (*b*) How much power is dissipated by the resistor and by each of the two internal resistances? (*c*) What is the power output at the terminals of the 12.0-V battery? What is the power input to the terminals of the 6.0-V battery? (*d*) Check to see that the total power produced is equal to the total power stored and dissipated.

SOLUTION (*a*) From $P = \mathcal{E}I$, the power produced by the source of emf in the 12.0-V battery is

$$P_{\mathcal{E}_1} = (12.0\text{ V})(1.0\text{ A}) = 12.0\text{ W}.$$

The 6.0-V battery is being charged, and its source of emf therefore absorbs power:

$$P_{\mathcal{E}_2} = (6.0\text{ V})(1.0\text{ A}) = 6.0\text{ W}.$$

(*b*) From $P = I^2R$, the power dissipated by each resistance is

$$P_R = (1.0\text{ A})^2(4.5\ \Omega) = 4.5\text{ W},$$
$$P_{r_1} = (1.0\text{ A})^2(1.0\ \Omega) = 1.0\text{ W},$$

and

$$P_{r_2} = (1.0\text{ A})^2(0.5\ \Omega) = 0.5\text{ W}.$$

(*c*) In Sec. 26-5, we found that the terminal voltage of the 12.0-V battery is 11.0 V, so the power output at the terminals of the 12.0-V battery is

$$P_1 = (1.0\text{ A})(11.0\text{ V}) = 11.0\text{ W}.$$

We also found that the terminal voltage of the 6.0-V battery is 6.5 V, so the power input to the terminals of the 6.0-V battery is

$$P_2 = (1.0\text{ A})(6.5\text{ V}) = 6.5\text{ W}.$$

Both P_1 and P_2 may also be calculated by combining the power of the source of emf with the power dissipated in the internal resistance. For the 12.0-V battery,

$$P_1 = 12.0 \text{ W} - 1.0 \text{ W} = 11.0 \text{ W},$$

and for the 6.0-V battery,

$$P_2 = 6.0 \text{ W} + 0.5 \text{ W} = 6.5 \text{ W}.$$

(*d*) The 12.0-V source of emf produces 12.0 W, the 6.0-V source stores 6.0 W, and the total power dissipated in the three resistances is 1.0 W + 4.5 W + 0.5 W = 6.0 W. As required by the principle of energy conservation, the total power generated is equal to the total power stored and dissipated.

EXAMPLE 26-9 Power Transfer in a Circuit

For the circuit of Fig. 26-19, equate the power generated by $\mathcal{E}$ to the power dissipated in r and R and thereby derive the circuit equation $\mathcal{E} - Ir - IR = 0$.

SOLUTION The power generated by the source of emf is $\mathcal{E}I$ and the power dissipated in the resistors is $I^2r + I^2R$. To conserve energy, we have

$$\mathcal{E}I = I^2r + I^2R,$$

which reduces to the circuit equation,

$$\mathcal{E} - Ir - IR = 0.$$

EXAMPLE 26-10 Electrical Properties of a Lightbulb

A lightbulb dissipates 10 W when connected across the terminals of a battery with an emf of 6.0 V and negligible internal resistance. (*a*) What is the current flowing through the bulb? (*b*) What is the resistance of the bulb?

SOLUTION (*a*) From $P = \mathcal{E}I$,

$$10 \text{ W} = (6.0 \text{ V})I,$$

so the current through the bulb is

$$I = 1.7 \text{ A}.$$

(*b*) From Ohm's law, the resistance of the bulb is

$$R = \frac{V}{I} = \frac{6.0 \text{ V}}{1.7 \text{ A}} = 3.5 \ \Omega.$$

DRILL PROBLEM 26-7

A wire of unknown resistance is used to connect the terminals of a 1.5-V source that has negligible internal resistance. If the power dissipated in the wire is 0.75 W, what is the resistance of the wire?
ANS. 3.0 Ω.

DRILL PROBLEM 26-8

In deriving Eqs. (26-12) and (26-13), we did not consider any changes in the kinetic energy of the charge carriers. Why?
ANS. v_d is constant.

SUMMARY

1. **Electric current**
   (*a*) Current is the rate of flow of charge through a cross section of a conductor:

$$I = \frac{dq}{dt}.$$

   (*b*) If there are n charge carriers per unit volume, each of charge magnitude e, moving at a drift speed v_d down a wire of cross-sectional area A, then the current in the wire is

$$I = neAv_d.$$

   (*c*) The current density J is the current per unit cross-sectional area A:

$$J = \frac{I}{A} = nev_d.$$

2. **Resistance and Ohm's law**
   (*a*) The resistance R of a conductor is defined as the potential difference V across the ends of the conductor divided by the current I flowing through it.

   (*b*) If the potential difference between the ends of a sample of material varies linearly with the current through the sample, the material is said to obey Ohm's law. The constant of proportionality between the potential difference and the current is the resistance R of the sample:

$$V = IR \qquad (R \text{ constant}).$$

   (*c*) If a material obeys Ohm's law, then the electric field at any point in the material is proportional to the current density at that point:

$$\mathbf{E} = \rho \mathbf{J},$$

   where ρ is the resistivity of the material. This equation may also be written as

$$\mathbf{J} = \sigma \mathbf{E},$$

   where $\sigma = 1/\rho$ is the conductivity of the material.

3. **Temperature dependence of resistivity**
For most applications, we can relate the resistivity ρ or the resistance R at a temperature T to the resistivity ρ_0 or the resistance R_0 at a temperature T_0 with

$$\rho = \rho_0[1 + \alpha(T - T_0)]$$

or

$$R = R_0[1 + \alpha(T - T_0)],$$

where α is the temperature coefficient of resistivity.

4. **The battery and electromotive force**
A source of emf is a device in which nonelectric energy is converted to electric energy. The quantitative measure of emf is the potential difference between the terminals of the source when no current is flowing through it.

5. **Single-loop circuits**
The single-loop circuit is analyzed by summing the potential differences across all elements in the loop and setting that sum equal to zero.

6. **Power in electric circuits**
(*a*) The power produced by a source of emf $\mathcal{E}$ is $\mathcal{E}I$, where I is the current leaving the positive terminal of the source. If current enters the positive terminal, then the source is storing energy at a rate $\mathcal{E}I$.
(*b*) The power dissipated in a resistor of resistance R is

$$P = IV = I^2R = \frac{V^2}{R},$$

where I is the current through the resistor and V is the potential difference across the resistor.

Questions

26-1. In the previous chapter you learned that the electric field inside a conductor must vanish. Then in this chapter you learned how to relate the electric field *inside* a wire to the current through it. Explain why these two results are consistent with one another.

26-2. Discuss how resistance and resistivity differ.

26-3. The resistances of most semiconductors and insulators decrease with temperature. Can you explain why?

26-4. The resistance of an ohmic conductor usually increases when it carries a high current. Why?

26-5. When the potential difference between the ends of a wire is doubled, the current through the wire increases by a factor of 2.8. What do you conclude about the wire?

26-6. When a wire is heated, its resistance varies because of the temperature dependence of its resistivity. However, temperature changes affect the dimensions of the wire as well. Using tables of temperature coefficients of expansion and resistivity, discuss which effect is more significant in determining resistance changes. Keep in mind also that both the length and the area of the wire increase with temperature.

26-7. Why do the filaments in incandescent lamps usually break just after the lamps are turned on?

26-8. Why do long-distance electric power transmission lines operate at very high voltages?

26-9. Does the sign of the potential difference across a resistor depend on the direction of the current flowing through it? Answer the same question for a battery.

26-10. Discuss how potential difference and emf differ.

26-11. Is the direction of the current through a battery always from the negative to the positive terminal?

26-12. The energy we can extract from a battery is always less than the energy that goes into it when it is being charged. Why?

26-13. Can the terminal voltage of a battery ever exceed its emf?

26-14. Does a 100-W household bulb burn at 100 W when it is connected to a 12-V car battery?

26-15. Batteries are often rated in ampere-hours. What does this rating mean?

26-16. Nichrome wire is often used as the heating element in ordinary appliances such as toasters. Discuss the characteristic(s) that make it suitable for such a purpose.

26-17. The size of the current that flows through a person's body is the most important factor in determining the seriousness of an electric shock. For example, a current of approximately 100 mA is usually fatal if it lasts for more than a few seconds. Given this, evaluate the relative dangers of a 12-V car battery capable of producing 100 A and a 1000-V power supply capable of producing 1 A.

Problems

Electric Current

26-1. A wire carries a current of 2.0 A. What is the charge that has flowed through its cross section in 1.0 s? How many electrons does this correspond to?

26-2. What is the current in a wire when 2.0×10^{24} electrons flow through it in 1 h?

26-3. Suppose the potential difference applied between two electrodes in a gas tube is large enough to ionize the atoms of the gas. If 6.0×10^{18} electrons and 4.0×10^{18} singly ionized ions move through a cross section of the tube in 1 s, what is the current in the gas?

26-4. (*a*) The density of gold is 19.3×10^3 kg/m^3, and its atomic mass is 197 g/mol. Using the fact that there is one free electron per gold atom, determine the number of free electrons per cubic meter in gold. (*b*) Make the same calculation for silver, whose density and atomic mass are 10.5×10^3 kg/m^3 and 107 g/mol, respectively. Silver also has one free electron per atom.

26-5. A current of 2.0 A flows through a silver ($n = 5.91 \times 10^{28}$ electrons/m^3) wire with a circular cross section of radius 1.5 mm. (*a*) What is the current density in the wire? (*b*) What is the drift speed of the electrons in the wire?

26-6. The current through a wire is given by $I(t) = I_0 e^{-\alpha t}$, where $I_0 = 2.0$ A and $\alpha = 6.0 \times 10^{-3}\ \text{s}^{-1}$. What is the total charge that passes through the wire between $t = 1.0 \times 10^{-4}$ s and $t = 4.0 \times 10^{-4}$ s?

26-7. The current density in a cylindrical wire of radius 2.5 mm is 300 A/m^2. In what time will Avogadro's number of electrons pass through the wire?

Resistance and Ohm's Law

Note: If there is no reference to temperature in a particular problem, you can assume that all conductors are at the temperature specified in Table 26-1.

26-8. A wire 50 m long and 2.0 mm in diameter has a resistance of 0.25 Ω. What is the resistivity of the material used to manufacture the wire?

26-9. A 5.0-m length of wire with cross-sectional area 0.015 cm^2 has a measured resistance of 0.088 Ω. Calculate the resistivity of the material used to make the wire. What is the material?

26-10. A tungsten wire and a copper wire have the same length. If these wires also have the same resistance, what must the ratio of their cross-sectional areas be?

26-11. Compare the lengths of a silver and a nichrome wire, each having a diameter of 0.10 mm and a resistance of 10 Ω.

26-12. Wires made of aluminum and copper have the same length, and the diameter of the copper wire is 1.0 mm. If both wires have the same resistance, what is the diameter of the aluminum wire?

26-13. Two cylindrical wires of the same material have diameters in the ratio 2:1. How do the resistances per unit length of the two wires compare?

26-14. A copper wire 5.0 m long and a silver wire 20.0 m long have the same resistance. If the cross sections of both wires are circular, what is the ratio of their diameters?

26-15. A bird stands on a frayed electric power line carrying 2.0×10^3 A. The resistance per unit length of the wire is 1.5×10^{-5} Ω/m, and the feet of the bird are 3.0 cm apart. What is the potential difference between the bird's feet?

26-16. Make a plot of the following potential difference-current data and determine the resistance of the sample.

V (mV)	*I* (mA)
1.00	0.26
2.00	0.53
3.00	0.78
4.00	1.03
5.00	1.30

26-17. When a wire 5.0 m long with a diameter of 0.050 mm is connected to a 12-V battery, a current of 4.0 A flows through it. What is the resistivity of the material used to fabricate the wire?

26-18. A current 0.10 A passes through a coil of platinum wire. If the resistance of the wire is 50 Ω, what is the potential difference between its ends? Suppose that the cross-sectional area of the wire is 1.0×10^{-8} m^2. What is the length of the wire?

26-19. A wire is made from a material with $n = 6.0 \times 10^{28}$ conduction electrons per cubic meter. The wire has a square cross section 1.0 mm on a side, it is 5.0 m long, and the resistance between its ends is 0.12 Ω. (*a*) What is the resistivity of the material? (*b*) If a potential difference of 0.36 V is applied between the ends of the wire, what is the resulting current? (*c*) What is the electric field in the wire? (*d*) What is the current density? (*e*) What is the drift speed of the conduction electrons in the wire?

26-20. The cross section of a copper wire is a square 2.0 mm on a side. (*a*) If the wire is 4.0 m long, what is its resistance? (*b*) What is the current in the wire when a potential difference of 0.020 V is applied between its ends? (*c*) What is the electric field in the wire?

26-21. A 0.75-Ω wire is melted and drawn out to four times its original length. What is its new resistance?

26-22. A wire is melted and drawn into a new wire whose length is three times that of the original. If the resistance of the new wire is 10 Ω, what was the resistance of the original wire?

Temperature Dependence of Resistivity

26-23. Compare the resistivities of (*a*) copper, (*b*) silver, and (*c*) nickel at 20 and 100°C.

26-24. A copper wire is at 20°C. To what temperature must it be heated in order to double its resistance?

26-25. A nichrome heating element operates at 110 V. When it is first turned on, its temperature is 20°C and 3.0 A flows through it. A short time later, the current reaches a steady value of 2.7 A. What is the operating temperature of the element?

26-26. When a wire is moved from ice water to boiling water, its resistance changes from 50.2 to 56.9 Ω. What is α for the material used in manufacturing the wire?

26-27. The resistance of a thin silver wire is 1.0 Ω at 20°C. The wire is placed in a liquid bath and its resistance rises to 1.2 Ω. What is the temperature of the bath?

26-28. A gold wire and a silver wire have the same dimensions. At what temperature will the silver wire have the same resistance that the gold wire has at 20°C?

EMF and Single-Loop Circuits

26-29. When the switch S in the circuit shown in the accompanying figure is open, the terminal voltage of the battery is 1.51 V. When the switch is closed, the terminal voltage drops to 1.40 V. What are the emf ℰ and the internal resistance r of the battery?

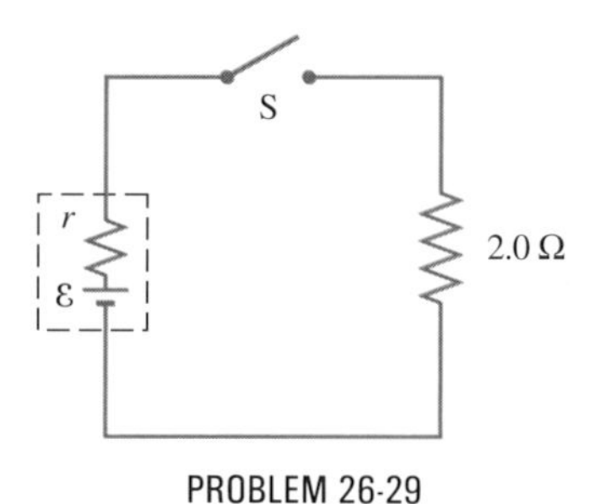

PROBLEM 26-29

26-30. Suppose that the terminal voltage of the battery in the circuit shown in the accompanying figure is 1.45 V. What is the internal resistance r of the battery if its emf is 1.50 V?

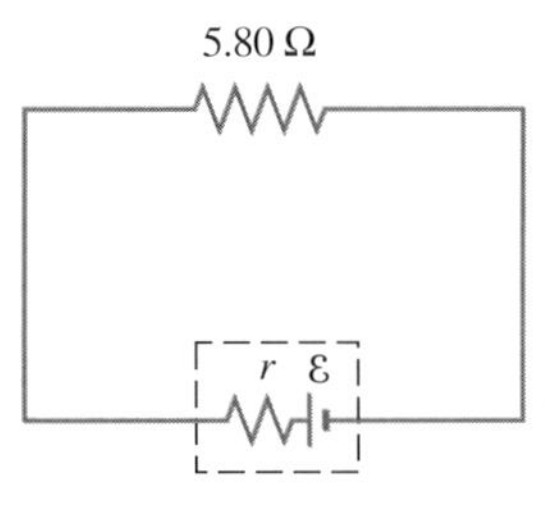

PROBLEM 26-30

26-31. The open-circuit terminal voltage of a car battery is 12.0 V. When the battery terminals are connected by a thick cable, the current that flows through the cable is 100 A. (*a*) What is the emf of the battery? (*b*) What is the internal resistance of the battery? (*c*) What is the current when the battery is connected to a 5.0-Ω resistor? (*d*) What is the potential difference between the terminals of the battery for this case?

26-32. The potential difference across the terminals of a battery is 9.0 V when there is a current of 2.5 A flowing through it from the negative to the positive terminal. When the current is 2.0 A in the reverse direction, the voltage across the terminals is 12.0 V. What are the internal resistance and the emf of the battery?

26-33. A battery with an emf of 12.0 V and an internal resistance of 0.10 Ω is charged by a 13.0-V battery of negligible internal resistance. (*a*) How should the two batteries be connected? (*b*) What is the current flowing between the batteries?

26-34. When a 5.0-Ω resistor is connected to a battery whose emf is 12.0 V, 2.3 A flow through the resistor. (*a*) What are the potential differences between the ends of the resistor and the terminals of the battery? (*b*) What is the internal resistance of the battery?

26-35. Find (*a*) the current flowing in the circuit shown in the accompanying figure, (*b*) the potential across each of the resistors, and (*c*) the potential across the terminals of the battery. Assume $\mathcal{E} = 12.0$ V and $r = 0.5$ Ω.

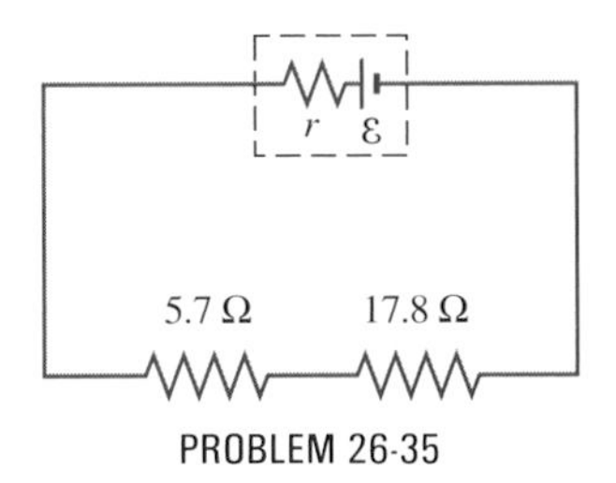

PROBLEM 26-35

26-36. (*a*) What is the current flowing in the circuit shown in the accompanying figure? (*b*) What is the potential difference between the terminals of each of the batteries? Assume $\mathcal{E}_1 = 10.0$ V, $r_1 = 0.2$ Ω, $\mathcal{E}_2 = 5.0$ V, and $r_2 = 0.4$ Ω.

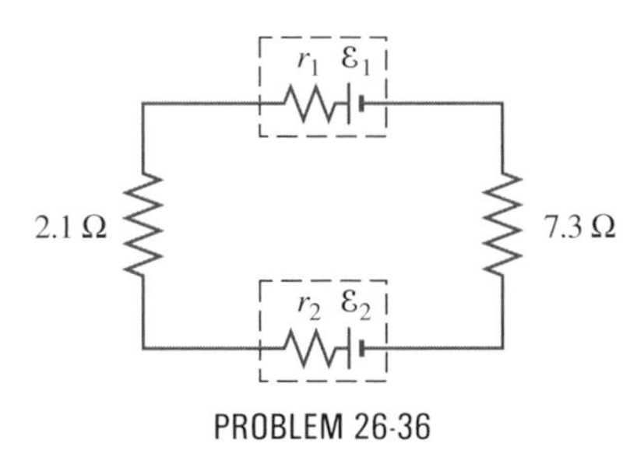

PROBLEM 26-36

Power in Electric Circuits

26-37. What is the power dissipated in the internal resistance of the battery of Prob. 26-31 when it is connected to the 5.0-Ω resistor? What is the power dissipated in the resistor? What is the power output of the battery?

26-38. In Prob. 26-33, what is the energy per second that is stored chemically in the battery being charged? How much time is needed to store 1.0×10^6 J in this battery?

26-39. In Prob. 26-34, what is the power output of the emf of the battery? What is the power dissipated in the internal resistance of the battery? What is the power dissipated in the 5.0-Ω resistor? Verify that energy is conserved here.

26-40. For the circuit of Prob. 26-35, determine (*a*) the power output of the battery and (*b*) the power produced by the chemical reaction in the battery. (*c*) Verify that energy is conserved in this circuit.

26-41. In the circuit of Prob. 26-36, how much chemical energy is drawn from the 10.0-V battery in 1 h? How much chemical energy is stored in the 5.0-V battery in 1 h?

26-42. What is the maximum potential that can be applied across a 100-Ω resistor rated at 1/2 W?

26-43. What is the current through a 40-W headlight that is connected to a 12-V automobile battery?

26-44. What is the resistance of a 100-W lightbulb designed to operate at 110 V? What current passes through the bulb?

26-45. A 500-W electric heater is designed to operate from a 110-V source. (*a*) What is the resistance of the heater? (*b*) What current does it draw? (*c*) If the line voltage drops to 90 V, determine the power dissipated in the heater, assuming that the resistance of the heating element does not change. Is this a good assumption?

26-46. A current of 25.0 A is drawn from a 12.0-V battery for 10 min. By how much is the chemical energy of the battery reduced?

26-47. A heating element draws 12 A from a 220-V source. (*a*) What is the resistance of the element? (*b*) What power is dissipated in the element?

26-48. (*a*) We are charged for electric power in kilowatt-hours. How much energy in joules does this unit correspond to? (*b*) If electric energy is 8 cents per kilowatt-hour, what would it cost to leave a 100-W lightbulb burning for 8 h?

26-49. A 12-V battery is rated at 300 A·h. What total energy can the battery deliver? (*Note:* An emf source rated at 1 A·h can produce (1 C/s) (3600 s) = 3600 C of charge before it needs to be recharged.)

26-50. What is the total energy stored in a 12-V, 80-A·h battery when it is fully charged? (See Prob. 26-49.)

General Problems

26-51. A copper wire with a square cross section is redrawn so that its cross section is circular. If the length of a side of the square section is equal to the diameter of the circular section, by what fraction is the resistance of the wire changed by the redrawing?

26-52. A wire is redrawn so that its new length is *N* times its original length. What is the ratio of the resistances of the new and original wires?

26-53. Suppose you have a collection of wires that are all made from the same material and have the same length, but have various cross-sectional areas. Show that the resistances of the wires are inversely proportional to their masses.

26-54. A small heater is designed to operate off a 12-V automobile battery. If the heater warms 250 cm^3 of water from 20 to 90°C in 5.0 min, how much current does it draw from the 12-V battery? Assume that all of the energy dissipated in the heater goes into the water.

26-55. A spherical shell with inner radius r_1 and outer radius r_2 is made from a material with resistivity ρ. Show that the resistance R between the inside and outside of the shell is

$$R = \frac{\rho}{4\pi}\left(\frac{1}{r_1} - \frac{1}{r_2}\right).$$

26-56. The accompanying figure shows a resistor that is made by forming a material of resistivity ρ into a hollow cylinder of length l, inner radius r_1, and outer radius r_2. (*a*) What is the resistance between the two ends of the cylinder? (*b*) What is the resistance between the inner and outer surfaces of the cylinder?

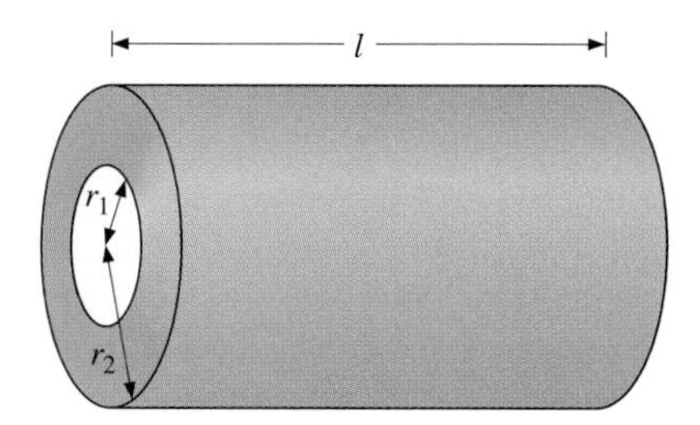

PROBLEM 26-56

26-57. When a carbon resistor and a nichrome resistor are connected as shown in the accompanying figure, the combination has a resistance of 2.0 kΩ at 0°C. What resistance at 0°C must each element have so that the resistance of the combination is independent of temperature?

PROBLEM 26-57

26-58. In the circuit shown in the accompanying figure, determine the current and the potential differences across all batteries and resistances. Assume that $\mathcal{E}_1 = 2.0$ V, $r_1 = 0.5\ \Omega$, $\mathcal{E}_2 = 4.0$ V, $r_2 = 0.5\ \Omega$, $\mathcal{E}_3 = 6.0$ V, and $r_3 = 0.2\ \Omega$. Verify that energy is conserved in the circuit.

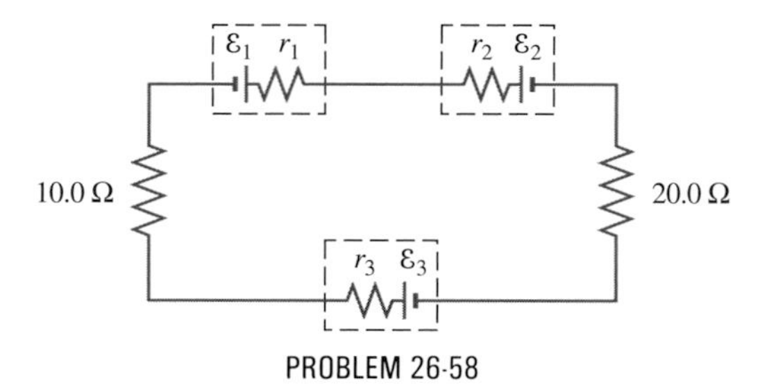

PROBLEM 26-58

Complicated circuits are analyzed using the rules developed here for direct current circuits.

CHAPTER 27 Direct Current Circuits

Preview

In this chapter we consider dc circuits and their applications. The main topics to be discussed here are the following:

1. **Kirchhoff's rules.** Kirchhoff's rules are developed and then used to analyze simple dc circuits.
2. **Resistors in series and in parallel.** We consider how a combination of resistors can be reduced to a simpler but equivalent system.

*3. **Household wiring.** The operation of simple household circuits is discussed. Grounding and circuit breakers are also considered.

4. **Electrical measurements in dc circuits.** Four instruments used to make electrical measurements in dc circuits are investigated.

In the previous chapter our study of electric circuits was limited to the single-loop circuit. Our analysis of this type of circuit was based on the principle that there is no net change in electric potential over a closed path. We now begin a study of more complicated circuits that can consist of several closed loops with many resistors and batteries. A simple example of a multiloop circuit is the two-loop circuit in Fig. 27-1.

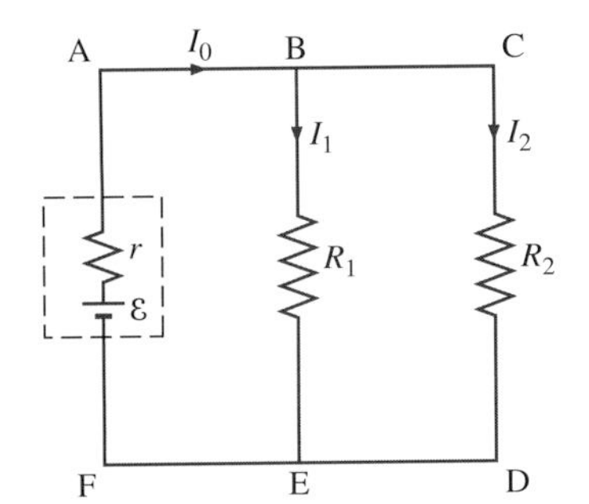

FIGURE 27-1 A two-loop circuit.

A printed circuit board. Kirchhoff's rules are needed to understand how the circuit works.

If the current passing through any element in a circuit does not change with time, the circuit is known as a *direct current* (*dc*) circuit. This is the type of circuit that will be analyzed here. In future chapters, you will also learn about circuits whose currents vary exponentially as well as sinusoidally with time. These circuits can also be analyzed with the rules we now consider.

27-1 Kirchhoff's Rules

Multiloop circuits are generally analyzed using a set of two rules developed by Gustav Robert Kirchhoff (1824–1887). One rule relates the current entering to that leaving a *junction* in a circuit. *A junction is any point in circuit where three or more wires meet.* For the circuit of Fig. 27-1, B and E are junctions. *The part of a circuit between two junctions is called a branch.* Hence BE, BCDE, and BAFE are branches of the circuit of Fig. 27-1. Kirchhoff's rules state that

1. The sum of the currents entering a junction is equal to the sum of the currents leaving that junction.
2. The net change in the electric potential in going around one complete loop in a circuit is equal to zero.

The first rule, which is also known as the *junction rule*, follows from the principle of conservation of charge. It simply states that charge does not accumulate or vanish at a junction. For example, current I_0 enters junction B of Fig. 27-1 along path AB, while I_1 and I_2 leave along BE and BC; Kirchhoff's first rule then tells us that

$$I_0 = I_1 + I_2.$$

Kirchhoff's second rule (also called the *loop rule*) is based on the fact that the electrostatic field is conservative. This results in no net change in the electric potential around a closed path. In fact, our analysis of single-loop circuits in the last chapter was based on this same principle. Before using the second rule, let's review the changes in electric potential that occur when we traverse the different components of a circuit.

When crossing a source of emf from the negative to the positive terminal (see Fig. 27-2*a*), the change in potential is positive, because the positive terminal of the source is at the higher potential. However, if the path traverses a source of emf from its positive to its negative terminal, as shown in Fig. 27-2*b*, the potential change is negative. The potential change in crossing a source of emf depends only on the orientation of the terminals relative to the direction of the path, and *not* on the direction of the electric current. *The change in electric potential is always positive if the path crosses a source of emf from the negative to the positive terminal and always negative in the opposite direction.*

We also encounter a change in potential when traversing a resistance. In this case, the direction of current flow is important in determining whether the change is positive or negative, since

A $\mathcal{E}$ B

Path

$\Delta V = V_B - V_A = +\mathcal{E}$

(*a*)

A $\mathcal{E}$ B

Path

$\Delta V = V_B - V_A = -\mathcal{E}$

(*b*)

FIGURE 27-2 Potential changes in traversing a source of emf.

positive charge always moves through a resistance in the direction of decreasing potential. As an example, consider the path shown in Fig. 27-3*a*, which is along the direction of the current. The potential at B must be less than that at A, because the current (positive charge) flows from A to B. The path therefore takes us through a potential difference of $-IR$. If our path is opposite to the current flow (see Fig. 27-3*b*), the potential change is $+IR$.

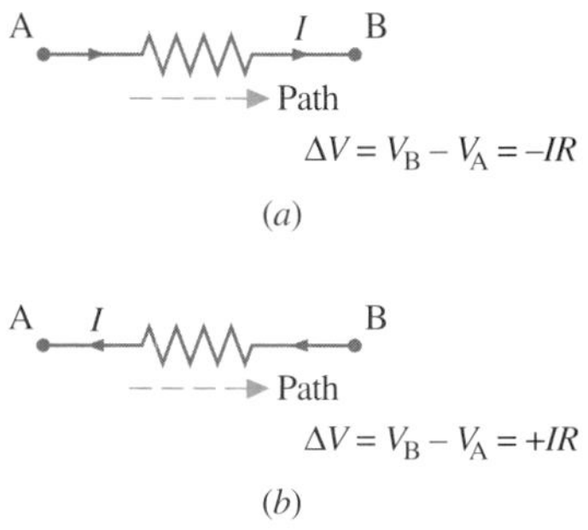

A I B
Path
$\Delta V = V_B - V_A = -IR$
(*a*)

A I B
Path
$\Delta V = V_B - V_A = +IR$
(*b*)

FIGURE 27-3 Potential changes in traversing a resistance. The sign of the change in potential depends on the direction of the current flowing through the resistance.

To illustrate the use of Kirchhoff's rules, let's consider the circuit of Fig. 27-4. In this case, all elements in the circuit are known, and we wish to determine the unknown currents (I_1, I_2, and I_3). Applying the junction rule at either C or F, we have

$$I_1 = I_2 + I_3,$$

where the *assumed* directions of the currents are those shown in the figure. In general, prior knowledge of the direction of a current is not needed to analyze a circuit. If by chance you should select a direction that is *opposite* to the actual flow of current, the proper application of Kirchhoff's rules will simply yield a *negative* value for that current. This indicates that the current flows in the opposite direction to that which was initially assumed. However, the *magnitude* of the calculated current will still be correct.

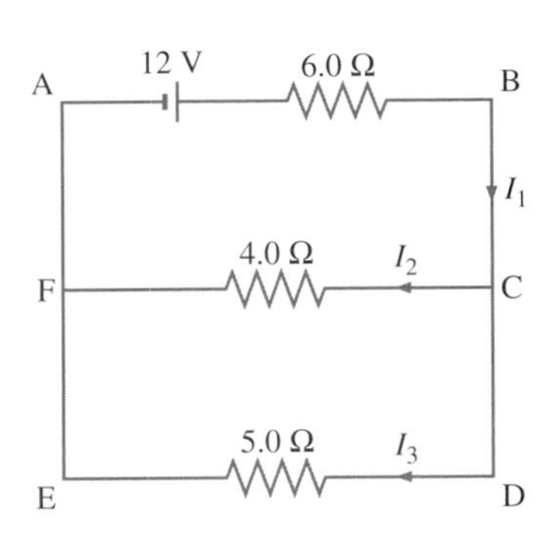

FIGURE 27-4 A two-loop circuit. Here the internal resistance of the battery is negligible.

For the complete loop ABCFA,* the loop equation gives

$$12\text{ V} - I_1(6.0\ \Omega) - I_2(4.0\ \Omega) = 0,$$

while for FCDEF,

$$I_2(4.0\ \Omega) - I_3(5.0\ \Omega) = 0.$$

These two equations contain three unknown currents. However, using the junction rule to substitute $I_2 + I_3$ for I_1 in the first equation, we obtain a system of two equations and two unknowns:

$$12\text{ V} - I_2(10.0\ \Omega) - I_3(6.0\ \Omega) = 0,$$

and

$$I_2(4.0\ \Omega) - I_3(5.0\ \Omega) = 0.$$

In solving these equations, we find that $I_2 = 0.81$ A and $I_3 = 0.65$ A. Consequently, $I_1 = I_2 + I_3 = 1.46$ A. The positive values of I_1, I_2, and I_3 indicate that these currents actually flow in the directions that were initially chosen for them.

Notice that we did not have to use the loop ABDEA of Fig. 27-4 to determine the currents. Its loop equation is

$$12\text{ V} - I_1(6.0\ \Omega) - I_3(5.0\ \Omega) = 0,$$

which, using $I_1 = I_2 + I_3$, can be rewritten as

$$12\text{ V} - I_2(6.0\ \Omega) - I_3(11.0\ \Omega) = 0.$$

This is nothing more than the sum of the other two loop equations and is therefore not independent of these equations. A general rule of thumb is that an N-loop circuit furnishes N independent loop equations. In analyzing such a circuit, you can find the N equations by making sure that each loop contains at least one element (a resistor or a battery) that is in no other loop. Once you have done this, you can proceed to solve the equations.

EXAMPLE 27-1 Analysis of a Multiloop Circuit

In the circuit of Fig. 27-1, assume that the source of emf is a 2.0-V battery with negligible internal resistance ($r = 0$). The currents through R_1 and R_2 are 1.0 A and 2.0 A, respectively. Determine (*a*) the current I_0 that flows through the battery and (*b*) the values of R_1 and R_2.

SOLUTION (*a*) At either B or E, the junction rule gives

$$I_0 = I_1 + I_2 = 1.0\text{ A} + 2.0\text{ A} = 3.0\text{ A}.$$

(*b*) Since we have a two-loop circuit, we can find two independent equations to solve for the unknowns R_1 and R_2. To find R_1, we apply the loop rule to the closed path ABEFA. Along AB (and EF), we do not traverse any circuit element so there is no potential change. From B to E, our path follows the direction of the current I_1 and takes us through the resistor R_1. The potential change across the resistor is therefore $-I_1 R_1 = -(1.0\text{ A})R_1$. Along the segment from F to A, the potential change in traversing the source of emf is +2.0 V (remember we are assuming $r = 0$ in this example). Thus we have

$$2.0\text{ V} - (1.0\text{ A})R_1 = 0,$$

and

$$R_1 = 2.0\ \Omega.$$

*The order of the letters tells us that we are going clockwise around the upper loop in the circuit of Fig. 27-4.

Around path BCDEB, the loop rule yields

$$-I_2 R_2 + I_1 R_1 = 0,$$

so

$$R_2 = \frac{I_1}{I_2} R_1 = \left(\frac{1.0\ \text{A}}{2.0\ \text{A}}\right)(2.0\ \Omega) = 1.0\ \Omega.$$

EXAMPLE 27-2 ANOTHER MULTILOOP CIRCUIT

(*a*) Determine the currents that flow through the 2.0-Ω resistor and the two batteries in the circuit of Fig. 27-5. (*b*) What are the potential differences between points C and D? between points A and E?

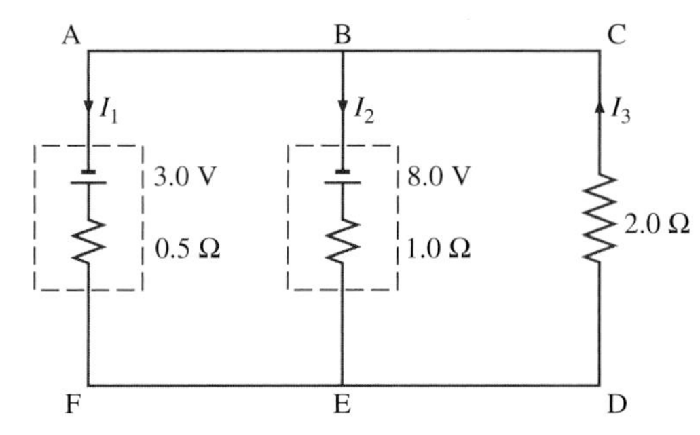

FIGURE 27-5 The multiloop circuit of Example 27-2.

SOLUTION (*a*) With the currents designated as shown in the figure, the junction rule yields at either B or E

$$I_3 = I_1 + I_2. \qquad (i)$$

With the loop rule, we have for ABEFA

$$8.0\ \text{V} - I_2(1.0\ \Omega) + I_1(0.5\ \Omega) - 3.0\ \text{V} = 0,$$

or

$$5.0\ \text{V} + I_1(0.5\ \Omega) - I_2(1.0\ \Omega) = 0. \qquad (ii)$$

Now applying the loop rule to BEDCB, we find

$$8.0\ \text{V} - I_2(1.0\ \Omega) - I_3(2.0\ \Omega) = 0,$$

and substitution for I_3 using Eq. (*i*) then gives

$$8.0\ \text{V} - I_1(2.0\ \Omega) - I_2(3.0\ \Omega) = 0. \qquad (iii)$$

There are two unknowns, I_1 and I_2, in Eqs. (*ii*) and (*iii*). Solving these equations, we obtain

$$I_1 = -2.0\ \text{A} \quad \text{and} \quad I_2 = 4.0\ \text{A}.$$

From Eq. (*i*), $I_3 = -2.0\ \text{A} + 4.0\ \text{A} = 2.0\ \text{A}$. The negative value of I_1 indicates that this current actually flows opposite to the direction assigned to it at the beginning of the problem. Figure 27-6 shows the currents and their actual directions.

You may have noticed that one path (ABEFA) used in analyzing this circuit is clockwise, while the other (BEDCB) is counterclockwise. When analyzing a circuit, *it is not necessary that all chosen paths have the same sense*. It is only important that the rules for finding potential differences across circuit elements as illustrated in Figs. 27-2 and 27-3 be followed for all paths.

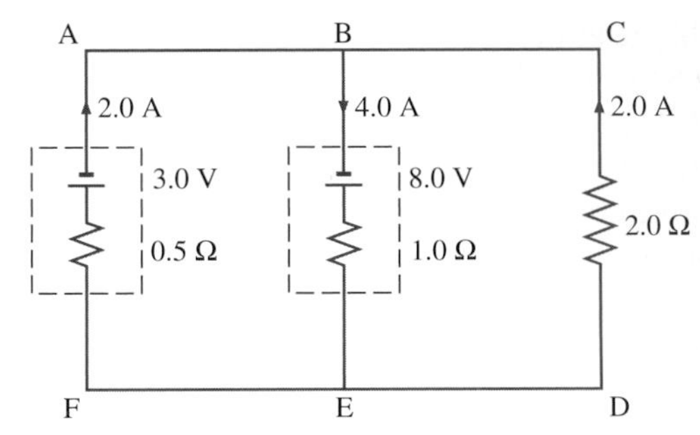

FIGURE 27-6 The actual currents and their directions for the multiloop circuit of Example 27-2.

(*b*) The potential difference between two points in a circuit can be determined by following any path in the circuit that leads from one point to the other. We begin our calculation by denoting the potential at point C of Fig. 27-6 as V_C. If we follow a direct path from C to D, we traverse the 2.0-Ω resistor in the opposite direction to the 2.0-A current. The change in potential in crossing this resistor and arriving at D is therefore (2.0 A) (2.0 Ω); hence the potential at D is

$$V_D = V_C + (2.0\ \text{A})(2.0\ \Omega),$$

so

$$V_D - V_C = 4.0\ \text{V}.$$

To find the potential difference between A and E, we start at A and follow the path ABE. This yields

$$V_E = V_A + 8.0\ \text{V} - (4.0\ \text{A})(1.0\ \Omega),$$

so

$$V_E - V_A = 4.0\ \text{V}.$$

The result $V_E - V_A = V_D - V_C$ is an expected one since points A, B, and C are at the same potential, as are points F, E, and D.

DRILL PROBLEM 27-1

For the circuit of Example 27-1, calculate the potential difference between points F and A. How is this related to the potential differences between points E and B and between points D and C? Use your results to determine the values of R_1 and R_2 and compare them with those found in Example 27-1.
ANS. $V_A - V_F = V_B - V_E = V_C - V_D = 2.0\ \text{V}$.

DRILL PROBLEM 27-2

Consider the circuit shown in Fig. 27-1. If $\mathcal{E} = 30\ \text{V}$, $r = 0.6\ \Omega$, $R_1 = 4.0\ \Omega$, and $R_2 = 6.0\ \Omega$, what are the values of I_0, I_1, and I_2? Use the junction rule together with the loop rule for paths ABEFA and BCDEB.
ANS. $I_0 = 10.0\ \text{A}$; $I_1 = 6.0\ \text{A}$; $I_2 = 4.0\ \text{A}$.

DRILL PROBLEM 27-3

Repeat the calculation of the previous drill problem by using the junction rule together with the loop rule for (*a*) paths CBEDC and CAFDC and (*b*) paths BAFEB and BCDEB.

DRILL PROBLEM 27-4

By applying the loop rule to paths ABEFA, BCDEB, and ACDFA of Fig. 27-5, we obtain three equations in I_1, I_2, and I_3. Can these equations be solved for unique values of these currents? Explain.
ANS. No; the three equations are not independent.

27-2 Resistors in Series and in Parallel

Resistors through which the *same current* flows are said to be *in series*. For example, the three resistors R_1, R_2, and R_3 in the circuit of Fig. 27-7*a* are in series since the current I flows through each one. By applying the loop rule to the closed path ABCDA, we find

$$-IR_1 - IR_2 - IR_3 + \mathcal{E} = 0.$$

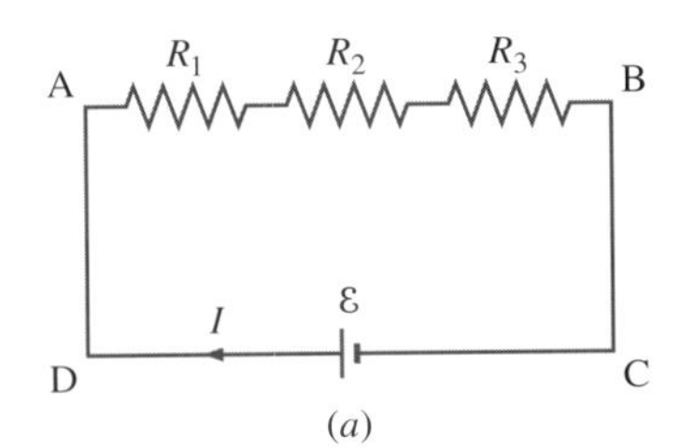

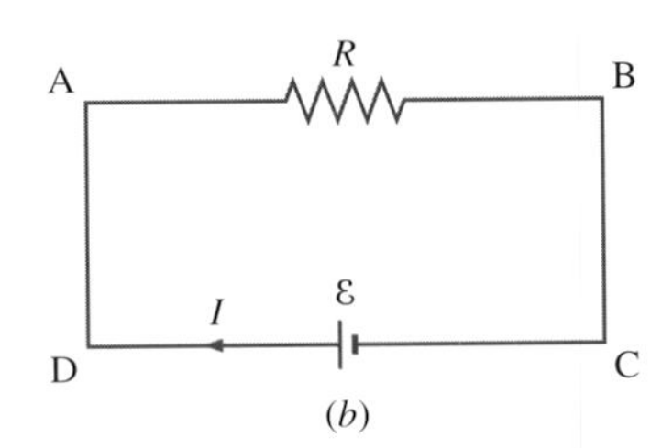

FIGURE 27-7 (*a*) A circuit with a battery and three resistors in series. (*b*) A circuit equivalent to that of part (*a*), where $R = R_1 + R_2 + R_3$.

The current is then

$$I = \frac{\mathcal{E}}{R_1 + R_2 + R_3} = \frac{\mathcal{E}}{R},$$

where

$$R = R_1 + R_2 + R_3.$$

Therefore the net resistance of the series resistors R_1, R_2, and R_3 is simply the sum of their individual resistances, and the circuit of Fig. 27-7*a* is equivalent to the circuit shown in Fig. 27-7*b*. This argument can be applied to any number of resistances in series. *The net resistance R_S of an arbitrary number of resistances in series is the sum of the individual resistances*; that is,

$$R_S = R_1 + R_2 + R_3 + \ldots \tag{27-1}$$

Notice that the net series resistance is always greater than any of the individual resistances.

Now suppose that R_1, R_2, and R_3 are connected as shown in Fig. 27-8*a*. The three resistances have the *same potential difference* across them and are said to be *in parallel*. The current I is split into I_1, I_2, and I_3, which flow through R_1, R_2, and R_3, respectively. We may write

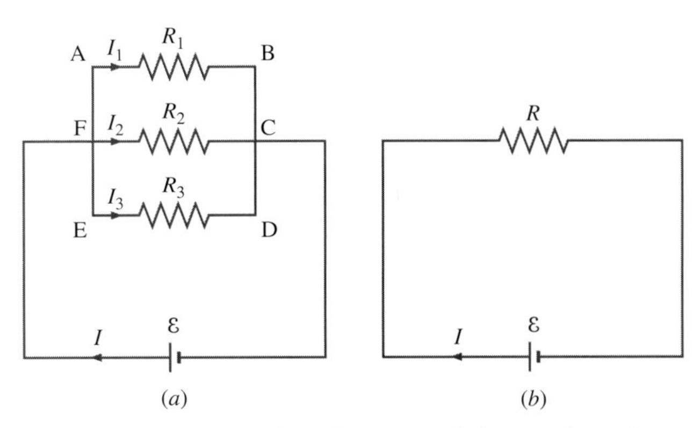

FIGURE 27-8 (*a*) A circuit with a battery and three resistors in parallel. (*b*) A circuit equivalent to that of part (*a*), where $1/R = 1/R_1 + 1/R_2 + 1/R_3$.

$$I = I_1 + I_2 + I_3 = \frac{\mathcal{E}}{R_1} + \frac{\mathcal{E}}{R_2} + \frac{\mathcal{E}}{R_3}$$

$$= \mathcal{E}\left(\frac{1}{R_1} + \frac{1}{R_2} + \frac{1}{R_3}\right) = \frac{\mathcal{E}}{R},$$

where

$$\frac{1}{R} = \frac{1}{R_1} + \frac{1}{R_2} + \frac{1}{R_3}.$$

Hence the net resistance of the three parallel resistances is equal to R, and the circuit of Fig. 27-8*a* is equivalent to the circuit shown in Fig. 27-8*b*. This argument is applicable to any number of resistances in parallel. *The net resistance R_P of an arbitrary number of resistances in parallel is related to the individual resistances by*

$$\frac{1}{R_P} = \frac{1}{R_1} + \frac{1}{R_2} + \frac{1}{R_3} + \ldots \tag{27-2}$$

You can see from this equation that the net parallel resistance is always less than any of the individual resistances.

EXAMPLE 27-3 Series and Parallel Resistors

What are the net resistances of a 1.00-Ω resistor and a 100-Ω resistor when they are in series? in parallel?

SOLUTION Let $r = 1.00\ \Omega$ and $R = 100\ \Omega$. The net resistance of the series combination is

$$R_S = r + R = 101\ \Omega.$$

The resistance of the parallel combination is found from

$$\frac{1}{R_P} = \frac{1}{r} + \frac{1}{R} = \frac{1}{1.00\ \Omega} + \frac{1}{100\ \Omega}$$

which yields $R_P = 0.99\ \Omega$.

Notice that $R_S \approx R$ and $R_P \approx r$. The results of this simple example provide us with the following useful information: For a combination of two resistances R and r such that $R \gg r$, the resistance of the series combination is approximately equal to R, the larger resistance, while the resistance of the parallel combination is approximately equal to r, the smaller resistance.

EXAMPLE 27-4 RESISTOR COMBINATIONS

Suppose that $R_1 = 10.0\ \Omega$, $R_2 = R_3 = 20.0\ \Omega$, and $\mathcal{E} = 25.0$ V in the circuits of Figs. 27-7*a* and 27-8*a*. For each circuit, calculate (*a*) the net resistance, (*b*) the current flowing through the source of emf, and (*c*) the current flowing through each resistor and the potential difference across each resistor.

SOLUTION (*a*) For the circuit of Fig. 27-7*a*, the net resistance is

$$R_S = 10.0\ \Omega + 20.0\ \Omega + 20.0\ \Omega = 50.0\ \Omega.$$

(*b*) The current flowing through the source of emf is therefore

$$I = \frac{25.0\ \text{V}}{50.0\ \Omega} = 0.500\ \text{A}.$$

(*c*) Since the three resistors are in series, this 0.500-A current flows through each resistor. The potential differences are

$$(0.500\ \text{A})(10.0\ \Omega) = 5.0\ \text{V} \quad \text{across } R_1,$$
$$(0.500\ \text{A})(20.0\ \Omega) = 10.0\ \text{V} \quad \text{across } R_2,$$

and

$$(0.500\ \text{A})(20.0\ \Omega) = 10.0\ \text{V} \quad \text{across } R_3.$$

The sum of these potential differences is 25.0 V, the value of the emf, in agreement with Kirchhoff's loop rule.

Now let's consider the circuit of Fig. 27-8*a*.

(*a*) The net resistance R_P of the three parallel resistors is found from

$$\frac{1}{R_P} = \frac{1}{10.0\ \Omega} + \frac{1}{20.0\ \Omega} + \frac{1}{20.0\ \Omega} = \frac{1}{5.0\ \Omega},$$

so

$$R_P = 5.0\ \Omega.$$

(*b*) The current that flows through the source of emf is now

$$I = \frac{25.0\ \text{V}}{5.0\ \Omega} = 5.0\ \text{A}.$$

(*c*) Since the resistors are in parallel, the same 25.0-V potential difference exists across each one. The currents that pass through the respective resistors are therefore

$$I_1 = \frac{25.0\ \text{V}}{10.0\ \Omega} = 2.5\ \text{A},$$
$$I_2 = \frac{25.0\ \text{V}}{20.0\ \Omega} = 1.25\ \text{A},$$

and

$$I_3 = \frac{25.0\ \text{V}}{20.0\ \Omega} = 1.25\ \text{A}.$$

Notice that $I = 5.0\ \text{A} = I_1 + I_2 + I_3$, which is the junction rule applied at F or C.

EXAMPLE 27-5 CIRCUIT ANALYSIS BY COMBINING RESISTANCES

Consider the circuit shown in Fig. 27-1. If $\mathcal{E} = 30$ V, $r = 0.6\ \Omega$, $R_1 = 4.0\ \Omega$, and $R_2 = 6.0\ \Omega$, what are I_0, I_1, and I_2? Use the rules for combining resistances to simplify the circuit. Compare your answers with those of Drill Prob. 27-2.

SOLUTION Because R_1 and R_2 are in parallel,

$$\frac{1}{R_P} = \frac{1}{R_1} + \frac{1}{R_2} = \frac{1}{4.0\ \Omega} + \frac{1}{6.0\ \Omega},$$

and the net resistance across the battery is

$$R_P = 2.4\ \Omega.$$

This resistance is in series with the internal resistance of the source of emf, as shown in Fig. 27-9*a*. The net resistance seen by $\mathcal{E}$ is therefore

$$R = 2.4\ \Omega + 0.6\ \Omega = 3.0\ \Omega.$$

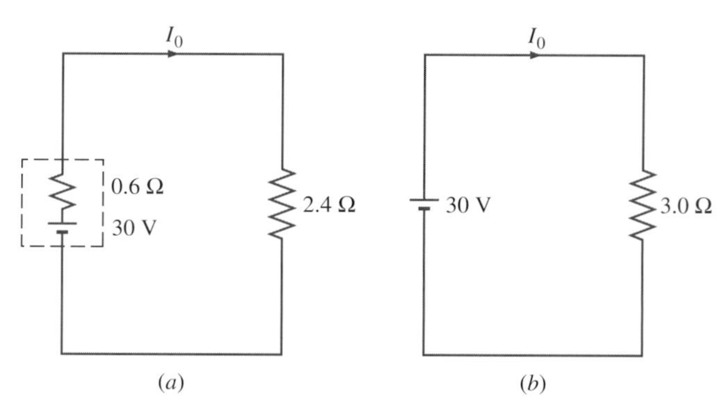

FIGURE 27-9 Simplification of the circuit shown in Fig. 27-1.

Figure 27-9*b* shows the simplest circuit that is equivalent to that of Fig. 27-1. Using this circuit, we find

$$I_0 = \frac{\mathcal{E}}{R} = \frac{30\ \text{V}}{3.0\ \Omega} = 10\ \text{A}.$$

The potential difference between points A and F in Fig. 27-1 is

$$V_A - V_F = 30\ \text{V} - (0.6\ \Omega)(10\ \text{A}) = 24\ \text{V},$$

which is also equal to $V_B - V_E$ and $V_C - V_D$. Thus

$$V_B - V_E = I_1 R_1 = 24\ \text{V},$$

and

$$I_1 = \frac{24\ \text{V}}{4.0\ \Omega} = 6.0\ \text{A}.$$

Finally,

$$V_C - V_D = I_2 R_2 = 24\ \text{V},$$

and

$$I_2 = \frac{24\ \text{V}}{6.0\ \Omega} = 4.0\ \text{A}.$$

EXAMPLE 27-6 ANOTHER APPLICATION OF THE RESISTANCE COMBINATION RULES

Simplify the circuit of Fig. 27-10*a* by using the rules for combining series and parallel resistors, then use the resulting equivalent circuits to determine the currents I_1 through I_5.

SOLUTION First we add the two 5.0-Ω resistors in series to obtain a 10.0-Ω resistor in parallel with the 10.0-Ω resistor between B and C. Upon combining these two resistors, we have the equivalent circuit of Fig. 27-10*b*, where the 3.0-Ω, the 5.0-Ω, and the 2.0-Ω resistors are in series. By combining these three resistors, we are left with the equivalent circuit of Fig. 27-10*c*. Since the 15.0-Ω and 10.0-Ω resistors are in parallel, the circuit can be further simplified to that of Fig. 27-10*d*.

We can now work backwards from the simplest circuit to determine the various currents. Starting with Fig. 27-10*d*,

$$I_1 = \frac{12\ \text{V}}{6.0\ \Omega} = 2.0\ \text{A} \qquad \text{and} \qquad V_A - V_D = 12\ \text{V}.$$

From Fig. 27-10*c*,

$$I_2 = \frac{V_A - V_D}{15.0\ \Omega} = \frac{12\ \text{V}}{15.0\ \Omega} = 0.80\ \text{A}$$

and

$$I_3 = \frac{V_A - V_D}{10.0\ \Omega} = \frac{12\ \text{V}}{10.0\ \Omega} = 1.2\ \text{A}.$$

From Fig. 27-10*b*,

$$V_B - V_C = I_3(5.0\ \Omega) = 6.0\ \text{V}.$$

Finally, we have with Fig. 27-10*a*,

$$I_4 = \frac{V_B - V_C}{10.0\ \Omega} = \frac{6.0\ \text{V}}{10.0\ \Omega} = 0.60\ \text{A}$$

and

$$I_5 = \frac{V_B - V_C}{10.0\ \Omega} = \frac{6.0\ \text{V}}{10.0\ \Omega} = 0.60\ \text{A}.$$

In the analysis of the last three examples, the resistance combination formulas of Eqs. (27-1) and (27-2) were used to reduce a network of resistances to a single equivalent resistance. However, resistances do not always occur as parallel and series combinations that can be reduced to a single resistance. In such cases, you should try to first simplify the circuit as much as possible using Eqs. (27-1) and (27-2), then complete the analysis with Kirchhoff's laws.

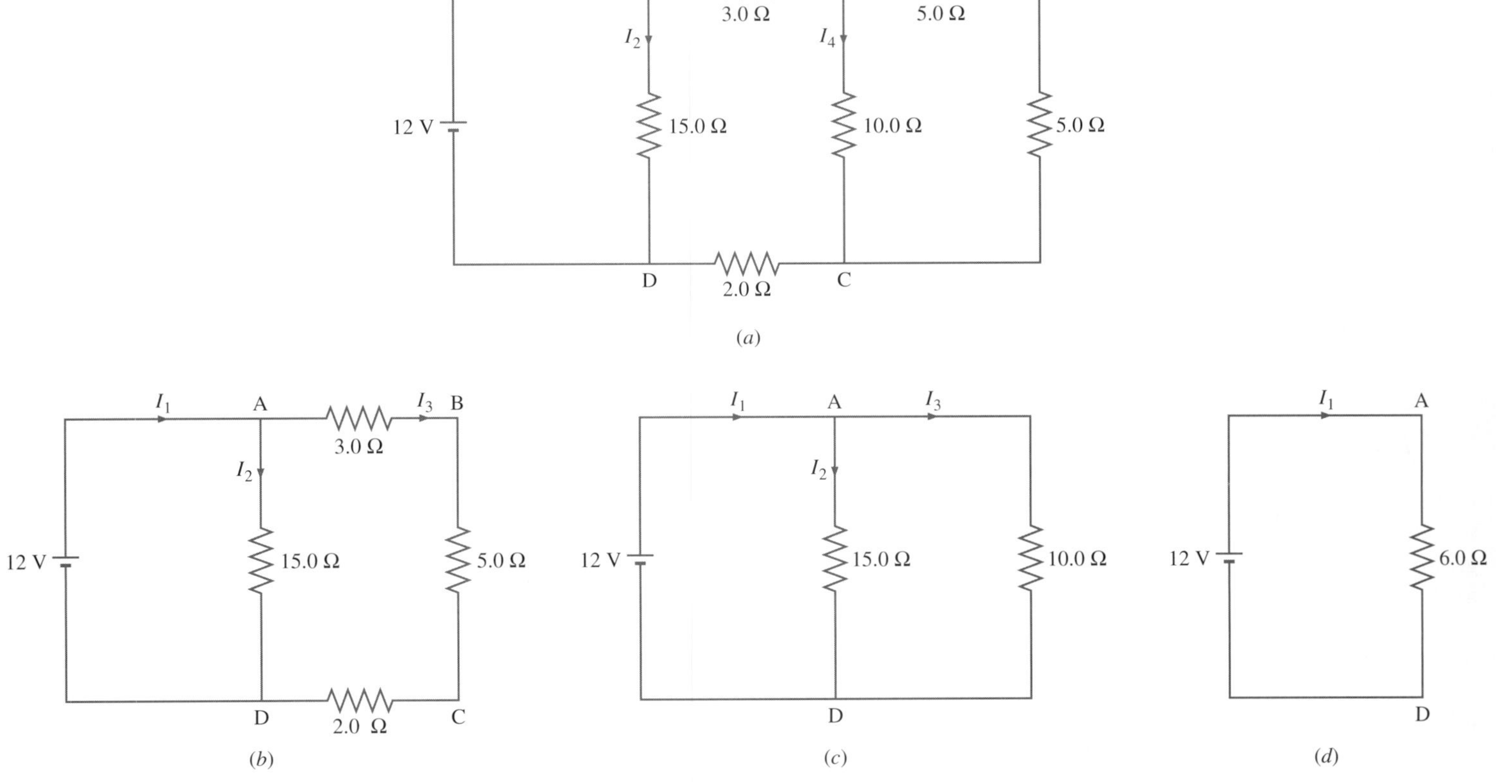

FIGURE 27-10 (*a*) The circuit of Example 27-6. (*b*)–(*d*) Equivalent but progressively simpler circuits.

DRILL PROBLEM 27-5

Calculate the net resistances of the resistor combinations shown in Fig. 27-11*a*, *b*, and *c*.
ANS. (*a*) 12.0 Ω; (*b*) 1.3 Ω; (*c*) 2.2 Ω.

2.0 Ω 6.0 Ω 4.0 Ω
(*a*)

3.0 Ω
6.0 Ω
4.0 Ω
(*b*)

1.0 Ω
3.0 Ω
5.0 Ω
2.0 Ω
(*c*)

FIGURE 27-11 Resistor combinations for Drill Prob. 27-5.

DRILL PROBLEM 27-6

(*a*) What is the net resistance seen by the 12.0-V battery in the circuit of Fig. 27-12? (*b*) What are the currents in the various branches?
ANS. (*a*) 8.50 Ω; (*b*) 1.42 A, 0.71 A, 0.71 A.

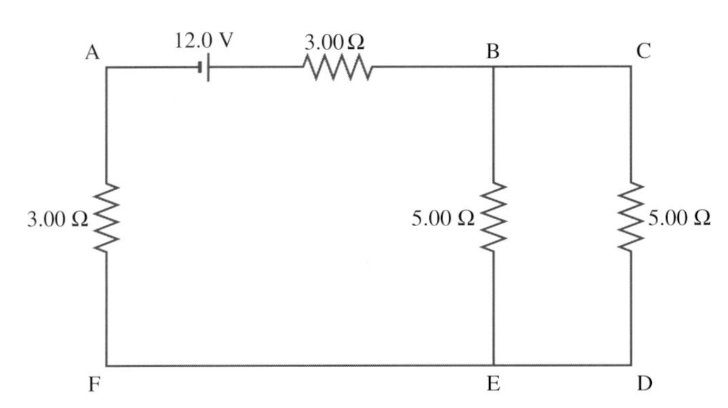

FIGURE 27-12 The circuit of Drill Prob. 27-6.

*27-3 HOUSEHOLD WIRING

Figure 27-13 depicts the wiring in a typical house. Two lines enter the house from a power pole, one at 120 V relative to ground, and the other grounded. The voltage actually alternates at 60 cycles per second, but for present purposes we can assume that it is steady (a dc voltage). Connected between the two power lines are sets of circuits within the house. To prevent the wires from overheating, a circuit breaker (or a fuse in an older house) is inserted as shown. The circuit breaker is set to open when the current through it reaches an upper limit, frequently 15 or 20 A. Once it opens, the current throughout the circuit is terminated.

The filament lamp was invented by Thomas Edison.

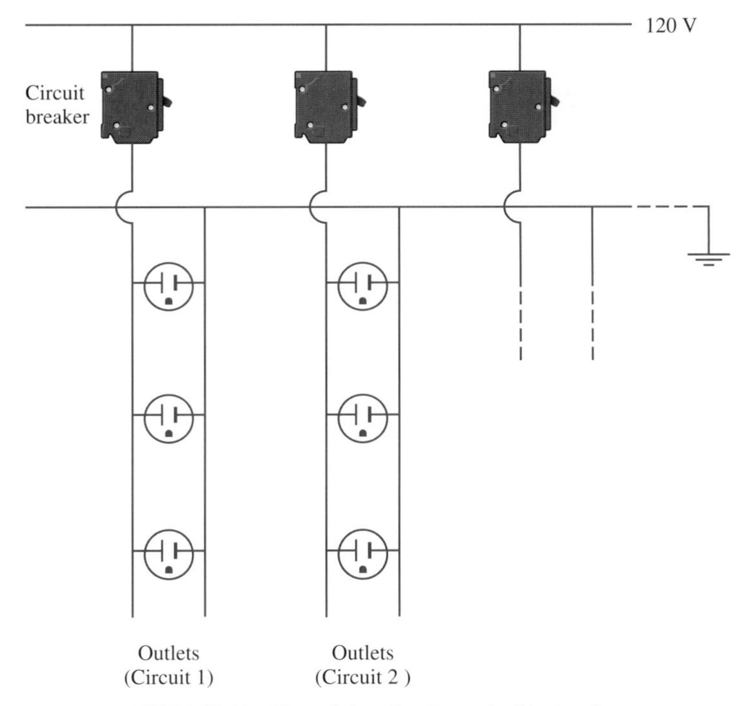

FIGURE 27-13 The wiring for household circuits.

The pairs of open points represent plugs in the wall. When appliances such as lamps, radios, and hair dryers are "plugged in," they are connected in parallel across the 120 V. A typical operating circuit is shown in Fig. 27-14. Since the appliances are connected in parallel, one appliance can be removed without affecting the current through the others. If too many appliances are connected in a particular circuit, the current through the circuit breaker will exceed the maximum allowed and the breaker will open. Of course, if the 120-V line inadvertently touches ground, the resulting large current will also cause the circuit breaker to open.

What happens if a person accidentally touches the 120-V line, as shown in Fig. 27-15*a*? This often occurs when a wire whose insulation is frayed touches the metal casing of an electric appliance (Fig. 27-15*b*). If the resistance between the boy's

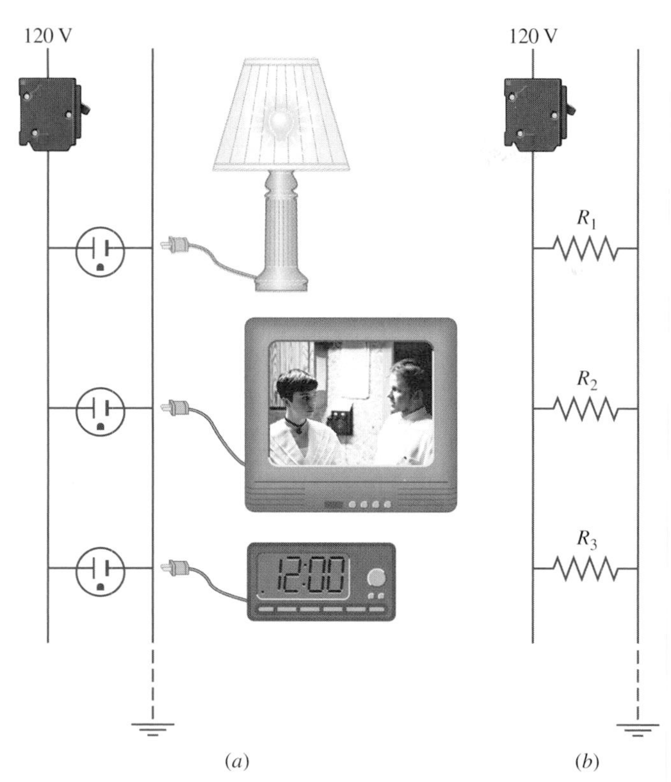

FIGURE 27-14 (*a*) A typical operating circuit in a house. (*b*) A schematic representation of the circuit in part (*a*).

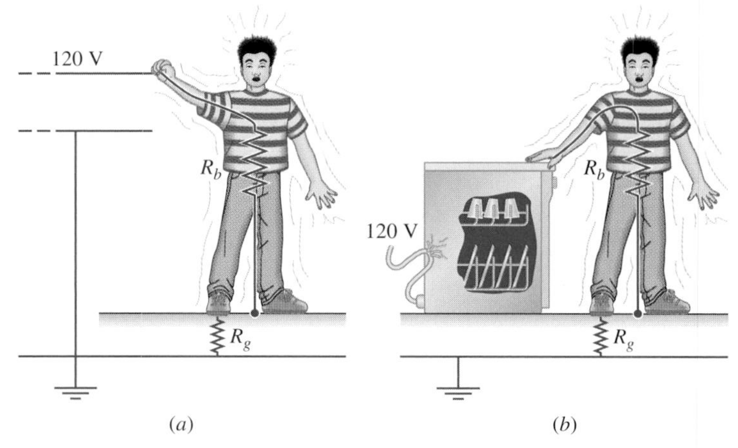

FIGURE 27-15 (*a*) When the 120-V line is touched, the resistance between the line and ground is $R_b + R_g$. (*b*) This accident can occur when a 120-V line with frayed insulation touches the metal case of an ungrounded appliance.

hand and his feet is R_b and the resistance between his feet and ground is R_g, the current through his body is

$$I = \frac{120\text{ V}}{R_b + R_g}.$$

If he is standing on a surface that is a good insulator such as a wooden floor, then R_g is so large and I so small that he will suffer only a mild shock. However, if he is standing on a wet floor that is in contact with metal drain pipes, R_g will be quite small and I can be large enough to be fatal.

As protection against accidental shock, metal cases are usually grounded. You have probably noticed that there is a third semicircular terminal on a wall plug. This terminal is connected to ground, usually through a water pipe. (See Fig. 27-16*a*.) An appliance with a three-prong plug is wired so that the prong inserted into the semicircular terminal is attached through the wire to the appliance case. The case is therefore grounded whenever the appliance is plugged in. If a wire at 120 V happens to touch the case, the wire is connected directly (or *shorted*) to ground, and the resulting current surge causes the circuit breaker to open as shown in Fig. 27-16*b*. The power to the appliance is then immediately shut off.

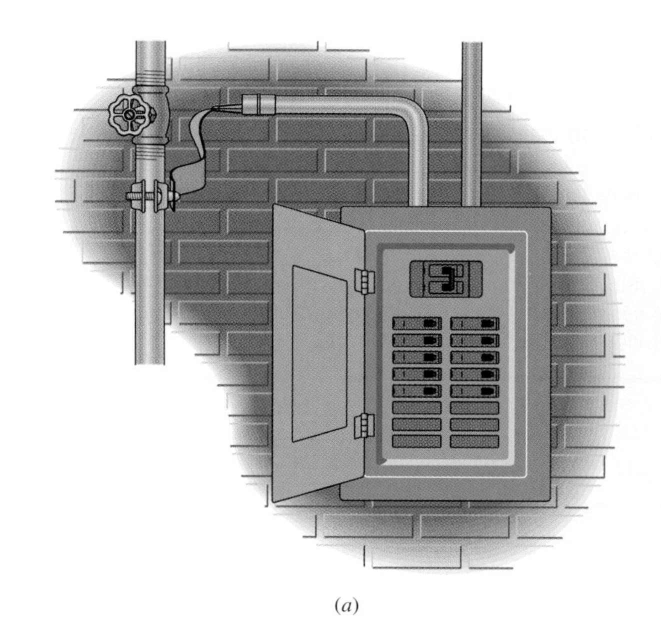

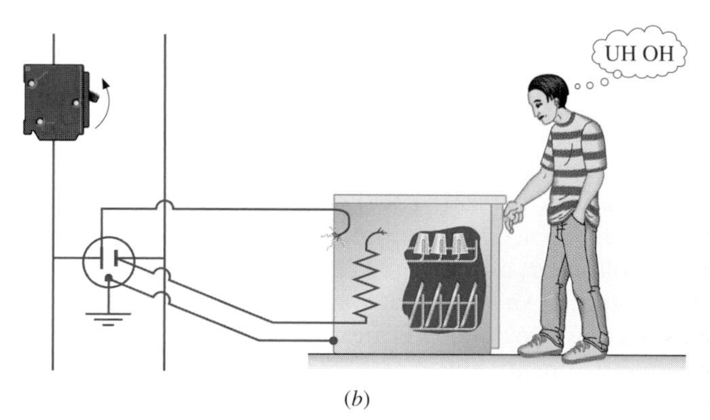

FIGURE 27-16 (*a*) Household circuits are usually connected to ground through a water pipe. (*b*) Because of the grounded case, the circuit breaker opens when the 120-V line comes into contact with the case. This serves as protection against electric shock.

Household appliances that use large amounts of power (for example, electric stoves and clothes dryers) generally operate with an input voltage of 240 V. This voltage is provided with the help of another wire that is −120 V with respect to ground. A voltage of 240 V between the two live terminals of an outlet is obtained by connecting the +120-V line to one terminal and the −120-V line to the other terminal. Since $P = IV$, an appliance that must consume 2000 W would require a current of

16.7 A if it is designed to operate at 120 V, but a current of only 8.3 A if it is designed for 240 V. Now the power dissipated in the wires of the circuit connected to the appliance is given by $P = I^2R$. Hence when the current is halved for the larger voltage, the power dissipated in the wires is one-fourth what it would be for 120 V. As a result, safety requirements to prevent overheating in the wires of a household circuit can be satisfied with smaller wires in a 240-V circuit.

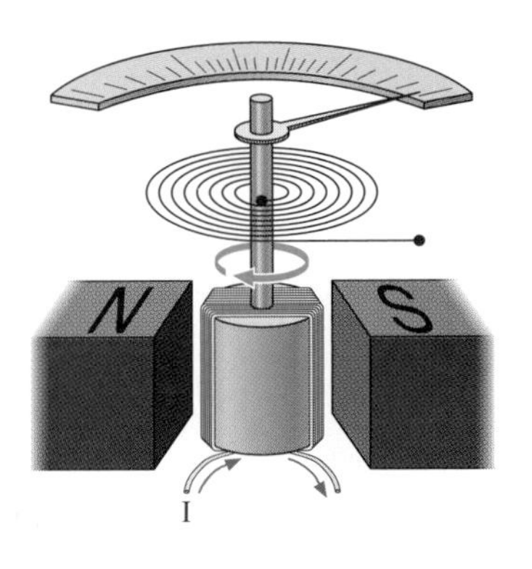

FIGURE 27-17 The galvanometer. When a current passes through the coil of wire, it rotates due to the torque of the magnetic field.

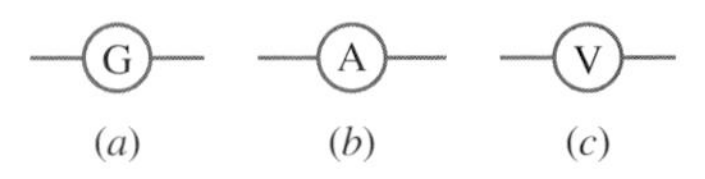

FIGURE 27-18 Circuit representations of (*a*) a galvanometer, (*b*) an ammeter, and (*c*) a voltmeter.

EXAMPLE 27-7 HOUSEHOLD APPLIANCES

A 600-W microwave oven and an 800-W heater are operating on a 120-V line that is protected by a 15-A circuit breaker. Will the circuit continue to operate if a 150-W lamp is connected to it? What happens when a 500-W hair dryer is added to the line?

SOLUTION The oven draws (600 W)/(120 V) = 5.0 A, and the current through the heater is (800 W)/(120 V) = 6.7 A. When the lamp is turned on, an additional (150 W)/(120 V) = 1.3 A flows through the circuit breaker. The total current is 5.0 A + 6.7 A + 1.3 A = 13.0 A, which is less than the 15.0-A maximum so the circuit will continue to operate. The hair dryer draws (500 W)/(120 V) = 4.2 A. When it is connected, the additional current is sufficient to trip the circuit breaker.

DRILL PROBLEM 27-7

How many 150-W lamps can be connected to a 120-V line that is protected by a 20-A circuit breaker?
ANS. 16.

27-4 ELECTRICAL MEASUREMENTS IN DC CIRCUITS

In this section we discuss four instruments used to make electrical measurements in dc circuits. They are (1) the ammeter, which measures current; (2) the voltmeter, which measures potential difference; (3) the potentiometer, which is a special type of voltmeter; and (4) the Wheatstone bridge, which measures resistance. An important component of all of these instruments is a sensitive device for measuring small currents. In most modern instruments, this measurement is made with electronic circuitry whose analysis is inappropriate for this textbook. However, we can base our study on the galvanometer,* a sensitive, current-measuring device in common use before the advent of electronic instruments. The galvanometer is basically a coil of wire between the poles of a magnet. When a current passes through the coil, the magnetic field exerts a torque on it and causes it to rotate. (See Fig. 27-17.) The amount of rotation is proportional to the current and may be read on a calibrated scale through the deflection of a needle. The circuit representations of a galvanometer, an ammeter, and a voltmeter are shown in Fig. 27-18.

*The galvanometer will be discussed in detail in Sec. 28-5.

The important characteristics of a galvanometer are its resistance and the current required to deflect its needle full-scale. Typically, a galvanometer has a resistance between 10 and 1000 Ω, and the current required to produce full-scale deflection is on the order of 10^{-4} to 10^{-6} A.

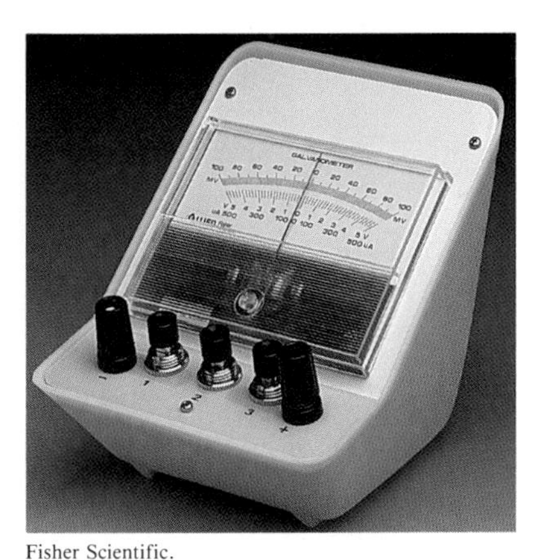

A student galvanometer. A current of 500 μA deflects the needle full-scale. Its internal resistance is 200 Ω.

Fisher Scientific.

AMMETER

When an ammeter is inserted in the branch of a circuit, as shown in Fig. 27-19*a*, it measures the current flowing along that branch. An ammeter basically consists of a galvanometer and a "shunt resistor" connected in parallel, as illustrated in Fig. 27-19*b*. The coil of the galvanometer has an internal resistance that we represent by R_g. Typically, $R_g \gg R$, the shunt resistance. By choosing the appropriate value for R, we can convert a galvanometer that measures small currents to an ammeter capable of measuring currents of several amperes.

Suppose we want full-scale deflection in the galvanometer to correspond to a current I_f in the branch of the circuit containing the ammeter. If a current $(I_g)_f$ deflects the galvanometer full-scale, then with a current I_f in the branch, a current $I_f - (I_g)_f$ flows through the shunt resistor. Figure 27-19*b* shows the branching of the currents in the ammeter. Since the shunt resistor and the galvanometer are connected in parallel,

$$(I_g)_f R_g = [I_f - (I_g)_f]R,$$

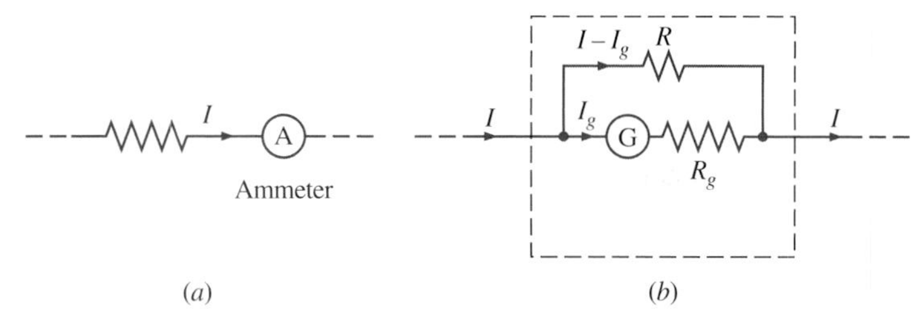

FIGURE 27-19 (*a*) An ammeter measures the current flowing through the branch of a circuit into which it is inserted. (*b*) An ammeter basically consists of a resistor in parallel with a galvanometer. Typically $R_g \gg R$. For full-scale deflection, $I = I_f$ and $I_g = (I_g)_f$.

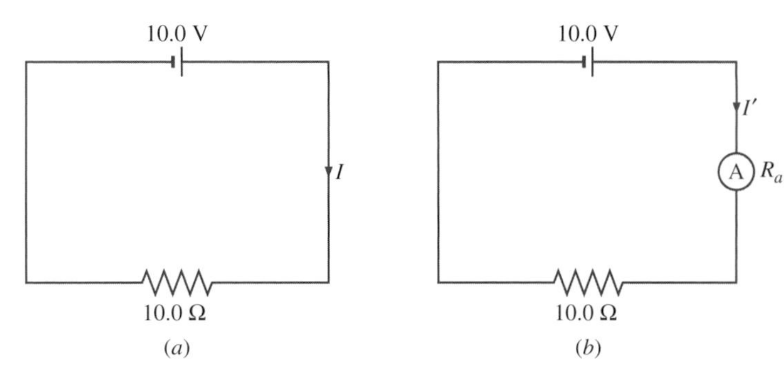

FIGURE 27-20 When an ammeter is inserted into a circuit, it changes the current flowing in the circuit because of its internal resistance. To minimize this effect, the internal resistance of an ammeter should be very low.

so the resistance of the shunt resistor must be

$$R = \frac{(I_g)_f R_g}{I_f - (I_g)_f}. \tag{27-3}$$

For a current $I < I_f$ in the branch, the current I_g through the galvanometer is less than $(I_g)_f$. We then have

$$I_g R_g = (I - I_g)R,$$

so

$$I = \frac{R + R_g}{R} I_g.$$

Thus the current in the branch is directly proportional to the current through the galvanometer, and the scale on the galvanometer can be calibrated directly in terms of the current I in the branch, with a range from 0 to I_f.

EXAMPLE 27-8 Designing an Ammeter

A galvanometer has an internal resistance of 20.0 Ω, and a 1.00-mA current deflects it full-scale. (*a*) What shunt resistance is needed to convert this galvanometer to an ammeter that reads full-scale when the current passing through it is 1.00 A? (*b*) What is the net resistance of the ammeter?

SOLUTION (*a*) From Eq. (27-3), the resistance of the shunt resistor is

$$R = \frac{(I_g)_f R_g}{I_f - (I_g)_f} = \frac{(1.00 \times 10^{-3}\ \text{A})(20.0\ \Omega)}{1.00\ \text{A} - 0.00100\ \text{A}} \approx 2.00 \times 10^{-2}\ \Omega.$$

(*b*) The net resistance R_a of the ammeter is found from

$$\frac{1}{R_a} = \frac{1}{2.00 \times 10^{-2}\ \Omega} + \frac{1}{20.0\ \Omega},$$

and to three significant figures $R_a = 2.00 \times 10^{-2}\ \Omega$. This resistance is much lower than the resistance of the galvanometer. In fact, *an ammeter must have a low resistance in order to make accurate measurements*. As an example, consider the circuit of Fig. 27-20*a* in which the current in the single branch is $I = (10.0\ \text{V})/(10.0\ \Omega) = 1.00$ A. Suppose that we wish to measure the current with an ammeter as shown in Fig. 27-20*b*. The insertion of the ammeter adds a resistance R_a to the circuit. As a result, the current in the branch is now

$$I' = \frac{10.0}{10.0 + R_a}\ \text{A}.$$

If the resistance of the ammeter were 2.0 Ω, the measured current would be

$$I' = \frac{10.0}{10.0 + 2.0}\ \text{A} = 0.83\ \text{A},$$

which is 17 percent lower than its value when the ammeter is not in the circuit. However, if $R_a = 0.10\ \Omega$, the ammeter would measure a current of

$$I' = \frac{10.0}{10.0 + 0.10}\ \text{A} = \frac{10.0}{10.1}\ \text{A} = 0.990\ \text{A},$$

which differs from the current without the ammeter by only 1.0 percent. Thus to minimize its effect on the current, an ammeter must have a very low resistance.

DRILL PROBLEM 27-8

(*a*) Given a galvanometer with an internal resistance of 40.0 Ω and a full-scale deflection of 8.00×10^{-4} A, design an ammeter that will read full-scale when the current is 10.0 A. (*b*) What is the resistance of the ammeter?
ANS. (*a*) Add a shunt resistance of $3.20 \times 10^{-3}\ \Omega$; (*b*) $3.20 \times 10^{-3}\ \Omega$.

DRILL PROBLEM 27-9

An ammeter is used to measure the current through the battery in the circuit of Fig. 27-20. If the current is to be within 5.0 percent of its value before the ammeter is inserted into the circuit, what is the maximum resistance that the ammeter can have?
ANS. 0.5 Ω.

Voltmeter

When a voltmeter is connected across two points in a circuit (see Fig. 27-21*a*), it measures the potential difference between those points. A voltmeter is constructed by adding a resistance

R in series with a galvanometer, as shown in Fig. 27-21b. Like the shunt resistor in an ammeter, this resistance limits the current through the galvanometer coil. Suppose, for example, that a galvanometer has an internal resistance of 20 Ω and a full-scale deflection current of 1.0 mA. The deflection of its needle will be off-scale if the voltage across the galvanometer exceeds $(1.0 \times 10^{-3}\ \text{A})\ (20\ \Omega) = 0.020$ V. Hence this galvanometer used by itself cannot measure potential differences larger than 0.020 V. However, in a voltmeter, most of the potential difference is across the series resistor because $R \gg R_g$, and the potential difference across the galvanometer stays under the 0.020-V maximum.

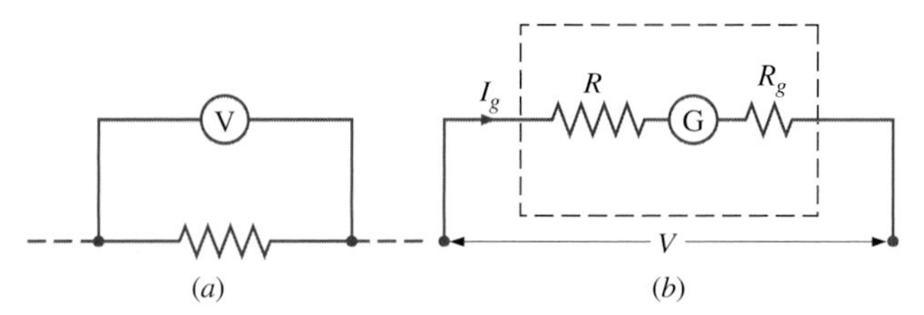

FIGURE 27-21 (a) A voltmeter that measures the potential difference between two points is connected across those points in a circuit. (b) A voltmeter basically consists of a resistor in series with a galvanometer. Typically, $R \gg R_g$.

Suppose we want a full-scale deflection current $(I_g)_f$ in the galvanometer when the voltmeter of Fig. 27-21b is connected across a potential difference V_f. Since R and R_g are in series, we have

$$V_f = (I_g)_f R + (I_g)_f R_g.$$

Thus the resistance we put in series with the galvanometer must be

$$R = \frac{V_f}{(I_g)_f} - R_g. \tag{27-4}$$

For a potential difference $V < V_f$, the current I_g through the galvanometer is less than $(I_g)_f$, and

$$V = I_g(R + R_g).$$

The potential difference is therefore proportional to I_g, and the scale of the galvanometer can be calibrated directly in volts, with a range from 0 to V_f.

EXAMPLE 27-9 Designing a Voltmeter

A galvanometer has an internal resistance of 10.0 Ω and a full-scale deflection of 1.00 mA. (a) What series resistance must be added to convert this galvanometer to a voltmeter that has a full-scale reading of 10.0 V? (b) What is the net resistance of the voltmeter?

SOLUTION (a) The required value of R is such that when a current of 1.00 mA flows through the galvanometer, the potential difference V across the voltmeter is 10.0 V. From Eq. (27-4),

$$R = \frac{V_f}{(I_g)_f} - R_g = \frac{10.0\ \text{V}}{1.00 \times 10^{-3}\ \text{A}} - 10.0\ \Omega,$$

which, to three significant figures, gives $R = 1.00 \times 10^4\ \Omega$.

(b) Since $R \gg R_g$, the net resistance of the voltmeter $(R + R_g)$ is, to three significant figures, equal to R, or $1.00 \times 10^4\ \Omega$. The voltmeter resistance is therefore much higher than the internal resistance of the galvanometer. More importantly, the voltmeter must be a high-resistance device if its effect on the electrical characteristics of the circuit into which it is inserted is to be negligible.

As an example, consider the circuit of Fig. 27-22a. The current I through the 50-Ω resistor is 0.100 A. The potential difference across this resistor is (0.100 A) (50 Ω) = 5.0 V, and the potential difference across the 100-Ω resistor is (0.100 A) (100 Ω) = 10.0 V. To measure the potential difference across the 50-Ω resistor, we connect the voltmeter as shown in Fig. 27-22b. Suppose that the voltmeter has a resistance $R_v = 500\ \Omega$. Then the net resistance of the parallel combination of R_v and the 50-Ω resistor is 45 Ω, and the total circuit resistance becomes 145 Ω. The current through the battery is no longer 0.100 A, but is instead

$$I' = \frac{15.0}{145}\ \text{A} = 0.103\ \text{A}.$$

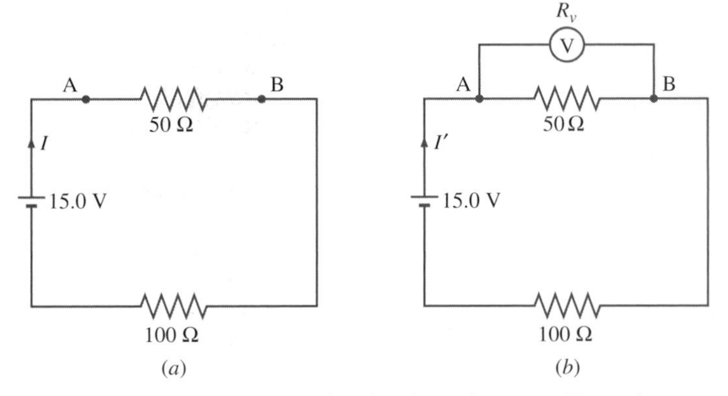

FIGURE 27-22 When used in a circuit, the voltmeter affects the potential difference between the points across which it is connected. To minimize this effect, the resistance of a voltmeter should be very high.

The voltage across the 50-Ω resistor as well as the voltmeter is therefore

$$(0.103\ \text{A})(45\ \Omega) = 4.7\ \text{V},$$

which is about 6 percent less than the 5.0-V potential difference across the 50-Ω resistor prior to the insertion of the voltmeter.

However, if the voltmeter has an internal resistance of $1.00 \times 10^4\ \Omega$, the net resistance in the circuit is 150 Ω, the current I' is 0.100 A, and the potential difference across the 50-Ω resistor becomes

$$15.0\ \text{V} - (0.100\ \text{A})(100\ \Omega) = 5.0\ \text{V},$$

which is the same as the voltage across the 50-Ω resistor without the voltmeter in place. *The higher the internal resistance of a voltmeter, the more accurate are its measurements.* The reason is that the voltmeter is always connected in parallel with some resistance r in a circuit. If its internal resistance is very large compared with r, the resistance of the parallel combination will be approximately equal to r itself (see Example 27-3), so the insertion of the voltmeter will have a negligible effect on the circuit.

DRILL PROBLEM 27-10

Consider the circuit of Fig. 27-22*b*. If the voltmeter measures the potential difference across the 50-Ω resistor to be 4.9 V, what must be the internal resistance of the voltmeter?
ANS. 1.7×10^3 Ω.

By a turn of the dial, the multimeter becomes a voltmeter, an ammeter, or an ohmmeter (a resistance-measuring device). Reproduced with permission.

*POTENTIOMETER

The *potentiometer* is a simple circuit used to measure potential differences. The current drawn by the potentiometer during measurements is so small that it acts like a voltmeter of essentially infinite resistance. Despite this fact, the potentiometer is used far less frequently than the voltmeter to make measurements in everyday circuits. The voltmeter can be used with much greater ease and convenience, and its resistance can usually be made large enough so that its effect on a circuit is negligible. Nevertheless, there are special applications for which effectively no current may be drawn from the device across which a potential difference must be measured. In such cases, a potentiometer, or an "infinite-resistance" meter, must be used for the measurement.

The essential features of the potentiometer are shown in Fig. 27-23. The battery of emf $\mathcal{E}_0$ is connected in series with a resistance that can be varied to produce a convenient voltage

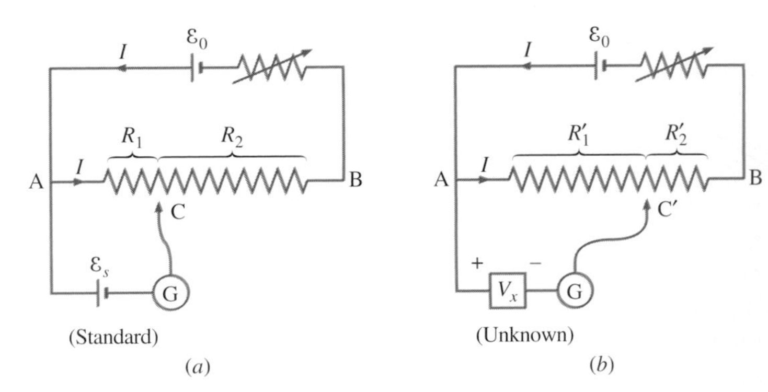

FIGURE 27-23 (*a*) The potentiometer is calibrated using a standard emf $\mathcal{E}_s$. (*b*) An unknown voltage V_x is measured by finding the contact point C′ that corresponds to null deflection of the galvanometer.

between A and B. Also connected between these two points is a resistor that has a contact point C so that its resistance can be divided into two parts, R_1 and R_2. In the common student potentiometer, the resistor is a long, uniform wire with a sliding contact that makes the connection at C. Other components of the circuit are a galvanometer G and a precise standard source of emf $\mathcal{E}_s$. The unknown potential difference to be measured is V_x.

Before being used to determine an unknown potential difference, the potentiometer must be calibrated. As shown in Fig. 27-23*a*, this is accomplished by moving the contact until there is no current through the galvanometer. Then

$$\mathcal{E}_s = IR_1, \qquad (i)$$

where I is the current flowing through the resistor from A to B.

Now an unknown voltage V_x is inserted in place of $\mathcal{E}_s$ and the contact is moved to a new point C′ where again no current flows through the galvanometer, as illustrated in Fig. 27-22*b*. In this case, we have

$$V_x = IR_1'. \qquad (ii)$$

The current I is the same in Eqs. (*i*) and (*ii*), so dividing Eq. (*ii*) by Eq. (*i*) and rearranging terms, we find

$$V_x = \mathcal{E}_s \frac{R_1'}{R_1}, \qquad (iii)$$

which gives the unknown V_x in terms of the standard $\mathcal{E}_s$ and the ratio of the two known resistances, R_1' and R_1. Since the resistance of a uniform wire is proportional to its length [see Eq. (26-8)], we have for the student potentiometer, $R_1' = kl_1'$, $R_1 = kl_1$, and

$$V_x = \mathcal{E}_s \frac{l_1'}{l_1}. \qquad (iv)$$

Hence with knowledge of the standard emf $\mathcal{E}_s$ and careful measurements of the lengths l_1 and l_1', the unknown potential difference V_x can be determined precisely (usually, with an accuracy of about 10^{-3} percent).

Note that when the potentiometer circuit is balanced, no current flows through $\mathcal{E}_s$ or V_x; consequently, their internal resistances are unimportant. Also, the emf $\mathcal{E}_0$ of the battery is of no concern, provided it is greater than either $\mathcal{E}_s$ or V_x so that the potentiometer can be balanced. Furthermore, the galvanometer does not have to be calibrated accurately because it has only to show a *null reading*, indicating that no current flows through the galvanometer.

*Wheatstone Bridge

The circuit of Fig. 27-24*a* is known as a *Wheatstone bridge*, named after its inventor Charles Wheatstone (1802–1875). The resistances are adjusted until there is no current flowing through the galvanometer. At that point, the four resistances are related

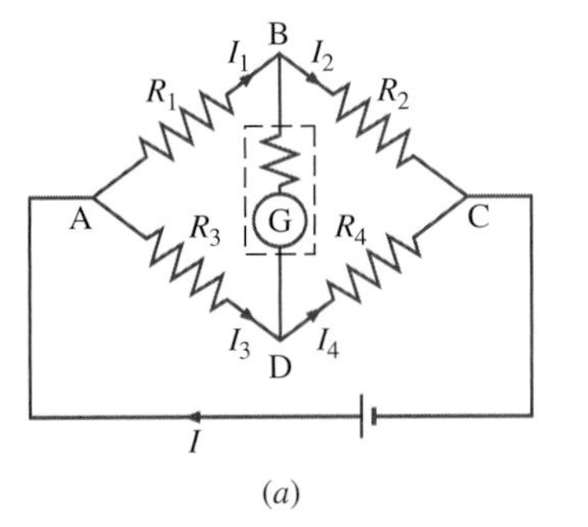

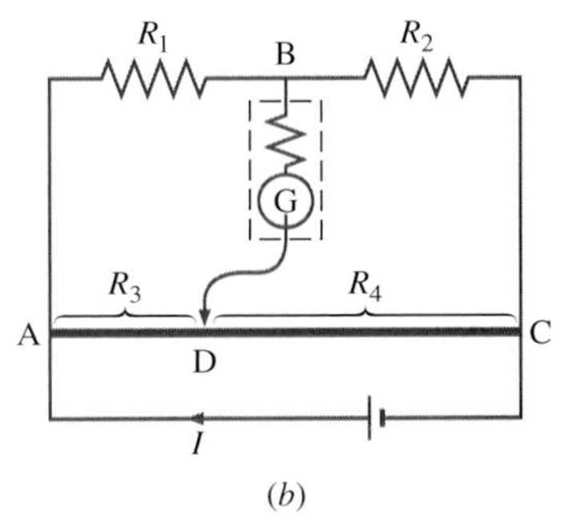

FIGURE 27-24 (*a*) The network of resistances in a Wheatstone bridge. (*b*) A practical form of the Wheatstone bridge.

by an equation that can be used to determine any one of them in terms of the other three. This relationship can be obtained as follows. With no current flowing through the galvanometer, points B and D must be at the same potential; that is

$$V_B = V_D.$$

We may then write

$$V_A - V_B = V_A - V_D,$$

or

$$I_1 R_1 = I_3 R_3. \qquad (i)$$

Similarly,

$$V_B - V_C = V_D - V_C,$$

so

$$I_2 R_2 = I_4 R_4. \qquad (ii)$$

Since there is no current flowing through the galvanometer,

$$I_1 = I_2 \qquad (iii)$$

and

$$I_3 = I_4. \qquad (iv)$$

Dividing Eq. (*i*) by Eq. (*ii*), we find

$$\frac{I_1}{I_2}\frac{R_1}{R_2} = \frac{I_3}{I_4}\frac{R_3}{R_4}, \qquad (v)$$

which, from Eqs. (*iii*) and (*iv*), reduces to

$$\frac{R_1}{R_2} = \frac{R_3}{R_4}. \qquad (vi)$$

A practical form of the Wheatstone circuit is shown in Fig. 27-24*b*. R_1 is an unknown resistance whose value we wish to determine, R_2 is a standard resistor whose resistance is accurately known, and R_3 and R_4 are the resistances of sections of a long, uniform wire AC to which is attached a movable contact D. This contact is adjusted until no current flows through the galvanometer. At this point, the Wheatstone bridge is said to be "balanced" and R_1 may be calculated from Eq. (*vi*), with R_3/R_4 given by the ratio of the wire lengths corresponding to AD and DC, respectively. Note that the condition for balance is independent of the battery, for it does not depend on either the battery's emf or its internal resistance.

DRILL PROBLEM 27-11

When the standard $\mathcal{E}_s$ in Fig. 27-23*a* is a mercury cell of emf 1.01860 V, a null reading is obtained at $l_1 = 19.23$ cm. With an unknown voltage V_x, the null reading is found to be at $l_1' = 8.27$ cm. What is the value of V_x?
ANS. 0.438 V.

DRILL PROBLEM 27-12

Suppose that the Wheatstone bridge of Fig. 27-24*a* is unbalanced and a current flows from B to D. If $I_1 = I_3 = 0.100$ A, (*a*) which is greater, R_1 or R_3? (*b*) If $R_1 = 0.99R_3$, and the internal resistance of the galvanometer is R_3, what is the current flowing between B and D?
ANS. (*a*) R_3; (*b*) 0.001 A.

SUMMARY

1. **Kirchhoff's rules**
   (*a*) The sum of the currents entering a junction is equal to the sum of the currents leaving the junction.
   (*b*) The net change in the electric potential around a closed loop in an electric circuit is zero.
2. **Resistors in series and in parallel**
   (*a*) The net resistance R_S of an arbitrary number of resistors in series is the sum of their individual resistances:

$$R_S = R_1 + R_2 + R_3 + \ldots$$

   (*b*) The net resistance R_P of an arbitrary number of resistors in parallel is related to their individual resistances by

$$\frac{1}{R_P} = \frac{1}{R_1} + \frac{1}{R_2} + \frac{1}{R_3} + \ldots$$

*3. **Household wiring**
Appliances in a household circuit are connected in parallel. A circuit breaker limits the total current in the circuit. To protect against dangerous shocks, the metal cases of many household appliances are grounded.
4. **Electrical measurements in dc circuits**
   (*a*) The ammeter is a low-resistance device that measures current.
   (*b*) The voltmeter is a high-resistance device that measures potential difference.
   *(*c*) The potentiometer measures potential differences without drawing current.
   *(*d*) The Wheatstone bridge is used to measure resistance.

QUESTIONS

27-1. Discuss the consequences of connecting identical batteries (*a*) in series and (*b*) in parallel across a given resistance.

27-2. Why are the batteries in a two-cell flashlight connected in series? Is there any advantage to arranging them in parallel?

27-3. Can the potential difference between the terminals of a battery ever be opposite in direction to the emf?

27-4. A high resistance and a low resistance are connected in parallel. Is the net resistance closer in value to the high or the low resistance?

27-5. Answer the previous question for the two resistances connected in series.

27-6. Is there any connection between Kirchhoff's second rule and $\oint \mathbf{E} \cdot d\mathbf{l} = 0$?

27-7. In order that they not quickly burn out, would you connect twenty 6-V bulbs in series or in parallel with a 120-V source?

27-8. A 50-W, 110-V bulb glows normally when connected across a 110-V source, while a 1000-W, 110-V bulb glows only dimly when connected across the same source. What does this tell you about the source?

27-9. (*a*) How would you connect two filaments to make a three-way lightbulb that burns at 50, 100, and 150 W? (*Hint:* Assume that one filament has twice the resistance of the other.) (*b*) Sketch how a three-way switch is connected to the filaments so that all three settings of the bulb can be used.

27-10. Are the lights in an automobile connected in series or in parallel?

27-11. Devise a method for accurately measuring the emf and internal resistance of a battery.

27-12. Explain why electricians move potentially live wires with the backs of their fingers.

27-13. Is the internal resistance of the galvanometer important in either the Wheatstone bridge or the potentiometer? (*Hint:* Consider the sensitivities of the instruments.)

27-14. The instructions to a string of Christmas lights you just bought say that when one bulb is burned out, the others will remain lit. What can you infer about how the bulbs are connected?

27-15. Why do lights on a car dim when the engine is started?

27-16. Explain why lights in a small workshop may dim when a large power saw is turned on.

27-17. Are fuses wired in series or in parallel with the devices they protect?

27-18. Why is an ungrounded electric appliance dangerous when used outside?

27-19. If the current flowing through a body determines the seriousness of an electric shock, why do we see high-voltage rather than high-current warnings?

27-20. For measuring potential differences, what are the advantages and disadvantages of using a potentiometer rather than a high-resistance voltmeter?

27-21. Discuss the similarities and differences between a voltmeter and ammeter.

27-22. What is the resistance of an ideal ammeter? of an ideal voltmeter?

27-23. If the battery of a Wheatstone bridge deteriorates and its terminal voltage drops, can the bridge still be used to determine unknown resistances accurately?

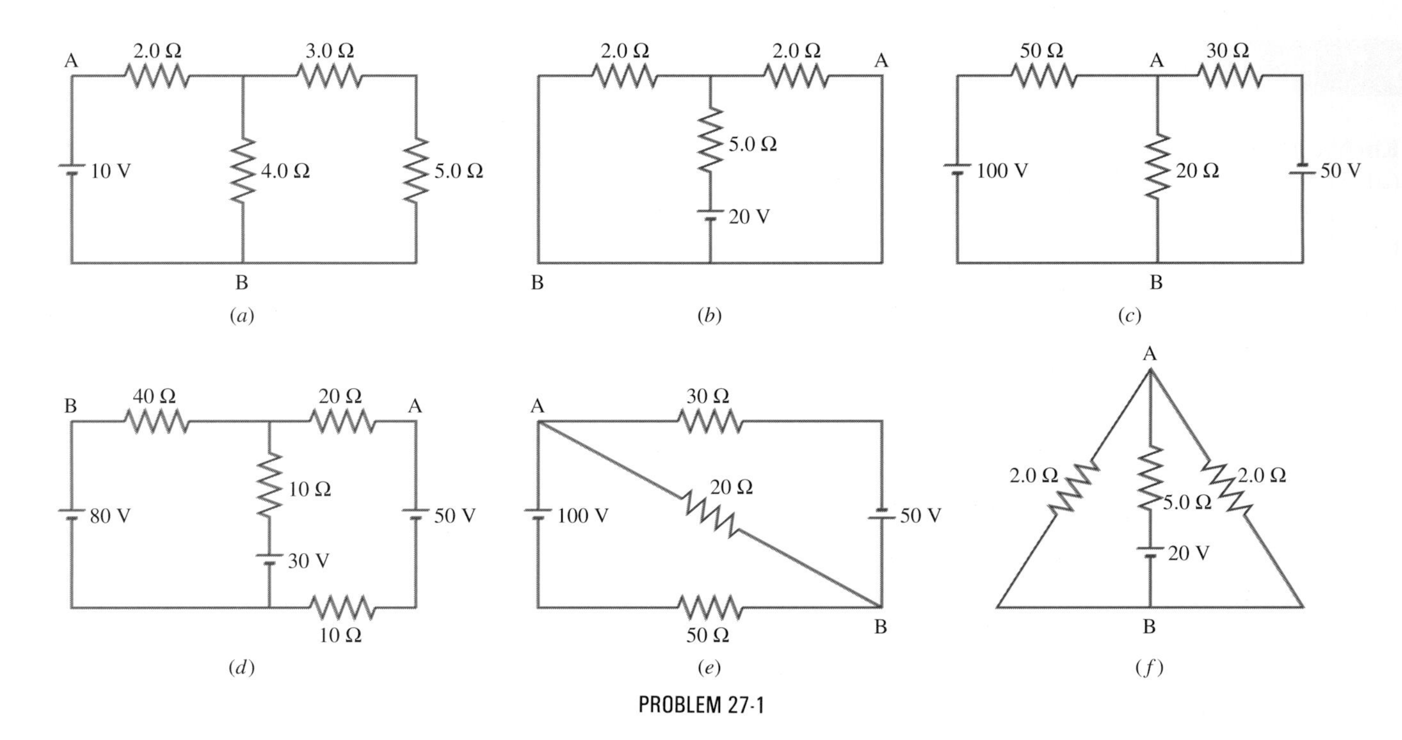

PROBLEM 27-1

Problems

Kirchhoff's Rules

27-1. Find the current in each branch of the circuits shown in the accompanying figure.

27-2. Calculate the potential difference between points A and B in each of the circuits of Prob. 27-1.

27-3. A current of 2.0 A flows from M to N through the section of circuit shown in the accompanying figure. (*a*) If 100 W is absorbed between M and N, what is the potential difference between these two points? (*b*) What is $\mathcal{E}$?

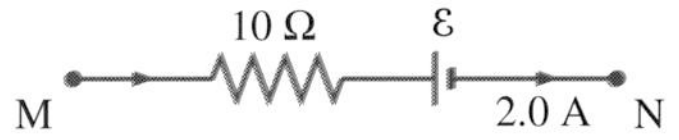

PROBLEM 27-3

27-4. For the circuit shown in the accompanying figure, calculate the potential difference between A and B and between A and C.

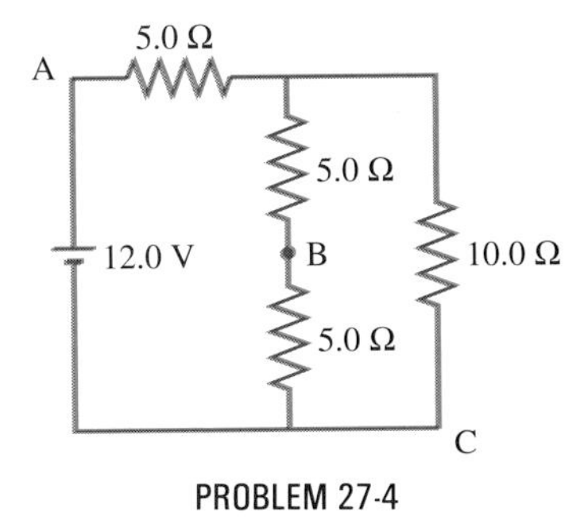

PROBLEM 27-4

27-5. Two sources of emf supply the currents indicated to resistances R_1 and R_2 in the accompanying figure. Determine all unknown branch currents and the voltages across R_1 and R_2.

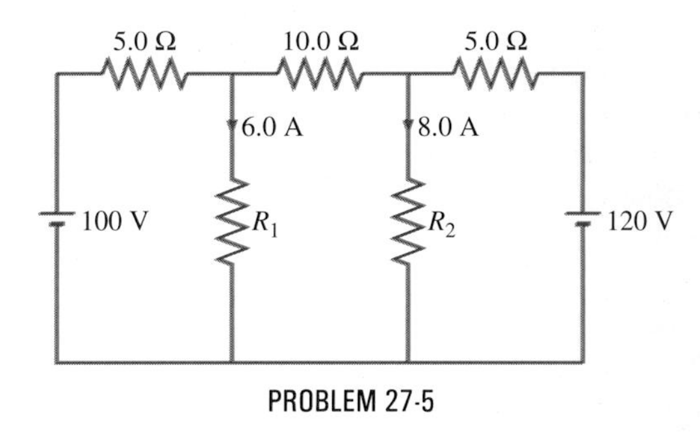

PROBLEM 27-5

27-6. For the circuit shown in the accompanying figure, calculate (*a*) the current through each resistor, (*b*) the power dissipated in each resistor, and (*c*) the power delivered by each source of emf.

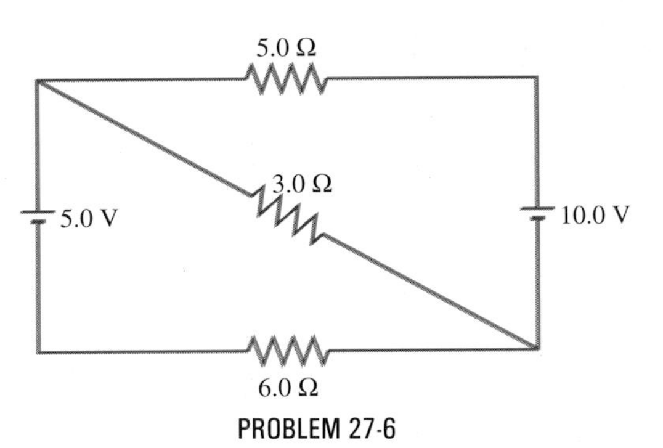

PROBLEM 27-6

27-7. If $V_B - V_A = 20$ V for the circuit branch shown in the accompanying figure, what is the current I?

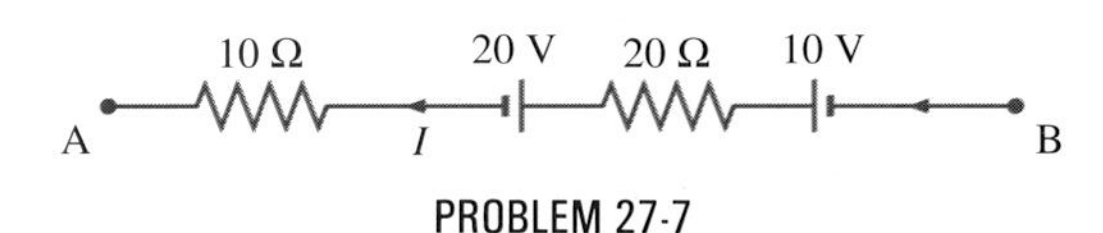

PROBLEM 27-7

27-8. For the circuit shown in the accompanying figure, calculate (*a*) the current through the 20-V source of emf, (*b*) the potential difference across the 20-Ω resistor, and (*c*) the current through the 40-Ω resistor.

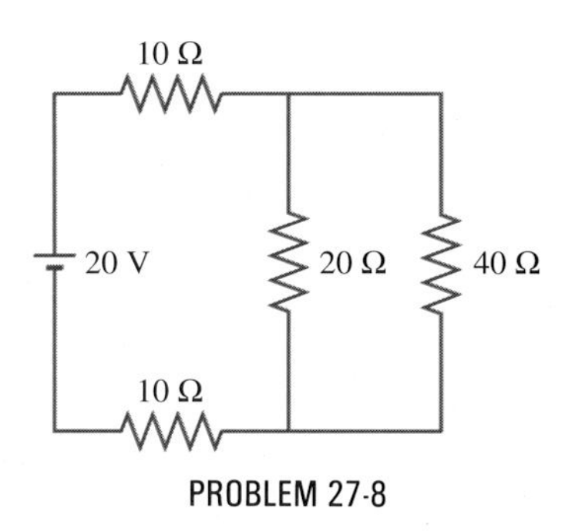

PROBLEM 27-8

27-9. The two resistors in the circuit shown in the accompanying figure dissipate 60 and 150 W, respectively, when placed across a 120-V source. (*a*) What is the value of each resistance? (*b*) What is the current output of the source?

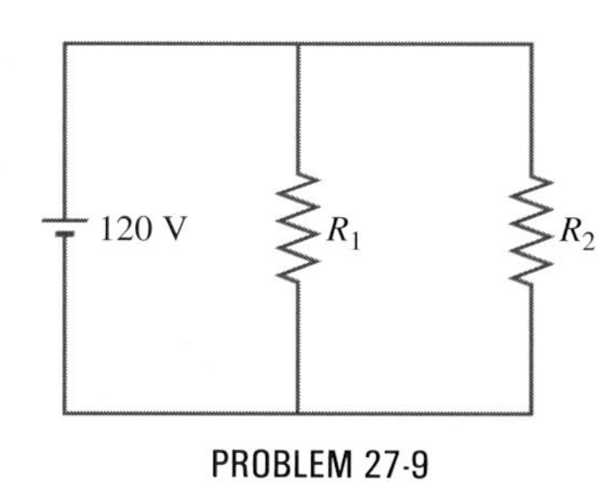

PROBLEM 27-9

27-10. Suppose the 120-V source in Prob. 27-9 is replaced by a 60-V source. (*a*) What is the power dissipated in each resistor? (*b*) What is the current output of the source?

27-11. In the network shown in the accompanying figure, the current through the 5.0-Ω resistor is 0.50 A. Calculate (*a*) the currents through the 15-V and the 50-V sources and (*b*) the potential difference $V_B - V_A$.

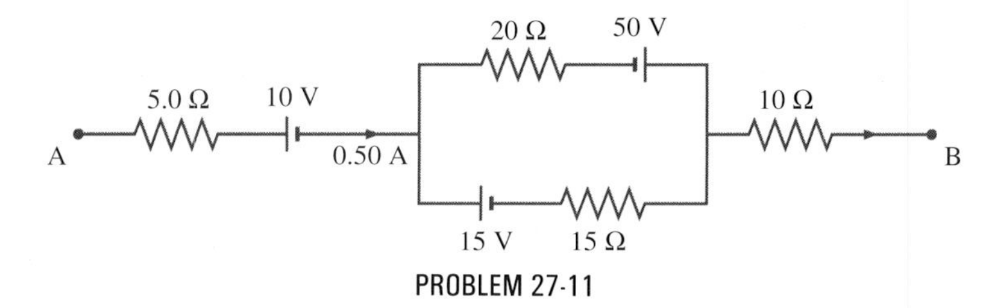

PROBLEM 27-11

27-12. (*a*) Find the current in each branch of the circuit shown in the accompanying figure. (*b*) Find the power produced (or absorbed) by each source of emf. (*c*) Find the potential difference across the terminals of each battery.

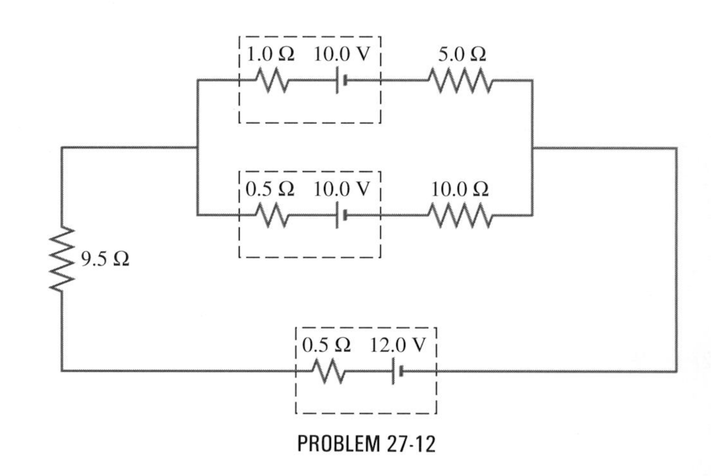

PROBLEM 27-12

27-13. Find the current in each branch of the circuit shown in the accompanying figure.

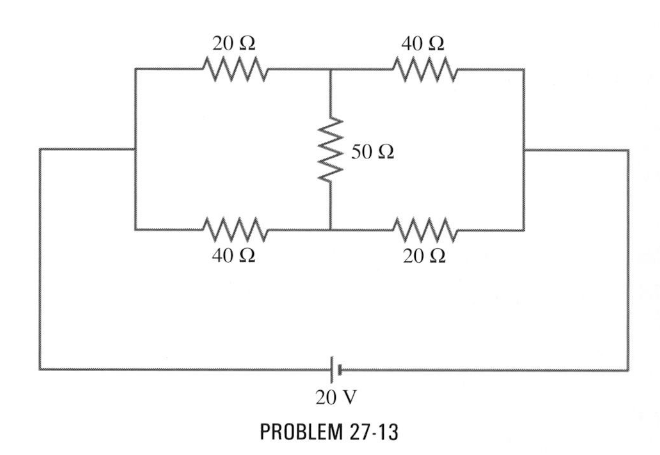

PROBLEM 27-13

27-14. For the circuit shown in the accompanying figure, determine the current through the 12.0-V battery and the potential difference between points A and B.

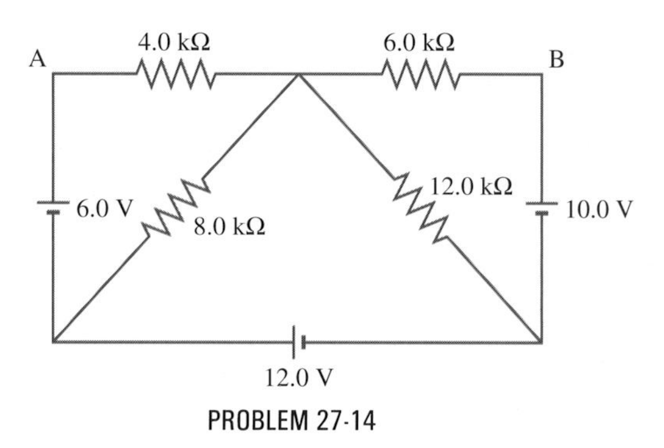

PROBLEM 27-14

27-15. Find (a) the currents in all branches of the circuit shown in the accompanying figure and (b) the power output of each battery.

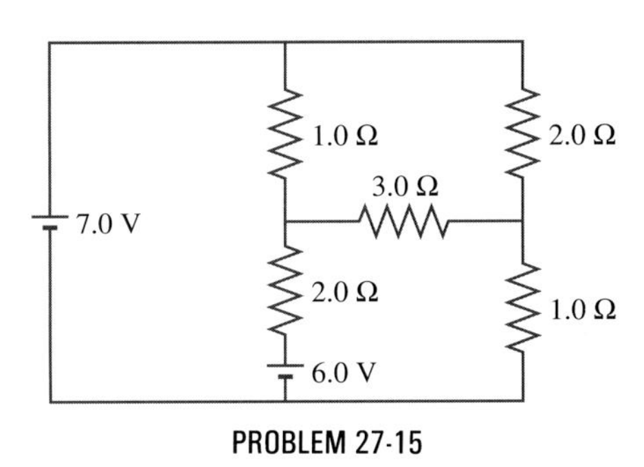

PROBLEM 27-15

27-16. What is the reading of the ammeter in the circuit shown in the accompanying figure? Express your answer in terms of $\mathcal{E}$ and R.

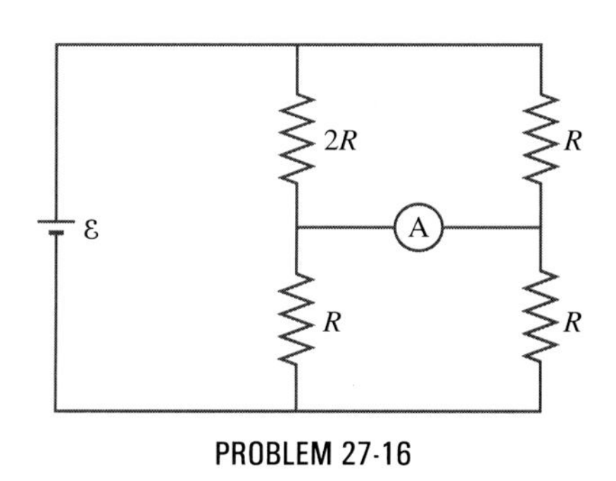

PROBLEM 27-16

Resistor Combinations

27-17. You are given ten 1.0-Ω resistors and a 10-V source. How would you connect the resistors so that the current flowing through the source is a maximum? a minimum? What are the values of the maximum and the minimum currents?

27-18. Find the equivalent resistance between A and B for the combinations of resistors shown in the accompanying figure.

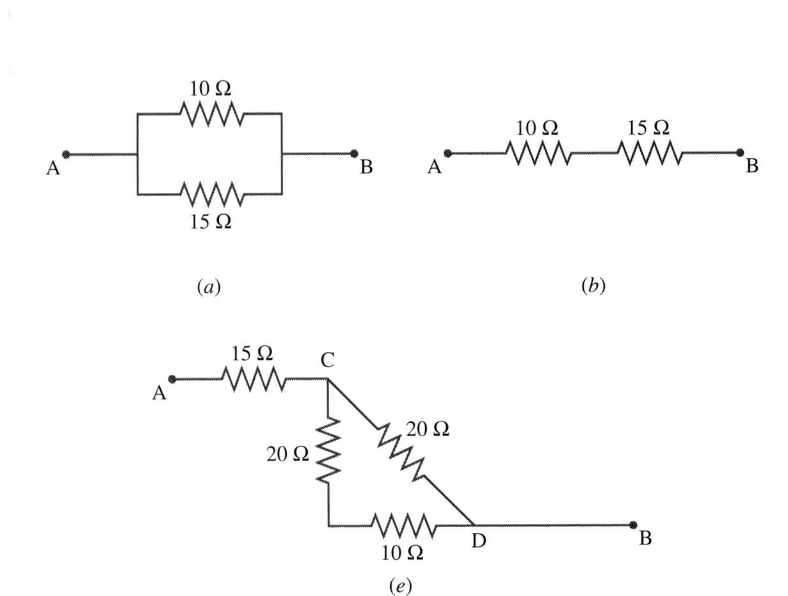

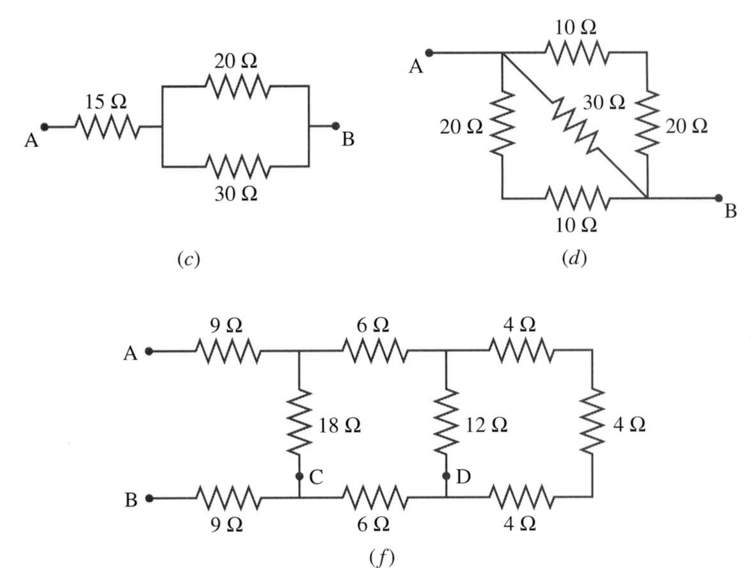

PROBLEM 27-18

27-19. Suppose that points C and D are connected by a wire in circuits (e) and (f) of Prob. 27-18. Determine the new values of the net resistances.

27-20. Show how a 500-Ω, 2-W resistor can be made using several 500-Ω, 1-W resistors.

27-21. What resistor R must be added in parallel with the 25.0-Ω resistor in the circuit shown in the accompanying figure so that the potential difference across the terminals of the 5.0-V battery is 4.9 V?

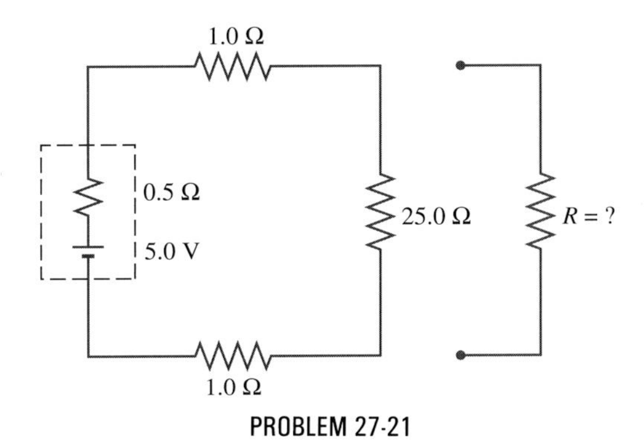

PROBLEM 27-21

27-22. (a) Combine resistances to find the current through the 12-V source of the circuit shown in the accompanying figure. (b) What is the current through the 10-Ω resistor? (c) What is the potential difference across the 60-Ω resistor?

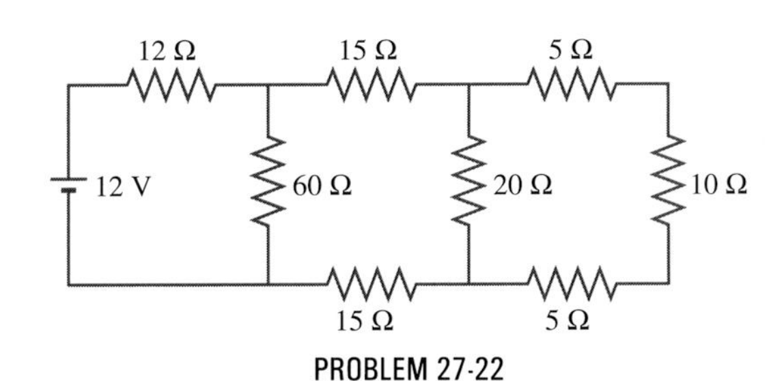

PROBLEM 27-22

27-23. Find the equivalent resistance between A and B for each of the networks shown in the accompanying figure.

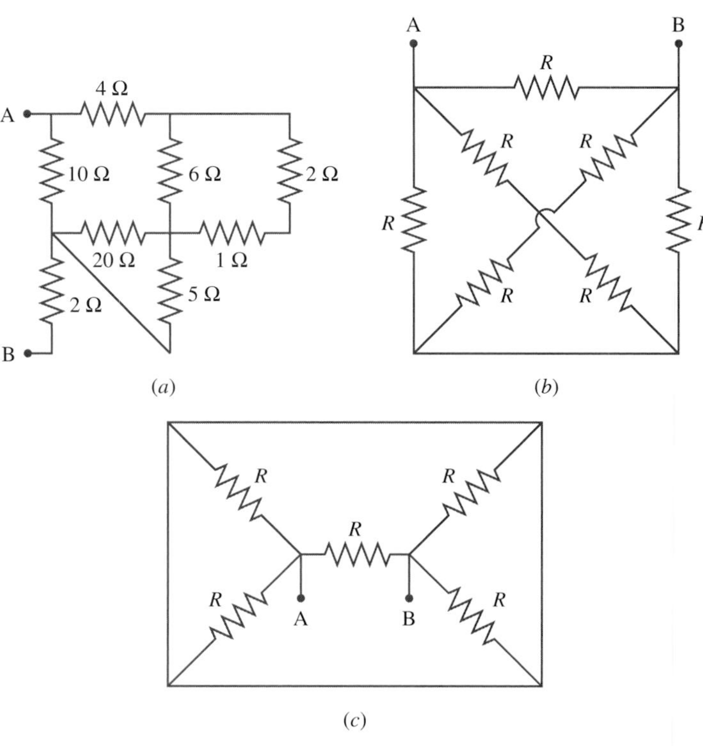

PROBLEM 27-23

27-24. If $r = 1.0\ \Omega$ in the network shown in the accompanying figure, what is the resistance between A and B?

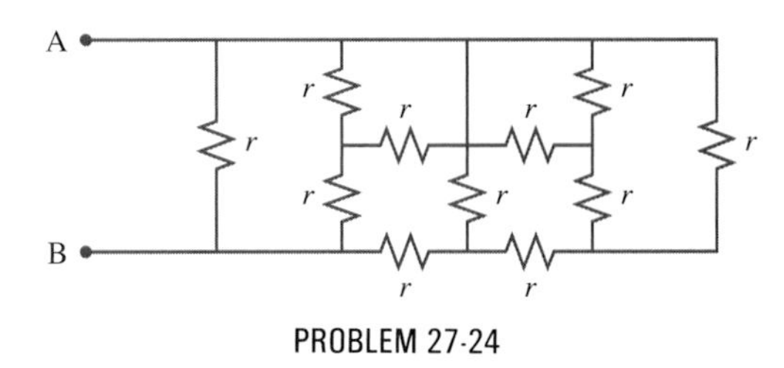

PROBLEM 27-24

27-25. Simplify the circuit shown in the accompanying figure by combining resistances and then calculate (*a*) the current through the 10-Ω resistor and (*b*) the potential difference across the 20-Ω resistor.

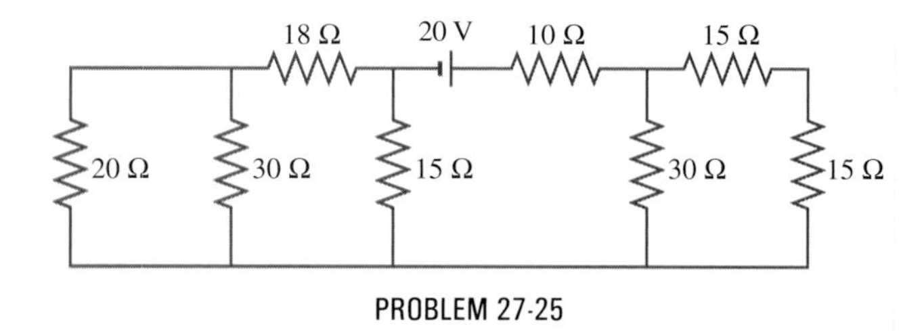

PROBLEM 27-25

27-26. (*a*) Combine resistances to simplify as much as possible the circuit shown in the accompanying figure. (*b*) Find the currents through the two batteries. (*c*) Determine the potential difference between A and B.

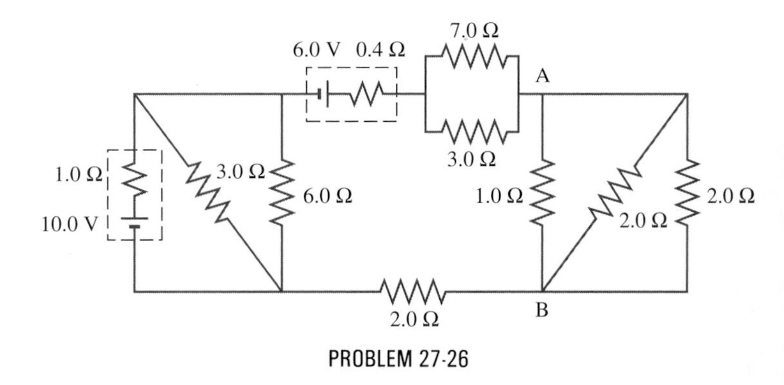

PROBLEM 27-26

Household Wiring

27-27. A 120-V power line is protected by a 15-A fuse. How many 100-W bulbs can be connected along this line without exceeding the rating of the fuse?

27-28. It is found that while six 200-W lightbulbs are lit when connected to a 120-V power line a seventh bulb causes the circuit breaker to open. Based on this information, what can you say about the capacity of the circuit breaker?

27-29. Suppose you connect a 1000-W toaster and a 400-W blender to a 120-V line, whose current is limited by a 15-A circuit breaker. (See the accompanying figure.) (*a*) Does the circuit breaker open? (*b*) What happens when a 150-W bulb is added to the line?(*c*) Next, a 500-W frying pan is added. What happens now?

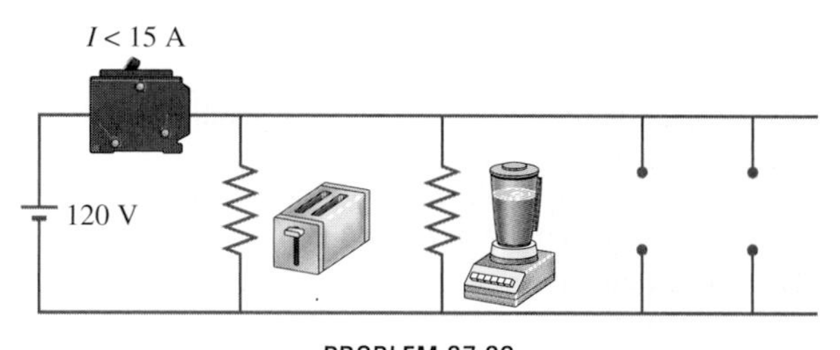

PROBLEM 27-29

Electrical Measurements

27-30. The resistance of a galvanometer is 100 Ω, and a current of 500 μA deflects it full-scale. (*a*) Show how the galvanometer can be used to construct a voltmeter that reads 50 V full-scale. (*b*) Show how the galvanometer can be used to construct an ammeter that reads 2.0 A full-scale.

27-31. A galvanometer has an internal resistance of 100 Ω, and 0.50 mA deflects it full-scale. (*a*) Show how the galvanometer can be used to construct an ammeter that reads from 0 to 2.0 A. (*b*) How can you use the galvanometer to make a voltmeter that reads 150 V full-scale?

27-32. A 12-V battery has an internal resistance of 0.50 Ω. (*a*) What is the reading of a voltmeter of internal resistance 200 Ω that is placed across the terminals of the battery? (*b*) Suppose the terminals of the battery are connected to a 200-Ω resistor. What will the voltmeter read when it is connected across the resistor?

27-33. Suppose that a Wheatstone bridge is balanced with $R_2 = 9.756\ \Omega$, $R_3 = 10.06\ \Omega$, and $R_4 = 15.73\ \Omega$. What is the value of R_1?

27-34. Suppose that the galvanometer used in a potentiometer has an internal resistance of 100 Ω and can detect a minimum current of 2.0×10^{-3} A. What is the maximum uncertainty in measuring an unknown voltage with this potentiometer?

27-35. The voltmeter used in the circuits shown in the accompanying figure has a resistance of 10,000 Ω. What is the voltmeter reading for each case?

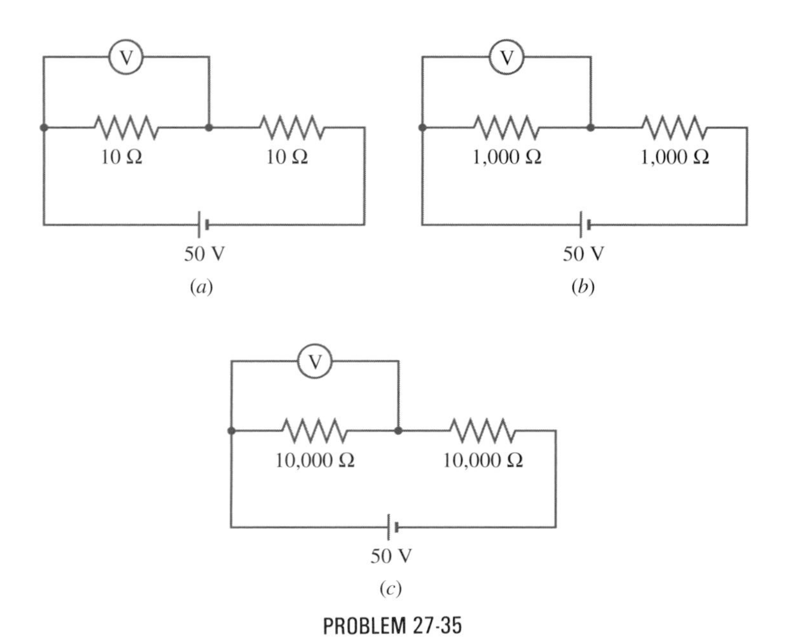

PROBLEM 27-35

27-36. A slide-wire potentiometer is balanced against a 1.0182-V standard cell when it is set at 39.8 cm along a 100-cm wire. When the standard cell is replaced by an unknown emf, the balance setting is at 14.2 cm. What is the value of the unknown emf?

27-37. The accompanying figure shows a Wheatstone bridge with R_1 and R_2 furnished by the sections AC and CB of a slide-wire resistor. When the bridge is balanced, the lengths of AC and CB are 27.3 and 72.7 cm, respectively. What is the value of R_x?

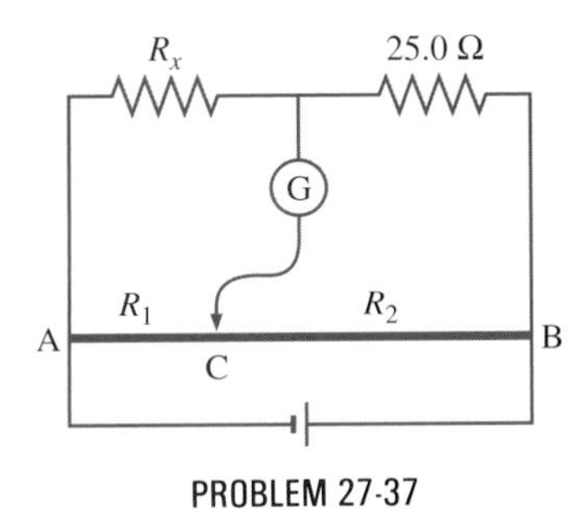

PROBLEM 27-37

27-38. A 200-Ω resistor and a 100-Ω resistor are connected in series across a potential difference of 80 V. (See the accompanying figure.) A voltmeter placed across the 200-Ω resistor reads 50 V. What does the voltmeter read when it is connected across the 100-Ω resistor?

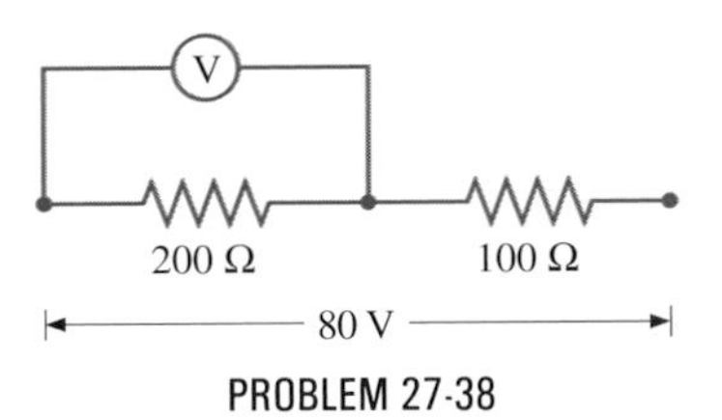

PROBLEM 27-38

27-39. An ohmmeter is a device used to measure resistance. A circuit diagram for a simple ohmmeter is shown in the accompanying figure. A galvanometer with a resistance of 100 Ω and a full-scale deflection of 500 μA is connected to a resistance r and a 1.5-V battery of negligible internal resistance. The value of r is adjusted so that the galvanometer deflects full-scale when the terminals are shorted. The galvanometer scale is then calibrated to read the value of the resistance R connected to the terminals. (*a*) What is the value of r? (*b*) What is the value of R when the deflection is 25, 50, and 75 percent full-scale? (*c*) Is the scale linear?

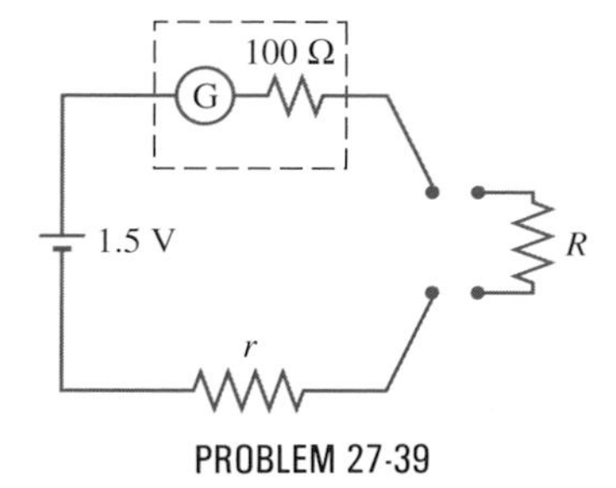

PROBLEM 27-39

27-40. A student measures an unknown resistance R by connecting it in series with a known resistance r, placing the combination across the terminals of a source $\mathcal{E}$, then measuring the voltage across each resistor. (See the accompanying figure.) If $r = 15.0\ \Omega$, $V_r = 5.0$ V, and $V_R = 12.0$ V, what is R?

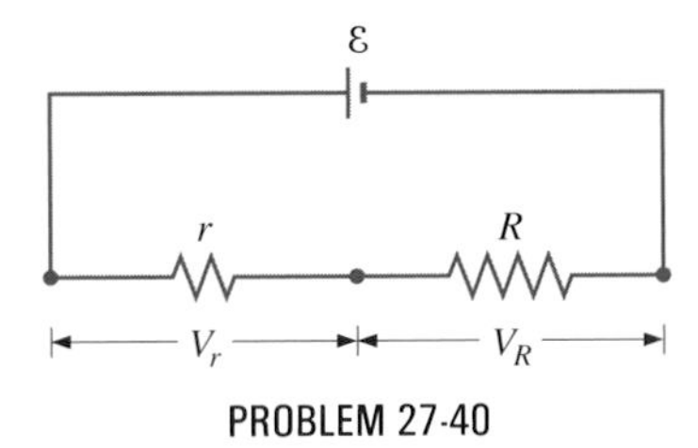

PROBLEM 27-40

27-41. Suppose that in the student potentiometer of Sec. 27-4, $\mathcal{E}_s = 1.437$ V, $l_1 = 10.76$ cm, and $l_1' = 12.23$ cm. What is V_x?

General Problems

27-42. The 2.0-Ω resistor of the circuit shown in the accompanying figure dissipates energy at a rate of 2.0 W. (*a*) What is the power output of each of the two identical sources of emf? (*b*) What is the power output at the terminals of each battery?

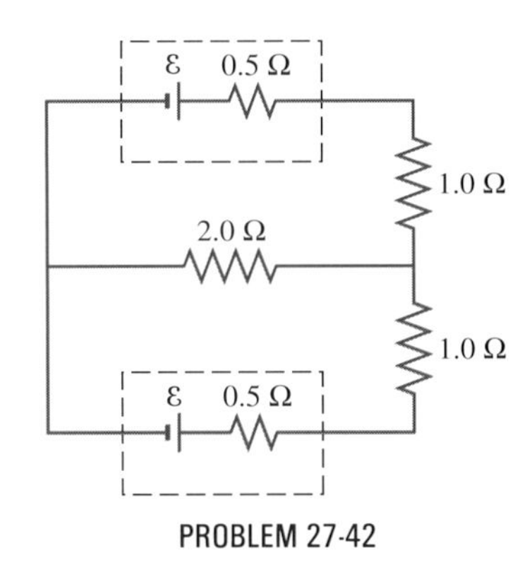

PROBLEM 27-42

27-43. A resistor R is connected, as shown in the accompanying figure, across two batteries of emf ℰ and internal resistance r. What is the relationship between R and r so that maximum power is delivered to R with (*a*) the switch S open and (*b*) the switch S closed? What is the maximum power in each case?

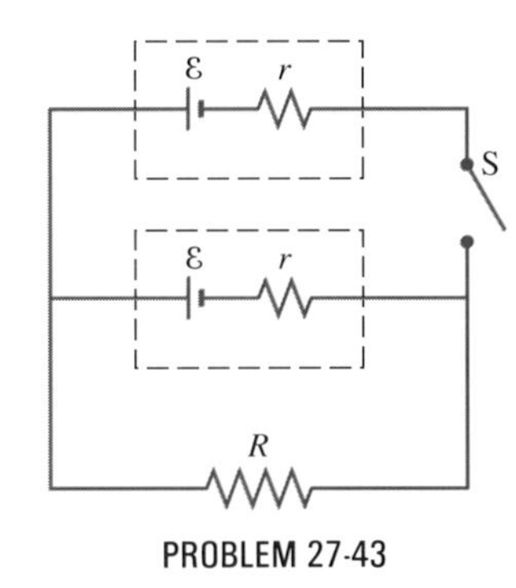

PROBLEM 27-43

27-44. If $I = 2.0$ A in the circuit shown in the accompanying figure, what are (*a*) the power dissipated in the two 5.0-Ω resistors and (*b*) the potential differences between A and B and between B and C?

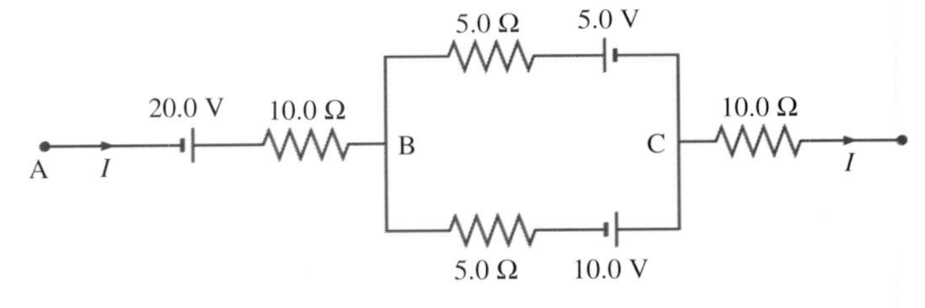

PROBLEM 27-44

27-45. Determine the equivalent resistance between A and B in the network shown in the accompanying figure. (*Hint:* Assume that a battery of emf ℰ is placed across A and B, then calculate the current I through the battery.)

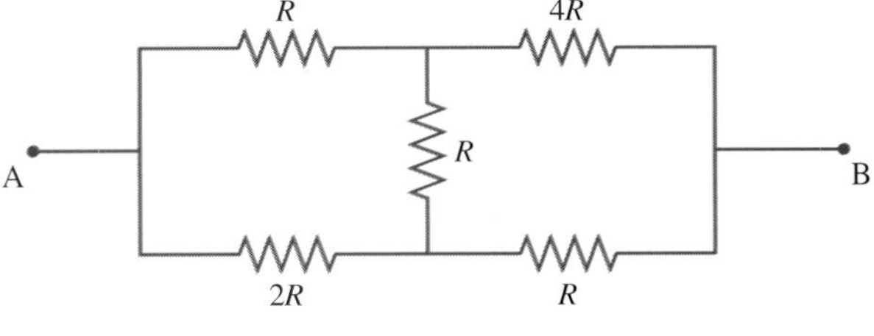

PROBLEM 27-45

27-46. Calculate I_1, I_2, and I_3 in the circuit shown in the accompanying figure.

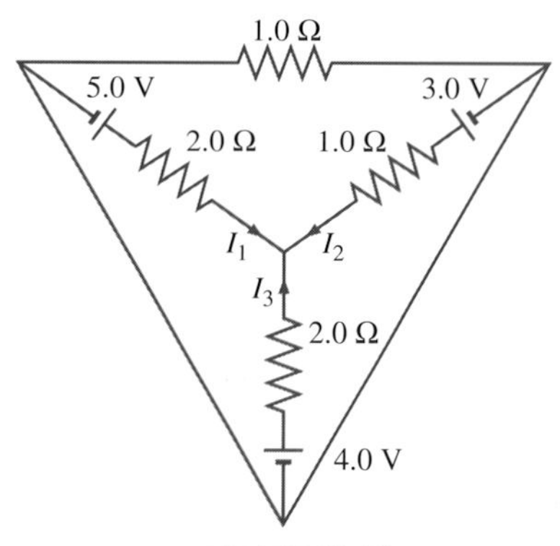

PROBLEM 27-46

27-47. (*a*) What is the current through the battery in the circuit shown in the accompanying figure? (*b*) What is the equivalent resistance between A and B for the network of resistors?

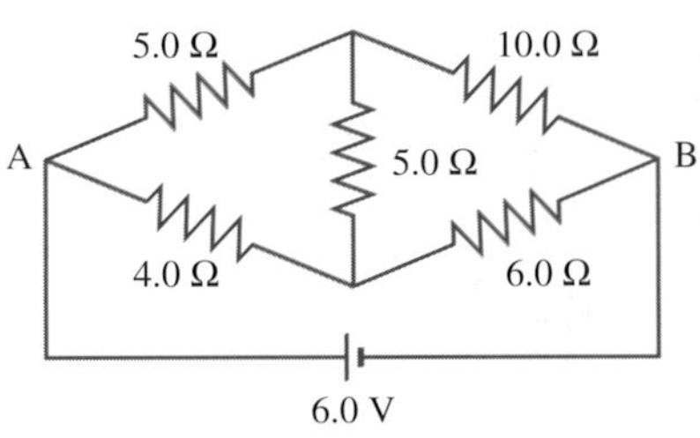

PROBLEM 27-47

27-48. A voltmeter of resistance R_V and an ammeter of resistance R_A are connected in order to measure an unknown resistance R_x (see accompanying figure). The actual resistance is $R_x = V/I'$, where I' is the current through R_x, and V is the reading of the voltmeter. The calculated resistance is $R_x' = V/I$, where I is the ammeter reading. Show that

$$\frac{1}{R_x} = \frac{1}{R_x'} - \frac{1}{R_V}.$$

If $V = 10.00$ V and $I = 10.00$ mA, by how much do R_x and R_x' differ when a 1.0-MΩ voltmeter is used?

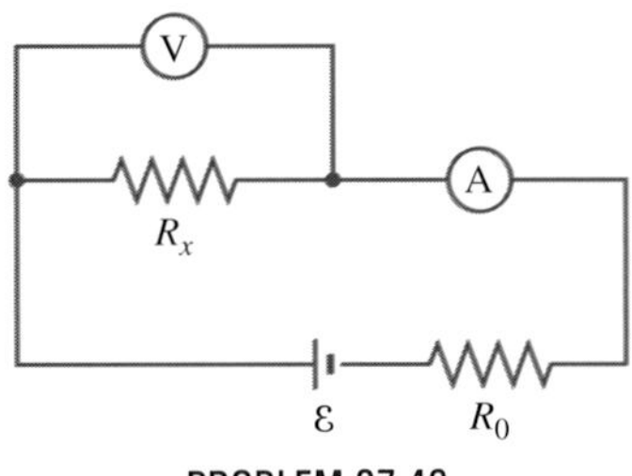

PROBLEM 27-48

27-49. Rather than using the circuit of the previous problem, suppose that you measure R_x by connecting the ammeter and voltmeter as shown in the accompanying figure. In this case, the calculated resistance is also $R_x' = V/I$, where V and I are the readings of the voltmeter and ammeter, respectively. Use the fact that the true resistance is $R_x = V'/I$, where V' is the voltage across R_x, to show that

$$R_x = R_x' - R_A,$$

where R_A is the ammeter resistance. What are the values of R_x and R_x' when $V = 1.50$ V, $I = 15.0$ mA, and $R_A = 0.20\ \Omega$?

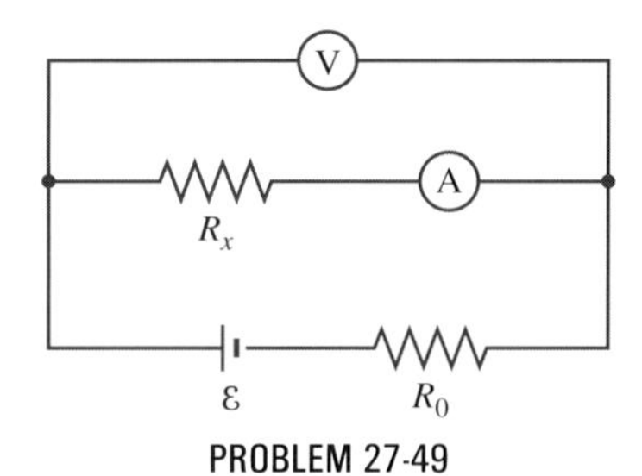

PROBLEM 27-49

27-50. Calculate the potential difference between A and B (see the accompanying figure) with (*a*) the switch S open and (*b*) the switch S closed.

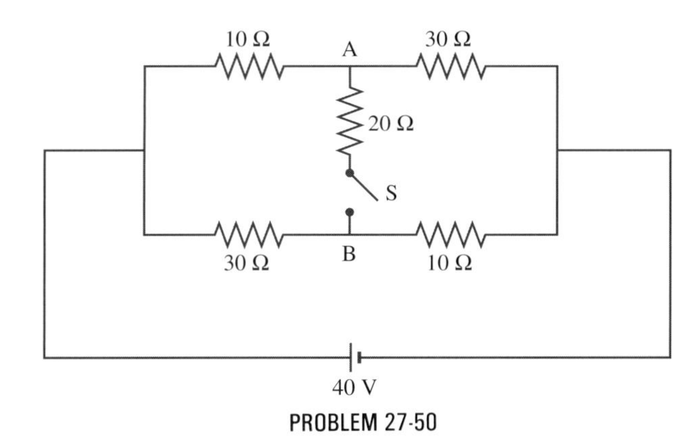

PROBLEM 27-50

27-51. What is the equivalent resistance between A and B for the circuit shown in the accompanying figure? The value of r is 2.0 Ω.

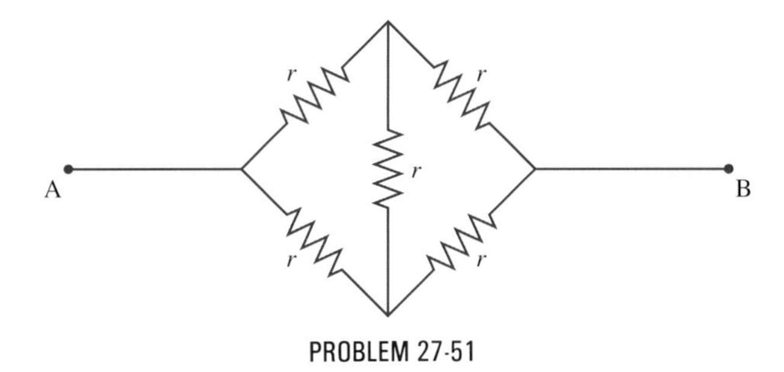

PROBLEM 27-51

27-52. (*a*) Determine the equivalent resistance seen by the battery in the accompanying figure and then find the power output of the battery. (*b*) Using the various equivalent circuits developed in solving part (*a*), determine the currents in the branches of the circuit. (*c*) What is the potential difference between A and B?

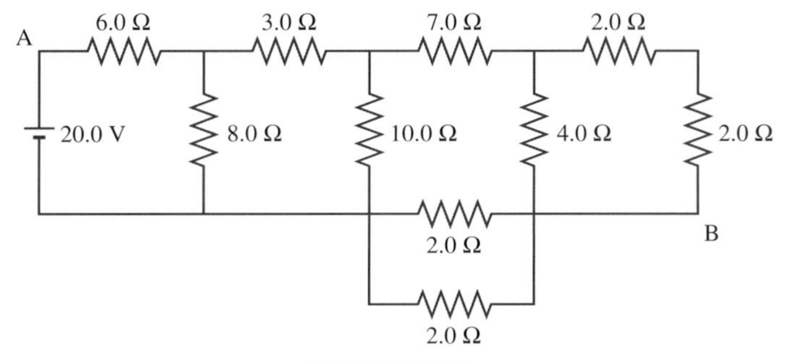

PROBLEM 27-52

27-53. The circuit of a multiscale ammeter is shown in the accompanying figure. The three settings of the switch give full-scale readings of 0.10, 1.0, and 10.0 A. If $R_g = 100\ \Omega$ and $I_g = 1.0$ mA for full-scale deflection, what are the values of R_1, R_2, and R_3?

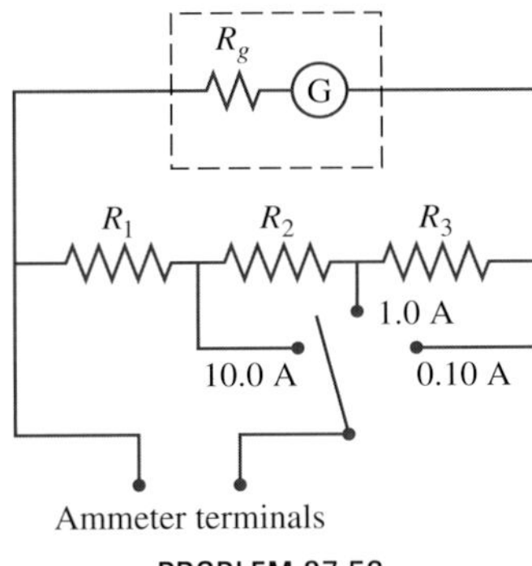

PROBLEM 27-53

27-54. The circuit of a multiscale voltmeter is shown in the accompanying figure. The three settings of the switch give full-scale readings of 3.0, 15.0, and 150 V. If $R_g = 100\ \Omega$ and $I_g = 1.0$ mA for full-scale deflection, what are the values of R_1, R_2, and R_3?

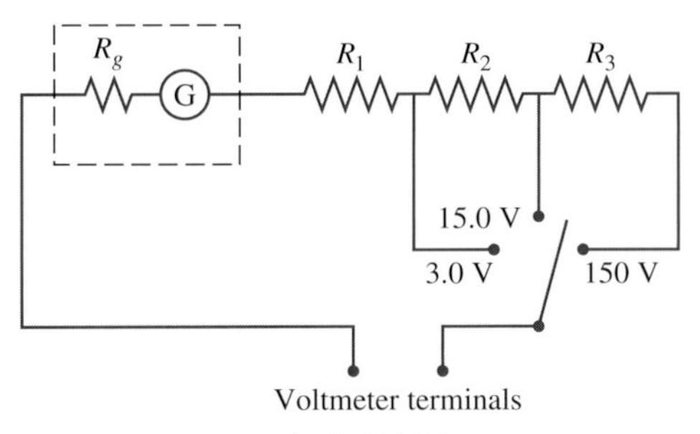

PROBLEM 27-54

27-55. Thevenin's theorem states that a network of resistors and sources of emf connected to a load resistor R can be replaced by a single source $\mathcal{E}_T$ and a single resistor R_T as shown in part (*a*) of the accompanying figure. The magnitude and polarity of $\mathcal{E}_T$ are found by removing R and measuring (or calculating) the potential difference across the resulting open circuit from A to B. The resistance R_T is given by $R_T = \mathcal{E}_T/I$, where I is the current from A to B when R is short-circuited. Determine the Thevenin equivalent circuits "seen" by the load resistor R in parts (*b*) and (*c*) of the figure. Calculate the current through R in each case.

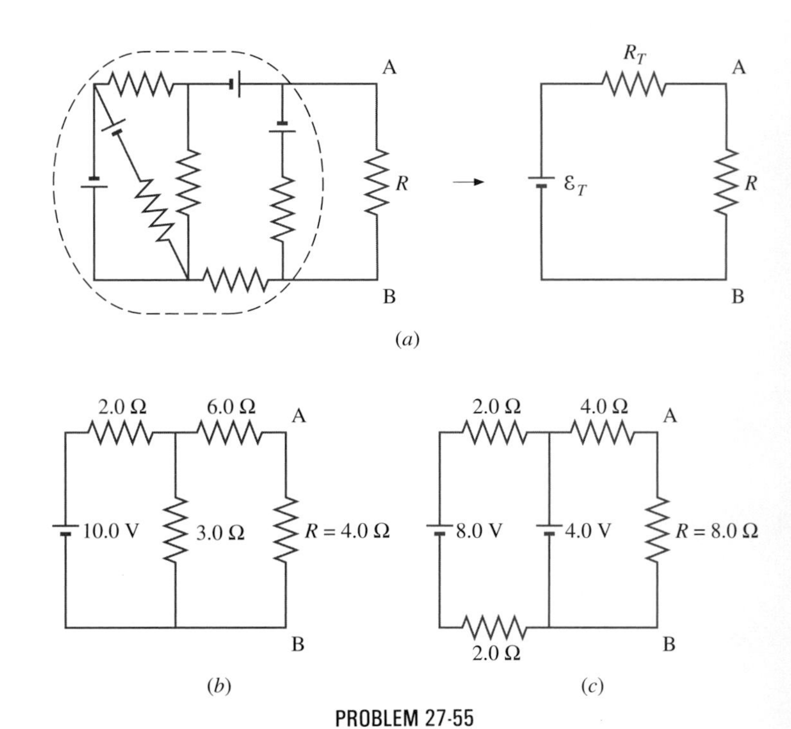

PROBLEM 27-55

An industrial electromagnet has a strong magnetic field.

CHAPTER 28 THE MAGNETIC FIELD

PREVIEW

In this chapter we define the magnetic field and study the forces it exerts on moving charged particles and current-carrying wires. The main topics to be discussed here are the following:

1. **The magnetic field and its force on charged particles.** The magnetic field is defined in terms of its force on a moving charged particle.
2. **Motion in a uniform magnetic field.** The circular orbits of charged particles moving in a uniform magnetic field are studied.
3. **Magnetic force on a current.** The force of the magnetic field on a moving charged particle is used to derive the force of the field on a current-carrying wire.
4. **A current loop in a uniform magnetic field.** The effect of a uniform magnetic field on a current loop is investigated. The magnetic dipole moment is defined in terms of the current loop.

*5. **Applications of the magnetic force law.** Applications considered are the galvanometer, the direct current motor, the Hall effect, the velocity selector, the mass spectrometer, and the cyclotron.

Most people are aware of many of the simple properties of bar magnets. For example, magnets always have two poles; one is designated as north (N), the other as south (S). Magnetic poles repel if they are alike (both N or both S), they attract if they are opposite (one N, the other S), and both poles of a magnet attract unmagnetized pieces of iron. A well-known example of a magnet is the compass needle. It is a thin bar magnet suspended at its center so that it is free to rotate in a horizontal plane. (See Fig. 28-1*a*.) Now the earth is also a magnet in the sense that it produces a magnetic field. This "earth magnet" is oriented such that its south pole is located near the geographic North Pole of the earth. Because opposite poles attract, a compass needle rotates until its north pole points toward the south pole of the "earth magnet" (that is, until it is directed toward the north), as illustrated in Fig. 28-1*b*.

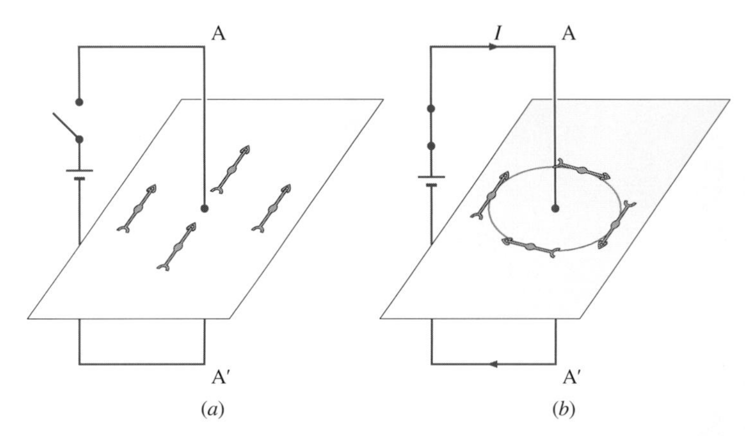

FIGURE 28-2 (*a*) With no current in wire AA′, the compass needles all point north. (*b*) With a current flowing as shown, the needles are deflected to point along the circle centered on AA′.

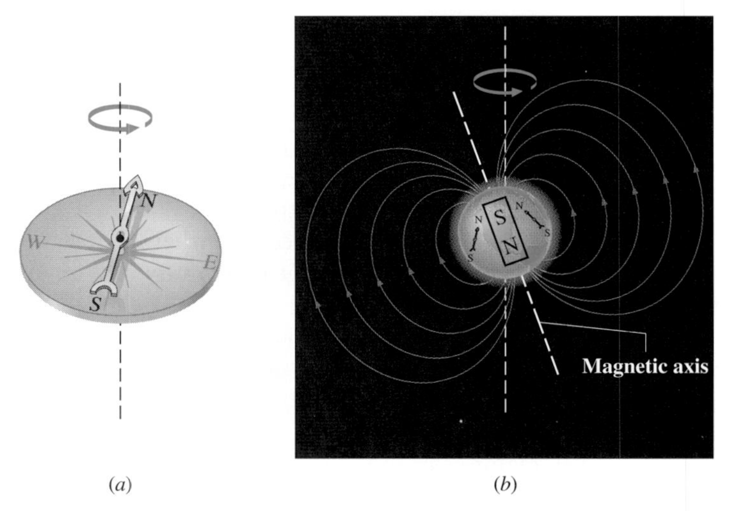

FIGURE 28-1 (*a*) The compass needle is a thin bar magnet that is free to rotate about an axis through its center. (*b*) The earth behaves magnetically as if a large bar magnet were located at its center as shown. The north pole of the compass needle points toward the south pole of the earth magnet, which is about 1200 km from the geographic North Pole. Also shown are the magnetic field lines of the earth.

Not so well known is the fact that current-carrying wires also participate in magnetic interactions. This was discovered in 1819 by Hans Oersted (1777–1851), who, while performing lecture demonstrations for some students, noticed that a compass needle moved whenever current flowed in a nearby wire. (See Fig. 28-2.) Further investigation of this phenomenon convinced Oersted that electric current somehow causes a magnetic force. He reported this finding to an 1820 meeting of the French Academy of Science.

Soon after this report, Oersted's investigations were repeated and expanded upon by other scientists. Among those whose work was especially important are Jean-Baptiste Biot (1774–1862) and Felix Savart (1791–1841), who investigated the forces exerted on magnets by currents; André Marie Ampère (1775–1836), who studied the forces exerted by one current on another; François Arago (1786–1853), who found that iron can be magnetized by a current; and Humphry Davy (1778–1829), who discovered that a magnet exerts a force on a wire carrying electric current.

The evidence from the various experiments led Ampère to propose that electric current is the source of all magnetic phenomena. To explain permanent magnets, he suggested that matter contains microscopic current loops that are somehow aligned when a material is magnetized. Today we know that permanent magnets are actually created by the alignment of spinning electrons, a situation quite similar to that proposed by Ampère. Given that Ampère's model of permanent magnets was developed almost a century before the atomic nature of matter was understood, it is remarkably accurate and insightful!

In this chapter we investigate the forces exerted on moving charges and current-carrying wires by magnetic fields. The source of the magnetic field will be examined in the next chapter. There are many important applications of the magnetic-force law. These include the electric motor, the particle accelerator, and the mass spectrometer.

28-1 The Magnetic Field and Its Force on Charged Particles

Let's imagine that we are observing the motion of a particle of known charge and mass that is in the vicinity of a permanent magnet. With measurements of the particle's acceleration and Newton's second law, we can determine the net force on the particle. If we subtract from this net force the effects of the gravitational field and any electric field that may be present, we will find that there yet remains an unidentified force on the charged particle—a force that must be connected in some way to the presence of the permanent magnet. Careful experimentation discloses four important properties of this force:

1. The magnitude of the force is proportional to both the charge q and the speed v of the particle. There is no force on a charged particle that is at rest or a particle that is uncharged.

2. The force and the velocity of the particle are always perpendicular to each other.

The magnet normally used in magnetic resonance imaging (MRI) has a field of 1.5 T.

3. For a fixed speed, the force at a particular point varies with the direction of the velocity. There is one line along which the particle can move in either direction without experiencing a force.

4. The forces on a positive charge and a negative charge moving with the same velocity are in opposite directions.

These observations can be accounted for if the magnet produces a **magnetic field** that exerts the force on the particle. All experiments, including our hypothetical one, are consistent with the following force law:

$$\mathbf{F}_m = q\mathbf{v} \times \mathbf{B}, \qquad (28\text{-}1)$$

where $\mathbf{F}_m$ is the force on the particle, q and $\mathbf{v}$ are the charge and velocity, respectively, of the particle, and $\mathbf{B}$ is the magnetic field.* The directions of $\mathbf{v}$, $\mathbf{B}$, and $\mathbf{F}_m$ are related by the right-hand rule, as summarized in Fig. 28-3. When you place the fingers of your right hand along $\mathbf{v}$ and turn them through the smaller angle θ between the vectors into $\mathbf{B}$, $\mathbf{F}_m$ points along your thumb if q is positive and opposite to it if q is negative. Also, from Eq. (2-11), the magnitudes of the three vectors satisfy

$$F_m = qvB \sin \theta. \qquad (28\text{-}2)$$

Since $\mathbf{v}$ and $\mathbf{F}_m$ are always perpendicular to each other, $\mathbf{F}_m$ does no work. Consequently, *the kinetic energy (and the speed) of a particle moving solely in a magnetic field must remain constant*.

The SI unit of the magnetic field is the *tesla* (T). This unit is named in honor of Nikola Tesla (1856–1943), an engineer who made significant contributions in the field of electric power generation. From Eq. (28-1),

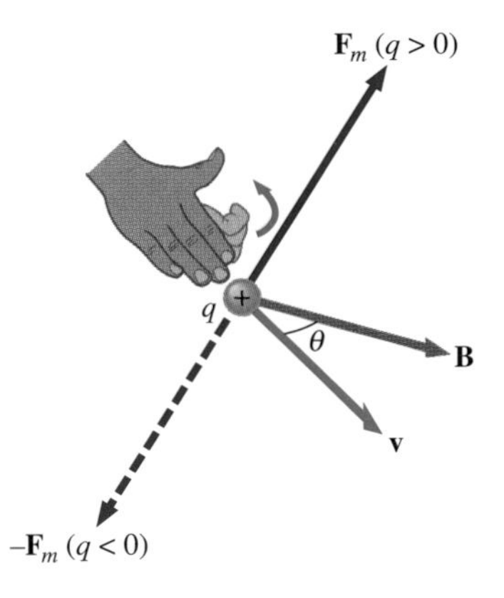

FIGURE 28-3 The right-hand rule gives the direction of the magnetic force $\mathbf{F}_m$ on a moving particle of charge q. If $q < 0$, $\mathbf{F}_m$ is directed opposite to the thumb.

$$1 \text{ T} = 1 \frac{\text{N}\cdot\text{s}}{\text{C}\cdot\text{m}} = 1 \frac{\text{kg}}{\text{C}\cdot\text{s}}.$$

In many applications, the cgs magnetic-field unit, the *gauss* (G), is used. The gauss is related to the tesla by

$$1 \text{ T} = 10^4 \text{ G}.$$

Conventional laboratory magnets produce fields up to approximately 2.5 T, while superconducting magnets generate fields over 10 times higher. And how strong is the magnetic field of the earth? It is minuscule compared with these laboratory fields. While varying with location, it is typically only about 0.5 G, or 5×10^{-5} T.

Like the electric field, the magnetic field is visualized with the help of field lines. The lines are drawn such that (1) at each point in space, a field line is tangent to the magnetic field passing through that point and (2) the spatial density of the field lines in a given region is proportional to the magnitude of the magnetic field in that region. Notice that these are also characteristics of electric field lines, as specified in Sec. 23-3. In the next chapter when the source of the magnetic field is considered, you will see that there are also major differences between electric and magnetic field lines.

In many examples involving the magnetic field, it is convenient to have the magnetic field lines perpendicular to the plane of the page. A field line that is *directed into the page* is represented in a figure by "x", while a field line *coming out of the page* is represented by "•".

EXAMPLE 28-1 AN ALPHA PARTICLE MOVING IN A MAGNETIC FIELD

An alpha particle ($q = 3.2 \times 10^{-19}$ C) moves through a uniform magnetic field whose magnitude is 1.5 T. The field is directed parallel to the positive z axis of the rectangular coordinate system of Fig. 28-4. What is the magnetic force on the alpha particle when it is moving (*a*) in the positive x direction with a speed of 5.0×10^4 m/s, (*b*) in the negative y direction with a speed of 5.0×10^4 m/s, (*c*) in the positive z direction with a speed of 5.0×10^4 m/s, and (*d*) with a velocity $\mathbf{v} = (2.0\mathbf{i} - 3.0\mathbf{j} + 1.0\mathbf{k}) \times 10^4$ m/s?

SOLUTION (*a*) If you place your fingers along $\mathbf{v}$ (the positive x direction) and turn them into $\mathbf{B}$ (the positive z direction), your thumb points along $\mathbf{F}_m$, which is in the negative y direction. (See

*The magnetic field is also known as the *magnetic induction* or the *magnetic flux density*. We will generally use the term *magnetic field*.

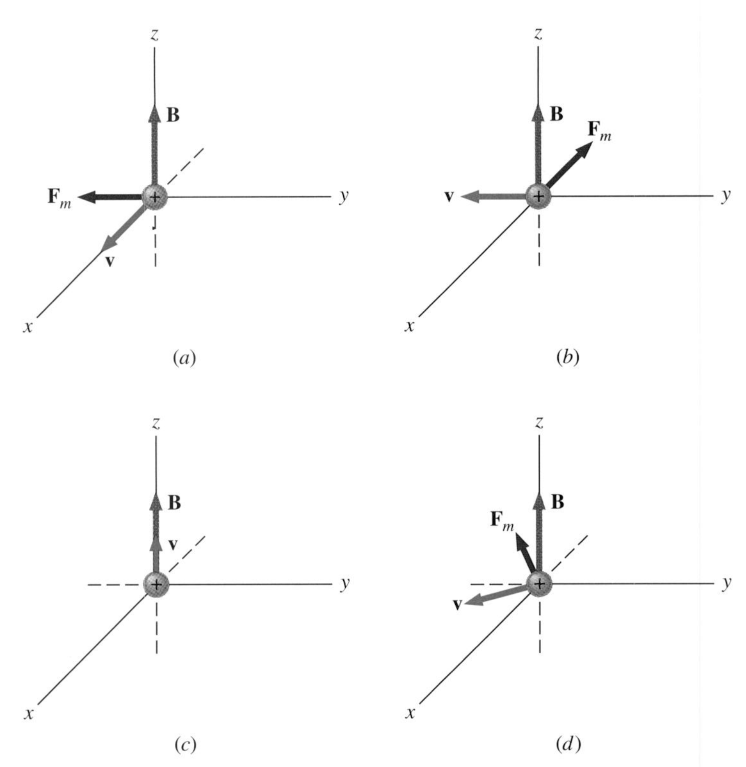

FIGURE 28-4 The magnetic forces on an alpha particle moving in a uniform magnetic field.

Fig. 28-4*a*.) Since **v** and **B** are perpendicular, the magnitude of the magnetic force on the alpha particle is

$$F_m = qvB \sin 90° = (3.2 \times 10^{-19}\text{ C})(5.0 \times 10^4\text{ m/s})(1.5\text{ T}) = 2.4 \times 10^{-14}\text{ N},$$

and

$$\mathbf{F}_m = -2.4 \times 10^{-14}\mathbf{j}\text{ N}.$$

(*b*) With the right-hand rule applied to the vectors of Fig. 28-4*b*, $\mathbf{F}_m$ is found to point in the negative *x* direction. Again, **v** and **B** are perpendicular, so

$$F_m = qvB \sin 90° = (3.2 \times 10^{-19}\text{ C})(5.0 \times 10^4\text{ m/s})(1.5\text{ T}) = 2.4 \times 10^{-14}\text{ N},$$

and

$$\mathbf{F}_m = -2.4 \times 10^{-14}\mathbf{i}\text{ N}.$$

(*c*) Since **v** and **B** are parallel (see Fig. 28-4*c*), $\mathbf{v} \times \mathbf{B} = 0$, and $\mathbf{F}_m = 0$.

(*d*) The vectors for this case are shown in Fig. 28-4*d*. Since neither the angle between **v** and **B** nor the plane of the two vectors is apparent, we use the distributive property of the vector cross product to determine the magnetic force:

$$\begin{aligned}\mathbf{v} \times \mathbf{B} &= [(2.0\mathbf{i} - 3.0\mathbf{j} + 1.0\mathbf{k}) \times 10^4] \times (1.5\mathbf{k})\text{ T}\cdot\text{m/s} \\ &= [3.0(\mathbf{i} \times \mathbf{k}) - 4.5(\mathbf{j} \times \mathbf{k}) + 1.5(\mathbf{k} \times \mathbf{k})] \times 10^4\text{ T}\cdot\text{m/s} \\ &= (-3.0\mathbf{j} - 4.5\mathbf{i}) \times 10^4\text{ T}\cdot\text{m/s}.\end{aligned}$$

Therefore

$$\mathbf{F}_m = q\mathbf{v} \times \mathbf{B} = (-14.4\mathbf{i} - 9.6\mathbf{j}) \times 10^{-15}\text{ N}.$$

28-2 Motion in a Uniform Magnetic Field

Consider a particle of charge q and mass m moving in a uniform magnetic field **B**. Initially, the particle's velocity vector **v** lies in a plane perpendicular to **B**, which is directed into the page, as shown in Fig. 28-5. To determine the subsequent motion of the particle, we make the following observations based on the magnetic force law, $\mathbf{F}_m = q\mathbf{v} \times \mathbf{B}$.

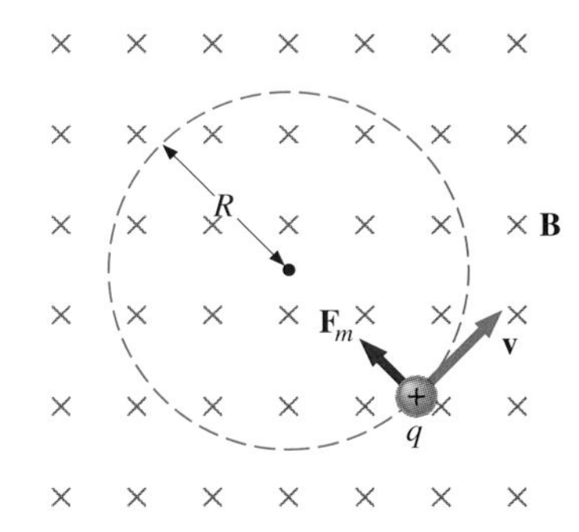

FIGURE 28-5 The motion of a charged particle that enters a uniform magnetic field with a velocity perpendicular to the field. The magnetic field is directed into the page.

First, by the right-hand rule, there is no magnetic force component along the direction of the field **B**. If our particle is initially moving in a plane perpendicular to **B**, it must remain in that plane.

Second, $\mathbf{F}_m$ and **v** are always mutually perpendicular, again from the right-hand rule. This also means that the speed of any charged particle in a magnetic field remains constant.

Finally, since **v** and **B** are perpendicular in this case, and v and B are both constant, the magnitude of the magnetic force,

$$F_m = qvB \sin 90° = qvB,$$

is also constant.

From these observations, *the charged particle moves in a plane at constant speed under the influence of a force of constant magnitude that always acts perpendicular to the velocity vector*. We have encountered motion of this type before in our study of mechanics—it is simply circular motion at constant speed. The radius R of the circle may be found with Newton's second law. Since the magnitude of the centripetal force on the particle is $F_m = qvB$,

$$qvB = m\frac{v^2}{R}$$

and

$$R = \frac{mv}{qB}. \tag{28-3}$$

The particle moves in the circle of radius R at constant speed v, so its period is given by

$$T = \frac{2\pi R}{v},$$

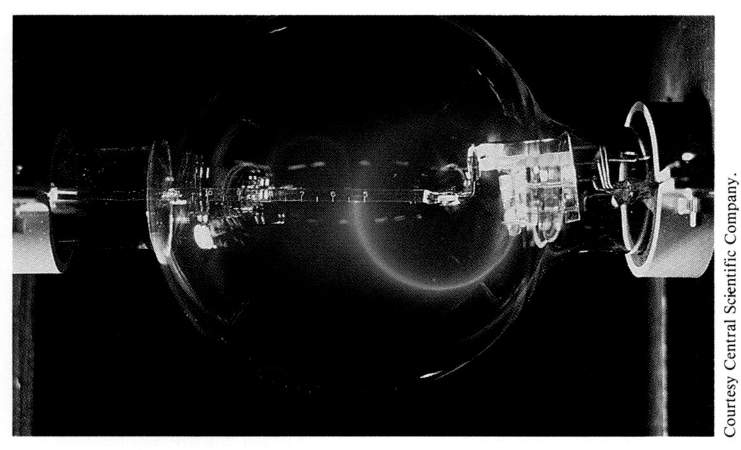

The circular path taken by an electron beam in a magnetic field. Low-pressure helium inside the tube gives off a green glow as the electrons collide with it, thereby producing an outline of the path.

which, when Eq. (28-3) is used to substitute for v, becomes

$$T = \frac{2\pi m}{qB}. \tag{28-4}$$

Notice that the period is independent of the radius R and the speed v. This fact is very important for the operation of a cyclotron, a device that will be discussed in Sec. 28-5.

One application of Eq. (28-3) is the determination of the momentum of a charged elementary particle. Typical circular paths of such particles moving in a uniform magnetic field are shown in the bubble chamber* photograph of Fig. 28-6. With measurements of the radii of these paths and knowledge of q and B, we can determine the momenta of the particles from Eq. (28-3).

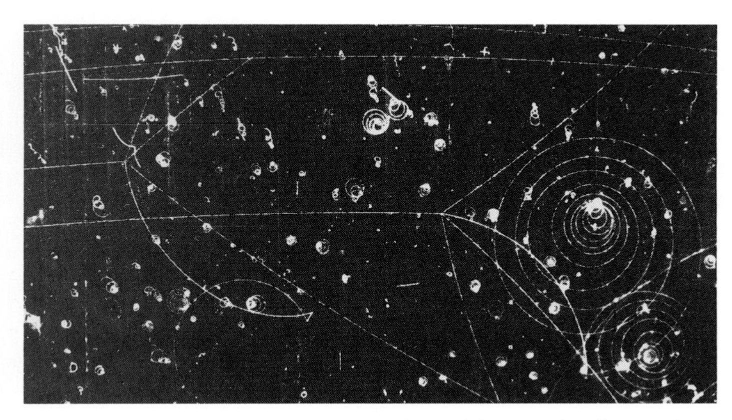

FIGURE 28-6 Circular paths of charged particles moving in a uniform magnetic field inside a bubble chamber.

EXAMPLE 28-2 **CIRCULAR ORBIT OF A PROTON IN A MAGNETIC FIELD**

A proton ($q = 1.6 \times 10^{-19}$ C, $m = 1.67 \times 10^{-27}$ kg) enters a uniform magnetic field of magnitude 1.5 T. The initial velocity of the proton is perpendicular to the field, and its magnitude is 2.0×10^7 m/s. Determine the radius and period of the proton's motion.

SOLUTION Since the proton moves in a plane perpendicular to the magnetic field, its path is a circle of radius (Eq. 28-3)

$$R = \frac{mv}{qB} = \frac{(1.67 \times 10^{-27}\text{ kg})(2.0 \times 10^7\text{ m/s})}{(1.6 \times 10^{-19}\text{ C})(1.5\text{ T})} = 0.14\text{ m}.$$

The time taken to traverse the circle once is, from Eq. (28-4),

$$T = \frac{2\pi m}{qB} = \frac{2\pi(1.67 \times 10^{-27}\text{ kg})}{(1.6 \times 10^{-19}\text{ C})(1.5\text{ T})} = 4.4 \times 10^{-8}\text{ s}.$$

DRILL PROBLEM 28-1

A uniform magnetic field of magnitude 1.5 T is directed horizontally from west to east. (*a*) What is the magnetic force on a proton at the instant when it is moving vertically downward in the field with a speed of 4.0×10^7 m/s? (*b*) Compare this force with the weight W of the proton.
ANS. (*a*) 9.6×10^{-12} N toward the south; (*b*) $W/F_m = 1.7 \times 10^{-15}$.

DRILL PROBLEM 28-2

An electron moving at 5.0×10^7 m/s enters a uniform magnetic field of magnitude 10 G with its velocity perpendicular to the field. What is the radius of the electron's path?
ANS. 28 cm.

We have just found that if the initial velocity **v** of a charged particle is perpendicular to the magnetic field **B**, the particle will move at a constant speed v in a circle whose radius R is given by Eq. (28-3). Of course, in many instances the initial velocity and the magnetic field are not perpendicular. For example, **B** may be directed along the z axis while **v** has components along all three rectangular axes, as illustrated in Fig. 28-7. For this particular case, there is no magnetic force in the z direction, so

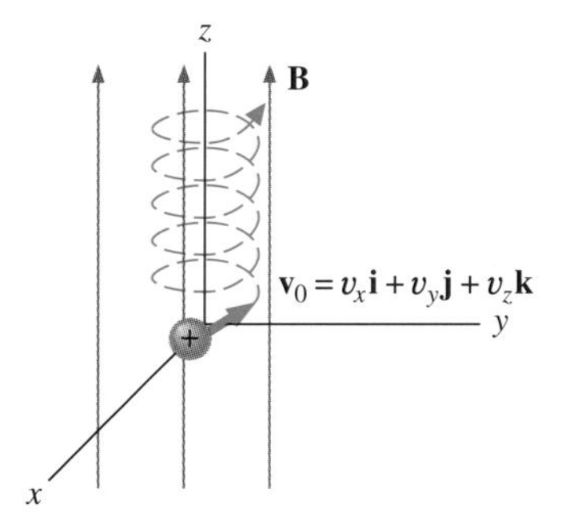

FIGURE 28-7 The helical path of a charged particle that has a velocity component parallel to the uniform magnetic field.

*A bubble chamber contains liquid (often hydrogen) heated under pressure nearly to its boiling point. If a charged particle enters the chamber immediately after the pressure is suddenly reduced, the ionization it produces in the liquid leads to bubble formation along its path, which may then be photographed.

$a_z = 0$ and v_z is constant. The velocity components v_x and v_y, however, are affected by the magnetic force $q\mathbf{v} \times \mathbf{B}$ and do change with time. In Drill Prob. 28-3, you will show that v_x and v_y vary such that the projection of the particle's path onto the xy plane is a circle whose radius is given by Eq. (28-3) with $v = \sqrt{v_x^2 + v_y^2}$. The combination of circular motion in the xy plane and motion at constant velocity in the z direction produces the "stretched-out" circle, or *helix*, shown in Fig. 28-7.

Prominences are visible during total solar eclipses as red protrusions at the edge of the solar disk. They consist of ionized gases trapped in the magnetic fields around magnetically active regions of the sun.

EXAMPLE 28-3 HELICAL MOTION OF A PROTON IN A MAGNETIC FIELD

A uniform magnetic field of magnitude 0.500 T is directed parallel to the z axis of Fig. 28-8. A proton enters the field at a point P on the x axis with a velocity $\mathbf{v}_0 = (2.00\mathbf{j} + 4.00\mathbf{k}) \times 10^6$ m/s. Where is the proton 6.89×10^{-7} s later?

SOLUTION The proton moves along a helix whose projected circle on the xy plane has a radius given by Eq. (28-3), with

$$v = \sqrt{v_x^2 + v_y^2} = \sqrt{(0^2 + 2.00^2)} \times 10^{12}\ \text{m/s} = 2.00 \times 10^6\ \text{m/s}.$$

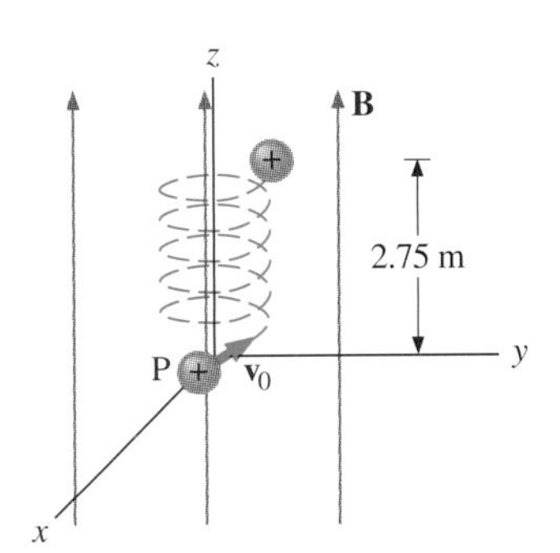

FIGURE 28-8 A helical path of a proton.

Thus the radius of the circle is

$$R = \frac{(1.67 \times 10^{-27}\ \text{kg})(2.00 \times 10^6\ \text{m/s})}{(1.60 \times 10^{-19}\ \text{C})(0.500\ \text{T})} = 4.18 \times 10^{-2}\ \text{m}.$$

The period of the motion is given by Eq. (28-4):

$$T = \frac{2\pi(1.67 \times 10^{-27}\ \text{kg})}{(1.60 \times 10^{-19}\ \text{C})(0.500\ \text{T})} = 1.31 \times 10^{-7}\ \text{s}.$$

So, in 6.89×10^{-7} s, the proton circulates $(6.89 \times 10^{-7}\ \text{s})/(1.31 \times 10^{-7}\ \text{s}) = 5.25$ times around the helix. Since the proton also moves along **B** at the constant velocity $v_z = 4.00 \times 10^6$ m/s during this time, it moves up the z axis a distance

$$z = (4.00 \times 10^6\ \text{m/s})(6.89 \times 10^{-7}\ \text{s}) = 2.75\ \text{m}.$$

To find the position of the proton in the xy plane, we note that during this 6.89×10^{-7} s time interval, the proton circulates 5.25 times around the helix. Since the proton enters the field at P, which is on the x axis, it must be on the y axis after 5.25 revolutions. Now the radius of its orbit projected on to the xy plane is 4.18×10^{-2} m, so the coordinates of the proton 6.89×10^{-7} s after it enters the field are $(0, 4.18 \times 10^{-2}, 2.75)$ m.

DRILL PROBLEM 28-3

A particle of charge q and mass m moves in a uniform magnetic field $\mathbf{B} = B\mathbf{k}$ with a velocity $\mathbf{v} = v_x\mathbf{i} + v_y\mathbf{j} + v_z\mathbf{k}$. (*a*) Show that $\mathbf{F}_m = q\mathbf{v} \times \mathbf{B}$ has no z component. (*b*) Show that $\mathbf{F}_m$ and $v_x\mathbf{i} + v_y\mathbf{j}$, the velocity in the xy plane, are perpendicular; then argue that the xy motion of the particle is a circle of radius $R = m\sqrt{v_x^2 + v_y^2}/qB$.

DRILL PROBLEM 28-4

A proton moves at a speed of 1.0×10^7 m/s in a plane perpendicular to a magnetic field of magnitude 0.75 T. (*a*) What is the radius of the proton's orbit? (*b*) How many revolutions does it make in 5.7×10^{-7} s?
ANS. (*a*) 14 cm; (*b*) 6.5.

28-3 MAGNETIC FORCE ON A CURRENT

Electric current is an ordered movement of charge. A current-carrying wire in a magnetic field must therefore experience a force due to the field. To investigate this force, let's consider the infinitesimal section of wire shown in Fig. 28-9. The length and cross-sectional area of the section are dl and A, respectively, so its volume is $A\,dl$. The wire is formed from material that contains n charge carriers per unit volume, so the number of charge carriers in the section is $nA\,dl$. If the charge carriers

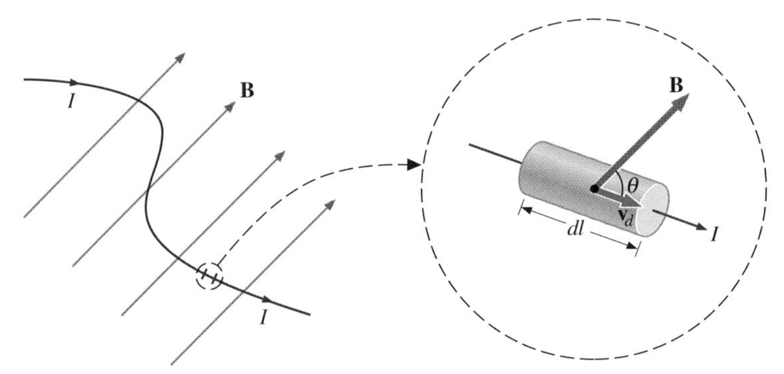

FIGURE 28-9 An infinitesimal section of current-carrying wire in a magnetic field.

move with a drift velocity $\mathbf{v}_d$, the current I in the wire is, from Eq. (26-2),

$$I = neAv_d.$$

The magnetic force on any single charge carrier is $e\mathbf{v}_d \times \mathbf{B}$, so the total magnetic force $d\mathbf{F}_m$ on the $nA\ dl$ charge carriers in the section of wire is

$$d\mathbf{F}_m = (nA\ dl)e\mathbf{v}_d \times \mathbf{B}.$$

With $d\mathbf{l}$ defined to be a vector of length dl pointing along $\mathbf{v}_d$, we can rewrite this equation as

$$d\mathbf{F}_m = neAv_d\, d\mathbf{l} \times \mathbf{B},$$

or

$$d\mathbf{F}_m = I\, d\mathbf{l} \times \mathbf{B}. \tag{28-5}$$

This is the magnetic force on the section of wire. Notice that it is actually the net force exerted by the field on the charge carriers themselves. The direction of this force is determined with the right-hand rule. In this case, you align your figures with the current and rotate them into the magnetic field. Your thumb then points along the force.

In order to determine the magnetic force $\mathbf{F}_m$ on a wire of arbitrary length and shape, Eq. (28-5) must be integrated over the entire wire; that is,

$$\mathbf{F}_m = \int_{\text{wire}} I\, d\mathbf{l} \times \mathbf{B}. \tag{28-6}$$

This integral must be evaluated with care, because the directions of both $d\mathbf{l}$ and $\mathbf{B}$ (and therefore of $d\mathbf{l} \times \mathbf{B}$) usually vary along the wire. If the wire section happens to be straight and if $\mathbf{B}$ is uniform, then Eq. (28-6) reduces to the much simpler equation

$$\mathbf{F}_m = I\mathbf{l} \times \mathbf{B}, \tag{28-7}$$

which is the force on a straight, current-carrying wire in a uniform magnetic field.

EXAMPLE 28-4 Balancing the Gravitational and Magnetic Forces on a Current-Carrying Wire

A wire of length 50 cm and mass 10 g is suspended in a horizontal plane by a pair of flexible leads. (See Fig. 28-10*a*.) The wire is then subjected to a constant magnetic field of magnitude 0.50 T, which is directed as shown. What are the magnitude and direction of the current in the wire needed to remove the tension in the supporting leads?

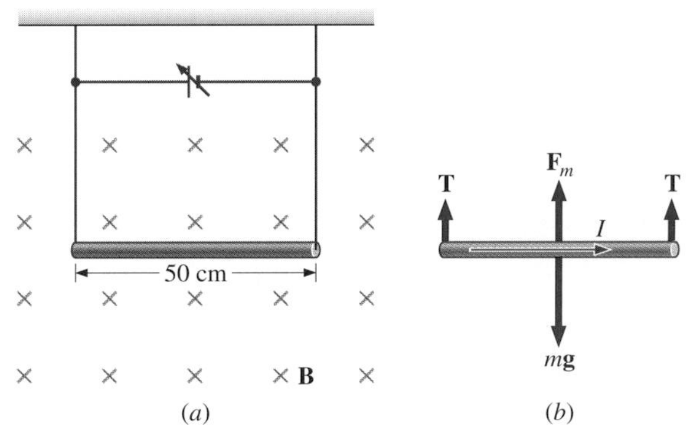

FIGURE 28-10 (*a*) A wire suspended in a magnetic field. (*b*) The free-body diagram for the wire.

SOLUTION From the free-body diagram of Fig. 28-10*b*, the tensions in the supporting leads go to zero when the gravitational and magnetic forces balance each other. If $\mathbf{F}_m$ is to point upward, then from the right-hand rule, the current I must flow from left to right along the wire. The magnitude of I may be determined by equating the two forces:

$$mg = IlB;$$

thus

$$I = \frac{mg}{lB} = \frac{(0.010\ \text{kg})(9.8\ \text{m/s}^2)}{(0.50\ \text{m})(0.50\ \text{T})} = 0.39\ \text{A}.$$

DRILL PROBLEM 28-5

A long, rigid wire lying along the y axis carries a 5.0-A current flowing in the positive y direction. If a constant magnetic field of magnitude 0.30 T is directed along the positive x axis, what is the magnetic force per unit length on the wire?
ANS. $-1.5\mathbf{k}$ N/m.

28-4 A Current Loop in a Uniform Magnetic Field

In this section we determine the force and torque on a current-carrying loop of wire in a uniform magnetic field. Figure 28-11 shows a rectangular loop of wire that carries a current I and has

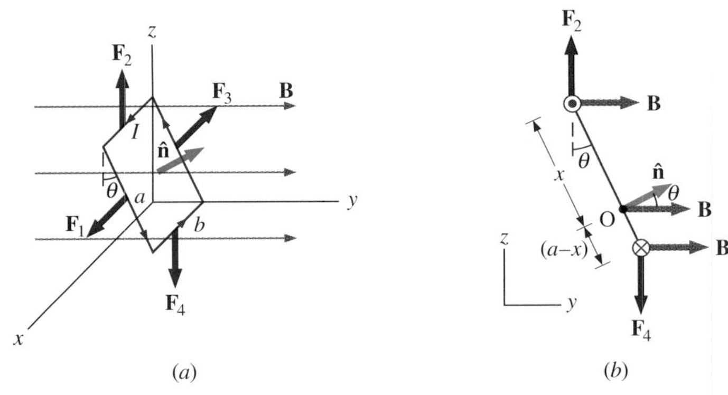

FIGURE 28-11 (*a*) A rectangular current loop in a uniform magnetic field is subjected to a net torque but not a net force. (*b*) A side view of the coil.

sides of length a and b. The loop is in a uniform magnetic field $\mathbf{B} = B\mathbf{j}$. The magnetic force on a straight current-carrying wire of length l is given by $I\mathbf{l} \times \mathbf{B}$. To find the net force on the loop, we have to apply this equation to each of the four sides. The force on side 1 is

$$\mathbf{F}_1 = IaB \sin(90° - \theta)\mathbf{i} = IaB \cos\theta\mathbf{i},$$

where the direction has been determined with the right-hand rule. The current in side 3 flows in the opposite direction to that of side 1, so

$$\mathbf{F}_3 = -IaB \sin(90° + \theta)\mathbf{i} = -IaB \cos\theta\mathbf{i}.$$

The currents in sides 2 and 4 are perpendicular to $\mathbf{B}$, and the forces on these sides are

$$\mathbf{F}_2 = IbB\mathbf{k}$$
$$\mathbf{F}_4 = -IbB\mathbf{k}.$$

We can now find the net force on the loop:

$$\begin{aligned}\sum \mathbf{F} &= \mathbf{F}_1 + \mathbf{F}_2 + \mathbf{F}_3 + \mathbf{F}_4\\ &= IaB\cos\theta\mathbf{i} + IbB\mathbf{k}\\ &\quad + (-IaB\cos\theta\mathbf{i}) + (-IbB\mathbf{k}) = 0.\end{aligned}$$

While this result ($\sum\mathbf{F} = 0$) has been obtained for a rectangular loop, it is far more general and holds for current-carrying loops of arbitrary shape; that is, *there is no force on a current loop in a uniform magnetic field.*

To find the net torque on the current loop of Fig. 28-11, we first consider $\mathbf{F}_1$ and $\mathbf{F}_3$. Since they have the same line of action and are equal and opposite, the sum of their torques about any axis is zero. Thus if there is any torque on the loop, it must be furnished by $\mathbf{F}_2$ and $\mathbf{F}_4$. Let's calculate their torques around the axis that passes through point O of Fig. 28-11*b* (a side view of the coil) and is perpendicular to the plane of the page. The point O is a distance x from side 2 and a distance $(a - x)$ from side 4 of the loop. The moment arms of $\mathbf{F}_2$ and $\mathbf{F}_4$ are $x\sin\theta$ and $(a - x)\sin\theta$, respectively, so the net torque on the loop is

$$\begin{aligned}\boldsymbol{\tau} = \boldsymbol{\tau}_1 + \boldsymbol{\tau}_2 &= -F_2 x\sin\theta\mathbf{i} - F_4(a - x)\sin\theta\mathbf{i}\\ &= -IbBx\sin\theta\mathbf{i} - IbB(a - x)\sin\theta\mathbf{i},\end{aligned}$$

which simplifies to

$$\boldsymbol{\tau} = -IAB\sin\theta\mathbf{i}, \tag{28-8a}$$

where $A = ab$ is the area of the loop.

Notice that this torque is independent of x; it is therefore independent of where the point O is located in the plane of the current loop. Consequently, the loop experiences the same torque from the magnetic field about *any axis* in the plane of the loop and parallel to the x axis.

A closed current loop is commonly known as a **magnetic dipole** and the term IA is known as its **magnetic dipole moment** $\mathcal{M}$. Actually, the magnetic dipole moment is a vector that is defined as

$$\boldsymbol{\mathcal{M}} = IA\hat{\mathbf{n}}, \tag{28-9a}$$

where $\hat{\mathbf{n}}$ is a unit vector directed perpendicular to the plane of the loop. (See Fig. 28-11*a*.) The direction of $\hat{\mathbf{n}}$ is obtained with a right-hand rule—if you curl the fingers of your right hand in the direction of current flow in the loop, then your thumb points along $\hat{\mathbf{n}}$. If the loop contains N turns of wire, then its magnetic dipole moment is given by

$$\boldsymbol{\mathcal{M}} = NIA\hat{\mathbf{n}}. \tag{28-9b}$$

In terms of the magnetic dipole moment, the torque on a current loop due to a uniform magnetic field (Eq. 28-8*a*) can be written simply as

$$\boldsymbol{\tau} = \boldsymbol{\mathcal{M}} \times \mathbf{B}. \tag{28-8b}$$

It can be shown that Eq. (28-8*b*) holds for a planar current loop of arbitrary shape. (See Prob. 28-61.)

DRILL PROBLEM 28-6

Potential energy of a magnetic dipole in a magnetic field. Using a calculation analogous to that of Sec. 25-6 for an electric dipole, show that the potential energy of a magnetic dipole in a magnetic field is given by

$$U = -\boldsymbol{\mathcal{M}}\cdot\mathbf{B}, \tag{28-10}$$

where U is zero when the angle between $\boldsymbol{\mathcal{M}}$ and $\mathbf{B}$ is 90°.

DRILL PROBLEM 28-7

A circular current loop of radius 2.0 cm carries a current of 2.0 mA. (*a*) What is the magnitude of its magnetic dipole moment? (*b*) If the dipole is oriented at 30° to a magnetic field of magnitude 0.50 T, what is the torque it experiences and what is its potential energy? (See Drill Prob. 28-6.)
ANS. (*a*) 2.5×10^{-6} A·m²; (*b*) 6.3×10^{-7} N·m, -1.1×10^{-6} J.

*28-5 Applications of the Magnetic Force Law

There are many scientific and engineering applications that depend on the magnetic force. We examine several of these applications in this section.

Galvanometer

As we discussed in Sec. 27-4, the basic component of nonelectronic ammeters and voltmeters is a current-measuring device known as the *galvanometer*. In the typical pivoted-coil galvanometer (Fig. 28-12), a flat coil of wire that carries the current to be measured encircles a fixed, soft-iron* core N times. The coil is suspended so that it can rotate around an axis that passes through the center of the coil and is perpendicular to the plane of the figure. The coil is placed between the poles of a magnet and a pointer is attached to detect its rotation. Because of the soft-iron core and the shape of the poles, the magnetic field is radial about the axis of the coil. It is therefore always parallel to the plane of the coil at the wires of the coil that are perpendicular to the plane of the page in Fig. 28-12. Thus the angle between the magnetic field and these wires is always 90°, and for any orientation, the magnetic torque on the coil is $\tau_m = NIAB$.

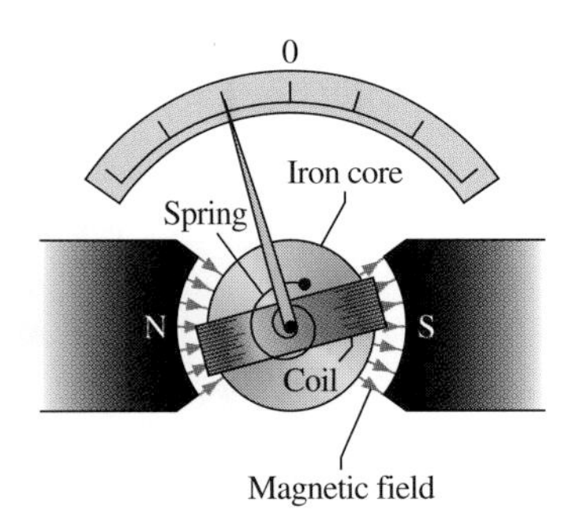

FIGURE 28-12 A pivoted-coil galvanometer. Notice the radial magnetic field lines.

The spring shown in the figure is attached to the coil and is designed to exert an opposing torque τ_s that is directly proportional to the angular displacement ϕ of the coil ($\tau_s = k\phi$). So when a current I passes through the coil, it will rotate to a position where $\tau_s = \tau_m$, or

$$k\phi = NIAB,$$

which gives

$$I = \frac{k}{NAB}\,\phi.$$

Since I and ϕ are directly proportional to each other, the angular scale may be calibrated to give the value of the current passing through the galvanometer. It can then be used as a current-measuring device.

*The effect of iron on a magnetic field will be discussed in Sec. 29-7.

Direct Current Motor

Direct current motors convert electric energy into mechanical energy. An important component of a dc motor is a planar coil that is placed between the poles of a strong magnet, as depicted in Fig. 28-13*a*. At the instant represented by Fig. 28-13*b*, the magnetic torque is causing the coil to rotate clockwise. This rotation decreases θ, so the torque decreases correspondingly until it vanishes at $\theta = 0°$ (Fig. 28-13*c*). Because it is moving, the coil continues to rotate past this position, whereupon the torque reverses direction (Fig. 28-13*d*). The coil then slows

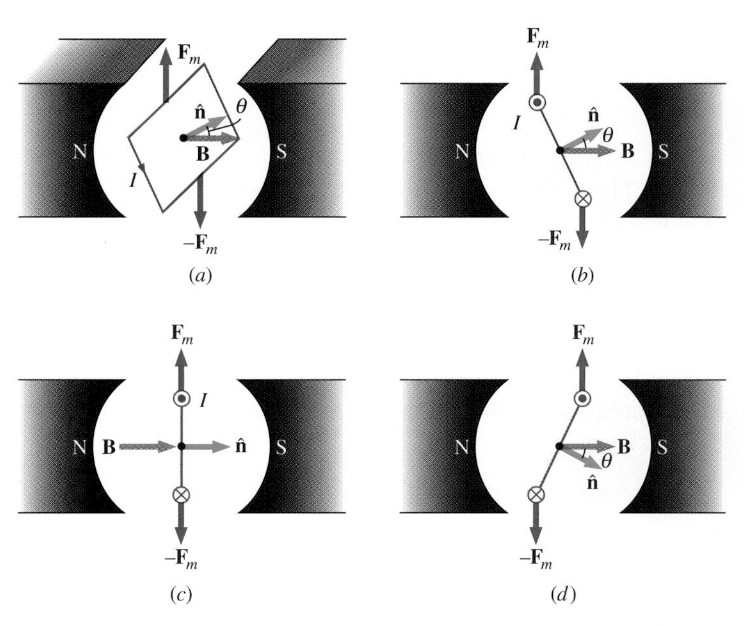

FIGURE 28-13 The various positions of a rotating rectangular coil in a dc motor. (*b*), (*c*), and (*d*) show side views of the coil.

down, stops, and begins to rotate counterclockwise. Upon reaching 0°, it again overshoots, stops, and reverses direction back toward 0°. If left unattended, these oscillations quickly become damped, and the coil stops in its equilibrium position at 0°.

In a simple dc motor, a *commutator*, two conducting half-rings insulated from each other, is connected between the coil and a power source. With the help of Fig. 28-14, you can show

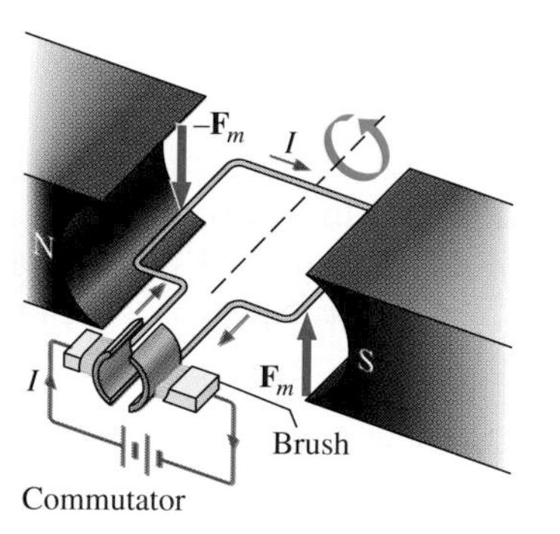

FIGURE 28-14 Whenever the brushes are in contact with the commutator, the power supply drives a current through the coil.

that the commutator makes the current (and therefore the torque) reverse direction whenever the coil passes through 0° or 180°. As a result, the coil is constantly turned counterclockwise and rotates steadily. This rotation is typically used to drive an external mechanism through an attached shaft.

A practical dc motor actually has many turns of wire distributed around a cylindrical soft-iron core. This coil-core arrangement, an example of which is shown in Fig. 28-15, is called an *armature*. With this distribution of current loops, the armature experiences a constant torque. The field of the electromagnet is produced by current from the same source that supplies the rotating coil.

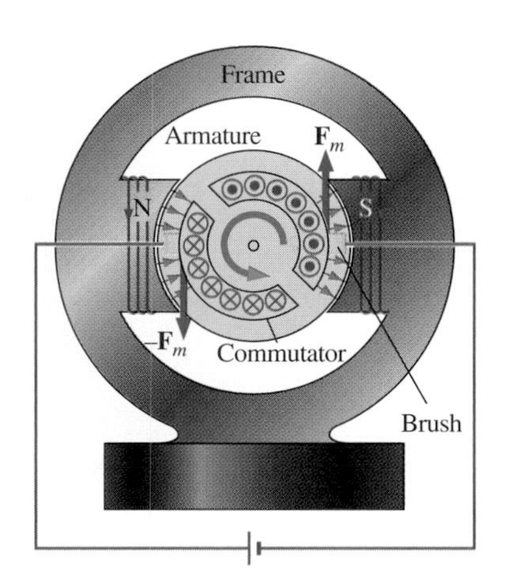

FIGURE 28-15 A representation of a practical dc motor. The currents through the armature coils and the electromagnet are supplied by the same source.

HALL EFFECT

In 1879 E. H. Hall (1855–1938) devised an experiment that can be used to identify the sign of the predominant charge carriers in a conducting material. From a historical perspective, this experiment is very important, for it was the first to demonstrate that the charge carriers in most metals are negative. We investigate the Hall effect by studying the motion of the free electrons along a metallic strip of width l in a constant magnetic field. (See Fig. 28-16*a*.) Since the electrons are moving from left to right, the magnetic force they experience pushes them to the bottom edge of the strip. This leaves an excess of positive charge at the top edge of the strip, resulting in an electric field $\mathbf{E}_H$ directed from top to bottom. The charge concentration at both edges builds up until the electric force on the electrons in one direction is balanced by the magnetic force on them in the opposite direction. Equilibrium is reached when

$$eE_H = ev_d B,$$

where e is the magnitude of the electron charge, v_d is the drift speed of the electrons, and E_H is the magnitude of the electric field created by the separated charge. Solving this for the drift speed, we find

$$v_d = \frac{E_H}{B}.$$

If the current in the strip is I, then from Eq. (26-2),

$$I = nev_d A,$$

where n is the number of charge carriers per unit volume and A is the cross-sectional area of the strip. Combining the equations for v_d and I, we have

$$I = ne\left(\frac{E_H}{B}\right)A.$$

The field E_H is related to the potential difference V_H between the edges of the strip by

$$E_H = \frac{V_H}{l}.$$

The quantity V_H is called the **Hall potential**. It can be measured with a potentiometer. Finally, with the equations for I and E_H combined, we obtain

$$V_H = \frac{IBl}{neA}, \tag{28-11}$$

where the upper edge of the strip of Fig. 28-16*a* is *positive* with respect to the lower edge.

What if the charge carriers are positive, as represented in Fig. 28-16*b*? For the same current I, the magnitude of V_H is still given by Eq. (28-11). However, the upper edge is now *negative* with respect to the lower edge. Therefore, by simply measuring the sign of V_H, we can determine the sign of the majority charge carriers in a material.

Hall-potential measurements show that electrons are the dominant charge carriers in most metals. However, there are a few metals, such as tungsten and beryllium, and many semiconductors whose Hall potentials indicate that the majority charge carriers are positive. It turns out that conduction by positive charge is caused by the migration of missing-electron sites (called *holes*) on ions. Conduction by holes is studied in courses on condensed-matter physics.

The Hall effect can be used to measure magnetic fields. If a material with a known density of charge carriers n is placed in a magnetic field and V_H is measured, then the field can be determined with Eq. (28-11). In research laboratories where the fields of electromagnets used for precise measurements have to be extremely steady, a "Hall probe" is commonly used as part of an electronic circuit that regulates the field.

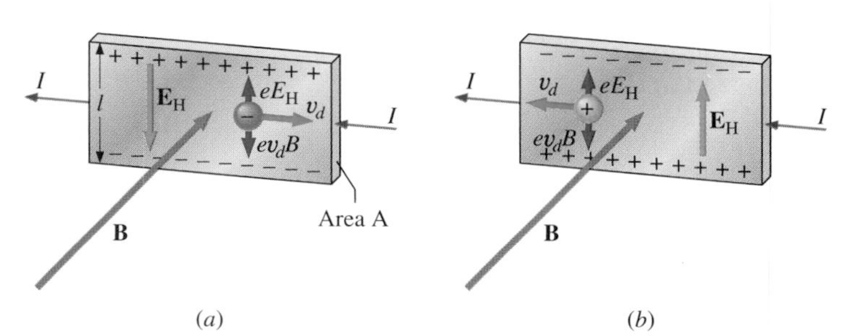

FIGURE 28-16 In the Hall effect, a potential difference between the top and bottom edges of the metal strip is produced when moving charge carriers are deflected by the magnetic field.

EXAMPLE 28-5 **THE HALL POTENTIAL IN A SILVER RIBBON**

Figure 28-17 shows a silver ribbon whose cross section is 1.0 cm by 0.20 cm. The ribbon carries a current of 100 A from left to right, and it lies in a uniform magnetic field of magnitude 1.5 T. Using $n = 5.9 \times 10^{28}$ electrons per cubic meter for silver, find the Hall potential between the edges of the ribbon.

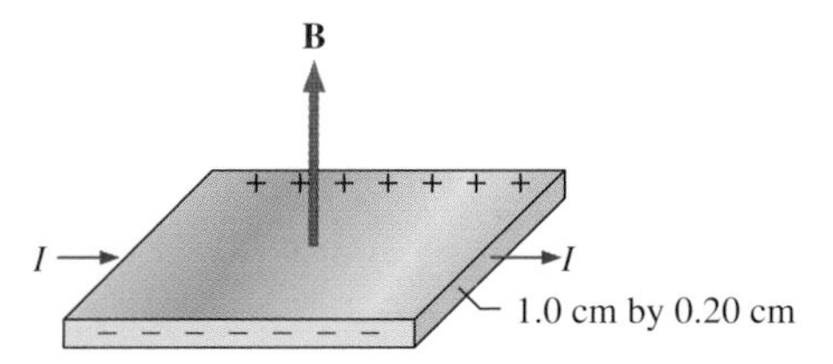

FIGURE 28-17 A silver ribbon in a magnetic field.

SOLUTION Since the majority charge carriers are electrons, the polarity of the Hall voltage is that indicated in the figure. The value of V_H is calculated with Eq. (28-11):

$$V_H = \frac{IBl}{neA} = \frac{(100\text{ A})(1.5\text{ T})(1.0 \times 10^{-2}\text{ m})}{(5.9 \times 10^{28}/\text{m}^3)(1.6 \times 10^{-19}\text{ C})(2.0 \times 10^{-5}\text{ m}^2)}$$
$$= 7.9 \times 10^{-6}\text{ V}.$$

As in this example, the Hall potential is generally very small, and careful experimentation with sensitive equipment is required for its measurement.

DRILL PROBLEM 28-8

A Hall probe consists of a copper strip ($n = 8.5 \times 10^{28}$ electrons per cubic meter) that is 2.0 cm wide and 0.10 cm thick. What is the magnetic field when $I = 50$ A and the Hall potential is (*a*) 4.0 μV and (*b*) 6.0 μV?
ANS. (*a*) 1.1 T; (*b*) 1.6 T.

VELOCITY SELECTOR

A *velocity selector* is used to obtain a beam of charged particles (ions) all moving with one particular velocity. The essentials of this device are shown in Fig. 28-18. Of the charged particles in region I, only those traveling along the line AA′ pass through the two small openings O_1 and O_2. They then encounter mutually perpendicular electric and magnetic fields **E** and **B** in region II.

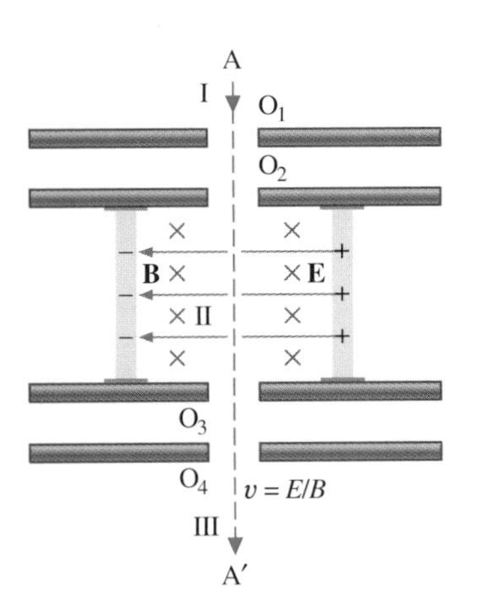

FIGURE 28-18 The velocity selector. The ions emerging at O_4 are all moving with the speed $v = E/B$.

Those particles that are not deflected by the combined fields continue along AA′ and pass through the small openings O_3 and O_4. For these particles, the net force is zero, so

$$qE = qvB,$$

and

$$v = \frac{E}{B}. \qquad (28\text{-}12)$$

Emerging from the velocity selector at O_4 is a beam of charged particles, all moving in the same direction with speed $v = E/B$, which can be set to the desired value by adjusting the fields.

DRILL PROBLEM 28-9

An ion passes through a velocity selector whose electric and magnetic fields are $E = 2.0 \times 10^4$ V/m and $B = 0.10$ T. Determine the speed of the ion.
ANS. 2.0×10^5 m/s.

MASS SPECTROMETER

The *mass spectrometer* is a device that separates ions according to their charge-to-mass ratios. One particular version, the *Bainbridge mass spectrometer*, is illustrated in Fig. 28-19. Ions produced at a source are first sent through a velocity selector, from which they all emerge with the same speed $v = E/B$. They then enter a uniform magnetic field $\mathbf{B}_0$ where they travel in a circular path whose radius R is given by Eq. (28-3). The radius is measured by a particle detector located as shown in the figure.

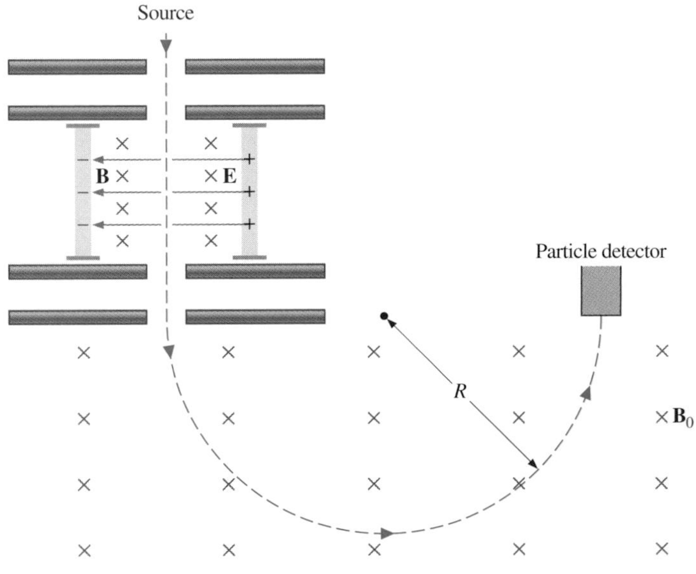

FIGURE 28-19 The Bainbridge mass spectrometer.

The relationship between the charge-to-mass ratio q/m and the radius R is determined by combining Eqs. (28-3) and (28-12):

$$\frac{q}{m} = \frac{E}{BB_0 R}. \tag{28-13}$$

Since most ions are singly charged ($q = 1.6 \times 10^{-19}$ C), measured values of R can be used with this equation to determine the masses of ions. With modern instruments, masses can be determined to one part in 10^8.

An interesting use of the spectrometer is as part of a system for detecting very small leaks in research apparatus. In low-temperature physics laboratories, a device known as a *dilution refrigerator* uses a mixture of He^3 and He^4 and other cryogens to reach temperatures well below 1 K. The performance of the refrigerator is severely hampered if even a minute leak between its various compartments occurs. Consequently, before it is cooled down to the desired temperature, the refrigerator is subjected to a leak test. A small quantity of gaseous helium is injected into one of its compartments while an adjacent, but supposedly isolated, compartment is connected to a high-vacuum pump to which a mass spectrometer is attached. A heated filament ionizes any helium atoms evacuated by the pump. The detection of these ions by the spectrometer then indicates a leak between the two compartments of the dilution refrigerator.

DRILL PROBLEM 28-10

With hydrogen as a source, singly charged ions are found in a Bainbridge mass spectrometer at R = 3.26 mm and R = 6.56 mm. If $E = 2.00 \times 10^5$ V/m and $B = B_0 = 0.800$ T, what do you conclude about the hydrogen sample?

ANS. It contains two isotopes of hydrogen, one with $m = 1.67 \times 10^{-27}$ kg ($_1H^1$) and one with $m = 3.36 \times 10^{-27}$ kg ($_1H^2$).

Cyclotron

The cyclotron was developed by E. O. Lawrence (1901–1958) to accelerate charged particles (usually protons, deuterons, or alpha particles) to large kinetic energies. These particles are then

Ernest Lawrence, the inventor of the cyclotron.

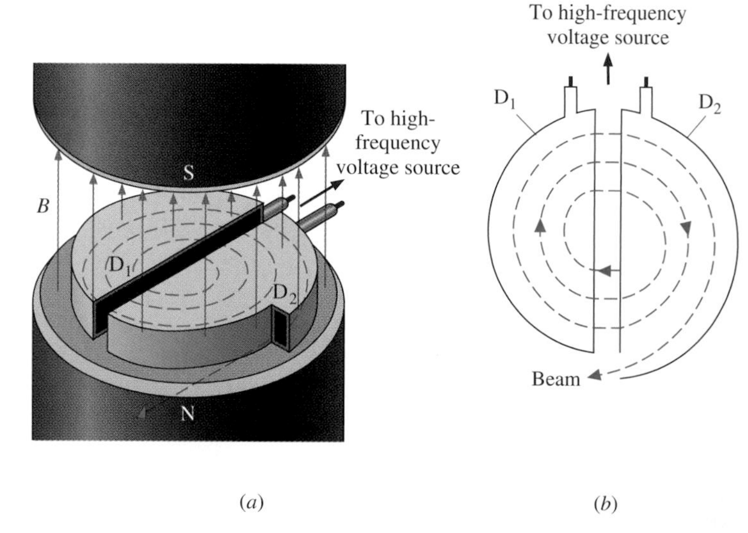

FIGURE 28-20 (*a*) The cyclotron. (*b*) A view of the orbital plane.

used for nuclear-collision experiments and to produce radioactive isotopes. The cyclotron is illustrated in the sketches of Fig. 28-20. The particles move between two flat, semicylindrical metallic containers D_1 and D_2 called "dees." The dees are enclosed in a larger metal container, and the apparatus is placed between the poles of an electromagnet that provides a uniform magnetic field. Air is removed from the large container so that the particles neither lose energy nor are deflected because of collisions with air molecules. The dees are connected to a high-frequency voltage source that provides an alternating electric field in the small region between them. Because the dees are made of metal, their interiors are shielded from the electric field.

Suppose a positively charged particle is injected into the gap between the dees when D_2 is at a positive potential relative to D_1. The particle will then be accelerated across the gap and enter D_1 after gaining a kinetic energy $q\overline{V}$, where $\overline{V}$ is the average potential difference the particle experiences between the dees. When the particle is inside D_1, only the uniform magnetic field **B** of the electromagnet acts on it, so the particle moves in a circle of radius (Eq. (28-3)

$$r = \frac{mv}{qB}$$

with a period (Eq. 28-4)

$$T = \frac{2\pi m}{qB}.$$

The period of the alternating voltage source is set at T, so while the particle is inside D_1 moving along its semicircular orbit in a time $T/2$, the polarity of the dees is reversed. When the particle reenters the gap, D_1 is positive with respect to D_2, and the particle is again accelerated across the gap, thereby gaining a kinetic energy $q\overline{V}$. The particle then enters D_2, circulates in a slightly larger circle, and emerges from D_2 after spending a time $T/2$ in this dee. This process repeats over and over until the orbit of the particle reaches the boundaries of the dees. At that point, the particle (actually, a beam of particles) is extracted from the cyclotron and used for some experimental purpose.

(*a*)

(*b*)

(*a*) One of the dees from a cyclotron. (*b*) The Fermilab particle accelerator at Batavia, Illinois. The main synchrotron ring is over 1.5 km in diameter.

The operation of the cyclotron depends on the fact that in a uniform magnetic field a particle's orbital period is independent of its radius and its kinetic energy. Consequently, the period of the alternating voltage source need only be set at the one value given by Eq. (28-4). With that setting, the electric field accelerates particles every time they are between the dees.

If the maximum orbital radius in the cyclotron is R, then from Eq. (28-3) the maximum speed of a circulating particle of mass m and charge q is

$$v_{\max} = \frac{qBR}{m},$$

and its kinetic energy when ejected from the cyclotron is

$$\frac{1}{2}mv_{\max}^2 = \frac{q^2B^2R^2}{2m}. \tag{28-14}$$

The maximum kinetic energy attainable with this type of cyclotron is approximately 30 MeV. Above this energy, relativistic effects become important, which causes the orbital period to increase with the radius. Up to energies of several hundred MeV, the relativistic effects can be compensated for by making the magnetic field gradually increase with the radius of the orbit. However, for very high energies, much more elaborate methods must be used to accelerate particles.

Particles are accelerated to very high energies with either *linear accelerators* or *synchrotrons*. The linear accelerator accelerates particles continuously with the electric field of an electromagnetic wave that travels down a long evacuated tube. The Stanford linear accelerator is about 3.3 km long and accelerates electrons and positrons to energies of 50 GeV. The synchrotron is constructed so that its bending magnetic field increases with particle speed in such a way that the particles stay in an orbit of fixed radius. The world's two highest-energy synchrotrons are located at the Fermi Laboratory in Batavia, Illinois, and at CERN, which is on the Swiss-French border near Geneva. These synchrotrons can accelerate beams of approximately 10^{13} protons to energies of about 10^3 GeV.

EXAMPLE 28-6 ACCELERATING ALPHA PARTICLES IN A CYCLOTRON

A cyclotron used to accelerate alpha particles ($m = 6.65 \times 10^{-27}$ kg, $q = 3.2 \times 10^{-19}$ C) has a radius of 0.50 m and a magnetic field of 1.8 T. (*a*) What is the period of revolution of the alpha particles? (*b*) What is their maximum kinetic energy?

SOLUTION (*a*) From Eq. (28-4), the period is given by

$$T = \frac{2\pi m}{qB} = \frac{2\pi(6.65 \times 10^{-27}\ \text{kg})}{(3.2 \times 10^{-19}\ \text{C})(1.8\ \text{T})} = 7.3 \times 10^{-8}\ \text{s}.$$

(*b*) We can determine the maximum kinetic energy from Eq. (28-14):

$$\frac{1}{2}mv_{\max}^2 = \frac{q^2B^2R^2}{2m} = \frac{(3.2 \times 10^{-19}\ \text{C})^2(1.8\ \text{T})^2(0.50\ \text{m})^2}{2(6.65 \times 10^{-27}\ \text{kg})}$$
$$= 6.2 \times 10^{-12}\ \text{J} = 39\ \text{MeV}.$$

DRILL PROBLEM 28-11

A cyclotron is to be designed to accelerate protons to kinetic energies of 20 MeV using a magnetic field of 2.0 T. What is the required radius of the cyclotron?
ANS. 0.32 m.

SUMMARY

1. **The magnetic field and its force on charged particles**
   If a particle of charge q has a velocity $\mathbf{v}$ in a magnetic field $\mathbf{B}$, the force on that particle due to the magnetic field is

   $$\mathbf{F}_m = q\mathbf{v} \times \mathbf{B}.$$

2. **Motion in a uniform magnetic field**
   (*a*) If a particle of mass m and charge q enters a uniform magnetic field moving perpendicular to the field at a velocity $\mathbf{v}$, the particle travels in a circle of radius

   $$R = \frac{mv}{qB}.$$

   The period of its orbit is

   $$T = \frac{2\pi m}{qB}.$$

   (*b*) If the particle's velocity is not perpendicular to the field, the particle moves along a helical path whose projection onto the plane perpendicular to the field is a circle.

3. **Magnetic force on a current**
   (*a*) If a current I flows through an infinitesimal segment of wire of length dl that is in a magnetic field $\mathbf{B}$, the magnetic force on that segment is

   $$d\mathbf{F}_m = I\, d\mathbf{l} \times \mathbf{B},$$

   where $d\mathbf{l}$ is a vector of length dl pointing along the current.

   (*b*) The net force on a finite segment of wire is

   $$\mathbf{F}_m = \int_{\text{wire}} I\, d\mathbf{l} \times \mathbf{B},$$

   which reduces to $I\mathbf{l} \times \mathbf{B}$ for a straight wire.

4. **A current loop in a uniform magnetic field**
   (*a*) If a current I flows through a planar loop of wire encircling an area A, the magnetic dipole moment of the loop is

   $$\mathcal{M} = IA\hat{\mathbf{n}},$$

   where $\hat{\mathbf{n}}$ is a unit vector whose direction is given by a right-hand rule.

   (*b*) When a magnetic dipole of dipole moment $\mathcal{M}$ is in a uniform magnetic field $\mathbf{B}$, the force on the dipole is zero, the torque on it is

   $$\boldsymbol{\tau} = \mathcal{M} \times \mathbf{B},$$

   and its potential energy is

   $$U = -\mathcal{M} \cdot \mathbf{B}.$$

*5. **Applications of the magnetic force law**
   (*a*) The essential component of a pivoted-coil galvanometer is a coil of wire placed in a magnetic field. The deflection of the coil is proportional to the current passing through it.

   (*b*) The operation of the dc motor is based on the fact that a magnetic field exerts a torque on a current-carrying coil placed in a magnetic field. A commutator reverses the direction of the current every half rotation, thereby causing the torque to always turn the coil in the same direction.

   (*c*) The Hall potential across a current-carrying ribbon is proportional to the current through the ribbon. The sign of the Hall potential depends on the sign of the dominant charge carriers in the ribbon.

   (*d*) When ions enter a velocity selector, only those with a speed

   $$v = \frac{E}{B}$$

   emerge from the selector.

   (*e*) The mass spectrometer is commonly used to measure the masses of ions.

   (*f*) The cyclotron is used to accelerate charged particles to high energies.

QUESTIONS

28-1. Discuss the similarities and differences between the electric force on a charge and the magnetic force on a charge.

28-2. (*a*) Is it possible for the magnetic force on a charge moving in a magnetic field to be zero? (*b*) Is it possible for the electric force on a charge moving in an electric field to be zero? (*c*) Is it possible for the resultant of the electric and magnetic forces on a charge moving simultaneously through both fields to be zero?

28-3. At a given instant, an electron and a proton are moving with the same velocity in a constant magnetic field. Compare the magnetic forces on these particles. Compare their accelerations.

28-4. Does increasing the magnitude of a uniform magnetic field through which a charge is traveling necessarily mean increasing the magnetic force on the charge? Does changing the direction of the field necessarily mean a change in the force on the charge?

28-5. Suppose we tried to define the magnetic field so that it was always parallel to the magnetic force. What difficulties would we encounter?

28-6. An electron passes through a magnetic field without being deflected. What do you conclude about the magnetic field?

28-7. If a charged particle moves in a straight line, can you conclude that there is no magnetic field present?

28-8. How could you determine which pole of an electromagnet is north and which pole is south?

28-9. Describe the error that results from accidentally using your left rather than your right hand when determining the direction of a magnetic force.

28-10. Discuss the possible dangerous effects of a very strong magnetic field.

28-11. Consider the magnetic force law $\mathbf{F}_m = q\mathbf{v} \times \mathbf{B}$. Are **v** and **B** always perpendicular? Are $\mathbf{F}_m$ and **v** always perpendicular? What about $\mathbf{F}_m$ and **B**?

28-12. Why can a nearby magnet distort a television picture?

28-13. Suppose an electron enters a region where there is either a uniform electric field or a uniform magnetic field. Will the motion of the electron indicate which type of field it is?

28-14. How can you determine the direction of a magnetic field by observing the deflection of a long, straight wire?

28-15. A magnetic field exerts a force on the moving electrons in a current-carrying wire. What exerts the force on the wire?

28-16. When the switch is closed in the circuit shown in the accompanying figure, in which direction will the wire between the poles of the magnet "jump"?

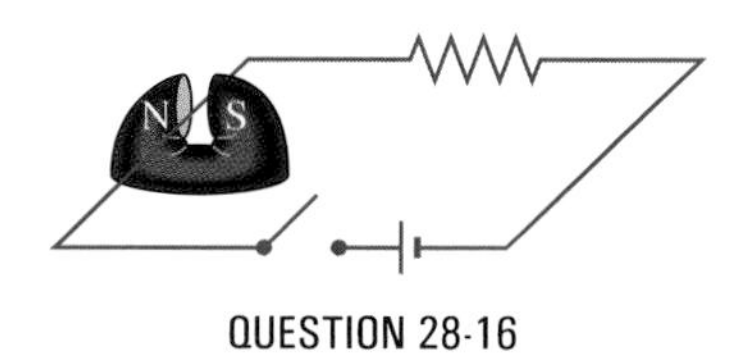

QUESTION 28-16

28-17. There are regions where the magnetic field of the earth is almost perpendicular to the surface of the earth. What difficulties does this cause in the use of a compass?

28-18. Although the magnetic force does no work on a charged particle, it does affect the particle's motion. Can you think of another force that is like the magnetic force in this respect?

28-19. Hall potentials are much larger for poor conductors than for good conductors. Why?

28-20. A Hall potential with the polarity shown in the accompanying figure is produced in a semiconducting ribbon. What is the sign of the dominant charge carriers in the ribbon?

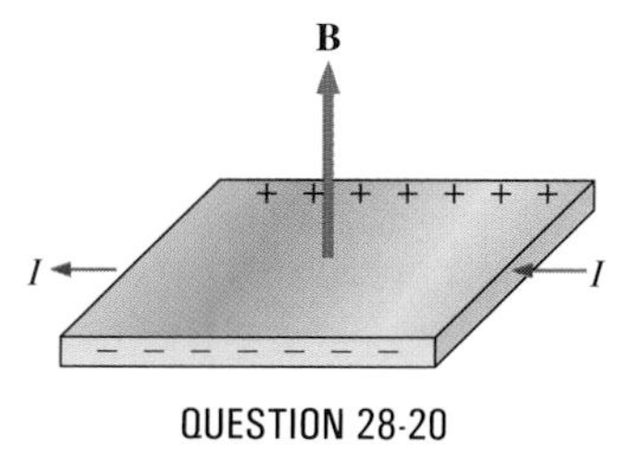

QUESTION 28-20

28-21. Describe the primary functions of the electric field and the magnetic field in a cyclotron.

28-22. Suppose the field of the permanent magnet in a galvanometer has lost some of its strength. What effect will this have on the current readings of the meter?

28-23. When the polarity of the voltage applied to a dc motor is reversed, does the direction of rotation of the armature change?

PROBLEMS

The Magnetic Field and Its Force on Charged Particles

28-1. (*a*) An electron moving eastward in a uniform magnetic field is deflected northward. What is the direction of the magnetic field? (*b*) If a proton moves northward in the same field, which way will it be deflected?

28-2. Charged particles enter magnetic fields as shown in the accompanying figure. For each case, determine the direction of the magnetic force on the particle at the moment represented by the figure.

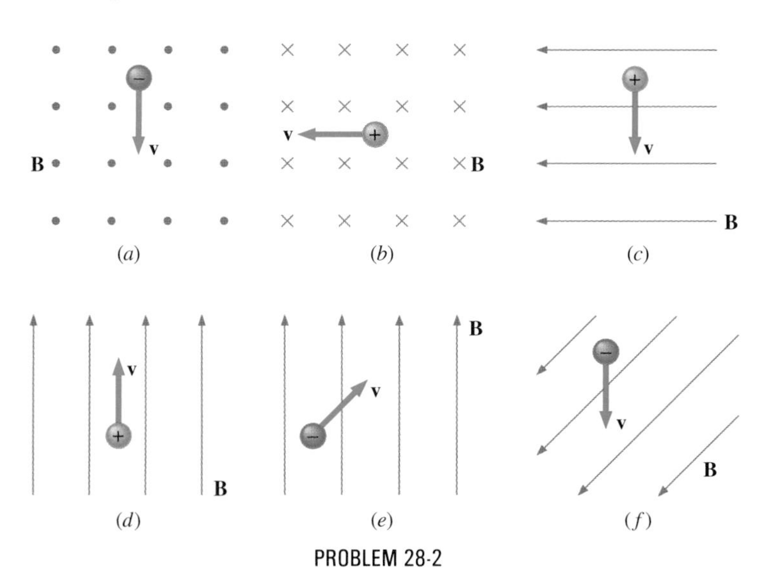

PROBLEM 28-2

28-3. Calculate the magnetic force on a hypothetical particle of charge 1.0×10^{-10} C moving with a velocity of $6.0 \times 10^4 \mathbf{i}$ m/s in a magnetic field $\mathbf{B} = 1.2\mathbf{k}$ T.

28-4. Repeat the previous problem, with the magnetic field now $\mathbf{B} = (0.4\mathbf{i} + 1.2\mathbf{k})$ T.

28-5. An electron is projected into a uniform magnetic field $\mathbf{B} = (0.50\mathbf{i} + 0.80\mathbf{k})$ T with a velocity $\mathbf{v} = (3.0\mathbf{i} + 4.0\mathbf{j}) \times 10^6$ m/s. What is the magnetic force on the electron?

28-6. The mass and charge of a water droplet are 1.0×10^{-4} g and 2.0×10^{-8} C, respectively. If the droplet is given an initial horizontal velocity of $5.0 \times 10^5 \mathbf{i}$ m/s, what magnetic field will keep it moving in this direction? Why must gravity be considered here?

28-7. Four different proton velocities are given in the accompanying figure. For each case, determine the magnetic force on the proton in terms of e, v_0, and B_0.

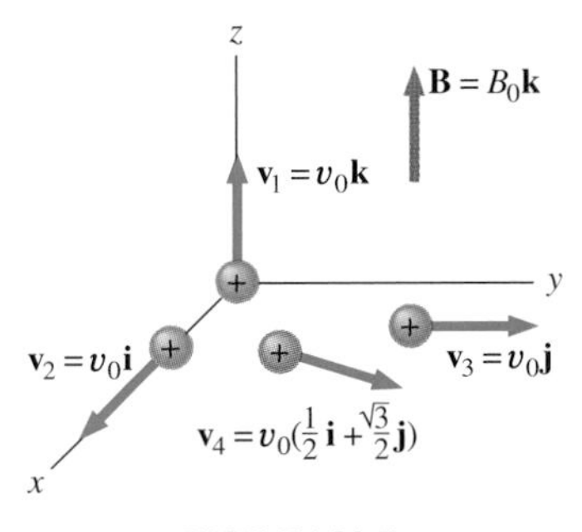

PROBLEM 28-7

28-8. An electron of kinetic energy 2000 eV passes between parallel plates that are 1.0 cm apart and kept at a potential difference of 300 V. What is the strength of the uniform magnetic field **B** that will allow the electron to travel undeflected through the plates? Assume **E** and **B** are perpendicular.

28-9. An alpha particle ($q = 3.2 \times 10^{-19}$ C, $m = 6.65 \times 10^{-27}$ kg) moving with a velocity $\mathbf{v} = (2.0\mathbf{j} - 4.0\mathbf{k}) \times 10^6$ m/s enters a region where $\mathbf{E} = (5.0\mathbf{i} - 2.0\mathbf{j}) \times 10^4$ V/m and $\mathbf{B} = (1.0\mathbf{i} + 4.0\mathbf{k}) \times 10^{-2}$ T. What is the initial force on it?

28-10. An electron moving with a velocity $\mathbf{v} = (4.0\mathbf{i} + 3.0\mathbf{j} + 2.0\mathbf{k}) \times 10^6$ m/s enters a region where there is a uniform electric field and a uniform magnetic field. The magnetic field is given by $\mathbf{B} = (1.0\mathbf{i} - 2.0\mathbf{j} + 4.0\mathbf{k}) \times 10^{-2}$ T. If the electron travels through the region without being deflected, what is the electric field?

28-11. An electron beam, an electric field of magnitude 3.0×10^5 V/m, and a magnetic field of magnitude 5.0×10^{-2} T are mutually perpendicular. If the electron beam is not deflected, (*a*) what is the relative orientation of **v**, **B**, and **E**? (*b*) What is the speed of the electrons in the beam?

Motion in a Uniform Magnetic Field

28-12. An alpha particle travels in a circular path of radius 25 cm in a uniform magnetic field of magnitude 1.5 T. (*a*) What is the speed of the particle? (*b*) What is its kinetic energy in electron-volts? (*c*) Through what potential difference must the particle be accelerated in order to give it this kinetic energy?

28-13. A particle of charge q and mass m is accelerated from rest through a potential difference V, after which it encounters a uniform magnetic field **B**. If the particle moves in a plane perpendicular to **B**, what is the radius of its circular orbit?

28-14. At a particular instant, an electron is traveling west to east with a kinetic energy of 10 keV. The earth's magnetic field has a horizontal component of 1.8×10^{-5} T north and a vertical component of 5.0×10^{-5} T down. (*a*) What is the path of the electron? (*b*) What is the radius of curvature of the path?

28-15. Repeat the calculations of Prob. 28-14 for a proton with the same kinetic energy.

28-16. What magnetic field is required in order to confine a proton moving with a speed of 4.0×10^6 m/s to a circular orbit of radius 10 cm?

28-17. An electron and a proton move with the same speed in a plane perpendicular to a uniform magnetic field. Compare the radii and periods of their orbits.

28-18. A proton and an alpha particle have the same kinetic energy and both move in a plane perpendicular to a uniform magnetic field. Compare the periods of their orbits.

28-19. A singly charged ion takes 2.0×10^{-3} s to complete eight revolutions in a uniform magnetic field of magnitude 2.0×10^{-2} T. What is the mass of the ion?

28-20. A particle moving downward at a speed of 6.0×10^6 m/s enters a uniform magnetic field that is horizontal and directed from east to west. (*a*) If the particle is deflected initially to the north in a circular arc, is its charge positive or negative? (*b*) If $B = 0.25$ T and the charge-to-mass ratio (q/m) of the particle is 4.0×10^7 C/kg, what is the radius of the path? (*c*) What is the speed of the particle after it has moved in the field for 1.0×10^{-5} s? for 2.0 s?

28-21. A proton, a deuteron, and an alpha particle are all accelerated through the same potential difference. They then enter the same magnetic field, moving perpendicular to it. Compute the ratios of the radii of their circular paths. Assume that $m_d = 2m_p$ and $m_\alpha = 4m_p$.

28-22. A singly charged ion moving in a uniform magnetic field of magnitude 7.50×10^{-2} T completes 10 revolutions in 3.47×10^{-4} s. Identify the ion.

28-23. Two particles have the same linear momentum, but particle A has four times the charge of particle B. If both particles move in a plane perpendicular to a uniform magnetic field, what is the ratio R_A/R_B of the radii of their circular orbits?

28-24. A uniform magnetic field of magnitude B is directed parallel to the z axis, as shown in Fig. 28-7. A proton enters the field with a velocity $\mathbf{v} = (4.0\mathbf{j} + 3.0\mathbf{k}) \times 10^6$ m/s and travels in a helical path with a radius of 5.0 cm. (*a*) What is the value of B? (*b*) What is the time required for one trip around the helix? (*c*) Where is the proton 5.0×10^{-7} s after entering the field?

Magnetic Force on a Current

28-25. What is the force per unit length on a long, straight wire carrying a current of 15 mA in a direction perpendicular to a magnetic field of magnitude 0.50 T?

28-26. A 5.0-m section of a long, straight wire carries a current of 10 A while in a uniform magnetic field of magnitude 8.0×10^{-3} T. Calculate the magnitude of the force on the section if the angle between the field and the direction of the current is (*a*) 45°, (*b*) 90°, (*c*) 0°, (*d*) 180°.

28-27. A 1.0-m segment of wire lies along the x axis and carries a current of 2.0 A in the positive x direction. Around the wire is the magnetic field $\mathbf{B} = (3.0\mathbf{j} + 4.0\mathbf{k}) \times 10^{-3}$ T. Find the magnetic force on this segment.

28-28. The accompanying figure shows a long, straight wire that carries a current of 20 A and is placed in a uniform magnetic field of magnitude 0.50 T. What is the magnetic force per unit length on the wire?

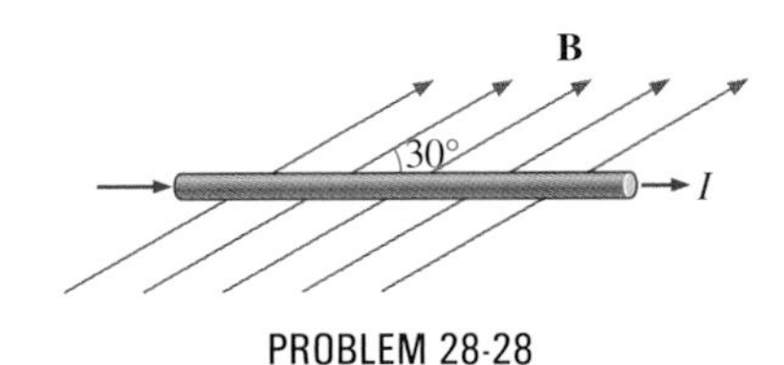

PROBLEM 28-28

28-29. An electromagnet produces a magnetic field of magnitude 1.5 T throughout a cylindrical region of radius 6.0 cm. A straight wire carrying a current of 25 A passes through the field as shown in the accompanying figure. What is the magnetic force on the wire?

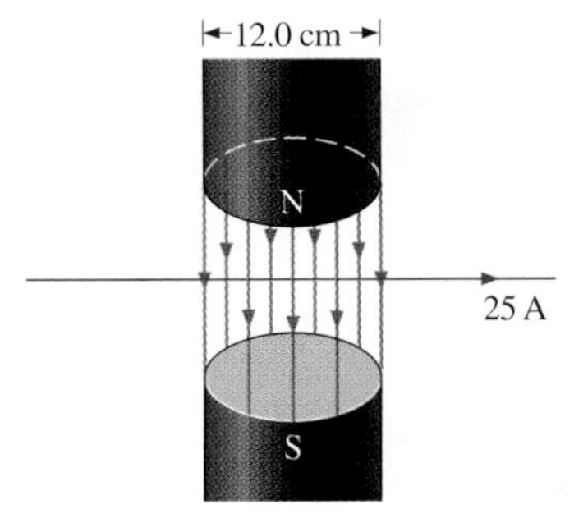

PROBLEM 28-29

28-30. For the cases shown in the accompanying figure, **B** is uniform. Determine the direction of the force on each wire.

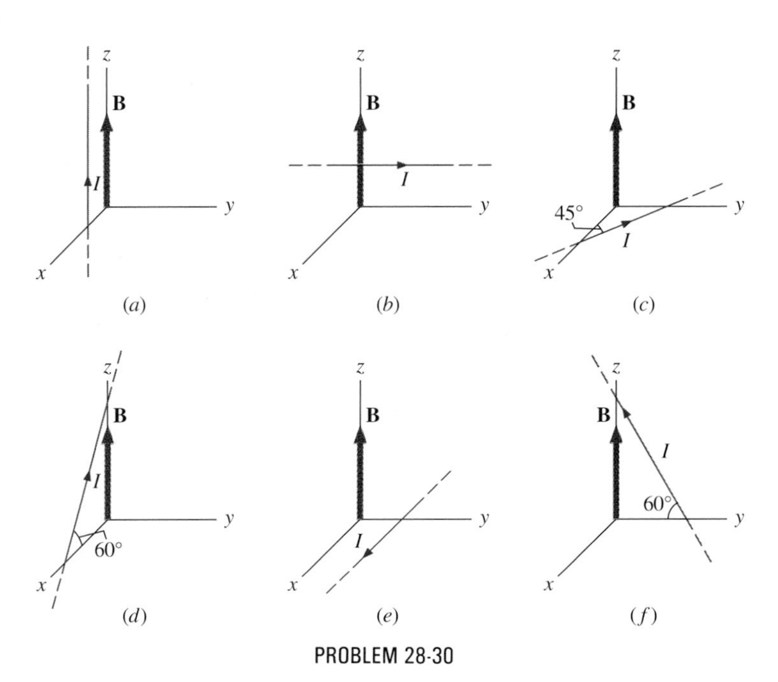

PROBLEM 28-30

28-31. A conducting wire is suspended by two cords, as shown in the accompanying figure. The wire has a linear density of 0.050 kg/m, and it is oriented perpendicular to a magnetic field of magnitude 2.0 T. What current must the wire carry if the tension in the supporting cords is to be zero? What is the direction of the current?

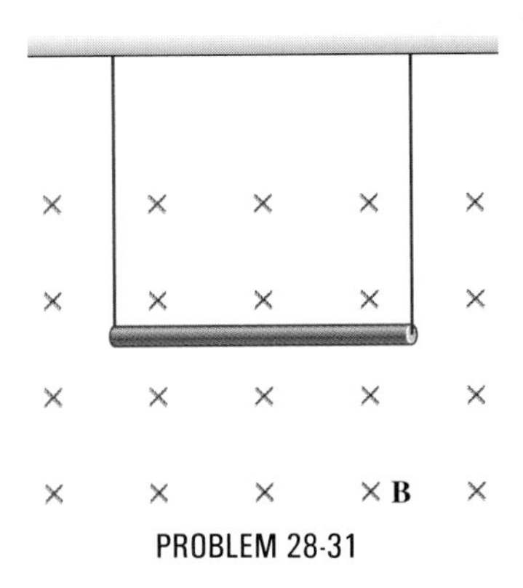

PROBLEM 28-31

A Current Loop in a Uniform Magnetic Field

28-32. The current loop shown in the accompanying figure lies in the plane of the page, as does the magnetic field. Determine the net force and the net torque on the loop if $I = 10$ A and $B = 1.5$ T.

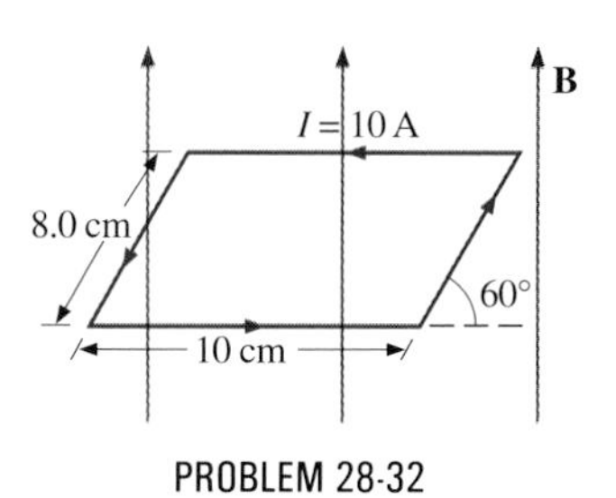

PROBLEM 28-32

28-33. A circular coil of radius 5.0 cm is wound with five turns and carries a current of 5.0 A. If the coil is placed in a uniform magnetic field of strength 0.50 T, what is the maximum torque on it?

28-34. A circular coil of wire of radius 5.0 cm has 20 turns and carries a current of 2.0 A. The coil lies in a magnetic field of magnitude 0.50 T that is directed parallel to the plane of the coil. (*a*) What is the magnetic dipole moment of the coil? (*b*) What is the torque on the coil?

28-35. A current-carrying coil in a magnetic field experiences a torque that is 75 percent of the maximum possible torque. What is the angle between the magnetic field and the normal to the plane of the coil?

28-36. A 4.0-cm by 6.0-cm rectangular current loop carries a current of 10 A. What is the magnetic dipole moment of the loop?

28-37. A circular coil with 200 turns has a radius of 2.0 cm. (*a*) What current through the coil results in a magnetic dipole moment of $3.0\ \mathrm{A \cdot m^2}$? (*b*) What is the maximum torque that the coil will experience in a uniform field of strength 5.0×10^{-2} T? (*c*) If the angle between $\mathcal{M}$ and **B** is 45°, what is the magnitude of the torque on the coil? (*d*) What is the magnetic potential energy of the coil for this orientation?

28-38. The current through a circular wire loop of radius 10 cm is 5.0 A. (*a*) Calculate the magnetic dipole moment of the loop. (*b*) What is the torque on the loop if it is in a uniform 0.20-T magnetic field such that $\mathcal{M}$ and **B** are directed at 30° to each other? (*c*) For this position, what is the potential energy of the dipole?

28-39. A wire of length 1.0 m is wound into a single-turn planar loop. The loop carries a current of 5.0 A, and it is placed in a uniform magnetic field of strength 0.25 T. (*a*) What is the maximum torque that the loop will experience if it is square? (*b*) if it is circular? (*c*) At what angle relative to **B** would the normal to the circular coil have to be oriented so that the torque on it would be the same as the maximum torque on the square coil?

28-40. Consider an electron rotating in a circular orbit of radius r. Show that the magnitudes of the magnetic dipole moment $\mathcal{M}$ and the angular momentum l of the electron are related by $\mathcal{M}/l = e/2m$.

Applications of the Magnetic Force Law

28-41. The coil of a pivoted-coil galvanometer has 50 turns and encloses an area of $5.0\ \mathrm{cm^2}$. The coil rotates in a radial magnetic field of 0.025 T, and the torsional constant of the restoring spring is $\kappa = 1.5 \times 10^{-8}\ \mathrm{N \cdot m/degree}$. Find the angular deflection of the coil when a 2.0-mA current flows through it.

28-42. The coil in a pivoted-coil galvanometer has 200 turns and an area of $3.0\ \mathrm{cm^2}$, and it rotates in a radial magnetic field of strength 0.025 T. If a current of 2.0 mA produces an angular deflection of 20°, what is the torsional constant of the restoring spring?

28-43. A strip of copper is placed in a uniform magnetic field of magnitude 2.5 T. The Hall electric field is measured to be $E_H = 1.5 \times 10^{-3}$ V/m. (*a*) What is the drift speed of the conduction electrons? (*b*) Assuming that $n = 8.0 \times 10^{28}$ electrons per cubic meter and that the cross-sectional area of the strip is $5.0 \times 10^{-6}\ \mathrm{m^2}$, calculate the current in the strip. (*c*) What is the *Hall coefficient* $1/nq$?

28-44. The cross-sectional dimensions of the copper strip shown in the accompanying figure are 2.0 cm by 2.0 mm. The strip carries a current of 100 A, and it is placed in a magnetic field of magnitude $B = 1.5$ T. What are the value and the polarity of the Hall potential in the copper strip?

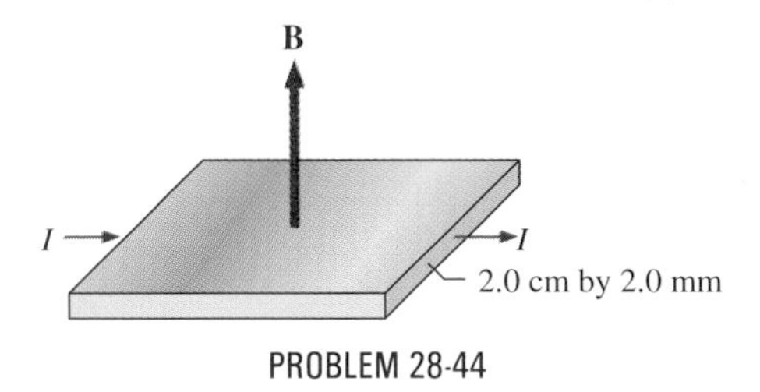

PROBLEM 28-44

28-45. The magnitudes of the electric and magnetic fields in a velocity selector are 1.8×10^5 N/C and 0.080 T, respectively. (*a*) What speed must a proton have to pass through the selector? (*b*) Also calculate the speeds required for an alpha particle and a singly ionized ${}_8O^{16}$ atom to pass through the selector.

28-46. A charged particle moves through a velocity selector at constant velocity. In the selector, $E = 1.00 \times 10^4$ N/C and $B = 0.250$ T. When the electric field is turned off, the charged particle travels in a circular path of radius 3.33 mm. Determine the charge-to-mass ratio of the particle.

28-47. A beam of singly ionized chlorine atoms composed of the two isotopes ${}_{17}Cl^{35}$ ($m = 5.85 \times 10^{-26}$ kg) and ${}_{17}Cl^{37}$ ($m = 6.18 \times 10^{-26}$ kg) is sent through a Bainbridge mass spectrometer. The fields in the velocity selector are $B = 0.500$ T and $E = 4.00 \times 10^4$ V/m, and the strength of the magnetic field that separates the ions is also 0.500 T. If the ions are bent through 180° as shown in Fig. 28-19, what is the separation between the two isotopes when they hit the detector?

28-48. The strengths of the fields in the velocity selector of a Bainbridge mass spectrometer are $B = 0.500$ T and $E = 1.20 \times 10^5$ V/m, and the strength of the magnetic field that separates the ions is $B_0 = 0.750$ T. A stream of singly charged Li ions is found to bend in a circular arc of radius 2.32 cm. What is the mass of the Li ion?

28-49. The strength of the magnetic field in a 20-MeV proton cyclotron is 1.5 T. (*a*) What is the maximum radius of the circulating protons? (*b*) What is the frequency of the voltage source used to accelerate the protons?

28-50. A physicist is designing a cyclotron to accelerate protons to one-tenth the speed of light. The magnetic field will have a strength of 1.5 T. Determine (*a*) the rotational period of the circulating protons and (*b*) the maximum radius of the protons' orbit.

28-51. A cyclotron with an alternating dee potential of frequency 1.2×10^7 Hz accelerates alpha particles to a maximum energy of 18 MeV. (*a*) What is the magnetic field in the cyclotron? (*b*) What is the maximum radius of the alpha particles' orbit?

28-52. The magnetic field in a cyclotron is 1.25 T, and the maximum orbital radius of the circulating protons is 0.40 m. (*a*) What is the kinetic energy of the protons when they are ejected from the cyclotron? (*b*) What is this energy in MeV? (*c*) Through what potential difference would a proton have to be accelerated to acquire this kinetic energy? (*d*) What is the period of the voltage source used to accelerate the protons? (*e*) Repeat the calculations for alpha particles.

General Problems

28-53. A long, rigid wire lies along the x axis and carries a current of 2.5 A in the positive x direction. Around the wire is the magnetic field $\mathbf{B} = 2.0\mathbf{i} + 5.0x^2\mathbf{j}$, with x in meters and $\mathbf{B}$ in millitesla. Calculate the magnetic force on the segment of wire between $x = 2.0$ m and $x = 4.0$ m.

28-54. The accompanying figure shows a copper rod of mass 250 g that is resting on a pair of conducting horizontal rails 1.0 m apart. A current of 40 A flows from one rail to the other through the rod. If the coefficient of static friction between the rod and rails is 0.50, what is the smallest magnetic field that will cause the rod to slide? (*Hint:* The magnetic field is neither vertical nor horizontal.)

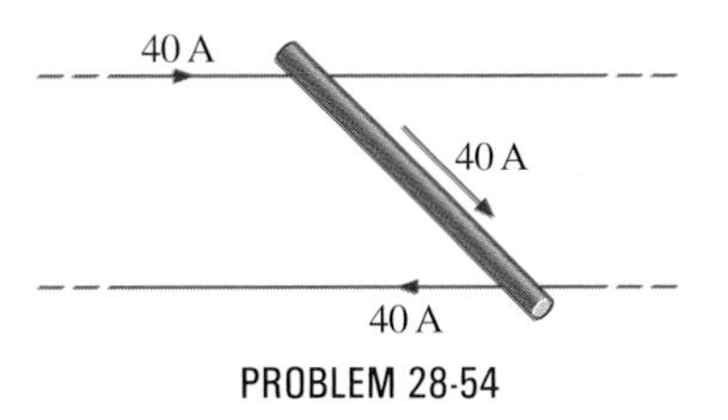

PROBLEM 28-54

28-55. Consider a cylindrical disk of copper that has a radius of 8.0 cm and a height of 0.25 cm. The density of copper is 8.96 g/cm^3, its atomic mass is 63.54 g/mol, and each atom has 29 electrons. Because electrons "spin," they each have a spin magnetic dipole moment of magnitude $\mathcal{M} = 9.3 \times 10^{-24}$ A·m^2. Suppose all magnetic dipoles are aligned parallel to the axis of the cylinder. (*a*) What is the net magnetic dipole moment of the cylinder? (*b*) What current around the circumference of the cylinder would give the same magnetic dipole moment?

28-56. Suppose you use a wire of length l to make a flat, circular coil of N turns. Since l is fixed, the larger the number of turns, the smaller is the area enclosed by the wire. (*a*) If a current I flows through the coil, what is its magnetic dipole moment? (*b*) Show that the maximum magnetic dipole moment is achieved by making $N = 1$ and that in this case, $\mathcal{M} = Il^2/4\pi$.

28-57. An insulating disk of radius r is charged uniformly with a total charge q, and it rotates around its central axis with constant angular velocity ω. Show that the magnitude of the magnetic dipole moment of the rotating charge is given by $\mathcal{M} = \omega q r^2/4$.

28-58. A circular loop of wire of area 10 cm^2 carries a current of 25 A. At a particular instant, the loop lies in the xy plane and is subjected to a magnetic field $\mathbf{B} = (2.0\mathbf{i} + 6.0\mathbf{j} + 8.0\mathbf{k}) \times 10^{-3}$ T. As viewed from above the xy plane, the current is circulating clockwise. (*a*) What is the magnetic dipole moment of the current loop? (*b*) At this instant, what is the magnetic torque on the loop?

28-59. (*a*) Show that the ratio of the Hall electric field E_H to the electric field E that drives the current is

$$\frac{E_H}{E} = \frac{B}{\rho n e},$$

where ρ is the resistivity of the material. (*b*) What is this ratio for copper at room temperature when $B = 0.75$ T?

28-60. In a particular cyclotron, protons are accelerated to an energy E_0. Show that the cyclotron can accelerate ions of atomic mass number A and charge Ze to the energy

$$E = \frac{Z^2}{A} E_0.$$

28-61. Show that $\boldsymbol{\tau} = \mathcal{M} \times \mathbf{B}$ is valid for any planar loop by replacing the given loop with a collection of infinitesimally thin rectangular elements, each carrying a current I, as shown in the accompanying figure.

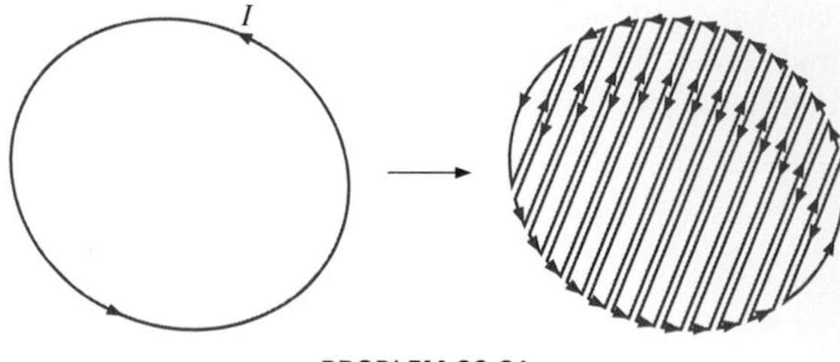

PROBLEM 28-61

A magnetically levitated train.

CHAPTER 29 Sources and Properties of the Magnetic Field

Preview

In this chapter we consider how the magnetic field is created by electric current; we also look at two fundamental properties of the static magnetic field. The main topics to be discussed here are the following:

1. **The Biot-Savart law.** This law gives us a method for calculating the magnetic field of an arbitrary distribution of electric currents. Examples of such calculations are presented.

*2. **Definitions of the ampere and the coulomb.** The mutual magnetic force between two long, parallel, current-carrying wires is used to define the ampere. This definition is then used to define the coulomb.

3. **Gauss' law for the magnetic field.** This law describes one of the two fundamental properties of the static magnetic field.

4. **Ampère's law.** The second fundamental property of the static magnetic field, represented by Ampère's law, is studied.

5. **Applications of Ampère's law.** We discuss how Ampère's law can be used to calculate the magnetic fields of certain spatially symmetric current distributions and consider some examples of these calculations.

6. **The solenoid.** We investigate the magnetic field of the solenoid, a tightly wound cylindrical coil.

*7. **Magnetic materials.** The behavior of materials in magnetic fields is considered, and a simple explanation of magnetism in matter is presented.

In studying the time-independent electric (or electrostatic) field we took the following steps. First we investigated the source of the field, which is electric charge. We found that charge both creates and interacts with electric fields. Next, by interpreting the Coulomb interaction between point charges using the field concept, we obtained a mathematical representation of the electrostatic field that in principle can be used to calculate the field of an arbitrary, fixed charge distribution. Finally, the inverse-square dependence of the field of a point charge on distance from that charge led us to two important results: (1) Gauss' law, which is expressed in terms of a surface integral of the electric field, and (2) the conservative nature of the electrostatic field, which is represented by a line integral of the field.

Our study of the time-independent magnetic field proceeds in a similar fashion. In the last chapter it was stated that the source of the magnetic field is moving charge or electric current. We also saw how moving charge and current interact with a magnetic field. Now we consider the basic law used to calculate the magnetic field produced by a current; it is called the Biot-Savart law. We will then study the characteristics of the field, which will also lead us to two important results: (1) Gauss' law for magnetism, which is expressed in terms of a surface integral of the magnetic field, and (2) Ampère's law, which is represented by a line integral of the magnetic field.

29-1 The Biot-Savart Law

Mass produces the gravitational field and also interacts with that field. Charge produces the electric field and interacts with that field. Since moving charge (or current) interacts with the magnetic field, we might expect that it also creates the field—and it does. The equation used to calculate the magnetic field produced by a current is known as the *Biot-Savart law*. It is an empirical law named in honor of two scientists who investigated the interaction between a straight, current-carrying wire and a permanent magnet. The Biot-Savart law states that at any point P (see Fig. 29-1), the magnetic field $d\mathbf{B}$ due to an element $d\mathbf{l}$ of a current-carrying wire is given by

$$d\mathbf{B} = \frac{\mu_0}{4\pi}\frac{I\,d\mathbf{l}\times\hat{\mathbf{r}}}{r^2}. \tag{29-1}$$

The constant μ_0 is known as the *permeability of free space* and is exactly $4\pi \times 10^{-7}$ T·m/A* in the SI system. The infinitesimal wire segment $d\mathbf{l}$ is directed along the current I, r is the distance from $d\mathbf{l}$ to P, and $\hat{\mathbf{r}}$ is a unit vector that points from $d\mathbf{l}$ to P.

Equation (29-1) is in many ways similar to

$$d\mathbf{E} = \frac{1}{4\pi\epsilon_0}\frac{dq\,\hat{\mathbf{r}}}{r^2},$$

*This is really the value for a vacuum. For magnetic phenomena, however, there is very little difference between air and a vacuum.

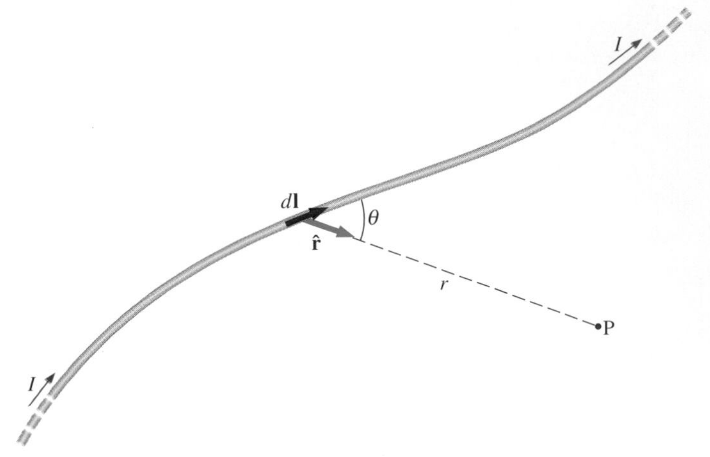

FIGURE 29-1 The current element $I\,d\mathbf{l}$ produces a magnetic field at point P given by the Biot-Savart law.

which gives the electrostatic field of a charge element dq. Both equations are expressed in terms of differential sources, a charge element dq for the electric field and a current element $I\,d\mathbf{l}$ for the magnetic field; both vary as $1/r^2$; both involve a constant of free space; and both are vector equations. However, the directional properties of the two fields are quite different. Whereas the electric field of dq points in the radial direction $\hat{\mathbf{r}}$, the magnetic field of $I\,d\mathbf{l}$ is *perpendicular* to $\hat{\mathbf{r}}$.

The direction of $d\mathbf{B}$ is determined by applying the right-hand rule to the vector product $d\mathbf{l}\times\hat{\mathbf{r}}$. The magnitude of $d\mathbf{B}$ is

$$dB = \frac{\mu_0}{4\pi}\frac{I\,dl\sin\theta}{r^2}, \tag{29-2}$$

where θ is the angle between $d\mathbf{l}$ and $\hat{\mathbf{r}}$. Notice that if $\theta = 0°$, $d\mathbf{B} = 0$. The field produced by a current element $I\,d\mathbf{l}$ has no component parallel to $d\mathbf{l}$.

The magnetic field due to a finite length of current-carrying wire is found by integrating Eq. (29-1) along the wire:

$$\mathbf{B} = \frac{\mu_0}{4\pi}\int_{\text{wire}}\frac{I\,d\mathbf{l}\times\hat{\mathbf{r}}}{r^2}. \tag{29-3}$$

Since this is a vector integral, contributions from different current elements may not point in the same direction. Consequently, the integral is often difficult to evaluate, even for fairly simple geometries.

EXAMPLE 29-1 Magnetic Field of a Thin, Straight Wire

Figure 29-2 shows a section of a thin, straight wire of length l that carries a current I. What is the magnetic field at a point P, which is located a distance R from the wire?

SOLUTION Let's begin by considering the magnetic field due to the current element $I\,d\mathbf{x}$ located at the position x. By the right-hand rule, $d\mathbf{x}\times\hat{\mathbf{r}}$ points out of the page for *any* element along the wire.

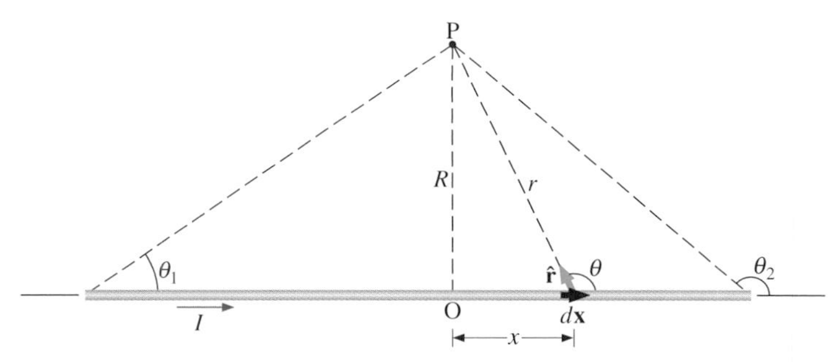

FIGURE 29-2 A section of a thin, straight wire. The independent variable θ has the limits θ_1 and θ_2.

At point P therefore, the magnetic fields due to all current elements have the same direction. This means that we can calculate the net field there by evaluating the scalar sum of the contributions of the elements. With $|d\mathbf{x} \times \hat{\mathbf{r}}| = (dx)(1)\sin\theta$, we have from the Biot-Savart law

$$B = \frac{\mu_0}{4\pi}\int_{\text{wire}} \frac{I \sin\theta\, dx}{r^2}. \qquad (i)$$

The integrand contains three variables (θ, x, and r), two of which must be expressed in terms of the third before the integral can be evaluated. With θ as the independent variable,

$$x = R\cot(\pi - \theta) = -R\cot\theta,$$
$$dx = R\csc^2\theta\, d\theta,$$

and

$$r = \frac{R}{\sin(\pi - \theta)} = \frac{R}{\sin\theta} = R\csc\theta.$$

Upon substituting these into Eq. (i), we find

$$B = \frac{\mu_0 I}{4\pi}\int_{\text{wire}} \frac{\sin\theta(R\csc^2\theta\, d\theta)}{(R\csc\theta)^2} = \frac{\mu_0 I}{4\pi R}\int_{\theta_1}^{\theta_2} \sin\theta\, d\theta, \qquad (ii)$$

where θ_1 and θ_2 correspond to the two ends of the section of wire. Evaluating Eq. (ii), we have

$$B = \frac{\mu_0 I}{4\pi R}\int_{\theta_1}^{\theta_2} \sin\theta\, d\theta = \frac{\mu_0 I}{4\pi R}(\cos\theta_1 - \cos\theta_2). \qquad (29\text{-}4)$$

An important limiting case of this calculation is the infinitely long, straight wire. Then $\theta_1 = 0$, $\theta_2 = \pi$ rad, and Eq. (29-4) reduces to

$$B = \frac{\mu_0 I}{2\pi R}. \qquad (29\text{-}5)$$

The field lines of the magnetic field of the infinite wire are circular and centered at the wire (see Fig. 29-3), and they are identical in every plane perpendicular to the wire. Since the field decreases with distance from the wire, the spacing of the field lines must increase correspondingly with distance. The direction of this magnetic field may be found with a form of the right-hand rule (illustrated in the same figure). If you hold the wire with your right hand so that your thumb points along the current, then your fingers wrap around the wire in the same sense as **B**.

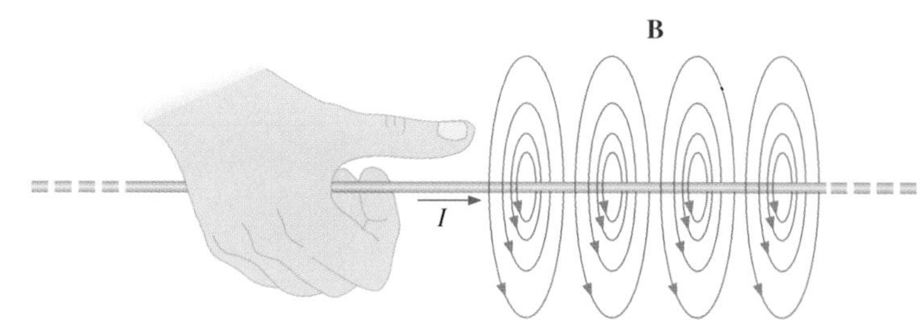

FIGURE 29-3 Some magnetic field lines of an infinite wire. The sense of **B** can be found with a form of the right-hand rule.

The shape of these field lines can be observed experimentally by placing small compass needles near the wire, as illustrated in Fig. 29-4a. When there is no current in the wire, the needles align with the magnetic field of the earth. However, when a large current is sent through the wire, the compass needles all point tangent to the appropriate circle. Iron filings sprinkled on a horizontal surface also delineate the field lines, as shown in Fig. 29-4b.

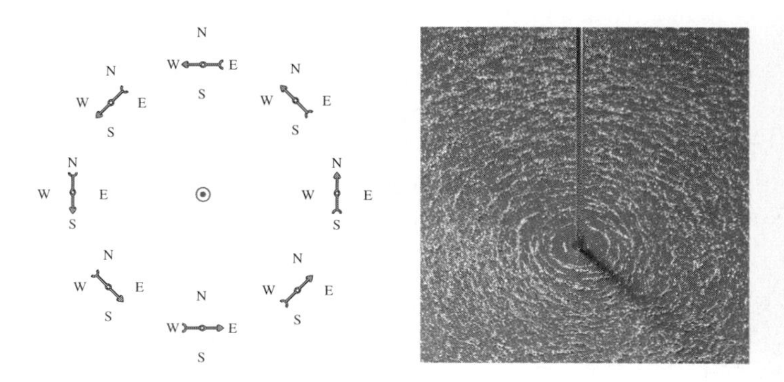

FIGURE 29-4 The shape of the magnetic field lines of a long wire can be seen using (a) small compass needles and (b) iron filings.

EXAMPLE 29-2 Axial Magnetic Field of a Circular Current Loop

The circular current loop of Fig. 29-5 has a radius R, carries a current I, and lies in the xz plane. Calculate the magnetic field due to the current at an arbitrary point P along the axis of the loop.

SOLUTION Let P be a distance y from the center of the loop. From the right-hand rule, the magnetic field $d\mathbf{B}$ at P produced by the current element $I\,d\mathbf{l}$ is directed at an angle θ above the y axis as shown. Since $d\mathbf{l}$ lies along the x axis and $\hat{\mathbf{r}}$ is in the yz plane, the two vectors are perpendicular, and

$$dB = \frac{\mu_0}{4\pi}\frac{I\,dl\sin 90^\circ}{r^2} = \frac{\mu_0}{4\pi}\frac{I\,dl}{y^2 + R^2}, \qquad (i)$$

where we have used $r^2 = y^2 + R^2$. Now consider the magnetic field $d\mathbf{B}'$ due to the current element $I\,d\mathbf{l}'$ which is directly opposite $I\,d\mathbf{l}$ on the loop. The magnitude of $d\mathbf{B}'$ is also given by Eq. (i), but it

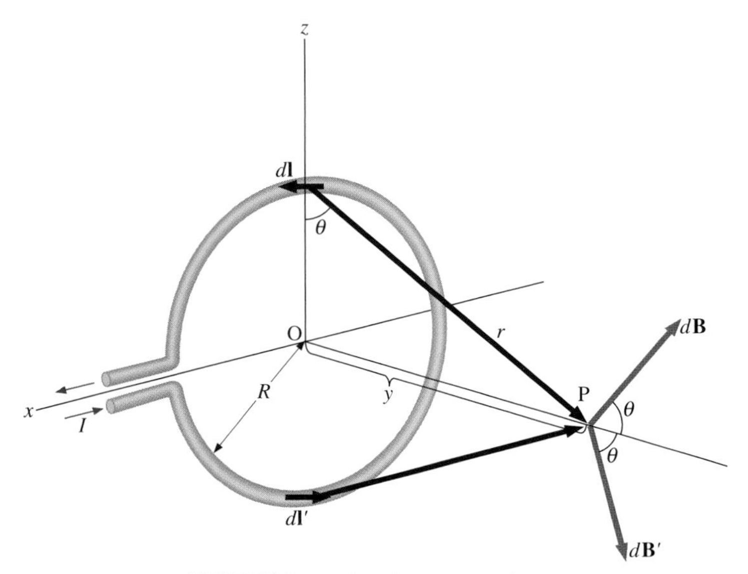

FIGURE 29-5 A circular current loop.

is directed at an angle θ below the y axis. The components of $d\mathbf{B}$ and $d\mathbf{B}'$ perpendicular to the y axis therefore cancel, and in calculating the net magnetic field, only the components along the axis need to be considered. Those components perpendicular to the axis of the loop sum to zero in pairs. Hence at point P,

$$\mathbf{B} = \mathbf{j}\int_{\text{loop}} dB\cos\theta = \mathbf{j}\,\frac{\mu_0 I}{4\pi}\int_{\text{loop}} \frac{\cos\theta\, dl}{y^2 + R^2}. \qquad (ii)$$

For all elements dl on the wire, y, R, and $\cos\theta$ are constant and related by

$$\cos\theta = \frac{R}{\sqrt{y^2 + R^2}}.$$

Now from Eq. (ii), the magnetic field at P is

$$\mathbf{B} = \mathbf{j}\,\frac{\mu_0 IR}{4\pi(y^2 + R^2)^{3/2}}\int_{\text{loop}} dl = \frac{\mu_0 IR^2}{2(y^2 + R^2)^{3/2}}\,\mathbf{j}, \qquad (29\text{-}6a)$$

where we have used $\int_{\text{loop}} dl = 2\pi R$. As discussed in Sec. 28-4, the closed current loop is a magnetic dipole of moment $\mathcal{M} = IA\hat{\mathbf{n}}$. For this example, $A = \pi R^2$ and $\hat{\mathbf{n}} = \mathbf{j}$, so the magnetic field at P can also be written as

$$\mathbf{B} = \frac{\mu_0 \mathcal{M}}{2\pi(y^2 + R^2)^{3/2}}. \qquad (29\text{-}6b)$$

By setting $y = 0$ in Eqs. (29-6), we obtain the magnetic field at the center of the loop:

$$\mathbf{B} = \frac{\mu_0 I}{2R}\,\mathbf{j}, \qquad (29\text{-}7a)$$

or

$$\mathbf{B} = \frac{\mu_0 \mathcal{M}}{2\pi R^3}. \qquad (29\text{-}7b)$$

The calculation of the magnetic field due to the circular current loop at points off-axis requires rather complex mathematics, so we'll just look at the results. The magnetic field lines are shaped as shown in Fig. 29-6. Notice that one field line follows the axis of the loop. This is the field line we just found. Also, very close to the wire the field lines are almost circular, like the lines of the long, straight wire.

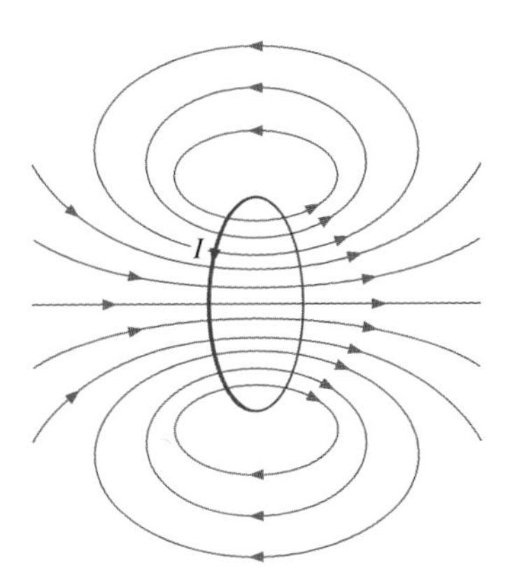

FIGURE 29-6 The magnetic field lines of a circular current loop.

EXAMPLE 29-3

Magnetic Field of the Electron in the Bohr Hydrogen Atom

The simple Bohr model of the hydrogen atom was described in Example 25-2. What is the magnetic field that the proton experiences because of the orbiting electron?

SOLUTION In that example, the kinetic energy of the electron was found to be

$$\frac{1}{2}m_e v^2 = \frac{e^2}{8\pi\epsilon_0 r},$$

so

$$v = \sqrt{\frac{e^2}{4\pi\epsilon_0 m_e r}} \qquad (i)$$

The current due to the motion of the electron is the charge per unit time passing any point in the electron's orbit. It is therefore given by $I = e/T$, where $T = 2\pi r/v$ is the orbital period. With the help of Eq. (i), we obtain

$$I = \frac{e^2}{\sqrt{16\pi^3\epsilon_0 m_e r^3}}. \qquad (ii)$$

The magnetic field at the center of the electron's orbit can now be determined by substituting Eq. (ii) into Eq. (29-7a). This yields

$$B = \frac{\mu_0 e^2}{\sqrt{64\pi^3\epsilon_0 m_e r^5}}. \qquad (iii)$$

The orbital radius for the ground state of hydrogen is $r = 5.3 \times 10^{-11}$ m. Using this together with the known values for μ_0, e, ϵ_0, and m_e in Eq. (iii), we obtain $B \approx 12$ T. This magnetic field is *very* strong! In fact, it is comparable to fields produced in the laboratory with strong superconducting magnets.

DRILL PROBLEM 29-1

Calculate the magnetic field due to a long, straight wire carrying a current of 2.0 A at a point 1.0 cm from the wire.
ANS. 4.0×10^{-5} T.

DRILL PROBLEM 29-2

A very long, straight wire carrying a current of 10 A is placed in a uniform magnetic field of magnitude $B = 1.0 \times 10^{-4}$ T. (See Fig. 29-7.) Where does the resultant magnetic field vanish?
ANS. 2.0 cm below the wire.

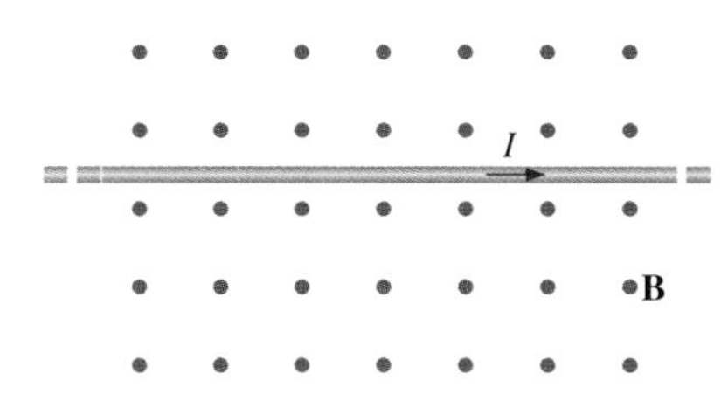

FIGURE 29-7 A long wire placed in a uniform magnetic field.

DRILL PROBLEM 29-3

A current of 100 A flows in a circular loop of radius 10.0 cm. (*a*) What is the magnetic field at the center of the loop? (*b*) What is the magnetic field at a point on the axis of the loop 5.0 cm from its center?
ANS. (*a*) 6.28×10^{-4} T; (*b*) 4.50×10^{-4} T.

*29-2 Definitions of the Ampere and the Coulomb

The definitions of both the ampere and the coulomb are based on the magnetic interaction between a pair of current-carrying wires. To investigate those definitions, let's consider the two very long, parallel wires of Fig. 29-8*a*. They are separated by a distance d and carry currents I_1 and I_2 in the same direction. Each current produces a magnetic field that exerts a force on the current in the other wire. As a result, the wires exert forces on one another.

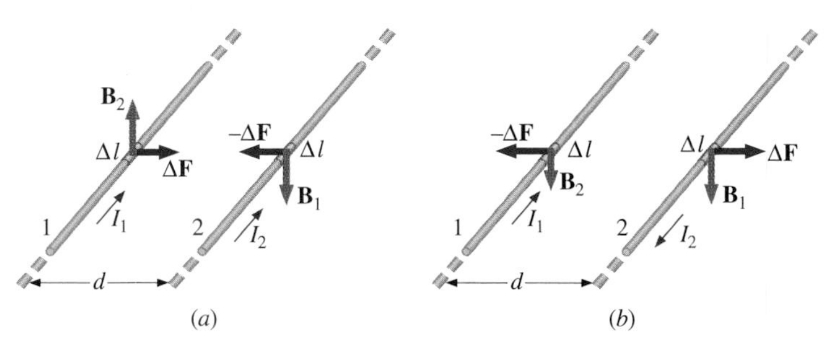

FIGURE 29-8 (*a*) Two long, parallel wires carrying currents I_1 and I_2 in the same direction attract one another. (*b*) If I_1 and I_2 flow in opposite directions, the wires repel.

A simple device used in undergraduate laboratories for measuring the forces between two parallel, current-carrying wires.

At wire 2, the magnitude of the field $\mathbf{B}_1$ due to I_1 is, from Eq. (29-5),

$$B_1 = \frac{\mu_0 I_1}{2\pi d}.$$

Its direction is shown in the figure. From Eq. (28-5), the force $-\Delta\mathbf{F}$ on a section of length Δl of wire 2 has magnitude

$$\Delta F = I_2 B_1 \,\Delta l = \frac{\mu_0 I_1 I_2 \,\Delta l}{2\pi d}.$$

By the right-hand rule, this force is directed toward wire 1. Therefore the magnitude of the force per unit length on wire 2 due to the field of wire 1 is

$$\frac{\Delta F}{\Delta l} = \frac{\mu_0 I_1 I_2}{2\pi d}. \tag{29-8}$$

If you consider the force on wire 1, you will find that it is equal and opposite to that on wire 2. Hence *parallel currents attract*. If the direction of one of the two currents is reversed, the forces also reverse; thus *antiparallel currents repel*, as illustrated in Fig. 29-8*b*.

The force between two parallel, current-carrying wires provides the basis for the definitions of the ampere and the coulomb. With μ_0 *assigned* the value $4\pi \times 10^{-7}$ N/A^2, Eq. (29-8) is used to define the ampere as follows:

Two infinitely long, parallel wires are placed exactly one meter apart in a vacuum with the same current I flowing in each. When I is set so that the magnetic force per unit length on each wire is exactly 2×10^{-7} N/m, I is exactly one ampere.

The coulomb is then defined in terms of the ampere:

If the current in a wire is one ampere, then one coulomb of charge passes through a cross section of the wire in one second.

The definition of the coulomb, along with Coulomb's law and a force measurement, could be used to determine ϵ_0.

However, the electric force between charges cannot be measured very accurately, so this method is not very precise. For this same reason the coulomb is not defined in terms of the electric force. The best value of ϵ_0 is actually determined with the speed of light c (2.99792458×10^8 m/s). It will be shown in Chap. 35 that c is related to μ_0 and ϵ_0 by

$$c^2 = \frac{1}{\mu_0 \epsilon_0};$$

so with μ_0 assigned the value $4\pi \times 10^{-7}$ N/A^2, this equation gives ϵ_0 to nine significant figures.

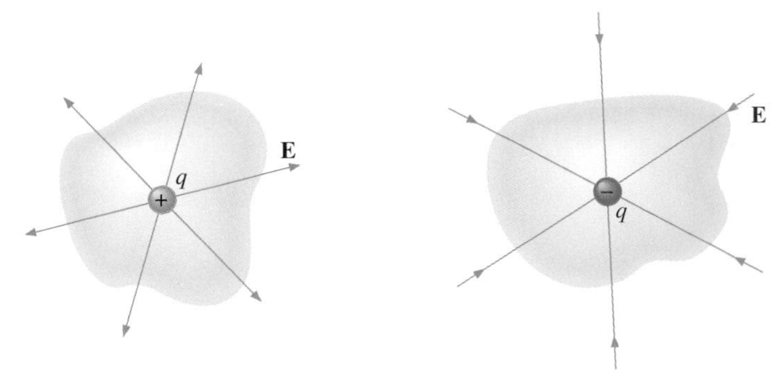

FIGURE 29-10 Since electric field lines begin and end at charges, there must be a nonzero flux through any closed surface bounding a net charge.

DRILL PROBLEM 29-4

The wire rings of Fig. 29-9 lie in parallel planes. The currents through the rings of Fig. 29-9*a* flow in the same direction, while those of Fig. 29-9*b* flow in opposite directions. For each case, determine whether the currents attract or repel.
ANS. They attract in part (*a*) and repel in part (*b*). *Parallel current loops attract; antiparallel current loops repel.*

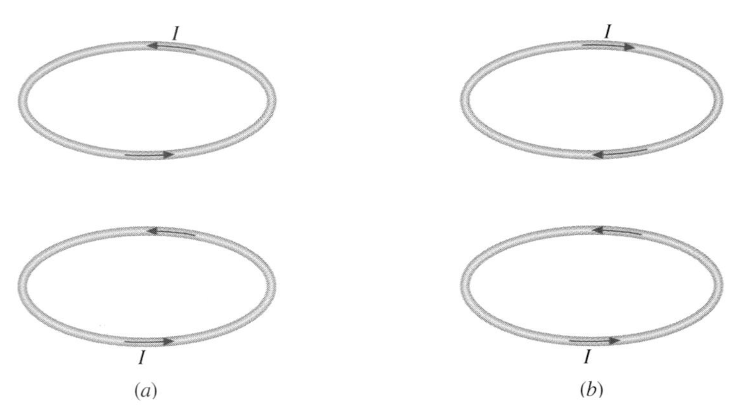

FIGURE 29-9 The wire rings of Drill Prob. 29-4.

29-3 Gauss' Law for the Magnetic Field

For an electric field, Gauss' law tells us that if a closed surface S contains a net charge Q_{in}, then the net electric flux crossing that surface is

$$\Phi_e = \oint_S \mathbf{E} \cdot \hat{\mathbf{n}}\, da = \frac{Q_{\text{in}}}{\epsilon_0}.$$

Inherent in the formulation of Gauss' law are two facts: (1) Electric charges are sources and sinks of the electric field, and (2) the electric field of an isolated point charge varies inversely with the square of the distance from the charge. As Fig. 29-10 shows, there must be a net electric flux passing through a closed surface that encloses an isolated charge. That flux is positive if the charge is positive and negative if the charge is negative. More generally, the net electric flux through a closed surface is positive/negative whenever the net charge enclosed by the surface is positive/negative.

Now let's consider how magnetic field lines differ from electric field lines. Figure 29-11 shows the field lines of a long wire. All of these field lines form closed loops. It turns out that this is a general property of the magnetic field because no point sources or sinks* of the field have been discovered. What then is the net flux across a closed surface in a magnetic field? Because magnetic field lines are closed, any field line that enters the surface must also leave that surface. Since flux is a measure of the net number of field lines crossing a surface, *the net magnetic flux through any closed (or Gaussian) surface must be zero*. This is a statement of *Gauss' law for the magnetic field*. Mathematically, this fundamental property of the magnetic field is represented by a surface integral of the field:

$$\oint_S \mathbf{B} \cdot \hat{\mathbf{n}}\, da = 0, \tag{29-9}$$

where the integral is over a Gaussian surface S, and $\hat{\mathbf{n}}$ is the unit normal to S at the area element da.

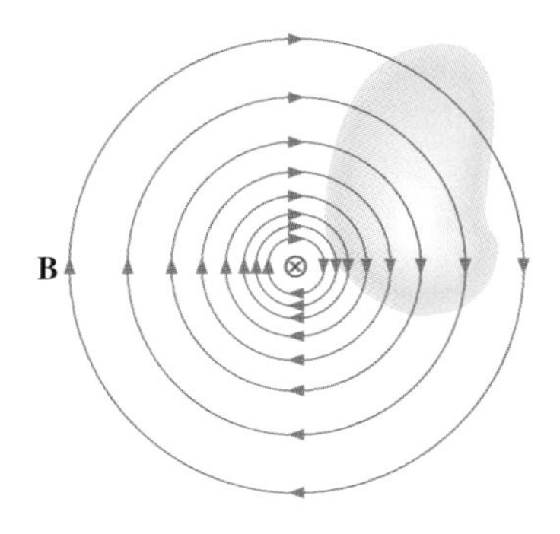

FIGURE 29-11 Magnetic field lines form closed loops, which results in a zero net magnetic flux through any closed surface.

*Scientists still search for the magnetic point source. If one is ever found, Eq. (29-9) will have to be altered so that the right-hand side is no longer zero. Presently, with no strong evidence to the contrary, we must accept the fact that "magnetic point charges" (often called *magnetic monopoles*) do not exist.

EXAMPLE 29-4 **Net Magnetic Flux Through a Closed Surface**

An infinite, current-carrying wire lies along the z axis as shown in Fig. 29-12a. Calculate the magnetic flux through the surface of the cube shown.

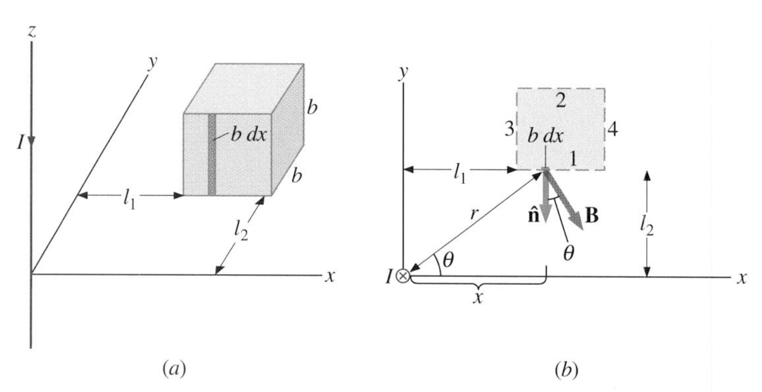

FIGURE 29-12 (a) A cubical surface and an infinite, straight wire. (b) The geometry for determining the flux through the front surface of the cube.

SOLUTION Since the magnetic field of the current-carrying wire is parallel to the top and bottom surfaces of the cube, there is no flux through either of these surfaces. For the front surface (1) shown in Fig. 29-12b, the flux through an area element $b\,dx$ is

$$d\phi_1 = B\cos\theta\, da = \frac{\mu_0 I}{2\pi r}(\cos\theta\, b\, dx). \qquad (i)$$

Now for the geometry shown,

$$\cos\theta = \frac{x}{r} \qquad \text{and} \qquad r^2 = x^2 + l_2^2,$$

so Eq. (i) can be rewritten as

$$d\phi_1 = \frac{\mu_0 Ib}{2\pi}\frac{x\,dx}{x^2 + l_2^2}.$$

Integrating this between $x = l_1$ and $x = l_1 + b$, we obtain

$$\phi_1 = \frac{\mu_0 Ib}{2\pi}\int_{l_1}^{l_1+b}\frac{x\,dx}{x^2 + l_2^2} = \frac{\mu_0 Ib}{4\pi}\ln\frac{(l_1+b)^2 + l_2^2}{l_1^2 + l_2^2}.$$

The fluxes through the other three surfaces can be found in the same way. The results are

$$\text{Back surface (2):} \quad \phi_2 = -\frac{\mu_0 Ib}{4\pi}\ln\frac{(l_1+b)^2 + (l_2+b)^2}{l_1^2 + (l_2+b)^2},$$

$$\text{Left surface (3):} \quad \phi_3 = -\frac{\mu_0 Ib}{4\pi}\ln\frac{l_1^2 + (l_2+b)^2}{l_1^2 + l_2^2},$$

$$\text{Right surface (4):} \quad \phi_4 = \frac{\mu_0 Ib}{4\pi}\ln\frac{(l_1+b)^2 + (l_2+b)^2}{(l_1+b)^2 + l_2^2}.$$

The total flux through the cubical surface is then

$$\Phi_m = \phi_1 + \phi_2 + \phi_3 + \phi_4.$$

Using ln AB = ln A + ln B and ln (A/B) = ln A − ln B, we find

$$\Phi_m = \frac{\mu_0 Ib}{4\pi}\ln\left[\frac{(l_1+b)^2 + l_2^2}{l_1^2 + l_2^2}\,\frac{l_1^2 + (l_2+b)^2}{(l_1+b)^2 + (l_2+b)^2} \times \frac{l_1^2 + l_2^2}{l_1^2 + (l_2+b)^2}\,\frac{(l_1+b)^2 + (l_2+b)^2}{(l_1+b)^2 + l_2^2}\right],$$

which reduces to

$$\Phi_m = \frac{\mu_0 Ib}{4\pi}\ln 1 = 0,$$

in agreement with Gauss' law for magnetism.

29-4 Ampère's Law

The second fundamental property of the static magnetic field is that, unlike the electrostatic field, it is nonconservative. This property involves a relationship between the magnetic field and its source, electric current. It is expressed in terms of the line integral of **B** and is known as *Ampère's law.* This law can also be derived directly from the Biot-Savart law. We now consider that derivation for the special case of an infinite, straight wire.

Figure 29-13 shows an arbitrary plane perpendicular to an infinite, straight wire whose current I is directed out of the page. The magnetic field lines are circles directed counterclockwise and centered on the wire. To begin, let's consider $\oint \mathbf{B}\cdot d\mathbf{l}$ over the closed paths M and N. Notice that one path (M) encloses the wire while the other (N) does not. Since the field lines are circular, $\mathbf{B}\cdot d\mathbf{l}$ is the product of B and the projection of $d\mathbf{l}$ onto the circle passing through $d\mathbf{l}$. If the radius of this particular circle is r, the projection is $r\,d\theta$, and

$$\mathbf{B}\cdot d\mathbf{l} = Br\,d\theta.$$

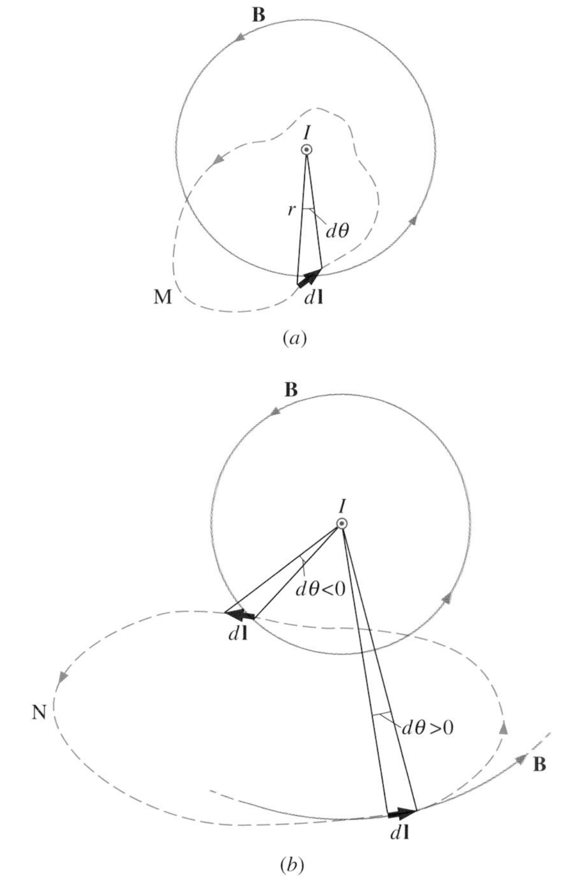

FIGURE 29-13 The current I of a long, straight wire is directed out of the page. The integral $\oint d\theta$ equals 2π and 0, respectively, for paths M and N.

With **B** given by Eq. (29-5),

$$\oint \mathbf{B}\cdot d\mathbf{l} = \oint \left(\frac{\mu_0 I}{2\pi r}\right) r\,d\theta = \frac{\mu_0 I}{2\pi}\oint d\theta.$$

For path M, which circulates around the wire, $\oint_M d\theta = 2\pi$, and

$$\oint_M \mathbf{B}\cdot d\mathbf{l} = \mu_0 I.$$

Path N, on the other hand, circulates through both positive (counterclockwise) and negative (clockwise) $d\theta$ (see Fig. 29-13*b*), and since it is closed, $\oint_N d\theta = 0$. Thus for path N,

$$\oint_N \mathbf{B}\cdot d\mathbf{l} = 0.$$

The extension of this result to the general case is Ampère's law: *Over an arbitrary closed path,*

$$\oint \mathbf{B}\cdot d\mathbf{l} = \mu_0 I, \tag{29-10}$$

where I is the total current passing through any open surface S whose perimeter is the path of integration.

To determine whether a specific current I is positive or negative, curl the fingers of your right hand in the direction of the path of integration, as shown in Fig. 29-14. If I passes through S in the same direction as your extended thumb, I is positive; if it passes through S in the direction opposite to your extended thumb, it is negative.

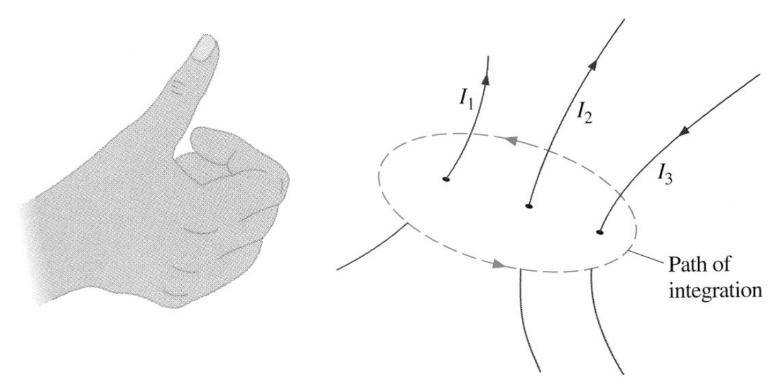

FIGURE 29-14 The signs of the currents in Ampère's law can be determined by a form of the right-hand rule. For this path, I_1 and I_2 are positive, and I_3 is negative.

EXAMPLE 29-5 AMPÈRE'S LAW

Use Ampère's law to evaluate $\oint \mathbf{B}\cdot d\mathbf{l}$ for the current configurations and paths of Fig. 29-15*a–f*.

SOLUTION The results are summarized in the following table.

Configuration	Current I	$\oint \mathbf{B}\cdot d\mathbf{l}$
(*a*)	3 A + 2 A = 5 A	μ_0(5 A)
(*b*)	−3 A − 2 A = −5 A	μ_0(−5 A)
(*c*)	−3 A − 2 A = −5 A	μ_0(−5 A)
(*d*)	3 A + 2 A = 5 A	μ_0(5 A)
(*e*)	4 A + 2 A = 6 A	μ_0(6 A)
(*f*)	3 A − 3 A = 0 A	0

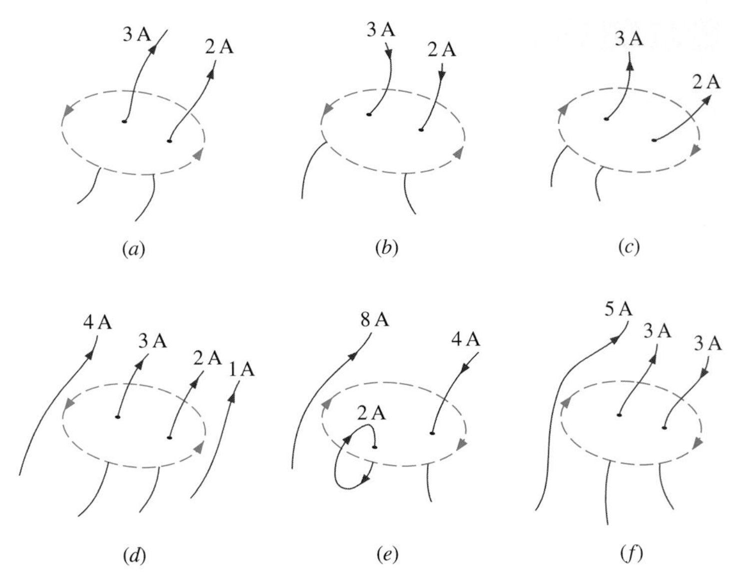

FIGURE 29-15 Current configurations and paths of Example 29-5.

DRILL PROBLEM 29-5

Use Ampère's law to evaluate $\oint \mathbf{B}\cdot d\mathbf{l}$ for the current configurations and paths of Fig. 29-16.
ANS. (*a*) 0; (*b*) μ_0(2 A); (*c*) μ_0(9 A); (*d*) μ_0(−3 A).

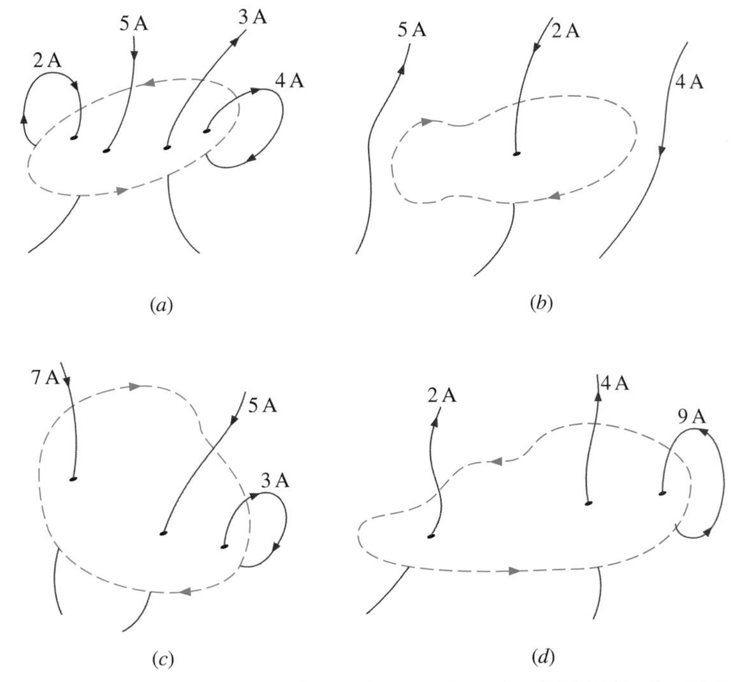

FIGURE 29-16 Current configurations and paths of Drill Prob. 29-5.

29-5 Applications of Ampère's Law

In Sec. 24-3 we used Gauss' law to calculate the electric fields of certain spatially symmetric charge distributions. The integral $\oint_S \mathbf{E}\cdot\hat{\mathbf{n}}\,da$ was first factored into a product of E and a surface area, then the resulting equation was solved for E. With Ampère's law we can follow an analogous procedure to calculate the magnetic fields of many spatially symmetric current distributions. In this case, the integral $\oint \mathbf{B}\cdot d\mathbf{l}$ is factored into a product of B and a path length, then the resulting equation is solved for B. The following examples illustrate this procedure.

EXAMPLE 29-6 Using Ampère's Law to Calculate the Magnetic Field of a Thin Wire

Use Ampère's law to calculate the magnetic field due to a steady current I in an infinitely long, thin, straight wire.

SOLUTION An arbitrary plane perpendicular to the wire, with the current directed out of the page, is shown in Fig. 29-17*a*. The *possible* magnetic field components in this plane, B_r and B_θ, are shown at arbitrary points on a circle of radius r that is centered on the wire. Since the field is cylindrically symmetric, neither B_r nor B_θ varies with position on the circle. Also from symmetry, the radial lines, if they exist, must be directed either all inward or all outward from the wire. This means, however, that there must be a net magnetic flux across an arbitrary cylinder concentric with the wire. (See Fig. 29-17*b*.) As this contradicts Gauss' law, the radial component of the magnetic field must be zero (Fig. 29-17*c*), and we can write

$$\mathbf{B} = B_\theta\hat{\boldsymbol{\theta}}.$$

B_θ can be determined by applying Ampère's law to the circular path shown. Over this path, $\mathbf{B}$ is constant in magnitude and is parallel to $d\mathbf{l}$; so

$$\oint \mathbf{B}\cdot d\mathbf{l} = B_\theta \oint dl = B_\theta(2\pi r),$$

and Ampère's law reduces to

$$B_\theta(2\pi r) = \mu_0 I.$$

Finally, since B_θ is the only component of $\mathbf{B}$, we can drop the subscript and write

$$B = \frac{\mu_0 I}{2\pi r}.$$

This, of course, agrees with the Biot-Savart calculation of Example 29-1.

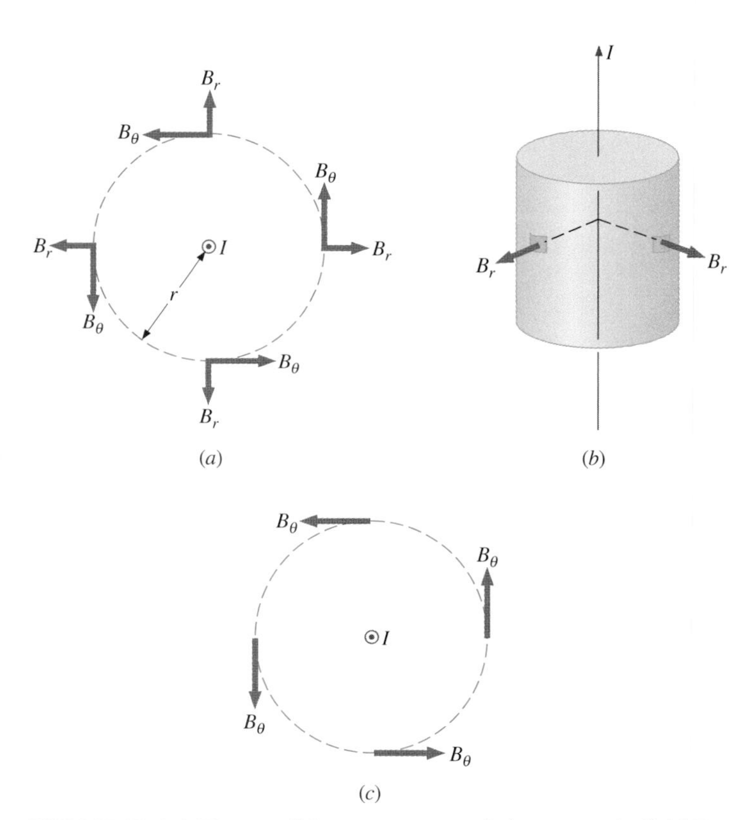

FIGURE 29-17 (*a*) The possible components of the magnetic field **B** due to a current I, which is directed out of the page. (*b*) If the field has a radial component, Gauss' law will be violated. (*c*) The field of a current-carrying wire must therefore be directed as shown.

EXAMPLE 29-7 Calculating the Magnetic Field of a Thick Wire with Ampère's Law

The radius of the long, straight wire of Fig. 29-18*a* is a, and the wire carries a current I_0, which is distributed uniformly over its cross section. Find the magnetic field both inside and outside the wire.

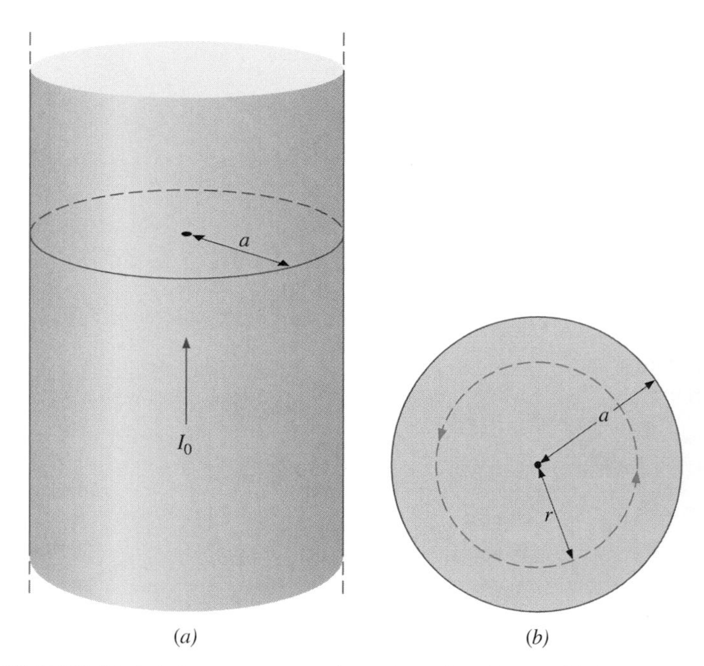

FIGURE 29-18 (*a*) A current-carrying wire of radius a. (*b*) A concentric circular path within the wire.

SOLUTION For the same reasons given in Example 29-6, the magnetic field lines must be circular. Thus for any circular path of radius r that is centered on the wire,

$$\oint \mathbf{B}\cdot d\mathbf{l} = \oint B\,dl = B\oint dl = B(2\pi r).$$

From Ampère's law, this equals the total current passing through any surface bounded by the path of integration.

Consider first a circular path that is inside the wire ($r \leq a$) such as that shown in Fig. 29-18b. Since the total current I_0 is distributed uniformly over the cross section of the wire, the current I passing through the area enclosed by the path is

$$I = \frac{\pi r^2}{\pi a^2} I_0 = \frac{r^2}{a^2} I_0.$$

Using Ampère's law, we obtain

$$B(2\pi r) = \mu_0\left(\frac{r^2}{a^2}\right) I_0,$$

and the magnetic field inside the wire is

$$B = \frac{\mu_0 I_0}{2\pi}\frac{r}{a^2} \qquad (r \leq a).$$

Outside the wire, the situation is identical to that of the infinite, thin wire of the previous example; that is,

$$B = \frac{\mu_0 I_0}{2\pi r} \qquad (r \geq a).$$

The variation of B with r is shown in Fig. 29-19.

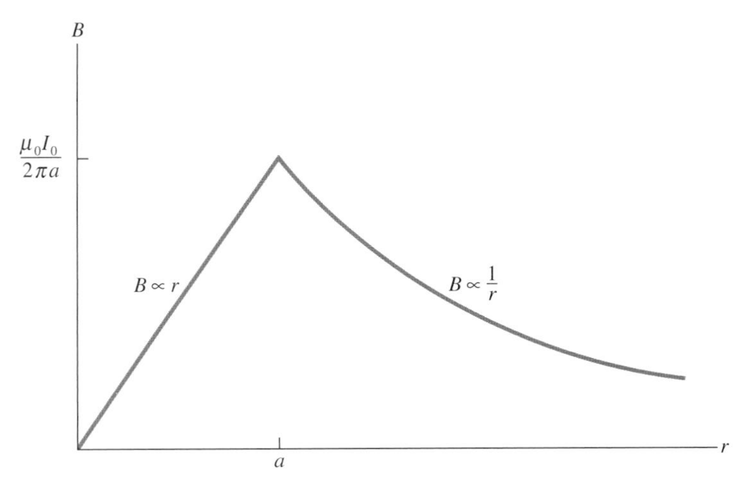

FIGURE 29-19 Variation of the magnetic field produced by the current I_0 in a long, straight wire of radius a.

DRILL PROBLEM 29-6

A hollow conducting cylinder of inner radius r_1 and outer radius r_2 carries a current I_0, which is distributed uniformly over its cross section. (See Fig. 29-20.) What is the magnetic field for (a) $r \leq r_1$, (b) $r_1 \leq r \leq r_2$, and (c) $r \geq r_2$?

ANS. (a) $B = 0$; (b) $B = \frac{\mu_0 I_0}{2\pi r}\frac{r^2 - r_1^2}{r_2^2 - r_1^2}$; ($c$) $B = \frac{\mu_0 I_0}{2\pi r}$.

FIGURE 29-20 The hollow conducting cylinder of Drill Prob. 29-6.

EXAMPLE 29-8 MAGNETIC FIELD OF A TOROID

A *toroid* is a donut-shaped coil closely wound with one continuous wire, as illustrated in Fig. 29-21a. If the current in the wire is I, what is the magnetic field both inside and outside the toroid? The toroid has N windings.

SOLUTION We'll begin by assuming that there is cylindrical symmetry around the axis OO′. Actually, this assumption is not precisely correct, for, as Fig. 29-21b shows, the view of the toroidal coil varies from point to point (for example, P_1, P_2, and P_3) on a circular path centered on OO′. However, if the toroid is tightly wound, all points on the circle become essentially equivalent (Fig. 29-21c), and cylindrical symmetry is an accurate approximation.

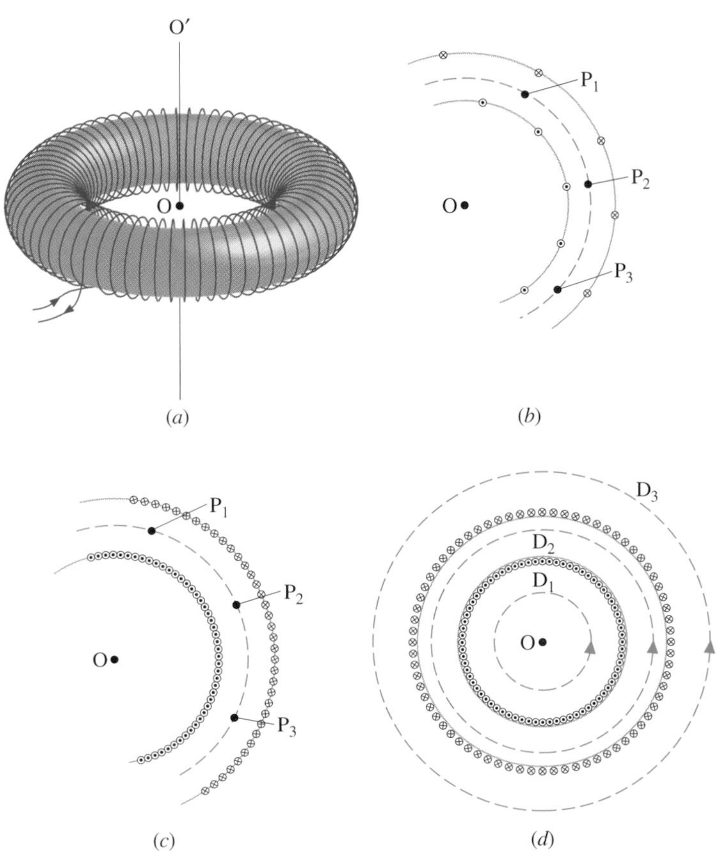

FIGURE 29-21 (a) A toroid. (b) A loosely wound toroid does not have cylindrical symmetry. (c) In a tightly wound toroid, cylindrical symmetry is a very good approximation. (d) Various paths of integration.

With this symmetry, the magnetic field must be tangent to and constant in magnitude along any circular path centered on OO′. This allows us to write for each of the paths D_1, D_2, and D_3 shown in Fig. 29-21*d*

$$\oint \mathbf{B}\cdot d\mathbf{l} = B(2\pi r).$$

Ampère's law relates this integral to the net current passing through any surface bounded by the path of integration. For a path that is external to the toroid, either no current passes through the enclosing surface (path D_1), or the current passing through the surface in one direction is exactly balanced by the current passing through it in the opposite direction (path D_3). In either case, there is no net current passing through the surface, so

$$\oint B(2\pi r) = 0,$$

and

$$B = 0 \text{ (outside the toroid).}$$

For a circular path within the toroid (path D_2), the current in the wire cuts the surface N times, resulting in a net current NI through the surface. We now find with Ampère's law

$$B2\pi r = \mu_0 NI,$$

and

$$B = \frac{\mu_0 NI}{2\pi r} \text{ (within the toroid).} \tag{29-11}$$

The magnetic field is directed in the clockwise direction for the windings shown. When the current in the coils is reversed, the direction of the magnetic field also reverses.

The magnetic field inside the toroid is not constant, as it varies inversely with the distance r from the axis OO′. However, if the central radius R (the radius midway between the inner and outer radii of the toroid) is much larger than the cross-sectional diameter of the coils, the variation is fairly small, and the magnitude of the magnetic field may be *approximated by*

$$B = \frac{\mu_0 NI}{2\pi R} \tag{29-12}$$

everywhere within the toroid.

DRILL PROBLEM 29-7

A toroid of central radius 20.0 cm has 2000 closely spaced windings wrapped in loops of radius 1.00 cm. The current in the coil is 5.00 A. Compute the magnetic field within the toroid at (*a*) the inner radius, (*b*) the central radius, and (*c*) the outer radius of the toroid.
ANS. (*a*) 1.05×10^{-2} T; (*b*) 1.00×10^{-2} T; (*c*) 0.95×10^{-2} T.

DRILL PROBLEM 29-8

Consider using Ampère's law to calculate the magnetic fields of a finite straight wire and of a circular loop of wire. Why is it not useful for these calculations?

29-6 THE SOLENOID

A long wire wound in the form of a helical coil is known as a *solenoid*. Solenoids are commonly used in experimental research requiring magnetic fields. A solenoid is generally easy to wind, and near its center, its magnetic field is quite uniform and directly proportional to the current in the wire.

The magnetic field outside two superconducting solenoids is delineated by iron nails aligned along field lines.

Figure 29-22*a* shows a solenoid consisting of N turns of wire tightly wound over a length L. A current I is flowing along the wire of the solenoid. We assume that the wire is so thin that in a length increment dy there are $(N/L)dy$ turns and therefore a current

$$dI = \frac{NI}{L}\,dy. \tag{29-13}$$

We first calculate the magnetic field at the point P of Fig. 29-22*b*. This point is *on the central axis of the solenoid*. The magnetic field $d\mathbf{B}$ due to the current dI in dy can be found with the help of Eq. (29-6*a*):

$$d\mathbf{B} = \frac{\mu_0 dI\,R^2}{2(y^2+R^2)^{3/2}}\,\mathbf{j} = \left(\frac{\mu_0 IR^2N}{2L}\,\mathbf{j}\right)\frac{dy}{(y^2+R^2)^{3/2}},$$

where Eq. (29-13) has been used to replace dI. The resultant field at P is found by integrating $d\mathbf{B}$ along the entire length

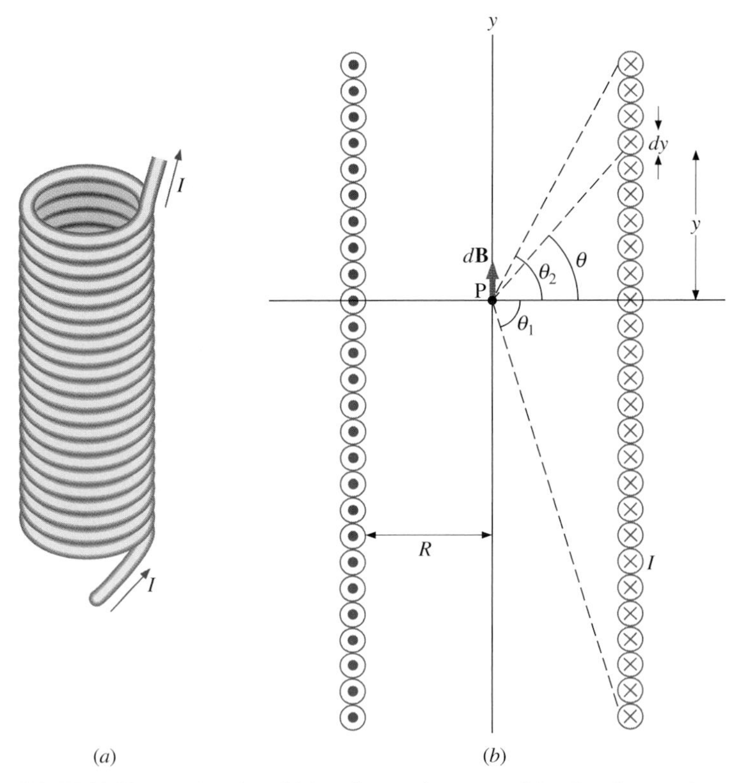

FIGURE 29-22 (*a*) A solenoid is a long wire wound in the shape of a helix. (*b*) The magnetic field at the point P on the axis of the solenoid is the net field due to all the current loops.

of the solenoid. It's easiest to evaluate this integral by changing the independent variable from y to θ. From inspection of Fig. 29-22*b*,

$$\sin\theta = \frac{y}{\sqrt{y^2+R^2}}.$$

Taking the differential of both sides of this equation, we obtain

$$\cos\theta\, d\theta = \left[-\frac{y^2}{(y^2+R^2)^{3/2}} + \frac{1}{\sqrt{y^2+R^2}}\right] dy = \frac{R^2\, dy}{(y^2+R^2)^{3/2}}.$$

When this is substituted into the equation for $d\mathbf{B}$, we have

$$d\mathbf{B} = \left(\frac{\mu_0 IN}{2L}\mathbf{j}\right)\cos\theta\, d\theta.$$

This can now be easily integrated between θ_1 and θ_2 (see Fig. 29-22*b*) to find the magnetic field at P:

$$\mathbf{B} = \left(\frac{\mu_0 IN}{2L}\mathbf{j}\right)\int_{\theta_1}^{\theta_2}\cos\theta\, d\theta = \frac{\mu_0 IN}{2L}(\sin\theta_2 - \sin\theta_1)\mathbf{j}. \tag{29-14a}$$

Finally, with the number of turns per unit length represented by $n\ (= N/L)$,

$$\mathbf{B} = \frac{\mu_0 nI}{2}(\sin\theta_2 - \sin\theta_1)\mathbf{j}, \tag{29-14b}$$

which is the magnetic field *along the central axis of a finite solenoid*.

Of special interest is the *infinite solenoid* for which $L \to \infty$. From a practical point of view, the infinite solenoid is one whose length is much larger than its radius ($L \gg R$). In this case, $\theta_1 = -\pi/2$ and $\theta_2 = \pi/2$. Then from Eq. (29-14*a*), the magnetic field *along the central axis of an infinite solenoid* is

$$\mathbf{B} = \frac{\mu_0 IN}{2L}\left[\sin\frac{\pi}{2} - \sin\left(-\frac{\pi}{2}\right)\right] = \frac{\mu_0 IN}{L}\mathbf{j} \tag{29-15a}$$

or

$$\mathbf{B} = \mu_0 nI\mathbf{j}. \tag{29-15b}$$

You can find the direction of **B** easily with a right-hand rule: Curl your fingers in the direction of the current, and your thumb points along the magnetic field.

The magnetic field of the infinite, tightly wound solenoid is further investigated in Supplement 29-1. There it is shown that the field has the following properties:

1. It is zero everywhere outside the solenoid.
2. It is directed parallel to the axis of the solenoid everywhere inside the solenoid.

We now use these properties, along with Ampère's law, to calculate the magnitude of the magnetic field at any location inside the infinite solenoid. Consider the closed path of Fig. 29-23. Along segment 1, **B** is uniform and parallel to the path. Along segments 2 and 4, **B** is perpendicular to part of the path and vanishes over the rest of it. Therefore segments 2 and 4 do not contribute to the line integral in Ampère's law. Along segment 3, $\mathbf{B} = 0$, and again there is no contribution to the line integral. As a result, we find

$$\oint \mathbf{B}\cdot d\mathbf{l} = \int_1 \mathbf{B}\cdot d\mathbf{l} = Bl.$$

Since there are n turns per unit length on the solenoid, the current that passes through the surface enclosed by the path is nlI; so, from Ampère's law,

$$Bl = \mu_0 nlI,$$

and

$$B = \mu_0 nI$$

within the solenoid. This agrees with what we found earlier for B on the central axis of the solenoid. Here, however, the location of segment 1 is arbitrary, so we have now found that this

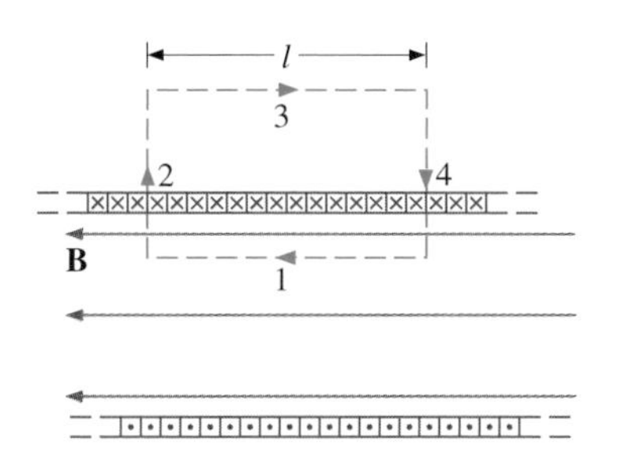

FIGURE 29-23 The path of integration used in Ampère's law to evaluate the magnetic field of an infinite solenoid.

equation gives the magnetic field *everywhere inside the infinite solenoid*.

The calculation of the magnetic field of a *finite*, tightly wound solenoid at points off the central axis requires mathematical methods that are not appropriate for the level of this textbook. The magnetic field lines found with this calculation are shown in Fig. 29-24. Near the center of the solenoid, the field lines are fairly evenly spaced and axial, indicating that the magnetic field in this region is almost axial and uniform. Close to the ends of the solenoid, however, the field lines bend considerably, as they leave one end of the solenoid, curl around in space, and enter the other end of the solenoid.

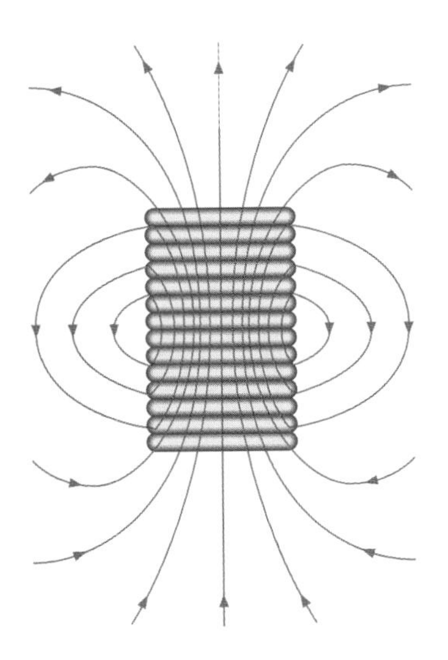

FIGURE 29-24 The magnetic field lines of a finite solenoid.

EXAMPLE 29-9 Magnetic Field Inside a Solenoid

A solenoid has 300 turns wound around a cylinder of diameter 1.20 cm and length 14.0 cm. If the current through the coils is 0.410 A, what is the magnitude of the magnetic field inside and near the middle of the solenoid?

SOLUTION Since the length of the solenoid is reasonably large compared with its diameter, the field near its middle is approximately uniform and given by $B = \mu_0 nI$. The number of turns per unit length is

$$n = \frac{300\text{ turns}}{0.140\text{ m}} = 2.14 \times 10^3\text{ turns/m}.$$

Now from $B = \mu_0 nI$, the magnetic field produced by a current of 0.410 A is

$$B = (4\pi \times 10^{-7}\text{ T}\cdot\text{m/A})(2.14 \times 10^3\text{ turns/m})(0.410\text{ A})$$
$$= 1.10 \times 10^{-3}\text{ T}.$$

DRILL PROBLEM 29-9

Near the center of a tightly wound solenoid of length 50 cm, the magnetic field is 2.4×10^{-3} T when a current of 1.0 A flows through its coils. How many turns of wire are there on the solenoid? Assume that the infinite-solenoid approximation is valid.
ANS. 9.6×10^2.

*29-7 Magnetic Materials

Why are certain materials magnetic and others not? And why do certain substances become magnetized by a field while others are unaffected? To answer such questions, we need an understanding of magnetism on a microscopic level. Within an atom, every electron travels in an orbit and spins on an internal axis. Both types of motion produce current loops and therefore magnetic dipoles. For a particular atom, the net magnetic dipole moment is the vector sum of the magnetic dipole moments of its constituent electrons. The values of the magnetic dipole moment $\mathcal{M}$ of several types of atoms are given in Table 29-1. Notice that some atoms have a zero net dipole moment, and that the magnitudes of the nonvanishing moments are typically $10^{-23}\text{ A}\cdot\text{m}^2$.

TABLE 29-1 **Magnetic Moments of Some Atoms**

Atom	Magnetic Moment ($10^{-24}\text{ A}\cdot\text{m}^2$)
H	9.27
He	0
Li	9.27
O	13.9
Na	9.27
S	13.9

In a handful of matter, there are approximately 10^{26} atoms and ions, each with its magnetic dipole moment. If there is no external magnetic field present, the magnetic dipoles are randomly oriented—there are as many pointing up as down, as many pointing east as west, and so on. Consequently, the net magnetic dipole moment of the sample is zero. But when the sample is placed in a magnetic field, these dipoles tend to align with the field [see Eq. (28-8*b*)], and this alignment determines how the sample responds to the field. On the basis of this response, a material is said to be either *paramagnetic*, *diamagnetic*, or *ferromagnetic*.

In a paramagnetic material, only a small fraction of the magnetic dipoles is aligned with the applied field. Since each dipole produces its own magnetic field, this alignment contributes an extra magnetic field, which enhances the applied field. When a ferromagnetic material is placed in a magnetic field, its magnetic dipoles also become aligned, but much more completely so than in the paramagnetic case. In fact, the alignment is generally so complete that the net field produced by the magnetic dipoles is usually stronger than the applied field. Diamagnetic materials are composed of atoms that have no net magnetic dipole moment. However, when a diamagnetic material is placed in a magnetic field, a magnetic dipole moment is induced in its atoms. This induced dipole moment is directed opposite to the applied field and therefore produces a magnetic field that opposes the applied field. We now consider each type of material in greater detail.

PARAMAGNETIC MATERIALS

For simplicity, we will assume our sample is a long, cylindrical piece that completely fills the interior of a long, tightly wound solenoid. When there is no current in the solenoid, the magnetic dipoles in the sample are randomly oriented and produce no net magnetic field. With a solenoid current, the magnetic field due to the solenoid current exerts a torque on the dipoles that tends to align them with the field. In competition with the aligning torque are thermal collisions that tend to randomize the positions of the dipoles. The relative importance of these two competing processes can be estimated by comparing the energies involved. From Eq. (28-10), the energy difference between a magnetic dipole aligned with and against a magnetic field **B** is $U_B = 2\mathcal{M}B$. If $\mathcal{M} = 9.3 \times 10^{-24}$ A·m² (the value for atomic hydrogen) and $B = 1.0$ T, then $U_B = 1.9 \times 10^{-23}$ J. At a room temperature of 27°C, the thermal energy per atom (see Sec. 18-4) is $U_T \approx kT = (1.38 \times 10^{-23}\text{ J/K})(300\text{ K}) = 4.1 \times 10^{-21}$ J, which is about 220 times greater than U_B. Clearly, energy exchanges in thermal collisions can seriously interfere with the alignment of the magnetic dipoles. As a result, only a small fraction of the dipoles is aligned at any instant.

The four sketches of Fig. 29-25 furnish a simple model of this alignment process. In Fig. 29-25*a*, before the field of the solenoid (not shown) containing the paramagnetic sample is applied, the magnetic dipoles are randomly oriented and there is no net magnetic dipole moment associated with the material. With the introduction of the field, there is a partial alignment of the dipoles, as depicted in Fig. 29-25*b*. The component of the net magnetic dipole moment that is perpendicular to the field vanishes. We may then represent the sample by Fig. 29-25*c*, which shows a collection of magnetic dipoles completely aligned with the field. By treating these dipoles as current loops, we can picture the dipole alignment as equivalent to a current around the surface of the material. (See Fig. 29-25*d*.) This fictitious surface current produces its own magnetic field, which enhances the field of the solenoid.

We can express the total magnetic field **B** in the material as

$$\mathbf{B} = \mathbf{B}_0 + \mathbf{B}_m, \tag{29-16a}$$

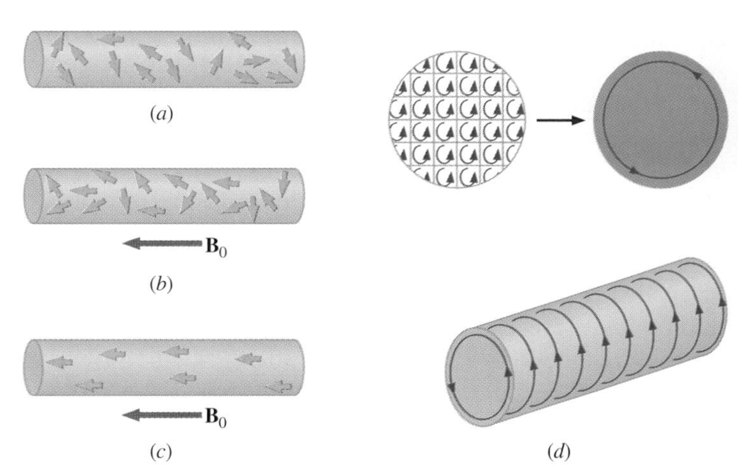

FIGURE 29-25 The alignment process in a paramagnetic material filling a solenoid (not shown). (*a*) Without an applied field, the magnetic dipoles are randomly oriented. (*b*) With a field, partial alignment occurs. (*c*) An equivalent representation of part (*b*). (*d*) The internal currents cancel, leaving an effective surface current that produces a magnetic field similar to that of a finite solenoid.

where $\mathbf{B}_0$ is the field due to the current I_0 in the solenoid, and $\mathbf{B}_m$ is the field due to the surface current I_m around the sample. Now $\mathbf{B}_m$ is usually proportional to $\mathbf{B}_0$, a fact we express by

$$\mathbf{B}_m = \chi \mathbf{B}_0, \tag{29-17}$$

where χ is a dimensionless quantity called the **magnetic susceptibility**.* Values of χ for some paramagnetic materials are given in Table 29-2. Since the alignment of magnetic dipoles is so weak, χ is very small for paramagnetic materials. By combining Eqs. (29-16*a*) and (29-17), we obtain

$$\mathbf{B} = \mathbf{B}_0 + \chi \mathbf{B}_0 = (1 + \chi)\mathbf{B}_0, \tag{29-16b}$$

*For a vacuum, $\mathbf{B}_m = 0$, so $\chi = 0$.

TABLE 29-2 **Magnetic Susceptibilities** Unless otherwise specified, values given are for room temperature.

Paramagnetic Materials	χ	Diamagnetic Materials	χ
Aluminum	2.2×10^{-5}	Bismuth	-1.7×10^{-5}
Calcium	1.4×10^{-5}	Carbon (diamond)	-2.2×10^{-5}
Chromium	3.1×10^{-4}	Copper	-9.7×10^{-6}
Magnesium	1.2×10^{-5}	Lead	-1.6×10^{-5}
Oxygen gas (1 atm)	1.8×10^{-6}	Mercury	-2.8×10^{-5}
Oxygen liquid (90 K)	3.5×10^{-3}	Hydrogen gas (1 atm)	-2.2×10^{-9}
Tungsten	7.8×10^{-5}	Nitrogen gas (1 atm)	-6.7×10^{-9}
Air (1 atm)	3.6×10^{-7}	Water	-9.1×10^{-6}

which, for a sample within an infinite solenoid, is

$$B = (1 + \chi)\mu_0 nI. \quad (29\text{-}18a)$$

This expression tells us that the insertion of a paramagnetic material into a solenoid increases the field by a factor $(1 + \chi)$. However, since χ is so small, the field isn't enhanced very much.

The quantity

$$\mu = (1 + \chi)\mu_0 \quad (29\text{-}19)$$

is called the **magnetic permeability** of a material. In terms of μ, Eq. (29-18a) can be written as

$$B = \mu nI \quad (29\text{-}18b)$$

for the filled infinite solenoid.

DRILL PROBLEM 29-10

Suppose a paramagnetic material with magnetic susceptibility χ fills the volume of a toroid. The toroid has N turns of wire and a central radius R. What is the net field inside the toroid when a current I is flowing through the wire?
ANS. $B = (1 + \chi)\mu_0 NI/2\pi R$.

DIAMAGNETIC MATERIALS

A magnetic field always induces a magnetic dipole in an atom. This induced dipole points opposite to the applied field, so its magnetic field is also directed opposite to the applied field. In paramagnetic and ferromagnetic materials, the induced magnetic dipole is masked by the much stronger permanent magnetic dipoles of the atoms. However, in diamagnetic materials, whose atoms have no permanent magnetic dipole moments, the effect of the induced dipole is observable.

We can describe the magnetic effects of diamagnetic materials with the same model developed for paramagnetic materials. In this case, however, the fictitious surface current flows opposite to the solenoid current, and the magnetic susceptibility χ is negative. Values of χ for some diamagnetic materials are also given in Table 29-2.

DRILL PROBLEM 29-11

Is a paramagnetic sample attracted or repelled by either pole of a bar magnet? What about a diamagnetic sample?

FERROMAGNETIC MATERIALS

The common magnet is made of a ferromagnetic material such as iron or one of its alloys. Experiment reveals that a ferromagnetic material consists of tiny regions known as *domains*. Their volumes typically range from 10^{-12} to 10^{-8} m^3, and they contain about 10^{17} to 10^{21} atoms. Within a domain, the magnetic dipoles are rigidly aligned in the same direction by a coupling among the atoms. This coupling, which is due to quantum mechanical effects, is so strong that even thermal agitation at room temperatures cannot break it. The result is that each domain has a net dipole moment.

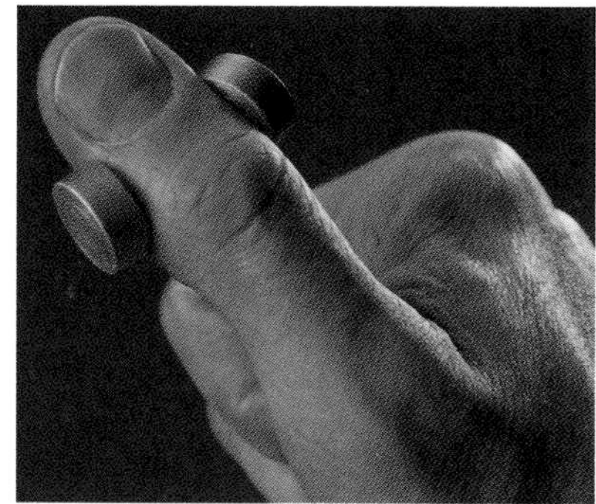

Human tissue has little effect on the force between two magnets.

Courtesy Central Scientific Company.

If the domains in a ferromagnetic sample are randomly oriented, as represented in Fig. 29-26a, the sample has no net magnetic dipole moment and is said to be *unmagnetized*. Suppose that we fill the volume of a solenoid with an unmagnetized ferromagnetic sample. When the magnetic field $\mathbf{B}_0$ of the solenoid is turned on, the dipole moments of the domains rotate

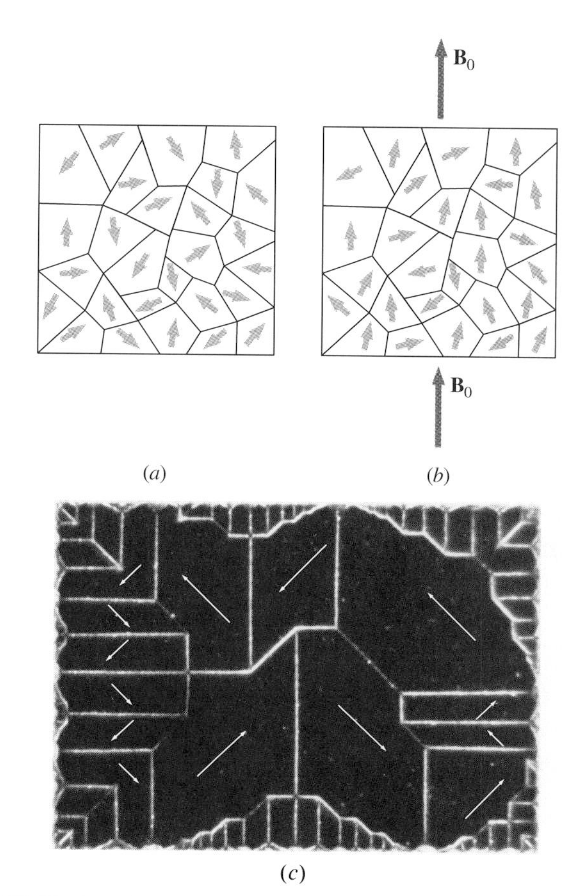

FIGURE 29-26 (a) The domains are randomly oriented in an unmagnetized ferromagnetic sample such as iron. The arrows represent the orientation of the magnetic dipoles within the domains. (b) In an applied magnetic field, the domains align somewhat with the field. (c) The domains of a single crystal of nickel. The white lines show the boundaries of the domains. These lines are produced by iron oxide powder sprinkled on the crystal.

so that they align somewhat with the field, as depicted in Fig. 29-26*b*. In addition, the aligned domains tend to increase in size at the expense of the unaligned ones. The net effect of these two processes is the creation of a net magnetic dipole moment for the ferromagnet that is directed along the applied magnetic field. Now this net magnetic dipole moment is much larger than that of a paramagnetic sample, and the domains, with their large numbers of atoms, do not become misaligned by thermal agitation. Consequently, the field due to the alignment of the domains is quite large.

Besides iron, only four other elements contain the magnetic domains needed to exhibit ferromagnetic behavior. They are cobalt, nickel, gadolinium, and dysprosium. Many alloys of these elements are also ferromagnetic. Ferromagnetic materials can be described using Eqs. (29-16) through (29-19), the paramagnetic equations. However, the value of χ for a ferromagnetic material is usually on the order of 10^3 to 10^4, and it also depends on the past history of the material. A typical plot of B (the total field in the material) versus B_0 (the applied field) for an initially unmagnetized piece of iron is shown in Fig. 29-27. Some sample numbers are: (1) for $B_0 = 1.0 \times 10^{-4}$ T, $B = 0.60$ T and $\chi = (0.60/1.0 \times 10^{-4}) - 1 \approx 6.0 \times 10^3$; (2) for $B_0 = 6.0 \times 10^{-4}$ T, $B = 1.5$ T and $\chi = (1.5/6.0 \times 10^{-4}) - 1 \approx 2.5 \times 10^3$.

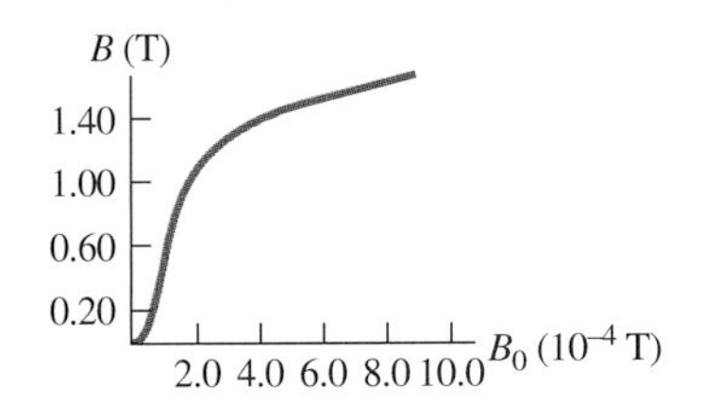

FIGURE 29-27 The magnetic field B in annealed iron as a function of the applied field B_0.

When B_0 is varied over a range of positive and negative values, B is found to behave as shown in Fig. 29-28. The same B_0 (corresponding to the same current in the solenoid) can produce different values of B in the material. *The magnetic field B produced in a ferromagnetic material by an applied field B_0 depends on the magnetic history of the material.* This effect is called *hysteresis*, and the curve of Fig. 29-28 is called a *hysteresis loop*. Notice that B does not disappear when $B_0 = 0$ (i.e., when the current in the solenoid is turned off). The iron stays magnetized, which means that it is a *permanent magnet*.

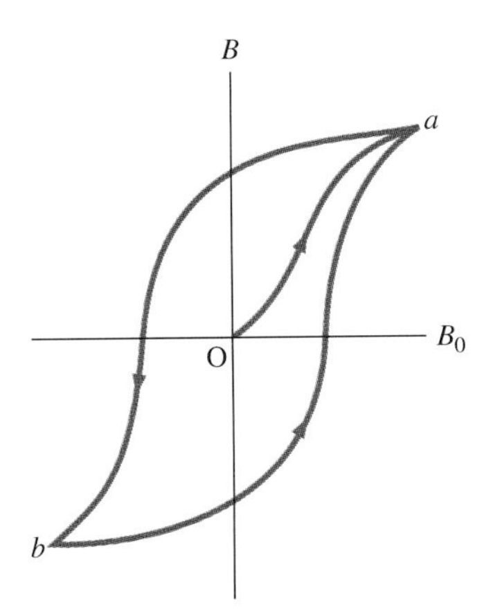

FIGURE 29-28 A typical magnetization curve for a ferromagnet. When the material is first magnetized, it follows the curve from O to a. When B_0 is reversed, it takes the path shown from a to b. If B_0 is reversed again, the material follows the curve from b to a.

Like the paramagnetic sample of Fig. 29-25*d*, the partial alignment of the domains in a ferromagnet is equivalent to a current flowing around the surface. A bar magnet can therefore be pictured as a tightly wound solenoid with a large current circulating through its coils (the surface current). You can see in Fig. 29-29 that this model is quite good. The fields of the bar magnet and the finite solenoid are strikingly similar. The figure also shows how the poles of the bar magnet are identified. To form closed loops, the field lines outside the magnet leave the north (N) pole and enter the south (S) pole, while inside the magnet, they leave S and enter N.

The interactions of magnetic poles are easily understood in terms of this model. A north and south pole situated as shown in Fig. 29-30*a* interact like two currents circulating in the same

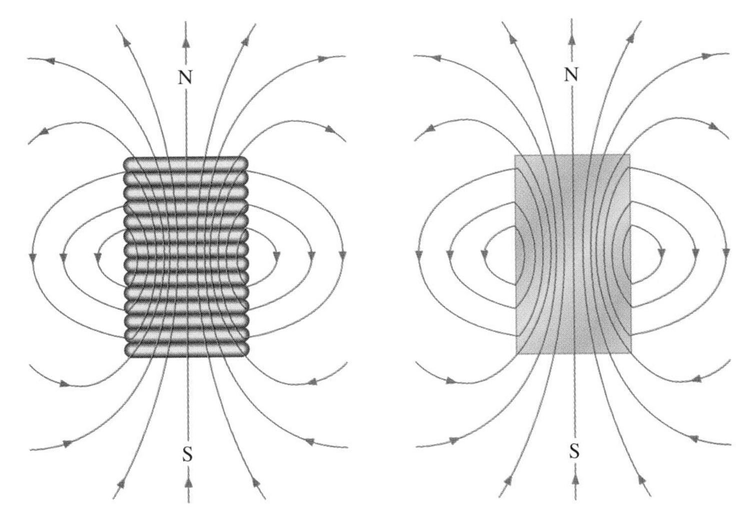

FIGURE 29-29 Comparison of the magnetic fields of a finite solenoid and a bar magnet.

The field lines of the bar magnet are indicated by the iron filings sprinkled around the magnet.

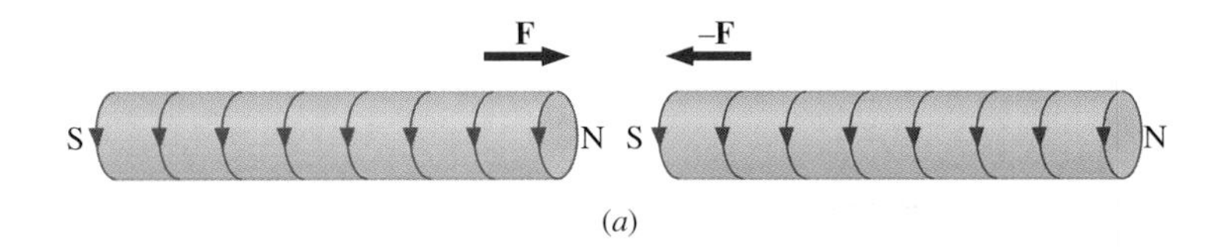

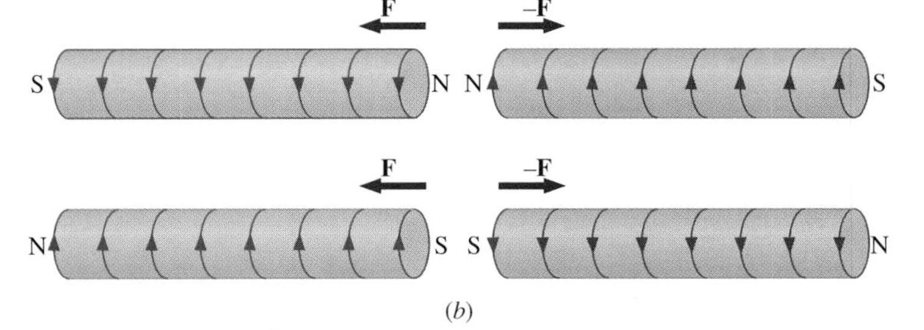

FIGURE 29-30 (*a*) Attraction and (*b*) repulsion of magnetic poles in terms of the solenoid model.

direction; the poles therefore attract. On the other hand, two north poles or two south poles interact like currents circulating in opposite directions (see Drill Prob. 29-4), so they repel, as indicated in Fig. 29-30*b*.

EXAMPLE 29-10 IRON CORE IN A COIL

A long coil is tightly wound around an iron cylinder whose magnetization curve is shown in Fig. 29-27. (*a*) If $n = 20$ turns per centimeter, what is the applied field B_0 when $I_0 = 0.20$ A? (*b*) What is the net field B for this same current? (*c*) What is the magnetic susceptibility in this case?

SOLUTION (*a*) The applied field B_0 of the coil is

$$B_0 = \mu_0 n I_0 = (4\pi \times 10^{-7}\ \text{T}\cdot\text{m/A})(2000/\text{m})(0.20\ \text{A}) = 5.0 \times 10^{-4}\ T.$$

(*b*) From inspection of the magnetization curve of Fig. 29-27, we see that, for this value of B_0, $B = 1.4$ T. Notice that the internal field of the aligned atoms is much larger than the externally applied field.

(*c*) Now from Eq. (29-16*b*), the magnetic susceptibility is found to be

$$\chi = \frac{B}{B_0} - 1 = \frac{1.4\ \text{T}}{5.0 \times 10^{-4}\ \text{T}} - 1 = 2.8 \times 10^3.$$

DRILL PROBLEM 29-12

Both a north and a south pole attract an unmagnetized piece of iron. Explain why. (*Hint:* How are current loops aligned by a magnetic field?)

DRILL PROBLEM 29-13

Repeat the calculations of Example 29-10 for $I_0 = 0.040$ A.
ANS. (*a*) 1.0×10^{-4} T; (*b*) 0.60 T; (*c*) 6.0×10^3.

29-8 THE STATIC MAGNETIC FIELD: A SUMMARY

We have now completed our investigation of the static magnetic field. The properties of this field are the following:

1. The fundamental source of the static magnetic field is the moving point charge or current element $I\,d\mathbf{l}$.
2. This source creates a magnetic field given by the Biot-Savart law:
$$d\mathbf{B} = \frac{\mu_0}{4\pi}\frac{I\,d\mathbf{l} \times \hat{\mathbf{r}}}{r^2}. \tag{29-1}$$
3. The magnetic force $\mathbf{F}_m$ on a particle of charge q moving with a velocity $\mathbf{v}$ in a field $\mathbf{B}$ is given by
$$\mathbf{F}_m = q\mathbf{v} \times \mathbf{B}. \tag{28-1}$$
The force of the magnetic field on a current element $I\,d\mathbf{l}$ is
$$d\mathbf{F}_m = I\,d\mathbf{l} \times \mathbf{B}. \tag{28-5}$$
4. The flux of the magnetic field through an arbitrary Gaussian surface S is zero:
$$\oint_S \mathbf{B}\cdot\hat{\mathbf{n}}\,da = 0. \tag{29-9}$$
5. The line integral of the static magnetic field around an arbitrary closed path satisfies
$$\oint \mathbf{B}\cdot d\mathbf{l} = \mu_0 I, \tag{29-10}$$
where I is the total current cutting through an arbitrary surface bounded by the path of integration.

SUMMARY

1. **The Biot-Savart law**
The magnetic field of a current-carrying wire is
$$\mathbf{B} = \frac{\mu_0}{4\pi}\int_{\text{wire}} \frac{I\,d\mathbf{l} \times \hat{\mathbf{r}}}{r^2}.$$

*2. **Definitions of the ampere and the coulomb**
(*a*) Suppose that two infinitely long, parallel wires are placed exactly 1 m apart in a vacuum, with the same current flowing in each wire. When the current is set so that the magnetic force per unit length on each wire is exactly 2×10^{-7} N/m, the current is exactly 1 A.
(*b*) If the current in a wire is 1 A, then 1 C of charge passes through a cross section of the wire in 1 s.

3. **Gauss' law for the magnetic field**
$$\oint_S \mathbf{B}\cdot\hat{\mathbf{n}}\,da = 0.$$

4. **Ampère's law**

$$\oint \mathbf{B} \cdot d\mathbf{l} = \mu_0 I.$$

5. **Applications of Ampère's law**
If $\oint \mathbf{B} \cdot d\mathbf{l}$ can be factored into a product of B and a path length, then Ampère's law can be used to calculate B.
6. **The solenoid**
(*a*) The magnetic field along the central axis (say, the y axis) of the finite solenoid is

$$\mathbf{B} = \frac{\mu_0 nI}{2}(\sin\theta_2 - \sin\theta_1)\mathbf{j}.$$

(*b*) Everywhere inside the infinite solenoid,

$$\mathbf{B} = \mu_0 nI\mathbf{j},$$

and everywhere outside, $\mathbf{B} = 0$.

*7. **Magnetic materials**
Materials are classified as paramagnetic, diamagnetic, or ferromagnetic, depending on how they behave in an applied magnetic field.

SUPPLEMENT 29-1 MAGNETIC FIELD OF THE INFINITE SOLENOID

In this supplement we show that the magnetic field of an infinitely long, cylindrical solenoid is parallel to the axis everywhere inside and vanishes everywhere outside the solenoid. We must assume that the windings are so tight that the current through the coils may be approximated as a sheet of current. (See Fig. 29-31*a*.) If n is the number of turns per unit length of the solenoid, and if a current I flows through the windings, then nI is the current per unit length on the sheet. To make full use of the symmetry of the configuration, we will employ the cylindrical coordinate system shown. The unit vectors are $\mathbf{k}$ (z direction), $\hat{\mathbf{r}}$ (radial direction), and $\hat{\boldsymbol{\theta}}$ (azimuthal direction), and the axis of the solenoid is the z axis.

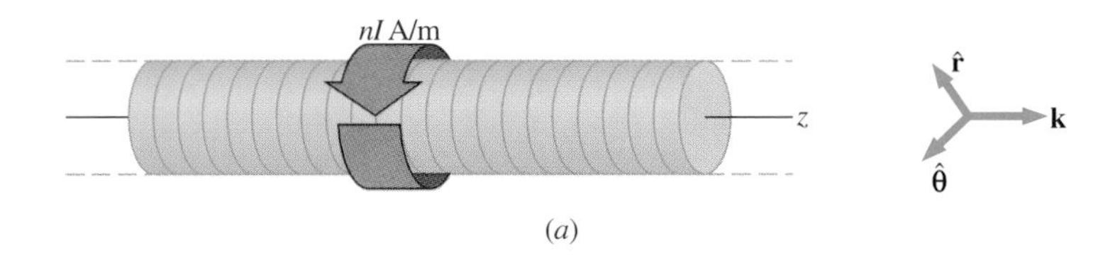

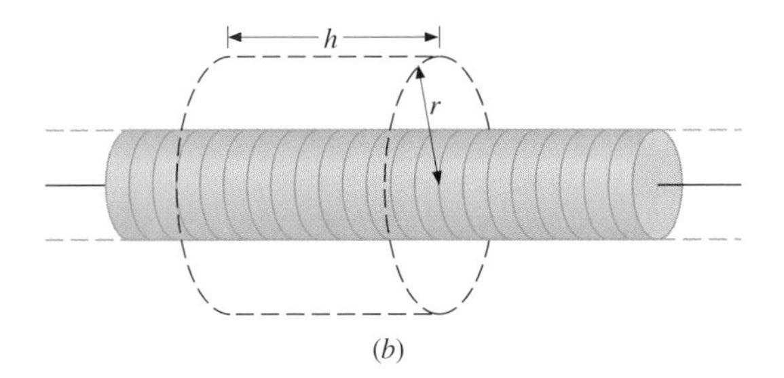

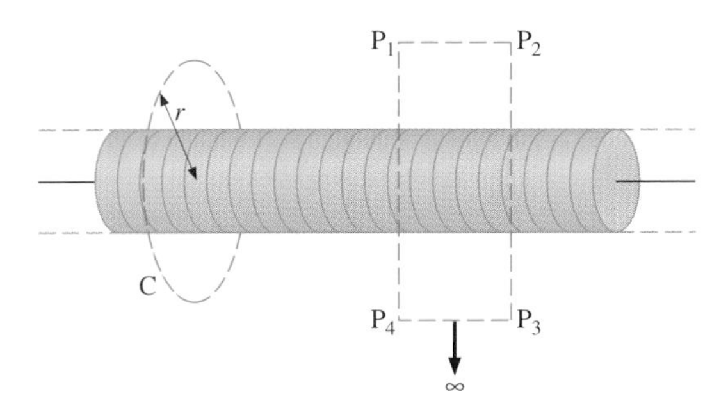

FIGURE 29-31 (*a*) An infinitely long solenoid is wound so tightly that the current through the coils may be approximated as a current sheet. (*b*) A cylindrical Gaussian surface surrounding the solenoid. (*c*) Two paths of integration for application of Ampère's law.

We begin by showing that the magnetic field has no radial component B_r either inside or outside the solenoid. Figure 29-31*b* shows a Gaussian cylinder that is concentric with the central axis of the solenoid. Its radius may be less than or greater than the radius of the solenoid. Only B_z contributes to the magnetic flux over the end caps. Since the solenoid is infinitely long, the magnetic field at any cross section of the cylinder perpendicular to its axis is the same. Therefore the flux entering one end is equal to that leaving the other end, and the total flux through the two end caps is zero. Only B_r contributes to the flux through the cylindrical Gaussian surface, and symmetry requires B_r to be constant over the surface. We now have from Gauss' law for magnetism

$$\oint_S \mathbf{B} \cdot \hat{\mathbf{n}}\, da = B_r(2\pi rh) = 0,$$

so $B_r = 0$ *both inside and outside the solenoid.*

Next we show that the magnetic field has no azimuthal component B_θ either inside or outside the solenoid. The circular path C of Fig. 29-31*c* is concentric with the central axis of the solenoid, and it lies in a plane perpendicular to that axis. The radius r of the circle can be larger or smaller than the radius of the solenoid. Along C, only B_θ contributes to $\oint \mathbf{B} \cdot d\mathbf{l}$. Furthermore, the system is cylindrically symmetric, so B_θ is constant over C, and

$$\oint \mathbf{B} \cdot d\mathbf{l} = B_\theta \oint dl = B_\theta(2\pi r). \qquad (i)$$

With the assumption of a current sheet, there is no current flow in either the positive or negative z direction. Consequently, no current crosses the flat surface enclosed by the circular path. Thus from Ampère's law and Eq. (*i*),

$$B_\theta(2\pi r) = 0,$$

so $B_\theta = 0$ *both inside and outside the solenoid.*

Notice that the current-sheet approximation is an essential part of this proof. In reality, there is a pitch to the coils, and some current passes through the surface bounded by C. Consequently, B_θ is not exactly zero outside the solenoid; however, it is very small there.

Finally we examine the third component, B_z, of the magnetic field. The path $P_1P_2P_3P_4$ of Fig. 29-31*c* lies in a plane that intersects the solenoid. There is no net current crossing the plane and $B_r = B_\theta = 0$ everywhere from the two previous steps. Ampère's law therefore becomes

$$\oint \mathbf{B} \cdot d\mathbf{l} = \int_{P_1}^{P_2} B_z\, dl + \int_{P_3}^{P_4} B_z\, dl = 0. \qquad (ii)$$

Now suppose the path is stretched so that P_3 and P_4 approach infinity while P_1 and P_2 stay close to the solenoid. Assuming that $B_z \to 0$ along P_3P_4 as this segment of the path recedes from the solenoid, we have from Eq. (*ii*)

$$\int_{P_1}^{P_2} B_z\, dl = 0,$$

so $B_z = 0$ *outside the solenoid.*

To summarize, we have found that the field due to the current through the coils of the infinite solenoid vanishes outside the solenoid. Everywhere within the solenoid, $B_r = B_\theta = 0$, so the field is directed along the axis of the solenoid.

QUESTIONS

29-1. For calculating magnetic fields, what are the advantages and disadvantages of the Biot-Savart law in comparison with Ampère's law?

29-2. Describe the magnetic field due to the current in two wires connected to the two terminals of a source of emf and twisted tightly around each other.

29-3. How would you orient two long, straight, current-carrying wires so that there is no magnetic force between them?

29-4. Is Ampère's law valid for all closed paths? Why isn't it normally useful for calculating a magnetic field?

29-5. How can the magnetic flux through an open surface be nonzero if the magnetic flux through a closed surface is always zero?

29-6. How are Gauss' law for electric fields and Ampère's law similar? How are they different?

29-7. Compare and contrast the electric field of an infinite line of charge and the magnetic field of an infinite line of current.

29-8. Is **B** constant in magnitude for points that lie on a magnetic field line?

29-9. Is the magnetic field conservative?

29-10. Why didn't we define a scalar potential for the magnetic field?

29-11. Is the magnetic field of a current loop uniform?

29-12. Based on your knowledge of the magnetic field lines of a circular current loop, make a plausibility argument as to why the field of an infinite solenoid is constant inside and zero outside.

29-13. What happens to the length of a suspended spring when a current passes through it?

29-14. Two concentric circular wires with different diameters carry currents in the same direction. Describe the force on the inner wire.

29-15. Is the magnetic field inside a toroid completely uniform? almost uniform?

29-16. Explain why $\mathbf{B} = 0$ inside a long, hollow copper pipe that is carrying an electric current. Is $\mathbf{B} = 0$ outside the pipe?

29-17. Two identical short solenoids are placed close together, as shown in the accompanying figure. The same current flows through each solenoid. Do the solenoids attract or repel each other?

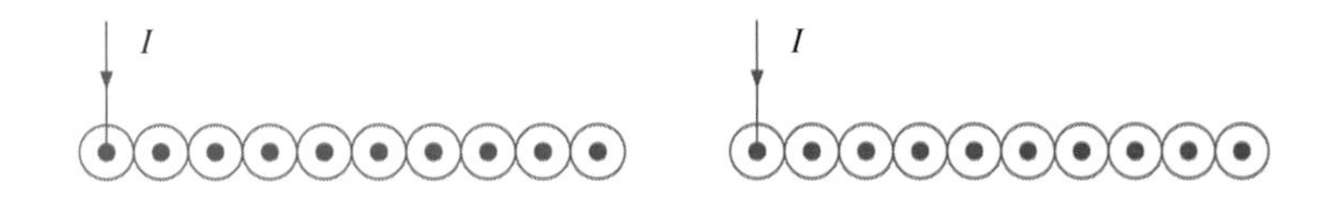

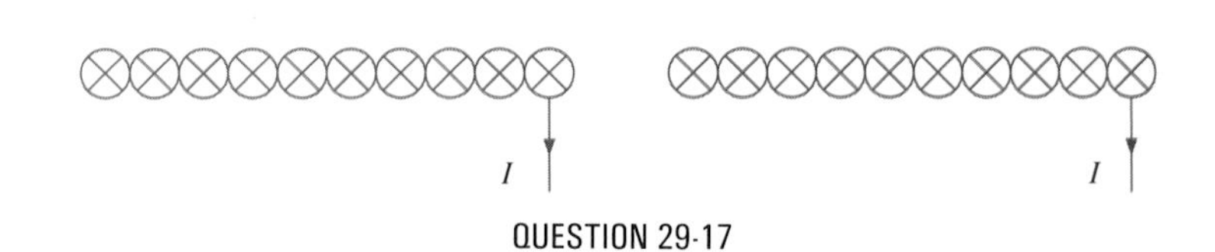

QUESTION 29-17

29-18. In Supplement 29-1, the current through the windings of an infinite solenoid was taken to be equivalent to a current sheet (Fig. 29-31*a*). The field outside the solenoid was then found to be zero. In reality, the windings are helical and the magnetic field outside the solenoid is not exactly zero. Explain.

29-19. Discuss the origin of the magnetic dipole moment in atoms.

29-20. Can an atom have a zero magnetic dipole moment?

29-21. Discuss the differences between paramagnetic, diamagnetic, and ferromagnetic materials.

29-22. A diamagnetic material is brought close to a permanent magnet. What happens to the material?

29-23. Compare what happens to a nail when it is attracted by the north pole of a permanent magnet with what happens when it is attracted by the south pole.

29-24. If you cut a bar magnet into two pieces, will you end up with one magnet with an isolated north pole and another magnet with an isolated south pole? Explain your answer.

PROBLEMS

The Biot-Savart Law and the Magnetic Fields of Currents

29-1. A typical current in a lightning bolt is 10^4 A. Estimate the magnetic field 1 m from the bolt.

29-2. The magnitude of the magnetic field 50 cm from a long, thin, straight wire is 8.0 μT. What is the current through the wire?

29-3. A transmission line strung 7.0 m above the ground carries a current of 500 A. What is the magnetic field on the ground directly below the wire? Compare your answer with the magnetic field of the earth.

29-4. A long, straight, horizontal wire carries a left-to-right current of 20 A. If the wire is placed in a uniform magnetic field of magnitude 4.0×10^{-5} T that is directed vertically downward, what is the resultant magnetic field 20 cm above the wire? 20 cm below the wire?

29-5. The two long, parallel wires shown in the accompanying figure carry currents in the same direction. If $I_1 = 10$ A and $I_2 = 20$ A, what is the magnetic field at P?

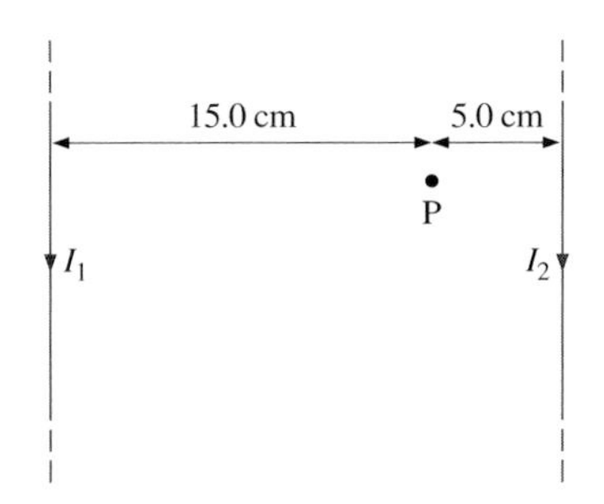

PROBLEM 29-5

29-6. The accompanying figure shows two long, straight, horizontal wires that are parallel and a distance $2a$ apart. If both wires carry a current I in the same direction, (*a*) what is the magnetic field at P_1? (*b*) at P_2?

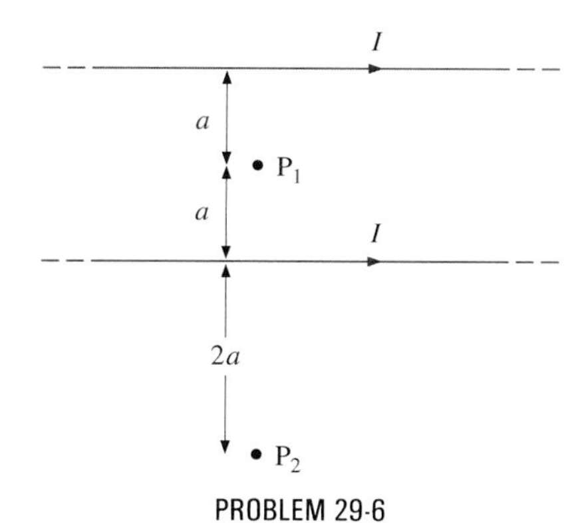

PROBLEM 29-6

29-7. Repeat the calculations of Prob. 29-6 with the direction of the current in the lower wire reversed.

29-8. Consider the wires of Prob. 29-6. At what distance from the top wire and between the wires is the net magnetic field a minimum? Assume that the currents are equal and flow in opposite directions.

29-9. The accompanying figure shows a long, straight wire carrying a current of 10 A. What is the magnetic force on an electron at the instant it is 20 cm from the wire and traveling parallel to the wire with a speed of 2.0×10^5 m/s? Describe qualitatively the subsequent motion of the electron.

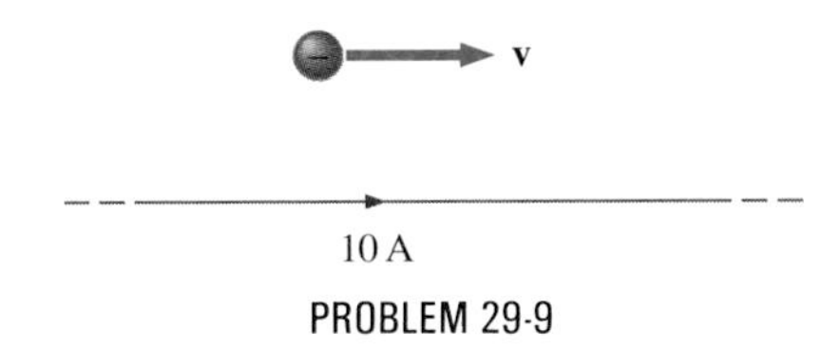

PROBLEM 29-9

29-10. When the current through a circular loop is 6.0 A, the magnetic field at its center is 2.0×10^{-4} T. What is the radius of the loop?

29-11. How many turns must be wound on a flat, circular coil of radius 20 cm in order to produce a magnetic field of magnitude 4.0×10^{-5} T at the center of the coil when the current through it is 0.85 A?

29-12. A flat, circular loop has 20 turns. The radius of the loop is 10.0 cm and the current through the wire is 0.50 A. Determine the magnitude of the magnetic field at (*a*) the center of the loop and (*b*) on the axis at 1.0, 2.0, and 10.0 cm from the center of the loop.

29-13. A circular loop of radius R carries a current I. At what distance along the axis of the loop is the magnetic field one-half its value at the center of the loop?

29-14. Field lines for the earth's magnetic field were shown in Fig. 28-1*b*. The field is due to current loops that circulate in the molten material of the earth's core. The field lines are similar to those of a circular current loop whose normal is inclined at 11° to the rotational axis of the earth. They emerge from the surface of the earth near the geographic South Pole and reenter the surface near the geographic North Pole. (*a*) As viewed from the North Pole, does this effective current loop flow clockwise or counterclockwise? (*b*) In New York City, the magnetic field of the earth has a downward vertical component of 6.0×10^{-5} T and a horizontal component of 1.7×10^{-5} T that points northward. What force does this field exert on a 15-keV electron at the instant it is moving horizontally from north to south?

29-15. Two flat, circular coils, each with a radius R and wound with N turns, are mounted along the same axis so that they are parallel and a distance d apart. What is the magnetic field at the midpoint of the common axis if a current I flows in the same direction through each coil?

29-16. For the coils of the previous problem, what is the magnetic field at the center of either coil?

29-17. Find the magnetic field at the center C of the rectangular loop of wire shown in the accompanying figure.

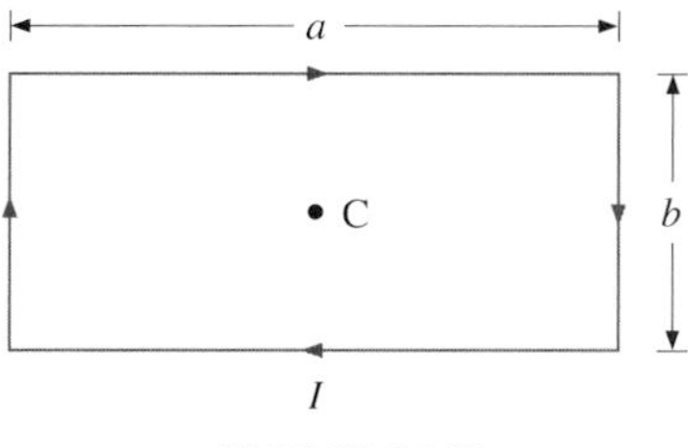

PROBLEM 29-17

29-18. What is the magnetic field at P due to the current I in the wire shown in the accompanying figure?

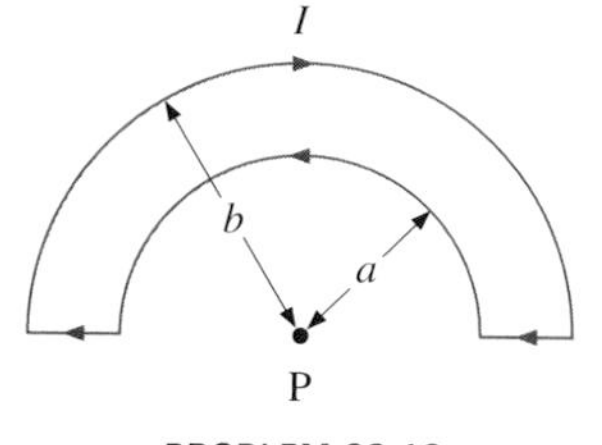

PROBLEM 29-18

29-19. The accompanying figure shows a current loop consisting of two concentric circular arcs and two perpendicular radial lines. Determine the magnetic field at the point P.

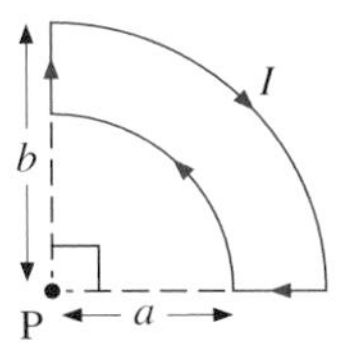

PROBLEM 29-19

29-20. Two long wires, one of which has a semicircular bend of radius R, are positioned as shown in the accompanying figure. If both wires carry a current I, how far apart must their parallel sections be so that the net magnetic field at P is zero? Does the current in the straight wire flow up or down?

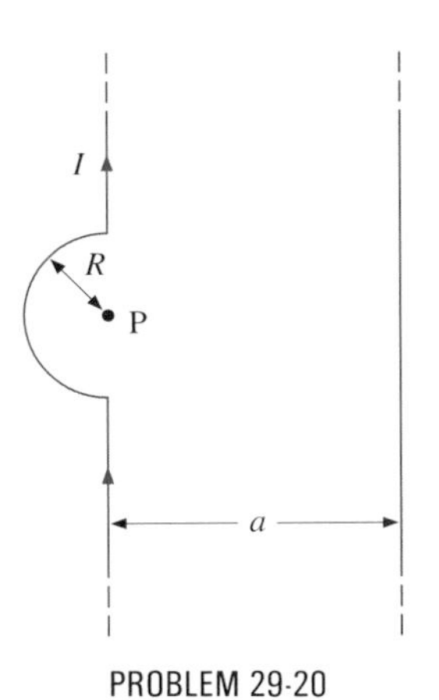

PROBLEM 29-20

29-21. Three long, straight, parallel wires, all carrying 20 A, are positioned as shown in the accompanying figure. What is the magnetic field at the point P?

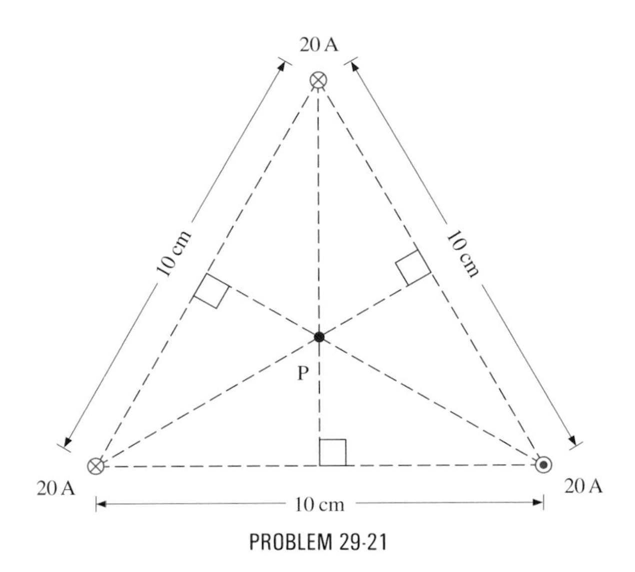

PROBLEM 29-21

Magnetic Forces between Current-Carrying Wires

29-22. Two long, straight wires are parallel and 25 cm apart. (*a*) If each wire carries a current of 50 A in the same direction, what is the magnetic force per meter exerted on each wire? (*b*) Does the force pull the wires together or push them apart? (*c*) What happens if the currents flow in opposite directions?

29-23. Two long, straight wires are parallel and 10 cm apart. One carries a current of 2.0 A, the other a current of 5.0 A. (*a*) If the two currents flow in opposite directions, what is the force per unit length of one wire on the other? (*b*) What is the force per unit length if the currents flow in the same direction?

29-24. Two long, parallel wires are hung by cords of length 5.0 cm, as shown in the accompanying figure. Each wire has a mass per unit length of 30 g/m, and they carry the same current in opposite directions. What is the current if the cords hang at 6.0° with respect to the vertical?

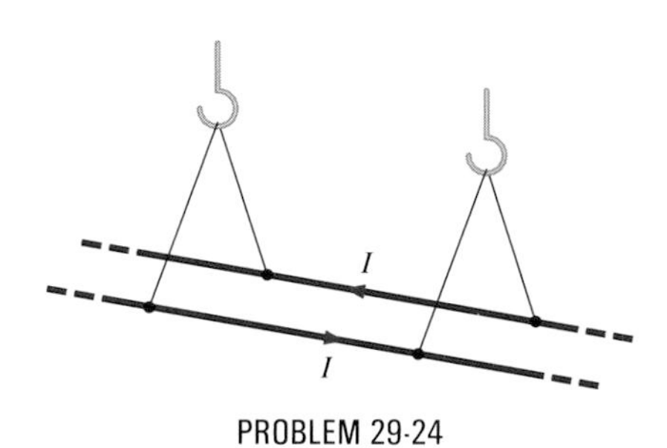

PROBLEM 29-24

Gauss' Law for the Magnetic Field

29-25. A friend proposes to build a magnet that will produce a spherically symmetric magnetic field given by $\mathbf{B} = f(r)\hat{\mathbf{r}}$. On the basis of what you have learned about magnetic field lines and Gauss' law for magnetism, how would you respond to this friend?

29-26. Verify the expressions for the magnetic fluxes ϕ_2, ϕ_3, and ϕ_4 in Example 29-4.

29-27. Consider the hypothetical magnetic field $\mathbf{B} = b_1 x^2 \mathbf{i} + b_2 yx\mathbf{j}$. Calculate $\oint_S \mathbf{B} \cdot \hat{\mathbf{n}}\, da$ over the cube of side 1.0 m shown in the accompanying figure. From this result find the relationship between the constants b_1 and b_2.

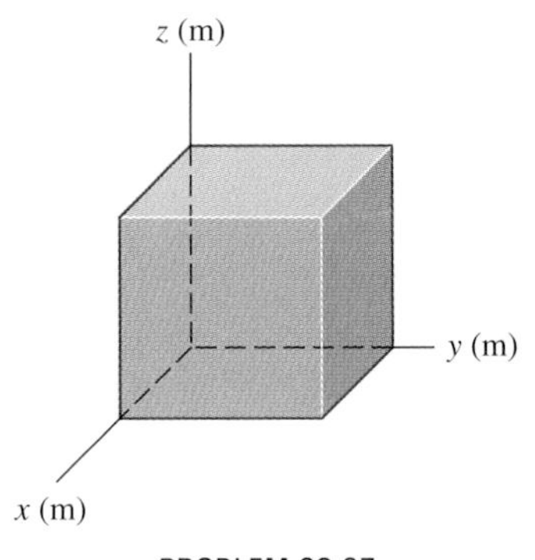

PROBLEM 29-27

Ampère's Law and Its Applications

29-28. A current I flows around the rectangular loop shown in the accompanying figure. Evaluate $\oint \mathbf{B}\cdot d\mathbf{l}$ for paths A, B, C, and D.

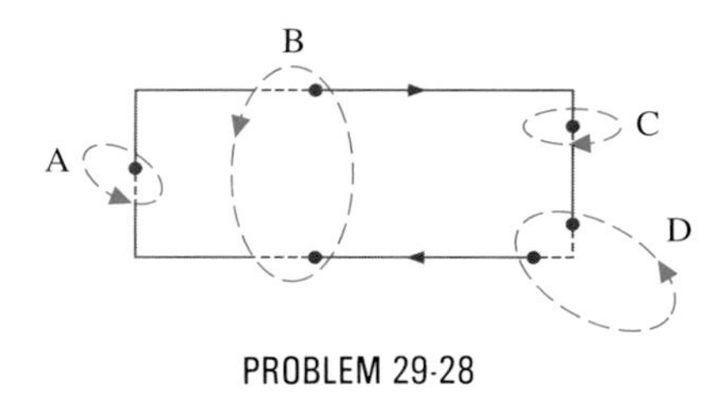

PROBLEM 29-28

29-29. Evaluate $\oint \mathbf{B}\cdot d\mathbf{l}$ for each of the cases shown in the accompanying figure.

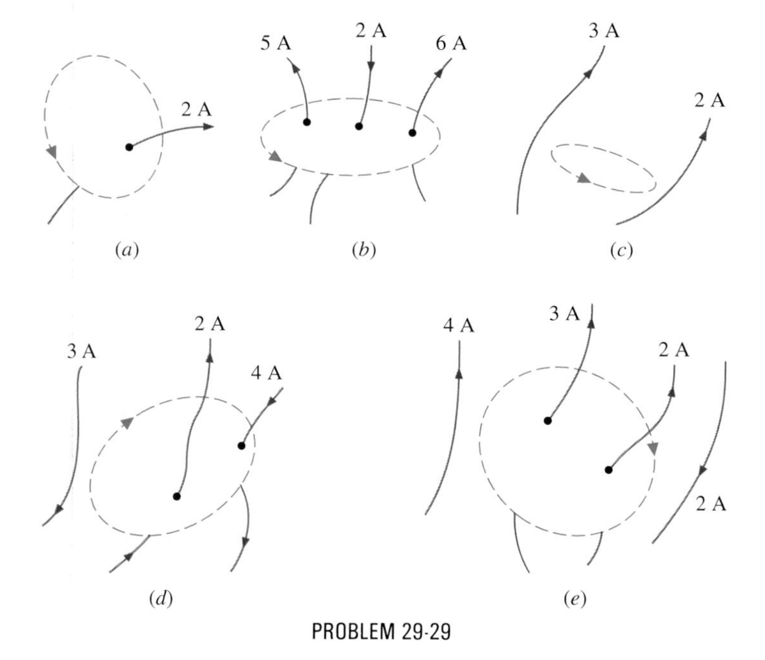

PROBLEM 29-29

29-30. The coil whose lengthwise cross section is shown in the accompanying figure carries a current I and has N evenly spaced turns distributed along a length l. Evaluate $\oint \mathbf{B}\cdot d\mathbf{l}$ for the paths indicated.

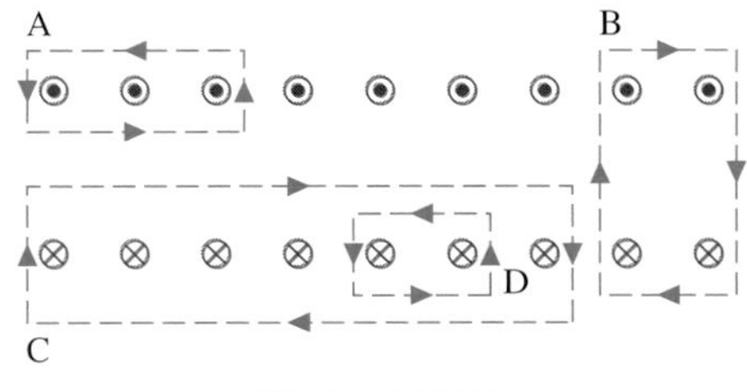

PROBLEM 29-30

29-31. How much current passes through the front face of the cube of Prob. 29-27?

29-32. A superconducting wire of diameter 0.25 cm carries a current of 1000 A. What is the magnetic field just outside the wire?

29-33. A long, straight wire of radius R carries a current I that is distributed uniformly over the cross section of the wire. At what distance from the axis of the wire is the magnitude of the magnetic field a maximum?

29-34. The accompanying figure shows a cross section of a long, hollow, cylindrical conductor of inner radius $r_1 = 3.0$ cm and outer radius $r_2 = 5.0$ cm. A 50-A current distributed uniformly over the cross section flows into the page. Calculate the magnetic field at $r = 2.0$ cm, $r = 4.0$ cm, and $r = 6.0$ cm.

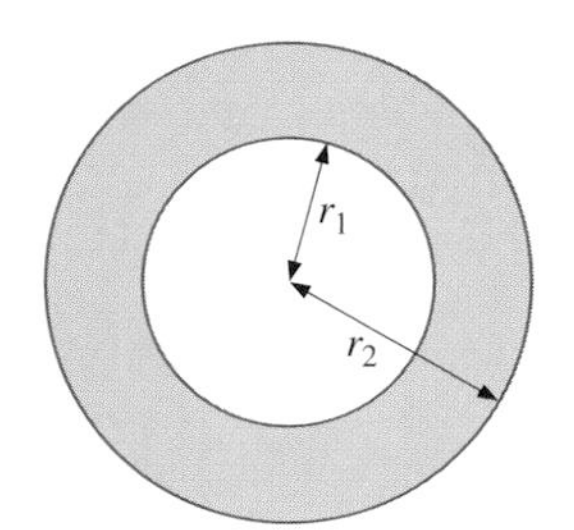

PROBLEM 29-34

29-35. A long, solid, cylindrical conductor of radius 3.0 cm carries a current of 50 A distributed uniformly over its cross section. Plot the magnetic field as a function of the radial distance r from the center of the conductor.

29-36. A portion of a long, cylindrical coaxial cable is shown in the accompanying figure. A current I flows down the center conductor, and this current is returned in the outer conductor. Determine the magnetic field in the regions (*a*) $r \le r_1$; (*b*) $r_2 \ge r \ge r_1$; (*c*) $r_3 \ge r \ge r_2$; and (*d*) $r \ge r_3$. Assume that the current is distributed uniformly over the cross sections of the two parts of the cable.

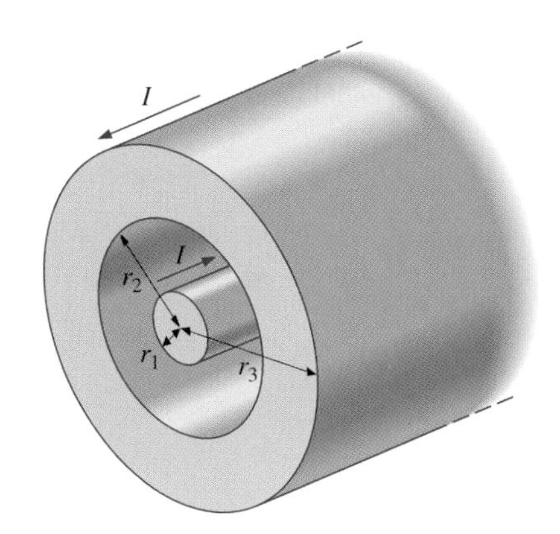

PROBLEM 29-36

29-37. Current flows along a thin, infinite sheet as shown in the accompanying figure. The current per unit length along the sheet is j in amperes per meter. (*a*) Use the Biot-Savart law to show that $B = \mu_0 j/2$ on either side of the sheet. What is the direction of **B** on each side? (*b*) Now use Ampère's law to calculate the field.

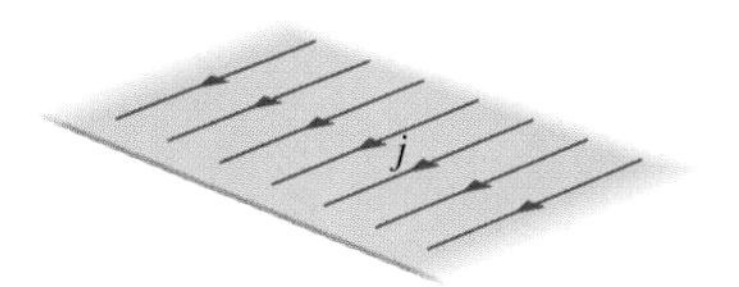

PROBLEM 29-37

29-38. (*a*) Use the result of the previous problem to calculate the magnetic field between, above, and below the pair of infinite sheets shown in the accompanying figure. (*b*) Repeat your calculations if the direction of the current in the lower sheet is reversed.

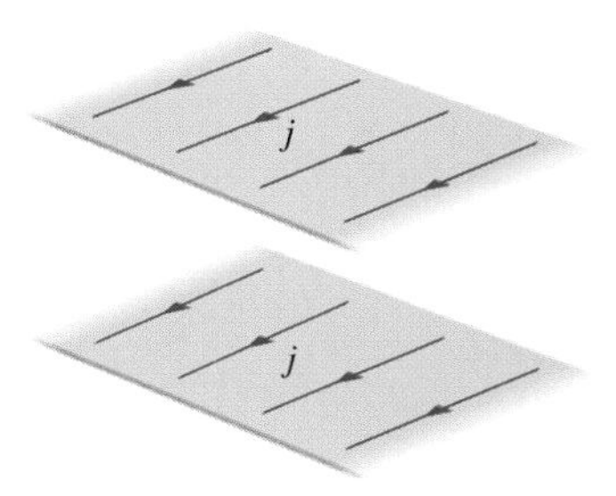

PROBLEM 29-38

29-39. We often assume that a magnetic field is uniform in a region and zero everywhere else. Show that in reality it is impossible for a magnetic field to drop abruptly to zero, as illustrated in the accompanying figure. (*Hint:* Apply Ampère's law over the path shown.)

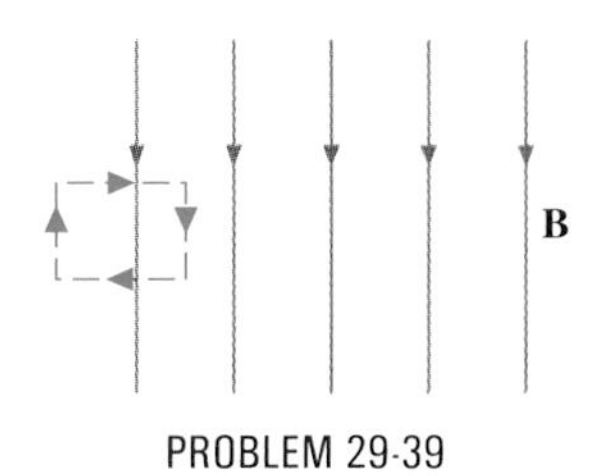

PROBLEM 29-39

Solenoids and Toroids

Unless stated otherwise, assume that the solenoids in the rest of the problems are infinite and that all solenoids and toroids are tightly wound.

29-40. A solenoid is wound with 2000 turns per meter. When the current is 5.2 A, what is the magnetic field within the solenoid?

29-41. A solenoid has 12 turns per centimeter. What current will produce a magnetic field of 2.0×10^{-2} T within the solenoid?

29-42. If the current is 2.0 A, how many turns per centimeter must be wound on a solenoid in order to produce a magnetic field of 2.0×10^{-3} T within it?

29-43. A solenoid is 40 cm long, has a diameter of 3.0 cm, and is wound with 500 turns. If the current through the windings is 4.0 A, what is the magnetic field at a point on the axis of the solenoid that is (*a*) at the center of the solenoid, (*b*) 10.0 cm from one end of the solenoid, and (*c*) 5.0 cm from one end of the solenoid? (*d*) Compare these answers with the infinite-solenoid case.

29-44. Determine the magnetic field on the central axis at the opening of a semi-infinite solenoid. (That is, take the opening to be at $x = 0$ and the other end to be at $x = \infty$.)

29-45. By how much is the approximation $B = \mu_0 nI$ in error at the center of a solenoid that is 15.0 cm long, has a diameter of 4.0 cm, is wrapped with n turns per meter, and carries a current I?

29-46. A solenoid with 25 turns per centimeter carries a current I. An electron moves within the solenoid in a circle that has a radius of 2.0 cm and is perpendicular to the axis of the solenoid. If the speed of the electron is 2.0×10^5 m/s, what is I?

29-47. A toroid has 250 turns of wire and carries a current of 20 A. Its inner and outer radii are 8.0 and 9.0 cm. What are the values of its magnetic field at $r = 8.1$, 8.5, and 8.9 cm?

29-48. A toroid with a square cross section 3.0 cm × 3.0 cm has an inner radius of 25.0 cm. It is wound with 500 turns of wire, and it carries a current of 2.0 A. What is the strength of the magnetic field at the center of the toroid?

29-49. How is the percentage change in the strength of the magnetic field across the face of a toroid related to the percentage change in the radial distance from the axis of the toroid?

29-50. Show that the expression for the magnetic field of a toroid reduces to that for the field of an infinite solenoid in the limit that the central radius goes to infinity.

29-51. A toroid with an inner radius of 20 cm and an outer radius of 22 cm is tightly wound with one layer of wire that has a diameter of 0.25 mm. (*a*) How many turns are there on the toroid? (*b*) If the current through the toroid windings is 2.0 A, what is the strength of the magnetic field at the center of the toroid?

Magnetic Materials

29-52. The magnetic field in the core of an air-filled solenoid is 1.50 T. By how much will this magnetic field decrease if the air is pumped out of the core while the current is held constant?

29-53. For a solenoid with a ferromagnetic core, $n = 1000$ turns per meter, and $I = 5.0$ A. If B inside the solenoid is 2.0 T, what is χ for the core material?

29-54. A 20-A current flows through a solenoid with 2000 turns per meter. What is the magnetic field inside the solenoid if its core is (*a*) a vacuum and (*b*) filled with liquid oxygen at 90 K?

29-55. The magnetic dipole moment of the iron atom is about 2.1×10^{-23} A·m². (*a*) Calculate the maximum magnetic dipole moment of a domain consisting of 10^{19} iron atoms. (*b*) What current would have to flow through a single circular loop of wire of diameter 1.0 cm to produce this magnetic dipole moment?

29-56. Suppose you wish to produce a 1.2-T magnetic field in a toroid with an iron core for which $\chi = 4.0 \times 10^3$. The toroid has a mean radius of 15 cm and is wound with 500 turns. What current is required?

29-57. A current of 1.5 A flows through the windings of a large, thin toroid with 200 turns per meter. If the toroid is filled with iron for which $\chi = 3.0 \times 10^3$, what is the magnetic field within it?

29-58. A solenoid with an iron core is 25 cm long and is wrapped with 100 turns of wire. When the current through the solenoid is 10 A, the magnetic field inside it is 2.0 T. For this current, what is the permeability of the iron? If the current is turned off and then restored to 10 A, will the magnetic field necessarily return to 2.0 T?

General Problems

29-59. For a wire element $d\mathbf{l}$, $I\,d\mathbf{l} = \mathbf{J}A\,dl = \mathbf{J}\,dv$, where A and dv are the cross-sectional area and volume of the element. Use this, the Biot-Savart law, and $\mathbf{J} = ne\mathbf{v}$ to show that the magnetic field of a moving point charge q is given by

$$\mathbf{B} = \frac{\mu_0}{4\pi}\frac{q\mathbf{v} \times \hat{\mathbf{r}}}{r^2}.$$

29-60. The infinite, straight wire shown in the accompanying figure carries a current I_1. The rectangular loop, whose long sides are parallel to the wire, carries a current I_2. What are the magnitude and direction of the force on the rectangular loop due to the magnetic field of the wire?

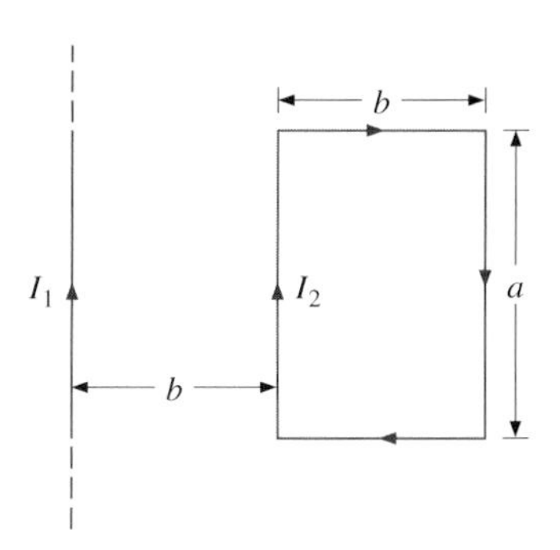

PROBLEM 29-60

29-61. A current I flows around a wire bent into the shape of a square of side a. What is the magnetic field at the point P that is a distance z above the center of the square (see the accompanying figure)?

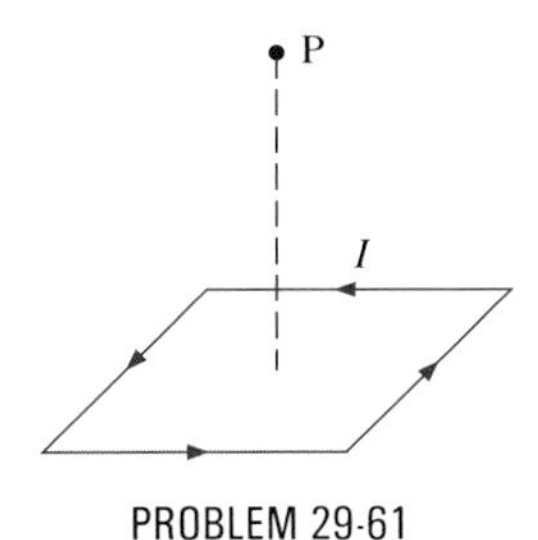

PROBLEM 29-61

29-62. A reasonably uniform magnetic field over a limited region of space can be produced with the *Helmholtz coil*, which consists of two parallel circular coils centered on the same axis. The coils are connected so that they carry the same current I. Each coil has N turns and radius R, which is also the distance between the coils. (*a*) Find the magnetic field at any point on the z axis shown in the accompanying figure. (*b*) Show that dB/dz and d^2B/dz^2 are both zero at $z = 0$. (*Note:* These vanishing derivatives demonstrate that the magnetic field varies only slightly near $z = 0$.)

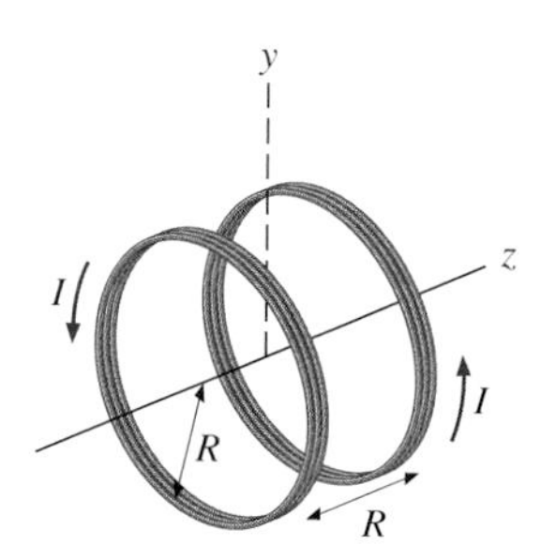

PROBLEM 29-62

29-63. A charge of 4.0 μC is distributed uniformly around a thin ring of insulating material. The ring has a radius of 0.20 m and rotates at 2.0×10^4 rev/min around the axis that passes through its center and is perpendicular to the plane of the ring. What is the magnetic field at the center of the ring?

29-64. A thin, nonconducting disk of radius R is free to rotate around the axis that passes through its center and is perpendicular to the face of the disk. The disk is charged uniformly with a total charge q. If the disk rotates at a constant angular velocity ω, what is the magnetic field at its center?

29-65. Consider the disk of the previous problem. Calculate the magnetic field at a point on its central axis that is a distance y above the disk.

29-66. Consider the axial magnetic field $B_y = \mu_0 IR^2/2(y^2 + R^2)^{3/2}$ of the circular current loop of Fig. 29-5. (*a*) Evaluate $\int_{-a}^{a} B_y\, dy$. Also show that

$$\lim_{a \to \infty} \int_{-a}^{a} B_y\, dy = \mu_0 I.$$

(*b*) Can you deduce this limit without evaluating the integral? (*Hint:* See the accompanying figure.)

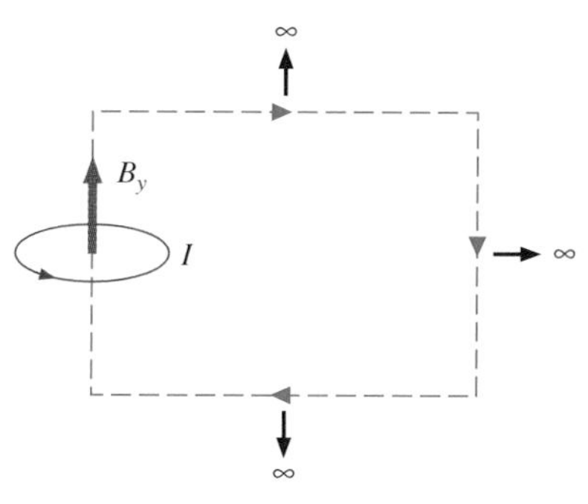

PROBLEM 29-66

29-67. The accompanying figure shows a flat, infinitely long sheet of width a that carries a current I uniformly distributed across it. Find the magnetic field at the point P, which is in the plane of the sheet and at a distance x from one edge. Test your result for the limit $a \to 0$.

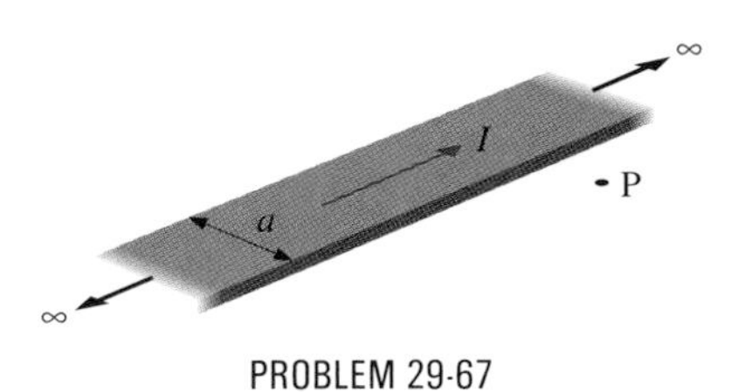

PROBLEM 29-67

29-68. The current density in the long, cylindrical wire shown in the accompanying figure varies with distance r from the center of the wire according to $J = cr$, where c is a constant. (*a*) What is the current through the wire? (*b*) What is the magnetic field produced by this current for $r \leq R$? for $r \geq R$?

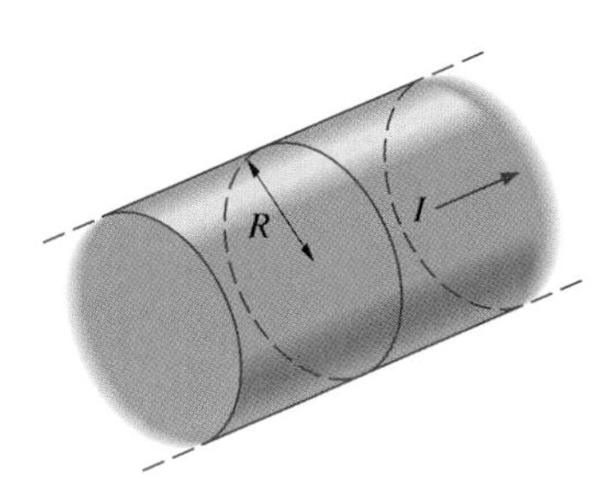

PROBLEM 29-68

29-69. A long, straight, cylindrical conductor contains a cylindrical cavity whose axis is displaced by **a** from the axis of the conductor, as shown in the accompanying figure. The current density in the conductor is given by $\mathbf{J} = J_0\mathbf{k}$, where J_0 is constant and $\mathbf{k}$ is along the axis of the conductor. Calculate the magnetic field at an arbitrary point P in the cavity by superimposing the field of a solid cylindrical conductor with radius R_1 and current density $\mathbf{J}$ onto the field of a solid cylindrical conductor with radius R_2 and current density $-\mathbf{J}$. Then use the fact that the appropriate azimuthal unit vectors can be expressed as $\hat{\boldsymbol{\theta}}_1 = \mathbf{k} \times \hat{\mathbf{r}}_1$ and $\hat{\boldsymbol{\theta}}_2 = \mathbf{k} \times \hat{\mathbf{r}}_2$ to show that everywhere inside the cavity the magnetic field is given by the constant

$$\mathbf{B} = \tfrac{1}{2}\mu_0 J_0 \mathbf{k} \times \mathbf{a},$$

where $\mathbf{a} = \mathbf{r}_1 - \mathbf{r}_2$ and $\mathbf{r}_1 = r_1\hat{\mathbf{r}}_1$ is the position of P relative to the center of the conductor and $\mathbf{r}_2 = r_2\hat{\mathbf{r}}_2$ is the position of P relative to the center of the cavity.

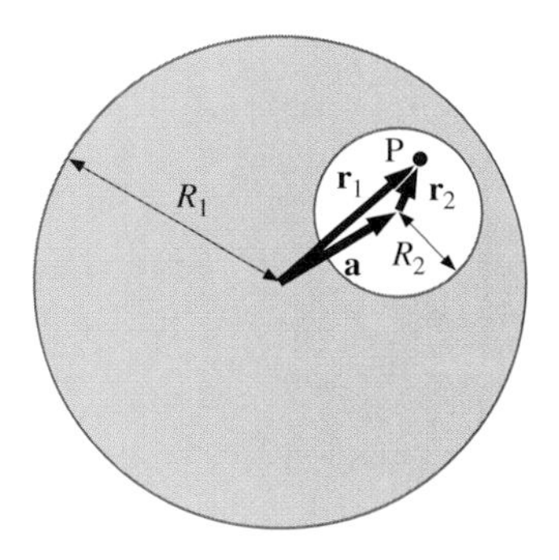

PROBLEM 29-69

29-70. A hypothetical current flowing in the z direction creates the field $\mathbf{B} = C[(x/y^2)\mathbf{i} + (1/y)\mathbf{j}]$ in the rectangular region of the xy plane shown in the accompanying figure. Use Ampère's law to find the current through the rectangle.

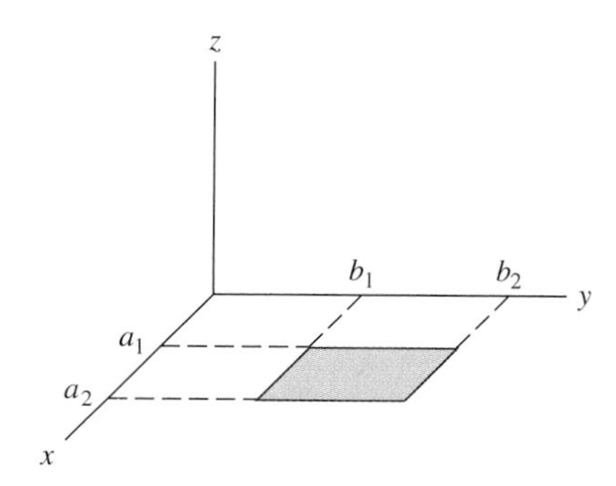

PROBLEM 29-70

Electric generators: an application of Faraday's law.

CHAPTER 30 FARADAY'S LAW

PREVIEW

In this chapter we investigate the electrical effects produced by a time-varying magnetic flux. The main topics to be discussed here are the following:

1. **Time-varying magnetic fields and Faraday's law.** We consider how an emf in a circuit is induced by changing magnetic flux due to a time-varying magnetic field.
2. **Induced electric field.** We discuss the nonconservative electric field induced by a time-varying magnetic field.
3. **Motionally induced emf.** The emf induced in a wire that is moving through a magnetic field is studied. Motionally induced emf's and emf's induced by time-varying magnetic fields are found to be related to changing magnetic flux by the same mathematical law. This relationship, which holds for all induced emf's, is called Faraday's law.

*4. **Applications of Faraday's law.** The applications considered are the alternating current source, eddy currents, and back emf's in dc motors.

Although we have been considering electric fields created by fixed charge distributions and magnetic fields produced by constant currents, electromagnetic phenomena are not restricted to these static situations. Most of the interesting applications of electromagnetism are, in fact, time-dependent. So that we can investigate some of these, we now remove the time-independent assumption that we have been making and allow the fields to vary with time. In this and the following chapter, you will see that there is a wonderful symmetry in the behavior exhibited by the time-varying electric and magnetic fields. Mathematically, this symmetry is expressed by an additional term in Ampère's law and by another of Maxwell's equations—Faraday's law.

30-1 Time-Varying Magnetic Fields and Faraday's Law

The first productive experiments concerning the effects of time-varying magnetic fields were performed by Michael Faraday in 1831. One of his early experiments is represented in Fig. 30-1. The wire loop is connected to a sensitive ammeter and the bar magnet is free to move in any direction. When the magnet is stationary, as shown in Fig. 30-1*a*, the ammeter needle does not deflect. This indicates that there is no current in the loop. However, when the magnet is moved toward the loop, as seen in Fig. 30-1*b*, the needle deflects to the right, indicating that current flows counterclockwise around the loop. Finally, when the magnet is moved away from the loop, as shown in Fig. 30-1*c*, a clockwise current flow is detected by a deflection of the ammeter needle toward the left.

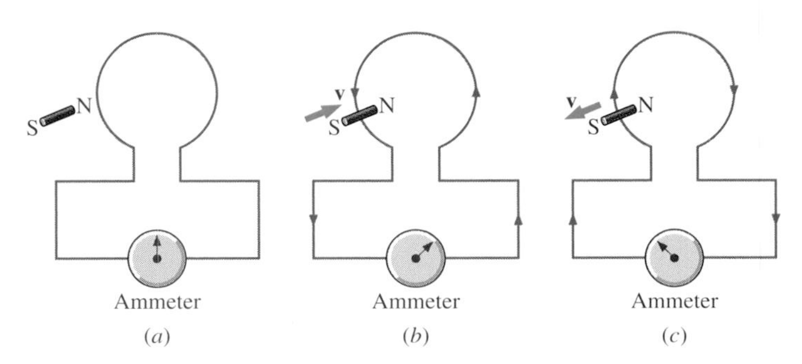

FIGURE 30-1 (*a*) A stationary bar magnet does not induce a current in the circuit. (*b*) and (*c*) A moving magnet produces a current whose direction depends on whether the magnet is moving toward or away from the coil.

Faraday also discovered that a similar effect can be produced using two circuits—a changing current in one circuit will induce a current in a second, nearby circuit. For example, when the switch is closed in circuit 1 of Fig. 30-2*a*, the ammeter needle of circuit 2 momentarily deflects, indicating that a short-lived current surge has been induced in that circuit. The ammeter needle quickly returns to its undeflected position, where it remains.

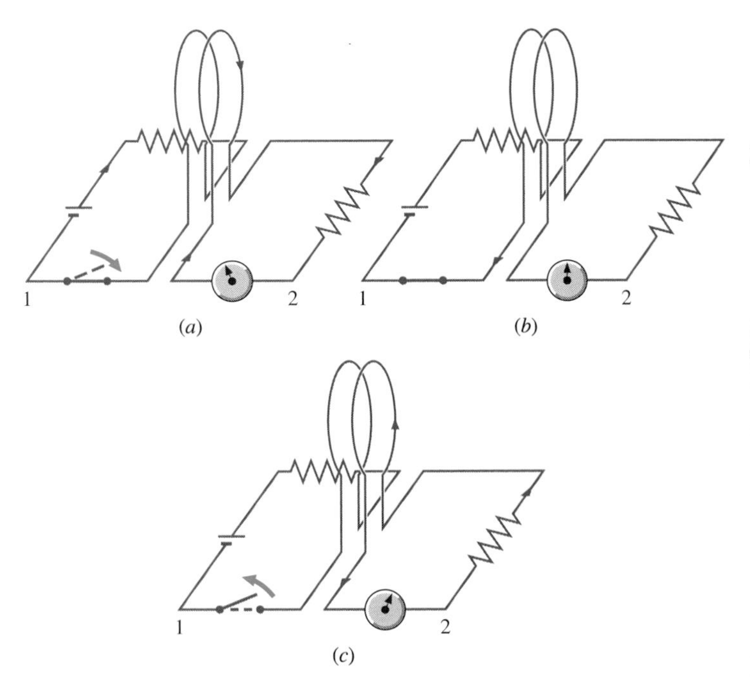

FIGURE 30-2 (*a*) Closing the switch of circuit 1 produces a current surge in circuit 2. (*b*) If the switch remains closed, no current is observed in circuit 2. (*c*) Opening the switch again produces a current in circuit 2 but in the opposite direction from before.

However, if the switch of circuit 1 is now suddenly opened, another short-lived current surge in the direction opposite from before is observed in circuit 2.

Faraday realized that in both experiments, a current flowed in the circuit containing the ammeter only when the magnetic field in the region occupied by that circuit was *changing*. As the magnet of Fig. 30-1 was moved, its magnetic field at the loop changed; and when the current in circuit 1 of Fig. 30-2 was turned on or off, its magnetic field at circuit 2 changed. Faraday was eventually able to interpret these and all other experiments involving magnetic fields that vary with time in terms of the following law, which we now call *Faraday's law*:

The emf $\mathcal{E}$ induced in a circuit by a changing magnetic field is equal to the negative time rate of change of the magnetic flux ϕ_m through any open surface S bounded by that circuit.

This means that if

$$\phi_m = \int_S \mathbf{B}\cdot\hat{\mathbf{n}}\,da,$$

then the induced emf is

$$\mathcal{E} = -\frac{d}{dt}\int_S \mathbf{B}\cdot\hat{\mathbf{n}}\,da = -\frac{d\phi_m}{dt}. \tag{30-1}$$

The negative sign describes the direction in which the induced emf drives current around a circuit. However, that direction is most easily determined with a rule known as Lenz's law, which will be discussed shortly.

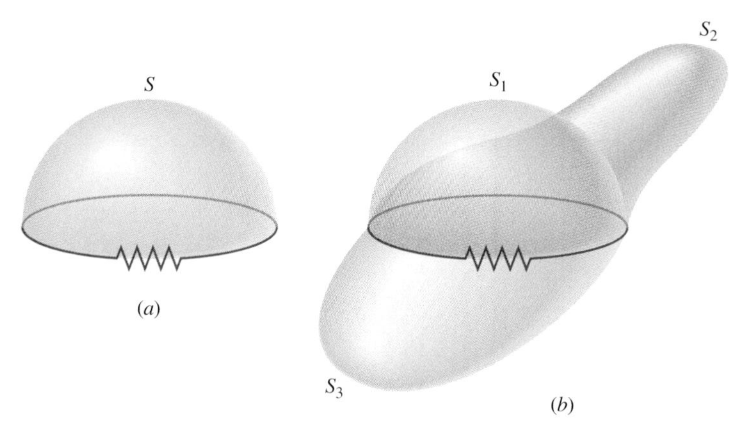

FIGURE 30-3 (*a*) A circuit bounding an arbitrary open surface *S*. The planar area bounded by the circuit is not part of *S*. (*b*) Three arbitrary open surfaces bounded by the same circuit. The value of ϕ_m is the same for all these surfaces.

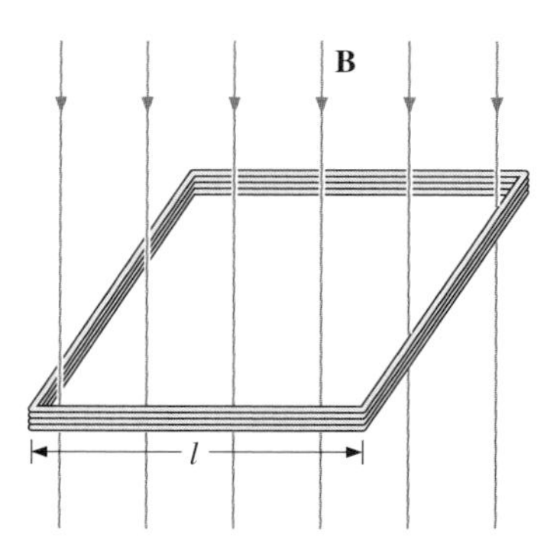

FIGURE 30-4 A square coil with *N* turns of wire.

Figure 30-3*a* depicts a circuit and an arbitrary surface *S* that it bounds. Notice that *S* is an *open surface*. It can be shown that *any* open surface bounded by the circuit in question can be used to evaluate ϕ_m. For example, ϕ_m is the same for the various surfaces $S_1, S_2, \ldots$ of Fig. 30-3*b*.

The SI unit for magnetic flux is the weber (Wb). Since the unit for *B* is the tesla,

$$1 \text{ Wb} = 1 \text{ T}\cdot\text{m}^2.$$

Occasionally, the magnetic field unit is expressed in webers per square meter (Wb/m^2).

In many practical applications of Eq. (30-1), the circuit of interest consists of a number *N* of tightly wound turns. (See Fig. 30-4.) In such cases, the net magnetic flux through the circuit is *N* times the flux through one turn, and Faraday's law is written as

$$\mathcal{E} = -\frac{d}{dt}(N\phi_m) = -N\frac{d\phi_m}{dt}. \tag{30-2}$$

The playback head of a tape deck. As the magnetized tape passes by a coil, the changing magnetic flux induces a current in the coil. This current is amplified and used to drive the speakers.

EXAMPLE 30-1 A SQUARE COIL IN A CHANGING MAGNETIC FIELD

The square coil of Fig. 30-4 has sides $l = 0.25$ m long and is tightly wound with $N = 200$ turns of wire. The resistance of the coil is $R = 5.0\ \Omega$. The coil is placed in a spatially uniform magnetic field that is directed perpendicular to the face of the coil and whose magnitude is decreasing at a rate $dB/dt = 0.040$ T/s. (*a*) What is the magnitude of the emf induced in the coil? (*b*) What is the magnitude of the current circulating through the coil?

SOLUTION (*a*) The flux through one turn is

$$\phi_m = BA = Bl^2,$$

so from Faraday's law,

$$|\mathcal{E}| = \left| -N\frac{d\phi_m}{dt} \right| = Nl^2\frac{dB}{dt}$$
$$= (200)(0.25\text{ m})^2(0.040\text{ T/s}) = 0.50\text{ V}.$$

(*b*) The magnitude of the current induced in the coil is

$$I = \frac{\mathcal{E}}{R} = \frac{0.50\text{ V}}{5.0\ \Omega} = 0.10\text{ A}.$$

DRILL PROBLEM 30-1

A closely wound coil has a radius of 4.0 cm, 50 turns, and a total resistance of 40 Ω. At what rate must a magnetic field perpendicular to the face of the coil change in order to produce Joule heating in the coil at a rate of 2.0 mW?
ANS. 1.1 T/s.

While the direction in which the induced emf drives current around a wire loop (called the *sense* of the emf) can be found from Eq. (30-1) through the negative sign, it is usually easier to determine this direction with *Lenz's law*, named in honor of its discoverer,* Heinrich Lenz (1804–1865). We state Lenz's law as follows:

The sense of the induced emf is such as to always *oppose* the change in magnetic flux that causes the emf.

So, to determine an induced emf $\mathcal{E}$, you first calculate the magnetic flux ϕ_m, and then obtain $d\phi_m/dt$. From Eq. (30-1), the magnitude of $\mathcal{E}$ is given by $\mathcal{E} = |d\phi_m/dt|$. Finally, you can apply Lenz's law to determine the sense of $\mathcal{E}$.

Let's apply Lenz's law to the system of Fig. 30-5*a*. We designate the "front" of the closed conducting loop as the region containing the approaching bar magnet, and the "back" of the

*Faraday, independent of Lenz, also discovered this law.

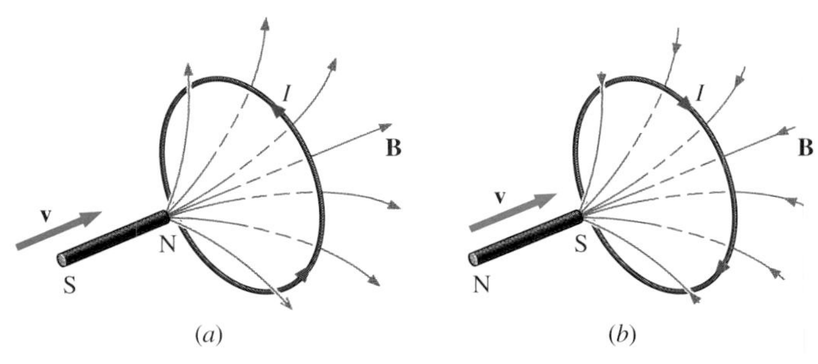

FIGURE 30-5 The change in magnetic flux caused by the approaching magnet induces a current in the loop.

loop as the other region. As the north pole of the magnet is moved toward the loop, the flux through the loop due to the field of the magnet changes because the number of field lines directed from the front to the back of the loop is increasing. A current is therefore induced in the loop. By Lenz's law, the direction of the current must be such that its own magnetic field is directed in a way to *oppose* the changing flux caused by the field of the approaching magnet. Hence the induced current circulates so that its magnetic field lines through the loop are directed from the back to the front of the loop. Alternatively, we can determine the direction of the induced current by treating the current loop as a magnet that *opposes* the approach of the north pole of the bar magnet. This occurs when the induced current flows as shown, for then the face of the loop nearer the approaching magnet is also a north pole.

Figure 30-5*b* shows the south pole of a magnet moving toward a conducting loop. In this case, the flux through the loop due to the field of the magnet changes since the number of field lines directed from the back to the front of the loop is increasing. To oppose this change, a current is induced in the loop whose field lines through the loop are directed from the front to the back. Equivalently, we can say that the current flows in a direction so that the face of the loop nearer the approaching magnet is a south pole, which then repels the approaching south pole of the magnet.

Another example illustrating the use of Lenz's law is given in Fig. 30-6. When the switch is opened, the decrease in current through the solenoid causes a decrease in magnetic flux through its coils, which induces an emf in the solenoid. This emf must oppose the change (the termination of the current) causing it. Consequently, the induced emf has the polarity shown and drives current in the direction of the original current. This may generate an arc across the terminals of the switch as it is opened.

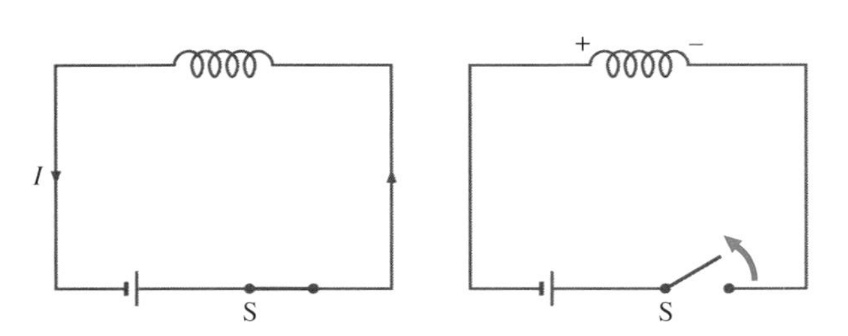

FIGURE 30-6 Opening switch S terminates the current, which in turn induces an emf in the solenoid.

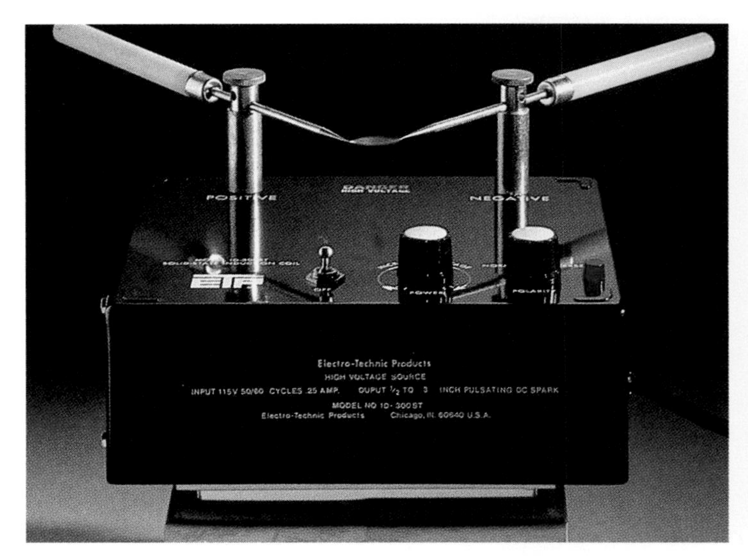

A potential difference between the ends of the sharply pointed rods is produced by inducing an emf in a coil. This potential difference is large enough to produce an arc between the sharp points.

DRILL PROBLEM 30-2

Using Lenz's law, verify the directions of current flow in Fig. 30-1.

DRILL PROBLEM 30-3

Find the direction of the induced current in the wire loop of Fig. 30-7 as the magnet (*a*) enters, (*b*) passes through, and (*c*) then leaves the loop.
ANS. To the observer shown, (*a*) the current flows clockwise as the magnet approaches, (*b*) decreases to zero when the magnet is centered in the plane of the coil, and (*c*) then flows counterclockwise as the magnet leaves the coil.

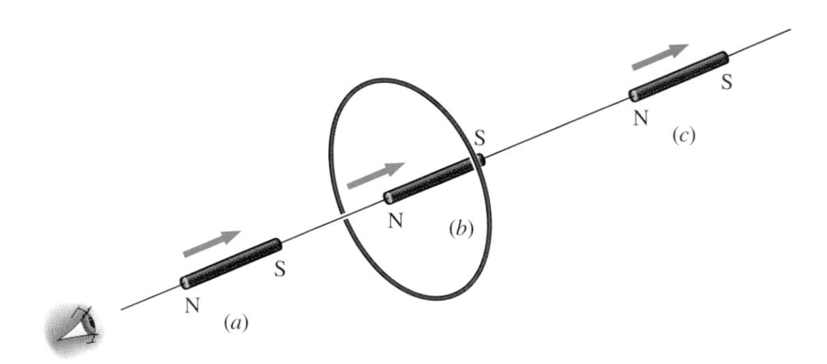

FIGURE 30-7 A bar magnet passing through a wire loop.

DRILL PROBLEM 30-4

Verify the directions of the induced currents in Fig. 30-2.

EXAMPLE 30-2 A Circular Coil in a Changing Magnetic Field

A magnetic field **B** is directed perpendicular to the plane of a circular coil of radius $r = 0.50$ m. (See Fig. 30-8.) The field is cylindrically symmetric with respect to the center of the coil, and its magnitude decays exponentially according to $B = 1.5e^{-5.0t}$, where B is in teslas and t is in seconds. (a) Calculate the emf induced in the coil at the times $t_1 = 0$, $t_2 = 5.0 \times 10^{-2}$ s, and $t_3 = 1.0$ s. (b) Determine the current in the coil at these three times if its resistance is 10 Ω.

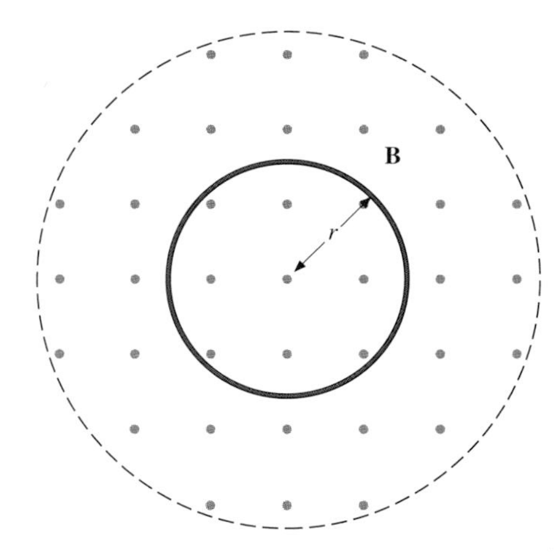

FIGURE 30-8 A circular coil in a decreasing magnetic field.

SOLUTION (a) Since **B** is perpendicular to the plane of the coil, the magnetic flux is given by

$$\begin{aligned}\phi_m &= B\pi r^2 = (1.5e^{-5.0t}\ \text{T})\pi(0.50\ \text{m})^2 \\ &= 1.2e^{-5.0t}\ \text{Wb}.\end{aligned}$$

From Faraday's law, the magnitude of the induced emf is

$$\mathcal{E} = \left|\frac{d\phi_m}{dt}\right| = \left|\frac{d}{dt}(1.2e^{-5.0t}\ \text{Wb})\right| = 6.0e^{-5.0t}\ \text{V}.$$

Since **B** is directed out of the page and is decreasing, the induced current must flow counterclockwise so that the magnetic field it produces through the coil also points out of the page. For all three times, the sense of $\mathcal{E}$ is counterclockwise; its magnitudes are

$$\mathcal{E}(t_1) = 6.0\ \text{V}; \qquad \mathcal{E}(t_2) = 4.7\ \text{V}; \qquad \mathcal{E}(t_3) = 0.040\ \text{V}.$$

(b) From Ohm's law, the respective currents are

$$I(t_1) = \frac{\mathcal{E}(t_1)}{R} = \frac{6.0\ \text{V}}{10\ \Omega} = 0.60\ \text{A};$$

$$I(t_2) = \frac{4.7\ \text{V}}{10\ \Omega} = 0.47\ \text{A};$$

and

$$I(t_3) = \frac{0.040\ \text{V}}{10\ \Omega} = 4.0 \times 10^{-3}\ \text{A}.$$

EXAMPLE 30-3 Changing Magnetic Field Inside a Solenoid

The current through the windings of a solenoid with $n = 2000$ turns per meter is changing at a rate $dI/dt = 3.0$ A/s. The solenoid is 50 cm long and has a cross-sectional diameter of 3.0 cm. A small coil consisting of $N = 20$ closely wound turns wrapped in a circle of diameter 1.0 cm is placed in the middle of the solenoid such that the plane of the coil is perpendicular to the central axis of the solenoid. Assuming that the infinite-solenoid approximation is valid at the location of the small coil, determine the magnitude of the emf induced in the coil.

SOLUTION Since the field of the solenoid is given by $B = \mu_0 nI$, the flux through each turn of the small coil is

$$\phi_m = \mu_0 nI\left(\frac{\pi d^2}{4}\right),$$

where d is the diameter of the coil. Now from Faraday's law, the magnitude of the emf induced in the coil is

$$\begin{aligned}\mathcal{E} &= \left|N\frac{d\phi_m}{dt}\right| = \left|N\mu_0 n\frac{\pi d^2}{4}\frac{dI}{dt}\right| \\ &= 20(4\pi \times 10^{-7}\ \text{T}\cdot\text{m/s})(2000\ \text{m}^{-1})\frac{\pi(0.010\ \text{m})^2}{4}(3.0\ \text{A/s}) \\ &= 1.2 \times 10^{-5}\ \text{V}.\end{aligned}$$

The jumping ring. When a current is turned on in the vertical solenoid, a current is induced in the metal ring. The resulting magnetic force between the currents causes the ring to jump off the solenoid.

30-2 Induced Electric Field

The fact that emf's are induced in circuits implies that work is being done on the conduction electrons in the wires. What can possibly be the source of this work? We know that it's neither a battery nor a magnetic field, for a battery does not have to be present in a circuit where current is induced, and magnetic fields never do work on moving charges. The answer is that *the source of the work is an electric field* **E** *that is induced in the* *wires. The work done by* **E** *in moving a unit charge completely around a circuit is the induced emf* $\mathcal{E}$; *that is,*

$$\mathcal{E} = \oint \mathbf{E} \cdot d\mathbf{l}, \tag{30-3}$$

where $\oint$ represents the line integral around the circuit. Faraday's law can be written in terms of the induced electric field as

$$\oint \mathbf{E} \cdot d\mathbf{l} = -\frac{d\phi_m}{dt}. \tag{30-4}$$

There is an important distinction between the electric field induced by a changing magnetic field and the electrostatic field produced by a fixed charge distribution. Specifically, the induced electric field is nonconservative because it does net work in moving a charge over a closed path, while the electrostatic field is conservative and does no net work over a closed path. Hence electric potential can be associated with the electrostatic field, but not with the induced field. The following equations represent the distinction between two types of electric field:

$$\oint \mathbf{E} \cdot d\mathbf{l} \neq 0 \quad \text{(induced)}$$
$$\oint \mathbf{E} \cdot d\mathbf{l} = 0 \quad \text{(electrostatic)} \tag{30-5}$$

Our results can be summarized by combining Eqs. (30-3) and (30-4). We then obtain

$$\mathcal{E} = \oint \mathbf{E} \cdot d\mathbf{l} = -\frac{d\phi_m}{dt}. \tag{30-6}$$

EXAMPLE 30-4 Induced Electric Field in a Circular Coil

What is the induced electric field in the circular coil of Example 30-2 (and Fig. 30-8) at the three times indicated?

SOLUTION By cylindrical symmetry, the induced electric field in the coil is constant in magnitude. Since **E** is tangent to the coil,

$$\oint \mathbf{E} \cdot d\mathbf{l} = \oint E\, dl = E 2\pi r,$$

which, when combined with Eq. (30-6), gives

$$E = \frac{\mathcal{E}}{2\pi r}.$$

The sense of $\mathcal{E}$ is counterclockwise, and **E** circulates in the same direction around the coil. The values of E are

$$E(t_1) = \frac{6.0\text{ V}}{2\pi(0.50\text{ m})} = 1.9\text{ V/m};$$

$$E(t_2) = \frac{4.7\text{ V}}{2\pi(0.50\text{ m})} = 1.5\text{ V/m};$$

and

$$E(t_3) = \frac{0.040\text{ V}}{2\pi(0.50\text{ m})} = 0.013\text{ V/m}.$$

When the magnetic flux through a circuit changes, a nonconservative electric field is induced, which drives current through the circuit. But what happens if $d\mathbf{B}/dt \neq 0$ in free space where there isn't a conducting path? The answer is that this case can be treated *as if a conducting path were present*; that is

> Nonconservative electric fields are induced wherever $d\mathbf{B}/dt \neq 0$, whether or not there is a conducting path present.

These nonconservative electric fields always satisfy Eq. (30-6). For example, if the circular coil of Fig. 30-8 were removed, an electric field *in free space* at $r = 0.50$ m would still be directed counterclockwise, and its magnitude would still be 1.9 V/m at $t = 0$, 1.5 V/m at $t = 5.0 \times 10^{-2}$ s, etc. The existence of induced electric fields is certainly *not* restricted to wires in circuits!

EXAMPLE 30-5 Electric Field Induced by the Changing Magnetic Field of a Solenoid

Figure 30-9*a* shows a long solenoid with radius R and n turns per unit length; its current decreases with time according to $I = I_0 e^{-\alpha t}$. What is the magnitude of the induced electric field at a point a distance r from the central axis of the solenoid (*a*) when $r > R$ and (*b*) when $r < R$? (See Fig. 30-9*b*.) (*c*) What is the direction of the induced field at both locations? Assume that the infinite-solenoid approximation is valid throughout the regions of interest.

SOLUTION (*a*) The magnetic field is confined to the interior of the solenoid where

$$B = \mu_0 n I = \mu_0 n I_0 e^{-\alpha t},$$

so the magnetic flux through a circular path whose radius r is greater than R, the solenoid radius, is

$$\phi_m = BA = \mu_0 n I_0 \pi R^2 e^{-\alpha t}.$$

The induced field **E** is tangent to this path, and because of the cylindrical symmetry of the system, its magnitude is constant on the path. Hence from

$$\left|\oint \mathbf{E} \cdot d\mathbf{l}\right| = \left|\frac{d\phi_m}{dt}\right|,$$

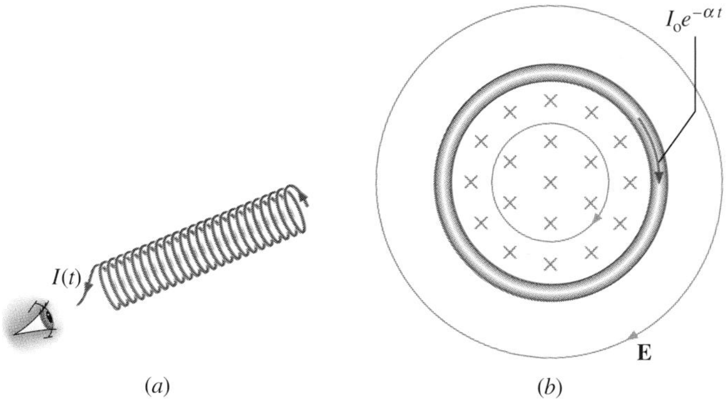

FIGURE 30-9 (*a*) The current in a long solenoid is decreasing exponentially. (*b*) A cross-sectional view of the solenoid from its left end. The cross section shown is near the middle of the solenoid. An electric field is induced both inside and outside the solenoid.

$$E(2\pi r) = \left| \frac{d}{dt} (\mu_0 n I_0 \pi R^2 e^{-\alpha t}) \right| = \alpha \mu_0 n I_0 \pi R^2 e^{-\alpha t},$$

and

$$E = \frac{\alpha \mu_0 n I_0 R^2}{2r} e^{-\alpha t} \quad (r > R). \qquad (i)$$

(*b*) For a path of radius r inside the solenoid, $\phi_m = B\pi r^2$, so

$$E(2\pi r) = \left| \frac{d}{dt} (\mu_0 n I_0 \pi r^2 e^{-\alpha t}) \right| = \alpha \mu_0 n I_0 \pi r^2 e^{-\alpha t},$$

and the induced field is

$$E = \frac{\alpha \mu_0 n I_0 r}{2} e^{-\alpha t} \quad (r < R). \qquad (ii)$$

You can see from Eqs. (*i*) and (*ii*) that E increases with r inside and decreases as $1/r$ outside the solenoid.

(*c*) The magnetic field points into the page as shown in Fig. 30-9*b* and is decreasing. If either of the circular paths were occupied by conducting rings, the currents induced in them would circulate as shown, in conformity with Lenz's law. The induced electric field must be so directed as well.

DRILL PROBLEM 30-5

Suppose that the coil of Example 30-2 is square rather than circular. Can Eq. (30-6) be used to calculate (*a*) the induced emf and (*b*) the induced electric field?
ANS. (*a*) Yes; (*b*) no; we lose our symmetry argument so E can no longer be factored from $\oint \mathbf{E} \cdot d\mathbf{l}$ when the integral is evaluated around the square coil.

DRILL PROBLEM 30-6

What is the magnitude of the induced electric field in Example 30-5 at $t = 0$ if $r = 6.0$ cm, $R = 2.0$ cm, $n = 2000$ turns per meter, $I_0 = 2.0$ A, and $\alpha = 200\ s^{-1}$?
ANS. 3.4×10^{-3} V/m.

DRILL PROBLEM 30-7

The magnetic field of Fig. 30-10 is confined to the cylindrical region shown and is changing with time. Identify those paths for which $\mathcal{E} = \oint \mathbf{E} \cdot d\mathbf{l} \neq 0$.
ANS. P_1, P_2, P_4.

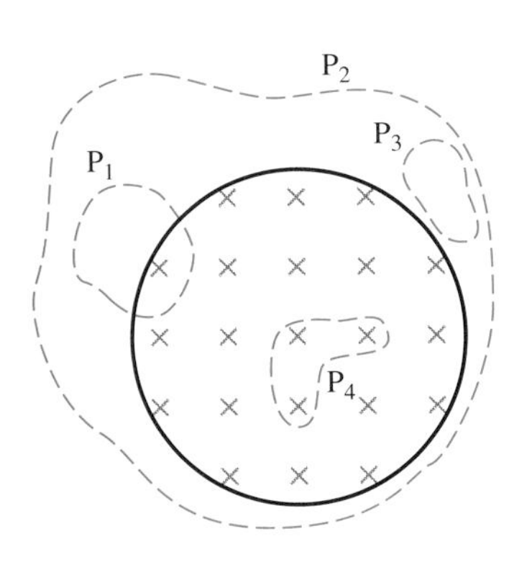

FIGURE 30-10 Closed paths and changing magnetic field in a cylindrical region.

DRILL PROBLEM 30-8

A long solenoid of cross-sectional area 5.0 cm^2 is wound with 25 turns of wire per centimeter. It is placed in the middle of a closely wrapped coil of 10 turns and radius 25 cm, as shown in Fig. 30-11. (*a*) What is the emf induced in the coil when the current through the solenoid is decreasing at a rate $dI/dt = -0.20$ A/s? (*b*) What is the electric field induced in the coil?
ANS. (*a*) 3.1×10^{-6} V; (*b*) 2.0×10^{-7} V/m.

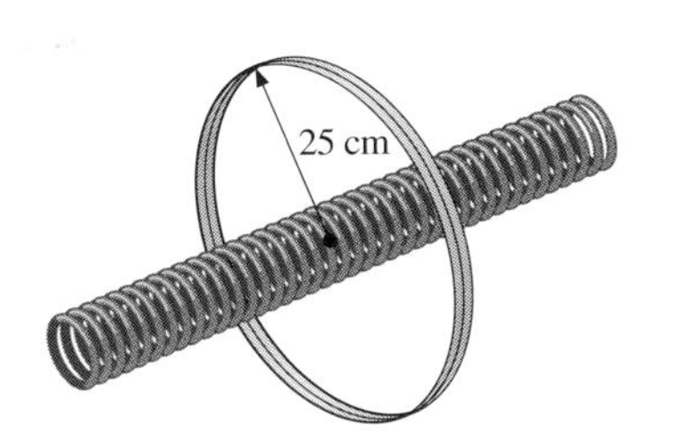

FIGURE 30-11 A concentric coil and solenoid.

30-3 MOTIONALLY INDUCED EMF

Magnetic flux depends on two factors: the strength of the magnetic field and the area through which the field lines pass. If either of these quantities varies, there is a corresponding variation in magnetic flux. So far, we've only considered flux changes due to a changing field. Now we look at the other possibility: a changing area through which the field lines pass. Two examples of this type of flux change are represented in Fig. 30-12. In Fig. 30-12*a* the flux through the rectangular loop increases as it moves into the magnetic field, and in Fig. 30-12*b* the flux through the rotating coil varies with the angle θ.

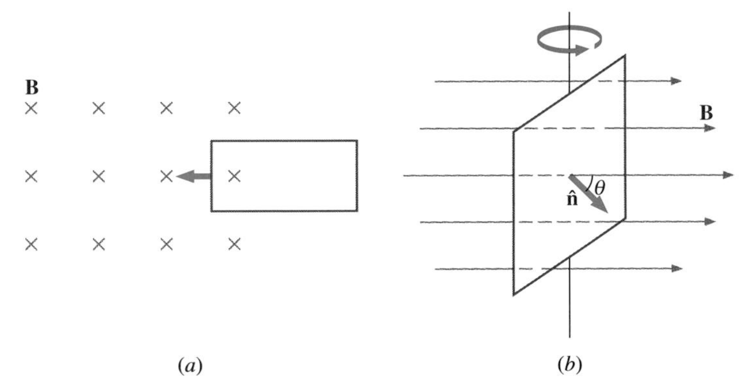

FIGURE 30-12 Two examples of flux changes due to changes in the areas through which the magnetic field lines cross.

It's interesting to note that what is perceived as the cause of a particular flux change actually depends on the frame of reference. For example, if you are at rest relative to the moving coils of Fig. 30-12, you would see the flux vary because of a changing magnetic field—in Fig. 30-12*a* the field would be moving from left to right in your reference frame, and in Fig. 30-12*b* it would be rotating. It is often possible to describe a flux change through a coil that is moving in one particular reference frame in terms of a changing magnetic field in a second frame where the coil is stationary. However, reference-frame questions related to magnetic flux are not appropriate for the level of this textbook. We'll avoid such complexities by always working in a frame at rest relative to the laboratory and explain flux variations as due to either a changing field or a changing area.

To study the effects of motional flux changes, we first consider the conducting rod NO of Fig. 30-13. It is moving to the right at a constant velocity **v** in a constant magnetic field **B**. Since the conduction electrons in the rod are also moving to the right, they experience a magnetic force $-e\mathbf{v} \times \mathbf{B}$ that drives them downward, resulting in an upward current I. (See Fig. 30-14*a*.) This causes a charge separation in the rod that produces an electric field **E** pointing downward. Equilibrium is quickly reached and the forces on the conduction electrons due to **E** and **B** balance, as shown in Fig. 30-14*b*. Then $I = 0$, $eE = evB$, and the potential difference between the ends of the rod is

$$V = El = Blv, \tag{30-7}$$

where l is the length of the rod.

Suppose we place the rod on the resistanceless rails of the circuit of Fig. 30-15. The rails are connected by a resistor of resistance R. We push the rod to the right with a force $\mathbf{F}_a$ so that it moves with a constant velocity **v**. As the rod moves, current

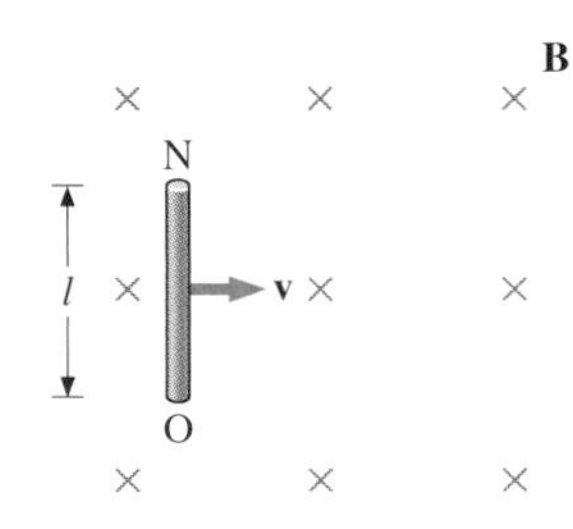

FIGURE 30-13 A conducting rod is moving in a magnetic field.

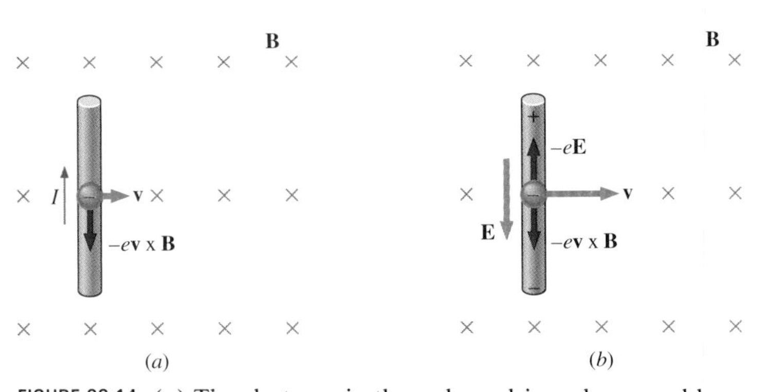

FIGURE 30-14 (*a*) The electrons in the rod are driven downward by the magnetic force. (*b*) In equilibrium, the electric force due to the charge separation is equal and opposite to the magnetic force.

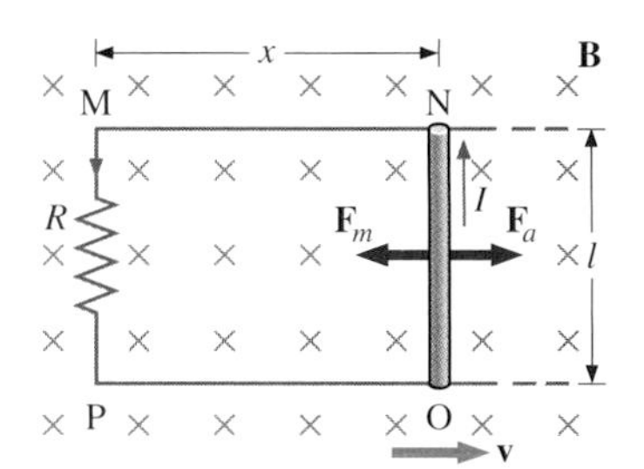

FIGURE 30-15 A conducting rod is pushed to the right at constant velocity. The resulting change in the magnetic flux induces a current in the circuit.

again flows upward from O to N but now continues counterclockwise around the completed circuit NMPO. The source of emf here is the applied force $\mathbf{F}_a$, which, in pushing the rod to the right, creates in conjunction with **B** a current flow in the circuit. You might find it helpful to picture the moving rod as a battery. Recall that the emf of a battery is the potential difference across its terminals when they are not connected to an external circuit. Analogously, the emf produced by $\mathbf{F}_a$ is given by the potential difference between the ends of the isolated moving rod. From Eq. (30-7), this *motionally induced emf* is

$$\mathcal{E} = Blv, \tag{30-8}$$

and the current induced in the circuit is

$$I = \frac{Blv}{R}.$$

From an energy perspective, $\mathbf{F}_a$ produces power $F_a v$, and the resistor dissipates power I^2R. Since the rod is moving at constant velocity, the applied force F_a must balance the magnetic force $F_m = IlB$ on the rod when it is carrying the induced current I. Thus the power produced is

$$F_a v = IlBv = \frac{Blv}{R}\, lBv = \frac{l^2B^2v^2}{R},$$

while the power dissipated is

$$I^2R = \left(\frac{Blv}{R}\right)^2 R = \frac{l^2B^2v^2}{R}.$$

In satisfying the principle of energy conservation, the produced and dissipated powers are equal.

Now let's look at the system in terms of changing magnetic flux. The area enclosed by the circuit MNOP of Fig. 30-15 is lx, and the magnetic flux through it is

$$\phi_m = Blx.$$

Since B and l are constant and the velocity of the rod is $v = dx/dt$,

$$\frac{d\phi_m}{dt} = Bl\,\frac{dx}{dt} = Blv,$$

which from Eq. (30-8) is also the emf induced in the circuit. Thus the magnitude of the induced emf satisfies $\mathcal{E} = |d\phi_m/dt|$. Furthermore, the sense of the induced emf satisfies Lenz's law, as you can verify by inspection of Fig. 30-15.

This calculation of motionally induced emf is not restricted to a rod moving on conducting rails. With $\mathbf{F} = q\mathbf{v} \times \mathbf{B}$ as the starting point, it can be shown that $\mathcal{E} = -d\phi_m/dt$ holds for *any* flux change caused by the motion of a circuit. We found in Sec. 30-1 that the emf induced by a time-varying magnetic field obeys this same relationship, which we called Faraday's law. Thus Faraday's law *holds for all flux changes*, whether they be produced by a changing magnetic field, by motion, or by a combination of the two.

Michael Faraday working in his laboratory at the Royal Institute, London.

We can calculate a motionally induced emf with Faraday's law *even when there is not an actual closed circuit*. We simply imagine an enclosed area whose boundary includes the moving conductor, calculate ϕ_m, then find the emf from Faraday's law. For example, we can let the moving rod of Fig. 30-16 be one side of the imaginary rectangular area represented by the dashed lines. The area of the rectangle is lx, so the magnetic flux through it is $\phi_m = Blx$. In differentiating this equation, we obtain

$$\frac{d\phi_m}{dt} = Bl\frac{dx}{dt} = Blv,$$

which is identical to the potential difference between the ends of the rod that we determined earlier.

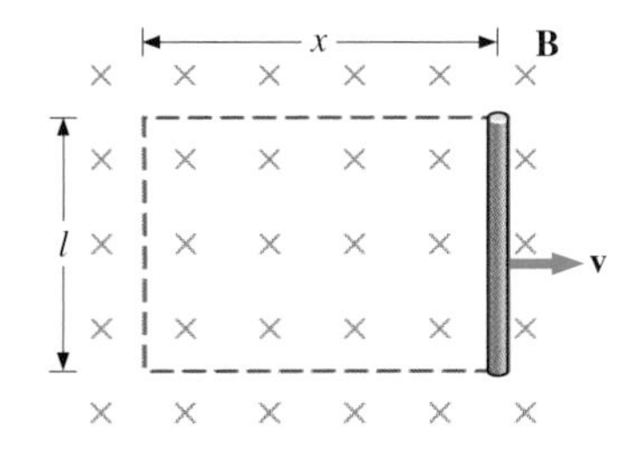

FIGURE 30-16 With the imaginary rectangle shown, Faraday's law can be used to calculate the induced emf in the moving rod.

EXAMPLE 30-6

A Metal Rod Rotating in a Magnetic Field

Figure 30-17*a* shows a metal rod OS that is rotating in a horizontal plane around O. The rod slides along a wire that forms a circular arc PST of radius r. The system is in a constant magnetic field $\mathbf{B}$ that is directed out of the page. (*a*) If you rotate the rod at a constant angular velocity ω, what is the current I in the closed loop OPSO? Assume that the resistor R furnishes all of the resistance in the closed loop. (*b*) Calculate the work per unit time that you do while rotating the rod and show that it is equal to the power dissipated in the resistor.

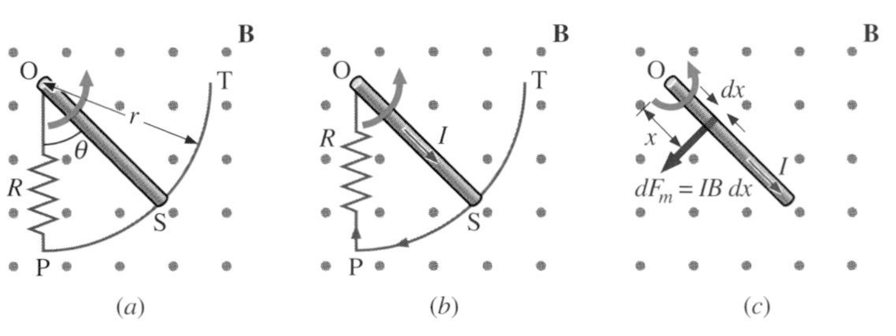

FIGURE 30-17 (*a*) A metal rod slides along a circular wire in a horizontal plane. (*b*) The induced current in the rod. (*c*) The magnetic force on an infinitesimal current segment.

SOLUTION (*a*) From geometry, the area of the loop OPSO is $A = r^2\theta/2$. Hence the magnetic flux through the loop is

$$\phi_m = BA = B\frac{r^2\theta}{2}.$$

Differentiating this and using $\omega = d\theta/dt$, we have

$$\mathcal{E} = \left|\frac{d\phi_m}{dt}\right| = \frac{Br^2\omega}{2},$$

which when divided by the resistance R of the loop yields for the magnitude of the induced current

$$I = \frac{\mathcal{E}}{R} = \frac{Br^2\omega}{2R}. \qquad (i)$$

As θ increases, so does the flux through the loop due to $\mathbf{B}$. To counteract this increase, the magnetic field due to the induced current must be directed into the page in the region enclosed by the loop. Therefore, as Fig. 30-17*b* illustrates, the current circulates clockwise.

(*b*) You rotate the rod by exerting a torque on it. Since the rod rotates at constant angular velocity, this torque is equal and opposite to the torque exerted on the current in the rod by the magnetic field. The magnetic force on the infinitesimal segment of length dx shown in Fig. 30-17*c* is $dF_m = IB\,dx$, so the magnetic torque on this segment is $d\tau_m = x\,dF_m = IBx\,dx$. The net magnetic torque on the rod is then

$$\tau_m = \int_0^r d\tau_m = IB\int_0^r x\,dx = \frac{1}{2}IBr^2.$$

The torque τ that you exert on the rod is equal and opposite to τ_m, and the work that you do when the rod rotates through an angle $d\theta$ is $dW = \tau\,d\theta$. Hence the work per unit time that you do on the rod is

$$\frac{dW}{dt} = \tau\frac{d\theta}{dt} = \frac{1}{2}IBr^2\frac{d\theta}{dt} = \frac{1}{2}\left(\frac{Br^2\omega}{2R}\right)Br^2\omega = \frac{B^2r^4\omega^2}{4R}, \quad (ii)$$

where we have used Eq. (i) to substitute for I. The power dissipated in the resistor is $P = I^2R$, which can be written as

$$P = \left(\frac{Br^2\omega}{2R}\right)^2 R = \frac{B^2r^4\omega^2}{4R}. \quad (iii)$$

Comparing Eqs. (ii) and (iii), we see that

$$P = \frac{dW}{dt}.$$

Hence the power dissipated in the resistor is equal to the work per unit time that you do in rotating the rod.

EXAMPLE 30-7

A Rectangular Coil Rotating in a Magnetic Field

A rectangular coil of area A and N turns is placed in a uniform magnetic field $\mathbf{B} = B\mathbf{j}$, as shown in Fig. 30-18. The coil is rotated about the z axis through its center at a constant angular velocity ω. Obtain an expression for the induced emf in the coil.

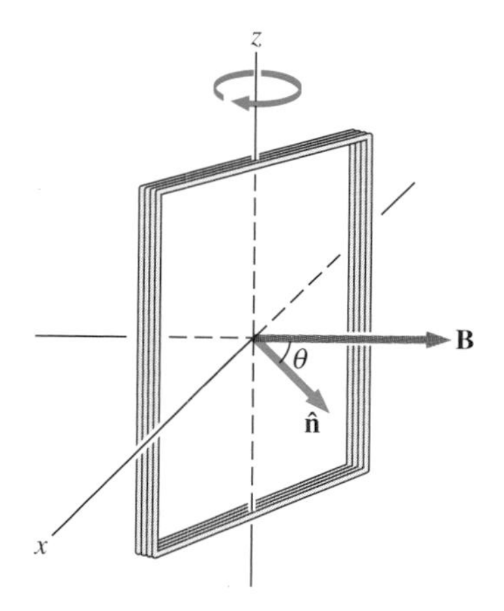

FIGURE 30-18 A rectangular coil rotating in a uniform magnetic field.

SOLUTION When the coil is in a position such that its normal vector $\hat{\mathbf{n}}$ makes an angle θ with the magnetic field $\mathbf{B}$, the magnetic flux through a single turn of the coil is

$$\phi_m = \int_S \mathbf{B}\cdot\hat{\mathbf{n}}\, da = BA\cos\theta.$$

From Faraday's law, the emf induced in the coil is

$$\mathcal{E} = -N\frac{d\phi_m}{dt} = NBA\sin\theta\,\frac{d\theta}{dt}.$$

The constant angular velocity is $\omega = d\theta/dt$. The induced emf therefore varies sinusoidally with time according to

$$\mathcal{E} = \mathcal{E}_0 \sin\omega t, \quad (30\text{-}9)$$

where $\mathcal{E}_0 = NBA\omega$.

DRILL PROBLEM 30-9

Figure 30-19 shows a rod of length l that is rotated counterclockwise around the axis through O by the torque due to $m\mathbf{g}$. Assuming that the rod is in a uniform magnetic field $\mathbf{B}$, what is the emf induced between the ends of the rod when its angular velocity is ω? Which end of the rod is at a higher potential?
ANS. $\mathcal{E} = Bl^2\omega/2$, with O at a higher potential than S.

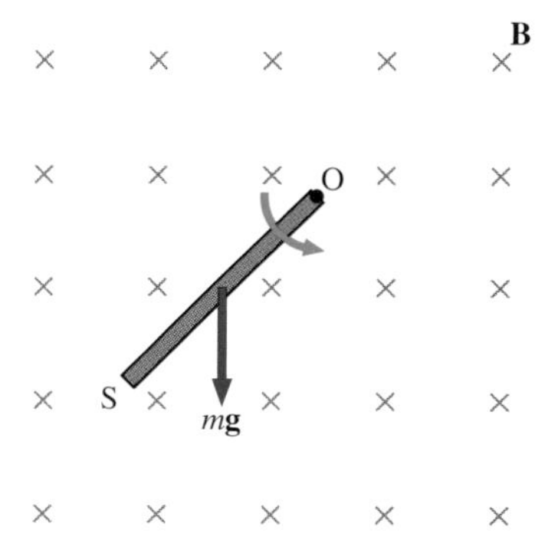

FIGURE 30-19 A rod rotating in a uniform magnetic field.

DRILL PROBLEM 30-10

A rod of length 10 cm moves through a 1.5-T magnetic field at a constant velocity of 10 m/s. What is the potential difference between the ends of the rod?
ANS. 1.5 V.

*30-4 Applications of Faraday's Law

A variety of important phenomena and devices can be understood with Faraday's law. In this section we examine three of these.

Alternating Current Source

The primary component of the alternating current source is a coil rotating in a magnetic field (see Example 30-7). The coil

By turning a coil in a magnetic field, the student is able to produce a current sufficient to light the bulb.

is connected to an external mechanical device (e.g., a steam turbine), which keeps the coil rotating at constant angular velocity thereby inducing an emf given by Eq. (30-9). This induced emf is coupled to an external circuit by attaching "slip rings" to the coil, as illustrated in Fig. 30-20. The slip rings rotate with the coil and make continuous contact with stationary brushes. These brushes serve as terminals that can be connected to an external circuit.

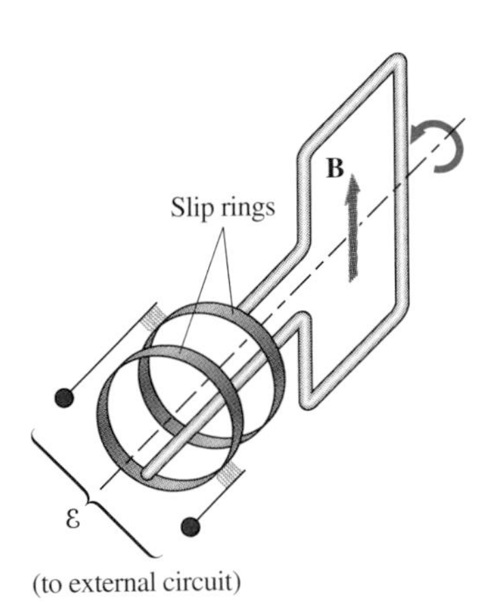

FIGURE 30-20 The "slip rings" couple the induced emf in the rotating coil to an external circuit.

Commercial electric power is produced with a complicated version of this generator. A large turbine is turned by high-pressure steam or by falling water. (See Fig. 30-21.) The turbine is connected to coils that rotate in a magnetic field,* thereby converting the mechanical energy of the driven turbine to electric energy. The generator supplies current to every device connected to it. The more devices there are—that is, the more lights, televisions, toasters, and motors connected to the power lines—the larger the current that must be supplied by the generator. By Lenz's law, the current flows in a direction such that a magnetic force acts on the coil to retard its turning; and the larger this current, the larger the retarding force. The power supplied to the turbine, and correspondingly the amount of fuel burned, must therefore increase with the load.

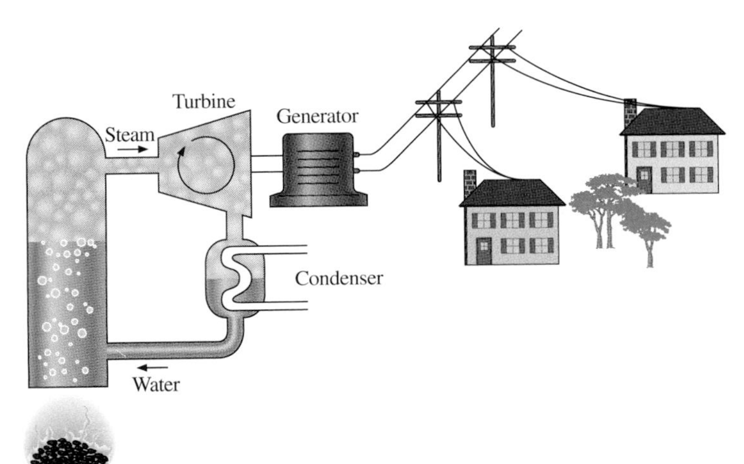

FIGURE 30-21 Representation of the production and distribution of electric power.

Eddy Currents

In Fig. 30-22a, a copper disk suspended from a plastic rod is shown swinging into a magnetic field. When the disk first enters the field, the magnetic flux through any closed path lying entirely in the copper increases. Consequently, emf's are induced around the closed conducting paths, and currents circulate in the disk. These circulating currents are called *eddy currents*. By Lenz's law, these currents flow in a sense such that the magnetic force on them opposes the motion of the disk into the field. If the disk swings far enough so that it starts to leave the field, the eddy currents reverse direction, and the magnetic force acts to prevent the disk from leaving. If the magnetic field is large enough, the magnetic force on the induced eddy currents can stop the disk before it moves out of the field.

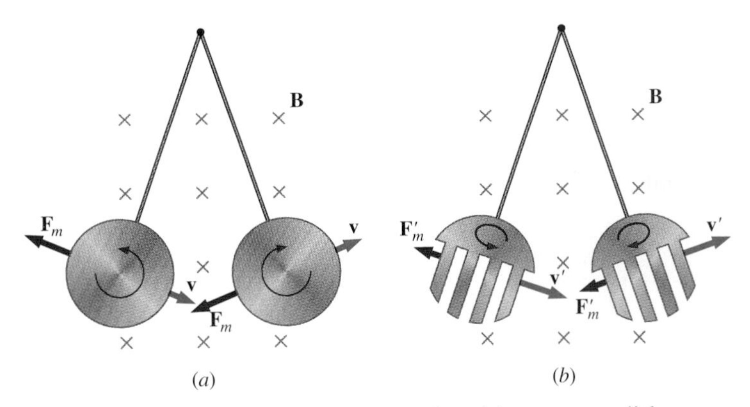

FIGURE 30-22 (a) Eddy currents are produced in a copper disk swinging through a magnetic field. (b) Slots cut in the disk reduce the eddy currents and allow the disk to swing more freely.

From an energy standpoint, the disk slows down because its kinetic energy is dissipated in the resistive heating of the eddy currents. These losses can be reduced by interrupting some of the conduction paths. For example, if slots are cut as shown in Fig. 30-22b, the effective resistance of the disk is increased, the eddy currents are decreased, and the energy dissipated is also decreased. As a result, the disk swings more freely.

Eddy currents in the iron cores of motors can cause troublesome energy losses. These are usually minimized by constructing the cores out of thin, electrically insulated sheets of iron. The magnetic properties of the core are hardly affected by the lamination, while the resistive heating is reduced considerably.

*Actually, most modern alternating current generators produce power by rotating a magnetic field through fixed coils.

DRILL PROBLEM 30-11

A thin, rectangular aluminum plate is dropped vertically into the region between the poles of a strong magnet, as shown in Fig. 30-23. This is repeated for an identical piece made of plastic. Compare the motions of the two falling objects.

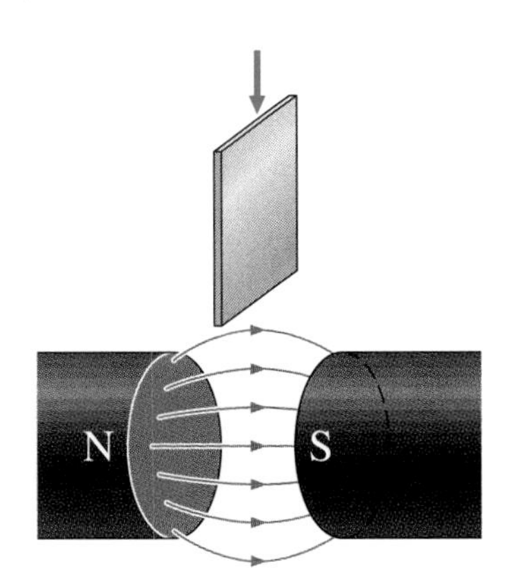

FIGURE 30-23 A rectangular plate is dropped between the poles of a magnet.

Back emf in the dc Motor

As the armature of a dc motor (see Sec. 28-5) rotates, an emf given by Eq. (30-9) is induced in each of its coils. The sum of these over all coils is called the **back emf** of the motor. Like the net magnetic torque on the armature, the back emf is independent of the angular position of the armature. It is also proportional to the angular velocity of the motor.

Here we'll only consider the *series-wound motor* whose armature coils and electromagnet coils are connected in series across an external power source. The circuit representation for this motor is given in Fig. 30-24. The resistances of the field and armature coils are R_f and R_a, respectively, $\mathcal{E}_i$ is the back emf induced in the rotating coil, and $\mathcal{E}_s$ is the emf of the external power source. By Lenz's law, the back emf acts to oppose the rotation of the coils; consequently, the sense of $\mathcal{E}_i$ is opposite to that of $\mathcal{E}_s$.

The current through the coils is given by

$$I = \frac{\mathcal{E}_s - \mathcal{E}_i}{R_f + R_a}. \tag{30-10}$$

Since $\mathcal{E}_i$ increases with the angular velocity of the armature, the faster a motor turns, the smaller the current it draws. As

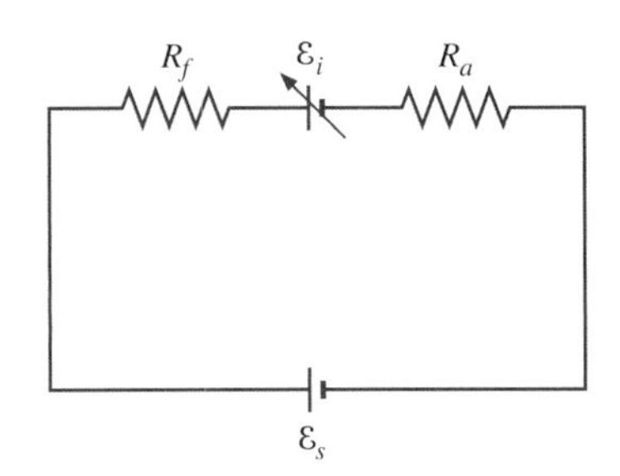

FIGURE 30-24 Circuit representation of a series-wound direct current motor.

an example, suppose that $R_f = 2.0\ \Omega$, $R_a = 1.0\ \Omega$, $\mathcal{E}_s = 120$ V, and the motor is turning at a rate such that $\mathcal{E}_i = 117$ V. Then the current drawn is

$$I = \frac{120\text{ V} - 117\text{ V}}{2.0\ \Omega + 1.0\ \Omega} = 1.0\text{ A}.$$

If the angular velocity of the motor is now decreased by a heavier load to the point where $\mathcal{E}_i = 99$ V, the current increases to

$$I = \frac{120\text{ V} - 99\text{ V}}{2.0\ \Omega + 1.0\ \Omega} = 7.0\text{ A}.$$

Finally, if the motor stalls, $\mathcal{E}_i = 0$ and

$$I = \frac{120\text{ V}}{2.0\ \Omega + 1.0\ \Omega} = 40\text{ A},$$

a current probably large enough to damage the coils if it is not quickly terminated.

Since the induced emf across the coils of the motor is $\mathcal{E}_i$ and the current through the coils is I, the power used to rotate the coils is $\mathcal{E}_i I$. This quantity is also the power output of the motor. Now, electric power is dissipated in the coil resistances as $I^2R_f + I^2R_a$, so from the principle of energy conservation, the electric power produced by the source must equal the power used to rotate the coils plus the power dissipated in the resistances; that is,

$$\mathcal{E}_s I = \mathcal{E}_i I + I^2R_f + I^2R_a.$$

Because the magnetic torque on a rotating coil is proportional to the current through it, slower rotation of the armature results in a larger torque. This is easily observed in ordinary shop machinery. For example, a rather small magnetic torque is required to keep a grinding wheel rotating freely; however, when the wheel slows as it is grinding metal, the torque on it increases correspondingly.

EXAMPLE 30-8 A Series-Wound Motor in Operation

The total resistance $(R_f + R_a)$ of a series-wound dc motor is 2.0 Ω. When connected to a 120-V source, the motor draws 10 A while running at constant angular velocity. (*a*) What is the back emf induced in the rotating coil? (*b*) What is the mechanical power output of the motor? (*c*) How much power is dissipated in the resistance of the coils? (*d*) What is the power output of the 120-V source? (*e*) Suppose the load on the motor increases, causing it to slow down to the point where it draws 20 A. Answer parts (*a*) through (*d*) for this situation.

SOLUTION (*a*) From Eq. (30-10), the back emf is

$$\mathcal{E}_i = \mathcal{E}_s - I(R_f + R_a) = 120\text{ V} - (10\text{ A})(2.0\ \Omega) = 100\text{ V}.$$

(*b*) Since the potential across the armature is 100 V when the current through it is 10 A, the power output of the motor is

$$P_m = \mathcal{E}_i I = (100\text{ V})(10\text{ A}) = 1.0 \times 10^3\text{ W}.$$

(c) A 10-A current flows through coils whose combined resistance is 2.0 Ω, so the power dissipated in the coils is

$$P_R = I^2R = (10\text{ A})^2(2.0\ \Omega) = 2.0 \times 10^2\text{ W}.$$

(d) Since 10 A is drawn from the 120-V source, its power output is

$$P_s = \mathcal{E}_s I = (120\text{ V})(10\text{ A}) = 1.2 \times 10^3\text{ W}.$$

Notice that we have an energy balance: 1.2×10^3 W = 1.0×10^3 W + 2.0×10^2 W.

(e) Repeating the same calculations with $I = 20$ A, we find $\mathcal{E}_i = 80$ V, $P_m = 1.6 \times 10^3$ W, $P_R = 8.0 \times 10^2$ W, and $P_s = 2.4 \times 10^3$ W. The motor is turning more slowly in this case, so its power output and the power output of the source are larger.

SUMMARY

1. **Time-varying magnetic fields and Faraday's law**
   (a) If the magnetic flux due to a time-varying magnetic field through any open surface S bounded by a circuit of N turns is ϕ_m, then the emf $\mathcal{E}$ induced in the circuit is given by Faraday's law:

   $$\mathcal{E} = -N\frac{d\phi_m}{dt}.$$

   (b) Lenz's law: The sense of the induced emf is such as to always oppose the change in magnetic flux that causes that emf.
2. **Induced electric field**
   If ϕ_m is the magnetic flux through any open surface bounded by an arbitrary closed path, there is an electric field **E** induced over that path such that

   $$\oint \mathbf{E}\cdot d\mathbf{l} = -\frac{d\phi_m}{dt},$$

   where the line integral is evaluated around the path.
3. **Motionally induced emf**
   (a) When the magnetic flux through a circuit of N turns changes because it is moving through a magnetic field, an emf is induced in the circuit such that

   $$\mathcal{E} = -N\frac{d\phi_m}{dt}.$$

   (b) All changes in magnetic flux, whether produced by a changing magnetic field, by motion through a magnetic field, or by a combination of these two, satisfy

   $$\mathcal{E} = -N\frac{d\phi_m}{dt}.$$

   This relationship is known as Faraday's law.

*4. **Applications of Faraday's law**
   (a) When a coil of N turns and area A rotates with an angular velocity ω around an axis that is perpendicular to a uniform magnetic field **B**, an emf

   $$\mathcal{E} = \mathcal{E}_0 \sin \omega t$$

   is induced in the coil, where $\mathcal{E}_0 = NBA\omega$.
   (b) Eddy currents are induced in a conductor when the magnetic flux through closed loops in the conductor changes because of its motion through a magnetic field. These currents circulate so as to impede the motion of the conductor.
   (c) When the armature of a dc motor rotates, a back emf is induced in each of its coils. This back emf opposes the current through the coils.

QUESTIONS

30-1. A stationary coil is in a magnetic field that is changing with time. Does the emf induced in the coil depend on the actual values of the magnetic field?

30-2. In Faraday's experiments, what would be the advantage of using coils with many turns?

30-3. A copper ring and a wooden ring of the same dimensions are placed in magnetic fields so that there is the same change in magnetic flux through them. Compare the induced electric fields and currents in the rings.

30-4. The circular conducting loops shown in the accompanying figure are parallel and coaxial. When the switch S is closed, what is the direction of the current induced in D? When the switch is opened, what is the direction of the current induced in D?

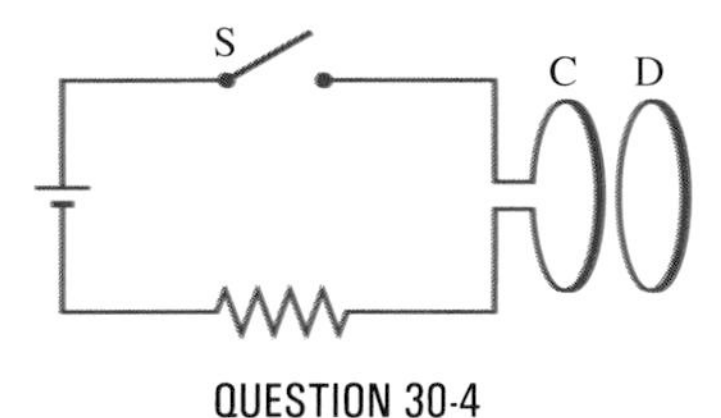

QUESTION 30-4

30-5. The north pole of a magnet is moved toward a copper loop, as shown in the accompanying figure. If you are looking at the loop from above the magnet, will you say the induced current is circulating clockwise or counterclockwise?

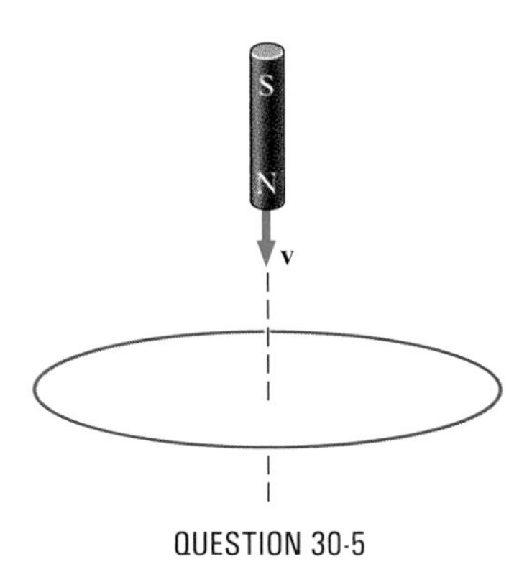

QUESTION 30-5

30-6. Discuss the factors determining the induced emf in a closed loop of wire.

30-7. Does the induced emf in a circuit depend on the resistance of the circuit? Does the induced current depend on the resistance of the circuit?

30-8. How would changing the radius of loop D in the figure for Ques. 30-4 affect its induced emf?

30-9. How would changing the radius of the coil concentric with the solenoid of Fig. 30-11 affect its induced emf? its induced current?

30-10. How can a magnetic field be used to make a charged particle at rest start moving?

30-11. Can there be an induced emf in a circuit at an instant when the magnetic flux through the circuit is zero?

30-12. Does the induced emf always act to decrease the magnetic flux through a circuit?

30-13. How would you position a flat loop of wire in a changing magnetic field so that there is no induced emf in the loop?

30-14. The accompanying figure shows a conducting ring at various positions as it moves through a magnetic field. What is the sense of the induced emf for each of these positions?

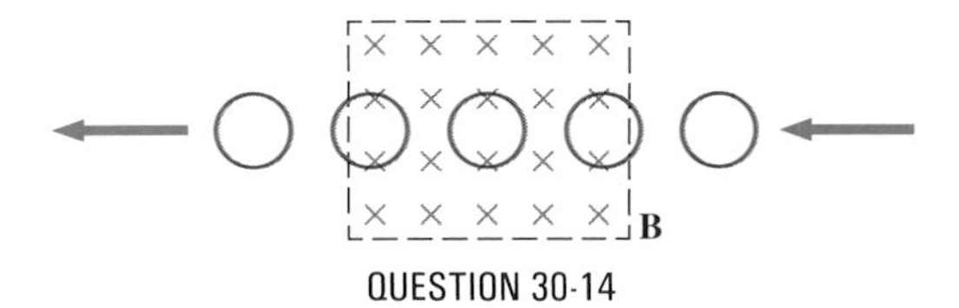

QUESTION 30-14

30-15. Show that $\mathcal{E}$ and $d\phi_m/dt$ have the same units.

30-16. A bar magnet falls under the influence of gravity along the axis of a long copper tube. If air resistance is negligible, will there be a force to oppose the descent of the magnet? If so, will the magnet reach a terminal velocity?

30-17. Around the North Pole, the earth's magnetic field is almost vertical. If an airplane is flying northward in this region, which of its wingtips is positively charged and which is negatively charged?

30-18. The work required to accelerate a rod from rest to a velocity **v** in a magnetic field is greater than the kinetic energy $mv^2/2$ of the rod. Why?

30-19. A wire loop moves translationally (no rotation) in a uniform magnetic field. Is there an emf induced in the loop?

30-20. The copper sheet shown in the accompanying figure is partially in the magnetic field. When it is pulled to the right, a resisting force pulls it to the left. Explain. What happens if the sheet is pushed to the left?

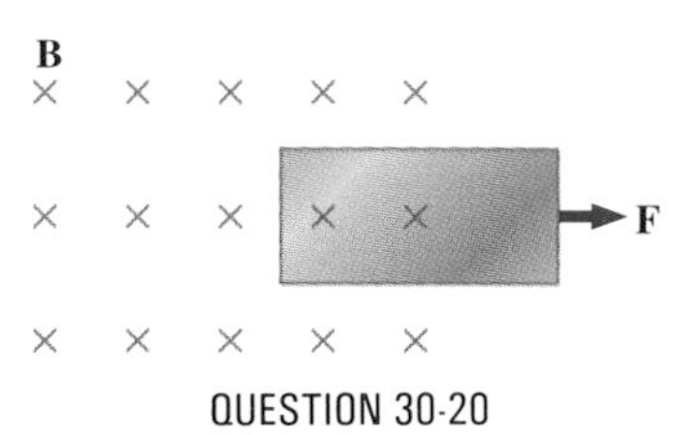

QUESTION 30-20

30-21. A conducting sheet lies in a plane perpendicular to a magnetic field **B** that is below the sheet. If **B** oscillates at a high frequency and the conductor is made of material of low resistivity, the region above the sheet is effectively shielded from **B**. Explain why. Will the conductor shield this region from static magnetic fields?

PROBLEMS

Changing Magnetic Field and Induced emf

30-1. Determine the direction of the induced current for each case shown in the accompanying figure on the right.

30-2. A single-turn circular loop of wire of radius 50 mm lies in a plane perpendicular to a spatially uniform magnetic field. During a 0.10-s time interval, the magnitude of the field increases uniformly from 200 to 300 mT. (*a*) Determine the emf induced in the loop. (*b*) If **B** is directed out of the page, what is the sense of the current induced in the loop?

30-3. The normal to the plane of a single-turn conducting loop is directed at an angle θ to a spatially uniform magnetic field **B**. (*a*) Show that the emf induced in the loop is given by $\mathcal{E} = (dB/dt)(A \cos\theta)$, where A is the area of the loop.

30-4. A 50-turn coil has a diameter of 15 cm. The coil is placed in a spatially uniform magnetic field of magnitude 0.50 T so that the face of the coil and the magnetic field are perpendicular. Find

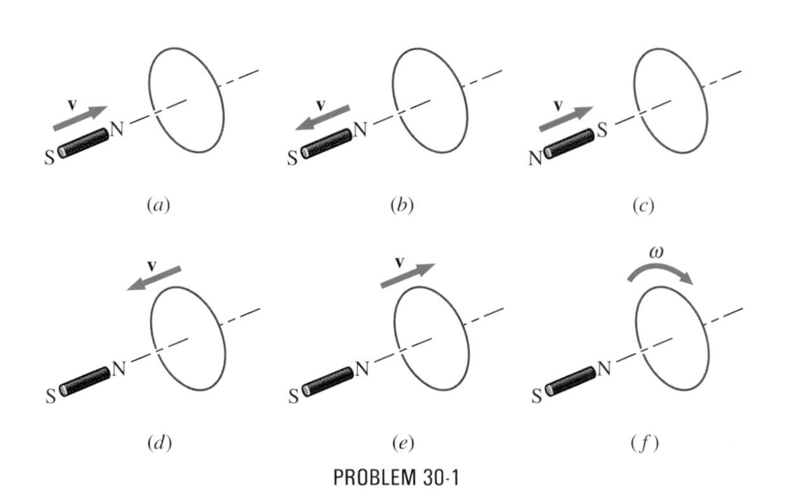

PROBLEM 30-1

the emf induced in the coil if the magnetic field is reduced to zero uniformly in (*a*) 0.10 s, (*b*) 1.0 s, and (*c*) 60 s.

30-5. Repeat your calculations of the previous problem with the plane of the coil making an angle of (*a*) 30°, (*b*) 60°, and (*c*) 90° with the magnetic field.

30-6. A square loop whose sides are 6.0 cm long is made with copper wire of radius 1.0 mm. If a magnetic field perpendicular to the loop is changing at a rate of 5.0 mT/s, what is the current in the loop?

30-7. The magnetic field through a circular loop of radius 10.0 cm varies with time as shown in the accompanying figure. The field is perpendicular to the loop. Plot the induced emf in the loop as a function of time.

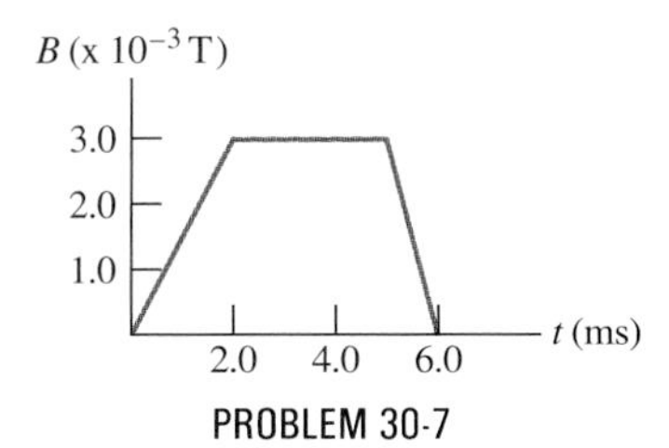

PROBLEM 30-7

30-8. When a magnetic field is first turned on, the flux through a 20-turn loop varies with time according to $\phi_m = 5.0t^2 - 2.0t$, where ϕ_m is in milliwebers, t is in seconds, and the loop is in the plane of the page with the unit normal pointing outward. What is the emf induced in the loop as a function of time? What is the direction of the induced current at $t = 0$, 0.10, 1.0, and 2.0 s?

30-9. The accompanying figure shows a single-turn rectangular coil that has a resistance of 2.0 Ω. The magnetic field at all points inside the coil varies according to $B = B_0 e^{-\alpha t}$, where $B_0 =$ 0.25 T and $\alpha = 200\ \text{s}^{-1}$. What is the current induced in the coil at $t = 1.0 \times 10^{-3}$, 2.0×10^{-3}, and 2.0 s?

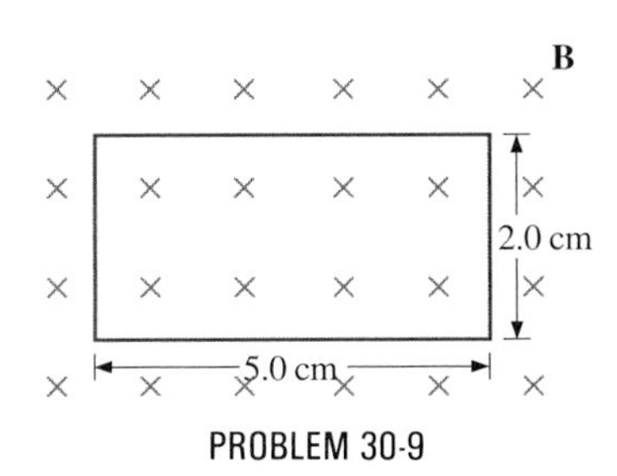

PROBLEM 30-9

30-10. How would the answers to Prob. 30-9 change if the coil consisted of 20 closely spaced turns?

30-11. The magnetic flux through the loop shown in the accompanying figure varies with time according to

$$\phi_m = 2.0e^{-3t} \sin 120\pi t,$$

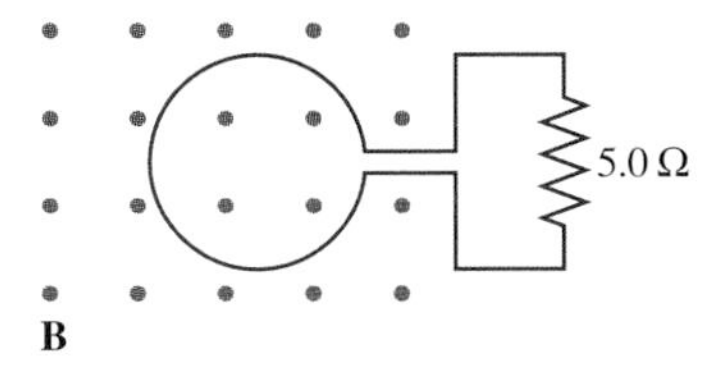

PROBLEM 30-11

where ϕ_m is in milliwebers. What are the direction and magnitude of the current through the 5.0-Ω resistor at (*a*) $t = 0$; (*b*) $t = 2.17 \times 10^{-2}$ s, and (*c*) $t = 3.0$ s?

30-12. A long solenoid with $n = 10$ turns per centimeter has a cross-sectional area of 5.0 cm² and carries a current of 0.25 A. A coil with five turns encircles the solenoid. When the current through the solenoid is turned off, it decreases to zero in 0.050 s. What is the average emf induced in the coil?

30-13. A rectangular wire loop with length a and width b lies in the xy plane, as shown in the accompanying figure. Within the loop there is a time-dependent magnetic field given by $\mathbf{B}(t) = C(x\mathbf{i} \cos \omega t + y\mathbf{k} \sin \omega t)$, with $\mathbf{B}(t)$ in tesla. Determine the emf induced in the loop.

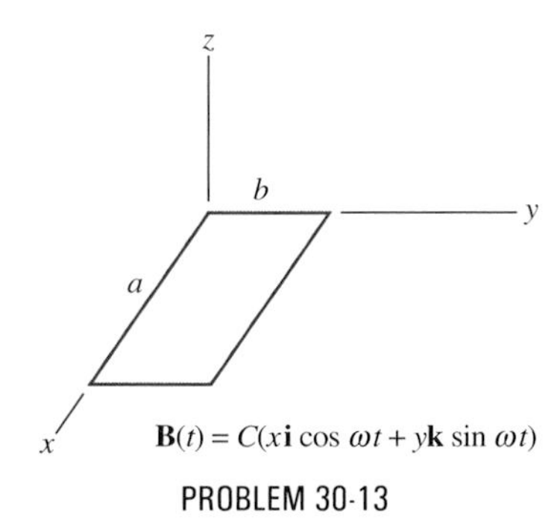

PROBLEM 30-13

30-14. The magnetic field perpendicular to a single wire loop of diameter 10.0 cm decreases from 0.50 T to zero. The wire is made of copper and has a diameter of 2.0 mm. How much charge moves through the wire while the field is changing?

30-15. The accompanying figure shows a long, straight wire and a single-turn rectangular loop, both of which lie in the plane of the page. The wire is parallel to the long sides of the loop and is 0.50 m away from the closer side. At an instant when the emf induced in the loop is 2.0 V, what is the time rate of change of the current in the wire?

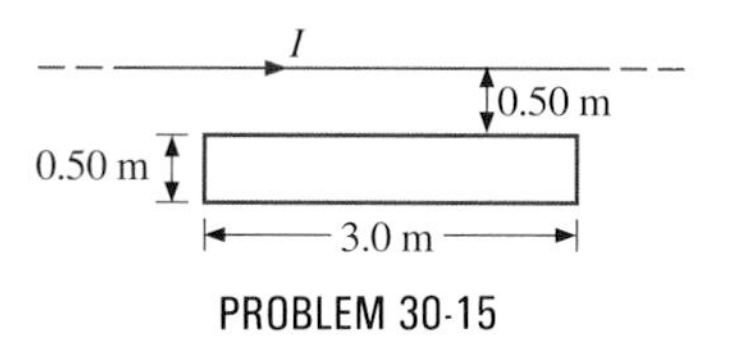

PROBLEM 30-15

Induced Electric Field

30-16. Calculate the induced electric field in the coil of Prob. 30-4. Assume that the magnetic field is cylindrically symmetric with respect to the central axis of the coil.

30-17. Assuming cylindrical symmetry with respect to the central axis of the loop of Prob. 30-7, plot the induced electric field in the loop as a function of time.

30-18. The current I through a long solenoid with n turns per meter and radius R is changing with time as given by dI/dt. Calculate the induced electric field as a function of distance r from the central axis of the solenoid.

30-19. Calculate the electric field induced both inside and outside the solenoid of the previous problem if $I = I_0 \sin \omega t$.

30-20. Over a region of radius R, there is a spatially uniform magnetic field $\mathbf{B}$. (See the accompanying figure.) At $t = 0$, $B = 1.0$ T, after which it decreases at a constant rate to zero in 30 s. (*a*) What is the electric field in the regions where $r \leq R$ and $r \geq R$ during that 30-s interval? (*b*) Assume that $R = 10.0$ cm. How much work is done by the electric field on a proton that is carried once clockwise around a circular path of radius 5.0 cm? (*c*) How much work is done by the electric field on a proton that is carried once counterclockwise around a circular path of any radius $r \geq R$? (*d*) At the instant when $B = 0.50$ T, a proton enters the magnetic field at A, moving with a velocity $\mathbf{v}$ ($v = 5.0 \times 10^6$ m/s) as shown. What are the electric and magnetic forces on the proton at that instant?

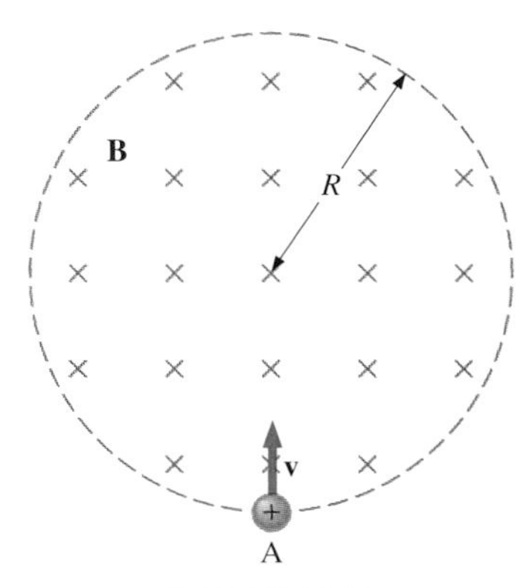

PROBLEM 30-20

30-21. The magnetic field at all points within the cylindrical region whose cross section is indicated in the accompanying figure starts at 1.0 T and decreases uniformly to zero in 20 s. What is the electric field (both magnitude and direction) as a function of r, the distance from the geometric center of the region?

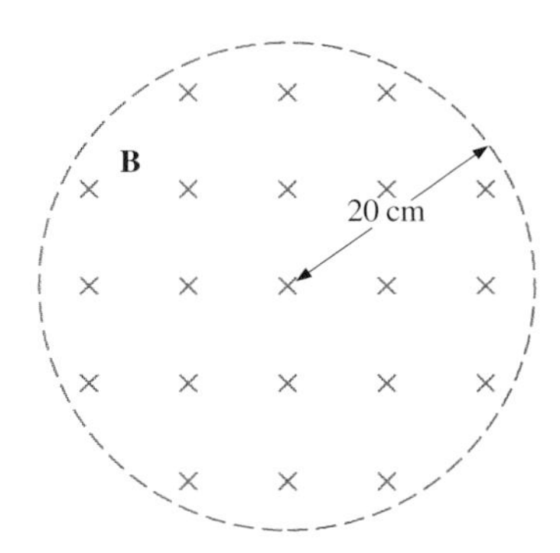

PROBLEM 30-21

Motionally Induced emf

30-22. An automobile with a radio antenna 1.0 m long travels at 100 km/h in a location where the earth's horizontal magnetic field is 5.5×10^{-5} T. What is the maximum possible emf induced in the antenna due to this motion?

30-23. The rectangular loop of N turns shown in the accompanying figure moves to the right with a constant velocity $\mathbf{v}$ while leaving the poles of a large electromagnet. (*a*) Assuming that $\mathbf{B}$ is uniform between the pole faces and negligible elsewhere, determine the induced emf in the loop. (*b*) What is the source of work that produces this emf?

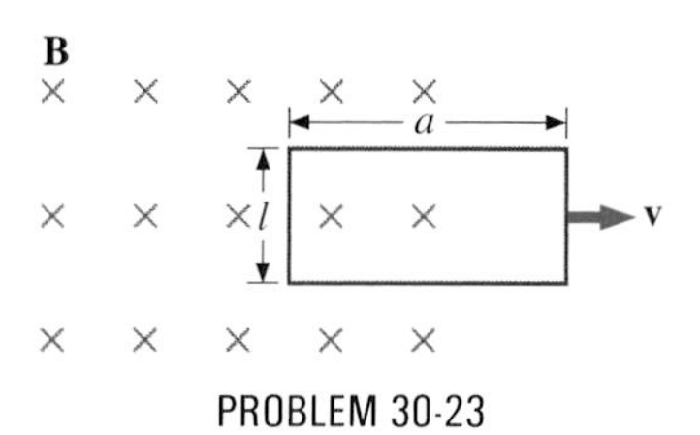

PROBLEM 30-23

30-24. Suppose the magnetic field of the previous problem oscillates with time according to $B = B_0 \sin \omega t$. What then is the emf induced in the loop when its trailing side is a distance d from the right edge of the magnetic field region?

30-25. A coil of 1000 turns encloses an area of 25 cm^2. It is rotated in 0.010 s from a position where its plane is perpendicular to the earth's magnetic field to one where its plane is parallel to the field. If the strength of the field is 6.0×10^{-5} T, what is the average emf induced in the coil?

30-26. In the circuit shown in the accompanying figure, the rod slides along the conducting rails at a constant velocity $\mathbf{v}$. The vector $\mathbf{v}$ is in the same plane as the rails and directed at an angle θ to them. A uniform magnetic field $\mathbf{B}$ is directed out of the page. What is the emf induced in the rod?

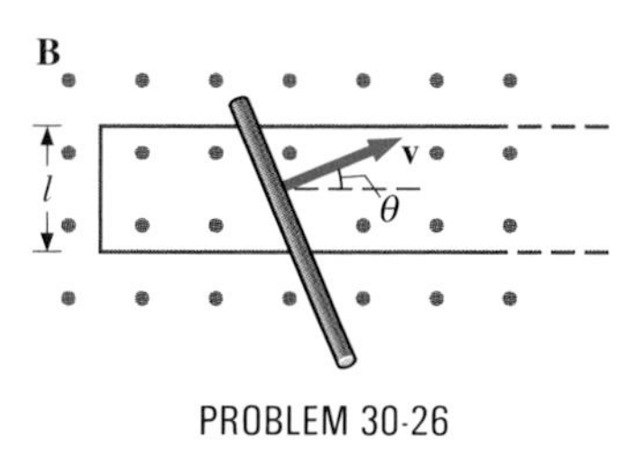

PROBLEM 30-26

30-27. The rod shown in the accompanying figure is moving through a uniform magnetic field of strength $B = 0.50$ T with a constant velocity of magnitude $v = 8.0$ m/s. What is the potential difference between the ends of the rod? Which end of the rod is at a higher potential?

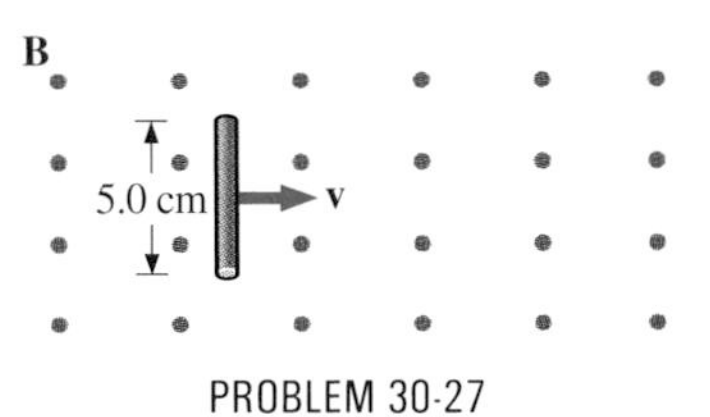

PROBLEM 30-27

30-28. A 25-cm rod moves at 5.0 m/s in a plane perpendicular to a magnetic field of strength 0.25 T. The rod, velocity vector, and magnetic field vector are mutually perpendicular, as indicated in the accompanying figure. Calculate (*a*) the magnetic force on an electron in the rod, (*b*) the electric field in the rod, and (*c*) the potential difference between the ends of the rod. (*d*) What is the speed of the rod if the potential difference is 1.0 V ?

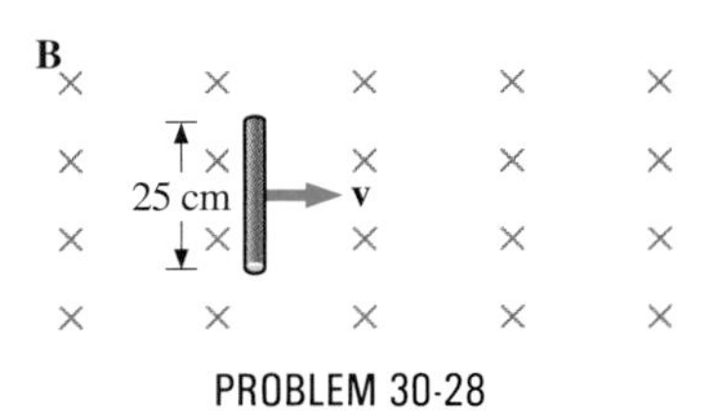

PROBLEM 30-28

30-29. In the accompanying figure, the rails, connecting end piece, and rod all have a resistance per unit length of 2.0 Ω/cm. The rod moves to the left at $v = 3.0$ m/s. If $B = 0.75$ T everywhere in the region, what is the current in the circuit when $a =$ 8.0 cm? when $a = 5.0$ cm? Specify also the sense of the current flow.

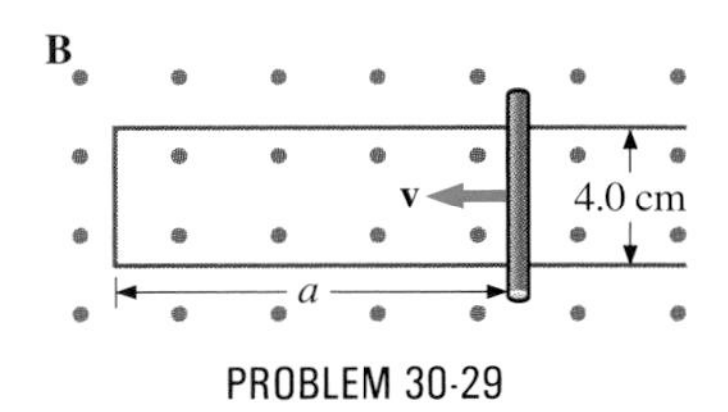

PROBLEM 30-29

30-30. The rod shown in the accompanying figure moves to the right on essentially resistanceless rails at a speed of $v = 3.0$ m/s. If $B = 0.75$ T everywhere in the region, what is the current through the 5.0-Ω resistor? Does the current circulate clockwise or counterclockwise?

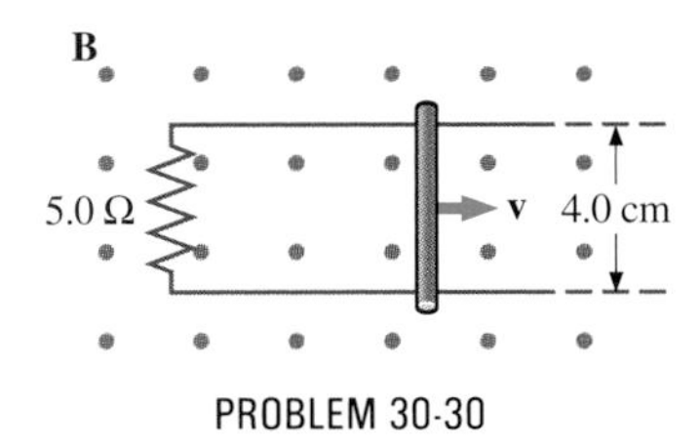

PROBLEM 30-30

30-31. The accompanying figure shows a conducting rod that slides along metal rails. The apparatus is in a uniform magnetic field of strength 0.25 T, which is directed into the page. The rod is pulled to the right at a constant speed of 5.0 m/s by a force **P**. The only significant resistance in the circuit comes from the 2.0-Ω resistor shown. (*a*) What is the emf induced in the circuit? (*b*) What is the induced current? Does it circulate clockwise or counterclockwise? (*c*) What is the magnitude of **P**? (*d*) What are the power output of **P** and the power dissipated in the resistor?

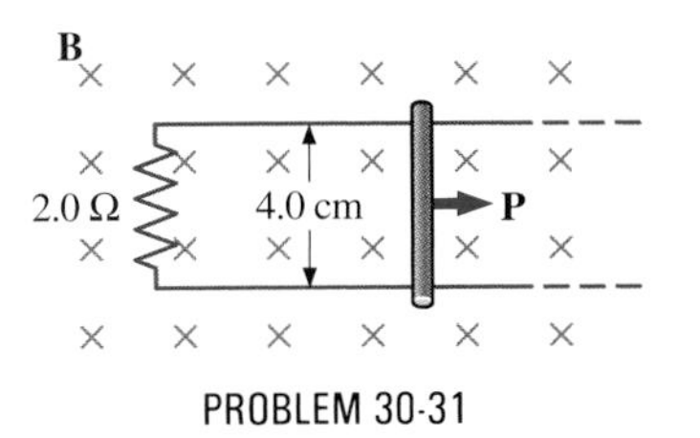

PROBLEM 30-31

30-32. A coil with 50 turns and area 10 cm^2 is oriented with its plane perpendicular to a 0.75-T magnetic field. If the coil is flipped over (rotated through 180°) in 0.20 s, what is the average emf induced in it?

30-33. A 20-turn planar loop of flexible wire is placed inside a long solenoid of n turns per meter that carries a constant current I_0. The area A of the loop is changed by pulling on its sides while ensuring that the plane of the loop always remains perpendicular to the axis of the solenoid. If $n = 500$ turns per meter, $I_0 = 12$ A, and $A = 20$ cm^2, what is the emf induced in the loop when $dA/dt = 100$ cm^2/s?

30-34. The conducting rod MN shown in the accompanying figure moves along parallel metal rails that are 25 cm apart. The system is in a uniform magnetic field of strength 0.75 T, which is directed into the page. The resistances of the rod and the rails are negligible, but the section PQ has a resistance of 0.25 Ω. (*a*) What is the emf (including its sense) induced in the rod when it is moving to the right with a speed of 5.0 m/s? (*b*) What force is required to keep the rod moving at this speed? (*c*) What is the rate at which work is done by this force? (*d*) What is the power dissipated in the resistor?

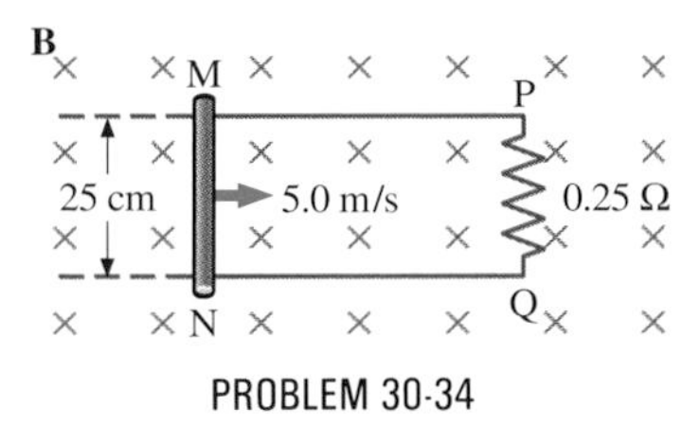

PROBLEM 30-34

Applications of Faraday's Law

30-35. Design a current loop that, when rotated in a uniform magnetic field of strength 0.10 T, will produce an emf $\mathcal{E} = \mathcal{E}_0 \sin \omega t$, where $\mathcal{E}_0 = 110$ V and $\omega = 120\pi$ rad/s.

30-36. A flat, square coil of 20 turns that has sides of length 15 cm is rotating in a magnetic field of strength 0.050 T. If the maximum emf produced in the coil is 30 mV, what is the angular velocity of the coil?

30-37. A 50-turn rectangular coil with dimensions 0.15 m × 0.40 m rotates in a uniform magnetic field of magnitude 0.75 T at 3600 rev/min. (*a*) Determine the emf induced in the coil as a function of time. (*b*) If the coil is connected to a 1000-Ω resistor, what is the power as a function of time required to keep the coil turning at 3600 rev/min? (*c*) Answer part (*b*) if the coil is connected to a 2000-Ω resistor.

30-38. The square armature coil of an alternating current generator has 200 turns and is 20.0 cm on a side. When it rotates at 3600 rev/min, its peak output voltage is 120 V. (*a*) What is the frequency of the output voltage? (*b*) What is the strength of the magnetic field in which the coil is turning?

30-39. A flip coil is a relatively simple device used to measure a magnetic field. It consists of a circular coil of N turns wound with fine conducting wire. The coil is attached to a ballistic galvanometer, a device that measures the total charge that passes through it. The coil is placed in a magnetic field **B** such that its face is perpendicular to the field. It is then flipped through 180°, and the total charge Q that flows through the galvanometer is measured. (*a*) If the total resistance of the coil and galvanometer is R, what is the relationship between B and Q? Because the coil is very small, you can assume that **B** is uniform over it. (*b*) How can you determine whether or not **B** is perpendicular to the face of the coil?

30-40. The flip coil of Prob. 30-39 has a radius of 3.0 cm and is wound with 40 turns of copper wire. The total resistance of the coil and ballistic galvanometer is 0.20 Ω. When the coil is flipped through 180° in a magnetic field **B**, a charge of 0.090 C flows through the ballistic galvanometer. (*a*) Assuming that **B** and the face of the coil are initially perpendicular, what is B? (*b*) If the coil is flipped through 90°, what is the reading of the galvanometer?

30-41. A 120-V, series-wound dc motor has a field resistance of 80 Ω and an armature resistance of 10 Ω. When it is operating at full speed, a back emf of 75 V is generated. (*a*) What is the initial current drawn by the motor? When the motor is operating at full speed, what are (*b*) the current drawn by the motor, (*c*) the power output of the source, (*d*) the power output of the motor, and (*e*) the power dissipated in the two resistances?

30-42. A small series-wound dc motor is operated from a 12-V car battery. Under a normal load, the motor draws 4.0 A, and when the armature is clamped so that it cannot turn, the motor draws 24 A. What is the back emf when the motor is operating normally?

30-43. A 120-V, series-wound dc motor draws 0.50 A from its power source when operating at full speed, and it draws 2.0 A when it starts. The resistance of the armature coils is 10 Ω. (*a*) What is the resistance of the field coils? (*b*) What is the back emf of the motor when it is running at full speed? (*c*) The motor operates at a different speed and draws 1.0 A from the source. What is the back emf in this case?

30-44. The armature and field coils of a series-wound motor have a total resistance of 3.0 Ω. When connected to a 120-V source and running at normal speed, the motor draws 4.0 A. (*a*) How large is the back emf? (*b*) What current will the motor draw just after it is turned on? Can you suggest a way to avoid this large initial current?

General Problems

30-45. A copper wire of length L is fashioned into a circular coil with N turns. When the magnetic field through the coil changes with time, for what value of N is the induced emf a maximum?

30-46. A copper sheet of mass 0.50 kg drops through a uniform horizontal magnetic field of 1.5 T, and it reaches a terminal velocity of 2.0 m/s. (*a*) What is the net magnetic force on the sheet after it reaches terminal velocity? (*b*) Describe the mechanism responsible for this force. (*c*) How much power is dissipated as Joule heating while the sheet moves at terminal velocity?

30-47. A circular copper disk of radius 7.5 cm rotates at 2400 rev/min around the axis through its center and perpendicular to its face. The disk is in a uniform magnetic field **B** of strength 1.2 T that is directed along the axis. What is the potential difference between the rim and the axis of the disk?

30-48. A short rod of length a moves with its velocity **v** parallel to an infinite wire carrying a current I. (See the accompanying figure.) If the end of the rod nearer the wire is a distance b from the wire, what is the emf induced in the rod?

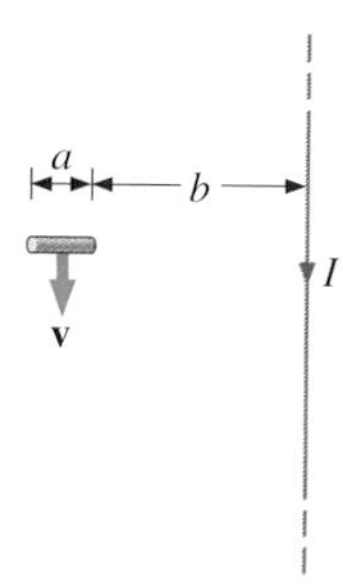

PROBLEM 30-48

30-49. A rectangular circuit containing a resistance R is pulled at a constant velocity **v** away from a long, straight wire carrying a current I_0. (See the accompanying figure.) Derive an equation that gives the current induced in the circuit as a function of the distance x between the near side of the circuit and the wire.

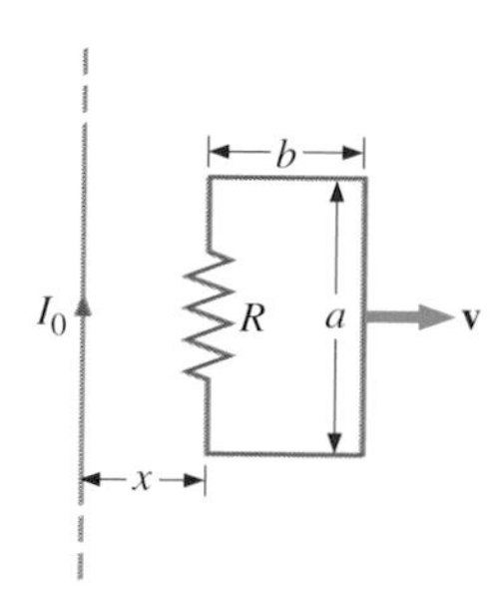

PROBLEM 30-49

30-50. Two infinite solenoids cross the plane of the circuit as shown in the accompanying figure. The radii of the solenoids are 0.10 and 0.20 m, respectively, and the current in each solenoid is changing such that $dB/dt = 50$ T/s. What are the currents in the resistors of the circuit?

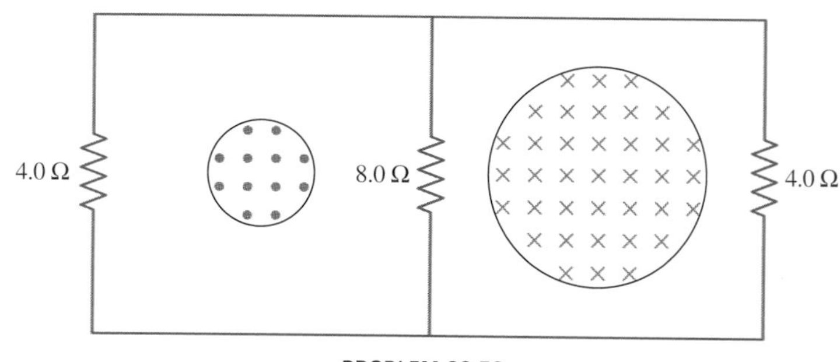

PROBLEM 30-50

30-51. An eight-turn coil is *tightly wrapped* around the outside of the long solenoid as shown in the accompanying figure. The radius of the solenoid is 2.0 cm and it has 10 turns per centimeter. The current through the solenoid increases according to $I = I_0(1 - e^{-\alpha t})$, where $I_0 = 4.0$ A and $\alpha = 2.0 \times 10^{-2}$ s^{-1}. What is the emf induced in the coil when $t = 0$, $t = 1.0 \times 10^2$ s, and $t \to \infty$?

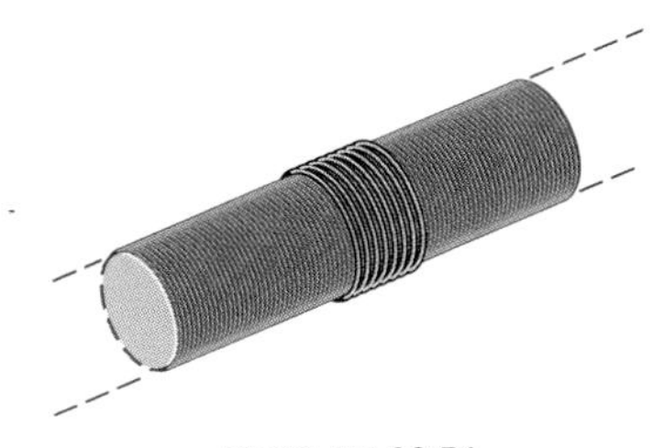

PROBLEM 30-51

30-52. The accompanying figure shows a long rectangular loop of width w, mass m, and resistance R. The loop starts from rest at the edge of a uniform magnetic field **B** and is pushed into the field by a constant force **F**. Calculate the speed of the loop as a function of time.

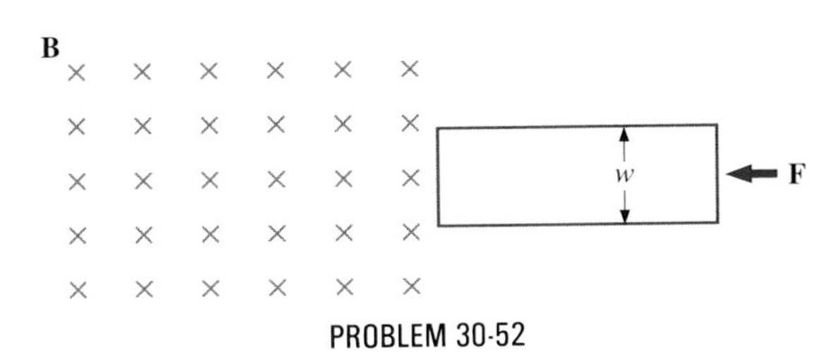

PROBLEM 30-52

30-53. A square bar of mass m and resistance R is sliding without friction down very long, parallel conducting rails of negligible resistance. (See the accompanying figure.) The two rails are a distance l apart and are connected to each other at the bottom of the incline by a resistanceless wire. The rails are inclined at an angle θ, and there is a uniform vertical magnetic field **B** throughout the region. (*a*) Show that the bar acquires a terminal velocity given by

$$v = \frac{mgR \sin\theta}{B^2 l^2 \cos^2\theta}.$$

(*b*) Calculate the work per unit time done by the force of gravity. (*c*) Compare this with the power dissipated in the Joule heating of the bar. (*d*) What would happen if **B** were reversed?

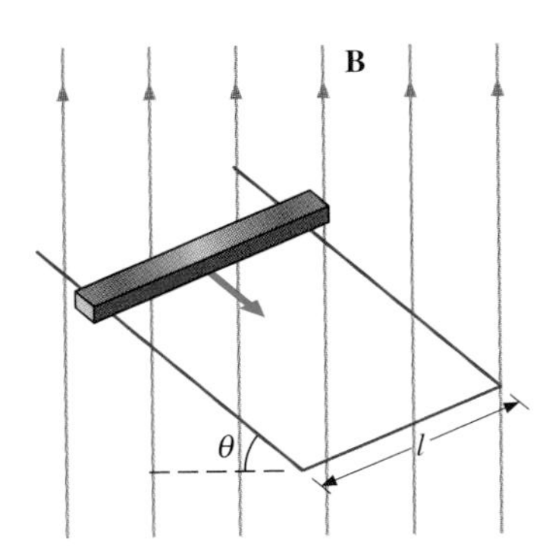

PROBLEM 30-53

30-54. The accompanying figure shows a metal disk of inner radius r_1 and outer radius r_2 rotating at an angular velocity ω while in a uniform magnetic field directed parallel to the rotational axis. The brush leads of a voltmeter are connected to the disk's inner and outer surfaces as shown. What is the reading of the voltmeter?

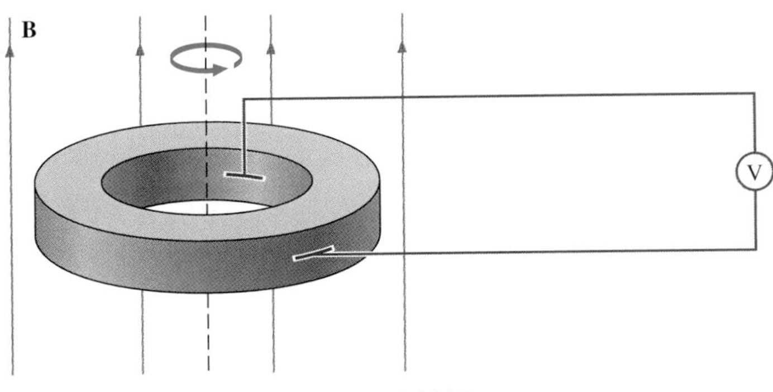

PROBLEM 30-54

30-55. A long solenoid with 10 turns per centimeter is placed inside a copper ring such that both objects have the same central axis. The radius of the ring is 10.0 cm, and the radius of the solenoid is 5.0 cm. (*a*) What is the emf induced in the ring when the current I through the solenoid is 5.0 A and changing at a rate of 100 A/s? (*b*) What is the emf induced in the ring when $I = 2.0$ A and $dI/dt = 100$ A/s? (*c*) What is the electric field inside the ring for these two cases? (*d*) Suppose the ring is moved so that its central axis and the central axis of the solenoid are still parallel but no longer coincide. (You should assume that the solenoid is still inside the ring.) Now what is the emf induced in the ring? (*e*) Can you calculate the electric field in the ring as you did in part (*c*)?

30-56. The current in the long, straight wire shown in the accompanying figure is given by $I = I_0 \sin \omega t$, where $I_0 = 15$ A and $\omega = 120\pi$ rad/s. What is the current induced in the rectangular loop at $t = 0$ and at $t = 2.09 \times 10^{-3}$ s? The resistance of the loop is 2.0 Ω.

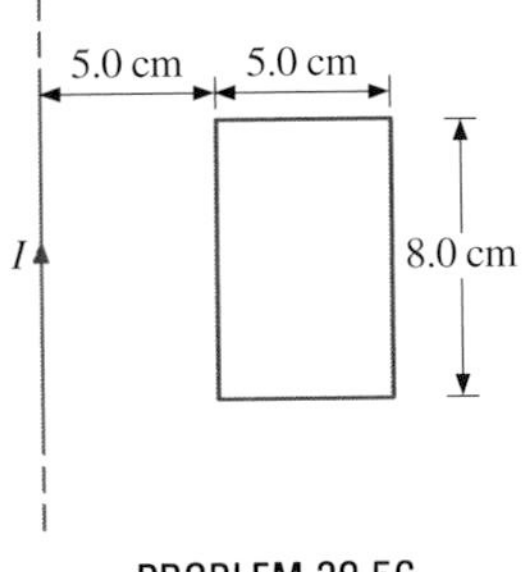

PROBLEM 30-56

The discoverers of Maxwell's equations.

CHAPTER 31 MAXWELL'S EQUATIONS AND ELECTROMAGNETIC ENERGY

PREVIEW

This chapter is devoted to Maxwell's equations and electromagnetic field energy. The main topics to be discussed here are the following:

1. **Induced magnetic field.** We explain how a changing electric field induces a magnetic field.
2. **Maxwell's equations.** Maxwell's equations are summarized and discussed.
3. **Electromagnetic field energy.** We obtain expressions for the energy stored in an electric field and in a magnetic field.
4. **The Poynting vector.** A vector that describes the transport of electromagnetic field energy is introduced.

In the previous chapter we found that a time-varying magnetic field induces a nonconservative electric field. With the many analogies that have been made between the electric and the magnetic fields, a question that naturally comes to mind is whether the opposite effect can occur. That is, when an electric field varies with time, does it induce a magnetic field? In this chapter you will see that indeed it does. With the equation describing how this induction occurs, we will then have Maxwell's complete set of equations.

31-1 Induced Magnetic Field

We investigate the induced magnetic field using the imaginary experiment of Fig. 31-1. A current I flows through the wires connecting two large, circular, parallel plates, thus charging them oppositely. The wires are long and straight, so we can assume that the system is cylindrically symmetric. With edge effects ignored, the electric field in the region between the plates is, from Eq. (23-7),

$$E = \frac{q}{\epsilon_0 A},$$

where A is the area of each plate and q the magnitude of its charge. The current is related to this field by

$$I = \frac{dq}{dt} = \epsilon_0 A \frac{dE}{dt}. \tag{31-1}$$

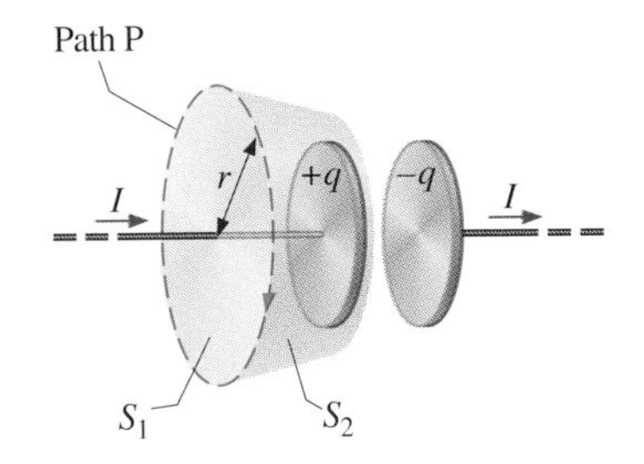

FIGURE 31-1 Two charging parallel plates. Path P bounds two different surfaces S_1 and S_2. The conduction current I only crosses S_1.

With cylindrical symmetry, Ampère's law for the circular path P of radius r and the planar surface S_1 it bounds gives

$$B(2\pi r) = \mu_0 I,$$

so

$$B = \frac{\mu_0 I}{2\pi r}$$

at any point on P. However, when Ampère's law is applied to path P and the curved surface S_2 (also bounded by P), we find

$$B(2\pi r) = 0,$$

since there is no current through S_2. This second result therefore tells us that $B = 0$ along P.

Ampère's law has given us two values for B along the same path! Does this mean there is something inherently wrong with Ampère's law? And which field, if either, is the correct one? The resolution of this apparent inconsistency came from Maxwell. He reasoned that Ampère's law, as given by Eq. (29-10), is valid only for *static* situations. He hypothesized that for *time-varying* conditions, a term involving the time rate of change of the electric field must be included in the right-hand side of Eq. (29-10). For the charging plates of Fig. 31-1, the additional term is the right-hand side of Eq. (31-1) multiplied by μ_0. The revised form of Ampère's law for this configuration is therefore

$$\oint \mathbf{B}\cdot d\mathbf{l} = \mu_0 I + \mu_0\epsilon_0 A \frac{dE}{dt}. \tag{31-2}$$

Applying this equation along the path P and over the surface S_2, we obtain

$$\oint \mathbf{B}\cdot d\mathbf{l} = B2\pi r = \mu_0\epsilon_0 A \frac{dE}{dt},$$

so

$$B = \frac{\mu_0\epsilon_0 A}{2\pi r}\frac{dE}{dt}.$$

Using Eq. (31-1) to substitute for I, we simplify this to

$$B = \frac{\mu_0 I}{2\pi r},$$

a result consistent with that of path P and surface S_1. This apparent paradox is therefore resolved with Maxwell's hypothesis.

Equation (31-2) can be generalized by simply replacing the term $A(dE/dt)$ by $d\phi_e/dt$, where ϕ_e, the electric flux, is

$$\phi_e = \int_S \mathbf{E}\cdot\hat{\mathbf{n}}\, da.$$

We then have Maxwell's revision of Ampère's law:

$$\oint \mathbf{B}\cdot d\mathbf{l} = \mu_0 I + \mu_0\epsilon_0 \frac{d\phi_e}{dt}. \tag{31-3}$$

Just as an electric field is induced when magnetic flux changes, so too is *a magnetic field induced when electric flux changes*. Because of Maxwell's contribution, Eq. (31-3) is commonly known as the *Ampère-Maxwell law*. It is the last of Maxwell's equations.

Verification of the Ampère-Maxwell law did not follow immediately after its formulation because it was impossible to distinguish experimentally the magnetic field produced by the changing electric flux from that produced by the current in the wires. Actually, the first verification was somewhat indirect. Maxwell showed that with time variations included, the electromagnetic field equations predicted the existence of combined electric and magnetic fields that move at the speed of light. The existence of this *electromagnetic radiation*, and hence the correctness of Maxwell's hypothesis, was verified by Heinrich Hertz (1857–1894) in 1890, which, unfortunately, was 11 years after Maxwell's death. Eventually, visible light, radio waves, microwaves, x-rays, and ultraviolet light were all found to be forms of electromagnetic radiation, distinguished only by their frequency of oscillation (see Chap. 35). A bridge connecting optics and electromagnetism was therefore discovered.

Since both terms on the right-hand side of the Ampère-Maxwell law contain μ_0, the units of $\epsilon_0(d\phi_e/dt)$ should be the same as those of I—and they are, for the unit of $\epsilon_0(d\phi_e/dt)$ is

$$\frac{\mathrm{C}^2}{\mathrm{N\cdot m}^2}\,\frac{\mathrm{N\cdot m}^2}{\mathrm{C\cdot s}} = \frac{\mathrm{C}}{\mathrm{s}},$$

which is an ampere. Because it is equivalent to a current, the quantity $\epsilon_0(d\phi_e/dt)$ is called the **displacement current**.

For the charged plates of Fig. 31-1, the flux through surface S_2 is given by

$$\phi_e = EA = \frac{q}{\epsilon_0}.$$

The displacement current between the plates is then

$$\epsilon_0\frac{d\phi_e}{dt} = \frac{dq}{dt} = I.$$

The displacement current between the plates is therefore equal to the conduction current in the wire. This allows us to retain the idea that the flow of current is continuous throughout the entire circuit. Keep in mind, however, that the conduction and displacement currents are quite different in nature! One represents the flow of electric charge, while the other represents a changing electric flux.

EXAMPLE 31-1 An Induced Magnetic Field

The plates of radius R in Fig. 31-2 are being charged at a constant rate $I = dq/dt$. (*a*) Assuming that the electric field is confined entirely to the region between the plates, what is the magnetic field at a distance r from the axis passing through the center of the plates? (*b*) If $I = 3.0\times10^{-3}$ A, what is the magnetic field at $r = R = 10$ cm?

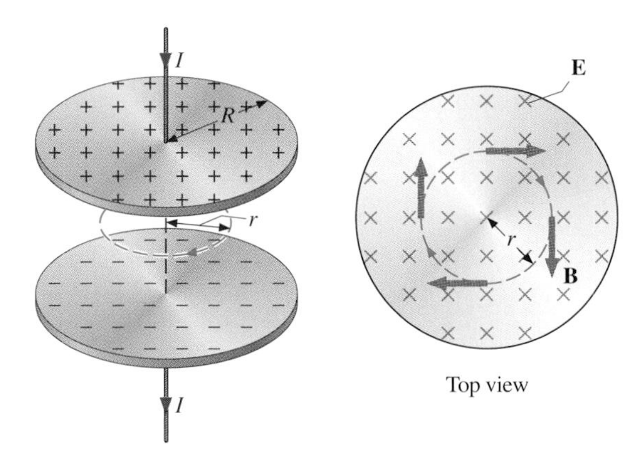

FIGURE 31-2 Circular plates being charged at a rate $I = dq/dt$. The path shown is concentric with the plates.

SOLUTION (*a*) To find the magnetic field, we apply the Ampère-Maxwell law using the circular path and its enclosed planar surface shown in the figure. The path is concentric with the axis through the centers of the plates and its radius r is less than R. Cylindrical symmetry forces B to be constant on the circular path, and the parallel plates make $\mathbf{E}$ constant over the surface bounded by the path. Thus

$$\oint \mathbf{B}\cdot d\mathbf{l} = B2\pi r, \qquad (i)$$

and

$$\phi_e = \int_S \mathbf{E}\cdot\hat{\mathbf{n}}\,da = E\pi r^2. \qquad (ii)$$

The conduction current *through the surface* bounded by the path is zero, so with Eqs. (*i*) and (*ii*), the Ampère-Maxwell law reduces to

$$B2\pi r = \mu_0\epsilon_0\pi r^2\frac{dE}{dt},$$

and the magnetic field along the path is

$$B = \mu_0\epsilon_0\frac{r}{2}\frac{dE}{dt} \quad (r \le R). \qquad (iii)$$

If $r \ge R$, Eq. (*i*) is unchanged while Eq. (*ii*) becomes

$$\phi_e = E\pi R^2, \qquad (iv)$$

since $\mathbf{E}$ is confined to the region where $r \le R$. Substitution of Eqs. (*i*) and (*iv*) and $I = 0$ into the Ampère-Maxwell law now yields

$$B2\pi r = \mu_0\epsilon_0\pi R^2\frac{dE}{dt},$$

and

$$B = \mu_0\epsilon_0\left(\frac{R^2}{2r}\right)\frac{dE}{dt} \quad (r \ge R). \qquad (v)$$

The electric field between the parallel plates is

$$E = \frac{q}{\epsilon_0 A} = \frac{q}{\epsilon_0\pi R^2},$$

so

$$\frac{dE}{dt} = \frac{1}{\epsilon_0\pi R^2}\frac{dq}{dt} = \frac{I}{\epsilon_0\pi R^2}, \qquad (vi)$$

where I is the current *in the wires* connected to the plates. Using this result in Eqs. (*iii*) and (*v*), we obtain for the magnetic field

$$B = \frac{\mu_0 r}{2\pi R^2}I \quad (r \le R), \qquad (vii)$$

and

$$B = \frac{\mu_0}{2\pi r}I \quad (r \ge R). \qquad (viii)$$

In Chap. 29, we determined the direction of the magnetic field due to a conduction current in a long, straight wire by using a right-hand rule (see Fig. 29-3). The direction of **B** here can be obtained by applying this rule to the displacement current. Since dE/dt is positive, the displacement current $\epsilon_0(d\phi_e/dt)$ is also positive and directed from the positive to the negative plate. Therefore **B** circulates as shown in Fig. 31-2.

(*b*) At $r = R = 10$ cm, the magnetic field is

$$B = \frac{\mu_0}{2\pi R} I = \frac{4\pi \times 10^{-7}\ \text{T}\cdot\text{m/A}}{2\pi(0.10\ \text{m})}(3.0 \times 10^{-3}\ \text{A})$$
$$= 6.0 \times 10^{-9}\ \text{T}.$$

Notice how small the induced magnetic field is! Generally, it is very difficult to detect because of its magnitude. In contrast, induced electric fields are observed with the simplest of experiments.

31-2 Maxwell's Equations: A Summary

Maxwell's equations are generally considered to be the culmination of the study of electricity and magnetism. Maxwell hypothesized that *all* electromagnetic field phenomena could be explained in terms of these four basic equations. Up to the present time, no verifiable experimental result has contradicted this hypothesis. His equations are as important and useful as any scientific principle ever discovered. Visible light, radio communications, electronics, electric power generation, and electric motors are just a few of the many devices and phenomena described by these equations.

We now present a brief summary of Maxwell's equations. As written, they really apply only in a vacuum. The effect of material media on electromagnetic fields was discussed briefly in Sec. 29-7 and will be considered further in Chap. 32.

1. *Gauss' law for electric fields*

$$\oint_S \mathbf{E}\cdot\hat{\mathbf{n}}\,da = \frac{Q_{\text{in}}}{\epsilon_0}. \qquad (24\text{-}4)$$

This equation represents the fact that the electric field lines of static charge distributions originate on positive charges and terminate on negative charges. Gauss' law is also a consequence of the inverse-square nature of the field of a point charge.

2. *Gauss' law for magnetic fields*

$$\oint_S \mathbf{B}\cdot\hat{\mathbf{n}}\,da = 0. \qquad (29\text{-}9)$$

This equation states that there are no point sources or sinks (monopoles) for the magnetic field, whose lines always form closed loops in space.

3. *Faraday's law*

$$\oint \mathbf{E}\cdot d\mathbf{l} = -\frac{d}{dt}\int_S \mathbf{B}\cdot\hat{\mathbf{n}}\,da. \qquad (30\text{-}4)$$

Faraday's law states that a nonconservative electric field is induced by a changing magnetic flux. If there is no changing magnetic flux, $\oint \mathbf{E}\cdot d\mathbf{l} = 0$, and **E**, which is then the Coulomb field produced by a static charge distribution, is conservative.

4. *Ampère-Maxwell law*

$$\oint \mathbf{B}\cdot d\mathbf{l} = \mu_0 I + \mu_0\epsilon_0 \frac{d}{dt}\int_S \mathbf{E}\cdot\hat{\mathbf{n}}\,da. \qquad (31\text{-}3)$$

This equation was originally formulated by Ampère without the changing-electric-flux term. The general form presented here is a result of Maxwell's modification. It states that magnetic fields are produced by electric current and by changing electric fields.

One final point: Although Maxwell's equations are fundamental laws that allow us to calculate the electromagnetic field at any point in space, they do not address the forces exerted on charged particles by the field. A charge q in an electromagnetic field is subjected to both the electric force $q\mathbf{E}$ and the magnetic force $q\mathbf{v} \times \mathbf{B}$. The net force on the charge is given by the *Lorentz force law*:

$$\mathbf{F} = q\mathbf{E} + q\mathbf{v} \times \mathbf{B}.$$

With the help of the Lorentz force law, we are able to determine the motion of a charged particle in an electromagnetic field.

31-3 Electromagnetic Field Energy

The energy stored in an electromagnetic field has two distinct parts, one associated with the electric field and one with the magnetic field. In a particular region of space, the electromagnetic energy does not necessarily remain constant, since it may be transported into or out of the region or it may be produced or dissipated within the region.

Let's first investigate the electric field energy stored between the parallel plates of Fig. 31-3. The plates are being charged as small increments of positive charge dq are moved from the negative to the positive plate. The distance between the plates is l and the area of each plate is A. At the instant represented in

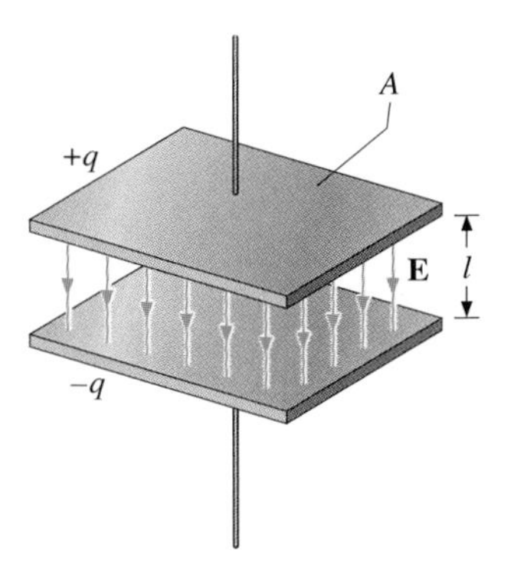

FIGURE 31-3 Energy is stored in the electric field between two charged parallel plates.

the figure, charge q has been transferred,* so the potential difference between the plates is

$$V = El = \frac{ql}{\epsilon_0 A},$$

where we have used $E = \sigma/\epsilon_0 = q/\epsilon_0 A$ to substitute for the electric field.

If an additional charge dq is transported from the negative to the positive plate, the work dW that must be done against the potential difference of the plates is

$$dW = V\,dq,$$

which, writing V as $ql/\epsilon_0 A$, becomes

$$dW = \frac{ql}{\epsilon_0 A}\,dq.$$

The total amount of work W needed to transfer a net charge Q between the plates is therefore

$$W = \int_0^Q \frac{ql}{\epsilon_0 A}\,dq = \frac{l}{\epsilon_0 A}\int_0^Q q\,dq = \frac{Q^2 l}{2\epsilon_0 A}.$$

Substituting E for $Q/\epsilon_0 A$ and (vol) for the volume lA of the region between the plates, we find

$$W = \tfrac{1}{2}\epsilon_0 E^2(\text{vol}).$$

The quantity W represents the work done to charge the plates, or equivalently, the work done to produce the electric field between the plates. This by definition is *the energy U_e stored in the electric field between the parallel plates*:

$$U_e = \tfrac{1}{2}\epsilon_0 E^2(\text{vol}). \tag{31-4}$$

The **electric-field energy density** u_e is by definition the electric field energy per unit volume. Hence

$$u_e = \frac{U_e}{(\text{vol})} = \frac{1}{2}\epsilon_0 E^2. \tag{31-5}$$

Although this equation was obtained for parallel plates, it does not contain any parameters related to that special case. Consequently, we might expect that it is more general than its derivation implies—and it is. Equation (31-5) represents the energy density of *any* electric field.

The energy stored in a magnetic field can be easily found by considering the long solenoid of Fig. 31-4. The solenoid has n turns per unit length, its cross-sectional area is A, and it carries a current i. To avoid having to consider the variation in the magnetic field near the ends of the solenoid, we consider a section of length l near the center of the solenoid where the magnetic field is given by $B = \mu_0 ni$. When the current through the

*In principle, the charging can occur by means of an outside agent lifting positive charge off the negative plate and carrying it to the positive plate. In practice, the charging is done by connecting one plate to a battery's negative terminal and one plate to its positive terminal.

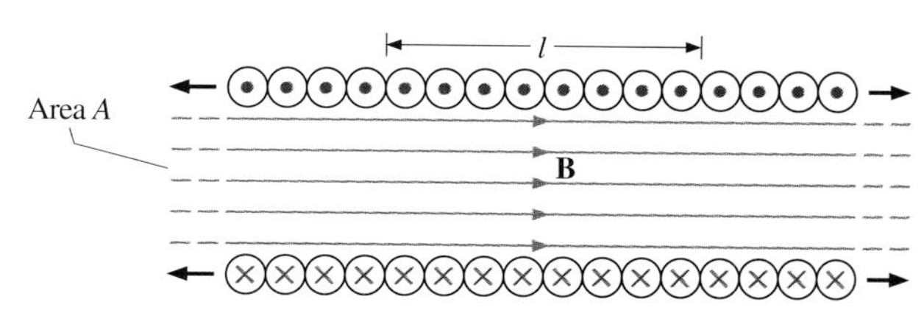

FIGURE 31-4 Energy is stored in the magnetic field of a solenoid.

solenoid changes, B changes, and an emf is induced. The magnetic flux ϕ_m through any single turn of wire in the section is

$$\phi_m = BA = \mu_0 niA.$$

Since there are $N = nl$ turns in the section, the magnitude of the induced emf between the two ends of the section is, from Faraday's law,

$$\mathcal{E} = N\frac{d\phi_m}{dt} = nl\frac{d}{dt}(\mu_0 niA) = \mu_0 n^2 lA\frac{di}{dt}.$$

The power input P to this section of the solenoid is

$$P = \mathcal{E}i = \mu_0 n^2 lA\frac{di}{dt}\,i.$$

Thus the energy U_m stored there when the current increases from zero to its final value I is

$$U_m = \int_0^\infty P\,dt = \int_0^\infty \mu_0 n^2 lAi\frac{di}{dt}\,dt$$
$$= \mu_0 n^2 lA\int_0^I i\,di = \frac{1}{2}\mu_0 n^2 lAI^2.$$

Substituting B for $\mu_0 nI$ and (vol) for the volume lA, we find that the *energy stored in the magnetic field of the section of the solenoid is*

$$U_m = \frac{B^2}{2\mu_0}(\text{vol}). \tag{31-6}$$

The **magnetic-field energy density** u_m is by definition the energy of the magnetic field per unit volume; therefore

$$u_m = \frac{U_m}{(\text{vol})} = \frac{B^2}{2\mu_0}. \tag{31-7}$$

This equation, like its electrical counterpart, $u_e = \epsilon_0 E^2/2$, is independent of the configuration used in its derivation—and it is also more general than its derivation implies. Equation (31-7) represents the energy density of *any* magnetic field.

EXAMPLE 31-2 Work Required to Charge a Conducting Sphere

How much work is required to place a total charge Q on a spherical conductor of radius R? Assume no energy is lost to Joule heating.

SOLUTION The work W required to charge the conductor is equal to the energy U_e stored in the electric field. From Example 25-15, the electric field of the spherical conductor is

$$E = 0 \qquad (r < R)$$

$$E = \frac{1}{4\pi\epsilon_0}\frac{Q}{r^2} \qquad (r \geq R),$$

which, with $u_e = \epsilon_0 E^2/2$, gives for the energy density

$$u_e = 0 \qquad (r < R)$$

$$u_e = \frac{Q^2}{32\pi^2\epsilon_0}\left(\frac{1}{r^4}\right) \qquad (r \geq R).$$

To find the total energy stored in the electric field, we integrate u_e from $r = 0$ to $r = \infty$:

$$U_e = \int u_e\,dv = \int_0^R u_e\,dv + \int_R^\infty u_e\,dv$$

$$= 0 + \frac{Q^2}{32\pi^2\epsilon_0}\int_R^\infty \frac{1}{r^4}(4\pi r^2\,dr) = \frac{Q^2}{8\pi\epsilon_0 R},$$

so the work required to charge the conductor is

$$W = \frac{Q^2}{8\pi\epsilon_0 R}.$$

EXAMPLE 31-3

The radius of the long, straight wire of Fig. 31-5 is a, and the wire carries a current I_0 distributed uniformly over its cross-sectional area. What is the magnetic field energy stored in the cylindrical volume of height h and radius b shown?

SOLUTION In Example 29-7, we found that the magnetic fields inside and outside a current-carrying wire are, respectively,

$$B = \frac{\mu_0 r I_0}{2\pi a^2} \qquad (r \leq a)$$

and

$$B = \frac{\mu_0 I_0}{2\pi r} \qquad (r \geq a).$$

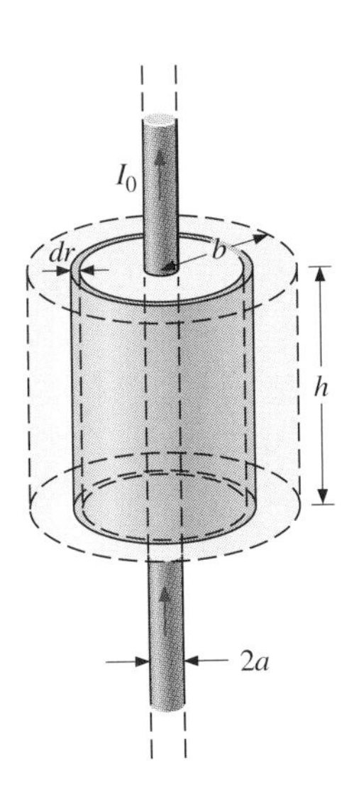

FIGURE 31-5 The electromagnetic energy stored in the cylindrical region is determined using the cylindrical shell of thickness dr shown.

Now from $u_m = B^2/2\mu_0$, the energy density of the magnetic field of the wire is

$$u_m = \frac{\mu_0 I_0^2}{8\pi^2 a^4} r^2 \qquad (r \leq a)$$

and

$$u_m = \frac{\mu_0 I_0^2}{8\pi^2 r^2} \qquad (r \geq a).$$

The total magnetic field energy stored in the designated volume is found by integrating u_m over that volume. Since u_m depends only on the radial distance r, we can use a cylindrical shell of height h and thickness dr as our volume element. Thus $dv = 2\pi rh\,dr$, and the total magnetic field energy is

$$U_m = \frac{\mu_0 I_0^2}{8\pi^2 a^4}\int_0^a r^2(2\pi rh\,dr) + \frac{\mu_0 I_0^2}{8\pi^2}\int_a^b \frac{1}{r^2}(2\pi rh\,dr)$$

$$= \frac{\mu_0 I_0^2 h}{16\pi} + \frac{\mu_0 I_0^2 h}{4\pi}\ln\frac{b}{a}$$

$$= \frac{\mu_0 I_0^2 h}{16\pi}\left(1 + 4\ln\frac{b}{a}\right).$$

DRILL PROBLEM 31-1

Two large, circular parallel plates of radius 20 cm are 2.0 mm apart. If the potential difference between them is 400 V, what is the energy stored in the electric field of the plates? Neglect edge effects.
ANS. 4.5×10^{-5} J.

DRILL PROBLEM 31-2

A tightly wound 5000-turn solenoid is 100 cm long and has a cross-sectional diameter of 2.0 cm. If a current of 2.0 A flows through the solenoid, how much energy is stored in the magnetic field in a 20-cm-long section near the middle of the solenoid?
ANS. 4.0×10^{-3} J.

DRILL PROBLEM 31-3

The magnitudes of electric and magnetic fields that can be easily produced in the laboratory are $E = 1.0 \times 10^5$ V/m and $B = 1.0$ T, respectively. Compare the energies stored in each of these fields within a 1.0-m^3 volume.
ANS. $U_e = 4.4 \times 10^{-2}$ J; $U_m = 4.0 \times 10^5$ J. Typically, the energy stored in a magnetic field is much larger than that stored in an electric field.

31-4 THE POYNTING VECTOR

To conclude our discussion of electromagnetic energy, we now ask how that energy is transported through space. What we find here will be very important for our study of electromagnetic radiation in Chap. 35. We start by considering the region of space

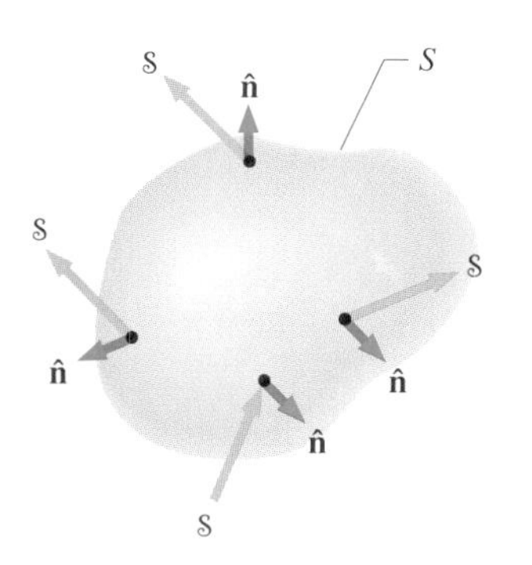

FIGURE 31-6 The electromagnetic field energy U stored in a given volume changes as electromagnetic energy is transported across the surface by the Poynting vector $\mathbf{S}$.

shown in Fig. 31-6. This region, which is free of matter, has a volume V and is bounded by a closed surface S. The outward unit normal at the surface is $\hat{\mathbf{n}}$. Although it won't be done here, the basic relationship that describes the transport of electromagnetic field energy can be derived directly from Maxwell's equations. That relationship is

$$\frac{d}{dt}\int_V \left(\frac{\epsilon_0 E^2}{2} + \frac{B^2}{2\mu_0}\right) dv + \oint_S \left(\frac{\mathbf{E}\times\mathbf{B}}{\mu_0}\right)\cdot\hat{\mathbf{n}}\, da = 0. \tag{31-8}$$

The term $\int_V (\epsilon_0 E^2/2 + B^2/2\mu_0)dv$ is a generalization of Eqs. (31-5) and (31-7). It represents the *total electromagnetic energy U stored in the volume V*. Since $\epsilon_0 E^2/2$ is the electric-field energy density u_e and $B^2/2\mu_0$ the magnetic-field energy density u_m, we can write

$$U = \int_V (u_e + u_m)\, dv. \tag{31-9}$$

The second term of Eq. (31-8) is a surface integral that represents the *rate at which electromagnetic energy flows across the surface S*. It is positive if energy leaves the region of volume V, and it is negative if energy enters the region. The rate dU/dt at which electromagnetic energy changes in the region is therefore equal to the negative of the rate at which electromagnetic energy crosses the boundary S of the region:

$$\frac{dU}{dt} = -\oint_S \left(\frac{\mathbf{E}\times\mathbf{B}}{\mu_0}\right)\cdot\hat{\mathbf{n}}\, da. \tag{31-10}$$

In the SI system, the units of these terms are joules per second or watts.

The quantity

$$\mathbf{S} = \frac{1}{\mu_0}(\mathbf{E}\times\mathbf{B}) \tag{31-11}$$

is called the **Poynting vector**, in honor of its discoverer, J. H. Poynting (1852–1914). *The direction of* $\mathbf{S}$ *at any point gives us the direction of electromagnetic energy transport, and its magnitude represents the rate at which electromagnetic energy crosses a unit area.* In the SI system, the Poynting vector has the unit watt per square meter. Equation (31-10) may now be written as

$$\frac{dU}{dt} = -\oint_S \mathbf{S}\cdot\hat{\mathbf{n}}\, da. \tag{31-12}$$

Notice that if $\oint_S \mathbf{S}\cdot\hat{\mathbf{n}}\, da = 0$, then

$$\frac{dU}{dt} = 0;$$

that is, the total electromagnetic field energy stored in a region of volume V remains constant if there is no energy flow across the bounding surface S of that region.

Although we have been considering energy flow across a closed surface, the Poynting vector can be used in the same manner to calculate the rate of energy flow across any surface. For an open surface S, we simply write this rate as $\int_S \mathbf{S}\cdot\hat{\mathbf{n}}\, da$.

Our treatment of electromagnetic field energy has not included the effects of energy sources or energy-dissipating elements such as sources of emf and resistive elements, respectively. Although a thorough investigation of these effects is beyond the level of this textbook, a brief look at the dissipative term is provided by Example 31-5.

The power that solar cells use to produce current is calculated from $\int \mathbf{S}\cdot\hat{\mathbf{n}}\, da$, where $\mathbf{S}$ is the Poynting vector of the electromagnetic radiation from the sun (mostly visible light).

EXAMPLE 31-4 The Poynting Vector for the Electromagnetic Field of a Charging Capacitor

Consider the parallel plates of Example 31-1. (*a*) Calculate the time derivative of the total energy U stored in their electromagnetic field while they are charging at a constant rate dq/dt. (*b*) What are the magnitude and direction of the Poynting vector $\mathbf{S}$ at all points on the cylindrical surface of radius R that bounds the region between the plates? (*c*) Calculate $\int_S \mathbf{S}\cdot\hat{\mathbf{n}}\, da$ over this cylindrical surface and compare it with dU/dt.

SOLUTION (*a*) The electromagnetic energy U stored in a volume V is, from Eq. (31-9),

$$U = \int_V \left(\frac{\epsilon_0 E^2}{2} + \frac{B^2}{2\mu_0}\right) dv,$$

and its time derivative is

$$\frac{dU}{dt} = \int_V \left(\epsilon_0 E \frac{\partial E}{\partial t} + \frac{1}{\mu_0} B \frac{\partial B}{\partial t} \right) dv. \qquad (i)$$

We begin with the term $\partial B/\partial t$. In Example 31-1, we found that $B \propto dq/dt$ for charging parallel plates. Since dq/dt is constant here, $\partial B/\partial t = 0$, and Eq. ($i$) reduces to

$$\frac{dU}{dt} = \int_V \epsilon_0 E \frac{\partial E}{\partial t}\, dv. \qquad (ii)$$

For parallel plates,

$$E = \frac{q}{\epsilon_0 A}$$

so

$$\frac{\partial E}{\partial t} = \frac{1}{\epsilon_0 A} \frac{dq}{dt}.$$

Since neither of these terms varies with position between the plates, both can be factored from the integral of Eq. (ii), leaving

$$\frac{dU}{dt} = \epsilon_0 E \frac{dE}{dt} \int_V dv = \epsilon_0 E \frac{dE}{dt} \pi R^2 h, \qquad (iii)$$

where we have replaced $\partial E/\partial t$ by dE/dt, because E is a function only of time.

(b) From Example 31-1, the magnitude of the magnetic field at the boundary of the cylindrical volume is

$$B = \mu_0 \epsilon_0 \frac{R}{2} \frac{dE}{dt},$$

and **B** is tangent to the cylindrical side, as shown in Fig. 31-7. Because **E** points from the positive to the negative plate, the Poynting vector is given by

$$\mathbf{S} = \frac{1}{\mu_0} \mathbf{E} \times \mathbf{B} = -\left(\frac{\epsilon_0 R E}{2} \frac{dE}{dt} \right) \hat{\mathbf{r}} \qquad (iv)$$

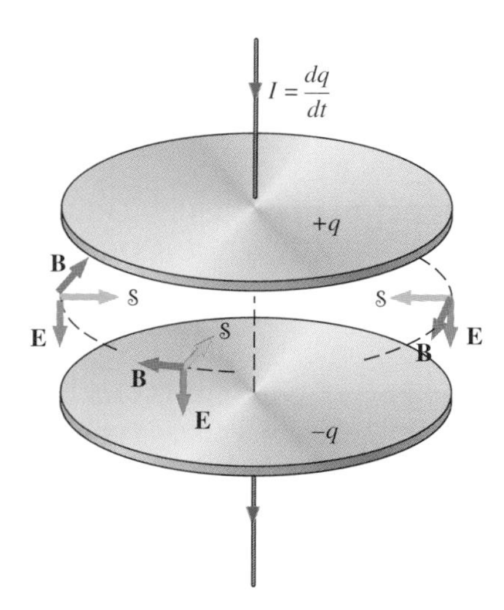

FIGURE 31-7 The directions of **E**, **B**, and **S** at the boundary of the region between two circular parallel plates.

everywhere on the boundary, where $\hat{\mathbf{r}}$ is directed radially outward from the central axis. Thus **S** is directed radially inward, and electromagnetic energy is flowing into the cylindrical region between the plates.

(c) At any given time t, the magnitude of **S** is constant over the cylindrical boundary. Since $\hat{\mathbf{n}} = \hat{\mathbf{r}}$ on that boundary,

$$\int_S \mathbf{S} \cdot \hat{\mathbf{n}}\, da = -\frac{\epsilon_0 R E}{2} \frac{dE}{dt} \int_S da = -\left(\frac{\epsilon_0 R E}{2} \frac{dE}{dt} \right) (2\pi R h)$$
$$= -\epsilon_0 E \frac{dE}{dt} \pi R^2 h. \qquad (v)$$

Finally, a comparison of Eqs. (iii) and (v) yields

$$\frac{dU}{dt} = -\int_S \mathbf{S} \cdot \hat{\mathbf{n}}\, da,$$

in agreement with Eq. (31-12).

EXAMPLE 31-5 The Poynting Vector and Joule Heating

Consider the section AA′ of the long, straight wire of Fig. 31-8. The length of the section is l, its radius is R, and when a current I flows through the wire, the potential difference between A′ and A is V. Determine the Poynting vector everywhere on the surface of the section, then integrate it over the surface to calculate the power flowing into the section.

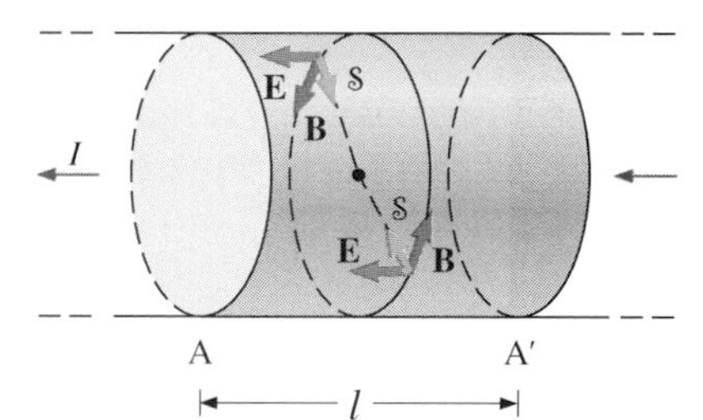

FIGURE 31-8 The electric and magnetic fields at the surface of a long, straight conducting wire. The direction of the Poynting vector is into the wire.

SOLUTION The potential difference and the electric field are related by

$$V = El, \qquad (i)$$

and the magnetic field at the surface of the wire is

$$B = \frac{\mu_0 I}{2\pi R}. \qquad (ii)$$

The directions of **E** and **B** are shown in the figure. Since $\mathbf{S} = (\mathbf{E} \times \mathbf{B})/\mu_0$, it points inward toward the axis of the wire section. With the help of Eqs. (i) and (ii), the magnitude of **S** is given by

$$S = \frac{1}{\mu_0} EB = \frac{IV}{2\pi R l}. \qquad (iii)$$

By integrating $\mathbf{S}$ over the surface of the section of wire (area = $2\pi Rl$), we find

$$\oint_S \mathbf{S}\cdot\hat{\mathbf{n}}\,da = S(2\pi Rl) = \frac{IV}{2\pi Rl}\,2\pi Rl = IV.$$

This is the electromagnetic field energy per unit time flowing across the surface into the wire. It is equal to the power dissipated by the section through Joule heating.

This result gives us an alternative way to look at the transfer of energy from a source of emf to a resistor. The source produces an electric field in the wire. This field in turn drives current through the wire, which creates a magnetic field around the wire. The two fields are directed so that the Poynting vector points into the surface of the wire. Hence energy is transferred from the source to the wire by means of the flow of electromagnetic field energy into the wire.

A solar-powered car.

SUMMARY

1. **Induced magnetic field**
   Wherever an electric field varies with time, there is an induced magnetic field. Magnetic fields are produced by currents and by changing electric fields according to the Ampère-Maxwell law:
   $$\oint \mathbf{B}\cdot d\mathbf{l} = \mu_0 I + \mu_0\epsilon_0\frac{d\phi_e}{dt}.$$

2. **Maxwell's equations**
   $$\oint_S \mathbf{E}\cdot\hat{\mathbf{n}}\,da = \frac{Q_{\text{in}}}{\epsilon_0}$$
   $$\oint_S \mathbf{B}\cdot\hat{\mathbf{n}}\,da = 0$$
   $$\oint \mathbf{E}\cdot d\mathbf{l} = -\frac{d}{dt}\int_S \mathbf{B}\cdot\hat{\mathbf{n}}\,da$$
   $$\oint \mathbf{B}\cdot d\mathbf{l} = \mu_0 I + \mu_0\epsilon_0\frac{d}{dt}\int_S \mathbf{E}\cdot\hat{\mathbf{n}}\,da$$

3. **Electromagnetic field energy**
   (*a*) The energy per unit volume stored in an electric field is
   $$u_e = \tfrac{1}{2}\epsilon_0 E^2.$$
   (*b*) The energy per unit volume stored in a magnetic field is
   $$u_m = \frac{B^2}{2\mu_0}.$$

4. **The Poynting vector**
   (*a*) The Poynting vector $\mathbf{S}$ is given by
   $$\mathbf{S} = \frac{1}{\mu_0}(\mathbf{E}\times\mathbf{B}).$$
   (*b*) The electromagnetic energy crossing a surface S per unit time is found by calculating the flux of $\mathbf{S}$ through S.

QUESTIONS

31-1. Why is it easy to demonstrate that changing magnetic fields produce electric fields but so difficult to demonstrate that changing electric fields produce magnetic fields?

31-2. Suppose that the charging plates of Fig. 31-2 are square rather than circular. What can you say about the shape of the induced magnetic field lines?

31-3. Which of Maxwell's equations (*a*) corresponds to Coulomb's law? (*b*) allows you to determine the magnetic field of a long, straight, current-carrying wire? (*c*) tells you that magnetic field lines form closed loops? (*d*) tells you that electric field lines do not necessarily form closed loops? (*e*) is used to calculate induced emf's? (*f*) includes a term known as the displacement current?

31-4. Can you directly measure the displacement current with an ordinary ammeter? How would an indirect measurement be made?

31-5. Speculate as to how Maxwell's equations will have to be altered if magnetic monopoles are ever found.

31-6. For circular field lines (either electric or magnetic), (*a*) is the field conservative? (*b*) is there a net flux through a closed surface?

31-7. For radial field lines (either electric or magnetic), (*a*) is the field conservative? (*b*) is there a net flux through a closed surface?

31-8. If the strength of the electric field in a given region of space is doubled, how are the electric energy density and the total electric energy in the region affected?

31-9. Answer Ques. 31-8 for the magnetic field.

31-10. Discuss what the Poynting vector represents.

31-11. At a given instant can the Poynting vector at a certain point in space be directed in more than one direction?

31-12. Is it possible for the Poynting vector to be time-dependent?

31-13. Describe power dissipation in a resistor in terms of the Poynting vector.

PROBLEMS

Induced Magnetic Field

31-1. Suppose that the plates shown in Fig. 31-2 are being charged at a constant rate of 2.0×10^{-2} C/s. The radius of each plate is 5.0 cm. Determine the direction and magnitude of the induced magnetic field at (*a*) 1.0 cm, (*b*) 2.0 cm, and (*c*) 20.0 cm from the central axis of the plates.

31-2. Suppose that the plates of Fig. 31-2 are *discharging* at a rate of 6.0×10^{-3} C/s. Determine the direction and magnitude of the induced magnetic field at the locations given in Prob. 31-1.

31-3. Two flat plates, each with an area of 150 cm^2, are placed close together and parallel to one another. If the electric field between the plates is changing at a rate $dE/dt = 2.5 \times 10^{10}$ V/m·s, what is the displacement current?

31-4. Two circular plates of radius 0.25 m are placed parallel to one another and 0.20 cm apart. If the potential difference between the plates increases at a constant rate $dV/dt = 1.5 \times 10^3$ V/s, what is the magnetic field between the plates at a distance of 0.15 m from the central axis of the plates? at a distance of 0.20 m?

31-5. The circular plates of Fig. 31-2 have a radius of 10.0 cm, and the electric field between them is given by $E = E_0 \sin \omega t$, where $E_0 = 2.54 \times 10^6$ V/m and $\omega = 120\pi$ rad/s. What is the induced magnetic field at (*a*) $r = 2.0$ cm and (*b*) $r = 15.0$ cm?

31-6. Show that the displacement current between two parallel circular plates can be written as $(\epsilon_0 \pi R^2/d)(dV/dt)$, where R is the radius of the plates, d is the distance between the plates, and V is the potential difference between the plates.

31-7. The potential difference between parallel plates of area A and separation d varies with time according to $V = V_0(1 - e^{-\alpha t})$. (*a*) Calculate the displacement current between the plates as a function of time. (*b*) If $V_0 = 200$ V, $\alpha = 0.25$ s^{-1}, $A = 100$ cm^2, and $d = 2.0$ mm, what is the value of the displacement current at $t = 1.0$ s?

31-8. The radii of two concentric, spherical conducting shells are R_1 and R_2, respectively. An alternating power source is connected between the shells such that their potential difference is given by $V = V_0 \cos \omega t$. What is the displacement current between the inner and outer shells?

Maxwell's Equations

31-9. Under static conditions, the electric field in the region $a < x < b$, $|y| < a$, $|z| < a$ is $\mathbf{E} = E(x)\mathbf{i}$. (*a*) If $E(x) = E_0$ and the net charge in the region is zero, is Gauss' law for the electric field satisfied? (*b*) What if $E(x) = E_0 e^{-\alpha x}$ and there is no net charge in the region? (*c*) Do these fields satisfy Maxwell's equations if they drop abruptly to zero for $|y| > a$ and $|z| > a$?

31-10. Consider the electric field of Prob. 31-9*a*, but now assume that there is also a magnetic field $\mathbf{B} = B\mathbf{k}$ in the same region. If $d\mathbf{B}/dt \neq 0$, is Faraday's law satisfied? Assume that there are no free charges or currents in the region.

31-11. In the region $a < x < b$, $|y| < a$, $|z| < a$, $\mathbf{B} = B(x)\mathbf{i}$. (*a*) If $B(x) = B_0$, is Gauss' law for the magnetic field satisfied? (*b*) What if $B(x) = B_0 e^{-\alpha x}$? (*c*) Does the existence of current in the region affect your answers to parts (*a*) and (*b*)? (*d*) Under time-varying conditions, does either of these fields satisfy the Ampère-Maxwell law?

31-12. Show that Maxwell's equations satisfy the principle of superposition. That is, show that if $\mathbf{E}_1$ and $\mathbf{B}_1$ satisfy the equations with sources Q_1 and I_1 and $\mathbf{E}_2$ and $\mathbf{B}_2$ satisfy the equations with sources Q_2 and I_2, then $\mathbf{E}_1 + \mathbf{E}_2$ and $\mathbf{B}_1 + \mathbf{B}_2$ satisfy the equations with sources $Q_1 + Q_2$ and $I_1 + I_2$.

Electromagnetic Field Energy

31-13. The electric-field energy density in a typical thundercloud is around 18 J/m^3. What is the electric field in the thundercloud?

31-14. What is the electric-field energy density corresponding to the dielectric strength of air (3.0×10^6 V/m)?

31-15. Two parallel circular plates of radius R are separated by distance d. Assuming the electric field between the plates is E, obtain an expression for the total energy stored in the field in terms of E, R, d, and the necessary physical constants.

31-16. How much work is required to charge an isolated conducting sphere of radius 20 cm to a potential of 4000 V?

31-17. The potential difference between two parallel plates is 4000 V. The area of each plate is 100 cm^2, and the plates are 2.0 mm apart. What is the energy stored in the electric field between the plates?

31-18. Parallel plates, each of area 100 cm^2, are separated by a gap of 2.0 mm. How much work must be done to give the plates equal and opposite charge of magnitude 2.0×10^{-7} C?

31-19. Suppose that the two large parallel plates shown in the accompanying figure are given equal and opposite charges of magnitude Q. The area of each plate is A, and the two plates are separated by a distance d. (*a*) What is the energy stored in the electric field between the plates? (*b*) If the plates are pulled apart

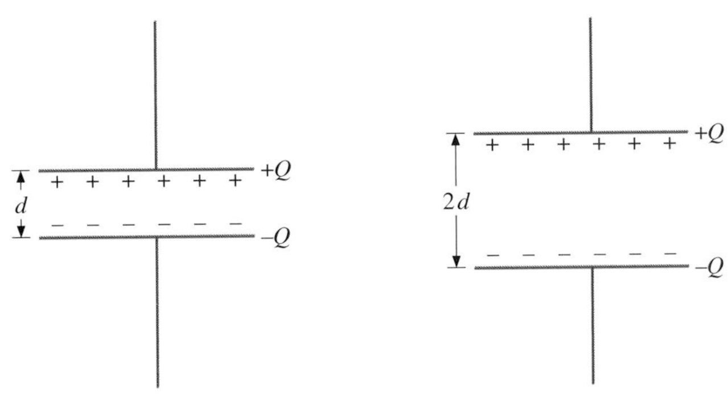

PROBLEM 31-19

until they are separated by $2d$, what is the energy stored in the electric field then? (*c*) What is the source of the increase in energy?

31-20. Suppose the plates of Prob. 31-19 are connected to opposite terminals of a battery, as shown in the accompanying figures. In this case, the potential difference between the plates remains constant at $\mathcal{E}$. (*a*) What is the energy stored in the electric field between the plates? (*b*) If the plates are pulled apart until they are separated by $2d$, what is then the energy stored in the electric field? (*c*) Account for the loss of energy. (*Hint*: Calculate the charges on the plates before and after they are pulled apart.)

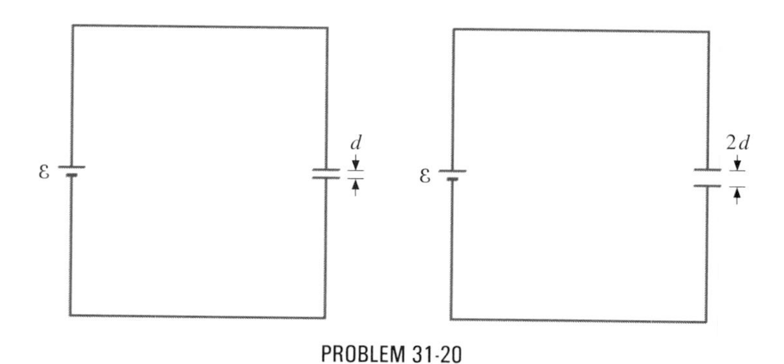

PROBLEM 31-20

31-21. A conducting sphere of radius R carries a charge q. (*a*) How much work is needed to bring an additional charge dq from infinity to the sphere? (*b*) Use the result of part (*a*) to show that the work needed to increase the charge on the sphere from zero to q is $q^2/8\pi\epsilon_0 R$. (*c*) Compare this result with that obtained in Example 31-2.

31-22. Two concentric spherical conducting shells of radii R_1 and R_2 are given equal and opposite charge of magnitude Q. (See the accompanying figure.) What is the electric field energy stored in the region between the shells?

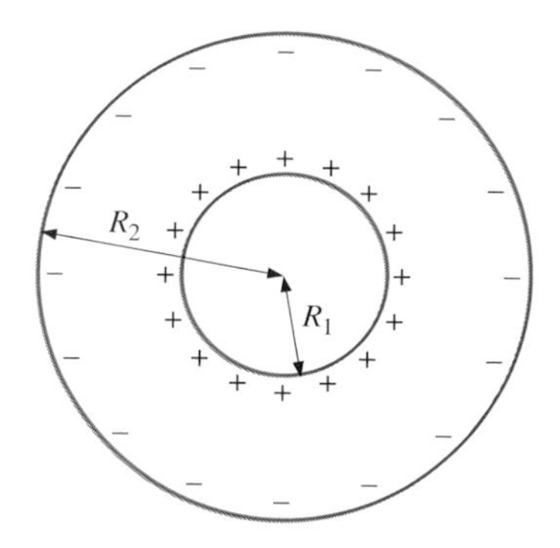

PROBLEM 31-22

31-23. Near its surface, the magnitudes of the earth's electric and magnetic fields are typically 100 V/m and 5.0×10^{-5} T. What is the energy density in each field?

31-24. A solenoid of length 90 cm and radius 2.0 cm is wound uniformly with 6000 turns. If the solenoid carries a current of 5.0 A, what is the energy stored in its magnetic field in a 15-cm section near the middle of the solenoid?

31-25. The toroid shown in the accompanying figure has a rectangular cross section with N uniformly spaced turns, and it carries a current I. What is the energy stored in its magnetic field?

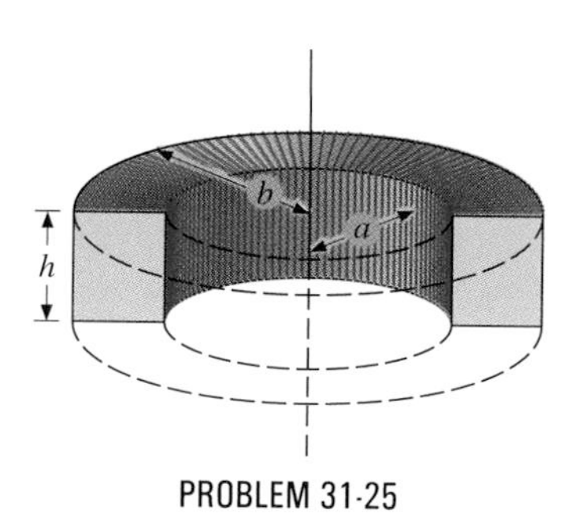

PROBLEM 31-25

31-26. Suppose that for the toroid of Prob. 31-25, $N = 1000$ turns, $I = 3.0$ A, $h = 2.0$ cm, $a = 8.0$ cm, and $b = 10.0$ cm. What is the energy stored in the magnetic field of the toroid?

31-27. A long, thin, straight wire carries a current of 10 A. Concentric with the wire is a cylindrical shell with an inner radius of 10 cm, an outer radius of 12 cm, and a height of 20 cm. What is the magnetic field energy stored within the cylindrical shell itself?

The Poynting Vector

31-28. Determine the directions of the Poynting vector for each of the cases shown in the accompanying figure.

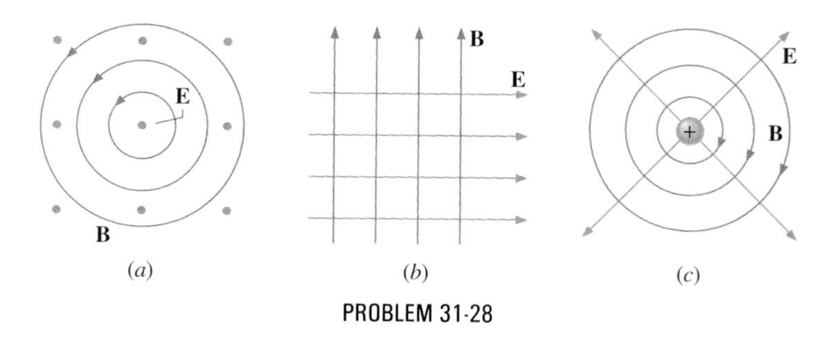

PROBLEM 31-28

31-29. A long solenoid of n turns per unit length and radius R carries a time-varying current $I(t)$. (*a*) What is the induced electric field at a point inside the solenoid at a distance r from its axis? (*b*) Compute the magnitude and direction of the Poynting vector at this point. Assume that $dI/dt > 0$.

31-30. A long, straight wire of radius 1.0 mm whose resistance per unit length is 2.5×10^{-3} Ω/m carries a current of 20 A. (*a*) Calculate the electric and magnetic fields at the surface of the wire. (*b*) Calculate the Poynting vector at the surface of the wire. (*c*) Determine the power flow into a 1.0-m section of the wire by integrating the Poynting vector over the surface area of the section. Does your result agree with what you find from Ohm's law?

31-31. A long, straight, cylindrical aluminum wire of radius 1.0 mm carries a current of 40 A. (*a*) What is the magnetic field at the surface of the wire? (*b*) What is the electric field there? (*c*) Determine the magnitude and direction of the Poynting vector at the surface of the wire. (*d*) How much power is transported by the Poynting vector across a section of the surface that is 1.0 m long? (*e*) Compare the power found in part (*d*) with that determined from Ohm's law.

General Problems

31-32. An electron with a velocity $\mathbf{v} = (2.5\mathbf{i} + 3.2\mathbf{j}) \times 10^5$ m/s enters a region where $\mathbf{E} = (2.8\mathbf{i} - 0.5\mathbf{j} + 1.6\mathbf{k}) \times 10^5$ V/m, and $\mathbf{B} = (-0.3\mathbf{i} + 0.9\mathbf{j} - 0.5\mathbf{k})$ T. Use the Lorentz force law to calculate the net force on the electron at the instant it enters the field.

31-33. A spatially uniform electric field **E** is produced over a cylindrical region of radius R. (See the accompanying figure.) At $t = 0$, $E = 1.0 \times 10^5$ V/m, and E decreases at a constant rate to zero in 30 s. (*a*) What is the magnetic field where $r > R$ and where $r < R$ during the 30-s interval? (*b*) Assume that $R = 10$ cm. At the instant when $E = 0.50 \times 10^5$ V/m, a proton enters the electric field at A moving with a velocity **v** ($v = 5.0 \times 10^6$ m/s) as shown. What are the electric and magnetic forces on the proton at that instant?

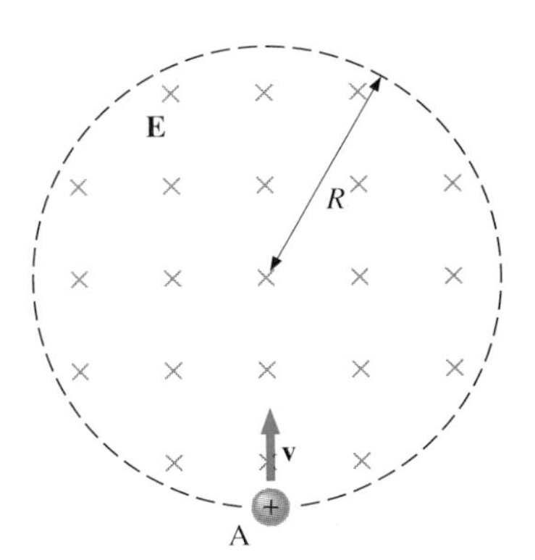

PROBLEM 31-32

31-34. An electric field is given by

$$\mathbf{E} = E_0\mathbf{k}\left(1 - \frac{r}{r_0}\right)\sin\omega t \qquad (r \le r_0),$$

$$\mathbf{E} = 0 \qquad (r > r_0).$$

Determine the induced magnetic field.

31-35. In this problem we consider *the electric field energy of an assembly of discrete point charges.* (*a*) Assume that three point charges are located as shown in the accompanying figure. Show that the work required to assemble these charges is given by

$$W = \frac{1}{4\pi\epsilon_0}\left(\frac{q_1q_2}{r_{12}} + \frac{q_1q_3}{r_{13}} + \frac{q_2q_3}{r_{23}}\right).$$

In Sec. 31-3, we interpreted this work to be the energy U_e stored in the electric field of the charges; that is,

$$U_e = \frac{1}{4\pi\epsilon_0}\left(\frac{q_1q_2}{r_{12}} + \frac{q_1q_3}{r_{13}} + \frac{q_2q_3}{r_{23}}\right).$$

(*b*) Generalize the result of part (*a*) to the case of N point charges and show that

$$U_e = \frac{1}{8\pi\epsilon_0}\sum_i\sum_j{}' \frac{q_iq_j}{r_{ij}}$$

where $\sum_j'$ indicates that the sum excludes the term with $j = i$.

(*c*) Rewrite the result of part (*b*) in the form

$$U_e = \frac{1}{2}\sum_i q_i \sum_j{}' \frac{1}{4\pi\epsilon_0}\frac{q_j}{r_{ij}}$$

and then show that

$$U_e = \frac{1}{2}\sum_i q_iV_i,$$

where V_i is the electric potential experienced by q_i due to the other $(N - 1)$ charges of the system.

(*d*) If charge is distributed continuously over a volume, we can write $dq = \rho dv$, where dv is the volume element surrounding the point where the charge density is ρ. Show that for a continuous charge distribution,

$$U_e = \frac{1}{2}\int \rho V\, dv.$$

(*e*) If the charge is distributed over the surface of a conductor, $\rho\, dv$ is replaced by $\sigma\, da$, where σ is the surface charge density at the element da. Use this in the result of part (*d*) to show that

$$U_e = \tfrac{1}{2}\epsilon_0E^2Ad$$

for parallel plates of area A that are separated by a distance d.

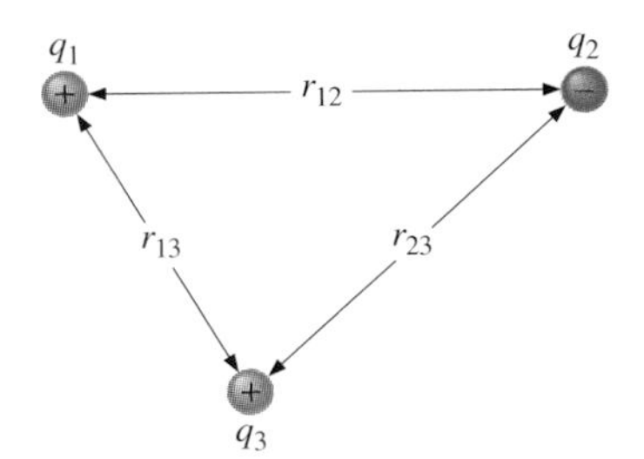

PROBLEM 31-34

31-36. Consider a spherical charge distribution for which the charge density is $\rho = \rho_0$ $(r \le R)$ and $\rho = 0$ $(r > R)$. (*a*) Show that the electrostatic field energy stored in the region $r \le R$ is

$$U_e' = \frac{1}{8\pi\epsilon_0}\frac{q^2}{5R},$$

where

$$q = \int \rho_0\, dv = \rho_0\left(\frac{4}{3}\pi R^3\right).$$

(*b*) Show that the electrostatic field energy stored in the region $r > R$ is

$$U_e'' = \frac{1}{8\pi\epsilon_0}\frac{q^2}{R};$$

then show that the total electrostatic field energy due to the charge distribution is

$$U_e = U_e' + U_e'' = \frac{1}{4\pi\epsilon_0}\frac{3q^2}{5R}.$$

31-37. Assume that the proton is a uniformly charged spherical volume of total charge $e = 1.6 \times 10^{-19}$ C and radius $R = 1.0 \times 10^{-15}$ m. Use the result of the previous problem to calculate the "electric self-energy" of the proton (the energy of the electric field within the proton). Compare this with the ionization energy of atomic hydrogen, which is 13.6 eV.

A strong electric field produced dielectric breakdown in this Plexiglas block. This caused the perforations seen in the photograph.

CHAPTER 32 CAPACITANCE AND INDUCTANCE

PREVIEW

This chapter is devoted to capacitance and inductance. The main topics to be discussed here are the following:

1. **Capacitance.** Capacitance is defined and discussed.
2. **Calculation of capacitance.** A general technique for calculating the capacitance of a pair of conductors is described and then applied to various configurations.
3. **Capacitors in parallel and in series.** We derive the rules for combining capacitors in parallel and in series.
4. **Energy stored in a capacitor.** The energy stored in the electric field of a capacitor is expressed in terms of the properties of the capacitor.
5. **Dielectrics and their molecular description.** The effects of placing a nonconducting material (a dielectric) between the plates of a capacitor are considered. These effects are then explained in terms of the properties of the molecules of the dielectric.
6. **Self-inductance.** Self-inductance is defined and discussed.
7. **Calculation of self-inductance.** A general technique for calculating self-inductance is described and applied to various configurations.
8. **Energy stored in an inductor.** The energy stored in the magnetic field of an inductor is expressed in terms of the properties of the inductor.

*9. **Mutual inductance.** Mutual inductance is introduced and explained.

Throughout the study of electricity and magnetism, we have drawn parallels between electric and magnetic phenomena. Our discussion of the similarities and differences between electricity and magnetism concludes with the introduction of two important components used in nearly all practical electric circuits: the capacitor and the inductor. These circuit elements have the common property that they possess the ability to store and release energy. In the case of the capacitor, this energy is stored in an electric field, while for the inductor, the energy is stored in a magnetic field.

In this chapter the basic properties of the capacitor and the inductor are examined. In the two following chapters we will consider the use of these two components in simple electric circuits. We will find that currents in circuits containing capacitors or inductors are time-dependent, even when the emf's of the sources are constant.

32-1 Capacitance

A *capacitor* is a device used extensively in electronic circuits to store and release electric energy. It consists of two conductors placed near to, but insulated from, each other. The two conductors of Fig. 32-1 represent an arbitrary capacitor. If initially uncharged, these conductors will become equally and oppositely charged when a quantity of charge $+Q$ is transferred between them. The charge separation produces an electric field and a potential difference between the conductors. The **capacitance** C of this capacitor is by definition

$$C = \frac{Q}{V}, \tag{32-1}$$

where Q and V are, respectively, the *magnitudes* of the charge and the potential difference between the conductors. *Capacitance is therefore a positive quantity.* We say that the capacitor has a charge Q, even though its total charge is zero. In SI units, Q is expressed in coulombs (C) and V in volts (V), so the unit of capacitance is a coulomb/volt (C/V), which is designated as a *farad* (F):

1 farad = 1 coulomb/volt.

A farad represents a very large capacitance. Capacitors used in electronic circuits typically have capacitances in the microfarad (μF) and even picofarad (pF) range, where

$$1\ \mu\text{F} = 10^{-6}\ \text{F} \qquad \text{and} \qquad 1\ \text{pF} = 10^{-12}\ \text{F}.$$

32-2 Calculation of Capacitance

The capacitance of a pair of conductors is usually calculated with the following approach:

1. Assume that the capacitor has a charge Q.
2. Determine the electric field $\mathbf{E}$ between the conductors. If there is symmetry in the system, you may be able to use Gauss' law for this calculation.
3. Find the potential difference between the conductors from

   $$V_B - V_A = -\int_A^B \mathbf{E}\cdot d\mathbf{l},$$

   where the path of integration leads from one conductor to the other. The magnitude of the potential difference is then $V = |V_B - V_A|$.
4. With V known, obtain the capacitance directly from $C = Q/V$.

As examples of this procedure, we now calculate the capacitances of three different arrangements of conductors, all of which are assumed to be in free space.

Parallel-Plate Capacitor

This capacitor, which is represented in Fig. 32-2, consists of two parallel conducting plates. The plates have a cross-sectional area A and are separated by a distance d. In Example 23-9, we found that if d is small, the electric field between the plates is uniform and given by

$$E = \frac{\sigma}{\epsilon_0},$$

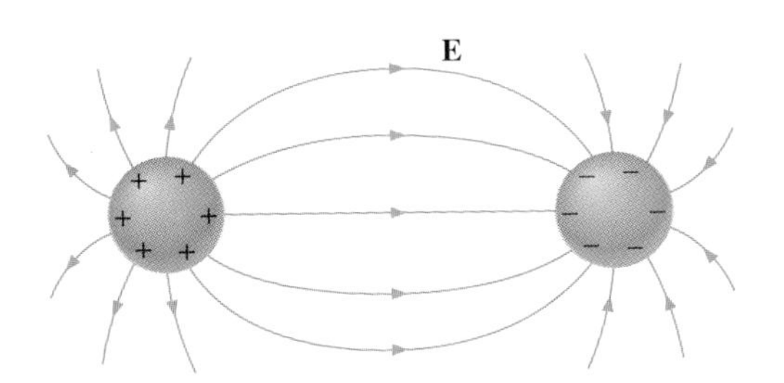

FIGURE 32-1 A system of two oppositely charged conductors that are insulated from each other is called a capacitor.

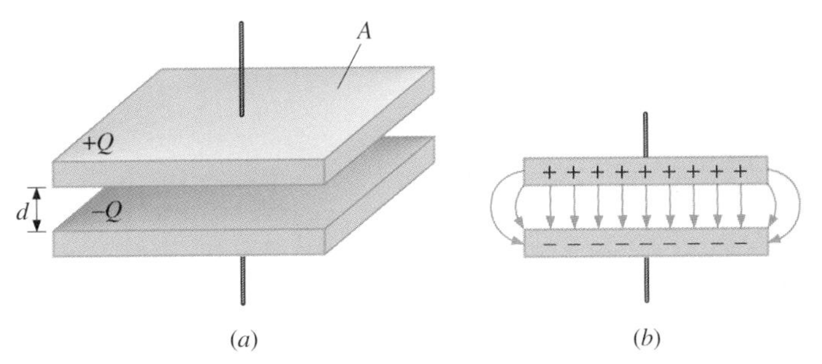

FIGURE 32-2 (*a*) A parallel-plate capacitor. (*b*) The electric field of the capacitor. In calculations, we ignore the fringing of the field near the edges.

where $\sigma = Q/A$ is the surface charge density on the plates. Since E is a constant, the magnitude of the potential difference between the plates is

$$V = Ed = \frac{Qd}{\epsilon_0 A}.$$

Therefore, from $C = Q/V$, the capacitance of the parallel-plate capacitor is

$$C = \frac{Q}{Qd/\epsilon_0 A} = \frac{\epsilon_0 A}{d}. \tag{32-2}$$

Notice that this expression is a function *only of the geometry* of this capacitor. In fact, this is true not only for the parallel-plate capacitor, but for all capacitors. The capacitance C *is independent of* Q *or* V. If Q changes, V changes correspondingly so that Q/V remains constant.

EXAMPLE 32-1 A 1-F Parallel-Plate Capacitor

Suppose you wish to construct a parallel-plate capacitor whose capacitance is 1.0 F. What is the area of the plates you must use if they are to be separated by 1.0 mm?

SOLUTION Rearranging $C = \epsilon_0 A/d$, we have

$$A = \frac{Cd}{\epsilon_0} = \frac{(1.0 \text{ F})(1.0 \times 10^{-3} \text{ m})}{8.85 \times 10^{-12} \text{ C}^2\cdot\text{m}^2/\text{N}} = 1.1 \times 10^8 \text{ m}^2.$$

Each square plate would have to be about 10,000 m across! Clearly, a capacitance of 1.0 F is unrealistically large for a capacitor of this type. However, with modern technological methods, it is possible to construct 1-F capacitors so small that they can be held in your hand. The necessary large plate area is obtained by simulating plates with granules of activated carbon. The effective surface area of just 1 g of this material is around 10^3 m^2.

DRILL PROBLEM 32-1

The capacitance of a parallel-plate capacitor is 2.0 pF. If the area of each plate is 2.4×10^{-4} m^2, what is the plate separation?
ANS. 1.1×10^{-3} m.

DRILL PROBLEM 32-2

Verify that Q/V and $\epsilon_0 A/d$ have the same dimensions.

Spherical Capacitor

Another set of conductors whose capacitance can easily be determined is the spherical capacitor of Fig. 32-3. It consists of two concentric spherical conducting shells of radii R_1 and R_2 that are given equal and opposite charges Q and $-Q$, respectively. From symmetry, the electric field between the shells is

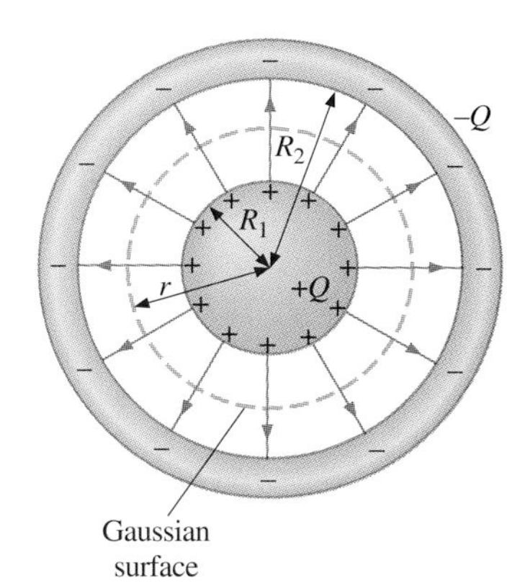

FIGURE 32-3 A spherical capacitor consists of two concentric conducting spheres.

directed radially outward. The magnitude of the field can be obtained by applying Gauss' law over a spherical Gaussian surface of radius r ($R_2 > r > R_1$) concentric with the shells. The enclosed charge is $+Q$, so

$$\oint_S \mathbf{E}\cdot\hat{\mathbf{n}}\, da = E(4\pi r^2) = \frac{Q}{\epsilon_0}.$$

The electric field between the conductors is therefore

$$\mathbf{E} = \frac{1}{4\pi\epsilon_0}\frac{Q}{r^2}\hat{\mathbf{r}},$$

and the magnitude of the potential difference between the conductors is

$$V = \int_{R_1}^{R_2} \mathbf{E}\cdot d\mathbf{l} = \int_{R_1}^{R_2}\left(\frac{1}{4\pi\epsilon_0}\frac{Q}{r^2}\hat{\mathbf{r}}\right)\cdot(\hat{\mathbf{r}}\, dr)$$
$$= \frac{Q}{4\pi\epsilon_0}\int_{R_1}^{R_2}\frac{dr}{r^2} = \frac{Q}{4\pi\epsilon_0}\left(\frac{1}{R_1} - \frac{1}{R_2}\right),$$

where the line integral is evaluated along a radial path from the inner to the outer conductor. Now using $C = Q/V$, we obtain for the capacitance of a spherical capacitor

$$C = \frac{Q}{V} = 4\pi\epsilon_0\left(\frac{R_1 R_2}{R_2 - R_1}\right). \tag{32-3}$$

As expected, C depends only on the geometry of the system.

EXAMPLE 32-2 Capacitance of an Isolated Sphere

Calculate the capacitance of a single, isolated conducting sphere of radius R_1. Compare your result with Eq. (32-3) in the limit as $R_2 \to \infty$.

SOLUTION Outside an isolated conducting sphere of charge Q and radius R_1, the electric field is

$$\mathbf{E} = \frac{1}{4\pi\epsilon_0}\frac{Q}{r^2}\hat{\mathbf{r}}.$$

We now assume that the other conductor of this capacitor is a concentric hollow sphere of infinite radius. The magnitude of the potential difference between the surface of the isolated sphere and infinity is

$$V = \int_{R_1}^{\infty} \mathbf{E}\cdot d\mathbf{l} = \frac{Q}{4\pi\epsilon_0}\int_{R_1}^{\infty}\frac{1}{r^2}\,\hat{\mathbf{r}}\cdot(\hat{\mathbf{r}}\,dr) = \frac{1}{4\pi\epsilon_0}\frac{Q}{R_1}.$$

The capacitance of the isolated sphere is therefore

$$C = \frac{Q}{V} = 4\pi\epsilon_0 R_1. \tag{32-4}$$

In the limit as the radius R_2 of the outer shell of a spherical capacitor goes to infinity, $R_2 - R_1 \approx R_2$, and Eq. (32-3) becomes

$$C = 4\pi\epsilon_0 R_1,$$

which is identical to Eq. (32-4). A single isolated sphere is therefore equivalent to a spherical capacitor whose outer shell is infinitely large. In principle, we can assign a capacitance to any isolated conductor in the same way. We assume that the "other plate" is a conducting sphere of infinite radius, determine V between the conductor and the infinite sphere when the charge on the conductor is Q, then take the ratio Q/V.

DRILL PROBLEM 32-3

The radius of the outer shell of a spherical capacitor is five times the radius of the inner shell. What are the dimensions of this capacitor if $C = 5.0$ pF?
ANS. $R_1 = 3.6 \times 10^{-2}$ m; $R_2 = 0.18$ m.

Cylindrical Capacitor

A cylindrical capacitor consists of two concentric cylinders, as shown in Fig. 32-4. The inner cylinder, of radius R_1, may either be a shell or completely solid. The outer cylinder is a shell of inner radius R_2. We assume that the length of each cylinder is l and that excess charges $+Q$ and $-Q$ reside on the inner and outer cylinders, respectively.

With edge effects ignored, the electric field between the conductors is directed radially outward from the common axis of the cylinders. Using the Gaussian surface shown, we have

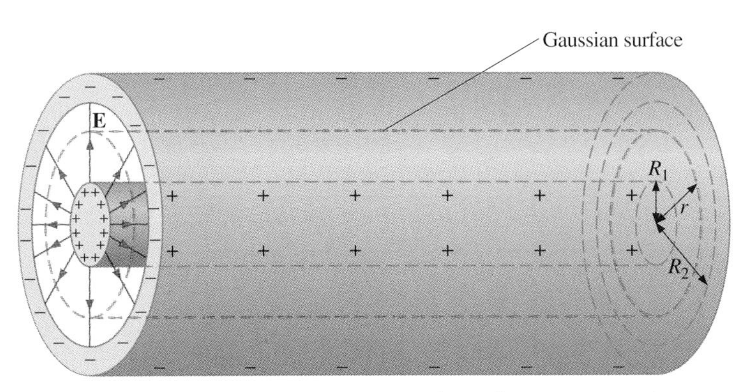

FIGURE 32-4 A cylindrical capacitor consists of two concentric conducting cylinders.

$$\oint_S \mathbf{E}\cdot\hat{\mathbf{n}}\,da = E(2\pi r l) = \frac{Q}{\epsilon_0},$$

so

$$\mathbf{E} = \frac{Q}{2\pi\epsilon_0 r l}\,\hat{\mathbf{r}}.$$

The potential difference between the cylinders is then

$$V = \int_{R_1}^{R_2}\mathbf{E}\cdot d\mathbf{l} = \frac{Q}{2\pi\epsilon_0 l}\int_{R_1}^{R_2}\frac{1}{r}\,\hat{\mathbf{r}}\cdot(\hat{\mathbf{r}}\,dr)$$
$$= \frac{Q}{2\pi\epsilon_0 l}(\ln r)\Big|_{R_1}^{R_2} = \frac{Q}{2\pi\epsilon_0 l}\ln\frac{R_2}{R_1},$$

and the capacitance of the cylindrical capacitor is

$$C = \frac{Q}{V} = \frac{2\pi\epsilon_0 l}{\ln(R_2/R_1)}. \tag{32-5}$$

As in the previous cases, C depends only on the geometry of the capacitor.

An important application of Eq. (32-5) is the determination of the capacitance per unit length of a *coaxial cable*, which is commonly used to transmit time-varying electric signals. This cable consists of two concentric cylindrical conductors separated by an insulating material. (Here we assume a vacuum between the conductors.) This configuration shields the electric signal propagating down the inner conductor from stray electric fields external to the cable. Current flows in opposite directions in the inner and outer conductors, with the outer conductor usually grounded. Now from Eq. (32-5), the capacitance per unit length of the coaxial cable is given by

$$\frac{C}{l} = \frac{2\pi\epsilon_0}{\ln(R_2/R_1)}.$$

In practical applications, it is important to select particular values of C/l. This can be accomplished with the appropriate choices of the radii of the conductors, and of the insulating material between them.

DRILL PROBLEM 32-4

When the capacitor of Fig. 32-4 is given a charge of 5.0×10^{-10} C, a potential difference of 20 V is measured between the cylinders. (*a*) What is the capacitance of the system? (*b*) If the cylinders are 1.0 m long, what is the ratio of their radii?
ANS. (*a*) 25 pF; (*b*) 9.2.

32-3 Practical Capacitors

Several types of practical capacitors are shown in Fig. 32-5. Common capacitors are often made of two small pieces of metal foil separated by a thin layer of insulation (to be discussed further in Sec. 32-6). The metal foils and insulation are encased

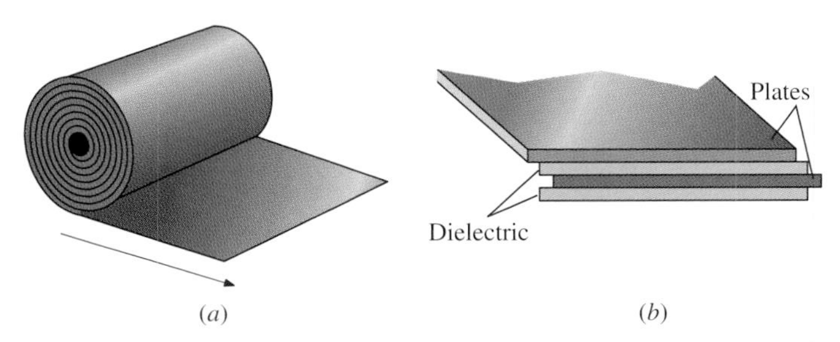

(a) A rolled capacitor. The effective area of the capacitor is made reasonably large by rolling the assembly as shown. (b) A magnified view of the end of the roll.

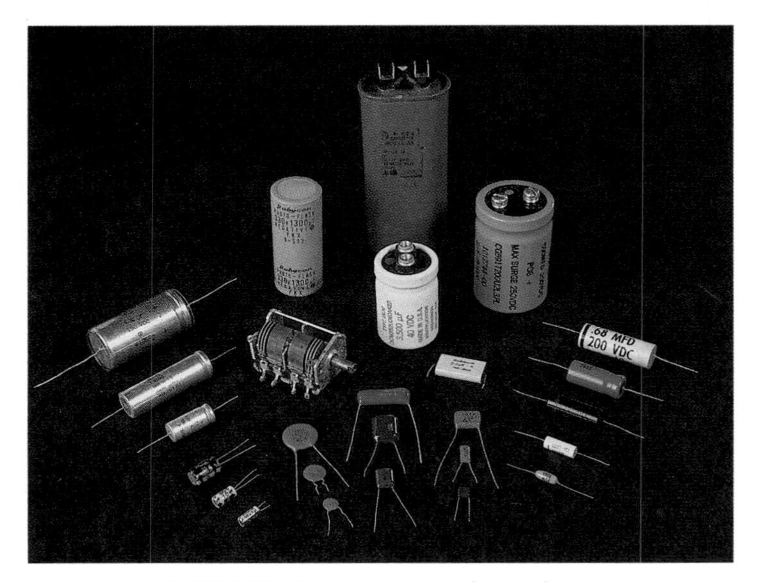

FIGURE 32-5 Common types of capacitors.

in a protective coating, and two metal leads are used for connecting the foils to an external circuit. Some common insulating materials are mica, ceramic, paper, and Teflon.

The electrolytic capacitor is also quite popular. It consists of an oxidized metal foil in a conducting paste. The thin oxide film at the surface of the foil acts as the insulation between the foil and the paste. The main advantage of an electrolytic capacitor is its high capacitance relative to the other common types of capacitors. However, one must be careful when using an electrolytic capacitor in a circuit, because it only functions properly when the metal foil is at a higher potential than the conducting paste. When the reverse occurs, electrolytic action destroys the oxide film. This type of capacitor cannot be connected across an alternating current source since half the time the voltage across its plates would have the wrong polarity.

A variable air capacitor is shown in Fig. 32-6. It has two sets of parallel metal plates, one of which is fixed, while the other is attached to a rotatable shaft. When the shaft is turned, the common cross-sectional area of the plates (and therefore the capacitance of the system) is changed.

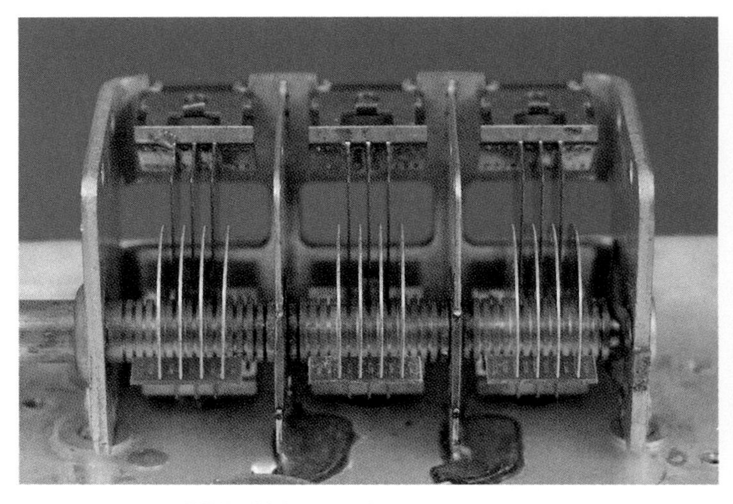

FIGURE 32-6 A variable air capacitor.

The symbols seen in Fig. 32-7 are circuit representations of various types of capacitors. We generally use the symbol shown in Fig. 32-7a. Notice its similarity to the geometry of a parallel-plate capacitor. If the capacitor is electrolytic, it is represented by the symbol seen in Fig. 32-7b, where the curved plate indicates the negative terminal. Finally, Fig. 32-7c denotes a capacitor whose capacitance can be varied (generally by mechanical adjustment of the geometry).

(a) (b) (c)

FIGURE 32-7 Various circuit representations of capacitors.

32-4 Capacitors in Parallel and in Series

Like resistors, capacitors may be connected in parallel or in series. Figure 32-8a shows a parallel combination of three capacitors that have been charged by a battery. Their capacitances are C_1, C_2, and C_3, and they contain charges Q_1, Q_2, and Q_3, respectively. Since the capacitors are in parallel, they all have *the same voltage V across their plates.* The net charge at the junction of the positive plates (the region enclosed by the dashed lines) is

$$Q = Q_1 + Q_2 + Q_3,$$

while $-Q$ is the net charge on the set of negative plates. The individual capacitances are given by

$$C_1 = \frac{Q_1}{V}; \qquad C_2 = \frac{Q_2}{V}; \qquad C_3 = \frac{Q_3}{V}.$$

A *single* capacitor that has the same capacitance C_p as this parallel combination must carry a charge Q and have a potential difference V between its plates. (See Fig. 32-8b.) Thus the equivalent capacitance is

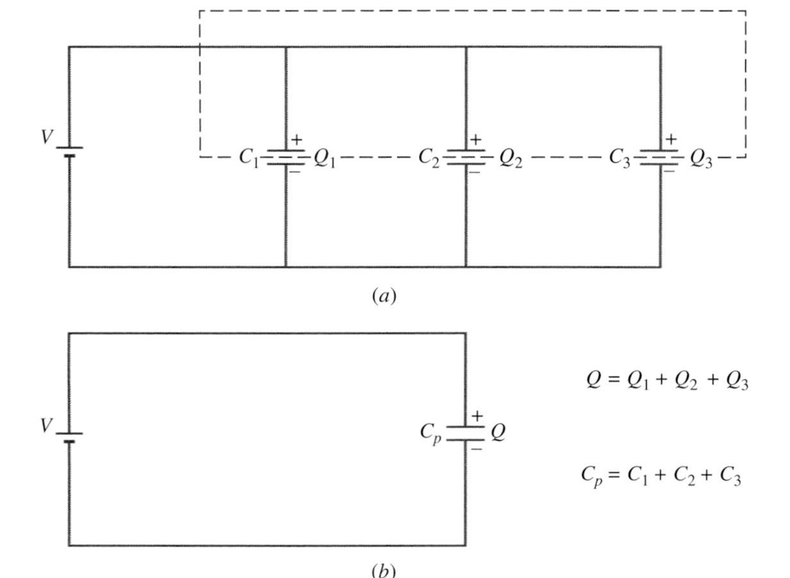

FIGURE 32-8 (*a*) Three capacitors in parallel. (*b*) A single capacitor with the equivalent capacitance of the parallel combination.

$$C_p = \frac{Q}{V} = \frac{Q_1 + Q_2 + Q_3}{V} = \frac{Q_1}{V} + \frac{Q_2}{V} + \frac{Q_3}{V}$$
$$= C_1 + C_2 + C_3.$$

By extending this argument to any number of capacitors, we find that the net capacitance of parallel capacitors is the sum of their individual capacitances:

$$C_p = C_1 + C_2 + C_3 + \ldots \tag{32-6}$$

Now let's consider a series combination of capacitors—for example, C_1 and C_2 in Fig. 32-9*a*. The region enclosed by the dashed lines is electrically isolated from the rest of the circuit by the dielectrics. This means that when the battery is connected to the uncharged capacitors, all it does is transfer charge from the lower plate of C_2 to the upper plate of C_1 until electrostatic equilibrium is reached. The charge on these plates must therefore be *equal* and *opposite* (say, $+Q$ on the upper plate of C_1 and $-Q$ on the lower plate of C_2). Because of the Coulomb interaction, the charge enclosed by the dashed lines is redistributed until $-Q$ resides on the lower plate of C_1 and $+Q$ on the upper plate of C_2. The *net* charge within that isolated region remains zero. So, at equilibrium, the two series capacitors become charged as shown in Fig. 32-9*b*. By extending this argument to any number of capacitors, we conclude that *all the capacitors of a series combination have the same charge.*

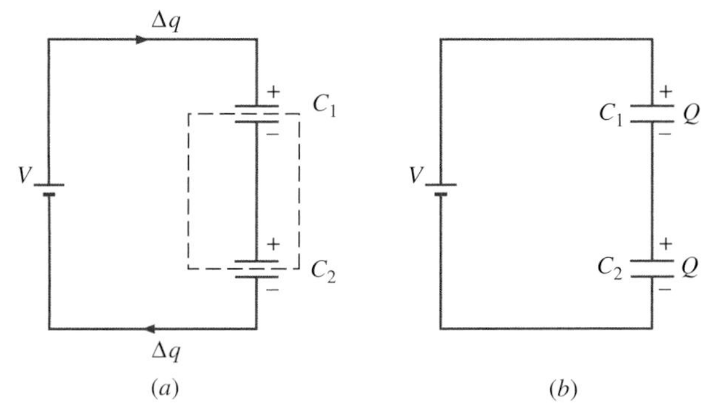

FIGURE 32-9 (*a*) Two series capacitors connected to a battery. The battery transfers charge from the lower plate of C_2 to the upper plate of C_1. (*b*) At equilibrium, the capacitors are equally charged.

Suppose that the voltages across each of the three series capacitors of Fig. 32-10 are V_1, V_2, and V_3, respectively; then

$$V = V_1 + V_2 + V_3.$$

Since all capacitors carry the same charge Q,

$$V_1 = \frac{Q}{C_1}, \qquad V_2 = \frac{Q}{C_2}, \qquad \text{and} \qquad V_3 = \frac{Q}{C_3}.$$

The single capacitor with the equivalent capacitance C_s of the series combination must carry a charge Q and have a potential difference V between its plates. (See Fig. 32-10*b*.) Thus

$$V = \frac{Q}{C_s} = V_1 + V_2 + V_3 = \frac{Q}{C_1} + \frac{Q}{C_2} + \frac{Q}{C_3},$$

and

$$\frac{1}{C_s} = \frac{1}{C_1} + \frac{1}{C_2} + \frac{1}{C_3}.$$

By extending this argument to an arbitrary number of capacitors in series, we find

$$\frac{1}{C_s} = \frac{1}{C_1} + \frac{1}{C_2} + \frac{1}{C_3} + \ldots \tag{32-7}$$

Notice that the formulas for calculating the net capacitances of parallel and series capacitors (Eqs. 32-6 and 32-7) are just the reverse of the formulas for the two equivalent resistor combinations (Eqs. 27-2 and 27-1).

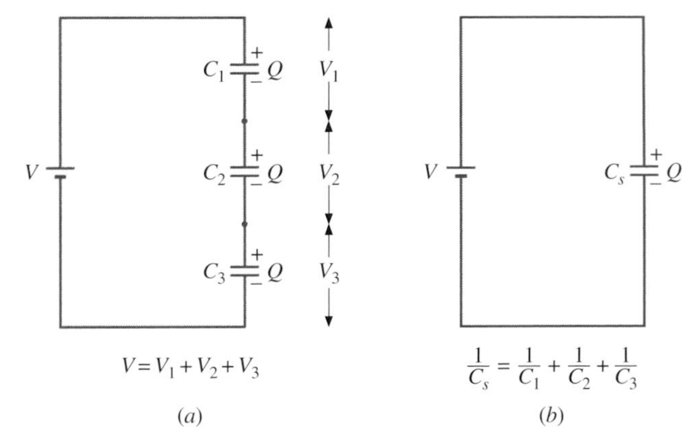

FIGURE 32-10 (*a*) Three capacitors in series. (*b*) A single capacitor with the equivalent capacitance of the series combination.

EXAMPLE 32-3 Net Capacitance of a Capacitor Combination

(*a*) Determine the net capacitance of the capacitor combination shown in Fig. 32-11*a*, where $C_1 = 12.0\ \mu F$, $C_2 = 2.0\ \mu F$, and $C_3 = 4.0\ \mu F$. (*b*) When a potential difference of 12.0 V is maintained across the combination, what are the charge on each capacitor and the voltage across each capacitor?

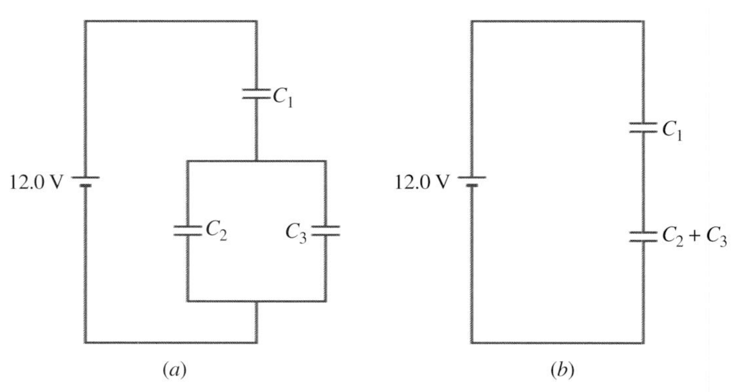

FIGURE 32-11 (*a*) A capacitor combination. (*b*) An equivalent two-capacitor combination.

SOLUTION (*a*) From Eq. (32-6), the equivalent capacitance of C_2 and C_3 is $C_2 + C_3 = 2.0\ \mu F + 4.0\ \mu F = 6.0\ \mu F$. This three-capacitor combination is therefore equivalent to two capacitors in series (Fig. 32-11*b*) whose net capacitance C may be found from

$$\frac{1}{C} = \frac{1}{12.0\ \mu F} + \frac{1}{6.0\ \mu F};$$

thus

$$C = 4.0\ \mu F.$$

(*b*) Consider the equivalent two-capacitor combination of Fig. 32-11*b*. Since the capacitors are in series, they have the same charge Q. In addition, the capacitors share the 12.0-V potential difference, so

$$\frac{Q}{12.0\ \mu F} + \frac{Q}{6.0\ \mu F} = 12.0\ V,$$

and

$$Q = 48\ \mu C.$$

We can now use this value of the charge to calculate the potential difference across C_1:

$$V_1 = \frac{Q}{C_1} = \frac{48\ \mu C}{12.0\ \mu F} = 4.0\ V.$$

Because C_2 and C_3 are in parallel, the potential difference across each is $V_2 = V_3 = 12.0\ V - 4.0\ V = 8.0\ V$, and the charges on the two capacitors are, respectively,

$$Q_2 = C_2V_2 = (2.0\ \mu F)(8.0\ V) = 16\ \mu C,$$

and

$$Q_3 = C_3V_3 = (4.0\ \mu F)(8.0\ V) = 32\ \mu C.$$

As expected, $Q_2 + Q_3 = Q$, the net charge on the parallel combination of C_2 and C_3.

DRILL PROBLEM 32-5

Determine the net capacitances of the networks of capacitors in Fig. 32-12*a*, *b*, and *c*. Assume that $C_1 = 1.0$ pF, $C_2 = 2.0$ pF, $C_3 = 4.0$ pF, and $C_4 = 5.0$ pF. Find the charge on each capacitor, assuming there is a potential difference of 12 V across each network.

ANS. (*a*) $C = 0.86$ pF, $Q_1 = 1.0 \times 10^{-11}$ C, $Q_2 = 3.4 \times 10^{-12}$ C, $Q_3 = 6.8 \times 10^{-12}$ C; (*b*) $C = 2.3$ pF, $Q_1 = 1.2 \times 10^{-11}$ C, $Q_2 = Q_3 = 1.6 \times 10^{-11}$ C; (*c*) $C = 2.3$ pF, $Q_1 = 9.0 \times 10^{-12}$ C, $Q_2 = 1.8 \times 10^{-11}$ C, $Q_3 = 1.2 \times 10^{-11}$ C, $Q_4 = 1.5 \times 10^{-11}$ C.

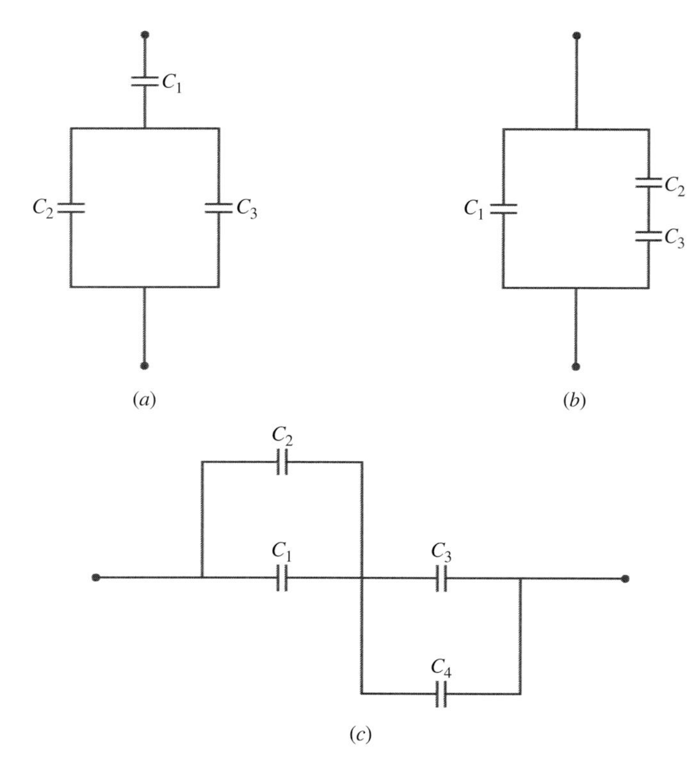

FIGURE 32-12 Capacitor networks.

32-5 Energy Stored in a Capacitor

A charged capacitor stores energy in the electric field between its plates. In Sec. 31-3 we found that the energy density in a region of free space occupied by an electric field E is given by

$$u_e = \tfrac{1}{2}\epsilon_0 E^2. \tag{32-8}$$

The amount of energy stored between the plates of a parallel-plate capacitor is therefore

$$U = \tfrac{1}{2}\epsilon_0 E^2(Ad),$$

where A is the cross-sectional area of the plates and d is their separation. For a parallel-plate capacitor, $E = V/d$ and $C = \epsilon_0 A/d$, so the energy stored in the electric field of the capacitor is

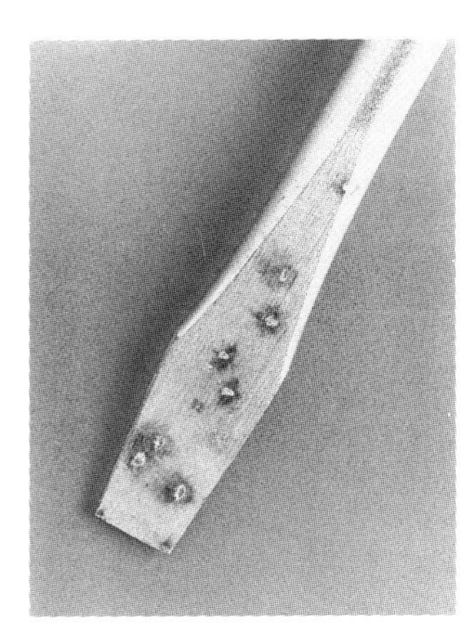

This screwdriver was used to discharge a large capacitor.

$$U = \frac{1}{2}\epsilon_0\left(\frac{V}{d}\right)^2(Ad) = \frac{1}{2}\left(\frac{\epsilon_0 A}{d}\right)V^2 = \frac{1}{2}CV^2. \qquad (32\text{-}9a)$$

Since $V = Q/C$, this equation may also be written as

$$U = \frac{1}{2}\frac{Q^2}{C}, \qquad (32\text{-}9b)$$

and

$$U = \tfrac{1}{2}QV. \qquad (32\text{-}9c)$$

While these expressions for the energy have been obtained for a parallel-plate capacitor, they hold for *any* type of capacitor. We can prove this by considering what happens when a capacitor is charged by a battery. Suppose that at some instant during the charging process, the charge on the capacitor is q and the potential difference between its plates is $V = q/C$. (See Fig. 32-13.) To move an infinitesimal charge dq from the negative plate to the positive plate (from a lower to a higher potential), an amount of work

$$dW = V\,dq$$

must be done on dq. This work becomes energy stored in the electric field of the capacitor. In order to give the capacitor a charge Q, the total work required is

$$W = \int_0^Q V\,dq = \int_0^Q \left(\frac{q}{C}\right)dq = \frac{1}{2}\frac{Q^2}{C},$$

which is Eq. (32-9b). Since the geometry of the capacitor has not been specified, this equation holds for *any* type of capacitor.

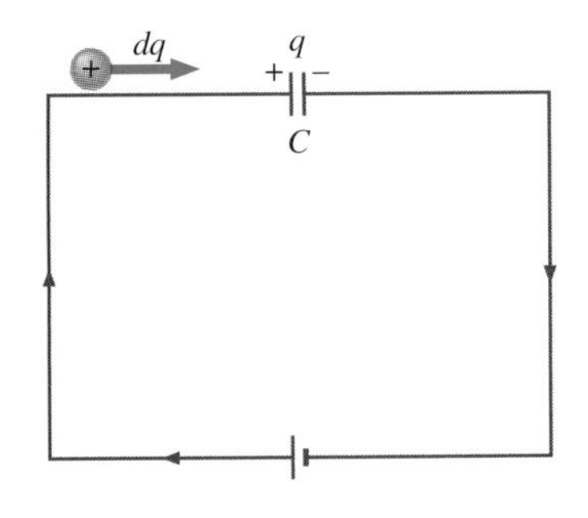

FIGURE 32-13 When an infinitesimal charge dq is moved from the negative to the positive plate of the capacitor, work $dW = V\,dq$ is done on dq.

EXAMPLE 32-4 ENERGY STORED IN A CAPACITOR

Calculate the energy stored in each of the capacitors of Fig. 32-11a after they are fully charged by the battery.

SOLUTION From Example 32-3, the voltage across the 12.0-μF capacitor (C_1) is 4.0 V, and there is a voltage of 8.0 V across both the 2.0-μF and the 4.0-μF capacitors (C_2 and C_3). Using $U = CV^2/2$, we find that the energies stored in the three capacitors are

$$U_1 = \tfrac{1}{2}(1.2 \times 10^{-5}\text{ F})(4.0\text{ V})^2 = 9.6 \times 10^{-5}\text{ J},$$
$$U_2 = \tfrac{1}{2}(2.0 \times 10^{-6}\text{ F})(8.0\text{ V})^2 = 6.4 \times 10^{-5}\text{ J},$$
$$U_3 = \tfrac{1}{2}(4.0 \times 10^{-6}\text{ F})(8.0\text{ V})^2 = 1.3 \times 10^{-4}\text{ J},$$

and the total energy stored in the system is

$$U = U_1 + U_2 + U_3 = 2.9 \times 10^{-4}\text{ J}.$$

To check this, we can calculate the energy stored in the single 4.0-μF capacitor that was found to be equivalent to the three-capacitor combination. There is a voltage of 12.0 V across it, so its stored energy is

$$U = \tfrac{1}{2}(4.0 \times 10^{-6}\text{ F})(12.0\text{ V})^2 = 2.9 \times 10^{-4}\text{ J},$$

in agreement with the previous result.

DRILL PROBLEM 32-6

The potential difference across a 5.0-pF capacitor is 0.40 V. (a) What is the energy stored in this capacitor? (b) The potential difference is now increased to 1.20 V. By what factor is the stored energy increased?
ANS. (a) 4.0×10^{-13} J; (b) 9 times.

DRILL PROBLEM 32-7

(a) Determine the energy stored in the spherical capacitor of Fig. 32-3 by calculating the energy of its electric field. (b) Use Eq. (32-3) to show that the energy can be written as $U = CV^2/2$.

32-6 DIELECTRICS

The insulating material placed between the plates of a capacitor is called a *dielectric*. To see what effect the insertion of a dielectric has on the characteristics of a capacitor, we consider the experiment shown in Fig. 32-14. Initially, a capacitor C_0 whose plates are separated by air is charged by a battery of emf V_0. When the capacitor is fully charged, the battery is disconnected. A charge Q_0 then resides on the plates and the potential difference between the plates is measured to be V_0. (See Fig. 32-14a.) Now suppose we insert a dielectric that fills the gap between the plates. In so doing, we find that the voltmeter reading has dropped to a *smaller* value V, which we write as

$$V = \frac{1}{\kappa}V_0,$$

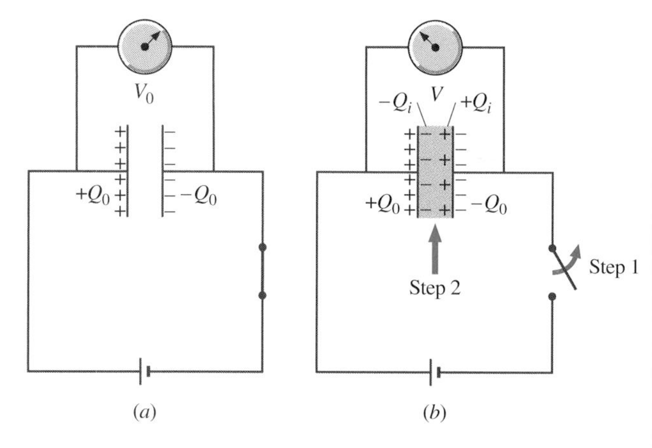

FIGURE 32-14 (*a*) When fully charged, the empty capacitor has a voltage V_0 and a charge Q_0. (*b*) The battery is disconnected (step 1), and then a dielectric is inserted into the charged capacitor (step 2). The voltage across the capacitor is found to decrease to $V = V_0/\kappa$. (The charges shown in the figure are discussed in Sec. 32-7.)

where the constant $\kappa > 1$. The value of κ depends on the material of the dielectric, and κ is known as the **dielectric constant** of the material. Once the battery is disconnected, there is no path for the charge to flow to or from the plates. Hence the insertion of the dielectric has no effect on the plate charge, which must remain at Q_0. From $C = Q/V$, we find that the capacitance with the dielectric in place is

$$C = \frac{Q_0}{V_0/\kappa} = \kappa\left(\frac{Q_0}{V_0}\right) = \kappa C_0. \tag{32-10}$$

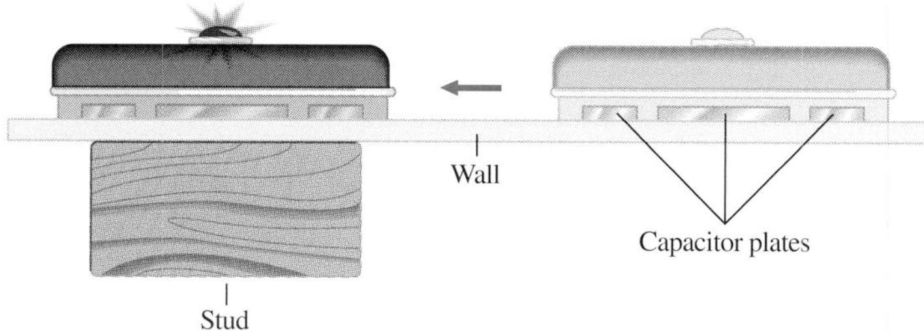

An electronic stud finder contains metal plates. The capacitance of its plates increases when it moves over a stud because of the dielectric constant of the wood.

Equation (32-10) was obtained for a situation in which a dielectric was inserted into an isolated capacitor. However, experiment shows that $C = \kappa C_0$ is valid in general; that is, *the capacitance of an empty capacitor is always increased by the factor κ when the space between the plates is completely filled by a material of dielectric constant κ.*

The electric field energy stored by a capacitor is also affected by the presence of a dielectric. For example, the energy of the empty capacitor of Fig. 32-14*a* is

$$U_0 = \frac{1}{2}\frac{Q_0^2}{C_0},$$

while the energy of the filled capacitor is

$$U = \frac{1}{2}\frac{Q^2}{C} = \frac{1}{2}\frac{Q_0^2}{\kappa C_0} = \frac{1}{\kappa}U_0.$$

As a dielectric is brought near an empty charged capacitor, the induced charges on the dielectric are attracted by the free charges on the plates. Consequently, the dielectric is "pulled" into the gap and work is done at the expense of the stored electric field energy, which is reduced. This is consistent with $U = (1/\kappa)U_0$, since $\kappa > 1$.

EXAMPLE 32-5 INSERTING A DIELECTRIC INTO AN ISOLATED CAPACITOR

An empty 20-pF capacitor is charged to a potential difference of 40 V. The charging battery is then disconnected and a piece of Teflon ($\kappa = 2.1$), which completely fills the gap between the plates, is inserted. (See Fig. 32-14.) What are the values of the (*a*) capacitance, (*b*) plate charge, (*c*) potential difference between the plates, and (*d*) stored energy with and without the dielectric?

SOLUTION (*a*) The original capacitance C_0 is 20 pF. With the dielectric, the capacitance increases to

$$C = (2.1)(20 \text{ pF}) = 42 \text{ pF}.$$

(*b*) Without the dielectric, the charge on the plates is

$$Q_0 = C_0V_0 = (20 \times 10^{-12} \text{ F})(40 \text{ V}) = 8.0 \times 10^{-10} \text{ C}.$$

Since the battery is disconnected before the dielectric is inserted, the plate charge is unaffected by the dielectric and remains at 8.0×10^{-10} C.

(*c*) The potential difference V_0 without the dielectric is 40 V. With the dielectric, the potential difference becomes

$$V = \frac{V_0}{\kappa} = \frac{40 \text{ V}}{2.1} = 19 \text{ V}.$$

(*d*) The stored energy without the dielectric is

$$U_0 = \tfrac{1}{2}C_0V_0^2 = \tfrac{1}{2}(20 \times 10^{-12} \text{ F})(40 \text{ V})^2 = 1.6 \times 10^{-8} \text{ J}.$$

With the dielectric inserted, the stored energy decreases to

$$U = \tfrac{1}{2}CV^2 = \tfrac{1}{2}(42 \times 10^{-12} \text{ F})(19 \text{ V})^2 = 7.6 \times 10^{-9} \text{ J}.$$

DRILL PROBLEM 32-8

When a dielectric is inserted into an isolated, charged capacitor, the stored energy decreases to 33 percent of its original value. (*a*) What is the dielectric constant? (*b*) How does the capacitance change?

ANS. (*a*) 3.0; (*b*) $C = 3.0C_0$.

32-7 Molecular Description of a Dielectric

The effect of a dielectric on capacitance may be understood by looking at its behavior on the molecular level. In general, molecules can be classified as either *polar* or *nonpolar*. There is a net separation of positive and negative charge in an isolated polar molecule (e.g., H_2O), while there is no charge separation in an isolated nonpolar molecule (e.g., O_2). Consequently, polar molecules have permanent electric dipole moments and nonpolar molecules do not.

Let's first consider a dielectric composed of polar molecules. In the absence of any external electric field, the electric dipoles are oriented randomly, as illustrated in Fig. 32-15*a*. However, if the dielectric is placed in an external field $\mathbf{E}_0$, the polar molecules align with the external field, as shown in Fig. 32-15*b*.* Opposite charges on adjacent dipoles within the dielectric neutralize each other so there is no net charge within the dielectric. (See the dashed circles of Fig. 32-15*b*.) However, this is not the case very close to the upper and lower surfaces (the regions enclosed by the dashed rectangles), where the alignment does produce a net charge. Since the electric field merely aligns the dipoles, the dielectric as a whole is neutral, and the surface charges induced on its opposite faces are equal and opposite. The *induced surface charges* Q_i and $-Q_i$ produce an additional electric field $\mathbf{E}_i$ within the dielectric, which *opposes* the external field, as illustrated in Fig. 32-15*c*.

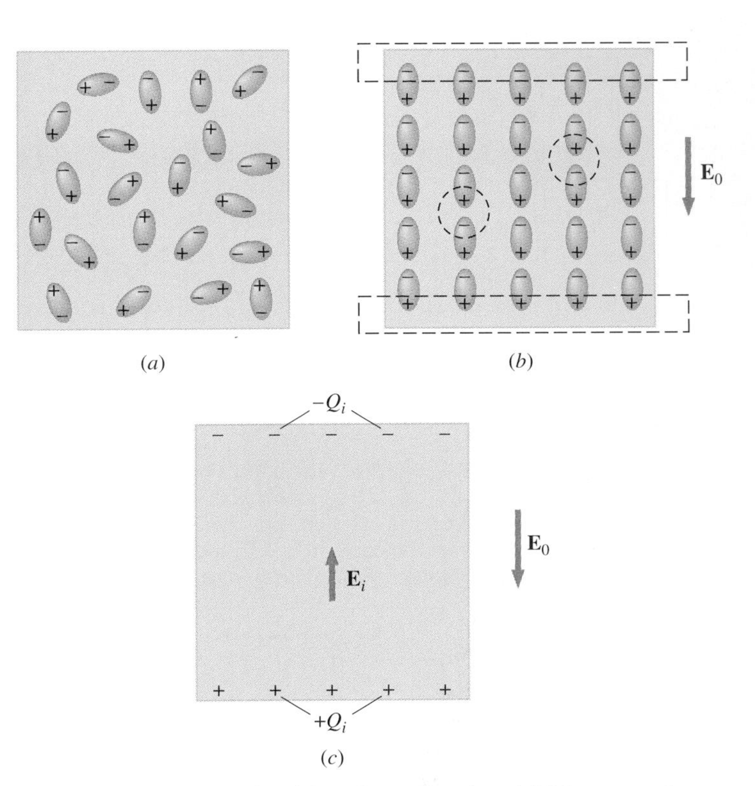

FIGURE 32-15 A dielectric with polar molecules. (*a*) No external field. (*b*) Application of an external field $\mathbf{E}_0$. (*c*) The induced field $\mathbf{E}_i$ produced by the surface charge Q_i of the dielectric.

The same effect is produced when the molecules are nonpolar. In this case, a molecule first acquires an *induced electric dipole moment* because the external field $\mathbf{E}_0$ causes a separation of its positive and negative charge. The induced dipoles of the nonpolar molecules align with $\mathbf{E}_0$ in the same way as the permanent dipoles of the polar molecules (shown in Fig. 32-15*b*). Hence the field within the dielectric is weakened regardless of whether its molecules are polar or nonpolar.

Now let's imagine filling the region between the parallel plates of a charged capacitor such as that of Fig. 32-16*a* with a dielectric. Then within the dielectric there is an electric field $\mathbf{E}_0$ due to the *free charge* Q_0 on the plates and an electric field $\mathbf{E}_i$ due to the induced charge Q_i on the surfaces of the dielectric. Their sum gives the net field within the dielectric:

$$\mathbf{E} = \mathbf{E}_0 + \mathbf{E}_i, \tag{32-11}$$

as shown in Fig. 32-16*b*. This net field can be considered to be the field produced by an *effective charge* $Q_0 - Q_i$ on the capacitor.

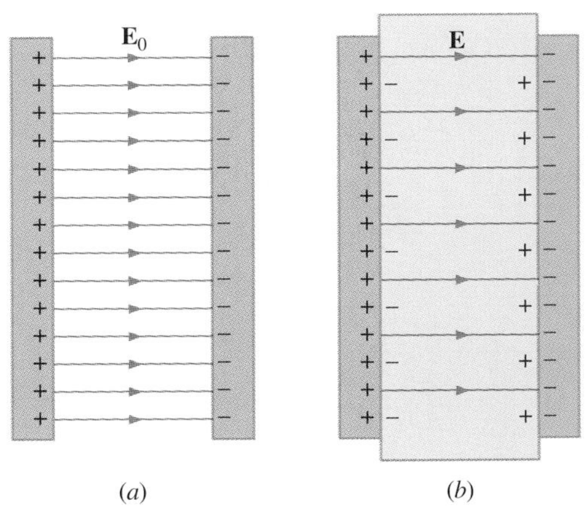

FIGURE 32-16 The electric field of (*a*) an empty capacitor and (*b*) a filled capacitor.

In most dielectrics, the net field $\mathbf{E}$ is proportional to the field $\mathbf{E}_0$ of the free charge. In terms of these electric fields, the dielectric constant of the material is defined as

$$\kappa = \frac{E_0}{E}. \tag{32-12}$$

Since $\mathbf{E}_0$ and $\mathbf{E}_i$ point in opposite directions, $E < E_0$, and $\kappa > 1$. Substitution of $\kappa = E_0/E$ into Eq. (32-11) yields

$$\mathbf{E}_i = \left(\frac{1}{\kappa} - 1\right)\mathbf{E}_0. \tag{32-13}$$

*In reality, the individual molecules are not aligned so perfectly with the field. At finite temperatures, they are constantly agitated by thermal energy. However, the *average* alignment is along the field.

Values of κ for a number of dielectrics are given in Table 32-1. Notice that κ is exactly 1.0 for a vacuum and very close to 1.0 for air under normal conditions—so close, in fact, that the properties of an air capacitor are essentially the same as those of a capacitor in a vacuum. The dielectric strengths of the materials are also given in Table 32-1. As explained for air in Example 25-16, the dielectric strength of an insulator represents the electric field at which the molecules of the material start to become ionized. When this happens, the material can conduct, thereby allowing charge to move through the dielectric from one capacitor plate to the other.

TABLE 32-1 **Dielectric Constants and Dielectric Strengths of Various Materials at Room Temperature**

Material	Dielectric Constant	Dielectric Strength (10^6 V/m)
Vacuum	1	∞
Dry air (1 atm)	1.00059	3
Glass	4–6	9
Teflon	2.1	60
Paper	3.7	16
Mica	6	150
Paraffin	2.3	11
Water	80	—

EXAMPLE 32-6 **ELECTRIC FIELD AND INDUCED SURFACE CHARGE**

Suppose that the distance between the plates of the capacitor of Example 32-5 (and Fig. 32-14) is 2.0×10^{-3} m and the area of each plate is 4.5×10^{-3} m^2. Determine (*a*) the electric field between the plates before and after the Teflon is inserted and (*b*) the surface charge induced on the Teflon.

SOLUTION (*a*) Since the voltage across the empty capacitor is $V_0 = 40$ V, the electric field between the plates is

$$E_0 = \frac{V_0}{d} = \frac{40\ \text{V}}{2.0 \times 10^{-3}\ \text{m}} = 2.0 \times 10^4\ \text{V/m}.$$

The electric field with the Teflon in place is, from Eq. (32-12),

$$E = \frac{E_0}{\kappa} = \frac{2.0 \times 10^4\ \text{V/m}}{2.1} = 9.5 \times 10^3\ \text{V/m}.$$

(*b*) The electric field of the filled capacitor is due to the effective charge on each side of the capacitor. As can be seen in Fig. 32-14*b*, this effective charge is the free charge on the plate minus the induced surface charge on the dielectric, or $Q_0 - Q_i$. Thus

$$E = \frac{Q_0 - Q_i}{\epsilon_0 A},$$

which yields

$$Q_i = Q_0 - \epsilon_0 AE = 8.0 \times 10^{-10}\ \text{C} - (8.85 \times 10^{-12}\ \text{C}^2/\text{N}\cdot\text{m}^2) \times (4.5 \times 10^{-3}\ \text{m}^2)(9.5 \times 10^3\ \text{V/m}) = 4.2 \times 10^{-10}\ \text{C}.$$

EXAMPLE 32-7 **INSERTING A DIELECTRIC INTO A CAPACITOR CONNECTED TO A BATTERY**

When a battery of emf V_0 is connected across an empty capacitor of capacitance C_0, the charge on the plates is Q_0. A dielectric of dielectric constant κ is inserted between the plates *while the battery remains in place*, as shown in Fig. 32-17. (*a*) What is the capacitance C, the voltage V across the capacitor, and the electric field E between the plates after the dielectric is inserted? (*b*) Obtain expressions for the free charge Q on the plates of the filled capacitor and the induced charge Q_i on the dielectric surface in terms of the original plate charge Q_0.

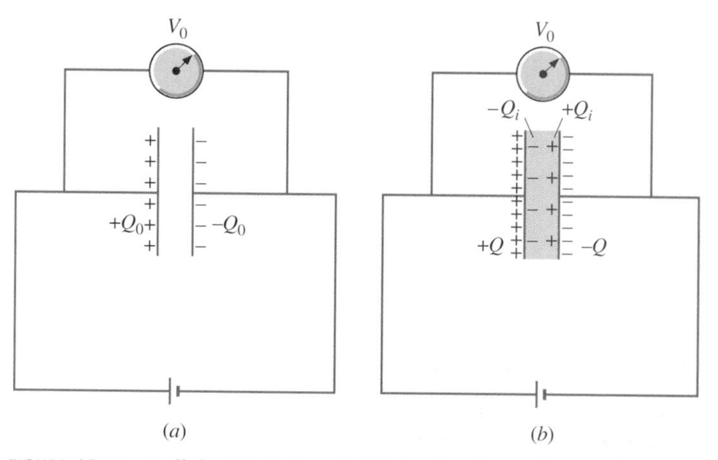

FIGURE 32-17 A dielectric is inserted into the charged capacitor while the battery remains connected.

SOLUTION (*a*) From Eq. (32-10), the capacitance of the filled capacitor is

$$C = \kappa C_0.$$

Since the battery is always connected to the capacitor plates, the potential difference between them remains at V_0 whether or not the dielectric is present; hence,

$$V = V_0.$$

The electric field of the filled capacitor is

$$E = \frac{V}{d} = \frac{V_0}{d} = E_0,$$

so the electric field has the same value before and after the dielectric is inserted.

(*b*) For the filled capacitor, the free charge on the plates is

$$Q = CV = (\kappa C_0)V_0 = \kappa(C_0 V_0),$$

where we have used $C = \kappa C_0$. Now the original plate charge was $Q_0 = C_0 V_0$, so

$$Q = \kappa Q_0.$$

The electric field E of the filled capacitor is due to the effective charge on each side of the capacitor, or $Q - Q_i$ (Fig. 32-17*b*).

Since this is the same as the field E_0 of the empty capacitor, we find

$$\frac{Q - Q_i}{\epsilon_0 A} = \frac{Q_0}{\epsilon_0 A},$$

so

$$Q_i = Q - Q_0.$$

Using $Q = \kappa Q_0$, we then obtain for the induced charge

$$Q_i = (\kappa - 1)Q_0.$$

DRILL PROBLEM 32-9

Using the results of Example 32-7, show that the energy U stored in the filled capacitor is related to the energy U_0 stored in the empty capacitor by $U = \kappa U_0$. The stored energy therefore increases with the dielectric in place if the voltage across the capacitor is kept constant. Compare this result with $U = U_0/\kappa$, which we found previously for an isolated charged capacitor.

DRILL PROBLEM 32-10

Repeat the calculations of Example 32-5 for the case in which the battery remains connected while the dielectric is placed in the capacitor.
ANS. (*a*) $C_0 = 20$ pF, $C = 42$ pF; (*b*) $Q_0 = 8.0 \times 10^{-10}$ C, $Q = 1.7 \times 10^{-9}$ C; (*c*) $V_0 = V = 40$ V; (*d*) $U_0 = 1.6 \times 10^{-8}$ J, $U = 3.4 \times 10^{-8}$ J.

32-8 SELF-INDUCTANCE

In Chap. 30, we discussed how an emf is induced in a circuit by a time-varying magnetic flux. In many of our calculations, this flux was due to an applied time-dependent magnetic field. Now the current flowing in a circuit produces its own magnetic field. Figure 32-18 shows some of the magnetic field lines due to the current in a circular loop of wire. If the current is constant, the magnetic flux through the loop is also constant. However, if the current I were to vary with time—say, immediately

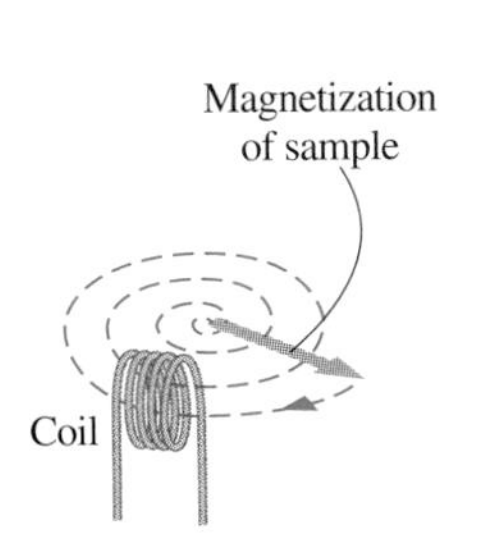

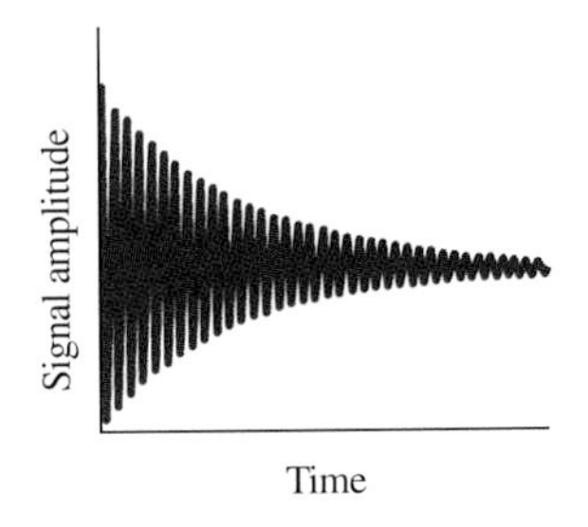

In magnetic resonance, signals are induced emf's in a receiver coil. A rotating magnetization vector, representing the motion of the magnetic moments of interest in the sample, is the source of a time-varying magnetic flux in the coil. This changing flux produces an induced emf.

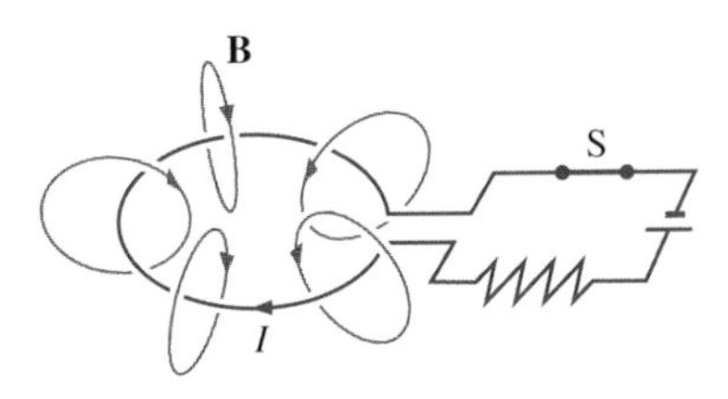

FIGURE 32-18 A magnetic field is produced by the current I in the loop. If I were to vary with time, the magnetic flux through the loop would also vary and an emf would be induced in the loop.

after switch S is closed—then the magnetic flux ϕ_m would correspondingly change, and Faraday's law tells us that an emf $\mathcal{E}$ would be induced in the circuit where

$$\mathcal{E} = -\frac{d\phi_m}{dt}.$$

Since the magnetic field due to a current-carrying wire is directly proportional to the current, the flux due to this field is also proportional to the current; that is,

$$\phi_m \propto I,$$

which we can also write as

$$\phi_m = LI,$$

where the constant of proportionality L is known as the **self-inductance** of the wire loop. If the loop has N turns, this equation becomes

$$N\phi_m = LI. \tag{32-14}$$

By convention, the positive sense of the normal to the loop is related to the current by the right-hand rule, so in Fig. 32-18 the normal points downward. With this convention, ϕ_m is positive in Eq. (32-14), so *L always has a positive value.*

For an N-turn loop, $\mathcal{E} = -Nd\phi_m/dt$, so the induced emf may be written in terms of the self-inductance as

$$\mathcal{E} = -L\frac{dI}{dt}. \tag{32-15}$$

In using this equation to determine L, it is easiest to ignore the signs of $\mathcal{E}$ and dI/dt, and calculate L as $|\mathcal{E}|/|dI/dt|$.

The SI unit for self-inductance is called the *henry* (H) in honor of Joseph Henry (1799–1878), an American scientist who discovered induced emf independent of Faraday. Notice from Eq. (32-15) that $1\ \text{H} = 1\ \text{V}\cdot\text{s/A}$.

Since self-inductance is associated with the magnetic field produced by a current, any configuration of conductors possesses self-inductance. For example, besides the wire loop, a long, straight wire has self-inductance, as does a coaxial cable.

A circuit element used to provide self-inductance is known as an *inductor*. It is represented by the symbol seen in Fig. 32-19, which resembles a coil of wire, the basic form of the

FIGURE 32-19 Symbol used to represent an inductor in a circuit.

FIGURE 32-20 Various inductors.

inductor. Figure 32-20 shows several types of inductors commonly used in circuits.

In accordance with Lenz's law, the negative sign in Eq. (32-15) indicates that the induced emf across an inductor always has a polarity that *opposes* the *change* in the current. For example, if the current flowing from A to B in Fig. 32-21*a* is increasing, the induced emf (represented by the imaginary battery) will have the polarity shown in order to oppose the increase. If the current from A to B is decreasing, then the induced emf will have the opposite polarity, again to oppose the change in current. (See Fig. 32-21*b*.) Finally, if the current though the inductor is constant, no emf is induced in the coil.

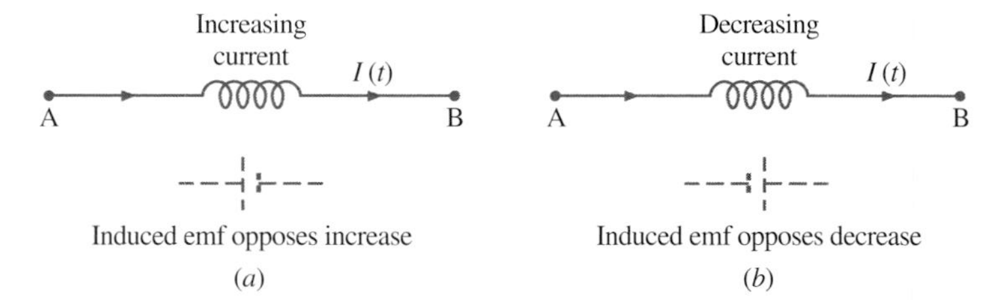

FIGURE 32-21 The induced emf across an inductor always acts to oppose the change in the current.

EXAMPLE 32-8 **SELF-INDUCTANCE OF A COIL**

An induced emf of 2.0 V is measured across a coil of 50 closely wound turns while the current through it increases uniformly from 0.0 to 5.0 A in 0.10 s. (*a*) What is the self-inductance of the coil? (*b*) With the current at 5.0 A, what is the flux through each turn of the coil?

SOLUTION (*a*) Ignoring the negative sign and using magnitudes, we have, from Eq. (32-15),

$$L = \frac{\mathcal{E}}{dI/dt} = \frac{2.0\text{ V}}{5.0\text{ A}/0.10\text{ s}} = 4.0 \times 10^{-2}\text{ H}.$$

(*b*) From Eq. (32-14), the flux is given in terms of the current by $\phi_m = LI/N$, so

$$\phi_m = \frac{(4.0 \times 10^{-2}\text{ H})(5.0\text{ A})}{50\text{ turns}} = 4.0 \times 10^{-3}\text{ Wb}.$$

DRILL PROBLEM 32-11

Current flows through the inductor of Fig. 32-21 from B to A. Is the current increasing or decreasing in order to produce the emf of Fig. 32-21*a*? of Fig. 32-21*b*?
ANS. (*a*) Decreasing; (*b*) increasing.

DRILL PROBLEM 32-12

A changing current induces an emf of 10 V across a 0.25-H inductor. What is the rate at which the current is changing?
ANS. 40 A/s.

32-9 CALCULATION OF SELF-INDUCTANCE

A good approach for calculating the self-inductance of an inductor consists of the following steps:

1. Assume that there is a current I flowing through the inductor.
2. Determine the magnetic field **B** produced by the current. If there is appropriate symmetry, you may be able to do this with Ampère's law.
3. Obtain the magnetic flux ϕ_m.
4. With the flux known, the self-inductance can be found from $L = N\phi_m/I$ (Eq. 32-14).

To demonstrate this procedure, we now calculate the self-inductances of two inductors.

CYLINDRICAL SOLENOID

Consider a long, cylindrical solenoid with length l, cross-sectional area A, and N turns of wire. We assume that the length of the solenoid is so much larger than its diameter that we can take the magnetic field to be $B = \mu_0 nI$ throughout the interior of the solenoid; that is, we ignore end effects in the solenoid. With a current I flowing through the coils, the magnetic field produced within the solenoid is

$$B = \mu_0\left(\frac{N}{l}\right)I,$$

so the magnetic flux through one turn is

$$\phi_m = BA = \frac{\mu_0 NA}{l}\,I.$$

Using Eq. (32-14), we find for the self-inductance of the solenoid

$$L = \frac{N\phi_m}{I} = \frac{\mu_0 N^2 A}{l}. \qquad (32\text{-}16a)$$

If $n = N/l$ is the number of turns per unit length of the solenoid, we may write Eq. (32-16a) as

$$L = \mu_0\left(\frac{N}{l}\right)^2 Al = \mu_0 n^2 Al = \mu_0 n^2(\text{vol}), \qquad (32\text{-}16b)$$

where (vol) $= Al$ is the volume of the solenoid. Notice that *the self-inductance of a long solenoid depends only on its physical properties* (such as the number of turns of wire per unit length and the volume), and not on the magnetic field or the current. *This is true for inductors in general.*

Rectangular Toroid

A toroid with a rectangular cross section is shown in Fig. 32-22. The inner and outer radii of the toroid are R_1 and R_2, and h is the height of the toroid. Applying Ampère's law in the same manner as we did in Example 29-8 for a toroid with a circular cross section, we find that the magnetic field inside a rectangular toroid is also given by

$$B = \frac{\mu_0 NI}{2\pi r},$$

where r is the distance from the central axis of the toroid. Because the field changes within the toroid, we must calculate the flux by integrating over the toroid's cross section. Using the infinitesimal cross-sectional area element $da = h\,dr$ shown in the figure, we obtain

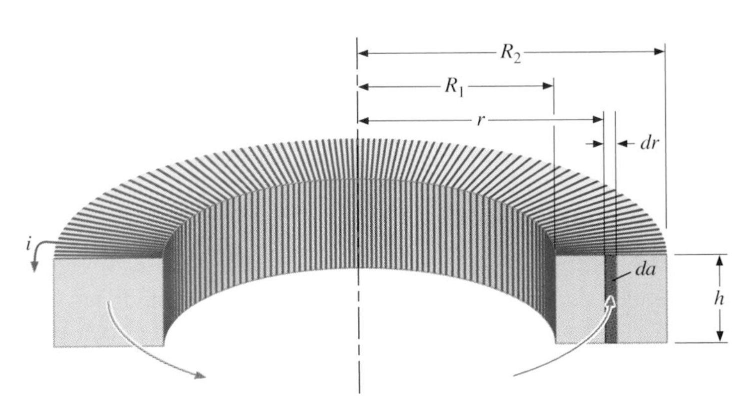

FIGURE 32-22 The rectangular toroid.

$$\phi_m = \int B\,da = \int_{R_1}^{R_2}\left(\frac{\mu_0 NI}{2\pi r}\right)(h\,dr) = \frac{\mu_0 NhI}{2\pi}\ln\frac{R_2}{R_1}.$$

Now from Eq. (32-14), we obtain for the self-inductance of a rectangular toroid

$$L = \frac{N\phi_m}{I} = \frac{\mu_0 N^2 h}{2\pi}\ln\frac{R_2}{R_1}. \qquad (32\text{-}17)$$

As expected, the self-inductance is a constant determined by only the physical properties of the toroid.

DRILL PROBLEM 32-13

(a) Calculate the self-inductance of a solenoid that is tightly wound with wire of diameter 0.10 cm, has a cross-sectional area of 0.90 cm^2, and is 40 cm long. (b) If the current through the solenoid decreases uniformly from 10 to 0 A in 0.10 s, what is the emf induced between the ends of the solenoid?
ANS. (a) 4.5×10^{-5} H; (b) 4.5×10^{-3} V.

DRILL PROBLEM 32-14

(a) What is the magnetic flux through one turn of a solenoid of self-inductance 8.0×10^{-5} H when a current of 3.0 A flows through it? Assume that the solenoid has 1000 turns and is wound from wire of diameter 1.0 mm. (b) What is the cross-sectional area of the solenoid?
ANS. (a) 2.4×10^{-7} Wb; (b) 6.4×10^{-5} m^2.

32-10 Energy Stored in an Inductor

The energy of a capacitor is stored in the electric field between its plates. Similarly, an inductor has the capability of storing energy—but in its magnetic field. In Sec. 31-3, we found that this energy can be found by integrating the magnetic energy density

$$u_m = \frac{B^2}{2\mu_0}$$

over the appropriate volume. Let's consider the long, cylindrical solenoid of the previous section. Again using the infinite-solenoid approximation, we can assume that the magnetic field is essentially constant and given by $B = \mu_0 nI$ everywhere inside the solenoid. Thus

$$U = u_m(\text{vol}) = \frac{(\mu_0 nI)^2}{2\mu_0}(Al) = \frac{1}{2}(\mu_0 n^2 Al)I^2.$$

With the substitution of Eq. (32-16b), this becomes

$$U = \tfrac{1}{2}LI^2. \qquad (32\text{-}18)$$

Although derived for a special case, this equation gives the energy stored in the magnetic field of *any* inductor. We can see this by considering an arbitrary inductor through which a changing current is passing. At any instant, the magnitude of the induced emf is $\mathcal{E} = L\, di/dt$, so the power absorbed by the inductor is

$$P = \mathcal{E}i = L\frac{di}{dt}i.$$

The total energy stored in the magnetic field when the current increases from 0 to I in a time interval from 0 to t can be determined by integrating this expression:

$$U = \int_0^t P\,dt' = \int_0^t L\frac{di}{dt'}i\,dt' = L\int_0^I i\,di = \frac{1}{2}LI^2,$$

which is Eq. (32-18).

EXAMPLE 32-9 Self-Inductance of a Coaxial Cable

Figure 32-23*a* shows two long, concentric cylindrical shells of radii R_1 and R_2. As discussed in Sec. 32-2, this configuration is a simplified representation of a coaxial cable. The capacitance per unit length of the cable has already been calculated. Now (*a*) determine the magnetic energy stored per unit length of the coaxial cable and (*b*) use this result to find the self-inductance per unit length of the cable.

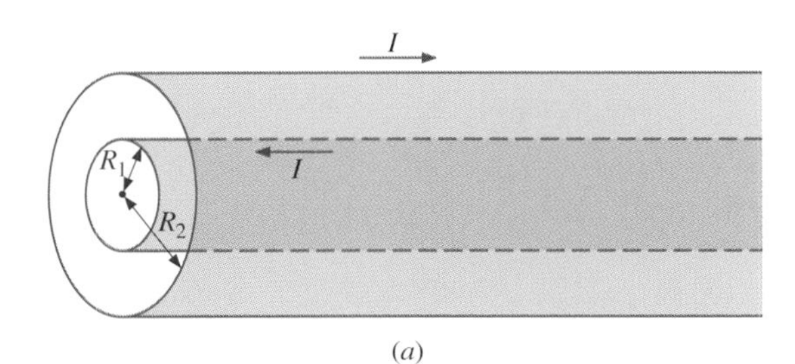

(*a*)

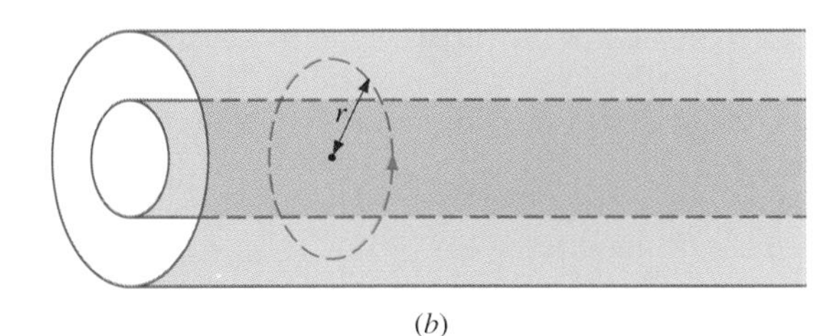

(*b*)

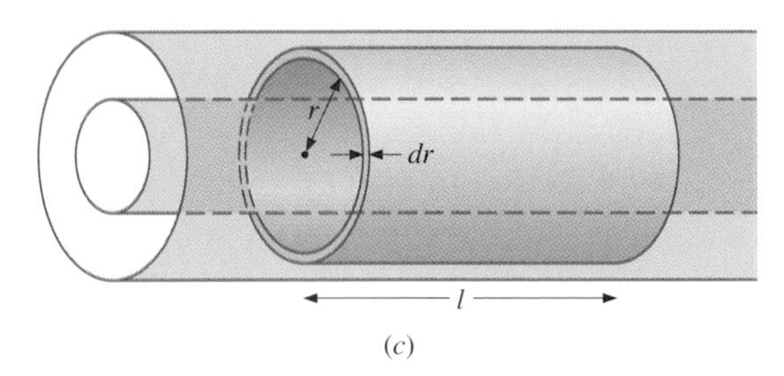

(*c*)

FIGURE 32-23 (*a*) A coaxial cable is represented here by two hollow, concentric cylindrical conductors along which electric current flows in opposite directions. (*b*) The magnetic field between the conductors can be found by applying Ampère's law to the dashed path. (*c*) The cylindrical shell is used to find the magnetic energy stored in a length l of the cable.

SOLUTION (*a*) The magnetic field between the conductors can be found by applying Ampère's law to the dashed circular path shown in Fig. 32-23*b*. Because of the cylindrical symmetry, B is constant along the path, and

$$\oint \mathbf{B}\cdot d\mathbf{l} = B(2\pi r) = \mu_0 I,$$

so

$$B = \frac{\mu_0 I}{2\pi r}.$$

In the region outside the cable, a similar application of Ampère's law shows that $B = 0$, since no net current crosses the area bounded by a circular path where $r > R_2$. This argument also holds when $r < R_1$, that is, in the region within the inner cylinder. All the magnetic energy of the cable is therefore stored between the two conductors. Since the energy density of the magnetic field is

$$u_m = \frac{B^2}{2\mu_0} = \frac{\mu_0 I^2}{8\pi^2 r^2},$$

the energy stored in a cylindrical shell of inner radius r, outer radius $r + dr$, and length l (see Fig. 32-23*c*) is

$$dU = u_m\,dv = u_m(2\pi r l)\,dr.$$

Thus the total energy of the magnetic field in a length l of the cable is

$$U = \int_{R_1}^{R_2} dU = \int_{R_1}^{R_2} \frac{\mu_0 I^2}{8\pi^2 r^2}(2\pi r l)\,dr = \frac{\mu_0 I^2 l}{4\pi}\ln\frac{R_2}{R_1}, \qquad (i)$$

and the energy per unit length is $(\mu_0 I^2/4\pi)\ln(R_2/R_1)$.

(*b*) From Eq. (32-18),

$$U = \tfrac{1}{2}LI^2, \qquad (ii)$$

where L is the self-inductance of a length l of the coaxial cable. Equating Eqs. (*i*) and (*ii*), we find that the self-inductance per unit length of the cable is

$$\frac{L}{l} = \frac{\mu_0}{2\pi}\ln\frac{R_2}{R_1}.$$

DRILL PROBLEM 32-15

How much energy is stored in the inductor of Example 32-8 after the current reaches its maximum value?
ANS. 0.50 J.

*32-11 Mutual Inductance

When two circuits carrying time-varying currents are close to one another, the magnetic flux through each circuit varies because of the changing current in the other circuit. Consequently, there is an emf induced in each circuit due to the changing current in the other. These emf's are therefore called *mutually induced emf's*. As an example, let's consider the two tightly

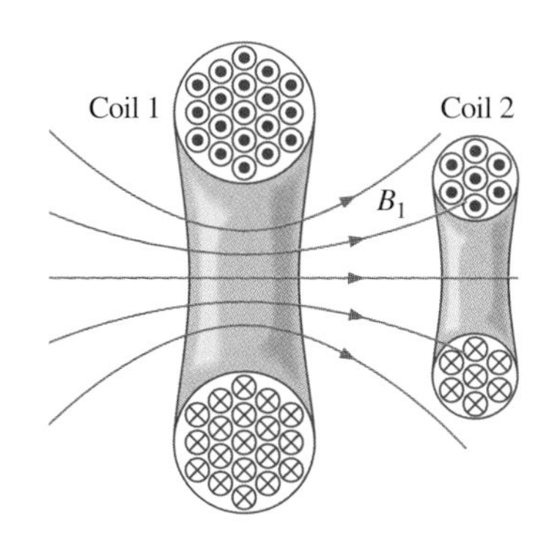

FIGURE 32-24 Some of the magnetic field lines produced by the current in coil 1 pass through coil 2.

wound coils of Fig. 32-24. Coils 1 and 2 have N_1 and N_2 turns and carry currents I_1 and I_2, respectively. The flux through a single turn of coil 2 produced by the magnetic field of the current in coil 1 is ϕ_{21}, while the flux through a single turn of coil 1 due to the magnetic field of I_2 is ϕ_{12}.

By definition, the **mutual inductance** M_{21} of coil 2 with respect to coil 1 is the ratio of the flux through the N_2 turns of coil 2 produced by the magnetic field of the current in coil 1 divided by that current; that is,

$$M_{21} = \frac{N_2\phi_{21}}{I_1}. \qquad (32\text{-}19a)$$

Similarly, the mutual inductance of coil 1 with respect to coil 2 is

$$M_{12} = \frac{N_1\phi_{12}}{I_2}. \qquad (32\text{-}19b)$$

The SI unit of mutual inductance is the henry, which is the same as that of self-inductance.

Like capacitance and self-inductance, mutual inductance is a geometric quantity. It depends on the shapes and relative position of the two coils, and it is independent of the currents in the coils. Now it can be shown that $M_{21} = M_{12}$, so we usually drop the subscripts associated with mutual inductance and write

$$M = \frac{N_2\phi_{21}}{I_1} = \frac{N_1\phi_{12}}{I_2}. \qquad (32\text{-}20)$$

The emf developed in either coil may be found by combining Faraday's law and the definition of mutual inductance. Since $N_2\phi_{21}$ is the total flux through coil 2 due to I_1, we have

$$\mathcal{E}_2 = -\frac{d}{dt}(N_2\phi_{21}) = -\frac{d}{dt}(MI_1) = -M\frac{dI_1}{dt}, \qquad (32\text{-}21a)$$

where we have used the fact that M is a time-independent constant. Similarly, we have

$$\mathcal{E}_1 = -M\frac{dI_2}{dt}. \qquad (32\text{-}21b)$$

You should be careful when using Eqs. (32-21), for $\mathcal{E}_1$ and $\mathcal{E}_2$ *do not necessarily represent the total emf's* in the respective coils. Each coil can also have an emf induced in it because of its self-inductance.

EXAMPLE 32-10 MUTUAL INDUCTANCE

Figure 32-25 shows a coil of N_2 turns and radius R_2 surrounding a long solenoid of length l_1, radius R_1, and N_1 turns. (*a*) What is the mutual inductance of the two coils? (*b*) If $N_1 = 500$ turns, $N_2 = 10$ turns, $R_1 = 3.10$ cm, $l_1 = 75.0$ cm, and the current in the solenoid is changing at a rate of 200 A/s, what is the emf induced in the surrounding coil?

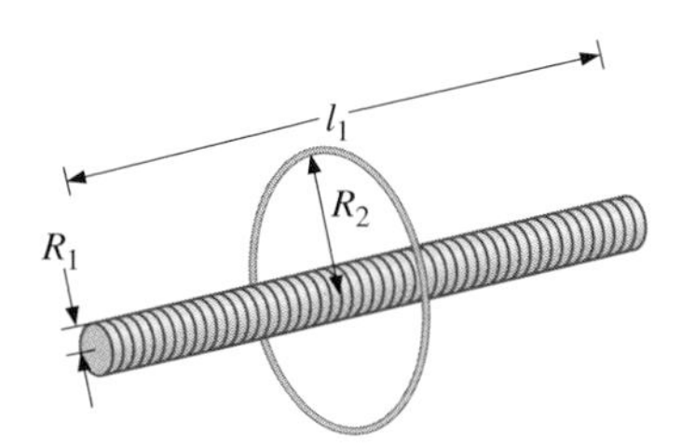

FIGURE 32-25 A solenoid surrounded by a coil.

SOLUTION (*a*) There is no magnetic field outside the solenoid, and the field inside has magnitude $B_1 = \mu_0(N_1/l_1)I_1$ and is directed parallel to the solenoid's axis. The magnetic flux ϕ_{21} through the surrounding coil is therefore

$$\phi_{21} = B_1\pi R_1^2 = \frac{\mu_0 N_1 I_1}{l_1}\pi R_1^2.$$

Now from Eq. (32-20), the mutual inductance is

$$M = \frac{N_2\phi_{21}}{I_1} = \left(\frac{N_2}{I_1}\right)\left(\frac{\mu_0 N_1 I_1}{l_1}\right)\pi R_1^2 = \frac{\mu_0 N_1 N_2 \pi R_1^2}{l_1}. \qquad (i)$$

Notice that M is independent of the radius R_2 of the surrounding coil. This is because the solenoid's magnetic field is confined to its interior. In principle, we can also calculate M by finding the magnetic flux through the solenoid produced by the current in the surrounding coil. This approach is much more difficult, because ϕ_{12} is so complicated. However, since $M_{12} = M_{21}$, we do know the result of this calculation.

(*b*) From Eq. (*i*) and the given values, the mutual inductance is

$$M = \frac{(4\pi \times 10^{-7}\ \text{T}\cdot\text{m/A})(500)(10)\pi(0.0310\ \text{m})^2}{0.750\ \text{m}}$$
$$= 2.53 \times 10^{-5}\ \text{H};$$

so from Eq. (32-21*a*), the emf induced in the surrounding coil is

$$\mathcal{E}_2 = -M\frac{dI_1}{dt} = -(2.53 \times 10^{-5}\ \text{H})(200\ \text{A/s})$$
$$= -5.06 \times 10^{-3}\ \text{V}.$$

DRILL PROBLEM 32-16

A current $I(t) = (5.0\ \text{A})\sin(120\pi\ \text{rad/s})t$ flows through the solenoid of part (*b*) of Example 32-10. What is the maximum emf induced in the surrounding coil?
ANS. 4.77×10^{-2} V.

SUMMARY

1. **Capacitance**
If the potential difference between an isolated pair of conductors is V when they carry equal and opposite charges $+Q$ and $-Q$, the capacitance of the pair is

$$C = \frac{Q}{V}.$$

2. **Calculation of capacitance**
The capacitance of a pair of conductors is calculated by assuming charges $\pm Q$ on them, determining the potential difference V between the conductors from the electric field of the charges, then dividing Q by V.

3. **Capacitors in parallel and in series**
(*a*) The net capacitance C_p of capacitors $C_1, C_2, \ldots, C_N$ connected in parallel is

$$C_p = C_1 + C_2 + \ldots + C_N.$$

(*b*) The net capacitance C_s of capacitors $C_1, C_2, \ldots, C_N$ connected in series is found from

$$\frac{1}{C_s} = \frac{1}{C_1} + \frac{1}{C_2} + \ldots + \frac{1}{C_N}.$$

4. **Energy stored in a capacitor**
The energy stored in the electric field of a capacitor is given by

$$U = \frac{1}{2}CV^2 = \frac{1}{2}QV = \frac{1}{2}\frac{Q^2}{C},$$

where C is the capacitance of the capacitor and Q and V are, respectively, the charge on and the voltage across its plates.

5. **Dielectrics and their molecular description**
(*a*) The capacitance of an empty capacitor is increased by κ when the space between its plates is completely filled by a dielectric with dielectric constant κ.
(*b*) When a dielectric is inserted between the plates of a capacitor, equal and opposite surface charge is induced on the two faces of the dielectric. The induced surface charge produces an electric field that opposes the field of the free charge on the capacitor plates.

6. **Self-inductance**
(*a*) If the magnetic flux through one turn of an N-turn coil is ϕ_m when it carries a current I, then the self-inductance of the coil is

$$L = \frac{N\phi_m}{I}.$$

(*b*) If the current through the coil is I, the emf induced in the coil is

$$\mathcal{E} = -L\frac{dI}{dt}.$$

7. **Calculation of self-inductance**
The self-inductance of an N-turn coil is calculated by assuming that a current I flows through the coil, determining the magnetic flux ϕ_m through one turn, then using $L = N\phi_m/I$.

8. **Energy stored in an inductor**
The energy stored in the magnetic field of an inductor is given by

$$U = \tfrac{1}{2}LI^2.$$

*9. **Mutual inductance**
(*a*) If ϕ_{21} is the magnetic flux through one turn of coil 2 due to the magnetic field of the current I_1 in coil 1, and ϕ_{12} is the magnetic flux through one turn of coil 1 due to the magnetic field of the current I_2 in coil 2, then the mutual inductance M of the two coils is

$$M = \frac{N_2\phi_{21}}{I_1} = \frac{N_1\phi_{12}}{I_2},$$

where N_1 and N_2 are the number of windings of coils 1 and 2, respectively.
(*b*) The emf's induced in the two coils are, respectively,

$$\mathcal{E}_1 = -M\frac{dI_2}{dt} \quad \text{and} \quad \mathcal{E}_2 = -M\frac{dI_1}{dt}.$$

QUESTIONS

32-1. Does capacitance depend on how much charge is on the plates? Does it depend on the voltage between the plates? Correlate your answers with the equation $C = Q/V$.

32-2. Would you place the plates of a parallel-plate capacitor closer together or farther apart to increase its capacitance?

32-3. The value of the capacitance is zero if the plates are not charged. True or false?

32-4. Two parallel-plate capacitors are identical except for the plate area. When they are connected in parallel across a battery, it is found that capacitor A has twice as much charge as capacitor B. Compare the plate areas of A and B.

32-5. Would capacitor A still have twice as much charge as B (see Ques. 32-4) if they were connected in series across the battery? If not, what would be the ratio of their charges?

32-6. Repeat Ques. 32-4 with the only difference between capacitor A and capacitor B being the plate separation rather than the area.

32-7. Is it feasible to apply the technique for calculating capacitance established in Sec. 32-2 to two concentric, oppositely charged cubes?

32-8. If the plates of a capacitor have different areas, will they acquire the same charge when the capacitor is connected across a battery?

32-9. Does the capacitance of a spherical capacitor depend on which sphere is charged positively or negatively?

32-10. The charged plates of a capacitor attract one another. In order to pull them farther apart, work is required. What becomes of the energy supplied by the work?

32-11. Will two identical capacitors store more energy if connected in series or in parallel?

32-12. Describe how you can measure ϵ_0 using a capacitor.

32-13. What would happen if a conducting slab rather than a dielectric were inserted into the gap between the capacitor plates?

32-14. Discuss how the energy stored in an empty, charged capacitor changes when a dielectric is inserted if (*a*) the capacitor is isolated so that Q doesn't change; (*b*) the capacitor remains connected to a battery so that V doesn't change.

32-15. Compare the rules for calculating the equivalent capacitances of capacitors in series and in parallel with the corresponding rules used for resistors.

32-16. Distinguish between dielectric strength and dielectric constant.

32-17. Water is a good solvent because it has a high dielectric constant. Explain.

32-18. Water has a high dielectric constant. Why then isn't it used as a dielectric material in capacitors?

32-19. Explain why a dielectric material experiences a force in a nonuniform electric field but not in a uniform field.

32-20. The dielectric constant of a substance containing permanent molecular electric dipoles decreases with increasing temperature. Explain.

32-21. Does self-inductance depend on the value of the magnetic flux? Does it depend on the current through the wire? Correlate your answers with the equation $N\phi_m = LI$.

32-22. Would the self-inductance of a 1.0-m-long tightly wound solenoid differ from the self-inductance per meter of an infinite, but otherwise identical, solenoid?

32-23. Discuss how you might determine the self-inductance per unit length of a long, straight wire.

32-24. The self-inductance of a coil is zero if there is no current passing through the windings. True or false?

32-25. Show that $LI^2/2$ has units of energy.

32-26. Show that $N\phi_m/I$ and $\mathcal{E}/(dI/dt)$, which are both expressions for self-inductance, have the same units.

32-27. A 10-H inductor carries a current of 20 A. Describe how a 50-V emf can be induced across it.

32-28. The ignition circuit of an automobile is powered by a 12-V battery. How are we able to generate large voltages with this power source?

32-29. When the current through a large inductor is interrupted with a switch, an arc appears across the open terminals of the switch. Explain.

32-30. How does the self-inductance per unit length near the center of a solenoid compare with its value near the end of the solenoid?

PROBLEMS

Note: Unless otherwise stated in the problems, assume that the capacitor gaps and inductor cores are air-filled.

Capacitance and Capacitors

32-1. The plates of a parallel-plate capacitor of capacitance 5.0 pF are 2.0 mm apart. What is the area of each plate?

32-2. How much charge flows through a 12-V battery when it is connected to a 50-μF capacitor?

32-3. A 60-pF capacitor has a plate area of 0.010 m^2. What is the plate separation?

32-4. The charge on a capacitor increases by 20 μC when the voltage across it increases from 30 to 40 V. What is its capacitance?

32-5. A set of parallel plates has a capacitance of 5.0 μF. How much charge must be added to the plates to increase the potential difference between them by 100 V?

32-6. How large must an isolated conducting sphere be in order to have a capacitance of 1.0 F?

32-7. Considering the earth to be a spherical conductor of radius 6400 km, calculate its capacitance.

32-8. If the inner sphere of a spherical capacitor has a radius of 0.50 cm, what must the radius of the outer sphere be to produce a capacitance of 1.5 pF?

32-9. If the capacitance per unit length of a cylindrical capacitor is 20-pF/m, what is the ratio of the radii of the two cylinders?

32-10. The inner conductor of a coaxial cable has a diameter of 1.0 mm, and the inner surface of the outer conductor has a diameter of 4.0 mm. What is the capacitance per unit length of the cable?

32-11. A parallel-plate capacitor has a capacitance of 20 μF. How much charge must leak off its plates before the voltage across them is reduced by 100 V?

32-12. The electric field between two parallel plates is 5.0×10^5 V/m. The area of each plate is 200 cm^2, and the plates are separated by 1.5 cm. What is the charge on each plate?

32-13. A capacitor is made from two flat parallel plates placed 0.40 mm apart. When a charge of 0.020 μC is placed on the plates, the potential difference between them is 250 V. (*a*) What is the capacitance of the plates? (*b*) What is the area of each plate? (*c*) What is the charge on the plates when the potential difference between them is 500 V? (*d*) What maximum potential difference can be applied between the plates without dielectric breakdown?

32-14. A parallel-plate capacitor is made from two square plates 25 cm on a side and 1.0 mm apart. The capacitor is connected to a 50-V battery. (*a*) What is the capacitance? (*b*) What is the charge on each plate? (*c*) What is the electric field between the plates? (*d*) With the battery still connected, the plates are pulled apart to a separation of 2.0 mm. Repeat the calculations of parts (*a*) through (*c*) for this case.

32-15. Answer part (*d*) of the previous problem if the battery is disconnected before the plates are separated to 2.0 mm.

32-16. A parallel-plate capacitor has a capacitance of 10 pF. (*a*) If the plates are 1.0 mm apart, what is the area of each plate? (*b*) If the charge on the plates is 2.0×10^{-3} μC, what is the potential difference across the plates?

32-17. Suppose that the capacitance of the variable capacitor of Fig. 32-6 can be manually changed from 100 to 800 pF by turning a dial, connected to one set of plates by a shaft, from 0 to 180°. With the dial set at 180° ($C = 800$ pF), the capacitor is connected to a 500-V source. After charging, the capacitor is disconnected from the source, and the dial is turned to 0°. (*a*) What is the charge on the capacitor? (*b*) What is the voltage across the capacitor when the dial is set at 0°?

32-18. What is the maximum charge that can be placed on a parallel-plate capacitor if each plate has an area of 5.2 cm^2?

Capacitors in Parallel and in Series

32-19. What is the maximum capacitance you can get by connecting three 1.0-μF capacitors? What is the minimum capacitance?

32-20. If the two combinations of the previous problem are connected to a 12-V battery, determine the charge on and voltage across each capacitor in each combination.

32-21. A 4.0-pF capacitor is connected in series with an 8.0-pF capacitor and a potential difference of 400 V is applied across the pair. (*a*) What is the charge on each capacitor? (*b*) What is the voltage across each capacitor?

32-22. Three capacitors with capacitances $C_1 = 2.0$ μF, $C_2 = 3.0$ μF, and $C_3 = 6.0$ μF are connected in series, and a potential difference of 500 V is applied across the combination. Determine the voltage across each capacitor and the charge on each capacitor.

32-23. The capacitors of the previous problem are connected in parallel, and the same 500 V is applied across the combination. Determine the voltage across each capacitor and the charge on each capacitor.

32-24. Find the equivalent capacitance between A and B for each of the sets of capacitors shown in the accompanying figure.

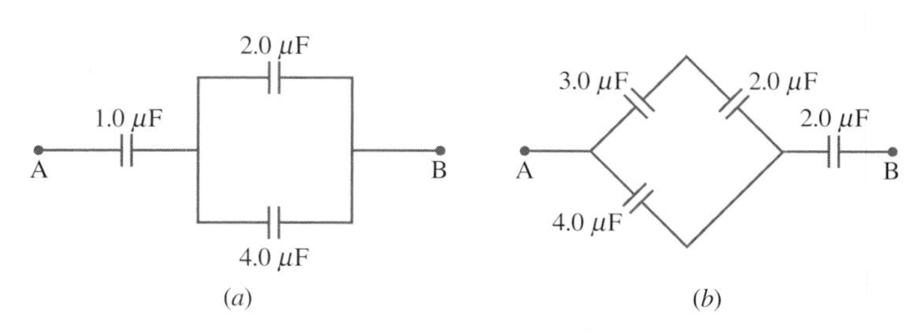

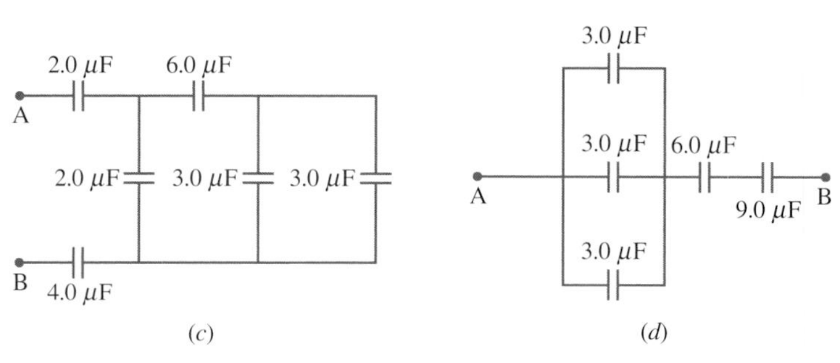

PROBLEM 32-24

32-25. The capacitors shown in the accompanying figure are all uncharged when 300 V is applied between A and B with the switch S open. (*a*) What is the potential difference $V_E - V_D$? (*b*) What is the potential at E after the switch is closed? (*c*) How much charge flows through the switch after it is closed?

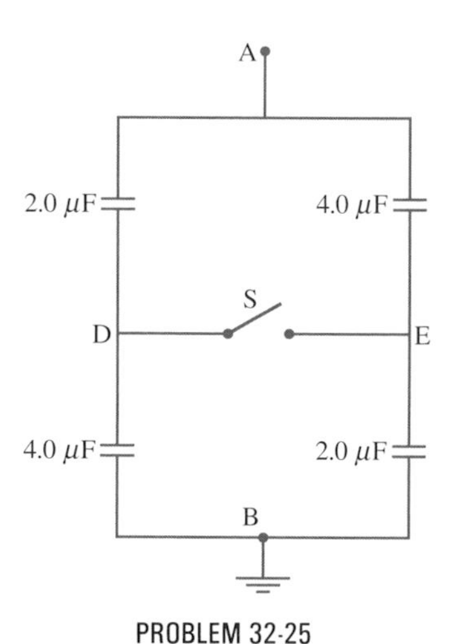

PROBLEM 32-25

32-26. How many 2.0-μF capacitors would you have to place in parallel if you wanted them to acquire a total charge of 1.0 C when connected to a 400-V source?

32-27. A 40-pF capacitor is charged to a potential difference of 500 V. Its terminals are then connected to those of an uncharged 10-pF capacitor. Calculate (*a*) the original charge on the 40-pF capacitor; (*b*) the charge on each capacitor after the connection is made; and (*c*) the potential difference across the plates of each capacitor after connection.

32-28. A capacitor C_1 carrying a charge Q is connected across an uncharged capacitor C_2. What is the resulting charge on and voltage across each capacitor?

32-29. A 2.0-μF capacitor and a 4.0-μF capacitor are connected in series across 1000 V. The charged capacitors are then disconnected from the source and connected to each other with terminals of like sign together. Find the charge on each capacitor and the voltage across each capacitor.

32-30. A 5.0-pF capacitor is charged to 100 V, and a 2.0-pF capacitor is charged to 200 V. (*a*) The positive plates of the two capacitors are connected together, and the negative plates are also connected together. What are the resulting charges on and voltages across the capacitors? (*b*) If plates of opposite sign are connected, what are the charges and voltages?

Energy Stored in a Capacitor

32-31. How much energy is stored in an 8.0-μF capacitor whose plates are at a potential difference of 6.0 V?

32-32. How much energy does a 2.0-pF capacitor with a charge of 6.0 pC store in its electric field?

32-33. A capacitor has a charge of 2.5 μC when connected to a 6.0-V battery. How much energy is stored in this capacitor?

32-34. A capacitor has square plates, 2.0×10^{-2} m to a side, which are separated by a distance of 1.0×10^{-3} m. If the energy stored in this capacitor is 1.0×10^{-10} J, what is the electric field between the plates?

32-35. How much energy is stored in the electric field of a metal sphere of radius 2.0 m that has a potential of 10 V?

32-36. How much energy is stored in the electric field of a metal sphere of radius 2.0 m that carries a charge of 3.0 μC?

32-37. What are the energies stored in the capacitor of Prob. 32-14 before and after the plates are pulled farther apart? Why does the energy decrease even though work is done in separating the plates?

32-38. What are the energies stored in the capacitor of Prob. 32-15 before and after the plates are pulled farther apart? Contrary to the previous problem, the work done here to separate the plates is equal to the change in stored energy. Explain why.

32-39. Consider the variable capacitor of Prob. 32-17. If friction is negligible, how much work is required to turn the dial from 180° to 0°?

32-40. Show that the plates of a parallel-plate capacitor attract each other with a force given by

$$F = \frac{q^2}{2\epsilon_0 A}.$$

Derive this by calculating the work necessary to increase the plate separation from x to $x + dx$ with the battery disconnected.

32-41. A parallel-plate capacitor with a capacitance of 5.0 μF is charged with a 12-V car battery, after which the battery is disconnected. Determine the minimum work required to increase the separation between the plates by a factor of 3.

32-42. Determine the decrease in energy when the capacitors of Prob. 32-27 are connected. Account for the lost energy.

32-43. (a) How much energy is stored in the electric fields of the capacitors shown in the accompanying figure? (b) Is this energy equal to the work done by the 400-V source in charging the capacitors?

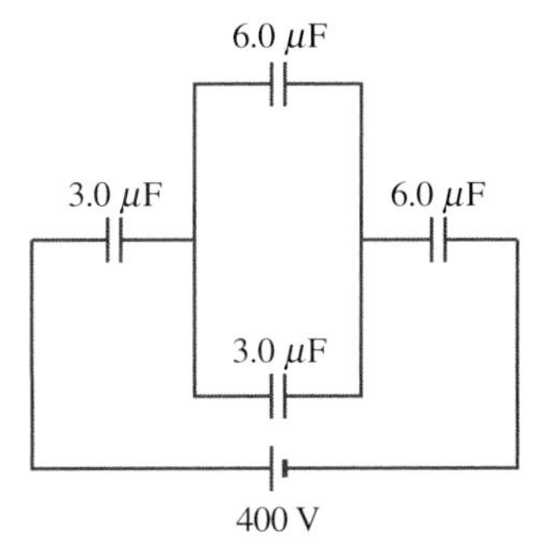

PROBLEM 32-43

Dielectrics and Their Molecular Description

32-44. What is the capacitance of two square plates 5.0 cm on a side that are separated by a 1.5-mm layer of paraffin?

32-45. An air capacitor is made from two flat parallel plates 1.0 mm apart. The inside area of each plate is 8.0 cm^2. (a) What is the capacitance of this set of plates? (b) If the region between the plates is filled with a material whose dielectric constant is 6.0, what is the new capacitance?

32-46. The electric field of an air-filled parallel-plate capacitor is 4.0×10^5 V/m. Without changing the charge on the plates, the air is replaced by an insulator of dielectric constant κ. This causes the electric field between the plates to drop to 1.5×10^5 V/m. (a) What is κ? (b) What is the induced charge density on the surfaces of the dielectric?

32-47. Two flat plates containing equal and opposite charges are separated by material 4.0 mm thick with a dielectric constant of 5.0. If the electric field in the dielectric is 1.5×10^6 V/m, what are (a) the charge density on the capacitor plates and (b) the induced charge density on the surfaces of the dielectric?

32-48. A parallel-plate capacitor with $A = 80$ cm^2 and $d = 1.0$ mm is given a charge of 2.0 μC. The electric field within the dielectric material filling the space between the plates is 8.0×10^6 V/m. (a) What is the dielectric constant of the material? (b) What is the charge induced on the dielectric surfaces?

32-49. For a Teflon-filled parallel-plate capacitor, $A = 50$ cm^2 and $d = 0.50$ mm. If the capacitor is connected to a 200-V battery, find (a) the free charge on the capacitor plates, (b) the electric field in the dielectric, and (c) the induced charge on the dielectric surfaces.

32-50. Two conducting plates, each of area 5.0 cm^2, are parallel to each other and 0.50 mm apart. The region between the plates is filled with a dielectric material with $\kappa = 12$. If the plates are connected across 400 V, find (a) the capacitance of the system, (b) the electric field in the dielectric, (c) the free charge on the plates, and (d) the induced charge on the dielectric surfaces.

32-51. A capacitor is made from two concentric spheres, one with radius 5.0 cm, the other with radius 8.0 cm. (a) What is the capacitance of this set of conductors if the region between them is filled with air? (b) If the region between the conductors is filled with a material whose dielectric constant is 6.0, what is the capacitance of the system?

Inductors and Self-Inductance

32-52. An emf of 0.40 V is induced across a coil when the current through it changes uniformly from 0.10 to 0.60 A in 0.30 s. What is the self-inductance of the coil?

32-53. The current shown in part (a) of the accompanying figure is increasing while that shown in part (b) is decreasing. In each case, determine which end of the inductor is at the higher potential.

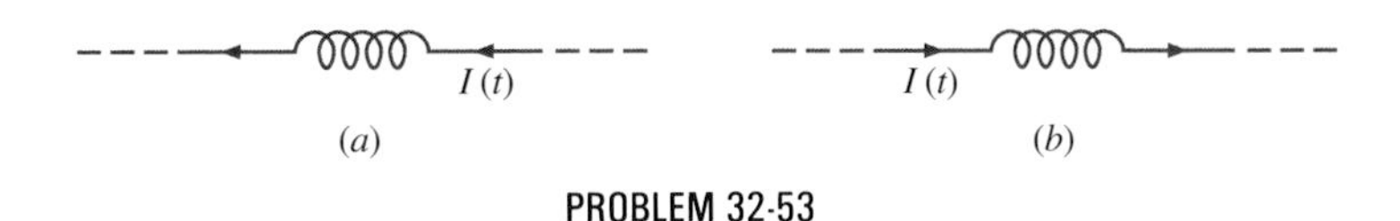

PROBLEM 32-53

32-54. What is the rate at which the current through a 0.30-H coil is changing if an emf of 0.12 V is induced across the coil?

32-55. A coil with a self-inductance of 2.0 H carries a current that varies with time according to $I(t) = (2.0 \text{ A}) \sin 120\,\pi t$. Find an expression for the emf induced in the coil.

32-56. A solenoid 50 cm long is wound with 500 turns of wire. The cross-sectional area of the coil is 2.0 cm^2. What is the self-inductance of the solenoid?

32-57. A coil with a self-inductance of 3.0 H carries a current that decreases at a uniform rate $dI/dt = -0.050$ A/s. What is the emf induced in the coil? Describe the polarity of the induced emf.

32-58. The current $I(t)$ through a 5.0-mH inductor varies with time as shown in the accompanying figure. The resistance of the inductor is 5.0 Ω. Calculate the voltage across the inductor at $t = 2.0$ ms, $t = 4.0$ ms, and $t = 8.0$ ms.

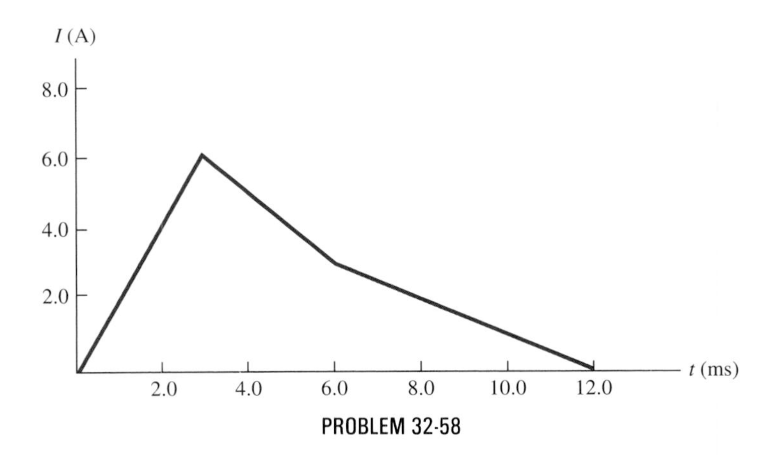

PROBLEM 32-58

32-59. A long, cylindrical solenoid with 100 turns per centimeter has a radius of 1.5 cm. (*a*) Neglecting end effects, what is the self-inductance per unit length of the solenoid? (*b*) If the current through the solenoid changes at the rate 5.0 A/s, what is the emf induced per unit length?

32-60. Suppose that the rectangular toroid of Fig. 32-22 has 2000 windings and a self-inductance of 0.040 H. If $h = 0.10$ m, what is the ratio of its outer radius to its inner radius?

32-61. What is the self-inductance per meter of a coaxial cable whose inner radius is 0.50 mm and whose outer radius is 4.00 mm?

Energy Stored in an Inductor

32-62. At the instant a current of 0.20 A is flowing through a coil of wire, the energy stored in its magnetic field is 6.0×10^{-3} J. What is the self-inductance of the coil?

32-63. What is the current flowing through the rectangular toroid of Prob. 32-60 when the energy in its magnetic field is 2.0×10^{-6} J?

32-64. Solenoid A is tightly wound while solenoid B has windings that are evenly spaced with a gap equal to the diameter of the wire. The solenoids are otherwise identical. Determine the ratio of the energies stored per unit length of these solenoids when the same current flows through each.

32-65. A 10-H inductor carries a current of 20 A. How much ice at 0°C could be melted by the energy stored in the magnetic field of the inductor?

32-66. A coil with a self-inductance of 3.0 H and a resistance of 100 Ω carries a steady current of 2.0 A. (*a*) What is the energy stored in the magnetic field of the coil? (*b*) What is the energy per second dissipated in the resistance of the coil?

32-67. A current of 1.2 A is flowing in a coaxial cable whose outer radius is five times its inner radius. What is the magnetic field energy stored in a 3.0-m length of the cable?

Mutual Inductance

32-68. When the current in one coil changes at a rate of 5.6 A/s, an emf of 6.3×10^{-3} V is induced in a second, nearby coil. What is the mutual inductance of the two coils?

32-69. An emf of 9.7×10^{-3} V is induced in a coil while the current in a nearby coil is decreasing at a rate of 2.7 A/s. What is the mutual inductance of the two coils?

32-70. Two coils close to each other have a mutual inductance of 32 mH. If the current in one coil decays according to $I = I_0 e^{-\alpha t}$, where $I_0 = 5.0$ A and $\alpha = 2.0 \times 10^3\ \text{s}^{-1}$, what is the emf induced in the second coil immediately after the current starts decaying? at $t = 1.0 \times 10^{-3}$ s?

32-71. A coil of 40 turns is wrapped around a long solenoid of cross-sectional area $7.5 \times 10^{-3}\ \text{m}^2$. The solenoid is 0.50 m long and has 500 turns. (*a*) What is the mutual inductance of this system? (*b*) The outer coil is replaced by a coil of 40 turns whose radius is three times that of the solenoid. What is the mutual inductance of this configuration?

32-72. A 600-turn solenoid is 0.55 m long and 4.2 cm in diameter. Inside the solenoid, a small (1.1 cm × 1.4 cm), single-turn rectangular coil is fixed in place with its face perpendicular to the long axis of the solenoid. What is the mutual inductance of this system?

32-73. A toroidal coil has a mean radius of 16 cm and a cross-sectional area of $0.25\ \text{cm}^2$; it is wound uniformly with 1000 turns. A second toroidal coil of 750 turns is wound uniformly over the first coil. Ignoring the variation of the magnetic field within a toroid, determine the mutual inductance of the two coils.

32-74. A solenoid of N_1 turns has length l_1 and radius R_1, and a second smaller solenoid of N_2 turns has length l_2 and radius R_2. The smaller solenoid is placed completely inside the larger solenoid so that their long axes coincide. What is the mutual inductance of the two solenoids?

General Problems

32-75. A metal plate of thickness t is held in place between two capacitor plates by plastic pegs, as shown in the accompanying figure. The effect of the pegs on the capacitance is negligible. The area of each capacitor plate and the area of the top and bottom surfaces of the inserted plate are all A. What is the capacitance of the system?

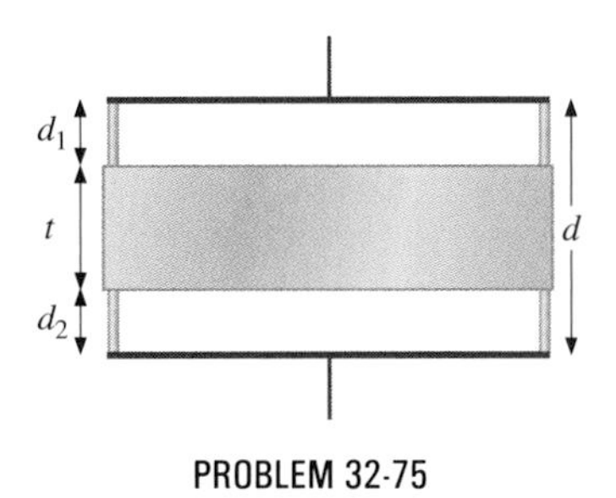

PROBLEM 32-75

32-76. A parallel-plate capacitor is filled with two dielectrics as shown in the accompanying figure. Show that the capacitance is given by

$$C = \frac{\epsilon_0 A}{d} \frac{\kappa_1 + \kappa_2}{2},$$

where A is the plate area and d is the separation of the plates.

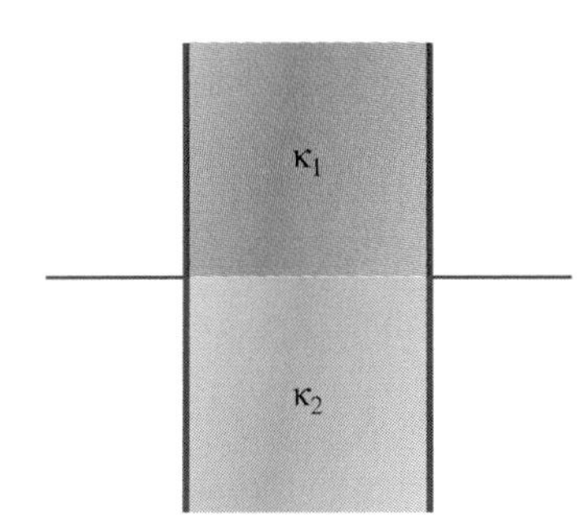

PROBLEM 32-76

32-77. A parallel-plate capacitor is filled with two dielectrics, as shown in the accompanying figure. Prove that the capacitance is given by

$$C = \frac{2\epsilon_0 A}{d} \frac{\kappa_1 \kappa_2}{\kappa_1 + \kappa_2}.$$

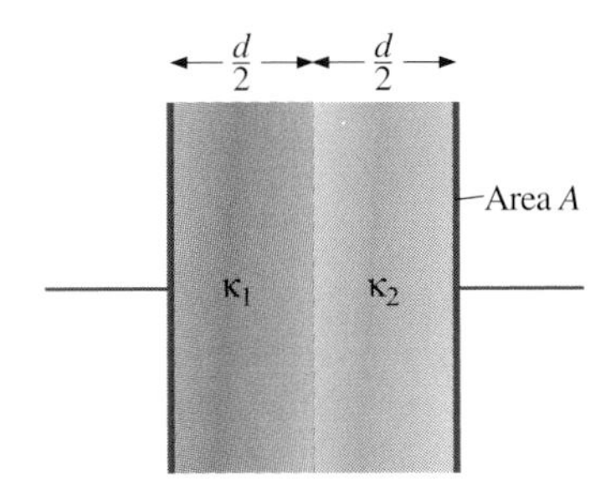

PROBLEM 32-77

32-78. The spherical capacitor of part (b) of Prob. 32-51 is connected to a 400-V source. Determine (a) the free charge on the spheres, (b) the electric field at $r = 7.0$ cm, and (c) the induced charge on the surfaces of the dielectric.

32-79. Show that the self-inductance per unit length of an infinite, straight, thin wire is infinite.

32-80. Two long, parallel wires carry equal currents in opposite directions. The radius of each wire is a, and the distance between the centers of the wires is d. Show that if the magnetic flux within the wires themselves can be ignored, the self-inductance of a length l of such a pair of wires is

$$L = \frac{\mu_0 l}{\pi} \ln \frac{d - a}{a}.$$

(*Hint:* Calculate the magnetic flux through a rectangle of length l between the wires and then use $L = N\phi/I$.)

·32-81. A small, rectangular single loop of wire with dimensions l and a is placed, as shown in the accompanying figure, in the plane of a much larger, rectangular single loop of wire. The two short sides of the larger loop are so far from the smaller loop that their magnetic fields over the smaller loop can be ignored. What is the mutual inductance of the two loops?

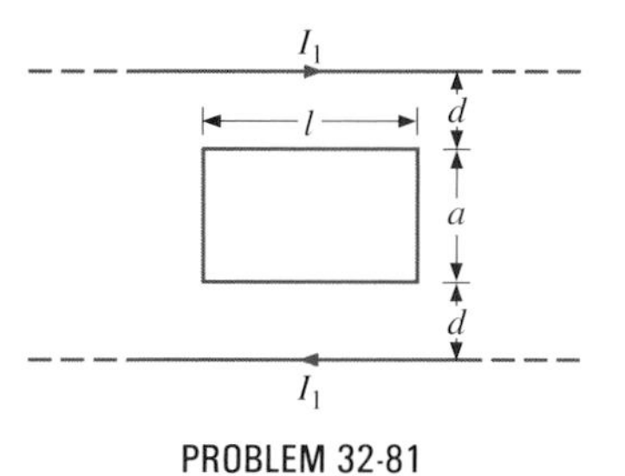

PROBLEM 32-81

32-82. Suppose that a cylindrical solenoid is wrapped around a core of iron whose magnetic susceptibility is χ. Using Eq. (29-18a), show that the self-inductance of the solenoid is given by

$$L = \frac{(1 + \chi)\mu_0 N^2 A}{l},$$

where l is its length, A its cross-sectional area, and N its total number of turns.

32-83. The solenoid of Prob. 32-56 is wrapped around an iron core whose magnetic susceptibility is 4.0×10^3. (a) If a current of 2.0 A flows through the solenoid, what is the magnetic field in the iron core? (b) What is the effective surface current formed by the aligned atomic current loops in the iron core? (c) What is the self-inductance of the filled solenoid?

32-84. A rectangular toroid with inner radius $R_1 = 7.0$ cm, outer radius $R_2 = 9.0$ cm, height $h = 3.0$ cm, and $N = 3000$ turns is filled with an iron core of magnetic susceptibility 5.2×10^3. (a) What is the self-inductance of the toroid? (b) If the current through the toroid is 2.0 A, what is the magnetic field at the center of the core? (c) For this same 2.0-A current, what is the effective surface current formed by the aligned atomic current loops in the iron core?

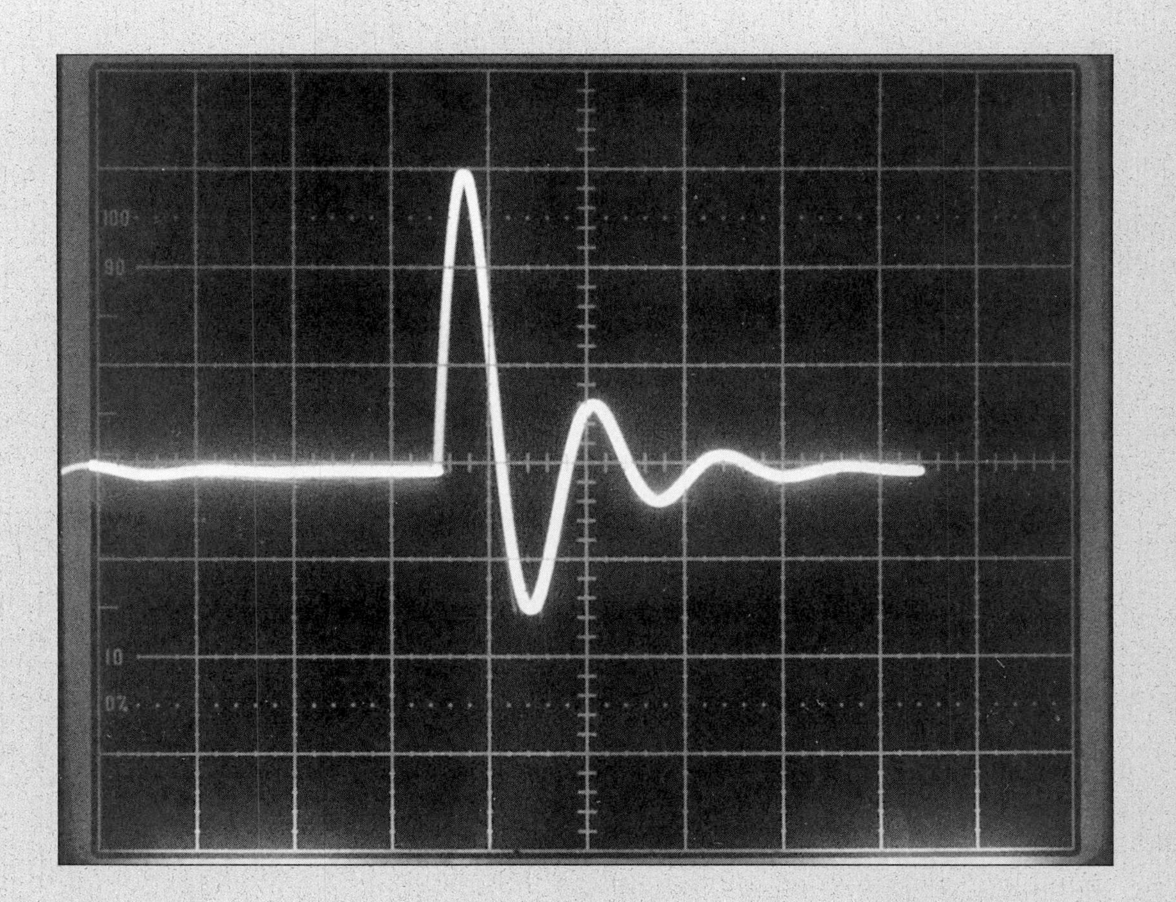

An oscilloscope pattern showing the decay in the oscillations of an RLC circuit.

CHAPTER 33 CAPACITORS AND INDUCTORS IN CIRCUITS

PREVIEW

In this chapter circuits containing capacitive and inductive elements are analyzed. The main topics to be discussed here are the following:

1. **RC and RL circuits.** With the help of Kirchhoff's rules, we investigate single-loop circuits containing a resistor and a capacitor and a resistor and an inductor. We calculate the currents through the elements, along with the voltages across the elements.

*2. **Freely oscillating circuits.** The LC circuit, both with and without resistance, is analyzed. An analogy to the mass-spring system, both with and without friction, is given.

In the previous chapter we studied in some detail the properties of capacitors and inductors. We now investigate some simple electric circuits containing these devices:

1. A capacitor connected to a resistor and a constant source of emf
2. An inductor connected to a resistor and a constant source of emf
3. Circuits 1 and 2 without the source of emf
4. A charged capacitor connected to an inductor, both with and without resistance in the circuit

The effects of a capacitor and an inductor on the electrical properties of a circuit are very similar. For example, circuits containing either a capacitor or an inductor carry time-varying rather than constant currents. In addition, both a capacitor and an inductor can produce electric currents due to the energy they store, either in an electric field (capacitor) or in a magnetic field (inductor).

The mathematical analysis of a circuit with capacitance is essentially the same as that of a circuit with inductance. We emphasize this by following a nearly identical sequence of steps when discussing each type of circuit. By comparing Sec. 33-1 ("RC Circuit") with Sec. 33-2 ("RL Circuit"), you will discover just how similar these two types of circuits are.

33-1 RC Circuit

A circuit with resistance and capacitance is known as an *RC circuit*. An RC circuit consisting of a resistor, a capacitor, a constant source of emf, and switches S_1 and S_2 is shown in Fig. 33-1*a*. When S_1 is closed, the circuit is equivalent to a single-loop circuit consisting of a resistor and a capacitor connected across a source of emf (Fig. 33-1*b*). When S_1 is opened and S_2 is closed, the circuit becomes a single-loop circuit with only a resistor and a capacitor (Fig. 33-1*c*).

Resistor and Capacitor across a Constant Source of emf

We begin with the RC circuit of Fig. 33-1*b*. Once S_1 is closed with S_2 open, the source of emf produces a current $I(t)$, and charge $q(t)$ starts to accumulate on the capacitor plates. Applying Kirchhoff's loop rule, we obtain

$$\mathcal{E} - IR - \frac{q}{C} = 0. \qquad (33\text{-}1a)$$

At $t = 0$, the capacitor is uncharged $[q(0) = 0]$, so there is no initial voltage across it. Thus from Eq. (33-1*a*), $I(0) = \mathcal{E}/R$. If there were no capacitance in the circuit, the current would remain at this value. However, the charge (and voltage) buildup in the capacitor causes the current to decay. Since $I = dq/dt$, Eq. (33-1*a*) can be written as

$$\mathcal{E} - R\frac{dq}{dt} - \frac{q}{C} = 0. \qquad (33\text{-}1b)$$

This is a first-order differential equation for $q(t)$. It is shown in Supplement 33-1 that with the initial condition $q(0) = 0$, the solution to Eq. (33-1*b*) is

$$q(t) = C\mathcal{E}(1 - e^{-t/RC}) = C\mathcal{E}(1 - e^{-t/\tau_C}), \qquad (33\text{-}2)$$

where

$$\tau_C = RC \qquad (33\text{-}3)$$

is the **capacitive time constant** of the circuit.

From Eq. (33-2), the charge $q(t)$ on the capacitor plates starts at zero, and when $t \to \infty$, it asymptotically approaches $C\mathcal{E}$. However, the quantity usually measured in the laboratory is the voltage $V_C(t)$ across the capacitor. Since $V_C(t) = q(t)/C$, we have from Eq. (33-2)

$$V_C(t) = \mathcal{E}(1 - e^{-t/\tau_C}). \qquad (33\text{-}4)$$

This voltage, which is plotted in Fig. 33-2*a*, starts at zero and asymptotically approaches $\mathcal{E}$. Notice from Fig. 33-1*b* that the voltage V_C across the capacitor and the emf $\mathcal{E}$ of the source oppose each other. Hence *as the capacitor charge approaches its maximum value (and V_C becomes equal to $\mathcal{E}$), the current decreases to zero.*

The energy stored in the electric field of a capacitor is

$$U_C = \tfrac{1}{2}CV_C^2;$$

so, as the capacitor becomes charged, its stored energy increases from zero and asymptotically approaches a maximum of $C\mathcal{E}^2/2$.

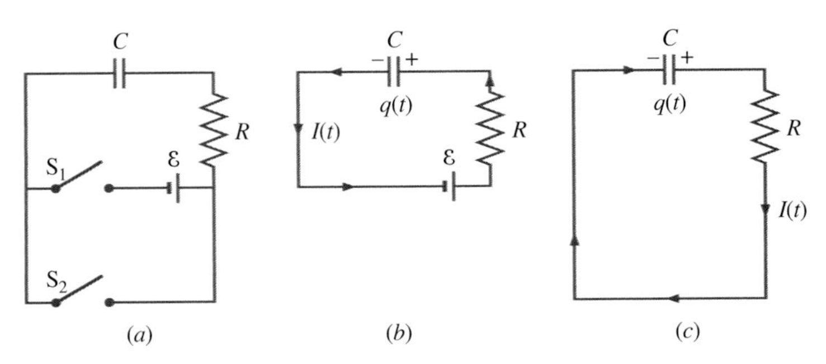

FIGURE 33-1 (*a*) An RC circuit with switches S_1 and S_2. (*b*) The equivalent circuit with S_1 closed and S_2 open. (*c*) The equivalent circuit after S_1 is opened and S_2 is closed.

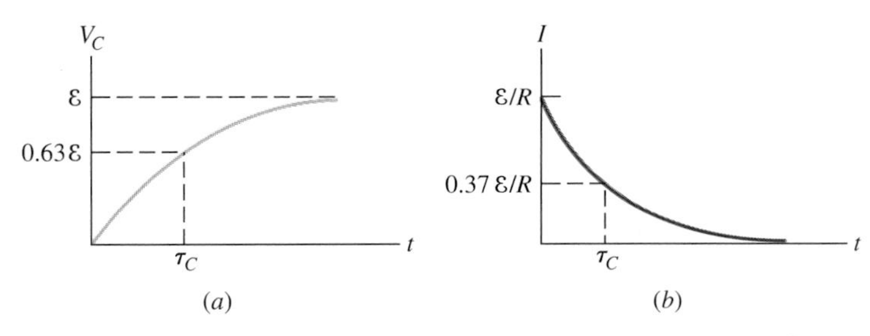

FIGURE 33-2 Time variation of (*a*) the voltage across the capacitor plate and (*b*) the electric current in the charging RC circuit of Fig. 33-1*b*.

The time constant τ_C tells us how rapidly the capacitor becomes charged. At $t = \tau_C$, the charge on the capacitor is, from Eq. (33-2),

$$q(\tau_C) = C\mathcal{E}(1 - e^{-1}) = 0.63C\mathcal{E},$$

which is 63 percent of the final value $C\mathcal{E}$. At this same instant, the voltage across the capacitor plates is also 63 percent of its final value $\mathcal{E}$. Notice that the smaller the capacitive time constant $\tau_C = RC$, the more rapidly the capacitor becomes charged.

We can find the time dependence of the current in the circuit of Fig. 33-1*b* using $I(t) = dq/dt$. From Eq. (33-2), we obtain

$$I(t) = \frac{\mathcal{E}}{R} e^{-t/\tau_C}. \tag{33-5}$$

This is plotted in Fig. 33-2*b*. At $t = 0$, there is no voltage across the capacitor, so $I(0) = \mathcal{E}/R$. The current decreases with time as the capacitor voltage increases, since this voltage opposes the emf of the source. Eventually, the current decreases to zero as the capacitor voltage approaches the emf of the source.

The time constant τ_C also tells us how quickly the current decays. At $t = \tau_C$,

$$I(\tau_C) = \frac{\mathcal{E}}{R} e^{-1} = 0.37 \frac{\mathcal{E}}{R} = 0.37I(0).$$

The current therefore decreases to about 37 percent of its initial value after one time constant. The shorter the time constant τ_C, the more rapidly the current decays.

Resistor and Capacitor Only

After enough time has elapsed so that the capacitor is essentially fully charged, the positions of the switches of Fig. 33-1*a* are reversed, giving us the circuit of Fig. 33-1*c*. At $t = 0$, the charge on the capacitor is $q(0) = C\mathcal{E}$. With Kirchhoff's loop equation, we obtain

$$\frac{q}{C} - IR = 0,$$

or

$$\frac{q}{C} + R\frac{dq}{dt} = 0. \tag{33-6}$$

Note that we have written $I = -dq/dt$ since the capacitor is discharging (that is, $dq/dt < 0$). The solution to Eq. (33-6) is also given in Supplement 33-1. It is, with the initial condition $q(0) = C\mathcal{E}$,

$$q(t) = C\mathcal{E}e^{-t/\tau_C}. \tag{33-7}$$

The capacitor therefore discharges exponentially with time. Since the voltage across the capacitor plates is $V_C(t) = q(t)/C$,

$$V_C(t) = \mathcal{E}e^{-t/\tau_C}, \tag{33-8}$$

which is also an exponential decay with time, as shown in Fig. 33-3.

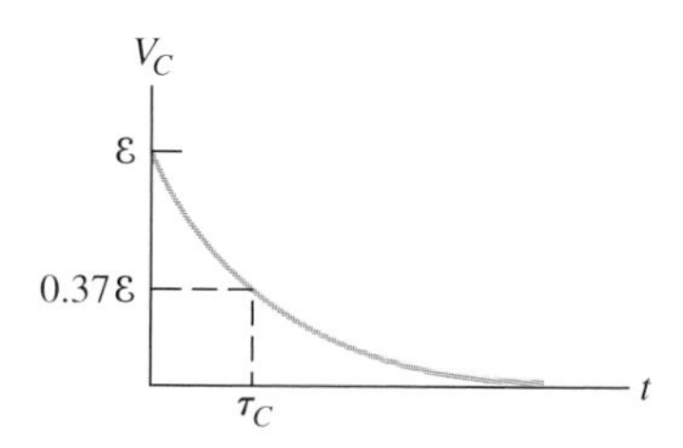

FIGURE 33-3 Time variation of the voltage across the capacitor plates in the discharging RC circuit of Fig. 33-1*c*. The electric current also decays exponentially.

The discharging capacitor produces a current in the circuit that is given by $I(t) = -dq/dt$:

$$I(t) = \frac{\mathcal{E}}{R} e^{-t/\tau_C}. \tag{33-9}$$

The current is initially $I(0) = \mathcal{E}/R$ and eventually decays to zero. During the same time, the capacitor is discharging according to Eq. (33-7). The energy, $C\mathcal{E}^2/2$, which was initially stored in its electric field, is dissipated by Joule heating in the resistance of the circuit.

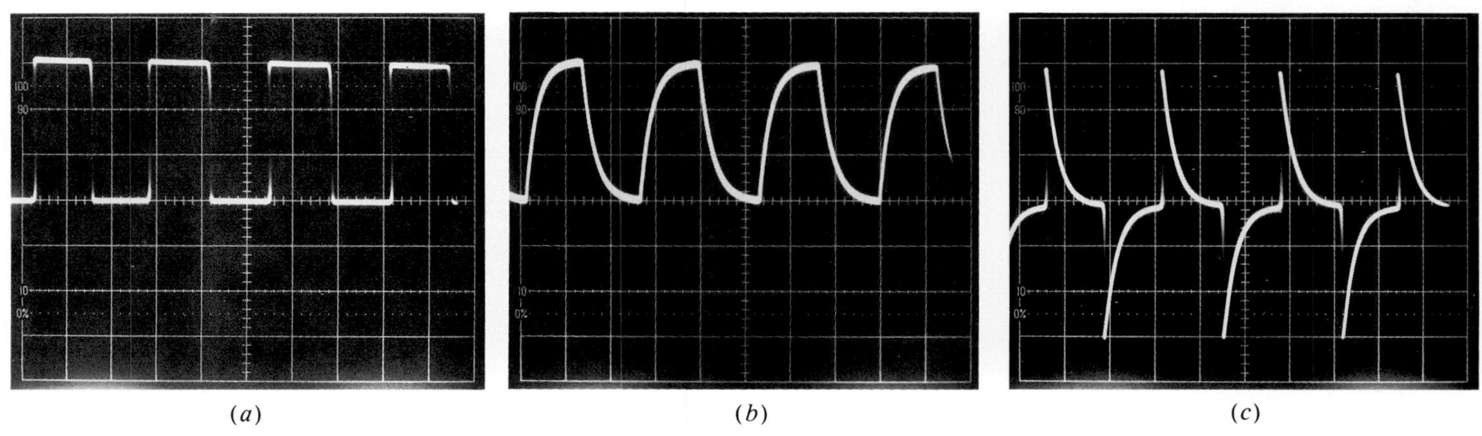

(*a*) (*b*) (*c*)

A generator in an RC circuit produces a square-pulse output in which the voltage oscillates between zero and some set value. Oscilloscope traces of: (*a*) the voltage output of the generator; (*b*) the voltage across the capacitor; (*c*) the voltage across the resistor.

EXAMPLE 33-1 A Charging Capacitor

In the circuit of Fig. 33-1*a*, $\mathcal{E} = 12$ V, $R = 60\ \Omega$, and $C = 2.0\ \mu$F. With S_1 closed and S_2 open (Fig. 33-1*b*), (*a*) what is the capacitive time constant of the circuit? (*b*) What are the charge on the capacitor plates and the current at $t = 0$, at $t = 1.5\tau_C$, and as $t \to \infty$?

SOLUTION (*a*) The capacitive time constant is

$$\tau_C = RC = (60\ \Omega)(2.0 \times 10^{-6}\ \text{F}) = 1.2 \times 10^{-4}\ \text{s}.$$

(*b*) The charge on the capacitor increases according to Eq. (33-2):

$$q(t) = C\mathcal{E}(1 - e^{-t/\tau_C}).$$

At $t = 0$, $(1 - e^{-t/\tau_C}) = (1 - 1) = 0$; so $q(0) = 0$, which is in agreement with the assumption that the capacitor is initially uncharged. At $t = 1.5\tau_C$ and as $t \to \infty$, we have, respectively,

$$q(1.5\tau_C) = C\mathcal{E}(1 - e^{-1.5}) = (2.0 \times 10^{-6}\ \text{F})(12\ \text{V})(0.78)$$
$$= 1.9 \times 10^{-5}\ \text{C},$$

and

$$q(\infty) = C\mathcal{E} = 2.4 \times 10^{-5}\ \text{C}.$$

The current in the circuit decays according to Eq. (33-5):

$$I(t) = \frac{\mathcal{E}}{R} e^{-t/\tau_C}.$$

At $t = 0$, $t = 1.5\tau_C$, and as $t \to \infty$, we obtain

$$I(0) = \frac{\mathcal{E}}{R} = \frac{12\ \text{V}}{60\ \Omega} = 0.20\ \text{A},$$
$$I(1.5\tau_C) = (0.20\ \text{A})e^{-1.5} = 4.5 \times 10^{-2}\ \text{A},$$

and

$$I(\infty) = 0.$$

EXAMPLE 33-2 A Discharging Capacitor

After the capacitor of Example 33-1 is fully charged, the positions of the switches are reversed so that the source of emf is removed from the RC circuit. (See Fig. 33-1*c*.) (*a*) How long does it take the capacitor charge to drop to half its initial value? (*b*) How long does it take before the energy stored in the capacitor is reduced to 1.0 percent of its maximum value?

SOLUTION (*a*) With the switches reversed, the charge on the capacitor decreases according to

$$q(t) = C\mathcal{E}e^{-t/\tau_C} = q(0)e^{-t/\tau_C}.$$

At a time t when the charge is one-half its initial value, we have

$$q(t) = 0.50q(0),$$

so

$$e^{-t/\tau_C} = 0.50,$$

and

$$t = -[\ln(0.50)]\tau_C = 0.69(1.2 \times 10^{-4}\ \text{s}) = 8.3 \times 10^{-5}\ \text{s},$$

where we have used the capacitive time constant found in Example 33-1.

(*b*) The energy stored in the capacitor is given by

$$U_C(t) = \frac{1}{2}C[V_C(t)]^2 = \frac{1}{2}C(\mathcal{E}e^{-t/\tau_C})^2 = \frac{C\mathcal{E}^2}{2}e^{-2t/\tau_C}.$$

If the energy drops to 1.0 percent of its initial value at time t, we have

$$U_C(t) = (0.010)U_C(0),$$

or

$$\frac{C\mathcal{E}^2}{2}e^{-2t/\tau_C} = (0.010)\frac{C\mathcal{E}^2}{2}.$$

Upon canceling terms and taking the natural logarithm of both sides, we obtain

$$-\frac{2t}{\tau_C} = \ln(0.010),$$

so

$$t = -\tfrac{1}{2}\tau_C \ln(0.010).$$

Since $\tau_C = 1.2 \times 10^{-4}$ s, the time it takes for the capacitor energy to decrease to 1.0 percent of its initial value is

$$t = -\tfrac{1}{2}(1.2 \times 10^{-4}\ \text{s})\ln(0.010) = 2.8 \times 10^{-4}\ \text{s}.$$

EXAMPLE 33-3 Energy in an RC Circuit

Consider the RC circuit of Fig. 33-1*b*. Determine the total energy produced by the battery and the total energy dissipated in the resistor during the entire process of charging the capacitor. Show that the difference in these two terms is equal to $CV_C^2/2$, the energy stored in the electric field of the capacitor.

SOLUTION The rate at which the battery produces energy is $\mathcal{E}I(t)$, where $I(t)$ is given by Eq. (33-5). The total energy produced is the integral of this rate over time. Thus

$$U_\mathcal{E} = \mathcal{E}\int_0^\infty I(t)\,dt = \mathcal{E}I(0)\int_0^\infty e^{-t/RC}\,dt = \mathcal{E}I(0)RC.$$

At $t = 0$, there is no voltage across the capacitor, so $I(0)R = \mathcal{E}$. Consequently, the total energy produced by the battery is

$$U_\mathcal{E} = C\mathcal{E}^2.$$

The energy dissipated in the resistor is $I^2(t)R$ integrated over time:

$$U_R = \int_0^\infty I^2(t)R\,dt = I^2(0)R\int_0^\infty e^{-2t/RC}\,dt = \frac{1}{2}C\mathcal{E}^2.$$

The energy stored in the capacitor is therefore

$$U_C = U_\mathcal{E} - U_R = \tfrac{1}{2}C\mathcal{E}^2,$$

which, since $V_C = \mathcal{E}$ when the capacitor is fully charged, is $CV_C^2/2$.

DRILL PROBLEM 33-1

The initial value of the current in the circuit of Fig. 33-1*c* is 0.10 A. After 2.0×10^{-2} s, the current drops to 0.030 A. (*a*) What is the capacitive time constant? (*b*) If $R = 5.0 \times 10^2\ \Omega$, what is the capacitance? (*c*) What is the initial charge on the capacitor?
ANS. (*a*) 1.7×10^{-2} s; (*b*) 3.3×10^{-5} F; (*c*) 1.7×10^{-3} C.

DRILL PROBLEM 33-2

A 5.0-MΩ resistor and an uncharged 20-μF capacitor are connected across a 60-V battery. At $t = 50$ s after the connection is made, what are (*a*) the energy stored in the capacitor, (*b*) the Joule heating in the resistor, (*c*) the rate at which the battery is delivering energy to the circuit, and (*d*) the rate at which energy is being stored in the capacitor?
ANS. (*a*) 5.6×10^{-3} J; (*b*) 2.7×10^{-4} W; (*c*) 4.4×10^{-4} W; (*d*) 1.7×10^{-4} W.

33-2 RL Circuit

A circuit with resistance and self-inductance is known as an *RL circuit*. An RL circuit consisting of a resistor, an inductor, a constant source of emf, and switches S_1 and S_2 is shown in Fig. 33-4*a*. When S_1 is closed, the circuit is equivalent to a single-loop circuit consisting of a resistor and an inductor connected across a source of emf (Fig. 33-4*b*). When S_1 is opened and S_2 is closed, the circuit becomes a single-loop circuit with only a resistor and an inductor (Fig. 33-4*c*).

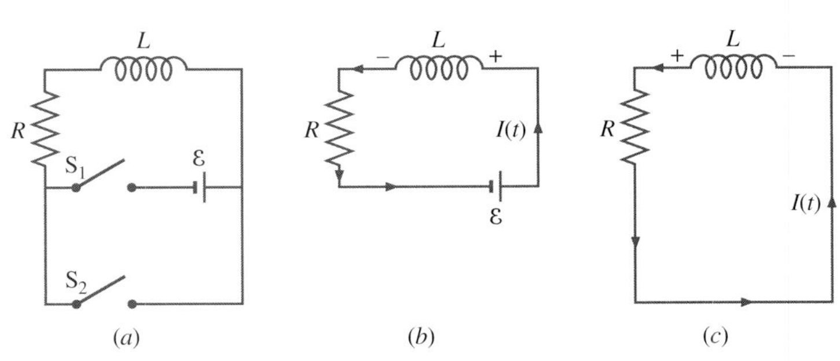

FIGURE 33-4 (*a*) An RL circuit with switches S_1 and S_2. (*b*) The equivalent circuit with S_1 closed and S_2 open. (*c*) The equivalent circuit after S_1 is opened and S_2 is closed.

Resistor and Inductor across a Constant Source of emf

We first consider the RL circuit of Fig. 33-4*b*. Once S_1 is closed with S_2 open, the source of emf produces a current in the circuit. If there were no self-inductance in the circuit, the current would rise immediately to a steady value of $\mathcal{E}/R$. However, from Faraday's law, the increasing current produces an emf $V_L = -L(dI/dt)$ across the inductor. In accordance with Lenz's law, the induced emf counteracts the increase in the current, and it is directed as shown in Fig. 33-4*b*. As a result, $I(t)$ starts at zero and increases asymptotically to its final value.

Applying Kirchhoff's loop rule to the circuit of Fig. 33-4*b*, we obtain

$$\mathcal{E} - L\frac{dI}{dt} - IR = 0, \qquad (33\text{-}10)$$

which is a first-order differential equation for $I(t)$. Notice its similarity to Eq. (33-1*b*)! If we replace $q(t)$ by $I(t)$, R by L, and $1/C$ by R in Eq. (33-1*b*), we obtain Eq. (33-10). Similarly, the solution to Eq. (33-10) can be found by making these substitutions in Eq. (33-2), the solution to Eq. (33-1*b*). This gives

$$I(t) = \frac{\mathcal{E}}{R}(1 - e^{-Rt/L}) = \frac{\mathcal{E}}{R}(1 - e^{-t/\tau_L}), \qquad (33\text{-}11)$$

where

$$\tau_L = L/R \qquad (33\text{-}12)$$

is the **inductive time constant** of the circuit.

The current $I(t)$ is plotted in Fig. 33-5*a*. It starts at zero, and when $t \to \infty$, $I(t)$ approaches $\mathcal{E}/R$ asymptotically. Now the induced emf $V_L(t)$ is directly proportional to dI/dt, or the slope of Fig. 33-5*a*. Hence, while at its greatest immediately after the switches are thrown, *the induced emf decreases to zero with time as the current approaches its final value of* $\mathcal{E}/R$. The circuit then becomes equivalent to a resistor connected across a source of emf.

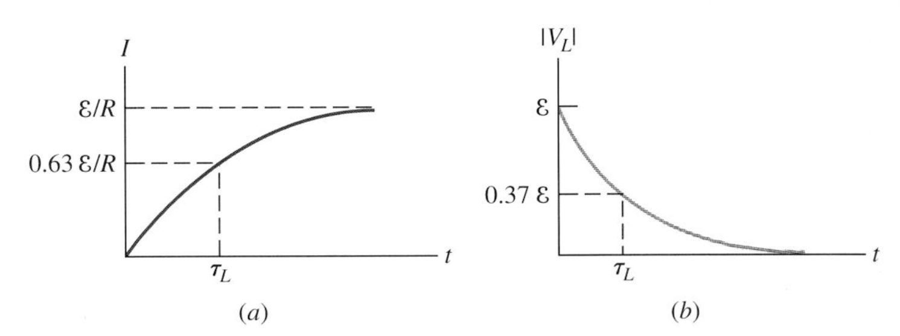

FIGURE 33-5 Time variation of (*a*) the electric current and (*b*) the magnitude of the induced voltage across the coil in the circuit of Fig. 33-4*b*.

The energy stored in the magnetic field of an inductor is

$$U_L = \tfrac{1}{2}LI^2;$$

so, as the current increases, the stored energy in the inductor increases from zero and asymptotically approaches a maximum of $L(\mathcal{E}/R)^2/2$.

The time constant τ_L tells us how rapidly the current increases to its final value. At $t = \tau_L$, the current in the circuit of Fig. 33-4*b* is, from Eq. (33-11),

$$I(\tau_L) = \frac{\mathcal{E}}{R}(1 - e^{-1}) = 0.63\frac{\mathcal{E}}{R},$$

which is 63 percent of the final value $\mathcal{E}/R$. The smaller the inductive time constant $\tau_L = L/R$, the more rapidly the current approaches $\mathcal{E}/R$.

We can find the time dependence of the induced voltage across the inductor of Fig. 33-4*b* using $V_L(t) = -L(dI/dt)$ and Eq. (33-11):

$$V_L(t) = -L\frac{dI}{dt} = -\mathcal{E}e^{-t/\tau_L}. \qquad (33\text{-}13)$$

The magnitude of this function is plotted in Fig. 33-5*b*. The greatest value of $L(dI/dt)$ is $\mathcal{E}$; it occurs when dI/dt is greatest, which is immediately after S_1 is closed and S_2 is opened. In the approach to steady state, dI/dt decreases to zero. As a result, the voltage across the inductor also vanishes as $t \to \infty$.

The time constant τ_L also tells us how quickly the induced voltage decays. At $t = \tau_L$, the magnitude of the induced voltage is

$$|V_L(\tau_L)| = \mathcal{E}e^{-1} = 0.37\mathcal{E} = 0.37V(0).$$

The voltage across the inductor therefore drops to about 37 percent of its initial value after one time constant. The shorter the time constant τ_L, the more rapidly the voltage decreases.

Resistor and Inductor Only

After enough time has elapsed so that the current has essentially reached its final value, the positions of the switches of Fig. 33-4*a* are reversed, giving us the circuit of Fig. 33-4*c*. At $t = 0$, the current in the circuit is $I(0) = \mathcal{E}/R$. With Kirchhoff's loop equation, we obtain

$$IR + L\frac{dI}{dt} = 0. \tag{33-14}$$

The solution to this equation is found by substituting I, L, and R for q, R, and $1/C$, respectively, in Eqs. (33-6) and (33-7); then

$$I(t) = \frac{\mathcal{E}}{R}e^{-t/\tau_L}. \tag{33-15}$$

The current starts at $I(0) = \mathcal{E}/R$ and decreases with time as the energy stored in the inductor is depleted (see Fig. 33-6).

The time dependence of the voltage across the inductor can be determined from $V_L = -L(dI/dt)$:

$$V_L(t) = \mathcal{E}e^{-t/\tau_L}. \tag{33-16}$$

This voltage is initially $V_L(0) = \mathcal{E}$ and it decays to zero like the current. The energy stored in the magnetic field of the inductor, $LI^2/2$, also decreases exponentially with time as it is dissipated by Joule heating in the resistance of the circuit.

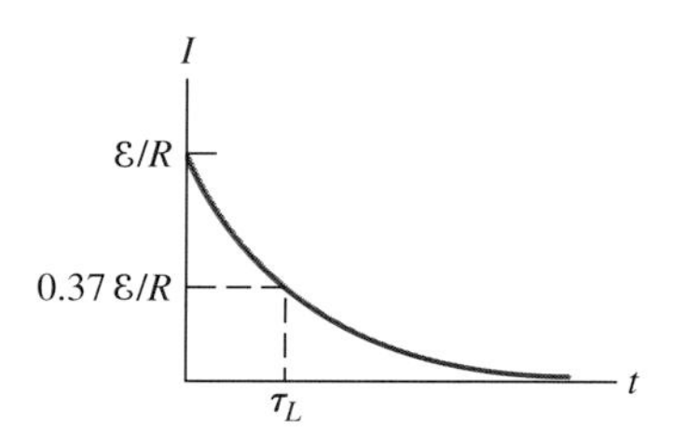

FIGURE 33-6 Time variation of the electric current in the RL circuit of Fig. 33-4*c*. The induced voltage across the coil also decays exponentially.

EXAMPLE 33-4 An RL Circuit with a Source of EMF

In the circuit of Fig. 33-4*a*, $\mathcal{E} = 2.0$ V, $R = 4.0\ \Omega$, and $L = 4.0$ H. With S_1 closed and S_2 open (Fig. 33-4*b*), (*a*) what is the time constant of the circuit? (*b*) What are the current in the circuit and the magnitude of the induced emf across the inductor at $t = 0$, at $t = 2.0\tau_L$, and as $t \to \infty$?

SOLUTION (*a*) The inductive time constant is

$$\tau_L = \frac{L}{R} = \frac{4.0\text{ H}}{4.0\ \Omega} = 1.0\text{ s}.$$

(*b*) The current in the circuit of Fig. 33-4*b* increases according to Eq. (33-11):

$$I(t) = \frac{\mathcal{E}}{R}(1 - e^{-t/\tau_L}).$$

At $t = 0$, $(1 - e^{-t/\tau_L}) = (1 - 1) = 0$; so $I(0) = 0$. At $t = 2.0\tau_L$ and as $t \to \infty$, we have, respectively,

$$I(2.0\tau_L) = \frac{\mathcal{E}}{R}(1 - e^{-2.0}) = (0.50\text{ A})(0.86) = 0.43\text{ A},$$

and

$$I(\infty) = \frac{\mathcal{E}}{R} = 0.50\text{ A}.$$

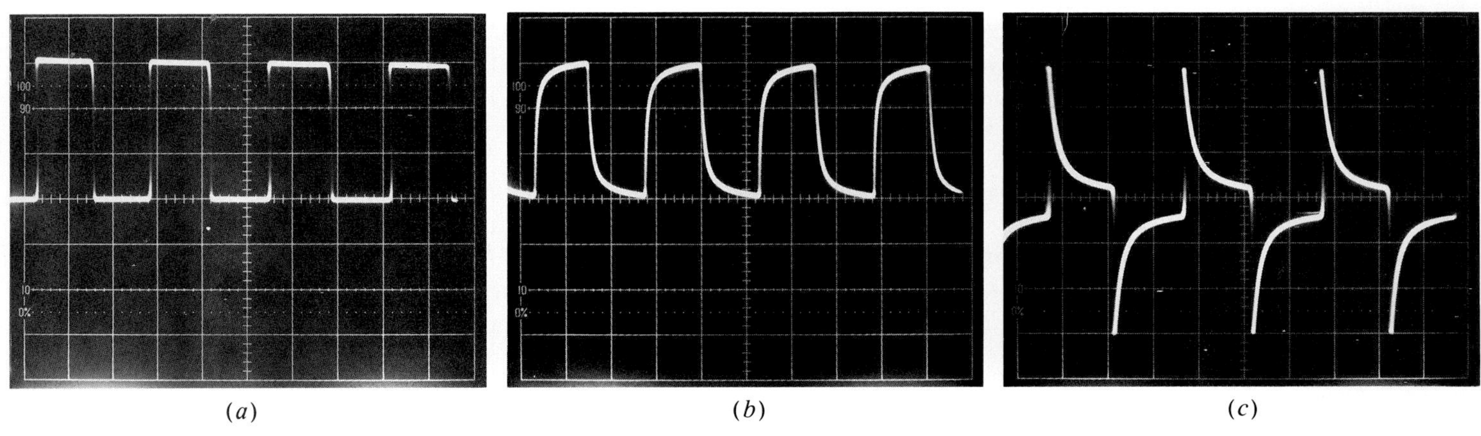

A generator in an RL circuit produces a square-pulse output in which the voltage oscillates between zero and some set value. Oscilloscope traces of: (*a*) the voltage output of the generator; (*b*) the voltage across the inductor; (*c*) the voltage across the resistor.

From Eq. (33-13), the magnitude of the induced emf decays as

$$|V_L(t)| = \mathcal{E}e^{-t/\tau_L}.$$

At $t = 0$, $t = 2.0\tau_L$, and as $t \to \infty$, we obtain

$$|V_L(0)| = \mathcal{E} = 2.0 \text{ V},$$
$$|V_L(2.0\tau_L)| = (2.0 \text{ V})e^{-2.0} = 0.27 \text{ V},$$

and

$$|V_L(\infty)| = 0.$$

EXAMPLE 33-5 An RL Circuit without a Source of emf

After the current in the RL circuit of Example 33-4 has reached its final value, the positions of the switches are reversed so that the circuit becomes the one shown in Fig. 33-4*c*. (*a*) How long does it take the current to drop to half its initial value? (*b*) How long does it take before the energy stored in the inductor is reduced to 1.0 percent of its maximum value?

SOLUTION (*a*) With the switches reversed, the current decreases according to

$$I(t) = \frac{\mathcal{E}}{R}e^{-t/\tau_L} = I(0)e^{-t/\tau_L}.$$

At a time t when the current is one-half its initial value, we have

$$I(t) = 0.50I(0),$$

so

$$e^{-t/\tau_L} = 0.50,$$

and

$$t = -[\ln(0.50)]\tau_L = 0.69(1.0 \text{ s}) = 0.69 \text{ s},$$

where we have used the inductive time constant found in Example 33-4.

(*b*) The energy stored in the inductor is given by

$$U_L(t) = \frac{1}{2}L[I(t)]^2 = \frac{1}{2}L\left(\frac{\mathcal{E}}{R}e^{-t/\tau_L}\right)^2 = \frac{L\mathcal{E}^2}{2R^2}e^{-2t/\tau_L}.$$

If the energy drops to 1.0 percent of its initial value at a time t, we have

$$U_L(t) = (0.010)U_L(0)$$

or

$$\frac{L\mathcal{E}^2}{2R^2}e^{-2t/\tau_L} = (0.010)\frac{L\mathcal{E}^2}{2R^2}.$$

Upon canceling terms and taking the natural logarithm of both sides, we obtain

$$-\frac{2t}{\tau_L} = \ln(0.010),$$

so

$$t = -\tfrac{1}{2}\tau_L \ln(0.010).$$

Since $\tau_L = 1.0$ s, the time it takes for the energy stored in the inductor to decrease to 1.0 percent of its initial value is

$$t = -\tfrac{1}{2}(1.0 \text{ s})\ln(0.01) = 2.3 \text{ s}.$$

DRILL PROBLEM 33-3

Verify that RC and L/R have the dimensions of time.

DRILL PROBLEM 33-4

(*a*) If the current in the circuit of Fig. 33-4*b* increases to 90 percent of its final value after 5.0 s, what is the inductive time constant? (*b*) If $R = 20\ \Omega$, what is the value of the self-inductance? (*c*) If the 20-Ω resistor is replaced with a 100-Ω resistor, what is the time taken for the current to reach 90 percent of its final value?
ANS. (*a*) 2.2 s; (*b*) 43 H; (*c*) 1.0 s.

DRILL PROBLEM 33-5

For the circuit of Fig. 33-4*b*, show that when steady state is reached, the difference in the total energies produced by the battery and dissipated in the resistor is equal to the energy stored in the magnetic field of the coil.

*33-3 Freely Oscillating Circuits

Circuits that do not have a source of emf can oscillate if they start with some energy stored in either a capacitor or an inductor. We'll consider two cases: (1) an LC circuit, which is an idealized circuit of zero resistance and contains an inductor and a capacitor; and (2) the more realistic RLC circuit with a capacitor, an inductor, and resistance.

LC Circuit

An LC circuit is shown in Fig. 33-7. If the capacitor contains a charge q_0 before the switch is closed, then all the energy of the circuit is initially stored in the electric field of the capacitor (Fig. 33-7*a*). This energy is

$$U_C = \frac{1}{2}\frac{q_0^2}{C}.$$

When the switch is closed, the capacitor begins to discharge, producing a current in the circuit. The current in turn creates a magnetic field in the inductor. The net effect of this process is a transfer of energy from the capacitor, with its diminishing electric field, to the inductor, with its increasing magnetic field.

In Fig. 33-7*b*, the capacitor is completely discharged, and all the energy is stored in the magnetic field of the inductor. At this

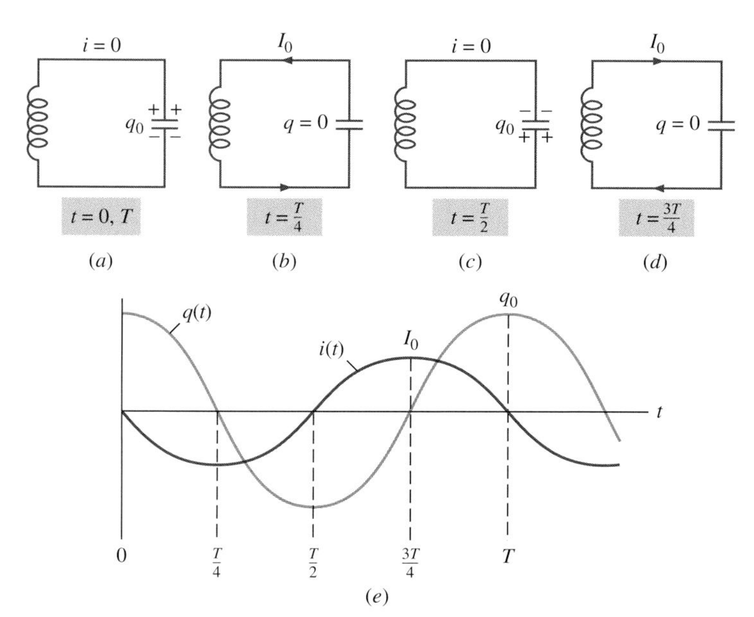

FIGURE 33-7 Electromagnetic oscillations in an LC circuit.

instant, the current is at its maximum value I_0, and the energy in the inductor is

$$U_L = \frac{1}{2}LI_0^2.$$

Since there is no resistance in the circuit, no energy is lost through Joule heating; thus

$$\frac{1}{2}\frac{q_0^2}{C} = \frac{1}{2}LI_0^2.$$

At an arbitrary time when the capacitor charge is $q(t)$ and the current is $i(t)$, the total energy U in the circuit is given by $q^2(t)/2C + Li^2(t)/2$. Because there is no energy dissipation,

$$U = \frac{1}{2}\frac{q^2}{C} + \frac{1}{2}Li^2 = \frac{1}{2}\frac{q_0^2}{C} = \frac{1}{2}LI_0^2. \tag{33-17}$$

After reaching its maximum I_0, the current $i(t)$ continues to transport charge between the capacitor plates, thereby recharging the capacitor. The electric field of the capacitor increases while the magnetic field of the inductor diminishes, and the overall effect is a transfer of energy from the inductor *back* to the capacitor. From the law of energy conservation, the maximum charge that the capacitor re-acquires is q_0. However, as Fig. 33-7*c* shows, the capacitor plates are charged *opposite* to what they were initially.

When fully charged, the capacitor once again transfers its energy to the inductor until it is again completely discharged, as shown in Fig. 33-7*d*. Then in the last part of this cyclic process, energy flows back to the capacitor and the initial state of the circuit is restored.

We have followed the circuit through one complete cycle. Its electromagnetic oscillations are analogous to the mechanical oscillations of a mass at the end of a spring. In this latter case, energy is transferred back and forth between the mass, which has kinetic energy $mv^2/2$, and the spring, which has potential energy $kx^2/2$. With the absence of friction in the mass-spring system, the oscillations would continue indefinitely. Similarly, the oscillations of an LC circuit with no resistance would continue forever if undisturbed.

The frequency of the oscillations in a resistance-free LC circuit may be found by analogy with the mass-spring system. For the former case, $i(t) = dq(t)/dt$, and the total electromagnetic energy U is

$$U = \frac{1}{2}Li^2 + \frac{1}{2}\frac{q^2}{C}.$$

For the mass-spring system, $v(t) = dx(t)/dt$, and the total mechanical energy E is

$$E = \tfrac{1}{2}mv^2 + \tfrac{1}{2}kx^2.$$

The equivalence of the two systems is clear. To go from the mechanical to the electromagnetic system, we simply replace m by L, v by i, k by $1/C$, and x by q. Now $x(t)$ is given by

$$x(t) = A\cos(\omega t + \phi),$$

where $\omega = \sqrt{k/m}$. Hence the charge on the capacitor in an LC circuit is given by

$$q(t) = q_0\cos(\omega t + \phi), \tag{33-18}$$

where the angular frequency of the oscillations in the circuit is

$$\omega = \sqrt{\frac{1}{LC}}. \tag{33-19}$$

Finally, the current in the LC circuit is found by taking the time derivative of $q(t)$:

$$i(t) = \frac{dq(t)}{dt} = -\omega q_0\sin(\omega t + \phi). \tag{33-20}$$

The time variations of q and i are shown in Fig. 33-7*e* for $\phi = 0$.

EXAMPLE 33-6 AN LC CIRCUIT

In an LC circuit, the self-inductance is 2.0×10^{-2} H and the capacitance is 8.0×10^{-6} F. At $t = 0$, all of the energy is stored in the capacitor, which has charge 1.2×10^{-5} C. (*a*) What is the angular frequency of the oscillations in the circuit? (*b*) What is the maximum current flowing through the circuit? (*c*) How long does it take the capacitor to become completely discharged? (*d*) Find an equation that represents $q(t)$.

SOLUTION (*a*) From Eq. (33-19), the angular frequency of the oscillations is

$$\omega = \sqrt{\frac{1}{LC}} = \sqrt{\frac{1}{(2.0 \times 10^{-2}\ \text{H})(8.0 \times 10^{-6}\ \text{F})}}$$
$$= 2.5 \times 10^3\ \text{rad/s}.$$

(*b*) The current is at its maximum I_0 when all the energy is stored in the inductor. From the law of energy conservation,

$$\frac{1}{2}LI_0^2 = \frac{1}{2}\frac{q_0^2}{C},$$

so

$$I_0 = \sqrt{\frac{1}{LC}}\, q_0 = (2.5 \times 10^3 \text{ rad/s})(1.2 \times 10^{-5} \text{ C})$$
$$= 3.0 \times 10^{-2} \text{ A}.$$

(*c*) The capacitor becomes completely discharged in one-fourth of a cycle, or during a time $T/4$, where T is the period of the oscillations. Since

$$T = \frac{2\pi}{\omega} = \frac{2\pi}{2.5 \times 10^3 \text{ rad/s}} = 2.5 \times 10^{-3} \text{ s},$$

the time taken for the capacitor to become fully discharged is $(2.5 \times 10^{-3} \text{ s})/4 = 6.3 \times 10^{-4}$ s.

(*d*) The capacitor is completely charged at $t = 0$, so $q(0) = q_0$. Using Eq. (33-18), we obtain

$$q(0) = q_0 = q_0 \cos \phi.$$

Thus $\phi = 0$, and

$$q(t) = (1.2 \times 10^{-5} \text{ C}) \cos (2.5 \times 10^3 t).$$

DRILL PROBLEM 33-6

The angular frequency of the oscillations in an LC circuit is 2.0×10^3 rad/s. (*a*) If $L = 0.10$ H, what is C? (*b*) Suppose that at $t = 0$, all the energy is stored in the inductor. What is the value of ϕ? (*c*) A second identical capacitor is connected in parallel with the original capacitor. What is the angular frequency of this circuit?
ANS. (*a*) 2.5 μF; (*b*) $\pi/2$ rad or $3\pi/2$ rad; (*c*) 1.4×10^3 rad/s.

RLC Circuit

When the switch is closed in the RLC circuit of Fig. 33-8*a*, the capacitor begins to discharge and electromagnetic energy is dissipated by the resistor at a rate i^2R. With U given by Eq. (33-17), we write

$$\frac{dU}{dt} = \frac{q}{C}\frac{dq}{dt} + Li\frac{di}{dt} = -i^2R,$$

which reduces to

$$L\frac{d^2q}{dt^2} + R\frac{dq}{dt} + \frac{1}{C}q = 0. \qquad (33\text{-}21)$$

This equation is analogous to

$$m\frac{d^2x}{dt^2} + b\frac{dx}{dt} + kx = 0, \qquad (14\text{-}23)$$

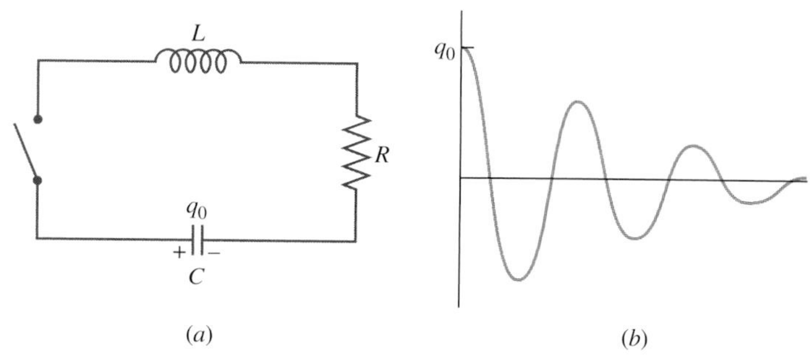

FIGURE 33-8 (*a*) An RLC circuit. Electromagnetic oscillations begin when the switch is closed. The capacitor is fully charged initially. (*b*) Underdamped oscillations of the capacitor charge.

the equation of motion for the damped mass-spring system. Recall that the solution to Eq. (14-23) can be underdamped ($\sqrt{k/m} > b/2m$), critically damped ($\sqrt{k/m} = b/2m$), and overdamped ($\sqrt{k/m} < b/2m$), as illustrated in Fig. 14-14. By analogy, the solution $q(t)$ to the RLC differential equation has the same feature. Here we look only at the case of underdamping. By replacing m by L, b by R, k by $1/C$ and x by q, and assuming $\sqrt{1/LC} > R/2L$, we obtain

$$q(t) = q_0 e^{-Rt/2L} \cos (\omega' t + \phi), \qquad (33\text{-}22)$$

where the angular frequency of the oscillations is given by

$$\omega' = \sqrt{\frac{1}{LC} - \left(\frac{R}{2L}\right)^2}.$$

This underdamped solution is shown in Fig. 33-8*b*. Notice that the amplitude of the oscillations decreases as energy is dissipated in the resistor. Equation (33-22) can be confirmed experimentally by measuring the voltage across the capacitor as a function of time. This voltage multiplied by the capacitance of the capacitor then gives $q(t)$.

DRILL PROBLEM 33-7

In an RLC circuit, $L = 5.0$ mH, $C = 6.0$ μF, and $R = 200$ Ω. (*a*) Is the circuit underdamped, critically damped, or overdamped? (*b*) If the circuit starts oscillating with a charge of 3.0×10^{-3} C on the capacitor, how much energy has been dissipated in the resistor by the time the oscillations cease?
ANS. (*a*) Overdamped; (*b*) 0.75 J.

SUMMARY

1. **RC and RL circuits**
   (*a*) In a charging RC circuit, the capacitor charge, the voltage across the capacitor, and the current are given by

$$q(t) = C\mathcal{E}(1 - e^{-t/\tau_C}),$$
$$V_C(t) = \mathcal{E}(1 - e^{-t/\tau_C}),$$
$$I(t) = \frac{\mathcal{E}}{R} e^{-t/\tau_C},$$

where $\tau_C = RC$ is the capacitive time constant of the circuit.

In a discharging RC circuit, the capacitor charge, the voltage across the capacitor, and the current are given by

$$q(t) = C\mathcal{E}e^{-t/\tau_C},$$
$$V_C(t) = \mathcal{E}e^{-t/\tau_C},$$
$$I(t) = \frac{\mathcal{E}}{R}e^{-t/\tau_C}.$$

(*b*) In an RL circuit containing a constant source of emf, the current and the voltage induced across the inductor are given by

$$I(t) = \frac{\mathcal{E}}{R}(1 - e^{-t/\tau_L})$$

and

$$V_L(t) = -\mathcal{E}e^{-t/\tau_L},$$

where $\tau_L = L/R$ is the inductive time constant of the circuit.

In an RL circuit without a source of emf, the current and the induced voltage across the inductor are given by

$$I(t) = \frac{\mathcal{E}}{R}e^{-t/\tau_L} \quad \text{and} \quad V_L(t) = \mathcal{E}e^{-t/\tau_L}.$$

*2. **Freely oscillating circuits**

(*a*) The total energy in an LC circuit is

$$U = \frac{1}{2}Li^2 + \frac{1}{2}\frac{q^2}{C}.$$

This energy is transferred in an oscillatory manner between the capacitor and the inductor with an angular frequency

$$\omega = \sqrt{\frac{1}{LC}}.$$

The charge on the capacitor and the current in the circuit are given by

$$q(t) = q_0 \cos(\omega t + \phi)$$

and

$$i(t) = -\omega q_0 \sin(\omega t + \phi).$$

(*b*) In an RLC circuit, electromagnetic energy is dissipated by the resistor at a rate i^2R, so

$$\frac{dU}{dt} = \frac{q}{C}\frac{dq}{dt} + Li\frac{di}{dt} = -i^2R.$$

The underdamped solution for the capacitor charge is

$$q(t) = q_0e^{-Rt/2L}\cos(\omega' t + \phi),$$

where

$$\omega' = \sqrt{\frac{1}{LC} - \left(\frac{R}{2L}\right)^2}.$$

SUPPLEMENT 33-1 SOLUTIONS TO EQUATIONS (33-1*b*) AND (33-6)

In this supplement we solve Eqs. (33-1*b*) and (33-6). The solutions to these two differential equations represent the time dependence of the charge on a capacitor in a charging RC circuit and in a discharging RC circuit, respectively.

Equation (33-1b): By rearranging terms in this equation, we obtain

$$\frac{dq}{dt} = \frac{\mathcal{E}}{R} - \frac{q}{RC} \qquad (i)$$

so

$$\frac{dq}{(\mathcal{E}/R) - (q/RC)} = dt. \qquad (ii)$$

At $t = 0$, the capacitor is uncharged $[q(0) = 0]$. Hence

$$\int_0^q \frac{dq'}{(\mathcal{E}/R) - (q'/RC)} = \int_0^t dt',$$

and

$$-RC\ln\left(\frac{\mathcal{E}}{R} - \frac{q}{RC}\right) + RC\ln\left(\frac{\mathcal{E}}{R}\right) = t,$$

which, because $\ln(A/B) = \ln A - \ln B$, simplifies to

$$\ln\left(1 - \frac{q}{C\mathcal{E}}\right) = -\frac{t}{RC}.$$

Finally, using the fact that when $\ln A = -x$, $A = e^{-x}$, we have

$$q(t) = C\mathcal{E}(1 - e^{-t/RC}).$$

Equation (33-6): Upon rearranging terms in this equation, we find

$$\frac{dq}{dt} = -\frac{1}{RC}q \qquad (iii)$$

so

$$\frac{dq}{q} = -\frac{1}{RC}dt. \qquad (iv)$$

At $t = 0$, the charge on the capacitor is $C\mathcal{E}$. Hence

$$\int_{C\mathcal{E}}^q \frac{dq'}{q'} = -\frac{1}{RC}\int_0^t dt',$$

and

$$\ln\left(\frac{q}{C\mathcal{E}}\right) = -\frac{1}{RC}t.$$

Finally, using the relationship between the natural logarithm and the exponential, we obtain

$$q(t) = C\mathcal{E}e^{-t/RC}.$$

Questions

33-1. Do Kirchhoff's rules apply to circuits that contain inductors and capacitors?

33-2. An uncharged capacitor is connected across the terminals of a battery. Does the charge that is eventually deposited on the plates of the capacitor depend on the internal resistance of the battery? Does the time required for the capacitor to become completely charged depend on the internal resistance?

33-3. Describe how you could use a capacitor to measure the resistance of a high-resistance voltmeter.

33-4. When a charging capacitor reaches $(1 - 1/e)$ of its final voltage, is the energy stored in the electric field of the capacitor also $(1 - 1/e)$ of its final value?

33-5. Explain the minus sign in $I = -dq/dt$, when q is the charge on the plates of a discharging capacitor and I is the current leaving the positive plate of the capacitor.

33-6. Describe how the currents through R_1 and R_2 in the accompanying figure vary with time after switch S is closed.

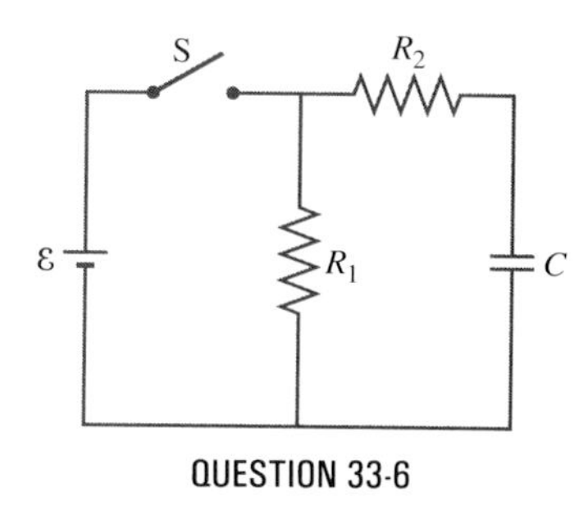

QUESTION 33-6

33-7. Does the time required for the current in the RC circuit of Fig. 33-1*b* to reach any fraction of its steady-state value depend on the emf of the battery?

33-8. Can a circuit element have both capacitance and inductance?

33-9. Use Lenz's law to explain why the initial current in the RL circuit of Fig. 33-4*b* is zero.

33-10. When the current in the RL circuit of Fig. 33-4*b* reaches its final value $\mathcal{E}/R$, what is the voltage across the inductor? across the resistor?

33-11. Does the time required for the current in an RL circuit to reach any fraction of its steady-state value depend on the emf of the battery?

33-12. An inductor is connected across the terminals of a battery. Does the current that eventually flows through the inductor depend on the internal resistance of the battery? Does the time required for the current to reach its final value depend on this resistance?

33-13. At what time is the voltage across the inductor of the RL circuit of Fig. 33-4*b* a maximum?

33-14. In the simple RL circuit of Fig. 33-4*b*, can the emf induced across the inductor ever be greater than the emf of the battery used to produce the current?

33-15. If the emf of the battery of Fig. 33-4*b* is reduced by a factor of 2, by how much does the steady-state energy stored in the magnetic field of the inductor change?

33-16. A steady current flows through a circuit with a large inductive time constant. When a switch in the circuit is opened, a large spark occurs across the terminals of the switch. Explain.

33-17. Describe how the currents through R_1 and R_2 in the accompanying figure vary with time after switch S is closed.

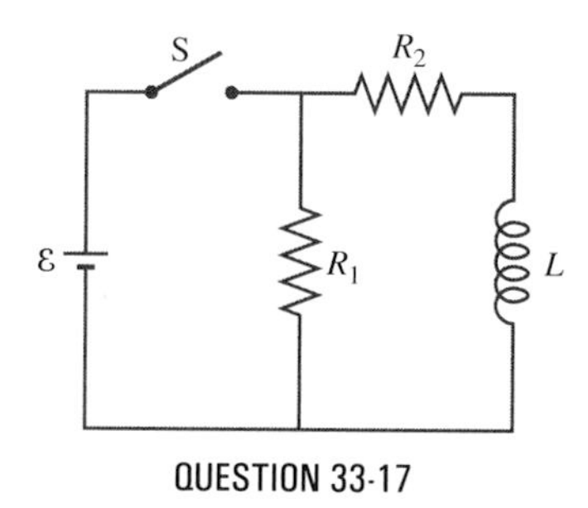

QUESTION 33-17

33-18. Discuss possible practical applications of RC and RL circuits.

33-19. When a wire is connected between the two ends of a solenoid, the resulting circuit can oscillate like an RLC circuit. Describe what causes the capacitance in this circuit.

33-20. In an LC circuit, what determines the frequency and the amplitude of the oscillations?

33-21. Describe what effect the resistance of the connecting wires has on an oscillating LC circuit.

33-22. Suppose you wanted to design an LC circuit with a frequency of 0.01 Hz. What problems might you encounter?

Problems

RC and RL Circuits

33-1. In the RC circuit of Fig. 33-1*b*, the capacitor is initially uncharged. If $\mathcal{E} = 12$ V, $C = 2.0\ \mu$F, and $R = 5.0$ kΩ, determine (*a*) $I(0)$, (*b*) $I(0.020\text{ s})$, (*c*) $I(\infty)$, (*d*) the rate at which energy is dissipated in the resistor at $t = 0.020$ s, (*e*) the energy stored in the capacitor at $t = 0.020$ s, and (*f*) the rate at which energy is stored in the capacitor at $t = 0.020$ s.

33-2. With the switch S open (see the accompanying figure shown at right), the voltage across the capacitor is 200 V. Determine the time t after the switch is closed at which the power dissipated in the resistor is 0.25 W.

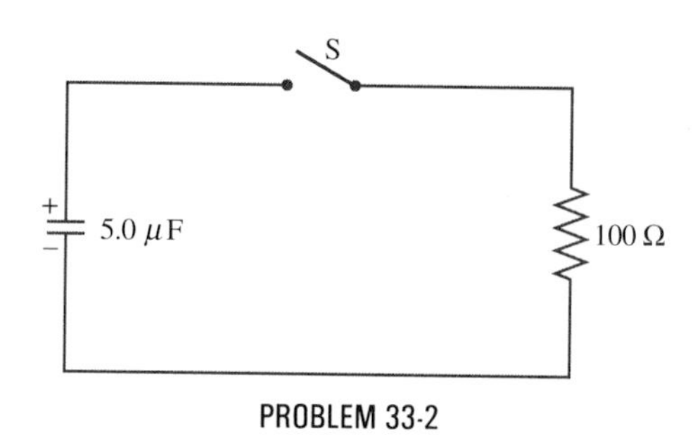

PROBLEM 33-2

33-3. When the capacitor in the RC circuit of Fig. 33-1c discharges, the difference in potential between its plates decreases from an initial voltage of 50 V to 5 V in exactly 4.0 s. (a) What is the time constant of the circuit? (b) What is the potential difference between the plates at $t = 10$ s?

33-4. How many time constants elapse before the capacitor in the RC circuit of Fig. 33-1b is charged to within 5.0 percent of its steady-state charge?

33-5. The capacitor of Fig. 33-1c is completely charged at $t = 0$. If $\mathcal{E} = 12$ V, $C = 2.0\ \mu$F, and $R = 5.0$ kΩ, determine (a) $q(0)$, (b) $I(0)$, (c) $q(0.010\text{ s})$, (d) $I(0.010\text{ s})$, (e) the rate at which energy is discharged from the capacitor at $t = 0.010$ s, and (f) the rate at which energy is dissipated in the resistor at $t = 0.010$ s.

33-6. The voltage across a slightly leaky capacitor with $C = 0.25\ \mu$F drops to half its initial value in 2.0 s. What is the resistance between the plates of the capacitor?

33-7. A capacitor is charged by connecting it to a 12-V battery. It is then disconnected from the battery and connected to a voltmeter with an internal resistance of 2.2 MΩ. After 4 s, the voltmeter reads 5 V. What is the capacitance of the capacitor?

33-8. How long does it take the energy stored in the electric field of a charging capacitor (see Fig. 33-1b) to reach half its maximum value? Express your answer in terms of the time constant of the circuit.

33-9. The capacitor in the circuit of the accompanying figure is completely charged by the 12-V battery. (a) What is the voltage across the capacitor? (b) If the switch S is opened, how much time elapses before the charge on the capacitor drops to 25 percent of its maximum value?

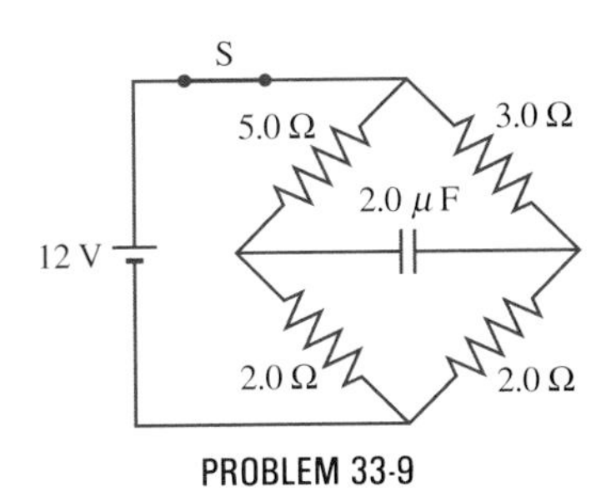

PROBLEM 33-9

33-10. In the circuit shown in the accompanying figure, the capacitor is uncharged when the switch S is open. (a) What are I_1, I_2, and I_3 immediately after the switch is closed? (b) What are these currents after the switch has been closed long enough for the capacitor to become completely charged?

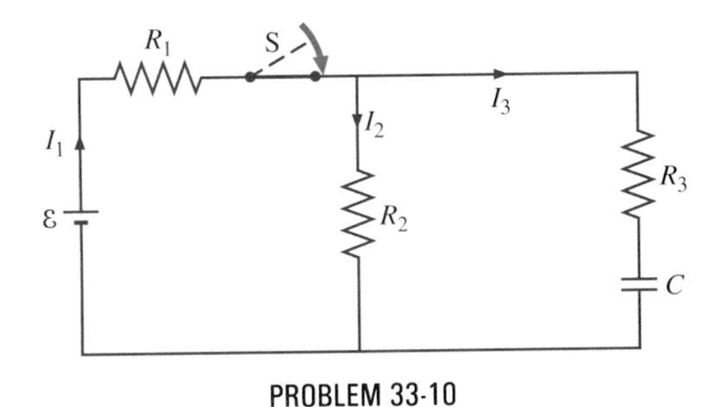

PROBLEM 33-10

33-11. A 5.0-μF capacitor with an initial stored energy of 1.0 J is discharged through a 1.0-MΩ resistor. (a) What is the initial charge on the capacitor? (b) What is the current through the resistor when the capacitor just starts to discharge? (c) Express the voltage across the resistor as a function of time. (d) Express the rate at which energy is dissipated in the resistor as a function of time.

33-12. In Fig. 33-4b, $\mathcal{E} = 12$ V, $L = 20$ mH, and $R = 5.0$ Ω. Determine (a) the time constant of the circuit, (b) the initial current through the resistor, (c) the final current through the resistor, (d) the current through the resistor when $t = 2\tau_L$, and (e) the voltages across the inductor and the resistor when $t = 2\tau_L$.

33-13. For the circuit shown in the accompanying figure, $\mathcal{E} = 20$ V, $L = 4.0$ mH, and $R = 5.0$ Ω. After steady state is reached with S_1 closed and S_2 open, S_2 is closed and immediately thereafter (at $t = 0$) S_1 is opened. Determine (a) the current through L at $t = 0$, (b) the current through L at $t = 4.0 \times 10^{-4}$ s, and (c) the voltages across L and R at $t = 4.0 \times 10^{-4}$ s.

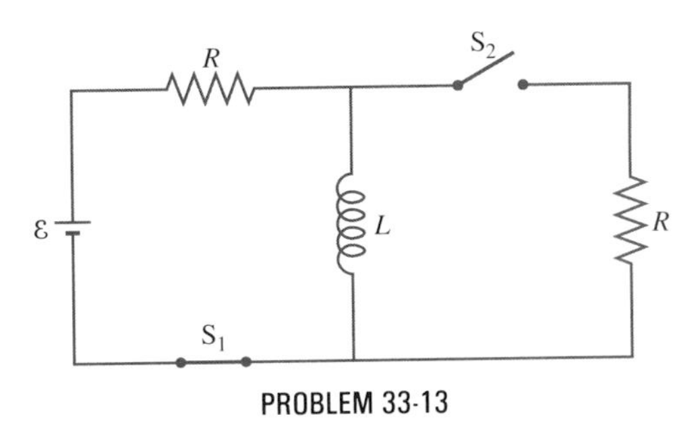

PROBLEM 33-13

33-14. The current in the RL circuit of Fig. 33-4b increases to 40 percent of its steady-state value in 2.0 s. What is the time constant of the circuit?

33-15. How long after the switch is thrown does it take the energy stored in the magnetic field of the inductor of Fig. 33-4b to reach half its maximum value? Express your answer in terms of the time constant of the circuit.

33-16. Determine dI/dt at the instant after the switch is thrown in the circuit of Fig. 33-4a, thereby producing the circuit of Fig. 33-4b. Show that if I were to continue to increase at this initial rate, it would reach its maximum $\mathcal{E}/R$ in one time constant.

33-17. The current in the RL circuit of Fig. 33-4b reaches half its maximum value in 1.75 ms. Determine (a) the time constant of the circuit and (b) the resistance of the circuit if $L = 250$ mH.

33-18. Consider the circuit shown in the accompanying figure. Find I_1, I_2, and I_3 when (a) the switch S is first closed, (b) after the currents have reached steady-state values, and (c) at the instant the switch is reopened (after being closed for a long time).

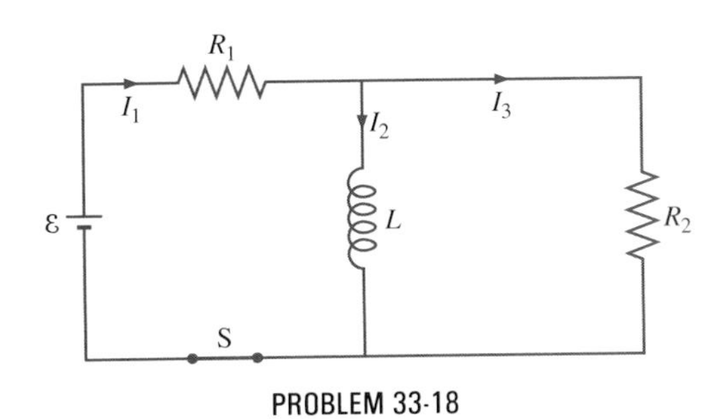

PROBLEM 33-18

33-19. For the circuit shown in the accompanying figure, $\mathcal{E} = 50$ V, $R_1 = 10\ \Omega$, $R_2 = 20\ \Omega$, $R_3 = 30\ \Omega$, and $L = 2.0$ mH. Find the values of I_1 and I_2 (*a*) immediately after switch S is closed, (*b*) a long time after S is closed, (*c*) immediately after S is reopened, and (*d*) a long time after S is reopened.

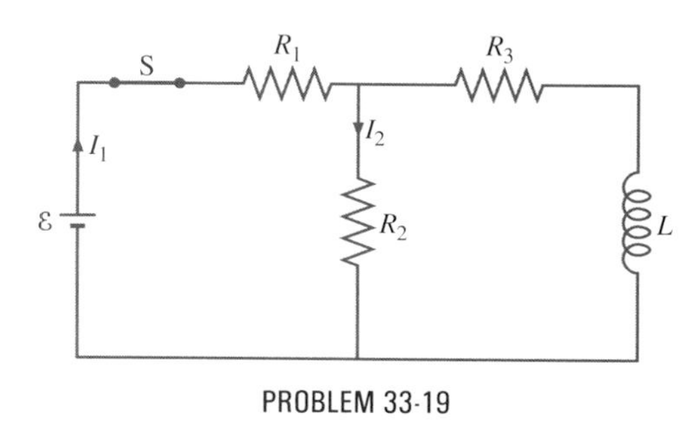

PROBLEM 33-19

33-20. For the circuit of the previous problem, find the current through the inductor 2.0×10^{-5} s after the switch is reopened.

33-21. Show that for the circuit of Fig. 33-1*c* the initial energy stored in the capacitor, $q(0)^2/2C$, is equal to the total energy eventually dissipated in the resistor, $\int_0^\infty I^2(t)R\,dt$.

33-22. Show that for the circuit of Fig. 33-4*c* the initial energy stored in the inductor, $LI^2(0)/2$, is equal to the total energy eventually dissipated in the resistor, $\int_0^\infty I^2(t)R\,dt$.

33-23. The initially uncharged capacitor of Fig. 33-1*a* is charged by setting the switches so as to obtain the circuit of Fig. 33-1*b*. (*a*) How much energy is supplied by the battery in the charging process? (*b*) How much of this energy is dissipated in the resistor? (*c*) How much of the energy is stored in the capacitor? (*d*) Does the answer to part (*a*) equal the sum of the answers to parts (*b*) and (*c*)?

Freely Oscillating Circuits

33-24. A 500-pF capacitor is charged to 100 V and then quickly connected to an 80-mH inductor. Determine (*a*) the maximum energy stored in the magnetic field of the inductor, (*b*) the peak value of the current, and (*c*) the frequency of oscillation of the circuit.

33-25. The self-inductance and capacitance of an LC circuit are 0.20 mH and 5.0 pF. What is the angular frequency at which the circuit oscillates?

33-26. What is the self-inductance of an LC circuit that oscillates at 60 Hz when the capacitance is 10 μF?

33-27. In an oscillating LC circuit, the maximum charge on the capacitor is 2.0×10^{-6} C, and the maximum current through the inductor is 8.0 mA. (*a*) What is the period of the oscillations? (*b*) How much time elapses between an instant when the capacitor is uncharged and the next instant when it is fully charged?

33-28. The self-inductance and capacitance of an oscillating LC circuit are $L = 20$ mH and $C = 1.0\ \mu$F, respectively. (*a*) What is the frequency of the oscillations? (*b*) If the maximum potential difference between the plates of the capacitor is 50 V, what is the maximum current in the circuit?

33-29. In an oscillating LC circuit, the maximum charge on the capacitor is q_m. Determine the charge on the capacitor and the current through the inductor when energy is shared equally between the electric and magnetic fields. Express your answer in terms of q_m, L, and C.

33-30. In the circuit shown in the accompanying figure, S_1 is opened and S_2 is closed simultaneously. Determine (*a*) the frequency of the resulting oscillations, (*b*) the maximum charge on the capacitor, (*c*) the maximum current through the inductor, and (*d*) the electromagnetic energy of the oscillating circuit.

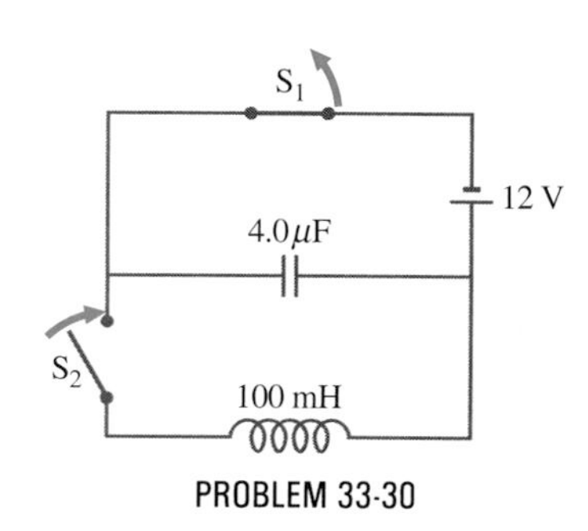

PROBLEM 33-30

33-31. An LC circuit in an AM tuner uses a coil with an inductance of 2.5 mH and a variable capacitor. If the natural frequency of the circuit is to be adjustable over the range 540 to 1600 kHz (the AM broadcast band), what range of capacitance is required?

33-32. In an oscillating RLC circuit, $R = 5.0\ \Omega$, $L = 5.0$ mH, and $C = 500\ \mu$F. What is the angular frequency of the oscillations?

33-33. In an oscillating RLC circuit with $L = 10$ mH, $C = 1.5\ \mu$F, and $R = 2.0\ \Omega$, how much time elapses before the amplitude of the oscillations drops to half its initial value?

33-34. What resistance R must be connected in series with a 200-mH inductor and a 10-μF capacitor if the maximum charge on the capacitor of the resulting RLC oscillating circuit is to decay to 50 percent of its initial value in 50 cycles? to 0.10 percent of its initial value in 50 cycles?

General Problems

33-35. With the switch S open, the capacitors shown in the accompanying figure are uncharged. When the switch is closed, they begin to charge. (*a*) What is the time constant of the circuit? (*b*) What is the voltage across each capacitor after one time constant? (*c*) What is the voltage across each resistor after one time constant?

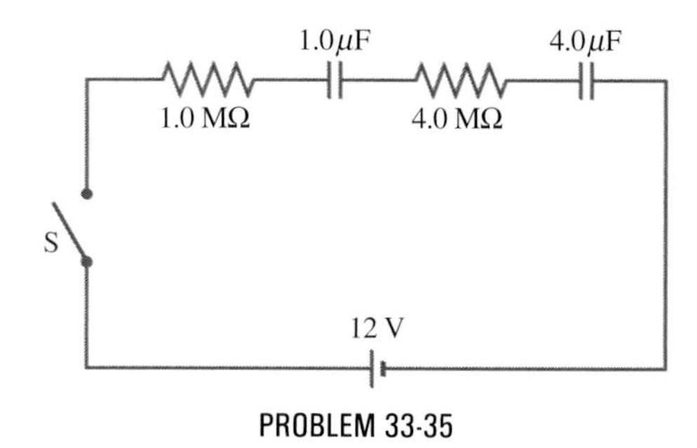

PROBLEM 33-35

33-36. When the switch S is closed, the initially uncharged capacitors shown in the accompanying figure start charging. (*a*) What is the time constant of the circuit? (*b*) What is the charge on each capacitor after one time constant? (*c*) What are the voltages across the resistor and the capacitors after one time constant?

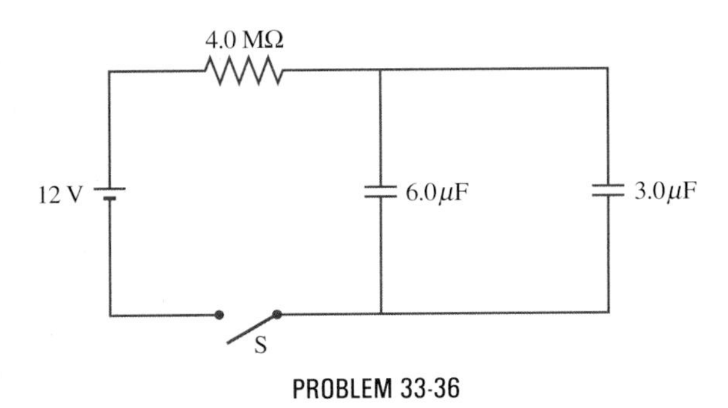

PROBLEM 33-36

33-37. The switch S of the circuit shown in the accompanying figure is closed at $t = 0$. Assuming the capacitors are initially uncharged, calculate (in terms of $\mathcal{E}$, C, and R) (*a*) the initial current through the battery, (*b*) the steady-state current through the battery, and (*c*) the steady-state charge on each capacitor.

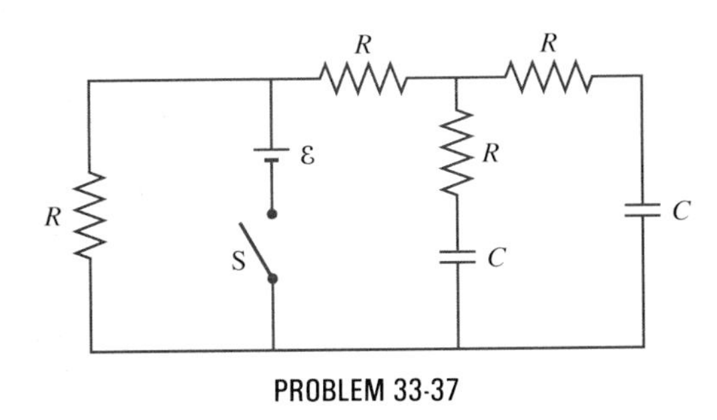

PROBLEM 33-37

33-38. Answer the questions of the previous problem for the circuit shown in the accompanying figure.

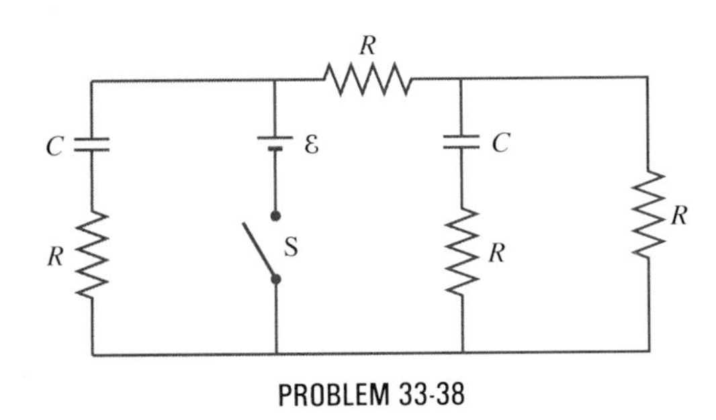

PROBLEM 33-38

33-39. The switch S is closed at $t = 0$, at which time the voltage across the charged capacitor C is equal to $\mathcal{E}$ and has the polarity indicated. (See the accompanying figure.) (*a*) Show that the current $I(t)$ is given by

$$I(t) = \frac{2\mathcal{E}}{R} e^{-t/RC}.$$

(*b*) Show that the voltage $V_C(t)$ across the capacitor is given by

$$V_C(t) = \mathcal{E}(1 - 2e^{-t/RC}).$$

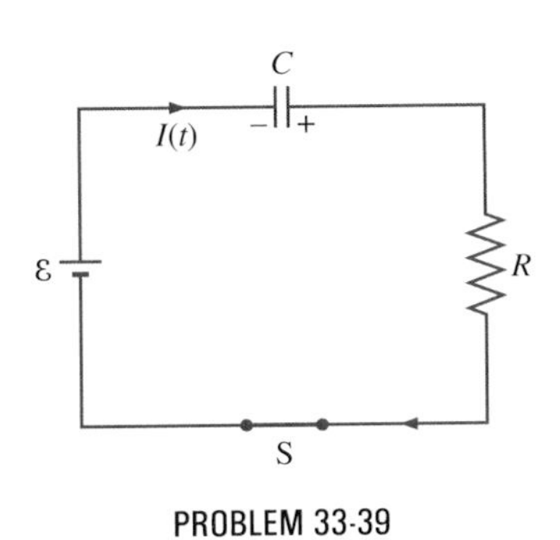

PROBLEM 33-39

33-40. The switch S of the circuit shown in the accompanying figure is closed at $t = 0$, at which time C_2 is uncharged and the voltage across C_1 is V. Show that the charges on the capacitors are given by

$$q_1(t) = \left(\frac{C_1 + C_2 e^{-t/\tau}}{C_1 + C_2}\right) C_1 V,$$

$$q_2(t) = \frac{C_1 C_2 V}{C_1 + C_2}(1 - e^{-t/\tau}),$$

where $\tau = RC_1C_2/(C_1 + C_2)$.

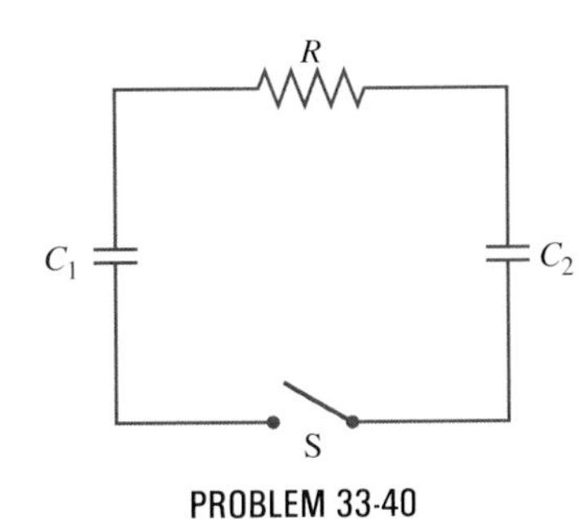

PROBLEM 33-40

33-41. The switch S of the circuit shown in the accompanying figure is closed at $t = 0$. Determine (*a*) the initial current through the battery and (*b*) the steady-state current through the battery.

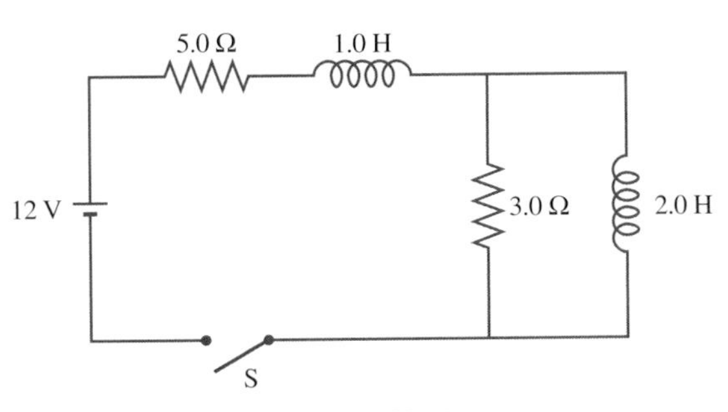

PROBLEM 33-41

33-42. In an oscillating RLC circuit, $R = 7.0\ \Omega$, $L = 10$ H, and $C = 3.0\ \mu$F. Initially, the capacitor has a charge of 8.0 μC and the current is zero. Calculate the charge on the capacitor (*a*) 5 cycles later and (*b*) 50 cycles later.

AC power is delivered through high-voltage transmission lines.

CHAPTER 34 ALTERNATING CURRENT CIRCUITS

PREVIEW

This chapter is devoted to simple ac circuits and selected applications of the diode and the operational amplifier. The main topics to be discussed here are the following:

1. **Four simple ac circuits.** We investigate circuits consisting of an ac source connected in turn to a resistor, a capacitor, an inductor, and a series combination of these three elements.
2. **Power in ac circuits.** We consider how power is dissipated and stored in ac circuits.
3. **Resonance.** An RLC series circuit operating at its resonant frequency is investigated.

*4. **The transformer.** The transformer and some of its applications are discussed.

*5. **Conversion of ac to dc: the diode.** We consider the properties of the diode and see how this device can be used to convert an ac voltage to a dc voltage.

*6. **The operational amplifier.** The operational amplifier and some of its applications are discussed.

Alternating current (ac) is the means by which electric power is delivered through transmission lines to our homes. Alternating current is produced by an alternating emf, which is generated in a power plant, as described in Sec. 30-4. The potential difference between the two sides of an outlet in a typical house alternates sinusoidally with a frequency of 60 Hz and an amplitude of either 156 or 311 V. The "voltage at a plug" can therefore be represented by $v = V_0 \sin \omega t$, where $V_0 = 156$ or 311 V* and $\omega = 120\pi$ rad/s.

Electric power is transmitted by alternating rather than direct current because of the practical advantages gained with the transformer. As you will learn in Sec. 34-4, the transformer is a device that can be used to change the amplitude of an alternating potential difference. This makes it possible to transmit power along lines at very high voltages, thereby minimizing resistive heating losses in the lines, then furnish that power to homes at much lower and safer voltages. Since constant potential differences are not affected by transformers, this capability is much more difficult to achieve with direct-current power transmission.

We will begin our study of ac circuits by using Kirchhoff's laws to analyze four simple circuits in which an alternating current flows. We have discussed the use of the resistor, the capacitor, and the inductor in circuits with batteries. These components will also be part of our ac circuits. However, since an alternating current is required, the constant source of emf is replaced here by an ac generator, a device that produces an oscillating emf.

34-1 Four Simple ac Circuits

We will assume that the power in the ac circuits to be discussed is furnished by an ac generator whose emf is given by

$$v(t) = V_0 \sin \omega t, \tag{34-1}$$

as shown in Fig. 34-1. The circuit representation of an ac generator is shown in Fig. 34-2. In the four ac circuits to be considered, (1) a resistor, (2) a capacitor, (3) an inductor, and (4) a series resistor-capacitor-inductor combination are connected in turn across an ac generator.

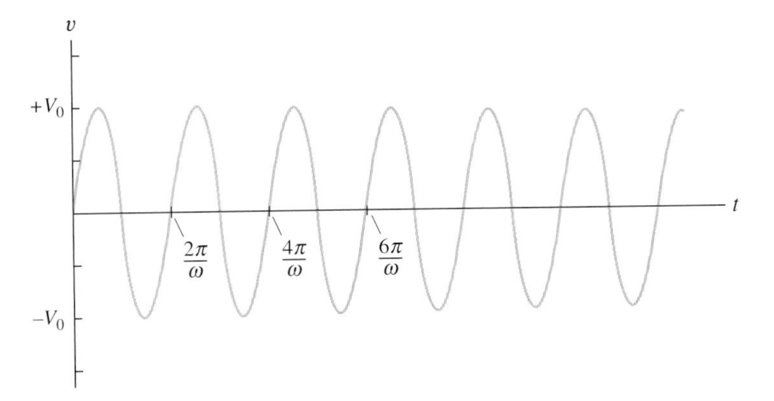

FIGURE 34-1 The output $v(t) = V_0 \sin \omega t$ of an ac generator.

FIGURE 34-2 Symbol used to represent an ac generator in a circuit diagram.

Resistor

From Kirchhoff's loop rule, the instantaneous voltage across the resistor of Fig. 34-3*a* is

$$v_R(t) = V_0 \sin \omega t, \tag{34-2}$$

and the instantaneous current through the resistor is

$$i_R(t) = \frac{v_R(t)}{R} = \frac{V_0}{R} \sin \omega t = I_0 \sin \omega t, \tag{34-3}$$

where $I_0 = V_0/R$ is the amplitude of the time-varying current. Plots of $i_R(t)$ and $v_R(t)$ are shown in Fig. 34-3*b*. Notice that both curves reach their maxima and minima at the same times; that is, *the current through and the voltage across the resistor are in phase*.

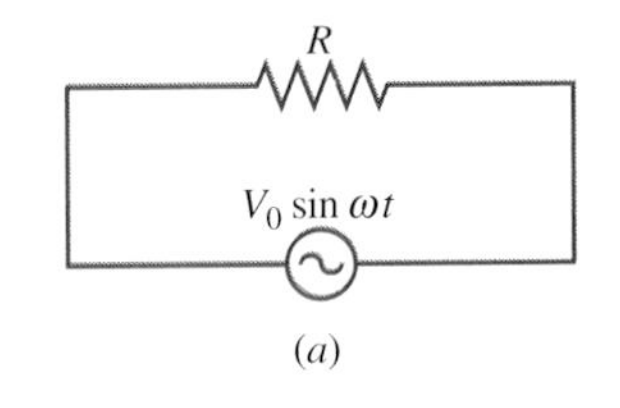

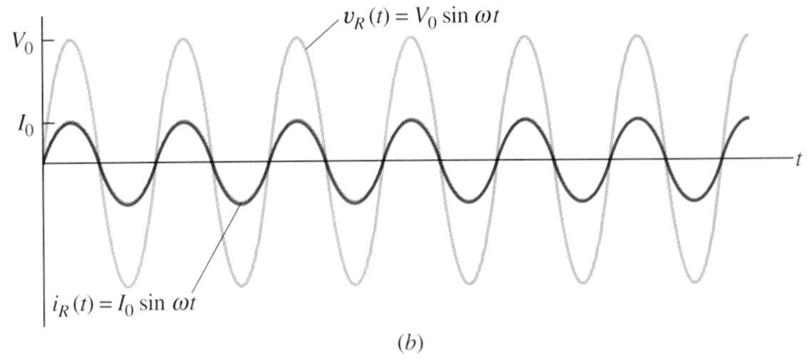

FIGURE 34-3 (*a*) A resistor connected across an ac generator. (*b*) The current $i_R(t)$ through and the voltage $v_R(t)$ across the resistor. Notice that the two quantities are in phase.

Graphical representations of the phase relationships between current and voltage are often useful in the analysis of ac circuits. Such representations are called *phasor diagrams*. The phasor diagram for $i_R(t)$ is shown in Fig. 34-4*a*. In this diagram, the arrow (or phasor) is rotating counterclockwise at a constant angular frequency ω, so we are viewing it at one instant in time. If the length of the arrow corresponds to the current amplitude I_0, the projection of the rotating arrow onto the vertical axis is $i_R(t) = I_0 \sin \omega t$, which is the instantaneous current.

*In Sec. 34-2, these voltage amplitudes will be shown to be equivalent to the 110- and 220-V ratings with which you are familiar.

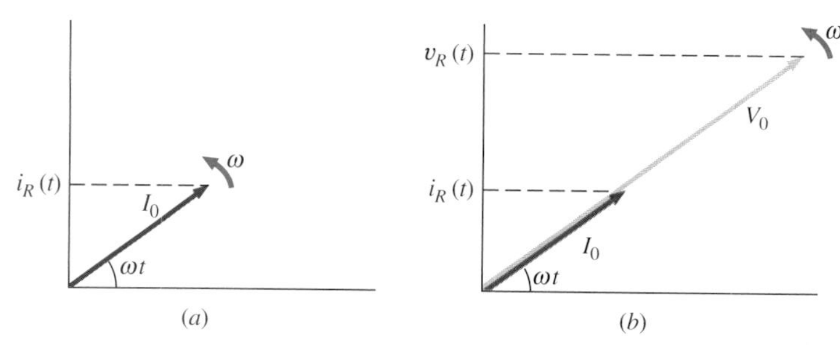

FIGURE 34-4 (*a*) The phasor diagram representing the current through the resistor of Fig. 34-3*a*. (*b*) The phasor diagram representing both $i_R(t)$ and $v_R(t)$.

Several different quantities can be depicted on the same phasor diagram. For example, both $i_R(t)$ and $v_R(t)$ are shown on the diagram of Fig. 34-4*b*. Since they have the same frequency and are in phase, their phasors point in the same direction and rotate together. The relative lengths of the two phasors are arbitrary because they represent different quantities.

Capacitor

From Kirchhoff's loop rule, the instantaneous voltage across the capacitor of Fig. 34-5*a* is

$$v_C(t) = V_0 \sin \omega t. \tag{34-4}$$

Consequently, the instantaneous charge on the capacitor is

$$q(t) = Cv_C(t) = CV_0 \sin \omega t.$$

Since the current in the circuit is the rate at which charge enters (or leaves) the capacitor,

$$i_C(t) = \frac{dq(t)}{dt} = \omega CV_0 \cos \omega t = I_0 \cos \omega t,$$

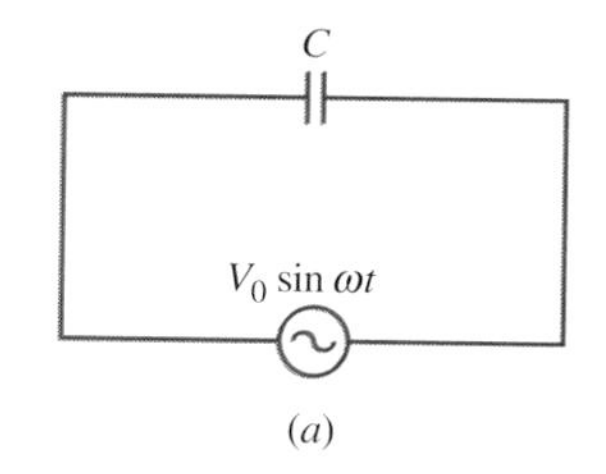

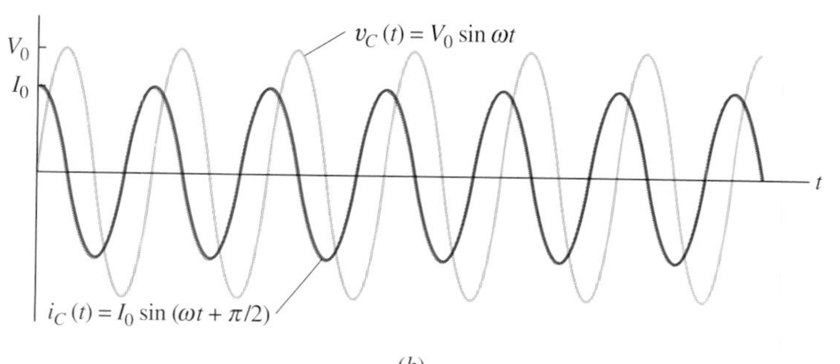

FIGURE 34-5 (*a*) A capacitor connected across an ac generator. (*b*) The current $i_C(t)$ through and the voltage $v_C(t)$ across the capacitor. Notice that $i_C(t)$ leads $v_C(t)$ by $\pi/2$ rad.

where $I_0 = \omega CV_0$ is the current amplitude. Using the trigonometric relationship $\cos \omega t = \sin(\omega t + \pi/2)$, we may express the instantaneous current as

$$i_C(t) = I_0 \sin\left(\omega t + \frac{\pi}{2}\right). \tag{34-5}$$

By dividing V_0 by I_0, we obtain an equation that looks very similar to Ohm's law:

$$\frac{V_0}{I_0} = \frac{1}{\omega C} = X_C. \tag{34-6}$$

The quantity X_C is analogous to resistance in a dc circuit in the sense that both quantities are a ratio of a voltage to a current. As a result, they have the same unit, the ohm. Keep in mind, however, that a capacitor stores and discharges electric energy, while a resistor dissipates it. X_C is known as the **capacitive reactance** of the capacitor.

A comparison of the expressions for $v_C(t)$ and $i_C(t)$ shows that there is a phase difference of $\pi/2$ rad between them. When these two quantities are plotted together, the current peaks a quarter cycle (or $\pi/2$ rad) ahead of the voltage, as illustrated in Fig. 34-5*b*. *The current through a capacitor leads the voltage across a capacitor by $\pi/2$ rad.*

The corresponding phasor diagram is shown in Fig. 34-6. Here the relationship between $i_C(t)$ and $v_C(t)$ is represented by having their phasors rotate at the same angular frequency, with the current phasor leading by $\pi/2$ rad.

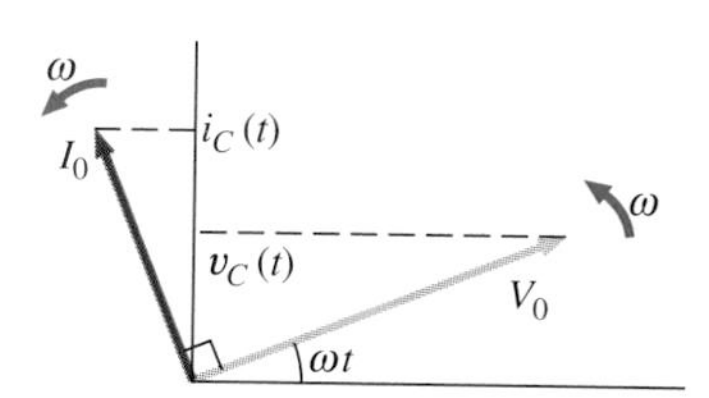

FIGURE 34-6 The phasor diagram for the capacitor of Fig. 34-5*a*. The current phasor leads the voltage phasor by $\pi/2$ rad as they both rotate with the same angular frequency.

Inductor

From Kirchhoff's loop rule, the voltage across the inductor of Fig. 34-7 is

$$v_L(t) = V_0 \sin \omega t. \tag{34-7}$$

Since $v_L(t) = L\, di_L(t)/dt$, we have

$$\frac{di_L(t)}{dt} = \frac{V_0}{L} \sin \omega t.$$

The current $i_L(t)$ is found by integrating this equation. Since the circuit does not contain a source of constant emf, there is no steady current in the circuit. Hence we can set the constant of integration, which represents the steady current in the circuit, equal to zero, and we have

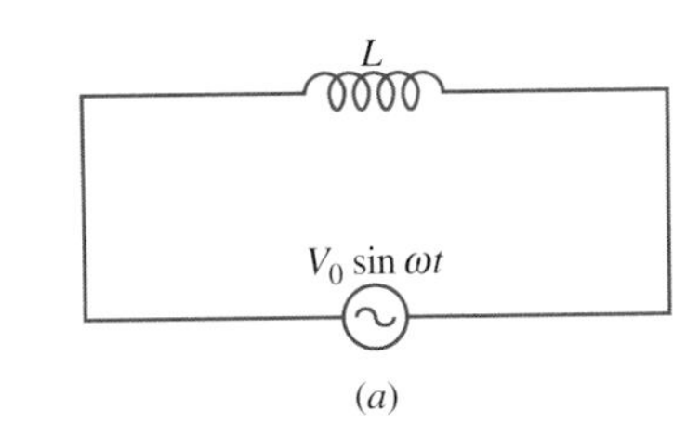

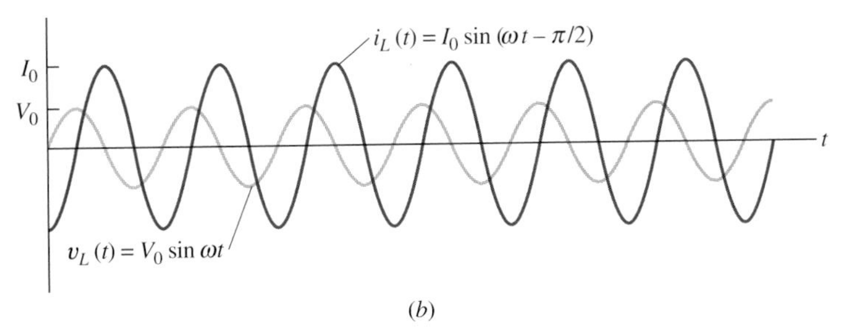

FIGURE 34-7 (*a*) An inductor connected across an ac generator. (*b*) The current $i_L(t)$ through and the voltage $v_L(t)$ across the inductor. Here $i_L(t)$ lags $v_L(t)$ by $\pi/2$ rad.

$$i_L(t) = -\frac{V_0}{\omega L}\cos \omega t = \frac{V_0}{\omega L}\sin\left(\omega t - \frac{\pi}{2}\right) = I_0 \sin\left(\omega t - \frac{\pi}{2}\right), \quad (34\text{-}8)$$

where $I_0 = V_0/\omega L$. The relationship between V_0 and I_0 may also be written in a form analogous to Ohm's law:

$$\frac{V_0}{I_0} = \omega L = X_L. \quad (34\text{-}9)$$

The quantity X_L is known as **inductive reactance** of the inductor and its unit is also the ohm.

There is a phase difference of $\pi/2$ rad between the current through and the voltage across the inductor. From Eqs. (34-7) and (34-8), *the current through an inductor lags the potential difference across an inductor by $\pi/2$ rad*. The phasor diagram for this case is shown in Fig. 34-8.

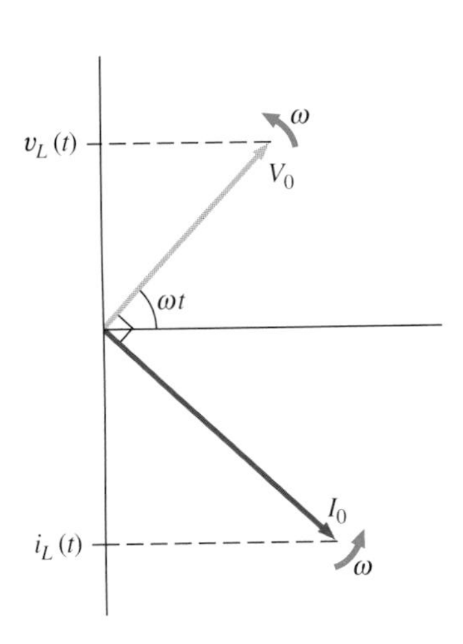

FIGURE 34-8 The phasor diagram for the inductor of Fig. 34-7*a*. The current phasor lags the voltage phasor by $\pi/2$ rad as they both rotate with the same angular frequency.

EXAMPLE 34-1 Simple AC Circuits

An ac generator produces an emf of amplitude 10 V at a frequency $f = 60$ Hz. Determine the voltages across and the currents through the circuit elements when the generator is connected to (*a*) a 100-Ω resistor, (*b*) a 10-μF capacitor, and (*c*) a 15-mH inductor.

SOLUTION The voltage across the terminals of the source is

$$v(t) = V_0 \sin \omega t = (10 \text{ V}) \sin 120\pi t,$$

where $\omega = 2\pi f = 120\pi$ rad/s is the angular frequency. Since $v(t)$ is also the voltage across each of the elements, we have

$$v(t) = v_R(t) = v_C(t) = v_L(t) = (10 \text{ V}) \sin 120\pi t.$$

(*a*) When $R = 100\ \Omega$, the amplitude of the current through the resistor is $I_0 = V_0/R = 10 \text{ V}/100\ \Omega = 0.10$ A, so

$$i_R(t) = (0.10 \text{ A}) \sin 120\pi t.$$

(*b*) From Eq. (34-6), the capacitive reactance is

$$X_C = \frac{1}{\omega C} = \frac{1}{(120\pi \text{ rad/s})(10 \times 10^{-6} \text{ F})} = 265\ \Omega,$$

so the maximum value of the current is

$$I_0 = \frac{V_0}{X_C} = \frac{10 \text{ V}}{265\ \Omega} = 3.8 \times 10^{-2} \text{ A},$$

and the instantaneous current is given by

$$i_C(t) = (3.8 \times 10^{-2} \text{ A}) \sin\left(120\pi t + \frac{\pi}{2}\right).$$

(*c*) From Eq. (34-9), the inductive reactance is

$$X_L = \omega L = (120\pi \text{ rad/s})(15 \times 10^{-3} \text{ H}) = 5.7\ \Omega.$$

The maximum current is therefore

$$I_0 = \frac{10 \text{ V}}{5.7\ \Omega} = 1.8 \text{ A},$$

and the instantaneous current is

$$i_L(t) = (1.8 \text{ A}) \sin\left(120\pi t - \frac{\pi}{2}\right).$$

DRILL PROBLEM 34-1

Repeat Example 34-1 for an ac source of amplitude 20 V and frequency 100 Hz.
ANS. (*a*) (20 V) sin $200\pi t$, (0.20 A) sin $200\pi t$;
(*b*) (20 V) sin $200\pi t$, (0.13 A) sin $(200\pi t + \pi/2)$;
(*c*) (20 V) sin $200\pi t$, (2.1 A) sin $(200\pi t - \pi/2)$.

Resistor, Capacitor, and Inductor in Series

The ac circuit shown in Fig. 34-9 is a series combination of a resistor, capacitor, and inductor connected across an ac source (an RLC series circuit) that produces an emf

$$v(t) = V_0 \sin \omega t.$$

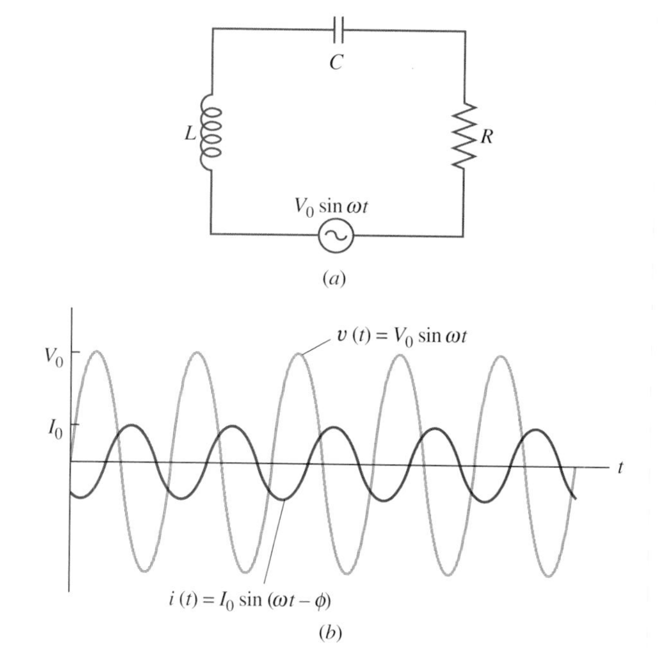

FIGURE 34-9 (*a*) An RLC series circuit. (*b*) A comparison of the generator output voltage and the current. The value of the phase difference ϕ depends on the values of *R*, *C*, and *L*.

Since the elements are in series, the same current flows through each. The relative phase between the current and the emf is not obvious when all three elements are present. Consequently, we will represent the current by the general expression

$$i(t) = I_0 \sin(\omega t - \phi), \tag{34-10}$$

where I_0 is the current amplitude and ϕ is the phase angle between the current and the applied voltage. Our task is to find I_0 and ϕ.

A phasor diagram involving $i(t)$, $v_R(t)$, $v_C(t)$, and $v_L(t)$ is very helpful for analyzing the circuit. As shown in Fig. 34-10, the phasor representing $v_R(t)$ points along the phasor for $i(t)$; its amplitude is $V_R = I_0 R$; the $v_C(t)$ phasor lags the $i(t)$ phasor by $\pi/2$ rad and has amplitude $V_C = I_0 X_C$; the phasor for $v_L(t)$ leads the $i(t)$ phasor by $\pi/2$ rad and has amplitude $V_L = I_0 X_L$.

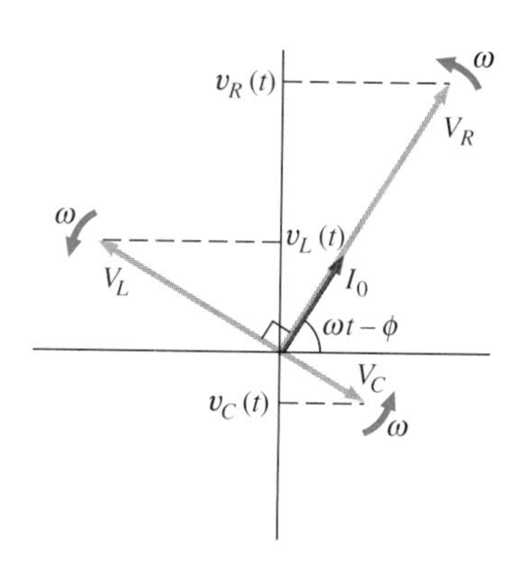

FIGURE 34-10 The phasor diagram for the RLC series circuit of Fig. 34-9*a*.

At any instant, the voltage across the RLC combination is $v_R(t) + v_L(t) + v_C(t) = v(t)$, the emf of the source. Since a component of a sum of vectors is the sum of the components of the individual vectors [e.g., $(\mathbf{A} + \mathbf{B})_y = A_y + B_y$], the projection of the vector sum of phasors onto the vertical axis is the sum of the vertical projections of the individual phasors. Hence if we add vectorially the phasors representing $v_R(t)$, $v_L(t)$, and $v_C(t)$ and then find the projection of the resultant onto the vertical axis, we obtain $v_R(t) + v_L(t) + v_C(t) = v(t) = V_0 \sin \omega t$.

The vector sum of the phasors is shown in Fig. 34-11. The resultant phasor has an amplitude V_0 and is directed at an angle ϕ with respect to the $v_R(t)$ [or $i(t)$] phasor. The projection of this resultant phasor onto the vertical axis is $v(t) = V_0 \sin \omega t$. We can easily determine the unknown quantities I_0 and ϕ from the geometry of the phasor diagram. For the phase angle,

$$\phi = \tan^{-1} \frac{V_L - V_C}{V_R} = \tan^{-1} \frac{I_0 X_L - I_0 X_C}{I_0 R},$$

which, with cancellation of I_0, becomes

$$\phi = \tan^{-1} \frac{X_L - X_C}{R}. \tag{34-11}$$

Furthermore, from Pythagoras' theorem,

$$V_0 = \sqrt{V_R^2 + (V_L - V_C)^2} = \sqrt{(I_0 R)^2 + (I_0 X_L - I_0 X_C)^2}$$
$$= I_0 \sqrt{R^2 + (X_L - X_C)^2}.$$

The current amplitude is therefore

$$I_0 = \frac{V_0}{\sqrt{R^2 + (X_L - X_C)^2}} = \frac{V_0}{Z}, \tag{34-12}$$

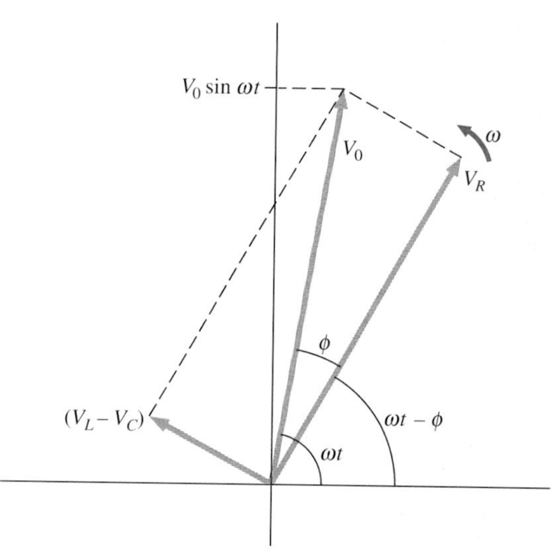

FIGURE 34-11 The resultant of the phasors for $v_L(t)$, $v_C(t)$, and $v_R(t)$ is equal to the phasor for $v(t) = V_0 \sin \omega t$. The $i(t)$ phasor (not shown) is aligned with the $v_R(t)$ phasor.

where

$$Z = \sqrt{R^2 + (X_L - X_C)^2} \tag{34-13}$$

is known as the **impedance** of the circuit. Its unit is the ohm, and it is the counterpart of resistance in a dc circuit.

Power capacitors are used to balance the impedance of the effective inductance in transmission lines.

EXAMPLE 34-2 AN RLC SERIES CIRCUIT

The output of an ac generator connected to an RLC series combination has a frequency of 200 Hz and an amplitude of 0.100 V. If $R = 4.00\ \Omega$, $L = 3.00 \times 10^{-3}$ H, and $C = 8.00 \times 10^{-4}$ F, what are (*a*) the capacitive reactance, (*b*) the inductive reactance, (*c*) the impedance, (*d*) the current amplitude, and (*e*) the phase difference between the current and the emf of the generator?

SOLUTION (*a*) From Eq. (34-6), the capacitive reactance is

$$X_C = \frac{1}{\omega C} = \frac{1}{2\pi(200\ \text{Hz})(8.00 \times 10^{-4}\ \text{F})} = 0.995\ \Omega.$$

(*b*) From Eq. (34-9), the inductive reactance is

$$X_L = \omega L = 2\pi(200\ \text{Hz})(3.00 \times 10^{-3}\ \text{H}) = 3.77\ \Omega.$$

(*c*) Substituting the values of R, X_C, and X_L into $Z = \sqrt{R^2 + (X_L - X_C)^2}$, we obtain for the impedance

$$Z = \sqrt{(4.00\ \Omega)^2 + (3.77\ \Omega - 0.995\ \Omega)^2} = 4.87\ \Omega.$$

(*d*) The current amplitude is

$$I_0 = \frac{V_0}{Z} = \frac{0.100\ \text{V}}{4.87\ \Omega} = 2.05 \times 10^{-2}\ \text{A}.$$

(*e*) From Eq. (34-11), the phase difference between the current and the emf is

$$\phi = \tan^{-1}\frac{X_L - X_C}{R} = \tan^{-1}\frac{2.77\ \Omega}{4.00\ \Omega} = 0.607\ \text{rad}.$$

DRILL PROBLEM 34-2

Find the voltages across the resistor, the capacitor, and the inductor in the circuit of Fig. 34-9 using $v(t) = V_0 \sin \omega t$ as the output of the ac generator.
ANS. $v_R = (V_0 R/Z) \sin(\omega t - \phi)$; $v_C = (V_0 X_C/Z) \sin(\omega t - \phi - \pi/2) = -(V_0 X_C/Z) \cos(\omega t - \phi)$; $v_L = (V_0 X_L/Z) \sin(\omega t - \phi + \pi/2) = (V_0 X_L/Z) \cos(\omega t - \phi)$.

34-2 POWER IN AC CIRCUITS

A circuit element dissipates or produces power according to $P = IV$, where I is the current through the element and V the voltage across it. Since the current and the voltage both depend on time in an ac circuit, the power

$$p(t) = i(t)v(t)$$

is also time-dependent and is known as the **instantaneous power**. A plot of $p(t)$ for various circuit elements is shown in Fig. 34-12. For a resistor, $i(t)$ and $v(t)$ are in phase and therefore always have the same sign. (See Fig. 34-3*b*.) For a capacitor or inductor, the relative sign of $i(t)$ and $v(t)$ varies over a cycle due to their phase difference. (See Figs. 34-5*b* and

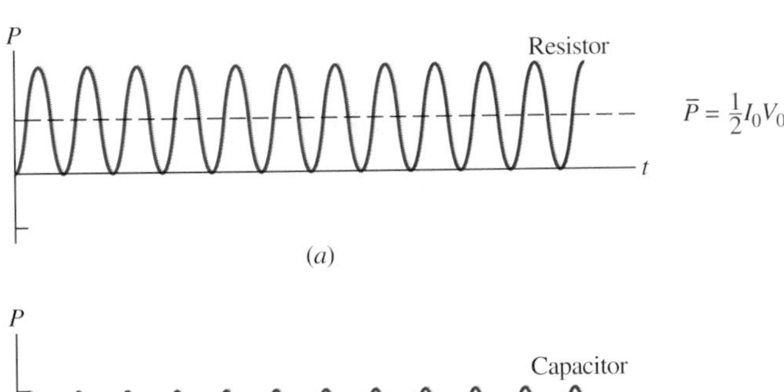

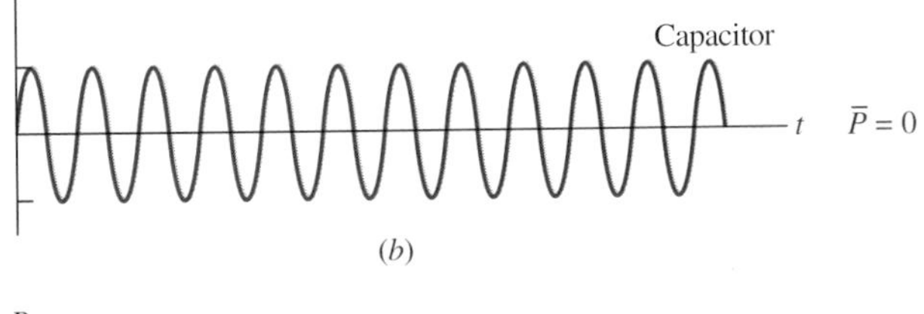

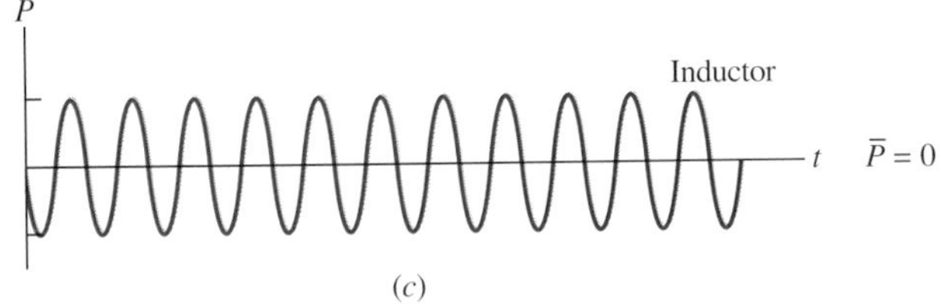

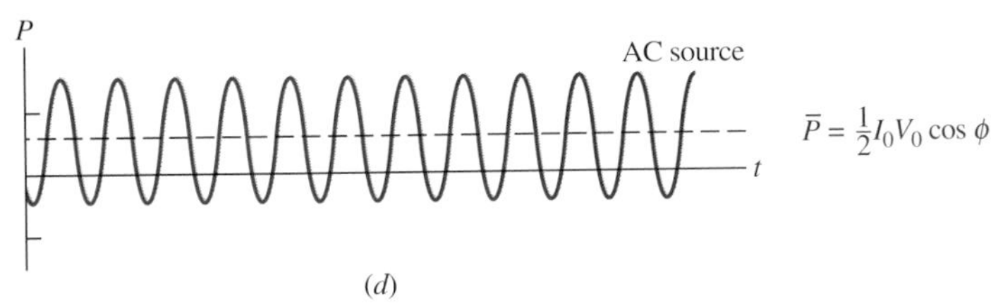

FIGURE 34-12 Graphs of instantaneous power for various circuit elements. (*a*) For the resistor, $\bar{P} = I_0V_0/2$, while for (*b*) the capacitor and (*c*) the inductor, $\bar{P} = 0$. (*d*) For the source, $\bar{P} = I_0V_0(\cos\phi)/2$, which may be positive, negative, or zero, depending on ϕ.

34.7*b*.) Consequently, $p(t)$ is positive at some times and negative at others, indicating that capacitive and inductive elements produce power at some instants and absorb it at others.

Because instantaneous power varies in both magnitude and sign over a cycle, it seldom has any practical importance. What we're almost always concerned with is the power averaged over time, which we will refer to as the **average power**. It is defined by the time average of the instantaneous power over one cycle:

$$\bar{P} = \frac{1}{T}\int_0^T p(t)\,dt, \tag{34-14}$$

where $T = 2\pi/\omega$ is the period of the oscillations. With the substitutions $v(t) = V_0 \sin \omega t$ and $i(t) = I_0 \sin(\omega t - \phi)$, this integral becomes

$$\bar{P} = \frac{I_0 V_0}{T}\int_0^T \sin(\omega t - \phi)\sin \omega t\,dt.$$

Using $\sin(A - B) = \sin A \cos B - \sin B \cos A$, we obtain

$$\bar{P} = \frac{I_0 V_0 \cos\phi}{T}\int_0^T \sin^2 \omega t\,dt - \frac{I_0 V_0 \sin\phi}{T}\int_0^T \sin \omega t \cos \omega t\,dt.$$

Evaluation of these two integrals yields

$$\frac{1}{T}\int_0^T \sin^2 \omega t\,dt = \frac{1}{2}$$

and

$$\frac{1}{T}\int_0^T \sin \omega t \cos \omega t\,dt = 0.$$

Hence the average power associated with a circuit element is given by

$$\bar{P} = \tfrac{1}{2} I_0 V_0 \cos\phi. \tag{34-15}$$

In engineering applications, $\cos\phi$ is known as the **power factor**.

For a resistor, $\phi = 0$, so the average power dissipated is

$$\bar{P} = \tfrac{1}{2} I_0 V_0.$$

A comparison of $p(t)$ and $\bar{P}$ is shown in Fig. 34-12*a*. To make $\bar{P} = (1/2)I_0 V_0$ look like its dc counterpart, we use the *root mean square* (or *rms*) values I_{rms} and V_{rms} of the current and the voltage. By definition, these are $I_{rms} = \sqrt{\overline{i^2}}$ and $V_{rms} = \sqrt{\overline{v^2}}$, where

$$\overline{i^2} = \frac{1}{T}\int_0^T i^2(t)\,dt \quad \text{and} \quad \overline{v^2} = \frac{1}{T}\int_0^T v^2(t)\,dt.$$

With $i(t) = I_0 \sin(\omega t - \phi)$ and $v(t) = V_0 \sin \omega t$, we obtain

$$I_{rms} = \frac{1}{\sqrt{2}} I_0 \quad \text{and} \quad V_{rms} = \frac{1}{\sqrt{2}} V_0. \tag{34-16}$$

We may then write for the average power dissipated by a resistor,

$$\bar{P} = \tfrac{1}{2} I_0 V_0 = I_{rms} V_{rms} = I_{rms}^2 R. \tag{34-17}$$

Alternating voltages and currents are usually described in terms of their rms values. For example, the 110 V from a household plug is an rms value. The amplitude of this source is $110\sqrt{2}$ V = 156 V. Since most ac meters are calibrated in terms of rms values, a typical ac voltmeter placed across a household plug will read 110 V.

For a capacitor and an inductor, $\phi = \pi/2$ and $-\pi/2$ rad, respectively. Since $\cos \pi/2 = \cos(-\pi/2) = 0$, we find from Eq. (34-15) that the average power dissipated by either of these elements is

$$\bar{P} = 0.$$

Capacitors and inductors absorb energy from the circuit during one half-cycle, then discharge it back to the circuit during the other half-cycle. This behavior is illustrated in the plots of Fig. 34-12*b* and *c*, which show $p(t)$ oscillating sinusoidally about zero.

The phase angle for an ac generator may have any value. If $\cos\phi > 0$, the generator produces power; if $\cos\phi < 0$, it absorbs power. In terms of rms values, the average power of an ac generator is written as

$$\bar{P} = I_{rms} V_{rms} \cos\phi. \tag{34-18}$$

For the generator in an RLC circuit,

$$\tan\phi = \frac{X_L - X_C}{R},$$

and

$$\cos\phi = \frac{R}{\sqrt{R^2 + (X_L - X_C)^2}} = \frac{R}{Z}.$$

Hence the average power of the generator is

$$\bar{P} = I_{rms} V_{rms} \cos\phi = \frac{V_{rms}}{Z} V_{rms} \frac{R}{Z} = \frac{V_{rms}^2 R}{Z^2}. \tag{34-19a}$$

This can also be written as

$$\bar{P} = I_{rms}^2 R, \tag{34-19b}$$

which designates that the power produced by the generator is dissipated in the resistor.

EXAMPLE 34-3 Power Output of a Generator

An ac generator whose emf is given by

$$v(t) = (4.00 \text{ V}) \sin(1.00 \times 10^4 t)$$

is connected to an RLC circuit for which $L = 2.00 \times 10^{-3}$ H, $C = 4.00 \times 10^{-6}$ F, and $R = 5.00\ \Omega$. (*a*) What is the root mean square voltage V_{rms} across the generator? (*b*) What is the impedance of the circuit? (*c*) What is the average power output of the generator?

SOLUTION (*a*) Since $V_0 = 4.00$ V, the rms voltage across the generator is

$$V_{rms} = \frac{1}{\sqrt{2}}(4.00\text{ V}) = 2.83\text{ V}.$$

(*b*) The impedance of the circuit is

$$\begin{aligned} Z &= \sqrt{R^2 + (X_L - X_C)^2} \\ &= \left\{(5.00\ \Omega)^2 + \left[(1.00 \times 10^4\text{ rad/s})(2.00 \times 10^{-3}\text{ H}) - \frac{1}{(1.00 \times 10^4\text{ rad/s})(4.00 \times 10^{-6}\text{ F})}\right]^2\right\}^{1/2} \\ &= 7.07\ \Omega. \end{aligned}$$

(*c*) From Eq. (34-19*a*), the average power transferred to the circuit is

$$\bar{P} = \frac{V_{rms}^2 R}{Z^2} = \frac{(2.83\text{ V})^2(5.00\ \Omega)}{(7.07\ \Omega)^2} = 0.800\text{ W}.$$

DRILL PROBLEM 34-3

An ac voltmeter attached across the terminals of a 45-Hz ac generator reads 7.07 V. Write an expression for the emf of the generator.
ANS. $v(t) = (10.0\text{ V})\sin 90\pi t$.

DRILL PROBLEM 34-4

Show that the rms voltages across a resistor, a capacitor, and an inductor in an ac circuit where the root mean square current is I_{rms} are given by $I_{rms}R$, $I_{rms}X_C$, and $I_{rms}X_L$, respectively. Determine these values for the components of the RLC circuit of Example 34-3.
ANS. 2.00 V; 10.0 V; 8.00 V.

34-3 RESONANCE

In the RLC series circuit of Fig. 34-9, the current amplitude is, from Eq. (34-12),

$$I_0 = \frac{V_0}{\sqrt{R^2 + (\omega L - 1/\omega C)^2}}.$$

If we are able to vary the frequency of the ac generator while keeping the amplitude of its output voltage constant, then the current will change accordingly. A plot of I_0 versus ω is shown in Fig. 34-13. In Chap. 14 we encountered a similar graph where the amplitude of a damped harmonic oscillator was plotted against the angular frequency of a sinusoidal driving force. (See Fig. 14-15.) This similarity is more than just a coincidence, as shown by the application of Kirchhoff's loop rule to the circuit of Fig. 34-9*a*. This yields

$$L\frac{di}{dt} + iR + \frac{q}{C} = V_0 \sin \omega t,$$

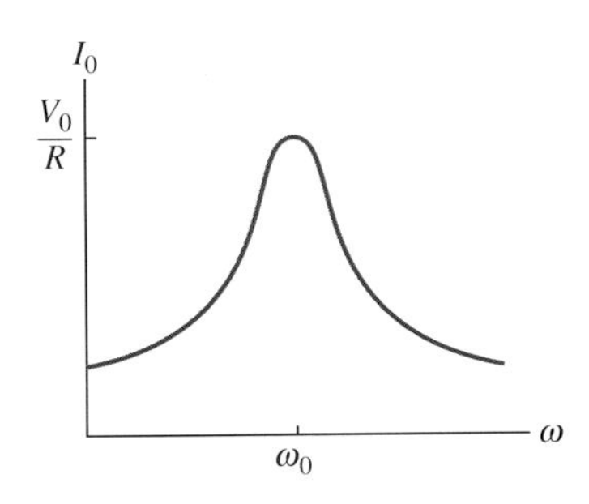

FIGURE 34-13 At an RLC circuit's resonant frequency, $\omega_0 = \sqrt{1/LC}$, the current amplitude is at its maximum value.

or

$$L\frac{d^2q}{dt^2} + R\frac{dq}{dt} + \frac{1}{C}q = V_0 \sin \omega t, \tag{34-20}$$

where $dq(t)/dt$ is substituted for $i(t)$. A comparison of Eqs. (34-20) and (14-24) clearly demonstrates that *the driven RLC series circuit is the electrical analog of the driven damped harmonic oscillator.*

The **resonant frequency** f_0 of the RLC circuit is the frequency at which the amplitude of the current is a maximum. By inspection, this corresponds to the angular frequency $\omega_0 = 2\pi f_0$ at which the impedance Z in Eq. (34-12) is a minimum, or when

$$\omega_0 L = \frac{1}{\omega_0 C},$$

and

$$\omega_0 = \sqrt{\frac{1}{LC}}. \tag{34-21}$$

This is the **resonant angular frequency** of the circuit. Substituting ω_0 into Eqs. (34-11) through (34-13), we find that at resonance,

$$\phi = \tan^{-1}(0) = 0,$$
$$I_0 = V_0/R,$$

and

$$Z = R.$$

Therefore at resonance an RLC circuit is *purely resistive*, with the applied emf and current in phase.

What happens to the power at resonance? Equation (34-19*a*) tells us how the average power transferred from an ac generator to the RLC combination varies with frequency. Notice that $\bar{P}$ also reaches a maximum when Z, which depends on the frequency, is a minimum; that is, when $X_L = X_C$ and $Z = R$. Thus *at resonance, the average power output of the source in an RLC series circuit is a maximum*. From Eq. (34-19*a*), this maximum is V_{rms}^2/R.

Figure 34-14 is a typical plot of $\bar{P}$ versus ω in the region of maximum power output. The **bandwidth** $\Delta\omega$ of the resonance peak is defined as the range of ω over which $\bar{P}$ is greater than

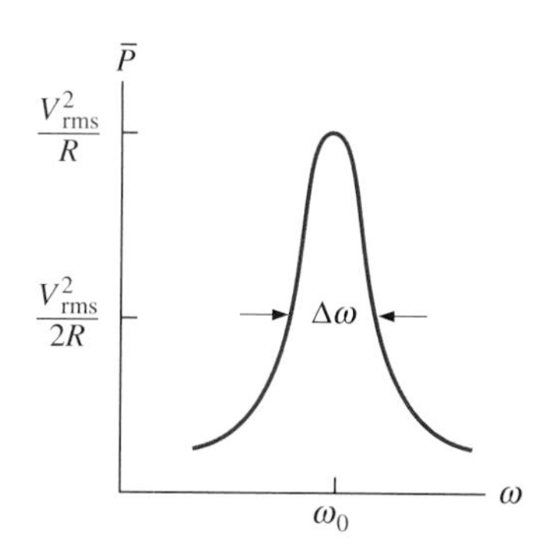

FIGURE 34-14 Like the current, the average power transferred from an ac generator to an RLC circuit peaks at the resonant frequency.

one-half the maximum value of $\bar{P}$. The sharpness of the peak is described by a dimensionless quantity known as the **quality factor** Q of the circuit. By definition,

$$Q = \frac{\omega_0}{\Delta\omega}, \tag{34-22a}$$

where ω_0 is the resonant angular frequency. A high Q indicates a sharp resonance peak. It can be shown (see Prob. 34-57) that Q is given in terms of the circuit parameters by

$$Q = \frac{\omega_0 L}{R}. \tag{34-22b}$$

Resonant circuits are commonly used to pass or reject selected frequency ranges. This is done by adjusting the value of one of the elements and hence "tuning" the circuit to a particular resonant frequency. For example, in radios and in television sets, the receiver is tuned to the desired station by adjusting the resonant frequency of its circuitry to match the frequency of the station. If the tuning circuit has a high Q, it will have a small bandwidth, so signals from other stations at even slightly different frequencies from the resonant frequency encounter a high impedance and are not passed by the circuit.

When a metal detector comes near a piece of metal, the self-inductance of one of its coils changes. This causes a shift in the resonant frequency of a circuit containing the coil. That shift is detected by the circuitry and transmitted to the diver by means of the headphones.

EXAMPLE 34-4 Resonance in an RLC Series Circuit

What is the resonant frequency of the circuit of Example 34-2? If the ac generator is set to this frequency without changing the amplitude of the output voltage, what is the amplitude of the current?

SOLUTION The resonant frequency is found from Eq. (34-21):

$$f_0 = \frac{1}{2\pi}\sqrt{\frac{1}{LC}} = \frac{1}{2\pi}\sqrt{\frac{1}{(3.00 \times 10^{-3}\text{ H})(8.00 \times 10^{-4}\text{ F})}}$$
$$= 1.03 \times 10^2\text{ Hz}.$$

At resonance, the impedance of the circuit is purely resistive, and the current amplitude is

$$I_0 = \frac{0.100\text{ V}}{4.00\ \Omega} = 2.50 \times 10^{-2}\text{ A}.$$

EXAMPLE 34-5 Power Transfer in an RLC Series Circuit at Resonance

(*a*) What is the resonant angular frequency of an RLC circuit with $R = 0.200\ \Omega$, $L = 4.00 \times 10^{-3}$ H, and $C = 2.00 \times 10^{-6}$ F? (*b*) If an ac source of constant amplitude 4.00 V is set to this frequency, what is the average power transferred to the circuit? (*c*) Determine Q and the bandwidth of this circuit.

SOLUTION (*a*) The resonant angular frequency is

$$\omega_0 = \sqrt{\frac{1}{LC}} = \sqrt{\frac{1}{(4.00 \times 10^{-3}\text{ H})(2.00 \times 10^{-6}\text{ F})}}$$
$$= 1.12 \times 10^4\text{ rad/s}.$$

(*b*) At this frequency, the average power transferred to the circuit is a maximum. It is

$$\bar{P} = \frac{V_{rms}^2}{R} = \frac{[(1/\sqrt{2})(4.00\text{ V})]^2}{0.200\ \Omega} = 40.0\text{ W}.$$

(*c*) The quality factor of the circuit is

$$Q = \frac{\omega_0 L}{R} = \frac{(1.12 \times 10^4\text{ rad/s})(4.00 \times 10^{-3}\text{ H})}{0.200\ \Omega} = 224.$$

We then find for the bandwidth

$$\Delta\omega = \frac{\omega_0}{Q} = \frac{1.12 \times 10^4\text{ rad/s}}{224} = 50.0\text{ rad/s}.$$

DRILL PROBLEM 34-5

In the circuit of Fig. 34-9, $L = 2.0 \times 10^{-3}$ H, $C = 5.0 \times 10^{-4}$ F, and $R = 40\ \Omega$. (*a*) What is the resonant frequency? (*b*) What is the impedance of the circuit at resonance? (*c*) If the voltage amplitude is 10 V, what is $i(t)$ at resonance? (*d*) The frequency of the ac generator is now changed to 200 Hz. Calculate the phase difference between the current and the emf of the generator.
ANS. (*a*) 159 Hz; (*b*) 40 Ω; (*c*) (0.25 A) sin $10^3 t$; (*d*) 0.023 rad.

DRILL PROBLEM 34-6

What happens to the resonant frequency of an RLC series circuit when the following quantities are increased by a factor of 4? (*a*) the capacitance, (*b*) the self-inductance, and (*c*) the resistance.
ANS. (*a*) Halved; (*b*) halved; (*c*) same.

DRILL PROBLEM 34-7

The resonant angular frequency of an RLC series circuit is 4.0×10^2 rad/s. An ac source operating at this frequency transfers an average power of 2.0×10^{-2} W to the circuit. The resistance of the circuit is 0.50 Ω. Write an expression for the emf of the source.
ANS. $v(t) = (0.14 \text{ V}) \sin (4.0 \times 10^2 t)$.

*34-4 The Transformer

Although ac electric power is produced at relatively low voltages, it is sent through transmission lines at very high voltages (as high as 500 kV). The same power can be transmitted at different voltages because power is the product $I_{\text{rms}} V_{\text{rms}}$. (For simplicity, we ignore the phase factor $\cos \phi$.) A particular power requirement can therefore be met with a low voltage and a high current, or a high voltage and a low current. The advantage of the latter choice is that it results in lower $I_{\text{rms}}^2 R$ ohmic losses in the transmission lines, which can be significant in lines that are many kilometers long.

So, the alternating emf's produced at power plants are "stepped up" to very high voltages before being transmitted through power lines; then they must be "stepped down" to relatively safe values (110 or 220 V rms) before they are introduced into homes. The device that accomplishes these important tasks is the *transformer*. As Fig. 34-15 illustrates, a transformer basically consists of two separated coils wrapped around a soft iron core. The primary coil has N_p turns and is connected to an alternating voltage $v_p(t)$. The secondary winding has N_s turns and is connected to a load resistor R_s. We'll assume the ideal case for which all magnetic field lines are confined to the core so that the same magnetic flux cuts each turn of both the primary and secondary coils. We will also neglect energy losses to magnetic hysteresis, to ohmic heating in the windings, and to ohmic heating of the induced eddy currents in the core. A good transformer can have losses as low as 1 percent of the transmitted power, so this is not a bad assumption.

Transformers are used to step down the high voltages in transmission lines to the 110 or 220 V used in homes.

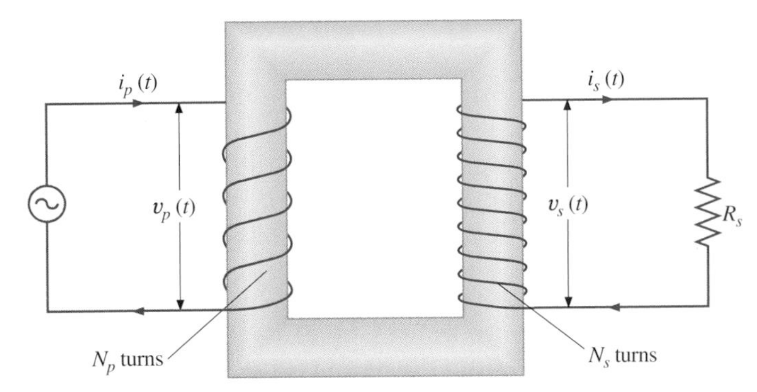

FIGURE 34-15 A transformer. The two coils are wrapped around a soft iron core.

To analyze the transformer circuit, we first consider the primary coil. The input voltage $v_p(t)$ is equal to the potential difference induced across the primary coil. From Faraday's law, the induced potential difference is $-N_p(d\phi/dt)$, where ϕ is the flux through one turn of the primary coil. Thus

$$v_p(t) = -N_p \frac{d\phi}{dt}.$$

Similarly, the output voltage $v_s(t)$ delivered to the load resistor must equal the potential difference induced across the secondary coil. Since the transformer is ideal, the flux through every turn of the secondary is also ϕ, and

$$v_s(t) = -N_s \frac{d\phi}{dt}.$$

Combining the last two equations, we have

$$v_s(t) = \frac{N_s}{N_p} v_p(t). \qquad (34\text{-}23)$$

Hence with appropriate values for N_s and N_p, the input voltage $v_p(t)$ may be "stepped up" ($N_s > N_p$) or "stepped down" ($N_s < N_p$) to $v_s(t)$, the output voltage.

From the law of energy conservation, the power introduced at any instant by $v_p(t)$ to the primary coil is equal to the power dissipated in the resistor of the secondary circuit; thus

$$i_p(t)v_p(t) = i_s(t)v_s(t),$$

which, when combined with Eq. (34-23), gives

$$i_s(t) = \frac{N_p}{N_s} i_p(t). \qquad (34\text{-}24)$$

If the voltage is stepped up, the current is stepped down, and vice versa.

Finally, we can use $i_s(t) = v_s(t)/R_s$ along with Eqs. (34-23) and (34-24) to obtain

$$v_p(t) = i_p\left[\left(\frac{N_p}{N_s}\right)^2 R_s\right],$$

which tells us that the input voltage $v_p(t)$ does not "see" a resistance R_s, but rather a resistance

$$R_p = \left(\frac{N_p}{N_s}\right)^2 R_s. \qquad (34\text{-}25)$$

Our analysis has been based on instantaneous values of voltage and current. However, the resulting equations are not limited to instantaneous values; they hold also for maximum and rms values.

EXAMPLE 34-6 A STEP-DOWN TRANSFORMER

A transformer on a utility pole steps the rms voltage down from 12 kV to 240 V. (*a*) What is the ratio of the number of secondary turns to the number of primary turns? (*b*) If the input current to the transformer is 2.0 A, what is the output current?

SOLUTION (*a*) Using Eq. (34-23) with rms values V_p and V_s, we have

$$\frac{N_s}{N_p} = \frac{240 \text{ V}}{12 \times 10^3 \text{ V}} = \frac{1}{50},$$

so the primary coil has 50 times the number of turns in the secondary coil.

(*b*) From Eq. (34-24), the output rms current I_s is

$$I_s = \frac{N_p}{N_s} I_p = (50)(2.0 \text{ A}) = 100 \text{ A}.$$

DRILL PROBLEM 34-8

A transformer steps the line voltage down from 110 to 9.0 V so that a current of 0.50 A can be delivered to a doorbell. (*a*) What is the ratio of the number of turns in the primary and secondary coils? (*b*) What is the current in the primary coil? (*c*) What is the resistance seen by the 110-V source?
ANS. (*a*) 12; (*b*) 0.042 A; (*c*) 2.6×10^3 Ω.

*34-5 CONVERSION OF AC TO DC: THE DIODE

In this and the next section we conclude our study of circuits by examining the electrical properties of two widely used semiconductor devices: the diode and the operational amplifier. We begin with the diode, which we discussed briefly in Sec. 26-2. Its circuit representation is shown in Fig. 34-16*a*. For reasons not discussed here, a diode effectively allows current to flow through it in only one direction, that indicated by the arrowhead in its representation. Figure 34-16*b* shows a plot of the current through a typical diode versus the voltage across it. If point A is at a higher potential than point B, current readily flows through the diode, which then acts like a conductor of very low resistance. However, if point B is at the higher potential, the current is very small and the effective resistance of the diode is very large. You can think of the diode as offering zero resistance to current flow in one direction and infinite resistance to flow in the other direction.

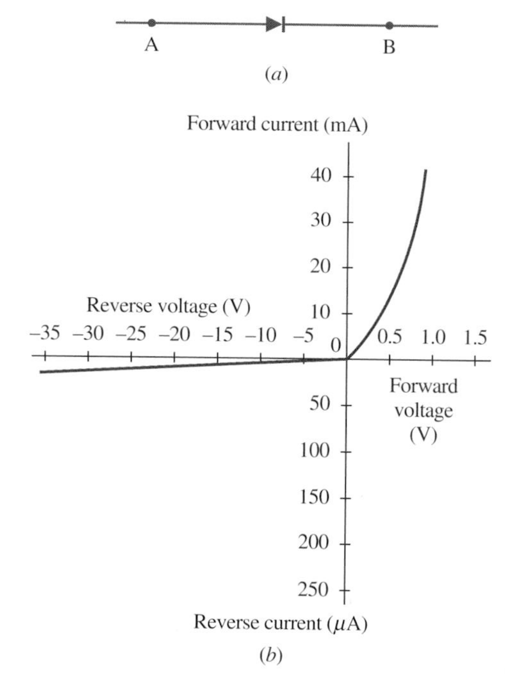

FIGURE 34-16 (*a*) Circuit representation of a diode. (*b*) A typical plot of the current through a diode as a function of the voltage across it. Notice that the forward current is in milliamps and the reverse current in microamps.

Diodes are commonly used in *rectifier circuits*, which convert an ac voltage to a voltage with a net polarity in one direction. In fact, the output voltages of rectifier circuits used in most electronic circuits are almost constant (dc). The simplest example of a rectifier is the *half-wave rectifier* of Fig. 34-17*a*. During the positive half-cycles of the ac generator's output $v(t) = V_0 \sin \omega t$ (see Fig. 34-17*b*), the diode behaves as a low-resistance conductor and allows the current to flow through it. The voltage $v_R(t)$ across the resistor is then the same as $v(t)$. However, during the negative half-cycles, the diode acts like an infinite resistance—no current then flows through the circuit, and $v_R(t)$ is zero. A plot of $v_R(t)$ versus time is shown in Fig. 34-17*c*. While this voltage is obviously not constant, it no longer reverses sign like $v(t)$.

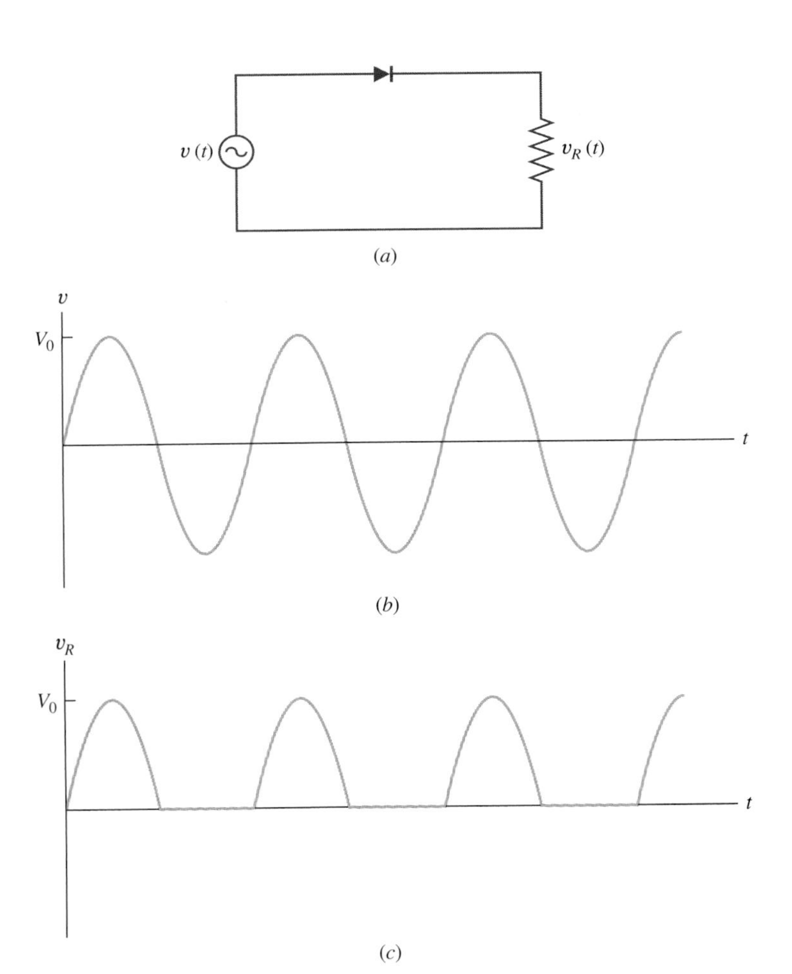

FIGURE 34-17 Half-wave rectification.

To further reduce the variation in $v_R(t)$, a capacitor may be added to the circuit as shown in Fig. 34-18*a*. During the positive half-cycles of $v(t)$, the capacitor becomes charged; it discharges during the negative half-cycles. The result is a "smoothing out" of $v_R(t)$, as illustrated in Fig. 34-18*b*.

A steadier dc output is obtained with *full-wave rectification*. A circuit that performs this function is shown in Fig. 34-19*a*. It consists of four diodes D_1 through D_4 and is known as a

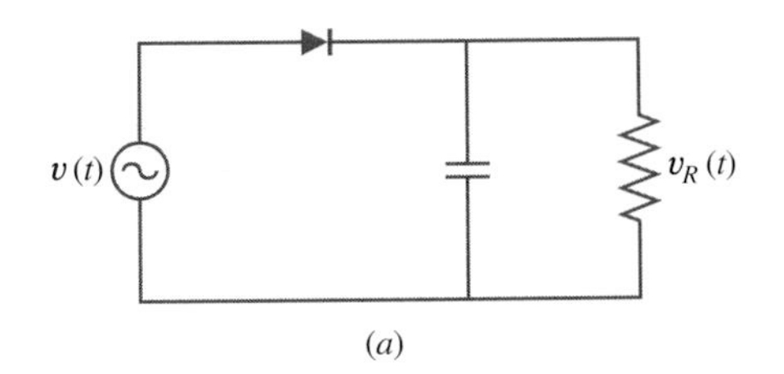

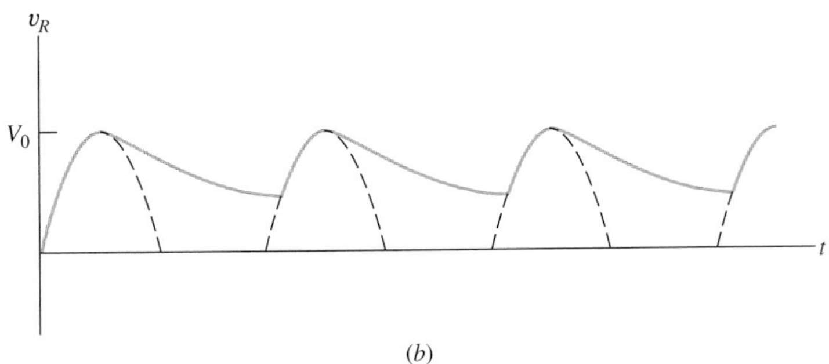

FIGURE 34-18 When a capacitor is used in half-wave rectification, its discharge "smooths out" the output voltage $v_R(t)$.

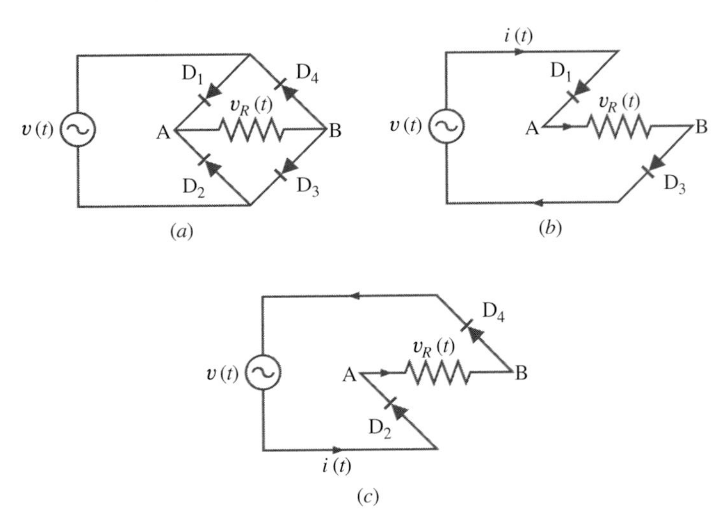

FIGURE 34-19 (*a*) A bridge rectifier that performs full-wave rectification. The output is the voltage $v_R(t)$ across the resistor. (*b*) The equivalent circuit during the positive half-cycle of $v(t)$. (*c*) The equivalent circuit during the negative half-cycle of $v(t)$.

bridge rectifier. The output voltage $v_R(t)$ is the potential difference across the ends AB of the resistor, while the input voltage is $v(t)$, which is produced by an ac source. During the positive half-cycles of $v(t)$, D_2 and D_4 block current while D_1 and D_3 conduct, so current flows through the rectifier as depicted in Fig. 34-19*b*. During the negative half-cycles, the current from the ac generator reverses direction, and the rectifier circuit is reduced to that of Fig. 34-19*c*. Because of the diodes, the current through the resistor always flows in the same direction, even though the current produced by $v(t)$ alternates. As a result, the voltage $v_R(t)$ across the resistor has the form shown in Fig. 34-20*b*. Again, the variation in $v_R(t)$ can be reduced by adding a capacitor across the resistor, as illustrated in Fig. 34-20*c*.

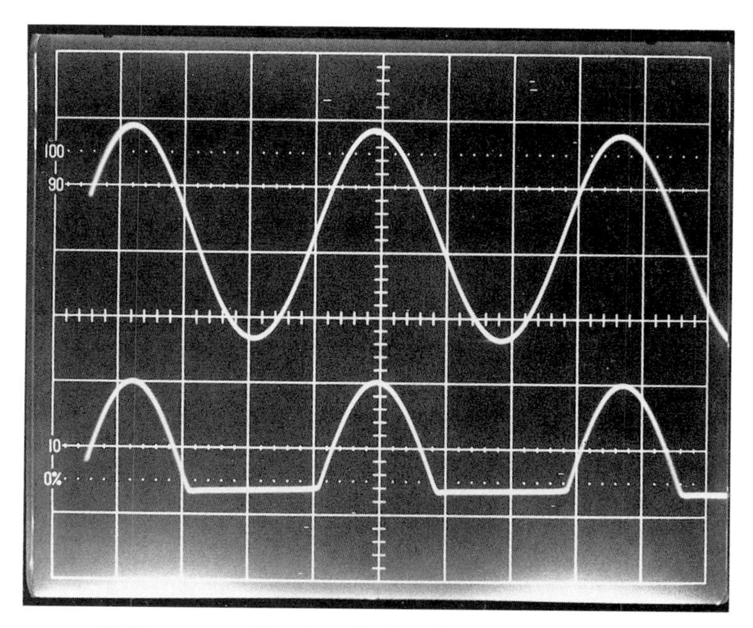

Half-wave rectification displayed on an oscilloscope.

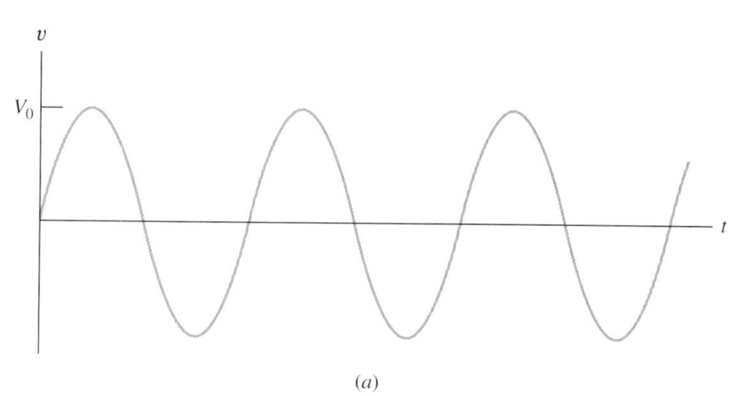

(a)

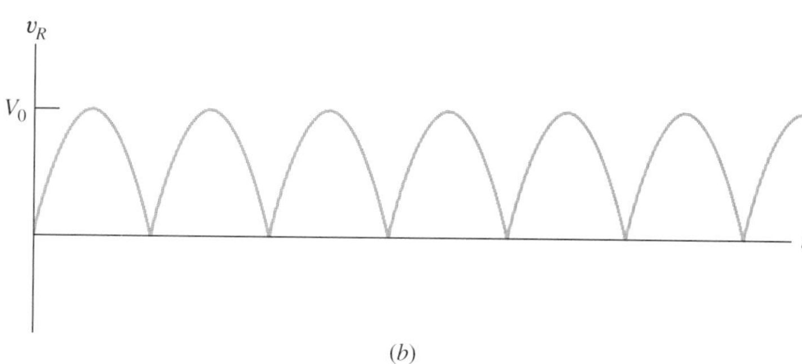

(b)

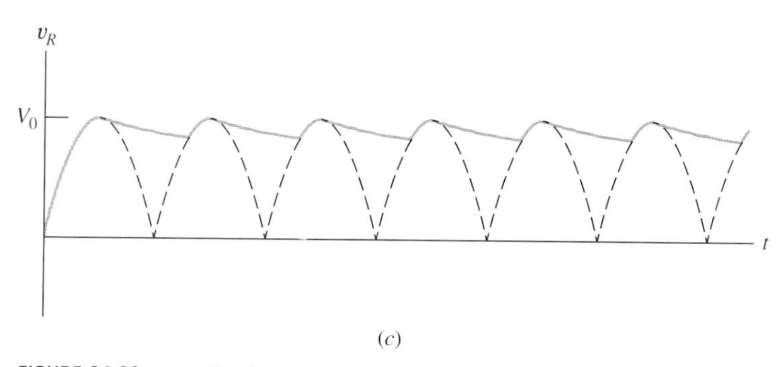

(c)

FIGURE 34-20 (a) The input voltage and (b) the output voltage of a full-wave rectifier. (c) The output voltage with a capacitor placed across the resistor.

*34-6 The Operational Amplifier

An *operational amplifier* ("op-amp") is an electronic circuit that can be used to amplify electric signals and perform mathematical operations such as addition and subtraction on these signals. The op-amp contains many components, including resistors, capacitors, diodes, and transistors. It is usually constructed or "integrated" on a minute piece (a "chip") of semiconducting material. The op-amp is then called an *integrated circuit (IC) operational amplifier*. Op-amps are used extensively in modern electronic circuits.

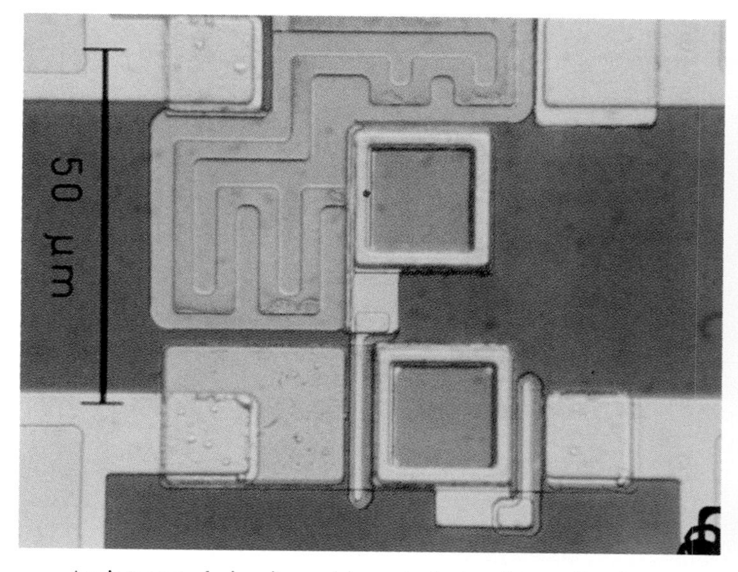

An integrated circuit used in optoelectronics applications.

A typical op-amp has eight terminals that are connected to an external circuit. For our purposes, we need discuss only five of these. They are the two *input*, one *output*, and two *power supply* terminals. The circuit representation of an op-amp with these five terminals is shown in Fig. 34-21. All input and output voltages are measured with respect to ground. Notice that the input terminals are marked "+" and "−." This convention does *not* refer to the polarities of the terminals. It indicates that an input signal at the "−," or *inverting*, terminal becomes inverted by the op-amp, while an input signal at the "+," or *noninverting*, terminal does not. A working op-amp has to be connected to two external dc sources at the power supply terminals. The voltages of these sources are equal in amplitude but opposite in sign. For example, if the V+ terminal is at +15 V, then the V− terminal must be at −15 V. In general, the power supply terminals are not shown in circuit diagrams.

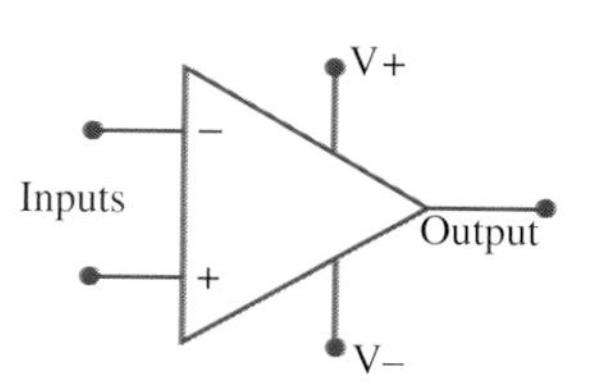

FIGURE 34-21 Circuit representation of an operational amplifier.

The operation of an op-amp is based on two important characteristics:

1. The potential difference between the inverting terminal and the noninverting terminal is very small. For simplicity, we assume that it is zero.

2. The currents at the inverting and noninverting terminals are very small. For simplicity, we assume that they are also zero.

Using these approximations, we now examine two of the basic applications of the op-amp—first as part of an *inverting amplifier*, then as part of a *summing amplifier*.

INVERTING AMPLIFIER

The noninverting terminal of an op-amp that is used as an inverting amplifier is connected to ground, as shown in Fig. 34-22. The input to the op-amp is the signal voltage v_i, which is applied through the *input resistor* R_i. The resistor R_f is called the *feedback resistor*, because it is used to "feed back" some of the output signal v_o to the input. The voltage at junction a is v_a.

Since the current flowing into the inverting terminal is by assumption negligible, the only currents at a are the ones passing through R_i and R_f. Thus from Kirchhoff's junction rule,

$$\frac{v_a - v_i}{R_i} + \frac{v_a - v_o}{R_f} = 0.$$

Now we are also assuming that the potential difference between the inverting and noninverting terminals is negligible; hence $v_a = 0$. We then obtain

$$-\frac{v_i}{R_i} - \frac{v_o}{R_f} = 0,$$

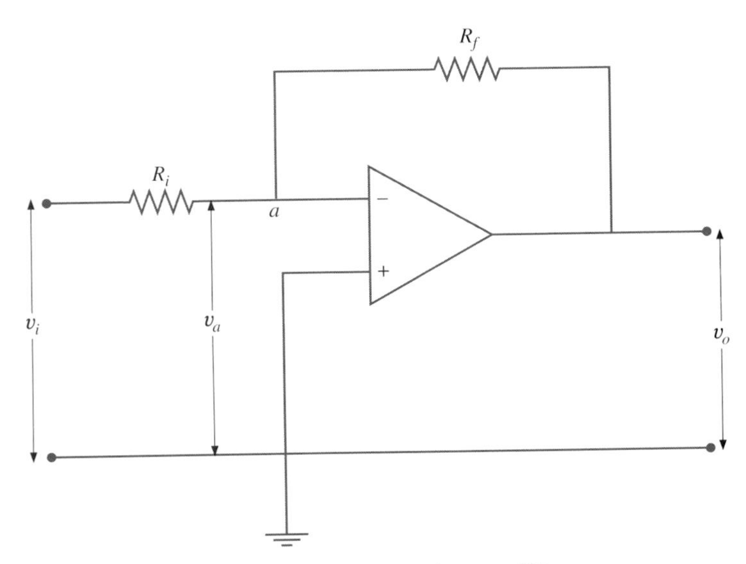

FIGURE 34-22 The inverting amplifier.

so the output of an inverting amplifier is

$$v_o = -\left(\frac{R_f}{R_i}\right)v_i. \tag{34-26}$$

You can see from this equation that the inverting amplifier performs the two operations responsible for its name. First, it amplifies the input v_i by the ratio R_f/R_i. This factor is known as the **gain** of the op-amp and can be controlled by choosing appropriate values for R_i and R_f. Secondly, as the negative sign indicates, the circuit inverts the input, leaving v_i and v_o out of phase by π rad.

EXAMPLE 34-7 AN INVERTING AMPLIFIER

In the circuit of Fig. 34-22, $v_i = (2.0\ \text{V}) \sin 20\pi t$, $R_i = 0.80 \times 10^3\ \Omega$, and $R_f = 1.2 \times 10^3\ \Omega$. (*a*) Write an expression for the output v_o of the circuit. (*b*) What is the gain of the circuit?

SOLUTION (*a*) From $v_o = -(R_f/R_i)v_i$, the output voltage is

$$v_o = -\left(\frac{1.2 \times 10^3\ \Omega}{0.80 \times 10^3\ \Omega}\right)(2.0\ \text{V}) \sin 20\pi t = -(3.0\ \text{V}) \sin 20\pi t.$$

A comparison of the output with the input is shown in Fig. 34-23.

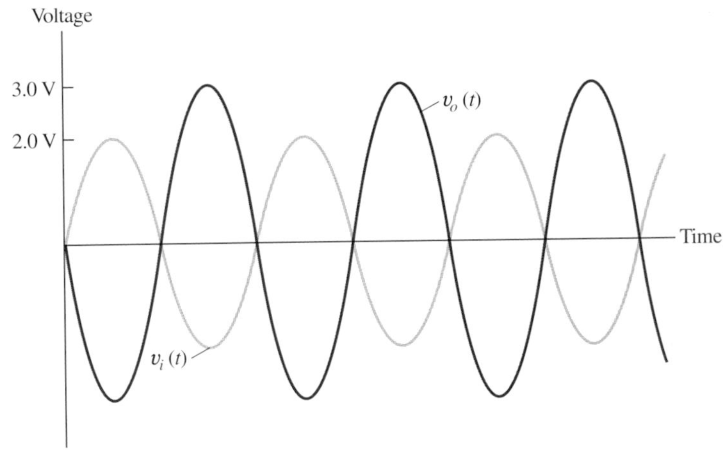

FIGURE 34-23 The input and output voltages of the inverting amplifier of Example 34-7.

(*b*) The gain of the circuit is

$$\frac{R_f}{R_i} = \frac{1.2 \times 10^3\ \Omega}{0.80 \times 10^3\ \Omega} = 1.5.$$

EXAMPLE 34-8 FEEDBACK RESISTANCE IN AN INVERTING AMPLIFIER

An op-amp used in an inverting amplifier circuit has an input resistance of 400 Ω. What feedback resistance must be used in order to obtain a gain of (*a*) 1.0, (*b*) 2.0, (*c*) 5.0, and (*d*) 0.50?

SOLUTION The gain of the op-amp is given by the ratio R_f/R_i. For a fixed input resistance, we control the gain by choosing an appropriate value of R_f:

(*a*) If $R_f/R_i = 1.0$, the op-amp simply inverts the input. Therefore, if $R_i = 400\ \Omega$, $R_f = 400\ \Omega$.

(*b*) If $R_f/R_i = 2.0$, $R_f = 2.0\,(400\ \Omega) = 800\ \Omega$.

(*c*) If $R_f/R_i = 5.0$, $R_f = 2000\ \Omega$.

(*d*) Finally, if $R_f/R_i = 0.50$, $R_f = 200\ \Omega$. In this case, the op-amp actually reduces the amplitude of the input.

DRILL PROBLEM 34-9

The output of an inverting amplifier is given by $v_o = (0.40\ \text{V})\cos(10\pi t - \pi/2)$. If $R_i = 3.0 \times 10^3\ \Omega$ and $R_f = 7.2 \times 10^3\ \Omega$, what is the input?
ANS. $(0.17\ \text{V})\cos(10\pi t + \pi/2)$.

DRILL PROBLEM 34-10

A noninverting amplifier. Consider the circuit of Fig. 34-24. By applying Kirchhoff's first rule at junction a, show that

$$v_o = \left(1 + \frac{R_f}{R_i}\right) v_i.$$

Notice that v_o and v_i have the same sign, so the input is not inverted. This circuit is known as a noninverting amplifier.

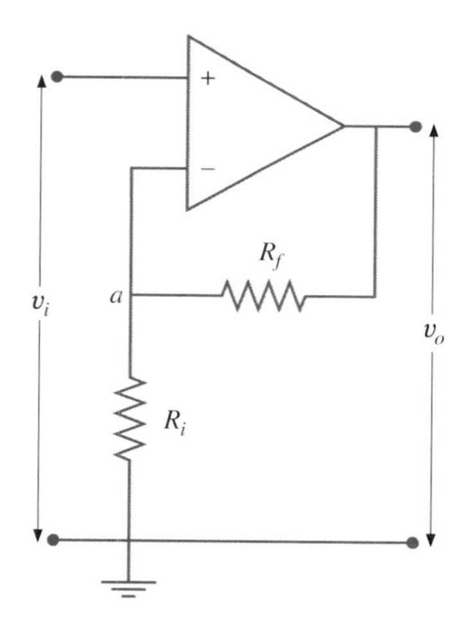

FIGURE 34-24 The noninverting amplifier.

SUMMING AMPLIFIER

Another common op-amp circuit is the summing amplifier. It not only amplifies but also sums several input signals. Like the simple amplifier, the summing amplifier may be inverting or noninverting. In this section, we study the inverting case. The noninverting summing amplifier will be considered in Drill Prob. 34-12.

The circuit for the inverting summing amplifier is shown in Fig. 34-25. Parallel inputs v_1, v_2, and v_3 are applied to the op-amp through resistors R_1, R_2, and R_3. The output is v_o. At junction a, we have from Kirchhoff's first rule

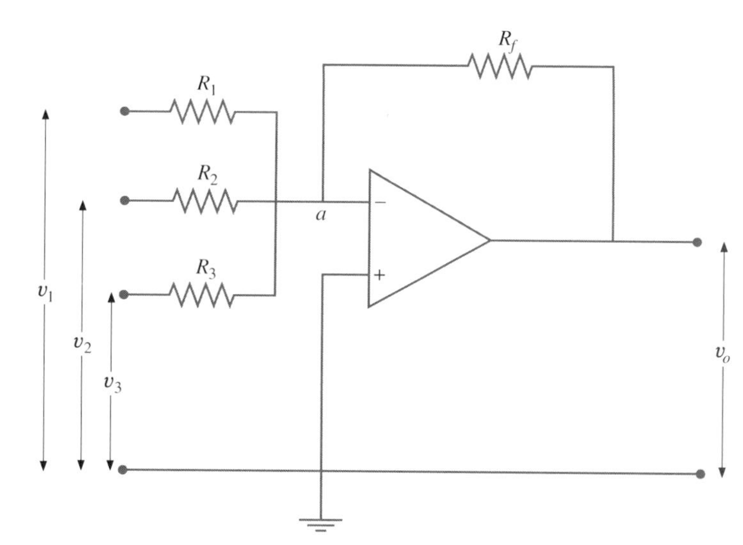

FIGURE 34-25 The summing amplifier (inverting).

$$\frac{v_1}{R_1} + \frac{v_2}{R_2} + \frac{v_3}{R_3} + \frac{v_o}{R_f} = 0,$$

where once again we assume that the current flowing into the noninverting terminal is negligible and that the inverting and noninverting terminals are essentially at the same voltage ($v_a = 0$). This equation may be rewritten as

$$v_o = -\left(\frac{R_f}{R_1}\right)v_1 - \left(\frac{R_f}{R_2}\right)v_2 - \left(\frac{R_f}{R_3}\right)v_3, \qquad (34\text{-}27)$$

which relates the output of the circuit to the different inputs. The inputs are individually amplified, then inverted and summed to yield the output of the inverting summing amplifier. By choosing appropriate values of the resistances, we can control the contribution of each input to the output.

If $v_2 = v_3 = 0$ (i.e., if there is only one input v_1), Eqs. (34-27) and (34-26) become identical and the circuit is an inverting amplifier. Also, if $R_f = R_1 = R_2 = R_3$, Eq. (34-27) reduces to

$$v_o = -v_1 - v_2 - v_3,$$

and the output is a simple inverted sum of the inputs.

EXAMPLE 34-9 A SUMMING AMPLIFIER

For the summing amplifier of Fig. 34-25, it is found that $v_o = -(2v_1 + v_2 + 3v_3)$. If $R_f = 6.0 \times 10^3\ \Omega$, what are R_1, R_2, and R_3?

SOLUTION By comparing the given expression for v_o with Eq. (34-27), we obtain

$$\frac{R_f}{R_1} = 2, \quad \frac{R_f}{R_2} = 1, \quad \text{and} \quad \frac{R_f}{R_3} = 3.$$

Since $R_f = 6.0 \times 10^3\ \Omega$, the values of the resistances are $R_1 = 3.0 \times 10^3\ \Omega$, $R_2 = 6.0 \times 10^3\ \Omega$, and $R_3 = 2.0 \times 10^3\ \Omega$.

DRILL PROBLEM 34-11

For the circuit of Fig. 34-25, $v_1 = (1.0\text{ V})\sin 200\pi t$, $v_2 = (2.5\text{ V})\sin 40\pi t$, and $v_3 = (1.5\text{ V})\sin 75\pi t$. If $R_f = 2.0 \times 10^4\ \Omega$, $R_1 = 5.0 \times 10^3\ \Omega$, and $R_2 = R_3 = 1.0 \times 10^4\ \Omega$, what is the output of the amplifier?
ANS. $v_o = -[(4.0\text{ V})\sin 200\pi t + (5.0\text{ V})\sin 40\pi t + (3.0\text{ V}) \times \sin 75\pi t]$.

DRILL PROBLEM 34-12

A noninverting summing amplifier. Consider the circuit of Fig. 34-26. Use Kirchhoff's first rule at junction a and the appropriate assumptions to show that

$$v_a = R_p\left(\frac{v_1}{R_1} + \frac{v_2}{R_2} + \frac{v_3}{R_3}\right), \qquad (i)$$

where

$$\frac{1}{R_p} = \frac{1}{R_1} + \frac{1}{R_2} + \frac{1}{R_3}.$$

Do the same at junction b, and then use Eq. (i) to show that

$$v_o = \left(1 + \frac{R_f}{R_i}\right)v_a = \left(1 + \frac{R_f}{R_i}\right)R_p\left(\frac{v_1}{R_1} + \frac{v_2}{R_2} + \frac{v_3}{R_3}\right).$$

Here the inputs are multiplied by constants and then summed without being inverted. For this reason, the circuit is called a noninverting summing amplifier.

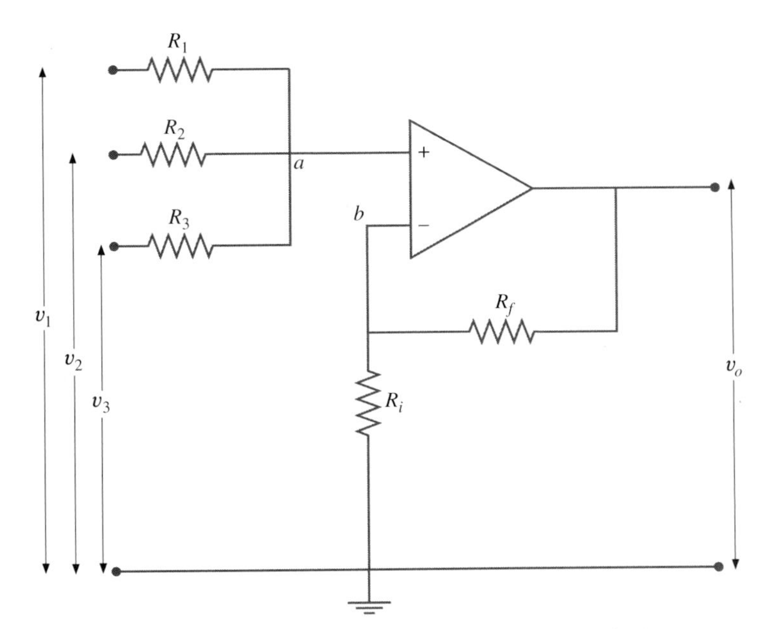

FIGURE 34-26 The summing amplifier (noninverting).

SUMMARY

1. **Four simple ac circuits**
   (*a*) The impedances of a resistor, a capacitor, and an inductor when used in an ac circuit oscillating at an angular frequency ω are R, $1/\omega C$, and ωL, respectively.
   (*b*) The net impedance of an RLC series circuit is

$$Z = \sqrt{R^2 + \left(\omega L - \frac{1}{\omega C}\right)^2}.$$

   (*c*) In the RLC series circuit, the current lags the voltage by a phase angle

$$\phi = \tan^{-1}\left(\frac{X_L - X_C}{R}\right).$$

2. **Power in ac circuits**
   (*a*) If $v(t) = V_0 \sin \omega t$ is the voltage across an element and $i(t) = I_0 \sin(\omega t - \phi)$ is the current through it, then the average power produced (or dissipated) by the element is

$$\bar{P} = \tfrac{1}{2} I_0 V_0 \cos \phi.$$

   For a resistor, $\phi = 0$, so $\bar{P} = I_0 V_0/2$, while for a capacitor and inductor, ϕ is $\pi/2$ and $-\pi/2$ rad, respectively, and in either case $\bar{P} = 0$.

   (*b*) In an ac circuit, the root mean square current and voltage are

$$I_{\text{rms}} = \frac{1}{\sqrt{2}} I_0 \quad \text{and} \quad V_{\text{rms}} = \frac{1}{\sqrt{2}} V_0,$$

   and the average power produced (or dissipated) by an element is

$$\bar{P} = I_{\text{rms}} V_{\text{rms}} \cos \phi.$$

3. **Resonance**
   (*a*) When an RLC series circuit is oscillating at the resonant angular frequency

$$\omega_0 = \sqrt{\frac{1}{LC}},$$

   its impedance is $Z = R$, and the average power output of its source is a maximum V_{rms}^2/R.
   (*b*) The bandwidth $\Delta\omega$ of the resonance peak is the range of ω over which $\bar{P}$ is greater than one-half the maximum value of $\bar{P}$. The sharpness of the peak is described by the quality factor

$$Q = \frac{\omega_0}{\Delta\omega} = \frac{\omega_0 L}{R}.$$

*4. **The transformer**

(*a*) The voltage across the secondary coil of a transformer is related to the voltage across the primary coil by

$$v_s(t) = \frac{N_s}{N_p} v_p(t),$$

where N_s is the number of turns in the secondary coil and N_p is the number of turns in the primary coil.

(*b*) The current through the secondary coil is related to the current through the primary coil by

$$i_s(t) = \frac{N_p}{N_s} i_p(t).$$

(*c*) The input resistance at the primary coil is related to the output resistance at the secondary coil by

$$R_p = \left(\frac{N_p}{N_s}\right)^2 R_s.$$

*5. **Conversion of ac to dc: the diode**

An ideal diode offers infinite resistance to current flow in one direction and zero resistance in the other direction.

*6. **The operational amplifier**

The operation of an op-amp is based on two important characteristics:

(*a*) The potential difference between the inverting terminal and the noninverting terminal is so small that it can be assumed to be zero.

(*b*) The currents at the inverting and noninverting terminals are small enough that they can be assumed to be zero.

QUESTIONS

34-1. What is the relationship between frequency and angular frequency?

34-2. Explain why at high frequencies a capacitor acts as an ac short while an inductor acts as an open circuit.

34-3. Show that the SI unit for capacitive reactance is the ohm. Show that the SI unit for inductive reactance is also the ohm.

34-4. For what value of the phase angle ϕ between the voltage output of an ac source and the current is the average power output of the source a maximum?

34-5. Discuss the differences between average power and instantaneous power.

34-6. Why do transmission lines operate at very high voltages while household circuits operate at fairly small voltages?

34-7. The average ac current delivered to a circuit is zero. Despite this, power is dissipated in the circuit. Explain.

34-8. Can the instantaneous power output of an ac source ever be negative? Can the average power output be negative?

34-9. The power rating of a resistor used in ac circuits refers to the maximum power dissipated in the resistor. How does this compare with the maximum instantaneous power dissipated in the resistor?

34-10. What is the impedance of an RLC series circuit at the resonant frequency?

34-11. In an RLC series circuit, can the voltage measured across the capacitor be greater than the voltage of the source? Answer the same question for the voltage across the inductor.

34-12. How can you distinguish the primary from the secondary coil in a step-up transformer?

34-13. Battery packs in some electronic devices are charged using an adapter connected to a wall socket. Speculate as to the purpose of the adapter.

34-14. Will a transformer work if the input is a dc voltage?

34-15. Why are the primary and secondary coils of a transformer wrapped around the same closed loop of iron?

34-16. Discuss the problems encountered if the capacitance used in a rectifier is (*a*) too small and (*b*) too great.

34-17. An operational amplifier provides a gain of 100. What parameter(s) could you adjust to double this gain?

34-18. The gain of an operational amplifier depends only on the ratio of the input to the feedback resistance. You could therefore obtain the same gain by using two low resistances or two very high resistances. Discuss what factors might determine your choice of low or high resistances.

PROBLEMS

Four Simple ac Circuits

34-1. Write an expression for the output voltage of an ac source that has an amplitude of 12 V and a frequency of 200 Hz.

34-2. Calculate the reactance of a 5.0-μF capacitor at (*a*) 60 Hz, (*b*) 600 Hz, and (*c*) 6000 Hz.

34-3. What is the capacitance of a capacitor whose reactance is 10 Ω at 60 Hz?

34-4. Calculate the reactance of a 5.0-mH inductor at (*a*) 60 Hz, (*b*) 600 Hz, and (*c*) 6000 Hz.

34-5. What is the self-inductance of a coil whose reactance is 10 Ω at 60 Hz?

34-6. At what frequency is the reactance of a 20-μF capacitor equal to that of a 10-mH inductor?

34-7. At 1000 Hz, the reactance of a 5.0-mH inductor is equal to the reactance of a particular capacitor. What is the capacitance of the capacitor?

34-8. A 50-Ω resistor is connected across the emf

$$v(t) = (160 \text{ V}) \sin (120\pi t).$$

Write an expression for the current through the resistor.

34-9. A 25-μF capacitor is connected to an emf given by

$$v(t) = (160 \text{ V}) \sin (120\pi t).$$

(*a*) What is the reactance of the capacitor? (*b*) Write an expression for the current output of the source.

34-10. A 100-mH inductor is connected across the emf of the previous problem. (*a*) What is the reactance of the inductor? (*b*) Write an expression for the current through the inductor.

34-11. The emf of an ac source is given by $v(t) = V_0 \sin \omega t$, where $V_0 = 100$ V and $\omega = 200\pi$ rad/s. Find an expression that represents the output current of the source if (*a*) a 20-μF capacitor, (*b*) a 20-mH inductor, and (*c*) a 50-Ω resistor is connected across it.

34-12. A 700-pF capacitor is connected across an ac source with a voltage amplitude of 160 V and a frequency of 20 kHz. (*a*) Determine the capacitive reactance of the capacitor and the amplitude of the output current of the source. (*b*) If the frequency is changed to 60 Hz while keeping the voltage amplitude at 160 V, what are the capacitive reactance and the current amplitude?

34-13. A 20-mH inductor is connected across an ac source with a variable frequency and a constant voltage amplitude of 9.0 V. (*a*) Determine the reactance of the circuit and the maximum current through the inductor when the frequency is set at 20 kHz. (*b*) Do the same calculations for a frequency of 60 Hz.

34-14. A 30-μF capacitor is connected across a 60-Hz ac source whose voltage amplitude is 50 V. (*a*) What is the maximum charge on the capacitor? (*b*) What is the maximum current into the capacitor? (*c*) What is the phase relationship between the capacitor charge and the current in the circuit?

34-15. A 7.0-mH inductor is connected across a 60-Hz ac source whose voltage amplitude is 50 V. (*a*) What is the maximum current through the inductor? (*b*) What is the phase relationship between the current through and the potential difference across the inductor?

34-16. What is the impedance of a series combination of a 50-Ω resistor, a 5.0-μF capacitor, and a 10-μF capacitor at a frequency of 2.0 kHz?

34-17. A resistor and capacitor are connected in series across an ac generator. The emf of the generator is given by $v(t) = V_0 \cos \omega t$, where $V_0 = 120$ V and $\omega = 120\pi$ rad/s; $R = 400$ Ω and $C = 4.0$ μF. (*a*) What is the impedance of the circuit? (*b*) What is the amplitude of the current through the resistor? (*c*) Write an expression for the current through the resistor. (*d*) Write expressions representing the voltages across the resistor and across the capacitor.

34-18. A resistor and inductor are connected in series across an ac generator. The emf of the generator is given by $v(t) = V_0 \cos \omega t$, where $V_0 = 120$ V and $\omega = 120\pi$ rad/s; $R = 400$ Ω and $L = 1.5$ H. (*a*) What is the impedance of the circuit? (*b*) What is the amplitude of the current through the resistor? (*c*) Write an expression for the current through the resistor. (*d*) Write expressions representing the voltages across the resistor and across the inductor.

34-19. In an RLC series circuit, the voltage amplitude and frequency of the source are 100 V and 500 Hz, respectively, $R = 500$ Ω, $L = 0.20$ H, and $C = 2.0$ μF. (*a*) What is the impedance of the circuit? (*b*) What is the amplitude of the current from the source? (*c*) If the emf of the source is given by

$$v(t) = (100 \text{ V}) \sin 1000\pi t,$$

how does the current vary with time? (*d*) Repeat the calculations with C changed to 0.20 μF.

34-20. An RLC series circuit with $R = 600$ Ω, $L = 30$ mH, and $C = 0.050$ μF is driven by an ac source whose frequency and voltage amplitude are 5000 Hz and 50 V, respectively. (*a*) What is the impedance of the circuit? (*b*) What is the amplitude of the current in the circuit? (*c*) What is the phase angle between the emf of the source and the current?

34-21. For the circuit shown in the accompanying figure, what are (*a*) the total impedance and (*b*) the phase angle between the current and the emf? (*c*) Write an expression for $i(t)$.

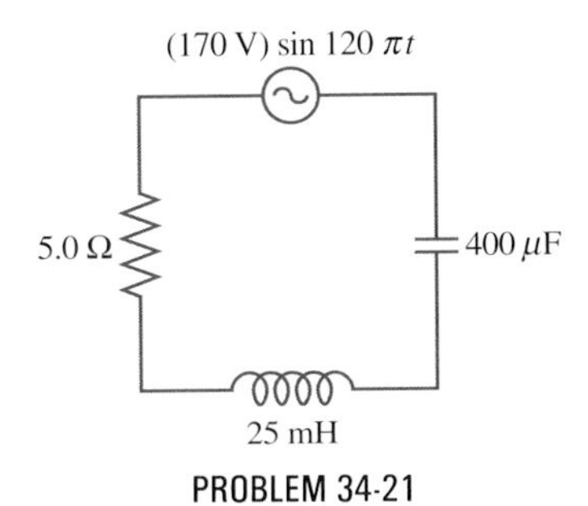

PROBLEM 34-21

34-22. What is the resistance R in the circuit shown in the accompanying figure if the amplitude of the alternating current through the inductor is 4.24 A?

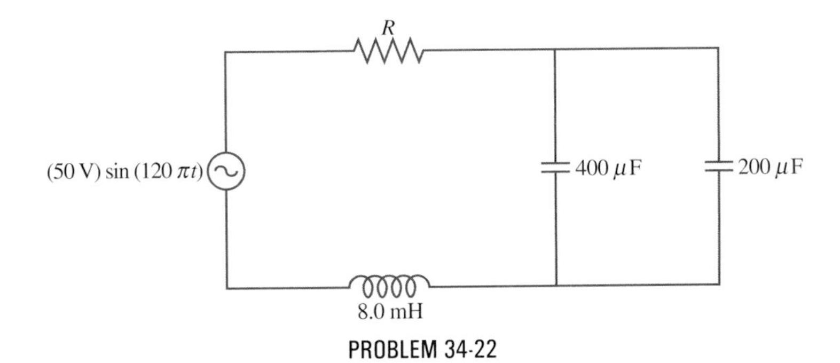

PROBLEM 34-22

Power in ac Circuits

34-23. Calculate the average power output of the source for all three circuits of Prob. 34-11.

34-24. Calculate the rms currents in Probs. 34-8, 34-9, and 34-10.

34-25. A 40-mH inductor is connected to a 60-Hz ac source whose voltage amplitude is 50 V. If an ac voltmeter is placed across the inductor, what will it read?

34-26. For the circuit of Prob. 34-19, find the average power output of the source and the average power dissipated in the resistor for both values given for the capacitance.

34-27. An ac source of voltage amplitude 10 V delivers electric energy at a rate of 8.0 W when its current output is 2.5 A. What is the phase angle ϕ between the emf and the current?

34-28. An RLC series circuit has an impedance of 60 Ω and a power factor of 0.50, with the voltage lagging the current. (*a*) Should a capacitor or an inductor be placed in series with the elements in order to raise the power factor of the circuit? (*b*) What is the value of the capacitance/self-inductance that will raise the power factor to unity?

34-29. An ac source of voltage amplitude 100 V and frequency 1.0 kHz drives an RLC series circuit with $R = 20\ \Omega$, $L = 4.0$ mH, and $C = 50\ \mu$F. (*a*) Determine the rms current through the circuit. (*b*) What are the rms voltages across the three elements? (*c*) What is the phase angle between the emf and the current? (*d*) What is the power output of the source? (*e*) What is the power dissipated in the resistor?

34-30. In an RLC series circuit, $R = 200\ \Omega$, $L = 1.0$ H, $C = 15\ \mu$F, $V_0 = 120$ V, and $f = 50$ Hz. What is the power output of the source?

Resonance

34-31. Calculate the resonant angular frequency of an RLC series circuit for which $R = 20\ \Omega$, $L = 75$ mH, and $C = 4.0\ \mu$F. If R is changed to $300\ \Omega$, what happens to the resonant angular frequency?

34-32. The resonant frequency of an RLC series circuit is 2.0×10^3 Hz. If the self-inductance in the circuit is 5.0 mH, what is the capacitance in the circuit?

34-33. (*a*) What is the resonant frequency of an RLC series circuit with $R = 20\ \Omega$, $L = 2.0$ mH, and $C = 4.0\ \mu$F? (*b*) What is the impedance of the circuit at resonance?

34-34. For an RLC series circuit, $R = 100\ \Omega$, $L = 150$ mH, and $C = 0.25\ \mu$F. (*a*) If an ac source of variable frequency is connected to the circuit, at what frequency is maximum power dissipated in the resistor? (*b*) What is the quality factor of the circuit?

34-35. An ac source of voltage amplitude 100 V and variable frequency f drives an RLC series circuit with $R = 10\ \Omega$, $L = 2.0$ mH, and $C = 25\ \mu$F. (*a*) Plot the current through the resistor as a function of the frequency f. (*b*) Use the plot to determine the resonant frequency of the circuit.

34-36. (*a*) What is the resonant frequency of a resistor, capacitor, and inductor connected in series if $R = 100\ \Omega$, $C = 5.0\ \mu$F, and $L = 2.0$ H? (*b*) If this combination is connected to a 100-V source operating at the resonant frequency, what is the power output of the source? (*c*) What is the Q of the circuit? (*d*) What is the bandwidth of the circuit?

34-37. Suppose a coil has a self-inductance of 20.0 H and a resistance of 200 Ω. What capacitance and resistance must be connected in series with the coil in order to produce a circuit that has a resonant frequency of 100 Hz and a Q of 10?

The Transformer

34-38. A step-up transformer is designed so that the output of its secondary windings is 2000 V (rms) when the primary is connected to a 110-V (rms) line voltage. (*a*) If there are 100 turns in the primary coil, how many turns are there in the secondary? (*b*) If a resistor connected across the secondary draws an rms current of 0.75 A, what is the current in the primary?

34-39. A step-up transformer connected to a 110-V line is used to supply a hydrogen-gas discharge tube with 5.0 kV (rms). The tube dissipates 75 W of power. (*a*) What is the ratio of the number of turns in the secondary to the number in the primary? (*b*) What are the rms currents in the primary and secondary coils? (*c*) What is the effective resistance seen by the 110-V source?

34-40. An ac source of emf delivers 5.0 mW of power at an rms current of 2.0 mA when it is connected to the primary coil of a transformer. The rms voltage across the secondary coil is 20 V. (*a*) What are the voltage across the primary coil and the current through the secondary coil? (*b*) What is the ratio of secondary to primary turns for the transformer?

34-41. A transformer is used to step down the 110 V from a wall socket to 9.0 V for a radio. (*a*) If the primary coil has 500 turns, how many turns does the secondary coil have? (*b*) If the radio operates at a current of 500 mA, what is the current through the primary coil?

34-42. A transformer is used to supply a 12-V model train with power from a 110-V wall plug. The train operates at 50 W of power. (*a*) What is the rms current in the secondary coil of the transformer? (*b*) What is the rms current in the primary coil? (*c*) What is the ratio of the number of primary to secondary turns? (*d*) What is the resistance of the train? (*e*) What is the resistance seen by the 110-V source?

34-43. A power plant generator produces 100 A at 15 kV (rms). A transformer is used to step the transmission line voltage up to 150 kV (rms). (*a*) What is the rms current in the transmission line? (*b*) If the resistance per unit length of the line is $8.6 \times 10^{-8}\ \Omega$/m, what is the power loss per meter in the line? (*c*) What would the power loss per meter be if the line voltage were 15 kV (rms)?

The Diode

34-44. If the diode shown in the accompanying circuit is ideal, what is the current through the 100-Ω resistor?

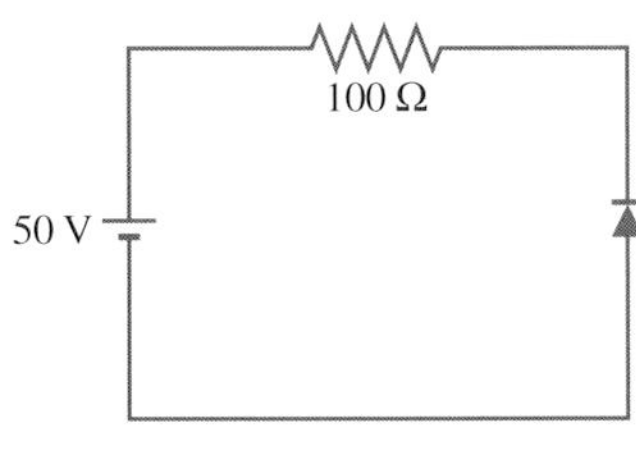

PROBLEM 34-44

34-45. Repeat the previous problem, with the direction of the diode reversed.

34-46. Assuming ideal diodes in the logic circuit shown in the accompanying figure, find V_0 when (*a*) $\mathcal{E}_1 = \mathcal{E}$, $\mathcal{E}_2 = \mathcal{E}_3 = 0$; (*b*) $\mathcal{E}_1 = \mathcal{E}_2 = \mathcal{E}$, $\mathcal{E}_3 = 0$; (*c*) $\mathcal{E}_1 = \mathcal{E}_2 = \mathcal{E}_3 = \mathcal{E}$; (*d*) $\mathcal{E}_1 = 0$, $\mathcal{E}_2 = \mathcal{E}_3 = \mathcal{E}$; and (*e*) $\mathcal{E}_1 = \mathcal{E}_2 = \mathcal{E}_3 = 0$.

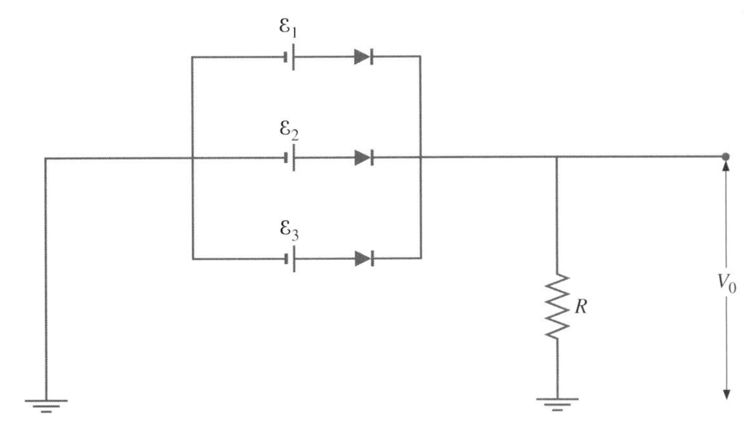

PROBLEM 34-46

34-47. Assuming that both diodes in the logic circuit shown in the accompanying figure are ideal, determine V_0 if (*a*) $\mathcal{E}_1 = \mathcal{E}_2 = 0$; (*b*) $\mathcal{E}_1 = 0$, $\mathcal{E}_2 = \mathcal{E}$; (*c*) $\mathcal{E}_1 = \mathcal{E}$, $\mathcal{E}_2 = 0$; and (*d*) $\mathcal{E}_1 = \mathcal{E}_2 = \mathcal{E}$.

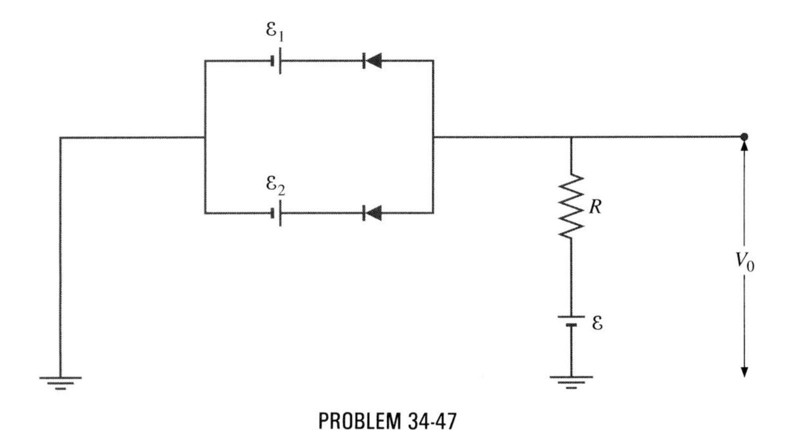

PROBLEM 34-47

34-48. The accompanying figure shows a "clipper circuit" to which is applied the given input voltage v_i. Show that the output voltage v_o varies as given in the figure.

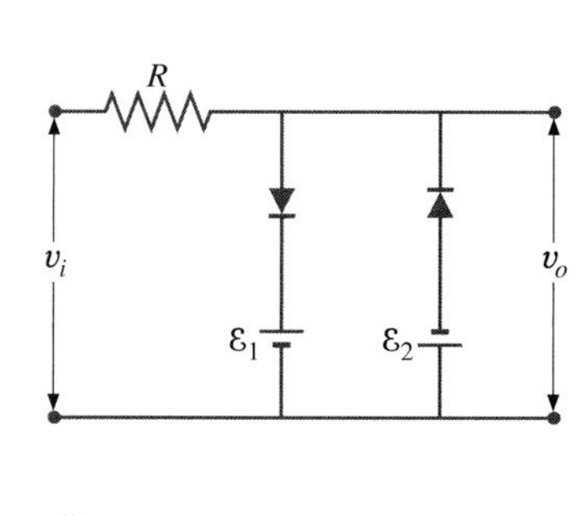

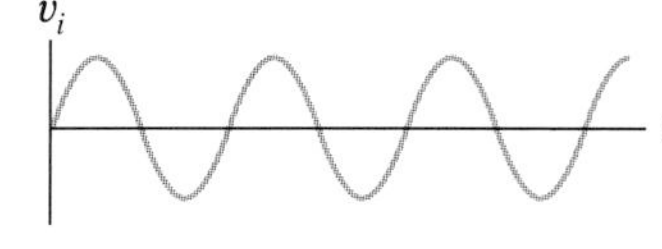

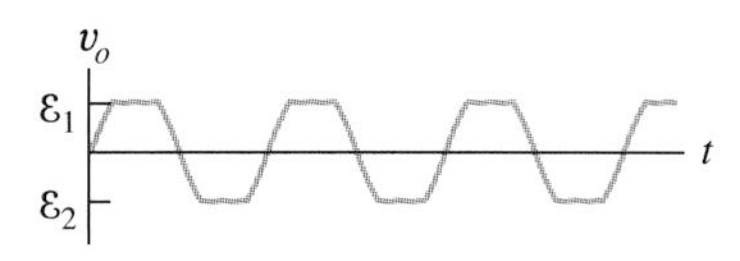

PROBLEM 34-48

Operational-Amplifier Circuits

34-49. An op-amp used in an inverting amplifier circuit has an input resistance of 250 Ω. What feedback resistance must be used in order to obtain a gain of 10?

34-50. The input resistance of the noninverting amplifier of Fig. 34-24 is 400 Ω. What feedback resistance is required to give the amplifier a gain of 10?

34-51. In the inverting amplifier of Fig. 34-22, $R_i = 1000\ \Omega$, $R_f = 100\ \Omega$, and $v_i = (2.0\ \text{mV}) \sin 120\pi t$. (*a*) What is the gain of the amplifier? (*b*) Write an expression that represents the output voltage v_o of the amplifier.

34-52. For the summing amplifier of Fig. 34-25, $R_1 = 200\ \Omega$, $R_2 = 200\ \Omega$, $R_3 = 300\ \Omega$, $R_f = 600\ \Omega$, and $v_1 = v_2 = v_3 = 2.0$ mV. What is v_o?

34-53. For the summing amplifier of Fig. 34-25, what values of R_1, R_2, and R_3 are needed to make the output voltage the negative sum of the input voltages?

General Problems

34-54. A coil with a self-inductance of 16 mH and a resistance of 6.0 Ω is connected to an ac source whose frequency can be varied. At what frequency will the voltage across the coil lead the current through the coil by 45°?

34-55. An RLC series circuit consists of a 50-Ω resistor, a 200-μF capacitor, and a 120-mH inductor whose coil has a resistance of 20 Ω. The source for the circuit has an rms emf of 240 V at a frequency of 60 Hz. Calculate the rms voltages across the (*a*) resistor, (*b*) capacitor, and (*c*) inductor.

34-56. An RLC series circuit consists of a 10-Ω resistor, an 8.0-μF capacitor, and a 50-mH inductor. A 110-V (rms) source of variable frequency is connected across the combination. What is the power output of the source when its frequency is set to one-half the resonant frequency of the circuit?

34-57. The equation $Q = \omega_0 L/R$ for an RLC series circuit can be derived using the following steps: (*a*) Find the equation that gives the values of ω for which $\bar{P}$ is exactly one-half its maximum value by setting Eq. (34-19*a*) equal to $V_{\text{rms}}^2/2R$. You should find that

$$\omega L - \frac{1}{\omega C} = \pm R.$$

(*b*) When this equation is multiplied by ω, it becomes

$$\omega^2 \mp \omega\left(\frac{R}{L}\right) - \left(\frac{1}{LC}\right) = 0.$$

Use the quadratic formula to find all solutions of this equation. Explain why only the positive root solutions,

$$\omega_1 = \frac{R}{2L} + \frac{1}{2}\sqrt{\frac{R^2}{L^2} + \frac{4}{LC}}$$

and

$$\omega_2 = -\frac{R}{2L} + \frac{1}{2}\sqrt{\frac{R^2}{L^2} + \frac{4}{LC}}$$

are meaningful physically.

(*c*) The bandwidth $\Delta\omega$ is $\omega_1 - \omega_2$. Use Eq. (34-21), justify ignoring $R^2/4L^2$ compared with $\omega_0^2 = 1/LC$ for a narrow peak, then show that

$$\omega_1 \approx \omega_0 + \frac{R}{2L},$$

$$\omega_2 \approx \omega_0 - \frac{R}{2L},$$

and

$$\Delta\omega \approx \frac{R}{L}.$$

(*d*) Finally, show that

$$Q \approx \omega_0 L/R.$$

34-58. The accompanying figure shows two circuits that act as crude *high-pass filters*. The input voltage to the circuit is v_{in} and the output voltage is v_{out}. Show that for the capacitor circuit

$$\frac{v_{out}}{v_{in}} = \frac{1}{\sqrt{1 + 1/\omega^2 R^2 C^2}},$$

and for the inductor circuit

$$\frac{v_{out}}{v_{in}} = \frac{\omega L}{\sqrt{R^2 + \omega^2 L^2}}.$$

Also show that for high frequencies, $v_{out} \approx v_{in}$, but for low frequencies, $v_{out} \approx 0$.

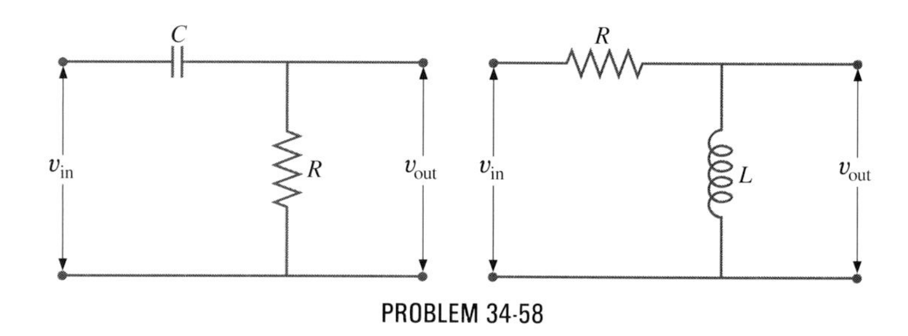

PROBLEM 34-58

34-59. The two circuits shown in the accompanying figure act as crude *low-pass filters*. The input voltage to the circuits is v_{in} and the output voltage is v_{out}. Show that for the capacitor circuit

$$\frac{v_{out}}{v_{in}} = \frac{1}{\sqrt{1 + \omega^2 R^2 C^2}},$$

and for the inductor circuit

$$\frac{v_{out}}{v_{in}} = \frac{R}{\sqrt{R^2 + \omega^2 L^2}}.$$

Also show that for low frequencies, $v_{out} \approx v_{in}$, but for high frequencies, $v_{out} \approx 0$.

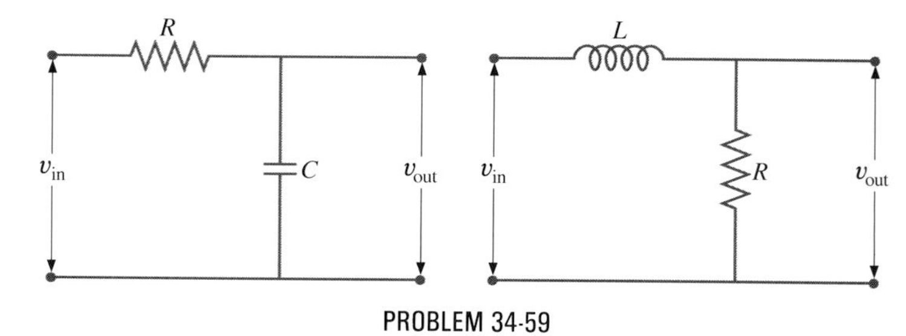

PROBLEM 34-59

PART V

ELECTROMAGNETIC WAVES AND OPTICS

A diamond sparkles because light entering the diamond undergoes a series of total internal reflections before emerging.

CHAPTER 35 Electromagnetic Waves and the Nature of Light

Preview

The nature of light and how it is related to electromagnetic fields are considered in this introductory chapter on optics. The primary topics to be discussed here are the following:

1. **Electromagnetic waves.** Using Maxwell's equations, we investigate the properties of electromagnetic waves and their relationship to light. We also discuss the electromagnetic spectrum.
2. **Speed of light.** Some of the first measurements of the speed of light are considered.
3. **Huygens' principle.** A geometrical method for determining the propagation of wavefronts is discussed.
4. **Laws of reflection and refraction.** Laws describing reflection and refraction at a planar interface are formulated. The phenomenon of total internal reflection is introduced.
5. **Polarization.** The polarization of light is explained in terms of the properties of the electromagnetic wave. The effect of a sheet of polarizing material on the intensity of a beam of unpolarized light is considered.

Maxwell's equations are often thought of as the culmination of the study of electricity and magnetism. Yet they are also the starting point for our next subject, *optics*, which is in essence a study of light. Our investigation of light will revolve around two questions of fundamental importance: (1) What is the nature of light? and (2) How does it behave under various circumstances? The answers to these questions can be found in Maxwell's field equations. These equations predict the existence of electromagnetic waves that travel at the speed of light. They also describe how these waves behave. Some examples of electromagnetic waves are visible light, radio and television signals, infrared and ultraviolet radiation, x-rays, and microwaves. Interestingly, not all light phenomena can be explained by Maxwell's theory. Experiments performed at the beginning of this century showed that light also has corpuscular, or particlelike, properties. These experiments and their ramifications will be discussed in Chap. 40.

In this chapter, we study the basic properties of light as electromagnetic radiation. Then, in Chaps. 36–38, we investigate the behavior of a beam of light when it encounters simple optical devices like mirrors, lenses, and apertures. Under many circumstances, the wavelength of light is negligible compared with the dimensions of the device, as in the case of ordinary mirrors and lenses. A light beam can then be treated as a ray whose propagation is governed by simple geometric rules. The part of optics that deals with such phenomena is known as *geometric optics*. However, if the wavelength is *not* negligible compared with the dimensions of the device (for example, a very narrow slit), the ray approximation becomes invalid and we have to examine the behavior of light in terms of its wave properties. This study is known as *physical optics*.

35-1 ELECTROMAGNETIC WAVES

If you have not studied electricity and magnetism (Chaps. 22–34) up to this point, you should read in place of this section the discussion of electromagnetic waves presented in Supplement 35-1. That material is sufficient to allow you to read the chapters on optics, although you will have to accept most of the properties of electromagnetic waves without proof. If you have already studied electricity and magnetism, you should read the more thorough discussion of electromagnetic waves presented here and skip the supplement.

Perhaps the most significant prediction of Maxwell's equations is the existence of combined electric and magnetic (or *electromagnetic*) fields that propagate through space. The traveling electromagnetic fields are often called *electromagnetic radiation*. To appreciate why electromagnetic radiation should even exist, let's imagine that there is a time-varying magnetic field $\mathbf{B}_0(t)$ in the region around point O of Fig. 35-1. We represent $\mathbf{B}_0(t)$ by one of its field lines. From Faraday's law, the changing magnetic field through a surface induces a time-varying electric field $\mathbf{E}_0(t)$ at the boundary of that surface. A field line representative of $\mathbf{E}_0(t)$ is shown. In turn, the changing electric field $\mathbf{E}_0(t)$ creates a magnetic field $\mathbf{B}_1(t)$ according to the Maxwell-Ampère law. And this changing field induces $\mathbf{E}_1(t)$, which induces $\mathbf{B}_2(t)$, and so on. We have then a self-continuing process that leads to the creation of time-varying electric and magnetic fields in regions farther and farther away from O. This phenomenon may be visualized as the *propagation of an electromagnetic wave* through space.

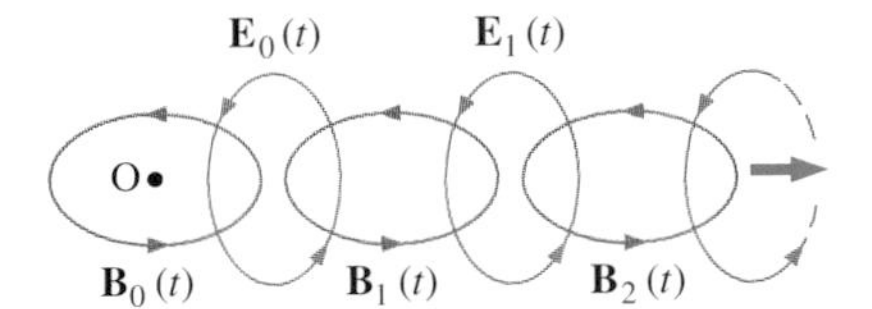

FIGURE 35-1 A plausibility argument for electromagnetic wave propagation. Starting with a time-varying magnetic field $\mathbf{B}_0(t)$ around point O, mutual induction between the electric and magnetic fields produces time-varying fields farther and farther away.

When we studied mechanical waves in Chaps. 16 and 17, we found that they traveled through a medium such as a string, water, or air. The properties of a particular type of wave were determined by the laws of Newtonian mechanics as applied to the medium responsible for the propagation of the wave. However, Maxwell's equations hold in free space; so electromagnetic waves, unlike mechanical waves, *do not require a medium for their propagation*.

A general treatment of the physics of electromagnetic waves is beyond the level of this textbook. We can, however, investigate the special case of an electromagnetic field that propagates as a wave through free space along the x axis of a given coordinate system. This field is a function of just the x coordinate and time. The y component of the electric field is then written as $E_y(x, t)$, the z component of the magnetic field as $B_z(x, t)$, etc. Since we are assuming free space, there are no free charges or currents, so we can set $Q_{\text{in}} = 0$ and $I = 0$ in Maxwell's equations. (See Sec. 31-2.)

We start by asking what Gauss' law for electric fields tells us about electromagnetic waves. Our Gaussian surface is the surface of a rectangular box whose cross section is a square of side l and whose third side has length Δx, as shown in Fig. 35-2. Also shown at the square faces at x and $x + \Delta x$ are the x, y, and z components of the electric field associated with the electromagnetic wave that is traveling along the x axis. Since the

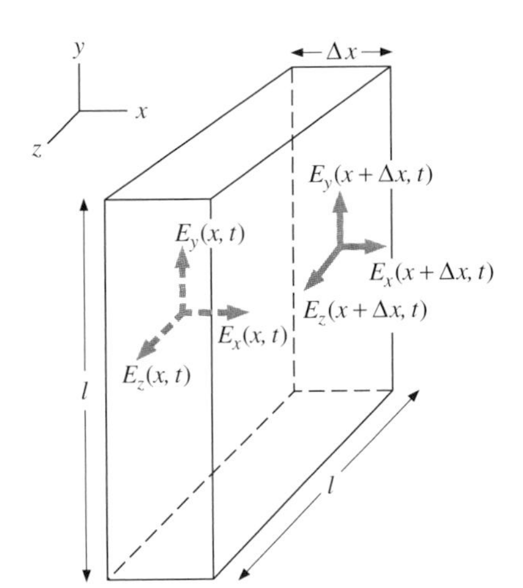

FIGURE 35-2 The surface of a rectangular box of dimensions $l \times l \times \Delta x$ is our Gaussian surface. The electric field shown at the square faces is associated with an electromagnetic wave propagating along the x axis.

y and z components of the electric field are independent of y and z, they do not contribute to the net electric flux through the surface. (Can you explain why?) Any net flux through the surface must therefore come from the x component of the electric field; that is,

$$\oint_S \mathbf{E}\cdot\hat{\mathbf{n}}\,da = -E_x(x,t)l^2 + E_x(x+\Delta x,t)l^2 = \frac{Q_{\text{in}}}{\epsilon_0}.$$

But $Q_{\text{in}} = 0$, so this component's net flux is also zero, and $E_x(x,t) = E_x(x+\Delta x,t)$. Thus if there is an x component of the electric field, it cannot vary with x. Such a uniform field component is inconsistent with the wave concept. Hence the component $E_x(x,t)$ cannot be part of an electromagnetic wave propagating along the x axis; that is, $E_x(x,t) = 0$ for this wave.

Since the magnetic field satisfies Gauss' law $\oint_S \mathbf{B}\cdot\hat{\mathbf{n}}\,da = 0$, regardless of whether free charges or currents exist, the same argument may be applied to this field. Thus our electromagnetic wave also has no magnetic field component in the x direction.

Without the components $E_x(x,t)$ or $B_x(x,t)$, both the electric and magnetic fields associated with the electromagnetic wave are transverse to the direction of propagation of the wave. *The electromagnetic field moves along the x axis with its electric and magnetic fields oscillating in the yz plane.* These transverse components can be investigated with the help of the two time-dependent Maxwell's equations—Faraday's law and the Ampère-Maxwell law. First, we apply Faraday's law over the front rectangular face of our surface, using the path shown in Fig. 35-3. Because $E_x(x,t) = 0$, we have

$$\oint \mathbf{E}\cdot d\mathbf{l} = -E_y(x,t)l + E_y(x+\Delta x,t)l,$$

which, assuming Δx is small and approximating $E_y(x+\Delta x,t)$ by

$$E_y(x+\Delta x,t) = E_y(x,t) + \frac{\partial E_y(x,t)}{\partial x}\Delta x,$$

becomes

$$\oint \mathbf{E}\cdot d\mathbf{l} = \frac{\partial E_y(x,t)}{\partial x}(l\,\Delta x).$$

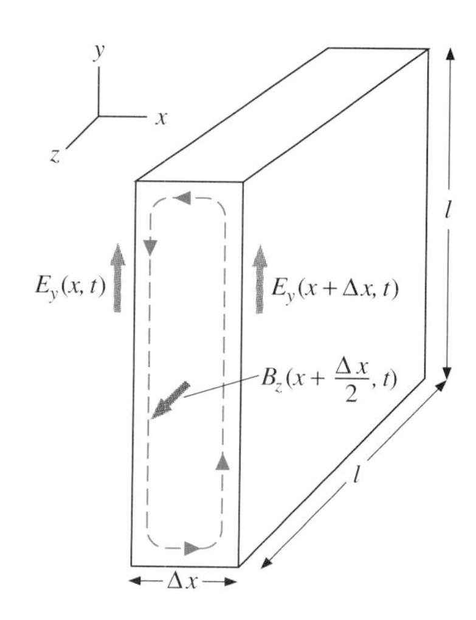

FIGURE 35-3 In applying Faraday's law to the front rectangular face, we traverse its edge in a counterclockwise direction to evaluate $\oint \mathbf{E}\cdot d\mathbf{l}$. Since Δx is very small, the magnetic field crossing the face at any point can be approximated by $B_z(x+\Delta x/2,t)$.

Since Δx is small, the magnetic flux through the face can be approximated by

$$\int_S \mathbf{B}\cdot\hat{\mathbf{n}}\,da = B_z\left(x+\frac{\Delta x}{2},t\right)(l\,\Delta x),$$

where we have used the value of the z component of the magnetic field at the center of the face. From Faraday's law,

$$\oint \mathbf{E}\cdot d\mathbf{l} = -\frac{d}{dt}\int_S \mathbf{B}\cdot\hat{\mathbf{n}}\,da,$$

so

$$\frac{\partial E_y(x,t)}{\partial x}(l\,\Delta x) = -\frac{\partial}{\partial t}\left[B_z\left(x+\frac{\Delta x}{2},t\right)\right](l\,\Delta x).$$

Canceling $l\,\Delta x$ and taking the limit as $\Delta x \to 0$, we are left with

$$\frac{\partial E_y(x,t)}{\partial x} = -\frac{\partial B_z(x,t)}{\partial t}. \qquad (35\text{-}1a)$$

When Faraday's law is also applied to the top face of the surface of Fig. 35-3, the resulting equation is

$$\frac{\partial E_z(x,t)}{\partial x} = \frac{\partial B_y(x,t)}{\partial t}. \qquad (35\text{-}1b)$$

Next we apply the Ampère-Maxwell law (with $I = 0$) over the same two faces of the rectangular box of Fig. 35-3. With the front surface, $\oint \mathbf{B}\cdot d\mathbf{l} = \mu_0\epsilon_0(d/dt)\int_S \mathbf{E}\cdot\hat{\mathbf{n}}\,da$ yields

$$\frac{\partial B_y(x,t)}{\partial x} = \mu_0\epsilon_0\frac{\partial E_z(x,t)}{\partial t}, \qquad (35\text{-}2a)$$

while for the top surface,

$$\frac{\partial B_z(x,t)}{\partial x} = -\mu_0\epsilon_0\frac{\partial E_y(x,t)}{\partial t}. \qquad (35\text{-}2b)$$

Notice that Eqs. (35-1) and (35-2) couple the electric and magnetic fields. Equations (35-1a) and (35-2b) tell us that if there is a component $E_y(x,t)$ propagating down the x axis, there is always a component $B_z(x,t)$ associated with it; and from Eqs. (35-1b) and (35-2a), we see that the component $E_z(x,t)$ is always accompanied by the component $B_y(x,t)$. Thus *an electromagnetic wave is never composed of an electric field by itself or a magnetic field by itself.*

Let's consider an electromagnetic wave composed of field components $E_y(x,t)$ and $B_z(x,t)$ that is propagating along the positive x axis with speed c. We know from Sec. 16-2 that for such a wave, the mathematical forms of the components are

$$E_y(x,t) = E_y(x-ct) \quad\text{and}\quad B_z(x,t) = B_z(x-ct).$$

Now for any function of the form $f(x,t) = f(x-ct) = f(\xi)$, where $\xi = x - ct$, we have

$$\frac{\partial f(x,t)}{\partial x} = \frac{df}{d\xi}\frac{\partial \xi}{\partial x} = \frac{df}{d\xi}\cdot 1 = \frac{df}{d\xi},$$

and

$$\frac{\partial f(x,t)}{\partial t} = \frac{df}{d\xi}\frac{\partial \xi}{\partial t} = \frac{df}{d\xi}(-c) = -c\frac{df}{d\xi}.$$

Using these two relationships in Eq. (35-1a) and (35-2b), respectively, we obtain

$$\frac{dE_y}{d\xi} = c\,\frac{dB_z}{d\xi},$$

and

$$\frac{dB_z}{d\xi} = \mu_0\epsilon_0 c\,\frac{dE_y}{d\xi}.$$

Substitution of $c\,dB_z/d\xi$ for $dE_y/d\xi$ in the second equation then yields

$$1 = \mu_0\epsilon_0 c^2,$$

so the speed of the electromagnetic wave in free space is given in terms of the permeability and the permittivity of free space by

$$c = \frac{1}{\sqrt{\mu_0\epsilon_0}}. \quad (35\text{-}3)$$

We could just as easily have assumed an electromagnetic wave with field components $E_z(x, t)$ and $B_y(x, t)$. The same type of analysis with Eqs. (35-1b) and (35-2a) would also show that the speed of an electromagnetic wave is $c = 1/\sqrt{\mu_0\epsilon_0}$.

The components of **E** and **B** associated with an electromagnetic wave must all satisfy the *wave equation*. For an electromagnetic wave propagating along an arbitrary direction, the wave equation [written here for the component $E_y(x, y, z, t)$] is

$$\frac{\partial^2 E_y(x, y, z, t)}{\partial x^2} + \frac{\partial^2 E_y(x, y, z, t)}{\partial y^2} + \frac{\partial^2 E_z(x, y, z, t)}{\partial z^2} - \frac{1}{c^2}\frac{\partial^2 E_y(x, y, z, t)}{\partial t^2} = 0. \quad (35\text{-}4)$$

This is the general form of the wave equation. In Drill Prob. 35-2, you will see how to derive the wave equation for the specific case of an electromagnetic wave traveling along the x axis.

The physics of traveling electromagnetic fields was worked out by Maxwell in 1873. He showed in a more general way than our simple derivation that electromagnetic waves *always* travel in free space with a speed given by Eq. (35-3). A fascinating aspect of this equation was that with $\mu_0 = 4\pi \times 10^{-7}$ N/A^2 and $\epsilon_0 = 8.85 \times 10^{-12}$ C^2/N·m^2, this speed came out to be 3.00×10^8 m/s, which agreed with the known value for the speed of light. Imagine the excitement that Maxwell must have felt when he discovered the equation! He had found a fundamental connection between two seemingly unrelated phenomena—electromagnetic fields and light.

Prior to Maxwell's work, experiments had already indicated that light was a wave phenomenon, although the nature of the waves was yet unknown. In 1801, Thomas Young (1773–1829) showed that when a light beam was separated by two narrow slits and then recombined, a pattern made up of bright and dark fringes was formed on a screen. This phenomenon was explained by assuming that light was composed of waves that added constructively at some points and destructively at others. Subsequently, Jean Foucault (1819–1868), with measurements of the speed of light in various media, and Augustin Fresnel (1788–1827), with detailed experiments involving interference and diffraction phenomena, provided further conclusive evidence that light was a wave phenomenon. So, light was known to be a wave and Maxwell had predicted the existence of electromagnetic waves that traveled at the speed of light. The conclusion seemed inescapable—light must be *a form of electromagnetic radiation*.

Now, Maxwell's electromagnetic waves weren't restricted to the frequencies of visible light. He showed that electromagnetic radiation with the same fundamental properties as visible light should exist at *any* frequency. And 10 years after Maxwell's death, Heinrich Hertz demonstrated experimentally that there were indeed electromagnetic waves outside the visible range of frequencies. A schematic representation of Hertz's apparatus is shown in Fig. 35-4. It consisted basically of a "spark coil" and an adjacent open wire loop. When the current through the primary of the spark coil is switched off, a large potential is induced in the secondary. This potential causes a dielectric breakdown of the air between the spherical electrodes connected to the secondary, resulting in sparks between them. Hertz found that corresponding sparks were produced between the electrodes of the adjacent wire loop. This result indicated that electromagnetic radiation had been generated by the sparks, and this radiation induced an emf in the open wire loop. The induced emf was large enough that it produced the observed sparks between the electrodes of the wire loop.

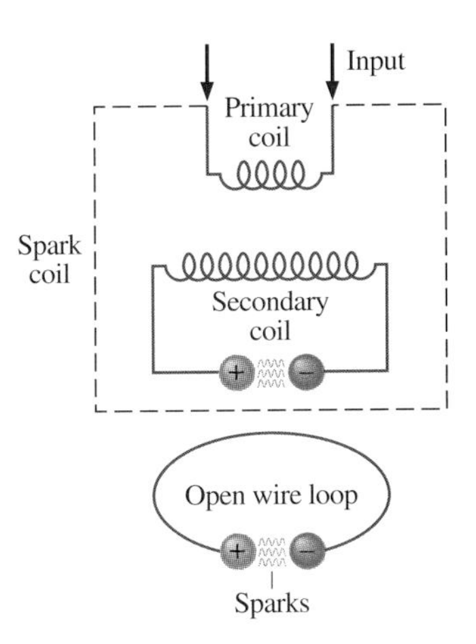

FIGURE 35-4 Hertz's spark-coil experiment was used to demonstrate the existence of electromagnetic waves outside the visible range of frequencies.

In addition, Hertz determined the speed of the waves to be 3.2×10^8 m/s, which, within experimental error, agreed with the accepted value for c. Finally, he showed that these nonvisible electromagnetic waves displayed some of the other known properties of light; specifically, they could be reflected, refracted, and polarized. With this added evidence, the validity of Maxwell's theory was no longer in doubt.

Electromagnetic waves are produced by accelerating charges. For example, the vibrating electrons in a conductor emit radiation at the frequencies of their oscillations. The electromagnetic field of a *dipole antenna* is shown in Fig. 35-5. The frequency of this radiation is the same as the frequency of the ac source that is accelerating the electrons in the antenna. At

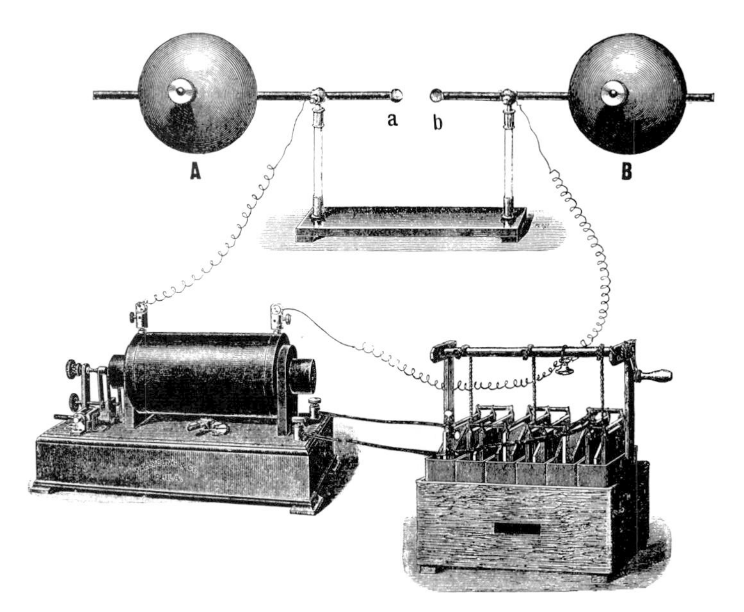

A sketch of Heinrich Hertz's apparatus used in his experiments on electromagnetic waves.

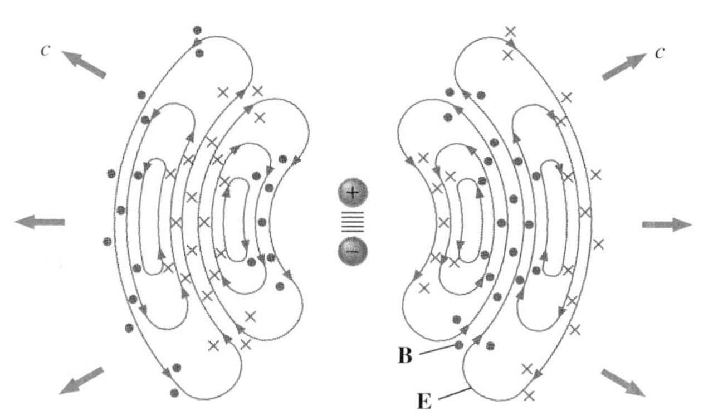

FIGURE 35-5 The oscillatory motion of the charges in a dipole antenna produces electromagnetic radiation. The electric field lines in one plane are shown. The magnetic field is perpendicular to this plane. This radiation field has cylindrical symmetry around the axis of the dipole. Field lines near the dipole are not shown. Far from the antenna the wavefronts are almost spherical and the radiation propagates like electromagnetic plane waves.

large distances from the antenna, this radiation essentially becomes plane waves (see Sec. 16-7).

The simplest electromagnetic waves are plane waves. When an electromagnetic plane wave moves along the x axis, its electric field oscillates in the yz plane and can therefore be resolved into a y component $E_y(x, t)$ and a z component $E_z(x, t)$. As we have seen, associated with these components are the magnetic field components $B_z(x, t)$ and $B_y(x, t)$, respectively. These pairs of electric-magnetic field components are themselves plane waves, which we will call the E_y-B_z wave and the E_z-B_y wave, respectively. A general electromagnetic plane wave propagating along the x axis can be represented as a linear combination of these two component-pair plane waves.

The E_y-B_z plane wave is shown in Fig. 35-6a. Its electric and magnetic fields may be written as

$$\mathbf{E} = E_0 \sin (kx - \omega t)\mathbf{j} \tag{35-5a}$$

and

$$\mathbf{B} = B_0 \sin (kx - \omega t)\mathbf{k}, \tag{35-5b}$$

where ω is the angular frequency, k is the wave number, and E_0 and B_0 are the amplitudes of the wave.

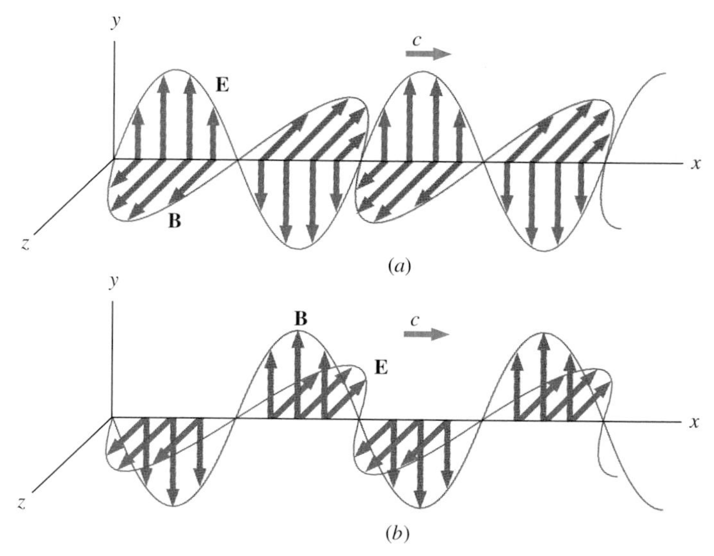

FIGURE 35-6 (a) An electromagnetic plane wave traveling down the positive x axis, with $\mathbf{E}$ oscillating parallel to the y axis and $\mathbf{B}$ parallel to the z axis. (b) This electromagnetic plane wave also travels down the positive x axis, but in this case, $\mathbf{E}$ oscillates along the z axis and $\mathbf{B}$ along the y axis. In both cases, the Poynting vector $\mathbf{S} = \mathbf{E} \times \mathbf{B}/\mu_0$ is directed along the positive x axis.

The relationships between the properties of an electromagnetic wave are the same as those of a mechanical wave; that is,

$$\omega = 2\pi f, \tag{35-6a}$$

$$k = \frac{2\pi}{\lambda}, \tag{35-6b}$$

and

$$c = f\lambda. \tag{35-6c}$$

In addition, there is a relationship between the electric and magnetic field amplitudes of the plane wave of Fig. 35-6a, which we can find by inserting Eqs. (35-5a) and (35-5b) into Eq. (35-1a). We then have

$$\frac{\partial}{\partial x}[E_0 \sin (kx - \omega t)] = -\frac{\partial}{\partial t}[B_0 \sin (kx - \omega t)],$$

so

$$kE_0 \cos (kx - \omega t) = \omega B_0 \cos (kx - \omega t).$$

Since $c = \omega/k$, this yields

$$E_0 = cB_0 \tag{35-7}$$

as the relationship between the electric and magnetic field amplitudes of the E_y-B_z plane wave. In Sec. 31-4, we discussed how the transport of electromagnetic energy is described by the Poynting vector $\mathbf{S}$. For the plane wave of Fig. 35-6*a*,

$$\mathbf{S} = \frac{1}{\mu_0}\mathbf{E} \times \mathbf{B} = \frac{1}{\mu_0}E_0 B_0 \sin^2 (kx - \omega t)\mathbf{i}.$$

The Poynting vector therefore lies along the direction of propagation of the wave. It gives us the instantaneous rate at which energy crosses a unit area perpendicular to the direction of wave propagation; this rate is called the **power flux**. Its SI unit is the watt per square meter (W/m^2).

At the frequencies of visible light, electromagnetic waves oscillate so rapidly that only the time-averaged power flux $\bar{\mathbf{S}}$ can be measured. For the plane wave of Fig. 35-6*a*, this is given by

$$\begin{aligned}\bar{\mathbf{S}} &= \frac{1}{T}\int_0^T \frac{E_0 B_0}{\mu_0}\sin^2 (kx - \omega t)\mathbf{i}\, dt \\ &= \frac{E_0 B_0}{2\mu_0 T}\int_0^T [1 - \cos 2(kx - \omega t)]\,\mathbf{i}\, dt,\end{aligned}$$

where T is the period of the wave. Since the average value of the cosine function over a cycle is zero, the integral of the second term vanishes and we are left with

$$\bar{\mathbf{S}} = \frac{E_0 B_0}{2\mu_0}\mathbf{i} = \frac{E_0^2}{2\mu_0 c}\mathbf{i} = \frac{cB_0^2}{2\mu_0}\mathbf{i}, \tag{35-8}$$

where we have used $E_0 = cB_0$. The magnitude of $\bar{\mathbf{S}}$ is the **intensity** I of the electromagnetic wave; it is

$$I = |\bar{\mathbf{S}}| = \frac{E_0 B_0}{2\mu_0} = \frac{E_0^2}{2\mu_0 c} = \frac{cB_0^2}{2\mu_0}. \tag{35-9}$$

Like other types of plane waves, the intensity of the electromagnetic plane wave is proportional to the square of the amplitude.

It's not surprising that energy is transported by light. This fact quickly becomes evident to anyone standing under the sun on a hot day. What we do learn from Maxwell is that the energy is transported by electromagnetic waves that, in the sun's case, travel through the vacuum of space to the earth.

The magnetron used in microwave ovens emits electromagnetic waves at a frequency of 2.45 GHz (microwaves). Their frequency matches a rotational vibration frequency of water molecules. Food is heated as its water molecules undergo resonance absorption of the microwave radiation.

The plane wave associated with the field components E_z and B_y is shown in Fig. 35-6*b*. It may be represented by

$$\mathbf{E} = -E_0 \sin (kx - \omega t)\mathbf{k} \tag{35-10a}$$

$$\mathbf{B} = B_0 \sin (kx - \omega t)\mathbf{j}. \tag{35-10b}$$

We place the minus sign in front of the right-hand side of Eq. (35-10*a*) so that the Poynting vector $\mathbf{E} \times \mathbf{B}/\mu_0$ points in the positive x direction. You can easily show that this plane wave also satisfies Eqs. (35-7), (35-8), and (35-9).

EXAMPLE 35-1 A Laser Beam

The beam from a small laboratory laser typically has an intensity of about 1.0×10^{-3} W/m^2. Assuming that the beam is composed of the plane waves of Fig. 35-6*a*, calculate the amplitudes of the electric and magnetic fields in the beam.

SOLUTION From Eq. (35-9), the intensity of the laser beam is

$$I = \frac{E_0^2}{2\mu_0 c},$$

so the amplitude of the electric field in the beam is

$$\begin{aligned}E_0 &= \sqrt{2\mu_0 cI} \\ &= \sqrt{2(4\pi \times 10^{-7}\ \text{N/A}^2)(3.0 \times 10^8\ \text{m/s})(1.0 \times 10^{-3}\ \text{W/m}^2)} \\ &= 0.87\ \text{V/m}.\end{aligned}$$

The amplitude of the magnetic field is then, from Eq. (35-7),

$$B_0 = \frac{E_0}{c} = 2.9 \times 10^{-9}\ \text{T}.$$

Electromagnetic waves transport momentum as well as energy. It can be shown with Maxwell's equations that the momentum per unit volume carried by these waves is

$$\mathbf{P} = \frac{\mathbf{S}}{c^2}. \tag{35-11}$$

Suppose that electromagnetic radiation is incident normally on a surface of area A, as shown in Fig. 35-7. In a time interval Δt, the portion of the electromagnetic wave in the volume

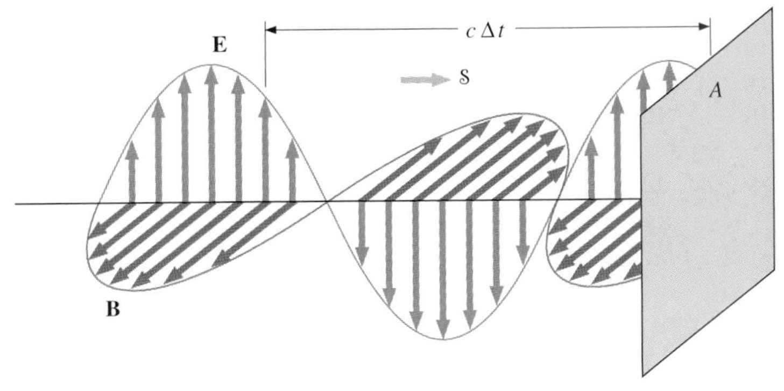

FIGURE 35-7 Electromagnetic radiation incident on a surface exerts a radiation pressure on that surface. In a time Δt, radiation in the volume $Ac\,\Delta t$ falls on the surface.

$A(c\,\Delta t)$ falls on the surface. If the radiation is completely absorbed by the surface, the average momentum it imparts to that surface in a time Δt is

$$\Delta\bar{\mathbf{P}} = \frac{\bar{S}}{c^2}(\text{vol}) = \frac{\bar{S}}{c^2}Ac\,\Delta t = \frac{\bar{S}A}{c}\,\Delta t.$$

The average force that the electromagnetic wave exerts on the surface is therefore

$$\bar{\mathbf{F}} = \frac{\Delta\bar{\mathbf{P}}}{\Delta t} = \frac{\bar{S}A}{c}.$$

The **radiation pressure** p of the wave is defined as the average force per unit area on the surface. Using the previous equation for the average force, we obtain

$$p = \frac{\bar{S}}{c} \quad \text{(total absorption).} \tag{35-12a}$$

If the wave is completely reflected by the surface, the momentum imparted to the surface is doubled because the momentum of the reflected wave is opposite to that of the incident wave. For this case, the radiation pressure becomes

$$p = \frac{2\bar{S}}{c} \quad \text{(total reflection).} \tag{35-12b}$$

Evidence of radiation pressure is found in the appearance of comets, which are basically chunks of icy material in which frozen gases and particles of rock and dust are embedded. When a comet approaches the sun, it warms up and its surface begins to evaporate. Some of the gases and dust form "tails" when they leave the comet. Notice in the photograph of Fig. 35-8*a* that a comet has *two* tails. The *ion tail* (Fig. 35-8*b*) is composed mainly of ionized gases. These ions interact electromagnetically with the "solar wind," which is a continuous stream of charged particles emitted by the sun. The force of the solar wind on the ionized gases is strong enough that the ion tail almost always points directly away from the sun. The second tail is composed of dust particles. Since the *dust tail* is electrically neutral, it does not interact with the solar wind. However, this tail is affected by the radiation pressure produced by the light from the sun. Although quite small, this pressure is strong enough to cause the dust tail to be displaced from the path of the comet.

(*a*)

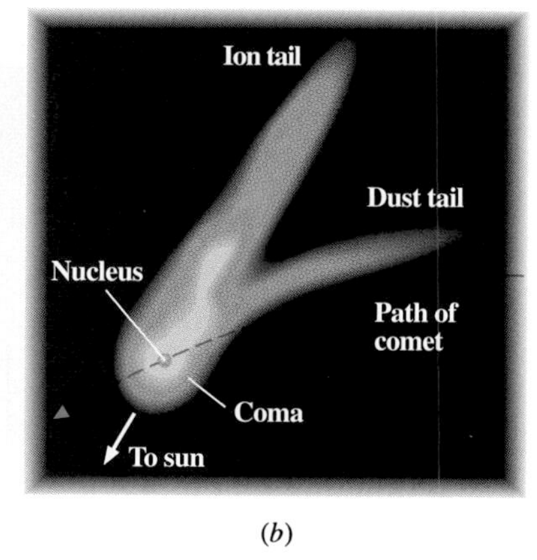

(*b*)

FIGURE 35-8 (*a*) Evaporation of material from a comet being warmed by the sun forms two tails. (*b*) The ion tail interacts with the solar wind and points away from the sun. The dust tail is curved slightly because of the radiation pressure of sunlight.

EXAMPLE 35-2 HALLEY'S COMET

On February 9, 1986, Comet Halley was at its closest point to the sun, about 9.0×10^{10} m from the center of the sun. The average power output of the sun is 3.8×10^{26} W. (*a*) Calculate the radiation pressure on the comet at this point in its orbit. (Assume that the comet reflects all the incident light.) (*b*) Suppose that a 10-kg chunk of material of cross-sectional area 4.0×10^{-2} m^2 breaks loose from the comet. Calculate the force on this chunk due to the solar radiation. Compare this force with the gravitational force of the sun.

SOLUTION (*a*) The intensity of the solar radiation is the average solar power per unit area. Hence at 9.0×10^{10} m from the center of the sun, we have

$$\bar{S} = \frac{3.8 \times 10^{26}\text{ W}}{4\pi(9.0 \times 10^{10}\text{ m})^2} = 3.7 \times 10^3\text{ W/m}^2.$$

Assuming the comet reflects all the incident radiation, we obtain from Eq. (35-12*b*)

$$p = \frac{2\bar{S}}{c} = \frac{2(3.7 \times 10^3\text{ W/m}^2)}{3.0 \times 10^8\text{ m/s}} = 2.5 \times 10^{-5}\text{ N/m}^2.$$

(*b*) The force on the chunk due to the radiation is

$$\begin{aligned} F_r &= pA = (2.5 \times 10^{-5}\text{ N/m}^2)(4.0 \times 10^{-2}\text{ m}^2) \\ &= 1.0 \times 10^{-6}\text{ N}, \end{aligned}$$

while the gravitational force of the sun is

$$\begin{aligned} F_g &= \frac{GMm}{r^2} \\ &= \frac{(6.67 \times 10^{-11}\text{ N}\cdot\text{m}^2/\text{kg}^2)(2.0 \times 10^{30}\text{ kg})(10\text{ kg})}{(9.0 \times 10^{10}\text{ m})^2} \\ &= 0.16\text{ N}. \end{aligned}$$

The gravitational force of the sun on the chunk is therefore much greater than the force of the radiation.

DRILL PROBLEM 35-1

A plane electromagnetic wave propagating in free space has an electric field of amplitude 9.0×10^3 V/m. (*a*) What is the amplitude of its magnetic field? (*b*) What is the intensity of this wave? (*c*) What is the radiation pressure that this wave exerts on an object that completely absorbs it?
ANS. (*a*) 3.0×10^{-5} T; (*b*) 1.1×10^5 W/m^2; (*c*) 3.6×10^{-4} N/m^2.

DRILL PROBLEM 35-2

One-dimensional wave equation. Differentiate Eq. (35-1*a*) with respect to x and Eq. (35-2*b*) with respect to t. (*b*) Use the resulting equations together with $c = 1/\sqrt{\mu_0 \epsilon_0}$ and the fact that $\partial^2 B_z(x, t)/\partial x\, \partial t = \partial^2 B_z(x, t)/\partial t\, \partial x$ to obtain the one-dimensional wave equation for E_y:

$$\frac{\partial^2 E_y(x, t)}{\partial x^2} - \frac{1}{c^2}\frac{\partial^2 E_y(x, t)}{\partial t^2} = 0.$$

35-2 ELECTROMAGNETIC SPECTRUM

Electromagnetic waves that we detect visually are known as *visible light*. Their wavelengths range from approximately 4.0×10^{-7} to 7.0×10^{-7} m. The visual sensation of color is produced by photoreceptor cells in the eye called *cones*. There are three sets of cones, each of which is especially sensitive over a particular range of wavelengths of the visible spectrum, as illustrated in Fig. 35-9. Light in the short-wavelength region primarily stimulates the set of cones responsible for the color sensation of blue; light in the middle of the visible spectrum primarily stimulates the cones responsible for the perception of green; and light in the long-wavelength region of the spectrum primarily stimulates the third set of cones and produces the sensation of red. Light in the appropriate wavelength band is designated as blue, green, or red, and these three colors are called the *primary colors*. By mixing different combinations of the primary colors, a wide range of different color sensations can be produced.* For example, the colors produced by a television screen are the result of the mixing of the three primary colors emitted by three different phosphors deposited on the surface of the screen. Figure 35-10 summarizes how the primary colors mix. A combination of red (R), green (G), and blue (B) light stimulates all three sets of cones in the eye and is perceived as white (W) light:

$$\mathrm{R + G + B = W}.$$

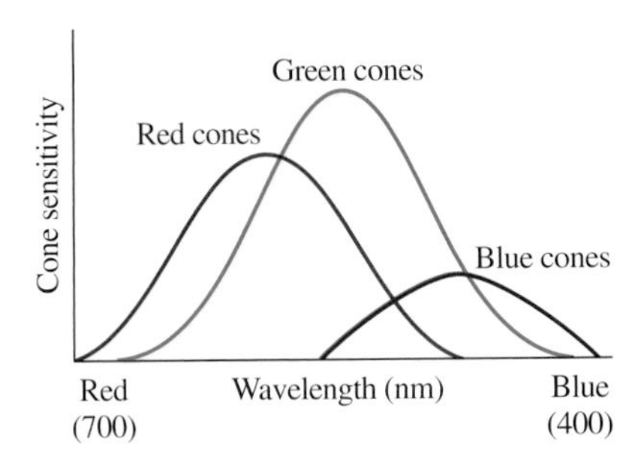

FIGURE 35-9 The three types of cones in the eye respond to different wavelengths of light as shown.

Other familiar combinations are blue plus green, which produces cyan (C); red and blue, which is seen as magenta (M); and red and green, which is perceived as yellow (Y). A wide variety of other colors can be produced by mixing combinations of nonprimary colors. For example, a mixture of magenta and yellow light gives

*We are considering the production of different colors by *addition*. Colors can also be produced by *subtraction*, which is the process responsible for the different colors we get when we mix paints. Subtraction will not be discussed here.

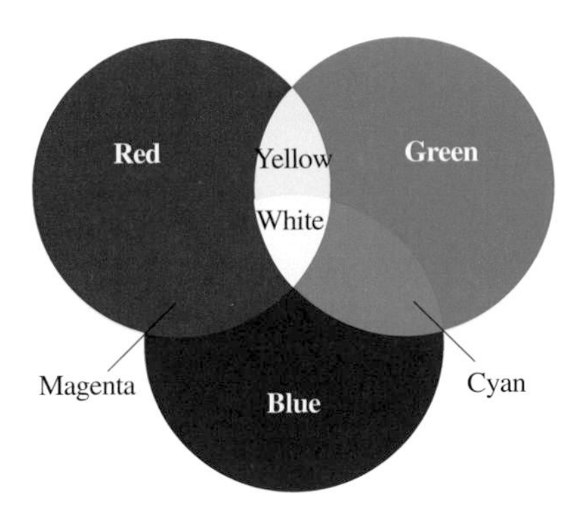

FIGURE 35-10 When different combinations of light beams of the three primary colors (red, blue, and green) are projected onto a white screen, the mixtures produce other colors.

$$\begin{aligned}\mathrm{M + Y} &= \mathrm{(R + B) + (R + G)}\\ &= \mathrm{(R + G + B) + R = W + R},\end{aligned}$$

which is seen as pink.

Light consisting of a single wavelength (in reality, a very narrow range of wavelengths) is called *monochromatic light*. The particular wavelength of the light determines which cones are stimulated and therefore the color of that light. For example, monochromatic light of wavelength 7.0×10^{-7} m is red, because it stimulates just the red cones; while monochromatic light of wavelength 5.9×10^{-7} m stimulates both red and green cones and is therefore yellow.

Occasionally, an overexposure to one color can desensitize the corresponding set of cones. Before color computer monitors became common, this was often experienced by programmers who sat for long periods in front of monitors with green displays. When they first looked away from their monitors, they found that white objects had a slight magenta tint. The reason for this is that continuous exposure to green light slightly desensitized the green cones in their eyes. The combination of the other two primary colors was therefore enhanced and provided a mild sensation of magenta.

Visible light is a very small part of the electromagnetic spectrum (Fig. 35-11). Radiation with wavelengths slightly longer than that of red light is called *infrared*. An example of infrared radiation is the heat we feel from a nearby burning object.

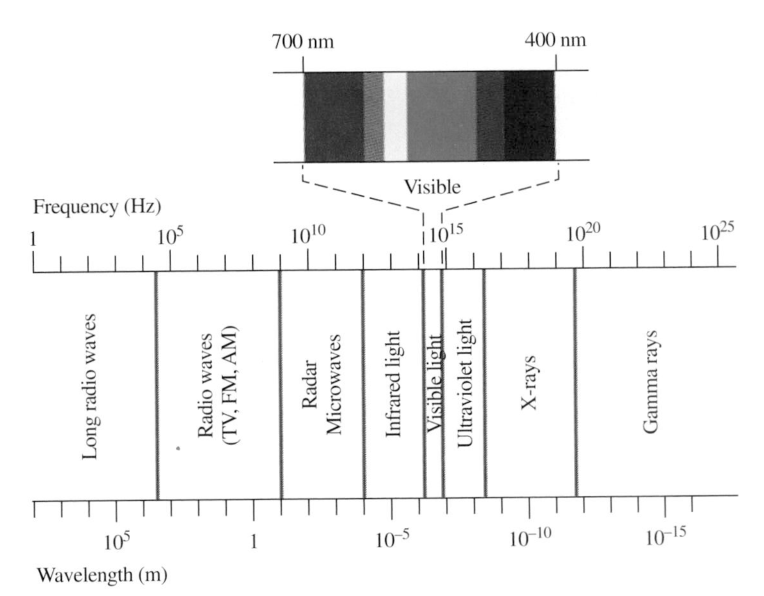

FIGURE 35-11 The electromagnetic spectrum.

Unlike visible light, much of the infrared part of the spectrum cannot penetrate the earth's atmosphere. For this reason, earth-orbiting satellites (Fig. 35-12) are used to detect infrared radiation from space. *Microwaves* and *radio waves* are also electromagnetic radiation. Their wavelengths are longer than that of infrared radiation. Beyond the short-wavelength end of the visible spectrum are *ultraviolet light*, *x-rays*, and *gamma rays*.

FIGURE 35-12 The InfraRed Astronomy Satellite (IRAS), launched in 1983.

We often use a unit of length called the *nanometer* (*nm*) to express wavelengths: 1 nm = 10^{-9} m. For example, a wavelength of 7.0×10^{-7} m is 700 nm. Other units used are the *micrometer* (μm), which is 10^{-6} m, and the *angstrom* (*Å*), which is 10^{-10} m.

35-3 Speed of Light

From everyday experience, you know that light travels very fast. At the start of a race at a track meet, you may have noticed that you see a puff of smoke from the starter's pistol before hearing its bang. Although the sound from the pistol travels at about 330 m/s, it still moves much more slowly than the light that carries the image of the firing.

The first measurement of the speed of light was made by the Danish astronomer Olaus Roemer (1644–1710) in 1675. He studied the orbit of Io, one of the four large moons of Jupiter, and found that it had a period of revolution of 42.5 h around its mother planet. He also discovered that this value fluctuated by a few seconds, depending on the position of the earth in its orbit around the sun. Roemer realized that this fluctuation was due to the finite speed of light and could be used to determine *c*.

Roemer found the period of revolution of Io by measuring the time interval between its successive eclipses by Jupiter. Figure 35-13*a* shows the planetary configuration when such a mea-

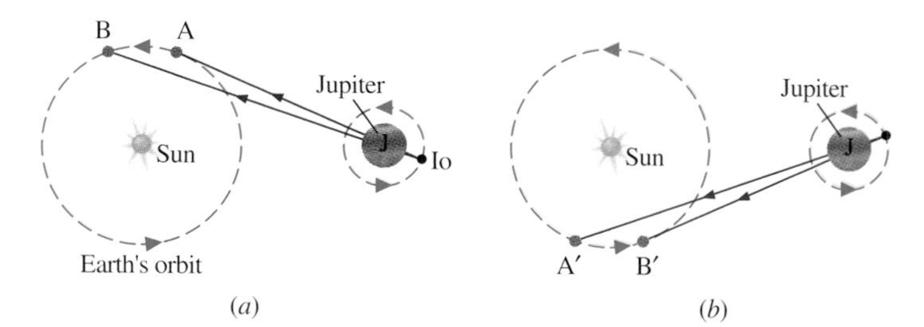

FIGURE 35-13 Roemer's astronomical method for determining the speed of light. Measurements of Io's period done with the configurations of parts (*a*) and (*b*) differ because of the difference between the distances BJ and B′J.

surement is made for the earth in the part of its orbit where it is receding from Jupiter. When the earth is at A, the earth, Jupiter, and Io are aligned. The next time this alignment occurs, the earth is at B, and the light carrying that information to the earth must travel to that point. Now imagine it is about six months later, and the planets are arranged as in Fig. 35-13*b*. The measurement of Io's period begins with the earth at A′ and Jupiter eclipsed by Io. The next eclipse then occurs when the earth is at B′, to which the light carrying the information of this eclipse must travel. Since B′ is closer to Jupiter than B, light takes less time to reach the earth when it is at the former location. The time interval between the successive eclipses of Io seen at A′ and B′ is therefore less than the time interval between the eclipses seen at A and B. By measuring the difference in these time intervals and with approximate knowledge of the distance between Jupiter and the earth, Roemer calculated that the speed of light was 2.0×10^8 m/s, which is 33 percent below the actual value.

The first successful terrestrial measurement of the speed of light was made by Armand Fizeau (1819–1896) in 1849. He placed a toothed wheel that could be rotated very rapidly on one hilltop and a mirror on a second hilltop 8 km away (Fig. 35-14). An intense light source was placed behind the wheel, so that when the wheel rotated, it chopped the light beam into a succession of pulses. The speed of the wheel was then adjusted until no light returned to the observer located behind the

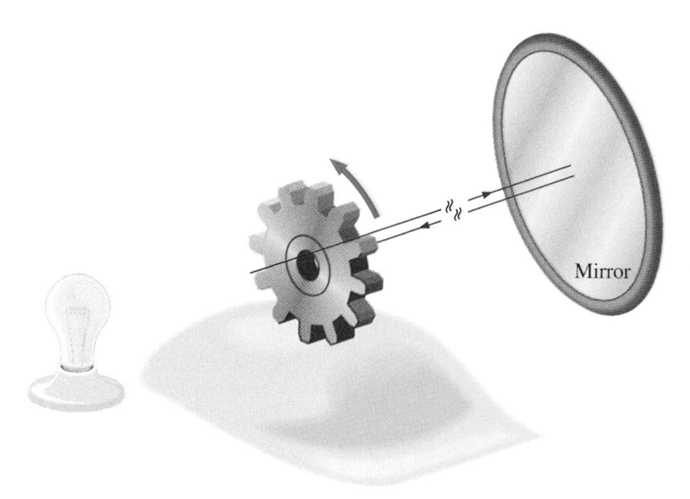

FIGURE 35-14 A representation of Fizeau's terrestrial method for measuring the speed of light.

wheel. This could only happen if the wheel rotated through an angle corresponding to a displacement of $(n + 1/2)$ teeth while the pulses traveled down to the mirror and back. Knowing the rotational speed of the wheel, the number of teeth on the wheel, and the distance to the mirror, Fizeau determined the speed of light to be 3.15×10^8 m/s, which is only 5 percent too high!

DRILL PROBLEM 35-3

Assume that the light pulse leaving an opening in Fizeau's wheel with 360 teeth is blocked upon its return by the adjacent tooth. If the wheel and mirror are 8.0 km apart and the wheel rotates at 1600 rev/min, what does Fizeau's measurement give for the speed of light?
ANS. 3.07×10^8 m/s.

Foucault modified Fizeau's apparatus by replacing the toothed wheel with a rotating mirror. In 1862 he measured the speed of light to be 2.98×10^8 m/s, which is within 0.6 percent of the presently accepted value. Albert Michelson (1852–1931) also used Foucault's method on several occasions to measure the speed of light. His first experiments were performed in 1878, and by 1926 he had refined the technique so well that he found c to be $299{,}796 \pm 4$ km/s.

The measurements of c have become so precise that its value is now taken to be exactly

$$c = 2.99792458 \times 10^8 \text{ m/s},$$

a number used to define the meter, as described in Chap. 1.

The quantity c represents the speed of light *propagating in free space*. In a transparent medium, the speed at which light travels is always less than c. The ratio of the speed of light c in free space to its speed v in a medium is known as the **index of refraction** n of the medium:

$$n = \frac{c}{v}. \tag{35-13}$$

The indices of refraction for various media are given in Table 35-1. Since light always travels more slowly than c in material media, n always exceeds 1. The index of refraction of air is so close to 1 that we will assume in most calculations that the speeds of light in air and in free space are the same.

TABLE 35-1 **The Indices of Refraction of Various Substances** The indices are given for yellow light ($\lambda = 589$ nm).

Medium	Index	Medium	Index
Vacuum (exactly)	1	Benzene	1.50
Air (STP)	1.00029	Crown glass	1.52
Ice	1.31	Sodium chloride	1.54
Water (20°C)	1.33	Polystyrene	1.55
Acetone	1.36	Carbon disulfide	1.63
Ethyl alcohol	1.36	Flint glass	1.65
Sugar solution (30%)	1.38	Sapphire	1.77
Carbon tetrachloride	1.46	Heavy flint glass	1.89
Fused quartz	1.46	Diamond	2.42
Sugar solution (80%)	1.49		

EXAMPLE 35-3 Comparing the Speeds of Light in Water and in Glass

Determine the ratio of the speed of light in water to the speed of light in glass for which $n_g = 1.50$.

SOLUTION From Eq. (35-13), the speeds of light in water and in glass are, respectively,

$$v_w = \frac{c}{n_w} \tag{i}$$

and

$$v_g = \frac{c}{n_g}. \tag{ii}$$

Dividing Eq. (i) by Eq. (ii) then yields

$$\frac{v_w}{v_g} = \frac{c/n_w}{c/n_g} = \frac{n_g}{n_w} = \frac{1.50}{1.33} = 1.13.$$

Hence light propagates more rapidly in water, the medium with the lower index of refraction.

35-4 Huygens' Principle

Figure 35-15 shows a photograph of a beam of light in air; the beam is incident on the surface of a semicircular glass slab. Part of the light is *reflected* back into the air, and the rest is *refracted* at the boundary and crosses into the glass. Reflection and refraction occur when light encounters a boundary between two transparent media.

The behavior of the light at the glass surface is similar to that of a mechanical wave in a string when it arrives at a boundary between two strings of different densities. (See Sec. 17-4.) Part

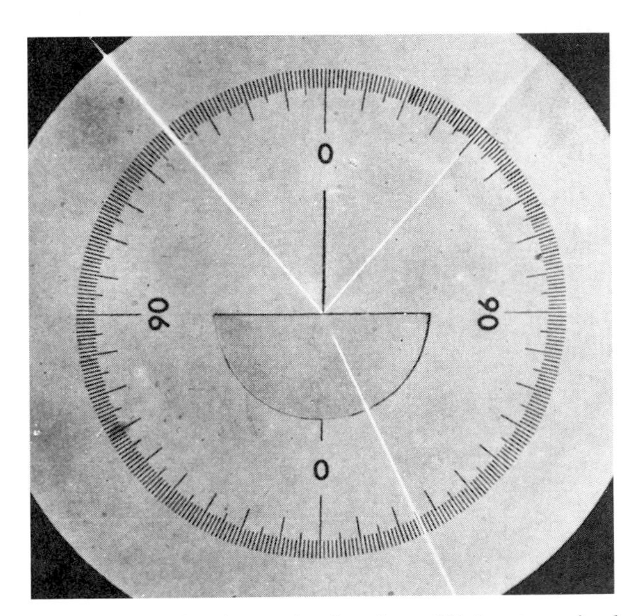

FIGURE 35-15 The reflection and refraction of light at an air-glass interface.

of the wave is reflected and the rest is transmitted. Since light is an electromagnetic disturbance, its reflection and refraction are explained at the most fundamental level in terms of Maxwell's equations. With these equations, we can determine what fractions of the incident light intensity are reflected and refracted, along with the directions of the reflected and refracted beams. There is, however, a much simpler theory, derivable from Maxwell's equations, that can be used to determine the directions of the beams but not their relative intensities. This theory was introduced by the Dutch physicist Christian Huygens (1629–1695) almost two centuries before the formulation of Maxwell's equations. Huygens stated that

> Every point on a wavefront can be treated as a source of secondary waves (called wavelets), which spread out uniformly at the speed of light in the medium of propagation. The tangents to the surfaces of these wavelets at a given instant then make up the new wavefront.

Figure 35-16*a* and *b* represents the propagation of a spherical and a plane wave as described by Huygens' principle. The initial wavefronts from which the wavelets emanate are represented by AA′. At a time t later, the radii of the wavelets are vt, and BB′ are the wavefronts formed from these wavelets. The fact that the wavelets are spherical is based on the assumption that the medium of propagation is *isotropic*; that is, the speed of light v in the medium is the same in any direction. All media will be assumed to have this basic property.

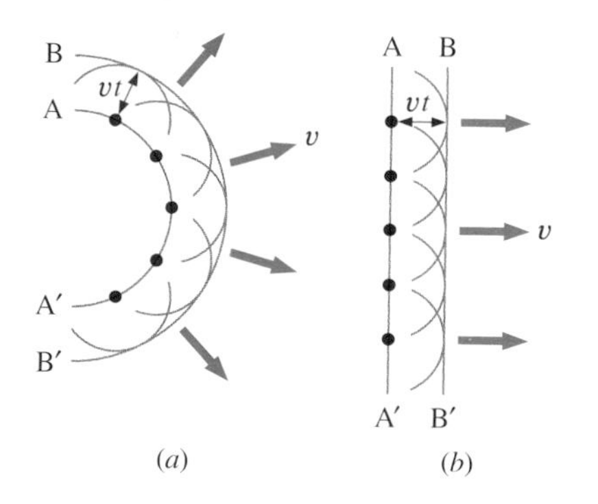

FIGURE 35-16 The propagation of (*a*) a spherical wave and (*b*) a plane wave according to Huygens' principle. The wavefronts serve as sources (the dots) for secondary wavelets.

Notice that there is no backward propagation of the wavelets in the Huygens construction. This seemingly arbitrary omission is justified by the Maxwellian theory, which predicts that the amplitude of the wavelets in the backward direction is zero.

35-5 Laws of Reflection and Refraction

Rays representing the reflection and refraction of a light beam at an interface between two media are shown in Fig. 35-17. The incident ray falls on the interface at an angle θ_1, the reflected ray leaves at θ_1', and the refracted ray leaves at θ_2. All three angles are measured with respect to the normal to the interface. The angles θ_1 and θ_1' are known as the **angle of incidence** and

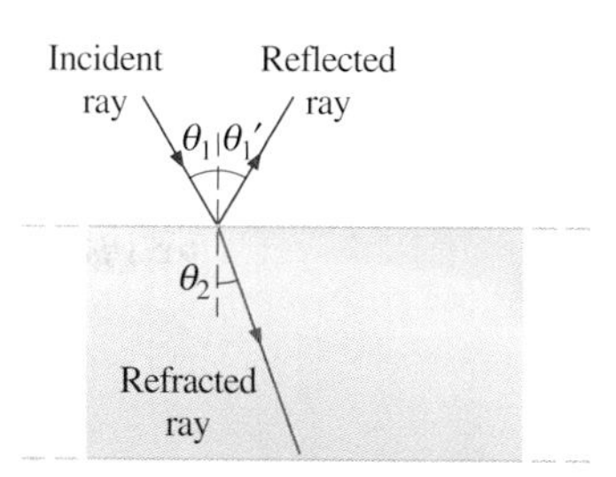

FIGURE 35-17 Reflection and refraction at an interface.

the **angle of reflection**, respectively, while θ_2 is called the **angle of refraction**. The relationship between the directions of the incident and reflected rays is a very simple one: *The angle of reflection equals the angle of incidence*; that is,

$$\theta_1 = \theta_1'. \tag{35-14}$$

This is the *law of reflection*. The direction of the refracted ray is given by *Snell's law* (or the *law of refraction*), named after its discoverer, Willebrod Snell (1591–1626). This law states that

$$n_1 \sin\theta_1 = n_2 \sin\theta_2, \tag{35-15}$$

where n_1 and n_2 are the indices of refraction of media 1 and 2, respectively.

According to Snell's law, a light beam that passes from a medium with a lower refractive index to one with a higher index is bent toward the normal ($\theta_1 > \theta_2$). Examples of this occur when light travels from air to glass or from air to water. The light from celestial objects is similarly refracted when it enters the earth's atmosphere from the vacuum of space. As a result, a celestial object appears to be at a slightly greater altitude above the horizon than it really is. This effect is most pronounced near the horizon where the apparent altitude of celestial objects is raised by about 0.5°. Snell's law also indicates that a light beam entering a medium with a lower refractive index is bent away from the normal. In fact, if the direction of the light ray in Fig. 35-18*a* is reversed, the ray retraces its own path. This is a general property of light known as *reversibility*.

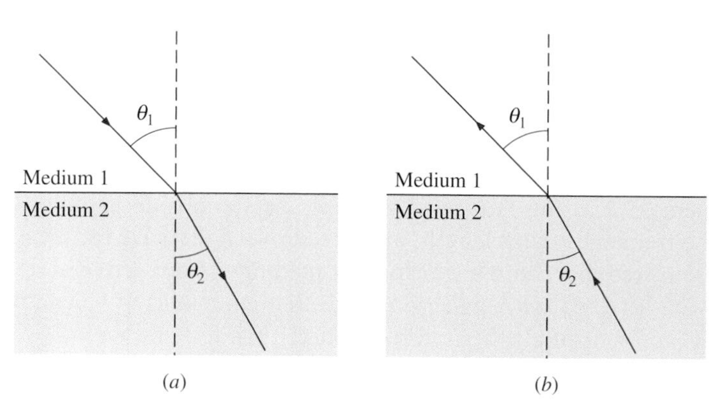

FIGURE 35-18 Snell's law and the reversibility of light. The incident ray is in (*a*) the medium with the smaller index of refraction, (*b*) the medium with the greater index of refraction.

Light from the submerged portion of the straw is refracted at the boundary between the water and the air. This refraction makes the straw appear bent.

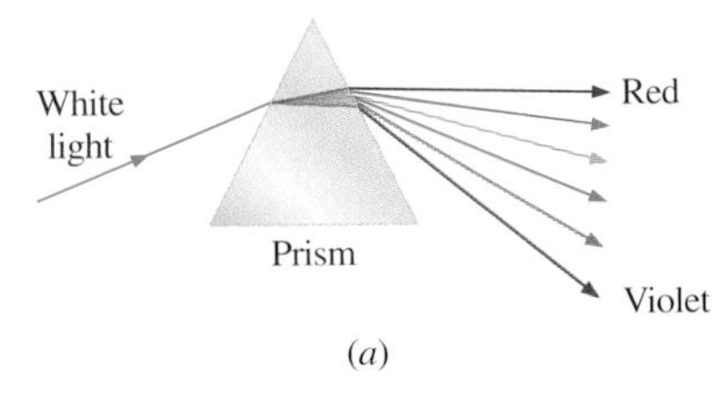

(*a*)

(*b*)

FIGURE 35-20 Dispersion of white light by a prism.

The index of refraction of a medium actually depends on the wavelength of the light. All indices given in Table 35-1 were measured with yellow light ($\lambda = 589$ nm) from a sodium lamp. Figure 35-19 shows how the refractive indices of various materials change with wavelength. In general, the refractive index of a substance increases as the wavelength decreases. Because the index of refraction varies with wavelength, a beam of white light is separated into its component wavelengths when it is refracted. This phenomenon is known as *dispersion*. The separation is so slight that it usually takes two refractions, such as we find with the prism of Fig. 35-20, before it is readily observable. Unless stated otherwise, we will neglect the effect of dispersion when studying refraction.

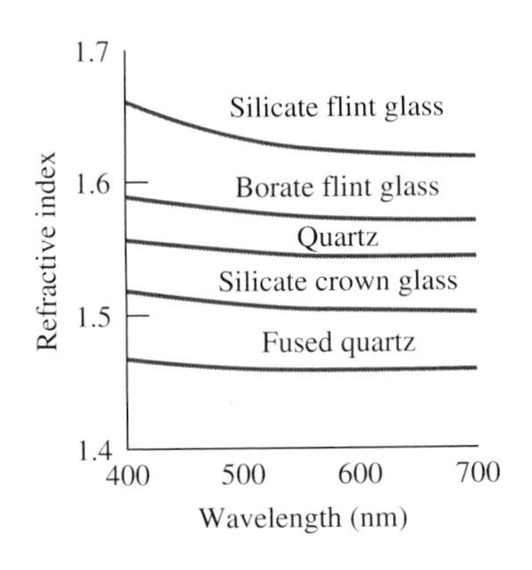

FIGURE 35-19 Variation of the index of refraction with wavelength for various materials.

The laws of reflection and refraction are verified in Supplement 35-2, using Huygens' principle. Also considered is how the frequency, wavelength, and speed of a light wave traveling from free space into a transparent medium of refractive index n are affected. If f_0 and λ_0 are the frequency and wavelength of monochromatic light in free space, then it is shown in Supplement 35-2 that

$$f = f_0, \tag{35-16a}$$

$$\lambda = \frac{\lambda_0}{n}, \tag{35-16b}$$

and

$$v = \frac{c}{n}, \tag{35-16c}$$

where f, λ, and v are the frequency, wavelength, and speed of the light in the medium. When monochromatic light crosses from free space into a transparent medium, its frequency remains the same, while its wavelength and speed are both reduced by the factor $1/n$.

EXAMPLE 35-4 REFRACTION AT AN AIR-GLASS INTERFACE

A light ray traveling in air strikes the surface of a rectangular block of glass with $n = 1.50$ at an angle of incidence of 60.0°, as shown in Fig. 35-21*a*. (*a*) Determine the paths of the reflected and refracted rays. (*b*) What happens to the refracted ray at the bottom surface of the glass block?

SOLUTION (*a*) The reflected and refracted rays are shown in Fig. 35-21*b*. From the law of reflection, the angle of reflection must equal the angle of incidence, or 60.0°. The angle of refraction θ_2 can be found using Snell's law:

$$(1.00)\sin 60.0° = (1.50)\sin\theta_2,$$

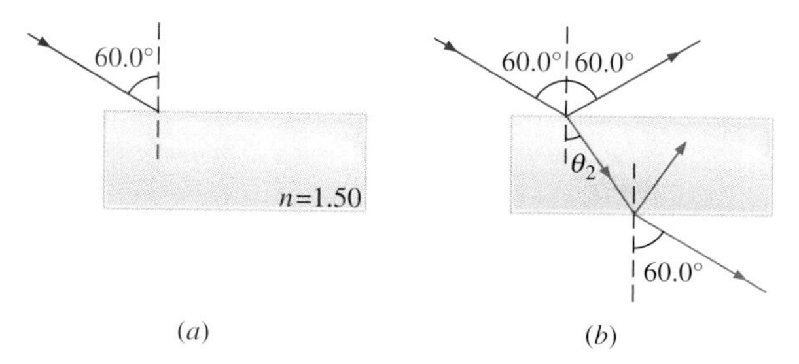

FIGURE 35-21 Reflection and refraction of a light ray incident on a glass block.

so

$$\theta_2 = 35.3°.$$

(*b*) The refracted ray strikes the opposite face of the block with an angle of incidence of 35.3° and is once again split into a reflected and refracted ray. The second refracted ray emerges into the air at an angle θ_1 found from

$$(1.50) \sin 35.3° = (1.00) \sin \theta_1,$$

so

$$\theta_1 = 60.0°.$$

The directions of the light ray when entering and leaving the block are therefore the same. However, the ray is laterally displaced and is diminished in intensity.

DRILL PROBLEM 35-4

Suppose that the light ray of Example 35-4 is composed of two wavelengths, 400 and 700 nm. Using the variation in refractive index with wavelength shown in Fig. 35-19 for silicate crown glass, determine the angle between the two refracted rays at the top surface. Which wavelength has the larger angle of refraction? Which wavelength has a greater lateral displacement in passing through the block?
ANS. 0.47°; 700 nm; 400 nm.

DRILL PROBLEM 35-5

What happens to the frequency, wavelength, and speed of light that crosses from a medium with index of refraction n_1 to one with index of refraction n_2?
ANS. The frequency does not change, while the wavelength and speed change by the factor n_1/n_2.

DRILL PROBLEM 35-6

A monochromatic light beam of frequency 6.00×10^{14} Hz crosses from air into a transparent material where its wavelength is measured to be 300 nm. What is the index of refraction of the material?
ANS. 1.67.

35-6 Total Internal Reflection

Figure 35-22*a* shows the reflection and refraction of a light ray at the interface between media 1 and 2 whose refractive indices are n_1 and n_2. The ray is initially in medium 1 and has an angle of incidence θ_1 and an angle of refraction θ_2. We assume that $n_1 > n_2$, so the ray is refracted away from the normal as shown. From Snell's law, we may write

$$\frac{n_1}{n_2} \sin \theta_1 = \sin \theta_2.$$

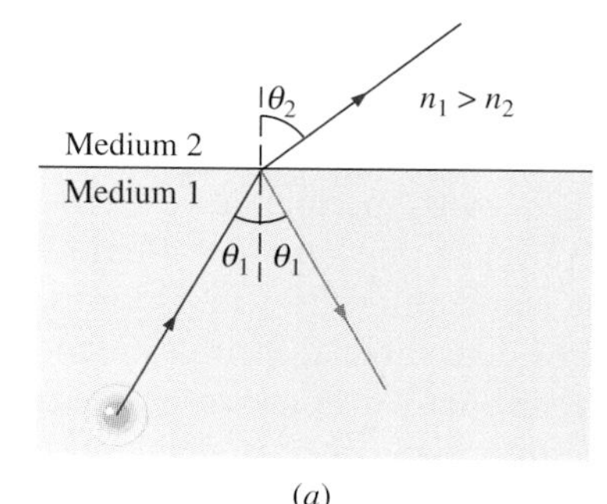

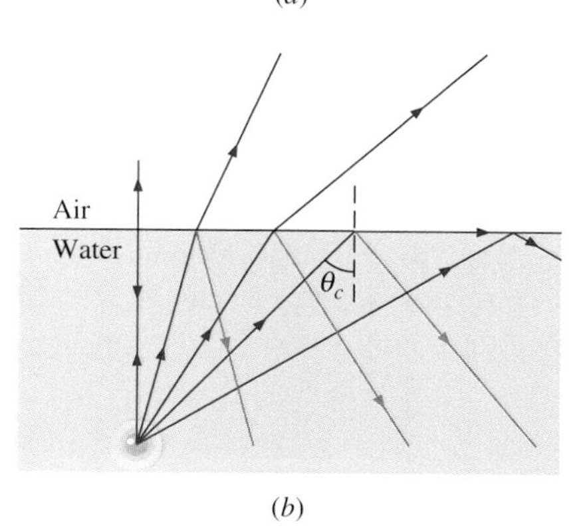

FIGURE 35-22 (*a*) A light ray from a source in the medium with a higher index of refraction is bent away from the normal when it is refracted at the interface. (*b*) When the angle of incidence at the air-water interface exceeds the critical angle, total internal reflection occurs.

The right-hand side of this equation is a sine function that has a range from 0 to 1. The left-hand side must therefore have the same range; that is,

$$1 \geq \frac{n_1}{n_2} \sin \theta_1 \geq 0.$$

Thus for Snell's law to apply to the refraction shown in Fig. 35-22*a*, the angle of incidence θ_1 must satisfy

$$\theta_c \geq \theta_1 \geq 0,$$

where

$$\frac{n_1}{n_2} \sin \theta_c = 1,$$

or

$$\theta_c = \sin^{-1} \frac{n_2}{n_1}. \tag{35-17}$$

The angle θ_c is called the **critical angle**. When the angle of incidence exceeds θ_c, no refracted beam is observed, and the incident beam is completely reflected at the boundary. This phenomenon, known as *total internal reflection*, only occurs when the light originates in the medium with the higher refractive index. Light rays emanating with increasing angles of incidence from a point source immersed in water are shown in Fig. 35-22*b*. When these angles are less than the critical angle, the rays are partially reflected and partially refracted at the boundary. At the critical angle, the refracted ray skims the water surface; for larger angles of incidence, total internal reflection occurs and all of the light is reflected back into the water.

A practical application of total internal reflection is the optical fiber. This is a thin filament of transparent material with a high index of refraction. Light that enters the filament is transported along its length by a series of total internal reflections, as illustrated in Fig. 35-23*a*. With special materials, losses in light intensity due to absorption can be kept to a minimum, allowing optical signals to be transmitted up to 200 km with enough intensity remaining to be detectable at their destination. Since rays from different parts of an object are scrambled by the multiple reflections along an optical fiber, a single optical fiber cannot transmit an image. To overcome this problem, bundles of closely packed fibers with their relative positions fixed are used. Each individual fiber transmits rays from a very small region of the object. When the light from all the fibers is then viewed or projected onto a screen, an image of the object is formed. These fiber bundles are very useful in medicine. Physicians often insert them into the body and view internal organs, thereby avoiding exploratory surgery. The fibers are also used for viewing when modern microsurgery is performed.

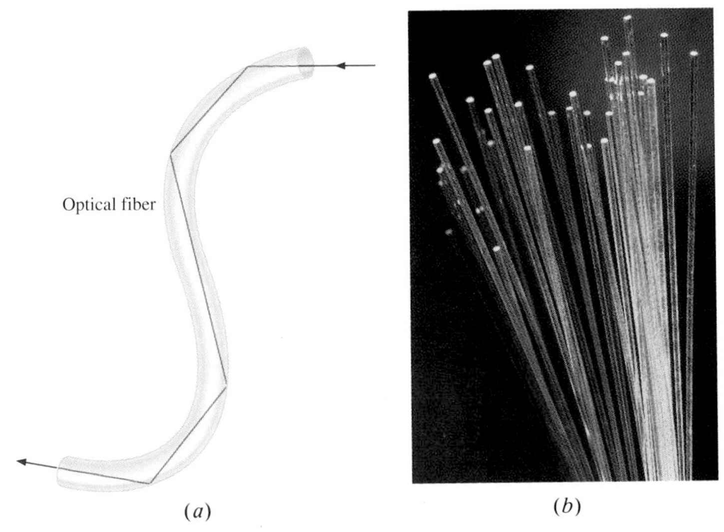

FIGURE 35-23 (*a*) Light travels down an optical fiber through a series of total internal reflections. (*b*) A bundle of optical fibers.

DRILL PROBLEM 35-7

What is the critical angle associated with an air-glass interface? with a glass-water interface? In each case, specify the medium of the incident light if total internal reflection is to occur. Use $n = 1.50$ for glass.
ANS. 41.8°; 62.5°; glass in both cases.

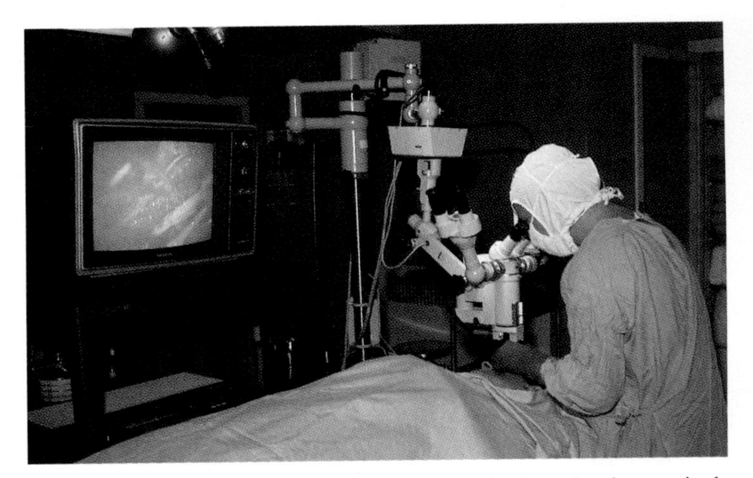

During this surgery, the image of an organ is viewed using optical fibers.

*35-7 THE RAINBOW

The formation of a rainbow is due to a combination of several of the optical phenomena we have discussed—refraction, dispersion, and total internal reflection. Figure 35-24*a* shows a ray of monochromatic light incident on a water droplet in the atmosphere. The ray is first refracted at the front surface of the droplet. It then travels to the back surface of the droplet where it is internally reflected back to the front surface. There the ray is once again refracted as it leaves the water and enters the air. The French philosopher and mathematician René Descartes

The colors of the visible spectrum are displayed in a rainbow.

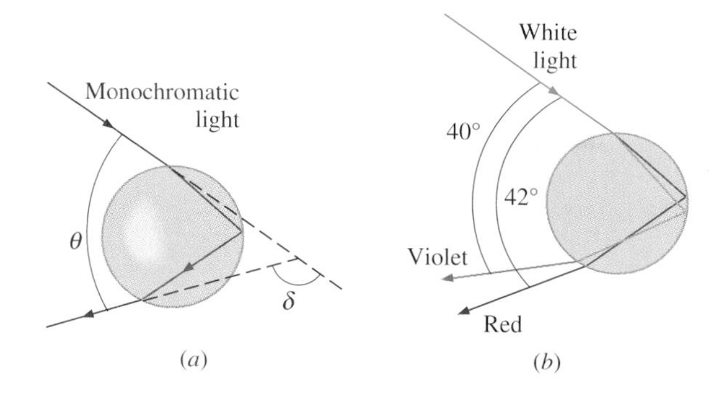

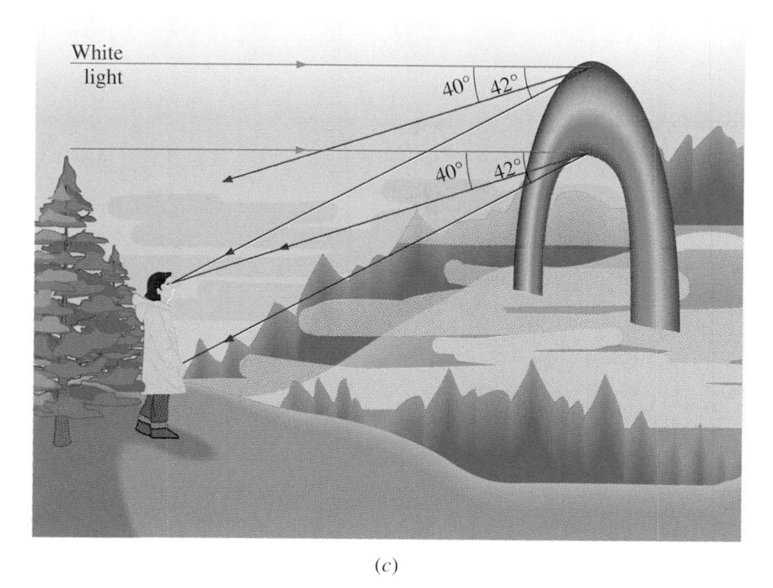

FIGURE 35-24 (*a*) Refraction of a monochromatic light ray by a spherical water droplet. The angle of deviation is δ, and $\theta = 180° - \delta$. (*b*) White light falling on a droplet is dispersed. Primarily violet light is refracted at $\theta = 40°$, while most of the refracted light at $\theta = 42°$ is red. (*c*) The formation of a rainbow when the sun is low on the horizon.

(1596–1650) showed that when a ray of one particular color falls on a droplet at the appropriate point such that its angle of deviation δ is a minimum, then all other rays of the same color striking the droplet surface near this same point will be refracted in nearly the same direction as that ray. Consequently, this color is enhanced when its light rays strike the droplet in the direction corresponding to minimum angular deviation. Since water is a dispersive medium with different indices of refraction for different colors, the direction of minimum angular deviation varies with color. That direction, expressed in terms of θ, where $\theta = 180° - \delta$, is 42° for red light and 40° for violet light (Fig. 35-24*b*). Between these two extremes are the directions for which the other colors of the visible spectrum are enhanced.

Figure 35-24*c* shows an observer looking skyward during a rainstorm. We'll assume that the sun is low on the horizon so that its white light rays are essentially horizontal. If the observer looks in the directions corresponding to $\theta = 42°$, she sees red light. If she looks in the directions corresponding to $\theta = 40°$, she sees violet light. She sees the other colors of the visible spectrum between 40° and 42° as the other "bands" of the rainbow.

35-8 Polarization

We conclude this chapter by discussing another property of light known as *polarization*. The polarization of a wave specifies the directions of its oscillatory motion. As an example, consider the harmonic wave of Fig. 35-25. It travels in the x direction along the string as the displacement of the string oscillates along the y axis. Because the oscillations occur along a single axis, this wave is *linearly polarized*. The plane that is common to both the oscillations and the direction of propagation is called the *plane of polarization*. For the wave of Fig. 35-25, this is the xy plane.

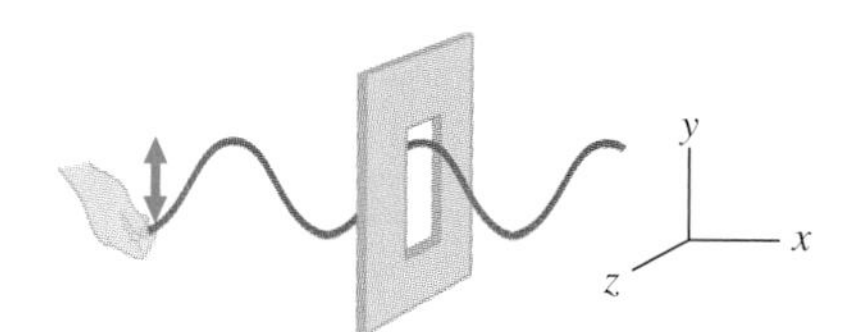

FIGURE 35-25 A traveling harmonic wave on a string. The wave is linearly polarized along the y axis.

By convention, the polarization of an electromagnetic wave is specified by its electric field. The plane wave shown in Fig. 35-6*a* is therefore linearly polarized in the y direction. If you could view the electric field vector of this wave as it directly approached you, you would see it always oscillating along the y axis, as illustrated in Fig. 35-26*a*.

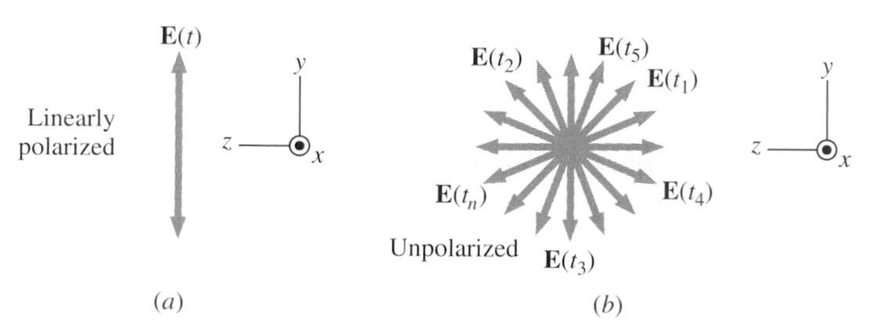

FIGURE 35-26 (*a*) The electric field vector of an electromagnetic wave polarized along the y axis is always oscillating along that axis. (*b*) In an unpolarized beam of light, the direction of **E** fluctuates randomly with time.

An ordinary beam of light consists of a very large number of waves emitted by the molecules of a light source. While each wave is polarized in a particular way, the overall orientation of the electric field vector is completely random. Although this vector points in a particular direction at any instant, that direction changes randomly with time (Fig. 35-26*b*). However, the average magnitude of **E** is the same for all directions. The light beam is said to be *unpolarized*.

It is possible to produce a linearly polarized beam from unpolarized light. To do so, we have to eliminate the vibrations in one direction from an unpolarized beam. One common way to do this is to send the light beam through a sheet of *Polaroid*. This material is a plastic containing long molecular chains that are aligned with one another. After immersion in an iodine solution, the material becomes conducting, but only along the molecular chains. When an unpolarized light beam passes through a sheet of Polaroid, the electric field component parallel to the chains stimulates electrons to move in that direction. As a result, this component is absorbed, and only the component perpendicular to the molecular chains is transmitted by the Polaroid. This process is represented in Fig. 35-27. The axis along which the transmitted light is polarized is called the *transmission axis* of the sheet.

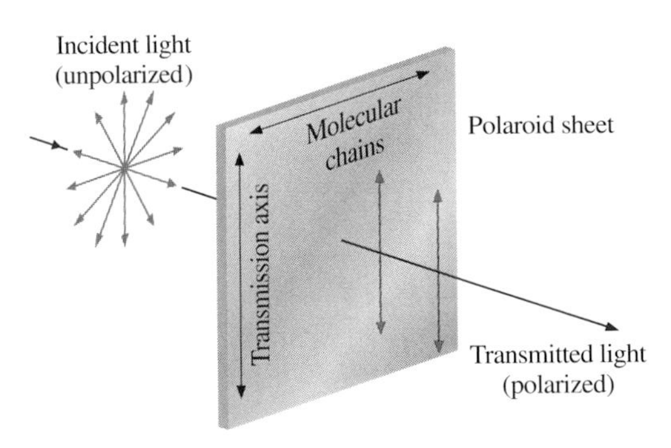

FIGURE 35-27 When unpolarized light is passed through a sheet of Polaroid, the light becomes linearly polarized along the transmission axis of the Polaroid.

How is the intensity of the unpolarized light affected by the Polaroid? To answer this question, we resolve the electric field vector into components parallel and perpendicular to the transmission axis. Since the electric field is rapidly fluctuating (about once every 10^{-8} s) along random directions, the components averaged over time are equal. The parallel component gets transmitted while the perpendicular component is absorbed. Thus when a beam of unpolarized light passes through a Polaroid sheet, its intensity is reduced by one-half.

Suppose that we have two polarizing sheets P_1 and P_2. If they are aligned with their transmission axes perpendicular, as shown in Fig. 35-28*a*, no light reaches the observer. However, if the axes are at an angle θ with respect to each other, as shown in Fig. 35-28*b*, the electric field vector of magnitude E_1 emerging from P_1 has a component $E_1 \cos\theta$ along the transmission axis of P_2. This component passes through P_2 and is detected by the observer. Since the intensity is proportional to the square of the electric field magnitude, the intensity I of the transmitted beam varies with θ according to

$$I = I_M \cos^2\theta, \tag{35-18}$$

where I_M is the *maximum value of the transmitted intensity*. This is *Malus' law*, named for Etienne Malus (1775–1812). Notice that when $\theta = 90°$, $I = 0$, the situation depicted in Fig. 35-28*a*.

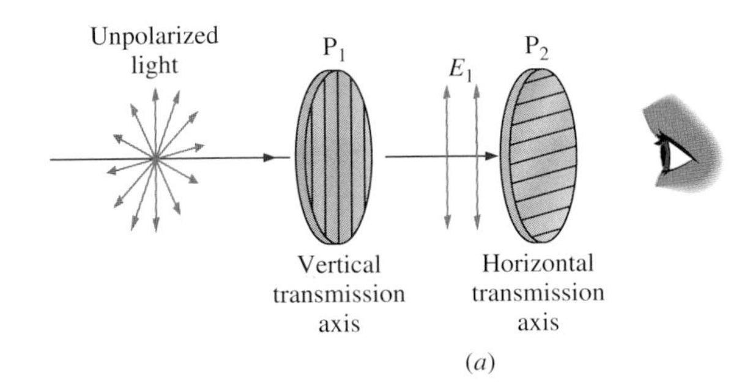

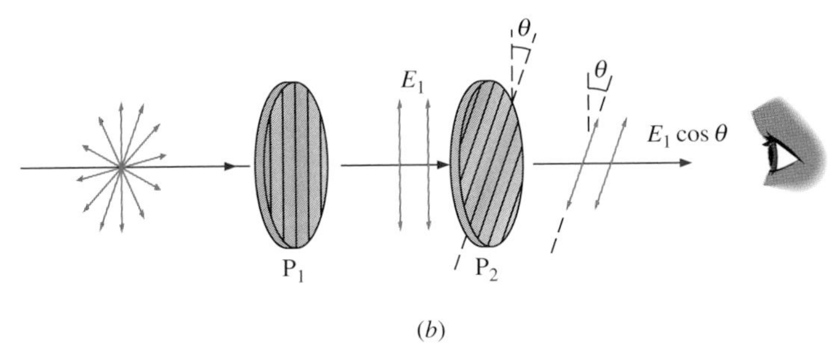

FIGURE 35-28 A combination of two polarizing sheets. (*a*) When the transmission axes of the sheets are mutually perpendicular, no light reaches the observer. (*b*) When the axes are at an angle θ with respect to each other, the intensity of the transmitted beam is $I = I_M \cos^2\theta$, where $I_M \propto E_1^2$.

EXAMPLE 35-5 INTENSITY OF A POLARIZED BEAM

Suppose that the intensity of the unpolarized beam incident on the polarizing sheets of Fig. 35-28*b* is I_0. What is the intensity of the light that passes through both sheets when (*a*) $\theta = 30°$ and (*b*) $\theta = 60°$? Express your answers in terms of I_0.

SOLUTION (*a*) The intensity of the light that passes through P_1 is $I_0/2$. This is also the maximum intensity I_M that can be transmitted through the combination. From Malus' law, we find for $\theta = 30°$ that the intensity passing through both sheets is

$$I = \frac{I_0}{2}\cos^2 30° = 0.38 I_0.$$

(*b*) At $\theta = 60°$, Malus' law gives

$$I = \frac{I_0}{2}\cos^2 60° = 0.13 I_0.$$

DRILL PROBLEM 35-8

For what value of θ will the transmitted beam of Fig. 35-28*b* have one-third the intensity of the beam incident on the first sheet?
ANS. 35°.

Light is also polarized when it is reflected off the interface between two media, with the amount of polarization depending on the angle of incidence of the light beam and the refractive indices of the media. Figure 35-29*a* shows a beam of unpolarized light that is incident on an interface between media with refractive indices n_1 and n_2. By definition, the *plane of incidence* is the plane containing the incident and refracted beams. In the figure it is the plane of the page. The light beam's electric field is resolved into components perpendicular to (the dots) and parallel to (the arrows) the plane of incidence, as illustrated in the figure. These two components are equal in the unpolarized incident beam. However, both the reflected beam and the refracted beam are polarized (at least partially) for all incident angles except 0° and 90°, and the perpendicular component of the electric field of the reflected beam is always greater than the parallel component.

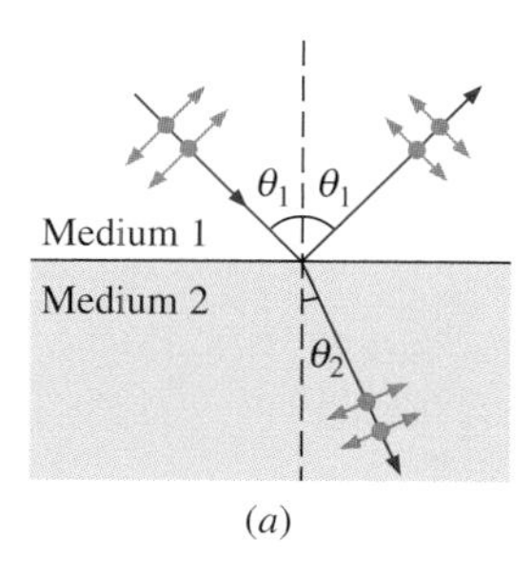

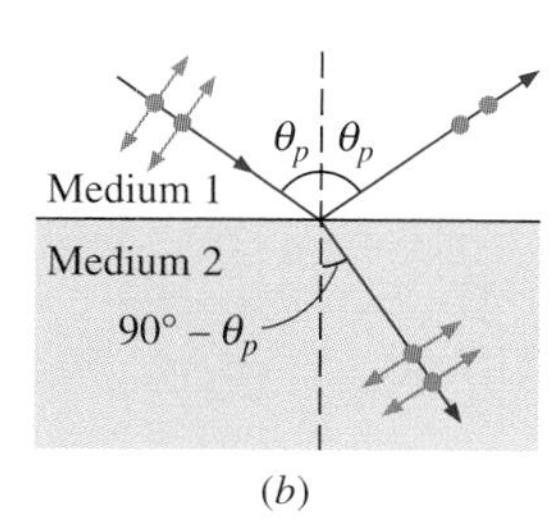

FIGURE 35-29 (*a*) When unpolarized light is incident on a reflecting surface, the reflected and refracted beams become partially polarized. (*b*) At Brewster's angle θ_p, the reflected beam is completely polarized parallel to the reflecting surface.

In 1812, David Brewster (1781–1868) discovered that when *the reflected beam is perpendicular to the refracted beam*, the electric field in the reflected beam is perpendicular to the plane of incidence; that is, *the reflected beam is linearly polarized perpendicular to the plane of incidence*. As Fig. 35-29*b* indicates, this occurs when the angle of refraction is the complement of the angle of incidence. Designating this angle of incidence as θ_p, we obtain from Snell's law

$$n_1 \sin\theta_p = n_2 \sin(90° - \theta_p) = n_2 \cos\theta_p,$$

so θ_p, which is called **Brewster's angle**, is given by

$$\theta_p = \tan^{-1}\frac{n_2}{n_1}. \tag{35-19}$$

For angles of incidence near Brewster's angle, the reflected beam's electric field is primarily perpendicular to the plane of incidence. Consequently, light reflected from a surface near Brewster's angle is strongly polarized. Brewster's angles for light in air reflected off water and glass surfaces are 53° and 56°, respectively. Much of the "glare" that bothers our vision on sunny days comes from light reflected at approximately these angles. For this reason, the lenses of many sunglasses are made of Polaroid whose transmission axis is oriented such that most of the polarized reflected light is absorbed. If you look at the surface of a body of water through such glasses, you will find the decrease in reflected light quite apparent.

The light from a liquid-crystal display is linearly polarized. What can you infer about the relative orientations of the Polaroid lenses of the "3-D" glasses?

When light falls on a system of particles, it is absorbed and then partially reradiated by the particles, a process commonly known as *scattering*. For example, light from the sky is sunlight that has been scattered by the gas molecules and dust in the atmosphere. It turns out that short-wavelength light is scattered more than light with longer wavelengths. Consequently, the sky appears blue (Fig. 35-30). The scattered light is also polarized. You can observe this by looking at the sky through Polaroid sunglasses. You'll find that maximum polarization occurs when the light is scattered through 90°. Interestingly, bees are able to detect polarized light and, in fact, use this ability to fly to and from their hives without getting lost.

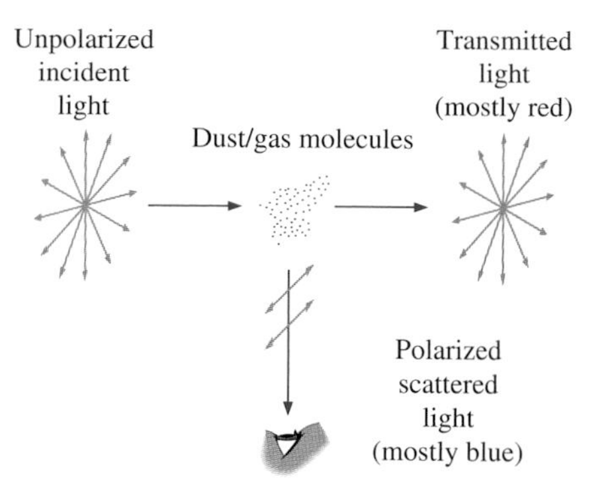

FIGURE 35-30 The sky appears blue because short-wavelength light (blue) is scattered by gas and dust in the atmosphere more strongly than long-wavelength light (red).

Finally, there are certain crystalline materials, such as calcite or quartz, which through refraction separate an unpolarized beam of light into two linearly polarized beams traveling in different directions (Fig. 35-31). The axes of polarization of

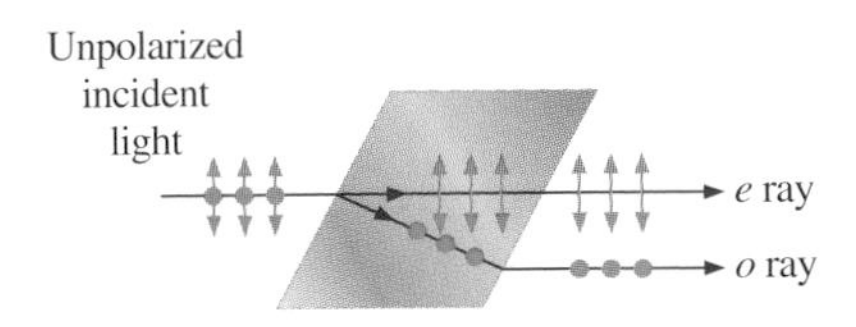

FIGURE 35-31 In certain crystalline materials, an incident beam of unpolarized light is separated upon refraction into two linearly polarized beams.

the refracted beams are mutually perpendicular. One beam, known as the *ordinary (o) ray*, travels with the same speed in all directions. The other, the *extraordinary (e) ray*, has a speed that depends on its direction of propagation.

This phenomenon is easily observed by placing a piece of this double-refracting crystalline material on top of a grid and then looking through the crystal. As shown in Fig. 35-32, two images of each line appear. Each image corresponds to one of the rays and hence to one of the polarization directions. You can observe the polarization by placing a sheet of Polaroid on top of the crystal and then rotating it. The two images will alternately appear and disappear as the Polaroid transmits and absorbs the appropriate polarization component.

FIGURE 35-32 Two images are produced by the calcite crystal as it splits the incident light into two polarized beams.

SUMMARY

1. **Electromagnetic waves**

Maxwell's equations predict the existence of combined electric and magnetic (or electromagnetic) fields that propagate through space. These electromagnetic waves travel in free space at a speed

$$c = \frac{1}{\sqrt{\mu_0 \epsilon_0}},$$

which is the measured speed of light in free space. Light is a form of electromagnetic radiation.

Electromagnetic plane waves obey the general wave relations

$$\omega = 2\pi f, \qquad k = \frac{2\pi}{\lambda}, \qquad c = f\lambda,$$

where ω, f, k, and λ are the angular frequency, frequency, wave number, and wavelength, respectively. Furthermore, the field amplitudes of the plane waves are related by

$$E_0 = cB_0.$$

Electromagnetic waves transport energy and momentum. The intensity of an electromagnetic plane wave is

$$I = \frac{E_0^2}{2\mu_0 c} = \frac{cB_0^2}{2\mu_0},$$

and the pressure of the wave on a surface is

$$p = \frac{\overline{S}}{c} \qquad \text{(total absorption)}$$

and

$$p = \frac{2\overline{S}}{c} \qquad \text{(total reflection)},$$

where $\overline{S}$ is the time-averaged value of the Poynting vector.

2. **Speed of light**

The speed of light in free space is taken to be exactly

$$c = 2.99792458 \times 10^8 \text{ m/s}.$$

Light always travels more slowly in a medium. The index of refraction n of the medium is defined as

$$n = \frac{c}{v},$$

where v is the speed of light in the medium.

3. **Huygens' principle**

Every point on a wavefront can be treated as a source of secondary waves (called wavelets), which spread out uniformly at the speed of light in the isotropic medium of propagation. The tangents to the surfaces of these wavelets at a given instant then make up the new wavefront.

4. **Laws of reflection and refraction**

Suppose that a light ray in medium 1 falls on the interface between medium 1 and medium 2 at an angle of incidence θ_1. If the reflected ray leaves at the angle of reflection θ_1' and the refracted ray leaves at the angle of refraction θ_2, then

$$\theta_1 = \theta_1' \qquad \text{(law of reflection)}$$

and

$$n_1 \sin\theta_1 = n_2 \sin\theta_2 \qquad \text{(Snell's law)}.$$

In total internal reflection, a light ray is completely reflected at an interface—there is no refracted beam. Total internal reflection only occurs under the following conditions:

(*a*) The incident ray is in the medium with the higher refractive index.
(*b*) The angle of incidence must be greater than the critical angle θ_c, where

$$\theta_c = \sin^{-1} \frac{n_2}{n_1}.$$

It is assumed in the expression for θ_c that the incident ray is in medium 1, so $n_1 > n_2$.

5. **Polarization**

The polarization of an electromagnetic wave is specified by its electric field. In an unpolarized light beam, the time-averaged magnitude of **E** is the same for all directions.

A linearly polarized beam may be produced by transmitting an unpolarized beam through a sheet of Polaroid. The axis along which the transmitted light is polarized is called the transmission axis of the sheet. The polarized beam has half the intensity of the unpolarized beam.

If unpolarized light is passed through two polarizing sheets whose transmission axes are oriented at an angle θ with respect to each other, the intensity of the transmitted beam varies with θ according to

$$I = I_M \cos^2 \theta,$$

where I_M is the maximum value of the intensity of the transmitted beam. This is Malus' law.

The polarization of light also occurs during reflection, scattering, and refraction by certain crystalline materials. In reflection, the amount of polarization depends on the angle of incidence. When the angle of incidence equals Brewster's angle, the reflected beam is completely polarized.

SUPPLEMENT 35-1 ELECTROMAGNETIC WAVES

The physics of traveling electromagnetic fields was worked out by James Clerk Maxwell (1831–1879) in 1873. He showed that electromagnetic waves travel in free space at a speed

$$c = \frac{1}{\sqrt{\mu_0 \epsilon_0}}, \tag{35-3}$$

where μ_0 and ϵ_0 are the permeability and permittivity of free space. These two constants are associated with the magnetic and the electric properties, respectively, of free space. A fascinating aspect of Eq. (35-3) was that with the known values of these constants ($\mu_0 = 4\pi \times 10^{-7}$ N/A^2 and $\epsilon_0 = 8.85 \times 10^{-12}$ C^2/N·m^2), c came out to be 3.00×10^8 m/s, *which agreed with the known value for the speed of light* in the 1870s!

Prior to Maxwell's work, experiments had already indicated that light was a wave phenomenon, although the nature of the waves was yet unknown. In 1801 Thomas Young (1773–1829) showed that when a light beam was separated by two narrow slits and then recombined, a pattern made up of bright and dark fringes was formed on a screen. This phenomenon was explained by assuming that light was composed of waves that added constructively at some points and destructively at others. Subsequently, Jean Foucault (1819–1868), with measurements of the speed of light in various media, and Augustin Fresnel (1788–1827), with detailed experiments involving interference and diffraction phenomena, provided further conclusive evidence that light was a wave phenomenon. So light was known to be a wave and Maxwell had predicted the existence of electromagnetic waves that traveled at the speed of light. The conclusion seemed inescapable—light must be *a form of electromagnetic radiation.*

Now, Maxwell's electromagnetic waves weren't restricted to the frequencies of visible light. He showed that electromagnetic radiation with the same fundamental properties as visible light should exist at *any* frequency. And 10 years after Maxwell's death, Heinrich Hertz demonstrated experimentally that there were indeed electromagnetic waves outside the visible range of frequencies. A schematic representation of Hertz's apparatus is shown in Fig. 35-4. It consisted basically of a "spark coil" and an adjacent open wire loop. When sparks were produced in the secondary of the spark coil, Hertz observed corresponding sparks between the electrodes of the adjacent wire loop. This result indicated that electromagnetic radiation had been generated by the sparks in the spark coil, and this radiation induced the observed sparks between the electrodes of the wire loop.

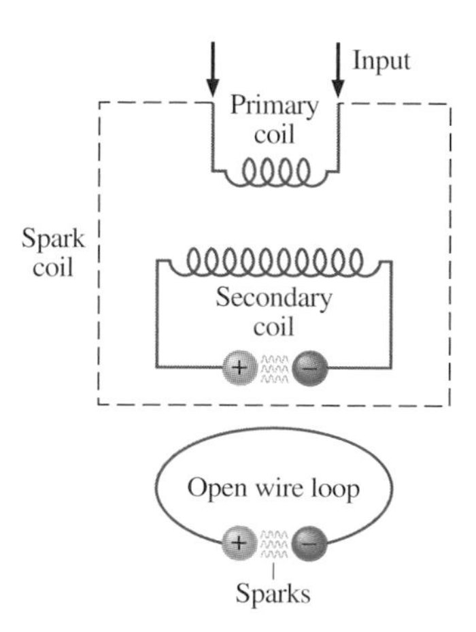

FIGURE 35-4 Hertz's spark-coil experiment was used to demonstrate the existence of electromagnetic waves outside the visible range of frequencies.

In addition, Hertz determined the speed of the waves to be 3.2×10^8 m/s, which, within experimental error, agreed with the accepted value for c. Finally, he showed that nonvisible electromagnetic waves displayed some of the other known properties of light; specifically, they could be reflected, refracted, and polarized. With this added evidence, the validity of Maxwell's theory was no longer in doubt.

Electromagnetic waves are produced by accelerating electric charges. For example, the vibrating electrons in a conductor emit radiation at the frequencies of their oscillations. The electromagnetic field of a straight-wire antenna is shown in Fig. 35-5. The frequency of this radiation is the same as the frequency of the ac source accelerating the electrons in the antenna. At large distances from the antenna, this radiation essentially becomes plane waves. (See Sec. 16-7.)

According to Maxwell, electromagnetic plane waves are a combination of an electric field **E** and a magnetic field **B** that oscillate perpendicular to each other and to the direction of wave propagation. The SI unit for **E** is the newton per coulomb (N/C), which is the same as

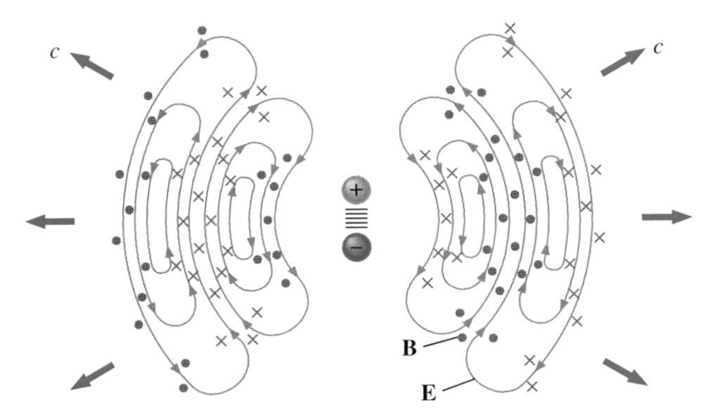

FIGURE 35-5 The oscillatory motion of the charges in a dipole antenna produces electromagnetic radiation. The electric field lines in one plane are shown. The magnetic field is perpendicular to this plane. This radiation field has cylindrical symmetry around the axis of the dipole. Field lines near the dipole are not shown. Far from the antenna the wavefronts are almost spherical and the radiation propagates like electromagnetic plane waves.

the volt per meter (V/m). The coulomb is the basic unit for electric charge and the volt is a joule per coulomb. The SI unit for **B** is the tesla (T) where $1\ \mathrm{T} = \mathrm{N \cdot s/m \cdot C}$.

Figure 35-6 shows a plane electromagnetic wave with angular frequency ω and wave number k traveling in the positive x direction. With amplitudes E_0 and B_0, the electric and magnetic fields of this plane wave are given by

$$\mathbf{E} = E_0 \sin(kx - \omega t)\mathbf{j}, \tag{35-5a}$$

and

$$\mathbf{B} = B_0 \sin(kx - \omega t)\mathbf{k}. \tag{35-5b}$$

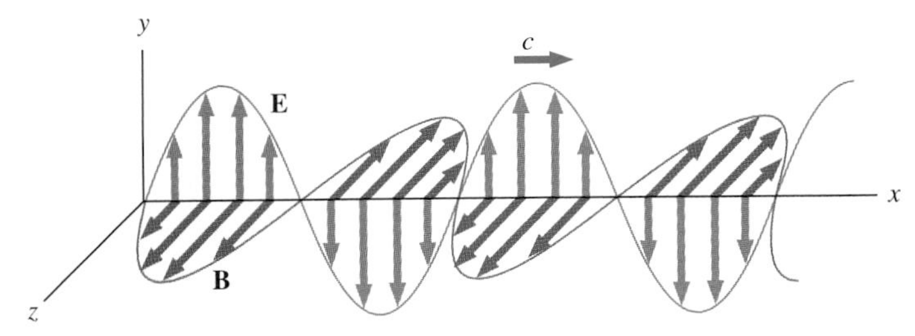

FIGURE 35-6 An electromagnetic plane wave traveling down the positive x axis, with **E** oscillating parallel to the y axis and **B** parallel to the z axis.

The relationships between the properties of an electromagnetic wave are the same as the relationships between the properties of a mechanical wave; that is,

$$\omega = 2\pi f, \tag{35-6a}$$

$$k = \frac{2\pi}{\lambda}, \tag{35-6b}$$

and

$$c = f\lambda. \tag{35-6c}$$

In addition, it can be shown that the field amplitudes are related by

$$E_0 = cB_0. \tag{35-7}$$

The transport of electromagnetic energy is described by the Poynting vector $\mathbf{S} = (1/\mu_0)\mathbf{E} \times \mathbf{B}$. Hence for the plane wave of Fig. 35-6,

$$\mathbf{S} = \frac{1}{\mu_0}\mathbf{E} \times \mathbf{B} = \frac{1}{\mu_0} E_0 B_0 \sin^2(kx - \omega t)\mathbf{i}.$$

Notice that the Poynting vector lies along the direction of propagation of the wave, which therefore transports energy. It gives us the instantaneous rate at which energy crosses a unit area perpendicular to the direction of wave propagation; this rate is called the *power flux*. Its SI unit is the watt per square meter (W/m^2).

At the frequencies of visible light, electromagnetic waves oscillate so rapidly that only the time-averaged power flux $\bar{\mathbf{S}}$ can be measured. For the plane wave of Fig. 35-6, this is given by

$$\begin{aligned}\bar{\mathbf{S}} &= \frac{1}{T}\int_0^T \frac{E_0 B_0}{\mu_0} \sin^2(kx - \omega t)\mathbf{i}\,dt \\ &= \frac{E_0 B_0}{2\mu_0 T}\int_0^T [1 - \cos 2(kx - \omega t)]\mathbf{i}\,dt,\end{aligned}$$

where T is the period of the wave. Since the average value of the cosine function over a cycle is zero, the integral of the second term vanishes and we are left with

$$\bar{\mathbf{S}} = \frac{E_0 B_0}{2\mu_0}\mathbf{i} = \frac{E_0^2}{2\mu_0 c}\mathbf{i} = \frac{cB_0^2}{2\mu_0}\mathbf{i}, \tag{35-8}$$

where we have used $E_0 = cB_0$. The magnitude of $\bar{\mathbf{S}}$ is the **intensity** I of the electromagnetic wave; it is

$$I = |\bar{\mathbf{S}}| = \frac{E_0 B_0}{2\mu_0} = \frac{E_0^2}{2\mu_0 c} = \frac{cB_0^2}{2\mu_0}. \tag{35-9}$$

Like other types of plane waves, the intensity of the electromagnetic plane wave is proportional to the square of the amplitude.

It's not surprising that energy is transported by light. This fact quickly becomes evident to anyone standing under the sun on a hot day. What we do learn from Maxwell is that such energy is transported by electromagnetic waves that, in the sun's case, are traveling through the vacuum of space to the earth.

EXAMPLE 35-1 A Laser Beam

The beam from a small laboratory laser typically has an intensity of about 1.0×10^{-3} W/m^2. Assuming that the beam is composed of plane waves like that shown in Fig. 35-6, calculate the amplitudes of the electric and magnetic fields in the beam.

SOLUTION From Eq. (35-9), the intensity of the laser beam is

$$I = \frac{E_0^2}{2\mu_0 c},$$

so the amplitude of the electric field in the beam is, from Eq. (35-7),

$$\begin{aligned}E_0 &= \sqrt{2\mu_0 cI} \\ &= [2(4\pi \times 10^{-7}\ \mathrm{N/A^2})(3.0 \times 10^8\ \mathrm{m/s}) \\ &\quad (1.0 \times 10^{-3}\ \mathrm{W/m^2})]^{1/2} = 0.87\ \mathrm{V/m}.\end{aligned}$$

The amplitude of the magnetic field is then

$$B_0 = \frac{E_0}{c} = 2.9 \times 10^{-9}\ \text{T}.$$

Besides energy, electromagnetic waves transport momentum. It can be shown that the momentum per unit volume carried by these waves is

$$\mathbf{P} = \frac{\mathbb{S}}{c^2}. \qquad (35\text{-}11)$$

Suppose that electromagnetic radiation is incident normally on a surface of area A, as shown in Fig. 35-7. In a time interval Δt, the portion of the electromagnetic wave in the volume $A(c\,\Delta t)$ falls on the surface. If the radiation is completely absorbed by the surface, the average momentum it imparts to that surface in a time Δt is

$$\Delta\bar{\mathbf{P}} = \frac{\bar{\mathbb{S}}}{c^2}(\text{vol}) = \frac{\bar{\mathbb{S}}}{c^2} Ac\,\Delta t = \frac{\bar{\mathbb{S}}A}{c}\Delta t.$$

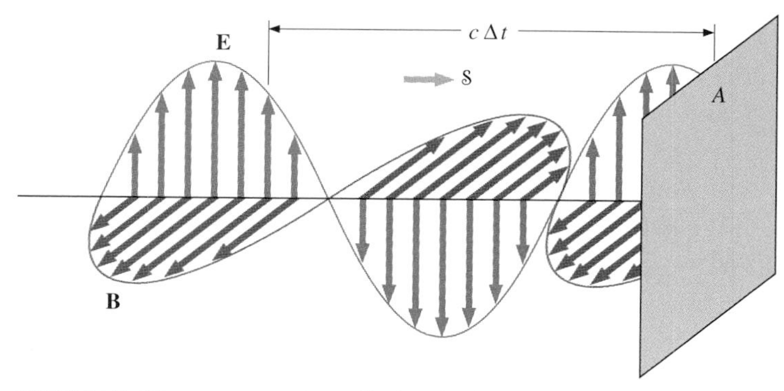

FIGURE 35-7 Electromagnetic radiation incident on a surface exerts a radiation pressure on that surface. In a time Δt, radiation in the volume $Ac\,\Delta t$ falls on the surface.

The average force that the electromagnetic wave exerts on the surface is therefore

$$\bar{\mathbf{F}} = \frac{\Delta\bar{\mathbf{P}}}{\Delta t} = \frac{\bar{\mathbb{S}}A}{c}.$$

The **radiation pressure** p of the wave is defined as the average force per unit area on the surface. Using the previous equation for the average force, we obtain

$$p = \frac{\bar{\mathbb{S}}}{c} \quad \text{(total absorption)}. \qquad (35\text{-}12a)$$

If the wave is completely reflected by the surface, the momentum imparted to the surface is doubled because the momentum of the reflected wave is opposite to that of the incident wave. For this case, the radiation pressure becomes

$$p = \frac{2\bar{\mathbb{S}}}{c} \quad \text{(total reflection)}. \qquad (35\text{-}12b)$$

Evidence of radiation pressure is found in the appearance of comets, which are basically chunks of icy material in which frozen gases and particles of rock and dust are embedded. When a comet approaches the sun, it warms up and its surface begins to evaporate. Some of the gases and dust form "tails" when they leave the comet. Notice in the photograph of Fig. 35-8*a* that a comet has *two* tails. The *ion tail* (Fig. 35-8*b*) is composed mainly of ionized gases. These ions interact electromagnetically with the "solar wind," which is a continuous stream of charged particles emitted by the sun. The force of the solar wind on the ionized gases is strong enough that the ion tail almost always points directly away from the sun. The second tail is composed of dust particles. Since the *dust tail* is electrically neutral, it does not interact with the solar wind. However, this tail is affected by the radiation pressure produced by the light from the sun. Although quite small, this pressure is strong enough to cause the dust tail to be displaced slightly from the path of the comet.

(*a*)

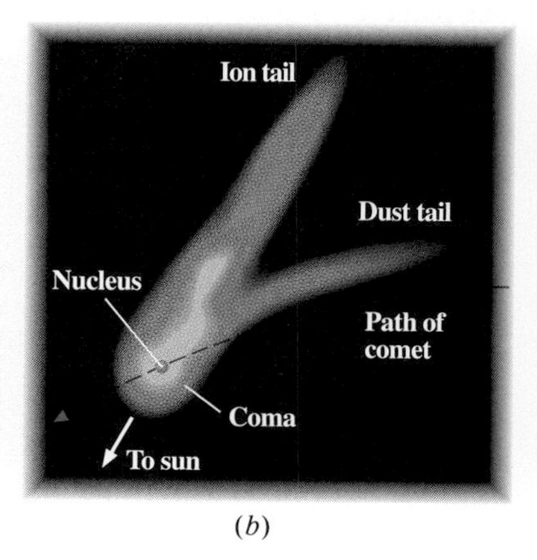

(*b*)

FIGURE 35-8 (*a*) Evaporation of material from a comet being warmed by the sun forms two tails. (*b*) The ion tail interacts with the solar wind and points away from the sun. The dust tail is curved slightly because of the radiation pressure of sunlight.

EXAMPLE 35-2 HALLEY'S COMET

On February 9, 1986, Comet Halley was at its closest point to the sun, about 9.0×10^{10} m from the center of the sun. The average power output of the sun is 3.8×10^{26} W. (*a*) Calculate the radiation pressure on the comet at this point in its orbit. (Assume that the comet reflects all the incident light.) (*b*) Suppose that a 10-kg chunk of material of cross-sectional area $4.0 \times 10^{-2}\ \text{m}^2$ breaks loose from the comet. Calculate the force on this chunk due to the solar radiation. Compare this force to the gravitational force of the sun.

SOLUTION (*a*) The intensity of the solar radiation is the average solar power per unit area. Hence at 9.0×10^{10} m from the center of the sun, we have

$$\bar{\mathbb{S}} = \frac{3.8 \times 10^{26}\ \text{W}}{4\pi(9.0 \times 10^{10}\ \text{m})^2} = 3.7 \times 10^3\ \text{W/m}^2.$$

Assuming the comet reflects all the incident radiation, we obtain from Eq. (35-12*b*)

$$p = \frac{2\bar{\mathbb{S}}}{c} = \frac{2(3.7 \times 10^3\ \text{W/m}^2)}{3.0 \times 10^8\ \text{m/s}} = 2.5 \times 10^{-5}\ \text{N/m}^2.$$

(*b*) The force on the chunk due to the radiation is

$$F_r = pA = (2.5 \times 10^{-5}\ \text{N/m}^2)(4.0 \times 10^{-2}\ \text{m}^2) = 1.0 \times 10^{-6}\ \text{N},$$

while the gravitational force of the sun is

$$F_g = \frac{GMm}{r^2}$$
$$= \frac{(6.67 \times 10^{-11}\ \text{N}\cdot\text{m}^2/\text{kg}^2)(2.0 \times 10^{30}\ \text{kg})(10\ \text{kg})}{(9.0 \times 10^{10}\ \text{m})^2}$$
$$= 0.16\ \text{N}.$$

The gravitational force of the sun on the chunk is therefore much greater than the force of the radiation.

DRILL PROBLEM 35-1

A plane electromagnetic wave propagating in free space has an electric field of amplitude 9.0×10^3 V/m. (*a*) What is the amplitude of its magnetic field? (*b*) What is the intensity of this wave? (*c*) What is the radiation pressure that this wave exerts on an object that completely absorbs it?
ANS. (*a*) 3.0×10^{-5} T; (*b*) 1.1×10^5 W/m²; (*c*) 3.6×10^{-4} N/m².

Supplement 35-2 Huygens' Principle and the Laws of Reflection and Refraction

In this supplement we see how the laws of reflection and refraction can be obtained using Huygens' principle. We also see how the frequency, wavelength, and speed of a light wave changes as it goes from one medium to another.

Law of Reflection

Figure 35-33*a* shows a wavefront $A_1A_2A_3A_4$ of speed v that is incident on a flat, reflecting surface. We assume that at $t = 0$, point A_1 of the wavefront is at the surface. To find the position of the wavefront at time t, we simply construct Huygens wavelets with radius vt—the surface of tangency to these wavelets then forms the new wavefront. Such a construction is shown in Fig. 35-33*b*. Point A_1 is now at A_1', A_2 has just reached the surface at A_2', and A_3 and A_4 are at A_3' and A_4', respectively. The section of the original wavefront that has reached the surface within this time is the reflected part of the wavefront $A_1'A_2'$. At the time $2t$, the Huygens wavelets form the wavefront shown in Fig. 35-33*c*, where the reflected part of the wavefront is now $A_1''A_2''A_3''$.

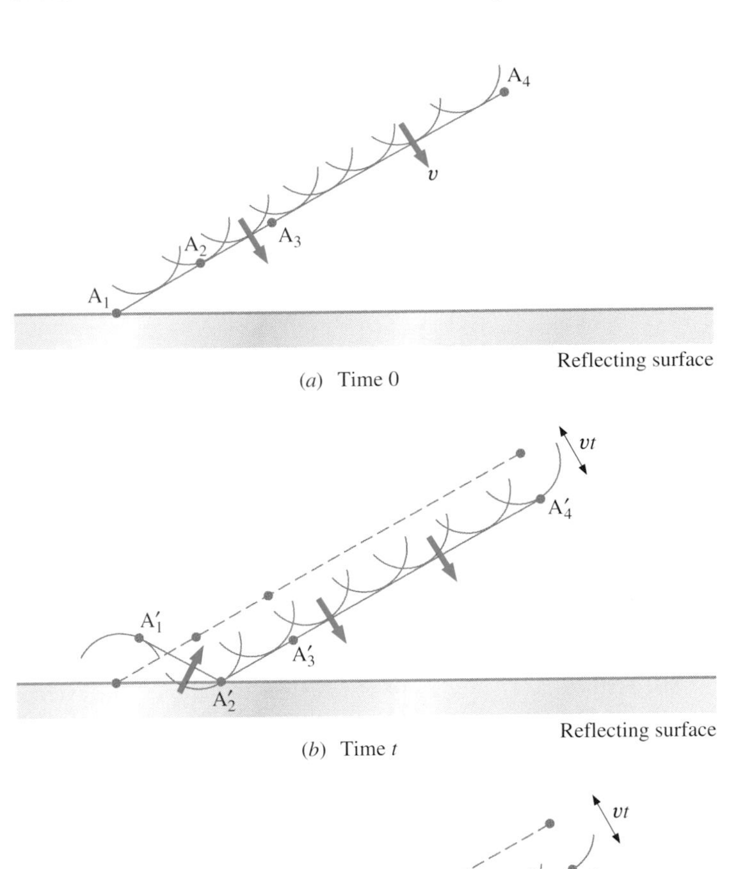

FIGURE 35-33 The reflection of a Huygens wavefront.

We can inspect more closely the reflection of the wavefront with the help of Fig. 35-34, which shows the positions of the wavefront at times 0 and t. During the time interval t, A_2 moves a distance vt to A_2', as indicated, while A_1 moves to the point A_1'. The point A_1' is located by first drawing a circular wavelet of radius vt centered at A_1 and then constructing the tangent from A_2' to the circular wavelet. Notice that the angles of incidence and reflection are represented by θ and θ'.

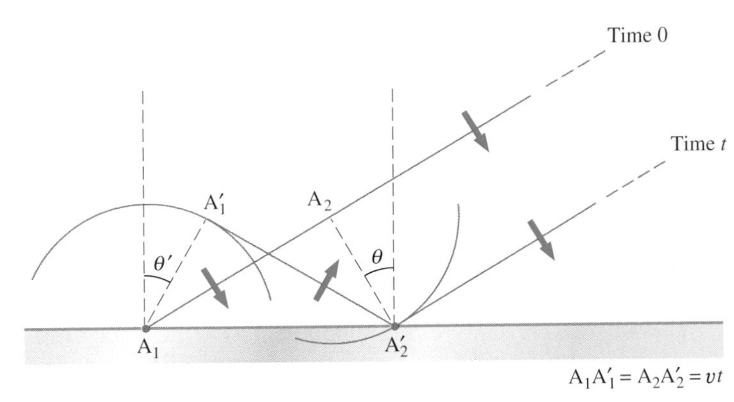

FIGURE 35-34 Geometry of the wavefront during reflection.

Now $A_1A_1'A_2'$ and $A_2'A_2A_1$ are right triangles, each with a side (A_1A_1' and A_2A_2') of length vt, and both with the common hypotenuse A_1A_2'. The two triangles are therefore similar and $\angle A_1'A_1A_2' = \angle A_2A_2'A_1$. ("$\angle ABC$" represents the angle between the lines AB and BC.) Moreover,

$$\theta' + \angle A_1'A_1A_2' = \frac{\pi}{2} \quad \text{and} \quad \theta + \angle A_2A_2'A_1 = \frac{\pi}{2};$$

thus

$$\theta = \theta',$$

which is the law of reflection.

Law of Refraction

Figure 35-35*a* shows a wavefront $A_1A_2A_3A_4$ incident on an interface between transparent media 1 and 2. At $t = 0$, point A_1 of the wavefront is at the interface. To find the position of the wavefront at time t, we again construct Huygens wavelets and then find the surface of tangency to these wavelets. We assume that the wave speeds in medium 1 and medium 2 are given by v_1 and v_2, respectively, and that $v_1 > v_2$. At time t, the wavelets in medium 1 have a radius $v_1 t$, while the wavelets in medium 2 have a smaller radius $v_2 t$. Because the radii are different, the surface of tangency to the wavelets is bent at the interface (Fig. 35-35*b* and *c*). The wavefront upon entering medium 2 therefore propagates in a different direction than in medium 1.

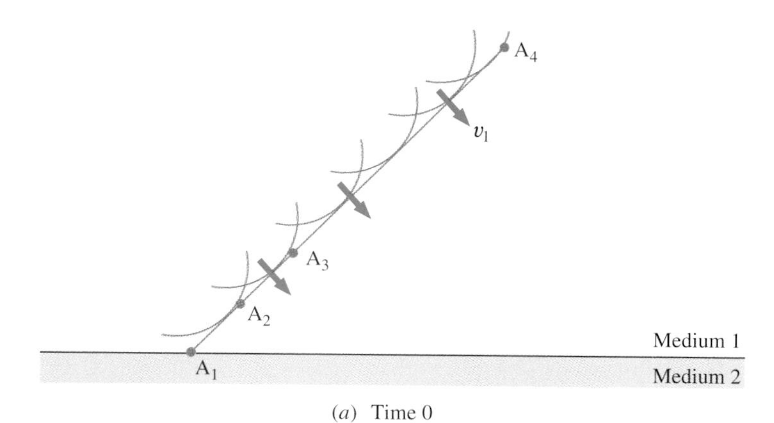

(*a*) Time 0

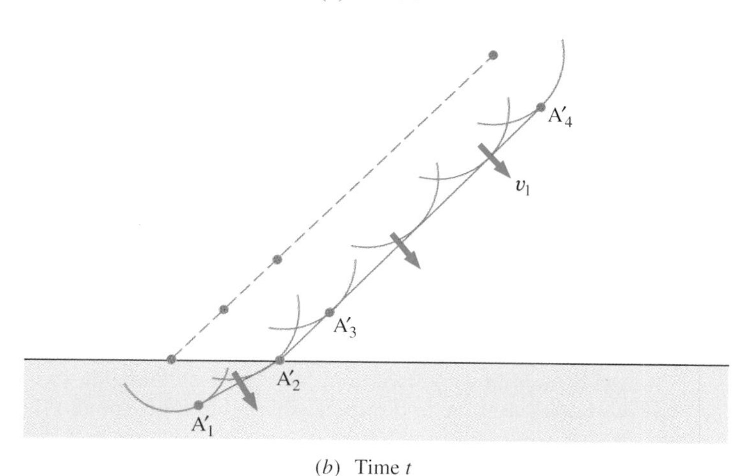

(*b*) Time *t*

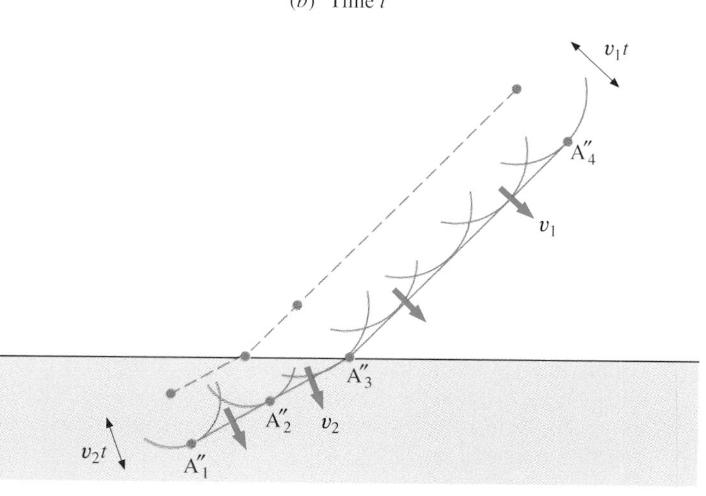

(*c*) Time 2*t*

FIGURE 35-35 The refraction of a Huygens wavefront.

We can inspect more closely the refraction of the wavefront with the help of Fig. 35-36, which shows the positions of the wavefront at times 0 and t. During the time interval t, A_2 moves a distance $v_1 t$ to A_2', as indicated, while A_1 moves to the point A_1', which is located by first drawing the circular wavelet of radius $v_2 t$ centered at A_1 and then constructing the tangent from A_2' to the wavelet. The angle of incidence is θ_1 and the angle of refraction is θ_2. Since $\angle A_2A_2'A_1 = \pi/2 - \theta_1$ and the triangle $A_1A_2A_2'$ is a right triangle, we have

$$\angle A_2A_1A_2' = \theta_1,$$

so

$$\sin\theta_1 = \frac{A_2A_2'}{A_1A_2'}.$$

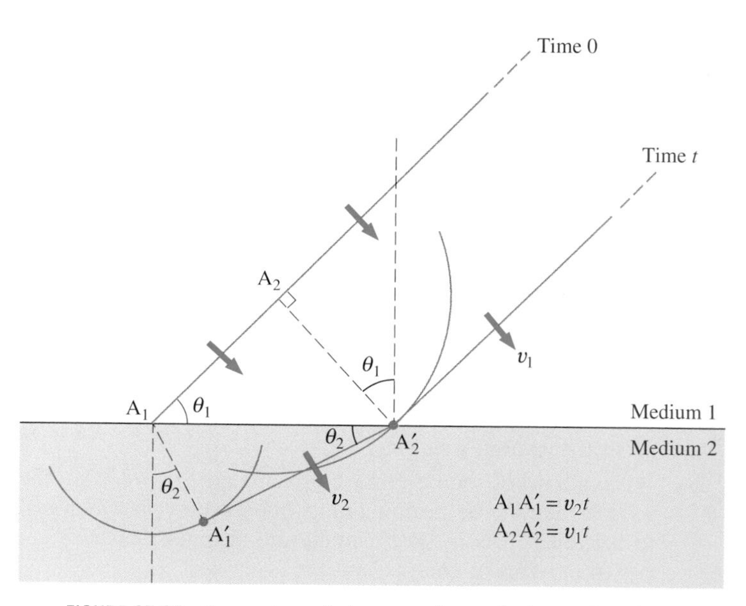

FIGURE 35-36 Geometry of the wavefront during refraction.

Substituting $A_2A_2' = v_1 t$ into this equation, we obtain

$$A_1A_2' = \frac{v_1 t}{\sin\theta_1}.$$

The same argument applied to the right triangle $A_1A_1'A_2'$ yields

$$A_1A_2' = \frac{v_2 t}{\sin\theta_2}.$$

Equating the two expressions for A_1A_2', we find

$$\frac{v_1 t}{\sin\theta_1} = \frac{v_2 t}{\sin\theta_2}.$$

Finally, substituting for v_1 and v_2 using Eq. (35-13) gives

$$n_1 \sin\theta_1 = n_2 \sin\theta_2,$$

which is the law of refraction, or Snell's law.

Wave Properties in Different Media

Here we consider what happens to the frequency, wavelength, and speed of a Huygens wave when it crosses an interface between two media. For simplicity, we assume that the wave is originally traveling in free space, then crosses into a medium of refractive index n. The frequency, wavelength, and speed of the wave in free space are represented by f_0, λ_0, and c; in the medium, they are f, λ, and v.

Implicit in the construction of the Huygens wavefront of Fig. 35-36 is the assumption that the wavefront stays intact as it crosses the interface between the two media. Consequently, in a given time interval, the number of wavefronts incident on the interface is the same as the number of wavefronts leaving the interface; or stated another way, *the frequency of a wave is unaffected when the wave crosses an interface between two media*:

$$f = f_0.$$

If the wave frequency is unchanged, then the period T must be the same in free space and in the transparent medium. When a wavefront in free space moves a distance $\lambda_0 = cT$, a wavefront in the medium 2 moves $\lambda = vT$; hence

$$\frac{\lambda}{\lambda_0} = \frac{vT}{cT} = \frac{v}{c} = \frac{1}{n},$$

so

$$\lambda = \lambda_0/n.$$

Finally from Eq. (35-13),

$$v = \frac{c}{n}.$$

To summarize, *when a light wave crosses from free space into a medium of refractive index n, its frequency remains the same while its wavelength and speed are decreased by a factor 1/n.*

Questions

35-1. Review the relationships between the speed, frequency, and wavelength of an electromagnetic wave.

35-2. Compare the speed, frequency, and wavelength of electromagnetic waves produced by an AM radio station with those produced by an FM station.

35-3. If the electric field of an electromagnetic wave is oscillating along the z axis and the magnetic field along the x axis, in what possible direction(s) is this wave traveling?

35-4. How does the power flux depend on the electric field of an electromagnetic wave? How does it depend on the magnetic field?

35-5. Why is the radiation pressure of an electromagnetic wave on a perfectly reflecting surface twice as large as the pressure on a perfectly absorbing surface?

35-6. When a bowl of soup is removed from a microwave oven, the soup is found to be steaming hot while the bowl is only warm to the touch. Discuss the temperature changes that have occurred in terms of energy transfer.

35-7. Certain orientations of a TV antenna give better reception than others for a particular station. Explain.

35-8. What property of light corresponds to loudness in sound?

35-9. What is the physical significance of the Poynting vector?

35-10. Is the visible region a major portion of the electromagnetic spectrum?

35-11. Can the human body detect electromagnetic radiation that is outside the visible region of the spectrum?

35-12. Radio waves are normally polarized and visible light is usually unpolarized. Can you explain why?

35-13. Compare the speed, wavelength, and frequency of radio waves and x-rays traveling in a vacuum.

35-14. Describe how the speed, wavelength, and frequency of an electromagnetic wave are affected when the wave travels from one medium to another.

35-15. Speculate as to what physical process might be responsible for light traveling more slowly in a medium than in a vacuum.

35-16. Is the angle of reflection always equal to the angle of incidence? Is the angle of refraction always less than the angle of incidence?

35-17. Explain why an oar that is partially submerged in water appears bent.

35-18. How can you use total internal reflection to estimate the index of refraction of a medium?

35-19. Why can you see objects below the surface of water better when you are wearing Polaroid sunglasses? Do ordinary sunglasses help in this respect?

35-20. How can you determine by viewing the sky whether or not sunglasses are made of Polaroid?

35-21. Why is space black when viewed from an orbiting space capsule?

35-22. If you lie on a beach looking at the water with your head cocked sideways, your Polaroid sunglasses don't work very well. Why not?

35-23. Suppose two polarizing sheets are positioned with their polarizing axes perpendicular to each other so that no light is transmitted through the combination. Now a third polarizing sheet is placed between the two original sheets, and it is found that light is transmitted through this combination. Explain how this can happen.

35-24. Why aren't sound waves polarized?

Problems

Electromagnetic Waves and the Electromagnetic Spectrum

35-1. Microwaves used in a physics experiment have a wavelength of 3.25 cm. What is their frequency?

35-2. Determine the wave number, angular frequency, and period of the microwaves of the previous problem.

35-3. What are the wavelengths of (*a*) x-rays of frequency 2.0×10^{17} Hz? (*b*) yellow light of frequency 5.1×10^{14} Hz? (*c*) gamma rays of frequency 1.0×10^{23} Hz?

35-4. For red light of $\lambda = 660$ nm, what are f, ω, and k?

35-5. The AM dial on a radio ranges from 540 to 1600 kHz and the FM dial ranges from 88 to 108 MHz. What is the range of wavelengths for AM? for FM?

35-6. A radio transmitter broadcasts plane electromagnetic waves whose maximum electric field at a particular location is 1.55×10^{-3} V/m. What is the maximum magnitude of the oscillating magnetic field at that location? How does it compare with the earth's magnetic field?

35-7. A radio station broadcasts at a frequency of 760 kHz. At a receiver some distance from the antenna, the maximum magnetic field of the electromagnetic wave detected is 2.15×10^{-11} T. (*a*) What is the maximum electric field? (*b*) What is the wavelength of the electromagnetic wave?

35-8. The electric field of a plane electromagnetic wave moving along the x axis is given by $\mathbf{E} = E_0 \cos(kx - \omega t)\mathbf{j}$ where $E_0 = 2.00 \times 10^{-2}$ V/m and $\omega = 4.78 \times 10^6$ rad/s. (*a*) Write an expression for the magnetic field associated with the wave. (*b*) What are the frequency and the wavelength of the wave? (*c*) What is its average Poynting vector?

35-9. A plane electromagnetic wave travels northward. At one instant, its electric field has a magnitude of 6.0 V/m and points eastward. What are the magnitude and direction of the magnetic field at this instant?

35-10. The electric field of a linearly polarized electromagnetic wave is given by

$$\mathbf{E} = (6.0 \times 10^{-3}\text{ V/m}) \sin\left[2\pi\left(\frac{x}{18} - \frac{t}{6.0 \times 10^{-8}}\right)\right]\mathbf{j}.$$

Write the equations for the associated magnetic field and Poynting vector.

35-11. The magnetic field of a plane electromagnetic wave moving along the z axis is given by $\mathbf{B} = B_0 \cos(kz + \omega t)\mathbf{j}$, where $B_0 = 5.00 \times 10^{-10}$ T and $k = 3.14 \times 10^{-2}\text{ m}^{-1}$. (*a*) Write an expression for the electric field associated with the wave. (*b*) What are the frequency and the wavelength of the wave? (*c*) What is its average Poynting vector?

35-12. A 150-W lightbulb emits 5 percent of its energy as electromagnetic radiation. (*a*) What is the average Poynting vector 10 m from the bulb? (*b*) What is the radiation pressure on an absorbing sphere of radius 10 m that surrounds the bulb?

35-13. A small helium-neon laser operates at 2.5 mW. What is the electromagnetic energy in a 1.0-m length of the beam?

35-14. At the top of the earth's atmosphere, the Poynting vector associated with sunlight has a magnitude of about 1.4 kW/m^2. (*a*) What are the maximum values of the electric and magnetic fields for a wave of this intensity? (*b*) What is the total power radiated by the sun? Assume that the earth is 1.5×10^{11} m from the sun and that sunlight is composed of electromagnetic plane waves.

35-15. For an electromagnetic plane wave such as that given by Eqs. (35-5*a*) and (35-5*b*) or Eqs. (35-10*a*) and (35-10*b*), show that the average energy density in the electric field is equal to the average energy density in the magnetic field.

35-16. Suppose that $\bar{S}$ for sunlight at a point on the surface of the earth is 900 W/m^2. If sunlight falls perpendicularly on a kite with a reflecting surface of area 0.75 m^2, what is the average force on the kite due to radiation pressure? How is your answer affected if the kite material is black and absorbs all sunlight?

35-17. A 2.0-mW helium-neon laser transmits a continuous beam of red light of cross-sectional area 0.25 cm^2. Assuming the beam does not diverge appreciably, calculate (*a*) its average Poynting vector, (*b*) the amplitude of its associated electric and magnetic fields, and (*c*) the average force it exerts on a reflecting surface.

35-18. A monochromatic point source radiates electromagnetic energy in all directions at an average rate of 300 W. (*a*) What is the average Poynting vector at a point 5.0 m from the source? (*b*) What are the amplitudes of the electric and magnetic fields at that point?

35-19. The average value of the Poynting vector for a beam of plane electromagnetic waves is 250 W/m^2. The waves fall perpendicularly on the face of a square plate that is 25 cm on a side. Half the waves are absorbed and half reflected. (*a*) What is the net average force on the plate? (*b*) How much energy does the plate absorb in 5.0 min? (*c*) Estimate the change in temperature of the plate due to this absorbed energy by assuming that it is made of aluminum and is 2.0 mm thick.

Speed of Light

35-20. Compare the time it takes for light to travel 1000 m on the surface of the earth and in outer space.

35-21. How far does light travel underwater during a time interval of 1.50×10^{-6} s?

35-22. From his measurements, Roemer estimated that it took 22 min for light to travel a distance equal to the diameter of the earth's orbit around the sun. (*a*) Use this estimate along with the known diameter of the earth's orbit to obtain a rough value of the speed of light. (*b*) Light actually takes approximately $16\frac{1}{2}$ min to travel this distance. Use this time to calculate the speed of light.

35-23. Cornu performed Fizeau's measurement of the speed of light using a wheel of diameter 4.00 cm that contained 180 teeth. The distance from the wheel to the mirror was 22.9 km. Assuming he measured the speed of light accurately, what was the angular velocity of the wheel?

Laws of Reflection and Refraction

Note: Unless otherwise specified, the indices of refraction of glass and water in the following problems should be taken to be 1.50 and 1.33, respectively.

35-24. A light beam in air has an angle of incidence of 35° at the surface of a glass plate. What are the angles of reflection and refraction?

35-25. A light beam in air is incident on the surface of a pond, making an angle of 20° with respect to the surface. What are the angles of reflection and refraction?

35-26. When a light ray crosses from water into glass, it emerges at an angle of 30° with respect to the normal of the interface. What is its angle of incidence?

35-27. A pencil flashlight submerged in water sends a light beam toward the surface at an angle of incidence of 30°. What is the angle of refraction in air?

35-28. Light rays from the sun make a 30° angle to the vertical when seen from below the surface of a body of water. At what angle above the horizon is the sun?

35-29. The path of a light beam in air goes from an angle of incidence of 35° to an angle of refraction of 22° when it enters a rectangular block of plastic. What is the index of refraction of the plastic?

35-30. What are the angles of reflection and refraction for the light ray shown in the accompanying figure?

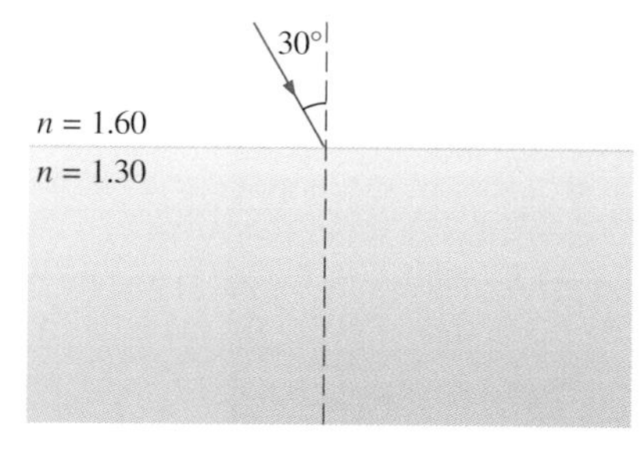

PROBLEM 35-30

35-31. A light beam is refracted by the interface between two transparent substances of refractive indices 1.7 and 1.4. The light originates in the medium with the higher index. (*a*) If the angle of incidence is 40°, what is the angle of refraction? (*b*) If the light beam originates instead in the medium with the lower index, what incident angle is necessary so that the beam is refracted at 40° to the normal?

35-32. Light of a certain frequency has a wavelength in water of 500 nm. What is its wavelength when it passes into benzene, with $n = 1.50$?

35-33. Light of wavelength λ_0 in vacuum has a wavelength of 450 nm in water and 411 nm in carbon tetrachloride. Determine λ_0 and n for carbon tetrachloride.

35-34. Suppose that when you look across the surface of the liquid in the tank shown in the accompanying figure, you can just see the lower right-hand corner C of the tank. What is the index of refraction of the liquid?

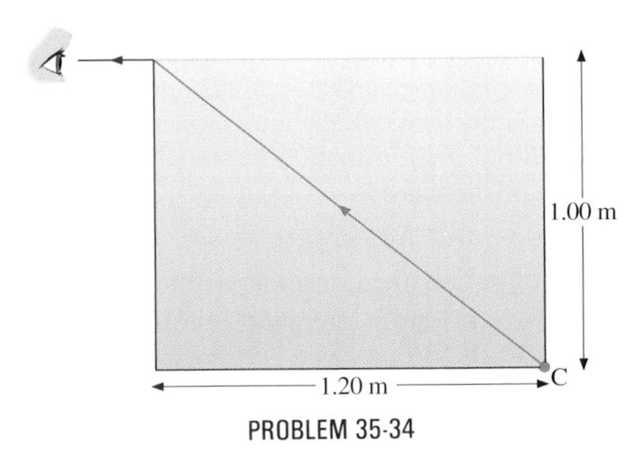

PROBLEM 35-34

35-35. For a certain material, the speed of light varies from a low (for violet) of 1.90×10^8 m/s to a high (for red) of 2.00×10^8 m/s. (*a*) What is the range of n over the visible spectrum? (*b*) A white light beam in air strikes the surface of the material at an angle of incidence of 30.0°. What are the angles of refraction for the violet and for the red light? Repeat this calculation for an incident angle of 75.0°.

35-36. The flat walls of an aquarium are made of glass with an index of refraction 1.5. If a ray traveling in the water strikes the water-glass interface at an angle of incidence of 30°, at what angle does the ray leave the outer surface of the wall?

Total Internal Reflection

35-37. The critical angle for the boundary between air and a certain liquid is 43°. What is the index of refraction of the liquid?

35-38. The index of refraction of carbon disulfide is 1.63. (*a*) What is the critical angle for a light ray going from the carbon disulfide to air? (*b*) from carbon disulfide to glass? (*c*) Is there a critical angle for a light ray going from air to carbon disulfide?

35-39. At what angle with respect to the vertical must a scuba diver look in order to see her friend on the distant shore? How is this angle related to the critical angle for a water-air boundary?

35-40. A point source of light is 2.0 m below the surface of a pond. What is the diameter of the largest circle on the surface through which light can emerge from the water?

35-41. Light rays fall normally on the vertical surface of the glass prism shown in the accompanying figure. (*a*) What is the largest value for ϕ such that the ray is totally reflected at the slanted face? (*b*) Repeat the calculation of part (*a*) if the prism is immersed in water.

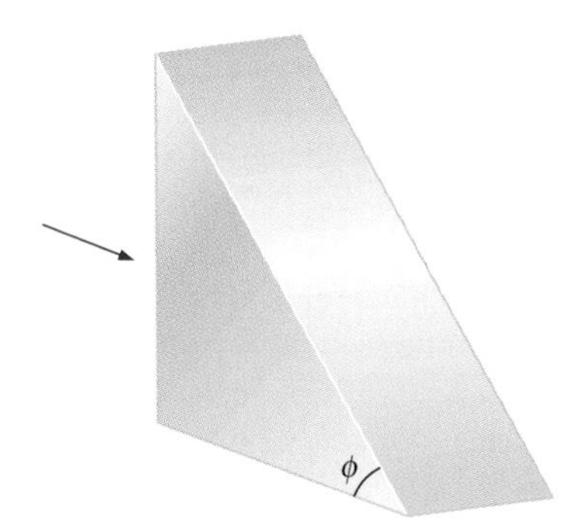

PROBLEM 35-41

35-42. A light ray is incident at an angle θ_i on the left face of a glass cube, as shown in the accompanying figure. The cube is completely surrounded by water. Find the maximum value for θ_i such that the light is totally internally reflected at the upper face of the cube.

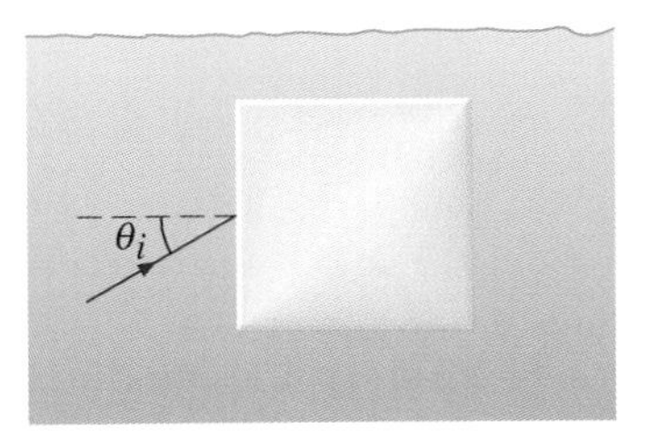

PROBLEM 35-42

Polarization

35-43. Two polarizing sheets P_1 and P_2 are placed together with their transmission axes oriented at an angle θ to each other. What is θ when only 25 percent of the maximum transmitted light intensity passes through them?

35-44. Suppose that in Prob. 35-43, the light incident on P_1 is unpolarized. At the determined value of θ, what fraction of the incident light passes through the combination?

35-45. What angle must the transmission axis of a polarizing sheet make with respect to the axis of polarization of a beam of linearly polarized light in order to reduce its intensity by 50 percent?

35-46. The light incident on polarizing sheet P_1 is linearly polarized at an angle of 30° with respect to the transmission axis of P_1. Sheet P_2 is placed so that its axis is parallel to the polarization axis of the incident light, that is, also at 30° with respect to P_1. (*a*) What fraction of the incident light passes through P_1? (*b*) What fraction of the incident light is passed by the combination? (*c*) By rotating P_2, a maximum in transmitted intensity is obtained. What is the ratio of this maximum intensity to the intensity of transmitted light when P_2 is at 30° with respect to P_1?

35-47. Three polarizing sheets are placed together such that the transmission axis of the second sheet is oriented at 25° to the axis of the first, while the transmission axis of the third sheet is oriented at 40° (in the same sense) to the axis of the first. What fraction of the intensity of an incident unpolarized beam is transmitted by the combination?

35-48. In order to rotate the polarization axis of a beam of linearly polarized light by 90°, a student places sheets P_1 and P_2 with their transmission axes at 45° and 90°, respectively, to the beam's axis of polarization. (*a*) What fraction of the incident light passes through P_1 and (*b*) through the combination? (*c*) Repeat your calculations for part (*b*) for transmission-axis angles of 30° and 90°, respectively.

35-49. What is Brewster's angle when light in air is reflected off the surface of a lake?

35-50. What is Brewster's angle when light in water is reflected off a glass surface?

35-51. It is found that when light traveling in water falls on a plastic block, Brewster's angle is 50°. What is the refractive index of the plastic?

General Problems

35-52. Consider a house whose residents use an average of 30 kW·h of electric energy per 24-h day. Suppose that on a typical day, there are 10 hours of sunlight at the location of the house, during which the average solar power flux is 300 W/m^2. The residents produce their electric power with solar panels that are 20 percent efficient; that is, they can convert 20 percent of the solar energy incident on them into electric energy. What total area of panels is required to supply the house with electric energy?

35-53. A spherical grain of dust of radius r and density ρ is floating in space above the earth's atmosphere where $\bar{S}$ for sunlight is about 1.4 kW/m^2. Assume that the dust particle absorbs all sunlight falling on it. (*a*) What is the average force on the grain of dust due to the sun's radiation? (*b*) What is the force on the grain due to the sun's gravity? (*c*) How small does r have to be before the radiation force is larger than the gravitational force? Calculate this value of r with $\rho = 2.0 \times 10^3$ kg/m^3.

35-54. A light ray falls on the left face of a prism (see the accompanying figure) at the angle of incidence θ for which the emerging beam has an angle of refraction θ at the right face. Show that the index of refraction n of the glass prism is given by

$$n = \frac{\sin \frac{1}{2}(\alpha + \phi)}{\sin \frac{1}{2}\phi},$$

where ϕ is the vertex angle of the prism and α is the angle through which the beam has been deviated. If $\alpha = 37°$ and the base angles of the prism are each 50°, what is n?

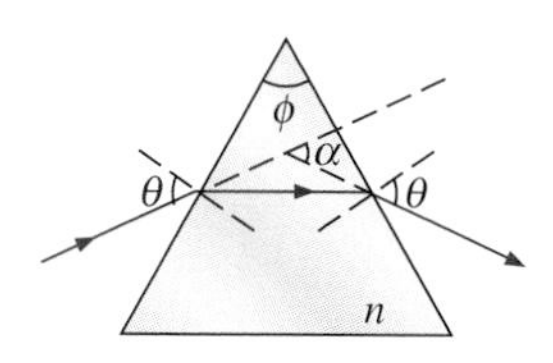

PROBLEM 35-54

35-55. If the apex angle in the previous problem is 20° and $n = 1.5$, what is the value of α?

35-56. Show that for small angles of incidence the light ray shown in the accompanying figure is displaced a distance

$$d = t\theta_1 \frac{n-1}{n}$$

when it emerges from the bottom of the transparent sheet.

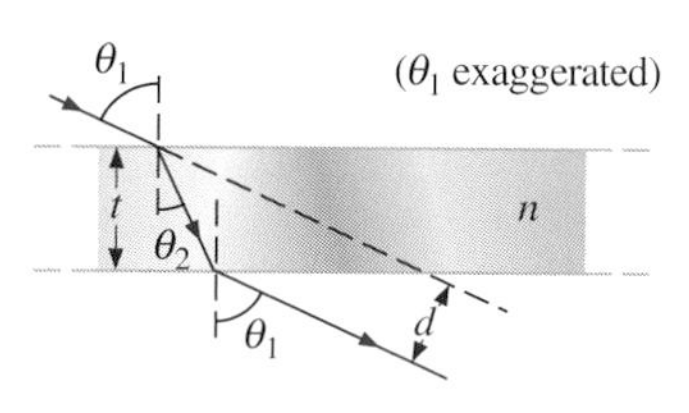

PROBLEM 35-56

35-57. Repeat the previous problem without using a small-angle approximation for the angle of incidence. Show then that

$$d = t \sin \theta_1 \left(1 - \frac{\cos \theta_1}{\sqrt{n^2 - \sin^2 \theta_1}}\right).$$

35-58. A layer of oil of index of refraction 1.48 floats on the surface of a pool of water. What is the maximum angle of incidence of a ray incident on the water-oil interface such that the ray will escape into the air? Does your answer depend on the index of refraction of the oil?

35-59. In the accompanying figure, the transmission axes of P_1 and P_3 are perpendicular. What value of θ will allow the maximum transmission of light through the combination? Make a plot of transmitted intensity versus θ. (What does this problem tell you about the configuration of Prob. 35-48?)

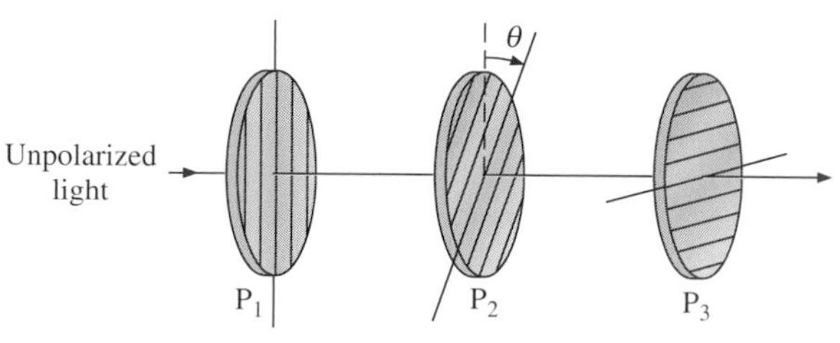

PROBLEM 35-59

35-60. Suppose a beam of light contains two linearly polarized waves whose electric fields are given by

$$\mathbf{E}_1 = E_1 \cos(kx - \omega t)\mathbf{j}$$

and

$$\mathbf{E}_2 = E_2 \cos(kx - \omega t + \phi)\mathbf{k}.$$

Suppose that $E_1 = E_2$ and $\phi = \pi/2$. (*a*) Show that at a fixed point x (use $x = 0$ for convenience) the net electric field component of the beam $\mathbf{E} = \mathbf{E}_1 + \mathbf{E}_2$ is a vector whose tip sweeps out a circle in the yz plane. The beam is said to be *circularly polarized*. (*b*) Repeat your calculation for a phase difference of $\phi = -\pi/2$. How do these two cases differ?

35-61. Repeat Prob. 35-60 for the case in which $E_1 \neq E_2$ and $\phi = \pi/2$. Show then that the tip of the electric field vector of the beam sweeps out an ellipse whose major axes lie along the y and the z axes. This beam is known as an *elliptically polarized* beam. Speculate as to what happens when $\phi \neq \pi/2$.

Kaleidoscopes use mirrors to produce multiple images such as these.

CHAPTER 36 FORMATION OF IMAGES

PREVIEW

In this chapter we consider how images are formed when light is reflected or refracted. The main topics to be discussed here are the following:

1. **Image formation by reflection.** The formation of images by plane mirrors and by spherical mirrors is discussed. We present a method for calculating the position of the image based on knowledge of the object position and the curvature of the mirror.
2. **Ray tracing.** We consider how to locate images by tracing light rays. Specific geometric rules of reflection used in this approach are listed.
3. **Image formation by refraction.** The formation of images by refraction at an interface between two transparent media is discussed. We present a method for calculating the position of the image based on knowledge of the object position, the curvature of the interface, and the refractive indices of the media.

Have you ever wondered why the warning "Objects are closer than they appear" is posted on rearview mirrors on the passenger side of most cars? Or how lenses can be used to magnify a speck of dust or to give us a better view of the planets? The answers to such questions can be found in the part of optics that deals with image formation through reflection and refraction. With the ideas to be developed in this and the following chapter, you will be able to determine the location of an image, its magnification, and its orientation. You will also understand some of the basic concepts used in the design of optical instruments.

The directions of propagation of the electromagnetic waves are represented here by rays whose behavior is analyzed with the laws of reflection and refraction. This ray approximation is valid when the wavelength of the light is much smaller than the dimensions of any aperture or barrier that the light might encounter. If the wavelength is comparable to such dimensions, then the effects of wave interference have to be considered, a topic discussed in Chap. 38.

36-1 Image Formation by Reflection

Most of our surroundings are visible because of reflection. For example, you can see this page because it is reflecting light into your eyes. In general, objects reflect light in all directions; this is known as *diffuse reflection*. It occurs when the surface of an object has irregularities whose dimensions are on the order of, or larger than, the wavelength of the incident light. However, if a surface is very smooth and has irregularities that are smaller than the wavelength, a reflected beam with a well-defined direction is produced. This process is called *specular reflection*. Reflection from a mirror or a highly polished surface is specular. Since we're interested in images formed by mirrors, we assume that all reflections are specular.

Images Formed by Plane Mirrors

A flat surface that is highly reflecting is called a *plane mirror*. It is commonly a silvered surface covered by a glass plate for protection. In Fig. 36-1*a*, a point source of light (the *object O*), is placed at a distance o in front of the reflecting surface. All rays diverging from O that are intercepted by the mirror obey the law of reflection. When the reflected rays are projected behind the mirror, they intersect at a single point I, which is a distance i behind the mirror. To an observer in front of the mirror, the reflected rays *appear* to diverge from I, which is known as the *image* of the source. This image is said to be *virtual*, because only the *projections* of the actual light rays pass through it. Images formed by plane mirrors are *always* virtual.

The relationship between the **object distance** o and the **image distance** i can be found by examining the two light rays shown in Fig. 36-1*b*. One ray is normal to the mirror and is therefore reflected back along its original path. The other makes an angle θ to the normal. When extrapolated behind the mirror, the two reflected rays also intersect at an angle θ at the image. With triangles OVP and IVP we find

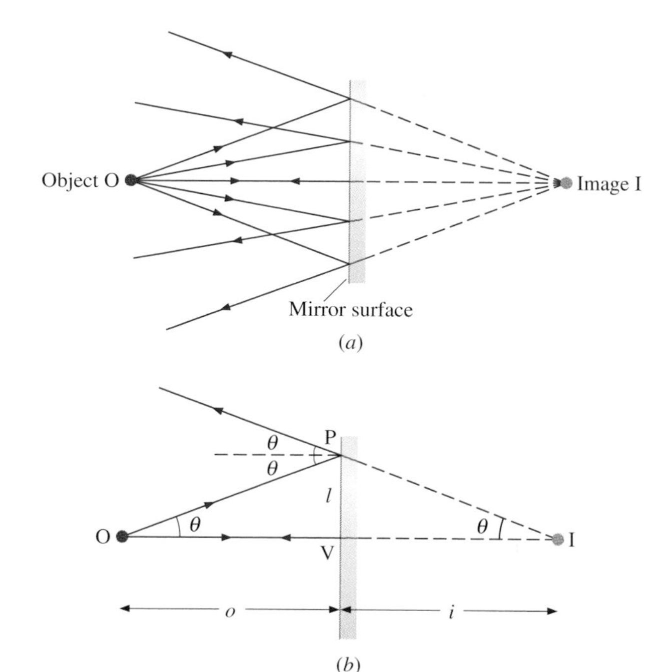

FIGURE 36-1 (*a*) The rays from object O reflected by the plane mirror converge to form an image I when they are extrapolated behind the mirror. (*b*) The image and the object are equidistant from the mirror surface.

$$\tan\theta = \frac{l}{o} = \frac{l}{i},$$

so o and i are equal. To denote that the image is behind the mirror and virtual, we represent its distance by $-i$ and write

$$o = -i \tag{36-1}$$

for the plane mirror.

In order to locate the image of an object of finite size, we simply draw two rays from any point on the object and find the point where their reflected rays intersect. This construction is shown in Fig. 36-2*a* for the two end points of the object. It's

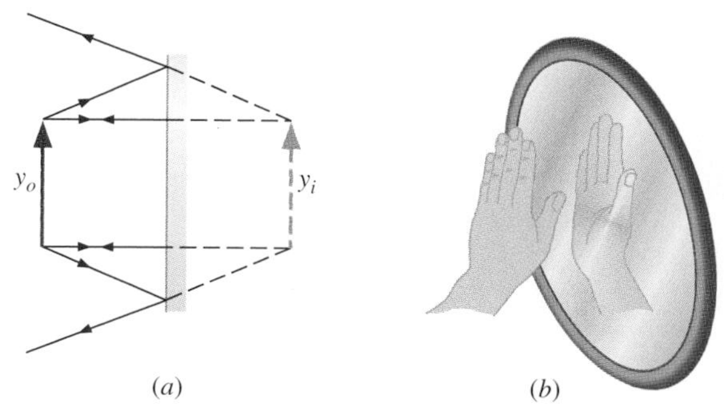

FIGURE 36-2 (*a*) The object and image sizes are the same for a plane mirror. (*b*) Left and right are interchanged upon reflection.

clear that for a plane mirror the heights of the image and object are the same. By definition, the ratio of image height to object height is the **lateral magnification** m. Thus the lateral magnification of a plane mirror is $m = 1$. While this is always the case for a plane mirror, the lateral magnification is seldom unity in other optical systems. You can also see from the figure that the image produced by a plane mirror is *upright*, or *erect*. However, left and right are interchanged by reflection, as illustrated by Fig. 36-2*b*.

Can you explain why the lettering on the fronts of many emergency vehicles is inverted?

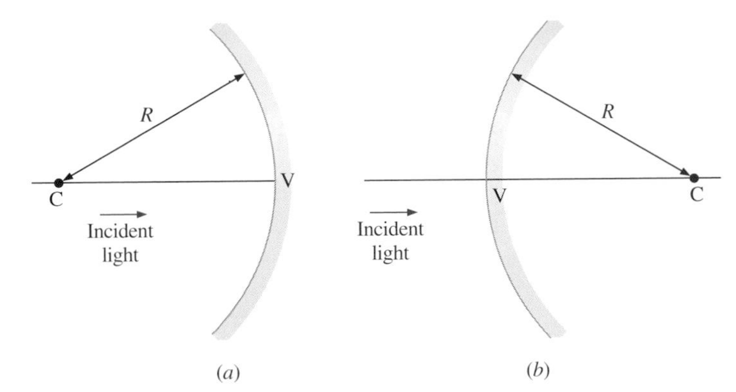

FIGURE 36-3 Two types of spherical mirrors: (*a*) concave and (*b*) convex.

Images Formed by Spherical Mirrors

Figure 36-3 shows two types of spherical mirrors: the *concave* mirror and the *convex* mirror. The **radius of curvature** R of each mirror is the radius of the imaginary sphere on whose surface the mirror lies. By definition, R is positive for a concave mirror and negative for a convex mirror. The *optical axis* passes through the *center of curvature* C of the sphere and intersects the mirror at the point V, which is known as the *vertex* of the mirror.

A curved mirror in a fun house.

In Fig. 36-4, an object O is placed on the optical axis of a concave mirror at a distance o from the vertex. Two rays that emanate from O and are reflected by the mirror are shown. One of these rays travels along the optical axis and is reflected back along the same path. (Why?) The other ray strikes the mirror at A and its angles of incidence and reflection are designated by θ. These two rays intersect at I, which is a distance i from the mirror.

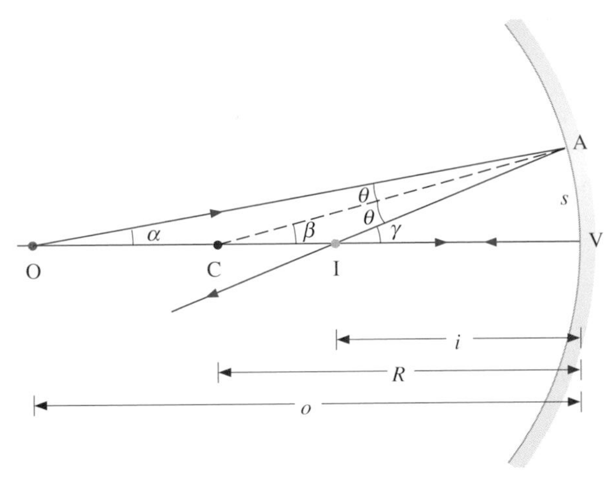

FIGURE 36-4 An image at I is formed at the intersection of the two reflected rays shown. In the paraxial-ray approximation, all the angles shown are assumed to be very small.

We assume that all angles shown in the figure are very small. This allows us to make small-angle approximations that greatly simplify our calculations. In the limit as these angles approach zero, the rays are (1) almost parallel to and (2) very close to the optical axis and are called *paraxial rays*. The relationships we will now develop are based on this *paraxial-ray approximation*. From geometry, the angles are related by

$$\beta = \alpha + \theta$$

and

$$\gamma = \beta + \theta.$$

With θ eliminated between these two equations, we obtain

$$\alpha + \gamma = 2\beta. \tag{36-2}$$

Since β is the angle that arc AV (of length s) subtends at the center of curvature,

$$\beta = \frac{s}{R};$$

and because all angles are assumed to be small, we also have

$$\alpha \approx \frac{s}{o}$$

and

$$\gamma \approx \frac{s}{i}.$$

Substituting the expressions for α, β, and γ into Eq. (36-2), we find

$$\frac{1}{o} + \frac{1}{i} = \frac{2}{R}. \tag{36-3a}$$

This *mirror equation*, because it is independent of any angle, is satisfied by *all* rays in the small-angle approximation. Hence all paraxial rays emanating from O intersect at the same point I, which is the image of O (Fig. 36-5a). With the mirror equation, we can locate the image if we know the position of the object and the radius of curvature of the mirror.

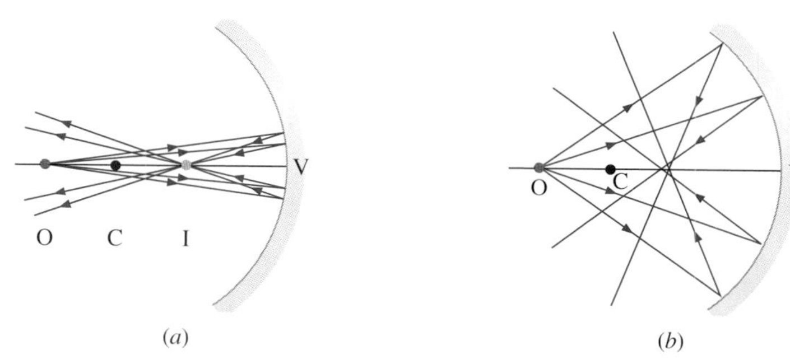

FIGURE 36-5 (a) All paraxial rays from O intersect at I. (b) The large-angle rays from O do not intersect at a single point. This results in a blurred image.

The assumption of small angles is very important here! Reflected rays that aren't paraxial do not even intersect at the same point, as illustrated in Fig. 36-5b. With nonparaxial rays, the image of the point object O is blurred.

How the image of an extended object is formed is shown in Fig. 36-6. All paraxial rays from a particular point on the object intersect at a common point on the image. Since rays actually converge at the image shown, the image is *real* and can be seen on a screen that is placed at I.

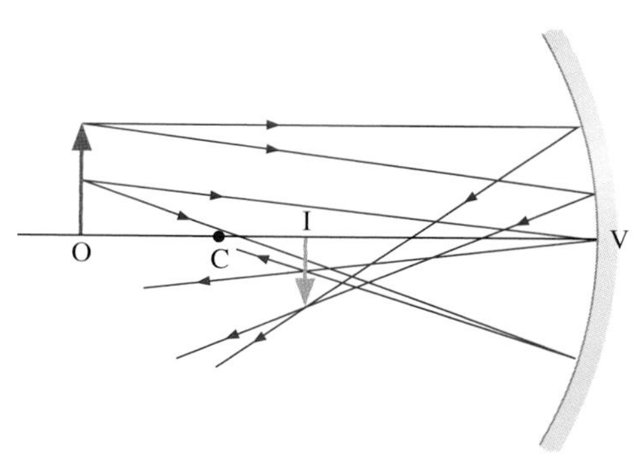

FIGURE 36-6 The image of an extended object. All paraxial rays from a point on the object intersect at the corresponding point on the image.

Figure 36-7 allows us to determine how the height of the image compares with the height of the object. In this figure, we have located the image by following the paths of two light rays from the top of the object. One ray passes through the center of curvature and is reflected back along its own path. The other ray has an angle of incidence θ at the vertex of the mirror and is therefore reflected at the same angle. From the geometry shown,

$$\tan\theta = \frac{y_0}{o} = \frac{y_i}{i},$$

so

$$\frac{y_i}{y_o} = \frac{i}{o}.$$

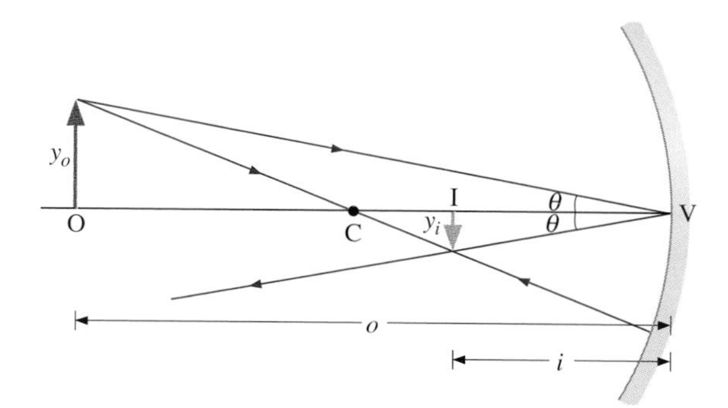

FIGURE 36-7 For spherical mirrors, the object and image heights are generally not the same. The ratio of image height to object height is known as the lateral magnification m.

Now y_i/y_0 is the lateral magnification because it is the ratio of image height to object height. We therefore have

$$m = -\frac{i}{o}, \tag{36-4}$$

where a minus sign has been inserted in order to indicate the orientation of the image. A positive value of m means that the image is upright, while a negative value tells us that it is inverted.

Although Eqs. (36-3a) and (36-4) have been derived for a concave mirror, they apply just as well to a convex mirror, for which R is negative. An example of image formation by a convex mirror is given in Fig. 36-8. Notice that the image is located behind the mirror and is formed by the extrapolation rather than by the actual intersection of light rays. It is therefore a virtual image.

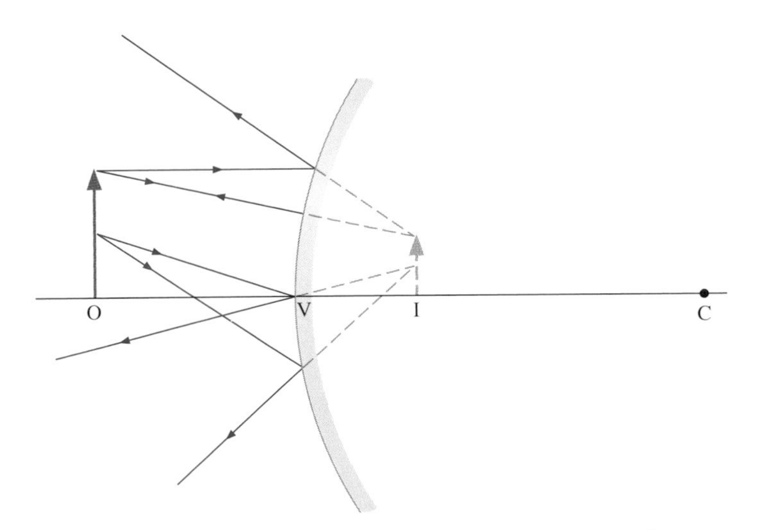

FIGURE 36-8 Formation of a virtual image by a convex mirror.

Diverging mirrors are frequently used for safety purposes because their shapes provide a wider field of view than that obtainable with a flat or a converging mirror.

The rays from an object at infinity that are incident on a mirror are parallel to the optical axis. With $o = \infty$ substituted into Eq. (36-3a), we obtain

$$\frac{1}{\infty} + \frac{1}{i} = \frac{2}{R},$$

so

$$i = \frac{R}{2}.$$

This image is located at the *focal point* F of the mirror. The distance of the focal point from the mirror is known as the **focal length** f. Since the image is at $R/2$,

$$f = \frac{R}{2}. \qquad (36\text{-}5)$$

With f replacing $R/2$, the mirror equation becomes

$$\frac{1}{o} + \frac{1}{i} = \frac{1}{f}. \qquad (36\text{-}3b)$$

The focusing of parallel rays by both a concave and a convex mirror is shown in Fig. 36-9.

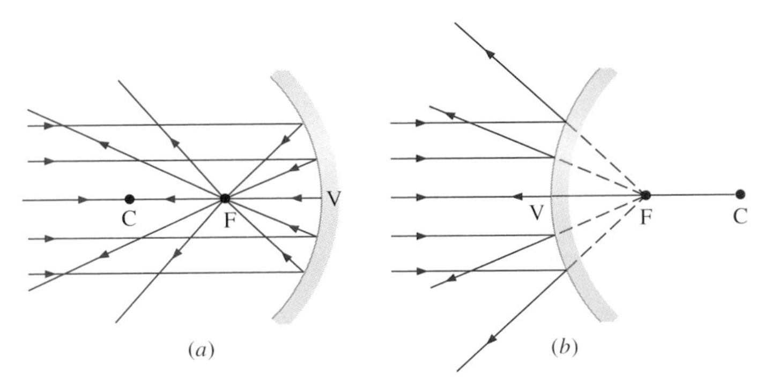

FIGURE 36-9 (a) Parallel rays reflected by a concave mirror converge at the focal point F. (b) For a convex mirror, the reflected rays also converge at the focal point when they are extrapolated behind the mirror.

Equations (36-3) through (36-5) are all applicable to both concave and convex mirrors, but before we can use them, we must specify a sign convention for o, i, and R. This same sign convention also works for refraction at a spherical surface, which will be discussed shortly. So in what follows, you can consider the "surface" as either reflective or refractive.

1. *Object distance:* $o > 0$ if the object is on the same side as the light that is incident upon the surface; $o < 0$ if the object is on the opposite side.*

2. *Image distance:* $i > 0$ if the image is on the same side as the light leaving the surface; $i < 0$ if the image is on the opposite side. Since a real image is formed by the intersection of actual reflected rays, it is characterized by a positive value of i. On the other hand, if $i < 0$, then the image is virtual.

*An object with a negative object distance is discussed in Chap. 37.

3. *Radius of curvature:* $R > 0$ if the center of curvature lies on the same side as the light leaving the surface; $R < 0$ if the center lies on the other side.

This convention is summarized in Fig. 36-10 for the two types of spherical mirrors. Since light rays cannot pass through a mirror surface, the incident and reflected rays are on the same side of the surface. Thus o, i, and R are all positive on the front side of a spherical mirror and negative on the back side.

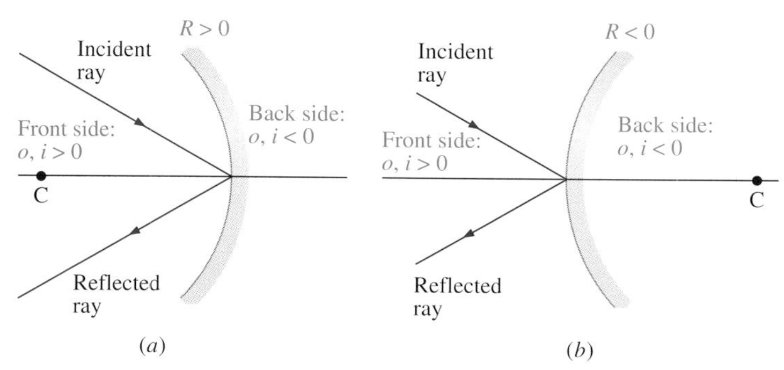

FIGURE 36-10 The sign convention for image formation by spherical mirrors: (*a*) concave mirror and (*b*) convex mirror.

EXAMPLE 36-1 Spherical Mirrors

A concave mirror has a radius of curvature of 16.0 cm. Determine the location, nature (real or virtual), orientation, and height of the image for (*a*) a 0.40-cm-high object on the axis at 40 cm from the mirror and (*b*) a 0.10-cm-high object 10 cm from the mirror. (*c*) Repeat your calculations for a convex mirror with the same radius of curvature.

SOLUTION (*a*) With our sign convention, $R = +16.0$ cm, $f = R/2 = +8.0$ cm, and $o = +40$ cm. From the mirror equation,

$$\frac{1}{40\text{ cm}} + \frac{1}{i} = \frac{1}{8.0\text{ cm}},$$

so the image distance is

$$i = +10\text{ cm}.$$

The image is therefore real and in front of the mirror. The lateral magnification is

$$m = -\frac{i}{o} = -\frac{10\text{ cm}}{40\text{ cm}} = -0.25,$$

which means that the image has a height of 0.10 cm and is inverted.

(*b*) With an object distance of +10 cm, we find

$$\frac{1}{10\text{ cm}} + \frac{1}{i} = \frac{1}{8.0\text{ cm}},$$

and

$$i = +40\text{ cm}.$$

A real image is formed 40 cm in front of the mirror. The lateral magnification is

$$m = -\frac{40\text{ cm}}{10\text{ cm}} = -4.0,$$

indicating that the image is inverted and has a height of 0.40 cm.

Notice that the object and image positions have simply been interchanged in parts (*a*) and (*b*). This is another example of the reversibility of light. *If an object at O produces an image at I, then an object at I in turn produces an image at O.* You can verify this fact quite easily from the mirror equation.

(*c*) For the convex mirror, $f = -8.0$ cm. If the 0.40-cm-high object is 40 cm away,

$$\frac{1}{40\text{ cm}} + \frac{1}{i} = \frac{1}{-8.0\text{ cm}},$$

so the image distance is

$$i = -6.7\text{ cm}.$$

The negative sign indicates that this image is virtual and located 6.7 cm behind the mirror. The lateral magnification is

$$m = -\frac{-6.7\text{ cm}}{40\text{ cm}} = +0.17;$$

hence the image is erect and has a height of 0.068 cm.

For a 0.10-cm-high object 10 cm from the mirror, we find

$$\frac{1}{10\text{ cm}} + \frac{1}{i} = \frac{1}{-8.0\text{ cm}},$$

and

$$i = -4.4\text{ cm}.$$

Once again, the image is virtual and behind the mirror. Since the lateral magnification is

$$m = -\frac{-4.4\text{ cm}}{10\text{ cm}} = +0.44,$$

the image is erect and has a height of 0.044 cm.

EXAMPLE 36-2 A Spherical Mirror of Unknown Curvature

An erect, virtual image twice the height of the object is formed 20 cm from the surface of a spherical mirror. What is the radius of curvature of this mirror? Is it concave or convex?

SOLUTION Since the image is virtual, $i = -20$ cm. The magnification is $m = +2.0$ because the image is erect. We can then find the object distance from

$$m = +2.0 = -\frac{-20\text{ cm}}{o},$$

which yields for the object distance

$$o = +10\text{ cm}.$$

From the mirror equation, we obtain

$$\frac{1}{10\text{ cm}} + \frac{1}{-20\text{ cm}} = \frac{2}{R},$$

so the radius of curvature of the mirror is

$$R = +40\text{ cm}.$$

Because R is positive, the mirror is concave.

DRILL PROBLEM 36-1

A concave mirror produces a real image at the same position as the object. What is the position of the object? What are the lateral magnification and orientation of the image?
ANS. At the radius of curvature; -1.0; inverted.

DRILL PROBLEM 36-2

A virtual image 5.0 cm from a spherical mirror is formed from an object 30 cm away from the mirror. Calculate the focal length of the mirror. Is the mirror concave or convex?
ANS. -6.0 cm; convex.

DRILL PROBLEM 36-3

Apply Eq. (36-3a) to a plane mirror by considering the limit $R \to \infty$. Compare your result with Eq. (36-1). Use Eq. (36-4) to determine the lateral magnification and orientation of the image.

DRILL PROBLEM 36-4

Show that as long as $o > 0$, the image of a convex mirror is always virtual.

36-2 Ray Tracing

An image can be found by tracing specific rays through the optical system. For reflection, four rays are normally used. They are drawn leaving a point on the object and reflecting off the mirror. Their point of intersection then gives us the corresponding point on the image. These four rays, which are shown in Fig. 36-11, satisfy the following rules:

1. An incident ray parallel to the optical axis is reflected so that it passes through the focal point (ray 1).
2. An incident ray that passes through the focal point is reflected and emerges parallel to the optical axis (ray 2).
3. A ray that passes through the center of curvature of the mirror returns along its original path (ray 3).

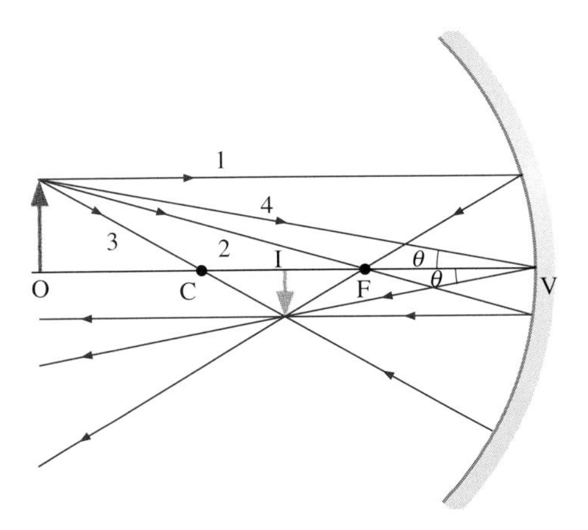

FIGURE 36-11 The four principal rays used in ray tracing.

4. The angle that a ray striking the vertex of the mirror makes with respect to the optical axis is the same before and after reflection (ray 4).

You don't need to trace all four rays to locate an image—any two will suffice, since you are only looking for their point of intersection. It is common practice to use the top of the object as the origin for the rays. The reflected rays then intersect at the top of the image and allow us to find its height. Various examples of ray tracing are given in Fig. 36-12 for both the concave and convex mirrors. Notice that extrapolations are necessary to find virtual images.

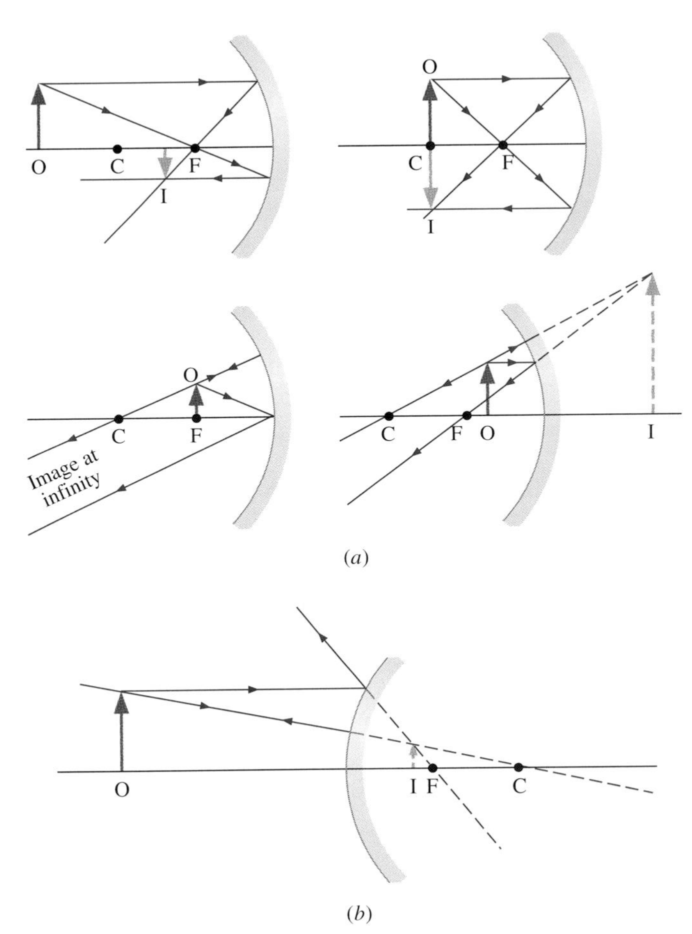

FIGURE 36-12 Ray tracing in spherical mirrors. While the nature of the image depends on the object position for a concave mirror [see (a)], the image of a real object for a convex mirror [see (b)] is always virtual and upright and has a smaller height than the object.

DRILL PROBLEM 36-5

Use the law of reflection to justify the four rules of ray tracing.

DRILL PROBLEM 36-6

Use ray tracing to find the locations and heights of the images in Example 36-1.

36-3 Image Formation by Refraction

When light from a source is refracted at a spherical interface between two transparent media, an image of that source is formed. Consider, for example, the point object of Fig. 36-13, which is in medium 1 at a distance o from a spherical boundary of radius R between media 1 and 2. To locate its image, we trace the paths of two rays emanating from the object. The first ray is the one that travels along the optical axis. Since it is normal to the boundary, it passes straight through. The second ray has an angle of incidence θ_1 and an angle of refraction θ_2. From the geometry shown in the figure,

$$\theta_1 = \alpha + \beta, \quad (36\text{-}6a)$$

and

$$\beta = \theta_2 + \gamma. \quad (36\text{-}6b)$$

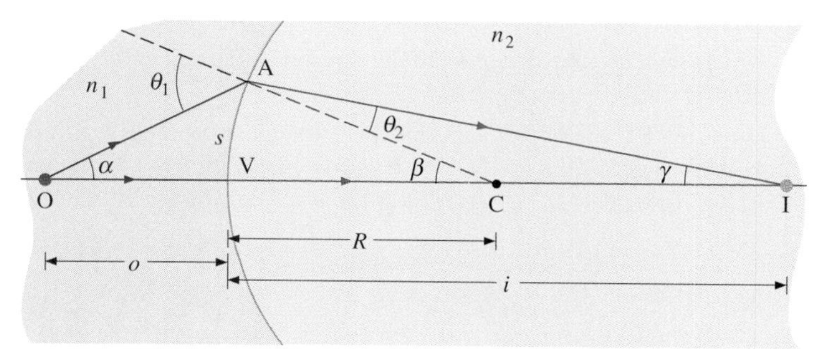

FIGURE 36-13 Formation of an image by refraction at an interface between two transparent media.

Refraction at a curved surface. Notice how much larger the numbers appear when the medium in the tube is water rather than air.

If the second ray is paraxial, all angles in these two equations are small. Thus $\sin\theta_1 \approx \theta_1$ and $\sin\theta_2 \approx \theta_2$, and Snell's law can be written as

$$n_1\theta_1 = n_2\theta_2. \quad (36\text{-}7)$$

Equations (36-6a) and (36-7) can be combined to express θ_2 in terms of α and β. Substituting the resulting expression into Eq. (36-6b) then yields

$$\beta = \frac{n_1}{n_2}(\alpha + \beta) + \gamma,$$

so

$$n_1\alpha + n_2\gamma = (n_2 - n_1)\beta. \quad (36\text{-}8)$$

Since the arc AV (of length s) subtends an angle β at the center of curvature, $\beta = s/R$. Also, in the paraxial approximation, $\alpha \approx s/o$ and $\gamma \approx s/i$. Using these expressions in Eq. (36-8) and canceling the common factor s, we are left with

$$\frac{n_1}{o} + \frac{n_2}{i} = \frac{n_2 - n_1}{R}. \quad (36\text{-}9)$$

Like Eq. (36-3a), its counterpart for the mirror, Eq. (36-9) is independent of the angles as long as they are small. *All paraxial rays emanating from the object intersect at the same point.* This point, which is a distance i from the refracting surface, is of course the image.

Equation (36-9) is used to locate the image formed by refraction in the same way that Eq. (36-3a) is used to find the image formed by a spherical mirror. *The rules developed for the mirror sign conventions also apply to refracting surfaces.* But do be careful in interpreting these rules! For a mirror, the rays incident upon and leaving the reflecting surface are always on the *same* side; for a refracting surface, the rays incident upon and leaving the surface are always on *opposite* sides, since light rays are refracted across the interface. The sign conventions for refraction at spherical surfaces are summarized in Fig. 36-14. *We will always assume that the incident light is in medium 1.*

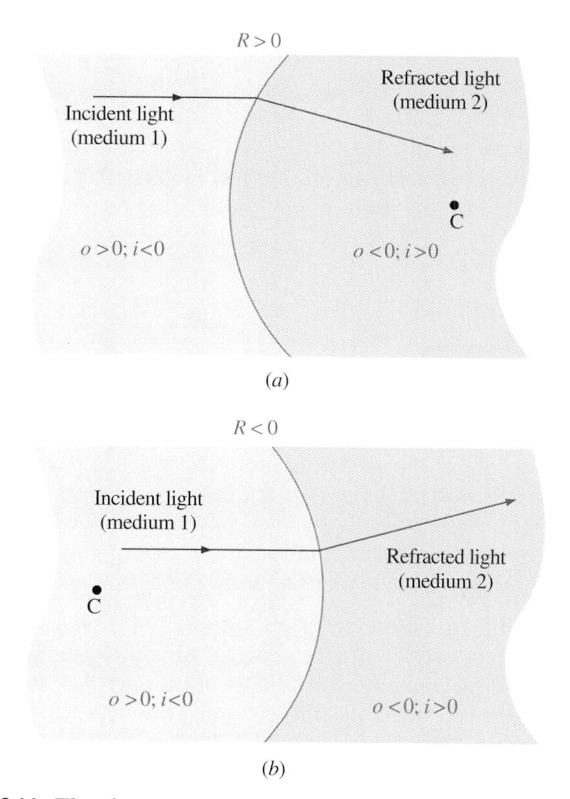

FIGURE 36-14 The sign convention for image formation by refraction at spherical boundaries: (a) convex refracting surface and (b) concave refracting surface.

The lateral magnification may be obtained with the help of Fig. 36-15, where two rays from the tip of an object of height y_o meet at the corresponding point on an image of height y_i. One ray passes through the center of curvature of the spherical surface so its direction is unchanged. The path of the second ray is obtained from Snell's law. With the paraxial approximation,

$$\sin\theta_1 \approx \frac{y_o}{o} \quad \text{and} \quad \sin\theta_2 \approx \frac{y_i}{i}.$$

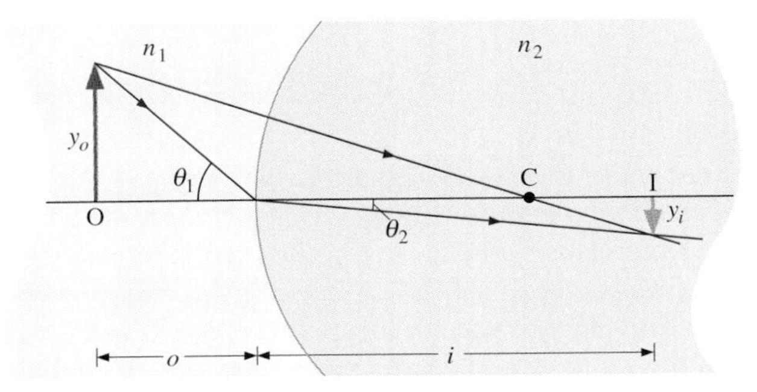

FIGURE 36-15 Ray diagram used to obtain the expression for lateral magnification by a spherical boundary.

Combining these equations with Snell's law then gives

$$n_1\frac{y_o}{o} = n_2\frac{y_i}{i}.$$

The lateral magnification m is the ratio of the image height to the object height, or y_i/y_o. We therefore obtain

$$m = -\frac{n_1 i}{n_2 o} \tag{36-10}$$

as the lateral magnification for refraction at a spherical interface between two transparent media. A minus sign has been inserted so that an inverted image is identified by a negative value of m.

EXAMPLE 36-3 Image Formation by Refraction

Find the positions of the image in the arrangement of Fig. 36-16 when an object is placed at (*a*) point A, which is 20 cm to the left of the spherical boundary and (*b*) point B, which is 5.0 cm to the right of the boundary. In both cases, determine the lateral magnification, nature, and orientation of the image. The radius of curvature of the refracting surface is 10 cm.

SOLUTION (*a*) Medium 1 is air with $n_1 = 1.0$, and medium 2 is glass with $n_2 = 1.5$. Since the center of curvature and the light leaving the surface are on opposite sides of the interface, $R = -10$ cm. The object is on the same side as the incident light, so $o = +20$ cm. Substituting these values into Eq. (36-9), we obtain

$$\frac{1.0}{20\text{ cm}} + \frac{1.5}{i} = \frac{1.5 - 1.0}{-10\text{ cm}},$$

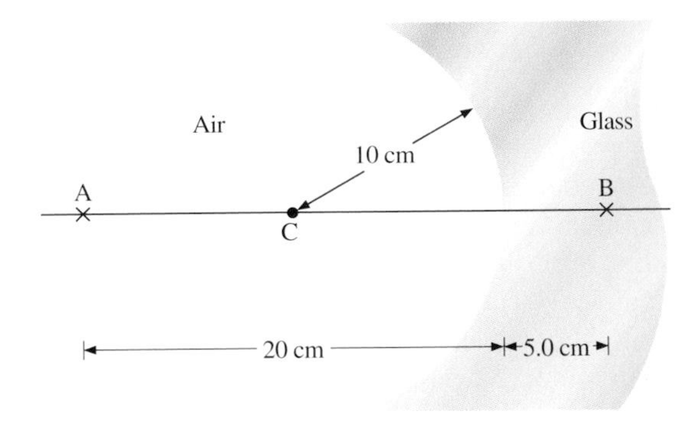

FIGURE 36-16 A spherical air-glass interface.

which yields $i = -15$ cm. The image is therefore virtual and on the air side of the boundary. From Eq. (36-10), the lateral magnification is

$$m = -\frac{(1.0)(-15\text{ cm})}{(1.5)(20\text{ cm})} = 0.50,$$

which tells us that the image is erect and half the height of the object.

(*b*) Now that the object is in the glass, the glass becomes medium 1, so $n_1 = 1.5$ and $n_2 = 1.0$. Also $R = +10$ cm [notice the change in sign from part (*a*)—why?] and $o = +5.0$ cm. From Eq. (36-9),

$$\frac{1.5}{5.0\text{ cm}} + \frac{1.0}{i} = \frac{1.0 - 1.5}{10\text{ cm}},$$

so the image distance is

$$i = -2.9\text{ cm}.$$

The image is therefore virtual and also on the glass side. The lateral magnification is

$$m = -\frac{(1.5)(-2.9\text{ cm})}{(1.0)(5.0\text{ cm})} = 0.86,$$

which indicates that the image is erect and slightly shorter than the object.

EXAMPLE 36-4 Coin in a Fountain

Figure 36-17 shows a coin resting at the bottom of a fountain whose depth is h. What is the apparent depth of the coin when viewed

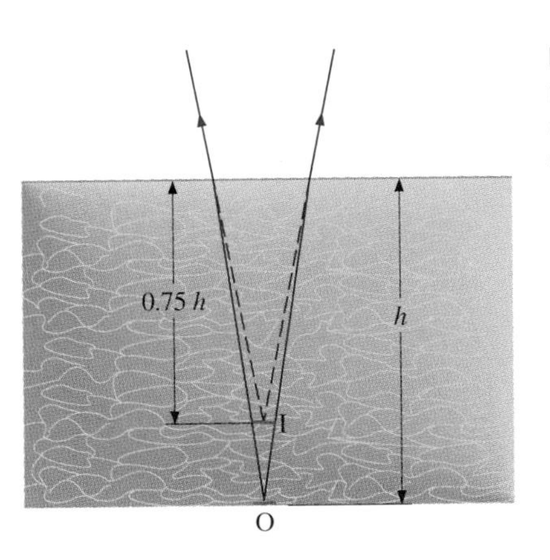

FIGURE 36-17 A coin rests at the bottom of a fountain (presently turned off) of depth h.

from directly overhead? If the coin has a diameter D, what is the diameter of the image?

SOLUTION The planar surface of the water is a spherical surface for which $R = \infty$. Equation (36-9) then reduces to

$$\frac{n_1}{o} + \frac{n_2}{i} = 0. \qquad (i)$$

The object distance is $o = h$, and since the object is in the water, $n_1 = 1.33$ and $n_2 = 1.0$. Substituting these values in Eq. (i), we have

$$\frac{1.33}{h} + \frac{1.0}{i} = 0,$$

and

$$i = -0.75h.$$

The image is therefore virtual and at a shallower depth than the actual coin.

From Eq. (i), the ratio of the image distance to the object distance for a planar refracting surface is

$$\frac{i}{o} = -\frac{n_2}{n_1}.$$

Substituting this expression into Eq. (36-10), we obtain for the lateral magnification

$$m = -\frac{n_1 i}{n_2 o} = -\left(\frac{n_1}{n_2}\right)\left(-\frac{n_2}{n_1}\right) = 1.$$

Hence, like a flat, reflecting surface, a flat, refracting surface *always has a lateral magnification of 1*. For this example then, the diameter of the image of the coin is D.

DRILL PROBLEM 36-7

An observer is viewing a small sea horse swimming in a spherical fishbowl, as shown in Fig. 36-18. Assuming that the bowl has a diameter of 60 cm and that the sea horse is 3.0 cm tall, find the location and height of the image. Neglect the refractive effects of the walls of the bowl.
ANS. −18 cm; 3.6 cm.

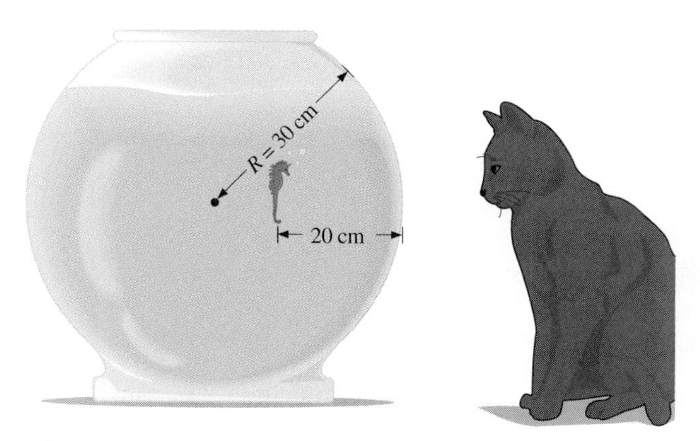

FIGURE 36-18 A small sea horse in a spherical fishbowl.

SUMMARY

1. **Image formation by reflection**
   (a) Plane mirrors: An object a distance o in front of a plane mirror has an image a distance i from the mirror such that

   $$o = -i.$$

   The image is virtual and has a lateral magnification $m = 1$.

   (b) Spherical mirrors: If an object a distance o from the vertex of a concave or a convex mirror has an image a distance i from the vertex, then in the paraxial-ray approximation,

   $$\frac{1}{o} + \frac{1}{i} = \frac{1}{f},$$

   where the focal length f is $R/2$, or one-half the radius of curvature of the mirror. The image may be real or virtual and has a lateral magnification $m = -i/o$. The image is erect if $m > 0$; it is inverted if $m < 0$.

2. **Ray tracing**
   The following rules are used in ray tracing to locate an image formed by reflection from a spherical mirror.
   (a) An incident ray parallel to the optical axis is reflected so that it passes through the focal point.
   (b) An incident ray that passes through the focal point is reflected parallel to the optical axis.
   (c) A ray that passes through the center of curvature of the mirror returns along its original path.
   (d) The angle that a ray striking the vertex of the mirror makes with respect to the optical axis is the same before and after reflection.

3. **Image formation by refraction**
   If an object in medium 1 a distance o from a spherical interface separating media 1 and 2 has an image at a distance i from the interface, then in the paraxial-ray approximation,

   $$\frac{n_1}{o} + \frac{n_2}{i} = \frac{n_2 - n_1}{R},$$

   where n_1 and n_2 are the refractive indices of the media and R is the radius of curvature of the interface. The lateral magnification is $m = -n_1 i/n_2 o$.

   The sign convention followed for both reflection and refraction at a spherical surface is expressed as follows:

(*a*) $o > 0$ if the object is on the same side as the light that is incident upon the surface; $o < 0$ if the object is on the opposite side.
(*b*) $i > 0$ if the image is on the same side as the light leaving the surface; $i < 0$ if the image is on the opposite side.
(*c*) $R > 0$ if the center of curvature lies on the same side as the light leaving the surface; $R < 0$ if the center of curvature lies on the other side.

QUESTIONS

36-1. What is the focal length of a plane mirror? What is the lateral magnification of a plane mirror?

36-2. The plane rearview mirror in a car has an adjustment for reducing the glare from headlights of trailing cars. This adjustment (usually made by shifting a small lever) changes the position of the mirror and much fainter images are seen. Explain how this mechanism works.

36-3. Discuss why the real image of an object in a concave mirror is always inverted.

36-4. Discuss why the virtual image of an object in a mirror is always upright.

36-5. Can a convex mirror produce an image that is larger than the object?

36-6. Is the wide-angle mirror on the passenger side of a car concave or convex?

36-7. How can you easily determine the focal length of a concave mirror? Why won't the same method work for a convex mirror?

36-8. Using ray diagrams, explain the advantage of a wide-angle side mirror over a plane mirror. Why aren't all the mirrors in a car wide-angle mirrors?

36-9. Is a shaving mirror convex or concave?

36-10. When you look at your reflection in a shiny round object such as a Christmas tree ornament, the image is quite distorted near the edge of the sphere. Why?

36-11. If you gaze at your image in the aluminum lid of a frying pan, you will notice that at certain distances the image may disappear when you are looking into the concave side, but it is always present when the lid is reversed. Explain.

36-12. What kind of spherical mirror (concave or convex) can be used to start a fire with sunlight? Where should the mirror be placed relative to the material to be ignited?

36-13. Can a virtual image be seen on a screen?

36-14. Can an image formed by a convex mirror be projected onto a screen without the help of additional mirrors or lenses? Answer the same question for a concave mirror.

36-15. In a flashlight, where is the lightbulb placed relative to the spherical reflector?

36-16. What is the radius of curvature of a plane refracting surface?

36-17. Can a virtual image be photographed? (*Hint:* Can you photograph your own reflection in a plane mirror?)

36-18. If photographic film is placed at the location of a virtual image, will it record the image?

36-19. If photographic film is placed at the location of a real image, will it record the image?

36-20. Does the focal length of a mirror change when it is immersed in water?

36-21. Explain why the sign on the ambulance on page 660 is painted in such a manner.

36-22. We can best view the inside of an aquarium filled with water by looking perpendicular to the glass wall. When viewed obliquely, objects inside appear distorted. Why is this the case?

36-23. Explain why a fish in a spherical fishbowl appears larger than it really is.

PROBLEMS

Image Formation by Reflection: Plane Mirrors

36-1. A 5.3-cm-long toothpick is placed with its length parallel to a plane mirror and 20.0 cm away from the mirror. How far is its image (*a*) behind the mirror? (*b*) from the toothpick?

36-2. The toothpick of the previous problem is now positioned so that its length is perpendicular to the mirror surface. The end that is closer to the mirror is 15.0 cm from the surface. Determine the distance between the images of the two ends of the toothpick.

36-3. A baby sits 120 cm in front of a large, flat mirror and watches her image. (*a*) How far away from her is the image located? (*b*) What is the size of the image? (*c*) If she waves at the image with her right hand, how does the image wave back?

36-4. If you walk toward a plane mirror at 1.5 m/s, how fast does your image appear to be approaching you?

36-5. Show that a plane mirror must be at least half your height before you can see your entire body in it.

36-6. A physicist gives his lady friend who is 5 ft 7 in. tall a wall mirror that is exactly half her height. How high from the ground must the mirror be hung so that she can see her entire image? Assume that the lady's eyes are 5 in. below the top of her head.

36-7. Unfortunately, the physicist of the previous problem is not a very good handyman. He hangs the mirror so that its top is 3 in. below the eye level of his friend. What fraction of her height can she see in this mirror?

36-8. The two plane mirrors shown in the accompanying figure are joined together at an angle ϕ such that a ray reflected by both mirrors leaves on a path parallel to its incident path as shown. What is the value of ϕ?

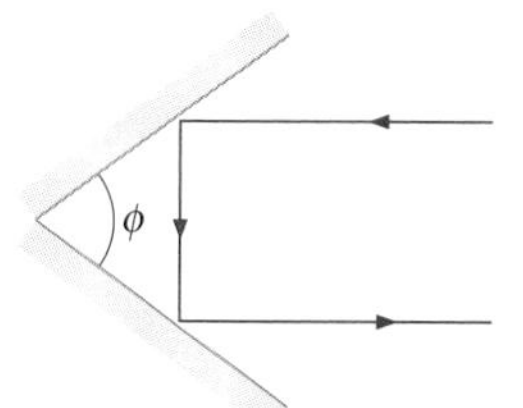

PROBLEM 36-8

36-9. The accompanying figure shows a person viewing the image of a point object in a 2.0-m-long plane mirror. Both the object and the observer are 4.0 m from the surface of the mirror, and the object is 3.0 m from the right end of the mirror. Over what range of values of d is the observer able to see the image in the mirror?

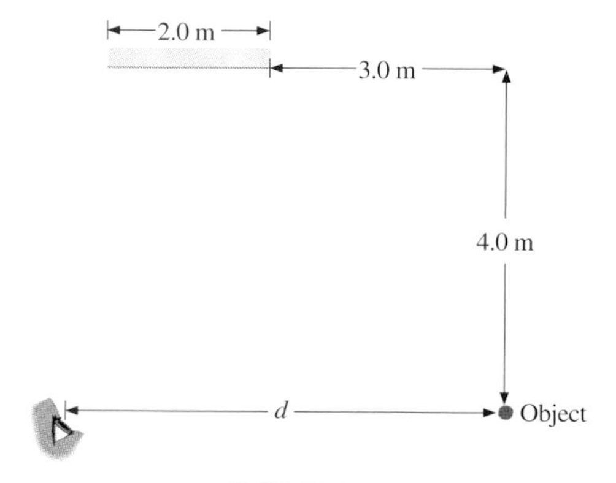

PROBLEM 36-9

Image Formation by Reflection: Spherical Mirrors

36-10. A concave mirror has a radius of curvature of 30 cm. (*a*) What is the focal length of the mirror? (*b*) Where is the image of an object placed 5.0 cm in front of the mirror?

36-11. A convex mirror has a radius of curvature of 24 cm. A virtual image of an object is produced 10 cm from the mirror. Where is the object located?

36-12. What is the radius of curvature of a concave reflecting surface that focuses parallel light rays to a point 20 cm in front of it?

36-13. An object placed 80 cm in front of a convex mirror has an image 32 cm behind the mirror. What is the focal length of the mirror?

36-14. The image of a distant car is located 15 cm behind a convex rearview mirror. What is the radius of curvature of the mirror?

36-15. A man is shaving in front of a concave mirror of radius of curvature 40 cm. The mirror is positioned so that the image of the man's nose is 2.0 times its actual size. How far is the man's nose from the mirror?

36-16. Suppose you are 25 cm from a reflecting spherical Christmas-tree ornament of diameter 8.0 cm. (*a*) Where is your image and what is its lateral magnification? (*b*) Construct the image with a ray diagram.

36-17. The face of a computer monitor is convex, with a radius of curvature of 100 cm. (*a*) How far from the screen must a programmer be sitting in order to see an image of himself that is half his size? (*b*) Where is the image located?

36-18. A concave mirror has a focal length of 50 cm. (*a*) Where must an object be placed so that its image will be erect and four times the size of the object? (*b*) Where will the image be?

36-19. The accompanying table shows object distance, object size, and mirror focal length (all in centimeters) for various situations. Use ray tracing to determine the image size and location for each case.

Object Distance	Object Size	Focal Length
40	2.0	20
10	1.0	20
30	1.5	−30
20	2.0	−40

36-20. Repeat the previous problem, given focal lengths of the opposite signs.

36-21. Complete the following table for spherical mirrors. All distances are in centimeters.

Mirror Type	Radius	Focal Length	Object Distance	Image Distance	Lateral Magnification	Real/Virtual Image
		−25	+40			
			+20	−30		
		+15		+30		
			+20		−2.0	
	+20			−40		
		−10			+2.0	

36-22. When looking into his convex wide-angle mirror, a driver sees the image of a truck. The image is 8.0 m from the mirror. If the radius of curvature of the mirror is 20.0 m, how far is the truck from the mirror?

36-23. A concave mirror of focal length 4.00 m is used to produce an image of the moon. The diameter of the moon is about 3.50×10^3 km, and its distance from the earth is approximately 3.86×10^5 km. What is the diameter of the image?

36-24. A concave mirror forms an image of a lightbulb filament on a wall 5.0 m from the mirror. The filament is 1.0 cm high, while its image is 50 cm high. (*a*) Where is the filament placed and (*b*) what is the focal length of the mirror?

36-25. A concave mirror of radius of curvature 80 cm is used to produce the image of a slide on a wall. The slide is illuminated by a bulb that shines light through it onto the mirror. (*a*) If the wall is 4.0 m from the mirror, where should the slide be placed? (*b*) If a picture on the slide is 2.0 cm high, what is the height of its image? (*c*) Is the image upright or inverted? (*d*) Use ray tracing to construct the *object*.

36-26. A dentist places his mirror 1.5 cm from a tooth and sees an enlarged image 4.0 cm behind the mirror. (*a*) Is the mirror concave or convex? (*b*) What is the focal length of the mirror? (*c*) What is the lateral magnification of the mirror? (*d*) Is the image inverted or upright?

36-27. Draw ray diagrams to locate either the object, image, or focal point for the situations shown in the accompanying figure.

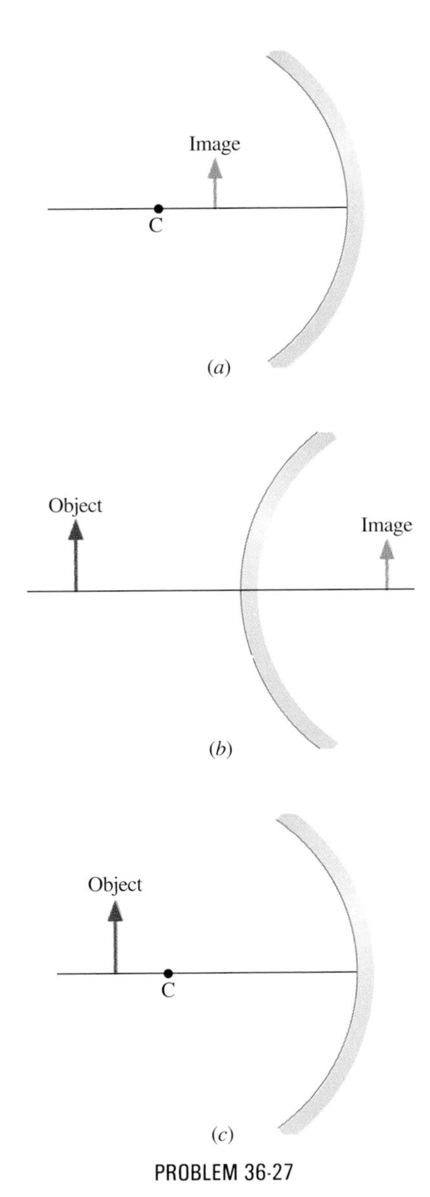

PROBLEM 36-27

36-28. Convex mirrors are often used to view the aisles in a store. (*a*) If a mirror has a focal length of -2.0 m, where is the image of a customer 15 m in front of the mirror? (*b*) If the customer is 1.5 m tall, how tall is her image? (*c*) Draw a ray diagram for this situation.

Images Formed by Refraction

36-29. One end of a long glass rod is ground into a convex spherical surface of radius 15 cm. An object is placed in front of this surface on the axis of the rod. Calculate the positions of the images when the object is (*a*) 10 cm, (*b*) 15 cm, and (*c*) 40 cm from the convex end of the rod. For each case, also determine the lateral magnification.

36-30. Repeat the previous problem, assuming that the rod is immersed in water.

36-31. Do Probs. 36-29 and 36-30, assuming that the spherical end of the rod is concave and has a radius of curvature of 15 cm.

36-32. Find the positions and lateral magnifications of the images due to the spherical surface of the rod of Prob. 36-29, assuming that the objects are at the given distances *within* the rod rather than outside it.

36-33. A barrel is filled with ethyl alcohol ($n = 1.36$) to a depth of 80 cm. How deep does the alcohol appear to be to someone who looks straight down at it?

36-34. A barrel is filled with an unknown liquid to a depth of 100 cm. To an observer looking straight down, the bottom of the barrel appears to be 65 cm below the surface of the liquid. What is the index of refraction of the liquid?

36-35. Suppose that you are 2.0 m below the surface of the water in a swimming pool and looking straight upward at a lightbulb that is 6.0 m above the surface. How far away from you does the lightbulb appear to be?

36-36. Complete the accompanying table for spherical refracting surfaces. All distances are in centimeters, and in all cases $n_1 = 1.0$ and $n_2 = 1.5$.

Surface Type	Radius	Object Distance	Image Distance	Lateral Magnification	Real/Virtual Image
	+50	+25			
			−20	2.0	
	−40			0.20	
		+40		1.0	

36-37. A concrete block 25 cm thick rests on the bottom of a filled swimming pool. How thick does the block appear to an observer looking at it from above the water?

36-38. An aquarium whose bottom is a mirror is filled with water to a depth of 100 cm. (See the accompanying figure.) A small sea horse floats motionless 20 cm below the surface. What is the apparent depth of the sea horse's image in the mirror as seen by an observer directly above?

PROBLEM 36-38

36-39. Parallel light rays fall on a transparent plastic sphere. If the rays focus to a point on the opposite side of the sphere, as shown in the accompanying figure, what is the index of refraction of the sphere?

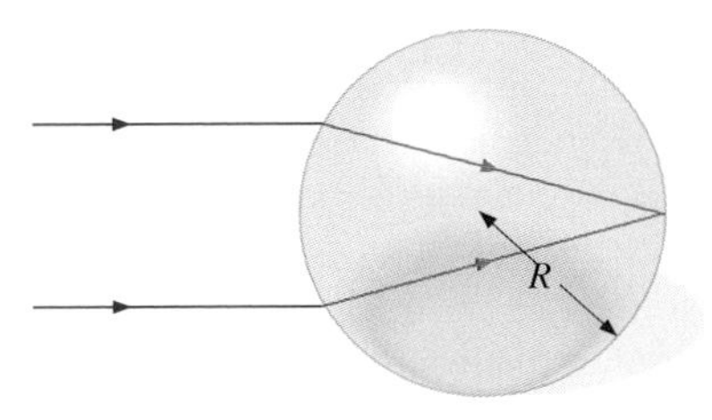

PROBLEM 36-39

36-40. The accompanying figure shows a glass rod of length 60 cm that has both ends ground into hemispherical convex surfaces, each of radius 10 cm. An object of height 2.0 cm is placed 30 cm from the left end. (*a*) Where is the image formed by the left surface? (*b*) What is the size of this image? (*c*) Use this image as an object for the right surface and find the position of the resulting image. (*d*) What is the size of this second image?

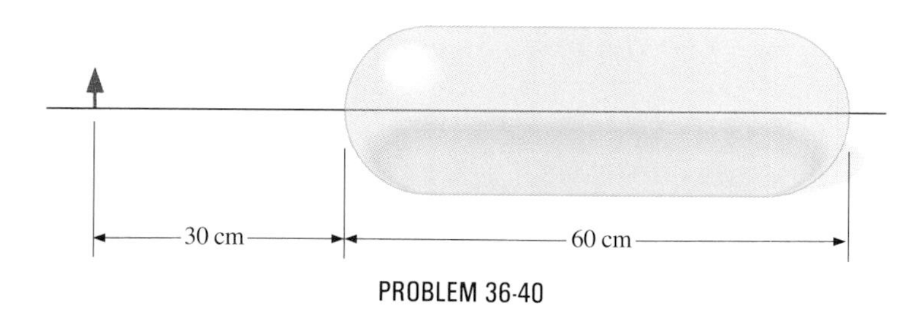

PROBLEM 36-40

36-41. A 4.0-cm-long goldfish is at the center of a spherical bowl of radius 40 cm, which is completely filled with water. (*a*) Where is the image of the fish for an observer looking into the bowl? (*b*) By how much is the image of the fish magnified? (*c*) Is its image inverted? Neglect any refraction due to the bowl.

36-42. A solid glass hemisphere of radius 10 cm is placed with its flat face on top of a table. A circular beam of light 2.0 cm in diameter aimed vertically downward enters the hemisphere centered on its axis. What is the diameter of the spot of light formed on the table?

36-43. The critical angle for a particular plastic material is 45°. A 2.0-cm-thick sheet of this plastic is placed on a table. How far below the upper surface of the plastic does an observer looking from above see a scratch on the surface of the table?

36-44. Enough light reflects off the concave end of a long rod of shiny, transparent material to form an image. An object is placed on the air side 20 cm from this end. The reflected image of the object is real and is 5.0 cm from the vertex. Where is the refracted image of the object?

General Problems

36-45. If the diameter of your pupils is 5.0 mm, what area of a plane mirror is used to reflect the rays from a point on your lips into one of your eyes? Assume that your lips and your eyes lie in a plane parallel to the mirror.

36-46. An "optical lever" is illustrated in the accompanying figure. It is used when very small rotational angles must be measured. A plane mirror is attached to the object whose rotation is to be determined, and a beam of light is reflected off the mirror to a distant scale. Even though the angle of rotation is small, the displacement of the beam on the scale is easily measurable because of the large distance D between the mirror and the scale. (*a*) Show that if the mirror rotates through an angle θ, the reflected beam rotates through an angle 2θ. (*b*) Also show that the displacement of the beam on the scale due to the rotation is $2D\theta$.

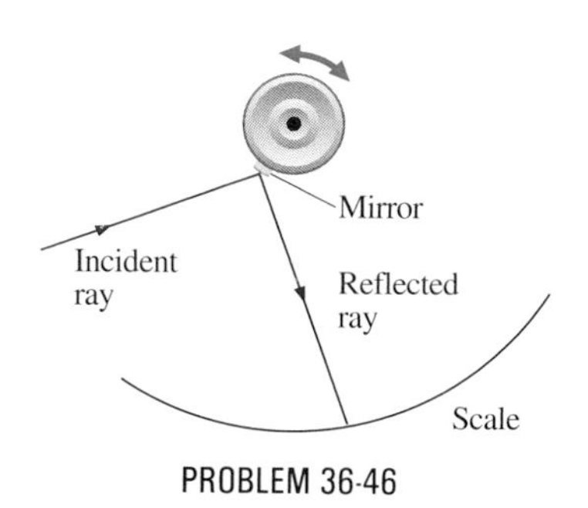

PROBLEM 36-46

36-47. A point object moves at a velocity v_0 along the optical axis toward a spherical mirror. Show that the velocity of its image is

$$v_i = -\left(\frac{f}{o-f}\right)^2 v_0.$$

36-48. Show that in order to produce an image of lateral magnification m with a mirror of focal length f, you must place the object a distance $d = (m - 1)f/m$ from the mirror.

36-49. A hemispherical plate is silvered on both sides. When used as a convex mirror, its lateral magnification is $+1/3$. For the same object distance, what is the lateral magnification when the plate is turned over and used as a concave mirror? (*Hint:* see Prob. 36-48.)

36-50. A small air bubble is inadvertently left during the manufacturing of a solid, spherical glass ball of diameter 50 cm. The bubble is located 15 cm from the surface of the ball. (*a*) Where is the image of the bubble for the observer shown in the figure? (*b*) By how much is the image magnified?

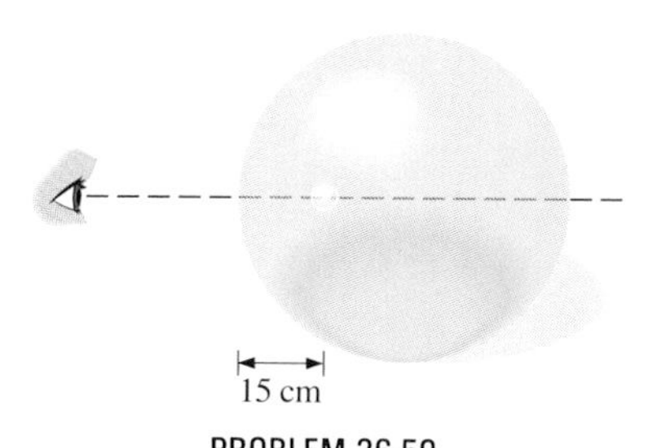

PROBLEM 36-50

36-51. If the spherical ball of Prob. 36-50 was resting at the bottom of a fountain, as shown in the accompanying figure, where would the image of the bubble be to an observer above the water?

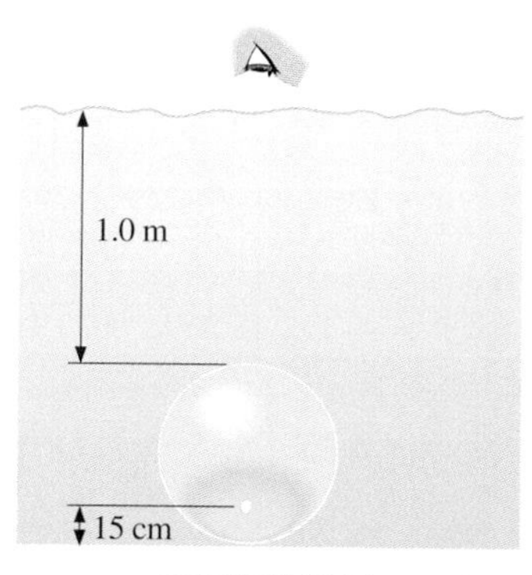

PROBLEM 36-51

A water droplet lens.

CHAPTER 37 LENSES AND OPTICAL INSTRUMENTS

PREVIEW

In this chapter we consider simple lenses and various optical instruments that consist of combinations of lenses and mirrors. The main topics to be discussed here are the following:

1. **Thin lenses.** A thin lens is defined and image formation by such a lens is explained. We then present two methods, one mathematical and the other geometric, for determining the location of an image based on knowledge of the location of the object and the properties of the lens.

*2. **Aberrations.** The effects on an image due to dispersion and nonparaxial light rays are discussed.

*3. **The eye.** A simple description of the eye is presented. Vision defects and methods for correcting those defects are discussed, and the relationship of the size of an object's image on the retina to the distance of the object from the eye is explained.

*4. **Optical instruments.** Three optical instruments—the magnifier, the microscope, and the telescope—are described.

In this chapter we study optical systems in which reflection and refraction may occur several times before an image is formed. In passing through a single lens, a light ray undergoes two refractions. Optical instruments such as microscopes and telescopes are generally made up of a combination of lenses and mirrors, so a ray may be reflected or refracted at many surfaces. Fortunately, locating images produced in such systems is not difficult with proper application of the laws of reflection and refraction.

An especially important and versatile component of many optical instruments is the *thin lens*. It is usually a piece of glass or plastic ground so that its two refracting surfaces are spherical and very close together. Much of this chapter is devoted to a study of the properties of the thin lens. As in the case of spherical mirrors, an image formed by a thin lens can be located if the object distance and the geometry of the lens are known. Finally, we discuss how the thin lens is used in various types of optical devices.

37-1 Thin Lenses

Figure 37-1 shows a transparent body of refractive index n and thickness t. The two ends of the body (which we call a "thick lens") are convex spherical surfaces, and the index of refraction of the surrounding medium is n'. Paraxial light rays from a point object O are refracted at the ends of the body and converge to form an image I. The radii of curvature of the left and the right surfaces are R_1 and R_2, respectively.* As shown in Fig. 37-2*a*, *refraction at the left surface produces an image I′ that acts as an object for the right surface. Refraction at the right surface then produces a final image I*, as shown in Fig. 37-2*b*.

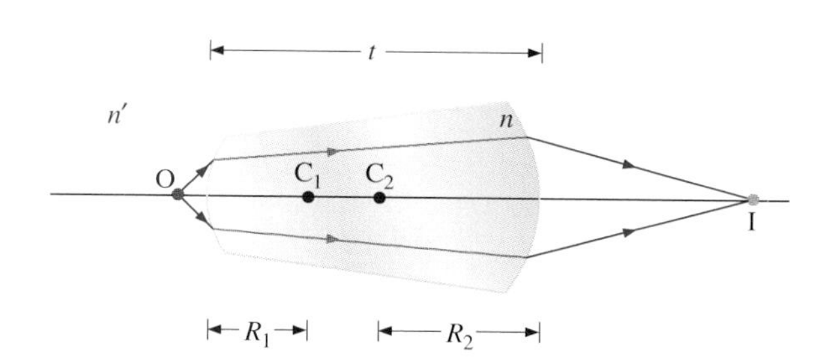

FIGURE 37-1 Example of image formation by a thick lens.

If I′ is a distance i' from the left surface, we have from Eq. (36-9),

$$\frac{n'}{o} + \frac{n}{i'} = \frac{n - n'}{R_1}. \tag{37-1}$$

*In our notation, R_1 always represents the radius of curvature of the *first* refracting surface encountered by the incident light; R_2 is the radius of curvature of the *second* surface at which refraction occurs.

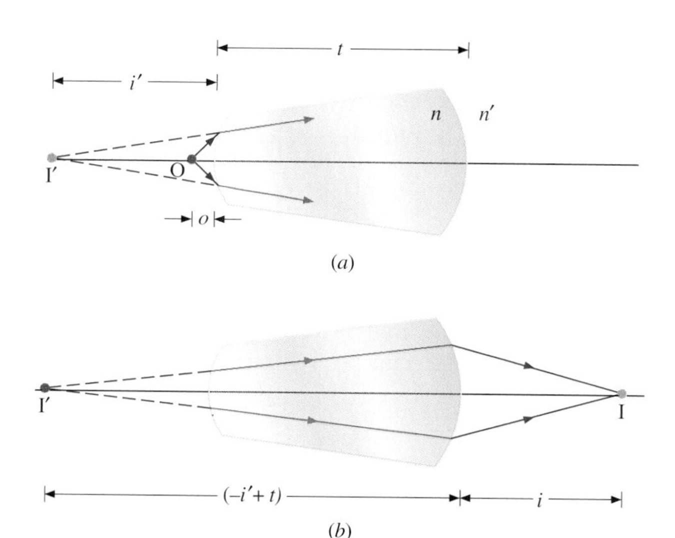

FIGURE 37-2 (*a*) An intermediate image I′ is formed by refraction at the left surface. (*b*) I′ acts as an object for the right surface. Refraction there forms the final image I.

For the case shown, the image is on the opposite side of the left surface as the light leaving that surface, so i' is negative. Its distance from the right refracting surface is therefore $-i' + t$. Since the image formed by the left surface acts as a real object for the right surface, Eq. (36-9) yields

$$\frac{n}{-i' + t} + \frac{n'}{i} = \frac{n' - n}{R_2}, \tag{37-2}$$

where i is the distance of the final image I measured with respect to the right surface.

As an example, suppose that $R_1 = 10.0$ cm, $R_2 = -20.0$ cm, $t = 65.0$ cm, $o = 5.00$ cm, $n' = 1.00$, and $n = 1.50$. Then at the left surface

$$\frac{1.00}{5.00 \text{ cm}} + \frac{1.50}{i'} = \frac{1.50 - 1.00}{10.0 \text{ cm}},$$

so

$$i' = -10.0 \text{ cm}.$$

Using this image as a real object for the right surface, we find

$$\frac{1.50}{(10.0 + 65.0) \text{ cm}} + \frac{1.00}{i} = \frac{1.00 - 1.50}{-20.0 \text{ cm}},$$

which yields $i = +200$ cm. Hence the final image is real and located in the surrounding medium at a distance of 200 cm from the right surface.

DRILL PROBLEM 37-1

Repeat the previous calculation, assuming that the spherical surfaces of the transparent body are concave, with $R_1 = -10.0$ cm and $R_2 = 20.0$ cm.

ANS. 21.7 cm from the right surface and within the body.

DRILL PROBLEM 37-2

Suppose that the image formed by refraction at the left surface of the transparent body of Fig. 37-2 is located on the same side as the light leaving the surface; now $i' > 0$. Verify that the expression $-i' + t$ still represents the distance of this image from the right surface.

We are interested in the thin lens, which is what the body of Fig. 37-1 becomes in the limit as its thickness t approaches zero. In this limit, Eq. (37-2) can be written as

$$-\frac{n}{i'} + \frac{n'}{i} = \frac{n' - n}{R_2}.$$

Adding this to Eq. (37-1), we obtain

$$\frac{n'}{o} + \frac{n'}{i} = \frac{n - n'}{R_1} + \frac{n' - n}{R_2},$$

which can be simplified to

$$\frac{1}{o} + \frac{1}{i} = \left(\frac{n}{n'} - 1\right)\left(\frac{1}{R_1} - \frac{1}{R_2}\right).$$

Usually the lens is surrounded by air, so $n' = 1$ and we have

$$\frac{1}{o} + \frac{1}{i} = (n - 1)\left(\frac{1}{R_1} - \frac{1}{R_2}\right). \tag{37-3}$$

The **focal length** f of a thin lens is defined as either the image distance i for an object at infinity ($o = \infty$), or the object distance o for an image at infinity ($i = \infty$). In either case, Eq. (37-3) yields

$$\frac{1}{f} = (n - 1)\left(\frac{1}{R_1} - \frac{1}{R_2}\right), \tag{37-4}$$

which is known as the *lensmaker's equation*. Recall that for a spherical mirror, the focal length is simply half the radius of curvature. For a thin lens, however, the focal length must be calculated using the lensmaker's equation.

Consider the thin lens shown in Fig. 37-3*a*. Here $R_1 > 0$ because the center of curvature of the first refracting surface is on the same side as the light leaving the surface; $R_2 < 0$ because the center of curvature of the second refracting surface is on the opposite side as the light leaving the surface. Hence since $n > 1$, $f > 0$. *A thin lens for which $f > 0$ is called a converging, or positive, lens.*

For the lens of Fig. 37-3*b*, $R_1 < 0$ and $R_2 > 0$, so $f < 0$. *A thin lens with $f < 0$ is called a diverging, or negative, lens.* You should convince yourself that if a lens is turned around (R_1 and R_2 are then interchanged), the sign and magnitude of f do not change.

Parallel light rays from a distant object on either side of a positive lens are refracted by the lens and converge to a point on the opposite side of the lens (Fig. 37-4*a*). These two points are called the *focal points F* of the lens. The focal points are situated symmetrically about the lens on the optical axis a distance f from the lens. Because of the reversibility of light, rays from a point object placed at either focal point of a converging lens pass through the lens as parallel rays, as shown in Fig. 37-4*b*. Like the converging lens, the diverging lens also has two symmetric focal points a distance $|f|$ from the lens. However, since light rays are diverged by the lens, only their extrapolated paths pass through these points, as shown in Fig. 37-5.

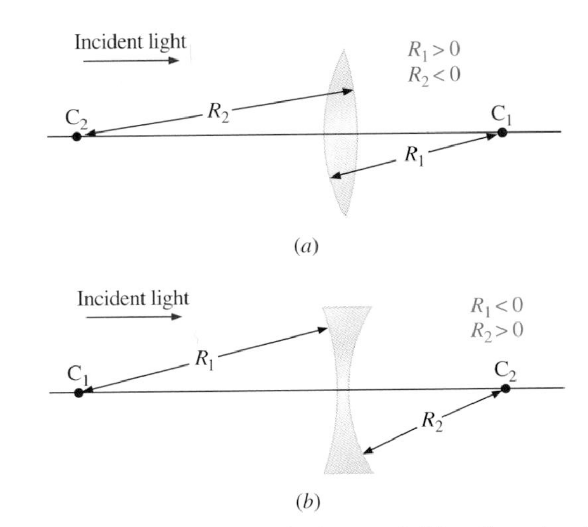

FIGURE 37-3 (*a*) A converging thin lens and (*b*) a diverging thin lens.

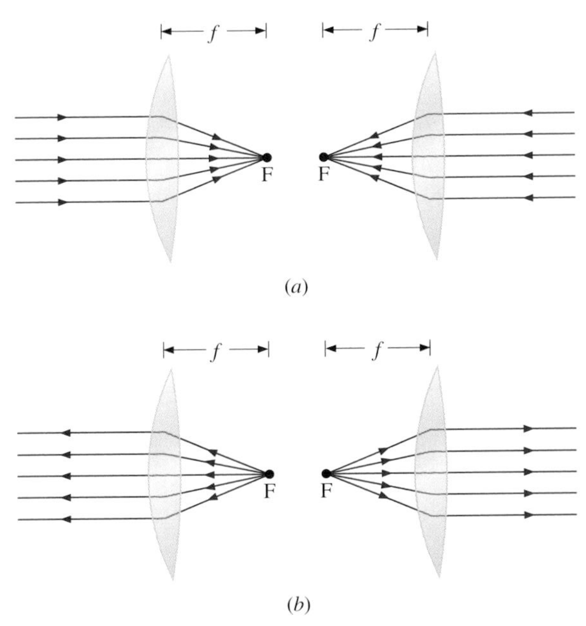

FIGURE 37-4 (*a*) When light rays parallel to the optical axis pass through a converging lens, they converge to the focal point on the opposite side of the lens. (*b*) Light rays from an object at a focal point pass through the lens and emerge as parallel rays.

By combining Eqs. (37-3) and (37-4), we obtain the *lens equation*,

$$\frac{1}{o} + \frac{1}{i} = \frac{1}{f}. \tag{37-5}$$

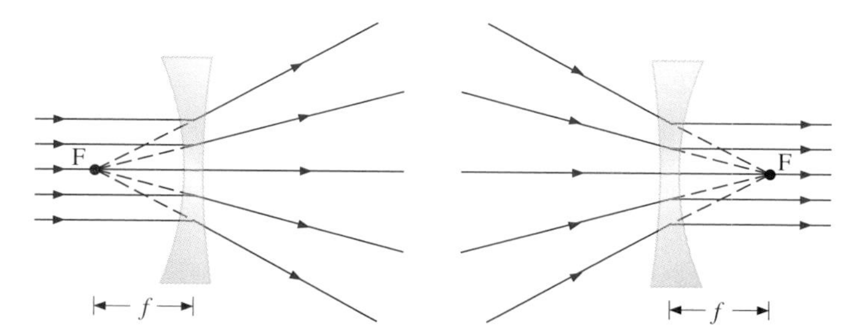

FIGURE 37-5 Extrapolations of the light rays shown pass through the focal points of the diverging lens.

Notice that it is identical to the mirror equation introduced in the previous chapter. For an object placed a distance o from a lens of focal length f, the lens equation allows us to find the distance i of the image from the lens.

The lateral magnification m of a thin lens is again the ratio of the image height to the object height. It can be related to the image and object distances with the help of Fig. 37-6, which shows two rays from an object of height y_o intersecting at an image of height y_i. From the similar triangles (colored green) on the left side of the lens,

$$\frac{y_o}{o-f} = \frac{y_i}{f},$$

so

$$\frac{y_i}{y_o} = \frac{f}{o-f} = \frac{1}{o(1/f) - 1}.$$

Using the lens equation to substitute for $1/f$, we obtain

$$\frac{y_i}{y_o} = \frac{1}{o(1/o + 1/i) - 1} = \frac{i}{o}.$$

The lateral magnification of a lens is therefore

$$m = -\frac{i}{o}, \tag{37-6}$$

where, once again, a negative sign has been inserted so that $m < 0$ indicates that the image is inverted.

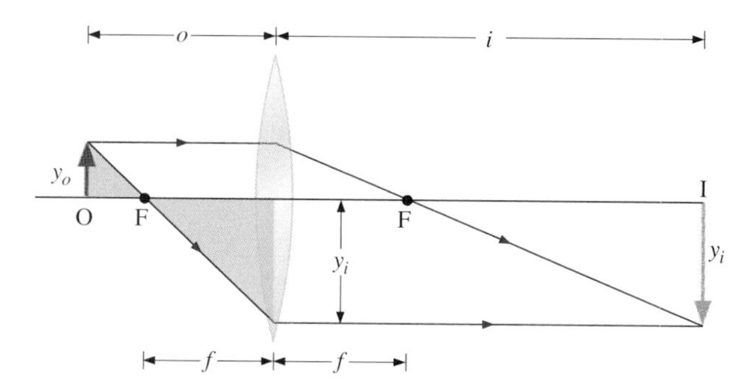

FIGURE 37-6 From the geometry of the rays shown, the lateral magnification of a thin lens is found to be $m = -i/o$.

The sign convention for refracting surfaces was given in Sec. 36-1. This convention is summarized in Fig. 37-7 for converging and diverging lenses. On the same side of the lens as the incident light (the front side), $o > 0$ and $i < 0$. On the same side of the lens as the refracted light (the back side), $o < 0$ and $i > 0$.

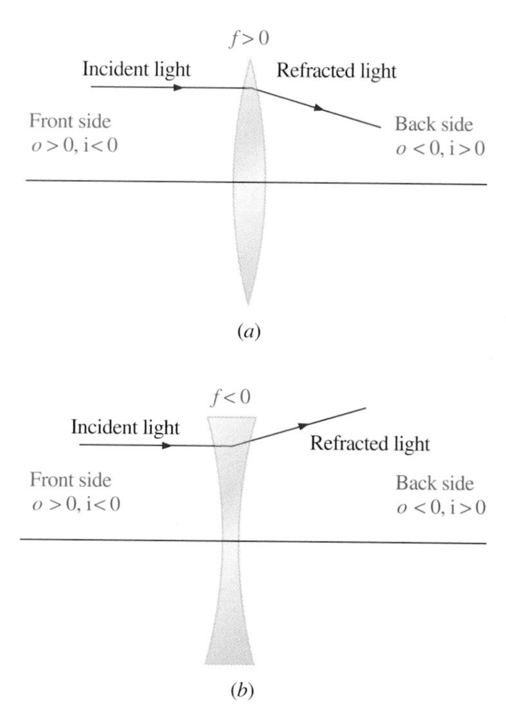

FIGURE 37-7 The sign convention for image formation by thin lenses: (*a*) converging lens and (*b*) diverging lens.

As demonstrated in Fig. 37-6, ray tracing can also be used to obtain an image formed by a thin lens. The three rays shown in Fig. 37-8 are used for this purpose:

1. An incident ray passing through a focal point is refracted so that it leaves parallel to the optical axis (ray 1).
2. An incident ray parallel to the optical axis is refracted so that it passes through a focal point (ray 2).
3. A ray passing through the center of the lens is not deflected (ray 3).

The reason ray 3 is not deflected is that the two sides of the lens near its center are essentially parallel. As Example 35-4 showed, parallel refracting surfaces do not change the direction of a ray. Furthermore, the displacement of the ray in crossing the lens is negligible for a thin lens.

When tracing rays 1 and 2 for a given optical system, you may have some question as to which of the two focal points is the appropriate one to use. To avoid any confusion, simply keep in mind that a light ray passing through a converging lens

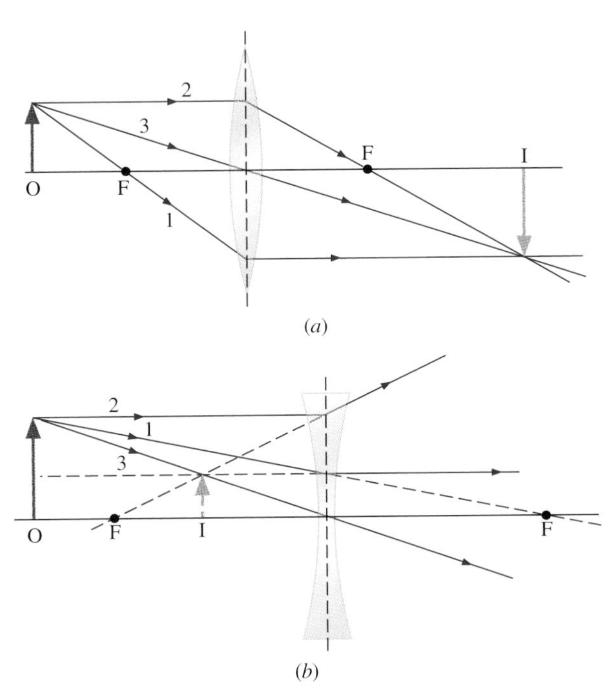

FIGURE 37-8 The three primary rays used in ray tracing for (*a*) a converging and (*b*) a diverging thin lens. Notice that it is the extrapolated rays that pass through the focal points of a diverging lens.

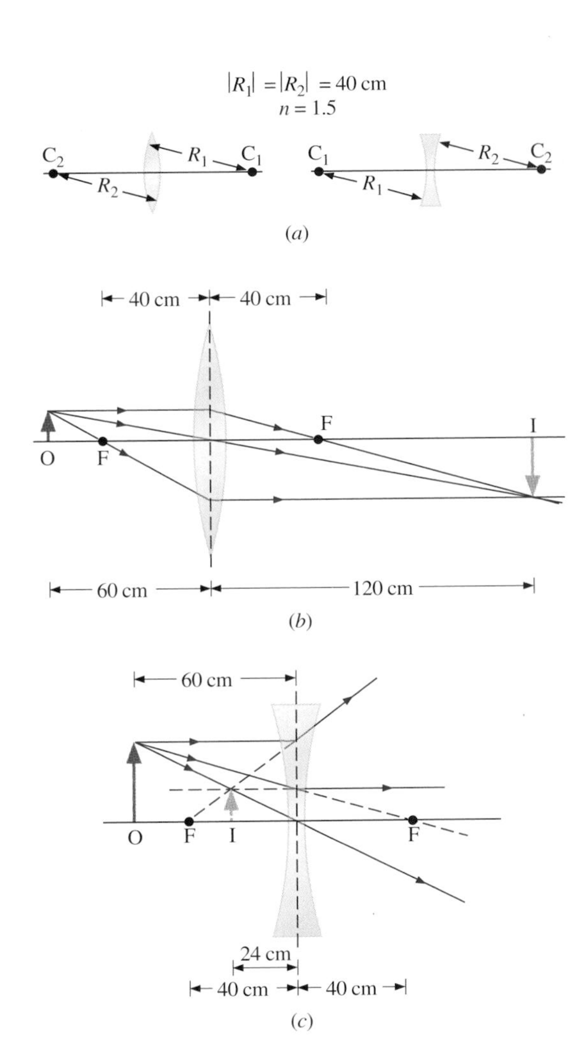

FIGURE 37-9 (*a*) The two lenses of Example 37-1. (*b*) Ray diagram for the converging lens. (*c*) Ray diagram for the diverging lens.

is always deviated *toward* the optical axis, while a light ray passing through a diverging lens is always deviated *away from* the optical axis. With this information, any ambiguity about which focal point is to be associated with a particular ray is eliminated.

EXAMPLE 37-1 Image Formation by Thin Lenses

The radii of all spherical surfaces on the two lenses of Fig. 37-9*a* are 40 cm, and the refractive index of the lens material is 1.5. (*a*) What are the focal lengths of the lenses? (*b*) An object is placed 60 cm in front of the converging lens. Where is the image located? Determine the nature, orientation, and lateral magnification of the image. Check your calculations with a ray diagram. (*c*) Describe the image if an object is placed 60 cm in front of the diverging lens. Construct a ray diagram for this case as well.

SOLUTION (*a*) We assume that the incident rays are traveling from left to right. For the converging lens, the center of curvature of the first refracting surface (the left surface in this example) and the light leaving that surface are on the same side of the surface, so $R_1 = +40$ cm. And since the center of curvature of the second refracting surface and the light leaving that surface are on opposite sides of the surface, $R_2 = -40$ cm. We now obtain, from the lensmaker's equation,

$$\frac{1}{f} = (1.5 - 1.0)\left(\frac{1}{40\text{ cm}} - \frac{1}{-40\text{ cm}}\right),$$

so the focal length of the converging lens is

$$f = +40\text{ cm}.$$

For the diverging lens, the same argument gives $R_1 = -40$ cm and $R_2 = +40$ cm. Then

$$\frac{1}{f} = (1.5 - 1.0)\left(\frac{1}{-40\text{ cm}} - \frac{1}{40\text{ cm}}\right),$$

and the focal length of the diverging lens is

$$f = -40\text{ cm}.$$

(*b*) With the lens equation, we have for the converging lens

$$\frac{1}{60\text{ cm}} + \frac{1}{i} = \frac{1}{40\text{ cm}},$$

which gives $i = +120$ cm. Because $i > 0$, the image is located on the right side of the lens (the same side as the light leaving the lens) and is real. The lateral magnification of the lens is

$$m = -\frac{i}{o} = -\frac{120\text{ cm}}{60\text{ cm}} = -2.0.$$

The image is therefore inverted and twice as high as the object. The corresponding ray diagram is shown in Fig. 37-9*b*.

(*c*) For the diverging lens, we use

$$\frac{1}{60\text{ cm}} + \frac{1}{i} = \frac{1}{-40\text{ cm}},$$

and find the image distance to be

$$i = -24 \text{ cm}.$$

The lateral magnification of the lens is then

$$m = -\frac{-24 \text{ cm}}{60 \text{ cm}} = 0.40.$$

Thus the diverging lens forms a virtual image that is 24 cm to the left of the lens, upright, and 0.40 times the height of the object. The ray diagram is shown in Fig. 37-9*c*.

EXAMPLE 37-2 IMAGE OF A VIRTUAL OBJECT

Figure 37-10*a* shows an image I′ that is formed by a converging lens (not shown). A diverging lens of focal length 50 cm is placed 20 cm to the left of I′ and directly in the path of the converging rays. Determine where the new image is formed. By how much is the image magnified? Check your calculations with a ray diagram.

SOLUTION The image I′ acts as an object O for the diverging lens. But notice from Fig. 37-10*b* that this object is on the side of the diverging lens opposite that of the incident light. This means that $o < 0$, and the object is *virtual*. Applying the lens equation, we obtain

$$\frac{1}{-20 \text{ cm}} + \frac{1}{i} = \frac{1}{-50 \text{ cm}},$$

so

$$i = +33 \text{ cm},$$

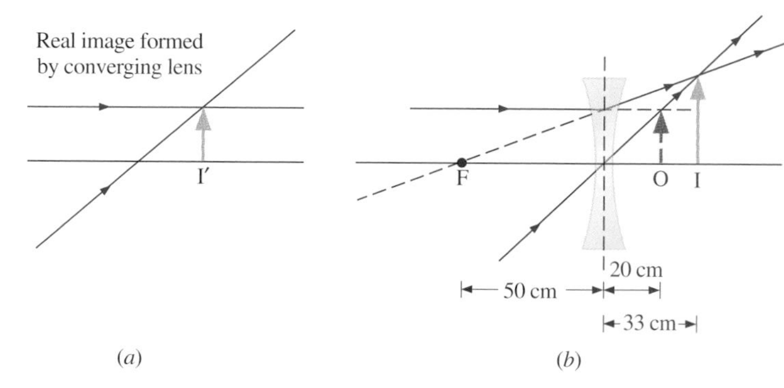

FIGURE 37-10 The image I′ acts as a virtual object O for the diverging lens, which forms a real image I.

and the final image I is real. The lateral magnification of the diverging lens is

$$m = -\frac{33 \text{ cm}}{-20 \text{ cm}} = +1.7.$$

The ray diagram is shown in Fig. 37-10*b*.

EXAMPLE 37-3 LENS COMBINATION

Figure 37-11*a* shows a diverging lens with $f_1 = -60$ cm and a converging lens with $f_2 = 60$ cm, which are 80 cm apart. An object is placed 100 cm to the left of the diverging lens. Determine the location of the image formed by this combination. By how much is the object magnified?

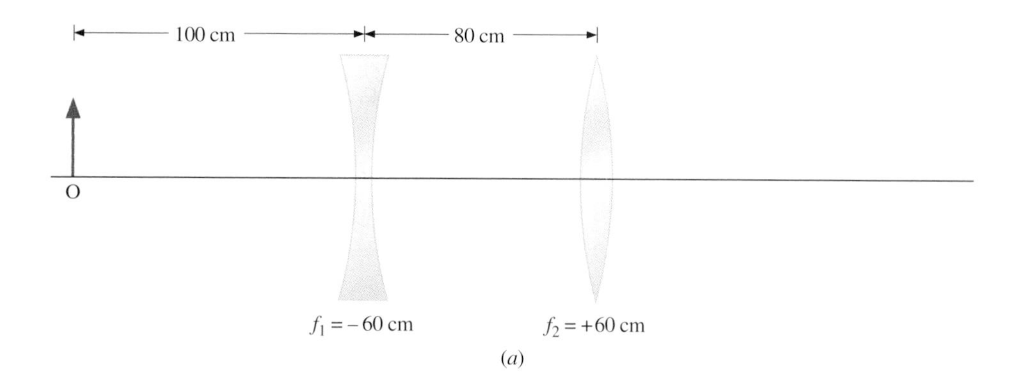

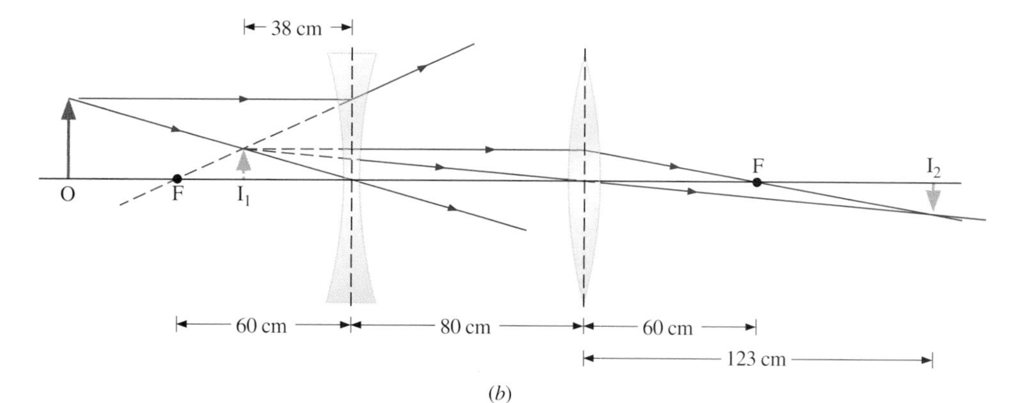

FIGURE 37-11 A lens combination.

SOLUTION Since light rays from the object strike the diverging lens first, we must begin by determining the location of its image I_1. From the lens equation,

$$\frac{1}{100\text{ cm}} + \frac{1}{i_1} = \frac{1}{-60\text{ cm}},$$

and

$$i_1 = -38\text{ cm}.$$

The negative sign indicates that the image is virtual and on the same side of the diverging lens as the object. The ray diagram of Fig. 37-11*b* shows I_1, which is 118 cm from the converging lens. Light rays appear to originate from this image so we treat it as a real object for the converging lens; hence

$$\frac{1}{118\text{ cm}} + \frac{1}{i_2} = \frac{1}{60\text{ cm}},$$

and

$$i_2 = +123\text{ cm}.$$

The final image I_2 is therefore real and 123 cm to the right of the converging lens. The ray diagram for the second lens is also shown in Fig. 37-11*b*.

The lateral magnification of the lens combination is the *product* of the lateral magnifications of the individual lenses. For the diverging lens,

$$m_1 = -\frac{-38\text{ cm}}{100\text{ cm}} = 0.38,$$

while for the converging lens,

$$m_2 = -\frac{123\text{ cm}}{118\text{ cm}} = -1.04.$$

Thus for the combination, the lateral magnification is

$$m = m_1 m_2 = (0.38)(-1.04) = -0.39.$$

The final image is therefore inverted and smaller than the original object.

DRILL PROBLEM 37-3

A converging lens has a focal length of 30 cm. Rays from a 2.0-cm high filament that pass through the lens form a virtual image at a distance of 50 cm from the lens. Where is the filament located? What is the height of the image?
ANS. +18.8 cm from the lens; 5.3 cm.

DRILL PROBLEM 37-4

When an object is placed 60 cm in front of a diverging lens, a virtual image is formed 20 cm from the lens. The lens is ground out of material with $n = 1.65$ such that its two spherical surfaces have the same radius of curvature. What is the value of this radius?
ANS. 39 cm.

*37-2 ABERRATIONS

The preliminary design of optical instruments is based on the simple equations we have been using to describe image formation in mirrors and lenses. Refinements in the design are generally necessary because these equations are derived assuming paraxial rays (rays that are almost parallel to and very close to the optical axis) and a single index of refraction for all wavelengths. For a lens, a more precise analysis of the path of a ray involves using Snell's law and taking dispersion into account. It is then found that all of the rays from a point object do not intersect at the same point to form a sharp image. The deviations of actual images from those predicted with our equations are called *aberrations*.

Spherical aberrations are a consequence of reflected or refracted rays not intersecting at one point because they are only approximately paraxial (Fig. 37-12*a*). These aberrations can be minimized by masking the outer portion of the lens or mirror, thereby using as little of the area at the center of the lens or mirror as possible—the smaller the area used, the more accurate the paraxial approximation becomes. For example, sharper images are obtained with cameras when the aperture is reduced.

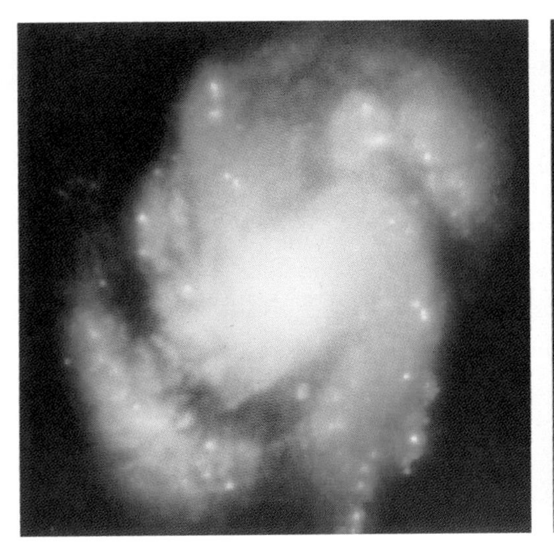

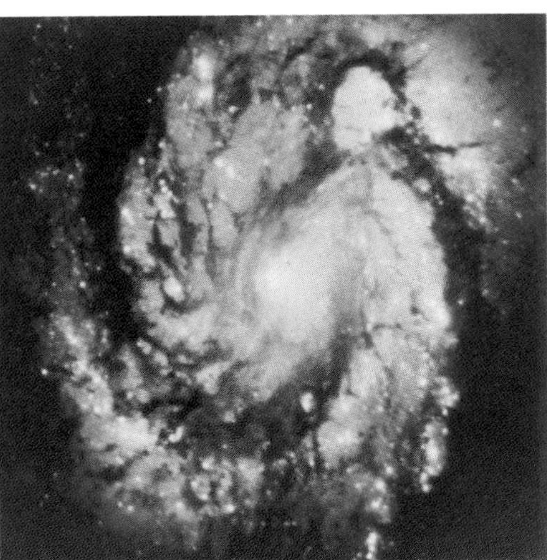

Because of imperfections in the main mirror of the Hubble telescope, its images were plagued by spherical aberration. In 1993 an extra lens was inserted by astronauts to correct this problem. Notice the dramatic improvements in the image of this distant galaxy.

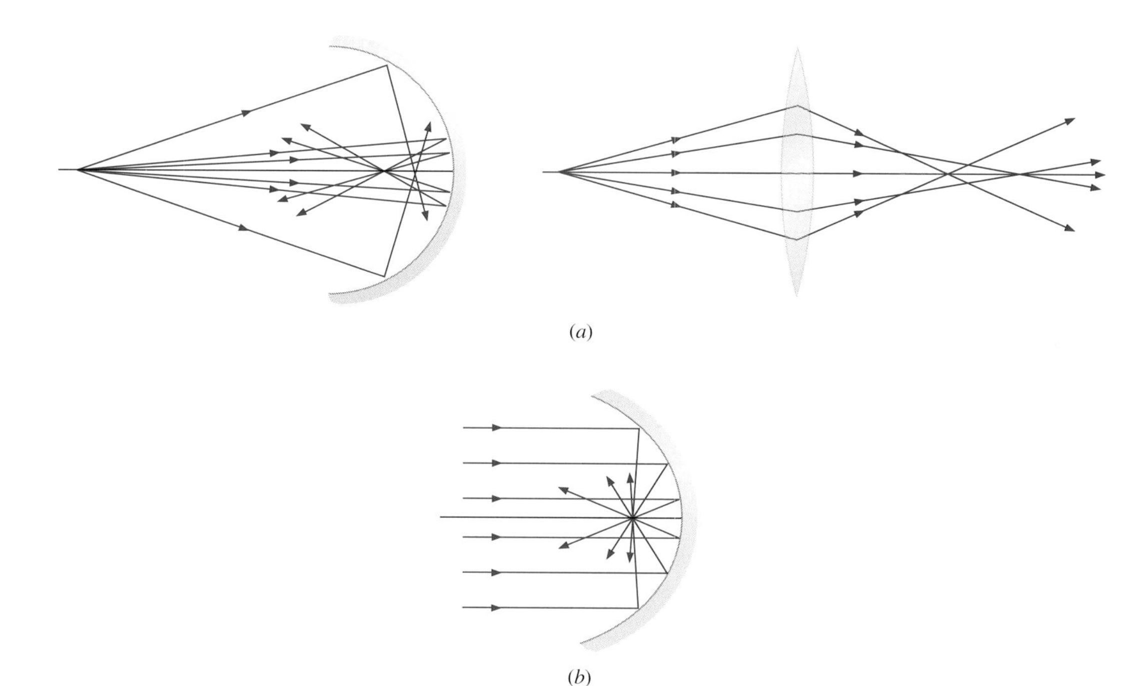

FIGURE 37-12 (*a*) Spherical aberration in a mirror or lens occurs with nonparaxial rays. (*b*) A parabolic reflecting surface helps in eliminating spherical aberration for distant objects.

The drawback here is that reducing the aperture also reduces the amount of light falling on the film. However, we can compensate for this by simply exposing the film longer. A similar effect occurs in the eye. When an object is viewed in bright light, the lens opening is reflexively reduced to protect the nerve endings in the eye. The rays entering the eye are then nearly paraxial, and spherical aberration is reduced. Conversely, the lens opening is enlarged in dim light, and images are not as clear.

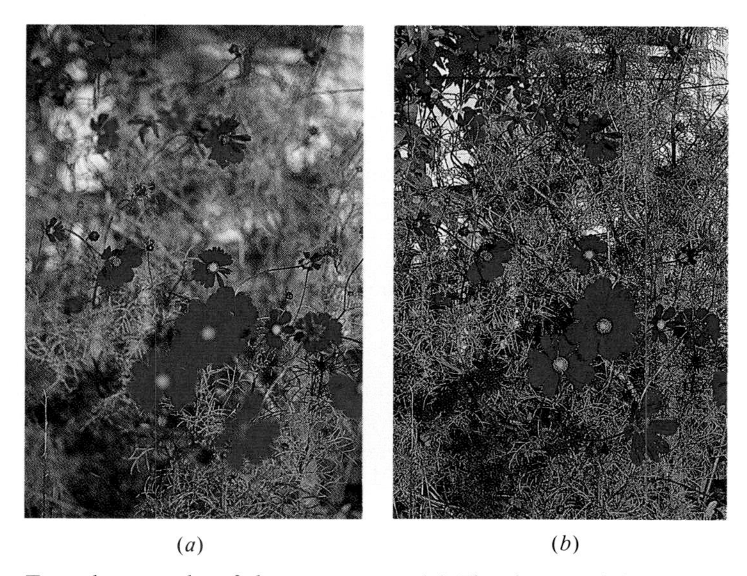

Two photographs of the same scene. (*a*) The shutter of the camera was opened wide so a large portion of the lens was used. (*b*) The shutter was set at a smaller opening so less of the lens was used. Notice that the image in part (*b*) is sharper than in part (*a*).

In many optical instruments, it is important to minimize the effect of spherical aberration. For distant objects, spherical aberration in a mirror may be eliminated by using a *parabolic* rather than a spherical surface. As illustrated in Fig. 37-12*b*, a parabolic surface reflects all parallel rays, no matter how far they are from the optical axis, through its focus. And because of the reversibility of light, rays from a source placed at the focus of a parabolic surface are all reflected parallel to the axis of the surface. This property is used in the design of searchlights that produce parallel light beams with parabolic reflectors.

Chromatic aberration is caused by the variation of the index of refraction with the wavelength of light. Since this only affects refraction, it exists in lenses but not in mirrors. Figure 37-13 shows how white light is dispersed when it passes through a lens. Notice how the position of the image changes with color. To minimize chromatic aberration, a lens is often constructed from several pieces of material with different indices of refraction. With the right combination, aberrations introduced by the different types of materials almost cancel one another.

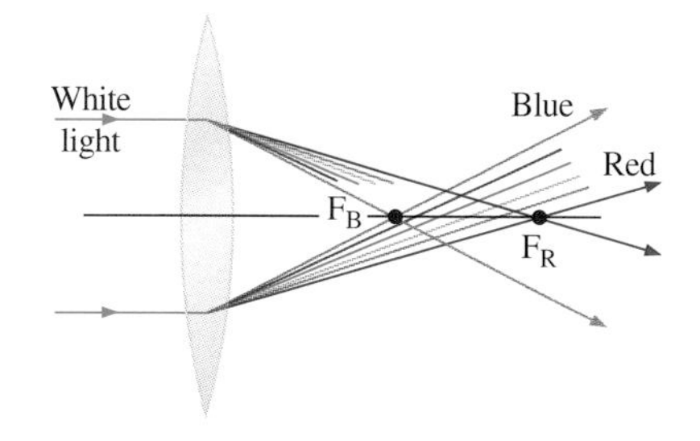

FIGURE 37-13 Chromatic aberration in a lens.

*37-3 The Eye

A sketch of the eye is shown in Fig. 37-14. Light enters the eye through an opening called the pupil, whose size is controlled by the iris. Through refractions at the cornea (the front surface membrane) and at the crystalline lens, this light forms an image on the retina. The image is detected by optical receptors, known as rods and cones, at the retina, and information is transmitted to the brain by the optic nerve. The rods are more sensitive than the cones, and they are responsible for sight in very dim light. However, the rods do not distinguish color, and the images they send to the brain are not very sharp. The cones, discussed earlier in Sec. 35-2, are responsible for color vision, and they produce the sharp images seen in bright light. The region between the cornea and the lens is filled with a liquid called the aqueous humor, and a thin jelly called the vitreous humor fills the volume between the lens and the retina. The indices of refraction of the humors are very close to that of water, while the crystalline lens has a refractive index of about 1.44. Because the refractive indices of the humors and the lens are almost identical, most of the refraction of the light entering the eye occurs at the cornea, whose index of refraction is about 1.38. Fine adjustments in focusing are made by the lens, as its shape (and hence its focal length) is changed by ciliary muscles located around the edge of the lens. This process is known as *accommodation*.

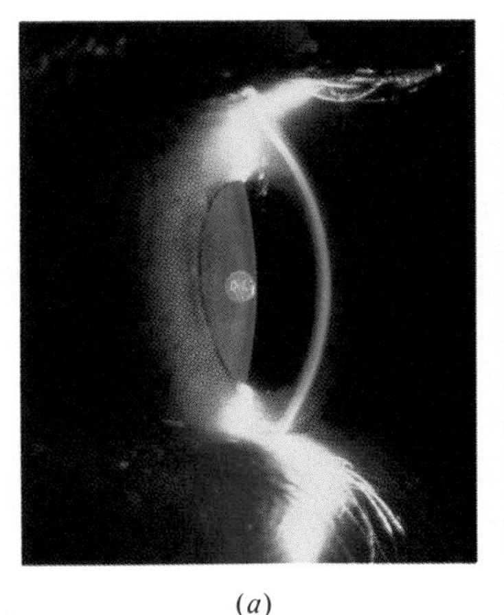

(*a*)

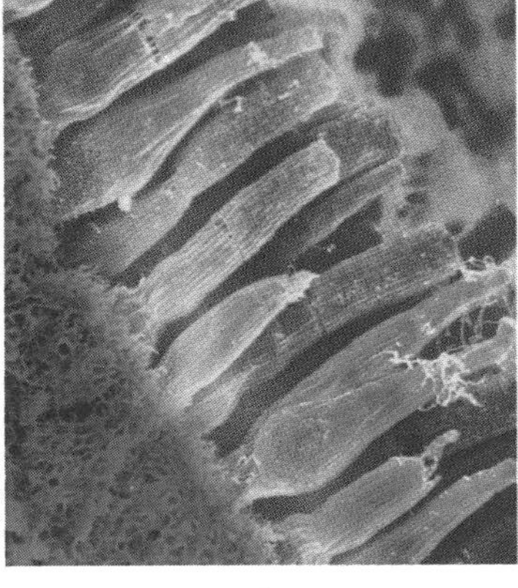

(*b*)

(*a*) A side view of the human eye. (*b*) Rods and cones magnified by a scanning electron microscope.

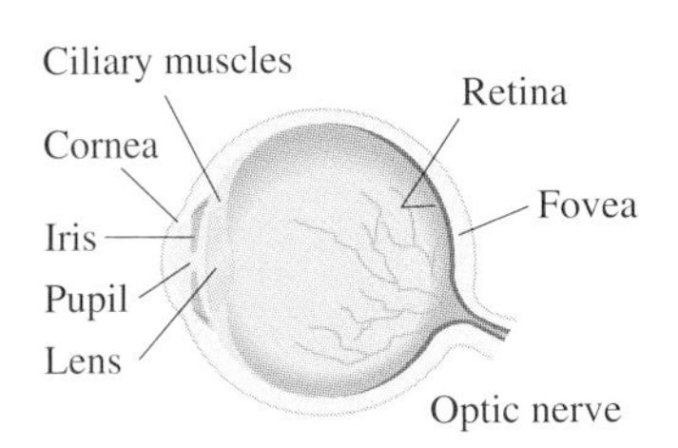

FIGURE 37-14 The human eye.

The extremes of the range of the object distance for which a clear image is formed on the retina are called the *far point* and the *near point*. The far point of the normal eye (corresponding to fully relaxed ciliary muscles) is at infinity. The near point depends on the eye's ability to accommodate. We generally use 25 cm, a normal reading distance, as the near point of the normal eye in calculations. Since the lens loses flexibility with age, our near points increase as we get older. A typical near point of the healthy eye of a 20-year-old is around 25 cm. This increases to about 50 cm when a person reaches 40 years old. The gradual recession of the near point is called *presbyopia*.

The most common defects in vision are caused by distortions in the shape of the eyeball. A normal eye focuses a distant object by relaxing the ciliary muscles. Now if the eyeball happens to be too long from front to back, the image is produced in front of the retina. This condition is known as *myopia* (or more colloquially, nearsightedness). The far point of a myopic eye is not at infinity, and in fact, is often quite close to the eye. If the eyeball is too short, a sharp image is formed behind the retina. This condition is called *hyperopia* (farsightedness). In this case, the image can still be focused on the retina through accommodation. However, the limited range of accommodation prevents the eye from focusing objects that are at the normal near point. The near point of a hyperopic eye is therefore more distant than that of a normal eye. Finally, when the cornea is not spherical, line objects in one direction are focused in a different plane than line objects in another direction. This condition is known as *astigmatism*.

Many vision defects are remedied with corrective lenses such as eyeglasses or contact lenses (Figs. 37-15 and 37-16). For myopia, a diverging lens is used to form the image of a distant object at the eye's far point. For example, if the far point of an eye is 300 cm, the corrective lens must focus objects at infinity

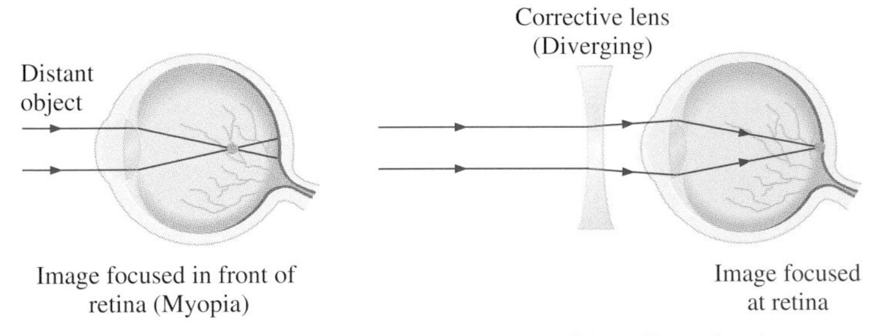

FIGURE 37-15 Myopia and its correction with a diverging lens.

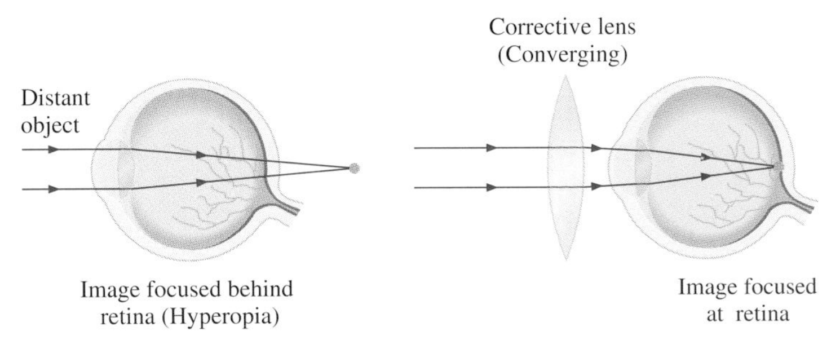

FIGURE 37-16 Hyperopia and its correction with a converging lens.

at $i = -300$ cm. The near point of either a presbyopic or a hyperopic eye is farther from the eye than the accepted distance of 25 cm. This defect is corrected with a converging lens that produces an image at the near point from an object located 25 cm from the eye (the normal reading distance). To correct astigmatism, line objects in different directions must be focused on the retina. This is accomplished with corrective lenses that themselves are not spherical.

EXAMPLE 37-4 Correcting Vision Defects

(*a*) A myopic eye has a far point of 200 cm. What is the focal length of the diverging lens needed to correct this defect? (*b*) The near point of a presbyopic eye is 60 cm. What is the focal length of the converging lens required to correct this defect?

SOLUTION (*a*) This corrective lens must produce an image at −200 cm from an object at infinity, so from the lens equation,

$$\frac{1}{\infty} + \frac{1}{-200 \text{ cm}} = \frac{1}{f},$$

and the diverging lens has a focal length of

$$f = -200 \text{ cm}.$$

(*b*) The lens must form an image at −60 cm of an object at 25 cm. The lens equation then gives

$$\frac{1}{25 \text{ cm}} + \frac{1}{-60 \text{ cm}} = \frac{1}{f},$$

and the focal length of the converging lens is

$$f = 43 \text{ cm}.$$

DRILL PROBLEM 37-5

If a hyperopic eye has a near point of 100 cm, what is the focal length of the corrective lens needed so that objects at 25 cm can be viewed comfortably?
ANS. +33 cm.

The closer an object is to us, the larger it appears to be. Its apparent size is determined by the size of its image on the retina, or equivalently, by the number of rods and cones stimulated by the light. As shown in Fig. 37-17, the size of the image on the retina depends on the angle subtended at the eye by the object. In fact, since the image distance (the distance from the lens to the retina) remains constant, *the size of the image is directly proportional to the angle that the object subtends at the eye*. The largest focused image of a given object formed on the retina by the unaided eye is therefore produced when the subtended angle is largest, that is, when the object is at the near point. With 25 cm as the distance of the near point from the eye, the maximum angle an object of height y (in centimeters) can subtend at the eye is

$$\theta_N \approx \frac{y}{25 \text{ cm}}.$$

To increase the size of the image on the retina, we use an optical device such as a microscope or a telescope to view the object. The image produced by such a device must be far enough from the eye (>25 cm) in order to be seen clearly. If this image subtends an angle θ at the eye, then by definition, the **angular magnification** M of the optical device is

$$M = \frac{\theta}{\theta_N}. \qquad (37\text{-}7)$$

The angular magnification of an optical device depends on its geometry and the properties of its components. In the following sections, we consider three such devices.

*37-4 The Simple Magnifier

The simple magnifier, which is a single converging lens, is a basic optical device used to magnify an image on the retina. When an object is placed at the focal point of the lens of Fig. 37-18*a*, the eye perceives the erect, virtual image formed by the lens. Since this image is at infinity, it can be easily focused

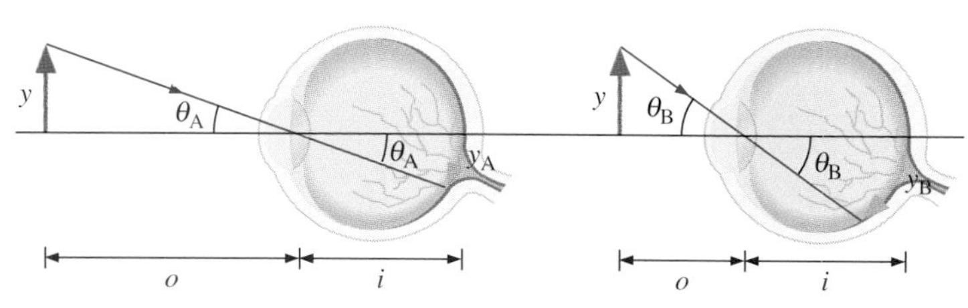

FIGURE 37-17 The apparent size of an object is determined by the size of its image on the retina. This is proportional to the angle that the object subtends at the eye.

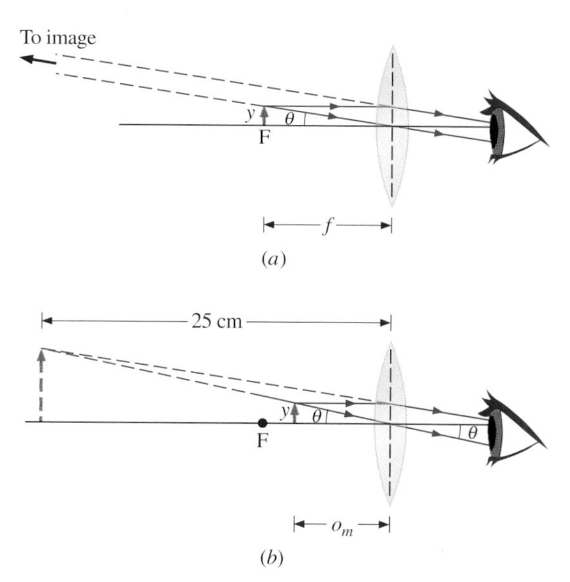

FIGURE 37-18 A simple magnifier with an image at (*a*) infinity and (*b*) the near point (see Example 37-5).

by the normal eye. If the focal length f of the lens is less than 25 cm, the angle θ subtended at the eye will be greater than $\theta_N \approx y/(25\text{ cm})$.* Hence there is an angular magnification. Notice that if the magnifier is removed, the eye cannot produce a focused image of the object on the retina since the object is closer than the near point.

The angular magnification depends on the focal length of the lens. From Fig. 37-18*a*, $\theta \approx y/f$, so from Eq. (37-7) the angular magnification of the simple magnifier with the image at infinity is

$$M \approx \frac{y/f}{y/(25\text{ cm})} = \frac{25\text{ cm}}{f}, \qquad (37\text{-}8a)$$

where f is expressed in centimeters. For example, if an object is placed at the focal point of a converging lens with $f = 10$ cm, $M = 25\text{ cm}/10\text{ cm} = 2.5$.

The eye seen through a magnifier.

EXAMPLE 37-5 Maximum Angular Magnification of a Simple Magnifier

(*a*) Is it necessary for an object to be placed at the focal point of a simple magnifier in order for its image on the retina to be enlarged? If not, what is the range of object distances allowed? (*b*) What is the corresponding range in the values of M?

SOLUTION (*a*) Since the eye sees the virtual image formed by the magnifier, it is only necessary that this image be *at the near point or beyond* in order to be focused on the retina. The nearest distance o_m that the object can be placed from the lens corresponds to the case shown in Fig. 37-18*b*, where the virtual image is at the near point of 25 cm, so $i = -25$ cm. Then from the lens equation,

$$\frac{1}{o_m} + \frac{1}{-25\text{ cm}} = \frac{1}{f},$$

which yields for the nearest object distance

$$o_m = \frac{(25\text{ cm})f}{25\text{ cm} + f}.$$

Therefore, the range of object distances for a magnifier is from $(25\text{ cm})f/(25\text{ cm} + f)$, when the image is at the near point, to f, when the image is at infinity. For example, if a magnifier has a focal length of 15 cm, the image of an object placed anywhere between 9.4 and 15 cm from the lens is magnified.

(*b*) Figure 37-18*b* shows that the virtual image subtends the same angle θ at the eye as does the object at o_m. Thus

$$\theta = \frac{y}{o_m}.$$

The maximum angular magnification of a simple magnifier is then

$$M = \frac{\theta}{\theta_N} = \frac{y/o_m}{y/25\text{ cm}} = \frac{25\text{ cm}}{o_m} = 1 + \frac{25\text{ cm}}{f}. \qquad (37\text{-}8b)$$

As the object is moved out toward the focal point of the lens, the angular magnification decreases to $(25\text{ cm})/f$, as given by Eq. (37-8*a*). The range in angular magnification M is therefore

$$1 + \frac{25\text{ cm}}{f} \geq M \geq \frac{25\text{ cm}}{f}.$$

*37-5 The Microscope

In its simplest form, a microscope is a combination of two converging lenses that produces highly magnified images of very small objects. The viewing lens is called the *eyepiece* or *ocular*,

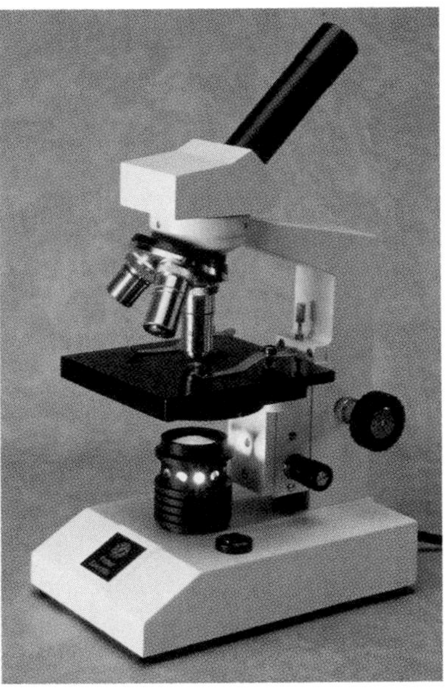

A modern microscope.

Fisher Scientific.

*The distance between the lens and the eye is assumed to be negligible.

and the lens closer to the object is known as the *objective* (Fig. 37-19).

Consider an object of height y. Its image on the retina of the unaided eye is largest when it is placed at the near point (25 cm) of the eye. Then the angle θ_N subtended at the retina is

$$\theta_N = \frac{y}{25 \text{ cm}}.$$

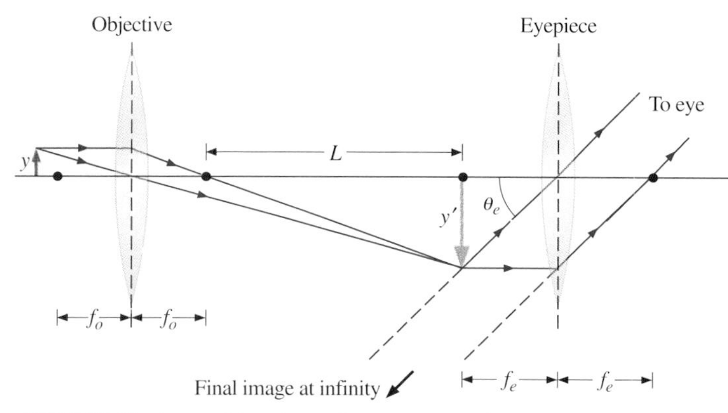

FIGURE 37-19 A combination of two converging lenses is used as a simple microscope. The viewed object is just beyond the focal point of the objective. The rays shown passing through the eyepiece are for construction purposes and do not represent continuations of the rays shown passing through the objective.

Now suppose we look at this object through a microscope by placing it just beyond the focal point of the objective, as shown in Fig. 37-19. The focal lengths of the objective and eyepiece are f_o and f_e, respectively, and L is the distance between the focal points of the lenses. The objective produces an enlarged real image of length y'. If the image is at the focal point of the eyepiece, this lens will act as a simple magnifier. It produces a final virtual image at infinity that subtends an angle at the retina given by

$$\theta_e = \frac{y'}{f_e}.$$

From the similar triangles immediately to the right of the objective lens of Fig. 37-19,

$$\frac{y}{f_o} = \frac{y'}{L},$$

so

$$\theta_e = \frac{yL}{f_o f_e}.$$

From Eq. (37-7), the angular magnification of the microscope is therefore

$$M = -\frac{\theta_e}{\theta_N} = -\frac{(25 \text{ cm})L}{f_o f_e}, \tag{37-9}$$

where the negative sign has been inserted to indicate that the image is inverted.

DRILL PROBLEM 37-6

The focal lengths of the objective and eyepiece in a microscope are 1.0 and 2.5 cm, respectively, and the two lenses are 15.0 cm apart. What is the angular magnification of the microscope?
ANS. −150.

*37-6 THE TELESCOPE

While a simple telescope has a lens arrangement similar to that of the microscope, the two instruments are very different in their applications. The microscope is used to examine minute objects that can be positioned close to the objective, while the telescope is used to view large, distant objects.

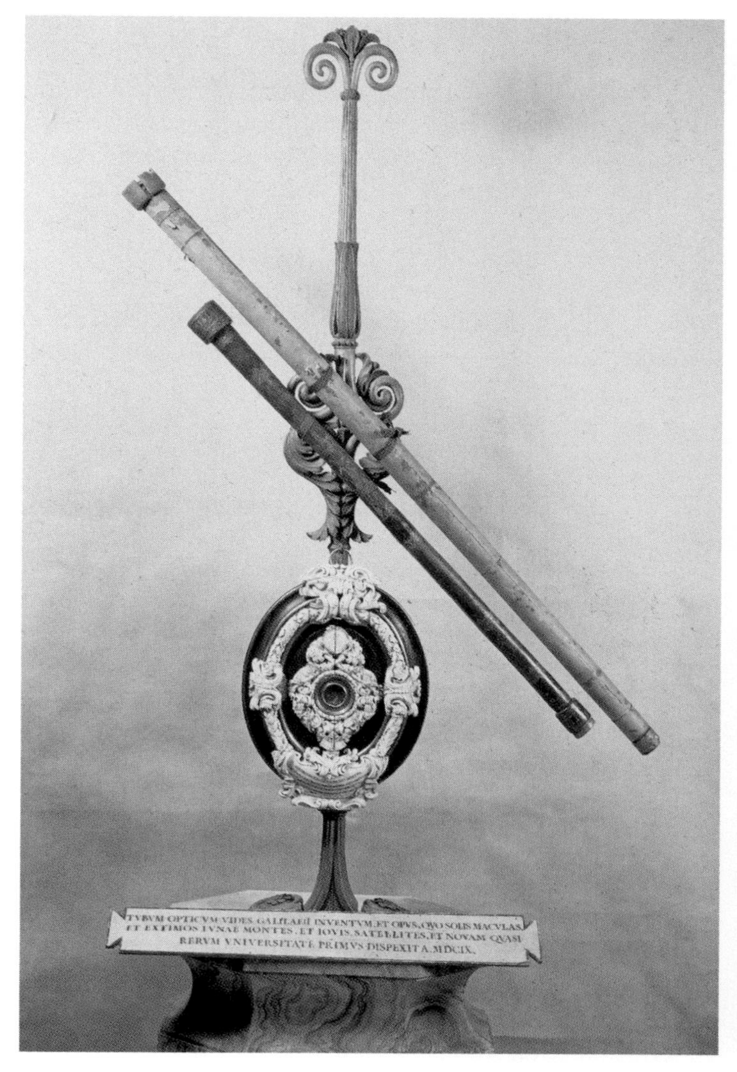

A telescope used by Galileo. With a telescope, Galileo studied the surface of the moon and discovered the moons of Jupiter.

Figure 37-20 shows a simple *refracting telescope* (or a *refractor*). The objective produces a real, inverted image of a distant object at its focal point. This location coincides with the focal point of the eyepiece, which then acts as a simple magnifier to

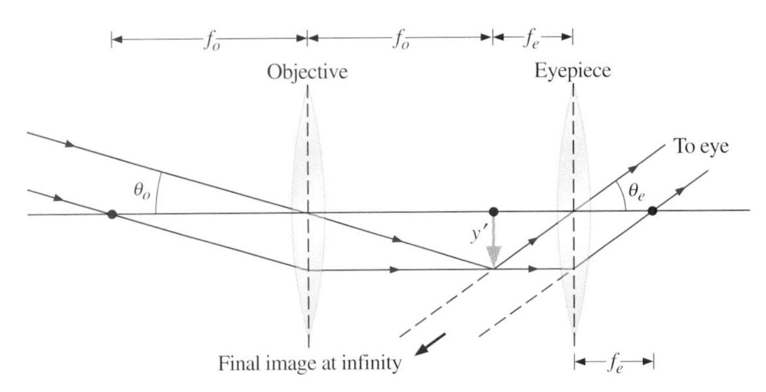

FIGURE 37-20 Like the microscope, a refracting telescope is a combination of two converging lenses. The focal points of the objective and the eyepiece are at a common location between the lenses. The rays shown passing through the eyepiece are for construction purposes.

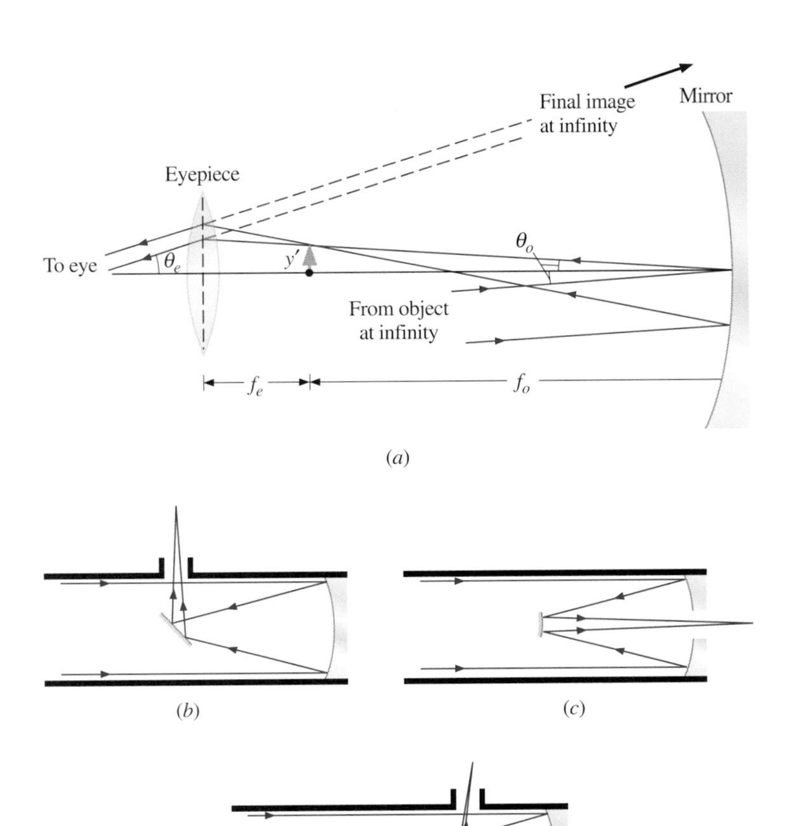

FIGURE 37-21 (*a*) A sketch of a prime-focus reflecting telescope. Other types of reflecting telescopes are (*b*) the Newtonian focus, (*c*) the Cassegrain focus, and (*d*) the coude focus. In all cases, light comes from an object at infinity on the left.

produce a final, enlarged image at infinity. When viewed without the telescope, the object subtends an angle θ_o at the retina. Its image seen through the telescope subtends an angle θ_e at the retina. For small angles,

$$\theta_o = \frac{y'}{f_o} \quad \text{and} \quad \theta_e = \frac{y'}{f_e}.$$

From Eq. (37-7), the angular magnification of the telescope is therefore

$$M = -\frac{\theta_e}{\theta_o} = -\frac{f_o}{f_e}. \tag{37-10}$$

Once again, a minus sign is included to designate that the image is inverted.

Another type of telescope is the *reflecting telescope* (or *reflector*). As Fig. 37-21*a* illustrates, it uses a spherical mirror rather than an objective lens to form an image. Since the eyepiece is placed in front of the mirror, it prevents some of the light falling on the telescope from reaching the mirror. Notice that the geometry of the reflected rays of this telescope is essentially the same as that for the refracted rays of the refracting telescope. Consequently, Eq. (37-10) also gives the angular magnification of the reflecting telescope. Sketches of various types of reflecting telescopes are shown in Fig. 37-21*b*, *c*, and *d*.

The "size" of a telescope is generally denoted by the diameter of the objective lens or the mirror. A larger telescope gathers more light and forms a brighter image and also has better "resolution" (discussed in the next chapter). For these reasons, astronomical telescopes are made as large as possible. The largest refracting telescope in the world is located at the Yerkes Observatory at the University of Chicago. Its objective lens has a diameter of 102 cm. Unfortunately, difficulties in producing large, high-quality lens surfaces and in supporting such massive pieces of glass limit the size of refractors. These problems are less severe for mirrors, and as a result, the largest astronomical telescopes are all reflectors. In the United States, a reflector with a 5.1-m-diameter mirror is located on Mount Palomar in California. Another noteworthy reflector is the 6.0-m telescope at Mount Pastukhov in Russia. The mirror alone of this telescope weighs approximately 42 tons! Finally, the Keck telescope (Fig. 37-22) atop Mauna Kea in Hawaii has a reflecting surface with a 10-m diameter. Since a single mirror of this size would warp under its own weight, the reflecting surface is

FIGURE 37-22 The mounting of the Keck telescope.

actually a composite of 36 closely packed hexagonal segments, all individually computer-controlled and aligned. This mirror actually weighs 2 tons less than the Palomar mirror, while having twice the diameter.

DRILL PROBLEM 37-7

A refracting telescope with angular magnification −20 has an objective with a focal length of 40.0 cm. (*a*) What is the focal length of the eyepiece? (*b*) What is the distance between the objective and the eyepiece?
ANS. (*a*) 2.0 cm; (*b*) 42.0 cm.

An image of Jupiter and its rings gathered by the Keck Observatory's 10-m primary mirror.

SUMMARY

1. **Thin lenses**
   (*a*) If an object is a distance o from a thin lens of focal length f, the image is located a distance i from the lens, where

$$\frac{1}{o} + \frac{1}{i} = \frac{1}{f}.$$

The image may be real or virtual. Its lateral magnification is $m = -i/o$. If $m > 0$, the image is erect; if $m < 0$, it is inverted.

(*b*) The focal length of a thin lens is given by the lensmaker's equation:

$$\frac{1}{f} = (n-1)\left(\frac{1}{R_1} - \frac{1}{R_2}\right).$$

(*c*) The following rules are used to construct the image formed by a lens by tracing the paths of light rays:

(i) An incident ray passing through a focal point is refracted so that it leaves parallel to the optical axis.

(ii) An incident ray parallel to the optical axis is refracted so that it passes through a focal point.

(iii) A ray passing through the center of the lens is not deflected.

*2. **Aberrations**
Spherical aberration occurs when reflected or refracted rays do not intersect at one point because they are only approximately paraxial. Chromatic aberration is caused by the variation of the index of refraction with the wavelength of light.

*3. **The eye**
   (*a*) Through refraction at the cornea and at the crystalline lens, an image of an object is formed on the retina of the eye.
   (*b*) The extremes of the range of the object distance for which a clear image is formed on the retina are called the far point and the near point.
   (*c*) The most common defects in vision are caused by distortions in the shape of the eyeball. Myopia occurs when the eyeball is too long from front to back; it is corrected with a diverging lens. Hyperopia occurs when the eyeball is too short; it is corrected with a converging lens. The gradual recession of the near point of the eye with age is called presbyopia; this is also corrected with a converging lens.
   (*d*) The size of an object's image on the retina is proportional to the angle that the object subtends at the eye.
   (*e*) The angular magnification of an optical device is

$$M = \frac{\theta}{\theta_N},$$

where θ is the angle subtended at the eye by the image produced by the device and θ_N is the angle subtended when the object is 25 cm from the eye.

*4. **Optical instruments**
   (*a*) The simple magnifier is a single converging lens used for magnification. If f is the focal length of the simple magnifier,

$$M = \frac{25\text{ cm}}{f}$$

for an image at ∞; and

$$M = 1 + \frac{25\text{ cm}}{f}$$

if the image is 25 cm from the eye.

(*b*) With the aid of the microscope, enlarged images of very small objects are formed on the retina. The angular magnification of the microscope is

$$M = -\frac{(25\text{ cm})L}{f_o f_e},$$

where f_o and f_e are the focal lengths, respectively, of the objective and the eyepiece and L is the distance between the focal points of the two lenses.

(*c*) A telescope is used to view large, distant objects. Its angular magnification is

$$M = -\frac{f_o}{f_e},$$

where f_o and f_e are the focal lengths, respectively, of the objective and the eyepiece.

QUESTIONS

Note: Unless stated otherwise in the questions and problems that follow, a "lens" is assumed to be a thin lens and the index of refraction of glass is assumed to be 1.50.

37-1. Why is a converging lens also known as a "positive" lens? Why is a diverging lens known as a "negative" lens?

37-2. Give the sign conventions for thin lenses and for spherical mirrors.

37-3. When a lens is turned around so that its other side faces the object, does the location of the image change?

37-4. Are the two foci of a lens always equidistant from the lens?

37-5. What happens to the focal length of a lens when it is immersed in water?

37-6. Describe how a virtual object for a lens can be produced.

37-7. Does a converging lens ever form a virtual image of a real object?

37-8. Does a diverging lens ever form a real image of a real object?

37-9. Are there any conditions for which a lens with convex surfaces becomes a diverging lens? for which a lens with concave surfaces becomes a converging lens?

37-10. How does the focal point of a converging lens for red light compare with its focal point for blue light? Answer the same question for a diverging lens.

37-11. How can you quickly measure the focal length of a converging lens? Could the same method be applied to determine the focal length of a diverging lens?

37-12. Devise a method for measuring the focal length of a diverging lens.

37-13. The thin piece of glass shown in the accompanying figure has parallel sides so that the radii of curvature of both surfaces are the same. What is the focal length of this "lens"?

QUESTION 37-13

37-14. How is the reversibility of light manifested in the lens equation?

37-15. Can the virtual image formed by a lens be photographed?

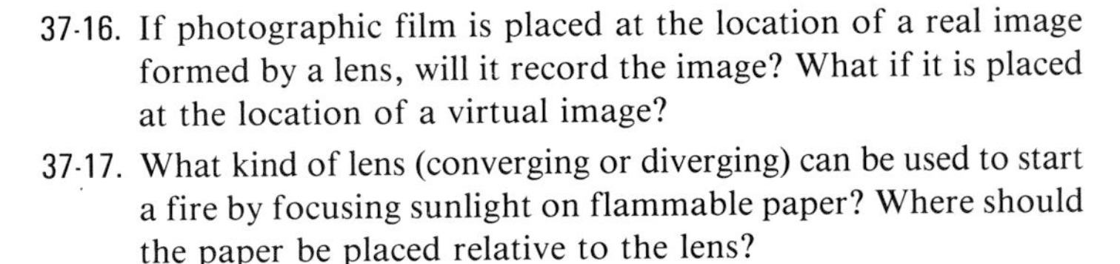

37-16. If photographic film is placed at the location of a real image formed by a lens, will it record the image? What if it is placed at the location of a virtual image?

37-17. What kind of lens (converging or diverging) can be used to start a fire by focusing sunlight on flammable paper? Where should the paper be placed relative to the lens?

37-18. Discuss what happens to the image of a nearby object when the central portion of a lens is obstructed.

37-19. Compare the images of a distant object formed by a converging lens and by an identical lens with a large hole drilled through its center.

37-20. A representation of a large astronomical telescope is shown in the accompanying figure. Notice that part of the light entering the instrument is obstructed by a small mirror used to redirect the rays to a viewer. Does this mean that only a portion of a star can be seen? How does the size of the obstruction affect the image?

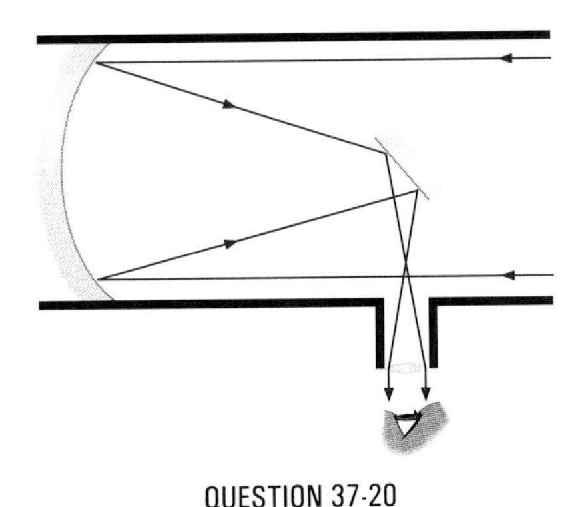

QUESTION 37-20

37-21. Compare angular magnification to lateral magnification.

37-22. Is the image formed on the retina upright or inverted?

37-23. Do eyeglasses affect the size of the image on the retina?

37-24. The accompanying figure shows a simple laboratory apparatus used to demonstrate the formation of images in the human eye. A glass lens can be placed at various positions (shown as A, B, and C) within a box. Water can be poured into the box, thus simulating the aqueous and vitreous humors. Suppose that when a converging lens is placed in position B and the box is filled with water, an image of a distant object is in sharp focus at the back of the box, as shown. This is the normal eye. Discuss whether the image would be behind, in front of, or remain focused at the back of the box for each of the following changes: (*a*) The lens is moved to position A; (*b*) the lens is moved to position C; (*c*) the water is removed from the box; (*d*) a liquid with a higher refractive index than water is poured into the box.

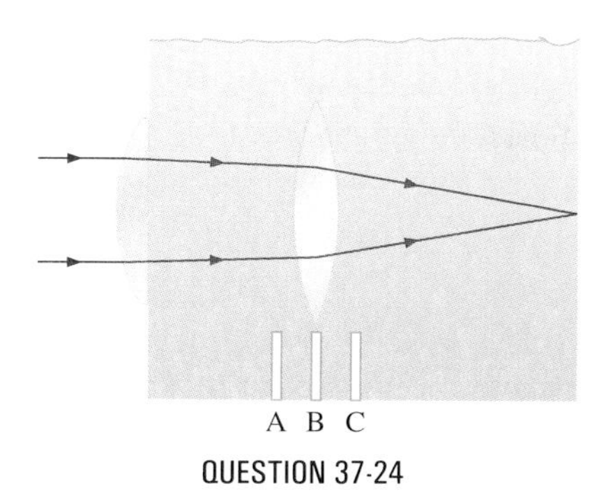

QUESTION 37-24

37-25. Suppose that by placing a corrective lens in front of the box of Ques. 37-24, we obtain a focused image on the "retina" even though the immersed lens is in position A. Is the corrective lens diverging or converging?

37-26. (*a*) Why are people with 20/20 vision farsighted under water? (*b*) Why might a nearsighted person have improved vision under water? (*Hint:* See Ques. 37-5.)

37-27. Poor underwater vision of the 20/20 eye can be corrected by wearing goggles. Explain.

37-28. Would you use the eyeglasses of a nearsighted or a farsighted person to focus sunlight?

37-29. Why is chromatic aberration an important consideration in lenses but not in mirrors?

37-30. Discuss the similarities and differences between a microscope and a telescope.

37-31. Does spherical aberration affect both lenses and mirrors?

37-32. Plastic bags filled with air are used as underwater lenses. What shape produces a converging lens? a diverging lens?

37-33. In terms of angular magnification, explain why objects are "closer than they appear" when viewed in the wide-angle mirror on the passenger side of a car.

PROBLEMS

Lenses

37-1. Where is the image of a fly that is itself 15 cm from the surface of a glass sphere of radius 20 cm? (See the accompanying figure.)

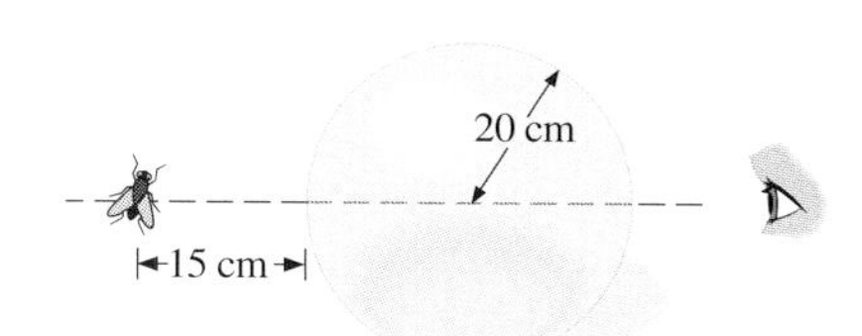

PROBLEM 37-1

37-2. Both ends of a glass rod 10 cm in diameter are ground to convex hemispherical surfaces, as shown in the accompanying figure. The overall length of the rod is 40 cm. If a small object is placed 20 cm to the left of the vertex of the left end of the rod, where is the image formed?

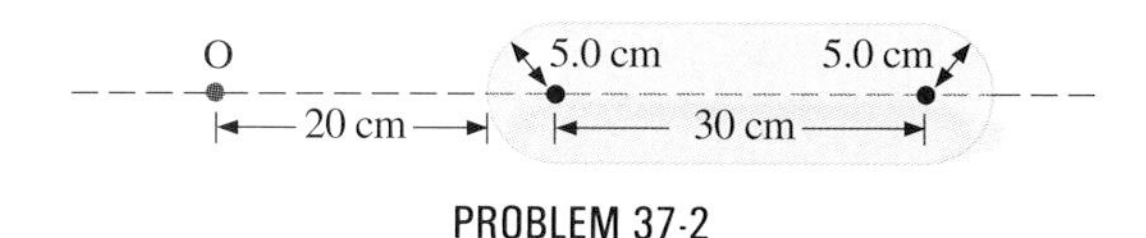

PROBLEM 37-2

37-3. Replace the left spherical surface of Prob. 37-2 by a convex spherical surface of radius 10 cm. Where is the image now formed? Assume that the section of the rod between the centers of the hemispherical surfaces is still 30 cm long.

37-4. The glass rod shown in the accompanying figure has one end ground flat and the other ground into a concave spherical surface of radius 10 cm. If an object is placed 10 cm to the left of the flat surface, where is its image?

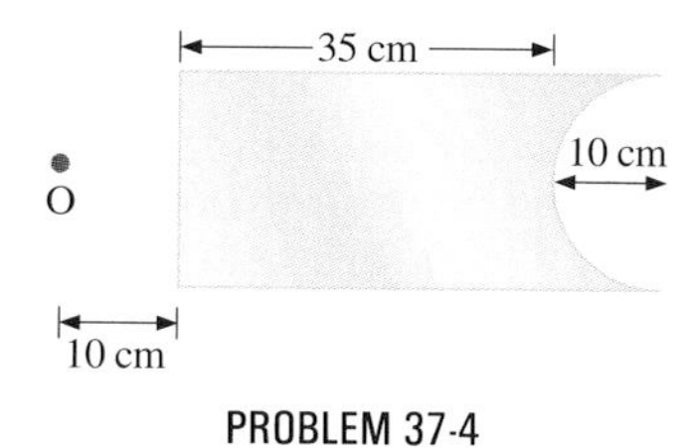

PROBLEM 37-4

37-5. A thin beam of parallel light rays is incident on a solid glass sphere of radius 6.0 cm, as shown in the accompanying figure. Where are the rays brought to a focus?

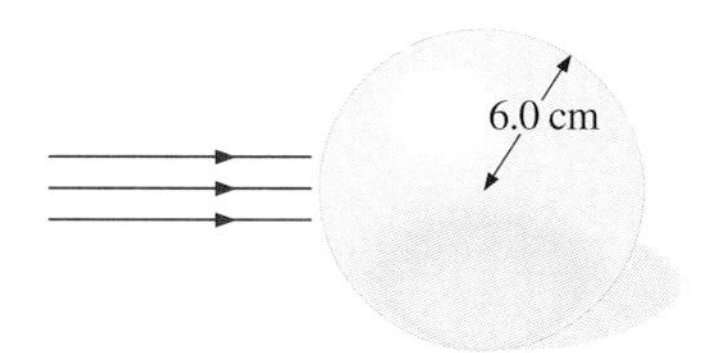

PROBLEM 37-5

37-6. Repeat Prob. 37-5 for the sphere immersed in water.

37-7. Sketch the lenses and calculate their focal lengths when they are composed of two surfaces ground as (*a*) identical convex surfaces of radius 20 cm, (*b*) a convex surface of radius 20 cm and a convex surface of radius 10 cm, (*c*) a convex surface of radius 20 cm and a flat surface, (*d*) identical concave surfaces of radius 20 cm, (*e*) a concave surface of radius 20 cm and a flat surface, (*f*) a convex surface of radius 20 cm and a concave surface of radius 10 cm, and (*g*) a convex surface of radius 10 cm and a concave surface of radius 20 cm.

37-8. A plano-convex lens has one flat side and one convex side, as shown in the accompanying figure. If the lens is made of plastic with $n = 1.6$ and has a focal length of 25 cm, what is the radius of curvature of the convex surface?

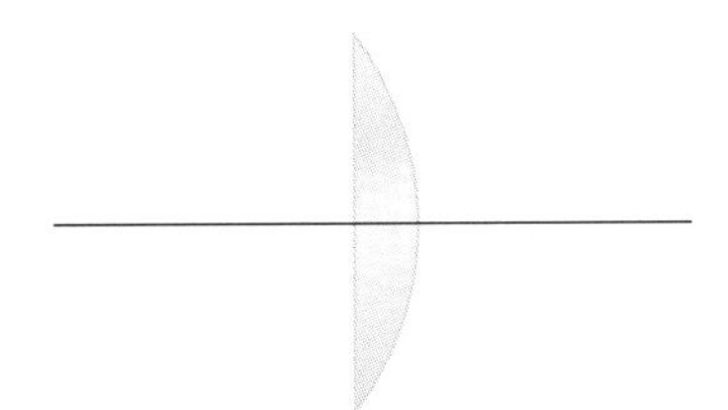

PROBLEM 37-8

37-9. Show that when a thin lens of refractive index n is used in a medium of refractive index n', the focal length of the lens is given by

$$\frac{1}{f} = \left(\frac{n}{n'} - 1\right)\left(\frac{1}{R_1} - \frac{1}{R_2}\right),$$

where R_1 and R_2 are the radii of curvature of the two spherical surfaces of the lens.

37-10. (*a*) Use the formula obtained in Prob. 37-9 to find the ratio of the focal length of a lens in air to the focal length of the same lens in water. (*b*) What is this ratio for a typical glass lens?

37-11. Show that the focal length of a thin lens is not changed when the lens is rotated so that the left and right surfaces are interchanged.

37-12. As an object is moved from the surface of a thin converging lens to a focal point, over what range does the image distance vary?

37-13. When a converging lens is immersed in water, the image of a distant object is focused 25 cm from the lens. What is the focal length of the lens in air? (See Prob. 37-9.)

37-14. Complete the following table for thin lenses made of glass. All distances are in centimeters.

Type of Lens	Focal Length	Object Distance	Image Distance	Real/Virtual Image	Lateral Magnification
	−20	+10			
	+20	+10			
		+20	+30		
			−30		+2.0
	+25				−.67
	−10	−30			
	−20	−10			

37-15. Verify your calculations of Prob. 37-14 using ray tracing.

37-16. Find the image location and lateral magnification [except in part (*f*)] using ray tracing for an object that is (*a*) 30 cm in front of a lens, with $f = +20$ cm; (*b*) 15 cm in front of a lens, with $f = +20$ cm; (*c*) 30 cm in front of a lens, with $f = -20$ cm; (*d*) 10 cm behind ($o < 0$) a lens, with $f = -20$ cm; (*e*) 30 cm behind a lens, with $f = -20$ cm; and (*f*) at infinity, with $f = -20$ cm.

37-17. A boy uses a converging lens to project sunlight onto a piece of paper. When the paper is placed 20 cm from the lens he finds that the rays become sharply focused. Later he examines an insect by placing the lens 10 cm from it. Compute the location and lateral magnification of the image of the insect.

37-18. By tracing the appropriate rays in the situations shown in the accompanying figure, determine the unknown quantity (object distance, image distance, or focal length).

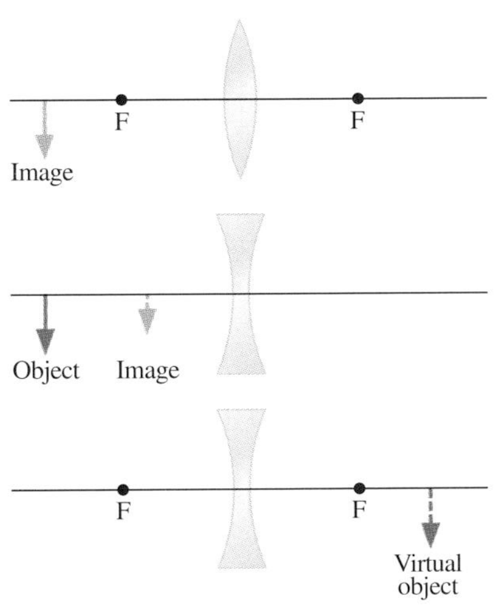

PROBLEM 37-18

37-19. A diverging lens is made of material with $n = 1.3$ and has identical concave surfaces of radius 20 cm. The lens is immersed in a transparent medium with $n = 1.8$. (*a*) What is now the focal length of the lens? (See Prob. 37-9.) (*b*) What is the minimum distance that an immersed object must be from the lens so that a real image is formed?

37-20. Equation (37-5) is called the *Gaussian form* of the lens equation. With x_o as the distance of the object from the first focal point and x_i as the distance of the image from the second focal point (see the accompanying figure), the *Newtonian form* of the lens equation is $x_o x_i = f^2$. Use Eq. (37-5) to derive this form.

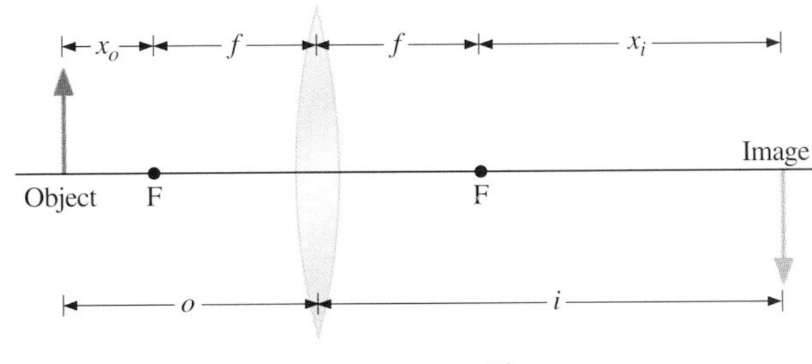

PROBLEM 37-20

37-21. An object on a photographic slide is 1.0 mm high. If you wanted to produce an image 30 cm high on a screen 5.0 m away from the slide, what type of lens would you use and where would you place it?

37-22. A diverging lens is held 20 cm above this page. If the image of the print is seen 10 cm beneath the lens, (*a*) what is the focal length of the lens? (*b*) What is the lateral magnification of the image? (*c*) Is the image real or virtual? upright or inverted?

37-23. Suppose that in Prob. 37-22 the diverging lens is replaced by a converging lens of focal length 25 cm. (*a*) Describe the image for this case. (*b*) Also use a ray diagram to obtain the image.

37-24. Each frame on a roll of movie film has a width of 70 mm. A projection lens of focal length 300 mm is used to produce the images of the frames on a screen 50 m away. (*a*) How far must the lens be from the film? (*b*) What is the width of the image on the screen?

37-25. A converging lens placed 25.0 cm from an object produces an image on a screen. When the object is moved 4.0 cm closer to the lens, the screen must be moved 3.0 cm farther away from the lens in order to retain a sharp image. What is the focal length of the lens?

37-26. So that sharp images of objects at various distances are formed on the film, a movable camera lens (known as a telephoto lens) is used to produce the images. If a camera with a telephoto lens of focal length 250 mm is used to take pictures of objects located between 3.0 and 50.0 m, over what distance must the lens be movable?

37-27. Two thin lenses of focal lengths f_1 and f_2 are placed in contact so that they have the same optical axis. (See the accompanying figure.) Show that this combination is equivalent to a single thin lens whose focal length is $f_1 f_2/(f_1 + f_2)$.

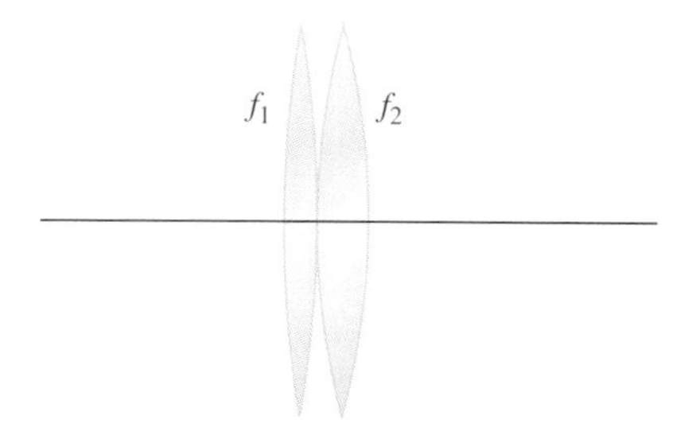

PROBLEM 37-27

37-28. Use the results of the previous problem to determine the overall focal length of the following combinations of two lenses in contact with each other: (*a*) two converging lenses with focal lengths 10 and 200 cm, (*b*) two diverging lens with focal lengths −40 and −80 cm, and (*c*) a converging lens with focal length 45 cm and a diverging lens with focal length −45 cm.

37-29. Suppose that a converging lens of focal length 20 cm is placed in contact with the lens combination of part (*b*) of Prob. 37-28. (*a*) What is the overall focal length of the combination? (*b*) Derive an equation similar to that of Prob. 37-27, but for three lenses in combination.

37-30. A converging lens with a focal length of 25 cm is placed 10 cm to the left of a diverging lens with a focal length of −15 cm. If an object is placed 10 cm to the left of the converging lens, where is the image formed by the combination? What is the overall lateral magnification?

37-31. An eyepiece consists of two positive thin lenses separated by a distance of 3.0 cm. Both lenses have a focal length of 6.0 cm. Where are the focal points of the eyepiece?

37-32. An object is placed 50 cm in front of a converging lens of focal length 25 cm, which is 100 cm in front of a plane mirror. (*a*) If you look through the lens toward the mirror, where will you see the image? (*b*) Is this image real or virtual? inverted or upright? (*c*) What is its lateral magnification?

37-33. An object is located 20 cm to the left of a converging lens with $f = 10$ cm. A second identical lens is placed to the right of the first lens and then moved until the image it produces is identical in size and orientation to the object. What is the separation between the lenses?

The Eye

37-34. (*a*) The far point of an eye is 150 cm. What corrective lens is needed for this eye to see distant objects? (*b*) Another eye has a near point of 75 cm. What corrective lens is needed so that this eye can be used for reading a book placed 25 cm from the eye?

37-35. A baseball umpire wears lenses with a focal length of 60 cm when he reads. What is his near point?

37-36. (*a*) Estimate the focal length of the lens system (the cornea, aqueous humor, lens, vitreous humor) of a human eye that focuses distant objects onto the retina. The distance from the lens to the retina is about 20 mm. (*b*) Compare this focal length with that of the lens needed to correct for distant vision in an eye whose far point is 40 cm.

37-37. In our estimates of the focal lengths of eyeglass lenses, we have ignored the distance between the lenses and the eyes. Recalculate the focal lengths of the corrective lenses of Prob. 37-34, assuming that the lenses are placed 2 cm in front of the eyes.

37-38. Although eyeglasses for myopia change the image location of the object being viewed, they do not alter the size of the image on the retina. That is, myopic people who wear glasses don't perceive the sizes of objects any differently than their friends with perfect vision. In Prob. 37-36, you found the focal length of the eye to be about 2 cm. In Prob. 37-37, the eyeglass lenses were assumed to be about 2 cm from the eyes. Use these two facts to explain why the corrective lens does not cause size distortion. (Imagine what difficulties such distortions would cause a person whose two eyeglass lenses were different!)

The Simple Magnifier

37-39. A magnifier produces an angular magnification of 3.3 for a normal relaxed eye. What is the maximum angular magnification when the magnifier is used with a normal eye?

37-40. With his eye relaxed, a philatelist is examining a stamp through a magnifier with $f = 10$ cm. (*a*) What is the angular magnification of the magnifier? (*b*) How far is the magnifier from the stamp under observation? (*c*) Wanting to study the stamp even more closely, the philatelist moves the magnifier until the image of the stamp is as large as possible and can still be seen clearly. What is the angular magnification of the magnifier now? (*d*) For this case, how far is the magnifier from the stamp?

37-41. A magnifier produces a maximum angular magnification of 4.0 when used by a teenager with a near point of 20 cm. (*a*) What is its maximum angular magnification when used by a middle-aged man with a near point of 50 cm? (*b*) Compare the sizes of the images on the retinas of the two observers.

The Microscope and the Telescope

37-42. A specimen subtends an angle of 5.0×10^{-3} rad when placed 25 cm from the unaided eye. What is its angular size when viewed through a microscope whose angular magnification is −160?

37-43. The angular magnification of a microscope is −1000. The focal points of its objective and its eyepiece are separated by 18 cm. If the focal length of the objective is 0.35 cm, what is the focal length of the eyepiece?

37-44. A compound microscope has three interchangeable objectives with focal lengths 2.0, 5.0, and 10.0 mm. The eyepiece has a 2.5-cm focal length, and the distance between the focal points of the objective and the eyepiece is 20 cm. What are the angular magnifications possible with this microscope?

37-45. A student uses a microscope with an objective of focal length 2.0 mm and an eyepiece of focal length 2.5 cm. The distance between the focal points of the two lenses is 20 cm. What is the angular magnification of the microscope?

37-46. The distance between the two lenses of a microscope is 19.00 cm. The focal lengths of the objective lens and the eyepiece are 0.50 and 2.50 cm, respectively. What is the angular magnification of the microscope?

37-47. A microscope has a barrel (the tube holding the two lenses at its opposite ends) 18.0 cm long and an eyepiece with a 1.5-cm focal length. What focal length must the objective have if the angular magnification of the microscope is −300?

37-48. Suppose you had a long cardboard tube that once held wrapping paper and that you also had two lenses of focal lengths 41.0 and 3.0 cm. (*a*) Assuming the lenses just fit inside the tube, describe how you could use these components to make a telescope. (*b*) What would be the angular magnification of the telescope?

37-49. The angular magnification of a telescope is −40, and the diameter of its objective lens is 8.0 cm. Suppose the telescope is aimed at a distant point source so that the source is on the axis of the telescope. What is the minimum diameter of the eyepiece required to collect all the light entering the objective lens from the source?

37-50. A telescope has an angular magnification of −120 and a barrel 1.20 m long. What are the focal lengths of the eyepiece and objective lenses?

37-51. The accompanying figure shows a sketch of a *Galilean telescope*, with both the object and its final image at infinity. Show that the final image is virtual and erect and that the angular magnification of the telescope is $-f_o/f_e$.

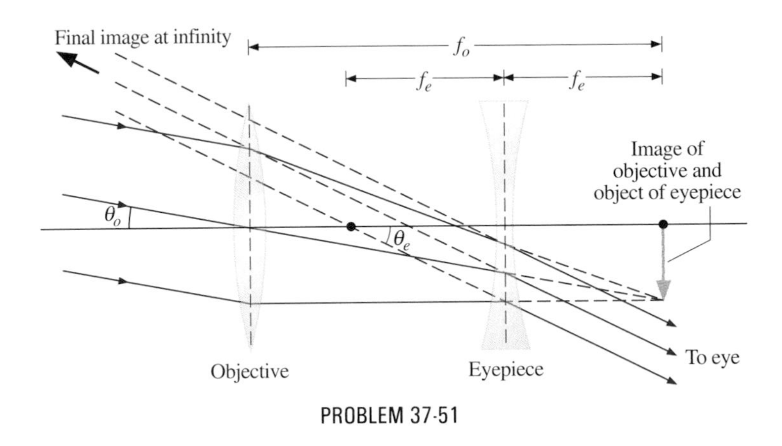

PROBLEM 37-51

37-52. The reflecting telescope of Fig. 37-21*a* is made using a mirror of radius of curvature 0.60 m and an eyepiece with focal length 2.0 cm. What is the angular magnification of the telescope?

37-53. The Hale reflecting telescope on Mount Palomar has a mirror of focal length 1680 cm. If the image of the mirror is observed with an eyepiece whose focal length is 1.20 cm, what is the angular magnification of the telescope?

General Problems

37-54. Suppose that the sphere of Prob. 37-1 is made of a transparent substance of unknown index of refraction. An image of the fly is observed 150 cm to the right of the right surface of the sphere. What is the index of refraction of the material?

37-55. Suppose you are looking straight through a glass pane of refractive index n at an object. The object is a distance d from the pane whose thickness is x. (See the accompanying figure.) (*a*) Show that you see the image of the object at $i = -x/n - d$. (*b*) Now argue that if the pane is thin, the image is at the same point as the object.

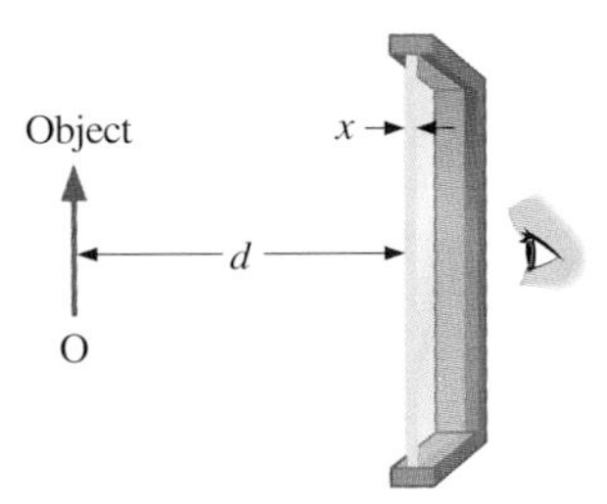

PROBLEM 37-55

37-56. An object and a screen are a distance d apart. (*a*) Find the two locations between the object and screen where a converging lens of focal length f can be placed to produce an image on the screen. (*b*) What is the magnification of the image for each location? (*c*) What is the maximum value of f if an image is to be formed?

37-57. Suppose that the plano-convex lens of Prob. 37-8 is part of an aquarium as shown in the accompanying figure. Calculate the position and lateral magnification of the image of a fish if it swims by at a distance of 10 cm from the lens.

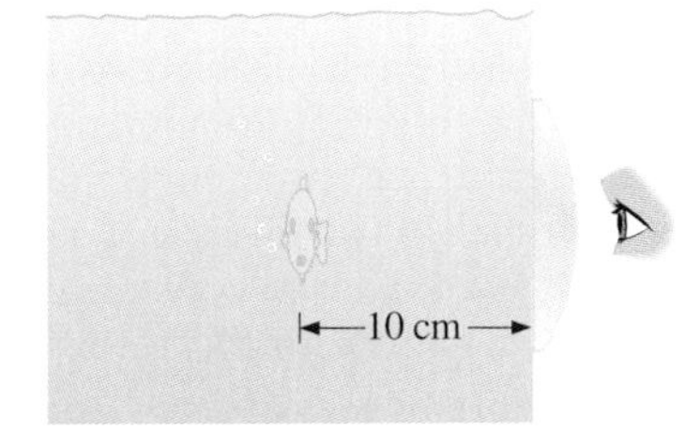

PROBLEM 37-57

37-58. A solid, transparent plastic ($n = 1.8$) sphere of radius 10 cm is floating in a pond. Assuming that the pond is deep enough so that light rays from the bottom that enter the sphere are essentially parallel, where is the image, as seen through the sphere, of a coin resting on the bottom?

37-59. Two converging lenses, each of focal length 30 cm, are placed 35 cm apart. An object is positioned 50 cm in front of one of the lenses. (*a*) Where is the image formed by the second lens located? (*b*) What is the overall lateral magnification? (*c*) Find the image with a ray diagram.

37-60. Consider the combination of a converging lens and a convex spherical mirror shown in the accompanying figure. The focal length of the lens is 25 cm, the radius of curvature of the mirror is 20 cm, and an object is placed 75 cm in front of the lens. Describe the image seen by an observer looking through the lens from its left side.

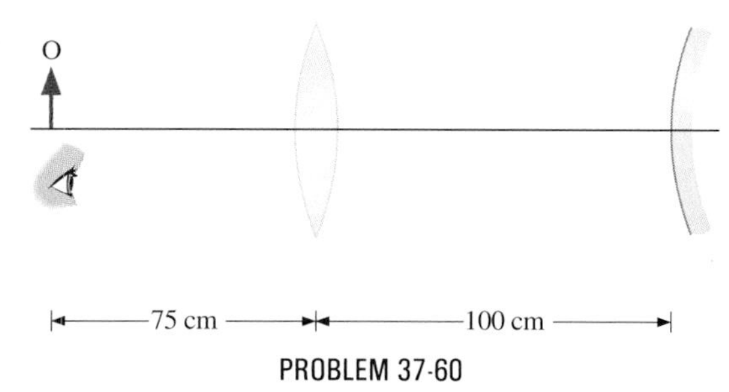

PROBLEM 37-60

37-61. Suppose you are a spectator at a football game watching a 2.0-m-tall quarterback from a seat 80 m away. At the same time, your friend is sitting in the living room watching a 15-cm image of the quarterback on a television screen 3.0 m away. For which of you is the image of the quarterback on the retina larger?

37-62. A nearsighted student wears contact lenses to correct for a far point that is 400 cm from her eyes. Her near point when she is not wearing the contact lenses is 20 cm. What is her near point when she is wearing the lenses?

37-63. A detective examines a fingerprint with a magnifier such that the image is at his near point, which is 30 cm. If the angular magnification for this position is 3.5, what is the focal length of the magnifier?

37-64. The objective and eyepiece of a microscope have focal lengths 1.0 and 3.0 cm, respectively. An object is in sharp focus when it is 1.1 cm from the objective, with the image at infinity. (*a*) What is the distance between the two lenses? (*b*) What is the angular magnification of the microscope?

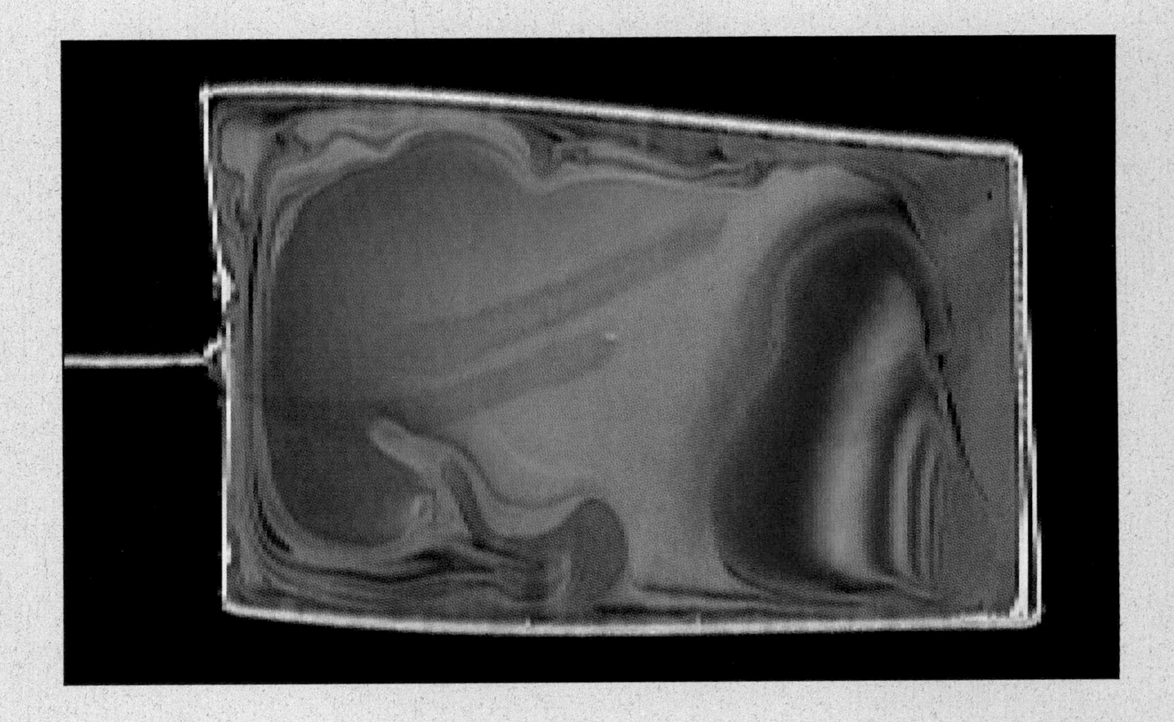

Interference pattern seen in a soap film.

CHAPTER 38 INTERFERENCE AND DIFFRACTION

PREVIEW

In this chapter we investigate phenomena that are a direct consequence of the wave nature of light. The main topics to be discussed here are the following:

1. **Double-slit interference experiment.** The interference of Huygens wavelets emanating from two closely spaced slits is described and explained. We derive the formulas used to calculate the positions of interference maxima and minima, along with the intensity distribution of the interference pattern.
2. **Multiple-slit interference patterns.** We investigate the intensity distribution of the interference pattern formed by Huygens wavelets emanating from *N* closely spaced slits.
3. **Interference in thin films.** We study the interference patterns produced by light reflecting off the two surfaces of a thin film.

*4. **The Michelson interferometer.** This device, which uses two interfering light beams to make very precise measurements of length, is considered.

*5. **Coherence.** We discuss the conditions that a light source must satisfy if its secondary Huygens wavelets are to be coherent.

6. **Diffraction.** The conditions necessary for Fraunhofer diffraction and for Fresnel diffraction are given.
7. **Fraunhofer diffraction by a single slit.** The intensity pattern produced by this type of diffraction is obtained, and the positions of the maxima and minima of the pattern are determined.

*8. **Fraunhofer diffraction by *N* slits.** The intensity pattern produced by this type of diffraction is given. The positions of the maxima and minima of this pattern are also found.

9. **Resolution.** We discuss the effects of diffraction at a circular opening on the ability of an optical instrument to distinguish the images of two objects.
10. **Diffraction grating.** This device, which is used to measure the wavelengths of light, is considered.

In geometric optics the propagation of light is described by rays that travel in straight lines and obey the laws of reflection and refraction. However, the rules of geometric optics are only valid when the wavelength of light is much smaller than the dimensions of the apertures or barriers involved. When this is not the case, phenomena arising from the wave nature of light are observed. Such phenomena fall under the general heading of physical optics.

38-1 Double-Slit Interference Experiment

We have already discussed how superposed mechanical waves can interfere constructively or destructively, depending on their phase difference. Since light is an electromagnetic wave, it must also exhibit interference effects. The first successful scientific investigations of optical interference were carried out by Thomas Young in 1801. In his famous experiment, Young passed sunlight through a pinhole on a board. The emerging beam fell on two pinholes on a second board. The light emanating from the two pinholes then fell on a screen where a pattern of bright and dark spots was observed. These fringes can only be explained if light is assumed to propagate as *waves*. Similar effects are found in mechanical waves. As an example, Fig. 38-1 shows circular water waves, generated by two synchronous vibrators in a ripple tank, adding and canceling at various locations.

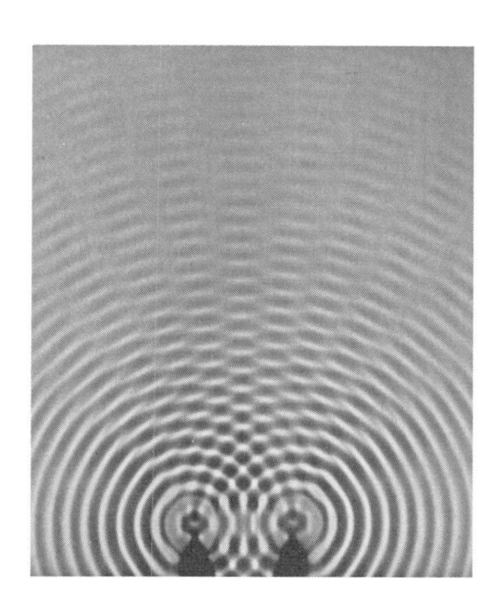

FIGURE 38-1 Photograph of an interference pattern produced by circular water waves in a ripple tank. Two thin plungers are vibrated up and down in phase at the surface of the water. Circular water waves are produced by and emanate from each plunger. The points where the water is calm (corresponding to destructive interference) are clearly visible.

We can analyze double-slit interference with the help of Fig. 38-2, which depicts an apparatus similar to Young's. Light from a monochromatic source falls on a slit S_0. The light emanating from S_0 is incident on two other slits S_1 and S_2 that are equidistant from S_0. A pattern of *interference fringes* on the screen is then produced by the light emanating from S_1 and S_2. All slits are assumed to be so narrow that they can be considered secondary point sources for Huygens wavelets. S_1 and S_2 are a distance d apart ($d \leq 1$ mm), and the distance between the screen and the slits is D (≈ 1 m), which is much greater than d.

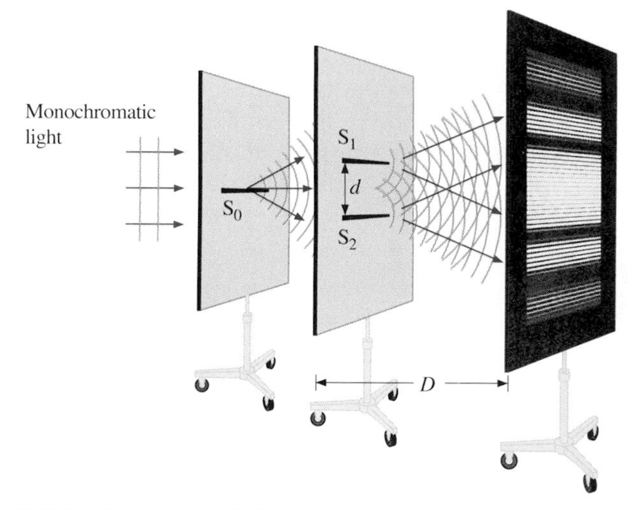

FIGURE 38-2 The double-slit interference experiment using monochromatic light and narrow slits. Fringes produced by interfering Huygens wavelets from slits S_1 and S_2 are observed on the screen.

Since S_0 is assumed to be a point source of monochromatic light, the secondary Huygens wavelets leaving S_1 and S_2 always maintain a constant phase difference (zero in this case because S_1 and S_2 are equidistant from S_0) and have the same frequency. The sources S_1 and S_2 are then said to be *coherent* (to be considered in detail in Sec. 38-5). Also, because S_1 and S_2 are the same distance from S_0, the amplitudes of the two Huygens wavelets are equal.

Figure 38-3 shows light rays from the two sources meeting at an arbitrary point P in space. From the principle of superposition, the resultant wave at this point is the sum of the two waves corresponding to the rays. If the two waves are to oscillate in phase at P, the difference $r_2 - r_1$ in their path lengths from their sources to P must be an integral number of wavelengths:

$$r_2 - r_1 = m\lambda \qquad (m = 0, \pm 1, \pm 2, \ldots). \tag{38-1}$$

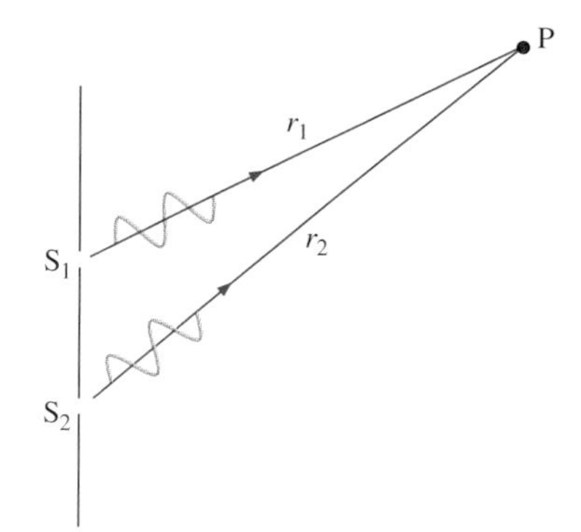

FIGURE 38-3 The path difference $r_2 - r_1$ between the Huygens wavelets from S_1 and S_2 at an arbitrary point P determines the intensity of the resultant wave there.

At any point where this equation is satisfied, the resultant wave has an amplitude twice that of its constituent waves, the light intensity is a maximum, and we have *constructive interference*. However, if

$$r_2 - r_1 = (m + \tfrac{1}{2})\lambda \qquad (m = 0, \pm 1, \pm 2, \ldots), \tag{38-2}$$

the resultant intensity is zero, which corresponds to *destructive interference*. Finally, at points not satisfying either Eq. (38-1) or Eq. (38-2), the intensities of the resultant waves lie somewhere between zero and the maximum. The addition of two harmonic waves for both constructive and destructive interference is illustrated in Fig. 38-4.

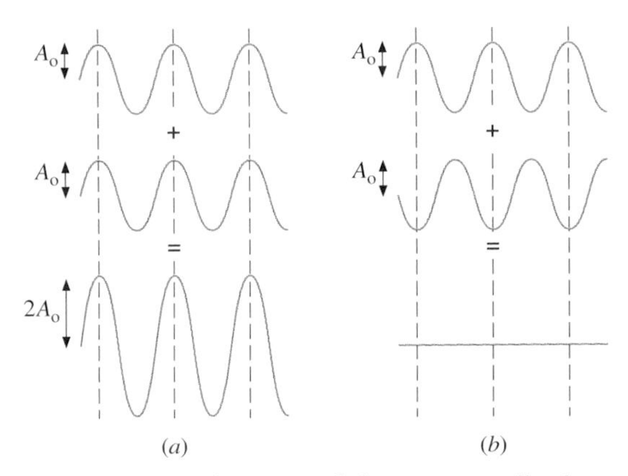

FIGURE 38-4 Two harmonic waves of the same amplitude and frequency undergoing (*a*) constructive interference and (*b*) destructive interference.

The wavefronts of the Huygens wavelets emanating from S_1 and S_2 are depicted by the blue curves in Fig. 38-5. The spacing between the solid blue curves (or the dashed blue curves) is the wavelength λ. Hence any adjacent solid and dashed curves are out of phase by π rad. Where solid (or dashed) curves intersect, the wavelets are in phase and we have constructive interference. Where a solid curve intersects a dashed curve, the wavelets are completely out of phase and destructive interference occurs.

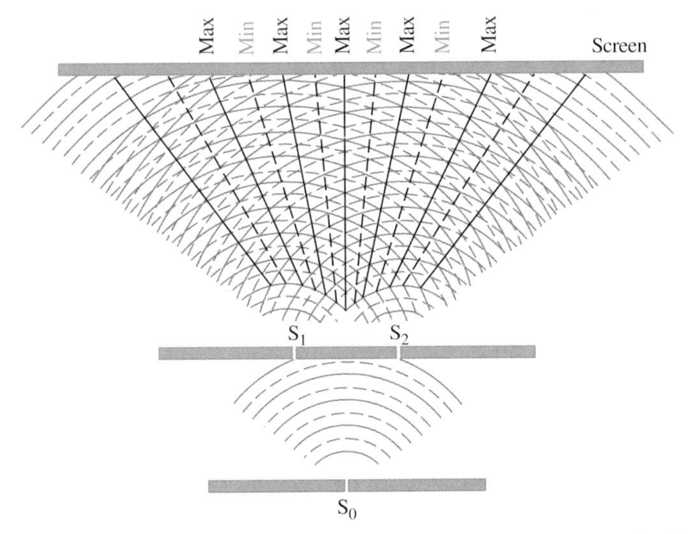

FIGURE 38-5 Interference of the Huygens wavelets in the double-slit experiment. The solid red lines represent points of constructive interference. The dashed red lines represent points of destructive interference.

The collections of points where constructive interference and destructive interference occur are represented by the solid and dashed red lines, respectively. Although the red lines appear to be straight in the figure, they are actually hyperbolic. By definition, a hyperbola is made up of points whose difference in distance from two fixed points (the "foci") is constant. This condition is satisfied by Eqs. (38-1) and (38-2). Hence the points of constructive and destructive interference may be represented by a series of hyperbolas whose foci are at S_1 and S_2.

Suppose that a screen is placed in front of the slits to catch the light. It is clear from Fig. 38-5 that constructive interference is observed wherever the solid red lines intersect the screen. The centers of the bright fringes are located at these points. Between the points of constructive interference are points of destructive interference where the dashed red lines intersect the screen. These correspond to the centers of the dark fringes. The centers of the bright and dark fringes are known as the *maxima* and *minima* respectively.

Figure 38-6*a* shows the light waves from S_1 and S_2 meeting at an arbitrary point P on the screen. Since $D \gg d$, the two light rays are approximately parallel with a path difference $r_2 - r_1 = d \sin \theta$, as designated in Fig. 38-6*b*. If P is located at

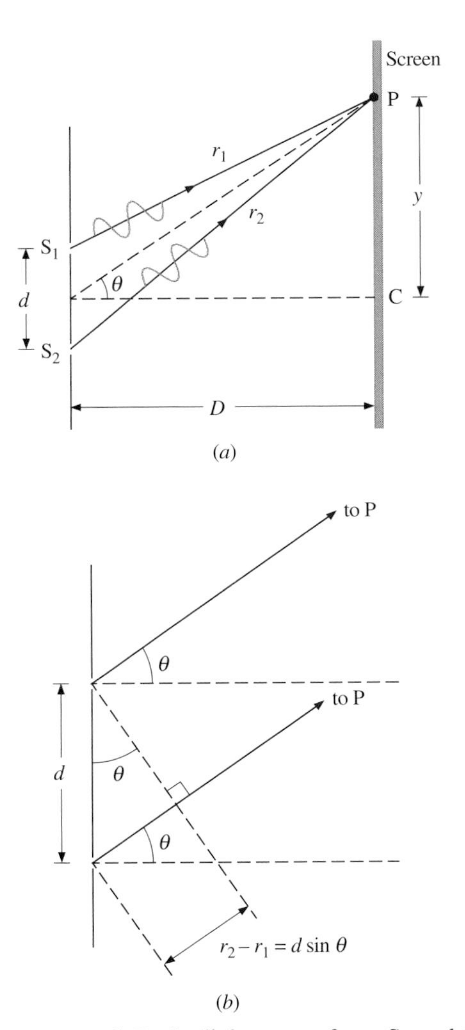

FIGURE 38-6 (*a*) To reach P, the light waves from S_1 and S_2 must travel different distances. (*b*) The path difference between the two rays is $d \sin \theta$.

a maximum, Eq. (38-1) gives for the positions of the interference maxima

$$d \sin \theta = m\lambda \qquad (m = 0, \pm 1, \pm 2, \ldots). \qquad (38\text{-}3a)$$

The bright fringe corresponding to the integer m is called the *mth-order* (or just *mth*) *bright fringe*. The bright fringe for $m = 0$ is known as the *central fringe*, and its center (point C in Fig. 38-6a) is called the *central maximum*. Higher-order bright fringes are situated symmetrically about the central fringe.

Typically, θ is small enough that $\sin \theta \approx \tan \theta = y/D$, where y is the distance from the central maximum. Equation (38-3a) may then be written as

$$y = \frac{m\lambda D}{d} \qquad (m = 0, \pm 1, \pm 2, \ldots). \qquad (38\text{-}3b)$$

At an interference minimum, the difference in the lengths of the paths of the two rays is $(m + 1/2)\lambda$, so

$$d \sin \theta = (m + \tfrac{1}{2})\lambda \qquad (m = 0, \pm 1, \pm 2, \ldots), \qquad (38\text{-}4a)$$

and the positions of the interference minima are given by

$$y = \left(m + \frac{1}{2}\right)\frac{\lambda D}{d} \qquad (m = 0, \pm 1, \pm 2, \ldots). \qquad (38\text{-}4b)$$

The first minima ($m = 0$) are adjacent to the central maximum on either side.

The assumption that parallel rays meet at the same point on the screen is approximated experimentally by positioning the screen and slits far apart compared with the distance between the slits. However, we can do even better by placing a converging lens between the slits and screen as illustrated in Fig. 38-7. Since the screen is in the focal plane of the lens, parallel rays leaving the slits are focused at the same point on the screen.

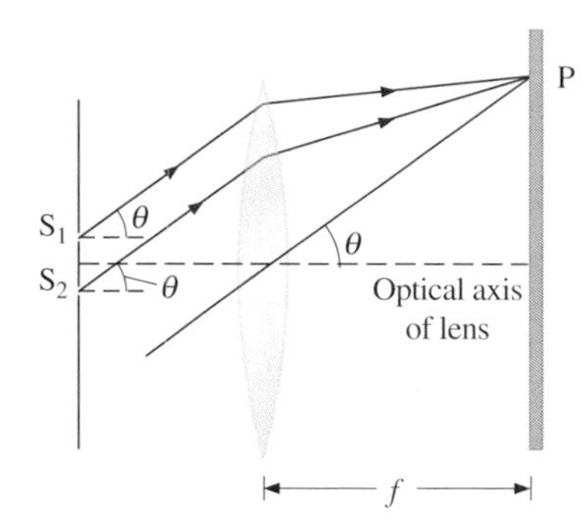

FIGURE 38-7 With a converging lens, parallel rays from S_1 and S_2 meet at the same point on the screen. By drawing a line through the lens center parallel to the rays, we find where they converge on the screen. Note that now $D = f$ in the interference equations.

How the light intensity varies over the screen is determined using the superposition principle. Since we are assuming that the waves from the slits are essentially planar when they reach the screen, we can write their electric field components at P as

$$E_1 = E_0 \sin \omega t$$

and

$$E_2 = E_0 \sin (\omega t + \phi),$$

where ϕ is the phase difference between the two waves at P. Since a path difference λ corresponds to a phase difference 2π rad, a path difference $d \sin \theta$ corresponds to a phase difference

$$\phi = \left(\frac{2\pi}{\lambda}\right)(d \sin \theta) = \frac{2\pi d}{\lambda} \sin \theta. \qquad (38\text{-}5)$$

The field of the resultant wave at P is given by

$$E = E_0 \sin \omega t + E_0 \sin (\omega t + \phi).$$

Using the identity

$$\sin L + \sin M = 2\left(\sin \frac{L + M}{2}\right)\left(\cos \frac{L - M}{2}\right),$$

we find

$$E = \left(2E_0 \cos \frac{\phi}{2}\right) \sin \left(\omega t + \frac{\phi}{2}\right). \qquad (38\text{-}6)$$

The electric field component of the resultant wave at P therefore oscillates at the same frequency as E_1 and E_2 and has an amplitude $2E_0 \cos \phi/2$. From Eq. (35-9), the intensity of this wave is

$$I = \frac{1}{2\mu_0 c}\left(2E_0 \cos \frac{\phi}{2}\right)^2 = I_0 \cos^2 \frac{\phi}{2}, \qquad (38\text{-}7)$$

where $I_0 = 2E_0^2/\mu_0 c$ is the intensity at $\phi = 0$. This point is located at the center of the pattern, because from Eq. (38-5) $\phi = 0$ corresponds to $\theta = 0$. Since θ (and therefore ϕ) changes with position on the screen, the intensity varies across the screen. A plot of the relative intensity versus $\sin \theta$ is shown in Fig. 38-8a. The interference maxima occur at points given by $\cos (\phi/2) = \pm 1$ and the interference minima occur where $\cos (\phi/2) = 0$. It can be shown (Drill Prob. 38-2) that these conditions are equivalent to Eqs. (38-3) and (38-4), respectively.

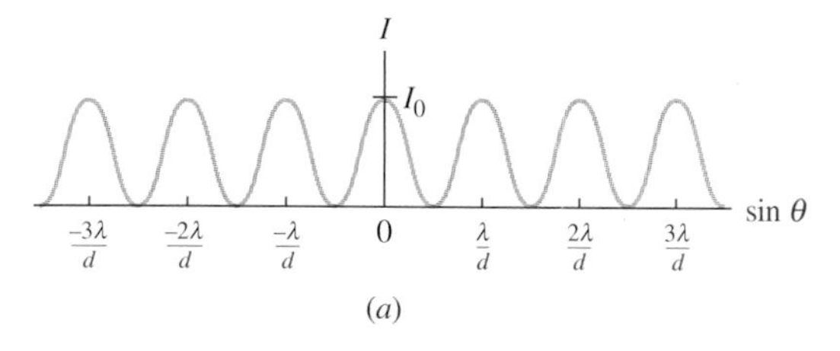

(a)

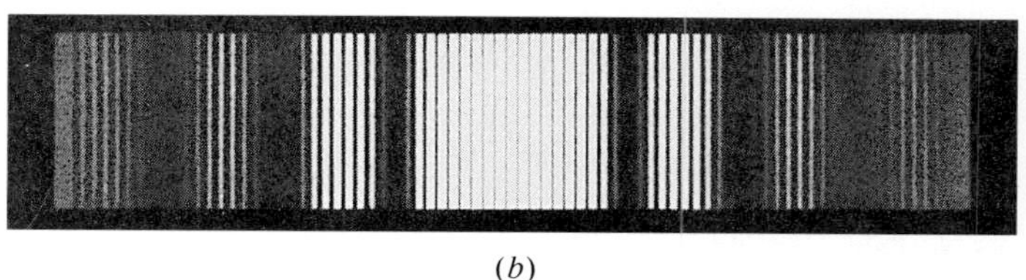

(b)

FIGURE 38-8 (a) Calculated intensity of the interference pattern in the double-slit experiment. (b) An actual intensity pattern.

Notice that all the fringes are predicted to have the same maximum intensity. However, this is not what is observed! As the photograph of Fig. 38-8*b* shows, the brightness of the fringes decreases with increasing order, and some of the fringes are missing. We will discuss these features in Sec. 38-8, where we consider the effects of the finite widths of the slits.

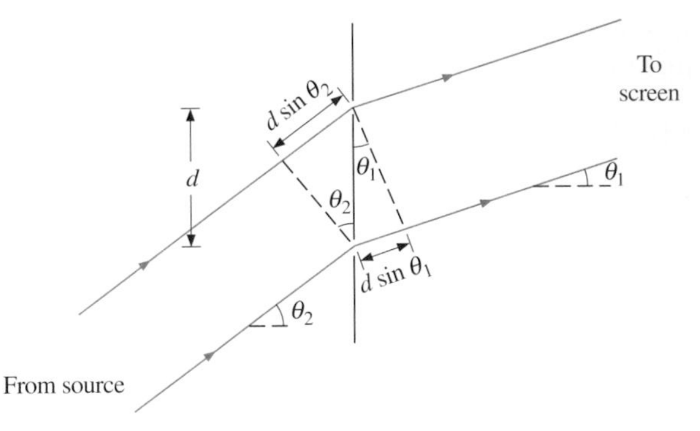

FIGURE 38-9 In this double-slit interference experiment, the light source is displaced off to one side.

EXAMPLE 38-1 The Double-Slit Experiment Using a Sodium Lamp

In a double-slit experiment, the source is a sodium lamp emitting yellow light of wavelength 589 nm. The separation of the slits is 0.20 mm and the screen is 2.0 m from the slits. Calculate the angular positions of the second and third maxima. What is their separation on the screen?

SOLUTION From Eq. (38-3*a*), we have for the second maximum

$$(2.0 \times 10^{-4}\ \text{m}) \sin \theta_2 = (2)(5.89 \times 10^{-7}\ \text{m}),$$

so the angular position of the second maximum is

$$\theta_2 = 0.34°.$$

A similar calculation yields for the third maximum $\theta_3 = 0.51°$.

The separation Δy of the maxima may be found by taking the difference between their positions as given by Eq. (38-3*b*):

$$\Delta y = (3 - 2)\frac{\lambda D}{d} = \frac{(5.89 \times 10^{-7}\ \text{m})(2.0\ \text{m})}{2.0 \times 10^{-4}\ \text{m}} = 5.9 \times 10^{-3}\ \text{m}.$$

Such close spacing of the fringes is typical of a double-slit experiment.

DRILL PROBLEM 38-1

In Young's experiment, the two slits are separated by 0.10 mm, and they are 0.50 m from the screen. If the distance between the central and the fifth maximum is found to be 10 mm, what is the wavelength of the light used?
ANS. 400 nm.

DRILL PROBLEM 38-2

Show (*a*) that the conditions $\cos \phi/2 = \pm 1$ and $d \sin \theta = m\lambda$ are equivalent and (*b*) that the conditions $\cos \phi/2 = 0$ and $d \sin \theta = (m + \frac{1}{2})\lambda$ are equivalent.

DRILL PROBLEM 38-3

Suppose that the point source in Young's experiment is displaced as shown in Fig. 38-9. Show that if the waves arriving at the slits can be assumed to be plane waves, the interference maxima appear on the screen at points given by

$$d(\sin \theta_2 - \sin \theta_1) = m\lambda.$$

38-2 Multiple-Slit Interference Patterns

The argument used to determine the interference maxima for two slits can be extended to N slits. As Fig. 38-10 shows, the path difference between adjacent sources to a point on the screen at an angle θ is $d \sin \theta$. If this path difference is $m\lambda$, the waves from all N sources are in phase and interfere constructively. *Therefore $d \sin \theta = m\lambda$ also gives the locations of the interference maxima for N slits.*

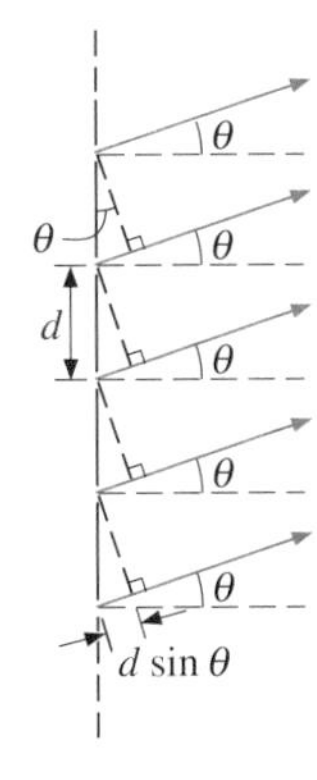

FIGURE 38-10 In a multiple-slit interference experiment, the path difference between adjacent slits is $d \sin \theta$. Constructive interference occurs at those points where the path difference between adjacent slits is $m\lambda$.

Determining the intensity at an arbitrary angle θ requires the addition of N harmonic waves, all with the same frequency and amplitude but with different phases. This can be done using phasor addition, a technique we introduced in the analysis of ac circuits (Chap. 34). The electric field component of a light wave,

$$E_1 = E_0 \sin \omega t,$$

is represented on the phasor diagram of Fig. 38-11*a* by a phasor of length E_0 that rotates counterclockwise at a constant angular frequency ω. As the phasor rotates, its projection onto the vertical axis is the field E_1. The phasor for a second electric field,

$$E_2 = E_0 \sin (\omega t + \phi),$$

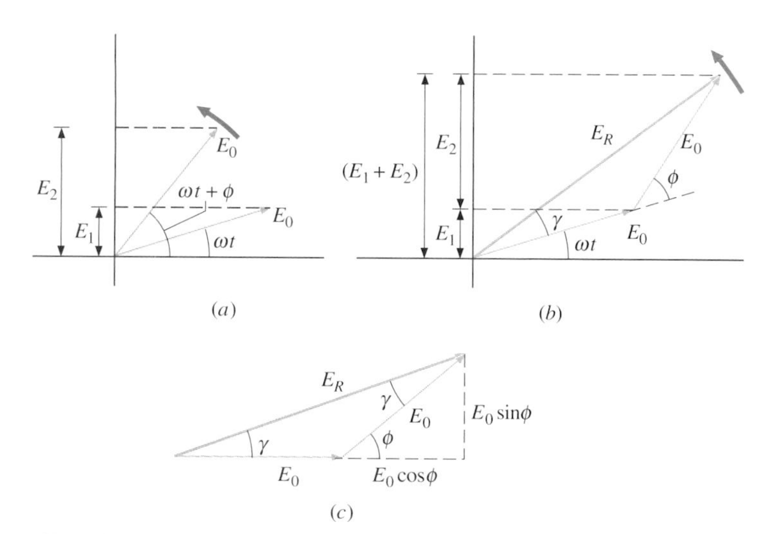

FIGURE 38-11 (a) The phasors for E_1 and E_2 rotate counterclockwise at the same angular frequency ω. The projections on the vertical axis are E_1 and E_2. (b) Summing the phasors as if they were vectors gives the resultant phasor for $E_1 + E_2$. (c) The addition of the phasors for $t = 0$.

is also shown. It is directed at an angle $(\omega t + \phi)$ with respect to the horizontal axis, and its projection onto the vertical axis is E_2.

The superposition of E_1 and E_2 is found by adding their phasors as if they were vectors, since the resultant phasor also rotates at an angular frequency ω and its projection onto the vertical axis is $E_1 + E_2$ (see Fig. 38-11b). *If the length of the resultant phasor is E_R, then the amplitude of the resultant field is E_R; if the angle between the phasors E_R and E_l is γ, then the resultant field's phase angle with respect to E_l is γ.* We can therefore represent the resultant electric field by

$$E = E_R \sin(\omega t + \gamma).$$

With the time chosen to be $t = 0$, the phasor addition of E_1 and E_2 is illustrated in Fig. 38-11c. From the geometry of the triangle,

$$E_R^2 = (E_0 + E_0 \cos\phi)^2 + (E_0 \sin\phi)^2$$
$$= 2E_0^2(1 + \cos\phi) = 4E_0^2 \cos^2\frac{\phi}{2};$$

thus the amplitude of the resultant electric field is

$$E_R = 2E_0 \cos\frac{\phi}{2}.$$

And because the triangle is isosceles,

$$\phi + (180° - 2\gamma) = 180°,$$

so the phase angle of the resultant field is

$$\gamma = \frac{\phi}{2}.$$

The resultant field is therefore given by

$$E = E_R \sin(\omega t + \gamma)$$
$$= 2E_0 \cos\frac{\phi}{2} \sin\left(\omega t + \frac{\phi}{2}\right).$$

Notice that this equation is identical to Eq. (38-6), which was obtained algebraically.

To determine the superposition of several waves of the same frequency, we add their phasors head-to-tail as if they were vectors. The length of the resultant phasor is the amplitude of the resultant wave, and the angle between the resultant phasor and the first phasor is the phase angle between the resultant wave and the first wave. An example of such a summation is shown in Fig. 38-12.

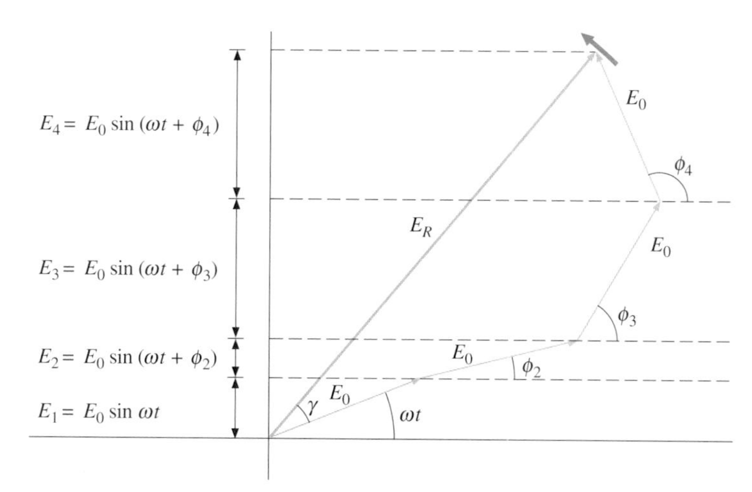

FIGURE 38-12 A summation of several phasors: $E_R \sin(\omega t + \gamma) = E_0 \sin \omega t + E_0 \sin(\omega t + \phi_2) + E_0 \sin(\omega t + \phi_3) + E_0 \sin(\omega t + \phi_4)$.

EXAMPLE 38-2 PHASOR ADDITION

Use the phasor addition method to determine

$$E = E_0 \sin \omega t + E_0 \sin(\omega t + 30°) + E_0 \sin(\omega t + 60°).$$

SOLUTION The phasor diagram for this sum is shown in Fig. 38-13 for $t = 0$. By measurement, the amplitude of the resultant field E_R is $2.7E_0$ and the phase angle is 30°, so

$$E = 2.7E_0 \sin(\omega t + 30°).$$

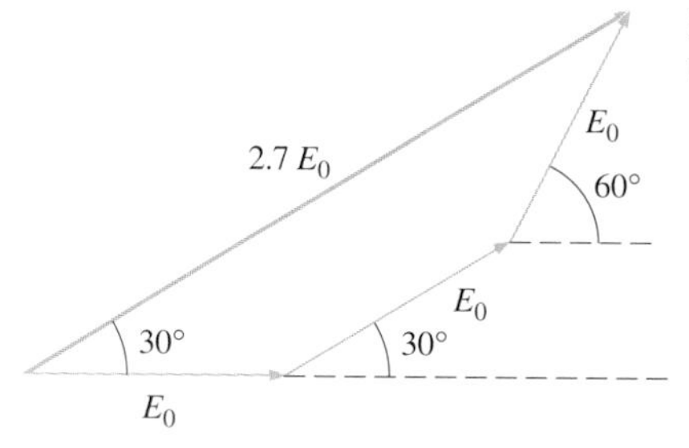

FIGURE 38-13 Addition of three phasors for $t = 0$.

Now let's use the phasor method to calculate the intensity distribution of the N-slit interference pattern. At an angle θ, the path difference between adjacent slits is $d \sin\theta$, so the phase difference between the Huygens wavelets from those slits is

$$\phi = 2\pi \frac{d \sin\theta}{\lambda}.$$

With the field of the light wave from the first slit represented by $E_0 \sin \omega t$, the resultant field from N slits is

$$E = E_0 \sin \omega t + E_0 \sin(\omega t + \phi) + E_0 \sin(\omega t + 2\phi) + \cdots + E_0 \sin[\omega t + (N-1)\phi].$$

The phasor diagram for this sum is shown in Fig. 38-14*a* for $N = 6$. The amplitude of the resultant field is E_R. The apex angle of every isosceles triangle is ϕ (Fig. 38-14*b*), so the angle subtended by the resultant is $N\phi$. Since the heads of the phasors are all the same distance r from the apex of the diagram, we have from Fig. 38-14*c*

$$\frac{1}{2} E_0 = r \sin \frac{\phi}{2}.$$

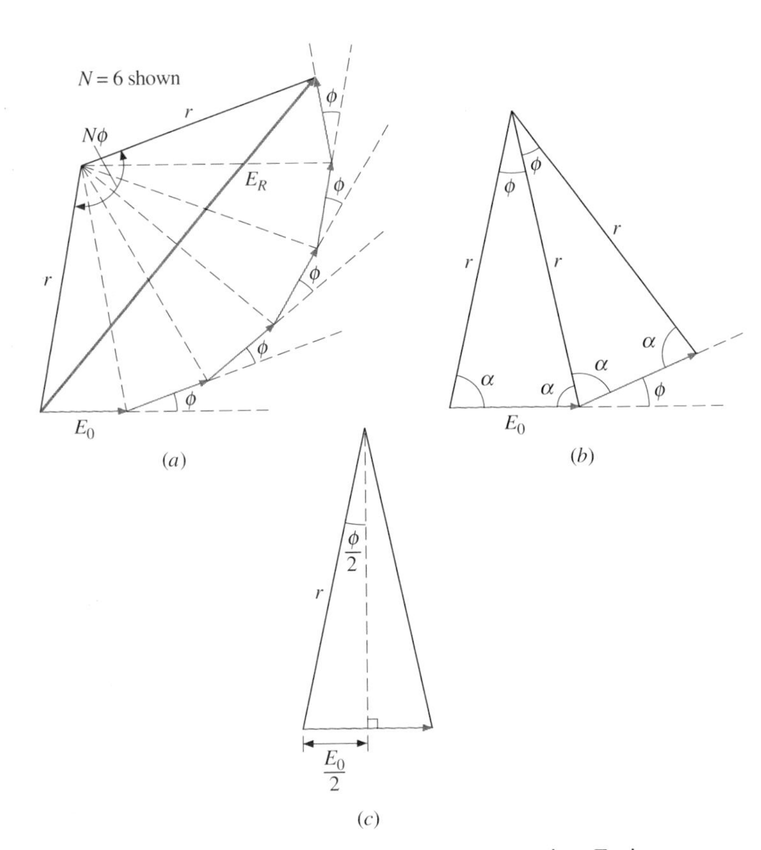

FIGURE 38-14 (*a*) The sum of N phasors representing $E_0 \sin \omega t + E_0 \sin(\omega t + \phi) + E_0 \sin(\omega t + 2\phi) + \cdots + E_0 \sin[\omega t + (N-1)\phi]$. (*b*) Since $\phi + 2\alpha = 180°$, the apex angle of each isosceles triangle is also ϕ. (*c*) Geometry of the isosceles triangle showing $E_0/2 = r \sin \phi/2$.

The same argument applied to the large triangle of Fig. 38-14*a* gives

$$\frac{1}{2} E_R = r \sin \frac{N\phi}{2}.$$

The ratio of the last two equations then yields

$$\frac{E_R}{E_0} = \frac{\sin N\phi/2}{\sin \phi/2},$$

so

$$E_R = E_0 \frac{\sin N\alpha}{\sin \alpha},$$

where

$$\alpha = \frac{\phi}{2} = \pi \frac{d \sin \theta}{\lambda}. \tag{38-8}$$

Thus the intensity of an N-slit interference pattern is, from Eq. (35-9),

$$I = \frac{1}{2\mu_0 c} E_R^2 = I_0 \left(\frac{\sin N\alpha}{\sin \alpha} \right)^2, \tag{38-9}$$

where $I_0 = E_0^2/2\mu_0 c$ is the intensity of one Huygens wavelet.

Plots of Eq. (38-9) for $N = 2$, 3, and 6 slits are shown in Fig. 38-15. Notice that the *principal maxima* are located at the same spots, given by $\alpha = \pi d \sin \theta/\lambda = 0, \pm\pi, \pm 2\pi, \ldots, \pm m\pi, \ldots$, in all three cases (and, in fact, for all values of N). This corresponds to $d \sin \theta = m\lambda$, which is the same result found at the beginning of this section for the centers of the bright fringes. At the principal maxima, $\sin \alpha = 0$ and $\sin N\alpha = 0$. To evaluate I from Eq. (38-9) at these points, we must apply L'Hôpital's rule to $\sin N\alpha/\sin \alpha$. We then find

$$\lim_{\alpha \to \pm m\pi} \frac{\sin N\alpha}{\sin \alpha} = \lim_{\alpha \to \pm m\pi} \frac{d(\sin N\alpha)/d\alpha}{d(\sin \alpha)/d\alpha} = N \frac{\cos(\pm Nm\pi)}{\cos(\pm m\pi)} = \pm N.$$

Therefore at any principal maximum,

$$I = I_0 \lim_{\alpha \to \pm m\pi} \left(\frac{\sin N\alpha}{\sin \alpha} \right)^2 = N^2 I_0.$$

The intensity of the N-slit pattern drops to zero whenever $\sin N\alpha = 0$, except at the points $\alpha = 0, \pm\pi, \pm 2\pi, \ldots$, where $\sin \alpha = 0$. Thus, $I = 0$ when

$$N\alpha = \pm\pi, \pm 2\pi, \pm 3\pi, \ldots, \pm(N-1)\pi, \pm(N+1)\pi, \ldots.$$

Consequently, there are $N - 1$ points where $I = 0$ between any two adjacent principal maxima. And since there must be a maximum between two adjacent minima, there are $N - 2$ *subsidiary maxima* between adjacent principal maxima. These subsidiary maxima are apparent from the plots of Fig. 38-15. They are not nearly as large as the principal maxima. You will investigate the relative intensities of the principal and subsidiary maxima in Prob. 38-69.

For a given number of slits N, an mth-order fringe begins at the point $N\alpha = \pm Nm\pi - \pi$ (where $I = 0$), peaks at $N\alpha = \pm Nm\pi$ (where $I = N^2 I_0$), then ends at $N\alpha = \pm Nm\pi + \pi$ (where once again $I = 0$). In terms of the parameter α, the width of the fringe is $\Delta\alpha = 2\pi/N$. We can use this with Eq. (38-8) to determine the angular width of a fringe:

$$\Delta\alpha = \Delta \frac{\pi d \sin \theta}{\lambda} = \frac{\pi d \cos \theta}{\lambda} \Delta\theta = \frac{2\pi}{N},$$

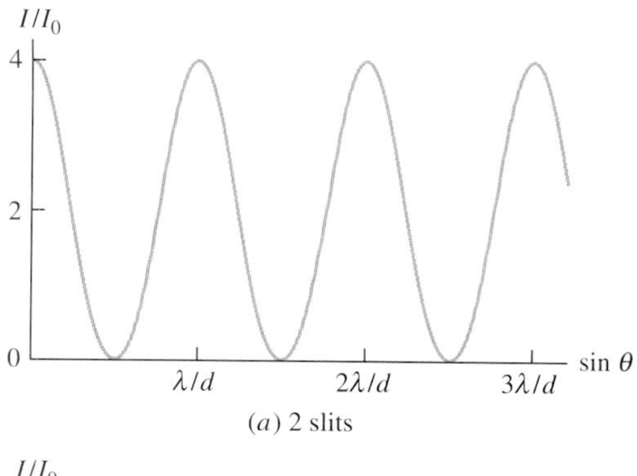

(a) 2 slits

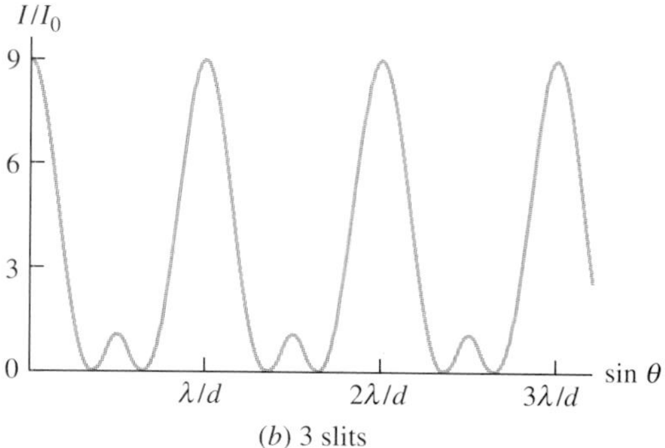

(b) 3 slits

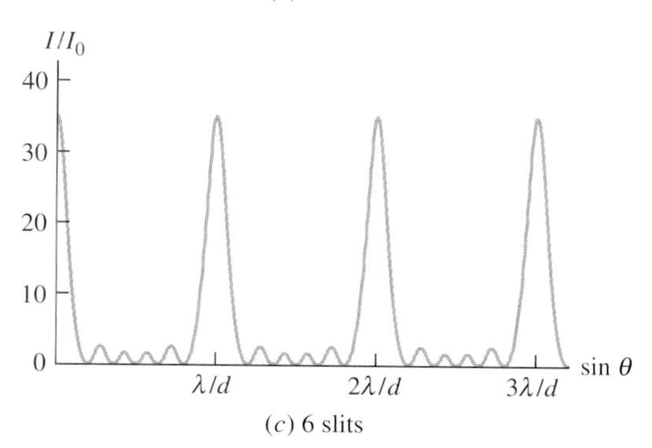

(c) 6 slits

FIGURE 38-15 The intensity distributions for 2-, 3-, and 6-slit interference patterns as given by Eq. (38-9).

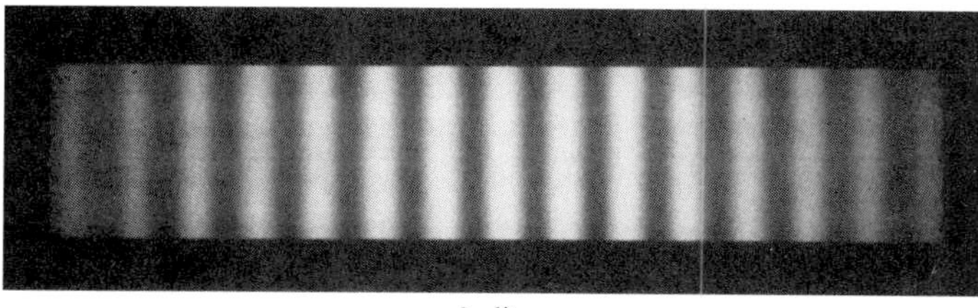
2 slits

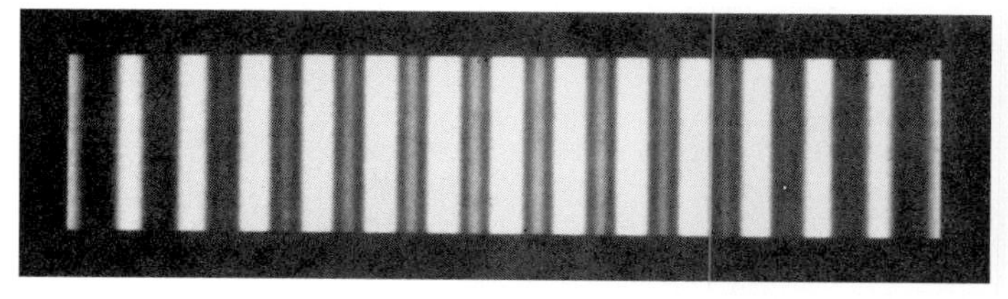
3 slits

4 slits

5 slits

FIGURE 38-16 Actual interference patterns for various numbers of slits.

which yields for the angular width of a principal maximum

$$\Delta\theta = \frac{2\lambda}{Nd\cos\theta}. \qquad (38\text{-}10)$$

The width of a fringe therefore decreases with the number of slits N.

Some photographs of actual interference patterns are shown in Fig. 38-16. You can see both the narrowing of the fringes with N and the subsidiary fringes. However, notice in the photographs that the intensities of the principal maxima decrease with m, while Eq. (38-9) predicts that all the principal maxima in a pattern should have the same intensity. Just as for the double-slit pattern, this discrepancy occurs because we have neglected the effects of the finite width of the slits.

EXAMPLE 38-3 Multiple-Slit Interference Using a Hydrogen Lamp

A hydrogen-gas lamp emits visible light at four wavelengths, $\lambda =$ 410, 434, 486, and 656 nm. (*a*) If light from a hydrogen lamp falls on N slits separated by 0.025 mm, how far from the central maximum are the third maxima when viewed on a screen 2.0 m from the slits? (*b*) By what distance are the second and third maxima separated for $\lambda = 486$ nm?

SOLUTION (*a*) We have $d = 2.5 \times 10^{-5}$ m, $D = 2.0$ m, and $m = 3$, so from Eq. (38-3*b*),

$$y_3 = 3\,\frac{\lambda D}{d} = 3\lambda\,\frac{2.0\text{ m}}{2.5 \times 10^{-5}\text{ m}}.$$

Inserting the given wavelengths into this equation, we find that $y_3 = 9.8$, 10.4, 11.7, and 15.7 cm.

(*b*) Again from Eq. (38-3*b*),

$$y_3 - y_2 = (3-2)\,\frac{\lambda D}{d}$$
$$= (4.86 \times 10^{-7}\text{ m})\left(\frac{2.0\text{ m}}{2.5 \times 10^{-5}\text{ m}}\right)$$
$$= 3.9 \times 10^{-2}\text{ m} = 3.9\text{ cm}.$$

DRILL PROBLEM 38-4

Monochromatic light of frequency 5.5×10^{14} Hz falls on 10 slits separated by 0.020 mm. What is the separation between the first and third maxima on a screen that is 2.0 m from the slits?
ANS. 10.9 cm.

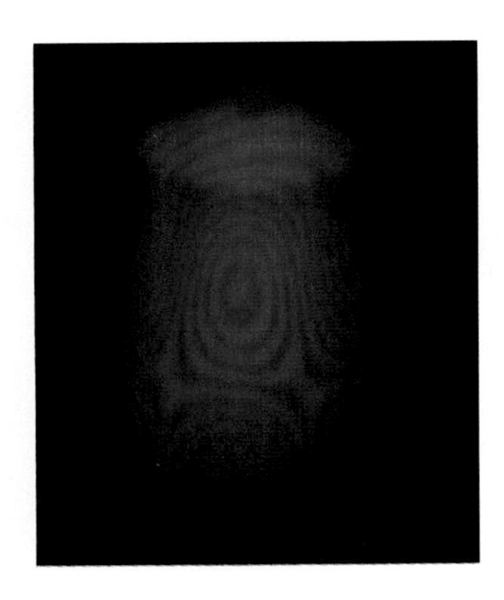

Vibrational modes of a bell exhibited with interferometric holography.

38-3 Interference in Thin Films

Interference patterns are also produced when light is reflected off thin, transparent films. Such patterns are commonly seen in reflections from soap bubbles and oil films. To study this phenomenon, we consider a transparent film of uniform thickness t and index of refraction n that is surrounded by air, as shown in Fig. 38-17. Because of partial reflections and refractions at the boundaries of the film, an incident monochromatic light ray emerges as two parallel rays. For *near-normal* incidence, their path difference is approximately $2t$. You might expect the light waves to interfere constructively if $2t = m\lambda$ and destructively if $2t = (m + 1/2)\lambda$, where m is an integer and λ is the wavelength of the light in the film. (Recall that $\lambda = \lambda_0/n$, where λ_0 is the wavelength in a vacuum.) However, *light waves undergo a phase change of π rad upon reflection at an interface beyond which is a medium of higher refractive index*. This effect is analogous to what we have seen for mechanical waves: a traveling wave on a string is inverted upon reflection at a boundary to which a heavier string is tied. (See Sec. 17-4.) For the thin film of Fig. 38-17, only the light reflected at the top surface has this phase change. The phase change is equivalent to adding an extra $\lambda/2$ to the path difference. Thus the *interference conditions for light incident normally on a thin film in air* are given by

$$2t = (m + \tfrac{1}{2})\lambda \qquad m = 0, 1, 2, \ldots \qquad \text{(constructive)},$$

and

$$2t = m\lambda \qquad m = 0, 1, 2, \ldots \qquad \text{(destructive)}.$$

Substituting λ_0/n for λ, we may rewrite these two equations as

$$2nt = (m + \tfrac{1}{2})\lambda_0 \qquad m = 0, 1, 2, \ldots \qquad \text{(constructive)}; \qquad (38\text{-}11a)$$

$$2nt = m\lambda_0 \qquad m = 0, 1, 2, \ldots \qquad \text{(destructive)}. \qquad (38\text{-}11b)$$

If light of a particular color is destroyed in the reflected beam, it is enhanced in the transmitted beam. For example, suppose that the blue portion of an incident beam of white light is not reflected. The transmitted beam then has a blue tinge, while the reflected beam is predominantly green and red and hence appears yellow. On the other hand, when a color is enhanced in the reflected beam, its complement is dominant in the transmitted beam.

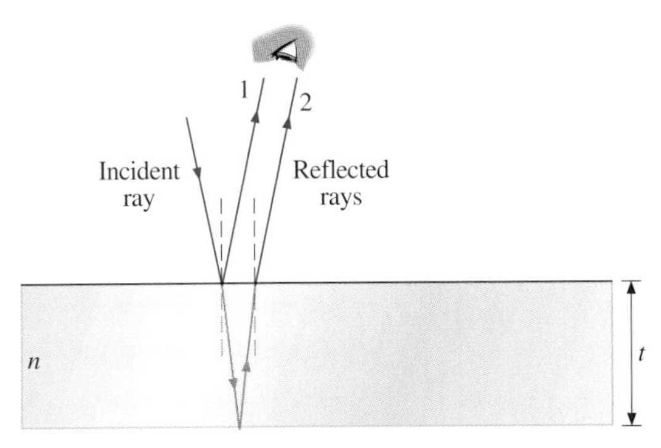

FIGURE 38-17 Interference occurs between the reflected rays of near-normal incident light on a thin, transparent film surrounded by air.

EXAMPLE 38-4 **Interference for a Soap Film**

White light is incident normally on a soap film ($n = 1.33$) with thickness 150 nm that is surrounded by air. What wavelength is missing from the reflected light?

SOLUTION We wish to find the wavelength for which destructive interference occurs. From Eq. (38-11b), those wavelengths satisfying

$$2(1.33)(150 \text{ nm}) = m\lambda_0$$

suffer destructive interference. The only wavelength in the visible region that satisfies this equation is $\lambda_0 = 400$ nm, which corresponds to violet light.

The interference conditions described by Eqs. (38-11) are only valid if the thin film is bounded on both sides by media whose indices of refraction are either *both greater or both less than* that of the film, for only then will there be a phase change upon reflection at one surface but not at the other. If the thin film happens to be situated between two media, one with a lower and the other with a higher refractive index than the film's, there will be phase changes either at both boundaries or not at all. Consequently, the constructive and destructive designations in Eqs. (38-11) must be interchanged. This situation is considered in the following example.

EXAMPLE 38-5 INTERFERENCE FOR A THIN FILM OF WATER

A film of water rests on a glass plate. For near-normal incidence, monochromatic light of wavelength 600 nm undergoes total destructive interference when reflected from the film. What is the minimum thickness of the water film?

SOLUTION Since the index of refraction of water is between that of glass and air, destructive interference in the film corresponds to

$$2nt = (m + \tfrac{1}{2})\lambda_0.$$

For *minimum* thickness of the water film, $m = 0$; thus

$$2(1.33)t = (0 + \tfrac{1}{2})(600 \text{ nm}),$$

and the thickness of the film is $t = 113$ nm.

The last example suggests how losses in the intensity of light passing through an optical instrument can be minimized. Since about 4 percent of the light incident normally on a lens is reflected, much of the intensity of the light entering a multilens device may be lost through partial reflections. This presents a serious problem, especially when the object being viewed is very faint (e.g., a distant star). To reduce this loss, a thin layer of transparent material whose thickness is appropriate for minimizing reflected light is deposited on the lens. In order to keep the film as thin as possible, the material should have a refractive index between those of air and glass. Magnesium fluoride (MgF_2), whose index is 1.38, is often used.

Suppose we wanted to minimize the reflection of green light ($\lambda_0 = 550$ nm) by depositing a thin film of MgF_2 on glass. The thickness t of this film would have to satisfy

$$2(1.38)t = (m + \tfrac{1}{2})(550 \text{ nm}),$$

so the thinnest possible film would have a thickness $t = 100$ nm. A lens that has been coated to minimize the reflection of green light will have a purplish tinge because of reflection of the other visible wavelengths, such as red and violet. You can often see this tinge on a camera lens.

DRILL PROBLEM 38-5

A thin film with $n = 1.30$ rests in air. What is the minimum thickness that this film must have if constructive interference is to occur for normally reflected red light ($\lambda_0 = 650$ nm)?
ANS. 125 nm.

DRILL PROBLEM 38-6

A transparent film in air has a thickness that is much smaller than the wavelengths of visible light. Will this film appear bright or dark when we shine white light on it? Explain.
ANS. Dark.

DRILL PROBLEM 38-7

Suppose that a camera lens is coated with a film of material with $n = 1.70$. What is the thinnest coating that will minimize the reflection of green light ($\lambda_0 = 550$ nm)? On the basis of your calculation, explain the statement "In order to keep the film as thin as possible, the material should have a refractive index between those of air and glass."
ANS. 162 nm.

An interesting and important type of thin-film interference occurs when the thickness of the film varies. An air film of variable thickness can be created by placing an object with a small cross section such as a fiber between the ends of a pair of juxtaposed microscope slides (Fig. 38-18). If the angle of the thin wedge is α, the thickness t of the air film at any point is approximately

$$t = x\alpha,$$

where x is the distance along the slides from the apex of the wedge. In this case, there is a phase change upon reflection at the lower surface but not at the upper surface of the air film. The conditions for interference are then

$$2x\alpha = (m + \tfrac{1}{2})\lambda_0 \qquad \text{(constructive)}$$

and

$$2x\alpha = m\lambda_0 \qquad \text{(destructive).}$$

Bright interference fringes therefore occur along the slides at $x = \lambda_0/4\alpha,\ 3\lambda_0/4\alpha,\ 5\lambda_0/4\alpha$, etc.

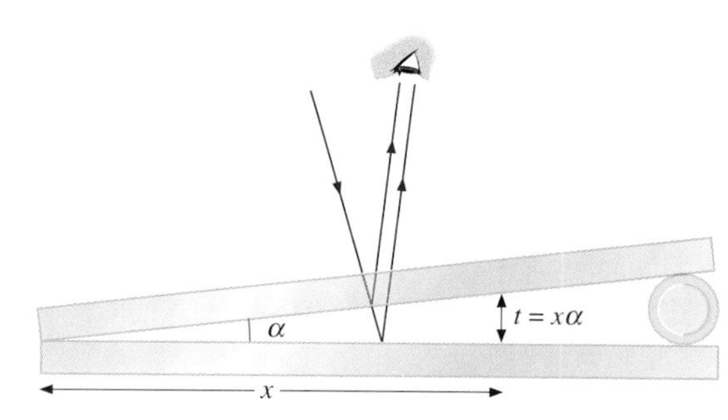

FIGURE 38-18 Interference in a thin air film of varying thickness.

EXAMPLE 38-6 NEWTON'S RINGS

A plano-convex lens is placed on a flat, glass surface to create a thin air film, as shown in Fig. 38-19. When the system is illuminated at normal incidence by monochromatic light, circular interference fringes are observed. Derive the equation for the radius of the mth bright fringe and the mth dark fringe. These fringes were studied and described by Isaac Newton and are called *Newton's rings*.

SOLUTION If R is the radius of the convex side of the lens, the distance x from the center of the lens is given in terms of the film thickness t and R by

$$x^2 = R^2 - (R - t)^2,$$

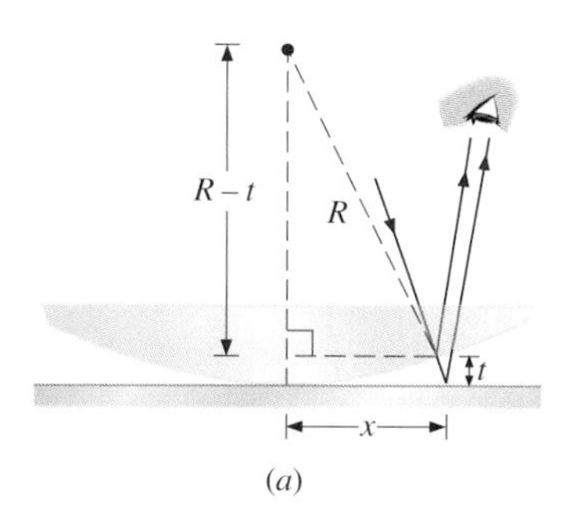

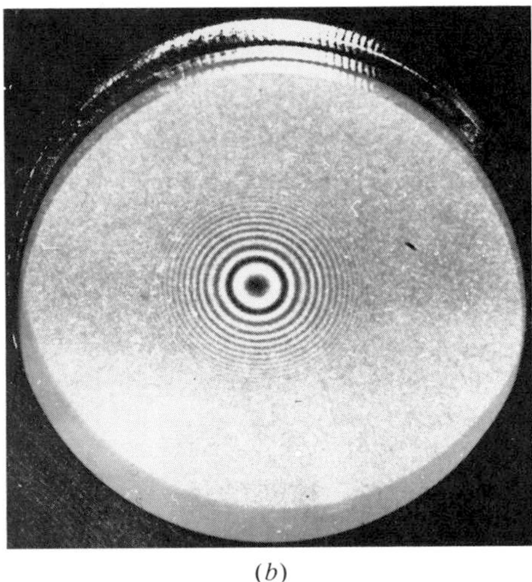

FIGURE 38-19 (*a*) A lens on a glass surface produces a thin air film of varying thickness. (*b*) The circular interference fringes are known as Newton's rings.

or

$$x^2 = 2Rt - t^2.$$

Since $R \gg t$, this is accurately approximated by

$$x^2 \approx 2Rt.$$

A phase change occurs at only one boundary, so constructive interference occurs when

$$2t = (m + \tfrac{1}{2})\lambda_0.$$

We can find the radius of the mth bright fringe by simply combining the last two equations; then

$$x_m = \sqrt{(m + \tfrac{1}{2})\lambda_0 R}.$$

Similarly, the radius of the mth dark fringe is

$$x_m = \sqrt{m\lambda_0 R}.$$

DRILL PROBLEM 38-8

(*a*) Why is the center of Newton's rings dark? (*b*) Why does the distance between adjacent rings decrease as the diameters of the rings increase?

*38-4 THE MICHELSON INTERFEROMETER

The Michelson interferometer* is a precision instrument that produces fringes by splitting a light beam into two parts and then recombining them after they have traveled different optical paths. Figure 38-20 depicts the interferometer and the path of a light beam from a single point on the extended source S, which is a ground-glass plate that diffuses the light from a monochromatic lamp of wavelength λ_0. The beam strikes the half-silvered mirror M, where half of it is reflected upward and half passes through the mirror. The reflected light travels to the movable plane mirror M_1, where it is reflected back through M to the observer. The transmitted half of the original beam is reflected back by the stationary mirror M_2 and then toward the observer by M. Since both beams originate from the same point on the source, they are coherent and therefore interfere. Notice from the figure that one beam passes through M three times and the other only once. To ensure that both beams traverse the same thickness of glass, a compensator plate C of transparent glass is placed in the arm containing M_2. This plate is a duplicate of M (without the silvering) and is usually cut from the same piece of glass used to produce M. With the compensator in place, any phase difference between the two beams is due solely to the difference in the distances they travel.

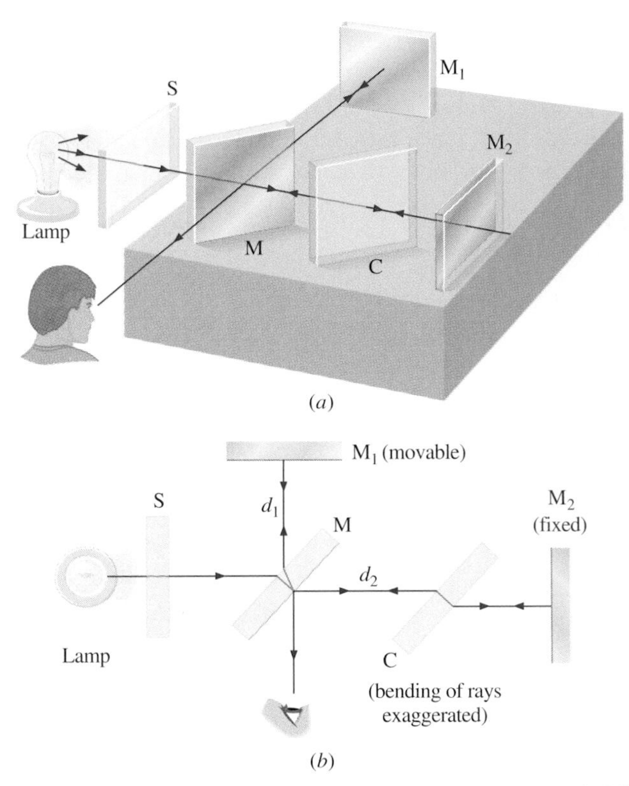

FIGURE 38-20 (*a*) The Michelson interferometer. The extended light source is a ground-glass plate that diffuses the light from a lamp. (*b*) A planar view of the interferometer.

*This instrument was invented by the American physicist Albert A. Michelson (1852–1931).

The path difference of the two beams when they recombine is $2d_1 - 2d_2$, where d_1 is the distance between M and M_1 and d_2 is the distance between M and M_2. Suppose this path difference is an integer number of wavelengths $m\lambda_0$. Then there is constructive interference and a bright image of the point on the source is seen at the observer. Now the light from any other point on the source whose two beams have this same path difference also undergoes constructive interference and produces a bright image. The collection of these point images is a bright fringe corresponding to a path difference of $m\lambda_0$. When M_1 is moved a distance $\lambda_0/2$, this path difference changes by λ_0, and each fringe will move to the position previously occupied by an adjacent fringe. Consequently, by counting the number of fringes passing a given point as M_1 is moved, an observer can measure minute displacements that are accurate to a fraction of a wavelength.

An important application of this measurement was the definition of the standard meter. As mentioned in Chap. 1, the length of the standard meter was once defined as the mirror displacement in a Michelson interferometer corresponding to 1,650,763.73 wavelengths of a particular fringe of krypton-86 in a gas discharge tube.

Fringes produced with a Michelson interferometer.

*38-5 Coherence

In our study of interference patterns created with slits, we assumed that the light illuminating the slits had been produced by a point source and was perfectly monochromatic. With these two assumptions, the slits became coherent sources that generated harmonic light waves with well-defined phase differences. However, real sources aren't points, nor are they monochromatic. Different atoms located at different spots radiate completely uncorrelated waves of various frequencies. As a result, the light signals leaving the slits are only "partially coherent."

Effects caused by a spread in the wavelength or frequency of the light are studied under the general classification of *temporal coherence* phenomena; effects that result from the finite size of the source are called *spatial coherence* phenomena. We will consider these two independent of each other; that is, we'll assume a point source when investigating temporal coherence and a monochromatic source when studying spatial coherence. But keep in mind that both may significantly influence a particular interference pattern, although in many cases one is much more dominant than the other.

Temporal Coherence

We will study temporal coherence by calculating the double-slit interference pattern formed when the monochromatic point source is replaced by a point source that emits light waves over a frequency range from $f_0 - \Delta f/2$ to $f_0 + \Delta f/2$. Interference only occurs for light waves from the two slits that have the same frequency. For example, if the source emits white light, the "blue waves" of frequency 6.5×10^{14} Hz from the two slits interfere, as do the "red waves" of frequency 4.7×10^{14} Hz, etc. However, a "blue wave" does not interfere with a "red wave," nor does a "green wave" interfere with a "yellow wave," etc. So what we observe on the screen is a sum of interference patterns with intensity distributions given by Eq. (38-7) for every frequency. *Each frequency component produces an interference pattern* that is calculated by adding sinusoidal electric fields and then squaring to get the intensity, as described in Sec. 38-1. *The net intensity pattern is then determined by summing the intensity patterns over all frequencies.*

Figure 38-21a shows the interference patterns for three different frequencies. These frequencies correspond to the visible wavelengths 400, 550, and 700 nm. Notice that the central fringe for each frequency component is centered at $\theta = 0$, a fact that is apparent from $\sin\theta = m\lambda/d$ (Eq. 38-3a). Consequently, the color of the intensity pattern around $\theta = 0$ is the same as that of the source. For example, the net intensity pattern for a white-light source is white near $\theta = 0$. Away from the center of the pattern, different frequency components peak at different spots. This results in a separation of colors for small m in the interference pattern. For larger values of θ, there is strong overlap among different-order fringes for the various colors,* and the fringes are difficult to separate once again.

Now let's see what happens when the source emits over a continuous frequency band. To keep the calculations relatively simple, all frequency components are assumed to be present with the same intensity. Then at a particular θ, the net intensity contribution dI from a frequency interval df is, from Eq. (38-7),

$$dI = (I_f\,df)\cos^2\frac{\pi d\sin\theta}{\lambda} = I_f\,df\cos^2\frac{f\pi d\sin\theta}{c}$$
$$= I_f\,df\cos^2\gamma f,$$

where $\gamma = \pi d\sin\theta/c$, and I_f is a constant representing the intensity per unit frequency. Integrating dI over the assumed frequency band, we have for the net intensity at θ:

$$I = I_f\int_{f_0-\Delta f/2}^{f_0+\Delta f/2}\cos^2\gamma f\,df.$$

*For example, notice in Fig. 38-21a the overlap of the $m = 5$ fringe of the 550-nm light, the $m = 7$ fringe of the 400-nm light, and the $m = 4$ fringe of the 700-nm light (designated by the arrow).

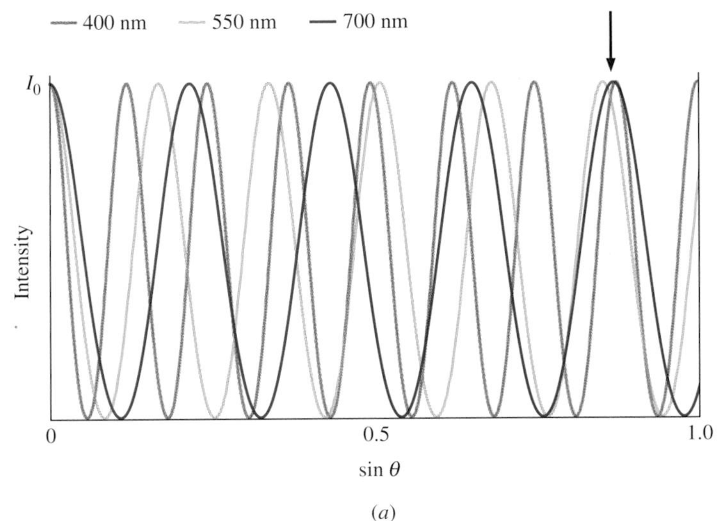

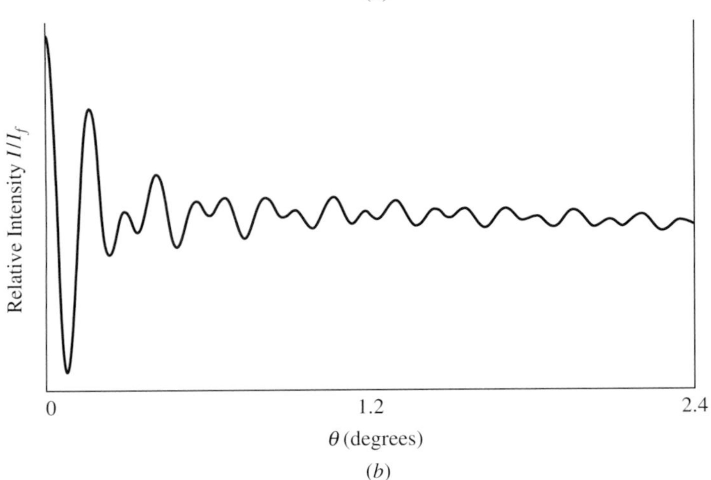

FIGURE 38-21 (*a*) The double-slit interference patterns for three wavelengths. (*b*) The double-slit interference pattern for a source that radiates with the same intensity over a frequency band Δf. Notice the disappearance of the fringes as θ gets larger.

Evaluating this integral and then simplifying the resulting equation with $\sin(A \pm B) = \sin A \cos B \pm \cos A \sin B$, we find

$$I = I_f\left(\frac{f}{2} + \frac{\sin 2\gamma f}{4\gamma}\right)\Bigg|_{f_0-\Delta f/2}^{f_0+\Delta f/2}$$

$$= I_f\left\{\frac{\Delta f}{2} + \frac{1}{4\gamma}\left[\sin 2\gamma\left(f_0 + \frac{\Delta f}{2}\right) - \sin 2\gamma\left(f_0 - \frac{\Delta f}{2}\right)\right]\right\}$$

$$= I_f\left[\frac{\Delta f}{2} + \frac{1}{2\gamma}(\cos 2\gamma f_0 \sin \gamma\, \Delta f)\right].$$

This intensity pattern is plotted in Fig. 38-21*b* using $d = 2.0 \times 10^{-4}$ m, $f_0 = 5.5 \times 10^{14}$ Hz, and $\Delta f = 3.0 \times 10^{14}$ Hz. For white light, the constant term $I_f \Delta f/2$ represents a uniform white background visible everywhere along the interference pattern. Superimposed on this background is a fluctuating intensity given by $I_f(\cos 2\gamma f_0 \sin \gamma\, \Delta f)/2\gamma$. At some points this term is dominated by red, at other points by blue, etc. It gives us the fringes we see in the white-light interference pattern. Since the size of this term decreases as γ (which is proportional to $\sin\theta$) increases, the fluctuations become less and less prominent as θ gets larger. Consequently, the fringes become increasingly obscured by the uniform white background at farther locations from the center of the pattern.

Do the outer fringes become so faint relative to the background that they are no longer visible? To answer this question, we have to quantify "no longer visible." The criterion we will use is that the fluctuating term has to be at least $1/\pi$ times* the constant term in order for a fringe to be observable; that is,

$$\frac{1/2\gamma(\cos 2\gamma f_0 \sin \gamma\, \Delta f)}{\Delta f/2} \geq \frac{1}{\pi}.$$

Since the cosine and sine functions are bounded by 1, this condition can only be satisfied if

$$\frac{1}{\gamma\, \Delta f} \geq \frac{1}{\pi}. \qquad (38\text{-}12a)$$

Suppose we are observing a point on the screen that is at an angle θ. The path difference from the two slits to this location is $d \sin\theta$. If the light travels this distance in a time Δt, then $c\,\Delta t = d \sin\theta$, and

$$\gamma = \frac{\pi d \sin\theta}{c} = \frac{\pi c\,\Delta t}{c} = \pi\,\Delta t.$$

Substituting this expression for γ into Eq. (38-12*a*), we obtain

$$\Delta f\, \Delta t \leq 1. \qquad (38\text{-}12b)$$

Thus *fringes are visible only when the product of the frequency bandwidth Δf of the source and the time Δt for light to travel the difference in path lengths from the slits to the screen ($d \sin\theta$) is less than or equal to 1.*

We can easily determine whether or not fringes will be visible at a particular point P on the screen if we know the frequency bandwidth Δf of the illumination. First, we calculate Δt with Eq. (38-12*b*) using the equality. This time interval is called the **coherence time**. Next we multiply Δt by c to get the **coherence length** $c\,\Delta t$ of the signal. Finally, if L is the difference in path lengths from the two slits to P, fringes are seen if $L \leq c\,\Delta t$; and they are not visible if $L > c\,\Delta t$.

EXAMPLE 38-7 VISIBILITY OF FRINGES

(*a*) Suppose that the slits of Example 38-1 are illuminated by white light from a point source. Take Δf to be 3.8×10^{14} Hz and determine how many orders of yellow (5.1×10^{14} Hz) fringes will be visible. (*b*) Repeat this calculation for a low-pressure Hg^{198} isotope lamp ($\lambda = 546$ nm) whose frequency bandwidth is 1.0×10^9 Hz.

*We use $1/\pi$ to keep the form of Eq. (38-12*b*) as simple as possible. Actually, this criterion slightly underestimates our ability to distinguish fringes.

SOLUTION (*a*) The coherence time is

$$\Delta t = \frac{1}{3.8 \times 10^{14}\ \text{Hz}} = 2.6 \times 10^{-15}\ \text{s},$$

so the coherence length is $(3.0 \times 10^8\ \text{m/s})(2.6 \times 10^{-15}\ \text{s}) = 7.9 \times 10^{-7}$ m. Since the path difference between the two slits and a point on the screen is $d \sin \theta = (2.0 \times 10^{-4}\ \text{m}) \sin \theta$, there are visible fringes if

$$(2.0 \times 10^{-4}\ \text{m}) \sin \theta \leq 7.9 \times 10^{-7}\ \text{m},$$

or

$$\sin \theta \leq 3.9 \times 10^{-3}.$$

From Eq. (38-3*a*), the order of a fringe is given by

$$m = \frac{d \sin \theta}{\lambda} = \frac{df \sin \theta}{c} = \frac{(2.0 \times 10^{-4}\ \text{m})(5.1 \times 10^{14}\ \text{Hz}) \sin \theta}{3.0 \times 10^8\ \text{m/s}}$$
$$= 340 \sin \theta,$$

so

$$\frac{m}{340} \leq 3.9 \times 10^{-3},$$

and

$$m \leq 1.3.$$

Thus only the central fringe and the $m = 1$ yellow fringe can be seen.

(*b*) In this case, the coherence time is

$$\Delta t = \frac{1}{1.0 \times 10^9\ \text{Hz}} = 1.0 \times 10^{-9}\ \text{s},$$

so the coherence length is

$$c\,\Delta t = (3.0 \times 10^8\ \text{m/s})(1.0 \times 10^{-9}\ \text{s}) = 0.30\ \text{m}.$$

Since this is considerably greater than the *maximum* path difference

$$L = (2.0 \times 10^{-4}\ \text{m}) \sin 90° = 2.0 \times 10^{-4}\ \text{m},$$

fringes are visible for all θ satisfying Eq. (38-3*a*).

DRILL PROBLEM 38-9

The coherence length of a typical commercial He-Ne laser is 6.0 m. What is the frequency bandwidth of the laser light?
ANS. 5.0×10^7 Hz.

Spatial Coherence

The problem with an extended source is that the interference patterns produced by the light emitted from different points on it are displaced relative to one another. For the source of Fig. 38-22, the interference pattern produced by S is centered at P, while the center of the pattern for S′ is at P′. (Drill Prob. 38-3 describes how P′ can be located.) It can be shown* that if *a disk-shaped monochromatic source of wavelength* λ *and radius R is located a distance L from double slits separated by d, it can be treated to a good approximation as a point source if*

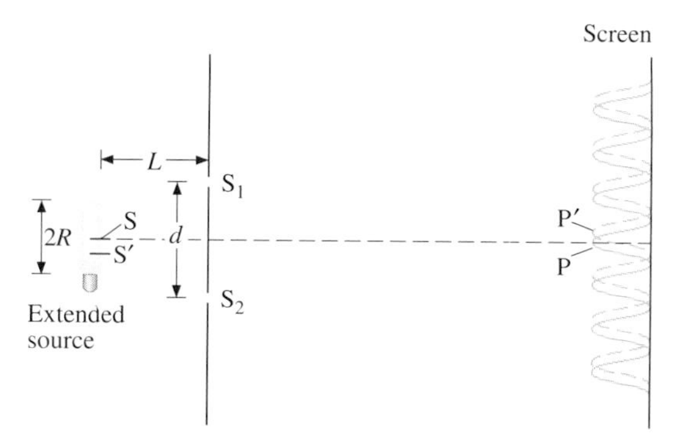

FIGURE 38-22 The two arbitrary points S and S′ on an extended source produce individual interference patterns P and P′, respectively.

$$R < 0.16 \frac{L\lambda}{d}. \tag{38-13}$$

For example, if the aperture S_0 in front of the sodium lamp of Example 38-1 (see Fig. 38-2) is a small round hole of radius R at 1.5 m from the slits, it can be treated as a point source if

$$R < 0.16 \frac{(1.5\ \text{m})(5.89 \times 10^{-7}\ \text{m})}{2.0 \times 10^{-4}\ \text{m}} = 7.1 \times 10^{-4}\ \text{m}.$$

EXAMPLE 38-8 A Sodium Streetlamp as a Point Source

The light from a sodium streetlamp of diameter 5.0 cm illuminates double slits that are separated by 0.30 mm. How far from the lamp must the slits be located if they are to produce sharp fringes?

SOLUTION Sharp fringes are produced if the source is effectively a point source. From Eq. (38-13), this is the case when

$$L > \frac{Rd}{0.16\lambda} = \frac{(2.5 \times 10^{-2}\ \text{m})(3.0 \times 10^{-4}\ \text{m})}{0.16(5.89 \times 10^{-7}\ \text{m})} = 80\ \text{m}.$$

Thus the slits must be at least 80 m from the streetlamp.

38-6 Diffraction

After passing through a narrow aperture, a wave propagating in a specific direction tends to spread out. For example, sound waves that enter a room through an open door can be detected even though the listener is in a part of the room where the geometry of ray propagation dictates that there should only be

*A very readable derivation of this equation can be found in Secs. 12-2 and 12-4 of *Optics* by Eugene Hecht.

silence (Fig. 38-23*a*). Sound waves also "bend" around obstacles. This allows the golfers of Fig. 38-23*b* to converse even though they are separated by several trees. The spreading and bending of sound waves are two examples of *diffraction*, a phenomenon exhibited by all types of waves.

(*a*)

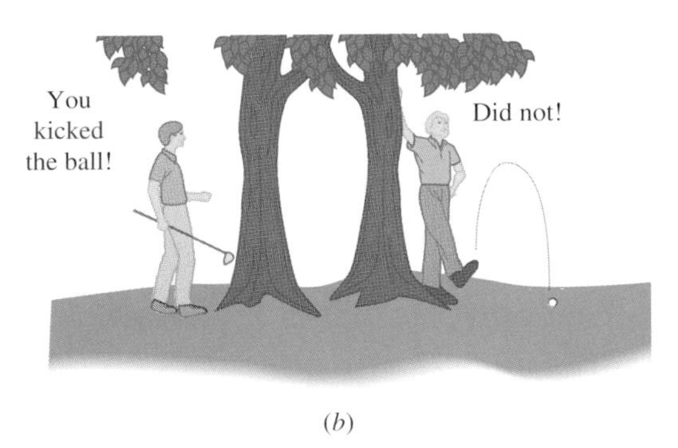

(*b*)

FIGURE 38-23 Because of the diffraction of sound waves, (*a*) the listener in the room is able to hear noise coming in through the door, and (*b*) the golfers who are separated by trees are able to converse.

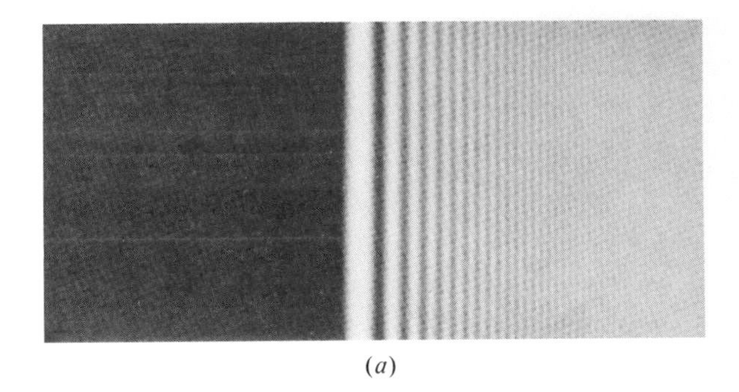

(*a*)

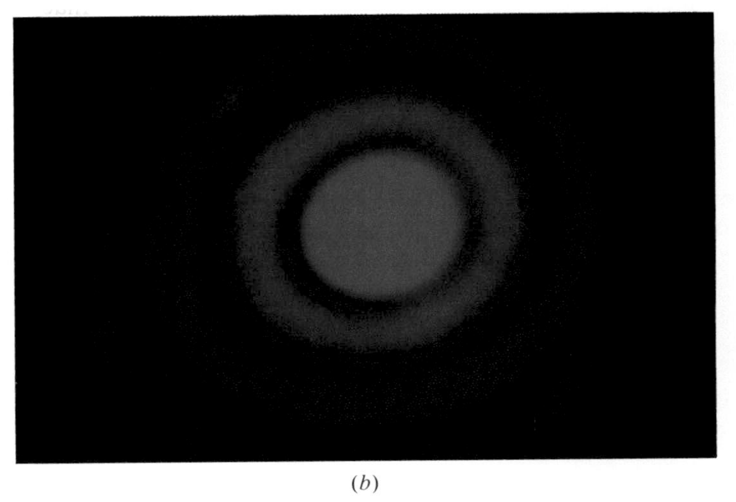

(*b*)

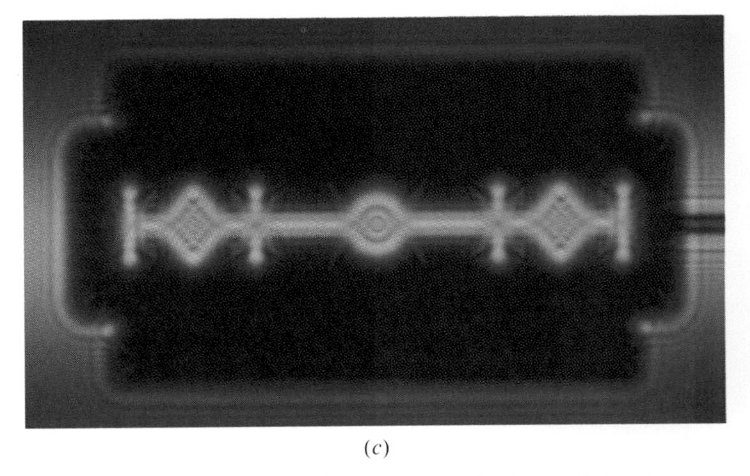

(*c*)

FIGURE 38-24 (*a*) The diffraction patterns of (*a*) a straight edge, (*b*) a small circular obstacle, and (*c*) a razor blade. In all three cases, plane waves fall on the diffracting obstacle and the diffraction pattern is observed on a screen placed behind and close to the obstacle.

The diffraction of sound waves is apparent to us because wavelengths in the audible region are approximately the same size as the objects they encounter, a condition that must be satisfied if diffraction effects are to be observed easily. Since the wavelengths of visible light range from approximately 390 to 770 nm, most common objects don't diffract light significantly. However, situations do occur in which apertures are small enough that the diffraction of light is observable. For example, if you place your middle and index fingers close together and look through the opening at a lightbulb, you'll see a rather clear diffraction pattern consisting of light and dark lines running parallel to your fingers. Other diffraction patterns are shown in Fig. 38-24.

When the light source and the observation point are each far enough from a diffracting object so that essentially plane waves fall on both the object and the observation point, we have *Fraunhofer diffraction*. If the plane-wave criterion is not met, then *Fresnel diffraction* occurs. Fraunhofer diffraction is reasonably simple to describe mathematically and will be discussed in the sections to follow.

38-7 Fraunhofer Diffraction by a Single Slit

Fraunhofer diffraction effects are calculated by dividing the wavefront at an aperture into an infinite number of evenly spaced sources of Huygens wavelets. The interference of these wavelets at the observation screen is then the diffraction pattern. Notice what was just said—*interference* of the wavelets gives the *diffraction* pattern. *Diffraction and interference are fundamentally the same phenomenon.* The distinction between the two is based on the number of Huygens wavelets involved. If the number is small, such as in the case of Young's double-slit experiment, we say that interference occurs. On the other hand, light patterns produced by Huygens wavelets whose sources are distributed *continuously* across a portion of a wavefront are said to produce diffraction. When the number of Huygens sources is large but finite, sometimes we speak of interference, other times of diffraction. For example, we have studied interference due to N slits; yet an optical device made up of N slits is commonly known as a diffraction grating.

Figure 38-25*a* shows a narrow slit of width b that is illuminated by coherent monochromatic light of wavelength λ. The incoming wavefronts are plane waves, and the wavefront across the slit is taken to be made up of N equally spaced sources separated by a distance b/N, each radiating with an amplitude ΔE_0. Eventually, we'll let N go to infinity and make the distribution of Huygens sources continuous across the slit. As shown in the figure, the light rays directed from each source to a particular point on the screen are parallel in Fraunhofer diffraction.

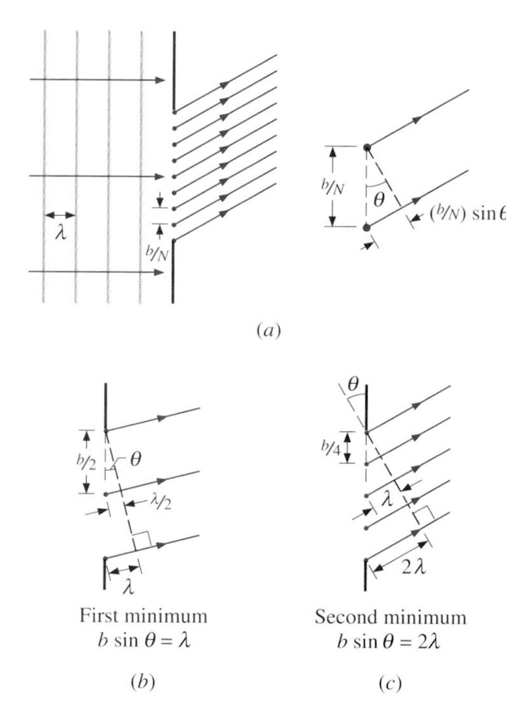

FIGURE 38-25 (*a*) The illuminated slit of width b is represented by N equally spaced sources where $N \to \infty$. The path difference between rays from adjacent sources is $(b/N)\sin\theta$. (*b*) Condition for the first minimum. (*c*) Condition for the second minimum. For clarity, the other sources that make up the slit in (*b*) and (*c*) are not shown.

We can determine the locations of the minima in the diffraction pattern by matching sources appropriately. Let's first assume that θ is such that the paths from the top and middle sources to the screen differ in distance by $\lambda/2$ (see Fig. 38-25*b*). Then the Huygens wavelets from these two sources arrive at the screen out of phase by π and cancel one another. Similarly, the wavelets from the two Huygens sources immediately below the original two cancel at the same point on the screen, as do the second set of sources down, and the third set, etc. Hence there is complete cancellation in pairs of sources at this location and a dark fringe is centered there. This point corresponds to

$$\frac{b}{2}\sin\theta = \pm\frac{\lambda}{2},$$

or

$$b\sin\theta = \pm\lambda.$$

By using this same argument with the slit divided into quarters (see Fig. 38-25*c*), sixths, eighths, and so on, we find the condition that gives us the centers of all of the dark fringes; it is

$$b\sin\theta = m\lambda \qquad (m = \pm1, \pm2, \pm3, \dots). \tag{38-14}$$

Between every pair of dark fringes is a bright fringe. The one between the $m = -1$ and $m = +1$ dark fringes is centered at $\sin\theta = 0$. At this point, the path length is the same for all rays from the slit. This is by far the brightest fringe. It is also twice as wide as the other bright fringes because $\Delta m = 2$ between its pair of bordering dark fringes, while $\Delta m = 1$ for all other pairs.

As in the double-slit experiment, we can focus parallel rays from the sources to the same point on the screen by placing a lens at a distance equal to its focal length in front of the viewing screen (Fig. 38-26). A second lens is also placed at a distance equal to its focal length in front of the point source. This produces an image of the source at infinity, thereby guaranteeing that plane waves fall on the slit.

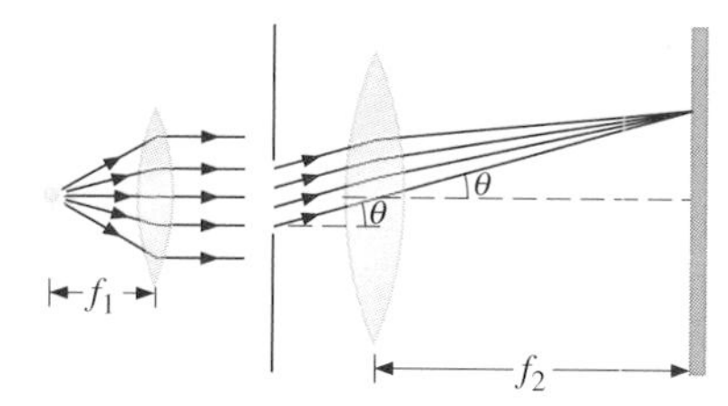

FIGURE 38-26 Fraunhofer diffraction can be observed by placing one converging lens at a distance equal to its focal length f_1 in front of the point source and a second converging lens at a distance equal to its focal length f_2 from the screen. Since the ray through the center of the lens is not deviated, the angle between this ray and the optical axis of the second lens is θ.

EXAMPLE 38-9 SINGLE-SLIT DIFFRACTION

Coherent light of wavelength 589 nm from a sodium lamp falls on a slit of width 0.250 mm. The resulting diffraction pattern is seen on a screen located 2.00 m from the slit. What are the widths of (*a*) the central peak and (*b*) the peaks adjacent to the central one?

SOLUTION (*a*) The first zeros of the diffraction pattern on either side of the central peak correspond to $m = \pm 1$ in Eq. (38-14). For these points,

$$\sin\theta_1 = \pm\frac{\lambda}{b} = \pm\frac{5.89\times 10^{-7}\text{ m}}{2.50\times 10^{-4}\text{ m}} = \pm 2.36\times 10^{-3}.$$

If $\pm y_1$ are the positions of these zeros relative to the center of the diffraction pattern, the width of the central peak is $2y_1$. Since θ_1 is small,

$$\sin\theta_1 \approx \frac{y_1}{D},$$

where D is the distance from the slit to the screen. Thus the width of the central peak is

$$2y_1 = 2D\sin\theta_1 = 2(2.00\text{ m})(2.36\times 10^{-3}) = 9.44\times 10^{-3}\text{ m} = 9.44\text{ mm}.$$

(*b*) For the second set of zeros, $m = \pm 2$, so

$$\sin\theta_2 = \pm\frac{2\lambda}{b} = \pm\frac{2(5.89\times 10^{-7}\text{ m})}{2.50\times 10^{-4}\text{ m}} = \pm 4.71\times 10^{-3},$$

and their positions are

$$y_2 = D\sin\theta_2 = (2.00\text{ m})(\pm 4.71\times 10^{-3}) = \pm 9.42\times 10^{-3}\text{ m} = \pm 9.42\text{ mm}.$$

The width of the peak between the $m = 1$ and the $m = 2$ zeros is therefore

$$y_2 - y_1 = 9.42\text{ mm} - 4.72\text{ mm} = 4.70\text{ mm},$$

or about half the width of the central peak.

DRILL PROBLEM 38-10

Suppose the slit size in Example 38-9 is changed until its width and that of its central diffraction peak are the same. What is the slit width then?
ANS. 1.53 mm.

Intensity of the Single-Slit Diffraction Pattern

To calculate the intensity of the diffraction pattern, we follow the phasor method used for N slits in Sec. 38-2. As shown in Fig. 38-25*a*, the path difference between waves from adjacent sources reaching the arbitrary point P on the screen is $(b/N)\sin\theta$, which is equivalent to a phase difference of $(2\pi b/\lambda N)\sin\theta$. The phasor diagram for the waves arriving at the point whose angular position is θ is shown in Fig. 38-27. The amplitude of the phasor for each Huygens wavelet is ΔE_0, the amplitude of the resultant phasor is E, and the phase difference between the wavelets from the first and last sources is

$$\phi = \frac{2\pi}{\lambda} b\sin\theta.$$

With $N\to\infty$, the phasor diagram approaches a circular arc of length $N\Delta E_0$ and radius r. Since the length of the arc is $N\Delta E_0$ for any ϕ, the radius r of the arc must decrease as ϕ increases (or equivalently, as the phasors form tighter spirals).

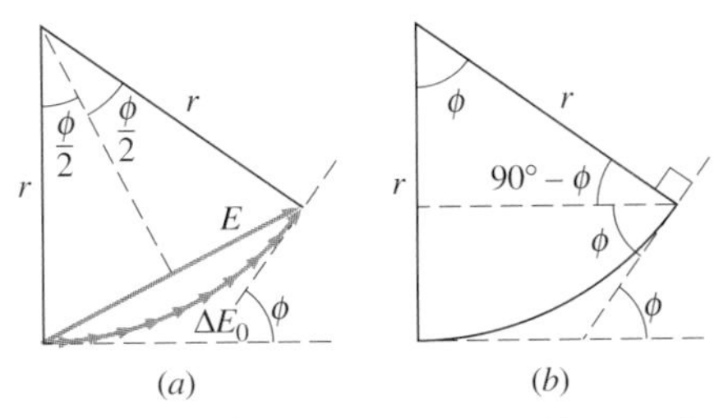

FIGURE 38-27 (*a*) Phasor diagram corresponding to the angular position θ in the single-slit diffraction pattern. The phase difference between the wavelets from the first and last sources is $\phi = (2\pi b/\lambda)\sin\theta$. (*b*) The geometry of the phasor diagram.

The phasor diagram for $\phi = 0$ (the center of the diffraction pattern) is shown in Fig. 38-28*a* using $N = 30$. In this case, the phasors are laid end to end in a straight line of length $N\Delta E_0$, the radius r goes to infinity, and the resultant has its maximum value $E_0 = N\Delta E_0$. The intensity of this maximum is then, from Eq. (35-9),

$$I_0 = \frac{1}{2\mu_0 c}(N\Delta E_0)^2.$$

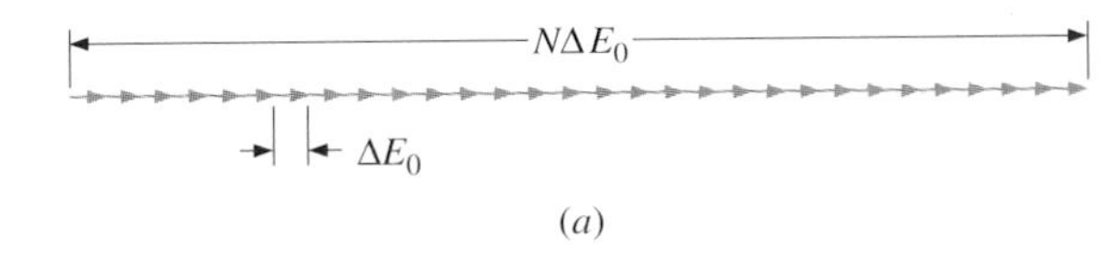

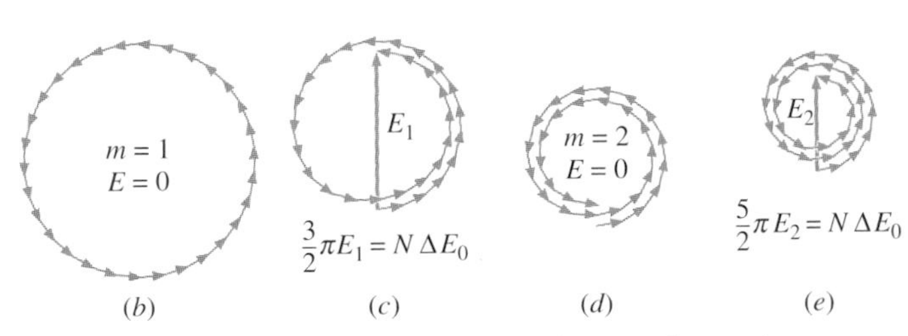

FIGURE 38-28 Phasor diagrams (with 30 phasors) for various points on the single-slit Fraunhofer diffraction pattern. Multiple rotations around a given circle have been separated slightly so that the phasors can be seen. (*a*) Central maximum, (*b*) first minimum, (*c*) first maximum beyond central maximum, (*d*) second minimum, and (*e*) second maximum beyond central maximum.

The phasor diagrams for the first two zeros of the diffraction pattern are shown in Fig. 38-28*b* and *d*. In both cases, the phasors add to zero, after rotating through $\phi = 2\pi$ rad for $m = 1$ and 4π rad for $m = 2$.

The next two maxima beyond the central maximum are represented by the phasor diagrams of Fig. 38-28*c* and *e*. In Fig. 38-28*c*, the phasors have rotated through $\phi = 3\pi$ rad and have formed a resultant phasor of magnitude E_1. The length of the arc formed by the phasors is $N\Delta E_0$. Since this corresponds to 1.5 rotations around a circle of diameter E_1, we have

$$\frac{3}{2}\pi E_1 = N\Delta E_0,$$

so

$$E_1 = \frac{2N\,\Delta E_0}{3\pi},$$

and

$$I_1 = \frac{1}{2\mu_0 c} E_1^2 = \frac{4(N\,\Delta E_0)^2}{(9\pi^2)(2\mu_0 c)} = 0.045 I_0,$$

where

$$I_0 = \frac{(N\,\Delta E_0)^2}{2\mu_0 c}.$$

In Fig. 38-28*e*, the phasors have rotated through $\phi = 5\pi$ rad, corresponding to 2.5 rotations around a circle of diameter E_2 and arc length $N\,\Delta E_0$. Thus

$$\frac{5}{2}\pi E_2 = N\,\Delta E_0,$$

and

$$I_2 = \frac{1}{2\mu_0 c} E_2^2 = \frac{4(N\,\Delta E_0)^2}{(25\pi^2)(2\mu_0 c)} = 0.016 I_0.$$

These two maxima actually correspond to values of ϕ slightly less than 3π rad and 5π rad. Since the total length of the arc of the phasor diagram is always $N\,\Delta E_0$, the radius of the arc decreases as ϕ increases. As a result, E_1 and E_2 turn out to be slightly larger for arcs that have not quite curled through 3π rad and 5π rad, respectively. The exact values of ϕ for the maxima are investigated in Prob. 38-74. In solving that problem, you will find that they are less than, but very close to, $\phi = 3\pi, 5\pi, 7\pi$ rad, ...

To calculate the intensity at an arbitrary point P on the screen, we return to the phasor diagram of Fig. 38-27. Since the arc subtends an angle ϕ at the center of the circle,

$$N\,\Delta E_0 = r\phi$$

and

$$\sin\frac{\phi}{2} = \frac{E}{2r},$$

where E is the amplitude of the resultant field. Solving the second equation for E and then substituting r from the first equation, we find

$$E = 2r\sin\frac{\phi}{2} = 2\,\frac{N\,\Delta E_0}{\phi}\sin\frac{\phi}{2}.$$

Now defining

$$\beta = \frac{\phi}{2} = \frac{\pi b\sin\theta}{\lambda}, \qquad (38\text{-}15)$$

we obtain

$$E = N\,\Delta E_0\,\frac{\sin\beta}{\beta}. \qquad (38\text{-}16)$$

This equation relates the amplitude of the resultant field at any point in the diffraction pattern to the amplitude $N\,\Delta E_0$ at the central maximum. The intensity is proportional to the square of the amplitude, so

$$I = I_0\left(\frac{\sin\beta}{\beta}\right)^2, \qquad (38\text{-}17)$$

where $I_0 = (N\,\Delta E_0)^2/2\mu_0 c$ is the intensity at the center of the pattern.*

For $\phi = 3\pi$ rad and 5π rad, $\beta = 3\pi/2$ rad and $5\pi/2$ rad. When these values of β are substituted into Eq. (38-17), we find that the intensities of the next two maxima beyond the central maximum are

$$I_1 = I_0\left(\frac{\sin 3\pi/2}{3\pi/2}\right)^2 = 0.045 I_0$$

and

$$I_2 = I_0\left(\frac{\sin 5\pi/2}{5\pi/2}\right)^2 = 0.016 I_0,$$

in agreement with what we found earlier in this section using the diameters and circumferences of phasor diagrams.

A plot of Eq. (38-17) is shown in Fig. 38-29, and directly below it is a photograph of an actual diffraction pattern. Notice that the central peak is much brighter than the others, and

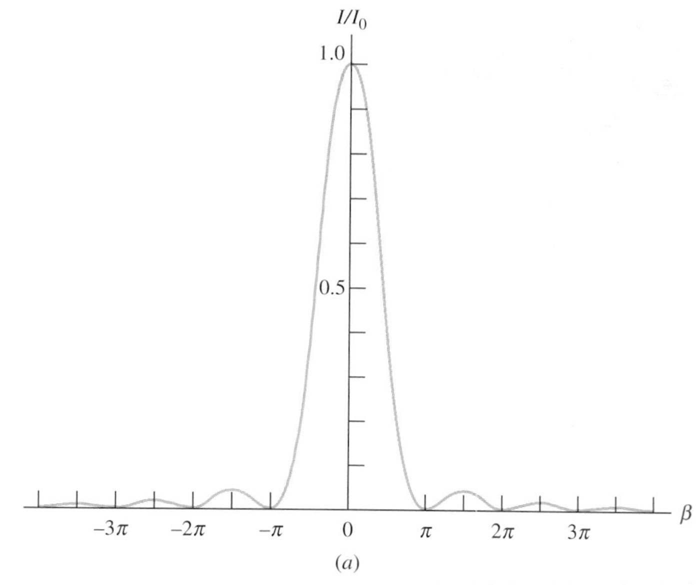

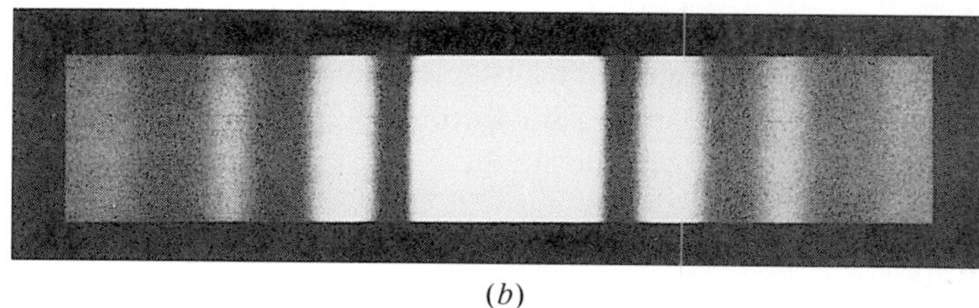

(*b*)

FIGURE 38-29 (*a*) The intensity distribution of a single-slit diffraction pattern. (*b*) The actual diffraction pattern.

*Note that from l'Hôpital's rule, $\lim_{\beta\to 0}(\sin\beta/\beta) = 1$, so $\lim_{\beta\to 0} I = I_0$.

that the zeros of the pattern are located at those points where $\beta = m\pi$ rad. This corresponds to

$$\frac{\pi b \sin\theta}{\lambda} = m\pi,$$

or

$$b \sin\theta = m\lambda,$$

which is Eq. (38-14).

If the slit width b is varied, the intensity distribution changes, as illustrated in Fig. 38-30. The central peak is distributed over the region from $\sin\theta = -\lambda/b$ to $\sin\theta = \lambda/b$. For small θ, this corresponds to an angular width $\Delta\theta \approx 2\lambda/b$. Hence an increase in the slit width results in a decrease in the width of the central peak. For a slit with $b \gg \lambda$, the central peak is very sharp, whereas if $b \approx \lambda$, it becomes quite broad.

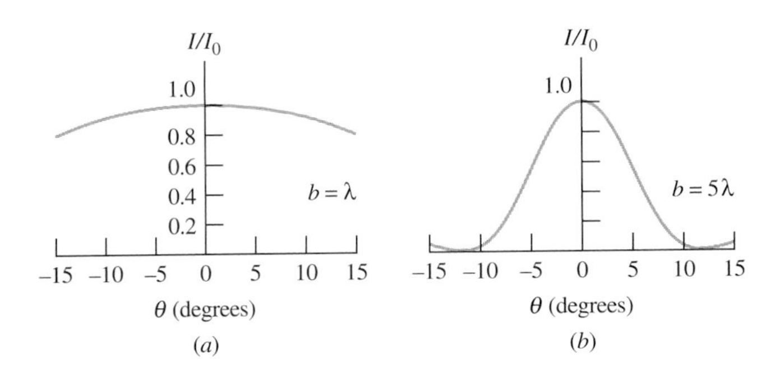

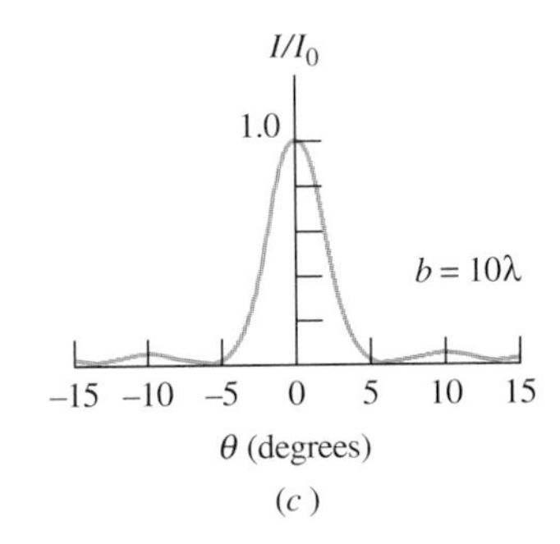

FIGURE 38-30 Single-slit diffraction patterns for various slit widths.

Apertures in everyday life have dimensions that are so much greater than the visible wavelengths that their diffraction patterns are significant only around $\theta = 0$; in other words, the light is essentially undiffracted. As a result, light, unlike sound, does not "bend" around ordinary obstacles or corners. For most objects, the geometry of ray optics quite adequately describes the formation of shadows and images.

*38-8 Fraunhofer Diffraction by N Slits

Because of the finite width of the slits, diffraction is always present in multislit interference experiments. Thus two effects are occurring simultaneously. First, there is interference among the Huygens wavelets from the different sources. This produces patterns like those shown in Fig. 38-15. Second, there is diffraction at each slit (see Fig. 38-29). In order to calculate the Fraunhofer diffraction pattern for any number of slits, we have to generalize the method we just used for the single slit. That is, we replace the wavefront across each slit by a uniform distribution of point sources that radiate Huygens wavelets and then sum the wavelets from all the slits. This gives the incident radiation at any point on the screen. Although the details of that calculation are complicated, the final result is quite simple:

> The diffraction pattern of N slits of width b that are separated by a distance d is the interference pattern of N point sources separated by d multiplied by the diffraction pattern of a slit of width b.

In other words, to find the diffraction pattern of N slits, we multiply the interference and diffraction patterns as given by Eqs. (38-9) (interference) and (38-17) (diffraction):

$$I = I_0\left(\frac{\sin N\alpha}{\sin\alpha}\right)^2\left(\frac{\sin\beta}{\beta}\right)^2. \tag{38-18}$$

Here I_0 is the intensity at $\theta = 0$ from any *one* of the slits, and

$$\alpha = \frac{\pi d \sin\theta}{\lambda},$$

and

$$\beta = \frac{\pi b \sin\theta}{\lambda}.$$

Equation (38-18) is plotted in Fig. 38-31 for $N = 2$, 3, and 6 slits and $d = 4b$. By comparing this figure with the photographs of actual patterns shown in Fig. 38-16, you can see that the decrease in the intensities of the principal maxima, the existence of the subsidiary maxima, and the narrowing of the fringes with N are all accurately predicted by Eq. (38-18).

A change in the slit width affects the intensities but not the locations of the interference maxima, which are determined from $\sin\alpha = 0$. With increasing θ, the intensity decreases because of the diffraction term $(\sin\beta/\beta)^2$. However, as $\theta \to 0$,

$$\left(\frac{\sin\beta}{\beta}\right)^2 \to \left(\frac{\beta}{\beta}\right)^2 = 1,$$

and

$$\left(\frac{\sin N\alpha}{\sin\alpha}\right)^2 \to \left(\frac{N\alpha}{\alpha}\right)^2 = N^2,$$

so I at $\theta = 0$ is

$$I(0) = N^2 I_0,$$

which is identical to what we found for N slits of negligible width.

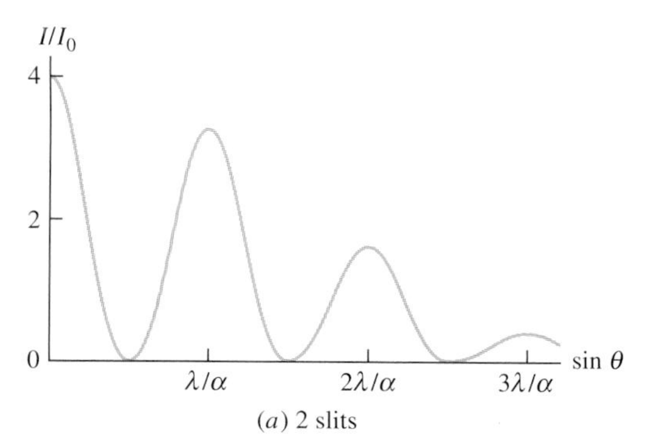

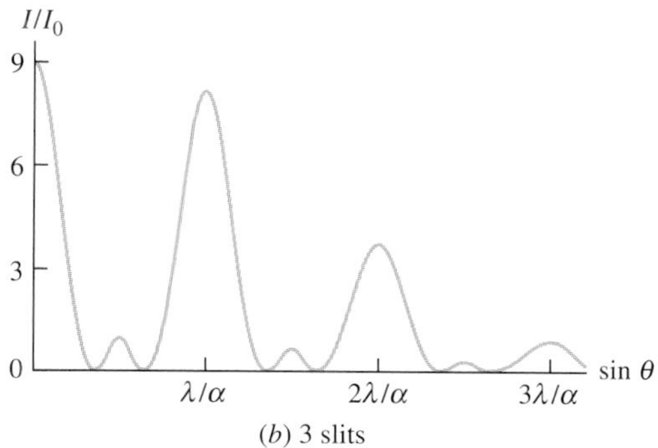

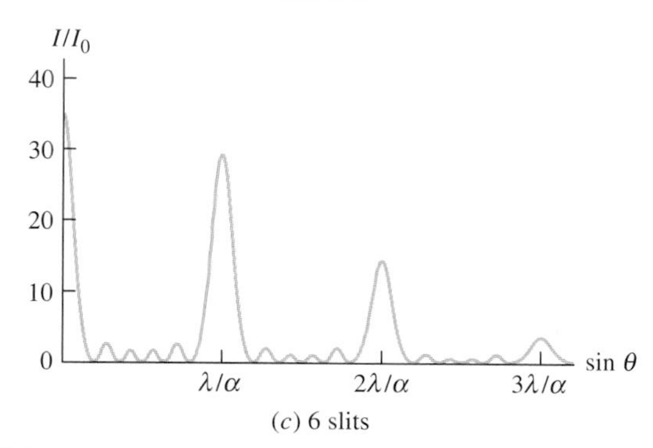

FIGURE 38-31 The Fraunhofer diffraction patterns for $N = 2$, 3, and 6 slits, with the effect of the slit width included. For all three cases, the distance between adjacent slits is four times the width of the slits.

EXAMPLE 38-10 TWO-SLIT DIFFRACTION

Suppose that in Young's experiment, slits of width 0.020 mm are separated by 0.20 mm. If the slits are illuminated by monochromatic light of wavelength 500 nm, how many bright fringes are observed in the central peak of the diffraction pattern?

SOLUTION From Eq. (38-14), the angular position of the first diffraction minimum is

$$\theta \approx \sin\theta = \frac{\lambda}{b} = \frac{5.0 \times 10^{-7}\ \text{m}}{2.0 \times 10^{-5}\ \text{m}} = 2.5 \times 10^{-2}\ \text{rad},$$

and from Eq. (38-3b), the spacing of the interference maxima is

$$\Delta y = \frac{\lambda D}{d} = \left(\frac{5.0 \times 10^{-7}\ \text{m}}{2.0 \times 10^{-4}\ \text{m}}\right) D = (2.5 \times 10^{-3})D.$$

Since the width of the central peak is $2D\theta$, the number of fringes in the central diffraction peak is

$$\frac{2D\theta}{\Delta y} = \frac{2D(2.5 \times 10^{-2}\ \text{rad})}{(2.5 \times 10^{-3})D} = 20.$$

DRILL PROBLEM 38-11

Use $\sin 2A = 2 \sin A \cos A$ to show that for $N = 2$, Eq. (38-18) reduces to the product of Eq. (38-7) and Eq. (38-17).

38-9 RESOLUTION

In most optical instruments, the light from an object passes through a circular aperture (usually a lens) to form an image. Because of diffraction, a point object is not focused to a point image, as was assumed in geometric optics. The image is instead a diffraction pattern with most of the intensity in the central peak. For a slit of width b, the first minimum of the Fraunhofer pattern occurs at $\sin\theta = \lambda/b$. We're interested here in a circular aperture, and the corresponding formula for the location of the first minimum turns out to be

$$\sin\theta = 1.22\,\frac{\lambda}{D}, \tag{38-19}$$

where D is the diameter of the aperture.

In order to determine if the images of two objects can be easily distinguished, we calculate the amount of overlap in their diffraction patterns. When the central maxima are sufficiently far apart, the overlap is small, and the two images are easily separable (Fig. 38-32a). However, if the central maxima are close, then we may only perceive a single merged image of the two objects, as shown in Fig. 38-32b. The minimum angular separation between the diffraction patterns such that the images can be resolved is given by *Rayleigh's criterion*. This states that when the central maximum of one diffraction pattern coincides with the first minimum of the other (as in Fig. 38-32c),

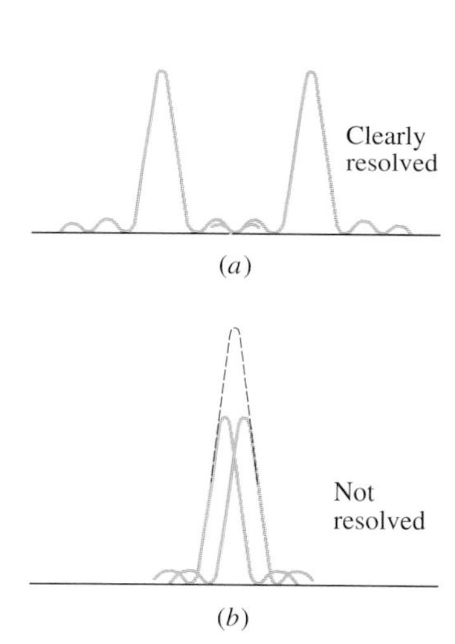

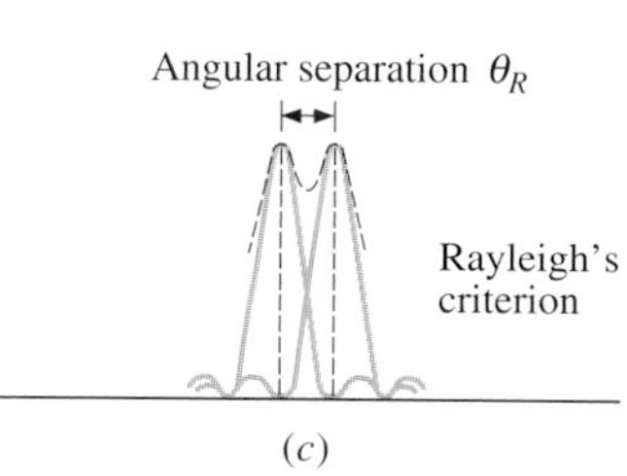

FIGURE 38-32 The resolution of the images depends on the amount of overlap in the diffraction patterns. (a) Little overlap and good resolution; (b) much overlap and no resolution; (c) Rayleigh's criterion, or the minimum angular separation for resolution.

the angular separation is just sufficient for their resolution. From Eq. (38-19), this minimum angular separation is

$$\theta_R = \sin^{-1}\left(1.22\,\frac{\lambda}{D}\right) \approx 1.22\,\frac{\lambda}{D}, \tag{38-20}$$

since $\lambda \ll D$.

Figure 38-33 shows that the angular separation of the centers of the diffraction patterns is equal to that of the objects. Thus Eq. (38-20) also gives the angular separation of two objects whose images are just resolvable by the optical instrument. If the objects are separated by l and are a distance L from the instrument, then $\theta_R \approx l/L$, and the objects must have a minimum linear separation of

$$l \approx 1.22\,\frac{L\lambda}{D} \tag{38-21}$$

if their images are to be resolved.

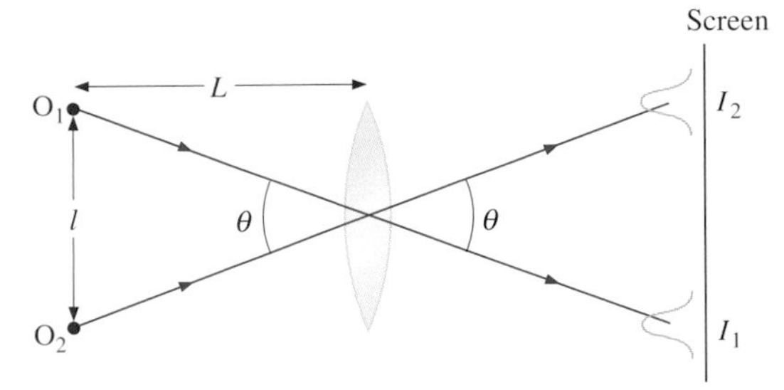

FIGURE 38-33 Since rays through the center of a lens are not deviated, the angular separations of the objects and of their diffraction patterns are the same.

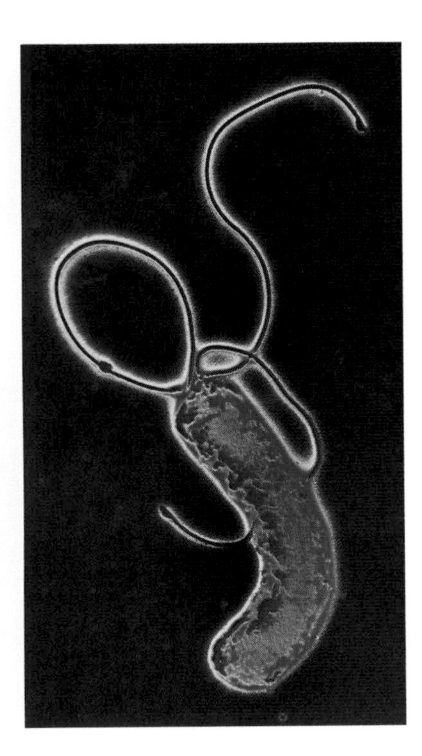

A magnified view (×5310) of the bacterium *Helicobacter*, taken with an electron microscope. We know from quantum mechanics that electrons have wavelike properties. Their wavelengths are typically much less than those of visible light. Hence very fine resolution can be obtained with an electron microscope.

EXAMPLE 38-11 Resolution of the Eye

The headlights of a car are 1.5 m apart. How far can you be from this car and still be able to resolve its headlights? Assume that the wavelength of the light is 500 nm, the diameter of your pupil is 5.0 mm, and $n = 1.33$ for the medium in the eye.

SOLUTION The situation is shown in Fig. 38-34. Within the eye, $\lambda = (500\text{ nm})/1.33 = 376\text{ nm}$. Using Rayleigh's criterion, we find

$$L \approx \frac{Dl}{1.22\lambda} = \frac{(5.0\times10^{-3}\text{ m})(1.5\text{ m})}{1.22(3.76\times10^{-7}\text{ m})} = 1.6\times10^{4}\text{ m}.$$

So this car has to be about ten miles away before its headlights become indistinguishable as separate entities to the eye.

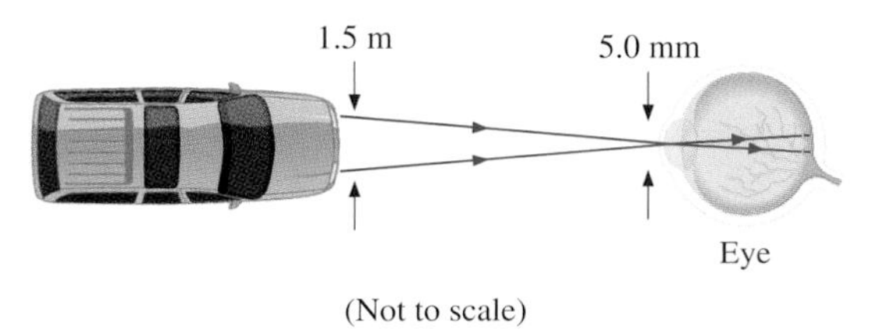

FIGURE 38-34 Resolution of a car's headlights.

As the car approaches, its headlights become more distinguishable.

DRILL PROBLEM 38-12

The Mount Palomar telescope, whose objective mirror has a diameter of 508 cm, is focused on the moon. (*a*) What is the minimum separation of two objects on the moon that can be resolved with the telescope? (*b*) Answer the same question if the objects are viewed through an eye with a pupil diameter of 4.0 mm. Assume that λ is 550 nm, and take the distance from the telescope to the area under observation to be 3.8×10^8 m.
ANS. (*a*) 50 m; (*b*) 48 km.

38-10 Diffraction Grating

The diffraction grating is a device commonly used in the measurement of the wavelengths of light emitted by various atoms. An atom of a particular element can be "fingerprinted" by the distribution of wavelengths (or *spectrum*) it radiates. Hence with measurements of the wavelengths emitted, a scientist is able to identify the elements contained in the sample producing the light.

The simplest type of diffraction grating is the multiple-slit configuration discussed in Sec. 38-3. It can be made by scratching equally spaced grooves on a glass surface with a diamond point, whose motion is accurately controlled by a precision positioning machine. A typical grating contains about 5000 lines per centimeter, so the distance d between these slits is $d = (1/5000)\text{ cm} = 2.0 \times 10^{-4}$ cm.

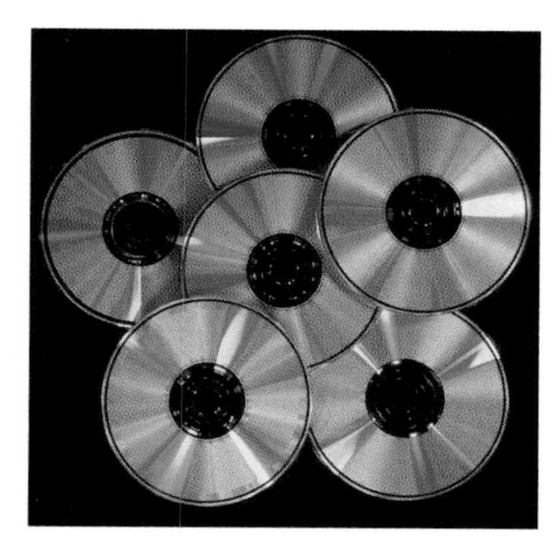

A compact disk with its evenly spaced grooves serves as a reflection grating.

For a given wavelength, the positions of the principal maxima in the N-slit interference pattern are given by $d \sin\theta = m\lambda$. A measurement of θ therefore provides us with a value for λ. As an example, consider a grating with 4800 lines per centimeter. Here

$$d = \frac{1}{4800} \times 10^{-2}\text{ m} = 2.08 \times 10^{-6}\text{ m}.$$

If the second maximum is found at $\theta = 30.2°$, we have

$$\lambda = \frac{d \sin\theta}{m} = \frac{(2.08 \times 10^{-6}\text{ m})(\sin 30.2°)}{2}$$
$$= 5.23 \times 10^{-7}\text{ m}.$$

A grating is usually one component in a *spectroscope*, a device used to measure radiation spectra. A photograph of a simple student spectroscope and its schematic representation are shown in Fig. 38-35. Light from the source is focused on the slit S_1 of the collimator C by the lens L_1. This slit is in the front focal plane of the lens L_2, so a parallel light beam emerges from L_2. The beam then falls on the grating G, where it is diffracted toward the telescope T. The light from the grating is focused to the back focal plane of L_3, where it can be viewed. By rotating the telescope through different angles, the observer is able to see the entire visible spectrum of the source. A scale at the telescope gives the angle θ, which is then used in Eq. (38-3*a*) to determine λ. Some spectra are shown in Fig. 38-36.

Photo courtesy of PASCO Scientific.

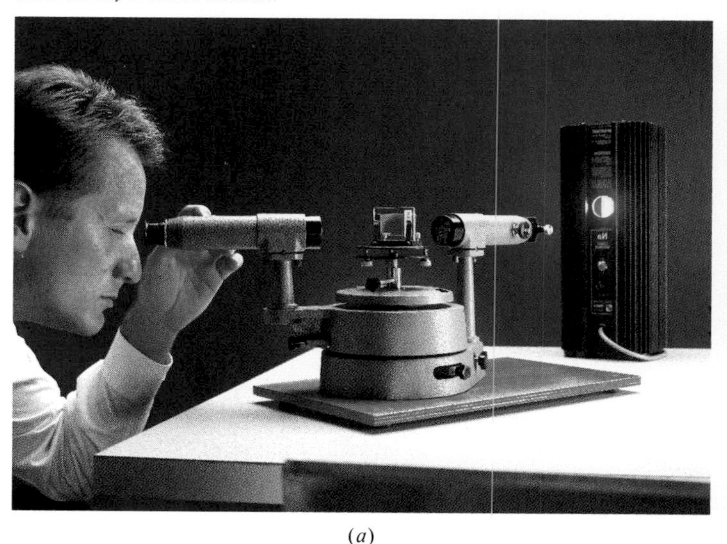

(*a*)

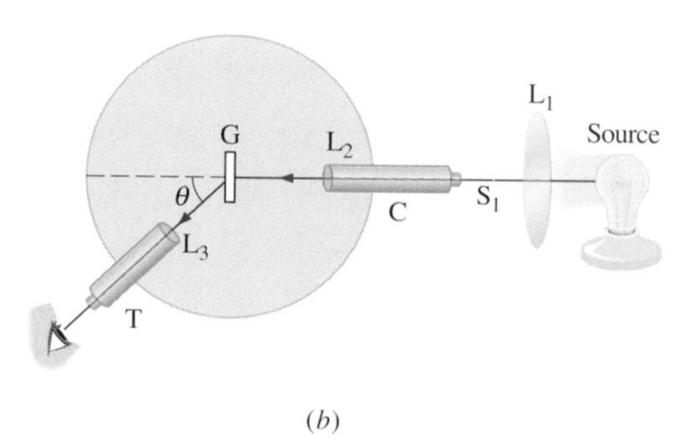

(*b*)

FIGURE 38-35 A student spectroscope and its schematic representation.

Our ability to analyze the spectrum of an atom is limited by how well we can resolve fringes corresponding to two wavelengths that are nearly equal. The difficulty arises because the fringes do have a width, so if they overlap too much, we cannot distinguish them. Once again, Rayleigh's criterion is used

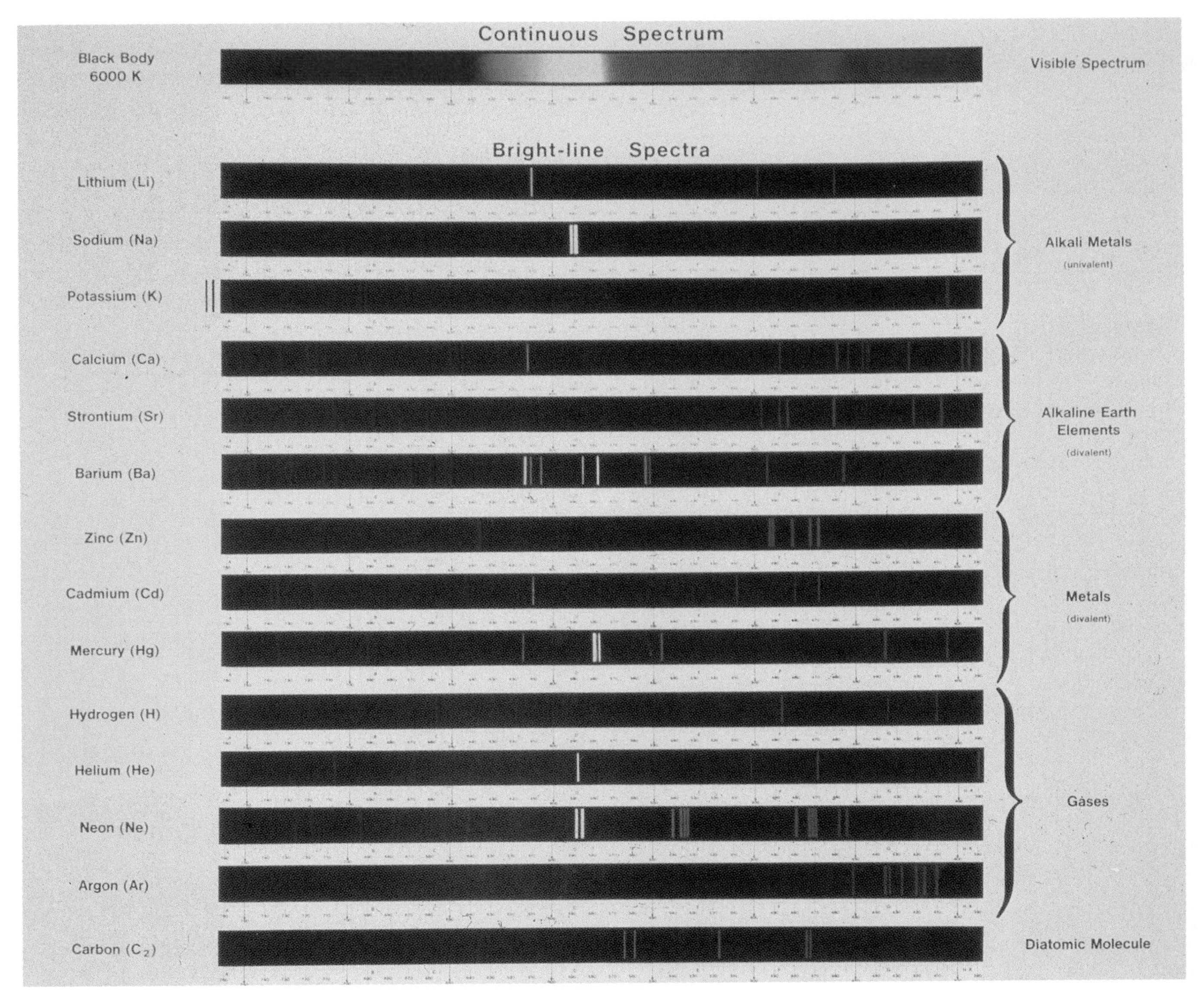

Sargent-Welch Scientific.

FIGURE 38-36 A spectrum chart.

to define the limiting condition for resolving two fringes. The two fringes shown in Fig. 38-37 become just resolvable when the maximum of one peak falls on top of the first minimum of the other peak. To quantify this condition, we have to compare the angular width of a fringe to the angular separation between the centers of the two fringes.

Since the fringes are assumed to be very close together, their angular widths are nearly equal and may be found from Eq. (38-10) using their average wavelength $\bar{\lambda}$. Representing this common angular width by $\Delta\theta_w$, we have

$$\Delta\theta_w = \frac{2\bar{\lambda}}{Nd\cos\theta}.$$

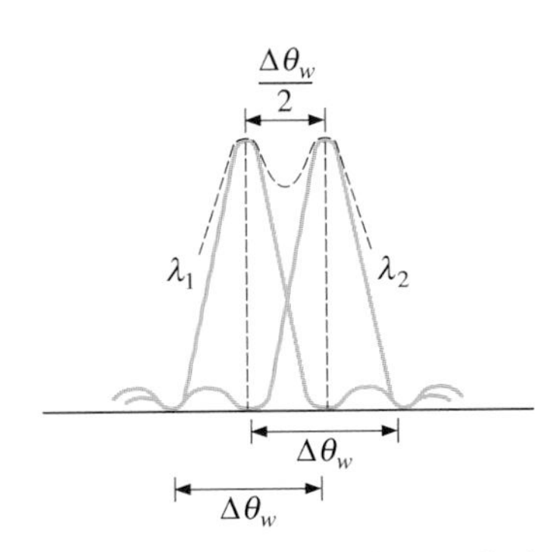

FIGURE 38-37 Rayleigh's criterion for the resolution of two fringes.

The angular separation $\Delta\theta$ of the centers of the two fringes can be obtained by taking differentials in Eq. (38-3*a*):

$$\Delta(d \sin \theta) = \Delta(m\lambda),$$

so

$$d \cos \theta \, \Delta\theta = m \, \Delta\lambda,$$

and

$$\Delta\theta = \frac{m \, \Delta\lambda}{d \cos \theta}.$$

From Rayleigh's criterion, the two fringes are just resolvable when $\Delta\theta = \Delta\theta_w/2$ (see Fig. 38-37). Thus

$$\frac{m \, \Delta\lambda}{d \cos \theta} = \frac{1}{2}\left(\frac{2\bar{\lambda}}{Nd \cos \theta}\right),$$

and

$$\frac{\bar{\lambda}}{\Delta\lambda} = mN.$$

The ratio $\bar{\lambda}/\Delta\lambda$ is called the **resolving power** R of the grating:

$$R = \frac{\bar{\lambda}}{\Delta\lambda} = mN. \qquad (38\text{-}22)$$

Notice that the value of R depends not only on the number of lines of the grating but also on the order of the fringes. Knowing R, we can determine whether or not two closely spaced fringes can be resolved. We gain resolution by using higher-order fringes. However, the disadvantage in doing so is that the intensities of the fringes decrease with order, so they become harder to see.

EXAMPLE 38-12 Resolving the Sodium Doublet

The yellow light of the sodium lamp is composed of two closely spaced fringes with wavelengths $\lambda = 589.00$ and 589.59 nm. These two fringes are called the *sodium doublet*. Can a grating of width 1.20 cm with 600 rulings per centimeter resolve these two fringes in first order? in second order?

SOLUTION Because the grating is 1.20 cm wide, the number of rulings it has is

$$N = (1.20 \text{ cm})(600 \text{ rulings/cm}) = 720 \text{ rulings}.$$

The resolving power required to distinguish the two fringes is

$$R = \frac{\bar{\lambda}}{\Delta\lambda} = \frac{(589.00 \text{ nm} + 589.59 \text{ nm})/2}{589.59 \text{ nm} - 589.00 \text{ nm}} = \frac{589.30 \text{ nm}}{0.59 \text{ nm}} = 998.$$

For first order, $R = (1)(720)$, which is insufficient to resolve the fringes. However, for second order, $R = (2)(720) = 1440$, which is large enough for their resolution.

SUMMARY

1. **Double-slit interference experiment**

   (*a*) The positions of the interference maxima are given by

   $$d \sin \theta = m\lambda$$

   or

   $$y = \frac{m\lambda D}{d} \qquad (m = 0, \pm 1, \pm 2, \ldots),$$

   where d is the distance between the slits and D is the distance between the slits and the screen.

   The locations of the interference minima are given by

   $$d \sin \theta = \left(m + \frac{1}{2}\right)\lambda$$

   or

   $$y = \left(m + \frac{1}{2}\right)\frac{\lambda D}{d} \qquad (m = 0, \pm 1, \pm 2, \ldots).$$

   (*b*) The intensity of the interference pattern varies with position on the screen according to

   $$I = I_0 \cos^2 \frac{\phi}{2},$$

   where $\phi = (2\pi d/\lambda) \sin \theta$ and I_0 is the intensity at $\phi = 0$.

2. **Multiple-slit interference patterns**

   (*a*) The positions of the N-slit interference maxima and minima are given, respectively, by

   $$d \sin \theta = m\lambda \qquad (m = 0, \pm 1, \pm 2, \ldots)$$

   and

   $$d \sin \theta = (m + \tfrac{1}{2})\lambda \qquad (m = 0, \pm 1, \pm 2, \ldots),$$

   where d is the distance between adjacent slits.

(*b*) The intensity of the interference pattern varies with position on the screen according to

$$I = I_0\left(\frac{\sin N\alpha}{\sin \alpha}\right)^2,$$

where I_0 is the intensity of one Huygens wavelet and $\alpha = \pi d \sin\theta/\lambda$.

3. **Interference in thin films**
   (*a*) Light waves undergo a phase change of π rad upon reflection at an interface beyond which is a medium of higher refractive index. There is no phase change if that medium has a lower index of refraction.
   (*b*) When a thin film of thickness t is surrounded by air, the interference conditions for light reflected nearly normally off the two surfaces are

$$2nt = (m + \tfrac{1}{2})\lambda_0 \qquad \text{(constructive)}$$

and

$$2nt = m\lambda_0 \qquad \text{(destructive)},$$

where n is the index of refraction of the film, λ_0 is the wavelength of the light in air, and $m = 0, 1, 2, \ldots$.
   (*c*) If the thin film is situated between two media, one with a lower and the other with a higher refractive index than the film's, the constructive and destructive interference conditions are interchanged.

*4. **The Michelson interferometer**
When the mirror in one arm of the interferometer moves a distance $\lambda/2$, each fringe in the interference pattern moves to the position previously occupied by the adjacent fringe.

*5. **Coherence**
   (*a*) Fringes in the double-slit experiment are visible only when the product of the frequency bandwidth Δf of the source and the time Δt for light to travel the difference in path lengths from the slits to the screen is less than or equal to 1.
   (*b*) If a disk-shaped monochromatic source of wavelength λ and radius R is located a distance L from double slits with separation d, it can be treated to a good approximation as a point source if

$$R < 0.16\,\frac{L\lambda}{d}.$$

6. **Diffraction**
Fraunhofer diffraction occurs when the light source and the observation point are each far enough from the diffracting object that essentially plane waves fall on both the object and the observation point. When these conditions are not met, Fresnel diffraction occurs.

7. **Fraunhofer diffraction by a single slit**
   (*a*) The positions of the minima in a single-slit diffraction pattern are given by

$$b\sin\theta = m\lambda \qquad (m = \pm 1, \pm 2, \pm 3, \ldots),$$

where b is the width of the slit. The maxima in this diffraction pattern are located approximately halfway between the minima.
   (*b*) The intensity of this diffraction pattern is given by

$$I = I_0\left(\frac{\sin\beta}{\beta}\right)^2,$$

where $\beta = \pi b \sin\theta/\lambda$ and I_0 is the intensity at $\theta = 0$.

*8. **Fraunhofer diffraction by *N* slits**
The diffraction pattern of N slits is the product of the interference pattern of N point sources and the diffraction pattern of a single slit.

9. **Resolution**
If two objects are a distance L from an optical instrument with a circular opening of diameter D, the objects must have a minimum linear separation of

$$l = 1.22\,\frac{L\lambda}{D}$$

if their images are to be resolved.

10. **Diffraction grating**
   (*a*) When light of wavelength λ falls on a diffraction grating whose slits are separated by d, fringes appear at positions given by the N-slit interference condition, $d\sin\theta = m\lambda$.
   (*b*) The resolving power $R = \bar{\lambda}/\Delta\lambda$ of a grating with N slits is

$$R = mN,$$

where m is the order of the fringes.

QUESTIONS

38-1. Why won't two small sodium lamps, held close together, produce an interference pattern on a distant screen?

38-2. What happens to the double-slit interference pattern if $d < \lambda$? if $d < \lambda/2$?

38-3. Why is monochromatic light used in the double-slit experiment? What would happen if white light was used?

38-4. At a point where two waves of unequal amplitude interfere destructively, the amplitude of the resultant wave is always smaller than the amplitude of either of the individual waves. True or false?

38-5. Describe what would happen to the interference pattern in Young's experiment if the two pinholes were made progressively larger.

38-6. Suppose that a double-slit apparatus is immersed in water. How would its interference pattern be affected?

38-7. Describe how the double-slit experiment can be used to correlate different colors with different wavelengths.

38-8. How would the double-slit interference pattern be affected if the separation between the two slits was gradually increased?

38-9. At a point of constructive interference in the double-slit interference pattern, the intensity is four times that of either of the constituent waves. Does this violate the law of energy conservation? Explain.

38-10. Suppose you could vary the wavelength of light emitted by your source over a continuous range of values. Describe how the appearance of the double-slit interference pattern would be affected as you gradually increased the wavelength through the range of the visible spectrum.

38-11. The vertical component of the vector sum of phasors representing oscillating fields gives the amplitude of the resultant field. Does the horizontal component have any physical significance?

38-12. By choosing the appropriate value of the constant phase difference between adjacent phasors, is it always possible to arrange N phasors ($N > 1$) of the same amplitude such that they sum to zero?

38-13. Compare interference and diffraction.

38-14. Explain how evenly spaced radio antennas can be used to produce a radio signal that propagates primarily in one direction.

38-15. (*a*) Explain why the center of the Newton's rings pattern is always dark. (*b*) When a thin film of kerosene spreads out on the surface of water, the thinnest part of the film looks bright in the interference pattern. What does this tell you about the refractive index of kerosene?

38-16. Why are interference fringes seen on thin films but not on thick films such as a block of glass?

38-17. A coated lens appears greenish-yellow when viewed with reflected light. What wavelengths are destroyed in the reflected light?

38-18. Describe how a Michelson interferometer can be used to measure the index of refraction of a gas (including air).

38-19. How does the motion of the molecules in a gas discharge tube affect the coherence length of the light emitted from the tube?

38-20. How does the width of the central peak of the single-slit diffraction pattern change as the width of the slit is varied?

38-21. If you and a friend are on opposite sides of a hill, you can communicate with walkie-talkies but not with flashlights. Explain.

38-22. What happens to the diffraction pattern of a single slit when the entire optical apparatus is immersed in water?

38-23. In our study of diffraction by a single slit, we assume that the length of the slit is much larger than the width. What happens to the diffraction pattern if these two dimensions are comparable?

38-24. A rectangular slit is twice as wide as it is high. Is the central diffraction peak wider in the vertical direction or in the horizontal direction?

38-25. Is higher resolution obtained in a microscope with red or blue light?

38-26. The resolving power of a refracting telescope increases with the size of its objective lens. What other advantage is gained with a larger lens?

38-27. The distance between atoms in a molecule is about 10^{-8} cm. Can visible light be used to "see" molecules?

38-28. Suppose light emerging from a spectroscope is dispersed into its various wavelengths. How can you tell whether a diffraction grating or a prism was used to separate the wavelengths of the light?

38-29. For a diffraction grating, what is the advantage of (*a*) closely spaced slits? (*b*) many slits?

38-30. Consider double slits separated by a distance d and a diffraction grating containing $1/d$ lines per meter. (*a*) How are their interference patterns alike? (*b*) How are they different?

38-31. In the study of spectra with a diffraction grating, what is the advantage of using a higher-order line? What is the disadvantage?

PROBLEMS

Double-Slit Interference Pattern

38-1. For 600-nm wavelength light and a slit separation of 0.12 mm, what are the angular positions of the first and third maxima in the double-slit interference pattern?

38-2. If the light source in Prob. 38-1 is changed, the angular position of the third maximum is found to be 0.57°. What is the wavelength of light being used now?

38-3. Red light ($\lambda = 710$ nm) illuminates double slits separated by a distance $d = 0.150$ mm. The screen and the slits are 3.0 m apart. (*a*) Find the distance on the screen between the central maximum and the third maximum. (*b*) What is the distance between the second and the fourth maxima?

38-4. Two sources are in phase and emit waves with $\lambda = 0.42$ m. Determine whether constructive or destructive interference occurs at points whose distances from the two sources are (*a*) 0.84 and 0.42 m, (*b*) 0.21 and 0.42 m, (*c*) 1.26 and 0.42 m, (*d*) 1.87 and 1.45 m, (*e*) 0.63 and 0.84 m, and (*f*) 1.47 and 1.26 m.

38-5. Two slits 4.0×10^{-6} m apart are illuminated by light of wavelength 600 nm. What is the highest-order fringe in the interference pattern?

38-6. Suppose that the highest-order fringe that can be observed is the eighth in a double-slit experiment where 550-nm wavelength light is used. What is the minimum separation of the slits?

38-7. In a double-slit experiment, the fifth maximum is 2.8 cm from the central maximum on a screen that is 1.5 m away from the slits. If the slits are 0.15 mm apart, what is the wavelength of the light being used?

38-8. The source in Young's experiment emits at two wavelengths. On the viewing screen, the fourth maximum for one wavelength is located at the same spot as the fifth maximum for the other wavelength. What is the ratio of the two wavelengths?

38-9. If 500- and 650-nm light illuminates two slits that are separated by 0.50 mm, how far apart are the second-order maxima for these two wavelengths on a screen 2.0 m away?

38-10. Two narrow slits 0.75 mm apart are illuminated by light of wavelength 500 nm. (*a*) What is the phase difference between the two interfering waves at a point 2.75 mm from the central maximum on a screen 3.0 m away? (*b*) What is the ratio of the intensity at this point to the intensity at the center of a bright fringe?

38-11. Determine what happens to the double-slit interference pattern if one of the slits is covered with a thin, transparent film whose thickness is $\lambda/[2(n-1)]$, where λ is the wavelength of the incident light and n is the index of refraction of the film.

Multiple-Slit Interference Patterns

38-12. With the help of a phasor diagram, find an expression for the resultant field when the following fields are added: $E_1 = E_0 \sin \omega t$, $E_2 = E_0 \sin(\omega t + 30°)$, and $E_3 = E_0 \sin(\omega t + 45°)$.

38-13. Consider an interference experiment using eight equally spaced slits. Determine the smallest phase difference in the waves from adjacent slits such that the resultant wave has zero amplitude. Verify your result with a phasor diagram.

38-14. Ten narrow slits are equally spaced 0.25 mm apart and illuminated with yellow light of wavelength 580 nm. (*a*) What are the angular positions of the third and fourth principal maxima? (*b*) What is the separation of these maxima on a screen 2.0 m from the slits?

38-15. Find the angular widths of the third- and fourth-order bright fringes of Prob. 38-14.

38-16. For a three-slit interference pattern, find the ratio of the peak intensities of a subsidiary maximum to a principal maximum.

38-17. What is the angular width of the central fringe for the interference pattern of (*a*) 20 slits separated by $d = 2.0 \times 10^{-3}$ mm? (*b*) 50 slits with the same separation? Assume that $\lambda = 600$ nm.

38-18. Fifty-one narrow slits are equally spaced and separated by 0.10 mm. The slits are illuminated by blue light of wavelength 400 nm. What is the peak intensity of the twenty-fifth subsidiary maximum in comparison with that of a primary maximum?

Interference in Thin Films

38-19. A transparent film of thickness 250 nm and refractive index 1.40 is surrounded by air. What wavelength in a beam of white light at near-normal incidence to the film undergoes destructive interference when reflected?

38-20. An intensity minimum is found for 450-nm light transmitted through a transparent film ($n = 1.20$) in air. (*a*) What is the minimum thickness of the film? (*b*) If this wavelength is the shortest for which the intensity minimum occurs, what are the next three lower values of λ for which this happens?

38-21. A thin film with $n = 1.32$ is surrounded by air. What is the minimum thickness of this film such that the reflection of normally incident light with $\lambda = 500$ nm is minimized?

38-22. Repeat your calculation of the previous problem with the thin film placed on a flat glass ($n = 1.50$) surface.

38-23. After a minor oil spill, a thin film of oil ($n = 1.40$) of thickness 450 nm floats on the water surface in a bay. (*a*) What predominant color is seen by a bird flying overhead? (*b*) What predominant color is seen by a seal swimming underwater?

38-24. A microscope slide 10 cm long is separated from a glass plate at one end by a sheet of paper. As shown in the accompanying figure, the other end of the slide is in contact with the plate. The slide is illuminated from above by light from a sodium lamp ($\lambda = 589$ nm), and 14 fringes per centimeter are seen along the slide. What is the thickness of the piece of paper?

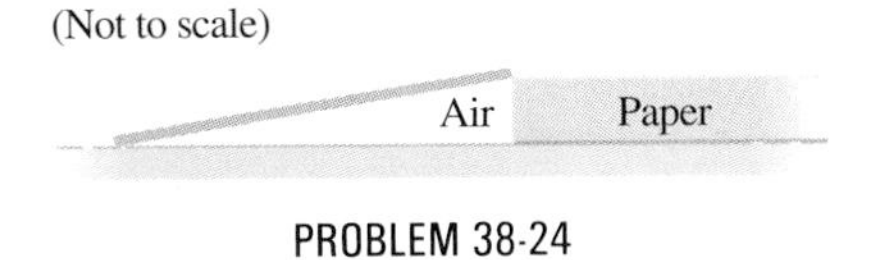

PROBLEM 38-24

38-25. Suppose that the setup of Prob. 38-24 is immersed in an unknown liquid. If 18 fringes per centimeter are now seen along the slide, what is the index of refraction of the liquid?

38-26. In a Newton's rings experiment, the radius of curvature of the lens is 40.0 cm and the light is monochromatic with a wavelength of 650 nm. What is the radius of the tenth bright ring?

38-27. The Newton's rings apparatus of the previous problem is immersed in a liquid, and the radius of the tenth bright fringe is measured to be 1.40 mm. What is the index of refraction of the liquid?

The Michelson Interferometer

38-28. When the traveling mirror of a Michelson interferometer is moved 2.40×10^{-5} m, 90 fringes pass by a point on the observation screen. What is the wavelength of the light used?

38-29. What is the distance moved by the traveling mirror of a Michelson interferometer that corresponds to 1500 fringes passing by a point on the observation screen? Assume that the interferometer is illuminated with the 606-nm spectral line of krypton-86.

38-30. A chamber 5.0 cm long with flat, parallel windows at the ends is placed in one arm of a Michelson interferometer. (See the accompanying figure.) The light used has a wavelength of 500 nm in a vacuum. While all the air is being pumped out of the chamber, 29 fringes pass by a point on the observation screen. What is the refractive index of the air?

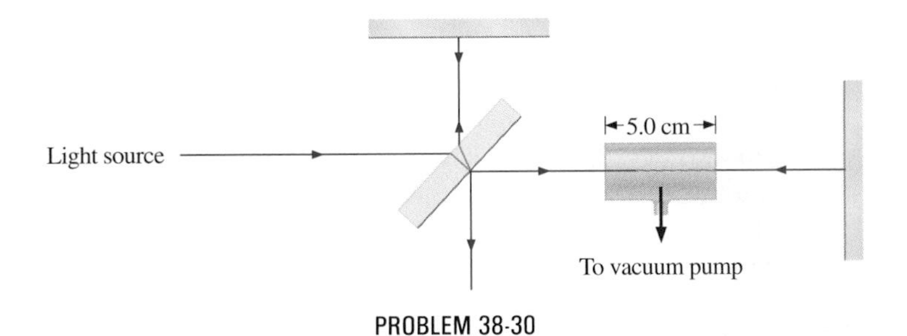

PROBLEM 38-30

Coherence

38-31. The emission from common laboratory lasers typically has a frequency bandwidth of about 3.0×10^6 Hz. (*a*) What is the coherence time of the light they emit? (*b*) What is the coherence length of the light? (*c*) Does this bandwidth have any practical effect on limiting the visibility of the fringes in Young's double-slit experiment?

38-32. An excited atom radiates light with a frequency bandwidth of 1.0×10^8 Hz. (*a*) If the average wavelength of the radiation is 600 nm, how does the frequency bandwidth compare with the average frequency? (*b*) What is the coherence length of the radiation? (*c*) If this light is used with a Michelson interferometer, how large can the path difference between the two arms be without destroying the fringe pattern? (*d*) Answer parts (*b*) and (*c*) for white light with a bandwidth of 3.0×10^{14} Hz.

38-33. A broadband filter is placed in front of a white-light source being used to illuminate double slits separated by 0.20 mm. The filter passes all light with wavelengths between 500 and 600 nm. (*a*) What is the frequency bandwidth of the transmitted light? (*b*) What is the coherence time of this light? (*c*) What is the coherence length? (*d*) Estimate the highest-order fringe at 550 nm that is visible in the interference pattern of the slits.

38-34. Yellow light ($\lambda = 589$ nm) emanates from a circular opening of radius 2.0 cm. How far away from double slits separated by 0.20 mm does the opening have to be before it qualifies as a point source for the slits?

38-35. A circular opening of variable diameter is placed in front of a large discharge tube producing red light ($\lambda = 650$ nm). Slits separated by 0.30 mm are positioned 4.0 m from the opening. As the size of the opening is decreased, at what diameter do easily visible fringes first appear?

Single-Slit Diffraction

38-36. What are the angular positions of the first and second minima in a diffraction pattern produced by a slit of width 0.20 mm that is illuminated by 400-nm light? What is the angular width of the central peak?

38-37. How far would you place a screen from the slit of the previous problem so that the second minimum is a distance of 2.5 mm from the center of the diffraction pattern?

38-38. The width of the central peak in a single-slit diffraction pattern is 5.0 mm. The wavelength of the light is 600 nm, and the screen is 2.0 m from the slit. What is the width of the slit?

38-39. For the pattern of the previous problem, determine the ratio of the intensity at 4.5 mm from the center of the pattern to the intensity at the center.

38-40. Consider a single-slit diffraction pattern for $\lambda = 589$ nm, $D = 1.0$ m, and $b = 0.25$ mm. How far from the center of the pattern are the centers of the first and second dark fringes?

38-41. How narrow is a slit that produces a diffraction pattern on a screen 1.8 m away whose central peak is 1.0 m wide? Assume $\lambda = 589$ nm.

38-42. (*a*) Assume that the maxima are halfway between the minima of a single-slit diffraction pattern. Then use the diameter and circumference of the phasor diagram, as described in Sec. 38-7, to determine the intensities of the fourth and fifth maxima in terms of the intensity of the central maximum. (*b*) Do the same calculation, using Eq. (38-17).

38-43. If the separation between the first and the second minima of a single-slit diffraction pattern is 6.0 mm, what is the distance between the screen and the slit? The light wavelength is 500 nm and the slit width is 0.16 mm.

38-44. Suppose that the central peak of a single-slit diffraction pattern is so wide that the first minima can be assumed to occur at angular positions of $\pm 90°$. For this case, what is the ratio of the slit width to the wavelength of the light?

38-45. Blue light of wavelength 450 nm falls on a slit of width 0.25 mm. A converging lens of focal length 20 cm is placed behind the slit and focuses the diffraction pattern on a screen. (*a*) How far is the screen from the lens? (*b*) What is the distance between the first and third minima of the diffraction pattern?

38-46. Consider the single-slit diffraction pattern for $\lambda = 600$ nm, $b = 0.025$ mm, and $D = 2.0$ m. Find the intensity in terms of I_0 at $\theta = 0.5°$, $1.0°$, $1.5°$, $3.0°$, and $10.0°$.

38-47. As an example of diffraction by apertures of everyday dimensions, consider a doorway of width 1.0 m. (*a*) What is the angular position of the first minimum in the diffraction pattern of 600-nm light? (*b*) Repeat this calculation for a musical note of frequency 440 Hz (A above middle C). Take the speed of sound to be 343 m/s.

Multiple-Slit Diffraction

38-48. The central diffraction peak of the double-slit interference pattern contains exactly nine fringes. What is the ratio of the slit separation to the slit width?

38-49. For a double-slit configuration where the slit separation is four times the slit width, how many interference fringes lie in the central peak of the diffraction pattern?

38-50. Light of wavelength 500 nm falls normally on 50 slits that are 2.5×10^{-3} mm wide and spaced 5.0×10^{-3} mm apart. How many interference fringes lie in the central peak of the diffraction pattern?

Resolution

38-51. The characters of a stadium scoreboard are formed with closely spaced lightbulbs that radiate primarily yellow light. (Use $\lambda = 600$ nm.) How closely must the bulbs be spaced so that an observer 80 m away sees a display of continuous lines rather than the individual bulbs? Assume that the pupil of the observer's eye has a diameter of 5.0 mm.

38-52. Can an astronaut orbiting the earth in a satellite at a distance of 180 km from the surface distinguish two skyscrapers that are 20 m apart? Assume that the pupils of the astronaut's eyes have a diameter of 5.0 mm and that most of the light is centered around 500 nm.

38-53. What is the minimum angular separation of two stars that are just resolvable by the 5.1-m Mount Palomar telescope? Use 550 nm for the wavelength of the light from the stars.

38-54. Suppose you are looking down at a highway from a jetliner flying at an altitude of 6.0 km. How far apart must two cars be if you are able to distinguish them? Assume that $\lambda = 550$ nm and that the diameter of your pupils is 4.0 mm.

38-55. Objects viewed through a microscope are placed very close to the focal point of the objective lens. (*a*) Show that the minimum separation s of two objects resolvable through the microscope is given by

$$s = \frac{1.22\lambda f_0}{D},$$

where λ is the wavelength of the light, f_0 is the focal length of the objective lens, and D is the diameter of the lens. (*b*) The "angle of acceptance" α of the objective lens is illustrated in the accompanying figure. Show that s is given in terms of α by

$$s = \frac{0.61\lambda}{\tan\alpha}.$$

(*c*) A typical value for α is 60°. What is the minimum separation s of two objects resolvable through the microscope if light of wavelength 550 nm is used?

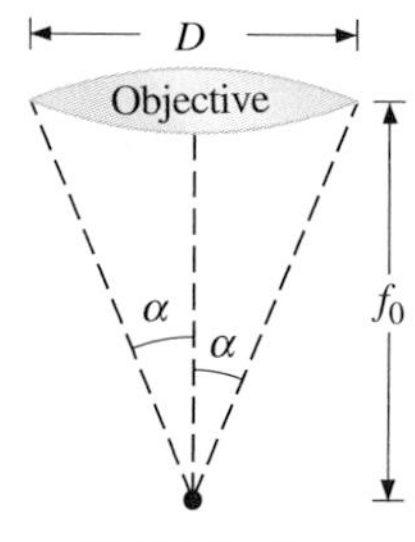

PROBLEM 38-55

38-56. Find the radius of a star's image on the retina of an eye if its pupil is open to 0.65 cm and the distance from the pupil to the retina is 2.8 cm. Assume that $\lambda = 550$ nm.

38-57. *Radio telescopes* are telescopes used for the detection of radio emission from space. Because radio waves have much longer wavelengths than visible light, the diameter of a radio telescope must be very large to provide good resolution. For example, the radio telescope in Arecibo, Puerto Rico (see the accompanying figure), has a diameter of 300 m and can be tuned to wavelengths as low as 4.0 cm. At this wavelength, what is the minimum angular separation of two stars that can be resolved by the telescope?

PROBLEM 38-57

38-58. A spy satellite orbits the earth at a height of 180 km. What is the minimum diameter of the objective lens in a telescope that must be used to resolve columns of troops marching 2.0 m apart? Assume that $\lambda = 550$ nm.

Diffraction Grating

38-59. How many lines per centimeter must be ruled on a 5.0-cm-wide grating if it is just able to resolve the two wavelengths 415.724 and 415.744 nm in first order?

38-60. Yellow sodium light consists of two closely spaced wavelengths, $\lambda_1 = 589.00$ nm and $\lambda_2 = 589.59$ nm. If this light falls normally on a diffraction grating with 5000 rulings per centimeter, how wide must the grating be to just resolve the two fringes in second order?

38-61. A grating with 5000 rulings per centimeter is used to produce two sets of fringes with $\lambda_1 = 570.720$ nm and $\lambda_2 = 570.730$ nm. (*a*) What is the highest order in the diffraction pattern of these fringes? (*b*) For this order, how wide must the grating be so that the two fringes are just resolved?

38-62. A diffraction grating produces a second maximum that is 8.97 cm from the central maximum on a screen 2.0 m away. If the grating has 600 rulings per centimeter, what is the wavelength of the light that produces the diffraction pattern?

38-63. A grating with 4000 lines per centimeter is used to diffract light that contains all wavelengths between 400 and 650 nm. How wide is the first-order spectrum on a screen 3.0 m from the grating?

38-64. A diffraction grating with 2000 rulings per centimeter is used to measure the wavelengths emitted by a hydrogen-gas discharge tube. (*a*) At what angles will you find the maxima of the two first-order blue fringes of wavelengths 410 and 434 nm? (*b*) The maxima of two other first-order fringes of the spectrum are found at $\theta_1 = 0.0972$ rad and $\theta_2 = 0.132$ rad. What are the wavelengths of these fringes?

38-65. For white light ($400 \text{ nm} < \lambda < 700$ nm) falling normally on a diffraction grating, show that the second- and third-order spectra overlap no matter what the grating constant d is.

38-66. How many complete orders of the visible spectrum ($400 \text{ nm} < \lambda < 700$ nm) can be produced with a diffraction grating that contains 5000 rulings per centimeter?

General Problems

38-67. In Young's double-slit experiment, the center of the third bright fringe is 1.80 cm from the center of the pattern. At which two positions does the intensity of this fringe drop to one-third of its maximum value?

38-68. White light falls on two narrow slits separated by 0.40 mm. The interference pattern is observed on a screen 3.0 m away. (*a*) What is the separation between the first maxima for red light ($\lambda = 700$ nm) and violet light ($\lambda = 400$ nm)? (*b*) At what point nearest the central maximum will a maximum for yellow light ($\lambda = 600$ nm) coincide with a maximum for violet light? Identify the order for each maximum.

38-69. (*a*) Show that in the N-slit interference pattern, the subsidiary maxima are located approximately at points where $\alpha = \pm 3\pi/2N$, $\pm 5\pi/2N, \ldots$. (*b*) Show that if N is large and θ is small, the intensity of a first subsidiary maximum (see Fig. 38-15) is 0.045 that of a principal maximum. Also show that the intensity of a second subsidiary maximum is 0.016 that of a principal maximum.

38-70. Different filters are placed in front of a white-light source, thereby producing monochromatic light of various wavelengths. The light falls normally on a flat, glass surface that is covered with a thin film of water. The reflected intensity is found to reach a minimum at $\lambda = 400$ nm and then rise to a maximum at $\lambda = 700$ nm. What is the thickness of the film?

38-71. A thin layer of transparent plastic ($n = 1.60$) is deposited on a glass ($n = 1.50$) surface. When illuminated from above by light ($\lambda = 589$ nm) from a sodium lamp, the surface appears dark. What are the two smallest values possible for the thickness of the plastic layer?

38-72. The sun is 1.5×10^{11} m from the earth and has a mean radius of 7.0×10^8 m. Does the sun qualify as a good point source for a pair of slits separated by 0.10 mm?

38-73. Microwaves of wavelength 10 mm fall normally on a metal plate that contains a slit 25 mm wide. (*a*) Where are the first minima of the diffraction pattern? (*b*) Would there be minima if the wavelength were 30 mm?

38-74. (*a*) By differentiating Eq. (38-17), show that the higher-order maxima of the single-slit diffraction pattern occur at values of β that satisfy $\tan \beta = \beta$. (*b*) Plot $y = \tan \beta$ and $y = \beta$ versus β and find the intersections of these two curves. What information do they give you about the locations of the maxima? (*c*) Convince yourself that these points do not occur exactly at $\beta = (n + 1/2)\pi$, where $n = 0, 1, 2, \ldots$, but are quite close to these values.

38-75. Three slits of width 0.025 mm are spaced 0.050 mm apart. Plane waves of wavelength 500 nm are incident normally on the slits. Calculate the ratios of the intensities of the $m = 1$ and $m = 2$ fringes to the intensity of the $m = 0$ fringe.

38-76. *Quasars*, or *quasi-stellar radio sources*, are astronomical objects discovered in 1960. They are strong emitters of radio waves and appear to be stars because of their great distances from our galaxy. The quasar 3C405 is actually two discrete sources that subtend an angle of 82 seconds at the earth. If this object is studied using radio emission at a frequency of 410 MHz, what is the minimum-diameter radio telescope that can resolve the images of the two sources?

38-77. Three closely spaced fringes of the calcium spectrum have wavelengths of 445.661, 445.588, and 445.477 nm. If a grating with rulings spaced a distance $d = 1.2 \times 10^{-4}$ cm apart is used to observe these fringes, what is the minimum width it must have if the fringes are to be resolved in first order?

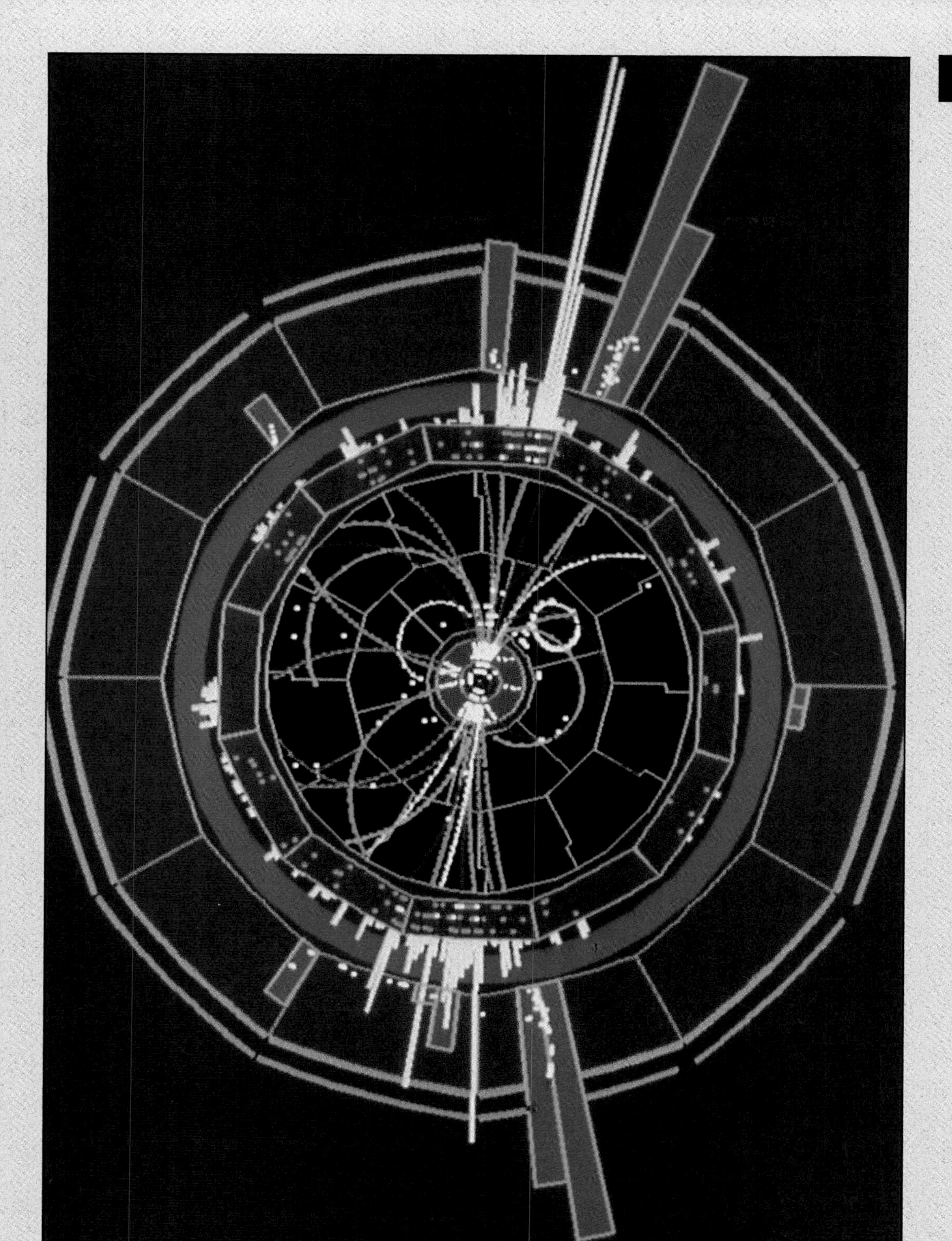

PART VI

MODERN PHYSICS

An application of Einstein's theory of special relativity: The Tokamak Fusion Test Reactor.

CHAPTER 39 THE SPECIAL THEORY OF RELATIVITY

PREVIEW

In this chapter we investigate the special theory of relativity. Since this is a theory about space and time, it has profound implications for all fundamental physical theories. The main topics to be discussed here are the following:

1. **Galilean transformation and invariance.** We examine the Galilean relationships between measurements of the position and time of the same physical event made in two inertial frames in relative motion. We also consider the invariance of physical theories under the Galilean transformation.
2. **Experimental basis for the special theory of relativity.** Observations that led to questions about the validity of the Galilean transformation are described.
3. **Postulates of the special theory of relativity.** Albert Einstein's two postulates for the special theory of relativity are presented and explained.
4. **Synchronization of clocks.** We discuss the importance of clock synchronization along with its inertial-frame dependence.
5. **Measurement of transverse lengths.** We consider the relationship between the measurements of the transverse length of the same object made in two inertial frames in relative motion.
6. **Time dilation.** The effect of motion on the rate of a clock is discussed.
7. **Measurement of longitudinal lengths.** The relationship between the measurements of the longitudinal length of the same object made in two inertial frames in relative motion is considered.
8. **Clock synchronization revisited.** A quantitative treatment of clock synchronization is given.
9. **Lorentz transformation.** We examine the Lorentz relationships between measurements of the position and time of the same physical event made in two inertial frames in relative motion.
10. **Velocity transformations.** The velocity of the same particle as measured in two inertial frames in relative motion is investigated.
11. **Relativistic momentum and energy.** Relativistic expressions for the momentum and energy of a particle are described. The principles of conservation of momentum and conservation of energy are considered in the context of the special theory of relativity.

The special theory of relativity was proposed in 1905 by Albert Einstein (1879–1955). With its acceptance came a profound change in the way space and time are perceived. The "commonsense" rules that we use to relate space and time measurements in the Newtonian world are just not correct in the Einsteinian world of speeds near that of light. For example, the special theory of relativity tells us that measurements of lengths and time intervals are not the same in reference frames moving relative to one another. A particle might be observed to have a lifetime of 1.0×10^{-8} s in one reference frame, but a lifetime of 2.0×10^{-8} s in another; and an object might be measured to be 2.0 m long in one frame and 3.0 m long in another frame.

Unlike Newtonian mechanics, which describes the motion of particles, or Maxwell's equations, which specify how the electromagnetic field behaves, special relativity is not restricted to a particular type of phenomenon. It is instead a "theory on theories" since its rules on space and time affect all fundamental physical theories.

We begin our study of special relativity by reviewing the evidence that led to its development. We then consider the postulates of the theory and investigate how they affect our perceptions of space and time. We also examine how Newton's second law must be modified to make it consistent with the postulates. This modification forces us to alter our perceptions of mass and energy as well, eventually leading us to Einstein's famous equation, $E = mc^2$.

Now the modifications of Newtonian mechanics does not mean that all you have learned about this subject is incorrect. As you will see, the equations of relativistic Newtonian mechanics differ from those of classical Newtonian mechanics only if the bodies under investigation are moving at relativistic speeds (i.e., speeds less than, but comparable to, the speed of light). In the macroscopic world that you encounter in your daily life, the relativistic equations reduce to the classical equations, and the predictions of classical Newtonian mechanics are in agreement with experiment.

39-1 Galilean Transformation and Invariance

In Chap. 4, we discussed how the space and time coordinates of an event observed in different reference frames are related in Newtonian mechanics. As a review, we now consider the particle P of Fig. 39-1, whose motion is analyzed relative to the inertial coordinate systems S and S′. If the origins of S and S′ coincide at $t = 0$, and if S′ is moving in the x direction at a constant speed u relative to S, then

$$x = x' + ut, \qquad y = y', \qquad z = z'. \tag{39-1a}$$

Implicit in these equations is the assumption that time measurements for observers in S and S′ are the same; that is

$$t = t'. \tag{39-1b}$$

These four equations are known collectively as the *Galilean transformation*.

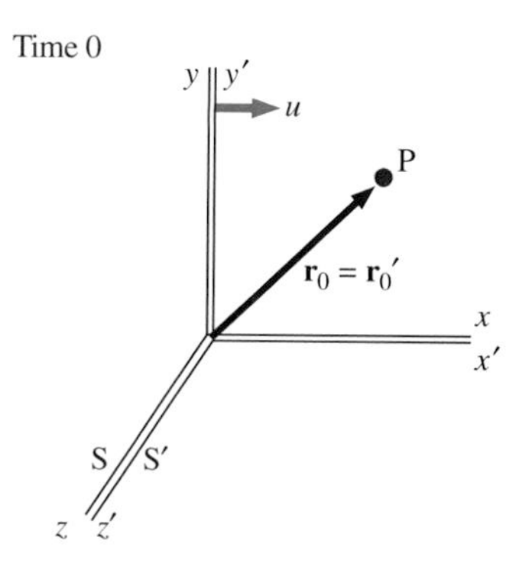

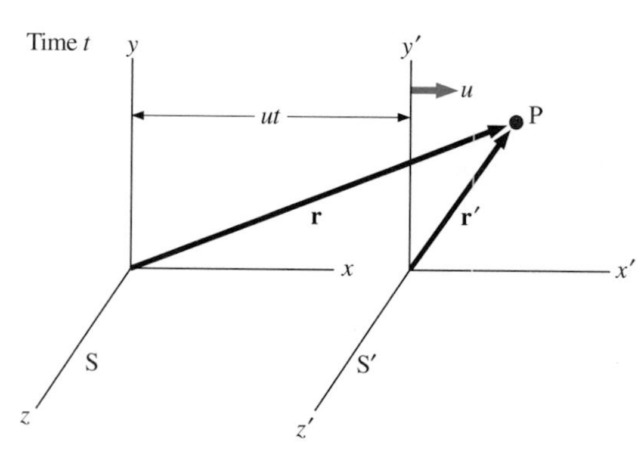

FIGURE 39-1 The motion of particle P can be analyzed relative to either of the inertial frames S or S′.

The Galilean velocity and acceleration transformation equations are found by differentiating Eqs. (39-1a) with respect to time. We then have

$$v_x = v_x' + u, \qquad v_y = v_y', \qquad v_z = v_z' \tag{39-2}$$

and

$$a_x = a_x', \qquad a_y = a_y', \qquad a_z = a_z'. \tag{39-3}$$

How are Newton's laws affected by the Galilean transformation? From Eq. (39-2), a reference frame moving at constant velocity relative to S′ is also moving at constant velocity relative to S. Thus observers in both S and S′ will identify the same frames as inertial. Newton's first law is therefore unaffected by the Galilean transformation.

To consider the effect of the Galilean transformation on the second law, let's consider a particular system of particles for which the interaction between any pair of particles depends solely on the distance between them.* In frame S, the distance between particle i and particle j is

$$|\mathbf{r}_i - \mathbf{r}_j| = \sqrt{(x_i - x_j)^2 + (y_i - y_j)^2 + (z_i - z_j)^2}.$$

In frame S′, the distance between particle i and particle j is

$$|\mathbf{r}_i' - \mathbf{r}_j'| = \sqrt{(x_i' - x_j')^2 + (y_i' - y_j')^2 + (z_i' - z_j')^2}.$$

*For example, the particles could be exerting gravitational forces on one another.

From the Galilean transformation, this may be written as

$$|\mathbf{r}_i' - \mathbf{r}_j'|$$
$$= \sqrt{[(x_i - ut) - (x_j - ut)]^2 + (y_i - y_j)^2 + (z_i - z_j)^2}$$
$$= \sqrt{(x_i - x_j)^2 + (y_i - y_j)^2 + (z_i - z_j)^2}.$$

Hence the distance between an arbitrary pair of particles is the same in S and S′; that is,

$$|\mathbf{r}_i - \mathbf{r}_j| = |\mathbf{r}_i' - \mathbf{r}_j'|.$$

Since the forces of interaction among the particles depend only on the distances between them, the net force on any particle is also the same in S and S′. Finally, assuming that mass is the same in all inertial frames and using $\mathbf{a} = \mathbf{a}'$ (Eq. 39-3), we conclude that Newton's second law has the same form in S and S′, or more generally, in all inertial frames.

Since forces between particles are the same in S and S′, we conclude that Newton's third law is also unaffected by the Galilean transformation.

Although we have just shown that Newton's laws are not affected by the Galilean transformation for a particular type of force, this fact is true for any force; that is, Newton's laws have the same form in all inertial frames. We say that the three laws are *invariant* under the Galilean transformation. Because of the invariance of Newton's laws, our view of the world using Newtonian mechanics is the same in all inertial frames. There does not exist a particular inertial frame with special properties that other inertial frames do not possess.

Now let's see if the basic laws of electromagnetism are invariant as well under the Galilean transformation. In our study of electricity and magnetism, we chose an arbitrary inertial frame and developed Maxwell's equations. These equations led us to the conclusion that the speed of light relative to our reference frame is $c = 1/\sqrt{\epsilon_0 \mu_0} = 3.00 \times 10^8$ m/s in any direction. But what about the propagation of light in an inertial frame moving relative to ours? Suppose a lightbulb at the origin of the reference frame S of Fig. 39-2*a* is flashed at $t = 0$, the same instant that the origin of a second reference frame S′ coincides with the origin of S. The frame S′ is moving in the x direction at a speed u relative to S. To an observer in S, the wavefront of the light pulse spreads out isotropically from the origin (see Fig. 39-2*b*). According to the Galilean transformation, the wavefront of the pulse observed in S′ moves at a speed $c - u$ along the $+x'$ axis and at a speed $c + u$ along the $-x'$ axis. Since $(c + u)t > (c - u)t$, an observer in S′ does not see a spherical wavefront (see Fig. 39-2*c*). Consequently, Maxwell's equations, inherent in which is the speed of light, are *not invariant* under the Galilean transformation.

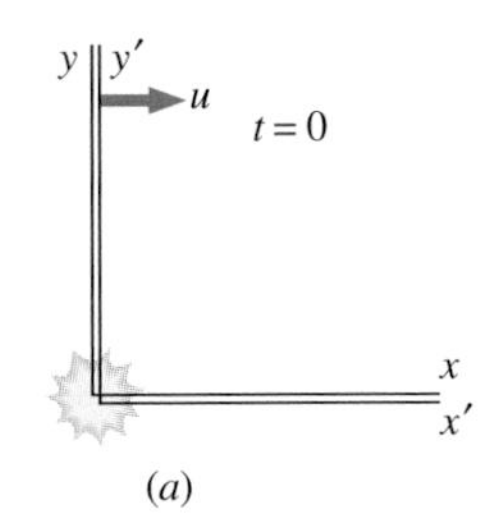

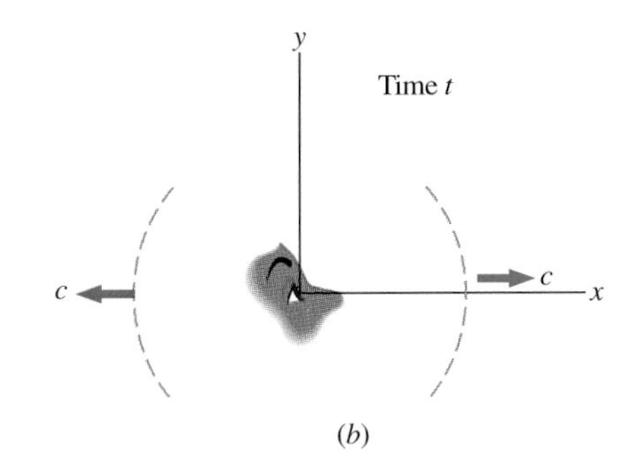

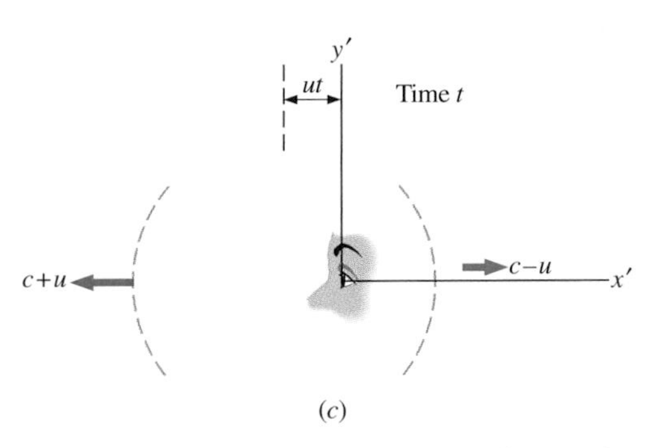

FIGURE 39-2 The Galilean transformation applied to electromagnetism. (*a*) The lightbulb is flashed when S and S′ coincide at $t = 0$. (*b*) The wavefront as seen in S is isotropic. (*c*) The wavefront as seen in S′ is not isotropic.

39-2 Experimental Basis for the Special Theory of Relativity

Scientists of the late nineteenth century were intrigued by the idea that Galilean invariance should exist for one set of laws of classical physics and not for another. Was it possible that either the Newtonian or Maxwellian equations were incorrect as written? There were many experiments performed and theories proposed in attempts to resolve this question. In the following paragraphs we discuss two of the most important of these.

Michelson-Morley Experiment

This experiment was performed in 1887 by Albert Michelson, the inventor of the interferometer discussed in Chap. 38, and Edward W. Morley (1838–1923). At that time, most physicists believed erroneously the universe was filled with an invisible substance known as *ether*, which supported the propagation of electromagnetic waves. Relative to the ether, light was thought to travel at a speed c in any direction. The existence of the ether seemed quite natural, because all mechanical waves were known to require a medium of propagation.

To analyze the motion of the earth through the ether, Michelson and Morley studied the interference fringes produced by a Michelson interferometer positioned with one of its arms parallel to the velocity vector **u** of the earth around the sun. The interferometer, including the light source, was assumed to be moving through the ether with the same velocity **u** (Fig. 39-3*a*). Relative to the ether, the incident and reflected light beams at mirrors M_1 and M_2 had to travel in the directions depicted in Fig. 39-3*b* in order to return to M. According to the Galilean transformation, the beam traveling from M to M_1 (a distance L) and back must have a speed $\sqrt{c^2 - u^2}$ in the earth/interferometer frame (Figs. 39-3*c* and *d*). It would therefore make the round trip in a time

$$t_1 = \frac{2L}{\sqrt{c^2 - u^2}}.$$

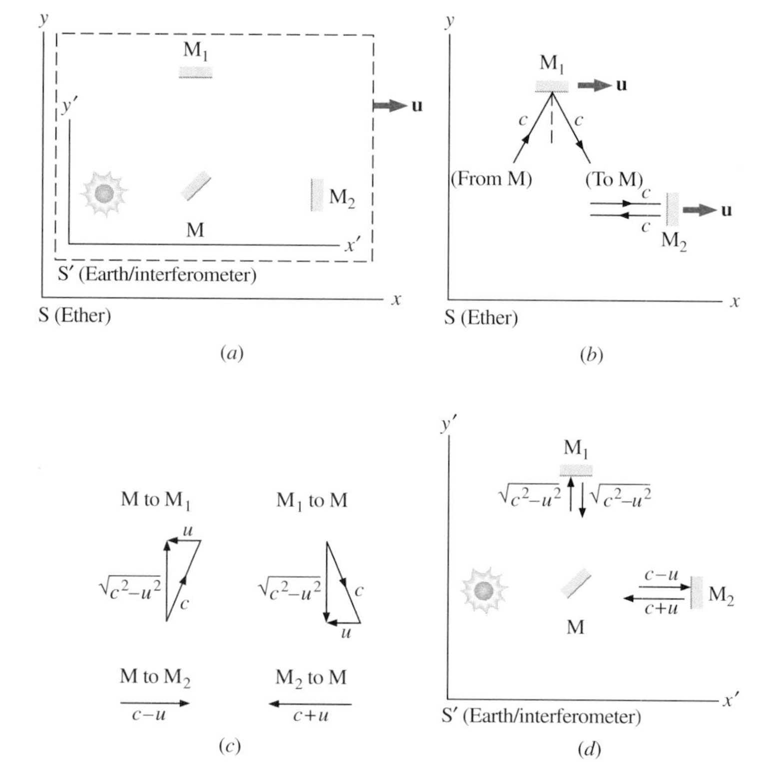

FIGURE 39-3 The Michelson-Morley experiment and the ether. (*a*) The frame of the earth/interferometer is assumed to have a velocity **u** relative to the ether. (*b*) The paths of the beams as seen in the ether frame. Light travels at a speed c in all directions in this frame. (*c*) The speeds of the beam in the earth/interferometer frame are determined using the Galilean transformation. (*d*) The beams and their speeds found in part (*c*) are shown in the earth/interferometer frame.

The other beam had to travel from M to M_2 (also a distance L) with a speed $c - u$ and from M_2 to M with a speed $c + u$, so the time for its round-trip time would be

$$t_2 = \frac{L}{c + u} + \frac{L}{c - u} = \frac{2Lc}{c^2 - u^2}.$$

The difference Δt in transit times for the two beams would then be

$$\Delta t = t_2 - t_1 = \frac{2Lc}{c^2 - u^2} - \frac{2L}{\sqrt{c^2 - u^2}}.$$

Since $u \ll c$, $1/(c^2 - u^2)$ and $1/\sqrt{c^2 - u^2}$ can be approximated by

$$\frac{1}{c^2 - u^2} = \frac{1}{c^2(1 - u^2/c^2)} \approx \frac{1}{c^2}\left(1 + \frac{u^2}{c^2}\right)$$

and

$$\frac{1}{\sqrt{c^2 - u^2}} = \frac{1}{c\sqrt{1 - u^2/c^2}} \approx \frac{1}{c}\left(1 + \frac{1}{2}\frac{u^2}{c^2}\right),$$

allowing Δt to be written as

$$\Delta t = \frac{Lu^2}{c^3}.$$

So, if the assumption about the motion of the earth through the ether were correct, the two beams would be out of phase at the detector by

$$\Delta\phi = \left(\frac{2\pi}{\lambda}\right)(c\,\Delta t) = \frac{2\pi L u^2}{\lambda c^2},$$

where λ is the wavelength of the light.

The experiment was repeated with the interferometer rotated by 90° to interchange the orientations of the arms relative to the earth's orbital velocity **u**. The same calculations then yielded

$$\Delta t' = -\frac{Lu^2}{c^3} \quad \text{and} \quad \Delta\phi' = -\frac{2\pi L u^2}{\lambda c^2}.$$

Because of the different phase shifts between the beams, the fringe patterns must be different for the two positions of the interferometer. As a result, an observer focusing at a given point of the pattern should see N fringes pass by during the rotation, where

$$N = \frac{\Delta\phi - \Delta\phi'}{2\pi} = \frac{2Lu^2}{\lambda c^2}. \tag{39-4}$$

With $L = 11$ m (the optical path used in the experiment), $\lambda = 590$ nm, and $u = 3.0 \times 10^4$ m/s, the earth's speed around the sun, Eq. (39-4) gives

$$N = \frac{2(11\text{ m})(3.0 \times 10^4\text{ m/s})^2}{(5.9 \times 10^{-7}\text{ m})(3.0 \times 10^8\text{ m/s})^2} = 0.37\text{ fringe},$$

a value large enough to be easily detected by the interferometer. However, no fringe shift was observed, indicating that the ether had to be at rest relative to the earth and that the speed of light was the same in all directions relative to the earth. The experiment was repeated at different times of the year with the earth at different points in its orbit, and it was also repeated with the interferometer oriented in different directions. In all cases, no fringe shift was observed within experimental error. This discovery that the speed of light is the same in all directions relative to the earth implied that either the ether does not exist or the ether is always stationary relative to the earth.

Aberration of Starlight

Because of the earth's motion around the sun, slight annual shifts in the positions of the stars relative to the earth can be observed. Such shifts are known as the *aberration of starlight*. This effect is most pronounced for a star directly above the sun-earth plane. Figure 39-4*a* shows the orientation of a telescope used to observe the light from such a star. If the earth is stationary relative to the star, then the telescope has to be held vertically in order for the starlight to traverse its entire length and reach the viewer. However if the velocity of the earth is **u** relative to the star, as shown, then the telescope has to be tilted slightly toward its direction of motion so that the starlight can travel all the way through the tube. In a time interval Δt, the telescope moves a horizontal distance $u\,\Delta t$ while the starlight moves a vertical distance $c\,\Delta t$. The tilt of the telescope is therefore

$$\alpha = \tan^{-1}\frac{u\,\Delta t}{c\,\Delta t} = \tan^{-1}\frac{u}{c} = \tan^{-1}\frac{3.0\times10^{4}\text{ m/s}}{3.0\times10^{8}\text{ m/s}}$$
$$= 1.0\times10^{-4}\text{ rad}.$$

As shown in Fig. 39-4*b*, this tilt is always in the direction of the earth's motion, and the star should appear to move in a circular path of angular diameter $2(1.0\times10^{-4}\text{ rad}) = 2.0\times10^{-4}$ rad. This is exactly what astronomers observe.

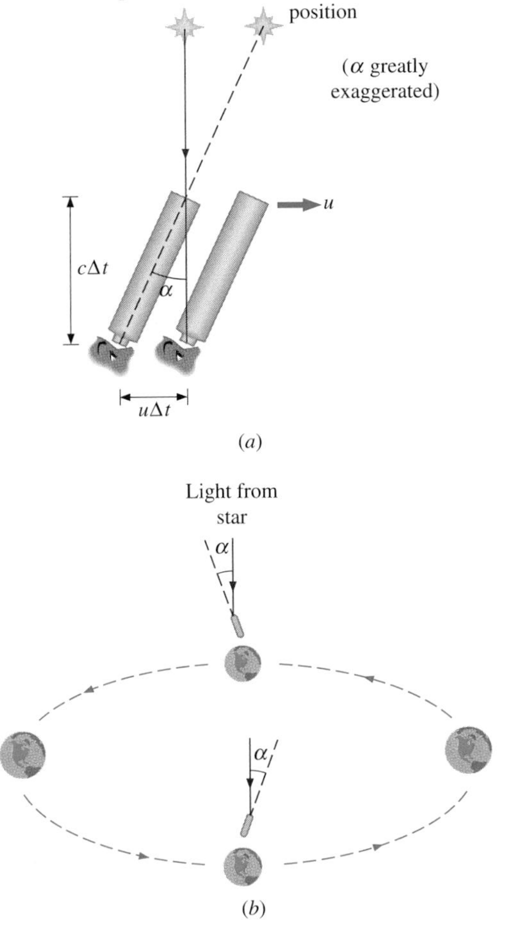

FIGURE 39-4 The aberration of starlight.

These observations imply that the earth is moving relative to the ether. If the earth and the ether moved together (as suggested by the results of Michelson and Morley), a beam of light from the star would be carried such that it would always fall vertically on the earth, as shown in Fig. 39-5. No aberration would then be observed.

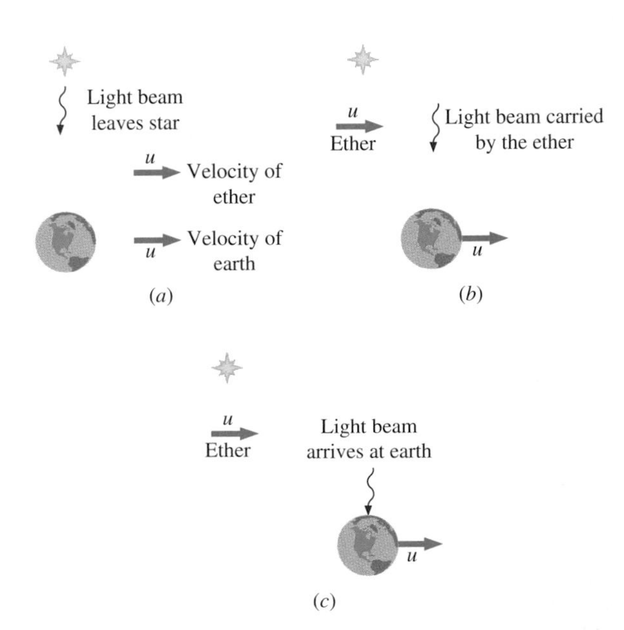

FIGURE 39-5 No aberration of starlight would occur if the earth and ether moved together.

Hence there was a basic discrepancy between the results of the Michelson-Morley experiment and the observation of stellar aberration. The former indicated that the earth was at rest relative to the ether, while the latter indicated that the earth and the ether were in relative motion. Which, if either, was correct? Many different ideas were developed to explain this inconsistency. Some were quite ingenious, but none withstood the ultimate test of experimental verification—until the special theory of relativity was proposed by Albert Einstein in his famous paper, "On the Electrodynamics of Moving Bodies."

39-3 Postulates of the Special Theory of Relativity

Einstein's special theory is based on the following two postulates:

1. The laws of physics are identical in all inertial frames.
2. Light travels in a vacuum with a speed c in any direction in all inertial frames.

The first postulate tells us that the mathematical form of any law must be identical in all inertial frames. If a physical law relating the variables x, y, and t is given by

$$\frac{dx}{dt} + 3y = 0$$

in frame S of Fig. 39-1, then relative to S′ the law must have the form

$$\frac{dx'}{dt'} + 3y' = 0.$$

This postulate denies the existence of a special or preferred inertial frame. The laws of nature do not give us a way to endow any one inertial frame with special properties. For example, there is no inertial frame that we can identify as being in a state of "absolute rest." We can only determine the relative motion of one frame with respect to the other.

Since Newton's laws of motion are invariant under a Galilean transformation, it might seem that these laws do not violate the first postulate. However, the transformation is based on the assumption of a single universal time ($t = t'$). In fact, as you'll soon see, the second postulate forces us to discard this notion of universal time—and consequently to modify Newton's second law.

While the first postulate may seem to be quite reasonable, the second one challenges our normal intuition. Consider a stationary lightbulb at the origin of S in Fig. 39-6. When it flashes, an observer in S sees a spherical pulse emanating from the origin with a speed c. Now suppose that the origin of the moving frame S′ coincides with that of S when the bulb is flashed. What would an observer in S′ detect? According to the second postulate, she also sees *a spherical pulse emanating from the origin of her reference frame*—even though S′ is moving to the right with a speed u relative to the bulb. This should seem somewhat strange to you! Common sense (and the Galilean transformation) would dictate that the observer in S′ see the light pulse move to the right with a speed $c - u$ and to the left with a speed $c + u$, as depicted in Fig. 39-2.

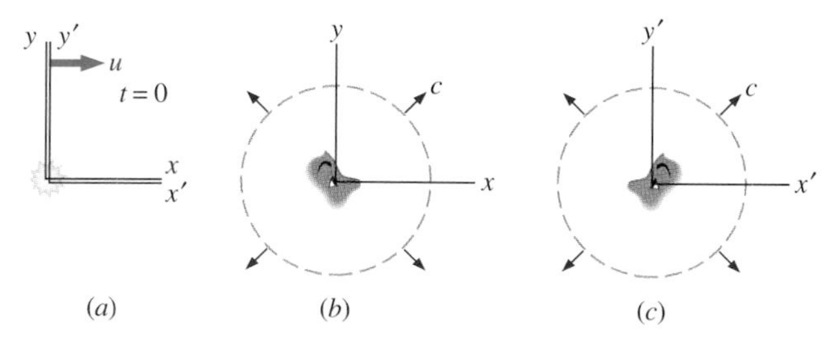

FIGURE 39-6 (*a*) The bulb flashes when the origins of the two frames coincide. (*b*) A spherical wavefront emanating from the origin is observed in each frame.

Einstein's second postulate provides a simple explanation of the result of the Michelson-Morley experiment. In determining the expected fringe shift, it was assumed that the speed of light depended on the relative motion of the interferometer and the ether. But this basic premise was wrong. The speed of light is the same in all inertial frames, so we should *expect* to obtain the null result. Einstein's theory also laid to rest the idea of an ether, for its presence would require the existence of a preferred frame.

The measured aberration of starlight is also consistent with Einstein's theory. This will be shown in Example 39-5 after we consider the relativistic velocity transformation, the counterpart of Eqs. (39-2).

We now examine the effects of Einstein's postulates on our perceptions of space and time.

39-4 Synchronization of Clocks

In Newtonian mechanics, it is assumed that all clocks can be synchronized, regardless of their relative motion. For example, an observer on the earth and an astronaut orbiting the moon in a space capsule can set their clocks to exactly 9:00 A.M. at the same instant. They would then agree on when an event occurs anytime in the future.

To see how this assumption holds up in the context of special relativity, we use the following hypothetical, or "thought," experiment on synchronization. An observer in S′ of Fig. 39-7*a* places two clocks, C_1' and C_2', along the x' axis, each at the same distance on opposite sides of a flashbulb. She triggers the bulb and the clocks are started when the wavefront representing the light pulse passes by. Since the clocks are equidistant from the bulb, they are started simultaneously and are therefore synchronized in the inertial frame S′.

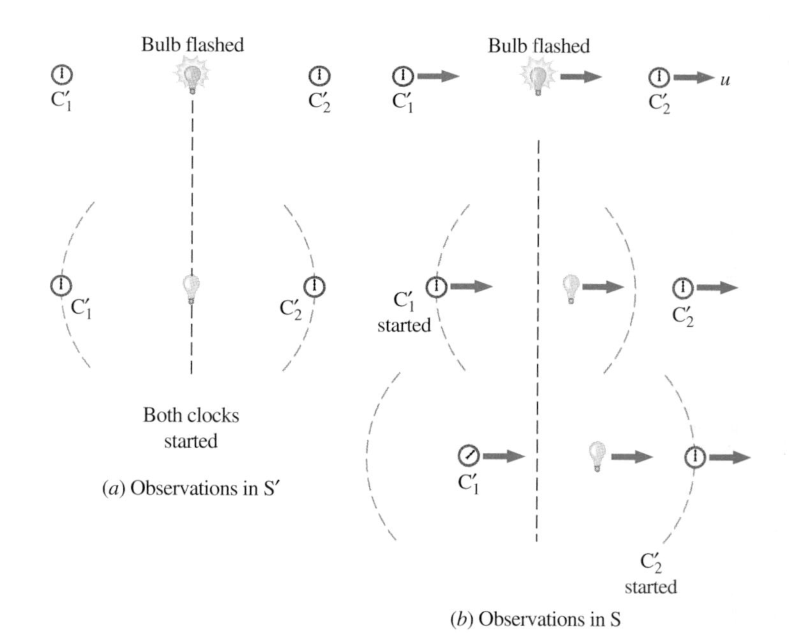

FIGURE 39-7 (*a*) In frame S′, the clocks C_1' and C_2' are stationary and equidistant from a flashbulb. The clocks are started simultaneously when the flash reaches them. (*b*) In frame S, relative to which S′ is moving to the right, the clocks are in motion. The light flash reaches C_1' first so the clocks are not started simultaneously in S.

How does someone in S, an inertial frame relative to which S′ is moving to the right with a speed u (Fig. 39-7b), observe the starting of the clocks? He sees the flash leave the bulb with a speed c in all directions (the second postulate). However, clock C_1' is moving toward the wavefront at a speed u, and C_2' is moving away from it at the same speed. Consequently, to the observer in S, light from the flash reaches C_1' before C_2'—and to him, C_1' is started before C_2'. Now he hasn't done anything incorrectly. The clocks are simply not synchronized in his frame. We can only conclude that *clocks synchronized in one inertial frame are not synchronized in a second inertial frame that is moving relative to the first frame*. We will return to this synchronization problem and treat it quantitatively in Sec. 39-8.

39-5 Measurement of Transverse Lengths

Figure 39-8 shows two meter sticks M and M′ that are at rest in the reference frames of two boys S and S′, respectively. A small paintbrush is attached to the top (the 100-cm mark) of stick M′. Suppose that S′ is moving to the right at a very high speed u relative to S, and the sticks are oriented so that they are perpendicular, or transverse, to the relative velocity vector of the boys. The sticks are held so that as they pass each other, their lower ends (the 0-cm marks) coincide.

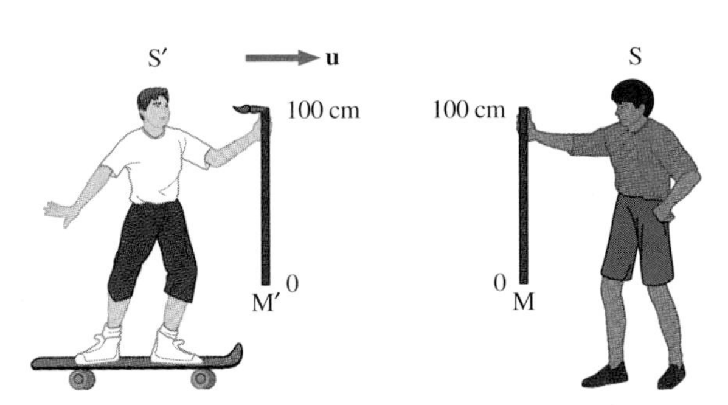

FIGURE 39-8 Meter sticks M and M′ are stationary in the reference frames of observers S and S′, respectively. When the sticks pass, a small brush attached to the 100-cm mark of M′ paints a line on M.

Let's assume that when S looks at his stick M afterwards, he finds a line painted on it, just below the top of the stick. Since the brush is attached to the top of the other boy's stick M′, S can only conclude that stick M′ is less than 1.0 m long. Now when the boys approach each other, S′, like S, sees a meter stick moving toward him with speed u. Since their situations are symmetric, each boy must make the *same* measurement of the stick in the other frame. So, if S measures stick M′ to be less than 1.0 m long, S′ must measure stick M to be also less than 1.0 m long; and S′ must see his paintbrush pass over the top of stick M and not paint a line on it. In other words, after the same event, one boy sees a painted line on a stick, while the other does not see such a line on that same stick!

Einstein's first postulate requires that the laws of physics (as, for example, applied to painting) predict that S and S′, who are both in inertial frames, make the same observations; that is, S and S′ must either both see a line painted on stick M, or both not see that line. We are therefore forced to conclude our original assumption that S saw a line painted below the top of his stick was wrong! Instead, we should have assumed that S finds the line painted right at the 100-cm mark on M. Then both observers will agree that a line is painted on M, and they will also agree that *both* sticks are exactly one meter long. We conclude then that *measurements of a transverse length must be the same in different inertial frames.*

Now we could just as well have assumed that the paintbrush was attached at the 100-cm mark of M and that it painted a line on M′. We would have then repeated the same argument with M and M′ and S and S′ interchanged. And the conclusion would have been the same—that the measurements of transverse lengths in different inertial frames agree.

39-6 Time Dilation

To study how time is affected by Einstein's postulates, we use the inertial frames S and S′ shown in Fig. 39-9a. In S′, a flashbulb and a clock are placed at the origin, and a mirror is positioned a distance l' above them. To an observer in S′, the bulb emits a light pulse that travels along the y' axis to the mirror, where it is reflected straight downward and back to the clock. With her clock, she measures the transit time of the pulse to be

$$\Delta t' = \frac{2l'}{c}.$$

In order to measure the transit time Δt in S, two observers with synchronized clocks located as shown in Fig. 39-9c are needed. Here the pulse travels a horizontal distance $u\,\Delta t$ while

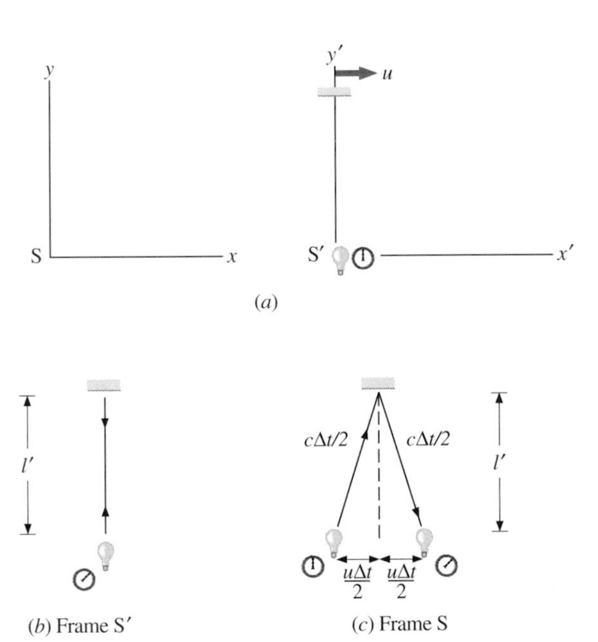

FIGURE 39-9 (a) The timing apparatus rests in S′, which is moving to the right relative to S with a speed u. (b) One round-trip of the light pulse in S′. (c) One trip of the light pulse in S.

covering a vertical distance $2l'$. (Remember, transverse lengths are the same in both frames.) Since the speed of the pulse is c,

$$c\,\Delta t = 2\sqrt{(l')^2 + (u\,\Delta t/2)^2},$$

which yields

$$\Delta t = \left(\frac{2l'}{c}\right)\frac{1}{\sqrt{1-\beta^2}},$$

where

$$\beta = \frac{u}{c}.$$

Finally, substituting $\Delta t'$ for $2l'/c$ we have

$$\Delta t = \frac{\Delta t'}{\sqrt{1-\beta^2}}. \qquad (39\text{-}5)$$

Since $\Delta t > \Delta t'$, the person in S′ measures a shorter time interval for this event than the people in S. In other words, the clock in S′ has recorded fewer "ticks" than the clocks in S. We therefore conclude that *moving clocks run slow*. This effect, known as *time dilation*, is real and is not caused by inaccurate clocks or improper measurements. Time-interval measurements of the same event differ for observers in relative motion.

A time interval like $\Delta t'$, which is measured by a single clock, is called a **proper time interval**. Since two clocks were used to determine Δt, it is not a proper time interval. If the roles of S and S′ were reversed in Fig. 39-9 so that two observers in S′ measured a time interval on a single clock stationary in S, then Δt would become the proper interval and $\Delta t' = \Delta t/\sqrt{1-\beta^2}$.

Since c is so large, timekeeping differences between clocks at rest and clocks moving at everyday speeds are completely insignificant. For example, if $u = 340$ m/s (the speed of sound in air), then

$$\Delta t = \frac{\Delta t'}{\sqrt{1-[(340\text{ m/s})/(3.0\times10^8\text{ m/s})]^2}} \approx \Delta t'(1 + 6.4\times10^{-13}).$$

A clock moving at the speed of sound would be slower by 1 s over a period of about 50,000 years!

Another aspect of Eq. (39-5) is that its right-hand side becomes imaginary if $u > c$. As this does not make sense physically, special relativity restricts all objects to *speeds less than c* in all inertial frames.

The dilation of time is independent of the nature of the particular clocks being used. Any moving device that can be used to keep time because of its regular beat (including a person's pulse) runs slow by the same factor $\sqrt{1-\beta^2}$. *Time dilation is an intrinsic property of time itself.*

Modern elementary-particle and cosmic-ray experiments furnish daily confirmation of the time dilation formula. Since the particles involved usually travel at speeds close to c relative to the earth, time dilation has a significant effect on how long they live in the laboratory before decaying. For example, in the rest frame of a beam of positive pions (π^+), one-half of that group on the average will decay in 1.8×10^{-8} s. (This is the **half-life** of the pion.) But suppose this beam of pions is produced by a particle accelerator, and, as measured in a laboratory, the beam is traveling at $0.99c$. Then relative to the laboratory, the half-life of the pion is

$$\Delta t = \frac{1.8\times10^{-8}\text{ s}}{\sqrt{1-(0.99)^2}} = 12.8\times10^{-8}\text{ s}.$$

This time difference has a significant effect on how far the pions will travel in the laboratory before decaying. If there were no time dilation, half the pions in the beam would have decayed after the beam had traveled only $(0.99c)(1.8\times10^{-8}\text{ s}) = 5.3$ m; yet, as verified experimentally, this actually occurs over a distance of $(0.99c)(12.8\times10^{-8}\text{ s}) = 38$ m.

Time dilation in elementary-particle experiments is quite easy to observe because the particles move so fast and have such short lifetimes. However, in the macroscopic world, time dilation effects are very difficult to measure. One such measurement was made in 1971 using two sets of cesium-beam atomic clocks. These clocks are accurate to approximately one part in 10^{10}, so extremely small differences in time measurements can be detected with the clocks. One set of clocks was flown around the earth in commercial jet airliners, and the other set was the reference clocks at the United States Naval Observatory. Equation (39-5) predicted that the moving clocks should lose $(275 \pm 21)\times10^{-9}$ s relative to the stationary clocks. The observed time loss was $(273 \pm 7)\times10^{-9}$ s, which, within the limits of uncertainty, is in agreement with the special theory of relativity.

EXAMPLE 39-1 THE TWIN PARADOX

Jean-Luc and Michel are 20-year-old twins living on the earth in the twenty-fifth century. Jean-Luc, who is a space traveler, flies to the star Altair (about 17 light-years away) at $u = 0.90c$, then returns to the earth at the same speed. What are the ages of the twins when they meet again?

SOLUTION Altair is 17 light-years away, so Jean-Luc makes the round-trip in a time (as measured by his brother on the earth)

$$\Delta t = \frac{34\text{ light-years}}{0.90c} = 38\text{ years}.$$

From Eq. (39-5),

$$\Delta t = 38\text{ years} = \frac{\Delta t'}{\sqrt{1-(0.90)^2}},$$

so the time $\Delta t'$ that passes in Jean-Luc's reference frame during the trip is

$$\Delta t' = (38\text{ years})\sqrt{1-(0.90)^2} = 17\text{ years}.$$

Thus when Jean-Luc returns to the earth, he is physically only 37 years old, while his twin brother Michel is 58 years old!

Now let's examine the problem from Jean-Luc's point of view. In his frame, Michel's clocks appear to run slow by the factor $\sqrt{1-\beta^2}$. Does this imply that it is Michel rather than Jean-Luc who is younger when they are reunited? Have we come across a contradiction in the special theory of relativity by studying the same situation from two different frames? The situation involving Michel and Jean-Luc is an example of what is known as the *twin paradox.* The resolution of this paradox is found in a closer examination of the two reference systems involved. With the slight acceleration of the earth ignored, Michel's frame is always inertial and is therefore appropriate for analyzing the entire motion of Jean-Luc and his rocket ship. Jean-Luc's frame, on the other hand, accelerates from rest (relative to the earth) to $0.90c$, then undergoes further changes in velocity in the round-trip. During the accelerations, Jean-Luc's frame is not inertial and consequently cannot be used in the time dilation formula. Predictions about events on the earth that are based on special relativity as applied in Jean-Luc's frame are just not valid. Therefore Jean-Luc is younger than Michel when they meet again.

EXAMPLE 39-2 **TIME DILATION AND PARTICLE DECAY**

The half-life of a charged K meson is 8.5×10^{-9} s. (*a*) To an experimenter in the laboratory where a beam of K mesons is moving at a speed $u = c/2$, how much time elapses before half of the K mesons decay? (*b*) How far does this beam travel in the laboratory before half of the particles decay?

SOLUTION (*a*) The half-life in the K meson's frame is $\Delta t' = 8.5 \times 10^{-9}$ s, so the experimenter measures a time

$$\Delta t = \frac{\Delta t'}{\sqrt{1-\beta^2}} = \frac{8.5 \times 10^{-9}\ \text{s}}{\sqrt{1-(\frac{1}{2})^2}} = 9.8 \times 10^{-9}\ \text{s}$$

for half of the mesons to decay.

(*b*) In this time Δt, the beam travels in the laboratory a distance

$$l = u\,\Delta t = \left(\frac{c}{2}\right)\Delta t = \frac{(3.0 \times 10^8\ \text{m/s})(9.8 \times 10^{-9}\ \text{s})}{2} = 1.5\ \text{m}.$$

DRILL PROBLEM 39-1

A group of charged K mesons is at rest in the laboratory. To an observer moving by them in a rocket ship at a relative speed of $u = c/2$, how much time elapses before half of the particles decay?
ANS. 9.8×10^{-9} s.

39-7 MEASUREMENT OF LONGITUDINAL LENGTHS

The inertial frame S′ of Fig. 39-10 is moving to the right at a speed u relative to the inertial frame S. At rest in S′ are a light source L′ and a mirror M′, which are placed a distance l' apart on the x' axis (see Fig. 39-10*a*). Since this distance is along the direction of the relative motion of the frames, it is a longitudinal length. A light pulse is sent from the source to the mirror, where it is reflected back to the source. A clock C′ placed at

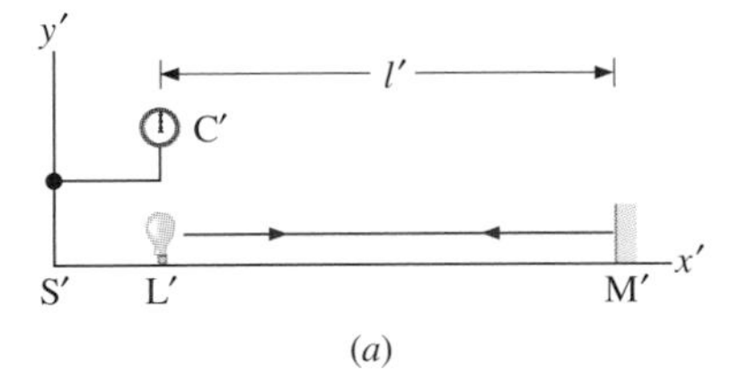

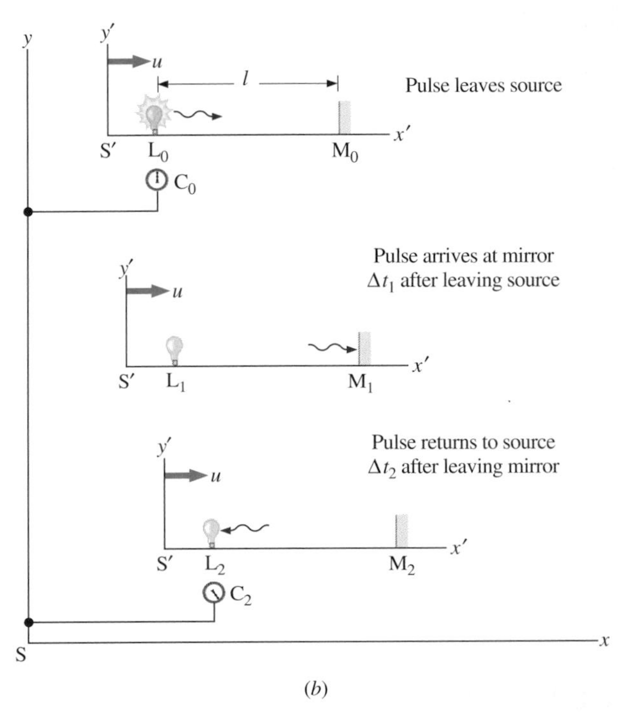

FIGURE 39-10 (*a*) A light source, clock, and mirror are at rest in frame S′. (*b*) In frame S, the instant that the pulse leaves the source is measured by clock C_0, which is in S. The instant that the pulse returns to the source is measured by C_2, which is in S.

the source is used to measure the transit time $\Delta t'$. As it is measured with a single clock, $\Delta t'$ is a proper time interval. It is given by

$$\Delta t' = \frac{2l'}{c}.$$

As measured in frame S, the light source and mirror are a distance l apart (Fig. 39-10*b*), and they both move to the right with a speed u as the light pulse travels between them. Two synchronized clocks C_0 and C_2 determine the instants when the pulse is sent and when it returns.

In the time Δt_1 that it takes the pulse to reach the mirror, the source and the mirror move from L_0 to L_1 and from M_0 to M_1, respectively. So while the light pulse travels a distance L_0M_1 at a speed c, the mirror travels a distance $(L_0M_1 - l)$ at a speed u. The time interval Δt_1 is therefore given by

$$\Delta t_1 = \frac{L_0M_1}{c} = \frac{L_0M_1 - l}{u},$$

which yields

$$L_0M_1 = \frac{lc}{c-u} = \frac{l}{1-\beta}.$$

During the pulse's return trip from M_1 to L_2 at a speed c, the source and the mirror travel from L_1 to L_2 and from M_1 to M_2 at a speed u. The time Δt_2 taken for this part of the trip is then

$$\Delta t_2 = \frac{L_2M_1}{c} = \frac{l - L_2M_1}{u},$$

which yields

$$L_2M_1 = \frac{l}{1+\beta}.$$

Over the entire trip, the net travel time of the pulse is

$$\Delta t = \Delta t_1 + \Delta t_2 = \frac{L_0M_1 + L_2M_1}{c} = \frac{l}{c}\left(\frac{1}{1-\beta} + \frac{1}{1+\beta}\right)$$

$$= \frac{2l}{c(1-\beta^2)}.$$

Finally, if we substitute $2l'/\Delta t'$ for c and apply the time dilation formula, we are left with

$$l = l'\sqrt{1-\beta^2}. \tag{39-6}$$

This *length contraction* equation tells us that *longitudinal distances, like time intervals, are not absolute.* The longitudinal dimension of an object measured in its own rest frame is referred to as its *rest length*. In any other inertial frame that is moving with respect to the rest frame, this length is always measured to be *shorter* than the rest length. The degree of contraction depends on the relative velocity of the object and the frame and can be found from Eq. (39-6).

Like time dilation, length contraction is insignificant at ordinary speeds. For example, if $u = 340$ m/s (the speed of sound in air),

$$l = l'\sqrt{1 - \left(\frac{340 \text{ m/s}}{3.0 \times 10^8 \text{ m/s}}\right)^2} \approx l'(1 - 6.4 \times 10^{-13}).$$

EXAMPLE 39-3 **LENGTH CONTRACTION**

An astronaut in a rocket ship measures the dimensions of his sleeping quarters to be 3.0 m × 2.0 m × 2.0 m. Observers on the earth see the ship moving by at a speed $u = 3c/5$ and in a direction parallel to the 3.0-m side of the room. What are the dimensions of the sleeping quarters as measured by the observers on the earth?

SOLUTION The length of the side parallel to the velocity of the ship is, from Eq. (39-6),

$$l = l'\sqrt{1-\beta^2} = (3.0 \text{ m})\sqrt{1 - (\tfrac{3}{5})^2} = 2.4 \text{ m}.$$

Since transverse lengths are the same in both inertial frames, the dimensions of the room as measured by the earth observers are 2.4 m × 2.0 m × 2.0 m.

DRILL PROBLEM 39-2

A spaceship leaves the earth and heads toward the star Alpha Centauri at a speed $u = 0.95c$. To observers on the earth, Alpha Centauri is 4.3 light-years away. What is the distance between the earth and this star as measured by a passenger on the ship?

ANS. 1.3 light-years.

39-8 Clock Synchronization Revisited

With what we've learned about time dilation and length contraction, we can now examine quantitatively the act of synchronizing clocks. Consider the two clocks C_1' and C_2' of Fig. 39-11, which are at rest a distance l' apart in the inertial frame S′. A flashbulb is placed exactly midway between them and is flashed. The clocks are synchronized in S′ by starting them at the instant $t' = 0$ when the light pulses from the bulb reach them.

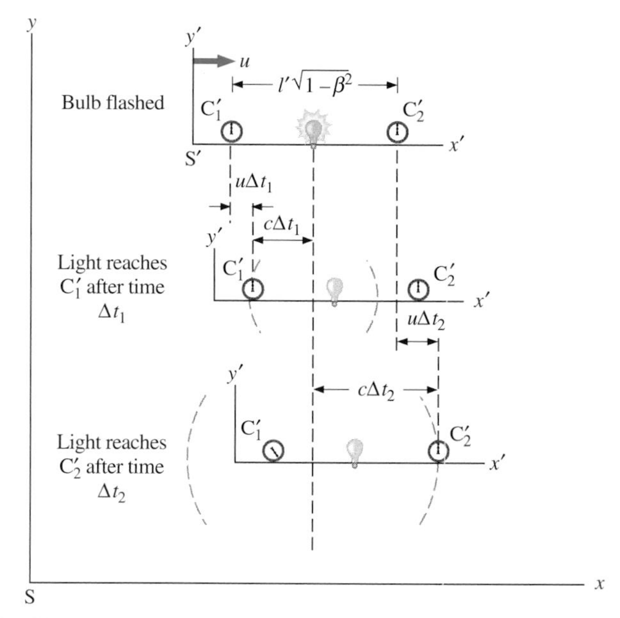

FIGURE 39-11 An observer in frame S sees clock C_1' started before C_2'. All sketches represent observations made relative to S.

Relative to the inertial frame S, S′ is moving to the right at a constant speed u. To an observer in S the clocks are a distance $l'\sqrt{1-\beta^2}$ apart with the lamp halfway between, and the pulses move in all directions at a speed c. Since C_1' is moving to the right at a speed u and the light pulse is approaching it at a speed c, the pulse will reach C_1' after a time interval Δt_1, where

$$c\,\Delta t_1 = \frac{l'\sqrt{1-\beta^2}}{2} - u\,\Delta t_1.$$

There is also a pulse moving to the right at a speed c "chasing" C_2', which is moving to the right at a speed u. This pulse arrives at C_2' after an interval Δt_2, where

$$c\,\Delta t_2 = \frac{l'\sqrt{1-\beta^2}}{2} + u\,\Delta t_2.$$

Solving these two equations for Δt_1 and Δt_2, we find

$$\Delta t_1 = \frac{l'\sqrt{1-\beta^2}}{2(c+u)} \quad \text{and} \quad \Delta t_2 = \frac{l'\sqrt{1-\beta^2}}{2(c-u)},$$

so

$$\Delta t_2 - \Delta t_1 = \frac{l'\sqrt{1-\beta^2}}{2}\left(\frac{1}{c-u} - \frac{1}{c+u}\right) = \frac{l'u}{c^2\sqrt{1-\beta^2}}.$$

The difference $\Delta t_2 - \Delta t_1$ in the time intervals represents how much *later* C_2' is started than C_1' *as measured in S*. (Remember, these two clocks are synchronized in S'.) Now to an observer in S, the clocks are not out of synchronization by this amount, because to him, they also *run slow*. From time dilation, while a time interval $\Delta t_2 - \Delta t_1$ has passed in S, a time interval $(\Delta t_2 - \Delta t_1)\sqrt{1-\beta^2}$ has passed in S'. *An observer in S therefore sees the hands on the two clocks in S' out of synchronization by an amount* $\delta t'$ (with C_1' ahead of C_2'), where

$$\delta t' = (\Delta t_2 - \Delta t_1)\sqrt{1-\beta^2}.$$

Finally, by substituting for $\Delta t_2 - \Delta t_1$, we obtain

$$\delta t' = \frac{l'u}{c^2}, \qquad (39\text{-}7)$$

where l' is the distance between the clocks as measured in S'.

EXAMPLE 39-4 **CLOCK SYNCHRONIZATION**

A space traveller has synchronized clocks at each end of his 100-m-long ship. Relative to the earth, the ship is moving by at a speed $u = 3c/5$. (*a*) What is its length as measured on the earth? (*b*) How much time elapses on the earth while the hands of the clocks move through 1 min? (*c*) Are the two clocks synchronized relative to the earth? If not, by how much are they out of synchronization? (*d*) Do the rates of the two clocks on the ship differ for observers on the earth?

SOLUTION (*a*) The rest length of the ship is 100 m. Observers on the earth measure a contracted length

$$l = l'\sqrt{1-\beta^2} = (100 \text{ m})\sqrt{1-(\tfrac{3}{5})^2} = 80 \text{ m}.$$

(*b*) With $\Delta t' = 60$ s, the time dilation formula yields for the time elapsed on the earth

$$\Delta t = \frac{\Delta t'}{\sqrt{1-\beta^2}} = \frac{60 \text{ s}}{\sqrt{1-(\frac{3}{5})^2}} = 75 \text{ s}.$$

(*c*) To observers on the earth, the hands on the clock in the rear of the ship are ahead of those in the front by an amount

$$\delta t' = \frac{l'u}{c^2} = \frac{(100 \text{ m})(3c/5)}{c^2} = 2.0 \times 10^{-7} \text{ s}.$$

(*d*) The rates at which clocks run differ only if they are in frames moving relative to one another. Since both clocks are on the spaceship, they run at the same rate according to observers on the earth.

DRILL PROBLEM 39-3

Two clocks are placed 200 m apart on the earth and synchronized. By how much are the clocks out of synchronization to space travelers who are moving by the earth at a relative speed of $4c/5$?
ANS. 5.3×10^{-7} s.

Albert Einstein in inertial frame S'.

39-9 LORENTZ TRANSFORMATION

At the beginning of this chapter, we used the Galilean transformation to relate the position and time of an event as observed in two different inertial frames. We assumed that there exists a single universal time (i.e., $t = t'$), so that time intervals measured by clocks in relative motion are the same. We now know from Einstein's theory that clocks in relative motion do *not* run at the same rate. Consequently, a basic premise of the Galilean transformation, and hence the transformation itself, is invalid.

The correct transformation must be consistent with the relativistic equations relating length and time-interval measurements in two inertial frames. To find this transformation, we use the inertial systems S and S' shown in Fig. 39-12, where S'

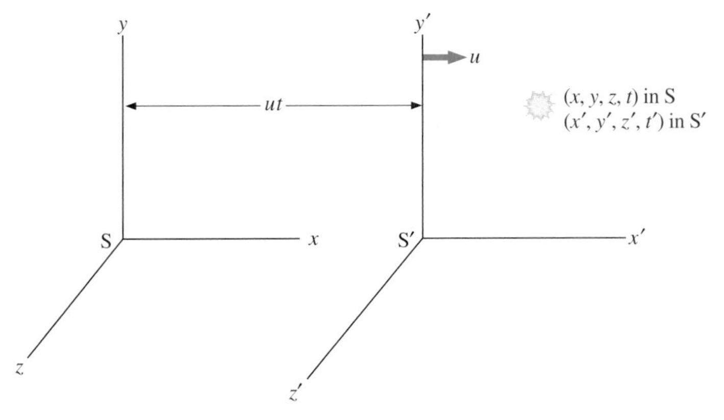

FIGURE 39-12 The position and time coordinates of an event as measured in frame S are (x, y, z, t). This same event is observed to occur at (x', y', z', t') in frame S'.

is moving in the x direction at a speed u relative to S. Our goal is to relate the position and time (x', y', z', t') of an event measured in S′ to the event's position and time (x, y, z, t) measured in S. We assume that clocks at the origins in the two systems are started $(t = t' = 0)$ when the origins coincide. These clocks are then used to synchronize all other clocks in their respective frames.

Suppose that an event occurs at $(x', 0, 0, t')$ in S′ and at $(x, 0, 0, t)$ in S, as depicted in Fig. 39-13. At the time t, an observer in S finds the origin of S′ to be at $x = ut$. With the help of a friend in S, she also measures the distance from the event to the origin of S′ to be $x'\sqrt{1-\beta^2}$. Thus the position of the event in S is

$$x = ut + x'\sqrt{1-\beta^2},$$

and

$$x' = \frac{x - ut}{\sqrt{1-\beta^2}}. \tag{39-8a}$$

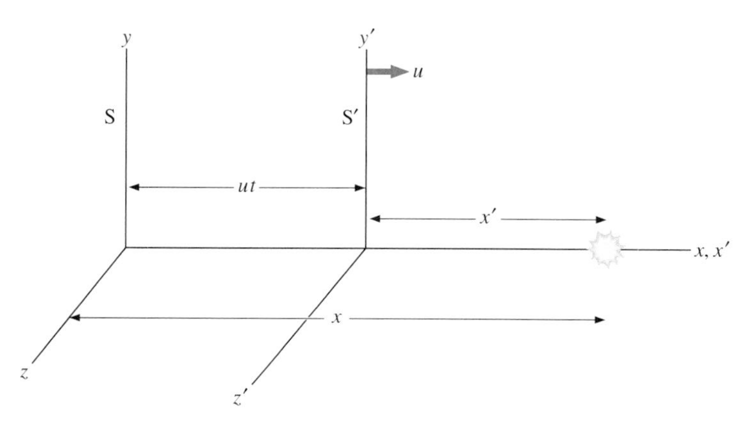

FIGURE 39-13 An event occurs at $(x, 0, 0, t)$ in S and at $(x', 0, 0, t')$ in S′. How x' is related to x and t gives one of the Lorentz transformation equations.

Since this equation provides a relationship between coordinates in the two frames, it is one of our new transformation equations. Notice that in the classical limit u/c (or β) $\to 0$, so $x' \to x - ut$. This is simply Eq. (39-1a), one of the Galilean transformation equations.

Now Einstein's theory does not affect lengths transverse to the relative velocity of the frames. The transformation equations for the other position coordinates are therefore

$$y' = y \tag{39-8b}$$

and

$$z' = z. \tag{39-8c}$$

We can find the transformation for time by imagining that there is a lightbulb at the origin of S′. Let's suppose that at the instant the origins of S and S′ coincide, the bulb is flashed. Relative to each frame, the light from the bulb emanates spherically with a speed c. After a time t, the radius of the sphere representing the leading wavefront in S is ct, and

$$x^2 + y^2 + z^2 - c^2t^2 = 0.$$

Similarly, in S′ the sphere representing the leading wavefront is described by

$$(x')^2 + (y')^2 + (z')^2 - c^2(t')^2 = 0.$$

Since the right-hand side of each equation is zero, we have

$$x^2 + y^2 + z^2 - c^2t^2 = (x')^2 + (y')^2 + (z')^2 - c^2(t')^2,$$

which reduces to

$$x^2 - c^2t^2 = (x')^2 - c^2(t')^2,$$

because $y = y'$ and $z = z'$. Finally, substituting for x' with Eq. (39-8a) and rearranging terms, we obtain

$$t' = \frac{t - ux/c^2}{\sqrt{1-\beta^2}}, \tag{39-8d}$$

the transformation for time. In the limit $u/c \to 0$, this equation also reduces to its Galilean counterpart, $t' = t$.

The set of equations (Eqs. 39-8a through d) that relate the position and time in the two inertial frames is known as the *Lorentz transformation*. They are named in honor of H. A. Lorentz (1853–1928) who proposed them. Interestingly, he justified the transformation on what was eventually discovered to be a fallacious hypothesis. The correct theoretical basis is, of course, Einstein's special theory.

The reverse transformations express the variables in S in terms of those in S′. You can find these by simply interchanging the primed and unprimed variables in Eqs. (39-8) and substituting $-u$ for u; then

$$x = \frac{x' + ut'}{\sqrt{1-\beta^2}} \tag{39-9a}$$

$$y = y' \tag{39-9b}$$

$$z = z' \tag{39-9c}$$

$$t = \frac{t' + ux'/c^2}{\sqrt{1-\beta^2}}. \tag{39-9d}$$

Hendrik Antoon Lorentz, winner of the Nobel Prize for Physics in 1902 for his work in explaining atomic phenomena.

Finally, we check the Lorentz transformation equations to see that they do account for time dilation, length contraction, and clock synchronization.

Time Dilation

Consider a clock fixed at x' in frame S′. Time intervals measured on it are proper times. To relate the time intervals $\Delta t'$ and Δt recorded by observers in S′ and S, respectively, for a given event, we write Eq. (39-9*d*) as

$$\Delta t = \frac{\Delta t' + (u\,\Delta x'/c^2)}{\sqrt{1-\beta^2}}.$$

Since the position of the clock in S′ is fixed, $\Delta x' = 0$, and

$$\Delta t = \frac{\Delta t'}{\sqrt{1-\beta^2}},$$

which is the time dilation equation.

Length Contraction

Consider a stick fixed in S′ with its ends at x_1' and x_2'. Because the stick is at rest in S′, $x_2' - x_1'$ is a proper length. The positions of these ends in S are found simultaneously (at time t) by two observers to be x_1 and x_2. The length of the stick in S is therefore $x_2 - x_1$. From Eq. (39-8*a*),

$$x_2' - x_1' = \frac{x_2 - ut}{\sqrt{1-\beta^2}} - \frac{x_1 - ut}{\sqrt{1-\beta^2}} = \frac{x_2 - x_1}{\sqrt{1-\beta^2}}.$$

With $l' = x_2' - x_1'$ and $l = x_2 - x_1$, we have

$$l = l'\sqrt{1-\beta^2},$$

which is the length contraction equation.

Clock Synchronization

Here we compare the readings of two clocks in S′ obtained at the same instant by observers in S. If the two clocks are $\delta x' = l'$ apart in S′, we have, from Eq. (39-9*d*),

$$\delta t = \frac{\delta t' + (u\,\delta x'/c^2)}{\sqrt{1-\beta^2}} = \frac{\delta t' + ul'/c^2}{\sqrt{1-\beta^2}} = 0.$$

The reason $\delta t = 0$ is that both clocks are read at the same instant in S. Solving for $\delta t'$, we find

$$\delta t' = -\frac{ul'}{c^2}.$$

This tells us how much the front clock *lags* the rear clock as observed in S, and it agrees with the clock synchronization equation.

39-10 Velocity Transformations

Like position, the velocity of a particle must be transformed when referenced in different frames. The particle's velocity components in S are dx/dt, dy/dt, and dz/dt, and in S′, which is moving along the x axis relative to S at a speed u, they are dx'/dt', dy'/dt', and dz'/dt'. To determine the transformation equation for $v_x' = dx'/dt'$, we first differentiate Eq. (39-8*a*) with respect to t'; this gives

$$v_x' = \frac{dx'}{dt'} = \frac{(dx/dt') - u(dt/dt')}{\sqrt{1-\beta^2}}.$$

From the time transformation equation (Eq. 39-9*d*),

$$\frac{dt}{dt'} = \frac{dt'/dt' + [(u/c^2)(dx'/dt')]}{\sqrt{1-\beta^2}} = \frac{1 + uv_x'/c^2}{\sqrt{1-\beta^2}},$$

and from the chain rule for differentiation,

$$\frac{dx}{dt'} = \frac{dx}{dt}\frac{dt}{dt'} = v_x\frac{dt}{dt'}.$$

When combined, these last two equations yield

$$\frac{dx}{dt'} = v_x\left(\frac{1 + uv_x'/c^2}{\sqrt{1-\beta^2}}\right).$$

Finally, with the expressions for dt/dt' and dx/dt' inserted into the equation for v_x', we find after some simplification that

$$v_x = \frac{v_x' + u}{1 + uv_x'/c^2}. \qquad (39\text{-}10a)$$

With similar calculations, the relativistic transformations for the other velocity components are found to be

$$v_y = \frac{v_y'\sqrt{1-\beta^2}}{1 + uv_x'/c^2} \qquad (39\text{-}10b)$$

and

$$v_z = \frac{v_z'\sqrt{1-\beta^2}}{1 + uv_x'/c^2}. \qquad (39\text{-}10c)$$

To obtain the reverse transformations, we simply interchange the primes and replace u by $-u$ in Eqs. (39-10); then

$$v_x' = \frac{v_x - u}{1 - uv_x/c^2} \qquad (39\text{-}11a)$$

$$v_y' = \frac{v_y\sqrt{1-\beta^2}}{1 - uv_x/c^2} \qquad (39\text{-}11b)$$

$$v_z' = \frac{v_z\sqrt{1-\beta^2}}{1 - uv_x/c^2}. \qquad (39\text{-}11c)$$

Notice that in the classical limit ($u/c \to 0$), both sets of velocity equations give $v_x = v_x' + u$, $v_y = v_y'$, and $v_z = v_z'$, which are the Galilean transformation equations.

EXAMPLE 39-5 **Relativistic Velocity Transformation**

The inertial frame S′ is moving in the x direction with a speed u relative to S. (*a*) A light source at rest in S′ sends a beam down the positive x' axis. Use the velocity transformation equations to show that the beam travels with a speed c relative to S. (*b*) Show that this is also the case if the beam is sent along the negative y' axis in S′. (*c*) Suppose that the stationary source in S′ is a star, and that the frame S is attached to the earth. At what angle relative to the vertical must an earth-bound observer aim his telescope in order to see the star? Compare this calculation with that made in Sec. 39-2.

SOLUTION (*a*) Relative to S′, $v_x' = c$, $v_y' = 0$, $v_z' = 0$. From the velocity transformations of Eqs. (39-10),

$$v_x = \frac{c+u}{1+uc/c^2} = c, \qquad v_y = 0, \qquad v_z = 0,$$

so

$$v = \sqrt{v_x^2 + v_y^2 + v_z^2} = \sqrt{c^2 + 0^2 + 0^2} = c.$$

(*b*) Now we have $v_x' = 0$, $v_y' = -c$, and $v_z' = 0$. Hence

$$v_x = \frac{0+u}{1+0} = u,$$

$$v_y = \frac{(-c)\sqrt{1-\beta^2}}{1+0} = -c\sqrt{1-\beta^2},$$

$$v_z = 0,$$

and

$$v = \sqrt{v_x^2 + v_y^2 + v_z^2} = \sqrt{u^2 + c^2(1-\beta^2)} = c.$$

As must be the case, the speed of light is c in both frames.

(*c*) As shown in Fig. 39-14, the angle relative to the vertical that an earth-bound observer must aim a telescope in order to see the star is given by

$$\alpha = \tan^{-1}\frac{v_x}{v_y} = \tan^{-1}\frac{u}{c\sqrt{1-\beta^2}}.$$

Now the speed of the star relative to the earth is essentially the same as the earth's orbital speed. Since this is much less than the speed of light, $\beta = u/c \approx 0$ and α is accurately approximated by

$$\alpha = \tan^{-1}\frac{u}{c},$$

which is the same as the equation found by analyzing the aberration of starlight with the Galilean transformation.

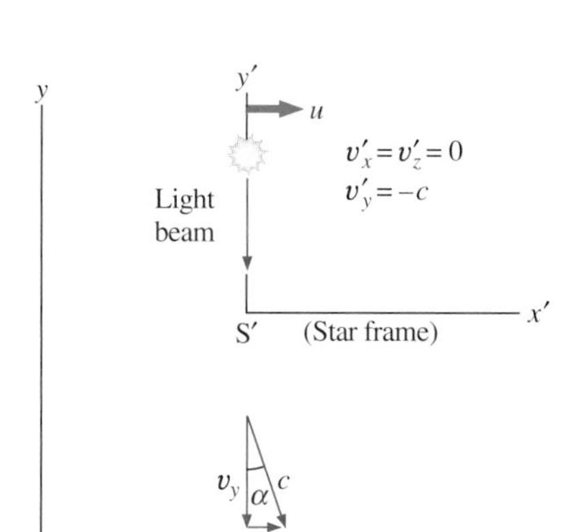

FIGURE 39-14 The aberration of starlight as analyzed using the Lorentz transformation.

EXAMPLE 39-6 Relative Velocity of Two Spaceships

Two spaceships, each with a speed $v = 4c/5$ relative to the earth, are moving directly toward one another (Fig. 39-15). What is the velocity of one ship with respect to the other?

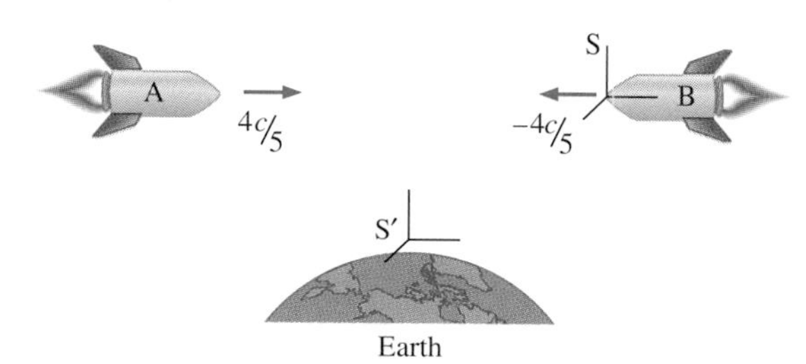

FIGURE 39-15 Spaceships moving toward each other. Each ship has a speed $4c/5$ relative to the earth.

SOLUTION Relative to ship B (frame S), the earth (frame S′) is moving with a velocity $(4c/5)\mathbf{i}$, so $u = 4c/5$.

Relative to the earth, ship A is moving with a velocity $(4c/5)\mathbf{i}'$, so its velocity components in S′ are $v_x' = 4c/5$, $v_y' = 0$, and $v_z' = 0$. From the velocity transformation, the velocity of A relative to B is

$$v_x = \frac{4c/5 + 4c/5}{1 + (4c/5)^2/c^2} = \frac{40}{41}c, \qquad v_y = 0, \qquad v_z = 0.$$

In this example, the Galilean transformation equations would give a relative speed of $8c/5$. However, as special relativity requires, the Lorentz transformation equations give a relative speed that is less than c. In this regard, you might also consult Drill Prob. 39-5.

DRILL PROBLEM 39-4

A neutral pion moves vertically downward at a speed $0.99c$ relative to the earth. The pion decays into two gamma rays (they are massless) that move off vertically in opposite directions, each at a speed c as observed in the pion's frame. What are the speeds of the two gamma rays relative to the earth?
ANS. c.

DRILL PROBLEM 39-5

A particle has a velocity $\mathbf{v}' = v_x'\mathbf{i}' + v_y'\mathbf{j}' + v_z'\mathbf{k}'$ relative to S′, and S′ moves in the positive x direction with a speed u relative to S. Use Eqs. (39-10) to show that

$$v^2 = v_x^2 + v_y^2 + v_z^2$$
$$= \frac{(v')^2 + 2v_x'u + (u^2/c^2)[c^2 - (v_y')^2 - (v_z')^2]}{c^2 + 2v_x'u + u^2(v_x')^2/c^2}c^2.$$

Use this equation to argue that if $v' < c$, then $v < c$, independent of u.

39-11 Relativistic Momentum and Energy

The Newtonian laws of momentum and energy conservation are both invariant under the Galilean transformation. As an example, let's consider momentum conservation for the interacting particles A and B of Fig. 39-16. Since their velocities in S and S′ are related through the Galilean transformation by

$$\mathbf{v}_A = \mathbf{v}'_A + \mathbf{u} \qquad \text{and} \qquad \mathbf{v}_B = \mathbf{v}'_B + \mathbf{u},$$

the particles' total momentum $\mathbf{P}$ in S is related to their total momentum $\mathbf{P}'$ in S′ by

$$\mathbf{P} = \mathbf{P}' + (m_A + m_B)\mathbf{u},$$

where $\mathbf{P} = m_A\mathbf{v}_A + m_B\mathbf{v}_B$ and $\mathbf{P}' = m_A\mathbf{v}'_A + m_B\mathbf{v}'_B$. Because m_A, m_B, and $\mathbf{u}$ are constants, and $t = t'$ under the Galilean transformation, we find by differentiation

$$\frac{d\mathbf{P}}{dt} = \frac{d\mathbf{P}'}{dt} = \frac{d\mathbf{P}'}{dt'}.$$

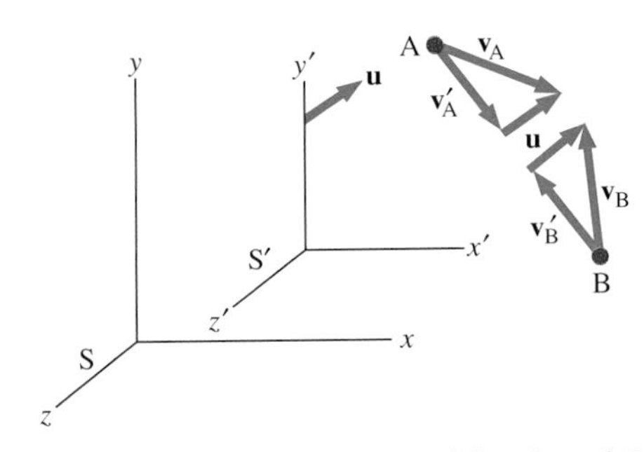

FIGURE 39-16 The interaction between particles A and B in the S and S′ frames. The velocities relative to the two frames are related by the Galilean transformation. Momentum as given by $\mathbf{p} = m\mathbf{v}$ is conserved in both frames.

Thus if momentum is conserved in S′ ($d\mathbf{P}'/dt' = 0$), it is also conserved in S ($d\mathbf{P}/dt = 0$); that is, momentum conservation is invariant under the Galilean transformation.

However, the special theory tells us that we should really transform the velocities using Eqs. (39-10) and (39-11). How does this affect the invariance of the momentum conservation law? We can answer this question in terms of the simple elastic collision depicted in Fig. 39-17, where we assume $m_A = m$ and $m_B = 2m$. In S′, the velocities of the two particles before the collision are $v'_{Ai} = v'$ and $v'_{Bi} = 0$. Using momentum conservation in S′, we find that after the collision (see Sec. 12-5), the velocities become

$$v'_{Af} = \left(\frac{m - 2m}{m + 2m}\right)v' + \left[\frac{2(2m)}{m + 2m}\right]0 = -\frac{1}{3}\,v',$$

and

$$v'_{Bf} = \left(\frac{2m}{m + 2m}\right)v' + \left(\frac{2m - m}{m + 2m}\right)0 = \frac{2}{3}\,v'.$$

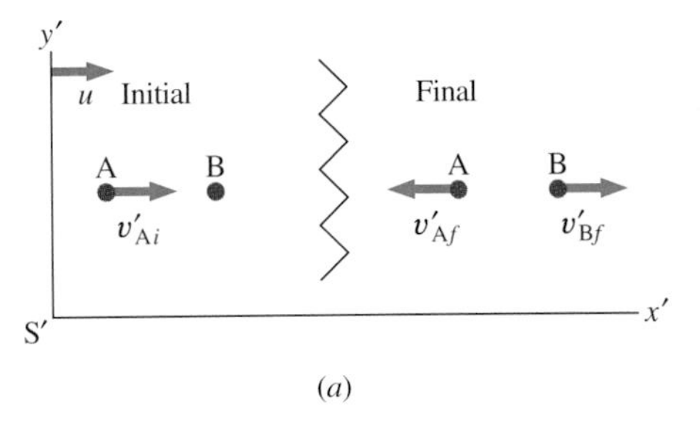

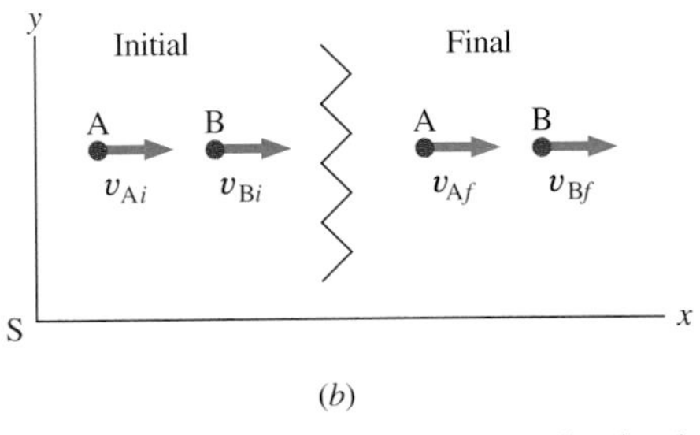

FIGURE 39-17 (*a*) An elastic collision in S′. The final velocities are found using the Newtonian law of momentum conservation. (*b*) Initial and final velocities of the particles in frame S found using the Lorentz transformation. Momentum is not conserved in this frame.

From the Lorentz transformation, the initial velocities in S are

$$v_{Ai} = \frac{v' + u}{1 + uv'/c^2} \qquad \text{and} \qquad v_{Bi} = \frac{0 + u}{1 + 0} = u;$$

while the final velocities are

$$v_{Af} = \frac{-\frac{1}{3}v' + u}{1 - uv'/3c^2} \qquad \text{and} \qquad v_{Bf} = \frac{\frac{2}{3}v' + u}{1 + 2uv'/3c^2}.$$

With these transformed velocities, let's compare the total initial and final momenta in frame S. The initial momentum is

$$m\left(\frac{v' + u}{1 + uv'/c^2}\right) + 2mu,$$

and the final momentum is

$$m\left(\frac{-\frac{1}{3}v' + u}{1 - uv'/3c^2}\right) + 2m\left(\frac{\frac{2}{3}v' + u}{1 + 2uv'/3c^2}\right).$$

We can verify that these momenta are not equal by simply substituting particular values for v' and u into the two equations. For example, $v' = 0.90c$ and $u = 0.90c$ yield an initial momentum of $2.794mc$ and a final momentum of $2.770mc$. Thus *the Newtonian momentum-conservation law is not Lorentz-invariant*. It can also be shown that *the Newtonian energy-conservation law is not Lorentz-invariant*. Now this does not mean that momentum and energy are not conserved for collisions in which the particles are moving at relativistic speeds. Instead, as reasoned by Einstein, the Newtonian expressions for momentum and energy are simply incorrect at these speeds. With the appropriate modifications of the expressions representing momentum and energy, their conservation laws do become invariant under the Lorentz transformation. The relativistically correct expressions for momentum and energy will now be discussed without a presentation of detailed arguments about their validity. As you study this material, do keep in mind that these expressions are consistent with the postulates of special relativity as well as experimental observations.

For a particle of mass m and velocity $\mathbf{v}$, the expression for its **relativistic momentum** is

$$\mathbf{p} = \frac{m\mathbf{v}}{\sqrt{1 - v^2/c^2}}. \tag{39-12}$$

At speeds for which $v/c \to 0$, this equation is accurately approximated by $\mathbf{p} = m\mathbf{v}$, the Newtonian formula. When v is comparable to c, the relativistic and Newtonian values of momentum differ considerably, as illustrated in Fig. 39-18.

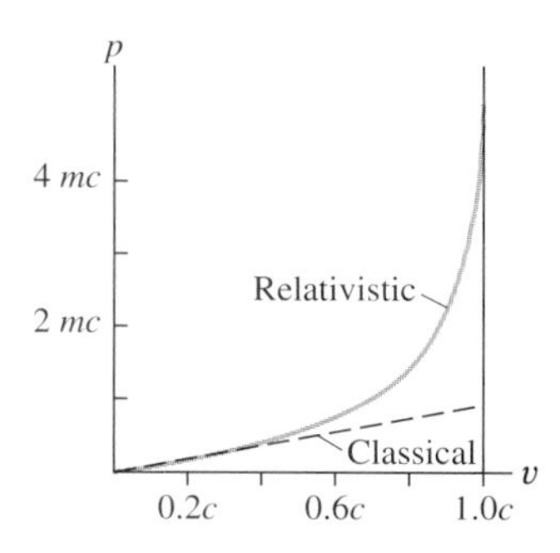

FIGURE 39-18 The classical momentum and relativistic momentum of a particle as functions of its speed.

The relativistic force and the time rate of change of a particle's relativistic momentum are related by

$$\mathbf{F} = \frac{d\mathbf{p}}{dt}, \tag{39-13}$$

which is identical to its Newtonian counterpart; however, $\mathbf{p}$ is now given by Eq. (39-12). Furthermore, Newton's second law is given by Eq. (39-13) and *not* by $\mathbf{F} = m\mathbf{a}$, an equation that is not relativistically invariant.

The relativistic form of the work-energy theorem is derived in Supplement 39-1. There it is shown that the work required to change the speed of a particle from 0 to v is

$$W = \frac{mc^2}{\sqrt{1 - v^2/c^2}} - mc^2.$$

The right-hand side of this equation is the **relativistic kinetic energy** T of the particle:

$$T = \frac{mc^2}{\sqrt{1 - v^2/c^2}} - mc^2. \tag{39-14}$$

Notice that $T \to \infty$ as $v \to c$. Consequently, as a particle approaches the speed of light, the work required to increase its speed approaches ∞. It is therefore impossible to accelerate a particle with mass to a speed c, as an infinite amount of work would be required.*

*Although we won't consider them at this point, there are particles with zero mass that move in free space at the speed of light c.

If the speeds of the particle at x_1 and x_2 are v_1 and v_2, respectively, the work-energy theorem becomes

$$\begin{aligned} W_{12} &= T_2 - T_1 \\ &= \left(\frac{mc^2}{\sqrt{1 - v_2^2/c^2}} - mc^2\right) - \left(\frac{mc^2}{\sqrt{1 - v_1^2/c^2}} - mc^2\right) \\ &= \frac{mc^2}{\sqrt{1 - v_2^2/c^2}} - \frac{mc^2}{\sqrt{1 - v_1^2/c^2}}. \end{aligned}$$

The quantity

$$E = \frac{mc^2}{\sqrt{1 - v^2/c^2}} \tag{39-15}$$

The Stanford linear accelerator for electrons is 3 km long. It accelerates electrons to an energy of 50 GeV, which, from Eq. (39-15), corresponds to a speed just less than the speed of light. What would the speed of a 50-GeV electron be if it obeyed the rules of classical Newtonian mechanics?

is the **total energy** of the particle. The work-energy theorem can therefore also be written as

$$W_{12} = E_2 - E_1. \tag{39-16}$$

From a comparison of Eqs. (39-14) and (39-15), we find that E and T are related by

$$E = T + mc^2. \tag{39-17}$$

Finally, by combining Eq. (39-15) with Eq. (39-12) and eliminating v, we obtain a useful relationship between E, p, and m:

$$E^2 - p^2c^2 = (mc^2)^2. \tag{39-18}$$

For a stationary particle, $E = mc^2$, which is known as the **rest-mass energy**. Hence from Eq. (39-17), *the total energy of a particle is the sum of its relativistic kinetic energy and its rest-mass energy*. The equation $E = mc^2$ implies that there is an equivalence between mass and energy; and in fact, mass can be converted to energy and vice versa during an interaction. Because c is numerically so large, there is a tremendous amount of energy associated with even a small mass. For example, if we could convert 1 kg of water completely into energy, we would have

$$E = mc^2 = (1.0 \text{ kg})(3.0 \times 10^8 \text{ m/s})^2 = 9.0 \times 10^{16} \text{ J}$$

of energy, which is enough to supply the United States with energy for roughly half a day!

According to Einstein, it is *the total energy E that is conserved in an interaction*, not the kinetic energy or the mass. If the mass changes by $-\Delta m$ during an interaction, there is a resultant change $+(\Delta m)c^2$ in the kinetic energy so that the total energy E remains constant. For example, if particles A and B interact to produce particles C and D ($A + B \rightarrow C + D$), then

$$(T_A + m_A c^2) + (T_B + m_B c^2) = (T_C + m_C c^2) + (T_D + m_D c^2),$$

and the change in kinetic energy is

$$(T_C + T_D) - (T_A + T_B) = [(m_A + m_B) - (m_C + m_D)]c^2.$$

In 1905, when Einstein proposed his ideas about mass and energy, scientists thought that mass was always conserved during interactions. This belief was supported by their studies of chemical reactions, which exhibited no measurable change in total mass.* However, when atomic nuclei were first split in the laboratory in the early 1930s, the interchange of mass and energy was confirmed. One of the first nuclear interactions studied was the fusion of a proton and a lithium nucleus into a product that quickly decays into two helium nuclei. This reaction is represented symbolically by

$$p + Li \rightarrow He + He.$$

The total mass of the reactants is $m_p + m_{Li} = 13.32759 \times 10^{-27}$ kg, and the total mass of the products is $2m_{He} = 13.29664 \times 10^{-27}$ kg. Thus there is a mass reduction of $\Delta m = (13.32759 - 13.29664) \times 10^{-27} \text{ kg} \approx 3.10 \times 10^{-29}$ kg. This shows up as kinetic energy $\Delta mc^2 = (3.10 \times 10^{-29} \text{ kg})(3.00 \times 10^8 \text{ m/s})^2 = 2.79 \times 10^{-12} \text{ J} = 17.5$ MeV carried by the He nuclei.

Today the conversion of mass into energy is a well-known phenomenon. Einstein is, in fact, probably best known in the nonscientific community for his equation $E = mc^2$. Here on the earth, nuclear power plants and weaponry are the most common applications of this remarkable formula. On a larger scale, it is not an exaggeration to say that our very existence depends on the conversion of mass to energy, for this is the principle behind energy production in the cores of stars. Protons there are constantly fused to form helium nuclei with a subsequent loss in mass. The average power generated by the sun is 4×10^{26} W. To account for this, about 4 million metric tons of matter has to be converted to energy each second! Fortunately, the sun is so massive that over its expected 10-billion-year lifetime, the total mass loss due to fusion is less than 1 percent of its total mass.

A fearsome example of the equivalence of mass and energy.

*As Prob. 39-58 demonstrates, there is a change in mass; however, it is just too small relative to the total mass to measure.

EXAMPLE 39-7 RELATIVISTIC MOMENTUM AND ENERGY

(*a*) Calculate the magnitude of the momentum of a proton moving at a speed of 2.00×10^8 m/s. (*b*) Calculate the total energy, the rest-mass energy, and the kinetic energy of the proton.

SOLUTION (*a*) For this proton,

$$\frac{1}{\sqrt{1 - v^2/c^2}} = \frac{1}{\sqrt{1 - (2.00 \times 10^8 \text{ m/s}/3.00 \times 10^8 \text{ m/s})^2}} = 1.34.$$

Its relativistic momentum is therefore

$$p = 1.34mv = 1.34(1.67 \times 10^{-27} \text{ kg})(2.00 \times 10^8 \text{ m/s}) = 4.48 \times 10^{-19} \text{ kg}\cdot\text{m/s}.$$

(*b*) From Eq. (39-15) the total energy of the proton is

$$E = 1.34mc^2 = 1.34(1.67 \times 10^{-27}\text{ kg})(3.00 \times 10^8\text{ m/s})^2$$
$$= 2.01 \times 10^{-10}\text{ J};$$

the rest-mass energy is

$$mc^2 = (1.67 \times 10^{-27}\text{ kg})(3.00 \times 10^8\text{ m/s})^2$$
$$= 1.50 \times 10^{-10}\text{ J};$$

and the kinetic energy is

$$T = E - mc^2 = (2.01 - 1.50) \times 10^{-10}\text{ J} = 5.1 \times 10^{-11}\text{ J}.$$

EXAMPLE 39-8 MOMENTUM AND ENERGY TRANSFORMATIONS

A particle is moving along the x' axis in the S′ frame at a velocity v'_x. (*a*) Use the relativistic velocity transformation to determine the transformation equations between S′ and S for momentum and energy. (*b*) Use the transformation equations to argue that momentum conservation and energy conservation are relativistically invariant.

SOLUTION (*a*) The momentum and total energy of the particle in frame S are

$$p_x = \frac{mv_x}{\sqrt{1 - v_x^2/c^2}} \quad \text{and} \quad E = \frac{mc^2}{\sqrt{1 - v_x^2/c^2}}.$$

Using the velocity transformation, we find

$$p_x = \frac{m(v'_x + u)/(1 + uv'_x/c^2)}{\sqrt{1 - [(v'_x + u)/(1 + uv'_x/c^2)]^2/c^2}}.$$

After some algebraic simplification, and with the help of Eqs. (39-12) and (39-15), this reduces to

$$p_x = \frac{p'_x + uE'/c^2}{\sqrt{1 - u^2/c^2}}, \qquad (i)$$

the *momentum transformation equation*.

Similarly,

$$E = \frac{mc^2}{\sqrt{1 - v_x^2/c^2}} = \frac{mc^2}{\sqrt{1 - [(v'_x + u)/(1 + uv'_x/c^2)]^2/c^2}},$$

which reduces to the *energy transformation equation*

$$E = \frac{E' + up'_x}{\sqrt{1 - u^2/c^2}}. \qquad (ii)$$

(*b*) Suppose p'_x and E' are conserved in S′. Then from Eq. (*i*), p_x, which is a function of these quantities, must be conserved in S; and from Eq. (*ii*), E is conserved in S.

EXAMPLE 39-9 DECAY OF A K° MESON

A K° meson at rest in S′ decays into two $\pi°$ mesons that move in opposite directions along the x' axis, as illustrated in Fig. 39-19. (*a*) Find the momenta and total energies of the two $\pi°$ mesons in S′. (*b*) If S′ is moving to the right at a speed $u = 4c/5$ with respect to S, what are the momenta and energies of the three particles in S? Do your calculations using the Lorentz velocity transformation. (*c*) Use the results of part (*b*) to verify that momentum and energy are conserved in S. For the K° meson $m_K = 8.85 \times 10^{-28}$ kg; for the $\pi°$ mesons $m_\pi = 2.40 \times 10^{-28}$ kg.

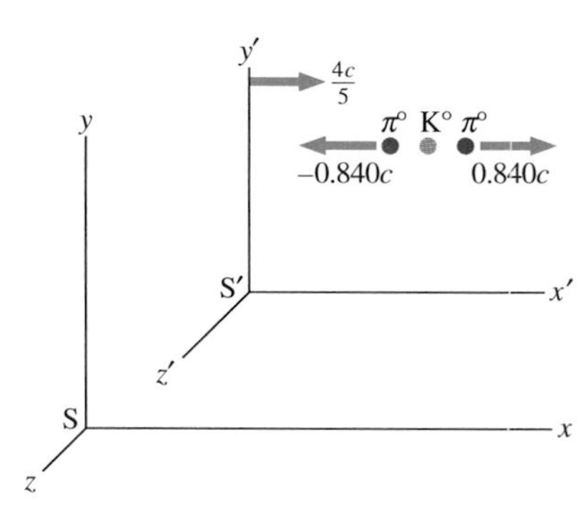

FIGURE 39-19 A K° meson is at rest in S′ when it decays into two $\pi°$ mesons that move off in opposite directions. Frame S′ is moving toward the right at speed $4c/5$ relative to S.

SOLUTION (*a*) To conserve momentum in S′, the two $\pi°$ mesons must move off with equal and opposite velocities since the K° meson is stationary. With the $\pi°$ meson velocity designated by v', the law of energy conservation in S′ gives

$$m_K c^2 = 2\left[\frac{m_\pi c^2}{\sqrt{1 - (v')^2/c^2}}\right],$$

so

$$(8.85 \times 10^{-28}\text{ kg})c^2 = 2\left[\frac{(2.40 \times 10^{-28}\text{ kg})c^2}{\sqrt{1 - (v')^2/c^2}}\right],$$

and

$$v' = \pm 0.840c.$$

The momenta and energies of the two $\pi°$ mesons are

$$(p'_\pi)_{1,2} = \frac{m_\pi v'}{\sqrt{1 - (v')^2/c^2}} = \frac{(2.40 \times 10^{-28}\text{ kg})(\pm 0.840c)}{\sqrt{1 - (0.840)^2}}$$
$$= \pm 1.11 \times 10^{-19}\text{ kg·m/s}$$

and the total energy of each $\pi°$ meson is

$$(E'_\pi)_{1,2} = \frac{m_\pi c^2}{\sqrt{1 - (v')^2/c^2}} = \frac{(2.40 \times 10^{-28}\text{ kg})c^2}{\sqrt{1 - (0.840)^2}}$$
$$= 3.98 \times 10^{-11}\text{ J}.$$

(*b*) From Eq. (39-10*a*), the velocity of the K° meson in frame S is

$$v_K = \frac{0 + 4c/5}{1 + 0} = 0.80c = 2.40 \times 10^8\text{ m/s},$$

so its momentum and energy in S are

$$p_K = \frac{m_K v_K}{\sqrt{1 - (v_K)^2/c^2}} = 3.54 \times 10^{-19}\text{ kg·m/s},$$

and

$$E_K = \frac{m_K c^2}{\sqrt{1 - (v_K)^2/c^2}} = 13.3 \times 10^{-11}\text{ J}.$$

Repeating these calculations for the two π° mesons, we obtain

$$v_{\pi 1} = \frac{0.840c + 0.800c}{1 + (0.840)(0.800)} = 0.981c = 2.94 \times 10^8 \text{ m/s},$$

$$p_{\pi 1} = \frac{m_\pi v_{\pi 1}}{\sqrt{1 - (v_{\pi 1})^2/c^2}} = 3.64 \times 10^{-19} \text{ kg}\cdot\text{m/s},$$

$$E_{\pi 1} = \frac{m_\pi c^2}{\sqrt{1 - (v_{\pi 1})^2/c^2}} = 11.1 \times 10^{-11} \text{ J}$$

for the first π° meson; and

$$v_{\pi 2} = \frac{-0.840c + 0.800c}{1 - (0.840)(0.800)} = -0.122c = -3.66 \times 10^7 \text{ m/s},$$

$$p_{\pi 2} = \frac{m_\pi v_{\pi 2}}{\sqrt{1 - (v_{\pi 2})^2/c^2}} = -0.09 \times 10^{-19} \text{ kg}\cdot\text{m/s},$$

$$E_{\pi 2} = \frac{m_\pi c^2}{\sqrt{1 - (v_{\pi 2})^2/c^2}} = 2.2 \times 10^{-11} \text{ J}$$

for the second π° meson.

(*c*) The momenta and energies of the three particles in S are tabulated as:

Particle	K°	$(\pi^\circ)_1$	$(\pi^\circ)_2$
Momentum (10^{-19} kg·m/s)	3.54	3.64	−0.09
Energy (10^{-11} J)	13.3	11.1	2.2

With round-off error taken into account, it is clear that momentum and energy are also conserved in frame S.

DRILL PROBLEM 39-6

Show that, like other relativistic expressions, the relativistic kinetic energy of Eq. (39-14) reduces to its Newtonian counterpart $mv^2/2$ in the limit $v/c \to 0$.

DRILL PROBLEM 39-7

Repeat the calculations of Example 39-7 for an electron moving at a speed of $0.60c$.
ANS. (*a*) 2.05×10^{-22} kg·m/s; (*b*) 1.02×10^{-13} J, 0.82×10^{-13} J, 0.20×10^{-13} J.

DRILL PROBLEM 39-8

Repeat the calculations of Example 39-9 using the momentum and energy transformation equations obtained in Example 39-8.

SUMMARY

1. **Galilean transformation and invariance**
The Galilean transformation equations relating coordinates in inertial frames S and S′ are

$$x = x' + ut, \qquad y = y', \qquad z = z', \qquad \text{and} \qquad t = t',$$

where it is assumed that the origins of the two frames coincide at $t = t' = 0$ and that S′ is moving along the x axis relative to S at a speed u. Newton's laws are invariant under this transformation; Maxwell's equations are not invariant under this transformation.

2. **Experimental basis for the special theory of relativity**
If there were a material medium (the ether) for the propagation of electromagnetic waves, it must be at rest relative to the earth, as demonstrated by the Michelson-Morley experiment. The aberration of starlight showed that the medium must be moving relative to the earth. This inconsistency is resolved by Einstein's special theory of relativity, which denies the existence of the ether.

3. **Postulates of the special theory of relativity**
(*a*) The laws of physics are identical in all inertial frames.
(*b*) Light travels in a vacuum with a speed c in any direction in all inertial frames.

4. **Synchronization of clocks**
Clocks synchronized in one inertial frame are not synchronized in a second inertial frame moving relative to the first frame.

5. **Measurement of transverse lengths**
Measurements of the transverse length of the same object are the same in two frames in relative motion.

6. **Time dilation**
(*a*) A proper time interval is a time interval measured on a single clock at rest in an inertial frame.
(*b*) If $\Delta t'$ is a proper time interval measured on a clock in inertial frame S′, then the corresponding time interval Δt measured in inertial frame S is

$$\Delta t = \frac{\Delta t'}{\sqrt{1 - \beta^2}},$$

where $\beta = u/c$ and u is the speed at which S′ is moving relative to S.

7. **Measurement of longitudinal lengths**
(*a*) The longitudinal length of an object measured in its own rest frame is the object's rest length.
(*b*) If l' is the rest length of an object measured in inertial frame S′, then the longitudinal length of the object measured in inertial frame S is

$$l = l'\sqrt{1 - \beta^2}.$$

8. **Clock synchronization revisited**
Two clocks synchronized in inertial frame S′ are out of synchronization in inertial frame S by

$$\delta t' = \frac{l'u}{c^2},$$

where l' is the distance between the clocks in S′ and u is the speed at which S′ is moving relative to S.

9. **Lorentz transformation**
The Lorentz transformation equations relating coordinates in inertial frames S and S′ are

$$x' = \frac{x - ut}{\sqrt{1-\beta^2}}, \qquad y' = y, \qquad z' = z,$$

and

$$t' = \frac{t - ux/c^2}{\sqrt{1-\beta^2}},$$

where it is assumed that the origins of the two frames coincide at $t = t' = 0$ and that S′ is moving along the x axis relative to S at a speed u. The reverse transformation is

$$x = \frac{x' + ut'}{\sqrt{1-\beta^2}}, \qquad y = y', \qquad z = z',$$

and

$$t = \frac{t' + ux'/c^2}{\sqrt{1-\beta^2}}.$$

10. **Velocity transformations**
The velocity transformation equations relating the velocity of a particle in the inertial frames S and S′ are

$$v_x' = \frac{v_x - u}{1 - uv_x/c^2},$$
$$v_y' = \frac{v_y\sqrt{1-\beta^2}}{1 - uv_x/c^2},$$
$$v_z' = \frac{v_z\sqrt{1-\beta^2}}{1 - uv_x/c^2},$$

and

$$v_x = \frac{v_x' + u}{1 + uv_x'/c^2},$$
$$v_y = \frac{v_y'\sqrt{1-\beta^2}}{1 + uv_x'/c^2},$$
$$v_z = \frac{v_z'\sqrt{1-\beta^2}}{1 + uv_x'/c^2},$$

where S′ is moving along the x axis relative to S at a speed u.

11. **Relativistic momentum and energy**
(*a*) The relativistic momentum of a particle of rest mass m moving with a velocity $\mathbf{v}$ is

$$\mathbf{p} = \frac{m\mathbf{v}}{\sqrt{1 - v^2/c^2}}.$$

(*b*) The force on a particle and the time rate of change of the particle's relativistic momentum are related by

$$\mathbf{F} = \frac{d\mathbf{p}}{dt}.$$

(*c*) The relativistic energy of a particle of rest mass m moving with a speed v is

$$E = \frac{mc^2}{\sqrt{1-\beta^2}}.$$

(*d*) The relativistic kinetic energy of a particle is

$$T = E - mc^2,$$

where mc^2 is the rest-mass energy of the particle.

(*e*) A useful relationship between E, p, and m is

$$E^2 - p^2c^2 = (mc^2)^2.$$

Supplement 39-1 Derivation of the Relativistic Form of the Work-Energy Theorem

Consider a particle moving along the x axis of an inertial frame. It is at rest at the origin, and it has a velocity v when its position is x. The work done on the particle between these two points is

$$W = \int_0^x F\,dx = \int_0^x \left(\frac{dp}{dt}\right) dx,$$

which, since $dx = (dx/dt)\,dt = v\,dt$, can be written as

$$W = \int_0^t \left(\frac{dp}{dt}\right) v\,dt.$$

From the chain rule, $dp/dt = (dp/dv)(dv/dt)$, so

$$W = \int_0^t \left(\frac{dp}{dv}\right)\left(\frac{dv}{dt}\right) v\,dt = \int_0^v v\left(\frac{dp}{dv}\right) dv.$$

From the expression for the relativistic momentum given in Eq. (39-12), we obtain

$$\frac{dp}{dv} = \frac{d}{dv}\left(\frac{mv}{\sqrt{1 - v^2/c^2}}\right) = \frac{m}{(1 - v^2/c^2)^{3/2}};$$

hence

$$W = \int_0^v \frac{mv}{(1 - v^2/c^2)^{3/2}}\,dv.$$

Integration of this last equation yields

$$W = \frac{mc^2}{\sqrt{1 - v^2/c^2}} - mc^2.$$

This expression is the relativistic form of the work-energy theorem.

QUESTIONS

39-1. Discuss how our lives would be different if the speed of light were 25 m/s.

39-2. Could you return from a space journey during which you traveled near the speed of light to find yourself younger than your children?

39-3. Could you return from a space journey during which you traveled near the speed of light to find yourself younger than you were when you began the journey?

39-4. A spaceship heads directly toward a laser at a speed of $0.90c$. At what speed does its pilot see the light from the laser approach the ship?

39-5. A cube rests at the origin of a reference frame with its sides parallel to the x, y, and z axes. Describe the dimensions of the cube as measured by observers in a frame moving at a relativistic speed (*a*) in the negative x direction, (*b*) in the positive x direction, (*c*) in the positive z direction, and (*d*) in the xy plane at 45° to the x axis.

39-6. Describe in detail what must be done to measure the length of an object moving at relativistic speeds.

39-7. Suppose that a spaceship could travel at such high speeds that a crew of space explorers could circumnavigate the entire universe in 30 years (ship time). Would they be able to tell their friends about their exploits when they returned to earth?

39-8. Discuss the situations in which two observers are needed to make a measurement.

39-9. The theory of relativity places an upper limit on the speed a particle can have. Are there also limits on the energy and momentum?

39-10. Two clocks positioned a meter apart along the equator are synchronized by observers at the equator. Are these clocks synchronized for observers at the North Pole?

39-11. Is it possible to observe two spaceships approaching each other at a relative speed greater than c? Is it possible to observe a spaceship approaching you at a relative speed greater than c?

39-12. Can you suggest any problems associated with defining a rigid body in a way that is consistent with the special theory of relativity?

39-13. Suppose you are standing at the curb watching a very fast express bus pass by. (*a*) If both you and a passenger on the bus are observing a clock on the bus, which of you is measuring a proper time interval? (*b*) Who measures the rest length of the bus? (*c*) Who measures the rest length between the streetlights?

39-14. Elementary-particle physicists usually express the masses of particles in MeV. Why can they use an energy unit to designate a mass?

39-15. Can a particle with zero mass be brought to rest in any inertial frame?

39-16. Why is the concept of simultaneity important for our measurements of length?

39-17. If the nuclear decay $A \to B + C$ is known to occur, what can you say about the mass of A compared with the mass of B plus the mass of C?

PROBLEMS

Note: In some problems, you will want to use one of the following approximations for small values of x:

$$1/\sqrt{1-x} \approx 1 + x/2 \qquad \text{or} \qquad \sqrt{1-x} \approx 1 - x/2.$$

Time Dilation

39-1. An unstable elementary particle travels at $0.90c$ relative to the laboratory. Calculate the ratio of its lifetime measured in the laboratory to its lifetime measured in its rest frame.

39-2. A passenger notes that he has spent 20 h on an airplane flying at 600 km/h. If he had not made this trip, how much older would he be?

39-3. One-half of a beam of pions (half-life = 1.8×10^{-8} s) is observed to decay in 3.0×10^{-7} s in the laboratory. What is the speed of the beam?

39-4. The star Alpha Centauri is about 4.3 light-years from the earth. With what constant speed must a spaceship travel from the earth to the star so that only one day passes by in the frame of the ship?

39-5. A clock moves past observers on the earth at a speed of $0.80c$. How much time will pass by for the earth observers while the moving clock ticks off 1 s?

39-6. An accelerator produces a beam of pions moving at a speed of $0.999980c$ relative to the laboratory. The half-life of the pions is 1.8×10^{-8} s. (*a*) If there were no time dilation, how far would the beam travel before half the pions decayed? (*b*) Since there is time dilation, how far does the beam actually travel before half the particles decay?

39-7. Two events occur at the same point in frame S′, which is moving by frame S at a relative speed of $0.60c$. In S′ the time interval between these events is 4.0 μs. (*a*) What is the time interval between these events as measured in S? (*b*) To observers in S, what is the distance between these events?

39-8. An unstable particle that is moving through a detector at $0.90c$ travels 2.0 m before decaying. How long does the particle live relative to its rest frame?

39-9. A pion produced by a cosmic ray at the top of the earth's atmosphere travels toward the earth at a speed of $0.99c$. Relative to its rest frame, the pion lives 3.0×10^{-8} s before decaying. (*a*) What is the time that this pion lives as measured with clocks on the earth? (*b*) How far does this pion travel with respect to the earth?

39-10. The half-life of the K^+ meson is 0.85×10^{-8} s. How fast is a beam of these mesons moving if half of them decay in 4.0×10^{-8} s in the laboratory?

Length Contraction

39-11. What length will observers measure for a spaceship moving by them at a speed of $0.80c$ if the passengers on the ship determine that its length is 100 m?

39-12. If the length of a stick is contracted to half its rest length, how fast is the stick traveling?

39-13. Plot the ratio (contracted length)/(rest length) versus u/c, where u is the speed of the object. Choose appropriate values for u/c so that your curve is clearly defined.

39-14. Do Prob. 39-4 using the concept of length contraction.

39-15. A meterstick moves by observers in the direction of its length at a speed of 2.8×10^8 m/s. What do these observers measure for the length of the stick?

39-16. A pilot measures the length of his spaceship to be 200 m. The ship is moving by the earth at a speed $u = 0.95c$ when a lamp at its tail emits a flash of light. (*a*) What is the time taken by the flash to reach the nose of the ship as measured by the pilot? (*b*) What is the time as measured by observers on the earth?

39-17. A cubical box with sides 30 cm long is placed in a rocket ship. The ship travels by the earth at a speed $u = 0.80c$, with one side of the box parallel to the ship's velocity. What do people on the earth measure the volume of the box to be?

39-18. Suppose that a proton is ejected from the sun toward the earth at a speed of $0.99c$. (*a*) Relative to the proton, how far away is the earth? (*b*) For the proton, how much time is required for the earth to reach it? (*c*) For earth observers, how much time is required for the proton to reach the earth?

39-19. Suppose there are human beings on a planet orbiting the star Sirius, which is about 9 light-years from the earth. A group of intrepid citizens of this planet decide to visit the earth and give themselves 12 years (ship time) to make the trip. At what speed must they travel?

39-20. Repeat Prob. 39-16, with the lamp in the nose of the ship and the flash detected in the tail.

39-21. Observers in S′ measure the rest length of a stick that is oriented at 30° to the x' axis to be 60 cm. If S′ is moving along the x axis with a speed $u = 0.80c$ with respect to S, (*a*) what is the orientation of the stick with respect to the x axis and (*b*) what is the length of the stick in S?

Clock Synchronization

39-22. A rocket ship is measured to be 200 m long by its pilot. There are synchronized clocks at each end of the ship, which is moving by the earth at a speed of $4c/5$. To observers on the earth, by how much are the two clocks out of synchronization?

39-23. To earth observers, two clocks that are synchronized on a rocket ship are 100 m apart and out of synchronization by 4.0×10^{-7} s. What is the speed of the ship relative to the earth?

39-24. Two marksmen on the ground are a distance l apart and have ray guns aimed upward. Relative to their frame, they fire their guns simultaneously at a rocket ship that is passing by at a relative speed u. (See the accompanying figure.) The ship passes so close to them that the rays burn holes in the ship at effectively the same instant that the guns are fired. (*a*) How far apart are the holes in the ship as measured by the marksmen? (*b*) Use the length contraction formula to determine how far apart passengers on the ship measure the holes to be. (*c*) To passengers on the ship, what is the difference in time between the firings of the ray guns? (*d*) What do the passengers measure for the distance between the guns? (*e*) Use the calculations of parts (*c*) and (*d*) to answer part (*b*).

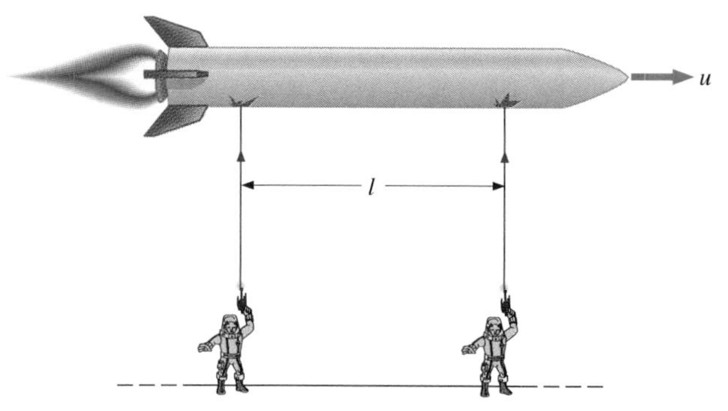

PROBLEM 39-24

Lorentz Transformation

Note: In the following problems, assume that the origins of S and S′ coincide at $t = t' = 0$.

39-25. An observer in inertial frame S measures the space and time coordinates of an event to be $x = 1000$ m, $y = 500$ m, $z = 500$ m, and $t = 4.0\ \mu$s. What are the space and time coordinates of the event in inertial frame S′, which is moving in the direction of increasing x at a speed $u = 0.80c$ relative to S?

39-26. The inertial frame S′ is moving in the direction of increasing x at a speed $u = 0.80c$ relative to inertial frame S. The space and time coordinates of an event in S′ are $x' = 1000$ m, $y' = 500$ m, $z' = 500$ m, and $t' = 4.0\ \mu$s. What are the space and time coordinates of the event in S?

39-27. Inertial frame S′ is moving in the direction of increasing x at a speed $u = 0.60c$ relative to inertial frame S. Event A occurs at the origin in S at a time $t = 0$, and event B occurs on the x axis at $x = 2000$ m and $t = 2.0\ \mu$s. (*a*) What are the space and time coordinates of the two events in S′? (*b*) You will find that in S′, B occurs before A, which is opposite to the order of the two events in S. Explain why the order can be reversed. (*Hint:* Is it possible for the two events to be cause and effect? You can answer this question by determining whether or not a signal can travel from A to B during the time interval between the instants when the two events occur.)

39-28. Inertial frame S′ is moving in the positive x direction relative to inertial frame S. Event A occurs in S on the x axis at x_A at a time t_A, and event B occurs on the x axis at x_B at a time t_B. (*a*) What speed must S′ have with respect to S so that both A and B occur at the same position in S′? (*b*) What speed is necessary for the events to occur at the same time in S′?

39-29. A stopwatch moves along the x axis at a speed $0.80c$. As it passes the origin, it reads zero. Use the Lorentz transformation to determine what time the stopwatch reads when it passes $x = 200$ m.

39-30. Do part (*b*) of Prob. 39-24 using the Lorentz transformation.

Velocity Transformation

39-31. A large spaceship moves away from the earth at a speed $u = 0.50c$. The ship contains a particle accelerator that produces protons that travel from the rear to the front of the ship. The protons travel at $v_x' = 0.70c$ relative to the ship. (*a*) What is the speed of the protons relative to the earth? (*b*) If the protons are produced in the front of the ship and travel to the rear at $v_x' = -0.70c$, what is the speed of the protons relative to the earth?

39-32. Two spaceships are moving in opposite directions, both at a speed of $0.90c$ as measured by an observer on the earth. What is the velocity of one ship with respect to the other?

39-33. The accompanying figure shows a liquid flowing down a long pipe from left to right at a speed u. (*a*) Using the fact that the speed of light in a stationary medium is c/n, where n is the refractive index of the medium, show that an observer in the laboratory measures the speed of light in the moving liquid to be

$$v = \frac{c/n + u}{1 + u/nc}.$$

(*b*) Show that this formula can be approximated by

$$v = \frac{c}{n} + \left(1 - \frac{1}{n^2}\right)u.$$

(*c*) Many years before the theory of relativity was developed, Fizeau found experimentally that v was given by $v = c/n + ku$. He called k the "dragging coefficient" and found an experimental value of 0.44 for water. What does the formula of part (*b*) give for k?

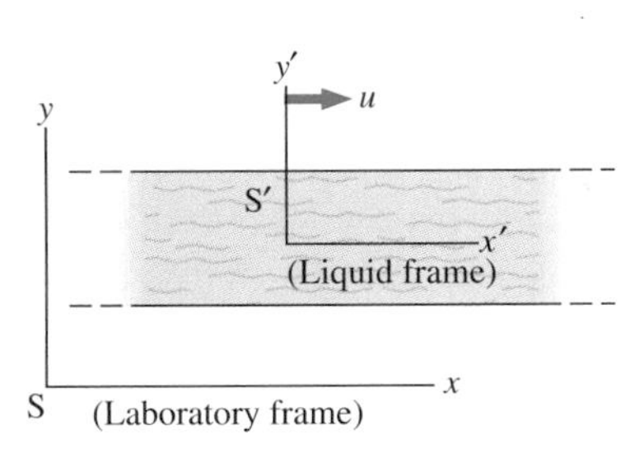

PROBLEM 39-33

39-34. A spaceship is moving directly away from the earth at a speed of 2.0×10^8 m/s. The pilot fires a probe back toward the earth at a speed of 2.5×10^8 m/s relative to the ship. With what speed does the probe approach the earth?

39-35. An astronomer on the earth observes galaxy X1 receding from the earth at a speed of $0.60c$. She also observes galaxy X2 receding in exactly the opposite direction at the same speed. What recessional speed would an astronomer in X1 measure (*a*) for our galaxy and (*b*) for X2?

39-36. The accompanying figure shows a spaceship moving parallel to the surface of the earth at a speed $u = 0.85c$. Relative to the ship, an electron is projected along the y' axis at $v_y' = -0.80c$. What is the velocity of the electron relative to the earth?

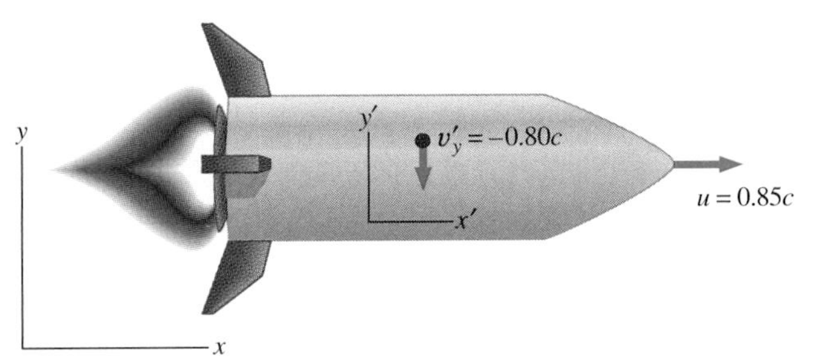

PROBLEM 39-36

39-37. The accompanying figure shows a spaceship moving by the earth at a speed $u = 0.90c$. In the ship, an electron is produced and moves off at 60° to the x' axis at a speed of $0.95c$. What is the speed of the electron relative to the earth?

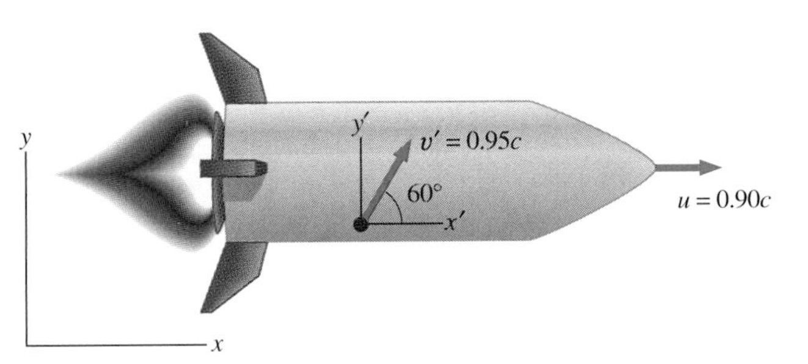

PROBLEM 39-37

Momentum and Energy

39-38. What are the momentum, energy, kinetic energy, and rest-mass energy of an electron moving at a speed of $0.80c$?

39-39. The mass of the hydrogen atom is sometimes written as 938.8 MeV/c^2. Determine this mass in kilograms.

39-40. At what speed are a particle's kinetic energy and rest-mass energy equal?

39-41. Suppose that electric energy costs eight cents per kilowatt·hour. If you could convert 1 kg of coal completely into electric energy, how much could you sell the energy for?

39-42. The positive pion's rest-mass energy is 140 MeV, and its half-life is 1.8×10^{-8} s. Consider a beam of these pions, each with an energy of 400 MeV. How much time elapses before half the pions in the beam decay?

39-43. What is the momentum of a proton whose energy is 1500 MeV?

39-44. An electron in an x-ray tube starts at rest and is accelerated from the cathode to the anode through a potential difference of 200 kV. When it arrives at the anode, what are (*a*) its kinetic energy in eV, (*b*) its energy in eV, and (*c*) its speed? (*d*) What is the electron's speed calculated classically?

39-45. In electron-positron annihilation, an electron and a positron (the electron's antiparticle) collide, and all of their mass is converted to electromagnetic radiation. If both particles are at rest when they annihilate, what is the total energy of the radiation? The mass of each particle is 9.1×10^{-31} kg.

39-46. In 1 year, the total consumption of electric energy in the United States is about 1.5×10^{19} J. If mass could be converted entirely into energy, how many kilograms of matter would be needed to supply this energy?

39-47. A possible source of energy for power plants of the future is the fusion of two deuterium nuclei producing a helium-three nucleus and a neutron. The mass of the deuterium nucleus is 3.3445×10^{-27} kg, the mass of the helium-three nucleus is 5.0083×10^{-27} kg, and the mass of the neutron is 1.6749×10^{-27} kg. (*a*) How much energy is released when one pair of deuterium nuclei fuses? (*b*) Assuming that the United States consumes 1.5×10^{19} J of electric energy per year, how much deuterium (in kilograms) is needed to supply this energy?

39-48. It takes 10.2 eV to excite the hydrogen atom from its ground state to its next-lowest energy state. By how much does the mass of 1 mol of hydrogen atoms (not H_2 molecules) in the ground state increase when half of these atoms are induced to make the transition to this second energy state? What is the percentage change in mass?

39-49. When a plutonium nuclear bomb explodes, the mass of the fuel decreases by 0.01 percent. (*a*) How much energy is released in the explosion of a bomb containing 25 kg of plutonium? (*b*) If the explosion takes place in 1.5 μs, what is the average power developed during the explosion? (*c*) Compare this with the average power consumed from fuel in the United States, which is about 8.0×10^{19} J/year.

General Problems

39-50. A wristwatch that keeps perfect time on the earth is worn by the pilot of a spaceship leaving the earth at a constant speed of 1.0×10^7 m/s. (*a*) When 5 s elapse on the watch, how much time elapses on the earth? (*b*) How many seconds does the watch lose in one day from the point of view of observers on the earth?

39-51. To observers on the earth, a star is d light-years away. A spaceship leaves the earth and heads toward the star. If the ship gets there in d years as measured by the pilot, what is the speed of the ship relative to the earth?

39-52. Relative to the earth, two spaceships travel in the same direction at speeds of $0.80c$ and $0.40c$, with the faster ship trailing the slower one by 20 light-seconds (the distance light travels in 20 s). Using the Lorentz transformation, determine how long it takes the faster ship to catch up to the slower one in the reference frames of (*a*) the earth, (*b*) the faster ship, and (*c*) the slower ship.

39-53. One of the high-speed trains of France is traveling between Paris and Lyons at 300 km/h. As the train speeds by, two friends standing alongside the track observe lightning bolts strike opposite ends of a passenger car simultaneously. The passengers on the train measure the resulting holes burned in the floor of the car to be 20 m apart. (*a*) What is the distance between the holes as measured by the observers on the ground? Is this length significantly different from 20 m? (*b*) Do the passengers of the train observe the lightning bolts to strike at the same instant? If not, by how much do the two times differ?

39-54. Do part (*b*) of Prob. 39-53 using the Lorentz transformation.

39-55. A π° meson ($mc^2 = 135$ MeV) moving in the laboratory at a speed of $0.90c$ decays into two gamma rays, both of which move off at speed c parallel to the velocity of the meson. (*a*) Determine the momentum and energy of the gamma rays in the rest frame of the meson. (*b*) Use these to determine the momentum and energy of the gamma rays in the laboratory. (*c*) Check to see that momentum and energy are conserved in the laboratory.

39-56. In S′ two events occur with space and time coordinates (x_1', y_1', z_1', t_1') and (x_2', y_2', z_2', t_2'). The same two events take place in S with space and time coordinates (x_1, y_1, z_1, t_1) and (x_2, y_2, z_2, t_2). (*a*) Show that

$$(\Delta x)^2 + (\Delta y)^2 + (\Delta z)^2 - c^2(\Delta t)^2 = (\Delta x')^2 + (\Delta y')^2 + (\Delta z')^2 - c^2(\Delta t')^2,$$

where $\Delta x = x_2 - x_1$, $\Delta t = t_2 - t_1$, etc. This demonstrates that the quantity

$$(\Delta x)^2 + (\Delta y)^2 + (\Delta z)^2 - c^2(\Delta t)^2$$

is invariant under the Lorentz transformation. (*b*) Explain why when $(\Delta x)^2 + (\Delta y)^2 + (\Delta z)^2 - c^2(\Delta t)^2 > 0$, the two events cannot be cause and effect.

39-57. The accompanying figure shows two particles of mass m that fuse to form a single particle of mass M. Use relativistic momentum and energy conservation to determine the mass M (in terms of m) and the velocity v of the particle formed.

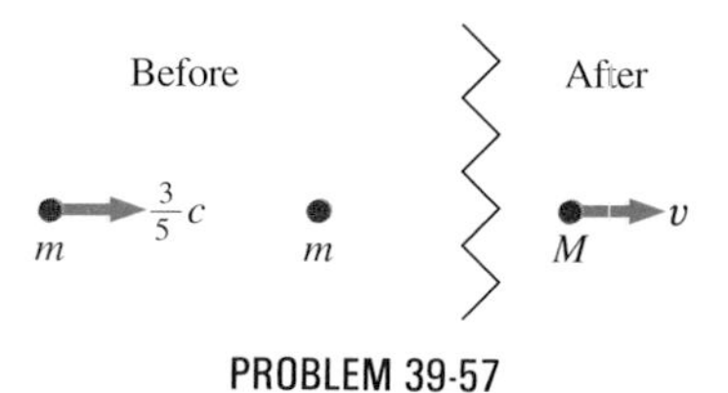

PROBLEM 39-57

39-58. Consider the reaction

$$2H_2 + O_2 \rightarrow 2H_2O$$

in which hydrogen and oxygen molecules combine to form water molecules. When 1 mol of water is formed, 2.9×10^5 J of energy is released. (*a*) Determine the reduction in mass that occurs in order to produce this energy. (*b*) What is the percentage change in the mass of the reactants? (*c*) Calculate the energy produced per molecule of water formed. Compare this with the energy produced per helium atom formed in the fusion reaction discussed in Sec. 39-11.

39-59. *Doppler shift for light.* A light source moves at a speed u along a straight line toward or away from an observer. If in its rest frame the source emits light of frequency f', show that the frequency f detected by the observer is

$$f = \sqrt{\frac{c \pm u}{c \mp u}}\, f',$$

where the upper signs correspond to the source approaching the observer and the lower signs correspond to the source receding from the observer. [*Hint:* In the frame of the source, the period $T' = 1/f'$ is a proper time interval. Also, the wavelength measured by the observer is $\lambda = (c \mp u)T$, where T is the time between crests in the observer's frame.]

APPENDIX A Conversion Factors

TABLE A-1 **Conversion Factors**

	Length					
	m	cm	km	in.	ft	mi
1 meter	1	10^{2}	10^{-3}	39.37	3.281	6.214×10^{-4}
1 centimeter	10^{-2}	1	10^{-5}	0.3937	3.281×10^{-2}	6.214×10^{-6}
1 kilometer	10^{3}	10^{5}	1	3.937×10^{4}	3.281×10^{3}	0.6214
1 inch	2.540×10^{-2}	2.540	2.540×10^{-5}	1	8.333×10^{-2}	1.578×10^{-5}
1 foot	0.3048	30.48	3.048×10^{-4}	12	1	1.894×10^{-4}
1 mile	1609	1.609×10^{5}	1.609	6.336×10^{4}	5280	1
1 angstrom	10^{-10}					
1 Bohr radius	5.292×10^{-11}					
1 fermi	10^{-15}					
1 light-year			9.460×10^{12}			

	Mass			
	kg	g	slug	u
1 kilogram	1	10^{3}	6.852×10^{-2}	6.024×10^{26}
1 gram	10^{-3}	1	6.852×10^{-5}	6.024×10^{23}
1 slug	14.59	1.459×10^{4}	1	8.789×10^{27}
1 atomic mass unit	1.661×10^{-27}	1.661×10^{-24}	1.138×10^{-28}	1
1 metric ton	1000			

	Time				
	s	min	h	day	yr
1 second	1	1.667×10^{-2}	2.778×10^{-4}	1.157×10^{-5}	3.169×10^{-8}
1 minute	60	1	1.667×10^{-2}	6.944×10^{-4}	1.901×10^{-6}
1 hour	3600	60	1	4.167×10^{-2}	1.141×10^{-4}
1 day	8.640×10^{4}	1440	24	1	2.738×10^{-3}
1 year	3.156×10^{7}	5.259×10^{5}	8.766×10^{3}	365.25	1

	Speed			
	m/s	cm/s	ft/s	mi/h
1 meter/second	1	10^{2}	3.281	2.237
1 centimeter/second	10^{-2}	1	3.281×10^{-2}	2.237×10^{-2}
1 foot/second	0.3048	30.48	1	0.6818
1 mile/hour	0.4470	44.70	1.467	1

	Force		
	N	dyne	lb
1 newton	1	10^{5}	0.2248
1 dyne	10^{-5}	1	2.248×10^{-6}
1 pound	4.448	4.448×10^{5}	1

TABLE A-1 continued

	Work, Energy, Heat					
	J	erg	ft·lb	eV	cal	Btu
1 joule	1	10^{7}	0.7376	6.242×10^{18}	0.2389	9.481×10^{-4}
1 erg	10^{-7}	1	7.376×10^{-8}	6.242×10^{11}	2.389×10^{-8}	9.481×10^{-11}
1 foot-pound	1.356	1.356×10^{7}	1	8.464×10^{18}	0.3239	1.285×10^{-3}
1 electron-volt	1.602×10^{-19}	1.602×10^{-12}	1.182×10^{-19}	1	3.827×10^{-20}	1.519×10^{-22}
1 calorie	4.186	4.186×10^{7}	3.088	2.613×10^{19}	1	3.968×10^{-3}
1 British thermal unit	1.055×10^{3}	1.055×10^{10}	7.779×10^{2}	6.585×10^{21}	2.520×10^{2}	1

	Pressure				
	Pa	dyne/cm²	atm	cmHg	lb/in.²
1 pascal	1	10	9.869×10^{-6}	7.501×10^{-4}	1.450×10^{-4}
1 dyne/centimeter²	10^{-1}	1	9.869×10^{-7}	7.501×10^{-5}	1.450×10^{-5}
1 atmosphere	1.013×10^{5}	1.013×10^{6}	1	76	14.70
1 centimeter mercury*	1.333×10^{3}	1.333×10^{4}	1.316×10^{-2}	1	0.1934
1 pound/inch²	6.895×10^{3}	6.895×10^{4}	6.805×10^{-2}	5.171	1
1 bar	10^{5}				
1 torr				1 (mmHg)	

*Where the acceleration due to gravity is 9.80665 m/s² and the temperature is 0°C.

TABLE A-2 **Other Useful Conversion Factors**

Area

1 square mile = 640 acres
1 acre = 43,560 ft²
1 barn = 10^{-28} m²

Volume

1 liter = 1000 cm³ = 3.531×10^{-2} ft³
1 U.S. fluid gallon = 4 U.S. fluid quarts = 8 U.S. fluid pints
1 U.S. fluid pint = 16 U.S. fluid ounces
1 U.S. fluid gallon = 231 in.³ = 3.786 liters

Angular Measurement

1 radian = 57.30° = 0.1592 rev
π radians = 180° = $\frac{1}{2}$ rev

Density

1 kilogram/cubic meter = 10^{-3} g/cm³ = 1.940×10^{-3} slug/ft³

Speed

1 knot = 1 nautical mi/h = 1.688 ft/s = 0.5145 m/s = 1.852 km/h
1 mile/minute = 60.00 mi/h = 88.00 ft/s

Power

1 foot-pound/second = 1.818×10^{-3} hp = 1.356 W = 1.356×10^{-3} kW
1 horsepower = 550 ft·lb/s = 745.7 W
1 British thermal unit/hour = 0.2161 ft·lb/s = 3.929×10^{-4} hp = 0.2930 W

APPENDIX B Some Fundamental Physical Constants*

Constant	Symbol	Value	Relative Uncertainty†
Avogadro constant	N_A	6.0221367×10^{23} mol^{-1}	0.59
Bohr radius	a_0	$5.29177249 \times 10^{-11}$ m	0.045
Boltzmann constant	k	1.380658×10^{-23} J/K	8.5
Deuteron rest mass	m_d	$3.3435860 \times 10^{-27}$ kg	0.59
Electron Compton wavelength	λ_C	$2.42631058 \times 10^{-12}$ m	0.089
Electron magnetic moment	μ_e	$9.2847701 \times 10^{-24}$ J/T	0.34
Electron rest mass	m_e	$9.1093897 \times 10^{-31}$ kg	0.59
Elementary charge	e	$1.60217733 \times 10^{-19}$ C	0.30
Gravitational constant	G	6.67259×10^{-11} m^3/s$^2 \cdot$kg	128
Neutron rest mass	m_n	$1.6749286 \times 10^{-27}$ kg	0.59
Permeability constant	μ_0	$1.25663706 \ldots \times 10^{-6}$ T$\cdot$m/A	Exact
Permittivity constant	ϵ_0	$8.85418781 \ldots \times 10^{-12}$ C^2/N$\cdot$m^2	Exact
Planck constant	h	$6.6260755 \times 10^{-34}$ J$\cdot$s	0.60
Proton magnetic moment	μ_p	$1.41060761 \times 10^{-26}$ J/T	0.34
Proton rest mass	m_p	$1.6726231 \times 10^{-27}$ kg	0.59
Rydberg constant	R_H	$1.0973731534 \times 10^{7}$ m^{-1}	0.0012
Speed of light in a vacuum	c	2.99792458×10^{8} m/s	Exact
Stefan-Boltzmann constant	σ	5.67051×10^{-8} W/m$^2\cdot$K^4	34
Universal gas constant	R	8.314510 J/mol$\cdot$K	8.4

*As specified by the Committee on Data for Science and Technology (CODATA) of the International Council for Scientific Unions. See E. Richard Cohen and B. N. Taylor, "Reviews of Modern Physics," **59**, 1121–1148 (1987).

†Parts per million. Defined constants have no error and are indicated by the notation "Exact."

APPENDIX C THE INTERNATIONAL SYSTEM OF UNITS

BASE UNITS

In the SI system, there are seven fundamental, or base, units corresponding to length, mass, time, electric current, thermodynamic temperature, amount of substance, and luminous intensity. The definitions of the base units and the years they were adopted are given in the glossary below.

Meter (m): The meter is the distance traveled by light in a vacuum in a time interval of 1/299,792,458 of a second (1983).

Kilogram (kg): The kilogram is the mass of the international prototype of the kilogram (1889).

Second (s): The second is the duration of 9,192,631,770 periods of the radiation emitted when the cesium-133 atom makes the transition between the two hyperfine levels of its ground state (1967).

Ampere (A): When two straight, parallel conductors of infinite length and negligible circular cross section are placed one meter apart in a vacuum, the ampere is that current maintained in both conductors such that a force of 2×10^{-7} newtons per meter of length is produced between the conductors (1948).

Kelvin (K): The kelvin is the fraction 1/273.16 of the thermodynamic temperature of the triple point of water (1967).

Mole (mol): The mole is the amount of substance that contains as many elementary entities as there are atoms in 0.012 kg of carbon-12 (1971).

Candela (Cd): The candela is the luminous intensity, in a given direction, of a source that emits monochromatic radiation of frequency 540×10^{12} Hz and that has a radiant intensity in that direction of 1/683 watt per steridian (1979).

SOME UNITS DERIVED FROM THE BASE UNITS

Quantity	Name of Unit	Symbol	Derived Unit
Frequency	hertz	Hz	s^{-1}
Force	newton	N	$kg \cdot m/s^2$
Pressure	pascal	Pa	$kg/m \cdot s^2$
Energy	joule	J	$kg \cdot m^2/s^2$
Power	watt	W	$kg \cdot m^2/s^3$
Electric charge	coulomb	C	$A \cdot s$
Electric potential	volt	V	$kg \cdot m^2/A \cdot s^3$
Electric resistance	ohm	Ω	$kg \cdot m^2/A^2 \cdot s^3$
Capacitance	farad	F	$A^2 \cdot s^4/kg \cdot m^2$
Inductance	henry	H	$kg \cdot m^2/A^2 \cdot s^3$
Magnetic field intensity	tesla	T	$kg/A \cdot s^2$

APPENDIX D Mathematical Formulas

Quadratic Formula

There are two values of x that satisfy $ax^2 + bx + c = 0$. They are

$$x = \frac{-b \pm \sqrt{b^2 - 4ac}}{2a}.$$

Pythagorean Formula

For a right triangle with perpendicular sides of length a and b and hypotenuse of length c,

$$a^2 + b^2 = c^2.$$

Trigonometric Functions

$\cos\theta = a/c$	$\sin\theta = b/c$
$\tan\theta = b/a$	$\cot\theta = a/b$
$\csc\theta = c/b$	$\sec\theta = c/a$

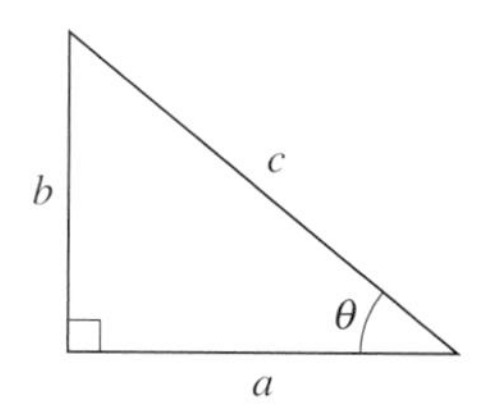

Law of Sines and Law of Cosines

1. Law of sines

$$\frac{a}{\sin\alpha} = \frac{b}{\sin\beta} = \frac{c}{\sin\gamma}$$

2. Law of cosines

$$c^2 = a^2 + b^2 - 2ab\cos\gamma$$

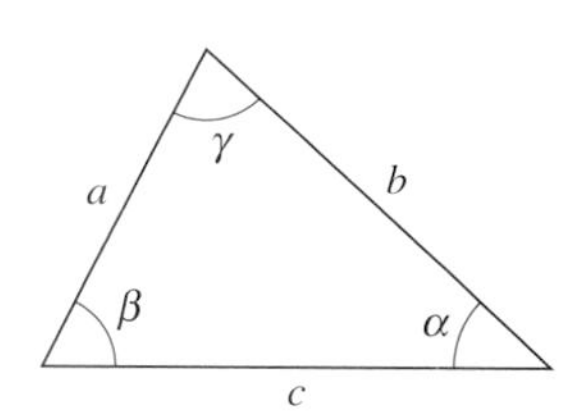

Properties of Simple Geometric Objects

1. Triangle of base b and height h

   Area $= \frac{1}{2}bh$

2. Circle of radius r

   Circumference $= 2\pi r$
   Area $= \pi r^2$

3. Sphere of radius r

   Surface area $= 4\pi r^2$
   Volume $= \frac{4}{3}\pi r^3$

4. Cylinder of radius r and height h

   Area of curved surface $= 2\pi rh$
   Volume $= \pi r^2 h$

5. Right circular cone of radius r and height h

   Area of curved surface $= \pi r\sqrt{r^2 + h^2}$
   Volume $= \frac{1}{3}\pi r^2 h$

Trigonometric Identities

1. $\sin\theta = 1/\csc\theta$
2. $\cos\theta = 1/\sec\theta$
3. $\tan\theta = 1/\cot\theta$
4. $\sin(90° - \theta) = \cos\theta$
5. $\cos(90° - \theta) = \sin\theta$
6. $\tan(90° - \theta) = \cot\theta$
7. $\sin^2\theta + \cos^2\theta = 1$
8. $\sec^2\theta - \tan^2\theta = 1$
9. $\tan\theta = \sin\theta/\cos\theta$
10. $\sin(\alpha \pm \beta) = \sin\alpha\cos\beta \pm \cos\alpha\sin\beta$
11. $\cos(\alpha \pm \beta) = \cos\alpha\cos\beta \mp \sin\alpha\sin\beta$
12. $\tan(\alpha \pm \beta) = \dfrac{\tan\alpha \pm \tan\beta}{1 \mp \tan\alpha\tan\beta}$
13. $\sin 2\theta = 2\sin\theta\cos\theta$
14. $\cos 2\theta = \cos^2\theta - \sin^2\theta = 2\cos^2\theta - 1 = 1 - 2\sin^2\theta$
15. $\sin\alpha + \sin\beta = 2\sin\frac{1}{2}(\alpha + \beta)\cos\frac{1}{2}(\alpha - \beta)$
16. $\cos\alpha + \cos\beta = 2\cos\frac{1}{2}(\alpha + \beta)\cos\frac{1}{2}(\alpha - \beta)$

Differentiation

1. $\dfrac{d}{dx}[af(x)] = a\dfrac{d}{dx}f(x)$
2. $\dfrac{d}{dx}[f(x) + g(x)] = \dfrac{d}{dx}f(x) + \dfrac{d}{dx}g(x)$
3. $\dfrac{d}{dx}[f(x)g(x)] = f(x)\dfrac{d}{dx}g(x) + g(x)\dfrac{d}{dx}f(x)$
4. $\dfrac{d}{dx}f(u) = \left[\dfrac{d}{du}f(u)\right]\dfrac{du}{dx}$
5. $\dfrac{d}{dx}x^m = mx^{m-1}$
6. $\dfrac{d}{dx}\sin x = \cos x$
7. $\dfrac{d}{dx}\cos x = -\sin x$

8. $\dfrac{d}{dx} \tan x = \sec^2 x$

9. $\dfrac{d}{dx} \cot x = -\csc^2 x$

10. $\dfrac{d}{dx} \sec x = \tan x \sec x$

11. $\dfrac{d}{dx} \csc x = -\cot x \csc x$

12. $\dfrac{d}{dx} e^x = e^x$

13. $\dfrac{d}{dx} \ln x = \dfrac{1}{x}$

14. $\dfrac{d}{dx} \sin^{-1} x = \dfrac{1}{\sqrt{1 - x^2}}$

15. $\dfrac{d}{dx} \cos^{-1} x = -\dfrac{1}{\sqrt{1 - x^2}}$

16. $\dfrac{d}{dx} \tan^{-1} x = \dfrac{1}{1 + x^2}$

Integration

Note: An arbitrary constant should be added to the right-hand side of each indefinite integral.

1. $\displaystyle\int af(x)\, dx = a \int f(x)\, dx$

2. $\displaystyle\int [f(x) + g(x)]\, dx = \int f(x)\, dx + \int g(x)\, dx$

3. $\displaystyle\int x^m\, dx = \frac{x^{m+1}}{m + 1} \quad (m \neq -1)$

4. $\displaystyle\int \sin x\, dx = -\cos x$

5. $\displaystyle\int \cos x\, dx = \sin x$

6. $\displaystyle\int \tan x\, dx = \ln|\sec x|$

7. $\displaystyle\int \sin^2 ax\, dx = \frac{x}{2} - \frac{\sin 2ax}{4a}$

8. $\displaystyle\int \cos^2 ax\, dx = \frac{x}{2} + \frac{\sin 2ax}{4a}$

9. $\displaystyle\int \sin ax \cos ax\, dx = -\frac{\cos 2ax}{4a}$

10. $\displaystyle\int e^{ax}\, dx = \frac{1}{a} e^{ax}$

11. $\displaystyle\int xe^{ax}\, dx = \frac{e^{ax}}{a^2} (ax - 1)$

12. $\displaystyle\int \ln ax\, dx = x \ln ax - x$

13. $\displaystyle\int \frac{dx}{a^2 + x^2} = \frac{1}{a} \tan^{-1} \frac{x}{a}$

14. $\displaystyle\int \frac{dx}{a^2 - x^2} = \frac{1}{2a} \ln \left|\frac{x + a}{x - a}\right|$

15. $\displaystyle\int \frac{dx}{\sqrt{a^2 + x^2}} = \sinh^{-1} \frac{x}{a}$

16. $\displaystyle\int \frac{dx}{\sqrt{a^2 - x^2}} = \sin^{-1} \frac{x}{a}$

17. $\displaystyle\int \sqrt{a^2 + x^2}\, dx = \frac{x}{2} \sqrt{a^2 + x^2} + \frac{a^2}{2} \sinh^{-1} \frac{x}{a}$

18. $\displaystyle\int \sqrt{a^2 - x^2}\, dx = \frac{x}{2} \sqrt{a^2 - x^2} + \frac{a^2}{2} \sin^{-1} \frac{x}{a}$

Taylor Series

1. $\displaystyle f(x) = f(a) + \left(\frac{df}{dx}\right)_{x=a} (x - a) + \frac{1}{2!} \left(\frac{d^2f}{dx^2}\right)_{x=a} (x - a)^2 + \frac{1}{3!} \left(\frac{d^3f}{dx^3}\right)_{x=a} (x - a)^3 + \cdots$

2. $\displaystyle (1 \pm x)^n = 1 \pm \frac{nx}{1!} + \frac{n(n - 1)x^2}{2!} + \cdots \quad (x^2 < 1)$

3. $\displaystyle (1 \pm x)^{-n} = 1 \mp \frac{nx}{1!} + \frac{n(n + 1)x^2}{2!} + \cdots \quad (x^2 < 1)$

4. $\displaystyle \sin x = x - \frac{x^3}{3!} + \frac{x^5}{5!} - \cdots$

5. $\displaystyle \cos x = 1 - \frac{x^2}{2!} + \frac{x^4}{4!} - \cdots$

6. $\displaystyle \tan x = x + \frac{x^3}{3} + \frac{2x^5}{15} + \cdots$

7. $\displaystyle e^x = 1 + x + \frac{x^2}{2!} + \cdots$

8. $\ln (1 + x) = x - \frac{1}{2}x^2 + \frac{1}{3}x^3 - \cdots \quad (|x| < 1)$

APPENDIX E Periodic Table of Elements

6	Atomic number
C	Symbol
12.011	Atomic mass*

Group I	Group II	Transition Elements										Group III	Group IV	Group V	Group VI	Group VII	Group O
1 H 1.0079																	2 He 4.0026
3 Li 6.94	4 Be 9.012											5 B 10.81	6 C 12.011	7 N 14.007	8 O 15.999	9 F 18.998	10 Ne 20.18
11 Na 22.990	12 Mg 24.305											13 Al 26.982	14 Si 28.086	15 P 30.974	16 S 32.06	17 Cl 35.453	18 Ar 29.948
19 K 39.098	20 Ca 40.08	21 Sc 44.956	22 Ti 47.90	23 V 50.94	24 Cr 51.996	25 Mn 54.938	26 Fe 55.847	27 Co 58.933	28 Ni 58.71	29 Cu 63.546	30 Zn 65.38	31 Ga 69.72	32 Ge 72.59	33 As 74.922	34 Se 78.96	35 Br 79.904	36 Kr 83.80
37 Rb 85.467	38 Sr 87.62	39 Y 88.906	40 Zr 91.22	41 Nb 92.906	42 Mo 95.94	43 Tc (98)	44 Ru 101.07	45 Rh 102.906	46 Pd 106.4	47 Ag 107.868	48 Cd 112.41	49 In 114.82	50 Sn 118.69	51 Sb 121.75	52 Te 127.60	53 I 126.90	54 Xe 131.30
55 Cs 132.905	56 Ba 137.33	57-71	72 Hf 178.49	73 Ta 180.95	74 W 183.85	75 Re 186.207	76 Os 190.2	77 Ir 192.22	78 Pt 195.09	79 Au 196.966	80 Hg 200.59	81 Tl 204.37	82 Pb 207.2	83 Bi 208.980	84 Po (209)	85 At (210)	86 Rn (222)
87 Fr (223)	88 Ra (226)	89-103	104 Rf (261)	105 Ha (262)	106 † (263)	107 † (262)	108 † (265)	109 † (266)									

Lanthanide Series	57 La 139.906	58 Ce 140.12	59 Pr 140.908	60 Nd 144.24	61 Pm (145)	62 Sm 150.4	63 Eu 151.96	64 Gd 157.25	65 Tb 158.925	66 Dy 162.50	67 Ho 164.930	68 Er 167.26	69 Tm 168.934	70 Yb 173.04	71 Lu 174.967
Actinide Series	89 Ac (227)	90 Th 232.038	91 Pa 231.039	92 U 238.029	93 Np (237)	94 Pu (244)	95 Am (243)	96 Cm (247)	97 Bk (247)	98 Cf (251)	99 Es (253)	100 Fm (257)	101 Md (258)	102 No (259)	103 Lr (260)

* Average value based on the relative abundance of isotopes on earth. For unstable elements, the mass of the most stable isotope is given in parentheses. † Unnamed.

APPENDIX F Answers to Odd-Numbered Problems

Chapter 1

1-1. L/T, L; m/s and cm/s; m and cm **1-7.** 9.47×10^{15} m **1-9.** 29.9 inHg **1-11.** 50 mi/h **1-13.** 38.8 m **1-15.** (*a*) 28.7°; (*b*) 4.71 rad **1-17.** 1.3×10^9 m, 3.3×10^6 m **1-19.** 0.070 m^3 **1-21.** 1.31 acres **1-23.** 1.4 g/cm^3 **1-27.** 1.2×10^{17} kg/m^3 **1-29.** (*a*) 3.27 m^2; (*b*) 4.5 ft; (*c*) 93.0 cm^2; (*d*) 3.7×10^{-6} kg/m^3 **1-31.** 1.46×10^{-3} m^3 **1-33.** 1.85×10^3 m, 6.07×10^3 ft, 1.15 mi **1-35.** (*e*) 8.7 mi

Chapter 2

2-3. 151 km at 61° south of west **2-5.** 18.4 at 300° **2-7.** $4a$, 0 **2-9.** $\mathbf{A} = 8.66\mathbf{i} + 5.00\mathbf{j}$, $\mathbf{D} = -16\mathbf{i} + 12\mathbf{j}$, $\mathbf{F} = 6.1\mathbf{i} + 6.1\mathbf{j} + 5.0\mathbf{k}$, $\mathbf{H} = -6.0\mathbf{i} - 8.0\mathbf{j} + 17.3\mathbf{k}$ **2-11.** (*a*) $11.7\mathbf{i} + 9.0\mathbf{j}$; (*c*) $-22.9\mathbf{i} + 8.0\mathbf{j}$; (*e*) $-9.1\mathbf{i} + 16.0\mathbf{j}$; (*g*) $17.3\mathbf{i} - 46.4\mathbf{j}$ **2-13.** (*a*) $(-3.0\mathbf{i} - 7.0\mathbf{j} + 3.0\mathbf{k})$ cm; (*b*) 8.2 cm; (*c*) 21.6 cm **2-15.** (*a*) $-10\mathbf{j}$ cm; (*b*) $-10\mathbf{i}$ cm; (*c*) $(5.0\mathbf{i} + 8.7\mathbf{j})$ cm; (*d*) $(8.7\mathbf{i} - 5.0\mathbf{j})$ cm; (*e*) $(-5.0\mathbf{i} - 8.7\mathbf{j})$ cm **2-17.** (*a*) $2.0\mathbf{i}$ m, $(-1.0\mathbf{i} + 1.73\mathbf{j})$ m, $(-1.0\mathbf{i} - 1.73\mathbf{j})$ m; (*b*) $(1.0\mathbf{i} + 1.73\mathbf{j})$ m; (*c*) $-2.0\mathbf{i}$ m; (*d*) $(3.0\mathbf{i} - 1.73\mathbf{j})$ m; (*e*) 0 **2-19.** (*a*) $(2.4\mathbf{i} + 2.4\mathbf{j})$ m **2-21.** 59° **2-23.** 55.6°, 64.9°, 45.0° **2-29.** $19.5\mathbf{k}$, $-55.2\mathbf{k}$, $127\mathbf{i} + 170\mathbf{j} - 83\mathbf{k}$, $146\mathbf{i} - 136\mathbf{j} - 12\mathbf{k}$ **2-31.** $(4.0\mathbf{i} - 3.0\mathbf{j})$ **2-33.** $-(8755\mathbf{i} + 5\mathbf{j})$ lb **2-35.** (*a*) $\mathbf{M}\cdot\mathbf{V} = 0$; (*b*) $M_x(x - x_0) + M_y(y - y_0) + M_z(z - z_0) = 0$; (*c*) $4x + 7y + 9z = 49$

Chapter 3

3-1. (*a*) $5.0\mathbf{i}$ m; (*b*) $-3.0\mathbf{i}$ m **3-3.** $-2.0\mathbf{i}$ km, $5.0\mathbf{i}$ km, $7.0\mathbf{i}$ km **3-5.** 0.414 s **3-7.** 3.4 m **3-9.** $-35.8\mathbf{j}$ m/s, $-2.7\mathbf{j}$ m/s, $-5.1\mathbf{j}$ m/s **3-11.** -7.0 cm/s, -1.0 cm/s **3-13.** -16.0 m/s, -8.0 m/s^2; -40.0 m/s, -8.0 m/s^2 **3-15.** (*a*) 1 mi/h·s = 3600 mi/h^2; (*b*) 5.4×10^4 mi/h^2 **3-17.** (*a*) $v(t) = 10.0t - 12.0t^2$, $a(t) = 10.0 - 24.0t$; (*b*) -28.0 m/s, -38.0 m/s^2; (*c*) 0.83 s; (*d*) 0, 0.83 s; (*e*) 1.15 m **3-19.** (*a*) 0, 0.50 m/s^2, -1.5 m/s^2; (*b*) 300 m, 0 **3-21.** v: +, 0, −, +; a: −, 0, 0, + **3-23.** 150 m **3-25.** (*a*) 525 m; (*b*) 180 m/s **3-27.** (*a*) -6.0 m, 6.0 m/s; (*b*) 12.0 m, 0; (*c*) 2.0 m **3-29.** $91g$ **3-31.** 15 mi/h **3-33.** -2.4 m/s^2 **3-35.** (*a*) 8.7×10^5 m/s; (*b*) 7.9×10^{-8} s **3-37.** (*a*) 4.17×10^{-5} m/s^2; (*b*) 1.42×10^{-2} m/s, 5.86×10^{-3} m/s **3-39** 87 mi/h **3-41.** -0.90 m/s^2 **3-43.** (*a*) v_0/a; (*b*) $v_0^2/2a$ **3-47.** (*a*) 11.6 m/s; (*b*) 3.6 m; (*c*) 16.1 m/s; (*d*) 3.79 s; (*e*) 29.1 m/s **3-49.** (*a*) 19.6 m/s; (*b*) 19.6 m **3-51.** (*a*) They are in free fall; (*b*) 0.90 s, 0.40 s **3-53.** (*a*) 39.2 m; (*b*) 8.1 m/s; (*c*) 35.8 m; (*d*) -9.8 m/s^2 **3-55.** (*a*) 34 m/s; (*b*) 149 m; (*c*) -64 m/s; (*d*) -9.8 m/s^2 **3-57.** (*a*) 6.37 s; (*b*) 59.4 m/s **3-59.** (*a*) m/s^2, m/$s^{2.5}$; (*b*) $At - \frac{2}{3}Bt^{1.5}$; (*c*) $\frac{1}{2}At^2 - \frac{4}{15}Bt^{2.5}$ **3-61.** (*a*) $0 \le t \le 5.0$ s: 3.2 m/s^2, $5.0 \text{ s} < t \le 11.0$ s: -1.5 m/s^2, $t > 11.0$ s: 0; (*b*) 6.4 m, 69.0 m, 116 m **3-63.** The music from the speakers is one-half beat behind the singer. **3-65.** They collide moving at 34 mi/h and 40 mi/h, respectively. **3-67.** 10 m **3-69.** 87 m, 93 m, 99 m, 105 m

Chapter 4

4-1. $(1.0\mathbf{i} - 4.0\mathbf{j} + 6.0\mathbf{k})$ m **4-3.** $(2.5\mathbf{i} + 4.7\mathbf{j} + 2.8\mathbf{k})$ m **4-5.** 2.99×10^4 m/s **4-7.** $-30\mathbf{i}$ cm/s **4-9.** $(13.7\mathbf{i} + 5.7\mathbf{j})$ m, $(-2.3\mathbf{i} + 5.7\mathbf{j})$ m, $(1.7\mathbf{i} + 12.6\mathbf{j})$ m **4-11.** $(3.7\mathbf{i} + 1.5\mathbf{j})$ m/s, $(-1.8\mathbf{i} + 4.6\mathbf{j})$ m/s, $(0.2\mathbf{i} + 1.8\mathbf{j})$ m/s **4-13.** (*a*) $c_1\mathbf{i} - 2c_2t\mathbf{j}$; (*b*) 0, $50\mathbf{i}$ m/s, $-9.8\mathbf{j}$ m/s^2 **4-15.** (*a*) $y = 6.3 - 0.67x$; (*b*) $3.0\mathbf{i} - 2.0\mathbf{j}$ cm/s **4-17.** $t = 0.50$ s, $p = 0.75$ m **4-19.** $[4.0t\mathbf{i} + (8.0 + 3.0t)\mathbf{j}]$ m/s, $[(5.0 + 2.0t^2)\mathbf{i} + (2.0 + 8.0t + 1.5t^2)\mathbf{j}]$ m **4-21.** (*a*) $4.0\mathbf{i}$ m/s^2; (*b*) $8.0\mathbf{j}$ m/s; (*c*) $(2.0t^2\mathbf{i} + 8.0t\mathbf{j})$ m **4-23.** $v_y(t) = 2cxv_x(t)$ **4-25.** (*a*) 0.45 s; (*b*) 6.7 m/s; (*c*) 8.0 m/s **4-27.** 350 m/s **4-29.** 1.78×10^3 m **4-31.** (*a*) 0.417 s; (*b*) 2.8 ft **4-33.** 28.9 ft/s **4-35.** 86.4 m away **4-37.** 1.07 m **4-41.** 24.9 m **4-43.** 5.0 m/s^2, 17.3 m **4-45.** 40 m/s^2 **4-47.** 7900 m/s **4-49.** 9.12 cm **4-51.** $-(1.85\hat{\mathbf{r}} + 2.50\hat{\boldsymbol{\theta}})$ m/s^2 **4-53.** (*a*) $(4.0t\mathbf{i} + 3.0t\mathbf{j} + 5.0t\mathbf{k})$ m; (*b*) $\mathbf{r}(t) = \mathbf{r}'(t) + (4.0t\mathbf{i} + 3.0t\mathbf{j} + 5.0t\mathbf{k})$ m; (*c*) $\mathbf{v}(t) = \mathbf{v}'(t) + [4.0\mathbf{i} + 3.0\mathbf{j} + 5.0\mathbf{k}]$ m/s; (*d*) $\mathbf{a}(t) = \mathbf{a}'(t)$ **4-55.** (*a*) 0.14 h; (*b*) 0.30 h; (*c*) 68° relative to the shore; (*d*) 7.4 km/h, 0.11 h; (*e*) 0.10 h, 0.30 km **4-57.** 76° north of west, 1.55 h **4-59.** 29.2 km/h at 14° north of east **4-61.** He misses by 5.8 m. **4-63.** $\sin\theta_1/\sin\theta_2 = v_1/v_2$ **4-65.** (*a*) $(8.0\mathbf{i} + 7.0\mathbf{j})$ m/s; (*b*) 41.2° relative to the horizontal; (*c*) 2.5 m **4-67.** (*a*) Yes, at $t = 4.17$ s; (*b*) no **4-69.** (*a*) 71.7 ft/s; (*b*) no, the ball clears by 5.6 ft; (*c*) no, the ball clears by 0.5 ft **4-71.** 29.1° relative to the horizontal, 32° relative to the sidelines with $v_0 = 73$ ft/s; $R = 142$ ft

Chapter 5

5-1. (*a*) 2.5 m/s^2, (*b*) 2500 N **5-3.** (*a*) $9.0\mathbf{i}$ m/s^2; (*b*) 113 m **5-5.** 3900 N **5-7.** $(1.6\mathbf{i} - 0.8\mathbf{j})$ m/s^2 **5-9.** $(-1.5\hat{\mathbf{r}} + 3.0\hat{\boldsymbol{\theta}})$ m/s^2, $(-3.8\mathbf{r} + 7.5\boldsymbol{\theta})$ N **5-11.** 254° **5-13.** $(-7.5\mathbf{i} + 9.0\mathbf{j})$ N **5-15.** $F/mg = 9.5$ **5-17.** 544 N **5-19.** (*a*) 147 N; (*b*) 25.5 N; (*c*) 15 kg; (*d*) 0; (*e*) 15 kg **5-21.** 3.9×10^{25} kg **5-23.** 3.37×10^{23} kg **5-25.** $10M_E$ **5-29.** $-g\hat{\mathbf{r}}/4$ **5-31.** $(0.60\mathbf{i} - 8.40\mathbf{j})$ m/s^2 **5-33.** (*a*) 14.1 m/s; (*b*) 601 N **5-35.** $Gm\lambda l/d(l + d)$ **5-37.** $\Delta g/g = -2.7 \times 10^{-3}$

Chapter 6

6-1. 133 N, 80 N **6-3.** 198 N, 394 N **6-5.** 2.0 m/s^2 **6-9.** 0.10 **6-11.** 687 N **6-13.** (*a*) 3.35 m/s^2; (*b*) 4.2 s **6-15.** $\mu_S = \tan\theta$ **6-17.** $4g/9$, $g/9$, $g/9$ **6-19.** (*a*) 10 kg; (*b*) 60 N; (*c*) 98 N; (*d*) 0 **6-21.** (*a*) 6.9 m/s^2; (*b*) 0.24 N **6-23.** 0.45 m/s^2, 360 N **6-25.** (*a*) 2000 N, 1000 N; (*b*) 4500 N **6-27.** 459 N, 1.83 m/s^2, 202 N **6-29.** $1.43g$ **6-31.** 2.3 m/s^2, 82 N **6-33.** 31.4 N, 7.84 N **6-35.** (*a*) No; (*b*) 5.5 m/s^2 **6-37.** 16.6 cm **6-39.** 20.7 m/s **6-41.** 12° **6-43.** 21.1 m/s **6-45.** $R = 1936$ ft **6-47.** 1.9×10^{27} kg **6-49.** 1 yr **6-51.** 35.4 N, 7.8 N **6-53.** $[(\mu_S + \mu_K)(m_1 + m_2)]/(1 - \mu_S)$ **6-55.** 3.1 m/s^2 **6-59.** 140 lb **6-61.** (*a*) 86°; (*b*) no; (*c*) it slides to the bottom of the hoop. **6-63.** 7.17×10^3 s **6-65.** (*a*) 0.049 kg/s; (*b*) 0.57 m

Chapter 7

7-1. 100 J **7-3.** (*a*) 2.4 J; (*b*) -2.4 J; (*c*) 0 **7-5.** (*a*) 1.79×10^3 J; (*b*) -1.79×10^3 J; (*c*) 0 **7-7.** -1.83 J **7-9.** 1.86×10^4 J **7-11.** (*a*) 7.73×10^5J; (*b*) 4.0×10^3 J; (*c*) 1.82×10^{-16} J **7-13.** (*a*) 2.6×10^3 J; (*b*) 6.4×10^3 J **7-15.** 2.8 m/s **7-17.** 50 J, 1000 J **7-19.** 1.28×10^4 N **7-21.** (*a*) 6.1 m; (*b*) 6.5 m/s **7-23.** (*a*) -1.44 J; (*b*) 0.33

7-25. $E_i/E_f = 145$ **7-27.** (*a*) 2.75 J; (*b*) $-3.0 \ln x$ J
7-29. -3.1×10^{10} J **7-31.** $-(a/x^2) + (2b/x^3)$
7-33. 9.70 m/s **7-35.** 6.6 m/s **7-37.** 38.7 m/s **7-39.** $2h$
7-41. (*a*) $\sqrt{6gR}$; (*b*) $6mg$ **7-45.** 2.78×10^3 m/s
7-49. (*a*) 1.07×10^4 m/s; (*b*) no **7-51.** -1.44 J **7-53.** 0.24
7-55. -1.89×10^3 J **7-57.** -38 J **7-59.** -106 J **7-61.** 122 W
7-63. 6.80×10^{-6} W **7-65.** (*a*) 8.55 J; (*b*) 8.55 W
7-67. 1.72×10^3 W **7-69.** $\sqrt{2h/g}$ **7-71.** (*a*) 5.30 m/s; (*b*) -55.5 J; (*c*) 5.30 m/s; (*d*) 1.12 m **7-73.** (*b*) $F(x) = -kx + 2\alpha Axe^{-\alpha x^2}$ **7-75.** (*a*) $d + v_0\sqrt{2dm/F}$; (*b*) $v_0, v_0 + \sqrt{2Fd/m}$
7-77. (*a*) 0, -4.0 N, 6.0 N, 0; (*b*) 9.0 m, 15.0 m; (*c*) 7.7 m, 16.3 m; (*d*) 6.9 m/s, 5.7 m/s, 8.0 m/s, 10.6 m/s
7-79. (*a*) 0.65 m; (*b*) 0.024 m **7-81.** (*a*) 40 hp; (*b*) 4.0×10^7 J, 4.0×10^7 J; (*c*) 80 hp, 4.0×10^7 J, 8.0×10^7 J

CHAPTER 8

8-1. 0.40 m from the 4.0-kg particle **8-3.** $0.29l$ from the sides
8-5. 1.5×10^9 m **8-7.** $(2L/3) + (R/3)$ from the free end of the stick **8-9.** $3L/4$ from the floor **8-11.** $\bar{X} = 0$, $\bar{Y} = 0.402L$
8-13. (*a*) 22.2 cm from the base; (*b*) 19.5 cm from the base; (*c*) fill to 45.8 cm **8-15.** $4R/3\pi$ from the flat side
8-17. (0, 10.8 cm) **8-19.** (0.57 m, 0.39A) **8-21.** $(3.0\mathbf{i} - 2.7\mathbf{j})$ m/s
8-23. (*a*) $(3.8\mathbf{i} + 0.12\mathbf{j} + 2.1\mathbf{k})$ cm, $(4.0\mathbf{i} + 2.4\mathbf{j} + 1.3\mathbf{k})$ cm/s, $(3.4\mathbf{i} + 0.36\mathbf{j} - 0.40\mathbf{k})$ cm/s^2; (*b*) 0, $(45\mathbf{i} + 18\mathbf{j} - 36\mathbf{k})$ dyne, $(6\mathbf{i} + 6\mathbf{j} + 6\mathbf{k})$ dyne, $(35\mathbf{i} - 15\mathbf{j} + 20\mathbf{k})$ dyne; (*c*) $(86\mathbf{i} + 9\mathbf{j} - 10\mathbf{k})$ dyne; (*d*) yes
8-25. (*a*) $(-4.00\mathbf{i} + 2.25\mathbf{j})$ m; (*b*) $-0.25\mathbf{j}$ m/s; (*c*) $(-4.00\mathbf{i} + 7.25\mathbf{j})$ m, $2.25\mathbf{j}$ m/s; (*d*) it doesn't affect them.
8-27. $1.25ml^2$ **8-29.** 5.12×10^{51} kg·m^2 **8-31.** $I_x = \frac{1}{6}Ma^2$, $I_y = \frac{1}{2}Mb^2$ **8-33.** 0.50 kg·m^2 **8-35.** $\frac{2}{9}MA^2$ **8-37.** $\frac{3}{5}MR^2$
8-39. (*a*) 0.33 m to the right of OO′; (*b*) 7.07 kg·m^2, 21.1 kg·m^2
8-41. (*a*) 0.78 ft from the base; (*b*) 1.60 ft from the base
8-43. 3.66×10^4 m from the point where the shell was shot.

CHAPTER 9

9-1. (*a*) 7.27×10^{-5} rad/s; (*b*) 1.99×10^{-7} rad/s
9-3. (*a*) 209 rad/s; (*b*) 2.09×10^3 rad $= (1.20 \times 10^5)°$
9-5. (*a*) -0.0582 rad/s^2; (*b*) 16.7 rev **9-7.** (*a*) 5.55 rad/s^2; (*b*) 66.7 rad/s **9-9.** (*a*) -4.4 rad/s^2; (*b*) 4.8 s
9-11. (*a*) 1.26×10^3 cm/s; (*b*) 7.89×10^4 cm/s^2
9-13. (*a*) 7.0 rad/s; (*b*) 22.5 rad; (*c*) 10 cm/s^2, 490 cm/s^2; (*d*) 10 cm/s^2, 63 cm/s^2 **9-15.** 310 N **9-17.** 48 N
9-19. (*a*) 230 N·m (ccw); (*b*) 230$\mathbf{k}$ N·m
9-21. $(4.0\mathbf{i} + 2.0\mathbf{j} - 16.0\mathbf{k})$ N·m **9-23.** 8.0 N·m
9-25. 44 N·m **9-27.** (*a*) 3.93 N·m; (*b*) 12 N·m
9-29. 0.42 rad/s^2, 0.21 m/s^2, 0.084 m/s^2 **9-31.** 20 s
9-33. (*b*) 2.85×10^3 N·m **9-35.** (*a*) 8.16 rad/s; (*b*) 8.00 rad/s
9-37. $(m_2 + M/2)L/(m_1 + m_2 + M)$ **9-39.** 0.78 m
9-41. (*a*) $3g/2$; (*b*) $3g/4$; (*c*) $mg/4$ **9-43.** 0.78 rad/s^2, 0.52 rad/s^2
9-45. (*a*) 38.2 kg·m^2; (*b*) 2.02×10^5 W **9-47.** (*a*) $\sqrt{3gl}$, $\sqrt{6gl}$; (*b*) $3.4g$, $6g$ **9-49.** (*c*) 5.4 mm

CHAPTER 10

10-1. 66.7 rad/s **10-3.** (*a*) 7.0 m; (*b*) 2.0 s; (*c*) 7.0 m/s, 2.0 s
10-5. $(3F/4M)(1 + r/R)$ **10-7.** $(T/M) - \mu_K g$, $(2/R)[(T/M) - \mu_K g]$ **10-9.** (*a*) Mg; (*b*) $\omega^2R^2/4g$
10-11. (*a*) $12v_0^2/49\mu_K g$; (*b*) 4.8 m **10-13.** (*a*) 600 J; (*b*) 20.4 m
10-15. 23.9 m **10-19.** $0.79Mg$ **10-21.** 11.3 m/s, 55°
10-23. (*a*) 35.4 cm/s^2, 47.2 rad/s; (*b*) 70.8 cm
10-25. $1.5d$ **10-27.** 3.85 m/s **10-29.** $f = \frac{2}{7}mg \sin\theta$, $N = (mg/7)(17 \cos\theta - 10) + mv_0^2/(R - r)$

CHAPTER 11

11-1. 80 kg **11-3.** 20 kg **11-5.** The 5.0-cm thread breaks when $m = 0.86$ g. **11-7.** 0.072 kg, 0.064 kg **11-9.** 822 N **11-11.** 25 N
11-13. No **11-15.** (*a*) 120 lb; (*b*) 36 in. from her head
11-17. 800 N, 721 N **11-19.** $M\sqrt{2Rd - d^2}/(R - d)$
11-21. $W[(l_1/l_2) - 1]$, $W(l_1/l_2) + mg$ **11-23.** 778 N, 778 N at 45°
11-25. 1.50×10^3 N, 1.62×10^3 N at 30° **11-27.** 172 N, 392 N
11-29. $A_x = -B_x = -200$ N, $A_y + B_y = 400$ N
11-31. (*a*) $\frac{1}{2}(w - h \tan\theta)$; (*b*) $\tan^{-1}(w/h)$ **11-33.** (*a*) 300 N; (*b*) 213 N **11-35.** 3.8×10^{-7} **11-37.** 6.24 cm **11-39.** 259 N
11-41. 7.2×10^4 N **11-43.** 9.3×10^{10} N/m^2 **11-45.** 0.50, 1.0 kg
11-47. 200 N, 200 N **11-49.** $F = Mg \tan\theta$, $f = 0$
11-51. (*a*) 419 N; (*b*) 942 N at 40° **11-53.** 34.3° (numerical solution) **11.55** (*a*) 1400 N; (*b*) 320 N; (*c*) 520 N
11-57. (*a*) 1.10×10^9 N/m^2; (*b*) 5.5×10^{-3}, 1.57×10^{-2}; (*c*) 11 mm, 31.4 mm

CHAPTER 12

12-1. $p_{16}/p_{10} = 1.6$ **12-3.** 2.0 kg **12-7.** He moves at -0.13 m/s
12-9. $-(0.2\mathbf{i} + 1.4\mathbf{j} + 2.0\mathbf{k}) \times 10^2$ m/s
12-11. $-(5.3\mathbf{i} + 2.7\mathbf{j} - 4.7\mathbf{k}) \times 10^3$ m/s
12-13. $(-0.80\mathbf{i} + 2.63\mathbf{j})$ m/s, 4.63×10^3 J **12-15.** 6.9 cm
12-17. 11.5 cm **12-19.** (*a*) $m_2d/(m_1 + m_2)$ from m_1; (*b*) same place **12-21.** 9.1 m; no **12-23.** (*a*) 10 N·s; (*b*) 1.0×10^4 N **12-25.** (*a*) -3.46 kg·m/s; (*b*) -3.46 N·s
12-27. 167 N, 250 N **12-29.** $(25\mathbf{i} + 10\mathbf{j} + 30\mathbf{k})$ N·s; $(10\mathbf{i} + 10\mathbf{j})$ kg·m/s, $(25\mathbf{i} + 10\mathbf{j} + 30\mathbf{k})$ kg·m/s, $(25\mathbf{i} + 10\mathbf{j} + 30\mathbf{k})$ kg·m/s **12-31.** -3.6×10^6 m/s, 2.4×10^6 m/s
12-33. (*a*) 26.7 cm/s; 13.3 cm/s, -13.3 cm/s; -6.7 cm/s, 6.7 cm/s; 13.4 cm/s, 33.4 cm/s; (*d*) -3.0 ft/s; 7.0 ft/s, -7.0 ft/s; -3.0 ft/s, 3.0 ft/s; -10.0 ft/s, 0 **12-35.** 42°, 48°, 22.4 m/s
12-37. 250 m/s **12-39.** 720 bullets/min **12-41.** 5.0 ft/s
12-43. $(55\mathbf{i} + 10\mathbf{j} - 25\mathbf{k}) \times 10^3$ m **12-45.** 2.9 m **12-47.** 0.16 m
12-49. (*a*) $2m_1^2m_2v_1^2/(m_1 + m_2)^2$; (*b*) $m_1/m_2 = 1$

CHAPTER 13

13-1. $-2.0\mathbf{k}$ kg·m^2/s **13-3.** $-\mathbf{k}$, 0, $\mathbf{k}$, 0
13-9. (*a*) $-(28.0\mathbf{i} + 8.0\mathbf{j} - 4.0\mathbf{k})$ kg·m^2/s; (*b*) $(-90\mathbf{i} + 60\mathbf{j} + 70\mathbf{k})$ N·m; (*c*) $(-90\mathbf{i} + 60\mathbf{j} + 70\mathbf{k})$ N·m
13-11. (*a*) 0.157 kg·m^2/s; (*b*) 0.628 kg·m^2/s; (*c*) 0.99 J, 3.96 J
13-13. $l_{orbit}/l_{poles} = 3.8 \times 10^6$ **13-15.** $l_{rot}/l_{orbit} = 8.3 \times 10^{-6}$
13-17. (*a*) 72°; (*b*) 7.39×10^{-2} kg·m^2/s, 44.6×10^{-2} kg·m^2/s; (*c*) no **13-19.** $-1.2\mathbf{j}$ kg·m^2/s **13-21.** $v_{CM} = 5.0$ m/s, $\omega = 18$ rad/s
13-23. 150 rev/min **13-25.** (*a*) 20.9 rad/s; (*b*) 229 J, or a factor of 3.3 **13-27.** 8.08×10^3 m/s, 6.26×10^3 m/s
13-29. 31.1 rev/min **13-31.** (*a*) 14 rad/s; (*b*) 0.014 J; (*c*) 10 rad/s; (*d*) 0.035 J; (*e*) work done by bug
13-33. (*a*) 6.4 rad/s; (*b*) 6.4 rad/s **13-35.** 1360 rad/s
13-37. 52 cm **13-39.** 2.5 rad/s, no **13-41.** (*a*) 109 cm above the feet; (*b*) 22.3 kg·m^2; (*c*) 2.0 rev/s; (*d*) nearly 3 rev

CHAPTER 14

14-1. 0.85 s **14-3.** 0.78 kg **14-5.** (*a*) 0.28 s; (*b*) 0.5%
14-7. 6.3 J **14-9.** (*a*) 49 N/m; (*b*) 5.0 cm; (*c*) 0.63 s; (*d*) 1.59 Hz; (*e*) 9.9 rad/s; (*f*) 1.57 rad; (*g*) 0.50 m/s; (*h*) 4.95 m/s^2; (*i*) 6.13×10^{-2} J **14-11.** (*a*) 0.25 m; (*b*) 0.314 s; (*c*) 0.785 rad; (*d*) 3.18 Hz; (*e*) 20 rad/s; (*f*) 5.0 m/s; (*g*) 100 m/s^2
14-13. (*a*) 1.26 m/s, 31.6 m/s^2; (*b*) 1.00 m/s, -18.9 m/s^2; (*c*) 94.7 N/m; (*d*) 0.118 J **14-15.** (*a*) 1.39 J; (*b*) 1.1 s

14-17. 9.78 m/s^2 **14-19.** (*a*) 0.39 rad/s; (*b*) 0.78 m/s; (*c*) 0.061 J; (*d*) 0.39 rad/s, 0.78 m/s, 0.122 J **14-21.** (*a*) 1.64 s; (*b*) 1.58 s **14-23.** 0.16 s **14-25.** 0.64 m from the handle **14-27.** 0.13 m **14-31.** 1.75 s, 1.71 m **14-35.** 50 Hz **14-37.** 0.38 Hz **14-39.** $(1/2\pi)\sqrt{9g/8l}$

CHAPTER 15

15-1. 6.48 kg **15-3.** $V_{FW} = 0.501$ m^3, $V_{SW} = 0.488$ m^3 **15-5.** 9.5×10^{17} kg/m^3 **15-7.** (*a*) 1.67×10^4 N/m^2; (*b*) 1.0×10^{-2} N; (*c*) 4.0 N **15-9.** (*a*) 0.228 atm; (*b*) 17.3 cmHg **15-11.** 488 m **15-13.** (*a*) 1087 atm; (*b*) 2.20×10^8 N **15-15.** $\Delta p = 1.1$ mmHg

15-17.

	(*a*)	(*b*)	(*c*)	(*d*)
Absolute (N/m^2)	1.54×10^5	4.8×10^4	1.05×10^5	7.5×10^4
Gauge (N/m^2)	5.3×10^4	-5.3×10^4	3.9×10^3	-2.67×10^4
cmHg	116	36	79	56
cmH$_2$O	1.58×10^3	4.9×10^2	1.07×10^3	7.6×10^2
torr	1160	360	790	560
pascal	1.54×10^5	4.8×10^4	1.05×10^5	7.5×10^4
atm	1.53	0.47	1.04	0.74
bar	1.54	0.48	1.05	0.75

15-19. 641 N **15-21.** 0.10 **15-23.** 1.12×10^3 kg/m^3 **15-25.** 311 cm^3 **15-27.** (*a*) Float; (*b*) sink; (*c*) one with 12-cm edges **15-29.** 2.5 m/s, 40 m/s **15-31.** 6.9 m/s **15-33.** 7.2 m/s **15-35.** 44 in./s **15-37.** 8.0 m/s, 4.5×10^4 N/m^2 **15-39.** 0.79×10^5 N/m^2 **15-41.** $\sqrt{2gh}$ **15-43.** 5.7 m **15-45.** (*a*) 2.48×10^5 N/m^2; (*b*) 17.1 m/s; (*c*) 0.34 m^3/s **15-47.** 2.0×10^{-2} m^3/s **15-49.** 12.3 m/s **15-51.** 15 m/s **15-53.** (*a*) $8Q$; (*b*) $Q/8$ **15-55.** (*a*) 4.1×10^4 N/m^2; (*b*) 7.3×10^4 N/m^2 **15-57.** 22.8 N/m^2 **15-59.** 1.65 cm **15-63.** 95 cm **15-65.** (*c*) For 50 m: 1.00×10^5 N/m^2 and 1.00×10^5 N/m^2; for 1000 m: 0.890×10^5 N/m^2 and 0.883×10^5 N/m^2 **15-67.** 30.8 m **15-69.** (*b*) 0.1203 kg **15-71.** 97 m/s

CHAPTER 16

16-1. 3.0 m/s, −2.0 m/s, −5.0 m/s **16-3.** No, because $y \to \infty$ as $x \to \infty$ **16-5.** (*a*) 1.5 cm/s **16-7.** 2.25 N **16-9.** 2.5×10^{-2} kg/m **16-11.** 16.0 N **16-13.** $v_2 = 1.2v_1$ **16-15.** (*a*) 140 m/s, 185 m/s; (*b*) 1.0×10^{-2} kg/m **16-17.** $\rho_B = 4.5\rho_A$ **16-19.** 2.10×10^9 N/m^2 **16-23.** 0.50 m, 3.14 m **16-25.** $y = 2.0 \sin \pi x/2$ (cgs units); no **16-27.** 17 to 0.017 m **16-29.** 0.5, 0.5, 2 **16-31.** 0.50 cm, 15.9 Hz, 5.24 cm, 0.063 s, 83.3 cm/s **16-33.** (*a*) Halved, doubled, same, same; (*b*) same, same, same, doubled; (*c*) decreased by 0.89, same, decreased by 0.89, same **16-35.** 14.5° or 165.5° **16-37.** (*a*) 0.40 m, 25 Hz, 10 m/s; (*b*) 3.14 m/s, 493 m/s^2 **16-39.** $s = (3.0 \times 10^{-2}) \cos (2.67\pi x - 40\pi t + \phi)$ (SI units), where $\phi = \pm 1.16$ rad **16-41.** 0.56 N/m^2 **16-43.** (*a*) $s = (4.0 \times 10^{-9}) \times \sin (55.4x - 1.88 \times 10^4 t + \phi)$, $\Delta p = (3.2 \times 10^{-2}) \times \cos (55.4x - 1.88 \times 10^4 t + \phi)$ (SI units) **16-45.** 1.0×10^{-3} W **16-47.** (*a*) $\frac{1}{4}$; (*b*) 4; (*c*) 1; (*d*) 1 **16-49.** 183 Hz **16-53.** 254 W **16-55.** 1.4×10^3 W/m^2 **16-57.** Earth 1, Venus 1.7, Mars 0.12, Saturn 1.0, Pluto 4.0×10^{-5} **16-59.** Same intensity, different energy **16-61.** 64 dB **16-63.** 2.0×10^{-7} W **16-65.** 31 m/s **16-67.** 404 Hz **16-69.** 405 Hz, 395 Hz **16-71.** (*a*) 418 Hz, (*b*) 385 Hz **16-73.** $dv/v = -0.01$; $dv/v = -0.25$, which is not accurate because dv is so large; more accurately, $\Delta v/v = -0.29$. **16-75.** 15.9 **16-77.** (*b*) $t = 8.3 \times 10^{-12}$ s, $T = 10^{-5}$ s, yes **16-79.** (*a*) 448 Hz; (*b*) 452 Hz

CHAPTER 17

17-3. 5.0 cm, 1.0 cm **17-5.** 0, $4I$ **17-7.** 3.8 cm **17-9.** $2.12 \sin (x + 0.50t + \pi/4)$ (SI units) **17-11.** $4.8 \sin (2.0x + 40t + 5\pi/24)$ (SI units) **17-13.** (*a*) $\phi = 0.32\pi$ rad; (*b*) $y_1 = 2.0 \sin (0.50x + 400t)$, $y_2 = 2.0 \sin (0.50x + 400t + 0.32\pi)$ (cgs units) **17-15.** 360 cm **17-17.** 8 Hz **17-19.** 1 Hz **17-23.** (*a*) $y(x,t) = 0.30 \cos (2.0x + 40t + \pi)$; (*b*) $y(x,t) = 0.30 \cos (2.0x + 40t)$ (SI units) **17-25.** $(6.67 \times 10^{-4}) \cos \pi(2.0x + 50t)$, $(2.67 \times 10^{-3}) \cos \pi(1.0x - 50t)$ (SI units) **17-27.** Inverted, inverted, upright, upright, inverted **17-29.** (*a*) 3.4; (*b*) $0.455A_{in}$, $-0.545A_{in}$ **17-31.** 1.3, 1, 1.13 **17-33.** $0.160 \cos (4.0x - 200t)$, $0.200 \cos (2.4x - 200t)$ (SI units) **17-35.** 2.0 cm, 0.50 cm/s, 0.50 Hz, 1.0 cm **17-37.** 0.80 m **17-39.** 1.0 cm **17-41.** 15, 30, 45, 60 Hz **17-43.** 160 m/s **17-45.** 35.5 N **17-47.** 18 Hz **17-49.** 0.873 kg **17-51.** (*a*) 5 and 6; (*b*) 360 N **17-53.** (*a*) 86.8 cm; (*b*) 21.7 cm **17-55.** 180 Hz **17-57.** 240 m/s **17-59.** 300 Hz **17-61.** Third and fifth, 0.85 m **17-63.** 1520 N **17-65.** 101 N **17-67.** $0.444\mu_{196}$, $0.198\mu_{196}$, $0.088\mu_{196}$ **17-69.** (*a*) 0.19 m; (*b*) 0.39 m **17-71.** (*a*) 0.588π rad; (*b*) 1.7 m; (*c*) 0.85 m

17-73. (*a*)

Reflected	Transmitted
100 m/s	71 m/s
200 Hz	200 Hz
0.50 m	0.35 m

(*b*) $y_{in} = (0.10) \cos \pi(4.0x - 400t)$
$y_{re} = (-0.017) \cos \pi(4.0x + 400t)$
$y_{tr} = (0.083) \cos \pi(5.7x - 400t)$ (SI units)

17-75. $0.93A$ **17-77.** (*a*) 5; (*b*) 1.20 m; (*c*) 1.20 m **17-79.** 2.19×10^5 N/m^2 **17-81.** 352 m/s

CHAPTER 18

18-1. (*a*) 55.5; (*b*) 55.5 **18-3.** 574°F **18-5.** 80.6°F, 300 K **18-7.** (*a*) 1096°H; (*b*) 640°H; (*c*) −292°H **18-11.** 3.3×10^{-7} cm

18-15.

p(mmHg)	150	250	80	425	320	750
T(K)	60	100	32	170	128	300

V(L)	0.50	0.80	0.38	0.10	1.17	1.50
T(K)	300	480	230	600	700	900

p(mmHg)	800	282	400	1070	150	1920
V(L)	1.20	3.40	2.40	0.90	6.4	0.50

18-17. 0.08156 atm·L/mol·K **18-19.** 0.19 atm **18-21.** 4.6% **18-23.** (*a*) 644 kg; (*b*) 64 kg **18-25.** +0.50 **18-27.** 7.6×10^{-14} J **18-29.** +50% **18-31.** $\Delta(mgh) = 3.25 \times 10^{-26}$ J compared with $\frac{3}{2}kT = 6.21 \times 10^{-21}$ J **18-33.** (*a*) 1/1/1; (*b*) $V_{H_2}/V_{N_2} = 3.74$, $V_{H_2}/V_{CO_2} = 4.69$ **18-35.** 516 m/s **18-37.** 1.0×10^4 N, 5.0×10^7 N/m^2 **18-43.** 413 K **18-45.** (*a*) 0.131 mol; (*b*) 0.076 mol, 0.055 mol **18-47.** (*a*) 9.60×10^4 N/m^2; (*b*) 1.48 m **18-49.** 13% **18-51.** (*a*) 1.14 kg/m^3 for moist air, 1.16 kg/m^3 for dry air; (*b*) higher humidity increases range slightly; (*c*) yes

CHAPTER 19

19-1. 6.21×10^{-21} J **19-3.** 12.8 J **19-5.** 108 cal **19-7.** (*a*) 3.02×10^5 J; (*b*) 6.30×10^5 J **19-9.** 13.8°C

19-11. 6.67 g **19-13.** (*a*) 24.8 J/mol·°C; (*b*) 24.3 J/mol·°C; (*c*) 51.4 J/mol·°C **19-15.** 0.875 J/s **19-17.** 1.52×10^5 J **19-19.** 40°C **19-21.** 1.08 kg **19-23.** 67 g **19-25.** 0.60 W/m·K **19-27.** 10°C **19-29.** 25°C, 22°C **19-31.** 1.996 m **19-33.** 118°C **19-35.** (*a*) 1.2×10^{-5} $(°C)^{-1}$; (*b*) steel **19-39.** 0.50 m, 1.00 m **19-41.** 0.060 cm^2 **19-43.** 1.8 **19-45.** By 60% **19-49.** 2.4 cm^3 **19-51.** (*a*) 1.2×10^{-3} m; (*b*) 1.2×10^{-4} m; (*c*) 0.96×10^{-4} m; (*d*) 6.8×10^{-6} m^2; (*e*) 1.0×10^{-5} m^3 **19-53.** 0°C **19-55.** 1.7×10^3 J

CHAPTER 20

20-1. 404 J **20-3.** 4.8×10^{-2} mol **20-5.** 1.9×10^3 N/m^2 **20-7.** 202 J, 404 J, 404 J, 606 J **20-9.** Isobaric; 1.8 **20-11.** 0, 250 J **20-13.** $pV \ln 4$ **20-15.** −1630 J **20-17.** 0, 159 J, −159 J **20-19.** (*a*) −150 J; (*b*) −400 J **20-21.** 0, no **20-23.** (*a*) 24 J; (*b*) 1.3×10^{-2}°C **20-25.** 6.8 m **20-27.** 0.060°C **20-29.** (*a*) 374 J; (*b*) 104 J; (*c*) 3740 J **20-31.** 7.57×10^3 J **20-33.** 10.95 L **20-35.** $\gamma = 1.41$, so diatomic **20-37.** 189 K **20-41.** Isothermal, independent of the type of gas **20-43.** (*a*) 241 K; (*b*) 383 K; (*c*) 1776 J; (*d*) 1776 J **20-45.** (*a*) 199 J; (*b*) 0.80°C **20-47.** 0, 11.6°C

CHAPTER 21

21-1. 500 J, 300 J **21-3.** (*a*) 150 J; (*b*) 350 J **21-5.** (*a*) 600 J; (*b*) 800 J **21-7.** (*a*) 69 J; (*b*) 11 J **21-9.** (*a*) 0.50; (*b*) 100 J; (*c*) 50 J **21-11.** 2.0 **21-13.** 50 J **21-17.** $\kappa = (1 - \epsilon)/\epsilon$ **21-19.** $\epsilon = 0.38$, so no for 45%, maybe for 35% **21-21.** 1.45×10^7 J **21-23.** (*a*) 546 K; (*b*) 137 K **21-25.** 0.067 J/K **21-27.** −61 J/K **21-29.** $-Q_h/T_h$, Q_c/T_c, 0 **21-31.** (*a*) 5.76 J/K; (*b*) 260 K **21-33.** (*a*) 170 J/K; (*b*) 290 J/K **21-35.** 3.78×10^{-3} W/K **21-37.** 430 J/K **21-39.** 80°C, 80°C, 6.70×10^4 J, 215 J/K, −190 J/K, 25 J/K **21-41.** $\Delta S_{H_2O} = 215$ J/K, $\Delta S_R = -208$ J/K, $\Delta S_U = 7$ J/K **21-43.** (*a*) 1200 J; (*b*) 600 J; (*c*) 600 J; (*d*) 0.50 **21-45.** (*a*) 750 K; (*b*) 338 K **21-47.** 46 W **21-53.** 75 J/K **21-55.** 5.0×10^3 J/K

CHAPTER 22

22-1. 1.93×10^5 C **22-3.** (*a*) 2.50×10^{10} electrons; (*b*) 1.1×10^{-10}% **22-5.** 8.5×10^8 electrons **22-7.** 8.8×10^{22} lb **22-9.** 54 N toward q_2 **22-11.** 0.052 N to the left **22-13.** 6.0×10^{-8} C, no **22-15.** 5.12 N **22-17.** $(1.35kq^2/a^2)(\mathbf{i} - \mathbf{j})$ **22-19.** $(6.8\mathbf{i} + 11.9\mathbf{j} - 24.2\mathbf{k}) \times 10^{-5}$ N **22-21.** $-(0.13\mathbf{i} + 0.28\mathbf{j})$ N **22-25.** $kQq[(x_2 - x_1)/x_1x_2]$ **22-27.** $\lambda ql\mathbf{i}/[4\pi\epsilon_0 a(a + l)]$ **22-29.** $2kq\lambda/R$ downward

CHAPTER 23

23-1. (*a*) 200 N/C up; (*b*) 2.0×10^{-6} N down **23-3.** 1.0×10^{-7} N/C **23-5.** 3.5×10^{16} m/s^2 **23-7.** 2.78×10^{-9} C **23-9.** (*a*) 1.15×10^{12} N/C; (*b*) 1.47×10^{-6} N **23-13.** (*a*) 2.61×10^5 N/C toward q_2; (*b*) 4.18×10^{-14} N toward q_1 **23-15.** -2.0×10^{-6} C at d and 4.0×10^{-6} C at $\sqrt{2}d$, with both on the same side of the origin **23-17.** $\dfrac{2\sqrt{2}kq}{a^2}(\mathbf{i} + \mathbf{j})$ **23-19.** (*a*) $\dfrac{0.29kq}{a^2}(\mathbf{i} - \mathbf{j})$, (*b*) $\dfrac{0.29kq^2}{a^2}(\mathbf{i} - \mathbf{j})$ **23-21.** 2.0×10^{16} m/s^2 **23-23.** 2.9×10^4 N/C **23-25.** 2.0 μC/m^2, 2.26×10^5 N/C **23-27.** $\dfrac{4kq}{a\sqrt{a^2 + L^2}}, \dfrac{kq}{a(a + L)}$ **23-29.** (*a*) $\dfrac{1}{2\pi\epsilon_0}\left(\dfrac{\lambda_y}{a}\mathbf{i} + \dfrac{\lambda_x}{b}\mathbf{j}\right)$; (*b*) $\left(\dfrac{\lambda_x + \lambda_y}{2\pi\epsilon_0 c}\right)\mathbf{k}$ **23-31.** $(Q \sin\theta)/4\pi\epsilon_0 R^2\theta$ **23-33.** (*a*) 14.2 cm; (*b*) 2.84×10^{-8} s **23-35.** 1.6×10^7 N/C **23-37.** (*b*) 4.81×10^{-19} C, 6.41×10^{-19} C, 3.19×10^{-19} C **23-39.** (*a*) $3.2 \times 10^{-15}\mathbf{i}$ N; (*b*) $1.92 \times 10^{12}\mathbf{i}$ m/s^2; (*c*) $7.84 \times 10^6\mathbf{i}$ m/s, 11.8**i** m **23-41.** (*a*) 2.0×10^{-8} s; (*b*) 1.78×10^{-3} m; (*c*) 7.07×10^{-3} m **23-43.** (*a*) $(1.02 \times 10^4\mathbf{i} + 1.86 \times 10^4\mathbf{j})$ N/C; (*b*) $(0.051\mathbf{i} + 0.093\mathbf{j})$ N **23-45.** 51.2° **23-47.** $-\dfrac{2k\lambda}{r}(\mathbf{i} + \mathbf{j})$ **23-49.** $(-6.0\mathbf{i} + 11.0\mathbf{j} + 2.0\mathbf{k}) \times 10^6$ m/s, $(-5.0\mathbf{i} + 8.0\mathbf{j} + 4.0\mathbf{k}) \times 10^{-2}$ m **23-51.** (*a*) $3.2 \times 10^{-15}\mathbf{i}$ N; (*b*) $1.92 \times 10^{12}\mathbf{i}$ m/s^2; (*c*) $(7.8\mathbf{i} + 3.0\mathbf{j}) \times 10^6$ m/s, $(11.8\mathbf{i} + 6.0\mathbf{j})$ m **23-53.** $7.5qE_0a$ **23-55.** $(\lambda_x\lambda_y/2\pi\epsilon_0)(\ln b/a)$ **23-57.** $\dfrac{2k\lambda_0}{r}\dfrac{\alpha - \alpha\cos\theta_0 e^{-\alpha\theta_0} + \sin\theta_0 e^{-\alpha\theta_0}}{1 + \alpha^2}$

CHAPTER 24

24-1. 4.4×10^4 N·m^2/C **24-3.** 3.77×10^4 N·m^2/C **24-5.** a^4 **24-7.** 0, $-2q/\epsilon_0$, q/ϵ_0, $-4q/\epsilon_0$, $-2q/\epsilon_0$, $3q/\epsilon_0$ **24-9.** q/ϵ_0 **24-11.** 3.54×10^{-7} C **24-13.** $\phi_e = 0$ **24-15.** (*a*) 9.77×10^{-3} C/m^3; (*b*) 5.65×10^5 N·m^2/C; (*c*) 5.65×10^5 N·m^2/C, 1.38×10^5 N·m^2/C; (*d*) 1.25×10^5 N·m^2/C **24-17.** 4.5×10^7 N/C **24-19.** (*a*) $-5.4 \times 10^6\hat{\mathbf{r}}$ N/C; (*b*) $-13.5 \times 10^6\hat{\mathbf{r}}$ N/C; (*c*) $-6.75 \times 10^6\hat{\mathbf{r}}$ N/C **24-23.** 0, $(\sigma R/\epsilon_0 r)\hat{\mathbf{r}}$ **24-27.** -1.77×10^{-9} C/m^2 **24-29.** (*a*) $E_0\pi R^2$; (*b*) 0 **24-31.** $Q/3\epsilon_0$ **24-33.** $(\alpha r^3/5\epsilon_0)\hat{\mathbf{r}}$, $(\alpha R^5/5\epsilon_0 r^2)\hat{\mathbf{r}}$ **24-35.** (*a*) 157 N; (*b*) 1.7×10^{32} m/s^2 **24-37.** $E = \dfrac{\rho a}{2\epsilon_0}$ $(|z| \geq a/2)$, $E = \dfrac{\rho z}{\epsilon_0}$ $(|z| \leq a/2)$

CHAPTER 25

25-1. 2.9 V, −2.9 V **25-3.** (*a*) -4.5×10^6 V; (*b*) −9.0 J **25-5.** 1.8 J **25-7.** −5.8 J **25-9.** 1.1×10^8 m/s **25-11.** (*a*) $-2.7 \times 10^4\mathbf{i}$ V/m, 0; (*b*) 0, 3.6×10^4 V; (*c*) $(7.2\mathbf{i} + 3.6\mathbf{j}) \times 10^4$ V/m, 0; (*d*) $(1.8\mathbf{i} + 2.7\mathbf{j}) \times 10^4$ V/m, 2.7×10^4 V **25-13.** $5.7kq/a$ **25-15.** $0.59kq/a$ **25-17.** $\dfrac{\sigma}{2\epsilon_0}\left(\sqrt{x^2 + b^2} - \sqrt{x^2 + a^2}\right)$ **25-19.** (*a*) -4.5×10^3 V, (*b*) -1.8×10^{-2} J **25-21.** (*a*) 8.0×10^{-18} J, 2.0×10^{-17} J; (*b*) 1.2×10^{-17} J; (*c*) -1.2×10^{-17} J **25-23.** 30×10^6 V **25-25.** 2.5×10^{-4} J, 1.6×10^{15} eV **25-27.** (*a*) 252 V; (*b*) 9.4×10^6 m/s **25-29.** (*a*) 1.0×10^5 V; (*b*) 0; (*c*) 1.8×10^5 V; (*d*) 0 **25-31.** (*a*) 1.0×10^9 J; (*b*) 390 kg **25-33.** −200**k** V/m **25-35.** $k\lambda L/x(x + L)$ **25-37.** 8.0×10^{-30} C·m **25-39.** (*a*) 7.2×10^{-3} V, $5.8 \times 10^6\mathbf{j}$ V/m; (*b*) 5.1×10^{-3} V, $(4.3\mathbf{i} + 1.4\mathbf{j}) \times 10^6$ V/m; (*c*) 0, $-2.9 \times 10^6\mathbf{j}$ V/m **25-41.** (*a*) 0; (*b*) 8.0×10^{-25} N·m; (*c*) 5.7×10^{-25} N·m **25-43.** 2.0 cm **25-47.** (*a*) 5.0×10^4 V/m; (*b*) 4.4×10^{-7} C/m^2, -4.4×10^{-7} C/m^2 **25-49.** (*a*) 4.0×10^{-6} C/m^2, -4.0×10^{-6} C/m^2; (*b*) 4.5×10^5 V/m; (*c*) 9.0×10^3 V **25-51.** outside: $E = \dfrac{\sigma a}{\epsilon_0 r}$, $V = \dfrac{\sigma a}{\epsilon_0}\ln\left(\dfrac{a}{r}\right)$; inside: $E = 0$, $V = 0$ **25-53.** 2.1×10^{-5} C **25-55.** (*a*) 3.2×10^{-10} C/m^2; (*b*) 7.1×10^{-11} C/m^2; (*c*) 6.4×10^{-12} C **25-57.** (*a*) 0; (*b*) $Q/4\pi\epsilon_0 r^2$; (*c*) 0; (*d*) 0; (*e*) $Q(b - a)/4\pi\epsilon_0 ab$ **25-59.** (*a*) 2.8×10^5 V/m; (*b*) 0; (*c*) 3.1×10^4 V/m; (*d*) 8.8×10^3 V, 5.0×10^3 V, 3.8×10^3 V; (*e*) -5.0×10^{-8} C, 5.0×10^{-8} C **25-63.** $(2kQe/m_eR)^{1/2}$ **25-65.** $-5.5kq^2/a$ **25-67.** (*a*) $(\sigma/2\epsilon_0)\left(\sqrt{R^2 + z^2} - z\right)$; (*b*) $(\sigma/2\epsilon_0)\left(1 - z/\sqrt{R^2 + z^2}\right)$ **25-69.** 2.3×10^7 m/s

CHAPTER 26

26-1. 2.0 C, 1.25×10^{19} electrons **26-3.** 1.6 A **26-5.** (*a*) 2.83×10^5 A/m^2; (*b*) 3.0×10^{-5} m/s **26-7.** 1.64×10^7 s **26-9.** 2.64×10^{-8} Ω·m, aluminum **26-11.** 4.94 m, 0.0785 m **26-13.** 1:4 **26-15.** 9.0×10^{-4} V **26-17.** 1.2×10^{-9} Ω·m **26-19.** (*a*) 2.4×10^{-8} Ω·m; (*b*) 3.0 A; (*c*) 0.072 V/m; (*d*) 3.0×10^6 A/m^2; (*e*) 3.1×10^{-4} m/s **26-21.** 12 Ω **26-23.** At 100°: (*a*) 2.58×10^{-8} Ω·m; (*b*) 2.07×10^{-8} Ω·m; (*c*) 1.06×10^{-7} Ω·m **26-25.** 290°C **26-27.** 73°C **26-29.** 1.51 V, 0.16 Ω **26-31.** (*a*) 12 V; (*b*) 0.12 Ω; (*c*) 2.34 A; (*d*) 11.7 V **26-33.** (*a*) (+) to (+); (*b*) 10 A **26-35.** (*a*) 0.50 A; (*b*) 2.85 V, 8.90 V; (*c*) 11.8 V **26-37.** 0.7 W, 27.4 W, 28.1 W **26-39.** 27.6 W, 1.1 W, 26.5 W **26-41.** 1.8×10^4 J, 9.0×10^3 J **26-43.** 3.33 A **26-45.** (*a*) 24.2 Ω; (*b*) 4.55 A; (*c*) 335 W **26-47.** (*a*) 18.3 Ω; (*b*) 2640 W **26-49.** 1.30×10^7 J **26-51.** +62% **26-57.** 0.89 kΩ, 1.11 kΩ

CHAPTER 27

27-1. (*a*) 2.1 A, 1.4 A, 0.71 A; (*c*) 1.9 A, 0.16 A, 1.8 A; (*f*) 1.7 A, 3.3 A, 1.7 A **27-3.** (*a*) 50 V; (*b*) 30 V **27-5.** 5.5 A, 0.5 A, 8.5 A; 72.5 V, 77.5 V **27-7.** −0.33 A **27-9.** (*a*) 240 Ω, 96 Ω; (*b*) 1.75 A **27-11.** (*a*) 1.57 A, 2.07 A; (*b*) −8.9 V **27-13.** 0.70 A, 0.39 A, 0.30 A, 0.087 A, 0.30 A, 0.39 A **27-15.** (*a*) 3.0 A, 1.0 A, 0, 2.0 A, 1.0 A, 3.0 A; (*b*) 21 W, 0 **27-17.** In parallel, 100 A; in series, 1 A **27-19.** (*e*) 15 Ω; (*f*) 25 Ω **27-21.** 225 Ω **27-23.** (*a*) 7 Ω; (*b*) $4R/7$; (*c*) $R/2$ **27-25.** (*a*) 0.57 A; (*b*) 2.3 V **27-27.** 18 **27-29.** (*a*) No (11.6 A); (*b*) nothing (12.9 A); (*c*) it opens (17.1 A) **27-31.** (*a*) 0.025 Ω in parallel; (*b*) 3.0×10^5 Ω in series **27-33.** 6.24 Ω **27-35.** (*a*) 25.0 V; (*b*) 23.8 V; (*c*) 16.7 V **27-37.** 9.39 Ω **27-39.** (*a*) 2900 Ω; (*b*) 9000 Ω, 3000 Ω, 1000 Ω; (*c*) no **27-41.** 1.633 V **27-43.** (*a*) $R = r$, $\mathcal{E}^2/4r$; (*b*) $R = r/2$, $\mathcal{E}^2/2r$ **27-45.** $37R/23$ **27-47.** (*a*) 1.0 A; (*b*) 6.0 Ω **27-49.** (*b*) $R_x = 99.80$ Ω, $R'_x = 100.00$ Ω **27-51.** 2.0 Ω **27-53.** 0.010 Ω, 0.091 Ω, 0.910 Ω **27-55.** 0.54 A, 0.33 A

CHAPTER 28

28-1. (*a*) Upward; (*b*) eastward **28-3.** $-7.2 \times 10^{-6}\mathbf{j}$ N **28-5.** $(-5.1\mathbf{i} + 3.8\mathbf{j} + 3.2\mathbf{k}) \times 10^{-13}$ N **28-7.** 0, $-ev_0B_0\mathbf{j}$, $ev_0B_0\mathbf{i}$, $(ev_0B_0/2)(\sqrt{3}\mathbf{i} - \mathbf{j})$ **28-9.** $(42\mathbf{i} - 19\mathbf{j} - 6.4\mathbf{k}) \times 10^{-15}$ N **28-11.** (*b*) 6.0×10^6 m/s **28-13.** $(2mV/qB^2)^{1/2}$ **28-15.** (*b*) 270 m **28-17.** $R_p = 1830R_e$, $T_p = 1830T_e$ **28-19.** 1.3×10^{-25} kg **28-21.** $1{:}\sqrt{2}{:}\sqrt{2}$ **28-23.** 1/4 **28-25.** 7.5×10^{-3} N/m **28-27.** $-(8{,}0\mathbf{j} - 6.0\mathbf{k}) \times 10^{-3}$ N **28-29.** 4.5 N **28-31.** 0.25 A from left to right **28-33.** 9.8×10^{-2} N·m **28-35.** 49° **28-37.** (*a*) 12 A; (*b*) 0.15 N·m; (*c*) 0.11 N·m; (*d*) −0.11 J **28-39.** (*a*) 0.078 N·m; (*b*) 0.099 N·m; (*c*) 52° **28-41.** 83° **28-43.** (*a*) 6.0×10^{-4} m/s; (*b*) 38 A; (*c*) -7.8×10^{-11} m^3/C **28-45.** (*a*) 2.25×10^6 m/s; (*b*) 2.25×10^6 m/s **28-47.** 6.6 mm **28-49.** (*a*) 0.43 m; (*b*) 2.3×10^7 Hz **28-51.** (*a*) 1.6 T; (*b*) 0.39 m **28-53.** 0.23**k** N **28-55.** (*a*) 1.2×10^3 A·m^2; (*b*) 5.7×10^4 A **28-59.** (*b*) 3.3×10^{-3}

CHAPTER 29

29-1. 2.0×10^{-3} T **29-3.** 1.4×10^{-5} T **29-5.** 6.7×10^{-5} T **29-7.** (*a*) $\mu_0 I/\pi a$; (*b*) $\mu_0 I/8\pi a$ **29-9.** 3.2×10^{-19} N **29-11.** 15 **29-13.** $0.77R$ **29-15.** $\mu_0 NIR^2/(R^2 + d^2/4)^{3/2}$ **29-17.** $\left(2\mu_0 I\sqrt{a^2 + b^2}\right)/\pi ab$ **29-19.** $\mu_0 I(b - a)/8ab$ **29-21.** 1.4×10^{-4} T **29-23.** (*a*) 2.0×10^{-5} N/m, repulsive; (*b*) 2.0×10^{-5} N/m, attractive **29-25.** Impossible, because $\oint_S \mathbf{B}\cdot\hat{\mathbf{n}}\, da \neq 0$ **29-27.** $b_2 = -2b_1$ **29-29.** (*a*) $8\pi \times 10^{-7}$ N/A; (*b*) $36\pi \times 10^{-7}$ N/A; (*c*) 0; (*d*) $8\pi \times 10^{-7}$ N/A; (*e*) $-20\pi \times 10^{-7}$ N/A **29-31.** $I = 0$ **29-33.** At $r = R$ **29-35.** $0 \le r < 3.0$ cm: $B = (1.1 \times 10^{-2}r)$ T, $r \ge 3.0$ cm: $B = (10^{-5}/r)$ T **29-41.** 13 A **29-43.** (*a*) 6.3×10^{-3} T; (*b*) 6.2×10^{-3} T; (*c*) 6.1×10^{-3} T; (*d*) 6.3×10^{-3} T **29-45.** 3.4% **29-47.** 0.012 T, 0.012 T, 0.011 T **29-49.** $\Delta B/B = -\Delta r/r$ **29-51.** (*a*) 5.0×10^3; (*b*) 9.6×10^{-3} T **29-53.** 320 **29-55.** (*a*) 2.1×10^{-4} A·m^2; (*b*) 2.7 A **29-57.** 1.1 T

29-61. $\dfrac{4\mu_0 Ia^2}{\pi(a^2 + 4z^2)(2a^2 + 4z^2)^{1/2}}$ **29-63.** 4.2×10^{-9} T

29-65. $\mu_0\sigma\omega\left(\sqrt{y^2 + R^2} - y - \dfrac{R^2}{2\sqrt{y^2 + R^2}}\right)$

29-67. $(\mu_0 I/2\pi a)\ln(1 + a/x)$

CHAPTER 30

30-1. As viewed from the magnet: (*a*) ccw; (*b*) cw; (*c*) cw; (*d*) ccw; (*e*) cw; (*f*) 0 **30-5.** (*a*) 2.2 V, 0.22 V, 0.0037 V; (*c*) 4.4 V, 0.44 V, 0.0074 V **30-7.** 0 to 2 s: -4.7×10^{-2} V; 2 to 5 s: 0; 5 to 6 s: 9.4×10^{-2} V **30-9.** 2.0×10^{-2} A, 1.7×10^{-2} A, 0 **30-11.** (*a*) 150 mA; (*b*) 46 mA; (*c*) 0.019 mA **30-13.** $-(Cab^2\omega\cos\omega t)/2$ **30-15.** 4.8×10^6 A/s **30-17.** 0 to 2 ms: $E = -7.5 \times 10^{-2}$ V/m; 2 to 5 ms: $E = 0$; 5 to 6 ms: $E = 15.0 \times 10^{-2}$ V/m **30-19.** Inside: $(\mu_0 nI_0\omega r/2)\cos\omega t$; outside: $(\mu_0 nI_0\omega R^2/2r)\cos\omega t$ **30-21.** $r < 0.20$ m: $E = \dfrac{r}{40}$ V/m; $r > 0.20$ m: $E = \dfrac{1.0 \times 10^{-3}}{r}$ V/m **30-23.** (*a*) $NBlv$; (*b*) the agent pulling the loop **30-25.** 15 mV **30-27.** 0.20 V, the lower end **30-29.** 1.9 mA, 2.5 mA, both ccw **30-31.** (*a*) 0.050 V; (*b*) 25 mA (ccw); (*c*) 2.5×10^{-4} N; (*d*) 1.3 mW **30-33.** 1.5 mV **30-35.** If single turn, area = 2.9 m^2 **30-37.** (*a*) (850 V) sin $120\pi t$; (*b*) (720 W) $\sin^2 120\pi t$; (*c*) (360 W) $\sin^2 120\pi t$ **30-39.** (*a*) $B = QR/2NA$, (*b*) maximize Q **30-41.** (*a*) 1.33 A; (*b*) 0.50 A; (*c*) 60 W; (*d*) 38 W; (*e*) 23 W **30-43.** (*a*) 50 Ω; (*b*) 90 V; (*c*) 60 V **30-45** $N = 1$ **30-47.** 0.85 V **30-49.** $I = \mu_0 I_0 abv/[2\pi Rx(x + b)]$ **30-51.** 1.0 μV, 0.14 μV, 0 **30-53.** (*b*) $m^2g^2R\sin^2\theta/B^2l^2\cos^2\theta$; (*c*) same as (*b*); (*d*) same result **30-55.** (*a*) 9.9×10^{-4} V; (*b*) 9.9×10^{-4} V; (*c*) 1.6×10^{-3} V/m for both; (*d*) 9.9×10^{-4} V; (*e*) no

CHAPTER 31

31-1. (*a*) 1.6×10^{-8} T; (*b*) 3.2×10^{-8} T; (*c*) 2.0×10^{-8} T **31-3.** 3.3×10^{-3} A **31-5.** (*a*) $(1.1 \times 10^{-10}$ T$)\cos\omega t$; (*b*) $(3.5 \times 10^{-10}$ T$)\cos\omega t$ **31-7.** (*a*) $(\alpha\epsilon_0 AV_0/d)e^{-\alpha t}$; (*b*) 1.7×10^{-9} A **31-9.** (*a*) Yes; (*b*) no; (*c*) no **31-11.** (*a*) Yes; (*b*) no; (*c*) no; (*d*) only if $\mu_0 I + \epsilon_0\mu_0(d\phi_e/dt) = 0$ **31-13.** 2.0×10^6 V/m **31-15.** $\epsilon_0\pi E^2R^2d/2$ **31-17.** 3.5×10^{-4} J **31-19.** (*a*) $Q^2d/2\epsilon_0 A$; (*b*) $Q^2d/\epsilon_0 A$; (*c*) work done in separating plates **31-21.** (*a*) $dW = q\,dq/4\pi\epsilon_0 R$ **31-23.** $u_e = 4.4 \times 10^{-8}$ J/m^3, $u_m = 9.9 \times 10^{-4}$ J/m^3 **31-25.** $(\mu_0 N^2I^2h/4\pi)\ln(b/a)$ **31-27.** 3.6×10^{-7} J **31-29.** (*a*) $(\mu_0 nr/2)(dI/dt)$; (*b*) $(\mu_0 n^2r/2)(IdI/dt)$, toward the solenoid axis **31-31.** (*a*) 8.0×10^{-3} T; (*b*) 0.34 V/m; (*c*) 2.1×10^3 W/m^2; (*d*) 13.5 W; (*e*) same **31-33.** (*a*) $-(1.85 \times 10^{-14}$ T/m$)(R^2/r)$, $-(1.85 \times 10^{-14}$ T/m$)r$; (*b*) 8.0×10^{-15} N, 1.5×10^{-27} N **31-37.** 1.44×10^4 eV

CHAPTER 32

32-1. 1.1×10^{-3} m^2 **32-3.** 1.5 mm **32-5.** 500 μC **32-7.** 712 μF **32-9.** 1:16 **32-11.** 2000 μC **32-13.** (*a*) 80 pF; (*b*) 3.6×10^{-3} m^2; (*c*) 0.040 μC; (*d*) 1200 V **32-15.** (*a*) 2.77×10^{-10} F; (*b*) 2.77×10^{-8} C; (*c*) 5.0×10^4 V/m **32-17.** (*a*) 0.40 μC;

(*b*) 4000 V **32-19.** 3.0 μF, 0.33 μF **32-21.** (*a*) 1.07×10^3 pC; (*b*) 267 V, 133 V **32-23.** 500 V; 1.0×10^{-3} C, 1.5×10^{-3} C, 3.0×10^{-3} C **32-25.** (*a*) 100 V; (*b*) 150 V; (*c*) 300 μC **32-27.** (*a*) 2.0×10^{-8} C; (*b*) 4.0×10^{-9} C, 1.6×10^{-8} C; (*c*) 400 V **32-29.** 889 μC, 1780 μC; 444 V **32-31.** 144 μJ **32-33.** 7.5 μJ **32-35.** 1.11×10^{-8} J **32-37.** 6.9×10^{-7} J, 3.5×10^{-7} J, charge is transferred to the battery **32-39.** 7.0×10^{-4} J **32-41.** 720 μJ **32-43.** (*a*) 0.13 J; (*b*) no, because of resistive heating in the wires **32-45.** (*a*) 7.1 pF; (*b*) 42 pF **32-47.** (*a*) 66 μC/m^2; (*b*) 53 μC/m^2 **32-49.** (*a*) 3.7×10^{-8} C; (*b*) 4.0×10^5 V/m; (*c*) 1.9×10^{-8} C **32-51.** (*a*) 14.8 pF; (*b*) 88.8 pF **32-53.** (*a*) Right end; (*b*) right end **32-55.** $(480\pi \text{ V}) \sin (120\pi t - \pi/2)$ **32-57.** 0.15 V **32-59.** (*a*) 0.089 H/m; (*b*) 0.45 V/m **32-61.** 4.2×10^{-7} H/m **32-63.** 0.010 A **32-65.** 6.0 g **32-67.** 7.0×10^{-7} J **32-69.** (*a*) 3.6×10^{-3} H **32-71.** 3.8×10^{-4} H in both cases **32-73.** 2.3×10^{-5} H **32-75.** $\epsilon_0 A/(d_1 + d_2)$ **32-81.** $(\mu_0 l/\pi) \ln(1 + a/d)$ **32-83.** (*a*) 10 T; (*b*) 8.0×10^3 A; (*c*) 0.50 H

CHAPTER 33

33-1. (*a*) 2.4 mA; (*b*) 0.32 mA; (*c*) 0; (*d*) 5.3×10^{-4} W; (*e*) 1.1×10^{-4} J; (*f*) 3.4×10^{-3} W **33-3.** (*a*) 1.74 s; (*b*) 0.16 V **33-5.** (*a*) 24 μC; (*b*) 2.4 mA; (*c*) 8.8 μC; (*d*) 0.88 mA; (*e*) 3.9×10^{-3} W; (*f*) 3.9×10^{-3} W **33-7.** 2.1 μF **33-9.** (*a*) 1.4 V; (*b*) 7.4 μs **33-11.** (*a*) 3.2×10^{-3} C; (*b*) 0.63 mA; (*c*) $(632 \text{ V})\, e^{-t/(5.0 \text{s})}$; (*d*) $(0.40 \text{ W})\, e^{-2t/(5.0 \text{s})}$ **33-13.** (*a*) 4.0 A; (*b*) 2.4 A; (*c*) 12 V; 12 V **33-15.** $1.23\, \tau_L$ **33-17.** (*a*) 2.53×10^{-3} s; (*b*) 99 Ω **33-19.** (*a*) 1.67 A, 1.67 A; (*b*) 2.27 A, 1.36 A; (*c*) 0, −0.91 A; (*d*) 0 **33-23.** (*a*) $\mathcal{E}^2 C$; (*b*) $\mathcal{E}^2 C/2$; (*c*) $\mathcal{E}^2 C/2$; (*d*) yes **33-25.** 3.2×10^7 rad/s **33-27.** (*a*) 1.6×10^{-3} s; (*b*) 3.9×10^{-4} s **33-29.** $0.71\, q_m$, $0.71\, q_m/\sqrt{LC}$ **33-31.** 3.96 to 34.8 pF **33-33.** 6.9×10^{-3} s **33-35.** (*a*) 4.0 s; (*b*) 6.07 V, 1.52 V; (*c*) 0.88 V, 3.53 V **33-37.** (*a*) $5\mathcal{E}/3R$; (*b*) $\mathcal{E}/R$; (*c*) $C\mathcal{E}$ **33-41.** (*a*) 0; (*b*) 2.4 A

CHAPTER 34

34-1. $(12 \text{ V}) \sin 400\pi t$ **34-3.** 265 μF **34-5.** 26.5 mH **34-7.** 5.1 μF **34-9.** (*a*) 106 Ω; (*b*) $(1.51 \text{ A}) \sin (120\pi t + \pi/2)$ **34-11.** (*a*) $(1.26 \text{ A}) \cos 200\pi t$; (*b*) $(-7.96 \text{ A}) \cos 200\pi t$; (*c*) $(2.00 \text{ A}) \sin 200\pi t$ **34-13.** (*a*) 2.5×10^3 Ω, 3.6 mA; (*b*) 7.5 Ω, 1.2 A **34-15.** (*a*) 18.9 A; (*b*) $\pi/2$ rad **34-17.** (*a*) 774 Ω; (*b*) 0.155 A; (*c*) $(0.155 \text{ A}) \cos (120\pi t + 1.03)$; (*d*) $(62 \text{ V}) \cos (120\pi t + 1.03)$, $(103 \text{ V}) \cos (120\pi t - 0.54)$ **34-19.** (*a*) 687 Ω; (*b*) 0.146 A; (*c*) $(0.146 \text{ A}) \sin (1000\pi t - 0.754)$; (*d*) 1086 Ω, 0.092 A, $(0.092 \text{ A}) \sin (1000\pi t + 1.09)$ **34-21.** (*a*) 5.73 Ω; (*b*) 0.509 rad; (*c*) $(29.7 \text{ A}) \sin (120\pi t - 0.509)$ **34-23.** (*a*) 0; (*b*) 0; (*c*) 100 W **34-25.** 35.4 V **34-27.** 50° **34-29.** (*a*) 2.38 A; (*b*) 47.6 V, 7.6 V, 59.8 V; (*c*) 47.7°; (*d*) 113 W; (*e*) 113 W **34-31.** 1.83×10^3 rad/s, no change **34-33.** (*a*) 1.78×10^3 Hz; (*b*) 20 Ω **34-35.** (*b*) 712 Hz **34-37.** 0.127 μF, 1057 Ω **34-39.** (*a*) 45 turns; (*b*) 0.68 A, 0.015 A; (*c*) 165 Ω **34-41.** (*a*) 41 turns; (*b*) 41 mA **34-43.** (*a*) 10 A; (*b*) 8.6×10^{-6} W/m; (*c*) 8.6×10^{-4} W/m **34-45.** 0.50 A **34-47.** (*a*) 0; (*b*) 0; (*c*) 0; (*d*) $\mathcal{E}$ **34-49.** 2.50×10^3 Ω **34-51.** (*a*) 0.10; (*b*) $-(0.20 \text{ mV}) \sin (120\pi t)$ **34-53.** $R_1 = R_2 = R_3 = R_f$ **34-55.** (*a*) 156 V; (*b*) 42 V; (*c*) 154 V

CHAPTER 35

35-1. 9.22×10^9 Hz **35-3.** (*a*) 1.5×10^{-9} m; (*b*) 5.9×10^{-7} m; (*c*) 3.0×10^{-15} m **35-5.** AM: 556 to 188 m; FM: 3.41 to 2.78 m **35-7.** (*a*) 6.45×10^{-3} V/m; (*b*) 395 m **35-9.** Down, 2.0×10^{-8} T **35-11.** (*a*) $\mathbf{E} = -(0.150 \text{ V/m}) \cos (kz + \omega t)\mathbf{i}$; (*b*) 1.50×10^6 Hz, 200 m; (*c*) $-2.98 \times 10^{-5}\mathbf{k}$ W/m^2 **35-13.** 8.3×10^{-12} J **35-17.** (*a*) 80 W/m^2; (*b*) 2.5×10^2 V/m, 8.2×10^{-7} T; (*c*) 1.3×10^{-11} N **35-19.** (*a*) 7.8×10^{-8} N; (*b*) 2.3×10^3 J; (*c*) 7.7°C **35-21.** 338 m **35-23.** 114 rad/s **35-25** 70°, 45° **35-27.** 42° **35-29.** 1.53 **35-31.** (*a*) 51.3°; (*b*) 51.3° **35-33.** 600 nm, 1.46 **35-35.** (*a*) 1.50 to 1.58; (*b*) 18.4° to 19.5°; (*c*) 37.7° to 40.1° **35-37.** 1.47 **35-39.** 49° (the critical angle) **35-41.** (*a*) 48°; (*b*) 28° **35-43.** 60° **35-45.** 45° **35-47.** 0.38 **35-49.** 53° **35-51.** 1.59 **35-53.** (*a*) $(1.5 \times 10^{-5})r^2$ N, (*b*) $(2.5 \times 10^{-2})\rho r^3$ N, (*c*) 3.0×10^{-7} m **35-55.** 10.2° **35-59.** $I = (I_0/8) \sin^2 (2\theta)$, so maximum transmission at $\theta = 45°$

CHAPTER 36

36-1. (*a*) 20 cm; (*b*) 40 cm **36-3.** (*a*) 240 cm; (*b*) same as object; (*c*) the image waves with her left hand **36-7.** 0.84 **36-9.** 6.0 to 10.0 m **36-11.** 60 cm from the mirror **36-13.** −53 cm **36-15.** 10 cm **36-17.** (*a*) 50 cm; (*b*) 25 cm behind the screen **36-19.** For o = 40 cm, i = 40 cm, m = −1. For o = 30 cm, i = −15 cm, m = 0.5. **36-21.** (for first, third, and fifth lines of table)

Mirror Type	Radius, cm	Focal Length, cm	Object Distance, cm	Image Distance, cm	Lateral Magnification	Real/ Virtual Image
Convex	−50	−25	40	−15	0.38	Virtual
Concave	30	15	30	30	−1.0	Real
Convex	20	10	8.0	−40	5.0	Virtual

36-23. 3.63 cm **36-25.** (*a*) At 44 cm; (*b*) 18 cm; (*c*) inverted **36-29.** (*a*) −22 cm, 1.5; (*b*) −45 cm, 2.0; (*c*) 180 cm, −3.0 **36-31.** Air: (*a*) −11 cm, 0.75; (*b*) −15 cm, 0.67; (*c*) −26 cm, 0.43. Water: (*a*) −10 cm, 0.92; (*b*) −15 cm, 0.89; (*c*) −34 cm, 0.75 **36-33.** 59 cm **36-35.** 10.0 m **36-37.** 19 cm **36-39.** 2.0 **36-41.** (*a*) i = −40 cm (at center of bowl); (*b*) 1.3; (*c*) No **36-43.** 1.4 cm **36-45.** 4.9 mm^2 **36-49.** −1 **36-51.** 1.03 m below surface

CHAPTER 37

37-1. 210 cm to the right of the sphere's center **37-3.** 10 cm to the right of the right surface **37-5.** 3.0 cm beyond the sphere **37-7.** (*a*) 20 cm; (*c*) 40 cm; (*e*) −40 cm **37-13.** 6.4 cm **37-17.** i = −20 cm, m = +2.0 **37-19.** (*a*) 36 cm; (*b*) 36 cm **37-21.** f = +1.656 cm placed 1.661 cm in front of slide **37-23.** i = −100 cm, m = 5.0 **37-25.** 10.6 cm **37-29.** (*a*) f = 80 cm; (*b*) $\dfrac{1}{f} = \dfrac{1}{f_1} + \dfrac{1}{f_2} + \dfrac{1}{f_3}$ **37-31.** 2.0 cm beyond the eyepiece on either side **37-33.** 40 cm **37-35.** 43 cm **37-37.** −1.48 m, 0.34 m **37-39.** 4.3 **37-41.** (*a*) 8.5; (*b*) Teenager's image is 1.2 times bigger **37-43.** 1.3 cm **37-45.** −1000 **37-47.** 0.87 cm **37-49.** 2.0 mm **37-53.** −1400 **37-57.** 10.7 cm to the left to the lens; 1.43 **37-59.** (*a*) 17 cm beyond second lens; (*b*) −0.64 **37-61.** Your friend sees an image twice as big. **37-63.** 12 cm

CHAPTER 38

38-1. 0.29°, 0.86° **38-3.** (*a*) 4.26 cm; (*b*) 2.84 cm **38-5.** 6 **38-7.** 560 nm **38-9.** 1.2 mm **38-11.** Bright and dark fringes interchange positions. **38-13.** 45° **38-15.** 4.64×10^{-4} rad for both (to three sig. figs.) **38-17.** (*a*) 3.0×10^{-2} rad; (*b*) 1.2×10^{-2} rad **38-19.** 700 nm **38-21.** 189 nm **38-23.** (*a*) Green (504 nm); (*b*) magenta (white minus green) **38-25.** 1.29 **38-27.** 1.26 **38-29.** 4.55×10^{-4} m

38-31. (*a*) 3.33×10^{-7} s; (*b*) 100 m; (*c*) no
38-33. (*a*) 1.0×10^{14} Hz; (*b*) 1.0×10^{-14} s; (*c*) 3.0×10^{-6} m; (*d*) 5 **38-35.** 2.8 mm **38-37.** 0.625 m **38-39.** 0.011
38-41. 2.1 μm **38-43.** 1.9 m **38-45.** (*a*) 20 cm; (*b*) 0.72 mm
38-47. (*a*) 6.0×10^{-7} rad; (*b*) 0.93 rad **38-49.** 8
38-51. 8.8 mm or less **38-53.** 1.3×10^{-7} rad **38-55.** (*c*) 194 nm
38-57. 1.6×10^{-4} rad **38-59.** 4.16×10^{3} lines/cm **38-61.** (*a*) 3; (*b*) 3.80 cm **38-63.** 0.32 m **38-67.** $y = 1.62$ mm, 1.98 mm
38-71. 184 nm, 368 nm **38-73.** (*a*) $\theta = \pm 23.6°$; (*b*) no
38-75. 0.41, 0 **38-77.** 0.733 cm

CHAPTER 39

39-1. 2.29 **39-3.** $0.998c$ **39-5.** 1.67 s **39-7.** (*a*) 5.0 μs; (*b*) 900 m **39-9.** (*a*) 2.13×10^{-7} s; (*b*) 63.3 m **39-11.** 60 m
39-15. 0.359 m **39-17.** 1.62×10^{-2} m^3 **39-19.** $0.60c$
39-21. (*a*) 43.9°; (*b*) 43.3 cm **39-23.** $0.77c$ **39-25.** 67 m, 500 m, 500 m, 2.2 μs **39-27.** (*a*) $x'_A = 0$, $x'_B = 2050$ m; $t'_A = 0$, $t'_B = -2.5$ μs **39-29.** 5.0×10^{-7} s **39-31.** (*a*) $0.89c$; (*b*) $-0.31c$
39-33. (*c*) 0.43 **39-35.** (*a*) $0.60c$; (*b*) $0.88c$ **39-37.** $0.995c$
39-39. 1.67×10^{-27} kg **39-41.** \$$2 \times 10^9$
39-43. 6.24×10^{-19} kg·m/s **39-45.** 1.02 MeV
39-47. (*a*) 5.22×10^{-13} J; (*b*) 1.9×10^5 kg
39-49. (*a*) 2.3×10^{14} J; (*b*) 1.5×10^{20} W; (*c*) $P_{Pu}/P_{US} = 5.9 \times 10^7$
39-51. $0.707c$ **39-53.** (*a*) $20(1 - 3.9 \times 10^{-14})$ m; (*b*) no, 1.9×10^{-14} s **39-55.** (*a*) $\pm 3.60 \times 10^{-20}$ kg·m/s, 1.08×10^{-11} J for both; (*b*) 1.57×10^{-19} and -8.26×10^{-21} kg·m/s, 4.71×10^{-11} and 2.48×10^{-12} J; (*c*) Total $P = 1.49 \times 10^{-19}$ kg·m/s, Total $E = 4.95 \times 10^{-11}$ J
39-57. $M = 2.12m$, $v = c/3$

CREDITS

Part I

Opener: Photo courtesy of Ringling Bros. and Barnum & Bailey Combined Shows, Inc.

Chapter 1
Opener: Comstock Inc./Russ Kinne. **Page 5:** (top left) Courtesy National Institute of Standards and Technology; (bottom left) Courtesy National Institute of Standards and Technology; (bottom right) Courtesy National Institute of Standards and Technology. **Page 6:** Science Photo Library/ Photo Researchers. **Page 7:** Allsport Photography (USA).

Chapter 2
Opener: Reuters/Bettmann. **Page 14:** Image Finders.

Chapter 3
Opener: Image Finders. **Page 34:** (top) Peticolas/Megna, Fundamental Photographs, New York; (bottom) Courtesy Central Scientific Company. **Page 35:** Dr. Michelangelo Fazio, University of Milan.

Chapter 4
Opener: Allsport Photography (USA). **Page 49:** Courtesy Central Scientific Company. **Page 51:** Comstock Inc./Comstock Inc.

Chapter 5
Opener: Allsport Photography (USA). **Page 64:** Dr. Michelangelo Fazio, University of Milan. **Page 65:** Courtesy Central Scientific Company. **Page 68:** Explorer/Photo Researchers. **Page 70:** National Optical Astronomy Observatories.

Chapter 6
Opener: Insurance Institute for Highway Safety. **Page 88:** Reprinted with special permission of King Features Syndicate. **Page 89:** Allsport Photography (USA). **Page 92:** NASA.

Chapter 7
Opener: J. P. Varin/Jacana/Photo Researchers. **Page 107:** Allsport Photography (USA). **Page 118:** Science Photo Library/Photo Researchers. **Page 119:** Allsport Photography (USA).

Chapter 8
Opener: Allsport Photography (USA). **Page 130:** © The Harold E. Edgerton 1992 Trust, courtesy of Palm Press, Inc. **Page 136:** Allsport Photography (USA).

Chapter 9
Opener: Cedar Point Photo by Dan Feicht. **Page 149:** Leonard Kamsler. **Page 150:** Image Finders. **Page 153:** © Joseph P. Sinnot, 1994, Fundamental Photographs, New York. **Page 155:** UPI/Bettmann.

Chapter 10
Opener: Photo courtesy of Ringling Bros. and Barnum & Bailey Combined Shows, Inc. **Page 171:** Allsport Photography (USA). **Page 175:** American Bowling Congress.

Chapter 11
Opener: Will & Deni McIntyre/Photo Researchers. **Page 186:** (upper right) THE HUMAN MACHINE by R. McNeill Alexander. Copyright © 1992 by Columbia University Press. Printed with permission of the publisher. **Page 188:** Allsport Photography (USA). **Page 191** © Michael J. Pettypool/ UNIPHOTO Picture Agency. **Page 192:** Allsport Photography (USA).

Chapter 12
Opener: Richard Megna, Fundamental Photographs, New York. **Page 206:** Comstock Inc./Comstock Inc. **Page 209:** Fisher Scientific. **Page 210:** Brookhaven National Laboratory. **Page 215:** NASA.

Chapter 13
Opener: Allsport Photography (USA). **Page 231:** Comstock Inc./ C. Davidson. **Page 234:** Courtesy Central Scientific Company.

Chapter 14
Opener: Comstock Inc./M & C Werner. **Page 247:** Howard Miller Clock Co. **Page 250:** Monroe Auto Equipment. **Page 251:** UPI/Bettmann.

Chapter 15
Opener: The Goodyear Tire and Rubber Company. **Page 261:** Allied-Gator, Inc. **Page 262:** Vanessa Vick/Photo Researchers. **Page 265:** Will & Deni McIntyre/Photo Researchers. **Page 266:** (bottom left) Photo by F. N. M. Brown, from *An Album of Fluid Motion*, by Milton Van Dyke (Parabolic Press), reproduced with permission of Thomas J. Mueller, University of Notre Dame; (top right) NASA. **Page 269:** Courtesy Central Scientific Company. **Page 270:** (bottom left) UPI/Bettmann; (bottom right) ONERA.

Part II

Opener: Karl Weatherly/Tony Stone Images

Chapter 16
Opener: Allsport Photography (USA). **Page 292:** Martin Bough, Fundamental Photographs, New York. **Page 293:** (left) Science Photo Library/ Photo Researchers; (right) PSSC PHYSICS, 2nd edition, 1965; D. C. Heath & Company with Educational Development Center, Inc., Newton, MA. **Page 294:** Comstock Inc./George Lepp. **Page 296:** Kulik Photographic. **Page 297:** Comstock Inc./Russ Kinne. **Page 298:** PSSC PHYSICS, 2nd edition, 1965; D. C. Heath & Company with Educational Development Center, Inc., Newton, MA. **Page 299:** (left) PSSC PHYSICS, 2nd edition, 1965; D. C. Heath & Company with Educational Development Center, Inc., Newton, MA; (right) NASA. **Page 300:** © 1995 by Sidney Harris.

Chapter 17
Opener: Susana Millman © 1994. **Page 309:** Science Photo Library/Photo Researchers. **Page 311:** Science Photo Library/Photo Researchers. **Page 315:** Richard Megna, Fundamental Photographs, New York. **Page 316:** Comstock Inc./Comstock Inc. **Page 318:** Alex Martin/Washington National Cathedral. **Page 319:** (left) T. D. Rossing; (right) Copyright © 1979 by The Metropolitan Museum of Art.

Part III

Opener: LTV Steel Company

Chapter 18
Opener: A. Pierce Bounds/UNIPHOTO Picture Agency. **Page 330:** NASA, Jet Propulsion Laboratory, Photo P40697. **Page 331:** © 1995 by Sidney Harris. **Page 333:** Wahl Instruments. **Page 334:** Science Photo Library/ Photo Researchers. **Page 337:** Comstock Inc./P. Greenburg. **Page 338:** M. J. Manuel/Photo Researchers. **Page 344:** Images From Ideas.

Chapter 19
Opener: NASA. **Page 351:** Fisher Scientific. **Page 352:** Pella Corporation. **Page 353:** (top) Image Finders; (bottom) National Optical Astronomy Observatories. **Page 354:** Kulik Photographic. **Page 355:** Mark Burnett/ Photo Researchers.

Chapter 20
Opener: W. M. Keck Observatory, California Institute for Research in Astronomy. **Page 363:** Science Photo Library/Photo Researchers.

Chapter 21
Opener: California State Railroad Museum Library. **Page 376:** Nuclear Energy Institute. **Page 380:** M. Vantz, CEI. **Page 386:** (left) Richard Hutchings/Photo Researchers; (right) © 1995 by Sidney Harris.

Part IV

Opener: Color-Pic, Inc.

Chapter 22

Opener: Physics Curriculum and Instruction. **Page 397:** Science Photo Library/Photo Researchers. **Page 398:** The Granger Collection, New York. **Page 399:** Dr. Michelangelo Fazio, University of Milan. **Page 400:** Burndy Library of the Dibner Institute for the History of Science and Technology.

Chapter 23

Opener: Gordon Gore. **Page 409:** Princeton University. **Page 416:** Fisher Scientific. **Page 417:** Courtesy Central Scientific Company.

Chapter 24

Opener: Science Source/Photo Researchers. **Page 427:** Dr. Michelangelo Fazio, University of Milan.

Chapter 25

Opener: Peter Menzel Photography. **Page 440:** Science Photo Library/ Photo Researchers. **Page 447:** Burndy Library of the Dibner Institute for the History of Science and Technology. **Page 450:** (left) Michael Burgess/ Science Photo Library/Photo Researchers; (right) Image Finders. **Page 451:** Carnegie Institute of Washington, Department of Terrestrial Magnetism.

Chapter 26

Opener: Spencer Grant/Photo Researchers. **Page 463:** Comstock Inc./Russ Kinne. **Page 464:** Marzano Photography, Los Angeles. **Page 467:** (top) Physics Curriculum and Instruction; (bottom) Sears Merchandise Group. **Page 471:** Lew Lause/UNIPHOTO Picture Agency.

Chapter 27

Opener: Reinstein/The Image Works. **Page 478:** Science Photo Library/ Photo Researchers. **Page 484:** U.S. Department of the Interior, National Park Service, Edison National Historic Site. **Page 486:** Fisher Scientific. **Page 489:** Fluke Corporation.

Chapter 28

Opener: Courtesy Ohio Magnetics, Inc. **Page 502:** Intermagnetics General Corporation. **Page 504:** (top) Courtesy Central Scientific Company; (bottom) Lawrence Berkeley Laboratory. **Page 505:** NASA. **Page 511:** Science Photo Library/Photo Researchers. **Page 512:** (top) University of Colorado; (bottom) Fermi National Accelerator Laboratory.

Chapter 29

Opener: Courtesy of German Information Center. **Page 521:** Richard Megna, Fundamental Photographs, New York. **Page 522:** Photo courtesy of PASCO Scientific. **Page 529:** GE Research and Development Center. **Page 533:** (top) Courtesy Central Scientific Company; (bottom) Ralph W. de Blois. **Page 534:** Courtesy Central Scientific Company.

Chapter 30

Opener: Lowell J. Georgia Photography/Photo Researchers. **Page 546:** Courtesy Brush Industries, Inc. **Page 547:** Courtesy Central Scientific Company. **Page 548:** Marzano Photography, Los Angeles. **Page 552:** Bettmann. **Page 553:** Courtesy Central Scientific Company.

Chapter 31

Opener: Dr. Michelangelo Fazio, University of Milan. **Page 569:** Joe Sohm/ UNIPHOTO Picture Agency. **Page 571:** Comstock Inc./Stuart Cohen.

Chapter 32

Opener: Earle Williams, High Voltage Research Laboratory, MIT. **Page 579:** (left) Paul Silverman, Fundamental Photographs, New York; (right) The Image Works. **Page 582:** Erik Borg. **Page 583:** Sears Merchandise Group. **Page 587:** The Image Works.

Chapter 33

Opener: Paul Rudo. **Page 599:** William Moebs. **Page 602:** William Moebs.

Chapter 34

Opener: UNIPHOTO Picture Agency. **Page 616:** The Image Works. **Page 619:** Photo courtesy of Fisher Research Laboratory, Los Banos, CA. **Page 620:** Comstock Inc./Robert Houser. **Page 623:** (left) William Moebs; (right) Courtesy of AT&T Bell Labs.

Part V

Opener: © Larry Ulrich.

Chapter 35

Opener: UNIPHOTO Picture Agency. **Page 639:** Image Select. **Page 640:** Courtesy Samsung Electronics Co., Ltd. **Page 641:** NASA. **Page 643:** NASA. **Page 644:** PSSC PHYSICS, 2nd edition, 1965; D. C. Heath & Company with Educational Development Center, Inc., Newton, MA. **Page 646:** (left) Richard Megna, Fundamental Photographs, New York; (right) Science Photo Library/Photo Researchers. **Page 648:** (bottom left) UNIPHOTO Picture Agency; (top right) Comstock Inc./Russ Kinne; (bottom right) Michael Giannechini/Photo Researchers. **Page 651:** Marzano Photography, Los Angeles. **Page 652:** Paul Silverman, Fundamental Photographs, New York. **Page 655:** NASA.

Chapter 36

Opener: From *Through the Kaleidoscope*, by Cozy Baker. **Page 664:** (top) Image Finders; (bottom) Charles Gupton, UNIPHOTO Picture Agency. **Page 666:** Image Finders. **Page 669:** Marzano Photography, Los Angeles.

Chapter 37

Opener: Image Finders. **Page 682:** NASA. **Page 683:** The Image Works. **Page 684:** (left) Lennart Nilsson, BEHOLD MAN, Little, Brown and pany; (right) Omikron/Photo Researchers. **Page 686:** (left) Comstock Inc./ M & C Werner; (right) Fisher Scientific. **Page 687:** Scala/Art Resources, NY. **Page 688:** W. M. Keck Observatory, California Association for Research in Astronomy. **Page 689:** W. M. Keck Observatory, California Association for Research in Astronomy.

Chapter 38

Opener: Physics Curriculum and Instruction. **Page 697:** PSSC PHYSICS, 2nd edition, 1965; D. C. Heath & Company with Educational Development Center, Inc., Newton, MA. **Page 699:** From M. Cagnet, M. Francon, and J. C. Thrierr, *Atlas of Optical Phenomenon*, Berlin; Springer-Verlag, 1962, reprinted by permission. **Page 703:** From M. Cagnet, M. Francon, and J. C. Thrierr, *Atlas of Optical Phenomenon*, Berlin; Springer-Verlag, 1962, reprinted by permission. **Page 704:** Physics Curriculum and Instruction. **Page 706:** Bausch & Lomb. **Page 707:** Marzano Photography, Los Angeles. **Page 710:** (top left) Road Runner and Wiley E. Coyote are trademarks belonging to Warner Bros. © 1995; (top right) From M. Cagnet, M. Francon, and J. C. Thrierr, *Atlas of Optical Phenomenon*, Berlin; Springer-Verlag, 1962, reprinted by permission; (middle right) Physics Curriculum and Instruction; (bottom) Ken Kay, Fundamental Photographs, New York. **Page 713:** From M. Cagnet, M. Francon, and J. C. Thrierr, *Atlas of Optical Phenomenon*, Berlin; Springer-Verlag, 1962, reprinted by permission. **Page 716:** (left) Science Photo Library/Photo Researchers; (right) Image Finders. **Page 717:** (left) Science Photo Library/Photo Researchers; (right) Photo Courtesy of PASCO Scientific. **Page 718:** Sargent-Welch Scientific. **Page 724:** Arecibo Observatory, National Astronomy & Ionosphere Center, Cornell University.

Part VI

Opener: CERN Photo

Chapter 39

Opener: Princeton University, Plasma Physics Laboratory. **Page 738:** AIP Emilio Segre Visual Archives. **Page 739:** Dr. Michelangelo Fazio, University of Milan. **Page 743:** Stanford Linear Accelerator Center/U.S. Department of Energy. **Page 744:** Los Alamos National Laboratory. **Page 750:** © 1995 by Sidney Harris.

INDEX

A

D

J

K

L

M

N

O

S

The Greek Alphabet

alpha	A	α	iota	I	ι	rho	P	ρ
beta	B	β	kappa	K	κ	sigma	Σ	σ
gamma	Γ	γ	lambda	Λ	λ	tau	T	τ
delta	Δ	δ	mu	M	μ	upsilon	Υ	υ
epsilon	E	ϵ	nu	N	ν	phi	Φ	ϕ, φ
zeta	Z	ζ	xi	Ξ	ξ	chi	X	χ
eta	H	η	omicron	O	o	psi	Ψ	ψ
theta	Θ	θ, ϑ	pi	Π	π	omega	Ω	ω

Prefixes for Powers of Ten

Power	Prefix	Abbreviation	Power	Prefix	Abbreviation
10^{-18}	atto	a	10^{1}	deka	da
10^{-15}	femto	f	10^{2}	hecto	h
10^{-12}	pico	p	10^{3}	kilo	k
10^{-9}	nano	n	10^{6}	mega	M
10^{-6}	micro	μ	10^{9}	giga	G
10^{-3}	milli	m	10^{12}	tera	T
10^{-2}	centi	c	10^{15}	peta	P
10^{-1}	deci	d	10^{18}	exa	E

Some Units and Abbreviations

ampere	A	joule	J
atmosphere	atm	kelvin	K
British thermal unit	Btu	liter	L
calorie	cal	meter	m
coulomb	C	mile	mi
day	day	minute	min
degree Celsius	°C	mole	mol
degree Fahrenheit	°F	newton	N
dyne	dyne	ohm	Ω
electron-volt	eV	pascal	Pa
erg	erg	pound	lb
farad	F	radian	rad
foot	ft	revolution	rev
gauss	G	second	s
gram	g	tesla	T
henry	H	unified atomic mass unit	u
hertz	Hz	volt	V
horsepower	hp	watt	W
hour	h	weber	Wb
inch	in.	year	yr